LIFE

AN INTRODUCTION TO BIOLOGY

Third Edition

LIFE

AN INTRODUCTION TO BIOLOGY

Third Edition

WILLIAM S. BECK
Harvard University

KAREL F. LIEM
Harvard University

GEORGE GAYLORD SIMPSON
Late, University of Arizona

 HarperCollins*Publishers*

SPONSORING EDITOR:	Glyn Davies
DEVELOPMENT EDITOR:	Barbara Conover
MANAGING EDITOR:	Lois Lombardo
PROJECT COORDINATOR:	Arlene Grodkiewicz
ART DIRECTION, TEXT DESIGN:	Michael Mendelsohn, M 'N O Production Services, Inc.
COVER DESIGN:	Jaye Zimet
COVER PHOTOS:	*Front:* Julia Sims/Peter Arnold, Inc. *Back:* John Shaw
PHOTO RESEARCH:	Lynn Goldberg Biderman/Visual Impact; Marina Gordon
PRODUCTION:	Kewal K. Sharma
COMPOSITOR, PRINTER, BINDER:	Arcata Graphics/Kingsport
COVER PRINTER:	New England Book Components

LIFE: An Introduction to Biology, Third Edition

Library of Congress Cataloging-in-Publication Data

Beck, William Samson, 1923–
    Life / William S. Beck, Karel F. Liem.—3rd ed.
       p. cm.
    Rev. ed. of: Life / George Gaylord Simpson. 2nd ed. 1965.
    Includes bibliographical references.
    1. Biology.   I. Liem, Karel F.   II. Simpson, George Gaylord,
1902–1984 Life.   III. Title.
QH308.2B42   1991
574—dc20   89–38704
                                                      CIP

ISBN 0–06–040603–8   (Student Edition)
ISBN 0–06–500009–9 (Teacher Edition)

90 91 92 93   9 8 7 6 5 4 3 2 1

CONTENTS IN BRIEF

DETAILED CONTENTS

PART 1 THE SCIENCE OF BIOLOGY   1

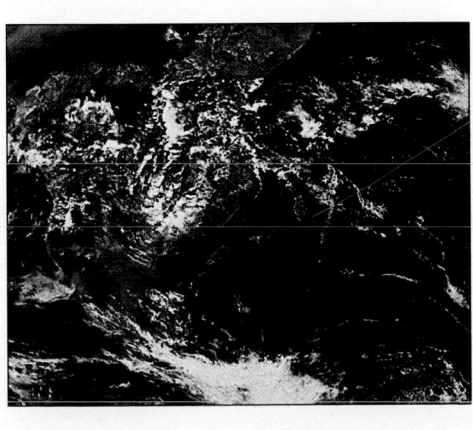

PART 2   BIOCHEMICAL AND CELLULAR BASIS OF LIFE   33

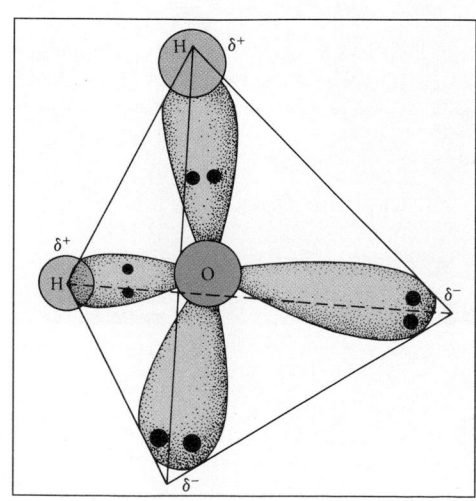

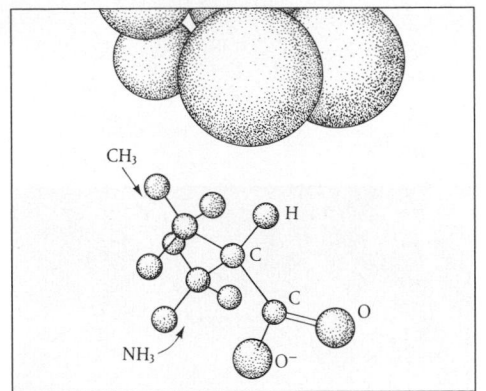

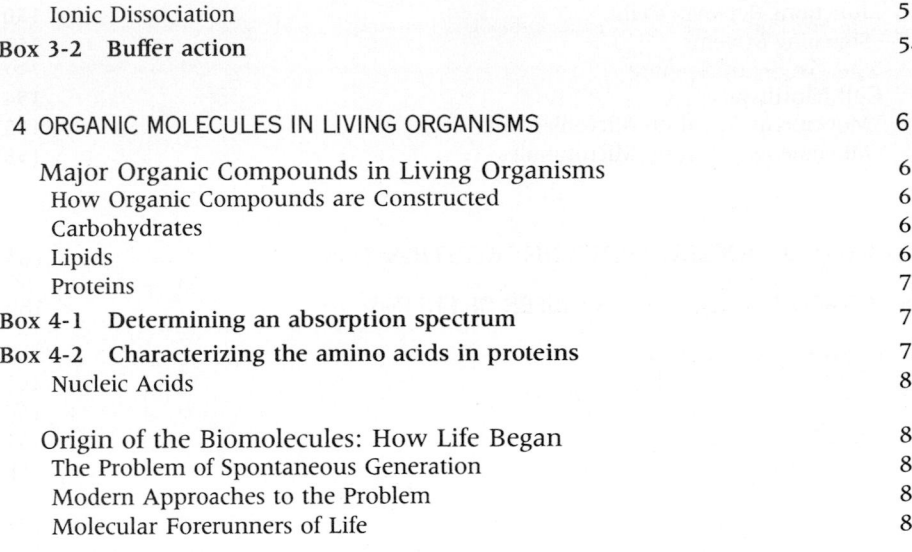
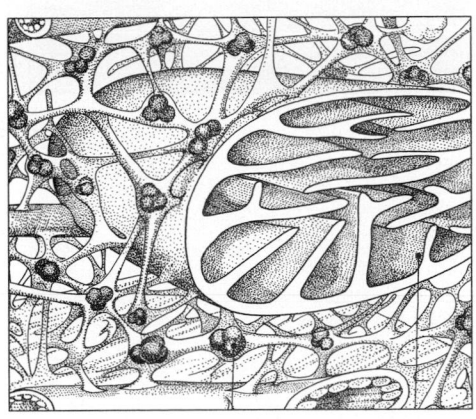
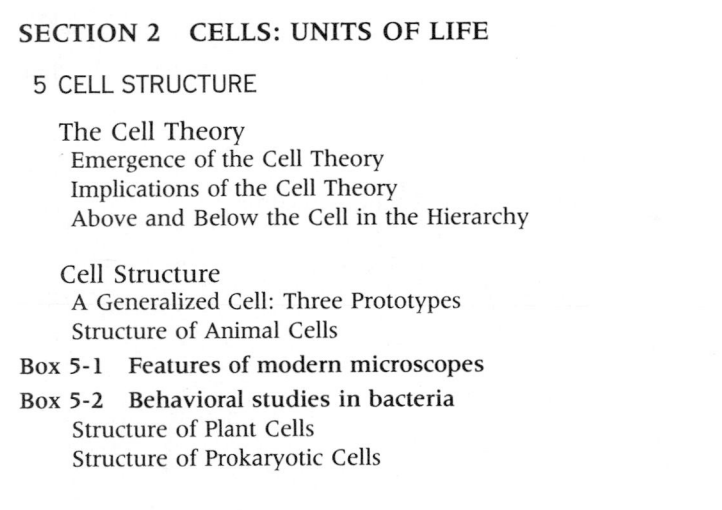
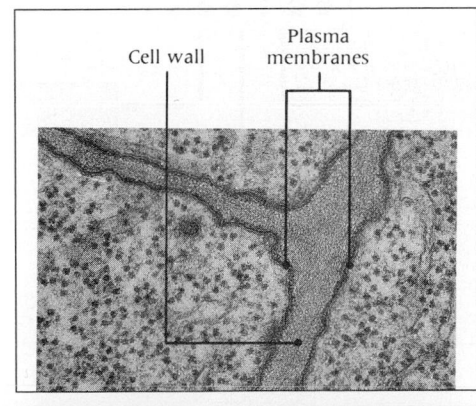
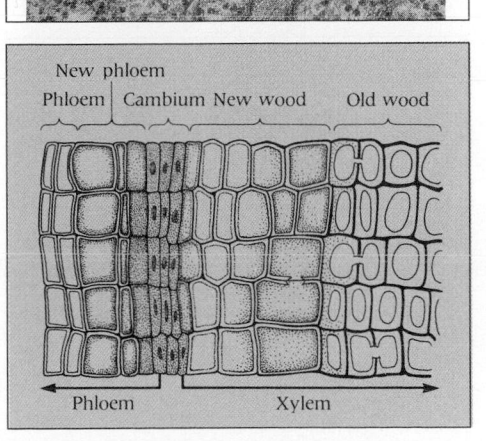

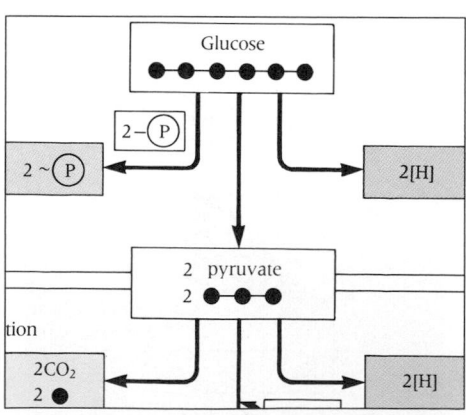

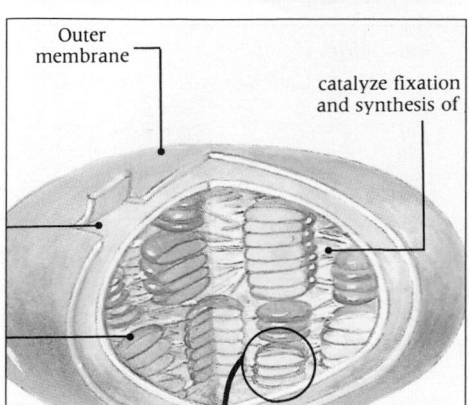

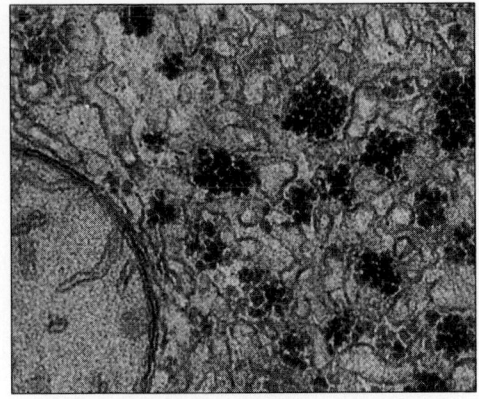

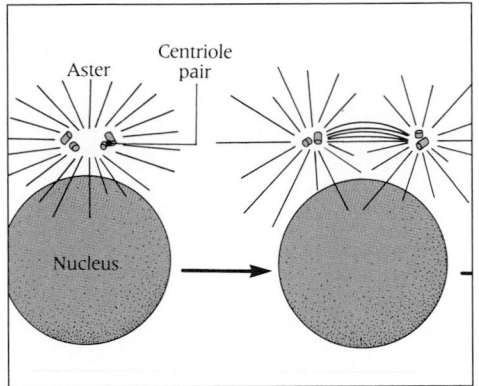

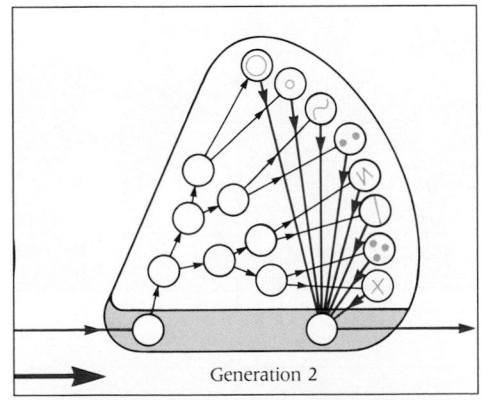

Generation 2

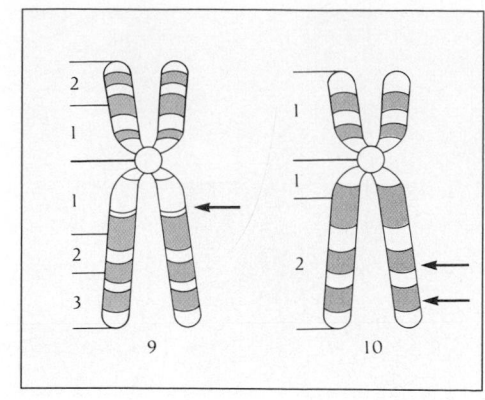

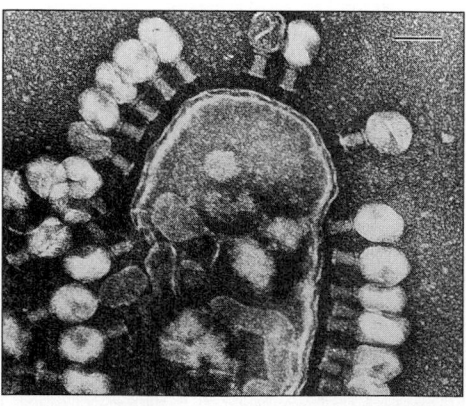

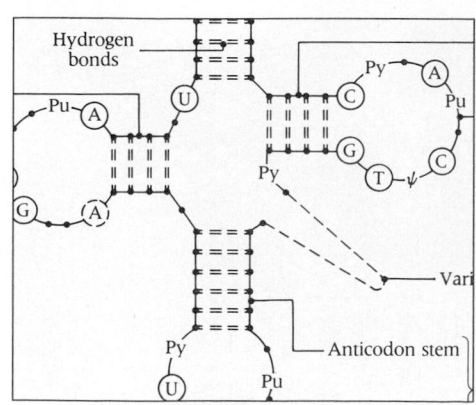

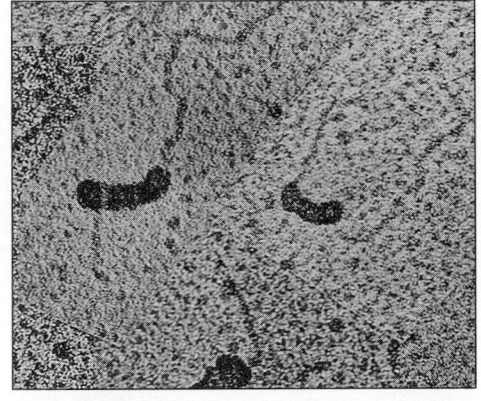

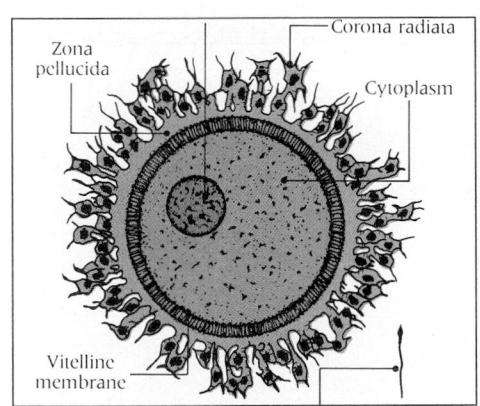

Corona radiata

Zona pellucida

Cytoplasm

Vitelline membrane

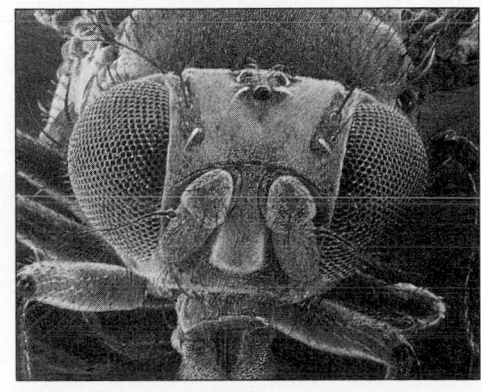

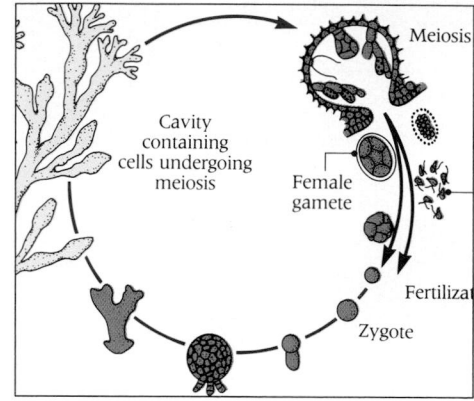

PART 3 HOMEOSTASIS

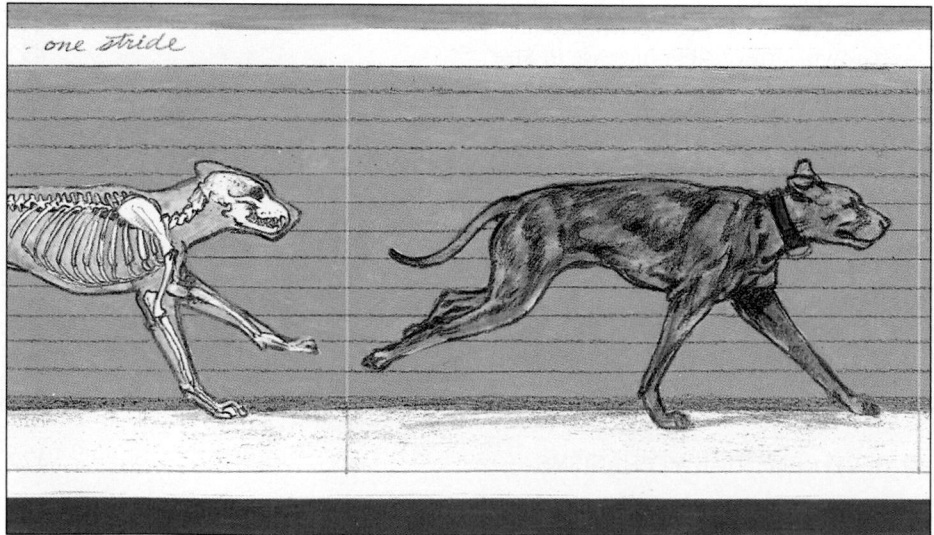

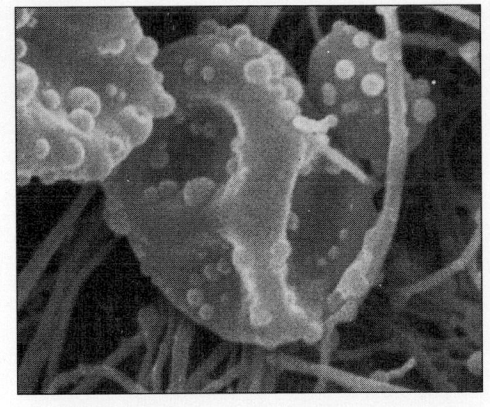

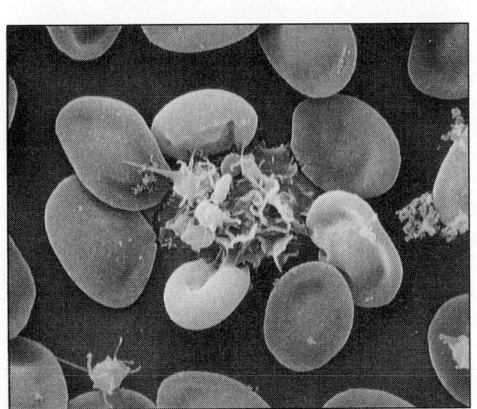

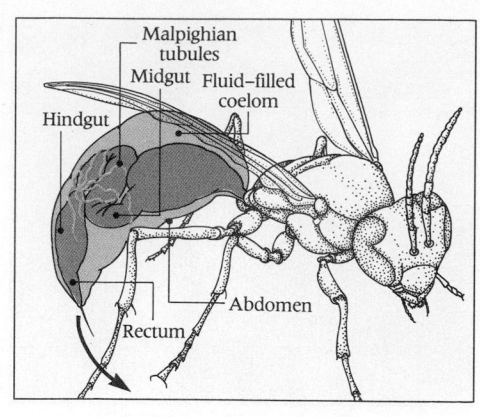

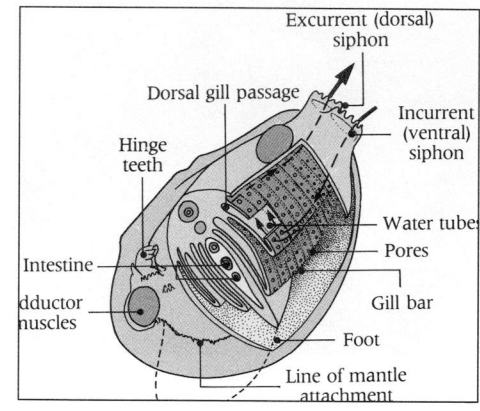

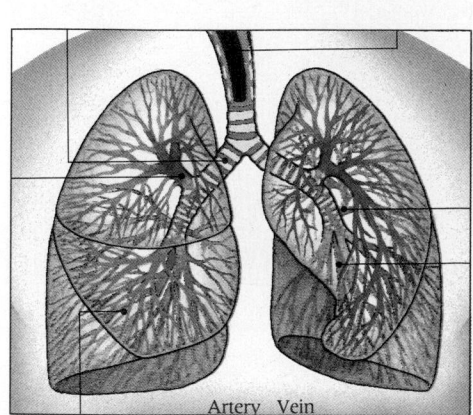

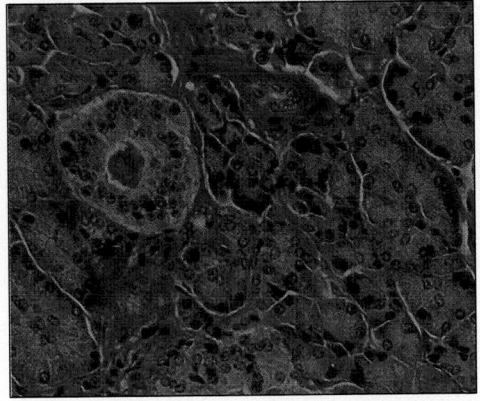

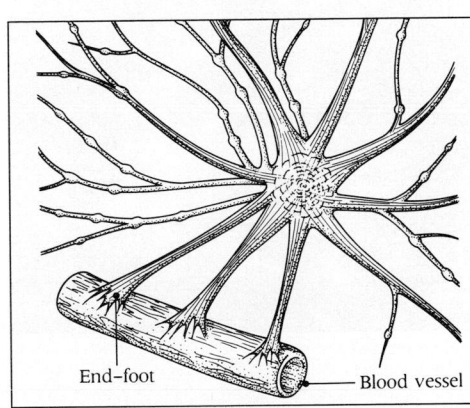

End-foot — Blood vessel

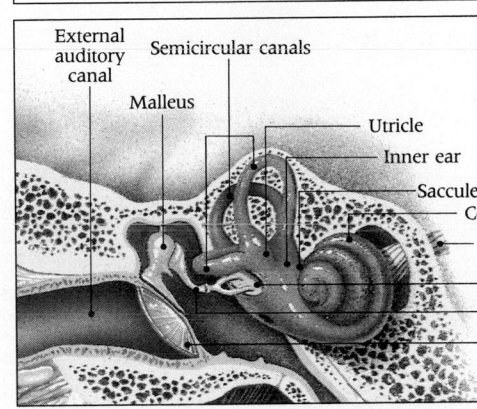

External auditory canal
Semicircular canals
Malleus
Utricle
Inner ear
Saccule
C

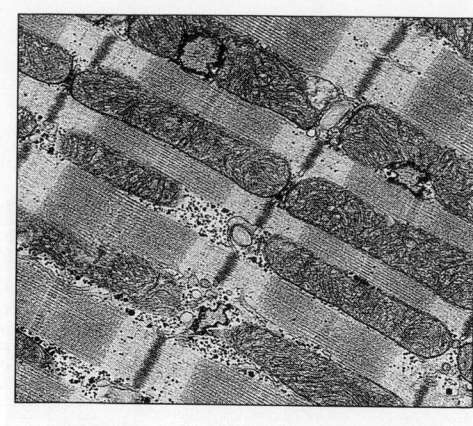

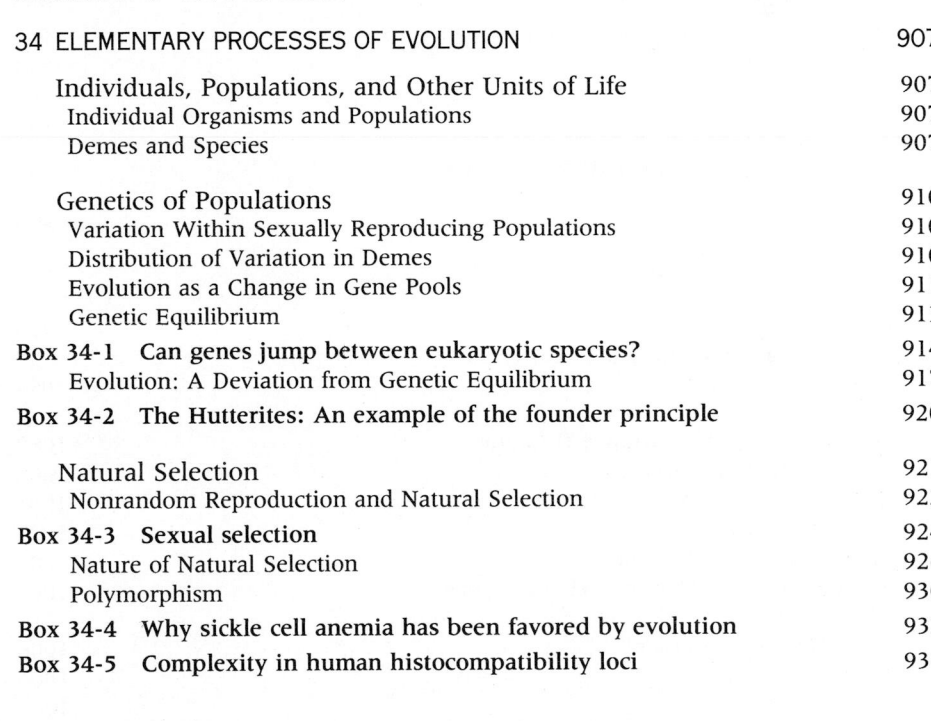

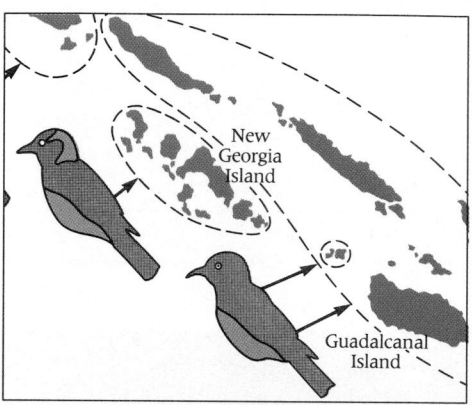

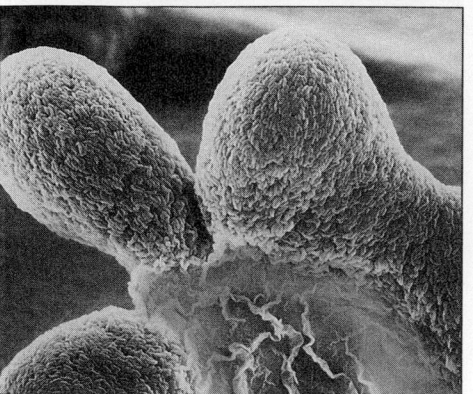

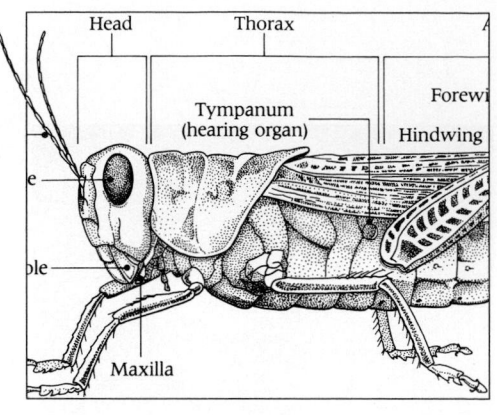

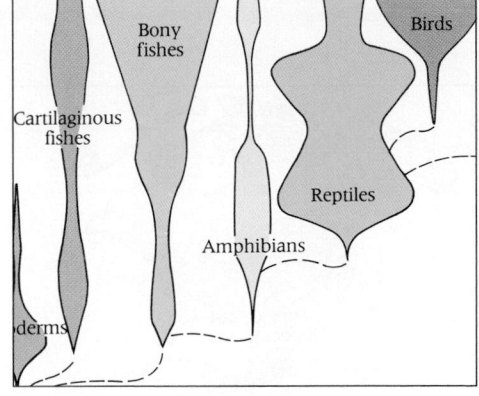

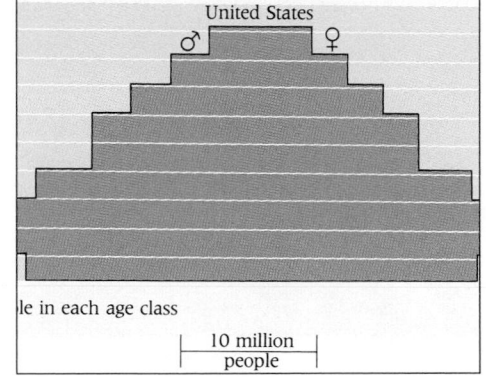

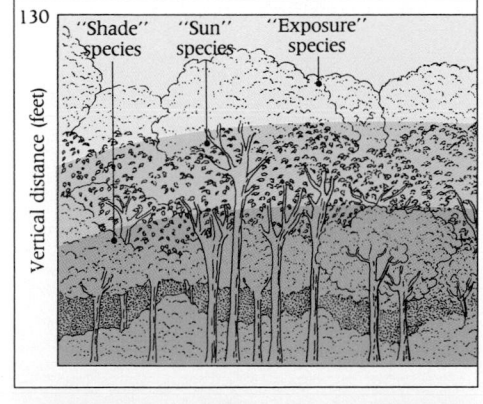

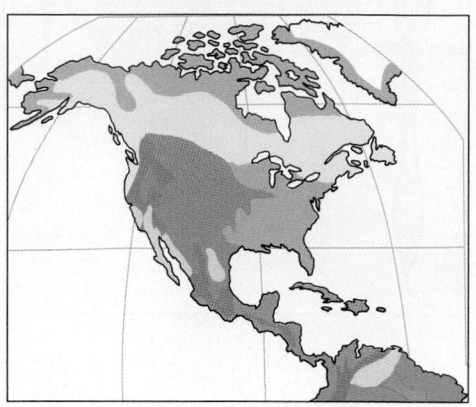

PREFACE

Of the many emotions stirred by the death of George Gaylord Simpson, one of the saddest was our regret that he did not live to see published the overdue third edition of *LIFE: An Introduction to Biology*, a book very close to his heart. He had hoped to see it (so he wrote just days before his death) and, though frustrated at the circumstances delaying it, he worked hard on early drafts, providing new text on certain topics and line-by-line critiques remarkable for their scholarship, elegance, and wit. He knew we treasured these notes of his and, accordingly, they grew longer and more captivating. It seemed at times that this cornucopia could itself make a glorious text book. Some day perhaps.

George Simpson was one of a kind—a seminal thinker in the fields of evolution and biogeography, a tenacious and original scientist of surpassing greatness, a distinguished teacher, and a profound and stylish man of letters. His many books ranged widely into history, philosophy, and themes of human destiny and they significantly influenced thought in this century. We miss him, and in dedicating this volume to his memory, (to paraphrase the words on Newton's tomb) we congratulate ourselves for having known him and learned from him.

The first edition of this influential text was published in 1957. Its authors were Simpson, Pittendrigh, and Tiffany. The second edition, by Simpson and Beck, appeared in 1965. It has taken until now to prepare the third edition. The hiatus was attributable mainly to three factors—George's long illness, the compelling demands of teaching and research, and, finally—a reason of possibly greater interest to our readers—the concerns we felt at times about the qualities an encompassing text should have in an era of rapid scientific progress and increasing diversity in college biology course offerings.

Partly to deepen thinking on these issues and partly to enlist a solid source of experience and wisdom in many areas of biology and its teaching, Simpson and I invited Karel F. Liem to become a coauthor. (Though a signatory to this Preface, Liem did *not* write that sentence—nor did he write this one: I was happy to welcome him as a collaborator.)

Before saying more about the values we have sought in this edition, we would like briefly to articulate the characteristics of contemporary biology as we see them —for it has been one of our aims, the

George Gaylord Simpson (1900–1984)
With bush baby, Galago alleni.

major one really, to offer a text that reflects the state of biology today, that tells it as it is.

Modern biology is an exciting, expanding arena that without question has attracted many of our very best minds. We do not say that lightly. There have been times—the late 1920s, for example—when physics drew the best of us. It was natural that it should have: that was where the action was. By any measure, from the number of papers published in such multifaceted journals as *Nature, Science,* or the *Proceedings of the National Academy of Sciences,* to the power we are acquiring to control and manipulate life, biology today is the queen of the sciences and young people know it.

Despite the mass and diversity of the new information pouring from the world's biological laboratories, there has been a clear trend toward the universalization and simplification of principles. Mechanisms discovered in one organism—for example, the genetic code—have turned out to be universal, or (significantly and interestingly) nearly universal. Newly discovered mechanisms by which hormones influence cells, it appears, resemble mechanisms underlying many disparate phenomena, including the transmission of nerve impulses across synapses and the generation of nerve impulses by light impinging upon the retina. New understandings of the effects of high pressure on chemical reactions have yielded insights on the functioning of deep sea animals that in turn have illuminated their geographical distribution and evolution. To those of us who have traveled this remarkable trail of discovery in recent decades (even as humble luggage carriers), the trend could hardly be more satisfying.

The convergence we speak of has gone so far as to bring together branches of biology long separated in the past. Advances in behavioral biology, for example, have joined it with ecology and evolutionary biology. Some sectors of molecular biology, so long an anathema to evolutionary biologists (including Simpson), have converged upon evolutionary biology itself. We do not argue that all of evolution is to be proximally explained in molecular terms. That would be nonsense. Nonetheless, molecular biologists have at last given evolutionary biologists plausible mechanisms for many aspects of adaptation—and newly discovered jumping genes, introns, and other surprising phenomena promise important insights into such major evolutionary events as the rise of new phyla. Turning the point around, evolution is responsible for all molecular events in living organisms today. We agree with Dobzhansky that nothing in biology makes any sense except in the context of evolution.

The authors of a book like this must make decisions in three daunting areas: (1) the topics to be covered and their sequences of presentation; (2) the proper level of scientific discussion and literary style; and (3) the values to be cultivated and commended to our readers. To plumb the wisdom of colleagues, we submitted what we wrote to reviewers, which included both specialists in various fields and generalists experienced in the teaching of biology. We benefitted from their advice and, with sincere gratitude, record their names below.

As expected, reviewers differed on certain of our decisions. One decision, however, that elicited little disagreement concerned our determination to continue an integrated sequence of presentation. The term "integration" denotes a sequence and philosophy of presentation in which broad topics, such as reproduction, are discussed in one place, with comparative consideration there of reproduction in plants, animals, and the other kingdoms—even though formal discussion of these kingdoms is still to come. The non-integrated approach (some have called it *dis*integrated) discusses plants, animals, and the rest of the kingdoms in discrete blocks—dealing, for example, with reproduction in five different places.

The previous editions of *LIFE* had a major impact on the teaching of biology in this country. This text (later claims to the contrary notwithstanding) pioneered the integrated approach to biology teaching, and received much acclaim for its readability, authority, and fidelity to the ringing opening declaration of the first edition's Preface:

"This book is based on strong convictions. We believe that there is a unified science of life, a general biology that is distinct from a shotgun marriage of botany and zoology, or any others of the special life sciences. We believe that this science has a body of established and working principles."

The third edition of *LIFE* keeps faith. We have chosen not only to continue the policy of integration in this edition, but to proclaim its good sense.

We believe that a strong integrative wind is blowing across the entire biological landscape. Aside from its scientific manifestations, there is an abundance of other evidence. For example, departments of zoology, such as that at the University of California at Berkeley, are being reorganized into departments of integrative biology. The American Society of Zoologists has become the Society of Integrative and Comparative Biology and has undertaken an urgent initiative aimed at promoting integrated biology teaching. Recent workshops of the National Science Foundation, Sigma Xi, and the American Institute of Biological Sciences have all advocated integrated teaching in the strongest terms. A workshop of the National Academy of Sciences will also deal with this issue in 1991. These thoughtful studies support the position that this book has taken since its first edition, and which it confidently takes now.

To be sure, there are arguments against integration. We would be happier if students reading Chapter 21 on reproduction were already versed in taxonomy and other aspects of organismic biology. But books can present only one topic at a time and we feel that a later chapter, say on plants (Chapter 41), is more meaningful to readers already knowledgeable about reproduction. We also hold that such broad matters as reproduction embody important principles that cut across the kingdoms, and that *these* are of transcending importance—both as basic principles and as stimulating evidence of the universality of mechanisms, itself a principle of utmost importance. In an effort to accommodate instructors whose courses are non-integrated, we have given early, brief descriptions of the major kingdoms—and, to continue the example of reproduction, have included some life cycles and other reproductive topics in later chapters on fungi, plants, etc. There are also enough cross-references to make it easy to cover all of the book's topics in any desired sequence and, as noted below, we have provided an alternate table of contents that is organized by kingdom.

It was of interest to us—and surely worth noting here—that the reviewing process propagated its own enigmas and riddles. We were pleased that most reviewers favored an integrated sequence. What we did not expect (but should have, perhaps) was the depth and breadth of their differences in certain other areas. If the truth be told, different reviewers sometimes praised and condemned the same passage, occasionally with amusing obversity. It became clear to us that some of our colleagues—a minority, we believe—with tastes unlike ours would place different weight than we did on certain biological concepts and their importance for beginners in the field.

A common point of contention related to the place of mechanistic arguments in a text like this. Here disagreement was wide and deep. Hardliners chastized us for not providing detailed data and "proof" for virtually every proposition, including many from evolutionary biology. We admire that viewpoint but this is not an encyclopedia or monograph. Others were on the other side, often denouncing us for including too many "details," especially biochemical details. We could not accept that view, not in this exciting moment in the history of biology when such ideas are bringing together disparate fields in new and exciting ways. An amino acid sequence may be a dreary exercise in some eyes, but amino acid sequences determine species, anatomy, and function. They also suggest evolutionary relationships and much else in biology that is profound and important. To back that opinion, we have shown *some* sequences—not all of them—and we have encouraged readers to consider them boldly for what they are—the building blocks of organisms that, like Lego toys, are put together in different ways with extraordinarily different consequences. It goes without saying that readers and teachers, using this text, can make their own decisions about what constitutes details by picking and choosing among the sections and chapters.

Some felt that occasional references to human biology do not belong in a general biology text. For a number of reasons we feel otherwise. Many students taking beginning biology are considering careers in medicine. We want these students—indeed, all students—to know that modern biomedical science rests firmly on a biological base, and that human biology is a part of biology. Since human biology includes considerations of disease as well as normalcy, the text offers discussions of cancer, AIDS, sickle cell anemia, infectious diseases, etc., when such discussion can usefully illuminate biological principles. The human body must not replace plants, animals, and other bedrock topics, but it should discreetly join them. It does so here. Indeed, we are impressed with the profound implications of modern biology for human life in all of its aspects, including not only medicine and pathology but agriculture and ecology. We hope our readers will feel that way too.

A WORD TO STUDENT READERS

To begin the study of biology is to embark on a fantastic voyage. In our long combined teaching experience, we cannot recall a time when our cherished field was more intellectually taut, vibrant, and interconnected. You will encounter important recurring concepts that a little thought will reveal to be quite familiar. For example, specificity in enzymes or receptors has the same basis as specificity in door keys. The information in DNA has many resemblances to the information in computer files. We know we are confronting you with many new facts and possibly repellent terms. Think about them and give them a chance. This is an exciting adventure of the human spirit—and real rewards await the diligent reader. It is also of great practical importance. We hope you will seize the day—and "go for it."

You may not believe it now, but the aim of biological research is clarity and simplification. Over and over again, you will see how clever men and women identified problems and solved them. Remember that almost everything we know in biology we discovered when someone did useful scientific work and reached conclusions that held up. Nothing was learned by witchcraft. To emphasize the point, we rejected the advice of reviewers who preferred "prettier" smoothcurved graphs, insisting instead on showing real data, the nitty-gritty of observation, which is often not very pretty. Pay attention to that. We want you to respect the pioneers who gathered those data, often with great difficulty. If you can, put yourself in their shoes—remember, they were once in your shoes. What would you have done in their place?

We also urge you to attack the reference lists with gusto. Press your library to locate hard-to-find books. If biology genuinely interests you, keep an eye on those superb weeklies *Science* and *Nature,* which are in every library. You will not understand everything in them, but the short reviews of new developments in the front of each issue are always lively and well written.

ACKNOWLEDGMENTS

It would be impossible to list all of the many people who assisted us in this undertaking. Scientific reviewers were legion. As already stated, all were helpful and stimulating. The following list includes most of those who read and criticized the text at various stages of its writing: I. Edward Alcamo, SUNY Farmingdale; Penny Amy, University of Nevada at Las Vegas; Karl Aufderheide, Texas A&M University; Edwin Battley, SUNY Stonybrook; Peter R. Bergethon, Boston University; Jerry J. Brand, University of Texas at Austin; Margaret Gould Burke, University of North Dakota; Ian M. Campbell, University of Pittsburgh; Anthony Chee, Houston Community College; David Gayle Davis, University of Alabama; Brian D. Earle, Cedar Valley College; H. W. Elmore, Marshall University; Alice Weaver Flaherty, Massachusetts Institute of Technology; David Fox, Columbia University; Darrell Galloway, The Ohio State University; Elizabeth Godrick, Boston University; Melvin H. Green, University of California at San Diego; Stephen C. Hedman, University of Minnesota; Paul Hertz, Barnard College; William Higgins, University of Maryland; Valerie M. Kish, Hobart & William Smith College; Robert M. Kitchin, University of Wyoming; Julius Kreier, The Ohio State University; Hallie M. Krider, University of Connecticut; Irene Kuter, Harvard Medical School; Edward McCrady, University of North Carolina; Paul M. Nollen, Western Illinois University; F. Scott Orcutt, Jr., University of Akron; Robert S. Platt, Ohio State University; Kenneth H. Roux, Florida State University; Raymond W. Scheetz, University of Southern Mississippi; Wayne Silver, Wake Forest University; Elizabeth R. Simons, Boston University; Nicholas Sturm, Youngstown State University; Gerard Summers, University of Missouri; Marshall D. Sundberg, Louisiana State University; Daryl Sweeney, University of Illinois at Urbana; Kenneth Wilkins, Baylor University; Charles Wimpee, University of Wisconsin at Milwaukee; Daniel E. Wivagg, Baylor University; and Philip R. Yant, University of Michigan.

In the interest of brevity, we can mention only a few of the others whose help was truly indispensable. Mary Ochs Hanlon did amazing things in preparing the manuscript and keeping order. Our editors at HarperCollins Publishers (especially Arlene Grodkiewicz) understood what we were about, believed in it, and earned our gratitude. We particularly appreciated the efforts of Michael Mendelsohn, a talented designer who by this time is surely a capable biologist.

Most of all, we want to thank some of our young academic colleagues who, while helping us in our labors, made them infinitely more enjoyable—especially the many undergraduate students who read and criticized chapters, and two present graduate students (and future stars): George Q. Daley, who wrote the essay questions; and Alice Weaver Flaherty, who critically reviewed the manuscript and wrote the summaries, glossary, and certain sections (tautology, memory, etc.). Her scientific breadth and editorial skill were immeasurably helpful to this book and its authors. What a pity that George and Alice never knew George Gaylord Simpson. We also thank Anne Roe Simpson for her enduring support. Friends, we salute you all. Finally, to Hanne T. Beck and Hetty K. Liem, so long our graceful and perdurable champions, go thanks beyond telling.

Karel, let's go get a beer.

WILLIAM S. BECK
KAREL F. LIEM

Here, in brief, are some of the book's concepts and features:

APPROACH

We have chosen again to hold with the central evolutionary theme of the previous editions, to tell an unfolding story that ascends the organizational hierarchy from atoms to human populations, and to try our best to convey the heady atmosphere of biology today—an uproarious (and, yes, evolving) discipline, in which each week brings both new answers and incisive new questions.

TEXT

The text is up-to-date and highly readable. Mechanisms are explained in a manner that makes them not only clear, but interesting and often exciting. We want readers to learn certain facts but we also want them to know *how* we know those facts—and in places we develop the historical background of a concept, in effect inviting the reader to decide what he or she would have done in the investigator's place—and where it is all going. We consider this an important part of the educational penumbra of a big text like this one.

ART PROGRAM

The book is richly endowed with new illustrations and photographs. Text and art are closely coordinated. The book is ornamented with paintings in part- and section-dividers that we find enchanting.

END-OF-CHAPTER ITEMS

Chapters end with carefully written, coherent summaries; clusters of quiz questions and essay questions; lists of key terms (boldfaced in the text and mostly defined in a notably full glossary at the end of the book); and a list of references, each annotated with critical comments. References include a thoughtfully selected mix of reviews of the *Scientific American* and *BioScience* type, popular and scholarly books and monographs, a scattering of papers of historical importance, and a few original papers reporting major discoveries. As noted below, we want students to dare these readings and we have done everything we could think of to make them appear as exciting and important as they truly are.

BOXED ESSAYS

One or more boxes in most chapters deal with interesting sidelights. Some delve more deeply into scientific issues. Others are—well—fun, and we enjoyed preparing them.

APPENDIX

The detailed classification of living organisms appearing in previous editions proved popular and was widely imitated. It is here updated. As far as we know, the present version is more detailed and interesting than any available in other texts and includes most of the many organisms mentioned in the text.

ALTERNATE CONTENTS

In the *Instructor's Manual,* we offer an alternate table of contents to readers and instructors who might prefer to consider topics in a nonintegrated sequence according to taxonomic groups. In this sequence, general (i.e., nonkingdom-oriented) topics, such as evolution and ecology, are listed under the same part and chapter headings as in the Detailed Contents on page VII. Kingdom-oriented topics originally appearing under these general headings have been transferred to new positions under appropriate kingdom headings.

ANCILLARY MATERIALS

A number of helpful aids have been developed in coordination with the text.

Instructor's Manual Newly written for this edition by Jay Templin of Widener University and Kenneth Wilkins of Baylor University, the manual includes outlines for each chapter as well as helpful teaching tips, list of resources, topics for discussions.

Testbank A testbank of 2500 questions and answers compiled by Jay Templin is available both in hard copy and on *Testmaster*—a computer program for IBM and MacIntosh systems.

Student Study Guide Also completely new for the third edition, a helpful study guide for students was written by Kenneth Wilkins of Baylor University. It includes not only objectives and outlines for each chapter, but also review summaries of the key concepts and self-tests to help students prepare for examinations.

HarperCollins Biology Laboratory Manual A completely new laboratory manual has been written by William Tietjen, of Bellarmine College. This lab text can be used independently and in various lab structures. All labs have been carefully chosen and class tested. Each lab begins with an introduction to the lab topic. Throughout each exercise, students will find cautions that alert them to problems or items of particular importance. In addition, boxes that demonstrate mathematical or statistical material are included when needed. Well-developed one- and four-color art included in each exercise helps clarify what the experiment is about.

Laboratory Manual Instructor's Guide This preparation guide includes hints for preparing labs, methods of building, equipment, learning objectives, projected times for each exercise, sample problems, help on caring for organisms, supply sources, and transparency masters.

Overhead Transparencies One hundred full-color transparencies of art from the text are available free to adopters. All illustrate key biological concepts discussed in the text.

HarperCollins Biology Encyclopedia The Biology Encyclopedia Laser Disk, produced in conjunction with Nebraska Interactive Video, Inc., offers the latest in visual technology. It contains transparencies, micrographs, slides, and film and video footage. Over 1500 images were provided by Carolina Biological Supply. The laser disk allows instant access to any image or footage, frame by frame or moving, simply by pushing a few buttons on a hand-held remote. The disk enhances the principles of biology covered in the text much more effectively than transparencies or videos.

ABOUT THE AUTHORS

LIFE: An Introduction to Biology, Third Edition, brings together two generations of teaching, research, and writing experience.

William S. Beck is Professor of Medicine at Harvard Medical School, Tutor in Biochemical Sciences at Harvard College, and Professor of Health Sciences & Technology at Massachusetts Institute of Technology. He is also Director of Hematology Research at the Massachusetts General Hospital. Dr. Beck received his M.D. from the University of Michigan and has divided his interests among medicine, biochemistry, and general biology. His research has centered on biochemical aspects of cell division and DNA synthesis, and the metabolism of vitamin B_{12}. A popular teacher and a widely published author of scientific articles and books (including *Modern Science and the Nature of Life*, *Human Design*, and *Hematology*), Professor Beck collaborated with George Gaylord Simpson on the second edition of *LIFE: An Introduction to Biology*.

Karel F. Liem is Henry Bryant Bigelow Professor of Ichthyology and Professor of Biology at Harvard University, where he is also Curator of Ichthyology. Dr. Liem, who received his Ph.D. from the University of Illinois, Urbana, received the 1987 Phi Beta Kappa Teaching Award, Harvard's most distinguished award in teaching. He is President of the American Society of Zoologists and Executive Editor of *The Journal of Experimental Zoology*. His research interests are evolutionary and functional morphology, systemic biology, respiratory physiology of lower vertebrates, and ecology.

George Gaylord Simpson (1902–1984) is remembered for his lasting contributions to the founding and further articulation of the modern evolutionary synthesis. Much of what is known and of what we understand of life's history and evolution has its original with Dr. Simpson. He served as Curator of Vertebrate Paleontology and Chairman of that Department at the American Museum of Natural History, and was Professor of Zoology at Columbia University. He later accepted an Alexander Agassiz Professorship from the Museum of Compatative Zoology at Harvard University, where he was also a professor of geology and biology. Dr. Simpson was the recipient of the AIBS Distinguished Service Award in 1978 and many other awards. His last teaching position was at the University of Arizona. A founder of the Society of Vertebrate Paleontology and the Society for the Study of Evolution, Dr. Simpson was the author of numerous books, including *The Meaning of Evolution* (1949), *Principles of Animal Taxonomy* (1961), and *Discoverers of the Lost World* (1984).

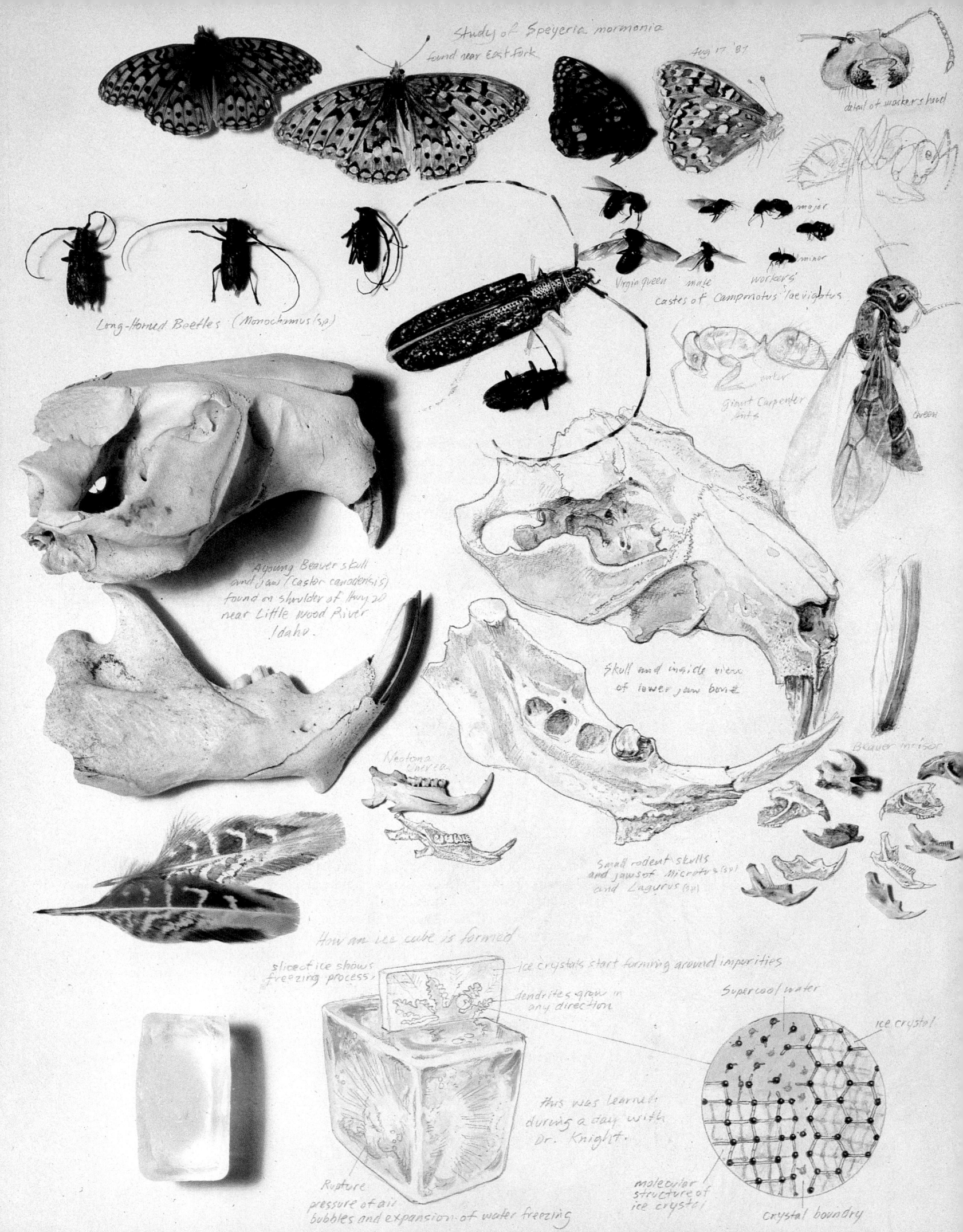

Study of *Speyeria mormonia*
found near East Fork
Aug 17 '87
detail of worker's head

Long-Horned Beetles (*Monochamus* sp)

Virgin queen   male   workers
castes of *Campanotus laevigatus*

Giant Carpenter Ants

worker

Queen

A young Beaver skull
and jaw (*Castor canadensis*)
found on shoulder of Hwy 20
near Little Wood River
Idaho.

Skull and inside view
of lower jaw bone

Beaver incisor

Neotoma cinerea

Small rodent skulls
and jaws of *Microtus* (sp)
and *Lagurus* (sp)

How an ice cube is formed

slice of ice shows
freezing process

Ice crystals start forming around impurities

dendrites grow in
any direction

Supercool water

ice crystal

this was learned
during a day with
Dr. Knight.

Rupture
pressure of air
bubbles and expansion of water freezing

molecular
structure of
ice crystal

crystal boundary

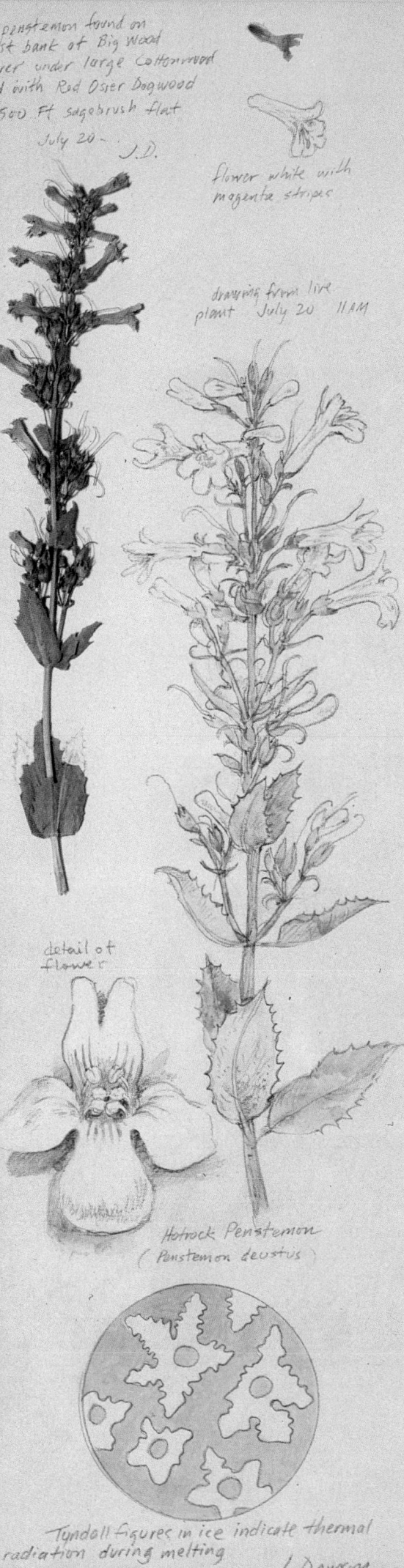

Penstemon found on
east bank of Big Wood
River under large Cottonwood
and with Red Osier Dogwood
5,500 Ft sagebrush flat
July 20 —
J.D.

flower white with
magenta stripes

drawing from live
plant July 20 11AM

detail of
flower

Hotrock Penstemon
(Penstemon deustus)

Tyndall figures in ice indicate thermal
radiation during melting
J. Dawson

PART 1

THE SCIENCE OF BIOLOGY

Modern biology, the science of life, has many facets and compartments. Though this book will speak of the details that distinguish species, a constant theme will be the universal principles underlying all of life. We introduce them in Part 1. They are illustrated by the differences between living and nonliving things. A spider web, for example, is not alive, yet it reflects the complexity and organization of the spider that made it. It is a product of its maker's behavior. It expresses the precision of the spider's adaptation to its environment. The spider can reproduce, and hence maintain a population of spiders throughout the ages, a population that endures although the individuals in it perish. The web is finally but preeminently an outcome of evolution which started long before there were any spiders.

These are the major themes of life and of this book, and we examine them in Chapter 1 as they are seen in a forest, a coral reef, and a polar icecap. We shall then organize the rest of the book in terms both of these main themes and of the levels of organization in nature, from molecules through cells, whole organisms, populations of single species, and communities of many species.

Chapter 2 considers the nature of science itself as a way of understanding and coping with the world in which we live. We shall briefly trace the expanding scope of science to its culmination when Darwin at last brought life into its dominion. Since the publication of The Origin of Species *in 1859, biology entered the mainstream of scientific thought, which until then had been shaped mainly by physics. Darwin's concepts of evolution and natural selection are the principles that bring order into the diverse facts of life.*

THE WORLD OF LIFE

The world of life
Biology, the science of life, concerns the immense diversity of organisms in the modern world and their ancestors in the distant past. Their elegance, beauty, and accessibility to diligent inquiry are constant sources of wonder and enjoyment.

Wherever we look, we view the world of life in one or another of its manifestations. In meadows and forests are familiar flowers, trees, birds, and insects—and less familiar soil bacteria, fungi, and worms. In pond water are microscopic organisms—one-celled creatures that without lungs take up oxygen and without stomachs digest food. In offshore shallows and dark ocean depths are strange worlds of fishes, corals, sea urchins, clams, and seaweeds. On farms and on ice floes, in zoos and in jungles are uncounted varieties of animals. We also see other people. And we see ourselves.

"KNOW THYSELF"

We are alive and we are related in complex and important ways to the rest of the living world. Our own lives are wholly dependent on other living things. We eat other organisms because our bodies require proteins and other materials that are manufactured only by living organisms. Most of us still shelter ourselves under timbers from trees and warm ourselves with fibers of cotton and wool. We live in communities of human beings and other living things that influence our climate, health, and well-being in ways we are only beginning to grasp. If merely to understand and control our own lives, we must ponder the nature of life—in other species and in ourselves.

Our connection with other living things is more than a simple relationship of neighbors in a community; we are related to other organisms as truly as we are related to our own sisters and brothers. Indeed, we share ancestors with every other animal and plant on Earth. We are, all of us, products of the same long and intricate history.

Without question, the best reason for studying **biology** is the admonition inscribed on the ancient Temple of Apollo at Delphi: "Know thyself." To know ourselves *well*, especially in this brilliant season of advance in the science of biology, we must examine all of life and life itself. One aim of biology—surely an aim of every branch of learning—is human betterment. In the years since this book was last revised, the urgency of that pursuit has grown.

As human beings have encroached further on other species of the Earth, thoughtful people everywhere have sensed with concern an intensifying human predicament, one that is deeply rooted in the issues of biology. It is the dilemma of an expanding human population in a world of limited resources, and it has created a surge of threats to the survival and well-being of all living creatures, ourselves included. Some are ancient perils in modern dress—war with an unprecedented potential for harm, pollution by newly concocted contaminants, serious diseases heretofore unheard of.

Against this dismal outlook is the stirring possibility that human knowledge, energy, and wisdom will prove equal to these looming challenges. We think that in very meaningful ways an exploration of biology can enhance that possibility—

1-1 The planet Earth
The Earth looked relatively small and isolated to the astronauts of the Apollo spacecraft during the first lunar mission.

and in surveying the field in these pages, we will attempt to exhibit its interconnectedness, vigor, and high promise.

A SMALL PLANET

An extraordinary event took place in the late 1960s. For the first time, the whole Earth was portrayed in a single photograph (Figure 1-1). This photograph, taken from lunar orbit during the Apollo 8 expedition, was stunning in its impact, because it demonstrated that we earthbound creatures are passengers on a small spaceship of limited size.

From their vantage point 200,000 miles away, the Apollo explorers could easily distinguish the regions of the Earth's surface. Continental landmasses were splotched with color—browns and reds where lands were dry, greens in forested regions, and whites where lands lay under snow. Ochre landmasses and blue seas were streaked with expansive white clouds floating in a shallow ocean of air.

The photograph spoke eloquently of Earth's relative insignificance in the vastness of space. Earth is but one of nine planets that orbit the sun along with myriad smaller bodies—asteroids, comets, moons, and meteors. Our sun is but one of the 100 billion stars making up the galaxy called the Milky Way. At least 10 billion other galaxies resembling our own are within range of our telescopes. At its closest point, our nearest neighbor, Venus, is 25 million miles from Earth. Moving at spaceship speed, travelers would need more than four months to get there. But even if humans were to reach Venus, they would find it a strange and forbidding place, with hot surface temperatures approaching 400°C (750°F), an atmosphere of 94% carbon dioxide, an unbroken cloud cover, and virtually no water or oxygen. To survive, human explorers would have to bring along the atmosphere and resources of Earth in elaborate life-support systems. Other planets of our solar system are even further from Earth and have even more inhospitable environments.

The message is plain: We are restricted to a planet of fixed size. It is our home. And, as Earth is finite, so are its resources—its land, water, and atmosphere—

1-2 (Opposite) The shifting continents of Earth
Approximate positions of the continental land masses at various times in Earth's history, from 540 million years ago to the present.

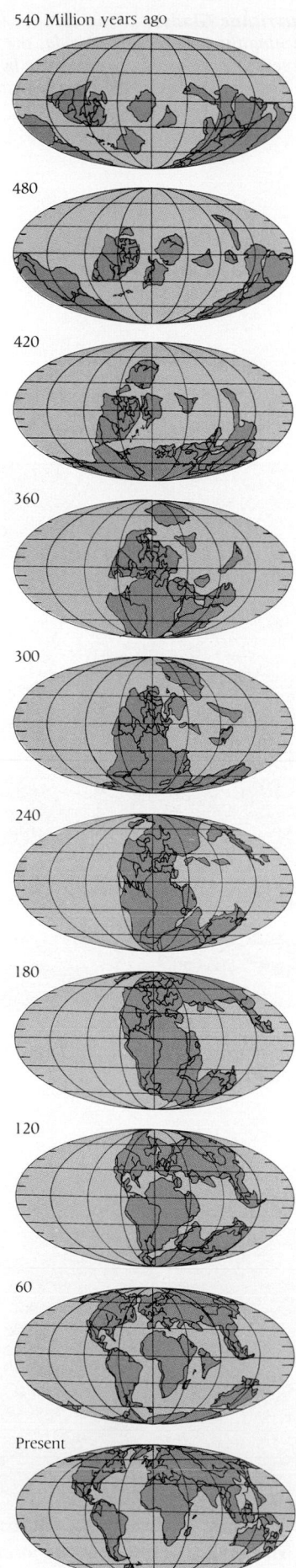

540 Million years ago

480

420

360

300

240

180

120

60

Present

and thus its capacity to support the living world confined to its surface. As with our imaginary spaceship to Venus, its life-support systems are of limited capacity. What we call the **biosphere**—the part of Earth in which life exists—includes only the outer crust of solid ground **(lithosphere)**, the surface waters **(hydrosphere)**, and a small portion of the air **(atmosphere).**

Earth has not always resembled the planet we know today. Indeed, there was a time when it did not exist at all. Though knowledge is still fragmentary, it is currently believed that the whole solar system—the sun, the Earth, the other planets and their moons—formed from nebular matter at about the same time. We can estimate *when* that happened only as an order of magnitude. It was about 4.5 to 5.0 eons ago. (In the dating of geological events, an **eon** is what Americans call a billion and the British call a thousand million years—or in the jargon of geologists, a thousand myrs.)

Earth's dense inner core is still hot and partly molten, but its hard, rocky outer crust formed early in Earth's history. Almost immediately, the crust began to be eroded by strong winds and rain. Sediments accumulated and water gathered in depressions to form rivers, lakes, and oceans. Life eventually appeared upon this evolving planetary surface.

It is intriguing to envision how Earth might have looked to an observer after its first million years. It was a scene of utter desolation: no trees on the land, no fishes in the sea, no living thing in the air, no sounds but those of the waves and the wind. Landmasses punctuated with jagged rocks rose abruptly from the seas. The gray, drab landscape resembled the surface of the moon, for color and diversification did not brighten the scene until microbes, plants, and animals had transformed the world.

The waters of Earth were fresh at first. The oceans became salty as freshwater rivers carrying salts dissolved from the land poured endlessly into the seas. Over the immense stretches of geological time, billions of tons of salt and other minerals were carried down to oceans, which became solutions of chemical compounds that eventually gave rise to the first stirrings of life. There is now good evidence that life existed over three eons ago.

The planet Earth and its atmosphere have an interesting history. The seas have changed constantly, in composition, shape, and position. Great mountain ranges have risen where seas once covered the land, and the ranges in turn have been reduced to sea level by erosion. Landscapes and seascapes, which appear so enduring to mortal eyes, are actually in constant turmoil. But the changes they undergo occur so slowly in terms of a human life they are all but imperceptible.

The science of geology became unsettled in recent years by evidence that the Earth's crust is composed of semirigid segments called plates that move slowly with respect to one another (Figure 1-2)—at the most a few inches a year. New ocean bed material is added to oceanic plates by the upwelling of molten rock from deep sources. To make room for new crust, old material is lost when one plate, sliding under another, finally reaches depths at which it melts. When plates collide, mountain ranges may rise. We believe that the Himalayan Mountains originated from the collision of the plate forming India with the central Asiatic plate (see Box 48-1).

From such movements in the past, continents have formed and split. When their segments drifted apart, separate smaller continents and oceans formed in the gaps between them. Thus, South America and Africa were once part of a supercontinent that split no more than 200 million years ago. As they drifted apart, the South Atlantic Ocean opened between them. Landmasses have changed position not only with respect to one another but also with respect to the North and South poles, and consequently to the various climatic zones. The details of these unimaginably massive changes and their effects on all of Earth's organisms are still being explored. Their study is an exciting area of scientific research.

Despite its isolation in space, the Earth has a critically important relationship with an extraterrestrial body 92.9 million miles away—the sun. The sun emits a constant stream of radiant energy, which we call **solar radiation,** that lights and warms the half of Earth that faces it. At every moment a new crescent of air, accompanying the twirling Earth beneath it, turns into the sunlit hemisphere as another segment rotates into the hemisphere of night. In the sunlit hemisphere, air is charged with the thermal energy that energizes the changing weather. Winds freshen by day. Clouds and winds form above sun-heated lands and oceans, some-

1-3 Hurricane Gladys
Approaching the west coast of Florida, the hurricane was photographed from space by the Apollo 7 astronauts on the morning of October 17, 1968.

times with hurricane intensity (Figure 1-3). All life on Earth is ultimately powered by the sun, and later we shall consider exactly how solar radiation governs the energy cycles of our planet.

Sunlight falls differently in the several latitudes of Earth. The resulting climates set certain broad limits on the possible varieties of terrestrial life (that is, land life as opposed to aquatic life). However, regional climates have not always been the same, and as a result, there have been major changes in the communities occupying each zone throughout life's history.

THREE VIEWS OF LIFE

We begin our study with a panoramic tour of the world of life. This view of life[1] takes in three "realms"—a forest, a reef, and a polar icecap—though we could have chosen forty. Despite their obvious differences, each domain abounds in living things that interact with their environment and struggle for survival in ways that differ only in detail. The point is clear: The basic patterns are everywhere the same.

A FOREST

In the beautiful ponderosa pine forests of the American Southwest (Figure 1-4), many trees are 300 to 400 years old and 150 feet tall. They stand well apart, as if intolerant of one another's shadow. They need full sunlight and, like all plants, special conditions of soil, slope, drainage, temperature, and rainfall. Seeds landing in areas that lack these requirements cannot grow into mature plants.

Plants are everywhere in the forest, yet they are not distributed randomly. Different **species** are found in different locations. Along the streams is one kind of tree community: alders, willows, and narrow-leaf cottonwoods. On exposed rocky slopes is another: piñon pines, tree junipers, and oak scrub. A cactus grows on a sandy, dry, southern slope. The forest floor, which receives only filtered sunlight, is carpeted with plants that create a green and flowery summer flourish, while an iris grows on the marshy bank of a pool. Each plant thrives where its own needs are best met.

On a calm summer day, the plants are motionless, seemingly lacking the animation we normally associate with life. But appearances deceive. Much *is* taking place within the plants. Roots are drawing water and other molecules from the soil. Gases are passing into and out of leaves, which are actively converting solar energy to chemical energy. Utilizing these materials and the energy they hold, leaves are

1-4 A ponderosa pine forest
These stately trees are conifers, the largest order of nonflowering seed plants.

1-5 A wild turkey strutting in the forest
These are called precocial birds because their young are able to walk and feed themselves shortly after hatching.

---

[1] This resonant phrase prompts an admiring salute to our preeminent coauthor, the late George Gaylord Simpson. He took the title of *This View of Life: The World of an Evolutionist*, a collection of his engaging essays, from these stirring words of Charles Darwin in *The Origin of Species:* "There is grandeur in this view of life, . . . having originally breathed into a few forms or into one; . . . whilst this planet has gone cycling on according to the fixed law of gravity, from so simple a beginning endless forms most beautiful and most wonderful have been, and are being, evolved."

fabricating foodstuffs that represent the starting point for every vital process of the forest and its residents.

This forest's inhabitants are numerous. Closer inspection reveals many barely visible varieties of plants and animals. Hundreds of insect species feed on everything from the nectar of flowers to the bodies of other insects. The soil swarms with mites, worms, bacteria, and molds, and the air is a conclave of spores, bacteria, and pollens.

There are **herbivorous animals** (animals that feed on plants). Squirrels tear up pine cones to get at their nutritious seeds, deer crop off the succulent lower leaves of trees, and wild turkeys (Figure 1-5) wander among the oaks and piñon pines, gorging themselves on acorns and pine seeds. Herbivores, however, are preyed upon by **carnivorous animals** (animals that eat other animals). Deer are killed and eaten by mountain lions; bobcats stalk the turkeys. Rabbits fall prey to hawks, coyotes, weasels, and other hungry predators.

Such a sequence, in which some organisms prey upon other organisms and are preyed upon by still other organisms, constitutes a **food chain.** The organisms comprising each link of the chain have the same objectives—the gathering of materials and energy. Although the ways in which these goals are met by plants, animals, and microorganisms are superficially different, they have basic features in common.

A CORAL REEF

Among the oldest and richest animal communities on Earth, coral reefs occur only in the clear well-lit shallows of tropical and subtropical seas (Figure 1-6A). Corals, among the major reef-building animals, demand certain conditions. These creatures,

1-6 A coral reef
(A) Corals create an underwater Eden. Its walls and intricate passages are populated with brilliantly colored fishes. (B) Rapidly multiplying colonies of finger coral, Porites divaricata, *construct huge calcareous structures. (C) Stony coral,* Pocillopora verrucosa, *spreads its tentacles to entrap plankton. It lives in mutually beneficial relationships with yellow-green algae, zooxanthellae. (D) Anthozoans belong to the phylum Coelenterata. Here anthozoans called sea anemones,* Actinaria, *trap a passing fish.*

of which there are hundreds of species, form the backbone of the reef. They live in all oceans, but they build reefs only when the water is warmer than 20°C (68°F).

Corals are sedentary relatives of the jellyfishes (phylum Coelenterata). They may look at first glance like dead stone, but they are living, polyp-like animals with hard external skeletons made of calcium carbonate (Figure 1-6B). Corals eat tiny organisms called algae, which enter their mouths after passing through the sieve of their flowerlike tentacles (Figure 1-6C). Some can even capture living fishes (Figure 1-6D). A reef is an enormous colony of interconnected individual corals. When a coral organism dies, its stony skeleton becomes part of an accumulating reef that in time can become an immense sculptured rampart.

The most majestic of all coral displays is the Great Barrier Reef, which runs roughly north and south for 1,250 miles off the northeastern coast of Australia (Figure 1-7), covering an area of 80,000 square miles. Much of the reef—perhaps 90% of it—lies below low tide, but in places platformlike patches of coral limestone rise above the surface and form seawalls against the full force of the Pacific Ocean. On some of these platforms, barely awash at low tide, winds and waves have piled enough coral and sand to form a coral island that is covered with palms and other vegetation. Seabirds roost or burrow there, and turtles come ashore to lay their eggs.

The reefs around the island create awesome undersea landscapes. In water of incredible clarity, brilliantly colored and bizarre living forms abound. Curiously, much of the color belongs to the hordes of algae and other microscopic organisms that live *in* the coral's bodies. Their role in the life of the corals is not yet fully understood, but we do know that many corals derive much of their energy from the sunlight algae capture and convert to chemical energy.

The diversity of living forms in a coral reef is astonishing. Embedded in clumps of coral are giant clams and tubes made by worms—spectacular creatures with colored tentacles. Soft corals (those without skeletons), sea anemones, and seaweeds find footing on the stony reef. Starfishes, sea cucumbers, and other animals lie on channel floors, and striped and spotted fishes circle and dart among the branches of the coral forests (Figure 1-8)—among them, batfish, squirrelfish, blue-spotted trout, and harlequin tuskfish. The waters are literally swarming with microscopic plants and animals.

A hole drilled into a coral reef reveals hundreds of feet of dead rock that was deposited by once-living corals and other reef organisms. A boring to a depth of 506 feet in the Great Barrier Reef disclosed skeletons of corals that lived thousands of years ago. These were identical with those living today. However, a boring on the Pacific island of Bikini brought up, from a depth of 2556 feet, skeletons laid down 20 to 25 million years ago by coral species quite different from those now growing on the reef. Reef corals of the type now existing cannot grow at depths greater than 250 feet. The presence of coral skeletons at depths where today's corals cannot live, the persistence of a given species for thousands of years, and the changing of a species over millions of years are three clues to the fact that life on Earth has undergone evolutionary change. We will return later to this great biological theme.

A POLAR ICECAP

On ancient maps, the vast landmass of Antarctica (Figure 1-9) was always labeled *terra incognita* (unknown land). Lying astride the South Pole, it is still the least-explored region of the Earth. Antarctica is a roughly circular group of land masses comprising 6 million square miles that is completely surrounded by the Antarctic Ocean, the merged southernmost parts of the Atlantic, Pacific, and Indian oceans.

The frozen polar continent consists essentially of an icecap atop a bedrock base (Figure 1-10), which was not adequately mapped until the advent of radar methods. Its coasts are shielded from human intruders by belts of pack ice, sometimes hundreds of miles wide, and hanging ice shelves. Island groups lie off the main landmass. Peaks in West and East Antarctica reach high above the sea—about 7,200 feet in the west and 14,500 feet in the east. The average thickness of the ice sheet is about 7,000 feet. Over 70% of Earth's fresh water is frozen into this ice sheet.

Beyond any question, Antarctica has the harshest climate on Earth. Its winter temperatures reach *minus* 57 to 63°C (−70 to 80°F)—the world record is −89.6°C (−130°F)—and severe winds blow incessantly toward the sea from the central

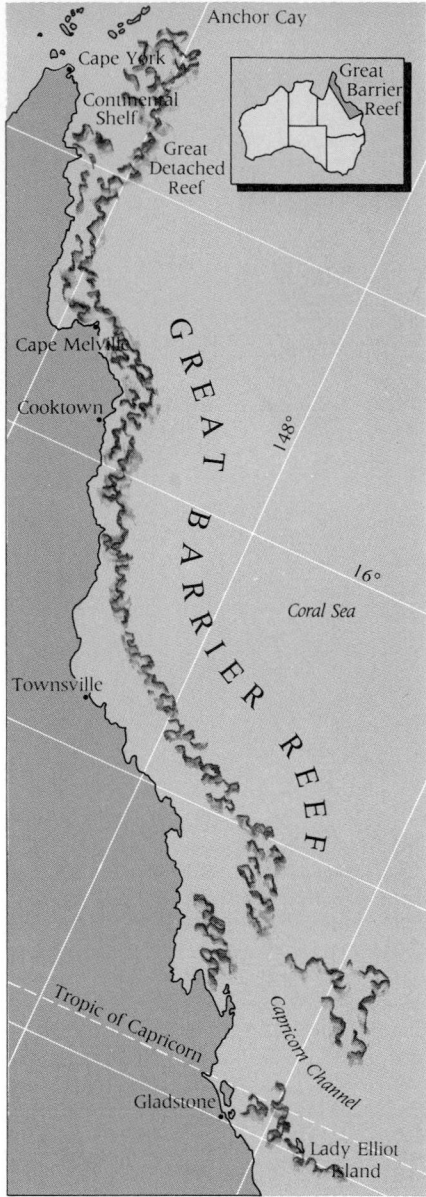

1-7 The Great Barrier Reef
The world's greatest display of coral is the Great Barrier Reef, which lies from 15 to 100 miles off the northeastern coast of Australia, running nearly 1250 miles in a roughly north-south direction.

1-8 Striped sweet-lips in the Coral Sea
These colorful coral reef fishes, Plectorynchus, are about 18 inches long.

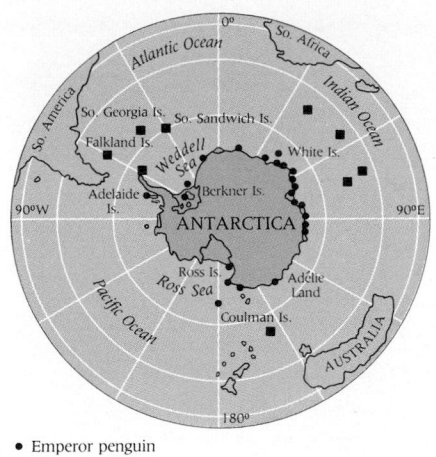

1-9 Antarctica

Sketch map of the Southern Hemisphere. Some of the islands where penguins breed are shown.

- Emperor penguin
- King penguin

1-10 (Right) Structure of the Antarctic icecap

Map of Antarctica, based on data obtained in the International Geophysical Year and the International Antarctic Glaciological Project. The graph below is a profile of the ice and bedrock made along the irregular line identified on the map by the letters A through M.

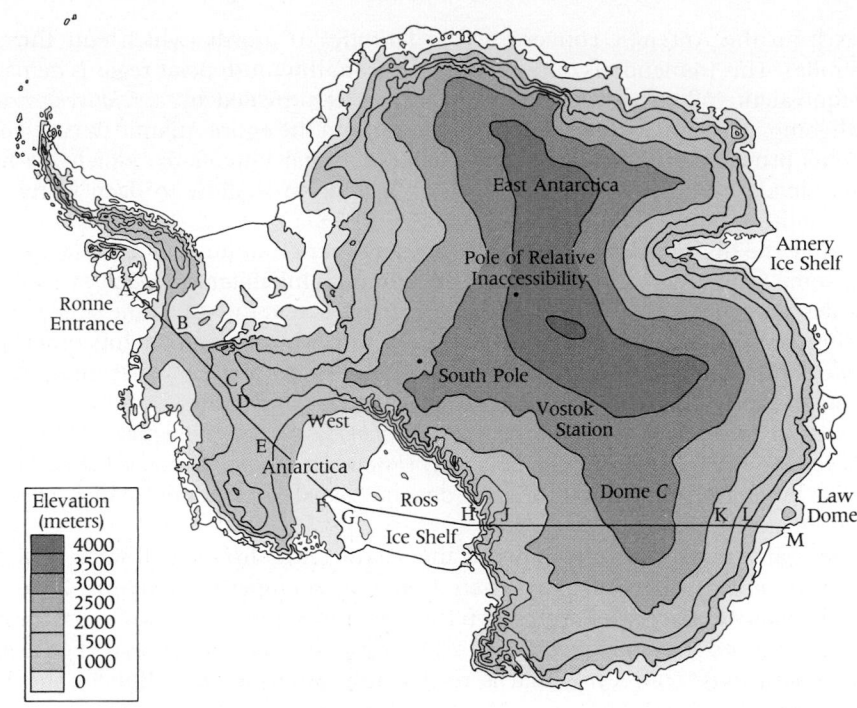

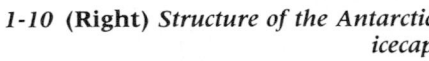

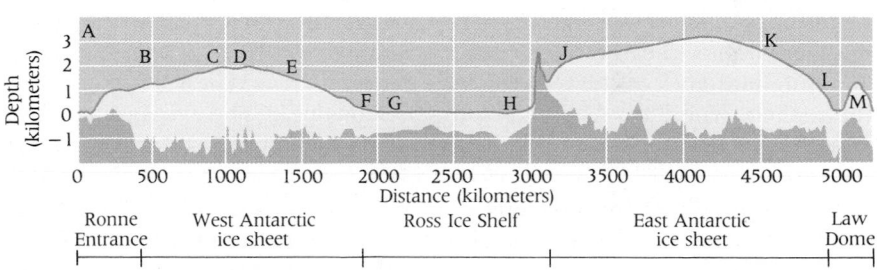

plateaus. One effect of the winds is to drive icebergs and melting shore ice out to sea. Each year, Antarctica discharges more than a trillion (10^{12}) tons of ice into the surrounding seas.

In the great expanse of encircling ocean, the waters are exceptionally rich in nitrates, phosphates, and other chemicals. These nutrients support the huge quantities of plankton living in the sea. (Plankton includes many species of passively floating or weakly motile algae—phytoplankton—and animals—zooplankton. Like other algae and plants, phytoplankton derive their energy from the sun.) The currents of this unusual sea are largely responsible for its abundant life (Figure 1-11). The sea's surface layer forms in spring and summer at the edge of the melting ice, which comes from winter pack ice, from icebergs breaking away from the coastal ice shelf, and from the Antarctic icecap itself. The water is, of course, very cold, but it does not sink because it comes from freshwater ice and thus is not as salty and dense as ordinary seawater.

In a zone called the Antarctic convergence (see Figure 1-11), surface waters flowing northward meet warmer southward-flowing waters. Surface waters sink, forming the Antarctic intermediate current, which has a higher salt content. It takes about seven years for the Antarctic intermediate current to travel the 4700

1-11 Patterns of water currents in the Southern Hemisphere

A profile of the Antarctic Ocean showing the upwelling of the warm, deep, nutrient-rich currents that come from the north. This water replaces cold, salty waters, which slide to the bottom, and surface waters, which are continuously diluted by fresh water from melting ice.

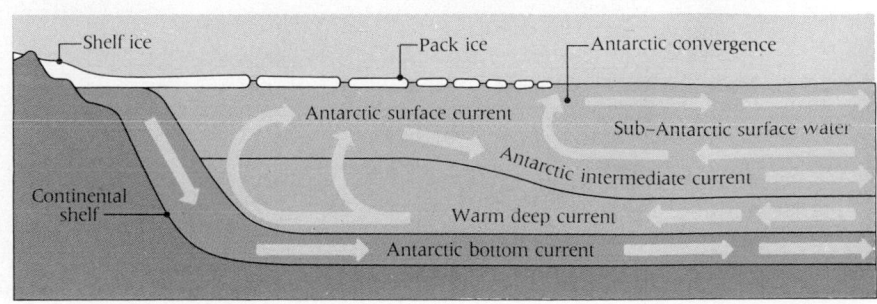

miles from the Antarctic convergence to latitude 20° north (which cuts through Australia). This tremendous outflow of water from the south polar regions demands an equivalent inflow to replace the loss. This is provided by a *relatively* warm, southgoing, deep current that originates throughout the entire Atlantic Ocean. These currents promote the plankton's remarkable seasonal migrations, which are both longitudinal (from place to place) and vertical (from shallow to deep water and back again).

Unlike the northern *arctic* regions, which have varied summer and winter vegetation, some year-round animal life, and even human inhabitants, *antarctic* vegetation includes only mosses and lichens. Some microbes live *in* the ice. Some single-celled organisms even live inside rocks. There are now no human inhabitants other than those manning military and scientific outposts. There is little year-round animal life on the icecap except for a few curious worms and insects—and a surprising number of birds. The emperor penguin (*Aptenodytes forsteri*; Figure 1-12) is the only year-round resident. Adélie penguins (*Pygoscelis adeliae*; Figure 1-13) and other smaller birds are summer visitors. All depend heavily on the sea, which contains an abundance of life.

The many varieties of penguins in the world range from small fairy penguins (*Eudyptula minor*) that weigh less than 2 pounds to emperors weighing 40 to 100 pounds. Many extinct types are known from their fossilized remains. One of these—*Pachydyptes ponderus*—weighed about 220 pounds! Of the species now living, each has a characteristic life cycle that is related to seasonal cycles, climate, and land and oceanic environments.

One of the true inhabitants of Antarctica, the Adélie penguin spends the winter along the edge of the pack ice and in nearby waters. In spring, which in the Southern Hemisphere begins in early October, they migrate to shore. With unerring accuracy each young bird goes to the rookery where it was born. When it is older, it goes to the rookery where it was breeding the previous year. The extraordinary ability of penguins to find their way home, often from far out at sea, has not yet been explained.

Couples that mated the previous year arrive separately—they did not spend the winter together—and get together again, usually in the same nesting location they had the year before. Ordinarily, the male arrives first. He reestablishes his territory—often after vociferous quarreling and fighting—and then begins nest building. The result is a circular structure about 15 inches in diameter. After he and his mate occupy the new nest, copulation begins in earnest, proceeding vigorously until early November.

Egg laying then commences, eggs arriving in pairs every few days. The female, who has been fasting for a month, then heads for the open sea. The male, who also has been fasting, stays home to incubate the eggs until his mate, now fed, returns in three weeks. By then, the male has been without food for three months and has lost 40% of his body weight. Now he seeks food, but he must take turns with his mate in incubating the eggs. When two hatchlings arrive, they must be kept warm, and both nestle under one parent or the other. In ten days, they are too large to fit under one parent, and observers never fail to smile at the sight of very large babies with their heads or rear ends sticking out from under mother or father. Soon the babies hungrily chase their parents—or any adult, though parents will feed *only* their own chicks.

1-12 (Left) *Emperor penguins*
Emperors, Aptenodytes forsteri, *the largest of all penguins, are twice the size of king penguins and 30 times the size of fairy (blue) penguins.*

1-13 (Below) *Adélie penguins*
Adélies, favorites of many penguin fanciers, were named by the French explorer Dumont d'Urville (1790 – 1842) for his wife.

1-14 (Left) *Adélie penguin rookery on the bare ground of Thorgeson Island*

1-15 (Right) *Oil-stained jackass penguin*
Blackfooted penguins, Spheniscus demersus, *bray like jackasses (hence, their second common name). Once numbering in the millions, only about 60,000 remain. This unfortunate creature will probably drown before it is poisoned by swallowed oil. If it comes ashore and preens, it will ingest even more oil.*

In the larger rookeries (Figure 1-14), penguins eliminate vast quantities of excrement. It has been said that anyone approaching a rookery from afar will always smell it first, hear it next, and see it last. Much excrement is also voided at sea; this is recycled through the marine food chain, first entering the phytoplankton, which utilize both the inorganic nutrients in seawater and the energy of sunlight in photosynthesis, then entering the small, floating or swimming herbivorous animals zooplankton and krill *(Euphausia superba),* which are eaten in turn by various crustaceans and fishes. Penguins may feed directly on krill, but fish is also a part of their diet. So is squid, which live on fish. Some animals prey on penguins—from penguin eggs to old penguins. In Antarctic regions, these predators are mostly skuas (relatives of gulls) and seals. Few of these species are wholly dependent on penguin diets.

In recent years, oil spills have fouled the waters of Antarctica and the landmass itself. The worst years of such pollution were 1945 to 1955, but it still continues, chiefly from supertankers in the dangerous waters off South Africa and South America. Many nearby islands have huge penguin colonies and large numbers of them have been oiled in slicks from spilled or jettisoned oil. Inevitably, an oiled penguin dies (Figure 1-15). In August 1974, the second largest spill on record, off the Strait of Magellan, spewed 16.8 million gallons of crude oil into the sea. Many thousands of penguins died soon thereafter. As far as we know, no human activity has yet caused any penguin species to become extinct.

COMMON ELEMENTS IN THE SKETCHES

Despite their diversity and complexity, forest, reef, and icecap have significant features in common.

- We see in each an extraordinary degree of *organization*. There is order and organization in the natural settings, in the living communities that occupy each setting, and in the individual organisms themselves.
- There is a ceaseless *exchange of materials and energy* in and between each of the organizations.
- The individual organisms in each domain and the communities of organisms both display a capacity for *self-regulation and control*.
- All the organisms *reproduce*.
- All are products of *evolutionary change*.
- There are characteristic *patterns of responsiveness and behavior*.
- Their lives exist within specific *geographical limits*.
- And finally, life and its physical settings—Earth's waters and lands—evolved over *many* years. In a word, they have a long *history*.

Let us briefly consider these shared elements. They are the buttresses of the science of biology and the matters with which this book is concerned.

ORGANIZATION

Each of the panoramas reveals a high level of organization. Earth, an organization of matter in space, is home for the populations of organisms that encompass the terrestrial world of life. (Here, terrestrial clearly means "living on Earth" rather than merely "living on land.") Forests, coral reefs, and polar icecaps, however desolate and remote, are organizations of living organisms and aggregates of the matter and processes we call the **environment.**

Furthermore, each living being is itself a highly organized entity. An organism may be a single bacterium or alga. Or, like a coral, pine tree, or human being, it may consist of a great many cells. Each is an organization of materials and functions dedicated to its own survival.

A feature of all living things—in a sense, the essence of organism—is a pattern of integration with multiple levels of organization. The cell is an organization that exists at a certain level of complexity, and we generally assert that the cell is the lowest level of organization at which life can unambiguously exist. If a single cell were always an independent organism, it would live in a community of other cells, joining them in certain activities, competing with them in others, and dying or dividing to form new offspring. Yet a cell may be part of a higher organization, a tissue. An organ, such as a brain or a leaf, is a whole made up of many tissues. This structure is at a more complex level of organization, existing and interacting in a community of other organs. In the same way, a whole animal or plant is at a still higher level of organization. Animals and plants interact with other animals and plants, not with individual organs or cells.

We may also descend the ladder—from the cell to structures such as the nucleus, particles within the nucleus, and particles within those particles. Eventually, we reach the level of molecules and atoms.

This organizational hierarchy suggests a useful plan for this book, and except for certain necessary digressions, we will orient our inquiry in this way. Thus, we will be climbing the organizational ladder outlined in Table 1-1. (The table, incidentally, mentions some of the sciences that deal with one or another of the levels. Clearly, biologists are working at virtually every rung of the ladder.)

TABLE 1-1
LEVELS OF BIOLOGICAL ORGANIZATION

Level of Organization	Related Fields of Science
Subatomic particles	Physics, chemistry, biophysics, radiobiology, radiology
Atoms	Physics, chemistry, biophysics, biochemistry
Molecules	Physics, chemistry, biophysics, biochemistry
Macromolecules	Physics, chemistry, biophysics, biochemistry, molecular biology
Cell organelles	Biochemistry, cytology
Cells	Biochemistry, cytology, microbiology, microbial genetics, molecular biology
Tissues	Histology, physiology, pathology
Organs and systems	Anatomy, physiology
Individual whole organisms	Anatomy, physiology, developmental biology
Populations (reproducing groups of organisms)	Ecology, systematics, behavioral science, genetics, evolutionary biology
Phyletic lineages (evolving populations)	Evolutionary biology, genetics, paleontology
Communities (associations of different populations and lineages)	Ecology, demography
The living world in space and time	Biology, "exobiology"

1-16 Ladybird beetles, **Hippodamia convergens,** *feeding on a leaf*
These colorful insects are common in forests, meadows, and gardens, where they are welcome because they feed on aphids and scale insects.

- Part 1 of the book concerns the science of biology, its history, and some of the issues that concern its students.
- Part 2 deals with the molecular and cellular basis of life. Although the cell is the first level at which life is present in the fullest sense, it can be understood best by looking *within* cells at their component parts.
- Part 3 surveys the functional systems and components of organisms.
- Part 4 explores whole organisms, their interactions in populations, and the interactions of populations. It also views the relations of human populations with other living organisms and with the physical environment—lithosphere, hydrosphere, and atmosphere. This is the level of the ecosystems, those fascinating, self-sustaining natural systems in which living organisms are collaborators and contestants.

EXCHANGE OF MATERIALS AND ENERGY

There is a constant flow of matter on Earth—from soil, water, and air to plant, from plant to animal, from animal to animal (perhaps through a longer or shorter *sequence* of animals), from dead plant and animal to bacterium and fungus, and back again to soil, water, and air. Each step of the process traps or liberates chemical energy. Such exchanges of materials and their content of chemical energy are termed metabolism. In forests and on reefs and icecaps, the exchanges of materials among the plants and animals are easily observed (Figure 1-16).

Energy is the capacity to do work. With minor exceptions, all energy used by organisms (including industrial humankind) comes from the sun—with one exception. The exception is the human use of atomic energy, which is increasing but is still negligible compared with the total use of solar energy by living organisms. In the forest, green leaves capture the radiant energy of sunlight and convert it by photosynthesis into useful molecules that incorporate both matter and energy. The first step synthesizes a sugar. When the sugar later breaks down, it contributes its stored energy to vital processes, and its materials become building blocks. The mountain lion that eats a deer that munched on leaves acquires energy from the sun that would not be available had a plant not captured and transformed it.

Every organism performs work and needs energy. A hawk obviously works when it flies. Less obviously, it also constantly expends energy when it is still. So does a bacterium or tree. Energy is needed not only to achieve organization but to maintain it. Organisms must spend energy or become disorganized, and disorganization leads inevitably to death.

SELF-REGULATION AND CONTROL

Maintenance of order, the opposite of disorder, is possible only by self-regulation and control. Those are capacities of the organismic level of integration, the first to possess machinery sufficient to resist the encroachments of a hostile environment. That machinery, which gives an organism a real if temporary measure of independence, operates in an astonishing number of ways.

- An organism exposed to cold generates heat in metabolism by breaking down glucose and other energy-containing molecules. Depletion of stored fuels induces hunger and stimulates food-seeking behavior and eating, which in turn restocks the fuel supply. The ability of many birds, fishes, and mammals to withstand frigid polar conditions is truly remarkable.
- Invasion by foreign matter is resisted. Bacteria are ingested and destroyed by highly efficient blood-borne cells (themselves unicellular living systems capable of propagation and long survival in laboratory cultures, but also components of larger multicellular organisms), *or* they are inactivated by chemical substances produced by the organism to destroy foreign materials.
- External physical danger is thwarted by voluntary movement like flight *or* by involuntary recoil reflexes *or* by use of a whole repertory of weapons, including electric shock, poisonous chemicals, thorns, teeth, claws, stench, and vocal cacophony.
- Damage to the body of a multicellular organism or to one of its cells is repaired.

In sum, every factor, however subtle, that can disturb the organism, whether unicellular or multicellular, is resisted by systems that are activated by the stimulus. Only in this way can an organism maintain its integrity.

Homeostasis

The precision of life's mechanisms for self-regulation has long impressed biologists. Harvard's illustrious professor of physiology Walter B. Cannon (1871–1945) called life's defenses against disorder the mechanisms of **homeostasis.** The homeostatic state—the state of relative constancy—is something we can observe in living organisms. Cannon realized that its successful preservation in a widely fluctuating environment must mean that certain agencies are operating to maintain it, whether or not we can observe them. Putting it more simply, nothing stays constant unless something works at keeping it that way. In the organism, these agencies are various systems in which a stimulus more or less automatically elicits a reaction that tends to restore equilibrium. One example is the pattern of events occurring when the body needs water: The stimulus (a slight increase in the chemical concentration of blood plasma) elicits thirst, water is drunk, and equilibrium is restored.

One can fit into the framework of homeostasis nearly every function of the living organism. The ingestion of food or the capture of energy from sunlight maintains energy reservoirs needed for other activities. Metabolism converts nutrients into available energy or building blocks. Excretory organs separate the unwanted by-products of metabolism, which they eliminate, from useful materials, which they retain. The status quo, especially of chemical composition, is thereby preserved. In death, homeostatic mechanisms have been overwhelmed.

The concept of homeostasis is a great central idea of biology. It permeates every aspect of biological organization. Homeostasis—an organism's efforts to restore its equilibrium when disturbed—gives it an *appearance* of purposive behavior, since these activities seem directed toward a future goal. It often looks as though it knows what it is doing. However, it does not.

Control Systems

Detailed study of homeostatic systems is the subject matter of such fields of biology as endocrinology, plant and animal physiology, immunology, neurophysiology, and many others. Each deals with a particular system or set of systems of self-regulation. Their common denominator is maintenance of control.

The nature of control processes has been examined thoughtfully in recent years. What interests us here is a universal pattern: An effect acting back upon a causal agency provides information on the consequences of the agency's previous actions and thereby influences its future behavior. Classic examples are the ordinary room thermostat that turns the heat off when the air temperature rises above a preset figure, and the helmsman of a boat who swings the rudder windward when he sees his vessel moving too far leeward. The controlling function of a helmsman consists in holding steady the course by swinging the rudder in a direction that will offset deviations from the preset course. The thermostat does the same, by turning the heat on and off. This is called negative feedback control. Clearly, control processes demand organization.

REPRODUCTION

All living things produce near-copies of themselves (Figure 1-17). The reproductive process is so accurate it is an axiom of biology that like begets like. Reproduction explains how whole communities of organisms can persist virtually unchanged for many thousands of years. We noticed this persistence on the coral reef, where species and communities that began millennia ago resemble those of today. Reproduction is responsible for the reef's very existence. All individual organisms die, but the species is sustained by reproduction.

Paradoxically, reproduction has another role beyond the replacement of lost individuals. We will see later that evolution depends on the occasional production of individuals with potentially useful variations—unusual traits that might better enable the individual to survive and reproduce (be more fit) in a changed environ-

1-17 White-tailed deer
These graceful animals, Odocoileus virginianus, *are much more plentiful in the eastern and central regions of North America than they were in colonial times.*

ment. Although a relatively rare event, the occurrence of variations is an essential part of the mechanism of evolutionary change (a topic that we will explore in Chapter 34).

EVOLUTIONARY CHANGE

Evolutionary change means changes that occur over long periods of time in populations of organisms. Since the time of Charles Darwin (1809–1882), few thinking people have doubted that all the millions of species of plants and animals have arisen from a remote common ancestor by a natural process operating across the eons. The reasons for that conclusion are many and complex. Indeed, it is their multiplicity and complexity that make them so convincing and so riveting. Evolution is the most pervasive theme in the drama of life. Evidence for that grand generalization will gradually unfold as we deal with life's many facets.

Vastness of Evolutionary Time

Several aspects of evolution were depicted in our portraits of the world of life. One is the vastness of time over which evolution has occurred to date. Every part of Earth's surface underwent profound change during this enormous time span, and change is still taking place. As we have seen, Earth's continents still drift about from place to place. Mountains and island chains still rise up and vanish. Oceans still broaden and contract.

Against this immense background of terrestrial and environmental change, living organisms have come, changed, and gone. Since billions of years are quite beyond our imagination, an analogy should help us comprehend how long life has existed on Earth. If Earth's 5-billion-year history is set equal to a single 24-hour day that started at midnight, life did not begin until about 1:30 P.M. Most of today's ocean basins appeared about an hour before midnight—at day's end. Our cave-dwelling ancestors were hunting their prey less than a second before midnight. The past 500 years is the day's last hundredth of a second. A human being's 50-year working life is but a thousandth of a second.

The spans of time over which evolutionary changes take place are so immense they dwarf the time interval available to any human observer. Hence, evolutionary biology confronts its students with an interesting dilemma: Very little of evolution can be *directly* observed in action. Humankind's time on Earth has indeed been small. We will later have the intellectual fun of learning how resourceful investigators bypassed that obstacle.

Adaptation and Diversification

Two important evolutionary principles are adaptation of organisms and diversification among organisms. Every kind of organism is adapted (that is, usefully organized) to survive and reproduce in a particular environment. It may seem trite to say that organisms are well fitted to live where they do live, but that statement has profound implications for evolutionary biologists. To say that a hawk is adapted to feed on small animals means more than saying that hawks can and do feed on them (Figure 1-18). It also implies that innumerable details of the hawk's organization must specifically serve that end. Often the details are far from obvious and require close study to be understood. Such considerations lead to a profound evolutionary question: "How did the hawk become so adapted?" We will answer it in due course.

Our portraits of life also emphasize the diversity of the participants and of their adaptations. Each species in each community has a way of life differing in some respect from that of any of its neighbors, and each is specifically adapted to its own way of life. Diversity is also a product of evolution. Evolution not only adapts each species to its own way of life, but its operations also parcel out the ways of life among the different species.

RESPONSIVENESS AND BEHAVIOR

Responsiveness means the many ways organisms react to changes in their environment. In plants, the response might be the modifications of cell membranes that take place as temperatures become cooler in the fall. Or the response might be the flowering of plants at certain times of the year. Because plant cells are confined by cell walls, plants are not able to move about in their environments, and therefore plants typically respond by changing their development and rates of growth (Figure 1-19). Generally, animals can move, and all have evolved complex patterns of behavior.

To biologists, **behavior** means the ways organisms act. It is not enough to say that an organism takes in matter and generates energy. A complete description of an animal's life would also have to point out that it moves, starts and stops, hunts, communicates, mates, and in one case reads and writes. The forest hawk is adapted to feed on small animals and behaves accordingly, swooping down and making off with them. Similarly, mountain lions attack, fleas hop, caribou migrate, and pigeons home.

It is now recognized that behavioral patterns are explainable only at the level of the whole organism. Consider a house cat. A detailed study of its metabolism, energy exchange, chemical composition, temperature regulation, and heredity would yield deep insights into the cat's inner workings—but it would not permit prediction of a fact known to every child: If you pull a cat's tail, it will scratch you.

One goal of behavioral research is a better understanding of human behavior. Although we should not assume that other animals are simplified humans, we should remember that human beings *are* animals who have evolved from nonhuman ancestors. Notable similarities exist between human beings and other animals. Behavioral biologists, as we shall see, study both the similarities and differences between human beings and other animals (Figure 1-20).

GEOGRAPHY OF LIFE

Our portraits of life imply that organisms vary geographically. No two places on Earth's surface are identical, and geographical differences lead inevitably to biological differences.

In the forest, the plant species growing on sunny hillsides differ from those on the shaded stream bank. Likewise, different species of plants and animals live on different parts of the reef. The differences relate to each species' own needs and to local differences in the environment.

It is significant for the argument that these several environments—forest, reef, and icecap—probably do not have a single species in common, except those introduced by humans. At first glance, this might be ascribed to the obvious differences in the environments. However, a faraway forest—say, in Australia—whose environment exactly matches that of an American forest may not share a single species with its American counterpart. Hence, the geography of life does not depend *only* on its relationships with the environment; it also depends on evolutionary history, especially on those factors that have promoted or impeded the spread of species across the Earth.

1-18 Goshawk with prey
These birds of prey, Accipiter gentilis, *are members of the large order Falconiformes, which includes vultures, eagles, hawks, buzzards, falcons, and kites. The goshawk's prey is a wood duck.*

1-19 Responsiveness
These plants are responding to low levels of light inside the room by modifying their growth pattern and growing toward the brightest available light source.

1-20 Aggression
Aggressive behavior patterns are common to most animals.

HISTORY OF LIFE

Evolution is not only the mechanism by which we explain a general phenomenon—life's variety. It is also a history. It is the actual sequence of events that have occurred in the time since life began.

Students of evolution have succeeded to a remarkable extent in reconstructing the sequence in which the various forms of life appeared and disappeared by studying their fossilized remains and their biochemical makeup. A whole catalog of intriguing questions was opened up by the discovery of this chronicle—questions, for example, on the *rate* of evolution and the *source* of biological novelties in the course of evolution. **Fossils,** which are the preserved remains or traces of early organisms, have provided the most helpful clues to this history. But the fossil record is fragmentary; perhaps only one out of 5,000 or 10,000 formerly existing species are represented by fossils. The origin of major groups of plants is notably unclear. Despite huge gaps in our knowledge, we have learned a good deal of evolutionary history. In this book, we shall probe its major episodes and ponder its meanings.

BIOLOGY AND HUMAN LIFE

Biology, we will find, does not deal exclusively with such classic settings as forests, reefs, and icecaps, where people are scarce and the pulse of human life is remote. It is preeminently the science of all living things, and human beings *are* living things, albeit very special ones. Needless to say, many issues of biology, from agriculture to zoology, are relevant to human life.

Many biological problems are raised by the conditions of human existence. Indeed, the villages, towns, and metropolises in which people live are communities as complex as coral reefs. In 1950, 64.0% of the 151 million Americans lived in urban areas. By 2050, some 82.4% of an estimated population of 309 million will be city dwellers. Clearly, urbanization is a major development in human history. It has important biological implications.

Several characteristics of the city merit comment in the context of this discussion. In various ways, the city parallels communities of living organisms in the wild. Like the forest, the city has different zones with different functions. Unlike forests, cities can be designed deliberately. People generally live in residential areas and travel via highways and other transportation systems to the business areas where they work. Parts of the city are connected by communication networks—telephone lines, radio and television transmissions, and a postal system. Elaborate supply

systems provide the food, water, and energy (electricity, natural gas, heating oil, gasoline) that people need each day. Various devices eliminate wastes produced by machines and individuals, in some cases transporting them to sewage treatment plants, in others dumping them thoughtlessly into rivers and oceans. Gaseous wastes from combustion engines and furnaces are discharged into the air (Figure 1-21); the result is smog and haze (Figure 1-22).

We are keenly aware of these things today. Many cities, grown far beyond their optimum size, have become severely overcrowded. As a result, a deteriorating environment often has too many cars on the highways (for example, in 1986 the streets of Boston were handling 80% more automobile traffic than they were designed for), too much waste in the sewers, too many sick in the hospitals, too much crime in the streets, and too many fumes in the atmosphere. Consequences of crowding can affect every aspect of human life. These considerations alone have stimulated in many an awareness of the nature and importance of relations between living things and their environment, the study of which is called **ecology.**

Complex social behavior is an important, if not unique, feature of human populations. People flock to cities seeking better and cheaper services. High-density living per se does not inevitably cause pollution in cities. That occurs when density increases faster than the maze of elements needed to cope with its dangers. Humanity is *part* of nature, but human society *dominates* much of Earth. The current ecological crisis arises because humankind has allowed itself to expand too rapidly and to degrade the environment. The large city of today is a poignant illustration of that quandary.

1-21 **(Left)** *Pollution of the air by industrial fuel combustion*

1-22 **(Right)** *Smog over Los Angeles*
Smog, a noxious mixture of gaseous, solid, and liquid pollutants, lies motionless over the city, irritating the eyes and throats of long-suffering residents and damaging their trees.

SUMMARY

A study of life begins with a view of the Earth, the resources of which support all living organisms. Within the complex world of living organisms can be found an astonishing variety of living forms and environments, ranging from the abundant life of forests and coral reefs to the austerity of the polar icecaps. Within the communities of organisms and the organisms themselves are hierarchical organizational levels, each accessible to different scientific approaches.

All living organisms share certain other features in addition to their hierarchical patterns of organization. Each organism continuously transfers matter and energy within itself and exchanges both with its surroundings. Each displays a capacity for homeostasis, or self-regulation. All organisms reproduce. All are products of evolutionary change. All display characteristic patterns of response and behavior and can live only in a specific range of environments. Finally, all organisms and their environments have evolved over long periods of geological time. Thus, they have a history.

It is instructive and sobering to consider the newest environment of living organisms, the city. In cities, human beings and individuals of many other species live together and interact. The activities and by-products of whole cities have profound effects on other biological environments, including rivers, the atmosphere, forests, farmlands, and oceans. The fact that many of these effects have been extremely detrimental is an impressive and important illustration of the delicacy of the complex biological systems and organizations that comprise life on Earth.

KEY TERMS

atmosphere
biology
biosphere
carnivorous animals

ecology
environment
food chain
herbivorous animals

homeostasis
hydrosphere
lithosphere
solar radiation

QUIZ QUESTIONS

1. _____ is (are) the ultimate source(s) of energy for carnivorous organisms in the food chain of a forest ecosystem.
   A. The decomposer species
   B. The herbivorous species
   C. The photosynthetic species
   D. The sun

2. Differences in climates in various areas of the Earth result largely from
   A. differences in the amount of sunlight striking at different latitudes.
   B. movements of plates of the Earth's crust.

C. the complexity of the food chain in a particular region.
   D. ocean currents.

3. Reproduction
   A. enables a species to persist over long periods of time.
   B. produces genetic variation that is important in survival of a species.
   C. is a universal feature of all living organisms.
   D. Choices A, B, and C are correct.

4. The mechanisms of _____ have been referred to as life's defenses against disorder.
   A. metabolism
   B. homeostasis
   C. adaptation
   D. photosynthesis

5. Evolution
   A. can be studied at the cellular level.
   B. can be studied by ecologists.
   C. is the process by which the Earth's diversity of life came about.
   D. A, B and C

ESSAY QUESTIONS

1. What major attributes are shared by all living organisms? Are any of these characteristics found in nonliving things? Give some examples.

2. Define homeostasis. Give an example of how an organism maintains internal homeostasis.

3. What is biological evolution?

4. Give some examples illustrating the operation and impact of biological processes and principles on human life in the modern world.

REFERENCES AND SUGGESTIONS FOR FURTHER READING

Benison, S., Barger, A. C., and Wolfe, E. (1987). *Walter B. Cannon: The Life and Times of a Young Scientist.* Harvard University Press, Cambridge, Mass.

The first of two volumes of a new biography of a seminal figure in the history of biology. It tells of his early life in science, from 1900 to 1917, and of that tumultuous period. One of the authors (A.C.B.) is a former student of Cannon's.

Cannon, W. B. (1939). *The Wisdom of the Body,* rev. ed. W. W. Norton, New York.

A famous and popular summary of the concept of homeostasis by its major proponent.

Darwin, C. (1950). *The Origin of Species by Means of Natural Selection.* [Original edition, 1859] Modern Library edition, Random House, New York.

One of the great books in the history of thought.

Deacon, G. (1984). *The Antarctic Circumpolar Ocean.* Cambridge University Press, New York.

An excellent and readable little book on the great ocean surrounding Antarctica, its explorations, and its physical and biological features.

Mayr, E. (1982). *The Growth of Biological Thought: Diversity, Evolution, and Inheritance.* Harvard University Press, Cambridge, Mass.

A masterful account of some of the central issues in the history of biology with an emphasis on evolutionary biology.

Medawar, P. B., and Medawar, J. S. (1983). *Aristotle to Zoos: A Philosophical Dictionary of Biology.* Harvard University Press, Cambridge, Mass.

A wise and witty ramble through many aspects of biology, infusing up-to-date scientific information with humor and insight.

Simpson, G. G. (1964). *This View of Life: The World of an Evolutionist.* Harcourt, Brace, New York.

Essays both brilliant and readable on what evolution means.

Simpson, G. G. (1976). *Penguins: Past and Present, Here and There.* Yale University Press, New Haven, Conn.

A delightful exegesis on an engaging animal.

Simpson, G. G. (1978). *Concession to the Improbable: An Unconventional Autobiography.* Yale University Press, New Haven, Conn.

The autobiography of a distinguished biologist.

Smith, J. M. (1986). *The Problems of Biology.* Oxford University Press, New York.

A provocative work that considers the nature of life and lucidly projects future avenues of investigation.

THE HISTORY AND SCOPE OF BIOLOGY

The history and scope of biology
Ancient ideas about the Earth and its organisms dominated thought until the dawn of science in the seventeenth century. Although this 1661 engraving depicts a new scientific idea—that the Earth revolves around the Sun—it includes an unscientific element, the signs of the Zodiac.

For centuries before the comparatively recent dawn of experimental science, whatever was known about living things was known to all. Somewhat later, biology's interpreters were the literary scholars, theologians, and philosophers of each age. In this chapter, we will briefly trace the path that thinking human beings followed in order to achieve the scientific abundance of the present day.

As we will see, the advent and growth of human knowledge were themselves major turning points in the history of life that set off whole new evolutionary trends.

THE ORIGIN OF SCIENCE

CONTRIBUTIONS OF THE ANCIENTS

The peoples of ancient Babylonia, Egypt, and Greece contributed much to a knowledge of the world and, in a sense, laid the foundations for modern science. But they lacked a world view in which science could flourish. The ancient Greeks, most notably Aristotle (384–322 B.C.), compiled perceptive descriptions of most of what was then known of animal life. However, they did not have a deep or systematic understanding of what they observed. With a few lucky exceptions, they were unable to formulate the right questions because they saw no need to temper their philosophies with scientific methods of reasoning.

EARLY SCIENCE

Science is based on observations and experiments that test each given *hypothesis,* which is a tentative explanation of the available facts. To be useful, a hypothesis must be potentially falsifiable. That means that it must be possible to discover facts or phenomena that might contradict it. A hypothesis remains valid only so long as such facts are not found.

A scientific theory is not an article of faith that in the mind of a believer could never be shown false by contrary facts. A wit once said, "I wouldn't have seen it if I hadn't believed it!" That nicely epitomizes the true believer's mode of thinking. To such thinkers, some beliefs do seem supportable by superficial evidence. But none can be falsified by a well-designed experiment. From ancient times to the present day, the growth of science has been characterized by the development of the **scientific method**—that is, by experimental or empirical methods.

The scientific world view emerged late in human history, within a single culture, that of western Europe, and over the brief span from the time of Nicolaus Copernicus (1473–1543) to the time of Charles Darwin (1809–1882). Its basis was a strict injunction: Questions and answers *must* be related to data and to observation rather than to ancient authority. In the quest for natural explanations of natural phenomena, hypotheses must be developed and then carefully tested.

Copernicus's theory that the Earth circles the sun *was* firmly grounded in precise observation, but it was violently disputed by those who sought knowledge from authority and philosophy. Hard data were not to be trusted; measuring instruments were suspect. A comparable argument occurred when Galileo (1564–1642), observing the sun through the newly perfected telescope, discovered sunspots. The authorities refused to believe that the sun could have such blemishes.

In its early years, the scientific method was applied mainly to physics, astronomy, and oceanic exploration. In a series of bold strokes, Europeans "discovered" the rest of the world—and its startlingly unfamiliar plants and animals. Natural history, still a useful and grandly evocative name for the natural sciences as a whole, had deteriorated after Aristotle. Now, with new data coming in from the world over, it recovered—but as an objective pursuit. Industrious workers began systematic collections of plants and animals and wrote detailed compendiums. Magnificently utilizing the new arts of printing and illustration, these early collectors—the first who could be called biologists in a modern sense—published huge tomes on natural history (Figure 2-1).

THE RISE OF CLASSIFICATION AND SYSTEMATICS

A flood of new discoveries increased the need to classify organisms, and many sought to establish systematic classifications that were not based on superficial resemblances or fancied properties. John Ray (1627–1705), an English botanist and a leading developer of the science of systematics, classified plants and animals and correctly grasped the nature of fossils. The preeminent Swedish physician and botanist Carolus Linnaeus (1707–1778) devised a formal system of *binomial* (two-name) nomenclature that is still in use. Although it has changed much since his day, modern classification began with Linnaeus (Figure 2-2). He and his successors brought to a pinnacle description and classification for its own sake.

Comte de Buffon (1707–1788) produced a voluminous and well-written natural history. Georges Cuvier (1769–1832) was among the first to include extinct animals in the system. These and other pre-Darwinian systematists tried very hard to develop a "natural classification" of living organisms. From our vantage point, it is clear that they never achieved one because their ideas of what is "natural" were based largely on philosophical notions unconnected with observable facts. For example, they often advocated making long lists of random observational data with the expectation that nature's laws would spring forth in many cases when the lists were long enough.

DARWIN AND THE DAWN OF MODERN BIOLOGY

In a real sense, the appearance of Charles Darwin's book *The Origin of Species by Means of Natural Selection* initiated the modern biological era. This great work was published in 1859, 28 years after the young Charles Darwin set sail on the famous voyage of the *Beagle* (see Box 2-1). The book accomplished three things.

- First, it proposed a theory of evolution, according to which all organisms arose from common ancestors by a natural historical process of change and diversification. It was the broadest generalization ever made about the relationships of living things.
- Second, it offered an explanation for the causes of evolution. The major point needing explanation was the *apparent purposiveness* of life, the fact that organisms seem so well designed for the functions they carry out. This supposed purposiveness, more than anything else, had given support to those who believed that life could not have arisen from natural causes and hence was not a suitable subject for scientific analysis.
- Third, it placed biology within the scope of science. Almost immediately, biological research began to expand explosively.

We should emphasize here what Darwin demonstrated: When logical and natural explanations for life's phenomena are sought, they can be found.

2-1 Illustrated natural history text
Early texts on natural history were splendidly illustrated. This page, taken from a text published in the nineteenth century, shows one of its two thousand engravings.

2-2 Carolus Linnaeus (1707–1778)
Elements of Linnaeus's system for naming and classifying organisms are still in use today. Linnaeus is shown here in Lapland costume. The "magic drum" at his side was probably used for carrying specimens.

THE NATURE OF SCIENCE

One reason people do scientific research is plain, ornery curiosity—the joy of pursuing knowledge for its own sake. Another is the practical hope of improving human welfare. Yet another, alas, is the interest some people have in destroying one another.

SCIENCE AND TECHNOLOGY

Whatever a scientist's motives may be, the upsurge of scientific activity in the modern world surely reflects an undeniable fact: Science gets results. It yields knowledge and it is utilitarian. The relation between these two aspects parallels the relation between *science* and *technology*. They depend on one another but are not the same. Science is humankind's response to curiosity, a way of gaining knowledge by explaining observed facts. Technology is our response to felt needs, the application of knowledge to human ends. As they say at M.I.T., "Science asks what exists; technology asks what you plan to do about it."

In this book, we will encounter many examples of science blossoming into practical technology. A notably stirring example is discussed in Chapter 17. Decades of pure research in the science of genetics abruptly spawned in the 1970s a new technology called *genetic engineering* or, more precisely, *recombinant DNA technology*. One of its pioneers is pictured in Figure 2-3.

The science of biology does indeed have many valuable technological applications, but the concern of this book is with the biological science on which technologies are based.

IMPORTANCE OF OBSERVATION

Observation is the critical part of any test of a proposed answer to a scientific question. No matter how easily repeated, observation is a foundation of scientific endeavor. But be clear on this: Observation is not the endeavor itself.

Observation is of little interest unless it can be related to other facts and ultimately explained. The isolated observation that a coral has a certain number of tentacles around its mouth becomes much more interesting when we recognize that this animal is one of a large category of similar creatures. This sort of generalization is fundamental in biology. The statement that all mammals have two sexes, male and female, is also a generalization. Once formulated, such generalizations lead scientists to search for the limits (do organisms other than mammals have two sexes?) and for contrary examples (are there mammals or other organisms with more or fewer than two sexes?).

Despite their importance, generalizations are descriptive, not explanatory. What we must do at another level is deal with the relationships between generalizations and real phenomena. Here we seek *explanation,* lawlike principles that expose causal forces. It is important to understand what "explanation" is (and is not), and we discuss it further in Box 2-2.

THE SPAN OF BIOLOGY

To this day, the identification and classification of organisms is one of the major preoccupations of biologists. The related field, **systematics,** has recently been revived by a change in perspective. When systematists moved from the classification of individuals to the classification of populations, biologists found a new significance in systematics. It is no longer mere pigeon-holing and labeling. It is now the study of the diversity of organisms coping with their environments.

Systematics has become so vast that no one can master more than a small part of it. The primary divisions are *botany* (the study of plants), *zoology* (the study of animals), and *protistology* (the study of the unicellular organisms called protists). Now that we recognize five kingdoms, we can add on this level *mycology* (the study of fungi) and *bacteriology.* Many specialties cover areas much smaller than kingdoms. No systematic zoologist works on the whole animal kingdom. He or she may be an *entomologist* (student of insects), *malacologist* (of mollusks), *ichthyologist* (of fishes), *herpetologist* (of amphibians and reptiles), *virologist* (of viruses), *ornithologist* (of birds), or be working in any of dozens of other kinds of *-ology.*

2-3 Barbara McClintock
McClintock was awarded the 1983 Nobel Prize in Medicine and Physiology. Her pioneering work on the genetics of corn plants anticipated today's recombinant DNA technology.

On December 27, 1831, the young English-man Charles Darwin (1809–1882; Figure 2A) sailed from Devonport, England, on board the *Beagle,* a 10-gun brig (Figure 2B). Captain Robert Fitzroy, only four years Darwin's senior, offered passage and a berth in his own cabin to any young man who would volunteer to go without pay as a naturalist and collector. Not yet 23, a student utterly lacking in distinction, an ardent hunter, fisherman, and collector of beetles, mollusks, and shells, young Darwin at first hesitated. But he was eager to escape from Cambridge, and in the end he accepted the invitation. The voyage of the surveying ship lasted five years. Darwin's encounters with such natural entities as flowers, bird beaks, and island beetles led in due course to a theory that shook the foundations of scientific thought.

The prevailing intellectual climate in 1831 was highly conservative. The great biologists of the time—such as Louis Agassiz and Richard Owen—leaned to the view that the successive life forms revealed by the fossil record were all separate creations, some of which had been extinguished by historical accidents. Nonetheless, the notion of evolution was stirring in people's minds. Darwin's own grandfather, Erasmus Darwin (1731–1802), was a celebrated poet, physician, and early evolutionist. His book *Zoonomia,* published in 1794, obliquely anticipated both the doctrines of inheritance of acquired characteristics, which we identify with Lamarck, and natural selection, later to be formulated by his grandson. Erasmus died seven years before Charles was born, but his ideas—and those of other early evolutionary thinkers—were being widely discussed during Charles's college years.

Charles Lyell (1797–1875), later to become Darwin's lifelong friend, opened the way for an evolutionary viewpoint by showing

2A Charles Darwin (1809–1882)
Darwin as he appeared around 1855 (at age 46) and in 1881 (at age 72).

ing that the planet must be very old, old enough to permit the slow, continuous changes we know as evolution. One of the few books Darwin took with him on the voyage was the first volume of Lyell's newly published *Principles of Geology.* The second volume reached him while he was on the *Beagle.* In these books, Lyell dismissed catastrophism, the theory that explained the extinction of species in the distant past by postulating a series of disasters, the most recent

of which was the flood described in the Bible. Species that survived these catastrophes were thought to have repopulated the world. Instead Lyell advocated the view that the slow, steady, cumulative effects of natural forces had produced continuous change in the course of Earth's history. To Darwin, the biological implications of this theory were very clear. If Earth had a long, continuous history and if no agencies other than existing physical forces were needed to explain the events inscribed upon the geological record, might not living organisms have had a similar history?

As the tiny vessel moved down the eastern coast of South America, through the Straits of Magellan and up the Pacific coast (Figure 2C), Darwin fished, hunted, observed nature, and traveled into the interior. No detail was too small to be fascinating and provocative. He seized the opportunity to study the plants and animals of the South American coast, and for five laborious years made careful notes and gathered priceless specimens of living species and fossil shells. On his long field trips, he encountered strange animals,

2B The H.M.S. Beagle *at anchor*
Only 90 feet in length, the Beagle *embarked on its long voyage of exploration with 74 people aboard.*

This fragmentation of systematics into specialties does, of course, reflect a present-day trend in every science. However, a contrary current is running too, and we will soon encounter abundant evidence of discoveries that apply to *all* living things and that tend therefore to erode the barriers between subdivisions.

In principle, biology should include any study of living (or formerly living) things. In practice, several branches of science that deal with living things are usually considered apart from biology. These include many aspects of *agriculture,* the study

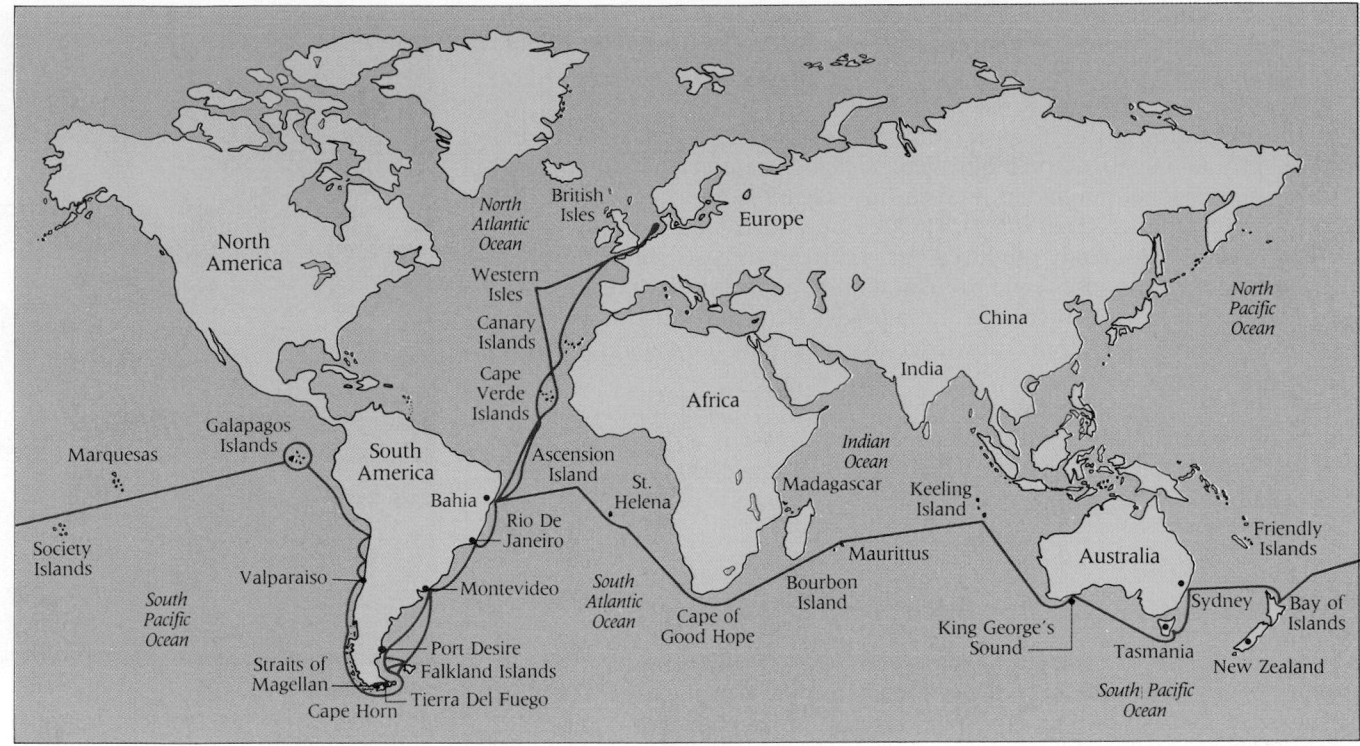

2C Map of the voyage of the **Beagle** *(1831)*

2D A giant tortoise
Darwin testing the speed of a giant tortoise on the Galápagos Islands.

unusual insects, lovely orchids, and new grasses and trees. From this stronghold of life grew Darwin's lifelong preoccupation with the problem of species.

Most interesting to him were the animals and plants of a small barren cluster of islands, the Galápagos, that lie directly on the equator some 600 miles off the coast of Ecuador. Once a refuge of buccaneers, the Galápagos were named for the islands' most striking inhabitants, the giant armored tortoises (*galápagos* in Spanish) that weigh up to 200 pounds and jostle through the underbrush like prehistoric monsters (Figure 2D). Remarkably, each island seemed to have its own type of tortoise: Local fishermen always knew which island was the home of any particular tortoise. There were also the drab little birds called finches, a group of 14 species, which differed from one another in body size, beak structure, and in the type of food they ate as they migrated from the arid coasts to the highlands during periods of drought. Their diversity played a large part in stimulating Darwin to formulate the principle of natural selection (see Chapter 34).

From his knowledge of geology, Darwin knew that these islands, clearly of volcanic origin, were much younger than the mainland, yet the plants and animals of the islands were as different from those of the mainland as were the inhabitants of the different islands from one another. Were the species on each island the product of a separate, special creation? "One might really fancy," Darwin wrote later, "that from an original paucity of birds in this archipelago, one species had been taken and modified for different ends." The birds had become transformed by their struggle for existence on their little islands into a series of types particularly suited to their particular niches. We will return to these historic birds in Chapter 36.

For years after his return, these questions continued to "haunt" him—in Darwin's word. We shall later consider the work of Darwin's mature years. It included his great book *The Origin of Species,* which through procrastination and scientific compunction he did not publish until 28 years after the *Beagle* had sailed from England.

of cultivated plants and domesticated animals in the context of food production; *anthropology,* the study of humankind, with emphasis on its social aspects; *psychology,* the study of thought and behavior; and *sociology,* the study of human communities.

One might add to this list *medicine,* the study of human disease and its treatment. However, it is an exciting fact that modern medical science is now in creative ferment *because* of recent advances in biology. We can agree that discussions of how to treat headaches or arthritis are facets of medicine that have no place in this book. But we must recognize that such topics as the nature of the cancer cell, the curious behavior of sickle hemoglobin, and the intriguing aspects of human sexuality traceable to chromosomal abnormalities are all aspects of modern medical science that are rooted in biological science.

As for **biotechnology,** this field explores the use of biological systems and techniques—recombinant DNA methods, gene splicing, transfer and cloning, cell hybridization, monoclonal antibodies, and so on—in the large-scale or industrial production of useful products (Figure 2-4). The area, now quickening in its sophistication and importance, has already produced bacteria that synthesize human insulin and other hormones. It promises soon to revolutionize agriculture by creating new kinds of soil organisms that can convert the nitrogen of the air to fertilizer. These achievements rest squarely on new knowledge of genes, how they work, and how their expression is regulated.

Each of the areas named above has become a complex field unto itself. But each has important relationships with biology. For these reasons, a sharp line cannot be drawn between biology and the other life sciences. Some perhaps would define biology as what is taught in biology classrooms.

LIVING ORGANISMS: ARE THEY MACHINES?

As we begin our journey of biological exploration, let us briefly consider a famous legacy of nineteenth-century biology: the eventual acceptance by the major scientific minds of that era that life can be explained in terms of **mechanism**—that is, in terms of physical and chemical laws.

According to **vitalism,** once the universally held contrary view, living substance differs fundamentally from nonliving substance. Its processes are not reducible to the mere interactions of material components, as the advocates of mechanism claimed. There is, said the vitalists, something special about the world of life.

We call concepts such as vitalism *anthropocentric* because they consider humankind the center and goal of all natural phenomena. An example of anthropocentrism was the early view that the sun rotates around the Earth—*our* Earth—not the other way around (Figure 2-5). Vitalists saw life as something beyond the reach of ordinary science. To think otherwise, they thought, would deny humankind its special status. This attitude sustained the view that living organisms contain a spark or a spirit that infuses life into nonliving materials.

The doctrine of **finalism** holds that the evolutionary history of life had a preordained pattern that was purposefully directed toward a future goal. Although vitalists were not always finalists, the two often went together, for in both views life is intrinsically beyond explanation. The *something* that transforms matter into life has been many things to many people. To the ancients, it was the psyche, the life principle, or the soul. To Aristotle, it was three souls. To later vitalists, it was variously called the *anima, élan vital, entelechy,* or *mystery of life.* To people of this persuasion, it was not a proper subject for scientific inquiry.

Early in the seventeenth century, René Descartes (1596–1650) made a famous distinction between mind and matter that pictured a dualistic world. The nonmaterial part was mainly concerned with intellectual matters, thinking, reasoning, and feeling. The strictly *mechanistic,* material part accounted for the remainder of life's functions. Living organisms, in this view, work like machines. Descartes gave such great impetus to mechanistic biology many began to view the soul as a superfluous hypothesis.

Biology was not purged of its spirits, however. The beginning of the twentieth century was a time of particularly vigorous debate between vitalists and mechanists. To the question "How does one explain human memory?" Henri Bergson (1859–1941) answered: *élan vital*—usually translated "vital force." A scientific explanation must be testable experimentally, but *élan vital* is unobservable. That made Bergson's explanation unscientific. Since we cannot know whether *élan vital* is there or not, there is no way to examine its properties. Even if that proposition were true in some sense, it would no more explain how an organism works than *élan locomotif* explains how a locomotive works.

Nonetheless, vitalism gained stature in some eyes with the emergence of what seemed like supportive experimental data. The German biologist Hans Driesch (1867–1941) demonstrated in newts the most astonishing restitutions of whole limbs following the amputation of a leg (see Chapter 19). From this, he concluded that in

2-4 Biotechnology
Biotechnology has made possible the production, in large quantities, of many biomolecules. These flasks contain solutions of interferon, a protein of animal origin, that was synthesized by bacteria following the application of recombinant DNA methods.

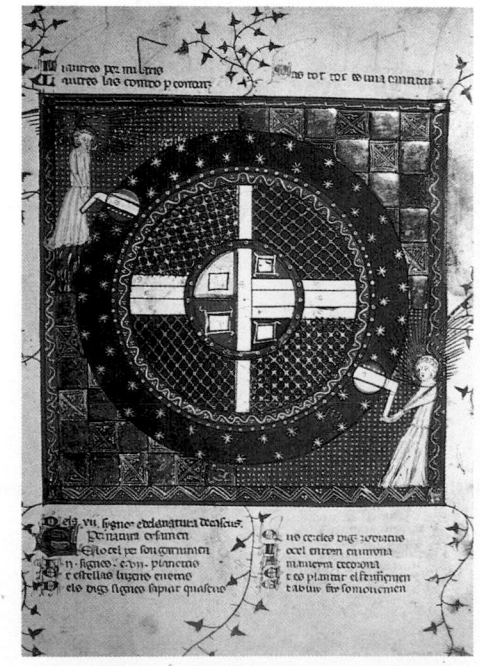

2-5 Anthropocentrism
This engraving from a fourteenth-century manuscript illustrates an anthropocentric view of the universe. Two angels are cranking the sky around a motionless Earth.

BOX 2-2

EXPLAINING EXPLANATION

Biological facts and generalizations are fully understood only when we have managed to explain them—and, as we shall see, many a biological phenomenon is still without explanation. Biology relies on explanations of at least three kinds.

- **Physiological** or **reductive explanations** relate biological structures and functions to events taking place at lower levels in the organizational hierarchy— and to external factors that may activate or otherwise influence events at these levels. Properties of membranes can be explained by properties of fatty acids; the rise of sap in trees is explained by the loss of water from leaves and by the cohesion, adhesion, and tension of water molecules; the circulation of blood is caused by the contractions of muscle fibers in the heart.
- **Functional** or **teleonomic explanations** relate biological structures and functions to their apparent purposes. We could explain digestion physiologically as the action of certain enzymes on foods. But we would explain it teleonomically as a step necessary in providing an organism with needed materials and energy. The teleonomic explanation of reproduction is that it perpetuates populations. And we would explain the structure of the eye, which so closely resembles a camera, by showing that its purpose is the same as a camera's—picking up images. This type of explanation is indispensable to the working biologist, who wants always to know: For what purpose does this structure or function exist?
- **Historical explanations** relate biological structures and functions to their evolutionary origins. The architecture of the human skeleton is explainable in part by the fact that our ancestors lived in trees and had to be adapted to jumping, swinging, and grasping. Migratory birds fly south in the winter because evolution has given them a genetic pattern that programs them to seek warmer climates at the onset of cool weather.

Finding the explanations of facts is a creative act—and many of the most creative scientists have been unable to tell how they do it. Some have declared that they arrived at certain important explanations by hunch or by sudden inspiration. It should be added that hunches and inspirations usually came to them only when they were actively seeking an explanation for certain facts. Alexander Fleming (1881–1955) discovered penicillin after observing that an unusual mold had killed the bacteria in a Petri dish. He would not have realized the significance of this observation had he not been already immersed in the general question of what causes bacterial colonies to die.

The first step in the development of a scientific explanation is the framing of a **hypothesis.** A hypothesis is a *possible explanation.* It may be based on a hunch or an inspiration. It may be the only conceivable explanation of the known facts. Whatever it is, it is not a hypothesis unless it can be tested—validated or invalidated—by an experiment or by further observations.

In testing a hypothesis, scientists look for exceptions to generalizations stated or implied in the hypothesis. They also make predictions—logical deductions of consequences that should occur—if the hypothesis is correct. They then see whether or not these phenomena actually do occur. The hypothesis that muscle contraction is due to the sliding of filaments past one another has a necessary implication. If this model is correct, certain changes should be seen in the bands and striations of muscle during a contraction. When it was found that these changes do in fact occur, the sliding filament explanation of muscular contraction was greatly strengthened. The larger the number of predictions or logical consequences a hypothesis yields, the better it is; for it can then be more thoroughly tested. A hypothesis also leads to new, useful, observed facts, and these lead to new hypotheses. When a hypothesis fails the test it must be modified or abandoned.

When a hypothesis passes the test, confidence in it grows. But passing a given test does not itself prove the validity of the hypothesis: It may have passed the test for other reasons. To be accepted, a hypothesis usually must be tested in more than one way and more than one time. At some point, we feel justified in accepting a given hypothesis as the most probable explanation. From then on, we proceed as if it were true and begin to speak of it as a *theory.* A theory is simply a hypothesis that has been tested often enough to convince scientists that it is probably correct. We speak of the ''acceptance'' of a theory, ''confidence'' in a theory, and ''probability'' of its correctness—but never of the ''proof'' of its correctness. If proof means the establishment of an eternal and absolute truth that is open to no possible exception or modification, then proof has no place in the natural sciences. A theory is always open to disproof. Good scientists must therefore be prepared to change, and sometimes discard, their most prized theories in the interest of advancing knowledge.

each organism there must reside a vital *entelechy* that keeps track somehow of the ''blueprints'' for the whole adult form. (Driesch resurrected the term *entelechy* from Aristotle, who used it to mean a potentiality that has become an actuality.) Driesch argued that the concept of mechanism cannot account for the origin of a whole organism from a single cell.

In response to this claim, biologists dispersed themselves along a broad spectrum of opinion ranging from thoroughgoing, hard-boiled materialism on one end to idealistic metaphysics on the other. Some natural scientists in the new century held that organisms depend for their autonomy on their *wholeness* and their marvelous integration. This position, sometimes called **organicism** or **holism,** is not unpalatable to materialists. To say that life depends upon the complex organization of its material parts is quite compatible with mechanism, whose advocates agree that there are no pieces of living matter, but only living organisms. But they disagreed when organicists began speaking of purpose and denying even the possibility of a physiochemical explanation of organismic integration.

Vitalistic views of the more flagrant kind were overturned in the nineteenth century largely because mechanistic theories were busily and successfully bearing fruit. Countless mechanistic theories, such as the discovery in 1628 by William Harvey (Figure 2-6) of the circulation of the blood, were successful because they invited experimental verification and accurately predicted events. The result was an accumulating body of knowledge, a self-validating and widening stream of scientific progress. In contrast, vitalistic theories led nowhere. Thus, vitalism withered—not so much because philosophical criticism was leveled against it but because, having argued either that nature is beyond explanation or that it is already explained by vital forces, vitalism was a barren guide to progress.

This is not to suggest that the debate has now been stilled. Today's working scientists are, with few exceptions, mechanists. Nevertheless, some organicists still proclaim the view that biology is irreducible to physics and chemistry. Were they maintaining merely that much of biology has *not yet* been reduced to physiochemical explanation, their position would be unassailable. But they hold that *in principle* such explanation is not possible.

Nothing so far discovered warrants this bleak view.

2-6 William Harvey (1578–1657)
A 1730 engraving.

IS THERE A LADDER OF NATURE?

One more question from biology's history deserves our early attention. It is the issue of how the world of life should be viewed.

We have spoken, in Chapter 1, of the *hierarchy of levels of organization* that rises from atoms and molecules, up through cells, organisms, and populations. Over the centuries this concept led to a seriously misleading idea. This is the notion that there is a *hierarchy of living organisms* rising from ''lower'' to ''higher'' forms. The old term **scala naturae,** which means ''ladder of nature,'' is still used to denote this concept. It came originally from the brilliant thought of Plato (422–347 B.C.).[1]

The idea of a *scala naturae* has echoed through the ages, radiating into religion, philosophy, and science, sometimes in strangely distorted ways. A form of this idea convinced Francis Bacon (ca. 1214–1294) that all things were created for human use. Immanuel Kant (1724–1804) concluded that there must be beings living on other planets that are farther up on the ladder.

The concept of the *scala naturae* strongly affected the history of biology as well, and its traces still linger. For instance, a conflict developed in the eighteenth century over whether the ''rule of continuity'' in nature meant that species of organisms are not real and distinct units. Were they not only artificially segmented steps on the ladder of nature? Buffon, the most distinguished French biologist of the time, eventually decided that species *are* objective realities and indeed the single and unchanging units in nature. However, Charles Bonnet (1720–1793) clung to the idea of continuity and denied that species exist. Around this time much thought was devoted to the ''missing link'' that would demonstrate the continuity of humankind with other animals and thus complete the chain of being. Bonnet held that the orang-utan is the missing link (Figure 2-7).

In those days, people assumed that the links of the chain of nature were immutable. In other words, after their creation by God, no new species were added and none were subtracted. Positions on the chain or ladder were rated from ''lower'' to ''higher'' on the basis of complexity in anatomy, function, or both.

Jean Baptiste Lamarck (1744–1829) is generally, but questionably, regarded as

2-7 The missing link?
The origin of humans was a hotly debated issue in the eighteenth century, as it is now. Charles Bonnet insisted that the orang-utan, pictured here, is the ''missing link.''

---

[1] Plato espoused a doctrine of *plenitude,* which says the universe is complete. Everything that can or should be *is.* Plato's student Aristotle added the idea that everything that *is* connects with everything else. There is continuity. Things intergrade, and they do so in a uniform and orderly way.

2-8 Jean Baptiste Lamarck (1744–1829)
Lamarck was one of the first biologists to adopt an evolutionary view of nature.

2-9 Acquired characteristics?
The giraffe was cited by Lamarck to illustrate the supposed inheritance of acquired characteristics. In his view, the continual reaching to the treetops lengthened the giraffe's neck, and the lengthened neck was passed on to offspring.

the first biologist to adopt a truly evolutionary view of nature (Figure 2-8). In his view, living animals display what he called the "order of nature," which runs from the "most imperfect" animals at the bottom to "the most perfect animals"— yes, humans—at the top. In the first of the two volumes of his *Philosophie Zoologique* (1809), he described the "animal chain" in detail and declared that species are not all as ancient as nature. They arose successively and are unchanging only for a time. He believed that the most imperfect animals, at the bottom of the chain, have arisen throughout time and are still now arising by spontaneous generation and that they have been, and are now still, climbing up the chain.[2] In his major work, *Natural History of Invertebrate Animals*, published in seven volumes from 1815 to 1822, Lamarck clarified somewhat the tangled threads of his thought. He held that the "power of life" would have followed a chain of being, a *scala naturae* (not his wording), if it had not been perturbed by environmental influences. These affected the characteristics of individual animals, and these acquired characteristics are heritable (Figure 2-9; also see Box 13-1).

"Lamarckism" accepted the concept of an ideal "ladder of beings." But it explained the fact that organisms do not really form a connected ladder or chain by a second apparently conflicting theory, which held that *acquired traits are heritable*. It will become apparent in Chapter 13 that they are not.

Cuvier rejected not only Lamarck's ideas on evolution but all other evolutionary concepts and anything resembling a *scala naturae*. It is an irony of history that although Cuvier rejected evolution, his censure of the views of Lamarck and his insistence on many groups of divergent organisms contributed significantly to an acceptance of evolutionary biology in later years.

After Cuvier, the perennial idea of *scala naturae* faded away—almost, but not quite completely. In the innumerable notes Darwin scribbled while developing his views on evolution, he wrote that he should never use the terms "lower" or "higher" in reference to organisms. Nevertheless, Darwin did use them in later publications, and some biologists use them still. When they do, they echo a discredited idea, for it is now clear that there never has been a *scala naturae*.

As we will learn in later chapters, all organisms living today have ancestries of equal length. So did all organisms living at any one time in the past. Their existence at a particular time, in a particular way, and in a particular place means that all were *adequately adapted* for life in their way, at that time, and in that place. A modern evolutionary biologist can agree with the poet Goethe (1749–1832), who wrote, "Every animal is an end in itself."

To refer to some organisms as higher or lower than other organisms in discussions of evolutionary biology makes no sense. We shall avoid this misuse of those terms in the present book, although it is sometimes tempting to use them. When contrasting, say, the structure of a sponge with that of an elegant land animal or flowering plant, it is easy to err by dismissing the former as "lower organisms" and exalting the latter as "higher organisms." If we err in this way, we ask to be forgiven.

---

[2] In his "additions" at the end of the second volume, Lamarck modified this view and seemed to be abandoning the *scala naturae*, strictly speaking, and even to be waffling about evolution. He ended by writing that nature forms an eternal circle, a perfect whole which completely fulfills the purpose for which it is destined "by its Creator," the "Sublime Auteur," which is God. At the time, these were perfectly reasonable ideas fully consonant with contemporary thought.

SUMMARY

Although ancient scholars accumulated many beliefs about the world around them, they did not systematically test these beliefs empirically to determine whether they were true. For a hypothesis to be scientific, it must be potentially falsifiable—that is, we must be able to conceive of possible results of a test which, if they occurred, would disprove the hypothesis. This criterion distinguishes scientific hypotheses from articles of faith, which are beliefs that no amount of evidence could invalidate.

Biology began as a branch of natural history. Like other sciences in their early stages, it was originally concerned more with collecting and classifying data than it was with formulating and testing hypotheses about the data. Darwin's proposed theory of natural selection marked the beginning of modern biology, providing a theoretical base for subsequent biological investigations. Biology, like all sciences, must work from observable facts and events, but facts by themselves are not useful unless they can be linked together in theories that provide explanations of the facts by invoking general laws. These laws, if they are firmly established, will allow some degree of prediction, and thus control of future events. In other words, they permit the development of a biological technology based on biological science.

Biology includes all fields that study the structure and function of living organisms. Many of the greatest advances in biology, however, have come from theories that explain living organisms in terms of the same chemical and physical laws that apply to *non*living mechanisms. The theory of natural selection is one such theory, applying to all systems that meet certain conditions, whether or not the systems are composed of living creatures. It dealt what

has proved to be a death blow to the doctrine of vitalism, the belief that living organisms are driven by a nonphysical "life force."

The theory of natural selection has also been influential in discrediting the idea of a *scala naturae,* or hierarchy of living things that rises from lower forms to higher forms. Although it is tempting to think of human beings as less primitive than mice or mushrooms, we must keep in mind that all organisms living today have evolutionary histories of equal length. Species that are not extinct have survived because they are adapted to the modern world, however similar to earlier forms they may appear. Thus, no species still prevalent today can truly be regarded as a lower form of life than any other species.

KEY TERMS

biotechnology	hypothesis	scientific method
finalism	mechanism	systematics
functional explanation	organicism	teleonomic explanation
historical explanation	physiological explanation	vitalism
holism	scala naturae	

QUIZ QUESTIONS

1. Many new species of organisms became known during the 1600s and 1700s. As a result, the biological field of _____ grew rapidly.
   A. natural history
   B. technology
   C. medicine
   D. systematics
   E. malacology

2. Most modern biological scientists are considered supporters of
   A. finalism.
   B. mechanism.

   C. vitalism.
   D. A and C

3. Which of the following is *not* an important element of Darwin's theory of evolution?
   A. inheritance of acquired traits
   B. presence of variation within a population of individuals
   C. natural selection
   D. all organisms arose from a common ancestor

4. The first step in the scientific method is
   A. development of a hypothesis.

   B. design of an experiment.
   C. observation.
   D. testing of an hypothesis.

5. Who of the following did *not* support some theory of evolution?
   A. Lamarck
   B. Cuvier
   C. Darwin
   D. A and B

ESSAY QUESTIONS

1. What is the relation between a hypothesis and a theory? Between a hypothesis and an experiment?

2. Characterize some of the major differences between the methods of ancient and modern biologists.

3. An unfortunate man died at the age of 50 with lung cancer that developed after years of heavy smoking. How would you "explain" his death? Is there more than

one way of explaining it? What does it mean to explain a phenomenon?

4. What is the difference between science and technology? Between vitalism and mechanism?

REFERENCES AND SUGGESTIONS FOR FURTHER READING

Barnett, S. A. (Ed.) (1958). *A Century of Darwin*. Harvard University Press, Cambridge, Mass.

The year 1958 was the centennial of the paper propounding the theory of natural selection, which was presented to the Linnean Society in London by Darwin and Wallace. Fifteen essayists here discuss its importance.

Beck, W. S. (1957). *Modern Science and the Nature of Life*. Harcourt, Brace, New York.

A brief account of the history of biology and the major issues in the philosophy of science.

Burkhardt, F., and Smith, S. (Eds.) (1985). *The Correspondence of Charles Darwin, Vol. I (1821–1836)*. Cambridge University Press, New York.

Fascinating insights into why Darwin remained with the *Beagle* despite the complaint in his letters after years at sea, "I hate every wave of the ocean."

Burkhardt, R. W., Jr. (1977). *The Spirit of System: Lamarck and Evolutionary Biology*. Harvard University Press, Cambridge, Mass.

A portrayal of how Lamarck viewed himself as a naturalist-philosopher. He emerges as a man possessed of a vision, but one he could not persuade colleagues to share.

Gould, S. J. (1977). *Ever Since Darwin: Reflections in Natural History*. W. W. Norton, New York.

Sparkling essays on various aspects of biology by an articulate colleague. "Science is not a heartless pursuit of objective information," he writes. "It is a creative human activity."

Hull, D. L. (1988). *Science as a Process: An Evolutionary Account of the Social and Conceptual Development of Science*. University of Chicago Press, Chicago.

A lively attempt to draw instructive parallels between the evolution of life and progress in science.

Mayr, E. (1988). *Toward A New Philosophy of Biology: Observations of an Evolutionist*. Harvard University Press, Cambridge, Mass.

A vigorous and learned defense of the author's view that biology is not reducible to physics and chemistry.

Monod, J. (1971). *Chance and Necessity: An Essay on the Natural Philosophy of Modern Biology*. Alfred A. Knopf, New York (paper edition, Vintage Books, New York).

A Nobel Prize winner sternly insists that life arose on Earth by chance, that humankind is alone in a dead universe, and that evolution is blindly driven by nothing more than the striving of organisms to survive.

PART 2
BIOCHEMICAL AND CELLULAR BASIS OF LIFE

All organisms—plant or animal, large or small—are composed of cells, either single cells or groups of cells. The discovery of this principle, the cell theory, was an outstanding achievement of the nineteenth century and ranks with the theory of evolution as a cornerstone of modern biology.

The study of life must start at levels below the cell, with the atoms and molecules that make up the structures of cells. These are discussed in Chapters 3 and 4. Like the organism of which it is a part—or a whole—each cell is complexly organized. A cell responds to stimuli and is capable of movement. It feeds and respires, expending energy to maintain its ordered state and execute other work such as secretion. And it can reproduce itself. In large organisms like ourselves, specialized parts like bones, muscles, and nerves are composed of special cells—bone cells, muscle cells, and nerve cells. Chapter 5 describes the organization of cells. Chapter 6 discusses cell membranes, a rapidly advancing field. Chapter 7 surveys certain cellular functions, including their organization into tissues of diverse functions.

Chapters 8 through 11 review the metabolic processes that extract the energy of the sun and provide other cells with energy-containing molecules and molecular building blocks. We turn then (in Chapters 12 through 20) to reproduction, heredity, and development, which account for life's continuity. We will review the evidence supporting the existence of genes and then discuss how they work. We will see that the molecular biology of the gene is one of the most explosively advancing areas of modern biological science, with enormous implications and promise for genetics, evolution, developmental biology, and many aspects of human life.

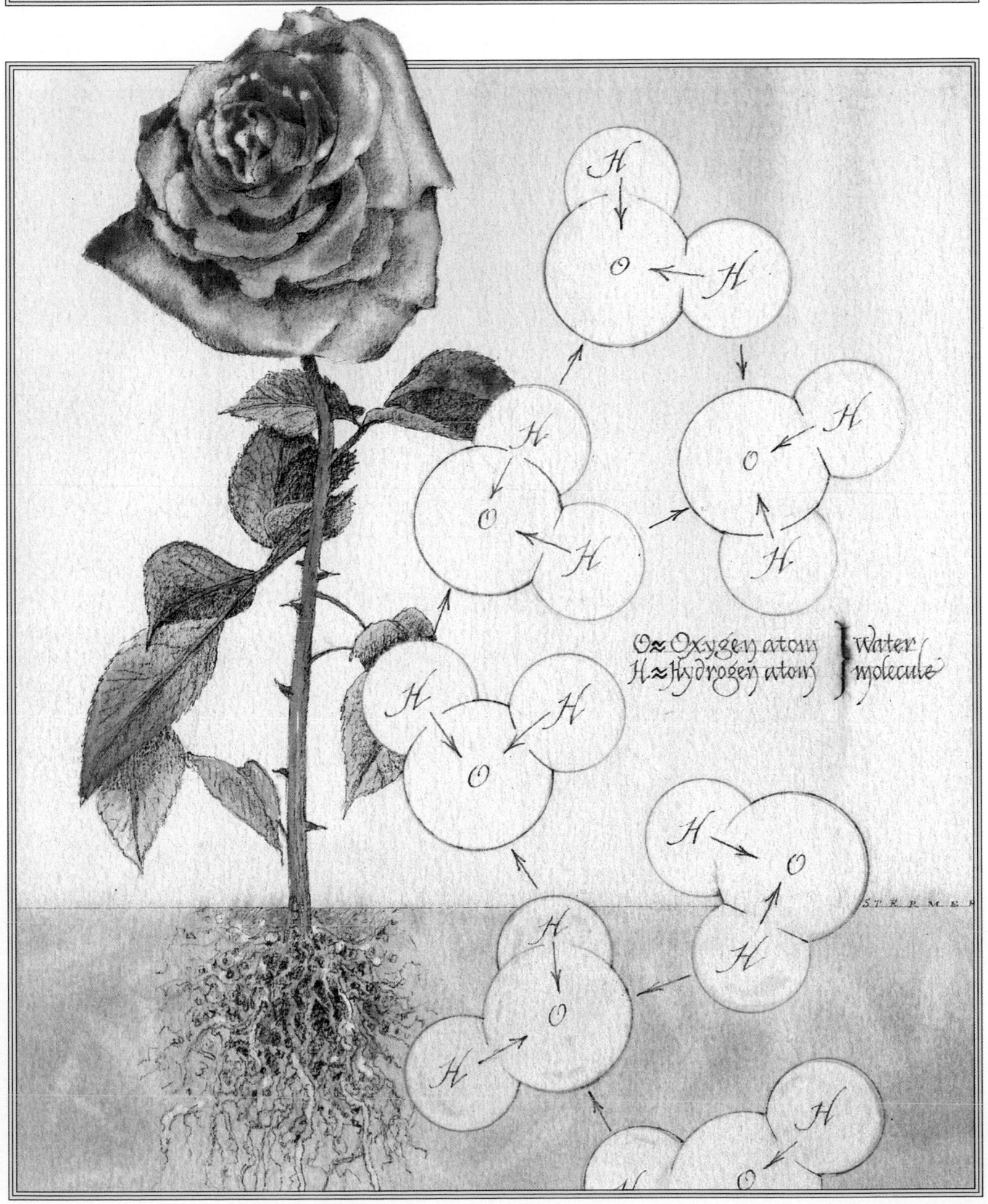

O ≈ Oxygen atom
H ≈ Hydrogen atom } Water molecule

ATOMS AND MOLECULES

Atoms and molecules
Having emerged on a planet made up of diverse chemical elements, living organisms are composed of many of those elements and the molecules into which they combine. Water, one of the most commonplace of molecules, has many exceptional properties of great biological importance. One such property, the lower density of ice compared to liquid water, explains why ice floats in a lake or ocean. This is a microscopic view under polarized light of ice crystals forming.

In the organizational hierarchy of living organisms, the cell is the lowest level at which life is unambiguously present. Within each cell are other structures; within each of these are component parts; within those, smaller parts. Eventually this downward escalator reaches the level of molecules and then of atoms—the fundamental building blocks of matter.

IMPORTANCE OF CHEMISTRY IN BIOLOGY

There are two classical approaches to the investigation of an organism: **morphology,** the study of its *structure*, and **physiology,** the study of its *function*. As we will see, biochemistry, the chemistry of living organisms, and biophysics, the physics of living organisms, elegantly link structure and function at the molecular level.

It is no longer possible to discuss biology without reference to these ideas. Partly for that reason, biology has become more than a descriptive science. Biochemistry has enormously enhanced the ability of biologists to uncover the mechanisms that make things happen.

New insights always raise new questions. We saw in the preceding chapter that there are no physical or chemical laws peculiar to living matter. Its atoms and molecules obey the same laws as the atoms and molecules of rocks, minerals, planets, and stars. Indeed, biochemical insights have made it harder to locate the frontier between living organisms and inanimate matter. To illuminate these issues we begin with biochemistry.

CHEMICAL STRUCTURES

SOME DEFINITIONS

An **element** of matter is defined as a single substance that cannot be changed by chemical means into a substance of a different kind. An **atom** is the smallest particle of an element that still retains that element's unique properties and the smallest unit that can enter into chemical combinations.

In view of the complexity of the material universe, the number of elements is surprisingly small. More than 100 are known, but only 92 occur naturally. Each element has a name and a corresponding symbol. About 35 are common and apparently essential for life, though many more elements can be found in living organisms in minute quantities. Some of the elements important in biology are listed in Table 3-1.

Within atoms are subatomic particles—the protons, neutrons, and electrons. These are much smaller than atoms and have distinctive properties of their own. Two or more atoms can combine to form a **molecule.** Some molecules contain many thousands of atoms. When molecules contain more than one kind of atom, they are also called **compounds.** The atoms of a compound are held together by specific forces called **bonds.** Atoms or molecules can *react* with each other and

TABLE 3-1
NAMES AND PROPERTIES OF THE MAJOR ATOMS OF BIOLOGICAL IMPORTANCE

Element	Symbol	Atomic Mass*	Number of Protons (Atomic Number)	Number of Neutrons in Neutral Form	Number of Electrons of Most Abundant Form
Hydrogen	H	1.0	1	0	1
Carbon	C	12.0	6	6	6
Nitrogen	N	14.0	7	7	7
Oxygen	O	16.0	8	8	8
Sodium	Na	23.0	11	12	11
Phosphorus	P	31.0	15	16	15
Sulfur	S	32.1	16	16	16
Chlorine	Cl	35.5	17	18	17
Potassium	K	39.1	19	20	19
Calcium	Ca	40.1	20	20	20

*The atomic weight of an element is the average mass of its atoms. Since an element may consist of two or more isotopes whose atoms differ in mass, the averages given are for the usual isotopic composition of each element. At one time, masses were stated in relation to the mass of oxygen, which arbitrarily was taken as 16.0. In 1960, an atomic mass unit was defined as 1/12 the mass of an atom of carbon 12, the predominant isotope of carbon. Isotopes are discussed below.

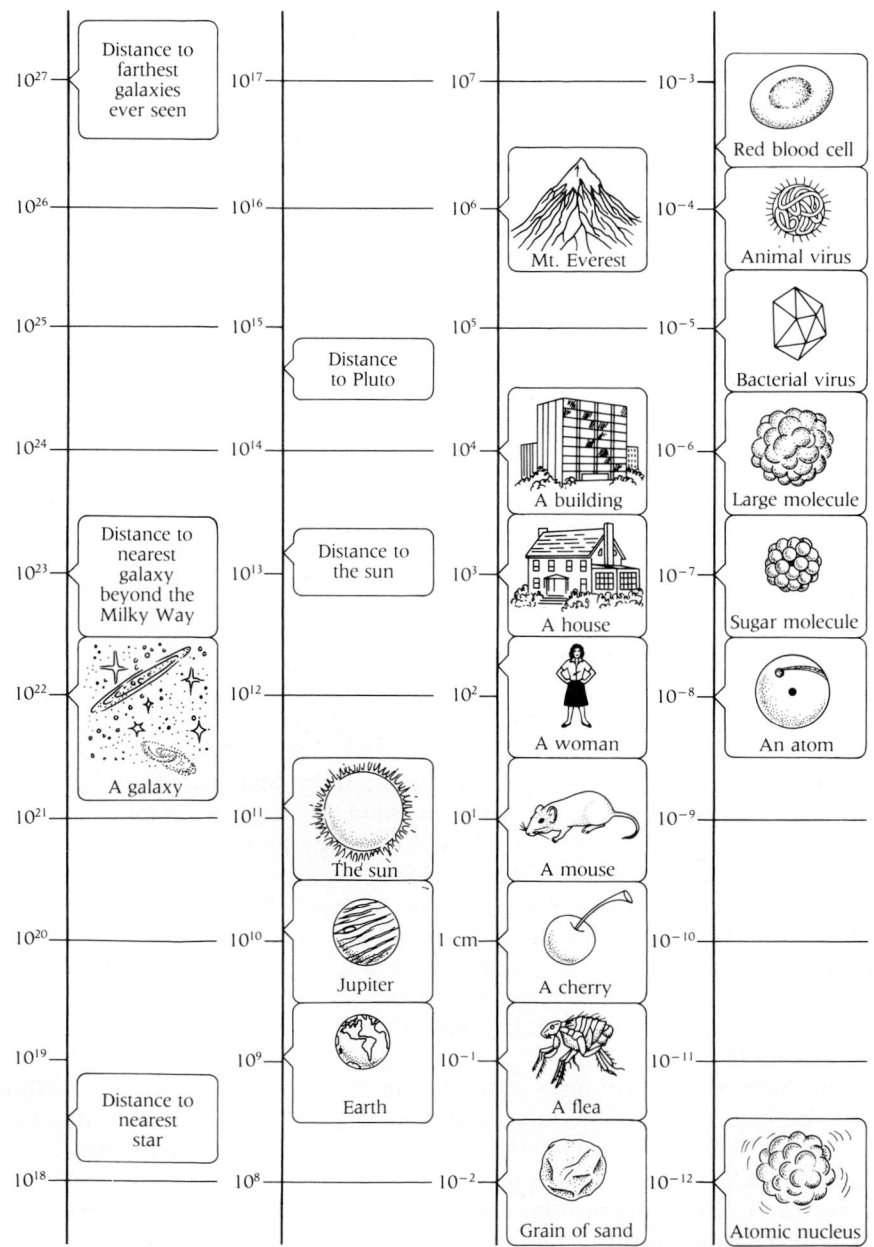

3-1 Relative linear dimensions of various objects
The scale ranges from 10^{-12} cm (the nucleus of an atom) to 10^{27} cm (the distance to the farthest galaxies yet seen). Note that the size of an atom has the same relation to that of a grain of sand as the size of a house to that of the Earth. The size scale is logarithmic, so that each step represents a tenfold change.

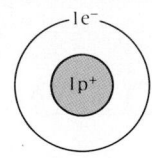

Hydrogen (H)
Atomic mass 1
Atomic number 1

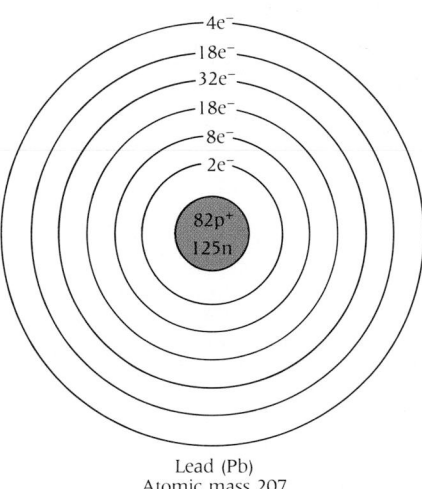

Lead (Pb)
Atomic mass 207
Atomic number 82

3-2 Atomic structures of hydrogen and lead
A hydrogen atom has only a single proton in its nucleus. Thus, its atomic mass is 1. Lead, by contrast, has 82 protons and 125 neutrons in its atomic nucleus, giving it an atomic mass of 207.

3-3 Isotopes of hydrogen and carbon
Hydrogen 1 (^1H) is by far the most common isotope of hydrogen. Carbon 12 (^{12}C) is the most common isotope of carbon. Of those shown, only tritium (^3H) and carbon 14 (^{14}C) are radioactive. Isotopes of an atom have similar chemical and biochemical properties, but different physical properties.

thereby produce different kinds of molecules. The simple (or empirical) formula for a molecule shows what atoms it contains. The formula for water is H_2O because water contains two atoms of hydrogen for every one atom of oxygen.

STRUCTURE OF ATOMS

Basic Particles

Each atom consists of a single nucleus and one or more electrons. The atomic nucleus contains protons and neutrons.[1] We can get some idea of the relative sizes of atoms and their nuclei—as well as of other structures in the universe—from the diagram in Figure 3-1. Note how small the nucleus is relative to the whole atom. Its diameter is roughly one ten-thousandth that of the atom itself—a little more than one foot wide if an atom were two miles across.

An atomic **nucleus** is certainly small, yet the mass or weight of the whole atom is concentrated almost entirely there. Hence, if nuclei could be packed together side by side, they would form matter of extremely high density and mass—perhaps of the order of 10^{12} grams per cubic centimeter. (Note that one cubic centimeter is equivalent to one milliliter.) As indicated in Box 3-1, 10^{12} grams is equal to about 1.1 million tons!

The **protons** and **neutrons** within the nucleus both have mass. We arbitrarily give each a mass value of 1. A proton also has a positive electric charge. A neutron, as its name implies, is electrically neutral. The third subatomic particle, the **electron,** is so small that its mass is assumed to be zero in most calculations. In fact, its mass is 1/1836 that of the lightest nucleus. That means that if a man weighed 150 pounds, or 2400 ounces, his electrons would contribute less than an ounce to the reading on the scale. Each electron bears a negative electric charge equal in magnitude to the positive charge of the proton. The total mass of the nucleus—that is, the number of protons and neutrons present—determines its **atomic mass** (also called *atomic weight*). For example, the simplest atom, hydrogen, has a nucleus consisting of a single proton; it contains no neutrons. Since the nucleus has a mass of 1, the atomic mass of hydrogen is said to be 1. (Atomic masses of other atoms of biological importance appear in Table 3-1.) In contrast, the nucleus of an atom of lead contains 82 protons and 125 neutrons. Hence, the atomic mass of lead is 207 (Figure 3-2). Note that the number of electrons equals the number of protons. This number, the number of protons, defines the **atomic number** of each element. The atomic number of hydrogen is 1; that of lead is 82.

Isotopes

Some elements occur in varying forms termed **isotopes**. Two isotopes of the same element have identical numbers of protons but different numbers of neutrons. Thus, they have the same atomic number but different atomic masses. As shown in Figure 3-3, carbon 12 (or, in the usual symbol, ^{12}C) and carbon 14 (^{14}C) are two isotopes of carbon. The former, the form in which carbon is most commonly found, has an atomic mass of 12. The latter has an atomic mass of 14. The isotope ^{14}C is **radioactive.** That means its nucleus is unstable, emitting energy as it changes to a more stable form. Both carbon isotopes have an atomic number of 6. Ordinary

---

[1] The atomic nucleus also contains other fundamental particles. Traditionally, their properties are matters for nuclear physics, not biology. However, that stronghold is under attack by biologists and should soon capitulate.

Hydrogen (^1H)

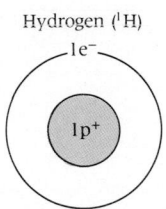

Deuterium (^2H)

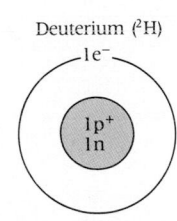

Tritium (^3H)

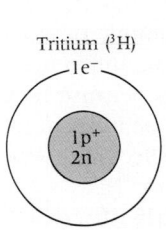

Carbon 12 (^{12}C)

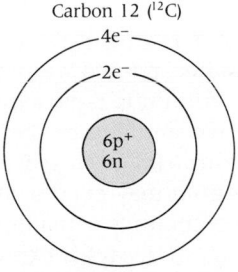

Carbon 14 (^{14}C)

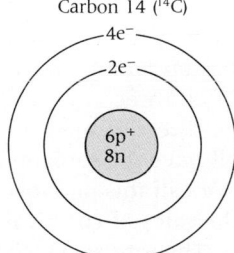

Scientists, and today many nonscientists (especially in Europe), use the metric system in all measurements of length, mass (or weight), and volume. We urge you to review these units and, if necessary, to commit their interrelationships to memory.

LENGTH

1 kilometer (km) = 1000 meters (or 10^3 m) = 0.62 mile, about 5/8 mile

1 meter (m) = 1/1000 kilometer (or 10^{-3} km) = 39.37 inches, just over a yard

1 centimeter (cm) = 1/100 meter (or 10^{-2} m) = 0.4 inch, just under 1/2 inch

1 millimeter (mm) = 1/1000 meter (or 10^{-3} m) = 0.04 inch

1 micrometer (μm) = 1/1000 millimeter (or 10^{-3} mm or 10^{-6} m)

1 nanometer (nm) = 1/1000 micrometer (or 10^{-3} μm or 10^{-9} m)

1 angstrom (Å) = 1/10 nanometer (or 10^{-4} μm or 10^{-10} m)

MASS

1 kilogram (kg) = 1000 grams (or 10^3 g) = 2.20 pounds

1 gram (g) = 1/1000 kilogram (or 10^{-3} kg) = 1/28 ounce

1 milligram (mg) = 1/1000 gram (or 10^{-3} g)

1 microgram (μg) = 1/1000 milligram (or 10^{-3} mg or 10^{-6} g)

1 nanogram (ng) = 1/1000 microgram (or 10^{-3} μg or 10^{-9} g)

VOLUME

1 liter (l) = 1000 milliliters (or 10^3 ml) = 1.06 quart

1 milliliter (ml) = 1/1000 liter (or 10^{-3} l) = 1 cubic cm $(cm^3)^*$

1 microliter (μl) = 1/1000 milliliter (or 10^{-3} ml or 10^{-6} l)*

* One ml of water weighs 1 g at 4°C. One μl of water weighs 1 mg at 4°C.

hydrogen (^{1}H) has an atomic mass of 1. One of its isotopes, known as deuterium, has an atomic mass of 2 (^{2}H). Another, termed tritium, is ^{3}H. The atomic number of all three is 1. Of the three, only tritium is radioactive. Despite these differences, all three isotopes of hydrogen behave similarly in chemical reactions because they all have the same number of electrons. However, reaction rates may be influenced. We call this an **isotope effect.** Tritium may exert a strong isotope effect and slow the rates of chemical reactions of compounds containing ^{3}H instead of ^{1}H.

The rate at which a radioactive atom decays is expressed as its **half-life,** which is the time it takes for half the radioactive atoms in a sample to convert to a stable form (or to another radioactive isotope). Half-lives of radioactive atoms vary widely. An isotope of nitrogen, ^{13}N, has a half-life of only 10 minutes; that of ^{14}C is 5730 years; ^{3}H, 12.3 years; ^{131}I, one isotope of iodine, 8.1 days; ^{133}I, another isotope of iodine, 20.9 hours; and ^{32}P, an isotope of phosphorus, 14.3 days. The half-life of ^{238}U, an isotope of uranium, is 4.5 billion years—close to the age of the Earth.

Radioactive and nonradioactive isotopes play at least two major roles in biological research. First, they are used as **tracers.** For example, if a molecule such as carbon dioxide (CO_2) contains ^{14}C instead of ^{12}C, it is radioactive. Nonetheless, compounds containing one isotope of carbon or the other behave almost identically in chemical reactions. Why? Because an atom's chemical reactivity is determined by how many electrons it possesses—and this number is the same in ^{14}C and ^{12}C. Therefore, if a radioactive sugar is found in the leaf of a plant exposed to $^{14}CO_2$, it can be safely concluded that the carbon atom of CO_2 became a carbon atom of the sugar—in other words, CO_2 is a precursor of that sugar.

Another role of isotopes has been in determinations of the age of old rocks and fossils (see Chapter 36).

Models of the Atom

Ordinary matter is electrically neutral—that is, it contains positive and negative electric charges in equal amounts. The story of the efforts of physicists over two centuries to fathom the structure of the atom is one of the great chapters in the history of science. Here we mention two modern theories of atomic structure. According to the early planetary atomic model—proposed in 1913 by Niels Bohr (1887–1962) of Copenhagen—the electrically negative electrons in an atom are attracted by the positively charged nucleus and move rapidly about it like planets about the sun in variously located orbital pathways, which occupy most of the atom's inner space. In fact, electrons travel only at certain fixed distances from the nucleus. The paths of the electrons in these orbits mark out a series of specific spherical, concentric **electronic shells** or **energy levels.** Each shell can hold only a fixed maximum number of electrons. In all atoms, the innermost or first shell can hold no more than 2 electrons. The second can hold up to 8 each, and the third shell can hold up to 18 electrons, but it is stable with 8 electrons.

The planetary atomic model also regards electrons as little spheres whose positions in space are exactly known. For a time, physicists accepted this. They eventually recognized, however, that electrons do travel in well-defined orbital shells, but are so omnipresent and unpredictable it is possible to speak only of the *probability* of their being here or there. For example, patterns of electron motion are now defined as a volume of space in which the electron is present, say, 90% of the time. This volume is the **orbital** of the electron. According to the electronic orbital model theory, electrons are viewed as a cloud of randomly moving particles (Figure 3-4). The orbital is a hollow shell of imaginary points defining a spherical layer within which the electron spends most of its time. Electrons in different orbitals differ in their amounts of energy. The more distant the orbitals, the more energy they possess. Orbitals thus define electron charge layers, volumes of space in which electrons can move and which can be distorted by other nearby atoms. In this model, the atom is more a vacant space than a small ball with a definite outer surface or boundary.

The picture conjures up a familiar image: a flock of small birds busily fending

3-4 The first three electron orbitals The density of colored shading shows the proportion of time an electron in that orbital spends in each part of the space surrounding the nucleus. These orbitals are superimposed in a single atom. Each orbital can contain up to two electrons. **(A)** The 1s orbital. **(B)** The 2s orbital. **(C)** A 2p orbital. Note that the orbitals of s electrons are spherical, while those of p electrons are dumbbell-shaped. There are three possible 2p orbitals, all oriented at right angles to one another.

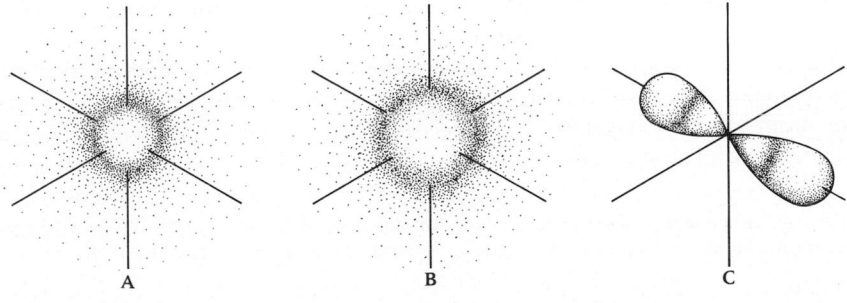

A   B   C

off a bird of prey. Like electrons they swoop around it, darting and looping, now near, now far, but like a nucleus the predator is always at their center. A rifle could fire a clip of bullets through the flock and never hit a bird—or, on the other hand, it just might hit one. In the same way, the "bullets"—actually, swiftly moving protons or other charged particles—in an atom-splitting *cyclotron* can pass through an atom and fail to hit even one electron. But if they do hit one and hit it hard enough, they will detach it from its orbit and cast it into "outer space." Unlike a bullet-pierced bird, however, such an electron is not damaged. It is merely freed to lead an unrestrained existence subject only to electrical influences. Many "bullets" can now be made to accomplish this kind of atomic disruption fairly dependably. Cosmic rays will do it, as will x-rays, emanations from radioactive substances, and streams of fast-moving electrons.

The hydrogen atom has a single electron in its first orbital. Since this shell can hold 2 electrons, hydrogen is said to have an *incomplete* shell. An atom of helium (He) does have 2 electrons in the first shell. Since this is the maximum possible number for this shell, it is said to be *complete*. Oxygen has 8 orbital electrons, 2 in the first shell and 6 in the second. Since the second shell can hold a maximum of 8 electrons, it also is incomplete (Figure 3-5). When atoms enter into chemical reactions, the number of electrons in the outer orbit increases or decreases.

Atoms whose electron shells are complete are notably stable. This means that they are **chemically inert,** that they rarely react with other atoms. Helium is such a substance.[2]

The atoms of all other elements have incomplete outermost shells and are therefore unstable—that is, they freely enter into chemical reactions. Electron-dot symbols provide a useful shorthand notation for representing the outermost electron shell.

$$\text{H} \qquad \text{He·} \qquad \text{:Är:} \qquad \text{:Ö·} \qquad \text{·C̈·}$$

$$\text{:F̈·} \qquad \text{:C̈l·} \qquad \text{Na} \qquad \text{Mg·}$$

Note that the inner electrons are not shown. When appropriate kinds of such atoms are brought together under favorable conditions, their incomplete outer electron shells may make them react with one another, as explained below.

The number of electrons that the atoms of a given element must gain or lose to complete its outer orbital shell is called its **valence.** Thus, oxygen has a valence of 2 and carbon a valence of 4. The valence of an atom determines how many bonds it can form when it reacts with other atoms.

CHEMICAL BONDS

Chemical reactions occur because atoms tend to utilize every opportunity to complete their outermost electron shell. A successful completion establishes a **chemical bond** that holds the atoms together in a **chemical compound.** An incomplete outer shell can become complete in several ways. Therefore, the resulting bonds are of several distinct types. Some are strong and some are weak.

Covalent Bonds

Atoms can become electronically stable by *sharing* electrons. When this occurs, electrons in the outermost shells move in orbitals that are shared by both atoms and we say that a **covalent bond** has been formed. If the bonded atoms share two electrons, a single covalent bond arises. When four electrons are shared, a double covalent bond forms. Covalent bonds are ordinarily the strongest kind of interatomic force. Atoms of elements of the halogen family, such as fluorine (F) and chlorine (Cl), have outer shells that are one short of the number of electrons needed to make the stable number of 8. One way in which the outer shell of a chlorine atom can be made stable is by reacting with another chlorine atom (Figure 3-6). The two atoms share electrons, each contributing one. Now each chlorine

[2] So are the other inert "noble gases": neon, 10 electrons in 2 shells; argon, 18 electrons in 3 shells; krypton, 36 electrons in 4 shells; xenon, 54 electrons in 5 shells; and radon, 86 electrons in 6 shells.

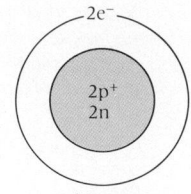

Helium (He)
Outer shell complete

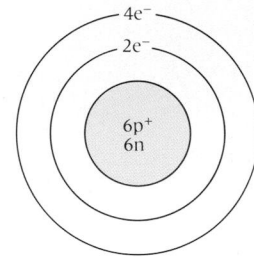

Carbon (C)
Outer shell incomplete by 4 electrons

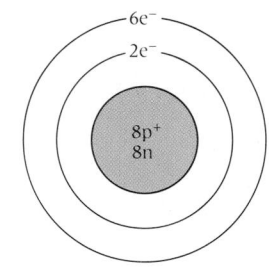

Oxygen (O)
Outer shell incomplete by 2 electrons

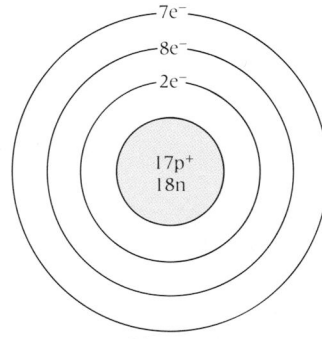

Chlorine (Cl)
Outer shell incomplete by 1 electron

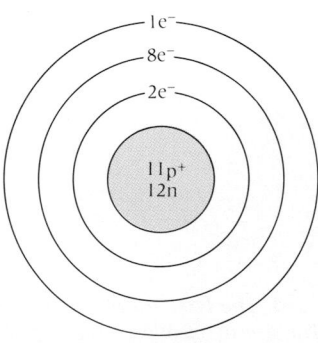

Sodium (Na)
Outer shell incomplete by 7 electrons

3-5 Atomic structures of various elements
The outer shell of helium is complete and contains two electrons. Because the shell is complete, the element is chemically inert. The outer shell of the other four atoms shown can contain a maximum of eight electrons each, but all contain a lesser number. Unlike helium, these elements are chemically reactive.

3-6 Formation of a covalent bond

(**A**) *Chlorine atoms lack one electron in their outer shells. They are more stable when they come together and share one pair of electrons, thereby completing their shells. The molecule Cl_2 results from this bond.* (**B**) *The shared electron pair can be represented as an overlap of the p orbitals of the two atoms.*

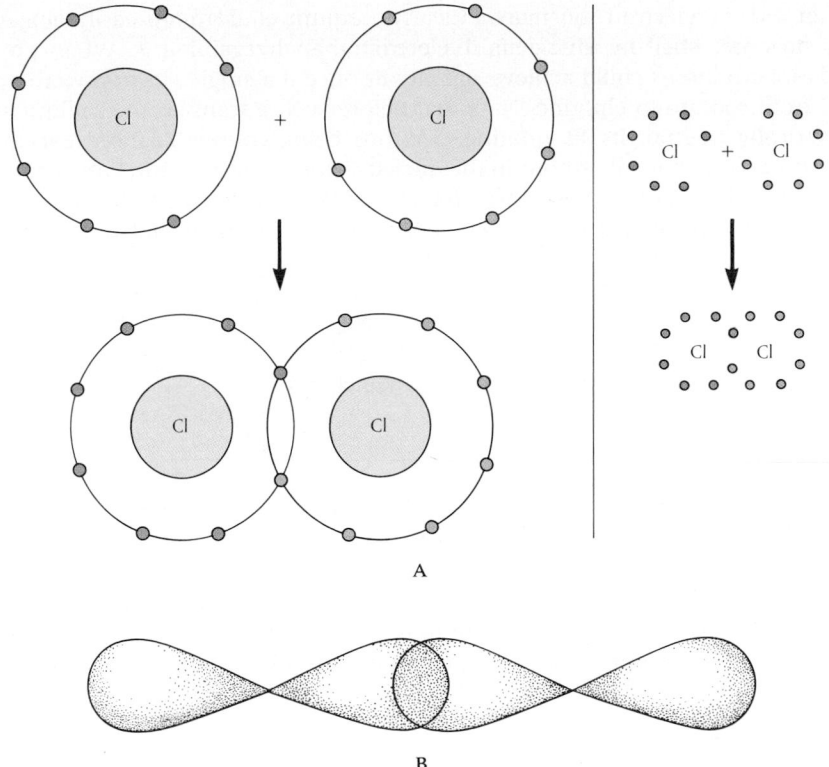

A

B

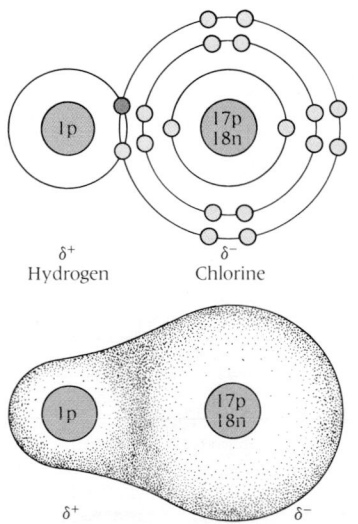

δ⁺
Hydrogen

δ⁻
Chlorine

δ⁺   δ⁻

3-7 A polar molecule

The two atoms of a molecule of hydrogen chloride or hydrochloric acid (HCl) share a pair of electrons. This forms a covalent bond that holds the atoms together. The electrons are attracted more strongly to the chlorine nucleus, which has 17 protons, than to the single proton of the hydrogen nucleus. Thus, the chlorine end of the molecule has a slightly negative charge (δ^-), while the hydrogen end, having lost its electron, has a slightly positive charge (δ^+).

atom behaves as if it actually possessed a full set of 18 electrons with a complete outer shell. The resulting combination—a chlorine *molecule* (Cl_2)—is stable. Hence, it is far less reactive than a chlorine *atom*. In the case of double covalent bonds (or triple ones), the atoms are held together by two (or three) pairs of electrons. The bonds are symbolized by double lines, as in $O{=}C{=}O$, the structural formula of CO_2, or by triple lines as in $H{-}C{\equiv}C{-}H$ (acetylene).

The bonds that form in Cl_2, or H_2 or O_2, are the results of two identical atoms bonding covalently. Such a bond is electrically symmetrical, each shared electron spending part of its time around one nucleus and part of its time around the other. When nonidentical atoms form a covalent bond, the shared electrons may be more powerfully attracted toward one atom than the other. This electrically *asymmetrical bond* is said to be **polar** (Figure 3-7). The contrasting *symmetrical* form is **nonpolar**. Covalent bonds between carbon and hydrogen are essentially symmetrical. In polar bonds electrons spend more time at one end of the bond than at the other. Molecules containing polar bonds are called polar molecules. They often contain one or more atoms of oxygen to which electrons are strongly attracted. Many of the special properties of water derive from the polar nature of water molecules. Another covalent bond of great biological importance is the one between carbon atoms.

Functionally, covalent bonds are notable for the large amount of energy needed to make them and to break them. This means that they hold atoms together very tightly. It should also be noted that when two or more atoms form covalent bonds with another central atom—as in the water molecule to be discussed later—these bonds are oriented at precise angles to one another. The angles are determined by the mutual repulsion of the outer electron orbitals of the central atom.

The covalent bond is so prevalent that G. N. Lewis (1875–1946), discoverer of its electronic basis, called it *the* chemical bond.

Ionic Bonds

An **ion** is an atom or a group of atoms that has acquired or lost one or more electrons. Consider sodium and chlorine.

The chlorine atom is unstable because it has 1 electron too few in its outer shell. The sodium atom is unstable because it has 7 electrons too few—or, to put it another way, 1 electron too many. Clearly, sodium could more easily achieve a complete outer shell by giving up 1 electron than by gaining 7. When present together, both atoms could achieve stability at once if a single electron were transferred from sodium to chlorine. Note that before such a transfer, the sodium atom is electrically neutral, its 11 orbiting electrons being counterbalanced exactly by the 11 positively charged protons in the nucleus. During the reaction with a chlorine atom, the sodium atom loses 1 negative charge when it loses an electron. Hence, it becomes positively charged: Its 11 protons are still present, but there are only 10 electrons.

$$Na = e^- + Na^+ \qquad (3.1)$$

Conversely, chlorine acquires a single negative charge.

$$Cl + e^- = Cl^- \qquad (3.2)$$

The charged forms, Na^+ and Cl^-, are ions. Negatively charged ions are called **anions,** and positively charged ions **cations.** The symbols Na^+, Mg^{2+}, or Cl^- denote ions and indicate the size and electrical character of the charge. Oppositely charged ions attract one another; the strong electrostatic attraction force that binds them together is an **ionic bond.** A group of ions bonded by this sort of electronic give-and-take is an **ionic compound.** For example, consider the reaction

$$Na^+ + Cl^- \rightarrow NaCl \qquad (3.3)$$

The product is the ionic compound NaCl, or sodium chloride.[3] Such an anion–cation combination is a **salt.** Salts are noted for their solubility in water and the ability of their solutions to conduct electric currents. Hence, they are **electrolytes.**

The formation of NaCl from Na^+ and Cl^- is an **electron-transfer** (or **ionic**) **reaction.** The total number of positive charges carried by the cation equals the total number of negative charges carried by the anion. Every transferred electron establishes one ionic bond. The number of bonds an ion can form with other ions is its ionic valence. The sodium ion has a valence of $+1$, the magnesium ion has a valence of $+2$, and the chloride ion has a valence of -1. In a sense an ionic bond is an extreme type of covalent bond. Electrons are shared in a highly asymmetrical manner, so that a majority of the bond's forces may be regarded as purely electrostatic—the attraction of electrical opposites. Note that ionic bonds are weaker than covalent bonds.

Hydrogen Bonds

Another type of chemical bond, the **hydrogen bond,** is much weaker than a covalent or ionic bond; but it is one of the most important molecular forces in the realm of biology. A hydrogen bond holds electrically charged groups under certain circumstances between an electronegative atom (the *acceptor* atom) and a hydrogen atom that is covalently bonded to another atom (the *donor* atom).

Consider the molecule of hydrogen fluoride (HF) in Figure 3-8. Note that the tiny hydrogen ion (H^+), lacking an electron, is a bare atomic nucleus resting on the surface of the larger and negatively charged fluoride ion. Since H^+ carries a positive charge, it is able to attract to itself another negatively charged fluoride ion. When one is found, a new ionic grouping occurs with the structure $(F^-H^+F^-)^-$ or HF_2^-, called the *hydrogen difluoride ion.* This ion is very stable. The bond holding it together is a hydrogen bond, indicated in a structural formula by a dotted line.

$$F^- \!-\! H^+ \cdots F^-$$

The bond is dotted to show that it is weaker than a covalent bond.

Interestingly, and for biochemists significantly, molecules of HF tend to *polymerize*—that is, they link up into chains—with such formulas as H_2F_2, H_3F_3, and so

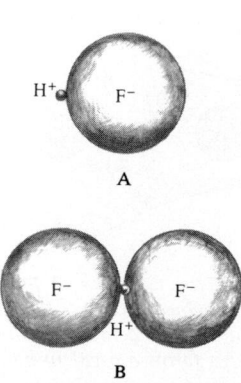

3-8 Bonds between hydrogen and fluorine atoms
(A) *A molecule of hydrogen fluoride or hydrofluoric acid.* **(B)** *A hydrogen difluoride ion, which has a net negative charge and contains a hydrogen bond.*

[3] In writing ionic compounds symbolically, one usually omits the electrical charges of the ions: NaCl is preferred to Na^+Cl^-.

on up through H_6F_6. These polymers arise because hydrogen bonds form between HF molecules.

$$H—F \cdots H—F \cdots H—F \cdots H—F$$

One small polymer, H_6F_6, is unusually stable because it can form an extra hydrogen bond that permits it to assume a stable ring structure.

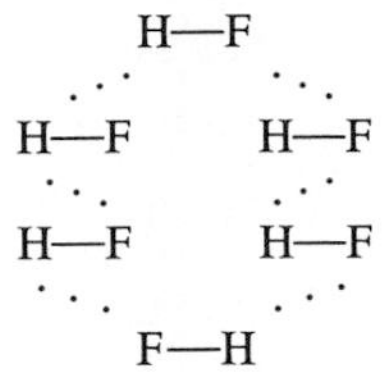

$$
\begin{array}{ccc}
 & H—F & \\
H—F & & H—F \\
H—F & & H—F \\
 & F—H &
\end{array}
$$

For biologists, the most familiar example of hydrogen bonding is the one occurring between neighboring water molecules. Because of its nearly tetrahedral arrangement, each water molecule tends to bond with four of its neighbors. Each water molecule has two attached hydrogen atoms and two unshared electron pairs. Thus, four hydrogen bonds can form between neighboring hydrogen and oxygen atoms.

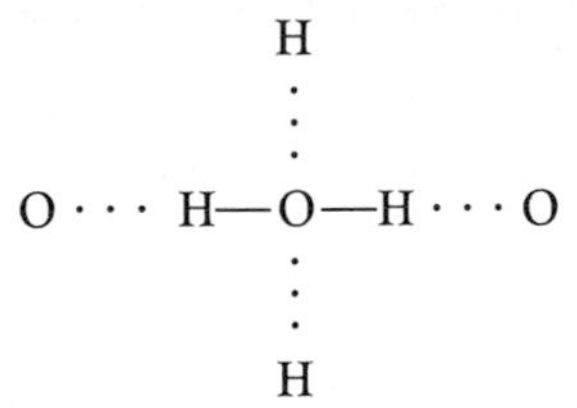

$$
\begin{array}{c}
H \\
\vdots \\
O \cdots H—O—H \cdots O \\
\vdots \\
H
\end{array}
$$

This distinctive tendency largely accounts for many of water's extraordinary properties.

A hydrogen bond can form only between certain elements. Its weakness is quite evident: It forms readily but it is easily broken—often to be replaced by another hydrogen bond. Despite this fact, large numbers of hydrogen bonds can act together to stabilize large molecules or groups of smaller molecules with considerable strength. That is why hydrogen bonds are so important in determining the properties of proteins, nucleic acids, and carbohydrates.

Weak Forces

Finally, there are the weak interactions between molecules that occur over very short distances as a result of mutual, nonspecific interactions of their electrons and nuclei. These feeble forces, termed **van der Waals forces,** are significant only when molecules are very close together.

Van der Waals forces combine with other weak bonds (hydrogen bonds) to help stabilize the three-dimensional structure of proteins and other large molecules. Roughly speaking, the relative strengths of three types of bonds in a *polar* medium are in the following ratio:

van der Waals forces	:	Hydrogen bonds	:	Ionic bonds	:	Covalent bonds
1	:	4.5	:	5	:	110

Certain weak forces called **hydrophobic interactions** occur when *nonpolar* molecules of low water solubility are surrounded by water molecules. The "attractions" caused by these forces actually result from repulsions caused by the solvent molecules. If surrounding solvent molecules "push away" nonpolar solute molecules, the latter may be combined almost as effectively as if they were attracted to one another. These forces also have an important role in stabilizing the molecular structures of proteins and other biomolecules.

TABLE 3-2

APPROXIMATE RELATIVE AMOUNTS
OF THE CHEMICAL ELEMENTS IN THE HUMAN BODY

Element	Symbol	Percent by Weight
Oxygen	O	65
Carbon	C	18
Hydrogen	H	10
Nitrogen	N	3
Calcium	Ca	2[*]
Phosphorus	P	1.1[*]
Potassium	K	0.35
Sulfur	S	0.25
Chlorine	Cl	0.15
Sodium	Na	0.15
Magnesium	Mg	0.05
Iron	Fe	0.0006
Iodine	I	0.00006
Cobalt, zinc, and other trace elements		Data incomplete

[*] Estimates differ widely.

MAJOR ELEMENTS OF LIVING ORGANISMS

All the materials in living organisms consist ultimately of chemical elements. The most abundant by far are oxygen, carbon, hydrogen, and nitrogen. As shown in Table 3-2, the "big four" account for 96% of the mass of the human body. A variety of other elements, including so-called trace elements and minerals, contribute the remaining 4%.

Interestingly, the four most abundant elements in the Earth's crust are oxygen, silicon, aluminum, and iron.[4] Clearly, these are not the same as the ones found in living organisms. We presume then that the elements in organisms must be uniquely fitted for their biological roles.

CARBON

Less than 200 years ago, it was widely believed that living organisms possess some sort of vital force that makes them fundamentally different from nonliving objects. When chemists finally examined living things, they discovered numerous substances that were previously unknown. The presence of these compounds—uric acid in urine, lactic acid in sour milk, citric acid in lemons, and countless others—seemed to bolster the notion of vitalism. These substances, it appeared, were unique because they were formed only by living organisms. They were called **organic** compounds to contrast them with the **inorganic** materials of lifeless rocks and minerals.

In 1826 the German chemist Friedrich Wöhler revolutionized thought on this question. At first by accident and later by design, he converted the "inorganic" substance ammonium cyanate (NH_4OCN) to the "organic" substance urea (NH_2CONH_2). This achievement and many that followed gradually demolished the imagined barrier between what were then viewed as the separate laws of inorganic and organic chemistry. We now know that the fundamental laws of both chemistries are identical. Because most molecules produced by living organisms contain carbon, the term "organic chemistry" came to mean the chemistry of carbon compounds. The study of compounds not containing carbon has been more or less lumped together under the heading "inorganic chemistry."

In living organisms, carbon atoms occur in thousands of different combinations, which are constantly being built up and broken down in the course of metabolism. In a sense, the biological itinerary of carbon atoms begins and ends with the simple molecule carbon dioxide (CO_2). At room temperature, CO_2 is a gas. At much lower temperatures ($-56.5°C$), it solidifies. Solid CO_2 is called dry ice. CO_2 is produced when pure elemental carbon burns—that is, when it is *oxidized*. It is also produced by those phases of the metabolism in which food molecules are oxidized.

---

[4] This statement takes no notice of the very abundant hydrogen in the waters of Earth. If hydrogen were taken into account, it would rank third, ahead of aluminum.

Gaseous CO_2 makes up about 0.03% of the atmosphere—a small percentage perhaps, but it accounts nevertheless for 2300 billion tons of CO_2. For land plants this is the main nonliving reservoir of carbon. The low percentage of CO_2 in the atmosphere is quite adequate for the world's plants under present conditions. Yet they would thrive even better at higher CO_2 levels. When the CO_2 in a greenhouse is artificially increased, tomatoes grow at twice their normal rate. During the Carboniferous period (more than 350 million years ago), when huge quantities of CO_2 were absorbed from the air into the dense vegetation that later became coal and other fossil fuels (Figure 3-9), the atmosphere had a much higher CO_2 level.

The oceans of the world comprise another major CO_2 reservoir. We can appreciate its enormous size by comparing the considerably larger amount of atmospheric CO_2 that dissolves in fresh water with the amount found in seawater. One liter of ice cold water dissolves less than 0.4 milliliters (ml) of atmospheric CO_2. A liter of seawater contains 34 to 56 ml of CO_2, which is 90 to 150 times the level of atmospheric CO_2. We will later see that high CO_2 levels are essential to the survival of the free-floating microscopic plants (phytoplankton) of the sea—critically important links in the oceanic food chain (see Chapters 43 and 45).

Gaseous CO_2 does not usually enter into chemical reactions directly. It dissolves freely in water, especially when under pressure (as we always discover when opening a soda bottle). In most of its biological reactions, dissolved CO_2 reacts with water to form the weak acid **carbonic acid** (H_2CO_3), which can react readily with other compounds.

The incorporation of dissolved CO_2 into the organic molecules of living organisms involves a complex reaction with water. In green plants, CO_2 and H_2O are combined into simple sugars in the remarkable process called photosynthesis (see Chapter 10). From the simple sugars, carbon is recombined into a huge number of other substances. Their breakdown in metabolism ultimately releases the organic carbon in large molecules, combining it again with oxygen to form CO_2.

HYDROGEN AND OXYGEN

Most compounds in living organisms contain carbon, hydrogen, and oxygen. Many contain only these elements.

Hydrogen is the lightest element. Free hydrogen (H_2) is a colorless, odorless gas that does not occur as such in most living organisms. Rather, hydrogen atoms are extensively combined with carbon, oxygen, and the other elements to form a huge variety of compounds. Compounds capable of dissociating to form a hydrogen ion (H^+) are called acids.

Free molecular **oxygen** (O_2) also contains two atoms. At moderate temperatures it is a gas. The great reservoir of O_2 is the atmosphere (some 21% by volume). Free O_2 dissolves in water, but much less readily than does CO_2. A liter of seawater dissolves only 8 ml of atmospheric O_2. In solution, oxygen remains in the form of O_2 and thus is available to aquatic organisms. In fact, O_2 must be dissolved in water to be used by living organisms, whether aquatic or terrestrial. Even air-living organisms can assimilate O_2 only if it is dissolved in water. This water, as discussed in Chapter 27, occurs in the thin layers of moisture lining the respiratory membranes. The low O_2 concentration in seawater relative to the atmosphere has many implications for marine organisms—implications for the pace of their lives and metabolism and the depths at which they live.

The principal biological role of oxygen in metabolism is to combine with carbon and hydrogen arising from the breakdown of organic compounds. These reactions result in carbon dioxide (CO_2) and water (H_2O).

$$C + O_2 \rightarrow CO_2 + \text{energy (heat)} \tag{3.4}$$

$$2H_2 + O_2 \rightarrow 2H_2O + \text{energy (heat)} \tag{3.5}$$

Energy (in the form of heat) is released in these reactions. Hence, there is much less energy in two molecules of H_2O than in two of H_2 and one of O_2.

Most organisms require oxygen and obtain it from water, oxygen-containing organic foods, and the atmosphere. The O_2 in the air can be used directly by most animals and plants. However, some primitive organisms live without O_2 and may even be killed by it.

Because an organism's O_2 consumption and CO_2 elimination are proportional

3-9 Coal
Hundreds of millions of years ago, huge amounts of CO_2 were absorbed from the air by the Earth's dense vegetation. The vegetation later became coal, a fuel that retains this carbon.

to its level of activity, it was once supposed that organic compounds like sugars and fats are "burned"—combined directly with O_2—in the body to furnish the body with energy. We know now that "burning" of "fuel" is not a very efficient method of mobilizing biologically useful energy. The chemical energy built into food molecules is released in other ways—by chemical reactions that use oxygen (see Chapters 8 and 9).

Plants, in the process of photosynthesis, utilize CO_2 (combining it with H_2O to form sugars) and release O_2. Animal metabolism utilizes O_2 and produces CO_2. Although some processes in plants utilize O_2, photosynthesis usually produces more O_2 than the plant needs. Therefore, green plants in the light generally contribute O_2 to the surrounding air or water. In the dark, where active photosynthesis cannot proceed, plants produce no extra O_2 although they continue to utilize O_2 and give off CO_2. That is why fish that share a pond with green plants can be suffocated at night: O_2 dissolved in the water is used up while none is produced.

NITROGEN

Nitrogen is the fourth most common element in living tissues, after oxygen, carbon, and hydrogen. It is a constituent of all proteins and nucleic acids, critically important classes of macromolecules.

A tremendous store of gaseous nitrogen (N_2) is the atmosphere, which is almost 80% N_2 (by volume). Yet animals and plants cannot use N_2 directly. Most plants can utilize nitrogen only in the form of inorganic compounds, such as ammonia (NH_3), nitrates (salts containing NO_3^-), or in some cases nitrites (salts containing NO_2^-). Animals acquire their nitrogen by eating plant or animal tissues.

The withdrawal of N_2 from the atmosphere and its incorporation into living organisms—termed **nitrogen fixation**—depends entirely on chemical processes that can convert free N_2 to ammonia, nitrates, or nitrites. These processes rarely involve a direct combination of elements, for a simple reason: Free N_2 is itself quite unreactive chemically. Such syntheses can occur when lightning strikes and, as we shall see, they probably took place on a primitive Earth which had an atmosphere containing no O_2. In today's world, compounds of nitrogen are formed mainly by a small number of unique living organisms that are able to convert the N_2 of the air into ammonia, nitrates, and nitrites. These organisms include some cyanobacteria and some bacteria. Certain kinds of nitrogen-fixing bacteria live in nodules in the roots of beans and related plants (legumes) (Figure 3-10). We discuss them in Chapter 46.

Once compounds of nitrogen have been formed, few organisms can completely decompose them and convert them back into free N_2. Some bacteria can do it; but the end products of nitrogen metabolism in most organisms are nitrogen-containing organic compounds. Plants and bacteria then convert these products into nitrates and nitrites. In this way, nitrogen atoms cycle repeatedly through the living world.

MINERALS AND TRACE ELEMENTS

The concentrations in living tissues of the major elements—carbon, hydrogen, oxygen, and nitrogen—are measured in grams per kilogram of tissue. Many other elements are required in smaller amounts by living organisms. One class of these elements are the **minerals:** sodium (not required by plants), magnesium, sulfur, potassium, chlorine, phosphorus, calcium, and several more. Their sources in most cases are the inorganic salts dissolved in water—in the soil, lakes, streams, and seas. Plants, and to an extent aquatic animals, usually acquire these salts directly from their watery environment. Land animals also acquire some of their salts directly—for example, human beings use salt shakers. But most of an animal's mineral requirement is satisfied by minerals present in certain of its foods. For example, milk is rich in calcium; bananas are a good source of potassium.

One class of mineral elements essential in the diet includes those that are required each day in large (gram) quantities. They serve as structural components of tissues (for example, calcium in bone) and as constituents of body fluids. Indeed, they are essential for the function of all cells.

A second class of essential elements occurs in organisms in much lower concentrations—milligrams or micrograms per kilogram of tissue. In the daily diets of humans, such elements are required in comparable quantities (Table 3-3).

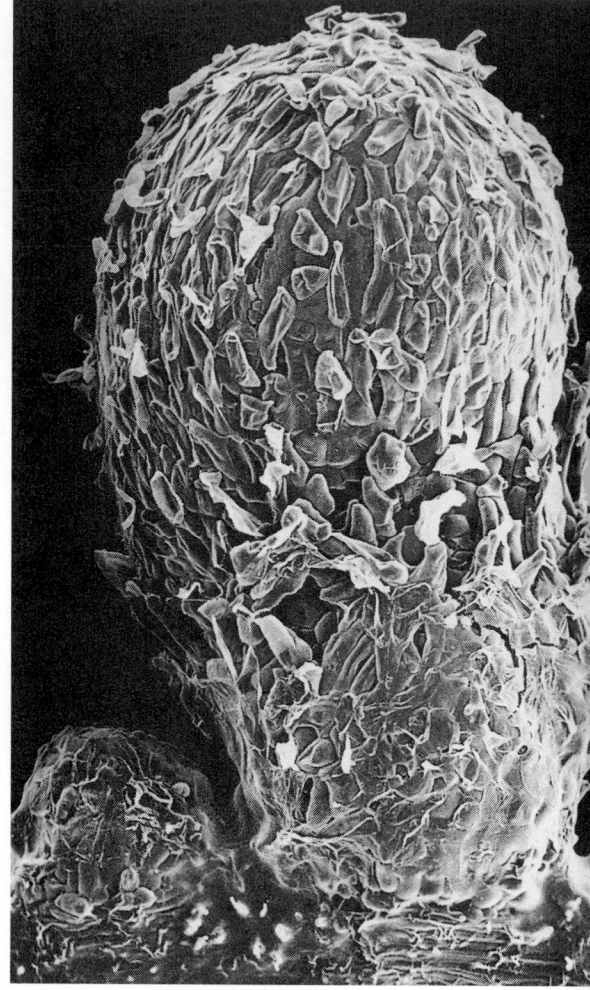

3-10 Nitrogen-fixing bacteria
A scanning electron micrograph of bacteria in a root nodule of a leguminous plant. Bacteria of this type are the principal sources of nitrogen for other organisms.

TABLE 3-3
TRACE ELEMENTS IN THE HUMAN DIET[*]

Element	Intake (mg/day)
Iron (males)	10
Iron (females)	18
Zinc	15
Manganese	2.5–5.0
Fluorine	1.5–4.0
Copper	2.0–3.0
Molybdenum	0.15–0.5
Chromium	0.05–0.2
Selenium	0.05–0.2
Iodine	0.15

[*] Recommended levels of trace elements in the daily diet of adult humans.

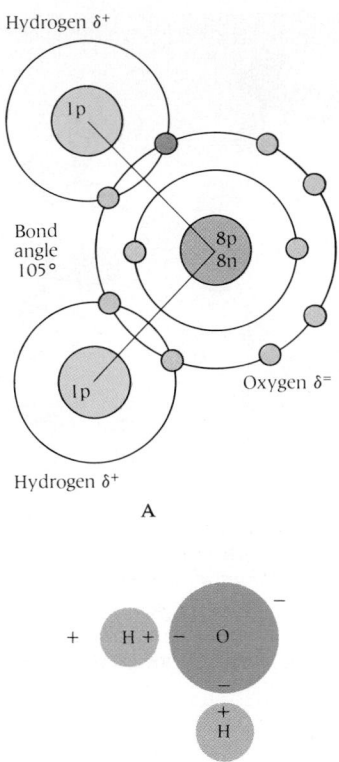

Hydrogen δ+

Bond
angle
105°

1p

8p
8n

1p

Oxygen δ=

Hydrogen δ+

A

+   H +   −   O   −

+
H

+

B

3-11 Water

(A) *A water molecule is bent with an angle of 105 degrees between the two hydrogen-oxygen bonds.* **(B)** *The distribution of charges in a single water molecule.*

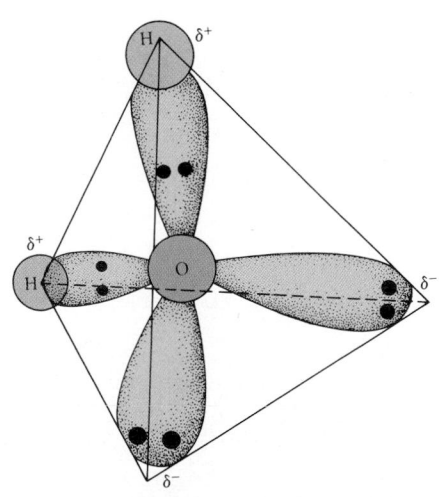

H   δ+

δ+

H   O   δ−

δ−

3-12 Structure of a water molecule

Its polarity is due to charge differences in different regions of the molecule. The regions containing the hydrogen atoms have a slightly positive charge (δ+); the opposite regions have a slightly negative charge (δ−). Note that when oxygen bonds covalently with two hydrogen atoms, two of the orbitals of its second-level electrons are formed by the shared electrons and have a slightly positive charge. The other two, called hybrid orbitals, have a slightly negative charge. The four orbitals are oriented to the four corners of a tetrahedron.

The first group are **macrominerals.** The second, the **microminerals,** were so difficult to measure in earlier days, they came to be known as *trace elements.* Remarkably, the trace elements present in living tissues include *all* of the chemical elements of the periodic table. Many have important functions as essential components of enzymes or of other proteins.

Macrominerals comprise about 99% of the mineral content of most organisms. Nevertheless, micronutrient elements have great significance despite their extremely low concentrations in tissues.

WATER

By far the principal contributor to the weight of all living tissues is water. In the human body, more than half of the weight of most tissues is due to water. More than 90% of the blood plasma and 60% of the red blood cells is water; muscle tissue is approximately 80% water, and water makes up considerably more than half—often more than three-quarters—of most other tissues. The only notable exceptions are certain relatively inert tissues, such as hair, horn, and the solid portion of bone. The spores of certain plants and bacteria have a low water content, as do seeds, but they are relatively inactive. When spores are transformed into cells that show active metabolism and growth, an increase in water content is an essential part of the transformation. Just that happens when seeds are watered and begin to sprout. Clearly, water is the indispensable matrix for the activities of living organisms.

Water is also a principal constituent of the environment in which organisms live. Although most of Earth's surface is covered with water and most of that water is liquid, large portions of it are in the form of ice and water vapor. The water of ice and vapor is also of great significance to the living world.

UNUSUAL PROPERTIES OF WATER

However odd it may seem, water is important to living organisms because it has extremely unusual properties. To the chemist familiar with other compounds, water is a strange substance the behavior of which is full of apparent anomalies. But it was these properties that gave water a key role in shaping the character of the physical and biological worlds and in helping to determine the course of their evolutionary development.

In a water molecule, one atom of oxygen is bonded covalently to each of two atoms of hydrogen. The shape of the water molecule is that of a triangle (Figure 3-11A). The powerful attraction of the oxygen nucleus tends to draw electrons away from the hydrogen protons, leaving the region around them with a net positive charge. The two pairs of unshared electrons tend to concentrate in directions pointing away from the O—H bonds (Figure 3-11B). If a tetrahedron is drawn about the oxygen atom, with the H nuclei at two of the corners, the unshared electrons (centers of negative charge) are concentrated along the lines leading to the other two corners. The water molecule is, therefore, an electrically polar structure (Figure 3-12).

In a group of water molecules clustered together, a positively charged region in one molecule tends to point toward a negatively charged region in one of its neighbors. Each of a water molecule's two regions with a concentration of negative charge attracts a proton of a neighboring water molecule. Each of its own protons attracts the oxygen of a neighbor (Figure 3-13). Thus each oxygen atom is the center of a tetrahedron of other oxygens. The linkage connecting the proton with a neighboring oxygen atom is a hydrogen bond. Hydrogen bonds also form readily between water molecules and the oxygen or nitrogen atoms of other molecules.

This ability to form hydrogen bonds prolifically accounts for most of the properties of water that make it so interesting to biologists. Among these properties are:

- Expansion on freezing
- A unique degree of cohesion and adhesion
- Extremely unexpected thermal properties
- Excellence as a solvent for polar molecules

Freezing

As their temperature decreases, most substances decrease in volume and therefore increase in density. (**Density** is defined as mass per unit volume; hence, if volume decreases and mass stays the same, density increases.) Remarkably, ice is less dense than cool water, the greatest density of liquid water occurring at 4°C. With further cooling or heating, the volume of a water sample increases. Consequently, its density decreases below or above 4°C.

A few other substances expand on solidifying, but the phenomenon is rare. Its profound importance for biology and geology will be seen again and again later in this book.

If ice were heavier than water, it would sink to the bottom on freezing. As Count Rumford showed about 150 years ago, it is possible to hold a piece of ice trapped at the bottom of a vessel filled with water, and then to heat, or even boil, the water at the top without melting the ice below. In his words:

> And so it would be with lakes, streams, and oceans were it not for the anomaly and the buoyancy of ice. The coldest water would continually sink to the bottom and there freeze. The ice, once formed, could not be melted, because the warmer water would stay at the surface. Year after year the ice would increase in winter and persist through the summer, until eventually all or much of the body of water, according to the locality, would be turned to ice. As it is, the temperature of the bottom of a body of fresh water cannot be below the point of maximum density; on cooling further the water rises; and ice forms only on the surface. In this way the liquid water below is effectually protected from further cooling, and the body of water persists. In the spring the first warm weather melts the ice and at the earliest possible moment all ice vanishes.

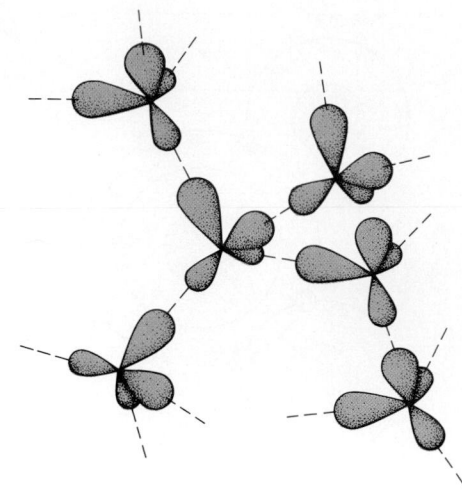

3-13 Hydrogen bonds in water
The positive regions of water molecules attract the negative regions of nearby water molecules, forming hydrogen bonds between them. These bonds, which continually break and re-form, account for many of the unusual properties of water.

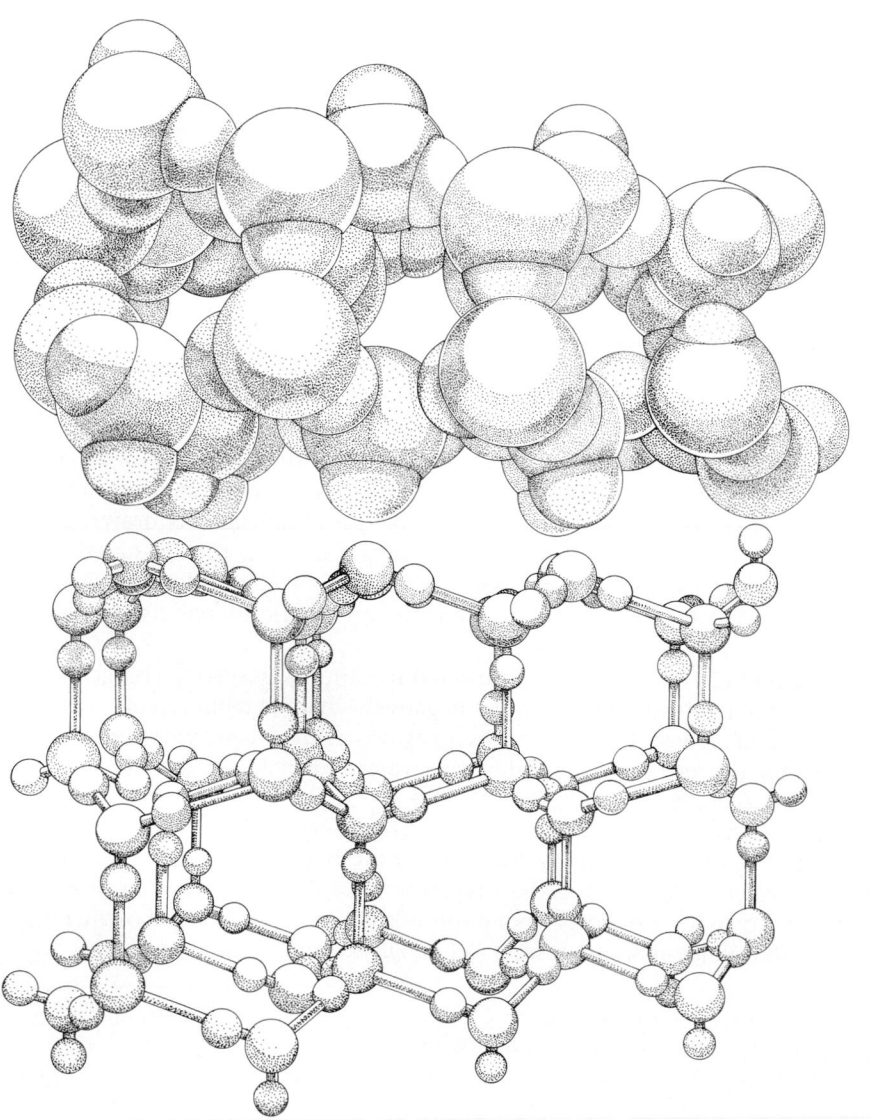

3-14 Ice crystal structure
The molecules above are shown with approximately correct shapes. Molecules are held in the lattice by hydrogen bonds. The large amount of open space gives ice its low density. To better show the geometry of the ice crystal lattice, the molecules below are drawn with small spheres representing oxygen atoms and smaller ones, hydrogen atoms.

Substance	Surface tension
Water at 0°C	75.6
Water at 20°C	72.75
Water at 40°C	69.56
Water at 100°C	58.9
Acetic acid	27.6
Acetone	23.7
Ammonia	41.3
Benzene	28.9
Chlorobenzene	33.2
Cyclohexane	25.3
Ethyl acetate	23.9
Chloroform	27.1
Ethyl alcohol	22.3
Ethyl ether	17.0
n–Hexane	18.4

3-15 Surface tensions
The surface tensions of water and several organic substances are compared here. The units are dynes/cm. Water's high surface tension is due to the hydrogen bonds forming between water molecules at and just below the surface.

3-16 Water strider, Gerridae
The water strider's ability to walk on water depends on the high surface tension of water.

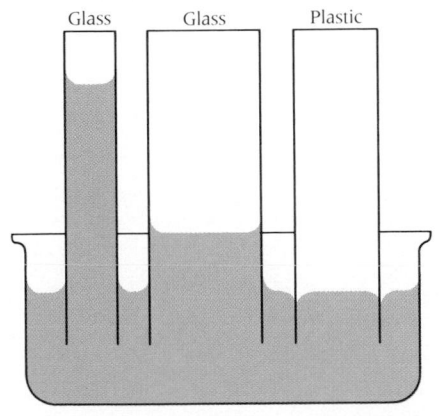

3-17 Capillary action
Water rises in thin glass tubes because polar water molecules are attracted to the charged groups on the glass surface and form hydrogen bonds with them. It does not rise as far in wider glass tubes because a smaller percentage of the water molecules is in contact with the glass surface. Water does not rise at all in plastic tubes because plastic surfaces are uncharged and therefore nonwettable.

This property of water is a direct result of the facility with which hydrogen bonds form between water molecules. As shown in Figure 3-14, these bonds give ice a loose and open structure—and hence a lower density—than it would have if the molecules were more closely packed. As a result, ice floats. As it melts, the open crystal structure is partially destroyed, and the molecules move closer together. Cool water is denser than ice. At temperatures higher than 4°C, the expansion due to increasing molecular agitation starts to overcome this effect and causes the density of water to decrease with increasing temperature.

Cohesion

As we have seen, one end of a water molecule (the hydrogen end) is more positively charged than the other (oxygen) end. Thus, the molecule is something like a bar magnet. In liquid water, the molecules align with the positive ends of some molecules abutting the negative ends of other molecules—and hydrogen bonds form between them. These bonds give water its considerable **cohesion**, a property defined as the holding together of *like* substances. When water molecules form hydrogen bonds with *other* molecules—such as cellulose in a plant cell—the phenomenon is termed **adhesion.**

The unique degree of molecular cohesion in water accounts for its **surface tension**. Water has high surface tension because hydrogen bonds form between water molecules at the surface and between them and the molecules just below the surface. Water has the highest surface tension of any known liquid, other than certain metals in the liquid state, and certain fused salts (Figure 3-15). When the surface of a liquid is increased, molecules which were formerly in the interior must be brought to the surface. To accomplish this, *work* must be done to oppose the attractive forces operating between these molecules and their neighbors. (Work is done when energy is applied in some task.) Thus, we speak of **surface energy**—the energy which must be expended to increase the surface tension.

Surface tension causes the surface of a liquid, like a stretched membrane or "skin," to contract in area as much as it can. Surface tension causes the water in a filled glass to rise above the rim; it makes it possible for insects like water striders to scuttle across the surfaces of ponds (Figure 3-16) and explains the tendency of water to creep upward into fine tubes (Figure 3-17), as it does in a plant stem. Surface tension is high in glass tubes because water molecules are attracted electrostatically to the charged groups on the glass surface. For that reason glass tubes are termed *wettable*, or *hydrophilic*. The surfaces of plastic or waxy tubes lack surface charges and are *nonwettable*, or *hydrophobic* (Figure 3-17).

Thermal Properties

The thermal properties of water have made it an ideal medium for the origin, development, and maintenance of life. Because of these properties, water plays a major part in keeping Earth's surface at a relatively constant temperature. Water also manages to regulate and moderate sudden temperature fluctuations *within* living organisms.

Heat capacity is the amount of heat required to raise the temperature of a certain quantity of a substance by 1°C without changing its phase (that is from solid to liquid or liquid to gas). Because of the high heat capacity of water, much more heat is necessary to increase the temperature of a given amount of water by a given number of degrees than is necessary to heat most other substances. Thus, water acts as a temperature buffer, maintaining temperature constancy more successfully than most other substances.

Water has a uniquely high heat capacity. The **specific heat of water**—that is, its heat capacity per gram—is by definition unity, or 1 (Figure 3-18). The standard *calorie* is the amount of heat required to raise the temperature of 1 gram of water 1°C (from 16.5° to 17.5°C). The specific heat of liquid ammonia, 1.23, surpasses that of water in this respect. The larger the specific heat, the smaller is the temperature rise produced by a given amount of heat. Consequently, the problem of temperature regulation is far simpler in systems composed chiefly of water or liquid ammonia than in any others. The processes of metabolism lead to the production of heat. An average man, weighing 60 kilograms, may in the course of an average day

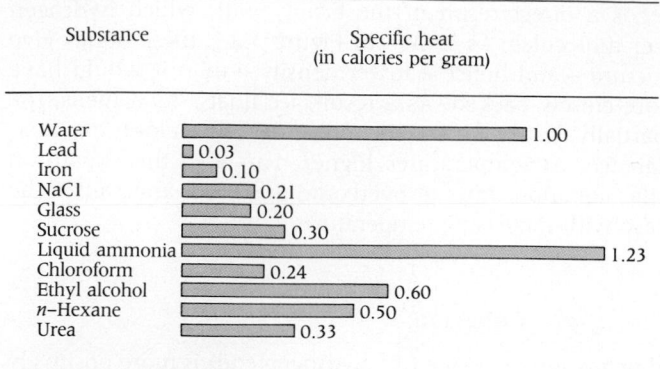

Substance	Specific heat (in calories per gram)
Water	1.00
Lead	0.03
Iron	0.10
NaCl	0.21
Glass	0.20
Sucrose	0.30
Liquid ammonia	1.23
Chloroform	0.24
Ethyl alcohol	0.60
n–Hexane	0.50
Urea	0.33

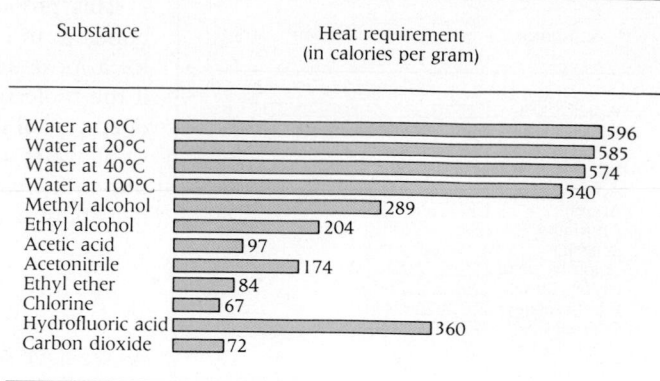

Substance	Heat requirement (in calories per gram)
Water at 0°C	596
Water at 20°C	585
Water at 40°C	574
Water at 100°C	540
Methyl alcohol	289
Ethyl alcohol	204
Acetic acid	97
Acetonitrile	174
Ethyl ether	84
Chlorine	67
Hydrofluoric acid	360
Carbon dioxide	72

produce 2500 kilocalories of heat, which in a closed system containing 60 kg of water would raise the temperature by more than 40°C. But by surface evaporation (and other cooling devices to be discussed later), the man can dissipate most of that heat. However, a far greater temperature rise would occur if his main body fluid was something other than water.

Hydrogen bonding is responsible for the high heat capacity of water. These bonds tend to restrict the movement of molecules. Since heat is in essence associated with molecular movement, a good deal of energy is needed to rupture the large number of hydrogen bonds, so relatively little of the energy applied may be left over to move the molecules more rapidly—that is, to increase the temperature.

The **heat of vaporization** is the amount of heat required to vaporize or evaporate a standard amount of liquid at its normal boiling point. The **heat of fusion** is the amount of heat required to convert a solid to a liquid at its melting point. Owing to its hydrogen bonding activity, water has very high heats of vaporization and fusion (Figure 3-19). This property of water is critically important in the regulation of climate and environmental temperature. Water enters the atmosphere by evaporation, and evaporation utilizes large amounts of heat because of the high heat of vaporization. Consequently, water tends to prevent rapid rises in temperature at times of sudden heat input (such as sunrise). Likewise, the high heat transfers involved in the freezing of water explain why large bodies of water seldom develop temperatures below the freezing point.

Solvent Properties

The special property of water that, perhaps more than any other, accounts for its significance in biology is the abnormally high value of what is called its **dielectric constant.** The dielectric constant involves the power of water to form hydrogen bonds and relates to the polarity of water molecules—to the presence of a positive charge at one end of a molecule and a negative charge at the other (see Figure 3-11B)—and to the *length* of the molecule.

Since water is a prolific hydrogen bond former, it freely combines with other water molecules, creating in effect a huge complex that is much larger than the individual water molecules within it. As a result, the distance between *free* positive and *free* negative charges is lengthened. Therefore, its dielectric constant is unusually high. A fluid with a high dielectric constant weakens the attraction between ions of opposite charge in an ionic bond. This makes the fluid a good solvent for polar molecules. In solid crystals of sodium chloride, for example, Na and Cl are united by ionic bonds. Their mutual attraction is only one-eightieth as great as when in water, where each ion is surrounded by water molecules that are attracted to it (Figure 3-20A). As a result, the Na^+ and Cl^- ions separate and drift away from the crystal into the surrounding water. In other words, the salt *dissolves* (Figure 3-20B).

The dielectric constant of water is extremely high compared to that of other solvents. Therefore, water, a highly polar substance—readily dissolves a large variety of polar molecules. They are of two types.

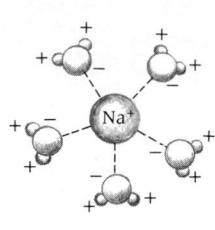 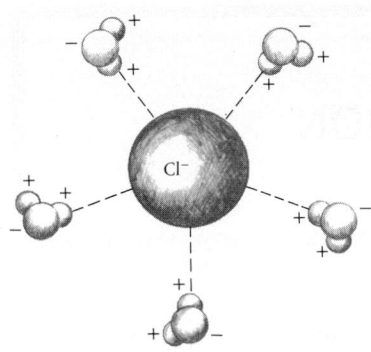 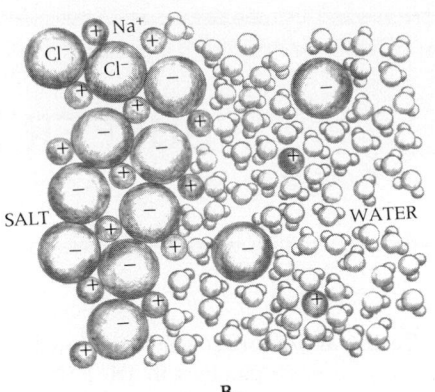

A

B

3-20 Salt in water

(A) When NaCl is added to water, each Na$^+$ and Cl$^-$ ion electrostatically attracts polar water molecules. Note that the oxygen end of a water molecule is attracted to Na$^+$, while the hydrogen end is attracted to Cl$^-$. Individual ions thus become hydrated—that is, surrounded by clusters of water molecules. Encased in these shells of water, the ions' attraction for one another diminishes greatly, and they move apart. In this way the crystal of NaCl dissolves. (B) Two hydrated ions move away from the rest of the crystal, which has not yet dissolved.

- Polar salts and ionic compounds readily form *aqueous* (watery) ionized solutions. These substances are nearly insoluble in chloroform or benzene, liquids that have low dielectric constants and are thus nonpolar.
- Polar nonionic compounds such as sugars and alcohols are soluble because water forms hydrogen bonds with them.

Substances that dissolve readily in water are **hydrophilic.** Those, like fats and oils, which seem to prefer their own kind to water, are **hydrophobic.**

IONIC DISSOCIATION

The electrostatic attraction between the sodium and chloride ions in NaCl accounts for the structural regularities of the NaCl crystal. But when solid NaCl dissolves in water, as we have just seen, Na$^+$ and Cl$^-$ ions leave the crystal and move randomly among the water molecules. This separation into free ions is termed **ionic dissociation.**

Dissociation of Salts in Water

In part, the solvent action of water is due to its high dielectric constant. Another phenomenon helps keep the Na$^+$ and Cl$^-$ ions apart once they have left the NaCl crystal.

Consider what happens when the ions of a salt dissolve in water. We may picture the simple ions as if they were tiny charged spheres. The positive charge on the sodium ion (a cation) tends to attract water molecules around it, the negative (oxygen) ends of the water molecules pointing toward the ion. The electric field around the ion is intense, and so the orienting force is very great. A whole cluster of water molecules becomes arranged around the ion, and the intense electric field causes them to pack very closely. Similarly, water molecules cluster around the anion, in this case with their positive (hydrogen) ends pointing toward the ion. These shells of water molecules around the ions produce electric fields of their own, and these fields are oriented in a way that opposes the fields arising from the ions themselves. The ions are, as it were, *kept apart* from one another by the water molecules clustered around them. Thus, water is an effective solvent for salts not only because of its high dielectric constant but also because water molecules tend to combine with ions to form **hydrated** ions.

Almost all chemical reactions of biological importance occur in aqueous solutions. Ionic compounds dissociate in water and yield free ions, as illustrated in the following examples.

$$NaCl \rightleftharpoons Na^+ + Cl^- \qquad (3.6)$$

Sodium chloride — Sodium ion — Chloride ion

$$CH_3COOH \rightleftharpoons CH_3COO^- + H^+ \qquad (3.7)$$

Acetic acid — Acetate ion — Hydrogen ion

$$NH_4OH \rightleftharpoons NH_4^+ + OH^- \qquad (3.8)$$

Ammonium hydroxide — Ammonium ion — Hydroxyl ion

The **buffering action** of a solution is defined as its ability to resist changes in $[H^+]$ that in its absence would otherwise result from the addition of acid or base. A buffer is the combination of substances that is responsible for this effect. As a general rule, the strongest buffering action is exhibited by ions of weak acids or weak bases. Strong acids and bases are almost completely dissociated in water. Hence they leave almost no reservoir of undissociated acid or base. Weak electrolytes that exhibit buffering action have an extremely important practical role within organisms, since cells can usually survive only within fairly narrow pH limits.

When HA dissociates to H^+ and A^-, a state of equilibrium exists:[*]

$$K = \frac{[H^+] [A^-]}{[HA]}$$

Solving for $[H^+]$:

$$[H^+] = K \frac{[HA]}{[A^-]}$$

[*] It should be noted that the discussion here concerns *concentrations* of H^+, A^-, HA, and so on. It would be more accurate to say that these equations apply rigidly only to *activity* values. Activities may be visualized as effective concentrations involving the behavior of both solute and solvent. We refer to concentrations instead because they are nearly correct and are more convenient.

Taking the negative logarithm of both sides:

$$-\log [H^+] = -\log K -\log \frac{[HA]}{[A^-]}$$

Substituting pH for $-\log [H^+]$ and pK for $-\log K$:

$$pH = pK -\log \frac{[HA]}{[A^-]}$$

If the sign is changed, we obtain the **Henderson-Hasselbach equation:**

$$pH = pK + \log \frac{[A^-]}{[HA]} \qquad (1)$$

or, in more general terms:

$$pH = pK + \log \frac{[\text{proton acceptor}]}{[\text{proton donor}]}$$

The most effective buffer action is exhibited by a *mixture of a weak acid (HA) and its salt (BA)*. Salts may be regarded as completely dissociated; hence, BA exists in solution as B^+ and A^-.

Let us assume that a weak acid HA dissociates only to the extent of contributing 10^{-5} mole of H^+ and A^- for each mole of HA present at the start. We say, then, that its pK value equals 5, bearing in mind that the

pK is the ionization constant of the weak acid. For our purposes, we may regard an acid this weak as entirely undissociated. That means that the amount of the ion A^- contributed by the acid HA is negligible compared with that coming from the salt BA, which dissociates completely. Hence, [BA] may be assumed to equal $[A^-]$. In view of the equivalence of [BA] and $[A^-]$, the Henderson-Hasselbach equation (Equation 1) may be rewritten in the form:

$$pH = pK + \log \frac{[\text{salt}]}{[\text{acid}]} \qquad (2)$$

Consider a mixture of a weak acid HA and its salt BA, both of which are present at concentrations of 0.1 M. Then:

$$pH = 5 + \log \frac{0.1}{0.1}$$

Therefore, the pH of the solution is 5.

What if 0.01 mole of base BOH is added to this solution? There forms an additional 0.01 mole of BA (from the combination of BOH with HA) and 0.01 mole of water; the latter may be ignored since its contribution to the total water present is negligible. The pH may be calculated again by inserting the increased salt concentration and decreased acid concentration in the equation:

Since the dissociated ions permit the passage of electric currents, these compounds are electrolytes.

Acids and Bases

In Equation 3.7, representing the dissociation of acetic acid, a hydrogen ion, H^+, is liberated. An **acid** is strictly defined as *any* compound that can liberate a hydrogen ion. A **base** has the ability to combine with the hydrogen ion furnished by an acid. Therefore, acids are commonly referred to as H^+ (proton) donors and bases as H^+ (proton) acceptors.

Note, in the following examples, that water can serve as both an acid and a base. Any substance capable of doing both is said to be **amphoteric.** Note also that acids and bases can be both electrically neutral molecules (for example, HCl and NaOH) or charged ions (for example, NH_4^+ and HPO_4^{2-}), since both can liberate or accept hydrogen ions.

$$pH = 5 + \log \frac{0.11}{0.09}$$

The pH is now 5.09, an increase of only 0.09 pH unit. The addition to the original buffer of 0.01 mole of a strong acid would decrease the pH by 0.09 pH unit, yielding a solution with a final pH of 4.91.

It is important to contrast this behavior with that of a strong acid like hydrochloric acid (HCl) at the same pH, namely 5. Since HCl dissociates completely, it would have to be present at a concentration of $0.00001M$ to yield a pH of 5. The addition of $0.01M$ base would bring the solution to a pH of almost 12, a change of 7 pH units, since there is no reservoir of undissociated acid. Even at pH 2 ($0.01M$ HCl), addition of $0.01M$ base brings the solution to neutrality, a change of 5 pH units. In sum, the addition of $0.01M$ base to a mixture of weak acid HA and its salt BA at pH 5 raises the pH by 0.09 units. The addition of $0.01M$ base to strong acid HCl at pH 5 raises the pH nearly 7 points. The former is a buffer; the latter is not.

In regions where weak acids and bases exert their buffering action, from about pH 2 to pH 12, this action offers an important means of maintaining constant pH. Near neutrality, where most living cells function, certain buffers are of paramount importance. The inorganic acids of greatest value in this regard are carbonic acid (H_2CO_3) and phosphoric acid (H_3PO_4).

Equations 1 and 2 have been derived in a way that makes them applicable to all weak electrolytes. They also describe the **titration** of all weak acids and bases. The titration curve of a monobasic weak acid is shown in Figure 3A. The pH at the midpoint of the titration curve, where 50% of the acid or base has been used up, equals the value of the **pK,** since here [salt]/[acid] is 1 and log [salt]/[acid] is zero. Buffering capacity

is always greatest at the pH that equals the pK value, diminishing at more acid or more alkaline values. Buffering capacity is determined largely by the *ratio* of salt to acid. When the ratio is 1, the pH is at the pK value of the system and maximal buffering is found (Figure 3B). This corresponds to the change in slope of the titration curve. The inflection point is the pK. The term pK' takes note of nonideal behavior.

For **polybasic acids**—that is, acids that can dissociate to yield more than one proton—each group has its own characteristic pK value which may be described by Equation 1. For example, phosphoric acid (H_3PO_4) has three protons and three different pKs. It can therefore buffer at three different regions of the pH scale:

H_3PO_4/BH_2PO_4	pK' = 2.1
$BH_2PO_4/BHPO_4$	pK' = 6.8
$BHPO_4/BOH$	pK' = 11.6

Figure 3B shows the titration curves of a number of buffer pairs of importance in the biochemistry laboratory.

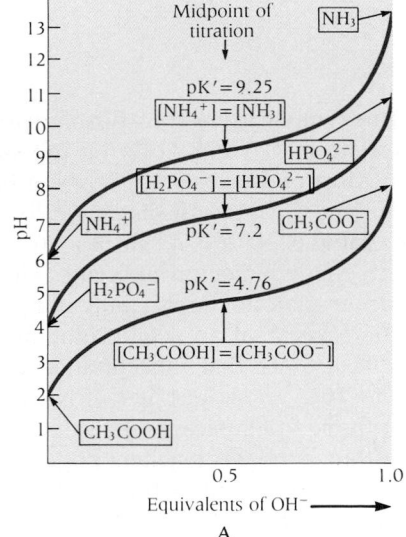

A

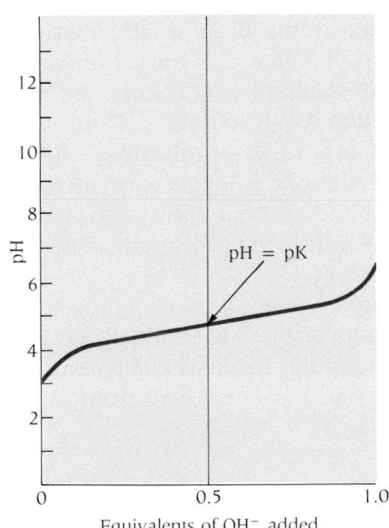

3A **Titration curve of a typical monobasic acid**

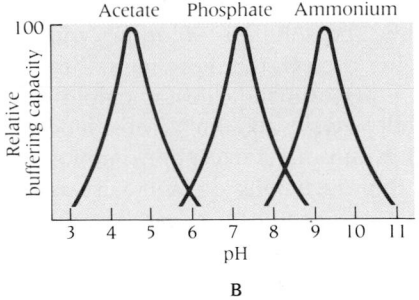

B

3B Buffer action
(**A**) *Acid-base titration curves of some acids, showing the major ions present at the beginning, midpoint, and end of a titration.* (**B**) *Relative buffering power plotted against pH. Note that the maximum buffering power occurs where the pH equals the pK.*

Acids	Bases
$HCl \rightleftharpoons H^+ + Cl^-$	$NH_3 + H^+ \rightleftharpoons NH_4^+$
$H_2CO_3 \rightleftharpoons H^+ + HCO_3^-$	$NaOH + H^+ \rightleftharpoons Na^+ + H_2O$
$NH_4^+ \rightleftharpoons H^+ + NH_3$	$HPO_4^{2-} + H^+ \rightleftharpoons H_2PO_4^-$
$H_2O \rightleftharpoons H^+ + OH^-$	$H_2O + H^+ \rightleftharpoons H_3O^+$

Acids and bases vary significantly in the *extent* to which they dissociate into free ions in aqueous solutions. Those that dissociate completely are called strong acids or bases. Weak acids and bases dissociate only to a limited and varying degree. For example, a strong acid such as HCl dissociates completely into H^+ and Cl^-. Virtually no HCl remains in the form of the intact ionic compound. The same is true of the strong base sodium hydroxide, NaOH. However, when acetic acid, which

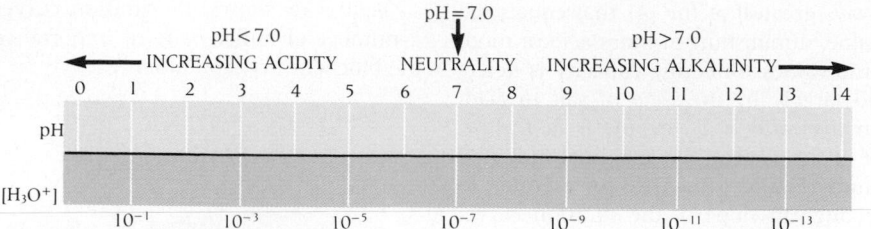

3-21 The pH scale
Along the bottom of the scale are the actual concentrations of hydrogen ions in moles per liter. Corresponding pH values appear at the top of the scale.

may be symbolized as HAc, dissolves in water, only a small fraction of the molecules dissociate to H^+ and Ac^-; the rest remain in the form of intact (undissociated) HAc. Most of the acids and bases of biological interest are weak. All salts, of both strong and weak acids, dissociate completely in water to free cations and anions. For example, NaCl is a salt of a strong acid (HCl); NaAc is a salt of a weak acid (HAc). Many biological molecules contain multiple acidic or basic groups that are in various states of dissociation at a given pH.

The *strength* of an acidic solution depends on the concentration of hydrogen ions in it. Since that concentration can vary enormously, from 1 mole of H^+ per liter to 10^{-14} mole per liter, it is convenient to represent the concentration on a logarithmic scale. Because the value is generally a fraction and hence has a negative logarithm, a more convenient positive scale is obtained simply by using the negative of the logarithm. This is the meaning of the symbol **pH** (Figure 3-21).

Pure water at ordinary temperatures dissociates to free H^+ and OH^-, but only to a slight extent. Only 10^{-7} moles per liter do so; therefore, 7 is the pH of water.[5] This is defined as **neutrality** (under conditions of standard temperature and pressure). Solutions with pHs below 7 are **acidic,** and the lower the pH, the more acidic the solution. Solutions with pHs above 7 are **basic** or **alkaline,** and the higher the pH, the more basic the solution. Since the scale is logarithmic, a change of 1 pH unit indicates a tenfold change in absolute concentration of H^+, or in acidity versus alkalinity. Most biochemical reactions in living cells occur at or near pH 7. But many exceptions are found to this rule.

Hydrogen ions do not exist as such in solution. As soon as a hydrogen ion (that is, a proton) is released or dissociates from a molecule—as in the dissociation of HCl—it is instantly picked up by another molecule. In dissociation reactions such as that of HCl, the other molecule is inevitably water. This results in the formation of a hydronium ion:

$$H^+ + H_2O \rightarrow H_3O^+ \text{ (hydronium ion)} \qquad (3.9)$$

Thus, the complete reaction for the dissociation of HCl would be

$$HCl + H_2O \rightarrow H_3O^+ + Cl^- \qquad (3.10)$$

For convenience it is customary to designate hydrogen ions as H^+, but it should be kept in mind that this does not match chemical reality.

There should be no confusion about the interchangeability of the terms "proton" and "hydrogen ion." We should recall that a hydrogen atom consists only of a nucleus of one proton with one electron orbiting around it (see Figure 3-2). Removal of that electron in the process of ionization leaves only the proton, which thus is, in fact, a hydrogen ion. While it is customary today to define acids and bases as proton donors or acceptors, measurements of the strength of particular acids and bases on the pH scale are expressed in terms of hydrogen ion (that is, proton) concentration. We shall use these terms interchangeably here. But we should recognize that what the pH scale actually measures is hydronium ion concentration.

[5] Since a mole of *any* substance (defined as its molecular weight in grams) contains 6.022×10^{23} molecules (Avogadro's number) and since 1 liter of water is equivalent to 52.72 moles of water (it would be instructive to consider how that figure was arrived at!), a liter contains 31.758×10^{24} molecules. Of these, only 31.758×10^{17}, or 1 in 527,200,000, undergoes dissociation.

Buffers

Since an organism ordinarily cannot tolerate significant alterations of its internal pH, several devices exist for keeping the pH nearly constant in the face of varying chemical circumstances. The main device is the **buffer.**

Changes in pH in a dynamic chemical system depend on the ratios of the concentration of H^+ acceptors, which are bases, to the concentration of H^+ donors, which are acids. For example, in a solution of H_2CO_3 (carbonic acid), the pH depends on the ratio $[HCO_3^-]/[H_2CO_3]$, where the brackets signify molar concentrations—that is, read $[HCO_3^-]$ as "the molar concentration of bicarbonate ion." To produce a pH of 7.4, the ratio must be 20 to 1. But since H_2CO_3 dissociates only slightly, such a high proportion of HCO_3^- cannot arise from its dissociation alone. If, however, the salt $NaHCO_3$ (sodium bicarbonate) is added to the solution, it dissociates completely into Na^+ and HCO_3^-. The concentration of HCO_3^- ions from this source can readily be adjusted to make the ratio $[HCO_3^-]/[H_2CO_3]$ equal 20 to 1 and thus to produce a pH of 7.4. In the absence of such a salt, the addition of strong base or acid would change the pH radically. However, in such cases, the presence of a completely dissociated salt of the weak acid minimizes—that is, buffers—these influences on the pH (see Box 3-2).

A biological buffering system is one that achieves and maintains a pH that is ideal for the organism in question. It operates whenever a *slightly ionized weak acid* (like H_2CO_3 or H_3PO_4) is mixed with one of its *salts* (like $NaHCO_3$ or NaH_2PO_4) which always ionizes completely.

SUMMARY

All organisms are constructed entirely of molecules, and thus of atoms. Hence, we must understand these constituents and their interactions in order to understand the structure and function of living things.

Atoms are composed of protons, neutrons, and electrons. The nucleus of an atom is composed of neutrons and protons. A cloud of electrons orbits the nucleus. The number of protons determines the identity of an element; the number of neutrons identifies each isotope of each element. Isotopes are both radioactive and nonradioactive. Biologists use them to "label" molecules. Procedures are then employed that make it possible to trace the fate of labeled molecules in the course of biochemical reactions.

The number of electrons in an atom—particularly in its outermost (highest) energy level—determines the chemical reactions in which the atom can take part. In chemical reactions, atoms bind to other atoms and form molecules. This is the result of a strong tendency for an atom to have a full number of electrons in its outer level. To accomplish this, electrons are lost or gained in chemical reactions, which in turn result in the formation of chemical bonds.

Chemical bonds are of four types: covalent bonds, ionic bonds, hydrogen bonds, and weak forces. Each has a distinctive role to play in the chemistry of living organisms.

The four most common atoms in biological systems are carbon, hydrogen, oxygen, and nitrogen. The most prevalent molecule is water.

Water has certain unusual properties shared by few other molecules. We are fortunate that it does, for without water's distinctive chemical and physical behavior, life as it is now constituted would be impossible. Water, unlike most liquids, expands when it freezes. It is cohesive and thus has a higher surface tension than most liquids. It also has a higher capacity for storing heat (a higher specific heat). For biologists, perhaps water's most significant property is its ability to dissolve other polar molecules and to exclude nonpolar molecules. Thus, it is the major transport vehicle for many of the materials essential for life.

Almost all chemical reactions of biological importance occur in aqueous solution. Water ionizes to a small degree, yielding a molar concentration of H^+ ions (protons) of 10^{-7}. We define pH as the negative exponent of the H^+ concentration. The pH of water is 7.

Salts, acids, and bases dissociate to their component ions to varying extents in water. The concentrations of these components in organisms are critically regulated and are crucial to their successful functioning. Buffers provide a sensitive method, indispensable to living organisms, for preventing excessive fluctuations in the acidity or alkalinity of their internal fluids.

KEY TERMS

acid
alkali
anion
atomic mass
atomic number
base
bond
buffer
cation
chemical reaction
cohesion
compound
covalent bond
density
electrolyte

element
half-life
heat capacity
heat of fusion
heat of vaporization
hydrogen bond
hydrophilic
hydrophobic
ion
ionic bond
ionic dissociation
isotope
mineral
molecule
neutrality

nitrogen fixation
nonpolar
organic
pH
polar
proton
radioactivity
salt
specific heat of water
surface energy
surface tension
valence
van der Waals forces

QUIZ QUESTIONS

1. Which of the following is the strongest type of attraction between atoms?
   A. covalent bonding
   B. hydrogen bonding
   C. ionic bonding
   D. van der Waals forces

2. Isotopes of a given element differ from one another in the number of _____ in their atoms.
   A. electrons
   B. neutrons
   C. protons
   D. ions

3. Which of the following is *not* a property of water that helps explain why water is so important to life on earth?
   A. Water expands when frozen.
   B. Water molecules attract each other.
   C. Water can absorb great quantities of heat with relatively small increase in its temperature.
   D. Nonpolar compounds dissolve well in water.

4. A solution with pH = 4 is _____ times more basic than a solution with pH = 2.
   A. 2
   B. 4
   C. 10
   D. 100

5. Carbon
   A. in the form of carbon dioxide gas, makes up about 21% of the atmosphere.
   B. enters plants in the form of carbon dioxide gas.
   C. is metabolized by animals and then released as a waste product called urea.
   D. is one of several microminerals important to living organisms.

ESSAY QUESTIONS

1. What is meant by the "polarity" of a water molecule? What are some of the unusual properties of water? How does the polar nature of these molecules account for these properties? How are these properties important for life?

2. What are the major types of chemical bonds? How are they formed and what holds them together? Which bonds are the strongest and which the weakest?

3. What is the difference between an acid and a base? What is the importance of acid-base balance in living organisms?

Define pH. How is the internal pH of living organisms maintained at a stable value?

4. What is an isotope? Name some of the isotopes important in biological research. Explain their importance. How are they used?

REFERENCES AND SUGGESTIONS FOR FURTHER READING

Cotton, F. A., Wilkinson, G., and Gaus, P. L. (1987). *Basic Inorganic Chemistry*, 2nd ed. John Wiley & Sons, New York.

New edition of a respected general text, of which one-third is devoted to molecular structure, chemical bonds, acids and bases, and related topics.

Douglas, B., McDaniel, D. H., and Alexander, J. J. (1983). *Concepts and Models of Inorganic Chemistry*, 2nd ed. John Wiley & Sons, New York.

Clear summaries of atomic structure, bonding, and aspects of thermodynamics.

Frank, H. S. (1970). The structure of ordinary water. *Science 169*, 635–641.

A key paper providing new insights into a fascinating substance.

Frieden, E. (1972). The chemical elements of life. *Scientific American 227*, January, 52–64.

A well-written, brief introduction to the atoms present in living organisms, with emphasis on the surprising diversity of trace elements.

Lehninger, A. (1982). *Biochemistry*, 2nd ed. Worth, New York.

A masterful survey of biochemical principles.

Maddox, J. (1989). Understanding hydrogen bonds. *Nature 339*, 173.

A brief note describing new insights into the nature of hydrogen bonds.

Moore, W. J. (1983). *Basic Physical Chemistry*. Prentice-Hall, Englewood Cliffs, N.J.

Validates the definition by G. N. Lewis that the scope of physical chemistry includes "everything that's interesting."

Pauling, L. (1960). *The Nature of the Chemical Bond*, 3rd ed. Cornell University Press, Ithaca, N.Y.

A classic work on chemical bonds of all kinds.

Stillinger, F. H. (1980). Water revisited. *Science 209*, 451–457.

More on the biophysical properties of water.

Stryer, L. (1988). *Biochemistry*, 3rd ed. W. H. Freeman, New York.

A deservedly popular textbook appears in a new edition. Striking color illustrations and an unusually complete index.

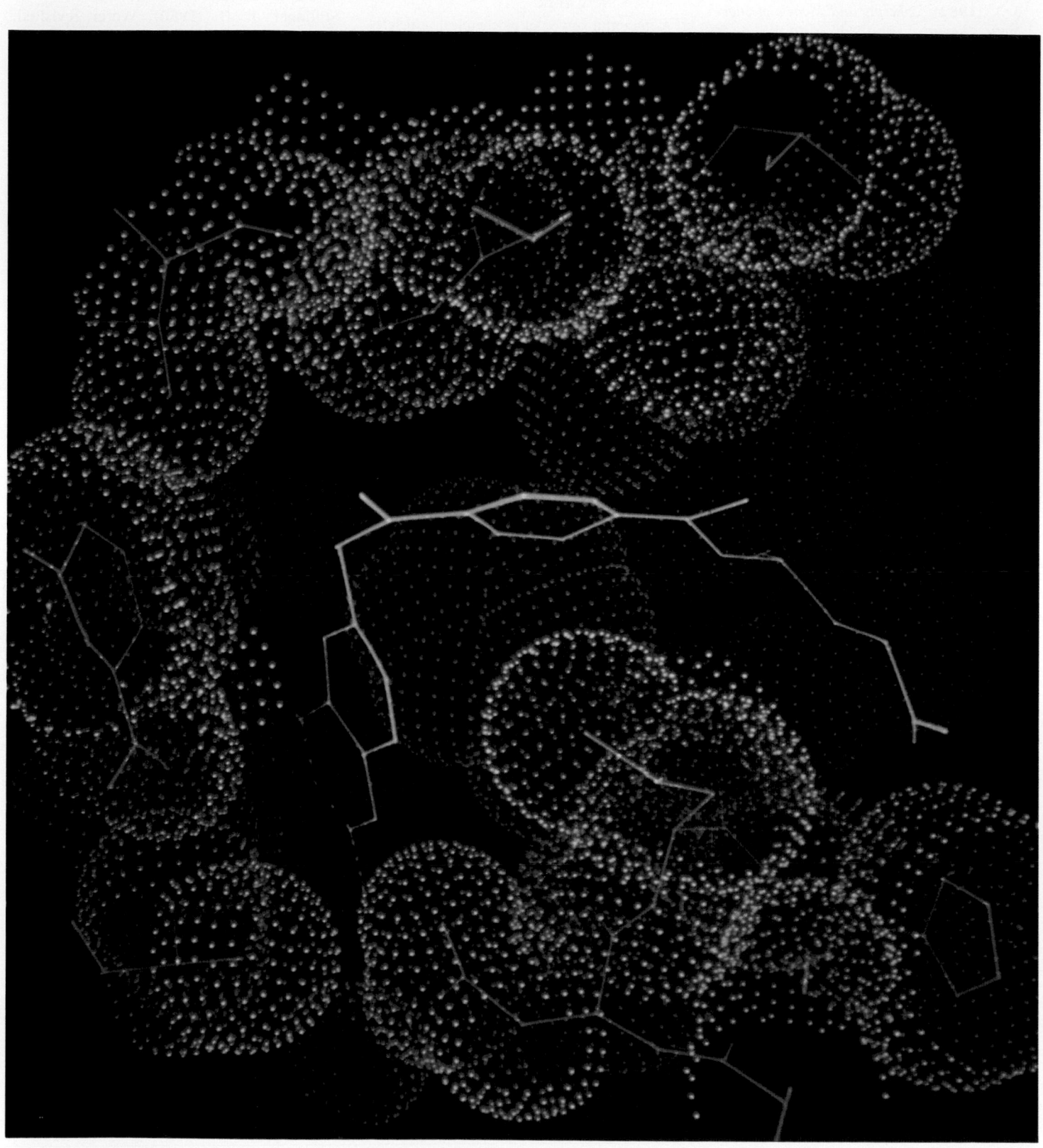

CHAPTER 4

ORGANIC MOLECULES IN LIVING SYSTEMS

Organic molecules in living systems
Knowing the rules governing the three-dimensional structure of proteins, biochemists can now devise computer images of individual proteins. This one shows how a molecule of the potent anti-cancer drug methotrexate, which closely resembles the vitamin folic acid, binds to the enzyme dihydrofolate reductase, which is essential to cell division. As a result, methotrexate powerfully inhibits the enzyme and ultimately blocks cell division.

The diversity of living organisms hides an astonishing inner unity. Although each living thing has unique features, all are composed of the same basic kinds of chemical substances—very large molecules called **macromolecules** and a variety of **small molecules**. The macromolecules include proteins, large carbohydrate molecules, lipids, and nucleic acids. The small molecules include amino acids, small carbohydrate molecules, fatty acids, nucleotides, and various vitamins and minerals. One of the surprising early discoveries of biochemistry was that the same kinds of amino acids are found in all forms of life.

Organisms as different as humans and *Escherichia coli* (a bacterial inhabitant of the animal intestine) have many common features at the molecular level. Both make use of an energy-rich molecule called adenosine triphosphate (ATP). In both species, genetic information flows from deoxyribonucleic acid (DNA) to ribonucleic acid (RNA) in essentially the same way. Even these critically important nucleic acids are fundamentally similar in the two distantly related species, differing mainly in the arrangements and sequences of their component nucleotides. That is like saying that a phonograph record of Beethoven's *Emperor Concerto* is fundamentally the same as a record of *Yankee Doodle*, except for the arrangement and shapes of the record grooves. In fact, the two records *are* fundamentally similar. They differ only in the nature of the information transferred—admittedly a major difference to some ears.

The list of molecular uniformities in the world of life includes the basic structure and composition of proteins (key building blocks and catalysts in all living organisms) and phospholipids (basic components of all membranes). There are more exotic examples. One would assume that cellulose, the major component of plant cell walls and hence of wood, is confined to plants. Yet cellulose is found in an odd group of marine animals called tunicates (see Chapters 43 and 44). One would also expect that hemoglobin, the protein that makes blood red, is confined to animals (since only animals have blood). But hemoglobin also occurs in the root nodules of some leguminous plants (see Box 34-1).

If we look more closely at diverse organisms, fundamental biochemical similarities become apparent in the small components of large, seemingly different macromolecules. Plant cellulose is a complex compound built up from small sugar molecules. Animal glycogen is another combination of nearly identical sugar molecules that are connected differently.

MAJOR ORGANIC COMPOUNDS IN LIVING ORGANISMS

The organic molecules of living organisms have four broad functions.

- Some are essential to body structure.
- Some serve as energy-rich fuels.

- Some transmit from one generation to another the genetic information that controls growth, differentiation, and biological specificity.
- Some operate primarily as catalytic agents in an organism's chemical processes.

Before we consider *how* they fulfill these functions, we must examine the compounds themselves.

HOW ORGANIC COMPOUNDS ARE CONSTRUCTED

The biologically important inorganic molecules are usually electrolytes, or ionic compounds. Examples are sodium chloride and calcium carbonate. In contrast most organic molecules are built with covalent bonds—for example, glucose. Indeed, the readiness with which they share electrons to form covalent bonds accounts for the particular biological fitness of carbon, hydrogen, oxygen, and nitrogen. Nonetheless, some organic compounds can undergo ionization. A molecule held together by covalent bonds may have certain localized groupings that ionize freely. We will deal with them later. Here let us familiarize ourselves with the covalently bonded portions of organic molecules.

Countless chemical compounds exist. Yet all are constructed from only a hundred-odd kinds of atoms. The uniqueness of each molecule thus depends upon the number, type, and three-dimensional arrangement of its component atoms.

The familiar chemical formula indicates the numbers and types of atoms present in each molecule. Carbonic acid, for example, is H_2CO_3. Much can be learned from such a formula. Consider the following simple examples: carbon dioxide, CO_2; carbon tetrachloride, CCl_4; methane, CH_4; chloroform, $CHCl_3$; formaldehyde, CH_2O; and ethane, C_2H_5. Among other things, the formulas show that a carbon atom has four combining sites. It can combine with four hydrogens because hydrogen has only one electron to share.[1] Oxygen has a valence (or covalence) of 2 (that is, it can form two covalent bonds); nitrogen's valence is 3; carbon's is 4.

Importance of Valence

The three-dimensional model of a molecule must be compatible with the valences of its component atoms. Since the valence of hydrogen is 1, the two hydrogens of H_2O cannot be attached to one another; although the valence of each would be satisfied, no means would remain for attaching the oxygen. We would have H_2 and O, not H_2O. Nor can the molecule have the structure H—H—O, because this assigns the central hydrogen atom a valence of 2. Therefore, the only possible structure for a water molecule is H—O—H.

Similarly, in methane (CH_4) each of the hydrogens must attach to the carbon and not to each other, as follows:

$$\begin{array}{c} H \\ | \\ H-C-H \\ | \\ H \end{array}$$

Likewise, the only possible arrangement of atoms in ethane (C_2H_5) is

$$\begin{array}{cc} H & H \\ | & | \\ H-C-C-H \\ | & | \\ H & H \end{array}$$

[1] Note that hydrogen can contribute its single electron in an ionic electron-transfer reaction or it can share its electron in the formation of a covalent bond. In the former case, we say that hydrogen has an ionic valence, or electrovalence, of +1. In the latter, we say that it has a covalence of 1.

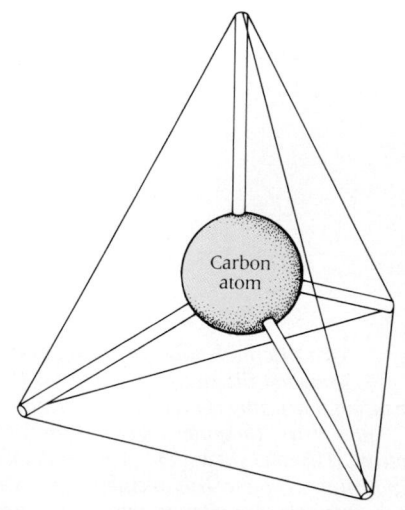

4-1 The three-dimensional geometry of a carbon atom
The four potential bonds point to the vertices of an imaginary symmetrical tetrahedron.

In carbon dioxide (CO_2), as we have seen, the two oxygens are attached to the carbon by two double bonds (O=C=O). However, when CO_2 is dissolved in water, one of the bonds can break and still leave the oxygen and carbon firmly held together by the other. The two molecules combine to form carbonic acid (H_2CO_3), which has the structure:

$$
\begin{array}{c}
\text{H}\!-\!\textbf{O} \\
| \\
\text{H}\!-\!\text{O}\!-\!\textbf{C}\!=\!\textbf{O}
\end{array}
$$

Carbonic acid

The H and OH of the water molecule are added to the O and C ends, respectively, of the broken bond.

Molecules can be represented on paper in several different ways. The **empirical formula** merely gives the ratio of atoms in a molecule (an example is H_2CO_3). The **structural formula** indicates the connections of atoms or atomic groupings in the molecule (an example is the above formula for carbonic acid). But the atoms within a molecule do not lie within a single plane as a printed structural formula might suggest. In three-dimensional reality, the four potential bonds of carbon point to the vertexes of an imaginary symmetrical tetrahedron, a pyramid whose three sides and base are identical equilateral triangles (Figure 4-1). In organic molecules, the spatial configuration of the atoms determines most of the properties of the substance. To obtain a truer picture of the structure of a molecule, chemists build models in which bond lengths and angles are built to scale.

Isomeric Compounds

A given empirical formula may permit more than one structural arrangement. Consider the formula C_2H_6O. The laws of valence hold that all the hydrogen atoms must be attached to oxygen or carbon, but two structures are still possible:

$$
\begin{array}{ccc}
\text{H} & & \text{H} \\
| & & | \\
\text{H}\!-\!\text{C}\!-\!\text{O}\!-\!\text{C}\!-\!\text{H} & \text{or} & \\
| & & | \\
\text{H} & & \text{H} \\
\text{Methyl ether} & &
\end{array}
\qquad
\begin{array}{c}
\text{H}\ \ \text{H} \\
|\ \ \ | \\
\text{H}\!-\!\text{C}\!-\!\text{C}\!-\!\text{O}\!-\!\text{H} \\
|\ \ \ | \\
\text{H}\ \ \text{H} \\
\text{Ethanol}
\end{array}
$$

When compounds have the same empirical formula and different structural formulas, they are **isomers** of one another.

There are two forms of isomerism. **Structural isomers,** which are compounds with the same empirical formula, have their constituent atoms linked in different ways. Methyl ether and ethanol are structural isomers. **Stereoisomers** are compounds that have the same empirical and structural formulas but that differ in the three-dimensional arrangement of the atoms. In one form of stereoisomerism, two molecules have the same structural formula but one is the *mirror image* of the other (Figure 4-2). This can occur whenever a molecule contains a carbon atom attached to four different atoms or groups of atoms. Such a carbon is called an **asymmetrical carbon.** Stereoisomers with a single asymmetrical carbon are sometimes called *epimers.* The prefixes D and L designate the two possible stereoisomers—as, for example, D-tartaric acid and L-tartaric acid.

It has long been known that one of a pair of stereoisomers will rotate a beam of polarized light to the right (*dextrorotation*) and the other rotates it to the left (*levorotation*). This observation had a prominent place in the history of biochemistry, the classical early work having been performed by Pasteur.[2] Such stereoisomers, also called *optical isomers,* are distinguished by the prefixes *d* and *l*. A solution containing equal amounts of *both* optical isomers would have no rotatory effect

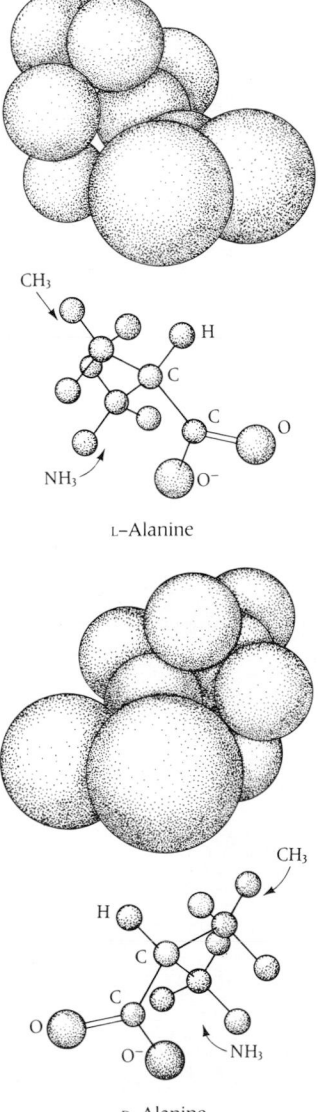

L–Alanine

D–Alanine

4-2 The two stereoisomers of the amino acid alanine

The structures of L-alanine and D-alanine are mirror images of each other. In the larger diagrams, atoms are drawn to scale. Structures of the isomers are more clearly seen in the skeletal models below. Note that amino groups (NH_2) are here shown in alternative form ($NH_3{}^+$).

---

[2] Pasteur observed in 1848 that the two stereoisomers of tartaric acid differ in their optical properties and form crystals that differ in the arrangements of their faces—so that one is a mirror image of the other. *Dextro*-tartaric acid (or *d*-tartaric acid) and *levo*-tartaric acid (or *l*-tartaric acid) have the same melting point, chemical behavior, density, and so on.

on polarized light. Such a mixture, termed a *racemic mixture* (or *racemate*), is called a *dl*-compound: Remember that the D and L prefixes denote absolute configurations of a molecule, not optical activity. The latter is indicated by *d* and *l*.

The terms *chiral* and *prochiral* are now used to describe molecular stereochemistry. A chiral molecule has handedness and thus is optically active. A prochiral molecule (or center within a molecule) lacks handedness and is optically inactive, but its identical substituents (for example, two —OH groups in a sugar) *are* distinguishable and therefore could become chiral in one step if one were altered chemically.

In sum, stereoisomers have identical chemical properties, but they can be distinguishable by three other properties:

- The effect of handedness (or chirality) on polarized light.
- The forms of their crystals.
- The actions upon them of specific enzymes.

When compounds capable of stereoisomerism are synthesized in the laboratory, equal numbers of dextrorotatory and levorotatory molecules are usually produced, canceling out each other's optical effects; and the mixtures are *optically inactive*. When such compounds are synthesized in organisms by stereospecific enzymes, they are always either right-handed or left-handed and never both. Therefore, they are *optically active*.

Functional Groups

Organic compounds would consist solely of carbon (and associated atoms) in chains and rings of varying sizes were it not for certain distinctive groupings of atoms that recur frequently in organic structural formulas. In the compounds just discussed, for example, we encountered the hydroxyl (—OH) group, which replaces one hydrogen in ethane, to make ethanol.

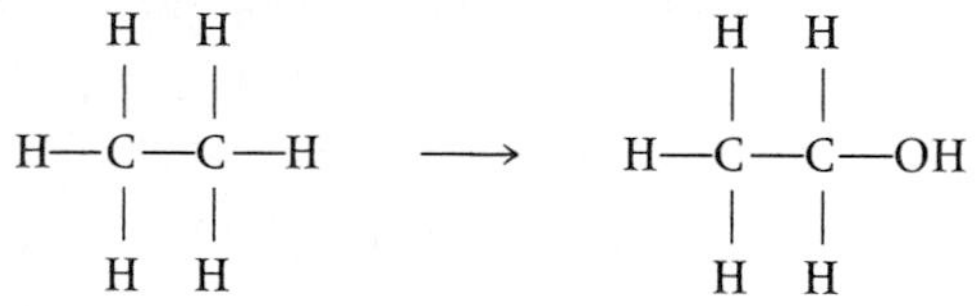

$$
\begin{array}{ccc}
\begin{array}{c}
\text{H} \quad \text{H} \\
| \quad | \\
\text{H—C—C—H} \\
| \quad | \\
\text{H} \quad \text{H} \\
\text{Ethane}
\end{array}
& \longrightarrow &
\begin{array}{c}
\text{H} \quad \text{H} \\
| \quad | \\
\text{H—C—C—OH} \\
| \quad | \\
\text{H} \quad \text{H} \\
\text{Ethanol}
\end{array}
\end{array}
$$

The chemical properties of such groups play a large role in determining the chemical behavior of the molecules of which they are a part. Hence, they are called **functional groups.**

The terminal functional group is essential in establishing the chemical behavior of the compounds in the following series:

$$
\begin{array}{ll}
CH_3\text{—}CH_2\text{—}CH_3 & \text{Propane} \\
CH_3\text{—}CH_2\text{—}CH_2\text{—}OH & \text{Propanol} \\
CH_3\text{—}CH_2\text{—}CHO & \text{Propionaldehyde} \\
CH_3\text{—}CH_2\text{—}COOH & \text{Propionic acid}
\end{array}
$$

Note that the last-named compound is an acid because it possesses a **carboxyl group** (—COOH), which dissociates in water to yield a hydrogen ion (H^+).

$$
CH_3\text{—}CH_2\text{—}COOH \rightleftharpoons CH_3\text{—}CH_2\text{—}COO^- + H^+ \tag{4.1}
$$

The carboxyl group is the hallmark of an **organic acid.** In Chapter 3 we noted that some organic molecules are electrolytes even though their carbon-to-carbon bonds are all covalent. They can possess one or more functional groups containing bonds that dissociate into ions in water. In some of the most biologically important organic molecules (for example, amino acids), there is one acidic functional group that liberates H^+ in water (from the carboxyl group —COOH) *and* one basic functional group that accepts H^+ (from the amino group —NH$_2$).

Figure 4-3 lists the most common functional groups of organic compounds. Note that some occur in series—for example, methyl, ethyl, and propyl. Functional

Name of Group	Structural Formula	Molecular Formula
Amino	$\begin{array}{c} H \\ \| \\ -N-H \end{array}$	$-NH_2$
Alkyl	$\begin{array}{c} H \quad H \quad\quad H \\ \| \quad \| \quad\quad \| \\ -C-C- \cdots -C-H \\ \| \quad \| \quad\quad \| \\ H \quad H \quad\quad H \end{array}$	$-C_nH_{2n+1}$
Methyl	$\begin{array}{c} H \\ \| \\ -C-H \\ \| \\ H \end{array}$	$-CH_3$
Ethyl	$\begin{array}{c} H \quad H \\ \| \quad \| \\ -C-C-H \\ \| \quad \| \\ H \quad H \end{array}$	$-C_2H_5$
Propyl	$\begin{array}{c} H \quad H \quad H \\ \| \quad \| \quad \| \\ -C-C-C-H \\ \| \quad \| \quad \| \\ H \quad H \quad H \end{array}$	$-C_3H_7$
Carboxyl	$\begin{array}{c} O \\ \| \| \\ -C-O-H \end{array}$	$-COOH$
Hydroxyl	$-O-H$	$-OH$
Aldehyde	$\begin{array}{c} O \\ \| \| \\ -C-H \end{array}$	$-CHO$
Keto (carbonyl)	$\begin{array}{c} O \\ \| \| \\ -C- \end{array}$	$-CO$
Sulfhydryl (thiol)	$-S-H$	$-SH$
Phenyl	$\begin{array}{c} H \quad H \\ \| \quad \| \\ C-C \\ // \qquad \backslash\backslash \\ -C \qquad\quad C-H \\ \backslash\backslash \qquad // \\ C=C \\ \| \quad \| \\ H \quad H \end{array}$	$-C_6H_5$

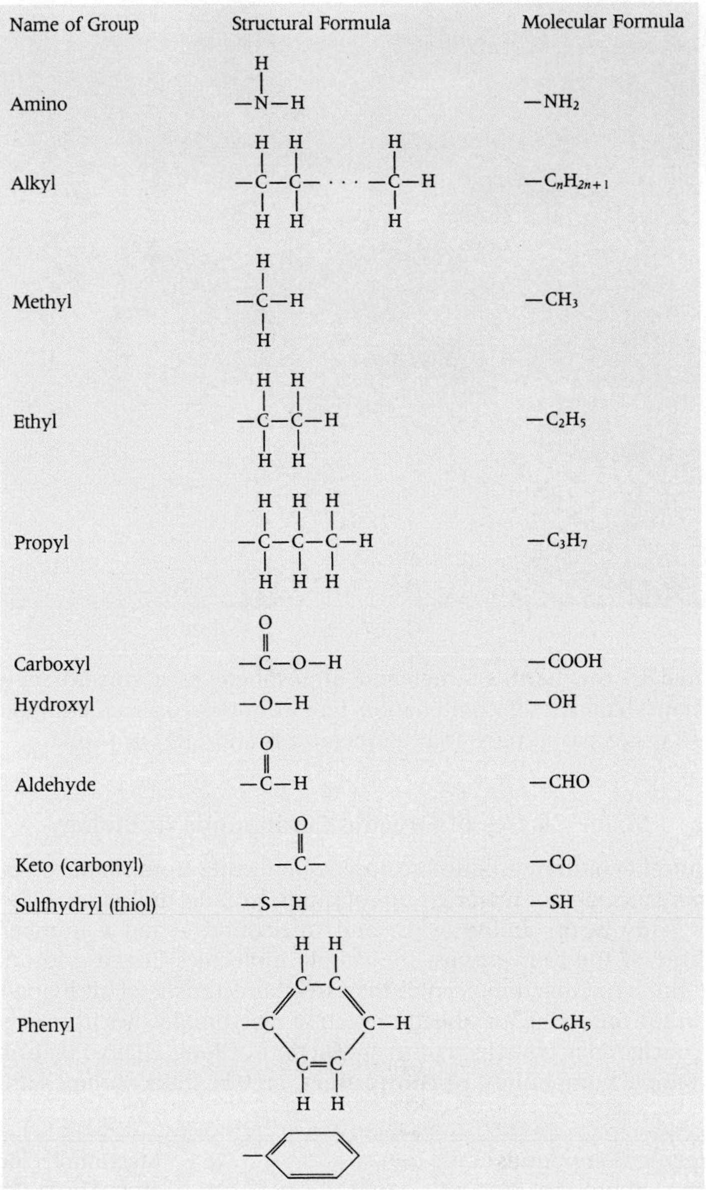

groups do not occur as free agents. As the open bonds in Figure 4-3 imply, functional groups are attached to other atoms.

Why Carbon?

We shall soon see that the origin of life on Earth was evidently preceded by a long series of reactions that gradually introduced chemical complexity. This process led in time to the major classes of organic molecules now found within living organisms. These are conveniently termed **biomolecules.** Before studying molecular evolution, we must become acquainted with these basic chemical structures.

One could ask, "Why are all the major biomolecules compounds of carbon?" Why weren't the earliest organisms made primarily of compounds of silicon (Si), another element of valence 4 and one that the Earth possesses in abundance far beyond that of carbon? Silicon can react with oxygen, as does carbon. Writers of science fiction have already contrived a world of siliceous beings in which *quartz,* or SiO_2, replaces the CO_2 of real life (Figure 4-4).

Part of the answer is that carbon atoms bond together to form chains more readily than silicon atoms do. Hence larger and more varied molecular structures can develop from carbon–carbon linkages. These include long and short chains, straight and branched chains, and ring structures. In addition, many (but by no means all) carbon compounds are readily soluble in water. The same is not true of silicon compounds. Most silicon has been long bound up in relatively insoluble minerals in the Earth's crust. Carbon, though present in many rocks as carbonate,

4-4 Quartz crystals
Quartz consists of molecules containing silicon and oxygen in covalent linkage.

has remained in the Earth's water and atmosphere as a constituent of reactive small molecules capable of combination into countless organic compounds. Also, of course, CO_2 is a gas at terrestrial temperatures and SiO_2 is a solid.

Major Classes of Organic Compounds in Biology

When chemical evolution produced complex molecules from simple ones, the result was the emergence of four major groups of simple organic molecules—monosaccharide sugars, fatty acids, amino acids, and nucleotides—and a number of minor ones. In three of the four groups, the simple molecules linked with one another to produce the large macromolecules that are characteristic of all living organisms. As shown in the following list, the formation of macromolecules from sugars, amino acids, and nucleotides consists in the production of long chains, or **polymers,** in which the simple compounds, or **monomers,** are the links of the chain.

Simple Organic Compounds in Living Organisms	Macromolecules Formed from These Compounds
Monosaccharide sugars	Polysaccharides
Fatty acids	Simple and complex lipids
Amino acids	Proteins
Nucleotides	Nucleic acids

Let us become familiar with the nature of these substances.

CARBOHYDRATES

The large class of compounds called **carbohydrates** is composed of the sugars, starches, and their many relatives. They consist of carbon, hydrogen, and oxygen, the last two ordinarily being present in the same proportions as in water. Therefore, most carbohydrates can be represented by the formula $C_x(H_2O)_y$. This relationship recalls the old incorrect idea that these compounds are combinations of carbon and water—**hydrates** of carbon, hence the name *carbohydrates*. In fact, the water molecule does not appear as such in the structural formula. Rather, a carbohydrate is a carbon chain bearing many **hydroxyl groups,** plus a single aldehyde or keto group.

Carbohydrates are of two kinds: (1) monosaccharides, or simple sugars; and (2) polysaccharides (also called glycans), which are long chains (polymers) of monosaccharides. A subgroup of polysaccharides is the oligosaccharides, which are short chains of 3 to 10 monosaccharides. Monosaccharides function chiefly as energy sources. Polysaccharides are storage forms of carbohydrates and, in plants, insects, and other species, they are vital structural elements.

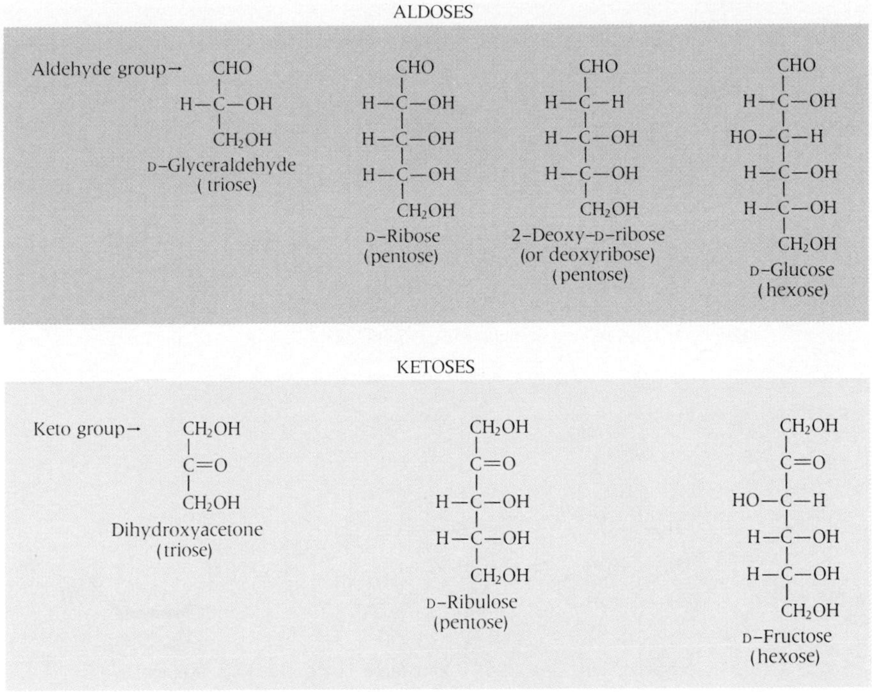

4-5 Important monosaccharides

Aldoses are sugars containing an aldehyde group. Ketoses contain a ketone group.

ALDOSES

```
Aldehyde group→   CHO          CHO          CHO          CHO
               H—C—OH       H—C—OH       H—C—H        H—C—OH
                 CH₂OH      H—C—OH       H—C—OH       HO—C—H
           D-Glyceraldehyde  H—C—OH       H—C—OH       H—C—OH
               (triose)       CH₂OH        CH₂OH       H—C—OH
                            D-Ribose     2-Deoxy-D-ribose  CH₂OH
                            (pentose)   (or deoxyribose)  D-Glucose
                                          (pentose)       (hexose)
```

KETOSES

```
Keto group→    CH₂OH        CH₂OH        CH₂OH
                C=O          C=O          C=O
               CH₂OH        H—C—OH       HO—C—H
         Dihydroxyacetone   H—C—OH       H—C—OH
              (triose)       CH₂OH       H—C—OH
                            D-Ribulose    CH₂OH
                            (pentose)   D-Fructose
                                         (hexose)
```

Monosaccharides

Monosaccharides cannot be split into simpler molecules by **hydrolysis** (the addition of water). They are named according to the number of carbon atoms in each molecule plus the -*ose* suffix, a characteristic of all carbohydrate nomenclature. Thus, there are trioses ($C_3H_6O_3$), tetroses ($C_4H_8O_4$), pentoses ($C_5H_{10}O_5$), hexoses ($C_6H_{12}O_6$), heptoses ($C_7H_{14}O_7$), octoses ($C_8H_{16}O_8$), and nonoses ($C_9H_{18}O_9$). Sugars containing an *aldehyde* group are known as **aldoses**; those containing a *keto* group are known as **ketoses**. We may refer to monosaccharides as aldopentoses, ketopentoses, aldohexoses, and so on.

The more important aldoses are the triose D-*glyceraldehyde*, the pentoses D-*ribose* and D-*deoxyribose*, and the hexose D-*glucose* (Figure 4-5).[3]

The more important ketoses are the triose *dihydroxyacetone*, the pentose D-*ribulose*, and the hexose D-*fructose* (see Chapter 8).

Many of these carbohydrates are dextrorotatory (they rotate polarized light to the right). It is not known why almost all naturally occurring sugars are dextrorotatory rather than levorotatory.

The structural formulas in Figure 4-5 portray the monosaccharides as straight-chain compounds. Many chemical properties of these sugars can be adequately explained only if the molecules are ring structures. Figure 4-6 compares the straight-chain structures of D-glucose and D-ribose and their ring formulations, which are termed D-glucopyranose and D-ribofuranose. As it turns out, both formulations are correct because the two forms interchange freely in solution, although the ring forms predominate. Biochemists write either of the two structures, depending on the circumstances.

But there is one more nomenclatural complication. When a straight-chain sugar is transformed into a ring form—that is, when it *cyclizes*, the first carbon in the chain becomes asymmetric. As a result, *two* isomeric ring structures can arise—one with an —OH group above the plane of the ring, one with the —OH below the plane of the ring. Respectively, these are designated α and β forms. The new asymmetric carbon is called the anomeric carbon atom and the α and β forms are called *anomers*. For example, glucose would have two ring forms: α-D-*glucopyranose* and β-D-*glucopyranose*.

---

[3] The very special importance of two of these sugars will be explained later in this chapter. Ribose and deoxyribose are parts of nucleic acids. Glucose, we will see later, is a critically important energy source.

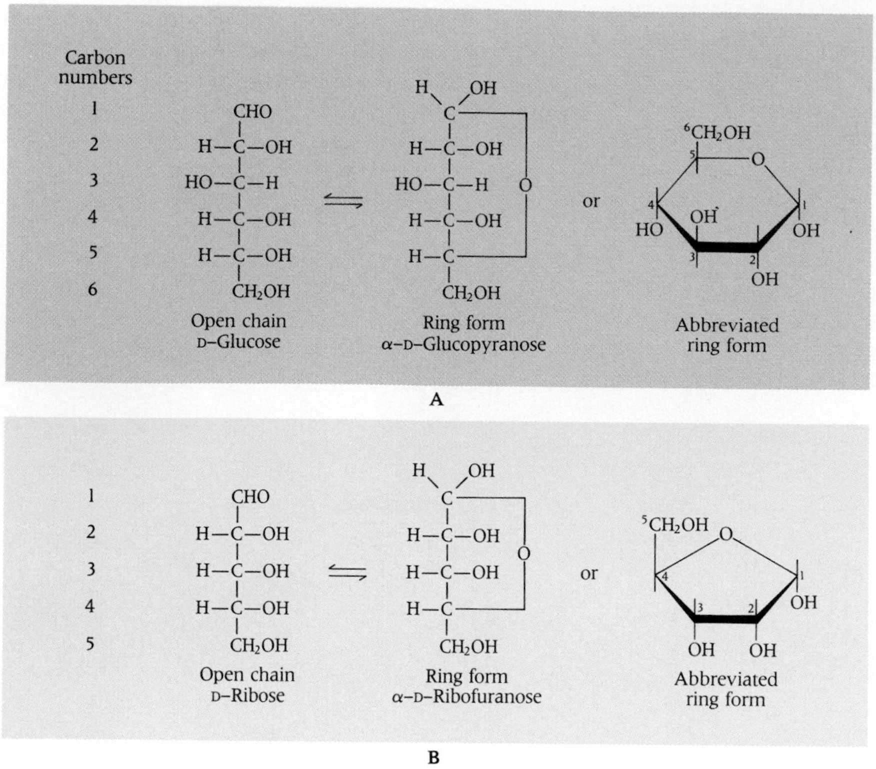

Carbon numbers

Open chain
D-Glucose — Ring form α-D-Glucopyranose — or Abbreviated ring form

A

Open chain
D-Ribose — Ring form α-D-Ribofuranose — or Abbreviated ring form

B

Disaccharides and Polysaccharides

The simplest polysaccharides are the **disaccharides,** which consist of two linked molecules of the same or different monosaccharides. Examples are maltose (which is glucose–glucose), lactose (glucose–galactose), and sucrose (glucose–fructose) (Figure 4-7). Sucrose, or ordinary table sugar, is abundant in plants. The world's supply comes mainly from sugarcane and sugar beets. Lactose occurs only in milk. Maltose arises in the course of starch breakdown.

The condensation of the two monosaccharide units in each disaccharide is a result of the elimination of one water molecule. A disaccharide bond forms when a hydrogen (H) from one carbohydrate molecule is caused to link with a hydroxyl group (—OH) of another carbohydrate molecule. A bond forms as the newly formed H_2O molecule is eliminated. The net reaction is the following **condensation reaction:**

$$2\ C_6H_{12}O_6 \underset{\text{Hydrolysis}}{\overset{\text{Condensation}}{\rightleftharpoons}} C_{12}H_{22}O_{11} + H_2O$$
Glucose Maltose Water (4.2)

Note that hydrolysis of a disaccharide reintroduces a water molecule and yields free monosaccharides.

The larger **polysaccharides** contain many monosaccharide units. An **oligosaccharide** is a small polysaccharide. Many include in their chains *amino sugars,* which are sugars in which one of the hydroxyl (—OH) groups is replaced by an amino (—NH$_2$) group. The most common amino sugars are D-galactosamine and D-glucosamine. The latter is prominent in chitin, a polysaccharide found in the exoskeletons of insects and crustaceans.

Many carbohydrates found in nature occur as polysaccharides. The various kinds differ in the nature of their recurring monosaccharide units, in the length of the chain, and in the degree of branching. The most common monosaccharide unit is D-glucose, but polysaccharides can also contain D-fructose and other monosaccharides. The most abundant glucose-containing polysaccharides are cellulose, glycogen, and other starches. All are polyglucoses, but they differ in molecular weight, branching structure, and solubility.

Glycogen, sometimes called animal starch, is the main storage form of carbohydrates in the animal body. Its molecular structure is highly branched.

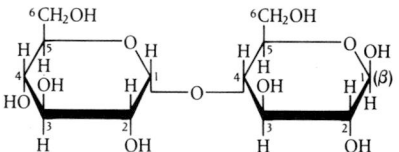

Maltose (β–form)

Lactose (α–form)

Sucrose

4-7 **Three important disaccharides**
Maltose consists of two glucose molecules, lactose of glucose and galactose, and sucrose of glucose and fructose. When two monosaccharides react to form a disaccharide, one molecule of H_2O is eliminated to form the bond.

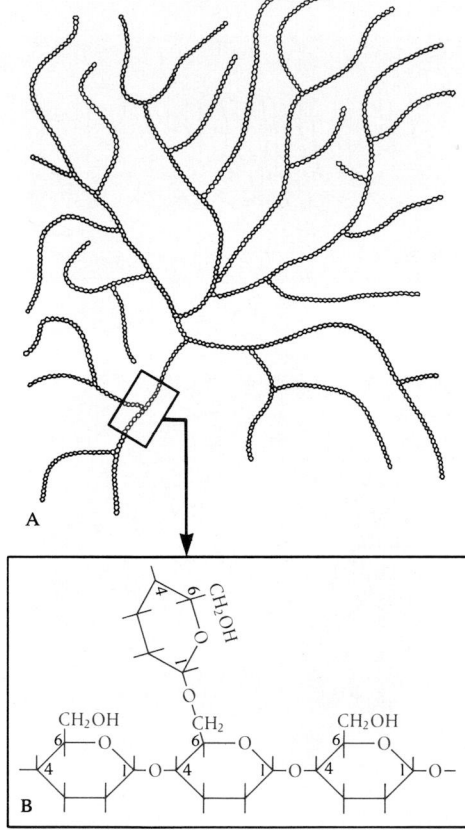

4-8 The general structure of starch and glycogen
(A) These polysaccharides consist of long, branched chains of covalently linked glucose molecules. Each bead in the drawing represents one glucose subunit. **(B)** An enlarged segment showing the molecular structure of a branch point.

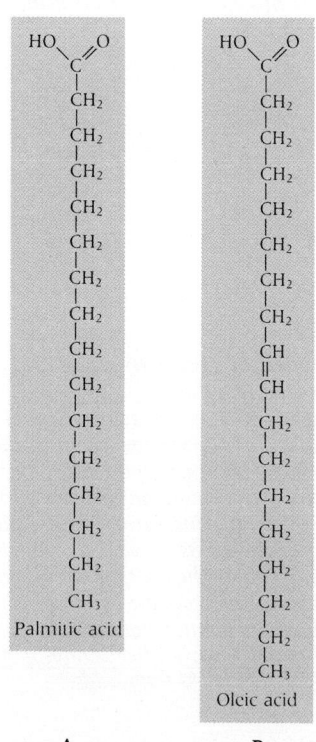

4-9 Structures of two fatty acids
(A) Palmitic, a saturated acid. **(B)** Oleic, an unsaturated fatty acid with one double bond.

Cellulose is the tough, insoluble constituent of the cell walls of plants—and, as noted earlier, of tunicates and a very few other animals. Its molecular chains—thousands of glucose units in length—are long, straight, and unbranched. Cellulose is said to contain more than 50% of the total organic carbon in the living world. Wood is 50% cellulose and cotton is 100% cellulose. Yet few organisms beyond the wood-boring shipworm *Teredo* and various microorganisms can digest cellulose. Termites can do it only because certain microorganisms (protists) living in their hindguts do it for them. Cows and sheep—as well as many undomesticated animals—also depend on microorganisms in their unusual stomachs (rumens) to digest the cellulose in their grassy diets.

Starch, branched like glycogen, has no structural role but it is a major storage carbohydrate in plants (Figure 4-8). Hence, it is a major food source of carbohydrates for animals. The potato consists mainly of starch.

LIPIDS

The **lipids** are a heterogeneous collection (fats, oils, waxes, and related compounds) of organic substances that are insoluble or sparingly soluble in water and freely soluble in organic solvents such as chloroform or ether. **Simple lipids,** or **triglycerides,** (also called **triglycerols**), contain only carbon, hydrogen, and oxygen, but the hydrogen and oxygen are not in a 2:1 ratio as in carbohydrates. Rather, it varies in different molecules.

The lipids have the following general functions:

- As structural components of membranes, they bridge the gap from water-soluble to water-insoluble phases without sharp discontinuities.
- As structural components of the animal body, they provide insulation and mechanical protection against injury, and they contour the body's secondary sexual characteristics.
- As intracellular storage depots, they are rich sources of energy.
- They facilitate the transport in body fluids of various molecules, including albumin and other proteins and cholesterol.
- Some have specialized roles, serving as vitamins or hormones.

Fatty Acids

Upon hydrolysis, triglycerides yield **glycerol** and **fatty acids.** Glycerol contains three —OH groups and is thus an alcohol. When the acidic —COOH groups of the fatty acids react with the three alcoholic —OH groups of glycerol, a triglyceride, or neutral fat, is formed in the following condensation reaction (we use R to symbolize any of several possible groups).

$$H_2C—O\boxed{H\ \ HO}—OC—R^1 \qquad\qquad H_2C—O—OC—R^1$$
$$HC—O\boxed{H+HO}—OC—R^2 \longrightarrow HC—O—OC—R^2 + 3H_2O$$
$$H_2C—O\boxed{H\ \ HO}—OC—R^3 \qquad\qquad H_2C—O—OC—R^3$$

Glycerol + Fatty acids → Triglyceride + Water (4.3)

Since R^1, R^2, and R^3 can be identical or different alkyl groups, it is clear that the possible number of different triglycerides is quite large.

The term *fatty acid* is usually taken to include the straight-chain members of the acetic acid series of organic acids. Almost all possess *even* numbers of carbon atoms. In other words, if $CH_3(CH_2)_nCOOH$ is the general formula for a fatty acid, *n* is ordinarily an even number. Some branched-chain fatty acids occur as minor components of lipids. The common fatty acids include palmitic acid, $CH_3(CH_2)_{14}COOH$ (or $C_{15}H_{31}COOH$) and stearic acid, $CH_3(CH_2)_{16}COOH$ (or $C_{17}H_{35}COOH$).

Fatty acids are divided into two classes depending on whether or not the carbon chain carries the maximum possible number of attached hydrogens. The **saturated fatty acids** have structures like the one in Figure 4-9A. In the **unsaturated**

fatty acids, double bonds join the carbon atoms that are not fully saturated with hydrogen (Figure 4-9B).

A well-known unsaturated fatty acid is oleic acid:

$$CH_3(CH_2)_7CH = CH(CH_2)_7COOH \text{ (or } C_{17}H_{33}COOH)$$

If the double bond in oleic acid were eliminated by the addition of two hydrogens, the product would be stearic acid, a saturated fatty acid. Table 4-1 lists some **polyunsaturated fatty acids,** in which there is more than one double bond. In general, unsaturated fatty acids predominate over saturated in biological systems.

Fatty acids with the longest carbon chains are often called *higher* fatty acids. The higher *saturated* fatty acids (lauric, myristic, palmitic, stearic, and other acids), in combination with glycerol, form the bulk of the body fat in most animals. However, the kinds of fatty acids present vary widely in different animal and plant species. The short-chain fatty acids are not widely distributed in nature. Certain fatty acids and other lipids are important components of cell membranes and membranous submicroscopic cellular components.

A fatty acid molecule has *two functional groups* with quite different properties: (1) a carboxyl group (—COOH) at one end (the "head" end); and (2) a long chain containing only carbon and hydrogen atoms (the "tail"). Thus, it has a dual personality. The carboxyl group is polar, water-soluble, and, therefore, *hydrophilic,* while the "tail" is *hydrophobic.* We say that molecules containing both hydrophilic and hydrophobic regions are **amphipathic.** If the polar head end were isolated from the rest of the amphipathic molecule it would be freely soluble in water. In contrast, the long tail, a nonpolar chain, would be poorly soluble in water but freely soluble in organic solvents. Their "two-faced" character accounts for certain characteristic properties of lipids. Fatty acid molecules in an immiscible mixture of water and oil orient themselves in a distinctive way at the interface (Figure 4-10A). The hydrocarbon chains project into the organic solvent and the carboxyl groups point toward the water layer. We shall see that exactly this arrangement is found in the lipid layers of biological membranes.

Soaps are the sodium or potassium salts of fatty acids. Sodium stearate is a well-known example. They are detergents because they surround oil droplets with their negatively charged water-soluble heads on the outside and their water-insoluble, oil-soluble tails on the inside mixed with the oil (Figure 4-10B).

Another important property of fatty acids derives from their three-dimensional structure. In this respect, saturated and unsaturated fatty acids have an interesting difference. A saturated fatty acid such as stearic acid (Figure 4-11) can have an infinite number of arrangements because each carbon–carbon bond has complete freedom of rotation. The extended form in Figure 4-11 is most common. However, as shown in the figure, unsaturated fatty acids like oleic and linoleic acids have a rigid kink in their chains because double bonds between carbons cannot rotate.

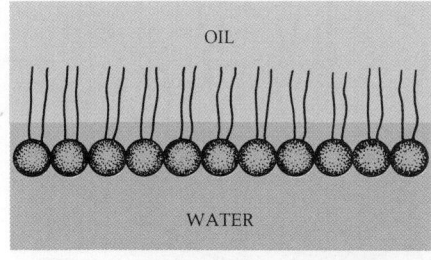

A

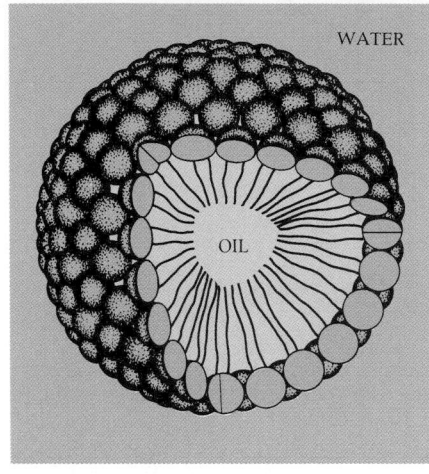

B

4-10 Orientation of fatty acids at an oil-water interface
(A) *At an oil-water interface, fatty acids orient their polar heads toward the water and their nonpolar tails toward the oil.* **(B)** *When the flat interface is disrupted, fatty acids can surround droplets of oil. Salts of fatty acids (soaps) do this effectively, thus causing the oil to be emulsified.*

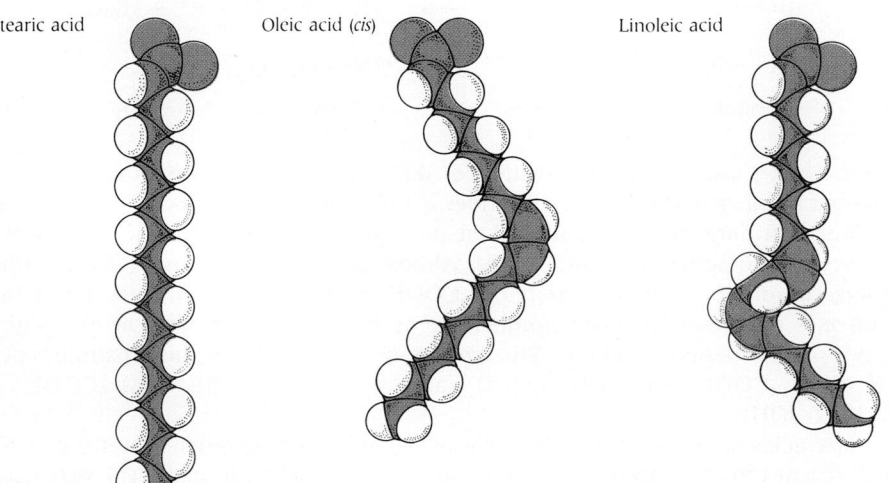

Stearic acid   Oleic acid (*cis*)   Linoleic acid

4-11 Models of some saturated and unsaturated fatty acids
These models of chains of saturated and unsaturated fatty acids show the influence of double bonds. Saturated fatty acids, such as stearic acid, have rotational freedom around every C—C bond. Unsaturated acids, such as oleic and linoleic acids, have C═C double bonds that cannot rotate. Their carbon chains thus have a fixed kink at each double bond, which makes them behave differently from saturated fatty acids in membranes. Clearly, they cannot be packed as tightly as saturated fatty acids.

TABLE 4-1
FATTY ACIDS

Common Name	Official Name	Number of Carbons	Molecular Formula
		SATURATED	
Butyric	*n*-Butanoic	4	C_3H_7COOH
Caproic	*n*-Hexanoic	6	$C_5H_{11}COOH$
Caprylic	*n*-Octanoic	8	$C_7H_{15}COOH$
Capric	*n*-Decanoic	10	$C_9H_{19}COOH$
Lauric	*n*-Dodecanoic	12	$C_{11}H_{23}COOH$
Myristic	*n*-Tetradecanoic	14	$C_{13}H_{27}COOH$
Palmitic	*n*-Hexadecanoic	16	$C_{15}H_{31}COOH$
Stearic	*n*-Octadecanoic	18	$C_{17}H_{35}COOH$
Arachidic	*n*-Eicosanoic	20	$C_{19}H_{39}COOH$
Behenic	*n*-Docosanoic	22	$C_{21}H_{43}COOH$
		UNSATURATED*	
One double bond			
Myristoleic	Δ^9-Tetradecenoic	14	$CH_3(CH_2)_3CH{=}CH(CH_2)_7COOH$
Palmitoleic	Δ^9-Hexadecenoic	16	$CH_3(CH_2)_5CH{=}CH(CH_2)_7COOH$
Oleic	Δ^9-Octadecenoic	18	$CH_3(CH_2)_7CH{=}CH(CH_2)_7COOH$
Two double bonds			
Linoleic	$\Delta^{9,12}$-Octadecadienoic	18	$CH_3(CH_2)_4CH{=}CHCH_2CH{=}CH(CH_2)_7COOH$
Three double bonds			
Linolenic	$\Delta^{9,12,15}$-Octadecatrienoic	18	$CH_3CH_2CH{=}CHCH_2CH{=}CHCH_2CH{=}CH(CH_2)_7COOH$
Eleostearic	$\Delta^{9,11,13}$-Octadecatrienoic	18	$CH_3(CH_2)_3CH{=}CHCH{=}CHCH{=}CH(CH_2)_7COOH$
Four double bonds			
Arachidonic	$\Delta^{5,8,11,14}$-Eicosatetraenoic	20	$CH_3(CH_2)_3(CH_2CH{=}CH)_4(CH_2)_3COOH$

* In official chemical nomenclature, the symbol Δ is used with a superscript numeral to denote the lowest-numbered carbon atom to which double bonds attach, the carboxyl carbon always being number 1. Thus Δ^9 indicates a double bond between the ninth and tenth carbons from the carboxyl (COOH) end of the molecular formula.

These features are of great significance for membrane structure.

We will see in later chapters that fatty acids have three major functions:

- They are building blocks of phospholipids and glycolipids.
- They serve as hormones and intracellular messengers.
- They are important energy sources.

Phospholipids

In one notable group of simple lipids, one fatty acid of a triglyceride has been replaced by the inorganic compound, phosphoric acid, and an additional organic compound. These are the **phospholipids.**

As shown in Figure 4-12, some components of the new substituent bear either positive or negative electrical charges—in contrast with the electrically neutral, long fatty acid residues still attached to the other two carbons. As a result, the molecule's head becomes even more polar and thus even more water-soluble. For this reason, phospholipids serve to bind water-soluble materials to water-insoluble materials. This is especially important in membranes.

A phosphorylated derivative of the membrane lipid phosphatidylinositol, which in the mid-1980s surged into scientific prominence, is phosphatidylinositol 4,5-bisphosphate. In this interesting compound, inositol triphosphate is attached to one carbon and arachidonate is attached to another. The hydrolytic release of these molecules, along with a molecule of diacylglycerol, gives phosphatidylinositol a molecule that plays a key role in a recently recognized cellular signaling system that is now under intensive study. Its substituent arachidonate is critically important in the metabolism of prostaglandins, which are discussed further in Box 28-1.

Steroids

The **steroids** are fat-soluble derivatives of a complex organic molecule consisting of four fused rings. As shown in Figure 4-13, this category includes *cholesterol* (present in the membranes of animal cells) *vitamin D*, a variety of *hormones*, and other substances.

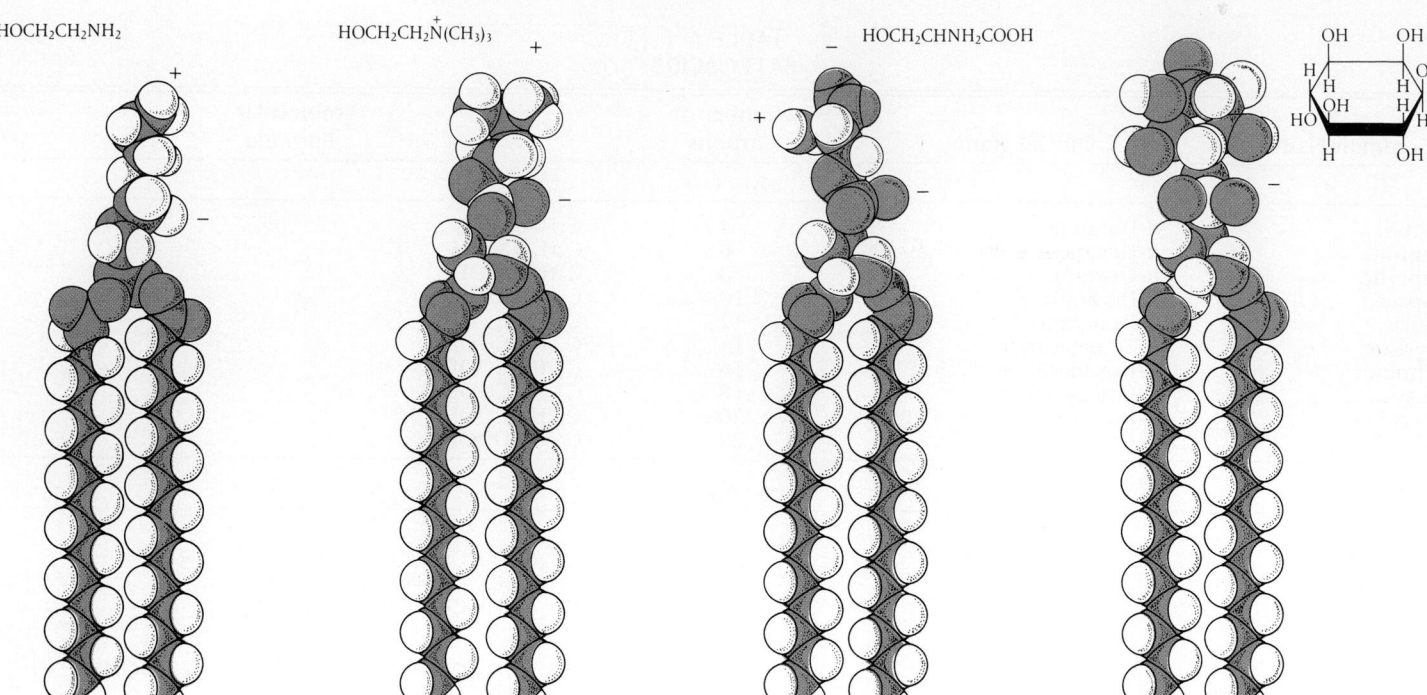

HOCH₂CH₂NH₂   HOCH₂CH₂N⁺(CH₃)₃   HOCH₂CHNH₂COOH

Phosphatidyl-
ethanolamine

Phosphatidyl-
choline

Phosphatidyl-
serine

Phosphatidyl-
inositol

PROTEINS

Proteins are the most abundant and most versatile organic molecules in living organisms, constituting 50% or more of their dry weight. To a large extent, the complexity and diversity of life itself depends upon the complexity and diversity of protein molecules.

In broad terms, proteins have the following functions.

- Key structural elements of living organisms—skin, connective tissue, cell membranes, hair, and so forth—are proteins.
- The agents that express genetic information are proteins (see Chapter 17).
- Enzymes, the specific catalysts of an organism's chemical reactions, are also proteins. We will be devoting several chapters to them.

Primary Protein Structure

Proteins are long, chainlike molecules of very high molecular weight and meticulously precise structure. They contain nitrogen as well as carbon, hydrogen, and oxygen. In addition, as we saw in Chapter 3, some proteins contain other elements, notably sulfur and the trace elements zinc, iron, and copper. When proteins are broken down by hydrolysis, the links in the chain are found to be low-molecular-weight components called **amino acids.** Only 20 different amino acids function as building blocks of all proteins of the living world.

Amino acids—to be more precise, α-*amino acids*—are organic acids with the general formula:

$$\underset{\underset{NH_2}{|}}{\overset{\overset{H}{|}}{R-\mathbf{C}-COOH}}$$

In formal nomenclature, the carbon immediately adjacent to the carboxyl carbon (boldface above) is designated the α carbon. The next one is β, the one after that γ, and so on through the Greek alphabet. The general formula given here is of an α-amino acid—that is, an amino acid in which the amino group attaches to the α

4-12 Structure of phospholipids
Models of four major phospholipids. They are named for the substituent bound to the phosphate of the head group. Phosphatidylinositol is a cell membrane component that participates in a cellular signaling system (see Chapter 28).

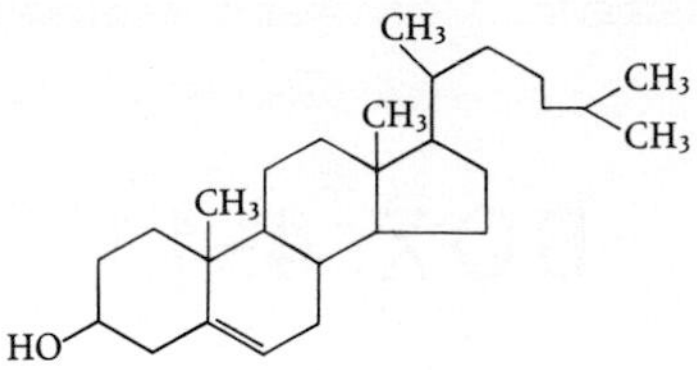

Cholesterol

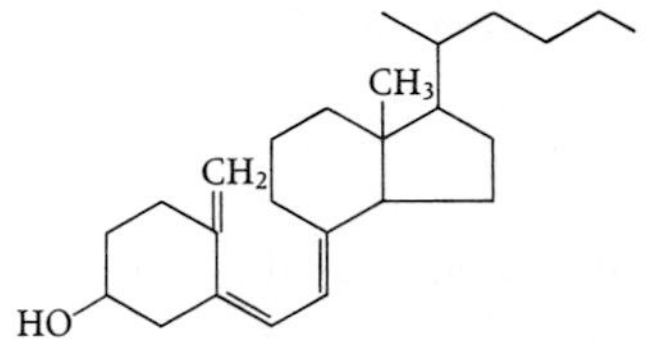

Vitamin D

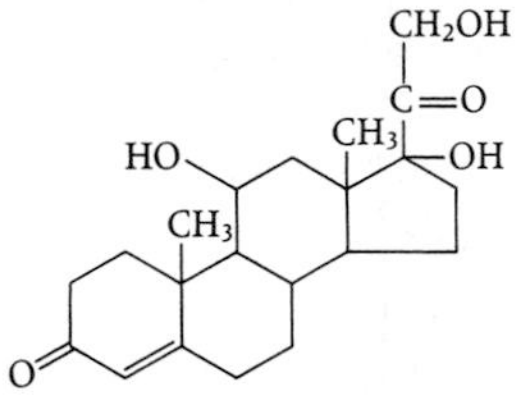

Cortisol

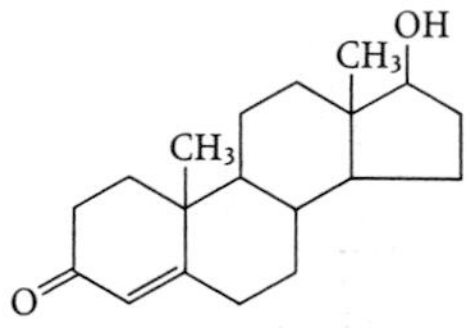

Testosterone

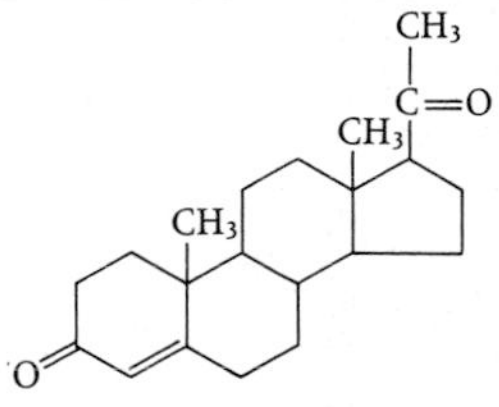

Progesterone

4-13 Some biomolecules derived from a steroid structure

Four fused rings of carbon atoms are found in the basic steroid structure underlying these compounds.

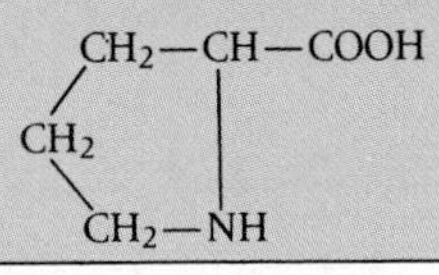

4-14 The R-groups associated with the 20 amino acids in almost all proteins

Major Amino Acids:
$$R-\underset{\underset{NH_2}{|}}{\overset{\overset{H}{|}}{C}}-COOH$$

Name	Abbreviations		Structure of R—	Category
	Three-letter	One-letter		
Alanine	Ala	A	CH_3-	Neutral
Arginine	Arg	R	$\underset{\underset{NH_2}{\vert}}{\overset{\overset{NH}{\|}}{C}}-NH-CH_2-CH_2-CH_2-$	Basic
Aspartic acid	Asp	D	$COOH-Ch_2-$	Acidic
Asparagine	Asp(NH₂)	N	$CONH_2-CH_2-$	Neutral
Cysteine	Cys	C	$SH-CH_2-$	Acidic
Glutamic acid	Glu	E	$COOH-CH_2-CH_2-$	Acidic
Glutamine	Glu	Q	$CONH_2-CH_2-CH_2-$	
Glycine	Gly	G	$H-$	Neutral
Histidine	His	H	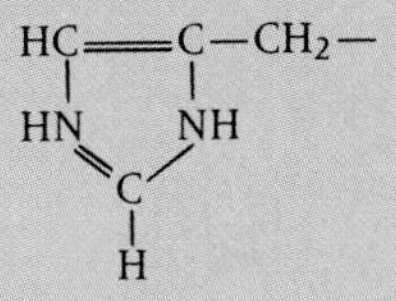 $\underset{\underset{\underset{H}{\vert}}{C}}{\overset{HC=C-CH_2-}{\underset{HN\quad NH}{\diagdown\diagup}}}$	Basic
Isoleucine	Ile	I	$CH_3-CH_2-\underset{\underset{CH_3}{\vert}}{CH}-$	Neutral
Leucine	Leu	L	$CH_3-\underset{\underset{CH_3}{\vert}}{CH}-CH_2-$	Neutral
Lysine	Lys	K	$NH_2-CH_2-CH_2-CH_2-CH_2-$	Basic
Methionine	Met	M	$CH_3-S-CH_2-CH_2-$	Neutral
Phenylalanine	Phe	F	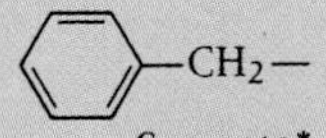 ⬡$-CH_2-$	Neutral
Proline	Pro	P	See note*	Neutral
Serine	Ser	S	CH_2OH-	Neutral
Threonine	Thr	T	CH_3CHOH-	Neutral
Tryptophan	Try	W	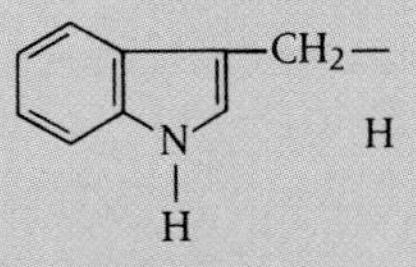 (indole)$-CH_2-$	Neutral
Tyrosine	Tyr	Y	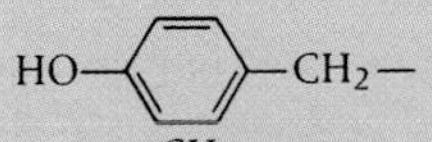 $HO-$⬡$-CH_2-$	Acidic
Valine	Val	V	$\underset{\underset{CH_3}{}}{\overset{\overset{CH_3}{}}{}}\!\!\diagdown CH-$	Neutral

*The structure of proline does not fit the prototype; it is
$$\underset{\underset{\underset{CH_2-NH}{\vert}}{CH_2}}{\overset{}{CH_2}}-\overset{}{CH}-COOH$$

When early biochemists attempted to analyze living tissues, they had to devise methods that use the same mild conditions of pH and temperature as are found in living organisms. One of these techniques is **spectrophotometry.**

Wavelengths of light are measured in nanometers (nm) or Angstrom units (Å) (see Box 3-1). We will later find (see Chapter 29) that the human eye can see light of wavelengths between 400 and 790 nm. Light of longer wavelengths (infrared light) or shorter wavelengths (ultraviolet or UV light) is invisible to the human eye, though the eyes of other species can see some of these wavelengths. To us, the longest visible wavelength appears as red and the shortest visible wavelength appears as violet. Intermediate wavelengths appear as the other colors of the spectrum.

When light passes through a solution of a given substance, the solution will absorb the energy of only certain wavelengths of light. The energy of other wavelengths is not absorbed. The molecular structure of the solute determines which energies are absorbed and which are not.

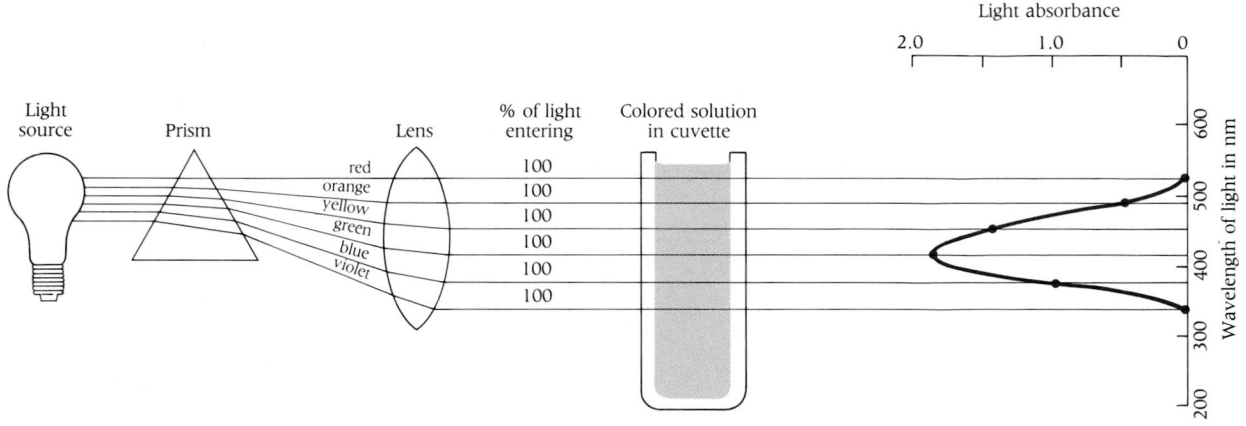

4A Basic elements of a spectrophotometer
A solution to be examined is placed in a cuvette. Incident light is split into its component wavelengths with a prism, and one selected single wavelength at a time is passed through the cuvette. The amount of light absorbed by the solution is measured by a photocell. A graph of light absorbed at various wavelengths constitutes an absorption spectrum. Absorption spectra depend on the chemical nature of the substance in solution and its concentration.

carbon (the carbon next to the COOH group). Amino acids making up proteins are all α-amino acids.[4] They are also almost all L-amino acids.

The distinguishing R group may be any of the 20 different structures shown in Figure 4-14. For example, when R is a hydrogen atom, the compound is glycine:

$$
\begin{array}{c}
H \\
| \\
H-C-COOH \\
| \\
NH_2
\end{array}
$$

The other R groups are more complex. Note that amino acids have two sets of standard abbreviations, which are used mainly in diagrams of amino acid sequences.

---

[4] A good many β-amino acids are found in nature but not in proteins. For example, β-alanine, an isomer of the commonplace protein constituent α-alanine, is a part of the structure of coenzyme A but not of proteins.

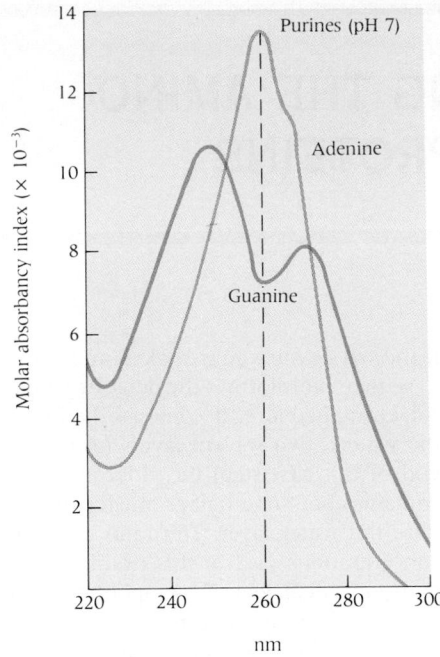

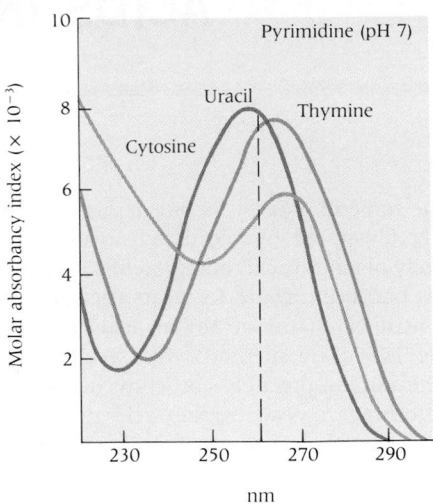

4B Absorption spectra of some purines and pyrimidines

The device that measures a solution's ability to absorb light at different wavelengths is a spectrophotometer (Figure 4A). White light from a source passes through a glass or quartz prism that diffracts the light into a spectrum of different wavelengths, each of which may be selectively passed through the solution in a special clear tube called a *cuvette*. A sensitive photocell detects how much light of each wavelength is transmitted by the solution. A graph plotting light absorbance versus wavelength is called an **absorption spectrum.** In the example in Figure 4A, "peak" absorption occurs at a wavelength of 420 nm. Interestingly, glass does not pass light wavelengths shorter than about 370 nm; therefore, spectrophotometry in the ultraviolet region requires that prisms, lenses, and cuvettes be made of optically pure quartz.

Absorption spectrums are useful in two major areas.

- First, the shape and character of an absorption spectrum—particularly in the infrared and UV regions—reveal molecular structure. Many biomolecules absorb maximally in the UV region of the spectrum. Nucleic acids and nucleotides, we will see, have absorption peaks near 260 nm (Figure 4B). Proteins have peaks around 280 nm. In these compounds, UV absorption is due to organic ring structures: *purines* and *pyrimidines* in nucleotides and *aromatic amino acids* such as tyrosine, tryptophan, and phenylalanine in proteins. Peptide bonds absorb at 220 nm.

- Second, the size of an absorption peak is proportional to the quantity of a substance in a solution, according to the Beer-Lambert Law. Hence, spectrophotometry permits precise quantitative analyses that are usually based on careful comparisons of peak heights in unknown solutions and in standard solutions containing known concentrations of the substance.

Such sequences are now so commonly printed that single-letter symbols have replaced three-letter abbreviations.

We see from the structural formula above that all amino acids possess an amino (NH_2) group on the α carbon—the carbon next to the carboxyl group carbon. An amino group is *basic*—that is, it can accept a hydrogen ion. The carboxyl group is *acidic*—that is, it can donate a hydrogen ion. Despite the presence of these two groups in *all* amino acids, some amino acids are said to be basic, others are acidic, and still others are neutral. These terms refer to properties of the R groups. Basic amino acids possess an additional amino (—NH_2) group—or a more complex substituent containing amino groups elsewhere in its R group. Acidic amino acids possess an additional carboxyl (—COOH) group. Neutral amino acids have only the single amino and carboxyl groups that define the amino acid class, and these cancel each other out electrically. None of the substituents in the R group of a neutral amino acid bear an electrical charge.

Amino acids join end to end to form a protein molecule. The carboxyl group of one amino acid combines with the amino group of the next to form a **peptide bond** (—CO—NH—) by removing a molecule of water:

The surge of progress in biochemistry is vividly illustrated by recent advances in the study of amino acids and proteins. Biochemists had been aware for years that proteins consist of chains of amino acids. Prior to the 1950s, no methods were available for determining precisely and conveniently the percentage of each amino acid in a given protein. Needless to say, there were also no methods for solving an even more difficult problem—the sequence of the amino acids in a given protein.

A major breakthrough took place when investigators found a way to determine the amino acid composition of a protein. To explain this method we must first look at the general principle of **chromatography,** which permits the separation and identification of substances, even when they are present in extremely small quantities or in complex mixtures. Although chromatography is done in many ways, it is always based on subtle differences among substances in their relative affinities for a *moving solvent* and a *stationary phase.*

Consider a simple experiment (Figure 4C). Equal volumes of chloroform and water (two solvents that do not mix with each other) are placed in a funnel and a substance is added (let us call it compound X) that is twice as soluble in water as in chloroform.

If the whole mixture is shaken and allowed to settle (chloroform is the denser of the two solvents; therefore, it comes to lie beneath the water), two solvent layers form. At the end of this first step, the chloroform layer contains about one-half as much compound X as the water layer. The ratio of the two concentrations (1:2 in this example) is the partition coefficient of compound X for a water-chloroform mixture.

Let us now place one volume of chloroform in each of two long series of consecutively numbered vessels. Next we transfer the water layer of vessel 1 (containing the amount of compound X it dissolved in the first step) to vessel 2 (containing only fresh chloroform). Meanwhile, one volume of fresh water is added to vessel 1, which already contains one volume of chloroform in which some compound X remains dissolved from the first step. If we shake vessels 1 and 2, the compound X *in each vessel* distributes itself between the two solvents exactly as predicated by the partition coefficient. Subsequent transfers are shown in Figure 4C. As each upper phase is transferred to the next vessel in the series, the first in the series always being added to fresh chloroform, fresh water is simultaneously added to vessel 1. In time, compound X will have moved to a small cluster of neighboring ves-

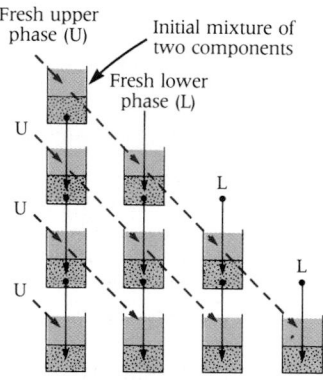

Fresh upper phase (U)

Initial mixture of two components

Fresh lower phase (L)

4C Transfer experiment
Each vessel holds water (above) and chloroform (below). A compound (X) is added to the vessel at top left (X is twice as soluble in water as in chloroform). After shaking and equilibration, twice as much X will be in the water as in the chloroform. The water layer is transferred to a new vessel (dashed arrow). The chloroform layer is also transferred (vertical arrow). Fresh water is added to vessels indicated by arrows. The process is repeated for the two vessels containing some X. After four transfers, all the X is concentrated in a few of the bottom vessels.

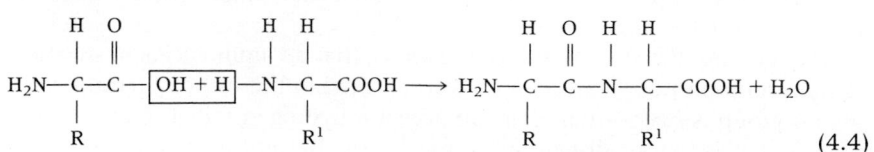

$$H_2N-C-C-\boxed{OH + H}-N-C-COOH \rightarrow H_2N-C-C-N-C-COOH + H_2O$$

(4.4)

Since each amino acid has lost part of a water molecule, each is now called an amino acid residue. Chains of amino acid residues, linked in such head-to-tail fashion, are called **peptides.** Two such residues joined together constitute a *dipeptide;* three, a *tripeptide;* and a large number (often many hundreds), a **polypeptide** (or a *polypeptide chain*). A protein molecule may have more than one polypeptide chain and have a molecular weight of hundreds of thousands.[5] A short peptide is called an **oligopeptide.**

[5] The great size of the average protein molecule is suggested by the following empirical formulas: casein of milk, $C_{700}H_{1130}O_{224}N_{180}S_4P_4$; gliadin of wheat, $C_{685}H_{1068}O_{211}N_{196}S_5$; human hemoglobin, $C_{3032}H_{4816}O_{872}N_{780}S_8Fe_4$.

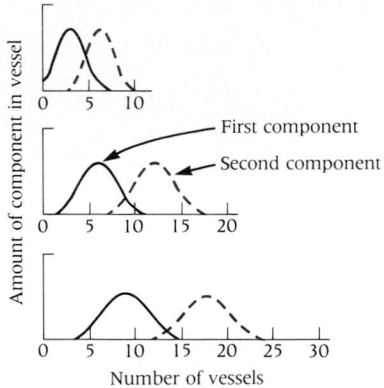

4D Separation of two components in extraction vessels

Two components with different partition coefficients are run through a series of transfers as in Figure 4C. The graphs show the distribution of the two components after 10, 20, and 30 transfers.

sels whose number can be predicted with astonishing accuracy from the partition coefficient alone.

If vessel 1 contained a second substance, compound Y, whose partition coefficient differed only slightly from that of compound X, we would soon see that it begins to separate from compound X after a number of transfers (Figure 4D). The distribution curve of the compound that is more soluble in the upper phase (in this case water) moves to the right more rapidly than that of the other compound.

All methods of chromatography operate on this principle. At each point in a long series of points, a molecule "decides" whether to stay in the stationary phase because it is more soluble there (as in paper chromatography) or more tightly bound there (as in ion exchange chromatography). Alternatively, it can cross into the moving phase if solubility or charge suits it better there. It would be hard to overestimate the importance of these simple methods for biological research.

It was the ingenious use of ion exchange chromatography that permitted the first suc-

cessful analysis of the relative percentages of each amino acid in a protein. In Figure 4E, each peak in the series of peaks represents one amino acid emerging from a column at a characteristic place. Prior to this operation, the protein was hydrolyzed completely so that each amino acid was liberated intact from the peptide chain. The area under each peak is proportional to its prevalence in the original protein. Astonishingly, this analysis can now be performed automatically on nanogram quantities of a protein.

A second major advance has been the development of methods for determining the *sequence* of amino acids in a protein. In one of these methods, the so-called Edman degradation, a chemical derivative is made of the final amino acid in the peptide chain. The "derivatized" terminal amino acid is then cleaved from the chain and identified. The next amino acid in the sequence is thus exposed and ready to be derivatized and identified in the same way. Using this method, the sequences of many proteins have been rapidly and accurately determined. Indeed, the whole procedure has been automated. It is now done by machines.

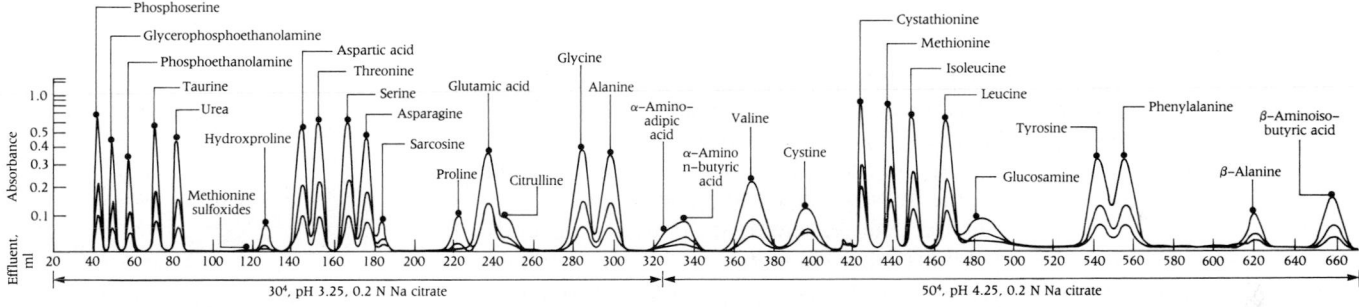

4E Pattern of amino acids emerging from an ion exchange column

The material added to the column was a sample of protein (ribonuclease) that had been hydrolyzed into *its component amino acids. Various buffers are poured through the column. As fractions emerge, the amino acid concentration in each is measured by a reaction that yields a color, the depth of which is proportional* *to amino acid concentration. The area under the peak for each amino acid is proportional to its prevalence in the original protein.*

Many sophisticated methods are now available to biochemists for characterizing and identifying proteins. We will mention two now.

One involves the measurement of light absorption by solutions of proteins (Box 4-1). It is simple to record the absorption spectrum of such a solution. The resulting curve has a distinctive shape for proteins—and, as noted later, a different shape for nucleotides, nucleic acids, and all the other biomolecules with a capacity to absorb light.

Another fundamental set of protein chemistry methods emerged from two Nobel Prize–winning studies of the 1950s. The first method determines the amino acids present and their relative amounts in a protein molecule. The second method determines their sequence (Box 4-2).

Insulin, a hormone produced by the animal pancreas, was the first protein to be fully sequenced. That early study revealed that the small insulin molecule (molecular weight 5733) contains two unequal-sized polypeptide chains held together by disulfide (—S—S—) bonds formed between the sulfhydryl (—SH) groups of opposed cysteine residues. This discovery showed with spectacular clarity the importance of —SH groups in establishing the final structure of the protein. The disulfide linkages formed between these —SH groups stabilize the structure (Figure 4-15).

The amino acid sequences of many proteins have now been determined. This body of work has yielded a fundamental principle of the greatest importance: Each protein's specific sequence of amino acids, which resembles a computer program in its precision, is responsible for the uniqueness and specificity of that protein and in turn for an enormous variety of living organisms.

Every peptide has a free —NH₂ group attached to its terminal amino acid residues, the so-called N-terminal amino acid, and a free —COOH group at the other, the C-terminal amino acid. Between these two ends, the long polypeptide chain of a protein molecule is folded and twisted into complex and unique configurations.

Higher Protein Structure

It would be difficult to overstate the importance of recent insights into the structure and chemistry of proteins. A protein molecule, with its long chains of amino acids arranged in diverse sequences, has an architecture ideally suited to wide-ranging, subtle variations. A useful terminology has been devised to specify the structure of a protein molecule at each of its levels.

- The **primary structure** of a protein is its linear amino acid sequence.
- Portions of the amino acid sequence are usually coiled into so-called **α-helixes.** Others, although folded, are uncoiled and are said to be *nonhelical,* or *extended.* The helical and nonhelical arrangement of a protein molecule constitutes its **secondary structure.** Helical coils are stabilized by strategically located bonds or cross-linkages that act to stabilize the folded structure. For example, there are hydrogen bonds between the —CO and —NH groups (of the primary, not the R groups) in successive turns of the helix (Figure 4-16). In a second common type of stable secondary structure, polypeptides are arranged in a **β-pleated sheet** (Figure 4-17). In this conformation, typical of the protein of silk, polypeptides aligned side by side are held together by hydrogen bonds.
- The **tertiary structure** is the three-dimensional pattern of foldings and turning of the amino acid sequence itself. Some say that the dividing line between secondary and tertiary structure is a matter of taste. Tertiary structure also depends upon the various types of stabilizing bonds shown in Figure 4-18. Here hydrogen bonds are chiefly between different R groups. Hydrophobic bonding also plays an important role.
- We speak of **quaternary structure** in those proteins with more than one polypeptide chain. It denotes the manner in which polypeptide chains are arranged in space. Such chains, the *subunits* of a protein, are held together by ionic bonds and other noncovalent (nonpolar) bonds.

Protein molecules in which the structural stabilizing bonds have been ruptured are said to be **denatured.** The classic example occurs when an egg is boiled. Heat denatures the egg white albumin as well as the various yolk proteins. Once this has occurred, it is usually impossible to reverse the process and **renature** the proteins. However, *some* proteins can be denatured reversibly. When the denaturing agent is removed, these proteins, usually the smaller molecular species, snap back into their original tertiary form. In Chapter 8, we discuss a famous example of this phenomenon (see Box 8-3).

The **conformation,** or overall shape, of a protein molecule depends upon the combined geometry of its secondary and tertiary structures. They in turn are consequences of its own unique primary structure. They also depend on the polarity, or lack of it, of R groups, which significantly influences protein folding.

Despite the seeming stability of protein structure, the protein molecule is a dynamic structure that is constantly undergoing remarkable structural fluctuations and inner motions. No longer do we consider the protein crystal structure demonstrated by an x-ray crystallographer as the rigid, unvarying design of a static structure. Rather, it is at best a representation of what might be called the "average structure." A crystal used in such an x-ray analysis might contain some 10^{20} protein molecules, yet it is highly probable that at any given instant *not one* of them exactly matches the average structure. We will encounter many examples of the dynamism of protein

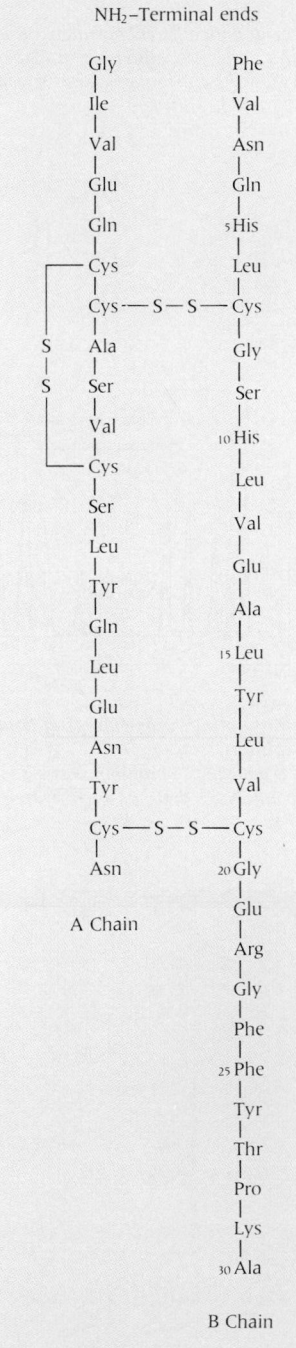

4-15 Amino acid sequence of insulin
In the insulin molecule, two short polypeptides are held together by disulfide (—S—S—) linkages.

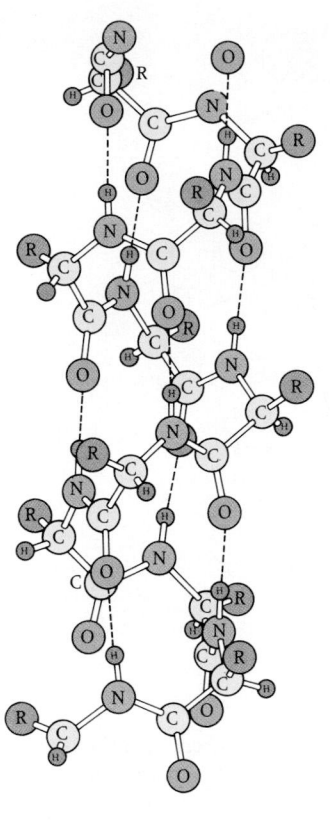

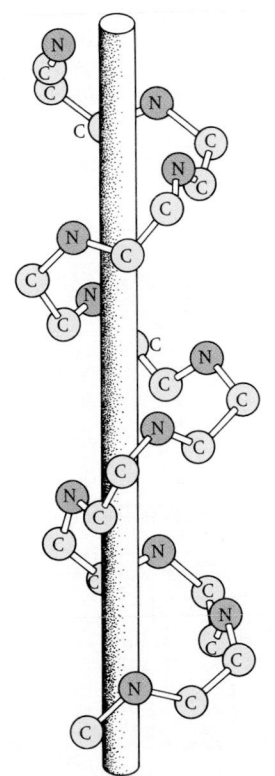

4-16 Helical structure of proteins
The α-helix is a hydrogen-bonded helical configuration of a single polypeptide chain that is present in many proteins. There are about 3.6 amino acids per turn of the helix. The sketch at the bottom emphasizes the helical arrangement by picturing the carbon and nitrogen backbone of the helix circling a rod.

structure. Perhaps the best known is the change in the conformation of enzyme structure in response to the enzyme's substrate, or various regulatory agents.

Larger proteins are usually classifiable as **fibrous** or **globular.** Fibrous proteins are long, threadlike molecules consisting of polypeptide chains arranged in a parallel fashion along a single axis to yield long fibers or sheets. Fibrous proteins tend to be physically tough structures that are insoluble in water. They are the main structural elements in the connective tissue of animals—the **collagen** of tendons, the **keratin** of hair, horn, leather, nails, and feathers (Figure 4-19A), and the **elastin** of elastic connective tissue.

Globular proteins (Figure 4-19B) have polypeptide chains that are folded into compact spherical or globular shapes. Most globular proteins are soluble in water. Nearly all of the many thousands of known enzymes are globular proteins, as are the antibodies, some hormones, and various blood proteins.

Whether they are fibrous or globular, large proteins usually consist of multiple polypeptide chains. Proteins possessing more than one chain are known as *oligomeric proteins;* their component chains, or subunits, are called *protomers.* Hemoglobin is a well-known oligomeric protein, which consists of four subunits. Interestingly, there are no covalent bonds between the subunits. Nevertheless, subunits are held tightly together by ionic bonds and nonpolar interactions with great stability and specificity.

Protein Molecules and Water Molecules

When a protein molecule is dissolved in water, the water molecules immediately adjacent to the surface of the protein molecule encounter a variety of environmental conditions not experienced by water molecules remote from that surface (Figure 4-20). These include positive and negative charges; polar and ionic side chains; hydroxyl, carbonyl, and amide groups; and a variety of hydrophobic or nonpolar nooks and crannies. As a result some of the water molecules are weakly bound to the protein's surface. We say that a protein molecule with water molecules bound to its surface is said to be **hydrated.**

Why is this worth mentioning? When watery solutions of proteins are cooled below the freezing point of water, some of the bound water molecules in the hydration "shell" remain unfrozen—even at temperatures of −70°C! This curious behavior is limited to solutions of macromolecules. It has important consequences for the new technology of cryobiology (the biology of cold). Rapid freezing of organs, tissues, cells, and foods tends to minimize damage by minimizing the formation of damaging ice crystals because much of the water remains unfrozen.

Conjugated Proteins

With the exception of hemoglobin, we have so far been discussing simple proteins. Many proteins are not simple but are **conjugated.** A conjugated protein is a protein that is attached to a nonprotein component—organic or inorganic. The attached substance is sometimes termed a *prosthetic group.*

The important conjugated proteins include the following classes:

- **Nucleoproteins,** such as those in chromatin, in which one or several molecules of protein are combined with a nucleic acid (DNA or RNA).
- **Glycoproteins,** such as γ-globulin, where the attached groups are polysaccharides.
- **Lipoproteins,** which are conjugated to triglycerides or other lipids by poorly understood linkages.
- **Metalloproteins,** such as ferritin, in which the attached group is a metal (in ferritin the metal is iron).
- **Chromoproteins,** such as hemoglobin and myoglobin, which have various chromophoric (i.e., colored) prosthetic groups.
- **Enzymes.** Some enzymes have coenzymes that are so tightly bound they are considered prosthetic groups. A few enzymes have loosely bound coenzymes that are not so considered. Coenzymes are essential participants in enzymatic reactions that serve as carriers of electrons or small chemical groups. We will discuss them in Chapter 8.

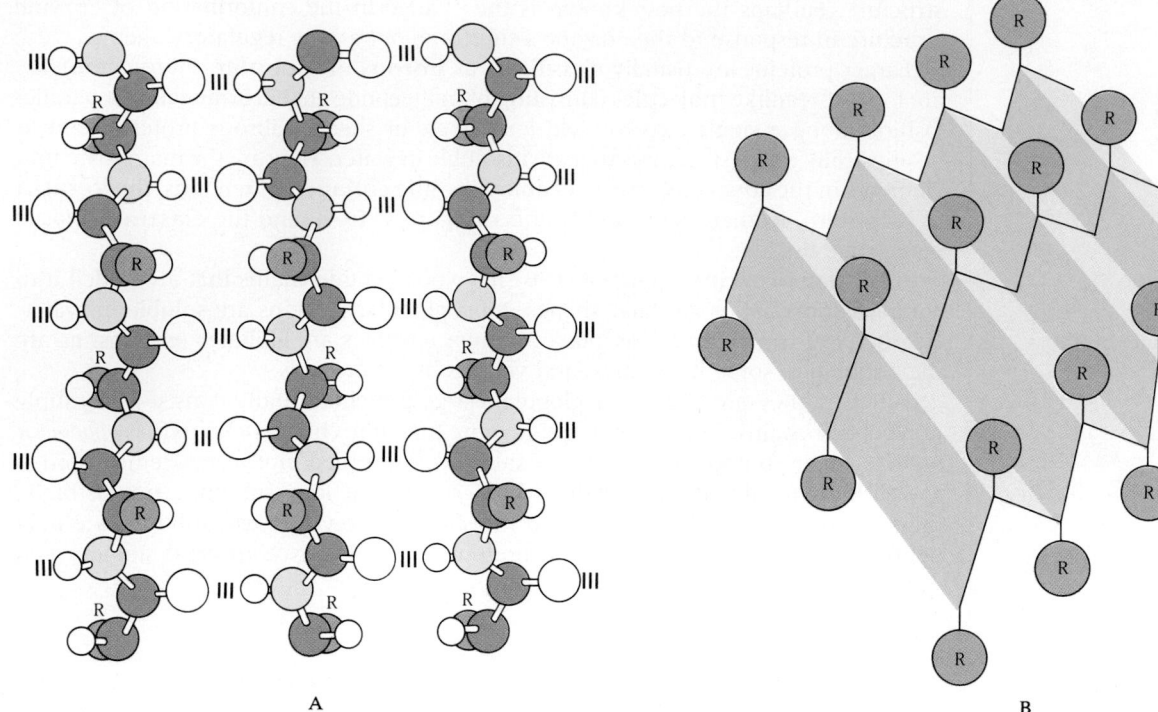

A   B

NUCLEIC ACIDS

The **nucleic acids** are the largest of the large biomolecules. Like polysaccharides and proteins, they are polymers made up of repeating small molecule units (or monomers). The units in nucleic acids are **nucleotides.** Thus, nucleic acids may properly be termed *polynucleotides*. In living organisms, nucleic acids, or polynucleotides, are often attached to proteins. This results in the formation of **nucleoproteins,** conjugated proteins with nucleic acid chains as prosthetic groups.

The two main kinds of nucleic acid are **deoxyribonucleic acid,** or **DNA,** and **ribonucleic acid,** or **RNA.** A major difference between DNA and RNA is the sugar present in their nucleotides. **Deoxyribose** occurs in the nucleotides of DNA. **Ribose** is in the nucleotides of RNA. We describe in Chapter 16 the nature of nucleotides and nucleic acids and how cells make them.

It is useful to keep in mind certain differences between proteins and nucleic acids, on the one hand, and the polysaccharides and lipids, on the other. Because of their structures, proteins and nucleic acids are **informational macromolecules** (Figure 4-21). Proteins can be viewed as long sentences written with a 20-letter alphabet.

In contrast, polysaccharides and lipids contain less, if any, information. The recurring building blocks of polysaccharides are either identical (as in starch, a polymer of glucose) or consist of only a few monotonously alternating sugars. One would be hard-pressed to write stirring prose with an alphabet containing only one or two letters. Likewise, lipids are low in information content compared to nucleic acids and proteins, since their fatty acid components include a small number of frequently repeating units.

ORIGIN OF THE BIOMOLECULES: HOW LIFE BEGAN

We have now been introduced to the major classes of biomolecules—the carbohydrates, lipids, proteins, and nucleic acids. The biomolecules we know today are all *products* of living organisms. Before life existed, such molecules obviously could not have been fabricated by living organisms. Yet these molecules *did* arise in the primitive oceans of Earth. Indeed, we believe that their **abiotic synthesis**—that is, their formation by processes independent of life—was a necessary first step on

4-17 β-pleated sheet protein structure
The β pleat is made of many polypeptide chains, or different parts of the same chain, held by intermolecular hydrogen bonds in a conformation resembling a corrugated sheet. (A) Ball-and-stick model of polypeptide chains in β conformation. (B) Diagram of three parallel polypeptide chains in β conformation, with imaginary pleated sheet between them.

Helical polypeptide chain

4-18 Bonds stabilizing the tertiary structure of a protein
All interactions shown (except those involving the N or C terminal amino acids) are between R groups. Top, left to right: Ionic bond; disulfide bond; hydrogen bond. Bottom, left to right: Interaction of nonpolar R groups; van der Waals forces; ionic bond. Of all of these interactions, only the disulfide bond is covalent.

FIBROUS PROTEINS GLOBULAR PROTEINS

Single chain

α–helical coil
(e.g. keratin)

Oligomeric
protein with
four chains
(e.g. hemoglobin)

Polypeptide
subunits

RNA

Supercoiling of α–helical
coils to form ropes

A

Portion of tobacco
mosaic virus particle,
a supramolecular
assembly with ~ 2200
polypeptide chains

B

4-19 Fibrous and globular proteins
*(A) Fibrous proteins often have α-helical
secondary structures. Several strands of such
helices are then supercoiled to form tough,
insoluble protein fibers. (B) Globular proteins
have primary amino acid chains that are folded
into compact, roughly spherical shapes. Most
globular proteins are soluble. Most enzymes are
globular proteins. The portion of a tobacco mosaic
virus has a "coat" made up of many globular
protein molecules.*

**4-20 Hydration of R groups on a protein
surface**
*Polar R groups weakly bind water molecules,
forming a shell of water around the protein which
may remain unfrozen even at temperatures far
below 0°C.*

the road to the eventual emergence of living organisms on Earth. How did these extraordinary events take place?

THE PROBLEM OF SPONTANEOUS GENERATION

From ancient times, people believed that life could arise spontaneously under certain peculiar circumstances: eels from the ooze of rivers; bees from the entrails of dead bulls; and maggots from the flesh of dead animals. So believed Aristotle (384–322 B.C.), and neither René Descartes (1596–1650), William Harvey (1578–1657), nor Isaac Newton (1642–1727) ventured to contradict him. The weight of centuries eroded belief in the spontaneous origin of maggots and mice; but the doctrine of spontaneous generation clung tenaciously to all theories of bacterial origin.

To shed light on this question, the Italian Abbé Lazzaro Spallanzani (1729–1799; Figure 4-22) performed a famous experiment. He found that whenever he sealed a broth from the air while it was boiling, it never developed bacterial growths. Therefore, he concluded, bacteria did not arise spontaneously from the broth, but came from air. To those who objected that he had ruined his broths by excessive boiling, he replied by breaking the seals of his flasks. Air rushed in and bacterial growth began. However, whatever Spallanzani did to purge contaminants—which he called the "seeds of life"—was viewed by believers as damaging to the "vital force" that in their opinion continuously gives rise to new life—to "life without parents" as one wit defined spontaneous generation.

Only after decades of controversy was the question resolved by the renowned studies of Louis Pasteur (1822–1895) in France and John Tyndall (1820–1893) in England. Pasteur (Figure 4-23) observed that eventually wine, milk, and meat broth are always overgrown by living bacteria. He conceded the logical possibility that the bacteria might have arisen anew from inanimate matter in these rich fluids. To him, the problem was a technical one: to repeat the work of those who claimed to have observed spontaneous generation, but to do so with sufficient care to exclude every possible hidden portal through which bacteria might enter. For those who contended that life did not enter from the outside, the proof had to go to the question of possible contamination. This was the main focus of Pasteur's extension of Spallanzani's earlier work.

In meticulous experiments, Pasteur showed that, following prolonged heating, a broth would contain growing microorganisms only after unfiltered air had been admitted to it. Only after he had gone to almost fanatical lengths to convince a stubbornly skeptical scientific community did it grudgingly relinquish its belief in spontaneous generation and accept the principle of **biogenesis**—the concept that all life comes from life.

Looking back on that period, we would have to agree that the hypothesis of spontaneous generation was entirely reasonable in its time. Despite our satisfaction at the downfall of that hypothesis, we must not forget that we now believe that life once did originate on Earth by a process that must be described as spontaneous generation. Pasteur's evidence against *latter-day* spontaneous generation brought the whole problem of life's origin to scientific attention for the first time.

MODERN APPROACHES TO THE PROBLEM

The problem of life's origin needed no solution as long as people believed that life could arise spontaneously on every side. Only when Pasteur showed that present-

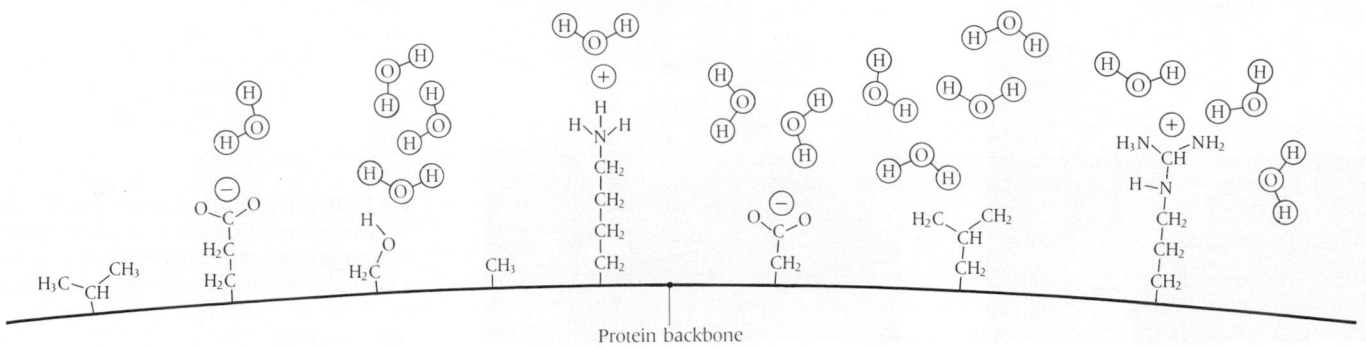

Protein backbone

day life must come from life did the scientific problem come into focus. As there were no witnesses, one can imagine only two kinds of direct evidence: (1) the discovery of fossil remains of the first organisms, and (2) a new appearance today of another "first organism" from nonliving matter, which, as noted below, would be an extremely unlikely event. Until recently, no such fossils had been found. However provocative clues have turned up. For example, primitive cyanobacteria have been found in rocks up to more than 3 billion years old. As for a new contemporary origin of life, it seems likely that if a fledgling new organism were not promptly eaten by an already existing organism, we would need to go through the same agonies Pasteur endured to prove it was not a contaminant.

Scientific consideration of the origin of life must thus be based on *indirect* evidence. Until recently, this meant armchair speculation.

Theoretical Approaches

Most biologists now agree that the earliest life forms could and did arise from nonliving matter by ordinary physical and chemical processes. Probably the major biomolecules came into existence in the primitive seas long before the advent of living, reproducing organisms. The infant world had its full complement of elements—hydrogen, carbon, nitrogen, and the rest. But these elements then existed largely in the form of simple molecules. The ocean was a huge, watery solution of chemical substances capable of interacting with each other and with chemical compounds in the atmosphere. The principles of thermodynamics and chemical kinetics (Chapter 8), both give clear theoretical hints as to how these simple molecules may have evolved into complex ones. In the ancient ocean, chemical interactions were not governed purely by chance, but by chance functioning within the strict laws of chemistry. Carbon atoms manifested their tendency to form long chains, and there began a period of chemical evolution that may have lasted 1.7 billion years, one-third of the Earth's history. The result was oceans of what J. B. S. Haldane called "dilute warm soup," a solution of many organic compounds that slowly grew more complex.

In the late 1920s, A. I. Oparin, a Soviet biochemist, first suggested that the primitive atmosphere contained gaseous methane (CH_4), ammonia (NH_3), hydrogen sulfide (H_2S), and water (H_2O) and that chemical and physical forces could have converted these molecules to simple organic compounds. Radiant energy of sunlight or the discharge of lightning could have caused these atmospheric gases to react with each other to form simple organic products that remained in solution in the sea. In time, Oparin suggested, they formed colloids (permanent suspensions of fine particles), which ultimately gave rise to the first living organisms. Oparin's hypothesis lacked supporting evidence and was not immediately accepted.

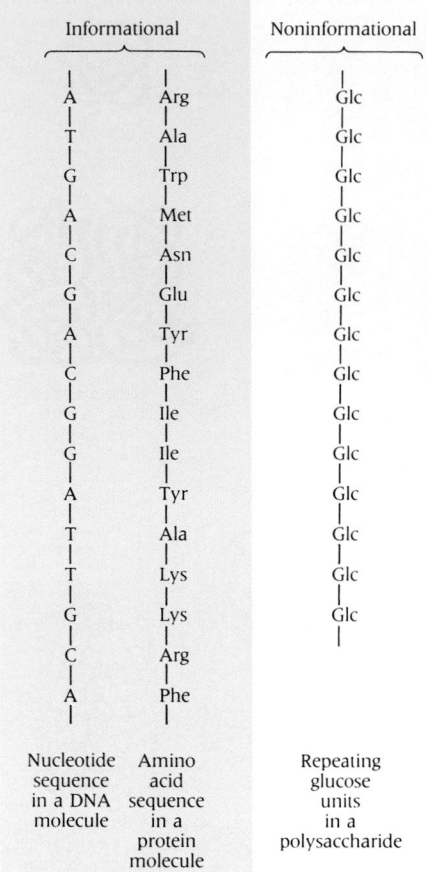

Informational		Noninformational
A	Arg	Glc
T	Ala	Glc
G	Trp	Glc
A	Met	Glc
C	Asn	Glc
G	Glu	Glc
A	Tyr	Glc
C	Phe	Glc
G	Ile	Glc
G	Ile	Glc
A	Tyr	Glc
T	Ala	Glc
T	Lys	Glc
G	Lys	Glc
C	Arg	
A	Phe	
Nucleotide sequence in a DNA molecule	Amino acid sequence in a protein molecule	Repeating glucose units in a polysaccharide

4-21 Not all macromolecules carry information
DNA and proteins have specific, highly complex sequences of subunits, whereas molecules such as polysaccharides contain monotonously repeating sequences. The letters A, T, C, and G symbolize the four bases of nucleic acid subunits. Ala, Arg, Trp, and so forth are abbreviations for amino acids. Glc stands for glucose.

4-22 (Left) Abbé Lazzaro Spallanzani (1729–1799)
Spallanzani was one of the first to dispute the notion of spontaneous generation. In a famous experiment, he demonstrated that bacteria come from the air, not from the broth within in a flask.

4-23 (Right) Louis Pasteur (1822–1895)
Pasteur showed that broth would contain living microorganisms only after unfiltered air was admitted to it. In this way, Pasteur invalidated the theory of spontaneous generation as it was then conceived.

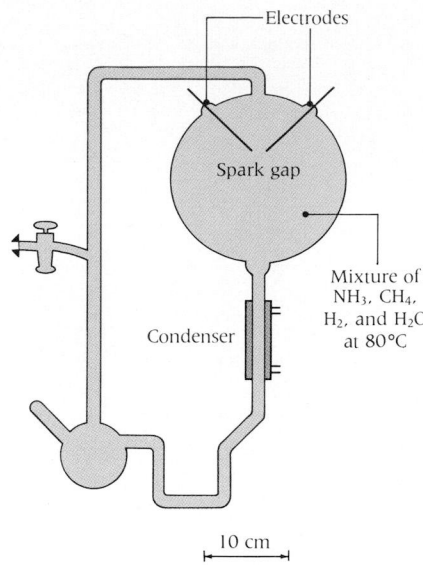

4-24 Apparatus used to simulate primitive atmospheric conditions
Organic compounds such as amino acids were synthesized under these conditions from a simple mix of gases thought to exist in Earth's early atmosphere.

Labels: Electrodes; Spark gap; Condenser; Mixture of NH_3, CH_4, H_2, and H_2O at 80°C; 10 cm

There has been continuing argument over exactly what gases the atmosphere contained during the period when life is believed to have begun. If it contained methane and ammonia and lacked oxygen, it was certainly different from today's atmosphere. Support for such a formulation recently came from laboratory experiments showing that the simple gases thought to be present in the primitive atmosphere *can* give rise to organic molecules.

In the first such experiment, in 1953, Stanley L. Miller and Harold C. Urey imaginatively tried to reproduce the conditions of an ancient world on a laboratory bench. They placed a gaseous "atmosphere"—a mixture of methane, ammonia, water, and hydrogen—above a watery "sea" in a special flask (Figure 4-24), warmed it to 80°C, and for a week subjected it to electrical sparks that simulated lightning. When they opened the flask and analyzed its contents, they found a remarkable collection of newly formed organic compounds, among them the α-amino acids glycine, alanine, aspartic acid, and glutamic acid, all common constituents of proteins. They also found several simple organic acids now known to occur in living organisms, including formic, acetic, propionic, lactic, and succinic acids.

They postulated that these organic compounds had arisen by sequences of single reactions. For example, hydrogen cyanide (HCN) was formed from methane and ammonia. Methane (CH_4) was converted by electrical discharge into a variety of other hydrocarbons which then reacted with HCN. After a further series of reactions, propionic acid, alanine, and other organic molecules appeared.

The many forms of radiant energy found capable of producing organic compounds in such gas mixtures include visible light, ultraviolet light, x-rays, gamma rays, electrical discharges, ultrasonic irradiation, and high-energy α- and β-particles. The many hundreds of organic compounds that have been formed in such experiments include examples of all the important molecules found in cells—the common amino acids, the nitrogenous bases (see Chapter 16) that are essential building blocks of nucleic acids, and many sugars.

It is doubtful that organic molecules are still being formed in this way in the world's oceans. Earth's surface cooled and the atmosphere changed substantially after the time when life began, hydrogen and carbon monoxide giving way to oxygen and carbon dioxide. This both slowed the formation of new organic molecules and increased the likelihood that such compounds would be oxidized away. Also, bacteria and other forms appeared that would be likely to consume any newly arisen organic molecules.

The experiments just described give a plausible explanation of the rise of small biomolecules. But we must also account for the polymerization of these simple molecules into the macromolecules that are such distinctive features of the chemistry of living organisms. They are critical components of organisms because two of them, as we have seen (see Figure 4-21), are carriers of biological information. These are the proteins and nucleic acids.

The origins of these structurally complex macromolecules seemed at one time to be the most formidable aspect of the scientific problem. It turned out, however, that in the experiments of Miller-Urey and others, polymerization of monomers occurs readily. For example, when Sidney Fox heated amino acids, some of them joined together to form protein-like macromolecules called **proteinoids.** Some of

4-25 Proteinoid microspheres
Such microspheres may form spontaneously when amino acids are heated. Some have enzymatic activity or form cell-like structures in water.

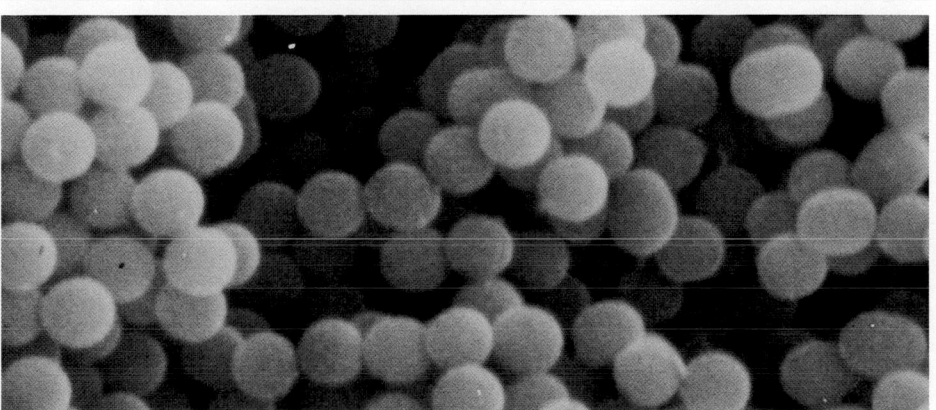

these products have molecular weights in the thousands and contain all of the amino acids common to modern proteins. Moreover, some have a variety of enzyme-like activities, and some have a distinct tendency to form cell-like structures upon contact with water (Figure 4-25). Similar successes have been achieved in the abiotic synthesis of polynucleotides.

MOLECULAR FORERUNNERS OF LIFE

Our purpose at this early point has been to introduce the major biomolecules and to show how they *may* have arisen on a primitive Earth that was without living organisms. The following view is favored by many investigators: Proteinoids evolved into modern proteins, and the tiny microspheres they formed eventually became cells. In other words, a variety of polymerizations led in time to the complex interdependencies that typify living cells.

Some scientists object to this view, arguing (1) that chemical changes in the primitive ocean were more likely to produce tar and sludge than finely honed biomolecules, and (2) that structures as complicated as nucleotides and nucleic acids would surely need to be produced by a preexisting organism. These objections can be summarized by a question: Is nucleotide and nucleic acid synthesis conceivable in a prebiotic world? A. G. Cairns-Smith has proposed that clay particles, which grow by a layering process, may have provided surfaces that collected the chemical precursors of biomolecules and accelerated their assembly and replication. The resulting "low-technology" product then evolved into "high-technology" form, a living organism. The answer to the question awaits evidence.

SUMMARY

The chemical characteristics of organic molecules, defined as molecules containing carbon, account for many of the properties of living organisms. Organic molecules are ordinarily held together with covalent, rather than ionic, bonds. As in inorganic compounds, the valences of the component atoms of an organic molecule dictate its molecular structure and geometry. The ability of a carbon atom to form up to four different covalent bonds means that certain carbon-containing compounds may have the same empirical formula but a different structural formula. Such compounds are said to be isomers of one another. Functional groups, which are small, distinctive groups of atoms that recur regularly in organic structural formulas, are crucial determinants of the chemical behavior of organic molecules.

The majority of the organic molecules found in living organisms can be categorized into four classes on the basis of their structural formulas and consequent chemical behavior. These classes are (1) the carbohydrates, including sugars (monosaccharides) and glycogen, starches, and cellulose (polysaccharides); (2) lipids, a heterogeneous group including fats, phospholipids, and steroids; (3) proteins; and (4) the nucleic acids, DNA (deoxyribonucleic acid) and RNA (ribonucleic acid), polymers of deoxyribonucleotides and ribonucleotides (respectively). The critical roles of these nucleotides in the mechanisms of genetics and heredity are considered in later chapters.

None of these molecular classes was present on Earth when the planet was formed. However, considerable evidence supports the hypothesis that early environmental conditions were different enough from those of today to favor the spontaneous formation first of small organic molecules and later of the macromolecules essential for the origin of life.

KEY TERMS

absorption spectrum
α-helix
amino acid
amphipathic
asymmetrical carbon
β-pleated sheet
carbohydrate
carboxyl group
cellulose
chromatography
conformation
conjugation
denaturation
deoxyribonucleic acid (DNA)
deoxyribose
disaccharide
empirical formula

fatty acid
functional group
glycogen
hydrolysis
informational macromolecule
isomer
lipid
macromolecule
monomer
monosaccharide
nucleic acid
nucleotide
organic acid
peptide
phospholipid
polymer
polypeptide

polysaccharide
primary structure
protein
quaternary structure
renaturation
ribonucleic acid (RNA)
ribose
saturated fatty acid
secondary structure
spectrophotometry
starch
stereoisomer
steroid
structural formula
tertiary structure
triglyceride
unsaturated fatty acid

QUIZ QUESTIONS

1. Which of the following categories of macromolecules includes representatives that frequently serve as energy storage molecules?
   A. proteins
   B. lipids
   C. carbohydrates
   D. A and B
   E. B and C

2. Although silicon is very abundant and has the same valence as carbon, carbon is better suited to be the predominant element in organic chemistry because
   A. carbon dioxide is a gas in the temperature range occupied by living organisms.
   B. carbon forms long chains more readily than silicon does.
   C. many carbon compounds readily dissolve in water.
   D. A and C
   E. A, B, and C

3. Polymerization reactions that produce many macromolecules are called _____ reactions.
   A. hydrolysis
   B. condensation
   C. carboxylation
   D. amination
   E. hydroxylation

4. Which of the following macromolecules is a protein?
   A. glycogen
   B. chitin
   C. triglyceride
   D. cholesterol
   E. keratin

5. The prosthetic group of a conjugated protein classified as a glycoprotein is
   A. a chromomorphic group.
   B. a polysaccharide.
   C. a nucleic acid.
   D. a metal ion.
   E. a lipid.

ESSAY QUESTIONS

1. What sort of a molecule is a carbohydrate? Distinguish between a monosaccharide, disaccharide, and polysaccharide and give some examples of each.

2. Characterize the general structure of an amino acid. What is meant by the "R group" of an amino acid? How are amino acids classified? What is the major role of amino acids in living organisms?

3. What is the difference between inorganic and organic molecules?

4. Describe the general structure of DNA and RNA. How do they differ?

5. If Pasteur and others proved that spontaneous generation does not occur, how can we account for the origin of life on Earth? Discuss some of the current theories on how living organisms arose billions of years ago. If a scientist claimed to have produced life from nonliving materials, what properties would the new "organism" have to have before it could be accepted as living?

REFERENCES AND SUGGESTIONS FOR FURTHER READING

Cairns-Smith, A. G. (1985). The first organisms. *Scientific American 252,* 90–101. *Seven Clues to the Origin of Life: A Scientific Detective Story.* Cambridge University Press, New York.

By a critic of the view that small molecules polymerized to form nucleic acids and proteins, which gradually evolved into cells. Here are two stimulating, nontechnical discussions of how, in the primordial seas of 4 billion years ago, clay surfaces, not "primordial soup," may have given rise to the molecules from which sprang the first living organisms.

Darnell, J. E., Jr. (1985). RNA. *Scientific American 253,* October, 68–78.

A brief review by a contributor to the field.

Dickerson, R. (1978). Chemical evolution and the origin of life. *Scientific American* September, 62–78.

A clear account of the most widely held view of how chemical evolution led to the rise of biomolecules and then of life.

Doolittle, R. F. (1985). Proteins. *Scientific American 253,* October, 88–99.

A useful current summary with fine illustrations.

Felsenfeld, G. (1985). DNA. *Scientific American 253,* October, 58–67.

A well-illustrated summary.

Karplus, M., and McCammon, J. A. (1986). The dynamics of proteins. *Scientific American 254,* April, 42–51.

An explanation of the fact that protein molecules are not rigid and are perpetually engaged in internal movements that can now be simulated with computers.

Margulis, L. (Ed.) (1970). *Origins of Life: Proceedings of the First Conference.* Gordon and Breach, New York.

The actual transcript of a lively conference in which workers of different disciplines debated the topic.

Oparin, A. (1938). *The Origin of Life.* Dover, New York.

A translation of the classic work that first enunciated the view that life arose spontaneously in the primordial seas.

Vance, D. E., and Vance, J. E. (Eds.) (1985). *Biochemistry of Lipids and Membranes.* Benjamin, Menlo Park, Calif.

A substantial, up-to-date volume summarizing a tumultuous field.

Vollhardt, K. P. C. (1987). *Organic Chemistry.* W. H. Freeman, New York.

Textbooks of organic chemistry, like the movies (and like the book you are reading), began years ago in black and white and are now in glorious color. This one is unusually appealing—for its clarity, appearance, and authority.

Weinberg, R. A. (1985). The molecules of life. *Scientific American 253,* October, 48–57.

A good brief review with fine color plates.

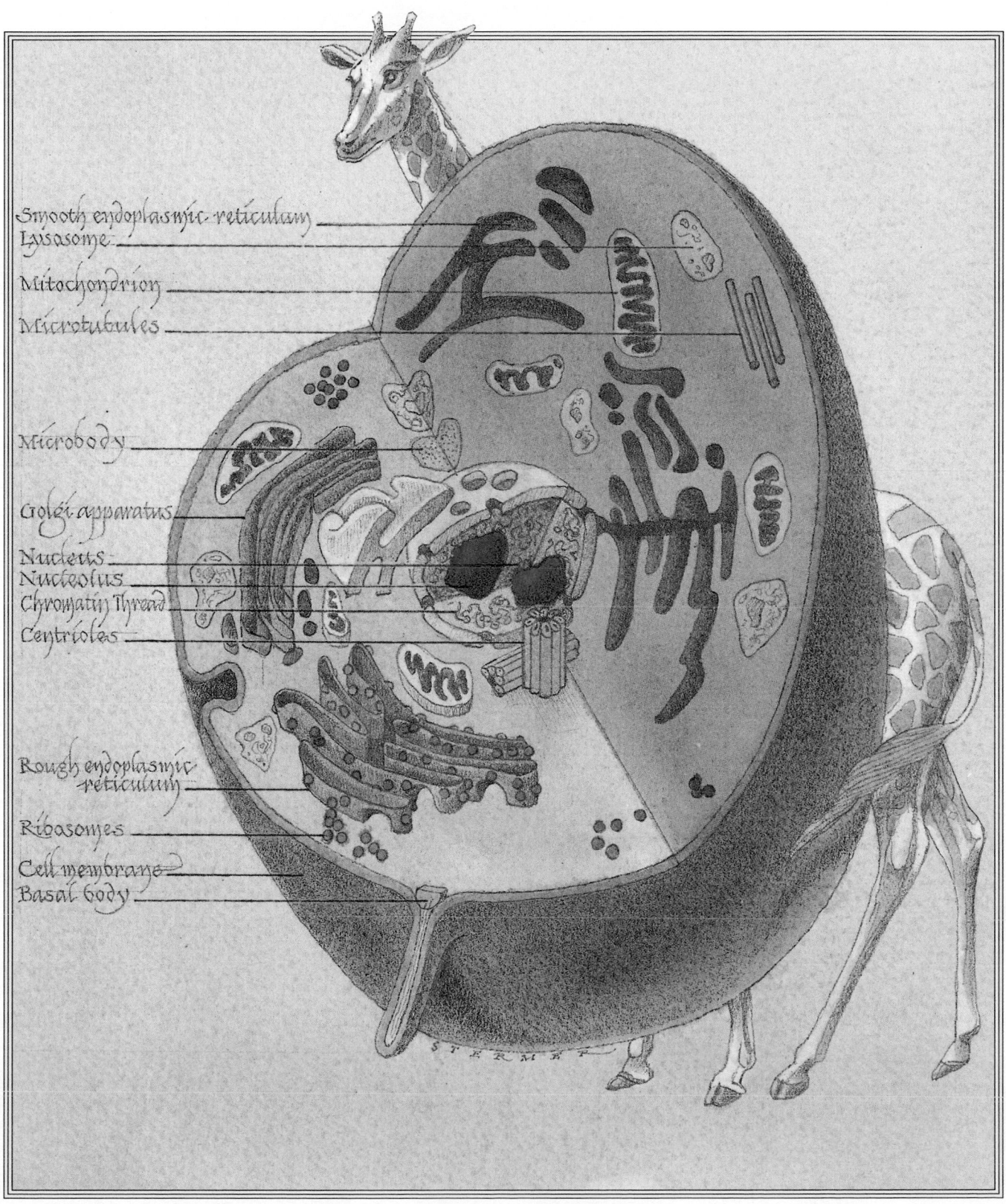

Smooth endoplasmic reticulum
Lysosome
Mitochondrion
Microtubules

Microbody

Golgi apparatus
Nucleus
Nucleolus
Chromatin Thread
Centrioles

Rough endoplasmic reticulum

Ribosomes
Cell membrane
Basal body

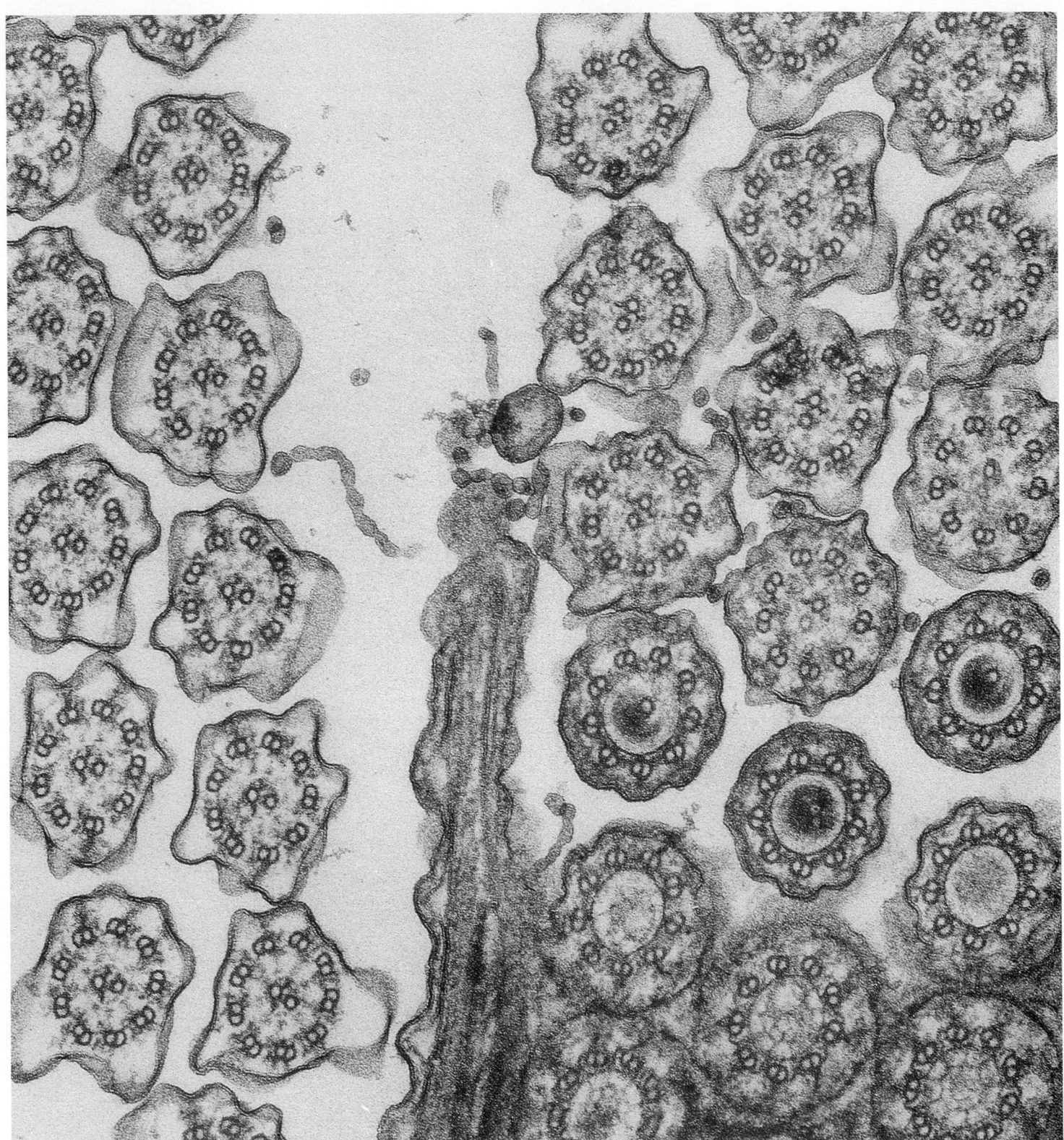

CELL STRUCTURE

Cell Structures
The electron microscope greatly enhances resolution of fine cell structure. This cross-section of cilia from the protist Tetrahymena *clearly reveals the "nine plus two" pattern of its microtubules. (× 90,000)*

The basic unit of life is the **cell,** the smallest organization of matter of which we can say, "This is alive." Each cell is a living system capable of regulating its vital processes and, with some exceptions, reproducing. Many cells are capable of organizing themselves into larger units called **tissues.**

The cell, then, is the minimum organization that in today's world is capable of all the processes we refer to collectively as "life." We say "today" because when life arose from inanimate matter billions of years ago, it was in a simpler form than are the cells we know. For simple as it may appear, the cell is exquisitely constructed and enormously complex.

THE CELL THEORY

All living organisms are composed of fundamentally similar entities called cells, each of which can arise only from a preexisting cell. That, in brief, is the **cell theory.** Like the theory of evolution, it is surely one of the great unifying principles of biology.

EMERGENCE OF THE CELL THEORY

Before microscopes were invented in the seventeenth century, there was no reason to suspect the existence of a world of living creatures too small to be seen by the naked eye. Even when the old lens grinders had produced reasonably powerful magnifying glasses of short focus, few scholars set about investigating microscopic living forms in a systematic way. The early microscopes were rather more like toys for the curious.

The Englishman Robert Hooke (1635–1703) was among the first to realize the importance of studying nature with instruments that could greatly extend the power of the eye. In 1665 Hooke published a remarkable treatise, his *Micrographia,* that launched the study of microscopic anatomy. Using a primitive microscope (Figure 5-1A), he carefully examined a thin slice of cork and observed many small cavities separated by walls (Figure 5-1B). We now know that these were the thickened walls of dead plant cells, the bodies of which had been lost long before. Unaware of what they were, Hooke called them "cells," meaning "little rooms" (Latin, *cella*).

The next major contribution came from the eccentric linen draper of Delft, Antonie van Leeuwenhoek (1632–1723). During his long life, Leeuwenhoek sent innumerable communications to the Royal Society of London describing the microscopic appearance of objects of all kinds. He observed swimming creatures in rainwater and wrote the first descriptions of bacteria, fungi, and protozoa. To doubters he offered affidavits from prominent citizens who had also seen the "wretched beasties," which he termed "animalcules." He did not think in biological terms, however.

The death of Leeuwenhoek ended the first inquisitive period of microscopic biology. Little more progress was made until 1838, 173 years after Hooke's description of cork! In that year the German botanist Matthias Schleiden (1804–1881) published a monograph on the microscopic anatomy of plants. A contemporary German physiologist, Theodor Schwann (1810–1882), recognized the similarity of Schleiden's plant cell nuclei to structures he had observed in animal nerve tissues. In 1839 he wrote, "We have seen that all organisms are composed of essentially like parts, namely, of cells."

Some historians of science today downgrade the contributions of Schleiden and Schwann. A careful reading of earlier literature discloses many statements directly or indirectly suggesting that cells are the units of life—and by 1835 the cell was generally regarded as an entity with a life and structure of its own. Therefore, when the time arrived for a unifying statement of the cellular nature of all living things, similar insights came to several individuals. Furthermore, Schleiden and Schwann made an extraordinary error. While recognizing the essential similarity of cells throughout the living world, they offered a most peculiar explanation of how cells arise in the first place. They likened the process to crystal formation—in disagreement with certain of their contemporaries, who believed correctly that new cells arise only by division of pre-existing cells. This feature of their work is usually forgotten.

A

IMPLICATIONS OF THE CELL THEORY

In 1845 Karl von Siebold (1804–1885) concluded that a protozoan is a simple animal consisting of one cell. By 1858 it was established by Rudolf von Virchow (1821–1902) that new cells can originate only by cell division. This insight deepened the significance of the cell theory by relating it to two other major insights that emerged at about the same time: the works of Pasteur showing that life does not originate by spontaneous generation, and those of Darwin on evolution.

Consider these relationships. Virchow's 1858 principle, that "all cells come from cells," was simply a more explicit statement of Pasteur's 1862 principle, that "all life comes from life." The relationship of the cell theory to the theory of evolution is implied in Virchow's sentence "The principle is thus established . . . that, throughout the whole series of living forms, whether entire animal or plant organisms . . . there rules an eternal law of *continuous development.*" Here Virchow was postulating an unbroken continuity of cell generations stretching back to the origin of life on Earth. It is only a step from this picture to the insight that all cells share a common ancestry, an essential element of modern evolutionary thought.

The transition from Hooke–Leeuwenhoek to Schleiden–Schwann was a step from simple observations of fact to an inductive generalization of great breadth. The transition from Schleiden–Schwann to Virchow was a leap from a simple generalization to a major explanatory theory.

The two parts of the cell theory—that all living things are composed of cells and that cells arise only from other cells—allowed the first scientific definition of living things.

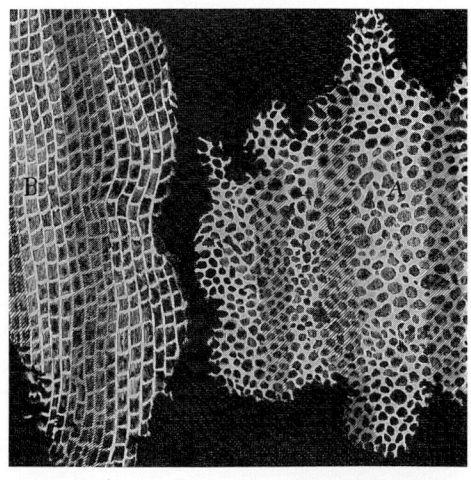

B

5-1 The work of Robert Hooke
(A) *Replica of Hooke's compound microscope.*
(B) *Hooke's drawing of the structure of cork (1665)*

ABOVE AND BELOW THE CELL IN THE HIERARCHY

Although the cell theory originally derived from investigations of *multicellular* organisms, it applies as well to *unicellular* forms of life. We should always bear in mind that even the simplest cells are highly complex organizations.

To better appreciate the significance of the cellular level in the hierarchy of biological organization, we should examine this hierarchy again.

The Organizational Hierarchy Reconsidered

The organizational hierarchy outlined in Table 1-1 appears in greater detail in Figure 5-2. The major **biomolecules** are ultimately derived from simple, low-molecular-weight precursors in the environment (CO_2, H_2O, and N_2). These precursors are converted by sequences of enzymatic reactions into biomolecules of intermediate molecular weight. These and their derivatives (termed *building blocks* in Figure 5-2) are then linked covalently to form the **macromolecules** of cells.

At the next higher level of organization, macromolecules of different kinds associate with one another to form **supramolecular complexes**—for example, lipoproteins, which are complexes of lipids and proteins; nucleoproteins, which are complexes of nucleic acids and proteins; glycoproteins, which are complexes of

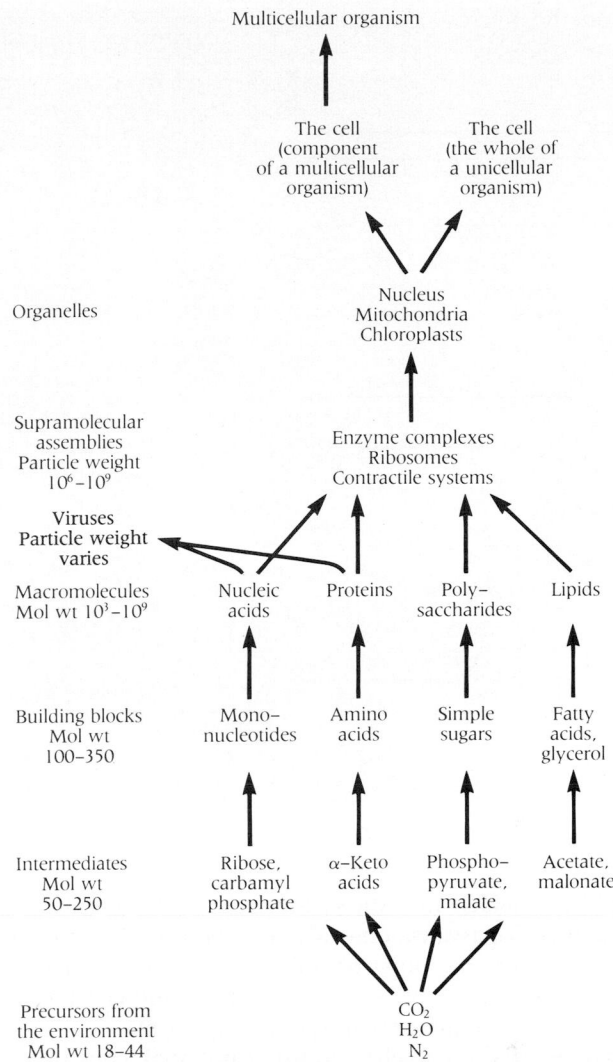

oligosaccharides and proteins; and others. The components of such complexes are held together by bonds of all kinds—covalent bonds and weak noncovalent forces. This noncovalent association of macromolecules into supramolecular complexes is nevertheless highly specific and, usually, quite stable.

At the next higher level of organization, various supramolecular complexes are assembled into subcomponents of the cell called **organelles.** These include nuclei, mitochondria, chloroplasts, and the rest. Together they make up the cell. Finally, we encounter the multicellular organism.

Unicellular and Multicellular Organisms

The plants and animals with which we are familiar are composed of many cells. An adult human contains about 5×10^{13} cells; an elephant and surely a giant sequoia tree contain much greater numbers. But these individual cells do not lead independent lives. They are not organisms. They are *parts* of organisms. Nonetheless there is a stage in the life history of every multicellular organism when it is a single cell—a single fertilized egg cell or a spore (an asexual reproductive cell).

Sometimes, as in the swimming spores of certain relatively simple organisms—for example, *Ulva,* a green alga known as sea lettuce—the single cell may lead an independent life for a time and behave very much like an independent organism. But the free-living single cell is still only a stage in a more complex life history.

In contrast, unicellular organisms are single cells living wholly independent lives. In the past, some of these were considered "plants" and others "animals" because in some ways they each resembled either multicellular plants or animals. We shall later discuss the differences between plants and animals (Chapter 42), the problems of classifying organisms (Chapter 38), and the recent debate over whether living

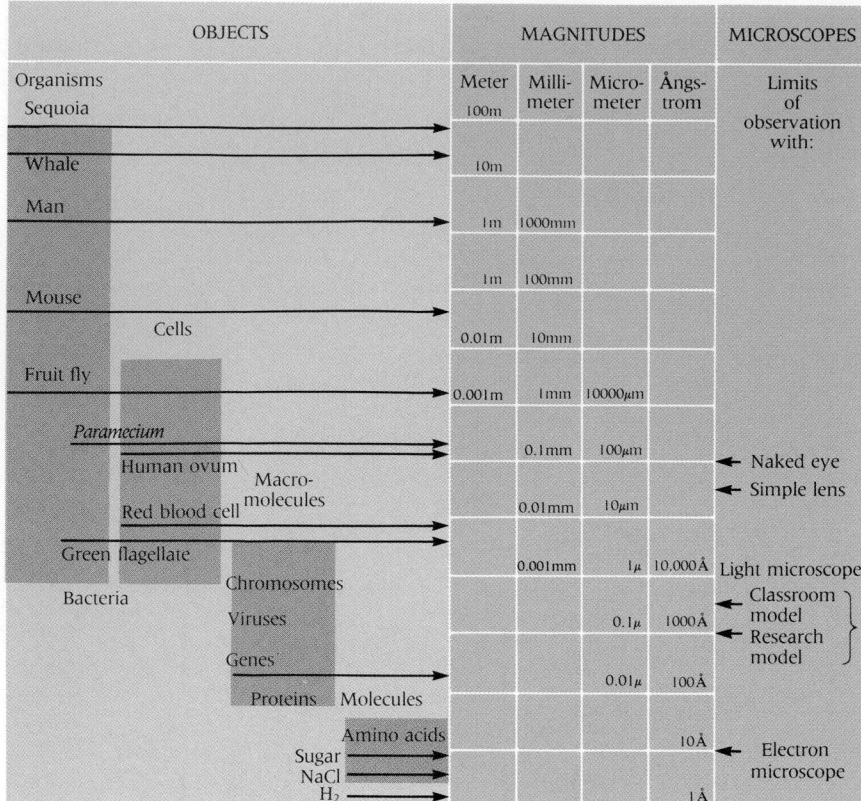

5-3 Biological objects, their magnitudes, and the microscopes that reveal them best

organisms should be divided into three, four, or five kingdoms. In this book we accept the five-kingdom classification—Monera, Protista, Plantae, Fungi, and Animalia—and we note that unicellular organisms, the subject at hand, occur in more than one kingdom. Indeed, the terms "plant" and "animal" are now considered quite inappropriate when applied to unicellular organisms.

It has been aruged that a unicellar protist is not analogous to a single cell of a multicellular organism but to the whole of such an organism. Most such unicellular organisms are capable of specific functions because they have intricate specialized parts resembling the mouths, eyes, and fins of multicellular animals.

A multicellular organism is a federation of specialized single cells. When we discuss multicellular organisms, the term **specialization** denotes a division of labor within the organism. To a large extent each cell of a multicellular organism depends on the presence and integrated functioning of the other cells—and thus upon the integrity and organization of the whole organism. The specialized cells of multicellular organisms acquire their distinctive properties during the process of **differentiation.**

Unlike a unicellular organism, a cell of a multicellular organism is not an independent unit. If such a cell grows and proliferates on its own, with a pattern of behavior that is not subservient to the whole organism's best interests, we call it a **neoplastic cell.**

Cell Sizes and Shapes

Cells vary greatly in size. Most cannot be seen without a microscope (see Figures 3-1 and 5-3), although a few varieties are visible to the naked eye. In a large, multicellular organism such as a human being, the diameter of an average cell is about 10 micrometers (μm)—or 0.01 millimeter (see Box 3-1). Some cells are quite large. For example, a single nerve cell in a large animal, although of small diameter, may reach a length of several feet. Eggs of animals are single cells before their development begins, and they are usually visible to the eye. Human eggs, which are rather small as eggs go, can also be seen without a lens but only as specks smaller than the periods on this page. The egg of an ostrich has a volume of about 1 quart. The egg of *Aepyornis,* an extinct bird of Madagascar, had a volume of more than 2 gallons (including its nutrient reserves). These are the largest cells yet encountered.

Cell size is limited by the need for the ratio of surface area to cell volume to be within certain limits. Cells obtain materials—gases and nutrients—through their surface membranes. Wastes leave by the same route. As cell diameter increases, the surface area increases as the square of the diameter and volume increases as the cube of the diameter. A critical size is soon reached beyond which the surface area can no longer sustain the needs of the cell mass within.

Which are the *smallest* cells known? That distinction probably goes to certain bacterial cells with diameters of only 0.35 to 0.40 µm—near the limit of resolution of an ordinary light microscope. Perhaps the smallest cells are of the bacterial genus *Mycoplasma*. While it is not possible to state the size of such cells with precision—they are variable in shape and dimension—they appear to be spherical cells with a diameter of about 0.35 µm. Its volume is 2.24×10^{-14} cubic centimeters. Put another way, a liter of packed mycoplasmas would contain over forty million billion (4×10^{16}) cells. About 8000 of these little organisms would fit inside a single human red blood cell. The approximate diameter of a carbon atom is about 1.4×10^{-10} meters. Hence these smallest living cells have diameters only 2500 times that of a carbon atom! Each mycoplasma consists of about 2.5×10^9 atoms that are organized into about 6×10^8 molecules, most of which are water. The solid part of the cell is made up of about 5×10^6 molecules of amino acids, sugars, fatty acids, and other biomolecules. These are organized into some 10^5 macromolecules, which are the cell's working machinery.

5-4 The diversity of cell shapes
This collection of animal cells is intended only to illustrate the great diversity of cell shapes. A comparable display could be made of cells from plants, fungi, protists, and monerans. **(A)** *Human muscle cells.* **(B)** *A fat storage cell.* **(C)** *Red blood cells.* **(D)** *White blood cell.* **(E)** *Intestinal epithelial cells (from axolotl, a larval salamander).* **(F)** *Cell from epithelial lining of human vagina.* **(G)** *Liver cells (mouse).* **(H)** *Intestinal epithelial cells (rat).* **(I)** *Human neuron.* **(J)** *Spermatozoon (human).* **(K)** *Ovum with sperm entering it (human).* **(L)** *Cells from human placenta with overlying syncytium (a multinucleate mass of protoplasm not separated by membranes) bearing cilia.* **(M)** *Cells from retina (human). All × 1000 except J, which is × 2000.*

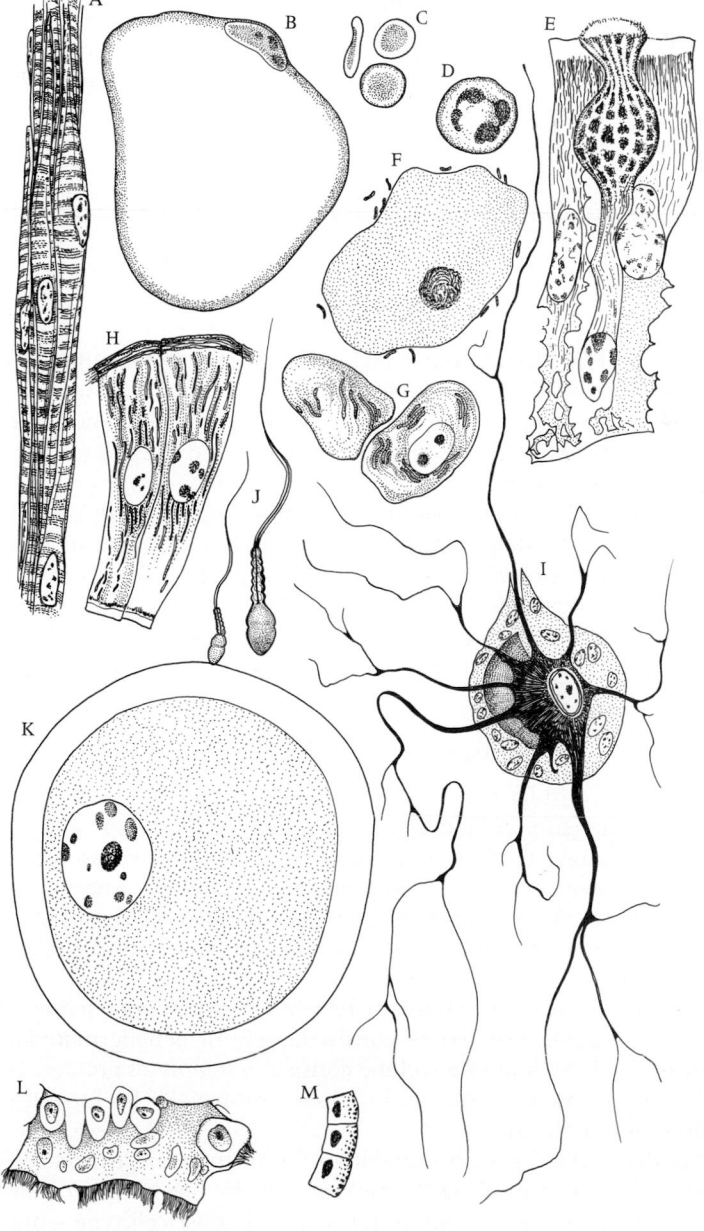

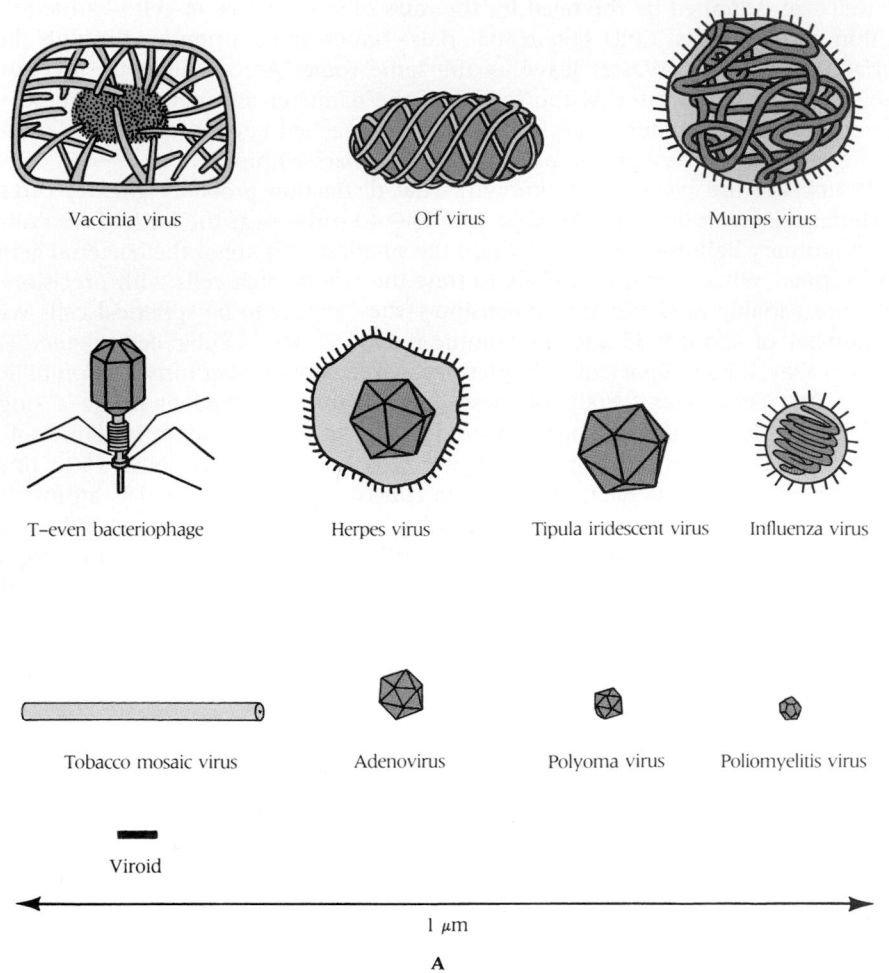

Vaccinia virus   Orf virus   Mumps virus

T–even bacteriophage   Herpes virus   Tipula iridescent virus   Influenza virus

Tobacco mosaic virus   Adenovirus   Polyoma virus   Poliomyelitis virus

Viroid

1 μm

A

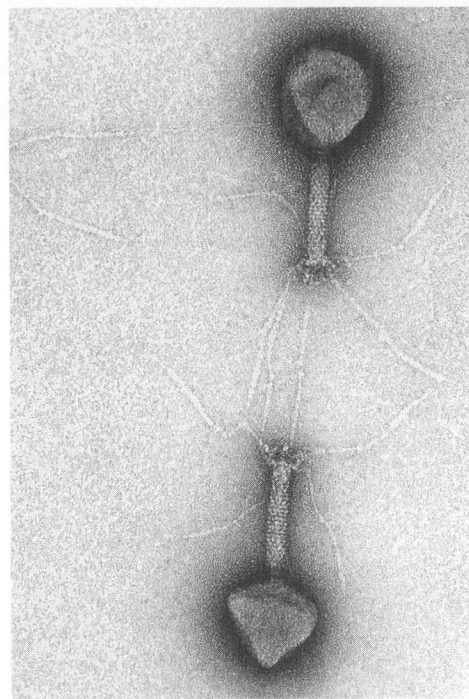

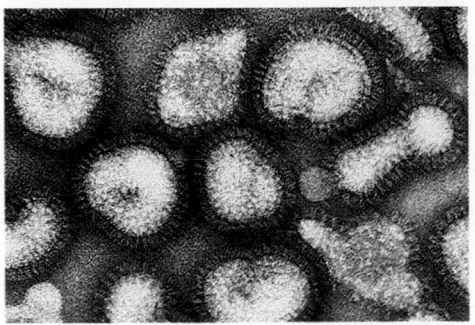

B

5-5 Relative sizes of viruses
(**A**) *Note the diversity of shapes within this group. The relative size of a viroid is shown in lower left.* (**B**) *Electron micrographs of a bacteriophage, × 110,000 (top); tobacco mosaic virus × 90,000 (middle); and influenza virus × 40,000 (bottom).*

One writer suggested that in our imaginations we should magnify this cell about 10 million times. It would then just fit into a small office with a bit of crowding. Each atom would now be a millimeter in size and each amino acid about the size of a marble. Proteins would be as large as baseballs and the cell's outer membrane as thick as a bathroom sponge.

It is impossible to generalize about cell shapes. Their variety is endless (Figure 5-4). In general, a cell's shape reflects its functions in an organism. Skin cells are flat and platelike. Nerve cells are elongated fibers suitable for transmitting impulses over long distances like telegraph wires. Muscle cells are elongated contractile structures. Red blood cells are biconcave discs with an unusually high ratio of surface area to cell volume that is appropriate to their role in O_2 and CO_2 transport.

Are Viruses and Plasmids Organisms?

Let us now briefly consider two special classes of subcellular entities, **viruses** and **plasmids,** which are considerably smaller than any known cells. A typical bacterial cell, for example, might have a diameter of 0.5 μm. A virus associated with that bacterium might have one-tenth that diameter—although some viruses are quite a bit larger (Figure 5-5). The nature of viruses, so familiar as causes of such human diseases as influenza, poliomyelitis, and viral pneumonia, has been endlessly debated. Are they or are they not living organisms?

Countless kinds of viruses exist. All are totally dependent upon living cells, though they are not cells themselves. They consist largely of genetic material—DNA or RNA—surrounded by a coat or envelope consisting mainly of protein. They possess genes, sometimes as many as several hundred (but usually far fewer). In contrast, cells contain many thousands.

The fact is that viruses are not capable of an independent existence. They enter cells as invaders, commandeering the cell's metabolic machinery to suit the virus's needs. Each virus type has an affinity for a specific host cell type—animal, plant, or bacterial. Once inside a host cell, the virus causes the cell to curtail its normal activities and to begin producing new viruses. In the course of this process, the host cell may be destroyed, although some viruses are able to linger within their host cells, sometimes for many generations, without destroying them. One would hardly term viral reproduction "self-reproduction," since it can occur only with the help of host cells. If the definition of *living* includes self-reproduction, viruses are not living. For this reason they are not included in any of the five kingdoms.[1]

Nevertheless, viruses do contain protein and either DNA or RNA. As we shall see, they also exhibit predictable genetic traits, despite the fact that in the isolated state they lack two of the major characteristics of living systems described in Chapter 1—self-reproduction and self-regulation. Their simplicity is deceptive, however. Their very interesting methods for duplicating themselves within cells are remarkable.

Plasmids are a less familiar class of subcellular components. Simpler in structure than viruses, they consist only of a bit of DNA that multiplies independently within the host cell (see page 418). They have no protein coat and evidently do not exist outside of cells. Like viruses, they are not themselves living organisms.

Plasmids are extremely common, occurring in almost every type of bacterium. However, they are not essential to bacterial cells, which in a normal environment can thrive without them. By mechanisms to be discussed later, plasmids determine the presence (or absence) of a rich array of a bacterial cell's traits—for example, the ability to resist the harmful effects of antibiotics. When certain bacteria contain certain plasmids, they resist attack by antibiotics (Chapter 16). When they are absent from these bacteria, the cells cannot resist antibiotics. (There are also types of antibiotic resistance that are unrelated to plasmids.) Thus, in the absence of antibiotics (that is, a normal environment), cells can get along very well without them. In recent years, plasmids have been widely studied because of their importance in recombinant DNA technology, or genetic engineering (Chapter 18).

Viroids and Prions

Two curious subviral particles merit brief mention. **Viroids** are small bits of "naked" nucleic acid (so far only RNA has been observed) with a molecular weight of only 10^5 (see Figure 5-5). The nucleic acid of the smallest known viruses has a molecular weight of about 10^6. Viroids lack a protein capsule. They have been clearly identified only in plant cells, where they replicate autonomously with the help of host cell enzymes and cause such picturesquely named plant diseases as chrysanthemum stunt, spindle tuber disease of potatoes, and at least eight others.

For a time it was believed that disease-producing DNA viroids might exist in animal cells and that such hypothetical subviral particles were the cause of scrapie, a usually fatal nervous system disease of sheep (Figure 5-6), and kuru, a strange degenerative human disease first observed in New Guinea and later attributed to the cannibalistic ceremony of eating human brains uncooked. However, it now appears that the causative agents in these diseases contain only protein without nucleic acid. To distinguish these agents from viruses or viroids, they were given the name **prion.** Prions may cause a number of chronic degenerative diseases of humans, sheep, and cattle, and are now under active study.

CELL STRUCTURE

THE GENERALIZED CELL: THREE PROTOTYPES

Until the mid-1950s, textbooks of biology contained an old and familiar drawing that portrayed the "generalized" cell as it appeared through the light microscope. Unchanged for perhaps 30 years and many times reprinted, the illustration presented an idealized composite picture of a typical cell (Figure 5-7). Cells come in many sizes and shapes. But for all their variety, they possess certain features in common.

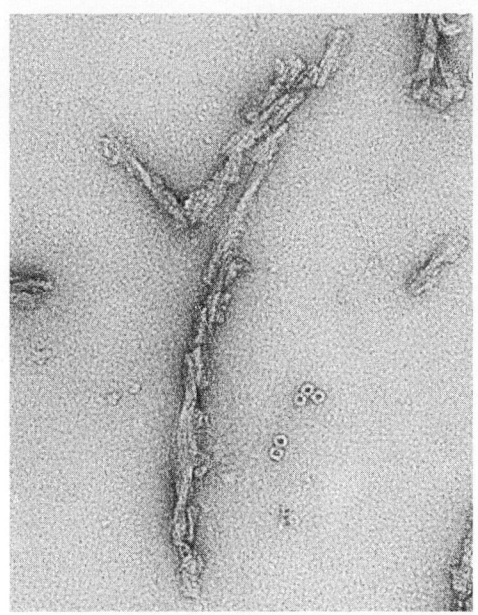

5-6 Prions
This electron micrograph shows fibrils isolated from the brain of a hamster infected with scrapie. The infectious particles, called prions, consist exclusively of protein and are 100 times smaller than the smallest virus. ($\times$ 147,000)

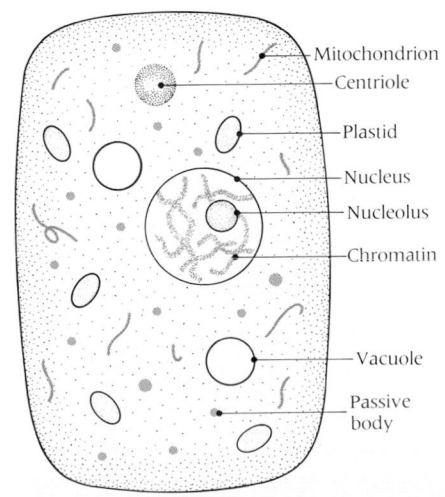

Mitochondrion
Centriole
Plastid
Nucleus
Nucleolus
Chromatin
Vacuole
Passive body

5-7 A "generalized cell" as it was conceived in the 1920s
Drawing combines features from several different cell types. With the advent of electron microscopy in the 1950s, diagrams such as this became outdated.

[1] Do not be confused by the fact that some vaccines—for example, the poliomyelitis vaccines of Salk and Sabin—are referred to as "killed" viruses or "live" viruses. These terms refer to the fact that some viruses have been chemically inactivated (that is, the virus can no longer replicate within a host cell) and others remain active.

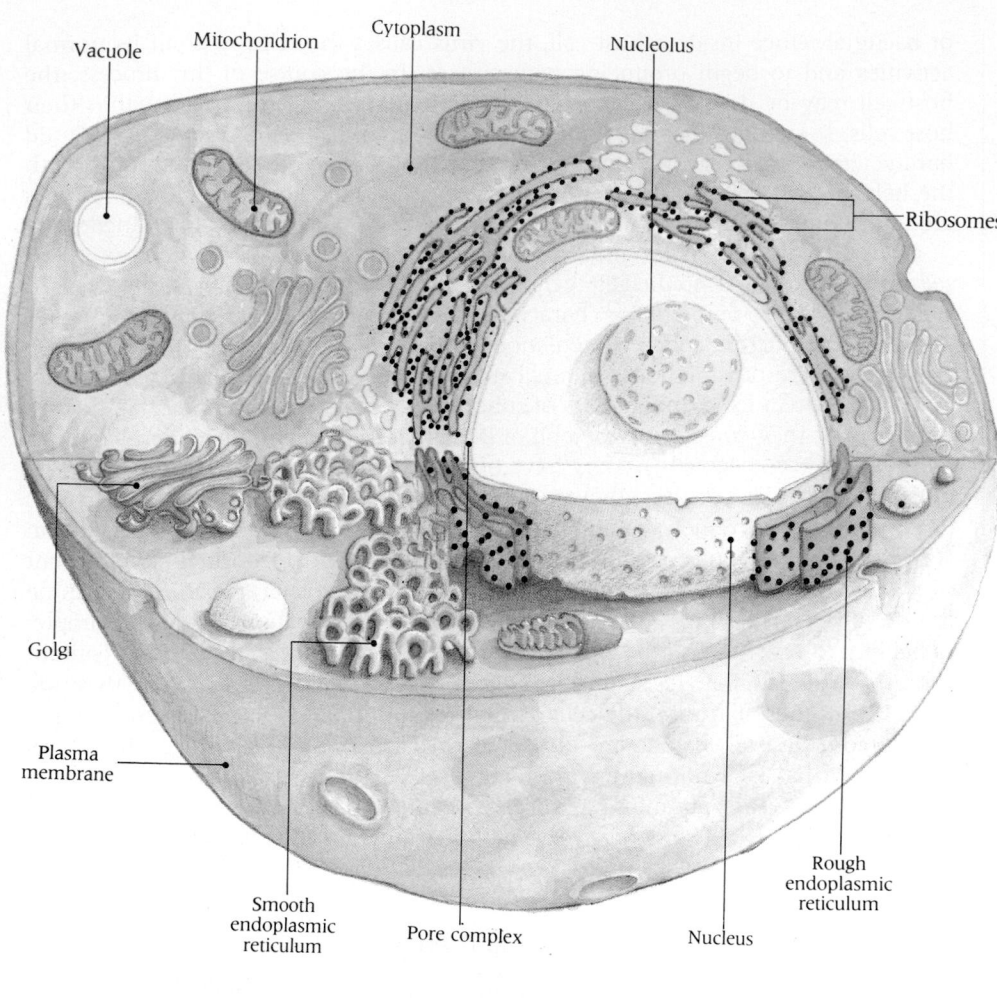

Vacuole Mitochondrion Cytoplasm Nucleolus

Ribosomes

Golgi

Plasma
membrane

Smooth
endoplasmic
reticulum

Pore complex

Nucleus

Rough
endoplasmic
reticulum

A

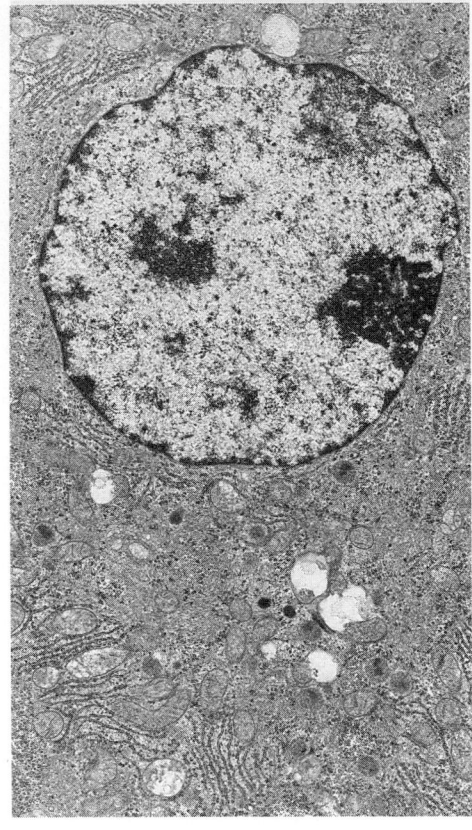

B

5-8 A contemporary "generalized" animal cell
A liver cell (hepatocyte). **(A)** *Drawing.*
(B) *Electron micrograph.* ($\times$ *10,000*)

These were displayed with the "generalized" cell, which showed a large central nucleus embedded in surrounding cytoplasm, the whole being encased in a membrane. Within the cytoplasm were various organelles—small nondescript bodies whose functions were barely known.

The electron microscope achieved much greater magnification than the light microscope (see Box 5-1). **Electron microscopy**—EM in the jargon of working scientists—disclosed such exquisite detail in the subcellular particles that a more contemporary and much more interesting diagram of a generalized cell soon emerged (Figure 5-8A). In fact, the cell in Figure 5-8A is an idealized *animal* cell, based largely on what was learned of a very well-studied animal cell—a liver cell, or hepatocyte (Figure 5-8B).

But to understand the broad scope of cell structure, we must look at two other kinds of "generalized" cells: a photosynthetic plant cell—the example in Figure 5-9 is a cell from a green leaf—and a bacterial cell—in this case the best known of them all, the colon bacillus *Escherichia coli*, or *E. coli* (Figure 5-10). It is readily apparent that the first two cell types have nuclei and that the bacterial cell does not.

Cells with nuclei are said to be **eukaryotic** (from Greek roots meaning "true nucleus"); cells without nuclei are **prokaryotic** ("before a nucleus"). These are the two great categories into which all cells are divided. Prokaryotic cells arose first in evolution. The eventual emergence of eukaryotic cells was one of evolution's greatest forward leaps. The cells of four of the five kingdoms are eukaryotic. Only the kingdom Monera (the bacteria and cyanobacteria) has prokaryotic cells.[2]

---

[2] We will consider in Chapter 38 the recent controversial claim of Carl R. Woese, of the University of Illinois, that a group of diverse but related bacteria are decisively different from true bacteria (Eubacteria) and thus represent still another kingdom, which he terms *Archaebacteria* ("old bacteria"). Their major distinctions from other prokaryotes are in the biochemical properties of their ribosomal RNAs, their cell membrane construction, and certain proteins.

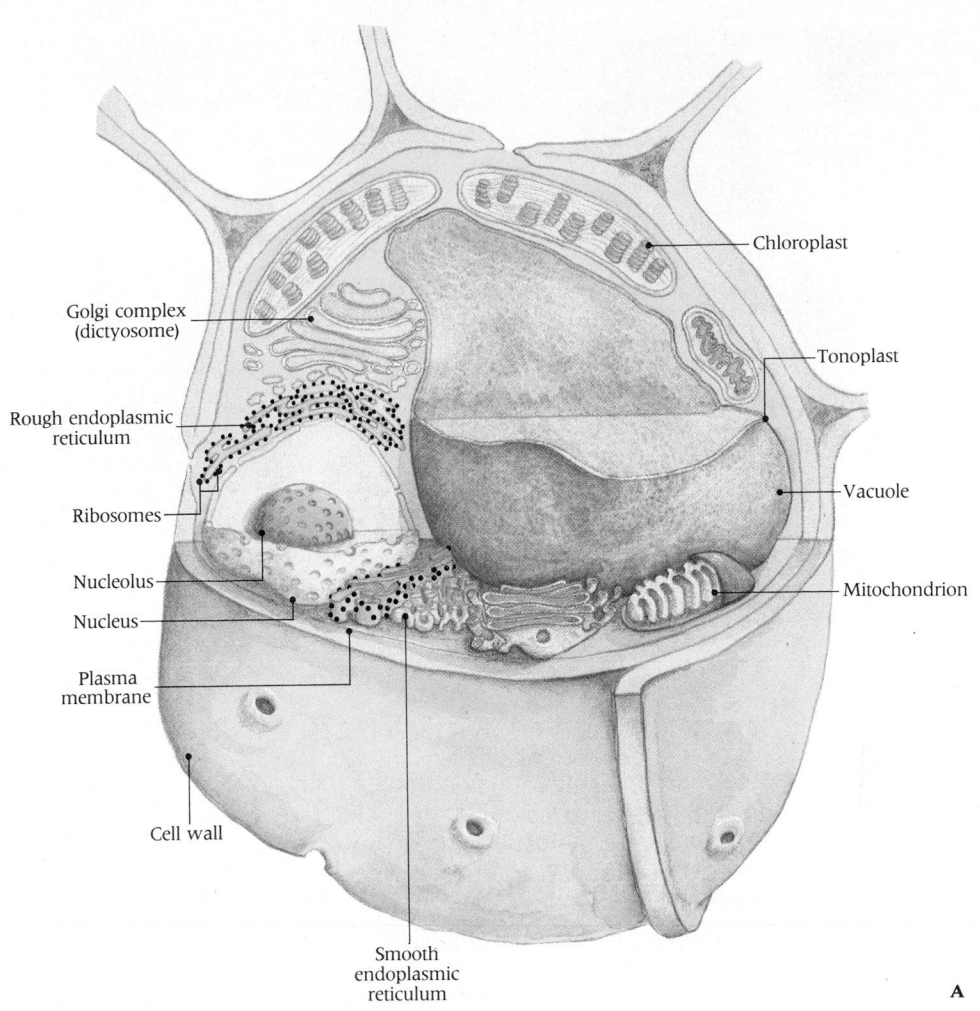

Chloroplast

Golgi complex
(dictyosome)

Tonoplast

Rough endoplasmic
reticulum

Vacuole

Ribosomes

Nucleolus

Mitochondrion

Nucleus

Plasma
membrane

Cell wall

Smooth
endoplasmic
reticulum

A

5-9 Plant cells from a green leaf
Details are discussed in the text. **(A)** *Drawing of a "generalized" cell.* **(B)** *Electron micrograph of an actual cell from a leaf of* Phleum pretense, *timothy.* (× 7000)

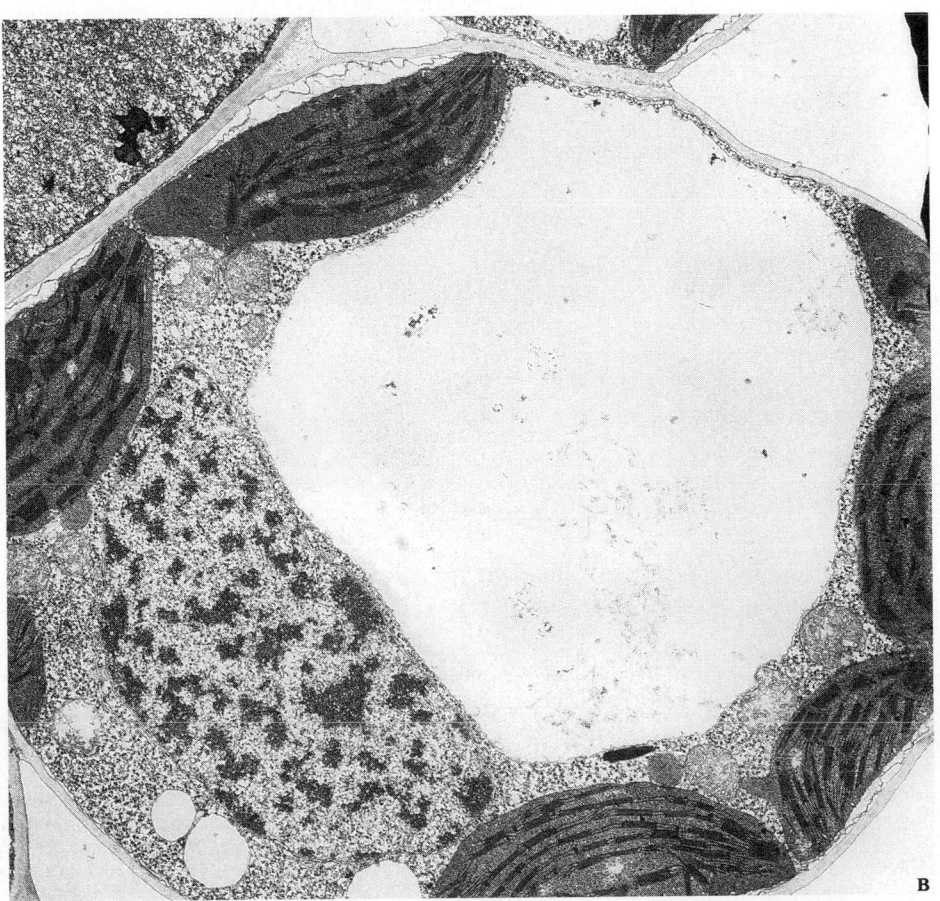

B

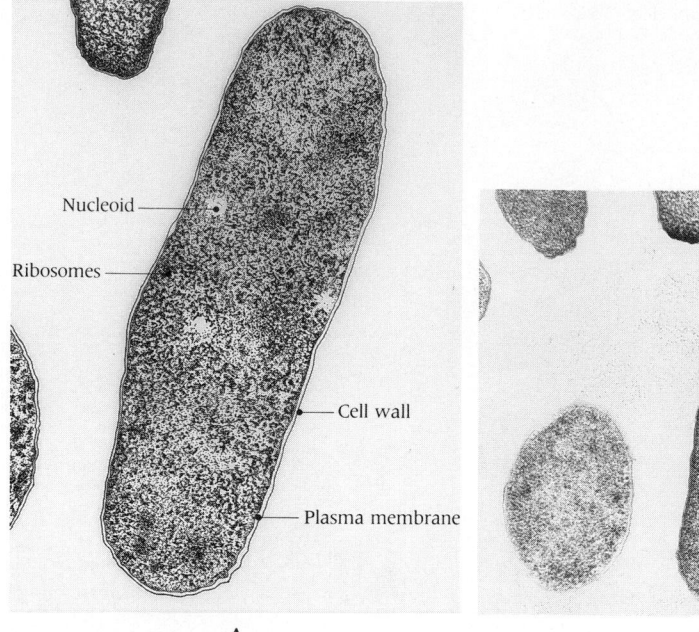

Nucleoid

Ribosomes

Cell wall

Plasma membrane

A

B

**5-10 "Generalized" bacterial cell
Escherichia coli**
Details are discussed in the text.
(A) Drawing. (B) Electron micrograph.
(× 2400)

STRUCTURE OF ANIMAL CELLS

Eukaryotic cells are larger and much more complex than prokaryotic cells. The volume of most eukaryotic cells is from 1,000 to 10,000 times greater than that of prokaryotes. This alone indicates the acquisition of many new functions and subcellular systems. It was also a necessary evolutionary prerequisite for the later emergence of multicellular organisms.

In sum, eukaryotic cells have the following features:

- A nucleus containing chromosomes and their DNA and surrounded by a double membrane (nuclear envelope)
- A surrounding mass of cytoplasm, where most of the cell's metabolic activities take place[3]
- A number of membrane-limited organelles within the cytoplasm
- A cell membrane (also called *plasma membrane*) at the cell surface that encases the cytoplasm

The Plasma Membrane

The surface membrane of a cell must be more than a simple container because certain substances enter and leave the cell freely while others do not. The behavior of molecules passing through plasma membranes clearly implies that the membrane has certain physical properties.

- It must form a closed container so that some of the substances pumped inward cannot leak out.
- It must constitute a true permeability barrier.
- It must have sidedness—that is, an inside and an outside.

If it can be seen at all with the light microscope, the **plasma membrane** (or **plasmalemma**) appears only as a thin boundary line (Figure 5-11). After many

[3] It was once taught that living organisms consist collectively of *protoplasm,* which T. H. Huxley had characterized in the nineteenth century as "the physical basis of life." Later, biologists gave the names *nucleoplasm* to the protoplasm inside the cell nucleus and *cytoplasm* to the protoplasm outside the nucleus. These were more convenient than scientific. "Protoplasm" is a misleading word for living matter if it is taken to mean a homogeneous substance of which a representative drop or piece can be studied. Protoplasm is really a complexly organized heterogeneous system, the smallest representative piece of which is the cell. Strictly speaking, it is no more meaningful to say that a cell is composed of "protoplasm" than it is to say that a radio is composed of "radioplasm." If we remember that *protoplasm* is merely a convenient word for the still incompletely understood substance of the cell, the term may be used—carefully.

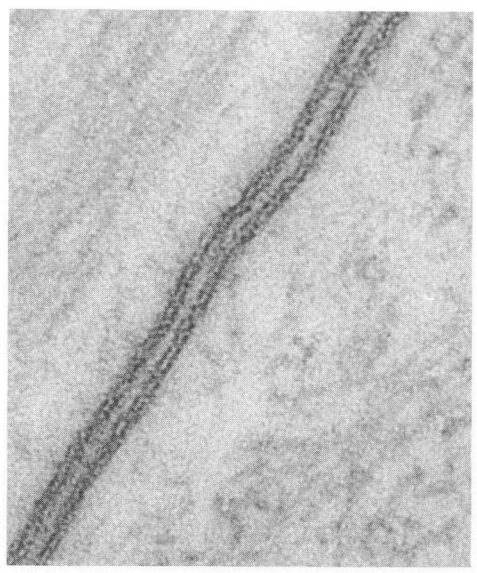

5-11 Plasma membrane
Two cells lie next to each other with their plasma membranes (two dark lines) separated by a small intercellular space. (× 20,000)

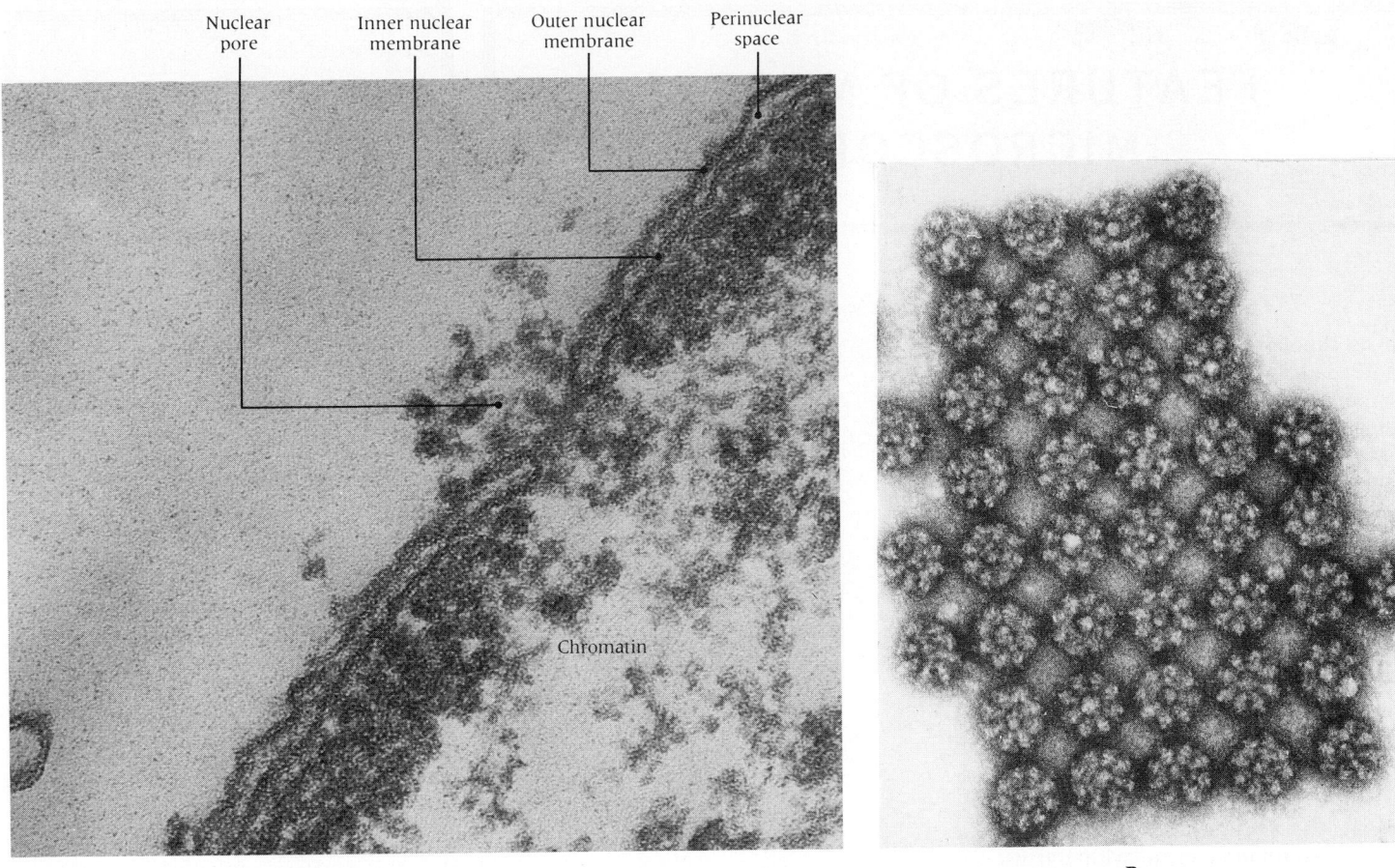

Nuclear pore   Inner nuclear membrane   Outer nuclear membrane   Perinuclear space

Chromatin

A

B

5-12 Structure of the nuclear envelope
(A) This electron micrograph shows the two nuclear membranes separated by the perinuclear space. The nuclear pore connects the chromatin with the cell cytoplasm. (× 160,000) (B) Nuclear pores seen from the top. Each pore reveals eight granules surrounding a central channel.

unsuccessful attempts to visualize it by electron microscopy, success finally came in the late 1950s to J. D. Robertson who used an unusually high powered electron microscope. It revealed an interesting and surprisingly fine structure. Membranes appeared to consist of a ''sandwich'' made up of outer, middle, and inner layers.

Plasma membranes have specialized properties in different cells and in different regions of the same cell. But this sandwichlike structure was found in all cells. A full discussion of its properties merits its own chapter (Chapter 6).

The Nucleus

By definition, a eukaryotic cell is a compartmentalized structure in which the intracellular space is divided into two regions: (1) the **nucleus,** which contains the genes and the structures involved in gene function; and (2) the **cytoplasm,** which contains the cell organelles, various membranes and particles, and the apparatus of protein synthesis. This nucleo-cytoplasmic compartmentation is ordinarily maintained by a specific eukaryotic membrane called the **nuclear envelope.**

The nucleus is usually the largest structure within the cell. Despite its prominence, the nucleus remains one of the least understood cell components. It is a roughly spherical body that is denser than the cytoplasm in which it is embedded. The surrounding nuclear envelope has a distinctive structure consisting of an outer and an inner membrane, each of which has the same bilayer organization as the plasma membrane. Between the outer and inner membranes of the nuclear envelope is the narrow **perinuclear space** (Figure 5-12A). The inner membrane resembles a relatively simple envelope, but it has complex functions. The outer membrane is intimately connected with the complex network of membranes in the cytoplasm called the endoplasmic reticulum. In their chemical composition, including especially their phospholipids, and in their bilayered structure, nuclear membranes are similar to membranes of the endoplasmic reticulum; however, there are enough differences to show that they are independent membrane systems.

At regular intervals the nuclear envelope is perforated by large **pores** (Figure 5-12B). At the margins of the pores, the outer membrane folds inward forming a continuous structure with the inner membrane, and creating a circular opening

In its simplest form, a microscope is little more than a metal tube with an **ocular lens** at the end near the eye and an **objective lens** at the end near the specimen. Such was the instrument used by Hooke in 1665 (see Figure 5-1).

The ability of a microscope to distinguish detail is termed its **resolving power.** As stronger lenses achieved higher magnifications, images began to show defects or aberrations—for example, distortion near the edges **(spherical aberration),** colored fringes **(chromatic aberration),** and blurring because of poor resolution. Spherical and chromatic aberrations were overcome by objective lenses made from combinations of two or more types of glass, and arranged so that defects in one offset those in the other. Superior resolution came with the use of improved **condensers** (combinations of lenses that focus light rays).

Light microscopes, such as the early simple ones, also have physical limits beyond which greater resolution is impossible. The limit is set by the wavelength of the light that is used. A light microscope's resolving power is limited to objects having a diameter larger than one-half of a given wavelength of visible light. Since the range of wavelengths of visible light (for humans) is be-

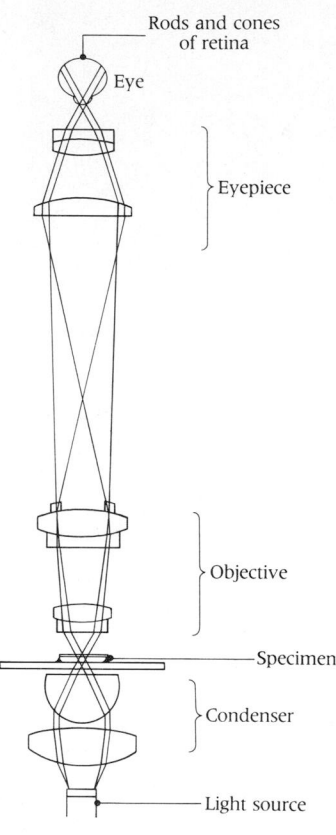

5B *Optical system of a bright-field light microscope*

tween 400 and nearly 800 nm, no object smaller than 190 nm in diameter can be resolved.

In recent years, many advanced microscopic techniques have evolved. They are in two categories, light microscopes and electron microscopes.

LIGHT MICROSCOPES

Bright-Field Microscope

This familiar microscope (Figure 5A) is by far the most common. Its optical system is sketched in Figure 5B. The image is observed directly by looking into the ocular lens. Coarse and fine controls permit up-and-down movements of the lens tube (or, in all modern microscopes, the specimen stage). This facilitates precise focus. The condenser can also be moved up and down so that rays from the light source, usually an electric lamp, converge at the specimen and then

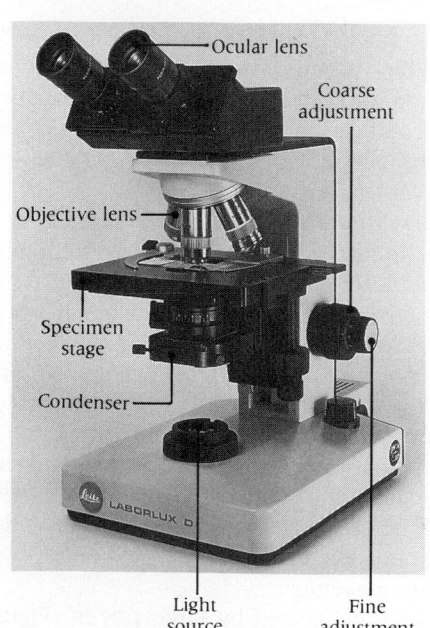

5A *Bright-field light microscope*

spread into a cone of light that completely fills the objective lens. The ocular lens merely magnifies the image produced by the objective lens. Light is transmitted through objects (Figure 5C). Their visibility is usually enhanced by staining.

Dark-Field Light Microscope

The visibility of many objects in a light microscope can be dramatically improved if an opaque disk is placed in the central axis of the condenser so that the light beam striking the specimen is in the shape of a hollow cone. Light scatter causes small objects to appear bright against a dark background (Figure 5D).

Phase Contrast Light Microscope

This instrument's optical system causes objects to be seen in bright or dark relief by virtue of slight variations in their refractive properties. Since many structures are visible without prior staining, living cells can be observed in action (Figure 5E).

Fluorescence Microscope

This is a light microscope that detects objects that have been rendered visible by special optical techniques that make them appear fluorescent.

Polarizing Microscope

This microscope uses polarized light to reveal structures that transmit light in a single plane. Only recently has it been successfully applied to biological systems. Major improvements in image quality have resulted from the use of video image processors—not only for polarizing microscopes, but for all light microscopes. In this technique, electronic subtraction of unwanted background greatly enhances image contrast.

ELECTRON MICROSCOPES

Electron Microscope

In place of visible light, this instrument transmits a beam of electrons whose wavelength is only 0.05 Å—in sharp contrast with the

5C Bright-field micrograph of diatom **Pleurosigma angulatum** *(× 100)*

5D Dark-field micrograph of diatom (× 100)

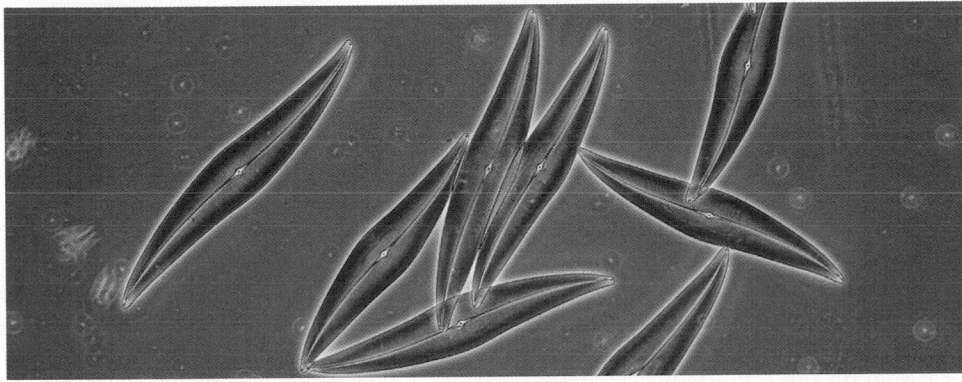

5E Phase contrast micrograph of diatom (× 100)

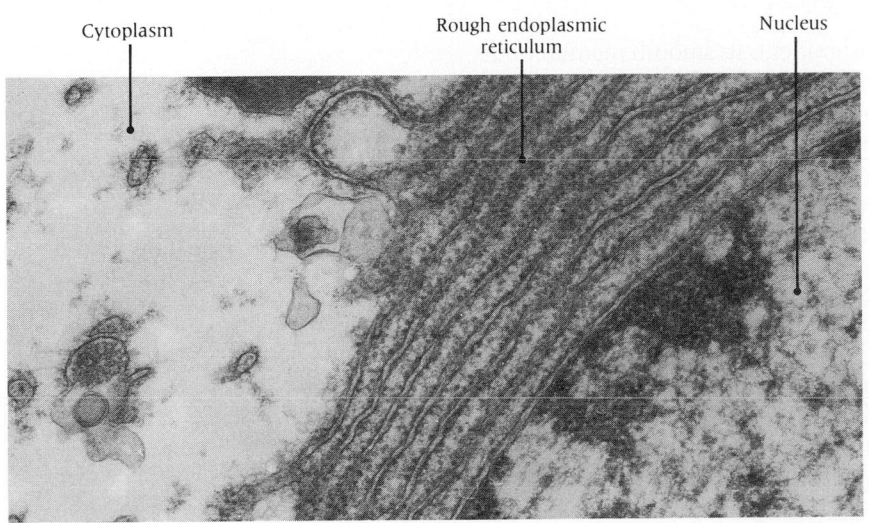

Cytoplasm   Rough endoplasmic reticulum   Nucleus

5F *Transmission electron micrograph of the structure of the nucleus and rough endoplasmic reticulum in a hepatocyte. (× 40,000)*

wavelengths of visible light, which ranges from 4000 to 8000 Å. In principle it resembles the light microscope, but it utilizes magnetic fields to focus electron beams instead of light-focusing optical lenses. Also, it projects its image upon a fluorescent screen or photographic plate instead of the human eye. In this way it achieves magnifications 100 times as great as those of the light microscope and far greater resolution (Figure 5F). Its main drawbacks are the elaborate procedures needed in the preparation of specimens and the expensive, bulky equipment.

High Voltage Electron Microscope

A huge variant of the electron microscope, the high voltage electron microscope is discussed on page 111. As seen in the example in Figure 5-30A, it can examine thicker specimens and thus give some sense of three-dimensional structure.

Scanning Electron Microscope

This remarkable instrument produces three-dimensional images of surfaces at magnifications between those obtained with light microscopes and transmission electron microscopes (Figure 5G). It depends on the scanning of the specimen surface by a rapidly moving focused electron beam. The electrons reflected from the surface are picked up by a detector that forms an image on a television screen.

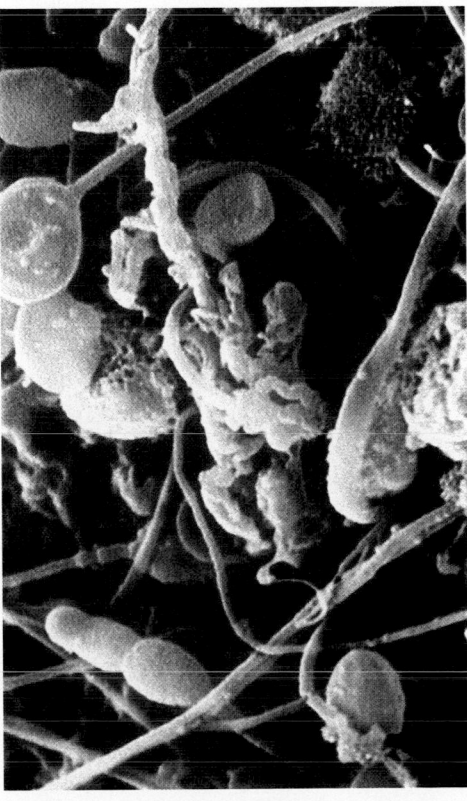

5G *Scanning electron micrograph of human sperm heads and tails (× 6000)*

about 70 nm in diameter. In the opening of some pores, a central granule or plug is found. Lying on each pore rim is a ring of eight symmetrically arranged granules. This is the **annulus.** The pore and the annulus together form the **nuclear pore complex.** It not only provides an opening through which the nucleus can communicate directly with the cytoplasm; it also regulates the diffusion of large molecules between the two compartments. In contrast with the cytoplasm, the interior of the nucleus is free of membranes.

Because it has a strong affinity for numerous dyes or stains, the fine structure of the nucleus can be brought into sharper contrast by dye treatment. The stained nucleus of a nondividing cell has a nonhomogeneous texture, the nuclear components obviously differing in their tendency to take up stain. The tangled network of strongly staining material was long ago called **chromatin** (Greek, *chroma,* "color"). The term now carries more specific biochemical or physiological meaning. The affinity of chromatin for basic dyes had for many years suggested to biologists that the nucleus must contain strongly acidic material. This material, we now know, is DNA, the molecular carrier of the genes—the fundamental units of heredity—and RNA, the other major nucleic acid form that has many functions in carrying out the instructions in each cell's genes. Chromatin strands condense during cell division into compact structures that appear as discrete chromosomes.

A small spherical body, the **nucleolus,** is often distinguishable within the nucleus by light microscopy. Many nuclei contain several nucleoli. Specific staining procedures reveal that the nucleolus contains substantial quantities of RNA. Recognition of this fact led to the discovery that the major function of the nucleolus is the production of the characteristic RNA of ribosomes, ribosomal RNA. As we will see, ribosomes and ribosomal RNA play essential roles in directing protein synthesis. This particular form of RNA is destined to be "shipped out" to the cytoplasm.

Endoplasmic Reticulum and Ribosomes

The **endoplasmic reticulum (ER)** is a continuous, interconnected, three-dimensional array of intracellular membranes. When the cell is viewed in cross section, the endoplasmic reticulum appears as systems of multiple canals defined by unit membranes. The space between the limiting membranes varies widely in different cell types and under different conditions. Sometimes the membranes seem to form tubules or canals 50–100 nm in diameter. Sometimes the spaces between membranes are much larger and are called **cisternae.** The whole system appears to provide channels of communication. It also comprises a cellular compartment in which specific reactions occur.

Two forms of endoplasmic reticulum can be distinguished: **rough** (abbreviated **RER**) and **smooth (SER)** (Figure 5-13). At high magnification the membranes of a typical rough reticulum are seen to be encrusted on the cytoplasmic side with dense particles about 25–30 nm in diameter. These particles are **ribosomes,** the ribonucleoprotein particles which are the sites of synthesis of all proteins (see Chapter 6).

SER is less well studied than RER, but equally prominent. Its smooth membranous surfaces are devoid of ribosomal granules. SER may appear in the same cell with RER. It is the site of synthesis of steroids and other lipids. In muscle cells, where it is known by the special name **sarcoplasmic reticulum,** it participates in muscle contraction (Chapter 31). In liver cells, it has a role in the detoxification of noxious substances, such as lipid-soluble poisons and drugs. These substances are converted by the SER to water-soluble molecules that are more readily removed from the body. Cells differ widely in their relative contents of RER and SER. Cells that are heavily involved in protein secretion (for instance, plasma cells secreting antibodies) are rich in RER, while other cells (such as those secreting steroid hormones) may be rich in SER.

Endoplasmic reticulum divides the cytoplasm into two compartments, the **cell sap** (or **cytosol**) and the **cisternal space** (Figure 5-14). The cytosol contains soluble enzymes, free ribosomes, and other factors needed in protein synthesis. The cisternal space is functionally continuous with the internal cavities of Golgi apparatus, lysosomes, and peroxisomes. Note that the ribosomes attached to RER are on the side of the membrane facing the cell sap.

Cytologists found that, when they gently homogenized cells, they were able to

5-13 Endoplasmic reticulum
Endoplasmic reticulum (A) seen by electron microscopy. (× 40,000) (B) A diagram of rough endoplasmic reticulum (RER) showing its three-dimensional branching structure. RER is studded with ribosomes and thus is the site of active protein synthesis.

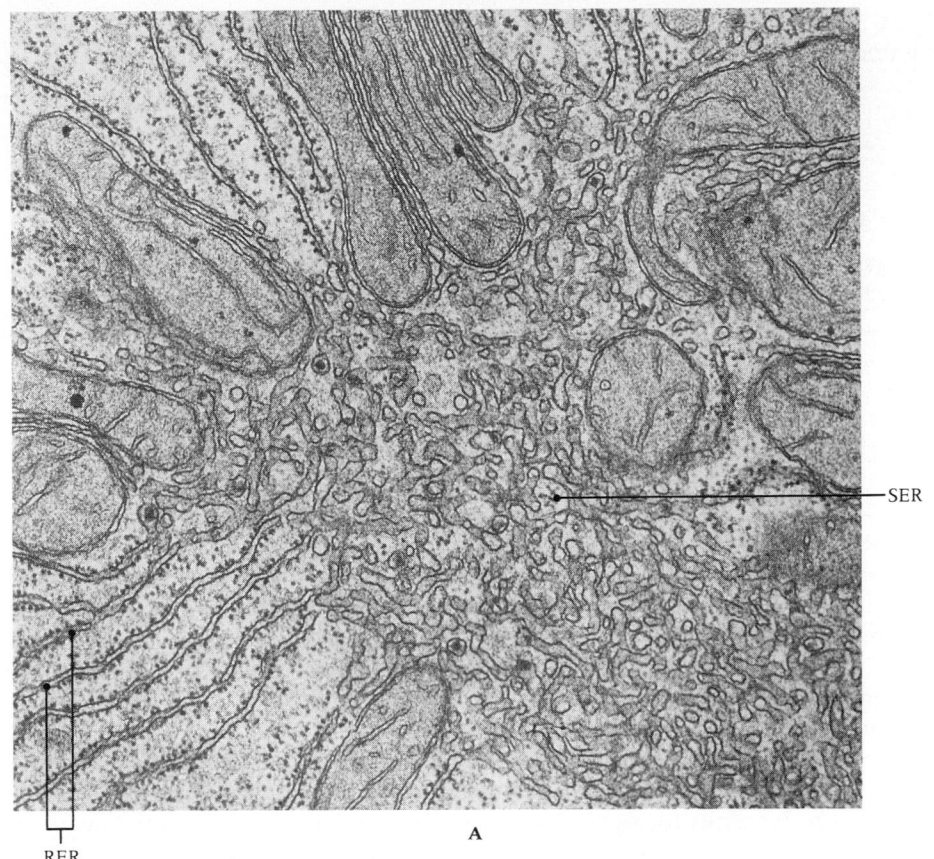

SER

RER

A

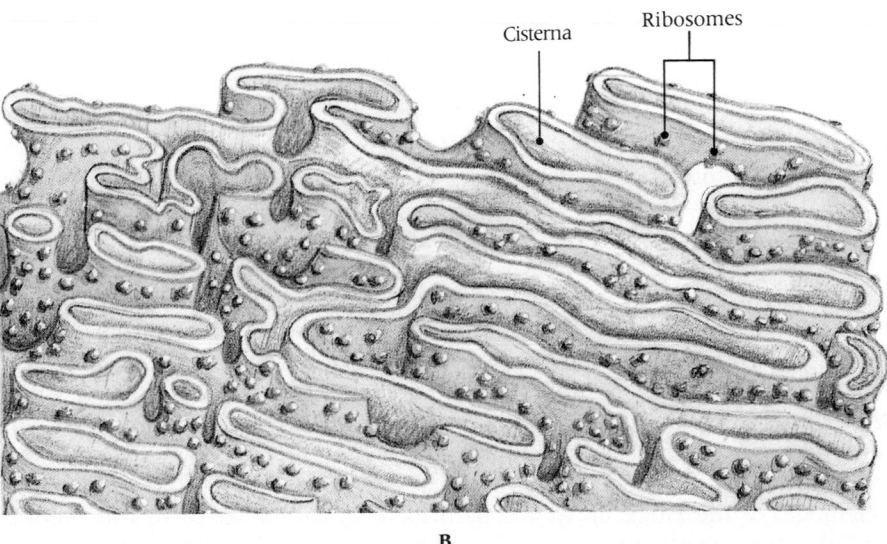

Cisterna   Ribosomes

B

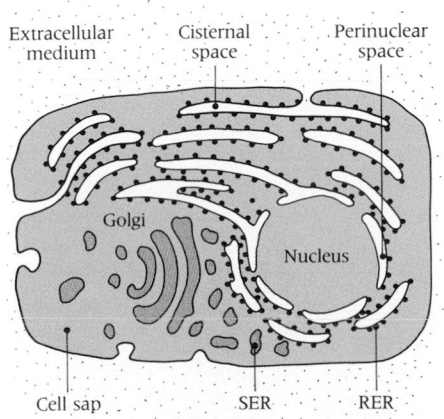

Extracellular medium   Cisternal space   Perinuclear space

Golgi

Nucleus

Cell sap   SER   RER

5-14 Relation between the cisternal space and the cell sap

separate many of the organelles in an ultracentrifuge. An **ultracentrifuge** is a device capable of spinning tubes of liquid suspensions at many thousands of revolutions per minute. This extremely rapid rotation produces forces up to 500,000 times the force of gravity. Hence an ultracentrifuge can readily separate different fractions of the suspension that have differing densities. Particles, whose density exceeds that of the suspending fluid, however slightly, fall or settle in the tube; those of lower densities rise or float (Figure 5-15).

Using the technique of differential centrifugation on fragmented cells, early investigators found that the lightest fraction, the one that settled out only after prolonged centrifugation at extremely high speeds, contained the RNA-rich particles we now call ribosomes. They were bound to fragments of endoplasmic reticulum and, before this fact was recognized through electron microscopy, were termed **microsomes.** When so-called microsomes were treated with detergents, the membranes of endoplasmic reticulum (whose high lipid content is responsible for their low density)

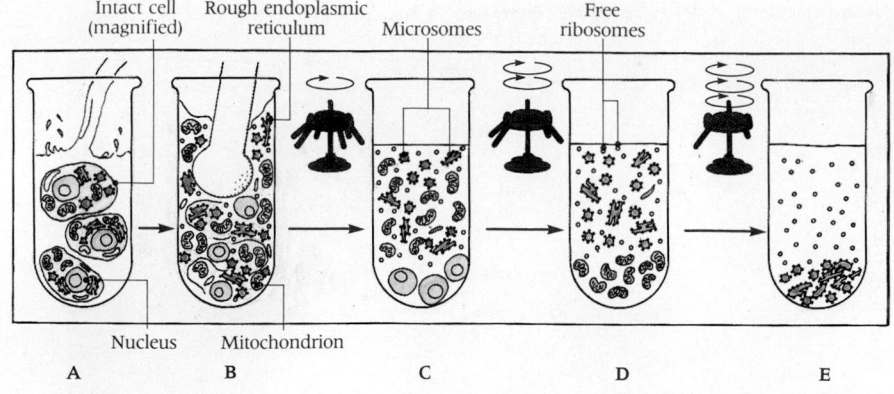

5-15 Differential centrifugation in the isolation of cellular organelles
(**A**) *Suspended cells.* (**B**) *The cells are broken up with a pestle, and the fragments are suspended in sucrose solution prior to centrifugation.* (**C**) *During the first centrifugation, heavier particles (nuclei) are sedimented.* (**D**) *When the supernatant fluid is removed and recentrifuged at a higher speed, the mitochondria are sedimented.* (**E**) *The remaining supernatant fluid is recentrifuged at a higher speed. This sediments microsomes, endoplasmic reticulum, plasma membranes, and Golgi complexes, leaving soluble molecules of the cytoplasm and free ribosomes above. Further centrifugation brings downs the ribosomes.*

dissolved away, leaving free ribosomes. In fragmented cells containing relatively little RER, differential centrifugation yields free ribosomes.

Ribosomes are discrete particles, about 25–30 nm in diameter. In eukaryotic cells, ribosomes contain equal amounts of RNA and protein by weight. Prokaryotic ribosomes are half as large and are two-thirds RNA and one-third protein. As noted above, eukaryotic ribosomes are formed in the nucleolus. When exposed to a lowered concentration of Mg^{2+} in the suspending medium, a ribosome breaks up into two subunits, one large and one small.

For reasons to be described later, most intact ribosomes occur in the cytoplasm in clusters of two to six, although some clusters contain many more. The clusters are called **polysomes** (or **polyribosomes**). In addition, some ribosomes float free in the cell sap. We shall later see that the polysome arrangement helps to promote protein synthesis.

The Golgi Apparatus

The **Golgi apparatus** was first observed in 1898 in the nerve cells of barn owls by the Italian cytologist Camillo Golgi (1843–1926). Electron microscopy showed that the Golgi apparatus, which occurs in all eukaryotic cells (and is termed a **dictyosome** in plants), consists of a half dozen or more flattened saclike structures that are stacked like dinner plates. They are made of smooth membranes (Figure 5-16A). Each saclike structure is termed a **cisterna.** Many of them form **buds** that pinch off from the larger ones. The Golgi membranes and ER appear not to be physically connected to one another.

The stack of cisternae in the Golgi apparatus has two different sides, or faces. The *cis* **face** is nearer the nucleus and next to a specialized patch of ER that lacks bound ribosomes and is called **transitional ER.** The *trans* **face** of the Golgi apparatus is at the opposite end.

Elements of the Golgi apparatus participate in the secretion of cell products. The Golgi apparatus is not a site of synthesis of secreted materials but, rather, a collecting, modifying, sorting, and packaging station. Proteins to be exported to the outside (secreted) are transferred from the endoplasmic reticulum in little vesicles termed **transitional vesicles** to the *cis* side of the Golgi apparatus, where they fuse with it (Figure 5-16C). The Golgi stack then refines "export proteins" by moving them from the *cis* to the *trans* face in little vesicles that pinch off of the lateral, bulbous parts of the Golgi complex. Some of these transport vesicles are termed **coated vesicles** because they contain an outer shell composed largely of the protein clathrin. Thus, the cisternae resemble the multi-stage plates of a distillation tower. Finally, vesicles filled with export products are fashioned by stages into larger **secretory vesicles** (or **granules**), which fuse with the outside plasma membrane before releasing their products to the outside. In this way, these vesicles promote the growth of the plasma membrane itself (Figure 5-16C).

Secretory vesicles occur conspicuously in specialized secretory cells like the granular cells of the duodenum or the pancreas, which secrete digestive enzymes, and liver cells (Figure 5-16B), which produce the albumin of blood plasma. In such cells, many secretory vesicles are converted into intracellular storage reservoirs for

5-16 The Golgi apparatus
(A) *The Golgi apparatus seen by electron microscopy.* **(B)** *Electron micrograph of a hepatocyte showing a Golgi complex characteristic of cells which, like this one, secrete protein prolifically. The circular black forms shown here are secretory vesicles packed with protein.* **(C)** *Drawing of the Golgi apparatus. Vesicles which bud off from the RER carry proteins synthesized there to the Golgi for processing. After processing is complete, new vesicles containing the final product bud off the other side of the Golgi. They may travel to the cell membrane, fuse with it, and release the protein into the extracellular fluid.*

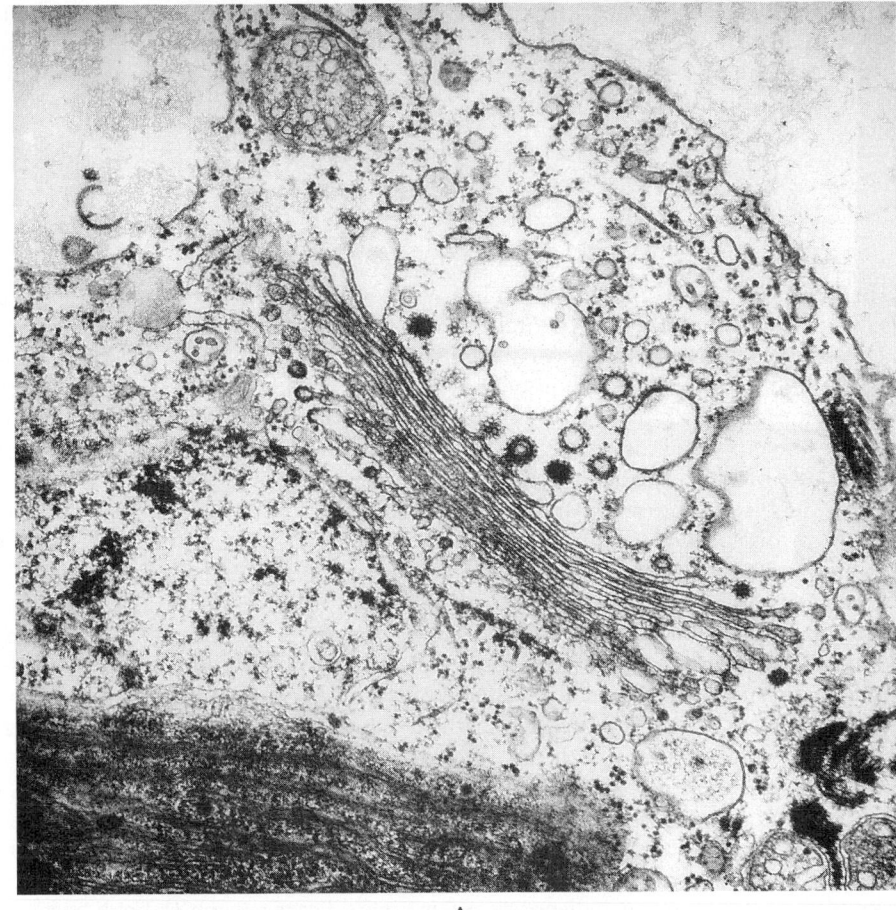

A

Golgi apparatus   Secretory vesicles

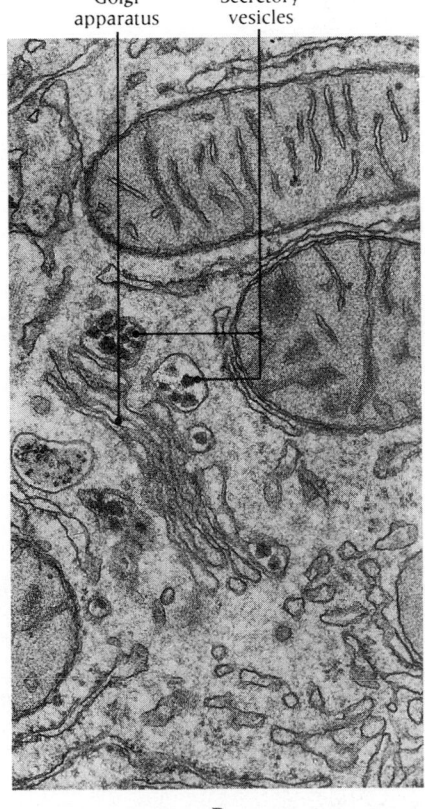

B

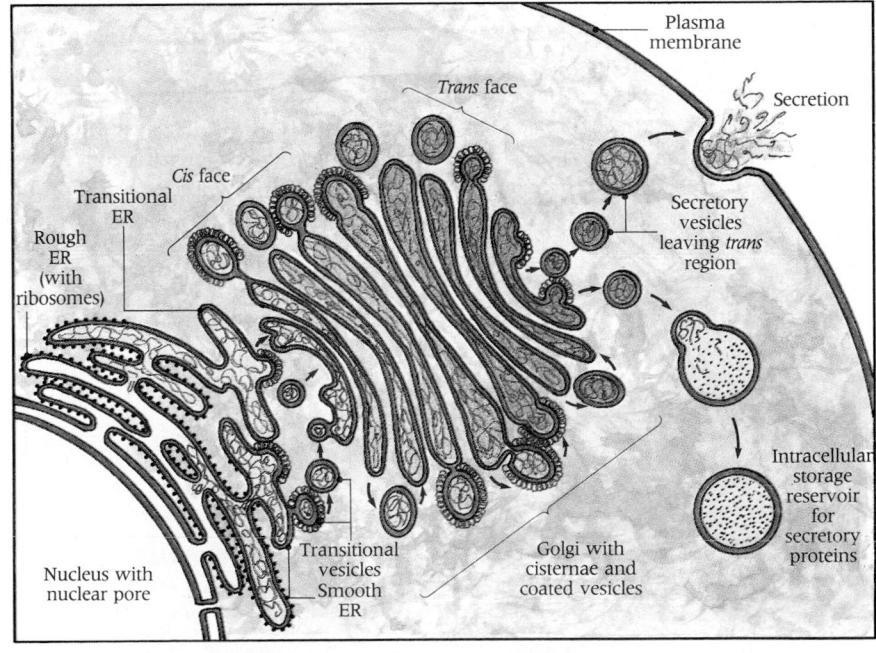

C

secretory proteins. Golgi apparatuses are prominent in such cells, but they also occur in nonsecretory cells.

The Golgi apparatus also forms a number of enzyme-containing **microbodies** that remain within the cell (see below). These include the lysosomes, which contain digestive enzymes; the peroxisomes (in plant and animal cells), which contain oxidative enzymes; and the glyoxysomes (only in plant cells), which contain enzymes that convert fat to carbohydrate. In addition, the Golgi apparatus participates in the biochemical processing of some of the proteins made in the RER by adding sugar groups to make glycoproteins (glycosylation) and sulfoproteins (sulfation), and by subjecting them to phosphorylation (addition of phosphate) and proteolytic cleavage.

Lysosomes and Other Microbodies

Lysosomes are single-membrane-bound vesicles, about 0.5 μm in diameter, that arise in the Golgi apparatus and contain many digestive (that is, hydrolytic) enzymes. These enzymes cleave macromolecules into smaller molecules that are disposed of metabolically. Ordinarily these powerful enzymes are kept isolated from the rest of the cytoplasm within their membranous envelope. Under certain conditions, however, the envelope ruptures and the enzymes pour out into the cell. Dissolution of all or part of the cell quickly follows.

From the viewpoint of cell survival, it is fortunate that the lysosome's contents are compartmentalized away from the rest of the cytoplasm. The rupture of lysosomes in injured cells has devastating effects on the cell. Indeed, the massive release of lysosomal enzymes accounts for the total removal of injured cells in a wound. It also is responsible for the normal loss of the tadpole's tail during development. Lysosomes also participate in the digestion of food particles within cells.

In human beings, the presence of defective lysosomes can lead to massive accumulations of materials awaiting digestion. This is the cause of the **lysosomal storage diseases.** An example is one of the glycogen storage diseases identified in 1963 by the Belgian investigator H. G. Hers; the first lysosomal storage disease to be recognized. The cells of people with this disorder lack a lysosomal enzyme, α-glycosidase, that normally degrades glycogen. As a result, the cells of liver, muscle, and other tissues contain large glycogen-filled lysosomes. Many storage diseases are known, among them such disorders as Gaucher's disease, in which lipids accumulate because of a defect in β-glucosidase; other lipids pile up in Niemann-Pick disease, in which sphingomyelinase is missing. In these disorders, the spleen and other body organs become choked with stored materials, grossly enlarged, and functionally impaired (Figure 5-17).

In white blood cells and other cells adapted to perform **phagocytosis,** defined as the ingestion of foreign particles such as invading bacteria (see Chapter 6), the particle becomes lodged in a plasma membrane-enclosed spherical vacuole in the cytoplasm, called a **phagosome** (Figure 5-18). Lysosomes quickly surround the vacuole, fuse with it, and discharge their corrosive enzymes into it. The ingested particles are rapidly digested.

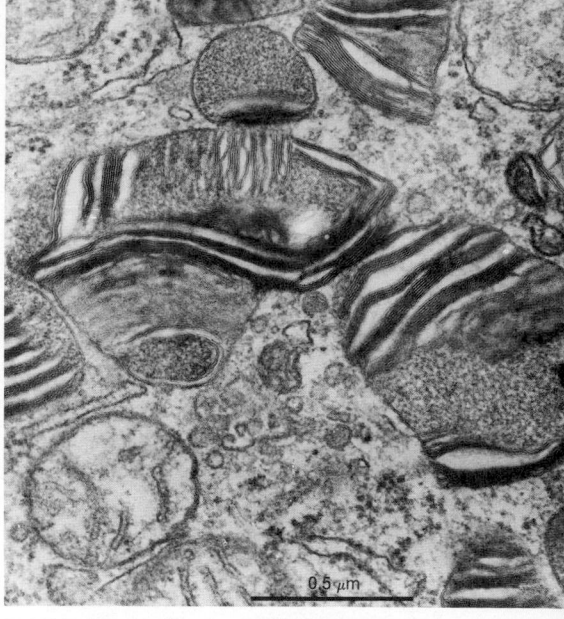

5-17 Lysosomal storage disease
Individuals lacking the lysosomal enzyme that breaks down glycolipids accumulate glycolipid in the lysosomes of their nerve cells. The resulting zebra bodies are characteristic of this storage disease. The nature of enzymes is discussed in Chapter 8.

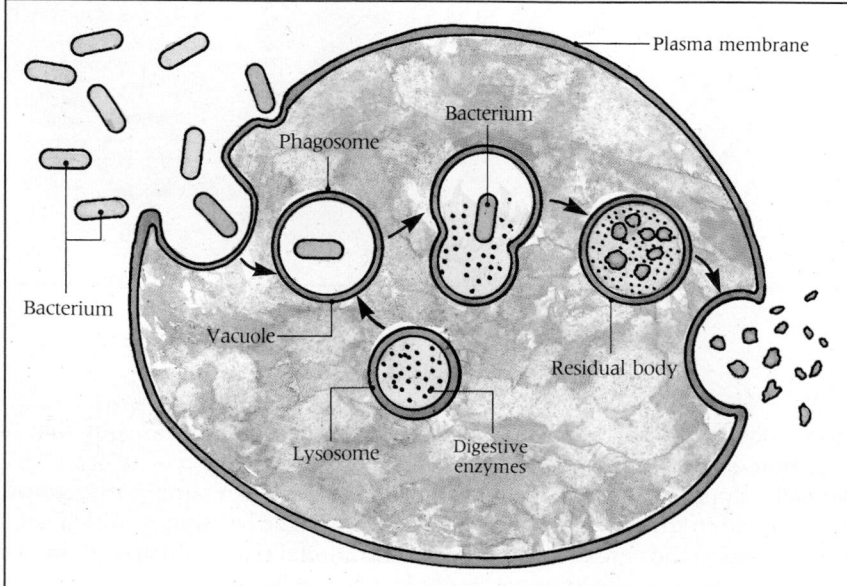

5-18 Phagocytosis
As a cell ingests a bacterium, the plasma membrane folds around it, forming a vacuole called a phagosome. A lysosome then fuses with the vacuole, and releases digestive enzymes which break down the bacterium. Useful products of this digestion diffuse into the cytoplasm. Undigestible material accumulates in the vacuole, which becomes a residual body. In some cases, the residual body migrates to the plasma membrane and ejects its contents.

Cell cytoplasm also contains a second class of microbodies, the **peroxisomes,** which are membrane-bound organelles containing enzymes that catalyze reactions involving hydrogen peroxide (H_2O_2). Two of these enzymes produce H_2O_2 (urate oxidase and D-amino acid oxidase) and one (catalase) destroys it, effectively converting H_2O_2 (a dangerous byproduct of the cell's biochemical machinery for attacking and killing ingested bacteria) into harmless O_2 and H_2O. Peroxisomes somewhat resemble lysosomes in structure (Figure 5-19). In green plants, they participate in the curious phenomenon called photorespiration (Chapter 10).

The name **liposome** has been given to interesting vesicles that are not produced in cells but on laboratory benches by biochemists (see Box 6-1).

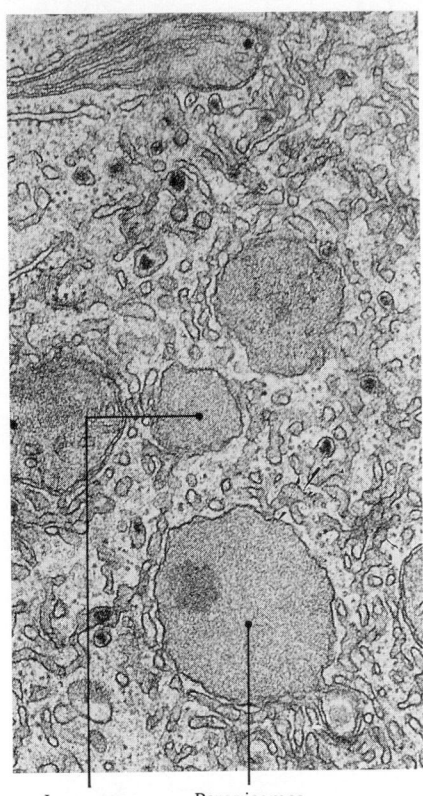

Lysosomes   Peroxisomes

5-19 Lysosomes and peroxisomes
An electron micrograph of peroxisomes shown in comparison to lysosomes. (× 30,000) In plants, peroxisomes participate in a phenomenon called oxidative photorespiration.

5-20 Mitochondria
(A) An electron micrograph showing the mitochondrial membranes and characteristic cristae. (× 60,000) (B) The surface of these vesicles, which are made from disrupted mitochondrial inner membranes, show knobs which are enzyme complexes crucial to the mitochondrion's ability to synthesize ATP. (× 600,000)

Mitochondria

The cytoplasm of nearly all eukaryotic cells contains bodies called **mitochondria,** among the largest of cell organelles, exceeded in size only by the nucleus—and by chloroplasts in plant cells (Figure 5-20). Although barely visible with the light microscope, mitochondria show intricate, fine structure in electron micrographs. Their numbers vary from cell type to cell type. This is a fact of some interest, since mitochondria are the cell's power plants—centers of cellular respiration and energy-yielding metabolism (see Chapter 9). A liver cell contains about 800 mitochondria. The few cell types containing no mitochondria (such as mammalian red blood cells) depend on less efficient means of energy production.

An average mitochondrion is about 1 μm in diameter. It has a bilayered outer membrane that is not connected with other membranes in the cell (Figure 5-20A). In many ways, this structure resembles the cell membrane but is less rigid. Special laboratory techniques reveal minute stalked, spherical bodies 8.5 nm in diameter, sometimes called "lollipops," that stud the inside surface of the inner membrane (Figure 5-20B). We will see in Chapter 9 that part of the enzymatic machinery that synthesizes the energy-rich molecule ATP in the course of metabolism is found within the spherical knobs and their stalks that protrude from the inside surfaces. The inner membrane also displays curious infoldings that form perpendicular plates called **cristae,** tubular or lamellar, which extend into the **matrix,** the gel-like interior of the mitochondrion (see Figure 5-20A).

The matrix contains many enzymes, some DNA, and ribosomes that are smaller than those found elsewhere in the cell. Mitochondrial DNA is circular; thus it differs from DNA found in the nucleus of the same cell. This difference supports the notion that during the early evolution of eukaryotic cells mitochondria may have arisen when primitive cells engulfed some sort of oxygen-utilizing bacteria,

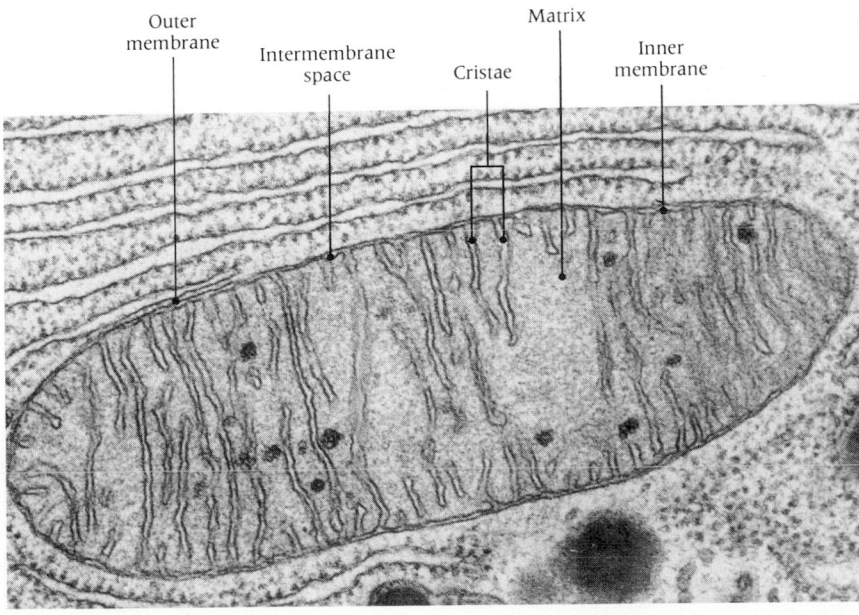

Outer membrane   Intermembrane space   Cristae   Matrix   Inner membrane

A

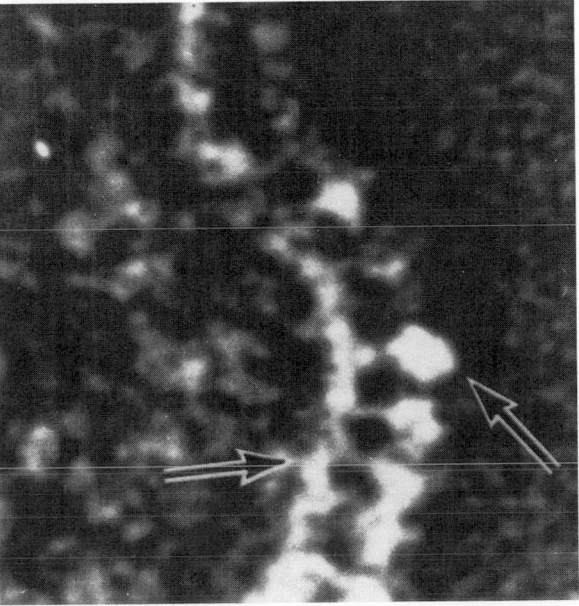

B

which thrived within the cell's cytoplasm in a symbiotic (mutually beneficial) relationship. Eventually, it is now believed, they become permanent features of eukaryotic cells.

Properties of Microtubules

Microtubules, a type of structure found in the cytoplasm of many kinds of cells, are composed of small globular protein molecules called **tubulins.** As shown in Figure 5-21A, two separate proteins, α-tubulin and β-tubulin, combine to form a **dimer**—or **heterodimer,** as it is properly termed because its subunits differ. The heterodimer is the basic building block of all microtubules.

In the microtubule, heterodimers pack in a helical array to form a hollow cylinder about 25 nm in diameter (Figure 5-21B). Cilia and flagella (see below) contain a bundle of microtubules called the axoneme. Many of these microtubules are doublets. At regular intervals the tubules of the doublets in the axoneme bind another protein called **dynein.** In some ways, this protein is analogous to the muscle protein **myosin** (Chapter 31) in that it forms projections that attach to one tubule of the outer doublet; and it splits ATP, the energy source for motility in these systems (Figure 5-22).

As we will see in Chapter 7 and elsewhere, the many functions of microtubules in eukaryotic cells include the following:

- They underlie ciliary and flagellary movements (discussed below).
- They participate in mitosis (Chapter 12).
- They account for rapid transport of materials through nerve fibers (Chapter 29).

Centrioles

Centrioles are small, paired cylindrical bodies lying in the cytoplasm near the nucleus, usually in the middle of the **centrosome,** a zone cleared of other particles. Electron microscopy shows that each centriole has a distinctive structure, consisting of a cluster of nine groups of microtubular triplets, each containing one complete microtubule and two partial ones that share subunits. The two centrioles of a pair always lie perpendicular to one another (Figure 5-23) and move to the poles of the mitotic spindle during cell division (see Chapter 12). Centrioles are found in most animal cells and a few plant cells.

Cilia and Flagella

Some cells are equipped on their surfaces with large numbers of **cilia,** which are motile, hairlike processes, or **flagella,** similar threadlike structures that are usually

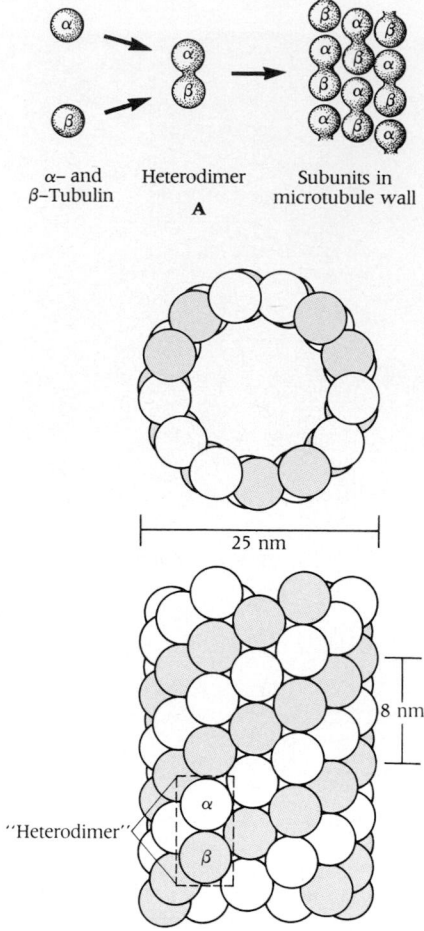

5-21 Microtubules
(A) Structural relationship between α- and β-tubulin. **(B)** Model of three-dimensional structure of a single microtubule. The tubulin polypeptides are aligned in 13 parallel rows, each composed of repeating αβ tubulin heterodimers.

5-22 The "dynein-walking" model
(A) Two adjacent outer doublets, before ATP addition and after ATP addition. The dynein arms of one subfiber of a microtubule "walk" along the adjacent doublet. **(B)** Sliding of microtubules during beating of a cilium. Outer doublets slide past each other and bend cilium.

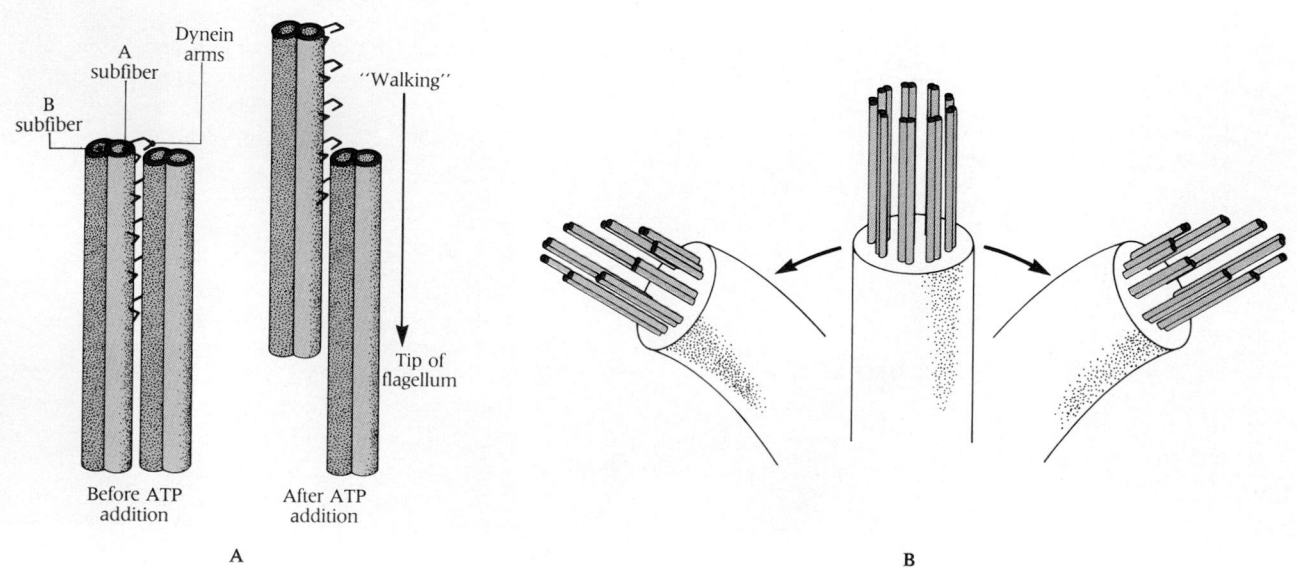

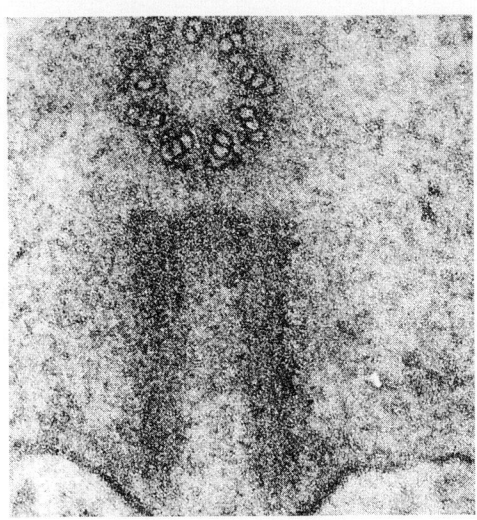

5-23 Two centrioles lying near nuclear envelope
The two bodies lie at right angles to each other in this electron micrograph. (× 60,000)

longer and less numerous. Cilia and flagella have two major roles: They propel cells through a fluid medium, and they create currents that move food particles toward food vacuoles and move such things as mucus in respiratory passages past the cell.

In most multicellular organisms, cilia move fluid past the surface of a cell that is anchored in place (Figure 5-24). They constantly sweep back and forth, in one direction like a stiff rod and in the other like a whiplash (Figure 5-25), and in a rotary screwlike pattern. The cilia of adjacent cells operate in a coordinated manner, and their aggregate motion resembles breeze-blown waves in a field of grain. It is easy to see how such a "meadow" of cilia, beating up to 20 times per second, can move considerable quantities of material.

In many free-swimming protists such as *Paramecium* and *Tetrahymena*, the cilia consist of numerous short, hairlike projections covering most of the exposed surface of the cell (Figure 5-26). For this reason, such organisms are called **ciliates.** Their movement is caused by the beating motion of the cilia, which literally row the protist through the water.

When cilia are longer and less numerous, they are called flagella. Prokaryotic and eukaryotic flagella are distinctly different structures. (Some features of bacterial flagella are discussed in Box 5-2.) For the moment, we are concerned with the flagella of eukaryotic cells (Figure 5-27). When a flagellum is attached to a movable cell, such as a sperm, it propels the cell through its liquid environment. Flagella are similar to cilia but are two or three times longer and more whiplike in their beat pattern. Unlike cilia, there are usually only a few flagella or a single one on each eukaryotic cell.

5-24 (Right) Beating cilia
Cilia on the surface of a cell are coordinated so that waves of beating pass across them, propelling fluid or debris past the cell.

Direction of Fluid Movement

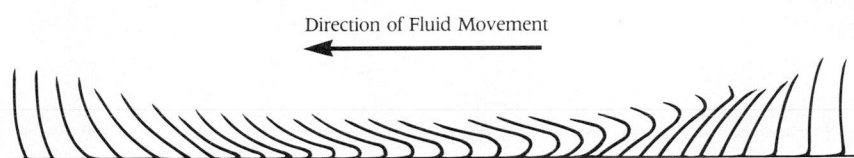

Each cilium or flagellum is anchored by a **basal body** (also called **kinetosome**) at its base (Figure 5-28). There are striking similarities between the structure of a basal body and other cell organelles. For example:

- Basal body structure has features in common with that of a cilium or flagellum. Both contain a cylindrical central axis called an **axoneme,** which consists of a sheaf of microtubules. The arrangement of the microtubules differs in each, however. Cilia and flagella have an outer circle of 9 *doublet* microtubules (two fused microtubules that on cross section look like a figure 8) surrounding two single microtubules in the center. This is the so-called **9 + 2 pattern** (Figure 5-29). This pattern is found in the cilia of eukaryotic cells and organisms of widely diverse kinds. Basal bodies have nine *triplet* microtubules arranged in a circle. There is no central microtubule (see Figure 5-28C).
- Basal body structure is identical to that of a centriole. Centrioles, like basal bodies, have 9 triplet microtubules arranged in a circle and with no central tubules.

Liquid

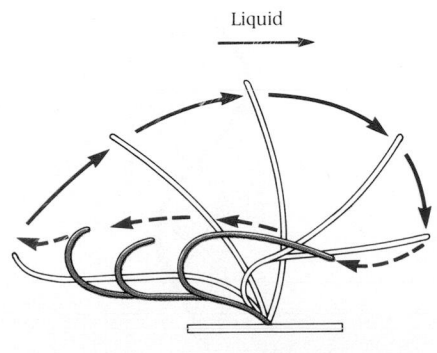

5-25 Ciliary motion
The beat of a typical cilium consists of a straight-armed effective stroke (white cilia) that moves a liquid (colored arrow), followed by a curling return stroke (grey cilia), often in a third dimension (out of the plane of the paper).

Unlike the structures at the bases of bacterial (prokaryotic) flagella, the basal body apparently has no active part in the generation of flagellar or ciliary motility. Rather, its role seems to be to anchor the beating cilium or flagellum against the stresses of the torque it produces. However, a recent study shows that the green alga *Chlorogonium* has two flagella able at times to detach themselves from their basal bodies. When they do, they continue to beat vigorously.

There is evidence that cilia and flagella (and other microtubular systems) operate by an ATP-dependent sliding filament mechanism similar to that of muscle (Chapter 31). In this mechanism, movement is caused by a cross-bridging cycle in which the hydrolysis of ATP induces conformational changes in the dynein links between the microtubules (see Figure 5-22).

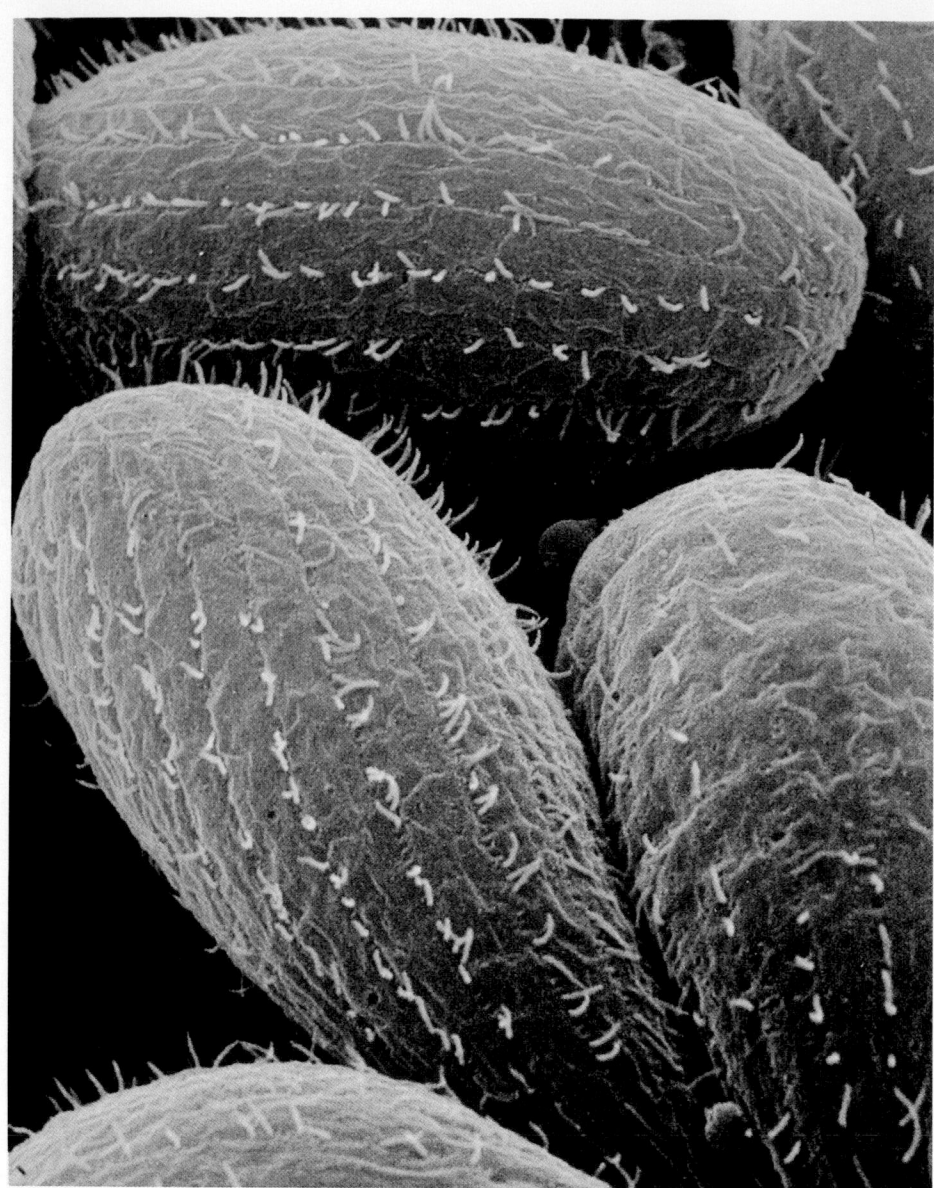

Organization of the Cytoplasm: The "Cytomatrix"

The advent of electron microscopy revealed that cytoplasm is a highly organized system containing not only membrane-enclosed vesicles and organelles but endoplasmic reticulum and ribosomes. However, the fluid medium in which all these organelles are suspended appeared quite structureless. Hence it came to be known by the noncommittal term **cytoplasmic ground substance.**

There were nevertheless good reasons to suspect that this material does have an internal structure. For example, its viscosity and elasticity make it seem much more like a gel than a fluid. Also, the organelles within it evidently cannot roam about freely as they would in an unstructured fluid. Their distribution is nonrandom, and thus it seemed likely that they are held in position by some sort of internal scaffolding.

Recent work finally revealed that cytoplasm is highly structured, containing throughout its volume an astonishing spongework or lattice of tiny filaments (Figure 5-30A). Recognition of this framework was made possible largely by the development in the 1960s of the high voltage electron microscope, current models of which stand 30 feet high and weigh more than 20 tons. Its chief advantage over the conventional electron microscope (if such a microscope could be called conventional) is its capacity to pass electrons through thicker specimens and thus to gain a three-dimensional perspective. With this equipment investigators discovered an irregular lattice of slender interconnected fiber strands throughout the cytoplasm—previously suspected but unseen—that are anchored to the outer cell membrane (Figure

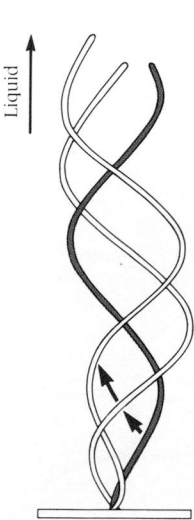

5-27 Flagellary motion
Successive waves propagated toward the tip of a flagellum move water (colored arrow) and thus propel the flagellated organism in the opposite direction.

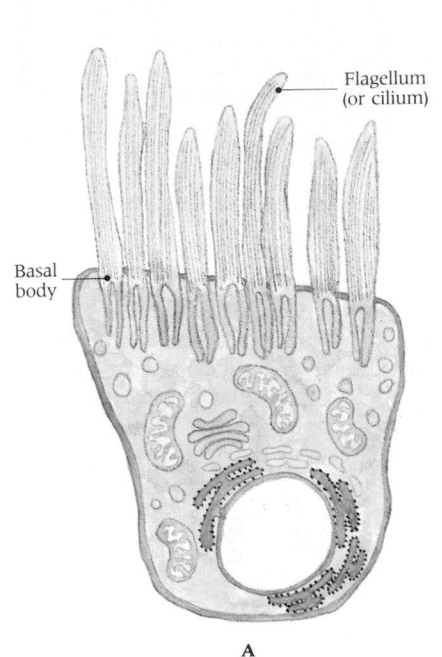

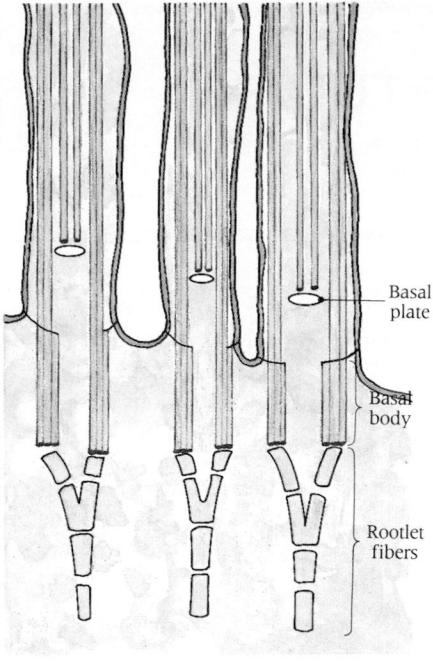

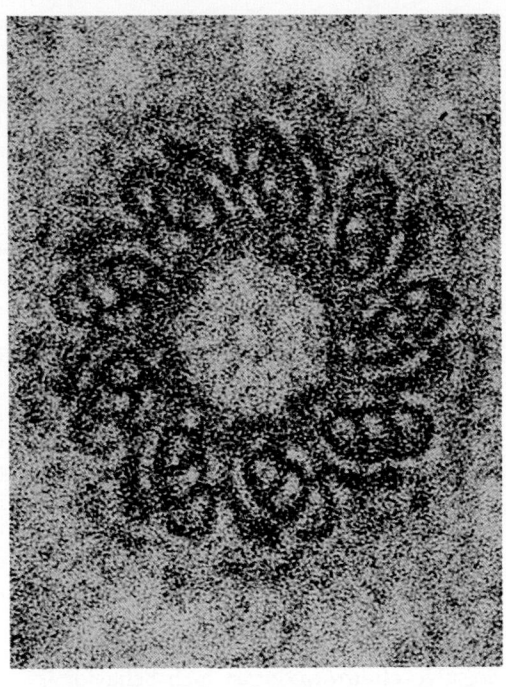

A
B
C

5-28 Basal bodies

(A) *Drawing of a cell showing the relation between basal bodies and flagella (or cilia).* **(B)** *Longitudinal section through three cilia showing basal bodies composed of microtubules (color).* **(C)** *Cross section of a basal body showing nine sets of three microtubules. Two microtubules of each group appear in the cilium of flagellum.* (× 60,000)

5-29 Structure of a cilium

(A) *Cross-section showing 9 + 2 pattern.* **(B)** *Electron micrograph of cilia in cross-section.* (× 77,000)

5-30B). This apparatus, the **microtrabecular lattice,** was eventually found in many eukaryotic cells.

The thin interlinked filaments of the microtrabecular lattice support the cell organelles and control their movements. The lattice divides the cell into two distinct phases: one, the protein-rich phase of the lattice itself; the other, the water-rich phase filling the spaces in the lattice. This fluid phase is a solution of small biomolecules such as glucose, amino acids, carbon dioxide, and oxygen.

As in all developing fields, there exists in the field of cell ultrastructure a degree of terminological confusion. With the discovery of multiple systems of filaments in cytoplasm, the term **cytoplasmic matrix** came into use to cover all of them collectively. Soon this was shortened to **cytomatrix.** The several filament systems include the microtrabecular lattice and, in addition, three other major filament systems that make up what has been termed the **cytoskeleton.** These systems— the microtubules, microfilaments, and intermediate filaments—form a network of larger fibers within the cell. What do these three structures do in cells?

Microtubules, as discussed above, are long, threadlike, polymeric molecules 25 nm in diameter that are constructed of many tubulin subunits arranged in the form of tubules. They help to maintain the cell's shape and participate in the movement of organelles in cell division and other mechanisms.

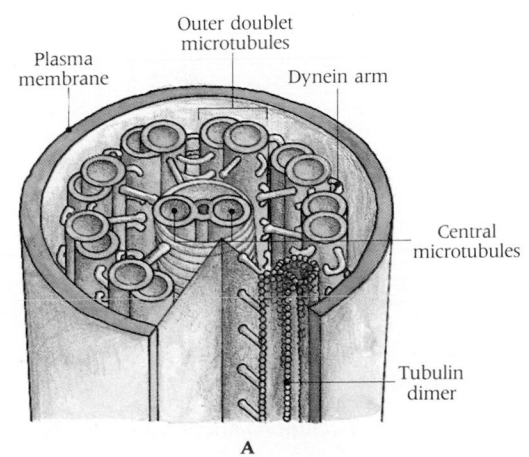

A

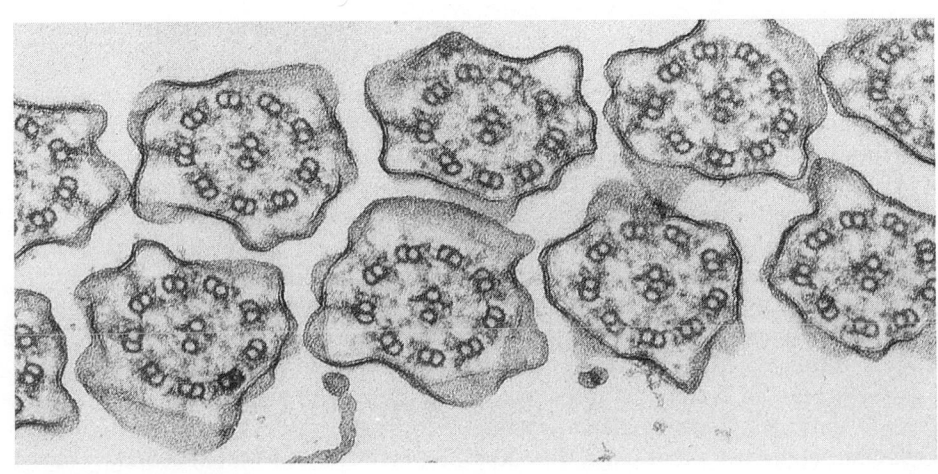

B

Motile bacteria, which are single-celled organisms at the simplest level of organization, are attracted or repelled by certain chemicals, and elegant new researches have begun to show how bacteria detect such stimuli and how they move in response to them. What remains unknown is the nature of the "nervous system" that links stimulus detection at one part of the cell with effector response in another part.

The capacity of bacteria for positive and negative **chemotaxis,** as such behavior is termed, has been recognized for a century. A breakthrough occurred in 1969 when Julius Adler, of the University of Wisconsin, found that bacteria sense chemical attractants and repellents by means of specific chemoreceptors. When Adler exposed the familiar motile bacterium *Escherichia coli* to various attractants, he found that the bacterial surface contains a small number of receptors (binding proteins) that detect chemical attractants and transport them into the cell. Mutants were then discovered that detect and bind the attractant but fail to transport it into the cell. These still perform chemotaxis. Other mutants that lack the surface receptor do not bind the attractant and do not display chemotaxis.

Next came ingenious studies by Howard Berg, then at the University of Colorado, on how a bacterium swims. Using a special mi-

croscope that automatically tracks an individual bacterium, Berg found that a swimmer's path consists of a random mix of straight runs interrupted by tumbles that change its direction (Figure 5H). Then he had a major surprise. Bacteria swim by rotating their flagella, which are joined to the cell by a hooklike device that is attached to a basal body, a set of rings embedded in

the cell envelope (Figure 5I). Unlike the eukaryotic flagellum, which operates by bending and whiplashing, the bacterial flagellum rotates about an axis like a ship's propeller. The most dramatic evidence for this conclusion was obtained by fixing the end of a flagellum to a glass slide by the use of specific antibodies. When a cell was tethered in this way, its flagellum was no longer free to ro-

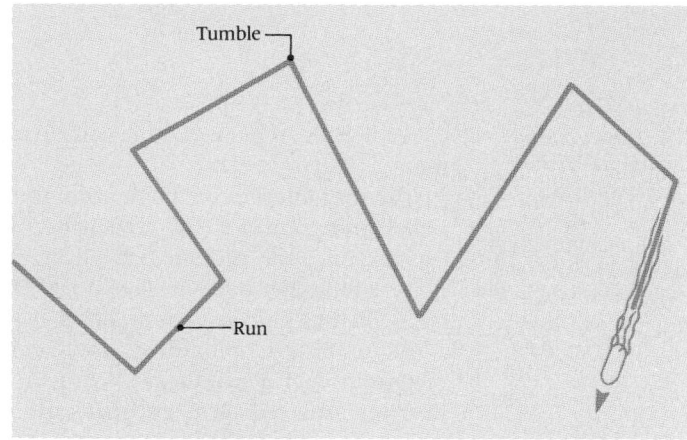

5H How bacteria swim
The normal path followed by a swimming E. coli *is a three-dimensional "random walk" (shown here in two dimensions). A series of almost straight "runs" is interrupted by "tumbles" in which the bacterium sets out in a new direction.*

Microfilaments are another set of threadlike elements that are 5–7 nm in diameter and occur in bundles. They are constructed mainly of the protein *actin*. Microfilaments facilitate a different set of movements, some of great importance in cell motility, cell division, and embryonic development. The beautiful filaments that can form from pure solutions of actin (Chapters 7 and 31) are the cytologist's microfilaments under a biochemist's name.

Intermediate filaments (0–15 nm in diameter) are less well understood. At least five types have been identified in different animal cells, each made of a different protein. Since no obvious role in motility or contraction has been demonstrated, investigators have grudgingly assumed that their purpose is purely structural.

The term *cytomatrix* encompasses all the cytoplasmic filaments (plus their associated proteins and water). Its discovery had a number of stimulating implications. First, it indicated that cytoplasm is a highly ordered system composed of slender cooperating elements that participate in all aspects of a cell's life. Second, it raised a basic biological question. Why do all eukaryotic cells have two parallel systems—

tate. However, the cell body rotated instead! The flagellar rotary motor utilizes as its energy source an intermediate in oxidative phosphorylation rather than ATP itself. Someone, hearing of this discovery, termed the flagellar motor "biology's first wheel."

In straight swimming runs, the flagella rotate counterclockwise, and all the flagella operate in unison. When the flagella switch to clockwise rotation, they fly apart, each fighting the other. Eventually one or another prevails. The result is a tumble. Significantly, the movements and path of a swimming bacterium in the presence of a chemical attractant are strongly biased by the sensory stimuli (Figure 5J). This accounts for chemotaxis.

When an individual bacterium is presented simultaneously with an attractant and a repellent, it must "decide" how to proceed. The "decision" that resolves the "conflict" depends on the relative concentrations of attractant and repellent. What is so very interesting is the notion that bacteria are able somehow to integrate multiple sensory stimuli. It is tempting to think of a bacterium as we think of a nerve cell, or neuron, which integrates excitatory and inhibitory inputs electrically, responding in an all-or-none fashion by generating an action potential. Perhaps the bacterium integrates different inputs and indeed couples receptors and flagella through changes in membrane potential. Answers to this and other questions await the results of work now in progress.

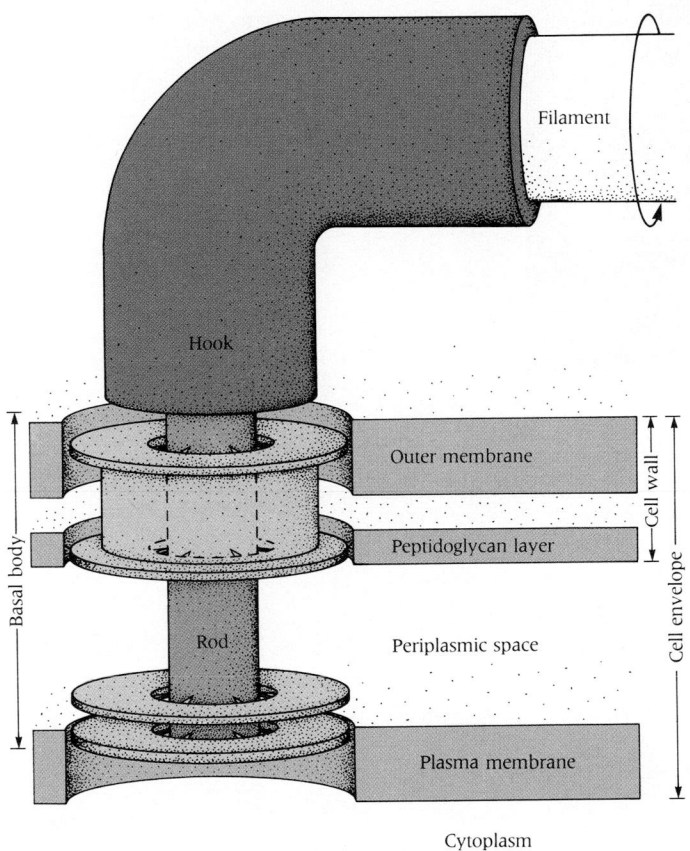

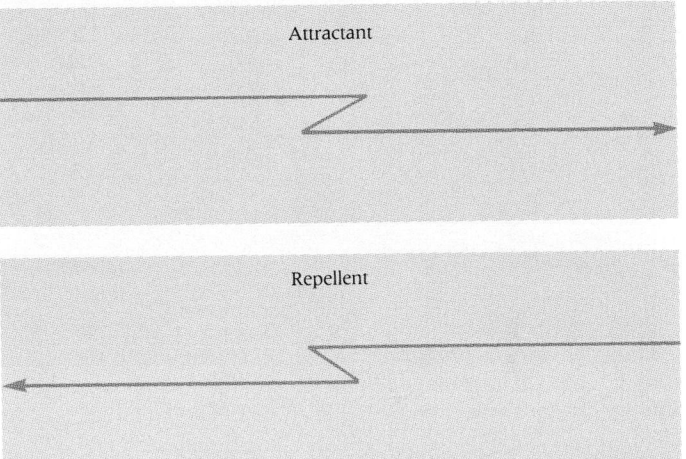

5J (Above) Chemotaxis in bacteria
The path of a cell in a gradient is biased by a change in the frequency of tumbling, which is inhibited when the cell is moving in the "right" direction (up-gradient in an attractant and down-gradient in a repellent) and promoted when the cell is going in the "wrong" direction. (Only runs that are in the "right" and "wrong" directions are shown).

5I (Left) Basal body of an E. coli flagellum
The hook and rotating filament insert into the cell envelope, which is composed of the cell wall and the basal body. The top ring is associated with the cell wall, the bottom ring with the periplasmic space and plasma membrane.

microfilaments and microtubules—for motility and support? Why do cells need two "separate but equal" mechanisms, each using a different protein, to do similar jobs? Perhaps one answer lies in subtle differences in sliding and polymerization mechanisms of the two systems that provide a larger repertoire of possible solutions to different problems occurring in the cell.

STRUCTURE OF PLANT CELLS

The plant cell typified by the green leaf cell in Figure 5-9 contains most of the distinctive organelles and structures seen in animal cells. These include the nucleus, mitochondria, Golgi apparatus (or dictyosome), endoplasmic reticulum, and ribosomes. In addition, plant cells contain important structures that are absent in animal cells, cell walls, and plastids (including chloroplasts). They also contain vacuoles, rarely prominent features of animal cells.

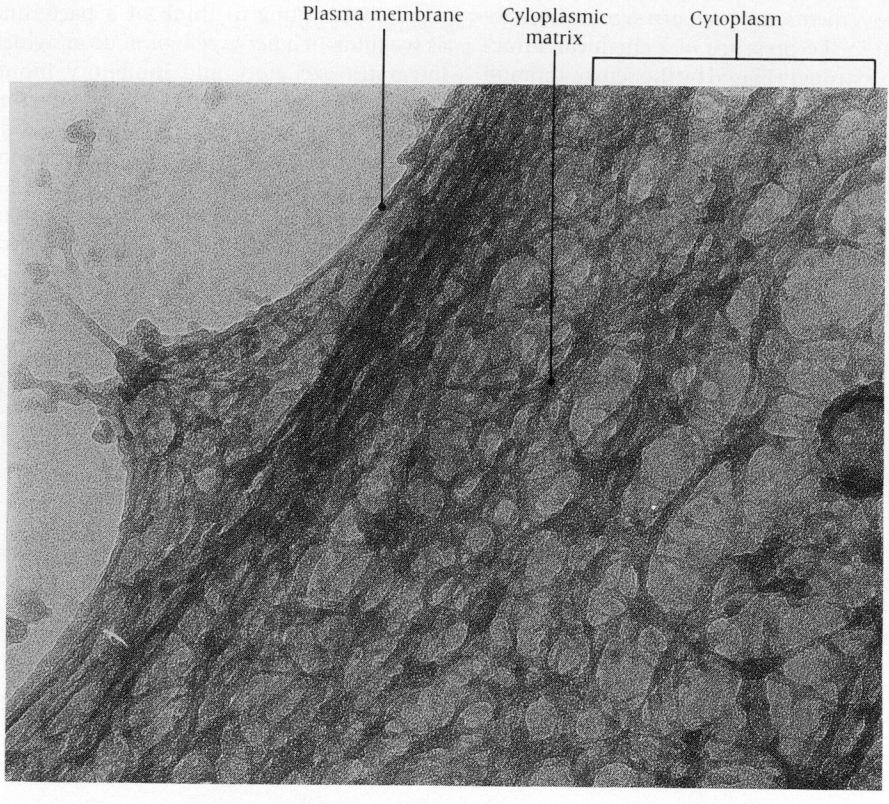

Plasma membrane Cyloplasmic matrix Cytoplasm

A

5-30 Cytoplasmic matrix
(A) *A web of proteins, here seen by high-voltage electron microscopy, fills the cytoplasm of many cells. (× 100,000)* **(B)** *Model of the microtrabecular lattice.*

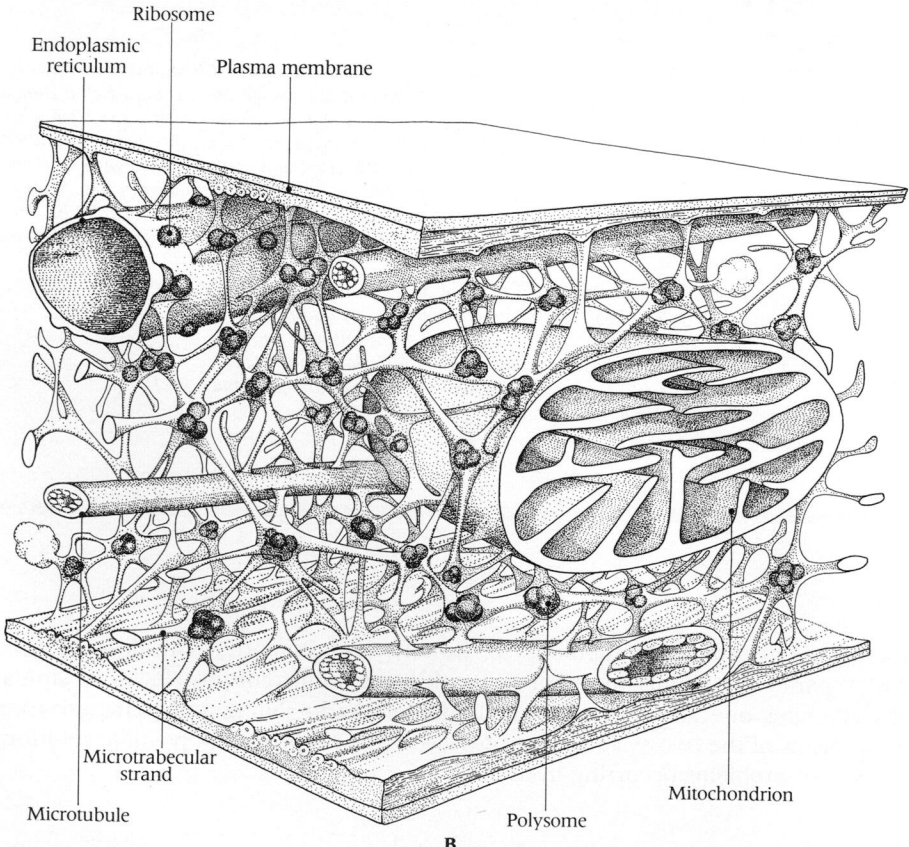

Endoplasmic reticulum   Ribosome   Plasma membrane

Microtubule   Microtrabecular strand   Polysome   Mitochondrion

B

The Cell Wall

Among eukaryotic organisms, plants, fungi, and the algal protists have cells surrounded by a more or less rigid **cell wall.** In all plants and many algal protists, the wall is composed mainly of cellulose, which is formed into microfibrils 10–25 nm wide (Figure 5-31). Most fungi have cell walls composed primarily of chitin.

The cell wall acts to keep the cell from rupturing from the increase in pressure

5-31 Structure of cell walls
Scanning electron micrograph of the cellulose microfibrils seen in the cell wall of Chaetomorpha melagonium. (× 50,000)

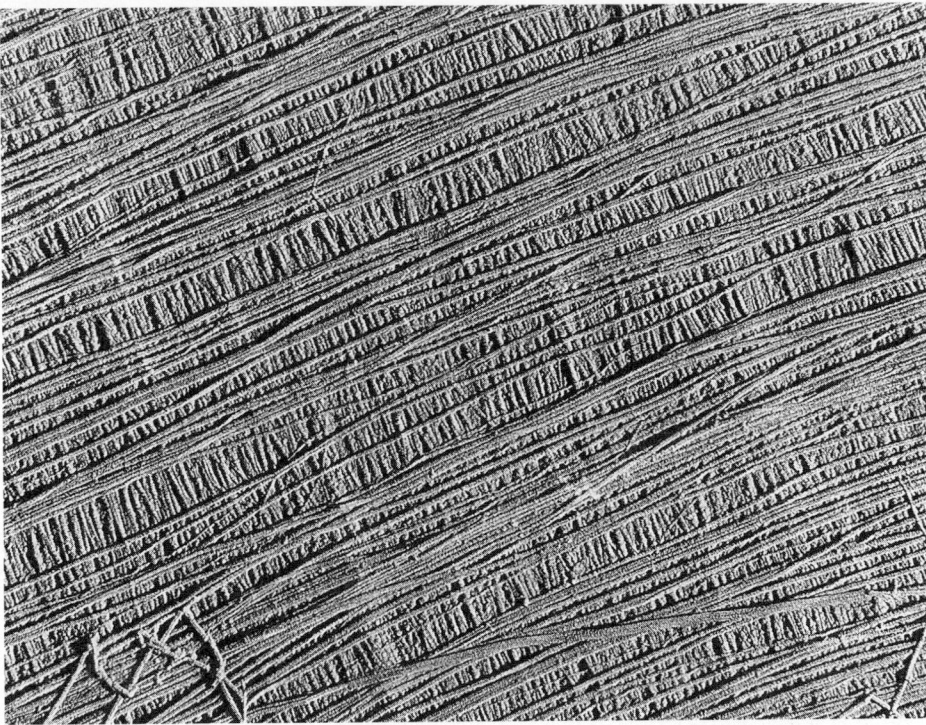

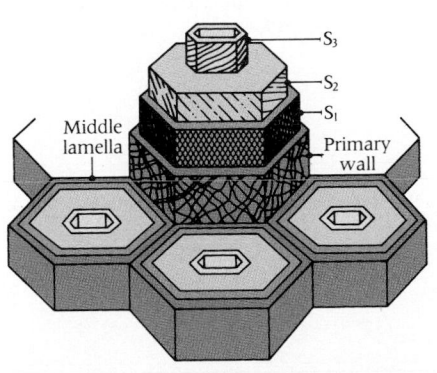

5-32 The layers of plant cell walls
The first layer is the middle lamella, a gluelike substance containing pectins that is shared by adjacent cells. The second, the elastic primary wall, contains cellulose and pectins. Fully grown cells often have a secondary cell wall, containing lignin and cellulose, that can itself contain three layers: S_1, S_2, and S_3.

that follows excessive water intake. This is directly attributable to a framework within the cell wall of *microfibrils* made of cellulose, which are embedded in a *matrix* of noncellulose carbohydrates (pectins and hemicellulose). Thus this rigid structure fixes the cell's size, shape, and position relative to other plant cells. In addition, plant cell walls play critical roles in secretion and digestion. They have tiny pores, 3.5 to 5.0 nm in diameter, that control the passage of materials into and out of the cell. Sheets of interconnected cell walls were what Hooke observed in his early microscopic glimpses of cork (see Figure 5-1).

The cell walls of many plants (and bacteria and fungi) can be experimentally removed with enzymes. The structure that remains is comparable to what remains when bacterial cells are grown in the presence of an agent—such as penicillin— that inhibits cell wall formation. In both cases, the cells that remain are enclosed only in their cell membranes. Such cells are called **protoplasts.** (Note that animal cells do not have cell walls and cannot form protoplasts.) Protoplasts have interesting osmotic properties, taking up water freely and becoming spherical. The point emphasized here is that the shapes of plant cells are molded by their cell walls.

The thickness of plant cell walls varies greatly from cell to cell. Careful electron microscopy and x-ray diffraction studies revealed at least two major layers in most plant cell walls (Figure 5-32). One is the **middle lamella,** a gluelike intercellular substance that forms as a common partition between two dividing cells and is composed mainly of acidic branched polysaccharides called *pectins*. Because of their branched structure, pectins trap many water molecules and readily form gels. For that reason purified pectins are added to induce fruit to "set" into jam or jelly. Pectins are broken down in fruit by the enzyme pectinase; this accounts for the softening of ripened fruit. The second cell wall layer is the elastic **primary wall** that each new cell forms on its side of the middle lamella. It consists chiefly of cellulose molecules, aggregated into the bundles of fiber cells called microfibrils. These in turn are clustered into variously arranged macrofibrils. The primary wall also contains pectins and certain other polysaccharides called **hemicelluloses.**

Many fully grown cells deposit yet another wall layer, the **secondary wall,** a rigid structure that contains cellulose and a hard, reinforcing polymer called **lignin.** (The density and durability of hardwoods are largely attributable to the high levels of lignin in their cell wall matrix.) Pectins are lacking. Secondary walls often have three distinct layers of their own—designated S_1, S_2, and S_3 (Figure 5-32)—that differ in the arrangements of their microfibrils. Secondary walls are most prominent in cells that give the plant structural strength and conduct water. Wood is composed mainly of secondary walls.

Plastids

Cells of plants (and algal protists) characteristically contain **plastids** (not to be confused with *plasmids*). They are membrane-encased bodies that in many ways resemble mitochondria. Like mitochondria, they contain a distinctive kind of DNA that differs from the DNA in the nucleus of the same cell.

In a sense, they may be looked upon as the fundamental organelles of the living world. Plastids that contain pigments are called **chromoplasts** (colored plastids); unpigmented ones are **leucoplasts** (colorless or white plastids). **Chloroplasts,** which are chromoplasts containing the green pigment chlorophyll, are the sites of photosynthesis. That is why they are critical to life. Some chromoplasts store pigments other than chlorophyll that bring color to many fruits and flowers. Among the leucoplasts are the **amyloplasts,** which store starch granules.

Compared to mitochondria, chloroplasts are quite large (from 4 to 6 and even 8 μm in diameter). There are a number—few to many—of chloroplasts per cell (Figure 5-33). A square millimeter of leaf contains some 500,000 chloroplasts. They are often disk-shaped.

Each choloroplast has a double membrane around it and a complex internal structure (Figure 5-34). Fine strands of chloroplast (nonnuclear) DNA, which, like mitochondrial DNA, is circular, permeate the mass of chloroplast **stroma.** Like mitochondria, chloroplasts contain small ribosomes similar to those of prokaryotes. A number of particlelike structures called *grana* are embedded within the stroma. Each granum consists of paired inner membranes that form regular stacks of disklike, flattened lamellae called **thylakoids,** which resemble stacks of coins. Chlorophyll and the photosynthetic apparatus are located in the thylakoid membranes. We discuss this remarkable photosynthetic membrane in Chapter 10. It is not unwarranted to suggest that the whole living world as it exists today depends, directly or indirectly, on the properties of the thylakoid membrane.

Vacuoles

Vacuoles, the most prominent feature of many plant cells, are bubblelike compartments in the cytoplasm that are bounded by membranes. The term **tonoplast** denotes an entire vacuole with its surrounding membrane. Vacuoles contain both nutrients and wastes in a watery solution called *cell sap.*

Vacuoles are small in young plant cells, but increase in size as plant cells mature (see Figure 5-9). Eventually, one or more large vacuoles push the cytoplasm against

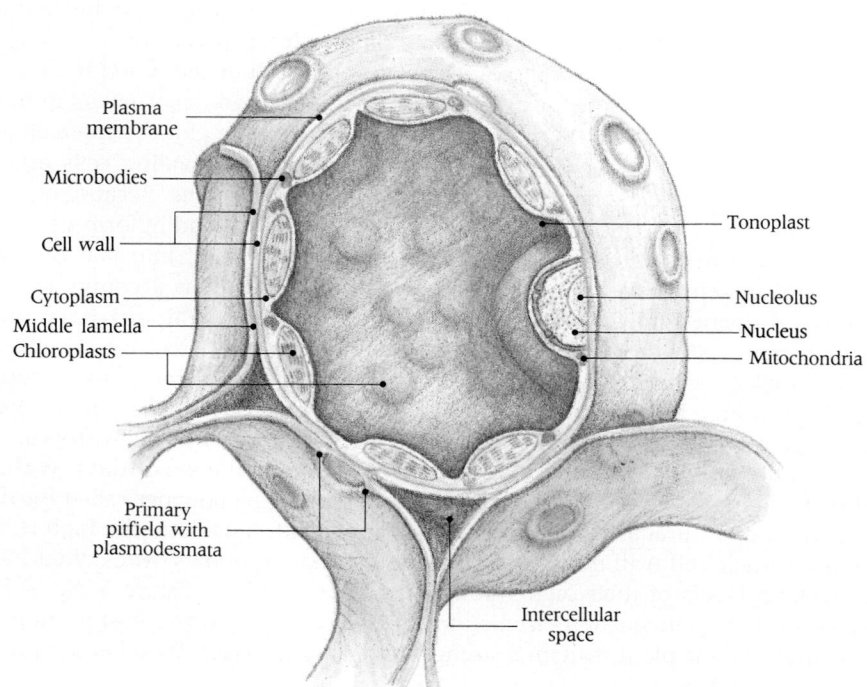

5-33 A chloroplast-containing cell
The chloroplasts are located in the cytoplasm along the margin of the cell. Most of the volume of the cell is occupied by the vacuole (tonoplast). See also Figure 5-9.

5-34 Chloroplasts
(A) *Internal structure of a chloroplast. Flattened individual lamellae called thylakoids are stacked to form grana.* **(B)** *An electron micrograph of a plant cell chloroplast.* (× 20,000)

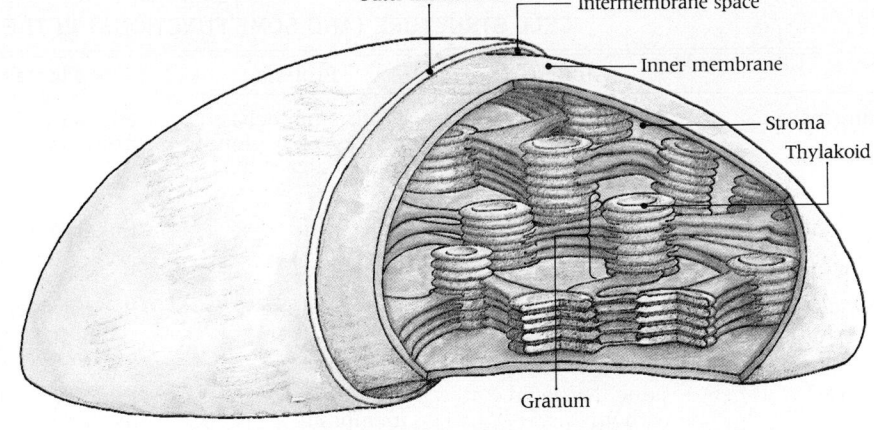

A

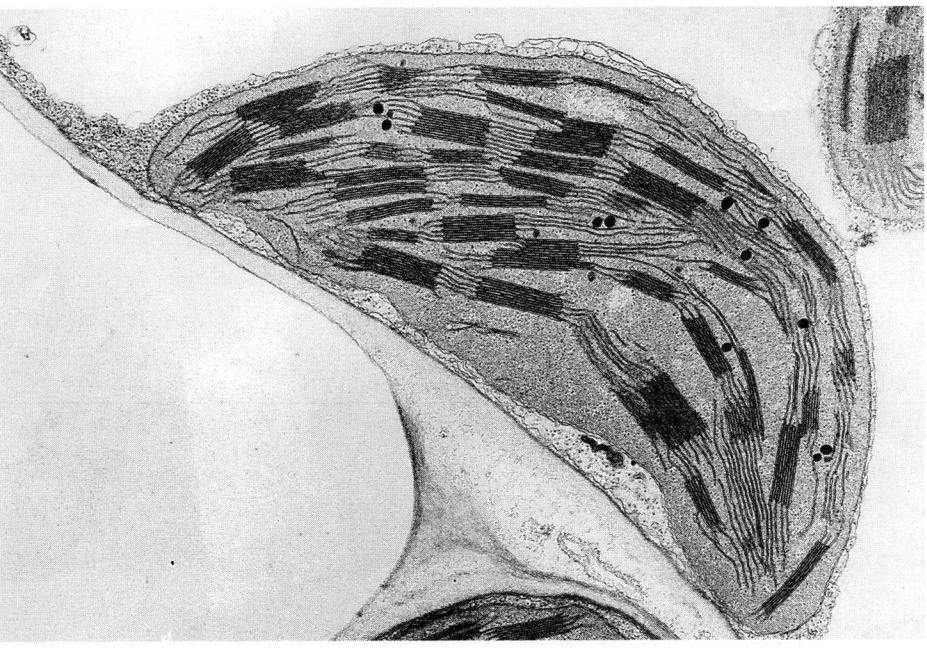

B

the cell membrane and wall, flattening it into a thin layer. The nucleus lies in the viscous cytoplasm outside the vacuole. As much as 90% of the cell's volume may be occupied by a large central vacuole.

Large vacuoles rarely occur in animal cells, but smaller ones of various sorts may be present. In some protist and animal cells, the outer cell membrane forms a pouch around a food particle, which later separates from the cell membrane and becomes a temporary food vacuole.

Mature plant vacuoles contain cell sap in which amino acids, sugars, and other important molecules are dissolved. Cell sap also contains familiar plant pigments, the **anthocyanins,** which account for the brilliant colors of plants and plant parts.

Another kind of vacuole is concerned with the maintenance of a proper water balance in some cells. These cells, found chiefly in many unicellular and simple multicellular animals, meet that problem by continually expelling excess water. Contractile vacuoles eject the water from the cell.

STRUCTURE OF PROKARYOTIC CELLS

Prokaryotes differ strikingly from eukaryotic cells (see Figure 5-10). They are small, seemingly simple cells surrounded by cell membrane. A cell wall is present but its biochemical makeup is different from that of eukaryotic cell walls. Prokaryotes differ most notably from eukaryotes in the following ways:

TABLE 5-1
CELL STRUCTURE (AND SOME FUNCTIONS) IN THE FIVE KINGDOMS*

	Monera	Protista	Plantae	Fungi	Animalia
Organisms	Bacteria, cyanobacteria, mycoplasmas, blue-green algae, etc.	Protists (amoebas, ciliates, etc.), slime molds, algae, etc.	Plants developing from embryos	Molds, mushrooms, yeasts, etc.	Animals developing from embryos
Cell organization	Prokaryotic	Eukaryotic	Eukaryotic	Eukaryotic	Eukaryotic
Cell membrane	Present	Present	Present	Present	Present
Cell wall	Present. Made of noncellulose polysaccharides	Present in many	Present. Made of cellulose polysaccharides	Present. Usually made of chitin	Absent
Nucleus	None. Nucleoid without surrounding membrane	Present. Nuclear membrane	Present. Nuclear membrane	Present. Nuclear membrane	Present. Nuclear membrane
Chromosomes	Single. Contains only DNA	Multiple. Some lack histones	Multiple. Contain DNA and protein	Multiple. Contain DNA and protein	Multiple. Contain DNA and protein
Endoplasmic reticulum	Absent	Usually present	Usually present	Usually present	Usually present
Mitochondria	Absent	Present	Present	Present	Present
Plastids	Absent	Present in some cells	Present	Absent	Absent
Ribosomes	Present (small)	Present	Present	Present	Present
Golgi apparatus	Absent	Present	Present	Present	Present
9 + 2 cilia or flagella	Absent	Present	Present	Present	Present
Endocytosis	Absent	Present	Present	Present	Present
Photosynthesis	Present in some	Present in some	Present in most	Absent	Absent

* Also see Table 37–2.

- They contain no membrane-surrounded nucleus and no membranous organelles such as mitochondria or endoplasmic reticulum.
- The organization of their DNA is not in the form of multiple threadlike, protein-containing chromosomes of the eukaryotic type. The single prokaryotic chromosome in each cell consists of a single molecule of double-helical DNA, densely coiled in the cell's nucleoid zone. No structural proteins of the type called histones are bound to this DNA.
- If prokaryotes contain chlorophyll, it is not in chloroplasts.
- If they have flagella, the 9 + 2 structure is lacking.
- The prokaryotic cell wall contains a rigid framework of noncellulose polysaccharide chains that are cross-linked with short peptide chains. The polysaccharides contain certain distinctive components—for example, amino sugars derivatives that do not occur in eukaryotic cell walls. One of its functions is to prevent bacterial protoplasts from swelling as they continue to take in water by osmosis. The cell membrane, however, is selectively permeable.
- Cell membranes generally resemble those in eukaryotes, although bacterial lipids are fewer in number and relatively simple. The cell wall is porous and allows most small molecules to pass through. It also contains enzymes that function in respiration and cell wall synthesis.
- The cell membrane often invaginates to produce membranous structures called *mesosomes*, which contain most of the cell's lipids, carotenoids, and several proteins. As discussed in Chapter 12, the mesosome participates in cell division.
- Prokaryotes often include storage granules containing polyglucose polymers. Other bacteria contain granules of stored lipid, or poly-betahydroxybutyric acid. When needed as fuel, these polymers are degraded to yield their smaller components in free form.
- Prokaryotic ribosomes are smaller than eukaryotic ones (other than those in mitochondria and chloroplasts), although each has a large and a small subunit. Like eukaryotic ribosomes, however, they are sites of protein synthesis.

The prokaryotes include the eubacteria, the cyanobacteria (blue-green algae), the spirochaetes, the rickettsiae, and the mycoplasma, or pleuropneumonialike organisms. Prokaryotes will be discussed in more detail in Chapter 38.

The classification of organisms into five kingdoms places all prokaryotes in the kingdom Monera. The other four kingdoms are composed of eukaryotic organisms. Classification is discussed in Chapter 37. Table 5-1 summarizes the basic cell structures found in each of the five kingdoms.

SUMMARY

The cell theory, one of the unifying forces of modern biology, states that all living organisms are composed of fundamentally similar cells, and that all cells are produced only by pre-existing cells.

Viruses are macromolecular complexes of nucleic acid and protein. Plasmids contain only DNA. Both should be distinguished from cells, for they are much smaller than cells and cannot reproduce independently outside of a host cell. Thus, cells are the smallest biological units that can be considered alive.

Most cells consist of a mass of cytoplasm surrounded by a semipermeable, phospholipid bilayer outer membrane. The cytoplasm is a complex, watery solution or suspension of proteins, carbohydrates, nucleic acids, and the other molecular building blocks discussed in Chapter 4.

Despite their basic similarities, cells can be divided into two different structural classes: eukaryotic cells, which have nuclei, and prokaryotic cells, which do not. Animals and plants are composed of eukaryotic cells, as are fungi and protists. Only monerans (bacteria and blue-green algae) are made up of prokaryotic cells.

This chapter focuses on the structure of eukaryotic cells. Within the cell, bound by the outer membrane, are other membrane-bound structures called organelles, which perform specialized functions for the cell. Chief of these organelles is the nucleus, which contains the cell's chromosomes and thus its DNA. Outside the nucleus, the cytoplasm is broken up by a continuous array of intracellular membranes called the endoplasmic reticulum (ER). Rough ER is studded with ribosomes, the ribonucleoprotein particles that are the sites of all protein synthesis, whereas smooth ER, which is not studded with ribosomes, is involved in lipid synthesis. The Golgi apparatus, a stack of flattened membranous sacs, is involved

in the secretion of cell products via secretory vesicles. It also forms enzyme-containing organelles such as lysosomes, which contain digestive enzymes. Mitochondria, with their intricate membranous substructure, are the centers for energy-yielding metabolism.

Not all eukaryotic organelles are membrane-bound. Ribosomes are not. Several organelles are composed of microtubules. These include centrioles, which are involved in cell division, and cilia and flagella, which are involved in cell motility. Microtubules, microfilaments, and intermediate filaments also exist outside membrane-bound organelles; they help to maintain the organization of the cytoplasm.

Some eukaryotes have additional features. Plants and algal protists have a cell wall, often containing cellulose, which keeps them from taking up too much water and bursting. These organisms also have plastids, most notably chloroplasts—the sites of photosynthesis. Large membrane-bound vacuoles filled with water, nutrients, and wastes are common in most mature plant cells. Many fungi have cell walls composed of chitin.

Prokaryotic cells are significantly different from eukaryotic cells. They have both cell membranes and cell walls, but their cell walls are biochemically unlike those of plant cells. They have no membrane-bound structures in their cytoplasm and, most notably, no nucleus. Their DNA is found not in protein-containing chromosomes, but as a single naked loop in the nucleoid region of the cytoplasm. Like eukaryotes, they have ribosomes and other membraneless organelles. But their ribosomes are smaller, and their flagella have a structure different from those in eukaryotes. In general, prokaryotes are relatively simpler than eukaryotes, although they are extraordinarily complex.

KEY TERMS

cell
cell theory
cell wall
centriole
centrosome
chemotaxis
chloroplast
chromatin
chromoplast
cilium
cristae

cytoplasmic ground substance
cytoplasmic matrix
cytoskeleton
differentiation
dimer
dynein
electron microscopy (EM)
endoplasmic reticulum (ER)
eukaryote
flagellum
Golgi apparatus

intermediate filament
lignin
lysosome
microfilament
microtrabecular lattice
microtubule
middle lamella
mitochondrion
myosin
nuclear envelope
nuclear pore complex

nucleolus
nucleus
organelle
peroxisome
phagocytosis
plasma membrane
 (plasmalemma)
plasmid

plastid
polysome (polyribosome)
prokaryote
protoplast
ribosome
rough endoplasmic reticulum
 (RER)
secretory vesicles

smooth endoplasmic reticulum
 (SER)
stroma
thylakoid
tissue
tubulin
vacuole
virus

QUIZ QUESTIONS

1. Which of the following is *not* characteristic of a eukaryotic plant cell?
  A. annuli associated with the nuclear envelope
  B. thylakoids inside of chloroplasts
  C. cristae inside of mitochondria
  D. dictyosomes associated with Golgi bodies
  E. ribosomes associated with endoplasmic reticulum

2. Which of the following is a correct description of a viroid?
  A. globular proteins surrounded by nucleic acid
  B. small bits of nucleic acid without a protein coat
  C. globular proteins lacking a nucleic acid coat
  D. small bits of protein associated with lipid molecules

3. Which of the following organelles is bounded by a double membrane?
  A. lysosome
  B. nucleolus
  C. ribosome
  D. plastid
  E. A and B

4. Endoplasmic reticulum is involved in all of the following *except*
  A. production of ribosomes.
  B. synthesis of proteins.
  C. transportation of molecules to the cell membrane for export.
  D. synthesis of lipids.

5. Which of the following is *not* a feature of prokaryotic cells?
  A. chlorophyll not located in chloroplasts
  B. flagella that lack the "9+2" pattern of fibrils
  C. only one chromosome located in the nucleus
  D. absence of organelles bounded by membranes
  E. presence of a cell wall

ESSAY QUESTIONS

1. How do viruses differ from cells? Contrast their life cycles, means of reproduction, and methods of acquiring nutrients.

2. What advantages does internal compartmentalization provide for cells? Why are all multicellular organisms eukaryotic?

3. What features are shared by mitochondria and prokaryotic cells?

4. Describe the components of the cytoskeleton. How does the function of the cytoskeleton compare with that of the bony skeleton in our bodies?

5. Contrast the structure and function of typical plant cells and animal cells. How do their distinctive features facilitate their particular roles?

6. What cellular requirements constrain cell size and shape?

REFERENCES AND SUGGESTIONS FOR FURTHER READING

Fields, B. N. (Ed.) (1985). *Virology*. Raven, New York.
  A comprehensive textbook.

Ford, B. J. (1985). *Single Lens: The Story of the Simple Microscope*. Harper & Row, New York.
  A survey of the construction and use of simple microscopes from the late sixteenth to the mid-nineteenth century. Interesting notes on the microscope's first great master, Antony van Leeuwenhoek.

Holtzman, E. (1989). *Lysosomes*. Plenum, New York.
  A useful review of the cytoplasmic organelles that were once viewed as little "intracellular stomachs."

Holtzman, E., and Novikoff, A. B. (1984). *Cells and Organelles*. 3rd ed. Holt Saunders/ Holt, Rinehart and Winston, New York.
  The latest edition of a respected text.

Kleinsmith, K. J., and Kish, V. M. (1988). *Principles of Cell Biology*. Harper & Row, New York.

A clearly written, up-to-date account of modern cell biology.

Porter, K. R. (1984). The cytomatrix: a short history of its study. *J. Cell Biology 99* (No. 1, Pt. 2), 3s–12s.

A brief historical review by an early worker in this field.

Prescott, D. M. (1988). *Cells: Principles of Molecular Structure and Function.* Jones and Bartlett, Boston.

A short textbook of cell biology emphasizing the molecular basis of cell structure and function.

Prusiner, S. B. (1984). Prions. *Scientific American 251,* October, 50–58.

A review of an intriguing subject by the person who discovered prions in 1982.

Rothman, J. E. (1985). The compartmental organization of the Golgi apparatus. *Scientific American 253,* October 74–89.

A clear summary.

Shively, J. M., et al. (1988). Functional inclusions in prokaryotic cells. *International Review of Cytology 113,* 35–100.

An up-to-date discussion.

Swanson, C. P., and Webster, P. L. (1985). *The Cell.* Prentice-Hall, Englewood Cliffs, N.J.

A useful textbook that covers the field well.

Weiss, L. (Ed.) (1983). *Histology: Cell and Tissue Biology.* Elsevier, New York.

One of the truly outstanding books on histology.

Wolfe, S. L. (1981). *Biology of the Cell.* 2nd ed. Wadsworth, Belmont, Calif.

An excellent text covering much more than cell structure.

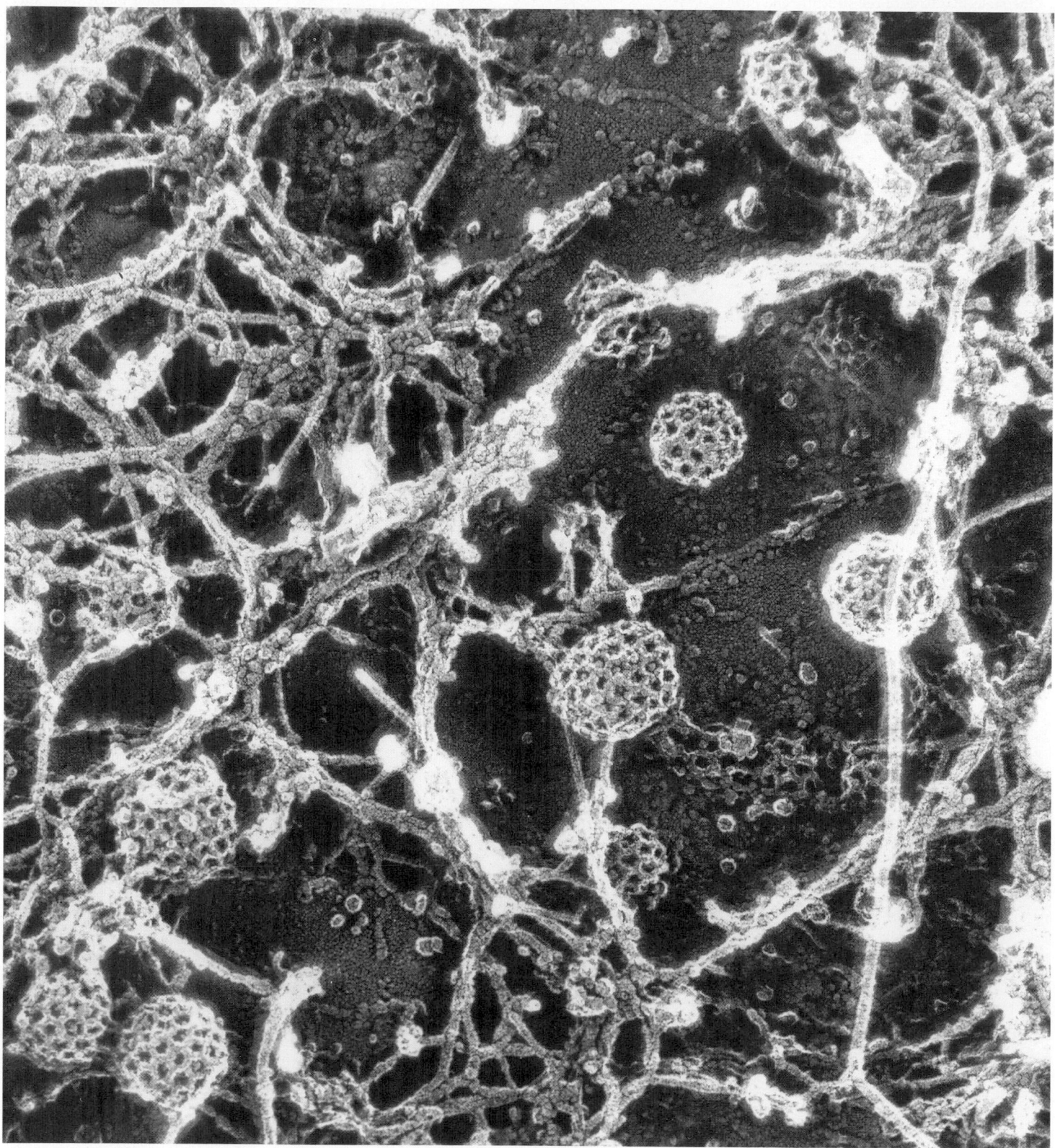

MEMBRANES AND CELL WALLS

Physical objects, whether living or inanimate, usually possess a specialized delimiting structure at the boundary where the object ends and the rest of the world begins. It may take the form of skin, membrane, wall, siding, veneer, sheath, whatever—but its basic functions are almost always the following:

- Protection
- Maintenance of physical integrity of the entity within the boundary
- Control of the composition of the entity

In eukaryotic cells, this enveloping structure is the plasma membrane. The eukaryotic cell also contains many membrane-limited compartments—nucleus, mitochondria, and so on (those who have counted them say there are 20 or more)—within its interior spaces. It is accurate to say that cell organization is based on a series of membrane systems (see Figure 5-14).

Knowledge of biological membranes emerged gradually in the 1940s and 1950s as biochemists slowly brought to light the properties of the membranes of mitochondria and chloroplasts. After investigators in the 1960s elucidated the basic architecture of membranes and its molecular basis, interest in membranes increased explosively.

This chapter summarizes present understanding of membrane structure and demonstrates how elegantly this design implements membrane functions. We will see that these arrangements are indispensable to life.

PRINCIPLES OF MOLECULAR MOVEMENT

Before we can usefully inquire how biological membranes function, we must consider general aspects of how molecules and fluids pass through membranes.

MOLECULAR MOVEMENT IN THE CYTOPLASM

Table 6-1 summarizes in gross terms the chemical makeup of tissues from two widely differing organisms, a sea urchin and a human infant. The predominance of water is the most obvious feature. Proteins are the next largest constituent. Then come lipids and carbohydrates, and a host of small molecules, including minerals.

Even though most cells are composed largely of water, it does not follow that cells can be considered as simple watery solutions of molecules. Rather it appears that the viscosity of cytoplasm varies from cell to cell, and even the most fluid portions of cytoplasm are more viscous than water—or most watery solutions. Moreover, viscosity varies greatly in different parts of the same cell.

The essence of a true solution is the random movement and uniform distribution of its constituent molecules. Yet a measure of order—by definition, nonrandomness—exists in cytoplasm. This depends to a large extent on the properties of large protein molecules. For one thing, these molecules may be individually globular in

Membranes and cell walls
This scanning electron micrograph shows a network of actin microfilaments in the cytoplasm of a macrophage. The spherical structures with polygonal networks on their surfaces are coated vesicles that arose when materials entered the cell by the process of endocytosis. The main protein in the coats of these vesicles is clathrin.
(× 30,000)

shape, but their aggregates can assume fibrous shapes that form the stabilizing cytoplasmic microtubules and microfilaments of the cytomatrix (see Chapter 5). This produces a cell interior in which fibers form a semisolid meshwork of small spaces containing watery solutions. Significantly, though large protein molecules can dissolve to form ordinary solutions, they can also form **colloidal systems** in the cell water.

If fine clay particles are mixed with water, the larger ones will settle out; but many of the smaller ones remain suspended indefinitely. Such a stable suspension is called a **colloid.** Particles of a colloid are larger than ordinary soluble molecules, but they are smaller than particles that would normally settle out of suspension under the force of gravity. Many large protein molecules are capable of forming colloidal suspensions.

Consider the behavior of gelatin, a typical large protein molecule. Ordinary gelatin dessert is a colloidal system of gelatin in water that can exist as a **sol** (liquid) or **gel** (solid or semisolid), depending on the temperature. It is a freely flowing liquid when it is dissolved in warm water, and it is a semisolid when it is cool. In general, all colloids can be either sols or gels. They are readily transformed from one state to the other—in so-called **sol–gel transitions**—by changes in temperature, ionic strength, and such other factors as Ca^{2+} concentration.

In sum, much of the cell mass is both solution and colloid.

This situation accounts for many of the cell's properties. The capacity of colloids to form semisolid gels explains in part how protoplasm, which may be 80% water, can maintain any structure whatsoever. Variations in viscosity in different parts of a cell also reflect reversible sol–gel changes. Significantly, different conditions may exist in different parts of the cell because of variations in local conditions. These conditions comprise the **microenvironment.**

RANDOM MOVEMENTS OF PARTICLES

Molecules move about continuously in both colloids and solutions. Let us briefly consider the physical patterns of these movements.

Diffusion

Molecular motion is easily demonstrated. If a bottle of perfume is opened, the fragrance permeates the room in minutes. Molecules move out of the bottle and through the air of the room, eventually reaching the observer's nose. In an ordinary room, convection currents (little breezes) may help to carry the molecules. But even if there were no convection currents, the normal random motions of the molecules of perfume would mix them with the molecules of air. Given enough time, the molecules of air and perfume would be uniformly mixed. Particles move faster in gases than in liquids and faster in liquids than in solids. They also move faster in hot substances than in cold.

Soon after the bottle is opened, perfume molecules would be detected far from the bottle. The difference in concentration of perfume molecules in the bottle and in the air a distance from the bottle is known as a **concentration gradient.** The concentration gradient shows that the concentration of perfume molecules becomes progressively lower until it approaches zero in the farthest part of the room.

If one were to follow the course of any single molecule, its random, seemingly senseless pattern would give no hint of the general behavior of the whole population of molecules. In fact, the motion of each individual molecule is quite random. That means that some actually move back into the bottle as others move out into the room. The net effect, however, is a movement of molecules from regions of high concentration to regions of low concentration. The reason is simple: There are more molecules in the high concentration zone that can wander randomly into the low concentration zone than the other way around.

The general tendency of molecules to spread from regions of higher concentration to those of lower concentration is called **diffusion.** When perfume molecules are evenly distributed throughout the room, a state of **equilibrium** has been reached.

Diffusion also accounts for the spread of molecules of **solute** (a substance dissolved in another substance or solvent) throughout the **solvent** (the medium that dissolves a solute)—for example, when a cube of sugar is dropped into a glass of water. Figure 6-1 shows the three stages of diffusion.

TABLE 6-1
CHEMICAL CONSTITUENTS OF
PROTOPLASM EXPRESSED AS A
PERCENTAGE OF TOTAL BODY MASS

	Sea Urchin	Human Infant
Water	78.3	66.0
Protein	15.2	16.0
Lipids	4.8	12.4
Carbohydrates	1.4	0.6
Ash*	0.3	5.0

* Ash includes all the elements not included in the other categories (iron, potassium, other metals, etc.).

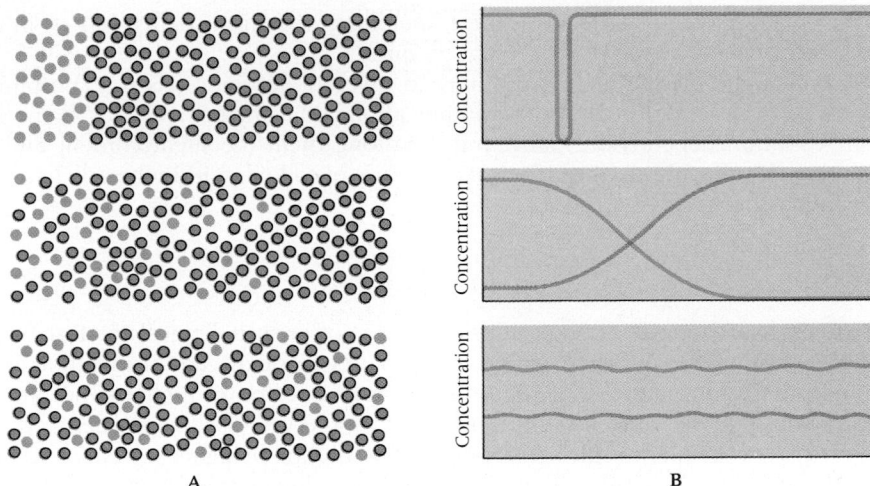

6-1 Diffusion
(**A**) *Diagrams tracing movements of two types of molecules that were initially separated. They mix by random motion until the solution is homogeneous. As colored molecules move to the right, the other molecules move to the left. While individual molecular motion continues after equilibrium is reached, the* net *movement of molecules slows until it becomes zero at equilibrium.* (**B**) *Graph showing the concentration of the two molecules across the container in their initial states, during equilibration, and at equilibrium.*

A

B

- Complete initial separation of solute and solvent molecules.
- Mixing of the two before equilibrium is established, with establishment of a concentration gradient of both molecular species across the vessel.
- Achievement of an equilibrium state at which solute and solvent are uniformly distributed. Even when solute molecules are uniformly distributed throughout the solvent, they continue to move randomly. This movement has no effect on the equilibrium.

In Chapter 8, we will consider the second law of thermodynamics and the concept of *entropy*, which is a formal expression of the degree of disorder in a system— that is, the greater the degree of disorder, the greater the entropy. A solution in which all dissolved molecules are distributed uniformly is less ordered than one in which all the sugar molecules are neatly arranged in one place and all the water molecules in another. We say that the equilibrated solution has a higher entropy level than existed before the solute molecules were arranged so randomly. As we will see, the second law holds that entropy (disorder) tends to increase spontaneously and energy must be invested to lower the entropy (increase the order) of a system. The point is this: Diffusion occurs spontaneously because an orderly array of solute molecules has a lower entropy than a dispersed array.

Molecules diffusing in a solvent do not have to push solvent molecules apart. There is plenty of room between the solvent molecules. On the other hand, molecules in a dense colloid may be so close together that new molecules entering by diffusion may have to force the colloidal particles apart. If a piece of dry gelatin is placed in water, it swells because water diffuses into the gelatin and separates its molecules. This sort of diffusion is called **imbibition**—that is, "drinking in." It is the familiar process that makes wooden doors stick in rainy weather, and makes bean seeds swell and burst their seed coats just before they sprout. Pressures produced by imbibition may become enormous (Figure 6-2). Imbibition by starch granules can develop pressures up to 15 tons per square inch. Ships loaded with dry rice have been split apart when water entering their holds was imbibed by the rice.

Semipermeability

Because every cell is surrounded by a membrane, the passage of substances through membranes is of great importance in a cell's life. Everything that enters or leaves a cell must pass through its outer membrane.

If a membrane, whether natural or artificial, offers no impediment to the movement of a substance through it, the membrane is said to be **permeable** (Latin, *per*, "through," and *meare*, "to pass") to that substance. If a membrane is permeable to some substances and less permeable or impermeable to others, it is considered semipermeable, or, in other terms, differentially or selectively permeable.

All cell membranes are semipermeable. They are relatively permeable to water, impermeable to colloids, and more or less permeable to various solutes. Hence the membrane can keep the colloidal components of cytoplasm from leaking out, can permit water to move freely into and out of the cell, and can act with discrimination

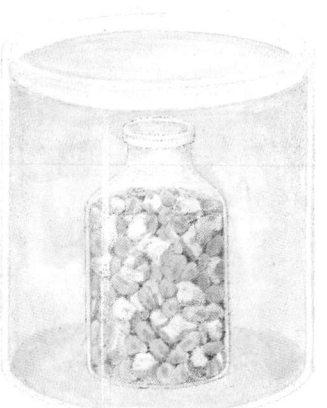

A

B

6-2 Imbibition
(**A**) *Kernels of dried field corn were placed in a bottle, which was immersed in water overnight.* (**B**) *By morning, the bottle was broken. Imbibition had caused the corn to swell, generating pressures strong enough to break the glass.*

in selectively passing particular solutes, including certain nutrients that go in and the secretions and waste products that go out.

Interestingly, membrane permeability to different substances is not constant. A substance may pass through the membrane at one time and be held back at another. Such variation depends on many things, among them the sugar content and pH of the surrounding fluids, the electrical properties of colloids within the cell, temperature, and other factors.

Osmosis

Osmosis is a special aspect of diffusion that is of great biological interest: It is a consequence of the diffusion of water molecules through certain membranes. In ordinary diffusion, particles of solute move from a region of higher concentration to a region of lower concentration. In osmosis, molecules of solvent (that is, water) move across a semipermeable membrane from a solution of higher solvent concentration on one side of the membrane to a solution of lower solvent concentration on the other side. We can demonstrate osmosis in the following way. If we take a bag made of a membrane permeable to water but not to sugar (such as a pig's bladder or a frog's skin), fill it with water, close it, and immerse it in a beaker of water, nothing noticeable happens. Equilibrium is being maintained because as many water molecules move out as move in. However, the membrane is impermeable to sugar, and if we put a sugar solution in the bag and immerse it in the water, the bag swells (Figure 6-3).[1] Water has diffused from the area of lower (zero) solute concentration on the outside of the membrane (bag) to the area of higher solute concentration inside.

A more quantitative experiment can be done with a simple device called an *osmometer*, illustrated in Figure 6-4. A tube capped with a semipermeable membrane (permeable to water and impermeable to sugar) is immersed in a beaker of water. The sugar molecules are retained in the tube, but water molecules move through the membrane from the beaker, where their concentration is higher, into the sugar solution, where their concentration is lower. The level of sugar solution rises in the tube. The simple explanation is found in what we know of diffusion. Since, as illustrated in Figure 6-3, water molecules are less concentrated in sugar solution than in pure water, more of them bump into the membrane on the water side than on the sugar side.

Osmotic pressure causes movement of water across cell membranes. The diffusion of water across the membrane into the tube generates **osmotic pressure,** which is defined as the pressure that must be exerted on a solution to stop the movement of water when solution and water are separated by a semipermeable membrane. If, in the set-up shown in Figure 6-4, the rise of sugar solution in the tube is opposed by a piston, water will continue to enter the sugar solution only until the pressure within the solution is high enough to cause water molecules to pass back and forth through the membrane in equal numbers. The amount of force on the piston needed to maintain equilibrium is a measure of the osmotic pressure. It is proportional to the number of solute particles per unit volume of solution. In practice, less cumbersome methods are used to measure osmotic pressure.

Osmotic pressure causes movement of water across cell membranes. The amount of osmotic pressure that can develop in a solution separated from water by a semipermeable membrane depends on the solute concentration, temperature, and other factors. Under standard conditions, this pressure has a characteristic value for each soluble substance. This is the **osmotic value** of the substance. If solutions A and B exert the same osmotic pressure, then they are **isotonic** relative to one another. If solution A has a higher solute concentration and exerts a higher osmotic pressure than solution B, then A is **hypertonic** compared to B and B is **hypotonic** compared to A. Figure 6-5 shows how these relationships influence conditions in cells. Note that a cell that is hypotonic with respect to the environment will lose water by osmosis and shrink; one that is hypertonic will take in water and swell.

In Chapter 8, we will discuss the critical difference between potential energy and kinetic energy. Briefly, potential energy is the capacity to do work that an object or particle has by virtue of its position or structural configuration. Kinetic

[1] **Dialysis** is defined as the diffusion of a solute through a semipermeable membrane under the conditions shown in Figure 6-3C.

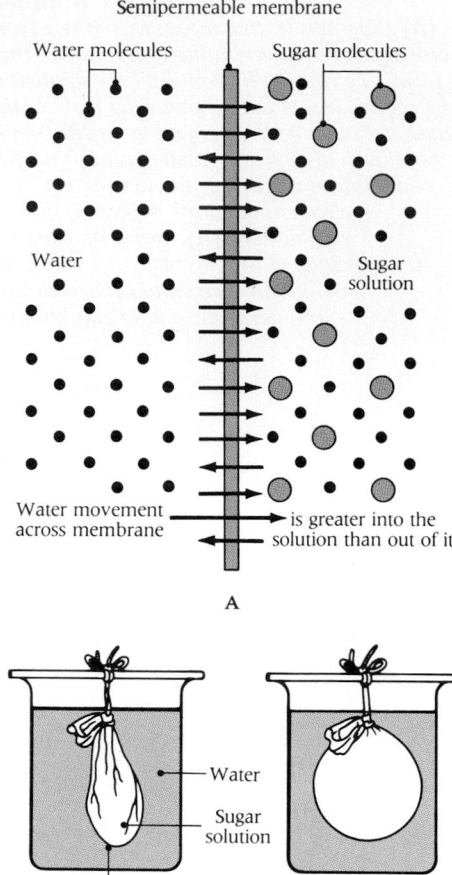

A

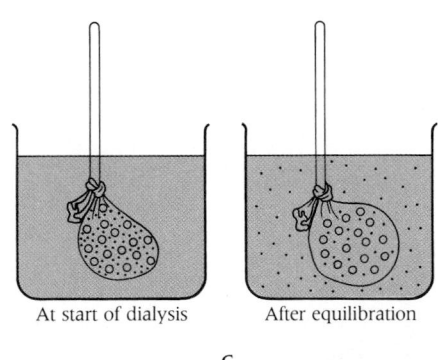

B

C

6-3 Semipermeability and osmosis
(A) Sugar molecules cannot cross this particular membrane because they are too big. Water molecules are less concentrated on the sugar-solution side of the membrane than on the sugar-free side. Therefore, the possibility that they will hit pores in the membrane, and thus cross it, is less on the sugar-solution side. As a result, more water enters the solution than leaves it. **(B)** The net movement of water into a sugar solution enclosed within a membrane container creates a pressure (osmotic pressure) that causes the container to swell. **(C)** Use of a semipermeable membrane to separate large and small molecules in the procedure called dialysis. The membrane used here can pass small molecules but not large ones.

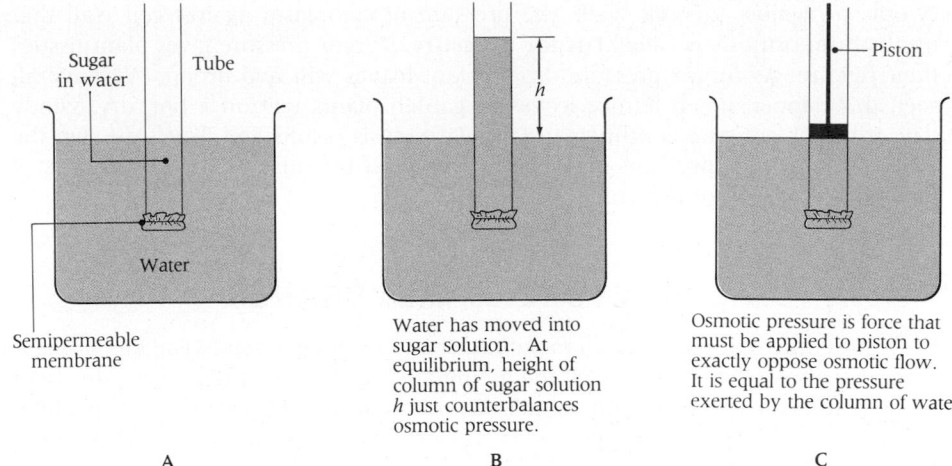

6-4 The principle of osmometry
A solution of sugar in water is placed in a tube covered with a semipermeable membrane and immersed in a beaker of water. (A) Initial state. (B) Final state.(C) A piston applies enough pressure to the water column to prevent the water flow. This pressure is equivalent to the osmotic pressure.

Sugar in water

Tube

Piston

Water

Semipermeable membrane

Water has moved into sugar solution. At equilibrium, height of column of sugar solution *h* just counterbalances osmotic pressure.

Osmotic pressure is force that must be applied to piston to exactly oppose osmotic flow. It is equal to the pressure exerted by the column of water.

A   B   C

energy is the energy an object has by virtue of its motion. The difference between potential and kinetic energy is of great biological importance. Just as water molecules at the top of Niagara Falls have far more potential energy (or **water potential,** as it is termed) than do those at the bottom because of their position, one could say that water molecules under pressure have a high water potential. Water always tends to move from an area of higher water potential to an area of lower water potential. Solute molecules always lower the water potential of the solution. This concept permits useful insights into osmosis. Clearly, a hypotonic solution has a higher water potential than a hypertonic solution. In osmosis, the movement of water molecules across semipermeable membranes continues until water potential is equal on the two sides of the membrane.

It would be unusual indeed to find a situation in the real world in which a simple solution was located on one side of a membrane and pure water was on the other side. Solutions in biological settings are always complex—that is, they contain several different solutes on both sides of the membrane. Hence, each solution has its own total osmotic value, which is the sum of all the osmotic values of substances in that solution.

Since cell membranes are semipermeable, water tends to move toward hypertonic environments. If the cytoplasm is either hypotonic or hypertonic with respect to the environment, this can pose a problem for the organism. Sometimes these effects are startling. If a cell of almost any water plant—a leaf of the aquatic plant *Elodea* is a good example—is soaked in a 15% sugar solution, its cell contents soon separate from the wall and form a mass in the center of the cell (Figure 6-6). This occurs because the sugar solution that has passed freely through the cell wall now encounters a selective semipermeable cell membrane. The large vacuole in the center of the cell originally contains a dilute solution with much lower osmotic pressure than that of the sugar solution on the other side of the membrane. The vacuole thus loses water and becomes smaller. The space between the cell membrane and the cell wall enlarges. The cell membrane and the protoplasm within it are eventually crowded into a ball at the center of the cell. This phenomenon is called **plasmolysis.** Plasmolyzed cells die unless the cell is transferred quickly enough from sugar solution to water. When that is done, it recovers.

Since the volume of such plant cells is kept constant by the rigid cell wall, the osmotic pressure exerted on the cytoplasm by the central vacuole ordinarily forces

6-5 Osmotic relationships of a cell
In an isotonic medium, water gain and water loss across the plasma membrane are balanced and the cell neither shrinks nor swells. In a hypertonic medium, there is a net loss of water from the cell. Therefore it shrinks. In a hypotonic medium, the cell gains water and swells.

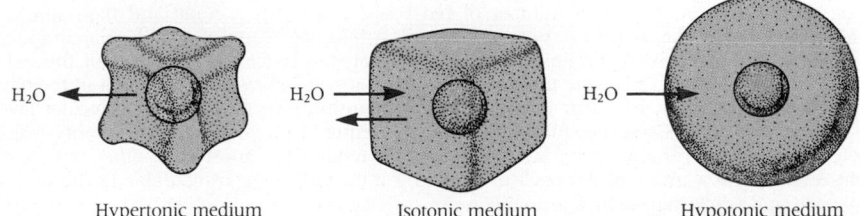

H_2O

H_2O

H_2O

Hypertonic medium   Isotonic medium   Hypotonic medium

cytoplasm against the cell wall. The pressure of cytoplasm against cell wall that results from osmosis is called **turgor pressure.** Turgor pressure gives plant tissues their rigidity. As turgor pressure drops, plant leaves wilt and droop. We have all seen this happen when lettuce leaves or garden plants wilt on a hot, dry, windy day. If turgor pressure continues to fall, plasmolysis results and leaves die. On the other hand, turgor pressure may be great enough to cause plant cells to burst if the cell wall were not present.

STRUCTURE OF MEMBRANES

In electron micrographs, the cell membrane has a distinctive sandwichlike, or bilayered, structure. In an average cell the membrane's outer and inner layers are about 2.5 nm thick, with a 3.5 nm middle layer; however, cell membranes do differ somewhat in different cell types. J. D. Robertson termed this whole aggregate a *unit membrane* to convey the idea that all cellular membranes are probably constructed from such units. A common error is to confuse the "sandwichlike" structure seen in electron micrographs of cell membranes and the bilayer structure to be discussed in the next few pages. The unit membrane concept proposed in 1959 by Robertson accounts for the two dark lines separated by a lighter interzone in electron micrographs of membranes. He suggested that the dark lines represent protein layers on the inner and outer membranes of a membrane model, to be discussed below. The less dense interzone, in this hypothesis, is formed by the hydrocarbon chains of the lipid bilayer.

The study of membrane structure has had a rich history. The year 1985 marked the sixtieth anniversary of the 1925 paper by E. Gorter and F. Grendel (respectively, a pediatrician and a chemist at the University of Leiden), which first proposed that biological membranes consist of a double layer of lipid molecules.[2] Since then, notions of membrane structure have been embodied in two famous models (a model in this context reflects a hypothesis that seeks to account for all the available data).

- The **Davson-Danielli model,** propounded in 1934 by Hugh Davson of University College, London, and James F. Danielli, then at University College, later at the State University of New York at Buffalo.
- The **fluid-mosaic model,** devised in 1972 by S. J. Singer and G. L. Nicolson, of the University of California, San Diego.

The Davson-Danielli hypothesis was correct in picturing the way phospholipid molecules are arranged in the bilayer and in adding proteins to the model, but it erred in assuming that membrane proteins form a uniform "coating" on both sides of the lipid bilayer (Figure 6-7). The fluid-mosaic model is a more sophisticated, interesting, and, incidentally, accurate conception that is the basis for most experimental work today.

MEMBRANE LIPIDS

The outer and inner layers of a cell membrane typically include more than 100 different lipid components, but despite their enormous diversity, membrane lipids occur in a limited number of categories:

- A matrix of cholesterol and phospholipid molecules (see Chapter 4).

---

[2] The old study of Gorter and Grendel is a splendid example of the way shrewd and confident investigators can make major discoveries with simple tools. What Gorter and Grendel did was to extract the lipids from a known number of red blood cells with acetone and then spread the extracted lipid on a water surface, where it formed a film consisting of a molecular monolayer. They found that the area of the lipid monolayer was always twice the total area of the red cell surface—even when red cells were tested from different species. They concluded that in the natural surface membrane, lipid is present in just sufficient quantity to form a double molecular layer—that is, a bilayer. It was a successful experiment that required no elaborate instruments—despite two technical errors turned up by later workers that tended to cancel each other out. A more recent reviewer, in a mood of enthusiasm, termed it a breakthrough comparable to the discovery by Watson and Crick that DNA is a double helix.

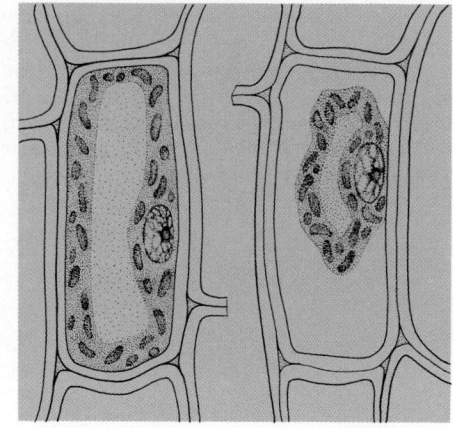

6-6 Water movements in a leaf cell of the water plant Elodea
Left: *Normal state with cell filling space within cell walls.* Right: *Concentration of solutes in the extracellular environment has been increased, making it hypertonic with respect to the cell interior. Some of the water molecules have diffused out of the cell, causing it to shrivel. This is called plasmolysis.*

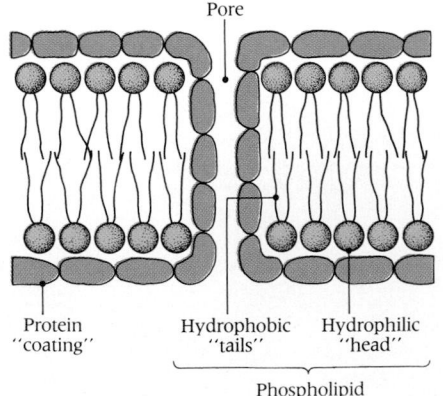

Pore

Protein "coating"   Hydrophobic "tails"   Hydrophilic "head"

Phospholipid

6-7 The Davson-Danielli model of the cell membrane
Two uniform layers of protein form a sandwich with the lipid bilayer in the middle. The entire structure is penetrated at various points by protein-lined pores. Note the orientation of the phospholipid molecules, with their nonpolar hydrophobic "tails" projecting into the interior and their polar hydrophilic "heads" near the surfaces.

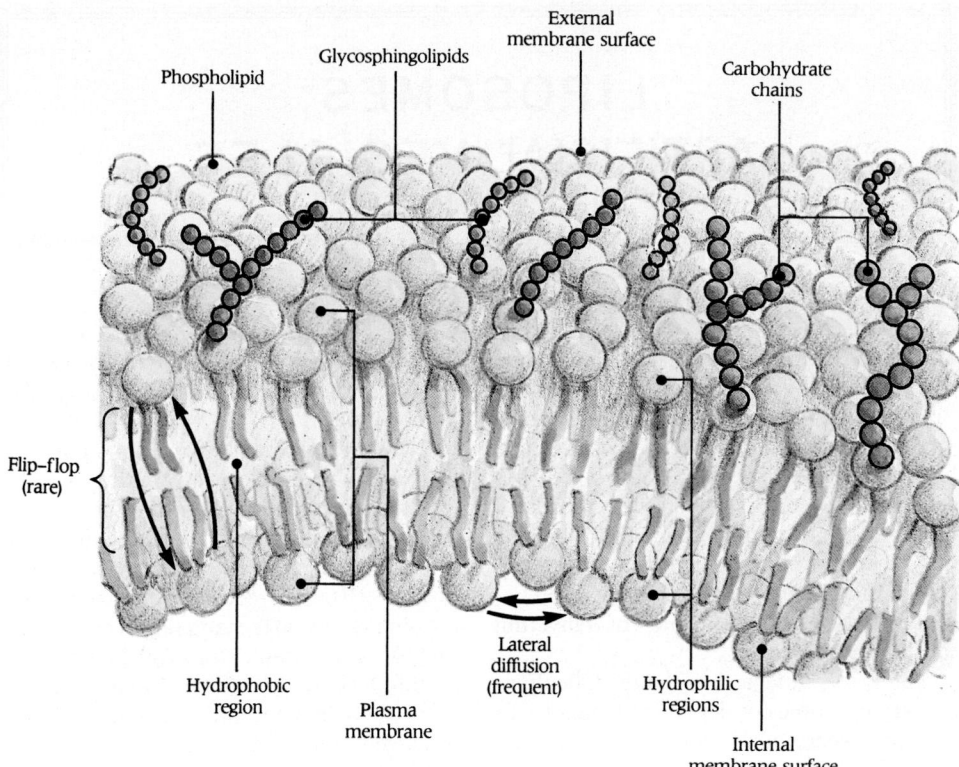

6-8 Lipid bilayer of cell membrane and associated glycolipids

The phospholipids in each leaf of the bilayer confront those of the other leaf tail to tail. Thus only the hydrophilic heads are exposed to the aqueous medium on the two membrane surfaces. This is the minimum-energy configuration for a suspension of lipids in water. Each monolayer is a two-dimensional fluid: lipid molecules can diffuse laterally as in a liquid film, but can rarely execute a ''flip-flop'' transition from one layer to the other. Lateral motion of two representative phospholipids is indicated by arrows.

Phospholipid

Glycosphingolipids

External membrane surface

Carbohydrate chains

Flip–flop (rare)

Hydrophobic region

Plasma membrane

Lateral diffusion (frequent)

Hydrophilic regions

Internal membrane surface

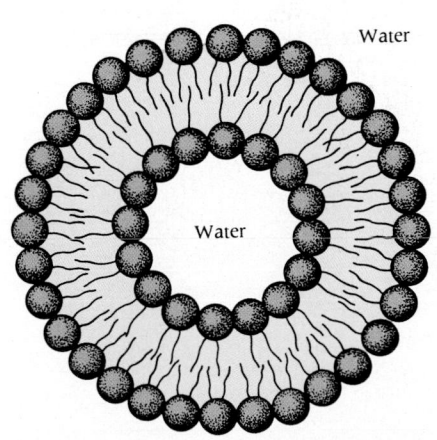

Water

Water

6-9 A liposome

Liposomes form spontaneously when phospholipids are mixed with water. Small membrane-lined vesicles in cells have a similar hollow spherical structure. They are stable and require little energy to maintain them. Compare with Figure 4–10.

- Glycosphingolipids, which have an outer end that is a branching structure of simple sugars (Figure 6-8) and a tail end that anchors them to the membrane surface. The root *-sphingo* identifies them as a special class of glycolipids that includes a structural component, sphingosine.
- Other glycolipids, which are less well understood. They are also confined to the outer monolayer.

Phospholipids are admirably suited to serve as membrane components because, we recall, they have two ends—a hydrophobic (''water-hating,'' or ''oily'') ''tail'' and a hydrophilic (''water-loving,'' or ''wettable'') ''head'' (see Figure 4-10). They are each positioned with their tails pointed inward (toward the inside of the membrane) and their heads pointed outward (toward the watery medium outside of the membrane) (see Figure 6-8). Note that the long axis of the molecules in both layers is roughly perpendicular to the plane of the bilayer.

Artificial phospholipid bilayers in water form closed **spherical vesicles**—called **liposomes**—with an outside and an inside (Figure 6-9). (Aspects of these structures are discussed in Box 6-1.) Closed vesicles form under these conditions because any exposed free edge of a bilayer that made contact with water would be in an energetically unfavorable state. Their energetic state is improved when the hydrophobic tails are sequestered on the inside of the bilayer. This tendency of phospholipid bilayers spontaneously to form a closed envelope with considerable mechanical strength is one of the properties that makes them so effective in biological systems.

Two other properties of bilayers have critical importance in the functions of membranes. The first relates to the hydrophobic conditions within the membrane. This makes the membrane nearly impermeable to many polar biomolecules, including amino acids, sugars, all the macromolecules, and a large variety of ions and other compounds that are soluble in water and insoluble in organic solvents. This property makes the bilayer an effective barrier. It would be a total obstacle, incompatible with life, were it not for the various gates and channels that cross the membrane.

The other important property of bilayers stems from the curious fact that they have many properties of a fluid. This means that the hydrophobic tails of the phospholipid molecules can wiggle about in a relatively nonresistant medium. In addition, each lipid monolayer is itself a two-dimensional fluid in which individual lipid molecules are free to ''creep'' sideways as though they were diffusing in a thin film—and they do so millions of times a second at a measurable rate of speed (120 μm per minute) (see Figure 6-8). Indeed, two neighboring phospholipid mole-

Microscopic vesicles (or sacs) called *liposomes* are composed of membranelike lipid layers surrounding a watery interior (see Figure 6-9). However, they are artifacts of the laboratory and not products of living cells. At first, biologists were interested in them largely because they seemed to be useful models of living membranes that might yield insights into how membranes function. More recently, they have attracted new attention because they offer novel means for delivering substances into cells.

For some time it was quite difficult to prepare liposomes of dependable quality. In the early years of liposome research—the 1960s—the best that anyone could do was to produce onionlike structures in which multiple lipid layers surrounded several very small interior compartments. Now it is possible to make vesicles with single bilayered membranes that are often large enough to enclose proteins, nucleic acids, and viruses. Under certain conditions, liposomes can be induced to enter cells. If they are carrying materials within them, these materials will also enter the cell—like the passengers in a Trojan horse.

To date, many hundreds of different substances have been entrapped in liposomes, including enzymes, glycolipids, antibodies, drugs, large DNA molecules, plasmids, and even chromosomes. However, progress has been less rapid in conveying these materials into cells. A few successes are on the record, but most of these are restricted to cells in tissue culture.

Workers in this field argue that liposomes have many potential future applications. For example, it may be possible one day to insert powerful drugs into infected cells or into cancer cells by means of liposomes that bypass normal cells. One recent report claimed impressive success in the targeting of an antiparasitic drug. The result was a 700-fold increase in the concentration of the drug in liver cells —the cells in which the infecting parasite is obliged to live during part of its life cycle.

Liposomes may also permit the introduction of enzymes into the body cells of individuals afflicted with diseases associated with the deletion of certain enzymes (see Chapter 15)—such genetic conditions, for example, as Tay-Sachs disease, in which a particular enzyme of lipid metabolism is missing or defective. So far this approach has been disappointing, as investigators have failed to obtain (and maintain) an intracellular concentration of the enzyme sufficient to correct the deficiency. Intense research continues, nonetheless, and success may yet be achieved.

Finally, if it becomes possible to introduce fragments of DNA containing new genes into cells, this maneuver could conceivably induce a permanent change in the genetic makeup of a recipient cell. In principle, that could offer a promising method of dealing with a genetic enzyme deletion. The successful introduction of DNA into cells via liposomes has been reported, but efficiency is still quite low. Many workers believe, nevertheless, that ways will be found to use liposomes as vehicles in a new and dependable technique of "gene therapy." We shall return to this subject later. Although satisfactory gene therapy is still in the future, we should record that it is already a subject of clamorous controversy, for it raises a prospect, disturbing to many, that human beings will soon be able to mischievously manipulate the heredity of their own and other species.

Addressing the issue of the prospects for useful applications, the reviewer George Poste recently wrote:

> I was highly skeptical (and remain so) about the feasibility of targeting liposomes to tumor cells . . . *in vivo* [that is, in the living organism]. My criticisms are directed only to those proponents of targeting who chose to ignore the substantial problems created by anatomic and physiologic factors *in vivo*. . . . Overselling always carries the danger that real opportunities may be lost because of skepticism elicited by earlier extravagant claims which have failed to come to fruition.

cules in the same monolayer can exchange places. They do so frequently. Only rarely (some claim it is only a few times a year) do phospholipids flip from one monolayer to the other. This fluidity, which is said to account for the dynamic state of lipids in the bilayer, is a fundamental tenet of the fluid-mosaic model of membrane structure. As we shall see, it is advantageous to the cell because it permits alteration of the membrane's properties in response to the cell's specific requirements.

Cholesterol is also an important membrane lipid. This molecule, which has a steroid nucleus, also has hydrophobic and hydrophilic ends. Most animal cells have a certain number of phospholipid molecules for every cholesterol molecule present. In many membranes, this ratio is around 2:1—but it varies widely. Cholesterol is one factor that enhances membrane fluidity. Oddly, there is no cholesterol in the membranes of prokaryotic cells or plant cells.

Of the **glycosphingolipids,** it can be said that these interesting structures are under intense investigation. Their widespread presence on the surfaces of animal cells has implicated them in cell growth and various regulatory functions. They are also part of the blood group systems (see Chapter 22) and apparently play roles in virus infections, cell-to-cell recognition, and cancer cell growth.

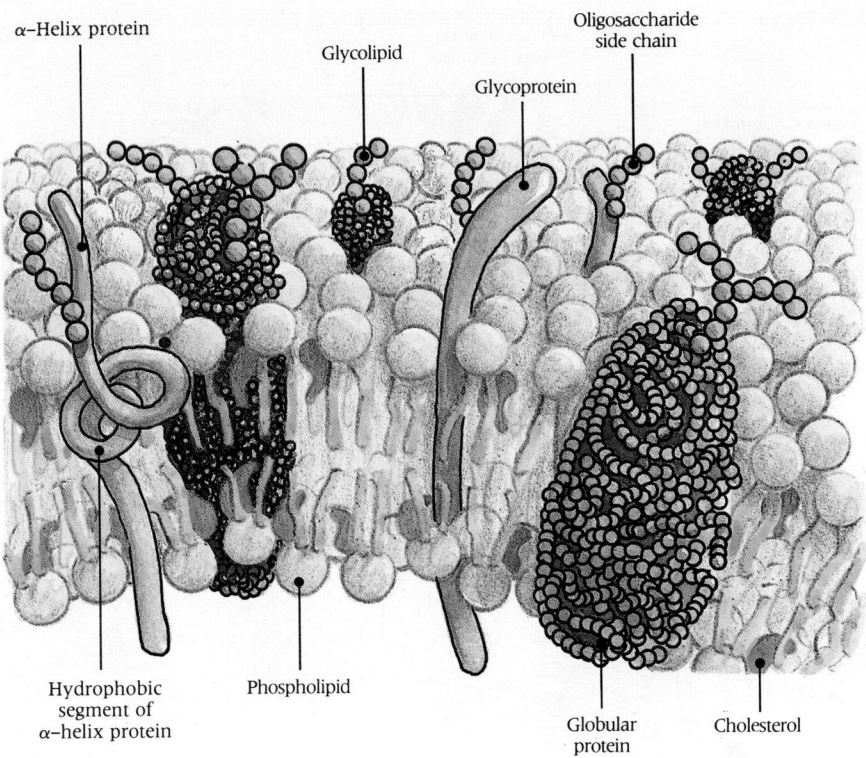

α–Helix protein
Glycolipid
Glycoprotein
Oligosaccharide side chain

Hydrophobic segment of α–helix protein
Phospholipid
Globular protein
Cholesterol

6-10 Plasma membrane

Model shows proteins, glycoproteins, glycolipids, and glycosphingolipids. Carbohydrate chains, here represented as strings of beads, may be bound to either lipids or proteins, but appear on the extracellular membrane surface only. Integral proteins are deeply imbedded in membrane structure. Peripheral proteins are attached to either surface. Cholesterol molecules float among the phospholipid molecules. Because they break up the close packing of the phospholipids, they increase membrane fluidity. At the front of the drawing are two transmembrane proteins: one is α-helical and one is globular.

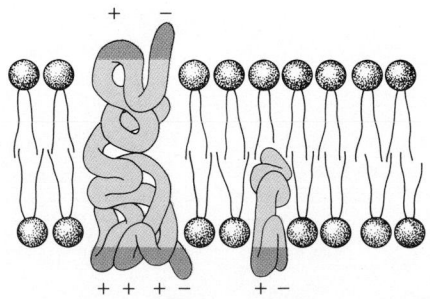

6-11 Orientation of proteins within membranes

Areas of the polypeptide chain that contain hydrophilic amino acids (dark color) tend to project outside the membrane. Areas of the chain where hydrophobic amino acids predominate (light color) tend to be folded within the inner hydrophobic part of the membrane.

MEMBRANE PROTEINS

Lipids form the structural matrix of a membrane, but the proteins embedded in it carry out most of its functions. It is largely the differing placement of the proteins that distinguishes the fluid-mosaic membrane model from its predecessor.

Membrane proteins are classifiable into two categories according to their structure. As shown in Figure 6-10, one type is a chunky globular protein that is embedded in the lipid bilayer like an iceberg in the sea. A second type of protein extends through the membrane as a single polypeptide chain in the form of a α-helix. A particular protein may even weave back and forth across the membrane several times with the intramembrane portion consisting of as many as seven α-helical regions. The ends of both types of proteins that are exposed to the extracellular environment may bear attached chains of monosaccharides and thus such proteins are glycoproteins.

One of the best studied of the α-helix proteins is **glycophorin,** the major glycoprotein of the red blood cell membrane. The carbohydrates of glycoproteins form patterns on cell surfaces, giving the cells a signaling system, rather like ship's flags, that permits each cell type to be recognized by other cells. A group of surface glycoproteins called the **MHC system** (major histocompatibility complex), a complex set of marker molecules, permits cells of one individual to recognize cells of another individual. These marker molecules are significant determinants of the success or failure of organ and tissue transplantation.

Another classification of membrane proteins is based on their positions. **Intrinsic** (or **integral**) **proteins** are deeply embedded in the lipid bilayer so that only a small part of the protein molecule reaches the inner or outer surface, or both surfaces. **Extrinsic** (or **peripheral**) **proteins** are not embedded in the bilayer but rest upon one surface or the other. An extrinsic protein is often bound to an intrinsic protein.

There is an instructive correlation between these positionings and the charges on membrane proteins. Those amino acids with charges (negative or positive) on their R groups or those whose R groups are —OH or —SH are polar or hydrophilic and those without them are nonpolar or hydrophobic. Predictably, hydrophilic amino acid residues are prevalent in the portions of a membrane protein's polypeptide chain that protrude into watery places—the cell's exterior or interior (Figure 6-11)—and hydrophobic residues predominate in portions of the membrane protein that are within the lipid bilayer. As a matter of fact, the locations of these residues in each protein's amino acid sequence has a dominant role in determining how

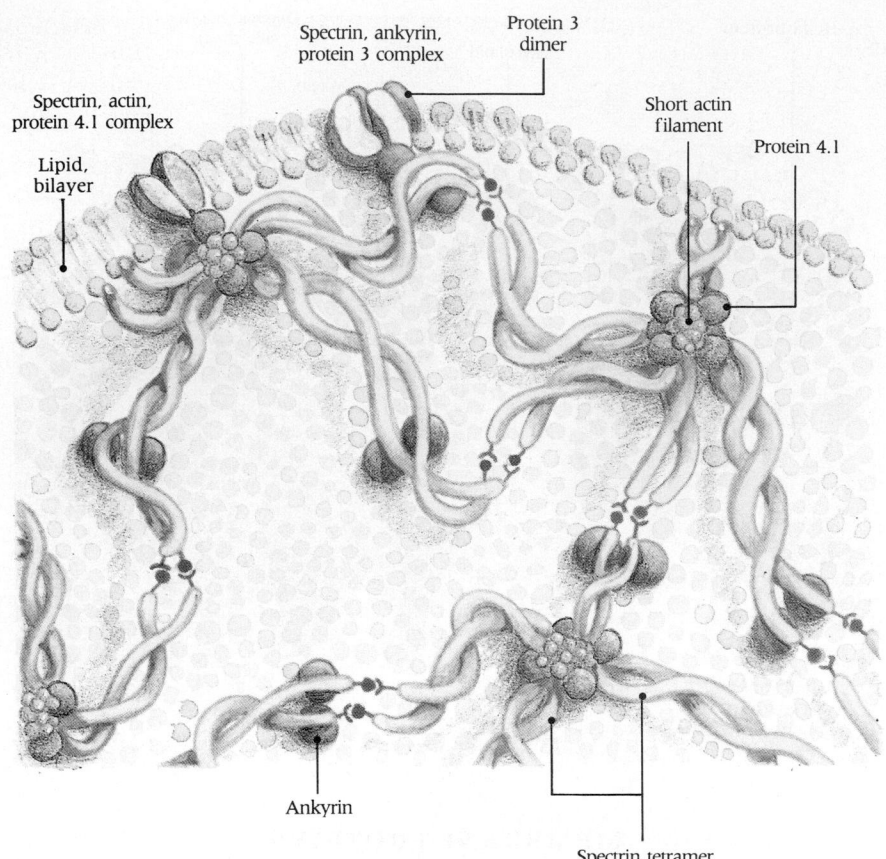

Spectrin, actin,
protein 4.1 complex

Lipid,
bilayer

Spectrin, ankyrin,
protein 3 complex

Protein 3
dimer

Short actin
filament

Protein 4.1

Ankyrin

Spectrin tetramer

6-12 Red blood cell membrane and its protein skeleton
Diagrams show scaffolding of membrane skeleton to underside (cytoplasmic side) of red cell membrane. Transmembrane proteins are anchored on the cytoplasmic side by fibrous proteins that help to maintain the characteristic biconcave shape of the cell. Note that spectrin binds directly to ankyrin, which is anchored to the membrane by a linkage to protein 3. Actin is also present. It binds to spectrin and the actin-spectrin complex binds to a membrane glycoprotein called glycophorin.

and where the protein is anchored in or on the membrane. Investigators have taken to referring to the *hydrophobicity* of the amino acid residues in a protein.

Some cells have a characteristic shape that is maintained by a membrane skeleton made up of various interconnected proteins located just beneath the outer cell membrane and anchored to specific membrane proteins. The best-studied case is the membrane skeleton of the red blood cell. This cell is a curious biconcave disk, the remarkable shape of which is essential to its critical oxygen-carrying mission (Figure 6-12).

Membranes differ widely in their protein content, ranging from 18% in the myelin membranes of nerve cells to 75% in the inner membrane of mitochondria. Certain membrane proteins, we will see, carry out transport functions within the membrane. About half the proteins in the membrane are free to move about. The rest are relatively immobile, probably by virtue of their attachments to the cell's membrane skeleton.

Many methods are now available for the study of membrane proteins. In the technique known as **freeze-fracturing,** one can examine the proteins by freezing and fracturing longitudinally through the lamellae of the lipid bilayer of the membrane. When the fragments are examined by electron microscopy (Figure 6-13), individual protein molecules appear as bumps on the fractured undersurfaces of the lipid bilayers. Observe how clearly these preparations reveal a membrane's individual intrinsic proteins.

PATTERNS OF MEMBRANE TRANSPORT

If internal cell composition is to stay constant, or nearly so, all molecular traffic between cell exterior and cell interior must be regulated. Control of this traffic is the job of the plasma membrane. This task has two major aspects: first, to prevent too much water from entering the cell under the influence of osmotic pressure; second, is to let necessary molecules (nutrients) into the cell and let unwanted molecules (wastes) out.

Many substances can cross the cell membrane in both directions. Three fundamentally different processes move them across: simple diffusion, facilitated diffusion, and active transport (Figure 6-14). We now consider these mechanisms.

6-13 Freeze-fracture preparation of a red blood cell membrane
When a frozen membrane cracks, it tends to split so that the two halves of the lipid bilayer are separated. Thus, it is possible to examine the several faces of a membrane, which are called PF and PS faces (for protoplasmic fracture and surface) and EF and ES (for exoplasmic fracture and surface). The left part of the photo shows EF—the outward face of the inner membrane. The right shows PS—the cell surface. The bumps are individual protein molecules. (× 40,000)

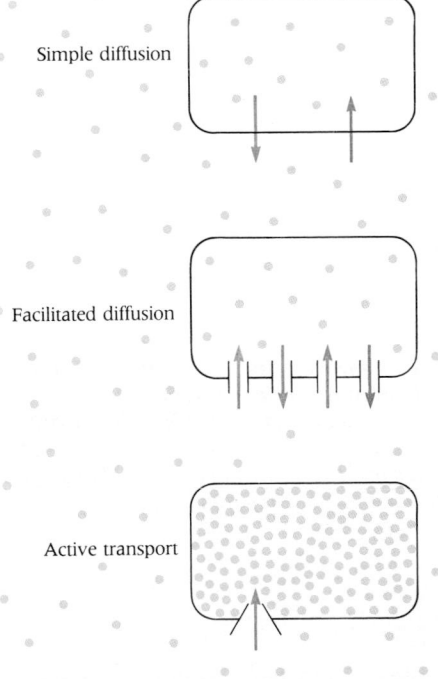

6-14 Three kinds of membrane transport
In simple diffusion and facilitated diffusion, molecules (dots) reach an equilibrium on both sides of the membrane. Active transport is an energy-dependent process, since molecules move against their concentration gradient and accumulate at a higher concentration on one side of the membrane.

SIMPLE DIFFUSION

Simple diffusion occurs when highly *nonpolar (hydrophobic) molecules* that are freely soluble in the membrane's lipid phase (like ether or ethylene) cross directly through the membrane by the random movements of diffusion. In such cases, all the features of diffusion are observed (see Figure 6-1). The final concentration of the substance is the same inside of the cell as outside. This represents the equilibrium state.

Water, carbon dioxide, and other *polar (hydrophilic) molecules* cannot use this entryway. Yet such molecules are not excluded by the membrane's lipid layers even though they are lipid-insoluble. Indeed, they enter cells freely. How is this to be explained? Possibly, water enters through tiny openings lined with membrane proteins bearing hydrophilic groups. These postulated openings are either part of the membrane's architecture or are momentary openings. But these "pores," if they exist, are too small to pass larger polar molecules—in either direction. This is just as well, because otherwise such critical molecules as disaccharides and amino acids would leak out.

FACILITATED DIFFUSION

The device with which membranes pass larger polar molecules—that is, larger than 0.3 nm, the width of a water molecule—is the carrier molecule. These are transport proteins within the membrane that "ferry" passenger molecules across (in either direction), traveling through what is termed a **membrane channel.** They operate in two ways: by promoting facilitated diffusion and by active transport. Both mechanisms enhance the selectivity and speed of membrane transport.

In part, **facilitated diffusion** involves diffusion because it is driven by the concentration of a substance, with molecules moving spontaneously from regions of higher concentration to regions of lower concentration. However, unlike simple diffusion, a carrier protein molecule in the membrane (termed a **transporter**) interacts with a passenger molecule (termed a **ligand**) on one side of the membrane (and they both must be in sufficient supply to have a fair chance of colliding). As a result, the ligand crosses the membrane, moving through a channel created by the three-dimensional shape of the transporter molecule.

Facilitated diffusion takes place at sites in the membrane that are occupied by such specialized protein molecules. The term *membrane channel* is applied to these sites, which should not be visualized as actual openings, pores, or holes through the membrane. A channel is a site where carrier molecules are concentrated. These molecules have a high degree of binding specificity, and thus they resemble, and indeed may be enzymes (Chapter 8). Since enzyme names almost always end in *-ase*, carrier molecules themselves are sometimes called *permeases*.

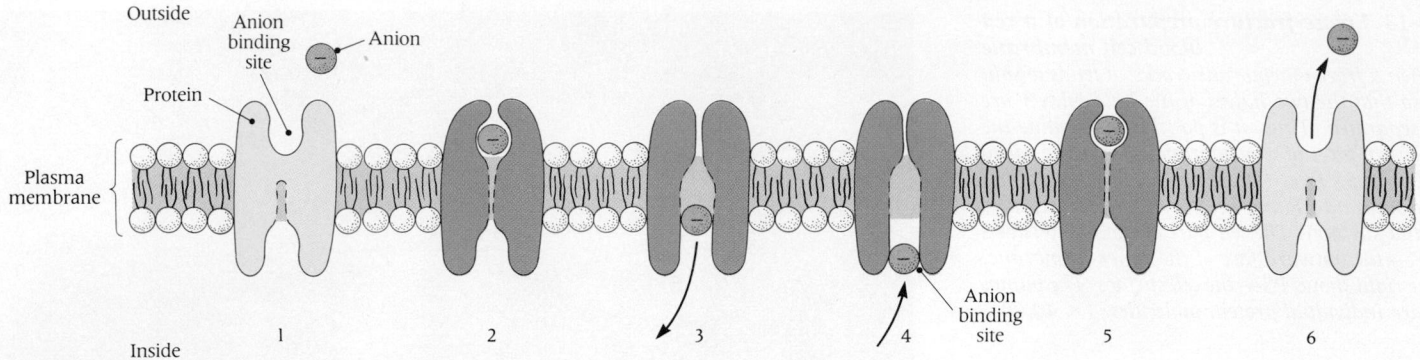

Facilitated diffusion has two distinctive characteristics.

- If the concentration of the passenger molecules is progressively increased, then the rate of transport into the cell will increase—but only up to a point. When all available carrier molecules are in action, the rate of transport cannot be increased further. We say then that the system has been *saturated.* Systems that operate by simple diffusion cannot be saturated.
- Facilitated diffusion, like simple diffusion, cannot move molecules against a concentration gradient—from regions of low concentration to regions of high concentration. That is a task for active transport.

The transport into the cell of various anions, such as chloride ion (Cl^-), provides a good example of facilitated diffusion. As we shall learn in Chapter 27, the movement of anions across the membrane is coupled with the transport of CO_2 by red cells. An *anion transport channel* in the red cell membrane contains a protein that binds an anion and then by changing confirmation moves the anion inward (Figure 6-15). Interestingly, this protein has been identified as *protein 3,* whose participation in the membrane skeleton of red cells is seen in Figure 6-12. It is the predominant integral glycoprotein of the red cell membrane. The model in Figure 6-15 is hypothetical. We still do not understand the details of anion transporter function. Another example of facilitated diffusion is the system that moves glucose from the animal intestine into its lining cells or from blood serum into body cells (Figure 6-16).

ACTIVE TRANSPORT

When a substance is transferred across a cell membrane in an "uphill" direction—that is, from a region of low concentration to one of high concentration—there must be an expenditure of energy. We term this process **active transport.** Active transport is accomplished by a **membrane pump** that depends for its operation on the high-energy molecule called ATP (Chapters 8 through 11). We know this because when ATP production in a cell is experimentally inhibited, there is a slow leak of ions and other molecules across the cell membrane until their levels are equal inside and outside the cell. The cell then dies.

Active transport works in both directions. In moving substances into the cell, it permits cells to perform what may be their most important task: to concentrate the nutrients and other needed molecules from the surrounding fluids, where they may be present in low concentrations. In moving substances out of the cell, active

6-15 Schematic diagram of anion transport system

In this hypothetical model, protein 3 catalyzes a one-for-one exchange of two different anions, probably by shuttling between two different conformations. In one, it binds an anion at the outside face of the membrane (1). This causes a conformational change that generates a tunnel through which the anion can cross the membrane (2). It also produces a new anion-binding site on the inside of the membrane (3). This allows an anion to bind on the outside (4), and reestablishes the original conformation. The second anion passes across the membrane in the opposite direction (5), and recreates the binding site on the outside of the membrane (6).

6-16 Sugar molecules move into the cell with the help of a carrier protein

(A) *Binding of sugar molecule to the carrier protein.* **(B)** *Change in protein structure to move the sugar through the lipid bilayer.* **(C)** *Release of sugar into the cytoplasm.*

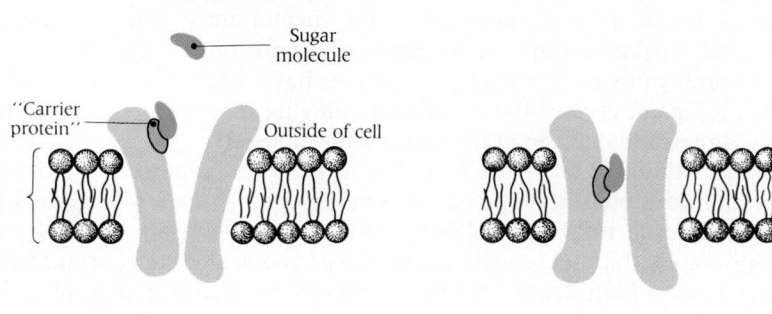
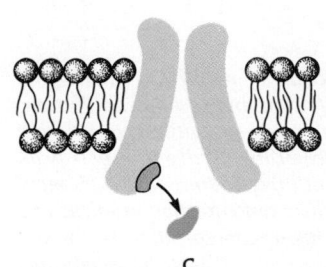

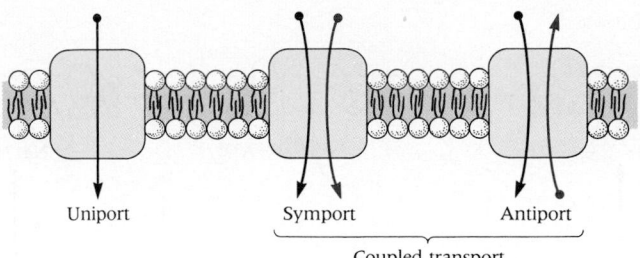

6-17 Schematic diagram of uniport, symport, and antiport membrane transport proteins. *The transported molecule is shown in black; the co-transported ion is shown in color*

Uniport   Symport   Antiport

Coupled transport

transport is essential in ridding cells of waste molecules that for one reason or another are unable to diffuse out by themselves.

Three active transport systems have been well studied.

■ The **sodium-potassium (Na⁺-K⁺) pump,** which transports Na⁺ out of the cell and K⁺ into it and is the best known example of active transport. The sodium-potassium pump has been studied mainly in red blood cells but it is present in most cells.
■ The **calcium pump,** which transports Ca^{2+} out of the cell.
■ The **proton pump,** which transports protons (H⁺) across cell membranes.

Each of these systems hydrolyzes ATP and thus is properly termed an **ATPase,** i.e., an enzyme that splits ATP. Let us consider them briefly.

We describe the sodium-potassium pump because of its great biological importance. As we have seen, most cells contain much more K⁺ than sodium ion (Na⁺) in their cytoplasm. A red blood cell contains 5 times as much K⁺ as Na⁺. Yet its fluid environment contains 40 times as much Na⁺ as K⁺. To a small extent, the membrane of the red blood cell is passively permeable to both types of ions. But if its permeability were uncontrolled, Na⁺ would leak in and K⁺ would leak out. To maintain its composition, therefore, the cell must constantly expel Na⁺ and accumulate K⁺ against an aggregate 40-fold concentration gradient. It has been shown that, in a human red blood cell, three Na⁺ ions pass outward for every two K⁺ ions passing inward, with each of these passages consuming one unit of ATP energy. The consistency of these ratios—three out for every two in—strongly suggests a connection between the two events.

Transport systems of this type, which carry different substances in *opposite* directions, are called **antiports** (Figure 6-17). Because three positively charged ions are expelled for every two positive ions taken up, this transport system creates an electrical charge difference—thus an electric potential—across the plasma membrane. Transport systems that move a molecule and an ion in the *same* direction (see below) are called **symports.** An example of a symport is shown in Figure 6-18. The simpler systems of facilitated diffusion (discussed above), which move a single molecule or ion across a membrane without coupling to the movement of another molecule or ion, are **uniports.**

Several ingenious models have been proposed to explain the workings of the

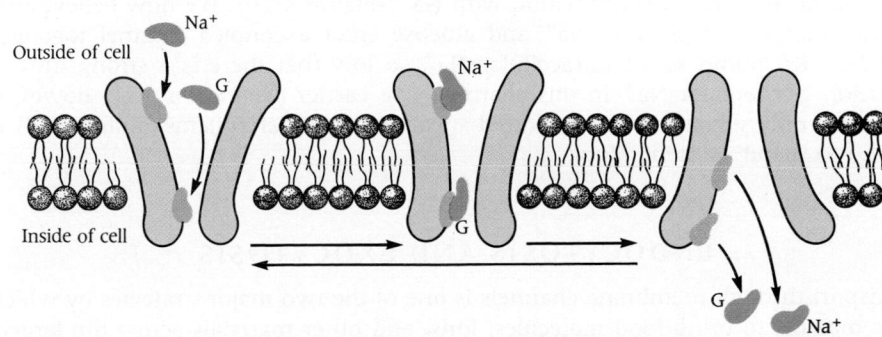

6-18 Facilitated diffusion of two substances using one carrier protein *The sequence left to right illustrates the binding–transport–release cycle involved in the passage of sodium ions and molecule G into the cell.*

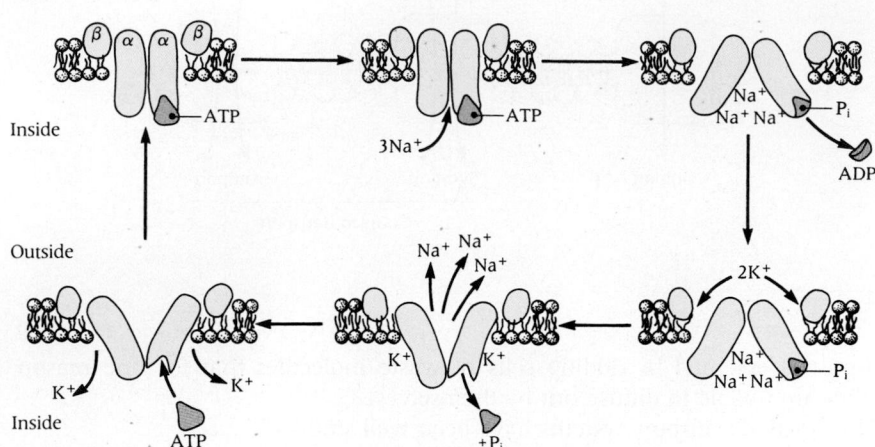

sodium-potassium pump. The one sketched in Figure 6-19 attributes the movement of K^+ and Na^+ ions across a membrane to a specific ATP-hydrolyzing enzyme (termed ATPase) in the membrane whose activity requires the presence of both Na^+ and K^+. Na^+ ions are bound to one of the subunits of the enzyme. ATP transfers a phosphate group to this subunit, thereby giving it an extra negative charge. This alters its conformation, in effect extruding Na^+ through an opened gate and bringing K^+ in on the same molecule, but at a different site. This model is not yet firmly established.

Another important active transport system of the surface membranes of many cells is the **calcium pump.** Many cellular functions depend critically on the cytoplasmic concentration of free calcium ions (Ca^{2+}). There are only two ways to raise their concentration. One is to release Ca^{2+} ions from intracellular storage depots in the endoplasmic reticulum. The other is to move Ca^{2+} ions from the extracellular fluid into the cell through its surface membrane. The second method involves a variety of highly regulated transport systems. Each consists of a cluster of specialized proteins embedded in the lipid bilayer that serves as a channel or pore. The molecular structure of these channels determines how many ions are allowed to enter by a remarkable opening-and-closing process known as the "gating" of the channel. Calcium channels are of special importance in nerve and muscle cells. In both these cell types, moving zones of altered electrical charge (termed **excitation waves**) pass down the cell membrane and operate the calcium channel "gate," thereby regulating the number of Ca^{2+} ions crossing the membrane and entering the cell.

In Chapters 9 and 10, we will encounter a **third active transport** system of great biological importance—the **proton pump.** This couples the transport of protons (H^+ ions) to the production of ATP and is responsible for the ability of a membrane to create a proton gradient across itself. This process, termed *chemiosmosis*, is the key mechanism in the major ATP-generating processes, oxidative phosphorylation and photosynthesis. We will also see that such a mechanism is a crucial part of the systems that acidify stomach contents and urine.

The systems just discussed are the three known active transport systems that are directly coupled to the hydrolysis of ATP. The transport of sugars and amino acids does not depend on energy provided by ATP. In these symport systems, molecular transfer is coupled—that is, the molecules are cotransported, moving through specialized channels in coordination with Na^+ (Figure 6-20). We now believe that in the transport of glucose, Na^+ and glucose enter a coupled channel together. The Na^+-K^+ pump keeps intracellular Na^+ so low that there is a strong inward diffusion gradient for Na^+ in this channel. The carrier protein for Na^+, however, functions only when the cotransported sugar (or, in other systems, amino acid) is bound to it and "carted" along.

ENDOCYTOSIS AND EXOCYTOSIS

Transport through membrane channels is one of the two major strategies by which cells manage to bring food molecules, ions, and other materials across the largely

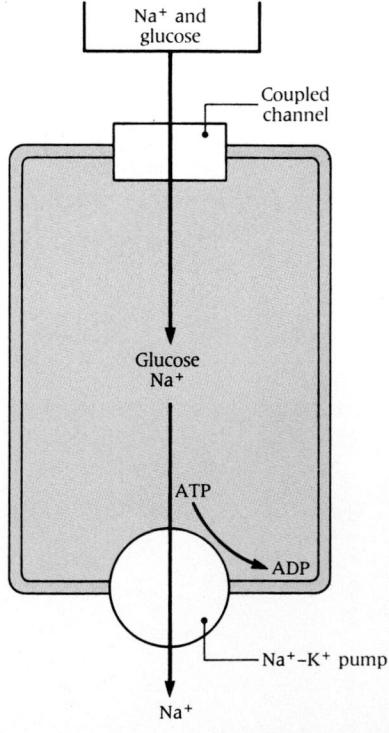

6-20 Active transport of glucose is coupled to Na⁺ transport
Glucose moves into the cell by facilitated diffusion, using the same carrier that transports Na⁺. This process is indirectly dependent on the active transport of Na⁺, using the ATPase pump found on the other side of the cell.

6-21 Three forms of endocytosis
Materials are taken into cells in three ways. Clockwise from top: (1) Phagocytosis; (2) Pinocytosis; (3) Receptor-mediated endocytosis.

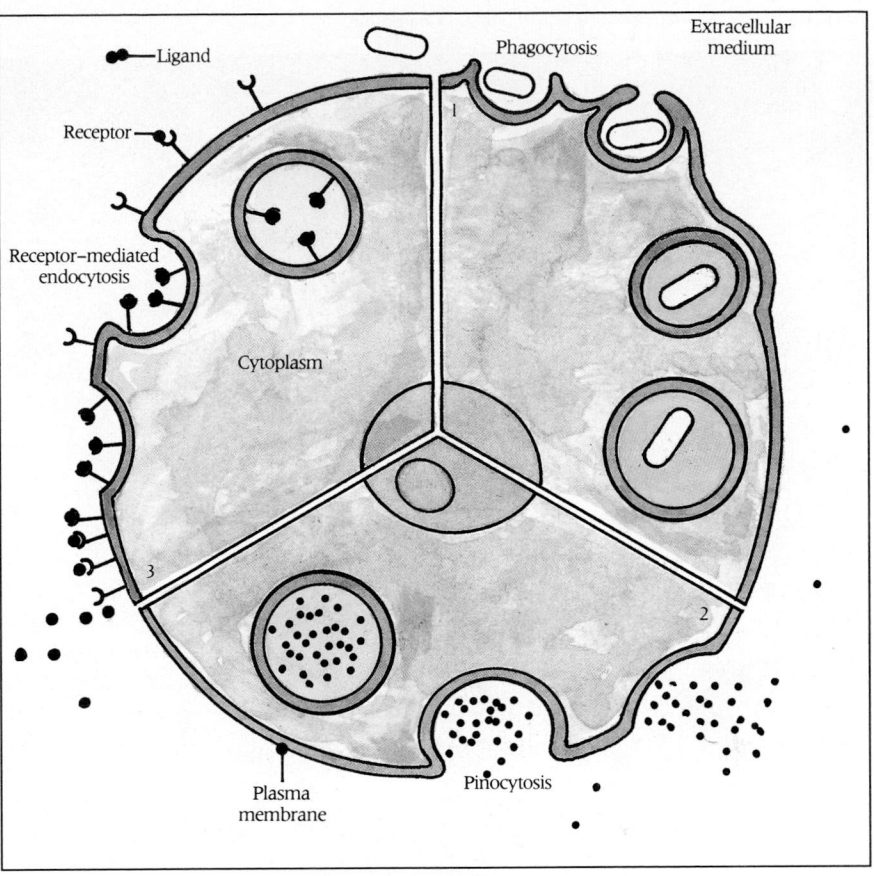

hydrophobic lipid layers of the cell membrane. A second tactic rests on the fact that the cell membrane can be pushed, pulled, and penetrated by various large objects. As a result, bits of membrane can pinch off from the rest of the cell membrane so that large particles coming in can be engulfed and particles going out can be exported. The former process is called *endocytosis;* the latter, *exocytosis.*

Endocytosis permits entry into the cell of molecules or particles that are too large for active or passive transport. It is the major method by which single-celled organisms, like amoebas, and simple multicellular animals acquire nutrients. It is also the mechanism that brings into the cell such important items as certain growth factors and hormones and, alas, viruses and poisons (or toxins). Like active and passive transport, endocytosis depends on the membrane's capacity for discrimination. Unfortunately, the cell surface is not always discriminating enough.

Long before biologists studied the daunting problems of passive and active transport across cell membranes, they had observed cells in the act of "eating." In 1882, Ilya Metchnikoff (1845–1916), a professor of zoology at the University of Odessa, saw wandering white blood cells engulfing bacteria and named the phenomenon **phagocytosis** (from *phagein,* Greek for "to eat," and *cytos,* "cell"). Later workers observed that the membrane of a cell occasionally closes up like a clenched fist and traps a small portion of the surrounding fluid in a vesicle or vacuole. This process was called **pinocytosis** (from *pinein,* "to drink"). It was soon recognized that pinocytosis occurs in nearly all cells—though, interestingly, not in that laboratory workhorse, the mammalian red blood cell. Thus we usually speak of two forms of endocytosis, depending on the nature of the materials taken in (Figure 6-21). When the process brings in a large molecular complex, a solid particle, or a bacterium or other organism, it is called *phagocytosis.* When the cell takes in a droplet of fluid, the process is called *pinocytosis.*

Pinocytosis itself has two forms. One is the nonspecific uptake of extracellular fluid in which a tiny droplet is surrounded by a bit of invaginated (in-pocketed or in-folded) cell membrane and internalized along with whatever ions or solutes it contains. The other, termed **receptor-mediated endocytosis,** is a highly specific mechanism by which the cell selects and ingests a single ingredient from the surrounding "extracellular soup."

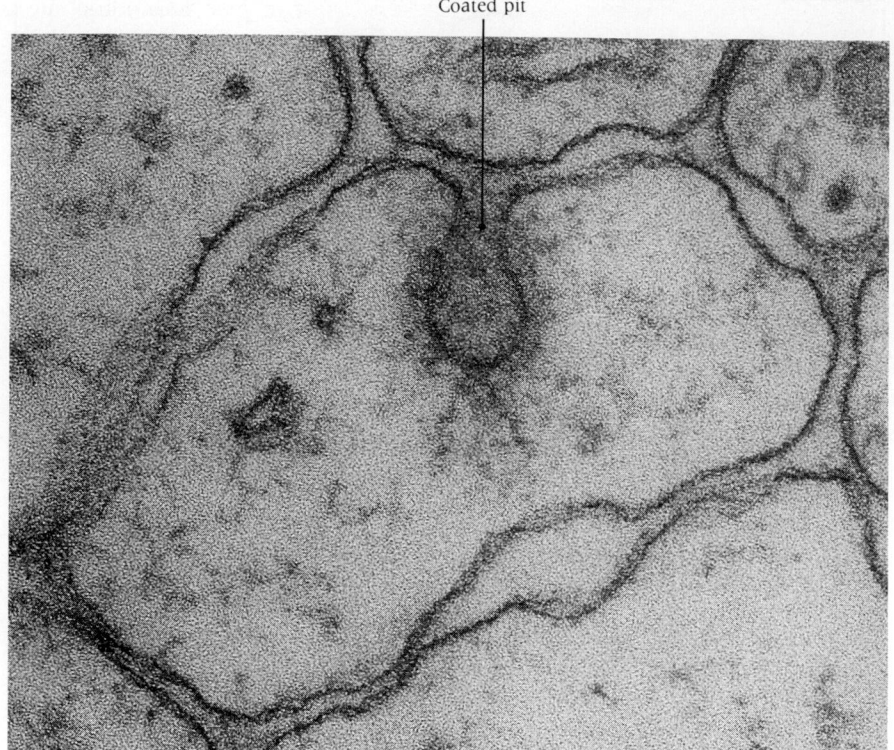

Coated pit

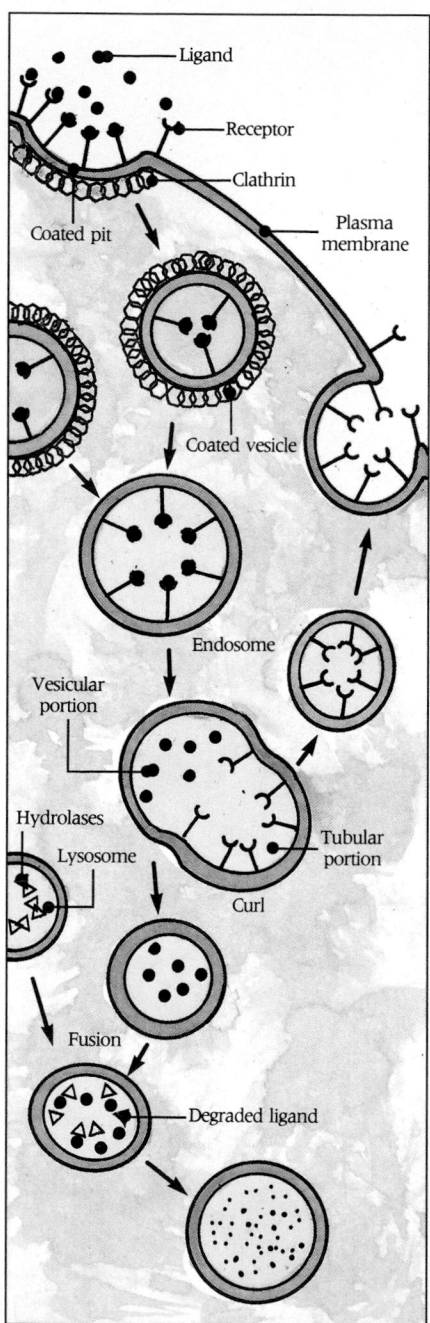

Ligand

Receptor

Clathrin

Coated pit

Plasma membrane

Coated vesicle

Endosome

Vesicular portion

Hydrolases

Lysosome

Tubular portion

Curl

Fusion

Degraded ligand

6-23 General pathway of receptor-mediated endocytosis

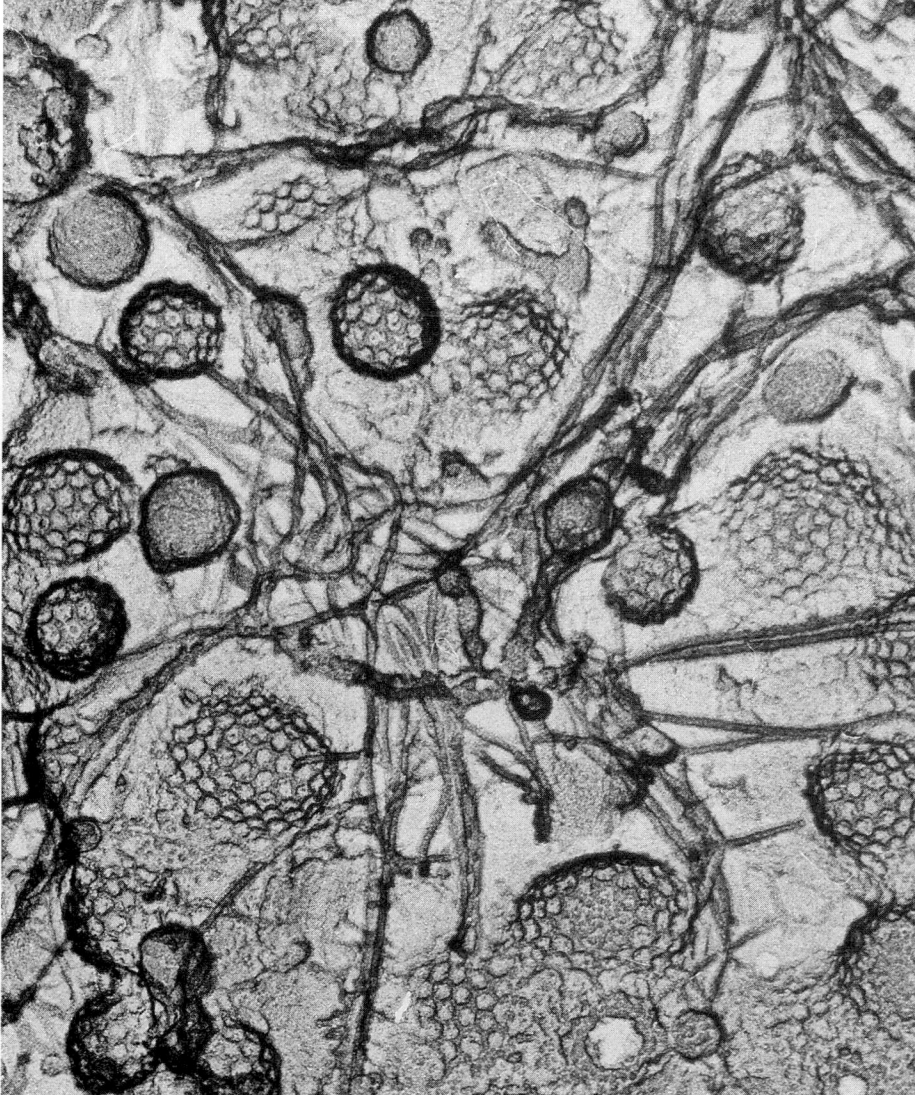

6-24 Structure of a coated vesicle
The cagelike structure of coated vesicles is seen in this view of the cytoplasmic side of the plasma membrane. Elongated filaments are also present throughout the cytoplasm. (× 28,000) Compare with the photograph opening this chapter.

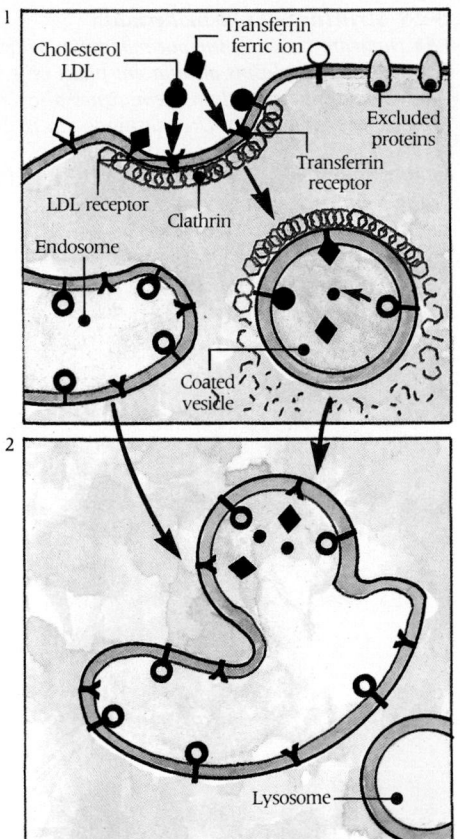

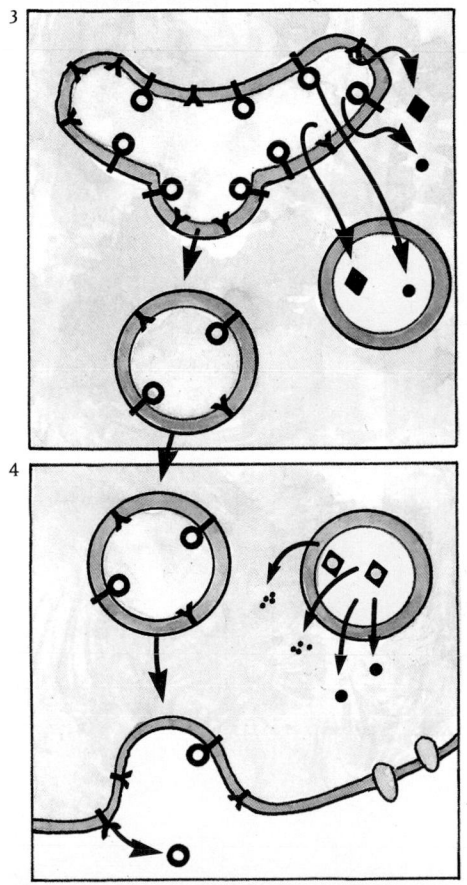

6-25 Uptake of cholesterol and ferric ions into the cell

Panels 1–4 show how cholesterol and ferric ions are brought into the cell and eventually enter the cytoplasm, while the receptors are recycled back to the plasma membrane. Compare with the general model in Figure 6-23.

Receptor-mediated endocytosis begins at a receptor, a specific binding site on the cell surface. There are many kinds of surface receptors. Each is a specialized protein molecule that binds a particular type of material in the surrounding fluid—even if it is present locally in very low concentrations. To a large extent, it is the specificity of the various receptors that accounts for the specificity of receptor-mediated endocytosis.

Extracellular particles or macromolecules destined to enter cells by receptor-mediated endocytosis are first bound to a surface receptor. (Materials that are specifically bound to receptors are referred to as their *ligands*.) The binding of a ligand to a receptor initiates endocytosis. Next, the ligand is brought into the cell by the invagination of the cell membrane (see Figure 6-21). A small saclike membrane forms around the incoming particle and carries it, along with a small portion of the external medium, into the cell interior.

Electron microscopy reveals that, in the early stages of endocytosis, a thick, dark, fuzzy coating appears on the cytoplasmic side of the cell membrane where it has indented inward to form a pit (Figure 6-22). Study of these structures, which are known as **coated pits,** identified the coating as molecules of a family of proteins called **clathrin.** The pit deepens, pinches off, and soon becomes a **coated vesicle** that internalizes both the passenger molecules (ligands) whose attachment to the surface receptors initiated the transport process and a small cluster of the receptors themselves (see Figures 6-21 and 6-23). When viewed by scanning electron microscopy (Figure 6-24), coated pits and vesicles resemble a cagelike structure made of a lattice of hexagons and pentagons. This pattern reflects the physical properties of clathrin polypeptide chains.

In scarcely more than a few minutes, the clathrin coat is disassembled and shed. What forms then, by the fusion of several recently coated vesicles, is an uncoated vesicle called an **endosome.** This is transformed into the awkwardly named structure **compartment for uncoupling of receptor and ligand,** or **CURL** in jargon. The pH within the CURL then decreases, and the more acidic conditions dissociate the receptor from its ligand. The former are restored to the cell surface; the latter fuse with cytoplasmic lysosomes and go from there to other parts of the cell. What has been called the **endocytic cycle** is now complete.

Two well-studied examples nicely illustrate receptor-mediated endocytosis in action (Figure 6-25). In the animal body, cholesterol arrives at the cell surface as part of the very large complex known as **low-density lipoprotein,** or **LDL.** This complex, its surface receptor, and their profound significance were discovered by Michael S. Brown and Joseph L. Goldstein, of the University of Texas, in a notably elegant series of experiments (and so thought the 1985 Nobel Prize committee). Iron comes bound to a large protein called **transferrin** (Figure 6-25).

Since pits and vesicles manage to exclude other proteins, some researchers suspect that clathrin is a molecular sorting device. Reports in 1985 demonstrated clathrin coatings around the *trans*-most cisternae of the Golgi apparatus. Clathrin appears also to coat primary lysosomes and newly formed secretory granules. This suggests a role for clathrin in both biosynthetic and endocytotic protein traffic.

Endocytosis working in reverse is called **exocytosis.** We have seen that many substances are exported by cells in secretory vesicles or special vacuoles (see Figure 5-18). When these vesicles reach the plasma membrane, they fuse with it and expel their contents to the exterior. Substances extruded in this way are digestive enzymes, hormones, or chunky particles of waste matter.

Other important functions of the surface membranes of cells include the structures that facilitate cell-to-cell communication (Chapter 7) and that play a role in development (Chapter 20); the hormone receptors that mediate hormone action (Chapter 28); the properties that lead to nerve impulse transmission (Chapter 29); and the surface structures that permit self-recognition, by which an organism can distinguish its own cells from those of another (Chapter 23).

CELL WALLS

PLANTS

As we saw in Chapter 5, plant cells do have a plasma membrane, but there are other distinctive features of their surface architecture that include a prominent cell

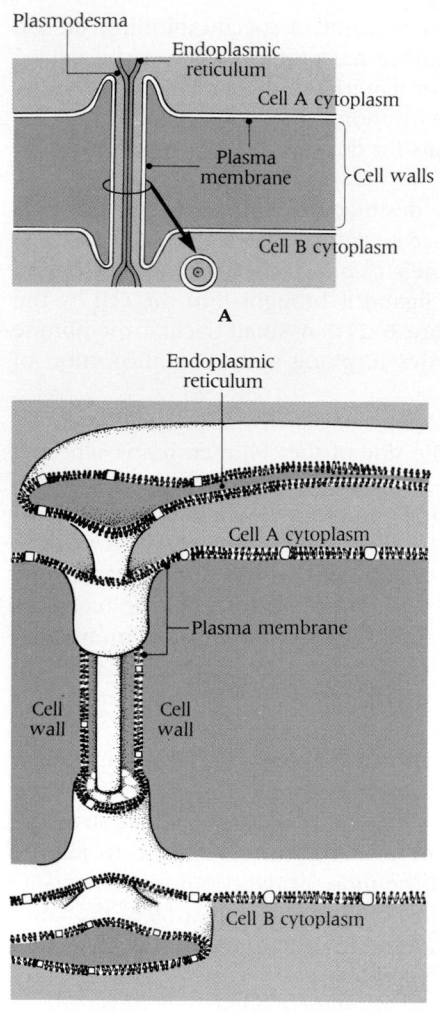

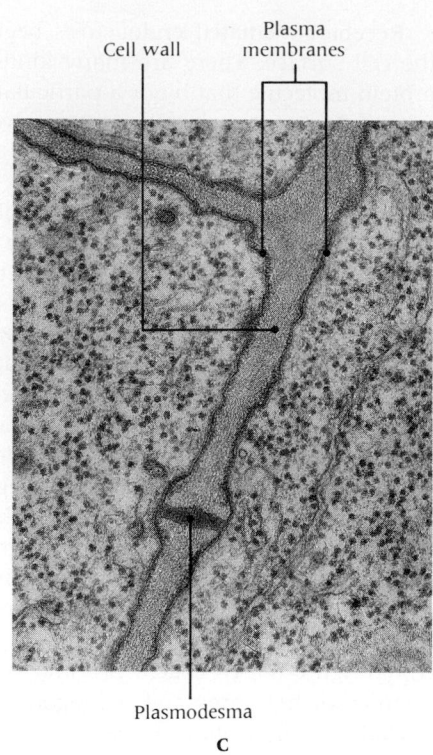

6-26 Structure of plasmodesmata
(A) *Diagram showing the plasmodesma, which connects the cytoplasms of adjacent plant cells. The endoplasmic reticulum membranes also traverse this space.* **(B)** *The plasmodesma is revealed in a cut-away view.* **(C)** *Electron micrograph of parts of three cells in the root tip of the bean,* Phaseolus vulgaris, *showing a plasmodesma.* (× 54,000)

wall consisting chiefly of cellulose and other polysaccharides. Cell walls are an evolutionary response to the problems of living in a hypotonic environment. Fungi, algae, and most bacteria also have thick outer walls that contain carbohydrates, but their structures differ from that of plant cell walls.

We have already considered the multi-layered design of the plant cell wall (see Figure 5-32). Plant cell walls are dotted with small holes connecting adjoining cells. These membrane-lined channels are called **plasmodesmata.** As shown in Figure 6-26, a plasmodesma is a concentric cylinder of plasma membrane and endoplasmic reticulum that traverses the wall between adjacent cells, interconnects their cytoplasms, and permits the passage of molecules below a certain size (molecular weight, 700 to 900). There is a small amount of bulk fluid flow from one cell to another through plasmodesmata, but diffusion through and across them is more prevalent. Plant cell walls (with secondary walls) are also dotted with structures called *pits,* patches of selectively permeable primary wall between adjacent cells.

Of the many differences between plant cells and animal cells, one of the most striking is the fact that plant cells are not bathed in an isotonic fluid medium. The interior of most plant cells is extremely hypertonic. Solutes accumulate at much higher concentrations than those of the surrounding watery habitat. This makes it possible to sustain the two distinctive properties—hydration (or water content) and turgor pressure (or hydrostatic pressure)—that make them rigid. Cells deprived of water lose their turgor and the plants they belong to wilt.

FUNGI AND BACTERIA

The cell walls of fungi do not contain cellulose, as do plant cell walls, but instead contain **chitin,** a polymer of the amino sugar glucosamine. As in plant cells, however, the cell wall surrounds a conventional protoplast that is sheathed in a cell membrane similar to that of other cell types.

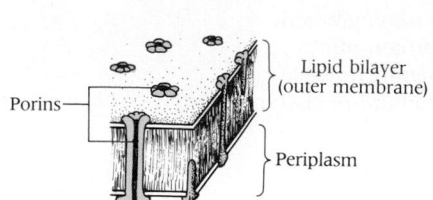

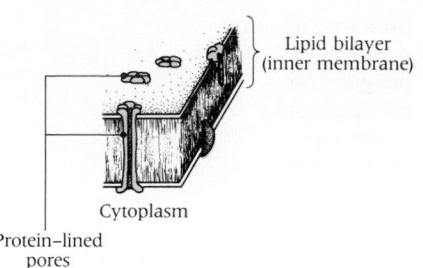

The cell walls of bacteria vary in composition depending on the bacterial species. All consist of rigid frameworks of cross-linked polysaccharide chains that have been described as one enormous molecule. The cell wall is mechanically responsible for maintaining the characteristic shape of a bacterium. Another function of this wall is as a permeability barrier that tends to keep such antibacterial molecules as antibodies and antibiotics from penetrating to the protoplast. Attached to the outside surface of this wall, again depending on the bacterial species, can be found an assortment of structures, substances, and devices. As we will see in Chapter 38, these include capsules, slime layers, and several kinds of the thin filaments.

The cell walls of some bacteria also contain an interesting protein called **porin.** In *E. coli,* porin molecules form channels that completely span the outer walls of the cell (Figure 6-27). The channels open and close by molecular mechanisms yet to be explained.

SUMMARY

All cells are separated from their environment by a plasma membrane composed of proteins, lipids, and a small amount of covalently attached carbohydrate. The properties of this membrane determine what molecules move into and out of the cytoplasm.

In the diffusion of molecules, whether across a membrane or through a solution, molecules move from an area in which they are initially more concentrated to one in which their initial concentration is lower. This results from the random motion of individual molecules. The laws of diffusion rest on the fact that more molecules in the high concentration zone can randomly move into the low concentration zone than can molecules in the low concentration zone move into the high concentration zone. If a semipermeable membrane is placed in the path of this diffusion, the molecules to which the membrane is impermeable cannot diffuse across it.

Most biological membranes are permeable to water. Osmosis, which is the diffusion of water molecules across such membranes, has great significance for cell structure and function. Most cells contain a high concentration of dissolved materials (solutes); hence, water tends to diffuse into them.

Cell membranes consist of a double layer (bilayer) of phospholipids (plus a few other lipids), which to a notable extent can slide laterally past each other in the plane of the membrane. In addition, membranes contain proteins of two classes: integral proteins are embedded in the lipid bilayer; peripheral proteins are bound to the lipid bilayer surface.

These proteins carry out many of the membrane's functions. Chief among them is facilitating the transport of molecules across the membrane. The transport of some molecules is not protein-mediated (lipid-soluble molecules may cross the membrane by simple diffusion through the lipid); however, larger molecules generally are transported across the membrane by protein-facilitated diffusion or active transport. In the latter case, which often transports molecules against a concentration gradient, the cell must expend energy to drive the transport mechanism. Many of the channels that transport ions or protons across the membrane operate in one direction only.

Endocytosis and exocytosis are other mechanisms by which materials can move, respectively, into and out of cells. In endocytosis, the cell membrane invaginates, engulfs liquid or other material, and then pinches off to form a vesicle full of that material on the other side of the membrane. Exocytosis reverses this process. Vesicles formed inside the cell fuse with the plasma membrane. As a result, they can release their contents into the exterior environment. This process is important in nerve impulse transmission.

While all cells have cell membranes, plants, fungi, and most bacteria also have rigid cell walls outside their cell membranes. The cell wall has little role in transport into or out of the cell but is important in maintaining the shape of the cell. By preventing its cell from swelling, it allows the cell to exist in a hypotonic environment without bursting.

KEY TERMS

active transport
chemiosmosis
colloid
colloidal systems
concentration gradient
Davson-Danielli model
diffusion
endocytosis

entropy
exocytosis
extrinsic (peripheral) protein
facilitated diffusion
fluid-mosaic model
gel
globular protein
glycoprotein

hydrostatic pressure
hypotonic
intrinsic (integral) protein
ligand
membrane pump
nonpolar (hydrophobic)
    molecule
osmosis

osmotic pressure
permeable
phagocytosis
phospholipid
pinocytosis
plasmodesmata

plasmolysis
polar (hydrophilic) molecule
proton gradient
proton pump
semipermeable
sodium-potassium pump

sol
sol-gel transitions
solute
solvent
transporter

QUIZ QUESTIONS

1. Which of the following means of transmembrane transport involves expenditure of energy by the cell *and* can move materials up a concentration gradient?
   A. active transport
   B. simple diffusion
   C. facilitated diffusion
   D. osmosis

2. Which one of the following is not a *correct* statement regarding the design of biological membranes?
   A. Carbohydrate molecules associated with membranes tend to be associated with protein molecules embedded in the membrane.
   B. Protein molecules can be located on the inner and outer surfaces of the membrane as well as traverse the entire thickness of the bilayer.
   C. Cholesterol is a type of lipid frequently found in membranes of animal cells.
   D. Hydrophobic ends of phospholipid molecules in the outer monolayer of the membrane are exposed directly to the extracellular environment.

3. Fish that live in a freshwater lake are _____ to their surroundings. Therefore, these fish
   A. hypotonic; face the problem of water loss.
   B. hypotonic; face the problem of water gain.
   C. hypertonic; face the problem of water loss.
   D. hypertonic; face the problem of water gain.
   E. isotonic; are in osmotic equilibrium with their surroundings.

4. Which of the following is (are) *not* a process(es) in which extracellular materials are brought into the cell?
   A. endocytosis
   B. exocytosis
   C. phagocytosis
   D. pinocytosis
   E. A and C

5. In addition to cell membranes, cells of all of the following kingdoms *except* _____ possess cell walls.
   A. Animalia
   B. Fungi
   C. Monera
   D. Plantae
   E. B and C

ESSAY QUESTIONS

1. Contrast the role of the plasma membrane of a cell with the integument (skin) of an animal.

2. A red blood cell is dropped into a beaker of distilled water (which is hypotonic relative to red cell cytoplasm). What happens? Would a plant cell have a different fate? Explain.

3. Contrast the means by which a cell absorbs material from its environment: simple diffusion; transport (facilitated, active); endocytosis; pinocytosis; phagocytosis. What structures of the cell are required for each process?

4. Trace the fate of a particle of food after ingestion by an amoeba by phagocytosis.

REFERENCES AND SUGGESTIONS FOR FURTHER READING

Bretscher, M. S. (1985). The molecules of the cell membrane. *Scientific American 253*, October, 100–108.

   A review of the bilayer plasma membrane, its proteins, and its functions.

Brodsky, F. M. (1988). Living with clathrin: its role in intracellular membrane traffic. *Science 242*, 1396–1402.

   A very stimulating update of the clathrin story.

Cevc, G., and Marsh, D. (1987). *Phospholipid Bilayers: Physical Principles and Models.* John Wiley & Sons, New York.

   A reasonably readable, high-level discussion of method and theory in a burgeoning field.

DeMello, W. C. (Ed.) (1987). *Cell-to-Cell Communication.* Plenum, New York.

   A review of ideas on how cells "talk to each other" via gap junctions and other channels.

Finean, J. B., Coleman, R., and Michell, R. H. (1984). *Membranes and Their Cellular Functions,* 3rd ed., Blackwell Scientific, Boston.

   An accurately written and attractively presented survey of membrane biology.

Gennis, R. B. (1989). *Biomembranes: Molecular Structure and Function.* Springer-Verlag, New York.

   Probably the best and most up-to-date treatment of membrane biology currently available. Readable and informative.

Gunning, B. E. S., and Overall, R. L. (1983). Plasmodesmata and cell-to-cell transport in plants. *BioScience 33*, 260–264.

A brief summary with useful diagrams and electron micrographs.

Orci, L., Glick, B. S., and Rothman, J. E. (1986). A new type of coated vesicular carrier that appears not to contain clathrin: its possible role in protein transport within the Golgi stack. *Cell 46*, 171–184.

Original paper demonstrating that protein transport through the Golgi stack is mediated by a novel type of clathrin-less vesicle.

Pearse, B. M. F., and Bretscher, M. S. (1981). Membrane recycling by coated vesicles. *Annual Rev. Biochem., 50,* 85–101.

A scholarly review of an important phenomenon that aroused surprise when it was discovered.

Racker, E. (1985). *Reconstitutions of Transporters, Receptors, and Pathological States.* Academic Press, Orlando, Fla.

A collection of lectures by a pioneer known for his wisdom and wit.

Singer, S. J., and Nicolson, G. L. (1972). The fluid mosaic model of the structure of cell membranes. *Science, 175,* 720–731.

The classic paper that introduced the fluid mosaic model of membrane structure.

Stein, W. D. (1986). *Transport and Diffusion Across Cell Membranes.* Academic Press, Orlando, Fla.

An enthusiastic introduction to current research in traffic across membranes.

Varner, J. E., and Lin, L.-S. (1989). Plant cell wall architecture. *Cell 56,* 231–239.

A brief and stimulating review.

Verner, K., and Schatz, G. (1988). Protein translocation across membranes. *Science 241,* 1307–1313.

Newly synthesized proteins must be translocated across one or more membranes if they are to reach their destinations. This exceptionally clear account explains the various mechanisms that make this possible.

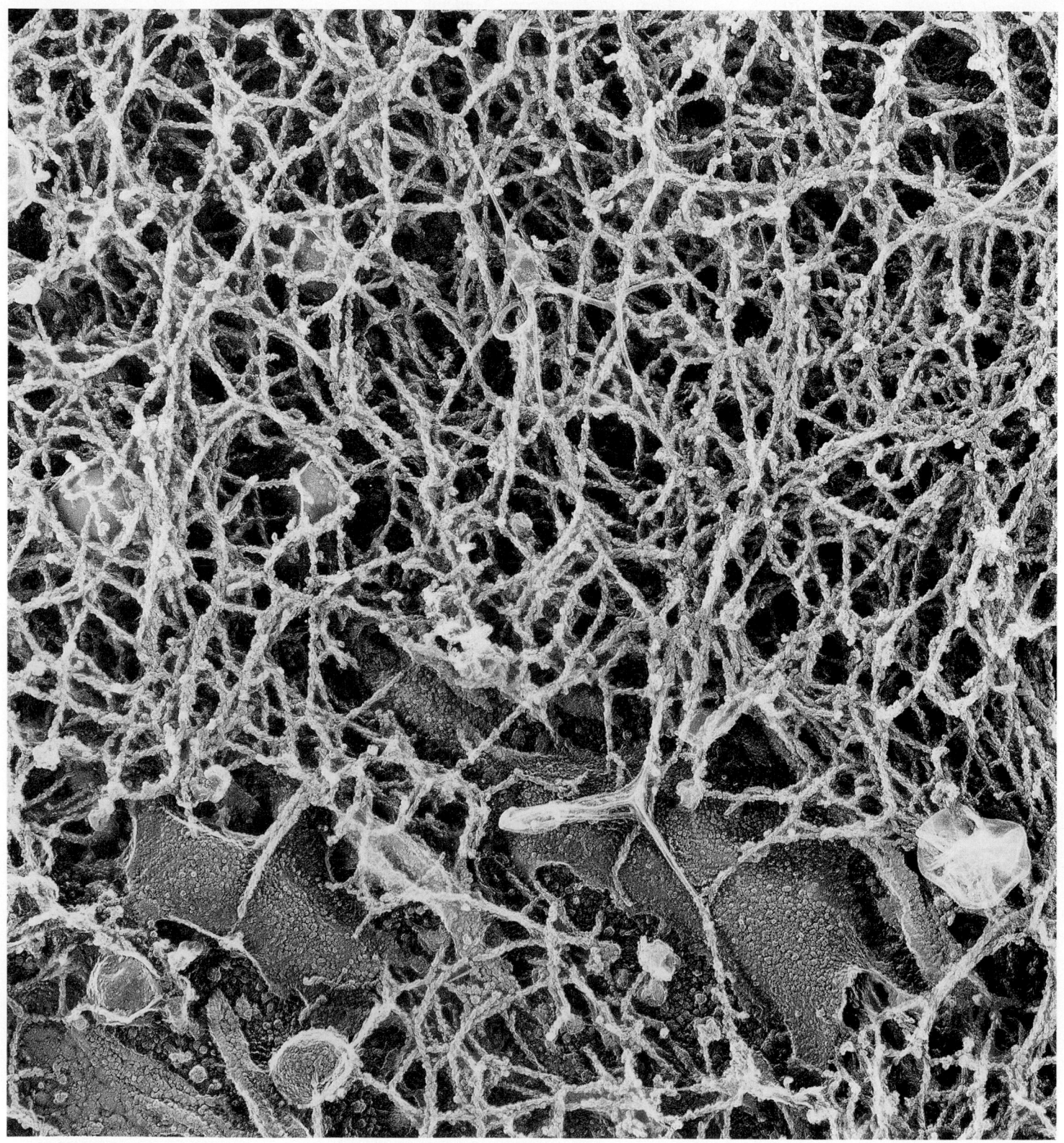

CELL FUNCTIONS

In this chapter we will consider how cells combine into tissues, how they communicate with one another, and how they move.

One aim of this chapter is to demonstrate again the universality of many biological mechanisms—among the different cell types of a given organism, among the different species, and among the five living kingdoms. Much that we learn of cell motility, for example, will acquire added meaning when we later explore the mechanisms of muscle contraction and nerve impulse transmission.

THE CELL AS A BUILDING BLOCK

We have already seen that some living organisms consist of only one cell. These include members of the kingdom Monera (the bacteria and blue-green algae, or cyanobacteria) and the unicellular members of kingdom Protista (the protists). The rest of the living world consists of multicellular organisms—animals (kingdom Animalia), plants (kingdom Plantae), and fungi (kingdom Fungi).

DIFFERENTIATION OF CELLS INTO TISSUES, ORGANS, AND SYSTEMS

A multicellular organism is made up of visibly different parts. Through the process of differentiation, its cells have become functionally and morphologically specialized. In other words, each cell has the same hereditary endowment—the original zygote. Yet they become different—they differentiate.

As a result, a cell in a root tip of a plant is different from one in a leaf; a leaf surface cell differs from a cell within the leaf. In the animal body, nerve cells, muscle cells, and blood cells all differ from each other. The search for an explanation of how this comes about is considered fully in Chapters 19 and 20.

Each cell type usually occurs with many others of its kind. An organized aggregation of like cells is called a **tissue.** In simple tissues, all cells are of the same type. Composite tissues contain two or more cell types in characteristic relationship with one another. Botanists term these tissues *complex*.

The most complex parts of organisms are made up of many kinds of tissues. The hand, for instance, is a functionally distinct **organ** of the body that includes skin, bones, muscles, nerves, blood, and other tissues. Organs cooperate and interact as parts of larger complexes called **systems.** The digestive system, for example, is a sequence of organs from the mouth through the esophagus, the stomach, the small intestine, and the large intestine to the anus. Each organ of the system is different, but all interact in the processes of nutrition—food ingestion, digestion, and waste elimination. Each organ in the digestive system is composed of many tissues; each tissue is composed of cells.

Cell functions
This stunning scanning electron micrograph shows a network of actin microfilaments inside a macrophage, a highly motile cell. The jagged edges of the fibers are due to bound myosin molecules, which were deliberately added as markers. The specificity of actin-myosin binding proves that the fibers contain actin. (× 30,000)

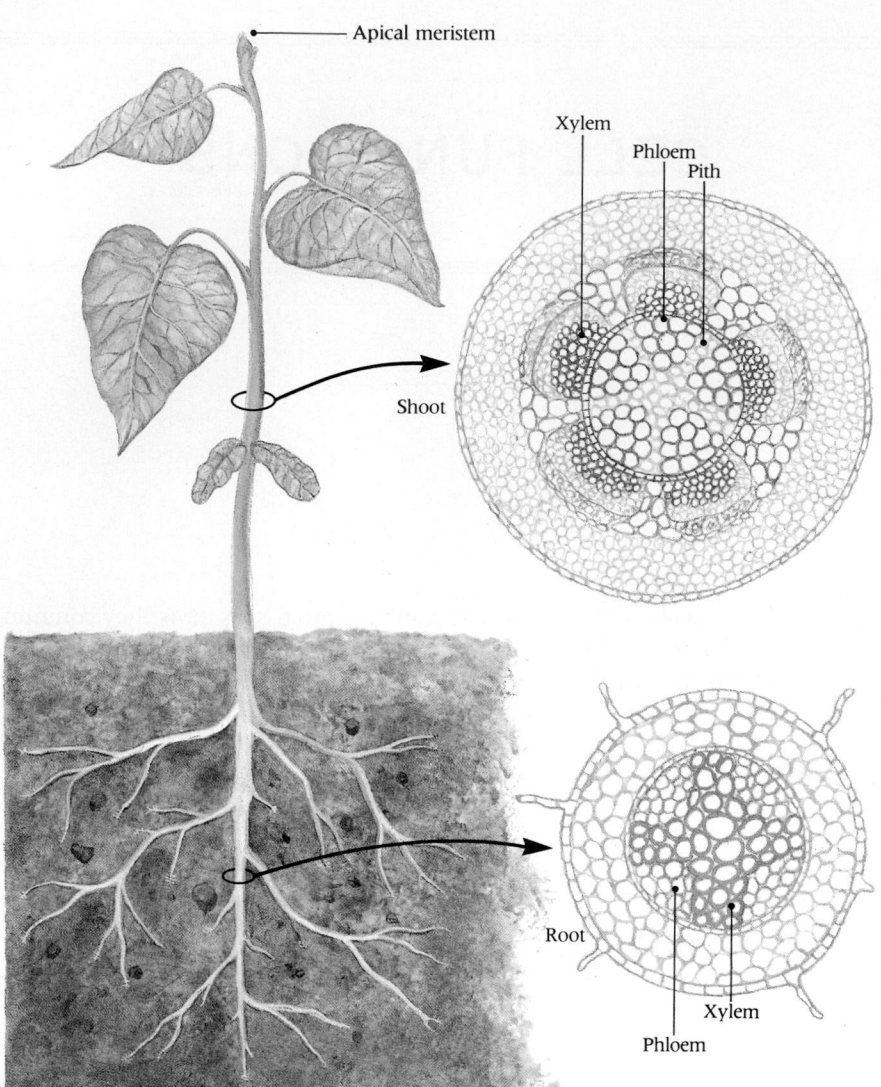

— Apical meristem

Xylem

Phloem

Pith

Shoot

Root

Xylem

Phloem

7-1 Cross section of root and stem of a green plant
Both structures show abundant phloem and xylem elements. Details are given in the text.

BRIEF OVERVIEW OF PLANT AND ANIMAL CELLS AND TISSUES

We here give necessarily brief notice to the various kinds of cells and tissues in plants and animals. Our purpose at this point is simply to display their immense variety, to define a few basic terms, and to introduce several key members from the cast of anatomical characters.

Plant Cells and Tissues

The more complex plants (mainly the green plants, with which we are most familiar) have bodies that consist of two organ systems—the underground **root system** and the aboveground **shoot system,** which includes stems and leaves (Figure 7-1).

Plants of this type are continually (or intermittently in the Temperate Zone) engaged in **primary growth,** which is the increase in length of shoots and roots. Such growth occurs in specialized regions called **meristems.** Meristems at the tips of roots and shoots are called **apical meristems.** Zones just above the points of leaf attachment (nodes) are called **intercalary meristems.** These are regions of active cell divisions and development. Growth that results in thickening of stems, branches, and roots is known as **secondary growth,** and it originates in the lateral meristems. Lateral meristems consist of a kind of meristematic tissue that is called **cambium.**

Growth in the meristematic tissues leads to the formation of three tissue systems: (1) protective, (2) fundamental, and (3) conductive.

Protective tissues are the outermost layers of cells that cover and protect under-

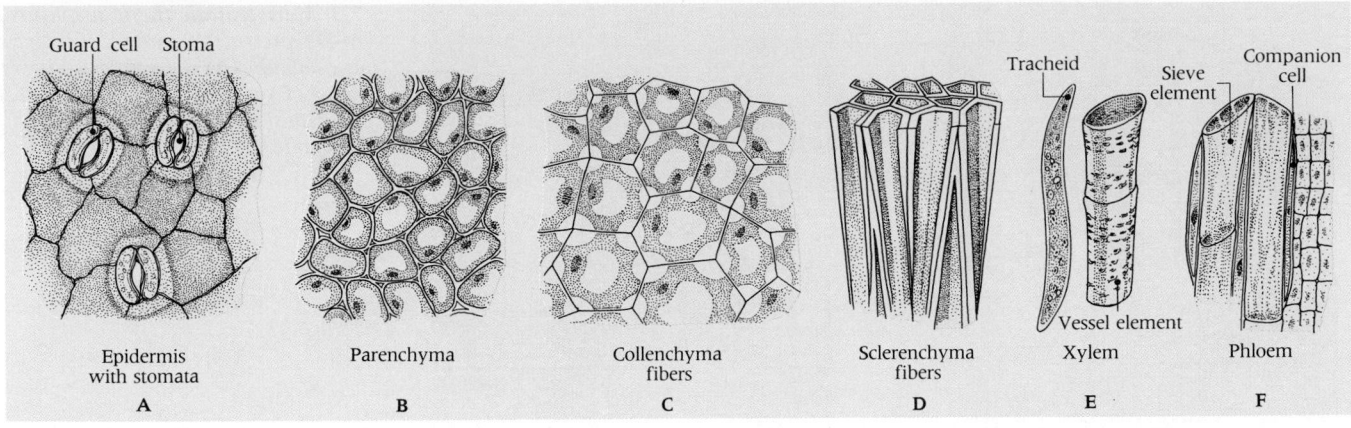

Guard cell   Stoma

Epidermis
with stomata
A

Parenchyma
B

Collenchyma
fibers
C

Tracheid

Vessel element
Sclerenchyma
fibers
D

Xylem
E

Sieve
element

Companion
cell

Phloem
F

*7-2 Cells in plant tissues display a wide
diversity of shapes*
Cell types are described in the text.

lying tissues in both organ systems. They consist chiefly of the **epidermis,** which is a sheet of flat cells that are fitted closely together with very little intercellular space (Figure 7-2A). Stems and roots undergoing secondary growth develop an outer layer **(periderm)** consisting of cork, a dead protective layer made of box-shaped cells (see Figure 5-1).

Three types of cells are collectively referred to as the **fundamental** (or **ground**) **tissues.** The **parenchyma,** is thin-walled and vacuolated (Figure 7-2B). Parenchymal cells often contain chloroplasts, the chlorophyll-containing organelles that are the sites of photosynthesis. They are by far the most plentiful of the fundamental tissue cells. The **collenchyma** and the **sclerenchyma** have thick walls that add greatly to the mechanical strength of a root, stem, or leaf (Figure 7-2C, D).

Conductive (or **vascular**) **tissues** are involved in transport functions within plants. There are two kinds—xylem and phloem. **Xylem** carries water and dissolved materials absorbed from the soil to the leaves. It also gives mechanical support to the plant. Two types of conducting cells make up the xylem of flowering plants (Figure 7-2E): tracheids and vessel elements. Both are elongated cells that begin their existence as unspecialized cellular derivatives of a meristem. As they mature, their walls become heavily thickened, often with conspicuous spiral bands.

Plants with xylem and phloem are referred to as *vascular* plants. All vascular plants have tracheids in their xylem; flowering plants have vessel elements in addition to tracheids. Both tracheids and vessel elements are dead at maturity. When they die, their cytoplasm disintegrates, leaving only the cell walls as systems of tubes extending from roots to leaves. Tracheids (Figure 7-2E) are narrower in diameter than vessel elements, and typically there are several openings in the end walls between a tracheid and the tracheids above and below it. Vessel elements (Figure 7-2E) are usually shorter than tracheids and have a larger diameter. They lose their end walls, thus forming a continuous open tubular vessel that extends along the axis of the plant.

Phloem is the major carrier to all parts of the plant of food materials manufactured in leaves and stored in root and stem materials. Mature **sieve cells** of phloem are elongated cylinders with walls less thickened than those of xylem cells (Figure 7-2F). Unlike xylem cells, phloem cells are connected to one another through perforations in their end walls called sieve plates. Phloem cells are usually associated with—or adjacent to—**companion cells** that actively secrete substances into the sieve elements, which are the vertical transport units of phloem. The nuclei of phloem cells disintegrate, but their cytoplasmic contents remain, fusing through the sieve plate pores and forming a protoplasmic highway along which food is efficiently carried. The manner in which growing meristematic tissues give rise to xylem, cambium, and phloem is illustrated in Figure 7-3. More will be said on the structure of xylem, phloem, and other plant tissues in Chapter 24, which surveys their transport functions.

In addition to xylem and phloem, transport in plants is possible because of the presence in plant cell walls of plasmodesmata. As discussed in Chapter 6, plasmodesmata are small plasma membrane-lined openings that penetrate the cell wall and permit direct communication between the cytoplasms of neighboring cells. The communication is subject only to a molecular size limit. Plant cell plasmodesmata have in common with animal cell gap junctions the capacity to transfer small mole-

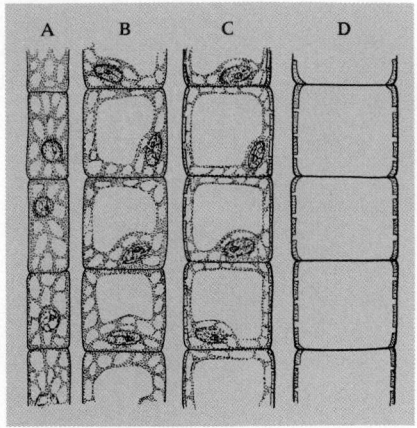

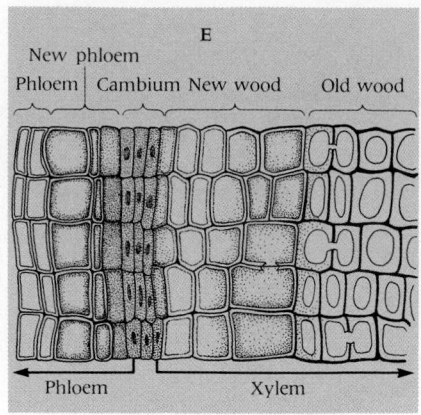

7-3 Meristematic tissue in plants
*A–D represent the sequence of changes that occur as cambium (**A**) differentiates into xylem (**D**). Panel (**E**) is a cross-section through a part of the stem showing that the cambium differentiates into both phloem and xylem tissues.*

cules from cell to cell. Note that gap junctions are regulated by Ca^{2+} ions, and plasmodesmata are not; gap junctions are constructed of transmembrane proteins, and plasmodesmata are not.

Animal Cells and Tissues

Four broad classes of tissues occur in virtually all multicellular animals, differing only in detailed structure and origin (Figure 7-4). These are (A) epithelial tissues, (B) connective tissues, (C) muscular tissues, and (D) nervous tissues. All the organs of animals are constructed of one or more of these basic tissue types.

Epithelial tissue, or **epithelium,** is a tissue that covers a surface. The external surface is only one of many surfaces in the animal body. Some internal surfaces are actually continuous with the outside; for example, the surface of the mouth. Other internal surfaces, like the linings of blood vessels, have no connections with the outside. They are surfaces, nonetheless, and are covered with epithelium.

Epithelial tissue consists of sheets of cells joined in continuous layers. Individual cells are closely joined to one another. In this respect, epithelium differs from connective tissue, in which much intercellular substance is found. It has one major character-

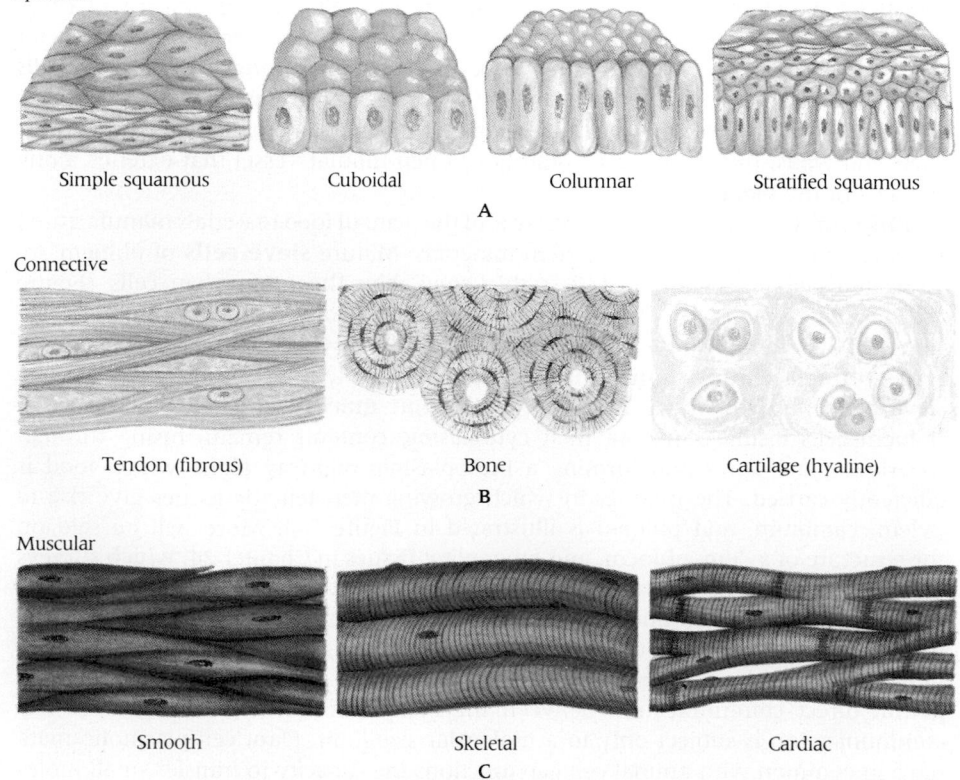

7-4 Major types of animal tissues
(A) The four types of epithelium. (B) Connective tissues. (C) The three types of muscle. See text for details.

istic: It forms the covering and lining membranes of the body and its organs, cavities, and passageways.

Epithelial cells are classified, according to the shape and structure of the individual cells, as **squamous, cuboidal,** and **columnar** (Figure 7-4A). If the cells are arranged in a single layer, the epithelium is called *simple;* if the cells are arranged in more than one layer, the epithelium is called *stratified.* An epithelial cell may be equipped on its free surface with cilia or with a fringelike structure composed of microvilli that light microscopists have traditionally called the *striated* or *brush border* (Figure 7-5).

Connective tissue bonds together and supports all other types of cells and tissues. It is characterized by a large amount of intercellular substance and a relatively sparse cell population. The nature of the intercellular substance provides the basis for a classification of connective tissue into three main types: connective tissue proper (e.g. tendon), bone, and cartilage (Figure 7-4B). Blood and lymph may be considered a form of connective tissue in which the intercellular substance is fluid. In most connective tissue proper, the intercellular substance is the proteinaceous, carbohydrate-containing extracellular matrix, which is soft, jellylike, and amorphous. In cartilage, the intercellular substance is rubbery. And, in bone, it is dense, hard, and mineralized. We should note that the deposition of generous amounts of intercellular material represents yet another strategy of cellular interaction—one that places cells relatively far apart, in contact with a common set of surroundings rather than with one another. In Chapter 31, we will see how extensively evolution has made use of different structural materials in variants of connective tissue.

Muscular tissue performs the mechanical task of body movement by physically shortening its cells and molecular components (see Chapter 31). There are two fundamental types of muscle: one is called *smooth* and the other *striated* (or skeletal) muscle (Figure 7-4C). Although this classification is made on the basis of microscopic appearance, it has a good deal of functional significance. Only striated muscle is under voluntary control. Smooth muscle operates involuntarily and contracts much more slowly.

A third type of muscle has distinctive structural features that reflect its unusual functional role. *Cardiac* (heart) *muscle* is an involuntary form of striated muscle that contracts automatically and rhythmically.

Nervous tissue is responsible for one of biology's most complex phenomena, the transmission of electrical impulses. The fundamental element is the **nerve cell,** or **neuron,** which has a central cell body containing the nucleus and two or more long protoplasmic processes known as *nerve fibers.* Nerve fibers that transmit nerve impulses away from the cell body are called axons. These often divide into many terminal branches and may be remarkably long. Nerve fibers that transmit impulses toward the cell body are called dendrites. They generally branch freely, close to the cell body. Two successive neurons are separated by a minute space called the synapse, which permits impulse conduction in one direction only. All the complexities of movement, perception, behavior, and conscious thought are based on the nerve's capacity to transmit electrical impulses and join with other nerve cells in the intricately organized computerlike association that is the nervous system (see Chapters 29 and 30).

CELL-TO-CELL COMMUNICATION

The organization of cells into tissues is an essential feature of multicellular organisms. This coming together means that cells in vast numbers have traded some of their independence for the many advantages resulting from functioning together as a coherent, unified whole.

Tissue cells communicate with one another in many ways—both subtle and obvious, direct and indirect. Perhaps the most conspicuous example of direct communication is the interaction of a sperm cell and an egg in the course of fertilization during sexual reproduction. That event clearly embodies a feature essential to any communication between cells—*recognition.* That, we now know, is made possible by specific surface structures on each of the cells.

Among the many varieties of cellular interactions, a number of distinct patterns have emerged. Direct communications include the following modes:

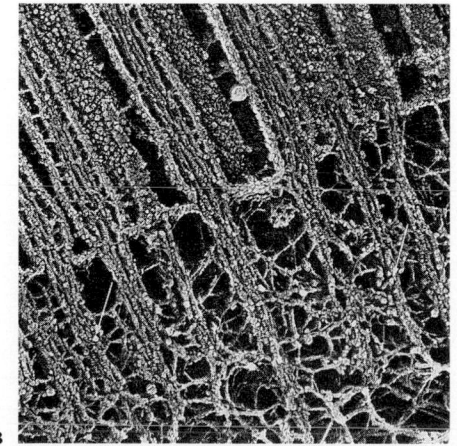

7-5 The intestinal brush border
(A) *Epithelial cells bearing numerous microvilli are stabilized by a network of filaments in the microvillus and in the cytoplasm.*
(B) *Freeze-etch view of brush border of an intestinal cell. (× 60,000)*

- *Direct contact.* Specificity in this type of interaction is influenced by the ability of cells to produce specific surface structures (or an extra-cellular matrix) that are capable of influencing other cells that contact their surfaces. This form of interaction is of great importance in the course of embryonic development (Chapters 19 and 20).
- *Binding.* Cells may bind to other cells and influence them through direct physical adhesion. This occurs, for example, when certain cells of the immune system bind to one another in the sequence of events leading to the synthesis of antibodies (Chapter 23).
- *Joining.* Cells may actually join their cytoplasms by means of localized connections, channels, or pores. In some cases, such junctions allow cytoplasmic mixing, with passage back and forth of cytoplasmic substance. In other cases there is traffic back and forth only of certain ions and molecules. This type of interaction is crucial as individual cells become tissues.

The major form of indirect communication is the following:

- *Secretion of messenger molecules.* Cells may secrete substances that are freely diffusible and that have the capacity to signal other cells that are near or far. Their signaling capacity depends on the anatomical circumstances, the amount of signaling substance secreted, and its stability. Hormones, the major example of signaling substances, also base their specificity on the specific structures on cell surfaces that permit them to recognize their targets (Chapter 28).

Many cell interactions are known to occur in animals and plants, but we should note that bacteria have this capacity also. One group of bacteria, the myxobacteria, actually gather into rudimentary multicellular organisms that are dependent on various forms of cell-to-cell interactions—including cell contact and the secretion of diffusible substances (Chapter 37).

JUNCTIONS BETWEEN CELLS

In most animal and plant tissues, the points at which constituent cells meet are termed **intercellular contacts,** or **junctions.** Three functional types are recognized.

- *Impermeable junctions* isolate cells from one another and thus preserve the distinctiveness of each cell's internal environment.
- *Adhering junctions* promote adhesion between cells and thereby reinforce tissue structure.
- *Communicating junctions,* by permitting cells to exchange nutrients and molecular information signals, promote functional coordination.

Electron micrographs and freeze-fracture preparations have revealed a good deal about the architecture of the several junctions that perform these functions. The following comments are concerned with animal cells. Some of the arrangements in plant cells (plasmodesmata and pits) were discussed in Chapter 6. Other aspects are discussed below.

Impermeable Junctions

Impermeable junctions, or **tight junctions,** are regions where the cell membranes of two adjacent cells appear to fuse into a bond that completely encircles each cell, as shown in Figure 7-6. This diagram and the several that follow are of the tight junctions existing between the lining cells of the small intestine. These cells must provide for the controlled absorption of intestinal contents by active transport and at the same time seal off channels of leakage between cells and other forms of unwanted molecular traffic. They also have unusual surface projections called **microvilli,** which facilitate absorption and display motility. Finally, these cells display most types of the known intercellular junctions.

In each tight junction, adjacent cell surface membranes are held together by interlinked rows of membrane proteins that adhere tightly to their counterparts in the next cell like opposing teeth in a zipper. As a result, there is no intercellular

7-6 Three-dimensional view of a tight junction between cells with microvilli
In this schematic diagram, a tight junction is located just beneath the microvilli where it acts to seal the cells together.

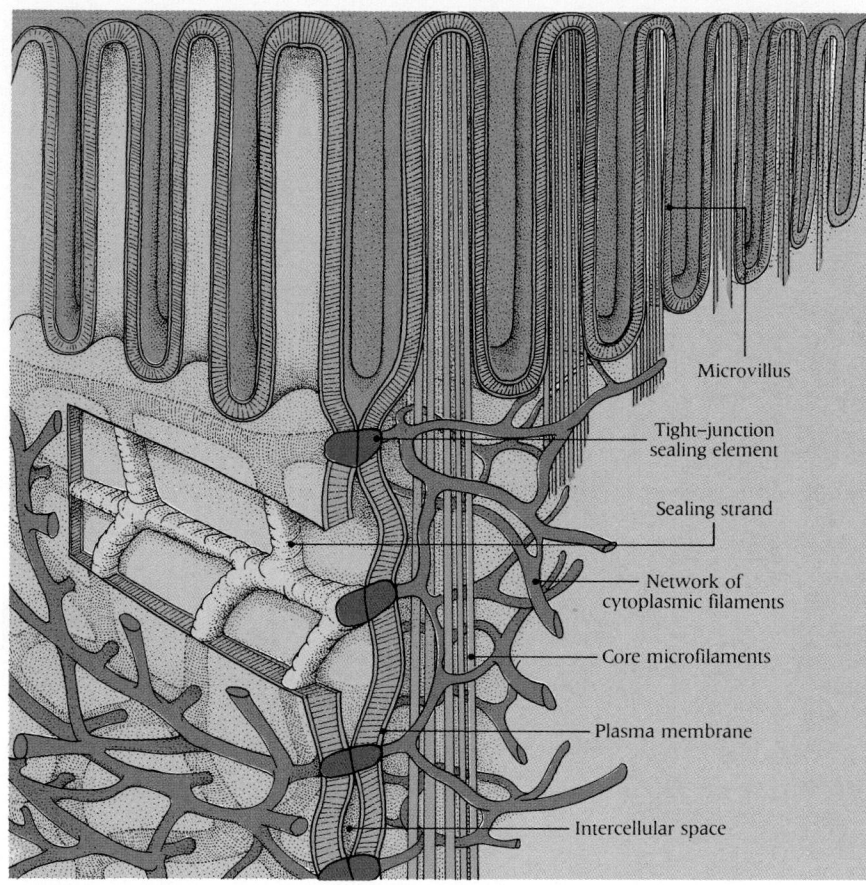

Microvillus

Tight–junction sealing element

Sealing strand

Network of cytoplasmic filaments

Core microfilaments

Plasma membrane

Intercellular space

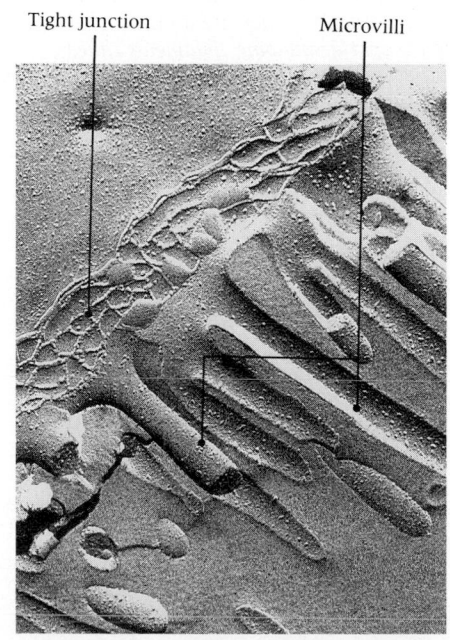

Tight junction

Microvilli

7-7 Structure of tight junctions
Freeze-fracture scanning electron micrograph showing a tight junction, which appears as ridges, made up of integral membrane protein molecules. Microvilli extend from the cell surface.
(× 42,000)

space. Tight junctions also serve as barriers to unlimited migration by membrane lipids and integral proteins (Chapter 6). Freeze-fracture micrographs show the zipper-like arrangement of proteins very nicely (Figure 7-7).

Adhering Junctions

The adhesion of neighboring cells is the responsibility of **adhering junctions** called **desmosomes,** which are composed of peripheral and integral membrane proteins, and characteristic filaments (termed *tonofilaments*). Desmosomes are of two types: spot desmosomes and belt desmosomes (Figure 7-8). These are accurately descriptive terms.

Spot desmosomes are disklike plaques or patches, just under the surface membranes of adjacent cells, that resemble "spot welds" between cells. The tonofilaments, fibers about 10 nm in thickness that contain a protein called **keratin,** radiate from the plaques into the surrounding cytoplasm on both sides of the junction. These fibers evidently anchor the desmosomes to the cytoskeleton of the underlying cytoplasm and to that of the adjacent cells (Figure 7-9).

Belt desmosomes are comparable structures in a bandlike arrangement that girdles the inside of the cell membrane with bundles of contractile filaments. These contain the protein *actin* rather than keratin. There is evidence that belt desmosomes, unlike spot desmosomes, are more involved with support of the cell than with the linking of adjacent cells.

Communicating Junctions

Perhaps the most interesting of the three types of cell junctions are the **communicating junctions** called **gap junctions,** which permit direct metabolic communication between cells (Figure 7-10). They are tiny pipes or channels that permit the passage of small molecules (molecular weight, less than 1200) between two cells and that under some conditions are regulated by local concentrations of Ca^{2+}, a high concentration tending to close the channels.

The walls of the channel of a gap junction are constructed of identical subunits

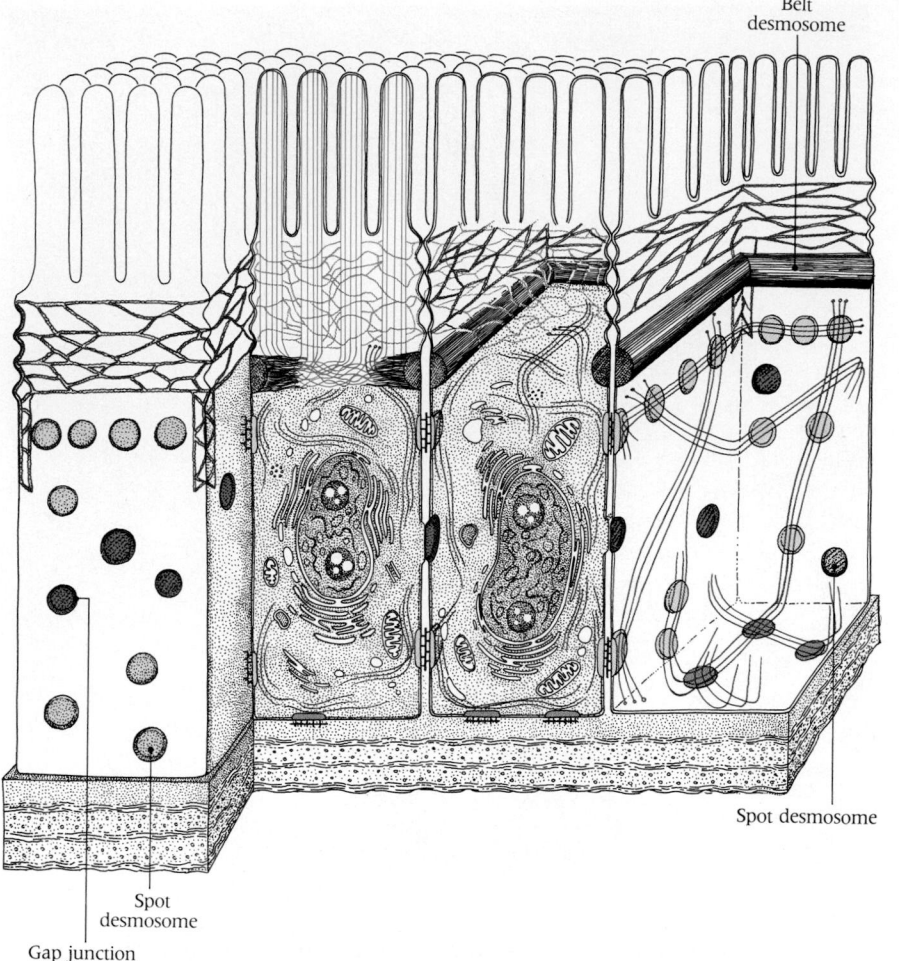

The belt desmosomes lack the thick plaquelike character of spot desmosomes.

Belt desmosome

Spot desmosome

Spot desmosome

Gap junction

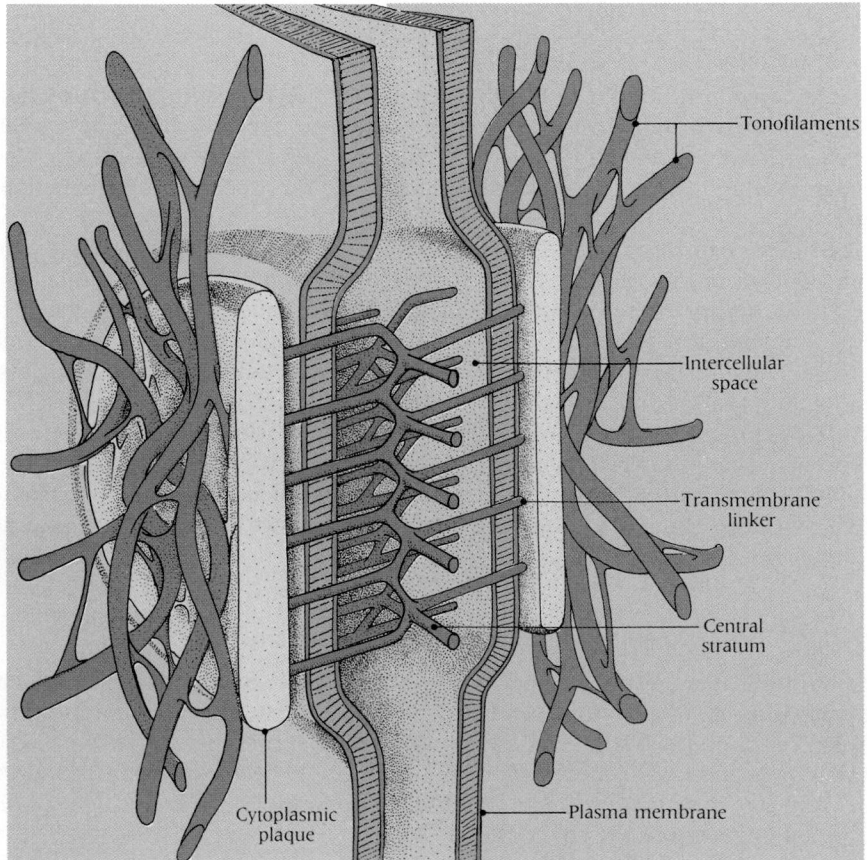

7-9 Three-dimensional view of a spot desmosome linking two cells
The tonofilaments form a tensile network in the cell interior. They are attached to plaques of the spot desmosome. Filaments called transmembrane linkers connect the plaques across the intercellular space.

Tonofilaments

Intercellular space

Transmembrane linker

Central stratum

Cytoplasmic plaque

Plasma membrane

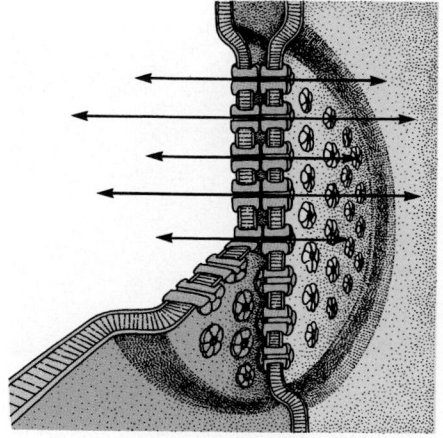

7-10 Model of a gap junction
Membrane proteins called connexins form channels that connect adjacent cells.

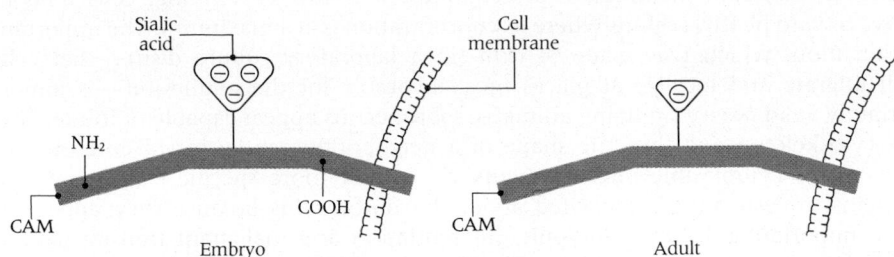

7-11 Signaling molecules of the cell membrane
Cell-adhesion molecules (CAMs) are often modified during development, thereby providing signals for cell changes. The CAM molecule has three regions, an NH₂ end, a COOH end, and a middle region containing sialic acid. In this case, the sialic acid part of the molecule differs in embryonic and adult cells.

of a protein called **connexin** that are packed in hexagonal array with their hydrophilic residues facing the central channel. The channel itself is termed a **connexon.**

We know that gap junctions are critically important in developing embryos and in nerve cells, but it is likely that their full significance has not yet been established. Their behavior in experimental systems can be astonishing. For example, when gland cells from a rat ovary were experimentally cultured with muscle cells from a mouse heart, gap junctions developed between the two drastically differing cell types. Subsequently, when one cell was stimulated by the experimenter, both would respond. This striking result was attributed to passage through the gap junctions of a signaling molecule, the ribonucleotide called **cyclic AMP.** We shall learn more of this important communicator molecule, one of several now known, in Chapter 28.

Comparable structures in plants, the plasmodesmata, were described in Chapter 6 (see Figure 6-26).

SIGNALING SYSTEMS

Cell Adhesion

One of the patterns of direct cellular interaction involves specific surface molecules. Such molecules—chiefly glycoproteins—serve as recognition signals between cells. Groups of surface molecules have been found that mediate cell-to-cell adhesion (without benefit of specific junctions), that appear for the first time at specific stages of embryonic life, and that regulate the movements of cells during development and the choice of sites where they come to rest. Some of these **cell-adhesion molecules,** or **CAM**s, appear to change as an embryo develops into an adult (Figure 7-11). New work in this area shows that CAMs structurally resemble antibodies (Chapter 23).

The adhesive attachment of cells to their surroundings influences cell shape and helps in maintaining cell function and tissue integrity. Such binding helps to anchor cells in place. It also provides signals—some researchers call them "positional signals"—that direct cellular traffic, growth, and development. Most cells possess several mechanisms that can bind them to the structures that surround them. For example, cells can bind to other cells and to so-called extracellular matrices.

An **extracellular matrix** is a meshwork of insoluble proteins and carbohydrates that is laid down by cells and that fills most of the intercellular spaces, more or less like cement between the bricks in a wall.

We now recognize that in different parts of the animal body, the matrix includes different combinations of proteins and carbohydrates called *collagen, proteoglycans, elastin, hyaluronic acid,* and various glycoproteins such as the *fibronectins* and *laminin.* The primary structure of many adhesive proteins in extracellular matrices includes the three-amino acid sequence arginine-glycine-aspartic acid. These sites in the protein molecule are believed to be the "sticking points" that bind to matching receptors on cell surfaces called **integrins.**

Consider, for example, one of the better understood classes of adhesive molecules—the high-molecular-weight glycoproteins called **fibronectins** (Latin *fibra,* "fiber," and *nectere,* "to bind or connect"). These are the sticky proteins that allow cells suspended in a culture medium to flatten out on a glass surface and adhere to it tenaciously. Note that fibronectins are concentrated at points where cells adhere (Figure 7-12). In another procedure for demonstrating their role in promoting adhesion, cultured cells can be shown to stick to fibronectin-coated beads and to spread upon their surfaces.

Not only do these proteins promote adhesion, but they stimulate **cell migration.**

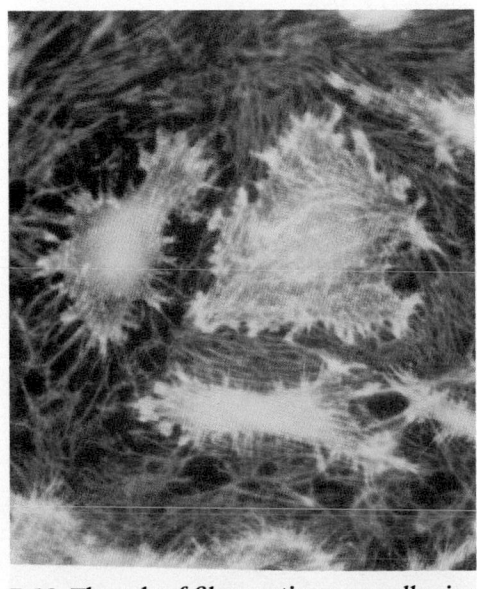

7-12 The role of fibronectin as an adhesive molecule
Micrograph showing that fibronectins are present in regions where cells attach to one another.

When the amount of adhesive protein present is low or limiting, cells tend to move toward nearby regions where its concentration is at a maximum. The important observation, which was made *in vitro* (in a laboratory culture dish)—that cells will migrate and localize at places most favorable for their adhesion—is almost certainly valid *in vivo* (in living animals). Fibronectins appear capable of influencing the cytoskeleton and thus the shape of a neighboring cell by establishing a "hot line" to its cytoplasmic microfilaments via one or more specific transmembrane proteins. Investigators are excited about the fibronectins because they appear to have important roles in embryonic differentiation and malignant transformation.

Another protein occurring on cell surfaces (and within cells) is so ubiquitous in the world of life it is named **ubiquitin.** Among its functions is to cause certain cells to "home"—that is, to find specific "roosting" sites in the body. In a recently discovered example, a "ubiquinated" cell surface (that is, a cell surface richly endowed with ubiquitin molecules) was responsible for the homing of lymphocytes in lymph nodes. More of that later.

A word of caution: One should not confuse cell adhesion and cell recognition. The former usually requires the latter, but the reverse is not true. Sperms recognize eggs before they adhere to them. While recognition is usually required in order for cells to adhere to each other, cells that recognize each other do not always adhere. Both cell adhesion and cell recognition may involve issues of specificity, adhesion usually requiring a measure of recognition, but many cases are known, especially from studies of embryonic development and immune mechanisms, in which cells that recognize each other do not adhere.

Molecular Messengers

Finally, we note only briefly the pattern of indirect cellular interaction—the secretion of molecular messengers, which are freely diffusible substances that can act on other cells at a remote location. When the secreted substances are hormones (the subject of the field of endocrinology), they act as so-called growth factors (see Chapter 28).

CELL MOTILITY

Motility is the ability of living systems to exhibit motion and to perform mechanical work at the expense of metabolic energy. Motility and mobility are often confused. The distinction is clear in the simplest motions observed in living cells with the light microscope: Brownian (random) motion of particles demonstrates their mobility under the influence of thermal agitation. On the other hand, motility occurs when forces transport the same particles much greater distances using metabolic energy.

The scope of motility, as it is presently understood, includes a huge variety of diverse phenomena. Here are a few examples.

- Flagellar movements of bacteria (prokaryotes)
- Flagellar and ciliary movement in eukaryotic cells
- Gliding in unicellular organisms (bacteria, cyanobacteria, diatoms)
- Saltatory motions (jerky hops and leaps) of particles in cytoplasm

7-13 Amoeboid movement is an example of nonmuscular movement
(A) *The extension and retraction of cytoplasm in the cell results in forward movement.*
(B) *Scanning electron micrograph of an amoeba showing two pseudopods surrounding a food particle. (× 200)*

Pseudopods

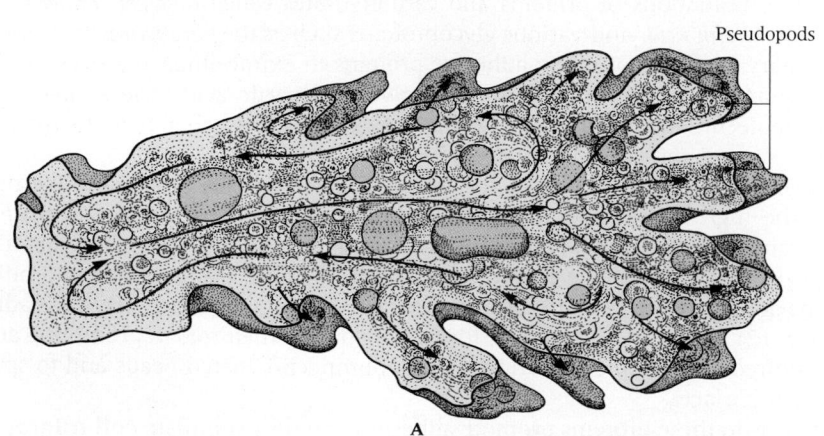

A

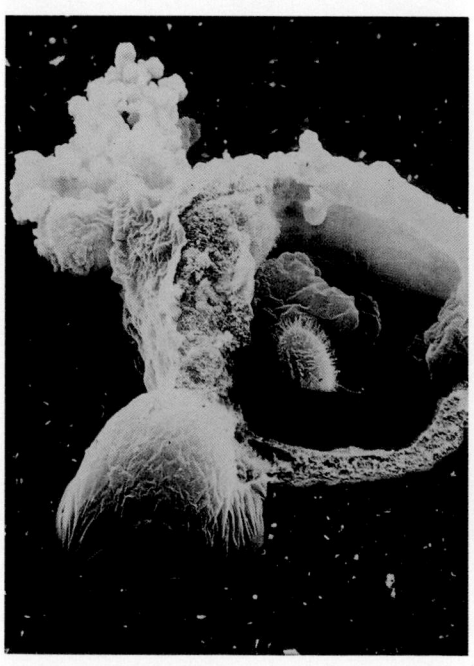

B

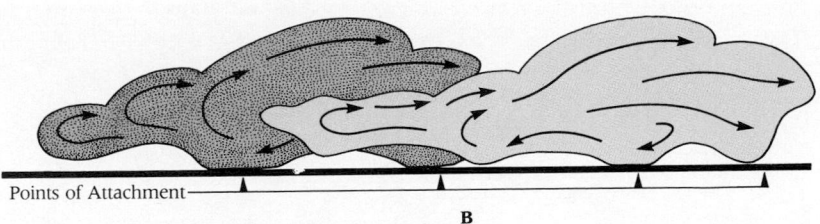

Points of Attachment

B

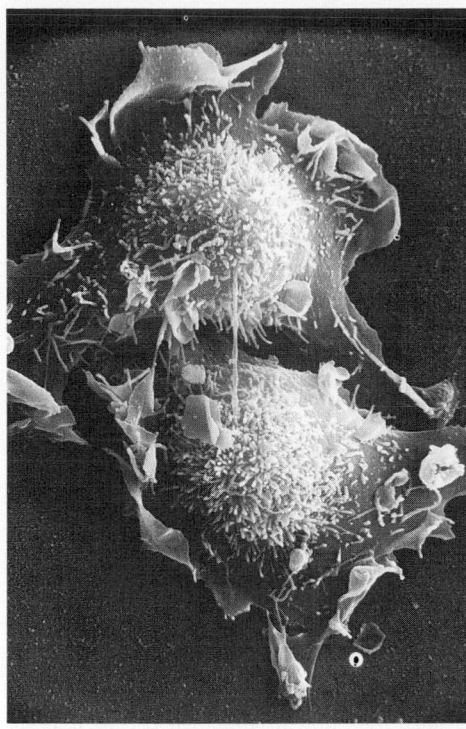

A

7-14 The ruffled membrane serves as a point of attachment in certain types of movement
(A) Scanning electron micrographs suggesting the movement of cells in tissue culture. (B) Some cells move by attachment and detachment from the underlying surface.

- Organelle movements such as deformations and translocations of chloroplasts, mitochondria, acrosomal filament extension, and the like
- Endocytotic cycle and related movements of endosomes
- Cytoplasmic streaming in protists and in plant, animal, and fungal cells
- Amoeboid movement (cell movement by cytoplasmic streaming)
- Motility of platelets (a blood cell involved in blood clotting)—their shape changes, transformations, and role in clot retraction
- Contractility of muscles
- Axoplasmic transport (transport of molecules through axons) in nerves
- Mitotic movements in plant and animal cell division (mitosis)
- Cytokinesis (division of parent cell cytoplasm into two daughter cells following mitosis) in plant and animal cells

These movements, many of which occur in almost all cells—prokaryotic and eukaryotic—are the result of the activities of one or both protein-based structures, microtubules and microfilaments (see Chapter 5). *Microfilaments* are extremely fine, unbranched fibers of varying length that are only 5–7 nm in diameter. *Microtubules* are unbranched strands, about 25 nm in diameter, that vary in length from a few nanometers to many micrometers. Both of these elements are constructed from distinctive proteins. Microfilaments are made of *actin*. Microtubules are composed of two proteins that together are termed *tubulin*. In the next sections we examine the role of microfilaments and microtubules in motility.

MOVEMENTS BASED ON MICROFILAMENTS

Muscle cells are manifestly motile—that is, they display movement and generate it. That other cells are motile too was established when Anton van Leeuwenhoek observed swimming sperm cells through his primitive microscope in the seventeenth century.[1]

Amoeboid Movements

Amoebas creep about freely and change their shapes. **Amoeboid movement** is characteristic of amoebas, as well as of other protists, but it also occurs in many cells of multicellular animals—for example, leukocytes, macrophages, platelets, and migrating embryonic cells.

The sequence of events in amoeboid movement is indeed striking (Figure 7-13). First the cell develops one or more bulges called **pseudopods** ("false feet"). A pseudopod attaches itself to a surface under the cell, and a **mantle** of the cytoplasm—that is, an outer layer, which is strikingly free of organelles—extends itself forward, the cytoplasm in the body of the cell streaming in the "intended" direction. The streaming process fills and enlarges the extended pseudopod. Soon the pseudopod contains most of the cell and it is no longer distinguishable from the cell body. The process of **contraction** retracts some pseudopods while at the same time new pseudopods form. As a result, the entire cell moves—at a few centimeters per hour.

A variant of this pattern is seen in the movement of **fibroblasts** across a glass culture dish. They move by a sequence of alternating attachments and detachments to the glass surface (Figure 7-14). The point of attachment is a delicate feathery

---

[1] The year 1976 was the tricentennial of the discovery of sperm. Although it was long known that semen causes conception, it was not known what factor in the semen is responsible. In reporting his discovery to the Royal Society of London, Leeuwenhoek wrote, "If your Lordship should consider such matters . . . offensive to the learned, I earnestly beg they be regarded as private, and either published or suppressed, as your Lordship's judgment dictates." Many argued that sperm cells are parasites unrelated to fertility.

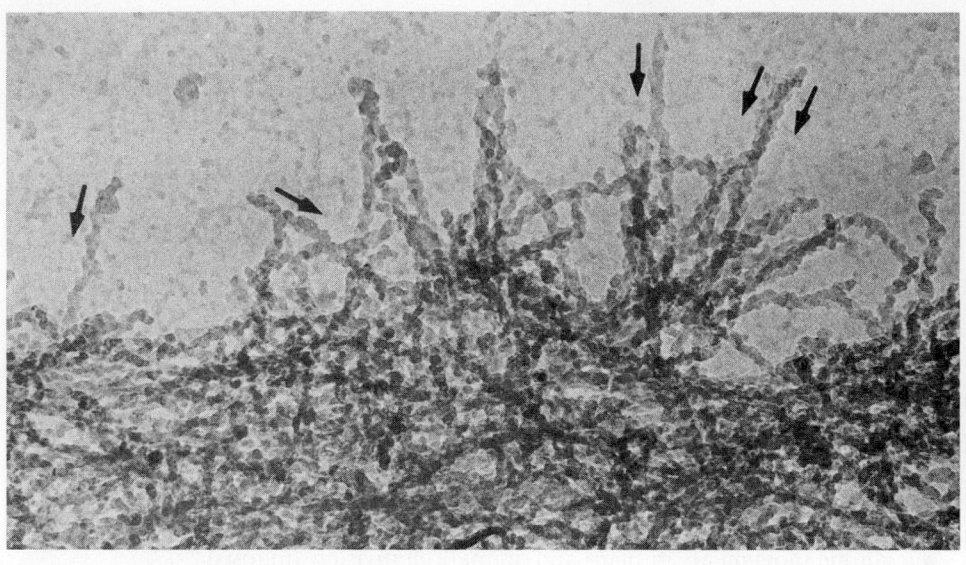

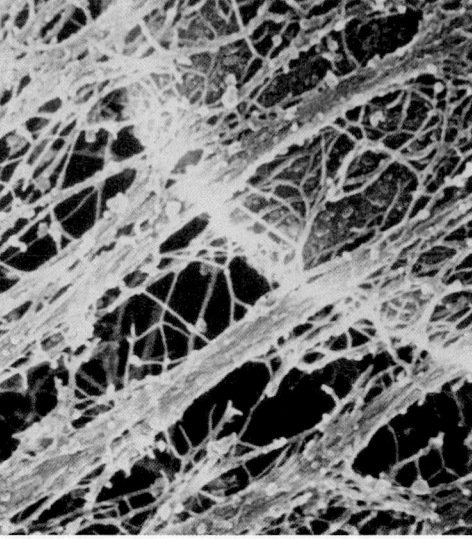

A

B

portion of an extended pseudopod called a **ruffle,** or **ruffled membrane.** Ruffling reflects the random arrangement of cross-linked actin filaments beneath the cell membrane.

Actin and Myosin

Biologists have been aware for decades of the presence in muscle cells of the two proteins called **actin** and **myosin.** Their central importance in muscle contraction was also recognized long ago. It thus came as something of a surprise when, in 1966, actin was discovered in plasmodia (a multinucleated mass of protoplasm occurring at certain stages of the life cycle) of the slime mold *Physarum*—structures that display amoeboid movement. In 1969, myosin was also identified in protists, fungi, and animal and plant cells. These were the first revelations that actin and myosin, so long familiar to muscle physiologists, occur as well in nonmuscle cells. Within a few years, they were identified in virtually all eukaryotic cells over the full evolutionary range.

Actin is not only universally present; it is abundant, comprising up to 15% of the total protein in actively motile cells like amoebas and blood platelets and 5 to 10% in migrating fibroblasts—a higher concentration than of any other cell protein! Often it is the major single protein of the cell. Actin has a singular property. Individual actin molecules, or monomers, can self-assemble into a variety of three-dimensional polymers called **actin filaments.** The capacity of actin to polymerize into filaments under precisely regulated conditions gives it an essential role in most forms of cell motility.

Under certain conditions the opposite ends of actin filaments are recognizably different in electron micrographs (Figure 7-15A). They are referred to as the **"barbed"** or (+) end and the **"pointed"** or (−) end. In the process of polymerization, monomers add on more rapidly to the (+) end.

In muscle, actin forms the **thin filaments**—elongated structures made up of globular actin monomers (Chapter 31). Myosin makes up the contrasting **thick filaments.** In nonmuscle cells, the actin filaments display somewhat different configurations in different cell types (Figure 7-15B), while thick filaments of myosin are rarely seen. In nonmuscle cells, myosin is revealed by special techniques to be closely associated with actin microfilaments. Not surprisingly, a sliding filament model similar to that now believed to account for muscular contraction has been offered to explain in part the motility of nonmuscle cells. Because the analogy with muscle is blurred by the absence of a precise structural framework in nonmuscle cells, the diversity of microfilament organization and interactions suggests that other mechanisms may also underlie the contractile movements of microfilaments. Nonetheless, the similarities between actin and myosin in muscle and nonmuscle cells are sufficient to justify the hypothesis that movement in both systems is the result of similar gliding mechanisms.

In effect, then, cells have little "musclelike structures" within them that can

7-15 Actin filaments play an important role in cell motility
(A) *Electron micrograph of actin filaments isolated from a neutrophil. The arrows indicate the orientation of pointed ends of the filaments.* (× 18,000) **(B)** *Scanning electron micrograph showing actin-containing filaments in a fibroblast. Also see the photograph opening this chapter.* (× 25,000)

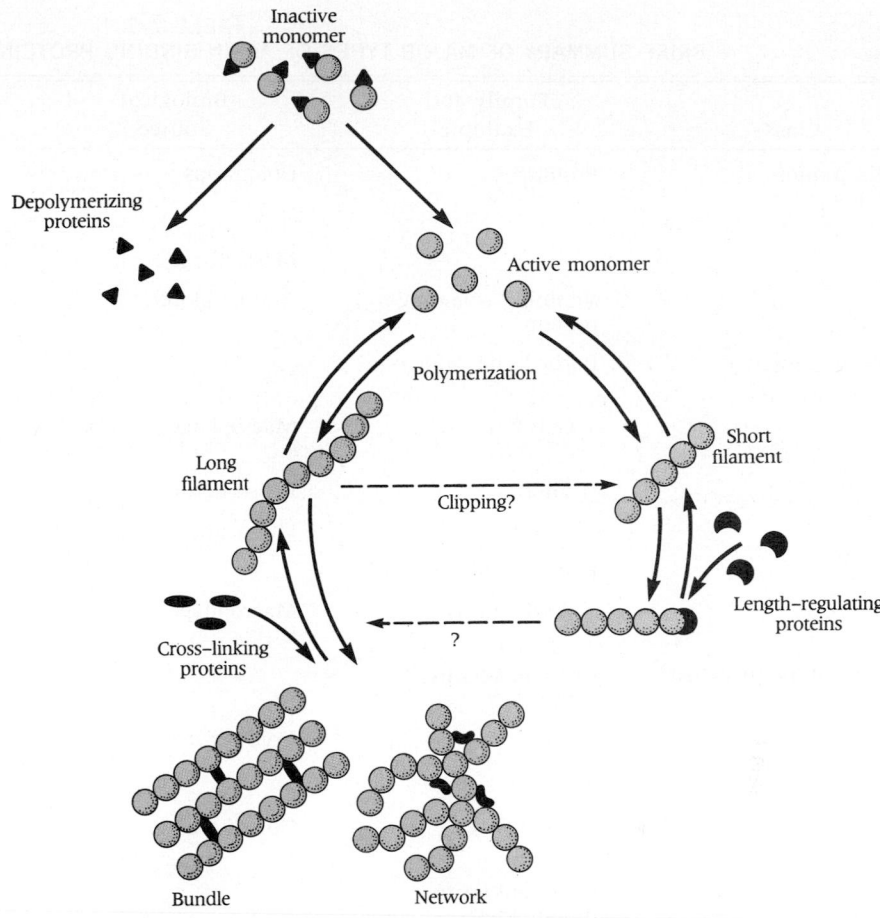

7-16 Actin-binding proteins regulate the assembly and complexity of actin filaments
Depolymerizing proteins, length-regulating proteins, and cross-linking proteins act to modify actin filaments.

produce local contractions. The result is cytoplasmic movement. The regulation of such movement appears to take place in the narrow zone at the cytoplasmic periphery, just below the cell membrane. The cell surface receives signals from the environment that provide information on the existence and location of food, light, moisture, poisons, or whatever. These signals are transmitted directly or indirectly through the cell membrane to the contractile machinery.

Actin also participates in other specialized movements of certain parts of the cytoplasm (see Chapter 12).

The Regulatory Role of Actin-Binding Proteins

Since polymerization of actin into filaments is a key event in cell motility, we must ask how this process is regulated. The answer is, in large part: by the many **actin-binding proteins** described in recent years—so many, in fact, that an editorialist in the journal *Nature* apologized to readers in late 1984 for the "seemingly endless proliferation of actin-binding proteins."[2]

The role of actin in cell motility is determined by the extent to which it polymerizes. Within cells, only a portion of the actin is usually polymerized. A repertory of actin-binding proteins governs the time, place, and extent of polymerization and organizes the resulting filaments into higher-order structures of increasing size and complexity. Thus these proteins are regulatory. Specific regulatory proteins probably exist for each step of the assembly process (Figure 7-16). As of 1989, it appeared that each cell contains dozens of regulatory proteins!

[2] As in all rapidly evolving fields of research, there is a severe nomenclature problem in this one. The favorite cells studied by investigators are brush-bordered intestinal epithelium, leukocytes and macrophages (motile animal cells), and the amoeba *Acanthamoeba* and the slime molds *Dictyostelium* and *Physarum* (motile "primitive" organisms). Newly discovered actin-binding proteins are given invented names by their discoverers (such as villin, fragmin, gelsolin, depactin) even though many of these picturesquely named proteins will turn out to be identical. Cases have already occurred in which the same protein was discovered and named three times! That problem is likely to continue because, as someone has remarked, "Scientists would rather use each others' toothbrushes than their nomenclature."

TABLE 7-1

BRIEF SUMMARY OF MAJOR TYPES OF ACTIN-BINDING PROTEINS WITH SELECTED EXAMPLES

Class	Family and Example	Biological Source	Molecular Weight	Role
Actin monomer binders	Profilins	Ubiquitous	13,000	Binds to end of actin monomers. Inhibits actin polymerization.
		Macrophages	15,500	Stabilizes actin monomer pool.
	Vitamin D–binding protein	Animal blood	58,000	
Capping proteins	Barbed-end cappers:			Prevents further polymerization and lengthening of filament.
	Gelsolin	Macrophages, brain, etc.	90,000	Followed by severing of filaments that shortens them.
	Villin	Brush border	95,000	
	Fragmin	Physarum	42,000	
	Pointed-end cappers:			
	Acumentin	Macrophages, leukocytes	65,000	
Cross-linking proteins	Cross-linkers to actin:			Forms filament bundles and filament networks.
	Spectrin	Erythrocytes	480,000	Acts as "spot welds" between actin filaments. Promotes gel formation.
	Filaments	Ubiquitous	540,000	Promotes perpendicular branching–actin fiber networks
	Cross-linkers to membranes:			Stabilizes cell structure.
	"Protein 110K"	Intestinal brush border	110,000	A myosin.

Despite the large number of reported actin-binding proteins, it is now clear that they will all eventually fit into a limited number of classes, each including several families. To grasp this classification, we should be aware that the actin monomer is a globular protein molecule with a molecular weight of 42,000. When it polymerizes, it forms a filament of molecular weight of 1.5–3.0 million. Actin-binding proteins are classified according to what part of the actin monomer or filament they bind to—and when. Table 7-1 summarizes what we now know of these classes and families and lists some of the individual proteins, their properties, sources, and functions. Note that some of these proteins—for example, gelsolin—are regulated by local changes in Ca^{2+} concentration. It binds and severs large filaments only when Ca^{2+} is high. Direct evidence that it binds the barbed end of the actin filament is seen in the electron micrographs in Figure 7-15A. Proteins not listed in Table 7-1 include some that bind to actin along the sides of the filament (tropomyosin) or cross-link to myosin. We will discuss these in connection with muscle contraction.

A major difficulty in this field, one that is inherent in all of molecular biology, is simply this: To what extent can events observed in the test tube be assumed to reflect events in the cell? It is easy to make actin polymerize *in vitro* into a filament network like that in Figure 7-17. Whether or not this is what happens in cells remains to be seen.

MOVEMENTS BASED ON MICROTUBULES

In Chapter 5 we surveyed the molecular structure of microtubules, which are components of many dissimilar cellular structures (Figure 7-18). We will later see (Chapter 31) that this mechanism, which is sketched in Figure 5-22, strikingly resembles the sliding filament mechanism of striated muscle contraction.

With the discovery of conditions that permitted the *in vitro* assembly of tubulin into microtubules, it was possible at last to study their biochemical properties. Microtubule assembly could now be demonstrated in preparations of brain homogenates

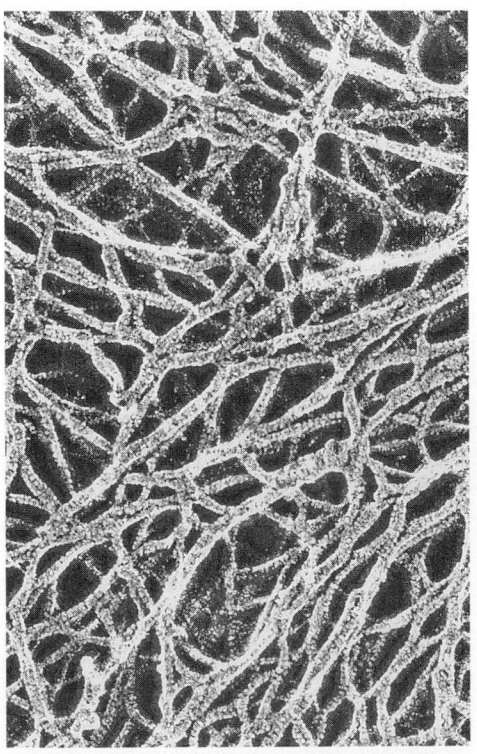

7-17 Formation of actin filaments in vitro
This electron micrograph shows purified actin molecules that form a network of filaments similar to that seen in the cell cytoplasm.

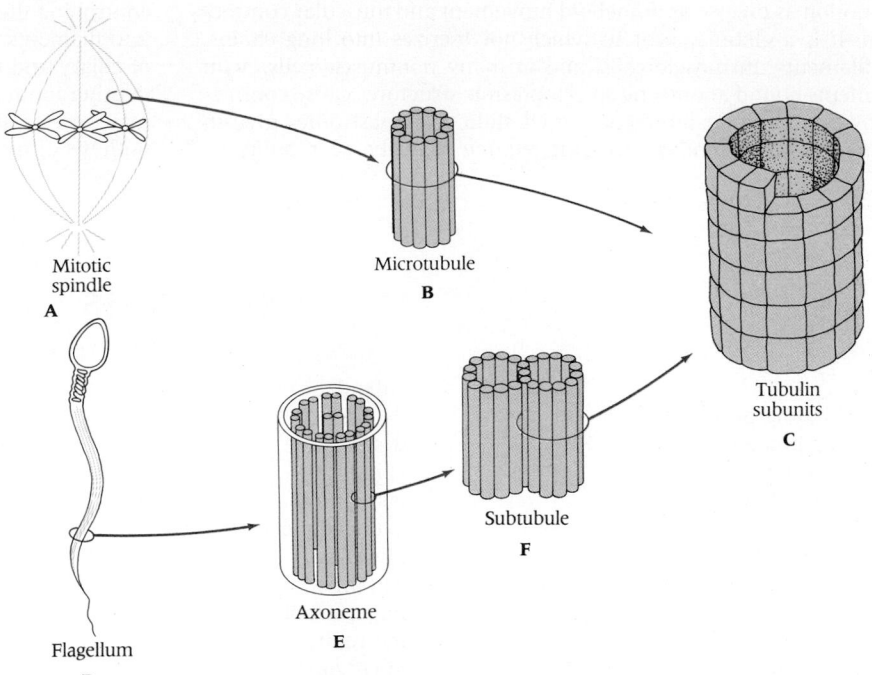

7-18 Microtubules are found in many dissimilar cellular structures
The sequence **A**, **B**, and **C**, shows, in progressively higher magnifications, the microtubular structure of a mitotic spindle fiber (Chapter 12). The sequence **D**, **E**, **F**, and **C** shows, in progressive magnifications, the microtubular structure of a sperm tail (Chapter 21).

Mitotic
spindle
A

Microtubule
B

Tubulin
subunits
C

Subtubule
F

Axoneme
E

Flagellum
D

containing Mg^{2+} ions, the ribonucleotide GTP, and a calcium chelator (a molecule that binds calcium so avidly that it effectively removes it from the reaction mixture). Tubulin was subsequently assembled from a number of other sources, including flagellar outer doublet microtubules.

By examining the incorporation of radioactive precursors into regenerating flagella, investigators determined that flagella are assembled principally at their distal (far) tips. Tubulin obtained from cytoplasmic microtubules also exhibits directional assembly. For example, addition of brain tubulin onto isolated basal bodies, centrioles, or axonemes resulted in microtubule polymerization predominantly onto the distal ends of these organelles.

Recent studies of microtubule assembly *in vitro* have shown that polymerization occurs at one end of the microtubule and depolymerization occurs at the opposite end. Thus the two ends of the microtubule have different critical concentrations for assembly, and, at polymerization equilibrium, the rate of tubulin addition onto one end of the microtubule would equal the rate of removal from the other end. The polarity of microtubules, as revealed in this directional polymerization, may permit them to function in directional intracellular movements. Of these movements, those exhibited during cell division (mitosis) have generated the most interest, and we will consider them later.

Our knowledge of motility has indeed come a long way since Leeuwenhoek first observed the movements of cilia and flagella in 1676.

SUMMARY

This chapter considers three phenomena essential to the survival and functioning of many cells: tissue and organ formation, cell-to-cell communication, and cell motility.

Tissue and organ formation involve differentiation of cells from unspecialized progenitors into groups of cells with highly developed individual structures and functions. This chapter presents a brief overview of different plant and animal tissue types, the functions of which will be discussed more thoroughly in later chapters.

Cell communication includes information exchange and encompasses many varieties of cellular interaction. Direct communication—that is, interaction that occurs via cell contact—often takes place at specialized areas called cell junctions. Impermeable junctions seal cells so tightly together that the intercellular space is obliterated and large molecules cannot slip between the joined cells. Adhering junctions do not anchor cells together as tightly. Communicating junctions differ from the other two in having channels

that allow direct cytoplasmic contact between the joined cells. Direct communication may also take place via cell adhesion without specific junctions. Molecules mediating this adhesion include certain glycoproteins, and receptors specific for them. Cell adhesion can stimulate many processes, including cell migrations during embryonic development.

Indirect communication (which will be discussed further in Chapter 28) involves the release of substances (hormones, growth factors, and so on) that can diffuse across great distances before they interact with other cells.

Motility is the ability of a cell to use metabolic energy to drive movement and perform mechanical work. All motility is due to one or both of two proteinaceous structures: microfilaments and microtubules. Microfilaments are made of the protein actin. Microtubules are made of two proteins that form a dimer called tubulin. Actin is found in almost all eukaryotic cells, and functions in types of motion as diverse as amoeboid movement and muscular contraction. It is a globular protein which polymerizes into long chains, or filaments. In muscle cells, and in many nonmuscle cells, actin filaments bound at one end to cytoplasmic structures cause contractions by molecular interactions that slide them past other myosin filaments. Actin-binding proteins regulate actin-based motility by controlling the extent to which the actin globules are polymerized into filaments. Microtubule-based movement, discussed in the case of ciliary and flagellar movement in Chapter 5, may also function in other intracellular movements. However, those related to mitosis may operate differently, according to the assembly-disassembly hypothesis of microtubule function.

KEY TERMS

actin
actin-binding protein
adhering junction
amoeboid movement
apical meristem
axon
belt desmosome
cambium
cell-adhesion molecule (CAM)
cell migration
collenchyma
columnar cell
communicating junction
conductive tissue
connective tissue

cuboidal cell
dendrite
desmosome
epidermis
epithelium
extracellular matrix
fibroblast
gap junction
impermeable junction
meristem
microvilli
myosin
organ
parenchyma
periderm

phloem
primary growth
protective tissue
pseudopod
sclerenchyma
spot desmosome
squamous cell
synapse
thick filament
thin filament
tight junction
tissue
vascular tissue
xylem

QUIZ QUESTIONS

1. Which of the following types of intercellular communication would be most appropriate in a situation where controlled passage of materials across a tissue was required?
   A. belt desmosomes
   B. gap junction
   C. spot desmosomes
   D. tight junction
   E. ruffled membrane

2. Mechanical support of plants is a role of _____ tissues.
   A. epithelial
   B. sclerenchymal
   C. parenchymal
   D. meristematic
   E. epidermal

3. Which of the following molecules is *not* an actin-binding protein?
   A. ubiquitin
   B. spectrin
   C. tropomyosin
   D. gelsolin
   E. protein 110K

4. Which one of the following statements is false?
   A. Cardiac muscle is a type of striated muscle.
   B. Smooth muscle lacks striations.
   C. Squamous cells are flattened.
   D. Large amounts of intercellular material occur in connective tissues.
   E. Growth in plant stem length occurs at the cambium.

5. Motility is a function of which of the following cell structures or functions?
   A. Flagella
   B. Pseudopodia
   C. Mitosis
   D. Muscle contraction
   E. All of the above

ESSAY QUESTIONS

1. Contrast the manner by which plant and animal cells communicate. Describe the participating structures.

2. What role might the cytoskeleton play in amoeboid movement? How do the roles of actin, myosin, and the cytoskeleton compare with those of the body's musculoskeletal system?

3. Contrast microtubule and microfilament structure and function.

4. Which of the following cell types are rich in actin? Which are rich in tubulin? Neurons; cardiac muscle; intestinal epithelial cells; macrophages.

5. What structures explain amoeboid movement?

REFERENCES AND SUGGESTIONS FOR FURTHER READING

Bretscher, M. S. (1984). Endocytosis: relation to capping and cell locomotion. *Science 224*, 681–686.

Review of the endocytotic cycle that is initiated with the formation of coated pits in the plasma membrane.

Bretscher, M. S. (1987). How animal cells move. *Scientific American 257*, December, 72–90.

Animal cells move by bringing bits of the plasma membrane into the cytoplasm by endocytosis and then recycling them in a directed way. The process is here explained.

Edelman, G. M. (1984). Cell-adhesion molecules: a molecular basis for animal form. *Scientific American 250*, April, 118–129.

A discussion of the molecules that mediate cell-to-cell adhesion, that appear during development, and that may regulate cell movements and thus the shape of a developing embryo and its organs.

Garrod, D. R., and Nicol, A. (1981). Cell behaviour and molecular mechanisms of cell-cell adhesion. *Biological Reviews 56*, 199–242.

A review of those cell-to-cell adhesive interactions that position cells within embryos.

Gibbons, I. R. (1988). Dynein ATPases as microtubule motors. *Journal of Biological Chemistry 263*, 15837–15840.

Recent views on the role of dynein.

Hollenbeck, P. (1989). Cell motility: Dissecting a molecular motor. *Nature 338*, 294–295.

Dynein and myosin have many similarities. This brief article discusses another similar protein, kinesin, which generates movement along microtubules in the direction opposite to dynein and promotes organelle movement.

Hynes, R. O. (1986). Fibronectins. *Scientific American 254*, June, 42–51.

Evidence that these adhesive proteins act as biological organizers by holding cells in position and guiding their migrations.

Lackie, J. M. (1986). *Cell Movement and Cell Behavior*. Allen & Unwin, Boston.

Probably the first general text on the broad subject of cell movement.

McIntosh, J. R., and Porter, M. E. (1989).

Enzymes for microtubule-dependent motility. *Journal of Biological Chemistry 264*, 6001–6004.

A brief review.

Pollard, T. D., and Cooper, J. A. (1986). Actin and actin-binding proteins. A critical evaluation of mechanisms and functions. *Annual Rev. Biochem. 55*, 987–1035.

A masterful summary of an exciting field.

Schliwa, M. (1986). *The Cytoskeleton. An Introductory Survey*. Springer-Verlag, New York.

A readable and comprehensive review of the filamentous networks in the cytoplasm of eukaryotic cells. Very good bibliographies.

Sloboda, R. D. (1980). The role of microtubules in cell structure and cell division. *American Scientist 68*, 290–298.

Discussion of the role of microtubules in spindle formation and function.

Weber, K., and Osborn, M. (1985). The molecules of the cell matrix. *Scientific American 253*, October, 110–120.

Good review with fine illustrations.

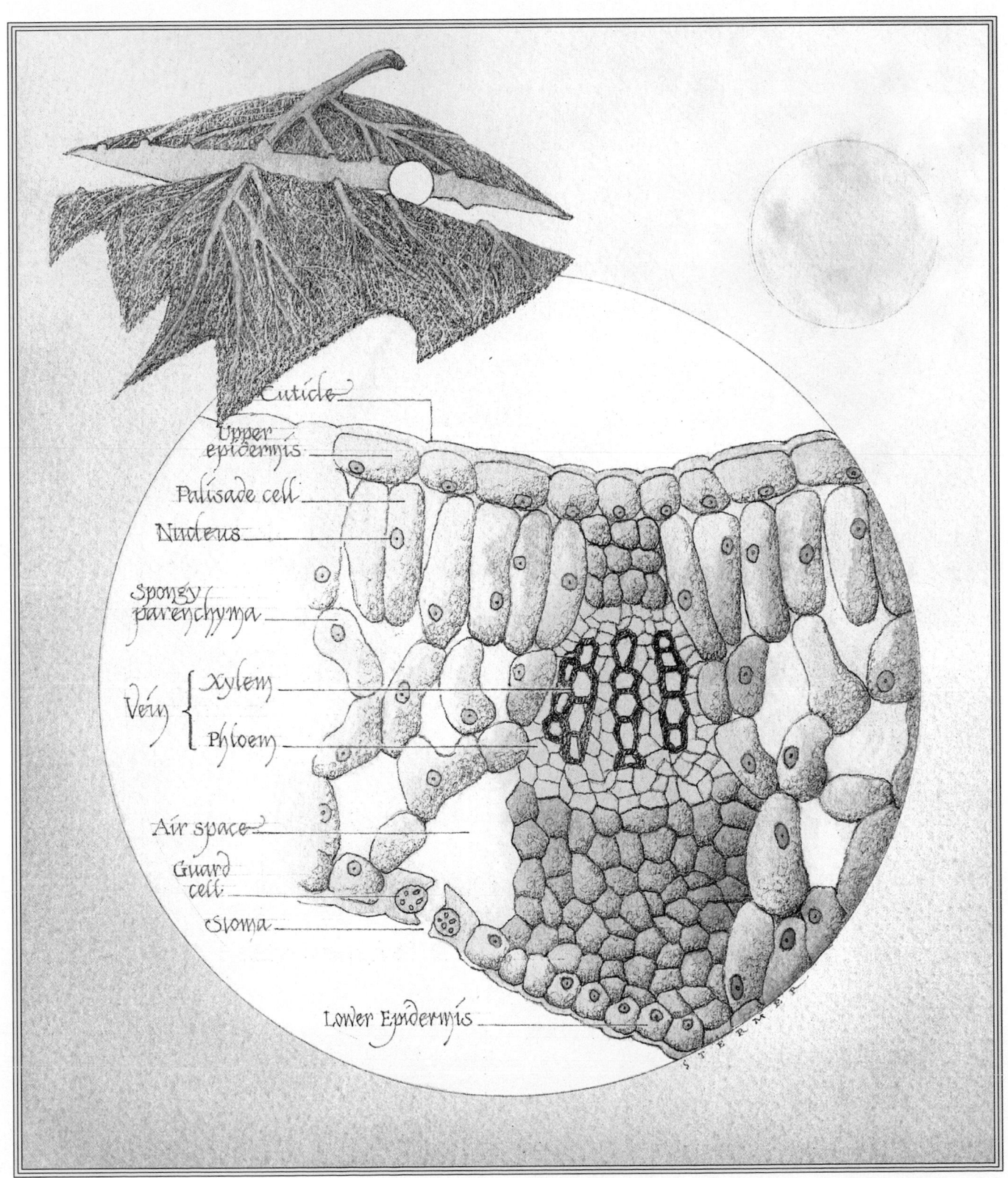

Cutide

Upper epidermis

Palisade cell

Nucleus

Spongy parenchyma

Vein { Xylem

Phloem

Air space

Guard cell

Stoma

Lower Epidermis

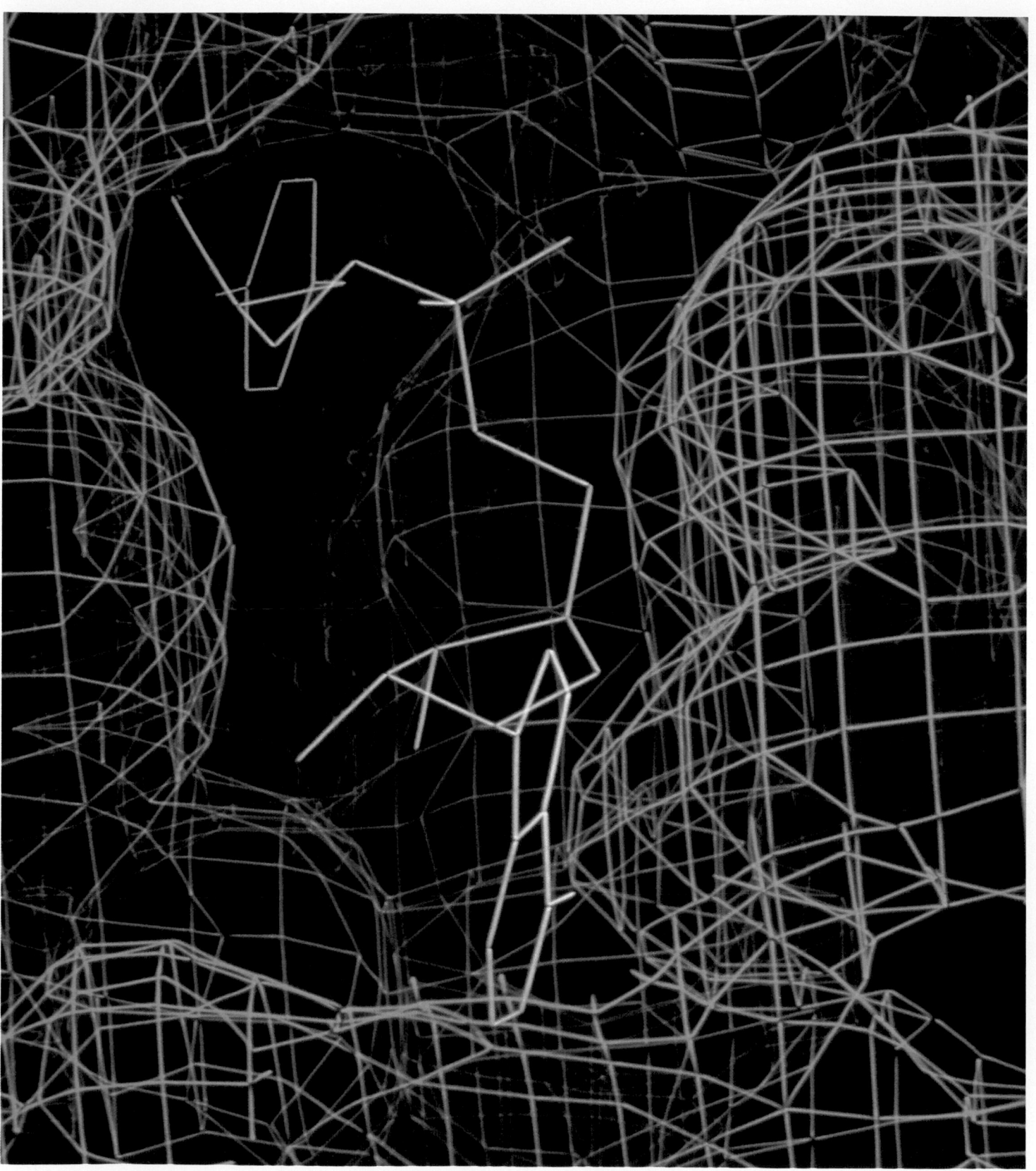

| CHAPTER 8 | METABOLISM AND THE TRANSFER OF ENERGY |

Metabolism and the transfer of energy
Enzymes are the remarkable catalysts that operate the machinery of metabolism. Modern biochemistry has probed deeply into their structure and mechanism. Pictured is a three-dimensional, computer-generated model of a portion of one enzyme, tyrosyl tRNA synthetase.

We introduced the major biomolecules in Chapter 4 and the structures and functions of cells in Chapters 5, 6, and 7. In a real sense, these were a cast of characters for the spectacular play about to unfold—a drama in which molecules change identities, cells defend themselves against hostile forces, and powerful agencies seem at times to violate the law.

IMPORTANCE OF ENERGY

Energy and work are defined easily enough in familiar terms: **Energy** is the capacity to do work, and **work** is something being accomplished. Nevertheless, the concept of energy is an elusive one. The definition tells what it does, not what it is. We will try now to clarify that definition.

ENERGY IN THE WORLD

The transformations of energy from one form to another and the efficiency of the processes that convert energy into work are central issues of physics, chemistry, and biology—as well as of commerce, agriculture, and human civilization in general. Energy constantly flows to the Earth's surface and environment and out again. The prime source is the sun.[1]

Plants capture a small fraction—only about 0.02%—of the solar radiation striking Earth. In the process of photosynthesis, they store solar energy in the form of energy-rich carbohydrates (Figure 8-1). These organic compounds in turn supply the energy needed for the entire world of life. The energy in these stores is released by processes of oxidation approximately as fast as it is being stored. Over millions of years, some of the energy-rich organic matter extant in the world was buried in huge deposits, where it was transformed into petroleum, coal, natural gas, and the other fossil fuels that provide most of the energy required by industrialized societies. Though long buried, these fuels are nevertheless products of the solar energy and photosynthesis of another time.

This natural energy cycle, which gives rise to a large portion of Earth's energy resources, lies at the heart of great issues of human affairs. Wars have been fought, empires have risen and fallen, and changes have been wrought in society as human beings gained and lost control over natural energy sources. The Industrial Revolution, for example, could proceed only after people learned how to transform the energy of coal into mechanical work with steam engines. The present energy crisis arises from the fact that Earth's deposits of fossil fuel may be used up in the relatively near future—just as living fuels (notably wood) were all but exhausted in parts of

[1] Other energy sources also exist. Small amounts of heat come from Earth's interior. The ebb and flow of tides are powered by the gravitation of Earth, the moon, and the sun; and the radiation that surrounds us in greater or lesser amount comes from cosmic radiation and the decay of radioactive materials.

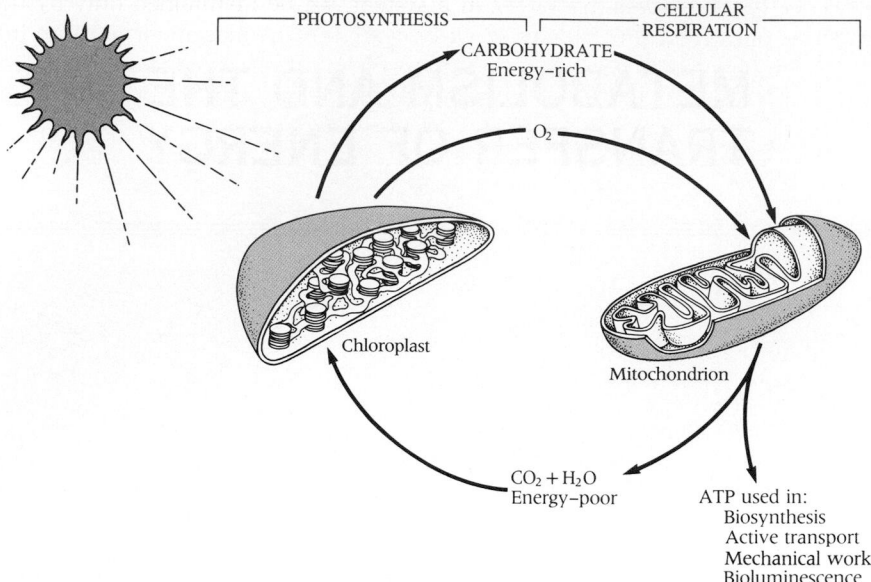

All life depends on the sun for energy. Photosynthesizing organisms trap solar energy, converting it to carbohydrate, which can then be transformed into ATP by cellular respiration. Prokaryotes can carry out the same kinds of energy transformations, but not in chloroplasts or mitochondria.

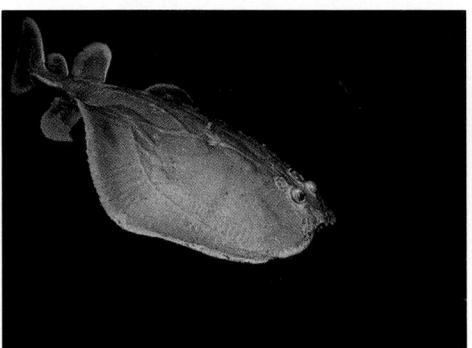

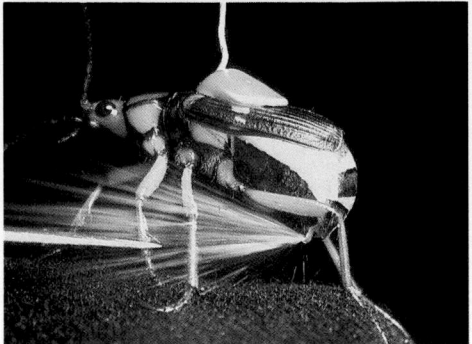

the world in the past. A shift in the course of human history comparable to the Industrial Revolution will doubtless follow an efficient conversion to nuclear, solar, tidal, hydroelectric, or geothermal sources of energy.

ENERGY AND ORDER

We advance science not only by discovering answers to problems, but by discovering the problems themselves. One of the great insights of modern physics was the perception that explanations are needed for order in the universe, since every process we know, left to itself, drifts toward disorder. These ideas are of fundamental importance to biology.

This inquiry began with reflections on the nature of heat and the motion of molecules. It was soon recognized that heat and molecular motion are one and the same thing—indeed, that heat *is* the motion of molecules. That is very different from stating that heat *causes* the motion of molecules. Further study gave rise to **thermodynamics,** a science that took its name from roots meaning heat and motion. It is a misnomer because thermodynamics concerns all forms of energy, not just heat. **Energetics** would be a better name.

Thermodynamic ideas can seem abstruse, but the basic principles are strikingly simple. They are concerned with such questions as these: What makes a chemical reaction take place? What explains the fact that heat flows from a hot body to a cold one and not the other way around? Why are many chemical and physical processes irreversible when allowed to run by themselves? Why do some but not all chemical reactions occur spontaneously? Consideration of such issues led to present understandings of order and randomness.

These are familiar themes of daily life. What, for example, is the difference between a Sunday newspaper and a pound of blank paper and a bottle of ink? Chemically, each includes the same sorts of molecules. Physically, they have the same weight and volume. The difference, obviously, is in the arrangement of the ink molecules—in other words, the order, the nonrandom spatial arrangement, of certain molecules. The opposite of high order is randomness.

We now recognize that what we call **information** depends upon a state of order as opposed to a state of randomness. A random splatter of ink on the paper would not make for interesting reading. If fingerprints were randomly patterned, they would give the police little information. The same is true for the ordered grooves of a phonograph record, the magnetic impulses of a computer disk, the ridges on a door key—and, as we shall see, the arrangements of amino acids in proteins and nucleotides in DNA (the molecules that make up the genes). Those ordered arrangements contain all of the information transmitted by heredity. Compelling evidence for that statement is seen in the physical similarity of identical twins. Clearly, the appearance of each twin was not a consequence of totally random processes.

8-2 Some uses to which living organisms put metabolic energy

(A) *The electric ray* (Torpedo nobiliana) *discharges 500 volt electric shocks that can kill or stun its prey.* (B) *Thousands of male fireflies,* Pteroptyx malaccae, *in a Malaysian mangrove tree flash synchronously and rhythmically through the night, providing a spectacular communal sexual display.* (C) *A bombardier beetle,* Brachinus crepitans, *in the act of discharging hot toxic secretions from its defensive organ. The beetle was provoked by a forceps pinching its front leg. Note how accurately it aims the spray.* (D) *A carnivorous Venus flytrap,* Dionaea muscipula, *seizing its prey.* (E) *A bacterium,* Escherichia coli, *rotates its flagellum.* (× 7000) (F) *Protists such as this amoeba,* Amoeba proteus, *are highly motile.* (× 160).

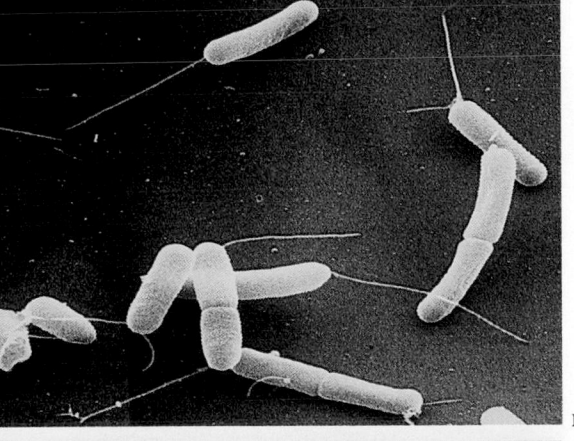

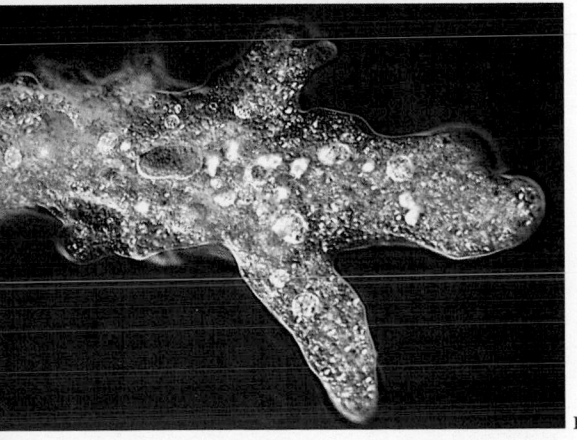

It was early recognized that order in a system can be maintained only by the expenditure of energy. Surely, energy was invested in the placement of those ink molecules on paper—by writers, editors, press operators, and so on. Other familiar examples surround us. If a garden is planted in a highly ordered arrangement of rows and designs, energy must be constantly expended to keep it from drifting toward disorder—unwanted weeds, ragged growth, and desiccation. Surely, disorder is the essence of today's environmental pollution.

This concept is of obvious relevance for biology. The ordered structure of a living organism is a highly unlikely arrangement for a group of molecules to find themselves in. Yet, the almost incredible complexity of that arrangement is provided for in a genetic "blueprint" within a single fertilized egg cell. When biologists recognized that the enormous information content of genes must reside in a highly specific and ordered molecular pattern, as in a key or phonograph record, a major advance took place.

How is this order maintained? It is not an easy question in view of the statistical aspect of many events. When we shuffle a deck of cards, it would not be impossible to end up with the top half of the deck containing all the red cards and the bottom half all the black ones. But such a result would be highly improbable. That would be a state of very high order and common sense tells us that the unordered random arrangement we usually get when we shuffle cards is the highly probable outcome of a shuffle.

Similarly, when we put an ice cube into a glass of water, the water becomes colder as its heat enters the ice and dissolves it. Although the ice is cold, it contains *some* heat, and the statistical probability is not zero that the ice cube could give off part of its heat and make the surrounding water *warmer* while the ice cube itself became colder. The fact that such a thing appears never to happen, that heat energy moves only in one direction, is one of the many phenomena explained by the laws of thermodynamics.

The larger the numbers in such situations, the smaller the probability of a highly ordered outcome. But that probability is never zero. Large numbers also permit certain predictions. If we heat a gas in a closed vessel, the pressure within it will increase. So states Charles' Law. Yet the gas consists of a great many individual molecules, which swarm in all directions, collide with each other, and describe zigzag paths along which they travel at great speeds. Although we can accurately predict the pressure change in the whole gas following the application of a certain amount of heat, it would be difficult indeed to predict the travels of an individual gas molecule. In fact, we must consider the behavior of the whole gas in terms of the statistical probability that all of its molecules will collectively behave in a predictable manner.

ENERGY IN BIOLOGY

We are all aware of energy considerations in the context of the inanimate world of machines, electricity, and steam. But biological transformations of energy in living organisms also influence human affairs. The whole of agriculture is a means for harnessing solar energy. Photosynthesis by green plants and their ingestion and oxidation by animals and other organisms to yield energy for the performance of work are the two major elements in the massive biological energy cycle. The amounts of energy circulated in this cycle far exceed the total energy exchanged by all existing machines. Indeed, the energy-transforming systems of living cells are much more efficient than anything yet constructed by human engineers. The study of energy transformations in living organisms has been termed **bioenergetics.**

The energy-transforming machinery of cells takes various forms. Some living organisms have acquired strange and fascinating devices for the transformation of energy (Figure 8-2). An electrical ray can deliver a shock of several hundred volts. When fireflies flash their lanterns, chemical energy is converted into light. The astonishing bombardier beetle can fire its "cannon"—which uses poisonous quinones as ammunition—by converting the chemical energy of hydrogen peroxide into the mechanical energy of pressure.

We can draw useful analogies between a living organism and a nonliving machine. (Remember the pitfalls of analogies! One system may resemble another in some ways, but not in others.) A comparison of the horse and the automobile is a good analogy to keep in mind.

- Both do a good deal of mechanical work.
- Both have maximum capacities for work that cannot be exceeded.
- Both require a continuous input of materials that serve as fuel—food for the horse, gasoline for the automobile.
- Both derive energy from organic molecules in this food or fuel, largely by the oxygen-consuming process of oxidation.
- Both transform energy of the food or fuel to several other forms—heat, sound, mechanical energy, and so on.
- Finally, food or fuel is not consumed completely. Waste products of the conversion are emitted by both the horse and the automobile—feces, urine, sweat, and CO_2 by the horse; exhaust, water, CO, and carbon particles by the car.

THE LAWS OF THERMODYNAMICS

Thermodynamics is a statistical science. It deals with the collective behavior of many atoms or molecules. Further, it deals only with relatively gross properties of matter in the bulk—like pressure, volume, temperature, and chemical composition—that are easily measured by simple methods.

First, a few definitions. In thermodynamics, the **universe** has three parts—the system, its boundary, and the surroundings (Figure 8-3). The **system** is all the matter within a defined region that is separated from the rest of the universe. It is the part of the universe under investigation. The **boundary** separates the system from the rest of the universe. Different types of boundaries separate the system from its surroundings. A **closed,** or **isolated, system** has a boundary that prevents any exchange of energy or matter between the system and its surroundings. In an **open** system, the boundary allows free exchanges of both energy (as work or heat) and matter. Living organisms are open systems.

Systems are characterized in terms of measurable quantities such as volume, temperature, and chemical composition. These are the **properties of the system.** When we know these properties, we know the **state of the system.**

Thermodynamics characterizes the behavior of a system by considering the **initial state** and **final state** of that system. Usually, we express its behavior by determining the *net change* in the properties of the system as it proceeds from its initial state to its final state. Such changes are conveniently designated by the Greek letter symbol Δ (pronounced "delta"). For example, H symbolizes heat content and ΔH means "change in heat content" between initial and final states. Remember that thermodynamics tells us about net changes in state properties. It says nothing about the paths that may have been taken to produce those changes.

The concept of **equilibrium** (to which we were introduced in Box 3-2) is related to the direction processes take. It is easy to define the equilibrium point in an isolated system: it is reached when the properties of the system are no longer changing. It is harder to ascertain an equilibrium point in open systems such as living organisms. Nonetheless, the concept of an equilibrium state *is* important in biological systems. In part, its importance lies in what it tells us about direction taken by spontaneously occurring processes.

According to thermodynamics, the macroscopic behavior of all matter in the universe conforms to its two major laws.[2]

- The **first law of thermodynamics** states that the total energy of a system and its surroundings remains constant. In other words, the energy of the universe is conserved.
- The **second law of thermodynamics** states that the **entropy** (randomness) of the universe is always increasing. This can be paraphrased by stating that the universe always moves in a direction of increasing randomness.

Concerning the first law, remember that the energy content of systems takes two major forms—**work** and **heat.** Work—the result of a force acting over a distance—is done when, for example, a carton of books is lifted or a pen is pushed across a paper. However, some of the energy of all systems takes the generally

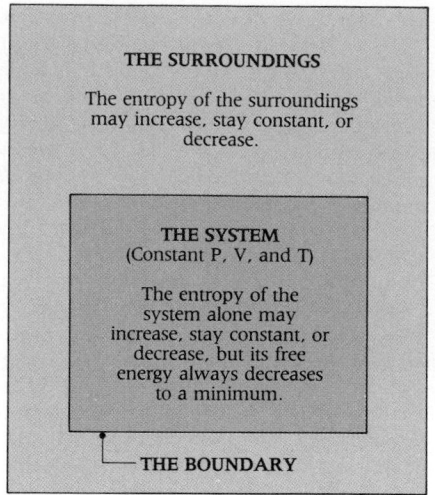

THE SURROUNDINGS

The entropy of the surroundings may increase, stay constant, or decrease.

THE SYSTEM
(Constant P, V, and T)

The entropy of the system alone may increase, stay constant, or decrease, but its free energy always decreases to a minimum.

THE BOUNDARY

THE UNIVERSE = SYSTEM + BOUNDARY + SURROUNDINGS

The entropy of the universe always increases to a maximum.

8-3 Thermodynamic description of the universe
The system, its boundary, and the surroundings make up the universe.

---

[2] There is a third law of thermodynamics which will not concern us here. It states that at a temperature of 0° Kelvin (absolute zero) the entropy of pure perfect crystals is zero.

less useful form of heat, a manifestation of the energy associated with random (disordered) molecular motion. The total energy of a system is the sum of the work energy and heat energy associated with each of its components. Whichever form the energy takes, it remains constant in amount. That is the first law.

The second law specifies the natural tendency for processes to go in one direction—for the water of Niagara Falls to go down and not up, for untended gardens to grow tangled and cluttered. We will shortly restate the law in formal terms, but we begin with the law's major mandate: the universe tends toward a state of maximum randomness. Thus events tend always toward an increase in entropy. It is true, of course, that many systems become *more* ordered with the passage of time. We will show that this can happen only when energy is applied to the system. That is why we are obliged to discuss the production and use of energy in living organisms in thermodynamic terms.

The First Law and Chemical Bond Energy

One implication of the first law of thermodynamics is that energy cannot be created or destroyed although it can be changed from one form to another. This is the principle of the conservation of energy.

For example, the mechanical energy of a waterfall may be converted into electrical energy by means of a turbine; electrical energy can be converted into light by a tungsten filament or heat on the kitchen stove; and so on. But never, through all of these transformations, can more energy be extracted than was invested originally. Nor will any energy be lost, although some may be transformed into forms that are useless in the performance of work (for instance, the heat of friction, the sound of random noise, and so forth).

Imaginary and Real Machines
This critical point deserves emphasis. We will recall that energy is either kinetic or potential. Kinetic energy is released energy that is associated with actual motion. Potential energy is just that—energy that is stored (potential) by virtue of the physical positions or arrangements of objects.

Consider an electric motor that is lifting a heavy object—say, a box of biology textbooks. While the box is moving, it has a certain amount of kinetic energy that is related to its mass and velocity. In addition, as it is raised in the gravitational field, it acquires potential energy that is related to its mass and its position. If all of the electrical energy used by the motor in lifting the box were utilized with 100% efficiency, the height of the box when it came to rest would depend only on the amount of electrical energy furnished the motor. In this hypothetical case, the potential energy of the box would now be equivalent to the electrical energy used to lift it.

Now couple the box to the crank of a generator (a device that transforms mechanical energy into electrical energy). What would happen if the box were allowed to fall from its perch? As it fell, its potential energy would rapidly diminish as it took the form of kinetic energy. If the coupling mechanism were perfect, the falling box would be doing work as it cranked the generator. Thus we would see, first, the conversion of potential energy into kinetic energy, and then the conversion of kinetic energy into electrical energy. In the ideal world of science fiction, this electrical energy could then be used to lift a similar box, the same distance, and the cycle could be repeated. If such a machine could be built—and it *cannot* be built—we would have constructed a perpetual motion machine. Why is such a machine impossible? Surely, it is not the first law that prohibits it, since that law's decree—total energy is conserved—has been obeyed.

The simple answer to this important question is that a real system of electrical motors, boxes to be lifted, and generators—no matter how carefully we built them—would squander some of its energy as heat. This heat loss is measurable if and only if we were to continuously add amounts of new electrical energy equivalent to the energy lost as heat. We could keep the cycle going indefinitely, but the cost would be a continuous source of fresh energy.

If energy lost as heat were not added back to the system, it would eventually run down. If a run-down system were enclosed in an isolated room so that none of the heat could escape, we would find that room's temperature higher at the end than when we started the engine. Calculations would show that all of the electrical energy initially supplied to the engine had been converted into heat energy. The first law is validated: energy was conserved.

The Energy of Chemical Bonds

We can readily apply this generalization to the energy transformations of living organisms. The horse and the automobile both convert the potential chemical energy of food or fuel to the kinetic energy of motion (as well as to heat, noise, and so forth). A process of oxidation in both systems releases energy from its potential state in the energy-donating molecules we call food or fuel. It is worth noting that the chemical compounds consumed by both animals and automobiles have the same source—the bodies (or the remains of bodies) of living organisms. Food, whether it be hay or porterhouse steak, is a collection of biomolecules—which, we recall, are carbon compounds. Gasoline is a hydrocarbon fuel refined from crude oil, which is a fossil fuel formed from the bodies of plants.

How did energy enter these compounds in the first place? Work is done in chemical reactions when atoms are joined by chemical bonds into stable molecules. Hence energy must be invested in forming the bonds (strong as well as weak) that hold a molecule together, just as energy has to be invested in raising a boulder to the top of a hill. To a very large extent, this energy first appeared on Earth as solar radiant energy.

The utilized solar radiant energy remains as potential energy within the electronic structure of molecules formed by photosynthesis. As explained in Chapter 3, a chemical reaction is in a sense solely concerned with electrons. A chemical reaction takes place if the electrons that maintain the structure of a chemical compound shift permanently from one configuration to another. If the potential energy of the first configuration is higher than that of the second, then the reaction gives off kinetic energy, usually in the form of heat; the second configuration is said to be more stable. Other forms of energy can also be liberated in chemical reactions. The chemical reactions in a storage battery generate electrical energy. The energy of exploding TNT is initially thermal and then becomes largely mechanical.

When a boulder rolls downhill after a little push, part of its original potential energy is converted to kinetic energy. When the boulder comes to rest, it is capable of doing less work. Similarly, when a chemical reaction converts potential energy of a system to kinetic energy, the system comes to rest at a lower potential energy level. The boulder is not changed chemically; only its position is changed. But the reactants in a chemical reaction are changed. Their physical properties are altered.

The molecular structure of glucose, for example, is rich in potential energy. Oxidation liberates some or all of this energy, the smaller molecular products (CO_2 and H_2O) containing lower energy levels than the original glucose molecule. The same thing happens in the burning of hydrocarbons in gasoline, coal, gas, and other fuels. When oxidation occurs by a slower or less spectacular process than burning, the liberated energy—or a significant portion of it—can be conserved in chemical forms more useful than heat. This is the central principle of energy metabolism in organisms.

The oxidative chemical reactions alluded to above result in the formation of the simple products CO_2 and H_2O plus the liberated energy that had been held within the more complex glucose molecule. The converse of this process is equally important. When the glucose molecule was built from CO_2 and H_2O by photosynthetic plants, solar energy was invested in constructing the covalent bonds. These fundamental relationships of the biological energy cycle are summarized in Figure 8-1.

Measuring Heat Energy

How can we measure the potential energy stored in chemical bonds? The first law suggests a simple answer. If the real-world lifting machine we described above converted all of its energy into heat, the change in temperature would provide a measure of the heat energy, which under the conditions specified was equivalent to the initial energy of the system. Such measurements are easily made in a device called a **calorimeter.** To perform **calorimetry,** we must first surround the room containing the machine with a water jacket containing a thermometer. Now *all* of the heat produced by the machine is absorbed by the water. At the end of the experiment, the temperature of the machine will be the same as it was at the beginning, but the water will be warmer.

The change in water temperature accurately indicates how much heat the machine produced. It reflects the internal energy of the system, which was equivalent to the electrical energy initially added. The heat content is called the **enthalpy** of the system (if we keep the system at constant pressure) and is represented by the symbol H. We would express the *change* in enthalpy as ΔH ("delta H"). Measurements of ΔH reflect the change in the internal energy of the system.

Now let us apply these concepts to chemical reactions, in which molecules are altered. Reactions start with molecules called **reactants** and end with molecules called **products.** Thus, a chemical reaction could be generalized as

$$\text{Reactants} \rightarrow \text{Products} \tag{8.1}$$

In the course of the reaction, chemical bonds are broken and formed. Calorimetry shows that most reactions are associated with changes in heat content. Some reactions give off heat and some absorb heat. If heat is given off as a reaction progresses, we conclude that the internal potential energy of the reaction products is lower than that of the reactants. Such reactions are said to be **exothermic.** Conversely, if heat must be added to make a reaction proceed, the internal energy of the products is higher than that of the reactants. Such reactions are **endothermic.**

The unit used to measure energy is the **Calorie,**[3] or kilocalorie (Calorie = 1000 calories = 1 kcal). The oxidation, for example, of one mole[4] of glucose ($C_6H_{12}O_6$) by molecular oxygen (O_2) to CO_2 and H_2O is an exothermic reaction yielding 673 kilocalories.

$$C_6H_{12}O_6 + 6O_2 \rightarrow 6CO_2 + 6H_2O$$
$$\Delta H = -673 \text{ kcal/mole} \tag{8.2}$$

An important convention governs all expressions of enthalpy. When a reaction is exothermic, enthalpy is given a minus sign. That means that heat is released into the surroundings. If a reaction is endothermic, enthalpy has a plus sign. That means that the reaction absorbs energy (as heat) from the surroundings.

The use of enthalpy (heat content) in the analysis of chemical reactions has one major advantage: it is easily measured in a calorimeter. Moreover, the value of enthalpy can be directly related (if pressure and volume are constant) to the internal energy change in a chemical reaction. We will see in a moment how knolwedge of that change (i.e., measurements of ΔH) can be entered into second law equations to yield estimates of the maximum energy available from a reaction. This is the **free energy** level—also known as the **Gibbs free energy** in honor of Josiah Willard Gibbs (1839–1903), a founder of the science of thermodynamics. The free energy of a system can produce work if a suitable linking mechanism is present (like the one in our electric motor system) to transfer the energy released in one reaction to another process that requires energy. We refer to such linkages as **coupled reactions.** Coupling is extremely important in bioenergetics.

Thermodynamics helps us predict whether physical or chemical processes can occur spontaneously. Although the first law is *not* the basis of these predictions, it is generally true that objects seek a state in which their internal energy is at its lowest possible level. Rocks roll downhill, not uphill, because their internal energy at the bottom of the hill is lower than at the top. In general terms, if the internal energy of the final state (or of the reaction products) is lower than the internal energy of the initial state (or the reactants), the spontaneous occurrence of the process is favored. For that reason, exothermic reactions generally tend to occur spontaneously. However, such an analysis is not sufficient to permit prediction of the spontaneous or natural direction of a process. That prediction derives from the second law.

---

[3] Many units are used to define energy. A calorie is the amount of energy needed to raise the temperature of 1 gram of water from 16.5°C to 17.5°C. Although biologists measure energy in calories, physicists find it convenient to use electron volts, ergs, and joules, and engineers prefer British thermal units and watt-hours. Conversion factors for all these units are given in physics texts. For example, a calorie has been standardized at 4.1855 joules. A joule is equal to a kg-m^2sec^{-2}.

[4] A *mole* of any substance is an amount equal to its molecular weight in grams. Thus, a mole of glucose is 180 grams of glucose.

The Second Law and Entropy

The second law allows us to predict both the tendency of a process to occur and the direction in which it will proceed. For these reasons, the second law has been somewhat more interesting to biologists than the first law. Because it predicts the direction of all events involving energy exchanges, the second law has been called "time's arrow."

Consider a simple experiment (Figure 8-4A). Two copper cubes, one warm and one cool, are placed side by side in an insulated box. The warm cube quickly cools and the cool cube warms. Soon both cubes are at the same intermediate temperature. We would say that this system is then in thermal equilibrium. The flow of heat energy out of the warm cube precisely equaled the flow into the cool cube. Both flows occurred spontaneously. If the two cubes had started at the same temperature, they would have remained at that temperature when placed in the box. No heat would have flowed from one to the other.

We have now proved something we already knew: Heat flows spontaneously from a hot object to a cold object and never the other way. Similarly, gas molecules flow freely from a zone of high pressure (like an inflated tire) to one of low pressure (the air outside the tire) and not the other way (Figure 8-4B). As discussed in Chapter 6, molecules of, say, sugar diffusing through a cup of warm coffee move from a more concentrated region (high order and low entropy) in the direction of random distribution (low order and high entropy). Finally, boulders roll spontaneously downhill but never uphill.

Spontaneous physical or chemical changes have a direction that is not accounted for by the first law. The direction is determined by the tendency of all systems to approach a state of equilibrium in which temperature, pressure, and other properties are uniform throughout the system. When the two cubes arrive at the same temperature, all the heat energy originally contained in them is maximally randomized—that is, uniformly distributed. It will never "unrandomize" by itself.

What does the second law tell us about this progression toward equilibrium? Several things. The following paraphrase of the second law is based on experience with heat engines: no process can result in absorption of heat from another body with total conversion of this heat into work. In other words, no heat engine is or can be 100% efficient. Though we can transform into heat *all* of the internal energy appearing as heat (under proper conditions this is the enthalpy), we cannot reconvert the same amount of heat into an equivalent amount of work or potential energy.

Why not? Because molecules in the random motion of heat have a certain level of kinetic energy, but this energy is undirected. If the motions of each molecule of the whole system were added up, the net motion would be zero. Since work is the application of force over a distance—and since the motion (or net distance traveled) in a heat energy system is zero, a system of heat-activated molecules cannot do work. To be sure, individual molecules do contain internal energy, but the random nature of that energy makes it unavailable for work.

Here is the link with the statistical concept of entropy. Any real process—whether a chemical reaction, the movement of electrons in a wire, or the movement of a piston in an engine—will promote collisions between particles and scatter many of them. Increased scatter increases both the loss of energy as heat and the randomness of particle motion, which is what we mean by entropy. That is why all real processes occurring in a given time interval inevitably increase the randomness or entropy of the universe. That is why the natural trend is toward increased entropy. One intriguing implication of this notion is that the universe is ultimately headed for a dismal fate—a state of complete randomness and disorder. This charming notion has been called the *entropic doom* or *heat death* of the universe.

Is it impossible then for a randomized system to unrandomize itself on its own, to become directed, unscattered, or ordered again? Since life depends on highly ordered and directed processes, and specific nonrandom sequences of events, how if this is not possible can life exist? The answer is: for a "life-machine" to exist, there must also exist an "anti-entropy machine." For order to return to a randomized system, processes must take place that can do real work in finite time intervals. These must also obey the second law. Hence, they can decrease the entropy of one system —a cell or an organism—only by increasing the entropy of the surroundings. One method of decreasing randomization is the linking of small random units like amino acids or nucleotides into large information-containing, nonrandom polymers like proteins or nucleic acids (Figure 8-5). To do this, however, requires energy.

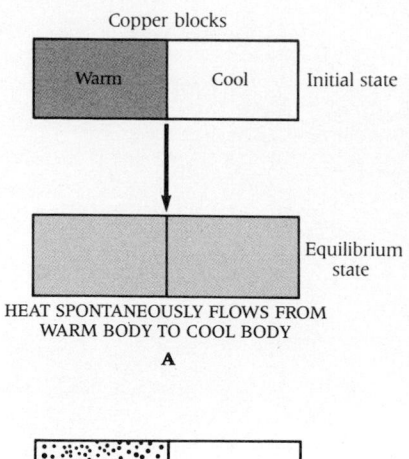

HEAT SPONTANEOUSLY FLOWS FROM WARM BODY TO COOL BODY

A

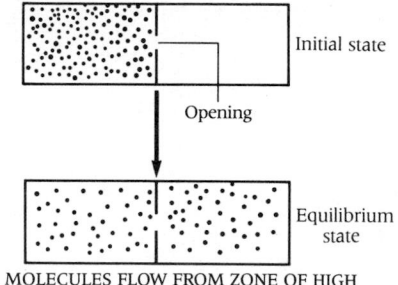

GAS MOLECULES FLOW FROM ZONE OF HIGH PRESSURE TO ZONE OF LOW PRESSURE

B

8-4 All systems tend toward a state of equilibrium
(A) *Heat spontaneously flows from the warm copper block to the cool block.* **(B)** *Gas molecules flow from a region of high pressure to a region of low pressure.*

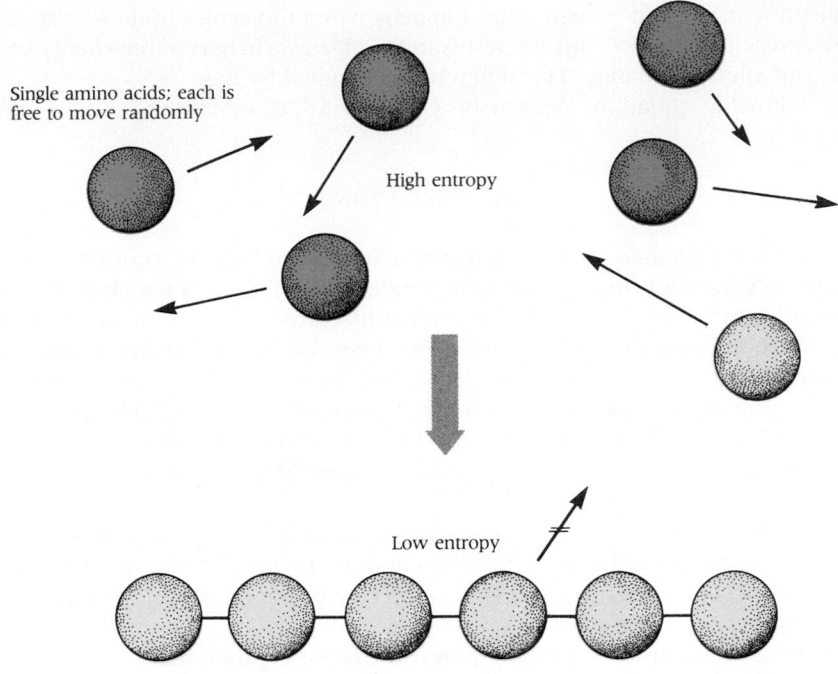

8-5 Entropy balance in a system
The formation of a protein from amino acids results in a decrease in the entropy of the system (an increase in the order). This decrease occurs when the amino acids are linked together to form a protein and are no longer free to move randomly. The formation of such a linkage requires energy.

Single amino acids; each is free to move randomly

High entropy

Low entropy

A peptide; movements of individual amino acids are inhibited

Precisely this happens at every level of the biological hierarchy shown in Table 1-1. At each level, entropy is decreased by mechanisms drawing energy taken from the surroundings—and ultimately from the sun. In these processes, the randomization of surroundings must accelerate because the entropy increase in the universe always exceeds the entropy decrease in the system.

Entropy can be measured and given a magnitude. The symbol for entropy in thermodynamics is S. A change in entropy is ΔS. Although ΔS for real events in the universe is always positive, ΔS for a given system may be positive, zero, or negative.

■ In systems reaching equilibrium (like those in Figure 8-4), entropy increases as particles become more random and ΔS has a positive value.
■ In systems *at* equilibrium and in certain reversible processes, there is no net change in the level of randomness and ΔS is zero.
■ In systems becoming more ordered (like the garden being tidied or the newspaper being published), particles become less random and ΔS has a negative value.

The Concept of Free Energy: Combining the First and Second Laws

Living organisms are open thermodynamic systems. Hence, there can be exchanges of work between the system and its surroundings that result in a decrease of entropy in the system. In such systems, simple measurements of ΔS will not tell us in which direction a process is going. If energy is entering the system from its surroundings, we cannot predict the direction of events from changes in entropy alone. But we *can* determine the direction of a process if we take into account both its changes in energy and changes in entropy. That would combine the first and second laws of thermodynamics.

Recognition of this fact led to the mathematical derivation of the thermodynamic function we call the free energy, or Gibbs free energy, symbolized by G (for Gibbs), which is related to the internal energy of a system as determined by measurements of enthalpy and entropy according to the following equation:

$$G = H - TS \qquad (8.3)$$

where H is the heat energy, or enthalpy, of the system, T is its temperature (in degrees Kelvin), and S is its entropy.

Our present concern is with what happens when molecules undergo change in chemical reactions; hence, our interest is in the *difference* between free energy values before and after the change. This difference is denoted by ΔG.

The following equation, written by Gibbs in 1878, establishes certain critical relationships:

$$\Delta G = \Delta H - T\Delta S \qquad (8.4)$$

where ΔG is the change in free energy of a system undergoing transformation at constant pressure and constant absolute temperature (T), ΔH is the change in heat content (defined above), and ΔS the change in entropy. Note that the properties of the surroundings do not enter into this expression. The equation states that when temperature and pressure are kept constant, the value of ΔG (especially whether that value is positive, negative, or zero) depends on the values of ΔH and ΔS. The value of ΔG is a useful predictor of whether a reaction can occur spontaneously: that can happen only when ΔG is negative. Consider the following cases:

- An exothermic reaction, we saw, is one in which ΔH is negative; because these increase entropy (ΔS becomes positive) and ΔG becomes negative. Thus, the reaction occurs spontaneously. This is the case in the oxidation of glucose (see below).
- An endothermic process (ΔH is positive) that increases randomness (ΔS is positive) will have a positive ΔG and thus will not occur spontaneously unless driven by added energy. This is the case in photosynthesis—the reverse of the oxidation of glucose.
- An endothermic reaction (ΔH is positive) will occur spontaneously if ΔG is negative. This occurs if ΔS is positive and the $T\Delta S$ term is large enough to overbalance the positive ΔH. This is the case with the spontaneously equilibrating particles of Figure 8.4.
- If there is no change in internal energy (ΔH equals zero) and the process increases the randomness of the system (ΔS is positive), then ΔG will be negative and again the process will occur spontaneously. This occurs when a solute mixes with a solvent.
- Finally, a system that undergoes no change in its internal energy or its entropy (ΔH and ΔS are both zero) is undergoing no change and is at equilibrium.

Bearing in mind the definition of ΔG, consider again the oxidation of glucose (Equation 8.2). We saw above that the ΔH of this reaction is -673 kcal/mole. The ΔG is -686 kcal/mole. The entropy factor has contributed 13 kcal/mole to the free energy change of the process. From a thermodynamic standpoint, this reaction is strongly favored to proceed spontaneously. The contribution to its spontaneity comes from both its exothermic character and the increased disorder of the final state. The negative ΔH indicates a significant loss of internal potential energy as the reactants are changed from glucose to CO_2 and H_2O. The positive sign of the $T\Delta S$ term indicates an increase in the entropy of the system due to the conversion of the nonrandom organization of a single glucose molecule into a more disordered array of $6CO_2$ molecules.

The value of ΔG is especially useful in bioenergetics. The total work that a system could potentially perform is the sum of internal energy released in the process (ΔH) and the energy represented in the $T\Delta S$ term, which obviously depends on the change in entropy. As mentioned earlier, this free energy coupled to other reactions or processes can perform work in the cell. The ability of reactions or processes to utilize energy from a reaction with a large negative ΔG is of fundamental importance to life.

We can summarize these fundamental ideas as follows:

- A reaction can occur spontaneously only when ΔG is negative.
- A reaction cannot occur spontaneously when ΔG is positive unless energy is applied to the system.
- When ΔG is zero, the system is at equilibrium and no net changes can occur. In fact, there may be changes in both directions in a reversible reaction, but they exactly offset each other: any conversion of reactants into products is exactly balanced by a conversion of products into reactants.

INTRODUCTION TO METABOLISM

Metabolism includes the following processes:

- All the chemical processes by which food and its derivatives are broken down to yield new building blocks and energy. This segment of metabolism is termed **catabolism.**
- All the chemical processes by which living cells and tissues are produced and built up. This is **anabolism.** Foods taken in by organisms never contain all the many kinds of molecules that are to be found in that organism's cells. This simple fact suggests an important conclusion. Most of the biomolecules present in organisms must be synthesized within them. The term **biosynthesis** denotes the syntheses of chemical compounds that occur in anabolic metabolic processes.
- All the regulatory mechanisms that govern these intricate systems.

It is easy to cite examples. Yeast cells are readily grown in cultures containing, as the only carbon source, the simple sugar glucose—plus various inorganic salts containing needed nitrogen and phosphorus. But the rapidly growing cells are found to contain a huge variety of large and small molecules—RNA, DNA, proteins, lipids, amino acids, and many others. One function of metabolism is to transform food molecules into those molecules that the organism needs. Since the process of taking molecules apart (catabolism) and reassembling the pieces (anabolism) requires energy, other major functions of metabolism are the extraction of energy stored in chemical compounds and the conversion of energy into useful forms.

CHEMICAL EQUILIBRIUM

Three questions can be asked about every chemical reaction: Will it occur? In which direction? And at what rate? Thermodynamics, we have just seen, deals with the first two questions. **Kinetics** is the branch of physical chemistry concerned with the third question.

Do not confuse the rate of a reaction with the extent of its progress toward completion. A man may be halfway home, but that doesn't tell us if he got there slowly or rapidly. Similarly, glucose can be oxidized slowly in a cell or rapidly in a calorimeter. The end products are the same but the reaction rates are different. In reversible reactions, the extent of the reaction depends upon the ratio between the rate constants of the forward and reverse reactions, k_1 and k_2. In a reversible second-order reaction—that is, a reaction whose rate is proportional to the concentration of two reacting substances—

$$A + B \underset{k_2}{\overset{k_1}{\rightleftharpoons}} C + D \tag{8.5}$$

the rate of the forward reaction equals k_1 [A][B] and that of the backward reaction is k_2 [C][D], brackets indicating molar concentrations.

We say that the reaction has attained **equilibrium** when the forward rate equals the backward rate—that is, when

$$k_1 [A][B] = k_2 [C][D] \tag{8.6}$$

At this point, [A], [B], [C], and [D] have reached fixed values, and the ratio [C][D]/[A][B] is constant. Expressing it mathematically, we get

$$\frac{[C][D]}{[A][B]} \frac{k_1}{k_2} = K_{eq} \tag{8.7}$$

The constant K_{eq} (which equals k_1/k_2) is the reaction's equilibrium constant. Box 8-1 shows how the **equilibrium constant (K_{eq})** relates to the change in free energy (ΔG) of a reaction.

Clearly, the final ratios of the concentrations of reactants to products would be the same in the reaction in Equation 8.5 whether we started with A and B or with C and D. We can now see why the reaction does not go to completion in one direction or the other. If it did do so in the forward direction, [A] and [B] would come to equal zero, and K_{eq} would equal infinity. Since $K_{eq} = k_1/k_2$, it could not equal infinity unless k_1 equaled infinity or k_2 equaled zero. This is not the case, since both reactions occur at finite rates.

A number of factors may intervene, however, to make such a reaction go to completion by preventing accumulation of the product. If one of the products, C or D, is an escaping gas or an insoluble precipitate, it is in effect removed from the system as fast as it is formed. As a result, its concentration drops to zero in the equilibrium equation—and the reaction continues until the initial reactants are consumed. This situation is known as "pulling a reaction to the right." That circumstance arises in living organisms in many settings—for example, when H_2CO_3 decomposes to H_2O and CO_2 (a gas) or when calcium ions (Ca^{2+}) precipitate in the form of insoluble bone salts.

It also occurs in the sequential steps of metabolic pathways, wherein successive enzymatic reactions utilize the products of the preceding reactions as substrates. This obviously pulls the preceding reaction to the right, since its products are thereby kept from accumulating. We presume that the high efficiency of this arrangement explains why sequential pathways have been so highly conserved in evolution.

HOW CATALYSTS AFFECT REACTION RATES

As we noted earlier, energy is the capacity to do work. Note that this definition does not state whether or not work is actually done. The energy of a boulder at the top of a hill is, as we said, potential energy. Even though the rock is at rest, it has the capacity to do work by rolling down the hill. If it does roll down the hill, then part of the potential energy becomes kinetic energy.

Unlike potential energy, kinetic energy always involves matter in motion. Work is accomplished when the movement is completed. Kinetic energy may take many forms; that of the falling boulder is mechanical energy, while light is kinetic energy in the form of moving particles called photons. Electricity is kinetic energy in which electrons move from atom to atom in a conductor. Heat is the random motion and kinetic energy of atoms and molecules. When a chemical reaction occurs, a movement of atoms within or between the reacting molecules converts the potential chemical energy to kinetic chemical energy.

Once any energy-yielding reaction is in progress, it is producing kinetic energy freely. However, special problems arise in connection with the initiation of the reaction. Just to get the reaction underway requires a small investment of energy. Similarly, an investment of energy is needed to start a reaction that converts compounds A and B to compounds C and D (Figure 8-6). This energy is called the **activation energy** (E_a).

Meaning of Activation Energy

All molecules above the temperature of absolute zero are in constant motion and thus have kinetic energy. Absolute zero is $-273.16°C$, or $-459.69°F$. At this temperature, which is also known as 0° Kelvin, all molecular motion would stop. However, every individual molecule in a large population of molecules does not possess the same amount of kinetic energy. If we picture a group of sugar molecules resting inside the small energy "trough" that keeps them from spontaneously uniting with O_2 and "sliding downhill" to CO_2 and H_2O (see Figure 8-6), we see that a few sugar molecules may somehow manage to acquire more energy than others in order to leap the energy barrier and slide down the energy-yielding downhill portion of the reaction curve. This occurs so rarely, however, as to be negligible. If, on the other hand, enough energy were supplied from outside the system, or if someone could lower the barrier, the reaction would be enormously accelerated. Such acceleration can be achieved in two ways: (1) by raising the temperature of the system, or (2) by adding a catalyst.

Effect of Temperature

In a chemical reaction in which molecules A and B react to form C and D, A and B are at first freely dispersed, moving about continuously and randomly. They react with each other only when they collide by chance. The dependence of chemical reactions on *molecular collisions* explains several of the basic laws governing chemical

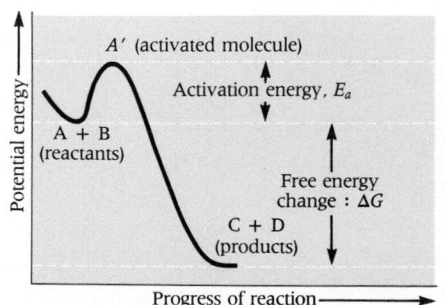

8-6 Conversion of potential energy into kinetic energy
In a chemical reaction in which A and B are converted to C and D, the input of activation energy (E_a) produces an activated molecule and initiates the conversion of substrate molecules into products with lower potential energy.

BOX 8-1

K_{EQ} IS RELATED TO ΔG

There is an important connection between the equilibrium constant (K_{eq}) and the change in free energy (ΔG). Let us explore this relationship. We can calculate the free energy change (ΔG) occurring during a chemical reaction by the use of an equation that is derivable from the law of chemical equilibrium. This deserves our attention because it will illuminate discussions of respiration and photosynthesis in Chapters 9 and 10.

Here is the same reaction shown in Equation 8.5.

$$A + B \rightleftharpoons C + D \qquad (1)$$

Let the reaction proceed under a set of conditions called the **standard state.** For biochemical processes, this means a pressure of one atmosphere, a temperature of 298.15°K, and a pH of 7.0. We then mix A and B and allow the reaction to proceed until there is no further change in the concentrations of reactants and products. This is the equilibrium state. The following relationship defines the free energy change ΔG:

$$\Delta G = \Delta G° + RT \ln \frac{[C][D]}{[A][B]} \qquad (2)$$

where ln denotes a natural logarithm to the base e (2.71828), R is the universal gas constant (1.987 kcal/degree · mole), T is the absolute temperature, and $\Delta G°$ is the standard free energy change, a term that will be defined below.

As we have seen, when a reaction is at equilibrium, a condition exists at which energy is at a minimum. That means that no further change is possible. Therefore, ΔG is 0. We can then say

$$0 = \Delta G° + RT \ln \frac{[C][D]}{[A][B]} \qquad (3)$$

This can be rearranged to

$$\Delta G° = -RT \ln \frac{[C][D]}{[A][B]} \qquad (4)$$

Since the apparent equilibrium constant K_{eq} for the reaction is as displayed in Equation 8.7,

$$K_{eq} = \frac{[C][D]}{[A][B]} \qquad (5)$$

we can substitute K_{eq} in Equation 4 above and obtain the general expression

$$\Delta G° = -RT \ln K_{eq} \qquad (6)$$
$$= -2.303 \, RT \log_{10} K_{eq} \qquad (7)$$

where ln is as defined above and $\log_{10}$ is a logarithm to the base 10. Conversely, K_{eq} is a function of the **standard free energy change,** a characteristic thermodynamic constant for any given chemical reaction. It tells us whether or not a reaction will proceed spontaneously under standard conditions. If $\Delta G°$ is negative, the reaction is **exergonic.** If $\Delta G°$ is positive, the reaction is **endergonic.**

The relation between K_{eq} and $\Delta G°$ is evident in the following table:

K_{eq}	$\Delta G°$ (kcal/mole)
0.001	+4.089
0.01	+2.726
0.1	+1.363
1.0	0
10.0	−1.363
100.0	−2.726
1000.0	−4.089

Equation 7 permits us under most circumstances to calculate the standard free energy change, $\Delta G°$, of any chemical reaction from its equilibrium constant, a value that can be easily estimated in the laboratory. Chemical reactions with a negative standard free energy change are exergonic, and they proceed spontaneously in the direction written; though the rate may be fast or slow. Reactions with a positive standard free energy change are endergonic, and they do not proceed spontaneously in the direction written. Instead, they proceed in the reverse direction.

It is important to understand the difference between $\Delta G°$, the so-called standard free energy change, and ΔG, the actual (or measured) free energy change. $\Delta G°$ is a constant for any given reaction at a given temperature. On the other hand, ΔG varies with the concentrations of the reactants and products. ΔG equals $\Delta G°$ only when all reactants and products are present in molar concentrations. In living cells, of course, concentrations are far lower.

reactions. All factors that increase the collision frequency increase the rate at which a reaction proceeds. Such factors include an increase in the concentrations of the reactants in solution and an increase in temperature, which always speeds molecular movement. Both conditions accelerate reactions by enhancing the probability of molecular collision.

Raising the temperature accelerates chemical reactions in part by providing the energy needed to surmount the *activation energy barrier*. Raising the temperature increases molecular movement and heightens the probability of molecular collisions.

As a result, the rate of a chemical reaction increases logarithmically as the temperature increases.

What Is a Catalyst?

Long before anyone understood how they worked, chemists and chemical engineers found that certain substances added in small quantities to reacting chemicals would dramatically speed up reactions. In many cases, the additions were of nickel, zinc, or platinum—finely powdered so that the particles had a large total surface area. These substances (and the countless others now used in science and industry) are called **catalysts.** One of the most closely guarded of industrial secrets is the nature of the catalyst being used in a given process. Whoever has the best catalyst often has the biggest bank account.

A catalyst accelerates the rate of a reaction by lowering its activation energy, thereby making it easier for the reaction to proceed. Catalysts are neither consumed nor permanently altered in the reactions they promote. Hence they are active in exceedingly small quantities. Note that catalysts affect only the *rate* of a reaction. They have no effect on the extent to which the reaction proceeds to completion—that is, on its equilibrium state. In other words, they do not cause reactions to occur that would not eventually occur by themselves.

Catalysts function by binding the reactants onto their surfaces, holding them close together and in correct orientation, often straining and weakening their chemical bonds. The resulting intermediate state, in which the catalyst has formed an unstable complex with the initial reactants, is brief and transient. Significantly, this complex has a lower activation energy barrier than do the reactants alone in the absence of catalysts (Figure 8-7). The complex rapidly decomposes to form the reaction products. When the reaction has been completed, the catalyst is liberated to act again.

Biological systems have gained control of reaction rates through the use of catalysts called enzymes.

ENZYMES

Among the reasons why living organisms do not tolerate temperature elevation as a method for accelerating their chemical reactions, the most obvious is that temperature elevation of more than a few degrees is devastating to living tissue, whose fragile structures are damaged by heat. Heat increases the rate of diffusion and tends to damage the complex structures of proteins. Even mild temperature increases would cause all chemical reactions to accelerate in an uncontrolled way. Hence the elaborate mechanisms by which organisms maintain internal homeostasis necessarily include methods for resisting large increases in their temperatures.

Many living organisms are isothermal, at least in critical portions of their bodies. That means that the temperature is about the same in different parts of a cell or in different cells of a tissue.[5] In this regard, there is an important difference between a horse and an automobile. Auto engines elevate the temperature in the combustion chamber by applying a hot spark, thereby initiating the combustion of gasoline. Horses have no such combustion chamber.

Chemical reactions in living organisms are accelerated by special catalysts called **enzymes.** Without enzymes, chemical reactions would take place much too slowly to support life. Enzymes comprise a special class of catalysts with the following properties.

- Until recently, it was believed that all enzymes are proteins (though not all proteins are enzymes). Evidence now exists that some RNA molecules have catalytic activity. Nonetheless, the great majority of enzymes are proteins.
- Enzymes operate under the mild conditions of temperature, pressure, and pH found in living cells.

[5] We will later learn that many animals are *not* isothermal, displaying considerable variations in body temperature. Nonetheless, all would be harmed by excessive heat.

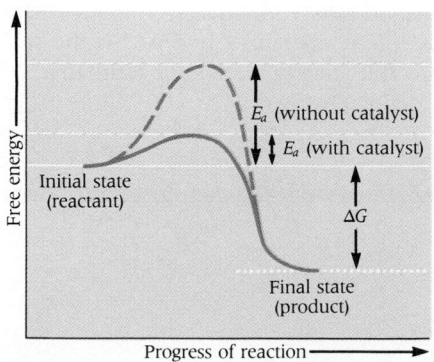

8-7 The effect of catalysts on chemical reactions
Catalysts lower the activation energy (E_a) needed to convert reactants to products. ΔG is the difference in free energy between reactants and products.

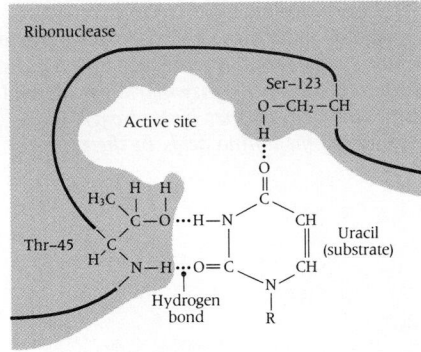

8-8 Substrate-enzyme interactions
Specific binding of a substrate to an enzyme involves formation of noncovalent bonds. In this case only two amino acids in the enzyme ribonuclease (Thr-45 and Ser-123) participate in the formation of hydrogen bonds with the substrate uracil in the active site. The numbers 45 and 123 denote the position of the amino acid in the enzyme protein primary structure. This particular threonine is 45th in the amino acid sequence; this serine is 123rd.

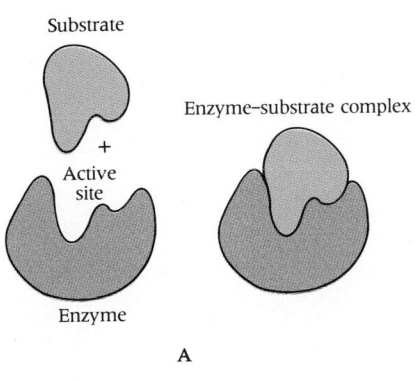

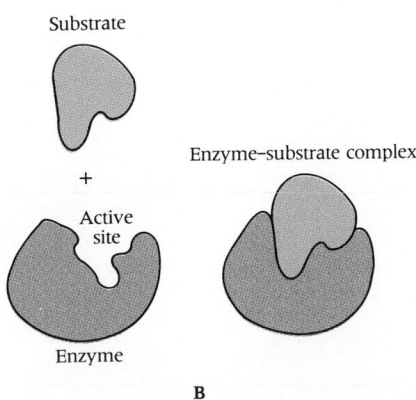

8-9 Two theories to explain enzyme specificity (A) *The lock-and-key theory proposes that the substrate precisely fits into the active site of the enzyme.* **(B)** *The induced fit model postulates that the substrate induces a conformational change in the active site so that its binding is ensured.*

- In contrast to such inorganic catalysts as "Raney nickel" or "platinum black," each enzyme demonstrates remarkable specificity as to the reactions it will or will not catalyze.
- Enzymes are outstandingly efficient—molecule for molecule, no inorganic catalysts approach enzymes in this regard. For example, one molecule of catalase, an enzyme that splits hydrogen peroxide (H_2O_2), can decompose 5 million molecules of H_2O_2 per minute.
- Certain enzymes are subject to sensitive regulation or control.

Although enzyme molecules, like all proteins, are synthesized only by living organisms, they may be readily extracted from tissues, purified, in some cases crystallized, and studied in isolation in test tube experiments. By means of such studies, scientists have dramatically expanded their knowledge of enzymes in recent years.

NOMENCLATURE

The original names given to enzymes in an earlier time (pepsin, trypsin, ptyalin, and so on) gave no hint of their biochemical actions. Such names are still convenient for everyday use, but a more rigorous and complex systematic nomenclature has become essential. Today, the names of most enzymes are derived by addition of the suffix -*ase* either to the name of the **substrate** (which is the reacting substance) or to a term describing the kind of reaction being catalyzed. In the former category are names for enzymes that split starch (Greek, *amylon*), the amylases; those that split fat (Greek, *lipos*), the lipases; those that act on urea, urease; and so on. In the latter category are various groups of enzymes with such names as oxidases, dehydrogenases, transferases, reductases, hydrolases, isomerases, and others. More formal and complex names (and coded numerical designations) are given to enzymes by a committee of the International Union of Biochemistry.

MECHANISM OF ACTION

An uncanny ability to distinguish between similar chemical structures is a major property of enzymes, though comparable specificity is also found elsewhere in biology—for example, in antibodies. The classic explanation for enzyme specificity is the famous **template** (or **lock-and-key**) **theory,** first proposed in 1884 by Emil Fischer (1852–1919) and Paul Ehrlich (1852–1915). A small patch of the very large enzyme surface is its **active site.** According to the template theory, the active site contains a fixed arrangement of chemical groups that together represent a precise, three-dimensional, negative image of a portion of the substrate molecule. In other words, they form a "lock" that will fit only a certain substrate "key." We know now that these chemical groups are often the side chains of amino acids, with whose charged polar and uncharged nonpolar substituents we are familiar (see Figure 4-14).

In very general terms, the template theory remains valid. An enzyme is a polypeptide, and the three-dimensional configuration of the amino acid side groups at the active site is a precise and specific one that usually binds only the molecules of a particular substrate (Figure 8-8)—though, as we shall see, it occasionally seems to make an error by binding other molecules of similar structure. (Interesting aspects of the binding of small molecules by proteins are discussed in Box 8-2.) The enzyme and substrate combine to form a short-lived **enzyme-substrate complex.** What happens to this complex?

According to the most current theory of enzyme activity, the **induced fit** model,[6] the shape—or conformation—of the active site of most enzymes changes when substrate is bound to it. Only then are certain reactive groups in the enzyme brought into the precise alignment essential for their catalytic action (Figure 8-9). In other words, the enzyme's active site is flexible—not rigid, as it was viewed in the early lock-and-key, or template, theory. Once a substrate molecule (or molecules) causes the enzyme to assume the proper alignment, it functions much like the old catalysts described above, making the substrate molecule (or molecules) more reactive, lower-

---

[6] The induced fit hypothesis of enzyme catalysis was first proposed in 1973 by Daniel E. Koshland, of the University of California, Berkeley, now the editor of the journal *Science*.

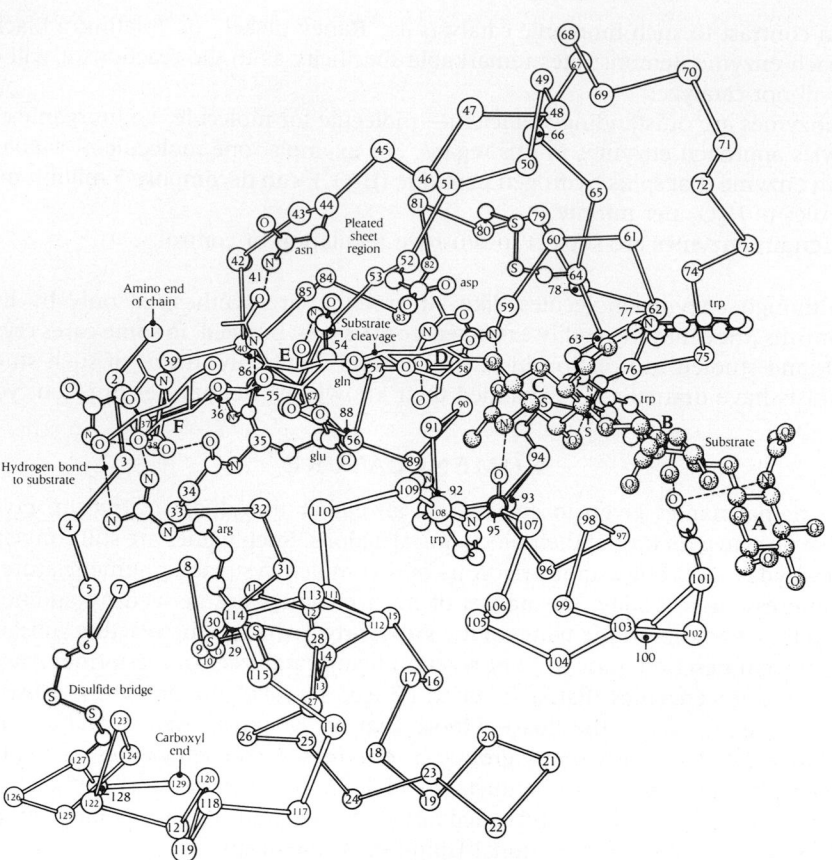

8-10 Three-dimensional structure of lysozyme as determined by x-ray diffraction *The substrate fits into the active site that runs horizontally across the enzyme. Substrate and enzyme are held together by hydrogen bonds involving specific amino acids in the protein.*

ing its activation energy enough to launch it into a chemical reaction. The enzyme is then liberated from the enzyme-substrate complex and restored to its original state. What is left behind are **reaction products.**

Perhaps glove-and-hand is a better analogy than lock-and-key, because a glove is not a three-dimensional image of a hand until the hand is put into it. An empty glove (free enzyme) may have any of several shapes; a proper fit is induced by a hand (substrate).

A Word About Enzyme Structure

We have learned a good deal in recent years about enzyme structure at the molecular level through the use of **x-ray diffraction.** In this procedure, the investigator prepares a protein crystal—a difficult procedure not yet achieved for many enzymes—and exposes it to an x-ray beam. By the use of computer analysis, the three-dimensional (3-D) structure of the protein can be deduced from the x-ray diffraction pattern. The model in Figure 8-10 shows the structure of muramidase (lysozyme), the first of several dozen enzymes to be successfully analyzed by this method.

We saw in Chapter 4 that the three-dimensional, or 3-D, structure of a protein is termed its *secondary* and *tertiary* structure. To the untutored eye, the structure in Figure 8-10 appears oddly irregular. It is anything but. Catalytic activity depends not only on the meticulous geometry of an enzyme's active site but on the intactness of the enzyme molecule's tertiary structure. The classic experiment that established this fact is described in Box 8-3. It showed, in sum, that when the native 3-D structure of an enzyme—in this case, ribonuclease—was destroyed (*denatured*) by the addition of compounds that break stabilizing disulfide and hydrogen bonds, a randomly arranged polypeptide chain forms that is totally devoid of catalytic activity. But when these conditions are withdrawn, the enzyme resumes its normal 3-D structure (we say it is *renatured*) and full catalytic activity is restored. Of course, this demonstration took place in a carefully controlled experimental setting. Under ordinary circumstances and for most proteins, full restoration of biological activity after denaturation is an exceedingly rare occurrence. Obviously, restoration is not possible if the denaturation process was too damaging—or if it altered the protein's primary structure. For example, we cannot yet undo the denaturing effects of

BOX 8-2

MANY PROTEINS BIND OTHER MOLECULES

We learned in Chapter 4 that the shape and toughness of fibrous proteins suits them to various structural functions. The globular proteins, which include enzymes, many hormones, antibodies, and transport proteins, have other roles. The major ones can be summarized as follows.

- *Enzymes*
  Catalysis of biochemical reactions

- *Hormones*[*]
  Transmission to other cells of "messages" that control rates of metabolic activity

- *Antibodies*
  Neutralization or inactivation of foreign substances

- *Transport proteins*
  Transportation of molecules, ions, or electrons from one place to another—across membranes, within membranes, within cells, or over long distances within the organism

A common feature of these classes of globular proteins is the ability to bind other molecules. The molecules being bound are called *ligands*. As in the case of enzymes, all proteins that bind ligands have a surface site—an active site or binding site—with a specific shape that can generate a specific array of forces. Proteins bind ligands through the mediation of the amino acid side chains at this site.

The fit is ordinarily perfect—in the geometrical sense and in terms of the forces arising when functional groups of the site interact with relevant sites on the ligand. Therefore the site can bind its ligand with great avidity. Because even a small change in the molecule is likely to spoil the fit and upset the interplay of forces, binding is both powerful and specific.

Like enzymes, nonenzymatic binding proteins often have an ability to modify the properties of their ligands. Consider the following:

- Hormones, whether proteins or not, influence metabolic rates because they can bind to specific sites in the cell called receptors, which are also proteins. As we have seen (Chapter 6) and will see again (Chapter 28), a receptor is usually part of a membrane—one that surrounds a cell or is within it. When it binds to a hormone, a receptor's properties are altered. This might have any number of effects, depending on which hormone it is. For example, a system of active transport through the membrane might be accelerated or slowed. That in turn would influence the rate of metabolism within a cell (or within an organelle).

- Antibodies are proteins of a type called immunoglobulins. They act by binding foreign substances called antigens (Chapter 23). Each antibody molecule bears two binding sites that precisely fit a specific antigen. Bound antigen is thereby inactivated.

- In a similar way, the transport proteins discussed in Chapter 6 "ferry" molecules or ions short distances across a membrane, often (as in active transport) transporting their ligands into areas of higher concentration. We have seen cases in which these proteins are thought to alter their passenger ligands—for example, glucose is altered as it is transported across some membranes.

- Finally, enzymes, our present concern, bind their substrates with great specificity. This alters the shape or conformation of the enzyme's active site (the induced fit mechanism). It also alters the ligand (substrate), subjecting it to the strains necessary for catalysis.

[*] Many hormones are not proteins (see Chapter 28).

---

heat in a hard-boiled egg, which converts the egg white proteins into an insoluble white mass.

Significance of the Enzyme-Substrate Complex

We saw in Figure 8-8 that the first step in the enzyme-catalyzed conversion of a substrate to its reaction products is the combination of the enzyme (E) with its substrate (S) to form a complex (ES). At this stage the substrate is activated. This is indicated by the symbol S'. The intermediate enzyme-substrate complex (now represented as ES') then gives rise to final products plus free enzyme again.

$$E + S \rightleftharpoons ES \rightleftharpoons ES' \rightarrow E + \text{products} \qquad (8.8)$$

Many observations support this theory. The slowest step of the reaction sequence is usually the first one, the formation of the ES complex from enzyme and substrate.

When the concentration of substrate is very low, some enzyme will have no substrate molecule to combine with and will remain free. In that circumstance, the whole sequence cannot operate at its maximum possible rate (Figure 8-11). The rate of reaction is mainly determined in this case by the amount of substrate present. But when substrate molecules are present in excess, all available enzyme combines with them, and the sequence proceeds at maximum rate. In this circumstance, the rate of the reaction is determined by the amount of enzyme present.

Temperature, as indicated above, also affects enzyme-catalyzed reactions. Most biological systems have optimal temperatures above which their enzymes will not perform. With rising temperatures, heat inactivates most enzymes (Figure 8-12) by destroying (denaturing) their secondary and tertiary structure. The relation between temperature and the rate of an enzymatic reaction or metabolic pathway is expressed in terms of a value called Q_{10}, which expresses the rate of increase for every 10°C rise in temperature. Very high temperatures (over 120°C) denature all proteins.

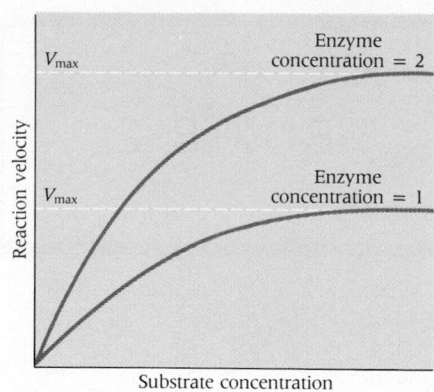

8-11 The effect of substrate concentration on enzyme-catalyzed reactions
The rate at which substrate is converted to product (reaction velocity) is dependent on the substrate concentration and reaches a maximum (V_{max}) that varies in direct proportion to the enzyme concentration.

Substrate Specificity and Enzyme Inhibition

The specificity of enzymes is exquisite and astonishing. The enzyme L-amino acid oxidase provides a useful example. We learned earlier that L-compounds and D-compounds are *stereoisomers*. L-amino acids and D-amino acids differ only in the three-dimensional arrangements of their R and NH_2 groups, thus:

$$(\text{R GROUP}) \quad H_3C \quad \overset{\displaystyle COOH}{\underset{\displaystyle H}{\overset{|}{C}}} \quad NH_2 \qquad\qquad H_2N \quad \overset{\displaystyle COOH}{\underset{\displaystyle H}{\overset{|}{C}}} \quad CH_3 \quad (\text{R GROUP})$$

D-Amino acid (D-alanine)   L-Amino acid (L-alanine)

This enzyme catalyzes the following reaction, which is an oxidative deamination:

$$\text{L-R}-\underset{\underset{\displaystyle NH_2}{|}}{CH}-COOH + FMN \rightarrow R-\underset{\underset{\displaystyle O}{\|}}{C}-COOH + NH_3 + FMNH_2 \qquad (8.9)$$

It is absolutely inactive with $\text{D-R}-\underset{\underset{\displaystyle NH_2}{|}}{CH}-COOH$ (or with various compounds in which the $-NH_2$ group has been modified). In other words, a mere rearrangement of groups within the substrate molecule is enough tampering to prevent this enzyme from catalyzing a reaction.

Enzymes differ in their degree of specificity. For example, **lipase,** which catalyzes the breakdown of a triglyceride to glycerol and its constituent fatty acids, is specific only for the ester linkages present in the triglyceride:

$$\begin{array}{l} H_2C-O-OC-R^1 \\ | \\ HC-O-OC-R^2 + 3H_2O \xrightarrow{\text{Lipase}} \\ | \\ H_2C-O-OC-R^3 \end{array} \quad \begin{array}{ll} H_2C-OH & R^1-COOH \\ | & + \\ HC-OH + & R^2-COOH \\ | & + \\ H_2C-OH & R^3-COOH \end{array} \quad (8.10)$$

Triglyceride   Water   Glycerol   3 Fatty acids

This, incidentally, is the reverse of the condensation reaction shown in Equation 4.3. For purposes of this discussion, the actual identities of the fatty acids R^1, R^2, and R^3 do not matter.[7] And yet other ester-splitting enzymes have even greater

---

[7] The three fatty acids of a typical triglyceride are not removed simultaneously. Rather they come off one after another. The two outer ester links are attacked preferentially; the one in the middle is more resistant to hydrolysis, so that di- and monoglycerides are produced as intermediate products. This is important in connection with the absorption of fatty food materials from the small intestine in animals (see Chapter 26).

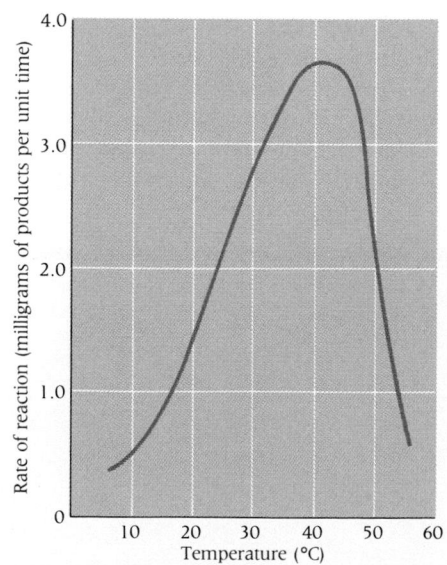

8-12 The effect of temperature on enzyme-catalyzed reactions
Under conditions of constant enzyme and substrate concentration, the reaction rate increases with rising temperature up to about 40°C. At higher temperatures the reaction rate declines, probably because of denaturation of the enzyme.

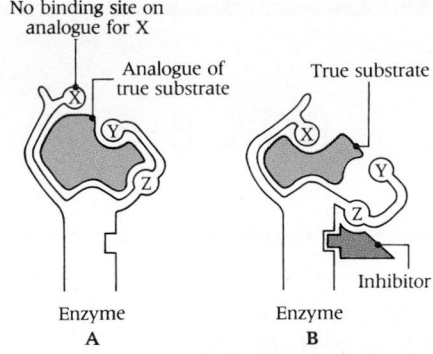

No binding site on analogue for X

Analogue of true substrate

True substrate

Inhibitor

Enzyme
A

Enzyme
B

8-13 Two mechanisms of enzyme inhibition
The enzyme contains three regions (X, Y, and Z) that make up the active site. **(A)** *In competitive inhibition, the active site is occupied by a molecule that closely resembles the substrate, but bonds only to sites Y and Z on the enzyme.* **(B)** *In noncompetitive inhibition, the substrate binds to site X, but not to Y and Z. These regions are not in proper position because inhibitor bound to another region of the enzyme has the effect of distorting the active site.*

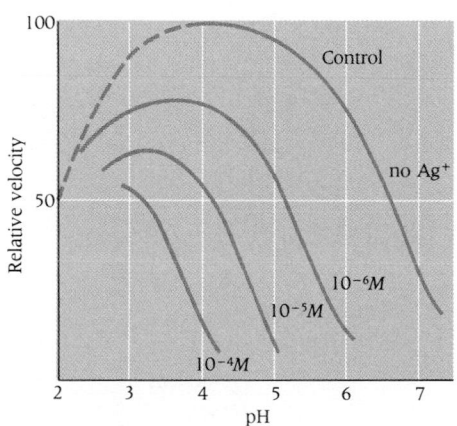

8-14 Noncompetitive inhibition of the enzyme saccharase by silver ions
Increasing the concentration of silver ions from $10^{-6}M$ to $10^{-4}M$ inhibits the activity of saccharase to an increasing extent.

specificity. Many attack only a single variety of substrate molecule. For example, acetylcholinesterase attacks only acetylcholine, a molecule of importance in the nervous system (see Chapter 29).

The phenomenon of **competitive inhibition** is an interesting and important consequence of the closely fitting complementary physical structures of enzymes and substrates. The enzyme succinic dehydrogenase catalyzes the oxidation of succinic acid as follows:

$$\begin{matrix} \text{COOH} \\ | \\ \text{CH}_2 \\ | \\ \text{CH}_2 \\ | \\ \text{COOH} \end{matrix} \xrightleftharpoons[\text{dehydrogenase}]{\text{Succinic}} \begin{matrix} \text{COOH} \\ | \\ \text{CH} \\ || \\ \text{HC} \\ | \\ \text{COOH} \end{matrix} + 2\text{H}^+ + 2e^- \qquad (8.11)$$

Succinic acid   Fumaric acid

If malonic acid, a molecule differing from succinic acid in lacking one —CH$_2$— group, is added to a solution of succinic acid and succinic dehydrogenase, the rate of the enzymatic oxidation sharply decreases. Yet malonic acid itself undergoes no chemical transformation. It attaches itself to the active site of succinic dehydrogenase and, by preventing access to the site by the normal substrate, prohibits enzyme activity (Figure 8-13A). In fact, malonic acid *competes* with succinic acid for a position on the active site. But it is a counterfeit substrate.

Another mechanism of enzyme inhibition is shown in Figure 8-13B. Here the inhibitor binds to a site on the enzyme other than the active site. As a result, it alters the shape of the enzyme molecule and thereby blocks its catalytic activity. This mechanism of inhibition is **noncompetitive**—that is, enzyme activity is inhibited irrespective of the availability of substrate. Examples are the action of *heavy metal ions*—such as silver ions (Ag$^+$)—on saccharase, a yeast enzyme (Figure 8-14). These inhibitory effects explain why such substances are poisons and why heavy metal pollution of rivers and soil is such a menace to all forms of life.

Certain enzymes may be inhibited or stimulated by low-molecular- weight substances called **effectors** that are bound to sites on the enzyme molecules at locations other than the active site. These are the *regulatory sites* (Figure 8-15). (When effectors act to inhibit an enzyme, it is a clear case of noncompetitive inhibition, since the inhibitory effector does not bind to the active site.) The most common form of this phenomenon is termed an **allosteric effect** (from Greek roots meaning ''other shape'') to emphasize the fact that the critical location on the enzyme is different from the one that is blocked in competitive inhibition by an analogue (close relative) of the substrate.

The interaction of enzyme and an allosteric effector bound at a distance from the active site causes a change in enzyme shape or conformation; the change is called an **allosteric transition.** The change produced by a negative effector decreases the affinity of enzyme for substrate. Therefore it is an **allosteric inhibitor.** The change produced by a positive effector increases the affinity of enzyme for substrate. Therefore it is an **allosteric activator.** In some cases the active site and the regulatory

8-15 Control of enzyme activity by allosteric effectors
(A) *The active site of the regulatory enzyme threonine deaminase binds the substrate threonine.* **(B)** *Binding of the allosteric effector, isoleucine, to the allosteric site changes the shape of the active site so substrate cannot bind to it.*

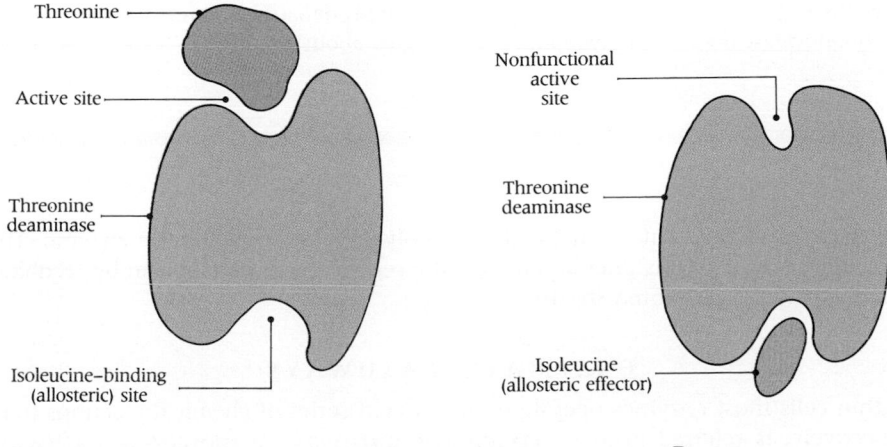

Threonine

Active site

Threonine deaminase

Isoleucine–binding (allosteric) site

Nonfunctional active site

Threonine deaminase

Isoleucine (allosteric effector)

A

B

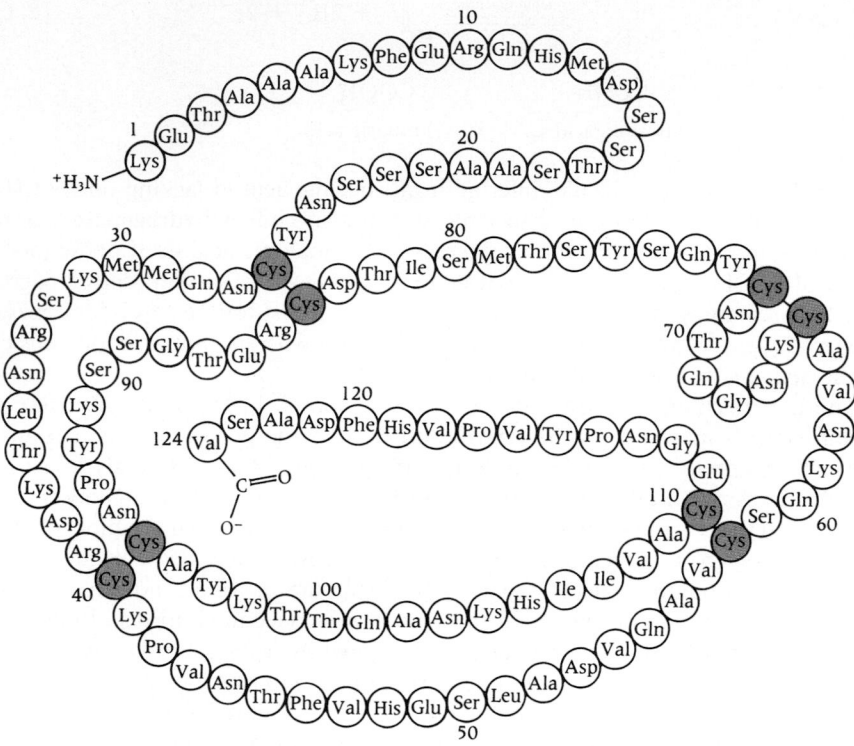

8A Structure of ribonuclease

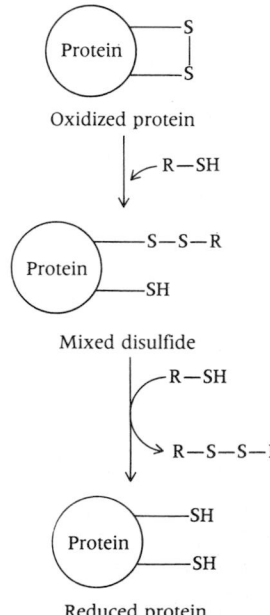

8B Mercaptoethanol (R–SH) is a reducing agent that breaks disulfide bonds.

We learned in Chapter 4 that all proteins have three levels of structure (and those with multiple subunits have four). In the 1960s a major contribution was made to our understanding of these levels by Christian Anfinsen at the National Institutes of Health (NIH). Anfinsen performed a simple experiment that brilliantly illuminated the critical relationship of the primary amino acid sequence to the overall conformation of a certain enzyme protein. The enzyme studied was ribonuclease, a small enzyme protein that hydrolyzes RNA.

Ribonuclease is a single polypeptide chain (Figure 8A) consisting of 124 amino acids and 4 disulfide bonds (—S—S—) that form stabilizing bridges between different parts of the chain. It was well known that an —S—S— bond is readily broken by a reducing agent such as mercaptoethanol, which converts it to two —SH groups (Figure 8B). The structure of ribonuclease is also stabilized by hydrogen bonds. When both kinds of stabilizing bonds are broken—the disulfide bonds with mercaptoethanol and the hydrogen bonds with urea—the result is a thoroughly denatured amino acid chain that wiggles about randomly (Figure 8C). Because

site may be on different subunits (that is, polypeptide chains) of the enzyme. The great importance of this phenomenon in the regulation of metabolism by feedback inhibition will be explored shortly.

ENZYMATIC PATHWAYS

Within cells most enzymes operate in an ordered series of chemical reactions that, collectively, is referred to as an **enzymatic pathway,** or **metabolic pathway.**

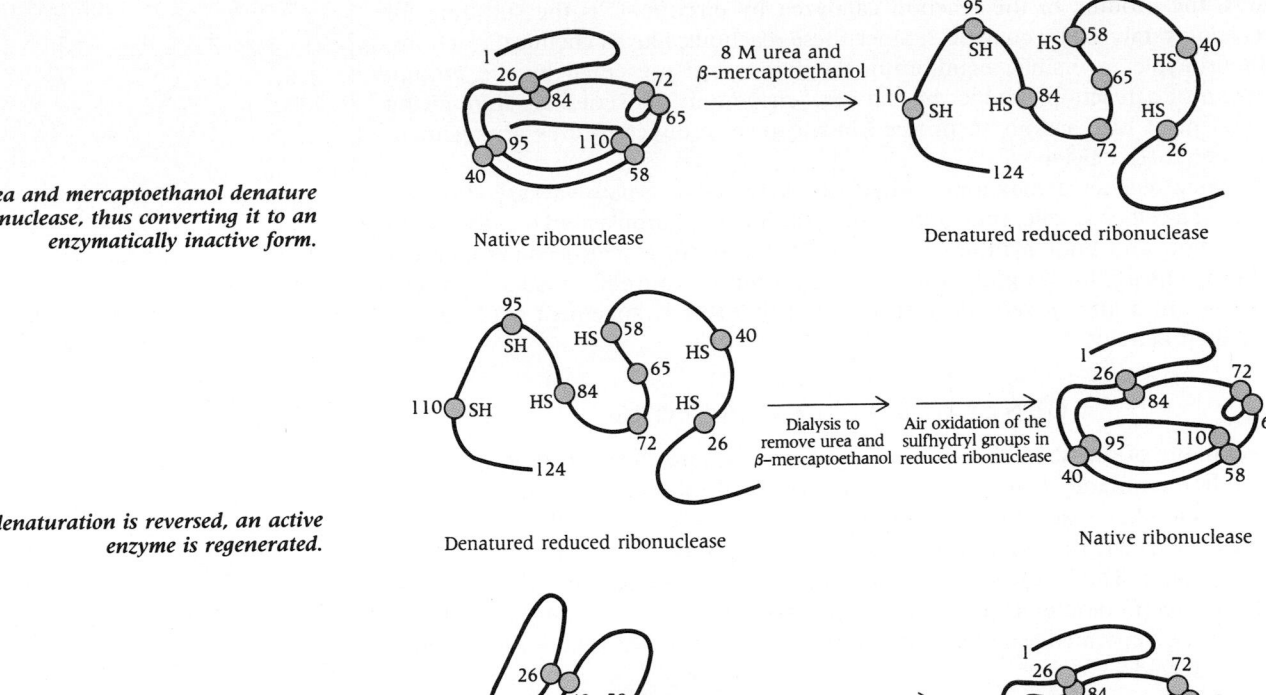

8C Urea and mercaptoethanol denature ribonuclease, thus converting it to an enzymatically inactive form.

Native ribonuclease

8 M urea and β–mercaptoethanol

Denatured reduced ribonuclease

8D When denaturation is reversed, an active enzyme is regenerated.

Denatured reduced ribonuclease

Dialysis to remove urea and β–mercaptoethanol

Air oxidation of the sulfhydryl groups in reduced ribonuclease

Native ribonuclease

8E A "scrambled" molecule is formed when disulfide bonds form in the wrong places.

"Scrambled" ribonuclease

Trace of β–mercaptoethanol

Native ribonuclease

it is a flexible, changing statistically random structure, it is called a **random coil** and is devoid of enzyme activity.

Anfinsen found that when he carefully removed the urea and mercaptoethanol by dialysis, the enzyme slowly regained its activity. The meaning of this chance finding seemed clear: After removal of the denaturing chemicals, the —SH groups of the denatured enzyme become reoxidized by air and the enzyme spontaneously refolds into its native catalytically active form (Figure 8D). Indeed, almost all of the original enzymatic activity is regained along with the protein molecule's original physical and chemical properties. The experiment clearly shows that the information needed to specify the complex three-dimensional structure of ribonuclease is in its amino acid sequence.

A rather different result was obtained when reduced ribonuclease (ribonuclease with free —SH bonds) is reoxidized while urea (the hydrogen-bond breaker) is still present. When urea is finally removed, the resulting product has only 1% of the enzyme activity of the native product. The reason is that the wrong —SH bonds paired up and disulfide bonds were formed in the wrong place (Figure 8E). There are 105 different ways to pair up 8 different SH bonds to form 4 —S—S— bonds. Only one of these combinations has enzyme activities. Tertiary structures based on any of the 104 wrong pairings have been termed *scrambled ribonuclease*. Anfinsen then found that the addition of a trace of mercaptoethanol would convert scrambled ribonuclease to native ribonuclease (Figure 8E).

Later studies of other proteins established the generality of this principle, a central one in molecular biology: Amino acid sequence specifies protein conformation. In 1964, Chris Anfinsen wrote:

It struck me recently that one should really consider the sequence of a protein molecule, about to fold into a precise geometric form, as a line of melody written in canon form and so designed by Nature to fold back upon itself, creating harmonic chords of interaction consistent with biological function. . . . The kinds of chords formed in a protein with scrambled disulfide bridges are dissonant, but given an opportunity for rearrangement by the addition of mercaptoethanol, they modulate to give the pleasing harmonics of the native molecule.

In such pathways, the product of the first enzymatic reaction in the series is the reactant in the second reaction, the product of which is the reactant in the third reaction, and so on.

This arrangement benefits the cell in a number of ways. First and foremost, it prevents waste. Consider the sequence:

$$A \underset{}{\overset{enz\ 1}{\rightleftharpoons}} B \underset{}{\overset{enz\ 2}{\rightleftharpoons}} C \qquad (8.12)$$

Here, B, the product of the reaction catalyzed by enzyme 1, is the substrate for the reaction catalyzed by enzyme 2. Since these reactions, like all chemical reactions, are in principle reversible, equilibrium conditions are prevented by the prompt utilization of a reaction product by the next enzyme in the series. By minimizing any tendencies of reactions to proceed in the reverse direction, this arrangement further enhances efficiency.

Efficiency reaches a maximum when the enzymes of a pathway are clustered together in a cell organelle, such as a mitochondrion or chloroplast. In the discussion of respiration and photosynthesis in Chapters 9 and 10, it will become apparent that the enzymes of these tightly controlled basic processes are embedded in organelle membranes in a precise geometric array—and that this arrangement critically enhances their activity.

COENZYMES AND COFACTORS

The technique of **dialysis,** shown in Figure 6-3C, exploits the properties of a semipermeable membrane. A solution of large and small molecules is placed in a bag made of cellophane or some other semipermeable membrane. When the bag is immersed in water, the small molecules pass through the membrane into the surrounding water. The large molecules are trapped inside the bag. In this way, low-molecular-weight biomolecules such as monosaccharides, amino acids, and nucleotides may be separated from macromolecules such as polysaccharides, proteins, and nucleic acids.

Early biochemists observed that when they exhaustively dialyzed tissue extracts containing certain active enzymes, the enzymes remaining in the bag lost their catalytic activity. They then found that activity could be fully restored merely by the addition of a little boiled tissue juice. This effect could not have been the result of the addition of fresh new enzymes, because boiled enzymes are ordinarily denatured. The dialyzed enzymes in the bag could also be reactivated by restoring to them the materials that had passed out of the bag—even after those materials had been boiled. The investigators concluded that tissues must contain molecules that are (1) essential for the catalytic activity of oxidative enzymes, (2) dialyzable and hence low in molecular weight, and (3) heat-stable and hence not proteins. The unfamiliar molecules were called **coenzymes;** the inactive enzymatic protein left in the bag, the **apoenzyme;** and the active combination of apoenzyme and coenzyme, the **holoenzyme.** Of course, these definitions apply only to those enzymes that have coenzymes. Not all enzymes do.

We define a coenzyme as an active and necessary participant in a reaction catalyzed by an apoenzyme, but one that is metabolically regenerated (by another coupled reaction) following its participation in the reaction so that it need be present only in small amounts. In Equation 8.9, the L-amino acid oxidase reaction, flavin mononucleotide (FMN) is a coenzyme. Another illustration is provided by the molecule called nicotinamide-adenine dinucleotide (NAD or, more precisely, NAD^+ to indicate its single positive charge; Figure 8-16) and its phosphate derivative NADP (or $NADP^+$). NAD^+ and $NADP^+$ are the coenzymes of many oxidative enzymes of the type called **dehydrogenases.** These enzymes catalyze the removal of hydrogen

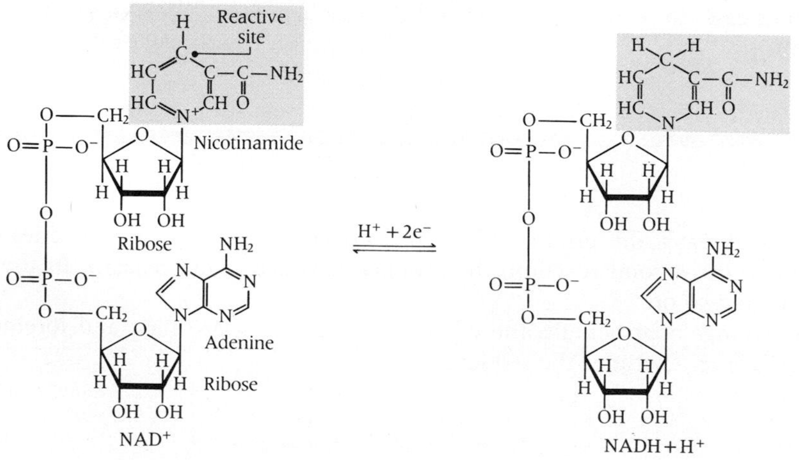

8-16 Oxidized and reduced forms of NAD
Nicotinamine adenine dinucleotide (NAD) is transformed from the oxidized state (NAD^+) to the reduced state ($NADH + H^+$) by the addition of a single proton and two electrons. Note the additional hydrogen atom at the reactive site of NADH. This hydrogen has stereospecifity.

ATP

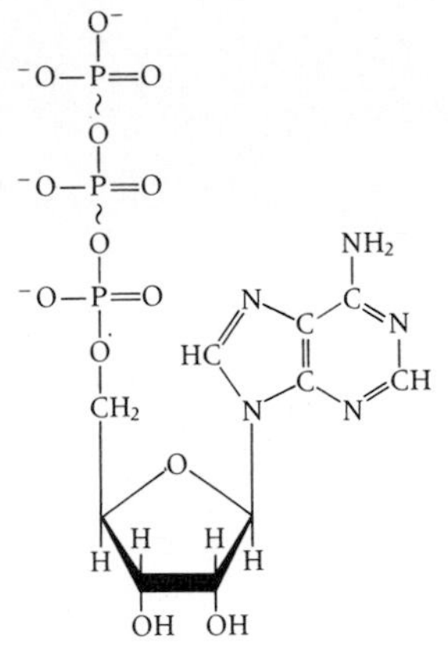

ADP

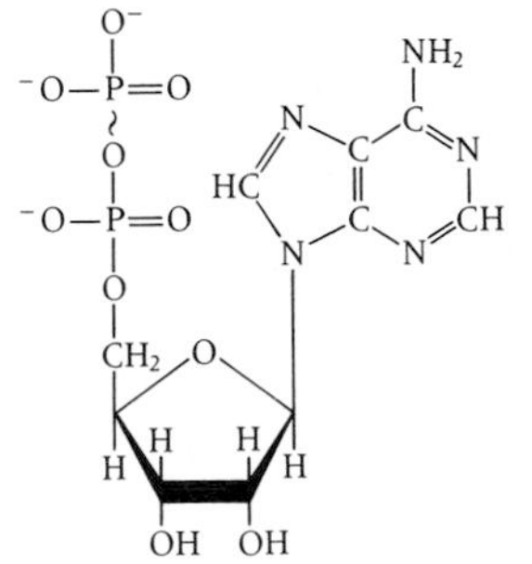

MgATP^{2-}

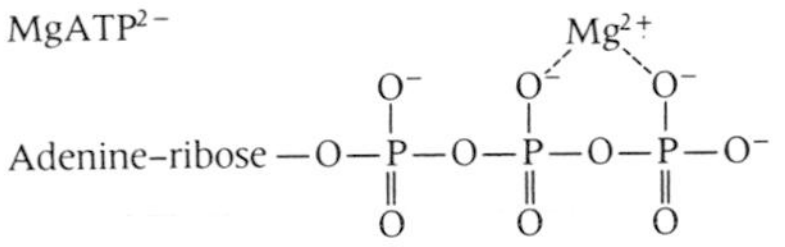

$$\text{Adenine–ribose}-\text{O}-\overset{\overset{\text{O}^-}{|}}{\underset{\overset{|}{\text{O}}}{\text{P}}}-\text{O}-\overset{\overset{\text{O}^{\angle}}{|}}{\underset{\overset{|}{\text{O}}}{\text{P}}}-\text{O}-\overset{\overset{\text{O}^-}{|}}{\underset{\overset{|}{\text{O}}}{\text{P}}}-\text{O}^-$$

Mg^{2+}

MgADP$^-$

$$\text{Adenine–ribose}-\text{O}-\overset{\overset{\text{O}^{\angle}}{|}}{\underset{\overset{|}{\text{O}}}{\text{P}}}-\text{O}-\overset{\overset{\text{O}^-}{|}}{\underset{\overset{|}{\text{O}}}{\text{P}}}-\text{O}^-$$

Mg^{2+}

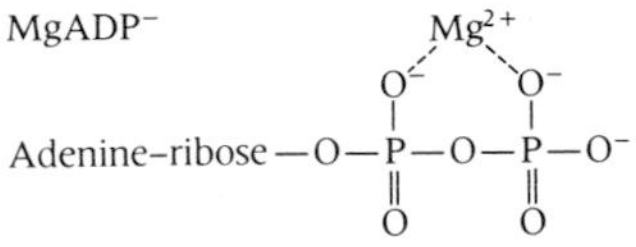

8-17 Structures of ATP and ADP
In the cell most of the ATP and ADP exists as its Mg^{2+} complex.

atoms from the substrate molecule—hydrogen removal with its accompanying electron being one form of oxidation. The function of NAD$^+$ and NADP$^+$ is to accept the freed electrons and pass them along to another substrate that is being reduced (that is, is accepting electrons). We may summarize these transactions in the following way:

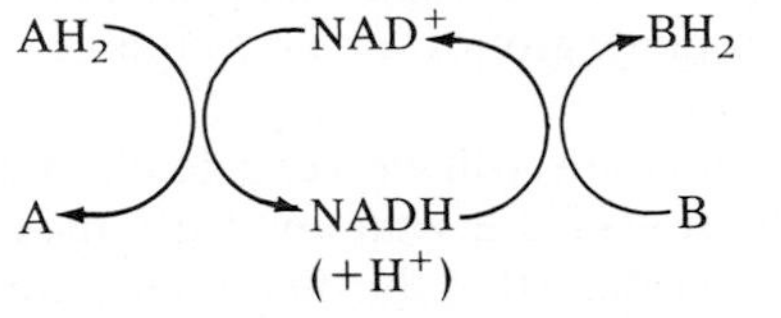

$$(8.13)$$

AH$_2$ symbolizes the initial substrate. It transfers its hydrogens to NAD$^+$ to form NADH (plus one H$^+$). The NADH is then recycled back to NAD$^+$ whenever it transfers its H to another molecule, here called B. The H of NADH (plus the H of one H$^+$) converts B to BH$_2$, regenerating NAD$^+$. Note that NAD$^+$ is not consumed in the sum of the two reactions. It is continually regenerated. Similar equations could be written for NADP$^+$ and NADPH.

Several conventions are used to designate oxidized and reduced forms of these coenzymes. The oxidized form has been symbolized as NAD$^+$, NAD, or NAD$_{ox}$; the reduced form as NADH plus H$^+$, NADH$_2$, or NAD$_{red}$. We will refer to them as NAD$^+$ and NADH (plus H$^+$) to remind ourselves that, in reactions involving these coenzymes, substrates being oxidized give up two electrons and two protons. One proton and both electrons are transferred to NAD$^+$ or NADP$^+$. The other proton is released into the solution, as follows:

$$\text{NAD}^+ + 2\text{H}^+ + 2e^- \rightleftharpoons \text{NADH} + \text{H}^+ \qquad (8.14)$$

In undergoing reduction, NAD$^+$ and NADP$^+$ accept only pairs of electrons. For brevity one H$^+$ may be subtracted from each side of the equation:

$$\text{NAD}^+ + \text{H}^+ + 2e^- \rightleftharpoons \text{NADH} \qquad (8.15)$$

Reduced FAD is always FADH$_2$.

Tissues also contain certain other nucleotides[8] that function as coenzymes. One group of paramount importance includes the ribonucleotides of adenine, notably adenosine diphosphate (ADP) and adenosine triphosphate (ATP) (Figure 8-17).

Adenine—ribose—Ⓟ
Adenosine monophosphate (AMP)
Adenine—ribose—Ⓟ ~ Ⓟ
Adenosine diphosphate (ADP)
Adenine—ribose—Ⓟ ~ Ⓟ ~ Ⓟ
Adenosine triphosphate (ATP)

ADP serves as an acceptor of phosphate groups—here symbolized Ⓟ—in certain reactions. The resulting product is ATP. In other reactions, ATP is a donor of a Ⓟ group, thereby regenerating ADP. The following scheme summarizes the situation:

$$\begin{array}{rl} & \text{X—Ⓟ} + \text{ADP} \rightleftharpoons \text{X} + \text{ATP} \\ & \text{ATP} + \text{Y} \rightleftharpoons \text{ADP} + \text{Y—Ⓟ} \\ \hline \text{Sum:} & \text{X—Ⓟ} + \text{Y} \rightleftharpoons \text{X} + \text{Y—Ⓟ} \end{array} \qquad (8.16)$$

---

[8] A **(mono)nucleotide** is a three-part molecule containing a nitrogenous base, a sugar (usually ribose or deoxyribose), and a phosphate. A **dinucleotide** is a six-part molecule consisting of two nucleotides attached through their phosphate groups. Some nucleotides are the monomeric units of DNA and RNA. There the bases are purines or pyrimidines. Other bases often occur in the nucleotides that function as coenzymes. NAD contains nicotinamide and the purine adenine.

Enzymes catalyzing reactions in which ATP is ultimately converted to ADP are often called adenosinetriphosphatases or ATPases. We encountered important ATPases in membrane transport systems (Chapter 6).

In these equations, X and Y symbolize any of a large group of biomolecules. Here again a coenzyme (ATP or ADP) acts as a **donor-acceptor** of a chemical group. In this case it is $\circled{P}$. In the case of NAD^+, it is H^+ plus two electrons. This is a recurring phenomenon of great significance in biochemistry. ADP and ATP, the donor-acceptors of phosphate groups, are profoundly important in the generation and storage of metabolic energy.

Nutritional aspects of vitamins will be considered later. However, we should note here that most of the so-called B vitamins are converted in the body to coenzymes. That is why they are essential in the diet. For example, the coenzyme NAD^+ derives from the vitamin nicotinamide; the coenzyme FAD (flavin adenine dinucleotide) contains riboflavin; and so on. The ill effects of vitamin deficiencies are direct consequences of shortages of the coenzymes needed in specific enzyme reactions.

Most coenzymes are bound to their apoenzymes as prosthetic groups. For example, the porphyrin derivatives called *cytochromes* are coenzymes that are tightly bound to their apoenzymes. We noted in Chapter 4 that proteins with tightly bound prosthetic groups are termed *conjugated*. In conjugated proteins (like these enzymes), another molecule is covalently attached to the protein.

Many enzymes require certain inorganic **cofactors.** These are also called *activators* to distinguish them from coenzymes (which function as donor-acceptors in group transfer reactions). The best-known inorganic activators are metals of the trace-element group (see Table 3-3), which are usually present as divalent cations—for example, Mg^{2+}, Ca^{2+}, Zn^{2+}, Cu^{2+}. In $\circled{P}$ transfers of the type discussed above, the positive charges of Mg^{2+} hold the negatively charged $\circled{P}$ groups in place (see Figure 8-17). Some enzymes require K^+ as a cofactor.

In sum, the catalytic machinery of an enzyme reaction may involve more than an enzymatic protein (a polypeptide made of amino acids alone). A coenzyme or a cofactor may also be necessary.

SPECIAL IMPORTANCE OF OXIDATIVE ENZYMES

Since oxidation is the key process by which chemical bond energy is extracted from molecules of food and fuel, it is appropriate to say a word about oxidation itself—and its opposite, reduction.

Historically, the term *oxidation* was first used for reactions in which O_2 is actually consumed—for example, the burning in air of free carbon.

$$C + O_2 \rightarrow CO_2$$
$$2C + O_2 \rightarrow 2CO \tag{8.17}$$

The burning of hydrogen in air forms water.

$$2H_2 + O_2 \rightarrow 2H_2O \tag{8.18}$$

When iron (Fe) is heated in oxygen, it forms iron oxide, Fe_2O_3, and other oxides such as FeO. Similar products form slowly when iron rusts in air—although these are usually associated with H_2O.

$$4Fe + 3O_2 \rightarrow 2Fe_2O_3 \tag{8.19}$$

Chemists soon recognized that fundamentally similar combinations may occur with elements other than oxygen. For example, carbon, hydrogen, and iron may "burn" in fluorine.

$$C + 2F_2 \rightarrow CF_4$$
$$H_2 + F_2 \rightarrow 2HF$$
$$2Fe + 3F_2 \rightarrow 2FeF_3 \tag{8.20}$$

Accordingly, the definition of an oxidation was broadened. An **oxidation** is now defined as any reaction in which an atom or molecule gives up electrons. For

TABLE 8-1
SUMMARY OF OXIDATION-REDUCTION REACTIONS

Oxidations [with respect to A]	Reductions [with respect to A]
Removal of electrons: $A \rightarrow A^+ + e^-$	Addition of electrons: $A + e^- \rightarrow A^-$
Addition of oxygen: $A + BO \rightarrow AO + B$	Removal of oxygen: $AO + B \rightarrow A + BO$
Removal of hydrogen: $AH + B \rightarrow A + BH$	Addition of hydrogen: $A + BH \rightarrow AH + B$
All three mechanisms: Release of energy[*]	All three mechanisms: Storage of energy

[*] In most cases.

example, in the oxidation of iron by both O_2 (Equation 8.19) and F_2 (Equation 8.20), iron undergoes the same transformation: it gives up electrons:

$$Fe \rightarrow Fe^{3+} + 3e^- \tag{8.21}$$

A **reduction** is the acquisition of electrons by an atom or molecule. Confusion sometimes arises from the fact that a "reduction" is accompanied by an acquisition of something (an electron). Remember that the gain of an electron (abbreviated e^-) reduces net electrical charge. In the reaction between Fe and O_2 in Equation 8.19, Fe is oxidized to Fe^{2+}, while oxygen is being reduced in a simultaneous reaction:

$$3O_2 + 12e^- \rightarrow 6O^{2-} \tag{8.22}$$

Thus the production of Fe_2O_3 requires both an oxidation and a reduction. The result is a net transfer of 12 electrons from Fe to O. Because there is no change in the total electrical charge, the reactions must occur simultaneously, one at the expense of the other. We refer to reactions in which one or more electrons pass from one molecule to another as **oxidation-reduction reactions.**

Oxidation may involve only a loss of electrons—as when Na is oxidized to Na^+. But often the traveling electron is accompanied by a proton (H^+), so that the final reaction may involve not only the addition of oxygen but the removal of hydrogen (Table 8-1). That is perhaps the most important form of biological oxidation. Since the ultimate source of oxidizing power is atmospheric O_2, it is fair to say that the final metabolic task of most aerobic organisms is to reduce O_2.

$$O_2 + 4e^- + 4H^+ \rightarrow 2H_2O \tag{8.23}$$

The electrons and protons that ultimately perform this critical task in many organisms arise from the oxidation of glucose (and other substrates):

$$C_6H_{12}O_6 + 6O_2 \rightarrow 6CO_2 + 6H_2O \tag{8.24}$$

Reduction of atmospheric oxygen is not accomplished directly. Many hydrogen transfers occur in metabolism before they are collectively transferred in a final sequence to oxygen, the final oxidant (electron acceptor). We earlier encountered an example of one of these transfers in the oxidative conversion of succinate to fumarate (Equation 8.11).

Electron removal is usually accompanied by a release of energy. Clearly, energy is released as heat and light in the oxidation of firewood by ordinary combustion. Energy release certainly accompanies oxygen addition or hydrogen removal. Conversely, reduction, or electron addition (which accompanies oxygen removal or hydrogen addition), is associated with a utilization of energy. Oxidation occurs as sugar molecules slide down the "energy hill"—like the one shown in Figure 8-3B. In short, oxidations are energy-yielding; reductions are energy-consuming.

These facts are fundamental to an understanding of the flow of energy in living organisms. Almost every biological energy transfer involves at some point in the metabolic pathway an oxidation-reduction process. Most are accomplished by the removal of electrons or hydrogen (dehydrogenation) from a substrate molecule. For example, sugar is oxidized with release of its energy into the cell by removal of its hydrogen and not by an addition of oxygen. The oxygen molecules ultimately utilized in the degradation of sugar are serving as electron (hydrogen) acceptors.

CONTROL OF CELL METABOLISM

The processes of metabolism occur in all cells. It should be remembered that metabolism is something more than the simple sum of all the individual chemical reactions within a cell. It is also the integrated activity by which a cell conducts business with its environment and thereby ensures its survival and the survival of its species.

As many metabolic pathways are in simultaneous operation in each cell, systems must exist to control and coordinate metabolic road traffic. Such control systems must have three important characteristics.

- First and foremost, they must control the rates of enzyme reactions.
- Second, they should prevent the waste of materials and energy.
- Third, they must function efficiently and uninterruptedly.

Until the mid-1950s, biochemists were engaged chiefly in identifying the steps of the sequences we call metabolic pathways (Chapters 9–11). Not until the pathways were mapped was it possible to search for factors controlling traffic along them. The stimulating revelations that followed disclosed the existence of several regulatory mechanisms. In general, they can be divided into two groups:

- Mechanisms controlling the catalytic activity of existing enzymes.
- Mechanisms controlling the rate of synthesis of individual enzymes.

FACTORS CONTROLLING ENZYME ACTIVITY

Although each reaction of the metabolic road map is catalyzed by a specific enzyme, the level of that enzyme's activity is an important determinant of the traffic volume along an entire pathway only when the reaction in question is a "bottleneck"—a step that fixes the rate of the sequence.

The slowest reaction is the **rate-limiting step** of a pathway. Fluctuations in its activity are of great importance. Usually these bottlenecks or pacemakers of metabolism are the *earliest* step in a sequence.

A classic example is seen in the synthetic pathway of pyrimidines, a component of nucleic acids. As shown in Figure 8-18, this short pathway begins with a condensation of two small molecules—carbamyl phosphate and the amino acid aspartic acid—by an enzyme called *aspartate transcarbamylase* (abbreviated *ATCase*). It ends with the synthesis of the nucleotide uridine-5'-phosphate, which is converted to uridine-5'-triphosphate (UTP), which then becomes cytidine-5'-triphosphate (CTP). When the steady-state level of CTP is high in a cell, it inhibits ATCase and thus slows further CTP synthesis—an example of negative feedback inhibition (discussed below). When CTP levels are low—when further CTP synthesis is needed—such inhibition of ATCase is relieved, and metabolic traffic down the pathway increases. In both situations, ATCase activity is the pathway's rate-limiting step.

Feedback Inhibition and Activation

In many major biosynthetic pathways, the catalytic activity of an early rate-limiting enzyme is inhibited by the major end product of the pathway. This phenomenon, called **feedback (end-product) inhibition,** blocks the pacemaker enzyme's activity when it is not needed.

Consider the example in Figure 8-19. A pathway in certain bacteria converts the amino acid L-threonine to another amino acid the bacteria need, L-isoleucine. However, if a generous supply of L-isoleucine is deliberately added to the culture, the bacteria promptly cease to make it for themselves. They do not resume its

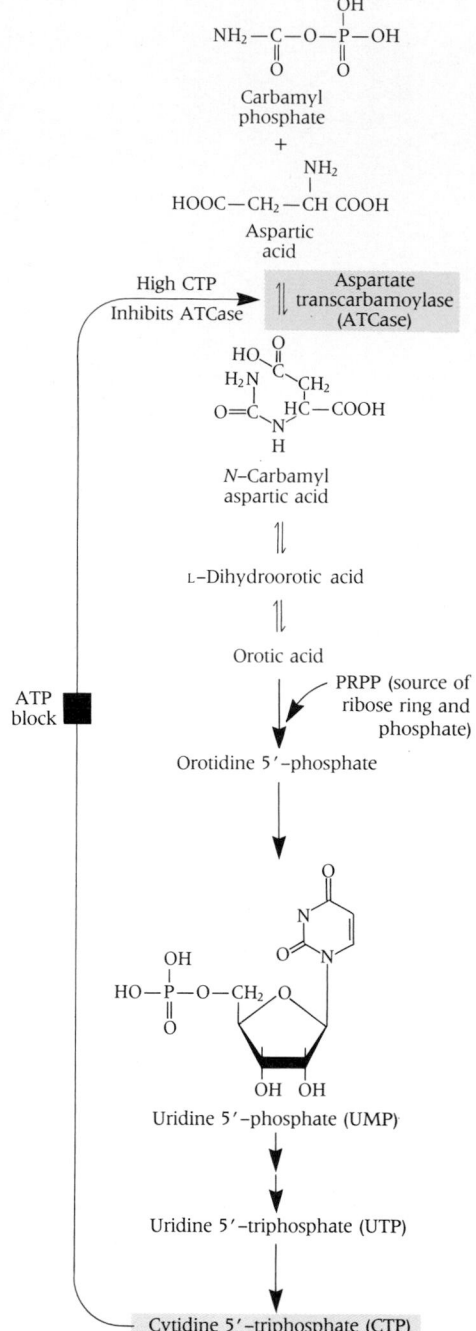

8-18 Regulation of CTP biosynthesis
Aspartate transcarbamylase (ATCase) is the rate-limiting enzyme in the CTP biosynthetic pathway. This enzyme is inhibited by high levels in the cell of CTP, the final product of the pathway. Such inhibition can be blocked by ATP.

8-19 Feedback inhibition of enzyme activity

The conversion of L-threonine to L-isoleucine requires five enzymatic reactions. The pathway is controlled by feedback inhibition, in which the final product of the pathway, isoleucine, inhibits the activity of the first enzyme, threonine deaminase.

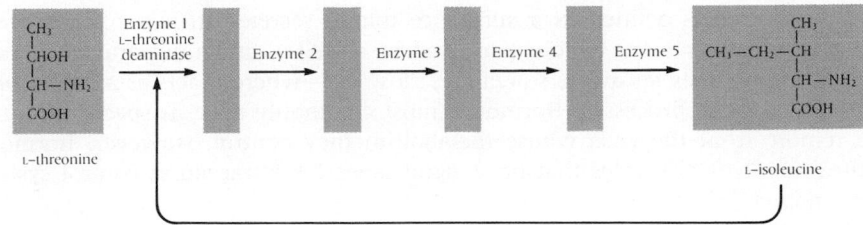

production until the L-isoleucine concentration in the medium drops. This temporary shutdown is mediated by a powerful inhibitory effect of L-isoleucine upon L-threonine deaminase, which, as shown in the figure, is the first enzyme in the biosynthetic pathway. The inhibitory action is quite specific. No other enzyme is inhibited and no other substance is an inhibitor—not even the stereoisomer D-isoleucine. Moreover, inhibition is instantaneous. Feedback inhibition occurs in all cells.

Inhibition of certain biosynthetic enzymes by certain end products makes it evident that these regulatory enzymes have two kinds of specificity: *substrate specificity* of the usual kind and *inhibitor specificity*. As suggested by Figure 8-19, there need be little chemical similarity between the inhibitory end product and the substrate for which the enzyme is targeted.

These are allosteric effects. As discussed above, the regulation of these enzymes depends on the activities of positive or negative allosteric effectors that interact with the enzyme's regulatory site, a site other than the catalytic active site. In so doing, they alter the shape of the enzyme in a way that influences the combination of the enzyme with its substrate (see Figure 8-13). Ordinarily, allosteric enzymes consist of more than one polypeptide chain or subunit.

Feedback inhibition by end product has a number of interesting features. Because the process is rapid and precise, a cell synthesizing a given substance will stop making that substance at once when the supply becomes adequate. It will start again when end-product concentration decreases. Moreover, control is automatic. It is efficient also because enzymes are constantly being synthesized and degraded in quantities appropriate to demand. Because only weak hydrogen or ionic bonds bind allosteric effectors to enzymes (no new covalent bonds are needed), the process requires little expenditure of energy. Thus the mechanism functions economically in preventing unwanted accumulations of biosynthetic products.

The concept of end-product inhibition evolved in part from a careful study of substrate-saturation curves, which describes the relation between the rate of an enzyme reaction and the concentration of its substrate (Figure 8-20). For ordinary, or nonregulatory, enzymes the curve follows a smooth rising course that gradually approaches a maximum in line B in the figure. We call curves of this shape *hyperbolic*. But for regulatory enzymes, the curve is characteristically *sigmoidal* or *S-shaped* in line A in the figure. This supports the idea that positive and negative effectors alter an enzyme's shape in a way that progressively alters its affinity for substrate molecules. In the presence of a positive effector, further binding of substrate is facilitated as substrate concentration increases. This phenomenon has been termed **cooperativity** because the occupation of a few active sites (at low substrate concentrations) alters the enzyme's conformation enough to increase the substrate affinity of the active sites still to be occupied. The presence of a negative effector tips the balance the other way.

Other Mechanisms

Other devices for regulating the activity of enzymes include (1) shifts in the availability of substrate, (2) hormones, (3) coupling mechanisms, and (4) the local presence of cofactors and various ions such as H^+.

Since the rate of an enzyme reaction is affected by the concentration of substrate until the enzyme is saturated with substrate (see Figure 8-20), the rate behavior of a metabolic pathway will vary with variations in the concentration of the initial substrate of the pathway. If glucose supply is limited by starvation, glucose metabolism is depressed. Conversely, an excess of glucose accelerates glycolysis to a point at which the rate-limiting enzyme becomes saturated with substrate.

8-20 Substrate-saturation curves for regulated and nonregulated enzymes

(A) A sigmoid curve is typical of a regulated enzyme, in which effector molecules alter the binding of enzyme to substrate. **(B)** A nonregulated enzyme is identified by a hyperbolic curve.

A **hormone** is defined as a substance that is secreted in living organisms—sometimes by cells in one region, sometimes by all cells—and transported to another region—sometimes far away, sometimes close by—where it acts as a regulator of certain metabolic processes. Hormones most commonly arise in special cells that are remote from the cells whose metabolism they control. However, hormonal control involves principles that have significance for intracellular control systems (see Chapter 28).

Since the rates of metabolic reactions are determined by their enzymes, it is natural to assume that hormones act by influencing enzyme behavior in some manner, direct or indirect. At least five mechanisms for hormonal control of an enzyme exist.

- Alteration of the activity levels of existing enzyme molecules
- Alteration of the rate of synthesis of new enzyme molecules
- Alteration of the permeability of cell membranes of the cell surface or of intracellular structures
- Direct participation by the hormone as a coenzyme in a reaction
- Competition by hormone with coenzyme for specific sites on the enzyme

Each of these mechanisms operates with different hormones.

Coupling mechanisms are inherent in the design of certain metabolic pathways. This is one of the very important lessons of the study of thermodynamics. Since the overall free-energy change for a series of reactions (such as a metabolic pathway) is equal to the sum of the free-energy changes of the individual steps, situations arise in which a thermodynamically unfavorable reaction (one that cannot occur spontaneously because ΔG is positive) can be driven by a thermodynamically favorable reaction (one that can occur spontaneously because ΔG is negative). Consider the following reactions:

$$
\begin{array}{lll}
(1) & A \rightleftharpoons B + C & \Delta G = +4 \text{ kcal/mol} \\
(2) & \underline{B \rightleftharpoons D} & \underline{\Delta G = -9 \text{ kcal/mol}} \\
(3) & A \rightleftharpoons C + D & \Delta G = -5 \text{ kcal/mol}
\end{array}
\qquad (8.25)
$$

Under ordinary conditions, reaction 1 cannot occur spontaneously, but reaction 2 can. Because the sum of the free-energy changes is negative, the sequence of the two reactions—reaction 3—can occur spontaneously. We say that these reactions are *coupled*. We will encounter many examples of energy coupling in our later studies of metabolism (Chapters 9–11). Indeed, the fundamental cycle shown in Figure 8-4 is such an example.

FACTORS CONTROLLING ENZYME SYNTHESIS

Induction and Repression

An important mechanism for controlling cell metabolism depends on factors controlling the rate at which cells synthesize the enzymes in each metabolic pathway. Two examples of variable rates of enzyme synthesis, induction and repression, led to much of our knowledge of this phenomenon.

Induction of enzyme synthesis is defined as the increased rate of synthesis of a single enzyme (or a few related enzymes) relative to the rate of synthesis of other proteins that results from the presence outside (or inside) of cells of substances called **inducers.** Inducers are identical or closely related to the substrate of the enzyme. For example, a culture of *E. coli* in a medium whose carbon source is glycerol contains only trace amounts of β-galactosidase, an enzyme that splits disaccharides known as *galactosides*. If one adds the galactoside lactose to the medium, there is a rapid 2000-fold increase in the rate of β-galactosidase synthesis (Figure 8-21). No compounds other than galactosides induce the synthesis of β-galactosidase. Enzymes such as β-galactosidase are called **inducible enzymes** to distinguish them from the **constitutive enzymes** normally present, which are noninducible.

Repression, like feedback inhibition, depends mainly upon the concentration of the final product of a reaction sequence. During repression, however, the end product inhibits the synthesis, not the catalytic activity, of an enzyme. Conversely, when the concentration of end product is low, synthesis of the enzyme is accelerated

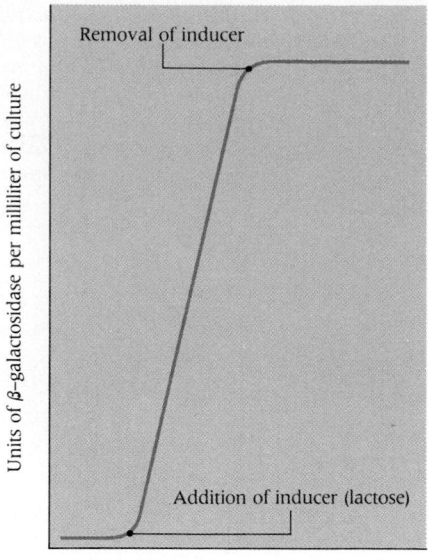

8-21 Induction of β-galactosidase synthesis
Addition of the inducer, lactose, to a culture of E. coli *leads to enhanced synthesis of the enzyme β-galactosidase.*

as a result of **derepression.** Derepression of the enzymes of L-isoleucine synthesis was illustrated in Figure 8-19.

Repression occurs in most of the major biosynthetic pathways. Like feedback control by allosteric inhibition, repression prevents the wasteful biosynthesis of enzymes. Feedback inhibition, however, is far more rapid and sensitive. In that case an already synthesized enzyme is temporarily inactivated. In a sense, repression is the coarse adjustment of a metabolic machine; inhibition is its fine adjustment.

Actions of Hormones in Regulating Enzyme Synthesis

Some hormones act by directly affecting enzyme activity. Others act by regulating the rates of synthesis of specific enzymes. Proof of this statement rests in part on a convincing demonstration that certain hormones are blocked from functioning after the administration of agents known to inhibit the synthesis of enzymes. One such agent is the antibiotic actinomycin D. This agent is known to enter the cell nucleus and form a complex with DNA in a way that ultimately prevents the synthesis of certain proteins. When the action of a hormone is blocked by actinomycin D, we can conclude that hormonal stimulation normally causes the production of new molecules of specific enzymes.

Enzyme Activation and Molecular Conversion

A final mechanism of metabolic control is implicit in the fact that many enzymes exist in tissues as inactive enzyme precursors that need to be converted to active enzymes. A familiar example is seen in the **zymogens,** inactive precursors of the digestive enzymes that are secreted into the stomach and intestine. For example, the inert molecule pepsinogen is secreted by the stomach. But this protein is useless in digestion until it is converted to the active protein-splitting enzyme pepsin (Chapter 26). This transformation, which is caused by certain enzymes as food leaves the stomach, involves the loss of 41 amino acid residues—and most of the helical structure of pepsinogen. It is an essential mechanism for one very good reason: If any cell synthesized pepsin molecules as such, they would destroy the cell that made them by hydrolyzing its proteins.

In the case of powerful enzymes—such as the protein-splitting enzymes of blood clotting (Chapter 22)—the existence of control mechanisms that delay "turning on" activity until the enzymes are in the proper physical locations is of obvious value to the organism.

SUMMARY

Energy, which is the capacity to do work, is continuously supplied to Earth by solar radiation. Photosynthetic organisms utilize this energy in the synthesis of energy-rich carbohydrates, which in turn are oxidized by all organisms as a source of energy to drive other metabolic reactions. Like the energy transfer reactions of chemistry and physics, these energy transfer reactions must obey the laws of thermodynamics.

These laws state that (1) the total energy of the universe is always constant, and (2) the entropy, or "randomness," of the universe always increases. The first law of thermodynamics implies that no energy is ever lost in the transfer. The second law implies that some of the energy of the transfer becomes useless for the performance of work. Determination of the change in free energy of a reaction permits a prediction of whether a reaction will occur spontaneously. When the change in free energy is zero, the reaction is at equilibrium.

Either chemical reactions require an input of energy to form new chemical bonds (endergonic reactions) or they release free energy when chemical bonds are broken (exergonic reactions). In general, endergonic reactions favor the presence of reactants; exergonic reactions favor the formation of products.

Metabolic reactions are characterized not only by thermodynamic parameters, but also by kinetic ones, kinetics being the study of how fast reactions proceed to equilibrium—that is, reaction rates.

Catalysts, by decreasing activation energy, increase reaction rates—without, however, either changing the extent to which the reaction proceeds to completion, or themselves being consumed in the reaction. Proteins called enzymes are the major catalysts in biological systems. Most are highly specific, catalyzing only one or a few types of reaction. This specificity is determined by the conformation of the enzyme's active site—the area which binds the reactants. A molecule with a structure similar to an enzyme's substrate may bind to the enzyme's active site and thus inhibit the enzyme. Some poisons work this way.

Some enzymes require organic coenzymes or inorganic cofactors for their catalytic function. Enzymes that catalyze oxidative reactions are especially important in metabolism, because those reactions provide the energy a cell needs to drive other reactions.

Nearly all biological reactions are enzymatically catalyzed. They are frequently controlled by mechanisms that act on these enzymes. These mechanisms fall into three classes: control of enzyme activity, enzyme synthesis, and pro-enzyme activation. Enzyme activity is often regulated by feedback inhibition, in which the product of a series of reactions can, when its concentration begins to increase, bind to and turn off a key enzyme in the series. In activation, a series of reactions is turned on when a precursor binds to and turns on a key enzyme. Other mechanisms, such as the presence of cofactors or certain hormones, can also control enzyme activity.

Enzyme synthesis can undergo induction or repression, resulting in increased and decreased rates of synthesis respectively. Often an increase in concentration of a reaction sequence's substrate induces synthesis of the enzymes involved, while an increase in the final product of the sequence represses synthesis of the relevant enzymes. Hormones may induce or repress synthesis of enzymes.

KEY TERMS

activation energy
active site
allosteric effect
apoenzyme
bioenergetics
calorie
catalyst
closed system
coenzyme
cofactor
competitive inhibition
conservation of energy
constitutive enzyme
cooperativity
coupled reaction
derepression
dialysis
effector
endergonic

endothermic
energetics
energy
enthalpy (H)
entropy (S)
enzymatic pathway
enzyme-substrate complex
enzyme
equilibrium
equilibrium constant (K_{eq})
exergonic
exothermic
feedback (end-product) inhibition
first law of thermodynamics
free energy (G)
holoenzyme
hormone
inducible enzyme

induction
kilocalorie
kinetics
metabolic pathway
mononucleotide
noncompetitive inhibition
open system
oxidation
oxidation-reduction reactions
rate-limiting step
reduction
repression
second law of thermodynamics
standard free energy change
substrate
thermodynamics
work
x-ray diffraction
zymogen

QUIZ QUESTIONS

1. Which of the following is *not* characteristic of an enzyme?
   A. Enzymes are proteins.
   B. An enzyme is specific as to the reaction it will catalyze.
   C. Enzymes can operate under wide ranges of temperature and pressure.
   D. Enzymes are usually named by adding the suffix -ase to the name of its substate.

2. Allosteric activator molecules bind at the _____ of an enzyme molecule.
   A. regulatory site
   B. active site
   C. exergonic site

   D. kinetic site
   E. catalytic site

3. Which of the following is *not* characteristic of an oxidation reaction?
   A. Oxidation reactions release energy.
   B. Oxidation reactions can involve removal of a hydrogen atom.
   C. Oxidation reactions can involve removal of electrons.
   D. Oxidation reactions can involve removal of an oxygen atom.
   E. None of the above.

4. In a chemical reaction, the total energy that is available for use in other reactions

is referred to as
   A. entropy.
   B. free energy.
   C. enthalpy.
   D. catalytic energy.
   E. energy of activation.

5. Regulation of cellular metabolism involves
   A. hormones.
   B. feedback inhibition.
   C. repression.
   D. B and C
   E. A, B and C

ESSAY QUESTIONS

1. Give examples of exergonic and endergonic biochemical reactions. Are all exergonic reactions exothermic and all endergonic reactions endothermic?

2. According to the second law of thermodynamics, all physical and chemical processes tend to increase the entropy of a system. The fact that living organisms are highly ordered and

complex systems seemingly defies the second law. Is that in fact the case? Given the constraints of the second law, how might one account for the origin and evolution of life?

3. Give examples of catabolic and anabolic processes. Characterize such processes in terms of their effect on the entropy of a system.

4. In metabolic systems, enzymes control the rates of chemical reactions. How do organisms control the activity levels of their enzymes?

5. What is the effect of an enzyme on the activation energy and rate of a reaction? What is the effect on temperature?

REFERENCES AND SUGGESTIONS FOR FURTHER READING

Fersht, A. (1985). *Enzyme Structure and Mechanism*, 2nd ed. W. H. Freeman, New York.

An enthusiastic, stimulating, and informative introduction to what the author calls the new golden age of enzymology.

Friedrich, P. (1985). *Supramolecular Enzyme Organization: Quaternary Structure and Beyond*. Pergamon Press, New York.

A book for nonspecialists that emphasizes the importance of associations between enzymes.

Kraut, J. (1988) How do enzymes work? *Science 242*, 533–540.

The author ascribes a rising interest in enzymes to the availability of mutant enzymes (see Chapter 17) with altered or defective active sites. These have permitted enlightening studies of structure-function relations. This paper reviews and amends our ideas on the enzyme-substrate complex.

Lehninger, A. L. (1982). *Principles of Biochemistry*. Worth, New York.

An abridged edition of *Biochemistry* (cited in Chapter 3), the late A. L. Lehninger's larger text. A fine summary.

Lipmann, F. (1971). *Wanderings of a Biochemist*. Wiley-Interscience, New York.

A master's autobiography. The work is richly laden with summaries of his many major contributions.

Royer, G. P. (1982). *Fundamentals of Enzymology: Rate Enhancement, Specificity, Control, and Application*. John Wiley & Sons, New York.

An authoritative summary.

Smith, E. B. (1982). *Basic Chemical Thermodynamics*. 3rd ed. Oxford University Press, New York.

A short, readable introduction to thermodynamics.

Smith, E. L., Hill, R. L., Lehman, I. R., Lefkowitz, R. J., Handler, P., and White, A. (1983). *Principles of Biochemistry*, 7th ed. (2 volumes).

A classic text, described by one reviewer as "the Rolls-Royce of biochemistry textbooks." Strong on fundamental aspects of enzymology and metabolism, its authors having contributed significantly to these fields.

Stokes, G. B. (1988). Estimating the energy content of nutrients. *Trends in Biochemical Sciences 13*, 422–424.

A stimulating presentation of a simple method for estimating the ATP yield from various compounds, explaining why more energy can be stored as depot fat than as an equal weight of glycogen.

HOW CELLS EXTRACT ENERGY FROM THE ENVIRONMENT

Details of the road maps of metabolism surpass the mastery of any one person. Although our interest is largely in the main highways and thoroughfares, we are obliged first to divide these megapathways into segments that can be understood in isolation and then to reassemble them into the larger picture. We begin with the pathways that produce the cell's major carrier of energy in chemical form—**adenosine triphosphate,** or **ATP.** The central role in energy exchanges of this remarkable nucleotide (see Figure 8-17) was established in 1941 by Fritz Lipmann and Herman Kalckar.

Over the eons, mechanisms evolved that couple the production of ATP to the utilization of oxygen. This is the principal method for the generation of ATP in heterotrophs and most autotrophs (these terms will be discussed shortly). Most of the phenomena of life are processes that utilize ATP. These phenomena serve the following three major purposes, among others:

- The performance of *mechanical work* in muscle contraction and other cellular movements
- The *active transport* of molecules and ions
- The *biosynthesis* of macromolecules and other biomolecules from simple precursors

We are aware of these processes much of the time, but we have no comparable awareness of the equally important systems that produce ATP except through our need for oxygen. Lack of oxygen is lethal to many cells because their need for ATP is constant.

CLASSIFICATION OF ORGANISMS ACCORDING TO ENERGY AND CARBON SOURCES

As noted, mechanisms arose in the course of evolution that couple the production of ATP to the utilization of oxygen. However, the atmosphere was without oxygen for a long time in the early years of Earth's history. Yet during that period life began and established itself. Obviously, mechanisms existed in early organisms that coupled the production of high-energy bonds to the utilization of something other than oxygen.

We can fairly accurately trace the evolution of these energy-yielding systems. Even in today's world, organisms exist that differ widely in their methods of extracting and utilizing metabolic energy. Here we will classify modern organisms according to the way they obtain and use energy and their sources of carbon. We will then attempt to deduce when they appeared during the course of evolution.

In the classification in Table 9-1, organisms are classified by their method of deriving *energy* from the environment into **phototrophs,** which obtain their energy by trapping light energy, and **chemotrophs,** which obtain it by the oxidation of foodstuffs.

How cells extract energy from the environment
The sun is the source of all of the energy that flows through the world of life.

TABLE 9-1

CLASSIFICATION OF ORGANISMS ACCORDING TO CARBON AND ENERGY SOURCES

Type of Organism	Carbon Source Class	Carbon Source	Energy Source Class	Energy Source	Electron Donor	Examples
Photolithotrophs	Autotroph	CO_2	Phototroph	Light	Inorganic compounds (H_2O, H_2S, S)	Photosynthetic plant cells, cyanobacteria, a few photosynthetic bacteria
Photoorganotrophs	Heterotroph	Organic compounds (as well as CO_2)	Phototroph	Light	Organic compounds	Nonsulfur purple bacteria
Chemolithotrophs	Autotroph	CO_2	Chemotroph	Oxidation reductions	Inorganic compounds (H, S, H_2S, Fe^{2+}, NH_2)	Hydrogen, sulfur, iron, and denitrifying bacteria
Chemoorganotrophs	Heterotroph	Organic compounds	Chemotroph	Oxidation reductions	Organic compounds (glucose)	All animals, most bacteria, nonphotosynthetic plant cells

Organisms can also be classified according to the nature of their *carbon sources.* Thus, as shown in Table 9-1, there are **autotrophs** (from the Greek for "self-feeder") and **heterotrophs** ("other-feeder").

■ Autotrophs are capable of making all their own molecular components (sugars, lipids, amino acids, and the rest) from simple small molecules: CO_2 as the sole carbon source, ammonia (NH_3) or nitrate (NO_3^-) as the nitrogen sources, and H_2O. Thus autotrophs are able to thrive in a purely inorganic medium.
■ Heterotrophs require preformed carbon compounds such as carbohydrates, lipids, amino acids, and other organic molecules—all products formed by other cells. In general, animals are heterotrophs and plants are autotrophs. Many bacteria, protists, and fungi are also heterotrophic.

Combining the two classifications, we call autotrophs that obtain energy from light **photosynthetic autotrophs,** or **photoautotrophs.** Autotrophs that obtain energy from oxidation-reduction reactions involving various chemical compounds in the environment are called **chemosynthetic autotrophs,** or **chemoauto-trophs.** Organisms are also divided into **lithotrophs** (Gr., *lithos,* stone), which utilize *inorganic* compounds as electron donors, and **organotrophs,** which use organic compounds as electron donors. Thus, to further compound the nomenclature, we can speak of **photolithotrophs** and **photoorganotrophs,** which are photo-trophs that, respectively, use inorganic and organic electron donors, and **chemo-lithotrophs** and **chemoorganotrophs,** which are chemotrophs that use one or the other electron donor. These terms are explained in summary fashion in Table 9-1. We will consider them again in Chapter 26.

An examination of these groups provides useful clues to the course of early evolution.

PHOTOSYNTHETIC AUTOTROPHS

The earliest organisms that arose on Earth were almost certainly heterotrophs, which depended on organic compounds that arose earlier by abiotic synthesis. Gradually, they depleted the supply of these compounds. During these early years, when the Earth's atmosphere contained no oxygen, a distinctive type of photosynthesis emerged in certain bacteria. It differed drastically from the photosynthesis of contemporary plants (Chapter 10). The metabolic processes of these early phototrophs were *anaerobic*—that is, requiring no oxygen—in contrast to *aerobic* processes. Indeed, most of these forms were so-called *obligate anaerobes;* that is, the presence of oxygen was actually detrimental. Because this system did not use H_2O as a reducing agent, it produced no O_2 and thus was anoxygenic. Rather, it used hydrogen, hydrogen sulfide (H_2S), or various other small organic compounds as reducing agents. Today this category is represented by only a few bacteria.

In all of these organisms, including the ones that survive today, sunlight is absorbed by a pigment called **bacteriochlorophyll.** Atmospheric CO_2 is converted into organic molecules according to the general scheme

9-1 Photosynthetic autotroph
This plant, called Prince's plume, grows along the Paria River at the boundary between Arizona and Utah.

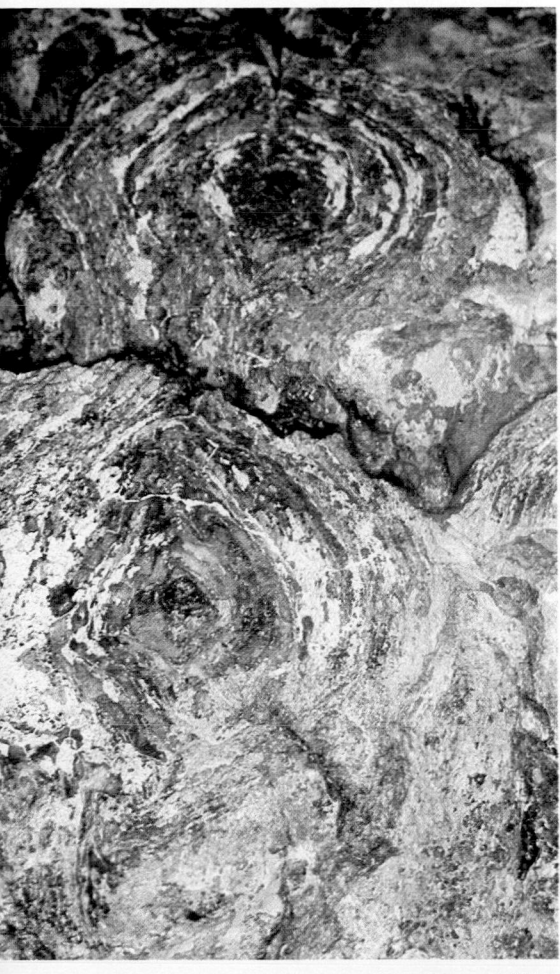

9-2 Stromatolites
Stromatolites are found in sedimentary rocks and consist of concentrically laminated masses of calcium carbonate and calcium magnesium carbonate that are of biological origin.

$$2H_2X + CO_2 \xrightarrow{hv} (CH_2O) + 2X + H_2O \qquad (9.1)$$

where H_2X is an unspecified reducing agent, (CH_2O) is part of a sugar molecule, hv is the standard term for the energy of one photon of light (h is Planck's constant and v is the frequency of radiation). (Radiation frequency is explained in Figure 10-4. See also Box 4-1.)

Those bacteria that used H_2S to reduce atmospheric CO_2 formed a carbohydrate as a product and elemental sulfur as a byproduct:

$$2H_2S + CO_2 \rightarrow (CH_2O) + 2S + H_2O \qquad (9.2)$$

Those that used molecular H_2 to reduce CO_2 formed methane as a product and water as a byproduct:

$$4H_2 + CO_2 \rightarrow CH_4 + 2H_2O \qquad (9.3)$$

Such bacteria, the descendants of which still thrive in swamps, marshes, and other lighted environments that are low in oxygen, have profoundly affected the Earth's atmosphere and its surface. The Earth's vast sulfur deposits are good examples.

Descendants of anaerobic photosynthetic autotrophs (Figure 9-1) became capable of photosynthesis of the more familiar kind—that releases oxygen (and thus is oxygenic). This mode is typical of cyanobacteria (blue-green algae) and green plants.

The organic compounds produced by early photosynthetic bacteria are believed to have supported complex communities of microorganisms. Traces of some of these communities remain in structures called **stromatolites** (Figure 9-2).

CHEMOSYNTHETIC AUTOTROPHS

There also arose the class of organisms called chemosynthetic autotrophs. These, too, are of two general types. Chemolithotrophs were (and still are) able to produce all needed organic compounds from CO_2 and use various reduced inorganic compounds as electron donors or reductants—and they can do this without light. Evidently they use what they can find.

For example, some chemolithotrophs oxidized ammonia (NH_3), which sufficed as the sole energy source. Others oxidized sulfur (as does the contemporary bacterium *Thiobacillus*), or ferrous iron, or other oxidizable constituents of the environment. Chemolithotrophs are not particularly numerous or important in the economy of nature today. Their study is instructive, nonetheless, because they delineate the evolutionary origins of such systems and they illustrate how diverse have been the biochemical approaches to the solution of the basic problems of energy and materials.

On the other hand, chemoorganotrophs are chemotrophs that do use organic compounds as a carbon source and as electron donors or reductants. These, then, are really heterotrophs. Some of them are anaerobes, which use a molecule other than oxygen as the ultimate electron acceptor—for example, sulfate, which is reduced to hydrogen sulfide (H_2S) by the bacterium *Desulfovibrio desulfuricans*. Many or most are aerobes. The category of chemoorganotrophs includes ourselves.

HETEROTROPHS

All heterotrophs require preformed organic compounds as food. Except in rare cases of cannibalism, this food is derived from the bodies (or their remains) of organisms of other species. A curious aspect of heterotrophic life is the extraordinary fussiness of many organisms about the food they will or will not eat.

Like autotrophs, heterotrophs can be grouped into various subclasses. For example, organisms vary widely in their method of utilizing nitrogen, a unique component of amino acids and hence of proteins. Some heterotrophs require preformed amino acids. Others can convert the gaseous N_2 of the air to amino acid nitrogen in the process called *nitrogen fixation*.

Similarly, organisms vary in their requirements for vitamins or growth factors. In general, if an organism requires a particular compound (such as a vitamin) in its food in order to survive and multiply, it is that the organism cannot itself synthesize that compound. On the other hand, if a compound plays a vital role in an organism's metabolism but is not required in its food, we must assume that the organism can and does synthesize it. Nutritional experiments thus can reveal which syntheses are performed in an organism. Humans are rather complex and exacting heterotrophs who require in their diets an energy source (usually but not necessarily carbohydrates), certain unsaturated fatty acids, at least 8 amino acids, and almost 20 vitamins.

When stated in general terms, the difference between autotrophy and heterotrophy seems clear-cut and definite. And the difference is clear-cut when we compare, say, trees and horses. Other organisms, however, reveal intergradations of autotrophy and heterotrophy and widely diverse biosynthetic patterns. The ancestors of all living organisms may have gone through autotrophic stages in which they could perform all necessary syntheses from free elements and simple inorganic compounds. The inability of present-day organisms to perform a particular synthesis would then imply an evolutionary loss of a biosynthetic ability possessed by remote ancestors. Although new syntheses have been acquired in some evolutionary lines, biosynthetic losses have probably been ruling factors in the evolution of biochemical systems. There can be little doubt, for example, that our remote ancestors could synthesize folic acid. However, we now need it in our diet because that synthetic capacity was lost somewhere along the line. The loss was not detrimental as long as environmental food sources were available to make up for the loss. However, such losses, as we shall see, profoundly affected the ecological relationships of heterotrophs.

PRODUCTION OF ENERGY-RICH BONDS

One of the most significant events in the evolution of energy-yielding systems was the appearance some 3 billion years ago of organisms capable of utilizing the sun's energy in the synthesis of sugars and other biomolecules. Some half billion years later, the descendants of the early phototrophs began producing oxygen as a byproduct of metabolism. This was the advance that led to the evolution of glycolysis and respiration and eventually added oxygen to Earth's atmosphere.

We will recall from the important cycle in which photosynthesis generates energy-rich carbohydrates which are utilized in cellular respiration (see Figure 8-4) that photosynthesis captures the kinetic energy of solar radiation and stores it as the potential energy of certain types of chemical bonds. We can summarize the net accomplishments of photosynthesis in one equation:

$$CO_2 + 2H_2O \xrightarrow{\text{light}} [CH_2O]_n + O_2 + H_2O \qquad (9.4)$$

where $[CH_2O]_n$ represents a carbohydrate. Note that in these and later equations the letter n denotes any (appropriate) number. For example, in $[CH_2O]_6$, n equals 6. This might represent glucose ($C_6H_{12}O_6$), one of several hexoses with the same ultimate formula.

Within an organism that has eaten a carbohydrate, the metabolic machinery of **glycolysis** and **respiration** converts the potential chemical bond energy of the carbohydrate into the kinetic energy needed to drive the organism's own life processes. These processes accomplish the opposite net reaction:

$$[CH_2O]_n + H_2O + O_2 \rightarrow CO_2 + 2H_2O \qquad (9.5)$$

We begin our review of metabolism with glycolysis and respiration, nature's inventions for breaking down carbohydrates and capturing their stored energy in the form of ATP. As we trace these pathways—first in broad terms and then in greater detail—we should keep clearly in mind that the energy-extracting processes that convert glucose to CO_2 and H_2O almost exactly reverse, if by different methods, the energy-investing processes of photosynthesis that convert CO_2 and H_2O to glucose.

In this chapter, and in the next on photosynthesis, we will emphasize the importance of several general principles of metabolism.

■ First, these systems—glycolysis-respiration and photosynthesis—are in the front line of the continuing struggle of living organisms against increasing entropy.

- The release of packets of free energy is coupled with enzymatic reactions that convert ADP and inorganic phosphate to the key compound ATP.
- Certain features are shared by the overall design of glycolysis and respiration, on the one hand, and photosynthesis, on the other. Neither system catalyzes the net chemical reactions shown in Equations 9.4 or 9.5 in a single step. Instead, both depend on multiple, small, discrete steps, each catalyzed by a different enzyme. These reaction sequences elegantly illustrate what we mean by *enzymatic pathways*.
- The presence of many small steps facilitates the arrangements discussed in Chapter 8 for controlling metabolism.
- In glycolysis-respiration, and in photosynthesis, the division of metabolic pathways into small steps facilitates the attainment of the three major objectives of intermediary metabolism: (1) the provision of energy in the form of ATP, (2) the generation of reducing power, and (3) the production of the material building blocks needed for the biosynthesis of cell constituents.

ATP: THE CURRENCY OF ENERGY EXCHANGE

The energy liberated as heat in exergonic reactions is no more useful to a cell as a prime energy source than is a bonfire under an automobile's engine.

What the cell needs is a method for trapping the energy of exergonic reactions before that energy is lost as heat. To do this successfully, it must be able to do three things:

- Couple energy-yielding reactions with energy-requiring reactions.
- Store it for later use.
- Transport energy from the place where it is produced to the places where it is utilized.

A cell does these things by investing energy in a special compound that serves the same function in metabolism that currency serves in the economy.

Nature's currency of energy exchange is a class of compounds containing **energy-rich bonds.** The most important are adenosine triphosphate (ATP) and adenosine diphosphate (ADP). In some cases, other nucleotides (for example, GTP) play the role of donor of energy-rich bonds; GTP is of notable importance in the biosynthesis of proteins. As shown in Chapter 8, the structures of ATP and ADP are as follows:

$$\text{ATP} \quad \text{A—R—}\textcircled{P}\text{—}\textcircled{P}\text{—}\textcircled{P}$$
$$\text{ADP} \quad \text{A—R—}\textcircled{P}\text{—}\textcircled{P}$$

where A denotes adenine, R denotes ribose, and $\textcircled{P}$ denotes a phosphate group. When ADP accepts a third phosphate, it becomes ATP. When ATP donates one of its phosphate groups, it becomes ADP.

The ATP molecule has a remarkable property. When its terminal phosphate group is split off by simple hydrolysis, under standard laboratory conditions, the amount of free energy liberated ($\Delta G°$) is about -7.3 kcal per mole. In a simple hydrolytic reaction, a bond is broken and the elements of water (H^+ and OH^-) are added to the ends of the broken bond. For example, in the hydrolysis of the X—$\textcircled{P}$ bond (where X is any group),

$$\text{X—}\textcircled{P} + H_2O \rightarrow \text{X—OH} + \text{H—}\textcircled{P} \tag{9.6}$$

At the same time, the potential energy in the original bond is released.

The actual ΔG for this hydrolysis for the reagent concentrations occurring within a cell is about -12 kcal per mole. In contrast, hydrolysis of most other phosphate ester bonds yields only 2 to 4 kcal per mole. When in 1941 Fritz Lipmann suggested that the two terminal phosphate bonds of ATP are distinctive ''energy-rich'' **phosphoanhydride bonds** and that they be designated by a clever symbol $\sim$, biochemists soon began calling them ''Lipmann squiggles.''

Thus the structure of ATP is properly represented as:

$$\text{A—R—}\textcircled{P}\text{\textasciitilde}\textcircled{P}\text{\textasciitilde}\textcircled{P}$$

It was soon shown that ADP contains an energy-rich phosphate bond.

$$A—R—\textcircled{P}\sim\textcircled{P}$$

Later work showed that certain bonds other than phosphate bonds can also be rich in energy—for example, the thioester bond in coenzyme A (CoASH) and various other compounds (see below).

Here was one of biochemistry's greatest insights, though some years were to pass before it was fully appreciated. In the words of Herman Kalckar:

> These important observations did not attract much attention at that time. . . . This may be attributed to the lack of appreciation of the importance of phosphorylation in cellular physiology. . . . However, Lipmann's forceful and imaginative formulation of phosphorylations exerted an ever-increasing influence [and they were] soon to become recognized as an essential process of the living cell.

The physiochemical reason why certain phosphoanhydride bonds become enriched in energy while others do not relates to the configuration of *pi* electrons in the bonds before and after transfer of the phosphate group. There is an energetic difference between the two kinds of configurations, and this approximates 7.3 kcal. The bond itself is not distinctive; rather there is a distinctive electron configuration across the bond. Certain kinds of enzymatic reactions can alter the electronic configuration of an ordinary phosphate bond, thus converting it to an energy-rich phosphate bond. *Dehydrations* and *oxidations* are the two major types of enzymatic reactions that can transform phosphate bonds in this way. It is important to remember that when we speak of the energy of a high-energy phosphate bond—or of any bond, for that matter—we are referring to the energy liberated when that bond is hydrolyzed.

Investigators soon perceived that when a phosphate group bearing an energy-rich bond is transferred from one molecule to another, it frequently takes its energy with it, and very little energy is lost in the process. As a result, when some low-energy compound (call it X) reacts with ATP, X can be converted to the high-energy compound, $X \sim \textcircled{P}$. In this sense, ATP is the currency of energy exchange among the cell's biochemical processes: Breakdown of ATP's energy-rich bonds is coupled with the creation of new bonds in the majority of biosynthetic pathways.

Almost alone, ATP carries the chemical energy arising in energy-yielding oxidations of foodstuffs to the sites of processes that are totally dependent on such chemical energy (Figure 9-3). These are the three major purposes of metabolism listed at the opening of this chapter.

In summary, ATP is the *charged form* of the energy-transporting system. ADP is the *discharged form*. In a continuous dynamic cycle, this energy-carrier system is charged during the oxidation of foodstuffs and discharged as cellular work is performed. We are now able to equate energy production with ATP generation. In most organisms, glucose is a prime source of metabolic energy, and we must now ask exactly how the breakdown of glucose leads to the synthesis of ATP.

QUANTITATIVE ASPECTS OF ENERGY METABOLISM

Let us consider a few simple quantitative relationships. As we have seen, glucose, which may be represented as $(CH_2O)_6$ or $C_6H_{12}O_6$, is oxidized according to the following overall reaction:

$$C_6H_{12}O_6 + 6O_2 \rightarrow 6H_2O + 6CO_2 + 686 \text{ kcal of free energy} \qquad (9.7)$$

Thermodynamics tells us that the amount of free energy liberated in the combustion of a given substance is always the same, no matter what the combustion method. When a mole of glucose (180 g) is oxidized by actual burning in a bomb calorimeter, 686 kcal of free energy is released. We may assume, therefore, that this much energy is ultimately liberated in the stepwise metabolic oxidation of glucose.

In 1937, Herman Kalckar in Denmark and V. A. Belitser in the Soviet Union observed that in the course of carbohydrate oxidation, *inorganic phosphate* is taken up and transformed into the terminal phosphate (now an *organic phosphate*) of ATP. Significantly, the uptake of oxygen and the conversion of inorganic phosphate

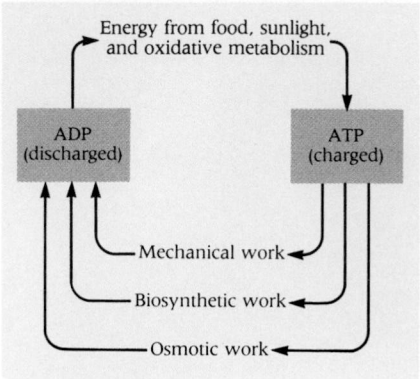

9-3 The ATP-ADP cycle
The energy of the terminal phosphate bond of ATP is used to perform a diverse array of work in the cell. ADP is converted back to ATP by oxidative metabolism.

(abbreviated P_i) to organic phosphate are *coupled reactions* (see Chapter 8). How did the investigators know that? The answer is: Except in the circumstances to be described below, approximately three inorganic phosphates are incorporated into three ADPs to form approximately three molecules of ATP for each atom of oxygen utilized. Coupling was implied by the 3 to 1 ratio—and by its regularity. When various chemical uncoupling agents were added to the system, especially 2,4-dinitrophenol, oxygen uptake continued normally, but phosphorylation of ADP to ATP no longer took palce.

This coupled process was given a coupled name: *oxidative phosphorylation*. Its discovery made biochemists aware for the first time that ATP generation depends upon oxidative metabolism. The 3 to 1 ratio suggested that the utilization of 12 atoms of oxygen in the total oxidation of glucose (as in Equation 9.7) should be accompanied by the production of 36 molecules of ATP. Accordingly, the equation should be expanded as follows:

$$C_6H_{12}O_6 + 6O_2 + 36ADP + 36P_i + 36H^+ \rightarrow$$
$$6CO_2 + 36ATP + 42H_2O \qquad (9.8)$$

This is the applicable equation in some tissues of many species. In others, for reasons to be explained below, the oxidation of glucose is theoretically supposed to yield 38 ATP molecules.

These figures, if correct, permit an interesting efficiency calculation. Since the terminal phosphate bond of a mole of ATP contains 7.3 kcal, 36 moles of ATP would represent an energy yield of 7.3 kcal $\times$ 36 moles or 263 kcal (38 moles of ATP would represent a yield of 7.3 kcal $\times$ 38 moles or 277 kcal). These are 38% and 40% of the 686 kcal contained in the original glucose—a very satisfactory yield. Does the cell actually do this? If so, how?

These questions place us in an interesting dilemma. It has been the traditional teaching that the cell *does* produce 36 (or 38) ATP molecules per molecule of glucose oxidized. That teaching was always supported by spreadsheet-like tables and charts (like those below) that summarized the bookkeeping of glucose metabolism: so many carbons, oxygens, and phosphates debited; so many ATPs and CO_2s credited. However, accumulating evidence of three kinds has put the neatness of this concept in question.

- It appears that *fewer* than 36 (or 38) ATPs are actually produced under the shifting, real-world conditions of cell metabolism.
- Certain previously overlooked factors are responsible for a lower-than-theoretical energy yield.
- Finally, the most disagreeable from a teacher's viewpoint, we are often unable to state the exact yield from glucose metabolism in a given cell or tissue.

Accordingly, we will present this remarkable and important story in a new way. First we will describe the transactions in each of the four stages of glucose metabolism that theory tells us are taking place. We may regard these as *maximum* possible yields; and their careful consideration should make an important point: there is value in such quantitative thinking, if only because it leads us to recognize discrepancies between theory and reality. Then, at the end of the chapter, we will describe these discrepancies—and, as best we can, the mechanisms underlying them.

THE MAJOR PATHWAYS

THE FOUR STAGES OF GLUCOSE METABOLISM

The individual reactions by which glucose is completely converted to CO_2 and H_2O occur in four clusters, or stages (Figure 9-4). As we identify the stages at which each of the several components in Equation 9.8 makes its entrance and exit, it is important to note carefully what happens to each carbon atom of the 6-carbon skeleton of glucose and at what point electrons (or hydrogen atoms) are transferred and $\sim\text{\textcircled{P}}$ groups are created from inorganic phosphates (P_is).

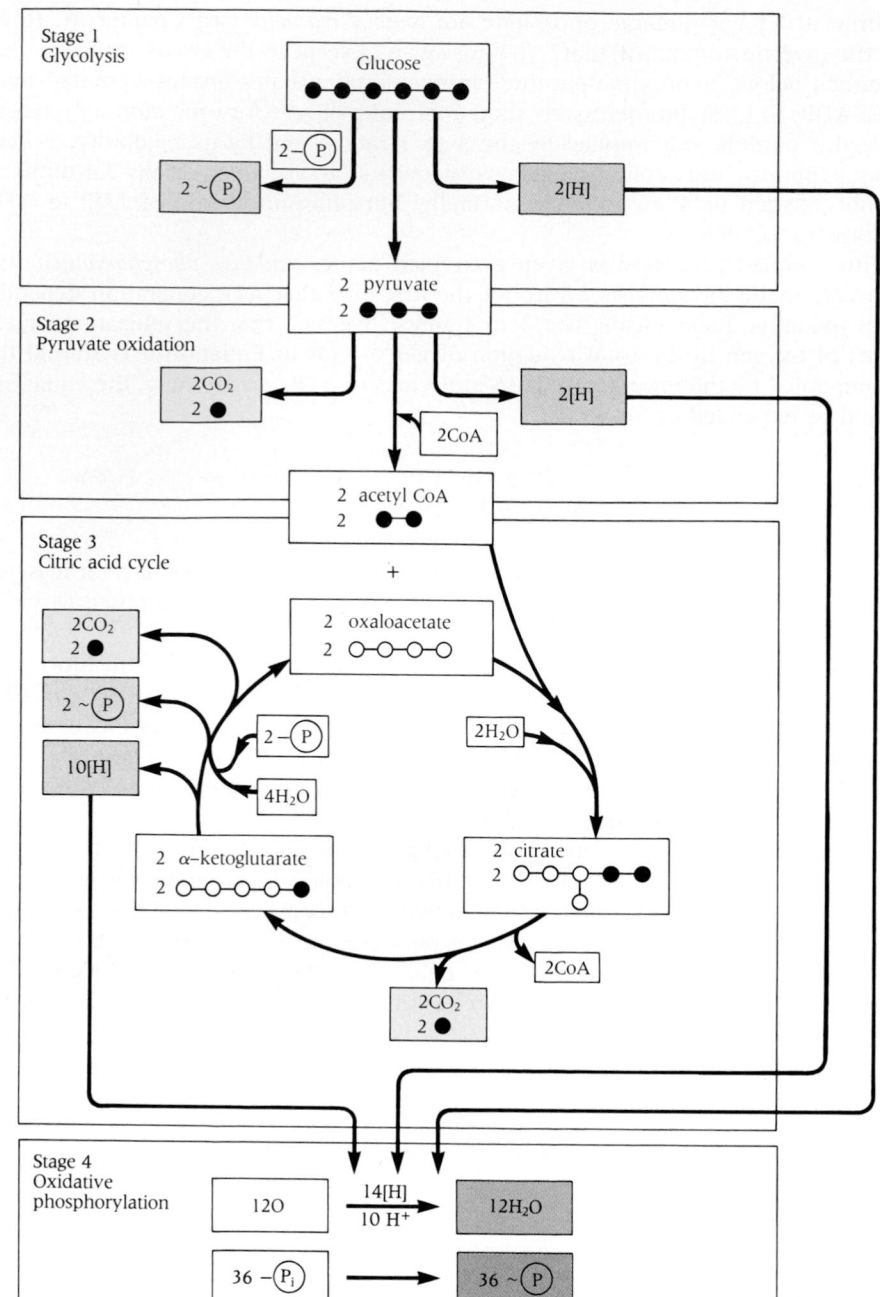

Stage 1
Glycolysis

Glucose

2 — (P)

2 ~ (P)

2[H]

2 pyruvate
2

Stage 2
Pyruvate oxidation

2CO₂
2

2[H]

2CoA

2 acetyl CoA
2

Stage 3
Citric acid cycle

+

2CO₂
2

2 oxaloacetate
2

2 ~ (P)

2 — (P)

2H₂O

10[H]

4H₂O

2 α–ketoglutarate
2

2 citrate
2

2CoA

2CO₂
2

Stage 4
Oxidative
phosphorylation

12O

$\dfrac{14[H]}{10\ H^+}$

12H₂O

36 — (Pᵢ)

36 ~ (P)

9-4 Four stages in the metabolism of glucose
During the breakdown of glucose, energy is released as ℗, hydrogen ions are transferred, molecular oxygen is consumed, carbon is released as CO_2, and water is produced.

This may seem at first glance an empty exercise. What difference does it make whether the 6-carbon glucose molecule splits into two 3-carbon fragments or three 2-carbon fragments—or whether the ATP arises in Stage 1 or Stage 4? The answer is that a good deal of important biology is reflected in these equations. What is being enacted here is evolution's answer to the twin problems of extracting energy from glucose and at the same time converting its carbon skeleton into small building blocks for later use in fabricating the biomolecules of the cell. As we go through these discussions, keep Figure 9-4 in view and carefully track each carbon atom, each pair of hydrogens arising in each oxidation, and each new ATP molecule.

Stage 1: Glycolysis

Stage 1 of glucose metabolism takes place in 10 steps, which are catalyzed by 10 different enzymes that are in solution in the fluid portion of the cytoplasm. These enzymatic reactions are listed in Table 9-2 and are shown in Figure 9-5. This stage,

TABLE 9-2
REACTIONS OF GLYCOLYSIS

Step	Reaction	Enzyme	ΔG°	Major Transformations
1.	Glucose + ATP → ADP + glucose 6-phosphate	Hexokinase	−4.0	Energy is invested at the outset. This energy is supplied by ATP, whose terminal ℗ is transferred to glucose to form glucose 6-phosphate. The overall reaction is exergonic, but some of the energy from ATP is conserved in the bond linking ℗ to the sugar molecule.
2.	Glucose 6-phosphate ⇌ fructose 6-phosphate	Phosphoglucose isomerase	+0.4	The molecule is rearranged, the 6-membered glucose ring becoming a 5-membered fructose ring. The reaction is an isomerization—that is, no carbons or other atoms are gained or lost. This reaction is reversible.
3.	Fructose 6-phosphate + ATP → fructose 1,6-diphosphate + ADP	Phosphofructokinase	−3.4	This reaction is similar to Step 1. Another ATP is invested, with the production of fructose 1,6-bisphosphate (the ℗'s are in the 1 and 6 positions). So far no energy has been recovered.
4.	Fructose 1,6-diphosphate ⇌ dihydroxyacetone phosphate + glyceraldehyde 3-phosphate	Aldolase	+5.7	The 6-carbon sugar is cleaved to two 3-carbon sugars.
5.	Dihydroxyacetone phosphate ⇌ glyceraldehyde 3-phosphate	Triosephosphate isomerase	+1.8	Since glyceraldehyde 3-phosphate (one of the 3-carbon sugars) is eventually used up, all of the other 3-carbon sugar (dihydroxyacetone phosphate) is ultimately converted (by isomerization) to glyceraldehyde 3-phosphate.
6.	Glyceraldehyde 3-phosphate + ℗ + NAD$^+$ ⇌ 1,3-bisphosphoglycerate + NADH + H$^+$	Glyceraldehyde 3-phosphate dehydrogenase	+1.5	Now the 3-carbon sugars (glyceraldehyde 3-phosphate) are oxidized. Hydrogens with their electrons are removed and NAD$^+$ is converted to NADH + H$^+$. For the first time, some energy has been extracted from the sugar. Some of this energy is used to attach a ℗ to what is now the 1 position of both molecules of glyceraldehyde 3-phosphate. The bond attaching this ℗ becomes an energy-rich bond.
7.	1,3-bisphosphoglycerate + ADP ⇌ 3-phosphoglycerate + ATP	Phosphoglycerate kinase	−4.5	The new energy-rich bond is transferred to a molecule of ADP to make a molecule of ATP (actually, 2 ATPs are made per original glucose molecule). This highly exergonic reaction pulls all the previous reactions to completion.
8.	3-Phosphoglycerate ⇌ 2-phosphoglycerate	Phosphoglyceromutase	+1.1	The remaining ℗ (at the 3 position) moves to the 2 position.
9.	2-Phosphoglycerate ⇌ phosphoenolpyruvate + H$_2$O	Enolase	+0.4	A molecule of H$_2$O is removed. This dehydration rearranges the energy distribution in the molecule and creates another energy-rich phosphate bond.
10.	Phosphoenolpyruvate + ADP ⇌ pyruvate + ATP	Pyruvate kinase	−7.5	The second new energy-rich bond is transferred to another molecule of ADP to form another molecule of ATP (actually, 2 more ATPs per original glucose). This reaction is also highly exergonic. Therefore the sequence is pulled decisively to completion.

called **glycolysis** because a sugar is split, requires no oxygen, as is evident in the *summary equation* for Stage 1.

$$C_6H_{12}O_6 + 2ADP + 2P_i + 2\ NAD^+ \rightarrow$$
Glucose

$$2C_3H_4O_3 + 2ATP + 2NADH + 2H^+$$
Pyruvate

(9.9)

We see that the following major transformations occur in the first stage of glucose metabolism:

- A 6-carbon glucose molecule is split with the eventual production of two molecules of pyruvate containing three carbons each.[1]
- No CO_2 is produced in this stage.
- Although two molecules of ATP were utilized early in the reaction sequence, enough new ATP was produced later (four molecules) to yield a net gain of two ATP molecules per molecule of glucose cleaved.
- An oxidative transfer of two hydrogens took place. Two molecules of NAD^+ participate in oxidative reactions (dehydrogenations) that yielded two molecules of NADH and two protons.

Production of ATP is only one part of the strategy of glucose metabolism. A second aspect is the extraction of energy from other compounds by oxidative reactions. This is accomplished by transfers of hydrogen to the coenzymes NAD^+ and, later, FAD. The result is the formation of NADH (and $FADH_2$), the reduced forms of these coenzymes. Hydrogens extracted from intermediate compounds undergoing oxidation take the form of NADH and $FADH_2$. These reduced forms are reoxidized in Stage 4, in which all of these transferred hydrogens are passed on to oxygen to form H_2O. Just as ATP provides a means of packaging free energy in energy-rich phosphate bonds (bundles containing 7.3 kcal per mole), NAD^+ permits the packaging of 52.7 kcal of free energy for every mole of NADH formed. The significance of these numbers will be explained in a moment.

The list of those who discovered the individual steps of glycolysis includes the principal architects of modern biochemical thought: Büchner, Harden, Embden, Meyerhof, Parnas, Carl and Gerty Cori, Warburg, Kluyver, and others. In developing the solution to this complex riddle, these investigators devised many of the fundamental techniques and concepts of modern biochemistry.

The Three Fates of Pyruvate

Biochemical understanding of glycolysis did not develop until the **fermentation** by yeast of glucose to ethanol (an alcohol) and CO_2 had been well studied.[2] Remarkably, the 10-step pathway of fermentation of glucose by yeast cells closely resembles the pathway of glycolysis by animal cells—down to pyruvate. The sequence of reactions from glucose to pyruvate is virtually identical in most organisms. However, the subsequent fate of pyruvate differs notably in various organisms (Figure 9-6).

In yeast and in most plant cells under conditions of low oxygen availability, the 3-carbon pyruvate molecule ($C_3H_3O_3$) is converted to 2-carbon acetaldehyde (CH_3CHO) by an enzyme called *pyruvate decarboxylase*. Acetaldehyde is then reduced to ethanol (CH_3CH_2OH) by alcohol dehydrogenase.

$$\text{Pyruvate} \rightarrow \text{acetaldehyde} + CO_2$$
$$\text{Acetaldehyde} + \text{NADH} + H^+ \rightarrow \text{ethanol} + NAD^+ \quad (9.10)$$

This is the short reaction sequence that through the ages has provided humankind with the CO_2 needed in baking and the alcohol essential for its cherished beverages.

A second possible fate of pyruvate is the one commonly occurring in animal cells, some plant cells, and many microorganisms. It is a simple reduction of 3-carbon pyruvate to 3-carbon lactate by the enzyme lactate dehydrogenase.

$$\text{Pyruvate} + \text{NADH} + H^+ \rightarrow \text{lactate} + NAD^+ \quad (9.11)$$

This reaction occurs in animal cells under two circumstances: (1) in muscle cells when the amount of available oxygen is limiting, as during intense muscular activity;[3]

---

[1] We use the terms *pyruvate* and *lactate* instead of *pyruvic acid* and *lactic acid* because, at pH 7, these acids are in the form of these anions.

[2] The term *fermentation* is used for many reactions by biochemists. The most familiar kind, which we mention here, occurs in organisms like brewer's yeast, which ferments glucose to ethanol and CO_2 rather than to lactic acid, as do "glycolyzing" cells. Alcoholic fermentation does not normally occur in the animal body.

[3] The rapid accumulation of lactic acid in the muscles of runners is a major factor leading to muscle fatigue and rigor (cramps).

9-5 Pathway of glycolysis
In this series of reversible reactions, one glucose molecule is converted to two pyruvate molecules, which then enter stage 2 reactions.

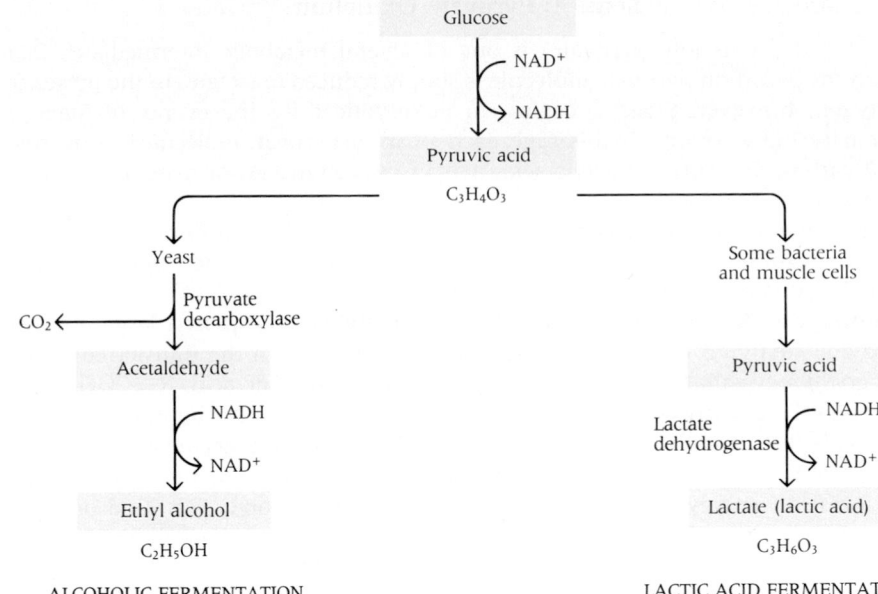

9-6 Two anaerobic routes of pyruvate metabolism
The electrons, produced in the conversion of glucose to pyruvate and stored in NADH, are used to make ethyl alcohol in yeast, or lactic acid in bacteria and animal muscle cells. Both pathways regenerate NAD⁺

and (2) in red blood cells, which require only a meager amount of ATP and therefore have no need to carry further the oxidation of pyruvate. Historically speaking, as yeast was to fermentation, muscle tissue was to glycolysis. But not until the late 1920s was it discovered that, except for the final transformations of pyruvate, the two pathways are identical.

In sum, glycolysis, Stage 1 of glucose metabolism, leads to the following overall or net changes:

- Conversion of one 6-carbon glucose molecule (a hexose) to two 3-carbon pyruvate molecules (a triose)
- Oxidation of two molecules of intermediary triose (glyceraldehyde 3-phosphate) with liberation of two pairs of hydrogen atoms, which simultaneously reduce two molecules of NAD^+ to NADH and H^+
- Utilization of two molecules of ATP in the early phosphorylations needed to launch the glycolytic sequence
- Synthesis of four molecules of ATP from four molecules of inorganic phosphate and four molecules of ADP. This compensates for the two ATPs utilized earlier and yields a net profit of two ATP molecules.

The entire glycolytic process can occur under anaerobic conditions—that is, in the absence of oxygen. This is possible because the only oxidative step (the dehydrogenation of glyceraldehyde 3-phosphate) can be offset by a later reduction of pyruvate to lactate or ethanol, the necessary hydrogens (electrons) coming from the NADH generated in the earlier oxidation. The conversion of pyruvate to lactate or ethanol regenerates NAD^+ by oxidizing NADH. Hence oxygen is not needed. It follows that glycolysis must proceed more rapidly in anaerobic cells than in aerobic cells. Pasteur, over a century ago, was the first to observe that the exposure of anaerobic cells to oxygen causes the rate of glucose consumption to decrease dramatically. This phenomenon is known as the **Pasteur effect.**

The mechanism of the Pasteur effect was not discovered until recently. The glycolytic enzyme *phosphofructokinase* (*PFK*) (see Figure 9-5) was shown to be allosterically regulated by ATP, AMP, and other substances. Under aerobic conditions, the large amount of ATP produced inhibits PFK and slows glycolysis. ATP production is low under anaerobic conditions, and the little ATP present is almost completely converted to ADP, which is degraded to AMP, which stimulates PFK and accelerates glycolysis.

Glycolysis is an obviously inefficient method of extracting energy from glucose—only two ATP molecules per molecule of glucose is certainly a poor yield. However, most cells take pyruvate in a third direction, which completes its oxidation in the next three stages of glucose metabolism. These stages lead to a far more satisfactory yield of ATP per glucose molecule utilized.

Stage 2: Pyruvate Oxidation

Pyruvic acid (or its ion, pyruvate) is one of several metabolic intermediates that oxygen the 3-carbon pyruvate molecule is simply reduced to lactate. In the presence of oxygen, however, Stage 2 begins. In eukaryotic cells, the events of Stage 2 occur in the mitochondria. In this stage, each 3-carbon pyruvate molecule is converted to a 2-carbon derivative of acetic acid, **acetyl coenzyme A,** or **acetyl CoA** (see Figure 9-4).

The eliminated carbon escapes as CO_2. This significant event disposes of one of the three carbons in pyruvate. In terms of the six carbons of the original glucose molecules, this represents one-third of them.

Coenzyme A (abbreviated **CoA**) is the essential coenzyme in a large number of enzyme-catalyzed group-transfer reactions. In all of them the transferred group is an *acyl group*—that is, the R—CO— portion of an organic acid, R—COOH (see Table 4-1). Like many coenzymes, CoA includes a vitamin in its chemical structure. In this case it is pantothenic acid. The active part of the CoA molecule is a sulfhydryl group (—SH). In acting as an acceptor-donor of acyl groups, the sulfhydryl of CoA forms an energy-rich bond that is not a phosphate bond. This kind of high-energy bond, which forms between the sulfhydryl group and the acyl group being transferred, is called a **thioester** bond (R—COS—CoA).

The summary equation for Stage 2 is as follows (bear in mind that this reflects the products of one glucose molecule):

$$2 \text{ pyruvate} + 2NAD^+ + 2CoA \rightarrow$$
$$2 \text{ acetyl CoA} + 2CO_2 + 2NADH + 2H^+ \quad (9.12)$$

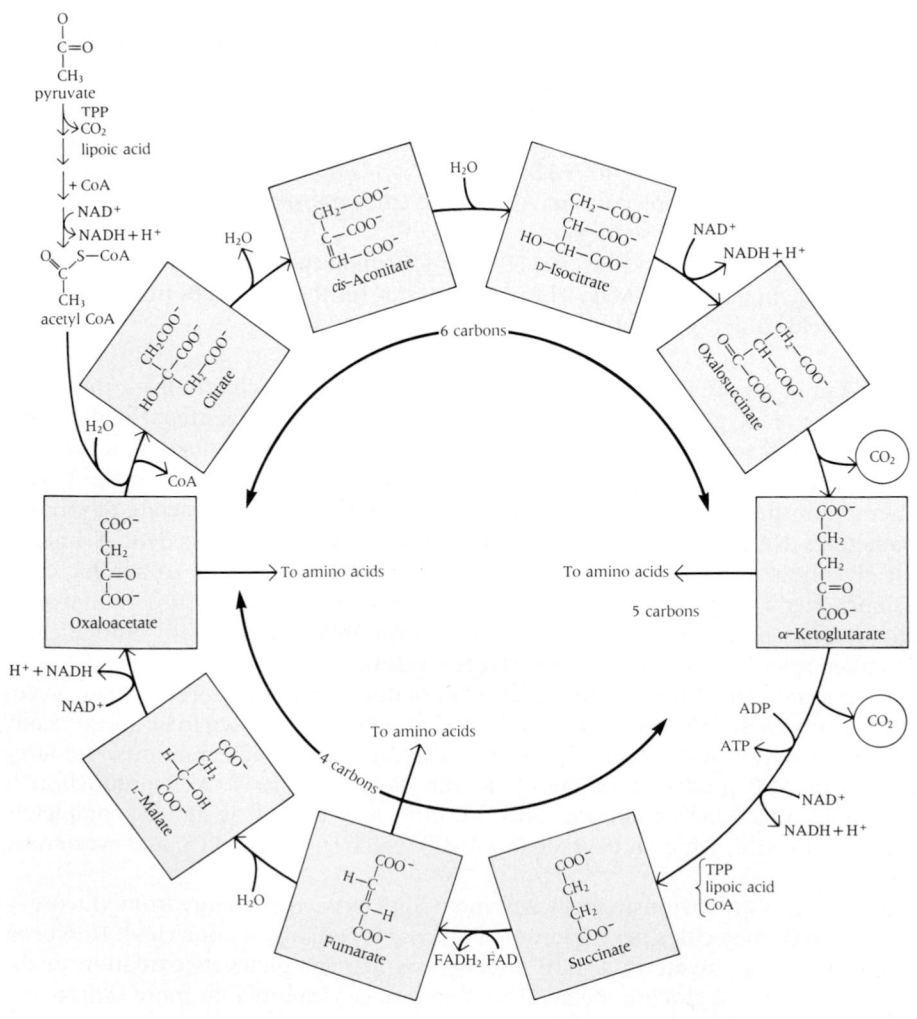

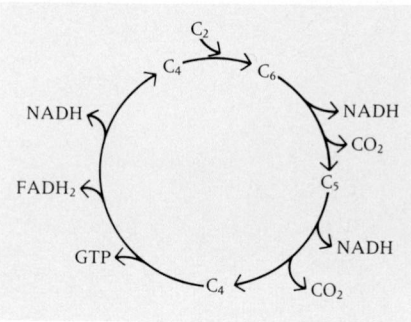

9-7 Pyruvate oxidation and the citric acid cycle
(A) *Pyruvate is oxidized to acetyl CoA using several coenzymes. The energy-rich acetyl CoA then enters the citric acid cycle, where its energy appears in GTP, NADH, and $FADH_2$.* **(B)** *A simplified scheme of the citric acid cycle indicating some of its byproducts.*

In sum, Stage 2 brings about the following net changes:

- A total of one oxidation per pyruvate molecule (or two per original glucose molecule). This leads to the removal of two pairs of hydrogens per original glucose molecule.
- Reduction of two NAD^+.
- Liberation of one carbon atom as CO_2 from each pyruvate molecule—or two CO_2 per original glucose molecule.
- Creation of two energy-rich thioester bonds in acetyl CoA.

As in Stage 1, no oxygen has been utilized (Equation 9.12). The conversion of pyruvate to acetyl CoA with the loss of only one carbon is a complicated process with many steps and four different coenzymes: NAD^+, CoA, lipoic acid, and thiamine pyrophosphate (TPP) (Figure 9-7A). Interestingly, all these steps are catalyzed by the pyruvate dehydrogenase complex, a multienzyme complex (molecular weight 7 million!) containing three different enzymes and several coenzymes (including lipoic acid and TPP) in close physical proximity. The complex is located in the mitochondrial matrix.

People deficient in the vitamin thiamine (also called vitamin B_1) may develop a devastating disease called *beriberi*. The major symptoms of this disease are muscular weakness, weight loss, and nerve and heart defects. In part, these symptoms are the result of impairment of the conversion of pyruvate to acetyl CoA.

Stage 3: Citric Acid Oxidation

Acetyl CoA, the major product of Stage 2, is another versatile "crossroads compound" (Figure 9-8). It enters not only into the **citric acid cycle** (synonyms: **tricarboxylic acid cycle, Krebs cycle**) but into the synthesis of fatty acids, porphyrins, steroids, and other biomolecules. It arises not only from glucose metabolism but from the breakdown of fatty acids and certain amino acids. That is why the citric acid cycle has been called the *final common pathway of metabolism*. Although most of the acetyl CoA normally comes from carbohydrate breakdown, the citric acid cycle cannot be considered a terminal pathway of carbohydrate metabolism alone. We saw in Figure 9-7A that a substantial part of the acetyl CoA formed escapes oxidation to CO_2 and H_2O because acetyl CoA is a biosynthetic building block.

In Stage 3, two 2-carbon acetyl CoA molecules (products of the metabolism of one glucose molecule) enter a metabolic cycle in which all of their carbons are ultimately converted to CO_2, according to the complicated summary equation.

$$2 \text{ acetyl CoA} + 4H_2O + 2GDP + 2P_i + 6NAD^+ + 2FAD \rightarrow$$
$$4CO_2 + 2GTP + 2CoA + 6NADH + 2FADH_2 + 4H^+ \quad (9.13)$$

This system requires **guanosine diphosphate (GDP),** a nucleotide resembling ADP, except that purine adenine is replaced by guanine.

9-8 Acetyl CoA is a crossroads compound in cell metabolism
Ethanol is formed from pyruvate only in yeast and several microorganisms.

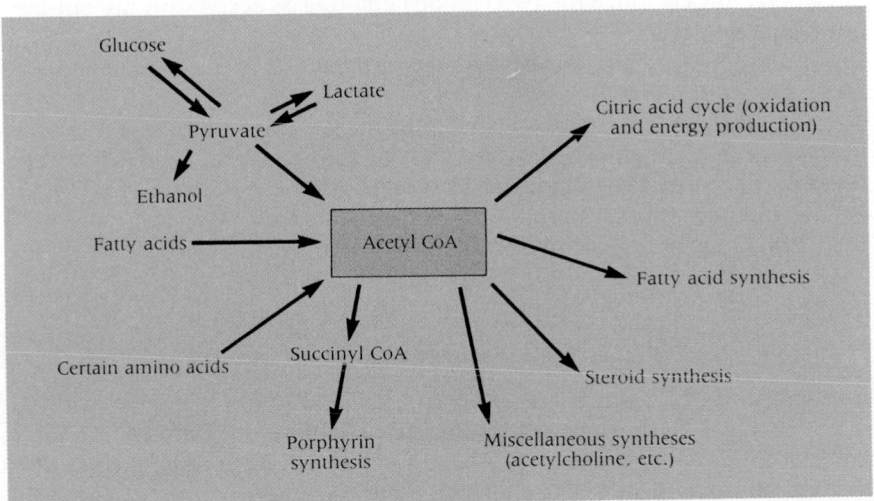

TABLE 9-3
CITRIC ACID CYCLE

Step	Reaction	Enzyme	Cofactors	Type	$\Delta G°$
1.	Acetyl CoA + oxaloacetate + H_2O → citrate + CoA	Citrate synthetase	CoA	Condensation	−7.5
2.	Citrate ⇌ *cis*-aconitate + H_2O	Aconitase	Fe^{2+}	Dehydration	+2.0
3.	*cis*-Aconitate + H_2O ⇌ isocitrate	Aconitase	Fe^{2+}	Hydration	−0.5
4.	Isocitrate + NAD^+ ⇌ α-ketoglutarate + CO_2 + NADH + H^+	Isocitrate dehydrogenase	NAD	Decarboxylation, oxidation	−2.0
5.	α-Ketoglutarate + NAD^+ + CoA ⇌ succinyl CoA + CO_2 + NADH + H^+	α-Ketoglutarate dehydrogenase complex	CoA, TPP, lipoic acid, FAD	Decarboxylation, oxidation	−7.2
6.	Succinyl CoA + Ⓟ + GDP ⇌ succinate + GTP + CoA	Succinyl CoA synthetase	CoA	Substrate-level phosphorylation	−0.8
7.	Succinate + FAD (enzyme-bound) ⇌ fumarate + $FADH_2$ (enzyme-bound)	Succinate dehydrogenase	FAD	Oxidation	~0
8.	Fumarate + H_2O ⇌ malate	Fumarase	None	Hydration	−0.9
9.	L-Malate + NAD^+ ⇌ oxaloacetate + NADH + H^+	Malate dehydrogenase	NAD^+	Oxidation	+7.1

The conversion to CO_2 of all the carbons entering this stage is accomplished in an ingenious way. First, as shown in Figure 9-7A, each 2-carbon acetyl CoA molecule combines with a 4-carbon molecule called **oxaloacetate** to form a new 6-carbon molecule called **citric acid**—or its ion, citrate. The citrate is then oxidized in stepwise fashion, first to a 5-carbon molecule, then to a 4-carbon molecule (Table 9-3). In the course of these oxidations, 2 carbon atoms are "knocked off" as CO_2 and 4 hydrogens are removed for each acetyl CoA molecule entering the cycle. These successive liberations of CO_2 occur in reactions that strikingly resemble those occurring in the Stage 2 oxidation of pyruvate.

The 4-carbon structure emerging from this eventful trip around the circuit is a brand new molecule of oxaloacetate—the same kind of molecule that initiated the cycle. It is free to start the cycle again by combining with a new molecule of acetyl CoA. The trick is in the cyclic pattern. All that is needed to keep it going (besides the necessary enzymes and coenzymes) is a constant supply of acetyl CoA.

This cycle, one of several that occur in metabolism, was discovered in 1937 by Sir Hans Krebs. It is often called the Krebs cycle, though in his papers Sir Hans modestly referred to it as the citric acid cycle.

In sum, Stage 3—the oxidations of the citric acid cycle—accomplishes the following (Figure 9-7B):

- They convert both the carbons of each acetyl CoA to CO_2. This is a *net* conversion of carbons to CO_2, since the two carbons entering as acetyl CoA are not the ones emerging as CO_2.
- Beyond the 2 GTP noted in the summary equation of Stage 3 (Equation 9.13), they produce no ATP.
- There is instead a substantial utilization of the hydrogens (electrons) extracted in the oxidations (i.e., dehydrogenations) of the first three stages. These take the form of reduced coenzymes (NADH and $FADH_2$).
- By the end of Stage 3, 14 of these hydrogen atoms are set to enter into the final oxidations of respiratory metabolism.

Stage 4: Oxidative Phosphorylation

After completion of 2 turns of the citric acid cycle, all 6 of the original carbons of glucose will have been converted to CO_2. In addition, 24 hydrogen atoms will have been transferred to the coenzymes that serve as hydrogen acceptors. Yet despite all this activity, only 4 new ATP molecules will have been harvested—2 in Stage

1 and 2 in Stage 3. If each molecule of glucose has potential energy enough to generate a maximum of 36 (or in some cases 38) ATP molecules, only one-ninth of that yield has been realized to this point.

It remains for Stage 4, **oxidative phosphorylation,** to finish the job. It has two tasks.

- To transfer 24 coenzyme-bound hydrogens to 12 atoms of oxygen so that free coenzyme is again available for hydrogen-transport duty
- To couple these hydrogen transfers with a conversion of inorganic phosphate to Ⓟ groups and transfer them to ADP to make ATP

Since 3 Ⓟ can in theory be formed for every 1 atom of oxygen utilized, these transactions would be expected to yield a total of 36 ATP molecules. The hydrogens from 8 of the 10 NADH molecules generate 3 ATP each to make 8×3, or 24, ATP. For each NADH molecule the following equation can be written:

$$\text{NADH} + \text{H}^+ + 3\text{ADP} + 3\text{P}_i + \tfrac{1}{2}\text{O}_2 \rightarrow$$
$$\text{NAD}^+ + \text{H}_2\text{O} + 3\text{ATP} \tag{9.14}$$

However, 2 NADH arising in glycolysis yields only 2 ATP each for a total of $24 + 4 = 28$ ATP. The explanation for this small discrepancy is that the 2 NADH arising in glycolysis (Stage 1) cannot pass as such from the cytosol where they are formed into the mitochondria where oxidative phosphorylation takes place. In addition, the hydrogens from 2 FADH$_2$ also generate only 2 ATP each for a total of 4 and a grand total of $28 + 4$, or 32.

TABLE 9-4
THEORETICAL YIELD OF HIGH-ENERGY BONDS AND HYDROGEN IN THE COMPLETE OXIDATION OF A MOLECULE OF GLUCOSE: A BALANCE SHEET

Reaction Sequence	ATP (or GTP) Formed (+) or Utilized (−)	Hydrogens Mobilized
Stage 1: Glycolysis—Glucose to Pyruvate (in Cytosol) (Equation 9.9)		
Phosphorylation of glucose	−1	
Phosphorylation of fructose 6-phosphate	−1	
Dephosphorylation of 2 molecules of 1,3-DPG	+2	
Dephosphorylation of 2 phosphoenolpyruvate	+2	
NADH formed in oxidation of 2 glyceraldehyde 3-phosphate		+2
Stage 2: Conversion of Pyruvate to Acetyl CoA (Inside Mitochondria) (Equation 9.12)		
NADH formed		+2
Stage 3: Citric Acid Cycle (Inside Mitochondria) (Equation 9.13)		
2 GTP formed from 2 succinyl CoA	+2	
NADH formed from oxidation of 2 isocitrate, 2 α-ketoglutarate, and 2 malate		+6
FADH$_2$ formed from oxidation of 2 succinate		+4
Stage 4: Oxidative Phosphorylation (Inside Mitochondria) (Equation 9.14)		
ATP formed from NADH formed in Stage 1 (each NADH yields 2 ATP, not 3)	+4	
ATP formed from NADH formed in Stage 2 (each yields 3 ATP)	+6	
ATP formed from NADH formed in Stage 3	+18	
ATP formed from FADH$_2$ formed in Stage 3	+4	
NADH and FADH$_2$ oxidized in Stage 4		−14
Net totals	+36	−0

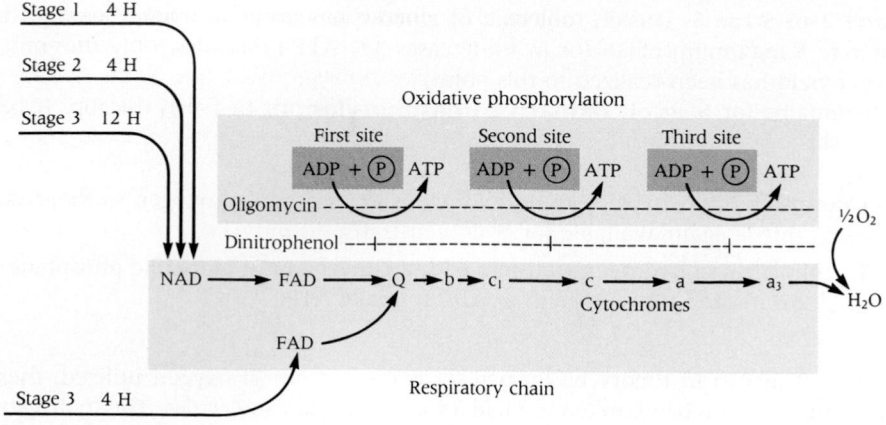

Stage 1 4 H
Stage 2 4 H
Stage 3 12 H

Oxidative phosphorylation

First site   Second site   Third site

ADP + Ⓟ ATP   ADP + Ⓟ ATP   ADP + Ⓟ ATP

Oligomycin

Dinitrophenol

NAD ⟶ FAD ⟶ Q → b → c_1 ⟶ c ⟶ a ⟶ a_3 ⟶ H_2O
                              Cytochromes

½O_2

FAD

Stage 3 4 H

Respiratory chain

9-9 An early view of oxidative phosphorylation
In this scheme electrons are transferred along a chain of carrier molecules, including cytochromes b, c_1, c, a, and a_3, to the final acceptor, oxygen. ATP synthesis is coupled to proton transfer. Dinitrophenol and oligomycin were known to block ATP synthesis.

Thus the summary equation for Stage 4 is as follows:

$$10NADH + 2FADH_2 + 10H^+ + 6O_2 + 32ADP + 32P_i \rightarrow$$
$$12H_2O + 32ATP + 10NAD^+ + 2FAD \quad (9.15)$$

Let us sum up the theoretical bookkeeping: 2 molecules of NADH were formed in Stage 1; 2 were formed in Stage 2; and 6 were formed in Stage 3, along with 2 molecules of $FADH_2$. This makes a total of 10 NADH and 2 $FADH_2$. Hence the total number of hydrogens flowing into the Stage 4 hopper is 7 pairs, or 14 (Table 9-4).

It is instructive to separate Equation 9.14, which shows the link between oxidation and phosphorylation for each NADH, into its exergonic component

$$NADH + H^+ + \tfrac{1}{2}O_2 \rightarrow NAD^+ + H_2O \quad (9.16)$$

and its endergonic component

$$3ADP + 3P_i \rightarrow 3ATP + 3H_2O \quad (9.17)$$

As noted above, the exergonic oxidation of 1 NADH liberates 52.7 kcal of free energy per mole. The endergonic formation of 3ATP conserves 7.3 kcal/mole per ATP, or 21.9 kcal free energy total. Thus the formation of 3ATP conserves at least 21.9/52.7—about 40%—of the total free energy decrease during oxidative phosphorylation.

The quest for understanding of the mechanisms of oxidative phosphorylation was one of the most elusive scientific challenges of this century, one that led imaginative investigators to wholly new ideas of how membranes function. The overall chemical reactions leading to the synthesis of ATP had been known for many years. We knew for some time, for example, that oxidative phosphorylation occurs on the cristae of the inner mitochondrial membrane of eukaryotic cells and that it involves a transfer of coenzyme-borne hydrogen atoms (or their chemically equivalent electrons, as explained in Table 8-1) from one carrier molecule to another through a respiratory chain of interlocking oxidation-reduction reactions.

According to this strictly chemical theory of two decades ago (Figure 9-9), each carrier is reduced as it receives hydrogen and is reoxidized as it transfers hydrogen to reduce the next link in the chain until the final hydrogen acceptor is reached—oxygen. Most of the carriers belong to a series of compounds called **cytochromes** (whose individual types are designated b, c, c_1, a, and a_3). As shown in Figure 9-10, these are conjugated proteins whose attached prosthetic groups are iron-containing porphyrins. Each cytochrome has a distinctive protein component and oxidation-reduction potential.

Details of many of the intermediate stages in these hydrogen transfers were carefully worked out, but the central question remained unanswered: How is the sequential transfer of hydrogen through the series of carrier molecules coupled to the synthesis of ATP?

Cytochrome

9-10 General structure of cytochromes
The centrally located iron atom within the porphyrin ring can be oxidized and reduced ($Fe^{3+} \rightleftarrows Fe^{2+}$), making these molecules ideally suited for transferring electrons. Compare this structural formula with those in Box 10-1.

The door to this central problem of bioenergetics was pried open a little in the mid-1920s by David Keilin of Cambridge University in a series of publications that described the chain of cytochromes that can transfer hydrogens from hydrogen donors at one end of the chain to an acceptor molecule of oxygen at the other. With a hand spectroscope, he established the presence of this chain in a wide variety of oxygen-utilizing organisms (animals, insects, bacteria). He also showed that the mix of cytochromes is quite similar from cell to cell and that the hydrogen transfer process is halted by cyanide and other poisons. This was a remarkable feat for an investigator with no formal training in biochemistry (he was a professor of parasitology). His only tools were a hand spectroscope, a prodigious appreciation of biological principles, and a tremendous imagination.

Prior to those years there had been a lengthy dispute about whether biological oxidations occurred by dehydrogenation or the direct oxidation of substrates by molecular oxygen (see Table 8-1)—the assumption being that it had to be one or the other. It was Keilin who realized that both were needed: Dehydrogenation at the beginning of the respiratory process; direct oxidation by molecular oxygen in the final step.

It took more than 50 years for biochemists to come to grips with the problem of how oxidation is coupled with the synthesis of ATP. In these days when it is popular to look for instant solutions, it is sometimes instructive to learn from the historical record what it takes to solve a major biological problem.

THE CHEMIOSMOTIC THEORY

According to the older chemical theory, oxidative phosphorylation occurs in much the same way as the ATP-yielding reactions of glycolysis and the citric acid cycle, whose individual steps are easily pinpointed as in Figures 9-5 and 9-8A and Tables 9-1 and 9-2. In this view, an energy-rich phosphate bond is formed and then transferred to ADP to make ATP. However, it was never possible to verify this hypothesis experimentally because biochemists were never able to isolate even one of the hypothetical energy-rich intermediate compounds.

A brilliant and imaginative proposal for coupling electron transfer to ATP synthesis was finally offered in 1961 by Peter Mitchell in England. Mitchell recognized that

9-11 The metabolism of mitochondria (A) *Electron micrograph of a mitochondrion, showing convoluted inner membrane.* (× 70,000) (B) *The inner mitochondrial membrane is the site of the major metabolic reactions that occur in mitochondria. Transport of precursors and products into and out of a mitochondrion depends upon membrane proteins called permeases.*

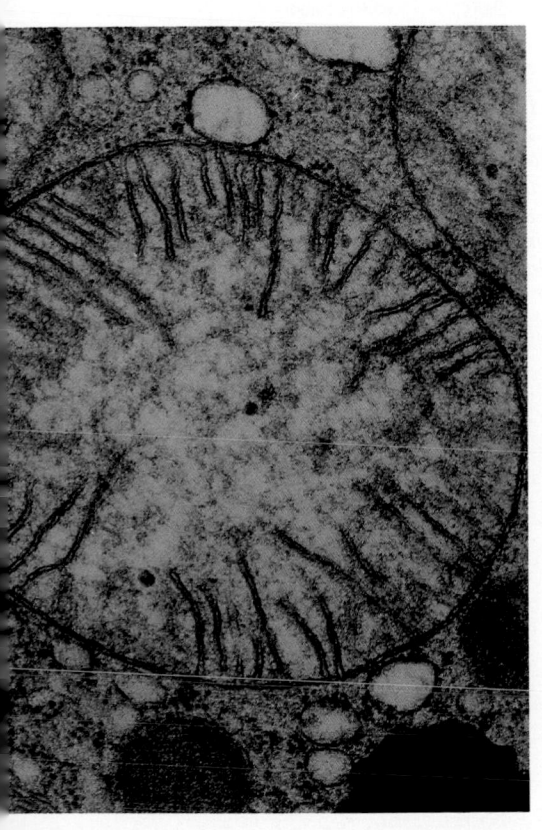

A

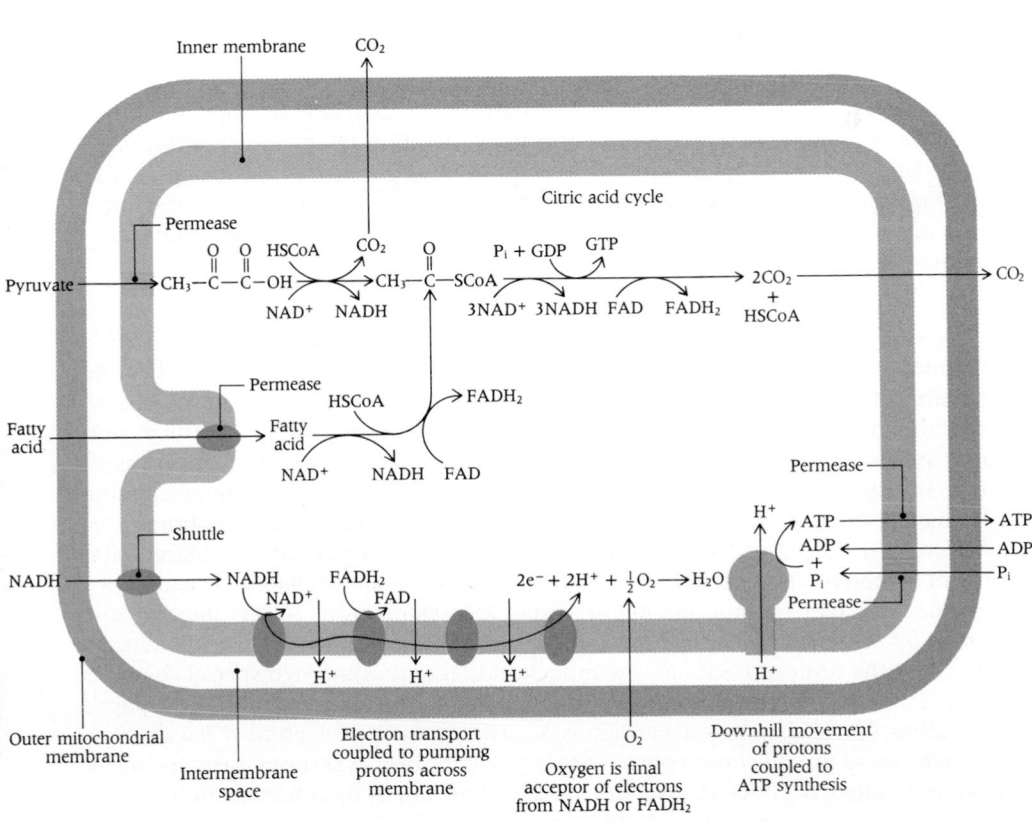

B

only intact mitochondria can make ATP. The driving force for ATP formation, he suggested, is not a chemical intermediate but a **proton gradient** that is maintained by a proton pump deriving energy from the hydrogens of NADH (and $FADH_2$). (The term *gradient* refers to a situation in which the concentration of a substance changes from one place to another.) It was Mitchell's idea that there is a difference in the concentration of protons (H^+ ions) between the two sides of the inner mitochondrial membrane—which, necessarily, must be intact.

Electron micrographs of a mitochondrion (see Figure 5-20) clearly show that the inner and outer mitochondrial membranes are structurally different entities that surround two spaces: the **intermembrane space** between the two membranes, and the **matrix**, or central compartment. Careful studies of pure isolated membranes have shown that each of the many enzymatic reactions occurring within a mitochondrion can be localized to a specific membrane or space.

- The outer membrane is freely permeable to most small molecules (molecular weight less than 10,000 daltons) including protons (H^+). It does not act as a barrier to the diffusion of many molecules. An outer membrane protein, **porin**, that forms channels through the membrane (see Chapter 6) largely accounts for its high permeability.
- The inner membrane is a far more selective permeability barrier, in part because it is rich in a lipid called **cardiolipin**, which makes the membrane less leaky to protons. The membrane also contains many intramembrane particles. Of these, four types are critically important in the transport of electrons from NADH or $FADH_2$ to O_2 and in the synthesis of ATP. Others are "permeases" that allow such large molecules as ATP to pass freely across mitochondrial membranes.
- The infoldings of the inner membrane, the cristae, greatly expand its surface area, thus enhancing its capacity to generate ATP (see Figure 5-20A). In some cells, the total area of the inner mitochondrial membranes per cell is 15 times that of the plasma membrane.

As shown in Figure 9-11, the inner membrane and matrix are the sites of most of the many enzymatic reactions associated with the oxidation of pyruvate (or fatty acid) to CO_2—and of the coupled synthesis of ATP from ADP and P_i.

- Coupling of pyruvate (or fatty acid) oxidation to the reduction of electron carriers NAD^+ and FAD to NADH and $FADH_2$ occurs in the matrix or on inner membrane proteins facing the matrix.
- Transfer of electrons from NADH and $FADH_2$ to O_2 occurs in the inner membrane and, as explained below, is coupled to the generation of an electrochemical potential across the membrane.
- Utilization of the energy in the transmembrane potential gradient for ATP synthesis occurs in protein complexes in the inner membrane called **F_0F_1-ATPase complexes.**

The exciting part of Mitchell's idea—that proton concentrations differ across the inner membrane—was that it was testable. Mitchell soon showed that the oxidations taking place in mitochondrial membranes actually do pump protons across the inner membrane (Figure 9-12). (This is the classic example of the proton pump mechanism mentioned in Chapter 6 as one of the three known forms of active transport.) As a result, more H^+ ions are present outside the membrane than inside. In other words, a proton gradient is created across the membrane.

A gradient develops because the various electron carriers are so placed that pairs of protons—the exact number is discussed later—are displaced outwardly for each electron pair moving down the respiratory chain from NADH. Since pH is defined by proton concentration, this results in a pH gradient across the membrane that drives the protons back into the mitochondrial matrix through special channels in the F_0F_1-ATPase complexes, the knoblike structures that resemble "lollipops" protruding into the matrix (see Figure 5-20B). The flow of protons back across the membrane provides the energy for the formation of ATP from ADP and P_i by a protein called **ATP synthase** in the sphere of the F_0F_1 complex. Indeed, this enzyme is also known as the H^+-ATPase discussed in Chapter 6. Despite the latter name, it is in fact a system for collecting the free energy released in the hydrolysis

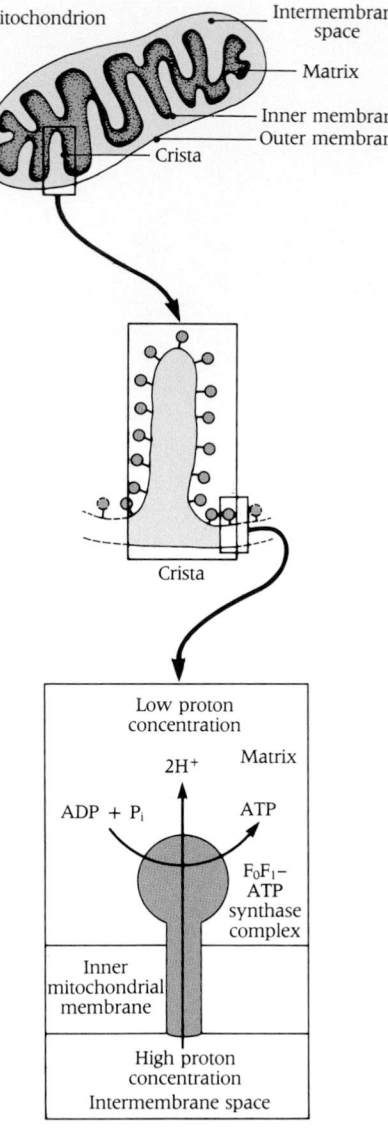

9-12 Oxidative phosphorylation is coupled to ATP synthesis
Protons are shunted into the intermembrane space, creating a proton gradient across the membrane. The potential energy established in this gradient is responsible for ATP synthesis, which is catalyzed by a membrane enzyme in the F_0F_1-ATP synthase complex.

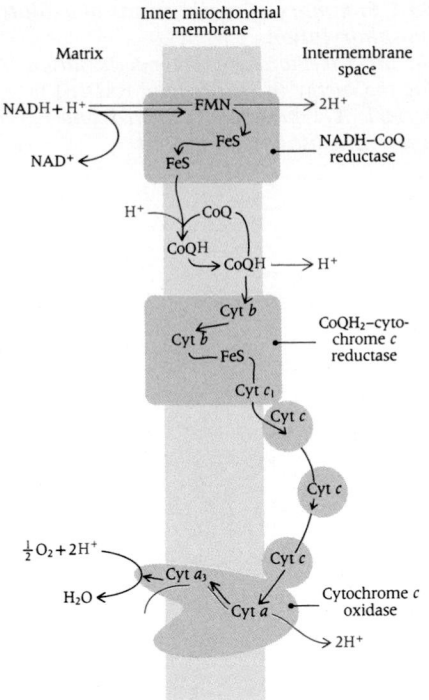

Inner mitochondrial membrane

Matrix | Intermembrane space

9-13 Flow of electrons along the inner mitochondrial membrane

For every pair of electrons transferred from NADH, three ATPs are synthesized. Electrons coming from FADH$_2$, generate only two ATP molecules.

of ATP and using it to move proteins up a concentration or an electrochemical gradient. The protons actually flow through a hydrophobic portion of the ATP synthase molecule. Careful experiments show clearly that this complex is the ATP-generating enzyme and that the generation of ATP is critically dependent upon the electrochemical proton gradient.

During glycolysis (Stage 1), two molecules of NAD$^+$ are reduced to NADH in the cell cytoplasm (see Figure 9-5). Electrons from this cytoplasmic NADH are shuttled into the electron transport chain of the inner mitochondrial membrane (see Figure 9-11). In the shuttle, the electrons are passed to a 3-carbon sugar (glycerol phosphate) that passes them on to the respiratory chain at the level of coenzyme Q (Figure 9-11). This point of entry bypasses one ATP-generating locus—hence, only two ATP are generated per electron pair.

In sum, electron transport in mitochondria, a system employing many types of carriers, is coupled to the pumping of protons. The movements of electrons from NADH (or FADH$_2$) to O$_2$ are catalyzed by electron carriers in four multiprotein complexes in the inner membrane. It is this electron transport that generates the proton gradient (see Figure 9-12). The electrons originally derived from glucose drive protons outward across the membrane. The inward return of protons by diffusion drives the formation of ATP. Interestingly, the terminal phosphate bond of ATP and the proton-motive force across a membrane are interconvertible forms of energy: A proton gradient can cause the synthesis of ATP from ADP and P$_i$; the hydrolysis of ATP to ADP and P$_i$ can be coupled to the pumping of protons.

The key mechanism in Mitchell's proposal is called **chemiosmosis.** In the years since 1961, many workers have shown that its basic postulates are correct, although some of the details remain controversial. Significantly, the theory also accounts for photosynthetic ATP production in the membranes of chloroplasts (see Chapter 10). In chemiosmosis, the crucial role is played by a membrane that divides one region from another. However, this membrane provides much more than shelter and a controlled internal milieu. It is an asymmetrical arrangement of carrier molecules across the membrane that establishes a proton gradient. Even the shape of the membrane is important: It must form a closed envelope if the proton gradient is to be maintained.

ADDING UP THE ENERGY YIELD: THEORETICAL AND REAL

The course of events as electrons flow along the respiratory chain is summarized in Figure 9-13, which shows how released energy is used to convert ADP to ATP. At the end of the chain, hydrogen atoms are accepted by oxygen to produce water. Each time a pair of electrons passes from NADH to oxygen, three molecules of ATP are formed. Each time a pair of electrons passes from FADH$_2$, which holds them at a slightly lower energy level than NADH, two molecules of ATP are formed.

Electrons flow freely along the respiratory chain only if ADP is available for conversion to ATP. Thus oxidative phosphorylation is regulated by the law of supply and demand. This coupling arrangement is an important mechanism for the control of metabolic rates.

It is significant that the light reactions of photosynthesis (see Chapter 10), also depend on an electron-carrying chain containing cytochromes and other carriers—even though these reactions are reductive rather than oxidative. Photosynthesis, as we will see, dehydrogenates water to form oxygen; the respiratory chain hydrogenates oxygen to form water. For those who have wondered why the term **respiration** is applied to these processes, the answer is in Equation 9.18. In some contexts, respiration means breathing, but in a biochemical setting the word denotes an exchange of gases, notably O$_2$ and CO$_2$.

The net results of the respiratory hydrogen cascade and its phosphorylations are precisely those we set as the goals of Stage 4 in Equation 9.14:

- Reduction of O$_2$ to form H$_2$O
- Oxidation of coenzymes to function again as hydrogen acceptors
- Generation of the bulk of the cell's ATP

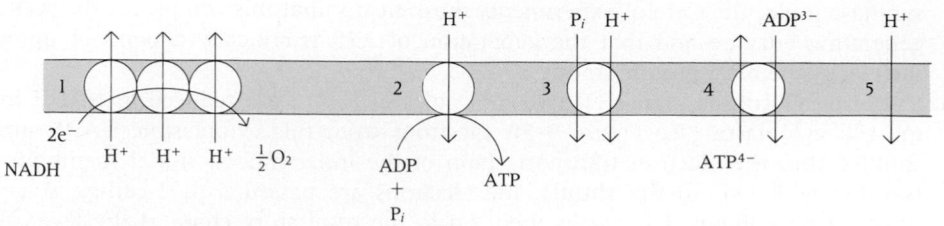

Outer mitochondrial compartment

Inner mitochondrial compartment

If we add up the summary equations of each of the four stages, we obtain a final encompassing equation that describes in theoretical quantitative terms what went into glycolysis and respiration and what came out.

$$C_6H_{12}O_6 + 6O_2 + 36ADP + 36P_i + 36H^+ \rightarrow$$
$$6CO_2 + 36ATP + 42H_2O \qquad (9.18)$$

The number of ATPs generated by the complete oxidation of a molecule of glucose is reckoned at 38 if the cost of various shuttles is overlooked. That is why some textbooks give that number while others state the ATP yield as 36. The discrepancy is attributable to the impermeability, noted above, of the mitochondrial membrane to the 2 NADHs generated outside the mitochondria in glycolysis. Before this shuttle was understood, it was assumed that every NADH molecule that is oxidized generates 3 ATP. These hypothetical 36 ATPs plus the 2 generated in glycolysis account for the 38 ATPs mentioned in earlier discussions.

As we mentioned earlier, the yields of ATP implied in Equation 9.18 represent the maximum number that would be realized were it not for certain newly recognized factors. These can be explained by reference to Figure 9-14, which summarizes the present understanding of how mitochondria utilize the energy in NADH (or FADH₂) to synthesize ATP molecules. At four of the five types of "sites," complications have been recognized.

- Site 1: The number of protons pumped per electron pair was always assumed to be 6—2 at each of 3 reduction loci (see Figure 9-11). Now we believe that 11 protons (4 at the first 2 loci and 3 at the third) are brought across the membrane by an electron pair coming from NADH. Since the lower-energy electrons from FADH₂ do not transport protons at Site 1, they would move only 7 protons to the outer compartment.
- Site 2: The exact manner in which energy released by proton flow is used to make ATP is not known. New data suggest that the passage of 4 protons releases enough energy to assemble 1 ATP. Thus, a mitochondrion could make 2.75 ATPs for each NADH and 1.75 ATPs for each FADH₂. If we try then to count ATPs per glucose, these values mean that we must reduce our total by 3. Furthermore, the proton gradient is used for things other than making ATP.
- Site 3: A transmembrane protein uses some proton gradient energy for active transport of P_i and other metabolites into the inner mitochondrial compartment. Thus we now must further reduce our estimates of ATP production.
- Site 5: Protons tend to leak back into the inner compartment. This movement is difficult to measure, but it is significant.

For all of these reasons, and perhaps others yet to be discovered, it is no longer possible to determine with precision the exact number of ATP molecules a cell can make from the energy in one glucose molecule. Although the attempt at accurate bookkeeping was useful in rationalizing a theoretical maximum that would apply if organisms functioned as we thought they should, there is equal value in discussing the function of mitochondria (and, in Chapter 10, chloroplasts) in terms of what we know about the mechanisms of ATP production. At least, this will illustrate the shifting nature of biological science—and the resulting necessity of adjusting our concepts to newer facts.

SUMMARY

Organisms may be classified as autotrophs or heterotrophs on the basis of their ability to fabricate their own biomolecules from simple precursors.

Autotrophs are capable of making all their own molecular components from such simple starting materials as CO_2 and various inorganic molecules. In photosynthetic autotrophs (notably plants), sunlight is the energy source that drives these synthetic reactions. Chemosynthetic autotrophs utilize energy extracted from oxidation-reduction reactions to drive various biosynthetic reactions. Heterotrophs, in contrast, cannot synthesize all of their molecular components and thus must obtain preformed organic molecules from their environment as components of their diets. Their synthetic reactions are driven by energy released from the catabolic breakdown of these dietary organic molecules.

Adenosine triphosphate (ATP) is a nucleotide synthesized by cells to store the energy released by these energy-yielding exergonic reactions. The phosphates of ATP are attached to the molecule by so-called high-energy bonds, which, when broken, release their energy in amounts sufficient to drive a wide variety of other endergonic chemical reactions.

In one sense, the story of the metabolism of glucose (and other basic nutrients) is an account of the ability of cellular metabolic pathways to generate ATP molecules. In fact, the complete oxidation of a single glucose molecule can in theory be coupled to the synthesis of up to 38 molecules of ATP. In another sense, the metabolism of glucose is an account of the release of molecular energy by various oxidations in the form of hydrogen atoms (electrons), which are trapped by the coenzymes NAD^+ and FAD to form the reduced coenzymes, NADH AND $FADH_2$. These are subsequently reoxidized when they pass their hydrogens to molecular oxygen to form H_2O. The energy released during hydrogen (electron) transfer is utilized to drive ATP formation. In actuality, the yield of ATP is less than the theoretical maximum for reasons explained in detail.

This complex process takes place in four stages, which are precisely understood in terms of their individual reactions, the energy needed to drive them, and their net energy yields.

In Stage 1, glycolysis, a glucose molecule is broken down in several steps to two molecules of pyruvate. Two molecules of ATP are synthesized in this stage. Pyruvate molecules can then have three different fates depending on the cell type and other circumstances: (1) in yeast, pyruvate is fermented to CO_2 and ethanol; (2) in many other organisms, it can be converted to a molecule of lactate; and (3) if molecular oxygen is present, pyruvate is converted to acetyl coenzyme A and then oxidized in the three final stages of glucose oxidation to form CO_2 and H_2O. Only the third fate results in the complete oxidation of glucose to CO_2 and H_2O.

In Stage 2, pyruvate oxidation, the major product is acetyl coenzyme A. This and the remaining stages take place within the mitochondrion, which is sometimes called the cell's "power plant."

In Stage 3, acetyl coenzyme A enters the citric acid cycle, in which a succession of oxidative reactions releases 2 CO_2 molecules, generates only 2 ATP molecules, and binds the 24 hydrogen atoms released in the various oxidative reactions of the first three stages to NAD^+ and FAD, which are converted to NADH and $FADH_2$.

In Stage 4, oxidative phosphorylation, the hydrogens bound to these reduced coenzymes are transferred to oxygen (to form H_2O). The hydrogen transfers are coupled to the synthesis of a theoretical maximum of approximately 36 or 38 ATP molecules. The chemiosmotic hypothesis explains how hydrogen transfers are coupled to ATP synthesis in the unique inner mitochondrial membrane, the site of the respiratory chain and oxidative phosphorylation. Protons pumped outside the membrane diffuse back across it through special channels. The passage of hydrogens into stalked spherical F_0F_1 complexes projecting into the matrix space causes them to efficiently synthesize ATP from ADP and inorganic phosphate. A number of recently discovered factors account for an actual ATP yield that is lower than theory would predict.

KEY TERMS

acetyl coenzyme A (acetyl CoA)
adenosine triphosphate (ATP)
aerobic
anaerobic
ATP synthase
autotroph
chemiosmosis
chemoautotroph
chemosynthetic autotroph

citric acid cycle
cytochrome
fermentation
glycolysis
heterotroph
inorganic phosphate
Krebs cycle
obligate anaerobe
oxidative phosphorylation

Pasteur effect
phosphoanhydride bond
photolithotroph
photoorganotroph
photosynthetic autotroph
phototroph
proton gradient
respiration
stromatolite

QUIZ QUESTIONS

1. Select the *true* statement:
   A. More ATP molecules are produced per molecule of glucose when glycolysis occurs in aerobic conditions.

   B. More ATP molecules are produced per molecule of glucose when glycolysis occurs in anaerobic conditions.

   C. Glycolysis produces equal numbers of ATP molecules regardless of the presence or absence of oxygen.

2. Which of the following reactions is endergonic?
   A. $ATP \rightarrow ADP + P_i$
   B. $NADH \rightarrow NAD^+ + H^+$
   C. $(CH_2O)_n + O_2 \rightarrow CO_2 + H_2O$
   D. A and B
   E. none of the above

3. Which of the following occurs only in aerobic conditions?
   A. glycolysis
   B. oxidative phosphorylation
   C. fermentation
   D. lactate respiration

4. The form in which carbon enters the citric acid cycle is
   A. citric acid.
   B. pyruvate.
   C. acetyl coenzyme A.
   D. oxaloacetate.
   E. NADH.

5. The form in which carbon leaves the citric acid cycle is
   A. acetaldehyde.
   B. ethanol.
   C. lactate.
   D. carbon dioxide.
   E. None of the above.

ESSAY QUESTIONS

1. When a poison such as cyanide impairs a cell's ability to generate ATP, what vital cell functions are lost?

2. Arrange the following types of organisms in a hypothetical evolutionary sequence: aerobic phototrophs, anaerobic phototrophs, aerobic heterotrophs, chemolithotrophs, chemoorganotrophs.

3. Characterize the four stages of glucose metabolism. What is the efficiency and net yield of ATP production in each? Why is glycolysis considered an anaerobic stage of glucose metabolism? At what stage is CO_2 liberated?

4. What stages of glucose metabolism are termed "respiration" and why?

5. What high-energy molecules are the major intermediates in the conversion of glucose energy into ATP energy?

REFERENCES AND SUGGESTIONS FOR FURTHER READING

Al-Awqati, Q. (1986). Proton-translocating ATPases. *Annual Review of Cell Biology 2,* 179–199.

Scholarly discussion of a mechanism—an ATP-splitting enzyme that moves protons across a membrane—that was entirely unsuspected until recently.

Baldwin, E. (1959). *Dynamic Aspects of Biochemistry,* 3rd ed. Cambridge University Press, New York.

This classic biochemistry text, exceptional for its readability, is strong in its treatment of coenzymes and their importance in energy metabolism.

Dickerson, R. E., and Geis, I. (1969). *The Structure and Action of Proteins.* Harper & Row, New York.

Another of the authors' beautifully illustrated monographs on protein structure and function.

Hinkle, P. C., and McCarty, R. E. (1978). How cells make ATP. *Scientific American 238,* March, 104–123.

Discussion of the chemiosmotic theory, according to which ATP is formed as a consequence of protons passing back through the inner mitochondrial membrane after having been pumped out by the respiratory chain. A bit more demanding than most *Scientific American* articles.

Kalckar, H. M. (1969). *Biological Phosphorylations: Development of Concepts.* Prentice-Hall, Englewood Cliffs, N.J.

A unique volume, by a major contributor, which traces the discovery of key principles of energy metabolism—glycolysis, the fate of acetyl CoA, and oxidative phosphorylation—by reprinting papers of the original discoverers. The works receive the author's comments from the vantage point of hindsight. Many of the papers are accompanied by pungent personal reminiscences.

Keilin, D. (1966). *The History of Cell Respiration and Cytochromes.* Cambridge University Press, New York.

An historical account by the investigator who in the 1920s observed that chains of cytochromes transfer electrons from one to the other in respiration.

Krebs, H. (1970). The history of the TCA cycle. *Perspectives in Biology and Medicine 14,* 154–170.

Here the discoverer of what we call the Krebs cycle and he calls the tricarboxylic acid cycle describes the history of that important contribution.

Krebs, H. (1981). *Otto Warburg: Cell Physiologist, Biochemist, and Eccentric.* Oxford University Press, New York.

An enjoyable account of the life of one founder of modern biochemistry by another.

McCarty, R. E. (1985). H^+-ATPases in oxidative and photosynthetic phosphorylation. *BioScience 35,* 27–30.

More on the ATPases of mitochondria (and chloroplasts and bacteria), which couple the synthesis and hydrolysis of ATP to the transmembrane fluxes of protons. Points out the striking differences between these ATPases and those discussed in Chapter 6 that actively transport ions across cell membranes.

Mitchell, P. (1979). Keilin's respiratory chain concept and its chemiosmotic consequences. *Science 206,* 1148–1159.
How the respiratory chain concept gave way to the chemiosmotic theory—by the latter's originator.

Racker, E. (1980). From Pasteur to Mitchell: a hundred years of bioenergetics. *Federation Proceedings 39,* 210–215.
Historical essay celebrating the first century of bioenergetics.

Slayman, C. L. (1985). Proton chemistry and the ubiquity of proton pumps. *BioScience 35,* 16–17.
One of a cluster of articles on various aspects of proton pumps.

Smith, E. L., Hill, R. L., Lehman, I. R., Lefkowitz, R. J., Handler, P., and White, A. (1983). *Principles of Biochemistry* (2 vols.), 7th ed. McGraw-Hill, New York.
A classic textbook by some of biochemistry's finest teachers.

Stryer, L. (1988). *Biochemistry.* 3rd ed. W. H. Freeman, New York.
Another excellent textbook.

CAPTURING THE SUN'S ENERGY: PHOTOSYNTHESIS

Capturing the sun's energy: photosynthesis
The Earth is carpeted with enormous masses of green plants that absorb solar energy and convert CO_2 to sugars. In this scene showing foliage in the Appalachian Mountains of Pennsylvania, the green leaves are actively engaged in photosynthesis. When the season ends, the leaves will yellow and fall from the trees.

The sun is the ultimate energy source for life on Earth, but only certain organisms are equipped to harvest solar energy. As we saw in Table 9-1, which classified organisms according to their energy and carbon sources, some organisms—the phototrophs—can utilize light as their energy source. Others—the chemotrophs—cannot.

In today's world, the principal organisms utilizing light energy are green plants. **Photosynthesis** is the process that captures light energy and uses it to build large organic molecules from atmospheric CO_2.

Knowledge of photosynthesis began in the seventeenth century with a simple experiment. Intending only to confirm his belief in the importance of water to life, the Belgian physician Jan Baptiste van Helmont (1577–1644) took a tub containing exactly 200 pounds of dry soil. In it he planted a 5-pound willow tree which he then irrigated with rainwater. Five years later, the willow had gained 164 pounds and the soil, dried, weighed only 3 ounces less than 200 pounds. The willow, he concluded, had been formed from rainwater.[1] His data were gathered carefully enough, but his conclusion was half wrong. The weight gain was as much due to CO_2 taken up from the air as to water taken from the soil. The small loss in soil weight was undoubtedly due to uptake of minerals by the willow.

The fact that a plant absorbs CO_2 from the air and releases O_2 was discovered by the English clergyman-chemist Joseph Priestley (1733–1804), who demonstrated that plants can "restore" air (Figure 10-1). A mouse left alone in a sealed chamber soon dies. A lone plant in another closed chamber also dies, but both survive when plant and mouse are placed in the same sealed chamber. We can now say that photosynthesis in the green plant absorbs the CO_2 exhaled by the mouse; the mouse in turn inhales the O_2 released by the plant. Interestingly, Priestley was not aware of the gases we call *carbon dioxide* and *oxygen*—he spoke of "good" air (or fresh air) and "fixed" air (by which he meant air "exhausted" by the burning of candles or the breathing of animals).[2] Nor did he realize that light was necessary for photosynthesis.

The Dutch physician Jan Ingenhousz (1730–1799) systematically repeated Priestley's experiments and showed conclusively that "fixed" air could be restored only by the interactions of sunlight and the leaves of plants (not roots, flowers, and fruits), especially the undersides of leaves.

---

[1] This faulty conclusion recalls a point often made by a later experimentalist, Rudolf Schoenheimer (1898–1941) of Columbia University, who first demonstrated with labeled compounds that metabolism is a "dynamic state" in which molecules are rapidly and reversibly transformed into one another. Schoenheimer said, "If you put a copper penny into a vending machine and get out a stick of gum, that does not prove that the machine can convert copper into chewing gum."

[2] Scientific historians usually cite August 1, 1774, as the birthday of oxygen because that is the day on which Priestley, using a 12 inch lens, focused the sun's rays on oxide of mercury (HgO) and caused it to liberate a gas in which "a candle burned . . . with a remarkably vigorous flame." But at least two years earlier, the Swedish chemist Carl Scheele, who died in 1786, had already isolated oxygen. Moreover, Scheele's discovery was a confirmation of a belief in its existence, whereas Priestley's discovery was unexpected.

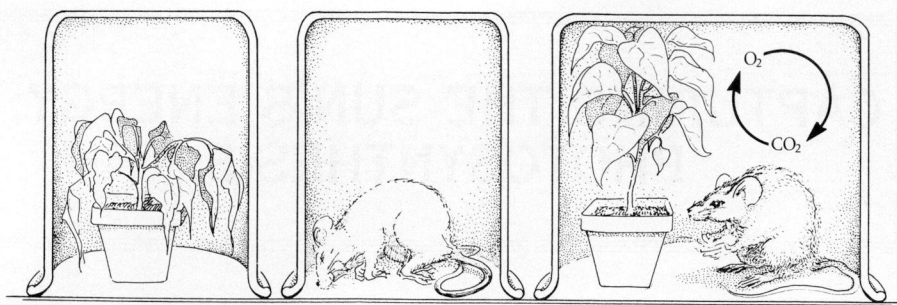

10-1 Priestley's experiment on restoration of air
Plants require CO_2 and animals require O_2 for life. If placed in separate sealed containers, a plant or an animal will die, but if both are in the same chamber, both could live.

The precise steps by which sugar molecules are assembled from CO_2 in green plants were only recently elucidated, in part from studies employing CO_2 molecules labeled with radioactive carbon (^{14}C).

FLOW OF ENERGY IN THE WORLD OF LIFE

The radiant energy that originates in thermonuclear reactions in the sun streams to Earth in the form of light (and other forms of radiation) and sustains all living organisms. Over the entire globe about 25% of the incoming solar radiation is scattered or reflected by the atmosphere back into space (Figure 10-2). This fraction of arriving solar energy is lost to the Earth. Another 25% is absorbed by the atmosphere. Of the remaining 50% that manages to reach the Earth's surface, some 45% of the total is absorbed as heat by lands or oceans; 5% (of the total) is reflected back into space; 24% is consumed in promoting water evaporation and related meterological (weather) phenomena. A very small fraction drives the wind and waves. An even smaller fraction is captured by the leaves of green plants. Indeed, the total solar energy reaching Earth per year is said to equal $173,000 \times 10^{12}$ watts. Of this, only 40×10^{12} watts is used in photosynthesis.

It is noteworthy that both the atmosphere and Earth's surface reflect a significant fraction of the incoming solar radiation. Bare ground and rocks reflect a much larger portion of the incident radiation than green fields and forests, which are more absorbent. Recently satellite engineers have acquired the capacity to estimate on a global scale the total mass of green plants—and, by inference, terrestrial photo-

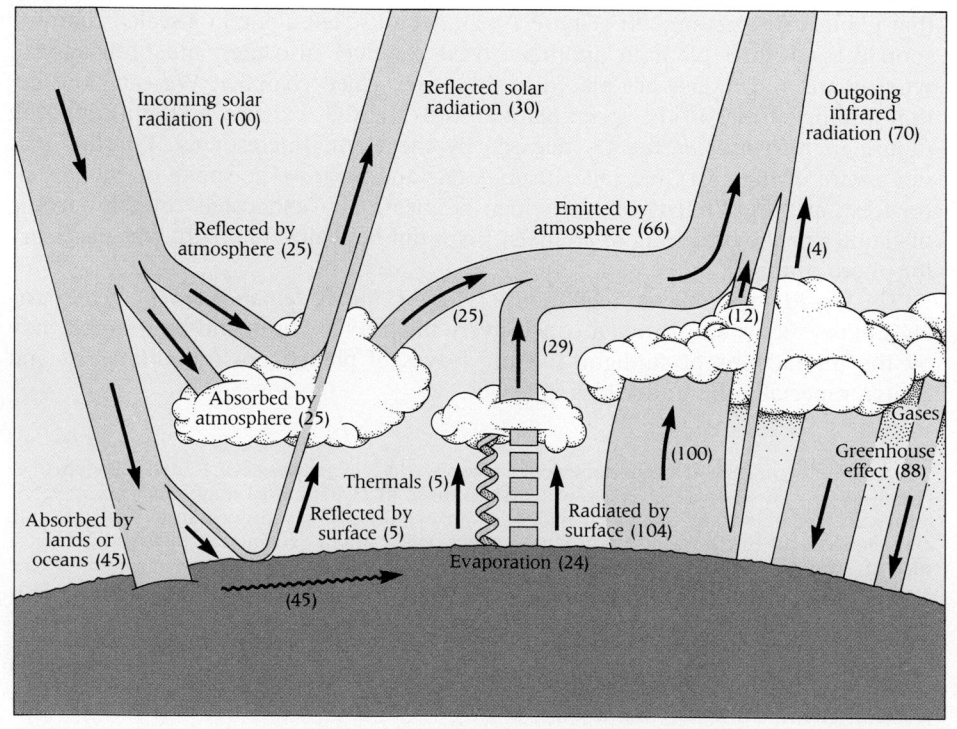

10-2 Distribution of the solar radiation reaching Earth
The diagram shows the disposition of incoming solar radiation. The numbers in parentheses are percentages represented by each arrow relative to total incoming solar radiation, which is 100%. The greenhouse effect (to be discussed in Chapter 46) arises because the Earth's atmosphere tends to trap heat near the surface.

10-3 Satellite-derived data showing the distribution of photosynthetically active radiation absorbed by terrestrial vegetation
The top diagram was made in August 1982 (late in the northern summer); the bottom one was made in February 1983 (winter). Brown/orange denotes low absorbance; dark green, medium absorbance; and red/violet high absorbance. Analysis of such maps yields estimates of green leaf densities and thus of terrestrial photosynthesis.

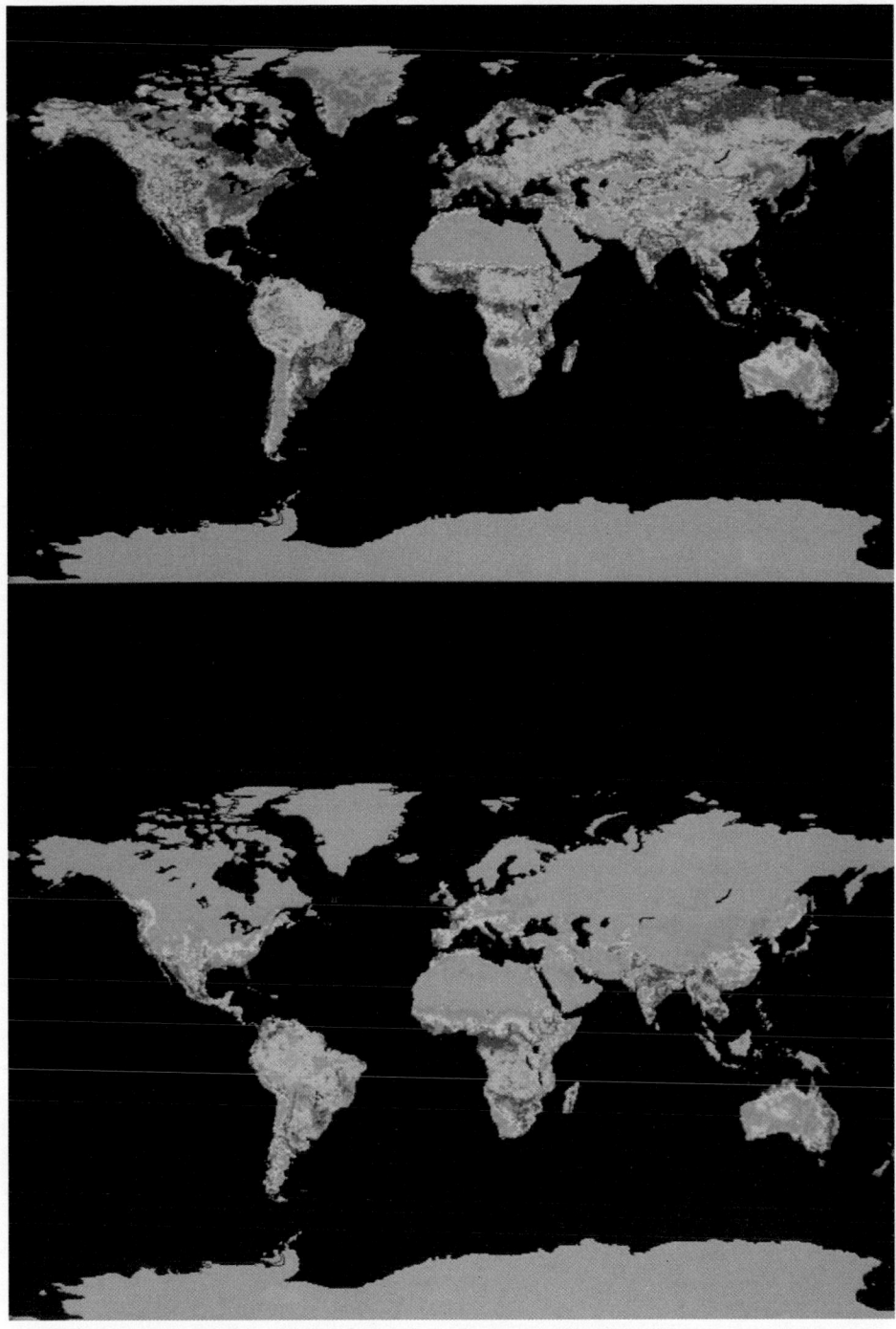

synthesis, with a high level of resolution. The example in Figure 10-3 clearly shows the areas of the world that absorb and those that reflect light energy—and how this varies with the seasons.

LIGHT AND SUNLIGHT

The **electromagnetic spectrum** (Figure 10-4) is defined as the total range of wavelengths of electromagnetic radiation—from the shortest cosmic rays to the longest radio waves. **Light** is only one small sector of this spectrum. However, light is unique in one respect: Its definition depends ultimately on the attributes of human vision. That is why it is also termed **visible radiation.**

The wavelength of light ranges from around 400 to nearly 800 nm—or, in alternative units preferred by physicists, 4000 to 8000 Å (see Box 4-1). At wavelengths

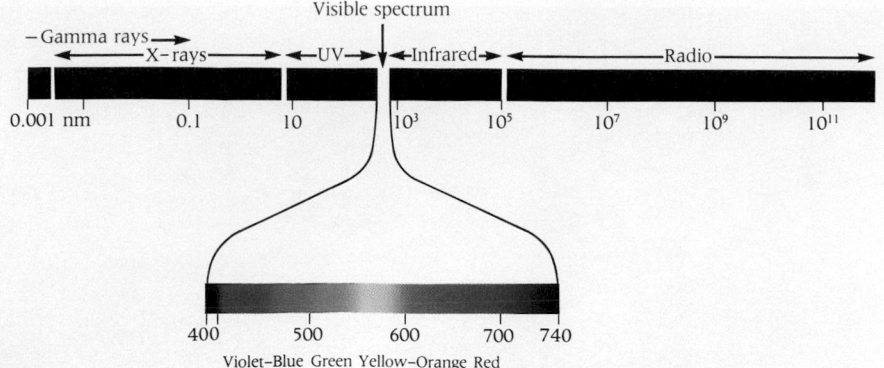

below 380 nm on the electromagnetic spectrum, radiant energy is known as **ultra-violet radiation** (or **UV light**). At even shorter wavelengths, **x-rays** (10 to 0.001 nm), then **gamma rays** (0.1 to 0.001 nm), and then **cosmic rays** (< 0.001 nm) exist. At the other end of the spectrum, radiant energy above 760 nm is known as **infrared radiation** (1000 nm to 0.1 mm). Then come **radio waves** (0.1 mm to >100 m).

The energy content of light—and other forms of radiation—is inversely proportional to its wavelength. Stating it precisely

$$e = \frac{hc}{\lambda}$$

where e is the energy of a photon in calories, c is the velocity of light in centimeters per second (2.998×10^{10}), λ is the wavelength of light in centimeters, and h is Planck's constant (1.583×10^{-34} cal/sec). This famous equation was developed by the German physicist Max Planck. Among other things, it tells us photons of short wavelength light (like blue light, wavelength 450 nm) have more energy than those of long wavelength light (like red light, wavelength 650 nm). As we saw in Box 4-1, the structure of some atoms and molecules causes them to absorb light of certain wavelengths. This is what accounts for the characteristic **absorption spectrum** of each light-absorbing chemical species. The explanation for this phenomenon lies in the features of atomic structure reviewed in Chapter 3. Since a specific amount of energy—not more and not less—is needed to boost electrons from a lower to a higher energy level, only photons of an energy that precisely equals the energy difference between the lower and higher levels can be absorbed. In other words, an atom or molecule absorbs certain light wavelengths and not others.

Our eyes perceive a mixture of all visible wavelengths as natural white or uncolored light. Ordinary sunlight is just such a mixture of radiations of different wavelengths or colors. A substance is perceived as having color when its molecular constituents absorb some wavelengths of light more than others. When sunlight strikes a colored object, some of its wavelengths are absorbed. The unabsorbed wavelengths are reflected back to our eyes, which thus receive more of some wavelengths than others. We perceive these dominant reflected wavelengths as color.

The sunlight reaching the ground continually varies in quantity and quality. Earth's atmosphere has already filtered the sunlight, absorbing most of its ultraviolet wavelengths and some of the infrared. High-energy ultraviolet photons are absorbed by ozone, oxygen, and CO_2 in the atmosphere; longer wavelength infrared photons are absorbed by water vapor and CO_2 in the air. In addition, an ever-changing amount of light is filtered out by clouds, smog, and other pollutants.

In summary, the sunlight reaching the ground consists of those photons that have managed to escape absorption or reflection. Of that sunlight, only about 25% consists of wavelengths capable of actively stimulating photosynthesis. Of the arriving light capable of stimulating photosynthesis, an exceedingly small fraction is actually utilized by green plants—only about two ten-thousandths of the solar radiation arriving on the Earth's surface.

TABLE 10-1

CLASSIFICATION OF PHOTOSYNTHETIC LIGHT-HARVESTING SYSTEMS

Type of Photosynthesis	Biological Groups	Pigments
Oxygenic photosynthesis (evolves oxygen)	*Photolithotrophs*[*]	
	Prokaryotes:	Chlorophyll *a* Phycobiliproteins Other minor pigments
	Cyanobacteria Prochlorophyta	
	Eukaryotes:	Chlorophyll *a* and *b*, or *c*, or *d* Carotenoids and xanthrophylls
	Chlorophyta Chrysophyta Plants	
Anoxygenic photosynthesis (does not evolve oxygen)	*Photoorganotrophs*[*]	
	Prokaryotes: Green sulfur bacteria Purple sulfur bacteria	Bacteriochlorophyll *a*, *b*, *c*, *d*, and *e*
	Eukaryotes: None	

[*] See Table 9-1.

OVERVIEW

Types of Photosynthesis

The bulk of photosynthesis is carried out by green plants. However, as we saw in the last chapter, photosynthesis occurs in organisms other than plants. As summarized in Table 10-1:

■ The more familiar type of photosynthesis is called **oxygenic** because it evolves oxygen. It occurs in higher plants and green algae, which are eukaryotes, and in certain prokaryotes, mainly the cyanobacteria (blue-green algae). These are the photolithotrophs of Tables 9-1 and 10-1.

■ Another type of photosynthesis is called **anoxygenic** because it does not evolve oxygen. It occurs only in prokaryotes—the so-called green sulfur and purple sulfur bacteria. These are the photoorganotrophs of Tables 9-1 and 10-1.

Note that, in plants, not all cells participate in photosynthesis. The woody portions, roots, and bark of many higher plants contain no chlorophyll. Only chlorophyll-containing cells are active in photosynthesis.

Quantitatively, anoxygenic photosynthesis is much less prevalent than oxygenic photosynthesis, though it was in the Earth's anaerobic early years that photosynthesis first evolved. The following discussion concerns oxygenic photosynthesis.

Key Events of Photosynthesis

In the most common form of photosynthesis (Figure 10-5), incoming **photons** (electromagnetic "particles" carrying energy but no mass) of sunlight are (in the customary metaphors) "harvested" or "trapped" by a **photosynthetic pigment.** As a result, an electron in the pigment is excited and caused to jump to a higher energy level. Under suitable conditions the excited electron is pried away from the pigment. It then passes to a series of electron carriers, which pass it to a final electron acceptor. As a result, ATP is formed, reducing power is generated as NADPH, and CO_2 is converted to sugar and other molecules.[3] Electrons in a low-energy state are returned to the pigment, restoring its original state.

---

[3] Note that NADP+ differs from NAD+ only in possessing an extra phosphate group (Chapter 9). However, the two coenzymes participate in distinctly different enzymatic reactions. NADP+ and its reduced form NADPH, we will see, have special connections with anabolic or biosynthetic pathways.

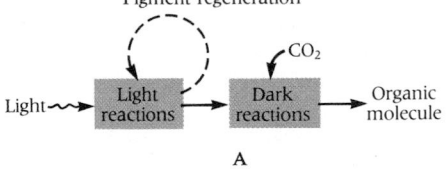

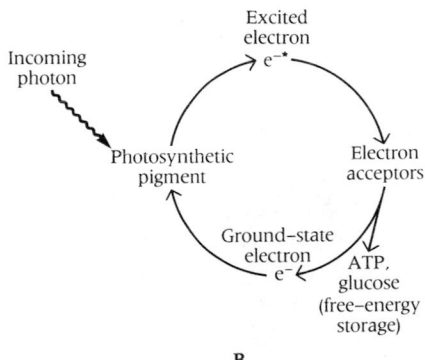

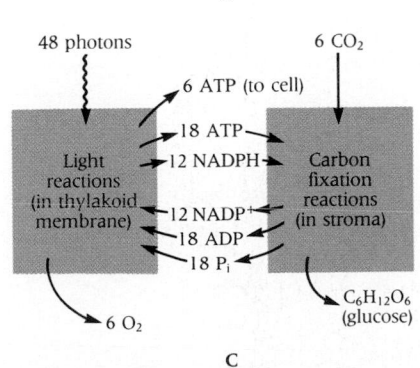

10-5 Summary of photosynthetic reactions
(A) *Solar energy is transformed into chemical bond energy in the presence of CO_2.* **(B)** *A photon is absorbed by a pigment molecule and the energy ultimately is converted to ATP and glucose.*
(C) *Solar energy is carried in the form of intermediates, ATP and NADPH, to the site of carbohydrate synthesis in the stroma.*

We have seen that among phototrophs various molecules are utilized as carbon sources (electron acceptors), and reductants (electron donors) (see Table 9-1). In green plants, the carbon source is CO_2 and the primary reductant is H_2O. The complex stepwise process of photosynthesis converts CO_2 and H_2O to sugar and other organic molecules:

$$6CO_2 + 6H_2O \xrightarrow{\text{Light}} 6O_2 + C_6H_{12}O_6 \qquad (10.1)$$

Equation 10.1 should be compared with Equation 8.1, which describes the fate of glucose in glycolysis and respiration.

The energy that makes glucose synthesis happen is supplied by the photons captured by chlorophyll. Thus the photosynthetic equation (Equation 10.1) can be rewritten as follows:

$$6CO_2 + 12H_2O + n\,h\upsilon \rightarrow C_6H_{12}O_6 + 6O_2 + 6H_2O \qquad (10.2)$$

where $h\upsilon$ is the energy of 1 photon of light (υ is the frequency of radiation, and h is Planck's constant). Note that in this version of the overall equation, water appears on both sides of the equation. This is a truer balance sheet for photosynthesis because the 6 H_2O among the products represents newly formed molecules; the 12 H_2O was starting material. The 6 O_2 among the products comes entirely from the 12 H_2O, and the oxygen in the 6 H_2O among the products comes from one of the oxygen atoms in the 6 CO_2 reactant. Of course, this net equation, like the others, hides the mechanisms.

The conversion of CO_2 into organic compounds is called **carbon fixation.** The organic compounds synthesized in the course of photosynthesis—notably carbohydrates—are rich in potential energy, which becomes usable in the form of ATP (see Figure 8-1).

We saw in Chapters 8 and 9 that 686 kcal of free energy is stored in one mole of glucose. Interestingly, the sequence in Equation 10.1 requires the capture of nearly 2000 kcal per mole of glucose formed. The difference between the energy utilized and that stored in sugars (about 1280 kcal per mole) is accounted for by energy liberated as heat during photosynthesis. Clearly, the process is strongly exergonic.

Other living organisms, including herbivores (plant-eating animals), carnivores (animals that eat other animals), and plants themselves have the ability to (1) release the energy stored in these food molecules by means of glycolysis and respiration, and (2) transfer it via the high-energy bonds of ATP and other molecules. Energy mobilized in this way is used in countless activities—in active transport, protein synthesis, and the myriad other tasks associated with the building and maintaining of new cells.

The entire food chain of life depends on plants. For that reason they are called the **primary producers.**

The Magnitude of Photosynthesis

Before turning from the energy cycle, we should note its immensity. The total amount of carbon fixed by all the plants growing on the Earth's surface is almost 10^{11} tons per year.[4] About one-third is accounted for by photosynthetic marine organisms. The total amount of carbon (as CO_2) in the Earth's atmosphere is about 7×10^{11} tons, and the total amount of carbon (largely as carbohydrates) in the Earth's vegetation is about 4.5×10^{11} tons.

It is instructive to compare these figures with data on the amount of solar energy reaching the Earth. Approximately 5×10^{24} kcal (or 1.73×10^{17} watts) of solar energy reaches the Earth's upper atmosphere annually. (This is one two-billionth

---

[4] The Earth's total carbon turnover due to photosynthesis is estimated in two ways: by averaging the yields of organic matter per unit area of field, forest, steppe, or ocean; and by determining the average utilization of incident solar energy by vegetation-covered areas. Both lead to a number approximating 10^{11} tons of carbon transformed annually from the inorganic to the organic state.

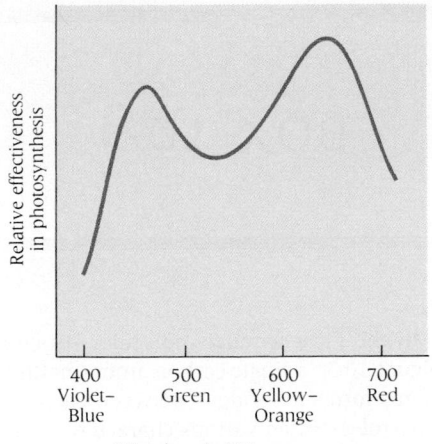

10-6 Action spectrum of visible light
Photosynthesis is at a maximum when light of red or blue wavelengths is used.

of the sun's total energy emission.) Of this incident solar energy, only about 3×10^{20} kcal is utilized in photosynthesis—less than one ten-thousandth of the incident solar energy!

It is instructive to compare this estimate of total annual biological energy flow with an estimate of the amount of energy expended by all the machines on Earth. Though obviously a difficult figure to fix and one that is increasing with time, it is believed to be more than 10^{19} kcal per year. We see that the annual flux of biological energy, which originates from the sun, is about 30 times the total energy flux in all machines.

HOW SUNLIGHT IS CAPTURED

The common form of photosynthesis, oxygenic, has three major stages:

- Stage 1: The trapping of quanta of light by the pigment chlorophyll in the thylakoid membranes of chloroplasts (see Figure 5-34); the use of absorbed light energy to remove electrons and protons from water (to form O_2); and the subsequent transfer of electrons through the thylakoid membrane to the ultimate electron acceptor $NADP^+$ (to form NADPH) in the stroma.
- Stage 2: The movement of protons down a concentration gradient from the lumen of the thylakoid to the stroma through a specialized protein in the thylakoid membrane that couples their movement to the synthesis of ATP from ADP and P_i. This chemiosmotic ATP-generating system resembles that operating in mitochondrial respiratory metabolism (Chapter 9). Stages 1 and 2 directly depend on light and thus are called **light reactions.**
- Stage 3: Utilization of the energy generated as ATP and NADPH to reduce atmospheric CO_2 to a sugar and other organic molecules. Reactions of this stage were once called dark reactions because glucose can be formed in the dark, given a supply of ATP and NADPH. However, plants almost never have a supply of ATP and NADPH in the dark and thus the term ''dark reactions'' is often misleading. This stage would be better termed the **carbon fixation** stage.

LIGHT-ABSORBING PIGMENTS

Absorption of Light by Chlorophyll

Light reaching a leaf may be absorbed (as is 60% of it), reflected (as is 10–40% of it), or transmitted through the leaf (as is a very small portion of it). Only light that is absorbed by molecules of the leaf can be utilized in photochemical reactions. As we have already seen, not all the absorbed light energy goes to this useful purpose. A portion of it alters the energy level of certain molecules and thereby accelerates chemical reactions. And some of it is reradiated as **fluorescence,** a process in which electromagnetic radiation (light) is emitted by the molecules that absorbed some inflowing energy.

To understand fluorescence, we should be aware that an electron stimulated by a photon of light rises to a higher energy level and stays there for only a very brief interval—in the range of 10^{-15} to 10^{-9} seconds. Such an electron is activated and ready to enter upon a chemical reaction. Had the electron been highly energized—as, for example, by a beam of x-ray or a gamma ray from an isotope such as ^{60}cobalt—it would have been torn away from its molecule, thus generating an ion. With ordinary levels of stimulation by light energy, the electron departs only if an electron acceptor (oxidizing agent) is at hand. If none is, it sinks back to its original energy level, emitting its excess energy as fluorescent light, which is at a longer wavelength (and thus has lower energy photons) than the excitatory incident light.

The light that is *not* absorbed, and therefore reflected and transmitted from leaves, consists mainly of green light. We suppose therefore that green light is utilized to a lesser extent than red or blue light, which is absorbed most efficiently. That this is true is easily proved by illuminating plants with pure light of different single specific wavelengths (termed **monochromatic light**) and measuring how efficiently each monochromatic illumination promotes photosynthesis. Figure 10-6 shows the

Two impressive evolutionary mechanisms, which we will encounter again and again, are evolution's tendency (1) to make the same inventions repetitively, and (2) to adapt existing materials to new and widely differing purposes.

The first mechanism is called **convergent evolution.** It is illustrated by the many instances in the course of evolution in which species of widely dissimilar ancestry came to resemble one another. An example is the similarity of the body shapes of whales, which are placental mammals, and fishes, which are not. The simple explanation of this phenomenon, which is discussed in Chapter 37, is that nature "selects" those features of organisms that work best in a given habitat or environment, irrespective of a species's evolutionary forebears.

The second mechanism, which has been called **historical opportunism** (Chapter 35), refers to frequent instances in which anatomical or biochemical structures were "lying around" in close proximity and an opportunity arose to combine them in new ways that would permit them to discharge novel functions. The story of how such opportunism led to the development of the mammalian ear is summarized in Box 35-1.

That this principle operates at the molecular level is convincingly illustrated by an interesting family of compounds—the **tetrapyrroles**—that are found in virtually all living organisms, from bacteria to human beings. In an impressive way, they illuminate the tendency of evolution to make the same inventions again and again and to adapt its earlier innovations to completely new and different purposes.

Tetrapyrroles are large ring structures made up of four smaller rings called **pyrroles.** A pyrrole ring contains four carbons and a nitrogen (Figure 10A). Presumably a very ancient organism acquired the ability to assemble four pyrroles into a tetrapyrrole (Figure 10B), deriving benefits from this new capacity that opened the door to new modes of life. Early organisms then began assembling tetrapyrroles in different ways. It is usual to classify contemporary tetrapyrrole molecules and their derivatives on the basis of three properties.

- The presence or absence of *carbon (methene) bridges* between adjacent individual pyrrole rings.
- The nature of the *metal atom* held in a complex at the center of the tetrapyrrole ring. The metal atom is held in place by the pincerlike process of *chelation.*
- The nature of the *substituent groups* attached as side chains to the pyrrole rings.

In the most familiar molecular structure (Figure 10C), a single carbon atom (methene group) forms a "bridge" between each pair of pyrroles. When various characteristic side chains are attached to this tetrapyrrole structure, it is called a **porphin ring.** But when

10A *Structure of a pyrrole ring*

Protoporphyrin IX

10B *Illustration of how four pyrrole rings are associated to form a tetrapyrrole, in this case the molecule is protoporphyrin IX*

results of such an experiment. This curve, termed an **action spectrum,** shows that maximum rates of photosynthesis are obtained with red and blue light.[5]

Chlorophyll is the principal pigment of leaves. Structurally, chlorophyll is a large molecule consisting of a porphyrin ring with a central atom of magnesium. Its long hydrophobic side chain anchors it to thylakoid membranes in the chloroplast.

Porphyrins are a widely occurring class of compounds built of four simpler nitrogen-containing **pyrrole rings.** Their extraordinary importance is discussed in Box 10-1. Note that the respiratory pigments of plants and animals—including the cytochromes—are also porphyrin derivatives. Heme and the cytochromes contain an atom of iron instead of magnesium. **Bacteriochlorophyll,** the major light-gathering pigment in many bacteria (see Table 10-1), is similar to chlorophyll except that one of the pyrrole groups has been reduced by 4 hydrogen atoms.

[5] Bear in mind the difference between an action spectrum and an absorption spectrum. The former is a graph (like that in Figure 10-6) showing the comparative effects or actions of different wavelengths of light on a biological system. The latter is a graph (like those in Box 4-1) showing the comparative extent to which different wavelengths of light have been absorbed as radiant energy passes through a medium.

four pyrroles are assembled in a slightly different way (Figure 10D)—with one of the four carbon bridges missing—the resulting structure is called a **corrin ring.**

Porphins form the structural framework of a large and diverse group of molecules of the most essential sort, as the following indicates:

■ When an atom of iron is bound into the center of the ring, the molecule is called

10C Porphin ring system showing methene bridges connecting the four pyrrole rings

10D Corrin ring system with three methene bridges

10E An iron atom at the center of heme

heme (Figure 10E). Heme is an essential component of all hemoproteins, which include hemoglobin and the cytochromes (see Figure 9-10).

■ When the metal is magnesium, the tetrapyrrole is **chlorophyll** or **bacteriochlorophyll** (see Figure 10-10).

■ When the metal is copper, the tetrapyrrole is **turacin,** a pigment found in bird's feathers.

■ When an atom of cobalt is inserted into a corrin ring, the result is a class of compounds that includes **cobalamin,** also known as **vitamin B$_{12}$.**

■ The light-sensitive molecule **phytochrome** is a plant tetrapyrrole that regulates seed germination and other processes in the lives of plants.

The biological properties of these molecules are determined not only by the side chains and metal present but by the attached protein. We learned in Chapter 4 that conjugated proteins carry an attached nonprotein prosthetic group. The **hemoproteins** are an important class of conjugated proteins, which includes many heme (iron–porphin–protein) complexes, among them the hemoglobins, myoglobins, cytochromes, and peroxidases. Patterns of alternating single and double bonds in the tetrapyrrole molecule confer brilliant colors upon all of these molecules. Hemoglobin and cobalamin are bright red; chlorophyll is green.

A survey of the metabolic roles played by these molecules provides the most convincing evidence that evolution utilized them in drastically different ways. For example, consider the following facts:

■ Chlorophyll's color makes it an excellent light trap.

■ Cobalamins utilize their cobalt atom as an appliance for accepting and donating hydride ions or small chemical units like methyl (—CH$_3$) groups.

■ Most of the activities of the hemoproteins are centered on their iron atom, but that atom has at least three modes of behavior in the several classes of hemoproteins. In hemoglobin, which has the task of carrying and delivering O$_2$ and CO$_2$, the iron remains in the Fe^{2+} state (reduced). In the cytochromes, it shuttles back and forth between the Fe^{3+} and Fe^{2+} states, functioning as an electron acceptor–donor in the respiratory chain. In the peroxidases, it remains in the Fe^{3+} (oxidized) state and participates in the enzymatic cleavage of hydrogen peroxide (H$_2$O$_2$), a powerful oxidant that became prevalent only when the Earth's atmosphere acquired O$_2$, threatening evolving life-forms until they began to defend themselves with peroxidase.

■ Biochemists have shown that an artificial heme molecule containing cobalt instead of iron works even better than natural heme in binding O$_2$. But iron is abundant on Earth and cobalt is not, and so evolution favored iron-porphins.

Biochemists have argued over whether corrins appeared on Earth before or after porphins. The evidence does seem to suggest that their antiquity greatly exceeds that of their more familiar—and plentiful—porphin relatives. The issue nicely illustrates the point: these two tetrapyrrole structures probably arose independently—at separate times and in different places. This structure was evidently an invention of such versatility that it probably arose on several occasions.

Until recently, it was believed that chlorophyll occurs exclusively in the form of the two major variants, **chlorophyll a** and **chlorophyll b,** plus a few minor ones. These differ only in small modifications of a porphyrin ring substituent: chlorophyll *a* has a methyl group (—CH$_3$) where chlorophyll *b* has a formyl group (—CHO). All eukaryotic phototrophs contain chlorophyll *a;* plants and some green algae have small amounts of chlorophyll *b* (Figure 10-7A). We must now modify that view to accommodate a surprising number of newly recognized chemical variants of chlorophylls *a* and *b*. Although their structures are still uncertain, some may be side-chain variants.

The action spectrum of photosynthesis (see Figure 10-6) becomes especially interesting when compared with the results of another experiment. If we determine the absorption spectrum of isolated chlorophyll—that is, how much of each wavelength of light it absorbs (see Box 4-1)—the results will be as shown in Figure 10-7B. Light absorbed by chlorophyll is mostly in the blue (400–475 nm) and red (625–700 nm) wavelengths. Predictably, chlorophyll transmits or reflects the wavelengths around the middle of the visible spectrum, the ones we see as green.

In many organic molecules with biological functions, the color of the molecules seems a coincidental result of a molecular structure that, in itself, is irrelevant to

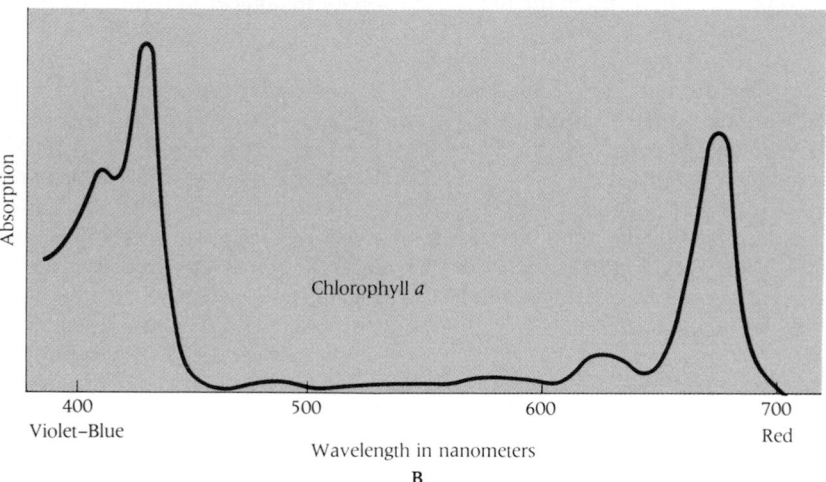

A

10-7 Chlorophyll is the major photosynthetic pigment
(**A**) Structures of chlorophylls a and b.
(**B**) Absorbance spectrum of chlorophyll a showing that it strongly absorbs red and blue light. The two major absorption peaks of chlorophyll b are at slightly higher wavelengths.

Chlorophyll *a*

| | | | |
400
Violet–Blue
500
600
700
Red
Wavelength in nanometers
B

Absorption

biological function. For example, hemoglobin is not a better oxygen carrier because it is red. But the color of chlorophyll is significant. It signifies the fact that certain wavelengths of the incident light are being absorbed and used, as suggested by the action spectrum of photosynthesis in Figure 10-6. Because of the alternating single and double bonds in chlorophyll, many electrons in the porphyrin ring and magnesium atom are free to move about. This is what makes chlorophyll (like other porphyrins) such a good light absorber.

The absorption of a photon of light briefly raises an electron in the chlorophyll molecule to an excited state (Figure 10-8). Under proper conditions, this can cause

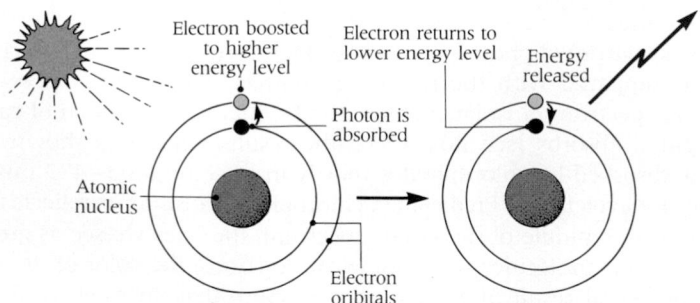

Electron boosted to higher energy level

Electron returns to lower energy level

Energy released

Photon is absorbed

Atomic nucleus

Electron orbitals

10-8 Absorption of energy by a chlorophyll molecule
Energy is released when the electron that has been raised to a higher energy level falls back to its original orbital. The drawing suggests that the energy-absorbing electron is oriented to a single atomic nucleus. In fact, these low and high energy orbitals are relative to many nucleii in a complex (pigment) molecule.

10-9 Absorption spectrum of carotenoids
The carotenoids are accessory pigments that absorb light in the blue region of the spectrum.

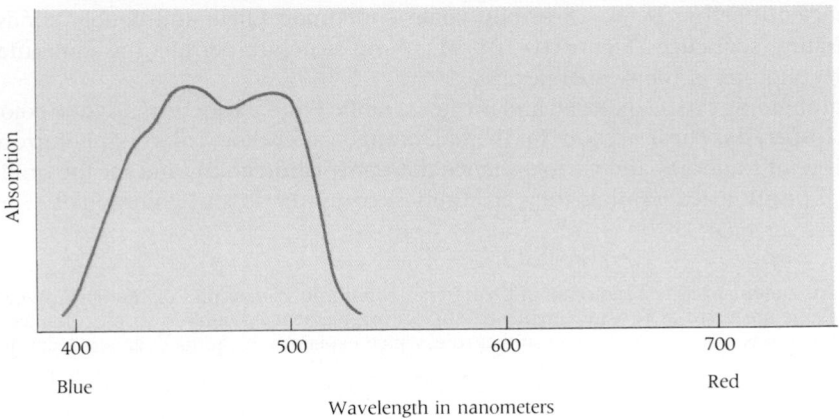

Absorption

400   500   600   700

Blue   Red

Wavelength in nanometers

chlorophyll to become an **electron donor**—or (Table 8-1) a reducing agent—that transfers electrons to other molecules. They reduce still other molecules, and so on down the chain.

Accessory Pigments

An action spectrum of photosynthesis shows the extent to which each wavelength of light actually stimulates photosynthesis. We do not get this information from an absorption spectrum of a pigment like chlorophyll, which indicates only to what extent each wavelength is absorbed.

The action spectrum of photosynthesis (Figure 10-6) is interesting because it is not identical to the absorption spectrum of chlorophyll. This suggests that other plant pigments may be absorbing light as well. This is precisely the case (Figure 10-9). These **accessory pigments** also absorb light that promotes photosynthesis.

A major group of accessory pigments is the **carotenoids,** which are localized to the thylakoid membranes. The most familiar is **β-carotene,** two 6-carbon rings

10-10 Structures of chlorophyll a, β-carotene, and retinal
Note that these three molecules have alternating double and single bonds that are responsible for their light-absorbing properties.

Chlorophyll a   β–Carotene   Retinal

connected together by an 18-carbon chain containing single and double bonds in alternating sequence (Figure 10-10). This long structure permits the molecule to absorb photons of many energies.[6]

Carotenoids can be oxidized and further modified into a variety of brightly colored **xanthophylls.** These account for the red, orange, and yellow colors of phototrophic bacteria, of tomatoes and carrots (hence the name carotenoid), and for the brilliant hue of autumn leaves after they are unmasked by the loss of chlorophyll (Figure 10-11).

[6] In the animal body, a molecule of β-carotene is split into 2 molecules of vitamin A, which is a metabolic precursor of **retinal** (Figure 10-10), an eye pigment essential in vision (Chapter 30). Predictably, it is also a light-absorber in that role, which explains why human vitamin A deficiency impairs vision.

10-11 Xanthophylls are responsible for the brilliant colors in autumn leaves

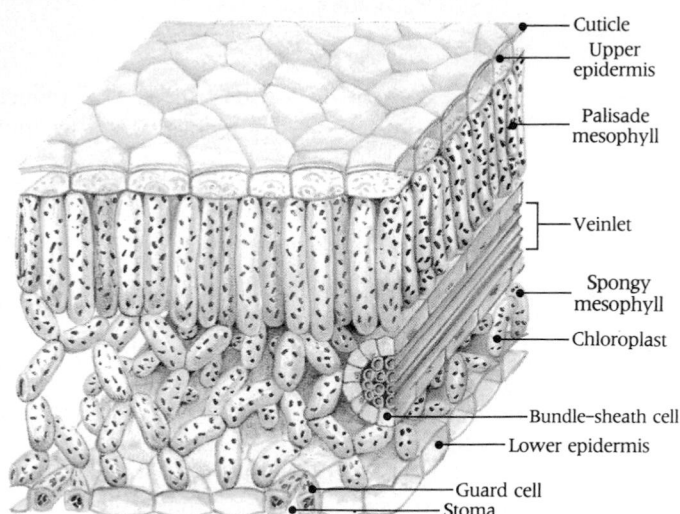

Cuticle
Upper epidermis
Palisade mesophyll
Veinlet
Spongy mesophyll
Chloroplast
Bundle–sheath cell
Lower epidermis
Guard cell
Stoma

A

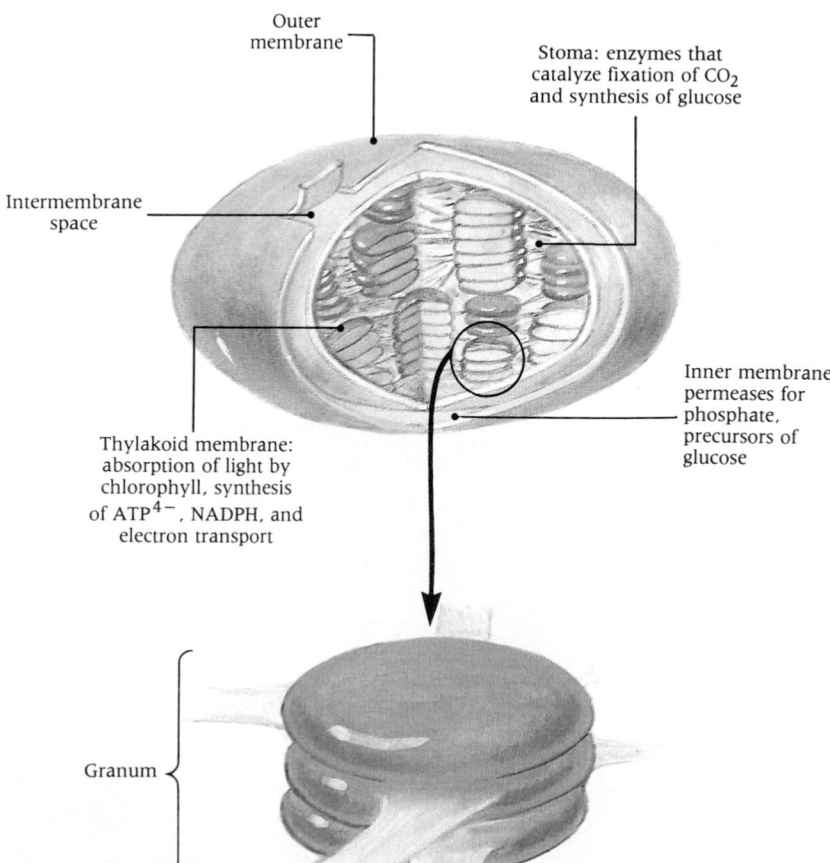

Outer membrane
Stoma: enzymes that catalyze fixation of CO_2 and synthesis of glucose
Intermembrane space
Inner membrane: permeases for phosphate, precursors of glucose
Thylakoid membrane: absorption of light by chlorophyll, synthesis of ATP^{4-}, NADPH, and electron transport
Granum

B

10-12 Structure of the chloroplast
(A) Cutaway diagram showing structure of a typical chloroplast and its components.
(B) Chloroplasts are localized in the mesophyll layer of the leaf. The thylakoid membranes are the site of the light reactions of photosynthesis.
(C) Electron micrograph of a chloroplast in a leaf mesophyll cell of corn. Thylakoids are stacked within the stroma. ($\times 16,000$)

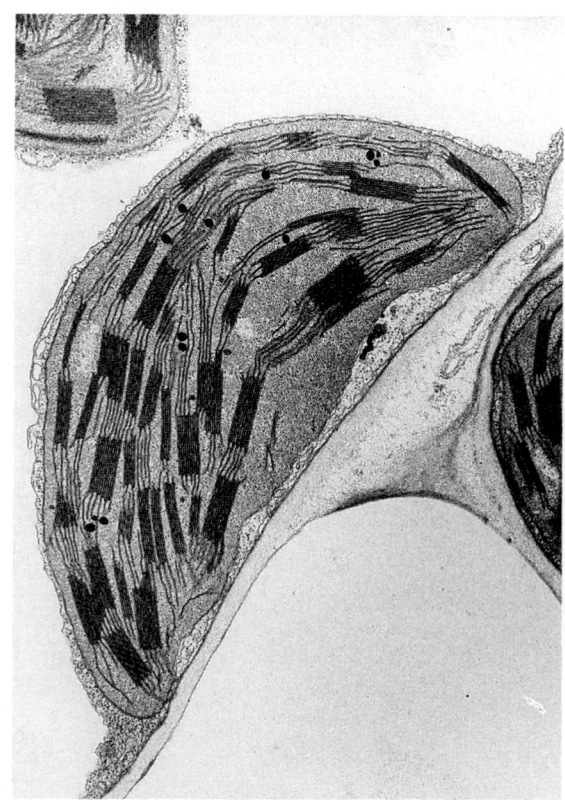

C

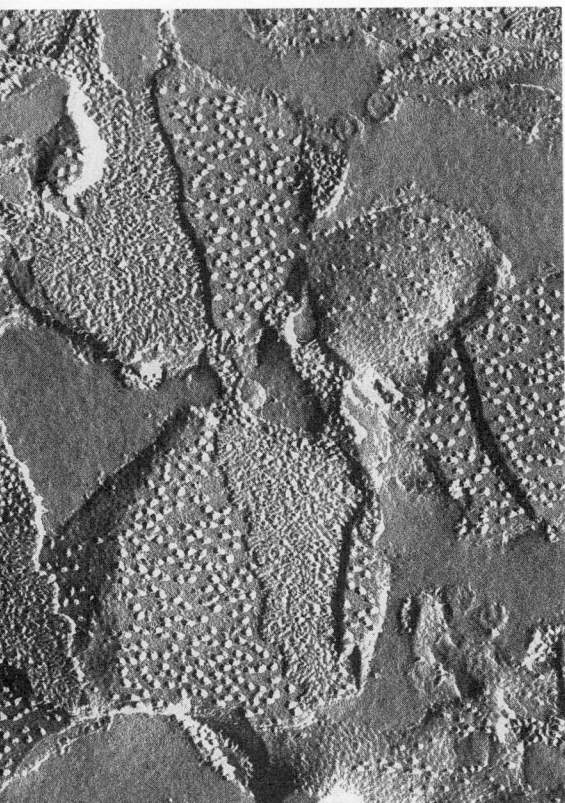

10-13 Scanning electron micrograph showing photosynthetic units embedded in thylakoid membrane

Application of the freeze-fracture technique to a chloroplast granum made possible this remarkable view of the arrangement of photosynthetic units within thylakoid membranes. The larger particles are photosystems containing pigment complexes. Smaller, more densely packed granules are probably enzymes of the photophosphorylation and electron transport systems. (× 70,000)

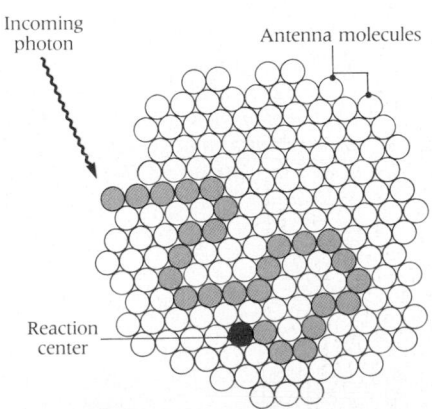

10-14 Transfer of energy in the antennae complex

Energy is transferred among antenna molecules until it is trapped in the reaction center.

In sum, the task of the carotenoids and other accessory pigments, wherever they occur in photosynthesizing cells, is to pass along their excited electrons according to the general scheme as shown in Figure 10-5.

THE CHLOROPLAST

To the naked eye, a leaf appears uniformly green, but a microscope reveals that the green color is confined to the chloroplasts in the leaf. We saw in Chapter 5 that the interior of the disk-shaped chloroplast, the **stroma,** is a dense, fluid-filled matrix in which green bodies called **grana** are located (Figure 10-12). The grana are shown by electron microscopy to be stacks of 10 to 20 flat photosynthetic membranes called **thylakoids** (Figure 10-12), which contain the chlorophyll and enzymes of the light-dependent reactions. (The enzymes of the carbon-fixing are in the stroma.) Except in the cyanobacteria (blue-green algae) and other photosynthetic bacteria, which have unstacked thylakoid membranes but lack chloroplasts, chlorophyll occurs only in the thylakoids of the grana. Collectively, the surface area of the thylakoid membranes is enormous. As noted earlier, a single cell on the upper leaf surface may contain 40 or 50 chloroplasts; thus a leaf may have 500,000 chloroplasts per square millimeter. A single mature chloroplast contains 40 to 60 grana.

In the chloroplast, chlorophyll and accessory photosynthetic pigments are located in discrete particles called **photosynthetic units** which are elegantly demonstrated by freeze-fracturing (Figure 10-13). Each photosynthetic unit in a thylakoid membrane contains about 300 pigment molecules, including chlorophyll *a*, chlorophyll *b*, carotenoids, and two manganese atoms (a biochemical rarity). All the pigments within a unit are capable of absorbing light, but one chlorophyll *a* molecule per unit is distinguished from the rest. This special molecule is the **reaction center** of the unit. It is distinguished from all other chlorophyll *a* molecules by the fact that it is bound to an integral membrane protein (molecular weight, 110,000) that confers special properties upon it. Only this chlorophyll molecule can mediate the transformation of light energy into chemical energy. All other pigment molecules are called **antenna pigments** because their arrangement—closely packed pigments that can very efficiently exchange energy—serves as an "antenna" for gathering light.

The membranes in fact contain two kinds of reaction centers—**photosystem I** and **photosystem II,** or **PS I** and **PS II.** They differ in responding to light of slightly different wavelengths. Thus they complement one another—and their interplay is of critical importance. The chlorophyll *a* molecule in the PS II reaction center is bound to a different protein molecule (molecular weight, 47,000). Electrons, we will see, are transported from PS II to PS I, each of which has a distinctive structure and spatial orientation within the molecular architecture of the thylakoid membrane.

The photosynthetic pigments of the photosystems are attached to proteins embedded within the thylakoid membranes. In PS I, 14 chlorophyll *a* molecules are bound to a large protein. In PS II, 3 molecules each of chlorophylls *a* and *b* are bound to a much smaller protein. The reaction center of PS I is a chlorophyll *a* complex called P_{700}. P stands for *pigment;* the subscript 700 refers to the wavelength (in nm) of optimally absorbed light. PS II also has a special chlorophyll *a* complex at its reaction center. For similar reasons it is called P_{680}.

The energy level of chlorophylls at the reaction center is lower than that of other chlorophylls. This accounts for their superior ability to trap light energy. When a chlorophyll molecule absorbs a photon of red or blue light,[7] one or another of its electrons is abruptly boosted to a higher energy level (see Figure 10-8). When this occurs in the antennalike network, the state of excitation (if not the electron itself) skips rapidly from one pigment molecule to another (Figure 10-14). It may or may not reach the reaction center; if it does, it is "trapped." That means that an excited electron in this particular chlorophyll molecule does not sink back to

[7] The efficiency of photosynthesis drops sharply with light of wavelengths greater than 680 nm, even though chlorophyll is able to absorb light at up to 700 nm. Photosynthesis by light at 700 nm is enhanced, however, if the incoming light includes shorter wavelengths (e.g., 600 nm). This is part of the evidence for the existence of two interacting photosystems.

its original energy level, emitting its extra energy as fluorescent light. Instead, the energized electron remains excited and escapes—moving on to a nearby electron acceptor molecule (oxidizing agent) that pries the electron away. This initiates a cascadelike electron flow from one carrier molecule to another.

The reaction center chlorophyll molecule, having lost an electron, has been oxidized. Until the lost electron is restored, it is precluded from raising another electron to an excited state.

THE LIGHT REACTIONS: GENERATION OF ATP AND NADPH

There are two modes of photosynthetic energy conversion that differ in the fates of the electrons excited by PS I.

- **Cyclic photophosphorylation** involves only one of the two photosystems (PS I). It occurs in all photosynthetic cells—but in plants it operates in tandem with noncyclic photophosphorylation. It is the sole system in prokaryotes. Probably the earliest photosynthetic system to arise in evolution, it may have been the only such system for a billion years or more.
- **Noncyclic photophosphorylation** involves both photosystems (PS I and PS II). It operates in green plants and certain bacteria.

Photophosphorylation is shorthand for light-induced ATP synthesis:

$$\text{ADP} + P_i + \text{light energy} \rightarrow \text{ATP} + H_2O \qquad (10.3)$$

Cyclic Photophosphorylation

This process, discovered in 1954 by Daniel I. Arnon, makes exclusive use of PS I (Figure 10-15). When P_{700}, the special chlorophyll at the reaction center, is excited by a photon of light, its energized electron is transferred to a carrier called **FeS** that contains iron and sulfur. At the same time, P_{700} is oxidized. From here, electrons flow through the membrane-bound chain of electron-transferring intermediates, as shown in Figure 10-15. The small release of free energy at each step keeps the energy release from occurring explosively. Note the striking resemblance of this sequence to the mitochondrial respiratory chain of oxidative phosphorylation (see Figure 9-12).

At the end of the chain, the now energy-poor electron is handed off by the last carrier, the copper-containing **plastocyanin (PC),** to the oxidized P_{700} molecule. Thus restored, it is ready to absorb another photon of light and start the sequence all over again. This is why we call the system *cyclic.*

We call it *photophosphorylation* because the sequence leads to a phosphorylation of ADP that generates one molecule of ATP. That ATP molecule is the solitary

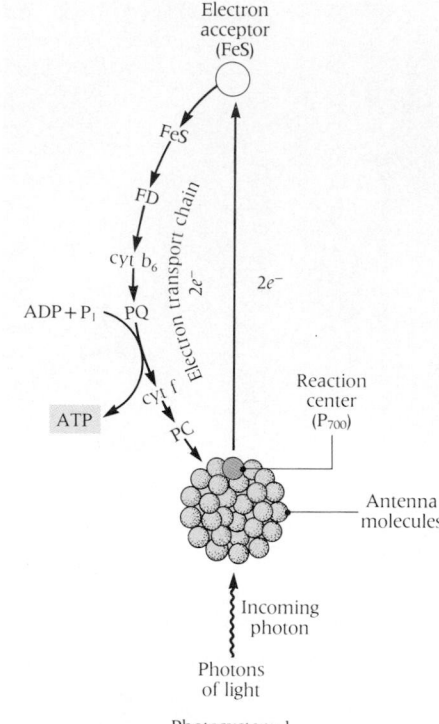

10-15 The flow of electrons in cyclic photophosphorylation
The electron transport chain is composed of a series of electron carrier molecules that shunt electrons in a cyclic direction through Photosystem I. Abbreviations: FeS (iron-sulfur protein), FD (ferredoxin), cyt b_6 and f (cytochromes), PQ (plastoquinone), PC (plastocyanin). The pumping of protons to form a gradient occurs at the step where ADP is converted to ATP.

10-16 The flow of protons across the thylakoid membrane
As electrons flow from PS II to PS I, protons are shunted into the thylakoid lumen, creating a region of high proton concentration relative to the stroma. Protons flow through the CF_0CF_1-ATP synthase complex, and the energy that is released is used to synthesize ATP.

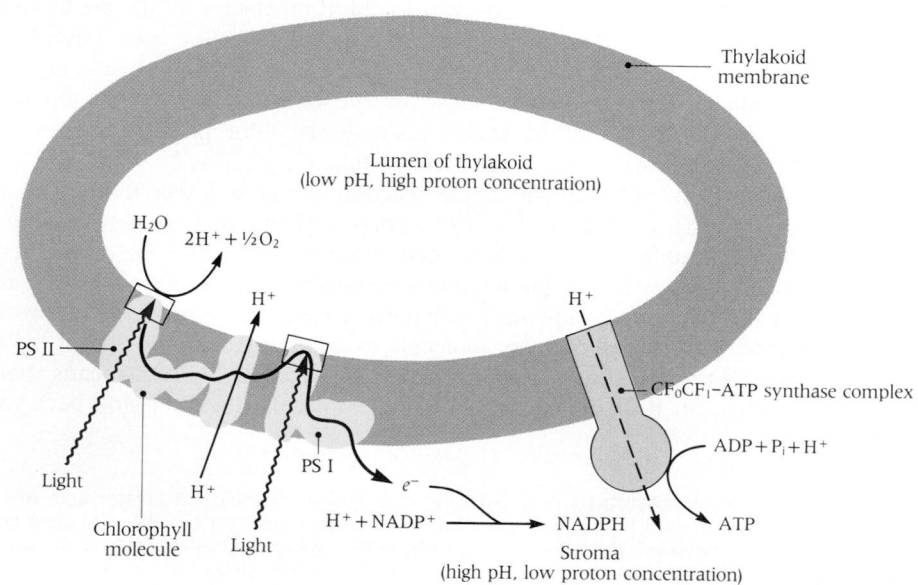

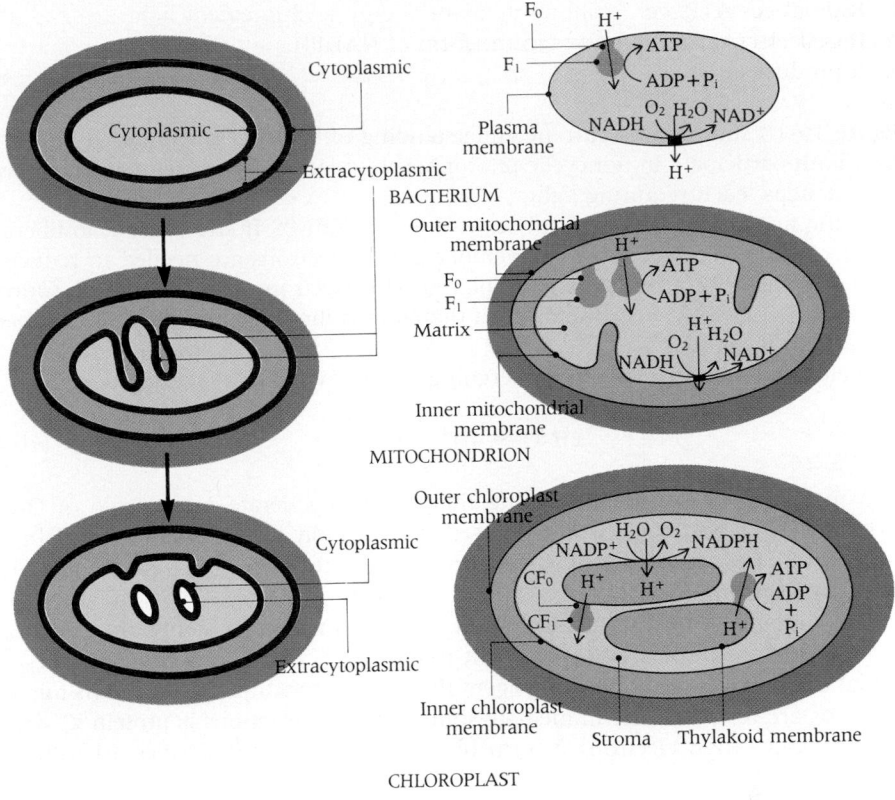

10-17 Membrane orientation and the direction of proton movement in bacteria, mitochondria, and chloroplasts
*The figure shows interesting differences in the membrane orientations in bacteria, mitochondria, and chloroplasts. **(A)** Cytoplasmic surfaces face colored areas. Surfaces facing uncolored areas are extracytoplasmic (outside of the cytoplasm). Thylakoid disks form by invagination of the inner chloroplast membrane. Therefore, inside surfaces of disks are extracytoplasmic and surfaces facing stroma are cytoplasmic (inside the cytoplasm). **(B)** In bacteria, mitochondria, and chloroplasts, F_0F_1 complexes always face the cytoplasmic surface. During electron transport, protons are always pumped out—that is, from the cytoplasmic to the extracytoplasmic membrane surface. During ATP synthesis, their direction reverses as they flow down a gradient through the F_0F_1 complexes. As discussed in Chapters 43 and 44, mitochondria and chloroplasts are thought to have evolved, respectively, from purple sulfur bacteria and cyanobacteria.*

product of this system. It is formed by the same chemiosmotic mechanism that operates in oxidative phosphorylation—the same, that is, except for one major difference. The flow of protons is in the opposite direction (Figure 10-16). Note in Figure 10-17 that because of the way thylakoid membranes are formed—by invagination of the inner chloroplast membrane—their inner surfaces are extracytoplasmic, facing away from the cell's cytoplasm. Protons are always pumped from the cytoplasmic to the extracytoplasmic surface of the membrane: In mitochondria they are pumped out of the matrix and into the intermembrane space; in chloroplasts they are pumped out of the stroma and into the inner thylakoid space. In both, their direction reverses during ATP generation.

Energy released in the electron transfer chain—notably by the carrier **plastoquinone (PQ)**—pumps protons across the thylakoid membrane to establish a proton gradient. As in oxidative phosphorylation (Figure 9-12), the return of protons through the membrane drives the **ATP synthase** enzyme complex—in chloroplasts termed **CF_0CF_1 complex** in analogy with the F_0F_1 complex of mitochondria. Here an ADP molecule is converted to ATP.

The following points summarize cyclic photophosphorylation:

- Its only product is ATP.
- It splits no water, evolves no O_2, and produces no NADPH.
- It yields no sugars or other organic molecules.

This sort of system may have been good enough for certain prokaryotes, but it would hardly serve the needs of plants and most prokaryotes, in which the ultimate purpose of photosynthesis is carbon fixation, not ATP generation. Plant cells do synthesize some ATP via the cyclic process, however.

Noncyclic Photophosphorylation

In 1957, Arnon discovered noncyclic photophosphorylation. This system has the following characteristics:

- It produces ATP.
- It generates reducing power in the form of NADPH.
- It produces free O_2.

Figure 10-18 summarizes current understanding of the system. Note that PS I and PS II both participate in noncyclic photophosphorylation. The scheme includes two distinct steps, each requiring light.

In the first step, 2 incoming photons interact with PS II and cause it to liberate a pair of energized electrons. (Remember that 2 electrons are needed to reduce 1 $NADP^+$.) These electrons must be replaced with electrons from an outside source. Ultimately, that source is water, which therefore is the fundamental reducing agent of plant photosynthesis.

Water performs this task by undergoing the following cleavage reaction:

$$2H_2O \rightarrow 4H^+ + O_2 + 4e^- \tag{10.4}$$

Two molecules of water undergoing this reaction generate 1 molecule of O_2, 4 protons ($4H^+$), and 4 electrons. Since the absorption of each photon of light by PS II results in the transfer of 1 electron, each PS II system must have the ability to lose 4 electrons.

This important reaction is brought about by a manganese-containing electron-seeking protein known as **Z.** The light-induced loss of 4 electrons leaves P_{680}, a potent electron-deficient oxidizing agent that cannot function again until its missing electrons are restored. The immediate source of these electrons is protein Z, which replenishes its own electrons in turn by splitting 4 water molecules according to Equation 10.4.

The energized electrons split from P_{680} are transported to the outer surface of the thylakoid membrane and the protons liberated from water are left in the lumen

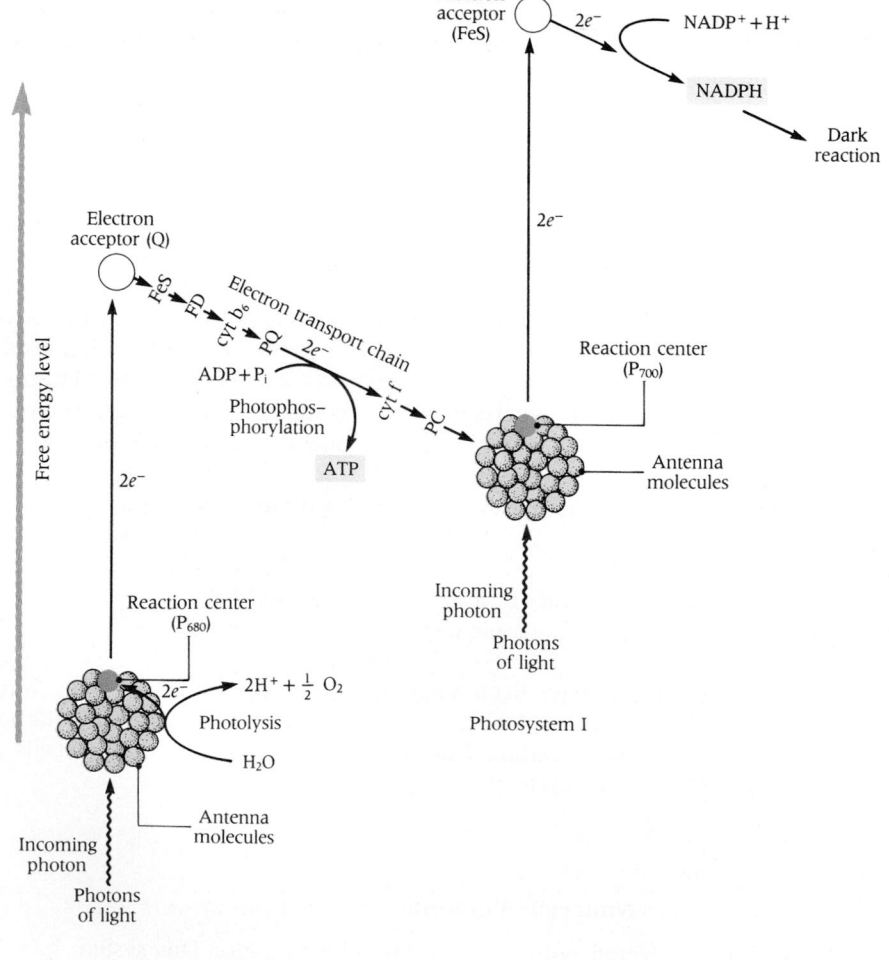

10-18 The flow of electrons in noncyclic photophosphorylation
Photons are absorbed by photosystem I and photosystem II, and electrons move from photosystem II to photosystem I to the final electron acceptor, $NADP^+$.

of the thylakoid. The electrons are transferred from PS II to PS I via a sequence of electron transport carriers that includes three of the same cytochromes used in cyclic photophosphorylation plus several other electron carriers. Once again, a transfer of electrons from plastoquinone—in the middle of the first electron transfer cascade—generates a proton-motive force across the thylakoid membrane. Once again, the electrons give up some of their energy at each step of the sequence in a now familiar pattern.

At the end of the first sequence, the transferred electrons reach plastocyanin (PC). There they remain, poised to recharge the P_{700} molecule in the second photosystem PS I following its light-induced ejection of electrons. This occurs in the second step of noncyclic photophosphorylation.

In that step, two incoming photons of light interact with P_{700} in PS I and energize its electrons, which pass to a second sequence of electron carriers across the membrane. The final electron acceptor at the end of this chain is a molecule of $NADP^+$, which is thereby reduced to NADPH. It is the reducing power of NADPH that drives the "dark reactions," reducing CO_2 as it is converted to sugar. The synthesis of ATP by the CF_0CF_1 complex is the major chloroplast function that depends upon the proton gradient. Two to three protons cross the membrane for each ATP formed.

Thus the sequence of electron transport in noncyclic photophosphorylation is as follows:

$$H_2O \rightarrow P_{680} \rightarrow \text{PS II transport chain} \rightarrow P_{700} \rightarrow \text{PS I transport chain} \rightarrow$$
$$NADP^+/NADPH \qquad (10.5)$$

So-called light reactions are associated with membranes because reaction centers and electron-transport systems must be held in an asymmetrical orientation capable of establishing a chemiosmotic proton gradient. Membranes also provide tough barriers that can sustain the difference in charge that has built up and can briefly store that energy.

CARBON FIXATION: CONVERSION OF CO_2 TO SUGAR

The Calvin Cycle

We turn now to the biosynthetic reactions of photosynthesis—the fixation of CO_2 and its conversion to glucose in a cyclical series of enzyme-catalyzed reactions known formally as the **reductive pentose phosphate cycle** and informally as the **Calvin cycle,** after its Nobel prize-winning discoverer, Melvin Calvin, of the University of California, Berkeley. Some texts, wishing to honor Calvin's coworkers, speak of the Calvin-Benson-Bassham cycle. These reactions are totally dependent on the ATP and NADPH generated in the light reactions.

10-19 The Calvin cycle
(A) The first step in the Calvin cycle involves the conversion of RuBP to PGA by the enzyme RuPB carboxylase. (B) The Calvin cycle regenerates RuBP and releases carbohydrate in the form of glyceraldehyde-3-phosphate. NADPH and ATP are required in order for this to occur. The structural formulas of many intermediates in this cycle appear in Figure 11-2. Note the 2 molecules of G-3-P emerging at the lower right. Their 6 carbons are converted in several steps to the hexose fructose 6-phosphate, which loses a phosphate and is converted to glucose. The presence of this phosphate explains the product mix in Equation 10.6, which includes a hexose-phosphate and only 17 P_is. If we consider glucose as the final hexose product, then 18 P_is would be formed.

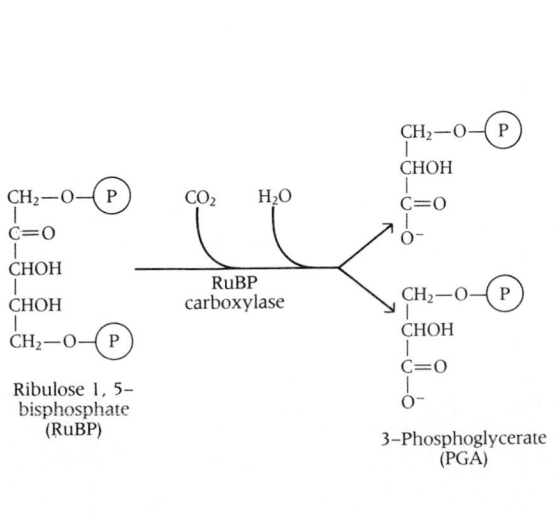

A

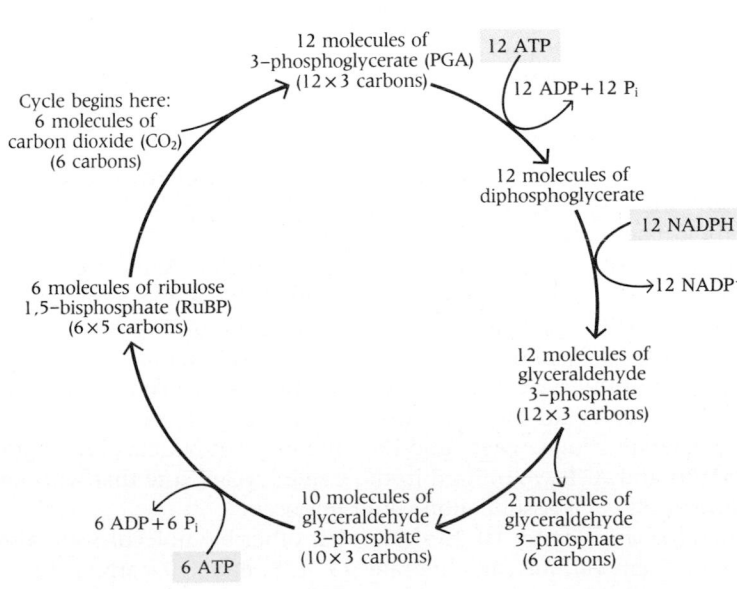

B

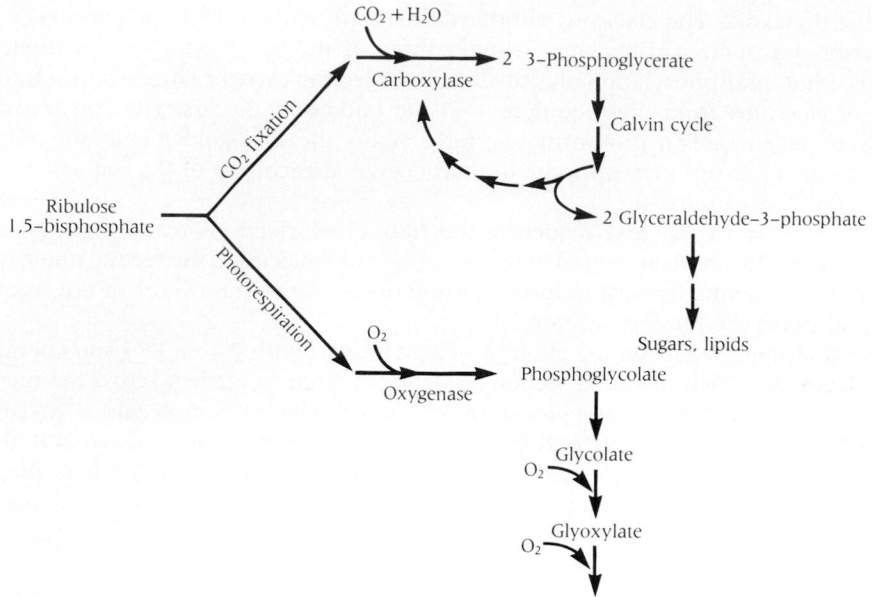

An understanding of how plants take in simple CO_2 molecules and then convert them to something as complex as sugar was achieved through the skillful use of CO_2 tagged with the radioactive carbon isotope ^{14}C. When $^{14}CO_2$ is supplied to photosynthetic cells in the presence of light, all intermediary compounds on the pathway to sugar are found to have become radioactive after separation by two-dimensional paper chromatography. They are then easily tracked with x-ray film and scintillation counters.

This approach to unraveling the metabolic pathway of CO_2 in plants began in the late 1940s. Experiments with the unicellular phototrophic green algae *Chlorella* and *Scenedesmus* showed that within seconds after the introduction of $^{14}CO_2$, radioactivity could be detected in sugars, sugar phosphates, amino acids, and organic acids in these cells.

One of the first molecules to become radioactive was the 3-carbon sugar phosphate **phosphoglyceric acid,** or **PGA** (Figure 10-19A). Significantly, the radioactivity in PGA was confined to its carboxyl (—COOH) carbon. Why is that significant?

The molecule that actually accepts CO_2 is **ribulose 1,5-bisphosphate,** or **RuBP.** Such additions of CO_2 are called **carboxylations.** The immediate reaction product is an unstable 6-carbon compound that promptly splits into two molecules of PGA. The enzyme that catalyzes the fixation of CO_2, **RuBP carboxylase,** comprises more than 15% of all the protein in chloroplasts. It is said to be the single most abundant protein on Earth. In laboratory jargon, this enzyme is called **Rubisco.**

This short pathway leads into the Calvin cycle, which takes place in the stroma of the chloroplasts. Figure 10-19B makes the following points:

- The first step is the carboxylation in which atmospheric CO_2 combines with RuBP under the catalytic influence of RuBP carboxylase. This is the locus of carbon fixation.
- The metabolic fate of the PGA carbons is complex. Some are converted to a hexose (fructose 6-phosphate) via a short pathway that essentially reverses the part of the glycolytic pathway in which fructose 6-phosphate is converted to glyceraldehyde 3-phosphate in animal cells (see Figure 9-5).
- One turn of the cycle also regenerates a new molecule of RuBP that can serve as CO_2 acceptor for the next turn of the cycle.
- The diagram shows where and how the major products of the light reactions NADPH and ATP are utilized in the Calvin cycle. Note that without these two compounds, CO_2 assimilation could not occur.
- The scheme in Figure 10-20 shows that other biomolecules are also formed— among them various carbohydrates (3, 4, 5, 6, and 7-carbon sugars, sucrose, and starch), lipids, and proteins.

Photorespiration

In most plants, photosynthesis is always accompanied by a curious and costly process called **photorespiration,** which takes place in light, consumes O_2, and oxidizes RuBP with the eventual release of CO_2. As shown in Figure 10-20, RuBP carboxylase functions as a carboxylase in catalyzing the carboxylation (that is, addition of CO_2) of RuBP. However, this enzyme can also catalyze a competing reaction in which it acts as an **oxygenase.** In this reaction, it catalyzes the oxidation (that is, by addition of O_2) of RuBP. In other words, CO_2 and O_2 compete with one another for a site on the enzyme.

When CO_2 concentrations are low, photorespiration predominates over photosynthesis. Plants that depend heavily on the Calvin cycle for CO_2 fixation cannot synthesize carbohydrates unless the atmospheric CO_2 concentration exceeds about 50 parts per million. At ordinary levels of atmospheric CO_2—about 340–345 parts per million—net levels of photosynthesis are curtailed by the coexistence of photorespiration.

The products of an addition of O_2 to RuBP consist of one molecule of the 3-carbon molecule PGA and one molecule of the 2-carbon molecule **phosphoglycolate** (Figure 10-20). This compound is hydrolyzed to glycolate and transported to peroxisomes, small organelles in the cytoplasm containing enzymes that generate and consume H_2O_2 (hydrogen peroxide). Here some of the glycolate is oxidized to CO_2 in a process that generates neither ATP nor NADPH. However, a complex series of reactions involving chloroplasts and mitochondria salvages some of the energy that would otherwise be lost, so that, in the end, only about one carbon of every three entering photorespiration is actually lost as CO_2.

In sum, photorespiration has the following effects:

- A key intermediate (phosphoglycolate) is squandered with release of some of its carbon as CO_2. Under certain environmental conditions (low CO_2, high temperature), the loss of fixed carbon is substantial, exceeding the rate of photosynthesis.
- Photosynthetic productivity is slowed in proportion.
- Some O_2 is utilized.

Why did we call photorespiration a curious process? For several reasons. First, it is unusual for an enzyme to catalyze two reactions. Second, it is even more unusual for the two functions to have opposing consequences. We believe that the existence of photorespiration is explained by the fact that the enzyme RuBP carboxylase-oxygenase evolved at a time when Earth's atmosphere contained high levels of CO_2 and little or no O_2. Under those conditions, the enzyme could function only as a carboxylase. Then, when photosynthesis evolved and added more and more O_2 to the atmosphere, the ratio of CO_2 to O_2 in the atmosphere declined. In a sense, early photosynthetic plants poisoned their own environment by producing O_2. But the enzyme, now confronted with two alternative substrates, began to oxygenate as well as carboxylate. With today's low ratio of CO_2 to O_2, the enzyme carboxylates most of the RuBP it encounters, but it does oxygenate some of it. Even though the enzyme is relatively ineffective in our modern environment, its importance is indicated by its conservation throughout evolution.

C_4 Photosynthesis

Carboxylation of RuBP is not the only way plants fix CO_2. In the 1960s, M. D. Hatch and C. R. Slack surprised workers in the field by showing that in some plant species—notably tropical plants, maize, sugar cane, and various other grasses—the first molecular product of CO_2 fixation is not the 3-carbon compound PGA but the 4-carbon organic acid oxaloacetate (Figure 10-21). Oxaloacetate, we recall, is a key intermediate in the citric acid cycle (Figure 9-7). Plants that utilize this pathway are called **C_4 plants** because some of the intermediates in their carbon fixation pathway are 4-carbon organic acids like oxaloacetate. These are to be contrasted with the more familiar **C_3 plants,** named for 3-carbon PGA. Both kinds of plants still depend on the Calvin cycle.

Among the major differences between the two systems are the following:

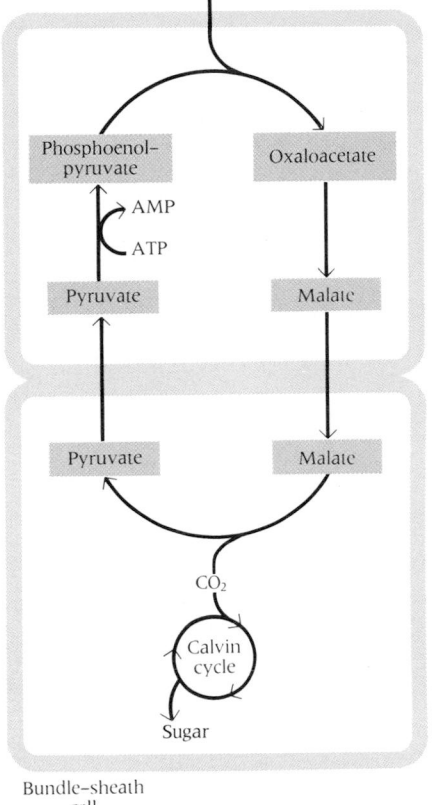

A

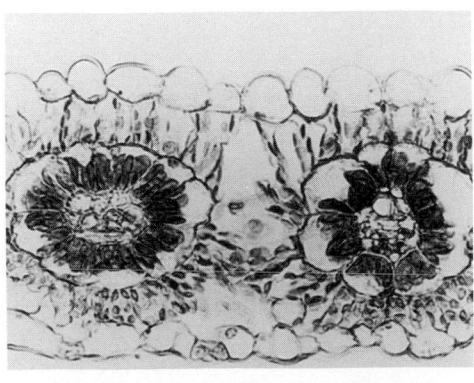

B

10-21 The C_4 photosynthetic pathway (*A*) *CO_2 is taken into mesophyll cells and becomes incorporated into 4-carbon acids, which are transferred to bundle-sheath cells, where the Calvin cycle extracts energy and converts it to sugar molecules.* (*B*) *Electron micrograph showing the arrangement of mesophyll cells and bundle-sheath cells in the leaf of a C_4 plant, corn.* (×200)

- In C$_3$ plants, CO$_2$ is fixed by combining with a 5-carbon acceptor molecule (RuBP). The immediate products are two 3-carbon compounds (3-PGA). A new 5-carbon acceptor molecule is regenerated in the Calvin cycle (see Figure 10-19).

- In C$_4$ plants, CO$_2$ is fixed by the enzyme phosphoenolpyruvate carboxylase, which combines it with a 3-carbon acceptor molecule to form a 4-carbon molecule (oxaloacetate). After a series of reactions, the newly formed molecule of oxaloacetate is reduced to malate, which then diffuses from the mesophyll cell, where it was formed, to a nearby bundle sheath cell. Here it loses a CO$_2$, which is picked up by RuBP carboxylase and then enters the Calvin cycle. The pyruvate left behind when malate lost a CO$_2$ diffuses back to the mesophyll cell, where, as a result of another cyclic reaction sequence, it generates a new 3-carbon compound (phosphoenolpyruvate or PEP) that can accept another CO$_2$, thereby reentering the C$_4$ pathway.

Other important differences between C$_3$ and C$_4$ are discussed below.

We saw above that the existence of photorespiration partially suppresses photosynthesis. This suggests that if photorespiration could be suppressed, photosynthesis would be more efficient. Interestingly, C$_4$ plants have evolved a mechanism for overcoming photorespiration that depends on local increases in CO$_2$ concentration at intracellular carboxylation sites. This accelerates the carboxylase reactions relative to the oxygenase reactions.

SIGNIFICANCE OF LEAF STRUCTURE

We deferred a discussion of the structure of leaves in Chapter 5 because the topic is most meaningfully considered in the context of photosynthesis. Indeed, leaf structure is a remarkable example of biological engineering in which structure and function are coupled. Consider the features of leaf anatomy in two major types of leaves.

10-22 Leaf shapes in nontropical plants

Poison sumac

Honey locust

Horse chestnut

Ginkgo

Maple

White oak

Sassafras

Beech

Willow

10-23 Tissue organization in a C₃ leaf

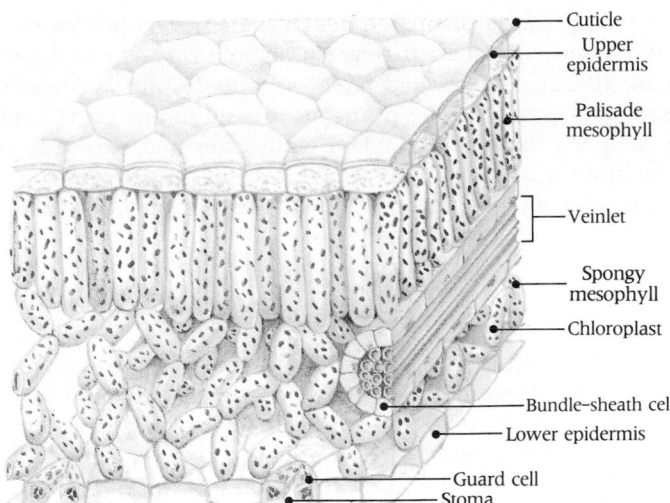

Cuticle
Upper epidermis
Palisade mesophyll
Veinlet
Spongy mesophyll
Chloroplast
Bundle–sheath cell
Lower epidermis
Guard cell
Stoma

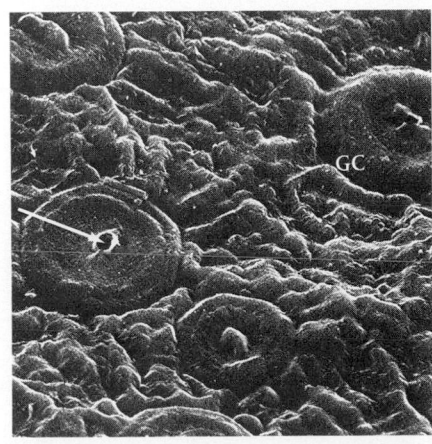

10-24 Structure of stomata
This is a scanning electron micrograph of the lower surface of a grape leaf, Mahonia aquifolium. *Each stoma consists of a pore (arrow) surrounded by two crescent-shaped guard cells (labeled GC). Changes in the turgidity of these cells causes stomata to open and close.*
(× 1300).

Leaves of Common Nontropical Plants

Each leaf bud arises during the growing season from a small mound of meristematic tissue. As the emerging leaf grows to maturity in the shape characteristic of its species (Figure 10-22), its cells differentiate into specialized tissues.

In many familiar green plants of the temperate zone, protective tissue takes the form of single cell layers, tightly fitted to the top and bottom leaf surfaces (Figure 10-23). This is the **epidermis.** Its cells produce the **cuticle,** which is transparent to light but is glossy and waterproof. The cuticle is composed of heavy deposits of a waxy lipid plus a substance called **cutin** (a complex polymer of cross-linked, long-chain fatty acids). The cuticle would also hinder the exchange of gases were it not for numerous small openings called **stomata** (singular, **stoma**) in the epidermis, especially on the leaf's lower surface (Figure 10-24). Openings and closings of each stoma are controlled by a pair of specialized **guard cells.** Stomata are usually open in the light and closed in the dark.

Osmosis controls the opening and closing of a stoma. Guard cells open it by accumulating ions (Figure 10-25). Water moves into the guard cells, causing them to become more turgid. Because of the distinctive pattern of cellulose fibers in their cell wall and the differences in wall thickness on each side, they do not swell evenly. Their ends and outside surfaces distend more easily than the surfaces next to the opening. As a result, the inner walls pull apart and the pore opens. Control of stomatal opening and closing and the role of stomata in the transport of water will be discussed in Chapter 24.

In considering what might cause guard cells to lower their osmotic potential, biologists once believed that light causes them to increase their internal solute concentration by promoting photosynthetic sugar production. It seemed an attractive theory because guard cells do contain chloroplasts, while other epidermal cells do not. However, further work revealed a better correlation between stomatal openings and the accumulation within guard cells of potassium ions (K^+): K^+ increases when stomata open and falls when they close. What then is the role of guard cell chloroplasts? Probably in these cells they function as a source of ATP, which energizes active transport of K^+ into the cells and diminishes their content of CO_2, a molecule that induces stomatal closure.

The stomatal pore is connected with the **intercellular spaces** in the part of the leaf that is sandwiched between the upper and lower epidermis. This region is largely the **mesophyll** ("middle leaf"), which includes the bulk of the leaf's photosynthetic tissue, most of its many-sided, thin-walled cells containing abundant chloroplasts (see Figure 10-23). Mesophyll has two layers: an upper **palisade mesophyll,** named for its thin, cylindrical, stand-on-end palisade cells, and a lower **spongy mesophyll,** an irregularly organized cell mass permeated by air spaces that communicate with the stomata. Palisade and spongy mesophyll are **parenchyma;** mesophyll cells are **parenchymal cells.**

The leaf also has a system of **vascular bundles,** each containing both xylem and phloem cells (Chapter 5), which run through the mesophyll surrounded by a thin sleeve of tightly packed **bundle-sheath cells.** As nature lovers know, trees have distinctive patterns of leaf veins, or venation, that are useful in species identification (see Figure 10-22). The vascular bundles in the leaf blade connect with those in the **leaf stalk** (or **petiole**), the **plant stem,** and ultimately the **root.** The leaf is constantly supplied through the xylem with water and minerals—and no leaf cell is very far from a vascular bundle. In turn, the leaf supplies the rest of the plant through its phloem with sugars manufactured by photosynthesis in its mesophyll cells.

Leaves with Kranz Anatomy

Early in this century, botanists first observed that the leaves of many tropical plants from dry habitats had anatomical patterns that differ strikingly from those just described. Not until the early 1970s did anyone recognize that leaves of this structure invariably display C_4 photosynthesis.

As shown in Figure 10-26, the main structural features of these leaves are the following:

- Vascular bundles are very close together, separated by only a few mesophyll cells. They are farther apart in C_3 plants.
- Leaf veins are surrounded by large bundle sheaths made up of cells containing many large chloroplasts.
- Outside the sheath is a sleeve of radially arranged mesophyll cells containing chloroplasts. This ringlike arrangement of photosynthetic cells inspired the term **Kranz anatomy,** from the German word for "crown" or "corona."
- Spongy mesophyll cells outside these rings have few chloroplasts.
- Mesophyll and bundle-sheath cells are both sites of photosynthesis. In C_3 plants, mesophyll is the primary site of photosynthesis.

Structural and metabolic patterns can be correlated in these leaves in an interesting way. For example, note in Figure 10-21 that, in C_4 plants, mesophyll cells, which are freely permeable to CO_2, are the sites at which the 4-carbon organic acid oxaloacetate is formed from carboxylation of phosphoenolpyruvate (PEP), the same molecule we encountered near the end of the glycolytic pathway (see Figure 9-5). As noted above, a cyclic sequence of reactions then regenerates a new acceptor molecule of PEP (and in so doing utilizes 2 ATPs per CO_2). An intermediate of this sequence, malate, is transported to a neighboring bundle-sheath cell, which is the site of RuBP carboxylation and the Calvin cycle.

Tropical plants on grasslands and savannahs are exposed to bright sunlight, intense heat, and frequent dry spells—and many tropical plants exhibit Kranz anatomy. Under such conditions, a leaf can conserve water only if the stomata close almost completely. However, this blocks a free exchange of gases—and CO_2 concentration within the leaf drops. In a C_3 plant, that would amplify photorespiration and drasti-

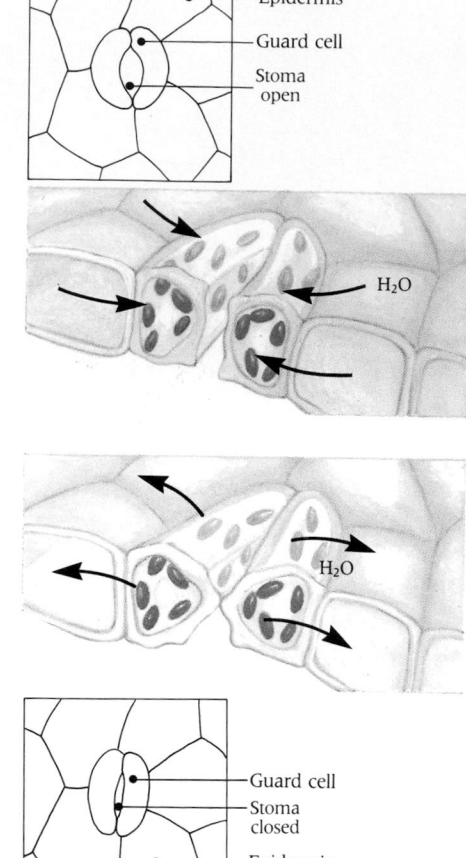

10-25 Control of stoma opening and closing
Water moves into guard cells, causing them to swell and leading to opening of the stoma. As water moves out of the cells, they return to their original shape and the stoma closes.

10-26 Structure of leaf with Kranz anatomy

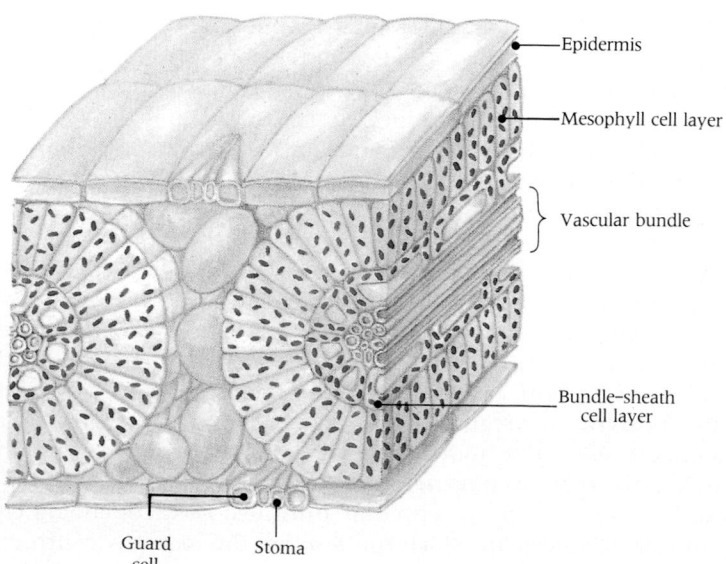

TABLE 10-2
COMPARISON OF C₃ AND C₄ PHOTOSYNTHESIS

Property	C_3 Plants	C_4 Plants
Primary CO_2 acceptor	RuBP	PEP
CO_2-fixing enzyme	RuBP carboxylase	PEP carboxylase
First product of CO_2 fixation	PGA	Oxoloacetate
Does Calvin cycle operate?	Yes	Yes
Locus of photosynthetic cells	Mesophyll	Mesophyll and bundle-sheath
Photorespiration	Sizable	Minimal

cally decrease photosynthesis. But Kranz-type leaves have C_4 photosynthesis, the efficiency of which is maintained at low levels of CO_2, which enters into mesophyll cells where it is fixed by combination with PEP. Closure of the stomata also minimizes CO_2 loss from bundle-sheath cells (the locus of the Calvin cycle). Kranz plants can convert CO_2 into sugar under conditions in which photorespiration would predominate over photosynthesis in C_3 plants that use only the Calvin cycle. Since photorespiration does not actively occur in C_4 plants, the plants thrive, actively synthesizing sugar from CO_2. The major differences between C_3 and C_4 photosynthesis are summarized in Table 10-2.

A remarkably detailed fossil leaf 5 to 7 million years old was recently found that clearly reveals Kranz anatomy (Figure 10-27). This discovery shows that C_4 photosynthesis was present in recent times, geologically speaking. Perhaps it evolved even earlier.

Variations

Two other metabolic patterns warrant brief mention. A small number of plant species have metabolic and structural features of both C_3 and C_4 photosynthesis. This has been termed **C_3–C_4 intermediate photosynthesis.**

Another pattern of C_4 metabolism is found in certain desert plants that are adapted to dry conditions and extremes of temperature—daytime heat and nighttime cold. These succulents, especially of the family Crassulaceae, are fleshy plants that avoid water losses by closing their stomata during the day and opening them at night. Unlike C_4 plants, they incorporate the CO_2 acquired at night into organic acids (mainly malic and isocitric acids) and store it in that form. During the day these acids release CO_2, which then participates in C_3 photosynthesis. All these activities occur within the same cell, in contrast with C_4 plants, which separate the C_4 pattern and the Calvin cycle.

This pattern is termed **CAM** (an abbreviation of **crassulacean acid metabolism**), only because it was first discovered in this family. At least 17 other families have CAM, not all succulents have it, and not all CAM plants are succulents. The

10-27 Microscopic view of a fossil leaf with Kranz anatomy

These fragments of a fossil grass leaf, Chloridoidae, recently found in northwestern Kansas clearly show well-developed Kranz anatomy. Note the prominent bundle sheath cells around a vascular bundle. This specimen, which is 5 to 7 million years old, shows that C_4 photosynthesis dates back to the Miocene.

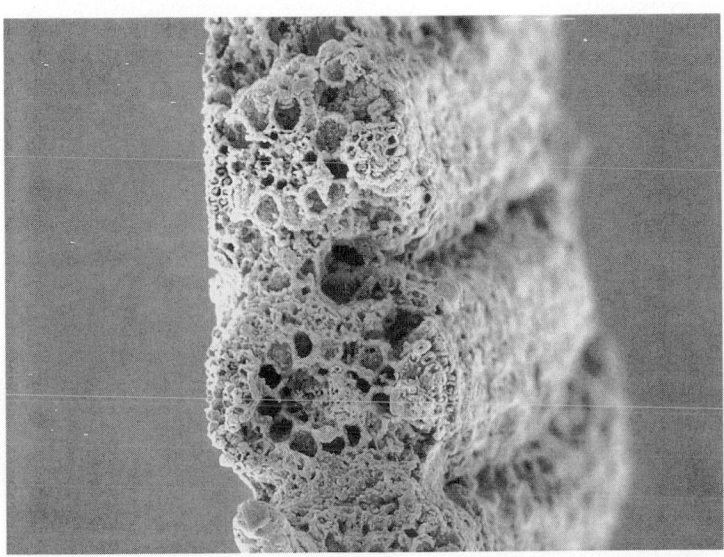
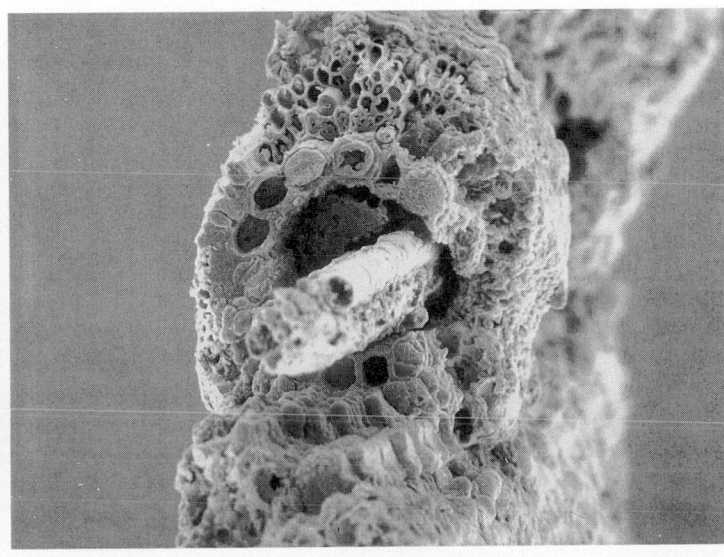

fact that CAM and C_4 evolved independently illustrates the tendency of evolution to arrive at multiple solutions to the challenges of the environment.

ENERGETICS REVISITED

The photosynthetic formation of glucose and O_2 from CO_2 and water can be capsulized in one of those overall equations (Equation 10.1), of which we have seen several. This equation tells what goes in and what comes out, but says nothing of how it all comes about.

The point is illustrated by these more detailed net equations for CO_2 fixation in C_3 and in C_4 plants, in which both pathways operate together.

$$\text{C}_3 \text{ Plants}$$
$$6CO_2 + 18ATP + 12NADPH + 12H_2O \rightarrow C_6H_{12}O_6\text{-phosphate} +$$
$$18ADP + 17P_i + 12NADP^+ \qquad (10.6)$$

$$\text{C}_4 \text{ Plants}$$
$$6CO_2 + 30ATP + 12NADPH + 12H_2O \rightarrow C_6H_{12}O_6 + 30ADP +$$
$$30P_i + 12NADP^+ + 18H^+ \qquad (10.7)$$

As the net equations make clear,

- C_3 plants require 18 ATPs to synthesize 1 glucose molecule. We saw in the caption of Figure 10-19 that if we write Equation 10.6 with *fructose 6-phosphate* as the product, 17 P_is are formed. If *glucose* is considered the final product, 18 P_is are formed. To synthesize a hexose, 6 rounds of the Calvin cycle are necessary because 1 carbon atom is reduced in each cycle (Figure 10-19); 12 ATPs are expended in phosphorylating 12 molecules of 3-phosphoglycerate to 1,3-bisphosphoglycerate, and 12 NADPHs are consumed in reducing 12 molecules of 1,3-bisphosphoglycerate to glyceraldehyde 3-phosphate. An additional 6 ATPs are needed to regenerate RuBB. Thus, 3 ATPs are needed per carbon fixed.
- C_4 plants require 30 ATPs to synthesize 1 glucose molecule. This reflects the need to expend 5 ATPs for each CO_2 fixed by the C_4 pathway (see Figure 10-21) as compared to 3 ATPs in C_3 metabolism. However, sunlight is plentiful and photorespiration sets no barrier, so this extra cost counts for little. As shown in Figure 10-28, C_4 plants perform photosynthesis considerably faster than do most C_3 plants under hot dry conditions, produce sugar faster, and grow faster.

Plant cells do rely on the conventional pathways of glycolysis and respiration to oxidize glucose to CO_2 and H_2O and generate needed ATP for cellular activities. The carbohydrates and other organic molecules of plants, into which energy derived from sunlight has been invested, are then utilized as food by animal cells, as well as by the plant cells themselves.

10-28 Efficiency of C_3 and C_4 photosynthesis compared
(A) *Effect of environmental CO_2 concentration on amounts of CO_2 fixed. At very low CO_2 concentrations, photosynthesis in a C_4 plant like corn is more efficient than in a C_3 bean plant. At high CO_2 concentrations, it is also more efficient. Note that the atmosphere ordinarily contains 330 ppm of CO_2.* **(B)** *Effect of environmental O_2 concentration on rate of photosynthesis. (Normal atmosphere is about 21% O_2). Corn photosynthesis is hardly affected at high O_2 levels. Bean photosynthesis is progressively decreased.* **(C)** *Effect of leaf temperature. Again, the bean, but not the corn, plant is affected. Clearly, C_3 plant photosynthesis is more efficient in cooler climates.*

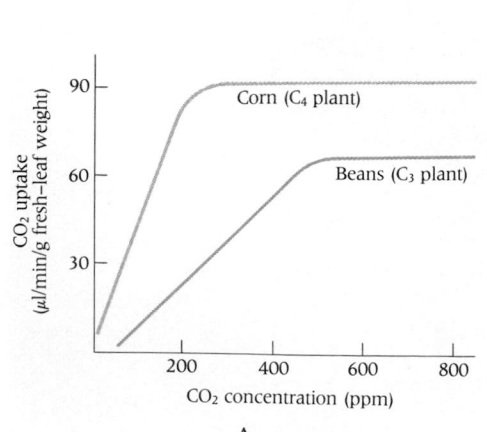

A

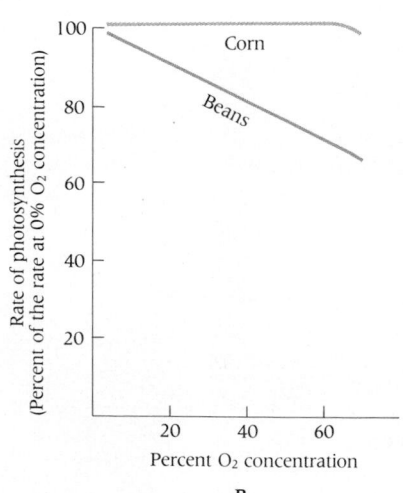

B

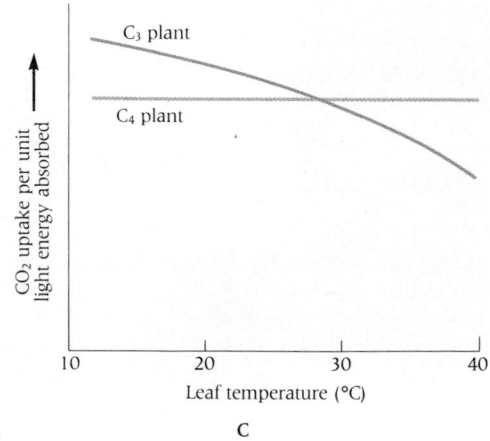

C

SUMMARY

Of the solar radiation reaching the Earth, only a small amount is absorbed by photosynthetic organisms—and only certain of its wavelengths can be used to drive photosynthesis.

The net result of photosynthesis is the synthesis of carbohydrates according to the equation: $6CO_2 + 6H_2O \rightarrow C_6H_{12}O_6 + 6O_2$. This synthesis is the primary source of energy-rich compounds for all organisms in the food chain. The process actually involves many individual enzymatic reactions, which operate in three pathways or stages: (1) the absorption of photons of light by the photosynthetic pigment chlorophyll and the liberation of electrons; (2) the transfer of electrons down a series of electron carriers to NADPH with the simultaneous synthesis of ATP and liberation of oxygen; (3) the utilization of ATP and NADPH to drive the fixation and reduction of CO_2 into carbohydrates. Stages 1 and 2 are called light reactions; stage 3 might occur in the absence or presence of light.

The light reactions depend on the presence of three classes of molecules within the thylakoid membranes of chloroplasts: light-absorbing pigments, electron carriers, and proteins associated with ATP synthesis. The ability of chlorophyll to donate electrons when excited by light stems from its porphyrin ring structure. Accessory light-absorbing pigments such as carotenoids produce excited electrons in response to light of wavelengths other than those which excite chlorophyll. Pigments and enzymes necessary for the light reactions are grouped in photosynthetic units of two types: photosystem I (PS I) and photosystem II (PS II).

The formation of ATP in photosynthesis proceeds either through cyclic or noncyclic photophosphorylation. The former predominates in prokaryotes. The latter, which is found only in eukaryotes, generates NADPH and O_2 as well as ATP. In cyclic photophosphorylation, which utilizes PS I, high-energy electrons flow through a chain of electron carriers, liberating small amounts of energy at each step, much as occurs in the mitochondrial respiratory chain of oxidative phosphorylation. A proton gradient arises across the inner chloroplast membrane, and by chemiosmosis the potential energy of the gradient drives the synthesis of ATP. Meanwhile, the now energy-poor electron is passed from the last electron carrier of the chain back to chlorophyll.

Noncyclic photophosphorylation utilizes both PS I and PS II—which together release two electrons. Because these electrons derive ultimately from the cleavage of H_2O to H^+, electrons (e^-), and O_2, noncyclic photophosphorylation causes the release of O_2. The electrons are energized in PS II and produce a strong oxidant that leads to the formation of O_2. Electrons are then carried to PS I. This results in the generation of a proton gradient, which drives the synthesis of ATP. Electrons are then passed to a second chlorophyll molecule in PS I. After being energized, they reduce $NADP^+$ to NADPH and produce another proton gradient.

The Calvin cycle uses the ATP and NADPH generated by the light reactions to incorporate atmospheric CO_2 into ribulose 1,5-bisphosphate (RuBP). This molecule splits to form two molecules of phosphoglyceric acid (PGA), some of which is then used to synthesize glucose. At the same time, photorespiration occurs. In that process, O_2 rather than CO_2 is added to RuBP and only one molecule of PGA is formed. This limits the efficiency of photosynthesis and thus is energetically wasteful. Some plants can avoid most photorespiration by fixing CO_2 to a three-carbon compound rather than to RuBP. This process is called C_4 photosynthesis. Although oxaloacetate rather than PGA is formed by such a fixation, it is converted through a series of reactions to PGA, which then enters the Calvin cycle and ultimately leads to the synthesis of glucose.

The structural organization of a leaf critically influences its ability to carry out photosynthesis. Leaves of nontropical plants are covered with a protective waxy coating, but the gas exchanges necessary for photosynthesis are made possible by small openings called stomata, which open and close as a function of the osmotic pressure in the cells around them. CO_2 and O_2 can therefore reach the middle of the leaf. Vascular tissue, containing xylem and phloem, bring H_2O and minerals to the leaf from the roots.

Many plants from dry tropical regions have Kranz anatomy, an organizational pattern different from that of nontropical plants. They exhibit C_4 photosynthesis and are better able to tolerate hot, dry conditions than plants that rely on C_3 photosynthesis. A variant of C_4 metabolism, called crassulacean acid metabolism (CAM), is yet another adaptation in succulent plants.

KEY TERMS

absorption spectrum
accessory pigments
action spectrum
antenna pigments
bacteriochlorophyll
β-carotene
C_3 plants
C_4 plants
Calvin cycle
carbon fixation
carboxylation
carotenoids
chlorophylls *a* and *b*
crassulacean acid metabolism
   (CAM)
cyclic photophosphorylation
dark reactions

electromagnetic spectrum
electron donor
guard cells
Kranz anatomy
light reactions
mesophyll
noncyclic
   photophosphorylation
oxygenase
oxygenic photosynthesis
P_{680}
P_{700}
palisade mesophyll
parenchymal cells
photon
photorespiration
photosynthesis

photosynthetic pigment
photosynthetic unit
photosystem I (PS I)
photosystem II (PS II)
primary energy producer
RuBP carboxylase/oxidase
   (Rubisco)
reaction center
ribulose 1,5-bisphosphate
   (RuBP)
spongy mesophyll
stomata (stoma)
stroma
thylakoids
xanthophylls

QUIZ QUESTIONS

1. Which of the following is *not* a photosynthetic pigment?
   A. xanthophyll
   B. carotenoid
   C. chlorophyll *a*
   D. plastoquinone
   E. A and B

2. Which of the following is *not* a product of noncyclic photophosphorylation?
   A. NADPH
   B. ATP
   C. carbon dioxide
   D. oxygen
   E. A and D

3. What is the role of phosphoenolpyruvate (PEP) in photosynthesis?
   A. PEP is an electron transport molecule in Photosystem II.
   B. PEP is the carbon dioxide acceptor in C4 photosynthesis.
   C. PEP is the electron donor in the reductive pentose phosphate cycle.
   D. PEP is the precursor molecule for xanthophyll.
   E. PEP is the carbon dioxide donor in C3 photosynthesis.

4. In C4 photosynthesis, the first molecular product of carbon dioxide fixation is
   A. oxaloacetate.
   B. phosphoenolpyruvate.
   C. malic acid.
   D. crassulacean acid.
   E. None of the above.

5. Chlorophyll molecules are located
   A. in the outer membrane of mitochondria.
   B. in the inner membrane of chloroplasts.
   C. in the stroma of chloroplasts.
   D. in the outer membrane of chloroplasts.
   E. in the thylakoid membrane.

ESSAY QUESTIONS

1. What biomolecules directly capture the energy of the sun?

2. What colors of monochromatic light best promote photosynthesis and why?

3. Contrast cyclic and non-cyclic photophosphorylation. Where is O_2 produced? Contrast "light reactions" and "dark reactions." Trace the steps in the incorporation of CO_2 into sugars.

4. How does light regulate the opening and closing of the stomata of leaves? What important role does the cell wall play in guard cells?

REFERENCES AND SUGGESTIONS FOR FURTHER READING

Anderson, J. M., and Andersson, B. (1982). The architecture of the photosynthetic membrane. *Trends in Biochemical Sciences 7*, 288–292.

A clearly written description of the intricate structure of photosynthetic membranes.

Cohen, I. B. (1976). Stephen Hales. *Scientific American 234*, May, 98–107.

The story of the English clergyman who measured the flow of water in plants and founded plant physiology—by a distinguished historian of science.

Dale, J. E. and Milthorpe, F. L. (1983). *The Growth and Functioning of Leaves.* Cambridge University Press, New York.

A collection of superior articles on aspects of the biology of leaves including structure, function and development and adaptations to various environments.

Danks, S. M., Evans, E. H., and Whittaker, P. A. (1983). *Photosynthetic Systems: Structure, Function, and Assembly.* John Wiley & Sons, New York.

A well-illustrated textbook on the broad topic of photosynthesis—its cellular aspects, the primary and secondary light reactions, and related metabolic pathways. A final chapter on the origin and assembly of chloroplasts is an engaging essay that raises evolutionary and philosophical issues.

Edwards, G., and Walker, D. (1983). *C3, C4: Mechanisms, and Cellular and Environmental Regulation, of Photosynthesis.* Blackwell/University of California Press, Berkeley.

Few textbook writers have the courage (or skill) to inject humor into their discussions—but these authors do. An informative and stimulating monograph.

Foyer, C. H. (1984). *Photosynthesis.* John Wiley & Sons, New York.

A thorough review.

Glazer, A. N. (1989). Light guides. Directional energy transfer in a photosynthetic antenna. *The Journal of Biological Chemistry 264*, 1–4.

A "minireview" illustrating the power of molecular biology in analyzing the details of photosynthetic mechanisms.

Govindjee, and Coleman, W. J. (1990). How Plants Make Oxygen. *Scientific American 262* (February), 50–58.

Stimulating and well-illustrated.

Gust, D., and Moore, T. A. (1989). Mimicking photosynthesis. *Science 244*, 35–41.

An interesting account of efforts to date to construct an artificial photosynthetic reaction center. Such studies increase our understanding of natural photosynthesis.

Halliwell, B. (1984). *Chloroplast Metabolism: The Structure and Function of Chloroplasts in Green Leaf Cells.* Clarendon (Oxford University Press), New York.

An excellent survey of mechanistic and technical aspects of chloroplast structure and function.

Knaff, D. B. (1988). The photosystem I reaction centre. *Trends in Biochemical Sciences 13*, 460–461.

A brief informative discussion.

Mansfield, T. A., and Davies, W. J. (1985). Mechanisms for leaf control of gas exchange. *BioScience 35*, 158–164.

A discussion of the interesting mechanisms which enable leaf stomata to optimize water loss with respect to carbon gain.

Mattoo, A. K., Marder, J. B., and Edelman, M. (1989). Dynamics of the Photosystem II reaction center. *Cell 56*, 241–246.

Latest word on a complex system.

Miller, K. R., and Lyon, M. K. (1985). Do we really know why chloroplast membranes stack? *Trends in Biochemical Sciences* *10,* 219–222.

A lucid discussion of a puzzling issue.

Oesterhelt, D. (1985). Light-driven proton pumping in halobacteria. *BioScience* 35, 18–21.

The simplest known proton pump is bacterio-rhodopsin, which forms patches in the plasma membrane of primitive halobacteria. The author discusses current views on how light activates this pump.

Sandved, K. B., and Prance, G. T. (1985). *Leaves.* Crown, New York.

Everything anyone needs to know about leaves, including information on leaf anatomy, physiology, and much else. Remarkable color photographs and fascinating lore about leafy plants.

Youvan, D. C., and Marrs, B. L. (1987). Molecular mechanisms of photosynthesis. *Scientific American 256,* June, 42–48.

What crystallography and molecular genetics reveal of the detailed events of photosynthesis.

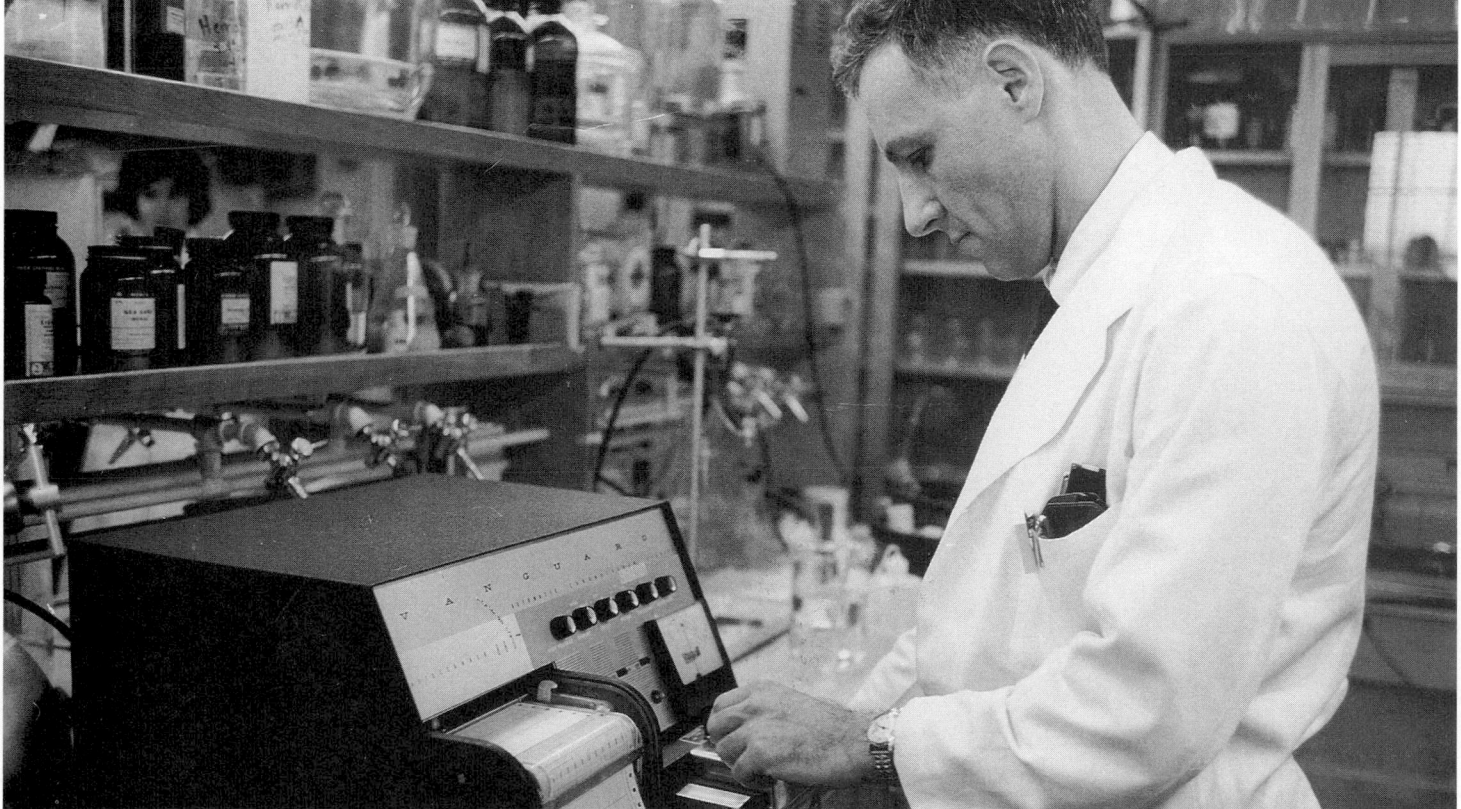

BIOSYNTHESIS OF CELL CONSTITUENTS

Biosynthesis of cell constituents
The labeling of biomolecules with radioisotopes greatly increased our knowledge of biosynthetic pathways. Here a biochemist works with two types of modern automated radiation detectors: a liquid scintillation counter (above); and a scanner (below) that detects radioactivity on paper chromatograms.

Three undertakings are common to all cells.

- Extraction of energy from nutrient molecules and synthesis of ATP and other compounds containing high-energy bonds
- Production of molecular building blocks
- Utilization of energy-rich bonds and molecular building blocks in the biosynthesis of cell constituents

In this chapter we view a panorama of energy-requiring biosynthetic processes in which cells utilize simple molecular precursors in fabricating the carbohydrates, lipids, and amino acids needed in the tasks of construction. All other cell functions—reproduction, repair, transport, and the rest—depend upon these processes.

Discussions of these matters fill long chapters in standard biochemistry textbooks. Our discussion is limited to selected examples chosen to illustrate basic principles.

HOW CELLS USE ENERGY IN BIOSYNTHESIS

The metabolic pathway leading to the biosynthesis of a particular class of biomolecules is almost never an exact reversal of the pathway followed in the metabolic breakdown of those same molecules. The two pathways may contain one or more identical steps that can proceed in either direction. But at least one step is usually dissimilar in the anabolic and catabolic pathways leading to and from the biomolecules in question. For example, a sequence of 12 well-known enzymatic reactions comprise the glycolytic pathway in which glycogen is degraded to lactic acid. However economical or logical it may seem for the synthesis of glycogen from lactic acid to proceed by a simple reversal of the 12 enzymatic steps, it appears that only 9 of the 12 enzymes are shared by the degradative and synthetic processes. In the synthetic pathway, the other 3 enzymatic reactions are replaced by 3 entirely different enzymatic reactions that participate only in the direction of synthesis.

The principle is illustrated in Figure 11-1. Note that the degradative (catabolic) pathways and synthetic (anabolic) pathways between proteins and amino acids, or between acetyl CoA and fatty acids, are dissimilar. This may seem a wasteful arrangement. In fact, it is essential, since the energetics of the catabolic pathway is impossible for anabolism. Energetically, catabolism is a "downhill" process, and synthesis proceeds "uphill." In many cases, the enzymes of corresponding catabolic and anabolic pathways differ in their intracellular locations and thus they are subject to differing regulatory influences. These factors allow them to operate independently and simultaneously.

This asymmetry has profound biological significance. For one thing, if the pathways of catabolism and anabolism were catalyzed by the same enzyme catalyzing the same reactions in opposite directions, no stable biological structure could exist. In freely reversible reactions, the equilibrium between precursor molecules (that is, molecules that are converted to other molecules) and product molecules would fluctuate excessively with every shift in the concentration of precursor molecules.

We saw in Chapter 9 that ATP is the major intermediary between those cellular processes that yield energy and those processes that consume energy. We know from thermodynamics that all chemical reactions are subject to a significant limitation: The total bond energies of the reaction products must always be less than the bond energies of the initial reactants. But, as we have seen, bond energies evade any limitation by coupling energy-requiring reactions or pathways with energy-releasing reactions—such as the release of the energy within the terminal energy-rich bonds of ATP.

In the language of thermodynamics, we would say that a biosynthetic reaction coupled to such an energy source always has a negative standard free energy change that keeps things moving in the direction of biosynthesis.

To this point we have considered four major mechanisms for the generation of ATP. They are as follows:

- Glycolysis
- Citric acid cycle
- Oxidative phosphorylation
- Photosynthetic light reactions

We will soon encounter a fifth mechanism of ATP generation:

- Fatty acid breakdown

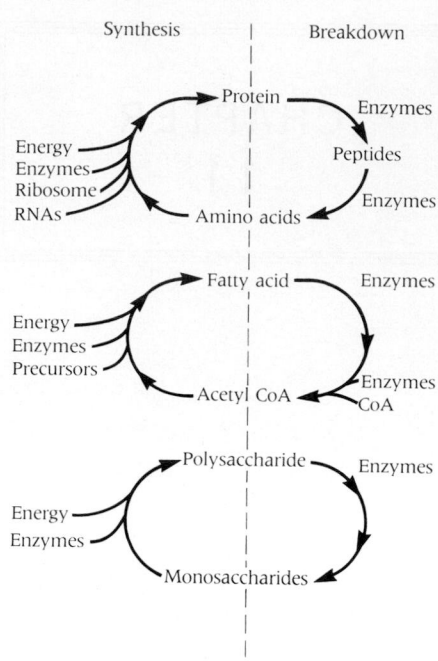

11-1 Synthesis and breakdown of cell molecules
The pathways generally followed in breakdown and synthesis of proteins, fatty acids, and polysaccharides are not the same. Note that energy is required to complete synthetic reactions.

MAJOR BIOSYNTHETIC PATHWAYS

We will now discuss the mechanisms by which cells synthesize key biomolecules: (1) ribose, (2) polysaccharides, (3) lipids, and (4) amino acids. The important topics of nucleotide and nucleic acid (DNA and RNA) biosynthesis and protein biosynthesis—both major utilizers of ATP—are deferred until Chapter 17 because they are more logically connected with the story of genes and their function.

A fundamental point can be made at the start: Most biosynthetic pathways begin with products of either the pentose phosphate shunt or the citric acid cycle.

CARBOHYDRATES
The Pentose Phosphate Shunt and Ribose Synthesis

The pentose phosphate shunt plays a special role in the metabolism of ATP and NAD^+ (and $NADP^+$). Both of these coenzymes are **ribonucleotides.** That means that they contain **ribose.** Thus their biosynthesis depends on the adequate functioning of this pathway, which is the major cellular source of ribose.

Four purposes of glucose metabolism are noteworthy in the present context.

- Production of the energy-rich bonds of ATP
- Provision of monosaccharide precursors for the production of such important polysaccharides as glycogen, starch, and cellulose
- Generation of reducing power in the form of NADH and NADPH
- Production of ribose, a 5-carbon sugar that is a constituent of RNA and its ribonucleotide precursors, and indeed of all ribonucleotides, including ATP. Ribonucleotides are, incidentally, the immediate precursors of deoxyribonucleotides, which are the precursors of DNA in which the sugar component is deoxyribose (see Chapter 4).

We learned in Chapter 9 how glucose is metabolized via the major pathways of glycolysis, respiration, and oxidative phosphorylation. We now consider an alternative pathway present in many cells for the oxidation of sugars. This is the sequence of reactions that carries out the last two of the four functions just listed. It is called the **pentose phosphate shunt** pathway. It is also known as the **hexose monophosphate** or **oxidative shunt** pathway. Here it will simply be called the *shunt.* In studying it, we will need once again to keep track of the number of carbons in key intermediates. This pathway is also critically involved in the fixation of CO_2 in photosynthesis (see Figure 10-21). In that situation, the reactions proceed in the reverse direction. (Remember that the incorporation of CO_2 into carbohydrates is a reductive process. Its elimination is oxidative.)

The shunt takes place in the cytoplasm. It begins with the phosphorylated derivative of the 6-carbon sugar, glucose, called glucose 6-phosphate (Figure 11-2). The conversion of glucose 6-phosphate to ribose includes several steps: in one, a single carbon of the glucose molecule is oxidized to CO_2, while in two steps $NADP^+$ is reduced to NADPH. With certain exceptions, only a small portion of the glucose that is catabolized travels through the shunt pathway. In normal cells, the bulk of it traverses the major glycolytic pathway from glucose to pyruvate. That sequence is usually called the **Embden-Meyerhof pathway,** after two of its discoverers, Gustav Embden and Otto Meyerhof.

Examine each step of the shunt pathway in Figure 11-2. An early intermediary product is ribose 5-phosphate. The ribose 5-phosphate continuously produced in

11-2 Fate of glucose in the shunt pathway

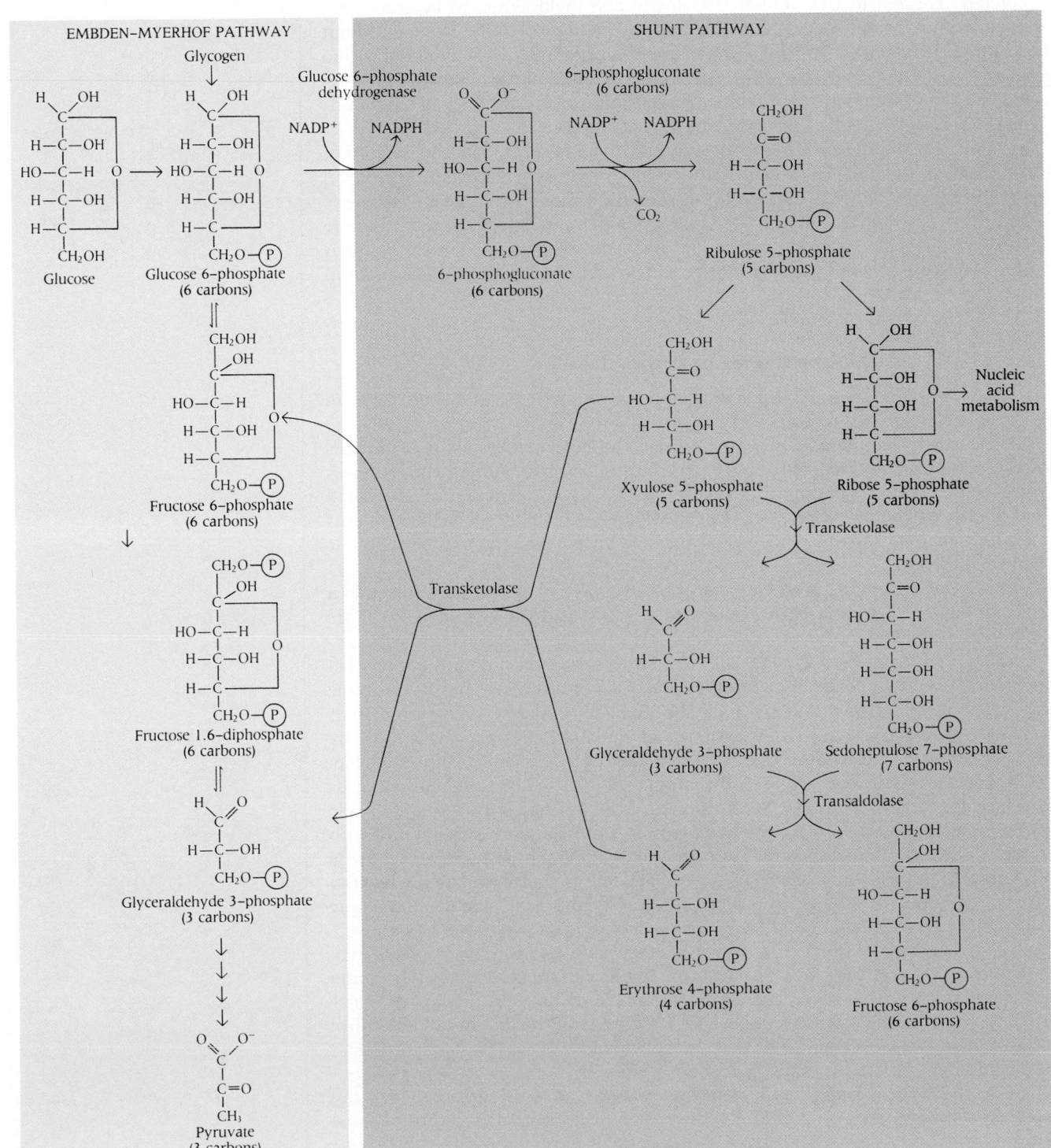

the shunt enters into further reactions that lead to the biosynthesis of ribonucleotides (including ATP and $NADH^+$). As noted above, these in turn lead to the production of deoxyribonucleotides.

The remainder of the shunt pathway includes a remarkable series of reactions in which the carbons of various 3-, 4-, 5-, 6-, and 7-carbon sugars are extensively scrambled. Since some of the final products of these interconversions are identical to various intermediates of the main glycolytic pathway, these are free to join the mainstream of traffic on its way to pyruvate. In this way, excess intermediates in the shunt can spill over and return to the main pathway. To an extent, the opposite is also true. The early steps of the Embden-Meyerhof pathway are notable for their reversibility (see arrows in Table 9-2). That means that some intermediates of glycolysis (including any that may have "spilled over" from the shunt) can be reconverted to glucose 6-phosphate, which can then enter or reenter the shunt and undergo further oxidative degradation.

Here, then, is another cycle. This one is capable of permitting glucose to be completely oxidized to CO_2 and H_2O—slowly and inefficiently, to be sure—but independently of the citric acid cycle. Although the volume of traffic passing through the shunt is light relative to traffic down the main pathway, it is essential to life because it is the indispensable source of ribose. In addition, this pathway serves a useful incidental function. It provides the cell with a serviceable—albeit round-about—"back-up" pathway for oxidizing glucose carbons to CO_2.

In sum, the shunt is a system for synthesizing ribose (a critically important 5-carbon sugar) and for generating NADPH, which serves as a hydrogen and electron donor in reductive biosyntheses. Some tissues have more active shunt pathways than others. Fat (adipose) tissue has a very high shunt activity; skeletal muscle has a very low activity. This is compatible with the fact that fat (but not muscle) requires large amounts of NADPH in the reductive biosynthesis of fatty acids from acetyl CoA.

Biosynthesis of Polysaccharides

The rather complicated topic of polysaccharide synthesis deserves our attention in this brief survey for two reasons.

First, from the standpoint of magnitude alone, the biosyntheses of polysaccharides (cellulose, starch, and glycogen) are by far the most substantial biosynthetic processes in the entire world, including as they do the woody substance of all trees and other plants. When this mass of metabolism is combined with the global biosynthesis of monosaccharides (mainly glucose), the two add up to a mountainous biosynthetic product.

Second, polysaccharide synthesis is a prime example of the principle illustrated earlier in Figure 11-1: The pathways of synthesis and pathways of breakdown are unambiguously different.

Consider glycogen. Even though cells are able to store large amounts of potential energy in energy-rich bonds, they must have access to a reserve supply of oxidizable nutrients that can be drawn upon rapidly during periods of need, such as in periods of hunger, starvation, or intense physiological activity. Lipids provide a crucially important reserve, but as an energy reservoir, glycogen is more readily available.

Glycogen forms in every cell of the animal body, but liver and muscle cells form it in the largest quantities. Structurally, it is a branched polysaccharide made from glucose monosaccharide units (see Figure 4-8). An insoluble compound with a molecular weight in the millions, glycogen serves admirably as a source of glucose molecules. They are linked in glycogen by special bonds (called **α-glycosidic bonds**) between carbon 1 of one glucose and carbon 4 of the next glucose. Branching results when an occasional bond forms instead between carbon 1 and carbon 6.

Glycogen synthesis is not merely a reversal of the pathway of glycogen breakdown. The discovery of that fact came in part from studies of rare individuals who could synthesize glycogen but were genetically incapable of breaking it down (Figure 11-3). It was a dramatic discovery because it was one of the first instances of firm proof of the principle that anabolic and catabolic pathways are not simply reversals of one another.

The pathway of glycogen synthesis begins with glucose 6-phosphate, which can be called a "crossroads compound" because each of its molecules confronts three or more available pathways (Figure 11-4). In one of those pathways, the enzyme

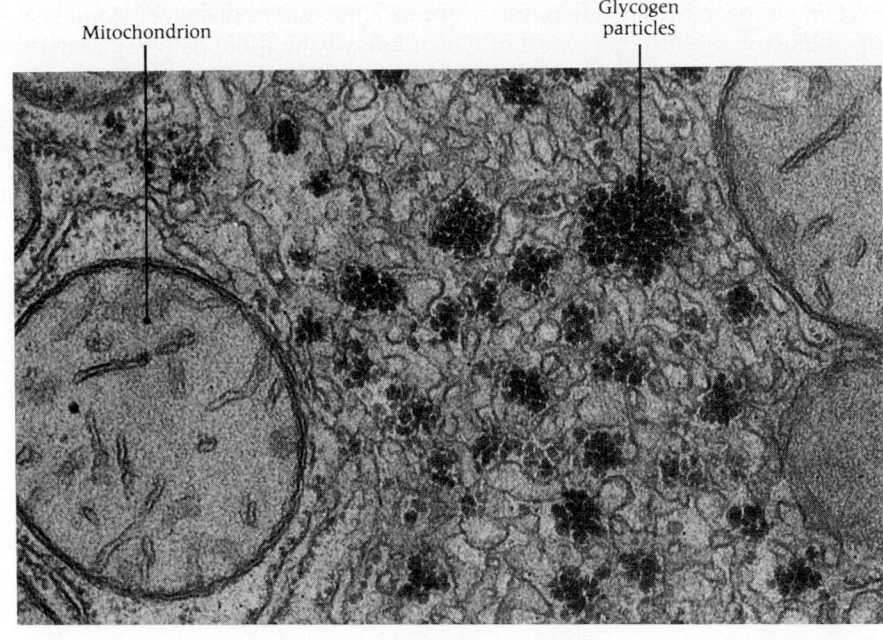

Mitochondrion   Glycogen particles

B

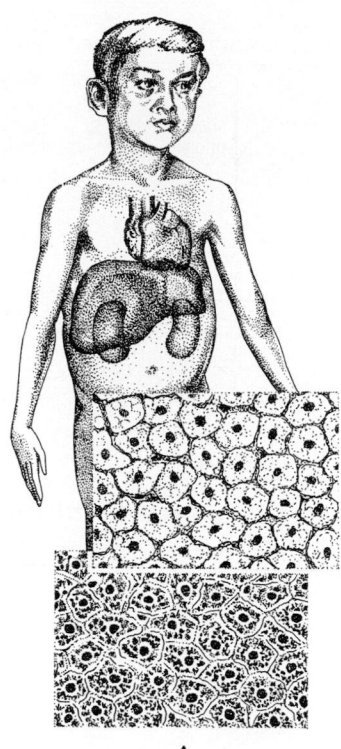

A

11-3 Glycogen storage disease
(A) *Excessive intracellular glycogen enlarges the liver, kidneys, heart, and other organs. Note that liver cells are packed with glycogen, as proved in the lower sections by the special stain, which colors glycogen red.* **(B)** *Electron micrograph showing the accumulation of glycogen within the lysosome of an individual unable to break down this polysaccharide.* (× 60,000)

phosphoglucomutase converts it to glucose 1-phosphate, which can then be polymerized to polyglucose, or glycogen.

However, the pathway of glycogen breakdown depends on **phosphorolysis.** Glycogen is split by phosphoric acid in the same way water brings about hydrolysis. Phosphorolysis is catalyzed by the enzyme **phosphorylase.**

$$(\text{glucose})_n + n\text{P}_i \xrightarrow{\text{Phosphorylase}} n \text{ (glucose 1-phosphate)} \qquad (11.1)$$

In this case n is a large number, $(\text{glucose})_n$ is glycogen, and P_i is inorganic phosphate. The reaction catalyzed by phosphoglucomutase is reversible; hence, this enzyme participates in both synthesis and breakdown.

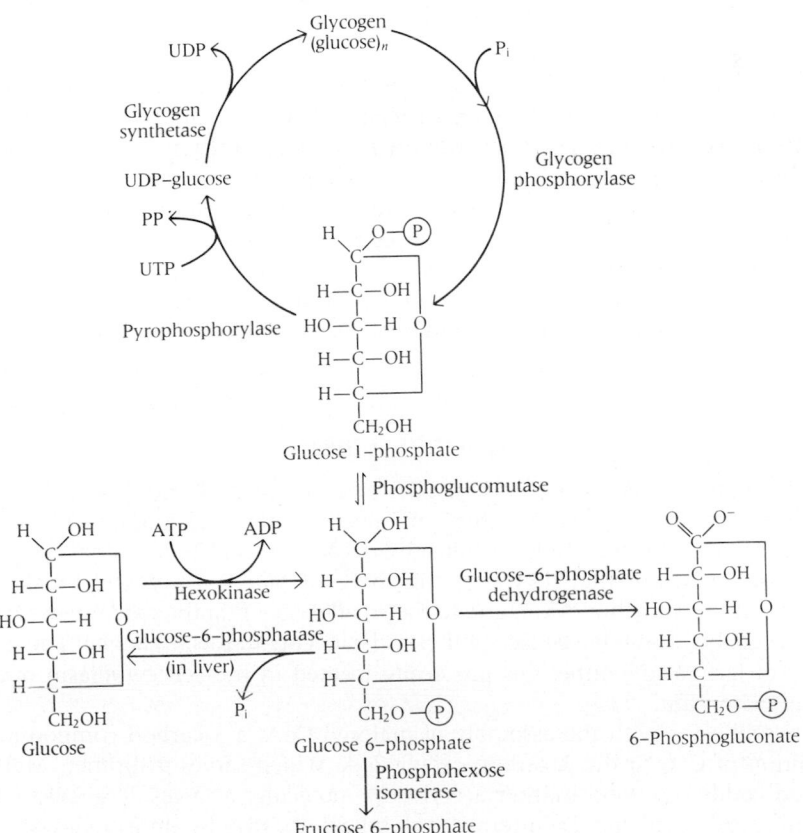

11-4 Summary of pathways of glycogen synthesis and breakdown and of the several possible fates of glucose

When glycogen breaks down, the major product (after phosphoglucomutase action) is glucose 6-phosphate, most of which follows the usual pathway to pyruvate and some of which enters the shunt (see Figure 11-4). Since glycogen is phosphorylated by P_i, only one molecule of ATP is needed to produce a molecule of fructose 1,6-diphosphate. In contrast, two molecules of ATP are needed to produce the same molecule from glucose. Therefore, when a cell relies on glycogen instead of glucose to provide its glucose 6-phosphate, there is a net gain of three molecules of ATP for each glucose unit of glycogen metabolized (see Table 9-4).

The pathways of starch synthesis and breakdown resemble the glycogen pathways (Figure 11-5). These pathways, of course, occur only in plant tissues.

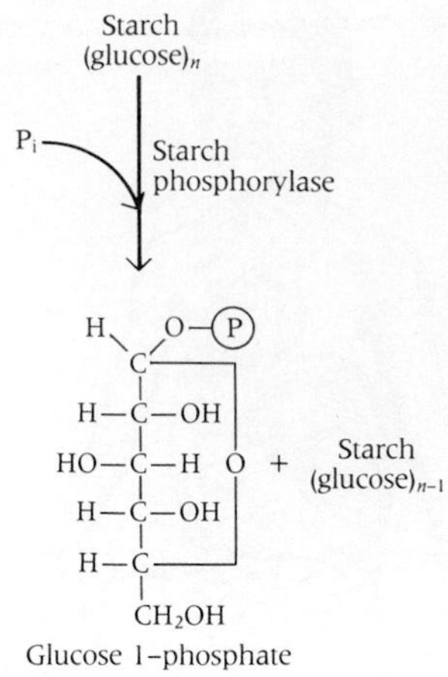

LIPIDS

Lipids provide still further evidence that critical biomolecules are built up and broken down by different mechanisms.

Fatty Acid Breakdown

Fatty acid breakdown occurs (after conversion of fatty acids to their CoA derivatives) by a stepwise removal of 2-carbon units in the form of acetyl CoA (Figure 11-6). The oxidation of fatty acids yields a rich harvest of energy. Consider the following:

■ Oxidation of palmitic acid, a 16-carbon fatty acid, yields a total of 131 ATP molecules! That is equivalent to 8.19 ATP molecules per original carbon.
■ Oxidation of glucose, a 6-carbon structure, yields a total of 36 ATP molecules. That is equivalent to 6 ATP molecules per original carbon.

Clearly, more energy is extracted per carbon atom from fatty acids than from carbohydrates.

The relationship of fatty acid breakdown to the interacting pathways of carbohydrate and protein catabolism is sketched in Figure 11-7. Note the central position of the citric acid cycle—and the integrated manner in which the separate pathways are tied together. We see that various amino acids can be converted to carbohydrates (this happens in starvation, when the body feeds on its proteins) and that carbohydrates can be converted to lipids (this happens in overnutrition, when the body converts excess nutrients to deposits of fat).

In many plants, a special metabolic system exists for the conversion of fats to carbohydrates during the germination of seeds. This occurs in cytoplasmic bodies called **glyoxysomes,** so named because their distinctive enzymatic pathway is the **glyoxylate cycle.** This cycle has some of the same steps as the citric acid cycle, but it differs in the fate of one 6-carbon intermediate compound, *isocitrate.* In the glyoxylate cycle isocitrate breaks down directly to yield succinate (4-carbon) and glyoxylate (2-carbon). In subsequent reactions, glyoxylate leads to the production of acetyl CoA, which can lead to the production of additional carbohydrates (see Figure 11-7). The prime function of this cycle is to promote the conversion of fat to sugar.

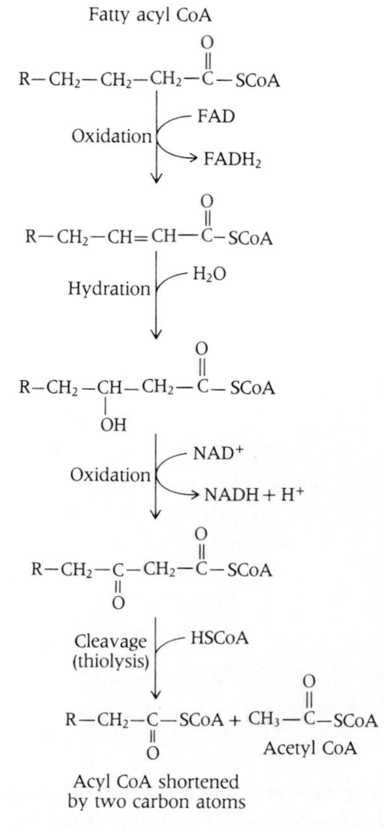

Fatty Acid Synthesis

Fatty acid synthesis begins with acetyl CoA (see Figures 9-8 and 11-7). For years it was believed that fatty acid synthesis involved nothing more than a reversal of the degradative pathway. However, later work showed that the two pathways are not only different, but that they occur in different places. Fatty acid breakdown takes place in the mitochondrion, and to the small extent that the oxidative enzymes can act reversibly, some fatty acid synthesis also occurs in this organelle. The major pathway of fatty acid synthesis is physically located in the cell cytoplasm outside of the mitochondria.

Synthesis begins with the assembly of malonyl CoA, a 3-carbon compound, by the addition of CO_2 to the 2-carbon acetyl CoA, which serves as primer. Malonyl CoA then condenses with another acetyl CoA molecule; as a result, a CO_2 is lost. The results, expressed simply in terms of numbers of carbons, are as follows:

11-5 *Simplified pathway of starch synthesis and breakdown*

11-6 *Pathway of fatty acid breakdown*

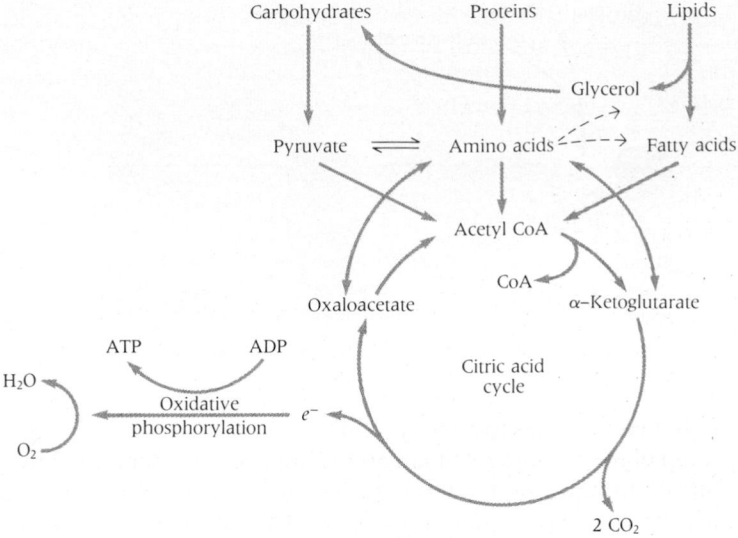

Primer: 2-carbon compd
                  (acetyl CoA)

First step: 2-carbon compd + 1-carbon compd = 3-carbon compd
               (acetyl CoA)   (CO_2)   (malonyl CoA)

Second step: 3-carbon compd + 2-carbon compd − 1-carbon compd
               (malonyl CoA)   (acetyl CoA)   (CO_2)

            = 4-carbon compd
              (butyryl CoA)   (11.3)

Butyryl CoA is a short-chain fatty acid derivative (of CoA). Thus, by this seemingly meandering pathway, the original 2-carbon primer is lengthened by two carbons. Butyryl CoA can now become the starting point of another 2-carbon addition involving a similar reaction with malonyl CoA. Subsequent 2-carbon additions by the same mechanism can bring the carbon chain length to 16 or 18 (see Table 4-2). In the following equation, 1 acetyl CoA is converted to 1 molecule of palmitic acid with the help of 7 malonyl CoA and 14 NADPH:

$$\text{Acetyl CoA} + 7\,\text{malonyl CoA} + 14\,\text{NADPH} + 7\,\text{H}^+ \rightarrow$$
$$\text{CH}_3(\text{CH}_2)_{14}\text{COOH} + 7\,\text{CO}_2 + 8\,\text{CoA} + 14\,\text{NADP}^+ + 6\,\text{H}_2\text{O} \quad (11.4)$$
(palmitic acid)

The intermediates in fatty acid synthesis in *E. coli* are linked to an **acyl carrier protein,** or **ACP,** which functions as though it were a larger CoA molecule. In most of the reactions in Equations 11.3 and 11.4, the CoA in a compound is replaced by ACP to make malonyl-ACP. For simplicity, we will assume that the intermediates remain as CoA derivatives. To a large extent, fatty acid synthesis takes place in the soluble portion of the cytoplasm, where participating enzymes have ready access to the substantial quantities of NADPH generated in the pentose phosphate shunt.

We have already noted the obvious advantages in having separate synthetic and degradative processes. However, in the case of fatty acid synthesis, the cell must pay a price for synthesizing fatty acids via a mechanism that repeatedly seems to detour through a step involving the 3-carbon compound malonyl CoA. It must utilize one extra ATP for each 2-carbon unit added to the chain. Thus, in the synthesis of one palmitic acid molecule, 7 ATP must be invested. The readiness with which we accumulate fat suggests that it is a price the body is willing to pay.

Synthesis of Membrane Lipids

We have seen that membranes consist chiefly of phospholipid bilayers in which various proteins are embedded (see Chapter 6). The chemical structures of phospholipids were discussed and portrayed in Chapter 4. We must now inquire how and

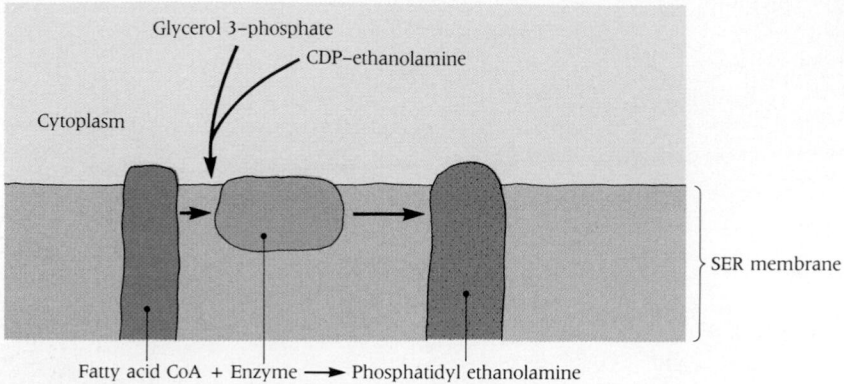

11-8 Biosynthesis of
phosphatidylethanolamine
Enzymes located in the membrane of the smooth
endoplasmic reticulum are responsible for the
synthesis of membrane phospholipids.

where phospholipids are put together by cells.

Investigation of these issues met many obstacles over the years. The low solubility of phospholipids in aqueous (watery) media, their tendency to form sheets and vesicles, their peculiar mix of highly hydrophobic and hydrophilic substituents— all these factors frustrated and puzzled investigators. The answer came with the recognition that membranes grow by expansion of existing membranes.

The biosynthetic pathway of one phospholipid in animal cells is summarized in Figure 11-8. Note that it takes place on the smooth endoplasmic reticulum (SER) membrane. (In bacteria, which have no endoplasmic reticulum, phospholipid synthesis takes place on the plasma membrane.) The initial substrate, a fatty acid CoA molecule, protrudes from the membrane into the cytoplasm. Then, as shown, it reacts with glycerol 3-phosphate and a nucleotide derivative of ethanolamine, CDP-ethanolamine, both soluble cytoplasmic molecules in whose synthesis ATP participated. A reaction sequence catalyzed by enzymes that are themselves anchored in SER membrane but protrude into the cytoplasm couples these two molecules to form the phospholipid phosphatidylethanolamine.

Probably all phospholipids are synthesized on SER membranes in eukaryotic cells. They then move to other membranes, transported it appears by special phospholipid exchange proteins.

THE SPECIAL PROBLEM OF NITROGEN

Amino acids, we will recall, differ from sugars and lipids in that they contain amino groups—nitrogenous substituents symbolized as —NH_2. Since proteins are chains of amino acids, proteins also contain nitrogen. Indeed, the average protein is 6.3% nitrogen by weight. The air we breathe is 80% gaseous nitrogen but inhaled nitrogen is biochemically inert. It can be transformed into amino acid nitrogen only by a few highly specialized cells, the nitrogen-fixing bacteria (see Chapter 46).

Amino Acid Breakdown

Like sugars and fatty acids, amino acids can be broken down to provide energy (in the form of ATP) and building blocks (in the form of acetyl CoA and other components of the Krebs cycle). As shown in Figure 11-7, the metabolic pathways of the three small biomolecules—sugars, fatty acids, and amino acids—are closely interrelated. Most (but not all) amino acids are capable of being metabolized to acetyl CoA. They may also be converted to pyruvate, which can then enter a pathway that ultimately generates glucose. Significantly, that pathway is not a mere reversal of glycolysis (which has three irreversible steps). The term **gluconeogenesis** denotes the production of glucose from noncarbohydrate precursors like amino acids or glycerol. Under ordinary nonemergency conditions, amino acid breakdown is a minor source of ATP.

The amino groups (—NH_2) liberated during amino acid breakdown are converted to *ammonium ions* (NH_4^+), some of which are reutilized in the biosynthesis of new amino acids. In most terrestrial vertebrates, the excess NH_4^+ is converted to *urea*, a small nitrogen-containing molecule, and then excreted in the urine. In birds and terrestrial reptiles, NH_4^+ is converted to *uric acid*, another nitrogen-containing compound, for excretion. In many aquatic animals, NH_4^+ is excreted itself. The three classes of organisms are called **ureotelic, uricotelic,** and **ammonotelic.**

$$
\begin{array}{c}
\text{COOH} \\
| \\
\text{CO} \\
| \\
\text{CH}_2 \\
| \\
\text{CH}_2 \\
| \\
\text{COOH}
\end{array}
+ \text{NH}_4^+ + \text{NADPH} + \text{H}^+ \rightleftharpoons
\begin{array}{c}
\text{COOH} \\
| \\
\text{CHNH}_2 \\
| \\
\text{CH}_2 \\
| \\
\text{CH}_2 \\
| \\
\text{COOH}
\end{array}
+ \text{H}_2\text{O} + \text{NADP}^+
$$

α–Ketoglutaric acid   Glutamic acid

Amino Acid Synthesis

The main importance of amino acids is their role as constituents of proteins. How are amino acids synthesized?

Note that amino acid metabolism does not involve a separate set of reactions unrelated to the major pathways discussed above. As we just saw, the metabolism of amino acids is closely linked with that of carbohydrates and fatty acids. One important type of enzymatic reaction plays a central role in the metabolism of amino acids because it provides a link between amino acids and the citric acid cycle (Figure 11-9). This reaction begins with α-ketoglutaric acid, which is called an α-keto acid because there is a keto group ($=$O) on the α-carbon (also called the number 2 carbon). This compound is reductively aminated—that is, it acquires an amino group. The product is glutamic acid (an α-amino acid). The equilibrium constant of the reaction is such that the reaction can occur in either direction. When it goes to the left, ammonia (NH_3 or NH_4^+) is liberated.

Why is this reaction important? Because in most organisms it provides the only means by which inorganic nitrogen (such as NH_3 or NH_4^+) can be incorporated into a non-nitrogenous organic molecule to form an amino acid. The *de novo* synthesis of amino acids depends almost entirely on the synthesis of glutamic acid from α-ketoglutaric acid, since the amino group, once it is introduced into glutamic acid, is readily transferred to other α-keto acids to form amino acids. This is accomplished by **transamination** (Figure 11-10). Its essential feature is an exchange of an α-amino group for an α-keto group. The reaction is catalyzed by an enzyme called **transaminase.** One of the two reaction products is an α-amino acid corresponding to the α-keto acid that entered into the reaction. This means that the glutamic acid–α-ketoglutaric acid system is a little shuttle that functions almost like a catalyst (that is, it is active in small quantities and never used up) in the synthesis of other amino acids. As a result, it participates indirectly in fixing ammonia into the α-amino acid (Figure 11-11). By means of this and related reactions, organisms can convert any of the three main α-keto acids formed in glycolysis and the citric acid cycle into the corresponding amino acids, as follows:

- The first, pyruvate, can be converted to alanine.
- The second, oxaloacetate, can be converted to aspartate.
- The third, α-ketoglutarate, can be converted to glutamate.

None of this is meant to imply that all amino acids are synthesized from intermediates of carbohydrate and fatty acid metabolism. The term **essential amino acids** refers to those amino acids that must be included in an organism's diet because it cannot synthesize them. Other amino acids are nonessential in the diet because they can be synthesized by an organism.

THE ASSEMBLY OF CELL STRUCTURES

The foregoing discussions illustrate how the major biomolecules are assembled from simple chemical precursors under the energizing impetus of ATP. The cell, however,

$$
\begin{array}{c}
\text{COOH} \\
| \\
\text{CH}_2 \\
| \\
\text{C}=\text{O} \\
| \\
\text{COOH}
\end{array}
+
\begin{array}{c}
\text{COOH} \\
| \\
\text{CH}_2 \\
| \\
\text{CH}_2 \\
| \\
\text{CH}-\text{NH}_2 \\
| \\
\text{COOH}
\end{array}
\xrightarrow[\text{Transaminase}]{\substack{\text{Pyridoxal}\\\text{phosphate}}}
\begin{array}{c}
\text{COOH} \\
| \\
\text{CH}_2 \\
| \\
\text{CH}-\text{NH}_2 \\
| \\
\text{COOH}
\end{array}
+
\begin{array}{c}
\text{COOH} \\
| \\
\text{CH}_2 \\
| \\
\text{CH}_2 \\
| \\
\text{C}=\text{O} \\
| \\
\text{COOH}
\end{array}
$$

Oxaloacetate   Glutamate   Aspartate   α–Ketoglutarate

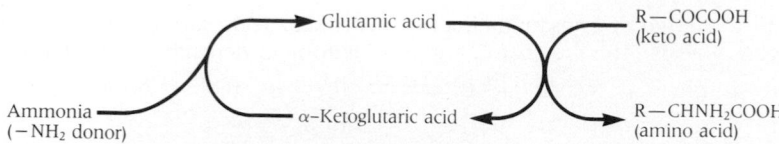

**11-11 The glutamic acid–
α-ketoglutaric acid shuttle**
*In many organisms, ammonia is reductively
converted to an amino group. By shuttling amino
groups between glutamic acid and α-ketoglutaric
acid, transaminases can generate other amino
acids. In terrestrial vertebrates, much NH_4^+ is
converted to urea, which is excreted.*

is not simply a little bag of randomly dispersed biomolecules. Nor is it a mere
container of enzymes. The biosynthesis of molecular components is but one stage
in the synthesis of an intact, structured, functioning cell.

There is a subsequent stage in the fabrication of cell components in which biomolecules are assembled into singular structures—membranes, microtubules, and ribosomes. As far as we know, these complex entities arise from individual molecules
of carbohydrate, lipid, and protein in ways that do not require the formation of
new covalent bonds between molecules, depending instead on such molecular interactions as hydrogen bonds, ionic bonds and nonpolar interactions.

Until recently, our knowledge of this stage—the final assembly of cell organelles—
was very sketchy. We knew only that mitochondria, chloroplasts, and other organelles replicate themselves in the course of cell division and that the identity, stability,
and function of each organelle are conferred by its component proteins and lipids—
and by the way they are put together. While the basis of these interactions is still
unclear, new work has shown that various "signals" tell a protein whether it belongs
in an organelle—that is, whether it is a resident or a visitor. These "signals" consist
of short amino acid sequences within certain proteins of the endoplasmic reticulum.

Such advances have suggested a **principle of self-assembly,** according to which
all the information needed to specify the three-dimensional structure of an organelle
resides within the amino acid sequences of its proteins. This principle, derived
from the classic work of Anfinsen on the renaturation of ribonuclease (see Box
8-3), holds that each linear chain of amino acids interacts with itself and with its
environment and then assumes a conformation of lowest free energy. The process
is spontaneous and requires no input of additional energy. The final conformation
displays functional and structural properties unique to that protein, or polypeptide,
including the ability to bind to other macromolecules in a specific way. As a result,
complex structures can and do self-assemble.

Current work shows that some proteins can self-assemble into functional structures, while others assemble only in the presence of additional proteins, which
are not components of the final structures. We shall say more about some of these
mechanisms after we have been introduced to the genetic role of DNA, which is
to direct the synthesis of protein molecules.

SUMMARY

The synthetic pathway of a class of biomolecules rarely proceeds
as an exact reversal of the pathway by which that class of molecules
is broken down catabolically. That fact allows cells to control synthesis and degradation separately.

In the biosynthesis of carbohydrates, lipids, and amino acids,
most biosynthetic pathways begin with compounds involved in
either the pentose phosphate shunt or the citric acid cycle, irrespective of the final product. Because the pentose phosphate shunt
produces ribose, it is important for the synthesis of ATP, NAD^+,
RNA, and DNA—all of which contain ribose or one of its derivatives.
The shunt begins with glucose 6-phosphate, from which ribose
5-phosphate is generated by the loss of CO_2. Other steps include
the synthesis of two NADPH molecules. Shunt metabolism can
result in the complete oxidation of glucose 6-phosphate to CO_2
and H_2O. Hence, in addition to its role in the synthesis of ribose,
the shunt provides a method of metabolizing glucose that is independent of the citric acid cycle.

Synthesis of the polysaccharide chains—glycogen, starch, and
cellulose—also begins with glucose 6-phosphate. In the synthesis
of glycogen, glucose 6-phosphate is converted to glucose 1-phos-
phate, which is then polymerized to glycogen. The depolymerization
of glycogen requires different enzymes.

Fatty acid breakdown, important steps of which occur in mitochondria, begins with the attachment of coenzyme A (CoA) to
the fatty acid. The CoA and two carbons are then removed as
acetyl CoA, leaving a fatty acid two carbons shorter—which can
be further degraded by repeating these steps. The acetyl CoA molecules produced can then enter the citric acid cycle. Fatty acid synthesis occurs in the cytoplasm and begins with the addition of CO_2
to acetyl CoA, forming a three-carbon compound called malonyl
CoA. Malonyl CoA (i.e., its ACP derivative) condenses with another
acetyl CoA, losing CO_2 in the process, to form a four-carbon compound. That compound can then react with another malonyl CoA
and lose another CO_2 to form a six-carbon compound, and malonyl
CoAs can continue to be added until the appropriate chain length
has been reached. The process requires NADPH as well as ATP.
The fatty acid CoA produced is used as a precursor for membrane
phospholipids. Their prototype, phosphatidylethanolamine, is synthesized by the reaction of a fatty acid in the SER membrane with
glycerol 3-phosphate and CDP-ethanolamine.

Amino acid breakdown and synthesis involves a distinctive feature owing to the fact that amino acids contain nitrogen. Most
amino acids can be broken down to acetyl CoA or pyruvate. The
amino groups (NH_2) liberated during this process are converted

to ammonia (NH_4^+), which may be reutilized in the synthesis of new amino acids, excreted directly, or converted to urea or uric acid and then excreted. In the synthesis of amino acids, α-ketoglutaric acid is reductively aminated by ammonia to form the amino acid glutamic acid. The amino group of glutamic acid is then transferred by transamination to other α-keto acids, forming new amino acids. The variety of amino acids formed depends on the nature of the α-keto acid that is transaminated. Most animals cannot synthesize every amino acid. Hence, they must obtain from dietary sources the ones they cannot synthesize themselves.

KEY TERMS

α-glycosidic bond
Embden-Myerhof pathway
essential amino acid
gluconeogenesis
glyoxylate cycle

glyoxysome
hexose monophosphate shunt
oxidative shunt
pentose phosphate shunt
phosphoglucomutase

phosphorolysis
principle of self-assembly
ribonucleotide
ribose

QUIZ QUESTIONS

1. Which of the following is *not* a pathway that relates to generation of ATP?
   A. pentose phosphate shunt
   B. photosynthesis
   C. oxidative phosphorylation
   D. Embden-Meyerhoff pathway
   E. amino acid synthesis

2. In which of the following processes is acetyl CoA *not* involved?
   A. transamination
   B. fatty acid synthesis
   C. Embden-Meyerhoff pathway
   D. fatty acid breakdown
   E. amino acid breakdown

3. An early step in the synthesis of amino acids involves amination of α-ketoglutarate. What is the form of the nitrogen in this reaction?
   A. uric acid
   B. ammonia
   C. molecular nitrogen
   D. NADPH
   E. transaminase

4. Synthesis of membrane lipids occurs
   A. inside of mitochondria.
   B. inside the nucleus.
   C. in association with endoplasmic reticulum.
   D. inside of chloroplasts.
   E. inside of Golgi bodies.

5. Which of the following pathways would increase its activity level during starvation?
   A. fatty acid synthesis
   B. amino acid synthesis
   C. fatty acid degradation
   D. amino acid degradation
   E. C and D

ESSAY QUESTIONS

1. Why do cells go to the great expense of using different pathways for the degradation (catabolism) and synthesis (anabolism) of biomolecules?

2. The metabolism of which of these molecules gives the highest yield of ATP molecules per carbon atom: glycogen, glucose, fatty acid? Why?

3. Where in the cell do the following processes occur: glycolysis, oxidative phosphorylation, hexose monophosphate shunt, citric acid cycle, fatty acid catabolism, fatty acid synthesis, synthesis of membrane lipids, light reactions of photosynthesis?

4. How do cells convert sugars to amino acids? Sugars to fatty acids?

REFERENCES AND SUGGESTIONS FOR FURTHER READING

Bender, D. A. (1985). *Amino Acid Metabolism*, 2nd ed. John Wiley & Sons, New York.

An excellent summary of current knowledge.

Bishop, W. R., and Bell, R. M. (1988). Assembly of phospholipids into cellular membranes: biosynthesis, transmembrane movement and intracellular translocation. *Annual Review of Cell Biology 4*, 579–610.

Bloch, K. (1965). The biological synthesis of cholesterol. *Science 150*, 19–28.

A Nobel Prize lecture that is wonderfully lucid.

Hadley, N. (1985). *The Adaptive Role of Lipids in Biological Systems*. John Wiley & Sons, New York.

A discussion of the great diversity of lipids and their importance in organismic function.

Stanbury, J. B., Wyngaarden, J. B., Frederickson, D. S., Goldstein, J. L., and Brown, M. S. (Eds.) (1983). *The Metabolic Basis of Inherited Disease*, 5th ed. McGraw-Hill, New York.

The editors of this distinguished work include two directors of the National Institutes of Health, two Nobel Prize winners, and five major contributors to our knowledge of genetically determined human diseases. Many chapters include concise reviews of biosynthetic pathways.

SECTION 4

HEREDITY AND GENETICS

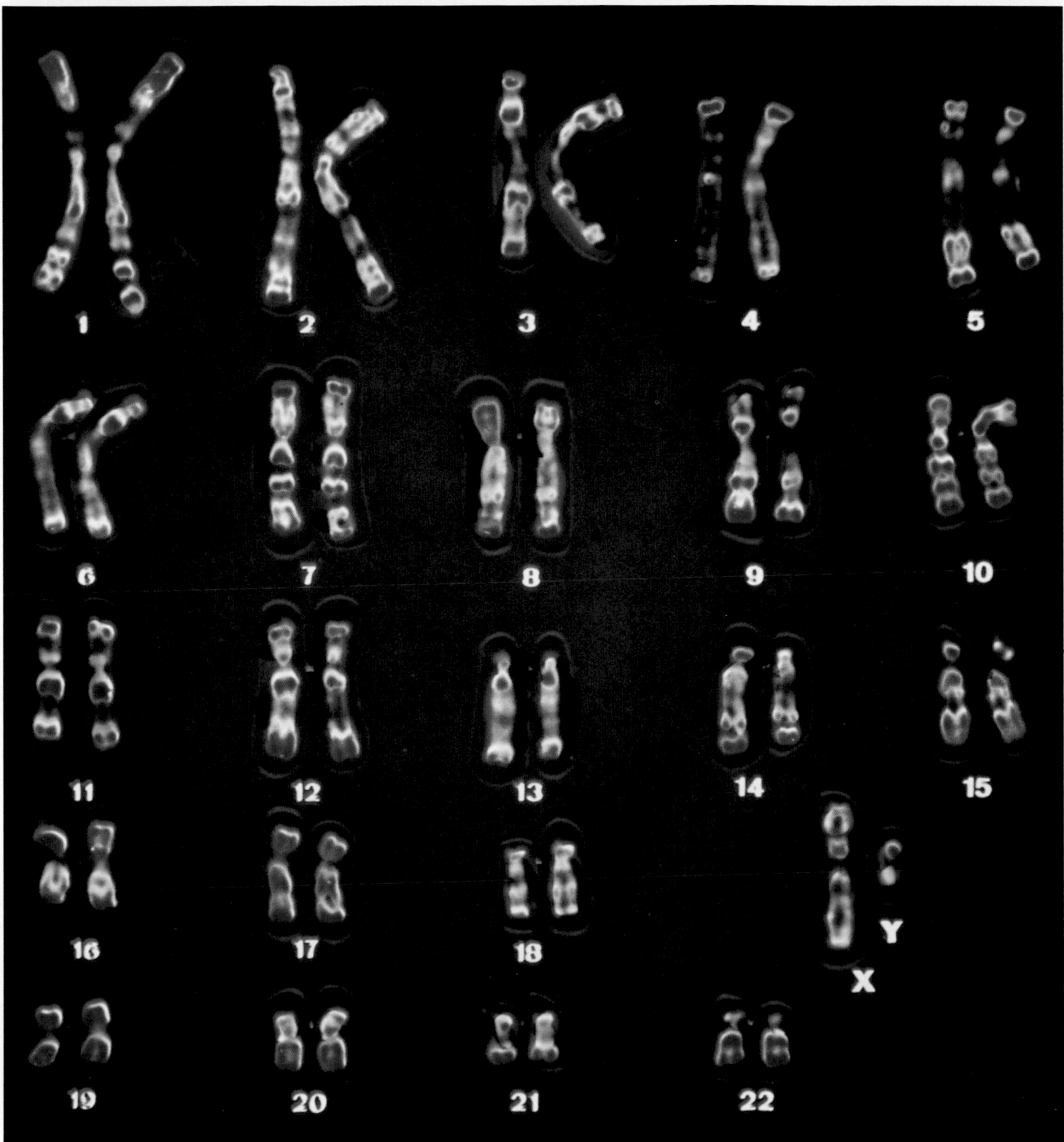

HOW CELLS REPRODUCE

How cells reproduce
*Mitosis, a remarkable mechanism of eukaryotic
cell division, has been described as the "ballet
of the chromosomes." Newer techniques make
it possible to examine the metaphase chromosomes
of an individual cell, classify them, and count
them. In this preparation, human chromosomes
treated with various dyes reveal distinctive
banding patterns that are useful in chromosome
identification.*

Reproduction, surely the most distinctive feature of living organisms, is an obvious necessity if a species is to survive, because individuals are not immortal. Reproduction occurs on two levels:

- Reproduction of individual cells, whether they be individual unicellular organisms or components of multicellular organisms
- Reproduction of multicellular organisms

The latter, incidentally, involves the reproduction of many individual cells. So does the growth and self-repair of multicellular organisms.

The ability to grow and reproduce is a basic attribute of cells. As we will later learn, the human body begins its existence as a single cell—a zygote—but it grows and develops through the processes of cellular reproduction and differentiation into an adult body consisting of more than 100 trillion (10^{14}) cells. To maintain the structure and function of its various tissues, the adult human body undertakes about a trillion (10^{12}) cell divisions every 24 hours—or 10^7 cell divisions per second. In the bone marrow alone, the cells that are precursors of red blood cells turn out 2.3 million (2.3×10^6) new red cells every second. Other tissues that turn over rapidly—white blood cells, epithelium lining the intestine, and skin cells—have similarly rapid rates of cell division. These considerations do not take into account abnormal cell divisions, such as occur in malignant tumors. One wonders how many cells there might be in a giant redwood, a whale, or a cricket.

This chapter concerns cellular reproduction. Organismic reproduction is discussed in Chapter 21.

Prokaryotic cells reproduce by dividing equally by **fission** (Figure 12-1A, B), or unequally by **budding** (Figure 12-2). Eukaryotic cells, with few exceptions, reproduce by **mitosis** (from *mitos,* meaning "thread"), a process that includes nuclear replication, or **karyokinesis** (Figure 12-3). Cytoplasmic replication, or **cytokinesis,** occurs in both prokaryotic and eukaryotic cells.[1]

Whole multicellular organisms duplicate themselves by one of the several modes of sexual or asexual reproduction. Although we will explore these later, let us briefly note how reproduction takes place in the four eukaryotic kingdoms. Despite the complexity of these life cycles, the replication of individual cells is essential to the reproduction of whole organisms.

- In *protists,* asexual reproduction is the usual mode, but sexual reproduction does occur.
- In *fungi,* reproduction is both sexual and asexual. Nearly all fungi can reproduce asexually by releasing spores provided by a single parent. Many are capable of

---

[1] Strictly speaking, *mitosis* denotes only nuclear division, though the term is often applied to cell division. *Karyokinesis,* which also signifies nuclear division, is a useful term because it avoids this ambiguity. We distinguish *karyokinesis* from *cytokinesis*. Cytokinesis usually follows karyokinesis promptly, but there are circumstances in which one occurs without the other.

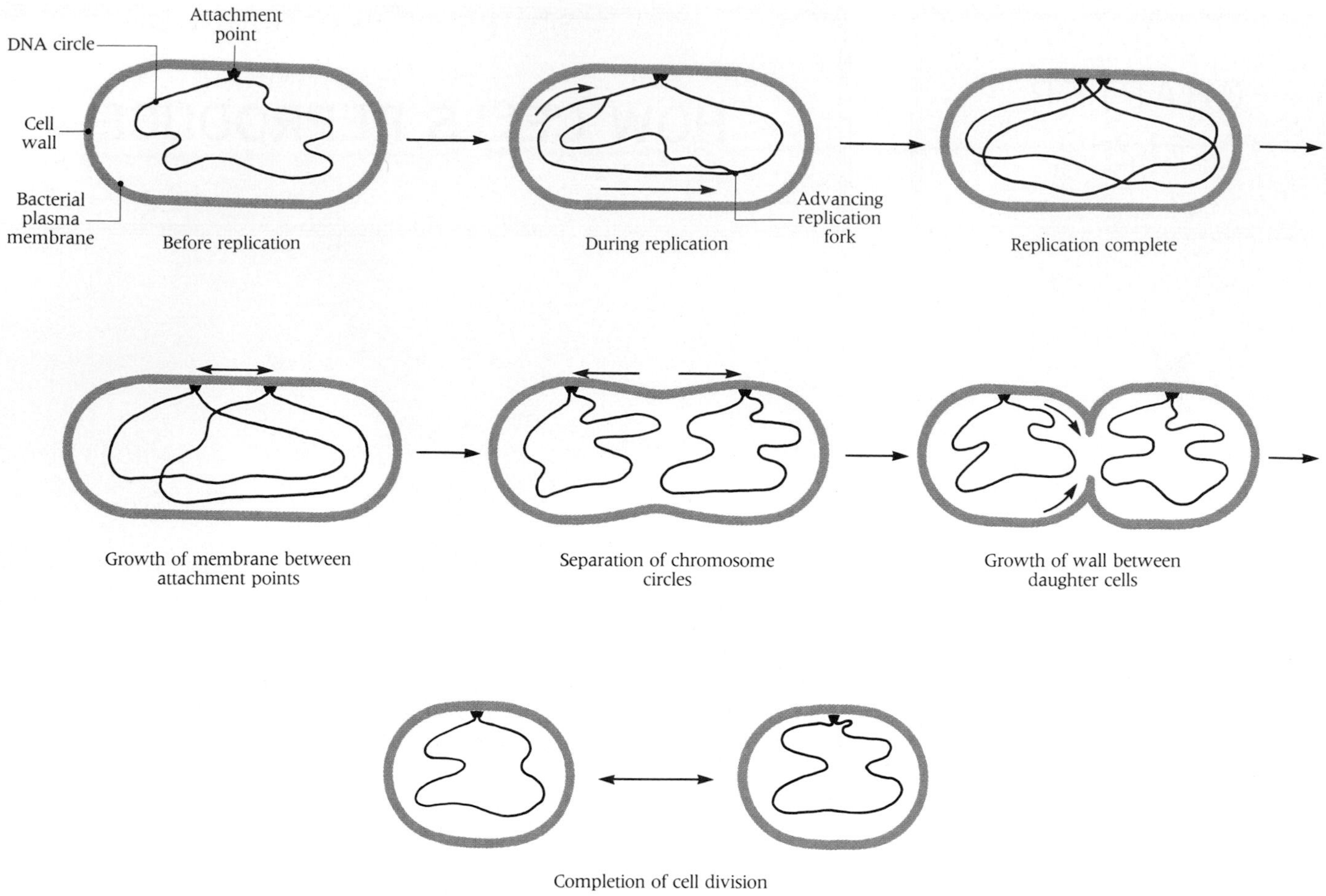

DNA circle — Attachment point

Cell wall

Bacterial plasma membrane

Before replication

During replication — Advancing replication fork

Replication complete

Growth of membrane between attachment points

Separation of chromosome circles

Growth of wall between daughter cells

Completion of cell division

A

sexual reproduction, but it differs from that of animals and plants, involving conjugation in which cells of different mating types come together but do not fuse. The resulting cells are haploid. Indeed, all fungal cells are haploid except the zygote, which is the only diploid stage in the life cycle.

- In *plants,* sexual reproduction, as in most eukaryotes, is also the rule with few exceptions. The life cycles of plants are highly distinctive and characteristic of this kingdom. Unlike animals, most of whose cells are diploid, and fungi, which are mostly haploid or dikaryotic, plants alternate haploid and diploid generations in an orderly sequence.

- In *animals,* sexual reproduction is almost the rule, though there are exceptions. Ordinarily, male and female **germ cells,** or **gametes,** combine, a male **sperm** cell entering a female **egg** cell (Figure 12-4). Fusion of the sperm nucleus with the egg nucleus produces a fertilized egg cell, or **zygote,** which contains nuclear material from both parents.

A zygote is a single cell, to be sure, but it is destined to undergo **development,** as it becomes a multicellular organism. For this to occur, two things must take place.

- First, a long series of replications by individual cells occurs.
- Second, a succession of transformations occurs in which daughter cells increasingly come to differ from their parent cells. This phenomenon, called **differentiation,** accounts for the fact that the fully developed animal has nerve cells, muscle cells, blood cells, epithelial cells, and the rest—even though all of them descended from the same zygote. Similarly, differentiation accounts for the transformation of a plant embryo into flowers, roots, stems, and other specialized structures.

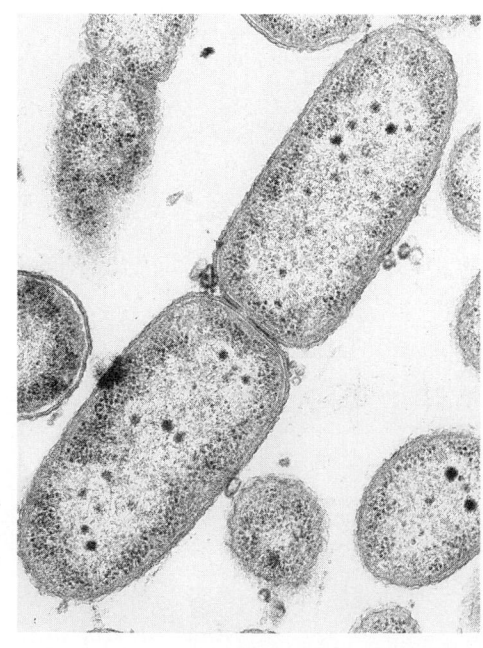

B

12-1 Division of bacterial cells by fission (A) DNA is attached to the plasma membrane of the bacterial cell. After the cell divides and its DNA is duplicated, each daughter cell contains the normal amount of DNA. (B) Electron micrograph showing fission in Pseudomonas aeruginosa. *The light areas in these cells contain DNA.* (× 50,000)

THE MEANING OF REPRODUCTION

Indeed, the capacity for reproduction is the feature that comes closest to giving us a definition of life. The endurance of life is not due to the capacity of individual organisms to repair and adjust themselves, although this does take place. It is a consequence of the capacity of populations of organisms to replace their worn-out members with vigorous new members—their offspring. Offspring, in turn, can repair and adjust for only so long before they must face their parents' fate. What survives over the millennia is the species. Thus reproduction is the bridge that spans successive mortal generations.

Two fundamental principles lie at the heart of reproduction:

- The first is the principle of **biogenesis:** Living things arise only from living things.
- The second is the principle of **heredity:** Like begets like. Human parents have human babies. Flies reproduce flies. Heredity requires the precise transmission of genetic information from parents to offspring.

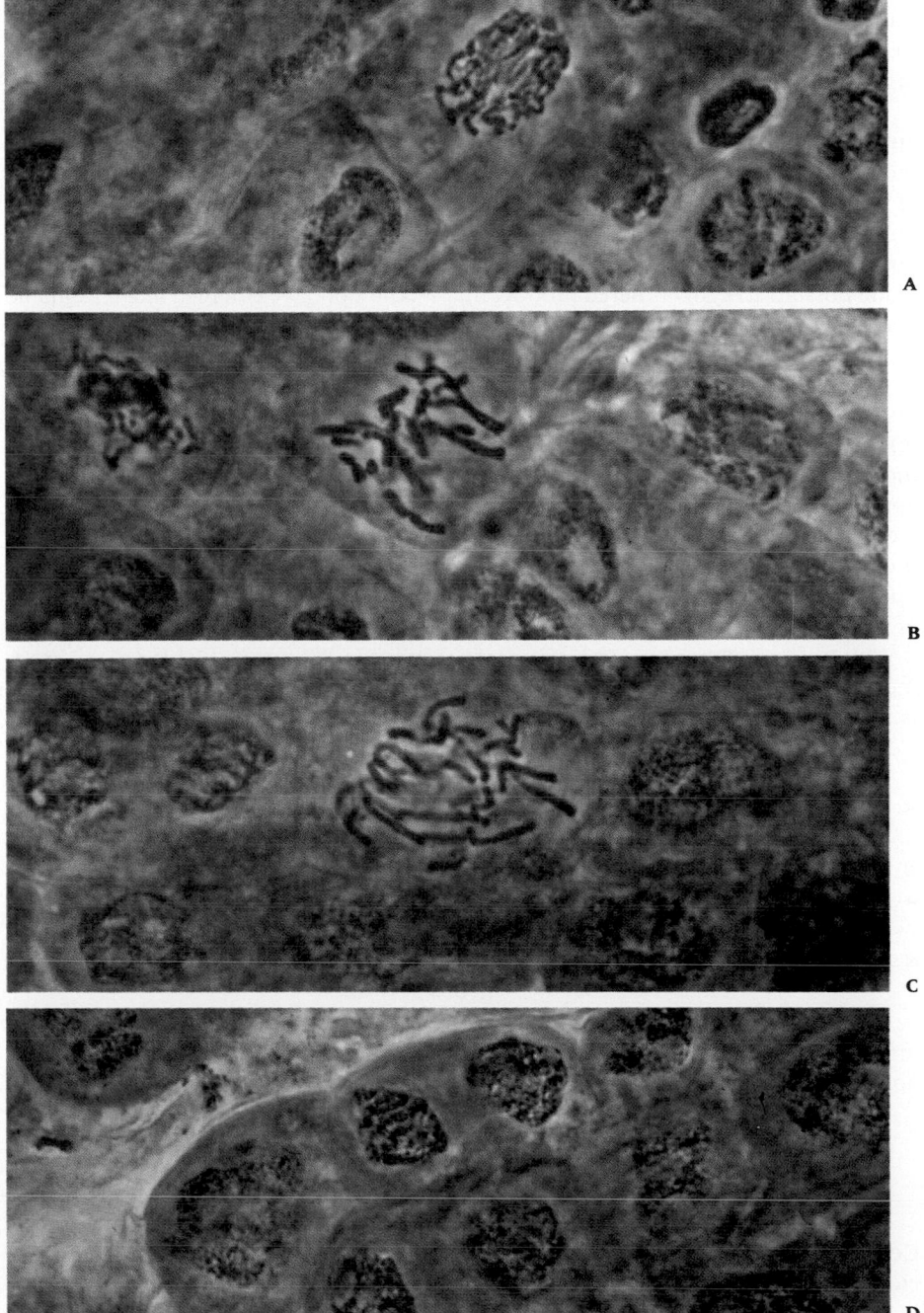

12-2 Reproduction by budding in prokaryotic cells
Scanning electron micrograph of Hyphomicrobium, *which reproduces by budding. (× 10,000)*

12-3 Mitotic division in plant cells
Cell division in the root tip of Narcissus *(A) Prophase. (B) Metaphase. (C) Early anaphase. (D) Telophase. During this sequence, chromosomes are duplicated and separated so that each new daughter cell receives a complete copy of the genetic material. (× 600)*

SIGNIFICANCE OF BIOGENESIS: THE CHAIN OF LIFE

We spoke in Chapter 4 of the old belief in the spontaneous generation, or abiogenesis, of living organisms from nonliving matter—and the experiments of Spallanzani, Pasteur, and Tyndall, which challenged that belief. As the concept of spontaneous generation faltered and the principle of biogenesis gained acceptance, it became clear that the biogenetic law, "All life from life," had to be restated more precisely. Rudolf Virchow (Figure 12-5) did so in 1855 when he declared, *"Omnis cellula e cellula"*—"All cells from cells." A cell, the unit of life, is always the product of another cell. Any new life must involve the formation of new cells from old. Even when the mature multicellular organism reproduces itself, a key event is the step in which certain cells produce gametes—the egg cells or sperm cells.

SIGNIFICANCE OF HEREDITY: THE TRANSFER OF INFORMATION

We learned in Chapter 8 of the constant drift of all matter toward a state of disorder, a tendency described by the second law of thermodynamics as increasing entropy. In view of this trend, the high order of living organisms can be maintained only through continuous investment of energy and work. But information—instructions on how, when, and where invested energy is to be applied—is required as well. Recognition of the need for information in the initiation and maintenance of living organisms was an important insight of modern biology.

It is revealing to compare the processes involved in the construction of a building and of a child. Certain principles govern both processes. For example, a complex building is unlikely to fall into existence spontaneously or by chance. The number of possible ways in which bricks and steel could be put together must be enormous. The actual arrangement embodied in a particular building is only one of all those possibilities. Chance was removed by the information in the architect's plan. This made possible a highly ordered product that would not have emerged in a reasonable period of time if construction workers went knocking things about in a random, uninformed way.

For the same reason, we reject the idea that a living organism—more complex by far than any building—could arise spontaneously from the materials of the nonliving world. Those materials—carbon, nitrogen, hydrogen, oxygen, and their molecular combinations—could be put together in many possible ways. Knowing approximately how many atoms of each element are present in the human body, we could conceivably estimate how many possible ways they could combine. Using the formulations of what is called *information theory,* we could then calculate how much information is needed to specify the particular combination that is a cell—or a multicellular organism like a human.

This information comes not from an architect's blueprint, but from the DNA molecules of the parent organism. The transfer of information from parent to offspring is the fundamental issue of the science of **genetics.**

In this and the next three chapters, we will learn:

- How the information that controls the development of an organism, its metabolism, and all else is coded in the nucleotide sequences of different DNA molecules
- How that information is duplicated (and conserved) during cellular reproduction
- How copies of the information are faithfully transmitted from generation to generation by genetic mechanisms
- How the information exerts control over a cell's metabolic machinery

We sketched the molecular structure of DNA in Chapter 4. We will lay the groundwork of genetics in this chapter by explaining cell division—and its remarkable variant, meiosis, essential in the reproduction of multicellular organisms.

PROKARYOTIC CELL DIVISION

Cell division, whether in prokaryotic or eukaryotic cells, displays certain unvarying elements. Prior to reproduction, a cell usually enlarges. At a critical size, it divides into two daughter cells, each usually about half the size of the parent cell. These

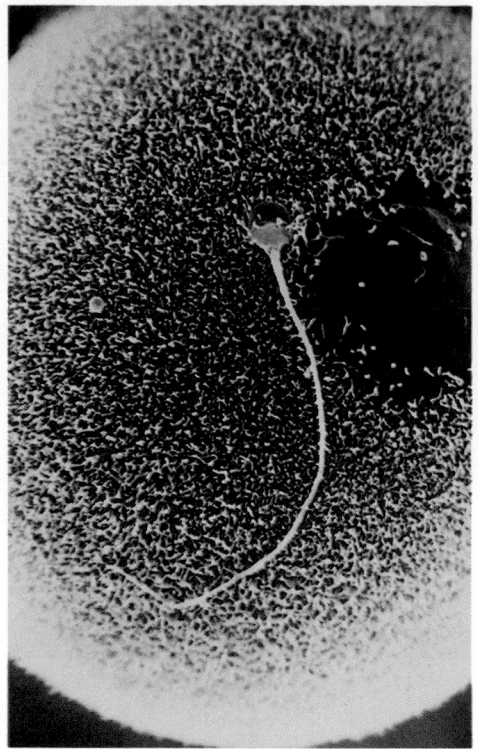

12-4 Sexual reproduction in animals
Scanning electron micrograph showing penetration of a human egg by a single sperm cell. ($\times$ 1000)

12-5 Rudolf Ludwig Karl Virchow (1821–1902)
In the company of former patients.

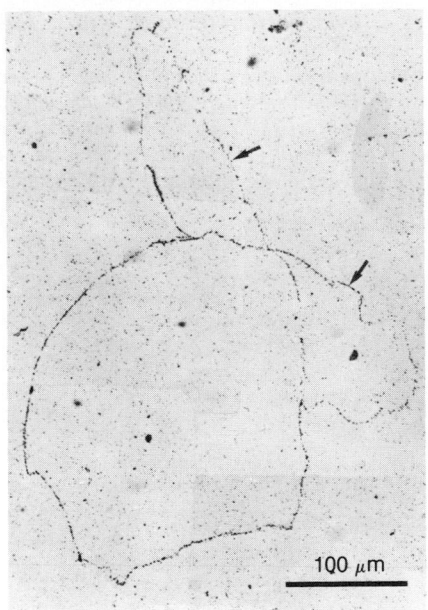

12-6 The bacterial chromosome
In this electron micrograph of an autoradiograph, radioactive DNA has exposed the photographic emulsion, revealing a closed circle of DNA undergoing replication. Compare this with the electron micrograph opening Chapter 16.

then grow to the size of their parent by assimilating various molecules from the environment and fashioning them into the biomolecules needed to fabricate cell substance.

There is a striking difference in the way prokaryotic cells (which lack nuclei) and eukaryotic cells (which have nuclei) accomplish this feat. On the face of it, prokaryotic cells have an easier task than eukaryotic cells. They are much smaller (often by a thousand fold or more). They are architecturally simpler, lacking a nucleus, mitochondria, endoplasmic reticulum, and many other of a eukaryotic cell's organelles (see Chapter 5). And they have much less DNA to duplicate. If all the DNA in a single *E. coli* cell were arranged end to end in a straight line, its length would be 1 mm. If the same were done with the DNA of an average animal cell, its length would be 1 meter.

The DNA of prokaryotic cells is usually called the **bacterial chromosome.** This is in analogy with the chromosomes of eukaryotic cells.

In true chromosomes, the negative charge on DNA is neutralized by positively charged protein molecules called **histones.** No such proteins have yet been found in prokaryotic cells.

A bacterial chromosome consists of a closed circle of DNA (Figure 12-6) that is found, not in a membrane-bound nucleus but in the **nucleoid,** a region that appears transparent in electron micrographs (see Figure 5-10). The nucleoid is attached to the cell membrane. Perhaps this circular arrangement protects the DNA from hydrolytic enzymes (called *exonucleases*) that can attack it only from a free end. (Eukaryotic nuclear DNA is not circular.[2]) In fact, the single DNA molecule of one *E. coli* cell is an enormous structure with a thickness of 2nm and a molecular weight of about 2.8 billion, equivalent to 4.2 million pairs of nucleotides. What is most remarkable is the way this very long molecule—when unwound, some thousand times as long as an *E. coli* cell—is packed into a small volume in the nucleoid. As noted above, the DNA of the prokaryotic chromosome, unlike eukaryotic DNA, is quite naked and vulnerable—that is, it is not bound to proteins.

The two major accomplishments of prokaryotic cell division are (1) the accurate duplication of the genetic information built into the DNA, and (2) the faithful distribution of copies to each new daughter cell. The method used to accomplish these ends looks deceptively simple (see Figure 12-1A, B).

A necessary preliminary step to chromosome duplication is the replication of its DNA (see Chapter 16). Throughout fission and chromosomal duplication, the DNA in prokaryotic cells remains attached to a patch of cell membrane. When chromosomal replication is complete, the new chromosome has its own independent point of attachment to the membrane. The cell wall grows inward between the attachment points of the two chromosomes, and the growing membrane beneath it pulls apart the two chromosomes and their points of attachment, dividing the cell into two daughter cells. By the time a full set of new cell walls has caused the cell to divide, the chromosomes are completely separated.

In sum, the cell cycle in prokaryotes consists of *DNA replication* followed by *cell division*. The factors that initiate DNA synthesis remain uncertain. In some organisms the entire process is completed in less than 20 minutes. Other prokaryotes take longer to complete the process. In contrast, the doubling times of many eukaryotic cells (in culture) are around 24 hours.

EUKARYOTIC CELL DIVISION

As noted above, cellular reproduction in eukaryotes (except for a very few protists) consists of two over-lapping phenomena: (1) **mitosis** (or **karyokinesis**), in which the nucleus reproduces itself; and (2) **cytokinesis,** in which the cytoplasm divides and the daughter nuclei are separated into two replicas of the parent cell.

Two aspects are noteworthy. One is the importance of the nucleus, the duplication of which (in mitosis) occurs before that of the rest of the cell (in cytokinesis). The

---

[2] However, the eukaryotic DNA in mitochondria and chloroplasts is circular. This is one of many features that these organelles share with prokaryotes—and it is one reason we believe that mitochondria and chloroplasts arose in evolution when prokaryotes invaded eukaryotic cells, survived within these cells, and remained in a mutually beneficial relationship.

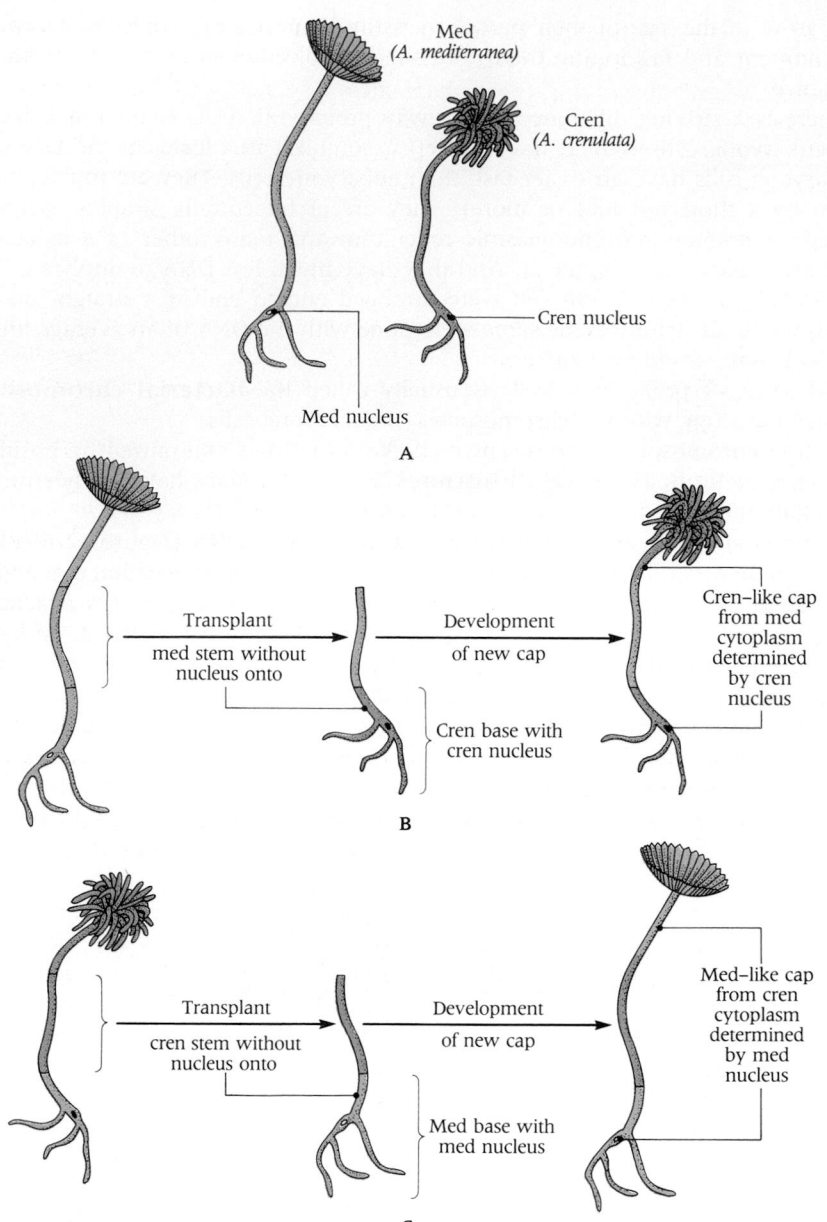

Med
(A. mediterranea)

Cren
(A. crenulata)

Cren nucleus

Med nucleus

A

Transplant
med stem without
nucleus onto

Development
of new cap

Cren–like cap
from med
cytoplasm
determined
by cren
nucleus

Cren base with
cren nucleus

B

Transplant
cren stem without
nucleus onto

Development
of new cap

Med–like cap
from cren
cytoplasm
determined
by med
nucleus

Med base with
med nucleus

C

other is the role of mitosis in producing daughter cells that conserve the chromosome number of the parent cell—a number that is characteristic for the species.

IMPORTANCE OF THE NUCLEUS

It was perhaps natural to suspect that the nucleus is a "control center" for cellular activities. However, it took a good deal of work to validate that assumption. For example, one can remove the nucleus from a cell (like an amoeba) that is large enough to permit such microsurgery. When that is done, the amoeba stops dividing and soon dies. If, however, a new nucleus is implanted within two or three days, the amoeba survives.

Many experiments have identified the nucleus as the repository of hereditary information. Studies of two marine protist species, the unicellular green algae *Acetabularia mediterranea* and *Acetabularia crenulata* (which we will call *med* and *cren*, respectively) were especially illuminating (Figure 12-7). Although they are single-celled organisms, these curious algae attain surprising complexity and size (up to 5 cm). Three parts are distinguishable in both species: a *cap*, a *stem* that contains chloroplasts, and a rootlike *base*. The nucleus lies in the base.

In a series of experiments in the 1930s, Joachim Hämmerling, of the Max Planck Institute for Marine Biology in Berlin, took the two *Acetabularia* species and cut away both stems and caps. The med and cren cells both regenerated a new stem and cap of characteristic type. If, however, the stem and cap were cut away, and the stem (lacking the cap) of one species was transplanted onto the cut base of the other species, interesting things happened. The new cap that regenerates is typical of the base. When a capless stem of med is grafted to a cren base, the newly regenerated cap is a cren cap (Figure 12-7B). The developmental "instructions" given the med stem must have come from the cren nucleus contained in the base.

CONSERVATION OF CHROMOSOME PATTERNS

How is the information within the nucleus accurately and dependably transmitted to two daughter cells when a cell divides? In eukaryotic cells, the gene-bearing structures are the linear bodies called *chromosomes,* of which a characteristic number (for the given species) occurs in each nucleus.

Chromosomes are best seen when cell division is interrupted at a certain stage of the cell cycle, prophase or metaphase (Figure 12-8). The figure also shows how photographs of these chromosomes can be cut up and rearranged to provide a **karyotype,** an orderly array of different chromosome types that can be carefully analyzed.

Keep in mind that the major function of mitosis is the duplication of chromosomes and the precise maintenance from one generation to the next of the parental chromosome number. Without such mechanisms, each cell division would alter the number of chromosomes handed down to each succeeding generation. Indeed, it was in part the realization that mitosis exactly preserves this number and distribution that, in 1902, led W. S. Sutton to conclude that chromosomes carry genetic information.

Most eukaryotic cells contain two full sets of chromosomes. Each chromosome in one set has a homologous partner in the other set: The two partners are **homologs.** Such cells are said to be **diploid.** Cells with multiple chromosome sets are called

TABLE 12-1
CHARACTERISTIC CHROMOSOME NUMBER IN VARIOUS SPECIES

Common Name	Binomial Name of Representative Species	Diploid Chromosome Number
	Animals	
Carp	*Cyprinus carpio*	104
Dog	*Canis familiaris*	78
Horse	*Equus calibus*	64
Cow	*Bos taurus*	60
Honeybee	*Apis mellifera*	56
Human	*Homo sapiens*	46
Orang-utan	*Pongo pygmaeus*	44
Monkey	*Macaca mulatta*	42
Rat	*Rattus norvegicus*	42
Mouse	*Mus musculus*	40
Alligator	*Alligator mississippiensis*	32
Hydra	*Hydra vulgaris*	32
Frog	*Rana pipiens*	26
Aardvark	*Orycteropus afer*	20
Flatworm	*Planaria torva*	16
House fly	*Musca domestica*	12
Fruit fly	*Drosophila melanogaster*	8
	Plants	
Fern	*Ophioglossum petiolatum*	1010
Horsetail	*Equisetum hiemale*	216
Sugarcane	*Saccharum officinarum*	80
Tobacco	*Nicotinana tabacum*	48
Corn	*Zea mays*	20
Snapdragon	*Antirhinum majus*	16
Barley	*Hordeum vulgare*	14
Garden pea	*Pisum sativum*	14
	Fungi	
Yeast	*Sacchoromyces cerevisiae*	36
Bread mold	*Neurospora crassa*	7*
Slime mold	*Dictyostelium discoideum*	7*

* Haploid number

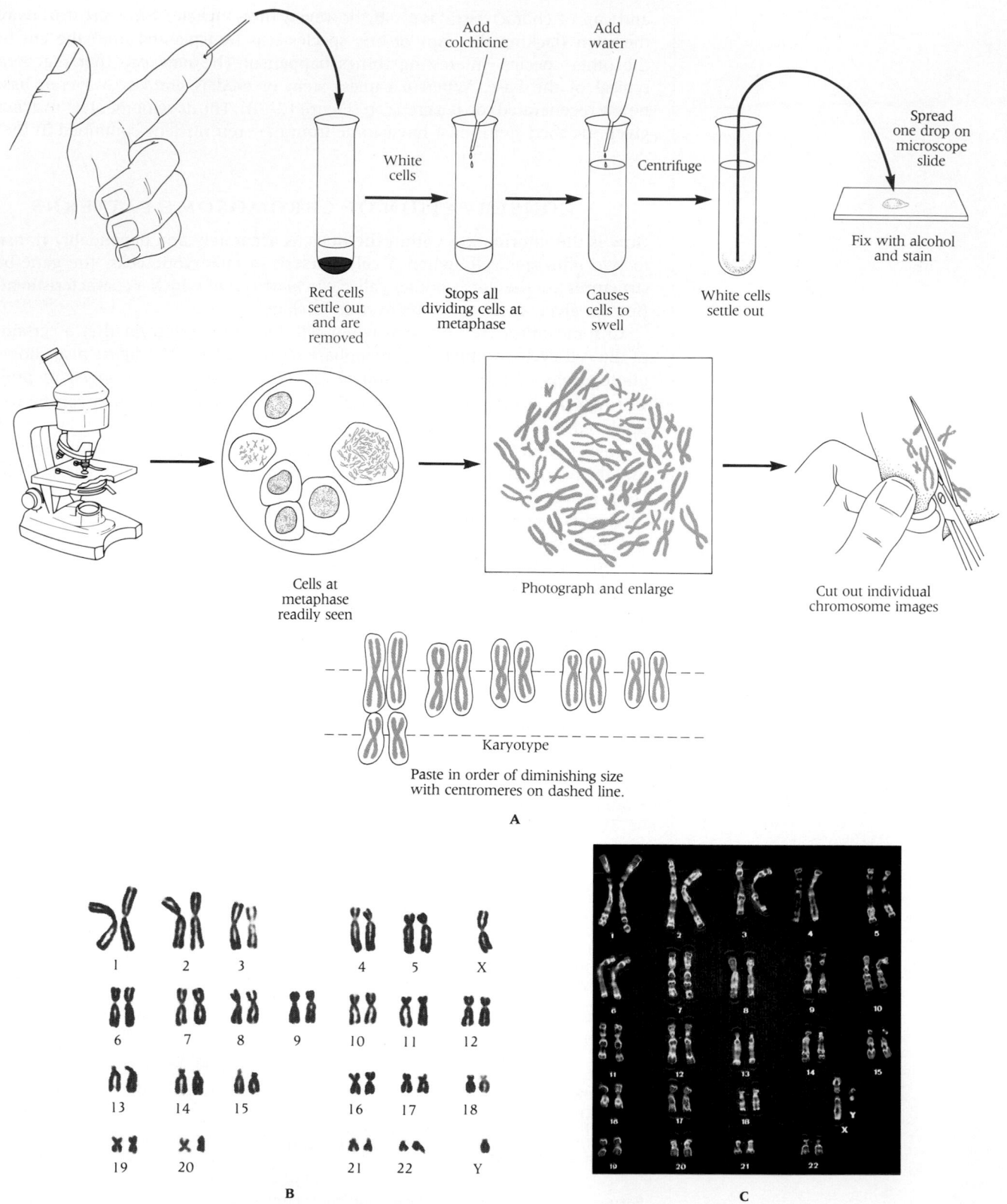

A

B

C

polyploid. The term **ploidy** refers to the number of chromosome sets in a nucleus. There is a characteristic number of chromosomes per cell in each species (Table 12-1). The number in a diploid human body cell is 46—or 23 pairs. We will later see that more than half of plant species are polyploid.

Haploid cells contain only one set of unpaired chromosomes. Ordinarily, each homolog of a pair in a diploid cell came from one of the parents—one from the male and one from the female. It follows that if cells in the animal or plant body are diploid, the germ cells that combined to form the original zygote must have

12-8 Demonstration and identification of human chromosomes
(A) Chromosomes are isolated from white blood cells in which mitosis has been arrested in metaphase by colchicine. Arrangement of chromosomes by size and centromere location yields a karyotype. (B) Karyotype of a human male. (C) Certain dyes bind to chromosomes, creating a banding pattern that can be used for chromosome identification.

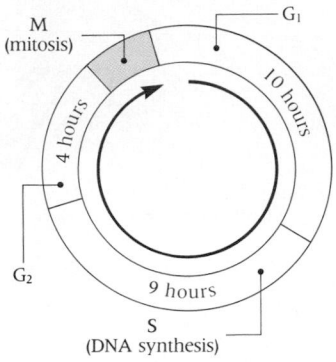

12-9 The eukaryotic cell cycle
During interphase the cell proceeds through a sequence of stages, designated G_1, S, and G_2. When a certain point is reached, a mitotic division (M) ensues.

been haploid. We shall learn in a moment how meiosis made them that way. In brief, mitosis preserves the diploid chromosome number; meiosis halves it.[3]

Keep in mind in the following discussions the difference between *chromosomes* (stainable nuclear bodies), *chromatin* (the combination of DNA and proteins—mostly histones—that make up the chromosomes); and *DNA* (a nucleic acid dimer consisting of two long chains of deoxyribonucleotides, the sequence of which contains all of the cell's genetic information in coded form).

THE CELL CYCLE

The **cell cycle** of dividing eukaryotic cells is the sequence of events from the beginning of one cell division to the beginning of the next. Its duration in typical actively dividing animal cells—like a red cell precursor in bone marrow or an epithelial cell in tissue culture—is about 24 hours. However, yeast cells and sea urchin embryo cells can divide in 2 hours.

The cell cycle has four phases, which are designated by the letters M, G_1, S, and G_2 (Figure 12-9).

- **M** stands for *mitosis;* it is the phase in which actual division of nucleus and cytoplasm occurs.
- Next there is **G_1,** the *first gap* and the one that comes after mitosis. The term gap denotes the absence of DNA synthesis.
- Next comes **S,** the *synthesis* stage, in which DNA is replicated and histones are synthesized.
- Finally, there is **G_2,** the *second gap*. This is the postsynthetic phase between DNA synthesis and the next mitosis.

Interestingly, not every cell is capable of dividing. In this regard, there are three cell populations—and most tissues contain some of each.

- Cells that are continuously dividing, going from one mitosis to the next (*cycling cells*). Examples are unicellular organisms and certain cells in regions of active growth—the tips of plant shoots and roots (meristems), blood cell precursors of animal bone marrow, and so on.
- Cells that temporarily leave the cell cycle and remain in a dormant state until environmental conditions stimulate their reentry into the cell cycle. We say these cells are in the **G_0** phase (read "G zero"). Examples are: the cells that divide vigorously in repairing wounds but are dormant until activated; the quiescent cells of plant roots.
- Cells that stop dividing after a certain number of divisions and then, after undergoing further differentiation, never divide again. Examples are neurons, which cease dividing soon after birth; mature white cells circulating in animal blood; and skeletal muscle cells. These cells are generally stopped in the G_0 phase.

Interphase

The "non-M" portion of the cell cycle, collectively termed **interphase,** includes three of the cell cycle's four stages: G_1, S, and G_2.

Viewed under a light microscope, the nucleus at interphase hides its detailed structure. The portions of the nucleus that stain darkly are the fibers of **chromatin.** Light microscopy reveals that in some areas the chromatin is tightly packed; in other areas it is in open networks. In older textbooks, the chromatin remaining tightly condensed during interphase was termed *heterochromatin,* and the chromatin in open networks during interphase was called *euchromatin.*

---

[3] We will later encounter certain organisms that normally have a haploid chromosome number in their body cells, at least during some stages of their life cycles. Also, some species have a chromosome number that reflects polyploidy. Most of these are the curious parthenogenetic and/or hermaphroditic forms to be discussed in Chapter 20. But some are familiar animal types. For example, most members of the butterfly genus *Polyommatus* have a diploid chromosome number of 45, but one species, *Polyommatus bellargus,* has 90, and another, *Polyommatus coridan,* has 180 chromosomes. These multiples of the original diploid number of 45 strongly suggest polyploidy.

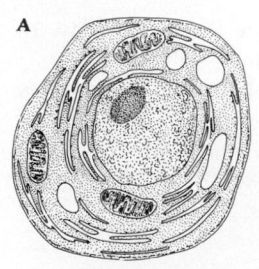

Interphase

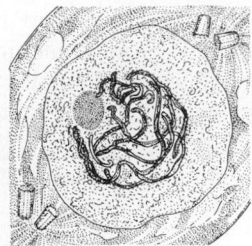

Early prophase

Mid–prophase

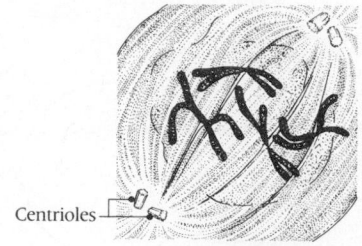

Centrioles

Late prophase

The first interphase stage is G_1. Its duration varies greatly. Slowly growing cells are generally cells with long G_1 stages—lasting many hours or days—whereas some rapidly dividing cells seem almost to lack a G_1 stage. In striking contrast, the other stages of the cycle are quite uniform in duration in different cell types. In most cases, they proceed in orderly sequence once G_1 is completed.

G_1 begins with the dismantling of the structural remnants of the last mitosis. It is a time of growth—of the cell and its organelles—and of preparation for the wave of DNA synthesis to come. Certain enzymes of the pathway of DNA synthesis now make their appearance, and protein and RNA synthesis proceeds actively.

The major event of the S phase—its defining feature—is DNA synthesis, in which replication of nuclear chromatin occurs. Enzymes strictly related to DNA synthesis increase in activity (see Chapter 16). The time between the completion of DNA synthesis and visible mitosis is occupied largely with RNA and protein synthesis and formation of structures (spindle fibers) that will be needed later.

The major events of the G_2 phase are preliminary preparations for the condensation of chromosomes, which take place in mitosis, and for the formation of the mitotic spindle, to be described below.

The major features of interphase are summarized as follows:

- The nuclear membranes are intact and nuclear structure is ill-defined, except for the nucleolus.
- DNA synthesis and chromatin replication occur in the S phase.

Mitosis (Karyokinesis)

It is impossible to observe mitosis without marveling at its satisfying logic and the elegance of its choreography—indeed, many writers have called it the "ballet of the chromosomes." Unlike interphase, mitosis is easily observed under a microscope.

It is convenient to think of the mitotic process as a sequence of stages or phases (see Figures 12-3, 12-10): prophase, metaphase, anaphase, and telophase. (Respectively, the prefixes mean "before," "between," "back," and "end.") We are still uncertain what factors induce an interphase cell to enter prophase and to start

12-10 Stages of mitosis
(A) (*above*) *Animal cell.* **(B)** (*below*) *Plant cell.*

Prophase

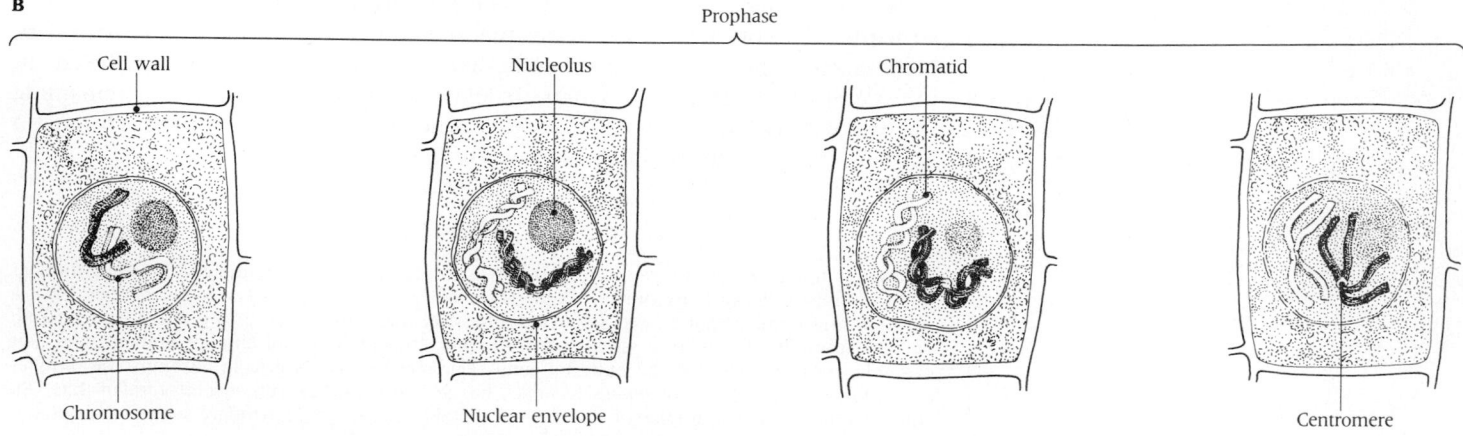

Cell wall · Nucleolus · Chromatid · Chromosome · Nuclear envelope · Centromere

Metaphase

Early anaphase

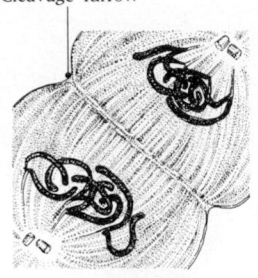

Cleavage furrow

Late anaphase

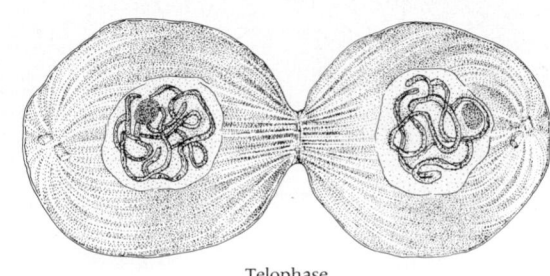

Telophase

dividing.[4] However that "decision" is made, the visible beginning of mitosis is marked by the onset of prophase.

Prophase

In prophase, the finely granular chromatin strands of the interphase nucleus condense into thickened, rod-shaped bodies within the nucleus. These are the **chromosomes.**

Chromosomes are made of the same chromatin that was either randomly dispersed or extended in the nondividing or interphase nucleus. Now, however, all of the chromatin is in a tightly packed configuration. Such sausagelike structures are obviously easier to move about without entanglement than long chromatin strands in the extended state.

In late prophase, each chromosome is composed of two roughly parallel longitudinal strands, called **chromatids,** that are joined together in a narrow region called the **centromere.** Because the centromere has a distinctive position in every chromosome pair, it divides each chromatid of the chromosome into "arms" of differing lengths that are characteristic for that chromosome pair. The two chromatid subunits are duplicates of one another and are termed **sister chromatids.** They are a result of the duplications of DNA and chromosomal proteins that took place during interphase.

Other preparations for division occur in prophase. The nuclear envelope and nucleolus start to disappear. As we saw in Chapter 5, most animal and some plant cells contain a pair of **centrioles.** During interphase, they position themselves just outside the nuclear envelope. As prophase begins, the centrioles separate and move apart toward the opposite poles of the nucleus (Figure 12-11). In animal cells, systems of radiating fibers become visible around each centriole by the time chromosome condensation has reached its maximum. These starlike arrays, termed **asters,** migrate to opposite poles of the cell, and spindle fibers start to form between them. The asters and spindle fibers are microtubules.

---

[4] The study of cell division in multicellular organisms has been sparked by the discovery of a variety of growth factors (see Chapter 28). The binding of growth factors such as platelet-derived growth factor (PDGF) or epidermal growth factor (EGF) to cell surface receptors may dispatch signals from the cell surface to the cell nucleus that ultimately generate instructions leading to cell division. Such growth factors apparently have no role in the division of unicellular organisms.

Metaphase

Early anaphase

Late anaphase

Early telophase

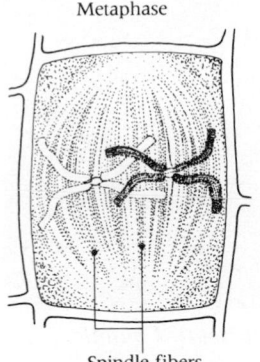

Spindle fibers

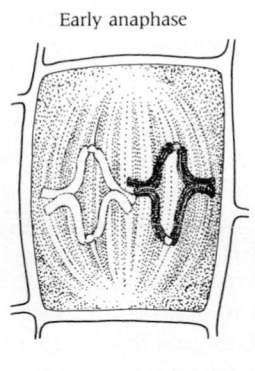

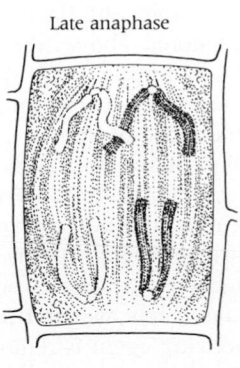

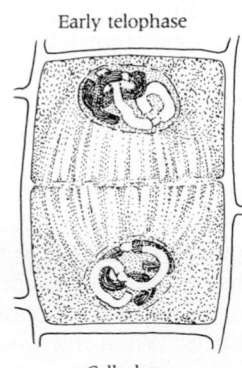

Cell plate

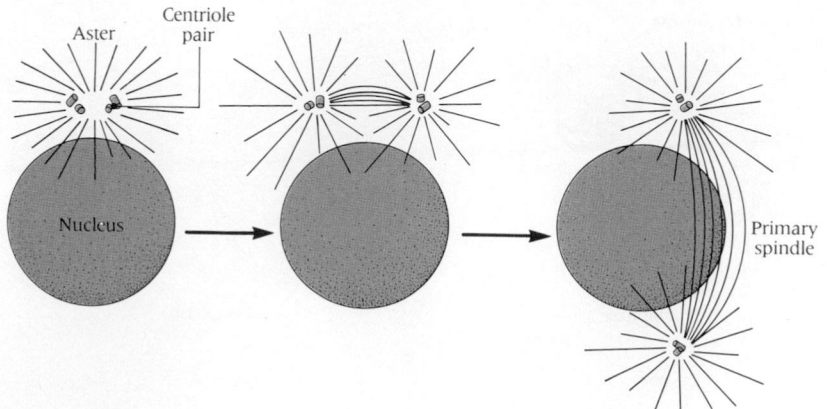

By late prophase the nucleolus has disappeared. Shortly thereafter, the nuclear envelope nearly disappears. This marks the end of prophase.

In summary, the major events in prophase include:

- Condensation of chromatin
- Cleavage of paired centrioles and centriolar migration toward poles
- Disappearance of the nucleolus
- Initial assembly of the spindle
- Breakup of the nuclear envelope

Metaphase

The **mitotic spindle** is completed soon after the beginning of metaphase. The spindle consists of microtubules, each about 25 nm in diameter, that vary in number from one (in certain fungi) to many hundreds or thousands. The spindle fibers are aligned in a roughly parallel array between the poles. Three-dimensionally, the spindle apparatus is shaped like two cones placed together base to base. The plane formed by the bases of the cones is the **equatorial plane** of the cell.

As discussed in Chapter 5, microtubules (of spindle fibers and other cell organelles) are composed of proteins that have many features in common with the contractile proteins of muscle (Chapter 31). In most eukaryotic cells, the spindle includes three classes of microtubules (Figure 12-12).

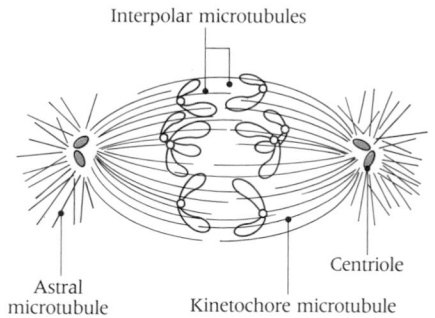

12-12 Anatomy of the mitotic spindle
Three classes of microtubules compose the spindle. Each has a unique position and performs a specific function.

- **Astral microtubules** radiate from the centrioles toward the peripheral regions of the cell.
- **Kinetochore microtubules** attach to the chromosomal kinetochores.
- **Interpolar microtubules** span the distance between the two spindle poles and make up the bulk of the spindle structure. Some extend from a pole to the equator. Others run from one pole past the equator into the opposite half-spindle. A few make the entire distance from pole to pole.

In animal cells, centrioles are at the centers of the mitotic poles. However, interpolar fibers radiate not from the centrioles themselves but from a diffuse zone of material around the centriole called the **microtubule-organizing center,** or **MTOC.** These are the centers from which spindle microtubules grow. Although certain specialized plant cells (for example, the flagellated spermatozoa of ferns) do have centriolelike organelles, as noted above, the cells of many plants lack centrioles. Nonetheless, plant cells still form mitotic spindles—hence we cannot say that centrioles are universally necessary. These **anastral spindles** lack astral fibers, however they closely resemble those of animal cells in most other respects.

At one moment the dividing cell contains no spindle fibers. Suddenly they are present in abundance. Where do they come from? The polarization microscope (see Box 5-1) reveals that spindle microtubules are constantly being formed by polymerization of subunits previously dissolved in the cytoplasm and constantly being depolymerized. Among the agencies capable of promoting depolymerization of the spindle microtubule are drugs, such as **colchicine,** that bind to spindle fibers. When this (or any) depolymerizing agent is removed, the spindle returns to normal. We should note that in 1952 Daniel Mazia successfully isolated intact

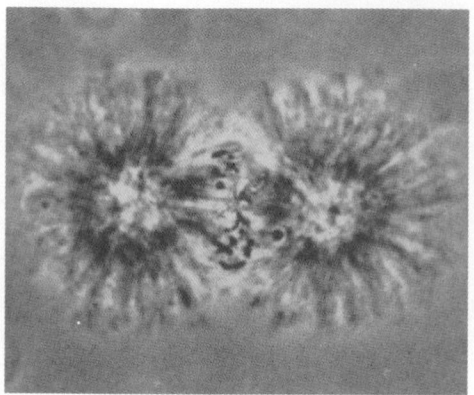

12-13 Micrograph of isolated mitotic spindle (× 1000)

12-14 Alignment of chromosomes in
metaphase

*Kinetochore microtubules connect each pole of
the kinetochore of each double-stranded
chromosome.*

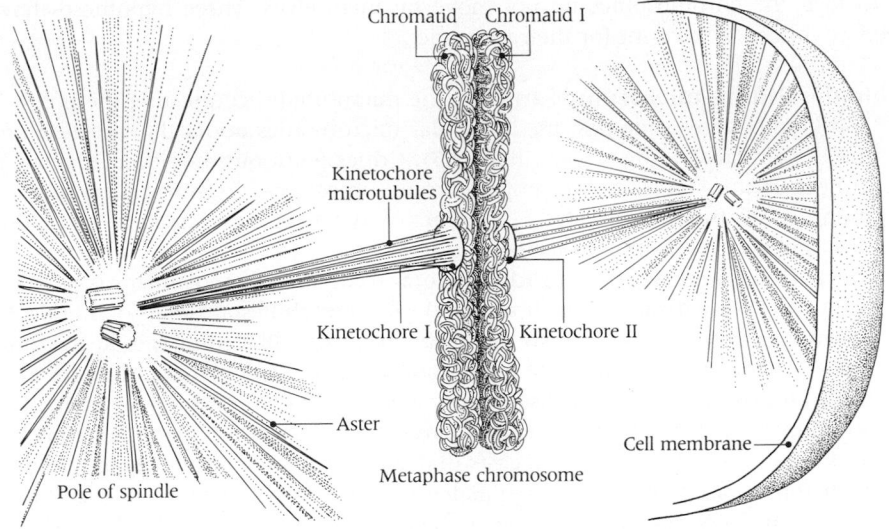

Chromatid   Chromatid I

Kinetochore
microtubules

Kinetochore I   Kinetochore II

Aster

Metaphase chromosome

Cell membrane

Pole of spindle

12-15 Microtubule assembly and
disassembly

*Chromosomes are thought to move toward the
poles as a result of subunit addition to the
chromatid end of microtubules, and simultaneous
loss of subunits from the pole end, which is
attached to the aster.*

Chromatid end
(assembly)

Pole end
(disassembly)

mitotic spindles. Despite their remarkable appearance (Figure 12-13), these preparations have not shed much light on the many unsolved problems of mitosis.

Next, a remarkable metaphase event takes place. The chromosomes, previously arranged in random fashion, move to the equatorial plane of the spindle and there line up in a random pattern. Then, with meticulous precision, they attach themselves to spindle microtubules (Figure 12-14). Each of the two sister chromatids of a metaphase chromosome has a **kinetochore,** a granule near the centromere, that serves as point of attachment and center for polymerization of spindle microtubules and locus of dynein, which powers spindle movement (see below).

Note that the centromere and kinetochore are two closely related parts of a metaphase chromosome. The centromere is a constricted region where the two chromatids are joined and held together. The kinetochore is a separate structure to which is attached the pole-to-chromosome spindle fibers. There is evidence that the DNA in the centromeric regions of each chromatid is distinctive in that it is enriched in adenine and thymine nucleotides, which occur in sequences of 6 to 10 nucleotides that are repeated millions of times. In some manner, centromeric DNA binds the proteins that cause a kinetochore to form.

Though the arms of the chromosomes may dangle from the equatorial plane, the kinetochores line up directly on it in metaphase. This chromosomal rearrangement is the high point of metaphase.

Note that the metaphase chromosomes arrange themselves in the equatorial plane—in striking contrast to what happens in the first metaphase of meiosis, which we discuss later in this chapter (see Figure 12-30). The two kinetochores now face in opposite directions toward the poles. The chromatids are ready to separate, and the stage is set for them to be pulled apart in anaphase. Meanwhile, the nuclear envelope has completely disintegrated.

The major events in metaphase are summarized below.

- Assembly (by polymerization of existing subunits) of the spindle microtubules to form the spindle
- Movement of chromosomes to the midpoint of the spindle
- Alignment of chromosomes end to end and linkage to spindle microtubules
- Disintegration and complete disappearance of the nuclear envelope

Anaphase

At the beginning of anaphase, each centromere divides. The chromatids of each chromosome, now free of each other, move apart quickly at a speed of 0.2–4 μm per minute. As they separate, the two arms of each chromatid characteristically form a V or a J, as though the entire structure were being towed against a current. A chromosome with a centrally positioned centromere resembles a V during anaphase; if the centromere is near the end of the chromosome, it resembles a J. At the same time, the poles move further apart by lengthening the pole-to-pole distance five- or sixfold.

In fact, the chromosomes do not move by themselves. Three hypotheses have been advanced to account for their movement.

- In the **sliding-microtubule model,** the microtubules connecting the poles and kinetochores slide past the interpolar microtubules under the direction of the dynein cross-bridges (see Chapter 5). Critics point out, however, that such a mechanism would tend to propel chromosomes away from the poles rather than toward them. Moreover, inhibitors of dynein ATPase, in some cases, do not impair chromosome movement.
- In the **depolymerization model,** chromosome-to-pole movement is based on the depolymerization (or shortening) of microtubules at their polar ends. In this model, the chromosomes are moved by the pull of shortening spindle microtubules. As the chromosomes are pulled poleward, the spindle microtubules are continuously shortening by disassembly at the pole end, even while assembly is taking place at the chromatid end (Figure 12-15). The simultaneous lengthening of the pole-to-pole distance suggests that microtubular subunits being removed from some spindle fibers are then added to the long pole-to-pole fibers. Thus one set of fibers may be shrinking as the other grows. This mechanism, which has been termed **microtubular treadmilling,** occurs in the test tube. Whether it operates in the cell has not been determined. It is difficult to visualize how microtubular depolymerization could exert a pull.
- In the **contractile-protein model,** spindle fibers are viewed as mere guidance tracks for chromosome movement while the motive force arises from other contractile proteins in the spindle—perhaps actin and myosin, both of which are present in the spindle in small amounts. This model also has shortcomings.

Whichever mechanism is correct, two matched sets of chromosomes have collected at the widely separated opposite ends of the cell by the end of anaphase (see Figure 12-10). The separated bodies may now be termed **daughter chromosomes.**

While these dramatic chromosomal movements are taking place, the process of cytokinesis—the physical division of the cytoplasm—begins.

In sum, the highlights of anaphase include:

- Movement of a full set of chromosomes toward each pole of the mitotic spindle, where the chromosomes collect
- Lengthening of the spindle fibers, pushing the poles apart
- Initiation of cytokinesis

Telophase
During telophase, the spindle vanishes and the chromosomes disappear from view as the chromatin again becomes extended, thin, and diffuse, with its affinity for stain diminished. When the chromosomes lose their individual identity at this point, they contain only one chromatid. But when condensation makes them visible again in the next prophase of the ensuing cell cycle, each chromosome has two chromatids. The reason: DNA replication occurs in the S stage of interphase.[5]

In telophase new nuclear membranes form around each of the two new daughter nuclei and new chromosomes regroup into the nuclei. Gradually there reappears a typical interphase nucleus containing one or more newly formed nucleoli. Telophase thus is prophase in reverse. Cytokinesis is usually completed in telophase.

In summary, the main events of telophase are the following:

- Chromosome visibility fades.
- Chromosomes still possess only one chromatid.
- A new nuclear envelope forms.
- One or more nucleoli reappear.
- Cytokinesis nears completion.

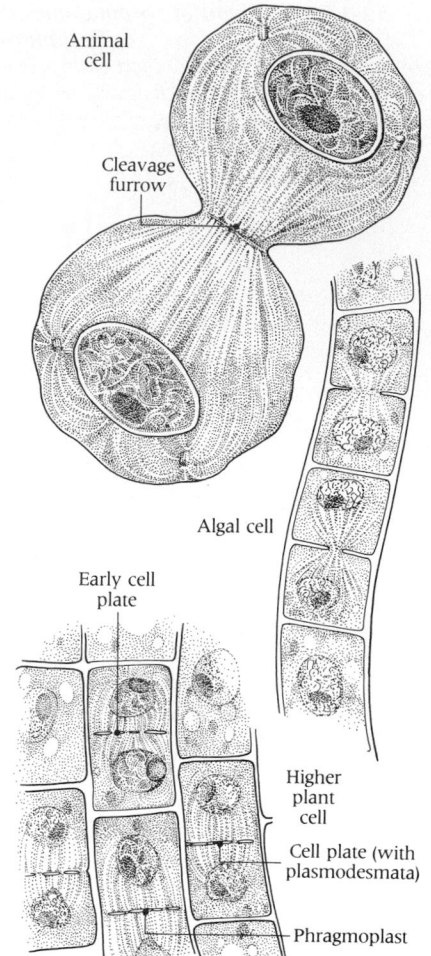

12-16 Diverse mechanisms of cytokinesis
In animal cells a cleavage furrow is formed by indentation of the plasma membrane. In higher plant cells, and many algal cells and fungi, new plasma membrane arises from cytoplasmic vesicles that fuse to form the cell plate. Patterns in algae vary.

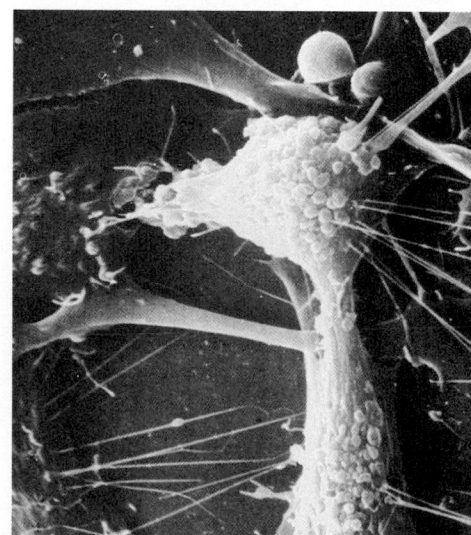

12-17 Formation of a cleavage furrow.
A cleavage furrow appearing at the end of mitosis results in the formation of two daughter cells. This scanning electron micrograph shows two animal cells in culture that are nearly separated. (× 6000)

---

[5] Biologists sometimes loosely say that chromosomes "split in two" in mitosis. This is unfortunate phrasing. If the duplex nature of the prophase chromosome resulted only from a longitudinal split of the earlier telophase chromosome, it could be repeated only so often. A cake can be cut into smaller pieces only so many times. Then very little is left! Clearly, replication of a chromosome involves the physical duplication of an existing one, and not the splitting of an existing one.

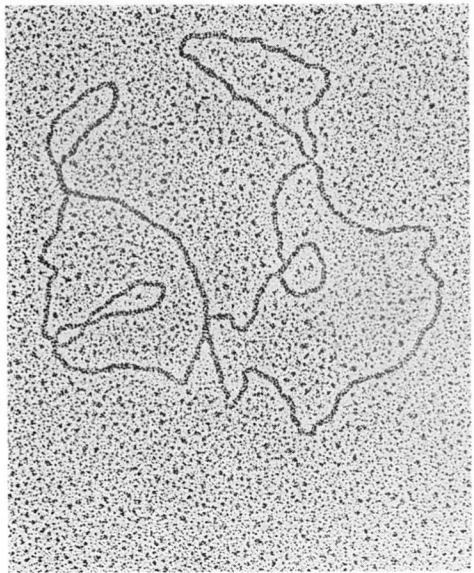

12-18 An isolated molecule of mitochondrial DNA
This electron micrograph shows a single molecule of mitochondrial DNA. Although twisted in several places, the molecule is in fact circular. (× 75,000) Compare with Figure 12-6, which shows circular prokaryotic DNA.

Mitosis in plant cells is quite similar to the process in animal cells. In some protists, the nuclear membrane never disappears; the spindle forms inside the nucleus.

In recent years, investigators have sought to identify critical metabolic events in each stage of the cell cycle, as it might be possible to design drugs that could selectively damage cancer cells by exploiting subtle differences in the cell cycles of normal and malignant cells. Although that hope has been only partially fulfilled— *some* drugs do this successfully in *some* cancers (for example, vincristine blocks mitosis in metaphase in certain leukemia cells and methotrexate blocks DNA synthesis in S stage in osteosarcoma, bone cancer, cells).

Cytokinesis

Karyokinesis (splitting of the nucleus) and cytokinesis (splitting of the cytoplasm) usually occur in sequence.[6] Interestingly, cytokinesis proceeds somewhat differently in animal and plant cells.

In animal cells, one sees a **division furrow** (or **cleavage furrow**) in the cell membrane that deepens until two separate daughter cells are formed (Figure 12-16). Its location is influenced by the position of the spindle midpoint. When the spindle is experimentally displaced by centrifugation, the locus of the furrow shifts. A developing division furrow is best seen with the scanning electron microscope (Figure 12-17).

It is still not known what mechanism triggers the onset of cytokinesis during the final stages of chromosome separation. The mitotic apparatus somehow initiates cytokinesis but is not responsible for its continuance. Once cytokinesis is launched, the spindle can be removed (or dissolved) without halting the division. A band of actin filaments, called a **contractile ring,** establishes the division furrow.

Cytokinesis is accompanied by duplication of cytoplasmic organelles, which divide autonomously, temporally independent of nuclear control. Chloroplasts and mitochondria derive from existing organelles by a process resembling fission or budding. Each of these organelles has DNA of its own that differs from nuclear DNA. Indeed, as shown in Figure 12-18, the circular structure of this DNA resembles that of prokaryotic DNA.

In centriolar reproduction, a new, small **procentriole** appears next to a mature centriole in the S stage. During prophase it elongates at a right angle to the old centriole (Figure 12-19). As a result, a full pair of centrioles exists at the end of prophase.

Organisms with rigid cell walls cannot form division furrows by a simple pinching-in process. They have solved this problem in different ways (see Figure 12-16). In some algae (and fungi), an inward growth of new cell wall and membrane erects a wall between daughter cells.

In plant cells, the process of cytokinesis is clearly different from that of animal cells, almost certainly because these cells are confined by rigid cell walls. The first sign of cytokinesis occurs in late anaphase with the appearance across the spindle equator of the group of vesicles that are thought to arise from the endoplasmic reticulum or the Golgi complex. They coalesce to form a structure known as a **phragmoplast** (see Figure 12-16). The vesicles contain materials for a future cell wall. The vesicles fuse together to form a special double-layered membrane sheet called a **cell plate**—a large but thin membrane-lined vesicle—across the cell's equator that eventually fuses with the plasma membrane and becomes a new cell wall at the point of cleavage. The new daughter cells remain connected by **plasmodesmata,** thin membrane-lined cytoplasmic passages through the cell wall.

In fungi and many algae (see Figure 12-16), new plasma membrane and cell walls grow inward from the outer walls until the edges meet. The new structure separates the daughter cells.

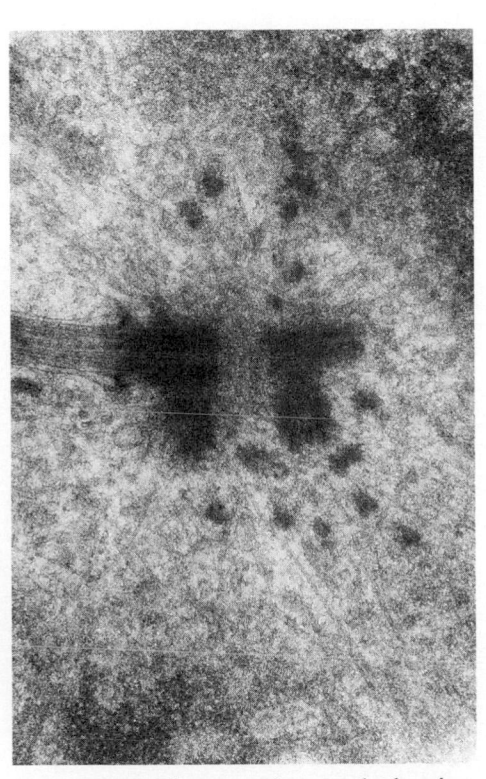

12-19 Electron micrograph showing duplicating centrioles (× 120,000)

---

[6] However, this seemingly self-evident statement is not always true. Mitosis without cytokinesis occurs commonly in green algae, fungi, green plants, and certain invertebrate animals. The result of such a process is a coenocyte, a large structure with many nuclei but few, if any, cellular partitions.

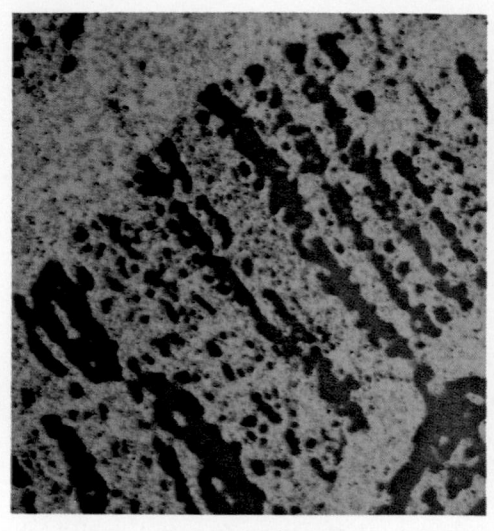

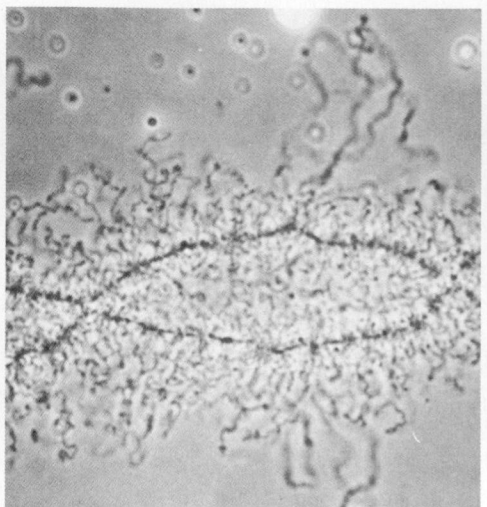

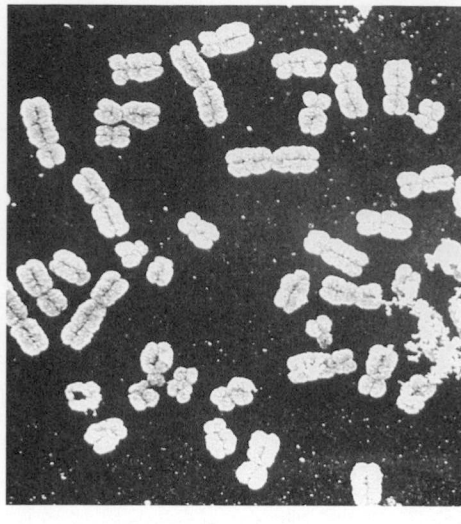

A   B   C

The major events of cytokinesis are summarized below:

- Duplication of cytoplasmic organelles
- In animals and other organisms without cell walls, formation of a division furrow
- In plants, formation of a cell plate

CURRENT CONCEPTS OF CHROMOSOME STRUCTURE

The 1970s brought a flood of reports on the fine details of chromosome structure. These advances also illuminated the curious interconversions of the extended chromatin of interphase and the tightly organized chromosomes that appear in prophase and disappear in telophase.

In devising a model of the eukaryotic chromosome, investigators were required to account for certain facts.

- Chromosomes of stained eukaryotic cells were found early by cytologists using light microscopy to display remarkable structural variety. Note the following in Figure 12-20: the common stubby **condensed chromosome** typical of dividing animal cells, which usually have characteristic banding patterns; the **lampbrush chromosome,** often seen in germ cells; and the huge banded chromosome, termed a **polytene chromosome,** which is commonly seen in some specialized cells of larval flies.
- DNA in the chromosome is associated with a considerable mass of proteins—among them, basic proteins of a type called **histones** and various **nonhistone proteins.** Both are believed to play a key role in the regulation of DNA's genetic functions. Histones are basic because they are rich in the basic amino acids lysine and arginine (see Figure 4-14). Their strong positive charge adapts them admirably to form ionic bonds with DNA (whose *A,* we should remember, stands for *acid*). The bonds between DNA and histones are ionic rather than covalent.
- There are five kinds of histones. The type termed **H1** is very rich in lysine; **H2A** and **H2B** are moderately rich in lysine; and **H3** and **H4** are rich in arginine. When isolated chromatin is placed in a solution of gradually increasing salt concentration, the various histone types are released in the sequence implied by their names.
- There is a diversity of nonhistone proteins.
- In electron micrographs, chromatin fibers stretched out at low salt concentrations resemble beads on a string (Figure 12-21). At higher salt levels (or in the presence of Mg^{2+}), fibers are considerably thicker.
- There is striking evidence in x-ray diffraction studies of some sort of repeating structure, a repeat "unit" occurring every 10 nm.

12-20 Diversity of chromosome structure (**A**) *Metaphase chromosomes displaying a highly condensed structure.* (**B**) *Lampbrush chromosome in a developing frog oocyte.* (**C**) *Polytene chromosome from a fruit fly showing the characteristic banding pattern.*

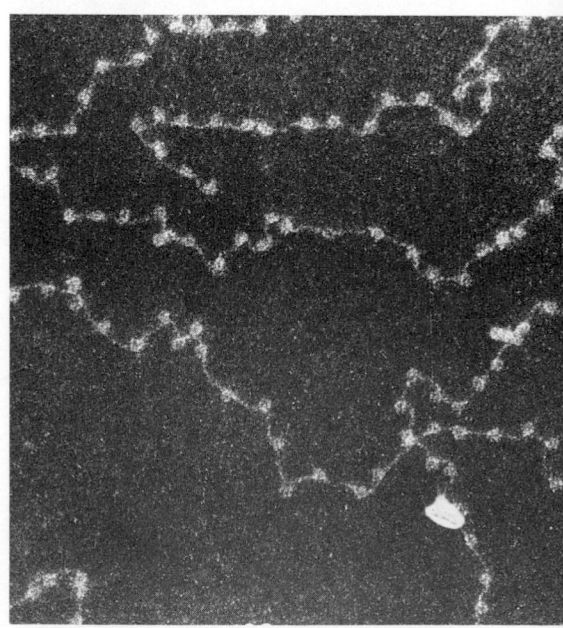

12-21 Electron micrograph illustrating beaded character of a chromatin fiber ($\times$ 150,000)

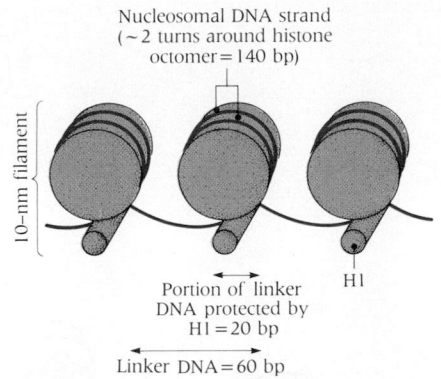

12-22 Structure of chromatin fiber
DNA is wound around a histone octomer to form a nucleosome. Chromatin is composed of an array of nucleosomes connected by linker DNA. Note that DNA lengths are expressed in basic pair (bp) units.

12-23 Anatomy of the nucleosome
The nucleosome is composed of a histone-containing octomer, around which is wrapped the DNA double helix. Linker DNA connects the nucleosomes and exposed portions are susceptible to cleavage by endonucleases (at arrows). Further exposure to endonucleases digests away all linker DNA, releasing single nucleosomes.

The Nucleosome

At first, these facts were given a simple explanation. The condensation of the chromosome was attributed to the coiling of DNA, itself a double helix, into a **superhelix** with a periodicity between turns of 10 nm. Histone was thought to be evenly distributed like the insulation on an electric wire. This model was soon discredited for lack of supporting data.

Opinion was radically altered in 1973 by the discovery of the **nucleosome.** In a key experiment, isolated chromatin was digested with DNA-cleaving enzymes called **endonucleases;** the products were pieces of regular length, each 200 base pairs (bp) long. In the age of molecular biology, lengths of DNA molecules are measured in **base pairs,** abbreviated **bp,** or **nucleotide pairs** abbreviated **np.** But if histones were removed first, endonuclease digestion produced DNA fragments of varying lengths. Periodicity was lost. Similar fragments were obtained with prokaryotic chromosomes, which lack histones. It began to appear that histone is not distributed evenly along the DNA strand.

In the current view, chromatin is visualized as a string of nucleosomes, nearly spherical bodies 10 nm in diameter (Figure 12-22). These are the basic repeating units. Nucleosomes are interconnected by strands of **linker DNA** that is about 60 bp long. Within a nucleosome, a strand of DNA 140 bp long is coiled nearly twice around a protein core consisting of a complex of 8 histone molecules that together form an **octomer** with the structure $(H2A)_2(H2B)_2(H3)_2(H4)_2$ (Figure 12-23). H1 histone remains outside the core—perhaps to protect a portion of the linker DNA (see Figure 12-22). This model nicely explains the endonuclease digestion patterns: The enzyme initially breaks DNA at an exposed place on the linker DNA while nucleosome DNA is protected by core histones (Figure 12-23).

Structure of Chromatin Fibers

Predictably, these long strings of nucleosome beads (10 nm in diameter) coil to form the higher-order structure of chromatin fibers (30 nm in diameter) that are found in both isolated chromatin and in chromatin within the nucleus. There is

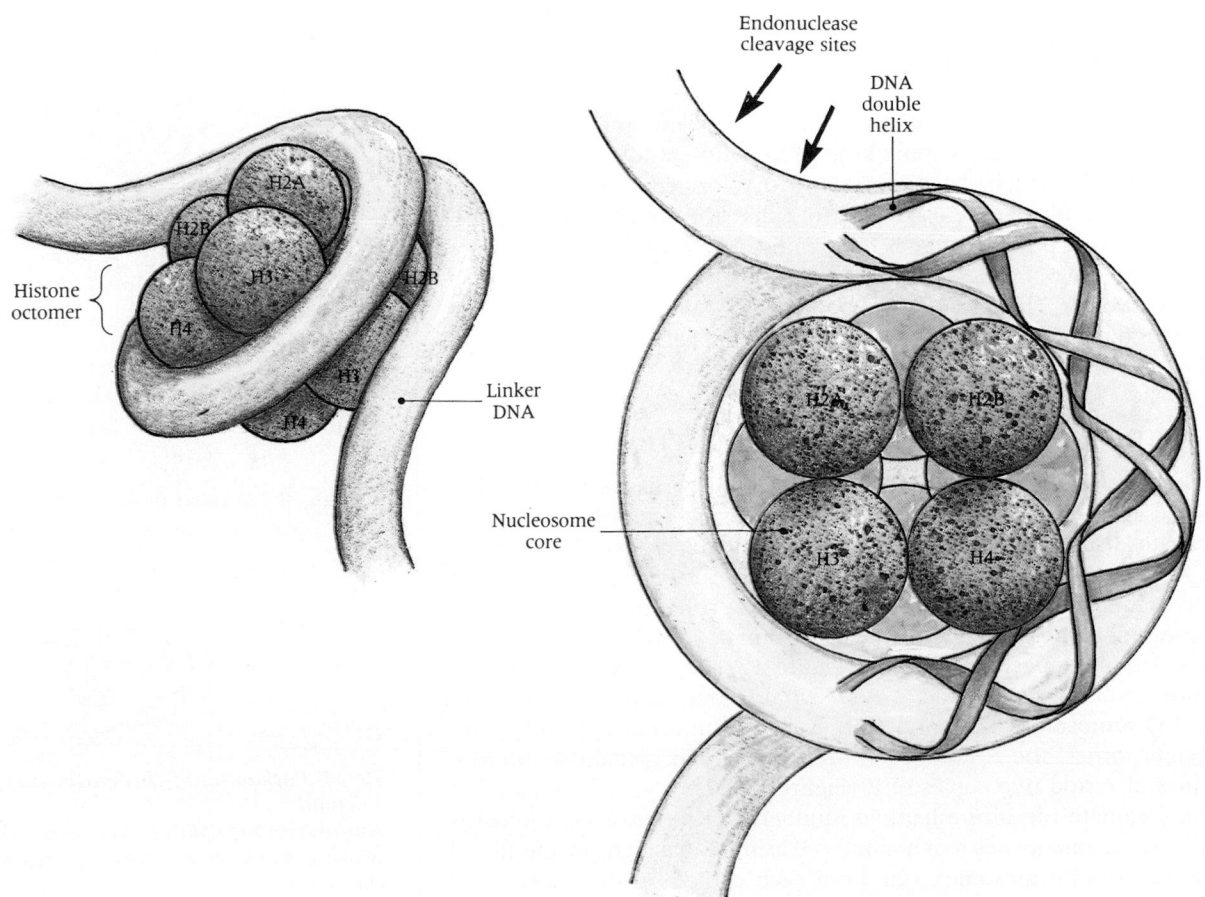

still considerable debate over the structure of these fibers. All investigators agree they are formed by the ordered folding of polynucleosome strings. In one model (Figure 12-24), nucleosomes are packed together into a helical DNA-protein complex called a **solenoid.** Within a solenoid a chromatin fiber is coiled into a helix containing six nucleosomes per turn.

Investigators, impressed with the appearance of the chromosome surface in the scanning electron microscope (Figure 12-25A) and the chromosome cross section (Figure 12-25B), proposed that chromatin fibers are organized in a radial loop, as shown in Figure 12-25C. Other models have been proposed, but this one seems unusually well supported. Whatever the details of the fiber's organization, it is a compact structure that likely must be altered before DNA can perform its genetic function.

BIOLOGICAL SIGNIFICANCE OF MITOSIS: A SUMMARY

Mitosis is a basic mechanism for cellular reproduction and transmission of hereditary information. Its major results are the following:

- Two cells come into being in place of one.
- By virtue of the orderly replication and distribution of the chromosomes on the spindle, each daughter cell is endowed with a precise copy of each and every chromosome.
- Every somatic cell in a multicellular organism has the same number and same kinds of chromosomes as every other cell.

The uniformity of chromosome quantity and quality among body cells is demonstrated by the procedure called **cloning,** in which a single cell from an adult plant (for example, a phloem cell of a carrot) develops into an entire new adult organism under suitable conditions. It would appear that each cell possesses a full complement of the genetic information specifying a whole organism.

This conclusion was confirmed in parallel experiments in animals by Gurdon and Brown, who succeeded after many trials in fertilizing the egg of a *Xenopus* frog with a cell nucleus isolated from an intestinal cell of a male tadpole. A normal adult developed!

MEIOSIS

Consider the cell of a hypothetical organism that underwent mitosis in Figure 12-10B. Assume that the organism is some kind of fly whose adult form is shown in Figure 12-26. The dividing cell had 4 chromosomes that formed 2 pairs, designated A^m and A^p, and B^m and B^p (Figure 12-27). (Note that the superscripts p and m denote paternal or maternal origin.) What is the significance of the fact that chromosomes occur in pairs in diploid cells?

All human cells have 46 chromosomes. As shown in Table 12-1, the dog has 78, garden pea 14, and the fruit fly 8. There is no correlation between an organism's complexity and its chromosome number.

PLOIDY AND THE SIGNIFICANCE OF FERTILIZATION

The meaning of this regularity lies in the nature of **sexual reproduction,** which joins two cells, one from each parent. These two special cells (sperm and egg) are called **germ cells** to distinguish them from ordinary body cells, which are called **somatic cells.** As mentioned earlier, the germ cells, also called **gametes,** differ from somatic cells in their chromosome count. Each gamete contains half the number of chromosomes present in the zygote—and in all the somatic cells of the multicellular organism. Clearly, the imaginary organism in Figure 12-10B, the somatic cells of which had 4 chromosomes (2 A's and 2 B's), would have to produce gametes containing only 2 chromosomes (1 A and 1 B). When a union of egg and sperm pooled these chromosomes, the resulting fertilized egg would contain 4 chromosomes—two copies of A and two copies of B (Figure 12-27).

The nucleus of a gamete contains a haploid number of chromosomes. Biologists denote the number of chromosomes in a haploid cell as *n.* The nucleus of a fertilized egg, with its two sets of chromosomes, one from each gamete, is diploid because

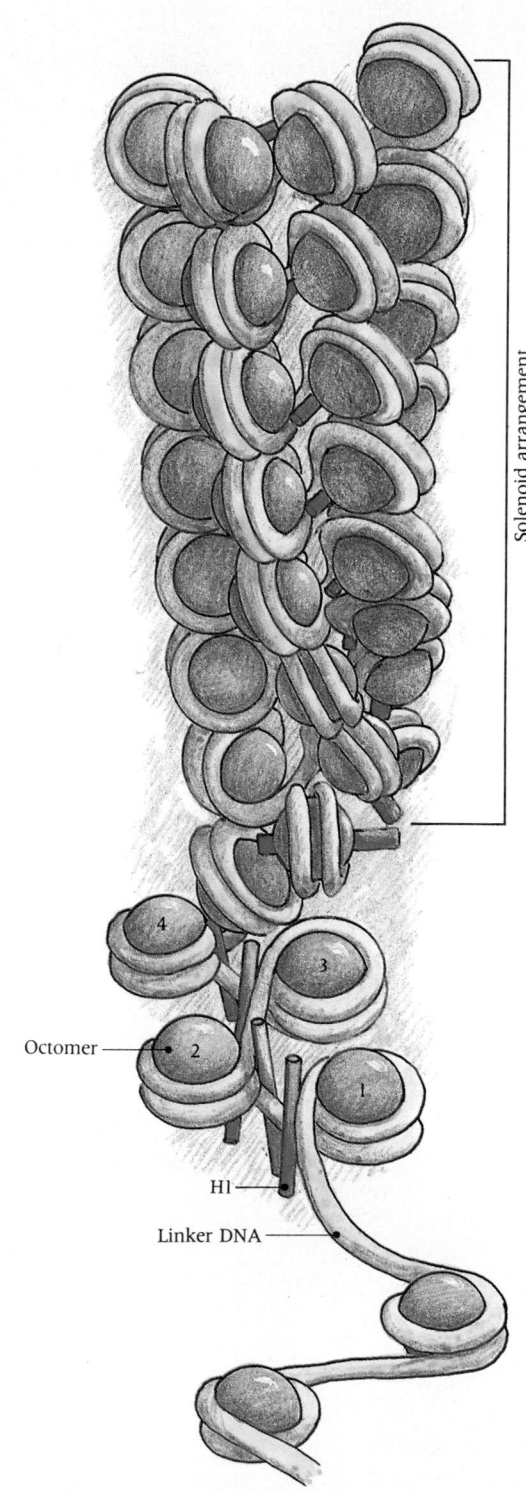

Solenoid arrangement

Octomer

4
3
2
1

H1

Linker DNA

12-24 Packaging of chromatin into a solenoid
Nucleosomes are tightly coiled to form a solenoid, which is a precursor of the condensed mitotic chromosome.

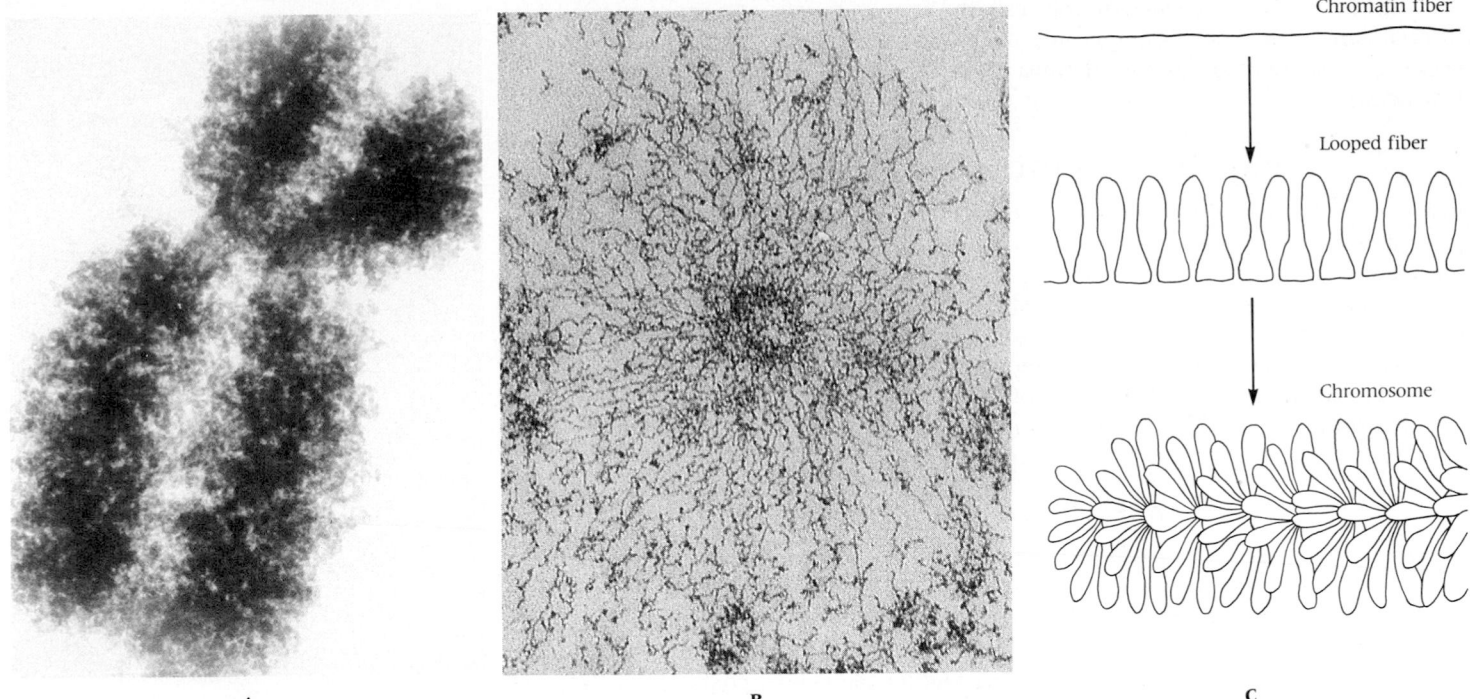

Chromatin fiber

Looped fiber

Chromosome

A   B   C

12-25 Higher order structure of the chromosome

(A) Electron micrograph showing the highly convoluted chromosome surface. (× 15,000) (B) In cross section, the chromosome appears to be radially arranged about a central core. (× 70,000) (C) In the radial loop model, the chromatin fiber is folded to form loops that arise from a central axis.

it has **2n** chromosomes. In each chromosome pair in a diploid nucleus, one chromosome derives from the male parent's sperm and the other from the female parent's egg. In Figure 12-27, we designated the chromosomes of paternal origin as A^p and B^p; those of maternal origin as A^m and B^m. The 2 A chromosomes make up a **homologous pair,** as do the two B chromosomes. A^p is homologous with A^m, and B^p is homologous with B^m—but neither A chromosome is homologous with either B chromosome.

Thus the diploid nucleus of a fertilized egg contains two complete sets of genetic information, one from the egg and one from the sperm. As the single-celled zygote

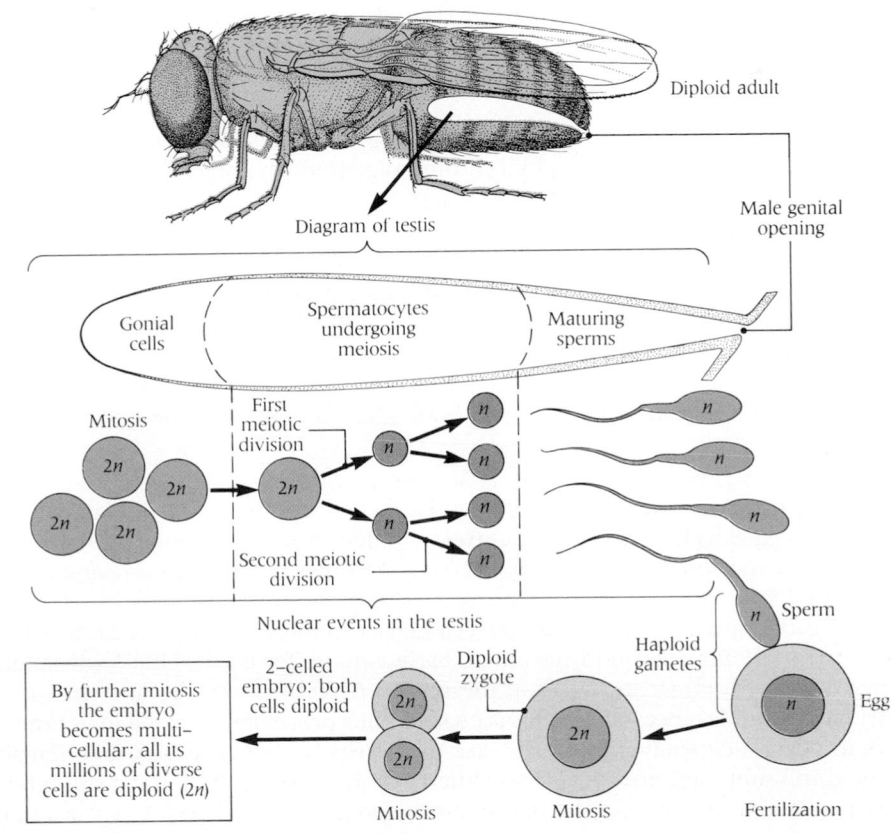

Diploid adult

Male genital opening

Diagram of testis

Gonial cells

Spermatocytes undergoing meiosis

Maturing sperms

Mitosis

First meiotic division

Second meiotic division

2n 2n 2n 2n 2n

2n

n n
n n
n n
n n

n
n
n
n

Nuclear events in the testis

Sperm

Haploid gametes

n Egg

Fertilization

Diploid zygote

2n Mitosis

2-celled embryo: both cells diploid

2n
2n

Mitosis

By further mitosis the embryo becomes multi-cellular; all its millions of diverse cells are diploid (2n)

12-26 The nuclear cycle in sexual reproduction

Haploid cells arise from diploid cells in the testis. Sperm are produced by meiosis and then contribute their genetic information to the next generation at fertilization.

develops as a result of repeated mitotic cell divisions, both sets are accurately copied in each interphase and the duplicates are distributed to new daughter cells. Thus every nucleus throughout the organism contains its own copy of all hereditary instructions.

HAPLOIDY ORIGINATES IN MEIOSIS

If a species is to retain a characteristic chromosome number and if all somatic cells of embryo and adult have a diploid chromosome number, somewhere in the life cycle of a sexually reproducing organism there must occur a distinctive form of cell division. That form was discovered in 1890: A unique form of nuclear division occurs in the production of gametes just before they mature into eggs and sperm cells. In this process, the diploid chromosome number ($2n$) typical of body cells is reduced to the haploid number typical of gametes (n). The special reductive division is called **meiosis** (from the Greek *meioun*, meaning "to diminish").

Where Meiosis Takes Place

Meiosis takes place during the production of gamete cells—the sperm and egg. Generally it occurs in specialized diploid cells of the gamete-producing organs—in animals, the testis or ovary, collectively termed **gonads;** in plants, the ovule, anther, or other structures (see Chapter 21).

Let us survey these processes in the male and female animal gonad.

Spermatogenesis

We have assumed that the hypothetical organism of Figure 12-10A is a fly. As shown in Figure 12-26 and 12-27, haploid gametes fuse to form a diploid zygote. Successive cell divisions distribute the chromosomes to all somatic cells, which are diploid, including the cells of the testis. Since the sperm cells produced by the testis are haploid, we can be sure that meiosis—a type of nuclear division that yields haploid cells—occurs at least in some cells of the testis.

These cells are shown schematically in Figure 12-28. Gamete production begins with **spermatogonia** (or **gonial cells**), which divide by ordinary mitosis. One daughter cell of a spermatogonial mitosis remains as a gonial cell; the other becomes a **primary spermatocyte.**

The entire meiotic process requires DNA replication in two divisions called **meiosis I** and **meiosis II.** Meiosis I is called the **reductional division;** meiosis II is called the **equational division.** The cell in which meiosis I takes place is diploid, having arisen by mitosis from a spermatogonial cell.

■ Meiosis I separates the chromosomal homologs, producing two haploid **secondary spermatocytes.** The haploid chromosome number is traceable to this reductional division.

■ Meiosis II separates the chromatids. By the mitotic division of two secondary spermatocytes, four cells are produced. These haploid **spermatids** then differentiate into haploid gametes called **spermatozoa,** or **sperm.** Since the secondary spermatocyte and its two daughter cells are all haploid, this is not a reductional division.

We call the process **gametogenesis** in both sexes, and spermatogenesis in males.

Oogenesis

Development of an **oocyte** into an egg cell in animals follows a similar pattern. However, egg development, or **oogenesis,** differs from spermatogenesis in that only one of the four emerging haploid cells becomes a functioning gamete (see Figure 12-28).

Unlike the sperm cell, the animal egg cell is immobile. It is the gamete that stores the nutrients needed to sustain an early embryo for a time after fertilization; hence it is much larger in size than the sperm. Growth of the egg and storage of nutrient (yolk substances) begin in prophase I. This process may take a long time—days to years—depending upon the size of the egg. In some species they remain arrested in prophase I until fertilized. Others remain in metaphase I and still others in metaphase II. A few, like the sea urchin's, complete meiosis before fertilization.

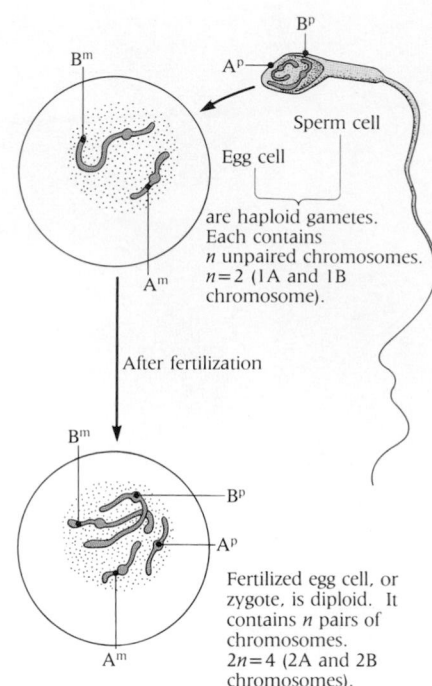

are haploid gametes. Each contains n unpaired chromosomes. $n=2$ (1A and 1B chromosome).

Fertilized egg cell, or zygote, is diploid. It contains n pairs of chromosomes. $2n=4$ (2A and 2B chromosomes).

12-27 Origin of diploidy at fertilization

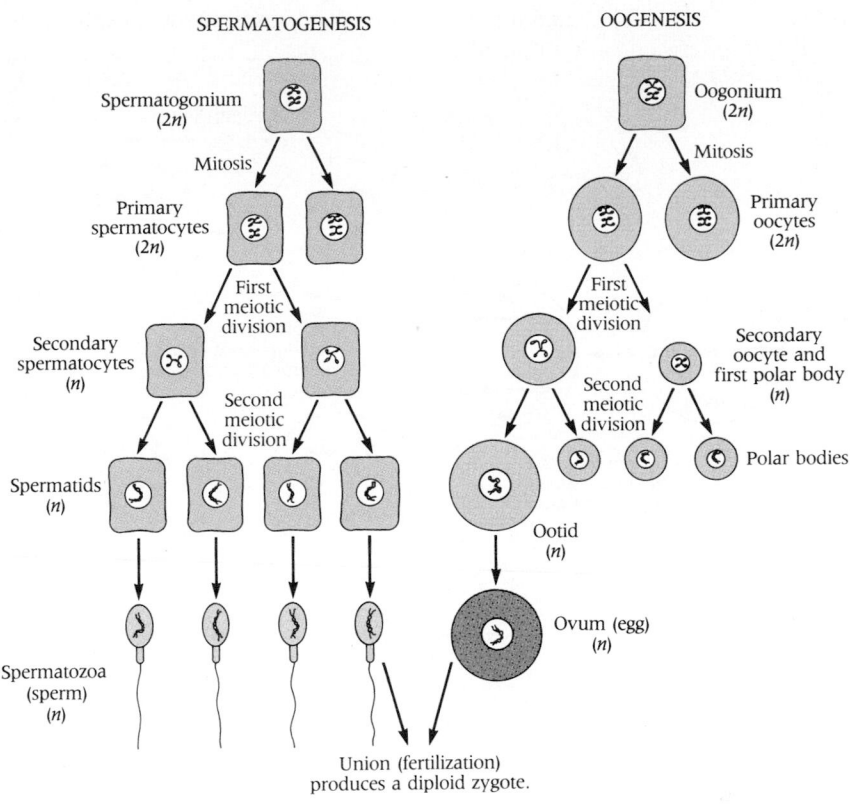

SPERMATOGENESIS   OOGENESIS

Spermatogonium (2*n*)   Oogonium (2*n*)

Mitosis   Mitosis

Primary spermatocytes (2*n*)   Primary oocytes (2*n*)

First meiotic division   First meiotic division

Secondary spermatocytes (*n*)   Secondary oocyte and first polar body (*n*)

Second meiotic division   Second meiotic division

Spermatids (*n*)   Polar bodies

Ootid (*n*)

Spermatozoa (sperm) (*n*)   Ovum (egg) (*n*)

Union (fertilization) produces a diploid zygote.

12-28 Gametogenesis in animals
In both spermatogenesis and oogenesis, haploid gametes are produced from diploid precursor cells. Note that four spermatozoa derive from each spermatogonium, but only one ovum is produced from each oogonium.

Egg proper

Zona pellucida of the egg   First polarbody of the egg   Follicle cells

12-29 Polar body
The unfertilized golden hamster egg shown in this phase-contrast micrograph has a single polar body (arrow). This indicates that the egg has completed meiosis I. (× 600)

Regardless of these differences, certain critical events after metaphase I are the same in all species. At anaphase I and telophase I, the cytoplasm divides unequally. As a result, one of the two nuclei present at prophase II is converted into a peculiar small structure called a **polar body,** which is little more than a vesicle containing the chromosomes lost during a meiotic division (Figure 12-29). The same thing happens during anaphase II and telophase II—when a second polar body is produced. During meiosis II, the first polar body either breaks up, remains quiescent, or divides, depending upon the species. As a result, the final product of oogenesis is a single large egg cell, called the **ovum,** with one, two, or three nonfunctioning polar bodies attached at one side, we emphasize again, is haploid.

Mechanisms in Meiosis

Meiosis and mitosis are compared in Figure 12-30 and in Tables 12-2 and 12-3 especially with regard to chromosome movements. Note that in mitosis a single replication of chromosomes is followed by one division of the nucleus. In meiosis, a single replication of chromosomes is followed by two divisions of the nucleus. That is why the chromosome number is halved in the daughter cells.

It is instructive to examine the amount of nuclear DNA in the different stages of meiosis and mitosis (Figure 12-31). If we denote as C (or $1C$) the amount of DNA in gametes, which are haploid, then the DNA level in diploid cells at G_1, before mitosis, would be twice this amount, or $2C$. The amount of DNA in G_2 cells, after DNA replication in S stage, would be $4C$. If this cell enters another mitosis, daughter nuclei will contain the normal $2C$ diploid level of DNA at the completion of division. However, if, instead, the cell enters a meiotic division, the $4C$ level of DNA will be reduced at the end of the two sequential meiotic divisions to the $1C$ haploid level. The DNA level is then restored to the $2C$ diploid value by fusion of two gamete nuclei during fertilization.

Cells enter a meiotic division after a series of ordinary mitotic divisions with their conventional interphases. Just before meiosis begins, interphase differs from an ordinary premitotic interphase in the time required for DNA replication in the S stage. In this premeiotic cell, S phase is quite prolonged. Indeed, the slowing of DNA replication in premeiotic cells may come on gradually in the several cell generations preceding meiosis.

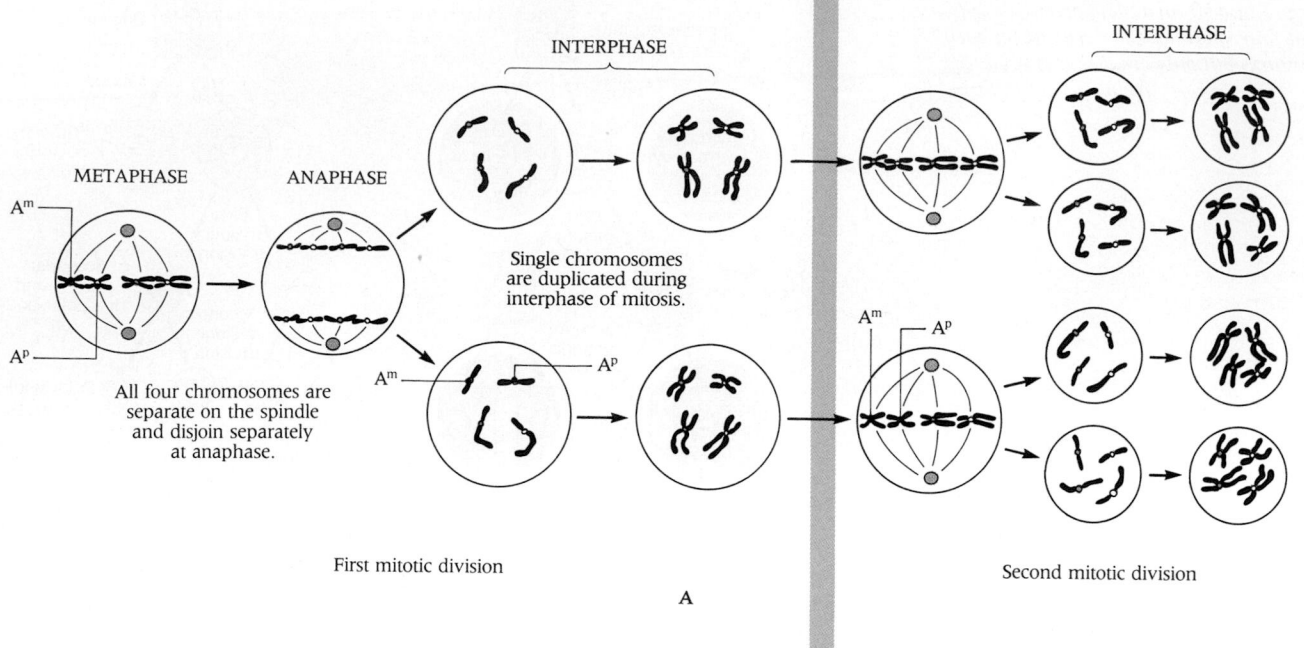

CHROMOSOME CYCLE IN MITOTIC DIVISION

INTERPHASE

INTERPHASE

METAPHASE

ANAPHASE

A^m

A^p

All four chromosomes are
separate on the spindle
and disjoin separately
at anaphase.

Single chromosomes
are duplicated during
interphase of mitosis.

A^m

A^p

A^m

A^p

First mitotic division

Second mitotic division

Products of 2 mitotic divisions are 4 diploid cells

A

CHROMOSOME CYCLE IN MEIOTIC DIVISION

INTERKINESIS

DIFFERENTIATION

Homologous chromosomes
segregate from each other;
only one member of each pair
enters the new nucleus.

A^p

A^p

A^p

A^p

METAPHASE I

ANAPHASE I

A^p

No chromosome duplication
during interphase between
the meiotic divisions.

A^p

A^m

A^m

A^m

A^m

First meiotic division

Second meiotic division

Products of 2 meiotic divisions are 4 haploid cells

B

Following the premeiotic G$_2$ stage, cells enter into the two consecutive divisions of meiosis. Each has four stages that would be similar to those of mitosis but for the events that result in haploidy. **Interkinesis,** a brief period of greater or shorter duration, may occur between the two meiotic divisions.

12-30 Mitosis and meiosis compared
(**A**) Mitosis. (**B**) Meisosis.

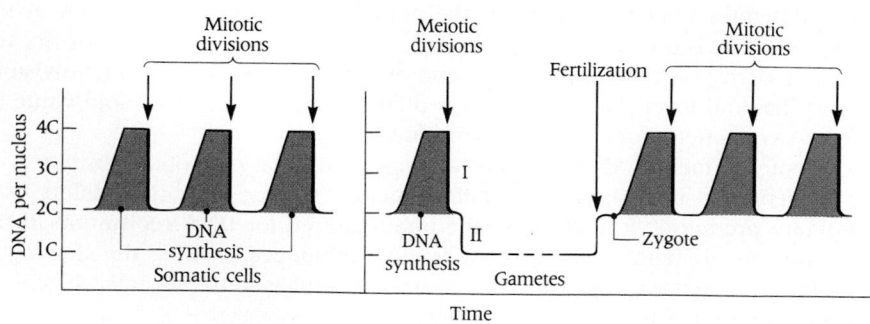

Mitotic
divisions

Meiotic
divisions

Fertilization

Mitotic
divisions

4C

3C

2C

1C

DNA per nucleus

I

II

DNA
synthesis

DNA
synthesis

Zygote

Somatic cells

Gametes

Time

12-31 Changes in nuclear DNA content in mitosis and meiosis
The base level (1C) represents the haploid amount of DNA.

TABLE 12-2
A COMPARISON OF MITOSIS AND MEIOSIS

Stage	Mitosis	Meiosis	
		First Division (I)	Second Division (II)
Prophase	Each chromosome appears as two identical chromatids. They move toward center of cell. Spindle forms. Nuclear envelope fragments. Nucleoli disappear.	Each chromosome appears as two identical chromatids. Homologous pairs synapse to form tetrads. Spindle forms. Nuclear envelope fragments.	Chromosomes reappear as double-stranded structures. Their number is now haploid. New spindle forms. Nuclear envelope fragments. Crossing-over occurs.
Metaphase	Double-stranded chromosomes align on equatorial plane and attach to spindle fibers. Chromatids separate by centromere division, and sister chromatids come under influence of opposite poles.	Tetrads align at spindle midpoint, but no centromere division occurs. Homologous chromosomes, each still consisting of two chromatids, attach to fibers from opposite poles.	Double-stranded chromosomes align at spindle center and attach to spindle fibers. This time, centromeres divide, two identical haploid sets in each of the dividing cells.
Anaphase	Chromosomes, now single-stranded, move toward opposite poles. Division furrow begins to form.	Chromosomes, still double-stranded, move toward opposite poles. Division furrow appears.	Each haploid set moves toward its own pole, and the second cleavage begins.
Telophase	Chromosomes fade from view. New nuclear envelopes form. Division of nucleus (karyokinesis) is usually accompanied by a division of the cytoplasm (cytokinesis) to yield two identical diploid daughter cells.	Spindle dissolves. New nuclear envelopes form. Two nonidentical cells with haploid chromosome number are formed, but each chromosome still consists of two identical (sister) chromatids.	Spindle dissolves. New nuclear envelopes form. Meiosis is complete, leaving a total of four haploid cells of two allelic types.

Prophase I

Prophase of meiosis I is more complex than mitotic prophase. It is divided into five stages, which have been named in two ways—by a newer nomenclature based on major activities of the chromosomes and by an older one based on structure of the chromosomes. Both systems are given in the following list, with functional terms preceding morphological ones:

- First stage: Condensation   Leptotene (*leptos*, "fine or thin")
- Second stage: Pairing   Zygotene (*zygon*, "yoke," or the letter *Y*)
- Third stage: Recombination Pachytene (*pachus*, "thick")
- Fourth stage: Coiling   Diplotene (*diplos*, "double")
- Fifth stage: Recondensation Diakinesis (*dia* and *kinesis*, "across" and "movement")

12-32 Chromosomal rearrangements during prophase I of meiosis

The only way to identify a cell's stage is by its appearance. The functional terms merely indicate what we believe chromosomes are doing rather than what they look like. The highlights of these stages are summarized in Figure 12-32.

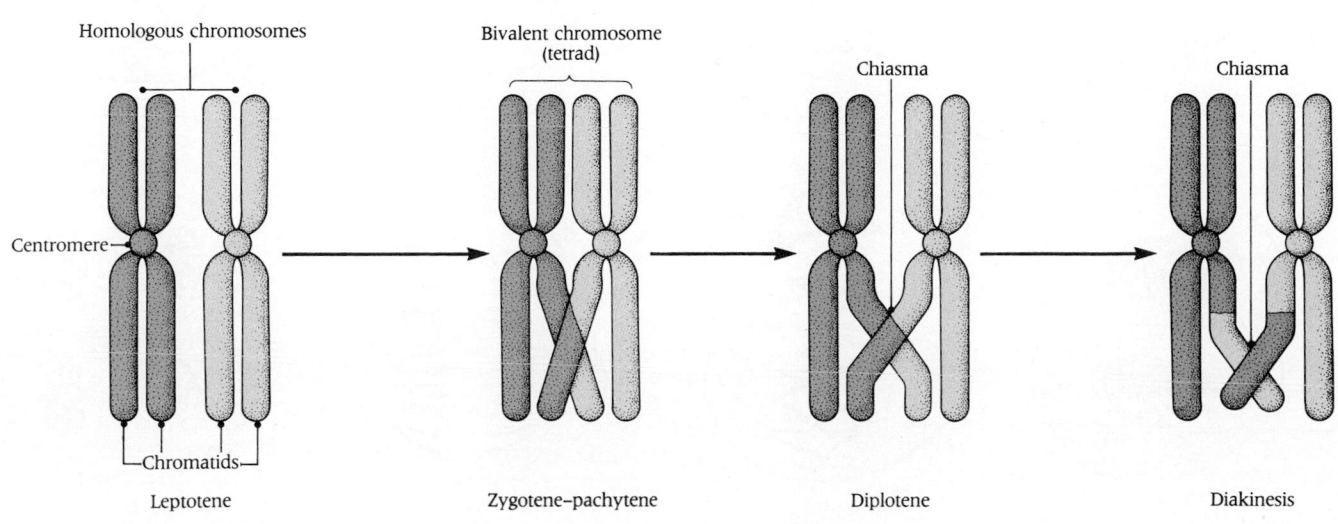

Leptotene   Zygotene–pachytene   Diplotene   Diakinesis

TABLE 12-3

MAJOR DIFFERENCES BETWEEN MITOSIS AND MEIOSIS

Feature	Mitosis	Meiosis
Number of division cycles	One	Two (I and II)
Change in chromosome number from parent to daughter cell	None	Number halves
Genetic identity of daughter cells	Identical	Several types present. Only four cells produced, but many genetic patterns may emerge.
Synapsis or tetrad formation	None	Occurs
Crossing-over	None	Occurs (in Prophase I)
General occurrence	In replication of somatic cells	In production of gametes
Duration	Brief (10–30 minutes in many cells)	May require several days to many weeks

12-33 Stages of meiosis

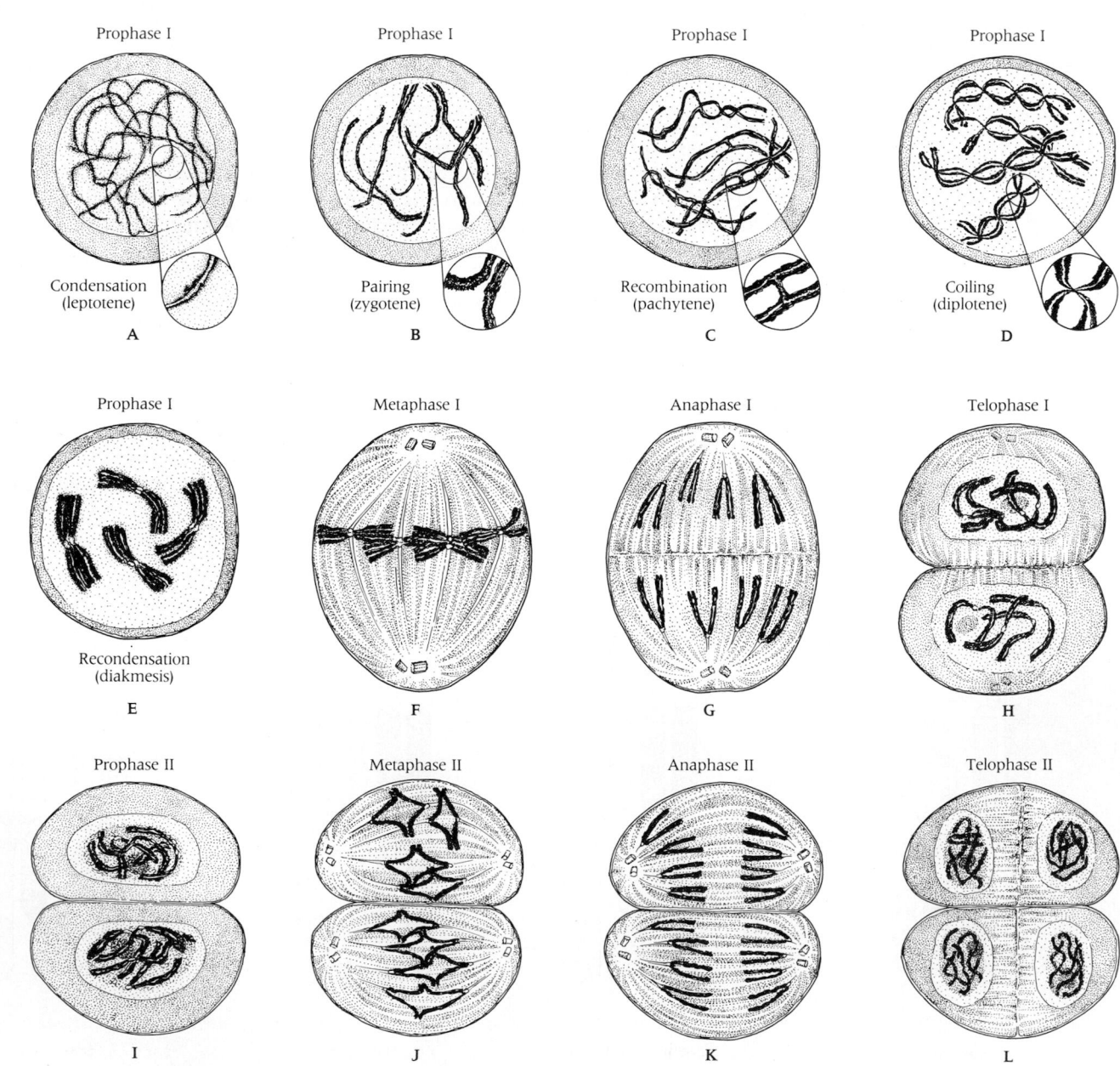

Prophase I	Prophase I	Prophase I	Prophase I
Condensation (leptotene)	Pairing (zygotene)	Recombination (pachytene)	Coiling (diplotene)
A	B	C	D

Prophase I	Metaphase I	Anaphase I	Telophase I
Recondensation (diakmesis)			
E	F	G	H

Prophase II	Metaphase II	Anaphase II	Telophase II
I	J	K	L

In the **condensation stage,** chromosomes become visible, appearing as long threads, often with beaded thickenings. The functional significance of these swellings, called **chromomeres,** is unclear. While this stage superficially resembles mitotic prophase, it yields chromosomes with recognizably different structures and arrangements. First, the two chromatids of each chromosome, which arose during premeiotic interphase, are so close together that they are hard to distinguish (Figure 12-33A). Hence the chromosomes appear as single rather than double threads. Second, both ends of each chromosome are in many cases attached to the nuclear envelope.

Next comes the **pairing stage.** We will recall that, in diploid cells, each chromosome is represented twice, one homolog of the pair deriving from the male parent, the other from the female. Thus they contain the same genes. Prior to the pairing stage, the two homologs of each pair are randomly distributed in the nucleus and often far apart. The condensation stage gives way to the pairing stage as the homologs begin to pair up (see Figure 12-33B).

Pairing, also called **synapsis,** precisely aligns the two homologs in a close chromomere-by-chromomere and gene-by-gene association. Although synapsis puts the homologs close together, they remain separated by a narrow space, about 0.2 μm wide—just wide enough to be visible in the light microscope. By electron microscopy the space is seen to contain an unusual structure called the **synaptonemal complex** (Figure 12-34), extending the length of the synapse and anchored to the nuclear envelope.

The **recombination stage** lasts much longer than the two previous stages. It begins when pairing is completed. In this stage, homologous chromosomes complete an exchange of segments that began in zygotene. In this way, they generate new gene alignments. This involves the process of **crossing-over** (see Figure 12-33C), which is described in Chapter 14. Since each chromosome of a synaptonemal pair contains two chromatids, a total of four chromatids are in close association in each of the paired structures during the recombination stage. Two terms are commonly applied to these synapsed chromosomes.

- When chromatids are being referred to, the synapsed chromosome pair is called a **tetrad** because four chromatids are present and visible.
- When chromosomes are being referred to, the same structure is called a **bivalent** because two homologous chromosomes make up each pair. The recombination stage ends as the synaptonemal complex appears to dissolve and homologs begin to separate. As the area between chromatids widens, all four chromatids of the tetrad become clearly visible.

These events mark the beginning of the **coiling stage** of meiotic prophase I (see Figure 12-33D). The homologous chromosomes become so widely separated that they actually appear to be repelling each other. However, they remain physically joined in several scattered locations called **chiasmata** (singular, **chiasma**). These points are sites of cross-overs, and they are part of the evidence that exchanges do take place between homologs in the recombination stage.

The synthesis stage may be lengthy. Indeed, in developing amphibian egg cells, this stage of prophase I may last a year or more. In humans, oocytes reach this stage in unborn females at about the fifth month of fetal life and remain arrested at this point until the individual reaches sexual maturity. Then, just before ovulation, one oocyte each month breaks arrest and continues the meiotic sequence. The time between the onset and completion of the synthesis stage in human females thus may extend for decades.

As this stage ends, the chromosomes again become condensed. This event marks the onset of **recondensation,** the final stage of meiotic prophase I (see Figure 12-33E). Chromosomes reach their greatest density in this stage. Perhaps for that reason, the chiasmata move toward the ends of the chromosomes. This process, called **terminalization,** may continue until many of the tetrads are held together by remaining chiasmata located at their tips. During this stage, nucleoli disappear and the cell envelope breaks up. Metaphase I follows.

Metaphase I

The remaining stages of meiosis are concerned primarily with division of the chromosomes into the haploid number. Near the end of prophase I, the spindle forms for

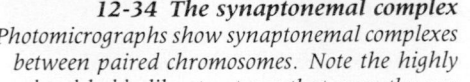

12-34 The synaptonemal complex
Photomicrographs show synaptonemal complexes between paired chromosomes. Note the highly ordered ladderlike structures that cross the gap between homologous chromosomes. (× 3500)

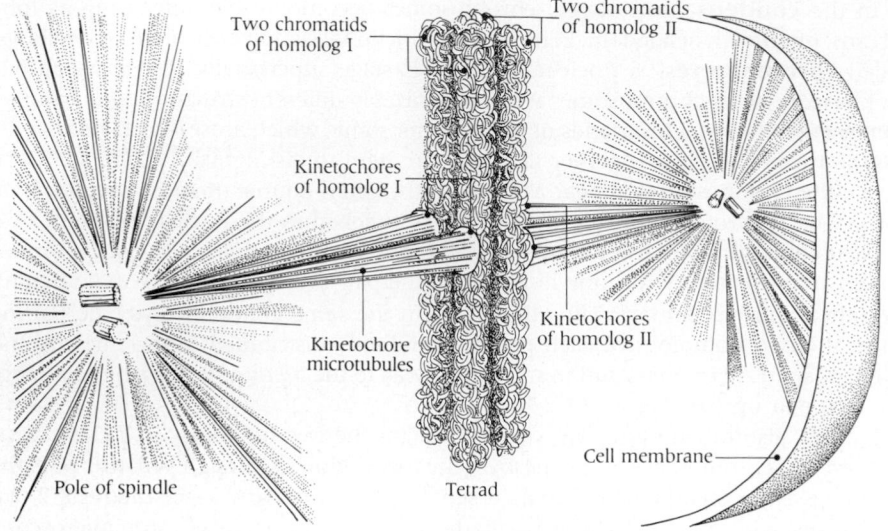

Two chromatids of homolog I

Two chromatids of homolog II

Kinetochores of homolog I

Kinetochores of homolog II

Kinetochore microtubules

Cell membrane

Pole of spindle

Tetrad

12-35 Alignment of tetrads in metaphase I of meiosis
Kinetochore microtubules link each chromatid with one pole of the cell. Compare this with Figure 12-14.

the first meiotic division. Centrioles, if present, complete their replication. This pattern is most typical of the meiotic divisions in spermatogenesis. In oogenesis, the centrioles disappear and cannot be detected in developing oocytes during meiosis. The spindle forms without the involvement of centrioles, essentially as it does in plants.

Breakdown of the nuclear envelope marks the beginning of metaphase I (see Figure 12-33F). The tetrads left scattered by the breakdown of the nuclear envelope move to the spindle midpoint. (Note again that pairing of homologous chromosomes into tetrads occurs in meiosis, not in mitosis.)

Another major divergence from the mitotic pattern—indeed, the most distinctive event of the divisions following prophase I—occurs as the tetrads become attached to spindle microtubules. Each homologous chromosome has two kinetochores, one for each of its two chromatids (Figure 12-35). Hence four kinetochores are present in each tetrad. The two kinetochores in each homolog connect to the same spindle pole at metaphase I. Both kinetochores of the other homolog attach to spindle microtubules connecting to the opposite pole. In other words, the two kinetochores of a homolog act together at metaphase I.

Anaphase I

The homologs separate as they move toward their respective poles at the beginning of anaphase I (see Figure 12-33G). Anaphase separates the chromosomes of every homologous pair and delivers each one to an opposing spindle pole. As a result, each pole receives a haploid number of chromosomes by the end of anaphase I. But because each chromosome is still double, containing two chromatids unlike the chromosomes in mitotic anaphase, a $2C$ amount of DNA is present at each of the cell's poles.

An important consideration now arises. If one member of each pair came originally from the male parent (A^p) and one came from the female parent (A^m), we may now inquire: Are all the chromosomes going to one pole derivatives of the male parent while all chromosomes going to the other pole are derivatives of the female parent? It obviously could happen that way because the members of each homologous pair separate in an entirely random manner. However, this arrangement would be highly unlikely when more than a few chromosome pairs are present.

In the example in Figure 12-36, A^m goes to one pole of the spindle in meiosis I and A^p to the other. But B^m may go to the same pole as A^m because the orientation of the B pair on the spindle is quite independent of that of the A pair. The alternative metaphase arrangements are equally probable: B^m may go to the same spindle pole with A^p as often as with A^m. The following combinations of chromosomes in the gametes are, therefore, all equally probable and equally frequent:

A^m and B^m
A^p and B^p
A^m and B^p
A^p and B^m

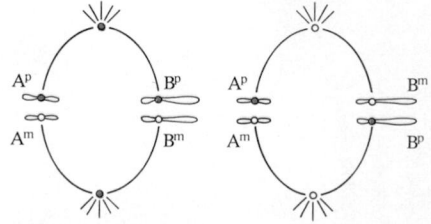

12-36 Assortment of chromosome pairs in metaphase
Maternal and paternal chromosomes of a given pair align randomly at metaphase. The two arrangements shown here occur in equal frequency.

The randomness of these combinations contributes to the genetic variability of the products of meiosis. In a human with 23 pairs of chromosomes, 2^{23} combinations of maternal and paternal chromosomes are possible. Even without recombination, the probability that two children of the same parents will receive the same combination of maternal and paternal chromosomes would be 1 chance out of 2^{23}, or 1 out of 70 trillion! The further variability introduced by recombination, which mixes chromatid segments randomly between maternal and paternal chromatids, makes it practically impossible for an individual to produce genetically identical offspring (except for identical twins).

Telophase I and interkinesis

Well-defined telophase I and interkinesis stages do not always occur between anaphase I and the second meiotic division. In most species, telophase I and interkinesis are transitory stages between the first and second meiotic divisions, in which the cells pause only briefly before entering prophase II (see Figure 12-33H). However, many variations are found in nature.

During this period, the single metaphase spindle is reorganized into two spindles that form in the regions of the telophase I spindle poles. Centrioles and asters, if present, also divide at this time, placing a single centriole at each pole of the two spindles. These events complete the cellular rearrangements leading to the second meiotic division.

The second meiotic division

After a brief or even nonexistent prophase II (see Figure 12-33I), the chromosomes left at the two poles by the first meiotic division move to the midpoints of the two newly formed spindles. If decondensation has occurred during telophase I or interkinesis, the chromosomes condense into tightly coiled rods as they move to the metaphase II spindle. As the chromosomes attach to the spindle, the two kinetochores of each chromosome rotate to face the opposite poles of the spindle and the chromatids attach to opposite spindle poles (see Figure 12-33J).

At anaphase II (see Figure 12-33K), the chromatids of each chromosome separate and move to opposite poles of the spindle. This delivers the haploid number of chromatids to each pole. Each pole now contains a $1C$ quantity of DNA, which is equivalent to one-fourth of the DNA present at G_2 in the cell that originally entered meiosis. At telophase II (see Figure 12-33L), new nuclear envelopes form around the four division products and the chromatids again fade from view.

The four new nuclei have different fates in different plant and animal species. In the males of animal species, all four nuclei are in separate cells, which differentiate into sperm cells, the functional male gametes (see Figure 12-28). In female animals, as noted, only one of the nuclei resulting from meiosis becomes functional as the egg nucleus. The remainder are compartmented at one side of the oocyte as polar bodies.

Complex plants undergo a similar developmental pattern. All four of the products of meiosis give rise to the "sperm" nuclei of pollen. Only one of the meiotic products in female plants survives to give rise, after several mitotic divisions, to an egg nucleus (see Chapter 21).

Biological Variations

We note in closing that we have been describing the familiar pattern seen in animals and many protists. As shown in Figure 12-37A, this has been termed **gametic,** or **terminal, meiosis,** to contrast it with two modes occurring in other eukaryotic organisms.

Sporic, or **intermediate, meiosis** (Figure 12-37B) occurs in complex plants and some fungi with alternating generations of haploid and diploid individuals.

Zygotic, or **initial, meiosis** occurs in most fungi and some protists (Figure 12-37C). Meiotic divisions occur immediately after fertilization, the zygote developing into four haploid cells. These eventually develop into haploid spores and ultimately into haploid individuals. A summary of patterns is shown in Figure 12-38.

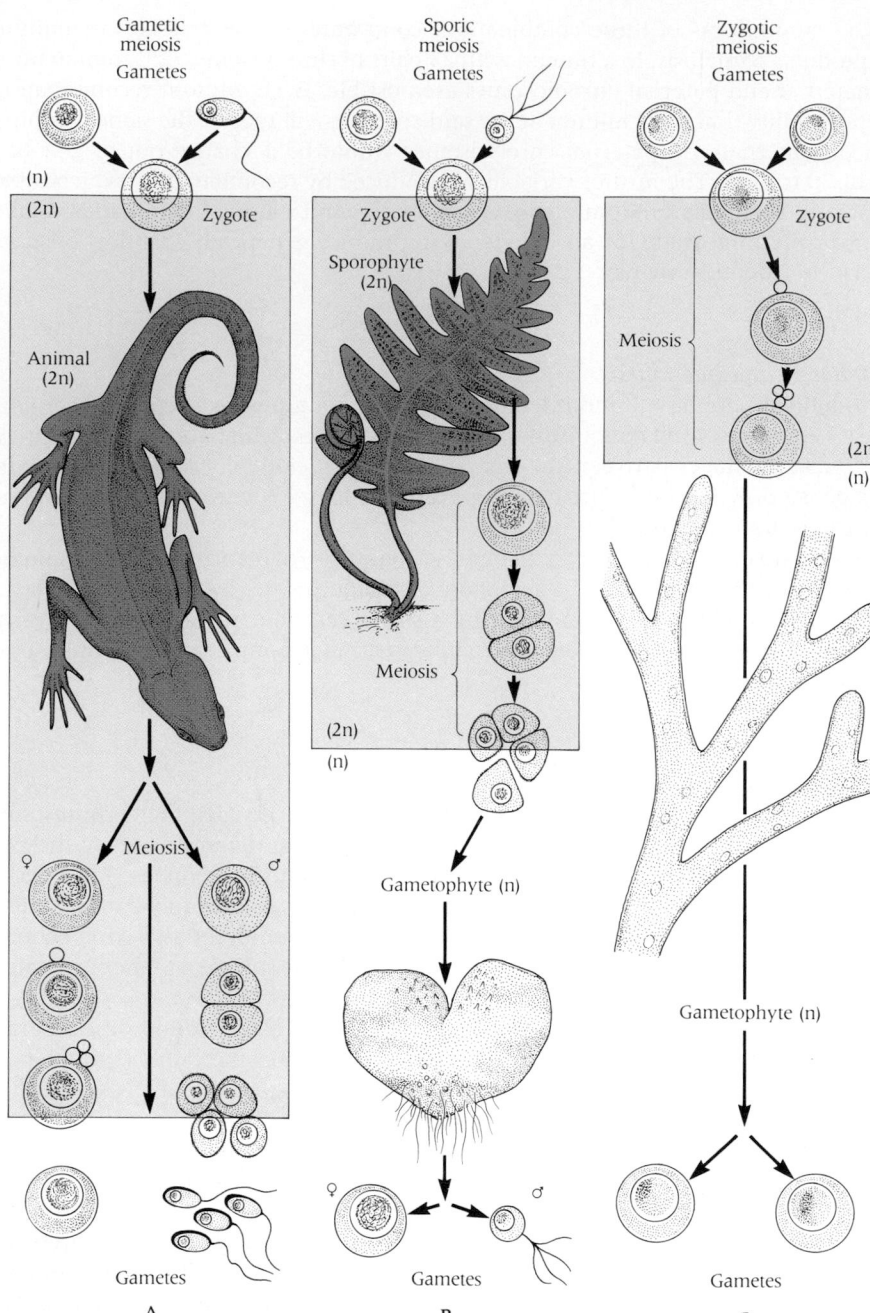

Figure labels within illustration:

Gametic meiosis / Gametes
(n)
(2n) Zygote
Animal (2n)
♀ Meiosis ♂
Gametes
A

Sporic meiosis / Gametes
Zygote
Sporophyte (2n)
Meiosis
(2n)
(n)
Gametophyte (n)
♀ ♂
Gametes
B

Zygotic meiosis / Gametes
Zygote
Meiosis
(2n)
(n)
Gametophyte (n)
Gametes
C

Abnormalities of Meiosis

A state in which the chromosome number or ploidy is normal is termed **euploidy.** Human beings, and presumably all species, commonly suffer from abnormalities of meiosis that result in cells containing one or more extra chromosomes. Abnormal states like those to be described are examples of **aneuploidy,** in which a cell has a chromosome number differing from the normal number for the species by a small number of chromosomes. Those particular disorders displaying high ploidy are forms of **polyploidy.** (However, as we will see later, polyploidy occurs normally in half of all plant species—indeed it plays a key role in the origin of new plant species.) The additional chromosome is presumably normal in terms of its genetic information and structure. What is abnormal is the chromosome number. Such aberrations are the result of errors of cell division that are called **nondisjunctions.**

A nondisjunction is a failure of the normal separation of homologous chromosomes during cell division. As we have seen, in meiosis duplicated chromosomes normally pair, or synapse, with their homologous partners and then separate, or disjoin. Failure to disjoin (that is, nondisjunction) yields one daughter cell with an extra chromosome and one lacking this chromosome (Figure 12-39).

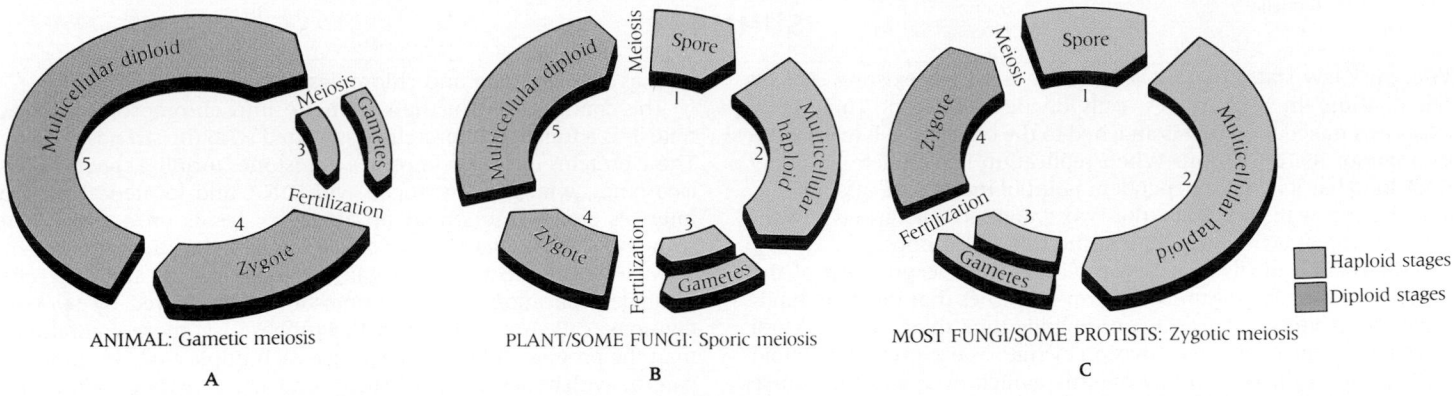

Haploid stages
Diploid stages

ANIMAL: Gametic meiosis

A

PLANT/SOME FUNGI: Sporic meiosis

B

MOST FUNGI/SOME PROTISTS: Zygotic meiosis

C

12-38 Summary of three types of meiosis

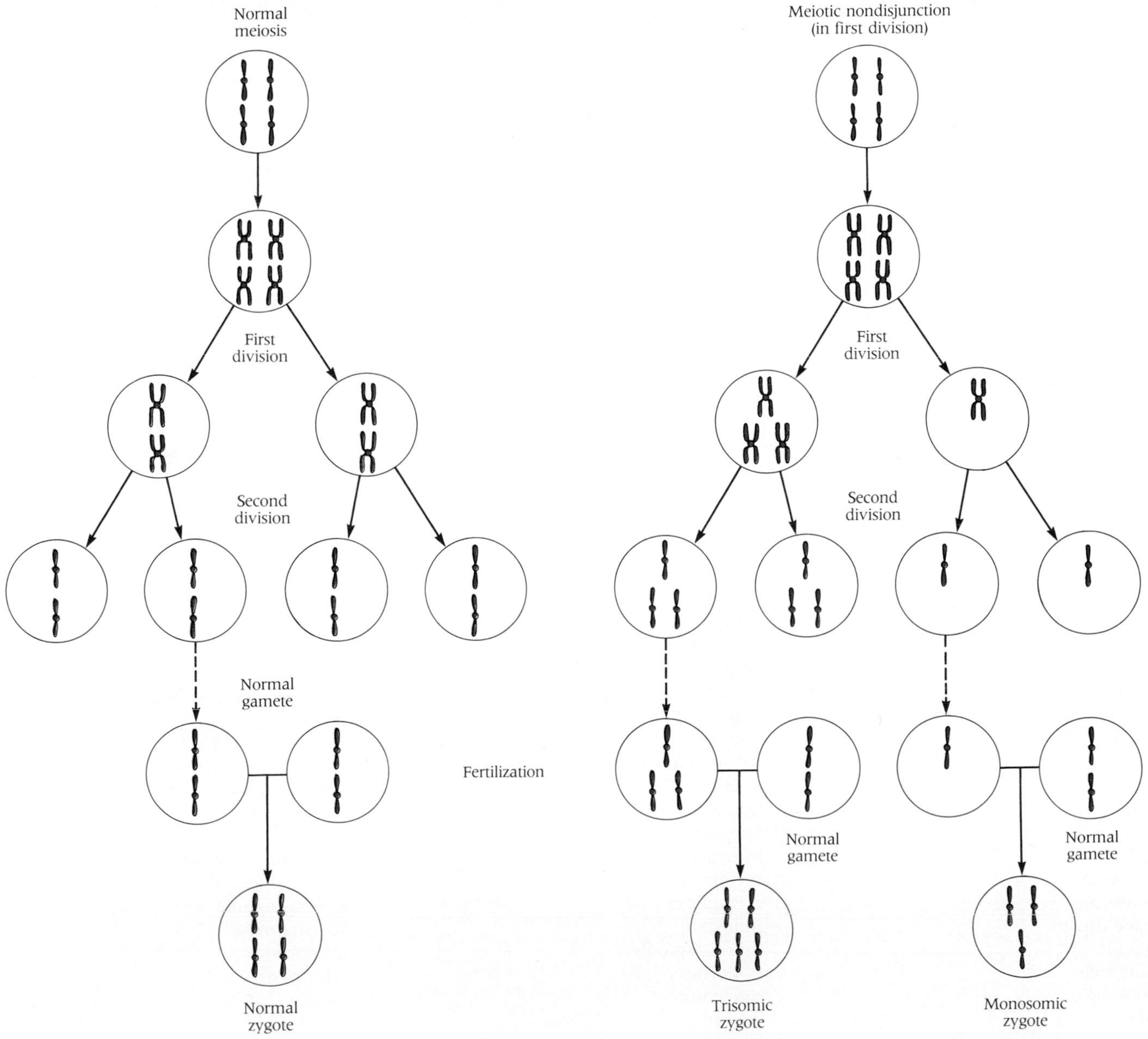

12-39 Meiotic nondisjunction
The failure of homologous chromosomes to separate properly during meiosis can lead to chromosomal imbalance in some of the daughter cells—for example, a trisomic gamete (with 2n + 1 chromosomes) or a monosomic gamete (with 2n − 1 chromosomes).

It is noteworthy that 40–50% of human embryos lost by miscarriage in the first 60 days of pregnancy have a chromosomal abnormality. These almost always involve changes only in chromosome number (aneuploidy) and hence are believed to result from nondisjunction.

Many embryos with chromosomal abnormalities die before birth. Nonetheless, many individuals do survive with various chromosomal abnormalities, and it is not unusual to observe aneuploidy in humans—and in organisms of all kinds. In humans, many of these are the sex chromosome disorders to be considered in Chapter 15.

Normal
meiosis

First
division

Second
division

Normal
gamete

Normal
zygote

Meiotic nondisjunction
(in first division)

First
division

Second
division

Fertilization

Normal
gamete

Normal
gamete

Trisomic
zygote

Monosomic
zygote

SUMMARY

Virchow's law states that all cells derive from pre-existing cells by cell division. In prokaryotes, cells divide by fission. Their DNA exists as a naked loop that is attached to the bacterial cell membrane throughout its replication. When replication is complete, the new DNA loop has its own independent point of membrane attachment, and the cell wall that splits the two daughter cells grows inward between the attachment points of the two DNA loops.

Eukaryotic cell division is complicated by the presence of the nucleus and the large number of chromosomes that must be duplicated. Most eukaryotic cells are diploid; thus they contain a full set of paired chromosomes. Division by mitosis preserves the diploid number of a cell. Division by meiosis, which occurs in the course of gamete formation, halves the chromosome number and produces a haploid cell. The eukaryotic cell cycle, the series of events from the beginning of one cell division to the next, can be divided into mitosis and interphase. Interphase includes growth of the cell and duplication of the DNA prior to the next mitosis.

Mitosis is a sequence of four stages: prophase, metaphase, anaphase, and telophase. In prophase, the duplicated chromosomes become visible, the nucleolus and nuclear envelope disappear, and the centrioles migrate to the poles of the cell.

Metaphase is distinguished by the formation of spindle microtubules that attach to the chromosomes and help them line up in the equatorial plane of the cell. In anaphase, the chromosomes divide at their centromeres and one member of each duplicated pair is pulled to each pole of the cell. Meanwhile, cytokinesis—division of the cytoplasm—begins and the poles begin moving apart. Telophase is the last stage of mitosis. The chromatids become diffuse and disappear from view, and new nuclear envelopes form around them. The completion of cytokinesis follows the end of the nuclear division. During cytokinesis, duplication of cytoplasmic organelles such as mitochondria and chloroplasts occurs.

The condensation of the chromatin into chromosomes during mitosis is a function of proteins associated with the strands of DNA. These proteins include complexes of histones forming spherical nucleosomes, which are wrapped with DNA and located at regular intervals along each strand of DNA, like beads on a string. The string of beads is then further coiled to form chromatin fibers.

Meiosis produces haploid gametes—the sperm and egg cells. A single replication of the chromosomes is followed by two cell divisions, called meiosis I and II. Prophase I is more complicated than the prophase of mitosis because each duplicated chromosome pairs up with its homolog to form a tetrad. Once this is accomplished, recombination occurs: the pairs of duplicated homologues exchange segments by the process of crossing over. In metaphase I, the tetrads line up along the cell's midpoint. Anaphase I is marked by the separation of the duplicated homologues, as each is pulled to the opposite pole of the cell. Although one homolog of each pair comes from each of the cell's original parents, the homologs assort independently of the parent from which they have originated. This randomness contributes to the genetic variability of the cells produced by meiosis, as does the process of recombination. Telophase I is followed by interkinesis, during which the single spindle at each pole reorganizes into two spindles.

The four spindles are put to use in meiosis II. The unpaired, doubled homologues move to the new spindle midpoints and are separated so that there are now four collections of chromatin within the cell, each containing only one set of unpaired, unduplicated homologues. During telophase II, new nuclear envelopes form around the four division products. The fate of these four nuclei vary in different species and cell lines, but normally at least one of them survives to become a haploid gamete.

KEY TERMS

anaphase
aneuploidy
aster
bacterial chromosome
base pairs (bp)
budding
cell cycle
cell division
cell plate
centriole
centromere
chiasma
chromatid
chromatin
chromomere
chromosome
cleavage furrow
contractile ring
crossing over
cytokinesis
diploid

egg
euploidy
fission
gamete
gametogenesis
germ cell
gonads
haploid
histone
homologous pair
interphase
karyokinesis
karyotype
kinetochore
meiosis
metaphase
mitosis
mitotic spindle
nondisjunction
nucleosome
oocyte

oogenesis
ovum
phragmoplast
ploidy
polar body
polyploidy
polytene chromosome
prophase
reductional division
sexual reproduction
sister chromatid
somatic cell
sperm
spermatid spermatocyte
spermatogenesis
spermatogonia
spermatozoa
synapsis
telophase
zygote

QUIZ QUESTIONS

1. The genetic diversity arising from meiosis results from
   A. crossing-over during prophase I.
   B. mixing of maternal and paternal chromatids during meiosis II.
   C. replication of DNA during interphase.
   D. A and B.
   E. A, B, and C.

2. The genophore of a prokaryotic cell
   A. is an elongate strand of DNA that is complexed with proteins.
   B. is a circular structure of DNA that is complexed with histones.
   C. is located in the nucleus.

   D. B and C.
   E. None of the above.

3. In which of the following stages of cell division is the nuclear envelope of a dividing eukaryotic cell intact?
   A. metaphase I
   B. anaphase II
   C. late prophase
   D. interphase
   E. A and C

4. Pairing of homologous chromosomes into tetrads occurs during
   A. prophase of meiosis I.
   B. metaphase of mitosis.

   C. telophase of meiosis II.
   D. interkinesis of meiosis.
   E. B and D.

5. For each diploid cell entering gametogenesis, _____ functional sperm cell(s) and _____ functional ovum (ova) is (are) produced.
   A. 4; 4
   B. 2; 4
   C. 2; 2
   D. 4; 1
   E. 1; 4

ESSAY QUESTIONS

1. How does cell division differ in prokaryotic and eukaryotic cells?

2. Outline the stages of mitosis. What is the role of the kinetochore? What might happen if a chromosome lost its kinetochore?

3. Contrast mitosis in plant and animal cells. What role do the centrioles play?

4. The drug colchicine binds to tubulin and causes depolymerization of microtubules. Cytochalasin B causes depolymerization of actin-based microfilaments. What effect would application of either drug have on mitosis in a typical plant cell?

5. The current concepts of chromatin structure pictures DNA and its associated proteins as "beads on a string." What experimental observations support this model?

6. Outline the stages of meiosis. What aspects mimic mitosis? What is a reduction division?

REFERENCES AND SUGGESTIONS FOR FURTHER READING

Alov, I. A., and Lyubskii, S. L. (1977). Functional morphology of the kinetochore. *International Review of Cytology Supp. 6*, 59–74.

A detailed summary.

Attardi, G. and Schatz, G. (1988). Biogenesis of mitochondria. *Annual Review of Cell Biology 4*, 289–333.

How mitochondria replicate during cell division.

Baserga, R. (1985). *The Biology of Cell Reproduction*. Harvard University Press, Cambridge, Mass.

This short book surveys the cell cycle and attempts to interpret its stages in molecular terms.

Bostock, C. J., and Sumner, A. T. (1978). *The Eukaryotic Chromosome*. North-Holland, New York.

A summary of newer views of chromosome structure.

Dunphy, W. G., and Newport, J. W. (1988). Unraveling of mitotic control mechanisms. *Cell 55*, 925–928.

Recent review of an interesting topic.

Edmunds, L. N., Jr. (Ed.) (1984). *Cell Cycle Clocks*. Marcel Dekker, New York.

A collection of fascinating essays.

Fincham, J. R. S., and Oliver, P. (1989). Meiosis: Initiation of recombination. *Nature 338*, 14–15.

New views on how recombination takes place.

Inoué, S. (1981). Cell division and the mitotic spindle. *J. Cell Biol. 91*, 132s–147s.

A consideration of the mechanisms of mitosis with emphasis on the nature and role of the mitotic spindle.

Margulis, L. (1981). *Symbiosis in Cell Evolution*. W. H. Freeman, New York.

A stimulating work that summarizes the many variations in mitosis in different organisms.

Mazia, D. (1974). The cell cycle. *Scientific American 203*, January, 54–64.

A classic article by an early student of the mechanisms of mitosis.

Mitchison, T. J. (1988). Microtubule dynam-

ics and kinetochore function in mitosis. *Annual Review of Cell Biology 4*, 527–549.

The role of the kinetochore is interpreted in light of recent advances in the structure and dynamics of microtubules.

Moens, P. B. (1977). The onset of meiosis. In: *Classic Papers in Genetics*, L. Goldstein, and D. M. Prescott (Eds). Academic Press, Orlando, Fla., pp. 93–108.

A classic paper on the factors underlying the onset of meiosis.

Nicklas, R. B. (1988). The forces that move chromosomes in mitosis. *Annual Review of Biophysics and Biophysical Chemistry 17*, 431–449.

The latest views on the mechanism of chromosome movement in mitosis.

Zimmerman, A. M., and Forer, A. (Eds.) (1981). *Mitosis/Cytokinesis*. Academic Press, Orlando, Fla.

A survey of current knowledge of mitosis and cytokinesis by 20 authors, each discussing one part of the eukaryotic division cycle.

GENETICS

Genetics

The study of heredity first became a science in the nineteenth century. Long before that, however, farmers all over the world knew intuitively how to modify and improve their animals by selective breeding. These unusual pigs from Denmark, products of generations of such breeding, are famous for their length, small size, smooth white hair, large ears, and refined heads.

Important new scientific ideas often emerge when widely separated lines of thought converge. The fusion of two such lines—from studies of cell structure and of heredity—culminated in the chromosome theory (Chapter 14). In this chapter, we consider an instance in which the coming together of a theory of heredity and the cell theory revealed the fundamental rules of inheritance.

Heredity has been pondered much longer than cells and their functions. Among the ancients, Aristotle had quite a bit to say about it, and eighteenth- and nineteenth-century scientists were preoccupied with the subject long before the emergence of the cell theory around 1840 (Box 13-1). We shall now see how fruitfully the two lines of study enhanced one another. Their convergence in 1902 launched a stream of biological insights and advances that is still at floodtide. Surely there is no area of contemporary biological research and application that does not depend in part on an understanding of genetics.

THE RELATION BETWEEN GERM CELLS AND SOMATIC CELLS

The realization that a multicellular organism derives from a single cell focused attention on the relation between germ cells and body cells (or somatic cells). As sketched in Figure 13-1, the different kinds of body cells (muscle, bone, nerve, and so on) are all descendants of a single zygote cell. But if each body cell is a descendant of the same zygote, are egg and sperm simply body cells of another kind? If not, where do gametes come from?

The question was answered by a gifted biologist of Freiburg, August Weismann (1834–1914). In 1885, Weismann, once a practicing physician, pointed out that the germ cells of each generation descend directly from the germ cells of the previous one. In other words, the somatic cells of each generation come from germ cells but do not give rise to them. It was Weissman's conclusion that the cells destined to become germ cells are physically set aside early in development, and only they can produce more germ cells. In his famous phrase, there is a "continuity of the germinal plasm." It is the "germinal plasm," or germ cells, that preserves heredity and ensures that offspring will resemble parents. We now know that this crucial separation is not between populations of cells but between different sets of the genetic instructions encoded in DNA.

Weissman also pointed out that there can be no influencing of the character of the offspring by environmental effects on somatic cells unless environmental forces interact with the germ plasm. He sought in many ways to refute with experimental data the ideas of Lamarck (see Chapter 2 and Box 13-1). He bred many generations of rats, cut off their tails at birth, and searched for offspring without tails. He searched in vain.

A one-way relationship between germ cells and somatic cells implies that acquired traits will not be inherited. It also makes impossible Darwin's pangenesis theory

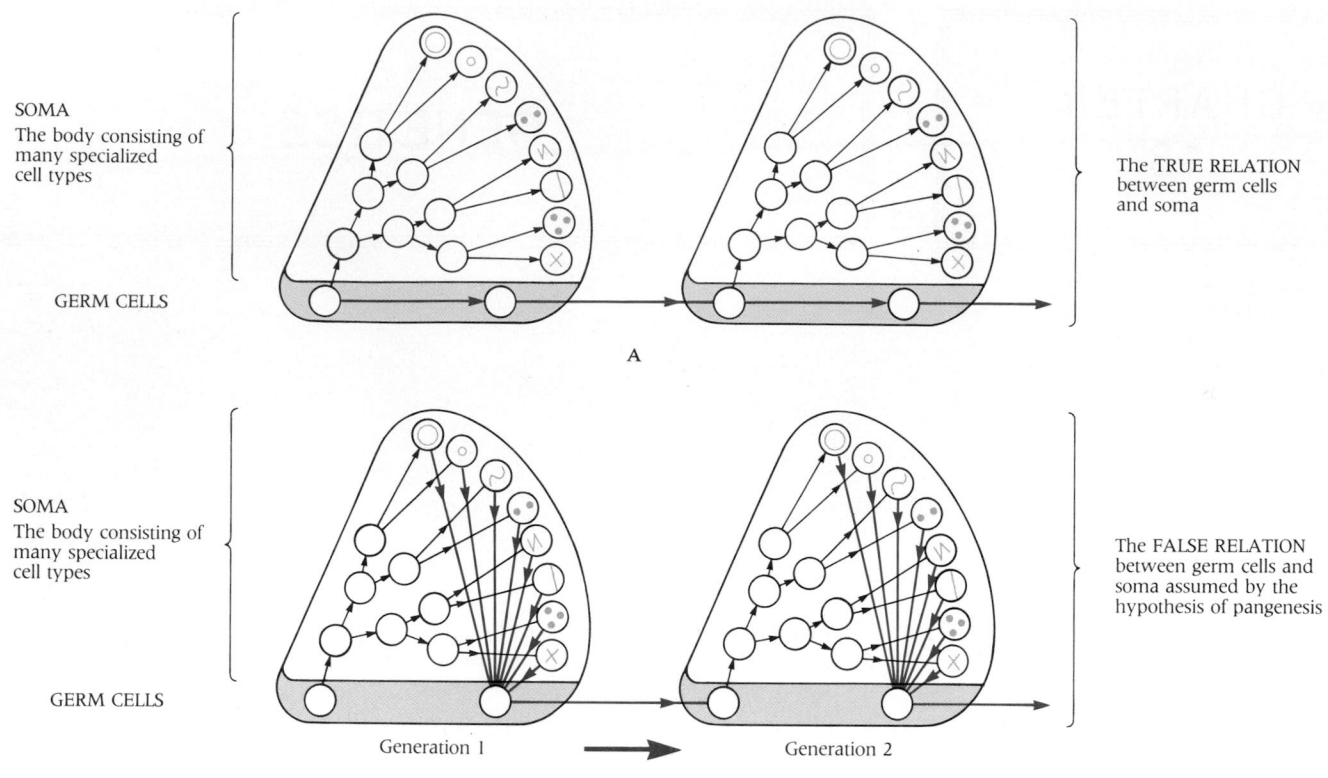

SOMA
The body consisting of
many specialized
cell types

GERM CELLS

The TRUE RELATION
between germ cells
and soma

A

SOMA
The body consisting of
many specialized
cell types

GERM CELLS

The FALSE RELATION
between germ cells and
soma assumed by the
hypothesis of pangenesis

Generation 1   Generation 2

B

(see Box 13-1) and all similar ideas that envision a two-way relationship between germ cells and somatic cells.

MENDELIAN PRINCIPLES OF HEREDITY

Johann Gregor Mendel (Figure 13-2) lived in the middle years of the nineteenth century (1822–1884). As a schoolboy he entered an Augustine monastery in the Moravian town of Brünn, Austria (now Brno, Czechoslovakia). In time the monastery sent him for two years (1851–1853) to study mathematics and science at the University of Vienna, where a professor, Franz Unger, stimulated his interest in evolution and heredity. Mendel then returned to Brünn—master of the monastery school, teacher of physics, sophisticated thinker, skillful chess player, and expert gardener.

Exploring a lifelong interest in heredity, Mendel observed that plants or animals may have offspring that differ from the parents in color, size, or some other **character** (a geneticist's synonym for *trait* or *characteristic*). Knowing that plants and animals arise from a single sperm and a single egg, he wondered if the appearance of variations among the offspring implies the existence of germ cells of various kinds. Recognizing that the question could be answered only by experimentation, Mendel began to cross-breed different strains of pea plants in the monastery garden. By careful experimentation and by the shrewd use of a tool quite new to biology— the statistical analysis of populations—he discovered the fundamental principles of the science we now call **genetics.**

Although Mendel's results were presented before the Natural Science Society of Brünn in 1865 and published the next year, their real significance was not fully grasped by other scientists (or even by Mendel himself) for many years. (Mendel, as some have said, was never a Mendelian!) Finally, around 1900, three independent experimenters—Karl Correns in Germany, Hugo De Vries in the Netherlands, and Erich von Tschermak in Austria—simultaneously discovered that Mendel's old results agreed with their own. Thereafter Mendel's fame and stature came to match Darwin's. The story of Mendel's neglect illustrates that a theory, however firmly it may rest on demonstrable facts, may be rejected until scientific progress generates a meaningful context for it.

13-1 The relationship between somatic cells and germ cells
(**A**) *Germ cells are direct descendants of specialized cells in the zygote and no germ cells arise from nonspecialized somatic cells.* (**B**) *An incorrect model, in which germ cells arise from various somatic cells.*

13-2 Johann Gregor Mendel (1822–1884)

BOX 13-1

IDEAS ABOUT HEREDITY BEFORE MENDEL

Early thought on heredity was entirely un-scientific. Consequently, it produced little beyond odd notions on strange matings between unrelated creatures and such myths as the one about the effects of a woman's first husband on any children she might have with a second husband.

It is convenient to review these early concepts under two headings.

BLENDING OF INHERITANCE

It was once believed that inheritance depended on an actual transfer of blood from generation to generation. Somehow the blood of an individual was derived from that of ancestors, as a result of mixing in proper proportions throughout many generations. Thus we spoke (and still speak) of "blood lines," "blue-bloodedness," and the like. According to this intuitive (and erroneous) notion, we inherit a half-and-half blend of our parents' blood, a quarter coming from each grandparent, and so on.

According to the concept of blending, combinations were created such as occur when one mixes paints. This lingering view was at the root of many early misconceptions about heredity. What we now call a giraffe was earlier called a *camelopard* and was supposed to have issued from the mating of a camel and a leopard, the leopard contributing its spots and the camel its long neck. Camels, it seems, were popular and versatile hybridizers in these early theories. One authority claimed that an ostrich was a cross between a camel and a sparrow—an odd couple, to say the least. Arabian scholars be-lieved sea cows to be crosses between humans and fishes, and Greek mythology was full of half-human, half-animal hybrids. We now know that only animals of the same or closely related species can mate to produce offspring. Even when two closely related species cross, as do the jackass and mare to yield a mule, the resulting offspring is usually infertile.

INHERITANCE OF ACQUIRED CHARACTERS

Another erroneous view was the persistent belief in the inheritance of acquired characters—or what has been termed **soft inheritance,** defined as inheritance in which the hereditary material transmitted from generation to generation is inconstant because it is subject to modification by the environment. Thus it was commonly believed that a man who exercises and becomes strong will pass on his large muscles to his children.

This theory was once much in vogue. Although it has gone under the name of *Lamarckism*—after Jean Baptiste Lamarck (1744–1829), the prolific French naturalist—the idea had been discussed by the ancient Greeks. In his famous work *Philosophie Zoologique* (1809), Lamarck wrote, "The environment affects the shape and organization of animals." Changes in the environment, he believed, can modify an animal's organs through use or disuse. These changes are then passed on to the offspring. The giraffe was Lamarck's most famous example. Animals, he declared, that stretched their necks reaching for the tender leaves at the treetops would have offspring with long necks. In time they became giraffes.

Another old idea was the theory of **pangenesis.** This view, prevalent before the advent of the chromosome theory of heredity, suggested that if a blacksmith's large muscles are inherited by his son, there must be some means for representing the state of his muscles—their *attributes,* to borrow a computer term—in whatever factors heredity transmits to his son. In ancient times Democritus speculated that particles called **pangens** come from all parts of the body. These particles appear in the semen in the male and in the blood in the female. During sexual intercourse, the male and female pangens combine.

In his later years, Darwin proposed what he called "a hypothesis of pangenesis" (though it wasn't clearly that), embodying the idea of inheritance of acquired characteristics. In it he defined **gemmules** much as Democritus had defined pangens. According to Darwin, they were particles—very small and hence invisible—that are released by somatic cells and that somehow became associated with germ cells. Hereditary material, in this view, is a mosaic of particles representing each kind of body cell.

These ideas are important historically because they were part of one of the first general theories of evolution, one that was later proved wrong. As we shall see, the mechanisms of heredity, as presently understood (despite some recent and controversial work to be discussed later) do not allow for soft inheritance, the inheritance of acquired characters.

MENDEL'S FIRST EXPERIMENTS WITH SINGLE CHARACTER DIFFERENCES

Mendel did most of his work with the common garden sweet pea, *Pisum sativum* (Figure 13-3A). For his purposes, this plant species had three advantages.

- First, it can self-pollinate—that is, it can fertilize its own egg cells with sperm from its own pollen in the same flower. It can also be cross-pollinated from the flower of one plant to flowers in different plants by an experimenter.
- Second, it has variants that differ in a clear-cut, discontinuous (either/or) manner—flowers are red or white, or axial or terminal; seeds are yellow or green,

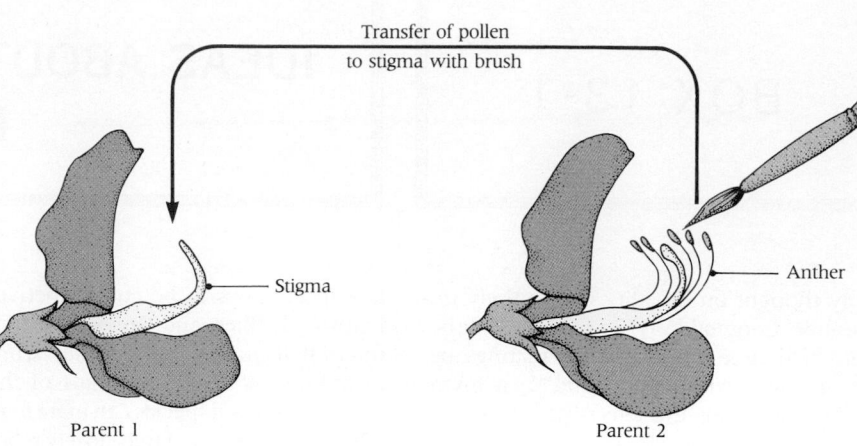

A

B

13-3 Mendel's model organism, the garden pea, **Pisum satirum**
(A) Appearance of the pea plant Mendel studied.
(B) Pea plants can be artificially cross-pollinated. They also self-pollinate.

or round or wrinkled; pods are inflated or constricted, or green or yellow; stems are long or short; and so on. (Interestingly, it was shown in 1990 that the wrinkled-seed character reported by Mendel is due to the presence of distinctive branching patterns in starch.)

■ Third, pure lines of pea plants, when self-pollinated, breed true for many generations—that is, plants with red flowers when crossed with others within the same population or when self-pollinated would give rise to offspring with red flowers and not some other color. These are termed **pure lines.**

In the two diagrams below, labels read: "Transfer of pollen to stigma with brush", "Stigma", "Anther", "Parent 1", "Parent 2".

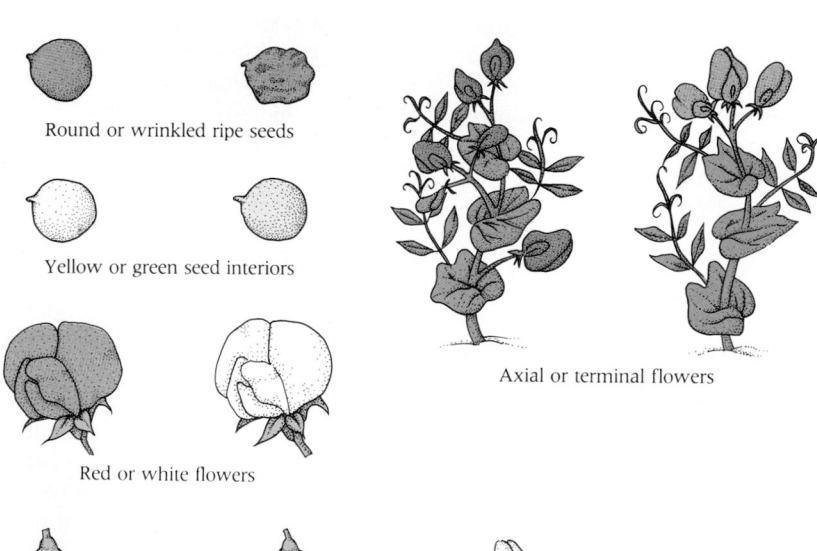

Round or wrinkled ripe seeds

Yellow or green seed interiors

Red or white flowers

Axial or terminal flowers

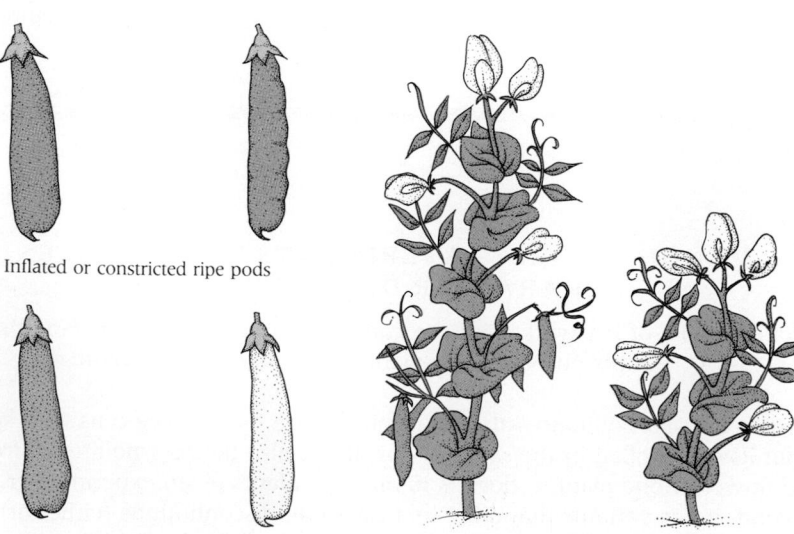

Inflated or constricted ripe pods

Green or yellow unripe pods

Long or short stems

13-4 The seven traits of pea plants studied by Mendel

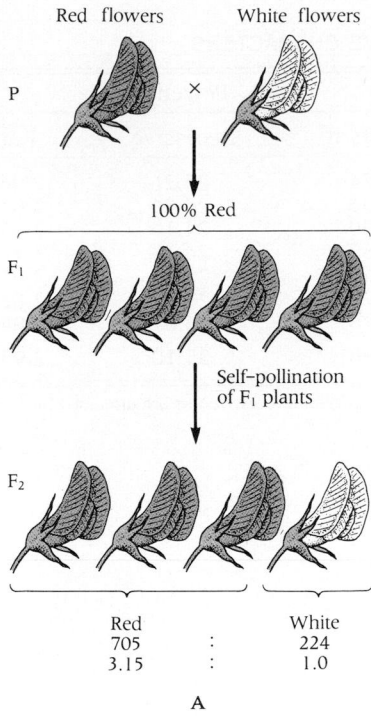

100% Red

F₁

F₂

Red	:	White
705	:	224
3.15	:	1.0

A

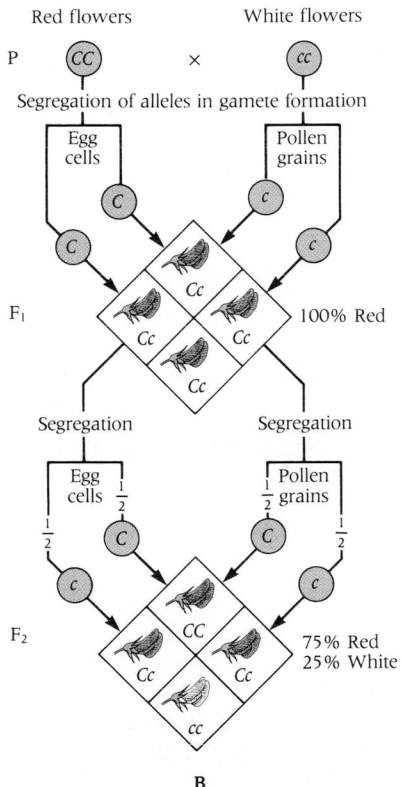

B

13-5 Crossing pea plants differing in a single character
(A) *The results of a cross between red-flowered and white-flowered plants, followed through two generations.* **(B)** *Genetic explanation of Mendel's results.*

Mendel recognized that only by analyzing such simple examples would he be likely to unravel the rules of heredity. His historic paper presented before the scientists of Brünn in February 1865 began with the opinion that his predecessors had failed to discover the laws of heredity because the test crosses they studied differed in too many characteristics.

Cross-pollination of any two pea plants is easily accomplished by an experimenter (or gardener). As shown in Figure 13-3B, the pollen-bearing anthers of one plant are clipped off to prevent self-pollination and pollen from another plant is transferred to the stigma of the other flower with a paint brush.

In his first experiments, Mendel crossed plant varieties that differed in only one trait—tall with dwarf plants, red-flowered with white-flowered plants, round-seeded with wrinkled-seeded plants, and so on, until seven character pairs had been examined (Figure 13-4). In all seven cases, results were essentially the same as those obtained in crosses between red-flowered and white-flowered types, which for brevity we will call red and white plants. Let us have a closer look at this cross.

When red flowers were crossed with white flowers, the results were surprising. The offspring—the members of the **first filial** (or **F₁**) **generation**—were all the same, perfect red copies of their red parent. The same result was obtained irrespective of whether the male (which contributes pollen) or the female (which contributes the ovum) member of the **parental generation** (called **P**) was the red parent.

When Mendel crossed two of the red F₁ offspring, transferring the pollen of one plant to another plant, he had another surprise. The next generation, the **second filial** (or **F₂**) **generation,** included red and white plants. How could this be explained? He repeatedly allowed red-flowered F₁ plants to self-pollinate until he had 929 F₂ seeds. He planted them and then counted the red- and white-flowered plants in this, the F₂ generation—one of the first and surely one of the most successful applications of numerical analysis in the history of biology. The results were impressive: In many trials, the F₂ generation contained almost exactly three reds for every one white (Figure 13-5A). The results, in summary, were as follows:

F₂ plants	Red : White
Actual numbers counted	705 : 224
Percent	75.9 : 24.1
Ratio	3.15 : 1.00

The important points to note are these.

- The F₁ generation consisted entirely of plants resembling only one of the parents.
- The F₂ generation consisted of plants resembling both parents.
- A parental trait found missing in the F₁ offspring appeared in one-fourth of the individuals in F₂.

Table 13-1 lists the actual data Mendel obtained in other experiments.[1] In every instance, the general result was the same: F₁ was all of one type, and F₂ contained both types, the types missing in F₁ making up about 25% of F₂. It was a result that could scarcely be ignored.

THE PRINCIPLE OF SEGREGATION: THE HYPOTHESIS OF PAIRED FACTORS

The most remarkable of Mendel's results was the reappearance of white-flowered plants in the F₂ generation. This, Mendel concluded, points to a separateness, or "atomicity," of heritable traits. Although F₁ plants do not have white flowers, they must nevertheless carry some sort of hereditary factor for them, because when F₁ plants are crossed among themselves, white flowers appear in one-fourth of their offspring. Mendel called these factors *elements.* This argument suggested two conclu-

---

[1] Statisticians showed several years ago that Mendel's results may have been "too good." That is, they gave better ratios than could be reasonably expected. Historians of science have suggested that Mendel's counts may have been biased by subconscious errors caused by his own expectations of what the results would be. In other words, Mendel may have had the final interpretation of his ratios clearly in mind before completing his experiments.

TABLE 13-1
MENDEL'S RESULTS IN CROSSES INVOLVING SEVEN PAIRS OF ALTERNATIVE CHARACTERS

Characters	F_1	F_2 Numbers of Plants		F_2 Percent		
		Dominant[*]	Recessive[*]	Dominant[*]	Recessive[*]	Ratio
Seeds: round vs. wrinkled	All round	5,474	1,850	74.74	25.26	2.96:1
Seeds: yellow vs. green	All yellow	6,022	2,001	75.06	24.94	3.01:1
Flowers: red vs. white	All red	705	224	75.90	24.10	3.15:1
Flowers: axial vs. terminal[†]	All axial	651	207	75.87	24.13	3.14:1
Pods: inflated vs. constricted	All inflated	882	299	74.68	25.32	2.95:1
Pods: green vs. yellow	All green	428	152	73.79	26.21	2.82:1
Stem length: tall vs. dwarf	All tall	789	277	73.96	26.01	2.84:1
Totals		14,949	5,010	74.90	25.10	2.94:1

[*] The dominant character is the one found in F_1; the recessive character is the one reappearing in smaller numbers in F_2. Dominance and recessiveness are discussed further in this chapter and in Chapter 14.
[†] Axial flowers occur all along the stem (axis) of a plant; terminal flowers cluster at the end of the stem (see Figure 13-4).

sions: first, that each plant carries at least two hereditary factors for each flower; second, that the factor for white is "dominated" by the factor for red when both are present in the same plant.

Let us restate Mendel's conclusions.

- A hereditary character is the result of some unknown factor (only later called a gene) that determines the presence of that trait.
- The factor exists in paired form, one member of the pair coming from one parent, the other coming from the other parent.
- These determinants must be physical entities or particles that remain intact throughout fertilization; there is no fusion of factors.
- Factors in each pair separate, or segregate, during the formation of germ cells (gametes), so that each gamete receives only one factor. As we will learn in Chapter 14, this is a consequence of meiosis, which produces gametes directly in animals and indirectly in plants.
- The factors for red flowers and white flowers are alternative forms of the same factor, the factor for red being dominant over that for white, which is recessive.

This hypothesis, many times validated by later workers, became known as Mendel's first law, or the **principle of segregation.**

Modern Terminology

So that we may discuss Mendel's work in modern terms, let us now introduce the terms later given to Mendel's hypothetical entities. The hereditary factors are **genes,** this term having been introduced in 1909 by the Danish biologist Wilhelm Johannsen. We shall later define the gene in contemporary terms.

Alternative forms of the same gene are called **alleles.** The alternative factors for red and white flowers are alleles of the flower color gene (Figure 13-6). Stated another way, the flower color gene in pea plants has two alleles, one for red flowers and one for white flowers. In the plant's cells, a flower color allele is present in the same location, or **locus,** on each of the two homologous chromosomes that bear the flower color gene. Similarly, the alternative genes for plant size (tallness and dwarfness) are alleles. The gene for tallness, however, is not an allele of the gene for redness. The existence of two alleles for a given trait—in these cases, flower color or plant size—suggests that a gene pair determines a given trait. This recalls the fact that body cells have a diploid chromosome number.

Bear in mind that any gene can have more than two alleles. Such variants may exist *in a population* in unlimited numbers. But no more than two alleles of any gene are found *in an individual.* One allele is located on each of the two homologous chromosomes present in diploid cells.

In simple cases such as those studied by Mendel, one allele is usually **dominant.** The other is **recessive.** It is customary in such simple cases to designate dominant and recessive alleles, respectively, by a capital and small letter (or letter combination). For example, *C* (for color) represents the red and *c* represents the white allele.

When the two alleles of a flower-color gene in a diploid cell are the same, we describe it as *CC* or *cc*. In such cases, the organism is said to be **homozygous** with respect to that gene. When the two alleles are different (*Cc*), the individual is **heterozygous** for that gene.

So that we may distinguish the outward appearance (which is visible and thus easy to determine) and the genetic pattern (which is impossible to determine by mere inspection), we speak of an organism's **phenotype** ("visible type") and **genotype** ("genic" or "genetic type"). The phenotype of all F_1 plants in our example is red; but the genotype of each plant includes the alleles for red and white. The terms homozygous and heterozygous refer to the two possible states of the genotype.

Modern Statement of Mendel's Hypothesis

Let us now restate Mendel's interpretation of the red–white cross in contemporary terms (see Figure 13-5B). The red and white plants of the P generation were both homozygous: the red-plant genotype was *CC* and the white-plant genotype was *cc*. When the parental plants produce gametes, the alleles segregate so that, in meiosis, the red plant produces only *C* gametes and the white plant only *c* gametes. When these gametes unite as a result of fertilization, only heterozygous F_1 plants with a *Cc* genotype result.[2] The phenotype of this plant, however, is red, and there is no way it can be distinguished by outward appearance from its red parent.

[2] It is incorrect to say that *Cc* and *cC* genotypes are identical, because the different sequences of the allele symbols imply different parental sources. However, *Cc* and *cC* genotypes are genetically equivalent in the effect they have on the phenotype.

When two F_1 reds (each of genotype Cc) are crossed, an F_2 offspring is produced. Before we can predict its genotype, we must ask what gametes are produced by the F_1 plants. The answer is that each Cc pair segregates. Therefore, two kinds of gametes are produced by each parental plant of the F_1 generation—C and c—and they are produced in equal numbers.

In the formation of F_1 plants, only one kind of fertilization was possible—a union of C and c that yielded Cc as the F_1 genotype. However, when two F_1 plants were mated, more than one type of fertilization was possible. In fact, four types were possible. Moreover, as shown in Table 13-2, all four are equally probable.

The checkerboards in Figures 13-5B and 13-7 are termed **Punnett squares,** after the English geneticist R. C. Punnett (1875–1967). To construct a Punnett square, the alleles present in one parent's gametes are written along one side; those from the other parent are written across the other side. The combination in each box illustrates all possible fertilizations. The second and third types of fertilization in Table 13-2 produce F_2 plants with genotypes Cc and cC. Since these are genetically identical, the scheme tells us that the F_2 generation should contain three kinds of genotypes in the following proportion:

Genotypes	CC : Cc : cc
Genotypic ratios	1 : 2 : 1
Phenotypes	red red white
Phenotypic ratios	3 red : 1 white

Because CC homozygotes and Cc heterozygotes are phenotypically the same (red), only two phenotypes appear in F_2. They appear in the ratio of 3 red : 1 white. The white-flowered cc homozygotes are the ones missing in F_1. They constitute one-fourth of F_2 plants.

The Hypothesis Is Tested

Having performed some crosses with pea plants differing in a single character—these are called **monohybrid crosses**—Mendel now needed firm evidence that the genotype of the red parent differed from that of the red F_1 offspring. He found the evidence in the **test cross.**

In a test cross, a plant of unknown genotype is crossed with a plant known to be homozygous for the recessive allele of a trait in question. This, of course, is the only form in which a recessive phenotype is recognizable. If, then one wished to determine whether the genotype of an unknown red plant is CC or Cc, it would be imperative to cross it with a white plant, which is necessarily of genotype cc. Possible outcomes are summarized in Figure 13-7.

■ A homozygous red (Cc) crossed with a homozygous white (cc) could produce only red offspring; but they would have to be Cc heterozygotes because the test parent was cc (Figure 13-7A). Although the appearance of red progeny from such a test would not totally prove that the red parent was homozygous, it would indicate a high level of probability.

■ A heterozygous red (Cc) crossed with a homozygous white cc would have a different outcome: Half the progeny would be red and half would be white (Figure 13-7B).

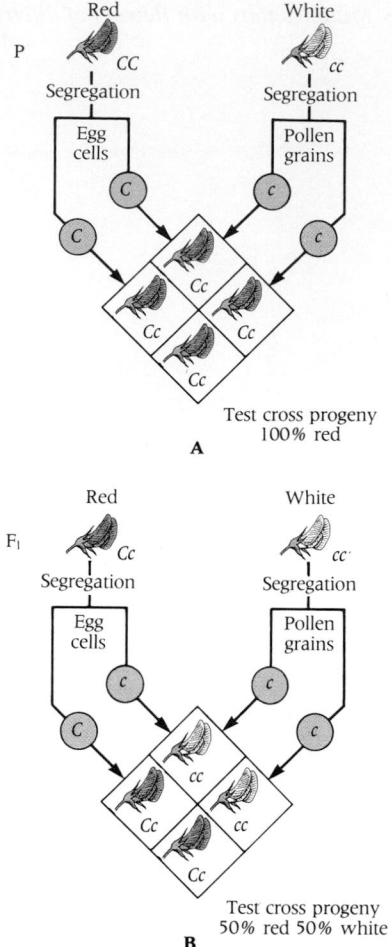

13-7 Principle of the test cross
*By using a homozygous recessive plant as one parent, the unknown genotype of the other parent can be determined by looking at the phenotype of the F_1 offspring. (**A**) Results when a phenotype of the unknown parent is CC. (**B**) Results when that parent's genotype is Cc.*

TABLE 13-2
POSSIBLE RESULTS OF A CROSS BETWEEN TWO HETEROZYGOTES WITH THE GENOTYPE Cc

Gametes from Female Parent		Gametes from Male Parent		Resulting F_1 Zygote	Frequency
C	×	C	=	CC	¼
C	×	c	=	Cc	¼
c	×	C	=	cC	¼
c	×	c	=	cc	¼

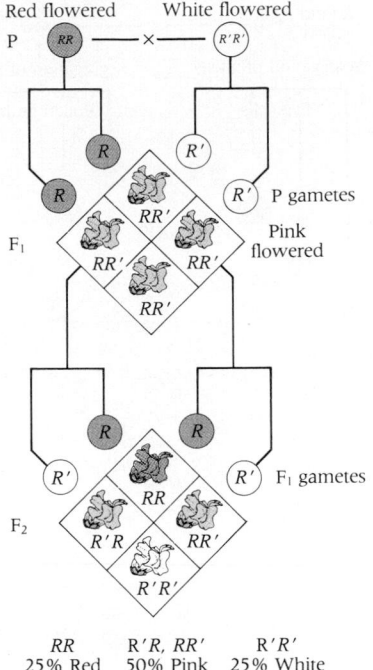

Red flowered   White flowered
P   RR ————×———— R'R'

R   R'

R   R'   P gametes

F₁   RR'   RR'
          RR'   RR'
               RR'   Pink
                        flowered

R   R

R'   R'   F₁ gametes

F₂   RR   RR'
          R'R   RR'
               R'R'

RR   R'R, RR'   R'R'
25% Red   50% Pink   25% White

13-8 Concept of incomplete dominance
When dominance is absent, each allele is designated by capital letter, with one distinguished by a prime superscript. In this cross of red-flowered and white-flowered snapdragons, the pink offspring have the genotype RR'.

The reason, shown in Figure 13-7, is that a white parent produces only c gametes, but a heterozygous red parent produces C and c in equal numbers. Hence Cc and cc fertilizations would be equally likely to occur and thus equally frequent. In summary, in a test cross with a homozygous recessive, it is possible to unmask the unknown half of a plant's genotype.

Mendel performed such a test cross with F₁ red plants and obtained offspring in a phenotypic ratio of approximately half red plants and half white plants. The appearance of any white-flowered progeny in this experiment would mean that the red-flowered parent must be heterozygous. Note, however, that a test cross such as Mendel performed reveals heterozygosity not merely by producing a few white offspring among the red. The test cross produced offspring in the particular quantitative relationship predicted by theory (that is, 1 red : 1 white). Thus was the Mendelian hypothesis made into a sound theory that has since been supplemented but never supplanted.

Incomplete Dominance

For characters such as flower colors, one allele may be fully dominant over the other. This is not always the case, however. Sometimes a heterozygous trait is halfway between the two homozygous traits. For example, a cross between a homozygous red snapdragon (*Antirrhinum*) and a homozygous white one yields F₁ offspring of an intermediate pink color (Figure 13-8), just as though red paint had been mixed with white paint. Does this prove Mendel's theory wrong?

Quite the contrary. For when two pink F₁ snapdragons are repeatedly crossed among themselves, the resulting F₂ generation contains red, pink, and white flowers in a ratio of 1:2:1 (Figure 13-8). This result adds even more weight to Mendelian theory.

Inheritance in which a heterozygous trait is visibly affected by both alleles is termed **intermediate inheritance.** The responsible genes are said to display **incomplete dominance.** In these situations, genotypic and phenotypic ratios in the F₂ generation are the same. Since complete dominance was lacking in the snapdragon, it was possible to recognize the phenotypes of heterozygotes, which clearly differ from the phenotypes of either homozygote.

In some cases of intermediate inheritance, a heterozygous phenotype is recognizable, but it is not intermediate between the two homozygous phenotypes—as pink is intermediate between red and white. In certain chicken strains, for example, a cross between black and white individuals produces offspring with the unusual appearance termed *blue Andalusian*. A cross between two blue Andalusians (when tested in large numbers) produces blacks, blue Andalusians, and whites in a ratio of 1:2:1.

Do not confuse intermediate inheritance with the outmoded theory of blending inheritance (see Box 13-1). Blending inheritance does not explain the snapdragon case because genes are involved. That is established by the fact that parental traits reappear in predictable ratios in the F₂ generation. Plainly, the validity of Mendel's hypothesis does not depend on whether one of a pair of alleles is dominant over the other.

As we shall see, certain traits are determined by more than one gene, a situation referred to as **polygenic inheritance.** In such cases, it may be very difficult to predict the phenotypic appearance of heterozygotes.

Codominance

Cases where both alleles produce an effect in the heterozygote are examples of **codominance.** A familiar example is found in the genetic determinants of human A and B blood types, which we will discuss again later (Chapter 22). Red blood cells have distinctive surface polysaccharides called **H antigens.** Whether a red cell is type A, type B, or type O depends on which sugars are attached to the ends of these antigen molecules. A single gene, I, determines a red cell's type (within the so-called ABO system). However, it has three alleles: I^A specifies type A, I^B specifies type B, and i specifies type O. This is an example of **multiple allelism,** a phenomenon recognized early in the history of genetics. Although only two alleles of a gene can occur in a diploid cell—and only one in a haploid cell—there can exist in a population of individuals many possible different allelic forms of a gene.

Interestingly, the alleles that specify blood types A and B are both dominant over the type O allele. That means that the heterozygous offspring of parents possessing the phenotype I^A and I^B will have the blood type AB. In other words, both alleles are codominant.

THE PRINCIPLE OF INDEPENDENT ASSORTMENT: DIFFERENCES OF TWO OR MORE CHARACTERS

Dihybrid Crosses

Having discovered the behavior of genes controlling single characters, Mendel set out to study a more complicated case: the simultaneous inheritance of two characters. Choosing the shape and color of pea seeds as test traits, he carried out a **dihybrid cross** using a strain producing round-yellow seeds and one producing wrinkled-green seeds. Each of these traits, when studied individually, obeyed the same rules as flower color; when studied together, however, they began to produce an interesting and complex picture.

Among the seven traits studied by Mendel were two, involving the shape and color of pea seeds. A seed can be round or wrinkled, and it can be green or yellow. Round (symbolized R) is dominant over wrinkled (r) and yellow (Y) is dominant over green (y). Therefore, a cross of homozygous round-yellow ($RRYY$) with homozygous wrinkled-green ($rryy$) yielded an F_1 generation that were all phenotypically round-yellow. But when two F_1 plants with round-yellow seeds were crossed with each other, two new phenotypes appeared in the F_2 generation in addition to the two original phenotypes (round-yellow and wrinkled-green). The new phenotypes were round-green and wrinkled-yellow.

How should one interpret this result? Mendel concluded that each gene is segregated to the gamete independently (Figure 13-9) and each pair of alleles is represented in the gametes by one of its members. Round-yellow plants ($RRYY$) produce RY (and not RR or YY) gametes. Wrinkled-green plants ($rryy$) produce ry gametes. The F_1 plants produced seeds that are phenotypically round-yellow and genotypically $RrYy$. Thus the F_1 generation is heterozygous for both pairs of alleles. It is a dihybrid, and, as shown in Figure 13-9, it would produce four gamete types: RY, Ry, rY, and ry. (Why not Rr, RR, Yy, or yy?)

When dihybrid F_1 plants are crossed among themselves to produce an F_2 generation, it is possible to predict the F_2 genotypes and phenotypes only if it is recognized that the four kinds of gametes form with equal frequency (Figure 13-10). Therefore, F_2 should include nine genotypes and four phenotypes: The expected phenotype frequencies would be 9 round-yellow : 3 round-green : 3 wrinkled-yellow : 1 wrinkled-green. Phenotypic ratios of 9:3:3:1 are characteristic of dihybrid crosses. Mendel's own results appear at the bottom of Figure 13-9. On the other hand, the genotypic ratio is 1:2:2:1:4:2:2:1. Confirm this in Figure 13-9.

One might have imagined that the gametes formed by an $RrYy$ could be limited exclusively to RY and ry if there were a rule mandating that dominant alleles had to be together in one gamete and recessives in the other. But when dihybrid crosses are actually carried out, the Rr pair of alleles segregate independently of the Yy pair, and F_1 produces two new kinds of gametes (Ry and rY) in addition to RY and ry. These new kinds of cells or individuals result from new gene combinations that were not present in the parents. Such results make it clear that in meiosis all four gamete types are produced with equal frequency (see Figure 13-9).

Note that the *dihybrid* test crosses were used by Mendel to judge the genotypes of dihybrids such as $RrYy$. As in the procedure shown in Figure 13-7, the tested individual is a homozygous recessive, in this case $rryy$, which produces only ry gametes.

Trihybrid and Higher Crosses

These concepts are readily applied to **trihybrid crosses**—and, indeed, to any cross with more than two characters. The Punnett square method is helpful in visualizing small numbers of possible combinations but cumbersome when more than two pairs of alleles are involved.

A far better method of predicting the kinds of offspring is based on a simple mathematical principle: If we know the probability that two independent events

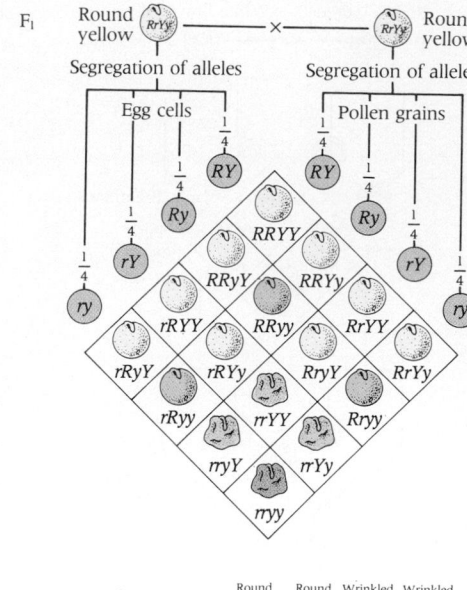

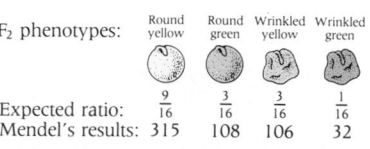

13-9 Possible gamete types from a dihybrid heterozygote
The shape and color genes are on different chromosome pairs, which assort independently at meiosis. Therefore each F_1 offspring produces four types of gametes in equal proportion. If parental chromosomes always stayed together, only two types of gametes would be produced. This is not supported by the data. Therefore this assumption is incorrect.

will occur, the probability that the two will occur together is the product of the separate probabilities.

To illustrate the *product rule*, consider a cross between two identical dihybrid round-yellow plants with the genotype *RrYy*. We could regard this single cross as two separate monohybrid crosses—one is *Rr* × *Rr*, the other *Yy* × *Yy*. We know that in such monohybrid crosses, we should expect progeny that are ¾ round and ¼ wrinkled and ¾ yellow and ¼ green.[3] To calculate the different combinations possible in the dihybrid cross, simply determine the product of each of the separate probabilities.

$$¾ \text{ round} × ¾ \text{ yellow} = \%_{16} \text{ round-yellow}$$

$$¾ \text{ round} × ¼ \text{ green} = ³/_{16} \text{ round-green}$$

$$¼ \text{ wrinkled} × ¾ \text{ yellow} = ³/_{16} \text{ wrinkled-yellow}$$

$$¼ \text{ wrinkled} × ¼ \text{ green} = ¹/_{16} \text{ wrinkled-green}$$

The result is the same 9:3:3:1 ratio that Mendel found experimentally.[4]

In a trihybrid (or higher) cross, the procedure is the same: We assume that three (or whatever the relevant number is) separate monohybrid crosses are taking place. Calculate, for example, how many wrinkled-yellow-tall offspring would occur in the progeny of two round-yellow-tall (*RrYyTt*) plants.

These classic experiments conclusively prove that in a mating of individuals differing in two characters whose genes are located in nonhomologous chromosomes, the genes do not necessarily remain in the same combinations found in the parents. The phenomenon of independent segregation of the two pairs of alleles is referred to as the **principle of independent assortment.** This is Mendel's second law.

MENDEL'S CONTRIBUTIONS RESTATED

What Mendel showed—and what the three botanists who rediscovered Mendel's work in 1900 independently verified—is briefly summarized below:

- When parents differ in a trait, the offspring often resemble one parent but not the other. This is the principle of dominance.
- When hybrids (heterozygotes) cross among themselves, they transmit with equal frequency either the dominant trait of one parent or the recessive trait of the other, but not both. This is the principle of segregation (or Mendel's first law).
- When parents differ in two or more pairs of traits, each pair may segregate independently of the others. As a result, all possible combinations of the two or more pairs occur in the gametes of the heterozygote, and their frequency is determined by chance. This is the principle of independent assortment (or Mendel's second law).

Mendel's work has stood the test of time and has provided the foundations of all modern ideas in the field of genetics. How curious, therefore, that his work went unnoticed for some 35 years following its publication. One reason may have been the total absence of contemporary ideas on the physical nature of cellular entities that might play the role Mendel had assigned to his hypothetical genetic determinants.

In the next chapter, we will consider the physical basis of heredity and consider the later extensions to Mendelian analysis.

[3] Note that *probability* can be expressed as a fraction, as a percentage, or as a decimal. All three conventions are used in different texts.

[4] Another useful basic rule of probability is the addition rule, which holds that the probability that either one of two mutually exclusive events will occur is the sum of their individual probabilities. For example, with two dice the probability of tossing two 3s or two 6s equals the sum of their individual probabilities: ¹/₃₆ + ¹/₃₆ = ¹/₁₈. The sum, ¹/₁₈, means that one or the other of these events will probably occur once in every 18 tosses.

SUMMARY

The information of heredity is passed through the generations from parents to offspring via the gametes (germ cells). Gametes ultimately give rise to all somatic (body) cells and to all future gametes.

In 1866, Gregor Mendel first proposed that individual traits inherited by organisms are governed by discrete factors (now called genes), which exist within higher plants and animals in pairs, each member of which derives from one parent. The forms of a gene are now known as alleles.

Different alleles of the same gene specify different versions of the same trait. A single gene may have many alleles distributed throughout a population, but an individual possesses only two of these alleles. If both alleles of a pair specify the same version of a trait, the organism is said to be homozygous with respect to that trait. If the alleles specify different versions of the trait, the organism is said to be heterozygous for that characteristic.

Mendel confirmed his so-called first law (the principle of segregation) in a cross of heterozygous yellow plants with homozygous green plants, which yielded a 1:1 ratio of yellow to green plants. This principle states that alleles segregate from each other during gamete formation, so that each gamete carries only one allele of each gene pair. The union of two gametes to form a zygote occurs randomly and is independent of which allele a gamete carries. In the zygote, the alleles continue their separate existence.

For some genes, one allele of the pair determines the phenotypic expression of a trait. Such an allele is said to be dominant; the other allele is recessive. A heterozygote has the same phenotype as a homozygote of the dominant allele, even though its genotype differs from that of the homozygote. Heterozygotes and homozygotes of this type can be readily distinguished from each other. A cross of two heterozygotes produces offspring with the phenotypic ratio of 25% recessive and 75% dominant. This ratio represents a probability statement: an offspring of heterozygous parents has a 25% chance of receiving both recessive alleles from them, a 50% chance of receiving one dominant and one recessive allele (and thus the dominant phenotype), and a 25% chance of being homozygous dominant. Not all heterozygotes have the phenotype of a dominant allele. In traits showing incomplete dominance, heterozygotes have an appearance distinct from each of the two homozygous phenotypes.

Mendel's second law (the principle of independent assortment) applies to crosses in which two or more traits are studied simultaneously. In dihybrid crosses, the gene pairs for two different traits are distributed to gametes independently of one another. Since the probability that a gamete will receive the recessive allele for one trait has no effect on the probability that it will receive the recessive allele for another trait, the frequency of phenotypes and genotypes produced in a dihybrid cross can be determined by the same probability analysis employed in a monohybrid cross.

KEY TERMS

allele
character
codominance
dihybrid cross
dominant
first filial generation (F$_1$)
gene
genotype
heterozygous

homozygous
incomplete dominance
independent assortment
intermediate inheritance
locus
monohybrid cross
multiple allelism
parental generation (P)
phenotype

polygenic inheritance
Punnett square
recessive
recombinants
second filial generation (F$_2$)
segregation
test cross

QUIZ QUESTIONS

1. Consider a cross in which a trait is inherited by complete dominance. What percentage of the F$_2$ from the mating of homozygous dominant and homozygous recessive individuals will possess the heterozygous genotype?
   A. 0 percent
   B. 25 percent
   C. 50 percent
   D. 75 percent
   E. 100 percent

2. Consider a cross in which a trait is inherited by incomplete dominance. What percentage of the F$_2$ from the mating of homozygous dominant and homozygous recessive individuals will possess the dominant phenotype?

   A. 0 percent
   B. 25 percent
   C. 50 percent
   D. 75 percent
   E. 100 percent

3. An independent assortment of traits from different parents occurs as a result of events during _____ of meiosis.
   A. prophase I
   B. prophase II
   C. metaphase II
   D. anaphase I
   E. telophase II

4. A test cross involves crossing an individual of unknown genotype with an individual of known genotype that

   usually is _____ for the trait in question.
   A. heterozygous
   B. homozygous recessive
   C. homozygous dominant
   D. codominant
   E. incompletely dominant

5. Which of the following individuals proposed a theory of evolution that was based on inheritance of acquired characteristics?
   A. Charles Darwin
   B. Gregor Mendel
   C. R. C. Punnett
   D. Jean Baptiste Lamarck
   E. A and D

ESSAY QUESTIONS

1. Which of Mendel's experimental results caused him to conclude that heritable factors (only later identified as genes) occur in pairs?

2. What is a test cross (sometimes termed a "back-cross")? What information does it provide about genotype of the organism being tested?

3. Define and contrast the concepts of polygenic inheritance, independent assortment, incomplete dominance, and codominance.

4. Do traits that exhibit incomplete dominance or codominance obey Mendel's first law (the principle of segregation) and second law (the principle of independent assortment)? Explain.

5. Determine the genotype and phenotype of the F_1 offspring of a cross between a homozygous tall, pink plant and a heterozygous tall, white plant. For this plant, height follows a simple Mendelian pattern of inheritance with tallness dominant over shortness. However, color is determined by two alleles, R (red) and W (white), which show incomplete dominance. What crosses should be performed to establish the genotype of the F_1 offspring?

REFERENCES AND SUGGESTIONS FOR FURTHER READING

Ayala, F. J., and Kiger, J. A., Jr. (1984). *Modern Genetics*, 2nd ed. Benjamin/Cummings, Menlo Park, Calif.

An excellent textbook.

Carlson, E. A. (1966). *The Gene: A Critical History*. W. B. Saunders, Philadelphia.

An interesting history.

Dunn, L. C. (1965). Mendel, his work, and his place in history. *Proc. Amer. Philos. Soc. 109*, 189–198.

An insightful and lucid essay.

Iltis, H. (1966). *Life of Mendel*. Haffner, New York.

A fine standard biography of Mendel that was first published in German in 1924 under the title *Gregor Mendel: Leben, Werk und Wirkung* (Springer, Berlin). The edition cited here is a translation by Eden and Cedar Paul that was published in 1932 and reissued in 1966.

Peters, J. A. (1959). *Classic Papers in Genetics*. Prentice-Hall, Englewood Cliffs, N.J.

A collection of reprints of the most significant papers in the history of genetics.

Snyder, L., Freifelder, D., and Hartl, D. (1985). *General Genetics*. Jones & Bartlett, Boston.

Snyder et al., Strickberger, and Suzuki et al. are good, up-to-date textbooks of genetics (of which there are many more on the shelves). Suzuki et al. presents excellent bibliographies.

Strickberger, M. (1985). *Genetics*. 3rd ed. Macmillan, New York.

Sturtevant, A. H. (1965). *A History of Genetics*. Harper & Row, New York.

A respected and authoritative work.

Suzuki, D. T., Griffiths, A. J. F., Miller, J. H., and Lewontin, R. C. (1986). *An Introduction to Genetic Analysis*. 3rd ed. W. H. Freeman, New York.

By now the significance of the fact that alleles are paired should be clear: Each parent contributes one member to the pair. Since gametes contain only one of each allele, allelic pairs must break apart in the course of gamete formation. So Mendel showed us.

The properties of genes clearly parallel the properties of chromosomes discussed in Chapter 12. Mendel was unaware of the existence of chromosomes when he performed his insightful experiments. Paradoxically, the pioneer students of chromosomes were unaware of Mendel's work. Only when De Vries, Correns, and von Tschermak independently rediscovered Mendel's work around 1900 did they begin to converge.

Even then, investigations of heredity and chromosomes persisted on separate tracks for the years 1900 to 1910. One reason for the delay was the propensity of chromosomes to fade from view between cell divisions. To many, this seemed to put their continuity and individuality in question.

THE CHROMOSOME THEORY

In 1902, a young American graduate student named W. S. Sutton was the first to connect Mendel's results with the movements of chromosomes in mitosis and meiosis. Similar ideas were expressed at about the same time by the German cytologist Theodor Boveri (1862–1915). In a classic paper titled "The Chromosomes in Heredity," Sutton theorized that the hereditary factors—the genes—are carried on the chromosomes. This was the first statement of the **chromosome theory.** It successfully explained the following:

- All of Mendel's results
- Why genes occur in pairs (because chromosomes do)
- Why each member of a gene pair is derived from one parent (because that is how chromosomes are derived)
- Why genes segregate at meiosis (because chromosomes do so in meiosis)

The striking parallels in the behavior of Mendel's genes and the chromosomes in meiosis are summarized in Figure 14-1.

CHROMOSOME MOVEMENTS ACCOUNT FOR MENDEL'S LAWS
Principle of Segregation

Here was a penetrating idea: Mendel concluded that Cc alleles in his F_1 red-flowered heterozygotes segregate to produce two types of gametes, C and c (see Figure 13-5B). The chromosome theory states that the two types of gametes ought to appear in equal numbers. Therefore an F_1 plant must contain a pair of chromosomes carrying the gene pair C and c. Figure 14-2A illustrates the relationship between the origin

The physical basis of heredity
Studies of the common fruit fly, Drosophila melanogaster, *have revealed many basic principles of genetics. This population of fruit flies is feeding in a laboratory dish.*

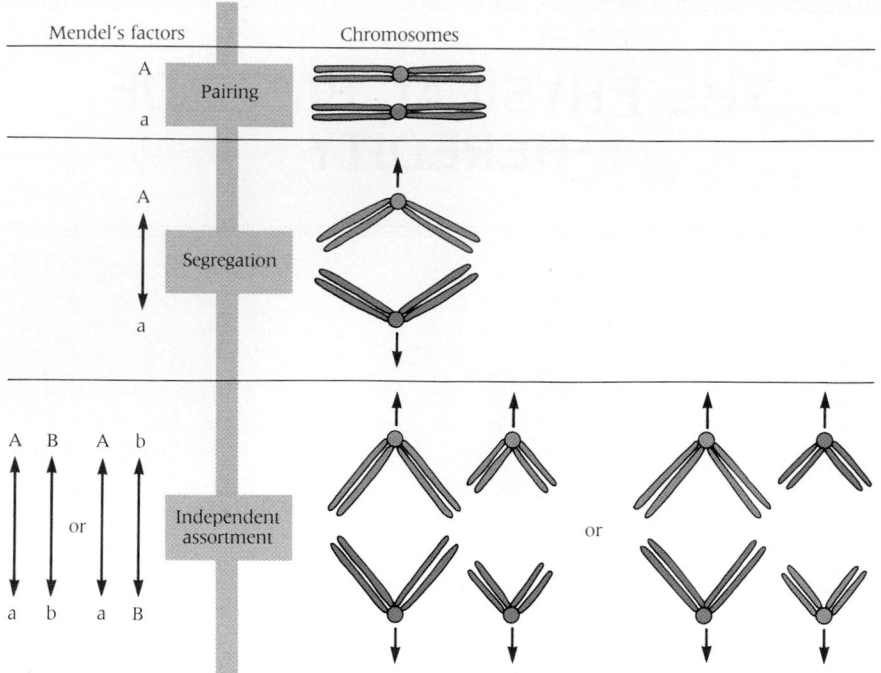

Mendel's factors | Chromosomes

Pairing

Segregation

Independent
assortment

14-1 Parallels in the behavior of Mendel's
factors and the chromosomes at meiosis
Colors represent homologous chromosome pairs.

of the F_1 red plant and the movement of its chromosomes during meiosis. It is apparent that the C and c alleles are on the large chromosome pair at meiotic metaphase I. Each member of the homologous chromosome pair duplicates, but the members go to opposite poles. Therefore, each telophase I nucleus contains a large double-stranded chromosome carrying either C or c. The two chromosome strands then separate in meiosis II to produce four F_1 gametes (Figure 14-2B).

In sum, each cell undergoing meiosis yields equal numbers of gametes bearing the C allele and the c allele. The segregation ratio will always be 1:1.

Principle of Independent Assortment

The Sutton–Boveri chromosome theory accounted for another of Mendel's discoveries—the independent assortment of multiple traits. Visualize two pairs of alleles (Rr and Yy) that are on different pairs of chromosomes—Yy on the larger chromosome pair and Rr on the smaller pair. In meiosis, the arrangement of the larger chromosomes on the equatorial plate at metaphase I is entirely independent of the arrangement of the smaller pair. Therefore, allele Y is as likely to end up in a gamete with r as it is with R. Hence, four types of gametes (YR, yr, Yr, and yR) are produced with equal frequency (Figure 14-3). Compare this arrangement with Figure 13-9, which shows a Punnett square illustrating the expected progeny from the cross $RrYy \times RrYy$. Note in the Punnett square the random and independent nature of the possible allelic combinations in the F_1 generation.

One of the early objections to the chromosome theory centered on doubts about the behavior of homologous chromosomes. Since in many species the chromosome pairs may appear superficially similar, this suggested the possibility that random pairing occurred in prophase I of meiosis—a member of one pair undergoing synapsis with a member of another pair. Decisive evidence against this possibility was found in species in which chromosomes do differ in size and shape.

In a famous experiment carried out in 1913, Elinor Carothers found an unusual species of grasshopper with testis cells that contain one pair of nonidentical homologous chromosomes termed a **heteromorphic pair** (Figure 14-4). She knew the chromosomes were homologous because they always paired up in meiosis I. One more oddity appeared in these cells: There was a single chromosome that had no homologous pairing partner. Recognizing that these curious chromosomes could serve as markers in an experiment designed to test whether different chromosomes do or do not segregate independently, Carothers studied anaphase nuclei to determine how often each of the two dissimilar chromosomes of the heteromorphic pair migrated to the same pole as the unpaired chromosome. If they migrated with unequal

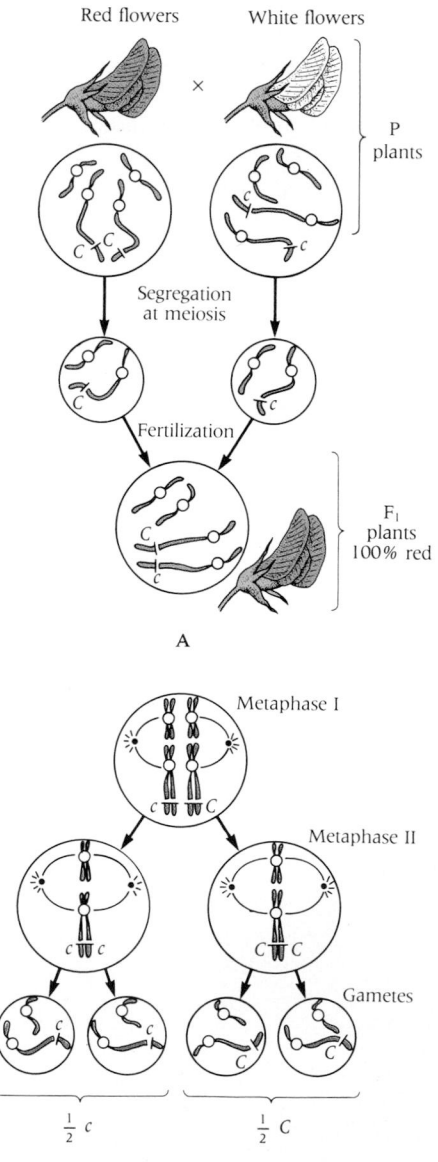

14-2 Chromosomal basis of the 1:1
segregation ratio
(A) Origin of the F_1 plant and its nucleus, which carries the alleles C and c on its large chromosomes. (B) Meiosis in an F_1 plant leads to equal numbers of c and C gametes.

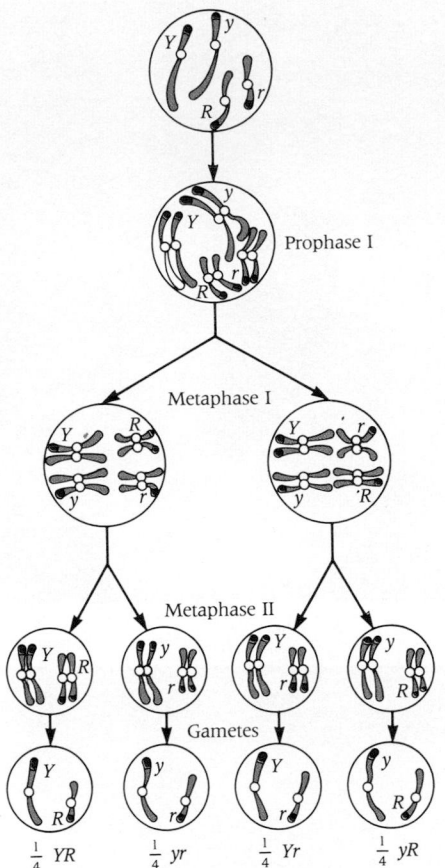

14-3 The route through meiosis of two heterozygous allele pairs

The figure traces the fate during meiosis of two heterozygous allele pairs (Yy and Rr), which are on separate chromosome pairs. In the end, the four gametes contain YR, yr, Yr, and yR with equal frequency.

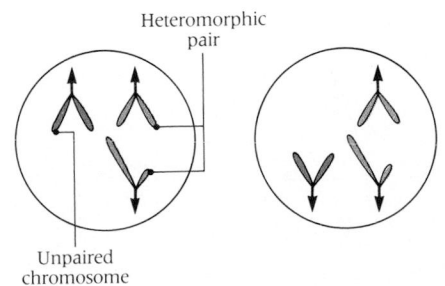

14-4 Evidence for independent assortment of chromosomes

In an unusual cell containing a heteromorphic chromosome pair and an unpaired chromosome, the unpaired chromosome appears with each member of the pair with equal frequency. Arrows indicate direction of movement toward the poles of the cell.

frequency, it would appear that chromosomes that are not homologues are not independently assorted. But if they migrated with equal frequency—as her studies clearly suggested—it would be established that chromosomes that are not homologues do assort independently.

STATISTICAL ASPECTS OF MENDEL'S THEORY

We spoke of the ratios in which different genotypes and phenotypes appear in the F_2 generation—the 3:1 ratio, the 1:2:1 ratio, the 9:3:3:1 ratio, and so on. These are called **Mendelian ratios.** Interestingly, the data of actual experiments (see Table 13-1) only approximate these ratios. For example, Mendel obtained 75.90% reds and 24.10% whites, rather than exactly 75.00% and 25.00% of each phenotype.

The reason for the small deviation of observed results from predicted results is both simple and important. An experimental sample is no more than a sample that gives an approximate representation of the larger population from which it is taken. In tossing a coin, we expect heads as often as tails. If we tossed it a million times, a 1:1 ratio would probably be closely approximated. If we tossed it only four times, four heads might very well appear instead of the expected ratio of 2 heads : 2 tails. If an observed or experimental ratio of heads to tails is to stay close to the expected ratio of 1:1, we need many tosses.

The same is true of crosses between F_1 red plants (*Cc*). If a geneticist makes this cross 100 times, the 100 eggs and 100 sperm that actually participate in producing offspring would be only small samples of the very large number of gametes produced by an F_1 plant. Therefore, the experimental results—like a poll before an election—would only approximate the true situation.

If an observed ratio differed from an expected ratio, such analysis could help us decide whether or not the deviation is significant. Statistical analysis can indicate how large a sample would be needed to give the hypothesis a meaningful test and can reveal that an observed result is not compatible with the hypothesis. In such a case, the hypothesis should be discarded and replaced with a new one—if the investigator can devise one. Conversely, statistics may strongly support the validity of a hypothesis. Further comments on these concepts, which have obvious importance for genetic studies, appear in Box 14-1.

CHROMOSOMAL GENETICS

Unlike much that preceded it, the chromosome theory was precise and quantitative. It rested on the observable behavior of chromosomes. And it permitted testable predictions. With the formulation of the Sutton–Boveri theory, modern genetics was born.

''FRUIT FLY GENETICS''

Early in the twentieth century, much experimental work in genetics was done with the common fruit fly, *Drosophila melanogaster*, a small gray or grayish-tan fly that hovers around garbage cans and fruit stands in late summer. More recently, other organisms from bacteria to human beings have gained favor as objects of genetic research. Although we shall later speak of these other organisms, and the advantages and disadvantages of their use, it is appropriate at this point to examine *Drosophila*. This organism was studied rewardingly by an American biologist, then at Columbia University, Thomas Hunt Morgan (1866–1945) and his young collaborators, C. B. Bridges, H. J. Muller, and A. H. Sturtevant.

Drosophila is a wonderfully serviceable experimental subject, which exists as a so-called **wild type** (a genotype or phenotype that occurs in nature or in comparable pedigreed laboratory strains) or in many variants, or **mutant** types (Figure 14-5).

Drosophila can be raised in large numbers in a small laboratory. Its life cycle is typical of that of many insects (Figure 14-6): The egg becomes a larval stage, also called the first **instar,** which in turn molts twice to become the second and third instars. A pupa then develops, within which the larval carcass is replaced by adult structures. An imago, or adult, then emerges, ready to mate within 12–14 hours. Indeed, *Drosophila* breeds prolifically, a new generation appearing every 12–14 days. The pea experiments that took Mendel seven years (and required a large garden)

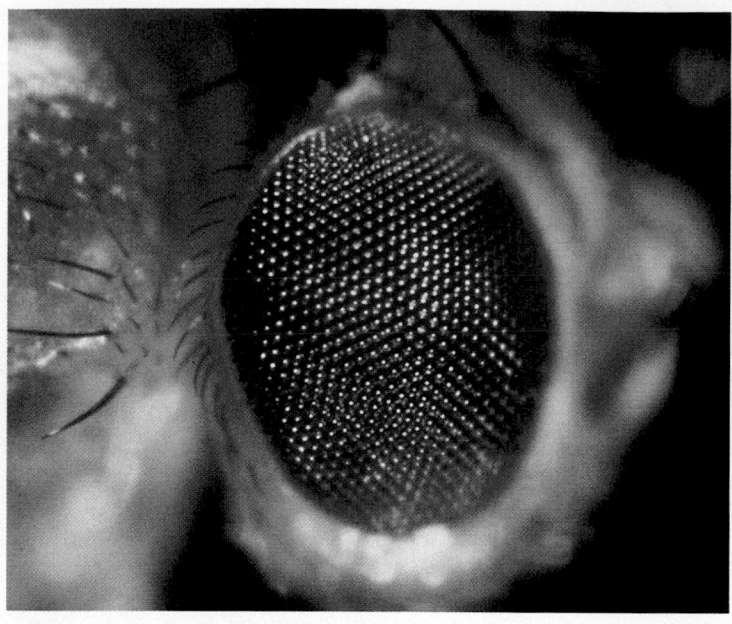

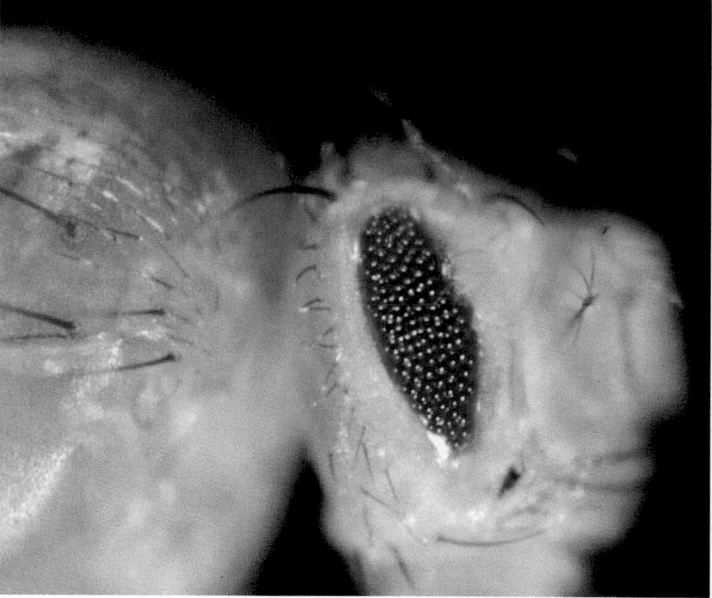

A   B

could be repeated with *Drosophila* in a few months (in a dozen small, easily tended bottles).

Drosophila also has other advantages that were not appreciated at first. It has very few chromosomes—only four pairs in the most studied varieties (Figure 14-7). Also, in the salivary gland cells the chromosomes are enormously larger, by some thousands of times, than in the cells of other tissues or in most other organisms (see Figure 12-20). The small number and large size of these chromosomes in *Drosophila* simplified their use in research, especially in studies seeking to correlate chromosome structure and heredity. Finally, the patterns of heredity of the fruit fly are so consonant with those of many other organisms, it seems reasonable to assume that whatever these patterns teach us of the principles of heredity in *Drosophila* is probably valid for most other organisms.

C

14-5 Drosophila *occurs in normal and many mutant forms*
(A) *Appearance of wild-type eye.* **(B)** *Eyeless mutant, showing reduced eye size.* **(C)** *Wild type. Note the red eyes.*

Notes on Genetic Symbols

Before proceeding further, it is appropriate to digress briefly on the conventions that govern the designation of genes and alleles.

Recognized genes are generally given standard names and symbols. The names usually consist of one or two words that may or may not be descriptive and often are whimsical. Note, for example, some of the curious gene names given some mutants ("speck body," "bobbed hairs," and so on). Symbols applied to certain organisms (such as pea plants) consist only of capital letters for dominant alleles and lower case letters for recessive alleles. Modern genetics developed a more sophisticated system for use with *Drosophila*, as described in the following:

- The alleles found in wild-type strains (that is, the common form found in nature) are designated with plus (+) signs. Wild-type alleles are usually dominant, but they are still marked with a plus sign when recessive.
- Genes are usually symbolized by the first letter of the name or by two-letter designations (where one letter would cause ambiguity). Thus the body color gene "ebony" has a wild-type allele e^+ and a mutant allele e. Writers may also refer to them simply as + and e, respectively.
- If the mutant allele is recessive to the wild type, its symbol is not capitalized. A mutant allele that is dominant is capitalized. For example, the following mutant alleles are dominant: B (bar, an eye shape), Cy (curly wings), and Pm (plum, a brownish eye color). The wild alleles are B^+, Cy^+, and Pm^+.
- As explained later in the text, a slash or line indicates linked allele locations on homologous chromosomes. For example, $abc/+++$, AbC/aBc, and $a++/+bc$ are combinations of three genes, written to indicate that the genes separated by a slash are on homologous chromosomes. Each of these combinations represents a different genotype, but all have the same phenotype.

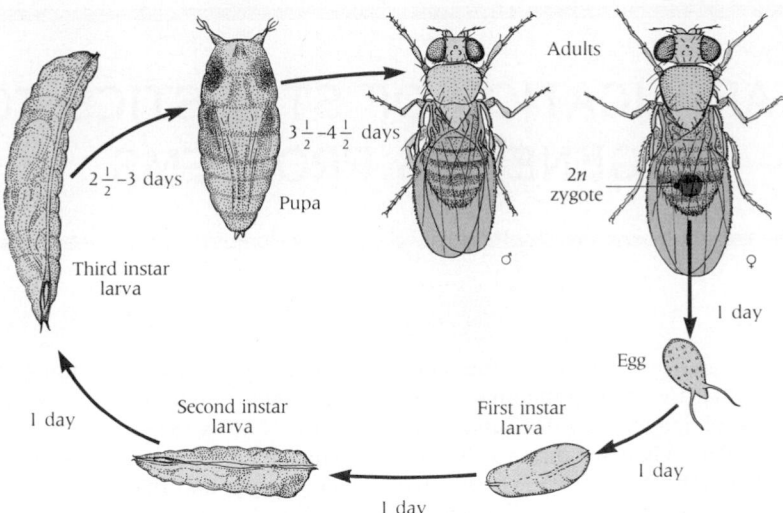

Adults

$3\frac{1}{2}$–$4\frac{1}{2}$ days

Pupa

$2\frac{1}{2}$–3 days

Third instar larva

1 day

Second instar larva

1 day

First instar larva

1 day

$2n$ zygote

♂   ♀

1 day

Egg

1 day

■ Linked alleles may be simply underlined together (for instance, <u>ABC</u>) without display of chromosome position.

Gene Linkage

An early major contribution of "fruit fly genetics" was the concept of **gene linkage,** which is defined as the association of genes on the same chromosome pair.

Independent assortment of genes observed by Mendel when he followed the inheritance of two allelic pairs is a particular case of a more complicated phenomenon. Two pairs of alleles do assort independently—but only when they are on separate chromosomes. Mendel did not realize it, but with one exception the pea plant genes he selected for study are on different chromosomes. Because of this, his experiments were largely uncomplicated by assortment patterns influenced by gene linkage.

Each chromosome carries many genes. The alleles on each chromosome are said to be linked. Where the chromosome goes, its genes go—unless, of course, crossing-over changes the chromosomal locations of the genes. Linked genes remain together through meiosis, in the gametes, and in the fertilized egg. Each "chromosome-load" of alleles segregates as a unit just as though it consisted of a single pair of alleles. We say that such genes are linked because they are inherited as a package.

When *Drosophila* are bred, four sets of traits usually appear in clusters. The clusters of responsible genes are called **linkage groups.** The fact that there are four such groups in the fruit fly has obvious significance: The value of *n*, the haploid number of the species, is 4. Diploid cells have four pairs of chromosomes. One of the four linkage groups includes a very small number of traits. Clearly, it belongs to the dotlike chromosomes, called pair IV (see Figure 14-7).

Of the hundreds of recognized genes in the fruit fly, we will choose as examples two allelic pairs, one affecting eye size, the other body color (see Figure 14-5B).

■ Pair 1: "Large eyes" (symbolized ey^+); and "eyeless" (symbolized ey), a term for reduced eye size. (Note: "large eyes" is dominant over "eyeless.")
■ Pair 2: "Tan" (symbolized b^+) and "black" (symbolized b). (Note: "tan" is dominant over "black.")

Imagine two fruit flies of known pure lines. One is homozygous for one recessive trait ("eyeless") and for one dominant trait ("tan"). Its genotype would be $eyeyb^+b^+$. The other is homozygous in the opposite dominant–recessive pattern: "large eyes" and "black." Its genotype would be ey^+ey^+bb. When we cross them (Figure 14-8), all members of the F₁ generation have the wild-type phenotype; the dominant traits are expressed. The F₁ genotype would be ey^+eyb^+b. What would it be if the F₁ fly is test crossed? The answer depends on the chromosomal locations of the two allelic pairs.

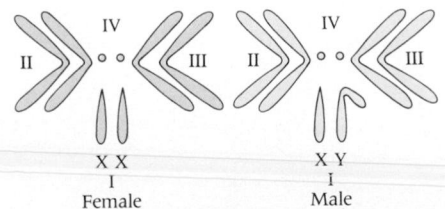

IV   IV

II >>○○<< III   II >○○< III

X X   X Y

I   I

Female   Male

***14-7 Chromosomes in* Drosophila**
The fruit fly has four pairs of chromosomes (I–IV). Pair I is the sex chromosomes, X and Y.

Genetics research often involves the study of **samples** that are chosen randomly as representatives of larger **populations.** Such sampling occurs when we mix a thousand small colored balls in a glass bowl and then take out a sample of, say, 20. The factors that determine whether or not a particular ball is in our sample are numerous and independent of each other. Hence the inclusion of the ball in the sample—indeed, the composition of the sample—is due to **chance,** which can be defined as the action of a multiplicity of independent causes. **Statistical tests** are methods of drawing valid conclusions from observations made on random samples. They help us to decide whether observed differences between random samples are due to chance or to other factors.

Consider a population of known composition—say, 1000 persons. **Frequency** is the number of individuals of a certain class in the whole population or sample—for instance, 500 males in a population of 1000. In a sample of 10 from that population, however, males and females may have frequencies of 4 and 6, respectively. Frequencies are often expressed as percentages—for example, 50% males. Statements of percentage frequencies in samples are useless if actual numbers of individuals are not specified.

When we pick an individual at random from a population of 500 males and 500 females, our expectation of picking a male is equal to our expectation of picking a female. The **probability** of picking an individual of a certain class by random sampling is defined as the relative frequency of individuals of that class in the population—that is, the frequency divided by the total population. Here the probability of picking a male is $500 \div 1000 = \frac{1}{2}$, or 0.5. The chance of picking a male is 1 out of 2, and the odds for male and female are even—$1:1$. If, in a population of 1000 organisms, 700 have a certain gene and 300 do not, the probability that any one will have the gene is $700 \div 1000$, or 0.7. The probability that it will not is 0.3. The sum of the probabilities of all classes always equals 1.0.

An understanding of the laws of probability is important in (1) understanding genetic mechanisms, (2) predicting the likelihood of certain results from a given cross, and (3) assessing how well an observed phenotypic ratio among progeny fits a postulated genetic mechanism. A good example of the role of probability laws is seen in studies of dihybrid ratios, the transmission of two pairs of genes that are on separate chromosome pairs. The law states, in essence, that the

probability of the simultaneous occurrence of two or more independent events is equal to the product of the probabilities that each will occur separately.

We can illustrate how the law works by tossing a coin. A coin tossed into the air is likely to land either "heads" or "tails" (if we neglect the highly improbable chance of its landing on an edge). We would therefore predict that the probability of a "head," or H, is 1 in 2, or $\frac{1}{2}$; the probability is the same for a "tail," or T. But if we toss one coin four times, we would not be surprised to get a ratio other than $2H:2T$. Indeed, we expect on occasion to see 4H, or 3H and 1H, or 1T and 3T, or 4T. However, if the number of tosses is very large, we would expect to approach a $1:1$ ratio. What if we toss two coins simultaneously for, say, 50 tosses? When this was actually done in an experiment, the result was HH12, HT27, TT11. We should now ask, "Are these the results we should have expected if each coin has an equal chance of landing either heads or tails?" In coin tossing as well as in breeding experiments—where given crosses yield certain progeny—it is necessary early in the game to formulate a hypothesis to predict or explain the results. We then ask, "Is an observed deviation from the predicted result

---

If they were on different chromosomes, the F_2 generation would display four different phenotypes in a ratio of $1:1:1:1$. As shown in Figure 14-8, just this is found. Hence, these particular alleles show independent assortment. They are free to segregate independently during anaphase I of the meiotic divisions leading to gamete production.

However, let us now follow two recessive traits, one ("black") affecting body color and one ("purple") affecting eye color.

- "Black" (symbolized b)
- "Purple" (symbolized p)

[Note: "purple" is recessive to the dominant wild-type red eye (p^+).]

As shown in Figure 14-9, crossing a purple eye–tan body fly (ppb^+b^+) with a red eye–black body fly (p^+p^+bb) yields F_1 flies that are once again all phenotypically wild type, though their genotype is p^+pb^+b. But this time when we test cross an F_1 fly, the phenotypic ratio in the F_2 is not $1:1:1:1$. The F_2 flies are phenotypically like the parents—with purple eye–tan body and red eye–black body appearing in

due to chance alone—or is it due to a faulty hypothesis?" If the latter, we must change the hypothesis.

To answer this question, we must make three assumptions about the coins and the conditions: (1) the coins are unbiased, (2) they are independent of each other, and (3) the result of one toss is unaffected by the preceding toss. Again, what kind of results should we expect from 50 tosses of two coins? Since each coin has 1 chance in 2, or a probability of ½, of coming up heads and the same probability of landing tails, the chance of HH—of *both* coins showing heads in the same toss—is ½ × ½, or ¼. Clearly, that result was not obtained. What should we expect with regard to HT? The chance of one coin landing H is ½; the chance of the other landing T is also ½. Now ½ × ½ = ¼, but the total array of possibilities thus arrived at (¼ HH + ¼ TT + ¼ HT) equals only 3/4. That leaves ¼ of the possibilities unaccounted for. However, the HT category should really be HT + TH, or 2 HT. Therefore, the probability of one head and one tail is 2(½ × ½), or ½.

The general statement for two independent events of known probability may be written as

$$a^2 + 2ab + b^2$$

where *a* represents the probability of a head and *b* the probability of a tail. If, as in this case, *a* and *b* each equal ½, then the value of this expression upon substitution becomes $(½)^2 + 2(½ × ½) + (½)^2$, or ¼ + 2/4 + ¼ = 1. Notice that the total of the probabilities is 1, and that $a^2 + 2ab + b^2$ is the expansion of the binomial $(a + b)^2$.

Our observations and expectations for a two-coin toss can be summarized as follows:

Class	Observed	Expected
HH	12	12.5
HT	27	25.0
TT	11	12.5
	50	50.0

How can we tell if the departure from expected results is too great to be accounted for by chance alone? The question implies that we do not expect to achieve exactly expected results very often. As Table 13-1 shows, even Mendel's results did not reflect the exact expected ratios (although they were surprisingly close!). Clearly, we need a mathematical tool to determine "goodness of fit." We have such a tool in the **chi-square (χ^2)** test.

The formula for calculating chi square is

$$\chi^2 = \sum \left[\frac{(o - c)^2}{c} \right]$$

where *o* = observed frequencies, *c* = calculated frequencies, and Σ indicates that the bracketed quantity is to be summed for all classes. Both *o* and *c* are calculated in actual numbers and not in percentages.

To calculate chi square for the coin tosses (which are expected to be 1:2:1), the data may be set up as below. Our question can now be restated as, "How often, by chance, can we expect a value of χ^2 equal to or greater

than 0.36?" The answer is obtained by consulting a table of chi-square values in a statistics book. There we will find that a χ^2 of 0.36 corresponds to a probability value of between 0.95 and 0.80. This means that, when expected results are 1:2:1, we can anticipate a deviation as large as or larger than we experienced in between 80 and 95% of repeated trials. Such a deviation, therefore, could be due to chance, and both our expectation and the assumptions on which it was based appear good. In other words, we have a good fit between observed results and our calculated expectations.

Properly used in genetics, the chi-square test can shed considerable light on the mechanisms operating in particular crosses. This test, one of many in the field of statistics, can be of tremendous value in deciding whether the observed traits in a breeding experiment were due to chance or whether they validated a hypothesis that predicted a definite ratio (for instance, 3:1 or 9:3:31). This is but one example of the power of statistical analysis. Many more formulas and tests are available that enhance the value and meaning of experimental data. For example, the **standard deviation** formula evaluates measurement or enumeration data; the concepts of **confidence limits** and **student's t test** facilitate the comparison of samples; and so on. All are discussed in textbooks on statistics.

Class	Observed o	Calculated c	Deviation $(o - c)$	Squared deviation $(o - c)^2$	$\frac{(o - c)^2}{c}$
HH	12	12.5	−0.5	0.25	0.02
HT	27	25.0	+2.0	4.00	0.16
TT	11	12.5	−1.5	2.25	0.18
Totals	50	50.0	0		$\chi^2 = 0.36$

a 1:1 ratio. No new combinations formed, as in the previous example. Note once again the importance of a test cross of an F_2 fly with a homozygous recessive.

Something different occurred in the second example. It was *gene linkage*. The two alleles remained together and did not segregate independently because they are on the same chromosome.

Arrangements of Linked Genes

When two genes are linked in an individual that is heterozygous for each gene, linkage may be of two kinds—a coupling conformation or a repulsion conformation.

- In a **coupling conformation,** the two dominant alleles are located on the same member of a chromosome pair, with the two recessives on the other.
- In a **repulsion conformation,** the dominant allele of one pair and the recessive allele of the other pair are located on one member of a chromosome pair, with the recessive of the first pair and the dominant of the second pair on the other member.

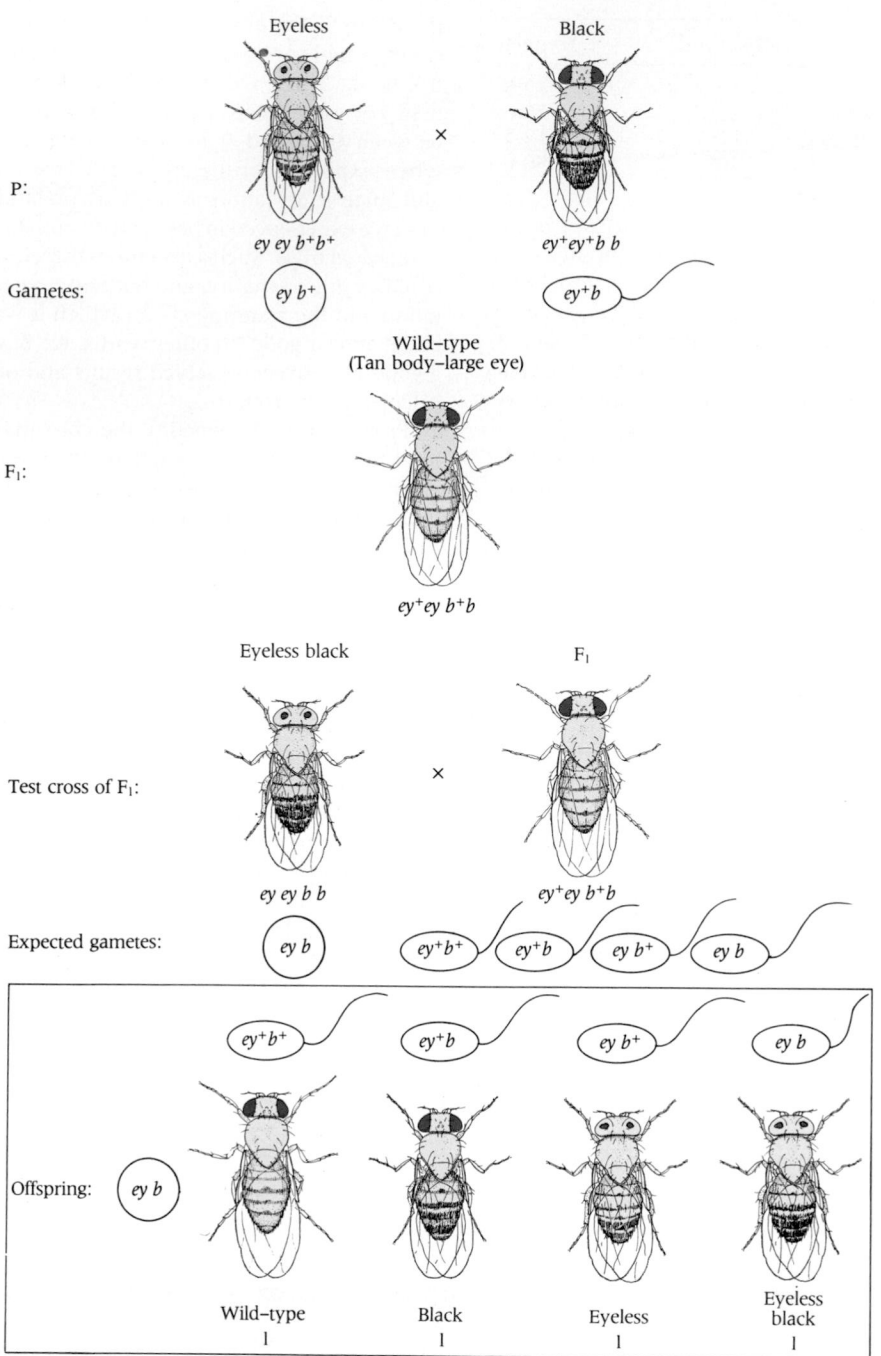

Using two hypothetical genes *Aa* and *Bb*, these arrangements would be as follows:

$$\frac{AB}{ab} \text{ and } \frac{Ab}{aB}$$

The coupling pattern on the left (two dominants on the same chromosome) is also called a *cis* arrangement; the repulsion pattern on the right (one dominant and one recessive on the each chromosome) is a *trans* arrangement.

In the practical world of plant and animal breeding, it is often important to know whether two dihybrids showing the same phenotype have a *cis* or a *trans* allelic arrangement, since the two types could yield very different progeny. How can these states be distinguished?

Predictably, the answer is by a test cross designed to determine which combination of alleles is together on a given chromosome. The approach rests on the fact that linked alleles stay together and do not assort independently.

It is necessary at this point to introduce symbols for displaying a chromosome pair on the printed page. It is convenient to use a slash or a bar. The alleles to the

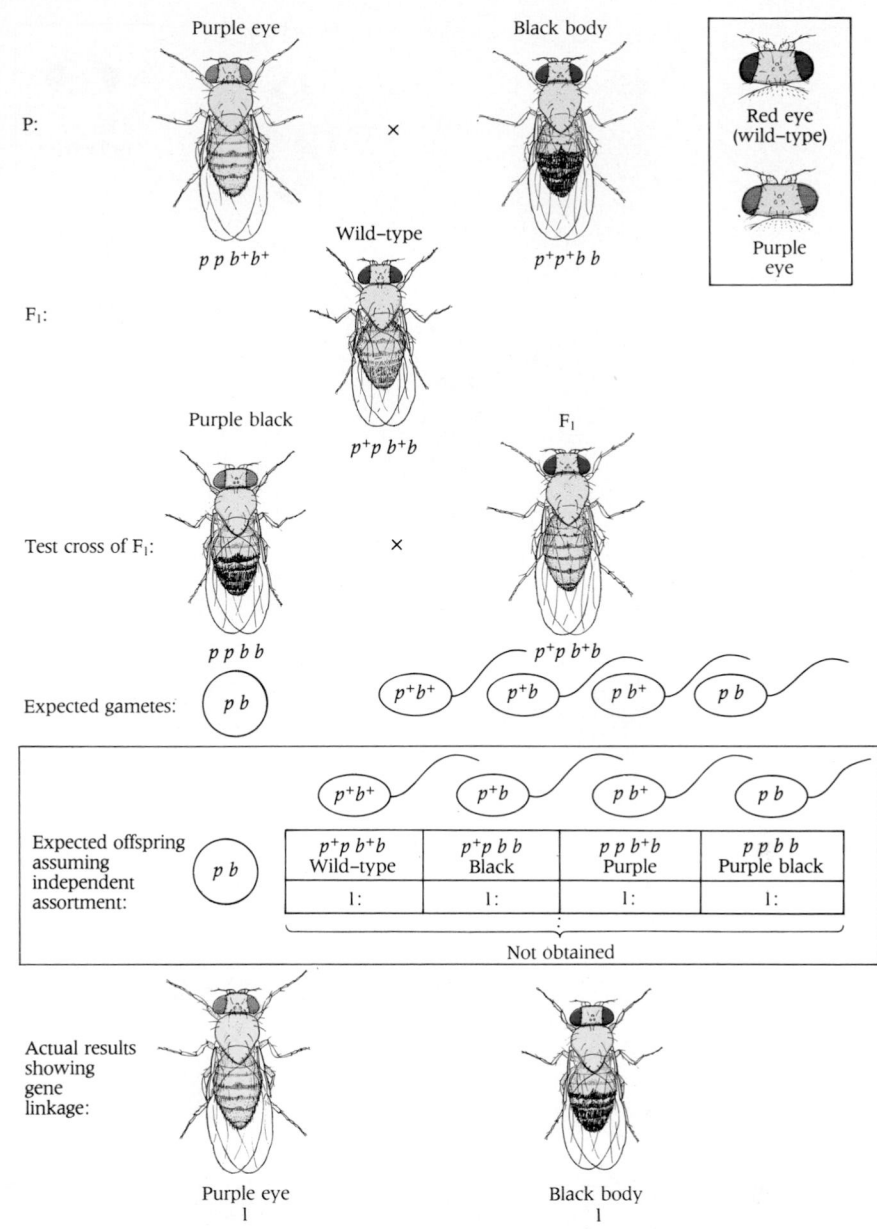

14-9 Linkage of purple and black genes
Linked genes do not assort independently. Four different types of gametes are produced by the F₁ fly; however a test cross with a homozygous recessive fly (purple fly) produces only two phenotypes among the offspring, in a 1:1 ratio.

left of the slash (or above the bar) are on one chromosome of the pair. Those to the right of the slash (or below the bar) are on the other chromosome. The two allelic combinations are linked and thus they remain together.

The cross illustrated in Figures 14-9 and 14-10 is between purple eye–black body flies. Each parental chromosome bears a recessive and a dominant allele: purple eye–tan body in one parent; red eye–black body in the other parent. These linked allelic combinations will stay together. In the F₁ generation all flies have a wild-type phenotype: red eye–tan body. In the test cross, an F₁ fly of wild-type phenotype is crossed with a double recessive, a purple eye–black body fly. The F₂ progeny include only two phenotypes in a 1:1 ratio. They are purple eye–tan body and red eye–black body—the same phenotypes as in the parents. The genotypes of the two F₂ classes are, respectively: $pp\ b^+b^+$ and p^+p^+bb, the same as the parental combinations. As predicted, the combinations did not come apart, and only the parental combinations were formed (purple eye–tan body and red eye–black body).

Crossing-over Changes Gene Positions on Chromosomes

Consider another example of linkage that will teach us another important lesson on how genes are packaged in a chromosome. Corn, a popular plant among geneticists—which brought a Nobel Prize to Barbara McClintock, one of the earliest corn geneticists—has the following genes affecting kernel color and texture (Figure 14-11). Note that capital letters indicate dominant alleles.

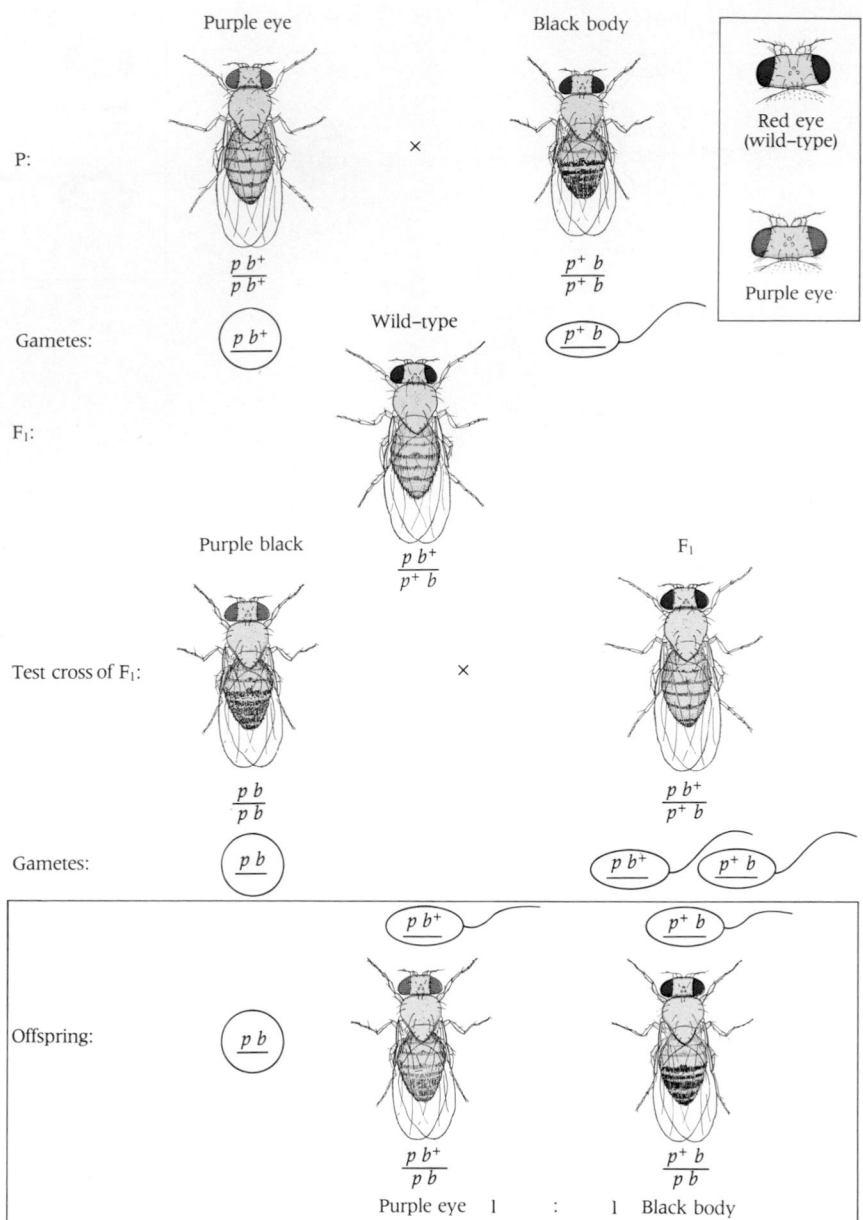

P:

Purple eye   ×   Black body

$\frac{p\ b^+}{p\ b^+}$ $\frac{p^+\ b}{p^+\ b}$

Gametes: $(p\ b^+)$ Wild-type $(p^+\ b)$

F₁:

$\frac{p\ b^+}{p^+\ b}$

Test cross of F₁:

Purple black   ×   F₁

$\frac{p\ b}{p\ b}$ $\frac{p\ b^+}{p^+\ b}$

Gametes: $(p\ b)$ $(p\ b^+)$ $(p^+\ b)$

Offspring: $(p\ b)$ $(p\ b^+)$ $(p^+\ b)$

$\frac{p\ b^+}{p\ b}$ $\frac{p^+\ b}{p\ b}$

Purple eye   1   :   1   Black body

14-10 Arrangement of linked genes in a heterozygote

In the F₁ fly, the purple and black genes are in repulsion conformation. A test cross produces a 1:1 ratio of purple to black. If these genes were in coupling conformation in the F₁, a test cross would produce a 1:1 ratio of wild to purple black.

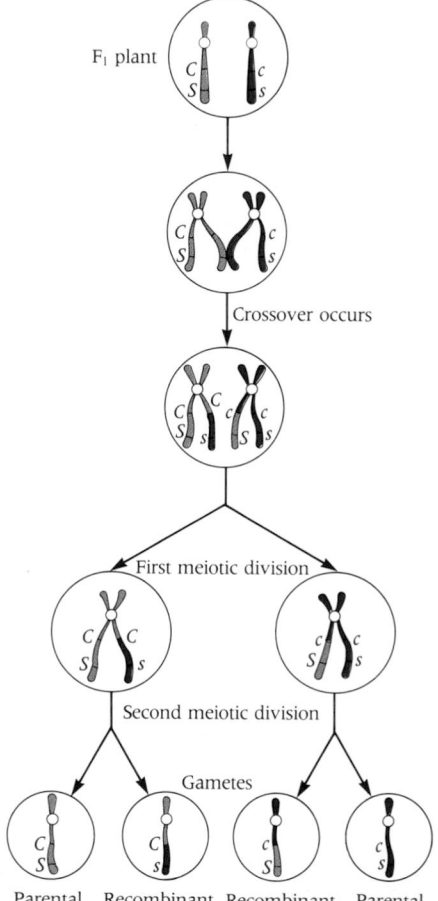

F₁ plant

Crossover occurs

First meiotic division

Second meiotic division

Gametes

Parental   Recombinant   Recombinant   Parental

	Alleles	Effect on kernel
Gene 1	C	Colored
	c	Colorless
Gene 2	S	Smooth (or full)
	s	Shrunken

Begin with the following cross:

P plants $CCSS \times ccss$

P gametes CS cs

F₁ plants $CcSs$

According to the rule of independent assortment, the F₁ plants should produce four types of gametes—*CS, cs, Cs, cS*—in equal numbers, each type comprising 25% of the gametes. The F₂ generation resulting from a F₁ × F₁ cross should have kernels in the ratio 9 colored-smooth : 3 colored-shrunken : 3 colorless-smooth : 1 colorless-shrunken (see Figure 13-9). Whether or not F₁ actually did produce gametes of all four kinds would be revealed by the phenotypic ratio in the F₂ generation.

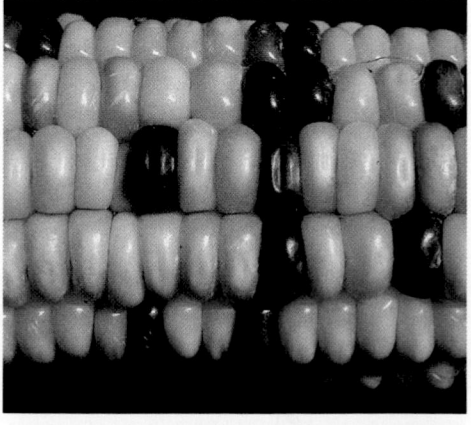

14-11 Phenotypic expression of corn kernels

The color and texture of corn kernels is represented by two different genes located on the same chromosome pair.

A more direct method of characterizing the gametes of the F_1 plants would be a test cross. If the F_1 plant (CcSs) to be tested were crossed with a "tester" plant known to be homozygous for both recessive genes (ccss) and thus capable of producing only cs gametes, the results would appear as in Table 14-1. Note the phenotypes of the test-cross offspring and their frequency. What can we make of these numbers?

The nature of the gametes produced by the tested plant is immediately apparent because it alone could have contributed the dominant alleles. If the Cc and Ss allelic pairs had assorted independently, the four possible phenotypes—colored-smooth, colored-shrunken, colorless-smooth, and colorless-shrunken—should occur in test-cross progeny in equal numbers (row 4 in Table 14-1). Such a result would mean that there is no linkage between Cc and Ss. But if Cc and Ss were on the same chromosome, they could not assort independently. In that case, the tested plant could produce only CS and cs gametes in equal numbers. Test-cross progeny would then be colored-smooth and colorless-shrunken in equal numbers (row 5).

The table shows (in row 6) the kinds of test-cross progeny actually obtained when this experiment was first carried out, in 1922, at the Cornell Agricultural Experiment Station by C. B. Hutchinson. Evidently, the two genes do not assort independently; the ratio among the F_2 phenotypes is close to 50% : 50%. We seem to be on the right track in attributing the results to linkage, but the results are not clear-cut.

Of the gametes produced by the CcSs plant, the commonest by far are CS and cs—the same as the ones produced by the parents. Thus they are **parental gametes.** New types (cS and Cs) are **recombinant gametes.** We cannot overlook the fact that a few recombinant gametes must have been operating in the test cross. Yet what we have said to this point about linkage would demand that test-cross progeny display the ratios given in either row 4 or row 5. Are the genes linked or are they not linked?

By contemplating the exceptions—the recombinant gametes—the science of genetics took an historic step. The sequence of discovery is instructive. First, it was found that some genes do not assort independently, as Mendel's second law says they should. Gene linkage was then uncovered. Next it was found in breeding experiments that supposedly linked genes failed to stay together with a small but measurable frequency. It became increasingly evident that alleles can change position on a chromosome. How they do it was soon discovered. The mechanism was given the explicit, if awkward, name **crossing-over.**

In the corn experiment detailed in Table 14-1, repeated test crosses showed that the alleles for kernel color and kernel texture are linked. However, exceptions were observed consistently. In this experiment, 1.75% of the F_2 generation had colored-shrunken kernels and 1.75% had colorless-smooth kernels. The discovery of how these exceptions come about was a profound advance, which gave support to emerging theories of genetics.

TABLE 14-1

USE OF THE TEST CROSS TO DETERMINE KINDS AND PROPORTIONS OF GAMETES PRODUCED BY THE F_1 CORN PLANT (CcSs)

	F_1 Plant CcSs (colored-smooth)				Tester Plant ccss (colorless-shrunken)
(1) Possible genotypes of F_1 gametes	CS	cs	Cs	cS	
(2) Possible genotypes of test-cross progeny (F_1 × tester)	CcSs	ccss	Ccss	ccsS	cs (only possible gamete genotype in tester plant)
(3) Phenotypes (i.e., traits of seeds) of test-cross progeny that correspond to genotypes shown	colored-smooth	colorless-shrunken	colored-shrunken	colorless-smooth	
(4) Ratios of phenotypes that would indicate no linkage between C and S	25%	25%	25%	25%	50% are recombinants
(5) Ratios of phenotypes that would indicate complete linkage between C and S	50%	50%	0%	0%	0% are recombinants
(6) Observed ratios indicating incomplete linkage between C and S*	[4030] 48.25%	[4030] 48.25%	[150] 1.75%	[150] 1.75%	3.5% are recombinants

* Numbers in brackets are actual numbers observed in Hutchinson's experiment. Percentages below are calculated from these data.

When investigators carefully scrutinized chromosome behavior early in the first meiotic division, they found the cause of the exceptions. As we saw in Chapter 12, when homologous chromosomes pair up in meiosis, the partners often exchange parts, literally crossing-over (Figure 14-12). Through-and-through breaks occur at corresponding parts of each chromosome during synapsis, and a piece of one chromosome is replaced with a piece of its homologue. All alleles on the exchanging parts travel, like passengers on a bus, from one chromosome to the other. Broken ends are joined and meiosis goes on. Crossovers of the kind that took place in the corn-breeding experiment are diagrammed in Figure 14-12.

Each chromosome, we recall, is a doublet made up of sister chromatids containing identical alleles. But when an individual chromatid of one chromosome crosses over with the chromatid of another chromosome, the chromosomes that emerge have nonidentical chromatids. These new recombinant chromatids, containing allelic combinations not found in the parents, faithfully replicate themselves when the cell divides.

Investigators studying this situation soon encountered a complication. If a chromatid can participate in one crossover, it can participate in two crossovers. This phenomenon, termed a **double crossover,** can return an allele to its original location (Figure 14-13). Detection of this phenomenon requires use of a marker gene as a reference point. As shown in the figure, the marker is at the *B* locus—between *A* and *C*. The occurrence of *AbC* in an offspring of parental combinations *ABC* and *abc* shows that a double crossover must have occurred to link *b* with *A* and *C*.

The corn-breeding experiment held yet another surprise. The F$_2$ generation regularly contained 1.75% of each "irregular" type. That yielded another vital concept. Why, we may ask, does colorless appear with smooth in only 1.75% of the F$_2$ offspring? Why not 5% or 10%? The answer was discovered between 1910 and 1915 by Morgan, who recognized that each chromosome is an elongated object on which genes are positioned like beads on a string. Each allele, he reasoned, must have a definite fixed position, or **locus,** on the chromosome. Experiments then revealed that crossing-over takes place constantly and with equal probability at all points along the chromosome. Therefore, a mating can provoke the formation of many genetic combinations—and crossing-over is a significant source of genetic variability in a population.

Crossing-over Permits Chromosome Mapping

If Morgan was right, the frequency of crossing-over between any two alleles should be proportional to the distance between them. The further apart two linked alleles are on a chromosome, the more often the chromosome will break between them. Morgan was right, and it soon became possible to map the loci of alleles on chromosomes. The work was pressed forward by A. H. Sturtevant, Morgan's graduate student.

Imagine a chromosome that carries genes *A* and *B,* which recombine with a frequency of 15%. Note that in such experiments it is customary to express results as percentages of recombination—that is, the percentage of progeny that are recombinants. Localizations on chromosome maps are expressed not in absolute units like micrometers or nanometers, but in arbitrary **map units.** One genetic map unit is the length of chromosome distance between gene pairs within which meiotic crossovers (recombinants) occur 1% of the time. A single map unit is sometimes called a **centimorgan,** in honor of Thomas Hunt Morgan.

Suppose there is another allele, *C,* on the chromosome. We would like to know two things: the sequence of the three genes and the distances between them. We could readily acquire that information by determining the frequency of recombinations between them.

The possible gene sequences are *A-B-C, A-C-B,* and *C-A-B.* Assume that we found the following crossover frequencies: between *A* and *B* is 15%, between *B* and *C* is 9%, and between *A* and *C* is 24%. The results imply the following gene arrangement:

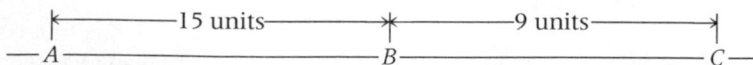

Had the recombination frequency between *A* and *C* been 6% (15 − 9) rather than 24% (15 + 9), we would have concluded that the sequence is *A-C-B.*

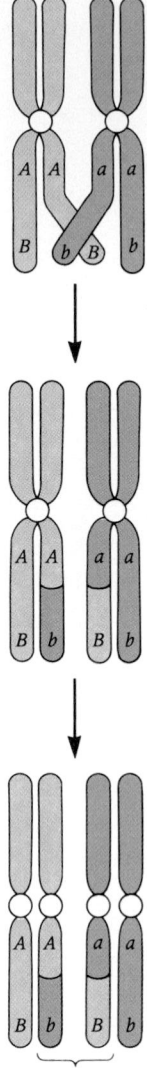

14-12 Crossing-over occurs during the tetrad stage of meiosis
The result of crossing-over is the production of two recombinant gametes that have a different combination of alleles than the parental gametes. Compare with Figure 12-32.

Recombinants

Three genes are required to determine whether a double crossover has occurred. The middle gene acts as a marker for this process.

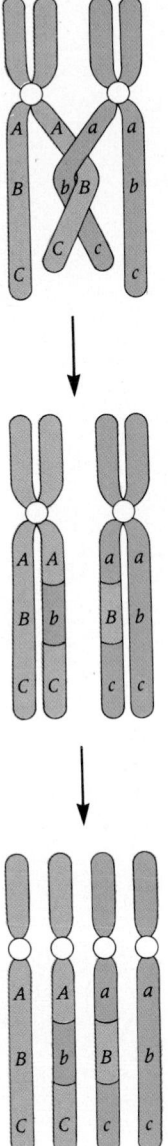

In summary, it is possible to construct a detailed **linkage map** of all the genes on a chromosome by systematically determining recombination frequencies between any gene and a pair of genes whose distance from one another is known.

The Morgan school recognized the significance of mapping studies, and by 1915 researchers had determined the location of more than 85 genes on one of the four chromosomes *in Drosophila*. Now we know the positions of hundreds of these genes (Figure 14-14). In later years, it became possible to confirm the gene order by microscopic studies of broken chromosomes. These permitted comparisons of missing chromosome regions with missing or altered genetic traits. All of this work gave strong support to the following fundamental concepts:

- First, nothing has been discovered to contradict the idea that genes are arranged sequentially on chromosomes.
- Second, each gene has a specific chromosomal locus that remains fixed.
- Third, a gene could now be defined as a locus on a chromosome that determines one or more traits. If two traits are linked so that recombination never occurs between them, it seemed reasonable to conclude that they are controlled by the same gene. The opposite case is even more compelling. If crossing-over does occur between them, however infrequently, two genes must be involved. Thomas Hunt Morgan defined the gene as the smallest unit of recombination.

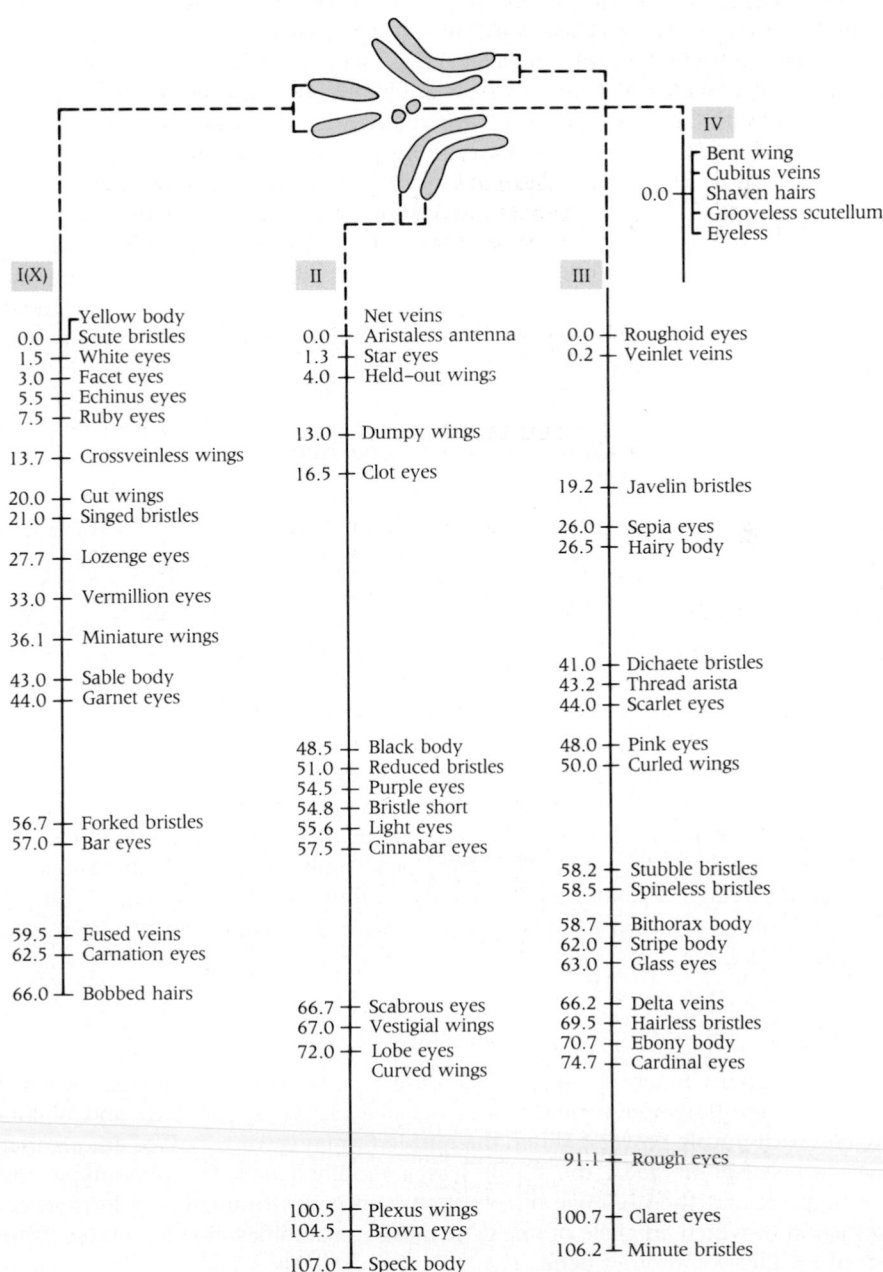

14-14 Linkage map of Drosophila chromosomes

The chromosome pairs are numbered I–IV. Each pair contains a specific set of genes. The distance between each gene locus is defined in map units.

However, we will soon see that modern understanding invalidates Morgan's definition. We now know that recombination can occur at different locations *within* a gene.

GENE INTERACTIONS

Most of the heritable traits we have considered thus far have been sharply discontinuous—in other words, distinct and readily separated from one another. Seeds were round or wrinkled, plants tall or dwarf, fly eyes purple or red. In these examples there were no "in-between" traits. Geneticists early learned, however, that most variations among organisms cannot be described in simple "yes/no" or "present/absent" terms. Rather, there is a gradation of traits.

For example, human height is determined genetically. But men and women in the real world are not tall or short; they are of many heights. Indeed, in any population this trait would display continuous variation. Many other traits are expressed in this way. As a matter of fact, the majority of the phenotypic characters of plants and animals display a broad spectrum of values. Other examples in human beings are intelligence, skin color, and eye color; in plants, size and seed color; and in mammals, coat color.

We have seen that some intermediate traits are attributable to incomplete dominance. But that is not the explanation in many cases. When that fact was recognized, the new science of genetics faced a serious crisis. Mendel himself was puzzled by the problem posed by traits that vary, not in yes/no fashion, but in continuous or graded fashion. When he did an experiment with bean plants, a cross between two F_1 hybrids yielded a whole series of different colors and not merely two types in the expected 3:1 ratio. Could it be that Mendel's principles were not always applicable? If not, their importance would be severely diminished.

Happily, investigators were soon able to explain continuous variation in terms that were in strict accordance with Mendelian principles. They found that some continuous variation is due to the additive effects of many different genes. This phenomenon, termed polygenic inheritance, occurs when two or more genes affect one phenotypic character. Such genes can interact in determining the character. In these cases, ordinary dominance does not occur and breeding experiments usually yield complicated phenotypic ratios.

Complementary Gene Action

Complementary gene action is probably the simplest type of interaction between two genes. Some single characters are determined by two genes. In these cases, the genes act together to produce a phenotypic trait that neither could produce alone.

A classic example is the genetics of the shape and size of the cock's comb (Figure 14-15). The dominant gene, *R*, yields a type of comb called "rose"; its recessive allele, *r*, produces what is termed a "single" comb. But another pair of allelic genes, *P* and *p*, is also involved. *P* produces a "pea" comb and *p* (like *r*) produces a "single" comb. The genotype of a cock with a single comb must be *rrpp*. A rose-combed cock could be *RRpp* or *Rrpp*. A pea-combed cock could be *rrPP* or *rrPp*. However, when *R* and *P* appear together (as in *RrPp*), the two genes collaborate to produce a "walnut" comb, a variety that neither could produce alone. Rose combs are found in Wyandotte chickens; pea combs occur in Brahma chickens. In a cross between a Wyandotte (*RRpp*) and a Brahma (*rrPP*), 9/16 of the F_2 offspring of two F_1 chickens would have walnut combs (*RrPp*). What kind of combs would the other 7/16 have?

Epistasis

In 1910, William Bateson obtained startling results when he crossed two pure-breeding white-flowered varieties of sweet pea (*Lathyrus odoratus*) and obtained progeny with purple flowers! When the purple F_1 plants were crossed among themselves (or self-pollinated), the result was a modified 9:3:3:1 phenotypic ratio. This suggests that the presence of two dominant genes resulted in a form of gene interaction in which an allele of one gene masks or modifies the phenotypic expression of an allele of another gene.

Pea   Rose

Walnut   Single

14-15 Comb types in chickens
The shape of the comb in chickens is due to the interaction of two genes.

14-16 A summary of modified dihybrid
Mendelian ratios resulting from gene
interaction

Type of gene interaction	A/B	A/bb	aa/B—	aa/bb
Four distinct phenotypes	9	3	3	1
Complementary gene action	9	7		
Dominant suppression by A of dominant gene B	13		3	
Recessive epistasis by aa of B/b genes	9	3	4	
Dominant epistasis by A of B/b genes	12		3	1
Duplicate genes	15			1

To denote this situation, Bateson coined the term **epistasis,** from roots meaning "standing upon." The masking gene is said to be epistatic to the masked gene, "standing upon" it or eclipsing it. When the basis of epistasis came to light, it was seen to represent no exception to Mendelian rules. Rather, it signified only that genes or allelic pairs never act in isolation.

Interacting nonallelic genes are dependent upon one another in the sense that neither can exert its phenotypic effect unless the other one does too. This sort of interaction is common when the expression of a phenotypic trait results from a series of enzymatic reactions, each of which is under the control of a different gene. Often an allele of the gene controlling the first reaction in the sequence masks the expression of one or more of the other genes. Bateson's purple pea plants appeared only when the dominant allele of gene *C* and the dominant allele of gene *P* were both present. If either was absent or inactive, flowers were white. Biochemists explained this phenomenon by showing that *C* governs the production of a chemical precursor of the purple pigment, anthocyanin. *P* governs the conversion of the precursor into anthocyanin. With either gene missing, anthocyanin cannot be produced, and the color remains white.

Modifier Genes

Still another form of gene interaction is seen with modifier genes. Mice have a gene that tends to produce white spotting, but its expression is modified by a whole series of other genes that are called **modifier genes** or **suppressor genes.** If many modifiers are present, the animal is almost completely white. The number of white spots in intermediate states depends on the number of modifiers acting.

Another example concerns human eye color. In the gene pair that controls the brown/blue phenotype, *B* is the dominant allele, *b* the recessive allele. Blue eyes occur only if *bb* is present. Either *BB* or *Bb* produces brown eyes. *B* determines the presence of a brown pigment (melanin) in the iris of the eye. However, many modifier genes influence the extent of pigment deposition and its precise location within the iris. This accounts for the common eye colors termed gray, hazel, green, and others.

We normally say that two blue-eyed parents will always have blue-eyed children because their alleles are recessive and homozygous (*bb*). However, blue-eyed parents can sometimes produce a brown-eyed child if one of the parents is *Bb* but is blue-eyed because of modifier genes.

A summary of the various types of gene interactions is shown in Figure 14-16.

14-17 Range of phenotypic variation due to polygenic inheritance

In this example the number of body segments in a population of millipedes (shown by the vertical bars) approximates a bell-shaped curve. The number of body segments is governed by polygenic inheritance patterns.

POLYGENIC INHERITANCE

The great majority of traits are influenced by many different genes, most of which are multiple alleles in the population (Chapter 13). Hence phenotypic characters frequently show a range of values, which when graphed generate a normal, or bell-shaped, curve (Figure 14-17). We term this situation **polygenic inheritance.**

Suppose a character such as plant height depends on two genes, A and B, each with two alleles, A^1 and A^2 and B^1 and B^2. Neither allele is dominant in either pair. Suppose further that A^1 and B^1 collaborate to produce tallness and A^2 and B^2 collaborate to produce shortness. The tallest plants would have the genotype $A^1A^1B^1B^1$; the shortest would have the genotype $A^2A^2B^2B^2$.

If plants of these two genotypes were crossed, half of the genes of the F_1 offspring would be talls (A^1 and B^1) and half would be shorts (A^2 and B^2). However, the plants themselves would be intermediate in height.

Suppose that now two intermediate F_1 plants were interbred. A Punnett square like that in Figure 14-18 indicates that five height classes should appear in the F_2 generation in the ratio summarized in Table 14-2. Experimental data would confirm the prediction. We conclude that when multiple genes affect the same character, each has a small but equal effect. Since segregation of the genes produces a range of quantitative phenotypic variations, they appear to act in an additive manner.

If the tallest and shortest plants were crossed and a great many genes were involved instead of two, the hypothesis of additive gene effects should permit the following predictions:

- F_1 would be phenotypically intermediate between the parental phenotypes, with no individuals as tall or as short as the tallest or shortest parent.
- F_2 would have many size classes that would tend to overlap and merge with one another.
- The extreme size classes of F_2 would be about equal to the tallest and shortest members of the original parental generation.
- Few individuals would be in the extreme size classes of F_2, and increasing numbers of individuals would be in classes nearer intermediate size.

A classic experiment by E. M. East (1879–1938) firmly validated these predictions. The variable trait studied was ear size in corn plants. The actual results are summarized in Figure 14-19. The parental plants had short and long ears. Note the distribution of lengths in the F_2. It would be easy to draw a smooth curve describing this distribution of ear sizes.

The concept of polygenic inheritance was first applied to human genetics in 1910–1913 by Gertrude and Charles B. Davenport, who studied skin color in natives of Bermuda and Jamaica, where marriages between white and black persons are common. It was assumed in those days that two major genes govern skin color, A and B, each with two alleles, A^1 and A^2 and B^1 and B^2. A pure white genotype would be $A^1A^1B^1B^1$. A pure black genotype would be $A^2A^2B^2B^2$. Intermediate skin color is the phenotype of the offspring of black-white crosses. The two gene–two allele model yields a Punnett square checkerboard that predicts five classes of skin color. Later workers showed that skin color is determined by a greater number of genes and that there is in fact a continuous gradation of skin color.

The continuous variations in natural populations almost always have distributions like those seen in the F_2 of experiments like East's. Most individuals are near the intermediate or average condition, and the number of individuals becomes smaller as deviation from the average becomes greater.

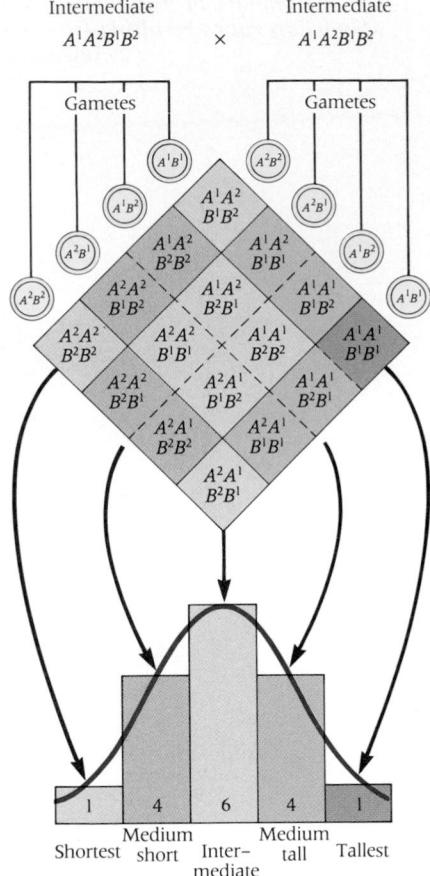

14-18 Plant height is determined by more than one gene
The predicted offspring of a cross between two plants of intermediate height are shown, assuming that only two genes are responsible.

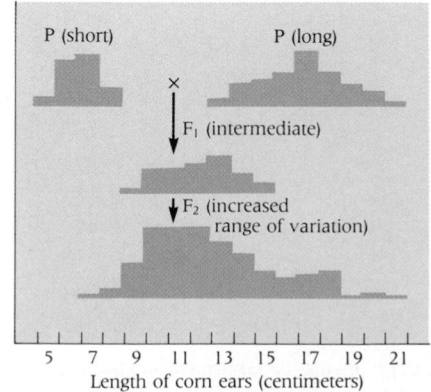

14-19 Polygenic inheritance of ear size in corn plants
The results show a broad variation in ear size in the F_2 generation, an indication that many genes are active in determining this trait.

TABLE 14-2
F_2 PROGENY FROM THE CROSS $A^1A^2B^1B^2 \times A^1A^2B^1B^2$

Genotypes	Expected Ratio	Phenotypes	Expected Ratio
$A^1A^1B^1B^1$	1	Tallest	1
$A^1A^1B^1B^2$	2 ⎫	Medium tall	4
$A^1A^2B^1B^1$	2 ⎭		
$A^1A^1B^2B^2$	1 ⎫		
$A^1A^2B^1B^2$	4 ⎬	Intermediate	6
$A^2A^2B^1B^1$	1 ⎭		
$A^1A^2B^2B^2$	2 ⎫	Medium short	4
$A^2A^2B^1B^2$	2 ⎭		
$A^2A^2B^2B^2$	1	Shortest	1

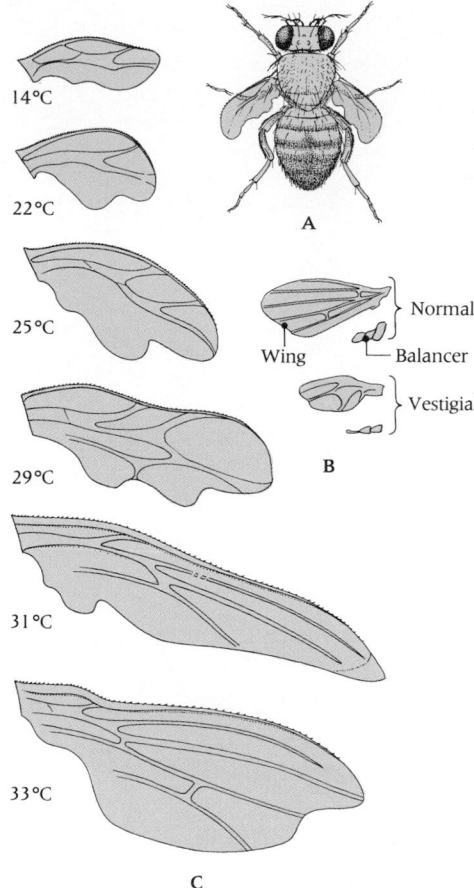

14-20 Action of the vestigial gene is influenced by the environment
(A) *Adult fly with vestigial wings. Compare with normal fly in Figure 14-5.* **(B)** *Relative sizes of wings and balancers in normal and vestigial flies. Balancers, also known as halteres, are rudimentary hind wings.* **(C)** *Temperature affects the degree of expression of the vestigial gene.*

Phenotypic expression
(each circle represents an individual)

Variable penetrance

Variable expressivity

Variable penetrance and expressivity

14-21 Penetrance and expressivity
Varying degrees of penetrance and expressivity are illustrated for a hypothetical trait such as pigment intensity. In each row all individuals have the same genotype with respect to this trait, although the degree of expression of the genotype is modified by other genes and the environment.

We should note, before leaving the topic of polygenic inheritance, that certain genes have been shown to have more than one phenotypic effect. For example, Manx cats have a recessive gene that determines tail-lessness and survival. Such genes are called **pleiotropic.** We shall return to pleiotropy later.

FACTORS INFLUENCING GENE FUNCTIONS

We have now seen that, by an assortment of methods, expression of a gene may be influenced by other genes. We might conclude that a gene's action is affected by its genic environment. It is also affected by its physical environment. To be sure, genes are within cells, but the intracellular environment is significantly influenced by the external environment.

ENVIRONMENTAL EFFECTS

Consider a famous example. *Drosophila* has a recessive gene with the odd name "vestigial" (*vg*). It is so named because, when expressed, the gene reduces the wings to mere vestiges of their normal size (Figure 14-20). Gene *vg* also affects other characters and its actions are affected by other genes. Interestingly, the action of this gene is strongly affected by the temperature at which the fly develops. At low temperatures (14–29°C), the wings of a fly homozygous for *vg* are rudimentary. At high temperatures (30–31°C), the wings are nearly as long as the wings of a fly with a normal wing-size gene.

The term **phenocopy** denotes an individual (like the *vgvg* fly) whose phenotype has been altered by the environment in a way that causes it to imitate a phenotype usually associated with another genotype. Although the induced modification resembles the phenotype of a known mutant, it is not heritable. When the true nature of a given character is not known, this situation is an obvious source of possible experimental error.

In inheriting a given gene, we acquire a **reaction range,** or a potentiality. Other factors may then determine whether the potentialities are to be realized. All phenotypes that actually occur, including those we consider abnormal, are necessarily within the reaction range of the underlying genotype.

The meaningful questions, then, about a phenotypic trait are not "Is it hereditary or environmental? Is it due to nature or to nurture?" Rather, we must ask, "What is the reaction range of the genotype, and what environmental factors may have led to expression of a particular position in the range?"

PENETRANCE AND EXPRESSIVITY

We suggested earlier that a dominant allele will always manifest itself phenotypically—or, in genetic terminology, a dominant allele will always express itself. But then we saw that genes can interact in various ways, so that expression of one gene can be altered by actions of another gene. This means that sometimes a dominant gene may not express itself phenotypically—or it may express itself in varying degrees of intensity. We speak, therefore, of the degree of a gene's penetrance and expressivity.

Penetrance is defined as the percentage of individuals carrying a gene (in a combination appropriate for its expression) that actually expresses the related phenotypic trait. If an allele always produces its expected phenotypic trait, its penetrance is said to be 100%. However, penetrance is often far below 100%.

Expressivity refers to the degree or extent to which a given genotype is expressed phenotypically—when it is penetrant. Note the distinction between expressivity and penetrance. The former denotes the ability of an allele to express itself (Figure 14-21).

Penetrance and expressivity are illustrated by a human gene that causes the whites of the eyes (the sclerae) to appear bluish. The gene usually behaves as a simple dominant. This means that anyone who has the gene, whether in heterozygous or homozygous form, also has the blue-sclera phenotype. But in fact only 90% of those who have the gene (as determined by family tree studies) show the trait. Moreover, expressivity is variable among those who display the trait—that is, the intensity of bluish coloration ranges from pale to dark blue. Low penetrance of the blue sclera gene in the other 10% means that these individuals will have white sclerae despite the presence of the blue-sclera gene.

MUTATIONS AND GENETIC VARIABILITY

It was implicit in Mendel's conclusions that genes be viewed as permanent and unvarying determinants guaranteeing that an offspring will resemble its parents. But if genes are stable, how do we account for the fact that countless variants, or **strains,** arise from a single species? Indeed, innumerable species have arisen in the course of time. As we shall see later, the theory of evolution rests on the proposition that species develop heritable variations that improve their chances for survival. The variant organism survives, but the variation has altered its character. How can these two views be reconciled? The answer is gene mutation.

In 1901, the Dutch biologist Hugo De Vries was studying inheritance in a variety of evening primrose (*Oenothera glazioviania*). He noticed that although Mendelian rules were generally followed, an occasional individual differed from both parents and indeed from anything ever seen among its ancestors. He postulated that the new trait was the phenotypic expression of an altered gene. Presumably, he reasoned, the altered gene would be transmitted to offspring like any other gene. He termed the hereditary gene alteration a **mutation** and the organism carrying it a **mutant** individual (properly, *mutant* is an adjective). It is an historic irony that only 2 of the 2000 changes reported by De Vries in the evening primrose were actually true mutations, according to later investigators. The other 1998 were the result of new gene combinations, or aneuploidy, rather than bona fide mutations. Despite this mistake, his concept of mutation was essentially accurate.

When a gene mutates, it abruptly changes in an all-or-none fashion. The change is permanent—until and unless the gene mutates again. As we know of genes by the results of their actions—the **test of progeny**—we would recognize a mutant gene by the appearance of altered traits in the progeny.

- When mutation occurs in germ cells, the test of progeny requires reproduction and the analysis of offspring for a generation or two. We know that new traits are heritable because they are transmitted to subsequent generations.
- When mutation occurs in somatic cells, the progeny are the daughters of the mutant cell, not the offspring of sexual reproduction. Somatic cell mutations are not transmitted to later generations.

Normally genes are faithfully copied when chromosomes replicate. If mutations occurred frequently, genes might become altered during every cell cycle and offspring cells might not resemble their parents. In 1927 H. J. Muller found that the frequency of mutations in *Drosophila* could be sharply increased by exposure of the gametes to x-irradiation; the higher the dose, the greater the number of induced mutations. Ultraviolet light and certain chemicals were also mutagenic. This work gave geneticists a potent new tool. It also led to the realization that the rare spontaneous mutation is the key to evolution. Mutation is biologically advantageous because it offers a constant supply of genetic variants, some of which permit plants and animals to adapt to a constantly changing environment. Mutation is, we should note, the basis of the variant genes we call *alleles*. Genes mutate from one allelic form to another.

Though rare, mutations occur at a regular and predictable rate in both germ cells and somatic cells. (A typical rate of spontaneous mutation in eukaryotes is 1 mutation at a given gene locus per 200,000 cell divisions.) For several reasons the frequency of mutation may seem lower than it actually is.

- There is a small but definite rate of **back mutation**—also called **reversion** and **reverse mutation.** (Mutation away from the common form is called a **forward mutation.**)
- Many mutations may produce alleles that are recessive and thus may not express themselves phenotypically until a suitable mating occurs.
- When they do express themselves, the new phenotypic traits may be so undesirable that the cell may be prevented from reproducing. Such mutations are said to be genetically lethal.
- Many nonlethal mutations cause phenotypic changes so subtle that they are

lost in the welter of nongenetic variations regularly caused by environmental influences. Such nongenetic variations are not transmitted to the offspring.

A mutation may not reveal itself phenotypically until the environment changes. For example, it was found during World War II that a new antimalarial drug, primaquine, produced severe and occasionally fatal blood disorders in a few individuals. Because the adverse reactions occurred mainly in blacks, suspicion arose that the disorder had a genetic basis. Investigation revealed that 10% of American blacks and a small number of Caucasians react unfavorably to primaquine and similar drugs. These sensitive individuals had inherited a specific enzyme deficiency that rendered them vulnerable to the drug's harmful effects. Their red blood cells contained abnormally low levels of the enzyme glucose 6-phosphate dehydrogenase. Presumably, individuals with this enzyme deficiency were getting along satisfactorily for generations until the environment was altered by the introduction of a new chemical agent. The mutation leading to the deficiency may have occurred ages ago. But it did not impair survival in the environment of that time. Only in the new environment did it express itself phenotypically—and injuriously. Clearly, it is not possible to declare unequivocally that a given mutation is advantageous, disadvantageous, or neutral unless we can predict the future environment.

Two modes of mutation occur.

- **Gene mutations,** in which a gene mutates from one allelic form to another.
- **Chromosome mutations,** in which changes occur in chromosome numbers or arrangements.

A growing list of human hereditary diseases is being attributed to mutations of both types, and the capacity of radiation to produce mutations and thus to threaten future generations has been a source of deepening concern to scientists studying the effects of nuclear fallout and of other environmental sources of radiation.

Though most of our knowledge of genes and chromosomes came from experiments with large multicellular organisms—both plants and animals—it became apparent in time that bacteria, fungi, and viruses also possess chromosomes. For this reason, each of these microorganisms is an admirable experimental subject for genetic research. In this role, they have notable advantages. They are usually haploid. Therefore, genetic expression is not influenced by dominance or recessivity. Microorganisms reproduce rapidly. *Escherichia coli,* a "workhorse" of modern genetics, can produce a new generation as speedily as every 20 minutes. In diploid cells, several generations may be necessary to detect the presence of a mutant gene. In a haploid bacterium, expression of a mutant gene occurs almost instantly.

Although bacteria do lack plainly visible features such as red eyes or pink flowers, their nutritional requirements and other chemical properties are genetically determined traits accessible to precise study. Moreover, they have useful properties that

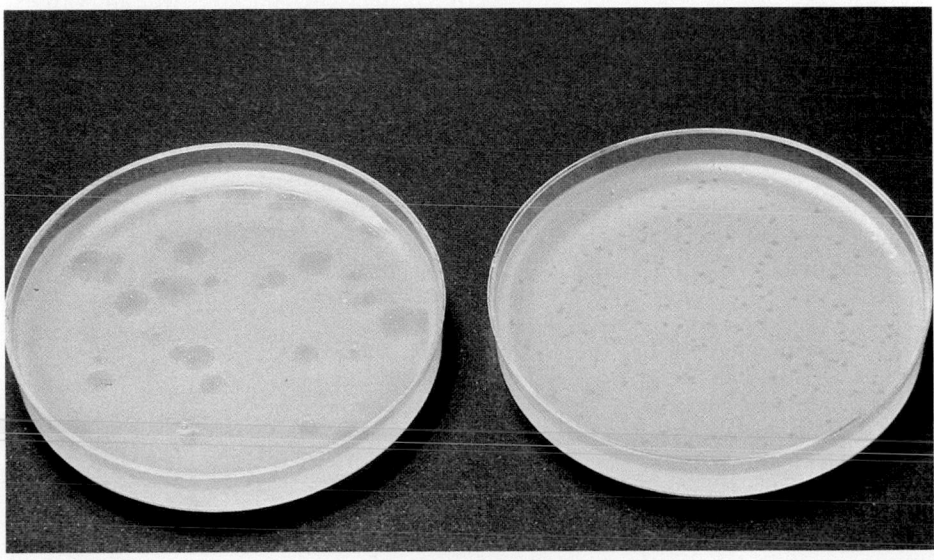

14-22 Virus infection of bacterial cells
When a virus infects a bacterial cell, it can destroy the cell, creating a hole, or plaque, in the bacterial "lawn." The left dish shows a number of plaques due to bacteriophage lambda. In the right dish, the bacteria are infected with T2 bacteriophage.

permit sensitive methods for the selection of small numbers of mutant cells from large populations.

As for viruses, which strictly speaking are not microorganisms, the most widely used types have been the bacteriophages (or phages, in laboratory jargon)—viruses that can live only by infecting certain bacterial cells. Of these, the best studied are the phages that infect *E. coli*. Their genes also control such subtle traits as host cell specificity, size, and shape of a plaque of lysed host cells (Figure 14-22). The figure shows that these traits are abruptly and visibly altered when viruses undergo mutation.

The genetics of microorganisms and viruses has been of immense scientific importance. We will shortly see why.

SUMMARY

The theory that genes are carried on chromosomes provided a physical basis for Mendel's principles of segregation and independent assortment. It was the striking parallels between the behavior of chromosomes during meiosis and the behavior of genes that led Sutton and Boveri to suggest that genes are carried by chromosomes. Thus, in meiosis alleles of a gene pair are always separated and distributed into different gametes because homologous chromosomes are always separated during meiosis, while alleles of genes on different, nonhomologous chromosomes are distributed randomly and independently to gametes. Because the process of distribution is random, statistical analysis permits predictions of the approximate results of a cross. The fruit fly *Drosophila melanogaster* was an important experimental subject in early studies of chromosomal genetics.

Linked genes (genes carried on the same chromsome) are important exceptions to the principle of independent assortment. Organisms have many thousands of genes and only a relatively small number of chromosomes, and the number of linkage groups corresponds to the number of chromosomes in that species. In a coupling or *cis* allele conformation, both dominant alleles of a linked gene pair are found on the same member of a chromosome pair; in a repulsion or *trans* conformation, the dominant allele of one gene and the recessive allele of the other are on the same member of the chromosome pair.

Crossing-over, or recombination, occasionally causes linked alleles to switch from a *cis* to a *trans* conformation (or vice versa). The frequency of this event is proportional to the distance between the two genes on a chromosome. This makes it possible to map the sequence of and relative distances between the genes on a chromosome.

Polygenic inheritance, in which two or more genes affect one phenotypic trait, is one reason why a population may show continuous variation in an inherited trait. Such genes may interact in several ways. Complementary genes interact to produce a phenotype that neither could produce alone. In epistasis, an allele of one gene masks the effects of an allele of another gene. This effect is similar to dominance, but occurs between alleles of different genes. Modifier genes may act singly or in concert to change the effect of other genes. As the number of genes involved in the expression of a single trait increases, it becomes more difficult to detect a discontinuous distribution of variation in that trait.

The expression of a gene is influenced not only by other genes, but by certain factors in the environment. This further complicates the relationship between a single gene and its phenotype. Penetrance measures the percentage of individuals carrying a gene that actually expresses the appropriate trait. Expressivity measures the degree to which a penetrant trait is expressed.

Much genetic variation is due simply to different combinations of preexisting alleles. Mutations (permanent changes in alleles) further extend the range of genetic variability. Mutations in somatic cells are not passed on to future generations, but mutations in germ cells are heritable. Spontaneous mutation rates are low, and to an extent can be reversed by back mutations. However, mutation rates can be increased by exposing organisms to high-energy radiation or certain chemicals (mutagens).

There are two general modes of mutation. In gene mutations, one allele is transformed into a new allele. In chromosome mutations, changes occur in chromosomal number or arrangement. Bacteria and viruses have been useful organisms in studies of the mechanisms of gene mutation.

KEY TERMS

back mutation	linkage group	polygenic inheritance
centimorgan	linkage map	population
chance	locus	probability
chromosome mutation	map unit	reaction rays
chromosome theory	Mendelian ratios	recombinant gamete
crossing-over	modifier gene	sample
epistasis	mutagen	standard deviation
expressivity	mutation	statistical test
forward mutation	penetrance	suppressor gene
gene linkage	phenocopy	test of progeny
gene mutation	pleiotropy	wild type

QUIZ QUESTIONS

1. If the phenotypic ratio of an F_1 generation were 9:7, then you would suspect that the trait was inherited by
   A. gene linkage.
   B. recombination.
   C. polygenic inheritance.
   D. epistasis.
   E. complementary gene action.

2. The presence of continuous phenotypic variation in an F_1 generation suggests that a character was inherited by
   A. gene linkage.
   B. recombination.
   C. polygenic inheritance.
   D. epistasis.
   E. complementary gene action.

3. Which of the following is *not* a source of variation in the genome of a species?
   A. recombination in somatic cells
   B. mutation of chromosomes in germ cells
   C. crossing-over
   D. gene linkage
   E. A and D

4. _____ explains how progeny can possess combinations of traits that neither parent possessed.
   A. The chromosome theory
   B. The law of independent segregation
   C. The law of independent assortment
   D. Polygenic inheritance
   E. A and D

5. The chromosome theory was proposed by
   A. W. S. Sutton.
   B. Thomas Hunt Morgan.
   C. Theodor Boveri.
   D. A and B
   E. A and C

ESSAY QUESTIONS

1. How does the chromosome theory provide a physical explanation of Mendel's laws of inheritance? At what stage of meiosis does one observe the mechanism of the first law (principle of segregation)? Which stage accounts for the second law (principle of independent assortment)?

2. What are the practical advantages of the fruit fly *Drosophila melanogaster* in genetic experiments? What can be learned from the fruit fly that cannot be learned from Mendel's pea plants?

3. Define gene linkage. What is its physical basis? Does gene linkage weaken Mendel's principle of independent assortment? Explain.

4. An allele for green skin (*G*) is dominant over an allele for grey skin (*g*), but the *G* allele demonstrates only 50% penetrance. What phenotypic ratio would appear in the F_1 generation from a mating between a homozygous green-skinned parent and a homozygous grey-skinned parent? (Assume 100% penetrance of the *g* allele.)

REFERENCES AND SUGGESTIONS FOR FURTHER READING

Allen, G. E. (1978). *Thomas Hunt Morgan: The Man and His Science*. Princeton University Press, Princeton, N.J.

An evocative portrait of a seminal figure in the history of genetics—and of the intellectual climate in which he thrived.

Ashburner, A., and Novitski, E. (Eds.) (1976). *The Genetics and Biology of Drosophila*. Academic Press, Orlando, Fla.

A definitive treatise.

Magasanik, B. (1988). Research on bacteria in the mainstream of biology. *Science 240*, 1435–1439.

A stimulating review by a distinguished authority on the importance of bacteria in modern research on genetics, biochemistry, and physiology.

O'Brien, S. J. (Ed.) (1984). *Genetic Maps*. Cold Spring Harbor Press, Cold Spring Harbor, N.Y.

A compendium summarizing detailed chromosome maps from 80 carefully studied species.

Rubin, G. M. (1988). *Drosophila melanogaster* as an experimental organism. *Science 240*, 1453–1459.

A review pointing out that *Drosphila* is now widely used not only in classical and molecular genetics but in many other fields, including developmental biology and neurobiology.

Sutton, W. S. (1903). The chromosomes in heredity. *Biol. Bull. 4*, 231–251.

A classic paper that has been reprinted in Peters, J. A. (1959). *Classic Papers in Genetics*, Pren-

tice-Hall, Englewood Cliffs, N.J. Worth reading.

Whitehouse, H. L. K. (1982). *Genetic Recombination: Understanding the Mechanisms*. John Wiley & Sons, New York.

An impressive survey by a respected authority. The work reviews recombination from the level of the nucleotide to the level of the chromosome.

Wilson, J. H. (Ed.) (1985). *Genetic Recombination*. Benjamin, New York.

A useful collection of 10 articles that appeared in *Annual Review of Genetics* and *Annual Review of Biochemistry* in the last decade. The main topics covered are meiotic, mitotic, and bacterial recombination. The level is advanced.

SEX AND HUMAN HEREDITY

Sex and human heredity
*Although we can describe the chromosomal
patterns associated with each sex, we are just
beginning to understand how the genes on these
chromosomes actually determine sex. Similarly,
we are increasingly familiar with the normal
and abnormal chromosomal patterns of human
beings—and thus we consider these related topics
in one chapter. These family photographs clearly
suggest that the transmission of parental traits
to offspring is as unerring in humans as it is
in* Drosophila *or pea plants.*

The evolution of sexual reproduction was a major advance because it greatly expanded opportunities for genetic variation. Prokaryotic organisms reproducing by fission can keep on dividing, duplicating themselves again and again. But for reasons discussed earlier, offspring of sexually reproducing organisms are never exact genetic duplicates of their parents. The widespread prevalence of sexual reproduction among living organisms suggests that this is the mode of reproduction that in the long run proved most advantageous.

In view of what we have learned about how genes determine phenotypic characters, we must now inquire whether genes determine the sex of an offspring, for surely that is a phenotypic character. We shall see that genes determine sex in a number of different ways.

SEX DETERMINATION

A major contribution of *Drosophila* research concerned the mechanism of sex determination. According to the chromosome theory of Sutton and Boveri, gene behavior (which can be judged only from breeding experiments) and chromosome behavior (which can be observed directly) coincide so strikingly it is apparent that genes are carried by chromosomes. The next step in validating this theory had to be the connection of specific genes with specific chromosomes and the localization of these genes within their chromosomes.

SEX CHROMOSOMES

In *Drosophila,* there is one chromosome pair that differs from all the others. Significantly, this pair looks different in males and females. Consider the metaphase chromosomes from a male and female *Drosophila* (see Figure 14-7). Four pairs of chromosomes are present: two large V-shaped pairs with centromeres in the middle (pairs II and III), a small dotlike pair (pair IV), and a pair that is visibly different in the two sexes (pair I). These are the **sex chromosomes,** or **gonosomes** (from *gonos,* meaning seed). The three pairs that are similar in both sexes are collectively designated **autosomes.**

In a female *Drosophila,* both sex chromosomes are rod-shaped, with centromeres near the ends. In the male, one sex chromosome is rod-shaped and one is J-shaped, with a centromere at the bend. The two female sex chromosomes are termed **X chromosomes.**[1] The male has one X chromosome and one **Y chromosome** (the

[1] This curious structure was first observed in 1891 by H. Henking who was studying meiosis in males of a species of Hemiptera (the true bugs). He noticed that one unpaired element moved to the poles in the first meiotic division along with the 11 paired chromosomes. He thought it was a nucleolus and called it an X body. Later work showed it to be a chromosome. Studies of another Hemipteran bug in 1905 by Edmond Wilson again revealed the unpaired chromosome. Wilson chose to call it the X chromosome. It is worth mentioning that the unpaired chromosome in the unusual male grasshoppers studied by Carothers (see Figure 14-3) was an X chromosome.

J-shaped member). A female can produce only one kind of gamete as regards its sex chromosome content: all carry an X chromosome. But a male can produce two kinds of gametes: one with an X chromosome and one with a Y chromosome. The female, thus, is **homogametic** and the male **heterogametic.** This is the pattern in many, but not all, species.

Again, the events at fertilization are determined by chance. As shown in Figure 15-1, half of a large number of fertilizations yield zygotes with two X chromosomes; half yield zygotes containing an X chromosome and a Y chromosome. The former become females, the latter males.

Sex determination, regarded as a supernatural phenomenon in past centuries, is thereby illuminated by the chromosome theory. Even the early geneticists, Mendel among them, realized that sex behaves as a Mendelian character, male and female offspring usually occurring in a 1:1 ratio. This is the classic monohybrid test-cross ratio predicted when any heterozygote (such as XY) is crossed with a homozygote (such as XX). Thus, sex (in most animals) is determined at the moment of fertilization.

DIFFERENCES AMONG SPECIES

Historically, those attempting to explain the determination of sex in Mendelian terms stumbled down a long and tortuous path. One source of difficulty was the early discovery of species employing mechanisms of sex determination differing from that of *Drosophila* and humans (Figure 15-2A).

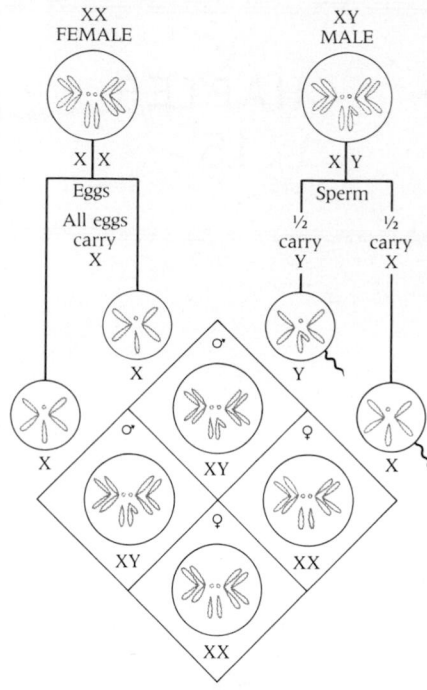

15-1 Sex chromosomes in Drosophila
In the fruit fly, females carry two X chromosomes and are homogametic, while males carry an X and a Y chromosome and are heterogametic. In this XX-XY system, a mating of male and female will produce females and males in equal numbers in the next generation.

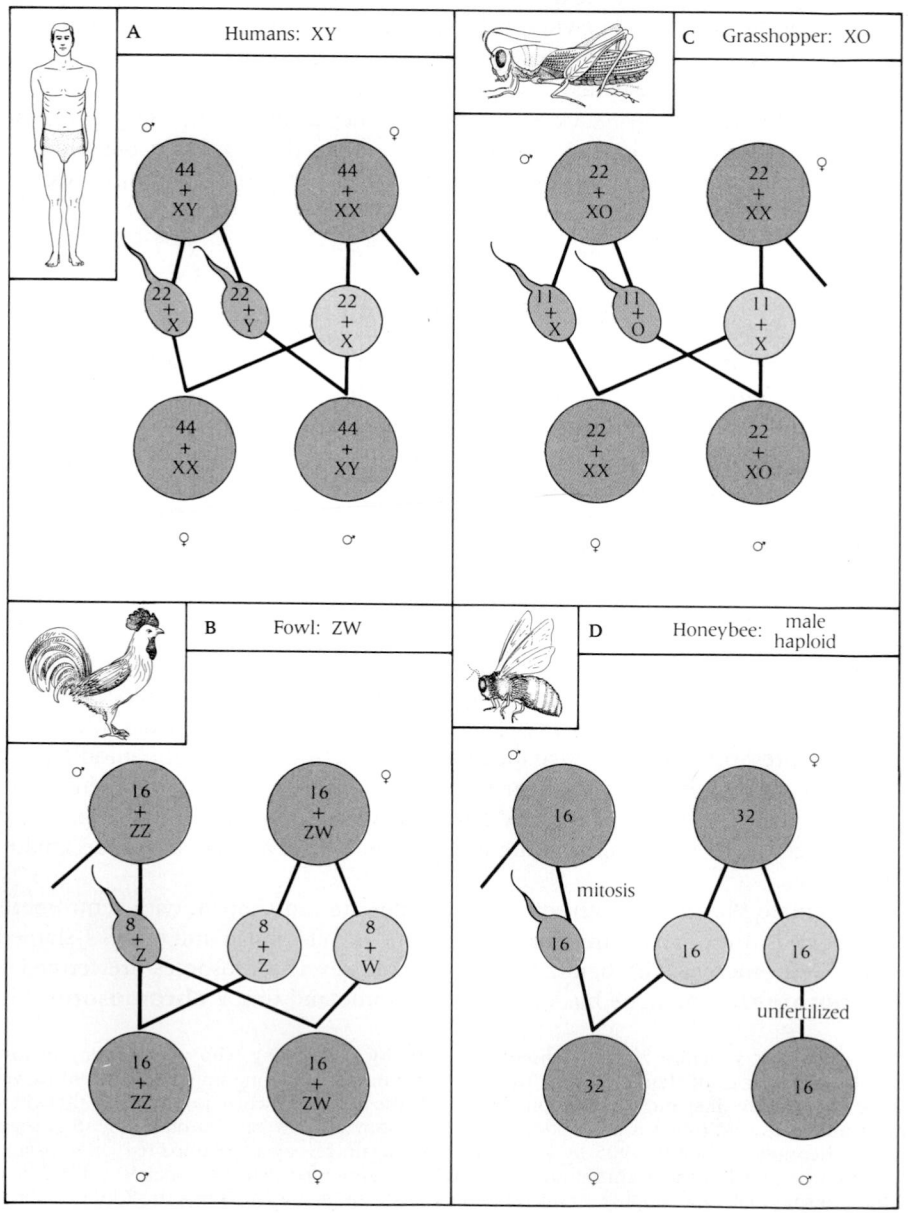

15-2 Sex determining mechanisms
(A) In the XX-XY system characteristic of humans and fruit flies, females produce gametes carrying only an X, while males produce both X and Y gametes. (B) The ZZ-ZW system found in birds produces females that are heterogametic (ZW) while males are homogametic (ZZ). (C) In the XX-XO system of grasshoppers, the males are XO and therefore produce two kinds of gametes, while the females produce only X gametes. (D) In honeybees, males are haploid (n), while females have the diploid chromosome number (2n).

- The **ZZ-ZW system** (Figure 15-2B) of birds, butterflies, and other species: the XX carrier is male and XY is female, the reverse of the pattern in *Drosophila*. To prevent confusion of the XX-XY system of *Drosophila* and this XX-XY system, the latter was renamed ZZ-ZW. In this system, the male is homogametic (ZZ) and the female heterogametic (ZW). Interestingly, this is the pattern in domestic strains of certain tropical fishes (e.g., guppies). However, wild strains of the same species have the opposite pattern: ZZ females and ZW males.
- The **XX-XO system** (Figure 15-2C) of grasshoppers, a species studied by Sutton: the Y chromosome is lacking. Females are XX and have 24 chromosomes. Males have 23 chromosomes and are XO (read "X-oh" not "X-zero"). Half of their sperm cells have one X chromosome; half have none.
- There are no sex chromosomes (Figure 15-2D) among the Hymenoptera (honeybees, ants, and so on). Males hatch from unfertilized cells and are haploid; females hatch from fertilized cells and are diploid. Hence, males have no fathers but have grandfathers on their mother's side!
- In some species with a single genotype, an offspring's sex may be determined by environmental factors, cytoplasmic elements, or hormones. For example, the marine annelid *Ophryotrocha* differentiates into a sperm-producing male as a young animal and then becomes an egg-laying female as it ages. In turtles, sex is influenced by environmental temperature.

Despite these variations, the XX-XY system of *Drosophila* is widespread and its study has been informative. It is also of special interest because it superficially resembles the human pattern. Humans have 22 pairs of autosomes that are the same in both sexes. The twenty-third pair consists of identical chromosomes, XX, in women, and of nonidentical chromosomes, X and Y, in men.

HOW DO GENES DETERMINE SEX?

Events in *Drosophila* raised the question: do X chromosomes carry the genes for femaleness and Y chromosomes the genes for maleness? Or, is an individual a female because Y is lacking or because two X chromosomes are present? We now know that sex is determined, not only by certain genes on sex chromosomes, but by interactions between genes on sex chromosomes and genes on various autosomes.

Classic studies in *Drosophila* showed that in occasional meiotic divisions one chromosome pair fails to separate properly. The result of such **nondisjunction** is a daughter cell with one chromosome too many and another daughter cell with one chromosome too few. When nondisjunction affects the sex chromosomes, gametes may contain one sex chromosome too many or too few—for example, an XX gamete in place of an X; XY rather than X; and O (meaning no sex chromosome) rather than X. If an XX gamete engages in fertilization, an XXX or XXY zygote would result.

We can examine such an occurrence in *Drosophila* using a Punnett square. Suppose that nondisjunction in a female results in an XX gamete and an O gamete (Figure 15-3). We draw the Punnett square as follows:

		Gametes from male parent	
		X	Y
Gametes from	XX	XXX	XXY
female parent	O	XO	YO

The progeny can include some offspring that appear normal and some that appear grossly abnormal.

- XXX progeny: abnormal, underdeveloped creatures that soon die
- XXY progeny: viable flies that resemble normal females
- XO progeny: capable of survival but as abnormal, sterile males
- YO progeny: incapable of development and soon die

These results suggest that in *Drosophila* sex is determined not by the presence or absence of Y but by the number of X chromosomes. Two X chromosomes generally produce a female. One X produces a male. At least one X is necessary for development.

NORMAL RESULTS AND THEIR INTERPRETATION
Normal meiosis, explaining origin of normal offspring

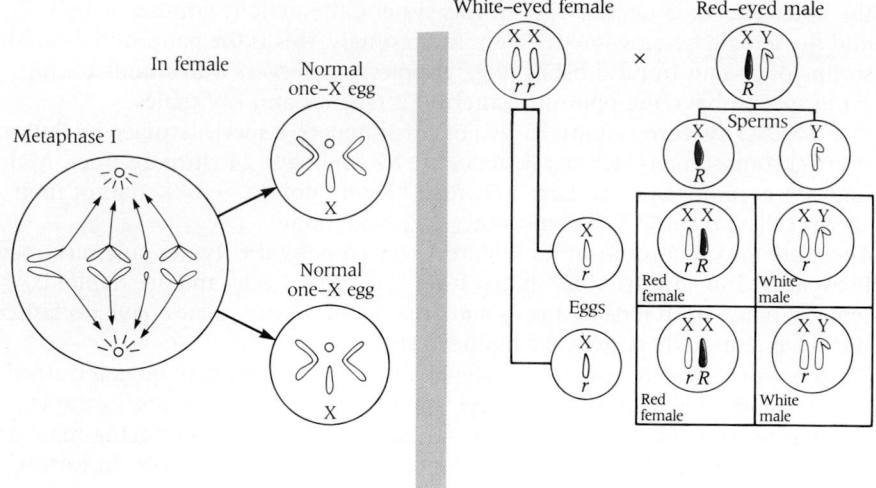

15-3 Nondisjunction of the X chromosome
The X chromosome carries the gene for eye color in Drosophila. A female that undergoes normal meiosis produces gametes containing a single X chromosome. If nondisjunction of the X chromosomes occurs during meiosis, the female will produce gametes that are XX or that lack an X chromosome (O). The results of matings between white-eyed females and red-eyed males can be used to determine whether nondisjunction has occurred.

EXCEPTIONAL RESULTS AND THEIR INTERPRETATION
Abnormal meiosis, explaining origin of abnormal offspring

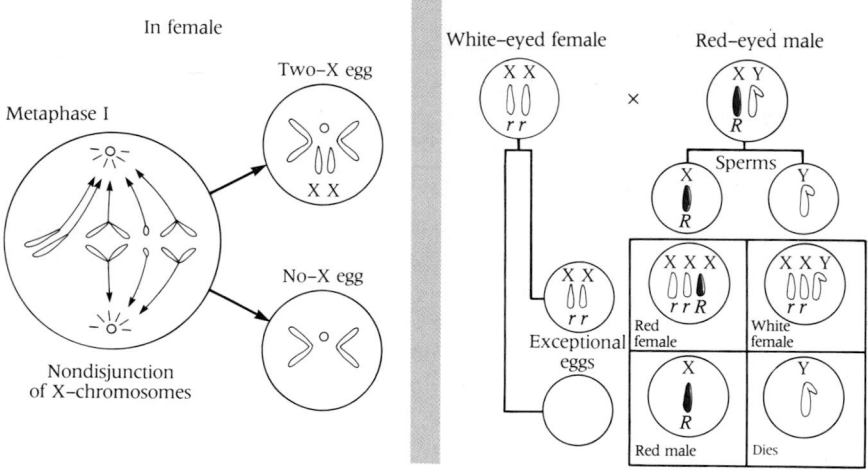

These and other variations were further characterized by determining the ratios of X chromosomes to autosomes (sometimes called A chromosomes). C. B. Bridges (in his doctoral thesis) observed that as the ratios of X chromosomes to A chromosomes decrease from 1.5 to 0.3, the degree of "femaleness" decreases (Table 15-1).[2] The normal male has an X:A ratio of 0.5 (one X chromosome for every two sets of autosomes); the normal female has an X:A ratio of 1.0 (two X chromosomes for every two sets of autosomes). Flies carrying two X chromosomes and three sets of autosomes (X:A = 0.67) are **intersexes** with phenotypic characteristics intermediate between those of the two sexes. Bridges called the XXX fly a **superfemale** because its X:A ratio is 3:2, or 1.5, a value higher than that of a normal female. An XXY fly would be a normal female because its X:A ratio is 1.0, the value typical of normal females.

Bridges's view, termed the *theory of genic balance,* was that the Y chromosome has no specific role in the actual determination of sex. It was strengthened by the discovery of the grasshopper system which completely lacks a Y chromosome. Later work, however, showed that the *Drosophila* pattern is not applicable to humans. We now believe that the Y chromosome in humans bears genes with male-determining properties—that maleness is determined by the presence of the Y chromosome

---

[2] Note the reference in Table 15-1 to *sets* of autosomes. The number of chromosome sets present in an organism is the same as its ploidy. An autosome set is the number of autosomes present in a haploid cell of an organism of a given species. In *Drosophila,* the normal diploid number is eight (four pairs); a female has three pairs of autosomes (two sets) and one pair of X chromosomes.

TABLE 15-1
PHENOTYPIC EFFECTS OF X:A RATIO IN FLIES

Number of X Chromosomes	Sets of Autosomes (A) (Three in a Set)	Ratio X:A	Sex Phenotype
3	2	1.5	Superfemale*
2	2	1.0	Normal female
2	3	0.67	Intersex
1	2	0.5	Normal male
1	3	0.33	Supermale*

* Superfemales and supermales are both weaker flies than their normal counterparts. There is nothing "super" about them except that their chromosomal ratio exceeds that characteristic of their sex. Note that a female develops when an X chromosome is present for each set of autosomes. A male develops when an X occurs for every two sets of autosomes. Departures from this norm yield anomalies.

and femaleness by its absence. Thus it is that an XXY *Drosophila* is a female and an XXY human a male.

New data reported in 1988 clearly indicates that a single gene on the human Y chromosome—termed the testis determining factor gene, or TDF—is directly responsible for male sexual development in the embryo. We will return to this later.

GENES ON THE X CHROMOSOME: SEX-LINKED CHARACTERS

THE DISCOVERY OF SEX LINKAGE

Most of the traits studied by Thomas Hunt Morgan in fruit flies followed simple Mendelian rules and were unrelated to the sex of the fly. But in 1909, while experimenting with the genetics of eye color, he encountered something peculiar.

The normal or wild-type *Drosophila* has bright red eyes that are controlled by the dominant allele *R*. One day Morgan observed a white-eyed male fly in the colony. Excited by this rare creature,[3] he crossed it with a normal red-eyed female and found that all members of F_1 had red eyes. That presumably meant that "white-eye" is controlled by a recessive allele. A cross between two F_1 siblings did yield an F_2 generation in which white eye reappeared in a ratio of three red eyes to one white eye. That seemed to confirm the recessiveness of white eye. But then Morgan noticed something exceedingly unusual that did not accord with Mendelian rules: all of the white-eyed flies were male! White-eyed females were not to be found (not until much later). The red-eyed flies occurred in a ratio of two females to one male. Moreover, the phenotypic pattern of F_2 depended on whether the white-eyed parent was male or female. To account for these confusing results, Morgan formulated an ingenious hypothesis that drew upon the work of Sutton, E. B. Wilson, and N. M. Stevens.

These results would be expected, he reasoned, if the allelic gene for eye color—dominant red (*R*) and recessive white (*r*)—were carried only on the X chromosome, while the Y chromosome lacked all eye-color genes (Figure 15-4). That he was right was shown by crosses between a white female (bearing an *r* on each X chromosome, according to the hypothesis) and a red male (bearing an *R* on its X chromosomes and no eye-color allele on its Y chromosome). In F_1, all male offspring were white-eyed because a male's X chromosomes always derive from its female parent. In this case, the female parent's X chromosomes carried *rr*. The male offspring received its Y chromosome from its male parent. A female offspring, on the other hand, received one X chromosome from its male parent and one from its female parent. All daughters must have been red-eyed because they all carried a paternal X chromosome. The male parent is red-eyed and red eye is dominant.

[3] Morgan discovered the original white-eyed male in early 1910—about the time his third child was born. In his biography of Morgan, Garland Allen tells of Morgan's first visit to his wife in the hospital. When she greeted him with the question, "How is the white-eyed fly?" he launched into his latest results with great excitement. It was very much later when he suddenly stopped and asked, "By the way, how is the baby?"

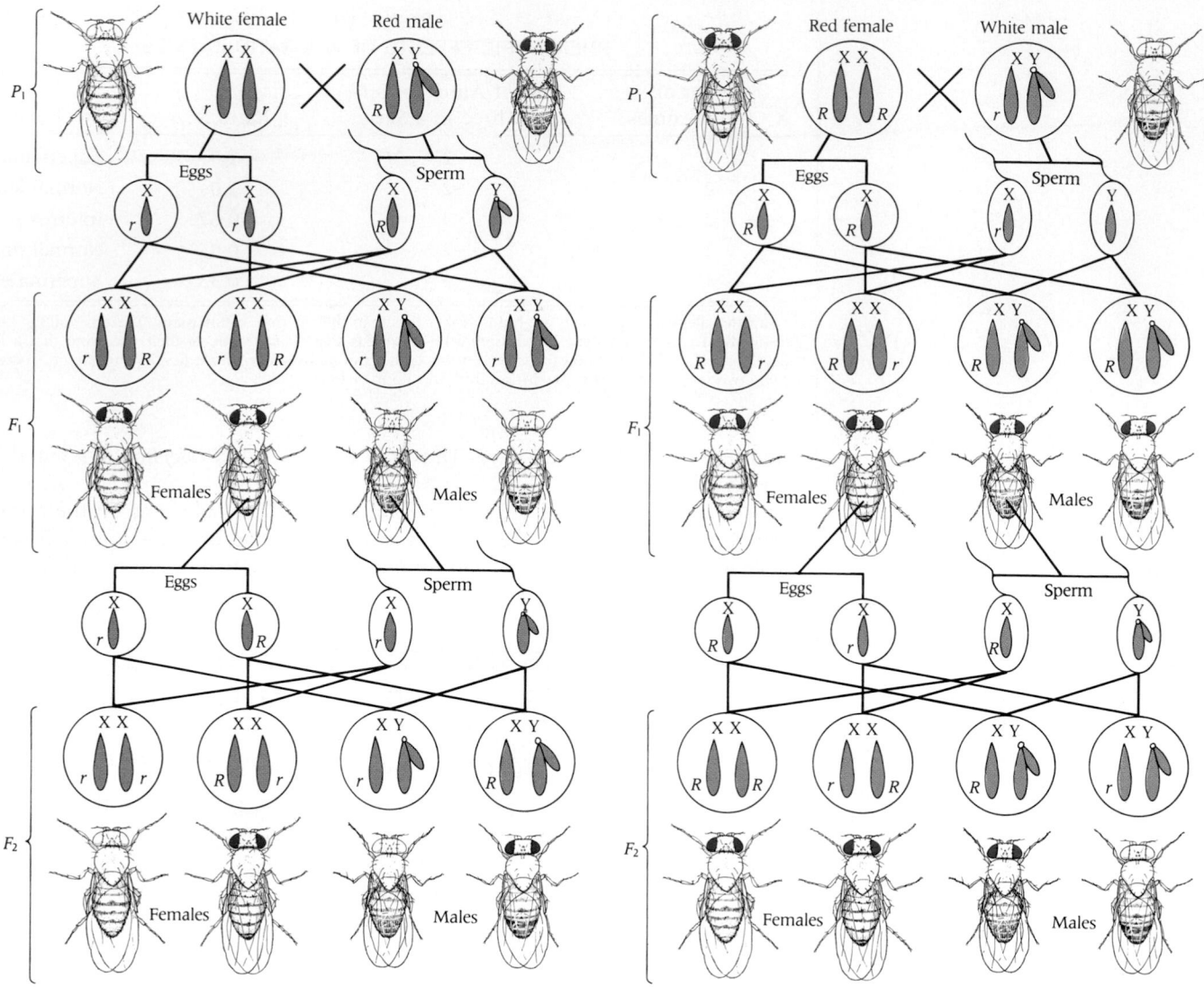

In summary:

P $X^r X^r$ $\times$ $X^R Y$
              white-eyed female   red-eyed male

F_1 $X^r X^R$ $\times$ $X^r Y$
              red-eyed female   white-eyed male

F_2 $X^r X^r$ $X^R X^r$ $X^r Y$ $X^R Y$
   white-eyed female   red-eyed female   white-eyed male   red-eyed male

Note that the dominant : recessive phenotypic ratio in the F_2 generation is 1 : 1, not 3 : 1. However, when the reciprocal cross is performed between a red-eyed female parent and a white-eyed male parent, the ratio is 3 : 1.

P $X^R X^R$ $\times$ $X^r Y$
              red-eyed female   white-eyed male

F_1 $X^R X^r$ $\times$ $X^R Y$
              red-eyed female   red-eyed male

F_2 $X^R X^R$ $X^r X^R$ $X^R Y$ $X^r Y$
   red-eyed female   red-eyed female   red-eyed male   white-eyed male

The key to the hypothesis is the assumption that eye-color genes are all on the X chromosome and absent on the Y chromosome. In such cases, it makes a considerable difference which parent has the dominant trait. On the other hand, it makes no difference which parent has the trait being studied in crosses involving genes on autosomal chromosomes.

This work showed, incidentally, that many genes on the X chromosome have nothing to do with sex determination. But the characters they determine are sex-linked because the genes are associated with a chromosome that is a part of the sex-determining system. As shown in Figure 14-14, the number of such characters is considerable.

NONDISJUNCTION AND EXCEPTIONAL OFFSPRING

Morgan's studies of eye color soon led to the discovery of another anomaly in eye-color inheritance, one that added weight to the conclusion that eye color genes are on the X chromosome. Bridges had observed that rarely (about once in 3000 times) a white-eyed daughter appeared in a cross between a white-eyed female (X^rX^r) and a red-eyed male (X^RY). This was a peculiar result, inasmuch as existing theory predicted that daughters with an R allele on the X chromosome from the male parent should be red-eyed heterozygotes. In crosses between these white-eyed females and red-eyed males, 4% of the female progeny were white-eyed like their female parents and 4% of the males were red-eyed like their male parents. The red-eyed males were also startling since the theory of **sex-linked inheritance** held that they should have received their X chromosomes from their female parents and thus should have X^r alone, with no companion gene from the Y chromosome.

Again Bridges convincingly attributed these exceptional offspring to nondisjunction. Consider the four types of zygotes arising in such a cross—this time in the context of sex-linkage. Again the female, which is white-eyed, produces two kinds of unusual gametes, XX (X^rX^r) and O. The male, which is red-eyed, produces ordinary X and Y gametes. When the red-eyed male parent fertilizes the white-eyed female parent (in which gametogenesis is disturbed by nondisjunction), the four kinds of zygotes are again the following:

Red-eyed male parent

		X^R	Y
White-eyed female parent	X^rX^r	$X^rX^rX^R$ Red-eyed female (dies)	X^rX^rY White-eyed female (fertile)
	O	X^RO Red-eyed male (sterile)	YO Male (dies)

As above, we can ignore two of these which are destined to die early: YO, the zygote with only a Y chromosome, and $X^rX^rX^R$, a female that is red-eyed like her normal XX (X^rX^R) sisters (see Figure 15-4).

The two fertilizations that concern us now are those producing a red-eyed male and a white-eyed female. Bridges showed that a red-eyed male arises when an egg receives an X chromosome (and with it an R allele) from its father, but no X chromosome from its mother. Like XY, XO becomes a male. As we have seen, the difference between a male and female fruit fly is not the presence of the Y chromosome in the male, but the X:A ratio of 0.5 when there is only one X chromosome (see Table 15-1). The white-eyed daughter was shown to be XXY. It is female because it has two X's. The presence of a Y chromosome does not cause it to be male. It is white-eyed because, as a result of nondisjunction, both Xs came from its white-eyed female parent.

Bridges's demonstration was important for several reasons.

- It was the first time abnormal gene segregation was correlated with a visibly abnormal pattern of chromosomes.
- It permitted the testable predictions that an exceptional white-eyed daughter will have a Y chromosome, unlike its normal XX sisters; and that an exceptional red-eyed son will lack a Y chromosome, unlike its normal XY brothers.
- It was a major conceptual advance. By showing that the chromosome theory permits successful predictions of the outcome of genetic breeding experiments it further validated the chromosomal location of genes.

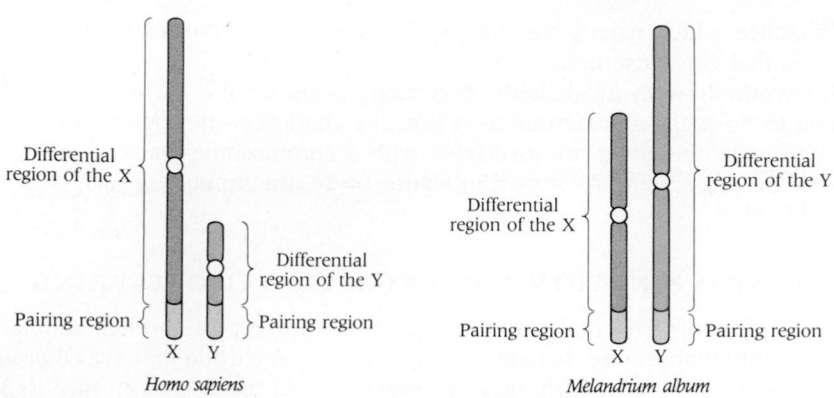

15-5 Homology of X and Y chromosomes
Pairing regions of X and Y chromosomes of humans and of the plant Melandrium album *are homologous and represent only a small part of each chromosome.*

This cytogenetic analysis of nondisjunction in *Drosophila* sex chromosomes (performed between 1913 and 1918) is a landmark in the history of genetics. Once again, seemingly contradictory data ended up confirming a theory and broadening its scope.

Y-LINKED INHERITANCE

Later work showed that certain human genes are located on a Y chromosome. **Y-linked inheritance** in these cases is very simple. Only men possess the Y chromosome. Therefore, only men possess Y-linked genes. Since all sons and no daughters receive a Y chromosome from their fathers, only sons receive the gene. In turn, they transmit it to all of their sons and to none of their daughters. If the gene always expresses itself phenotypically whenever it is present, the trait occurs only in males.

Only a few traits have been proved to be Y-linked. In humans, the trait of hairy ear rims is suspected to be Y-linked—and certainly the genes that determine "maleness" are Y-linked.

Restriction of a trait to males does not prove it is determined by a Y-linked gene. Some traits, like a bass voice, occur only in males, but are determined by autosomal genes that may be present in both sexes.

HUMAN SEX-LINKED CHARACTERS

Two factors complicate genetic analysis in human families. One is that breeding experiments are not feasible. In effect, the geneticist "takes what he gets" in the way of offspring, making inferences when possible and often encountering blind alleys. Second, patterns of inheritance differ for genes in different positions on the sex chromosomes. In a human male, only a small region at the ends of the X and Y chromosomes is truly homologous (Figure 15-5).

Despite these difficulties, many sex-linked traits are recognized in human heredity. Most are associated with abnormalities, or disease states. X-linked genes can be dominant or recessive, although the latter are the more common. Clues to the presence of X-linked dominant genes include the following (Figure 15-6A):

- All sons of affected males are normal.
- All daughters of affected males show the dominant phenotype.
- On the average, half the sons and half the daughters of affected females are affected (Figure 15-6B).

X-linked recessive mutations are more difficult to assess (Figure 15-6C).

- Affected (phenotypically abnormal) males have phenotypically normal offspring (unless mated with female carriers) and parents and are related to each other through females.
- If the female parent is homozygous, all the daughters of an affected male will be phenotypically normal but they will carry the gene in a masked (heterozygous) form.

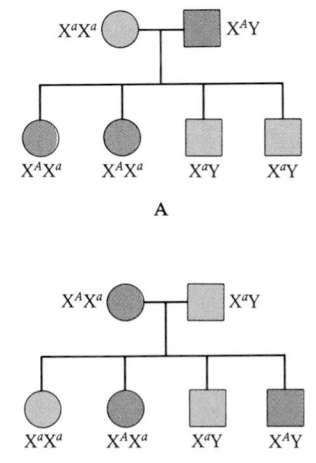

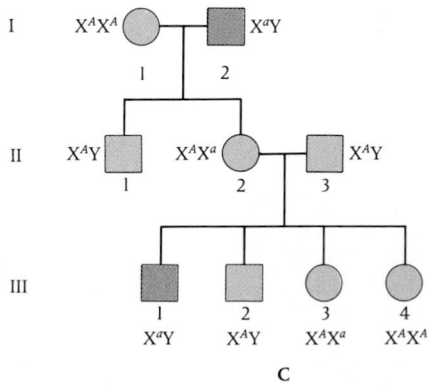

15-6 Sex-linked genes in humans
*In this diagram, circles represent females, squares are males, and shaded symbols are affected individuals. (**A**) Expression of an X-linked dominant trait when the male parent is affected. (**B**) Females expressing an X-linked dominant trait are heterozygous; the trait appears in half the offspring. (**C**) An X-linked recessive trait in the male disappears in the next generation and reappears in the grandsons.*

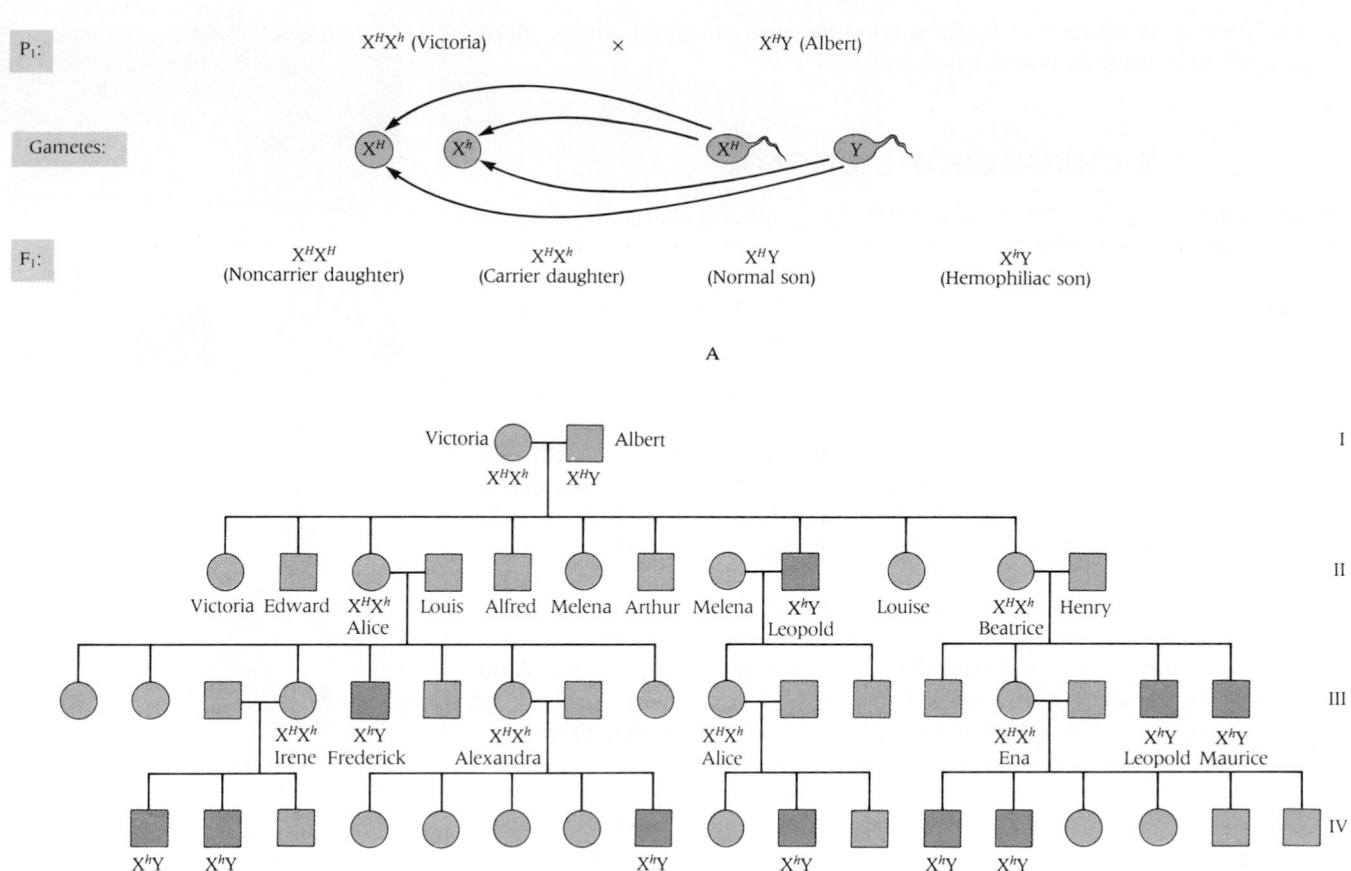

P₁: X^HX^h (Victoria) × X^HY (Albert)

Gametes: X^H X^h X^H Y

F₁:
X^HX^H (Noncarrier daughter) X^HX^h (Carrier daughter) X^HY (Normal son) X^hY (Hemophiliac son)

A

Victoria X^HX^h Albert X^HY I

Victoria Edward X^HX^h Alice Louis Alfred Melena Arthur Melena X^hY Leopold Louise X^HX^h Beatrice Henry II

X^HX^h X^hY Irene Frederick X^HX^h Alexandra X^HX^h Alice X^HX^h Ena X^hY X^hY Leopold Maurice III

X^hY X^hY Waldemar Henry X^hY Alexis X^hY Trematon X^hY X^hY Asturias Gonzalo IV

B

15-7 Hemophilia in European royalty
The hemophilia allele (h) *appeared first in Queen Victoria. Analysis of the family tree shows clearly that this gene follows the sex-linked pattern of inheritance.*

■ Half of the sons of a female "carrier" will be affected phenotypically.
■ A phenotypically normal female, with an affected father and half of her sons affected, can expect half of her daughters to transmit the gene.

One consequence of these considerations is evident: there are far more phenotypically affected males than females.

Surely the most famous examples of human sex-linked heredity are found in the family trees of people with hemophilia, a disease in which a defect of blood clotting causes a bleeding tendency. The genealogy of a hemophiliac family, like the illustrious one in Figure 15-7, reveals that only males exhibit the disease although it is transmitted from generation to generation by females.

The gene responsible for hemophilia is carried on an X chromosome. It is also recessive. If the X chromosome of the male XY pair is affected, the disease becomes manifest because a normal X chromosome is lacking. If one X chromosome of the female XX pair is abnormal, there is no sign of the disease because a normal X chromosome is present. However, half of the sons of such a female will be abnormal because she is a carrier. All the children of a normal (noncarrier) mother and a hemophiliac father will be normal; the sons receive their only X chromosome from their mother, who has two normal X's. Half the male children of a carrier mother and a normal father will be afflicted. Only one combination—a carrier mother and a hemophiliac father—can yield a hemophiliac female. This is an uncommon event, but it does occur.

Glucose-6-phosphate dehydrogenase deficiency is a genetic disorder that is responsible for a severe adverse reaction to the antimalarial drug primaquine (as well as to other drugs that are oxidizing agents). In many cases, it is an X-linked recessive trait in which affected males inherit the defect from carrier mothers. Some forms of this deficiency are not sex-linked.

Still another example is red-green colorblindness. Typically, the mother of a colorblind boy has normal vision. Nevertheless, he inherits an abnormal gene from her. His single X chromosome bears that abnormal gene. A male needs only one gene to be colorblind. The female has two X chromosomes. She would be colorblind only if she received an abnormal gene from both parents—that is, if she were a homozygous double recessive. This explains why this form of colorblindness affects about 2% of all men but only 0.01% of all women.

X CHROMOSOME DOSAGE

The dosage of a given gene refers to the number of copies of that gene in a cell. The dosage of a given gene can be increased or decreased. If, for example, nondisjunction of the X chromosome yields an XXX offspring, all genes on the X chromosome will be present in increased dosage. We are now concerned with the effect of gene dosage on the level of gene expression.

REGULATION

Nondisjunctional increase in the dosage of certain autosomes is associated with elevated levels of certain specific proteins—often with serious somatic defects. A familiar example is Down's syndrome, in which body cells contain three type 21 chromosomes (trisomy) and the individual displays physical abnormalities and mental retardation.

In contrast, the expression of genes located on X chromosomes in many instances seems independent of the number of X chromosomes present. It is well known, for example, that women (XX) and men (XY) have similar levels of certain blood proteins known to be determined by X-borne genes. This is true as well of X-determined traits in the XO system of grasshoppers—even when nondisjunction leads to an excess of X chromosomes.

Biochemical studies in *Drosophila* confirm that expression of some X-linked genes is unaffected by the number of gene copies present. For example, some alleles of the white locus produce the same eye colors in males and females, whereas other alleles produce more eye pigment in females than in males.

To explain the former cases, geneticists postulated the existence in insects of **dosage compensation genes.** When such genes are present in homozygous dosage they can inhibit the phenotypic expression of other X-linked genes. As a result, the dosage of X-linked genes in a female is not reflected in the phenotype. The molecular basis of this effect is unknown.

A more dramatic mechanism regulates expression of X-linked characters in mammals and many other animals. All X chromosomes in excess of one are inactivated. Indeed, they are (or it is) converted to a useless small clump of darkly staining heterochromatin known as a **Barr body** (and sometimes called **sex chromatin**) (Figure 15-8) that was first described in interphase somatic cells of females by Murray L. Barr of the University of Western Ontario in 1949. Any genes on this shrunken X chromosome are inactivated. The number of Barr bodies in a cell exactly equals the number of inactivated X chromosomes. For example, in cells with an abnormal number of X chromosomes—say, XXX or XXXX—only one X chromosome remains functional. The others are all converted into Barr bodies. Since fruit flies have dosage compensation genes, they do not inactivate their X chromosomes. Hence, they do not have Barr bodies.

MOSAICISM AND THE LYON HYPOTHESIS

If one of two X chromosomes normally present in each normal female somatic cell is inactivated, which one is it? In 1961, Mary Lyon in England and Liane Russell in the United States independently suggested that the X chromosome destined to be inactivated is randomly chosen in each cell. Early in embryonic development, according to this proposal (now called the **Lyon hypothesis** or **single-active-X hypothesis**), each cell of the female body randomly, independently, and permanently inactivates (or in scientific jargon "lyonizes") a single X which may be either the maternally derived or the paternally derived X chromosome (Box 15-1)[4].

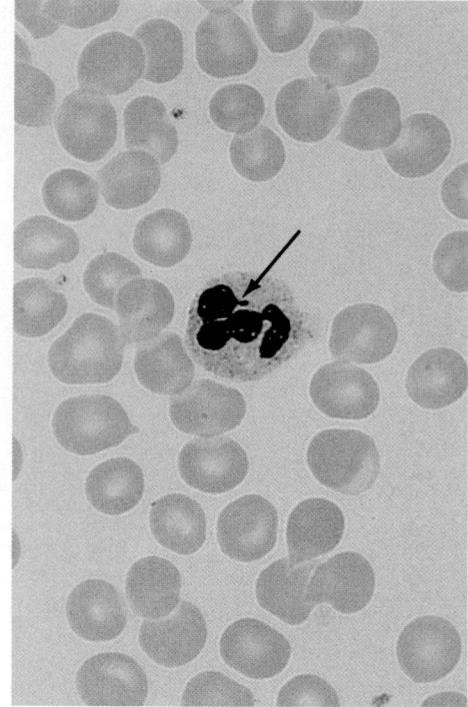

15-8 "Drumstick" in a white blood cell of a human female
In stained white blood cells (neutrophils), the Barr body appears as a drumstick-shaped appendage (arrow) of the oddly shaped nucleus. Because the nucleus is multi-lobed, or polymorphous, these cells are often called "polys." (×1500)

---

[4] Note that inactivation of one of the two X chromosomes occurs only in somatic cells. If the single X chromosome in a female gamete were inactivated, the chromosome pattern in a zygote after fertilization would not be normal.

BOX 15-1

THE LYON HYPOTHESIS AND THE ORIGIN OF TUMORS

In cancer, abnormal cells proliferate in an uncontrolled manner to the detriment of the whole body. Unless the resulting malignant tumor (defined as a mass of cancer cells) can be controlled by treatment—and often it cannot be—the outcome may be tragic. A major obstacle to scientific progress in this area is an ethical one: investigators cannot directly study mechanisms of tumor formation in human subjects. The only acceptable alternatives are investigations in animals—or indirect studies in humans.

A novel indirect approach cleverly takes advantage of a naturally occurring cell "label" to determine whether a tumor originated in a single cell or in a large group of cells. When available, such information provides useful clues to the initiating event. For example, one would expect a cancer that arises from a rare event like a spontaneous mutation to have a clonal, or single cell, origin. If, on the other hand, the tumor arose primarily from the activities of an infectious virus, one would anticipate a multicellular origin.

This problem can now be investigated by studying tumors that happen to arise in peo-ple with cellular mosaicism, whose bodies consist of two or more genetically distinct cell populations. Figure 15A schematically illustrates a normal tissue consisting of two types of cells, A or B, which occur with equal frequency. Note that each cell is either an A cell or a B cell. Consider a tumor of clonal origin. By definition, it began in one cell—say, an A cell. That means that all cells in that tumor will be of A type since they will all be descendants of the single progenitor A cell. They are, in fact, a clone of the single progenitor cell. If, on the other hand, a tumor is found to contain cells of both A and B types, it must have had a multicellular origin.

The most useful cell markers for studies of this kind depend upon the X chromosome inactivation envisioned in the Lyon hypothesis. This approach can be used for any tumor that happens to arise in a female who is heterozygous for a gene on the X chromosome. The X-linked enzyme, glucose-6-phosphate dehydrogenase (G-6-PD), is such a marker. According to the Lyon hypothesis, only one of the two X chromosomes is active in a given somatic cell. Therefore, females who have the usual wild-type G-6-PD gene (type B) on one X chromosome and the most common variant (type A) on the other, would have two cell populations in their bodies—one producing type A G-6-PD and the other, type B G-6-PD (Figure 15B). Tumors of clonal origin contain either type B or type A G-6-PD. In contrast, tumors of multicellular origin contain both enzyme types. Many tumors have now been studied in this way.

The use of G-6-PD markers is well illustrated by a recent investigation of one kind of leukemia (chronic myelocytic leukemia), a malignant proliferation of white blood cells. In this study, both B and A enzymes were found in the normal body tissues of 12 women who were heterozygous for G-6-PD and who had this type of leukemia. However, only one type of G-6-PD was found in their leukemic white blood cells. Conversely, in type B/type A heterozygotes without leukemia, the normal (nonleukemic) white blood cells invariably show both B and A enzymes. Thus, the finding of single-enzyme types strongly favors a clonal origin of this leukemia. We may conclude that the disease began in one cell.

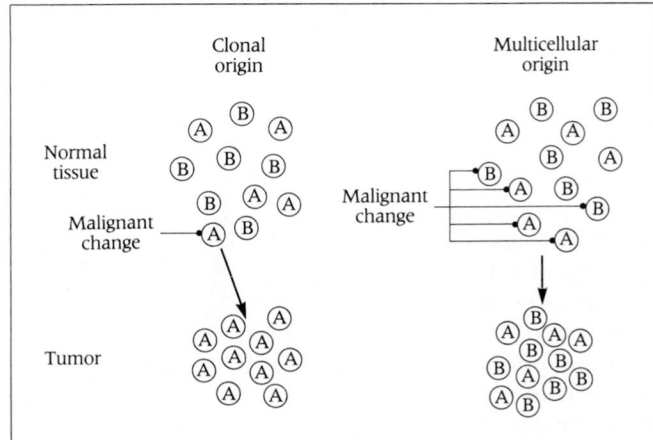

15A The concept of cellular mosaicism
A normal tissue consisting of cell types A and B can give rise to tumors of clonal origin (containing either type A or type B cells) or tumors of multicellular origin (containing both A and B cell types).

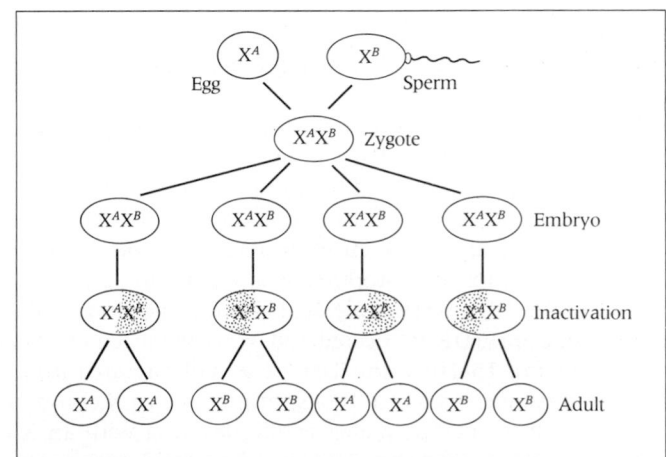

15B The Lyon hypothesis and the origin of tumors
The G-6-PD gene is X-linked and can be used as a marker to determine tumor origins. A female exhibiting two forms of this gene (X^A and X^B) will be expressing either one form or the other in any given cell. This can be used to determine the clonal or multicellular origin of tumors. (Note that the inactivated X is not shown.)

- The randomness of the inactivation means that in each cell there is only a 50% chance that any given X chromosome will be eliminated.
- The independence of its inactivation means that events in each cell are uninfluenced by events in neighboring cells.
- Its permanence means that once a cell inactivates an X chromosome, the pattern is duplicated in all of its subsequent progeny.

If one X chromosome in every cell of the female body is "lyonized," all of a woman's body cells can be divided into two groups, one in which the maternal X chromosome survives and one in which the paternal X chromosome is expressed. As a result, the retina of the eye of a woman heterozygous for X-linked colorblindness contains colorblind light receptors that alternate with normal ones in a scattered patchwork pattern.

Such individuals are called **mosaics.** Geneticists define mosaicism as the coexistence within one organism of two cell populations that differ in genotype even though they arose from the same zygote. Many examples are known. The enzyme glucose 6-phosphate dehydrogenase is carried on the X chromosome. Box 15-1 explains how this fact helps us determine whether a tumor originated in one cell or many. Similarly, an X-linked condition, congenital ectodermal dysplasia, in which the skin lacks sweat glands, displays a patchwork of normal and abnormal skin areas in heterozygous females (Figure 15-9).

The random inactivation of X chromosomes suggests an explanation for the fact that the sex of a mammal is determined by the presence or absence of the Y chromosome, rather than the ratio of X chromosomes to A chromosomes (autosomes) as in *Drosophila*. Because each mammalian cell undergoes "lyonization," it can have only one X chromosome; thus it is either an XO female or an XY male. Thus each cell has an equal number of X-linked genes, whereas in *Drosophila* females have twice the number of X-linked genes as males and dosage compensation is necessary.

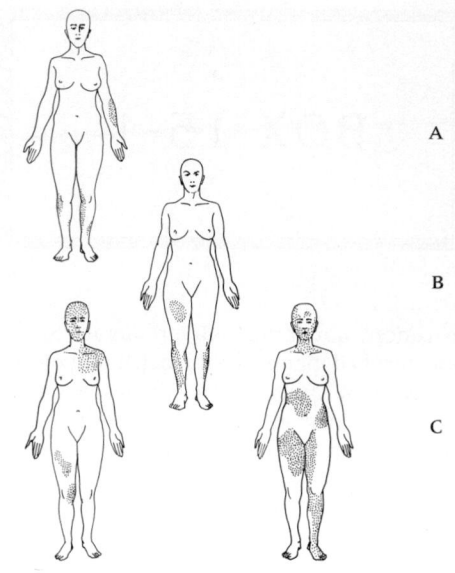

15-9 X chromosome inactivation
In the three generations shown (A, B, and C), each woman is believed to be heterozygous for the X-linked condition, congenital ectodermal dysplasia, which results in absence of sweat glands in some areas of the body (shaded). Note that identical twins in generation C exhibit a different distribution of sweat glands.

SEX ABNORMALITIES

Sex determination is the process that directs the zygote to become a male or female. **Sexual differentiation** is the process that carries out the genetic instructions and ensures that the character of the sexual organs is male or female. Occasional errors occur in both processes.

NEW INSIGHTS ON SEX DETERMINATION

In 1988, David C. Page of M.I.T. and coworkers reported the important discovery that a single gene on the Y chromosome determines the sex of a human embryo (Figure 15-10A).

The gene, termed **testis-determining factor gene,** or **TDF,** leads to the synthesis of a specific protein that binds to DNA or RNA and somehow causes embryonic development to proceed in the direction of maleness. The investigators were aided by studies of rare individuals whose sex did not match their sex chromosome pattern. One such person had two X chromosomes but was male because he also had 0.05% of a Y chromosome. Another was female in spite of having 99.8% of a Y chromosome. All who were male had the tiny portion of the Y chromosome in which the gene for TDF is located; all who were female lacked that region. In some cases (Figure 15-10B), the TDF gene had migrated from the Y chromosome to the X chromosome. As a result, offspring with two X chromosomes, one carrying the TDF gene, developed as males, while offspring with an X chromosome and a Y chromosome that lacked the TDF gene developed as females.

The discovery of the TDF gene in humans and various placental mammals touched off a predictable hunt for the corresponding gene in other animals. Surprisingly, a report in late 1988 conveyed the news that an almost identical gene in marsupials is located not on the Y chromosome but on an autosome. Yet the Y chromosome is male-determining in marsupials. Perhaps in these animals sex determination requires cooperation between the TDF gene and another gene that *is* on the Y chromosome.

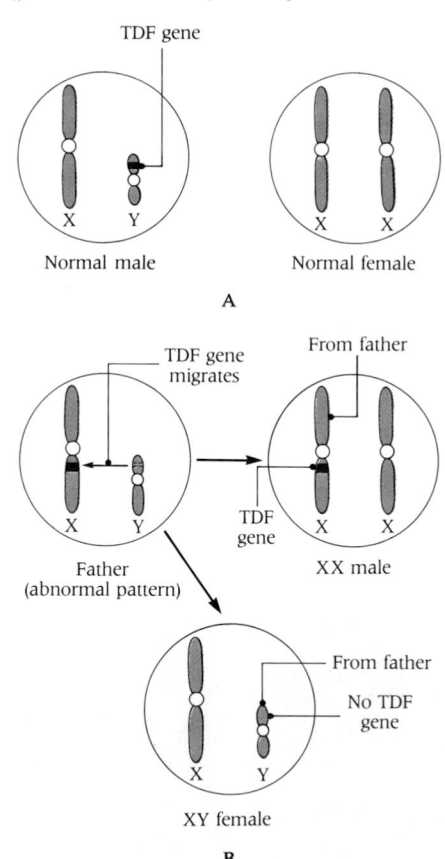

15-10 Evidence that the TDF gene determines maleness
(A) Sex chromosomes in normal male and female. The TDF gene is on the short arm of the Y chromosome. (B) If the TDF gene migrates from the Y chromosome to the X chromosome in a male, his offspring will show discrepancies between their sex and their sex chromosomes, including XX males and XY females.

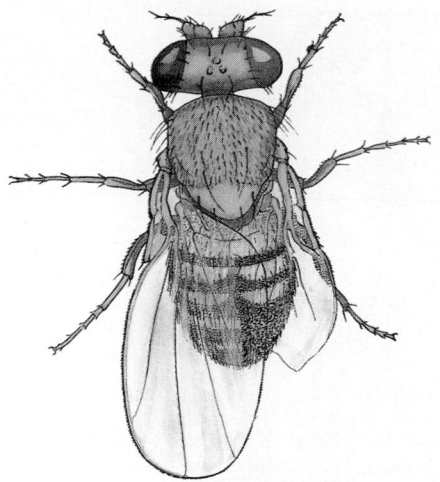

15-11 Gynandromorph in Drosophila
The right side of this fruit fly is male and shows a mutant eye color and wing length. The left side is female and possesses a normal eye and wing.

ABNORMAL SEXUAL DIFFERENTIATION: GYNANDROMORPHS

Gynandromorphs (*gyne,* female, *andros,* male, and *morphe,* form) are rare occurrences, in which parts of the body are female and parts are male.

The nondisjunctional loss of an X chromosome in *Drosophila* very early in the embryonic life of a female is illustrated in Figure 15-11. If the loss occurs in the first division of the zygote, all body tissues that derive from the bereft cell of the pair will carry only one X chromosome. The emergent adult is a bilateral gynandromorph—male on one side of the body and female on the other. If the mishap occurs later in zygote development, small patches of male tissue will be scattered against a female background.

The zygote of the fly in Figure 15-11 was originally heterozygous for X-linked genes controlling eye color and wing length. Normally, such a fly would have red eyes and long wings because these alleles, respectively *R* and *M,* are dominant, but loss of the X chromosome carrying *R* and *M* from a primitive embryonic cell led to the appearance of dominant phenotypic traits only on the left side of the body, where retention of both X chromosomes causes femaleness. The right or male side of the body has a white eye and miniature wing! Clearly, the gynandromorphic fly is a mosaic—half XX and half XO.

HUMAN ABNORMALITIES

Equally fascinating examples of sexual mosaicism are encountered in our own species. Nondisjunctional abnormalities produce a variety of intersexes. These are abnormal forms that are not clearly male or female. Perhaps arbitrarily, we do not consider them to be gynandromorphs, in which whole body regions may be male or female. Rather they are individuals whose sex is ambiguous for one reason or another. Many incongruous combinations occur. One may have small testes and ovaries. Another may have a testis on one side and an ovary on the other. Still another may have a mixed gonad containing male and female elements. Not surprisingly, most have severe psychological difficulties.

Such disorders reflect aberrations of various kinds. One is the XXY male, which can arise in two ways: (1) nondisjunction in a female gamete followed by fertilization of the resulting XX egg by a normal Y sperm; and (2) nondisjunction in a male gamete followed by fertilization of a normal X egg with the resulting XY sperm. XXY males have Klinefelter's syndrome, a phenotype in which testes are undeveloped, breasts are overdeveloped, limbs are lengthened, body hair is sparse, and intelligence is sometimes impaired (Figure 15-12). Almost all are sterile.

The second most common anomaly, occurring once in 3000 female births, is the XO female, in whom the second X chromosome is totally deleted. This phenotypic pattern is known as Turner's syndrome (Figure 15-13). These unfortunate individuals are usually of short stature with many body abnormalities, including sterility owing

15-12 Kleinfelter's syndrome
Individuals with this anomaly are males with an extra X chromosome. Thus they are XXY.

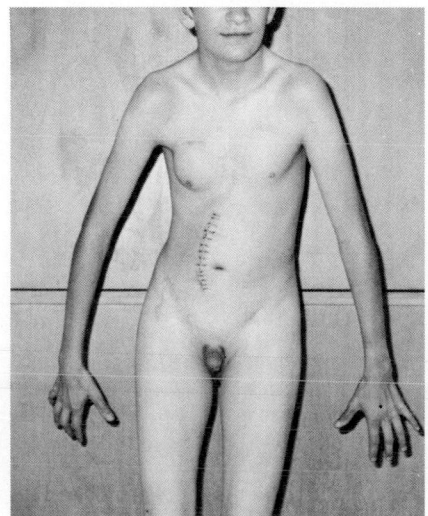

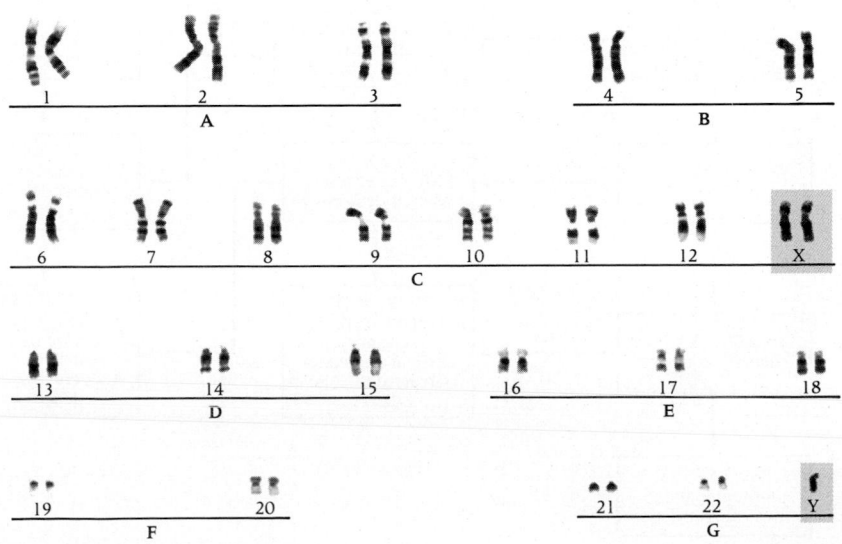

to failure of the ovaries to develop. Predictably—because they lack one X chromosome—the cells of women with Turner's syndrome have no Barr bodies.

In 1961, cytogeneticists first observed an XYY male, presumably the result of nondisjunction of an X and a Y chromosome during spermatogenesis in the subject's maternal grandfather. Many XYY males have turned up since then. They have normal sexual development and are fertile. Recent reports have claimed that the incidence of XYY genotype among institutionalized male criminals is 2.5 times the incidence in a control group. If further study confirms this finding, human society may have to alter its view of criminal and aggressive behavior. However, critics have questioned these studies and the matter remains undecided.[5]

Most of the human X and Y abnormalities observed so far are summarized in Figure 15-14. A large number of these variants are associated with mental deficiency, physical abnormality, and infertility. Many exhibit aneuploidy—that is, an abnormal total number of chromosomes. (The arrows in Figure 15-14 are not meant to suggest that one abnormal state gives rise to another.)

Finally, cases have been described of **sex chromosome mosaicism,** in which at least two populations of cells with different cell chromosome patterns are found in the same individual. For example, an XY/XXY mosaic would have one cell line with an XY pattern and another with XXY. These arise from a mitotic error in an early division in a zygote that had originally had either a normal or an abnormal sex chromosome pattern.

When confronted with an individual of indeterminate sex—or a male or female with an abnormal sex chromosome pattern—the physician attempts to establish the sex according to four different criteria.

- The **genetic sex,** which is determined by the presence or absence of Barr bodies and by a detailed study of the sex chromosomes present
- The **genital sex,** which is suggested by the appearance and structure of the genital organs
- The **somatic sex,** which is indicated by the secondary sexual characteristics (facial hair, breasts, etc.)

[5] Many have criticized the ethics of investigators wishing to screen infants for the XYY pattern in order to follow their development and then, if necessary, control criminal tendencies with psychiatric counseling (if that is possible). The criticism is that this unfairly stigmatizes these individuals in view of the scientific uncertainty surrounding the meaning of XYY.

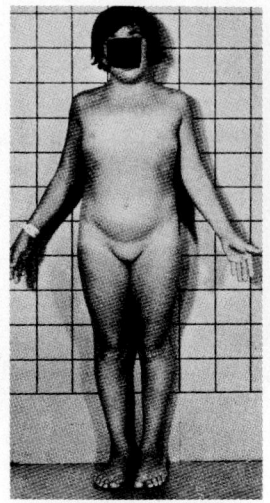

15-13 Turner's syndrome
Individuals with this anomaly are females lacking one of the X chromosomes. Thus they are XO. They are of short stature and often have a distinctive webbing between the neck and shoulders.

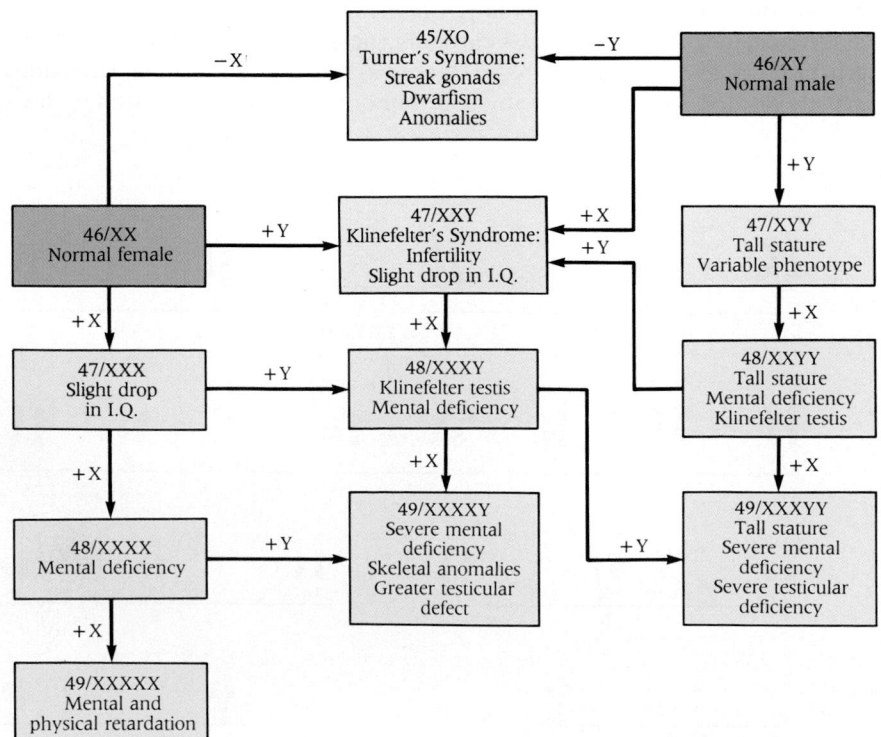

15-14 Abnormalities of human sex chromosomes
Lines connecting each box indicate additions of X or Y chromosomes by nondisjunction. Within each box, the total number of chromosomes is given, followed by the sex chromosome complement.

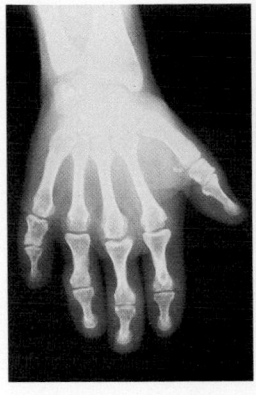 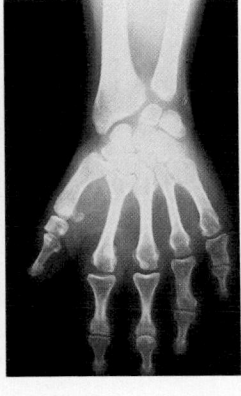 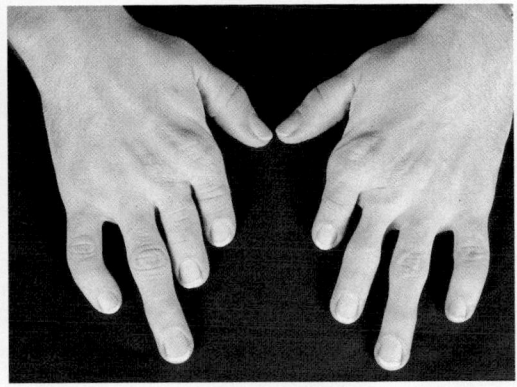 

A   B

15-15 Heritable human traits
(A) *Brachydactylism (short fingers) is due to a dominant gene. The first bones of the fingers are normal length, but the second and third bones are abnormally short.* **(B)** *The tongue-rolling ability is transmitted by a dominant gene.*

■ The **social** or **psychological sex,** which is the gender with which the individual has lived and grown up

Many of these individuals display striking inconsistencies between the findings in several of these areas. Obviously careful evaluation is essential in planning hormonal, surgical, or psychiatric treatment.

SEX-INFLUENCED CHARACTERS

As we noted in discussing Y-linked inheritance, characters determined by autosomal genes may be influenced by the sex of the organism. These are called **sex-influenced characters.** Such influence is usually mediated by hormones produced by the sex organs. For example, in sheep the allelic pair that determines the presence or absence of horns is autosomal. The horn gene, H, is dominant; the hornlessness gene, h, is recessive. But only the male heterozygote (Hh) is horned. The female heterozygote is hornless.[6]

Similarly, the genes controlling human secondary sexual characteristics (growth, development of breasts and sex organs, body hair distribution, etc.) are autosomal. But sex hormones modify their phenotypic expressions in the two sexes. Hence, they are sex-determined, rather than sex-linked. Hormones, as we shall see, are the most important environmental influence acting on genes.

HUMAN HEREDITY

The list of known heritable characters is not restricted to disease states caused by mutant genes. They include as well the blood groups and such curiosities as short fingers (brachydactylism; Figure 15-15A), the capacity to roll the tongue longitudinally (only 7 of 10 people can do it; Figure 15-15B), and the ability to distinguish certain tastes.

Nevertheless, disease states have had the most study. That fact is evident in current lists of genetically controlled human traits, which now exceed 3000. Although it is difficult to do formal genetic studies on human subjects, there is no reason to doubt that the genes responsible for these characters behave according to Mendelian principles.

TECHNIQUES OF STUDY
Formal Genetics

Because of the difficulties inherent in human genetics research, the recognition that a human disease seems genetically determined does not ensure that an investigator can elucidate its genetic mechanism. Although the basis of the genetics is the "test of progeny," geneticists have had to turn to the following less direct approaches.

---

[6] A comparable case in humans involves the sex-determined autosomal allele that causes premature baldness. It is active in the female homozygote. In the presence of testosterone (a male sex hormone), the allele behaves as a dominant and thus is expressed in the heterozygote over the wild-type gene.

- Analysis of pedigrees (genealogies), or family trees, to define patterns such as sex-linkage
- Study of identical twins, and the relationship of phenotype and the environment
- Study of homogeneous populations from isolated areas (such as remote villages and deep Alpine valleys) where outsiders rarely enter
- Analysis of consanguineous marriages—marriages between near relatives—in which there is an increase in the numbers of recessive homozygotes

The genetic traits of humans reside in the large number of genes in each cell—variously estimated at 50,000 or more. It is presumed that everyone has a substantial number of mutant genes. When we consider that there are polygenically determined characters as well as characters determined by single genes, we realize that the number of human characters is very large.

Detailed knowledge of human genes is meager but growing. We presume that much of what we have learned from other species applies to human beings. But many problems arise. For example, although a dominant human gene expresses itself in a heterozygote and a recessive gene only in a homozygote, our ability to recognize these expressions in a given individual depends on many things.

In considering the formal genetics of a specific human character, an investigator must try to answer certain questions.

- Is the character hereditary?
- If so, is it autosomal or sex-linked, dominant or recessive?
- How common is the gene in the general population?
- How often does it arise by mutation?
- What are its linkage relations?
- When a genetic basis for a disease is suspected, do relatives of an affected person have a higher incidence of the disease than the general population?
- Are environmental factors ruled out as causes of observed frequencies?

In human genetics, the cases most accessible to analysis involve only a single gene. In the simplest case, a rare autosomal dominant gene with complete penetrance, the mutant allele would be inherited from only one parent. The probabilities are that half the siblings and half the offspring of an affected individual would also be affected—but in a small human family, all or none might be affected. An affected individual would be heterozygous for that gene pair. It is through him or her that the trait is transmitted from generation to generation.

The most difficult case is the autosomal recessive. A phenotypically affected individual would have to inherit the mutant allele from both parents—and thus would be homozygous for that gene pair. Such families are usually discovered when an affected child is encountered. The chance is small that two unrelated parents will carry the same relatively uncommon mutant allele. But when a carrier of the mutant allele marries a relative, the chance increases. Thus, parental consanguinity is virtually the rule in some rare genetic diseases.

HUMAN BLOOD GROUPS

Perhaps the best-studied genetic systems in human beings are those determining the several major blood groups. The **ABO** system of blood groups—one of the two most important in the context of blood transfusion (the other is the **Rh** or **CDE** system)—was the first blood group system recognized. All people fall into one of four blood groups—A, B, AB, and O. Each inherited blood group is associated with a particular glycoprotein **antigen**[7] on the surfaces of all red blood cells. Note in Table 15-2 that group A individuals have antigen A on their red cells; those in group B have antigen B; those in group AB have both antigens A and B; and those in group O have neither antigen A nor antigen B.

It is remarkable that someone with one of these red cell antigens should have a serum **antibody** that is specific for the antigen not present on the red cells.

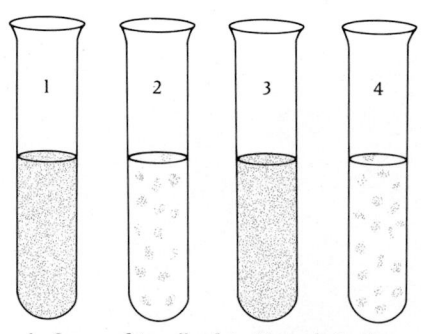

1. Serum of A, cells of A → No agglutination
2. Serum of A, cells of B → Agglutination
3. Serum of B, cells of B → No agglutination
4. Serum of B, cells of A → Agglutination

15-16 Tests for human blood groups
Incompatibility of blood group and serum antibodies is shown by clumping of red blood cells.

---

[7] Antigens are substances that induce certain cells in the animal body to form antibodies. These are proteins that react specifically with antigen molecules, usually precipitating and inactivating them. Most antigens are foreign substances. Thus antibody synthesis, the phenomenon called immunity, is an important defense mechanism (See Chapter 23).

TABLE 15-2
THE HUMAN ABO BLOOD GROUP SYSTEM

Blood Group (Phenotype)	Possible Genotypes	Red Cell Antigen(s)	Reaction with:		Antibody in Blood Plasma
			Anti-A	Anti-B	
O	ii	None	−	−	Anti-A, anti-B
A	$I^A I^A$ or $I^A i$	A	+	−	Anti-B
B	$I^B I^B$ or $I^B i$	B	−	+	Anti-A
AB	$I^A I^B$	A and B	+	+	None

Note: + indicates agglutination; − indicates no agglutination.

Since they are of different types, no reaction takes place in the body between cell antigen and serum antibody. But in a test tube, added antibodies do combine with the correct red cell antigen. The combination causes clumping, or **agglutination,** of the red cells (Figure 15-16); hence the antibodies are called **agglutinins.** The figures show how antibodies are used in agglutination tests to determine the ABO group of an unknown blood sample.

Of immediate interest here is the inheritance of the blood groups. They are determined by three alleles at a single gene locus, symbolized by I^A, I^B, and i. Many other blood group systems exist in addition to the ABO and Rh systems, among them the MNS system, the Kell system, Duffy, Lutheran, and dozens of others. All are genetically determined. Surprisingly, the biological roles of all blood groups remain unknown.

Knowledge of blood groups and their genetics is now an indispensable aid in the solution of several practical problems.

- Scientifically planned and safe blood transfusions are now possible.
- Disorders such as Rh disease, which occurs when the red cells of newborn Rh positive offspring are attacked by an antibody developed in the Rh negative mother, can now be successfully treated.
- Blood grouping of large populations has provided markers useful in the study of various anthropological problems—for example, population movements, common origins, and evolution of genes.
- Demonstration of blood groups in fresh or even dried blood or other materials is indispensable in establishing the human origin of the specimen and in identifying the individual from whom it came, of obvious importance in police work and forensic pathology.
- In cases of disputed paternity, blood grouping of mother, child, and putative father can often either exclude paternity or give a statistical probability that a given man is the father (Table 15-3). It is almost always possible, by the use of multiple genetic markers, to establish that a man is not the father of a child.

TABLE 15-3
GENETICS OF THE ABO SYSTEM: PHENOTYPES POSSIBLE IN CHILDREN

Phenotype of One Parent	Phenotype of Other Parent	Possible Phenotypes of Children	Impossible Phenotypes of Children
A	A	A, O	AB, B
A	B	A, B, AB, O	None
A	AB	A, B, AB	O
A	O	A, O	AB, B
B	B	B, O	A, AB
B	AB	A, B, AB	O
B	O	B, O	A, AB
AB	AB	A, B, AB	O
AB	O	A, B	O, AB
O	O	O	A, B, AB

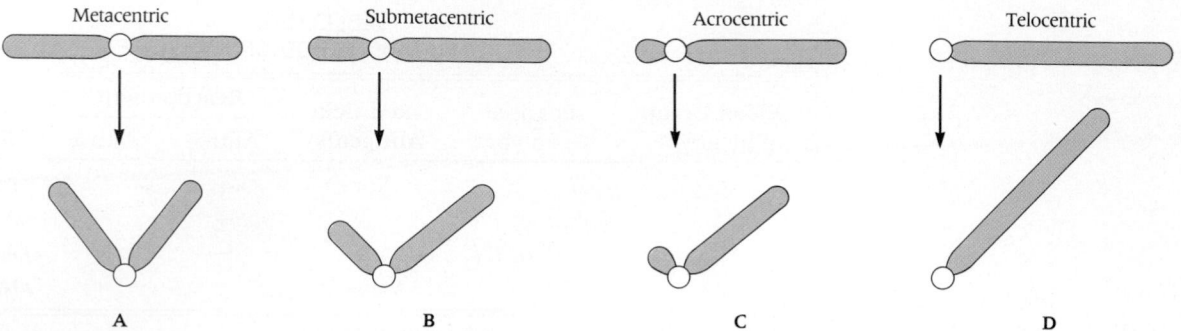

It is often possible to say that a certain man is probably the father. It is rarely possible to prove beyond all doubt that he is the father. But DNA fingerprinting can now do this.

CYTOGENETICS: KARYOTYPE ANALYSIS

Astonishingly, the correct number of human chromosomes was not known until 1956 when a simple technique was developed for observing them in the metaphase stage and arranging them in the orderly pattern known as the **karyotype** (see Chapter 12). Before that, researchers thought they had counted 48 chromosomes in human cells.

The chromosomes in a "metaphase plate" karyotype are classified by length and position of centromeres, which determines "arm" length (Figure 15-17).

- Those with median centromeres and two arms of equal length are **metacentric.**
- Those with submedian centromeres with one longer and one shorter arm are **submetacentric.**
- Those with centromeres near the chromosome ends and very unequal arm lengths are **acrocentric.**
- Those with centromeres at the ends with only one arm are **telocentric.**

Some chromosomes have, at one end, a small bulblike extension called a **satellite.**

On the basis of such criteria, human chromosomes are classified into seven groups, designated A to G. The pairs are numbered 1 to 22. The 23rd pair is XX in women and XY in men (see Figure 12-8B). Their properties are given in Table 15-4.

Karyotypes are studied today whenever genetic disease is suspected (and in certain malignant disorders such as leukemia). Abnormalities, which are surprisingly common, can affect chromosome structure and chromosome number. The major mechanisms responsible for these anomalies are summarized in Table 15-5.

The chromosome rearrangements include **deletions, inversions,** and **translocations.** All depend on prior breakage of chromosomes, found in 1–2% of metaphases

15-17 Chromosome classification
The position of the centromere determines chromosome type. Arrows represent movement of each chromosome toward the pole of the cell. **(A)** *Metacentric chromosome.* **(B)** *Submetacentric chromosome.* **(C)** *Acrocentric chromosome.* **(D)** *Telocentric chromosome.*

TABLE 15-4
PROPERTIES OF HUMAN CHROMOSOMES

Group	Chromosome Number	Size	Centromere Position	Special Features
A	1–3	Large	Metacentric	Each is distinctive
B	4–5	Large	Submetacentric	4 larger than 5
C	6–12, X	Medium	Submetacentric	6 resembles X
D	13–15	Medium	Acrocentric	13 has prominent satellite on short arm
E	16–18	Short	Metacentric/Submetacentric	Each has small satellite on short arm
F	19–20	Short	Metacentric	
G	21–22, Y	Very short	Acrocentric	21 and 22 have satellite on short arm

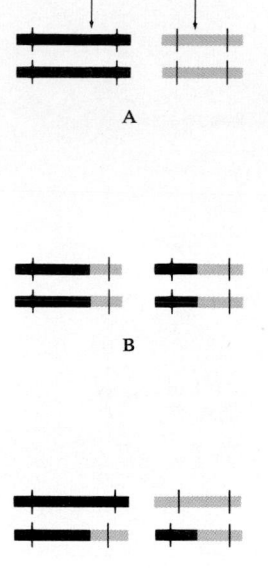

15-18 Chromosome translocations
(**A**) *Nonhomologous chromosomes (one pair black, the other pair colored) break at the arrows.* (**B**) *and* (**C**) *Once broken, parts of each nonhomologous pair rejoin, creating new arrangements of genetic material.*

TABLE 15-5
MECHANISMS UNDERLYING CHROMOSOME ABNORMALITIES

Category	Illustration			Remarks
Changes in Chromosome Structure				
Rearrangements				
Inversions	*abc*	→	*acb*	May unbalance meiosis and cause sterility
Translocations	*abcdef* *uvwxyz*	→	*abcdyz* *uvwxef*	May cause sterility
Deletions	*abcd*	→	*acd*	Nonreversible; lethal if homozygous
Duplications	*abc*	→	*abbc*	Important in evolution as source of new genetic materials and functions
	abc	→	*abcbc*	
	abc	→	*abccb*	
Changes in Chromosome Number				
Abnormal euploidy				Change in ploidy (multiples of haploid number)
Aneuploidy				Change in total number

in normal persons. Spontaneous breaks apparently reunite and "heal up." (How this repair process takes place is described in Chapter 17.)

When several breaks occur simultaneously, the broken ends that reunite may not be those that broke apart originally. Chromosomal segments may be thereby exchanged from one chromosome to another (Figure 15-18). This is a translocation. When such a rearrangement produces a large chromosome and a chromosomal fragment from two medium-sized chromosomes, with loss of chromosomal material, the result is a chromosome deletion.

Breakage may be promoted by external agencies such as virus infections and radiation. Some chemicals and drugs, such as lysergic acid (LSD) and cyclamates, increase breakage. Such agents induce breakage at random locations on any chromosome. The rearrangements induced by breakage may disturb meiosis and produce zygotes that cannot survive. Hence, rearrangements may cause sterility.

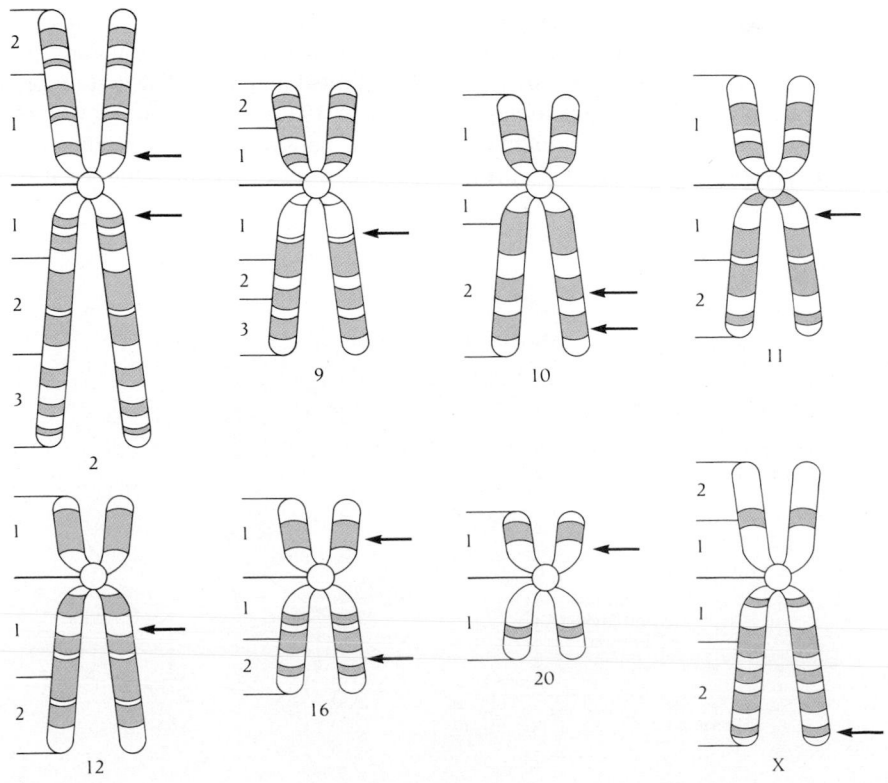

15-19 Fragile sites on human chromosomes
Arrows illustrate areas in eight human chromosomes that are fragile sites susceptible to chromosome breakage. The chromosomes are identified by number.

A new form of chromosome breakage was recognized recently: the **fragile site,** an area of "weakness" at a specific location on one or another chromosome that breaks only under certain conditions. Like other anomalies, fragile sites can be observed under the light microscope but, unlike them, they involve only a small portion of the genetic material, so that their phenotypic effects are much harder to assess. Many distinct fragile sites are known (Figure 15-19), but only one of them is definitely known to have adverse consequences. However, that one, the "fragile X syndrome," occurs in a significant proportion of all cases of inherited mental retardation in males. The weak spot is located at the very end of the long arm of the X chromosome. Interestingly, this structural defect can be observed only after cultivating cells in a medium deficient in folic acid, a vitamin whose coenzyme derivatives function normally in the pathways of purine and thymine synthesis— and thus of DNA synthesis.

Changes in chromosome number (Table 15-5) are of two types. **Abnormal euploidy** is defined as changes in chromosome counts in multiples of the normal monoploid (or haploid) number (n). Such changes include **triploidy,** wherein all chromosomes are present in triplicate ($3n$), **tetraploidy** ($4n$), and so on.

Aneuploidy ("not euploid") includes changes affecting selected portions of a chromosome set. Thus it causes odd chromosome numbers, usually classified as **hyperdiploid** ($>2n$) and **hypodiploid** ($<2n$). The most common of these are trisomy (that is, 3 present instead of 2) of chromosome 16 in group E or of chromosome 21 in group G and, as noted above, the XXY pattern (Klinefelter's) and the XO pattern (Turner's). Of these anomalies, only one is common: trisomy of chromosome 21 in group G. It occurs in 1.8 per 1000 births and accounts for the unfortunate disorder called Down's syndrome (Figure 15-20). For reasons unknown, the incidence is sharply higher among children born to older mothers. For example, its frequency is 3.5 per 1000 at age 35, 10 per 1000 at 40, and 22 per 1000 at 45.

It is important not to confuse chromosomal abnormalities with gene mutations, which are due to changes in a nucleotide in the DNA sequence. These subtle molecular changes cannot be seen, even with the most powerful microscopes.

Chromosome structure is similar in all races of mankind and human chromosomes have much in common with those of the chimpanzee and gorilla, though these apes have 48, not 46, chromosomes. At a critical stage in human evolution the number of chromosomes was decreased, presumably by translocation.

MAPPING METHODS

At present, only a small fraction of the human genes have been located on their chromosomes, most of them only approximately, although the rate of progress is accelerating rapidly. Yet even these still modest achievements have already made significant contributions to basic biology and to clinical medicine. The first (approximate) location of a human gene dates as far back as 1911. E. B. Wilson of Columbia University, comparing the distinctive inheritance pattern of color blindness with the transmission of the X and Y chromosomes, deduced that the gene for it must

15-20 Down's syndrome
One of the most common chromosomal abnormalities, Down's syndrome (also called trisomy-21) is associated with an extra chromosome 21. Affected individuals typically have a broad, flattened face and narrow eyelids.

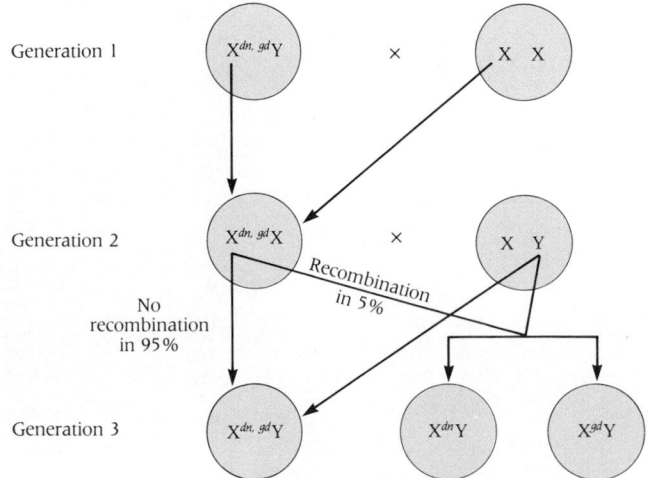

15-21 Mapping human chromosomes
One of the X chromosomes of the grandfather has both the colorblind gene (dn) and the G-6-PD gene (gd), and the woman in generation 2 receives one of her X's from her father. Any grandsons possessing either the dn gene or the gd gene (but not both) represent recombinants within the X chromosome. This occurs about 5% of the time and so the map distance between these two genes is estimated to be about 5 map units.

15-22 Human chromosome mapping

Genes for hundreds of diseases have been assigned to specific human chromosomes. A few are shown here. Chromosome numbers are indicated.

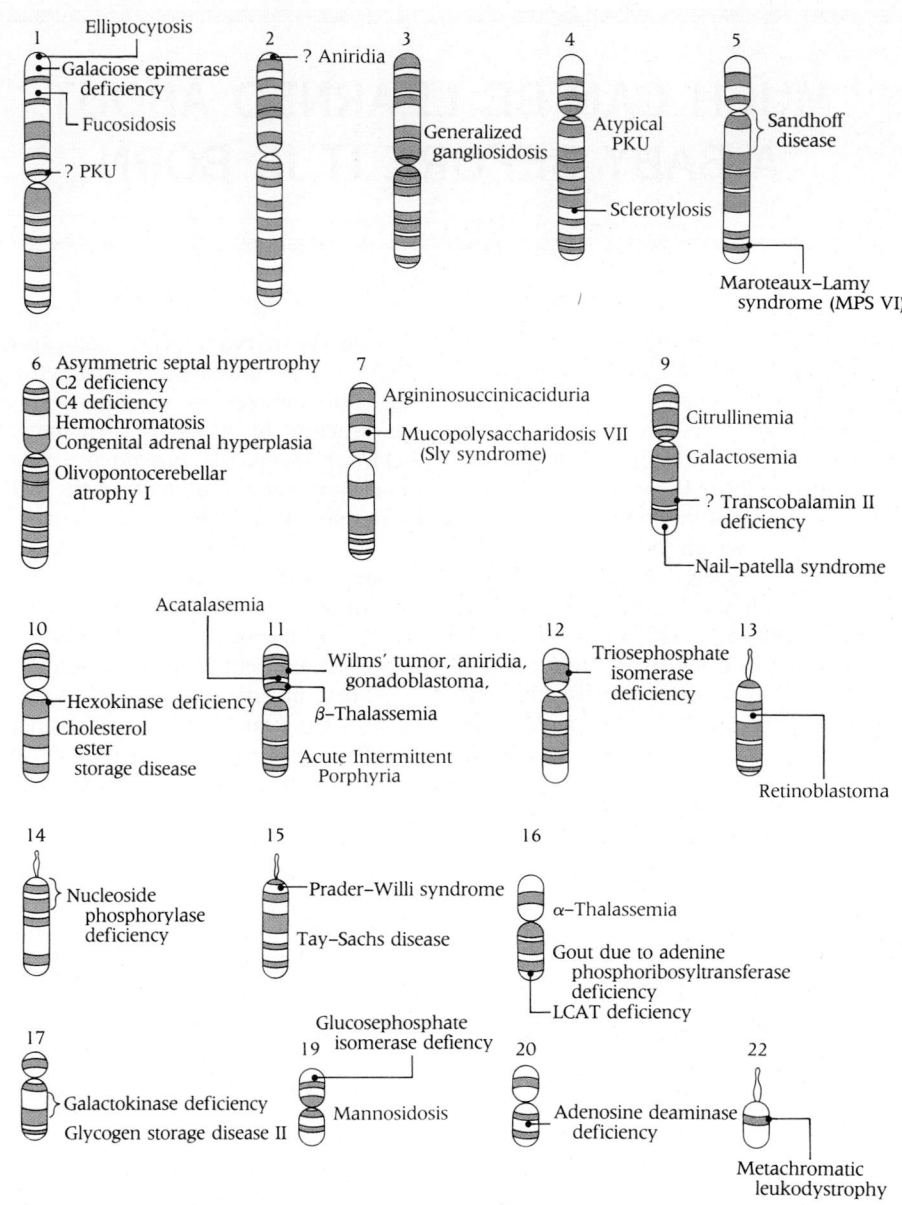

be located on the X chromosome. Thus, the assignment of specific genes to particular human chromosomes began with the study of sex-linked genes.

Ingenious investigators have found ways of doing such studies in selected cases. A recent study by Victor McKusick of Johns Hopkins University concerned several families that included women who were heterozygous for both colorblindness and glucose-6-phosphate dehydrogenase (G-6-PD) deficiency, the two well-known X-linked disorders we discussed earlier (Figure 15-21). Before attempting to judge the positions of the gene for colorblindness (*dn*) and that for G-6-PD deficiency (*gd*), the chromosome mapmaker would have to determine whether both aberrant genes are on the same X chromosome or on different X chromosomes. McKusick did that by examining the women's fathers. A father who phenotypically displayed both traits would have to have both genes on his single X chromosome. His heterozygous daughter would have this doubly affected X chromosome. If he showed just one of the traits, his daughter would have received from him an X that bore only that mutant gene. The other mutant gene would have to be on her other X chromosome, which came from her mother. Armed with knowledge of the fathers' patterns, McKusick could recognize recombinations if they occurred—that is, cases in which the sons had phenotypes different from those of their grandfathers. As it turned out, about 5% of the sons of these doubly heterozygous mothers showed trait combinations differing from those of their grandfathers and hence were "recombinants." On the basis of this "grandfather method," it was estimated that the *dn* and *gd* genes are about five "map units" apart on the X chromosome.

Advances in molecular biology have led to many practical benefits. One is our growing ability to diagnose genetic disorders in babies prenatally—that is, before birth. Some of these are serious diseases that affect millions of people throughout the world.

Structural malformations, such as anencephaly (in which the brain is partially or completely missing) or spina bifida (in which the vertebral column is open and the spinal cord is protected only by membranes), are now well correlated with high levels of a protein called α-fetoprotein in amniotic fluid and maternal blood. Enzyme deficiencies due to gene mutations are also identifiable in cultures of fetal cells obtained by amniocentesis, a safe and simple procedure (Figure 15C). Chromosome abnormalities, such as Down's syndrome, can now be detected during the mid-trimester of pregnancy by the analysis of fetal cells shed into the surrounding amniotic fluid.

Many disorders, however, are not readily detectable in fetal cells from amniotic fluid because the relevant genes are expressed phenotypically only in particular cell types. The best examples are conditions affecting the structure or synthesis of hemoglobin. Sickle cell anemia is a consequence of a common structural abnormality of the β-chain of human hemoglobin (see Chapter 22). A disabling disorder of wide occurrence, β-thalassemia, is a genetically determined underproduction of β-chains. It has been possible to diagnose these diseases prenatally by special biochemical methods since 1974.

At first, investigators performed diagnostic tests on fetal blood samples, since the β-chain gene is expressed—although at low levels—by the fourth or fifth month of pregnancy. This approach required (1) safe techniques for obtaining fetal blood samples (from placental veins) uncontaminated by maternal blood (Figure 15C), and (2) reliable methods for detecting β-chain production in tiny amounts. That these objectives were achieved is evident in experience surveys: in many cases 95% of blood samplings are technically successful, while the diagnostic error rate is less than 1%. The fetal loss rate—a worrying matter in the early years—fell steadily and is now below 5%. Such a risk is usually acceptable in families facing a one in four chance of having a severely affected child. These procedures, along with abortion and counseling, have sharply reduced the incidence of β-thalassemia in many areas.

With the advent of recombinant DNA techniques, it was recognized that β-thalassemia and sickle cell anemia might be diagnosed by examination of DNA from the cells in amniotic fluid rather than from fetal blood studies or from samples of chorionic villi from the placenta. Unlike methods that directly assess hemoglobin synthesis in fetal red cells, this approach seeks out alterations in DNA structure that are related to the condition being tested for. DNA analysis has two advantages: it is extremely sensitive and any fetal cell displays the defect. Hence, investigations can be done earlier in the pregnancy.

These considerations led to the first use

Step by step, other genes on the X chromosome are being related to these two. The *gd* gene, two genes for colorblindness (*dn* and *pn*), and the hemophilia gene (*h*) are now known to be quite close together. Other studies have shown that the overall length of the X chromosome is more than 200 map units. That length is sufficient to allow many pairs of genes, not closely linked on the chromosome, to become separated by a crossover, thereby producing X chromosomes carrying all sorts of recombinations of the traits being studied.

Other new methods for mapping human chromosomes include recombinant DNA methods, somatic cell hybridization, and various chromosome transfer techniques. Despite these advances, the human autosome map remains sketchy. However, plans are now afoot to sequence the DNA in the entire human genome—a project that will take years to complete and will cost billions of dollars. As shown in Figure 15-22, the evidence now available concerns many genetic diseases and a relatively small number of normal genes. Chromosome 1, for example, contains the genes for the Rh and Duffy (symbol Fy) blood group systems and for the PKU gene, which is discussed later in the chapter. Chromosome 11 carries the genes for insulin and for one of the polypeptide chains in hemoglobin.

PRENATAL DIAGNOSIS

Recent advances have made it possible to study the karyotype prior to birth through studies of the fluid that bathes the developing fetus within the amniotic cavity. The sac surrounding the fetus is lined by two layers of cells, the **chorion** and the

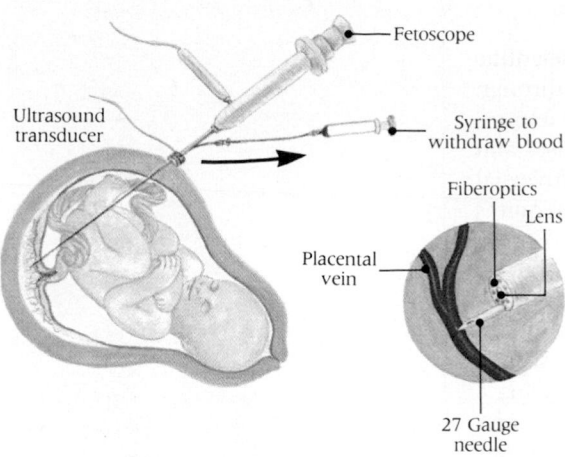

15C Obtaining a fetal blood sample
Ultrasound is employed to locate the placenta and fetus, and the needle is inserted using a lens and fiber optic system for visualization (see inset). Blood is drawn and prepared for genetic analysis.

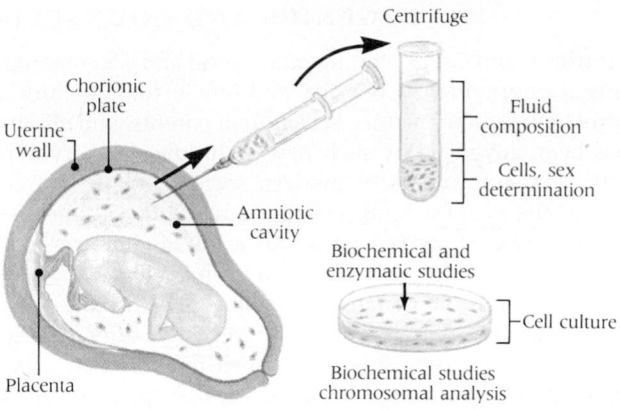

15D Amniocentesis
A sample is taken of the fluid surrounding the fetus and the fetal cells are tested for genetic defects. The optimum time for carrying out this procedure is about the sixteenth week of gestation.

of direct DNA analysis in prenatal diagnosis—by Yuet Wai Kan and Andree M. Dozy—in 1978. Their ingenious method employed restriction enzymes, which cleave DNA strands at specific nucleotide sequences. Thus it was possible to detect defects in 10 μg or less of DNA from fetal cells in amniotic fluid or their cultures (Figure 15D). At first, the effort was aimed at detecting the defective genes for sickle cell anemia, or β-thalassemia. This was difficult and a more reliable assay was required.

A remarkable finding provided the answer. Investigators had recently identified highly polymorphic regions of DNA in various parts of the genome. Hundreds, if not thousands, of such differing loci are scattered throughout the DNA. Two such regions flank the genes for the α-chain of hemoglobin. One or two are linked to the β-chain gene cluster. Family studies showed that when the β-chain is abnormal, as it is in β-thalassemia and sickle cell anemia, the more easily detected flanking sequences are often abnormal too. These sequences—and their changes—are like genetic fingerprints. The power of restriction enzyme mapping is such that a single base change in a restriction enzyme recognition site can now be detected easily.

The usefulness of this approach, called linkage analysis, has been limited by the fact that the sickle gene is linked to a polymorphic DNA sequence in only 60 to 70% of the cases—and by its variation in different populations.

amnion, and is filled with **amniotic fluid.** Suspended in the fluid are viable cells shed from the fetus's skin, respiratory tract, and other tissues. Removal of amniotic fluid by needle puncture, called **amniocentesis,** became a very safe procedure in the 1960s. Now one can study the karyotype in cells of an unborn infant in search of genetic diseases known to "run in the family." Such studies also permit diagnosis of the sex of a fetus from the presence of Barr bodies. Other prenatal studies that can be performed with cells obtained by amniocentesis are described in Box 15-2.

SOME IMPLICATIONS

Although still in an early stage, genetics has become a recognized branch of medicine and human geneticists are actively applying the insights of genetic science.

Such applications are of several types.

■ Cataloguing human genes
■ Establishing linkage groups and chromosome maps
■ Counseling of families with abnormal children on the probability that future children will be affected
■ Evaluation of consanguineous marriages for the likelihood of defective offspring

Let us briefly consider some of these issues.

EUGENICS AND COUNSELING

Eugenics (from Greek, *eus,* meaning good and *gen,* meaning born) refers to scientific efforts at improving the physical and intellectual qualities of human beings through control of hereditary factors, selection of parents, and elimination of defective fetuses. It has been advanced by such new techniques as karyotype analysis of fetuses and by the changing mores of modern society. It does, of course, raise controversial issues of the deepest kind, because any effort to "improve" our offspring requires someone to decide which particular array of characters is preferred.

Nonetheless, eugenics can play a useful role and its possible importance—along with knowledgeable genetic counseling, which aims only at assessing risks—may be illustrated with an example. The recessive abnormality phenylketonuria (PKU) occurs once in every 10,000 newborn infants (Caucasians and Orientals) with a carrier rate of 2 per 100.[8] It is much less frequent in blacks. A couple producing a child with this disorder would naturally wish to know the probability that their next child will also be affected. Often they assume that "lightning never strikes twice in the same place," that it is unlikely that a 1 in 10,000 chance would be realized again in their next child. Acting on this assumption, the couple has another child who, of course, has a high probability of being affected—since the ratio 1 : 10,000 refers to the incidence of PKU in a general population, not among the offspring of two parents, each of whom has been shown (by their first child) to possess the recessive PKU allele. Given the parental genotype, the chance that another child will be affected is 2500 times greater than it would have been had nothing been known of the parents' genetic constitution.

In contrast, consider the problem of Down's syndrome. The parents of a child with this disorder are of normal genotype. The abnormality of the affected child is the consequence of an abnormality of chromosome distribution—an autosomal nondisjunction in the mother—that occurs for unknown reasons. The frequency of the condition is a fraction of 1% in the children of women below 35 years of age, but it then rises steeply with advancing maternal age. Removal of amniotic fluid and karyotype analysis in high risk women can reassure many that their child will be unaffected. It will, at the same time, lead to recognition of those with the abnormal chromosome constitution.

THE QUESTION OF ABORTION

What can be done with the information acquired in this way? The liberalization of abortion laws has permitted therapeutic abortion in those cases in which the likelihood is great of a severely defective fetus. For example, the prenatal diagnosis of trisomy-21 (Down's syndrome) can be made with certainty.

Many believe that abortion is advisable in such cases. Extremely difficult questions arise, however, over the wisdom of sanctioning the abortion of all fetuses in which genetic defects have been recognized prenatally. It is far from clear where a line should be drawn. For example, should one terminate pregnancy in a serious X-linked recessive disorder if the fetus is female and thus only a carrier of the defective gene? If so, one would be terminating the development of a healthy fetus only because in later life she might have a greater chance of bearing an affected son. Another difficult question that arises: should a fetus with a treatable genetic disorder (e.g., PKU) be aborted to ensure that the defective allele is not passed on to future generations?

That choice is left to the individuals themselves. We do, however, counsel them about risks, the nature of the problem, and prospects for useful treatments and their costs.

Clearly, modern human genetics has raised profound questions for the future of humanity. If abortion is widely performed, will increasingly arbitrary standards be applied in decisions concerning the fitness of a given fetus and the desirability of its survival? Is there a moral justification for such life-death decisions? Will pressures develop for standards of desirability in genetically determined human

---

[8] This disorder must now be tested for by law in all newborn babies. Those with PKU appear outwardly normal at birth, but if the disorder is untreated, the infant during the first year of life gradually develops mental retardation, tremors, skin lesions, and other serious signs and symptoms. It can be treated with relative success by dietary restriction of the amino acid phenylalanine.

characteristics? If so, how will such standards be established and by whom? What should be the legal rights of the unborn fetus? Are parents to be held legally liable to their child in the future for allowing it to be born with a severe genetic handicap that could have been predicted?

Here is a cluster of profound problems that may be solved one day by the generation whose members are readers of this book. They are significant issues that affect our future and they are well worth your serious attention.

SUMMARY

The differentiation of individual organisms of a species into two different sexes is a notable illustration of the role of genes in development. In many species, the chromosomal endowments of males and females differ slightly, one of the two sexes possessing a pair of chromosomes that differ from one another in size and shape. These are the sex chromosomes. All other chromosomes are called autosomes. Many systems of sex determination exist. For example, sex can be determined by the ploidy of the zygote or by nongenetic, environmental factors.

The role of sex chromosomes varies in different species. In humans, the presence of a single gene normally found on the Y chromosome is sufficient to induce maleness, regardless of the number of X chromosomes present. In *Drosophila*, sex is determined not by the Y chromosome but by the ratio of X chromosomes to autosomes.

Not all sex-specific traits are determined by genes on the sex chromosomes. Often the gene for such a trait is autosomal, but it is expressed differently in males and females. Although genes on sex chromosomes (these genes are said to be sex-linked) may have roles unrelated to sex determination, the inheritance of sex-linked traits depends on the sex of the parent displaying the trait, as well as the sex of the offspring. In humans, alleles on the Y chromosome are expressed only in males. Recessive alleles on the X chromosome are often expressed in males, since a male receiving a recessive allele will be hemizygous, while most females receiving the allele will be heterozygous.

The dosage of a gene is the number of copies of that gene in each cell. The expression of many X-linked genes is independent of dosage. In humans, levels of the products of such genes are the same in males and females because all X chromosomes in excess of one are converted to inactive Barr bodies. Which of the two X chromosomes is preserved varies randomly from cell to cell within a normal human female. Thus, women heterozygous for X-linked traits are mosaics, some of their cells expressing one allele, the remainder expressing the other allele.

The study of human heredity is complicated by incomplete family trees, long generation times, and barriers to controlled breeding experiments. One of the best-studied human genetic systems is that determining surface antigens of red blood cells. Gross chromosomal abnormalities, detected by karyotype analysis, include translocations and deletions secondary to chromosomal breakage, as well as changes in chromosome number. However, this technique does not reveal either the incidence of gene mutations or the locations of genes. Whether or not a given gene is present, however, can be determined by identifying the gene's product. Often this can be done prenatally. The exact locus of a point mutation can be approximated by chromosomal mapping techniques with remarkable accuracy. DNA sequence studies and other techniques of molecular biology permit more precise positional assignments. Many difficult issues are raised by the discovery of a genetically abnormal fetus.

KEY TERMS

abnormal euploidy
acrocentric
agglutination
amniocentesis
amnion
amniotic fluid
aneuploidy
Barr body
deletion
dosage compensation genes
fragile site
genital sex
genetic sex

hemizygous
heterogametic
homogametic
intersex
inversion
karyotype
Lyon hypothesis
metacentric
mosaic
nondisjunction
sex chromosome
sex-influenced character
sex-linked inheritance

social (psychological) sex
somatic sex
superfemale
telocentric
testis-determining factor (TDF)
translocation
triploidy
X chromosome
XX-XO system
Y chromosome
Y-linked inheritance
ZZ-ZW system

QUIZ QUESTIONS

1. Sex-linked genes
   A. are located on the Y chromosome.
   B. are located on the X chromosome.
   C. are found on both X and Y chromosomes.
   D. carry information only for traits that are characteristic of that sex.
   E. C and D

2. In which of the following groups of animals are females homogametic?
   A. birds
   B. butterflies
   C. grasshoppers
   D. A and B
   E. None of the above

3. In the XX-XY system of sex determination, the sex of a fruit fly is determined by
   A. the presence of an X.
   B. the absence of an X.
   C. the presence of a Y.
   D. the absence of a Y.
   E. the ratio of X chromosomes to autosomes.

4. Which of the following is *not* a sex-linked condition?
   A. congenital ectodermal dysplasia
   B. hemophilia
   C. deuteranopia
   D. Down syndrome
   E. A and D

5. All of the following except _____ are sources of change in the genome.
   A. chromosomal translocation
   B. nondisjunction during meiosis
   C. gene mutations
   D. aneuploidy
   E. amniocentesis

ESSAY QUESTIONS

1. In humans, the presence of an intact Y chromosome determines the male phenotype. Does this mean that the trait of maleness is dominant? Explain.

2. At an infertility clinic, a phenotypical male is found to have Barr bodies in cells scraped from a mucous membrane. What pattern of sex chromosomes is present? A female patient at the same clinic lacks Barr bodies in her cells. Explain.

3. What genetic mechanisms give rise to chromosomal aberrations?

4. A man with a lifelong disorder and his normal wife have six children (three boys and three girls). All the girls (but none of the boys) have the same ailment as their father. Is the inherited allele most likely to be: autosomal recessive; autosomal dominant; X-linked dominant; X-linked recessive; or Y-linked? Explain your answer.

REFERENCES AND SUGGESTIONS FOR FURTHER READING

Hartl, D. (1983). *Human Genetics.* Harper & Row, New York.

A serviceable textbook pitched somewhat beyond the beginner's level.

Hoagland, M. (1981). *Discovery: The Search for DNA's Secrets.* Houghton Mifflin, Boston.

One of the discoverers of the role of mRNA vividly and compellingly offers a primer filled with the excitement and joy of discovery. An elegant delight.

Kevles, D. J. (1985). *In the Name of Eugenics: Genetics and the Uses of Human Heredity.* Alfred A. Knopf, New York.

A readable attempt to develop historical understanding of the eugenics movement written in the hope that it will guide us through the uncharted territory of genetic engineering.

Margulis, L., and Sagan, D. (1986). *Origins of Sex: Three Billion Years of Genetic Recombination.* Yale University Press, New Haven, Conn.

A stimulating work advancing a novel thesis on the evolution of sex.

Marx, J. L. (1988). Multiplying genes by leaps and bounds. *Science 240,* 1408–1410.

News item about a new gene amplification technique permitting investigators to synthesize millions of copies of a single DNA sequence in hours. It has already proved invaluable in studies of human heredity, molecular biology, and evolutionary biology.

Marx, J. L. (1989). Detecting mutations in human genes. *Science 243,* 737–738.

A brief discussion of the importance of new detection methods in assessing the genetic risks to humans of environmental chemicals and radiation.

Michod, R. E., and Levin, B. R. (Eds.) (1988). *The Evolution of Sex. An Examination of Current Ideas.* Sinauer/Blackwell Scientific, Boston.

Lucid essays on a central problem of evolutionary biology: the prevalence of maintenance of sexual reproduction. Explains the competing theories and suggests ways of choosing them.

Pengelley, E. T. (1978). *Sex and Human Life,* 2nd ed. Addison-Wesley, Reading, Mass.

A discussion of human sexuality that covers physiological as well as social issues.

Roberts, L. (1988). Genome project. *Science 242,* 1123. Carving up the human genome. *Science 242,* 1244–1246.

Two short notes on plans for determining the whole sequence of human DNA.

Stanbury, J. B., Wyngaarten, J. B., Frederickson, D. S., Goldstein, J. L., and Brown, M. S. (Eds.) (1983). *The Metabolic Basis of Inherited Disease,* 5th ed. McGraw-Hill, New York.

A classic work containing authoritative articles on human genetics and the known human genetic diseases.

Suzuki, D. T., and Knudtson, P. (1989). *Genethics: The Clash Between the New Genetics and Human Values.* Harvard University Press, Cambridge, Mass.

A provocative discussion of the medical potential of genetic engineering and the unease felt by many about human genetic manipulation.

Vogel, F., and Motulsky, A. G. (1987). *Human Genetics: Problems and Approaches.* Springer-Verlag, New York.

More a reference book than a textbook. A remarkable resource.

Weatherall, D. J. (1986). *The New Genetics and Clinical Practice.* Oxford University Press, New York.

The change in human genetics from a descriptive to a molecular science has occurred so rapidly that the discipline was left with an outmoded terminology and conceptual base. This work, by a distinguished colleague, revises the subject satisfyingly and concisely.

White, R., and Caskey, C. T. (1988). The human as an experimental system in molecular genetics. *Science 240,* 1483–1488.

This review of progress in the study of human genetics amply demonstrates the proliferation of new insights into the treatment and prevention of disease. This work also confronts society with challenging new ethical dilemmas.

White, R., and Lalouel, J. M. (1988). Chromosome mapping with DNA markers. *Scientific American 258,* February, 40–48.

On the use of new markers called restriction-fragment length polymorphisms to indicate the location on a chromosome of various genes.

Wilson, J. H. (Ed.) (1985). *Genetic Recombination.* Benjamin, Menlo Park, Calif.

A useful collection of 10 articles that appeared in *Annual Review of Genetics* and *Annual Review of Biochemistry* in the last decade. The main topics covered are meiotic, mitotic, and bacterial recombination. The level is advanced.

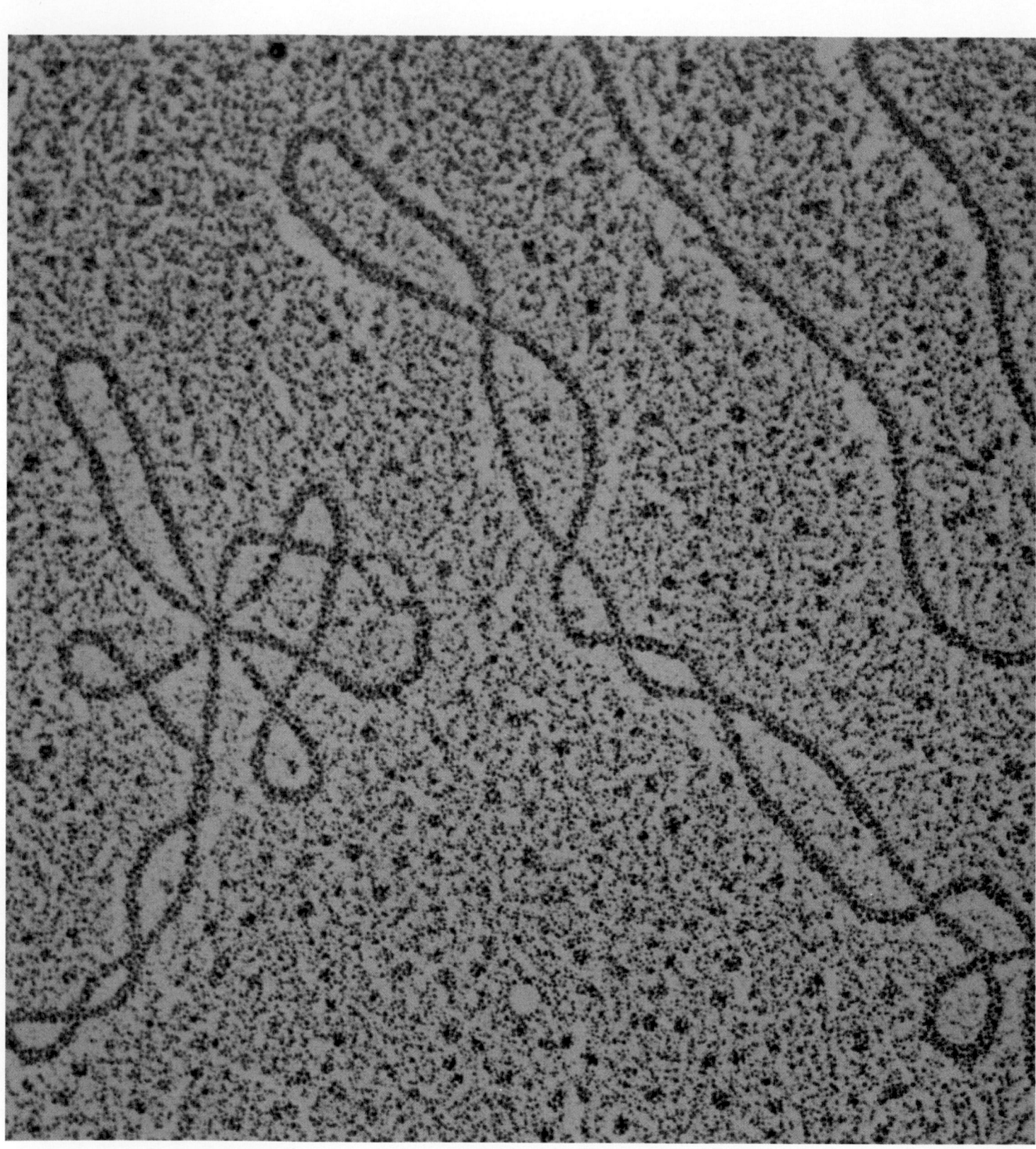

THE NATURE OF GENES

By the end of the 1930s, the first chapter of a powerful new science was all but written. The science was genetics—and the critical first chapter concerned the rules governing transmission of hereditary traits from generation to generation. Mendel's discoveries had been confirmed and extended by connecting the rules of heredity with the properties and behavior of chromosomes. The role of genes had been recognized and the difference between genotype and phenotype was understood. Finally, sophisticated mapping methods had placed individual genes in exact locations on many chromosomes.

These matters are usually placed collectively under the heading formal genetics or classical genetics to contrast them with molecular genetics, which came later and which now commands our attention.

WHAT IS A GENE?

While revealing nothing about the gene's physical nature, the data of formal genetics had nonetheless disclosed six of the gene's properties.

- Substantial stability from generation to generation
- A variable degree of mutability (despite extraordinary stability)
- A content of information that governs a cellular trait
- Alternative forms, or alleles, that produce distinct phenotypic effects
- Capacity for self-replication during cellular reproduction
- A specific locus along a linear sequence in or on a chromosome

The time had come to inquire, "What is a gene?"

The modern approach to that question, which we now consider, added broad new vistas to the science of genetics. Much of the research behind our current understanding of the nature of genes has been carried out since the 1960s. This work can illuminate fundamental issues in every biological domain.

We are concerned in this chapter with the following questions[1]:

- Of what does the genetic material consist?
- What are the physical and chemical properties of this material?
- How does it arise in metabolism?
- What form does genetic information take in genes?
- How is genetic information duplicated when cells divide?

As we discuss these questions, we should begin to reflect on how patterns of genetic information arose in the first place. The significance of the processes consid-

The nature of genes
The workhorse of prokaryotic genetic research has been the bacterium Escherichia coli. *In this remarkable electron micrograph, we see several isolated chromosomes from this organism. (×92,500) Like those of other bacteria, they are circular. To this time, more than 900 different genes have been located on the* E. coli *chromosome, which contains four million base pairs. Thus, it is huge compared to the plasmid (shown at the beginning of Chapter 18), which contains only a few thousand base pairs.*

[1] In Chapter 17, we will consider how a genetic "blueprint" causes a cell to display this or that phenotypic trait and what, in physical and chemical terms, occurs when a gene mutates. Chapter 18 will show how these activities are regulated.

ered in this chapter will be appreciated fully only in the broader context of biological evolution.

PHYSICAL NATURE OF GENES

SOME EARLY EVIDENCE

Early in the study of genetics it seemed that genes must be very small. The reasoning was as follows. The chromosome number characteristic of each species may be few, dozens, or hundreds, but mapping results show that single chromosomes contain many genes. Hence, the number of genes is very large and a gene must be considerably smaller than a chromosome. Thus, the mass of a gene is equal to (or smaller than) the mass of a chromosome divided by the number of genes it contains. Evidence that the gene is even smaller than this simple calculation implies came later.

Gene Size

A number of ingenious approaches permitted early indirect estimates of the size and number of genes in a cell nucleus. One technique employed a method similar to one that could be used to determine the area of a target at the far end of a dark room. If we fired a volley of small shot at such a target—and if we knew that the shot distributes itself evenly with one pellet per square inch in the target area—we could easily calculate the area of the target. If, for example, we found that the volley had hit the target an average of two times every firing, we could justifiably conclude (after repeated firings) that the target area is approximately two square inches. If we hit the target, on the average, only once every two times we fired a volley, we could conclude that the target area is about one-half square inch.

The size of a *Drosophila* gene was estimated in a similar way. Volleys of x-ray particles were used instead of bullets. The density of x-ray particles in a beam of radiation can be calculated with great accuracy. Early workers assumed that whenever a gene is hit it undergoes mutation, and thus produces a recognizably different fly in the next generation. These and other studies showed that a gene is very small, with an apparent length of 10 to 100 nm. Such approaches suggested that the four chromosomes of *Drosophila* contain at least 5,000 to 15,000 genes.

Gene Composition

The effort to determine what genes are made of began with intensive efforts to see them through the microscope. But even the most powerful electron microscope could not discern individual genes. Chromosomes in some tissues like the salivary glands of *Drosophila* are exceptionally large and display clearly defined cross-bands (see Figure 12-20C). Others show a linear sequence of the small bead-like swellings called chromomeres in the prophase of mitosis and meiosis. It was natural for early workers to suspect that these bands and swellings were individual genes, but crossing-over studies showed that both sorts of structures contain more than one gene.

Every clue made the gene seem a unique and remarkable entity. The stability of its information content through the generations suggested that its physical construction must be reasonably complex. If a single gene were made of many small molecules, it was argued, it would be difficult to imagine how its structural integrity could be maintained. If its information content had to depend on the average behavior of many small molecules, the risk would be high that their individual motions would rapidly and inevitably increase the disorder (entropy) of the system. An alternative view, according to such theorists as Schrödinger and Delbrück, was that the gene—whatever it might be—consists not of a collection of small molecules, but of one or a few enormous molecules. In that way, its stability, order, and permanence would be precisely determined and unvarying. This point is illustrated for amino acids and peptides in Figure 8-5. Genes would not be the by-products of statistical laws. It was a brilliant idea.

GENES ARE USUALLY (BUT NOT ALWAYS) MADE OF DNA

We now know that most genes are composed of **deoxyribonucleic acid,** or **DNA.**[2] The genes of some viruses are composed of **ribonucleic acid,** or **RNA.** In 1869, Friedrich Miescher (1844–1895), a young physician working in Tübingen, isolated a substance he called "nuclein" from the nuclei of white blood cells obtained from pus scraped from surgical bandages. Remarkably, considering the primitive methods then available, he recognized three properties of the molecules of this new and novel compound: they were acidic, rich in phosphorus, and quite large. He then turned to another starting material, sperm cells of the salmon from the headwaters of the Rhine, fishes for which the river in those days was celebrated. He chose these cells because their nuclei are unusually large. The nuclein from white cells was in fact a complex of DNA and protein, as was the material isolated from sperm cells. However, Miescher was able to convert the sperm cell material to pure DNA.

[2] The Germans call it DNS (Deoxyribonucleins) and the French call it ADN (acide deoxyribonucléique).

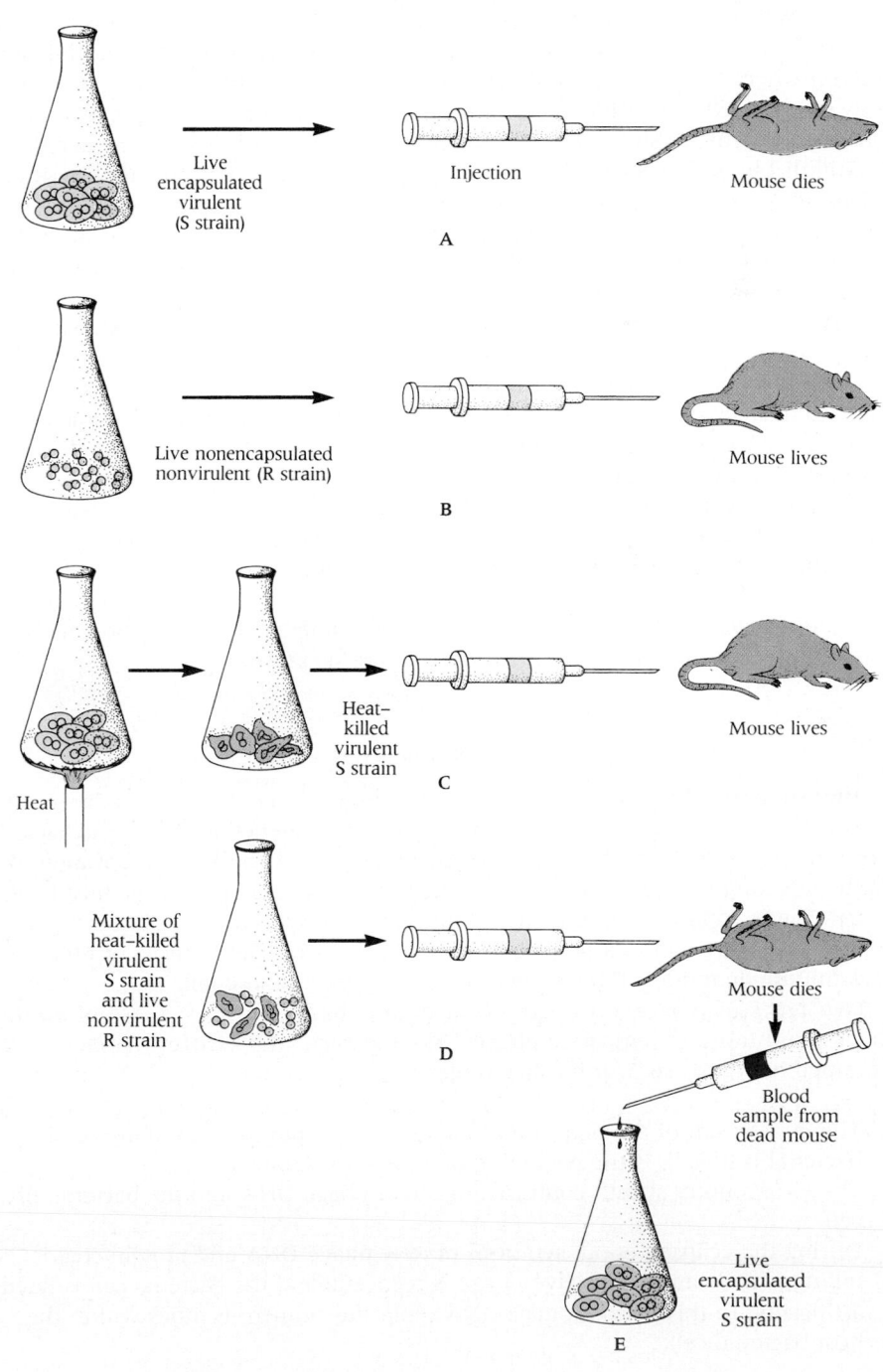

16-1 Discovery of the transforming principle

(A) Mouse dies following injection of live virulent cells (S strain). (B) Injection of live nonvirulent cells (R strain) into the mouse has no effect. (C) Injection of heat-killed virulent cells into the mouse has no effect. (D) When a mixture of heat-killed virulent cells and live nonvirulent cells is injected the mouse dies. (E) Blood samples taken from the dead mouse reveal live virulent cells, indicating that some material had been transferred from the dead virulent cells to the live cells, endowing these cells with the ability to cause pneumonia. This material was called the transforming principle and was later identified as DNA.

Live encapsulated virulent (S strain)   Injection   Mouse dies
A

Live nonencapsulated nonvirulent (R strain)   Mouse lives
B

Heat   Heat-killed virulent S strain   Mouse lives
C

Mixture of heat-killed virulent S strain and live nonvirulent R strain   Mouse dies
D

Blood sample from dead mouse

Live encapsulated virulent S strain
E

Like Mendel, his contemporary, Miescher failed to impress the scientific community. Indeed, this work was barely noticed for almost 75 years. For long years, most scientists believed that genes were made of protein. Since proteins are composed of varying combinations of 20 amino acids, they seemed to offer a very large number of possible configurations and thus rich opportunities for the storage in some form of genetic information.

Its presence in chromosomes suggested (but did not prove) that DNA is the message-carrying genetic material. However, a series of epoch-making discoveries that began in the 1940s soon provided the needed proof. It was the converging results of four major experimental approaches that established DNA as the genetic material. These concerned: (1) transformation, (2) viral replication, (3) transduction, and (4) conjugation. Let us examine them briefly.

Transformation of Bacteria

In 1928 Fred Griffith, a physician working in the Pathology Laboratory of the British Ministry of Health in London, obtained some startling results for which he had no explanation. His observations concerned the pneumonia-causing bacterium, *Streptococcus pneumoniae* (also called pneumococcus), which occurs in two forms recognizable by their distinctive colonies in culture: smooth (S) and rough (R). The difference between them is due to the different polysaccharides in the bacterium's capsule. Only smooth cells are virulent (i.e., able to produce pneumonia in animals). Rough cells are nonvirulent. Since each form "breeds true," capsule properties are hereditary.

Griffith knew that S cells injected into mice produced fatal pneumonia; R cells did not (Figure 16-1). But when living R cells (which alone were harmless) were first mixed with heat-killed S cells before injection into mice, all the mice died. He concluded that somehow the R cells had been transformed. Griffith died before the answer was revealed.

That discovery was made sixteen years later when O. T. Avery, C. M. MacLeod, and M. McCarty of the Rockefeller Institute showed that R cells could be transformed to S cells by exposing them to a cell-free extract of S cells. When cells so transformed were cultivated, their offspring were S cells. Therefore, the altered character was hereditary. The Avery group, however, discovered that the active **transforming principle** in the cell-free extracts was the DNA of the S cells.

Thus it appeared that donor cell DNA somehow entered the chromosome of a recipient cell, where it behaved like a new gene. For the first time DNA was shown to be the conveyor of genetic information. This landmark discovery, made when Avery was 67 years old, all but proved that genes consist of DNA—and, significantly, that genic DNA differs from the product it determines. Once again, however, few noticed this work. Of those who did, many were skeptical.

Viral Replication

Further support for the idea that DNA is the carrier of hereditary information came from studies in the late 1940s and early 1950s of the genetics of viruses and bacteria. The virus studied with particular intensity was one of the **bacteriophages** that live by infecting *E.coli* and that often disintegrate, or lyse, their bacterial host (Figure 16-2). These "phages," as they are commonly known, are to molecular genetics what *Drosophila* was to early genetics (see Figure 5-5). They consist of little more than minute quantities of DNA, an outer protein coat, and a tail.

Two types of replication cycles occur among bacteriophages: those of virulent phages, and those of temperate phages. The life cycle of a **virulent phage** is diagrammed in Figure 16-3. It has five stages.

- The attachment of the phage tail to a specific receptor, one of many on the bacterial cell wall, is the beginning of a virus infection.
- The virus's outer sheath contracts, injecting phage DNA into the bacterial host cell.
- During the eclipse period, synthesis of new phage DNA and protein coat is taking place, and no infective phage is recoverable if the bacterial cell is lysed artificially. In this period, phage DNA replicates numerous times within the host bacterial cell.

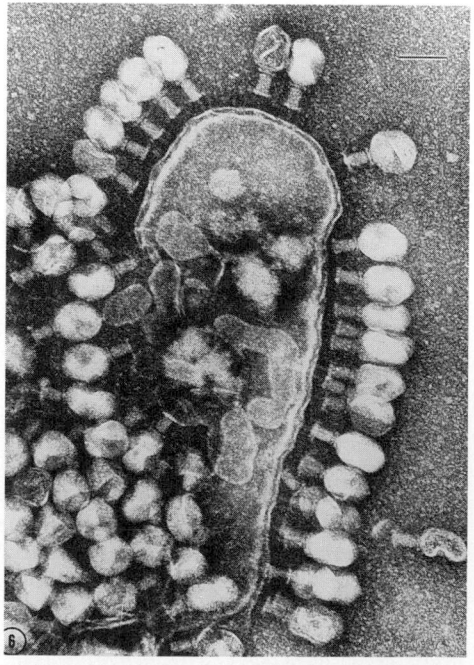

16-2 Bacteriophage T4 particles attached to outer surface of E. coli
An electron micrograph showing multiple bacteriophages attached to receptors on the surface of the cell's outer membrane. Some of the viruses have injected their DNA into the cell, as revealed by their empty heads. The entire replication cycle takes only 20 minutes. (×40,000)

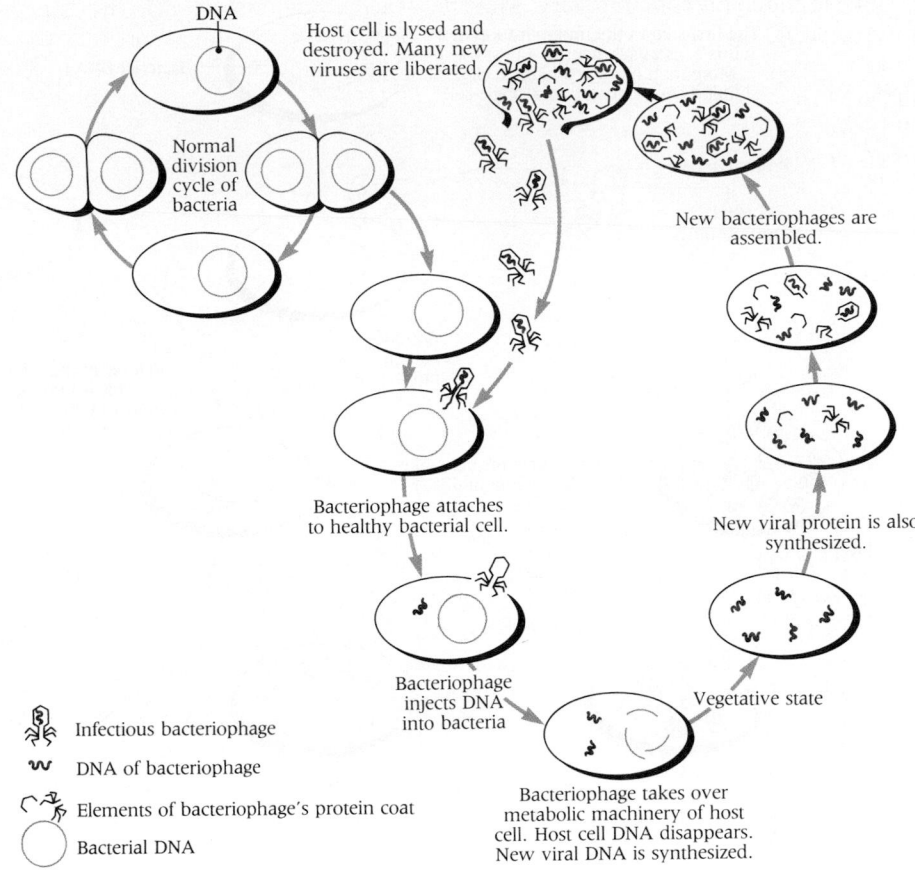

16-3 Life cycle of a virulent phage
When phage DNA enters the host cell, complete copies of the virus are produced. These eventually cause lysis of the cell and release of a new generation of viruses.

DNA

Host cell is lysed and destroyed. Many new viruses are liberated.

Normal division cycle of bacteria

New bacteriophages are assembled.

Bacteriophage attaches to healthy bacterial cell.

New viral protein is also synthesized.

Infectious bacteriophage

DNA of bacteriophage

Elements of bacteriophage's protein coat

Bacterial DNA

Bacteriophage injects DNA into bacteria

Vegetative state

Bacteriophage takes over metabolic machinery of host cell. Host cell DNA disappears. New viral DNA is synthesized.

- Assembly of phage DNA into new protein coats begins toward the end of the eclipse period. Phage DNA directs the production of protein coats and the assembly of some 50 to 200 new infective viral particles.
- Lysis of the host cell and release of several hundred new infective phage particles occurs within a short time after infection (only 13 minutes for phage T1, 22 minutes for phage T2). A phage-produced enzyme, lysozyme **(muramidase)**, brings about lysis of the host cell wall and release of mature phages.

Virulent bacteriophages helped establish that genes are made of DNA. In an ingenious experiment (Figure 16-4), Alfred D. Hershey and Martha Chase used

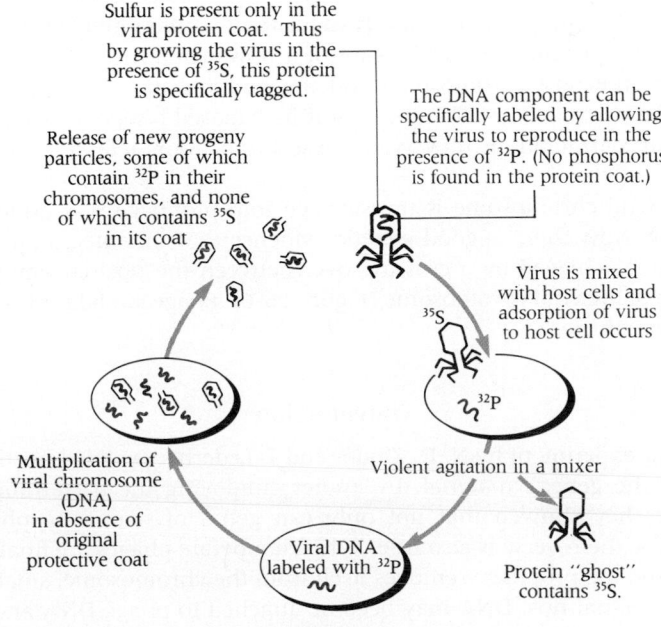

Sulfur is present only in the viral protein coat. Thus by growing the virus in the presence of ^{35}S, this protein is specifically tagged.

The DNA component can be specifically labeled by allowing the virus to reproduce in the presence of ^{32}P. (No phosphorus is found in the protein coat.)

Release of new progeny particles, some of which contain ^{32}P in their chromosomes, and none of which contains ^{35}S in its coat

Virus is mixed with host cells and adsorption of virus to host cell occurs

^{35}S

^{32}P

Violent agitation in a mixer

Multiplication of viral chromosome (DNA) in absence of original protective coat

16-4 The Hershey-Chase experiment
Bacteriophage DNA is labeled with ^{32}P and the protein coat with ^{35}S. Following infection, only the DNA enters the host cell to serve as the genetic material that guides the production of new viruses.

Viral DNA labeled with ^{32}P

Protein "ghost" contains ^{35}S.

CHAPTER 16   THE NATURE OF GENES

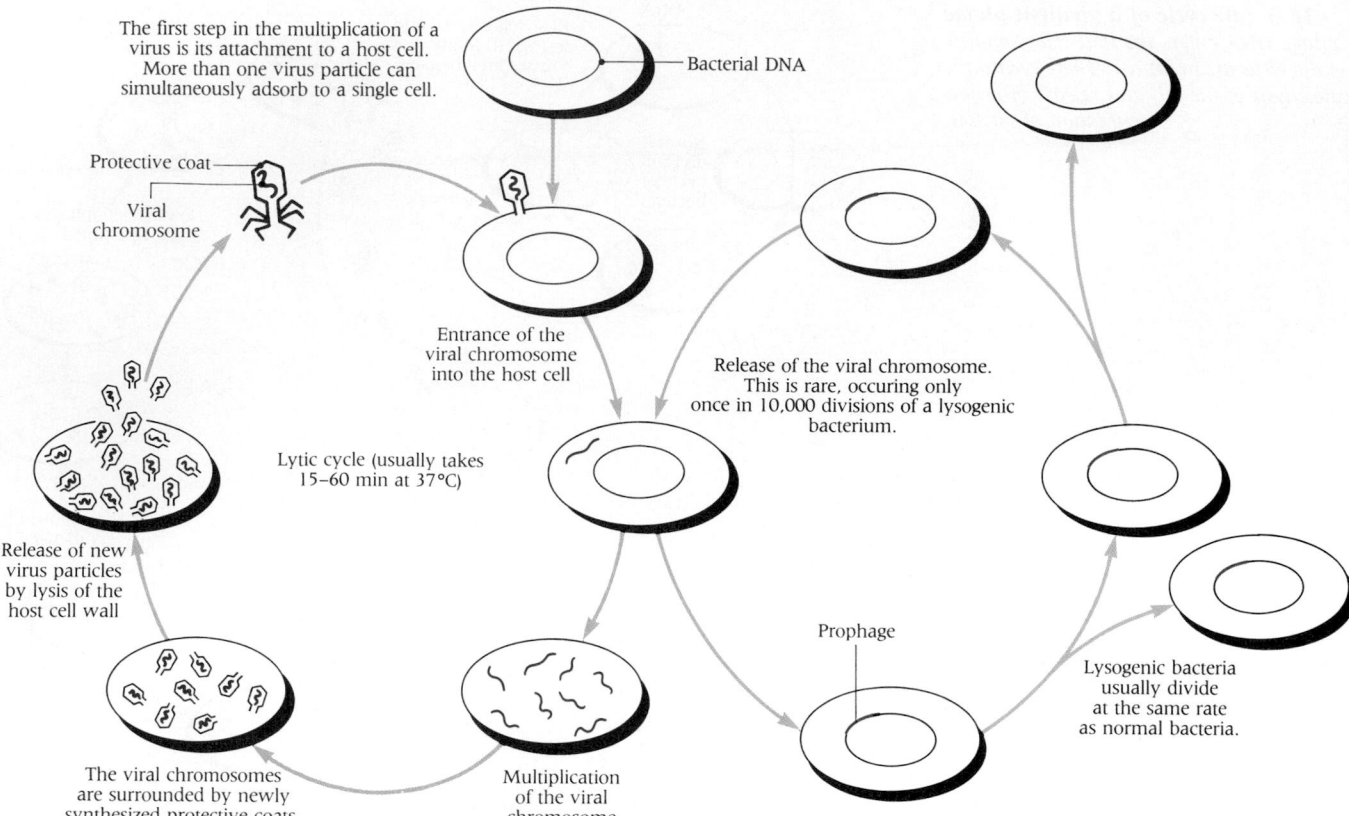

The first step in the multiplication of a virus is its attachment to a host cell. More than one virus particle can simultaneously adsorb to a single cell.

Bacterial DNA

Protective coat

Viral chromosome

Entrance of the viral chromosome into the host cell

Release of the viral chromosome. This is rare, occurring only once in 10,000 divisions of a lysogenic bacterium.

Lytic cycle (usually takes 15–60 min at 37°C)

Release of new virus particles by lysis of the host cell wall

Prophage

Lysogenic bacteria usually divide at the same rate as normal bacteria.

The viral chromosomes are surrounded by newly synthesized protective coats.

Multiplication of the viral chromosome

the radioactive isotopes ^{32}P and ^{35}S to prove that when phages infect a host cell, only their DNA enters the bacterial cell. The protein coat of the bacteriophage remains outside. Inside the host cell, it is the virus's DNA that takes over the cell's metabolic machinery, literally commandeering it to copy phage DNA again and again and to follow its instructions for producing new viruses. The host cell is finally destroyed in the end. This experiment confirmed the conclusion of the pneumococcus transformation experiment that DNA alone carries genetic information—and was the death knell for the theory that genes are made of protein.

Unlike virulent phages, **temperate phages** do not ordinarily lyse their host. Temperate phage DNA becomes integrated into the DNA of the bacterial host, where it becomes a **prophage** (Figure 16-5). A prophage replicates synchronously with the host DNA. Moreover, its genes recombine with host cell genes. All progeny of the bacterial host cell will contain this bacterial-plus-phage DNA.

Bacteria containing prophages are **lysogenic.** It is often difficult to know when a bacterium is lysogenic: we can be sure of it only when the virus chromosome is released from the host chromosome and the multiplication of new viruses commences. In 1954, André Lwoff found that if he exposed lysogenic bacteria to ultraviolet light or x-irradiation, they would lyse in an hour, releasing many virulent phages.

How the viral chromosome is transformed into prophage was quite mystifying until recently. Now there is good genetic evidence that the integration of the viral chromosome is achieved by a crossing-over between the host chromosome and a circular form of the viral chromosome (Figure 16-6). Phage lambda (λ) is an example of a temperate phage.

Transduction

A classic 1951 experiment by N. D. Zinder and J. Lederberg yielded further evidence that DNA is the genetic material. In studies employing the bacterium *Salmonella typhimurium,* they showed that not only can genes of temperate phages act as bacterial genes, the reverse is also true. When temperate phages are finally activated to replicate and the prophage removes itself from the chromosome, small fragments of adjacent bacterial host DNA may become attached to phage DNA and be carried

16-5 Life cycle of a temperate phage
When the DNA of a temperate phage enters the host cell, it may follow either of two pathways. (Left) In the lytic cycle, the virus DNA rapidly multiplies and new viruses are released upon lysis of the host cell. (Right) In the lysogenic cycle, the virus DNA is incorporated into the bacterial chromosome and is called a prophage.

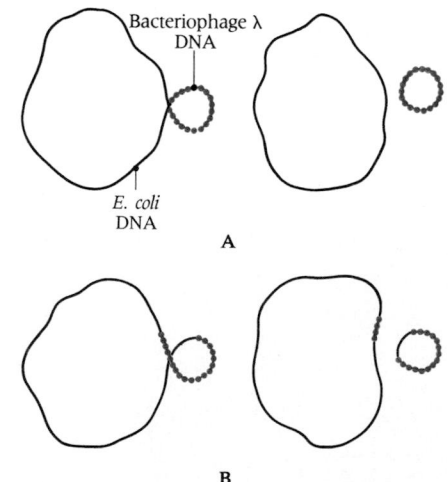

Bacteriophage λ DNA

E. coli DNA

A

B

16-6 Integration of virus DNA into bacterial DNA
(A) Sequence illustrates that circular virus DNA is the form required for exchange of genetic information between bacterial chromosome and virus. (B) Sequence shows a faulty attachment resulting in some exchange of DNA between prophage and bacterial chromosome.

along. If a temperate virus carrying a bit of bacterial host DNA in this manner infects a new host cell, it injects both viral and bacterial DNA into this new host. Sometimes the injected bacterial DNA recombines with the new host's genes. As a result, it may confer a new property on the new host. The process was called **transduction** (Box 16-1).

This work incidentally provided a practical new laboratory tool. A method of repairing genetic defects by gene transfer is today a major goal of cell biologists, one that raises visions of medical miracles. The problem is to pick out the few cells that acquire a new gene from the many that do not. For that reason, investigators have concentrated on genes that confer a selective advantage on recipient cells, thereby permitting them to grow under conditions detrimental to cells lacking that gene. The first undisputed success in this approach to gene transfer was achieved in 1977 by R. Axel, S. Silverstein, and M. Wigler of Columbia University who transferred a gene (*tk*) for the enzyme thymidine kinase, which they isolated from a virus called herpes simplex, into animal cells lacking the *tk* gene. They could tell that some cells had acquired a new gene because they had acquired a new ability to grow in a medium that required them to convert thymidine to thymidine phosphate. This demanded the presence of thymidine kinase. Through a virus-mediated process, the cell had acquired new instructions for producing the enzyme needed to metabolize thymidine.

Conjugation: The Mating of Bacteria

One more body of work implicating DNA as the genetic material rested on the surprising finding that sexuality exists among bacteria. We learned in Chapter 12 that bacteria reproduce by simple fission. If this were their only reproductive method, then the only way they could acquire new gene combinations would be by mutation, since the recombinations arising from sexual reproduction would be precluded.

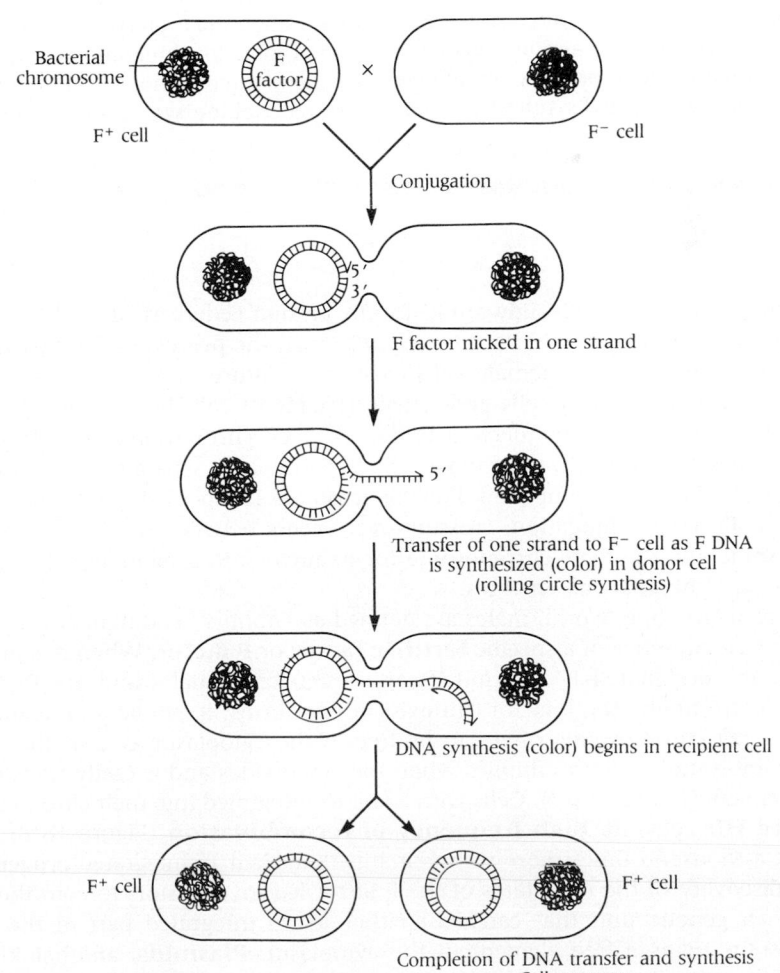

16-7 Conjugation in bacteria
F factor DNA of donor F$^+$ (male) cells transfers to an F$^-$ cell. (The size of the DNA is exaggerated here.) One F factor DNA strand is cleared and replication occurs in a recipient F$^-$ (female) cell.

BOX 16-1

HOW VIRUSES TRANSFER HOST DNA FROM CELL TO CELL

The discovery of transduction was announced in 1951 by Joshua Lederberg, then a young investigator at the University of Wisconsin, and Norton D. Zinder, his graduate student. It was one of those elegant discoveries that cause some people to speak of the beauty of science, yet it stemmed from a lucky accident.

In the months following André Lwoff's first paper on the induction of lysogenic bacteria, bacterial mating had thus far been demonstrated only in *E. coli.* It was the aim of Zinder and Lederberg to see if mating could be induced in two strains of *Salmonella typhimurium,* an intestinal bacterium that gives mice something resembling typhoid fever. Each of the strains lacked the capacity to synthesize one particular amino acid that it needed for growth, but each strain could make the amino acid the other strain was unable to make. The two amino acids were methionine and threonine. One strain, we will call it *met⁻thr⁺*, could synthesize threo-

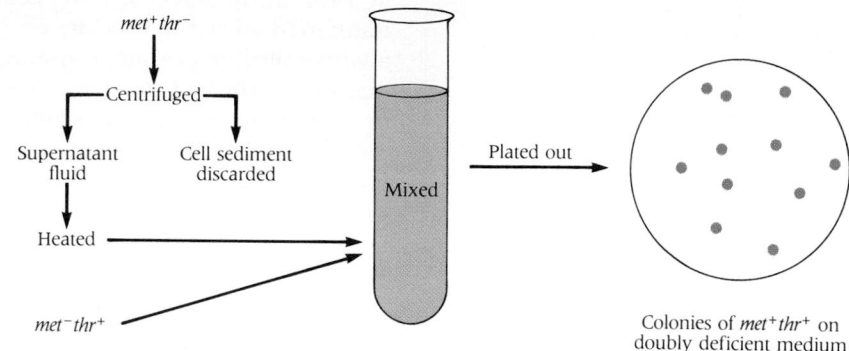

16A The production of double prototrophs is not based on conjugation
Modification of the original Zinder-Lederberg experiment in which only cell-free extract of one auxotroph is added to a culture of the other auxotroph. Colonies of double prototrophs develop on doubly deficient medium. This shows that their occurrence is not based on conjugation.

nine but was unable to synthesize methionine from simple precursors. Hence it was prototrophic for threonine and auxotrophic for methionine—and it could not grow unless methionine was added to the culture

medium. The other strain, *met⁺thr⁻*, was prototrophic for methionine but auxotrophic for threonine. In each experiment, the *met⁻thr⁺* and *met⁺thr⁻* strains were first grown separately. The idea was to place them

---

But in the 1940s and 1950s, Edward L. Tatum, Joshua Lederberg, and others established a primitive sexuality in bacteria. They termed the process **conjugation.**

For example, male and female cells exist in a culture of *E. coli.* As shown in Figure 16-7, male (donor) cells and female (recipient) cells become attached and form a little cytoplasmic connecting bridge. A male chromosome then begins to move through the bridge to the female cell. Often the cells separate before chromosome transfer has been completed. For this reason, a complete diploid cell is rarely formed. Following conjugation, crossing-over occurs between the female parental chromosome and the male chromosome (or fragment). Recombination occurs and a new kind of haploid cell emerges.

In this microscopic world, males are defined as "donors" and maleness is determined by the presence of a specific **fertility factor** or **F factor.** When it is present, the cells are designated F⁺. Recipients are designated females and are F⁻ (no F factor). Interestingly, the F factor can exist in two forms: it can be an integral part of the *E. coli* chromosome, or it can be free in the cytoplasm as a small circular free chromosome, which multiplies when the cell divides and is easily transferable to F⁻ recipients (Figure 16-7). Cells with F factors integrated into their chromosomes are called **Hfr** cells, for **high frequency of recombination** (Figure 16-8). They produce a thousand times more recombinants than their unintegrated progenitors.

The discovery of the two states of the F factor led investigators to coin the term episome, a genetic unit that can exist either as an integrated part of the main chromosome or as a free element in the cytoplasm. **Plasmids,** another kind of cytoplasmic genetic element, do not integrate into the main chromosome ring. As

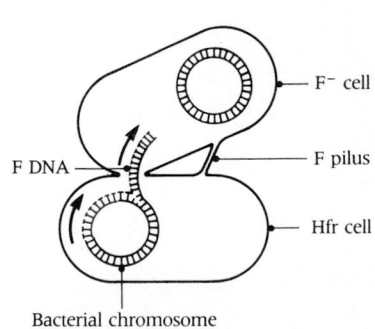

16-8 Conjugation in Hfr E. coli
Transfer of the bacterial chromosome of an Hfr cell to an F⁻ cell. The chromosome is led by a segment of F factor DNA.

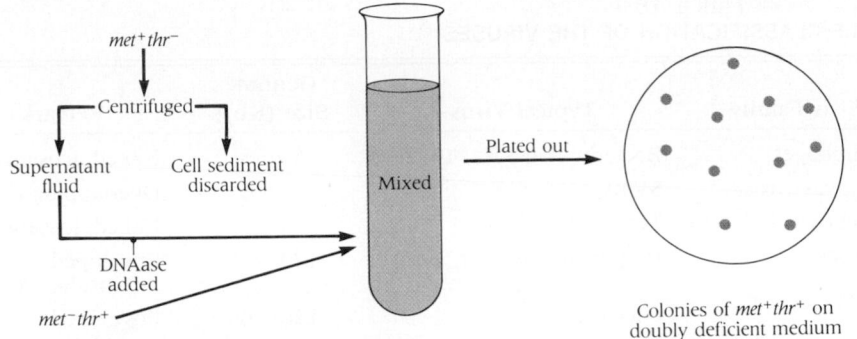

met⁺thr⁻ → Centrifuged → Supernatant fluid / Cell sediment discarded

DNAase added

met⁻thr⁺

Mixed

Plated out →

Colonies of met^+thr^+ on doubly deficient medium

16B The production of double prototrophs is not the result of transformation
Modification of the original Zinder-Lederberg experiment in which DNAase is added to the cell-free extract of one auxotroph. The continued appearance of double prototroph colonies on doubly deficient medium shows that they are not the result of transformation.

together. It was predicted that if mating occurred some bacterial offspring should appear that could make both methionine and threonine—that were met^+thr^+—and that would grow on a medium containing no added amino acid.

In the first experiment, the investigators mixed the two cultures and then plated them out on a medium lacking methionine and threonine. To their pleasant surprise, many colonies began to grow, and they felt certain they had demonstrated mating in *Salmonella.* But when they tried to mate strains that differed by more than one trait, the results were puzzling. If mating were actually taking place between these strains, new cells should appear that combined genes from the two parents in various proportions. Some would be more like one parent and others would be more like the other. But the observed progeny all resembled one parent, except for the single trait by which they were isolated and which was supplied by the other parent. Only one trait was being transferred at a time and this suggested, not mating, but something else. Random mutation seemed unlikely because the change appeared much more frequently in mixed cultures of the parent strains than in unmixed controls. Other experiments then ruled out (1) conjugation, by showing first that direct contact between the two parental strains was not necessary to get double prototrophs and second that a cell-free extract of one auxotroph could

substitute for it (Figure 16A); and (2) transformation, by showing that free DNA was not involved (Figure 16B).

A key experiment yielded the answer. After finding that recombinant bacteria continued to appear when all contact was barred between parental strains, Zinder and Lederberg found that an agent—a so-called "met⁺ factor"—could pass through filters of a type known to pass viruses but hold back bacteria. The factor was not naked DNA—and step by step Zinder and Lederberg proved that it had to be a temperate bacteriophage, later named P22. The *Salmonella* strains turned out to be lysogenic and the phage particles they harbored could occasionally be induced to come forth. They carried with them a small assortment of host genes that went along into other bacterial strains which they infected lysogenically. Like all phages, these are particular about which host cells they will infect. That is why the reciprocal experiment—in which a cell-free filtrate of met^-thr^+ was used in an effort to convert met^+thr^- cells to met^+thr^+—failed to produce prototrophs.

Since that time other investigators have succeeded in transferring all sorts of genetic traits from one cell to another by transduction. These traits have included drug resistance, motility factors, and factors related to immunogenicity (i.e., the capacity to elicit antibodies). Transduction has been demonstrated in many bacterial species. No one, however, has yet succeeded in mating members of the *Salmonella* group.

we shall see, episomes and plasmids are the key elements of genetic engineering.

With the recognition that the material transferred during bacterial conjugation is DNA, the evidence favoring DNA as the genetic material became irresistible. New data accumulated rapidly in support of this generalization.

SOME VIRAL GENES ARE MADE OF RNA

There is an exception to the generalization that DNA is the genetic material. In several types of viruses the genes consist of RNA. Table 16-1 lists the major classes of viruses. Note that they include the causal agents of such dreaded human diseases as poliomyelitis and AIDS, and agents that induce such plant diseases as tobacco mosaic. They also include certain forms that have become powerful tools of the molecular biologist, among them SV40, a valuable adjunct to plasmid research, and retroviruses.

Tobacco mosaic virus (TMV) has a place in history. In the 1950s, the application to a tobacco leaf of pure RNA isolated from TMV was shown to cause new whole TMV viruses to be produced within the leaf cells. Purified TMV protein on the other hand elicited no new viruses. Prior to this experiment, real doubt existed as to whether RNA would function as a carrier of genetic information.

We will account for the function of RNA genes—after we explain how DNA genes work. Before we do either, we must say more about the DNA and RNA molecules themselves. We have spoken of DNA as the message-carrying molecule or "information tape" that remains unchanged from generation to generation in

TABLE 16-1
A BRIEF CLASSIFICATION OF THE VIRUSES

Genetic Material	Character of Strands*	Host	Virus Family	Typical Virus	Genome Size (Kb)†	Remarks
DNA	Double	Bacteria	Coliphages	Bacteriophages T2, T4, T6	Varies	Naked, icosahedral
		Primates	Papovavirus	SV40	5.8	Overlapping genes
		Vertebrates	Adenovirus	Types 12, 18	35–40	Naked; icosahedral
			Herpes	Herpes simplex	120–200	Enveloped, icosahedral
			Pox	Vaccinia	120–300	Enveloped, complex
	Single	Bacteria	Coliphages	Coliphage φX174	5.4	
				Phage lambda	47	
		Vertebrates	Parvovirus	Canine parvovirus	1–5	Naked; icosahedral
RNA	Double	Vertebrates	Reoviruses	Diarrhea virus	18–30	Naked; icosahedral
	Single					
	−		Rhabdovirus	Vesicular stomatitis virus	12–15	
	+	Primates	Picornavirus	Poliomyelitis virus	7	
		Bacteria		Phage MS2	3.5	
	−	Mammals	Orthomyxovirus	Influenza virus	14	
	−	Mammals	Paramyxovirus	Newcastle disease virus	15	
		Plants	Plants	Tobacco mosaic virus	6.4	
RNA/DNA		Vertebrates	Retrovirus	Rous sarcoma, HIV	5–10	

* A plus (+) strand is a molecule of viral RNA that is functional—that is, capable of serving as mRNA (see Chapter 18). A minus (−) strand, its complementary sequence, cannot function as mRNA. Similarly, a DNA strand complementary to plus strand viral mRNA is a minus strand. A plus strand of RNA (i.e., mRNA) can be produced only on a template of minus strand RNA or DNA.

† In general 1 kb (kilobase) of DNA has a molecular weight of about 7000.

bacteria, in viruses, and presumably in all organisms. How can a single kind of molecule carry all the genetic information in all these many different cases? The answer is that all DNA is not the same.

ON DNA AND RNA

Nucleic acids are among the largest molecules in living organisms. Like proteins, they are polymers made up of repeating low molecular weight monomeric units. Unlike proteins, which may be viewed as long strings of words written with a 20-letter alphabet, nucleic acids contain words written with a 4-letter alphabet. *Word* and *letter* are not idle metaphors. The information content of a real word is determined by its letters and their sequence. Exactly the same statement can be made about proteins and nucleic acids. Both are long polymers constructed from components that, with minor exceptions, are identical in all living species.

TABLE 16-2
MOLECULAR COMPONENTS OF DNA AND RNA*

	DNA	RNA
Bases		
Purines	Adenine (A, Ade) Guanine (G, Gua)	Adenine (A, Ade) Guanine (G, Gua)
Pyrimidines	Cytosine (C, Cyt) Thymine (T, Thy)	Cytosine (C, Cyt) Uracil (U, Ura)
Sugars	Deoxyribose (d, dR)	Ribose (r, R)
Phosphoric acid	Phosphoric acid (P)	Phosphoric acid (P)

* Letters in parentheses are standard abbreviations. The 3-letter abbreviations for bases have been officially recommended, but are rarely used.

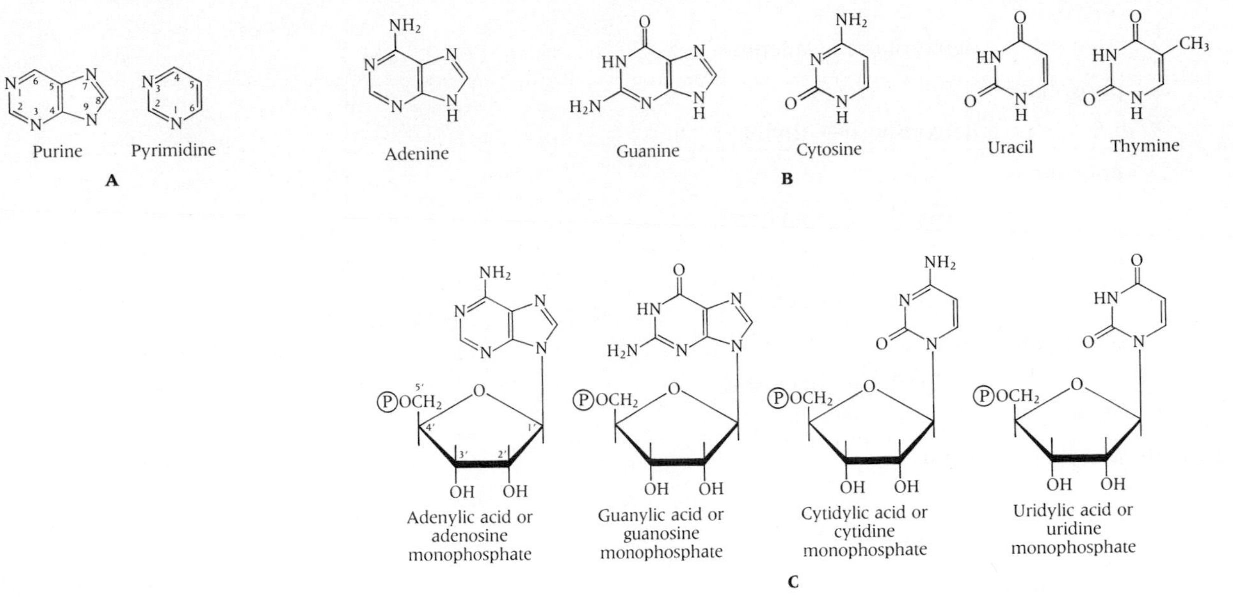

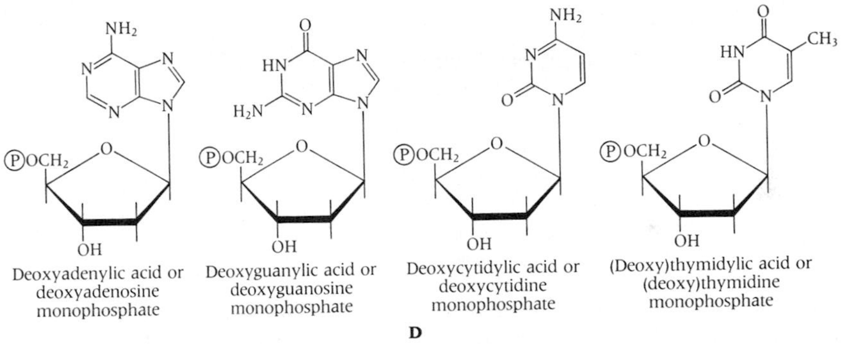

16-9 Chemistry of nucleic acids
(A) *Purines and pyrimidines are ring structures; the position of each carbon or nitrogen is identified by number.* **(B)** *The basic ring structures are modified by the addition of chemical groups to produce the five nitrogenous bases.* **(C)** *Ribonucleotides contain the sugar ribose and a phosphate group attached to the base.* **(D)** *Deoxyribonucleotides contain deoxyribose, which links the phosphate group to the nitrogenous base. Note that primed numbers (for example, 2') are used to identify positions in the sugars.*

NUCLEOTIDES

The four monomeric units of nucleic acids (or "letters") are **nucleotides.** Hence, a nucleic acid is a **polynucleotide.** In cells, a nucleic acid, or polynucleotide, is usually attached to a protein; the complex is a **nucleoprotein.** Nucleotides have importance beyond their role as links in nucleic acid chains. As we learned in Chapter 8, many coenzymes (e.g., ATP, NAD$^+$, NADP$^+$, and FMN) are nucleotides.

In a nucleotide, three simpler molecules are linked: an organic **nitrogenous heterocyclic base** (that is, a ring structure made up of carbon and nitrogen atoms), a **sugar,** and a **phosphate group.** As shown in Table 16-2, nucleic acid nucleotides (as opposed to coenzyme nucleotides) include bases called **purines** (**adenine** and **guanine**), and **pyrimidines** (**cytosine** and **uracil** or **thymine**). The nucleotide sugars differ in a small but decisive way. The sugar in RNA nucleotides is **ribose;** the sugar in DNA nucleotides is **deoxyribose.** Both are five-carbon aldehyde-type sugars, or aldopentoses. The difference between them is small but significant: deoxyribose lacks the oxygen atom attached to the number 2 carbon of ribose (Figure 16-9).

There is another crucial difference between RNA and DNA. Three of the four bases—adenine, guanine, and cytosine—occur in both RNA and DNA. However, the fourth base in RNA is the pyrimidine uracil. In DNA it is thymine, a methyl derivative of uracil. Respectively, these bases are abbreviated as **A, G, C, U,** and **T** in one convention and **Ade, Gua, Cyt, Ura,** and **Thy** in another.

POLYNUCLEOTIDE CHAINS

The structure of a DNA strand may be represented as a long chain of the following kind, the sequence of the bases differing in different DNAs:

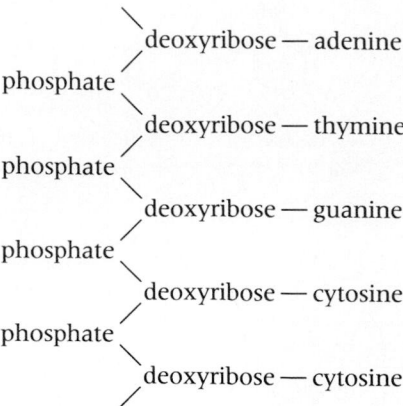

```
                deoxyribose — adenine
    phosphate
                deoxyribose — thymine
    phosphate
                deoxyribose — guanine
    phosphate
                deoxyribose — cytosine
    phosphate
                deoxyribose — cytosine
```

And RNA may be represented thus:

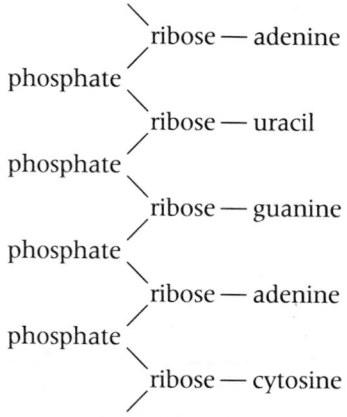

```
                ribose — adenine
    phosphate
                ribose — uracil
    phosphate
                ribose — guanine
    phosphate
                ribose — adenine
    phosphate
                ribose — cytosine
```

To understand the remarkable properties of these chains, we must be aware of certain aspects of the chemical structures of purines and pyrimidines (see Figure 16-9). Purines have two rings and pyrimidines one. Thus a purine is somewhat more space-consuming than a pyrimidine. Note, too, that one purine and one pyrimidine have an amino ($—NH_2$) group at the top of the ring (on carbon 6); the other purine and pyrimidines have a keto ($=O$) group in this position.

Prior to its discovery, there was a hint of DNA's helical structure in data showing that the ratio of 6-amino bases to 6-keto bases is always 1:1 (Table 16-3). The

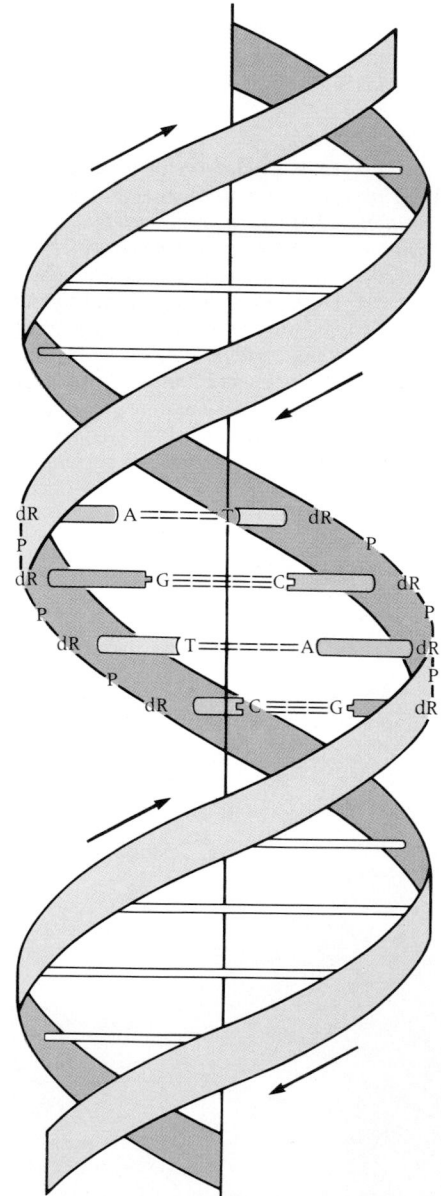

16-10 DNA double helix
The two strands of the helix are held together by hydrogen bonds between the complementary purine and pyrimidine base pairs. The backbone consists of alternating deoxyribose (dR) and phosphate (P) elements linked by covalent phosphodiester bonds. Note that the two polynucleotide chains run in opposite directions.

TABLE 16-3
BASE COMPOSITION OF DNA MOLECULES FROM DIFFERENT SPECIES

Source of DNA	Percentage Base Composition				$\dfrac{A + T}{G + C}$	$\dfrac{A + G}{C + T}$
	Adenine	Thymine	Guanine	Cytosine		
Plants						
Wheat germ	27.3	27.1	22.7	22.8	1.19	1.01
Animals						
Bovine (cow) sperm	28.7	27.2	22.2	20.7	1.30	1.00
Rat bone marrow	28.6	28.4	21.4	20.4	1.36	0.98
Sea urchin	32.8	32.1	17.7	17.3	1.88	1.01
Human	30.9	29.4	19.9	19.8	1.52	1.03
Herring	27.9	28.2	19.5	21.5	1.37	1.04
Marine crab*	47.3	47.3	2.7	2.7	17.52	1.00
Bacteria						
Escherichia coli	24.7	23.6	26.0	25.7	0.93	1.03
Sarcina lutea	13.4	12.4	37.1	37.1	0.35	1.02
Clostridium perfringens	36.9	36.3	14.0	12.8	2.73	0.99
Viruses						
Bacteriophage T7	26.0	26.0	24.0	24.0	1.08	1.00
Bacteriophage λ	21.3	22.9	28.6	27.2	0.79	0.94
Bacteriophage φX174	24.6	32.7	24.1	18.5	1.35	0.76

* Satellite DNA (defined in Chapter 17) comprises 10–30% of total DNA.

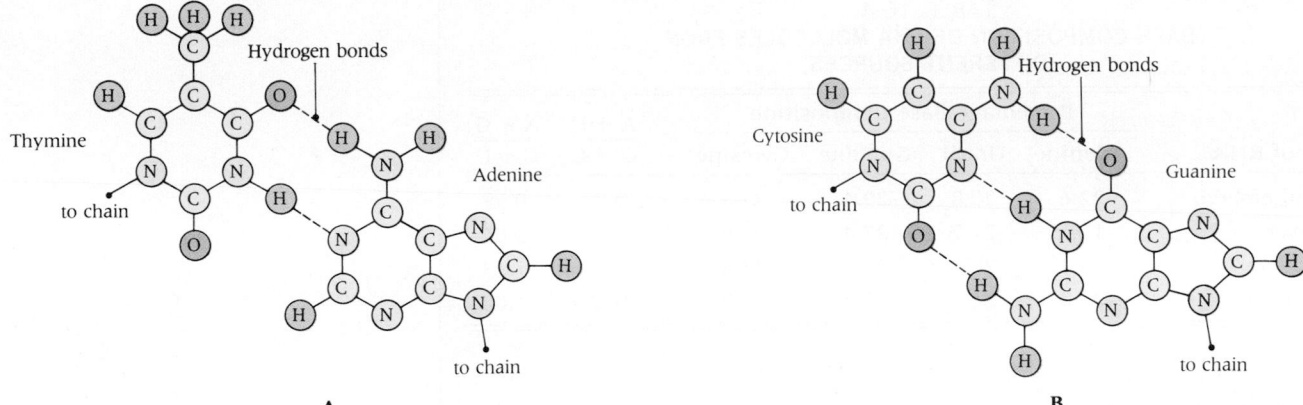

A   B

16-11 Pairing of complementary bases
In order to maintain constant spacing in the double helix, a purine must always pair with a pyrimidine: (A) adenine with thymine and (B) guanine with cytosine. Hydrogen bonds are shown as dashed lines between the bases. Note how precisely the paired bases fit together.

table reveals that these ratios are often not exactly 1 : 1. That is because experimental methods of measuring the percentages of A, T, G, and C in DNA have a small inherent error. Despite this error, the approximation of (A + G)/(T + C) to 1 in all DNAs studied (except those of some viruses) is obvious and impressive. These observations led Erwin Chargaff of Columbia University to the important conclusion that the amount of adenine in DNA equals the amount of thymine—and that, in the same way, guanine equals cytosine. It was an early clue to the existence of complementary base-pairing.

In 1953, after studying x-ray diffraction patterns of pure DNA, James D. Watson and F. H. C. Crick brilliantly deduced that the DNA molecule is a **double-stranded helix,** resembling a long twisted ladder (Figure 16-10), with two parallel nucleotide chains winding around an empty cylindrical space. The sides of the ladder consist of alternating deoxyribose and phosphate groups. The rungs between the sugars are paired bases. Base pairing is the key to the importance of the Watson-Crick discovery, for it was found that bases can be paired only in a certain way: adenine pairs with thymine, and cytosine pairs with guanine (Figure 16-11). Observe in these diagrams the prominent role of the 6-amino and 6-keto groups in forming hydrogen bonds.

Thus the structure of DNA as it predominantly occurs in nature must be represented as follows:

deoxyribose—adenine = = = =thymine—deoxyribose

phosphate   phosphate

deoxyribose—thymine = = = = adenine—deoxyribose

phosphate   phosphate

deoxyribose—guanine ≡ ≡ ≡ ≡cytosine—deoxyribose

phosphate   phosphate

deoxyribose—cytosine ≡ ≡ ≡ ≡guanine—deoxyribose

phosphate   phosphate

deoxyribose—thymine = = = = adenine—deoxyribose

The two sugar-phosphate chains are so arranged that the space between them can accommodate only base pairs consisting of a "large" purine and a "small" pyrimidine. Pairs consisting of two purines or two pyrimidines would not fit. Moreover, a base with a 6-keto group must stand opposite a base with a 6-amino group since only such a pair can form the hydrogen bonds required to hold the two strands together. Note that one hydrogen bond forms between the 6-amino and 6-keto groups. Other hydrogen bonds form between other amino and keto groups. The net result is 3 hydrogen bonds between G and C and 2 hydrogen bonds between T and A. That is why the two DNA strands of the DNA molecule are held together more tightly in DNAs with high levels of (G + C) relative to (A + T).

TABLE 16-4
BASE COMPOSITION OF RNA MOLECULES FROM DIFFERENT SOURCES

Source of RNA	Percentage Base Composition				$\dfrac{A+U}{G+C}$	$\dfrac{A+G}{C+U}$
	Adenine	Uracil	Guanine	Cytosine		
Sea urchin embryo	22.6	20.8	29.4	27.2	0.77	0.99
Bovine liver	17.1	21.7	27.3	33.9	0.63	1.04
Bovine kidney	19.7	20.2	26.7	33.4	0.66	1.13
Escherichia coli	25.3	21.2	28.8	24.7	0.87	1.00

Interestingly, these rules were found to be just as valid for pairings between a base in a DNA strand and a base in an RNA strand as for pairings between bases in two DNA strands. In other words, a DNA cytosine can pair with an RNA guanine, and a DNA adenine will pair with RNA uracil (the "pairing equivalent" of DNA thymine). If the base sequence of a DNA strand has any significance at all, then the sequence in its complementary strand, at least potentially, has the same significance since the two sequences are "images" of one another although the two strands run in opposite (antiparallel) directions. Complementary base-pairing is among the most significant biochemical phenomena of living organisms. This is what enables the DNA molecule to perform its preeminent functions: accurate self-duplication and accurate transmission of genetic information.

It is seen in Table 16-3 that the DNAs of closely related species have similar base compositions.[3] The situation is different in RNA. Since RNA may exist as a single-stranded nonhelix, it may not display complementary base-pairing. Hence the G:C and A:U ratios vary widely from one RNA to another (Table 16-4). Also, RNA molecules from different tissues of one species may differ. Some RNA molecules bend back on themselves like cloverleafs or hairpins. In that configuration, bases on one arm of the hairpin are complementary to bases on the other arm. Hydrogen bonds between the pairs keep the hairpin from turning into a single noodlelike structure of random shape.

DNA BIOSYNTHESIS

How is DNA synthesized in the cell? In 1957, four years after the revelation of the double-strandedness of DNA, Arthur Kornberg of Stanford University discovered **DNA polymerase,** an enzyme that catalyzes the polymerization of a new DNA strand from its four component **deoxyribonucleotides,** as follows:

[3] One DNA in the table is of notable interest. The DNA of the marine crab (*Cancer*) includes a component, termed satellite DNA, that consists almost exclusively of A and T. Its significance is discussed in Chapter 17.

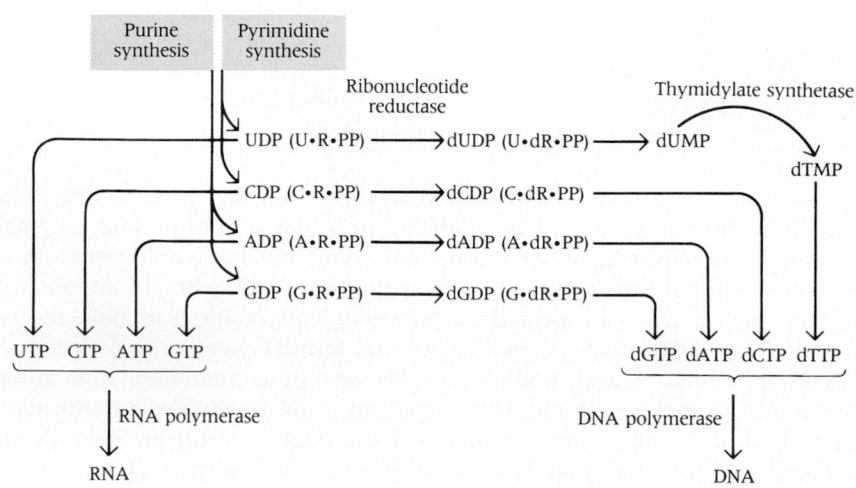

16-12 Synthesis of DNA and RNA and their nucleotide precursors
The diphosphate forms of the four ribonucleotides are precursors for the enzymatic synthesis of deoxyribonucleotides by action of ribonucleotide reductase. Uracil in dUMP is converted to thymine in dTMP by thymidylate synthetase. (Terms in parentheses show components of nucleotides.)

$$\left.\begin{array}{l}\text{dATP}\\\text{dCTP}\\\text{dGTP}\\\text{dTTP}\end{array}\right\} + \underset{\text{(template)}}{\text{DNA}} \xrightarrow{\text{DNA polymerase}} \underset{\text{(new)}}{\text{DNA}} + \underset{\text{(pyrophosphate)}}{\text{PP}_i} \qquad (16.1)$$

As for the four deoxyribonucleotide precursors themselves, we earlier saw that ribose produced in the pentose phosphate shunt pathway gives rise to the ribose of purine and pyrimidine ribonucleotides. When the four major ribonucleotides have been synthesized in the cell and converted to their diphosphate forms (UDP, CDP, ADP, and GDP; Figure 16-12), the enzyme **ribonucleotide reductase** converts them to deoxyribonucleotides. These molecules (see Figure 16-9) are converted to triphosphates, the necessary form for interaction with DNA polymerase. Ribonucleoside diphosphates are also converted to triphosphates, which are the substrates of another enzyme, **RNA polymerase.** This enzyme utilizes DNA as its template and synthesizes RNA in the complementary image of its DNA template.

Four important rules govern DNA synthesis.

First, polymerase can make new DNA only when copying an existing DNA molecule. The required DNA strand acts as an information-containing **template** that specifies which base is to be inserted next, according to rules of Watson-Crick base-pairing. If one DNA strand carries information in its base sequence, the opposing strand will carry the same information—in complementary form (Figure 16-13).

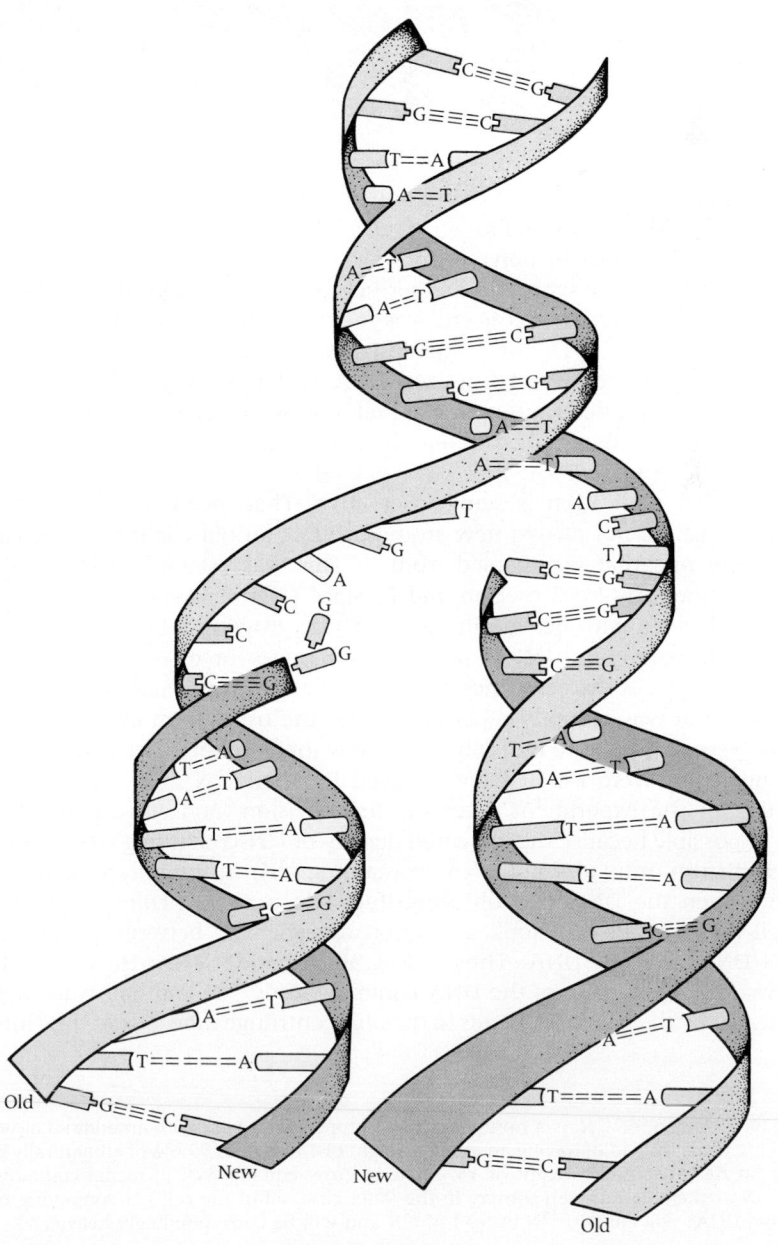

16-13 Replication of DNA
Each strand of the double helix serves as a template for the replication of a new DNA strand. After replication, the two DNA molecules (each a new and an old strand) return to a helical structure.

Original DNA molecule
(all ^{15}N labeled)

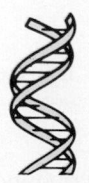

Once–replicated "hybrid" DNA
(each molecule contains one ^{15}N
and one ^{14}N strand)

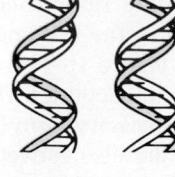

Twice–replicated DNA
(half the molecules "hybrid"
and half containing only ^{14}N)

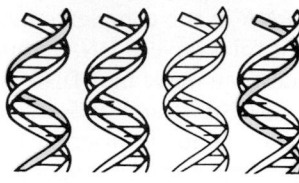

Thrice–replicated DNA
(one quarter of the molecules
"hybrid" and three quarters
containing only ^{14}N)

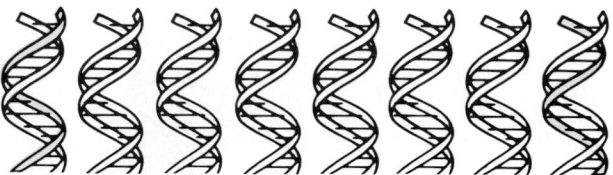

In general, polymerization of deoxyribonucleotides into DNA does not occur unless all four nucleotides are available.

Second, the product of polymerization is a faithful copy of its template. Early students of polymerase worried whether newly synthesized product DNA is in fact a copy of its template. That concern was met by comparing the base compositions of newly synthesized DNA and template DNA. When template DNA had a specific biological property, newly synthesized DNA had too. When biochemists finally devised methods for determining the actual base sequence of DNA (to be described later) they obtained the final proof that the newly synthesized DNA is a complementary image of template DNA.

Third, DNA replication is semiconservative. That means that each new DNA molecule consists, not of two new strands, but of an old chain that has paired up with a new one that was copied from it. This was established in an ingenious 1958 experiment by M. Meselson and F. Stahl (Figure 16-14). All of the DNA of *E. coli* was labeled with heavy nitrogen (^{15}N) by growing it for several generations in a medium containing ^{15}N as the only nitrogen source—so that all of the ^{14}N normally present in the cells, including that of DNA, was replaced.[4] Then, the ^{15}N in the medium was abruptly replaced by ^{14}N, the usual form of nitrogen, and the bacteria were permitted to multiply in the new medium for exactly two generations. DNA molecules were isolated and assayed for their ^{15}N and ^{14}N content at the beginning of the experiment, after the first division, and at various later times. This was possible because the increased density of ^{15}N-containing DNA easily distinguished it from ordinary light DNA (containing ^{14}N) in the ultracentrifuge (Figure 16-15). When the DNA was ultracentrifuged in a cesium chloride gradient after one cell division, all of it took a position intermediate between positions typical for ^{14}N-DNA and ^{15}N-DNA. Thus it was all "hybrid" DNA. However, after two cell divisions in ^{14}N, half of the DNA contained only ^{14}N and half was "hybrid"— as indicated by the two DNA bands in the ultracentrifuge tube, one at the "intermedi-

---

[4] Heavy nitrogen, or ^{15}N, is a nonradioactive isotope of nitrogen with an additional neutron in its nucleus. Thus, instead of having an atomic weight of 14, as does 99.6% of all naturally occurring nitrogen, it has an atomic weight of 15. Bacteria grow equally well in media containing either ^{14}N or ^{15}N as the sole nitrogen source. In the latter case, all of the cell's N-containing molecules (including DNA) will contain ^{15}N instead of ^{14}N and will be correspondingly heavier.

16-15 Proof of semiconservative replication of DNA

Details of the Meselson-Stahl experiment are presented. Compare with Figure 16-14.

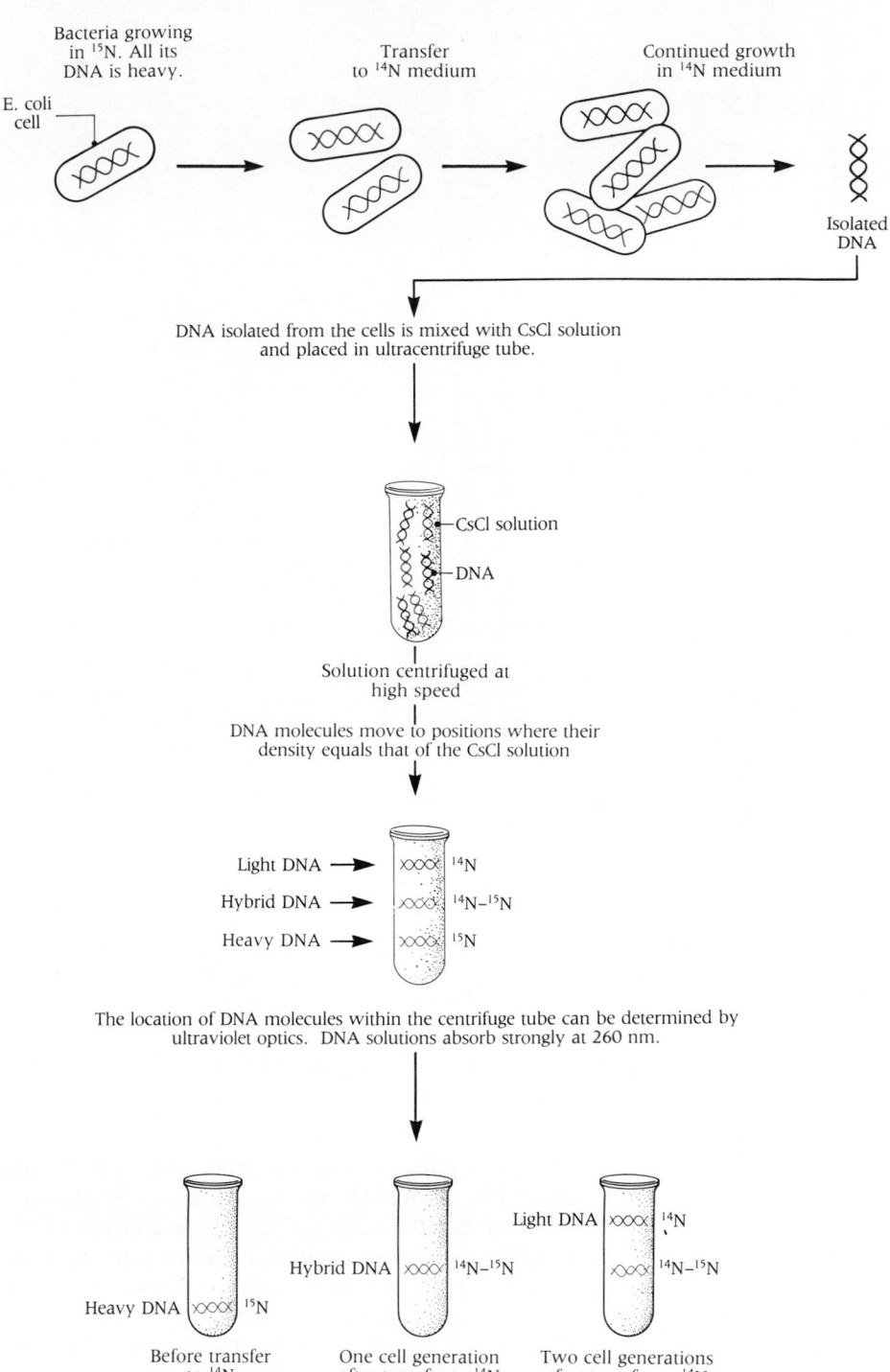

DNA isolated from the cells is mixed with CsCl solution and placed in ultracentrifuge tube.

CsCl solution

DNA

Solution centrifuged at high speed

DNA molecules move to positions where their density equals that of the CsCl solution

Light DNA → ^{14}N

Hybrid DNA → $^{14}N-^{15}N$

Heavy DNA → ^{15}N

The location of DNA molecules within the centrifuge tube can be determined by ultraviolet optics. DNA solutions absorb strongly at 260 nm.

Heavy DNA ^{15}N

Before transfer to ^{14}N

Hybrid DNA $^{14}N-^{15}N$

One cell generation after transfer to ^{14}N

Light DNA ^{14}N

$^{14}N-^{15}N$

Two cell generations after transfer to ^{14}N

ate" position, and one at the "all ^{14}N" position. Only these two types of DNA molecules were ever found: DNA containing only ^{14}N and DNA containing both ^{14}N and ^{15}N (hybrid molecules)—precisely what would be anticipated if the DNA double helix unwinds during replication and each of the strands serves as the template for the synthesis of a new strand. In a second experiment, Meselson and Stahl further supported this conclusion by isolating the DNA from cells that had undergone one division in the presence of ^{14}N. By heating the isolated DNA molecules, they were able to separate the two strands of the double helix—this is called "denaturing" a DNA molecule. When the individual strands were ultracentrifuged in a cesium chloride gradient, half had the density of ^{14}N-containing strands and half had the density of ^{15}N-containing strands. This elegant result provided final proof that in cell division one DNA strand goes to each of the daughter cells where it serves as a template for the synthesis of a complementary new strand.

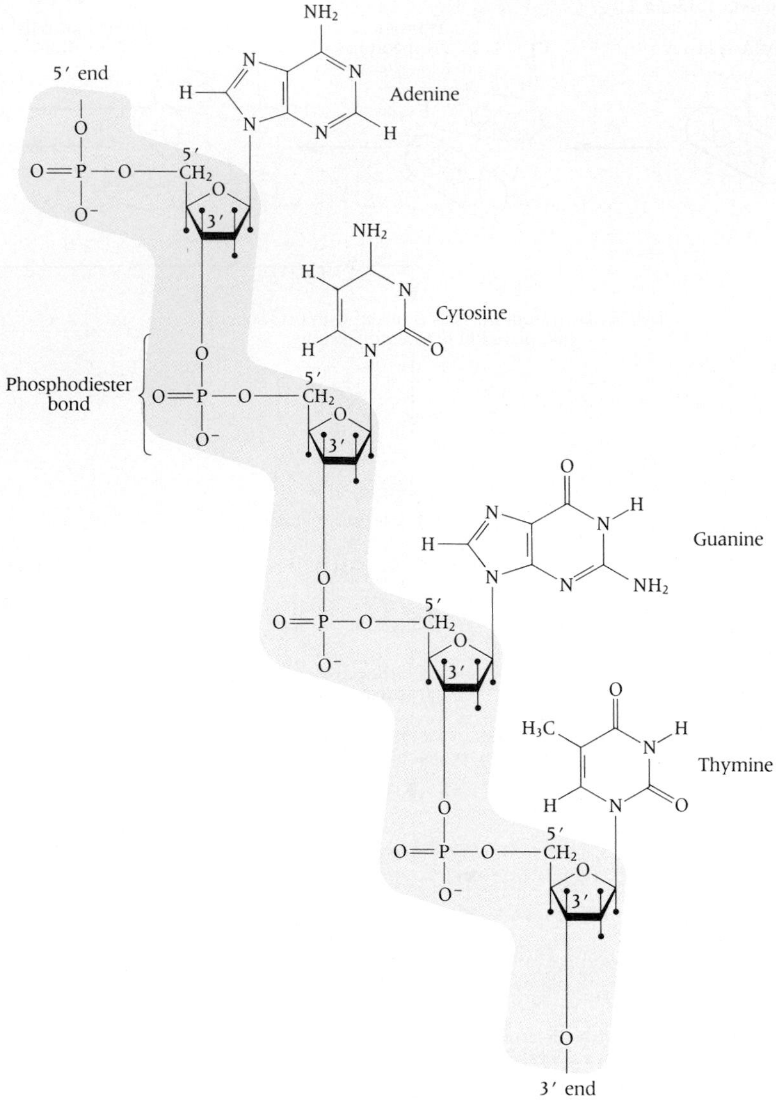

Adenine

Cytosine

Phosphodiester bond

Guanine

Thymine

Fourth, the elongating DNA strand grows in one direction—from the 5' end toward the 3' end. As we have seen, strands of DNA (and RNA) consist of nucleotides linked together by **phosphodiester bonds,** each end of which attaches to the sugar—deoxyribose or ribose—of its nucleotide neighbors. On one side, the bond attaches to the 5' carbon of the sugar; on the other side, it attaches to the 3' carbon (Figure 16-16). Hence, every DNA strand has a 5' end and a 3' end (Figure 16-17) and there is directionality in each strand. It is conventional in printed diagrams to place the 5' end at left and the 3' end at right. For example, the printed sequence ATCG signifies (5')ATCG(3'), thus:

A   C   G   T

5' end   P—dR—P—dR—P—dR—P—dR—P—   3' end

Nucleotides or sequences nearer to the 5' end than the 3' end are said to be "upstream."

It was once thought that all DNA is double-stranded. We now know that several small viruses—among them φX174, S13, F1, and others—contain DNA that exists normally as a single-stranded circle (see Table 16-1). An early clue to the single-strandedness of virus DNAs was the nonequivalence of the A and T, and of the G and C contents (see Table 16-3). Despite this circumstance, replication of single-stranded DNA does involve base pairings (Figure 16-18). In an intermediate stage, a complementary strand (termed a **minus strand**) forms on the original **plus strand,** in which the sequence is preserved but in complementary form. Thus, the DNA is temporarily double-stranded. This is termed the replicative form of the

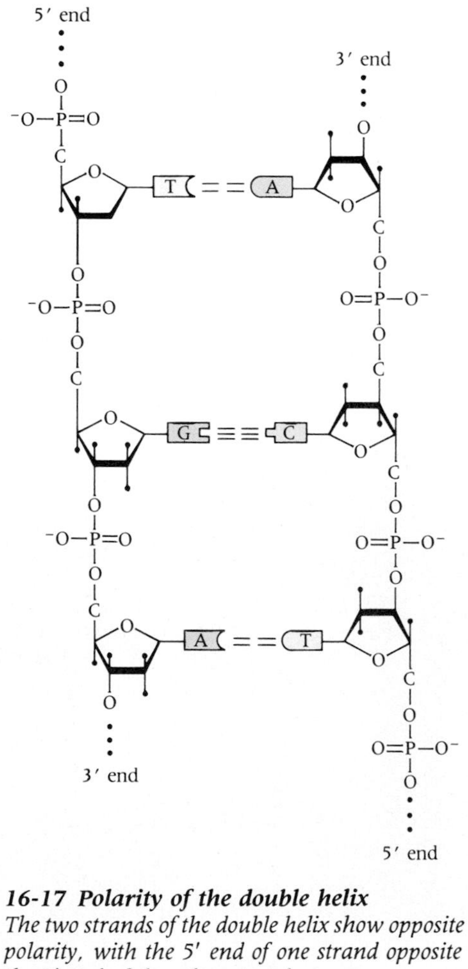

16-17 Polarity of the double helix
The two strands of the double helix show opposite polarity, with the 5' end of one strand opposite the 3' end of the other strand.

16-18 Replication of single-stranded DNA
Plus-strand DNA serves as the original template for replication of a single-stranded DNA molecule. Synthesis of the minus strand (color) creates a new template that is used to generate new plus-strand DNAs.

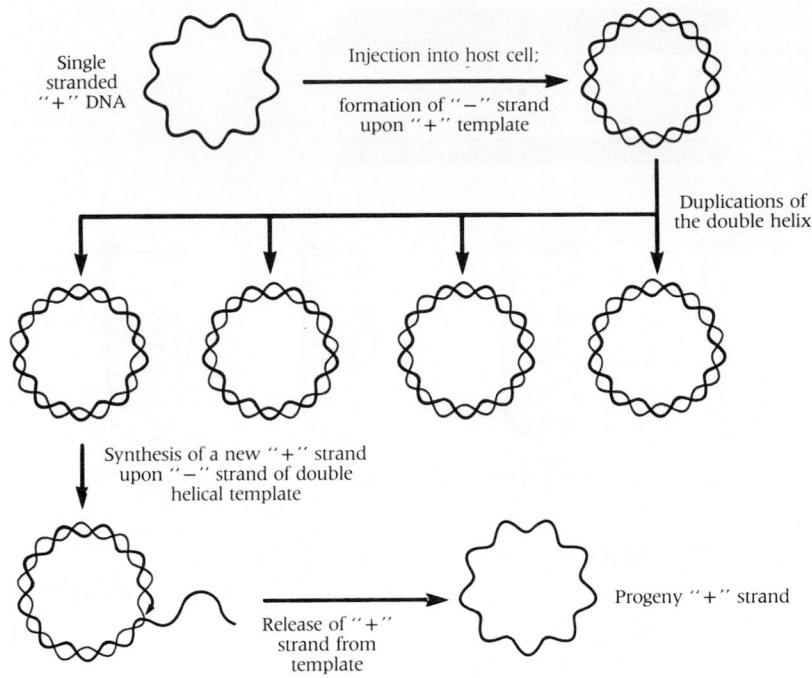

Single stranded "+" DNA

Injection into host cell; formation of "−" strand upon "+" template

Duplications of the double helix

Synthesis of a new "+" strand upon "−" strand of double helical template

Release of "+" strand from template

Progeny "+" strand

DNA, because it can copy itself repeatedly. The complementary minus strand serves as template for a large number of new single-stranded plus strands, which are then released.

One of the greatest advances in our understanding of these mechanisms—and those to be described in Chapters 17 and 18—followed the discovery of methods for determining the actual sequence of individual nucleotides, or bases, in DNA. One of the methods is summarized in Figure 16-19. Today, DNA sequencing—much of it automated—is as easily done as genetic crosses.

COMPLICATIONS ARISE

Kornberg discovered DNA polymerase in *E. coli* in 1957. At first, it seemed the long-sought explanation of DNA replication had been found, especially since eukaryotic cells appeared to have a similar polymerase and to synthesize DNA in the same manner. Alas, it was not to be. Kornberg's polymerase was ultimately given less importance when it was found that most cells contain three or four DNA polymerases. The original one is now called **DNA polymerase I** or **pol I.**

Later work revealed some important and surprising aspects of the DNA replication process. First, the two DNA strands are made differently (Figure 16-20). Cells rely primarily on **RNA polymerase** to make an **RNA primer,** and on **DNA polymerase III** to elongate both chains of the helix. The latter enzyme works in only one direction along the chain, however—the 5' to 3' direction. The two DNA strands are held together in an opposite orientation—that is, the direction of one is 5' to 3' and that of the other is 3' to 5' (see Figure 16-17). One is called the **leading strand** (3' to 5'); the other the **lagging strand** (5' to 3'). Because of this arrangement, a **replication fork** develops between the two strands. Assembly of a daughter strand proceeds smoothly when the leading strand is the template. Along the lagging strand, however, the new strand must be made in discontinuous chunks. These were named **Okazaki fragments** after their talented discoverer. They are ultimately joined together by the enzyme **DNA ligase.**

Because of the structural intricacy of an intertwined double helix, a replicating DNA molecule must twist and rotate. This is made possible by a class of specific enzymes called **DNA topoisomerases,** which transform the DNA strands from one structural form to another. The number of these enzymes is large—well more than a dozen. The following are the best understood of them (Figure 16-21):

■ An **unwinding enzyme** (or **helicase**), which solves a tricky mechanical problem by unwinding the DNA helix at the replication fork, using energy from ATP

16-19 **Method for sequencing DNA**
*Each end of the DNA is radioactively labeled.
The molecule is denatured and the strands
separated. The one to be sequenced is chemically
cleaved at the base G, A and G, T and C, and
C (test tubes 1–4). The resulting DNA fragments
are separated by size using gel electrophoresis,
with the smallest pieces moving the farthest in
the gel. The positions of the radioactive fragments
are determined by exposure on x-ray film, and
the sequence is read from the shortest to the
longest, generating the base sequence of the
original strand.*

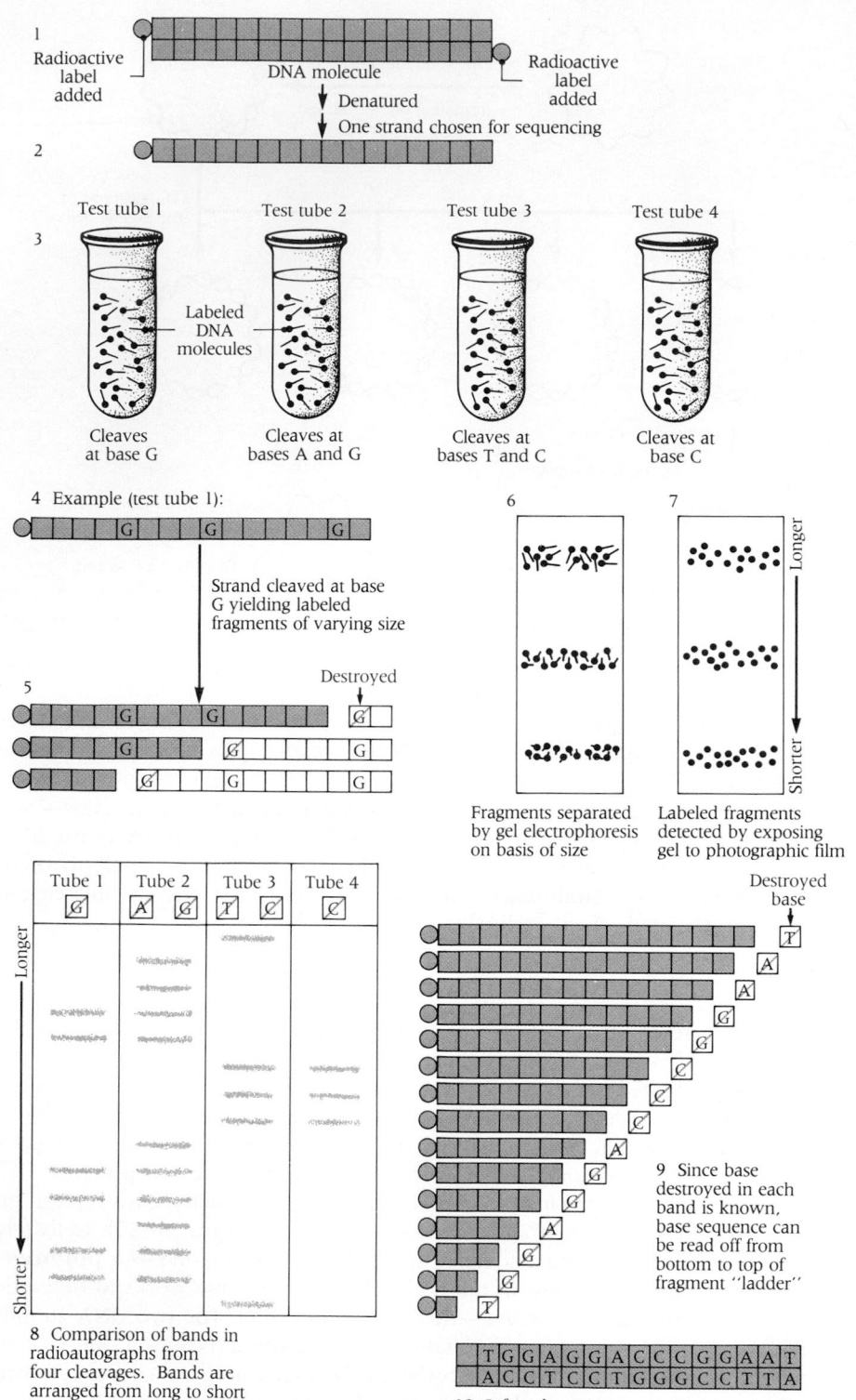

1 Radioactive label added — DNA molecule — Radioactive label added
Denatured
One strand chosen for sequencing
2

Test tube 1   Test tube 2   Test tube 3   Test tube 4

3   Labeled DNA molecules

Cleaves at base G   Cleaves at bases A and G   Cleaves at bases T and C   Cleaves at base C

4 Example (test tube 1):

Strand cleaved at base G yielding labeled fragments of varying size

5   Destroyed

6   7   Longer / Shorter

Fragments separated by gel electrophoresis on basis of size

Labeled fragments detected by exposing gel to photographic film

| Tube 1 | Tube 2 | | Tube 3 | | Tube 4 |
| G | A | G | T | C | C |

Longer / Shorter

8 Comparison of bands in radioautographs from four cleavages. Bands are arranged from long to short

Destroyed base
T
A
A
G
G
C
C
C
A
G
G
A
G
G
T

9 Since base destroyed in each band is known, base sequence can be read off from bottom to top of fragment "ladder"

| T | G | G | A | G | G | A | C | C | C | G | G | A | A | T |
| A | C | C | T | C | C | T | G | G | G | C | C | T | T | A |

10 Inferred sequence

- A **DNA ligase,** which links up the discontinuous Okazaki fragments
- A **DNA gyrase,** which twists DNA helices into superhelices by placing strategic nicks in the DNA chain (all the while utilizing ATP energy)

Far from the simple zipper-style polymerization of the uncomplicated DNA molecule that was visualized in the 1960s, we must now grapple with the contortions and rotations of supercoiled molecular duplexes (Figure 16-22). That has proved elusive and difficult. Three models of the mechanism appear in Figure 16-23. The first occurs in a few viruses. The second (unidirectional growth) may occur only occasionally; the third (bidirectional growth) is probably the most widespread.

16-20 Leading strand and lagging strand synthesis

(A) *As the parent DNA molecule unwinds, a replication fork is created so that DNA synthesis can occur.* **(B)** *Unlike the leading strand, which is replicated as a continuous piece of DNA, the lagging strand is synthesized as short pieces called Okazaki fragments. DNA synthesis is initiated with an RNA primer that is later removed so that the DNA pieces can be joined by DNA ligase.*

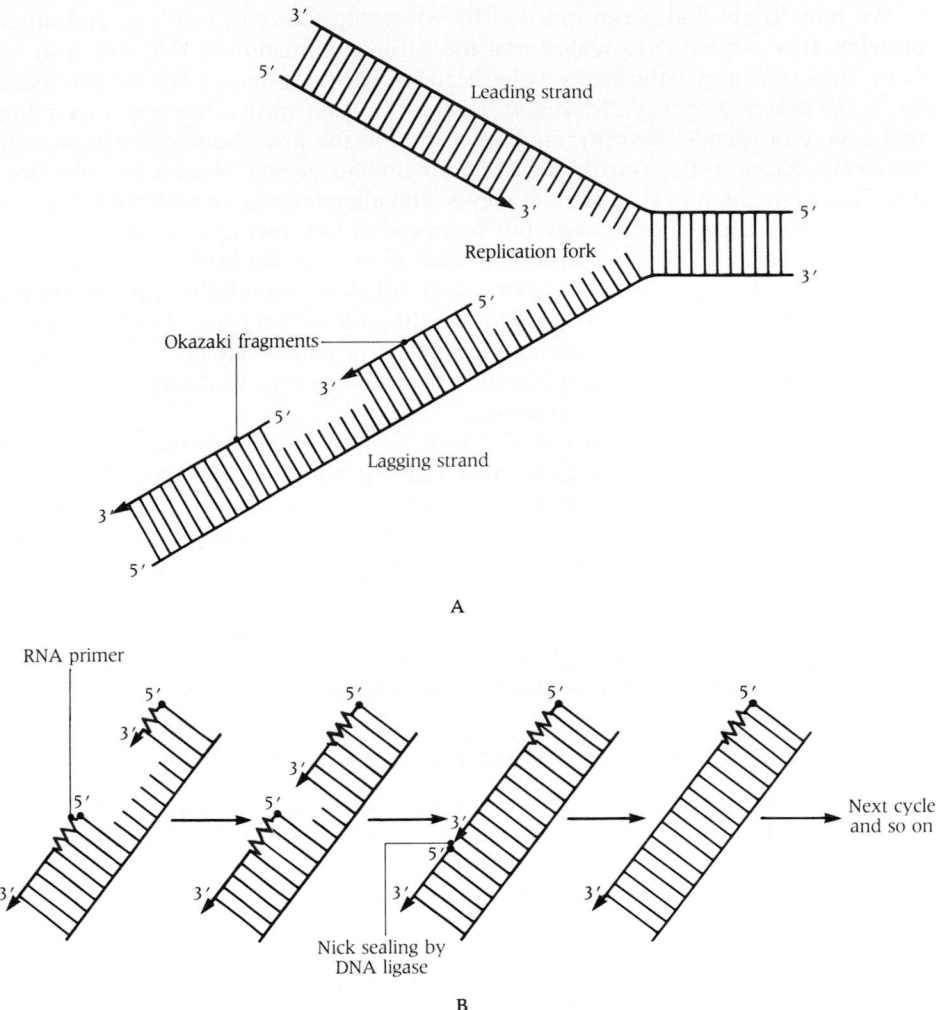

HOW GENES FUNCTION

We learned in Chapter 13 how genes are transmitted to the offspring, how they are recombined and altered, and how the regularities found in breeding experiments are thereby explained. We now inquire how the existence of a trait is determined by the presence of a gene. How, in a word, do genes act?

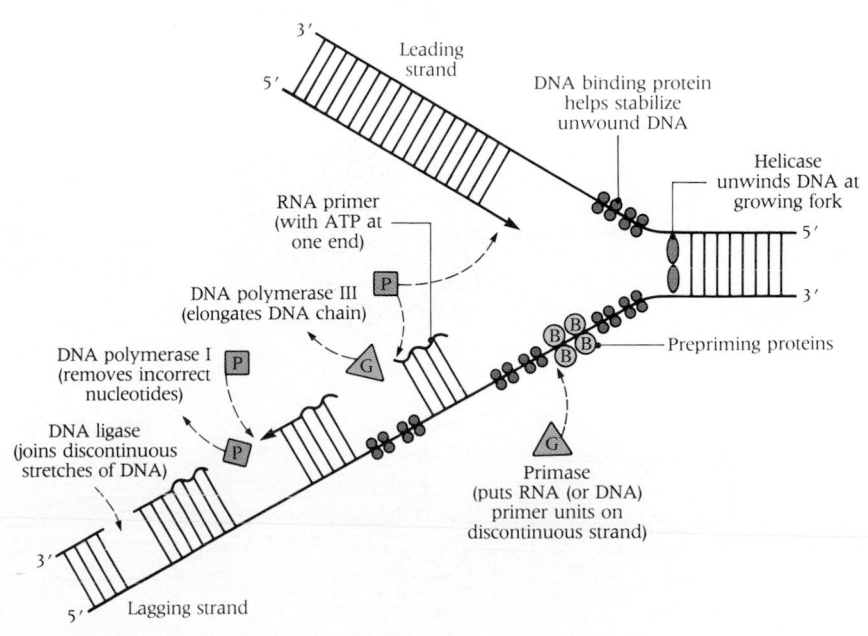

16-21 Multiple enzyme participation in DNA replication

Current model of DNA replication shows roles for more than twenty enzymes.

We now know that genes function by governing the cell's enzymes and other proteins. It is surprising to realize that the earliest intimation of this idea is to be found in old writings of the Scottish physician Sir Archibald Garrod. His extraordinary book *The Inborn Errors of Metabolism*, written in 1908 (just a few years after the rediscovery of Mendel's work), established him as the first "human geneticist." In the years before 1908, Garrod studied two human genetic disorders—albinism, total lack of pigment in skin, hair, and eyes, and alkaptonuria, in which the affected person's urine turns frighteningly but harmlessly black owing to its high content of a substance called homogentisic acid. Garrod argued that both conditions were caused by the absence of specific enzymes and called these conditions **inborn errors of metabolism.** By coming so very close to the truth—that genes direct organisms to manufacture particular enzymes and that defective genes fail to do so—his conclusions must be ranked along with Mendel's laws and Avery's transforming principle as an historic case of scientific prescience.

Not until the 1960s was it revealed how a gene determines the structure (or rate of synthesis) of an enzyme (or other protein) and precisely how DNA serves as keeper of the information that governs the structure of proteins. Imaginative work in the 1960s literally "cracked the code" and revealed exactly how this information is carried in the DNA molecule. Then, in quick succession, it was learned how the coded message is communicated to the site of protein synthesis, and how the amino acid sequences, which alone make proteins specific, are established. It was also learned, among other things, which enzymes are missing in the diseases studied so long ago by Garrod—albinism and alkaptonuria (Figure 16-24).

THE ONE GENE, ONE ENZYME HYPOTHESIS

Living organisms can be viewed as self-regulating metabolic machines dedicated to two enterprises vital for growth and maintenance: energy extraction and molecular biosynthesis.

The essential nature of metabolism may be illustrated quite simply:

$$A \xrightarrow{e_1} B \xrightarrow{e_2} C \xrightarrow{e_3} D \xrightarrow{e_4} E \xrightarrow{e_5} F$$

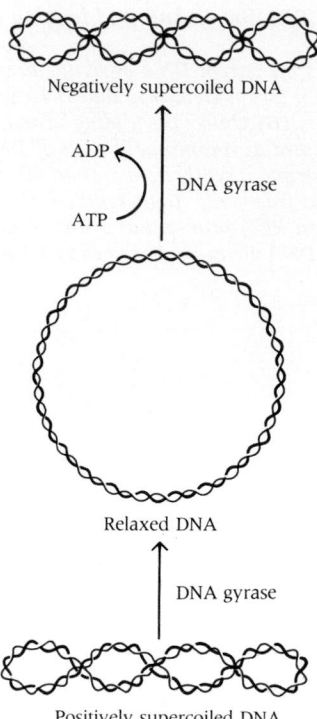

16-22 Supercoiled forms of DNA
The relaxed (nonsupercoiled) form of DNA can be converted to either positive or negative supercoils by the enzyme DNA gyrase. ATP is required to produce negative supercoils.

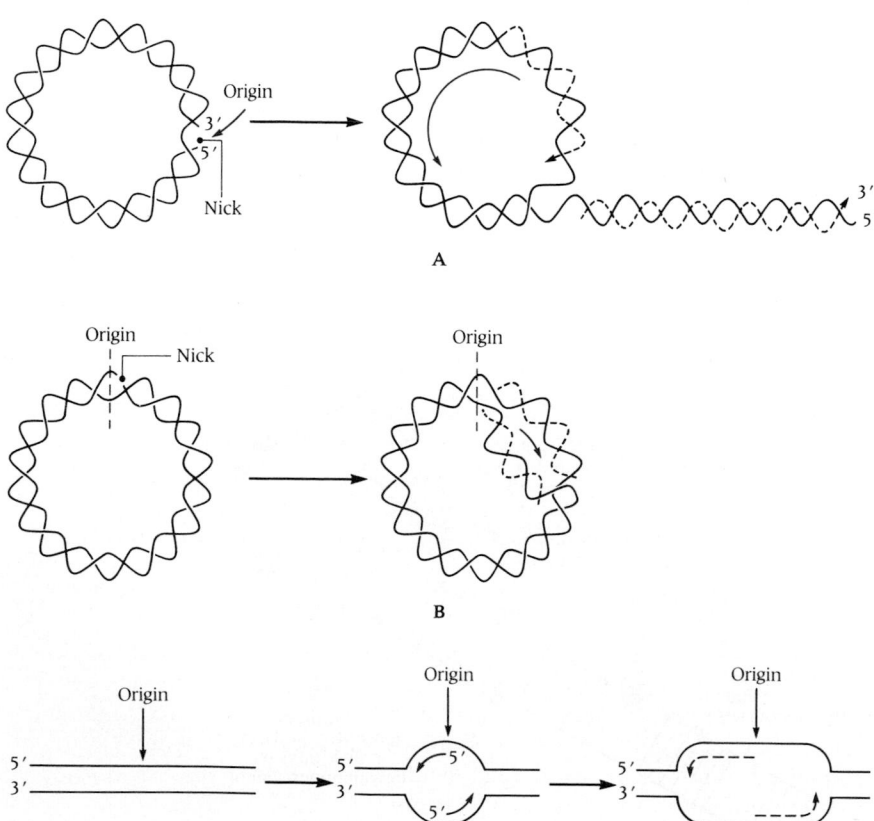

16-23 Three models of DNA replication
(A) *Rolling circle model found in some viruses.* **(B)** *Unidirectional replication of a circular DNA molecule.* **(C)** *Bidirectional replication, which is the most common form. In all cases the newly synthesized DNA appears as a dashed line.*

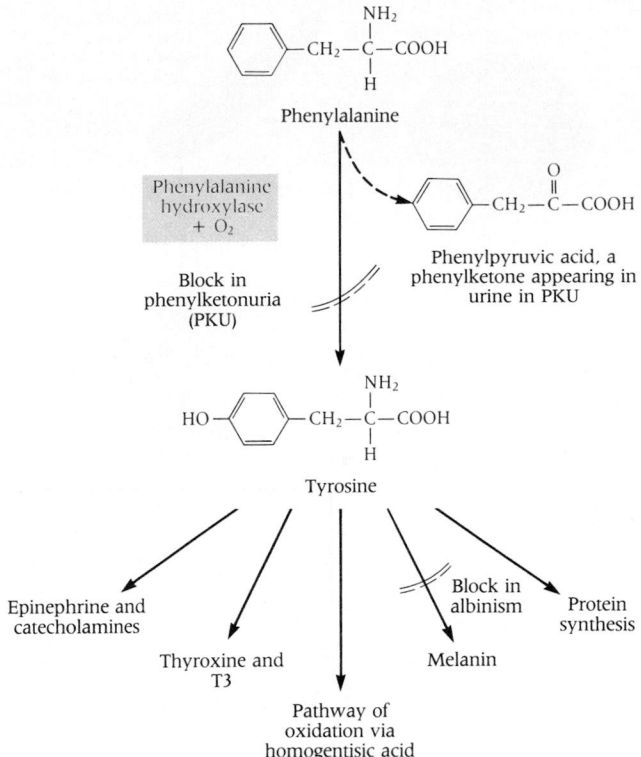

16-24 Pathways of phenylalanine and tyrosine metabolism
The diagram shows locations of defective enzymes causing metabolic blocks in mutants with phenylketonuria and albinism.

Here the upper case letters symbolize chemical compounds. *A* represents a nutrient molecule—say, glucose or a fatty acid—taken in as food. *F* is a compound needed for some essential purpose. The intermediate compounds, *B, C, D,* and *E,* are products and precursors in an orderly sequence of chemical reactions, each of which is catalyzed by a specific enzyme. These are designated e_1, e_2, and so forth.

We may regard every phenotypic trait as a consequence of one or another such sequence. When, for example, human skin has a brown color it became that way because a pigment, melanin, was synthesized in skin cells from simpler materials by a sequence of enzyme reactions like those above.

The Nobel Prize-winning work of George Beadle and Edward Tatum in the 1940s led to the first great simplification in our notions of how genes act. It was nicely epitomized by its name: the **one gene, one enzyme hypothesis.** Beadle and Tatum did their famous experiment, not with fruit flies, but with a common red (actually salmon-colored) mold of bread, *Neurospora crassa.*

They picked this organism for several reasons.

- It reproduces rapidly.
- It can easily be grown in vast quantities.
- Its complex life cycle, which we discuss briefly here, differed from that of experimental organisms traditionally studied by geneticists.
- It is haploid for much of its cycle. Therefore, mutations occurring in its single chromosome set can be detected promptly. No homologous chromosomes are present to mask a mutant gene with a dominant normal allele.
- Finally, it can grow on a simple medium containing only sugar as carbon and energy source plus one vitamin (biotin) and a few minerals and trace elements.

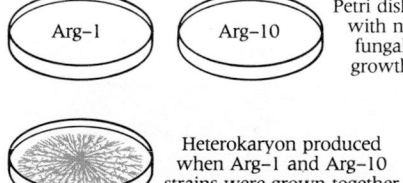

Petri dishes with no fungal growth

Heterokaryon produced when Arg–1 and Arg–10 strains were grown together

16-25 X-ray-induced mutations in Neurospora
Nutritional mutations can be created in this fungus using x-rays. Arg-1 and Arg-10 are mutant strains that cannot grow on minimal medium because each lacks the amino acid arginine. Fusion of cells from each strain generates a heterokaryon, which contains the combined genetic information from both cells. Each supplies an enzyme missing in the other. Hence, normal growth on minimal medium is again possible.

In view of these properties, Beadle and Tatum assumed that if a mutation arose that kept the mold from synthesizing a needed molecule (*F* in the above sequence), it would then be necessary to add a supply of *F* (or a metabolic precursor of it) to the culture medium. Such a mutant would be **auxotrophic** because, through a gene mutation, it acquired a nutritional requirement not present in the nutritionally independent **prototrophic** or wild-type strain.

Beadle and Tatum first collected *Neurospora* spores. Then they increased the mutation rate of the spores by exposing them to x-rays. This treatment produced

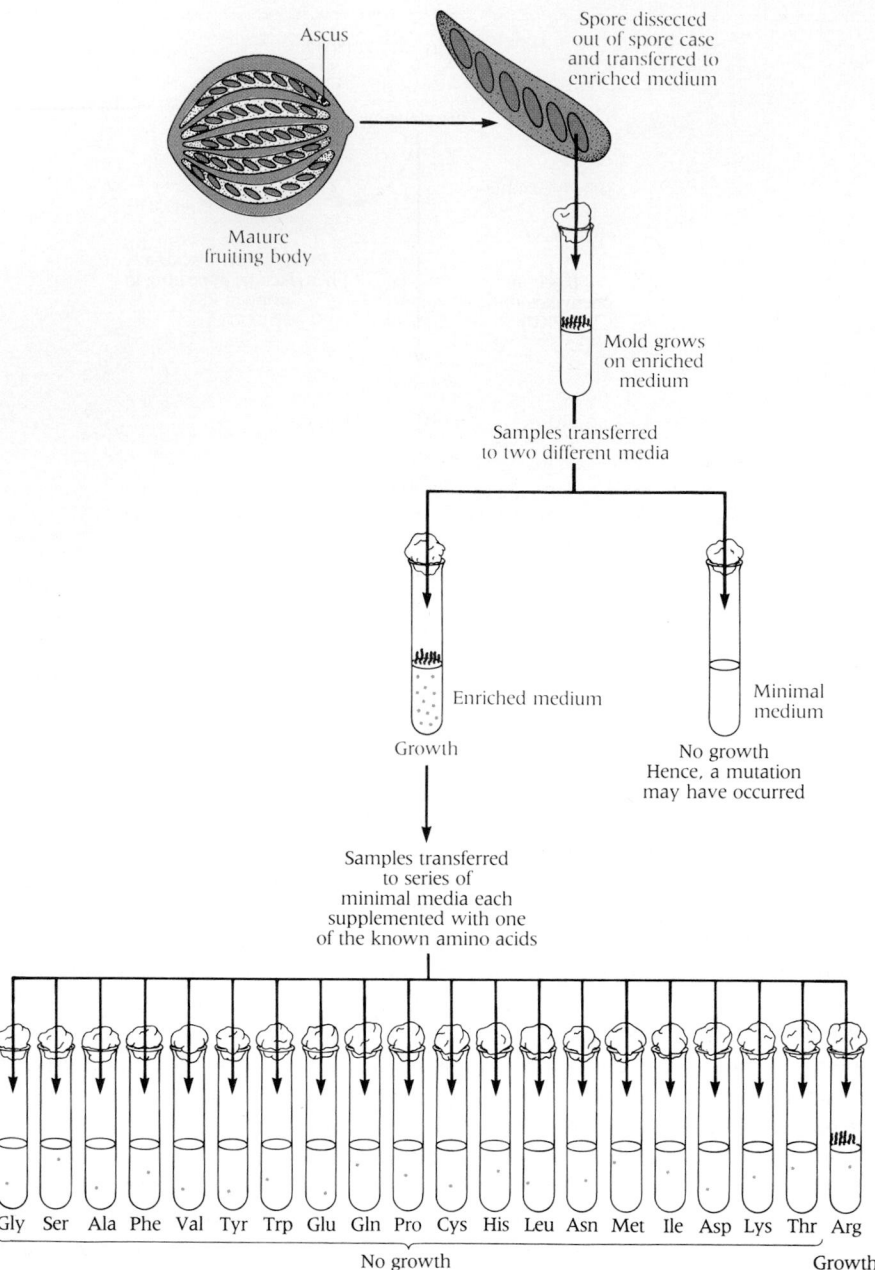

Ascus

Spore dissected out of spore case and transferred to enriched medium

Mature fruiting body

Mold grows on enriched medium

Samples transferred to two different media

Enriched medium

Growth

Minimal medium

No growth
Hence, a mutation may have occurred

Samples transferred to series of minimal media each supplemented with one of the known amino acids

| Gly | Ser | Ala | Phe | Val | Tyr | Trp | Glu | Gln | Pro | Cys | His | Leu | Asn | Met | Ile | Asp | Lys | Thr | Arg |

No growth   Growth

16-26 Defining nutritional mutants in Neurospora
Individual spores are selected and tested for their ability to grow on minimal medium supplemented with a single amino acid. In this case, the mutant is able to grow only when arginine is supplied, indicating that the biochemical pathway for arginine synthesis has been interrupted due to a defect in an essential enzyme.

some mutant cells that were no longer capable of growing on a simple **minimal medium** (Figure 16-25). However, they did grow normally in an enriched medium (one that was supplemented with amino acids). Subcultures of such mutants were then studied individually to determine which capacity they lacked. The general approach is illustrated in Figure 16-26.

Consider a typical experiment with organisms that were fully capable of synthesizing the amino acid arginine. The following are the final steps of that biosynthetic pathway:

$$\text{Nutrient compound} \xrightarrow{} \xrightarrow{e_1} \text{ornithine} \xrightarrow{e_2} \text{citrulline} \xrightarrow{e_3} \text{arginine}$$

The actual biochemical steps catalyzed by the three enzymes symbolized here as e_1, e_2, and e_3 are shown in Figure 16-27.

Many of the mutants thus produced were unable to thrive in a medium containing only sugar. Significantly, these mutants were anything but identical. In fact, they could be divided into three classes on the basis of their growth response after addition to the medium of ornithine, citrulline, or arginine.

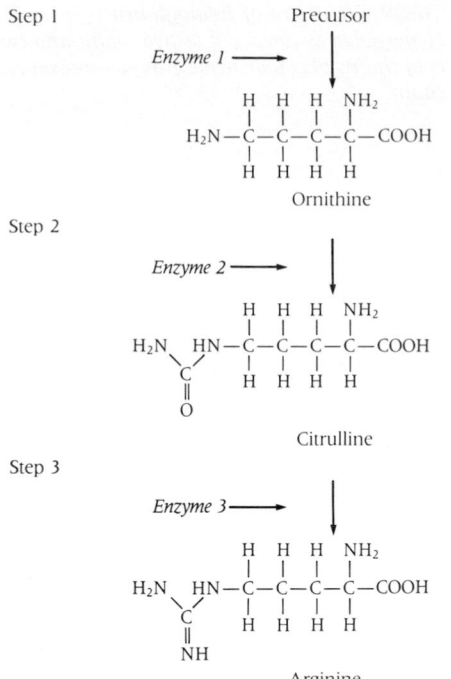

Step 1

Precursor

Enzyme 1 ⟶

$$H_2N-\underset{\underset{H}{|}}{\overset{\overset{H}{|}}{C}}-\underset{\underset{H}{|}}{\overset{\overset{H}{|}}{C}}-\underset{\underset{H}{|}}{\overset{\overset{H}{|}}{C}}-\underset{\underset{H}{|}}{\overset{\overset{NH_2}{|}}{C}}-COOH$$

Ornithine

Step 2

Enzyme 2 ⟶

$$H_2N \quad HN-\underset{\underset{H}{|}}{\overset{\overset{H}{|}}{C}}-\underset{\underset{H}{|}}{\overset{\overset{H}{|}}{C}}-\underset{\underset{H}{|}}{\overset{\overset{H}{|}}{C}}-\underset{\underset{H}{|}}{\overset{\overset{NH_2}{|}}{C}}-COOH$$
$$\underset{O}{\overset{||}{C}}$$

Citrulline

Step 3

Enzyme 3 ⟶

$$H_2N \quad HN-\underset{\underset{H}{|}}{\overset{\overset{H}{|}}{C}}-\underset{\underset{H}{|}}{\overset{\overset{H}{|}}{C}}-\underset{\underset{H}{|}}{\overset{\overset{H}{|}}{C}}-\underset{\underset{H}{|}}{\overset{\overset{NH_2}{|}}{C}}-COOH$$
$$\underset{NH}{\overset{||}{C}}$$

Arginine

16-27 Arginine biosynthesis
Three enzymatic reactions participate in arginine synthesis. Beadle and Tatum postulated that the synthesis of each enzyme is governed by a different gene.

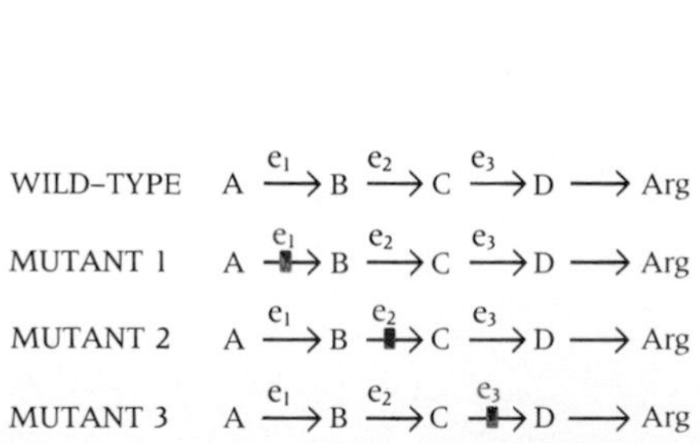

Mutant	Growth Response in Medium Containing:			Locus of Block in Synthetic Sequence
	Ornithine	Citrulline	Arginine	
1	+	+	+	e_1
2	−	+	+	e_2
3	−	−	+	e_3

Their irradiated mutant organisms still needed a supply of arginine, but radiation had evidently eliminated their ability to make it for themselves from commonplace nutrients such as glucose.

To Beadle and Tatum this result suggested the following conclusions:

■ Mutant 1: The absence (or nonfunction) of enzyme e_1 was implied by normal growth in a medium containing only ornithine as carbon source. This showed that the ornithine → citrulline → arginine sequence was intact and functioning.

■ Mutant 2: Normal growth in citrulline and poor growth in ornithine showed that in this mutant the citrulline → arginine step is intact and the ornithine → citrulline step is defective. This indicated that e_2 was missing or defective in mutant 2. Accumulation of ornithine in these cells supported that inference.

■ Mutant 3: A defect in the citrulline → arginine sequence is implied by the mutant's inability to grow on ornithine or citrulline, its ability to grow on arginine and its tendency to accumulate citrulline.

When Beadle and Tatum analyzed cells from the three mutant lines for their component enzymes they found that the enzymes we have called e_1, e_2, and e_3 were absent or present exactly as predicted from the nutritional patterns (Figure 16-28). When they sustained mutant growth by supplementing their mediums with the product of a missing reaction, new generations of offspring were obtained that had the same enzyme deficiency as their parents. The enzyme deficiencies were genetically determined! In order to survive a mutant strain would be in perpetual need of an external supply of the missing product.[5]

The implications of this classic experiment are unmistakable: A single gene governs the synthesis of a single enzyme. This was the one gene, one enzyme hypothesis.

Indeed all proteins are synthesized under the direct control of specific genes. Later workers renamed the hypothesis the one gene, one protein hypothesis.

When a protein contains more than one polypeptide chain subunit, the chains are synthesized separately and only then do they aggregate to form the complete protein. Often, the two or more polypeptide chains needed to form a complete protein are controlled by neighboring genes on a chromosome. Sometimes, as in the case of the two polypeptide chains of hemoglobin (the α and β chains) the two responsible genes are on different chromosomes. If a mutant gene impairs the synthesis of one subunit of a protein that happens to be an enzyme, the net effect may be the production of a defective or nonfunctional enzyme. This occurred in the Beadle-Tatum experiment. In such cases, "one gene, one polypeptide" is synonymous with "one gene, one enzyme."

THE CONTRIBUTION OF SICKLE CELL HEMOGLOBIN

Incisive studies of one protein, hemoglobin, opened the door further. The major protein in red blood cells, hemoglobin consists of equal amounts of polypeptide chains of two types, termed α and β chains (Figure 16-29). Each has a molecular weight of about 16,100. Since hemoglobin's molecular weight is 64,500, it is evidently a tetramer, $\alpha_2\beta_2$. The α chains in normal human hemoglobin contain 141 amino

WILD-TYPE $\quad A \xrightarrow{e_1} B \xrightarrow{e_2} C \xrightarrow{e_3} D \longrightarrow Arg$

MUTANT 1 $\quad A \xrightarrow{e_1} B \xrightarrow{e_2} C \xrightarrow{e_3} D \longrightarrow Arg$

MUTANT 2 $\quad A \xrightarrow{e_1} B \xrightarrow{e_2} C \xrightarrow{e_3} D \longrightarrow Arg$

MUTANT 3 $\quad A \xrightarrow{e_1} B \xrightarrow{e_2} C \xrightarrow{e_3} D \longrightarrow Arg$

16-28 Mutants of Neurospora with defective enzymes
Based on nutritional mutant studies, Beadle and Tatum were able to establish the sequence of defective enzymes in the biosynthetic pathway for arginine.

[5] This suggests an explanation for certain of the nutritional needs of humans and other complex organisms. For example, vitamin B_1 (thiamine) is as necessary in human metabolism as in *Neurospora* metabolism. But the normal (wild-type) *Neurospora* can make it from simpler compounds and humans cannot. Hence, humans require thiamine in their diets; they are in fact thiamine-requiring auxotrophs. Since a vitamin is an essential substance that an organism cannot make for itself, thiamine is a vitamin for humans and for a *Neurospora* mutant—but not for wild-type *Neurospora*. We may reasonably speculate that our remote evolutionary ancestors could make thiamine but that, through mutation, the enzymatic machinery for its synthesis was lost. Had there been no organisms in the environment to make it for the first mutants, they too might have perished.

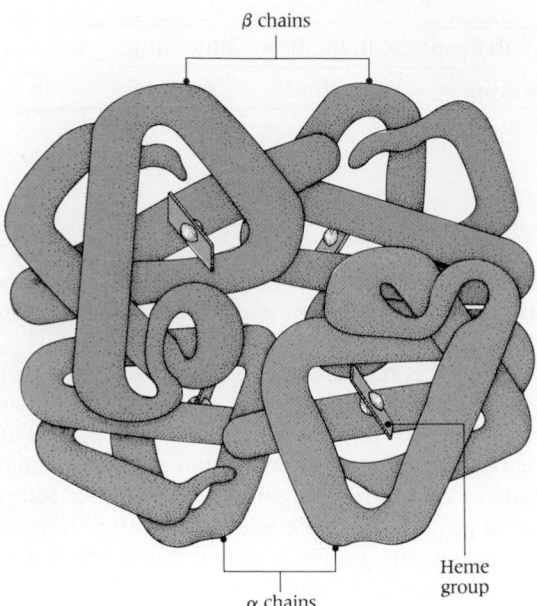

β chains

α chains

Heme group

16-29 Structure of hemoglobin
Hemoglobin is composed of two alpha and two beta chains plus four heme groups—one on each chain.

acids; the β chains contain 146 amino acids. If the β-chain gene undergoes mutation, both β chains (and neither α chain) are affected in a homozygous offspring; half of the β chains (and none of the α chains) are affected in heterozygotes. Clearly, synthesis of each chain is ordered by a different gene.

In 1948, Linus Pauling and his student Harvey Itano were probing the cause of sickle cell anemia, an hereditary blood disease that chiefly affects black people. They found it to involve an abnormal hemoglobin molecule that undergoes physical deformation, or "sickling," when oxygen is in short supply. The deformed red blood cells then obstruct blood flow. Pauling and Itano termed the abnormal molecule **hemoglobin S.** Normal hemoglobin was renamed **hemoglobin A.**

They knew that hemoglobin S is abnormal when they observed its abnormal rate of migration in an electrophoresis apparatus, a simple device for measuring the rate at which dissolved proteins (or other electrically charged molecules) migrate toward a positive or negative electrode (whichever is of opposite charge) (Figure 16-30). Since migration rates in this device depend on net electrical charge on the moving molecules, their surprising finding meant that hemoglobin S must have a different electric charge than hemoglobin A. Calculations revealed that it has, in fact, one less negative charge than hemoglobin A. Since the charge on a protein is the sum of the charges on its component amino acids (see Figure 4-14), it was clear that hemoglobin S must have a different amino acid composition than hemoglobin A.

In 1956, Vernon Ingram of M.I.T. found the answer. The abnormal hemoglobin of sickle cell anemia is due to the replacement of one amino acid in the β chains. Instead of ending with the normal sequence:

Val-His-Leu-Thr-Pro-Glu-Glu-Lys-. . .
1　　2　　3　　4　　5　　6　　7　　8

each abnormal molecule has the following sequence at its N-terminal end:

Val-His-Leu-Thr-Pro-*Val*-Glu-Lys-. . .
1　　2　　3　　4　　5　　6　　7　　8

Valine has replaced glutamate in position 6. As shown in Figure 4-14, glutamate possesses a negatively charged carboxyl group; valine does not. This accounts for the slow movement of the abnormal hemoglobin in an electrical field, the property that had led Pauling and Itano to its discovery. Geneticists quickly showed that the hemoglobin S molecule is transmitted as a simple Mendelian trait, with homozygotes (SS) exhibiting severe **sickle cell anemia** and heterozygotes (SA) a milder disorder known as **sickle cell trait,** in which only half the β chains are abnormal.

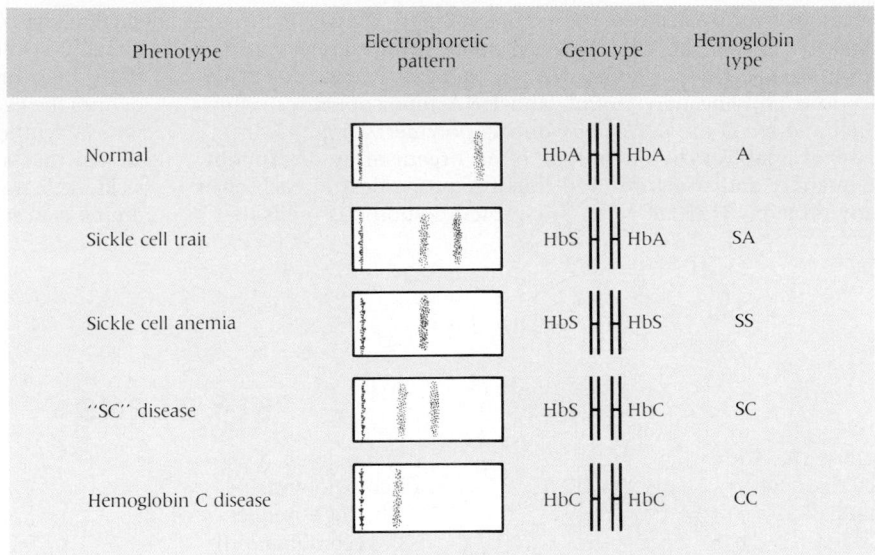

Phenotype	Electrophoretic pattern	Genotype	Hemoglobin type
Normal		HbA ‖ HbA	A
Sickle cell trait		HbS ‖ HbA	SA
Sickle cell anemia		HbS ‖ HbS	SS
"SC" disease		HbS ‖ HbC	SC
Hemoglobin C disease		HbC ‖ HbC	CC

Many aspects of this story are striking. To begin with, it was now clear that a gene mutation results in a specific and localized change in the amino acid sequence of a single protein. Thus, it gave support to the one gene, one polypeptide hypothesis of Beadle and Tatum. It seemed likely that the total amino acid sequence of hemoglobin is ordered by the genes governing the synthesis of two polypeptides—the α and β chains. These findings also suggested that the only business of genes is to order amino acid sequences. This concept was soon validated by studies of other proteins showing precise relationships between the order of mutable sites in their genes and the order of corresponding amino acid replacements. We conclude that each amino acid in a polypeptide chain is there because a specific portion of a gene orders it to be there.

SUMMARY

Most genes are composed of deoxyribonucleic acid (DNA); only those of certain viruses consist of ribonucleic acid (RNA). The total length of DNA in a prokaryote like *E. coli* is about 1.4 mm. Eukaryotic cells contain much larger amounts, ranging to about 1000 mm in human cells. However, only a small fraction of the material in each chromosome encodes genetic information. Most eukaryotic DNA is found in chromosomes, but small amounts occur in mitochondria and choroplasts.

Experimental approaches that established DNA as the genetic material related to bacterial transformation, viral replication, transduction, and conjugation. Transformation occurs when one strain of bacteria acquires genetic characters from a second strain after being exposed to a cell-free extract containing DNA from the second strain. Studies of viral replication showed that when bacteriophages infect a host cell and reproduce within it, only their DNA enters the cell. This DNA usually causes the metabolic machinery of the host cell to produce new viruses. In some cases, viral DNA becomes incorporated into host cell DNA. In transduction, the reverse process occurs: viruses may pick up pieces of host cell DNA and, when they subsequently infect a second cell, may transfer that DNA (and its genetic information) to the new host cell. Transduction is now widely used to introduce foreign genes into cells. Conjugation is the direct exchange of genetic information between two bacteria. It is thus a primitive form of sexual reproduction. The material transferred between conjugating organisms is DNA.

Nucleic acids are linear polymers of nucleotides, which are molecules containing an organic nitrogenous base (a purine or a pyrimidine), a sugar (ribose in RNA, deoxyribose in DNA), and phosphate in covalent linkage. It is the sequence of bases in polynucleotide strands that encodes genetic information. The bases found in nucleic acids include the purines adenine and guanine, and the pyrimidines cytosine and thymine or uracil. Thymine is present only in DNA; uracil only in RNA. Complementary pairing of bases in nucleic acids means that adenine in one polynucleotide strand can share hydrogen bonds only with thymine (or uracil) in a counterpart strand; similarly, cytosine can pair only with guanine.

DNA, unlike RNA, is a double helix. Its two polynucleotide strands are held together by hydrogen bonds between complementary base pairs. The two strands run in opposite (antiparallel) directions. Although RNA is a single-stranded molecule, it frequently doubles back upon itself, forming hydrogen bonds between short sequences of complementary bases. Each RNA strand is complementary to the DNA strand from which it was originally transcribed.

When DNA is synthesized from its four deoxyribonucleotide precursors (dATP, dGTP, dCTP and dTTP), its base sequence is copied from that of a preexisting molecule of DNA that serves as a template. This makes possible the preservation of genetic information from generation to generation. DNA replication is semiconservative. That is, each new DNA molecule consists of one strand from a preexisting DNA molecule and a second strand that has been newly synthesized in the complementary image of the old strand. More than 20 enzymes participate in the replication of a DNA molecule. These include DNA polymerases, which add nucleotides to a growing new strand, and DNA topoisomerases, which uncoil and recoil the DNA helix during polymerization.

Genes establish the phenotype of an organism by determining the quantity and structure (and thus the properties) of each of its many proteins. The one gene, one protein hypothesis holds that a single gene controls the synthesis of each protein. Those proteins containing more than one polypeptide chain are determined by more than one gene, one gene specifying each subunit. A gene defines the exact amino acid sequence of the protein or polypeptide whose synthesis it controls. Studies of hemoglobin first demonstrated that a gene mutation, in this case the one producing the sickle cell trait, results in a specific and localized change in the amino acid sequence of that protein.

KEY TERMS

adenine (A, Ade)
bacteriophage
conjugation
cytosine (C, Cyt)
deoxyribonucleic acid (DNA)
deoxyribonucleotide
deoxyribose
DNA gyrase
DNA ligase
DNA polymerase
DNA topoisomerase
episome
fertility factor (F factor)
guanine (G, Gua)

hemoglobin A
hemoglobin S
Hfr (high frequency of recombination)
lagging strand
leading strand
lysogeny
minimal medium
nitrogenous base
nucleic acid
nucleoprotein
nucleotide
Okazaki fragments
one gene, one enzyme hypothesis

plasmid
prophage
purine
pyrimidine
ribonucleic acid (RNA)
ribonucleotide reductase
ribose
RNA polymerase
RNA primer
template
thymine (T, Thy)
transduction
uracil (U, Ura)

QUIZ QUESTIONS

1. Which of the following is *not* a chemical component of deoxyribonucleic acid?
   A. uracil
   B. a pentose
   C. a phosphate group
   D. selected pyrimidines
   E. selected purines

2. Which of the following regarding DNA is *not* true?
   A. DNA is double-stranded in most organisms that possess DNA.
   B. The amount of thymine equals the amount of adenine.
   C. Complementary base pairs are joined by hydrogen bonds.
   D. The "side rails" of the DNA double helix are made of alternating

   molecules of phosphate and ribose.
   E. Complementary base pairs consist of one purine and one pyrimidine.

3. Each gene codes for
   A. one enzyme.
   B. one protein.
   C. one polypeptide.
   D. one nucleotide.
   E. one nucleoprotein.

4. Replication of DNA in organisms that possess double-stranded DNA is semi-conservative. This means that
   A. the old DNA molecule passes intact from the parental cell to a daughter cell.

   B. the new DNA molecule has one old strand and one new complementary strand.
   C. the new DNA molecule is entirely new and the old DNA molecule remains with the parental cell.

5. Important experiments that provided evidence that the chemical identity of the genetic material is DNA were conducted by
   A. Tatum and Lederberg.
   B. Avery, MacLeod, and McCarty.
   C. Griffith.
   D. A and B
   E. A, B, and C

ESSAY QUESTIONS

1. How do DNA and RNA differ in structure?

2. In addition to DNA polymerase III, what other enzymes are essential in the synthesis of a new strand of DNA?

3. What information did studies of sickle cell hemoglobin provide regarding the relationship of genetic mutation to protein structure?

4. If RNA is copied from a DNA template by RNA polymerase, what is the polarity of the nascent RNA chain relative to the template strand of DNA?

REFERENCES AND SUGGESTIONS FOR FURTHER READING

Adams, R. L. P., Knowler, J. T., and Leader, D. P. (1986). *The Biochemistry of the Nucleic Acids*. 10th ed., Chapman & Hall, New York.

A clear account of gene structure and expression.

Dickerson, R. E. (1983). The DNA helix and how it is read. *Scientific American 249,* December, 94–111.

A typically well-illustrated review.

Kelly, T., and McMacken, R., (Eds.) (1986). *Mechanisms of DNA Replication and Recombination*. Alan R. Liss, New York.

A collection of scholarly essays summarizing the state of knowledge.

Kornberg, A. (1984). DNA replication. *Trends in Biochemical Sciences 9,* 122–124.

Another authoritative discussion by an eminent investigator.

Kornberg, A. (1989). *For the Love of Enzymes: The Odyssey of a Biochemist*. Harvard University Press, Cambridge, Mass.

In this unique volume, Kornberg offers fascinating insights into the great works he performed and the life and thought of the man behind them.

McCarty, M. (1985). *The Transforming Principle: Discovering that Genes are Made of DNA*. W. W. Norton, New York.

A readable and moving historical reminiscence by one of the discoverers. Inexplicably, this monumental contribution had little impact on contemporary genetics and was never recognized by the Nobel Prize committee.

Radman, M., and Wagner, R. (1988). The high fidelity of DNA duplication. *Scientific American 259,* August, 40–46.

An explanation of why there are so few mistakes in the copying of DNA in the countless cell divisions occurring down through the generations.

Rawlings, C. J. (1986). *Software Directory for Molecular Biologists: A Complete Guide to the Selection of Computer Software for the Management and Analysis of Molecular Sequences*. Macmillan, New York.

For those interested in learning what computers are contributing to molecular biology, this volume provides a fascinating glimpse of the programs now available for the analysis of nucleotide sequences of DNA.

Sanger, F. (1988). Sequences, sequences, and sequences. *Annual Review of Biochemistry 57,* 1–28.

Delightful autobiographical essay by the astute investigator who won a Nobel Prize for sequencing a protein (insulin) and then won another for sequencing DNA.

Stanier, R. Y., Ingraham, J. L., Wheelis, M. L., and Painter, P. R. (1986). *The Microbial World,* 5th ed. Prentice-Hall, Englewood Cliffs, N.J.

Good coverage of genetic exchange and recombination in microorganisms.

Wang, J. C. (1982). DNA topoisomerases. *Scientific American 247,* July, 94–109.

A coherent explanation of a challenging topic.

MECHANISMS OF GENE ACTIVITY

Mechanism of gene activity
The expression "nurture and nature" suggests the dual control of an organism's traits. Both factors are apparent in these tidy Dutch tulip fields. Nurture is represented by abundant sunshine, water, and fertilizer. Nature consists in the genes that determine species, structure, and function. All of these tulips live in the same bountiful environment. Yet some are yellow and some are red, because each strain has a different set of color genes. This chapter describes how genes determine genetic traits.

We have already considered the theoretical argument that the physical substance of genes is more likely to consist of a few large molecules than a large collection of small ones in which random thermal motion would probably prevent necessary stability. With the recognition that the genetic substance is DNA—a large "chain" made up of "links" of four kinds—it was a small step to the realization that the vast numbers of genes in the living world differ only in the sequences of those four units. By analogy, all existing books in English may be said to differ only in the sequences of the letters of the alphabet. Only a code based on nucleotide sequences could account for the bewilderingly large numbers of genes and, hence, of the many specific proteins in living organisms.

HOW DNA CONTROLS PROTEIN SYNTHESIS

How does the information-containing nucleotide sequence of DNA determine the amino acid sequence of a protein? The question arouses curiosity because proteins contain 20 kinds of amino acids, while DNA contains only 4 kinds of nucleotides. To answer it we must trace the flow of genetic information from a nucleotide sequence in DNA to the final steps of protein synthesis in prokaryotic and eukaryotic cells.

FLOW OF GENETIC INFORMATION IN CELLS: THE "CENTRAL DOGMA"

In the following "road map" of the pathways of information transfer in eukaryotic cells, arrows indicate the direction of transfer.

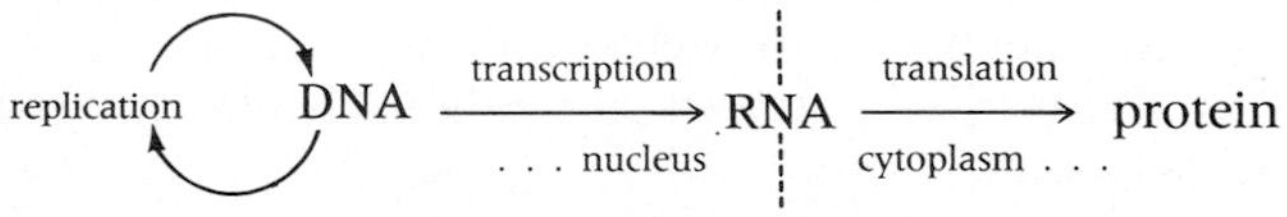

$$\text{replication} \ \circlearrowleft \ \text{DNA} \xrightarrow[\text{. . . nucleus}]{\text{transcription}} \text{RNA} \xrightarrow[\text{cytoplasm . . .}]{\text{translation}} \text{protein}$$

The map portrays three key events.

■ First, DNA itself repeatedly undergoes **replication** using a preexisting DNA strand as a template. This, we recall, occurs in the S phase of eukaryotic cell division—just before mitosis.

■ Second, DNA exercises its main function in **transcription.** This occurs along the DNA in chromosomes (Figure 17-1A). In this process, DNA directs the synthesis of a type of RNA called **messenger RNA** (abbreviated **mRNA**). Each nucleotide in mRNA is complementary to a nucleotide in one of the DNA strands. Hence, mRNA is a complementary image of the parent DNA molecule that effectively preserves its information content (the nucleotide sequence) in a new and usable form. Transcription is an apt term. As a mode of information transfer,

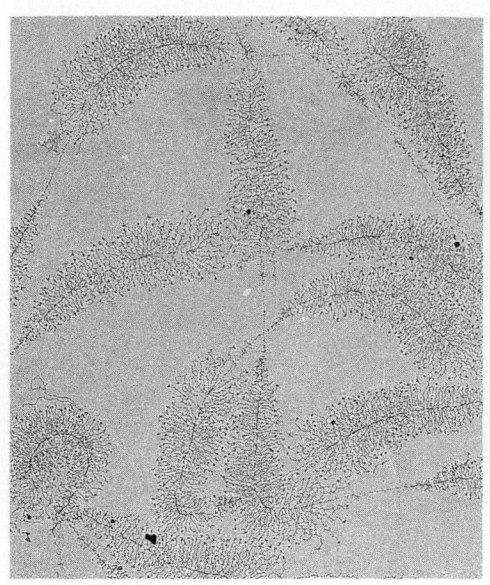

A   B

it resembles the stamping out of positive phonograph records by a negative master record.

■ The third step is **translation,** so called because in this step a 4-letter nucleic acid language is translated into a 20-letter protein language. In this step, which occurs on ribosomes (Figure 17-1B), a specific mRNA molecule guides the synthesis of a specific protein, or of one of its component polypeptide chains if the whole protein (like hemoglobin) contains more than one polypeptide.

In a word, then, the amino acids of all proteins are directly determined by RNA templates previously constructed upon DNA templates. The above scheme has been called "the central dogma" of molecular biology. Its highlights are as follows:

■ DNA is the primary information carrier.
■ The direction of information transfer is from DNA to RNA to protein; that is, the last two arrows point in one direction—away from DNA.
■ Only DNA self-replicates. This means that DNA is never made on an RNA or protein template (or almost never: exceptions come later). Likewise, RNA sequences are never copied on protein templates. Rather, RNA is synthesized from a DNA template. Information that has passed from DNA and RNA to protein has reached the end of the line and is not used again.

Transcription of the DNA Sequence: Messenger RNA

How is the information in nuclear DNA transferred to a protein being synthesized in the cytoplasm? The answer began to emerge in 1961 with the discovery that in *E.coli* (which is a prokaryote and thus lacks a nucleus) DNA directs the synthesis of mRNA. Soon thereafter a similar mechanism was found in eukaryotic cells. An enzyme—**RNA polymerase**—catalyzes the synthesis of mRNA only when template DNA is present, as follows:

$$\left.\begin{array}{l} n\ ATP \\ n\ CTP \\ n\ UTP \\ n\ GTP \end{array}\right\} \xrightarrow[\text{DNA template}]{\text{RNA polymerase}} RNA + nPP_i$$

Only one DNA strand governs the synthesis of mRNA. This is the **sense strand,** or **template strand.** The other remains untranscribed. Each DNA nucleotide in the sense strand attracts a complementary ribonucleotide. That is, a G in DNA attracts a C, which then enters the RNA polymer in the proper place; A attracts U (the ribonucleotide equivalent of T), and so on (Figure 17-2). The DNA thus gives form and sequence to the mRNA molecule being assembled one nucleotide at a

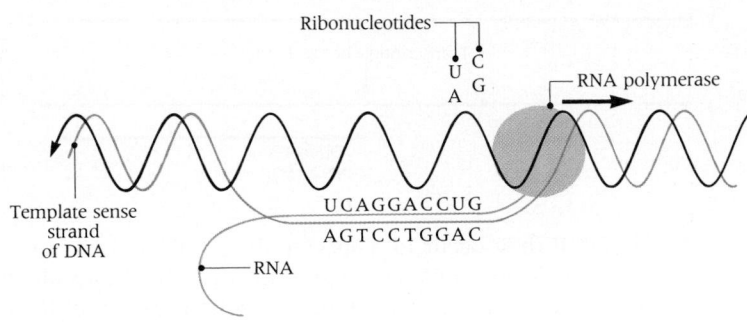

17-2 Synthesis of RNA on a DNA template
The template strand of DNA is transcribed into an RNA molecule by RNA polymerase and free ribonucleotides.

Ribonucleotides

U
C
G
A

RNA polymerase

Template sense
strand
of DNA

UCAGGACCUG
AGTCCTGGAC

RNA

time. DNA sequences that contain information specifying the structure of an mRNA molecule—and thus of a protein—are called **structural genes.**

DNA directs the synthesis of several kinds of RNA.

- **Messenger RNA (mRNA),** which directs protein synthesis
- **Transfer RNA (tRNA),** a set of adaptor molecules that individually carry amino acids to the sites of protein synthesis
- **Ribosomal RNA (rRNA),** a constituent of the ribosome, the organelle on which protein synthesis takes place
- **Heterogeneous nuclear RNA (hnRNA),** a class of RNA molecules that in eukaryotes are the immediate precursors of mRNA

Only mRNA is "stamped" with the genetic information needed to guide protein synthesis. In the fullest sense of the word, it is a messenger sent by DNA.

Until recently, there was no knowledge of the actual sequences of nucleotides (or their bases) in different DNAs. We knew the total amounts or relative percentages of A, T, G, and C. However, years were to pass before methods were devised for "sequencing" DNA. In one of two major approaches (Figure 16-19), the DNA base sequence is read simply by interpreting a ladder of bands in four electrophoretic lanes.[1] Today it is commonplace to see published DNA sequences like the one in Figure 17-3. Scientists are rapidly learning to recognize the special meanings of many subsequences within these sequences (now on file in computer databases the world over).

RNA transcription occurs in three major steps.

- Initiation of transcription
- Elongation of the growing RNA molecule
- Termination of transcription

The mechanism of **initiation** clearly illustrates that investigators are learning to read DNA sequences. Within the cell, RNA polymerase molecules constantly impinge upon DNA. However, these random encounters lead to tight binding between

17-3 Partial base sequence of DNA of a bacterial virus
This is a small portion of the DNA sequence from bacteriophage φX174, which has a single circular strand of only 5375 nucleotides for part of its life cycle.

[1] The sequencing procedure shown in Figure 16-19 is that of Allan Maxam and Walter Gilbert. It relies on different chemical methods that disrupt the DNA strand next to one or the other of its four bases.

5′ end ... ACAGACCTATAAACATTCTGTGCCGCGTTTCTTTGTTCCTGAGCATGGCACTATGTTTACTCTTGCGCTGGTTCGTTTTCCGCCTACTGCGACTAAAGAG

GGAAGTATCTTTAAAGTGCGCCGCCGTTCAACGGTATGTTTTGTCCCAGCGGTCGTTATAGCCATATTCAGTTTCGTGGAAATCGCAATTCCATGACTTA

ATGTTTTCCGTTCTGGTGATTCGTCTAAGAAGTTTAAGATTGCTGAGGGTCAGTGGTATCGTTATGCGCCTTCGTATGTTTCTCCTGCTTATCACCTTCT

TTGACTTGCTGACTTTGTGACCAGTATTAGTACCAACGCTTATTCATGCGCAAGAACGTTTAGTGGTCTTCCGCCAAGGACTTACTTACCCTTCGGAAGT

GTTGCAGTGGATAGTCTTACCTCATGTGACGTTTATCGCAATCTGCCGACCACTCGCGATTCAATCATGACTTCGTGATAAAAGATTGAGTGTGAGGTTA

TTTCAGACTTTGTACTAATTTGAGGATTCGTCTTTTGGATGGCGCGAAGCGAACCAGTTGGGGAGTCGCCGTTTTTAATTTTAAAAATGGCGAAGCCAAT

TATTTCTCGCCACAATTCAAACTTTTTTTCTGATAAGCTGGTTCTCACTTCTGTTACTCCAGCTTCTTCGGCACCTGTTTTACAGACACCTAAAGCTACA

GGACTAATCGCCGCAACTGTCTACATAGGTAGACTTACGTTACTTCTTTTGGTGGTAATGGTCGTAATTGGCAGTTTGATAGTTTTATATTGCAACTGCT

TTGTTTCAGTTGGTGCTGATATTGCTTTTGATGCCGACCCTAAATTTTTTGCCTGTTTGGTTCGCTTTGAGTCTTCTTCGGTTCCGACTACCCTCCCGAC

TAACGGCCCGCATGCCCCTTCCTGCAGTTATCAGTGTGTCAGGAACTGCCATATTATTGGTGGTAGTACCGCTGGTAGGTTTCCTATTTGTAGTATCCGT

AACGTCTACGTTGGTTTCATGGTTTGGTCTAACTTTACCGCTACTAAATGCCGCGGATTGGTTTCGCTGAATCAGGTTATTAAAGAGATTATTTGTCTCC ... 3′ end

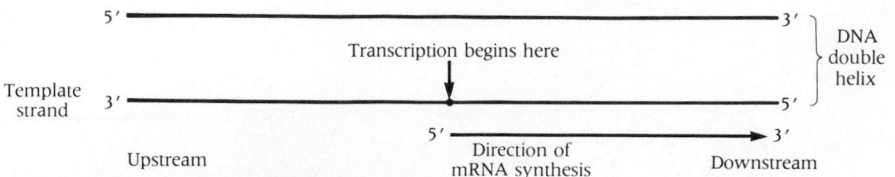

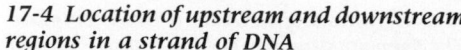

17-4 Location of upstream and downstream regions in a strand of DNA
Note the locations of the 3' and 5' ends of the template strand of DNA. Synthesis of mRNA occurs in a 5' → 3' direction.

enzyme and DNA only if they occur in a specific region of the DNA molecule. To simplify discussion of the location of such regions in the DNA molecule, the terms *upstream* and *downstream* are used to refer to locations nearer the 3' or 5' end, respectively, of the DNA strand or nearer the 5' or 3' end, respectively, of the mRNA strand (Figure 17-4). The DNA regions that bind RNA polymerase are called **promoters.** They are just upstream from the **initiation site,** which includes the first DNA base to be copied into RNA by the RNA polymerase and thus marks where transcription begins (Figure 17-5). Within two promoter regions there are

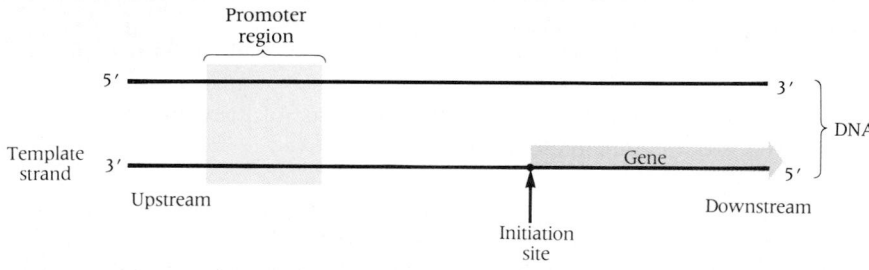

17-5 Relative positions of promoter and transcription initiation site in DNA
Note that the promoter is upstream from the gene being transcribed.

sequences of six bases that are very similar in different species. One is about 10 bases upstream from the initiation site (Figure 17-6).

Some 11 to 15 bases farther upstream in prokaryotic cells is another sequence that, with some variation, appears almost universally (Figure 17-7A). This is the **−35 region,** named for its location (in number of bases upstream from the initiator, whose first base is arbitrarily designated +1). It induces tight binding between polymerase and DNA. Note that each gene cluster has its own promoter sequence.

The arrangement is more complicated in eukaryotic cells. As shown in Figure 17-7B, the DNA sequences necessary for transcription are found in at least three regions close to the RNA initiation site: the so-called **TATA box** (a widely occurring regulatory sequence containing only adenine and thymine), and two regions located upstream between positions −110 and −40, one of which often includes the so-called **CAAT box,** a sequence containing cytosine, adenine, and thymine. Removal of these regulatory regions decreases the ability of a cloned gene to be transcribed by RNA polymerase. However, it is still not clear how these several eukaryotic promoter sites participate in the regulation of mRNA synthesis.

We have spoken of RNA polymerase as though it were a single enzyme. Prokaryotic cells do have only one RNA polymerase that makes all types of RNA. However,

17-6 Promoter sequences of various viral and bacterial genes
The two boxes, respectively 35 and 10 bases upstream from the initiation site, contain promotor sequences that bind RNA polymerase. The "consensus sequences" of many prokaryotes are shown at the top.

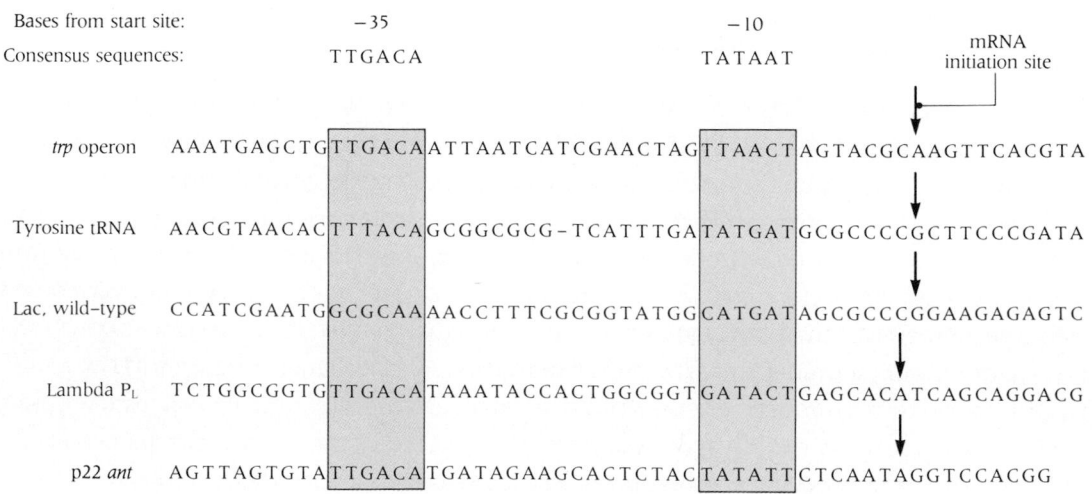

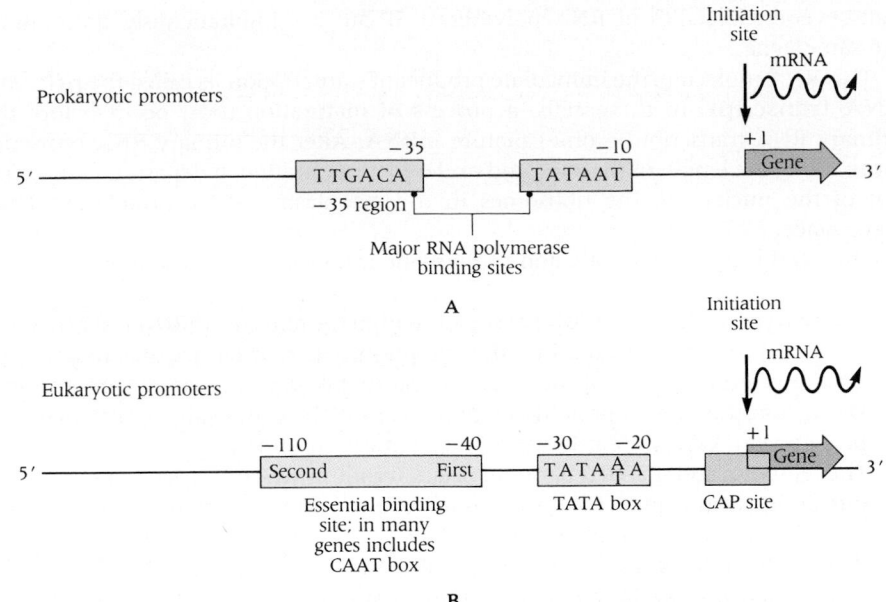

17-7 Transcriptional control elements (A) *Prokaryotic promoters include two 6-base sequences at 10 and 35 bases upstream from the mRNA initiation site.* (B) *At least three DNA sequences occur in eukaryotes, including the TATA box and two upstream elements between −110 and −40, which often includes the CAAT box.*

eukaryotic cells have three RNA polymerases (termed RNA polymerases I, II, and III) that make the different types of RNA in the cell nucleus. We are here concerned with RNA polymerase II, which is involved in the synthesis of mRNA.

The two DNA strands unwind and separate in the course of mRNA synthesis. As noted, only the sense strand serves as template in mRNA synthesis. Obviously, if both DNA strands were templates, a single gene would produce two different mRNAs with complementary nucleotide sequences. Evidence that this does not occur comes from genetic studies, which reveal that one gene leads to the synthesis of only one polypeptide chain, and from direct analyses of RNA-DNA hybrids.

When a molecule of RNA polymerase II is bound to an appropriate promoter site on the DNA molecule, transcription begins with the insertion of an ATP or GTP. In contrast with the pattern seen with nucleotides incorporated into the RNA chain during the elongation phase, the *initiating* ATP or GTP retains the three phosphate groups that made it ATP (or GTP).

In the **elongation** phase of transcription, complementary ribonucleotides line up along one of the two unwinding DNA strands of the structural gene. RNA polymerase II then links them to form a single-stranded RNA molecule (Figure 17-8). The RNA strand continues to elongate in the 5' → 3' direction until the polymerase reaches a specific short sequence in the DNA template. This position, called a **terminator sequence,** is read as a stop signal. Transcription is terminated and the completed RNA molecule (and then the RNA polymerase molecule) is released from the DNA. Note that while mRNA is being synthesized it does not remain hydrogen bonded to the DNA. Rather it dissociates from it completely except in the region of the moving RNA polymerase II molecule. This means

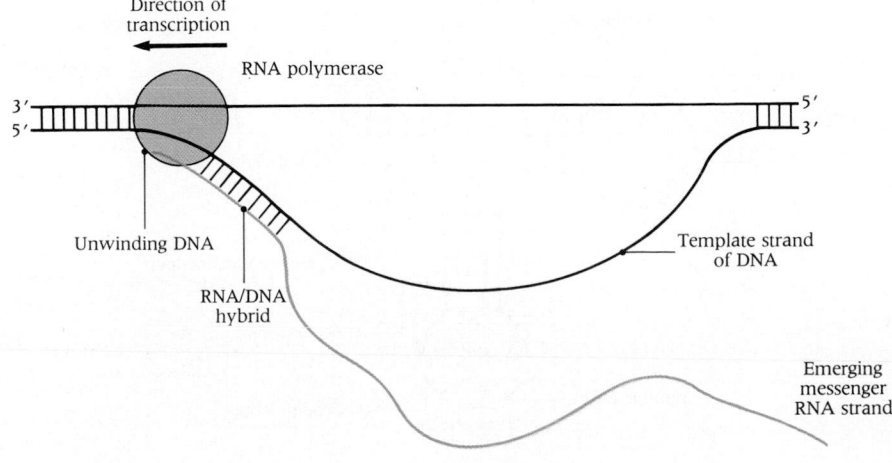

17-8 Messenger RNA dissociates from the DNA template strand during transcription *The strand of mRNA elongates as it emerges from the complex of RNA polymerase and template DNA. As DNA unwinds, a short RNA/DNA hybrid forms and unforms as the RNA chain is expelled.*

that several molecules of RNA polymerase II can be simultaneously transcribing the same gene.

This RNA molecule, the immediate product of transcription, is called the **primary RNA transcript.** In these cells, a process of maturation must occur before the primary RNA transcript becomes mature mRNA. After the primary RNA transcript forms along the length of the gene and undergoes maturation, it departs and migrates out of the nucleus to the ribosomes in the cytoplasm, where protein synthesis takes place.

The synthesis of eukaryotic and prokaryotic mRNA differs in three ways.

- In eukaryotic cells, chromosomal DNA is within the nucleus; mRNA is synthesized there and then transported into the cytoplasm where it participates in protein synthesis. Prokaryotic cells have no nuclear membrane to separate chromosomal DNA from the sites of protein synthesis and mRNA can begin to participate in protein synthesis while it is still being synthesized on DNA.
- The temporal and spatial separation of eukaryotic mRNA synthesis and translation into amino acid sequences allows this molecule to undergo a number of biochemical modifications, such as splicing.
- Prokaryotic mRNA has a very short half-life—seconds to minutes. Eukaryotic mRNA may survive in the cytoplasm for many hours.

Transfer RNA and the Process of Translation

How is a particular nucleotide sequence of mRNA translated into a specific amino acid sequence? The answer: protein synthesis takes place on the mRNA molecule, which serves as a blueprint after reversibly associating itself with ribosomes—in the cytoplasm in eukaryotes, near the DNA in prokaryotes. In eukaryotes, each mRNA strand usually becomes associated with clusters of 3 to more than 40 ribosomes called **polyribosomes** or **polysomes.**

Interestingly, free amino acids in the cell have no affinity whatsoever for mRNA molecules. Instead, the amino acids must be brought to the ribosome and its associated mRNA by adaptor molecules, which do have an affinity for mRNA. These adaptors are made of the type of RNA called **transfer RNA (tRNA).** A different tRNA (in some cases more than one) exists for each of the amino acids.

In 1965, Robert W. Holley and co-workers managed to separate the different tRNAs and succeeded in establishing the entire nucleotide sequence of one of them— the tRNA from yeast that is specific for the amino acid alanine (Figure 17-9A). For this achievement, he received the Nobel Prize in 1968. Although the molecule contains only 76 nucleotides—most other RNA molecules contain thousands—it took Holley and his many graduate students seven years of work and consumed a full gram of alanine tRNA that had to be purified from 300 pounds of yeast cells.

17-9 The nucleotide sequence of two transfer RNAs
(A) *Yeast alanine tRNA.* **(B)** *E. coli phenylalanine tRNA. Color indicates unusual bases.*

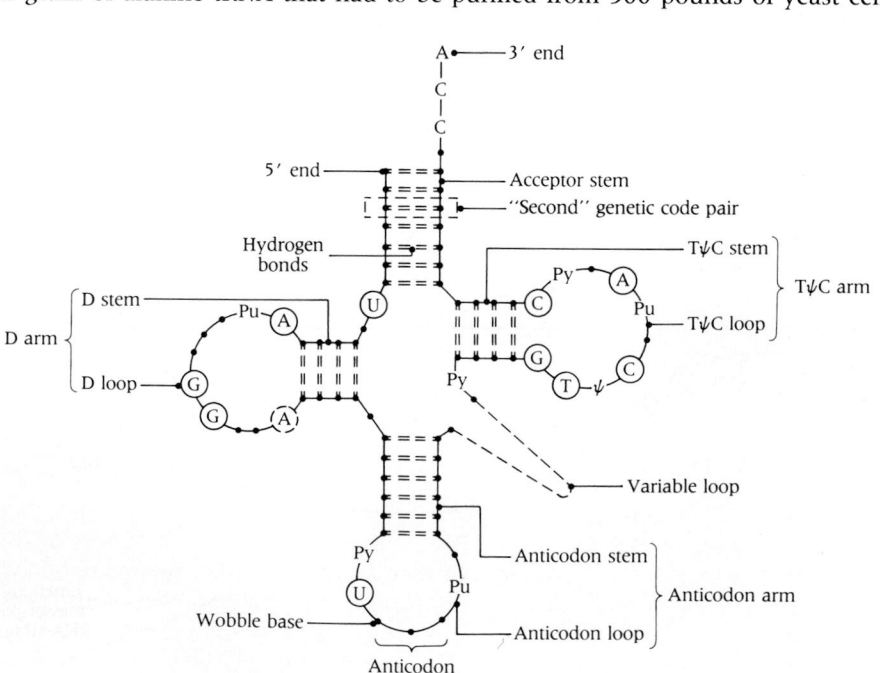

17-10 Cloverleaf structure of transfer RNA of prokaryotes and eukaryotes
In this two-dimensional view, three arms are clearly visible. The anticodon arm binds messenger RNA during protein synthesis. Circled bases occur in all tRNAs.

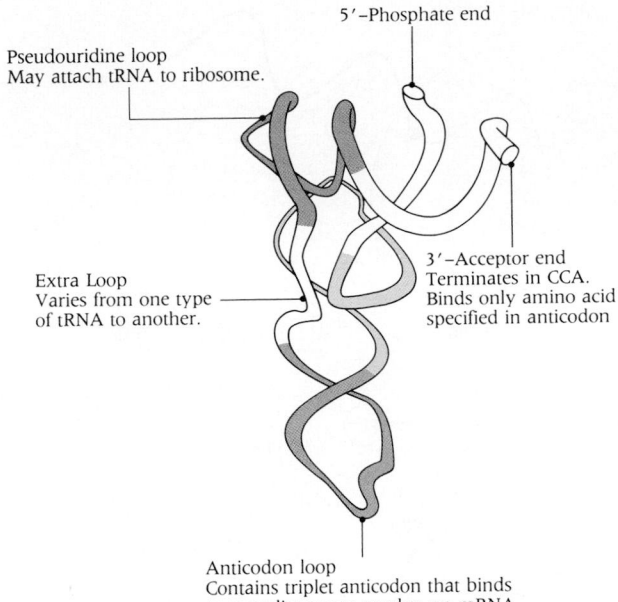

5'–Phosphate end

Pseudouridine loop
May attach tRNA to ribosome.

3'–Acceptor end
Terminates in CCA.
Binds only amino acid
specified in anticodon

Extra Loop
Varies from one type
of tRNA to another.

Anticodon loop
Contains triplet anticodon that binds
to complimentary condon on mRNA

A year later, the sequence of the 81 nucleotides of serine tRNA was reported by others. The two tRNA sequences were generally similar, but there were subtle and critical differences. Since this pioneering work, many tRNAs have been fully sequenced (Figure 17-9B).

Hundreds of tRNAs have now been purified and sequenced from different cell types and species. Although differences exist among them, the following four features are seen in all. These are well illustrated by the tRNAs in Figure 17-9 (which in proper terminology are now symbolized tRNA^{ala} and tRNA^{phe}) and the summary diagram in Figure 17-10.

- Each molecule has a cloverleaf shape held together by hydrogen bonds between paired complementary bases that find each other with unerring accuracy. The result is a characteristic array of stems, loops, and arms. In 1973, Alexander Rich of M.I.T. worked out the three-dimensional structure of phenylalanine tRNA (Figure 17-11).

- tRNAs contain a number of unusual bases—that is, purines and pyrimidines differing slightly from the usual ones, A, G, C, and U. In Holley's sequence of 77 bases (in alanine tRNA), nine are in this category (Figure 17-9A). Many were made unusual by the presence of one or more attached methyl (—CH$_3$) groups that had been added after the RNA chain was formed. In Figure 17-9, these are abbreviated m$_1$G (methyl G), m$_2$G (dimethyl G), and so on. Note that tRNA contains at least one thymine, a pyrimidine previously thought to occur only in DNA. The role of the unusual bases, which occur only in tRNA, is not yet known. Perhaps they influence the attachment of tRNA to ribosomes.

- The 3' end of every tRNA molecule terminates in the same three-base sequence: CCA. Before an amino acid can be attached to its specific tRNA, it must be activated by a family of some 20 specific enzymes called **aminoacyl-tRNA synthetases,** which catalyze a two-step reaction sequence. As shown in Figure 17-12, ATP energy is utilized to activate a specific amino acid and attach it to the proper tRNA molecule at its acceptor end—the end that terminates with CCA—to form an amino acid-tRNA complex. The complex then moves into position on the mRNA bound to the ribosomes (Figure 17-13). The activation of amino acids by ATP is one of the important points at which energy is needed to make proteins.

- In 1988 Y.-M. Hou and P. Schimmel at M.I.T. finally considered a long-pending question: how does each type of tRNA recognize the particular amino acid for which it is specific? They found that each tRNA molecule contains two bases that encode that information. When, for example, these two bases (the G-U pair near the end of the acceptor stem) were deleted from tRNA^{ala}, this tRNA molecule no longer recognized alanine (see Figures 17-9A and 17-10). When

NH$_2$
|
R — C — COOH
|
H

Aminoacyl–tRNA synthetase (specific)

ATP

NH$_2$
|
R — C — CO — AMP   Activated amino acid
|
H   + PP$_i$

HO — A
|
C
|
C

tRNA (specific)

NH$_2$
|
R — C — CO — O — A
| |
H   C
|
C

Amino acid–tRNA
complex

+ AMP

17-12 Activation of an amino acid and formation of amino acid-tRNA complex
Each amino acid is activated by ATP and a specific synthetase. It then attaches to the CCA end of a specific tRNA to form an amino acid-tRNA complex.

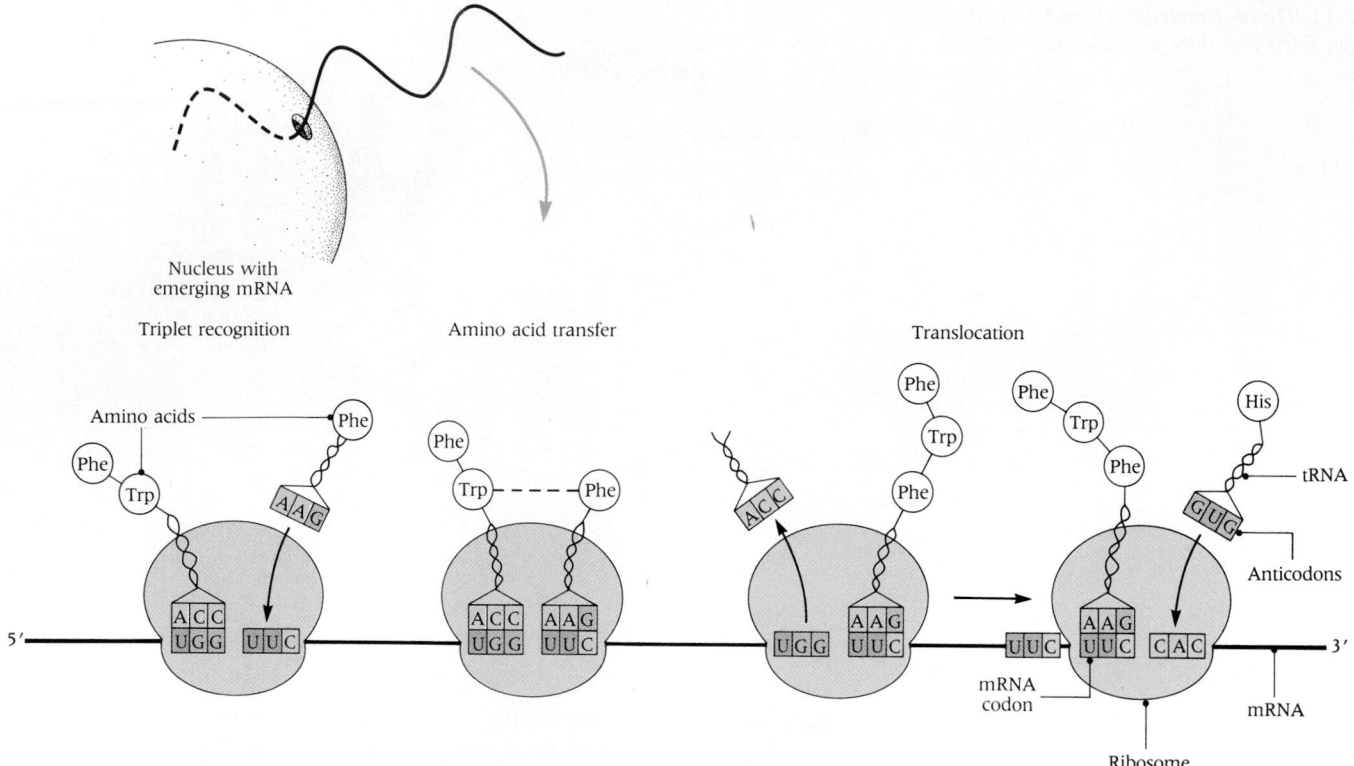

Nucleus with
emerging mRNA

Triplet recognition   Amino acid transfer   Translocation

17-13 *How proteins are made on the
ribosome in eukaryotes*
*Amino acids bound to tRNA approach the
ribosome and are incorporated into the growing
polypeptide chain. Coded information in the
messenger RNA ensures the correct amino acid
sequence.*

G and U were transferred to a molecule of tRNA^gly that normally binds glycine, the tRNA bound alanine. This two-base codescript, called a **paracodon,** has already been called biology's second genetic code. The first genetic code is discussed in the next section.

All tRNAs share a fifth feature, a critical role in translation, which we discuss now.

Translation and the Genetic Code

Each kind of tRNA may be compared to a bricklayer who is qualified to lay only one kind of brick. To get many amino acids into the right sequence many amino acid-bearing tRNAs must line up along the mRNA in the correct sequence. How this is done was unknown until investigators broke the genetic code.

In 1961, M. W. Nirenberg and J. H. Matthaei reported an exciting observation, made possible by the discovery of methods for producing synthetic mRNAs, the nucleotide content of which could be dictated by the experimenter. First, they prepared a synthetic polynucleotide containing only one kind of base, uracil (U). Thus it was called poly-U. When it was added to a test tube containing a cell-free system capable of synthesizing protein—one that already contained all necessary ingredients (amino acids, ATP, GTP, tRNAs, ribosomes, and enzymes) except mRNA—the system accepted it as mRNA and synthesized a "protein" containing only one repeating amino acid, phenylalanine! Evidently the base U (or some combination of Us) was the "code word," or **codon,** that caused phenylalanine to be inserted in a growing polypeptide chain.[2]

There were good reasons for believing that the correct code word for phenylalanine is UUU—and indeed, that the code words for all amino acids are such **triplets.** The nucleic acid alphabet has only 4 letters, but code words are needed for some 20 amino acids. If 2-letter code words were written from a 4-letter alphabet there could be only 4^2, or 16 possible codons—not enough to specify 20 amino acids.

---

[2] An author of this book was fortunate to be present at the International Congress of Biochemistry in Moscow in 1961 when Nirenberg first announced that UUU encodes phenylalanine. This announcement, coming less than a year after the famous "U2 incident" in which an American aircraft was shot down over the Soviet Union, excited biochemists in the corridors of Moscow University, who, predictably, began to refer to that year's great event as the "U3 incident."

TABLE 17-1
THE GENETIC CODE

Codon [5′ → 3′]	Amino acid*	Codon [5′ → 3′]	Amino acid*
UUU	Phe	AUU	Ile
UUC	Phe	AUC	Ile
UUA	Leu	AUA	Ile
UUG	Leu	AUG	Met
UCU	Ser	ACU	Thr
UCC	Ser	ACC	Thr
UCA	Ser	ACA	Thr
UCG	Ser	ACG	Thr
UAU	Tyr	AAU	Asn
UAC	Tyr	AAC	Asn
UAA	Stop	AAA	Lys
UAG	Stop	AAG	Lys
UGU	Cys	AGU	Ser
UGC	Cys	AGC	Ser
UGA	Stop	AGA	Arg
UGG	Trp	AGG	Arg
CUU	Leu	GUU	Val
CUC	Leu	GUC	Val
CUA	Leu	GUA	Val
CUG	Leu	GUG	Val
CCU	Pro	GCU	Ala
CCC	Pro	GCC	Ala
CCA	Pro	GCA	Ala
CCG	Pro	GCG	Ala
CAU	His	GAU	Asp
CAC	His	GAC	Asp
CAA	Gln	GAA	Glu
CAG	Gln	GAG	Glu
CGU	Arg	GGU	Gly
CGC	Arg	GGC	Gly
CGA	Arg	GGA	Gly
CGG	Arg	GGG	Gly

* Amino acid abbreviations are given in Table 3-3.

But a 3-letter word would offer 4^3 or 64 possible codons—more than enough to encode 20 different amino acids.

Imaginative experiments—done first with synthetic mRNAs of known composition and later by studying the binding of specific, pure tRNAs to nucleotide triplets of known sequence—revealed that each amino acid is encoded by a different three-nucleotide codon (Table 17-1). Note that some amino acids are specified by more than one codon. For that reason, we say in the language of cryptographers that the code is **degenerate.** This simply means that some code words are synonymous.

When the data were all in, it was clear that the first two nucleotides of the codon triplet are the most important. Variation or "wobble" in the third nucleotide is often permitted. Note in Table 17-1 that three of the 64 triplets do not code for amino acids. These are **stop signals** that terminate a polypeptide chain. They act as punctuation marks along the DNA strand, separating the sequences that comprise individual genes. All evidence indicates that the genetic code is universal, with few exceptions. The DNA of *Tetrahemena* and *Paramecium* show such exceptions, and the genetic code of mitochondrial DNA also shows interesting departures from the universal code.

Now reconsider protein synthesis in light of what we have just learned. The amino acid sequence of a new protein molecule is determined by the nucleotide sequence of the mRNA on the ribosome or polysome. Each tRNA contains a critically important nucleotide triplet, the **anticodon,** that is strategically located in the anticodon loop (see Figure 17-10). This unpaired anticodon triplet is complementary to the particular codon in the mRNA strand that specifies a given amino acid. Because some "wobble" is permitted in the third base of the triplet, some of the rare or unusual bases of tRNA occasionally appear in an anticodon triplet. For example, the anticodon of alanine tRNA (see Figure 17-9A) is CGI. I is the unusual base, hypoxanthine, which for base-pairing purposes resembles A.

The complementary mRNA codon for alanine is GCU (Table 17-1). The anticodon of phenylalanine tRNA is AAG. The complementary codon for phenylalanine is UUC. As in all cases involving complementary base-pairing, the tRNA anticodon and the mRNA codon are oriented in opposite directions in regard to 5' → 3' polarity when they become associated. Since codon base sequences are usually written in the 5' → 3' direction, anticodon triplets should be written in the opposite direction. For example, the anticodon that recognizes the codon UUC is written AAG, not GAA:

Codon   (mRNA)   (5') — U U C — (3')
Anticodon (tRNA)   (3') — A A G — (5')

Next, a tRNA molecule—for example, one carrying the anticodon AAG—is pulled in by a UUC codon in an imaginary mRNA molecule consisting (for simplicity) of alternating codons for phenylalanine (UUC) and tryptophan (UGG). The anticodon pairs up with its complementary codon and is bound to it by hydrogen bonds. As shown in Figure 17-13, several phenylalanines and a tryptophan have already linked together. A new histidine is about to be added. Amino acids drawn into alignment in this way are joined together by **peptide bonds** to form a growing chain whose amino acid sequence is determined by the length of the mRNA strand and the base sequences of its triplets. That sequence, as we have seen, is strictly determined by the number and sequence of bases in the section of DNA that comprised the gene.

Single ribosomes attach themselves to an mRNA strand to make a polysome (Figure 17-14). They then move, one after another, along the mRNA strand, "reading" its coded information. Appropriate amino acids brought by their tRNAs are positioned at each step of the way, and the polypeptide chain rapidly grows longer. The synthesis of a single protein molecule takes only a few seconds. As a ribosome drops off at the end of the mRNA strand, a new one attaches at the other end. When a full chain of amino acids has formed, the resulting polypeptide peels away from the polysome and moves to other regions of the cell where it functions as an enzyme, structural element, or secretory product. After producing several polypeptide chains, a single mRNA molecule eventually breaks down. New polypeptides form continuously so long as the gene is active and continues to produce new mRNA.

17-14 Assembly of a polypeptide chain on a polysome
During protein synthesis many ribosomes are attached to the mRNA simultaneously, thereby creating polypeptide chains of varying lengths along the mRNA molecule. Synthesis begins near the 5' end of the mRNA and proceeds toward the 3' end.

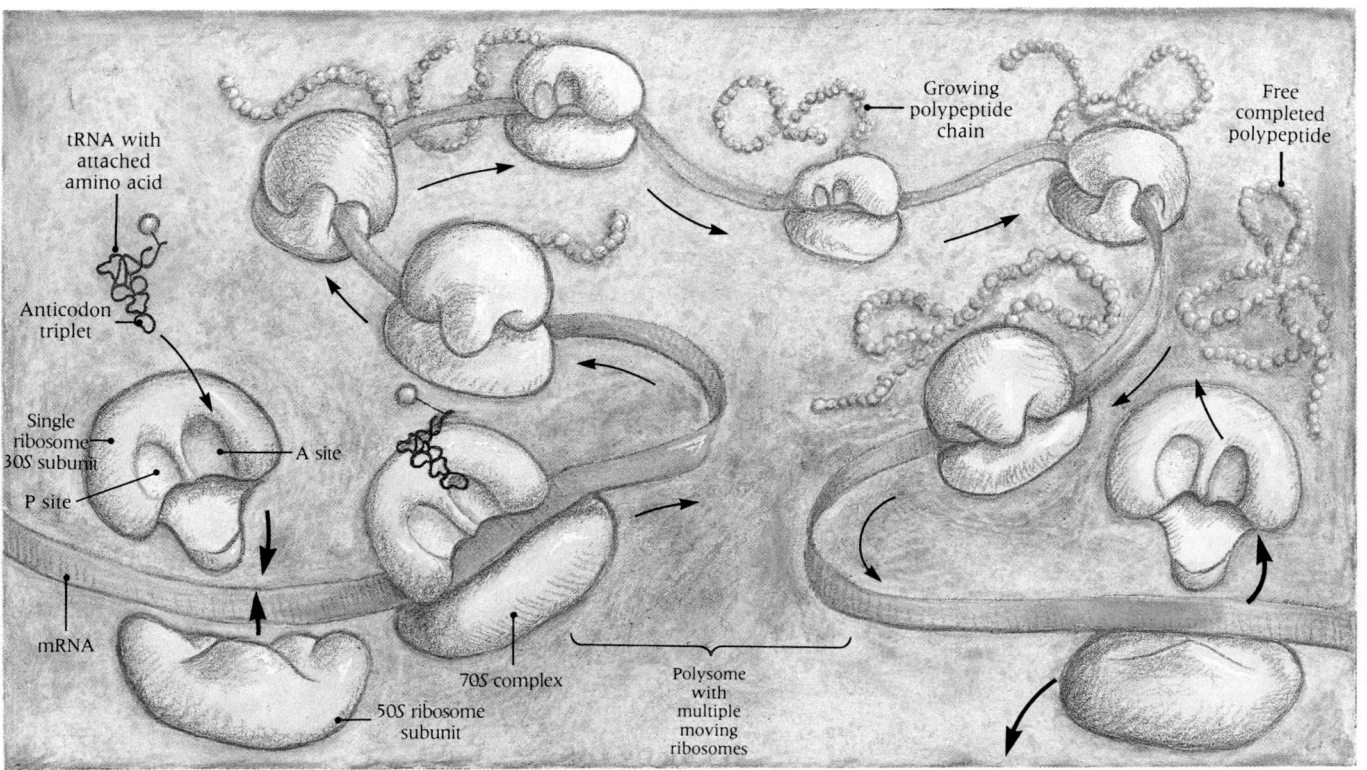

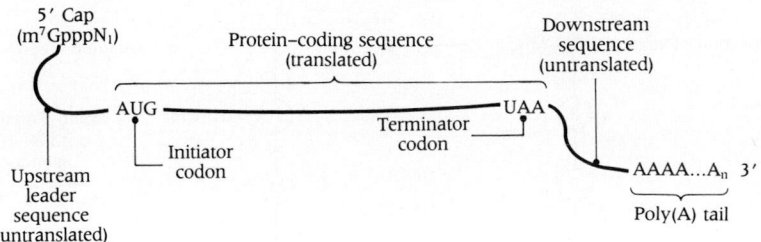

17-15 Structure of eukaryotic messenger RNA

The protein-coding region is flanked by noncoding sequences that are required for protein synthesis. These include the 5' cap, the poly(A) tail, and the initiator and terminator codons.

STRUCTURE OF MESSENGER RNA

In light of new insights into the genetic code, sequence studies were eagerly carried out on mRNA from many species. The results permitted the following generalizations, summarized in Figure 17-15, concerning the structure of mRNA:

- Beyond the expected sequences encoding protein synthesis, most mRNAs contain untranslated sequences that are similar, if not identical.
- The first untranslated segment is a long stretch of adenines at the 3' end in most eukaryotic RNAs. This terminal **poly(A) tail,** as it is known, contains 20−30 A's in invertebrates and fungi and 200 or more in vertebrates. Its role remains uncertain.
- Further upstream in the mRNA sequence (toward the 5' end), but before the start of the translated sequence, there is a stretch of 90−100 untranslated nucleotides in the middle of which is the sequence *AAUAA*. This sequence causes an enzyme to cleave the mRNA molecule 11 to 30 bases downstream. The poly(A) tail attaches to the newly created 3' end. Next upstream comes the **terminator codon** (which may be UGA, UAA, or UAG and stops translation).
- Further upstream is the translated coding sequence, which begins at its 5' end with the **initiator codon** that starts the translation process and is the first codon to be translated. This codon—AUG in both eukaryotes and prokaryotes—specifies the amino acid *methionine*. Almost all mRNAs in eukaryotes are **monocistronic**—that is, they encode a single polypeptide chain (Figure 17-16). Many (but not all) mRNAs in prokaryotes are **polycistronic**—that is, they encode multiple polypeptides, as indicated by their many start and stop signals. (The term **cistron** denotes a structural gene, a region of DNA encoding a single polypeptide, or functional RNA molecule.)
- Finally, a characteristic structure called the **5' cap** appears at the 5' end (after a **leader sequence** consisting of 10−50 untranslated nucleotides). This is a modification that affects the 5' end of the eukaryotic mRNA molecule. As noted above, synthesis of mRNA always begins with the insertion of an ATP or GTP, which retains its three phosphate groups. These undergo a curious alteration resulting in the formation of a 7-methylguanosine group at the 5' end of the RNA chain (Figure 17-17). The riboses of the first and sometimes the second nucleotides at the 5' end are methylated. Together, these changes are the cap of an mRNA molecule, and their addition is called capping. The methylated cap structure is believed to participate in the initiation of protein synthesis by facilitating the binding of mRNA to ribosomes, though its precise role remains unclear.

17-16 Monocistronic and polycistronic mRNAs

In eukaryotes, mRNAs usually contain information specifying only one gene, and are termed monocistronic. Prokaryotic mRNAs often contain several protein-coding regions and are termed polycistronic.

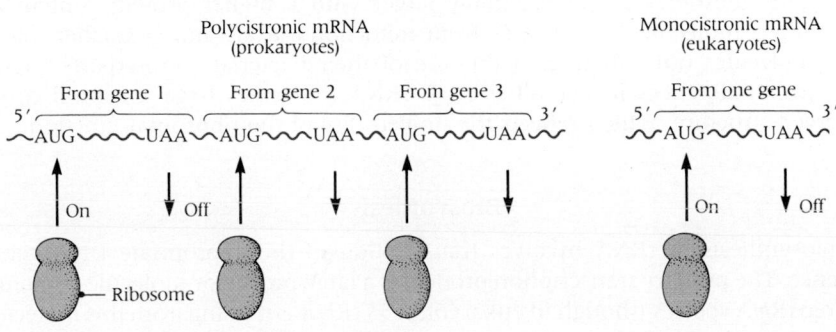

Methylation of guanine R—base → 5' Guanine—CH₃

OH OH

Methyl attached to N⁷ of guanine

Methylation of ribose R—base → Base O—CH₃

Methylation of ribose R—base → Base O—CH₃

17-17 Structure of the 5' cap
The 5' end of eukaryotic mRNA contains a methylated guanine, and two methylated riboses (R, ribose; Ⓟ, phosphate including its oxygens).

RIBOSOMES

As we have seen, amino acids are linked together in a sequence determined by mRNA on ribosomes. These complex arrays of protein and rRNA molecules have been intensively studied by investigators seeking to learn how these particles function at the molecular level.

Ribosomal Subunits

Ribosomes are composed of two subunits, one approximately twice the size of the other (Figure 17-18). In the extensively studied ribosomes of *E.coli*, the smaller subunit is named 30*S*, according to its sedimentation coefficient. (Sedimentation coefficients, expressed in *S*, or Svedberg units, relate to the behavior of particles in an ultracentrifuge and are approximate measures of particle size.) A 30*S* subunit contains 21 proteins and a single molecule of 16*S* rRNA. The larger (50*S*) subunit contains 32 proteins and two rRNA molecules (23*S* and 5*S*). The proteins vary in size from about 50 to 500 amino acids; their amino acid sequences have all been determined. Each protein is present in a single copy per ribosome, with the exception of one of the large subunit proteins, for which there are four copies. The overall mass ratio of RNA to protein is about 2:1. Cations, notably Mg^{2+} and polyamines, play an important role in maintaining the integrity of the ribosomal structures. Functionally active bacterial ribosomes can be reconstituted from their isolated RNA and protein components. Interestingly, the secondary structures of rRNA molecules— that is, the way in which the RNA is folded into loops by complementary base pairing—is virtually identical in all species studied.

Certain differences are found between prokaryotic and eukaryotic ribosomes. Eukaryotic ribosomes are considerably larger with a higher protein content and larger rRNA molecules. Ribosomes from mitochondria are much smaller, having rRNA molecules only about half the size of their bacterial counterparts. Despite these gross differences in overall size, all rRNA molecules have a central core of conserved structure, which reflects the universality of the ribosomal function.

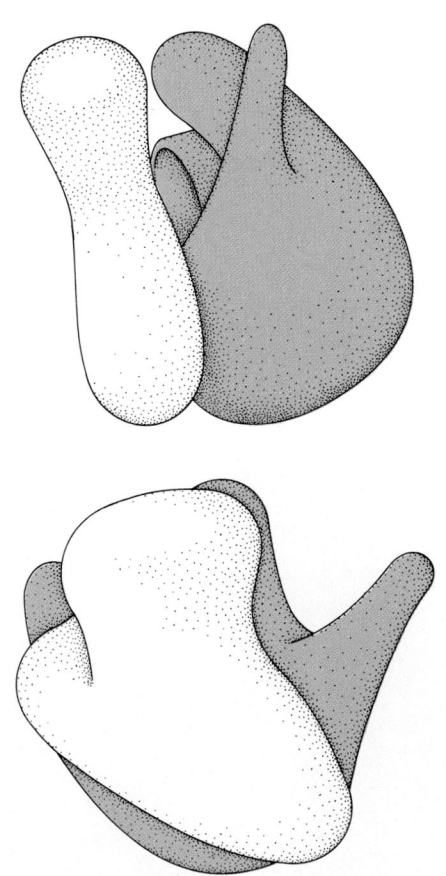

17-18 Three-dimensional model of the bacterial ribosome
This model of the E. coli *ribosome, based on electron micrographs, shows that the subunits are shaped unevenly, and that they fit together in a specific manner.*

Biosynthesis

The biosynthesis of rRNA involves transcription of the appropriate DNA genetic sequence. The primary transcription product is a large precursor molecule containing all three rRNA species (though in eukaryotes, 5*S* rRNA is missing from this molecule).

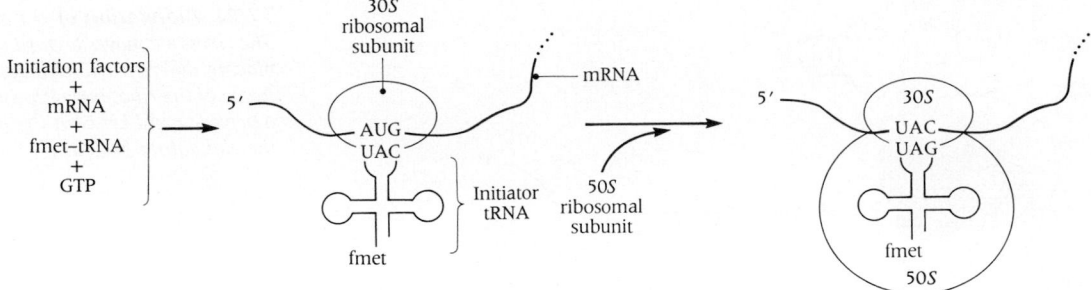

17-19 Initiation of protein synthesis in prokaryotes

The association of an initiator tRNA, with its bound formylated methionine, and the initiator codon of mRNA, is necessary before protein synthesis can begin. The large 50S subunit of the ribosome then attaches to the complex.

The precursor is subsequently cleaved in several stages to yield mature rRNA molecules. In prokaryotes, the ribosomal proteins are added while the rRNA is still being transcribed. In eukaryotes, rRNA is synthesized in the nucleolus and the completed precursor is then transported to the cytoplasm. The ribosomal proteins are themselves synthesized via mRNA on ribosomes in the cytoplasm.

MECHANISMS OF PROTEIN SYNTHESIS

Once ribosomal subunits are assembled, they participate in protein biosynthesis. Like transcription, translation includes three phases: initiation, elongation, and termination.

Initiation

The synthesis of a protein is initiated by a special type of tRNA, called **initiator tRNA** (Figure 17-19). Prokaryotes have two tRNAs that are specific for the amino acid methionine. One is an ordinary tRNA, termed tRNA^{met}, that inserts methionine within the growing peptide chain like any other amino acid. The other, a special tRNA, is termed fmet-tRNA (or tRNA^{fmet}) because after the methionine is attached to the tRNA, it is converted to *N*-formylmethionine (Figure 17-20). The formyl group blocks the amino group of methionine and thus prevents the addition of another amino acid via a new peptide bond. In prokaryotes, tRNA^{fmet} serves as an **initiator tRNA** that binds to the AUG initiator codon at the beginning of the protein-coding sequence, and binds with the mRNA to the 30S subunit. One end (the 3′ end) of the 16S RNA is also involved in the formation of this complex. A 50S subunit then joins the complex, forming a complete 70S ribosome. A similar mechanism exists in eukaryotes, except that the initiating methionine is not formylated.

Studies of the synthesis of the globin portion of hemoglobin showed that certain small proteins serve as initiation signals. In the globin system, one such **initiation factor,** termed eIF-2, serves as an inhibitor of protein synthesis initiation. But when it is phosphorylated (by a cAMP-dependent protein kinase) it is inactivated and initiation is allowed to proceed.

17-20 Formation of the tRNA-N-formylmethionine complex

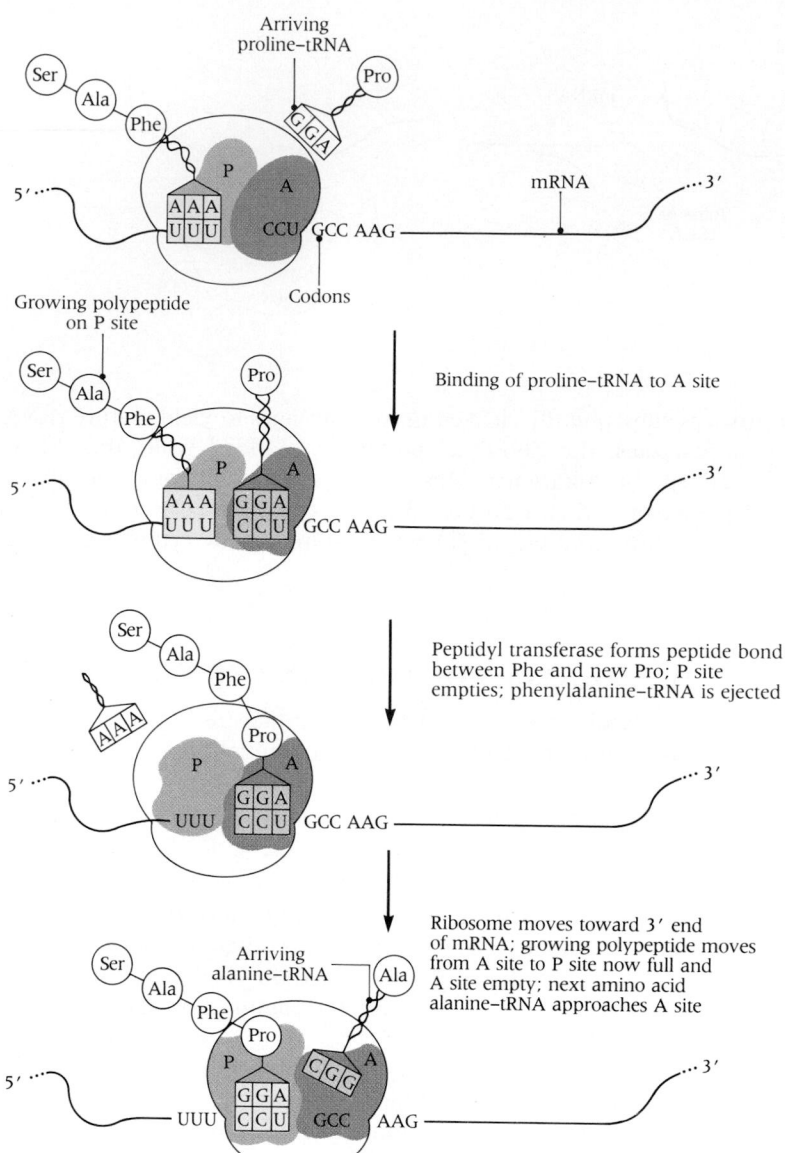

17-21 Elongation of the polypeptide chain
*The ribosome moves toward the 3' end of mRNA,
placing different triplet codons in the P-site and
A-site of the ribosome. Peptidyl transferase forms
a peptide bond between the new amino acid and
the elongating protein.*

Elongation

At this stage, the initiator aminoacyl tRNA occupies the **P-site** (peptidyl site) on
the ribosome, while a second tRNA binding site (the **A-site,** or aminoacyl site) is
free to accept the next coded aminoacyl tRNA molecule (Figure 17-21). The assembly
and elongation of the growing peptide chain includes several steps. First, incoming
amino acids bound to their tRNAs are bound to a ribosome at a ribosomal locus
called the A-site. The previously bound initiator tRNA occupies the ribosome's
P-site. The binding of amino acid-bearing tRNAs requires GTP and various **elongation
factors.**

In the course of elongation, amino acids are joined to the peptide chain by
peptide bonds formed by the enzyme **peptidyl transferase.** As this occurs, the
ribosome moves along the mRNA molecule. In a polysome, translation of the mRNA
can take place on several ribosomes simultaneously (Figure 17-22). Since the initiator
codon is at the 5' end of the mRNA, the ribosome moves from the 5' to the 3'
end of the mRNA molecule. Note that protein synthesis can occur on free ribosomes.
This usually produces proteins that are retained in the cell. Proteins that are to be
secreted by the cell are synthesized on the membranes of the rough endoplasmic
reticulum.

In the final step of elongation, **translocation,** the mRNA molecule moves, advancing one triplet unit. Once an aminoacyl tRNA is in the A-site, the initiator
amino acid (or at later stages the growing polypeptide chain) is transferred from

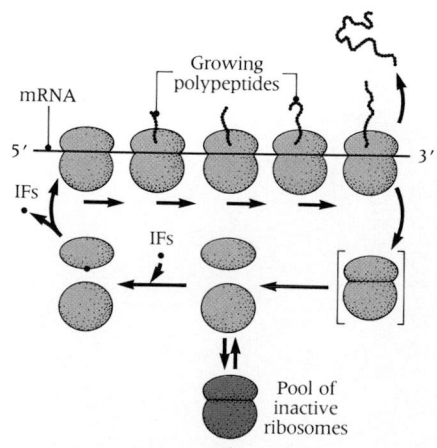

**17-22 Translation of an mRNA molecule by
several ribosomes**
*These simultaneous translations cause the
formation of a polysome. IFs are initiation factors
that influence the association and dissociation
of ribosomal subunits and affect mRNA binding.
Compare with Figure 17-14.*

17-23 Control of ribosomal protein synthesis

In E. coli, *certain ribosomal proteins are able to bind to their polycistronic mRNAs, thereby shutting off synthesis of that protein and slowing down translation.*

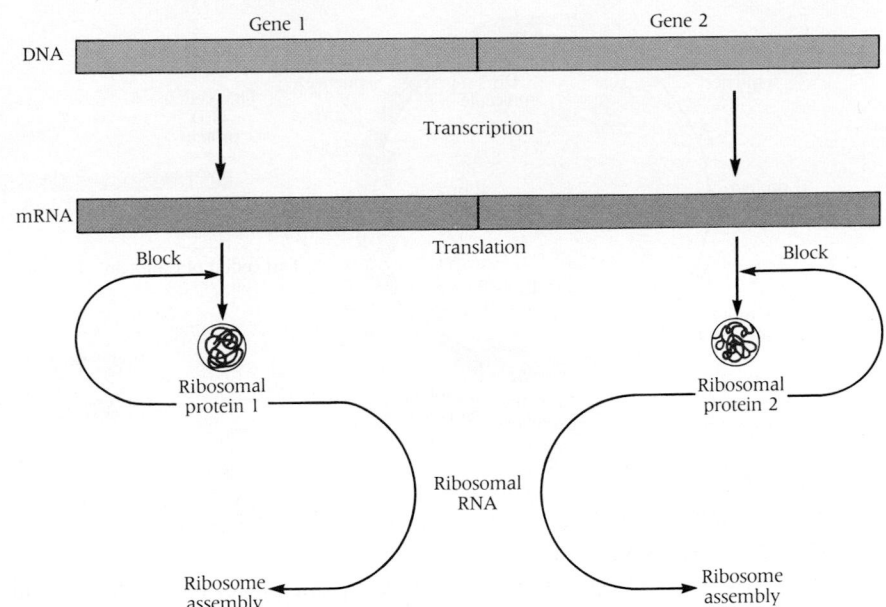

the P-site tRNA to the A-site aminoacyl tRNA. After peptide transfer has taken place, the peptide is attached to the A-site tRNA, and an "empty" tRNA molecule is at the P-site. The peptidyl tRNA complex is translocated to the P-site in order to free the A-site. It is probable that the empty tRNA occupies a third ribosomal site (exit or **E-site**) before finally being ejected from the complex.

Termination

Protein synthesis is terminated when a termination codon (UAA, UAG, or UGA) is reached. This happens because no tRNA contains an anticodon capable of binding to these triplets. Various **release factors** then cause the release of the finished polypeptide chain, the mRNA, and the empty tRNAs from the ribosome (see Figure 17-22).

In prokaryotes, there is a coupling of transcription and translation. As shown in Figure 17-23, a feedback loop couples transcription and translation by regulating the synthesis of certain ribosomal proteins. One of the proteins produced by the ribosomal protein genes can bind either to the RNA of ribosomes (rRNA) or to the mRNA directing its synthesis.

A mechanism as complex as protein synthesis has many control points. The major systems that regulate this process will be discussed in Chapter 18.

OVERLAPPING GENES

The triplet character of the genetic code permits some interesting bookkeeping, especially in viruses, where the circular DNA molecule is small and the total number of proteins scant. The 3:1 ratio suggests that a virus DNA circle containing, say, 9000 nucleotides could encode proteins containing no more than 3000 amino acids altogether. That is why molecular biologists were puzzled when the *E. coli* bacterio-phage φX174 was found to contain 15% too few nucleotides to encode all the amino acids present in its nine proteins.

The problem was solved by DNA and protein sequencing. When the amino acid sequences of the nine proteins, named A to J (Figure 17-24A), were compared with the DNA sequence (Figure 17-24B), it was surprising to see that parts of the DNA coding sequences for two of the proteins (B and E) are contained *within* the DNA sequences for two other proteins (A and D). In other words, the structural genes overlap. The **reading frame** of the E sequence is shifted one nucleotide to the right. The term reading frame refers to the fact that a sequence of triplet codons is read in sequence from some specific starting point. If the starting point is shifted, we say the reading frame is shifted.

The significance of this concept is easily explained by an analogy. Suppose the first letter (the T) were lost from this sentence of three-letter words:

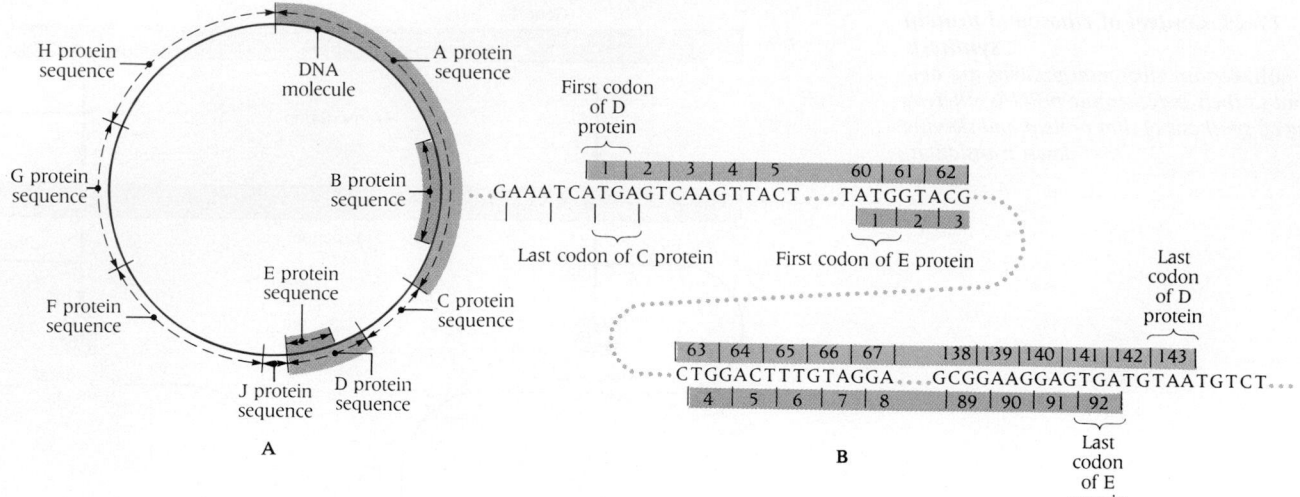

17-24 Arrangement of coding sequences in φX174 DNA
(A) The sequences for the B and E proteins are contained within the A and D sequences, respectively. **(B)** Partial sequence of the DNA in the regions of the A, B, D, and E genes.

THE CAT ATE THE RAT

This would cause the reading frame to be shifted by one place. The sentence would then read:

HEC ATA TET HER AT

The same sort of thing happens when an extra nucleotide, say a G, is inserted as follows:

THE CGA TAT ETH ERA T

These may be nonsense as English sentences. But if these three-letter sequences were codons they would now produce amino acid sequences completely different from the original sentence. The existence of a reading frame in the genetic context indicates that sequences are "read" from a fixed point in the gene.

Going back to the example of φX174 DNA, the reading frames are different for each of the two coding sequences. Hence, the amino acid sequences are read off differently (see Figure 17-25). Note that the E sequence gene overlaps the D sequence gene—indeed, the 92-amino acid E sequence is entirely within the D sequence.

Some small eukaryotic proteins are apparently encoded by overlapping genes. An example of such a **polyprotein** is pro-opiomelanocortin, a protein that contains within it several polypeptide hormones (see Chapter 28).

MOLECULAR BASIS OF GENE MUTATION

One of the most satisfying consequences of the deciphering of the genetic code was the long-awaited resolution of the molecular mechanism of gene mutations.

17-25 Effect on amino acid sequence when overlapping triplet codes are read in different frames
The mRNAs for both D protein and E protein have the same base sequences, but are read in different frames. Note that the E protein frame is shifted by one base.

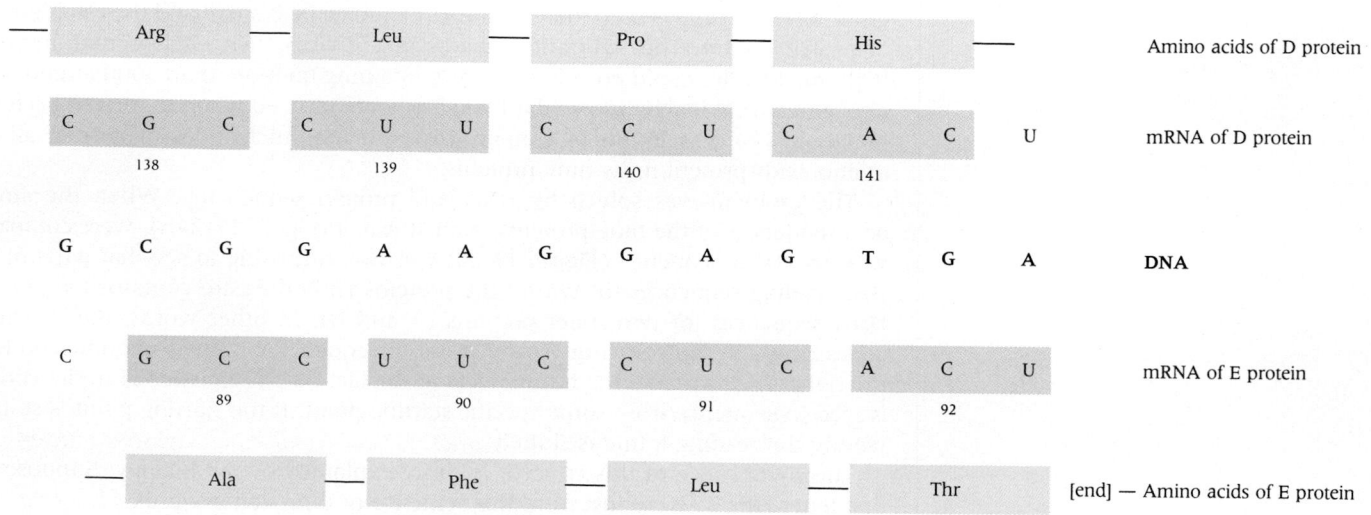

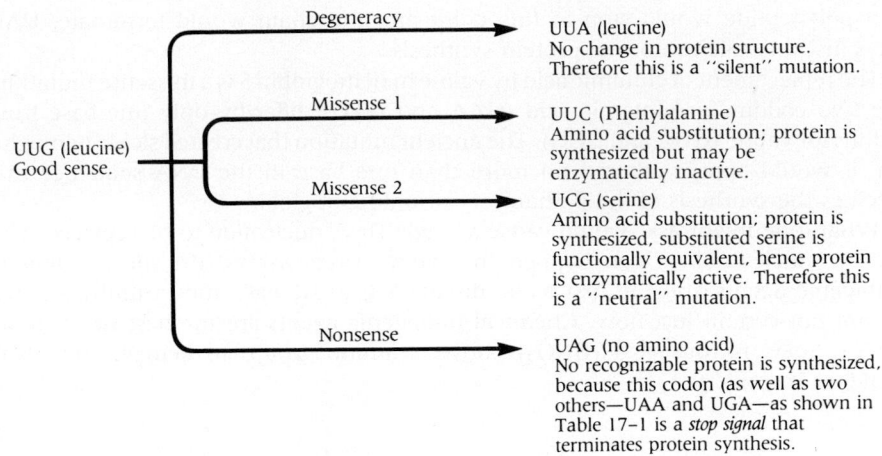

17-26 Consequences of a single alteration in a codon

A single change in the codon UUG can have different consequences, depending on the base that is affected.

Degeneracy → **UUA (leucine)**
No change in protein structure. Therefore this is a "silent" mutation.

Missense 1 → **UUC (Phenylalanine)**
Amino acid substitution; protein is synthesized but may be enzymatically inactive.

Missense 2 → **UCG (serine)**
Amino acid substitution; protein is synthesized, substituted serine is functionally equivalent, hence protein is enzymatically active. Therefore this is a "neutral" mutation.

UUG (leucine)
Good sense.

Nonsense → **UAG (no amino acid)**
No recognizable protein is synthesized, because this codon (as well as two others—UAA and UGA—as shown in Table 17–1 is a *stop signal* that terminates protein synthesis.

As discussed in Chapter 14, mutations fall into two categories. There are **chromosome abnormalities,** which are changes in the structure of chromosomes or their numbers (aneuploidy) (see Table 15-5). And there are **gene mutations,** which are changes in the sequences of base pairs in genetically active DNA—including DNA segments that encode protein structures and various regulatory sequences (enhancers, promoters, and so on) that influence the activity of structural genes. In a gene mutation, an mRNA with an altered base sequence is made from the mutant DNA sequence.

Most mutations result from defects in the sequence of bases in DNA. A **point mutation** involves alteration of a single base; a **deletion** or **insertion** mutation has a base sequence in which more than one base has been altered. Three types of defects occur.

- A simple switch—for example, of C–G to T–A or vice versa
- An insertion (or deletion) of a single base pair
- An insertion (or deletion) of a group of base pairs

Sequence defects in gene mutations are conveniently divided into **coding errors** and **reading errors.**

CODING ERRORS

Possible consequences of an altered base sequence are summarized in Figure 17-26, which considers UUG, a codon for leucine, and traces what would happen if its bases were altered in different ways. If UUG were changed to UUA, another codon for leucine, the resulting protein would be unchanged. But if UUG were changed to UCG, a codon for serine, or to UUC, a codon for phenylalanine, a protein molecule would be synthesized that contained a "wrong" amino acid. These would be **missense mutations.** We may illustrate their nature by returning to our analysis of "THE CAT ATE THE RAT," the phrase discussed earlier. If an error occurred that was confined to one word, as in

THE CAT ATE THE FAT

sense would be preserved but it would be missense. In

THE CAT ATE THR RAT

sense is largely preserved because the single error is localized. Even the deletion or addition of a three-letter sequence need not ruin the message, as in

THE CAT ATE RAT . . . or   THE CAT ATE HEC THE RAT

But it might ruin the message, as in

THE ATE THE RAT . . . or   THE CAT ATE NOT THE RAT

As shown in Figure 17-26, the protein resulting from a missense mutation may or may not be functionally active.

But if a mutation converted UUG to UAG, the resulting triplet would encode no amino acid. This was termed a nonsense mutation until Sidney Brenner of Cambridge University recognized that such triplets are stop signals. Further synthesis of a polypeptide would stop at this point and the chain would terminate. UAG plays just this role in normal protein synthesis.

The replacement of glutamic acid by valine in hemoglobin S is a missense mutation. The two codons for glutamic acid (GAA and GAG) differ by only one base from codons for valine (GUA and GUG). The ancient mutation that created sickle hemoglobin, it would appear, altered no more than one base in the DNA sequence that specifies the synthesis of the β chain of normal hemoglobin.

What, one might ask, would cause a single DNA nucleotide to be replaced with another nucleotide? It can happen in several ways. X-rays, the most potent of mutagenic agents and the first to be discovered, cause base substitutions, though we are not certain just how. Chemical mutagenic agents are the best understood. For example, nitrous acid (HNO_2) converts amino groups to keto groups—and therefore it converts C to U:

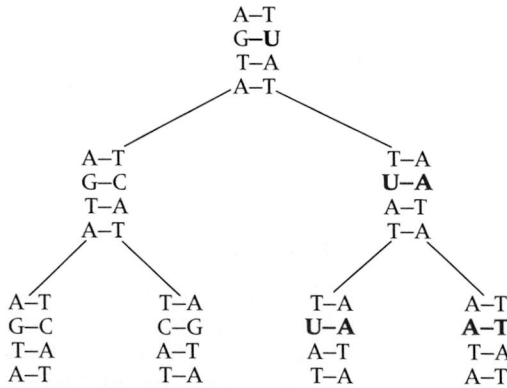

Cytosine   Uracil

If a DNA molecule were exposed to HNO_2, the following events could occur. First, a C within one strand could be converted to a U:

```
A–T                    A–T
G–C      HNO₂            U
T–A    ─────────→      T–A
A–T                    A–T
```

Then after chain separation, new complementary chains are synthesized. The new chains then separate and generate new complementary strands of their own:

```
                        A–T
                        G–U
                        T–A
                        A–T
              ┌──────────┴──────────┐
            A–T                    T–A
            G–C                    U–A
            T–A                    A–T
            A–T                    T–A
        ┌────┴────┐            ┌────┴────┐
      A–T      T–A          T–A      A–T
      G–C      C–G          U–A      A–T
      T–A      A–T          A–T      T–A
      A–T      T–A          T–A      A–T
```

Note that the change, or error, in base pairing is perpetuated: mutant DNA contains an A–T pair in place of a G–C pair.

Many mutagenic agents are chemical analogues of purines and pyrimidines. For that reason, they trick the mechanisms of DNA synthesis into making a pairing error. In view of the ease with which mutations can be induced with chemical agents, the relative rarity of natural or spontaneous mutations may seem surprising. If mutagenic agents now around us had existed throughout biological history, organisms sensitive to their mutagenic action might have long ago become extinct.

Our understanding of the molecular basis of mutation explains, at least in part, why most mutant alleles are genetically recessive, while wild-type alleles are dominant. In a homozygous recessive (let us call it *aa*), the phenotypic character commonly results from the failure of a mutant allele to produce any functional protein or enzyme. Heterozygotes (*Aa*) have one "good" allele; therefore, normal gene product is present. Because the wild-type allele occurs only once in heterozygotes, they may contain fewer of the relevant protein molecules than homozygotes (*AA*), which have two wild-type alleles. Often, the heterozygous phenotype is not half-way between the two homozygous phenotypes, because enough good enzyme molecules

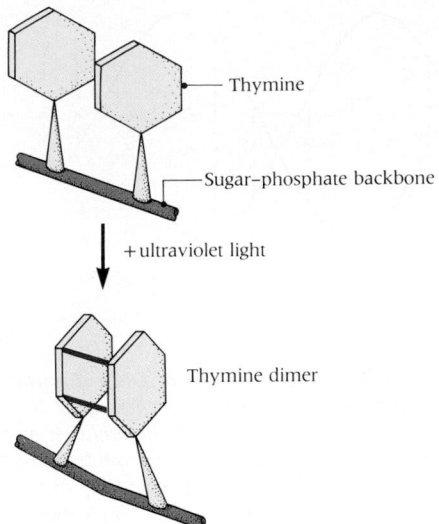

Thymine

Sugar–phosphate backbone

+ ultraviolet light

Thymine dimer

17-27 Formation of thymine dimer by ultraviolet radiation

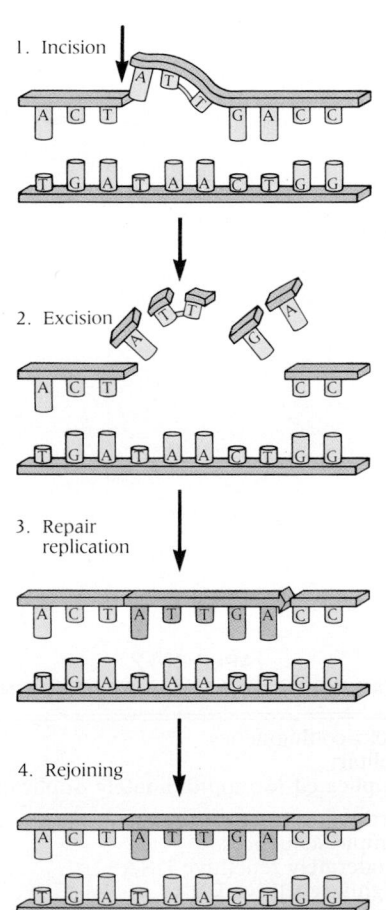

1. Incision

2. Excision

3. Repair replication

4. Rejoining

17-28 Pathway for excision repair
An endonuclease recognizes a thymine dimer and nicks the DNA at this point (1). The thymine dimer is removed by another enzyme (2) and the gap is repaired using the other strand as a template (3). The newly replicated DNA is joined to the original strand by DNA ligase (4).

are present to keep the metabolic reaction going at a rate approaching normal, even though the number of enzyme molecules is reduced.

DAMAGED DNA CAN BE REPAIRED

Survival of a species—and of an individual—obviously depends on the stability of genomes. Yet, damage to DNA can take many forms. For example, the presence in DNA of an incorrect or altered base, or of any abnormality that distorts the helix or prevents accurate base pairing, will tend to interfere with replication, protein synthesis, or recombination. Damage to DNA can cause the death of a cell. In eukaryotes, such damage sometimes causes a cell to become a cancer cell.

Fortunately, DNA stability and the precision of its replication are generally preserved by various **repair enzymes** that constantly correct lesions caused by harmful agents. In bacteria, significant damage to DNA molecules can even induce an emergency response whereby increased quantities of the repair enzymes are rapidly synthesized and caused to function as an orchestrated ensemble.

Various DNA repair systems have been recognized for years, but only recently have they been studied in detail. They are of three types.

- Enzymes that excise damaged base pairs
- Enzymes that remove lesions directly
- Post-replication repair systems

Let us consider some of these systems—and the lesions they repair.

The best known DNA lesion is the **pyrimidine dimer** that follows exposure of cells to ultraviolet radiation (Figure 17-27). A dimer forms when two adjacent pyrimidines (thymines or cytosines) become linked by the formation of an abnormal four-carbon ring. As a result, the dimer's two bases are pulled out of alignment. The hydrogen bonds linking their complementary bases are broken and the DNA backbone is distorted. This prevents the correct pairing of two bases on each side of the dimer. The presence of a single pyrimidine dimer is enough to interrupt both transcription and replication.

The process that removes the dangerous dimers is called **excision repair** (Figure 17-28). DNA strands containing dimers are attacked by a specific enzyme and repair is completed by two enzymes, DNA polymerase and DNA ligase, which respectively replace the excised nucleotides and ligate (tie together) the sugar-phosphate backbone, using the intact complementary strand as template.

Interestingly, there are countless specific enzymatic remedies for particular lesions of individual bases. If, for example, a single base is replaced by an incorrect one or if a base is altered, the incorrect base can be efficiently removed by enzymes, called *glycosylases,* that are specific for particular kinds of damaged bases. The resulting gap is filled by the correct base and patched.

Another mechanism, **post-replication repair,** operates only after replication of the damaged DNA is complete. In a cell exposed to ultraviolet radiation, DNA might replicate itself normally along the undamaged stretches of the parental template strands. However, when the replication fork (the point where the two parental strands are unwound and exposed to the replicating enzymes) reaches an unexcised pyrimidine dimer, correct base pairing is impossible along the damaged strand. The synthesis of one new strand is interrupted, to be reinitiated at some point in the undamaged region of the template beyond the dimer. As a result, there are post-replication gaps in the daughter strands. The missing information must somehow be recovered if the gap is to be repaired correctly. As shown in Figure 17-28, the required base sequence is available nearby—in the sister duplex (double strand) formed at the same replication fork. A recombination event substitutes one strand of the sister duplex for the missing bases in the post-replication gap, thereby supplying a template with the appropriate sequence of bases opposite the dimer. Thus, the gaps left in daughter strands by dimers that reach the replication fork before being excised are filled by **recombinational repair;** the dimer left in one parental strand is handled by excision repair.

Presumably mutation (especially of ultraviolet origin) involves damage too complex to be repaired. This is only now beginning to be understood. It now seems likely that in all organisms thousands of alterations occur every second but are

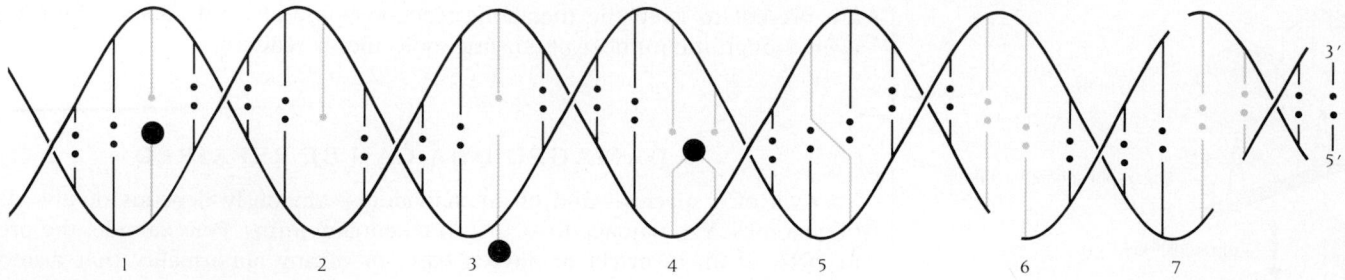

rapidly corrected by these excellent repair systems. Figure 17-29 summarizes several of the forms that damage to DNA can take.

READING ERRORS

In a second class of gene mutations, the error is in the reading of the code. In one such case, exposure to the acridine compounds used in dye synthesis causes DNA to contain one nucleotide too few or too many. The result is an abnormally shifted reading frame. Here we refer to the dislocating effect of a gain or loss in nucleotides.

When a mutation results in the insertion or deletion of one or two bases in the DNA sequence, the cell completely loses the ability to synthesize the protein determined by that gene. When three bases are inserted, a protein molecule may be synthesized even though it may contain one or more erroneous amino acids. When three bases are deleted the protein sequence will lack one amino acid and its synthesis might or might not be blocked. The insertion or deletion of one or two bases throws the sequence of bases off by one or two places, shifting the reading frame by one or two places and thereby causing major distortion of the message. In contrast, insertion or deletion of three bases will shift the reading frame three places—the length of a single codon—thereby preserving the proper sequence except for a local jumbled segment too small to prevent the synthesis of a near-perfect specific protein. These results add weight to the view that a module of three bases encodes one amino acid.

A final point concerns the overlapping genes. We have generally assumed that if a codon is sufficiently altered to undergo mutation a single protein will be affected. However, if gene overlap is widespread, the alteration of a single nucleotide could affect more than one protein. This would greatly amplify the genetic and evolutionary effects of mutation.

DNA ARRANGEMENTS AND REARRANGEMENTS

In a few years, sequencing DNA has generated an astonishing wealth of information on the organization of DNA in cells (Table 17-2).

REPEATED DNA SEQUENCES

Although we have seen that there are notable differences between the DNA of prokaryotic and eukaryotic cells, it seemed reasonable to extend our concept of the gene and its code from prokaryotic to eukaryotic cells. Indeed, investigators did suppose that the precision and uniqueness of the DNA base sequence are as important in one as in the other.

It came as a surprise, therefore, when biochemists discovered that much eukaryotic DNA consists of **repeated sequences**—multiple repetitive copies of the same (or similar) base sequences—in contrast with those sequences that are not repeated and are termed **unique** or **single-copy DNA.** In an ultracentrifugal density gradient, a small portion (usually less than 5%) of eukaryotic DNA readily separates from the bulk of the DNA because of a difference in density. Before this was understood, this fraction was termed **satellite DNA.** Nothing of the sort happens with prokaryotic DNA. For several years no one knew what satellite DNA is or how it differs from the bulk of the DNA. The curious discovery was then made that it consists of short sequences of nucleotides that are extensively repeated. Not all repetitive DNA

17-29 The several forms of DNA damage
The double helix shows: (1) a chemically altered base; (2) a deleted base; (3) the addition of a bulky foreign molecule; (4) a pyrimidine dimer; (5) cross-linked strands; (6) a broken strand; and (7) breaks in both strands. Such lesions disrupt base pairing or template continuity.

TABLE 17-2
CLASSIFICATION OF EUKARYOTIC DNA

Protein-coding genes
    Solitary
    Duplicated (or approximately duplicated)

Repetitive DNA sequences
    Simple sequences
    Moderately repetitive DNA
    Highly repetitive DNA

Mobile genetic elements
    Transposons
    Reverse transcription copies

RNA-coding genes
    (most tandemly duplicated)

Introns

Unclassified spacer DNA

has a distinctive buoyant density in the centrifuge tube that causes it to appear as satellite DNA. That depends on which bases are repeated.

Further studies suggested that repetitive eukaryotic DNA can be conveniently divided into three classes.

- **Simple sequence DNA** contains protein-encoding sequences but is not exclusively composed of them. For example, many sequences several hundred base pairs in length may be interspersed throughout the genetically active chromosomal DNA. These repetitive sequences are transcribed into RNA and are relatively stable in terms of sequence and location.
- **Moderately repetitive DNA** may be dispersed throughout the genome, or clustered in tandem repeats at the centers (centromeres) and ends (telomeres) of the chromosomes. The precision of repetition is often imperfect so that members of a "family" of repeated DNA sequences are best described as closely related, rather than identical. For the most part this class is composed of dispersed sequences of unknown function that are repeated up to 105 times. In addition, certain functional genes can be placed in this group. These include tandemly repeated genes for rRNA, tRNA, histones, actin, β-globin, and immunoglobulins.
- **Highly repetitive DNA** sequences are repeated a million or more times; some are dispersed while most are clustered at centromeres and telomeres. It is this group of clustered sequences that is known as satellite DNA. Most satellite DNA is not transcribed into RNA. However a small fraction of it may be essential to some genomic functions.

Sequences of repetitive DNA may constitute as little as 20% of all the DNA in the cells of some species and as much as 90% in others. Most repetitive DNA is of the moderately repetitive type. The wide distribution of repeated DNA, its persistence through millions of years of evolution, and the fact that at least some of it is expressed, or transcribed into RNA, all suggest that the repetition is important to the function and survival of the cell. We should bear in mind when we consider the regulation of gene expression in Chapter 18 that repetitive DNA occurs in eukaryotes, but not in prokaryotes.

A number of interesting patterns have emerged from comparisons of dispersed repeated sequences in eukaryotes of different phyla. Sea urchins have several thousand repeat families, some with as few as 100 members. Mammals, in contrast, have relatively few families, each containing many thousands, sometimes hundreds of thousands, of members. The *Alu* **family** (named for the restriction enzyme used to demonstrate it) occurs in many mammals. In humans it consists of between 3×10^5 and 5×10^5 repeats of a full 300 base-pair sequence and various smaller portions of it and comprises 3 to 6% of the total human DNA. (Human DNA contains about 6×10^9 base pairs—and 50,000 to 100,000 genes—per cell.)

Although many functions have been postulated for repeated DNA sequences, their functions remain unknown. One theory views them as functionless remnants of once active genes—now called **pseudogenes**—that arose with the emergence of mammals.

Repetition of DNA sequences is not the same as repetition of genes. Many functional genes are repeated and we refer to this as dosage repetition. Current evidence points to several broad classes of gene duplication.

- **Dosage repetition** is an increase in the number of gene copies per cell. Its most common forms are **duplication,** which is the precise doubling of the genome preparatory to cell division, and **endoduplication,** which leads to polyploidy.
- **Gene amplification** occurs when a cell selectively increases the copy number of a particular gene to a greater extent than it increases the copy number of the genes in the rest of the genome. It results from the repeated replication of the DNA in a limited portion of the genome. It occurs primarily in genes whose products are needed in larger amounts than could be supplied by a single gene—for example, in genes encoding tRNA and rRNA. Gene amplification was first described in the oocytes of certain amphibians (Figure 17-30). Some frog oocytes contain 400,000 times as much rRNA as a normal frog liver cell, despite the fact that the amount of DNA is about the same in both cell types. If the oocyte had the normal number of rRNA genes, it would take about 1000 years to

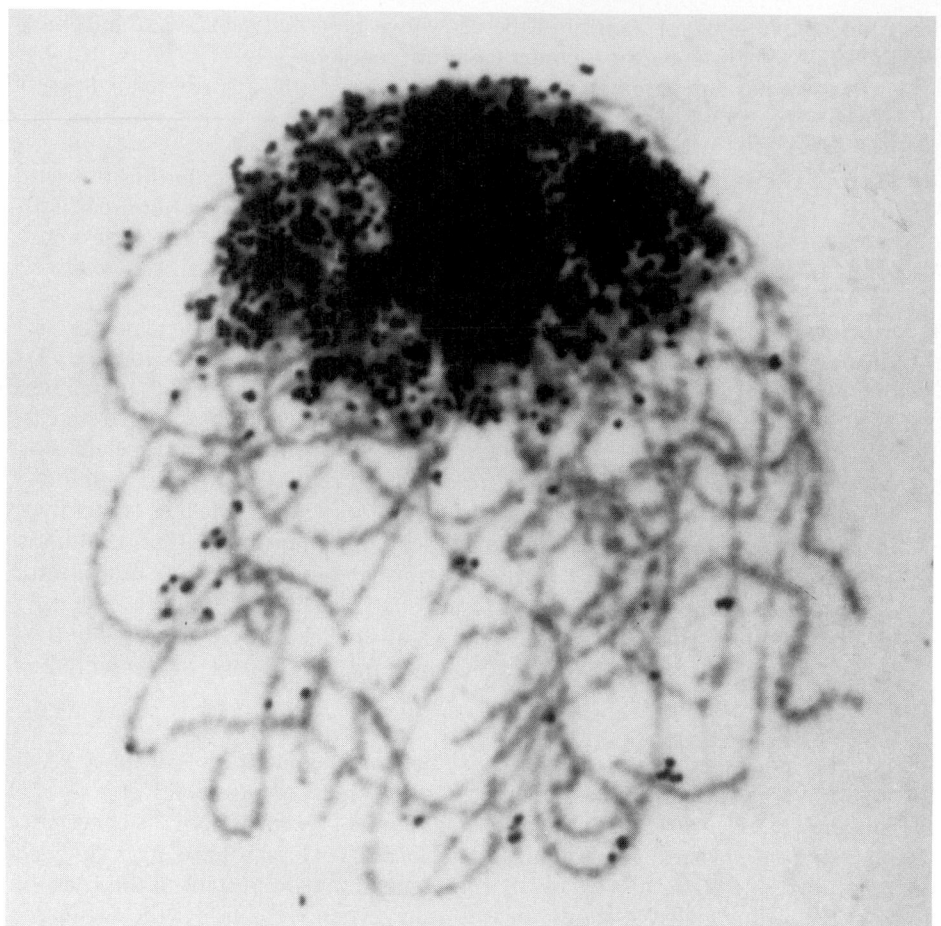

17-30 Electron micrograph showing amplification of genes encoding rRNA
This is a radioautograph of a Xenopus oocyte which has been incubated in radioactive uridine, a precursor of RNA. Dark grains of exposed photographic emulsion (reflecting the presence of newly formed radioactive RNA) are concentrated in the nucleolus, the main locus of rRNA synthesis. (×75)

synthesize the amount of rRNA in an oocyte. By amplifying these genes some 1000- to 2000-fold, the oocyte can synthesize this amount of rRNA in 6–9 months. The resulting ribosomes are stored in the egg because after fertilization no new rRNA, and hence no new ribosomes, are formed until much later.

■ **Variant repetition** is defined as genes (and their products) that are related but not identical. The genes for antibody proteins are in this class.

The majority of repeated DNA sequences do not appear to encode proteins. Only further work will reveal their role if it is other than the specification of gene products. What we have here is a remarkable and previously unappreciated situation. There exists in eukaryotic cells a great deal of DNA that codes for no gene products and thus has no known function. Significantly, DNA sequences of whatever kind, coding or noncoding, all share one notable function. They are capable of duplicating themselves—and we will later consider the provocative suggestion that this alone is the "selfish" purpose of such DNA. Whatever its function, once a DNA sequence has been incorporated into the eukaryotic genome there is no good way of getting rid of it.

INTERVENING DNA SEQUENCES

In 1977, molecular biologists studying the organization of eukaryotic genes made a startling discovery. In many structural genes, the DNA sequences that encode

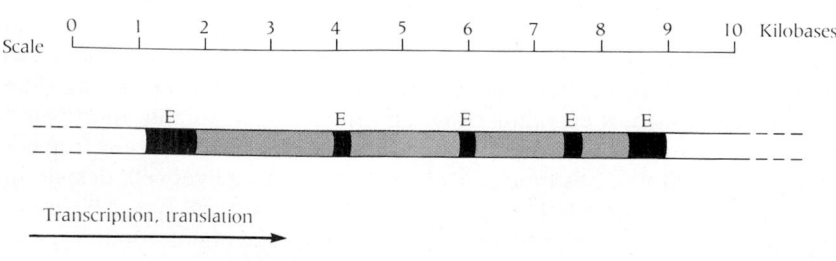

17-31 The cytochrome b gene
This gene is a complex of at least five exons (E), with four intervening noncoding sequences called introns.

proteins are interrupted by stretches of noncoding DNA that specify no gene product whatever. This means that it is not possible to demonstrate a one-to-one correspondence between the codon sequence of DNA and the amino acid sequence of its protein product.

As shown in Figure 17-31, the gene for cytochrome *b* consists of five actively coding sequences interrupted by four intervening noncoding sequences of variable length.[3] A second example is the gene for the enzyme dihydrofolate reductase, which includes five intervening sequences. The entire gene contains only 568 base pairs in the coding sequence, yet it is 32,000 base pairs long!

In sum, eukaryotic genes are often organized as discontinuous sequences of coding DNA with intervening noncoding sequences. The whole gene thus may be much longer than necessary for the coding of an amino acid sequence. The portions of DNA that actually encode an amino acid sequence were termed **exons** and the intervening sequences **introns.** Introns are almost never found in prokaryotes.

It was soon recognized that in eukaryotes all DNA bases, coding and noncoding, are transcribed into the newly formed mRNA, the so-called **primary transcript,** or **pre-mRNA** (Figure 17-32). Introns are copied right along with exons, but some sort of signal initiates **processing** and **splicing,** a term implying that the

---

[3] Cytochrome *b*, it will be recalled, is an electron carrier of oxidative phosphorylation in the inner mitochondrial membrane (see Chapter 9) and of the light reactions of photosynthesis in the chloroplast thylakoid membrane (see Chapter 10).

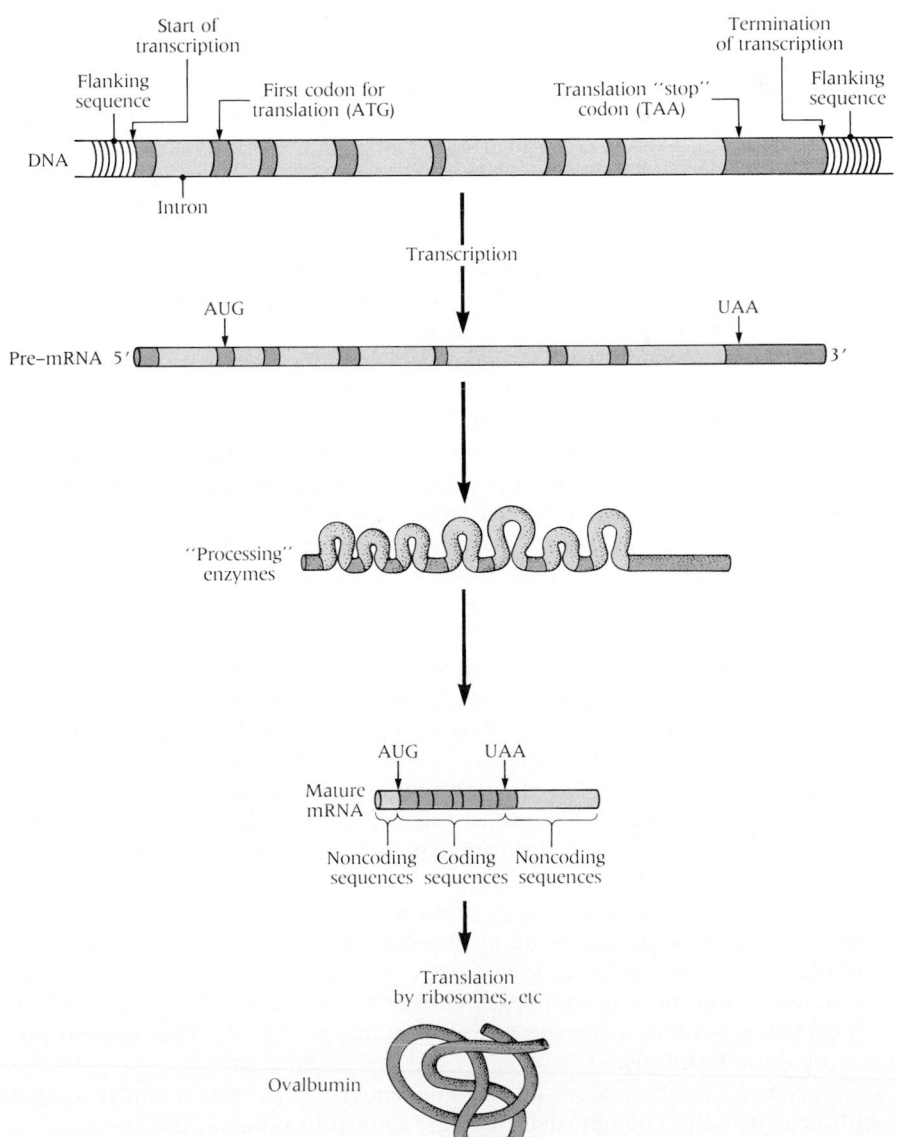

17-32 Conversion of pre-mRNA to mature mRNA

A gene, containing both introns and exons, is transcribed into the primary transcript, or pre-mRNA, with flanking sequences determining the start and stop signals for transcription. Introns are enzymatically removed and the remaining RNA pieces are spliced together to generate a mature mRNA molecule. This is translated in the example to ovalbumin.

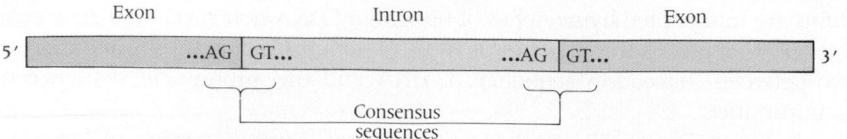

Exon   Intron   Exon

Consensus sequences

17-33 RNA splicing
Many exon-intron junctions contain the AG–GT consensus sequence, which is recognized by splicing enzymes during RNA maturation.

base sequences on either side of the intron must be joined together after the intron has been removed. A remarkable enzymatic reaction efficiently removes all intron sequences from primary transcript RNA.

Reflection on how this happens was influenced by another interesting observation. We stated above that nuclear DNA specifies the synthesis of four types of RNA: mRNA, rRNA, tRNA, and hnRNA. The last of the four, heterogeneous nuclear RNA, or hnRNA, was recognized when it was found that nuclei synthesize a group of variously sized RNA molecules that do not resemble any of the other three RNA types. It is a mixture of primary transcript molecules, which become mRNA after processing and splicing.

In these tailoring steps, introns are clipped out with exquisite accuracy. When free ends are rejoined the final product is **mature mRNA.** This post-transcriptional processing of primary transcript RNA is called **RNA maturation.** Its product, a mature mRNA molecule, is a continuous uninterrupted sequence of codons, all of which are active in translation. A similar maturation process operates in the synthesis of tRNA and rRNA.

The splicing process is quite remarkable: Introns—some many hundreds of bases long—are clipped out enzymatically from both ends of the RNA. The process works toward the middle and rather resembles the burning of a candle at both ends.

The base sequences at both exon-intron junctions were found to be similar in many genes (Figure 17-33). For example, at the junction at the 5′ end, the first two intron bases are often GT and a pyrimidine-rich cluster occurs near the 3′ end of the intron. We call such conserved sequences **consensus sequences.** Significantly, when these sequences are altered experimentally, intron removal and RNA splicing were impaired. Hence, these **splice junctions** are regarded as signals that activate splicing. The splicing mechanism generally depends on enzymes (a ligase and a nuclease) that have recently been isolated and purified. In one surprising case, pre-rRNA molecules catalyze their own splicing.

Interestingly, the otherwise bacterialike DNA of mitochondria turned out to have genes containing introns, now considered the hallmarks of "higher" or eukaryotic organisms. Thus, there is an inconsistency. If mitochondria are remnants of lower organisms that became established in the cells of higher organisms, mitochondrial DNA would not be expected to contain introns. We know that mitochondria are now no longer wholly independent because their own distinctive DNA codes for only a fraction of their proteins. Curiously, there is a striking similarity between certain introns in nuclear and mitochondrial DNA. A provocative recent theory about this will be discussed in Box 34-1.

We are left in an uncomfortable position. Only a small percentage of eukaryotic DNA (1 or 2% in humans) encodes proteins—and these are the only bits of eukaryotic DNA we understand at the moment. The rest consists mainly of introns and repeated sequences of unknown function. Although there is still no conclusive explanation of why many eukaryotic genes contain introns, biologists are speculating freely as new findings come rolling in. After finding genes that contain coding and noncoding sections, researchers spoke of introns as "silent regions." The silence is now being broken by the hum of function. Here are some examples of these findings.

- Recent evidence suggests that exons may encode special regions of protein molecules called **domains,** which are localized folding patterns that produce small globular units that may be endowed with special functions, such as fitting into membranes, and forming complexes with other molecules. When domains are connected by peptide sequences encoded by introns, the resulting protein can acquire these functions.
- In some cases, mutations along intervening sequences have been found to interfere with gene activity. Though such cases are limited in number, they point to possible effects of intervening sequences upon the coding portions of genes.

The origin and significance of introns pose a fascinating problem for workers in the field. Are they ancient features of genes, which in some cases were selectively or randomly lost (completely so in prokaryotes)? Or do they represent relatively recent insertions into previously intact coding sequences? It is clear that some introns have been in place over a very long period of evolutionary time. Introns might offer evolutionary advantages to eukaryotes by facilitating the development of variants of proteins or even new proteins. Introns might also be related to the phenomenon of biochemical microheterogeneity, which occurs in many proteins, among them antibodies (see Box 8-2 and Chapter 23). Though still unproved, the idea is that introns permit cells to generate a large variety of proteins. This argument rests on new evidence indicating that the coding sections of DNA might move about within the gene, thereby generating families of proteins that differ slightly.

This brings us to another complication.

MOVABLE DNA SEQUENCES

After recognizing noncoding intervening sequences in DNA, biologists asked, "Is it possible that widely separated coding sequences (exons) are shuffling about within chromosomes?" Such rearrangements, could they occur, would represent a new kind of recombination. They would also imply that eukaryotic genomes are less stable than classical genetic analysis had led us to believe. Movement of DNA coding sequences would surely work more efficiently in the presence of introns than in their absence, since intervening sequences would provide some maneuvering space along the DNA molecule. What excited everyone was the notion that such "jumps" could conceivably link together two different proteins to make a new composite protein.

The idea that genes might be movable dates back to the corn-breeding experiments of Barbara McClintock that began in the early 1930s (see Chapter 2).[4] McClintock's conclusion that gene segments in corn (*Zea mays*) rearrange themselves within and between chromosomes—that genes are "mobile"—aroused little excitement until it was found a few years ago that bacterial DNA sequences can be altered by insertions of short sequences, termed **transposable elements.** Here was a case in which chromosomes were rearranged in a completely unexpected way. A dramatic example was provided by bacterial genes specifying resistance to such antibiotics as penicillin or kanamycin, which can be carried on plasmids that have no homology with the bacterial chromosome. Such a resistance gene, when present in a cell's plasmid, can move into the chromosome—even though the two DNA segments (chromosome and plasmid) are not homologous. Such DNA segments, called **transposons,** introduced a new principle: mutation by insertion.

A transposon moves genes to new and unrelated sites. For years, the evidence for such movable DNA sequences in eukaryotes rested entirely on breeding experiments—the test of progeny. However, in the late 1970s, investigators began to get their hands on eukaryotic transposable elements, which are widespread—in plants, yeast, *Drosophila*, and other animals. Remarkably, these elements were found to have similar structural organization wherever they originated.

Investigators soon began searching for transposable genetic elements in eukaryotic cells by looking for their characteristic features in eukaryotic genomes. As shown in Figure 17-34, a typical eukaryotic transposon includes a *target site* in which 5 to 10 bases are repeated, a *long terminal repeat* (LTR) of 250 to 600 bases at each end of the transposon that is within it, and *inverted repeats* of 10 to 50 bases at the outer end of each LTR. Any gene within the transposon is located between the repeats. Amazingly, there is striking similarity in this general structure in bacteria and eukaryotes, although their mechanisms of transposon movement may differ.

In 1976, biologists identified a new class of *Drosophila* genes that are present in many copies—from 10 to 100 in each cell. Of the three classes of repetitive DNA listed earlier in this chapter, one resembles this new class of *Drosophila* genes which are moderately repetitive and often dispersed throughout the genome. It is now believed (from mapping studies) that what dispersed these elements was their capacity to move from one chromosomal location to another. As a result of their mobility,

[4] For her pioneering work, performed more than 50 years earlier, Barbara McClintock received the Nobel Prize in 1983, at the age of 81. Her photograph appears as Figure 2-3.

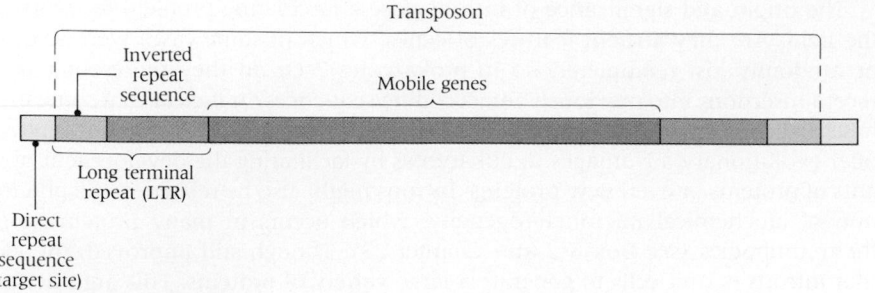

17-34 Structure of transposons
A transposon contains one or several genes which are flanked by repeating sequences of various kinds. These sequences are of different lengths and sequence, depending on the transposon, and are important in moving the transposon to other regions of the genome.

their chromosomal loci vary greatly in different *Drosophila* strains. They are even found in different places in different individuals of the same strain—in sharp contrast with ordinary patterns of gene distribution in individuals of the same species, which are constant. The dispersed repeated sequences have been described as "nomadic." Occupying no fixed position, they wander about in the genome.

There are four general types of chromosomal rearrangements (Figure 17-35): (A) insertions; (B) transpositions; (C) amplifications (repeated localized copying); and (D) deletions. Only the first two involve the movement of mobile genetic elements. The chromosomal rearrangements they promote thus give rise to mutations. Gene rearrangements, like those occurring in the development of antibody-producing cells (see Chapter 23), are now recognized as normal features of eukaryotic development. Although there is little similarity between the base sequences of different families of movable elements, the ones that have been thoroughly studied to date—in widely diverse species—have many features in common.

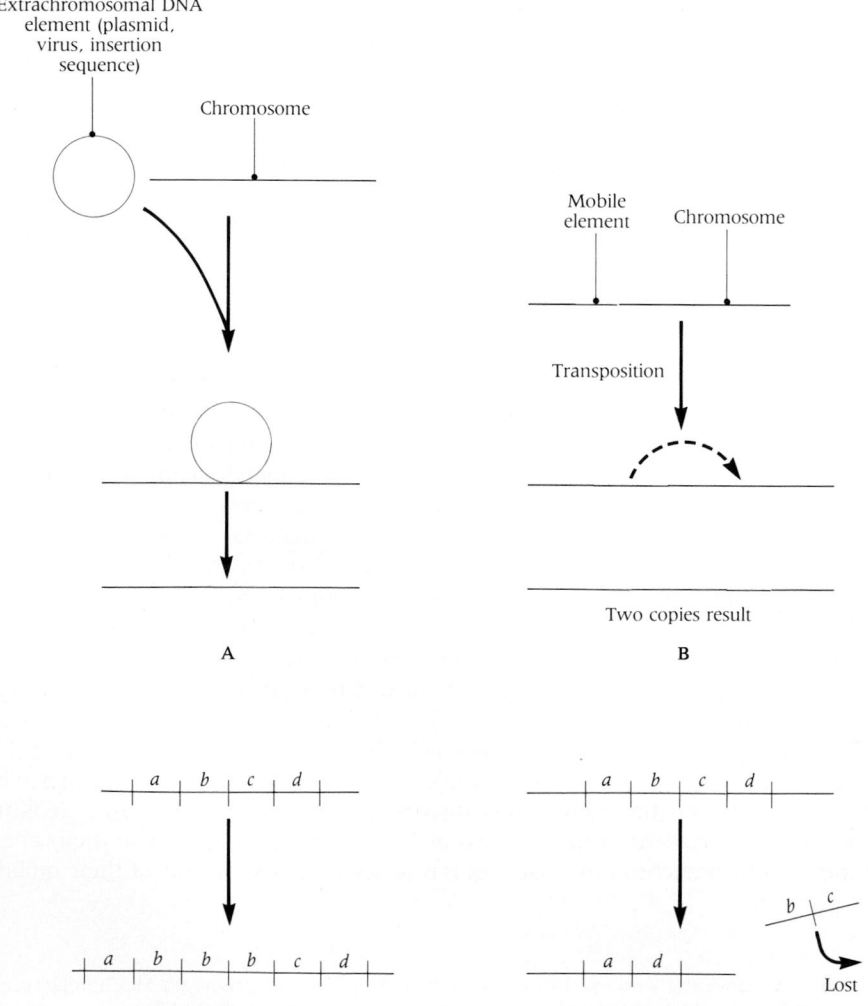

17-35 Types of DNA rearrangements
(A) *Insertion involves the introduction of foreign DNA from a plasmid or virus into a chromosome.* **(B)** *Transposition is the movement of sequence information from one site to another.* **(C)** *Amplification is the localized copying of a particular DNA sequence.* **(D)** *Deletion involves the loss of a sequence at a specific site.*

- They are 5000 to 7500 base pairs in length with certain characteristic sequence repeats at each end (see Figure 17-34).
- They are not known to carry genes coding for any specific cell component or characteristic, even though they direct the synthesis of large quantities of RNA transcripts.
- They can turn genes on and off as they move about the genome.
- They insert themselves into DNA by differing mechanisms.

DNA movements are also being recognized now between the organelles of eukaryotic cells. For example, certain base sequences have been found to occur in corn plants in both the mitochondrial DNA and chloroplast DNA of the same and other species. Since such apparent movement of DNA diminishes the separate identity of each organelle genome, the term *intracellular DNA promiscuity* has been applied to this phenomenon.

We are just at the beginning of our understanding of transposons. However, even now three conclusions can be stated with assurance.

- The DNA of the eukaryotic genome is much more mobile and unstable than most biologists had assumed.
- This situation provides an entirely new set of mechanisms whereby genes can be rearranged and diversity generated—both in the development of individual organisms (see Chapter 20) and in their evolutionary origin (see Chapter 34).
- It is likely that investigators will be able to use transposons to introduce functional new genes into the germ lines of plants and animals.

Movable genetic elements are a topic of intense current interest and controversy, as investigators try to fathom their role. As generators of genetic diversity, they may have evolutionary significance. The recent discovery of such a fundamentally different mechanism of genetic variation came at a time when most biologists thought that all or most of the important molecular mechanisms of genetics were known—with only details left to be gathered up. Are other novel basic mechanisms waiting to be discovered?

SUMMARY

Eukaryotic cells synthesize four kinds of RNA: messenger RNA (mRNA), transfer RNA (tRNA), ribosomal RNA (rRNA), and heterogeneous nuclear RNA (hnRNA). Much of the RNA arising in transcription is hnRNA, a small portion of which finds its way to the cytoplasm where it becomes mRNA. The rest is degraded in the nucleus. Molecules of several rRNA varieties become structural components of ribosomes; tRNA molecules bind specific amino acids and serve as adaptors in the assembly of polypetides along molecules of mRNA.

In transcription, the base sequence of DNA specifies the base sequence of mRNA, using the same base-pairing rules that apply when a DNA strand serves as template in DNA replication (except that uracil base-pairs like thymine). Gene transcription is initiated when RNA polymerase binds to a promoter sequence upstream to the gene. RNA polymerase then links ribonucleotides to form a single RNA strand, the base sequence of which is complementary to a gene sequence in one of the two DNA strands—the template strand. Elongation of the mRNA chain, which proceeds in a 5' to 3' direction, is regulated by various upstream promoter sequences (TATA box, etc.) that differ in eukaryotes and prokaryotes. Transcription stops when RNA polymerase reaches a terminator base sequence at the end of the gene, which releases the newly synthesized mRNA molecule and the RNA polymerase from the DNA molecule. Aside from the sequences which encode proteins (exons),

most eukaryotic mRNAs have many untranslated intervening sequences (introns), a methylated cap at the 5' end, an initiator codon (which specifies the amino acid methionine), and a poly(A) tail at the 3' end.

Molecules of tRNA have certain distinctive structural features: a cloverleaf shape (which contains the critical anticodon loop); many methylated and other unusual bases; a constant "acceptor" sequence (CCA) at the 3' end which binds the amino acid specific for that tRNA molecule; and a recently discovered "second genetic code" consisting of two bases that determine which amino acid is bound by each of the 20 specific tRNAs.

The nearly universal genetic code utilizes a three-base sequence for each amino acid. Hence 4^3 or 64 possible codons are available to encode 20 amino acids. Of these, 61 actually do so. Hence, the code is degenerate—that is, more than one codon specifies most amino acids. Each tRNA bears an anticodon triplet complementary to the mRNA codon for the amino acid of that tRNA. By lining up on a ribosome-bound mRNA molecule in a sequence determined by the sequence of mRNA triplet codons, tRNA molecules and their bound amino acids align correctly, thereby permitting accurate translation of the genetic "message" in mRNA.

Prokaryotic ribosomes, which sediment in an ultracentrifuge at 70*S*, are formed from a large (50*S*) and a small (30*S*) subunit. Eukaryotic ribosomes sediment at 80*S* and contain 40*S* and 60*S* subunits. Both classes of ribosomes contain many different proteins and one, two, or three rRNA molecules, which contain no genetic information. Ribosomes participate in the three stages of protein synthesis: initiation, elongation, and termination. Initiation occurs when an initiator tRNA (carrying formylmethionine in prokaryotes; methionine in eukaryotes) attaches to the initiator codon of ribosome-bound mRNA. The two bind to a small ribosomal subunit; a large subunit then joins them to form a complete ribosome.

Each ribosome has two rRNA binding sites: the P (peptidyl) site and the A (aminoacyl) site. Amino acid-tRNA molecules first occupy the A-site. This allows (in the presence of elongation factors and GTP) the formation of a peptide bond with the growing chain held on the P-site. Then the growing chain moves to the P-site. In a process called translocation, the tRNA carrying the growing chain moves to the P-site and advances the mRNA molecule to the next codon and ejects empty tRNA from the P-site so that another cycle can begin. When a termination codon on the mRNA is reached, chain elongation stops; and the mRNA, tRNA, and ribosomal subunits dissociate from the completed polypeptide.

When an mRNA molecule is translated, the resulting amino acid sequence depends on the reading frame within which its codons are read—that is, the point at which codon-anticodon pairing begins. Because codons are triplets, every base sequence has three reading frames. Some viruses utilize a reading frame shift to generate two polypeptides with different sequences from the same gene. Such genes are said to overlap.

Genetic mutations result from changes in DNA base sequences. Most are point mutations—a change, insertion, or deletion of a single base. They result in coding errors, in which a single changed codon inserts the wrong amino acid, and reading errors, in which one or two inserted or deleted bases change the reading frame and thus the coding of every amino acid downstream of the change. Many potentially harmful mutations are prevented by various DNA repair mechanisms that can reverse DNA damage.

The DNA of eukaryotes contains many puzzling repetitive DNA sequences. Most of the moderately and highly repetitive DNA is not transcribed and its function is unknown. Some of it, however, represents multiple copies of genes encoding protein or rRNA. In gene amplification, a cell selectively increases the copy number of particular genes whose products are needed in large amounts.

Many eukaryotic DNA sequences are interrupted by non-coding sequences, called introns. Introns are transcribed into the primary mRNA transcript, but later excised, leaving only the coding sequences (exons) in the mature mRNA. Consensus sequences at exon-intron junctions are the signals guiding intron excision.

Certain DNA sequences, called transposons are able to move from one chromosomal locus to another in prokaryotes and eukaryotes. Some, but not all, carry genes; others turn genes on and off as they move about their genomes. Such rearrangements are normal features of eukaryotic development. Transposon sequences may be useful in future attempts to introduce new genes into a genome.

KEY TERMS

amino acyl-tRNA synthetase
anticodon
central dogma
codon
consensus sequence
degenerate code
downstream
elongation factor
excision repair
exon
gene amplification
gene mutation
initiation factor
initiator codon
intron
messenger RNA (mRNA)
missense mutation
monocistronic

nonsense mutation
peptidyl transferase
point mutation
polycistronic
post-replication repair
pre-mRNA
primary RNA transcript
promoter
pyrimidine dimer
reading error
reading frame
recombinational repair
release factor
repair enzyme
replication
ribosomal RNA (rRNA)
RNA polymerase
RNA processing

satellite DNA
sense strand
simple sequence DNA
splice junction
splicing
structural gene
template
template strand
terminator codon
terminator sequence
transcription
transfer RNA (tRNA)
translation
transposable element
transposon
triplet
upstream

QUIZ QUESTIONS

1. Which of the following is the correct sequence of the events referred to as the central dogma?
   A. replication, transcription, translation
   B. duplication, translation, transcription
   C. translation, replication, transcription
   D. transcription, translation, replication
   E. None of the above

2. Which of the following is *not* involved in the process of translation?
   A. mature messenger RNA
   B. transfer RNA
   C. heterogeneous RNA
   D. activated amino acids
   E. ribosomes

3. Which of the following is the correct 5' to 3' order in which the listed sequences of eukaryotic DNA occur?
   A. TATA box, -35 region, initiation site, terminator
   B. -35 region, TATA box, initiation site, terminator
   C. initiation site, -35 region, TATA box, terminator

D. initiation site, promoter, TATA box, terminator
E. None of the above

4. The base sequence known as the anticodon is part of a _____ molecule.
   A. DNA

B. messenger RNA
C. heterogeneous RNA
D. transfer RNA
E. ribosomal RNA

5. Which of the following is not a means of introducing variation into the genome?
   A. shifted reading frame

B. substitution of a base
C. gene amplification
D. activity of transposons
E. B and D

ESSAY QUESTIONS

1. Where in the cell is DNA found? Where in the cell do transcription and translation take place?

2. How does RNA synthesis differ in prokaryotic and eukaryotic cells? Contrast prokaryotic and eukaryotic mRNA structure.

3. Describe the structure and properties of tRNA molecule. What role does tRNA play in translation? How does the system for recognition of tRNA molecules by amino-acyl tRNA synthetases constitute a "second genetic code"?

4. Describe the structure of mRNA. What

are untranslated sequences? How do we know that the terminal poly-A tail is added after RNA transcription?

5. What are transposons? How do they differ from viruses? How might their presence facilitate genetic variation, and hence evolution?

REFERENCES AND SUGGESTIONS FOR FURTHER READING

Crick, F. (1988). *What Mad Pursuit*. Basic Books, New York.

A delightful account of his life in science by one of the discovers of the double helix. Full of anecdotes and good advice.

Darnell, J. E., Lodish, H., and Baltimore, D. (1986). *Molecular Cell Biology*. Scientific American Books, W. H. Freeman, New York.

A masterful text by three outstanding molecular biologists.

Friedberg, E. C. (1985). *DNA Repair*. W. H. Freeman, New York.

A clearly written presentation of molecular mechanisms of DNA repair in prokaryotes and eukaryotes.

Howard-Flanders, P. (1981). Inducible repair of DNA. *Scientific American 245*, November, 72–80.

Discussion of the fact that prokaryotic DNA repair genes are regulated by a single repressor.

Judson, H. F. (1979). *The Eighth Day of Creation: The Makers of the Revolution in Biol-* ogy. Simon and Schuster, New York.

A stimulating and impressive historical narrative of the rise of molecular biology. A good read.

Kornberg, A. (1980). *DNA Replication*. W. H. Freeman, New York.

A clearly written text on DNA replication and repair by the discoverer of DNA polymerase.

Lin, E. C. C., Goldstein, R., and Syvanen, M. (1984). *Bacteria, Plasmids, and Phages: An Introduction to Molecular Biology*. Harvard University Press, Cambridge, Mass.

A general introduction of microbiology with emphasis on the bacteria and their associated plasmids and phages.

Rees, A. R., and Sternberg, M. J. E. (1984). *From Cells to Atoms: An Illustrated Introduction to Molecular Biology*. Blackwell Scientific, Boston.

Something of a coffee-table book with excellent diagrams and clear, concise explanations of various mechanisms.

Schimmel, P. R., Söll, D., and Abelson, J. N. (Eds.) (1980). *Transfer RNA: Structure, Properties, and Recognition, and Biological Aspects*. Cold Spring Harbor Laboratory, Cold Spring Harbor, N.Y.

A two volume collection of papers on all aspects of tRNA.

Watson, J. D. (1968). *The Double Helix*. Atheneum, New York.

This eye-opening personal account of the discovery that DNA is a double helix is must reading for anyone interested in modern biology.

Watson, J. D., Hopkins, N. H., Roberts, J. W., Steitz, J. A., and Weiner, A. M. (1987). *The Molecular Biology of the Gene*. 4th ed. Benjamin/Cummings, Menlo Park, Calif.

The fourth edition of Watson's fine book, first published in 1965. Described by one reviewer as a celebration of the crowning achievements of biology in the second half of the twentieth century.

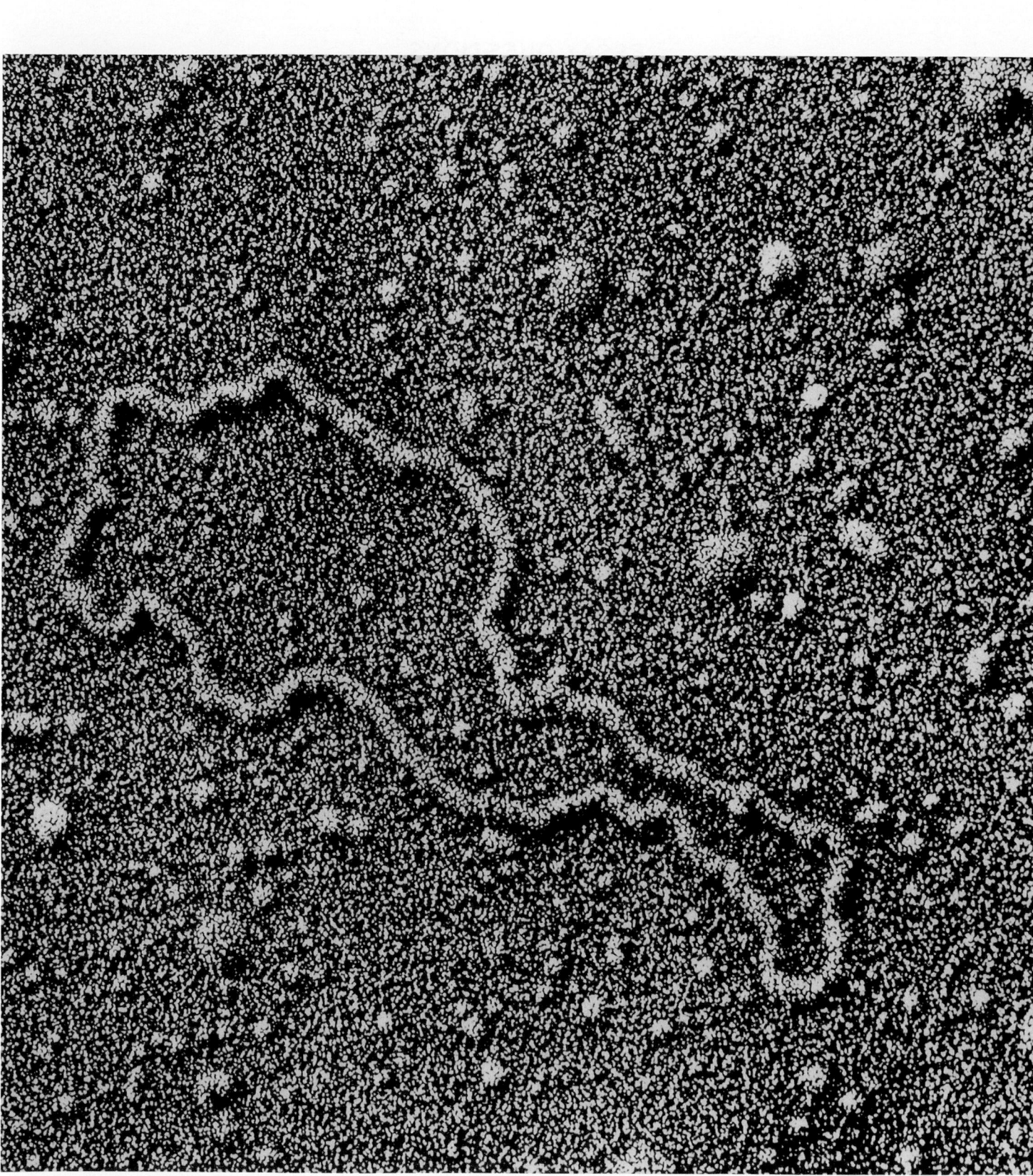

REGULATION OF GENE ACTIVITY

Regulation of gene activity
The study of heredity has advanced in this century from breeding experiments to investigations and manipulations of the genes themselves. Here is an electron micrograph of plasmid DNA isolated from the bacterium Escherichia coli. *Compare this structure with that of the larger* E. coli *chromosome shown at the beginning of Chapter 16. Plasmids are critically important elements in recombinant DNA technology.* (×210,000)

We have seen that DNA controls the traits and activities of cells, and thus of whole organisms, by specifying the structures of proteins and controlling their rates of synthesis. We also saw that a mutant gene produces either a protein with an erroneous amino acid sequence or no protein at all.

New questions now arise.

■ Is a gene always at work, constantly stamping out proteins at full throttle?
■ Is a gene somehow prohibited from resting and for the moment not encoding proteins?
■ Does the absence of a protein necessarily signify the absence or alteration of the gene that ordinarily determines that protein?

All three questions are important and each can be answered *no*. The functioning and survival of a cell require more than an outpouring of structurally correct ingredients. Coordination of gene expression is also necessary.

This chapter describes the intricate control mechanisms ensuring that proteins are synthesized only in the amounts needed. Although gene regulation is accomplished differently in prokaryotes and eukaryotes, all of the diverse control systems regulate either gene transcription into mRNA at one or another of its steps, or translation of mRNA sequences into amino acid sequences (see Chapter 17).

REGULATION IN PROKARYOTES

In the 1950s, Francois Jacob and Jacques Monod of the Pasteur Institute in Paris advanced our understanding of the regulation of certain enzymes of *E. coli*. The enzymes were those catalyzing the early metabolism of lactose—they became known as the *lac* operon—but their regulatory behavior turned out to be typical of many enzymes.

Lactose, we recall, is a disaccharide consisting of a glucose and a galactose molecule. Structurally, it is a **galactoside.** As shown in Figure 4-7, the glycosidic bond between the two sugars of a disaccharide can take either of two steric positions, termed α-form and β-form. By studying the fate of the β-form lactose molecule (a β-galactoside), Jacob and Monod found that three enzymes catalyze the critical steps of the early pathway of its metabolism.

■ First, **permease** brings lactose into the cell.
■ Second, **β-galactosidase** hydrolytically cleaves lactose into glucose and galactose (Figure 18-1).
■ Third, **transacetylase** attaches an acetyl group to one of the metabolic intermediates in the pathway.

Three different genes encode these three enzymes: gene Z encodes β-galactosidase, gene Y encodes permease, and gene A codes for the transacetylase molecule (Figure

18.2). Since each determines an amino acid sequence—and thus the structure of a protein—they were called **structural genes.** This is the kind of gene we have been discussing thus far. The investigators soon found that genes of another kind, which they called **regulatory genes,** determine the rates of protein synthesis.

INDUCTION AND REPRESSION

We now explore an example of prokaryotic gene control that allows these single cells to adjust to changes in their nutritional environment, thus optimizing their main function—growth and division. Consider a culture of *E. coli* growing in a medium in which the sole carbon source is succinic acid. The organisms contain only a trace of β-galactosidase, the enzyme that splits β-galactosides. However, addition of lactose (or some other β-galactoside) to the culture medium is followed within minutes by a several thousand-fold increase in the rate of synthesis of that enzyme. A high rate of enzyme synthesis continues as long as the bacterial cells are exposed to β-galactoside. When it is withdrawn, the rate of enzyme synthesis returns to its original low level. Apparently, these cells have mechanisms for turning genes on and off.

The "turning-on" mechanism is called **induction** (see Figure 8-21.) Induction is defined as an increase in the rate of synthesis of a single enzyme protein (or group of enzymes) following exposure of a cell to an **inducer** (Figure 18-2B). Such enzymes are said to be **inducible.** Lactose is, of course, a substrate for the induced β-galactosidase. But many active inducers may chemically resemble the usual substrate of an enzyme without being identical to it and thus are themselves inactive as substrates.

Repression of enzyme synthesis is a related "turning-off" mechanism. Repression is the inhibitory effect of the product of a reaction sequence on the synthesis of an early enzyme of that sequence (Figure 18-2C,D). We saw in Chapter 8 that synthesis of the first enzyme in the pathway of L-isoleucine synthesis is repressed by L-isoleucine, the end product of that pathway (see Figure 8-19). When the end product is present at a high concentration, synthesis of this key enzyme is repressed. Conversely, when the concentration of end product is low, synthesis of the enzyme accelerates. That is called **derepression.** Induction and repression are instances in which cells with the ability to synthesize a protein either do not start doing so (induction) or do not stop doing so (repression) until they receive the proper signal.

THE OPERON MECHANISM

Their studies of these systems revealed the basis of variable rates of enzyme synthesis—and brought a Nobel Prize to Jacob and Monod in 1965. Their postulated explanation of the relationship between induction and repression (Figure 18-2) was imaginative—and later work confirmed and supplemented it.

- Structural genes produce the mRNA that orders synthesis of the enzymes. In the *lac* operon, the three structural genes—Z (β-galactosidase), Y (permease), and A (transacetylase)—together produce a single polycistronic mRNA molecule. (Recall that *cistron* is a newer term for structural gene; see Chapter 17.)

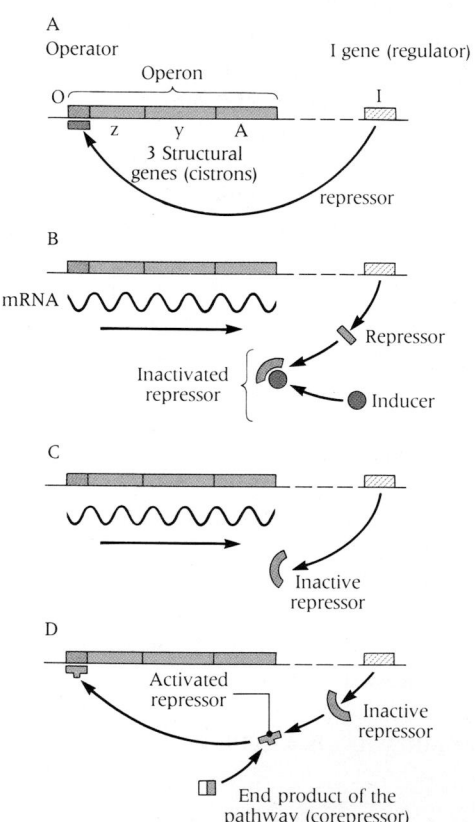

18-2 Regulation of the lac operon
(A) *Under normal conditions, the repressor protein, encoded by the I gene, binds to the operator sequence and blocks transcription of the Z, Y, and A structural genes. A promoter sequence (not shown here) is just upstream from the operator, overlapping it slightly (see Figure 18-3).* **(B)** *In induction, the inducer, lactose, inactivates the repressor and transcription proceeds.* **(C)** *Normally, the I gene makes an inactive repressor, and transcription proceeds.* **(D)** *In repression, the inactive repressor become activated upon binding to a corepressor molecule, and transcription is blocked.*

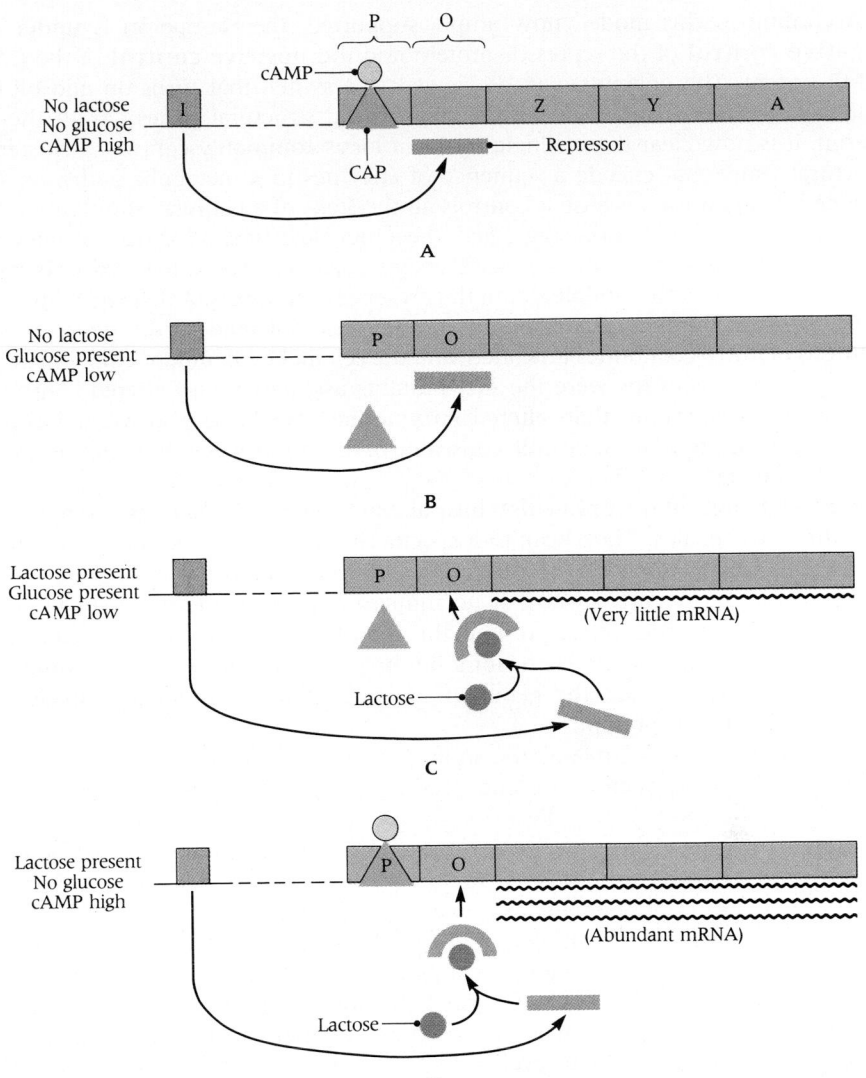

18-3 Regulation of the lac operon by glucose and cyclic AMP

(A) Cyclic AMP binds to the CAP protein, and this complex prevents RNA polymerase from binding to the promoter, P, thereby blocking transcription. **(B)** In the absence of cyclic AMP, the repressor continues to block transcription. **(C)** In the presence of lactose, the inactive repressor permits some mRNA to be made. **(D)** If RNA polymerase is permitted to bind to the promoter, in the presence of lactose, large amounts of mRNA are made. The molecular structure of cyclic AMP is shown in Figure 28-6.

- The rate of mRNA synthesis is controlled by a separate sequence located in a nearby region of the chromosome, away from the structural genes. This is the **operator,** or **O** region. The operator controls the three genes of the lactose system as a concerted unit, and is continuously active.
- A regulatory gene, the **I gene,** is also located nearby on the chromosome. It encodes a specific regulatory protein called the **repressor.**
- If no inducer is present in the culture medium (Figure 18-2A), repressor diffuses through the cell and binds to the operator, O, and it does so with extraordinary specificity. As long as the repressor is bound to the operator, the genes of the lac operon are not transcribed.
- However, if an inducer is present, it binds to the repressor and inactivates it. Transcription goes on at a high rate.
- Other regulatory sequences in the upstream control region of the *lac* operon include two **promoter** sites that bind RNA polymerase—P_i, promoter of the I gene, and P_{lac}, promoter of the O-Z-Y-A unit.
- The **CAP site** (CAP stands for **catabolite activator protein**) binds a CAP, which starts transcription (Figure 18-3). Discovery of CAP explained a puzzling fact: in cells grown in a medium containing glucose, synthesis of many sugar-metabolizing enzymes (including β-galactosidase) is markedly suppressed. Apparently, glucose or a breakdown product of glucose (a *catabolite*) prevents synthesis of the mRNA encoding these enzymes. Study of this phenomenon, termed **catabolite repression,** led to the realization that when cyclic AMP (cAMP) levels are low (as they are when glucose is plentiful) the CAP protein does not function and minimal *lac* mRNA is synthesized. When glucose is absent and another sugar is present, transcription is initiated and the sugar is metabolized.

According to this model, now amply supported, the *lac* operon is under the **negative control** of the repressor protein and the **positive control** of the CAP-cAMP system. The operator may be viewed as a switch that turns on and off the synthesis of mRNA transcribed from neighboring structural genes. As in the *lac* operon, it is now clear that a single operator locus commonly controls a cluster of structural genes that encode a sequence of enzymes in a metabolic pathway. The scheme is economical because it controls all the steps of a pathway simultaneously.

When, through mutation, the *I* gene becomes defective, all three enzymes are synthesized at full throttle. In such cases, enzyme synthesis is said to be **constitutive** because the *lac* operon is indifferent to the presence of inducers. A defective repressor molecule—the product of the mutant *I* gene—cannot bind to the operator, the synthesis of mRNA continues unabated, and the cell makes abundant enzyme constitutively. Such organisms were the first mutants associated with altered control of enzyme synthesis rather than altered enzyme activity. Jacob and Monod coined the term **operon** for the whole unit, consisting of structural genes, operator, promoter sites, and I gene.

Repressor molecules are proteins whose ability to bind to the *lac* operon is regulated by inducer molecules. They bear two specific binding sites, one for the operator DNA sequence and one for the inducer. The inducer combines with and inactivates the repressor (see Figure 18-2B). If no inducer is present, the repressor binds to the operator. If an inducer is present, the repressor combines with it, causing a change in the shape of the repressor's binding site that decreases its affinity for the operator. As a result, the genes of the operon are transcribed actively—as long as an inducer is present.

The hypothesis also explains the inhibition of transcription by a substance in the cell or medium, such as an end product of a biosynthetic pathway. In this case (see Figure 18-2D), the effector molecule that binds to the repressor plays a role opposite to that of an inducer. We call such a repressor-binding effector molecule a **corepressor.** Certain repressors do not function when they are uncombined with corepressors. Consider, for example, the amino acid tryptophan, which is a product of the biosynthetic pathway coded by the *trp* operon. Tryptophan molecules themselves are the corepressors that combine with the *trp* repressor protein. Only when such binding has occurred can this repressor bind to the operator and inhibit transcription of the genes encoding the enzymes that make tryptophan (Figure 18-4). In other words, a high level of end-product acts to repress its own further synthesis.

In summary, the alternate forms of the basic operon hypothesis can be summarized as follows:

- In induction: repressor + inducer → inactive repressor
                  (operon is transcribed)
- In repression: repressor + corepressor (end product) → active repressor
                  (operon is not transcribed)

The operon hypothesis makes it clear why the end products of biosynthetic pathways have been called corepressors: they, along with repressor molecules, cause repression of mRNA transcription and thus of enzyme synthesis. Repression occurs only in repressible pathways; these include most of the major biosynthetic pathways. Under ordinary conditions, most repressible enzymes are partially repressed—that is, they are synthesized at rates well below the maximum rate possible. In these circumstances, operators are partially saturated with corepressor-activated repressors.

Functionally, repression resembles feedback inhibition (see Chapter 8) in that it prevents wasteful endogenous synthesis (in the cell) of materials when there is an ample exogenous supply (outside of the cell). Feedback inhibition is rapid and sensitive and is the chief regulator of small-molecule production. Repression coordinates the competing processes of protein synthesis in a cell. By providing enzyme proteins in the right combinations and at the right time, it prevents what could otherwise be a lethal overactivity of the structural genes. If, for example, the structural gene for β-galactosidase were fully derepressed, the amount of the single enzyme protein synthesized could represent 8% of the cell's protein. A few such derepressed syntheses could wreck a cell's economy by permitting certain pathways to operate in either a wasteful or an inadequate manner. Repression is thus the coarse adjustment of a metabolic machine, and feedback inhibition the fine adjustment. Through these

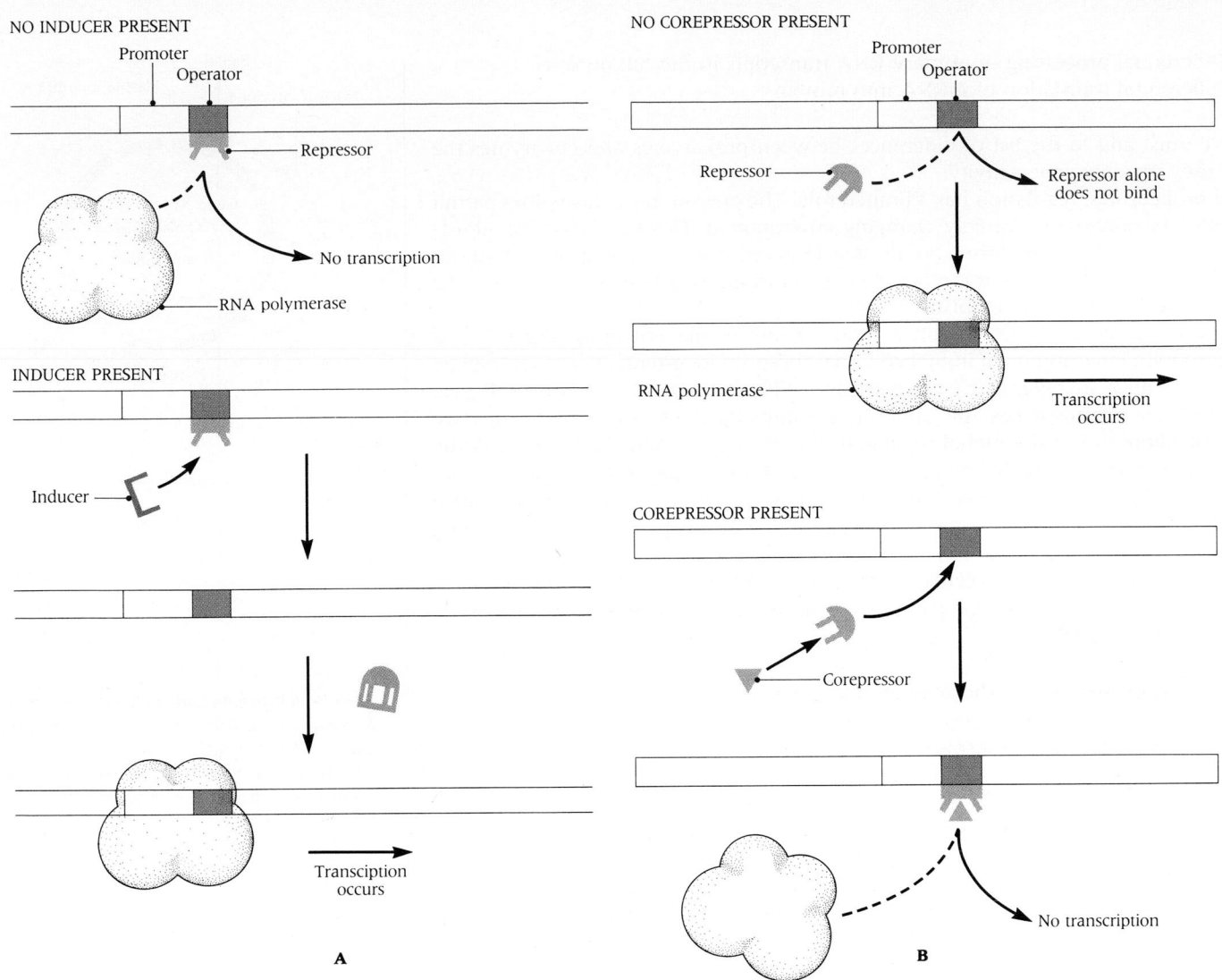

NO INDUCER PRESENT

Promoter
Operator

Repressor

No transcription

RNA polymerase

INDUCER PRESENT

Inducer

Transcription occurs

A

NO COREPRESSOR PRESENT

Promoter
Operator

Repressor

Repressor alone does not bind

RNA polymerase

Transcription occurs

COREPRESSOR PRESENT

Corepressor

No transcription

B

18-4 Induction and corepression in the bacterial operon

(*A*) *In an inducible system, transcription is blocked by the binding of repressor to the operator as long as the inducer is absent. When inducer is added, it pries the repressor from the DNA, allowing RNA polymerase to bind and transcription to proceed. (B) In corepressible systems, absence of corepressor is accompanied by transcription, since the shape of the repressor keeps it from binding to the operator. Added corepressor binds to the repressor, changing its conformation. Now it binds to DNA and stops transcription.*

systems, bacteria are capable of making rapid, short-term responses to changes in the environment.

The complexity and sensitivity of the mechanisms regulating prokaryotic gene transcription has only recently been appreciated. Not only are there negative and positive controls, but there are also many operons that are doubly regulated by both types of controls. In these systems, the rate of mRNA synthesis is at its maximum when a negatively controlling element is made ineffective by an inducer and a positively controlling factor (like CAP) is available to bind to the DNA.

REGULATION IN EUKARYOTES

The regulation of gene activity in eukaryotes is a hot topic in molecular biology today. An operon system like the *lac* operon in *E. coli* is rare or nonexistent in eukaryotes. Genetic crosses have revealed no operator genes and specific repressors have not been found.

To sharpen our inquiry, let us restate the question. How does a specialized eukaryotic cell selectively express only a small set of the tens of thousands of genes at its disposal? Unlike the situation in prokaryotes, regulation in eukaryotes depends upon multiple mechanisms that facilitate selective or differential gene functioning. In other words, these mechanisms selectively control the rates of protein synthesis by different genes in different cells. Furthermore, these mechanisms operate at every stage in the formation and utilization of mRNA. Thus there are several levels of control.

- Differential transcription, perhaps the most significant category of control mechanisms

- Differential processing of primary RNA transcripts in the cell nucleus
- Differential translation of mRNA into protein

We must add to the list of differences between prokaryotes and eukaryotes the way they regulate gene activity.

In prokaryotes, regulation has a limited role. The operon mechanism does permit sensitive responses to a rapidly changing environment. However, ensuing adjustments are for the short term, briefly affecting cell division and growth without producing permanent alterations in gene expression. This is in keeping with the short life cycles of these organisms.

In eukaryotes, regulation is on a more or less permanent basis. There is less concern with environmental influences. That these are long-term changes is implied by their critical involvement in growth and differentiation. Every body cell of a multicellular organism has the same gene complement its zygote had. Thus we must attribute to transcriptional regulation the diversity of adult cell types and the differences in their enzyme content. A liver cell, for example, contains many more enzymes than a fibroblast. Yet both cells contain the same genes. That the course of differentiation is guided by programs of selective or differential gene functioning is made clear in Chapter 20.

How do eukaryotic cells regulate rates of gene activity? To answer this question we must consider mRNA production and utilization and attempt to answer three questions (Figure 18-5).

- What signals influence the activity of a given gene?
- At what steps is control exerted?
- What molecular mechanisms of control occur at each step?

SIGNALS FOR GENE CONTROL

Hormones

In systems as precise as those regulating transcription, very little happens on a random basis. If a regulatory system puts the transcription throttle up or down it is fair to assume it is doing so in response to a signal. That assumption was substantiated by the discovery that many hormones are such signaling devices (Chapter 8). In effect, their levels are the signals that inform the system when to regulate, how to regulate, and by how much to regulate gene activity.

Hormones are substances that regulate the rates of specific metabolic events (see Chapter 28). Some hormones act directly upon various "target" enzymes, affecting their rates of catalytic activity. This mechanism is quite independent of gene activity. But other hormones control metabolism by regulating the rates of synthesis of specific target enzymes.

The best known case is that of the giant salivary gland chromosomes of *Drosophila*, which result from an unusual circumstance: in these organisms, cells of metabolically active organs like the salivary gland increase in size rather than number. Hence, chromosomes of these cells are enlarged by thousands of times by accumulations of RNA and protein. Thus, they can be easily observed by light microscopy. During development, visible enlargements, called **puffs,** occur along these chromosomes. Studies show that these are active sites of mRNA and protein synthesis. As shown in Figure 18-6, radioactive uridine (an RNA precursor) is actively incorporated in puff regions; and this effect is prevented by **actinomycin D,** a potent inhibitor of mRNA synthesis. The main proteins in puffs are DNA-binding proteins. Puffs develop in response to the hormone ecdysone, which causes fly larvae to molt. If the ecdysone is withdrawn, the puff disappears immediately. Other puffs are caused by such environmental effects as heat shock. Although puffs form at sites of active mRNA synthesis—and thus are visible signs of intense gene transcription—for some reason, puffs do not form in some active regions of these curious chromosomes.

Hormones are of many types, chemically speaking. The two largest categories are **steroid hormones** and **protein hormones.** Hormones differ widely in the kinds of cells they influence and the number of genes they affect in each target cell. For example, some steroid hormones from the adrenal cortex affect many kinds of cells and dozens of genes in each cell. The same is true of the protein hormones insulin and growth hormone.

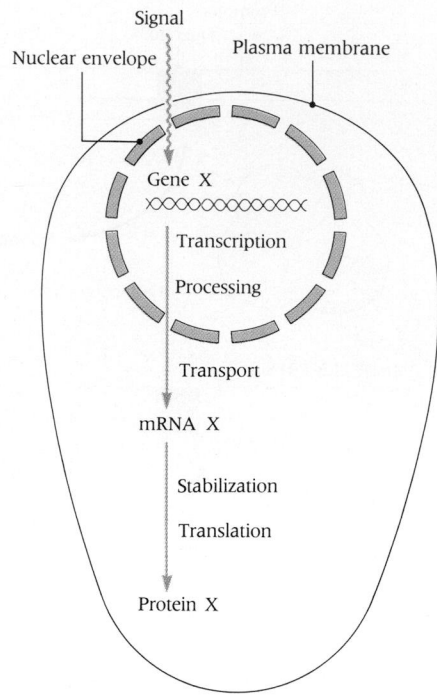

18-5 Mechanisms controlling gene activity
Three questions are to be answered in studying control: What signals a change in gene expression? At what level does the cell respond? What is the mechanism of the response?

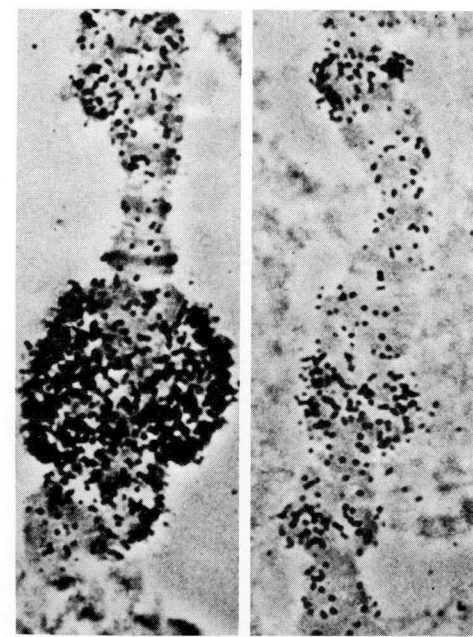

18-6 Puffing in polytene chromosomes
Uptake of radioactive uridine into RNA (black dots) is localized to the puff region (left), and is eliminated in the presence of an inhibitor of mRNA synthesis (right). (×2000)

Growth Factors and Other Signals

Another category of signaling agents includes a large number of the proteins known as **growth factors,** which we mentioned in Chapter 12. These agents, which include nerve growth factor, epidermal growth factor, and others, are arousing intense interest because new work implicates them in growth (both normal and malignant), differentiation, and various disease states. They, too, act by signaling the systems that regulate transcription. They also stimulate cell division.

These signaling molecules—both hormones and growth factors—operate in two ways. Those in one group act by binding to surface receptors on a target cell and dispatching a **second messenger** to the gene site. Many protein hormones and growth factors are in this group. Those in the second group enter the cell and do not require a second messenger to affect cell function. Most steroid and thyroid hormones are in this category (see Chapter 28).

Cell-to-cell contacts also act as signals that can regulate gene transcription (see Chapter 7). Though still poorly understood, they are under urgent study because there is evidence that a cell's "decision" to differentiate via one pathway or another may depend on information obtained via cell-to-cell contacts—for example, by the passage of small molecules from cell to cell through gap junctions.

STEPS AT WHICH CONTROL IS EXERTED

The synthesis and processing of mRNA and the mechanisms of translation, described in Chapter 17, provide a large number of "switches" that can serve as targets of regulatory mechanisms. While it is still difficult to pinpoint the mechanism of eukaryotic gene control, current work suggests that the following stages of gene expression are major control points.

Synthesis of mRNA in the Nucleus

Many examples of gene control at this level are recognized, among them the synthesis of egg-white proteins in the chick oviduct (signaled by the steroid hormone estrogen) and of the α and β chains of hemoglobin (signaled by the hormone erythropoietin). In these cases, labeling methods reveal sharp increases in the amounts of mRNA synthesized when the signal is *go*.

Control at the mRNA synthesis step depends on **transcriptional signals,** which are short DNA segments that powerfully influence the activity of neighboring structural genes. Although one or two occur downstream from the structural genes, most are upstream. Like the promoters of prokaryotes, these sequences lie upstream of the structural genes (see Figure 17-7). In eukaryotes, transcriptional control elements are of three types (Figure 18-7).

■ The **TATA box,** so named by whimisical molecular biologists for its TATA sequence (although that may vary a bit), is about 30 bp (base pairs) upstream from many eukaryotic genes. Apparently essential for transcription initiation, it is analogous to a similar sequence in prokaryotic DNA, in which the sequence is TATAAT or something similar. The TATA box helps fix the start site of transcription by positioning RNA polymerase (see Figure 17-6).

■ A second upstream sequence, CCAAT, also called the **CAAT box,** influences rates of transcription as well (see Figure 17-7).

■ Finally, DNA segments even farther upstream than the CAAT box have been implicated in gene control. These more remote segments, termed **enhancers,** appear to determine the efficiency of transcription.

18-7 Eukaryotic transcriptional control elements revisited
Control of transcription resides in three principal regions located upstream from the 3' end of the gene. These include the TATA box, a cluster of promotors that includes the CAAT box, and the enhancer region. Compare with Figures 17-6 and 17-7.

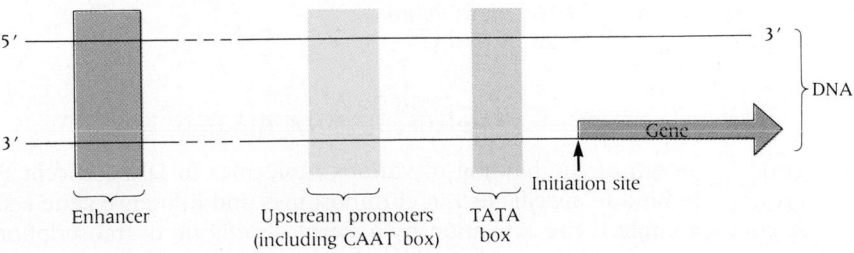

In addition, we now recognize a variety of **transcription factors** in prokaryotes and eukaryotes that have major roles in regulating mRNA synthesis. One recent essayist has predicted that this will be the next major arena of molecular biology.

Processing of the Primary RNA Transcript

Ever since hnRNA was identified as a precursor of mRNA, selective RNA processing seemed a likely locus of gene control—especially since processing includes a number of post-transcriptional modifications of the RNA molecule (methylation, capping, addition of poly(A), and splicing). Several dozen genes of mammals and *Drosophila* are now known to be controlled by differential RNA processing and there is reason to consider this a widely used control device.

Stabilization or Destabilization of mRNA in the Cytoplasm

The preceding two control mechanisms are of greater importance than is stabilization or destabilization of cytoplasmic mRNA. In other words, nuclear events are more important determinants of eukaryotic gene activity than are cytoplasmic events. Nonetheless, other factors affecting the stability of mRNA do play a role. One is obvious: it is the fact that in eukaryotes, mRNA is bound to protein and in prokaryotes it is naked. This doubtless accounts for the long half-life of eukaryotic messenger ribonucleoproteins and the short half-life of naked prokaryotic mRNAs.

Other factors control the rate of mRNA transport from nucleus to cytoplasm and the life of the mRNA molecule. Both can vary widely. For example, the half-life of the mRNA for casein (the abundant milk protein) is 5 hours in the absence of the hormone prolactin and 92 hours in its presence.

Translational Regulation

The last chapter revealed that the translation process includes many steps that would be useful intersections for a regulatory traffic control system. However, most of the translational controls actually observed—in both eukaryotes and prokaryotes—operate during initiation of protein synthesis. These include mechanisms regulating the activity or availability of mRNAs or a variety of initiation factors.

While translation could conceivably also be regulated during the elongation or termination phases, particularly through control of the availability or activity of the elongation and termination factors, there is little evidence supporting regulation of this type. In both prokaryotes and eukaryotes, elongation and termination evidently proceed at constant rates once translation is initiated. This probably represents an economy in the use of ribosomes, since controls inhibiting the elongation or termination stages of protein synthesis, in contrast to initiation, would shunt ribosomes into inactive complexes and reduce the quantity available for protein synthesis.

MOLECULAR MECHANISMS OF CONTROL

When we speak of "mechanism," we are asking, "How does it work?" Much more is known about transcriptional control mechanisms than of translational control mechanisms. The following molecular mechanisms of translational control have been established or suspected:

- Site-specific DNA binding
- Alterations in chromosome structure
- Altered levels of DNA methylation
- Nonspecific effects of histones

Site-specific DNA Binding

Little is known of the binding of various molecules to DNA, except that various proteins do bind to specific loci in chromosomes and influence gene activity levels. A good example is the activation by a steroid hormone of transcription of a gene

18-8 Electron micrograph showing the binding of a hormone-receptor complex to DNA

This remarkable preparation shows steroid hormone receptors (dark particles) bound to strands of DNA. The size and shape of the bound protein clusters indicate that they contain many subunits ($\times 100,000$).

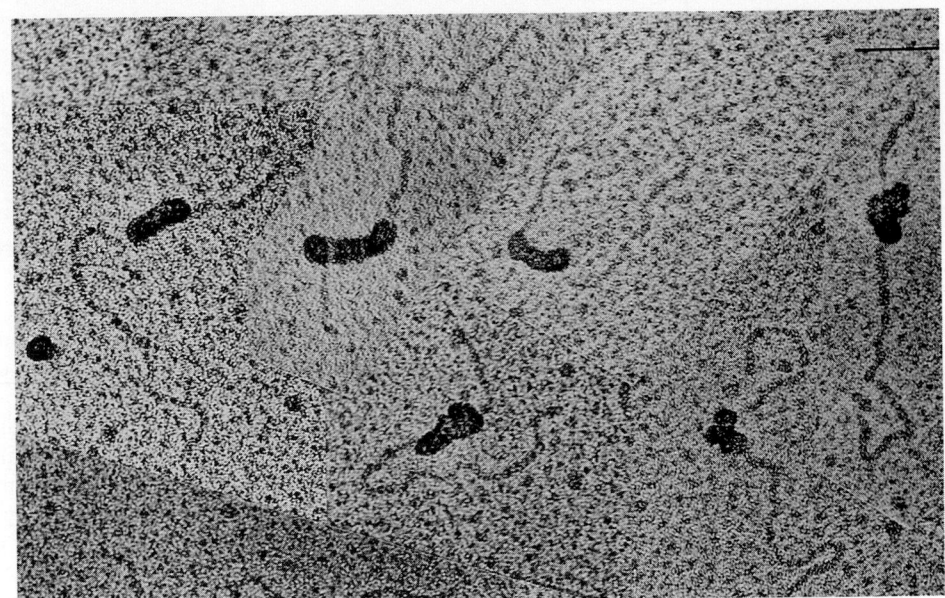

derived from a retrovirus called MMTV.[1] (As will be explained in Box 18-2, this is a DNA copy of a viral RNA that was once integrated into the genome of the cell or its precursors.) For this gene to be transcribed, the steroid hormone must bind to its receptor within the cell. The resulting steroid-receptor complex must then bind to a specific locus on the DNA molecule (Figure 18-8).

In other cases, specific proteins (rather than steroids) bind to regulatory sites on the DNA molecule and influence transcriptional activity.

Alterations of Chromatin Structure

The condensed character of chromatin in eukaryotic cells—and the central role of the nucleosome in chromatin structure (see Chapter 12)—strongly suggested that some sort of "unwinding" or "loosening" of chromatin packing might be a necessary preliminary to gene activity.

In the electron microscope, all DNA in eukaryotic cells looks essentially the same. From this perspective, DNA containing active genes cannot be distinguished from inactive DNA sequences. But the two are different and the differences are in fine structure—beyond the microscope's reach. For example, we learned in Chapter 12 that **DNase I** is a nuclease that can cut either DNA strand. It does so more readily in active chromatin regions than in inactive regions. Further, when a gene is active, a region of great sensitivity to DNase is found just upstream from the gene. The increased DNase sensitivity of such **hypersensitive regions** signifies a more open or relaxed structure that allows better access to the DNase "scissors."

Discovery of these hypersensitive regions raised an obvious question: are these regions related to DNA promoter sequences upstream from active genes? It is interest-

---

[1] For mouse mammary tumor virus.

18-9 The molecular basis of chromatin coiling

Loss of a nucleosome causes a local supercoil, the energy of which can cause the local DNA strands to untwist and separate ("melt") or to become a left-handed helix called Z-DNA (see Box 18-1).

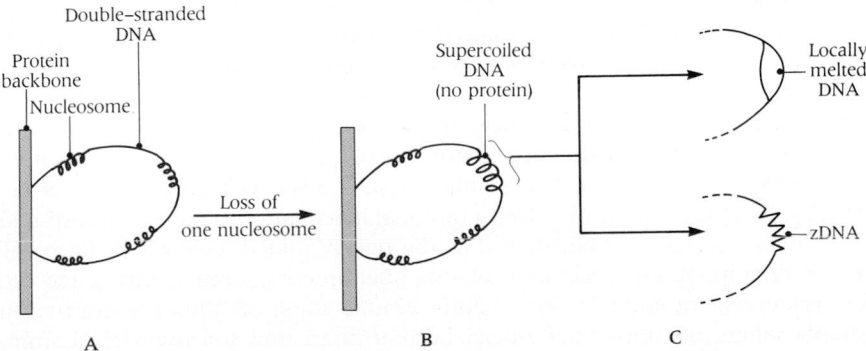

The sketches of DNA in Figure 18A and Figure 16-11 reveal a "right-handed" helix. Until recently, no one knew why DNA rotates to the right instead of the left. Then, in 1979, Alexander Rich and associates at M.I.T. encountered a form of DNA with a structure unlike any seen before. They found it while probing structural details of conventional DNA molecules, a form known as **B-DNA.**[*]

Some dismissed Rich's observation as a chemical oddity, but Rich, ever a believer that any structure (usual or unusual) has biological meaning, began seeking a function for the new molecule. When he found that it occurs only in specific regions of chromosomes, he began to suspect that it has a role in the regulation of gene expression. What was so peculiar about this new form of DNA?

[*] B-DNA, the biologically important form of DNA, possesses two dimensions, a major one is a 3.32 Å rise per base pair; a minor one is a 33.2 Å pitch per helix turn. So-called A-DNA differs only in its lower number of associated water molecules.

Knowledge of the structure of DNA and other macromolecules is most reliable when it is based on the data of x-ray diffraction. But this method requires good DNA crystals, and DNA does not yield very good crystals. Hence it is not possible to see single atoms of DNA in x-ray diffraction patterns and crystallographers could only guess at some of the fine structural details. Rich thought he could settle these issues by synthesizing some oligonucleotides, short chains of deoxyribonucleotides that can be crystallized. He made a group of tetramers (chains of four nucleotides) in which the bases were mostly G and C, and crystallized them, hoping to see individual atoms in x-ray diffraction studies. For atoms to be distinguishable, a crystal must diffract to a resolution of 1 Å or less. The effort was successful: the crystals that were obtained diffracted to 0.9 Å. Nearly two years were needed to solve the structure. When the pattern emerged, it was startling.

The polymer was the above-mentioned new form of DNA with a left-handed rather than right-handed helix (Figure 18B). However, it was not simply a left-handed version of ordinary DNA. Rich named the new DNA structure **Z-DNA.**

Soon after they detected the Z-DNA structure in synthetic polymers, the investigators showed that it occurs naturally. At this point, some considered Z-DNA a trivial finding. Others reasoned that life is opportunistic. Here was a stable form of DNA. Nature is likely to use it. The job was to find out how.

Z-DNA so far had been found only in highly concentrated salt solutions, which do not occur normally in nature. But it was found that Z-DNA can occur at normal salt concentrations if the C–G polymer is methylated. As we have seen, methylation of C–G base pairs in eukaryotic DNA may regulate gene transcription. When these sequences are methylated, genes are inactive, a hint that Z-DNA may have a regulatory role.

One way to find Z-DNA in biological systems was to make an antibody to it to see if it binds specifically to any naturally occurring DNA. This would signify the presence of Z-DNA in nature. This experiment led to

ing that two control elements may be involved: the TATA box is found between a gene and a hypersensitive site and the CAAT box is located within a hypersensitive region.

Several hypotheses seek to explain how these remote hypersensitive sequences could affect gene function. One hypothesis is that these sequences have some influence on chromatin structure. As we have seen, DNA in chromatin is clothed in proteins, mainly histones. The DNA-protein complex yields a structure resembling a string of beads (see Figures 12-21 and 12-22): histone octomers and the DNA coiling around them form the nucleosomes. DNA coiled around the histones may be less accessible to enzymes, including RNA polymerases, than is the linker DNA found between nucleosomes. One explanation for the effect of remote DNA segments on transcription efficiency is that they position nucleosomes so as to expose the initiation site to polymerase. Alternatively, the coiling or supercoiling of the DNA might bring together separate DNA regions that interact to bind the polymerase.

Another theory postulates that loss of a nucleosome (Figure 18-9), or some other change in the state of the nucleosomes, results in structural changes in DNA near transcribed genes. One such change could be the conversion of a normal right-handed DNA helix to the left-handed type known as **Z-DNA** (Box 18-1).

Finally, there is evidence that transcriptional inactivation is reflected morphologically in the conversion of chromatin to the tightly folded, condensed form called heterochromatin. A vivid example of this phenomenon occurs during red blood cell development in animals (see Chapter 22). As these cells mature, transcription gradually tapers off until their nuclei become inert and inactive. (In mammals,

an important observation. Z-DNA does occur in salivary gland polytene chromosomes of *Drosophila* (see Figure 11-16C). About 100 cells in each salivary gland continue to replicate their DNA but do not divide. About 1000 to 2000 copies of each chromosome arrange themselves in perfect alignment. As a result, the chromosomes display bands. Dark bands running down the chromosomes are separated by light areas called interbands. Nearly 90% of the DNA is in the bands. Application of Z-DNA antibodies to *Drosophila* giant salivary gland chromosomes had dramatic results. Z-DNA antibodies stained only the interband regions (Figure 18C).

Thus Z-DNA may be one of the elements regulating gene transcription. When genes are "turned on," regions of Z-DNA which previously were methylated lose their methyl groups. Perhaps the response of Z-DNA to torsional stress (twisting) is to open up in a way that amplifies gene function. Current work seeks to explore these possibilities.

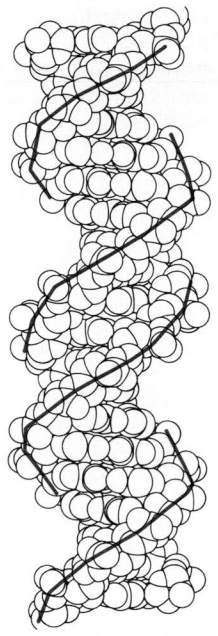

18A B-DNA structure showing right-handed helix
In this space-filling model of DNA, the double helix is not as easy to see. However, the turn of the helix (color) is clearly right-handed.

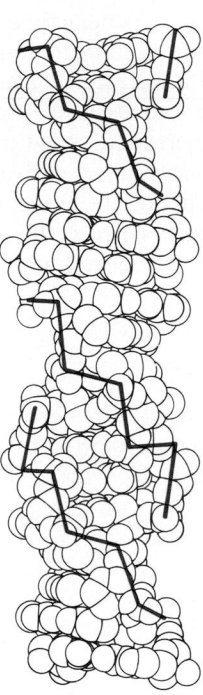

18B Z-DNA structure showing left-handed helix
In this uncommon form of DNA, the sugar-phosphate backbone zigzags down the molecule.

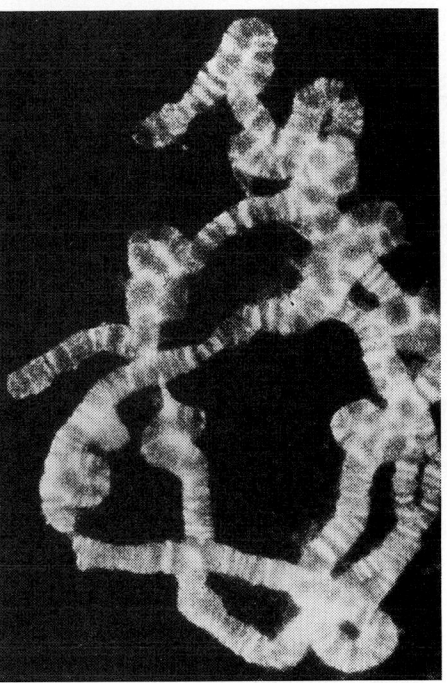

18C Micrograph showing antibodies to Z-DNA bound to interbands in polytene chromosomes (×600)

these shriveled nuclei are ejected from the cell.) This reduction in transcriptional activity is accompanied by the gradual condensation of the chromatin into a tightly packed mass.

Condensation of one of the two X-chromosomes in females into a block of tightly folded chromatin, the Barr body, also reflects transcriptional inactivation. According to the Lyon hypothesis, this inactivation is a form of dosage compensation that equalizes the number of active X-chromosomes in the nuclei of mammalian males and females.

Altered Levels of DNA Methylation

About 5% of the cytosine residues in mammalian DNA are in the form of 5-methyl-cytosine. Significantly, methylation of cytosine residues occurs after DNA has been synthesized. Recent work suggests a role for DNA methylation in regulating or blocking transcription of DNA. Actively transcribing genes are generally undermethylated. Perhaps demethylation is necessary for gene expression—especially during development and differentiation. In this connection, Z-DNA—DNA with a left-handed helix—may be significant (see Box 18-1).

The reliance on DNA methylation as a key element in the control of gene expression is limited to vertebrates. Methylated genes have not been detected in *Drosophila* or other invertebrates. Experimental methylation of *Drosophila* genes is purely suppressive, condemning them (in the words of one writer) to transcriptional silence. On the other hand, many plant cells contain exceptionally high levels of 5-methylcytosine in their DNA, and transcription of such methylated DNA does take place. Conceivably, replacement of cytosine with this more radiation-resistant derivative gives these cells increased tolerance to sunlight.

In sum, there is an inverse correlation between the level of methylation in the vicinity of a gene and its transcription rate. But are changes in DNA methylation causes or consequences of eukaryotic gene expression? A reliable method of manipulating methylation levels experimentally may now be possible. Cells exposed to the cytidine analogue 5-azacytidine (Figure 18-10) incorporate it into DNA in place of cytidine. It decreases DNA methylation because, unlike cytidine, it cannot be methylated. Significantly, 5-azacytidine appears to promote gene activity.

Several medical research groups have administered 5-azacytidine to human beings with β-thalassemia, a common genetically determined impairment of hemoglobin β-chain synthesis (see Chapter 15). If these patients could synthesize fetal hemoglobin (hemoglobin F) in its place, they would be notably benefited (see Chapter 22). Normally, hemoglobin F synthesis ceases at birth. Administration of 5-azacytidine did increase hemoglobin F levels in red blood cells, apparently by activating the genes that determine its synthesis. If this interpretation is correct, this is the first case in which the activity of a human gene was directly influenced by a drug.

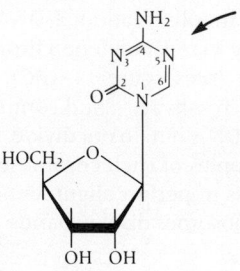

18-10 Molecular structure of 5-azacytidine
This analog of cytidine contains a nitrogen atom at the 5-position (arrow). Cytidine has a carbon atom at this position.

Nonspecific Effects of Histones

Histones are basic proteins that avidly form salts with acidic DNA molecules (see Chapter 12). Histone molecules, because of their unusual properties, might inhibit the template activity of specific segments of the DNA to which they were bound. This would be a way of restricting gene action in a specific, or selective, manner. However, inhibition of transcription by histones turned out to be a nonspecific dose-dependent interaction between these positively charged proteins and DNA. In other words, the more histones that are present, the more transcription is reduced.

A close look at the biochemical properties of the histones makes it seem unlikely that they could operate as specific gene regulators. Eukaryotic cells have thousands of genes, but only five kinds of histones (see Figure 12-22). The way histones bind to DNA—a nonspecific electrostatic attraction—also argues against specificity in the interaction. Such attractions are so strong they overshadow the possible modifying effects of interactions between histones and specific DNA sequences.

These features also make it unlikely that chemical modifications of histones would bring gene specificity to histone-DNA interactions. Many proteins in actively transcribed chromatin are more highly phosphorylated and acetylated than those in inactive chromatin. However, such modifications simply influence histones to do their nonspecific tasks more or less effectively. Also, some of these modifications involve the very diverse nonhistone proteins. It is true that in some cases histones

TABLE 18-1
MAJOR DIFFERENCES BETWEEN MECHANISMS REGULATING GENE EXPRESSION IN PROKARYOTES AND EUKARYOTES

Property	Prokaryotes	Eukaryotes
Chromosomes	Not membrane-limited; nonhistone DNA-binding proteins	In membrane-limited nucleus; includes histones
Nature of structural genes	Contiguous DNA sequences that are colinear with mRNA	Often interrupted by introns
RNA polymerases	One polymerase synthesizes all forms of RNA in cell	Three polymerases synthesize different forms of RNA
Synthesis of mRNA	Often polycistronic; short half-life; not protein-bound	Monocistronic; long half-life; protein-bound
Translation of mRNA	Coupled with transcription	Occurs only after processing of mRNA and transport to cytoplasm

in actively transcribing DNA do seem to be more highly modified. Perhaps, histone acetylation "opens out" nucleosome structure.

In contrast to the nonspecific effects of histones on transcription, there is evidence that the so-called nonhistone proteins may specifically regulate the activity of single genes. Some research suggests that a steroid hormone-receptor complex somehow interacts with the nonhistone proteins of some target genes, with subsequent gene activation taking place. Of course, prokaryotes lack histones.

In sum, the relationship of chromatin condensation to transcriptional regulation by the histone and nonhistone proteins is uncertain. Condensation could reflect histone activity, which is probably a generalized control to regulate the accessibility of DNA for transcription. Alternatively, condensation could be a part of a more specific gene control mechanism exerted by the nonhistone proteins.

These discussions have called further attention to fundamental differences between gene regulation in prokaryotes and eukaryotes (Table 18-1).

WHEN GENES ARE COMPOSED OF RNA

When viruses infect cells the viral genes expropriate the host's metabolic machinery and turn it to the task of making more viruses. But some viruses contain no DNA (see Table 16-1). Their genes are composed of single-stranded RNA. Can RNA molecules act as genes just as DNA molecules do? If so, how?

REPRODUCTIVE CYCLES OF RNA VIRUSES

Some RNA viruses contain a plus (+) strand and others a minus (−) strand. (Remember: a plus strand can function as mRNA; a minus strand, its complementary sequence, cannot.) To this point we have held that in the ordinary circumstances of cellular reproduction RNA molecules never serve as templates for the formation of new RNA molecules. That is the clear meaning of the "central dogma" of molecular biology (Figure 18-11). Now we will substitute "hardly ever" for "never."

Obviously, replication of RNA viruses of this type must involve a special polymerase that can form new RNA strands upon parental templates made of RNA. The

18-11 Information flow in cells and viruses
(A) In prokaryotes and eukaryotes, there is a unidirectional flow of information from DNA to RNA to protein. This scheme is called the "central dogma" of molecular biology. (B) The genome of DNA viruses consists of DNA. (C) In RNA viruses, the genome is composed of single-stranded RNA, called the plus (+) or the minus (−) strand. (D) In retroviruses, an RNA genome gives rise to double-stranded DNA which generates new RNA. Examples of each virus type are shown in parentheses.

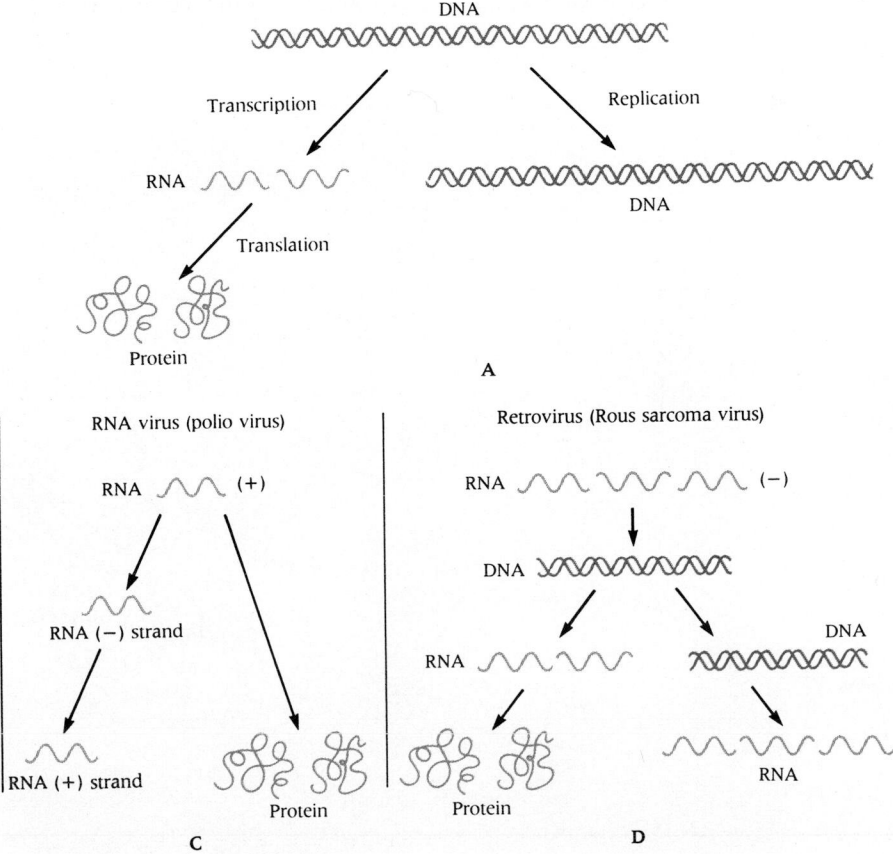

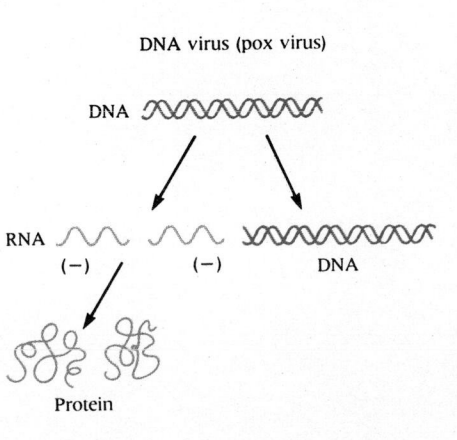

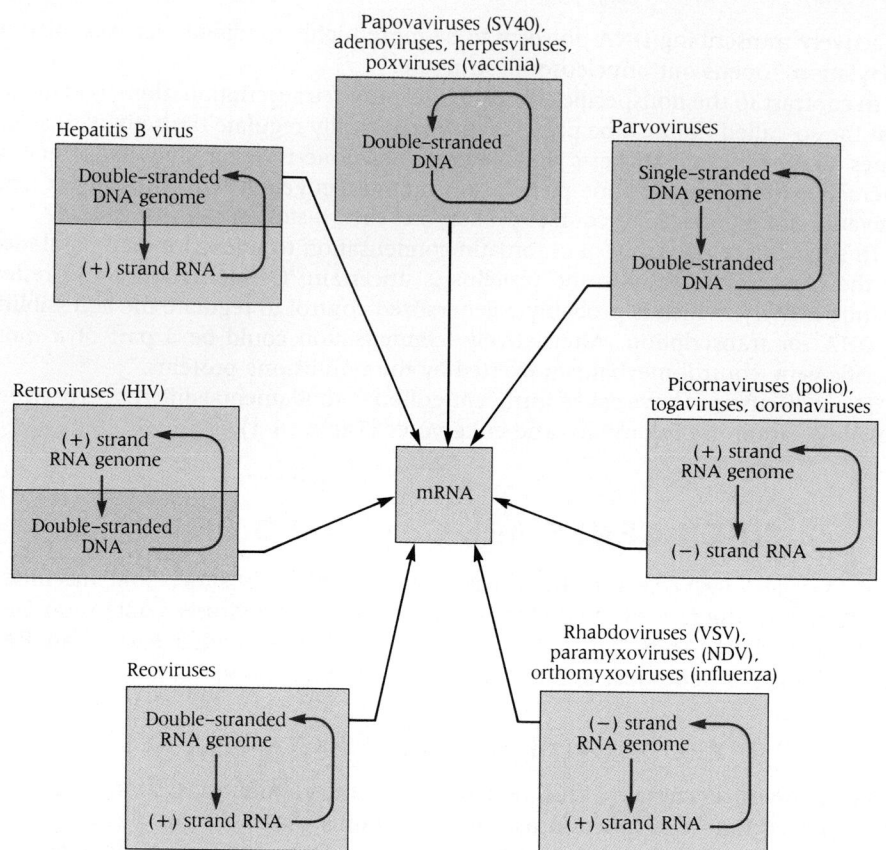

Papovaviruses (SV40), adenoviruses, herpesviruses, poxviruses (vaccinia)

Double–stranded DNA

Hepatitis B virus

Double–stranded DNA genome

(+) strand RNA

Parvoviruses

Single–stranded DNA genome

Double–stranded DNA

Retroviruses (HIV)

(+) strand RNA genome

Double–stranded DNA

mRNA

Picornaviruses (polio), togaviruses, coronaviruses

(+) strand RNA genome

(−) strand RNA

Reoviruses

Double–stranded RNA genome

(+) strand RNA

Rhabdoviruses (VSV), paramyxoviruses (NDV), orthomyxoviruses (influenza)

(−) strand RNA genome

(+) strand RNA

18-12 A classification of viruses by replication mechanism and relation of viral genome to mRNA
Viruses are assigned to seven groups according to genome nucleic acid (DNA or RNA), strandedness (double or single), and strand sense (− or +). Arrows inside boxes show information flow during replication. Arrows to viral mRNA begin at the template for mRNA synthesis. Compare with Table 16-1.

familiar RNA polymerases that make mRNA on DNA templates are DNA-dependent RNA polymerases. The special polymerase that makes possible the replication of RNA viruses is an RNA-dependent RNA polymerase that catalyzes the formation of a complementary RNA strand upon a single-stranded RNA template. It is usually called **RNA synthetase** (or **RNA replicase**). Clearly, the basic mechanism for copying parental nucleic acid base sequences is similar in principle in both DNA and RNA viruses.

18-13 "Flower" model of one of the three genes in bacteriophage MS2
This two-dimensional model of the coat-protein gene resembles a flower. Because the gene is made of RNA, uracil replaces thymine in the sequence.

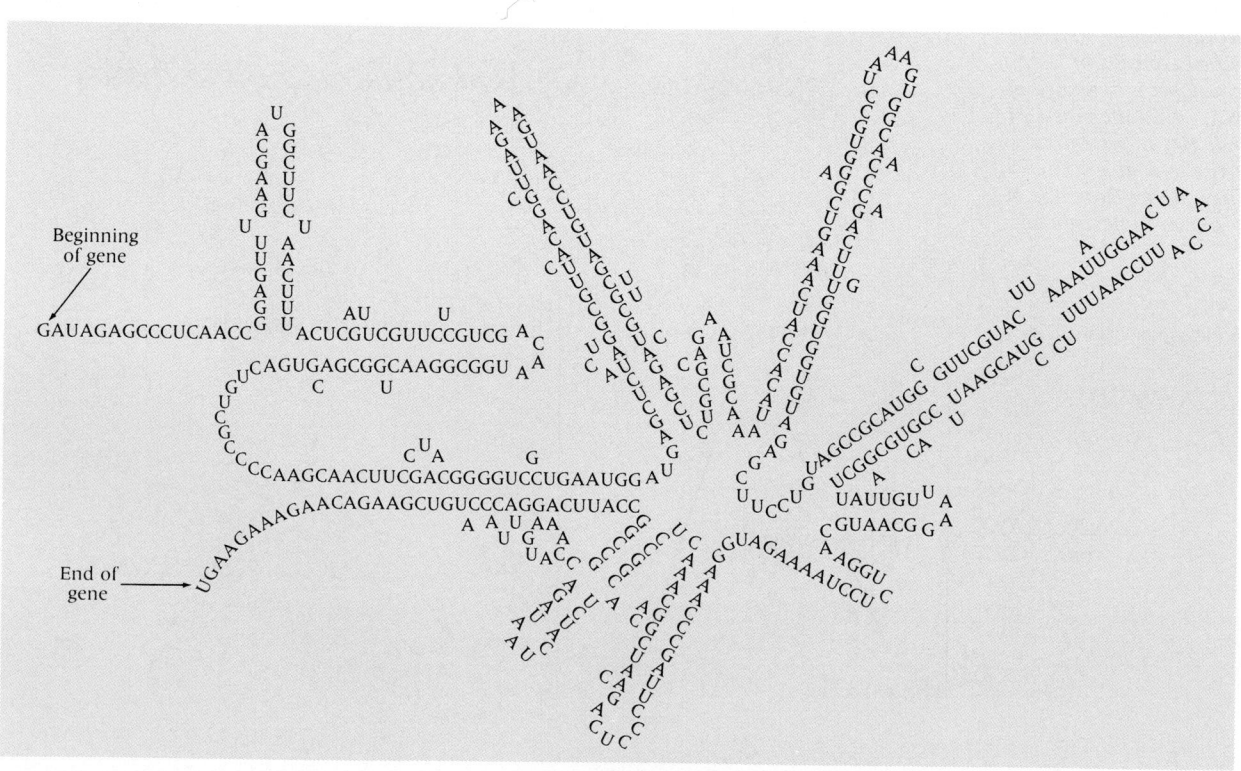

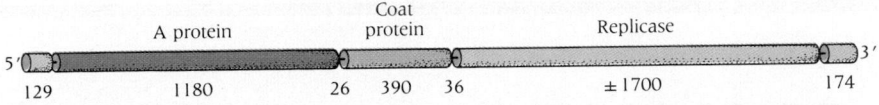

Once again, complementary base-pairing is the device that accurately duplicates a specific base sequence. RNA synthetase first appears in the cell just after the viral RNA enters the host cell and attaches to its ribosomes. New complementary minus strands of RNA are formed. These are promptly put to work as templates for new plus strands of RNA; exact copies of the original parental RNA.

Figure 18-12 summarizes the replication mechanisms of all known viruses. Note again, as shown earlier in Table 16-1, that RNA viruses include varieties with genomes of (+)-single-stranded RNA, (−)-single-stranded RNA, and double-stranded RNA.

The entire base sequence is now known for several RNA viruses. The sequence and proposed two-dimensional structure of a single gene of the bacteriophage known as MS2 is shown in Figure 18-13. Since the genes are made of RNA, not DNA, the sequence contains U instead of T. Several remarkable features are known about the MS2 virus. This virus, like other RNA phages, is very small—indeed, it is one of the simplest viruses known. Its "chromosome," only about 3600 nucleotides long, contains a mere three genes (Figure 18-14). Two genes encode two structural proteins—a small coat protein (CP gene), containing 129 amino acids, and an attachment protein (A gene), with 330 amino acids. Since the two proteins together contain 459 amino acids, the genes encoding them must contain three times that number of nucleotides, or 1377 nucleotides. The third gene (REP) encodes the enzyme necessary for self-replication of the RNA, the RNA replicase containing 580 amino acids. This accounts for another 1740 nucleotides, for a running total of 3117 (1740 plus 1377). This leaves approximately 500 nucleotides that do not encode a protein sequence. These presumably include termination codons.

In examining the nucleotide sequence of MS2, one can only guess how the genes are arranged when they are not at work. Presumably, the molecule folds up into hairpin-like loops, forming a structure that resembles a flower. Note the resemblance of the "flower" structure of this DNA sequence to the "cloverleaf" structure of tRNA (see Figure 17-10). Probably, the flower structure of the gene must be open before genetic information in each of the three sites (CP, A, and REP) can be read out during transcription.

THE MEANING OF RETROVIRUSES

In 1970, Howard Temin of the University of Wisconsin first suggested that certain cancer-causing viruses evolved from movable genetic elements in cells. Temin and David Baltimore of M.I.T. simultaneously showed that these viruses have RNA as their genetic material and that their duplication requires that the RNA first be copied into DNA by an RNA-dependent DNA polymerase aptly called **reverse transcriptase** that catalyzes a transfer of information in a direction opposite to that outlined in the "central dogma" (DNA to RNA).[2] Since this is the reverse of what had been thought to be the normal flow of information from DNA to RNA, the RNA tumor viruses became known as **retroviruses.**

Retroviruses cause the host cell to develop a new genetic trait, which is passed from cell to daughter cell. (When retroviruses carry a class of genes called *oncogenes,* or when they influence cellular *proto-oncogenes,* they cause malignant growth, as discussed in Box 18-2.) This effect of a retrovirus upon its host cell suggests that the virus endows the host cell with new or altered DNA molecules, just as though DNA synthesis in these cases were occurring on an RNA template. In this unusual life cycle (Figure 18-15), genomic RNA in an infecting virus is transcribed into a DNA intermediate (variously called **complementary DNA, copy DNA,** or **cDNA**), which becomes integrated into the DNA of the host cell, where it acts as a **provirus.**

The integrated proviral DNA has two possible destinies. It can be transcribed by ordinary cellular RNA polymerase into new viral RNA—and it does this without

[2] For this pioneering work, Temin and Baltimore (and Dulbecco) shared a Nobel Prize in 1975.

Researchers long suspected that cancer might be due to hypothetical "cancer genes." In recent years, two experimental strategies finally confirmed this suspicion: (1) studies of viruses that cause cancer in animals; and (2) studies of **oncogenes** (cancer-causing genes) in the DNA of cancer cells. (Oncogene comes from the Greek, *onkos,* meaning mass or tumor. Oncology is the scientific study of malignant tumors.)

This work revealed that the human genome contains a set of genes—more than 20, but less than 100—that are responsible for malignant tumors. These genes, which are potential initiators of these tumors, are called **proto-oncogenes** because they are precursors of the actual oncogenic determinants, the oncogenes. Proto-oncogenes are part of the genetic endowment of all normal cells. Their discovery was of immeasurable importance, because it foreshadowed new understanding of the long-sought molecular mechanisms by which cancers arise.

That viruses can cause malignant tumors was proved in 1910 by Peyton Rous of the Rockefeller Institute when he found that cell-free fluid from chicken tumors called **sarcomas** produced new sarcomas when injected into other chickens. Though a landmark experiment, it met with skepticism and apathy. Rous abandoned these studies—and not until decades had passed did other workers confirm the existence of the Rous sarcoma virus and show it to be an RNA retrovirus. Today this virus is easily seen with the electron microscope (Figure 18D). In 1966, at the age of 85, Peyton Rous received the Nobel Prize.

Many tumor viruses are known today. Some are DNA viruses; some are RNA viruses. Some are oncogenic only in animals that are not their natural hosts. Others elicit the changes, collectively termed **malignant transformation** only in cultured cells. Curiously, some tumor viruses that are powerfully oncogenic in animals do not cause transformation in cell cultures. Tumor viruses cause malignant transformation by virtue of their ability to integrate their genetic information into host-cell DNA, thereby forming transforming genes, or oncogenes. DNA and RNA tumor viruses create oncogenes in different ways: For DNA viruses, the oncogenes are integrated parts of the vi-

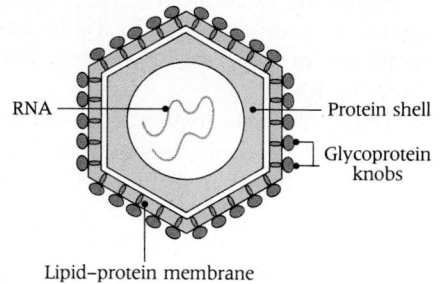

18D *Schematic diagram of the Rous sarcoma virus*

rus genome; and for RNA retroviruses, the oncogenes arise from normal or slightly modified cellular genes, called proto-oncogenes.

Our concern here is with retrovirus-induced oncogenes. The most compelling evidence for the genetic origin of cancer came from studies of these viruses, whose genes are carried in RNA but copied into DNA by reverse transcriptase early in viral replication (Figure 18E). DNA so produced is inserted, or integrated, into the host cell's chromosomal DNA. From then on, the host cell uses its customary machinery to express the integrated viral genes in the usual manner. These events can initiate malignant cell growth in at least two ways.

The integration of viral DNA into host-cell DNA is potentially mutagenic. It can damage cellular genes and can influence their expression by bringing them under the sway of powerful viral regulatory signals. This mechanism, an example of **insertional mutagenesis,** may be oncogenic.

Some (not all) retroviruses carry oncogenes in their own genomes—and the expression of these genes can lead to malignant growth. It seemed curious that the oncogenes of retroviruses contribute nothing to viral replication. Therefore, their presence in viral genomes posed a puzzle. The puzzle was solved with the discovery that retroviral oncogenes are not viral genes at all but wayward copies of cellular genes that were acquired in the course of viral replication by transduction. They are passengers in the viral genome (see Box 16-1). Transduction is probably a rare accident without benefit for the virus—and it is not limited to oncogenes. Its disclosure was of great importance because it led to the discovery of cellular genes,

the presence and actions of which appear crucial to all forms of oncogenesis.*

About one-third of viral oncogenes encode protein kinases that end up on the host cell's plasma membrane and phosphorylate the tyrosine residues of other proteins. This provides a persuasive explanation for malignant transformation: by phosphorylating numerous cellular proteins (in their tyrosine residues) a single enzyme could rapidly alter many aspects of cellular structure and function. However, protein phosphorylation is not the only way in which retroviral oncogenes act. As more and more of the proteins they encode were revealed, a diverse pattern emerged. Some are protein kinases, others are not. One protein is a component of a growth factor normally released by platelets, called platelet-derived growth factor (PDGF) (see Chapter 28). Some attach in the nucleus of the host cell, some in the cytoplasm, some at the plasma membrane (Figure 18F). Perhaps the only conclusion we can draw from this diversity is that cell growth is regulated by many integrated factors involving many regions of the cell. If that network is disturbed at any point, malignant growth might ensue.

As noted above, these are cellular genes, not viral genes in disguise, that can be found in every member of every vertebrate species. It seems likely that in their normal guise proto-oncogenes help in the normal control of cell growth and development.

Why, then, are transduced forms of proto-oncogenes oncogenic? There are two general answers to this question. Transduction may have unleashed the genes from their usual controls and the resulting abnormal expression of otherwise normal genes might be harmful. Alternatively, mutation during or after transduction could change the structure of the genes and the proteins they encode, thus giving rise to abnormal functions.

Two more points are of great current interest.

Chromosomal abnormalities are common in cancer cells. In the cells of some cancers,

* Note, incidentally, that oncogenes are named with three-letter symbols—for example *src, myc, abl, ras,* and so on. An oncogene in a virus is denoted with the prefix v- (e.g., *v-src*). The equivalent normal proto-oncogene in the cell is designated with the prefix c- (i.e., *c-src*).

parts of chromosomes undergo translocation—so that new abnormal chromosomes are formed by fusion with the parts of other chromosomes. The best known of these is the **Philadelphia chromosome** (symbolized **Ph^1**), found in the blood cell precursors of most individuals with chronic myelocytic leukemia, a malignant disorder of bone marrow. The small Ph^1 chromosome results from fusion of a piece of chromosome 9 to a piece of chromosome 22 (Figure 18G). Although the structural pattern of Ph^1 had been recognized for years, it was found only recently that the proto-oncogene c-*abl* lies at the break-point of chromosome 9 and a locus called break-point cluster region, or *bcr*, lies at the break-point of chromosome 22. Translocation moves the chromosome 9 piece onto chromosome 22 and results in the production of a "fusional" or chimeric mRNA that produces a novel protein kinase that differs from the one normally specified by the c-*abl* gene,

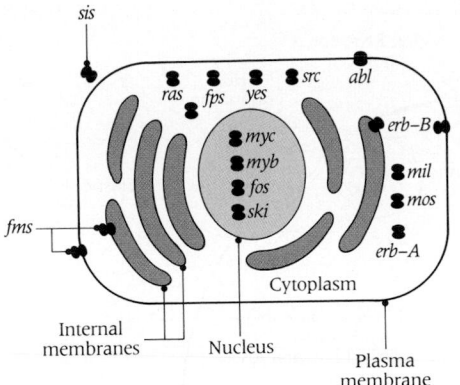

18F Diverse cellular locations of oncogene products

The protein products of oncogenes assume a variety of different locations within the host cell. In these examples, labeled with their oncogene names, protein products are lodged within the nucleus, cytoplasm, and various cellular membranes. In these locations, the proteins function in a way that alters the growth and differentiation of the cell.

strongly implicating this protein in the production of the malignant state called leukemia.

A number of retroviruses have been discovered that cause human cancers—unlike the majority of viruses that cause cancer only in laboratory animals. One such virus, discovered in 1980, causes a rare leukemia called human T-cell leukemia—and the virus was called HTLV-I.

Interestingly, the virus that causes acquired immunodeficiency syndrome, or AIDS, is HTLV-III (later renamed HIV, for human immunodeficiency virus). Though different from HTLV-I, it does transform T lymphocytes and often leads to the development of an unusual cancer called Kaposi's sarcoma.

Will knowledge of oncogenes facilitate the diagnosis and treatment of human cancer? The future holds the answer.

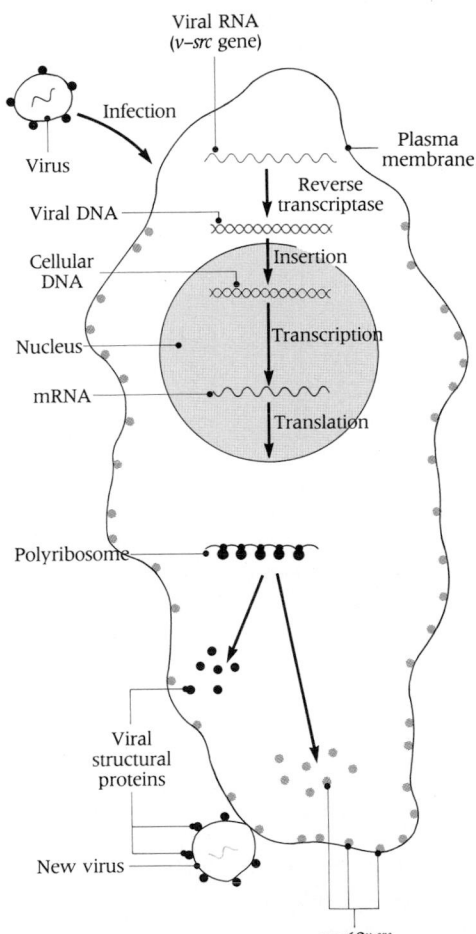

18E How molecular events in the life cycle of a retrovirus can cause cancer

The diagram shows how the insertion of a viral gene into the chromosome of the host cell leads to the production of new viral proteins like the one shown here (pp60^{v-src}), which can convert the cell into a cancer cell. In this example, this protein is encoded by the src oncogene of the Rous sarcoma virus. It attaches to the inner surface of the plasma membrane and catalyzes the modification (by phosphorylation) of various critical cell proteins. Compare with Figure 18–15.

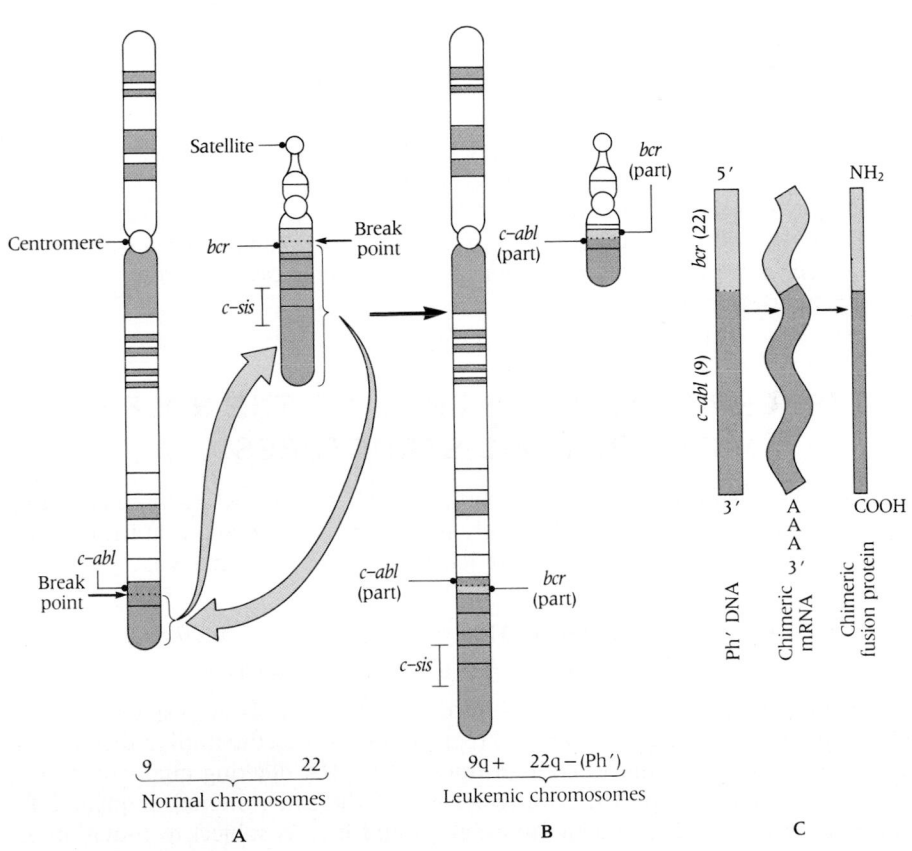

18G How translocation produces the Philadelphia chromosome

(A) *Normal chromosomes 9 and 22. The translocation break-points are within the c-abl proto-oncogene on 9 and within the bcr gene on 22.* **(B)** *Abnormal chromosomes in chronic myelocytic leukemia. Translocation rejoins the broken DNA sequences to form a Ph^1 chromosome.* **(C)** *The new DNA sequence is transcribed into a chimeric mRNA, which is translated into a chimeric fusion protein.*

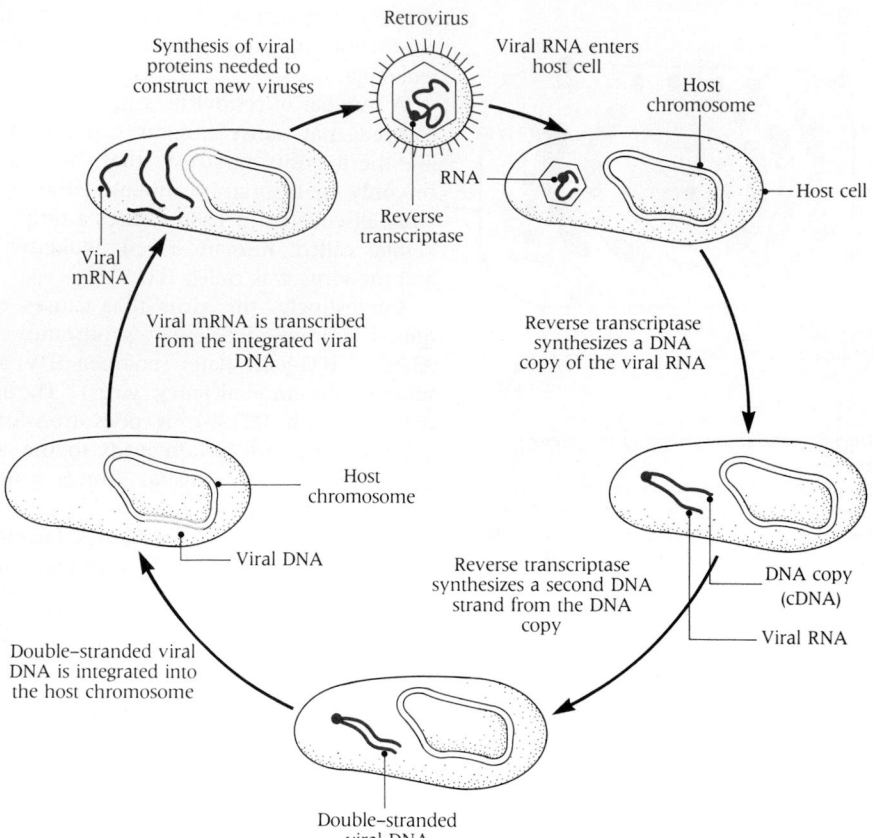

Synthesis of viral proteins needed to construct new viruses

Retrovirus

Viral RNA enters host cell

Host chromosome

RNA

Reverse transcriptase

Host cell

Viral mRNA

Viral mRNA is transcribed from the integrated viral DNA

Reverse transcriptase synthesizes a DNA copy of the viral RNA

Host chromosome

Viral DNA

Reverse transcriptase synthesizes a second DNA strand from the DNA copy

DNA copy (cDNA)

Viral RNA

Double–stranded viral DNA is integrated into the host chromosome

Double–stranded viral DNA

18-15 Reproductive cycle of a retrovirus
The single-stranded RNA that makes up the genome of this virus is transformed into a DNA copy (cDNA), which acts as a template to make double-stranded DNA. New viral RNA is made from this DNA.

killing the host cell. Or it can be replicated like a normal cellular gene and under certain conditions trigger malignant cell growth (Box 18-2).

Strikingly, the sequences of proviral DNAs (frequently present, in birds, mammals and other eukaryotes) are sometimes found in germ cells, and in such cases in all of the animal's offspring. These sequences can, in such cases, become part of a species' genetic endowment.

EXTRACHROMOSOMAL HEREDITY: THE ROLE OF CYTOPLASMIC STRUCTURES

We have spoken of nuclear genes and the functions of DNA (and RNA) in transmitting information as though these were the only mechanisms of hereditary transmission. They are not: some eukaryotic genes are located outside of the nucleus.

SELF-REPLICATING ORGANELLES: A CASE OF NON-MENDELIAN INHERITANCE

Cytoplasmic structures such as mitochondria and chloroplasts are capable of self-duplication. The mechanism of organelle replication was obscure until mitochondria and chloroplasts were found to contain traces of a DNA differing qualitatively (in base ratio and density) from the nuclear DNA of the same animal or plant cell. Nonnuclear DNA contains genetic information and it is as subject to mutation as nuclear DNA. Organelles containing nonnuclear DNA also contain ribosomes, tRNA, and the enzymes necessary to make proteins.

The DNA and ribosomes of mitochondria are in many ways similar to their prokaryotic counterparts. These and other features are compelling evidence that mitochondria (and other cytoplasmic organelles) arose in evolution from autonomous bacterial parasites that began living (perhaps in a vacuole) within the cytoplasm and eventually entered into a symbiotic relationship with the cell.

Mitochondria and chloroplasts of present day eukaryotes contain much less DNA than must have been present when they first arose in evolution. Chloroplasts contain DNA molecules that are one tenth the size of the DNA molecules of any known

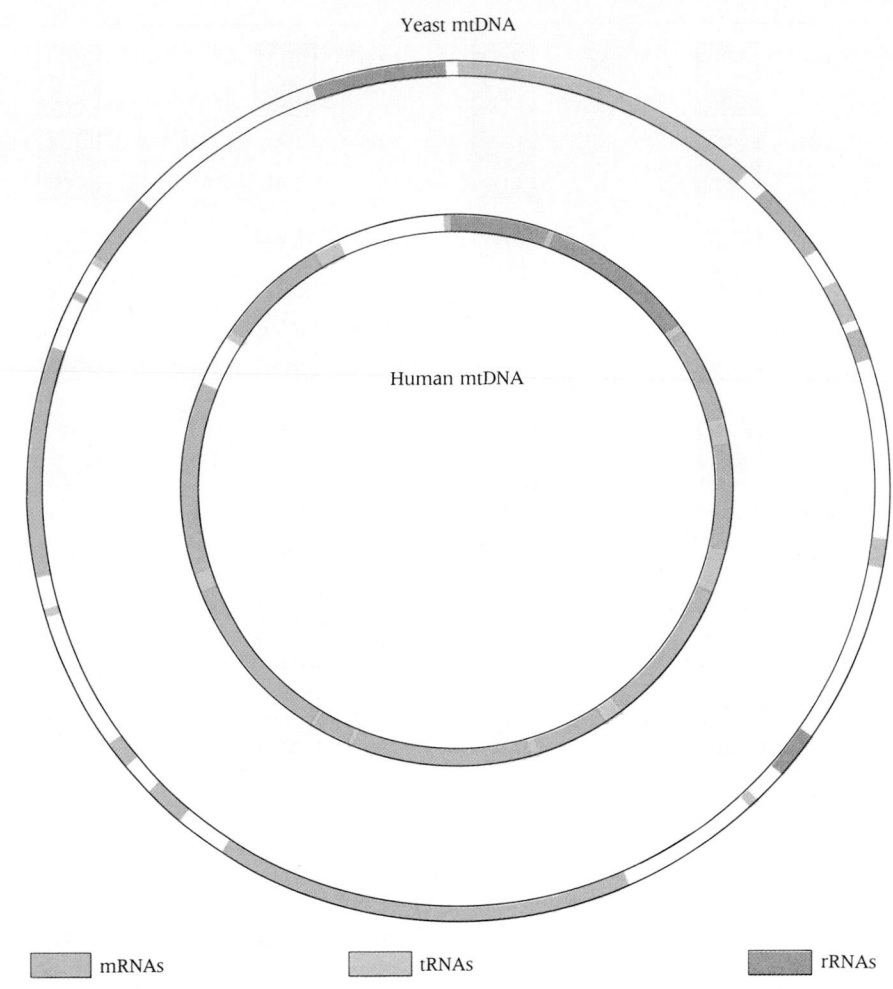

18-16 Maps of human and yeast mitochondrial DNAs
This generalized scheme shows the arrangement of genes encoding transfer RNAs, ribosomal RNAs, and messenger RNAs in each circular DNA molecule. Note the striking differences between the two mtDNAs.

Human mtDNA

mRNAs   tRNAs   rRNAs

bacteria. Mitochondrial DNA, or mtDNA, is even smaller. As noted earlier, the DNA molecules of both mitochondria and chloroplasts are circular.

The two best studied mitochondrial genomes are those of yeast and human mitochondria. The genomes of both have been mapped (Figure 18-16), yeast mtDNA has been partially sequenced, and human mtDNA has been fully sequenced. The two approaches, mapping and sequencing, yield insights and surprises.

- The yeast mitochondrial chromosome is five times larger than the human mitochondrial chromosome. Yeast mtDNA is approximately 78 kb in length, human mtDNA about 16.5 kb (or, more precisely, 16,569 base pairs).
- Maps of both genomes include genes for various tRNAs, rRNAs, enzymes (e.g., ATPase, cytochrome oxidase) and various other proteins—eleven proteins in human mtDNA, a smaller number in yeast mtDNA.
- One particular protein subunit of the enzyme ATPase is encoded by nuclear DNA in human cells and by mitochondrial DNA in yeast cells.
- Mammalian mtDNA has no introns or ordinary termination codons.
- Yeast mtDNA genes do contain introns. (The gene for a subunit of cytochrome oxidase contains nine introns!) Most of the yeast mtDNA sequence does not encode proteins and is not transcribed into RNA.
- Surprisingly (as discussed in Chapter 17), the genetic code of human mtDNA differs slightly from the universal code (Figure 18-17). This would have to be accounted for by any theory seeking to explain the evolution of eukaryotic cells.
- Chloroplast DNA is much larger than mtDNA (consisting of about 145,000 bp), encoding many more proteins than mtDNA.

We assume that many of the original functions of mtDNA were taken over by host cell nuclear DNA. Indeed, the DNA of a mitochondrion is scarcely long enough to encode the proteins in the mitochondrion's ribosomes, let alone the many proteins

Phe	UUU UUC	Ser	UCU UCC	Tyr	UAU UAC	Cys	UGU UGC
Leu	UUA UUG		UCA UCG	Ter	UAA UAG	Trp	UGA UGG
Leu	CUU CUC CUA CUG	Pro	CCU CCC CCA CCG	His	CAU CAC	Arg	CGU CGC CGA CGG
				Gln	CAA GAG		
Ile	AUU AUC	Thr	ACU ACC	Asn	AAU AAC	Ser	AGU AGC
Met	AUA AUG		ACA ACG	Lys	AAA AAG	Ter	AGA AGG
Val	GUU GUC GUA GUG	Ala	GCU GCC GCA GCG	Asp	GAU GAC	Gly	GGU GGC GGA GGG
				Glu	GAA GAG		

18-17 The genetic code of human mitochondrial DNA
The code used by human mitochondria differs from the universal code in that UGA encodes tryptophan, not termination, AUA codes for methionine, not isoleucine, and AGA and AGG serve as termination signals, not codons for arginine. Compare with Table 17-1.

of oxidative phosphorylation. This strongly suggests that many of the proteins in mitochondria are encoded by nuclear DNA, and are synthesized in the conventional way on nonmitochondrial ribosomes, and then imported into the mitochondrion. (This occurs abundantly in chloroplasts.) Such a scheme is shown in Figure 18-18. Recent data confirm that vital parts of the organelle are encoded by nuclear DNA. That explains why mitochondria do not divide or grow outside of cells and survive only within the cytoplasm. Some evidence suggests that several mitochondrial genes actually migrated to the nucleus in the course of evolution.

PLANT VARIEGATION AND OTHER INSTANCES OF MATERNAL INHERITANCE

Inheritance of information in mtDNA genes is managed differently than in nuclear genes. This was an inescapable conclusion when biologists finally recognized that cell organelles can give rise to new organelles and possess "nonnuclear genes." Here was the long-sought explanation for many well-known examples of **extrachromosomal inheritance** in variegated green leaves, *Paramecium, Chlamydomonas, Neurospora,* and yeast—that seemed to contradict Mendelian rules.

Variegation is defined as the appearance in otherwise normal green leaves of pale areas and blotches (Figure 18-19). Investigators found that immature chloroplasts, termed **proplastids,** are more numerous in the large maternal egg cells than in the small paternal sperm cells. Some have held that in the fusion that produces a zygote, the large egg contributes almost all of the cytoplasm and the sperm contributes almost none. In this quantitative view, zygote chloroplasts are more likely to descend from the proplastid chloroplast precursors of the maternal gamete than from the paternal gamete.[3]

In breeding experiments, plant variegation adheres to the pattern that would be expected if a cytoplasmic element derived from the maternal plant determines the trait (Figure 18-20). That element is the chloroplast—or its proplastid precursor. If the maternal line is green, the progeny are green. If the maternal line is pale,

[3] Another notion holds that, in certain regions of a leaf, proplastids are kept from developing into chloroplasts. In this view, variegation is due to differing properties of clusters of somatic cells.

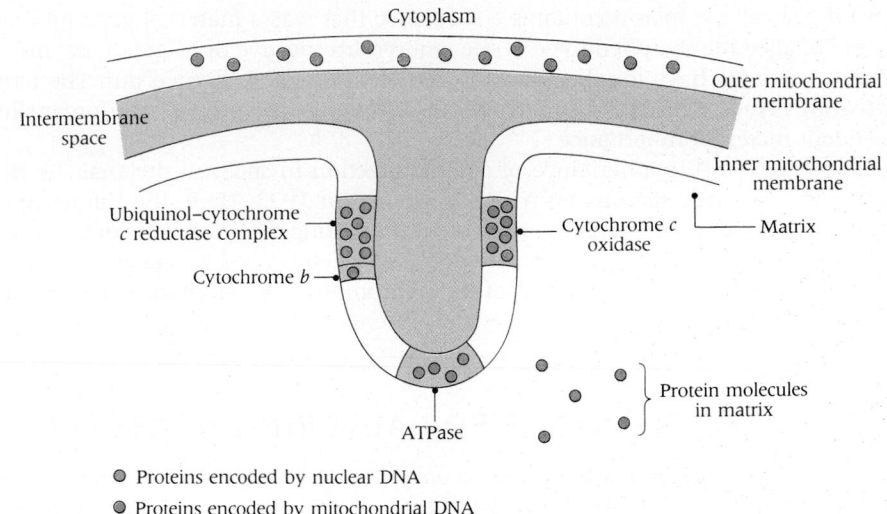

18-18 Cooperation of nuclear and mitochondrial genes
The inner mitochondrial membrane contains proteins encoded by mtDNA and by nuclear DNA.

Cytoplasm

Outer mitochondrial membrane

Intermembrane space

Inner mitochondrial membrane

Ubiquinol–cytochrome *c* reductase complex

Cytochrome *c* oxidase

Matrix

Cytochrome *b*

ATPase

Protein molecules in matrix

○ Proteins encoded by nuclear DNA
● Proteins encoded by mitochondrial DNA

18-19 Plant variegation
A Caladium *leaf exhibits variegation, a pattern in which white areas are interspersed with green areas. Note the clarity of the leaf veins.*

18-20 Explanation of plant variegation mechanism (Right)
This experiment, originally performed in Mirabilis jalapa, *shows the influence of the many chloroplasts in the egg cell (and none in the pollen grain) on the zygote.*

the progeny are pale. If a leaf loses some of its chloroplasts (and thus is pale in spots), no nuclear gene can initiate their formation *de novo*. Moreover, chloroplast genes can undergo mutation. An altered or mutant chloroplast is responsible for variegation. Since variegation is heritable, we must consider it an instance of extra-chromosomal inheritance.

Effects of maternal cytoplasm are common in nature. In many cases, however, a pattern of maternal inheritance may be deceptive. In some cases, the critical factor may be a substance in the maternal gamete cytoplasm that was a product of nuclear genes. For example, a cytoplasmic substance in the egg of a multicellular organism might influence phenotypic expression during early embryonic develop-

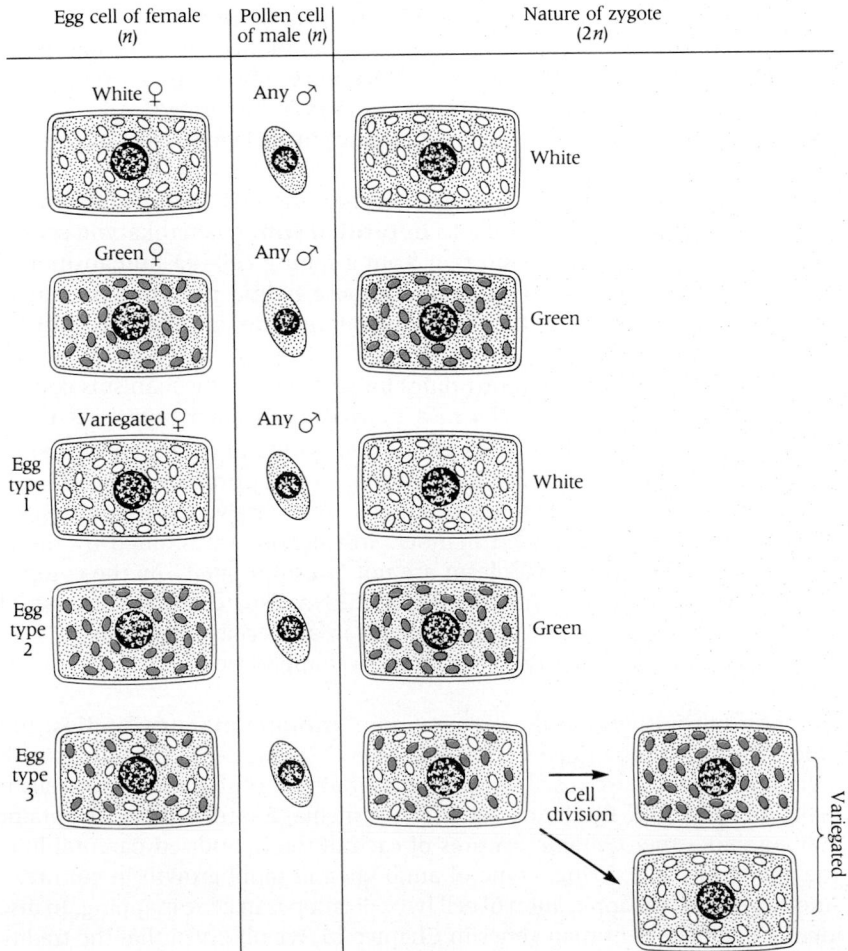

Egg cell of female (*n*)	Pollen cell of male (*n*)	Nature of zygote (*2n*)
White ♀	Any ♂	White
Green ♀	Any ♂	Green
Egg type 1 — Variegated ♀	Any ♂	White
Egg type 2		Green
Egg type 3		Cell division → Variegated

ment. If egg cell cytoplasm contains a substance that was a maternal gene product capable of affecting the phenotype of the embryo irrespective of its genotype, many cell divisions may have to take place before the substance is diluted out. The term **maternal effect** is used for such cases to distinguish them from true organelle-dependent maternal inheritance.

Another case is the inheritance of coiling direction in shells of the snail *Limnea* (Figure 18-21) made famous by A. H. Sturtevant in 1923. Here, the influence of the maternal cytoplasm is a permanent one. The coiling direction is a consequence of an embryo's early cleavage pattern—and this is determined by maternal nuclear genes. Thus, it is not an example of extrachromosomal inheritance, which can occur only when true organellar genes are present.

OTHER MECHANISMS FOR ALTERING HEREDITY

We are taught that heredity can be altered only by gene and chromosome mutations, but modern molecular biology has uncovered several other means of altering a cell's hereditary pattern. We have already considered a few of them, including:

- Transformation of bacteria
- Viral transduction
- Bacterial conjugation

To these may be added three more recently recognized methods for transferring genetic information from cell to cell: cell hybridization, chromosome-mediated gene transfer, and recombinant DNA technology or genetic engineering.

Let us briefly consider these innovations.

CELL HYBRIDIZATION

It was long known that viruses can fuse two or more cells into a new cell with more than one nucleus. It occurred to Henry Harris and John Watkins of Oxford in 1965 that viruses might fuse cells of different kinds into hybrid forms. When they tried to do this, they found that both living and killed viruses could fuse cells taken from different species, and even from different vertebrate orders. Some interspecific hybrids (from more than one species) could even be formed.

When cells are exposed to virus in this procedure (Figure 18-22), they clump together, their membranes fuse at the contact points between the cells, and the cytoplasms of two or more cells coalesce. When the cells are of different kinds, the resulting multinucleate cell is called a **heterokaryon.** A heterokaryon containing two nuclei from a human cell, and two from a mouse cell—a human-mouse hybrid!—appears in Figure 18-23. A human-mouse hybrid in mitotic metaphase is shown in Figure 18-24. Note that such hybrid cells do not develop into hybrid organisms.

Such cells offer fascinating opportunities for study of the mechanisms controlling gene activity. When the two nuclei in a heterokaryon enter mitosis at the same time they often fuse into a large single nucleus containing the chromosomes of both parent cells in various proportions. At mitosis, the chromosome combinations are determined by the parental contributions. Amazingly, these composite cells can still function in an integrated manner. Instructions transmitted by the genes of one species to the hybrid cytoplasm are not "mistranslated" by the cytoplasmic machinery of the other species. In other words, mRNA dispatched to hybrid cytoplasm does not induce production of "false" proteins in appreciable amounts.

Cell hybridization has had two major practical applications.

- The most impressive is in the preparation of **monoclonal antibodies.** In this process, which is described in detail in Box 23-2, short-lived cells that make a desired antibody are fused with an immortal line of mutant tumor cells to make a cell hybrid. By the use of selective media, a pure cell line is obtained that combines the desirable features of each of the hybridized parental lines—the ability to make a single type of antibody and rapid growth in culture.
- Another important application of cell hybridization is in gene mapping. In discussing the mapping of human genes in Chapter 15, we observed that the traditional

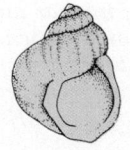 

Dextral   Sinistral

18-21 Two directions of shell coiling
Coiling of the shell of the snail, Limnea, is inherited through maternal nuclear genes.

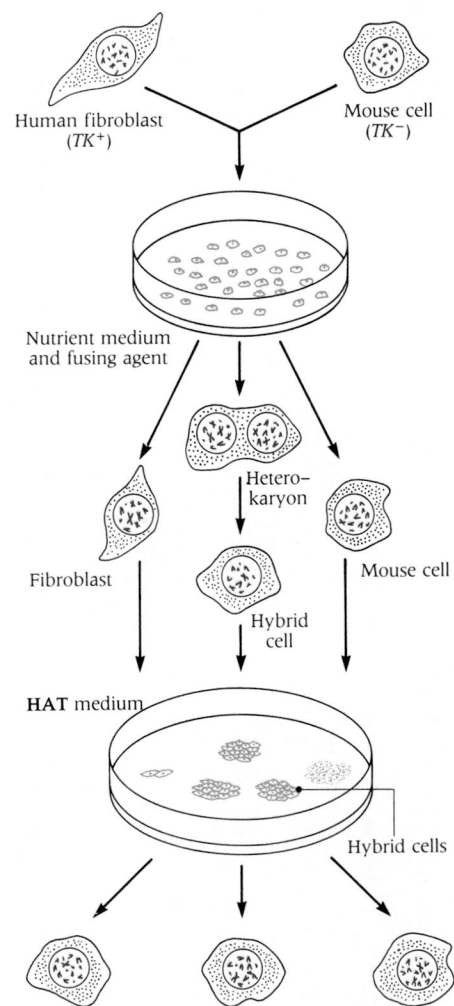

Human fibroblast (*TK*⁺)

Mouse cell (*TK*⁻)

Nutrient medium and fusing agent

Hetero-karyon

Fibroblast   Mouse cell

Hybrid cell

HAT medium

Hybrid cells

18-22 Somatic-cell hybridization of human and mouse cells
This technique permits the transfer of genes from one cell to another. Heterokaryons contain the TK gene (TK⁺) and thus are selected for in HAT medium.

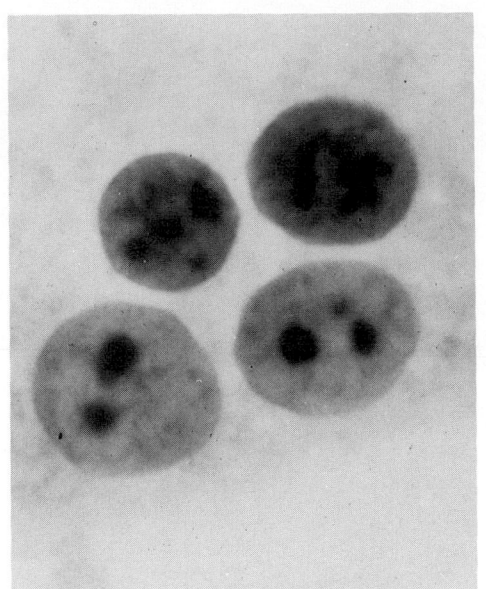

18-23 *Micrograph showing a human-mouse heterokaryon containing two nuclei, one from each "parent" cell* ($\times 1000$)

approach, studies of crossover frequencies, has given way to newer methods. One of the best involves study of the progressive and preferential loss of human chromosomes in human-mouse hybrid cells. If the human chromosome set includes a genetic marker, it is possible to correlate in each line of hybrids the presence or absence of that gene with the presence or absence of a given chromosome. In the example in Table 18-2, 5 hybrids were found to contain (or lack) 4 genes in several different combinations. The patterns suggest that: (1) genes 1 and 3 are linked because they are always present or absent together; (2) they are on chromosome 2 because their presence correlates with the presence of this chromosome; and (3) the location of gene 4 cannot be assigned from these results. Precise chromosome identification is easily accomplished by banding techniques (see Chapter 12) and other methods.

CHROMOSOME-MEDIATED GENE TRANSFER

A second method for the mapping of human genes depends on chromosome-mediated gene transfer. In 1973, investigators showed that purified metaphase chromosomes can serve as vectors for the transfer of genetic information into cultured cells. In an early experiment, cultures of wild-type Chinese hamster cells were halted in metaphase of mitosis. Cell membranes of these donor cells were disrupted, causing them to release chromosomes that were purified and added to recipient mouse cells deficient in hypoxanthine phosphoribosyltransferase (HPRT), an enzyme of purine synthesis. This enzyme converts a free purine (hypoxanthine) to a ribonucleotide. Recipient cells were then added to a medium that allows only those cells expressing HPRT to grow. Cells that receive and express complementing genes in this way will proliferate in such a selective medium, thus giving rise to transformed cell lines. In these experiments, one transformant appeared per 10^7 recipient cells.

This technique is also useful in gene mapping. The transfer process often results in breakage of the donor genome. This alters gene arrangements on chromosomes and permits intrachromosomal mapping.

It is instructive to compare the amount of donor genomic material transferred in cell hybridization and chromosome-mediated gene transfer with that in bacterial transformation (better named DNA-mediated gene transfer). A whole genome is transferred in cell hybridization, an intact chromosome or subchromosomal segment in chromosome-mediated gene transfer, and mere DNA fragments in the kilobase range in DNA-mediated gene transfer. Transfer of genes by such methods has been called **parasexuality,** which Haldane defined as "an alternative to sex."

GENETIC ENGINEERING: A NEW TECHNOLOGY BASED ON RECOMBINANT DNA

As discoveries like those described above tumbled out of the world's laboratories, a new technology emerged that permits almost incredible manipulations of genetic material. Workers in the field of **recombinant DNA technology** or **genetic engineering,** can now accomplish the following:

■ Individual genes can be produced in quantity, thus extending knowledge of gene structure and function. Now this technique employs an automated device called the "gene machine." Its products have been termed "designer genes"!

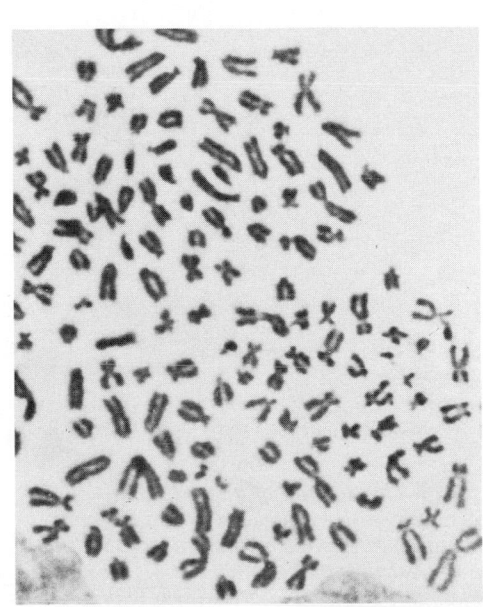

18-24 *Micrograph of a human-mouse heterokaryon in mitotic metaphase*
Arrows indicate several human chromosomes in a fluorescently stained human-mouse hybrid cell. ($\times 10,000$)

TABLE 18-2
HOW CELL HYBRIDIZATION ASSIGNS GENES TO SPECIFIC CHROMOSOMES

		Hybrid cell lines				
		A	B	C	D	E
Human genes	1	+	−	−	+	−
	2	−	+	−	+	−
	3	+	−	−	+	−
	4	+	+	+	−	−
Human chromosomes	1	−	+	−	+	−
	2	+	−	−	+	−
	3	−	−	−	+	+

- Individual genes can be transferred from eukaryotic organisms (plants and animals) into bacteria—and vice versa.
- Valuable proteins such as insulin or interferon can be manufactured by growing large quantities of the bacteria that have acquired the animal cell DNA that encodes these proteins.

We have seen that a bacterium can accept segments of new DNA and incorporate this foreign DNA into its own chromosome. To the genetic engineer, this is a desirable property of bacteria. The incorporated foreign DNA is henceforth replicated and transcribed along with the cell's own native DNA. To dramatize this situation, Har Gobind Khorana of M.I.T. synthesized the first artificial gene and proved its competence by introducing it into *E. coli*. These cells then replicated, transcribing and translating the gene like one of their own.

This procedure, in which a gene from any organism is transplanted into a microorganism, which then produces many gene copies, is called **gene cloning** (Figure 18-25). The DNA of an animal cell structural gene is not spliced directly into the bacterial genome, but into a **plasmid**, a small circular DNA molecule that exists outside the main DNA molecule of bacteria. Plasmid DNA is cut into pieces by one of the so-called **restriction enzymes**, which cleave DNA at specific sites determined by the short stretch of bases. For example, the restriction enzyme known as EcoRI will cleave the following short sequence of double-stranded DNA whenever it is encountered:

$$(5')..GAATTC..(3') \xrightarrow{\text{EcoRI}} (5')..G(3') \ + \ (5')AATTC..(3')$$
$$(3')..CTTAAG..(5') \qquad\qquad (3')..CTTAA(5') \ + \ (3')G..(5')$$

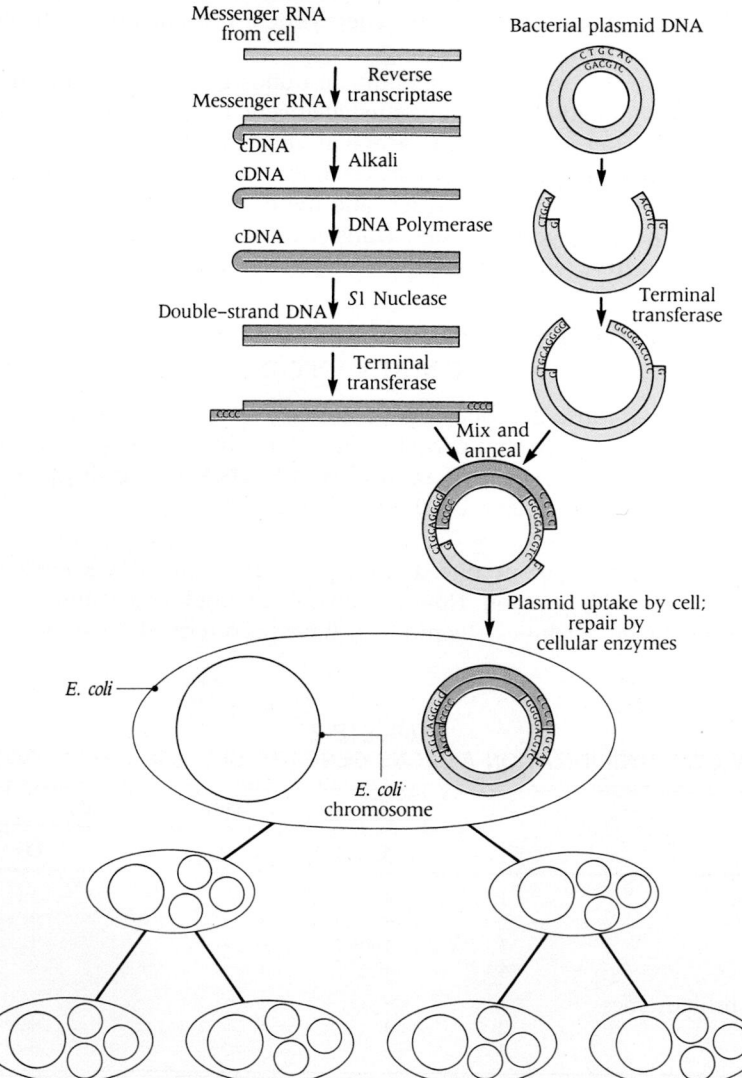

18-25 The method of gene cloning
A specific mRNA is isolated from a cell and transformed into a double-stranded DNA that is equivalent to the gene that encodes the mRNA. Plasmid DNA from bacteria is cut with a restriction enzyme (Pst) and the two DNA types are mixed together. The resulting recombinant plasmid is taken up by a bacterial cell and replicated to create many copies of the gene encoding the original mRNA.

Such enzymes recognize (and cleave) distinctive short sequences, each of which may occur in a large DNA molecule several times. All fragments produced by a given enzyme have the same self-complementary end. Thus a single fragment can circularize by base-pairing, or it can combine with another fragment. In the latter case, if fragments are from different sources, a recombinant DNA molecule is produced. DNA ligase links the free ends to produce a new intact recombinant plasmid. The plasmids are added to a culture of bacteria made permeable by osmotic pressure. Some plasmids enter the bacteria and are replicated when the bacterial DNA replicates.

The two DNA molecules making up a recombinant plasmid are termed **vector** (or **vehicle**) **DNA** and **insert** (or **passenger**) **DNA.** A vector is a DNA molecule that can self-replicate in its host. When an insert is joined to a vector, it is usually replicated along with the vector. Two classes of vectors have been used in *E. coli*—plasmids and certain viruses.

Plasmid vectors are chosen that give a transformed cell a selective advantage in the test tube. For example, some vectors carry antibiotic resistance genes. If antibiotic is in the culture, only transformed cells will grow.[4] The frequency of transformation is low, but a transformed cell is generally produced by the uptake of a single vector-insert DNA molecule. Transformed cells can be propagated indefinitely.

Human diseases associated with defective genes one day may be treated by the insertion of a normal competent gene into defective DNA. The powerful techniques of genetic engineering are a source of concern to some people, who fear the creation of dangerous new organisms (Box 18-3).

AGAIN, WHAT IS A GENE?

So far we have managed to avoid giving a complete answer to the question "What is a gene?" Its original definition was based on the data of formal genetics. Though not directly observable, the gene was defined as a fundamental particle that produces a clear-cut effect on an organism's phenotype, an effect that is heritable. Then it was learned that genes are in chromosomes and that they can be identified and mapped if they have distinctive alleles and if they are separable by crossing-over studies. This gave rise to the operational definition given earlier, according to which we identify a gene not by what it is, but by what it does.

Today we know that a gene is a segment of a DNA molecule. To specify a polypeptide chain, a structural gene must have a coding length of about 1000 base pairs. We also know that a mutation can result from a change in a single nucleotide of the DNA sequence.

Because ambiguity surrounds definitions of the gene as a unit of recombination, or as a unit that specifies protein structure, or a unit of mutation, some molecular biologists tried to replace the term "gene." Seymour Benzer coined the terms cistron, muton, and recon. A **cistron** is approximately equivalent to the classical geneticist's gene. It is the functional unit that determines a polypeptide chain. A mutational unit within a DNA molecule is a **muton.** It is the smallest part of a gene (a single nucleotide pair) that can be involved in a mutational event. A **recon** is a region of the gene that is too small to recombine genetically.

Later molecular biologists coined the term **orphon** for a newly discovered type of gene that leaves the bosom of its multigene family and comes to rest in isolation elsewhere in the genome. This new bit of evidence of the promiscuity of genetic material has generated intensive speculation about its evolutionary significance. Thus, there is a growing list of gene forms with *-on* suffixes: introns, exons, orphons, and transposons mentioned earlier plus, of course, the codons and operons.

We should bear in mind that in most discussions of the genes of plants and animals we are still speaking of particles that are recognized exclusively from studies of phenotypic traits and their heredity. Most of the molecular mechanisms described in this chapter were first worked out in viruses and bacteria. Many unforeseen complexities arose as investigators moved on to eukaryotic forms. For example, a

---

[4] We now realize that the transfer of plasmids is one of the reasons why the harmful bacteria that infect the human body can suddenly become resistant to a wide range of antibiotics. From the human viewpoint, then, these wandering pieces of genetic material can do vast damage.

Recent work has enabled molecular biologists, virtually at will, to transfer individual genes from eukaryotic organisms to bacteria. To some this triumph of recombinant DNA research seems both arrogant and dangerous—arrogant because in the view of these critics mere human beings are "tampering" with evolution, and dangerous because we are risking the unintended production of an Andromeda strain that might devastate human life in a deadly epidemic of some kind. The questions at issue were these.

- Might bacteria carrying foreign genes pose a serious threat to existing forms of life?
- Are investigators in this field tampering with the "natural flow" of evolution?
- Might these capabilities lead to potentially dangerous attempts to manipulate human heredity?

Discussion of these questions in the late 1970s split the scientific community into contentious camps. It is impossible, argued the critics, to predict the properties of newly modified or novel organisms. Therefore the consequences of their introduction into our biosphere are also unpredictable. Most would probably be innocuous. Some, by design and selection, could have great value for various human purposes. But others,

through inadvertence, might pose a major hazard to animal or plant life.

Although recombinant DNA technology now allows transfer of animal genes into plants, most of this work involves microorganisms such as *E. coli*, several strains of which are indigenous to the human body. Those who fear this technology contend that it is impossible to contain microorganisms within the four walls of a laboratory even under the most stringent conditions. Once released, and once in an appropriate ecological environment, the new organisms could not be recalled. They would be with us, perhaps forever.

Scientists on the other side argued that in order to be virulent (disease-producing), bacteria require many traits and not just single new genes. The strains of *E. coli* used in this work do not live in humans and do not cause human disease. Admittedly, their relatives are common inhabitants of the human colon, but the experimental bacteria are poorly adapted to life there. In a sense, recombinant DNA workers themselves have been the guinea pigs in this arena. To date there have been no reports of illness associated with infection by a strain of bacteria containing cloned DNA.

Is there a large inherent risk in inserting DNA into the bacterial genome? Surely bacteria have had countless opportunities over

the millennia to incorporate into themselves human and other alien DNAs from decaying cells and bodies. Had this happened, we would now be able to recognize short stretches of animal DNA in *E. coli* DNA. But we do not. Therefore such recombinants, if they occurred, have not survived.

These arguments favor the view that no monster is ever likely to be created by recombinant DNA methods. Probably, it is not possible to create a strain that would escape the laboratory and "head for town." As for the proposition that natural evolution should be left to itself, we might keep in mind that this is the same evolutionary process that has already brought us smallpox, bubonic plague, typhoid fever, and cancer. Is medicine's assault on these diseases a form of warfare against nature? Perhaps so, but few would deny its benefits. In fact, we have already done a fair amount of beneficial genetic "tampering" by domesticating and selecting animals and plants, grafting and hybridizing plants, and so on.

Concern has subsided somewhat in recent years in the face of evidence that nothing frightful has yet happened. Rigorous safety guidelines now in effect manage adequately to find a middle road between protecting people from unbidden hazards and providing them with the benefits of fundamental new knowledge.

---

virus consists merely of a stretch of nucleic acid and a protein coat; in a bacterium, a single circular DNA molecule carries genetic information. However, the chromosomes of plants and animals are more complex and, as we have seen, their DNA contains noncoding sequences and the enigmatic repeated sequences known collectively as satellite DNA. They also contain, in addition to DNA, the histones and nonhistone proteins that may have roles in mechanisms that regulate the production of mRNA.

Few indeed are the traits in eukaryotes that can be rigorously traced back, step by step, to a sequence of codons. In view of the fast-paced progress in molecular biology we have every right to assume that this will be possible in the near future.

SUMMARY

All cells possess intricate control mechanisms which ensure that proteins are synthesized economically in the amounts needed. Most of our knowledge of the molecular bases of these mechanisms centers on prokaryotes, notably *E. coli*. Many enzymes in these organisms are synthesized at a rate influenced by the availability of external nutrient molecules called inducers and corepressors. These control the rate of enzyme synthesis by regulating the rate at which the structural genes of these enzymes produce their mRNA templates.

Induction, an increase in the rate of synthesis of a single enzyme protein, follows exposure of a cell to an inducer. This inducer is often the substrate of the enzyme whose synthesis it induces or a chemical analogue of the substrate. Corepressors are usually the final product of a biosynthetic reaction sequence in which that enzyme takes part. In the operon mechanism, inducers and corepressors act by binding to regulatory proteins that are synthesized by nearby regulatory genes. In the well-studied *lac* operon of *E. coli*, an operator gene controls transcription by its nearby structural genes. A so-called I gene encodes a repressor protein, which binds to the operator locus, thereby inhibiting structural gene transcription. Repressor binding is inhibited by inducer molecules (hence transcription is activated) and it is facilitated by corepressors (in which case transcription is prevented).

Operon systems do not function in gene regulation in eukaryotes. Here, extracellular signals such as hormones and growth factors can exert control, often by binding to membrane receptors and generating second messengers. Regulation by these and other signaling mechanisms can occur at three levels: by changing the rates of mRNA transcription, by the processing of primary mRNA transcripts, or by influencing the translation of mRNA into protein. Control of mRNA transcription, clearly the most important, depends on the presence of short DNA segments called transcription signals which are upstream of their associated structural genes. They include so-called TATA boxes, CAAT boxes, and enhancers.

Four molecular mechanisms appear to control transcription: (1) site-specific DNA binding by various proteins; (2) factors affecting the loosening or unwinding of the tightly packed chromatin (a necessary preliminary to gene transcription); (3) decreases in the levels of methylated cytosine in the DNA of a gene increases its transcription (but only in vertebrates); and, (4) though the inhibition of transcription occurring when histones bind DNA is nonspecific, nonhistone proteins may specifically regulate the expression of certain genes.

Some viral genes are composed not of DNA, but of single-stranded RNA. Such viruses contain RNA-dependent RNA polymerases (RNA replicases) rather than the DNA-dependent RNA polymerases (RNA polymerase) typical of organisms with DNA genes. A class of RNA viruses called retroviruses replicate their RNA by copying it into DNA by means of reverse transcriptase and RNA-dependent DNA polymerase. This DNA is then integrated into the host cell DNA where it can be transcribed into the RNA and protein necessary to make new viruses. It can also be passed on to daughter cells along with the normal genome of the host cell. Such incorporated DNA may cause certain cellular genes called proto-oncogenes to become oncogenes, which are capable of triggering malignant transformations (cancer) in these cells.

Not all eukaryotic genes are located on nuclear chromosomes. Chloroplasts and mitochondria contain small amounts of their own DNA, which in many ways closely resembles prokaryote DNA. This gives support to the theory that such organelles were originally free-living bacterial parasites that entered a permanent symbiotic relationship with their host cells.

Since, in most species, nearly all of the cytoplasm of a zygote comes from the maternal egg, extrachromosomal inheritance is non-Mendelian. Only the maternal line contributes the genes found in chloroplasts and mitochondria. True extrachromosomal inheritance must be distinguished from effects of maternal cytoplasm, which are the products of nuclear genes.

The cell's heredity can be altered by mechanisms other than mutation. To the naturally occurring phenomena of transduction, transformation, and conjugation have been added three recent experimental techniques for transferring genetic information from cell to cell: (1) cell hybridization allows fusion of cells from different species, the resulting hybrids having immensely important practical roles in making monoclonal antibodies (see Chapter 23) and in gene mapping; (2) chromosome-mediated gene transfer provides another technique used in gene mapping; and (3) recombinant DNA technology allows single genes to be cloned via plasmids inserted into bacteria or other cells, which then replicate and mass produce them. Cultures of cells bearing recombinant genes are now used as synthetic "factories" for valuable proteins such as insulin and growth hormone. This remarkable and promising technology, also called genetic engineering, ranks as one of the century's greatest advances in the practical applications of biological science. In recent years, many genetically engineered products (hormones, growth factors, etc.) have become available.

KEY TERMS

actinomycin D
B-DNA
β-galactosidase
cDNA
CAP site
catabolite activator protein
catabolite repression
cistron
complementary DNA
constitutive enzyme
corepressor
derepression
enhancer

extrachromosomal inheritance
gene cloning
genetic engineering
inducer
inducible enzyme
induction
insertional mutagenesis
lac operon
long terminal repeat (LTR)
maternal effect
oncogene
operator
operon

parasexuality
passenger DNA
permease
plasmid
Philadelphia chromosome
promoter
proto-oncogene
provirus
puff
recombinant DNA technology
regulatory gene
repression
repressor protein

restriction enzyme
retrovirus
reverse transcriptase
RNA replicase

RNA synthetase
structural gene
transacetylase
transcriptional signal

transcription factor
vector DNA
vehicle DNA
Z-DNA

QUIZ QUESTIONS

1. Which of the following is *not* a component of an operon of a prokaryote?
   A. operator
   B. promoter
   C. -35 sequence
   D. CAP site
   E. CAAT box

2. At which of the following steps in the transfer of genetic information from nuclear DNA into polypeptides can gene activity be regulated?
   A. synthesis of mRNA
   B. processing of the primary RNA transcript in the nucleus
   C. translation

   D. in the cytoplasm during movement of mRNA to sites of polypeptide synthesis
   E. All of the above

3. Which of the following enzymes do *not* play a role in the production of new RNA molecules in the various RNA viruses?
   A. RNA-dependent RNA polymerase
   B. DNA-dependent RNA polymerase
   C. RNA synthetase
   D. DNA polymerase
   E. reverse transcriptase

4. Extranuclear chromosomes occur within various organelles, including

   A. endoplasmic reticulum.
   B. chloroplasts.
   C. lysosomes.
   D. A and B
   E. B and C

5. Which of the following processes transfers the greatest relative amount of the genome from one cell (or virus) to another cell?
   A. bacterial transformation
   B. chromosome-mediated gene transfer
   C. DNA-mediated gene transfer
   D. cell hybridization
   E. bacterial conjugation

ESSAY QUESTIONS

1. What is the "central dogma" of molecular biology? At what stages in the flow of information it describes does regulation of gene expression occur? How does regulation operate at each stage?

2. How does gene regulation in prokaryotes differ from that in eukaryotes? Would

   prokaryotic genes function in eukaryotic cells—and vice versa?

3. Does the existence of RNA viruses cast doubt on the "central dogma" of molecular biology? How do the life cycles of the following RNA viruses differ: a $(+)$ strand RNA virus; a $(-)$ strand RNA virus;

   a retrovirus? Outline the fates of the viral genomes upon infection, expression, and replication.

4. What aspects of mitochondrial DNA suggest a kinship of mitochondria and prokaryotes? What aspects of their genetic content inform us that they have diverged over evolution?

REFERENCES AND SUGGESTIONS FOR FURTHER READING

Angier, N. (1988). *The Search for the Oncogene.* Houghton Mifflin, Boston.

   A gripping account of real-world academic molecular biology and its human face.

Cech, T. R., (1985). Self-splicing RNA: implications for evolution. *International Review of Cytology 93,* 3–22.

   A review by the investigator whose startling 1982 discovery of the self-splicing ribosomal RNA of *Tetrahymena* revolutionized thinking on the nature of enzymes, the origin of life, and the meaning of introns.

Cohen, S. N. and Shapiro, J. A. (1980). Transposable genetic elements, *Scientific American 242,* February, 40–49.

   Clear exposition of how a trail of research led

to understanding of how genetic elements may be shifted from place to place in the genome.

Fedoroff, N. V. (1984) Transposable genetic elements in maize. *Scientific American 250,* June, 84–98.

   Summary of the evidence showing that the transposable genes discovered 40 years earlier by Barbara McClintock also exist in bacteria, animals, and other plants.

Fedoroff, N. V. (1989). About maize transposable elements and development. *Cell 56,* 181–191.

   A stimulating review.

Jaenisch, R. (1988). Transgenic animals. *Science 240,* 1468–1474.

   An account of what has been learned from

experiments in which foreign genes are introduced into the germ line of animals.

Keller, E. F. (1983) *A Feeling for Organism: The Life and Work of Barbara McClintock.* W. H. Freeman, New York.

   A biography of a remarkable biologist.

Krimsky, S. (1985). *Genetic Alchemy: The Social History of the Recombinant DNA Controversy.* The MIT Press, Cambridge, Mass.

   Though its title hints at science fiction, this is a carefully documented account of an extraordinary episode—a unique case in which scientists chose to stop doing certain experiments because of possible negative social consequences. Worth reading.

Marx, J. L. (1989). How DNA viruses may

cause cancer. *Science 243*, 1012–1013.

A summary of current views on an important question.

McClintock, B. (1984). The significance of responses of the genome to challenges. *Science 226*, 792–801.

The Nobel Prize address of the discoverer of transposable genetic elements of maize. The mobile genes she discovered 40 years earlier have since been identified in bacteria, other plants, and animals.

Multiple authors (1989). Special issue. The New Harvest: Genetically Engineered Species. *Science 244*, 1275–1325.

This special issue of *Science* includes reviews of the roles of genetic engineering in altering plant and animal species and the prospects for human gene therapy.

Novick, R. P. (1980). Plasmids. *Scientific American 243*, December, 102–127.

Well-illustrated survey of these accessory genetic elements, which are best known as carriers of antibiotic resistance and as useful tools for the genetic engineer.

Razin, A., Cedar, H., and Riggs, A. D. (Eds.) (1984). *DNA Methylation: Biochemistry and Biological Significance.* Springer-Verlag, New York.

An authoritative review.

Rich, A., and Kim, S. H. (1978). The three-dimensional structure of transfer RNA. *Scientific American 238*, January, 52–62.

A stimulating review by the investigators who determined the three-dimensional structure of tRNA.

Rich, A., Nordheim, A., and Wang, A. H. J. (1984). The chemistry and biology of left-handed Z-DNA. *Annual Review of Biochemistry 53*, 791–846.

A scholarly review.

Roberts, L. (1989). Ethical questions haunt new genetic technologies. *Science 243*, 1134–1136.

Provocative discussion of ethical issues inherent in genetic engineering.

Smith, H. O. (1979). Nucleotide sequence specificity of restriction endonucleases. *Science 205*, 455–462.

Nobel Prize address of one of the discoverers of restriction enzymes.

Temin, H. M. (1985). Reverse transcription in the eukaryotic genome: retroviruses, pararetroviruses, retrotransposons, and retrotranscripts. *Mol. Biol. Evol. 2*, 455–468.

A scholarly review by one of the discoverers of reverse transcription. Although addressed to workers in the field, the paper is clearly written and well worth tracking down. Excellent bibliography.

Schmid, C. W., and Jelinek, W. R. (1982). The Alu family of dispersed repetitive sequences. *Science 216*, 1065–1070.

A discussion of a curious family of related DNA sequences, some 300 base pairs in length, that are inserted at hundreds of thousands of different chromosomal locations.

Varmus, H. (1987). Reverse transcription. *Scientific American 257*, September, 56–64.

Well-illustrated discussion of viruses and other organisms which convert RNA to DNA and how they do it.

Varmus, H., (1988). Retroviruses. *Science 240*, 1427–1435.

A lively review summarizing the life cycle of retroviruses, their role in oncogenesis and the causation of AIDS and other diseases, and their usefulness as research tools.

Zubay, G. (1987). *A Practical Guide to Molecular Cloning.* Benjamin/Cummings, Menlo Park, Calif.

A good how-to book.

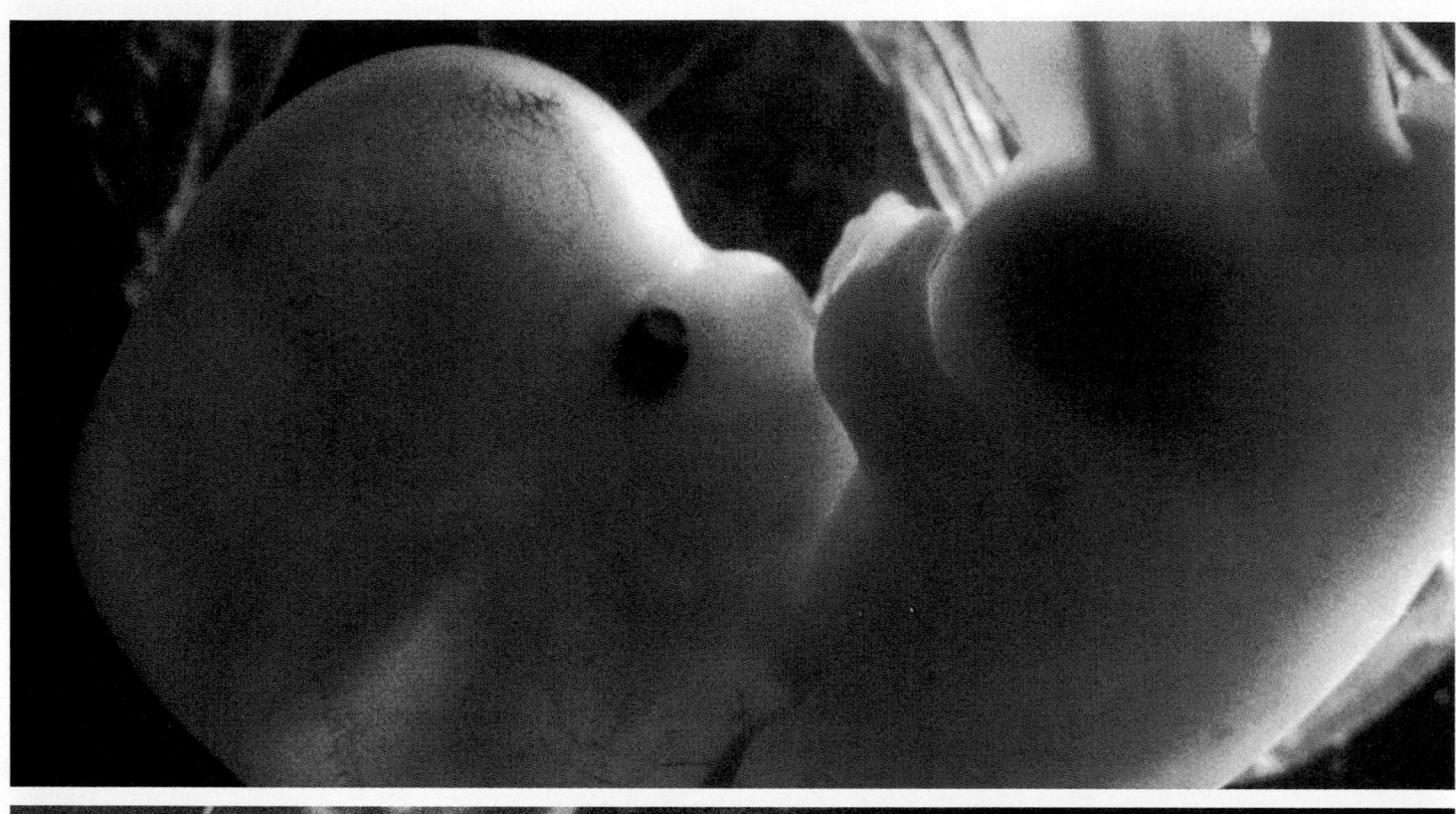
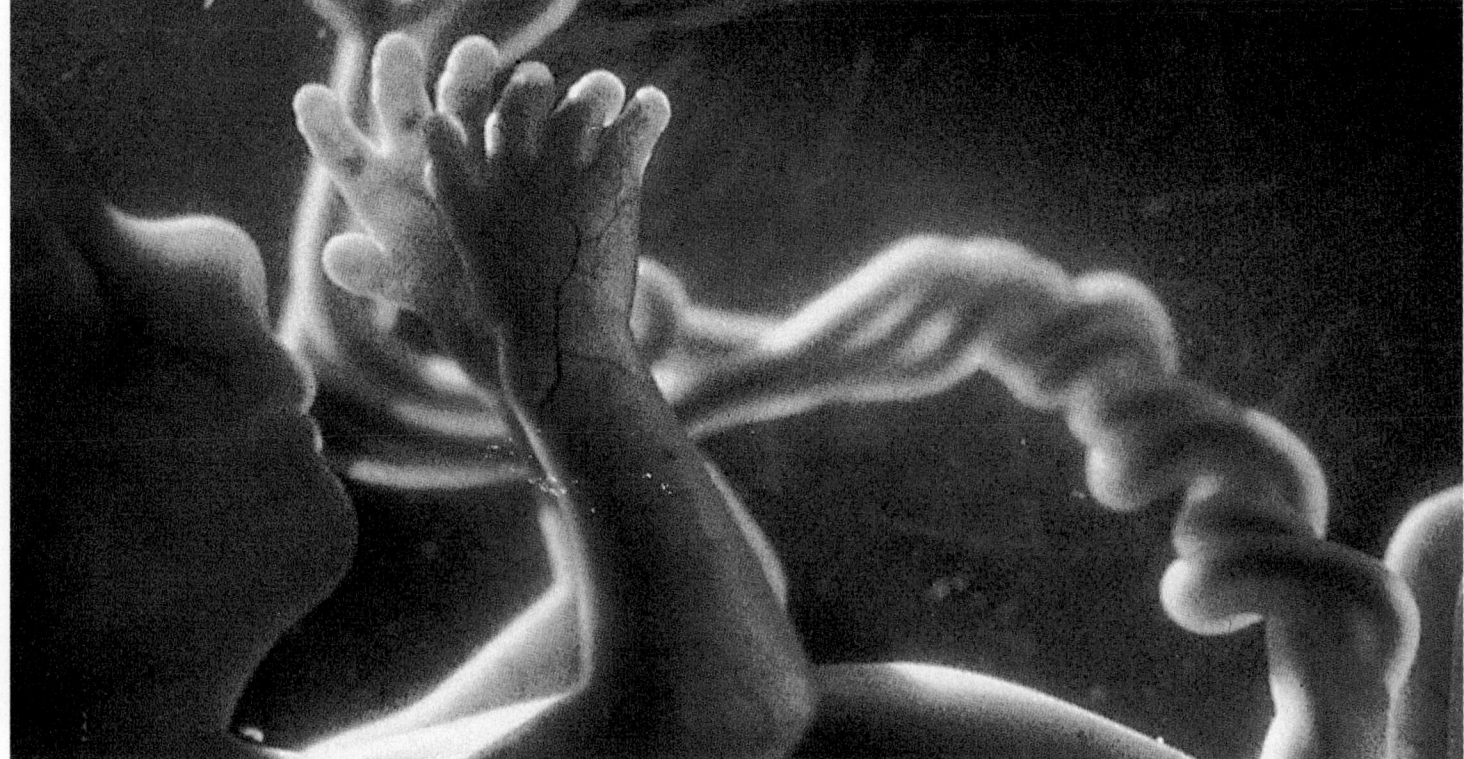

DEVELOPMENT OF FORM IN THE ANIMAL BODY

Development of form in the animal body
These photographs symbolize the remarkable process of embryonic development. The top photograph shows a human embryo at about 7 weeks. It is 3 cm long. Most of the organ systems and appendages have been formed and the nose and retina of the eye are visible. The large dark mass in the body cavity is the liver. The umbilical cord connects with the placenta. The lower photograph is of a human fetus at 4 months.

We cannot yet fully explain what leads a fertilized egg to produce daughter cells that in later generations become as diverse as the cells of bone and muscle, blood and nerve, phloem and mesophyll. How do human zygotes develop into human beings, frog zygotes into frogs, and rose zygotes in roses? Such questions are the concerns of **embryology,** which lately has been known by the broader name **developmental biology.**

As the mechanisms of gene expression yield to determined investigators, a major theme emerges: how can we rephrase the "classical questions" of embryology in the language of molecular biology?

That theme echoes through this chapter and the next. It is our aim here to consider developmental biology from both viewpoints—old and new. We begin with animal development. (Plant development will be discussed in Chapter 21.) First, we will briefly describe the early stages of animal development in classical terms. Second, we summarize present thinking on mechanisms—on how these events come about. Along the way we will encounter unsolved problems aplenty.

PROCESSES OF DEVELOPMENT: SOME DEFINITIONS

Before we consider the sequence of changes or stages that occur in the early development of the embryo, let us briefly note the processes that underlie embryonic development. Here we will merely define key terms and begin to sort out some of the problems that arise.

DEVELOPMENT: PROGRESSIVE CHANGES IN THE LIFE CYCLE

The terms **development** and **ontogeny** refer to the progressive changes that occur during a whole life history. Organisms of every kind have a life cycle. In sexually reproducing organisms, it generally pursues the following sequence:

fertilized egg → embryo → larva → adult → fertilized egg

A **larva** is an independent, small form that feeds actively and typically has a mode of life quite different from that of an adult. A tadpole is a frog larva, a caterpillar is a butterfly larva (Figure 19-1). Unfortunately, this term is used in different ways. The larva of an invertebrate like the sea urchin is an embryonic stage that develops directly into an adult. The larva of a moth or butterfly develops from an earlier embryonic stage. It is itself an adult form that cannot reproduce.

Many animals—including mammals—do not pass through a larval stage. In these forms, the embryo develops into a newborn form that at birth is essentially a sturdy little replica of an adult. Note, too, that in perennial plants and many animals, adults pass into a **senescent phase**—old age—after an extended period of reproductive life.

The processes of development, marked by progressive and cumulative changes at all levels of biological organization, has two notable features.

- First, it leads to irreversible structural change. Changes caused by nondevelopmental processes such as muscle contraction or nerve impulse conduction lead to transitory and reversible structural changes.
- Second, developmental processes are gradual and cumulative, spanning long segments of the organism's life history.

It is often difficult to draw a sharp line between developmental and nondevelopmental processes. Often a continuity exists between slow developmental processes and rapid nondevelopmental processes. For example, biosynthetic activity is largely nondevelopmental, yet the long-term control of what materials are synthesized (and how much of each) is a central aspect of developmental change. At another extreme, there is continuity between development and the slow course of evolution.

Development—the processes by which a multicellular organism arises from a zygote—may be viewed as a precisely controlled program of coordinated serial changes in the levels of gene expression over time. In considering these changes, it is useful to distinguish the individual component processes, which include determination, differentiation, growth, and morphogenesis. They are not independent of one another, nor do they occur in sequence. One or the other may predominate at various times in development. Sometimes all proceed together.

DETERMINATION: REGIONAL SPECIALIZATION

The vocabulary of developmental biology includes a cluster of terms related to the degree of "potency" possessed by a cell: they refer to the number of options open to the cell in its subsequent development.

- A **totipotent** cell has the full developmental capacity of the original zygote and thus can follow any or all of the possible developmental pathways.
- A **pluripotent** (or **multipotent**) cell has many possible developmental pathways open to it, but they are fewer in number than are the options of a totipotent cell.
- A **unipotent** cell has only one possible developmental pathway available to it. As we will see, development is marked by a progressive tendency of pluripotent cells (totipotent in plants) to become unipotent.

When the properties of pluripotent embryonic cells become defined in a way that causes them to head inextricably along a particular course of development, we say they are determined or committed—and the terms **determination** and **commitment** are often used interchangeably. The crux of determination is that a cell's fate is now predictable. A determined cell is irreversibly committed to a particular fate before it displays any outward changes. **Differentiation** is the visible or overt effect of determination or commitment.

As the embryo develops, stages are reached at which a local region or area of the embryo—a group of cells, a single cell or even a part of a cell—begins to differ from the rest of the embryo. At first, the difference is barely detectable; later it becomes decisive as the area develops into a recognizable organ or body component. A small nondescript protuberance or region will appear, for example, that later becomes a limb or an eye (Figure 19-2). This process of local specialization is due to differentiation. Investigators have learned to identify these sites before they manifest themselves, and they have developed techniques for removing or transplanting these areas for experimental purposes.

Critical questions are raised by the processes of determination. How are the few cells singled out from the many? Which of their properties sets them apart? At what point do they acquire their special characteristics, since all cells of an embryo descend from a single cell, the **zygote.**

DIFFERENTIATION: CELL SPECIALIZATION
The Concept of Differentiation

There is a key difference between the cells and tissues of an early embryo and those of an adult: the adult cells have already differentiated into a great many

19-1 Larval stage of development
The larvae of butterflies and moths appear to us as caterpillars. This brightly colored larva will eventually develop into an anise swallowtail butterfly, Papilio zelicoan. *The two orange structures on its head are everted stink glands.*

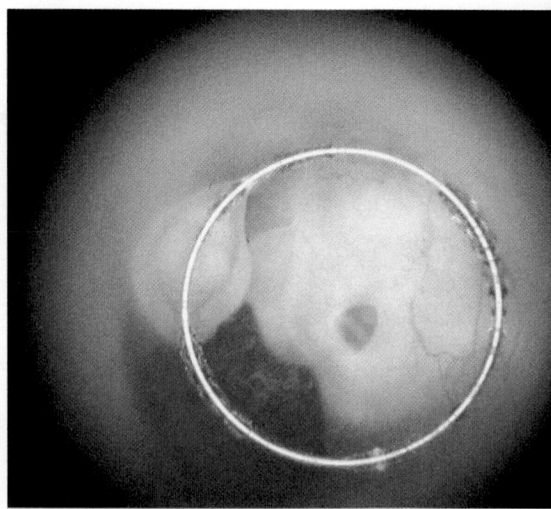

19-2 Micrograph showing early development of the human eye

19-3 Photograph showing annual rings of a hemlock tree

varieties. Early embryonic cells must not only undergo division, they must somehow be transfigured in the course of time from nonspecific cells to highly specific cells. The emergence of specialized structural or biochemical properties is what we mean by **differentiation.**

One should distinguish differentiation from determination. Differentiation is the actual appearance of new properties (structural, biochemical, functional, or whatever). It is preceded by determination. A cell that is to become a muscle cell is determined long before it is visibly transformed or differentiated into one.

Differentiation occurs in both plants and animals, although each group displays its own peculiarities. As we will later learn, a tree depends for its annual growth on undifferentiated cells that in each growing season differentiate rapidly into an assortment of specialized cells. Enduring signs of this process are seen in the annual rings of tree trunks (Figure 19-3). If plants of mature age did not retain many undifferentiated cells, they would not be able to grow.

The story is different in animals. The progeny of the animal zygote preserve both species and parental characters while at the same time acquiring new forms and functions. Thus, animal embryonic cells differentiate into liver cells, nerve cells, and the other types of body cells as development proceeds. Nonetheless, as we will later see, some cells in the adult animal body retain to a degree the character and multipotentiality of embryonic cells.

One important sign that differentiation has occurred is an accumulation of gene products. For example, a cell differentiating into a red blood cell accumulates hemoglobin. In many cases, it is more practical to define differentiation in this way than in terms of structural refinements.

Regeneration and Dedifferentiation

Once cells have differentiated, they ordinarily cannot become unspecialized again and revert to an undifferentiated form. However, such creatures as the flatworm *Planaria* or the coelenterate *Hydra* retain a capacity in adult life to develop whole new organs when called upon to do so. An adult salamander can regenerate an entire limb when its original limb has been amputated (Figure 19-4). In this remarkable process, the adult cells of the amputation stump lose many or all of the gene products that arose in their differentiation. These cells then begin to proliferate just like early embryonic cells (Figure 19-5). Muscle and nerve cells lose their specialized features and change into a mass of nonspecific cells, called the **blastema,** from which the new limb differentiates.

To some extent, the cells of all animals have regenerative powers. A whole liver regenerates after a portion of the liver of a rat is removed. And the processes of healing after a bone fracture, a skin wound, or a muscle tear are expressions of a developmental potential that appears to have been retained in the fully differentiated cell.

19-4 (Left) Normal (N) and regenerating (R) salamander limb
Appearance of normal forelimbs and regenerating stumps 14 days after limb amputation.

19-5 (Right) Appearance of cells in regenerating salamander limb
Note blastema cells (BC) and nerve (N).

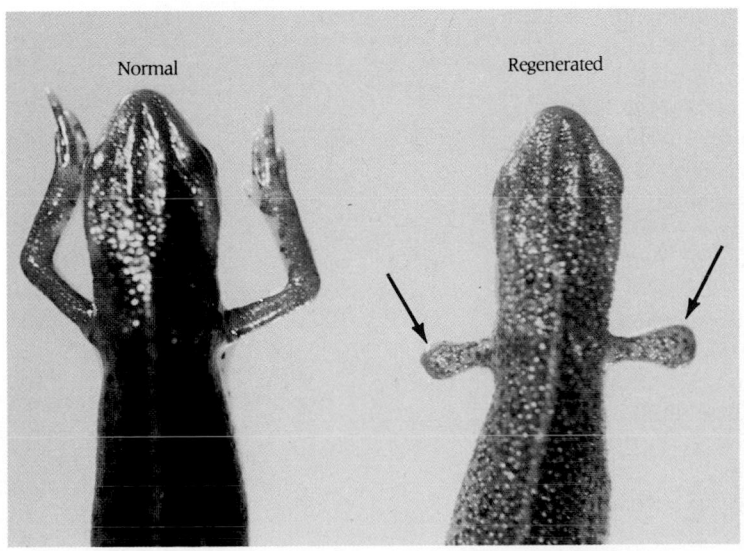

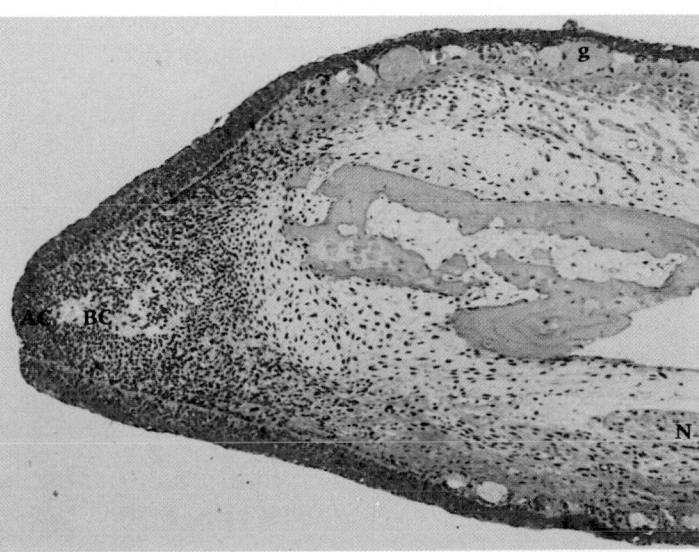

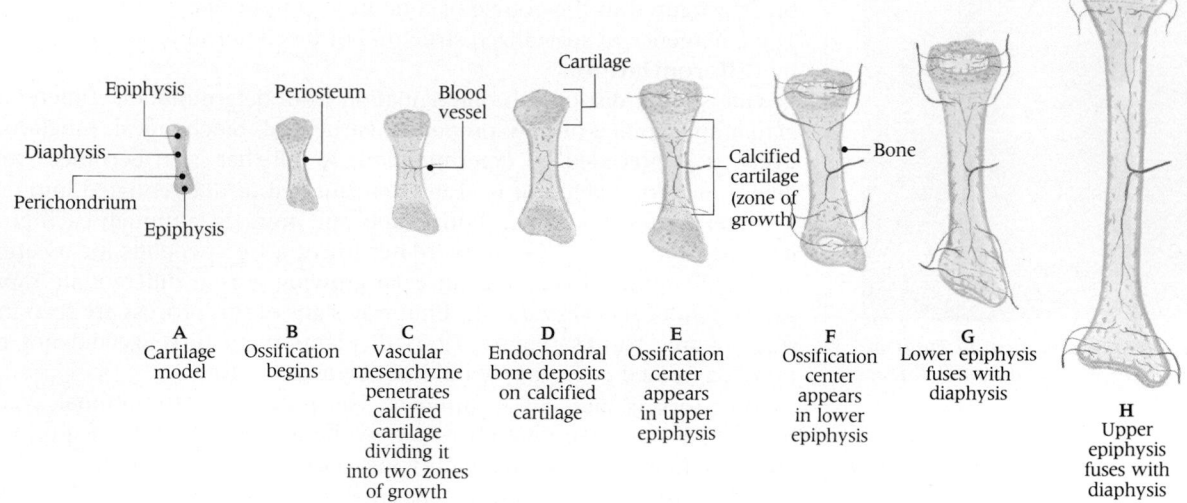

A
Cartilage
model

B
Ossification
begins

C
Vascular
mesenchyme
penetrates
calcified
cartilage
dividing it
into two zones
of growth

D
Endochondral
bone deposits
on calcified
cartilage

E
Ossification
center
appears
in upper
epiphysis

F
Ossification
center
appears
in lower
epiphysis

G
Lower epiphysis
fuses with
diaphysis

H
Upper
epiphysis
fuses with
diaphysis

GROWTH: IRREVERSIBLE INCREASE IN SIZE

An offspring must grow in the course of becoming an adult. Adult plants and animals may be thousands or millions of times larger than the zygote from which they came. Even the two daughter cells of a dividing protist must grow to reach their parent's size before either can divide again.

Development must include **growth,** which we define as an irreversible increase in the size of an organism. Usually growth results from an increase in the total number of cells in the organism. It is intriguing to realize that a newborn human baby, for example, contains more than 10^{13} cells. Individual organs in the baby— such as heart or brain—contain many billions of cells. The blood alone contains 10^{12} red cells. All of these cells, remember, came from a single ancestral zygote.

To an extent, growth can also result from an increase in the size of individual cells or of many cells. This, too, occurs in human growth. The infant's heart, for example, has the same numbers of cells as the adult's heart—even though it is only about 6% of its size and weight. Thus, a heart grows by increasing the size of its cells. Increase in the number of cells in a tissue is called **hyperplasia.** Increase in the size of an organ because of enlarged cells is called **hypertrophy.**

The Problem of Regulation

Growth obviously requires strict regulation that must operate at two levels—first, regulation of the growth of the whole body; second, regulation of the growth of individual body parts.

Characteristic patterns are seen when the growth of organisms of different types is measured at regular time intervals. In organisms lacking a larval stage, the sequence is quite familiar to us. Embryonic growth begins slowly but at birth growth is proceeding rapidly. It continues so into adolescence. Then it slows down and stops in adulthood.

The remarkable fact that in many organisms growth stops at a certain predictable point demands explanation. In mammals, the major mechanism for stopping bone growth involves peculiar structures called **epiphyses** within the limb bones and vertebrae (Figure 19-6). As long as these bones are growing, they maintain separate growth centers at each end, where new bony tissue is added between the center and the main portion of the bone. Eventually the end piece fuses with the rest of the bone. From then on, further bone growth is impossible.

Rigorous control is also exerted on the growth of organs and other body parts. In humans, body weight continues to increase to the age of 25 and beyond. However, the growth of various body parts follows different timetables. Figure 19-7A compares growth rates for the whole human body and some of its organs. The rapid increase in brain weight matches that of heart weight up to the age of 4. Then the brain stops growing. This growth pattern corresponds with the increase in head size. Figure 19-7B shows how strikingly growth rates differ among the several body parts.

19-6 Ossification and growth of bone
(A) The epiphyses are located at each end of the limb bone, which at this early stage is made of cartilage. (C-G) Growth zones at the epiphyses are responsible for increasing bone length. (H) When the final length is reached, the epiphyses fuse with the bone and growth stops.

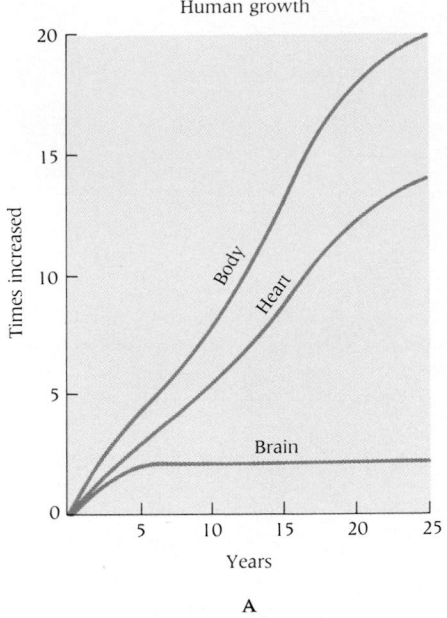

Human growth

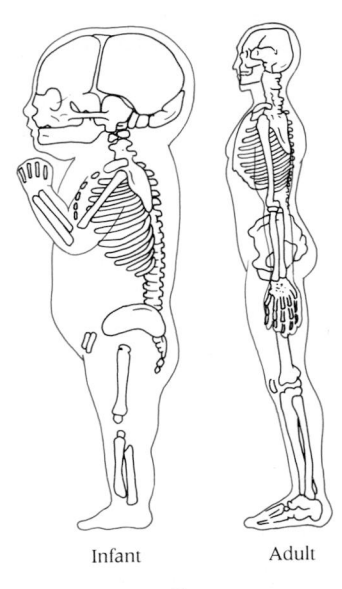

Infant   Adult

B

19-7 Relative growth of body parts in humans
(A) *The heart and especially the brain have a relatively smaller size increase than that of the body.* **(B)** *Major differences in the relative sizes of various body parts are strikingly evident in infant and adult.*

The growth of whole organisms and their body parts is regulated by hormones and growth factors. Both types of substance are currently under intensive study, and one brief illustration here will intimate their importance. When a portion of an animal's liver is removed, the remaining liver tissue regenerates. Significantly, the mass of regenerating cells increases only until the original liver size is restored. Here we have an instance in which liver cells that had long ago stopped growing actively resume growth under highly controlled circumstances. How do such cells ''know'' they should start dividing again, as they did in their embryonic days? How do they ''know'' when to stop growing? This puzzling phenomenon is apparently due to the production of one or more growth factors—probably by the stump of the liver—which act to promote liver growth, but only to the limit of its original mass. A long (and lengthening) list of such factors is now recognized. Fittingly, the discoverers of one of these factors, a protein called nerve growth factor (see Chapter 28), were recognized with the 1986 Nobel Prize in Medicine and Physiology.[1] We will return to this important topic.

The Problem of Cancer

Cancer is a pattern of abnormal growth exhibited by certain cells within a larger community of cells. By analogy, cancer cells are ''outlaws'' which show contempt for the welfare and orderly life of the larger community—that is, the whole organism—invading and plundering neighboring cells in an uncontrolled, or malignant, manner. We are able to recognize cancer cells reliably only by observing this behavior pattern. If a cancer cell is removed from an organism and grown in culture, its malignant character may not be recognizable until it is transplanted back into an organism and its uncontrolled growth observed. In the animal or plant body, abnormally growing cells may exist in a mass called a **tumor** (Figure 19-8). When a tumor consists of **malignant cells**—that is, cancer cells—it often spreads (usually via the bloodstream) throughout the animal's body by a process called **metastasis.** When a tumor forms a mass, but the cells do not metastasize throughout the body, the tumor is said to be **benign.**

Several difficulties have impeded scientific efforts to control and understand cancer.

- First, any attempt to establish some mechanism as a cause of cancer obliges us to find that agency operating in cancer cells and absent or inactive in noncancer cells. This search went unrewarded for many years, but a surge of new developments in the study of oncogenes (see Box 18-2) suggests that the prize may be near at hand.
- Second, it is likely that cells exhibiting the behavior pattern of cancer do not all do so for the same reason.
- Third, scientists have not yet succeeded in identifying a common metabolic feature that would distinguish cancer cells from normal cells. Throughout their search, scientists have accumulated many data, and differences have been found. For example, an enzyme normally present in white blood cells is lacking in the blood cancer, leukemia. Another example is the typically high rate of glycolysis in almost all cancer cells. These are exciting findings and, in the vernacular of journalism, may be ''a clue to the cause of cancer.'' Unfortunately, it is not at all clear whether such abnormalities are causes or consequences of the cancer with which they are associated.

MORPHOGENESIS: CREATION OF PATTERN AND SHAPE

A fourth developmental process, **morphogenesis,** involves the genesis of pattern and shape. An adult human has more cells than a human zygote. More importantly, the adult has a characteristic pattern of structure, shape, and function that arose from programmed foldings and mass movements of cells and groups of cells during the course of development.

The processes underlying these events rest entirely on a series of coordinated interactions of many differentiating and growing systems. Morphogenesis is the result of differentiation and growth.

---

[1] They were Rita Levi-Montalcini of Rome and Stanley Cohen of St. Louis.

Different morphogenetic mechanisms operate in animal and plant development.

■ In animals, the principal morphogenetic mechanism involves programmed rearrangements of individual cells and cell groups. Cells that originate in one embryonic region may not differentiate to their final form until they have migrated to another region. For example, the cells of three diverse organs—the medulla of the adrenal gland, the melanin-producing cells of the skin, and the cells of the sympathetic nervous system—all arise in a structure of the early embryo called the neural crest, which lies on either side of the developing spinal cord. These diverse cell types differentiate only after they have moved to other locations in the embryo.

■ In plants, cell movements are restricted by the rigidity of cell walls and extracellular cementing substances. Cell development in plants thus must occur where cells are formed. This is what happens, for example, when a single short cell, just above the intercalary peristem in a growing oat shoot, undergoes a single division and differentiates into the two guard cells of a stomatal apparatus. The final differentiation product is located at or near the side of the original differentiating cell.

These considerations suggest an important conclusion concerning the mechanism of development: A full account must reckon with the complex interactions that occur at several levels of organization. At the very least an understanding of development must explain the following:

■ How such molecular processes are spatially organized.
■ How these molecular processes are regulated.
■ How these processes are temporally arranged, including information on their rate, duration, and sequence.
■ How the products of each reaction are combined into the distinctive patterns of cells and tissues.

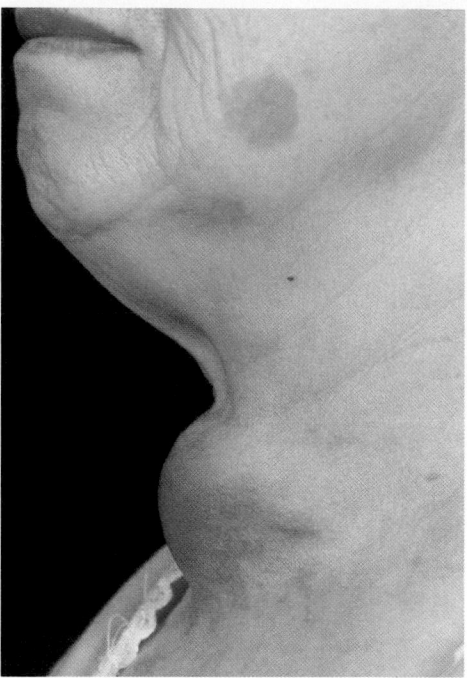

19-8 Tumor of thyroid gland

THE COURSE OF DEVELOPMENT

Animal development occurs in six stages.

■ **Gametogenesis** involves the maturation of parental germ cells.
■ **Fertilization** is the union of male and female gametes to form a zygote.
■ **Implantation** (in mammals) is the attachment of the early embryo to the uterine wall.
■ **Cleavage** involves division of the zygote into many cells, with no overall increase in size. Redistribution of the egg cytoplasm, including various regulatory molecules, occurs throughout the embryo.
■ **Gastrulation** entails the migration of embryonic cells and subsequent formation of two or three primary cell layers.

TABLE 19-1
ORGANISMS FREQUENTLY STUDIED BY DEVELOPMENTAL BIOLOGISTS

Broad Category	Organism	Genus
Chordates		
Cephalochordates	Amphioxus	*Branchiostoma*
Mammals	Human	*Homo*
	Mouse	*Mus*
Amphibians	Frog	*Rana*
	Toad	*Xenopus*
	Salamander	*Triturus*
Birds	Chick	*Gallus*
Invertebrates		
Insects	Fruit fly	*Drosophila*
Roundworms	Nematode	*Caenorhabditis*
Coelenterates	Hydra	*Hydra*
Echinoderms	Sea urchin	*Strongylocentrotus*
	Starfish	*Asterias*
Protists	Slime mold	*Dictyostelium*
Fungi	Bread mold	*Neurospora*

- **Neurulation** involves formation of notochord and neural crest; in **organogenesis,** primary cell types combine in various ways.

As in other areas of biology, certain organisms have become the experimentalist's "workhorses" (Table 19-1). Note that the list includes several types of vertebrates and invertebrates. In addition, important work has been done on the bread mold and slime mold.

MATURATION OF MALE AND FEMALE GERM CELLS

The animal life cycle begins with gametogenesis—the development of spermatozoa and ova (also called sperm and eggs). We saw in Chapter 12 that meiosis operates similarly in both sexes, at least as far as chromosome behavior is concerned. Chromosomes are reduced from the diploid to the haploid number. However, there are interesting and relevant differences between the sexes. As we have already seen, at least one distinctive aspect of gamete formation in females—the preservation of the cytoplasm—can significantly affect later embryonic development.

In sperm production, or **spermatogenesis,** the cells that give rise to spermatozoa divide symmetrically twice during meiosis. In the end each original diploid **spermatogonium** becomes four haploid **spermatids.** These in turn mature into four spermatozoa (or sperm). Cell size (and cytoplasmic mass) decrease drastically with each of these steps (Figure 19-9). A mature sperm consists of little more than a small nucleus containing condensed DNA, a wiggling tail or flagellum, and many mitochondria (Figure 19-10).

19-9 (Right) *Spermatogenesis*
Spermatogonia are continually replenished by mitosis. They then undergo meiosis and after two divisions produce spermatids. At each stage, the amount of cytoplasm decreases. The flagellum first appears on spermatids. Note that all cells are interconnected until the stage at which mature sperm are produced.

19-10 *Mature human sperm*
Micrograph showing large number of sperm in freshly ejaculated semen.

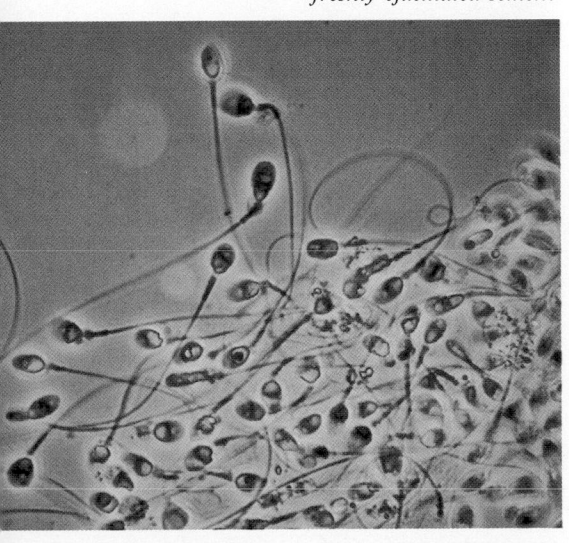

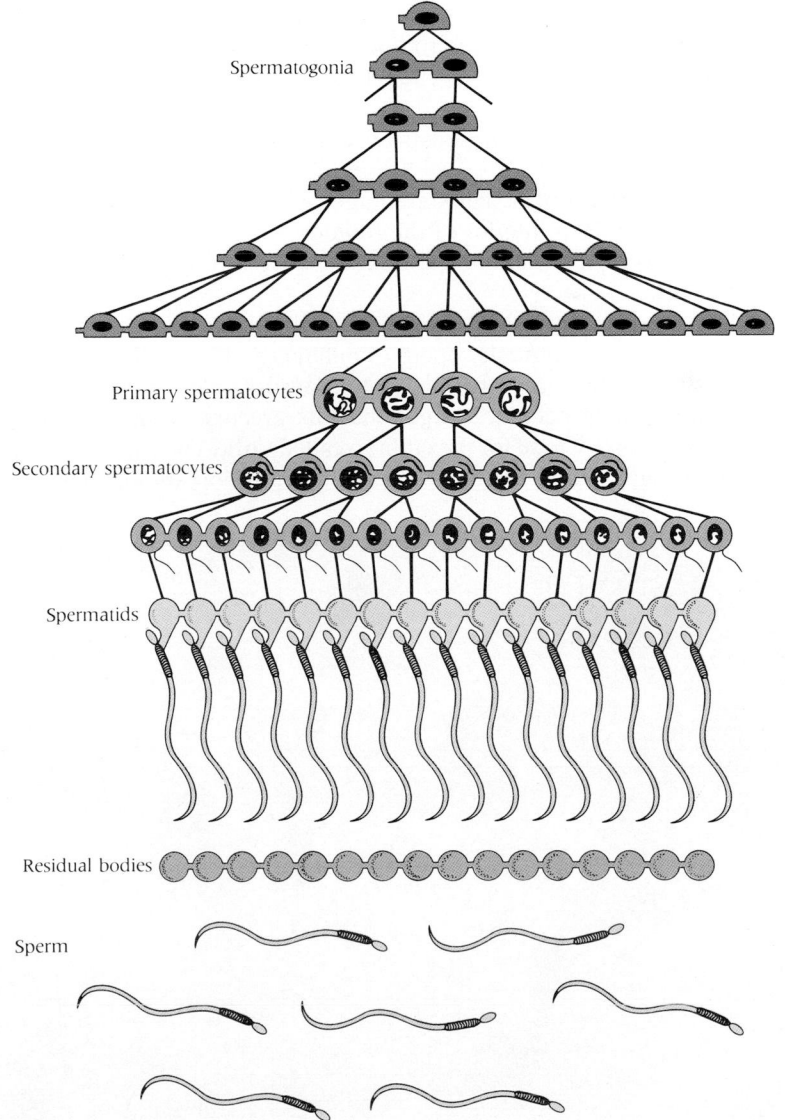

Spermatogonia

Primary spermatocytes

Secondary spermatocytes

Spermatids

Residual bodies

Sperm

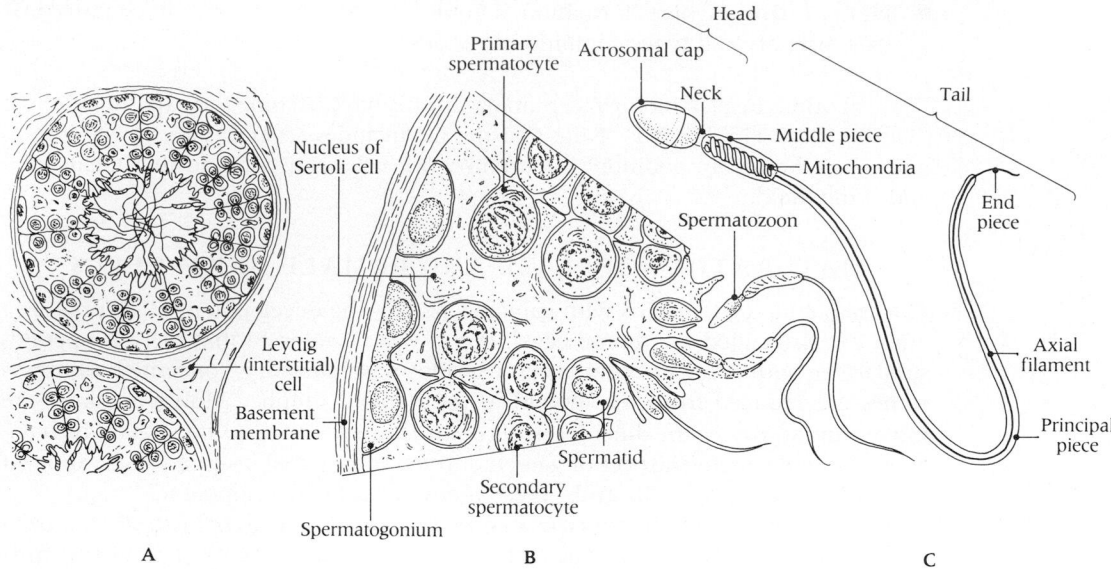

19-11 Structure of the seminiferous tubule
(**A**) *Low power view of tubule in cross section showing the location of Leydig cells.* (**B**) *High power view showing close association of Sertoli cells and cells involved in sperm production.* (**C**) *More detailed view of structure of mature spermatozoan.*

We will say more about spermatogenesis in Chapter 21, but we note here that it occurs in the testes, which in mammals are located in an extraabdominal sac called the scrotum in many but not all male mammals. (Among the exceptions are elephants and whales.) Spermatogenesis is inefficient at high body temperatures but is favored at the cooler temperatures of exposed scrotal sacs. When the testes fail to descend from the abdomen during development, spermatogenesis may be severely impaired. The testis contains many lobules, within each of which is one or several convoluted seminiferous tubules. Spermatogenesis occurs within these tubules, which contain three types of functioning cells (Figure 19-11).

- **Spermatogonia** are the primordial cells that in spermatogenesis differentiate into spermatozoa.
- **Sertoli,** or **sustentacular, cells** are large cells extending from the base of the epithelium to the deep interior of the tubules.
- **Leydig,** or **interstitial, cells** lie between adjacent tubules and secrete male sex hormones.

Egg production, or **oogenesis,** operates differently. Each original diploid **oogonium** gives rise in the end to only one haploid **ovum** (or **egg**). Unlike a spermatozoan, an ovum usually becomes larger than its precursor cell as it acquires the cytoplasmic collections of food for the embryo called **yolk.** This nutrient-rich mixture of carbohydrates, lipids, and proteins varies in composition among different species.

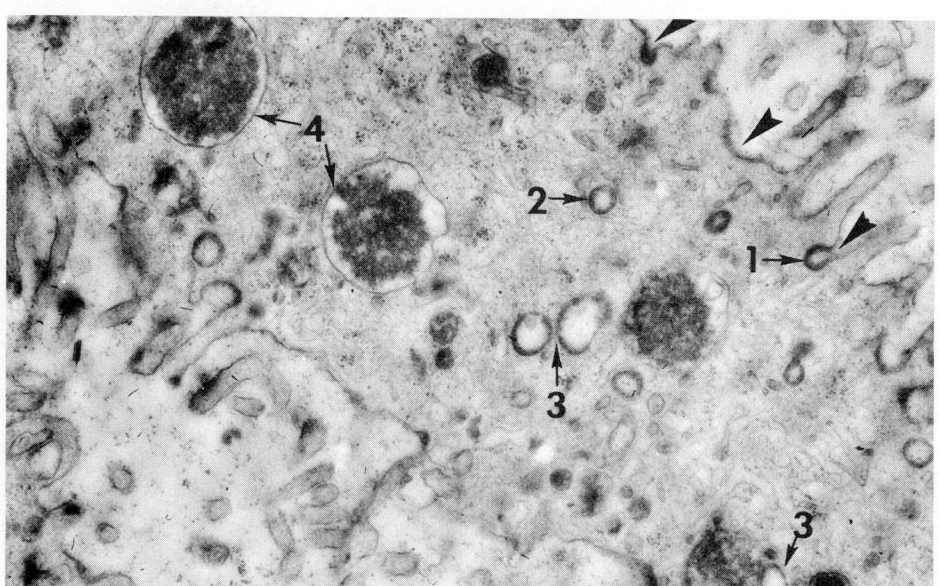

19-12 Electron micrograph showing the formation of yolk platelets in an oocyte of a milkweed bug
The arrowheads in the upper right-hand corner point to regions where pinocytosis is occurring at the cell surface. The numbered arrows point to sequential stages in which the pinocytic vesicles invaginate, bud off from the cell surface, and fuse together to form yolk platelets. ($\times$ 23,000)

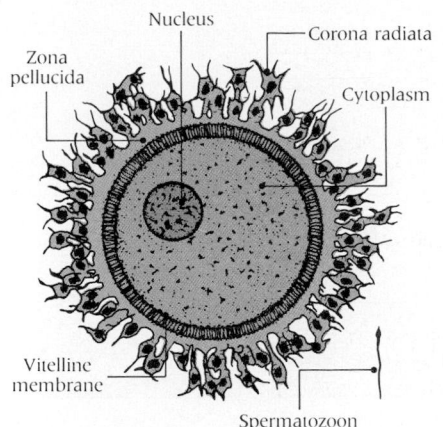

19-13 Structure of a mature ovum
A thin perivitelline space (not visible here) surrounds the vitelline membrane. Note the size of the ovum relative to that of a spermatozoon (sperm).

In birds, a huge amount of yolk makes up the bulk of the cell. In other species, including most mammals, the amount of yolk is meager. In many species, yolk components are synthesized elsewhere in the mother's body, brought to the developing egg, and assimilated by in-pocketings of the outer membrane.

In many species, yolk components synthesized in the liver of the mother's body are brought to the developing egg and assimilated through its cell surface by pinocytosis. For example, the liver produces large amounts of **vitellogenin,** a complex containing the phosphoprotein **phosvitin** and the lipoprotein **lipovitellin.** Vitellogenin crosses the plasma membrane of an oocyte in tiny pinocytotic vesicles, which eventually fuse to form nutrient-containing cytoplasmic bodies called **yolk platelets** (Figure 19-12).

Only one of the cells formed in meiosis II is destined to become a functional ovum. The other three do not mature into ova. Instead they form small, nonfunctional **polar bodies** that collect at one side of the oocyte and eventually degenerate (see Chapter 12). This unequal division diverts most of the cytoplasm to the single large egg cell that emerges from oogenesis (Figure 19-13).

Properties of the Egg

It is useful to distinguish **oviparous** animals, in which eggs are laid and development takes place outside of the mother's body, from the **viviparous** group, in which eggs are retained within the body and embryonic development occurs internally. There is a third intermediate type, termed **ovoviviparous,** in which development takes place internally, but there is no placenta whereby the mother can provide nourishment to the embryo.

In addition to nutrient-rich yolk, yolk platelets contain the hydrolytic enzymes—phosphatases, lipases, and proteases—necessary for the future utilization of these nutrients. As discussed in Chapter 5, there are many species differences in egg size and design. Size seems to depend on how long the food supply in the yolk must last.[2] In marine animals, in which the embryo or larva begins to feed itself early, eggs are small and there is little yolk. The yolk that is present is evenly distributed throughout the egg. In striking contrast, the embryos of birds must live on the yolk until they hatch. Hence in these eggs the yolk volume is vast relative to the size of the embryo. (Despite its size, a bird's egg is a single cell.)

The yolk content of the eggs of mammals resembles that of marine animal eggs. Mammalian embryos need little food reserves since their nourishment is soon to come from the mother's body through a remarkable structure peculiar to mammals, the **placenta,** which we describe and illustrate later. Mammalian eggs range in

[2] Other factors influencing egg size are: differences in the degree of complexity of animals at the time of hatching, the course of post-hatching development, and the extent of parental care. These factors are considered in Chapter 33.

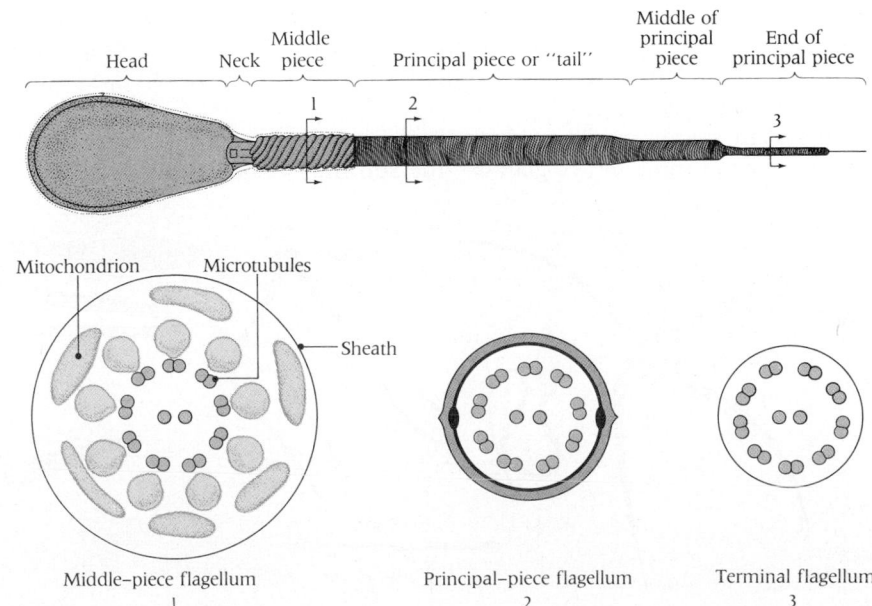

19-14 Structure of sperm and its flagellum
The major regions of the sperm are the head, neck, middle piece, and tail. Mitochondria are found only in the mid-piece. The 9 + 2 arrangement of microtubules occurs throughout the tail (flagellum). Note that the sperm shown above is greatly shortened for clarity.

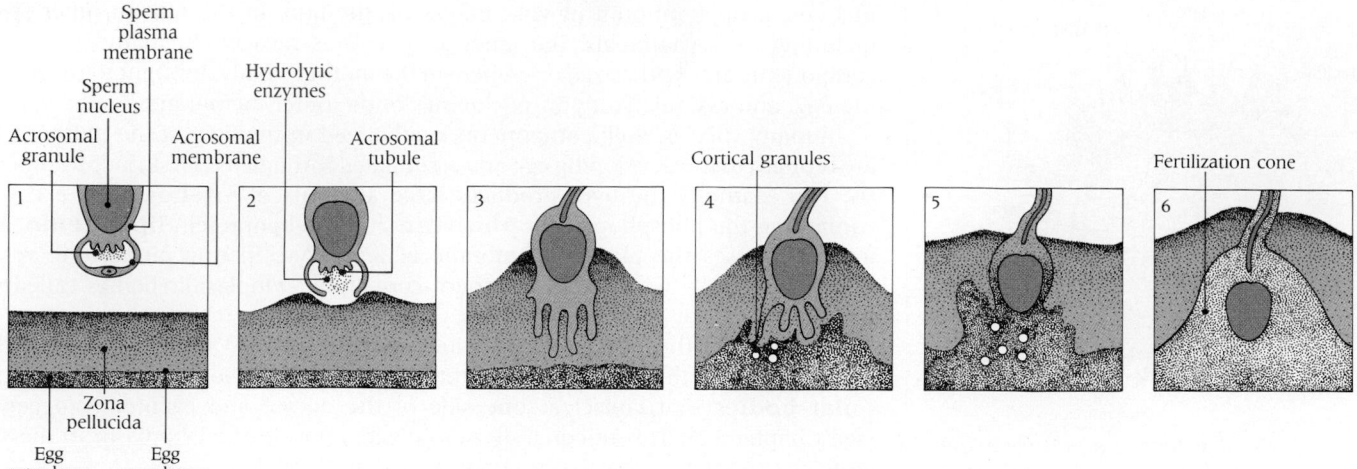

Sperm
plasma
membrane
Sperm
nucleus
Hydrolytic
enzymes
Acrosomal
granule
Acrosomal
membrane
Acrosomal
tubule
Cortical granules
Fertilization cone

Zona
pellucida

Egg
cytoplasm
Egg
membrane

size from 0.02 to 0.25 mm in diameter. The human egg, shown in Figure 19-13, is 0.14 mm in diameter—smaller than the period that ends this sentence and just visible to the naked eye. Note, however, how much larger it is than a sperm cell.

Properties of the Sperm

The three portions of a sperm cell—**head, middle piece,** and **tail** (Figure 19-14)—correspond roughly to its three functions: (1) activating and genetic functions, (2) metabolic functions, and (3) motility.

A sperm cell initiates development by doing two things. It activates an egg and it furnishes a nucleus containing a haploid number of paternal chromosomes. These functions are carried out by the head, which consists of a nucleus containing chromosomes in a tight package of unusual density, and the **acrosome,** which lies beneath a caplike membrane. The acrosome facilitates sperm contact with the egg plasma membrane by releasing powerful hydrolytic enzymes (Figure 19-15).

Before a sperm can penetrate an egg, it must reach it. This is made possible by a tail that propels the sperm at a velocity of 1 mm per minute. This is only 60 mm or 2.5 inches per hour. Like most cilia and flagella, the sperm tail is composed of two central microtubules surrounded by a ring of nine doublet tubules (see Figure 19-14). The microtubules are attached to a basal body or granule in the neck of the sperm.

The middle piece is the area about which we know the least. Mitochondria are located there, often in a spiral configuration around the microtubules. Presumably they power the beating tail. Other than the nucleus (and a centriole), the sperm contributes little to the egg. In many species, neither the tail nor the middle piece (with its mitochondria) enter the egg during fertilization. This explains why the mitochondrial genome is inherited only from the mother.

Sperm of different species show fascinating variations of size and shape (Figure 19-16). Sperm of some insects are 12 mm long—visible to the naked eye. A human sperm is about 60 μm long.

Mammalian sperm are released as a suspension in **seminal fluid,** the thick fluid secretions of three sets of glands—the **seminal vesicles,** the **prostate,** and

19-15 Principal steps in sperm-egg interaction during fertilization
(1) The sperm head approaches the egg.
(2) Contact is made and hydrolytic enzymes (black dots) are released from the acrosome into the zona pellucida, the egg's evelope, and its penetration begins. (3) The acrosome tubule approaches the egg plasma membrane.
(4) Contact between the two membranes is made.
(5) Cortical granules release enzymes. (6) The membranes fuse and the sperm nucleus enters the egg.

19-16 Variations in sperm morphology in different animal species
Note that these examples are not drawn to scale.

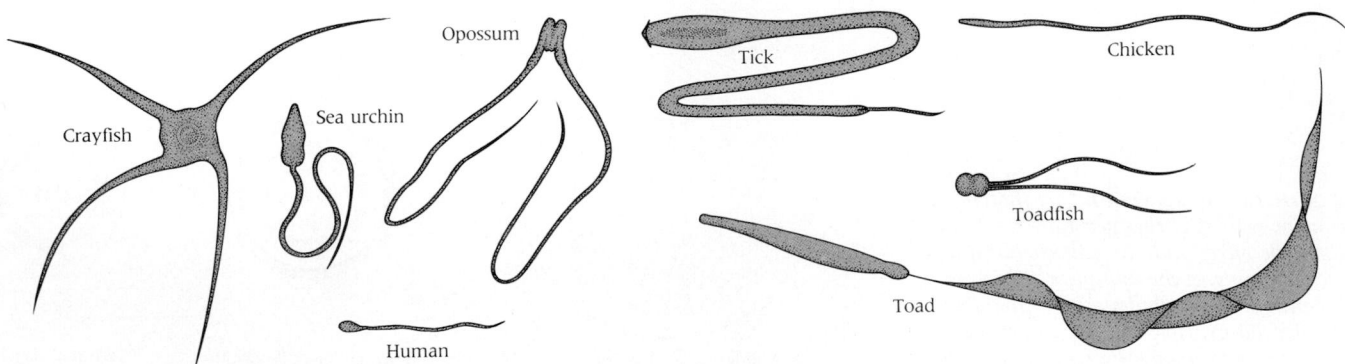

Opossum
Tick
Chicken
Crayfish
Sea urchin
Toadfish
Toad
Human

Cowper's glands. The mixture of sperm cells and seminal fluid is called **semen.** Seminal fluid has several functions.

- It contains sugars (chiefly fructose) that are essential for the energy metabolism of motile sperm cells.
- It physically transports sperm cells before and during ejaculation.
- It contains buffers that protect sperm cells from the relatively low pH of the female reproductive tract.
- It lubricates the passages through which sperm cells travel in the male reproductive system.

The amount of semen released at one time varies from species to species. In humans, the volume is 2 to 4 ml, about 10% of which consists of sperm. Each milliliter contains about 100 million cells. In humans with sperm counts below 20 million per milliliter, fertilization becomes less likely. Low sperm counts are caused by various diseases or drugs. For instance, anticancer drugs that impair DNA synthesis block spermatogenesis.

Relatively few sperm cells actually reach the egg. Of the nearly 400 million spermatozoa deposited in a human vagina during copulation, only a few hundred thousand find their way into the mouth of the uterus (cervix) before they die. A low sperm count reduces this number even further. The chance that fertilization will take place is limited also by the short lifespan of sperm cells. In the human female, most sperm cells survive only about 30 hours.

FERTILIZATION

The union of male and female gametes occurs outside the body in most fishes and amphibians. In all other vertebrates it takes place internally following copulation. The events of fertilization occur in three stages.

- Penetration of the egg by the sperm (Figure 19-17)
- Activation of the egg
- Fusion of gamete nuclei within the egg

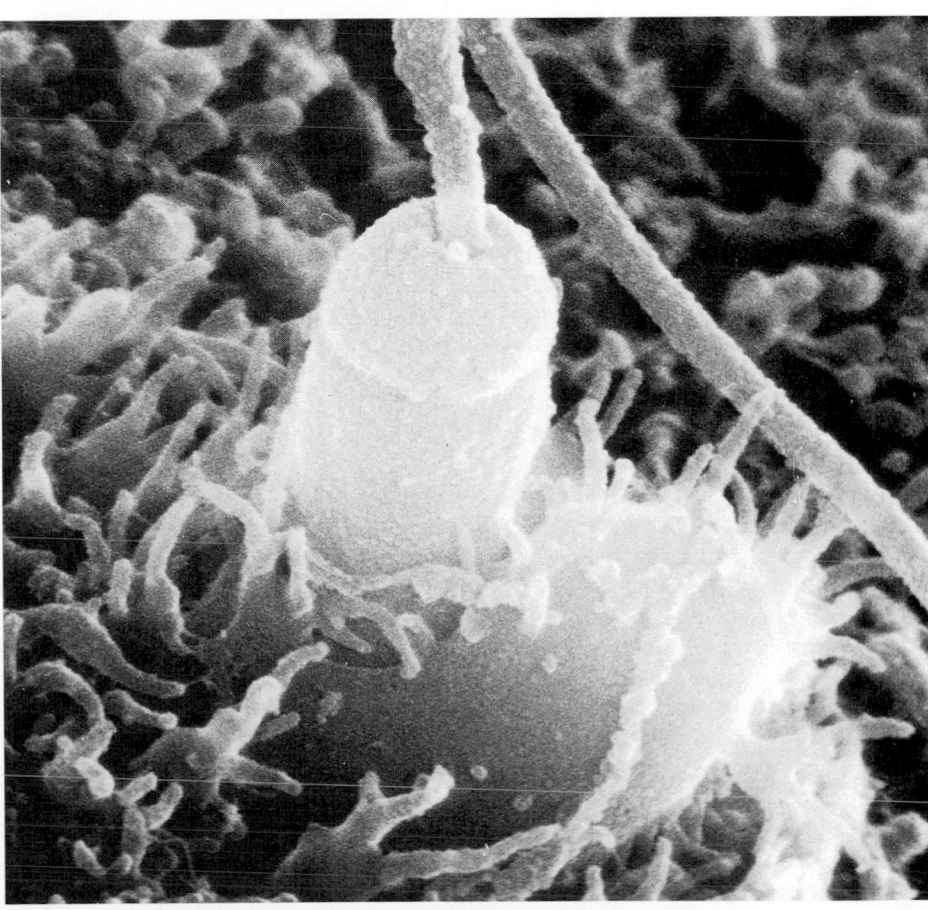

19-17 Sperm penetrating egg
A scanning electron micrograph showing a sea urchin sperm that has penetrated nearly half way into an egg. Note the fertilization cone and the surrounding microvilli on the egg's surface.
(× 6000)

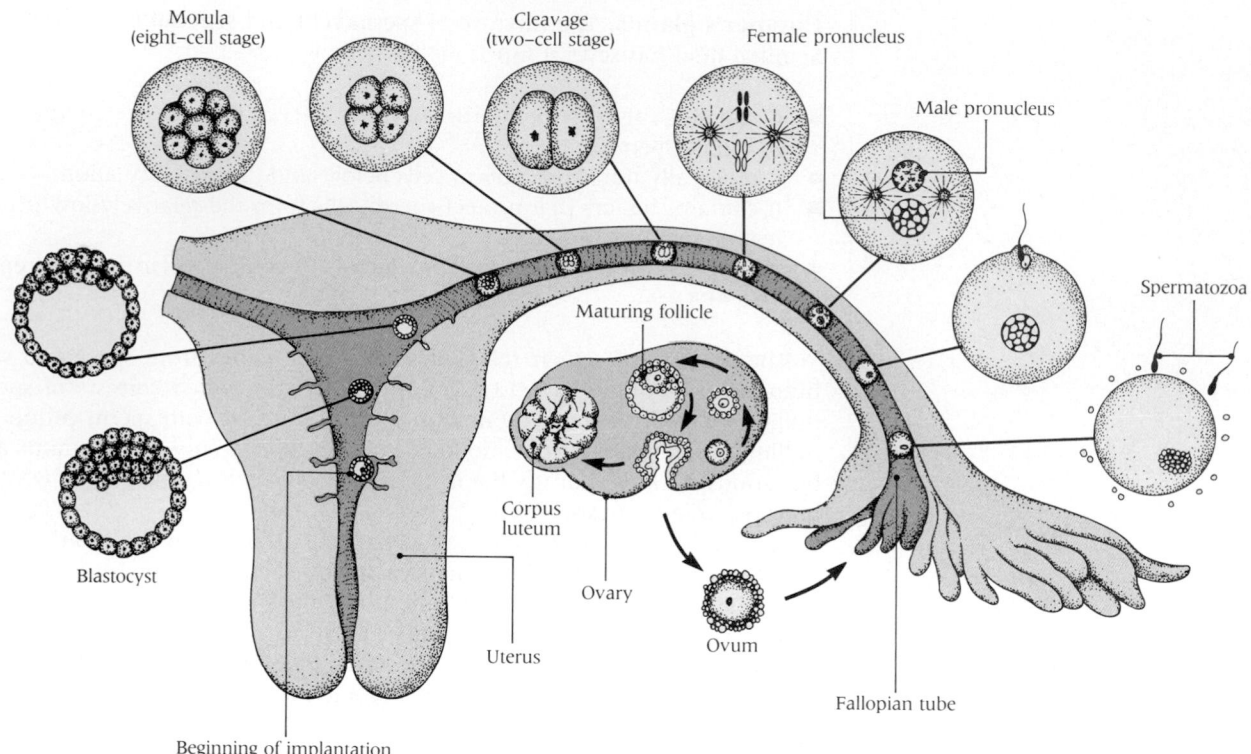

Morula (eight–cell stage)
Cleavage (two–cell stage)
Female pronucleus
Male pronucleus
Spermatozoa
Maturing follicle
Blastocyst
Corpus luteum
Ovary
Ovum
Uterus
Fallopian tube
Beginning of implantation

As discussed in Chapter 12, in the mammalian oocyte there is usually a long pause between the beginnings of the first meiotic division, which in humans starts in the third month of embryonic life and then pauses in metaphase, not to be completed until an ovulation occurs after puberty (see Figure 12-33). Only after entry of a sperm cell into the egg does the second meiotic division take place.

Mechanism of Penetration

After copulation, a relatively few active spermatozoa wander aimlessly from the vagina into the uterus and then into the Fallopian tubes (Figure 19-18). An ovum, surrounded by a ragged sheath of granulosa cells, is rolled and pushed toward the sperm by cilia of the cells lining the tube and by the downward peristalsis of its muscular layers. Even though a few hundred spermatozoa thrash about in the current, the ovum may not encounter one, in which case, fertilization does not occur. This is the pattern in humans and other mammals.

Much of what we know of the entry of sperm into egg at fertilization has come from studies of sea urchins, such as *Strongylocentrotus purpuratus,* and other invertebrates. The moment a sperm touches the **zona pellucida** (outer zone) of an egg, the acrosome in the sperm head becomes extended into a long, thin **acrosomal process** as its actin monomers polymerize (see Figures 19-15 and 19-17). The sperm expands the part of the acrosomal membrane that will fuse with the egg plasma membrane and, in effect, "reaches out" to make contact with the egg. Following contact, a threadlike filament tightly binds the sperm to the egg surface. In organisms that utilize external fertilization, it is obviously important that this binding be highly selective. For example, such selective binding is crucial in seawater that may contain sperm cells from several different species. The molecular basis of this species specificity is believed to reside in a sperm protein called **bindin.** The bindin from sperm of different species of sea urchin binds preferentially to surface receptors in the outer **vitelline envelope** encasing eggs of the same species.

In the marine worm, *Hydroides,* in which the fertilization mechanism has been studied exhaustively, the acrosome is instantly activated by contact between sperm and egg envelopes (see Figure 19-17). This is termed the **acrosome reaction.** As the acrosome opens, it liberates hydrolytic enzymes that digest a canal or passageway in the egg envelope, through which a sperm nucleus can enter.

Next there is fusion of the surface membranes of egg and sperm and lifting off of an outer egg membrane to which extraneous sperm may be attached. This estab-

19-18 Fertilization and passage of ovum into the uterus
The ovum is released from a follicle in the ovary and enters the Fallopian tube, where fertilization occurs. The fertilized egg completes cleavage as it travels toward the uterus, reaching the blastocyst stage just before implantation into the uterine wall. The diagram shows various stages undergone by a single egg.

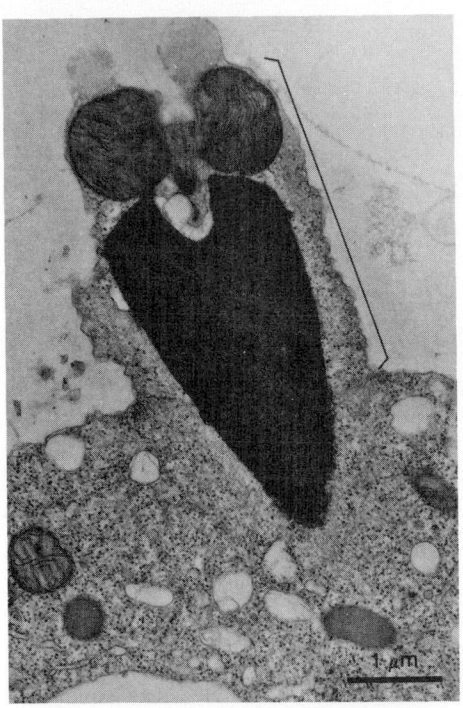

19-19 Electron micrograph illustrating fertilization cone
The cone is formed from egg cytoplasm that surrounds the entering sperm nucleus.
(× 12,000)

lishes a physical barrier that helps to prevent **polyspermy** (fertilization of one egg by more than one sperm). Establishment of this barrier depends on a set of events called the **cortical reaction** (see Figure 19-15). It depends on the presence in the egg of **cortical granules** that discharge their contents into the space between the sperm and egg plasma membranes. As a result, the zona pellucida envelope is converted into a tough **fertilization membrane** that effectively bars the entry of additional sperm cells into the egg. This is called the **zona reaction.** The establishment of this barrier is slow and time-consuming—yet the effective prevention of polyspermy clearly requires that a barrier be put in place rapidly following the entry of one sperm cell. Such a mechanism exists, though it is very short-lived: it is a rapid change in the electrical properties. As a result of this **depolarization** event, the egg's plasma membrane does not fuse with that of a second sperm cell. Another result is an increase in intracellular Ca^{2+} in the egg, the calcium ions arising from stores of bound calcium.

Recent work has revealed the presence in the zona pellucida of several important glycoproteins. One of these proved to be essential in the binding of a sperm head to the egg's plasma membrane; hence it was called the **sperm receptor.** Indeed, experiments showed that sperm will bind tightly to a glass bead that has been coated with purified sperm receptor protein.

Shortly after fertilization the egg cytoplasm creeps up around the entering sperm nucleus to form a **fertilization cone** (Figure 19-19). The next event is fusion of the **pronucleus** of each cell to form a single diploid nucleus. Now inside the egg, the sperm pronucleus moves toward the egg pronucleus, which completes its second meiotic division.

Activation of the Egg

We have spoken of the "activation" of an egg by an entering sperm. What does this mean at the molecular level? It is convenient to speak of two kinds of reactions to fertilization (Figure 19-20).

- The early responses to fertilization occur within the first minute and include depolarization of the egg plasma membrane, the cortical reaction, and an increase in intracellular Ca^{2+}.
- The late responses follow soon after, and include a rise in intracellular pH (a consequence of the release of H^+ ions into the surrounding medium), a five- to tenfold increase in protein synthesis, a sharp increase in membrane transport, the fusion of egg and sperm nuclei, and the initiation of DNA synthesis—all major events that presage the brisk biosynthetic activities that lie ahead.

Both sets of responses are part of the process of **egg activation.**

Studies in sea urchins and other marine invertebrates show that the unfertilized egg contains many ribosomes and a large amount of mRNA. Within minutes of fertilization, all of this metabolic machinery is "turned on" in preparation for the rapid cell divisions about to take place. One of the sperm's primary roles is to initiate this chain of events. We shall consider how the sperm accomplishes this in Chapter 20.

19-20 Early and late responses to fertilization
Egg activation involves a variety of events that take place sequentially before cleavage begins. The responses known to occur in sea urchin eggs are shown here. Note that the time scale is logarithmic.

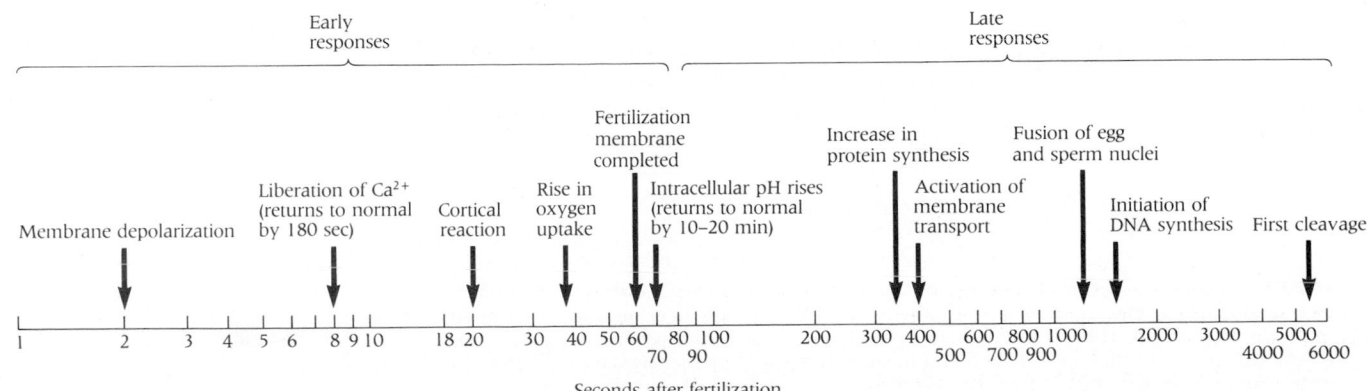

In some species (e.g., some toads and sea urchins), artificial stimuli (such as a needle prick or a shift to an abnormal pH) can mimic the whole process of fertilization and cause an egg to complete meiosis and start development. Evidently, a sperm is not essential for development. A haploid egg cell can go ahead in some instances and complete all or nearly all of its development. This development of a haploid embryo is called **parthenogenesis.** Haploid salamanders have been produced in this way. Haploid embryos can also be raised by surgically removing the egg nucleus just before fertilization. The haploid nucleus of the fertilizing sperm proves quite adequate to initiate and control development. A haploid egg in such a case can become diploid simply by duplicating its chromosomes without a cell division— or by reviving a polar body that has been destined for elimination.[3]

Some subsequent development has been observed in an egg cell artificially deprived of both gamete nuclei. To be sure, such eggs soon die, but the fact that they develop at all tells us that the cytoplasm carries some genetic information— probably in the form of long-lived mRNA and proteins.

Fusion of Nuclei

Nuclear fusion, or **syngamy,** takes place as the two haploid gamete nuclei join to become a single diploid zygote nucleus (see Figure 19-19). This process is triggered by the activation of the egg. Experimental insertion of a sperm nucleus into an egg in the absence of activation does not lead to syngamy.

Actual development of the fertilized egg (now a zygote) begins with syngamy. The stage is set for embryonic development following completion of the fertilized egg's second division.

Unsolved Problems

As noted above, sperm do not ordinarily penetrate the eggs of other species. This generalization applies as well to experimental manipulations in tissue culture dishes. The membranes of gametes have unique properties that permit them to fuse with one another but not with cells other than gametes. Similar properties—for example, the presence of bindin—appear responsible for species specificity in fertilization.

Thus sperm-egg interaction is usually restricted to the gametes of a given species and hybrid fertilizations occur only rarely. Other factors that tend to prevent the fertilization of eggs of one species by sperm of another species include species behavior patterns that usually keep mating from occurring. Even if individuals of different species do mate and development actually begins, it usually fails to progress because of mismatch in chromosome structure and number.

Of the many problems in the field of fertilization research, two are currently attracting much attention.

- One is *in vitro* **fertilization.** At the present time, fertilization of an egg by sperm in a laboratory culture dish has met with success in several species, including rabbits, cows, mice, and humans—as indicated by the development of live young following transfer of fertilized eggs into recipient females. The first successful *in vitro* human fertilization was achieved in England in 1974 (Box 19-1).

- The other is **sperm capacitation,** the conditioning sperm receive as they proceed through the female reproductive tract. Changes occurring in them during this journey are essential for later penetration of the egg. One involves the removal of certain components derived from semen that are tightly bound to the sperm surface. The need for capacitation is well established in rabbits, mice, and other species. Its significance in human reproductive physiology remains unclear.

---

[3] Some parthenogenesis occurs naturally, with complete development in the absence of fertilization. We shall later see that some male insects (e.g., aphids) normally develop from unfertilized eggs and thus are haploid. There is a remarkable population of all-female parthenogenetic lizards in the Caucasus region of the Soviet Union and in some fishes a sperm is necessary for activation but nuclear fusion does not occur. The male chromosomes are discarded!

BOX 19-1

ISSUES RAISED BY *IN VITRO* HUMAN FERTILIZATION

In July 1978, a baby girl named Louise Brown was born in England as a result of the first successful *in vitro* human fertilization. Through the efforts of the same research team, a boy was born next. Subsequently, children conceived *in vitro* were born in Australia, India, and the United States—and in June 1982 an American woman bore her second child to be conceived in this way. The births of these babies generated worldwide expressions of acclaim and concern. The acclaim was for Patrick Steptoe and Robert Edwards, the British investigators who had developed this technique for circumventing the childlessness of women with blocked oviducts. The concern arose over ethical questions that some people feared might arise from this ''unnatural'' pattern of human reproduction.

Natural fertilization takes place in the upper third of the oviduct. The early embryo then commences a series of mitotic divisions, cleaves into a morula, and becomes a fluid-filled blastocyst. Throughout this period, the embryo floats freely in the fluids of the female generative tract, traveling down the oviduct in the course of several days and finally entering the uterus. This is the preimplantation period of pregnancy. It ends with implantation—the adherence of the blastocyst to the endometrium and the invasion of its wall.

By this time, the embryo has reached the 12- to 16-cell stage. Implantation occurs about 6 days after fertilization.

When the oviducts are blocked, sperm cannot reach the egg. The result is infertility. Two approaches to the relief of this situation have been attempted. One is to repair the damaged oviducts by microsurgery. The other is to conduct the fertilization outside the body in a glass vessel.

The idea of extracorporeal fertilization using gametes isolated from an infertile couple, followed by placement of the resulting embryo in the woman's normal uterus, was first suggested in 1937. The efforts of investigators led to successful experimental pregnancies in mice and finally, in 1959, in rabbits.

Edwards began work on the human egg in 1963. He was joined by Steptoe in 1968 and their persistence for a decade and a half led to the birth of Louise Brown in 1978. Their procedure had three steps.

- An ovum that has completed meiotic maturation is recovered from a carefully selected subject.
- The mature ovum is fertilized *in vitro* and the embryo is subsequently cultured through part of its preimplantation development.
- The embryo is placed in the uterus of the woman from whom the ovum was obtained. This procedure differs from that usually performed in laboratory and domestic animals in that the embryo is placed in the uterus of the ovum's donor, rather than in the uterus of a surrogate mother.

The value of this procedure is chiefly in cases in which infertile couples desire to produce their own child rather than to adopt. However, there are countless opportunities for failure—e.g., in the isolation of the ovum, in the development of an embryo after exposure to sperm *in vitro*, and in implantation. At this writing, the chance that a live baby will be successfully produced once a woman has been selected for the procedure is about 1 in 30.

Investigators have extensively discussed ethical aspects of this procedure (which is not to be confused with surrogate motherhood) and have widely agreed on its acceptability. No normal fertilized embryos are discarded and the risks of the procedure to the potential offspring are minimal. We can expect that the procedure will become increasingly available to infertile couples throughout the world.

CLEAVAGE: ONE CELL BECOMES MANY CELLS

The next stage of embryonic development begins when the zygote divides. In this stage, termed **cleavage,** rapid cell division begins within hours of the formation of a diploid nucleus (see Figure 19-18). With further divisions the number of cells increases in geometric progressions—2, 4, 8, 16, 32 (Figure 19-21). A human zygote reaches the 2-cell stage about 30 hours after fertilization (Figure 19-22). The next three divisions occur at 12-hour intervals.

Interestingly, cleavage alters the transit time of the various stages of the cell cycle (see Chapter 12). Cleaving cells move rapidly between the S phase and mitosis, spending little time in the G_1 and G_2 phases.

In this early stage, mitotic divisions rapidly increase the number of cells but, interestingly, embryo size remains unchanged for a time because early cell divisions yield smaller and smaller individual cells. The original zygote is much larger than a typical adult body cell, but by the gastrula stage cleavage has decreased the embryo's cells to the size of the adult's cells.

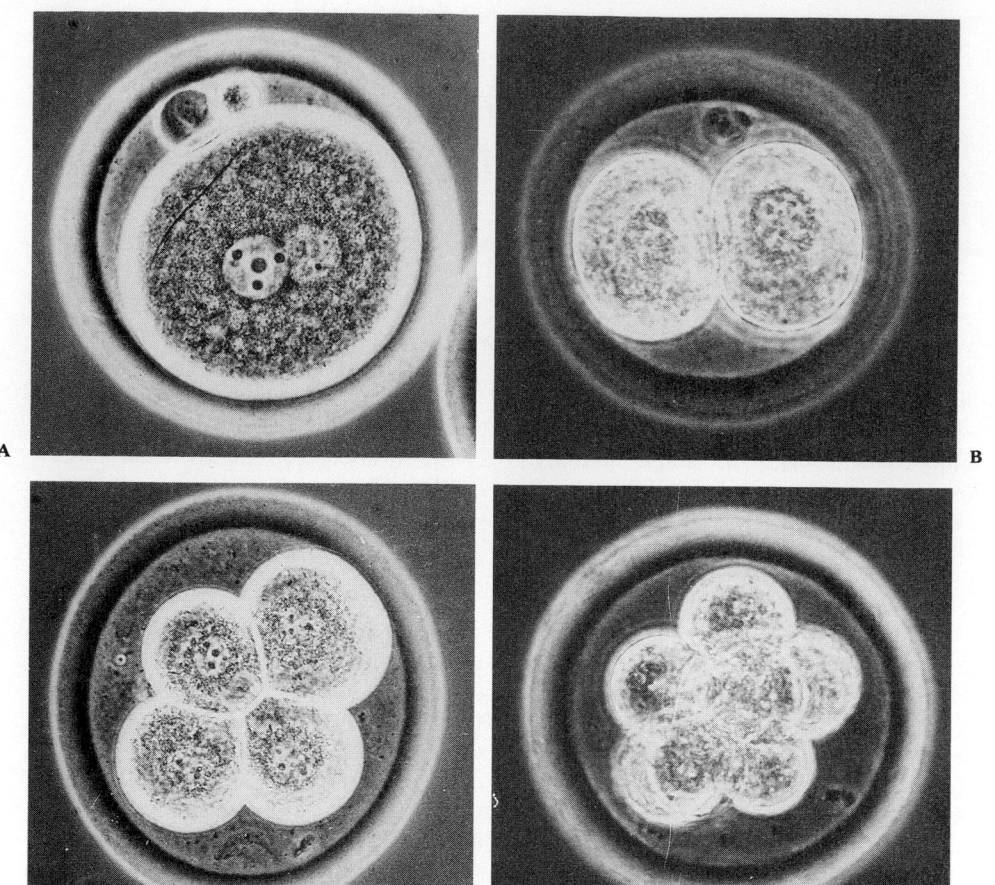

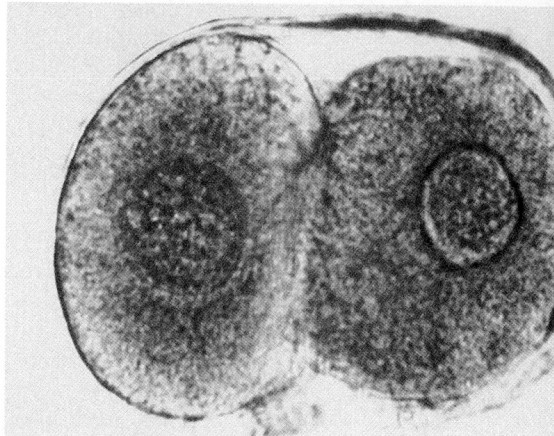

19-21 (Left) Cleavage stages in development
Development of hamster eggs as seen by phase-contrast microscopy. **(A)** Fertilized egg at the pronuclear stage, about 3 hours after sperm entry. (× 1000) The cell contains an egg pronucleus and a sperm pronucleus. **(B)** Two-cell stage, 20 hours after fertilization. **(C)** Four-cell stage. **(D)** Eight-cell stage. (B–D × 650)

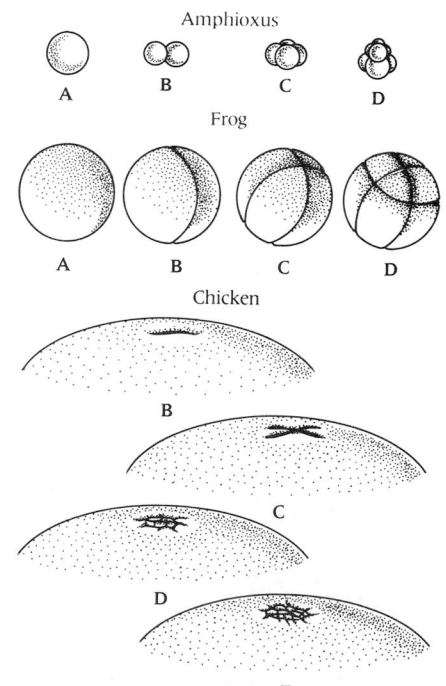

19-22 (Above) Cleavage in human eggs
The embryo is at the 2-cell stage, about 60 hours old.

Types of Cleavage Patterns

Despite wide variations in cleavage patterns in the animal kingdom, certain features are generally shared. In most species, early cleavage produces a raspberry-shaped mass of about 16–32 cells (see Figure 19-21). This is a **morula** (from the Latin word for mulberry). Each cell in the morula is called a **blastomere.** As the number of cells increases, the embryo becomes a **blastula** (termed **blastocyst** in mammals), containing a cavity, or **blastocoel.** In many species the embryo is now hollow.

Careful study of animal embryos of different major groups, particularly those listed in Table 19-1, has revealed notable variation. Characteristic patterns in the timing of divisions and their spatial orientation in each species and major group have emerged from these studies. For example, early embryos develop **polarity,** with a so-called **animal pole** and a **vegetal pole.** The cells of the animal half of the early embryo tend to be smaller than those of the vegetal half, though variations exist. For this reason, embryologists have concentrated on the following representative forms:

- Invertebrates (aquatic): sea urchin and starfish
- Amphibians (vertebrates): frog, toad, and salamander
- Birds (vertebrates): chick
- Mammals (vertebrates): mouse and human

We cannot pursue each of these models in extensive detail, but we can note certain striking regularities. We will find that a major determinant of the cleavage pattern is the amount of yolk present (Figure 19-23).

When a sea urchin embryo has 8 cells, all the daughter cells or blastomeres are approximately equal in size (Figure 19-24). We call such a pattern, in which the whole zygote divides up into equal-sized cells, an **equal holoblastic cleavage.** Sea urchin eggs contain little or no yolk.

19-23 The effect of yolk on cleavage
Cleavage is complete in the Amphioxus egg, which has little yolk. The frog's egg contains considerable yolk on one side (unstippled), and because of it, the cleavage plane across the egg cuts above it. In the chicken egg, which has a large yolk, cleavage is limited to a small area. The letters denote corresponding stages in all three animals.

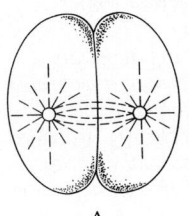 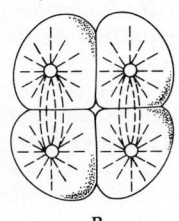 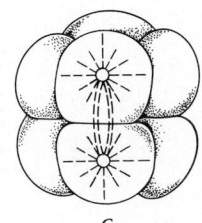 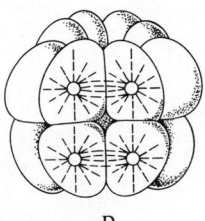 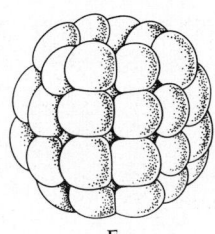 

A   B   C   D   E   F

19-24 Equal holoblastic cleavage in the sea urchin

In this type of cleavage, the zygote divides into cells that are the same size. The size of each cell decreases as cleavage proceeds, so that at the end of cleavage the many-celled embryo is no larger than the zygote after the first cleavage division.

In an 8-cell frog embryo, half of the cells are smaller than the other half (Figure 19-25). This is an **unequal holoblastic cleavage.** Eggs that give rise to such embryos contain more yolk in one hemisphere than in the other.

In birds, reptiles, and some fishes (where egg yolk is plentiful), cleavage is restricted to a tiny disk at one end of the egg, the **blastodisc.** This pattern is termed **meroblastic** (Figure 19-26). In a chicken egg, for example, the zygote includes the whole yolk mass. Cleavage planes could not pass through such a large structure and as a result, embryonic development is restricted to the blastodisc.

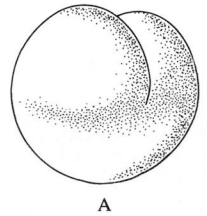 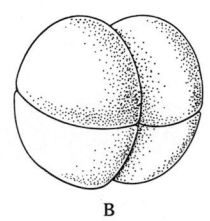 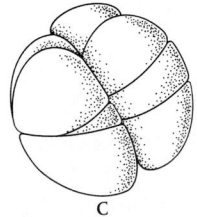 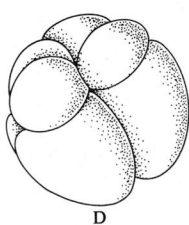 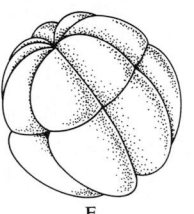 

A   B   C   D   E   F

19-25 Unequal holoblastic cleavage in the salamander

Cells of unequal size are produced as a result of this type of cleavage.

Patterns in mammalian embryos resemble those of reptiles except that the amount of yolk is meager. Cleavage is holoblastic, but blastomeres differ in size from the start. Their number does not increase by a regular doubling sequence but by a more complex progression.

Sea urchins and frogs form blastulas that are hollow spheres. Birds and mammals form a blastocoel that is more disklike (Figure 19-27).

When Does Differentiation Begin?

Experiments with embryos undergoing cleavage tell us something about when differentiation begins. It is possible to shake apart the individual cells and observe their subsequent behavior. In sea urchins and amphibians, each of the separated cells can go on to produce a complete embryo. Indeed, a whole sea urchin can come from a single cell of the 4-celled stage. We would be justified, therefore, in calling such cells totipotent. However, in worms, mollusks, and other species, the first cleavage divides the zygote into cells with unequal capacities for development. These would be multipotent cells.

We conclude from such results that the cells of the early cleavage stages are not irreversibly differentiated. To an extent they are all still alike, and in many instances each cell is capable of complete development—just like the original zygote. This may seem surprising. Should not cleavage be expected to produce partially differentiated cells?

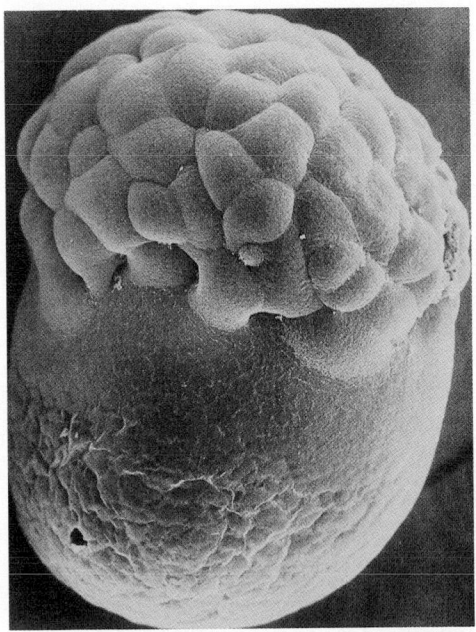

19-26 Micrograph illustrating the meroblastic cleavage pattern of an embryonic fish, Medaka

19-27 Comparison of blastulas derived from three cleavage types

In both the sea urchin and frog, the blastula is a hollow ball of cells. In birds it is a disk of cells that rests on the surface of the yolk.

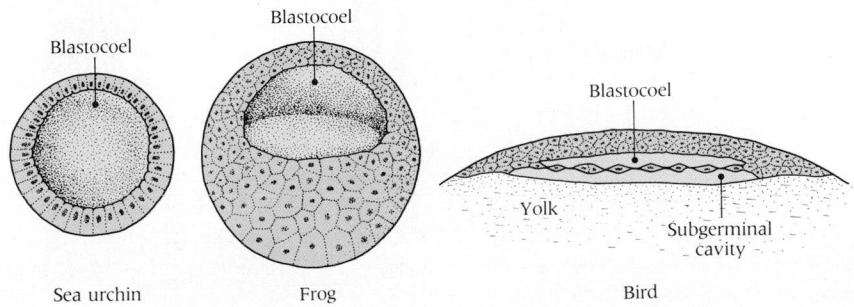

CHAPTER 19   DEVELOPMENT OF FORM IN THE ANIMAL BODY   465

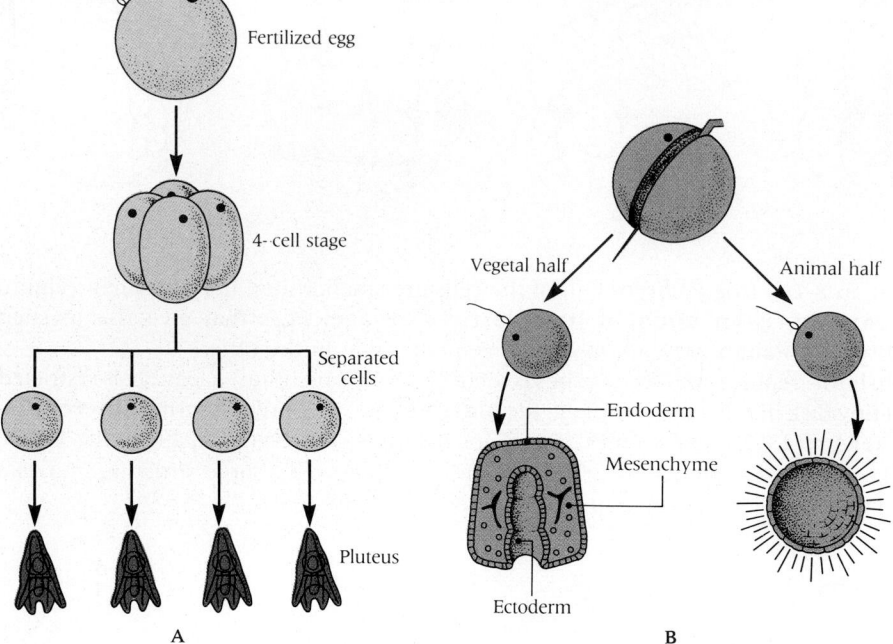

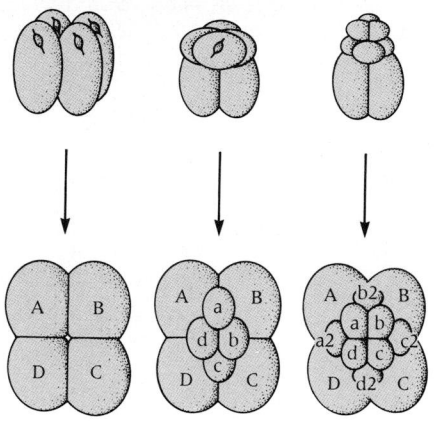

19-28 (Left) Evidence of cytoplasmic differentiation in the zygote
(A) Cleavage of a sea urchin egg along the primary axis produces four cells that, if separated and permitted to develop normally, produce normal pluteus larvae. (B) If the egg is divided across the primary axis into animal and vegetal halves, development does not proceed normally, since neither half contains the full complement of differentiation factors. The vegetal half reaches only the gastrula stage; animal-half development stops at the blastula stage.

19-29 Spiral cleavage in a snail
Two views of the 2-, 8-, and 12-cell embryos are shown. In the top row the orientation of mitotic spindles is evident in several of the cells. The bottom row illustrates the same cells seen from the top and labeled for clarity.

Part of the answer lies in the locus of the plane of cleavage. The first divisions of the zygote cleave the egg along what might be termed its primary axis. Up to the 4-celled stage, isolated individual cells are the same in that each can develop into a complete embryo (Figure 19-28A). If, however, a zygote is experimentally cleaved across the primary axis, the separated halves fail to develop into normal embryos (see Figure 19-28B).

The many natural variations of cleavage strategy include a group in which the planes of the earliest cleavage series do not describe 90° angles. When there is asymmetry of all three spatial axes the pattern becomes evident by the third division and we term it **spiral cleavage** (Figure 19-29). This cleavage pattern is characteristic of many invertebrates.

After differentiation has occurred in the cells of a developing embryo, an experimentally isolated single cell can no longer develop into a whole organism. Nor do the remaining cells develop normally if a cell is removed.

Further development of species with eggs of little yolk generally occurs in one of two ways. The immature embryo either becomes self-feeding, or it is nourished directly by the mother. As noted earlier, self-feeding embryos are called larvae. They are common in many invertebrate animal groups.

The interesting matter of twinning is considered in Box 19-2.

19-30 Relation between the fetal and maternal circulations
The placenta is the link between the mother and fetus. Nutrients and wastes exchange between the capillaries of the two physically separated bloodstreams.

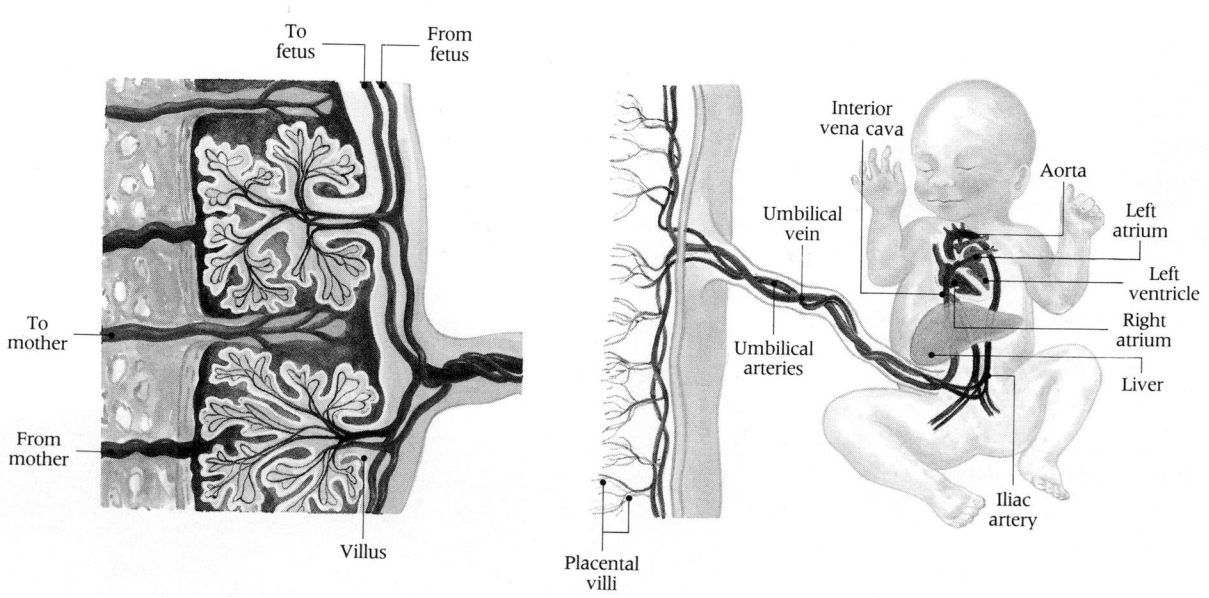

BOX 19-2

TWINNING

The term **twinning** originally meant the simultaneous (or nearly so) birth of two human infants by one mother. By extension, the term came to be applied to multiple births in other animals. Confusion arose as a result.

We now classify twinning in mammals into three categories.

- In one type, twinning occurs in species that usually produce only one ovum at a time. These monovulatory species include humans and other primates, as well as horses and many other ungulates.
- Twinning can also occur in species that normally produce two ova in each cycle. These diovulatory species include marmosets and armadillos.
- Twinning may also result in offspring of a single pregnancy in polyovulatory species that normally produces more than two ova per cycle, including marsupials, rodents, etc.

Regularly in at least one mammalian group, armadillos, and rarely in other species including humans, a single fertilized egg may give rise to four, or even eight identical embryos. This condition, termed **polyembryony,** is not the same as the condition arising in polyovulatory species.

Embryos developing from separate cells that are derived from the same morula possess exactly the same genetic material. This is one of the ways human **identical twins** arise. Evidently this occurs when an early embryo—supposedly at the 32- or 64-cell stage—for some unknown reason becomes separate embryos. If separation occurs again, identical quadruplets result.

Human twins are not always identical. Sometimes two eggs are liberated from the ovaries at the same time, are fertilized by separate sperm, and develop simultaneously. Twins developed in this way are no more and no less genetically similar than separately born brothers and sisters. They are called **fraternal twins.** When ingenious workers separated the blastomeres of 2-cell rabbit embryos and implanted them in foster mothers, they achieved a similar result—two separate embryos, each with a separate placenta and membrane sac, which developed normally. Identical twins must be of the same sex; fraternal twins may or may not be.

Triplets may be all fraternal, all identical, or two identical and one fraternal. More modern terms for identical and fraternal twins are **monozygotic** and **dizygotic** twins, respectively.

Multiple births are relatively rare in humans though the frequency of natural multiple births varies considerably in different ethnic groups. In the United States, twins occur in about 1 birth in 88. Of these, a third are identical twins and two-thirds are fraternal. Triplets occur about once in 8,000 births, and quadruplets once in 700,000. A tendency to produce twins or other multiple births may be hereditary; hence, the proportions may be higher in certain families. This rule applies only to dizygotic twins.

Curiously, dizygotic or fraternal twinning appears to be seasonal. In Finland, the peak "twin season" (for conception) is July. The nadir is January. There is evidence that continuous summer sunlight stimulates a brain structure (hypothalamus) with resulting multiple ovulation (see Chapter 30). There are no data that establish a seasonal incidence in New York State as far as we know.

IMPLANTATION

In mammals, the new embryo must now adapt to its environment in the uterus. In the human female, this environment has been well prepared by hormones, which have caused the lining of the uterus (the **endometrium**) to acquire a new layer of decidual cells rich in proteins, glycogen, and lipids. These are most abundant just before **implantation,** or nidation, of the fertilized egg. In this process, the new embryo, now well along in the cleavage stage, buries or implants itself among the decidual cells.

During the first week after implantation, the embryo lives on the nutrients stored in decidual cells. As the local blood supply increases, the embryo erodes its way

19-31 Gastrulation in the sea urchin
*Stages of development from blastula (**A**) to gastrula (**D**) are illustrated. Invagination produces the archenteron; the blastopore will become the anus.*

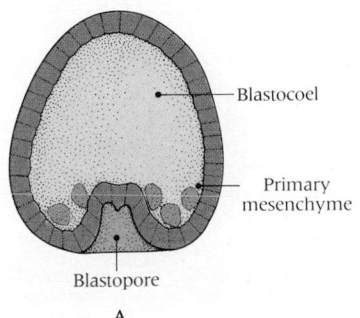

Blastocoel

Primary mesenchyme

Blastopore

A

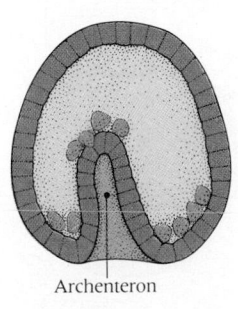

Archenteron

B

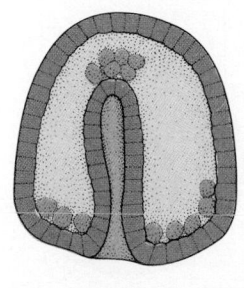

C

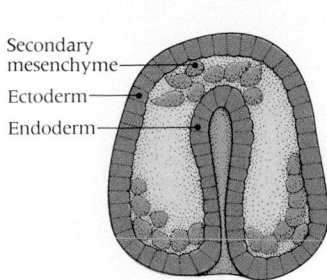

Secondary mesenchyme

Ectoderm

Endoderm

D

into one or more small blood vessels. Implantation is completed when erosion of a maternal blood vessel establishes a parasitic relationship between the embryo, and the uterus (Figure 19-30).

Efforts to cultivate embryos *in vitro* have generally failed because the embryo eventually needs nutrition and other support that can come only from the mother's bloodstream. A fertilized mouse egg may develop in culture to the stage at which it normally implants. But development rarely proceeds further *in vitro*. Indeed, a zygote does not develop even that far.

In true mammals (Eutheria), implantation is followed by the development of the placenta (see Figure 19-30), a disklike organ that attaches to the uterine wall and progressively eliminates the barrier between maternal and embryonic circulatory systems.

GASTRULATION: EMERGENCE OF THE BODY PLAN

The next major development after the blastula stage is the process of **gastrulation,** in which cells migrate into the blastocoel. As a result the embryo becomes a more complex multilayered structure. It is clear even in these early stages of embryogenesis that echinoderms, amphibians, birds, and mammals follow different developmental pathways.

The sea urchin presents the simplest case. Though the embryo may appear symmetrical on superficial inspection, it has polarity, with a vegetal hemisphere of larger cells destined to become the gut and skeleton and an animal hemisphere of relatively smaller cells that will form the rest of the organism.

After the blastula stage, a small pocket develops in the cells at the vegetal end, which begin to migrate into the blastocoel (Figure 19-31). This infolding of one region of the blastula within another is called **invagination.** As the in-pocketing process continues the embryo enlarges and a new cavity called the **archenteron** develops. It will function as the alimentary canal in the larva and it opens to the outside through the **blastopore,** which later becomes the anus. As a result of invagination, a single-layered zygote becomes a "two-ply" embryo, with two cell layers, called **germ layers,** in its wall. The outer layer is the **ectoderm** ("outer skin"); the inner layer is the **endoderm** ("inner skin"). Between the layers are **primary mesenchyme** cells. These and their decendants—secondary mesenchyme—play a directive role in gastrulation. A very similar process occurs in Amphioxus (Figure 19-32).

The process is a little different in eggs with more yolk, such as those of amphibians (Figure 19-33). When these embryos are about 50 hours old, one side of the blastula develops a blastopore. Cells around the blastopore begin to move toward the opening. They pass through it and crowd in toward the inside of the embryonic mass. The most active locus of these movements is the **dorsal lip** of the blastopore, the cells of which vigorously invaginate into the cavity of the blastula. The critical organizational role of certain dorsal lip cells, called the **chordamesoderm,** is discussed in Chapter 20.

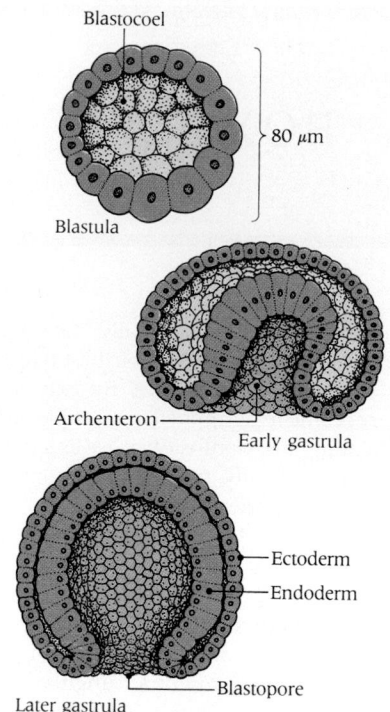

19-32 Gastrulation in Amphioxus
The hollow blastula invaginates to produce a two-layered late gastrula. Note that endoderm cells are larger than those of ectoderm.

19-33 Gastrulation in the frog
*In the blastula stage **(A)**, cells are stationary. As gastrulation begins, cells move from the animal pole inward to form three primary germ layers **(B–D)**. The blastocoel is lost and the yolk plug forms from cells in the vegetal pole.*

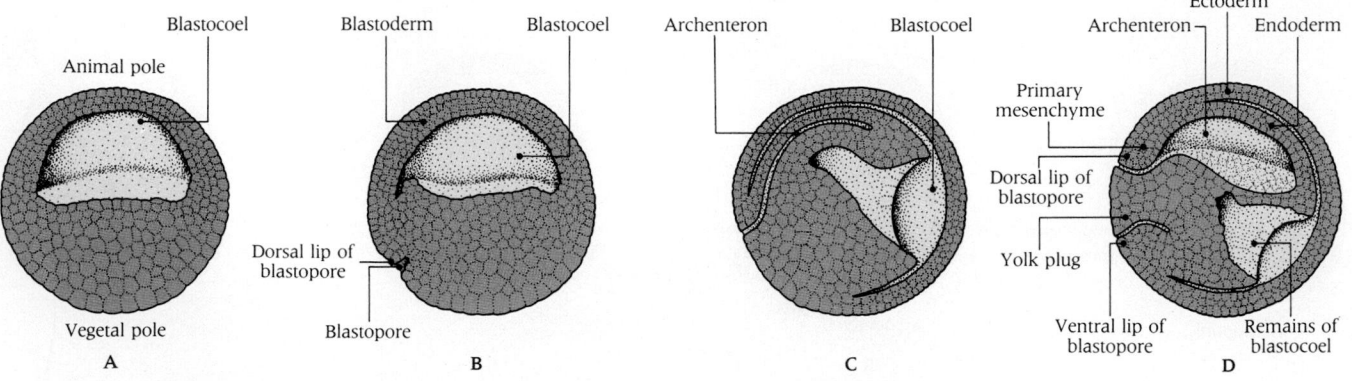

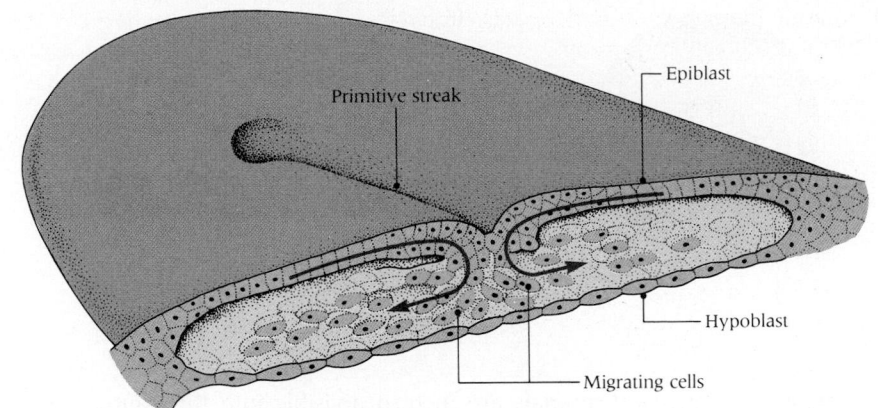

19-34 Gastrulation in the chick
Arrows indicate the direction of cell migration into and through the primitive streak to form mesoderm. The epiblast refers to the ectoderm and the hypoblast is a layer of cells separating the embryo from the yolk.

Primitive streak

Epiblast

Hypoblast

Migrating cells

An important event occurs then. The frog embryo begins to develop a "three-ply" wall. A similar process occurs in the chick embryo, which is awash in yolk. Its blastopore becomes an elongated groove called the **primitive streak** (Figure 19-34).

In all vertebrates, the development during gastrulation of a third cell layer between the ectoderm and endoderm leads to the appearance of the **mesoderm** ("middle skin"), which grows rapidly, gains in complexity, and soon ceases to be structurally simple.

At the completion of gastrulation in the frog the original blastocoel has been obliterated and the archenteron has formed (see Figure 19-33). Soon, a cavity called the **coelom** develops within the mass of mesoderm.

An important distinction is made between species in which the embryonic coelom develops within the mesodermal mass and those in which it develops from endodermal or ectodermal pouches. To understand the meaning of the coelom and its derivation, we should be aware that animal bodies have one of two architectural plans.

■ Animals may exhibit **radial symmetry** (i.e., with body parts arranged regularly around a central line rather than on two sides of a plane). Only two major phyla are radially symmetrical, the coelenterates (hydra, jellyfishes, sea anemones, etc.) and the comb jellies.
■ Animals with **bilateral symmetry** (i.e., the body has two similar halves on either side of a central plane) include the remaining animal phyla, some 28 or so, depending on how they are divided (Figure 19-35).

19-35 Body symmetry in animals
(A) *Parts of a radially symmetrical animal are arranged around a central axis like the spokes of a wheel.* **(B)** *Animals that are bilaterally symmetrical can be roughly divided into right and left halves.*

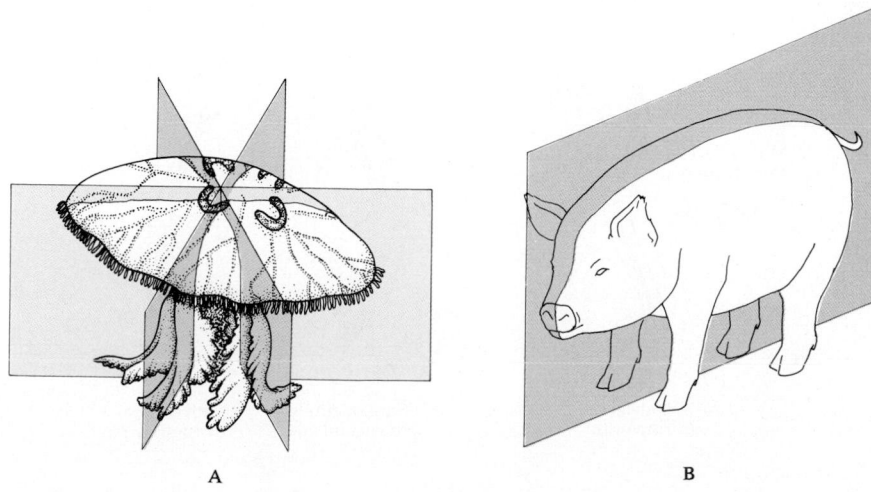

A

B

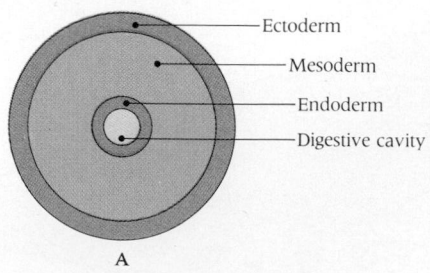

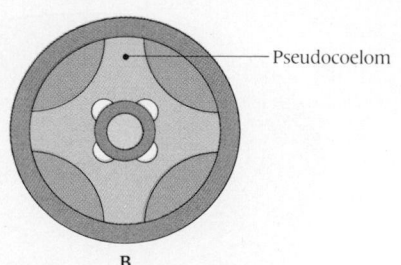

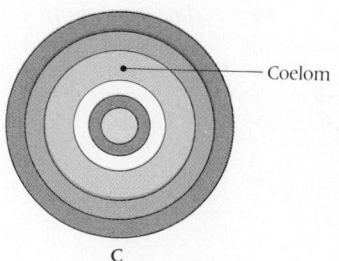

Bilaterally symmetrical animals are in turn divisible into three groups (Figure 19-36).

- **Acoelomates** lack a coelom.
- **Pseudocoelomates** have a body cavity but lack a true coelom.
- **Coelomates** have a true coelom.

What is the importance of the coelom? Gastrulation leads to the development of three tissue layers in all bilaterally symmetrical animals. In coelomates, the mesoderm opens up to form a space that expands and eventually becomes the main body cavity in which internal organs are suspended. A **pseudocoelom** is an internal cavity that does not develop within the mass of mesodermal tissues, but within

19-36 Acoelomate, pseudocoelomate, and coelomate body plans
(A) Animals with an acoelomate body lack a body cavity in the mesoderm layer.
(B) Pseudocoelomate animals have a body cavity that is only partially bounded by mesoderm.
(C) In coelomate animals, the body cavity is lined entirely by mesoderm.

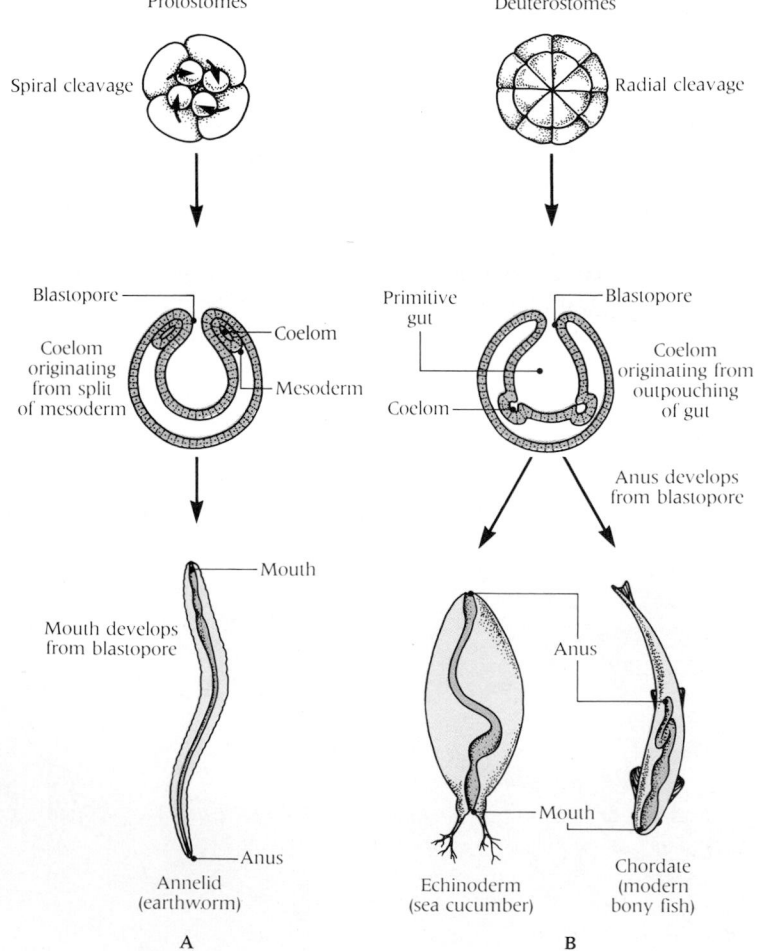

19-37 Developmental fates of the blastopore
(A) In protostomes such as the earthworm, the mouth develops from the blastopore. **(B)** In deuterostomes, the blastopore gives rise to the anus. The differences are of great importance in evolutionary theory.

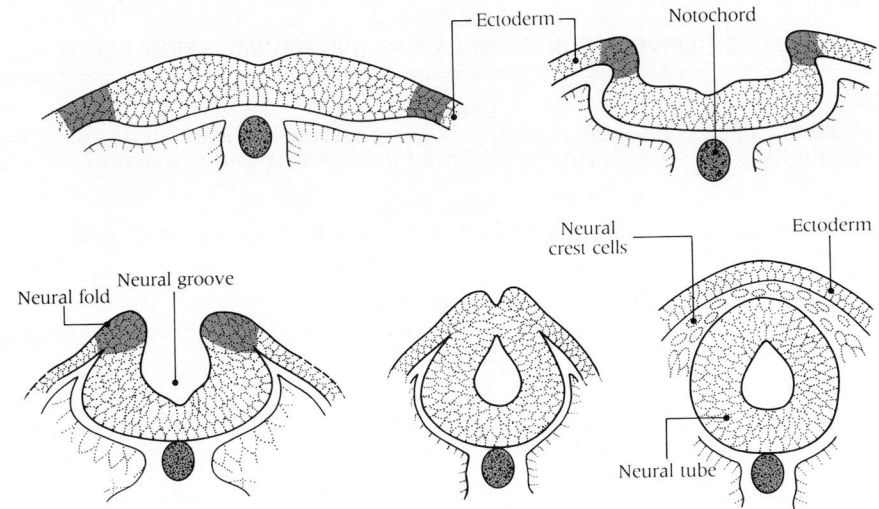

19-38 Neural tube formation in the frog embryo

After gastrulation, ectodermal cells rise to form the neural groove and neural fold, eventually fusing to produce the neural tube. Neural crest cells arise from the ectoderm. Note the position of the notochord.

ectoderm and endoderm.

One further level of classification is of interest. In coelomates, the blastopore has one of two fates (Figure 19-37). This structure enlarges as its cells proliferate and begin to move about.

- In the animals called **protostomes** ("first mouth"), the blastopore eventually turns into the animal's mouth. (This pattern also occurs in acoelomate animals.) The protostomes include the annelids, mollusks, and arthropods.
- In the animals called **deuterostomes** ("second mouth"), the blastopore becomes the anus at the end of the intestine; the mouth arises later as a secondary opening at the opposite end of the intestine. Deuterostomes include the echinoderms, chordates, and several other phyla.

As we will see in Chapter 43, several phyla of deuterostomes arose more recently than the protostome phyla. However, their evolutionary divergence occurred more than 680 million years ago.

In sum, the main feature of gastrulation is a rearrangement of cells by gross morphogenetic movements. A major result is the formation of distinctive cell layers. New cell surfaces come together for further developmental refinements. If morphogenetic cell movements are the key feature of gastrulation, cell interactions are the key to what comes after.

NEURULATION ORGANOGENESIS

Each of the three germ layers formed during gastrulation is a precursor of certain specific portions of the adult body (Table 19-2).

- The endoderm develops into the lining of the alimentary canal, the glands and other associated organs, and the lining of the respiratory system.
- The mesoderm develops into muscle (including the muscular walls of the alimentary and respiratory canals), connective tissues, bones, blood vessels and blood, inner skin layers, and many internal organs such as the heart, kidneys, and gonads (testes and ovaries).
- The ectoderm gives rise to the outer layer of the skin, the nervous system, and parts of sense organs associated with it.

In the course of these developments, the chordate embryo passes through a critical stage marked by the appearance of the notochord and the hollow neural

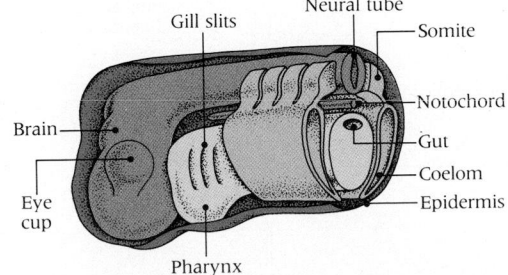

19-39 Neurula stage of development

This view of the front part of an embryo shows the relative locations of the neural tube and notochord. The nervous system displays outpocketings that will develop into the brain and eye.

TABLE 19-2
TISSUES DERIVED FROM GERM LAYERS AND NEURAL CREST CELLS

Ectoderm
Epidermis of the skin, nails, and hair; sweat glands in the skin; all nervous tissue; receptor cells in the sense organs; epidermis of the mouth, nostrils, and anus

Endoderm
Epidermis lining the gut, trachea, bronchi, lungs, urinary bladder, and urethra; liver; pancreas; thyroid gland

Mesoderm
All muscles; blood; connective tissue (including bone); kidneys; testes and ovaries; epithelia lining the body cavities

Neural crest cells
Gill arch elements; jaws; sheaths around motor nerve cells; medulla of adrenal gland; pigment cells (melanocytes) of skin and iris; sympathetic ganglia; ciliary muscles of eye; many bones and cartilages

tube. These events signal the stage of **neurulation.** Note that the chordate phylum includes the vertebrates as one of its subphyla (see Chapter 43). The evolutionary advent of neurulation marked the arrival of the chordates.

When gastrulation is completed and the blastopore is closing, two ridges rise on opposite sides of the dorsal (that is, back) surface of the embryo. These are the **neural folds.** Between them is the **neural groove.** Just below the neural groove, mesoderm in the midline forms the **notochord.** The embryo at this stage is called a **neurula** (Figures 19-38 and 19-39).

19-40 Derivation of the major tissue types
The central evolutionary role of neural crest cells is evident from the many characteristic features of vertebrates that derive from them.

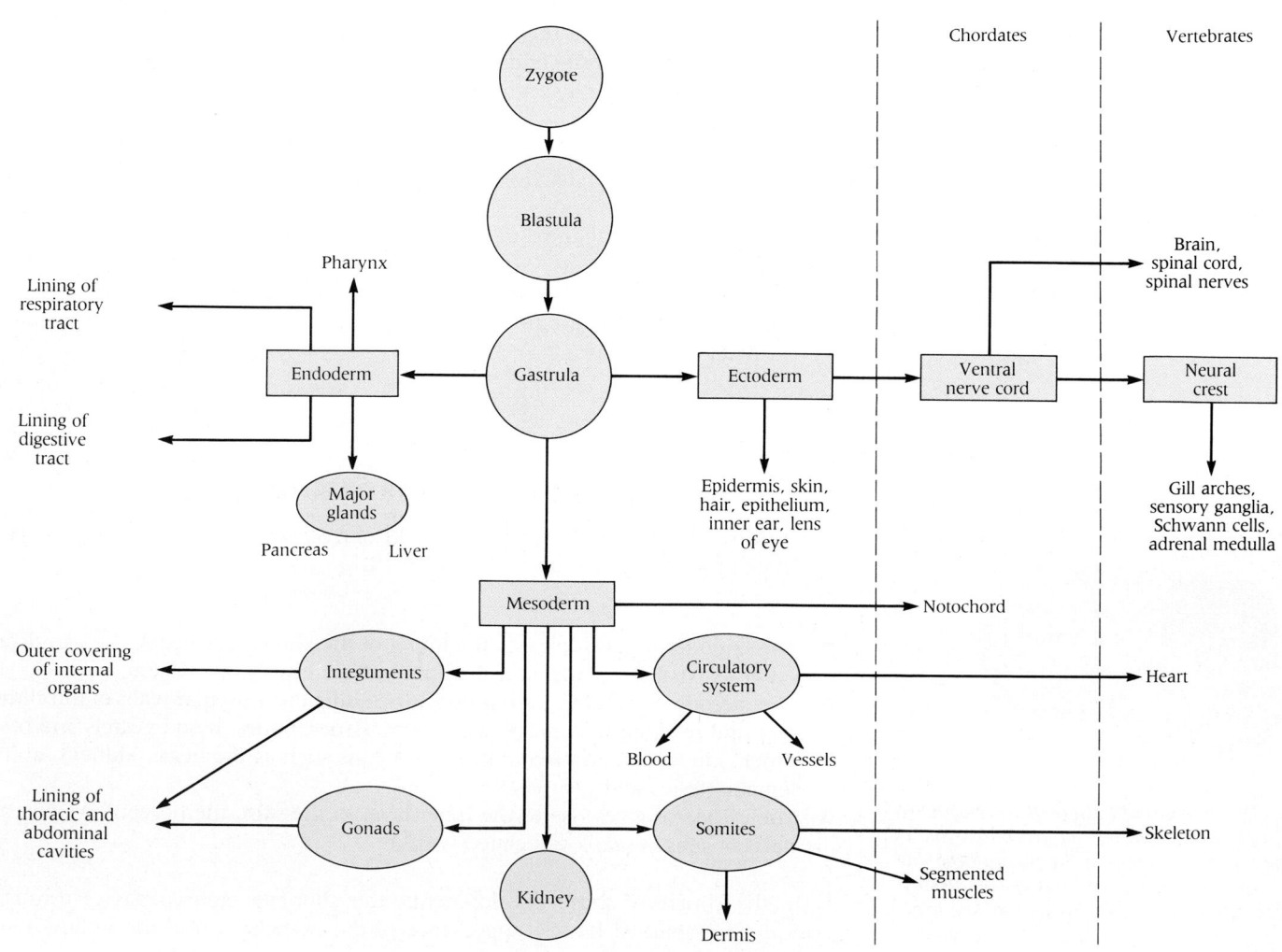

A groove deepens between the neural folds and soon the folds come together to form the neural tube, which is the precursor of the central nervous system. As shown in Figure 19-38, the young embryo contains a group of cells known as the neural crest. These cells are of great importance in the further development of the embryo. Their presence is a unique feature of the vertebrates (Figure 19-40).

Neural crest cells become detached from the tube and migrate great distances, becoming differentiated only when they reach their final destination. This specialization process is presumably influenced by the surrounding tissue. Some cells form sheaths around the motor nerve cells, which are growing out into the rapidly developing muscles. Other cells become sensory nerve cells. Still others become cartilage, or parts of the adrenal glands, eyes, or skin. Neural crest cells wander far more widely than all other embryonic cells and in a sense appear to break the tidy rules that once attributed all subsequent development to orderly dispositions of the three primitive germ layers. Though of ectodermal origin, neural crest cells evidently have potential and will travel.

In Chapter 21, we will trace the later life cycle of the animal embryo and will compare the principles of development in animals and plants.

SUMMARY

The development of a zygote into a multicellular animal requires a precisely controlled program of coordinated changes in levels of gene expression extending over a period of time. In the process of determination, in which regional specialization of tissues within an embryo occurs, multipotent embryonic cells (cells with many possible developmental paths open to them) become irreversibly committed to certain fates. Determination takes place before developing cells have visibly differentiated and before they have acquired the specialized structural or biochemical properties identifying them as mature or developed cells. In response to tissue injury, some cells in adult animals of certain species can dedifferentiate, multiply, and regenerate a damaged organ or limb.

All development includes growth, an irreversible increase in the number and size of the cells of an organism. Growth must be rigorously regulated if an organism is to function normally. Malignant growth (cancer) occurs when a cell and its progeny escape such control.

Growth and differentiation operating together result in morphogenesis, the development of pattern and shape. Morphogenesis also depends critically on cell migration.

The meiotic divisions that produce an egg differ from those producing sperm in that egg cytoplasm divides unequally. Spermatogenesis produces four small sperm cells from each progenitor cell. Meiosis in an egg progenitor produces only one large egg and three tiny, nonfunctional polar bodies. The size of the egg in different species is influenced by the amount of nutrient reserves the developing embryo is likely to need.

A sperm must be highly motile, and its small volume of cytoplasm is packed with mitochondria which power the single flagellum of the sperm. Despite their motility, only a few hundred thousand of the many millions of sperm released during ejaculation reach the egg. Sperm count is an important determinant of male fertility.

When a single sperm cell touches the surface of an egg cell, the acrosome in the sperm head elongates, and its membrane fuses with the plasma membrane of the egg. Rapid depolarization of the egg's plasma membrane after entry of the sperm prevents the egg membrane from fusing with another sperm. Next, an outer egg membrane forms a tough barrier that further blocks the entry of additional sperm. Sperm penetration activates the metabolic machinery of the egg. Zygote formation is completed when the haploid nuclei of sperm and egg fuse.

Cleavage, the rapid cell division which follows nuclear fusion, soon transforms the zygote into a raspberry-shaped mass of cells. The amount of yolk present is a major determinant of the cleavage pattern, which shows distinctive differences in zygotes of different species. In zygotes with little yolk, cleavage produces cells of approximately equal size; where yolk is plentiful, cleavage is largely restricted to one end of the egg. The cells of the early cleavage stage are not irreversibly differentiated. However, individual cells have by this time lost the ability to develop into a whole organism. The locus and orientation of early cleavage planes is crucial to the development of a normal embryo.

In mammals, the embryo at the blastula stage must implant into the uterine wall if it is to survive. It then gastrulates by a process of cell migration into a multilayered structure containing ectoderm, endoderm, and, in many species, mesoderm. Architectural patterns laid down during this stage are important for taxonomic classification. They include the development of radial or bilateral symmetry, the presence or absence of a coelom, and the fate of the blastopore.

Each of the three germ layers created during gastrulation is transformed by organogenesis into early forms of adult body parts. In vertebrates, organogenesis is accompanied by neurulation, the formation of the notochord and neural tube. Neural crest cells formed above the neural tube migrate widely in the embryo and are of crucial importance in further development.

KEY TERMS

acrosome
animal pole
bilateral symmetry
blastocoel
blastodisc
blastopore
blastula
cleavage
coelom
commitment
cortical reaction
determination
deuterostome
differentiation
ectoderm
egg activation
endoderm

fertilization
gametogenesis
gastrulation
holoblastic cleavage
implantation
larva
mesoderm
monozygotic twins
morphogenesis
morula
multipotent
neural groove
neurula
notochord
ontogeny
oogenesis
organogenesis

oviparous
ovoviviparous
ovum (egg)
placenta
pluripotent
polyspermy
primitive streak
protostome
pseudocoelomate
radial symmetry
spermatogenesis
totipotent
unipotent
vegetal pole
viviparous
yolk
zygote

QUIZ QUESTIONS

1. Which of the following is a correct sequence of early stages of embryonic development?
   A. zygote, blastula, morula, gastrula, neurula
   B. morula, gastrula, neurula
   C. gastrula, blastula, neurula, morula
   D. zygote, morula, blastula, neurula, gastrula

2. In which of the following stages is(are) the three germ layers differentiated?
   A. blastula   C. morula
   B. gastrula   D. neurula

   E. zygote   G. C and E
   F. A and C   H. B and D

3. Which of the following does not develop from the mesoderm?
   A. blood
   B. epidermis
   C. bone
   D. muscle

4. Which of the following types of cleavage is associated with embryos having large quantities of yolk?

   A. equal holoblastic
   B. unequal holoblastic
   C. meroblastic
   D. None of the above

5. Which of the following groups of organisms is a radially symmetric deuterostome?
   A. echinoderms
   B. chordates
   C. arthropods
   D. mollusks
   E. All of the above

ESSAY QUESTIONS

1. How does gametogenesis differ between the male and female species in mammals? Contrast the structures and the cell types involved.

2. What mechanisms prevent polyspermy (fertilization of an egg by more than one sperm cell)? Why is this important? How do identical and fraternal twins arise?

3. Define totipotency, multipotency, and unipotency in relation to cellular commitment and differentiation. What cells are totipotent; multipotent; unipotent?

4. What are the principal features of the following stages of development: cleavage, gastrulation, neurulation?

Contrast meroblastic and holoblastic cleavage.

5. How is the endoderm formed in embryos of: amphioxus, frog, sea urchin, and chicken?

REFERENCES AND SUGGESTIONS FOR FURTHER READING

Browder, L. W. (Ed.) (1985). *Oogenesis. Developmental Biology: A Comprehensive Synthesis, Vol. 1.* Plenum, New York.

Most books on eggs follow their development after fertilization. This is one of a surprisingly small number of books about oogenesis—the development of eggs *before* fertilization. A well produced and useful volume.

Browder, L. W. (1984). *Developmental Biol-*

ogy, 2nd ed. Saunders/Holt, Rinehart and Winston, New York.

A basic text that is enjoyable to read. Fine illustrations.

Buss, L. W. (1988). *The Evolution of Individu-*

ality. Princeton University Press, Princeton, N.J.

Original ideas on the evolutionary basis of development, with stimulating thoughts on how gastrulation arose in evolution; the advantages of maternal control of development, sex, ploidy, programmed cell death; the commonness of heterochrony, and of the common patterns of early ontogeny.

Edelman, G. M. (1989). Topobiology. *Scientific American 260* (May), 76–88.

Topobiologists, a term coined by Edelman, ask: How does a one-dimensional genetic code specify a three-dimensional animal? How do genetic changes lead to large evolutionary changes in animal morphology? These questions here get thoughtful answers.

Gans, C., and Northcutt, R. G. (1983). Neural crest and the origin of vertebrates: a new head. *Science 220*, 268–274.

An important contribution to understanding of the importance of neural crest cells in the evolution of vertebrates.

Goldberg, R. B. (1988). Plants: novel developmental processes. *Science 240*, 1460–1467.

The author explains that, although morphologically simple, plants have genetic processes equivalent in complexity to those of animals.

Loomis, W. F. (1986). *Developmental Biology*. Macmillan, New York.

A well-organized and brief summary of current concepts.

McNamara, K. J. (1986). A guide to the nomenclature of heterochrony. *Journal of Paleontology 60*, 4–13.

This discussion of the various aspects of heterochrony defines very clearly the differences between progenesis, neoteny, hypermorphosis, and acceleration.

DEVELOPMENTAL MECHANISMS

Developmental mechanisms
Advances in developmental biology have made it possible for investigators to produce the mutant Drosophila *which has four wings instead of two, as seen in this scanning electron micrograph.*

How, in the course of development, do cells endowed with all of the zygote's genes come to differ from one another in molecular composition, structure, and function? We alluded to this question in discussing how eukaryotes selectively regulate gene expression (see Chapter 18). We tackle it now in the context of development.

To begin with, genes are not segregated in a developing embryo, with one set of genes going to a future leg and another to a future eye. Indeed, there is compelling evidence that this is not the case. The chromosome set is visibly complete in different adult tissues. And cells that, seemingly, have proceeded for a time along one pathway of differentiation will develop into something entirely different if transplanted to another part of the embryo.

In plants, many somatic cells can give rise to whole new plants without participation of the usual reproductive cells. That is why plants can be propagated by cuttings. Both poorly-differentiated and well-differentiated plant cells have this capability. Many animal cells also have the capacity to regenerate cells and organs, as we saw in Chapter 19.

These phenomena confirm that all fully differentiated cells of an adult organism are endowed with all the genes needed to form any tissue. They also deepen the mystery of how different cells with the same genome can assume radically different forms and functions.

MOLECULAR STRATEGIES

Studies of development early suggested the likelihood that only a part of the genome is functioning in differentiated cells. The idea was that different portions of nuclear DNA may somehow be kept from functioning in different types of cells.

IMPORTANCE OF THE NUCLEUS

Indirect Evidence

We have already considered transplantation experiments showing that the style and presence of the umbrellalike cap in *Acetabularia* is dependent on the species of the nucleus (see Chapter 12). A review of the description of that experiment would be useful.

The German embryologist Hans Spemann (1869–1941) first showed that the nucleus is essential in eukaryote development. He constricted a fertilized newt egg with a thread so that only a thin bridge of cytoplasm connected the halves. One of the halves contained the single nucleus (Figure 20-1). Only the half cell containing the nucleus cleaved normally. When he relaxed the constriction enough to permit one of the nuclei arising during cleavage to escape into the other half, an identical twin of the first newt began developing. The second embryo grew more slowly, presumably because it had a late start, but in the end it was identical to the first.

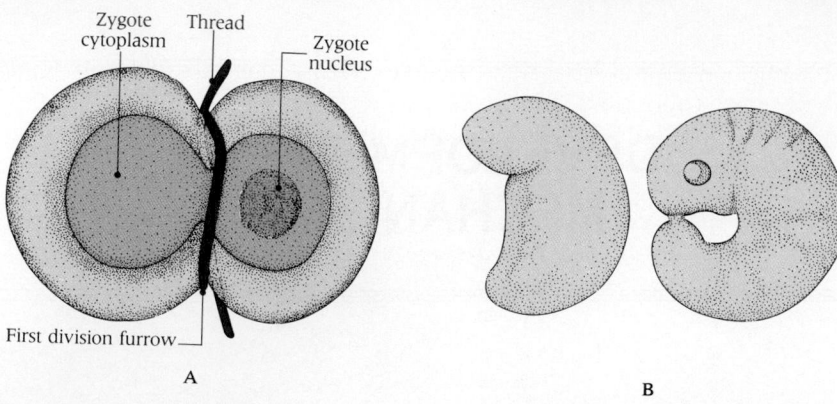

20-1 Importance of the nucleus in development
(A) *A zygote is constricted during the first cleavage division, leaving the nucleus in the right half. Only this half develops normally. If a nucleus of the developing blastula slips into the uncleaved half, cleavage of this half begins.* **(B)** *The result of delayed nucleation of the left half is indicated by the lag in development of the left embryo.*

Remarkably, a normal embryo develops in the retarded half even if the constriction is not relaxed until four divisions have occurred in the cleaving half. By this time, the developing half is composed of 16 cells. These results demonstrated two facts.

- Each nucleus at the 16-cell stage sufficiently resembles the original nucleus to support full development.
- Early cleavages do not restrict nuclear capabilities.

Spemann's early evidence that nuclear capabilities remain intact through the 16-cell stage led in time to more decisive studies in which the nucleus of one cell (embryonic or differentiated) was transplanted to an enucleated cell (one whose own nucleus had been removed). By ingeniously transferring the nucleus of an embryonic frog cell into an enucleated oocyte (Figure 20-2), R. Briggs and T. J. King showed in 1952 that any nucleus from a late blastula can replace the original egg nucleus. In 77% of the cases, an enucleated egg receiving such a nucleus develops normally into a tadpole. Briggs and King concluded that at least to the late blastula stage, no cell nucleus in the embryo has changed enough to keep it from supporting full development.

As we will see, John Gurdon later proved this statement to be just as true for nuclei taken from an adult *somatic* frog cell.

Direct Evidence

Such nuclear transplantation experiments could give only indirect hints of what is going on within a nucleus. They gave no clue whether the key changes of differentiation consist in alterations of gene structure or of gene activity rates—or perhaps neither.

New work marking a milestone of developmental biology soon illuminated the

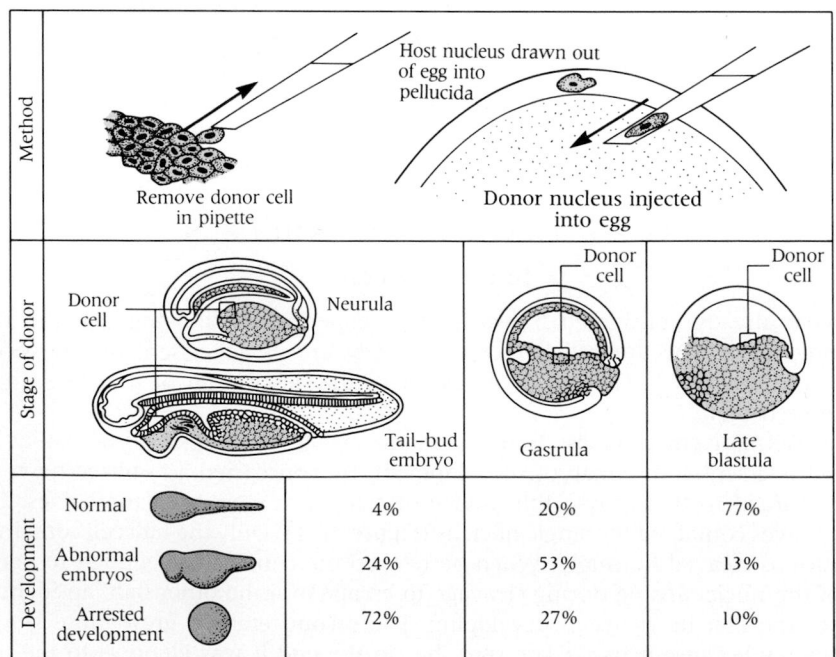

20-2 The Briggs-King nuclear transplantation experiment
The nucleus is removed from an oocyte and a donor cell derived from the tail-bud stage, neurula, gastrula, or late blastula is injected into the oocyte cytoplasm. Nuclei taken from the late blastula stage support the development of a greater proportion of normal frog embryos than do nuclei derived from later stages of development.

question. The new approach began with studies of the significance of the mechanisms regulating gene expression, which we surveyed in Chapter 18. Recall that in eukaryotes these include devices regulating transcription or translation by means of a variety of molecular mechanisms.

- Proteins bind in a site-specific manner to DNA.
- Chromatin undergoes structural changes.
- Patterns of DNA methylation are variable.
- Nonspecific "gene masking" by histones takes place.
- Gene amplifications result in an increase in the number of gene copies present and consequently the number of gene products synthesized per nucleus.
- Chromosomes are eliminated in certain species. For example, in the worm *Ascaris*, 70% of the chromosomal DNA present in the zygote is eliminated in the course of differentiation.

All of these mechanisms facilitate shifts in gene activity at specific moments in the course of development. Together they offer a medley of mechanisms for selectively programming gene action.

GENE REGULATION IN DIFFERENTIATION

Not long ago it seemed likely that we would never understand eukaryotic gene expression as well as we then understood prokaryotic gene expression. There were several good reasons for that view.

- Classical genetic studies are more difficult in eukaryotes than in prokaryotes.
- Most animals and plants have long generation times.
- The mammalian genome is thousands of times larger than that of a bacterium.
- Diploidy complicates the isolation of recessive mutations.
- Eukaryotes cannot easily be studied in the numbers necessary for reliable statistical analysis. Try plating out 10^6 mice on a petri dish!

In the 1970s, recombinant DNA technology eliminated many of these problems in a stroke. It is now possible to isolate any gene from any organism. Sequence studies soon revealed that many eukaryotic genes consist of noncontiguous coding sequences (exons) separated by noncoding intervening sequences (introns). Moreover, as we have seen, DNA sequencing methods revealed the entire nucleotide sequence of many genes and quickly led to the recognition of certain features common to all.

Further Evidence of Nuclear Transplantation

Despite these advances, it became obvious that DNA sequence studies alone would not explain the vagaries of gene expression, notably its on-and-off character. This problem required functional studies of active genes. A giant step was taken in this direction when gene transplantation methods were applied in the study of development.

In one approach, whole nuclei and purified DNA (containing genes of various kinds) were injected directly into anucleate oocytes[1] (i.e., cells in which the nuclei had been destroyed or removed) of *Xenopus laevis*, a South African clawed frog well suited to life in the laboratory. In this way, John Gurdon of Oxford University tried to answer a central question: If diverse specialized adult cells still contain the same complete genome of the original zygote, is cell differentiation accomplished, not by the loss of some genes and the retention of others, but by patterns of differential regulation of the same full gene set?

That the latter is the case was shown by transplantation of the nucleus of a *Xenopus* somatic (diploid) cell, such as a tadpole intestinal epithelial cell, into an

---

[1] The oocyte, we recall, is a precursor in oogenesis. Oocytes grow and accumulate in mother frogs. Following hormonal stimulation they mature into eggs. Oocytes are useful research objects because they are large (1.2 mm in diameter), resistant to manipulation, and they can live for weeks in culture media.

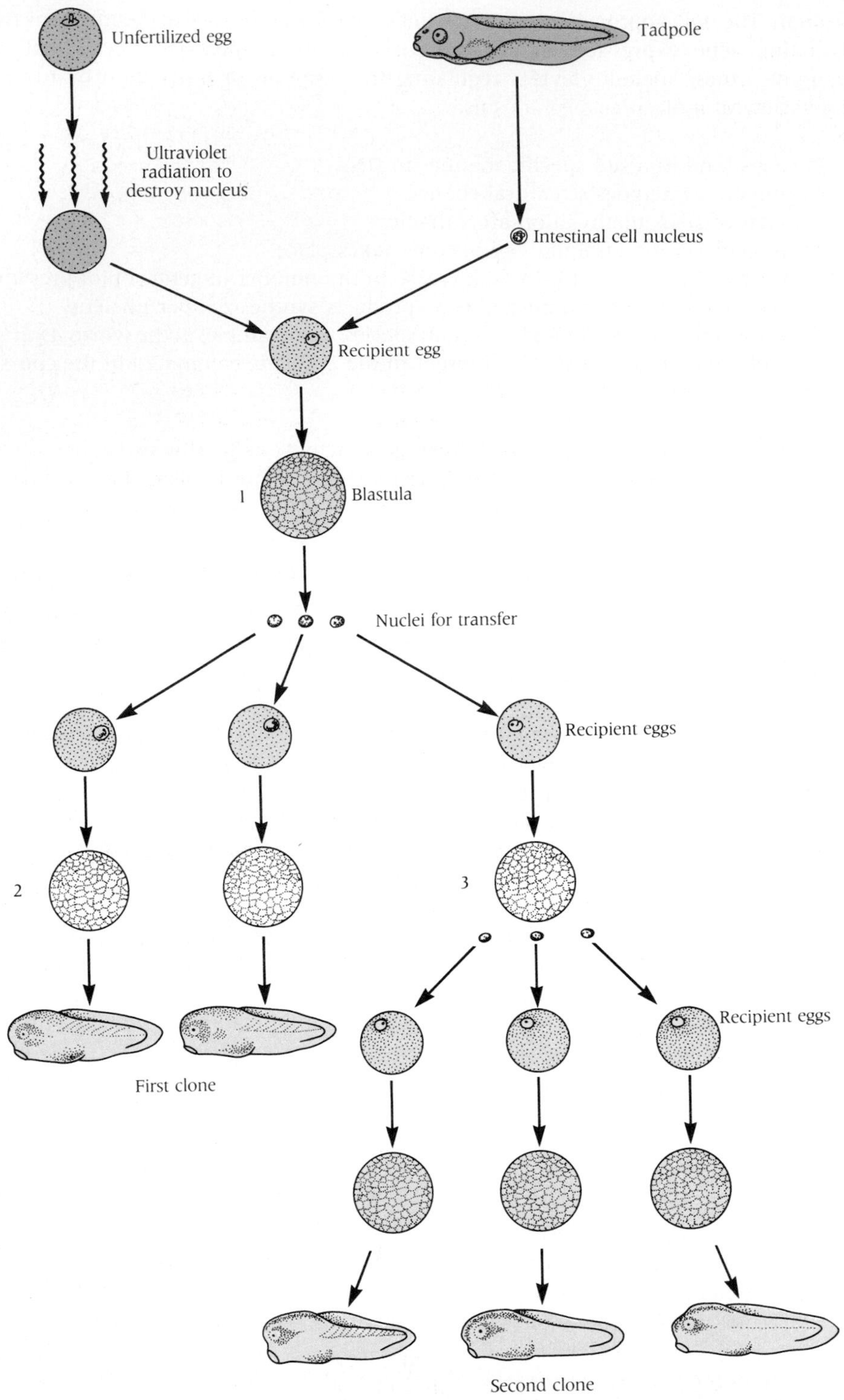

Unfertilized egg

Tadpole

Ultraviolet
radiation to
destroy nucleus

Intestinal cell nucleus

Recipient egg

1 Blastula

Nuclei for transfer

Recipient eggs

2

3

First clone

Recipient eggs

Second clone

20-3 The nucleus of a differentiated cell can support normal development
The nucleus of a tadpole intestinal cell is transplanted into an unfertilized and enucleated Xenopus *egg. Nuclei are then removed from cells at the blastula stage (1) and transplanted into unfertilized eggs, which subsequently develop into normal blastulas (2) and later into normal tadpoles and toads. Nuclei from blastulas can be transplanted into new eggs, giving rise to another generation of frogs (3).*

20-4 Tadpole stage of Xenopus laevis development

egg cell, the (haploid) nucleus of which had earlier been destroyed (Figure 20-3). This process is termed **cloning.**[2] In successful experiments, the diploid egg cell resulting from this procedure gave rise in time to a normal, fully formed, feeding-stage tadpole and then to a frog, one whose chromosomes came entirely from the donated intestinal epithelial cell nucleus. All the genes necessary for the frog's complex life cycle are present in a single specialized somatic cell.

This experiment also revealed something else: nuclear changes occurring in differentiation are not irreversible. The pattern of gene expression typical of the trans-

[2] Cloning can be defined in two ways, and so confusion may arise. In this experiment, the nucleus of a somatic cell containing the full genome of one individual—a frog in Gurdon's experiments—is implanted into an enucleated cell and yields a copy of the organism donating the nucleus. The cloning process connected with recombinant DNA technology (see Chapter 18) employs fragments of passenger DNA representing only a fraction of the genome of the organism from which they are derived.

planted nucleus must have been reprogrammed in a way that permitted new cell specialization patterns.

When the transplanted nucleus was taken from an adult gut cell, it nonetheless gave rise to a swimming tadpole with specialized cells of all sorts—including blood, muscle, and eye lens (Figure 20-4). Each of these cells actively expresses the genes for a characteristic protein—hemoglobin, myosin, and crystallin, respectively. The questions now needing answers were: Are gene-regulating substances responsible for the switch-over? If so, what are they and when do they arise in development?

Differentiated cell functions like these become apparent only after an egg has undergone some 10 to 15 cell divisions. If gene-regulating substances are present in the unfertilized oocyte, one might hope to find them, since frog oocytes can be cultured without dividing for many days after the insertion of a nucleus.

Are the genes of a transplanted nucleus reprogrammed (i.e., turned on or off) in the oocyte? Transplanted somatic cell nuclei do not divide, but they actively synthesize RNA—far more actively, in fact, than they did in the somatic cell from which they were taken. Indeed, three major RNA types—mRNA, rRNA, and tRNA— are briskly transcribed from the DNA of these nuclei in preparation for intense protein synthesis associated with early development. Hence, these genes are in fact reprogrammed.

In searching for the responsible oocyte ingredients, Gurdon carefully studied the proteins actually synthesized under direction of genes in the injected nucleus. He did so by comparing the proteins synthesized by *Xenopus* oocytes and by *Xenopus* kidney cells. Some proteins were found to be synthesized by both types of cells. Gurdon called these the **housekeeping proteins** that are required by all cells to build cytoskeletons, membranes, ribosomes, and other organelles. However, he found other proteins that are specific to either oocytes or to kidney cells. These, in Gurdon's terminology, are **luxury proteins,** encoded by genes expressed only in particular tissues.

To determine if it is only the level of gene expression that changes in differentiation, he injected *Xenopus* kidney cell nuclei into the oocytes of the newt *Pleurodeles waltlii,* an entirely different species whose proteins are easily distinguished from those of *Xenopus* by two-dimensional electrophoresis (Figure 20-5). Each newt oocyte was allowed to accumulate newly synthesized mRNA in its cytoplasm. It was then exposed

20-5 Reprogramming oocyte genes
Oocytes are removed from the newt Pleurodeles *(1a–c.).* Xenopus *kidney cells are cultured and their nuclei are injected into a newt oocyte (2a–d,3). After several days the oocyte contains mRNAs in the cytoplasm (4). Radioactively labeled amino acids are added (5), and protein synthesis proceeds for 6 hours (6). Radioactive proteins are extracted (7), and then separated by electrophoresis. The resulting radioautograph (8), reveals that* Xenopus *oocyte genes, normally inactive in kidney cells, are turned on by factors in the newt oocyte cytoplasm.*

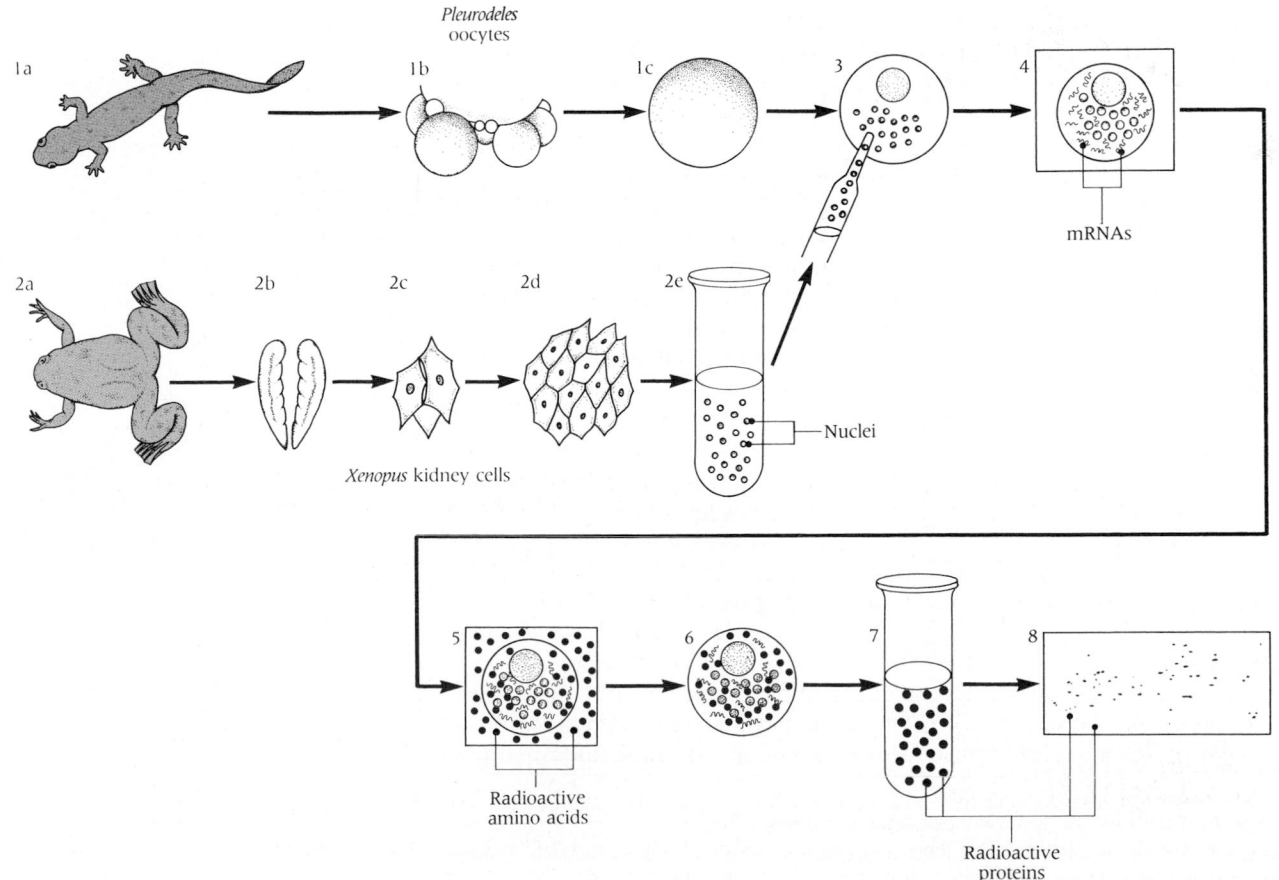

Pleurodeles oocytes

Xenopus kidney cells

mRNAs

Nuclei

Radioactive amino acids

Radioactive proteins

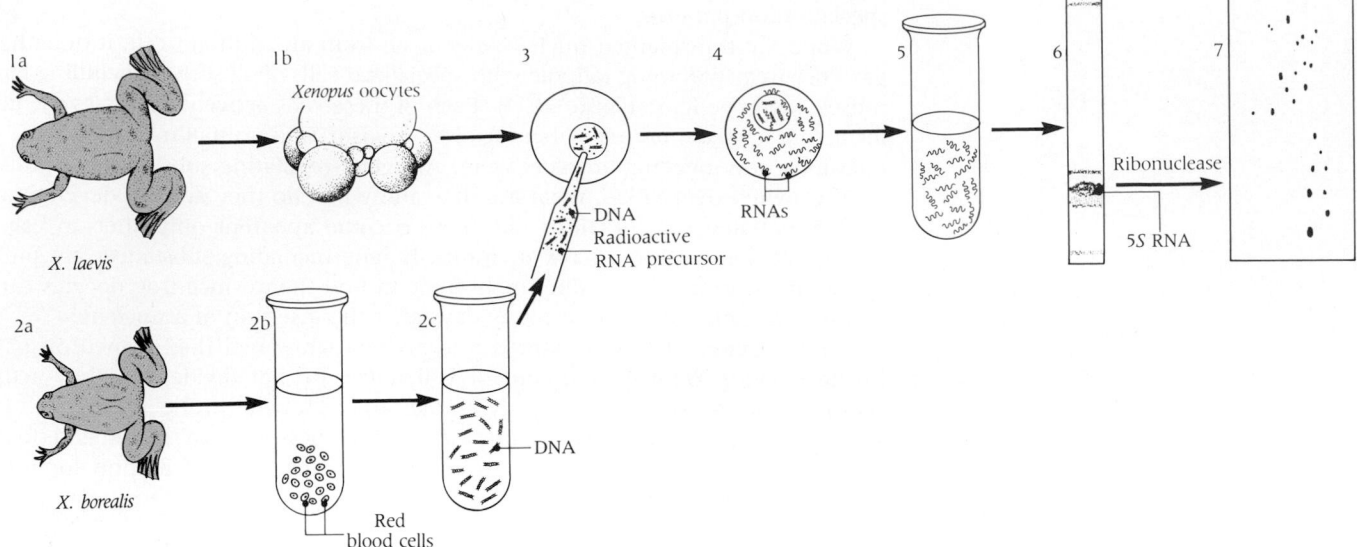

to radioactive amino acids to label all new proteins specified by the new mRNAs. Electrophoresis of the resulting radioactive proteins showed that the genes encoding several *Xenopus* housekeeping proteins had been expressed. But the genes for kidney-specific proteins were not expressed! More important, some of the newly synthesized proteins were characteristic of the *Xenopus* oocyte. These oocyte-specific frog proteins had not been synthesized in the kidney nucleus before transplantation. Hence their genes must have been activated by the cytoplasm of the newt oocyte. In other words, transplanted nuclei had been reprogrammed by normal components of newt oocytes to conform to an oocyte-specific pattern of gene expression. (Note that toads and newts are closely enough related to permit toad genes to "read" the regulatory signals provided by newt cytoplasm.)

The Evidence of Gene Transplantation

Nuclear transplantation seemed an unpromising method for further studying such cytoplasmic signaling substances. A more promising plan was to study the activation (or inactivation) of a single gene by observing changes in its activity after insertion into an oocyte (Figure 20-6). Accordingly, Gurdon injected into an oocyte many copies of a single gene (in the form of pure DNA) in a way that allowed gene products to be synthesized correctly and abundantly.

The results were striking.

- There was highly accurate transcription of genes normally encoding 5*S* rRNA[3] in the frog *Xenopus borealis* when its DNA was injected into *Xenopus laevis* oocytes. To appreciate this experiment, we should recall from Chapter 17 that in addition to 18*S*, 5.8*S*, and 28*S* rRNAs, which come from large molecules of so-called pre-rRNA, eukaryotic ribosomes also contain a short (120-nucleotide) 5*S* rRNA, the source of which is unrelated to the pre-rRNA gene. Almost all eukaryotic cells contain 5*S* rRNA genes, and their product, 5*S* rRNA, is virtually identical in these cells (and in mitochondria and chloroplasts). However, most animals have several sets of 5*S* rRNA genes, the nucleotide sequences of which differ slightly.[4] These and other small differences made it possible to recognize the 5*S* rRNA molecules of the two species.

- In the reverse experiment, *X. laevis* oocytes injected with purified *X. borealis* DNA synthesized RNA with the properties of *X. borealis* 5*S* rRNA. It is striking that RNA polymerase III, the enzyme that synthesizes 5*S* rRNA in *X. laevis* oocytes, could recognize the initiation and termination signals of the injected DNA, which define the stretch of DNA to be transcribed. The abundance of correct RNA products rapidly transcribed from the injected genes indicated that

20-6 DNA-injection experiment
(1) Oocytes are isolated from Xenopus laevis. *(2) Red blood cells from a related frog (X. borealis) provide DNA that contains the gene for 5*S *RNA (3). This DNA is injected into the oocyte along with radioactive RNA precursors, and radioactive RNA is synthesized (4). (5) Following extraction from the cell, the radioactive RNA is analyzed by electrophoresis and the 5*S *RNA is identified (6). Ribonuclease digestion of this RNA reveals a pattern of fragments that match those of 5*S *RNA from X. borealis (7), thus confirming that the oocyte is capable of transcribing RNA from the injected foreign DNA fragments.*

[3] Note again that the symbol *S* stands for Svedberg unit, a measure of sedimentation behavior in an ultracentrifuge. This in turn is a rough measure of molecular size.

[4] In addition, many other types of small RNA molecules are found in eukaryotic cells, though no functions have yet been assigned to them.

TABLE 20-1
COMPARISON OF RESERVES IN *XENOPUS* OOCYTE AND TYPICAL SOMATIC CELL

Component/Property	Oocyte	Somatic Cell
Cell diameter, μm	1500	10–50
Ribosomes, pg/cell	4×10^6	18
Nuclear DNA, pg/cell*	12	6
Mitochondrial DNA, pg/cell	3×10^3	0.06
5S rRNA and tRNAs, pg/cell	5×10^4	3
Yolk, % dry weight	45	0

Note: A pg (picogram) equals 10^{-12} grams (see Box 2-1).
* DNA levels are higher in oocytes than in somatic cells because amphibian oocytes contain many extrachromosomal nucleoli, each containing the DNA that encodes the massive stores of rRNA.

the polymerase had correctly transcribed only those parts of the DNA directly encoding 5S rRNA. Thus it does appear that oocytes contain regulatory substances capable of reprogramming the expression of genes in injected nuclei.

We call these molecules **cytoplasmic determinants.** They are molecules in the egg. After the egg is fertilized and begins to divide, they influence the developmental fates of embryonic cells. What could these substances be? The cytoplasm of unfertilized eggs (mature oocytes) is a storehouse of these items (Table 20-1).

- Developmental instructions (maternal mRNAs and various RNA transcripts of repetitive DNA sequences)
- Machinery of protein synthesis (ribosomes, tRNAs, etc.)
- Extranucleolar DNA that encodes all forms of RNA (mRNA, tRNA, hnRNA) other than rRNA.
- Food reserves (yolk, glycogen, lipids, etc.)
- Proteins that will be needed during cleavage (histones, tubulin, DNA and RNA polymerases, etc.)
- Machinery for generating energy (mitochondria)

All are used until their functions are taken over by the embryo's own genes.

The presence of this rich inventory of possible "determinants" in mature oocytes suggested that cell determination might begin even before the egg is fertilized. Indeed, Paul Gross found that an unfertilized sea urchin egg does contain such cytoplasmic determinants in inactive form, which become active when the egg is fertilized and divides. Only then do they begin to influence gene expression.

Recent work has identified at least three classes of cytoplasmic determinants, though there are almost certainly many more.

- One class consists of the **maternal mRNAs** that direct the synthesis of histones.
- A second determinant that influences the formation of mesoderm in frog embryos is a close relative of **transforming growth factor-β** (or **TGF-β**), which we will later see (in Chapter 28) is a protein already well known as a regulator of cell growth and inflammation.
- A third is a remarkable protein called **transcription factor IIIA** (or **TFIIIA**), which activates the genes for 5S ribosomal RNA.

The TFIIIA story is an exciting chapter of the current scene that bears brief telling. Haploid *Xenopus* cells contain some 20,000 copies of one form of 5S rRNA genes—the so-called "oocyte" 5S rRNA gene—and a smaller family of the very slightly different "somatic" 5S rRNA genes. In oocytes, both kinds of 5S rRNA genes are expressed. After gastrulation, only somatic 5S rRNA genes are expressed. Why should this be? What turns on oocyte 5S rRNA genes in oocytes and turns them off in somatic cells? Donald Brown of the Carnegie Institution of Washington demonstrated that a specific nonhistone protein called transcription factor IIIA (symbolized TFIIIA) is responsible. Only when it binds to a control sequence in the middle of the 5S rRNA gene can the gene be transcribed. Interestingly, TFIIIA has a lower binding affinity for the oocyte 5S rRNA gene than for the somatic 5S rRNA gene. However, TFIIIA is present in great excess in oocytes and thus all of its 5S rRNA genes are active. Inactivation of the oocyte gene occurs when the amount of TFIIIA becomes low relative to gene number (Table 20-2). We now know that,

TABLE 20-2
RELATION OF TFIIIA ABUNDANCE TO TRANSCRIPTION RATES

Stage of Development	Number of Free TFIIIA Molecules per $5S$ rRNA Gene	Ratio of Transcription Efficiency (Somatic Gene/Oocyte Gene)*
Oocyte	10^7	4
Mid-blastula	10	50
Gastrula	2	1000
Somatic cell	0.2	1000

*Ratio of somatic $5S$ rRNA to oocyte $5S$ rRNA synthesized per gene. In *X. laevis,* there are 50 times as many oocyte $5S$ rRNA genes as somatic $5S$ rRNA genes.

remarkably, TFIIIA also binds to the product of the oocyte-specific $5S$ RNA gene—the $5S$ RNA itself. The complex of TFIIIA and $5S$ RNA is stored as particles, which tend to turn off $5S$ RNA synthesis. In other words, early in oogenesis, when there is an excess of TFIIIA over $5S$ RNA, most of the factor binds to $5S$ RNA genes and *activates* transcription. Later in oogenesis, TFIIIA is sequestered in a $5S$ RNA storage particle, which *inactivates* transcription. We see that the availability of this factor (and others like it) has a critical role in determining which gene is transcribed in different tissues and at which times. This suggests that one driving force in development may be the selective activation and inactivation of specific genes.

LOCALIZATION AND ASYMMETRIC CELL DIVISION

Studies of mRNAs and other cytoplasmic determinants of gene transcription revealed a significant fact: most of these molecules are unevenly distributed in the egg. As a result, the two daughter cells of the first cleavage have them in differing amounts. If these molecules are in fact macromolecules and thus "stay put" in the cells that contain them—it can be concluded that they are distributed in different concentrations to cells formed during cleavage. This circumstance underlies an important principle of developmental biology: maternal factors or cytoplasmic determinants that are localized or segregated into particular cells during cleavage can, in principle, specify the fates of the recipient cells.

We now realize that a number of factors (Figure 20-7) can account for the differing fates of the cells arising in early cleavages. In brief, they are the following:

- A cytoplasmic determinant localized near one pole of an egg remains in that polar location during the course of cleavage (Figure 20-7A).
- A determinant that is uniformly distributed throughout the egg becomes progressively localized during early cleavage stages (Figure 20-7B).
- Unequal or asymmetric cell divisions occur, with resulting partitioning of cytoplasmic determinants (Figure 20-7C).
- External influences may cause determinants to become localized within the fertilized egg (Figure 20-7D).
- Incident light can also cause polarization (Figure 20-7E).

The localization of determinants is a first step in a series of events that ultimately lead to overt cell differentiation or differential gene expression. However, it has been said that determinants cause a bias, not a commitment. What do we mean by commitment?

COMMITMENT: HOW DO CELLS "DECIDE" TO DIFFERENTIATE?

Consider a system used to study differentiating muscle cells. Fragments of embryonic muscle-forming tissue are easily dissociated into individual muscle precursor cells called **myoblasts.** When the resulting cell suspension is cultured, the myoblasts settle to the bottom of the dish and fasten themselves to the glass (Figure 20-8A). There they grow into a continuous sheet. A dramatic event then takes place as seemingly nondescript growing immature cells are transformed into large elongated **myotubes** containing hundreds of nuclei that are lined up within a common cytoplasm. Soon long fibers appear, which begin in a few days to contract spontaneously like a mature muscle. As shown in Figure 20-8B, microscopy reveals the typical

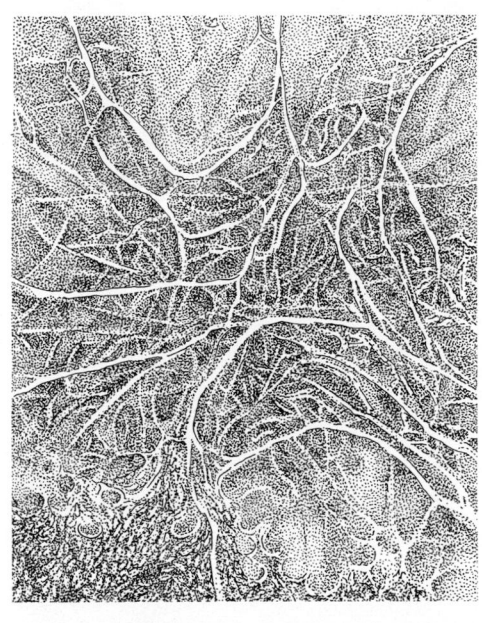

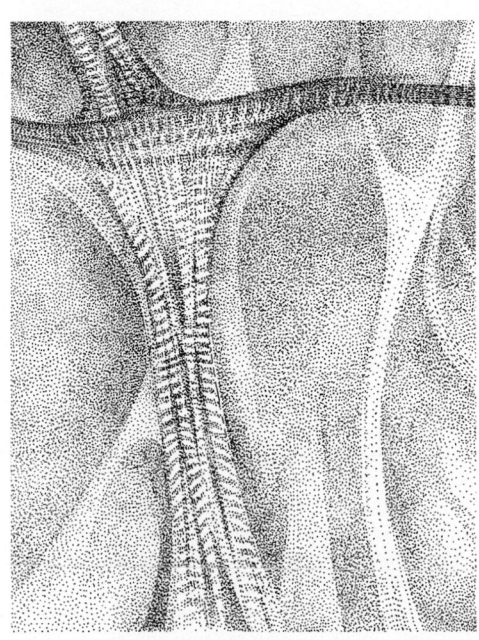

20-7 (Right) *Cytoplasmic localization and asymmetric cell division*
(A) *Localization of cytoplasmic factors in the pole plasm of* Drosophila *eggs is retained during cleavage.* **(B)** *In ctenophores, as development proceeds there is progressive localization of cytoplasmic materials to generate ciliated cells or light-producing cells.* **(C)** *Asymmetric cell division in nematodes results in the unequal partitioning of cytoplasmic factors and cell specialization.* **(D)** *The point of sperm entry determines the dorso-ventral polarity of the embryo. The black dots represent prospective muscle tissue and mark the dorsal surface of the embryo.* **(E)** *Light determines polarity of seaweed eggs, in which the root forms opposite the direction of light.*

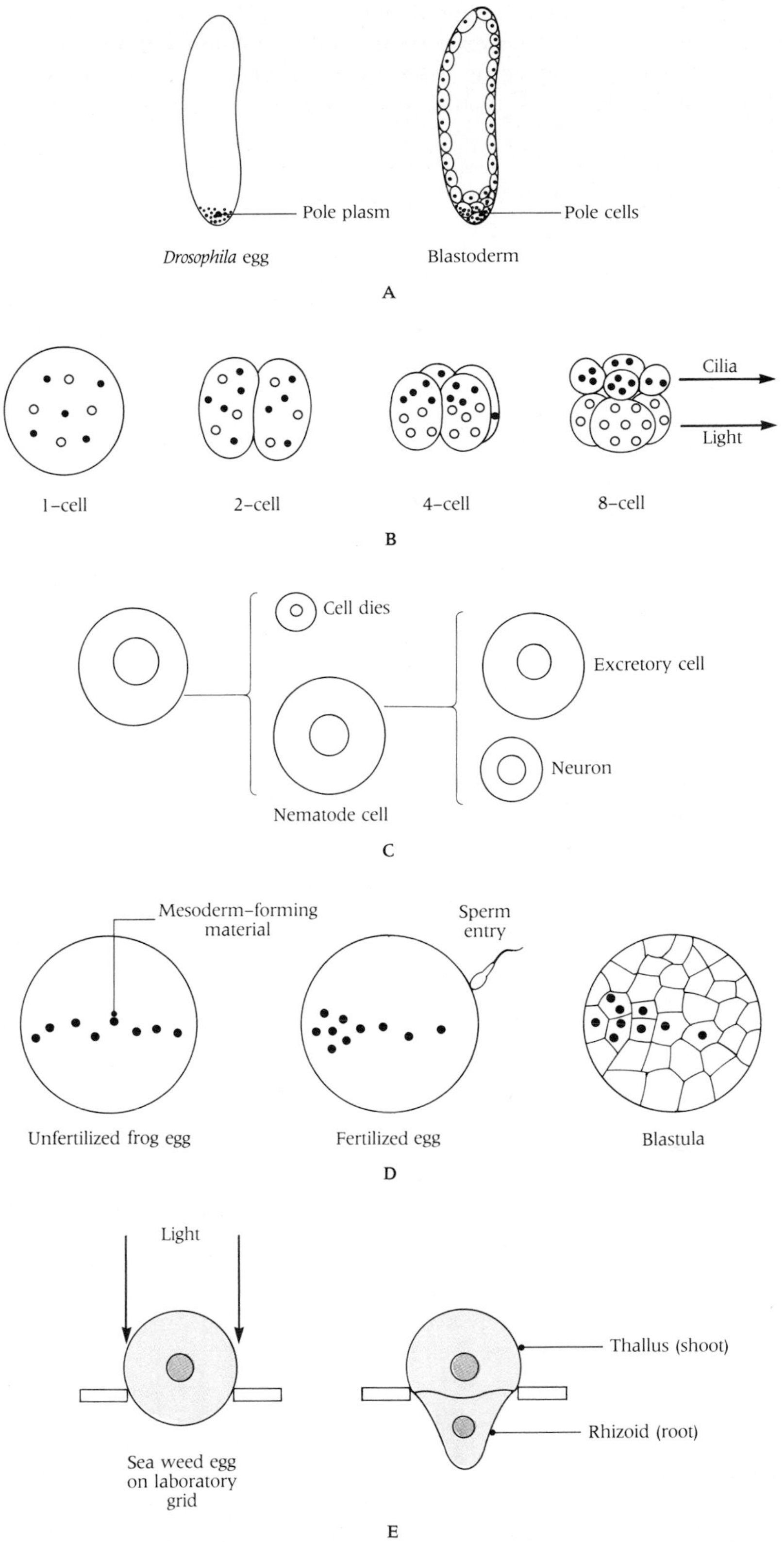

A

Pole plasm — Pole cells

Drosophila egg — Blastoderm

A

Cilia — Light

1–cell — 2–cell — 4–cell — 8–cell

B

Cell dies — Excretory cell — Neuron

Nematode cell

C

Mesoderm–forming material — Sperm entry

Unfertilized frog egg — Fertilized egg — Blastula

D

Light

Sea weed egg on laboratory grid — Thallus (shoot) — Rhizoid (root)

E

B

20-8 Muscle cells in culture

cross-striated pattern of differentiated muscle cells (see Chapter 7). In addition, biochemical analysis shows they contain the contractile proteins of muscle.

Here is a strong hint that when some sort of stimulus initiates a coordinated differentiation program a single cellular event triggers the program. Pending further clarification, this event is termed **commitment.** The cell is now committed—unipotent rather than pluripotent.

Three major questions now arise: When do the cells descending from a totipotent zygote acquire a restricted developmental potential? Is this determination a reversible event? And what cellular event initiates the differentiation program? To sort out the biochemical essentials of commitment, one can introduce specific metabolic inhibitors into cultures of cells exposed to a differentiation stimulus. Surprisingly, myoblast differentiation proceeds normally even when DNA synthesis is almost completely blocked. However, requirements for mRNA and protein synthesis appear absolute.

DIFFERENTIATION AND THE CAPACITY FOR CELL DIVISION

Many normal cells have lost the ability to divide. Examples are neurons, muscle cells, and the mammalian red blood cell, precursors of which lose their nucleus and other organelles (Figure 20-9). In fishes and birds, the nucleus of the red blood cell is retained but it is dormant, no longer able to support cell division.

Red blood cells are, of course, highly specialized structures designed for efficient transport of oxygen. The more specialized a cell becomes, the less able it is to divide. In contrast, the less sophisticated and unspecialized cells of connective tissue are ready to divide at any time.

Study of a dormant cell nucleus—one that cannot support cell division—supports the idea that factors in the cytoplasm influence gene activity. When a human macrophage (a large scavenger cell that can engulf foreign particles) is hybridized with a chicken red cell (a nucleated cell) by the methods described in Chapter 18, the previously inactive red cell nucleus begins making mRNA—like the macrophage nucleus. Evidently, cytoplasmic components contributed by the macrophage somehow derepress a long-silent portion of the red cell's genome.

NEW METHODS FOR TRACKING GENES DURING DEVELOPMENT

In recent years, development has been studied in specific organisms with great profit. To illustrate this work—and to delineate several important principles—we briefly discuss three (of many) such models. Interestingly, each was a favorite study object of biologists in other areas before it was chosen for developmental work (see Table 19-1). The familiar mouse, *Mus domesticus* or *Mus musculus,* has been studied by biologists of all kinds. The nematode *Caenorhabditis elegans* was a favorite object of neurobiological study. The fruit fly *Drosophila melanogaster* was and is a workhorse of chromosomal genetics.

TRANSGENIC MICE: MAKING MUTANTS BY GENE TRANSFER

In 1960 Beatrice Mintz of the Institute for Cancer Research in Philadelphia boldly attacked the problem of gene regulation in development by assembling artificial early embryos from the cells of different embryos. Her technique yielded whole mature mice containing two genetically distinguishable cell populations. Built-in cell markers enabled her to trace their fates during development.

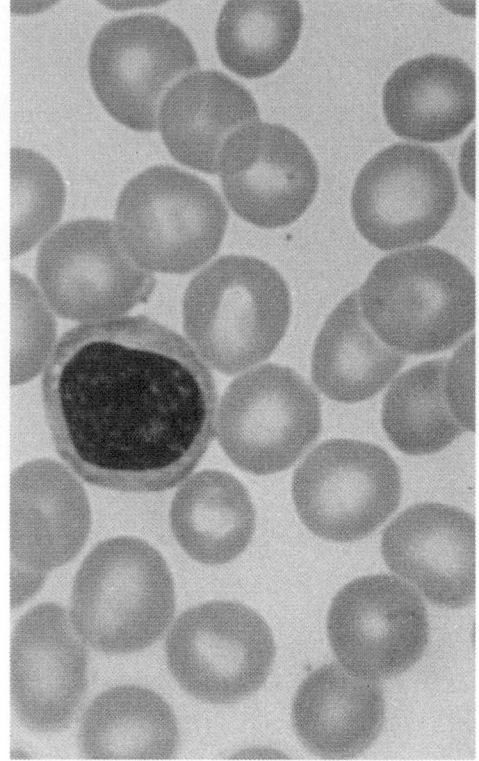

20-9 Stained smear of human blood
The large white blood cell (a lymphocyte) contains a nucleus. The surrounding red blood cells are nonnucleated. (×2000)

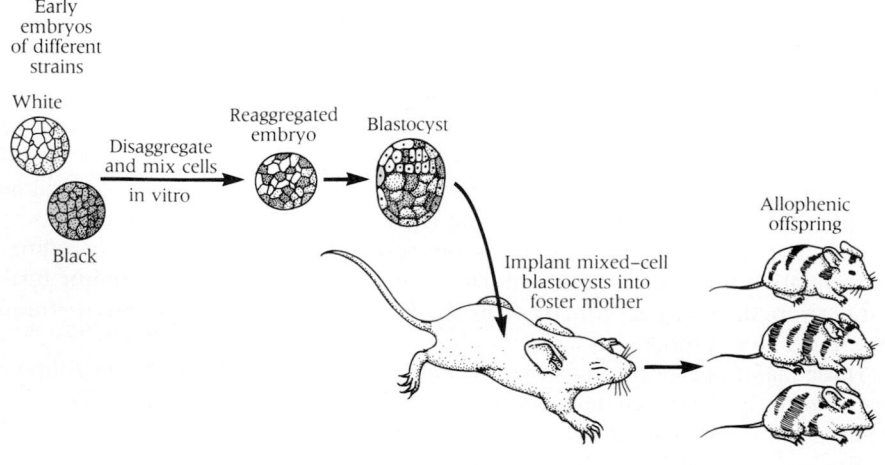

20-10 The production of transgenic mice
Embryos of two different strains (one with white hair and the other with black hair) are mixed and the reaggregated embryos implanted in a foster mother. The offspring have black and white hair in bands. The design of each coat depends on the location of individual cells in the embryo. Each precursor founder cell will give rise to either white or black hair, depending on its genetic makeup and segmental location.

BOX 20-1

"MANUFACTURING" MICE WITH SIX PARENTS

Chimeras are multicellular organisms derived from more than one zygote. They are genetic mosaics because they contain more than one genetically distinguishable cell population. The Mintz technique of producing chimeras by aggregating the cells of early-stage mouse embryos reached some sort of pinnacle in a 1978 experiment of C. L. Markert and R. M. Petters. After noting that all adult chimeras thus far produced—or "manufactured" in the authors' terminology—were derived from the cells of two embryos, the investigators produced the first adult chimeric mice derived from three. Such creatures would have six genetic parents!

In manufacturing "hexaparental" chimeric mice, strains were chosen with distinctive coat color phenotypes—black, white, and yellow. Embryos at the 4-cell or 8-cell stage were obtained from the oviducts of pregnant females. After enzyme treatment to remove their outer gelatinous coats, black,

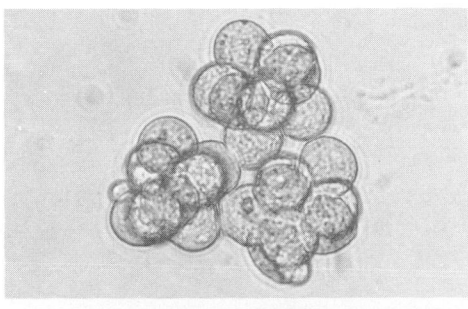

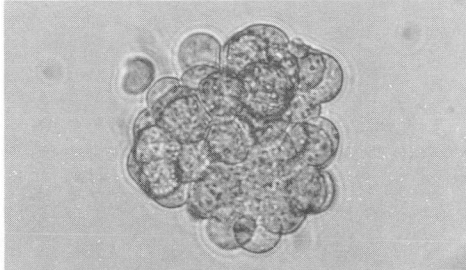

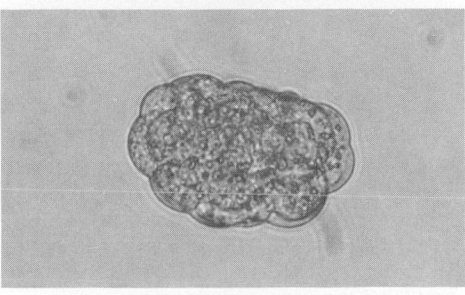

20A Reaggregated embryos of mixed cell types

white, and yellow embryos were placed close together in a culture plate for 24 hours. Embryos newly formed by cellular aggregation (Figure 20A) were transferred to females for gestation. The degree of genetic mosaicism in the resulting offspring could be determined 7 to 10 days after birth by examination of hair color patterns.

Preliminary tests ensured that embryos of all three genotypes (black, white, and yellow) were similar in their developmental ability. Pairwise combinations (indicated by the symbol ↔) were then constructed from the three genotypes. Many yellow ↔ white, black ↔ white, and black ↔ yellow chimeras were produced and found to develop equally and therefore to make equivalent contributions to any chimera derived from the aggregation of three embryos.

In most cases, three embryos aggregating into a single embryo produced an embryo that was more than triple the usual size. By the end of the culture period, the new embryos had developed small blastocoels (Figure 20B). Forty of these surviving triple embryos were transferred to two females, 20 to each female. Both foster mothers gave birth to offspring that included all three pairwise combinations: black ↔ yellow, yellow ↔ white, and black ↔ white. In addition, one individual emerged that was a mosaic of all three colors. This triple chimera (black ↔ yellow ↔ white) is shown in Figure 20C.

The clearly visible patches of pigmented fur on this hexaparental mouse evidently had a clonal origin from single cells arranged linearly from head to tail. White fur represents the smallest proportion of the coat (about 10%); yellow and black fur occupies the remaining areas of skin more or less equally.

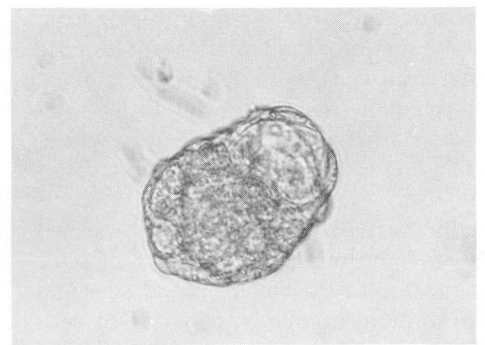

20B Blastocoel of new embryo

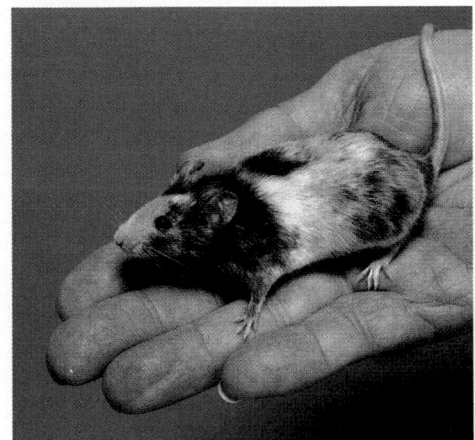

20C Triple chimera

The presence of black, yellow, and white patches of fur demonstrates unequivocally that this allophenic mouse arose from three different embryos.

During the embryonic development of the mouse the egg first undergoes a series of cleavages until about 64 cells have formed. Morphogenetic movements then shape these 64 cells into a hollow ball—the blastocyst—with a concentration of about 15 cells, the **inner cell mass (ICM),** on one side. The wall of the blastocyst, the **trophoblast,** gives rise exclusively to tissues of the placenta. The ICM is the forerunner of the embryo proper and also of several extraembryonic structures such as the yolk sac and amnion. How many of these ICM cells actually contribute to the tissues of the embryo rather than to extraembryonic tissues was an undecided issue of some importance. The determinative event that selects a few cells of the ICM to be the exclusive progenitor cells of the embryo must occur prior to the differentiation of primitive ectoderm and endoderm, probably when the ICM is composed of about 15 cells, or shortly thereafter.

These results are consistent with the hypothesis that three cells of the ICM are set aside for the formation of the embryo at the time the ICM is composed of 10 to 15 cells. The successful manufacture of a chimeric mouse from three genetically marked embryos also suggests that three cells in the ICM are allocated to form the adult organism. If this conclusion is correct, the vast majority of cells in the blastocyst contribute nothing to the tissues of the adult.

Mintz expected that such embryos could not develop normally. Many believed that the developmental fates of mammalian embryonic cells were fixed at an early stage. But when Mintz actually fabricated her anomalous embryos, a surprise lay ahead.

In one famous experiment (Figure 20-10), she placed together in a dish the morula stages of embryos of two different mouse strains stripped of their gelatinous coats. Cells from the embryos comingled and aggregated to form a new embryo which was carefully implanted in a foster mother that had been mated with an infertile male to prepare her hormonally for pregnancy (see Chapter 21). Remarkably, the mothers gave birth to normal-looking animals that were **mosaics** of cells derived from the two original embryos. Mintz called them **allophenic** mice because they had the characteristics (phenotypes) of two different genetic types (Figure 20-11). Others later produced such mice of even more extraordinary parentage (see Box 20-1).

In the creation of allophenic mice of two "parental" strains, the cells of embryos containing as many as 32 cells are still pluripotent. The proportions of the two parental cell types will vary from animal to animal and even from tissue to tissue within the same animal. When these cells are labeled, they are found to be extensively rearranged. Yet to the 32-cell stage the cells of the original embryos could still differentiate into tissues of all kinds. This work thus contradicted the view that developmental paths in the mammalian embryo are fully determined at a very early stage. No matter how the system was manipulated the end result was always a normal mouse.

Another surprise concerned the lineages of the cells that form a particular tissue. When the embryo cells of a pure black mouse were mixed with those of a pure albino mouse, the result was a black-and-white mouse (Figure 20-11). The surprise was in the way the pigment cells of each color were arranged. Coat color patterns usually consisted of transverse stripes that were sharply defined on the mouse's back and blurred on its abdomen. By observing the patterns produced by genetically different cells, Mintz could estimate the number of cells that underwent the initial commitment to a specific developmental pathway—these are called **founder cells**— and the approximate time at which that commitment occurred.

She concluded, for example, that coat pigment cells (melanocytes) were destined to form 17 stripes on each side and that each stripe derived from a clone of pigment cells arising from a single cell. In other words, all the melanocytes of a mouse coat came from only 34 segmentally distributed founder cells that were committed to become pigment-producing cells early in development. When she looked at patterns formed by other cell types, she found that every kind of tissue had a unique story, not only in the number of clones, but in the resulting patterns. This tracing of cell lineages in the allophenic mice led to several important conclusions.

- It strongly supported the hypothesis that commitment to a particular developmental pathway occurs very early in embryonic life.
- It showed that relatively few cells make the initial commitment.
- It revealed that groups of cells, not single cells, become the "founders" of developing structures.

Next Mintz and Illmensee examined the developmental consequences of specific mutations. They induced mutations in cultured cells and then introduced the mutant cells into a living embryo as had been done in making allophenic mice. The pluripotent cells in the first experiments were from a **teratocarcinoma,** a tumor arising from undifferentiated embryonic cells. We know these tumor cells are pluripotent because they give rise to all kinds of cells, including germ cells, within the tumor mass. They also produce malignant tumors when injected into mice after birth. But following their insertion into early mouse embryos, normal differentiation took place. Normal-looking mice were eventually born that were mosaics of cells of teratocarcinoma and embryonic origins.

This work soon led to an even bolder approach in which DNA was injected into mouse eggs. Foreign genes introduced directly into the germ lines of mice create new strains of so-called **transgenic** mice. At first aimed only at studying expression of the transferred genes themselves, the technique was soon used with foreign DNAs of all kinds. A new principle was quickly established: foreign DNA injected into eggs can cause a mutation by inserting itself into the recipient's own

20-11 An allophenic mouse

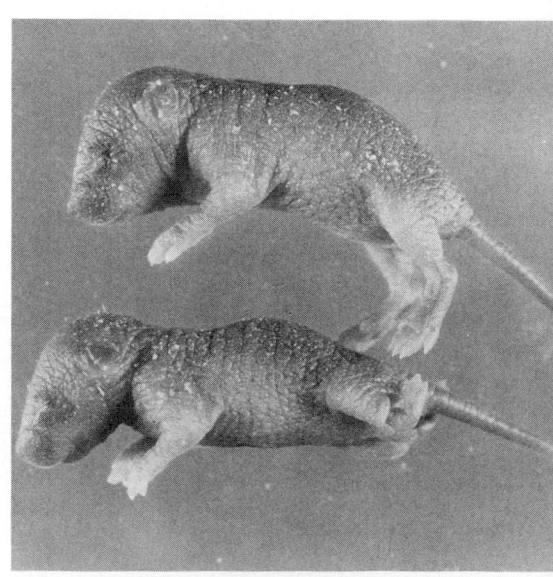

20-12 Mouse with limb deformity induced by insertional mutagenesis
These are two 4-day-old mice born of heterozygous parents. The upper animal is normal. The lower animal is a litter mate displaying the mutant phenotype—defective forelimbs and hind limbs.

20-13 Effect of growth hormone in mice
The photograph shows two male mice, which are 10-week-old siblings. The mouse on the left has a new genetically engineered gene consisting of a promoter fused to the rat growth hormone structural gene. This animal weighs 44 grams. His sibling without the gene weighs 29 grams. The gene is passed on to offspring, which also grow large. Mice that express the gene grow 2–3 times as fast as controls and reach a size up to twice normal.

genome. Hence the process was an example of **insertional mutagenesis** (see Chapter 18).

Many insertions produce interesting developmental abnormalities. In an early demonstration of the power of this method, Philip Leder, of Harvard Medical School, produced an insertional mutant termed *limb deformity* (Figure 20-12) with foreign DNA containing a modified oncogene called *myc* (see Box 18-2).

These successes have led to experiments employing DNA of every imaginable source, including the following:

- Normal tissue-specific genes from other species
- Mutant genes of all sorts
- Retroviral genes
- Oncogenes
- Regulatory genes

In one dramatic case, the injected DNA contained the promoter of a mouse gene that was chemically fused to a cloned structural gene encoding rat growth hormone, from which the regulatory region had been deleted. Of 21 mice that developed from these eggs after implantation in foster mothers, 7 carried the fusion gene and 6 of these grew significantly larger than their littermates (Figure 20-13).

The scientific and practical potential of this methodology needs no comment. It has obvious implications for investigations of cancer, development, gene regulation, and hormone action—and for methods of accelerating growth in farm animals, of correcting genetic disease, and of producing valuable gene products. Leder was actually granted a patent in 1988 for one of his transgenic mice.

CONTRIBUTIONS OF A TINY NEMATODE

In the 1970s Sidney Brenner of Britain's Medical Research Council took the small worm whose nervous system he had been studying and placed it at center stage of developmental biology. The nematode, *Caenorhabditis elegans*—or *C. elegans* as it is usually known (Figure 20-14)—has several irresistible assets from the investigator's viewpoint.

- It is only 1 mm long and is easily bred in large numbers.
- It is a self-fertilizing hermaphrodite producing both eggs and sperm (Figure 20-15). Though diploid, its capacity for self-fertilization facilitates study of recessive genes.
- It has a brief life cycle, fertilized eggs developing into mature worms in 3.5 days. A mere 14 hours after fertilization, the embryo hatches as a feeding larva. In the next day and a half the larva passes through four stages (molts). The worm emerging at the end of this developmental sequence is sexually active and can begin laying eggs within an hour.
- Its invariant anatomy is fully understood at the ultrastructural level.

20-14 The nematode, Caenorhabditis elegans **(×160)**

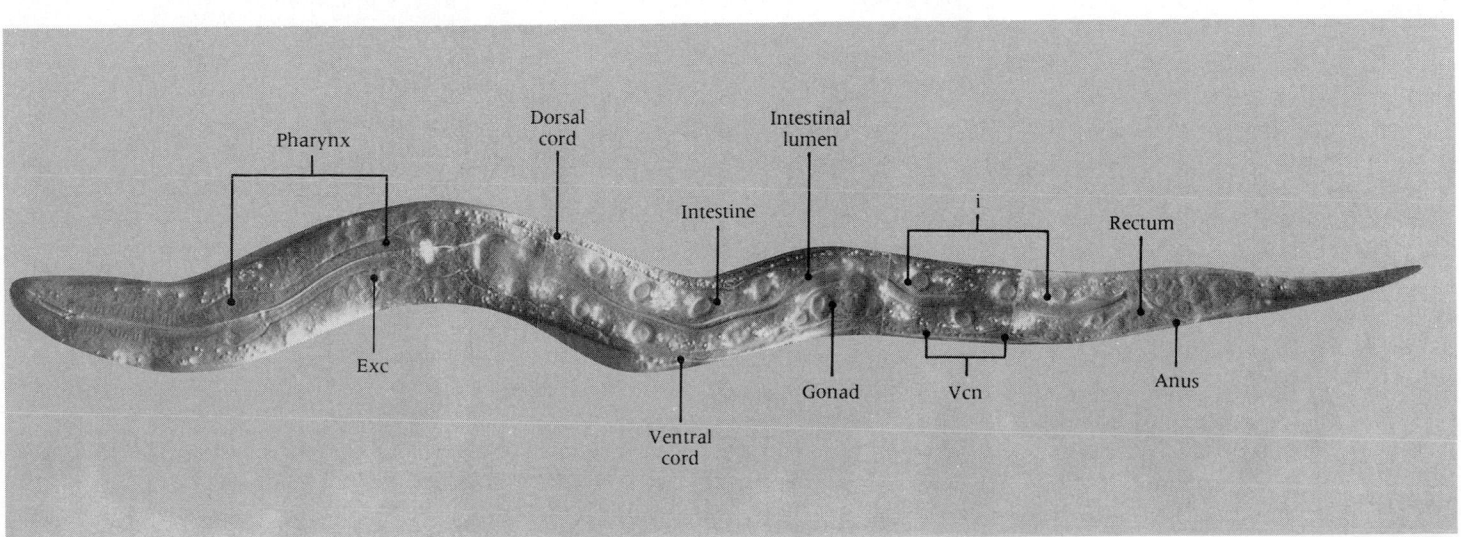

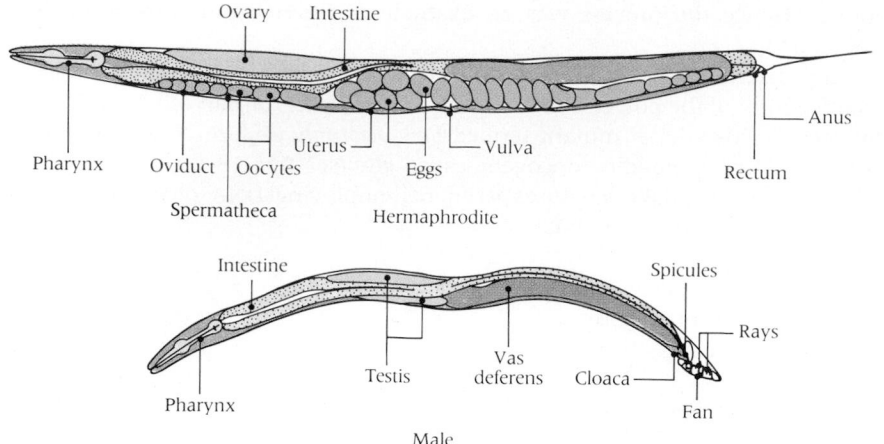

20-15 The sexes of C. elegans
The self-fertilizing hermaphrodite (top) produces both eggs and sperm. The male is slightly smaller and produces only sperm.

- It is transparent. Therefore an observer can watch cells divide in a living worm (Figure 20-16).

- Finally, and of great importance, the adult worm contains exactly 959 cells—the square root of the number of cells in *Drosophila*. Of these, 302 constitute its nervous system. We now know the complete developmental history of every one of these cells. By the use of a highly focused laser beam, it is possible to destroy almost any cell in the developmental sequence. The resulting worm will lack all of the descendents of that cell—except in rare cases where another cell fills in.

The elusive goal of this work—an understanding of how genes influence cells to assemble into an organism remains unmet. Much has been learned, however.

A diagram showing the origin and lineage of all 959 cells of *C. elegans* is sketched in Figure 20-17A. This diagram, a triumph of modern developmental biology, is called a **fate map.** It illustrates several points. As shown in summary form in Figure 20-17B, the initial cleavage of the zygote is asymmetric, producing a larger cell (termed AB) and a smaller cell (P1). (Note that every cell of *C. elegans* is numbered!) By the 16-cell stage the progeny of P1 have taken off on a pathway to the germ line. Other cells arising in the first four divisions are called *founder cells* because they are progenitors of major lineages and specific cell types. The major founder cells are AB and C (which give rise to a portion of the hypodermis or skin, nerve, and muscle cells), MS (glands, other muscle, and nerve cells), E (intestine), and D (still more muscle cells). Only a few lineages are labeled in Figure 20-17A. Note that the pharynx is made up of cells from both the AB and

20-16 Developmental sequence in C. elegans
These remarkable photographs track the appearance of specifically identifiable cells.
(a) The fertilized egg just before the first cleavage. The male and female pronuclei have not yet fused completely. (b) The beginning of gastrulation. Ea and Ep are intestinal precursor cells. D is a mesodermal precursor, and P4 is a germ cell precursor. (c) By the 16-cell stage, P4 (marked by star) is still distinguishable. (d,e) By 430 minutes, individual cells are more difficult to discern. (f) The white arrow points to the mouth and the two black arrows to germ cells.

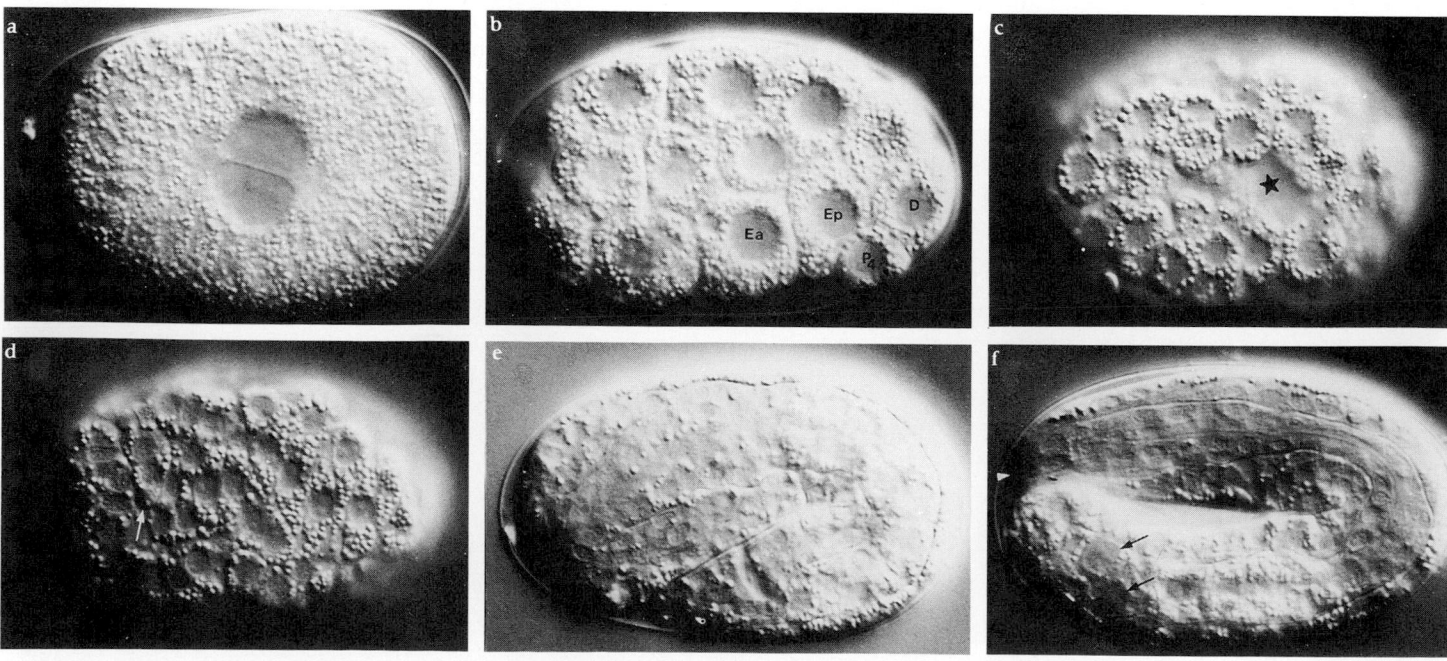

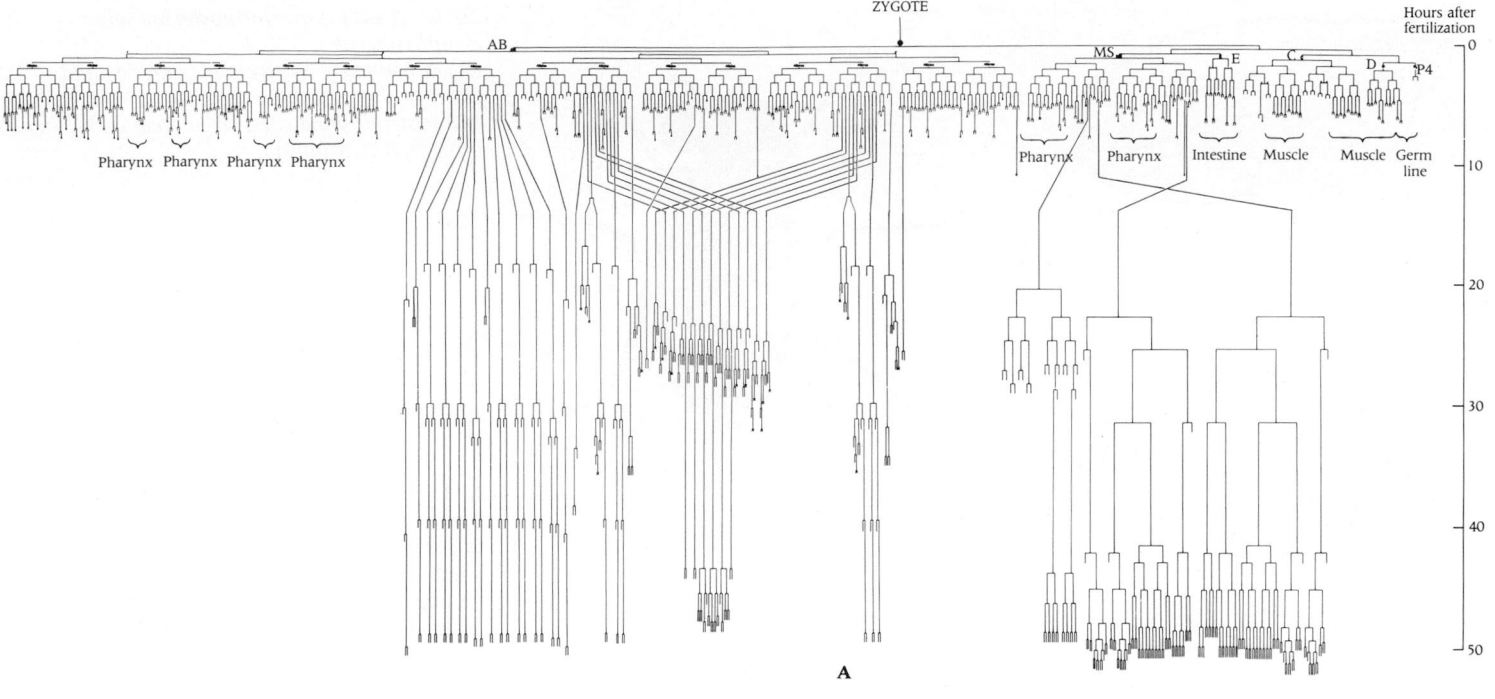

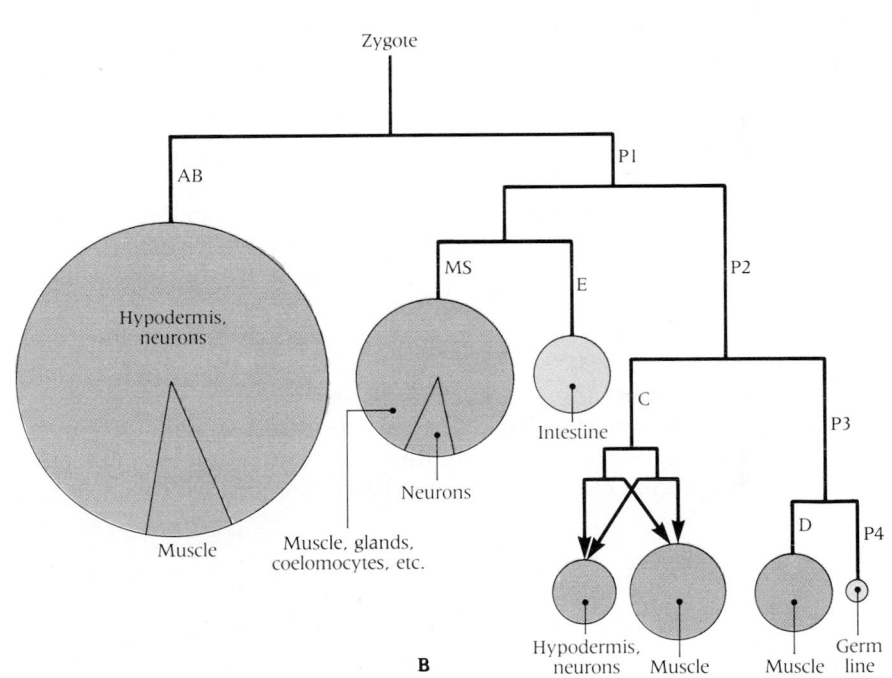

20-17 Fate map showing origin of all 959 cells in C. elegans

(A) *A complete cell lineage tree of an adult hermaphrodite. The vertical axis, representing time, spans 50 hours. Horizontal lines represent cell divisions. Vertical lines ending in X represent cell deaths (131 total). Only a few lineages are labeled.* **(B)** *Summary showing origin of founder cells AB, MS, E, C, D, and P_4 and their subsequent fate.*

MS lineages. Most of the unlabeled cells give rise to neurons or hypodermal cells (which lie just below the skin).

Among the lessons of this work is the observation that *cell death* is a characteristic feature of development. Cells are evidently programmed to die—and about one-eighth of the total do. Most are in the worm's developing nervous system. How this occurs is not yet known. Investigators have learned to induce mutations in individual cells of the lineage. In one set of mutants, cell death does not occur.

Clearly, a foundation has been laid for a wave of progress in the analysis of development—of how one remarkable creature puts itself together.

DROSOPHILA DEVELOPMENT: THE ''HOMEO BOX''

The history of *Drosophila* genetics is full of chance occurrences. Whenever a new gene was encountered, it was described, mapped, and recorded in textbooks. However, study of gene regulation was avoided because other organisms were much more

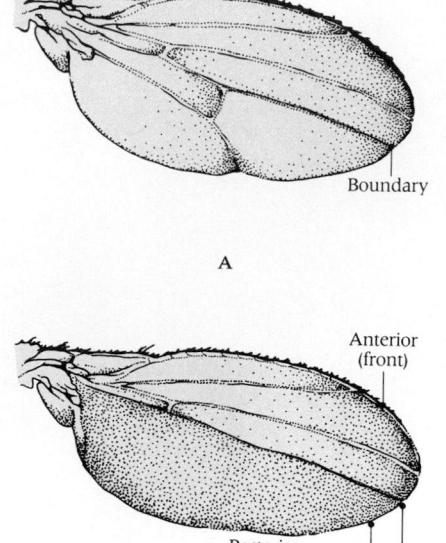

A

B

Boundary

Anterior (front)

Posterior (rear)

Boundary

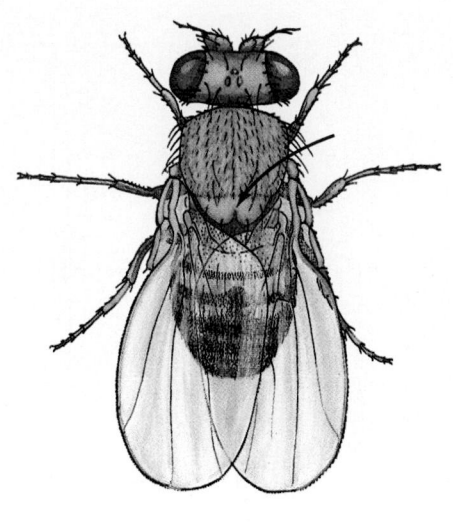

20-18 (Left) *Anterior/posterior wing compartments*
(A) *A boundary separates the anterior and posterior wing compartments.* **(B)** *The anterior and posterior wing edges differ from each other in normal* Drosophila.

20-19 (Right) *The* engrailed *mutation in* **Drosophila**
This mutation is characterized by a notch (at arrow) in the scutellum of the fly (a piece of the upper thorax), as well as a change in the posterior wing edge that makes it resemble the anterior wing edge.

accessible technically. As a result, several important advances were delayed. Consider this example.

In normal *Drosophila*, cells never grow across the antero-posterior boundary of the wing, which lies between the third and fourth wing veins (Figure 20-18A). The anterior (front) wing edge normally differs in appearance from the posterior (rear) wing edge (Figure 20-18B). However, there is a mutant gene called *engrailed*, or *en* (Figure 20-19). Its phenotypic expression was first recognized in 1929 as a posterior wing edge that had come to resemble an anterior wing edge. This was explained within the year: the wing normally develops in two compartments, each the descendents of a small group of founder cells. In the wings of flies with the *en* mutation, cells occur in patchwise fashion, with clones of cells arising in the posterior half invading the anterior half. No further study was undertaken until 1975 when it was found, among other things, that *en* is a regulatory gene that is responsible for the differences between anterior and posterior compartments. Its meaning emerged only after the recent discovery of **homeotic mutations**—mutations that cause cells to switch from one developmental fate to another.

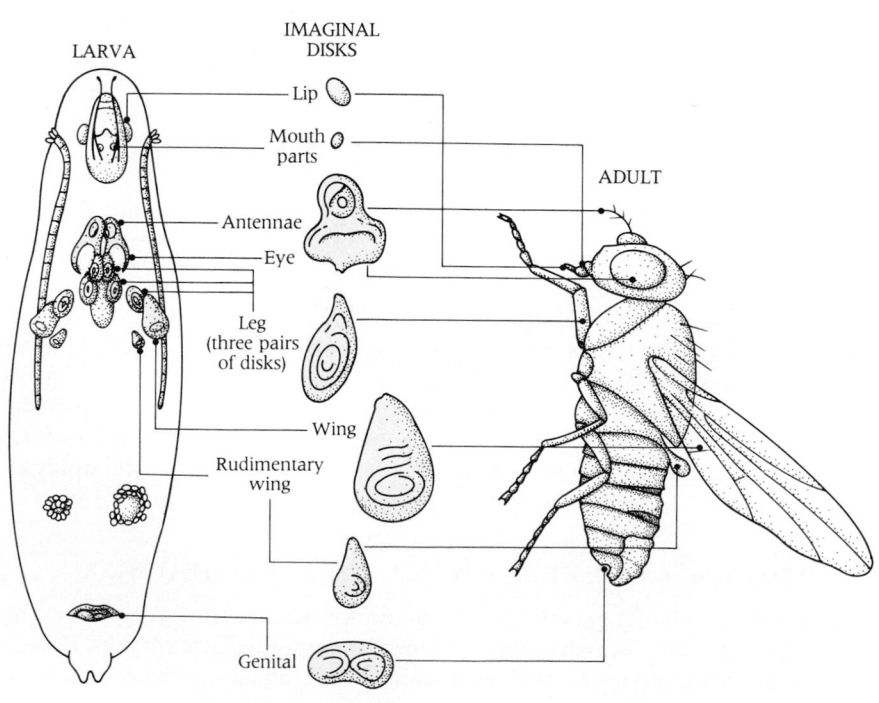

LARVA

IMAGINAL DISKS

Lip

Mouth parts

ADULT

Antennae

Eye

Leg (three pairs of disks)

Wing

Rudimentary wing

Genital

20-20 *Imaginal disks in* **Drosophila**
As the larva develops, groups of cells are set aside as imaginal disks. These give rise to the specific adult structures shown.

20-21 Segment patterns of larva and adult Drosophila

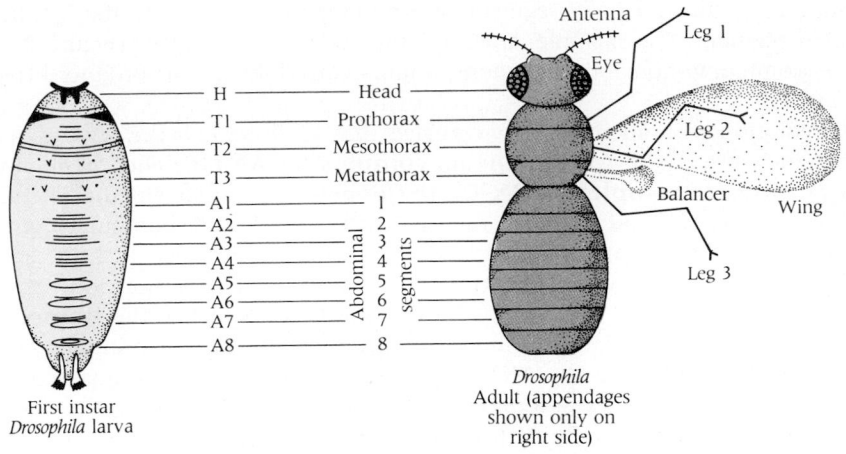

First instar
Drosophila larva

Drosophila
Adult (appendages
shown only on
right side)

During early embryogenesis, nuclei remain totipotent for a time. Then a change takes place that causes cell nuclei to fall into two categories. One set of cells develops into a larva. The other set will shape a second organism that will emerge from the larval carcass as an adult fly. This development is anticipated in the larva, which sets aside small packets of ectodermal cells called **imaginal disks** (often called simply **disks**) (Figure 20-20). The fate of each disk is precisely programmed. One, for example, becomes an eye. We know that because removal of that disk and transplantation to another larva results in an adult fly with an extra eye! The other disks and their fates are shown in Figure 20-20.

We must also recognize that an insect's body is subdivided into segments (Figure 20-21). In *Drosophila,* several segments—probably six—fuse to form the head. The thorax (chest) is composed of three segments, each carrying a pair of legs, and the abdomen of eight segments. In segmented creatures such as annelid worms or myriapods (millipedes, etc.), most body segments are very similar. Such similar segments are called **metameres.** Metameric segmentation, the basis of development in these organisms, consists mainly in the repetitive production of these body units. It also occurs in insect development but is less obvious because of the later appearance of differences between the segments.

Significantly, cells from one segment do not invade another during insect development. Thus each segment is a separate developmental unit derived from only a few founder cells. Within each segment are found the discrete compartments mentioned above, which are defined by the lineages of their constituent cells. Such compartments are responsible for distinguishing the anterior portion of structures from the posterior and the dorsal from the ventral. Although their presence is not obvious from the embryo's anatomy, it is clearly revealed by cell lineage studies showing that the *Drosophila* embryo is spatially organized into curious repeating units—the anterior and posterior developmental compartments of each body segment. These account for the striking zebralike, alternating pattern of encircling bands of cells on the surface of the young embryo. *Engrailed* is one of a number of genes upon which segmentation depends.

Homeotic genes specify segmental identity—that is, the kinds of segments that are formed. Homeotic mutations cause dramatic and visible modifications in the insect's segmented body plan by altering the developmental program of the cells of particular imaginal disks—usually by causing a disk to exchange one developmental pathway for another. Hence, we conclude that homeotic genes are involved in the control of the embryo's spatial organization or architecture. Instead of developing

20-22 Homeotic mutation in which antennae are replaced by legs
This is an extreme example of a homeotic mutation within the Antennapedia *gene complex.*

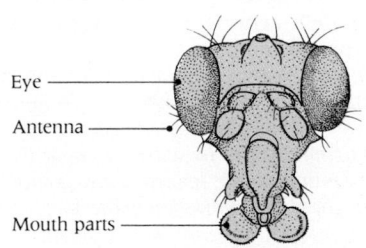

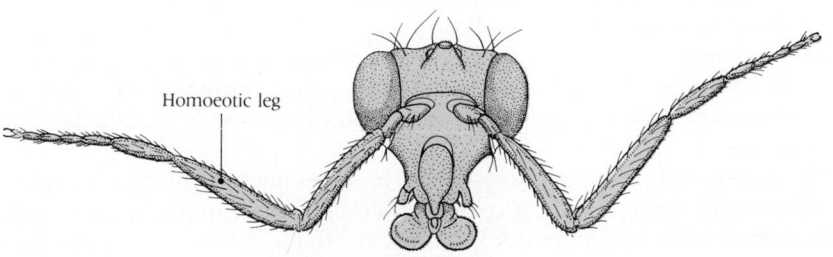

its normal features, a body segment of a mutant fly may acquire the features of another segment. For example, antennae may be replaced by legs (Figure 20-22). Or, by combining homeotic mutations, a four-winged fly may be produced (Figure 20-23).

Exploration of homeotic genes revealed that in *Drosophila* they occur in two large gene clusters, the **Antennapedia complex** (or **ANT-C**) after antenna-foot, and the **bithorax complex** (or **BX-C**). The former is concerned with the determination of thoracic and head segments, the latter with thoracic and abdominal segments.

Cloning of these complexes in 1984 led to a major discovery: a short segment of DNA (about 180 base pairs long) is a component of several homeotic genes in both complexes. Walter Gehring of Basel named the sequence the **homeo box** because of its presence in many homeotic genes. There are, for example, at least nine homeo boxes within the ANT-C and BX-C complexes. The first gene outside of the two complexes that was found to contain a homeo box sequence was *engrailed*, which is now known to distinguish the posterior and anterior compartments in each segment. The exceptional photograph in Figure 20-24 is of an experiment showing the spatial distribution of cells expressing the *engrailed* gene. This is further evidence of the presence of compartments within segments: the *engrailed* gene is expressed only in posterior compartment cells.

Homeo boxes were soon found in other species. A large fraction of the 180 bases found in these homeo boxes is identical to that of the *Drosophila* homeo box and the protein product of the gene shows a 75% homology with that of the fly. This degree of sequence conservation exceeds that of most protein-coding structural genes.

Research in this area is now following three trails.

- Attempts are now being made to prove that the presence of homeo boxes in other species does indeed mean that they represent equivalents of the fruitfly's homeotic genes.
- *Drosophila* genes associated with homeo boxes are being catalogued.
- The function of homeo boxes is being investigated.

It must be said that at this writing firm proof of the role of homeo boxes in development is still wanting. Yet a wealth of evidence supports such a role. Their postulated function is one of regulating the expression of other genes, such as those that determine the features characteristic of each body segment in *Drosophila*. Current thinking is that the protein encoded by the homeo box enables a number of gene regulators to recognize and bind to the gene under their control. P. H. O'Farrell at the University of California, San Francisco, recently found that the homeo box region of the *engrailed* gene protein product can bind to specific DNA sequences.

20-23 Homeotic mutations can produce a four-winged fly

EMBRYONIC INDUCTION

We must now consider an issue unique to developmental biology: It concerns the fact that a cell's fate depends not only on what sort of cell it is but on where it is in the embryo.

THE PRIMARY ORGANIZER

The issue of how a cell's fate is determined was posed dramatically in the 1920s by Hilde Mangold and Hans Spemann, who skillfully transplanted parts of amphibian embryos to other embryos of the same or different species. They identified a region in the undifferentiated cell mass of an early embryo that serves as an **organizer** for the remainder of the cell mass. This is the **dorsal lip** of the blastopore. The dorsal lip includes a group of cells called the **chordamesoderm.** They are the first to move through the blastopore—from outside to inside—at the beginning of gastrulation.

Within the embryo, chordamesoderm becomes mesoderm and notochord. By the processes of induction, these cells promote a key event of neurulation: formation of the neural tube from neural folds in the overlying ectoderm. We say this event

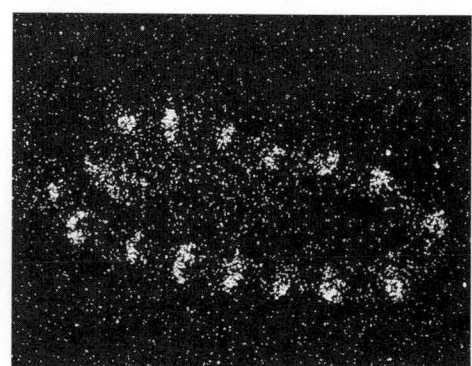

20-24 Cells expressing the engrailed gene in Drosophila
The bright areas on a photographic emulsion indicate the locations of these cells in an early embryo. They were made radioactive by hybridizing their DNA with a radioactive DNA probe containing the homeo box sequence. Each spot in the photo marks the posterior compartment of a future body segment.

20-25 Spemann's experiment showing induction by an organizer
Chordamesoderm in the dorsal lip is transplanted into the side region of a second frog embryo. This region develops a second notochord, neural tube, and other organs that were induced by the transplanted tissue.

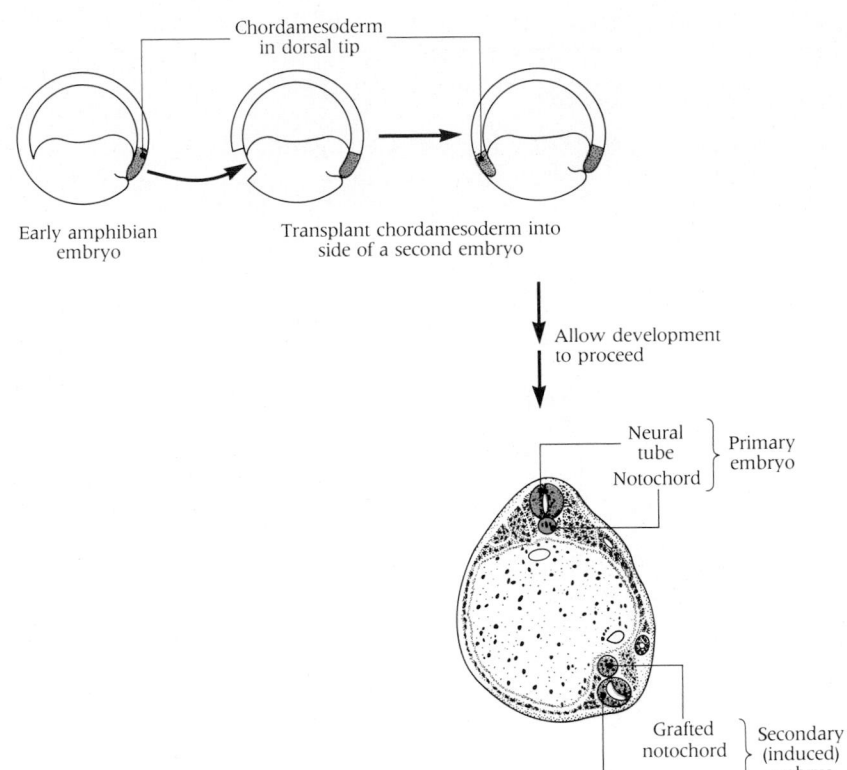

Chordamesoderm
in dorsal tip

Early amphibian
embryo

Transplant chordamesoderm into
side of a second embryo

Allow development
to proceed

Neural
tube   } Primary
Notochord   embryo

Grafted
notochord } Secondary
              (induced)
Neural tube } embryo

Postneurula embryo

is induced by organizer tissue, because that tissue must be nearby for the process to proceed, yet none of it becomes an actual part of the nervous system. If organizer tissue is removed surgically just before a neural tube is to be formed, the neural tube fails to develop. If the organizer is transplanted to another area of ectoderm, a neural tube develops there instead of in the normal position.

Spemann called the dorsal lip of a developing embryo the "primary" organizer because, when transplanted to a young enough embryo, it induces a whole second embryo to form like a Siamese twin (Figure 20-25).

Further insights into normal development have come from experiments in which pieces of an embryo were stained and followed during development—or were transplanted to new locations. Such observations have generated fate maps showing what organs will eventually be produced by various parts of the gastrula. We see in Figure 20-26A the location of a cell mass that would normally give rise to an eye. If these gastrula cells were transplanted to various locations on an older embryo, their future development would be determined by their new environment. Transplanted to the head region they develop into an eye and brain; transplanted behind the mouth, they develop into gills; when placed in the tail end of the embryo, they develop into the kidney duct, among other things. But if the same tissue is transplanted from an older embryo (Figure 20-26B), its fate is found to have become determined. No matter where it is transplanted, it differentiates into an eye.

In sum, what these cells differentiate into is controlled at an early stage by their location in the embryo. Their differentiation at these stages is not completely controlled by genes inherited from the zygote. All the cells have the same genes. At some point, however, the versatility of differentiation declines. A stage is reached when the fate of these cells is settled. At that point, transplanted cells develop independently of their new location. Cells destined to form an eye will become an eye no matter where they are. The extraordinary process by which the eye develops in embryogenesis is perhaps the most elegant of all illustrations of both morphogenetic movements to new locations and embryonic induction.

Presumably the organizer acts by producing specific factors or agents like those mentioned below in connection with cell interactions in tissue culture. In 1933, Johannes Holtfreter found that heat-inactivated dorsal lip tissue is as effective an inducer as living dorsal lip. The massive research that ensued showed that almost any killed tissue (from embryo or adult) can mimic the action of the dorsal lip on

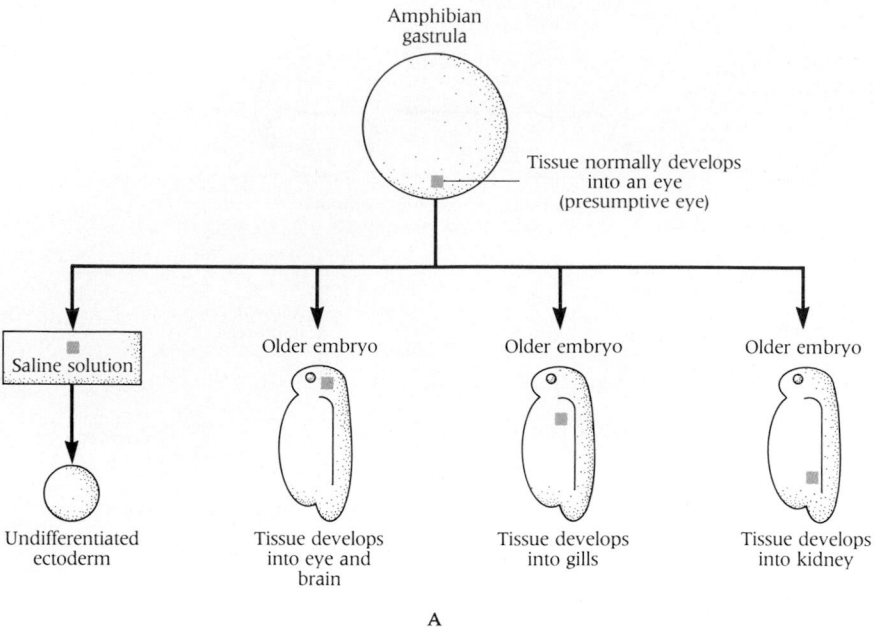

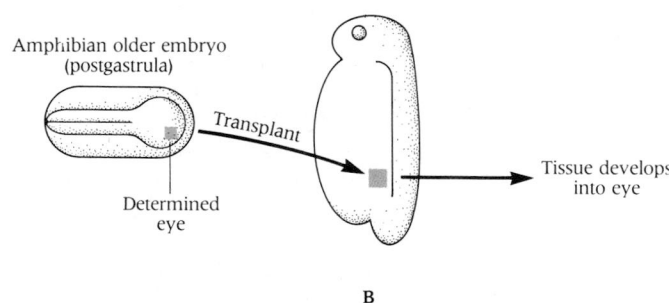

20-26 Determination versus presumptive fate of embryonic cells
(A) When taken from a gastrula, tissue that normally becomes an eye is not yet determined. If cultured in saline solution it remains undifferentiated. If transplanted into different sites of older embryos, it differentiates into to the tissue characteristic of that site. (B) Tissue taken from older embryos is determined, since after transplantation it develops only into the determined organ, in this case an eye.

embryonic ectoderm. Many chemical substances—nucleoproteins, steroids, etc.—were also effective. It appears that both chordamesoderm and diverse chemicals act by triggering some inherent capacity of embryonic cells to form a specific tissue, such as neural tissue. The "how" remains to be solved.

THE CONCEPT OF COMPETENCE

Work of this kind revealed that the host embryo must be capable of responding to the organizer. If cells normally destined to become epidermis are removed from a frog embryo at an early gastrula stage and transplanted into another embryo in a region destined to become nervous tissue, the grafted cells will become nervous tissue. If the same experiment is done in a late gastrula—only two days older—grafted tissue becomes epidermis as it would normally. The transient capability of cells to differentiate, depending on their location, is called **competence.**

Long before 1935, when Spemann received his Nobel Prize, Warren Lewis reported that the dorsal lip did not affect embryonic development. Unfortunately for him (and for science), he had transplanted chordamesoderm to older "incompetent" embryos. The transplanted cells differentiated into the cell and tissue types that had been destined to form without any influence of the surrounding host tissue. We now recognize that the ectoderm is not capable of reacting to induction—that is, it is not competent—until gastrulation occurs. Somewhat later it loses its competence completely.

We do not yet know the mechanism underlying these events. The major characteristic of embryonic induction is the transfer of information from one cell to another. Direct cell-to-cell contact would be one method. Another would involve the passage of molecules between cells, perhaps through gap junctions (see Chapter 7). If this is the case they must be small molecules, since macromolecules would not pass freely between cells. These molecules are still to be identified.

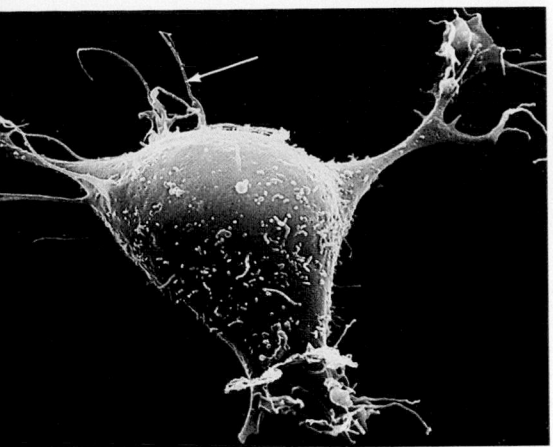

20-27 Filopodia
A scanning electron micrograph of a fibroblast showing many small filopodia (arrow) extending from the cell surface. (×9000)

CELLULAR STRATEGIES

Two phenomena peculiar to developmental biology and essential in the events associated with embryonic induction are programmed movements of cells and cell groups, sometimes over long distances, and changes in cell shapes.

MORPHOGENETIC MOVEMENTS

Cell and tissue migrations that help to form the shape and structure of an embryo are termed **morphogenetic movements.** We saw an example of such movements in the migration of surface cells through the blastopore in the course of gastrulation.

Cells also move long distances in the adult. White blood cells move into and through the tissues. Cancer cells move from one tissue to another. Morphogenetic movements, however, tend not to be movements of individual cells. Rather they are the spreadings and foldings of groups and sheets of cells.

We explored in Chapter 7 two topics fundamental to this subject: cell-to-cell communication and its dependence on junctions between cells and signaling systems; and cell motility and its dependence on microfilaments and microtubules. Those discussions would be usefully reviewed at this point.

Contact Inhibition and Contact Guidance

Our ideas about cell movements in the body derive largely from studies of fibroblast movement in a culture plate. Moving fibroblasts have a ruffled border on the advancing cell surface that bears many fine contractile processes called **filopodia** (Figure 20-27). When the filopodia of a moving fibroblast make contact with another fibroblast, the ruffled membrane becomes paralyzed, and movement of the cell in that direction stops (Figure 20-28). Localized intercellular connections called **tight junctions** (see Chapter 7) often develop at sites where the filopodia of migrating cells make contact. Nerve cells in culture display a similar pattern—but only when their filopodia touch another nerve cell, not a solid object or a fibroblast.

We call this phenomenon **contact inhibition** of movement. (It is to be distinguished from contact inhibition of growth, which is the inhibition of mitosis in cell cultures as a result of crowding). The mechanism of contact inhibition of movement remains unknown. It might result from an electrical connection between cells. Cytoplasmic contractility could be altered by local membrane changes that allow ions to flow from one cell to another.

An interesting issue is raised by these long-distance cellular movements: what guides cells in the proper direction? At present, three mechanisms seem to play important roles.

- One is a phenomenon called **contact guidance.** Cells do not move unless they adhere to a surface, and adherence must be reversible if moving cells are to cover any distance. The surface may actually guide the direction of cell movement.
- A second guidance mechanism depends on the presence of fibrils in the extracellular matrix that contain the protein **fibronectin.** Fibronectin affects the adhesion of cells to surfaces (see Chapter 7). It also influences and guides their migration, according to the principle that cells will migrate and localize to places where their adhesion is favored. When an antibody capable of inactivating fibronectin is injected into a frog blastula, the cell migrations necessary to convert a blastula into a gastrula are prevented.
- Migrations are also influenced by the organization of the **basal lamina,** a thin sheet of the extracellular matrix that separates two cell layers of different origin (Figure 20-29).

Migrating mesenchymal cells tend to move away from zones where they are abundant. When they encounter bare undersurfaces, they form local tight junctions and remain stationary. When filopodia find a suitable surface to which they can attach and contact, they can pull a whole mesenchyme sheet in their direction. Eventually these migrating tissues form a variety of adult tissues, including connective tissues (see Table 19-2).

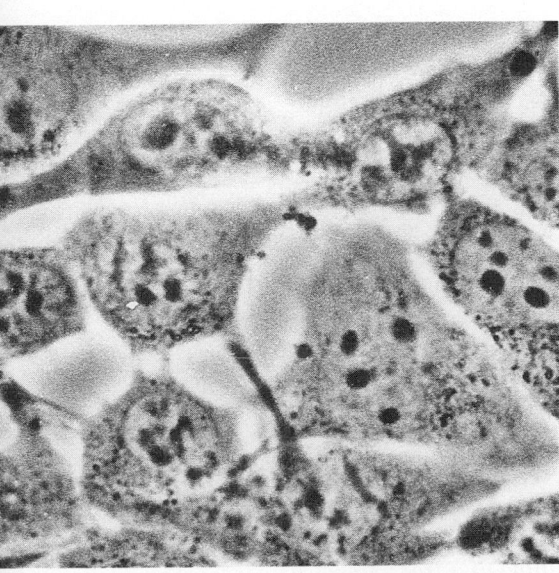

20-28 Contact inhibition
Phase contrast photomicrograph of fibroblasts in a crowded tissue culture. When the cells collide into one another, they stop moving. This phenomenon is called contact inhibition. (×1280)

Cell Reaggregation and Homing

One way to study the role of cell surfaces in cell-to-cell adhesion is to separate attached cells and then watch them come together again. A living sponge, pressed through a sieve, can be separated into individual cells and small cell clusters. If the cells are kept in a salt water suspension, they move together and reconstruct a sponge precise in every detail—including internal structures characteristic of the species. Reassociation is highly species-specific. If cells of two sponge species are mixed, they reaggregate, not into one mixed sponge, but into two sponges, each composed of cells of one species.

Something like this happens during the life cycle of the cellular slime mold *Dictyostelium discoideum* (see Chapter 39). At one stage in its life cycle, this slime mold consists of unicellular organisms called **myxamoebae.** Eventually the myxamoebae aggregate into a large protoplasmic mass that differentiates further into subsequent developmental stages. This coming together of myxamoebae is an example of **chemotaxis**—defined as the oriented movement of an organism under the influence of a chemical agent. The agent in this case is *cyclic AMP*, or *cAMP*, a substance that plays a messenger's role in many mechanisms in living organisms (see Chapter 28). In slime molds the cAMP that brings cells together may also regulate gene expression as these cells differentiate.

Johannes Holtfreter, a pioneer in the study of animal cell reassociation, found that when epidermis and neural tube cells are combined, the cells segregate so as to preserve a specific pattern, with epidermal cells on the outside and neural tube cells on the inside. Unlike sponge or slime mold cells, many animal cells recombine without concern for species—mouse cells readily join with chick cells. However, the process is highly tissue-specific.

A number of proteins, called **cell adhesion molecules,** or **CAMs,** promote the reaggregation of disaggregated embryonic cells. Studies with specific antibodies show that different CAMs exist, each possessing a pattern of specificity for particular cell types. One of these is termed N-CAM because it is associated with the development of the nervous system.

The mechanism by which cells recognize their neighbors is not settled. Recognition probably has to do with specific surface structures. Of course, disaggregation and reaggregation of the type demonstrated in cultures does not occur in the embryo. However, animal embryonic cells do sort themselves out on the basis of cell or tissue specificity.

When individual cells migrate through the embryo, they find their target locations with astonishing accuracy. When investigators injected pigment cells into the blood of unpigmented chickens, they could be traced with ease. In no case did a pigment cell lodge and then proliferate anywhere except where pigment cells are normally found. It appears that a type-specific targeting capability is characteristic of normal embryonic cells.

CELL INTERACTIONS

The Principle of Self-assembly

We have already seen examples of the principle of self-assembly. At the molecular level it is illustrated by two α and β chains spontaneously aggregating to form a tetrameric hemoglobin molecule (see Chapter 22). At the macromolecular level, it is seen when proteins and lipids come together to form cell membranes or organelles. Self-assembly doubtless has much to do with the aggregation of cells into tissues. Here we encounter higher orders of complexity because a single tissue can include a diversity of cell types (Figure 20-30).

The spontaneous self-assembly of large molecules into a stable configuration means that in a given environment it is thermodynamically more stable than a mixture of separate components. However, the assembly of cells into tissues involves cell interactions that we do not yet understand. Let us consider the approaches to the problem now being made.

Cell Interactions in Tissue Culture

Tissue culture is a useful tool in the study of cell interactions. In one experiment, tissues are dissociated into cells by the proteolytic enzyme trypsin. When epithelium

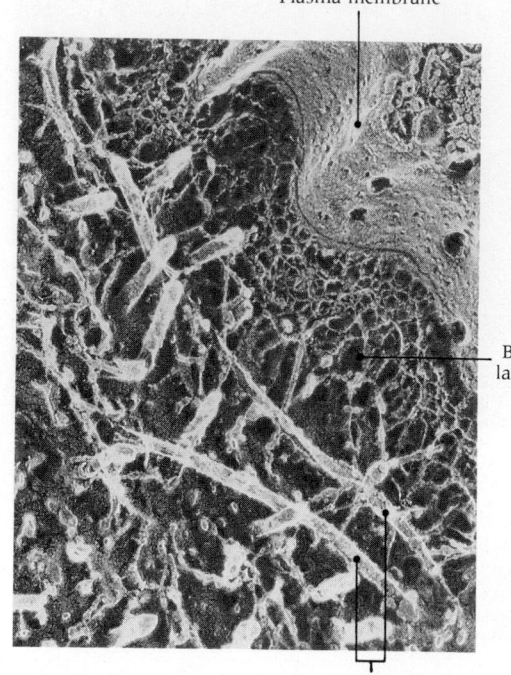

20-29 Basal lamina
Scanning electron micrograph of a quick-frozen, deep-etched preparation of skeletal muscle revealing the basal lamina which lies just outside the plasma (cell) membrane. Within the muscle (upper right) are the myofilaments. The extracellular space is filled with collagen fibrils. (×18,000)

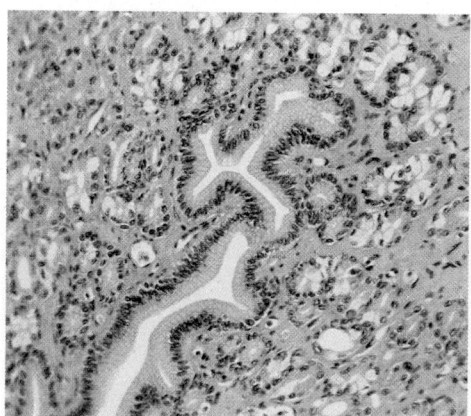

20-30 Histological tissue section of the lining of a frog's stomach
Note the variety of cell types, which include different kinds of epithelium (mainly columnar and glandular), connective tissue cells, and endothelium (lining the capillaries). All are coherently organized into a complex, functioning tissue. (×100)

from an embryonic mouse salivary gland is separated from mesenchyme, neither the salivary gland epithelium nor salivary gland mesenchyme differentiates into anything resembling a salivary gland when cultured alone. But when the two cell types are mixed, credible-looking salivary gland tissue appears in the culture plate. Salivary gland epithelium cultured with mesenchyme cells from kidney does not give rise to glandlike structures. If a filter with pores <1 nm in diameter is placed between the epithelium and salivary gland mesenchyme, recognizable salivary gland tissue again appears.

It thus appears that a specific factor influences epithelium to differentiate correctly. Direct cell-cell contact is not required. Filter pores are too small to allow whole cells to pass through, although it took years for investigators to realize that cell processes (threadlike extensions of the cell surface) can penetrate many filters— and to design special filters to prevent it.

As noted above, cultured cells of embryonic chick myoblasts are notable for their capacity to organize into contracting muscle fibers (see Figure 20-8). Even when a whole organism, with its multiple tissue types, is experimentally dismantled by trypsin treatment into a mixture of individual cells, under the right culture conditions the cells will reassemble.

Clearly, cells have not only the internal controls we have been discussing, but they have, in addition, subtle ways of interacting with the environment and with each other. The factors passing through filters (or trapped in them) may reflect one of these interactions.

Role of Collagen

Many believe that **collagen** is a key factor in tissue interactions. A fibrous protein that gives strength to connective tissue, collagen is synthesized by fibroblasts. The direct product of fibroblast activity is a fundamental unit called **tropocollagen** or **TC** (Figure 20-31). TC molecules have "heads" and "tails" and include three helical peptide chains. Under certain conditions (ion composition, temperature, and so on), TC molecules aggregate spontaneously to form collagen fibrils—a good example of "self-assembly."

Chick myoblasts in culture differentiate only in the presence of medium that was previously exposed to fibroblasts. Such a medium is said to be "conditioned." This implies a possible role for collagen in the assembly of cells into tissues. Exposure to conditioned medium is necessary only in the first 24 hours during which the cells become attached to the bottom of the plate. Probably it is the collagen in conditioned medium that orients random suspensions of cells into the specific patterns necessary for differentiation into recognizable units.

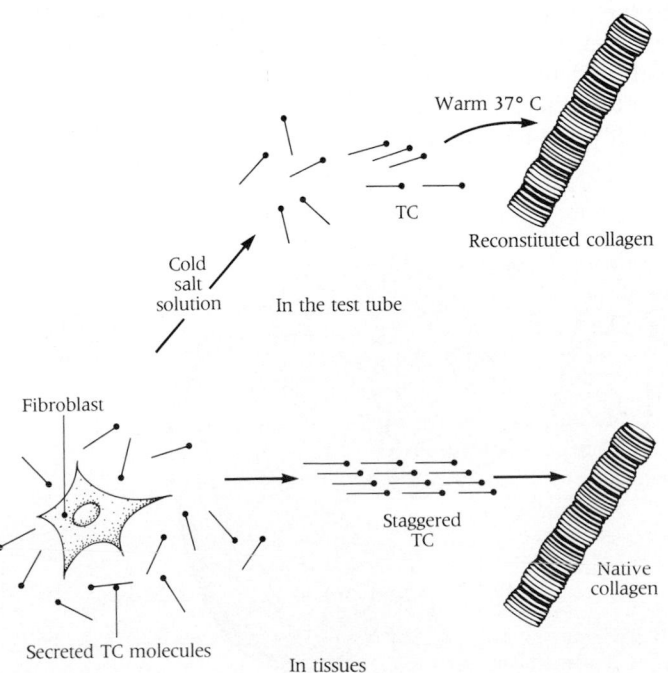

20-31 Relation between tropocollagen and collagen

Tropocollagen secreted by fibroblasts contains three polypeptide chains that are organized in a "head" to "tail" arrangement. Collagen fibrils are composed of staggered arrays of tropocollagen molecules that produce a pattern of cross-striations. Differing aggregation patterns in tissue and test tube produce different striation patterns.

In sum, development constantly changes the arrangements of cells, their patterns and movements. As a result, every body cell confronts a continuously changing environment in which cells impinge upon and influence each other. We are just beginning to understand how they do these things.

ONTOGENY AND PHYLOGENY

In some ways the course of embryonic development seems devious and indirect. Some have called it illogical. For example, human embryos develop tails, which then disappear (Figure 20-32). They also develop gill-like pouches, which are in part transformed into various structures, including the ear canal. At this stage the embryo looks rather like a fish, although the resemblance is superficial. The human embryo does not really have actual gills with real gill slits as do fishes—and fins are lacking. But it seems odd that a developing human being should go through a superficially fishlike state and have structures, even temporarily, that are so little related to human adult anatomy. Until evolution was recognized, such facts were inexplicable.

SUPPOSED RECAPITULATION

The tailed, gill-pouched stage of human development resembles not an adult fish but an embryonic fish. K. E. von Baer (1792–1876) noted this fact even before evolution was understood. He saw it as an example of the **biogenetic law,** which is really a descriptive generalization.

According to this generalization, at earlier stages embryos resemble those of other animals once considered "lower" in the scale of nature, and now deemed more like those of related or ancestral groups. As development proceeds, the embryos of different animals become more and more dissimilar. In its very earliest cleavage stages, a human embryo is rather like that of a starfish. In later stages, it comes to resemble the embryos of fish, amphibians, or reptiles. Still later, it is like the embryos of other mammals. Before birth, it becomes clearly human and unmistakably distinct from all other species.

Early evolutionists, especially E. H. Haeckel (1834–1919), restated the biogenetic law as the **principle of recapitulation,** according to which ontogeny repeats phylogeny. **Ontogeny** refers to the development of single individuals. **Phylogeny** refers to the evolutionary development of a population through many generations. In this formulation, successive stages of individual development correspond with successive adult ancestors in the line of evolutionary descent. The vaguely fishlike stage of the human embryo was believed to represent the stage when our adult ancestors were fishes. Von Baer had more correctly generalized the facts, but at a time when the principles underlying those facts could not be understood.

It is now clear that ontogeny does not repeat phylogeny. Ontogeny repeats ontogeny, with variations. Evolution is not a sequence of adults giving rise to later modified adults. What is passed on from one generation to the next is a developmental mechanism. It is the mechanism that changes in evolution.

20-32 Example of the biogenetic principle
A comparison of the embryos of shark and human reveals the presence of gill pouches in the neck region as well as the similarity of their circulatory systems.

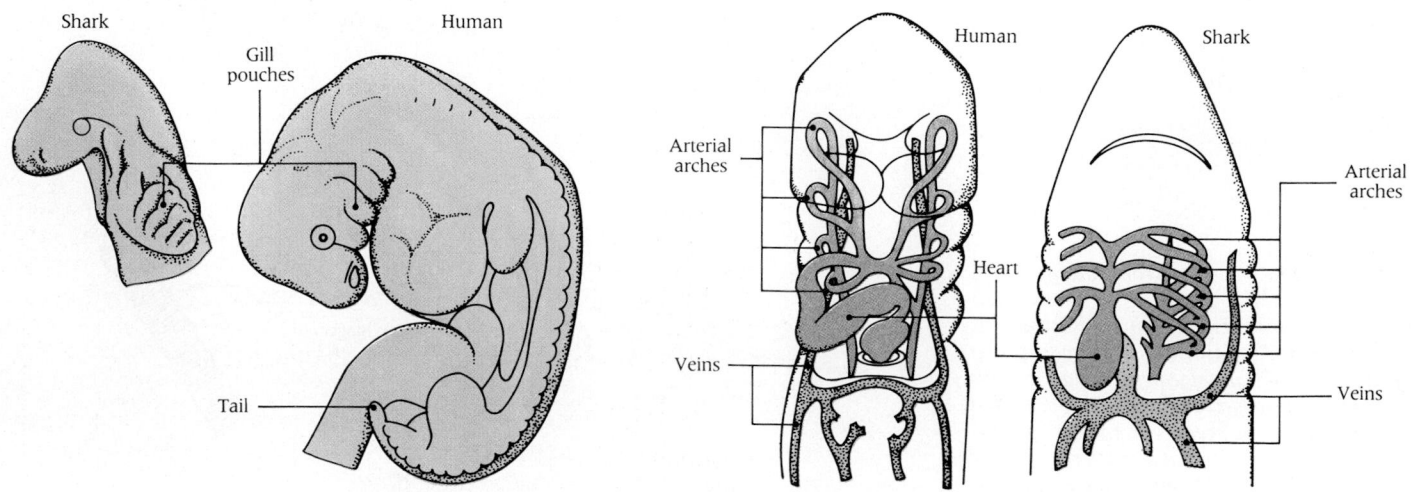

20-33 Recapitulation in the development of fish tail skeletons

In the development of fish tail skeletons, the tail develops into a structure (stage 1) resembling that of the adult of a distant ancestor (Pholidophorus). In stage 2 of development, a pattern emerges that resembles a more closely related evolutionary ancestor (Amia).

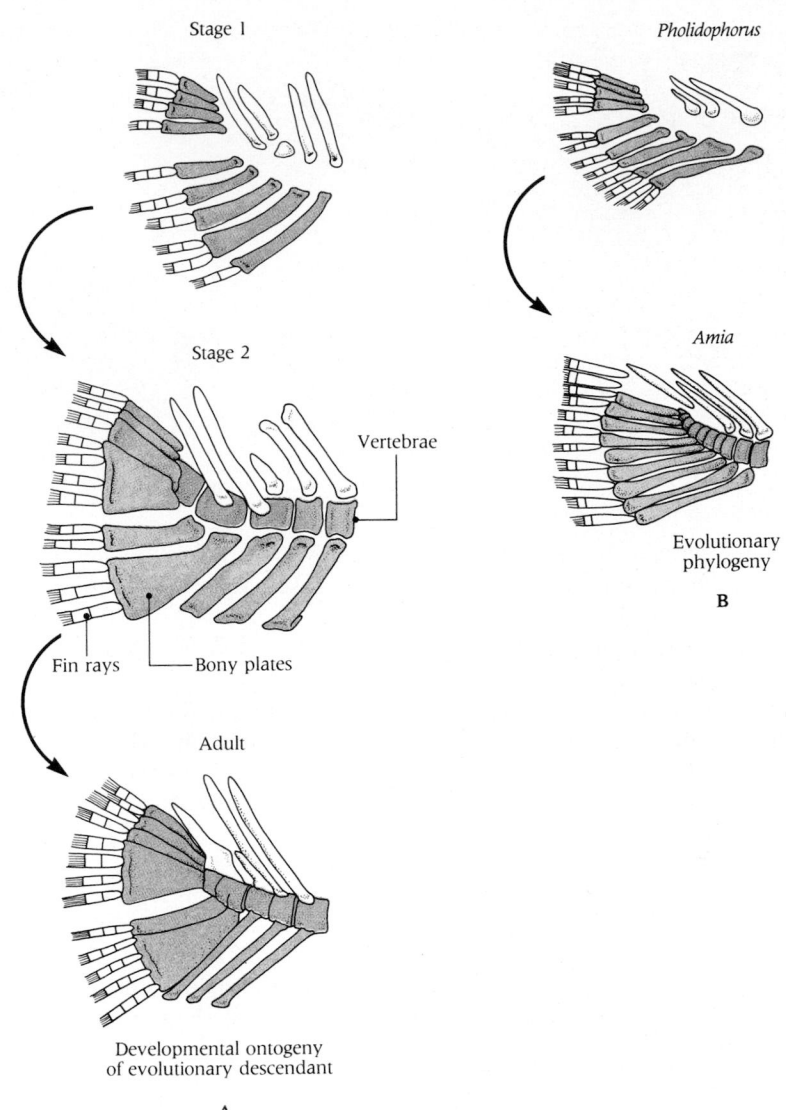

Stage 1

Pholidophorus

Stage 2

Vertebrae

Fin rays — Bony plates

Amia

Evolutionary phylogeny

B

Adult

Developmental ontogeny
of evolutionary descendant

A

HETEROCHRONY

To understand how differences in the forms of organisms evolved over time, it is necessary to ask what genetic programs control development and how they changed in the course of time. Today investigators looking into those questions are on two tracks. One revives the venerable approach of comparing the developmental pathways followed by different species. The other tries to identify the genes—such as homeotic genes—that control developmental "decisions" and thus are critical points at which evolutionary change can occur.

In closing the topic of developmental mechanisms, we cite one interesting example of the first approach that concerns variations seen in the timing of developmental processes, termed **heterochrony.** New forms evolved whenever there were changes in the relative rates of somatic trait development and gonadal maturation. If the developmental appearance of a somatic feature is accelerated with respect to gonadal maturation, a formerly adult trait will become a juvenile trait in the animal's descendents. This is recapitulation in the classic sense.

The evolution of fish tail skeletons is a good example (Figure 20-33). In forms arising in evolution (descendents), the adult tail is supported by a symmetrical skeleton with fully ossified (bony) vertebrae bearing large bony plates to which fin rays are movably attached. During development (Figure 20-33A), this adult condition is reached via two intermediate juvenile stages, each of which resembles the adult stage of an evolutionary ancestral form (Figure 20-33B).

- In stage 1 of development, the tail differentiates as an asymmetrical upturned structure lacking ossified vertebrae that closely resembles conditions in the adult of its most distant ancestor (*Pholidophorus*).

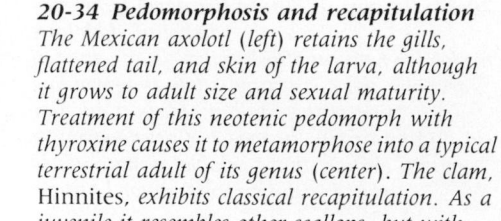

20-34 Pedomorphosis and recapitulation
The Mexican axolotl (left) retains the gills, flattened tail, and skin of the larva, although it grows to adult size and sexual maturity. Treatment of this neotenic pedomorph with thyroxine causes it to metamorphose into a typical terrestrial adult of its genus (center). The clam, Hinnites, exhibits classical recapitulation. As a juvenile it resembles other scallops, but with further maturation it attaches to the bottom and achieves the oysterlike form of its shell.

■ In stage 2 of development, the vertebrae ossify and the bony plates enlarge to resemble the pattern in the adult of a more closely related ancestor (*Amia*).

Thus, the phylogeny or ancestral history is recapitulated or repeated during the ontogeny of the tail skeleton of the descendent.

A second heterochronic mechanism is possible when maturation of the gonads is retarded while a somatic adult trait continues to develop. This phenomenon, termed **hypermorphosis,** often causes major size increases. It is a common evolutionary trend. The giant antlers of the Irish elk were results of hypermorphosis (see Chapter 43).

Timing shifts in developmental processes, heterochrony, can also lead to **pedomorphosis,** in which traits characteristic of juvenile ancestors are retained by adult descendents. Pedomorphosis can occur in two ways. Most often, somatic development is retarded with respect to the reproductive maturation. This is called **neoteny.** A famous example is the salamander axolotl, which remains aquatic and with its bushy gills and body fin looks like an overgrown larva (Figure 20-34). Yet it is sexually mature. In some neotenic salamanders injections of thyroxine, one of the thyroid hormones (see Chapter 28), causes the gills to disappear gradually and the body to lose its larval characteristics as in Figure 20-34. This suggests that (1) neoteny in these animals may be due to hypothyroidism, and (2) development of sexual maturity does not require thyroid hormone.

A second form of pedomorphosis is **progenesis,** which results from a different process. Here the gonads mature early. As a result, maturity is reached in a small-size juvenile body. Insect evolution, for example, may have been initiated by progenesis among millipedes. When a millipede hatches, it has only three pairs of legs and a limited number of body segments (Figure 20-35). If such a larva were to become sexually mature, an organism very similar to a primitive insect would result (Figure 20-36).

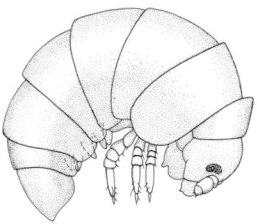

20-35 Progenesis
The newly hatched larva of the millipede, Glomeris, has only three pairs of legs and a limited number of body segments. The evolution of insects may have involved progenesis, in which the larval stage of the millipede attained sexual maturity and thus developed no further, creating an organism with fewer body segments.

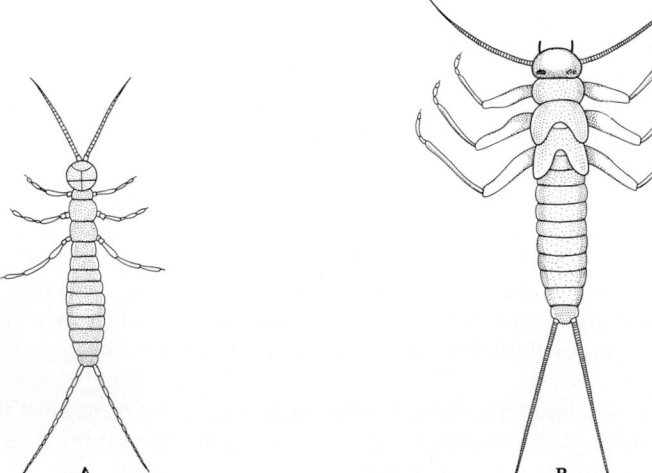

20-36 A proposed archetypic insect body pattern
The development of primitive insect patterns could result from progenesis. (A) An adult primitive wingless insect called Campodea, or bristle tail, about 4 mm long. (B) A late aquatic nymphal stage of Perla, a stone fly, about 3 cm long.

What are the biological implications of these timing switches?

- Pedomorphosis is a genetically convenient method of responding to environmental changes, since an existing, well-integrated developmental system need be altered only slightly.
- Neoteny occurs when it becomes more advantageous to remain in a hospitable larval environment than enter a harsh adult environment.
- Progenesis appears to be a response to environments in which a high reproductive rate or small size is advantageous.

The evidence of developmental biology often makes it possible to trace the evolutionary changes that have occurred in important structures—an example (one of many) is the evolutionary history of the mammalian ear, which is discussed in Box 35-1—and to reconstruct the detailed evolutionary history of modern organisms. A classic example is the brief appearance of pharyngeal gill pouches in the early embryogenesis of humans and other mammals (see Figure 20-32). This is strong evidence that the early ancestors of land vertebrates were aquatic forms.

SUMMARY

Attempts to understand the complicated processes of development have stimulated intense recent interest and many imaginative experiments. Although nuclear transplantation experiments have yielded important indirect evidence for the role of genes in directing embryonic development, functional studies of gene regulation are essential to a fuller understanding of what makes embryonic cells with identical genomes differentiate into a multiplicity of cell types.

Cytoplasmic determinants appear crucial in this regulation. When, for example, the nucleus of an adult skin cell is transplanted into an enucleated egg cell, there develops not more skin cells but a whole normal embryo. In part this is explained by the presence in oocyte cytoplasm of gene inducers that stimulate specific genes, rather than enhancing gene transcription generally. To date, three classes of oocyte cytoplasmic determinants have been discovered in various experimental animals: (1) maternal mRNAs; (2) transforming growth factor-β (TGF-β), a protein earlier found to have a role in growth and inflammation; and (3) transcription factor IIIA (TFIIIA), a nonhistone protein that selectively activates the gene for a variant of 5S rRNA found only in oocytes and then, by binding to 5S RNA, inactivates the gene.

Uneven spatial localization of determinants in the oocyte, unequal cell divisions of the zygote, and various external influences result in a series of steps that ultimately lead to overt cellular differentiation in the embryo. In general, the more specialized a cell becomes, the less able it is to undergo cell division.

Although the factors committing a cell line to a specific developmental fate are still unclear, powerful new methods now allow the tracking of genes during development. In one technique, foreign genes injected into eggs are often incorporated into the genome of the recipient cell, with the production of a mutant, transgenic organism in which the changes induced by an individual injected gene can be studied. In another, artificial embryos are assembled from the cells of different mouse embryos in a way that permits the fates of individual cells to be traced during development. In the nematode *Caenorhabditis elegans*, the origins and fates of every one of its cells have been mapped. Such maps reveal that programmed cell death is a characteristic feature of development. The discovery of homeotic mutations, mutations that cause gross changes

in body architecture by making cells switch developmental fates, has also enhanced understanding of the regulatory genes orchestrating differentiation.

A cell's fate depends not only on what kind of cell it is, but also on its location in the embryo. In the process of induction, the differentiation of a tissue is influenced by nearby organizer tissue. Transplantation experiments demonstrate that if tissue is moved from one part of an embryo to another early enough in development, it differentiates into an organ appropriate to its new location. Presumably organizer tissue mediates the induction by secreting specific substances. As embryonic development continues, cells lose the ability to respond to induction (competence). After that, they develop into the organ appropriate to their original site, regardless of where they have been transplanted.

Cell movements and changes in cell morphology shape the developing organism. Cell movements occur only when cells adhere to a surface. Depending on the nature of the surface, a moving cell may stop migrating (contact inhibition) or be directed to a specific location (contact guidance). Such interactions can be studied by dissociating embryonic cells and allowing them to reaggregate. Proteins called cell adhesion molecules (CAMs) promote reaggregation. Other proteins, notably collagen and fibronectin, influence cell shape changes and differentiation.

Embryos appear to pass through stages roughly resembling their evolutionary ancestors. This fact led early biologists to postulate that ontogeny (development of the individual) recapitulates phylogeny (evolutionary development of the species). However, the relationship between evolution and embryonic development is not simple. Heterochrony (variations in the timing of developmental processes) provides one mechanism for such recapitulation. If the developmental appearance of a somatic feature (such as gill slits) is accelerated, what was formerly an adult trait will become a juvenile trait or even an embryonic one in evolutionary descendents. Heterochrony may also produce descendents in which certain characteristics of juveniles are retained in adulthood. If advantageous, such heterochronic variations in development may have important evolutionary consequences.

KEY TERMS

basal lamina
cell adhesion molecule
chemotaxis
chimera
collagen
commitment
competence
contact guidance
contact inhibition

cytoplasmic determinant
fate map
founder cell
homeo box
heterochrony
homeotic mutation
housekeeping protein
morphogenetic movement
mosaic

neoteny
ontogeny
organizer
pedomorphosis
phylogeny
tight junction
transcriptional factor IIIA (TFIIIA)
transgenic animal

QUIZ QUESTIONS

1. _____ studied the problem of gene regulation of development by assembling artificial early embryos from cells of different embryos.
   A. Hans Spemann
   B. Beatrice Mintz
   C. Paul Gross
   D. Philip Leder
   E. Walter Gehring

2. Which of the following is not a strategy by which cells communicate with each other?
   A. contact guidance
   B. tight junctions
   C. contact inhibition
   D. chemical signaling
   E. self-assembly

3. *Caenorhabditis elegans* is an ideal model species for study of development because
   A. it has a short life cycle.
   B. the developmental history of every cell of the body is known.
   C. its body is transparent.
   D. A and B
   E. A, B, and C

4. _____ induces ectoderm to differentiate into structures of the nervous system.
   A. The "homeo box"
   B. Chordamesoderm
   C. Fibroblasts
   D. A hybridization probe
   E. A teratocarcinoma

5. _____ refers to a situation in which traits characteristic of juvenile ancestors are retained by adult descendants.
   A. Hypermorphosis
   B. Paedomorphosis
   C. Recapitulation
   D. Chemotaxis
   E. Insertional mutagenesis

ESSAY QUESTIONS

1. What is meant by the term "housekeeping proteins"? Given some examples. What proteins are said to be "tissue specific"?

2. What is meant by "commitment"?

3. Define "homeotic mutation." What is a homeo box and what is its function? What common properties do homeotic genes share? Suggest a plausible mechanism by which a zebra acquires its stripes.

4. What is "embryonic induction"? What role does it play in neurulation? Contrast the concepts of embryonic induction, competence, and commitment.

5. What is meant by the phrase "ontogeny recapitulates phylogeny"? Is this a true statement in the context of human development?

REFERENCES AND SUGGESTIONS FOR FURTHER READING

Cold Spring Harbor Symposia (1985). *Molecular Biology of Development.* Cold Spring Harbor Symposia on Quantitative Biology, Vol. 50. Cold Spring Harbor, N.Y.

A massive compilation of current research.

Davidson, E. H. (1986). *Gene Activity in Early Development.* 3rd ed. Academic Press, Orlando, Fla.

A multifaceted review of current knowledge of how genetic information is used in embryogenesis.

Dawid, I. B., and Sargent, T. D. (1988). *Xeno-pus laevis* in developmental and molecular biology. *Science 240,* 1443–1448.

A summary of the contributions of students of *Xenopus laevis,* the prime system for the study of embryogenesis in vertebrates.

De Pomerai, D. (1985). *From Gene to Animal. An Introduction to the Molecular Biology of Animal Development.* Cambridge University Press, New York.

Useful summary of current understanding.

De Robertis, E. M., and Gurdon, J. B. (1980). Gene transplantation and the analysis of development. *Scientific American 241,* December, 74–82.

Evidence that purified genes microinjected into an amphibian oocyte can be accurately "read."

DeRobertis, E. M., Oliver, G., and Wright, C. V. E. (1990). Homeobox genes and the vertebrate body plan. *Scientific American 263,* July, 46–52.

A recent, well-illustrated discussion of the gene group, so similar in fruit flies, frogs, and mammals, that controls the body plan.

Desplan, C., Theis, J., and O'Farrell, P. H.

(1985). The *Drosophila* developmental gene, *engrailed*, encodes a sequence-specific DNA binding activity. *Nature 318*, 630–635.

The first hint of mechanism in homeo box research.

Fjose, A., McGinnis, W. J., and Gehring, W. J. (1985). Isolation of a homeo box-containing gene from the *engrailed* region of *Drosophila* and the spatial distribution of its transcripts. *Nature 313*, 284–289.

An important paper from the Basel laboratory that first recognized what we now call the homeo box. Here it is shown that like many of the homeotic genes of the bithorax and *Antennapedia* complex, the *engrailed* gene has a homeo box sequence and a periodic pattern of expression in *Drosophila* embryos. This gene has since been cloned.

Gehring, W. J. (1987). Homeo boxes in the study of development. *Science 236*, 1245–1252.

A recent review of the author's pioneering research.

Gordon, J. W. (1989). Transgenic animals. *International Review of Cytology 115*, 171–230.

A summary of lessons learned from these remarkable preparations.

Hamburger, V. (1988). *The Heritage of Experimental Embryology: Hans Spemann and the Organizer*. Oxford University Press, New York,

The author, the same age as this century, was a graduate student in Spemann's laboratory in Freiburg. Here he offers a small masterpiece of explanation and recollection.

Howard, K. (1989). Developmental biology: *Drosophila* back to front. *Nature 338*, 618–619.

Brief discussion of the question: why does the fate of a cell depend on its position in the embryo?

Kenyon, C. (1988). The nematode *Caenorhabditis elegans*. *Science 240*, 1448–1453.

In *C. elegans*, patterns of cell division, differentiation, and morphogenesis can be observed at the single-cell level in intact, living animals. This is a stimulating review of what has been learned.

Kessin, R. H. (1981). Conservatism in slime mold development. *Cell 27*, 241–243.

Summary of studies implicating cyclic AMP in the aggregation step in the development of the cellular slime mold *Dictyostelium discoideum*.

Marx, J. (1984). *Caenorhabditis elegans:* getting to know you. *Science 225*, 40–42.

News story reporting that the complete cell lineage of *C. elegans* has been worked out, an important and portentous achievement that would aid future study of the role of genes in development.

Strome, S. (1989). Generation of cell diversity during early embryogenesis in the nematode *Caenorhabditis elegans*. *International Review of Cytology 114*, 81–124.

A recent review of progress in an active field.

Wolffe, A. P., and Brown, D. D. (1988). Developmental regulation of two 5S ribosomal RNA genes. *Science 241*, 1626–1632.

The developmental regulation of two kinds of *Xenopus* 5S rRNA genes (oocyte and somatic types) is due to differences in the stability of certain protein-protein and protein-DNA interactions. Here is the latest report by the discoverer of this important system.

Wolpert, L. (1978). Pattern formation in biological development. *Scientific American 239*, October, 154–164.

This clear treatment of the concept of pattern formation in the development of animals explains how positional information helps to determine the developmental fates of cells.

HOW ORGANISMS REPRODUCE

A walk through a New England forest in winter almost persuades one that life has perished, never to return. Trees stand gaunt and bare. Plants look dead. Even rodents and insects—if any are to be found—seem lifeless.

But spring brings resurrection. Life returns. New plants sprout from seeds and spores. Trees erupt in leafy grandeur. Insects emerge from numberless tiny eggs. As if from nowhere, birds return, nest, and mate. Hibernators awaken and the freshening season works its yearly magic. What we are witnessing, of course, is not death and resurrection but the dormancy and awakening of the changing seasons. Each spring nurtures new reenactments of the cycles in which life, repeating ancestral patterns, follows life.

Reproductive cycles are a universal attribute of life. They impel all living things, wherever they live and for however long. The life of every organism has a beginning. Its "birth" may consist in the fission of a single parent, the germination of a seed, the cracking of an egg, or the expulsive contractions of a distended uterus. Development follows birth, and at sexual maturity reproduction takes place once again. In sum, this cycle consists of the repetitive sequence:

reproduction → development → developed organism → reproduction → . . .

In this chapter, we consider the reproduction of organisms—with emphasis on eukaryotes. (The reproduction of cells was discussed in Chapter 12.) However whimsical it may seem, we will argue that the major role of the developed individual is to serve as a medium of reproduction. In this view, the individual is a drudge that appears between the acts. The acts, of course, are episodes of reproduction. An individual's other activities and functions are directed primarily at facilitating its reproductive moments.

GENERAL FEATURES OF THE REPRODUCTIVE CYCLE

In a sense, the story of reproduction is the story of biology. Because it has many forms and guises throughout the kingdoms of the living world, the only adequate definition of reproduction is a general one: the production of offspring by organisms.

NECESSITY OF REPRODUCTION

There would be no further need for reproduction if living organisms became immortal. If survival were the goal of every individual organism, then that goal would be achieved in an ultimate way by immortality. If, however, immortal organisms retained their old reproductive habits the world would become a very crowded place. Fanciful perhaps, but these considerations make clear the necessity of both death and reproduction. Moreover, as we shall see, without death and reproduction, there could be no evolution.

How organisms reproduce
Aspects of the process of reproduction, the most characteristic feature of all living systems, are illustrated here in a wide variety of organisms.

Organisms differ widely in their ways of reproducing. These variations seem related to longevity. Organisms that devote a large fraction of their time and energy to reproduction (as do insects) usually generate offspring at a rapid rate. But their survival as individuals is short. On the other hand, those that conserve their time and energy, giving meager amounts of both to reproduction (as do humans), in most cases survive longer as reproductive rates are slowed. It would be unwarranted to infer a direct cause-and-effect relationship between longevity and reproductive

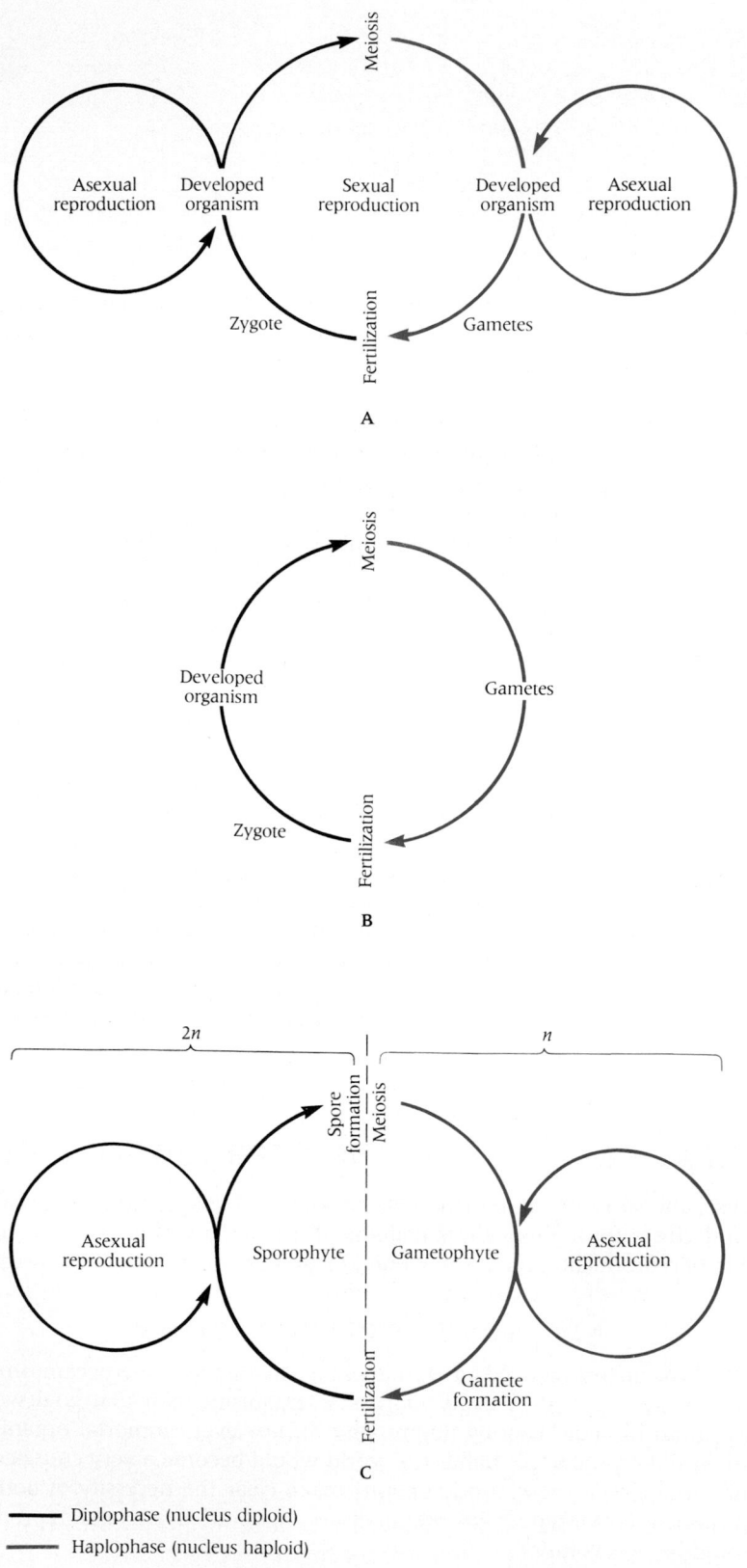

21-1 Generalized reproductive cycles
(A) In the generalized reproductive cycle, organisms can reproduce either asexually or sexually. (B) In the sexual reproductive cycle of most animals, gametes are produced to form a genetically distinct organism. (C) In the plant life cycle, plants form gametes in the gametophyte generation and spores in the sporophyte generation. Asexual reproduction may or may not occur, depending on the plant species.

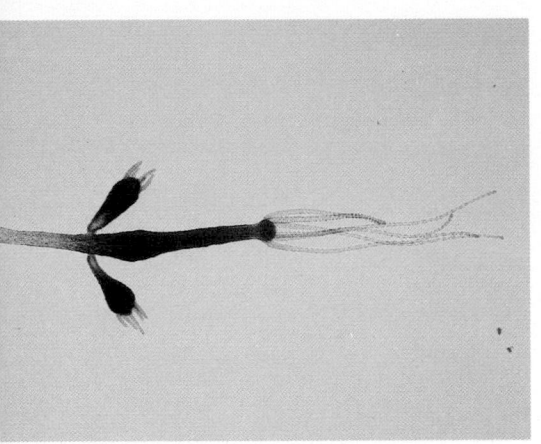

21-2 Vegetative reproduction in Hydra
Two buds with tentacles are forming on the body of this animal. Eventually each bud will separate from the parent and become a new individual.

output. Rather, these are **coadapted characters,** the importance of which relates to what ecologists call life history strategies (see Chapter 45). It appears, nonetheless, that in the course of evolution each species invests the amount of time and energy in reproduction that is appropriate to its circumstances.

SEXUAL AND ASEXUAL REPRODUCTION

We saw the significance of the difference between sexual and asexual reproduction in Chapter 15. It is summarized in Figure 21-1.

- **Asexual reproduction** involves no gametes and no fusion of nuclei. In many cases, it is little more than the division—by fission in prokaryotes, by mitosis and cytokinesis in eukaryotes—of a single parent. Hence, such reproduction is usually uniparental (one parent). For reasons of size and scale, uniparental asexual reproduction must be confined to prokaryotes, protists (unicellular eukaryotes) such as *Amoeba,* and certain plants and fungi. However, asexual reproduction is not always uniparental, as we shall see.
- **Sexual reproduction** involves a fusion of two haploid nuclei from different germ cells (gametes) that yields a single-celled diploid zygote. Hence, we define sexual reproduction as a generally biparental (two parents) process that usually leads to recombination of DNA from two different organisms. We say "usually" because there are exceptions.

Forms of Asexual Reproduction

There are at least four major modes of asexual reproduction.

- Fission, or mitosis and cytokinesis
- Vegetative reproduction
- Sporulation
- Parthenogenesis

Fission—or **binary fission** as it is sometimes called—is the reproductive mode of single-celled prokaryotic organisms. **Mitosis** (and **cytokinesis**) is the reproductive mode of many (but not all) unicellular eukaryotic organisms (see Chapter 12). In both processes, the organism enlarges and the cell wall or plasma membrane constricts to form two daughter cells, each of which regains the size and configuration of the original parent. An intriguing aspect of this reproductive system is that today's offspring are in part yesterday's parents. Parental protoplasm survives in the offspring, augmented by new protoplasm.

Vegetative reproduction is a process in which organisms sprout offspring from their bodies as appendages, which in time separate and take up an independent existence. Its mechanisms may be complex, but the principle is simple: part of the parent develops an offshoot that becomes a separate organism. Inevitably, parent and offspring are genetically identical. The process, often called **budding,** occurs in animals (Figure 21-2), fungi, and plants, but is less common in animals.

Many plants have special organs of vegetative reproduction—for example, the tubers of potatoes (Figure 21-3), bulbs of onions and tulips, bulblike organs of gladioli, and runners of strawberries. Vegetative reproduction also operates when plants are propagated by cuttings, a special form of which is **grafting** (the attachment of a cutting from one plant to the growing stem or root of another). Graft and host become part of the same physiological system and come to act like a single plant though there has been no exchange or mixture of chromosomes of the combined plant. When, for example, a peach branch is grafted to a plum tree, the grafted branch will continue to bear peaches while the rest of the tree bears plums.

Sporulation is a form of asexual reproduction employed by many multicellular organisms. Plants and fungi generally form gametes in one phase of their life cycle and **spores** in another. Like gametes, spores are haploid cells. Unlike gametes, however, they can undergo mitosis. Indeed, they can form whole multicellular organisms. The original individual plant may be termed a **sporophyte** (a plant that makes spores). The organism developing from dividing spores is called a **gameto-phyte** (a plant that makes gametes) (see Figure 21-1). As we shall see, such plants

21-3 A potato tuber is an organ of vegetative reproduction

TABLE 21-1

TERMINOLOGY OF REPRODUCTIVE BIOLOGY: A GLOSSARY

Term	Definition
Phases of Life Cycle:	
Diplophase	Phase of a life cycle, occurring after fertilization but before meiosis, in which cells are diploid.
Haplophase	Phase of a life cycle, occurring after meiosis but before fertilization, in which cells are haploid.
Gametophyte	A haploid (n), gamete-producing phase in the life cycle of a plant with alternation of generations. This phase, which alternates with diploid ($2n$) sporophyte phase, is prominent and independent or reduced and parasitic.
Sporophyte	A diploid ($2n$), sexual spore-producing phase (in which meiosis occurs) in the life cycle of a plant with alternation of generations. Alternates with haploid (n) gametophyte phase; may be prominent and independent or reduced and parasitic.
Reproductive Cells:	
Gamete	A haploid (n) reproductive cell that arises from meiosis. In fertilization, a gamete nucleus fuses with that of another gamete of opposite sex or type to produce a diploid zygote.
Spore	In plants, fungi, and some protists, sexual spores are haploid reproductive cells that arise from meiosis (meiospores) and are capable of undergoing mitosis. In fungi, asexual spores arising from mitosis (mitospores) are cast off somatic cells that can act as initial cells capable of developing into an adult (by mitosis) without fusion with another cell.
Types of Reproductive Cycles:	
Gametic meiosis	Haploid gametes formed by meiosis in diploid individuals, which fuse to form a diploid zygote that develops by mitosis into another diploid individual. It occurs in most animals and some protists. Developed organism is in diplophase of the cycle. Only the gametes are haploid cells.
Zygotic meiosis	Zygote divides immediately by meiosis to form four haploid cells that divide by mitosis to produce more haploid cells or a multicellular individual that forms gametes. Occurs in *Chlamydomonas* and other algae. Zygote is only diploid cell.
Sporic meiosis	Diploid individual (sporophyte) produces haploid spores by mitosis that do not function as gametes but undergo mitosis to form a multicellular haploid individual (gametophyte) that produces gametes that fuse to form diploid zygotes which develop into diploid individuals. Termed alternation of generations. Characteristic of plants and many algae.

switch back and forth between asexual and sexual modes of reproduction. (We offer a special glossary of the terminology of reproductive biology in Table 21-1.)

Parthenogenesis is a mode of asexual reproduction in which a haploid organism develops from an unfertilized egg that is activated by means other than sperm entry. It can be artificially induced in the laboratory (see Chapter 19), but in some animals—for example, rotifers, several genera of fishes, certain amphibians and lizards, and in some of the microscopic multicellular organisms found in stagnant water—males are unknown and parthenogenesis occurs naturally.

Parthenogenesis occurs cyclically in a number of species—for example, when they find themselves for a brief time in such a congenial environment that it is to their advantage to breed rapidly to make the most of it. An example is the tiny aphid that lands on a suitable plant in early spring (Figure 21-4). To exploit its favorable location before its enemies arrive, the aphid reproduces at a furious rate by parthenogenesis.

In animals, parthenogenesis can occur in two ways: by automixis and by apomixis.

■ In **automixis,** the egg undergoes normal meiotic reduction and diploidy is restored either by fusion of two of the four pronuclei, or by fusion of two genetically identical cleavage nuclei.

- In **apomixis,** offspring genetically identical to the parent arise without meiosis. Although a very rare event, it can give rise to a parthenogenetic strain. This mechanism probably led to weevils that are obligatorily parthenogenetic and the cyclically parthenogenetic aphids.

Parthogenesis occurs not only in animals but also in plants. About 80% of all plants display some form of asexual reproduction; about 50% are primarily or exclusively asexual. Botanists use the term apomixis to include both vegetative reproduction and **agamospermy,** which is the production of viable seeds in the absence of fertilization. In addition, the absence of meiosis results in the production of an egg with an unreduced chromosome number. This then develops into an embryo that is genetically identical to the parent plant. This mode of reproduction combines the advantages of vegetative reproduction (rapid spread of genetically identical organisms) with the advantages of seed production (resistance to drying, dormancy, and the availability of dispersal mechanisms). The dandelion (*Taraxacum officinale*), so common in our lawns, is an agamospermous plant.

21-4 Aphids, **Longistigma caryae,** *feeding on a plant*
By alternating sexual and asexual phases in their life cycle, aphids enjoy the benefits of both reproductive models.

Sexuality and Its Advantages

If asexual reproduction is so simple and rapid, why is it not the universal method of reproduction? The answer: because it does not result in the fusion of genetic elements from two organisms which generates diversity among offspring.

Before exploring the large advantages of sexual reproduction, we should be aware that it also has certain disadvantages. For one, genetic recombination, a major consequence of sexuality, can have the deleterious effect of breaking up the gene complexes of coadapted characters. That would be bad for organisms living in a constant environment. Second, sex has certain costs. Consider a species with no paternal care of the young. A female of that species which produced only female offspring parthenogenetically would be spared the burden of useless male offspring. These asexual females would be at an advantage when compared with sexual females producing equal numbers of male and female offspring—and whose genomes were repeatedly being diluted by meiosis and syngamy.

Why then do so many animals and plants reproduce sexually? One of the reasons may be that sexual reproduction provides counterbalancing advantages when two gamete nuclei fuse within a single cell, which then becomes a new organism. Gametes in most plants and animals are of two kinds: the smaller, motile male gametes, or sperm, and larger, nonmotile female gametes, or eggs. In some protists and plants, the two cells whose nuclei fuse in sexual reproduction are not visibly different from one another or from ordinary cells of the species. There are no specialized germ cells. As a matter of fact, in one view the gametes of protists are their nuclei. Even though male and female are indistinguishable, the term "gamete" is apt because such cells have gone through meiosis, have a haploid chromosome number, and are ready for fertilization.

Plants and animals have evolved many methods for reproducing sexually. Still, the essential features are universal. Whether sexual reproduction occurs in protists, in plants, or in animals, it has the same evolutionary and biological significance. That significance rests squarely on the fact that in sexual reproduction half the chromosomes from each of two individuals unite in a single new individual. In Chapter 12 we saw that during meiosis, parental gene combinations undergo recombination or reshuffling. From the evolutionary viewpoint, that is invigorating.

Consider, for example, two organisms homozygous for two gene pairs: *AABB* and *aabb*. If each can reproduce only by asexual means, the offspring will always be *AABB* and *aabb*, respectively, unless a rare mutation converts one of the alleles to something new. But if the two organisms were to reproduce sexually, the very next generation would be hybrids (*AaBb*). If these were to mate together, the next generation of offspring would be richly varied—*AABB, AABb, AAbb, AaBB, AaBb, Aabb, aaBB, aaBb,* and *aabb*. Crossing-over and recombination would add further variation. Sexual reproduction thus yields arrays of even richer variants because real organisms contain many thousands of gene pairs.

Organisms that reproduce asexually often exist in a "comfortable" environment that places them under little stress. However, we will soon see how asexually reproducing organisms sometimes turn to sexual reproduction when the "going gets tough." Examples are aphids in the autumn when food supplies dwindle and *Hydra* and various algae when ponds dry up and conditions deteriorate. This is a strategy that seeks to exploit the greater opportunities for variation in sexual reproduction and thus its possible genetic advantage in a changing, and menacing, environment. Sexual reproduction also tends to eliminate "deviant" chromosomes that are not able to pair.

Biological Distribution of Sex

Sexual reproduction is nearly universal in the animal kingdom and occurs in half of the plant kingdom. Even among bacteria, which we ordinarily think of as reproducing by asexual fission, a form of sexual reproduction can occur, as we saw in Chapter 16. This surprising discovery of the 1950s rested on two kinds of evidence.

First, it was found that gene exchange can and does occur between two strains of *E. coli*. In studies using mutant strains requiring one or another nutritional supplement, culturing two different mutants together produced an offspring containing two genes, each previously present in a different strain. The basis of such gene exchange is **conjugation.** Its existence was the second line of evidence revealing sex in bacteria (Figure 21-5; see also Figures 16-7 and 38-13).

Sexual union also occurs in other prokaryotes—mainly bacteria and perhaps cyanobacteria (blue-green algae), which lack true nuclei and mitotic and meiotic apparatuses. Some biologists object to calling this form of chromosomal fusion sexual reproduction. They say it is not the same as the joining of nuclear material from two individuals that occurs in eukaryotes during **syngamy**—fusion of the nuclei and cytoplasm of two haploid gametes in true sexual fertilization. Nevertheless, conjugation in prokaryotes and syngamy in most multicellular eukaryotes have the same biological significance. In both, genetic material arising from two parents is extensively recombined.

To simplify discussion, we will use the term **sex** to mean meiosis followed by nuclear fusion—events that occur only when there is a fully developed nuclear, chromosomal, and mitotic system. Such systems are found only in eukaryotes, the group that includes all multicellular organisms and protists. Eukaryotes have the structures and systems necessary for sexual reproduction and a great many reproduce sexually, though as noted above half of all plant species reproduce asexually.

Why are there only two sexes in sexually reproducing organisms? Because two sexes are enough to accomplish genetic recombination. More than two would surely complicate the genetic basis of sex determination—to say nothing of the processes of sexual union.

The two-sex generalization is not in conflict with the fact that in some species both kinds of sex organs are normally found in the same organism. Such organisms are called **hermaphrodites.** Hermaphrodites are "simultaneous" as in most land and freshwater snails, or "sequential," when an individual is first male and then female or vice versa, as in some ferns. Many plants have both male and female sex organs, though botanists rarely use the term hermaphroditic, preferring **perfect** for flowers with both parts, and **imperfect** for those lacking one part or the other. The terms **complete** and **incomplete** denote, respectively, the presence or partial absence of any of the four major parts (see Figure 21-16).

A perfect flower produces both seeds and pollen. This condition is commonest among flowering plants. Some plants (like corn and oak and birch trees) are **monoecious,** which means that male and female organs occur in the same plant but in different flowers. Others (like the avocado and willow tree) are **dioecious;** their male and female organs occur in different individual plants.

Hermaphroditism also occurs in animals—normally and pathologically (see Chapter 15). It is most commonly found in sponges and mollusks and in parasites such as tapeworms and flukes—creatures that might have difficulty finding one another if they had to rely on the ordinary pattern of sexual union. In these forms, the same individual produces male and female gametes and engages in self-fertilization.

Why isn't this the universal method of reproduction? The main reason is that the potential for variability among offspring is lower when the gene pool is not regularly fortified by the contributions of other individuals. Another likely reason is that the existence of two sexes divides the labor in advantageous ways. Females can specialize in making eggs, finding places to deposit them, and, in some species, nursing and protecting the young. Males specialize in finding and fertilizing females,

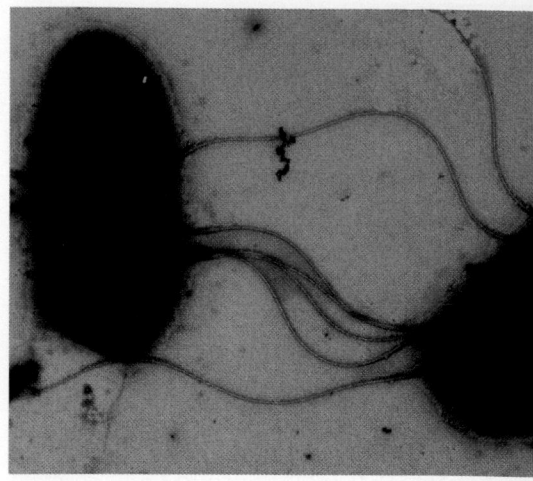

21-5 Conjugating E. coli
Here are two bacteria of opposite sex, or mating type. DNA is being transferred through long, hairlike channels from the male cell to the female.

scattering their genes widely, and in many species locating and protecting territory in which the young will be born.

Reproductive Strategies: A Brief Guide

Countless reproductive techniques have evolved. Before exploring them further, we should note certain general principles.

It is convenient to start with human reproduction, which resembles that of most other mammals, differing only in detail. It is a sexual cycle in which gametogenesis is followed by fertilization (see Figure 21-1B). Then comes development, meiosis, and gametogenesis again as the cycle repeats in the next generation.

Note that meiosis and fertilization divide the life cycle of a sexually reproducing organism into two parts.

- **Haplophase** occurs after meiosis but before fertilization. The chromosome number in the germ cells is haploid (there are n chromosomes).
- **Diplophase** occurs after fertilization but before meiosis in the offspring. All cells have the diploid number of chromosomes ($2n$ chromosomes).

Eukaryotes display three types of life cycles, as summarized in Figure 21-6 and Table 21-1. Let us list them and then explore them further.

- In **gametic meiosis,** haploid gametes are formed by meiosis that takes place in a diploid individual (see Figure 21-6A). They fuse to form a diploid zygote that develops by mitosis into another diploid individual. This type of life cycle is seen in most animals (including humans), some fungi (e.g., the water molds), and the brown alga, *Fucus.* Only the gametes are haploid cells (and thus in haplophase).
- In **sporic meiosis,** the diploid individual (sporophyte) produces haploid spores by meiosis (see Figure 21-6B). These spores do not function as gametes but undergo mitosis to form a multicellular haploid individual (gametophyte). This individual eventually produces gametes that fuse to form diploid zygotes, which later differentiate into diploid individuals. This cycle, termed **alternation of generations,** is characteristic of plants and many algae.
- In **zygotic meiosis,** the zygote immediately undergoes meiosis to form four haploid cells that divide by mitosis to produce more haploid cells or a multicellular individual that eventually gives rise to gametes by differentiation (see Figure 21-6C). This type of life cycle occurs in *Chlamydomonas* and several other algae, and fungi. The zygote is the only diploid cell.

Gametes are, in a real sense, haplophase organisms. They have only one set of chromosomes, but it is a complete set that permits an individual gamete to live and carry out its metabolic functions. Human gametes are short-lived cells that unite into a zygote or die. But, as we have seen, some organisms exist in which haploid cells arise by meiosis and lead longer lives—to the point of becoming independent multicellular organisms. Male bees, for instance, are haploid, as are the gametophytes of plants.

A specific life-cycle diagram like the one in Figure 21-7, which shows the reproductive cycle of the protist *Chlamydomonas,* can be readily generalized to illustrate the reproductive cycles of all living organisms. The broader scheme illustrates again the importance of fertilization and meiosis. Now we can place asexual reproduction in the scheme of things. When a plant gametophyte reproduces, it is on the asexual sidepath of the haplophase half of the cycle. Most protists repeatedly go through such an asexual cycle (by fission), but once in a while they go round the sexual cycle. Then come more asexual cycles. All sorts of variations occur in the sequences of asexual and sexual cycles. Occasionally, both cycles go on simultaneously in different offspring.

Fully developed organisms thus can develop in three ways.

- Development can occur in diplophase only, as in humans.
- Both diplophase and haplophase can be part of the life cycle, as in many plants and some animals (e.g., bees).
- In the rarest possibility, development takes place in haplophase only (although it does occur regularly in certain protists, plants, and fungi).

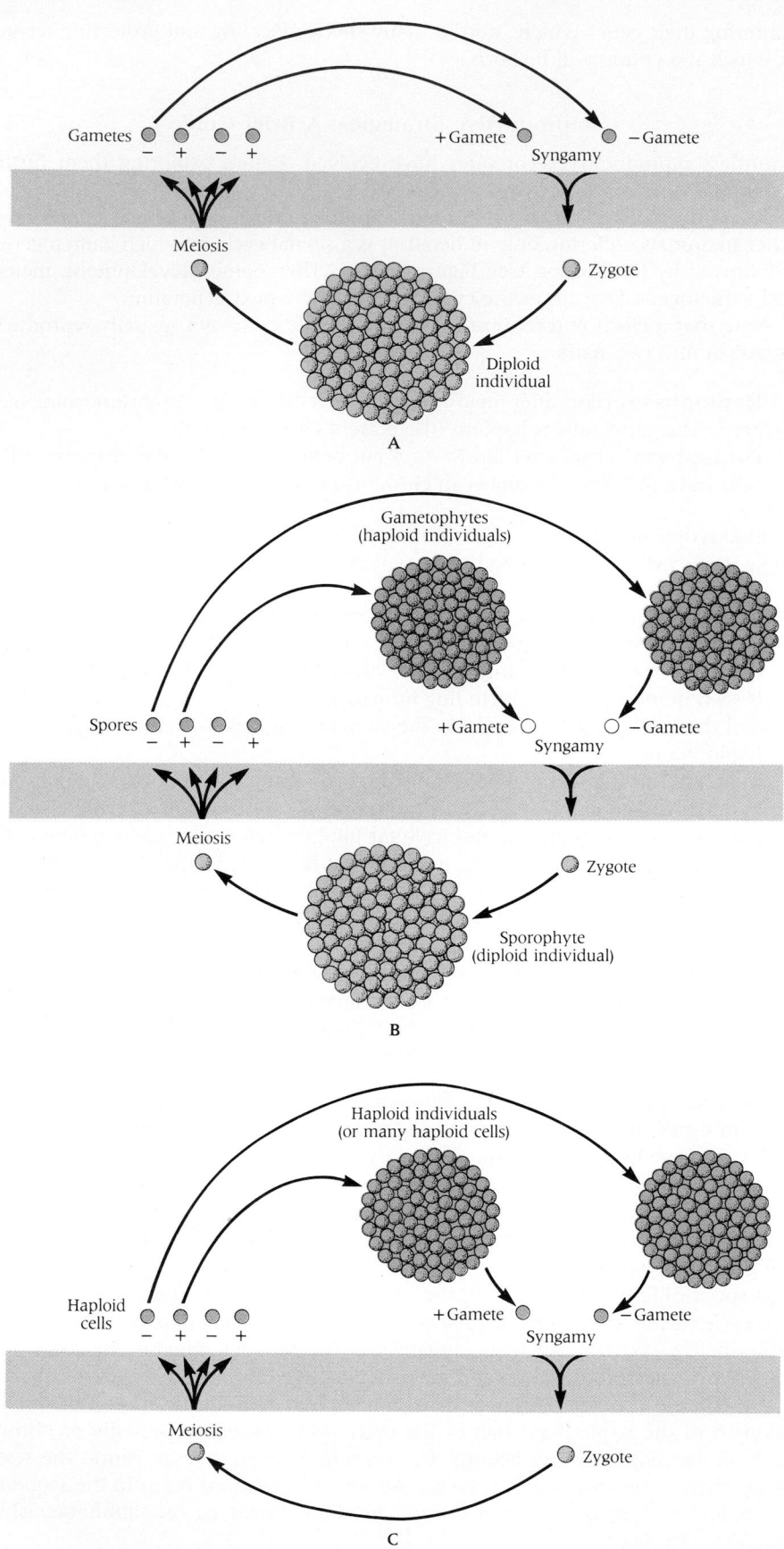

21-6 Principal types of life cycles
(**A**) *Gametic meiosis.* (**B**) *Sporic meiosis.*
(**C**) *Zygotic meiosis. Compare this figure with Figures 12-37 and 12-38.*

We see that many developed organisms, whether in diplophase or haplophase, can reproduce asexually as well as sexually. Figure 21-6 depicts the three known types of reproductive cycles. Note again that in all three of nature's reproductive stratagems, meiosis decreases the chromosome number from diploid to haploid.

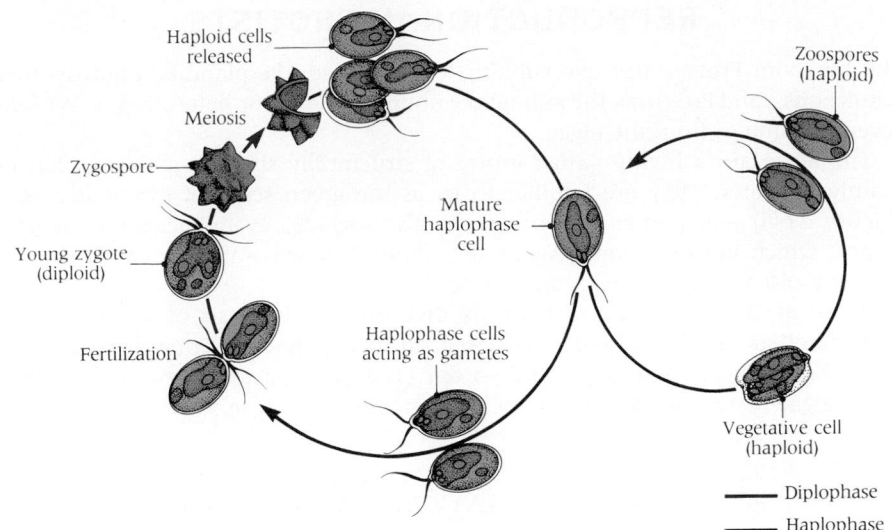

21-7 Reproductive cycle of Chlamydomonas

Most of the life cycle usually involves vegetative reproduction (asexual) during haplophase (right side of the diagram). Sometimes haploid cells act as gametes and fuse to form a diploid zygote (left). Although shown here at the same size, the gametes are often smaller than the mature haplophase cell.

Fertilization is the process in which fusion of haploid gametes returns the life history to the diplophase.

The main value of this important diagram lies in its broad depiction of the fundamental events occurring within a given reproductive cycle. This makes it easier to compare the many known kinds of cycles.

The balance of this chapter considers a variety of reproductive cycles in greater detail—first in protists, then in plants, fungi, and animals.

CLASSIFICATION OF ORGANISMS: A PREVIEW

Living organisms are divided into five **kingdoms.** We noted that fact in Chapter 5 and consider it fully in Chapter 37. They are as follows:

- Monerans (kingdom Monera): bacteria and blue-green algae (also called cyanobacteria)
- Protists (kingdom Protista)
- Plants (kingdom Plantae)
- Fungi (kingdom Fungi)
- Animals (kingdom Animalia)

The last three kingdoms reflect the three great ecological stratagems for larger organisms: production (plants), absorption (fungi), and consumption (animals).

Vast numbers of species have been described and named in all five kingdoms, yet many problems surround the classification of organisms. For example, one group of organisms, the algae, were once considered protists by some authorities or plants by others. There has even been disagreement about which organisms are to be called algae.

The greatest division is not between plants and animals, but between two kinds of microorganisms—the prokaryotic Monera and the eukaryotic Protista. In a sense, the five kingdoms encompass three great levels of life.

- Prokaryotes (bacteria of the kingdom Monera)
- Eukaryotic microorganisms (protists)
- Eukaryotic larger forms (plants, fungi, and animals)

Brief descriptions of the kingdoms accompany the following discussions, and tabular summaries may also be consulted for background as the reader makes his way through the balance of this chapter.

- Table 5-1 lists characteristic cell structures (and some functions) of the five kingdoms.
- Table 37-2 lists other characteristics of the five kingdoms.
- The *Appendix* lists the major phyla and other subdivisions of all five kingdoms.

REPRODUCTION IN PROTISTS

The kingdom Protista has two subkingdoms—Algae, the plantlike photosynthetic autotrophs,[1] and Protozoa, the animallike nonphotosynthetic heterotrophs. We select several examples from the algae.

The algae are a highly varied group of structurally simple organisms that live mainly in water. They are familiar to us as the green scum of freshwater ponds (Figure 21-8)—or part of it, since scum also includes cyanobacteria (blue-green algae), which are monerans. Algae also abound in sea water, in damp soil, and on the moist surfaces of rocks and wood.

Algae are classified into at least six divisions on the basis of various criteria. Their familiar names are generally based on their pigments. Thus there are green algae, brown algae, red algae, and so on. (The cyanobacteria, once called "blue-green algae," are now considered to be monerans, not algae.)

A GREEN ALGA: *CHLAMYDOMONAS*

Chlamydomonas, a single-celled, photosynthetic green alga, swims actively by beating its two long flagella and darts rapidly about in stagnant fresh water pools. As shown in Figure 21-7, *Chlamydomonas* spends most of its life cycle in haplophase (on the right side of the diagram), reproducing asexually by mitosis—both daughter cells remaining within the wall of the original cell. In some species, the daughter cells are promptly released; in others, they divide again (and again) within the walls of their parent cell until 4, 8, or 16 cells are eventually freed, each with its own wall and flagella. These haploid forms are called **zoospores,** a general term for motile, asexually reproducing cells.

Under certain environmental conditions (low nitrogen in the medium, etc.), gametes arise within the zoospore population in equal numbers of two mating types—some call them male and female, others prefer (+) and (−). These are slightly smaller than typical adult cells, but resemble them otherwise. Some green algae have male and female gametes that are similar in size, form, and structure. These gametes are termed **isogamous.** Others have unequal sized **(anisogamous)** male and female gametes, and still others are **oogamous,** with a motile sperm and a large, nonmotile egglike female gamete (or oogamete).

Every so often, two haploid gamete cells come together—much like gametes undergoing syngamy. The result is a diploid zygote, called a **zygospore.** Gamete cells that do not fuse increase in size and become indistinguishable from other adult cells.

The zygote promptly loses its flagella and sinks to the bottom, where it acquires a thick protective wall and becomes dormant. It does not develop further but instead undergoes meiosis, becoming four haploid cells, each of which grows two new flagella. They are released as motile haploid zoospores, which are free to resume their former asexual life.

Chlamydomonas, perhaps the most primitive of algae, spends most of its life in the haplophase. The diplophase lasts only long enough to permit meiosis or to withstand adverse conditions. The dominance of haplophase is characteristic of primitive plants and was undoubtedly the ancestral condition in the evolution of sexual reproduction.

SEA LETTUCE AND BROWN ALGAE

In other algae, the haplophase is often overshadowed by the diplophase portion of the reproductive cycle.

Consider the alga *Ulva*, commonly called sea lettuce (Figure 21-9). This organism produces prominent haploid gametophytes and diploid sporophytes, and it is difficult to distinguish them on superficial examination. In life histories of this kind, a multicel-

21-8 Algae-covered pond
Green algae are the most varied of all the algae. Most of the many thousands of species are aquatic, often occupying quiet ponds like this one.

---

[1] We will see in Chapter 38 that the term "alga" is not always used as a formal classification. Indeed, several of the various algae are not related to one another.

A

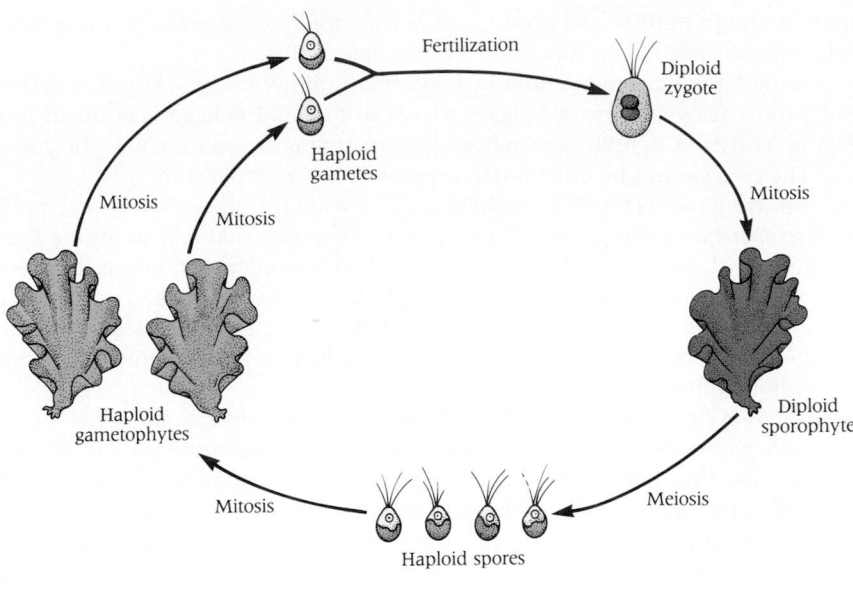

Fertilization

Diploid zygote

Haploid gametes

Mitosis

Mitosis

Mitosis

Mitosis

Haploid gametophytes

Diploid sporophyte

Mitosis

Meiosis

Haploid spores

B

21-9 Reproductive cycle of Ulva
*(A) A patch of Ulva off the Maine coast
(B) This green alga, commonly called sea lettuce,
reproduces by forming both diploid and haploid
cells during its life cycle.*

lular haploid stage alternates with a multicellular diploid stage. Hence the pattern reflects an alternation of generations. The sporophyte generation is the parent of the gametophyte generation, which is parent of the next sporophyte generation.

$$\text{Sporophyte} \xrightarrow{\text{meiosis}} \text{spore} \longrightarrow \text{gametophyte} \longrightarrow \text{gamete} \xrightarrow{\text{fertilization}}$$
$$\text{zygote} \longrightarrow \text{sporophyte} \ldots$$

In *Ulva*, the diploid sporophyte resembles a glistening flat ''leaf'' two cells thick and up to a meter long. Within it, specialized cells called sporocytes arise by differentiation and undergo meiosis to form haploid spores with four flagella. These well-equipped spores swim away, lose their flagella and begin to divide by mitosis, eventually forming another broad thin ''leaf.'' However, this one is a haploid gametophyte that looks very much like a diploid sporophyte. Indeed, they can be distinguished only by their counting chromosomes or observing their fate. The haploid gametophyte produces motile gametes (obviously haploid), which are male or female. Both types have flagella, but the female gamete is larger and of a slightly different color. Two gametes fuse to form a zygote, which rests and then begins the mitotic divisions that produce a new leaflike diploid sporophyte. Gametes that cannot find

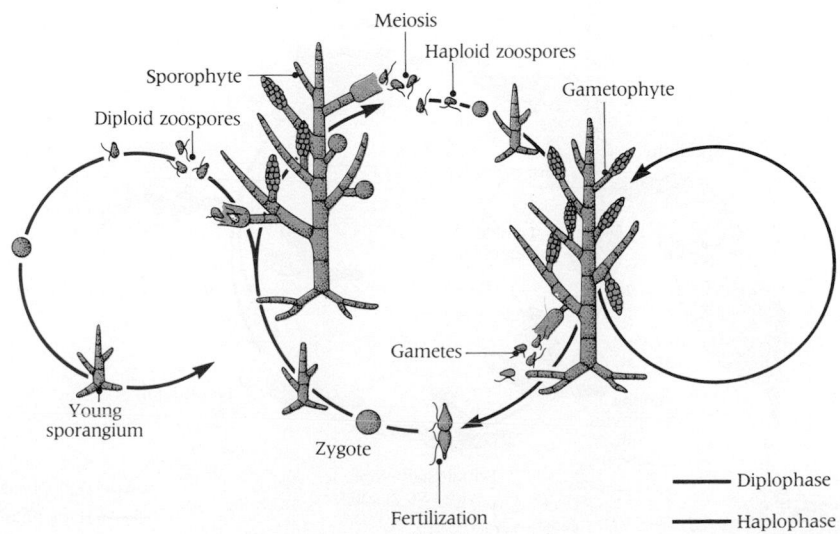

Meiosis

Haploid zoospores

Sporophyte

Diploid zoospores

Gametophyte

Gametes

Young sporangium

Zygote

Fertilization

—— Diplophase

—— Haplophase

21-10 Reproductive cycle of Ectocarpus
*In the life cycle of this brown alga, the
gametophyte and sporophyte generations are
equally prominent and very similar in
appearance. Vegetative reproduction
occurs in both.*

partners undergo mitosis and produce new gametophytes directly. In other words, gametes of this species can function much as spores do.

The same basic pattern is found in *Ectocarpus,* a brown algae classed as seaweed (though most seaweeds are red algae) that is so fine and delicate few might notice it (Figure 21-10). A definite sporophyte appears that is as prominent as the gameto-phyte. The two cannot be differentiated without microscopic study.

In sum, we have seen two reproductive patterns in the algae. In one, exemplified by *Chlamydomonas,* some gametophytes produce gametes that fuse to form a zygote. The zygote, with or without a resting period, directly undergoes meiosis to produce spores, which in turn produce new gametophytes. In the entire life cycle, only one cell, the zygote, is the diploid sporophyte. In the pattern exemplified by *Ulva* and *Ectocarpus,* gametes fuse to form a zygote, which then divides by mitosis to form a fully developed multicellular sporophyte.

Many life-cycle patterns differ from those of *Chlamydomonas* and *Ulva.* In some cases, both gametophyte and sporophyte generations are multicellular, with one phase (usually the sporophyte) being larger and more prominent than the other. For example, in *Laminaria,* a brown alga commonly known as kelp, the sporophyte is much more conspicuous than the gametophyte, becoming six feet in length or longer (Figure 21-11). In *Fucus,* another brown alga (Figure 21-12), the emphasis on the diplophase achieves some kind of limit. There is no gametophyte generation. Hence, this life cycle is similar to that of animals.

REPRODUCTION IN PLANTS

To this point, our excursions into the plant kingdom were considerations of the structure of plant cells and tissues in Chapters 5–7, of photosynthesis in Chapter 10, and, implicitly or explicitly, of genetics in Chapters 13–18. We discuss the wide world of plant life in Chapter 41. But if we are to recognize the sweep of reproductive biology, we must place the story of reproduction in plants alongside the story of reproduction in other kingdoms. So that we may do so, we here briefly introduce the plant kingdom (Table 21-2).

The plant kingdom is divided into **bryophytes** (nonvascular plants that include liverworts, hornworts, mosses, etc.) and **tracheophytes** (vascular plants that include ferns and seed plants). Needless to say, diversity in the plant kingdom is enormous.

FEATURES DISTINCTIVE TO PLANT REPRODUCTION

The major difference between the reproductive cycles of plants and animals is in their methods of gamete production. The method in animals is meiosis, in which chromosome number is reduced from diploid to haploid. In plants, gametes arise from meiosis so rarely it is fair to say that the absence of meiotic gametogenesis is a more universal feature of plant life than is photosynthesis. Meiosis does take

21-11 Reproductive cycle of Laminaria
(A) *The diplophase dominates the cycle with the production of a large sporophyte (six feet long). The gametophyte is a small filament of cells and occurs as two distinct sexes.* **(B)** *A pile of* Laminaria saccharina, *also called oarweed or poor man's weather glass.*

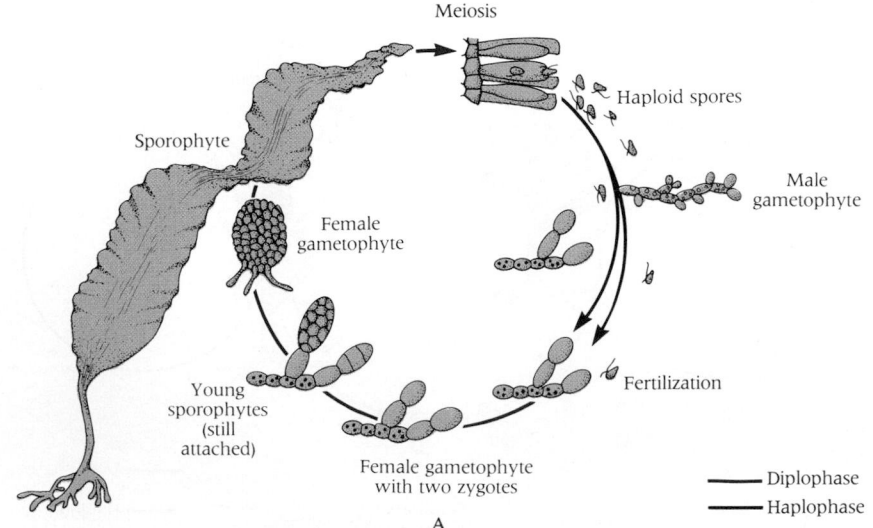

21-12 Reproductive cycle of Fucus
In this life cycle, the gametophyte generation is restricted to the gametes, and the sporophyte generation is completely dominant.

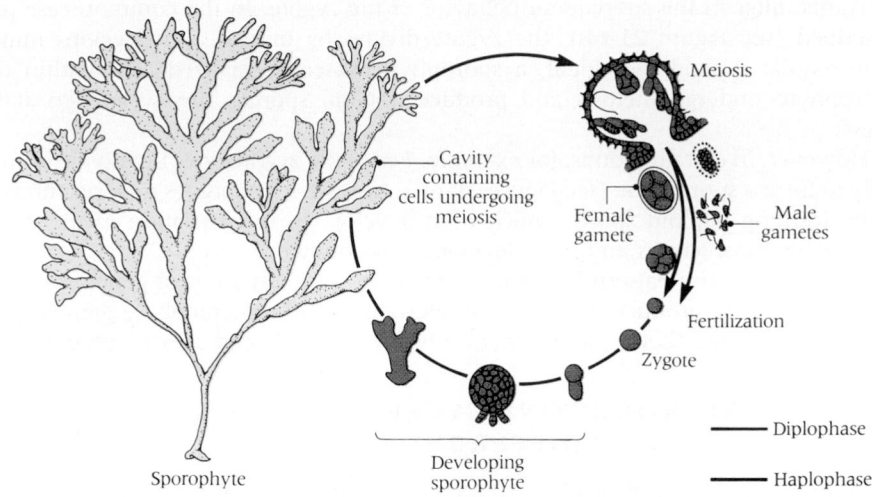

Cavity
containing
cells undergoing
meiosis

Meiosis

Female
gamete

Male
gametes

Fertilization

Zygote

Sporophyte

Developing
sporophyte

— Diplophase

— Haplophase

place in plants, but the resulting haploid cells do not immediately become gametes. They are, in fact, **spores.** Some call them meiospores to contrast them with mitospores, which arise by mitosis.

Gametophytes and Sporophytes

Before gamete production can take place, spores begin to divide by mitosis—haploid though they are. The result is a fully developed organism. Plants, it seems, can exist as conspicuous and fully developed organisms in both the haplophase and diplophase of the reproductive cycle. These are, respectively, the gametophyte, which is the developed haplophase organism, and the sporophyte, which is the developed diplophase organism.

The gametophyte arises from spores and produces gametes. The sporophyte produces spores and develops from a fertilization between gametes.

Variations on a Theme

In Figure 21-6, which outlines the main kinds of life cycles in animals and plants, we see that each plant cell undergoing sporic meiosis produces four haploid spores. Successive mitotic divisions of these cells can produce a multicellular haploid individual, a gametophyte. It in turn produces gametes—by mitosis, not meiosis—and when two of these fuse a zygote is produced.

TABLE 21-2
PROMINENT CHARACTERISTICS OF MAJOR PLANT GROUPS: A BRIEF SUMMARY

Group	Habitat	Method of Obtaining Energy	Relative Size	Structures Present			
				Roots, Stems, Leaves	Vascular Tissue	Seeds	Flowers
Bryophytes	Terrestrial moist, aquatic	Photosynthesis	Small	−	−	−	−
Ferns	Terrestrial, fresh water	Photosynthesis	Small to large	+	+	−	−
Seed plants:							
Gymnosperms	Terrestrial	Photosynthesis	Large	+	+	+	−
Angiosperms	Terrestrial, a few fresh water marine	Photosynthesis, a few are saprophytic	Small to large	+	+	+	+

Plants differ in the subsequent behavior of the zygote. In the common case just described (see Figure 21-6B), the zygote divides by mitosis and develops into a multicellular diploid individual, a sporophyte. Later, specialized cells within the sporophyte undergo meiosis and produce haploid spores. The cycle then starts anew.

However, in certain groups, for example, *Fucus*, the zygote does not divide mitotically to form a sporophyte (see Figure 21-6C). Instead, it undergoes meiosis immediately, forming haploid spores which then develop. In this variation of the cycle, there is no conspicuous and fully developed sporophyte.

In summary, the pattern in plants is described as alternation of generations—a diplophase spore-producing generation alternating with a haplophase gamete-producing generation, that is, an alternation between sexual and asexual phases.

REPRODUCTIVE CYCLES IN PLANTS OF DIFFERENT KINDS

To sense the diversity of plant reproductive patterns we consider typical examples from three widely differing plant categories—mosses (which are bryophytes), ferns, and nonflowering and flowering seed plants. We will see, among other things, that the cycles differ mainly in the extent to which each allows its gametophyte or sporophyte to develop into a visibly conspicuous multicellular organism. More will be said on this subject in Chapter 41.

Mosses

Mosses, among the simplest of green land plants, appear to have inherited from ancestral algae a reproductive cycle that is quite different from those of the more complex ferns and seed plants. The moss organism we all know, the one with little leaflike structures, is a haplophase gametophyte (Figure 21-13). It produces eggs and sperm—from the same plant in some species, from different male and female plants in other species.

Mosses usually live on land, but they favor moist environments. Fertilization is possible only when there is a thin coating of water through which the flagellated sperm cells can swim toward the eggs. Eggs are formed and retained in a structure called the **archegonium** (Figure 21-13A). Sperm are formed in and released from

21-13 Reproductive cycle of a moss
(**A**) *The prominent green gametophytes produce brown, parasitic sporophytes. Sperm reach the eggs by moving through a film of water.*
(**B**) *Fern on a moss-covered stump.*

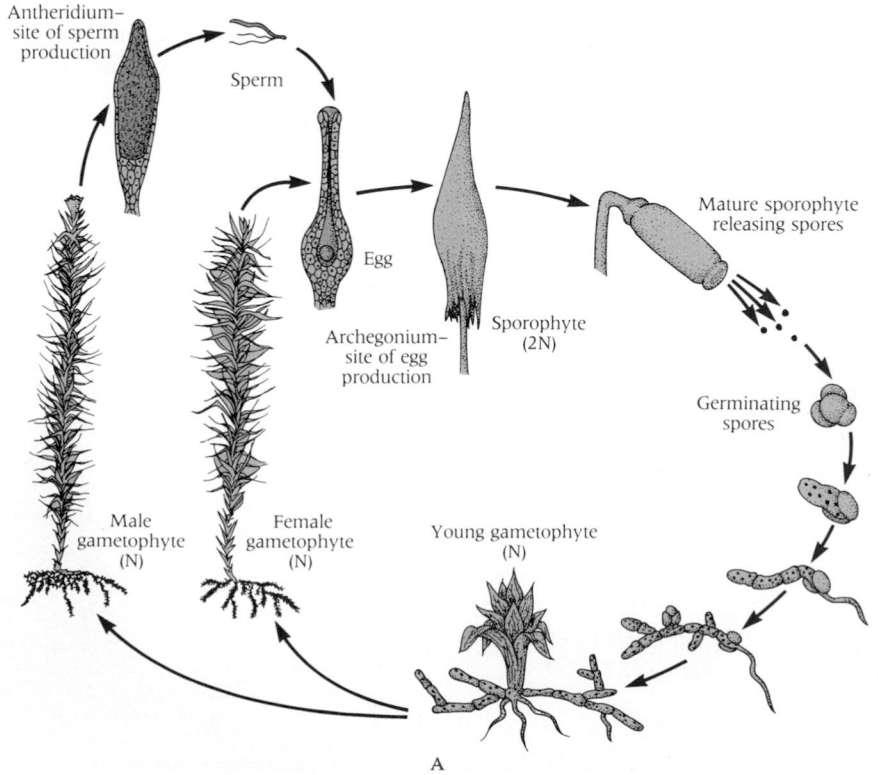

Antheridium–site of sperm production

Sperm

Egg

Archegonium–site of egg production

Sporophyte (2N)

Mature sporophyte releasing spores

Germinating spores

Young gametophyte (N)

Male gametophyte (N)

Female gametophyte (N)

A

B

21-14 Reproductive cycle of a fern
The sporophyte dominates the life cycle, while
the gametophyte is a small structure with male
and female organs. 1, Meiosis in the sporophyte
produces a haploid spore; 2, young gametophyte;
3, mature gametophyte; 4 and 6, antheridia
(male sex organs); 5 and 7, archegonia (female
sex organs); 6a, sperm released from antheridia;
7a, egg within archegonia fertilized by sperm,
8; 9, diploid zygote at base of archegonium; 10,
embryonic sporophyte within the archegonium,
11; 12, young sporophyte still attached to parent
gametophyte; 13, mature sporophyte; 14, clusters
of sporangia; 15, single sporangium; 15a, release
of haploid spores from sporangium.

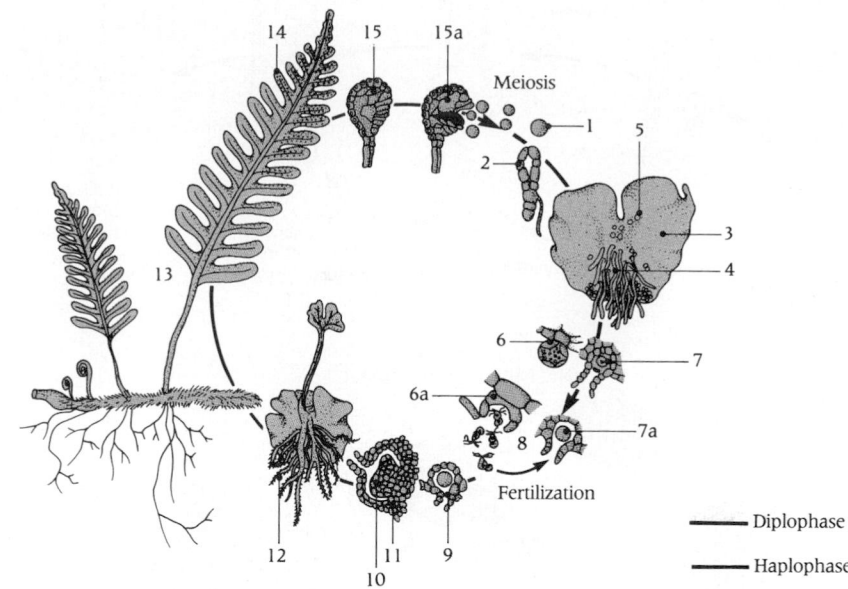

a structure called an **antheridium.** After fertilization of egg by sperm, the zygote
remains in place, developing into an embryonic sporophyte and then a mature
one with a capsule at the upper stalk end. Meiosis within the capsule yields spores,
which are liberated and widely dispersed. Those that find congenial conditions
settle down and develop into green gametophytes.[2]

Ferns

The familiar leafy fern plant is the diploid sporophyte (Figure 21-14). The undersides
of fertile leaves bear **sporangia,** small bodies which in meiosis produce spores
(Figure 21-15). Sporangia are numerous and often aggregated into clusters called
sori (singular: **sorus**). In time, a sporangium splits open, scattering mature spores.
The probability that a given wind-blown spore will land in a suitable location is
small. Hence, evolution has favored ferns that produce many millions of spores.
Those landing in moist environments germinate and develop into multicellular photosynthetic, haploid gametophytes that are small (5–10 mm across) and often heart-
shaped (Figure 21-14).

The gametophytes produce sperm and eggs in antheridia and archegonia, respectively, the former often developing first. To reach the eggs, which remain in the
archegonia, sperm cells must swim through a film of water. After fertilization, the
zygote develops into an embryonic sporophyte. For a time, the sporophyte is a
parasite, deriving food and water from the gametophyte. But soon the fern sporophyte
develops its own roots, stem, and leaves and grows into a large autonomous plant
that contrasts strikingly with the small, haploid gametophyte.

Nonflowering Seed Plants: Gymnosperms

The most complex and most recent plants in the evolutionary sense include the
majority of the familiar green plants that blanket the world. These are vascular
plants. Seed plants were the latest group to evolve and include the most complex
and specialized forms. This spectacularly successful category owes its prosperity to
two major evolutionary innovations—pollen and seeds. Seed plants are believed
to have arisen from fernlike ancestors. As in the ferns, the sporophyte is the conspicuous and dominant form. Gametophytes are often few in number compared with
those of ferns and they are microscopic in size.

**21-15 Fern showing many sporangia along
the leaf margin.**

[2] This description does not apply to all mosses. In a few species, sporophytes are unknown
and sexual reproduction has been lost.

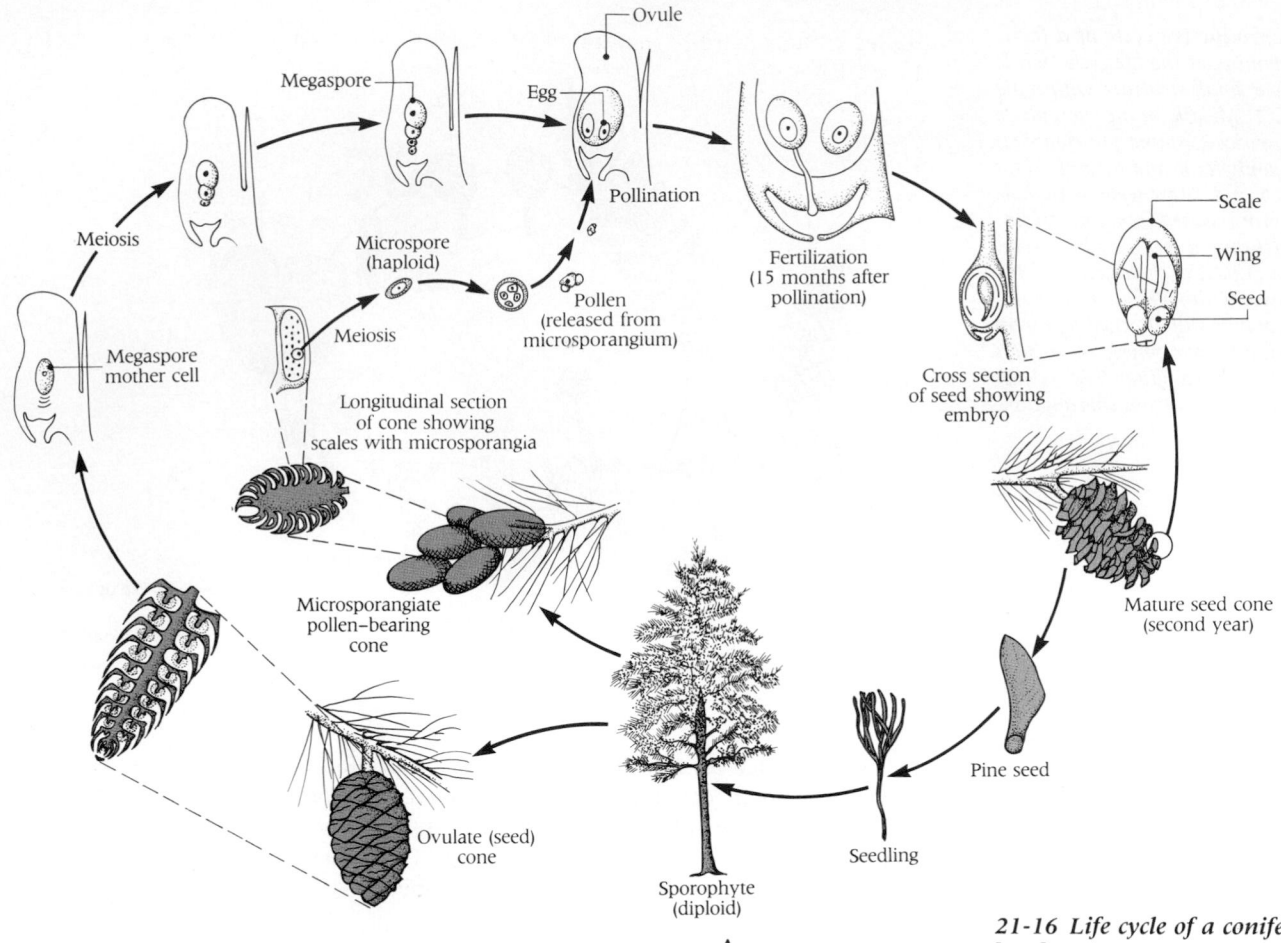

A

21-16 Life cycle of a conifer, Pinus lambertiana

(A) The tree is the sporophyte that gives rise to two different types of cones. The male cone produces microspores. The female cone produces megaspores. After fertilization, seeds form from the ovules in the cone. Each seed contains an embryo that may ultimately grow into a new sporophyte. (B) The tip of a pine shoot showing new plant growth (top), several pollen cones, and one seed cone.

As noted in Table 21-2, seed plants are divided into **gymnosperms,** which are nonflowering, and **angiosperms,** which are flowering.

Seed plants are far better adapted to life on land than ferns and mosses. They have evolved a life cycle that eliminates both the free-living—photosynthetic and nonparasitic—gametophyte stage and the swimming sperm cells. Water-dependent free-living gametophytes and swimming sperm are the two stages of the moss and fern life cycles that are most susceptible to drying out. In their place, seed plants evolved new kinds of small, parasitic gametophytes that are kept moist by structural enclosures.

The pine tree life cycle (Figure 21-16A) is a good example of nonflowering seed plant reproduction. Pine trees produce two kinds of spores, **megaspores** and **microspores.** The megaspores are produced in megasporangia, which are found on female (ovulate) cones (Figure 21-16). **Cones,** or **strobili** (singular, strobilus), are shortened stems with many spore-producing appendages. The appendages, or scales, of the female pine cone each have two **sporangia** (megasporangia) where meiosis occurs to produce megaspores. There is usually only one functional megaspore per megasporangium.

Unlike the spores of mosses and ferns, the megaspore is never shed from the megasporangium. This simple change has profound consequences. The megaspore can develop into a megagametophyte (female gametophyte) of several thousand cells while being protected from drying out by the megasporangium and the cone. The megagametophyte develops archegonia with eggs, an egg is fertilized, and the resulting zygote develops into an embryonic sporophyte, all within the megasporangium. An embryonic sporophyte inside the remains of the megagametophyte, which is still inside the megasporangium, is called a **seed.** The megagametophyte serves as a nutritive tissue; and the megasporangium, together with an additional tissue (the integument), forms the seed coat. The collective term for the megasporangium plus integument is **ovule;** each ovule can potentially become one seed. But the story is not complete until the sperm reaches an egg that is produced by a gametophyte inside an ovule in a cone that may be many meters off the ground.

Pines produce a second kind of cone, called a male, or pollen, cone (Figure 21-16B). Like all cones, it is a shortened stem with many spore-bearing appendages.

B

Spores (microspores) are produced by meiosis in sporangia (microsporangia) on the surface of the scales (leaves) of the cone. Each tree produces many pollen cones, each cone has many microsporangia, and each microsporangium produces many microspores; the total number of microspores produced by a large pine tree is enormous. Before being released from the microsporangium, each microspore begins developing into a microgametophyte (male gametophyte). Only a few mitotic divisions occur, so the microgametophyte never contains more than a few nuclei. This development takes place while the microgametophyte is still enclosed in the microspore wall, a thickened structure that prevents water loss. In this way, immature microgametophytes are protected against dessication. They are also small enough to be dispersed by the wind when they are released from the cones. The immature microgametophyte within a microspore wall is a **pollen grain.**

Pine pollen grains are wind-blown from pollen cones to seed cones of the same or perhaps a different tree. By chance, some pollen grains are blown between the open scales of the seed cones and land near the ovules. An opening in the integument (the micropyle) allows pollen grains to come to rest near the megasporangium. Here the pollen grain germinates (grows out of the microspore wall) and grows a short pollen tube through the megasporangium to the megagametophyte. The pollen grain has now become a mature microgametophyte and is capable of passing a sperm nucleus to the egg cell.

The evolution of pollen grains made it possible for plants to bring a microgametophyte to a megagametophyte that is still within the megasporangium. Such retention of the megaspore (and the magagametophyte that develops from it) was a critical step in the evolution of seeds, but this cannot occur unless the microgametophyte reaches the megagametophyte. Thus all seed plants produce pollen, and pollen is never produced by nonseed plants like mosses and ferns.

The pine microgametophyte is so much reduced, in the evolutionary sense, that its few cells are insufficient to form an antheridium and it produces only one sperm nucleus. In addition, there is a year's delay between the arrival of a pollen grain at the seed cone (pollination) and the fusion of a sperm nucleus with an egg nucleus (fertilization). Note that pollination and fertilization are related, but they are two different events. As we shall now see, fertilization in angiosperms usually occurs soon after pollination. This is only one of many ways that flowering plants differ from other seed plants.

Flowering Seed Plants: Angiosperms

The many adaptations of the flowering plants, or angiosperms—most notably the **flower**—make them well suited to life on land.

In a typical flower, there are four main kinds of structures—sepals, petals, stamens, and pistils—each derived (evolutionarily) from one or more leaves and each attached to an expanded stem tip called a receptacle (Figure 21-17). Flowers may have all of these structures, or may lack one or more of them.

Sepals protect the bud. In some flowering plants they become colored and function like petals; in others, they are green or absent.

Petals attract pollinating animals and thus are often brightly colored. They are usually absent in wind-pollinated plants.

Stamens have evolved from leaves with microsporangia on them. The leaf is usually reduced (by evolutionary loss of photosynthetic tissue) to a stalk (the **filament**) with four microsporangia (collectively called the **anther**) at its tip (see Figure 21-17). Meiosis occurs in the anthers to produce microspores. Each microspore divides once to form a tube cell and a generative cell. This pair of cells constitutes the immature pollen grain, or gametophyte. At some point, the generative cells divide to form two sperm cells. The mature pollen grain, with its tube cell nucleus and two sperm cells, is the mature male gametophyte. The three-celled pollen grain leaves the anther and is carried, usually by wind or insects, to a sticky stigma—either of the same flower or of another one.

The **pistil** is also derived from a leaf (or leaves) with sporangia on it (them), in this case megasporangia. As in pines, each megasporangium with its integument is termed an ovule. Unlike pines, flowering plant ovules are enclosed in **locules,** chambers filled with moist air which are the evolutionary result of leaves folding up around the ovules. The ovules and the seeds that will develop from them are thus protected from drying out. Each folded leaflike structure is a **carpel;** because

21-17 Structure of a typical flower
All of the flower parts are leaves that have been modified into other structures. The number of sepals, petals, stamens, and pistils varies in different species. Also see Figure 41-14.

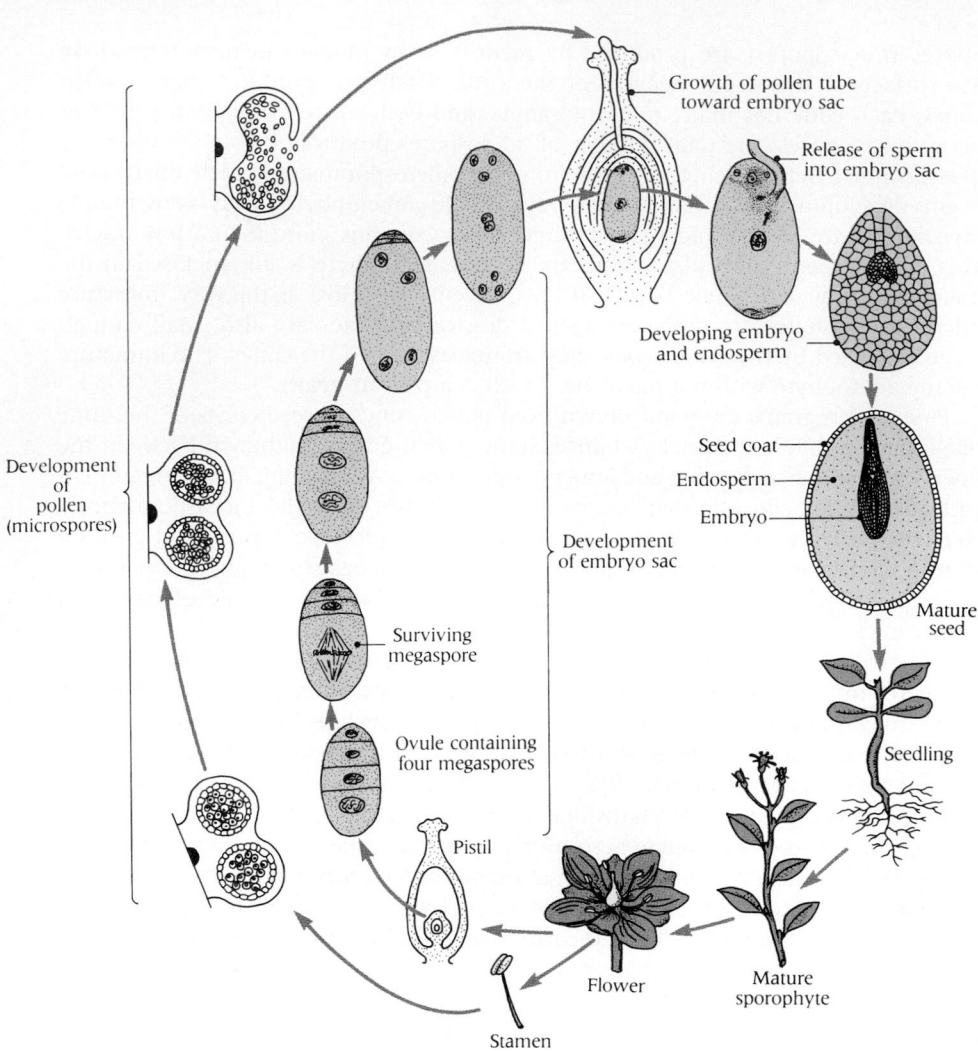

Growth of pollen tube toward embryo sac

Release of sperm into embryo sac

Developing embryo and endosperm

Development of pollen (microspores)

Development of embryo sac

Seed coat

Endosperm

Embryo

Mature seed

Surviving megaspore

Ovule containing four megaspores

Seedling

Pistil

Flower

Mature sporophyte

Stamen

of the way it evolved, each carpel has one locule. One or more carpels, more or less united at the center of a flower, comprise the pistil (Figures 21-17 and 21-18). Thus a flower may have a pistil with one carpel (as in Figure 21-18) or several carpels. Each carpel may have one to many ovules in its locule. The part of the pistil containing the ovules is called the **ovary.** The rest of the pistil is formed into a sticky, pollen-trapping **stigma** that is borne by a **style** (see Figure 21-17).

Within the ovules (megasporangia with integuments), meiosis occurs to make megaspores. Remarkably, only one cell per megasporangium undergoes meiosis. Four megaspores are produced, but only one megaspore is functional.[3] As in all seed plants, the megaspore is never released from the megasporangium.

The surviving megaspore in each ovule undergoes three mitotic divisions and forms a female gametophyte with eight nuclei. This structure is an **embryo sac.** Only one of its haploid nuclei becomes an egg. The eight nuclei assume the positions shown in Figures 21-18 and 21-19. The designated egg nucleus associates with two others (called synergid nuclei) that may be evolutionary vestiges of the flask-shaped archegonium around the egg in moss and fern gametophytes (see Figures 21-13 and 21-14).

Having reached the stigma, the pollen grain begins to germinate. A **pollen tube** emerges from each pollen grain, grows down the style, and enters the ovary. There it penetrates an ovule and the female gametophyte within. The pollen sperm cells move down to the end of the tube, whereupon the nuclei of the two sperm cells pass into the embryo sac.

One sperm nucleus from the pollen tube fertilizes the egg to form a zygote, which begins to develop into an embryo sporophyte while still within the ovule

[3] There are exceptions to this scheme among lilies and other plants.

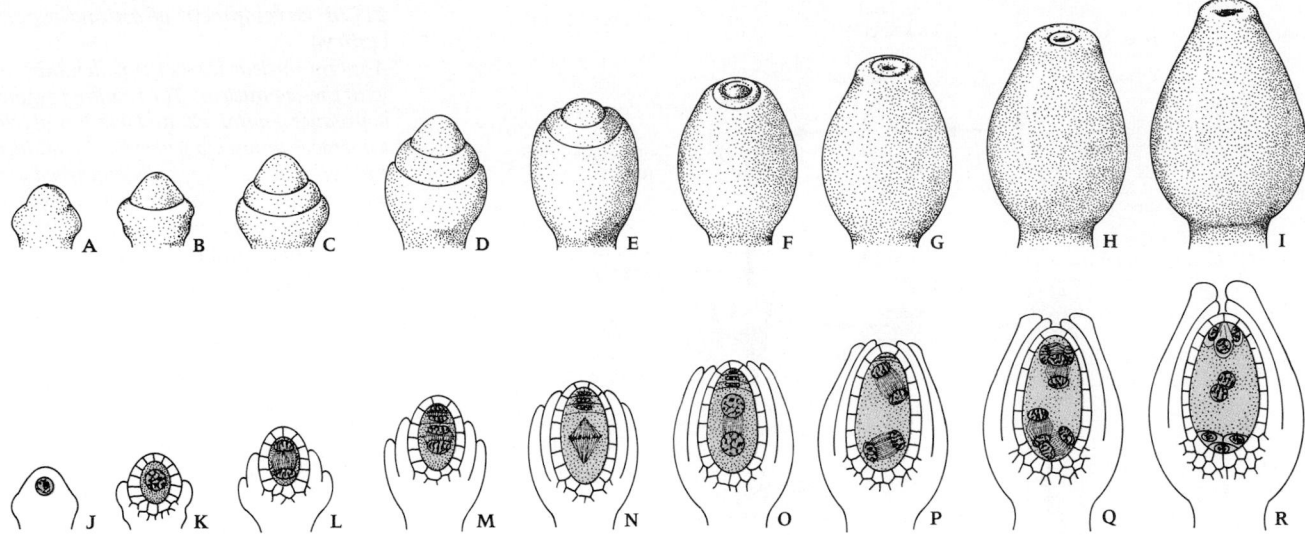

21-19 Development of an ovule
A–I, growth of ovule as seen from outside. J and K, magaspore mother cell; L and M, division producing four megaspores; N, three megaspores are compressed and one is enlarged, its nucleus in division; O, binucleate stage of embryo sac; P, four nuclei present; Q, eight nuclei present; R, mature female gametophyte with egg and two synergid nuclei at the top, two centrally located polar nuclei, and three other cells at the bottom.

in the ovary. In a second act of fertilization, the second sperm nucleus fuses with a fusion nucleus arising from two polar nuclei of the embryo sac. Since this structure is already diploid, the entrance of the haploid sperm nucleus makes it triploid. Now it has three times as many chromosomes as a haploid cell. This interesting cell, without parallel in any reproductive cycle outside of the flowering plants, develops into a multicellular mass called the **endosperm.** No real differentiation occurs in the endosperm. It consists merely of tissue rich in foodstuffs that nourish the growing sporophyte embryo. The starchy, oily, or proteinaceous endosperm forms the bulk of the food grains—corn, wheat, and other cereals—so important in the nutrition of humans and other animals.

Note on the embryonic development of plants
It is appropriate now to consider the story of plant embryogenesis.

Unlike animals, plants do not display continuity of the germ plasm discussed earlier (see Figure 13-1). The reproductive organs and gametes are not formed until well into the developmental sequence. Any somatic cell of the plant, therefore, can give rise to reproductive structures.

Following pollination and fertilization, the newly formed diploid zygote is found (together with the triploid endosperm) within the confines of the ovule, which will become the seed.

In angiosperms, careful microscopic studies of developing embryos have shown that immediately after fertilization, the organization of the zygote changes drastically. Although there is considerable variation in different species, a selected example is given in Figure 21-20. After the first nuclear division, for example, the two cells of the embryo are of unequal size. These two cells are already structurally differentiated and have different developmental fates. The larger so-called **basal cell,** which is nearest the **micropyle** (the opening in the ovule through which the pollen tube entered), divides to form the linear **suspensor** that holds the embryo. The smaller **terminal cell** divides to form a mass of cells that become the embryo proper. When about 50 cells have been formed in this globular mass, three tissues start to differentiate.

- A surface layer of cells appears that will form the plant's **epidermis.**
- A zone of small cells forms at the pole of the embryo opposite the suspensor. These elongated cells will form the shoot **apical meristems.** The shoot meristems grow upward and form leaves and lateral branches.
- Another mass of cells at the opposite end of the embryo forms the root meristems, which grow downward to form root structures.

Soon, the embryo loses its radial symmetry and becomes bilaterally symmetrical. Then the first embryo leaves appear. These are the **cotyledons** (Figure 21-21).

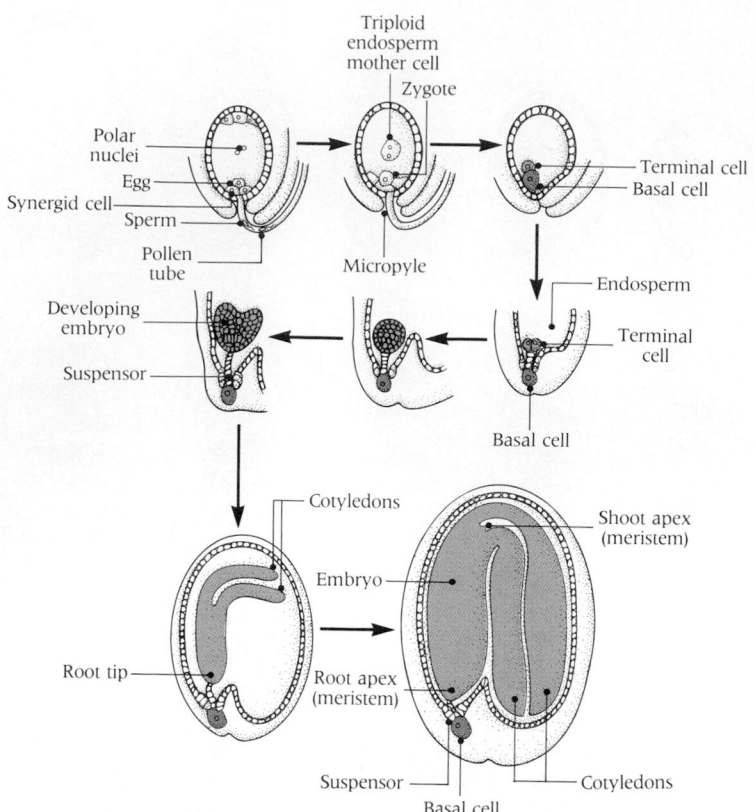

21-20 Development of an angiosperm embryo
A sperm nucleus leaves the pollen tube and fuses with the egg nucleus. The resulting zygote divides to produce a basal cell, and a terminal cell, which undergoes mitosis to generate the multicellular embryo. The embryo is distinguished by meristematic tissue at the root and shoot ends, as well as by cotyledons, which form the first leaves of the germinated seed.

Both meristems continue dividing throughout the life of the plant. In the developing seedling (Figure 21-21), the apical meristem is situated at the tip of the **epicotyl,** the portion of the central stem that rises above the cotyledons. The stem below the cotyledons is called the **hypocotyl.** An embryonic root at the lower end of the hypocotyl is called the **radicle.**

In gymnosperms, the zygote nucleus divides repeatedly after fertilization (Figure 21-22). But in many cases only after eight mitotic divisions, when 256 nuclei have accumulated, do cell walls begin to form. Differentiation follows, with cells farthest from the micropyle leading the way. Soon the cells begin to differ in size. The larger cells near the micropyle differentiate into a suspensor. The smaller cells at the opposite end of the embryo develop into the apical meristem of the root and

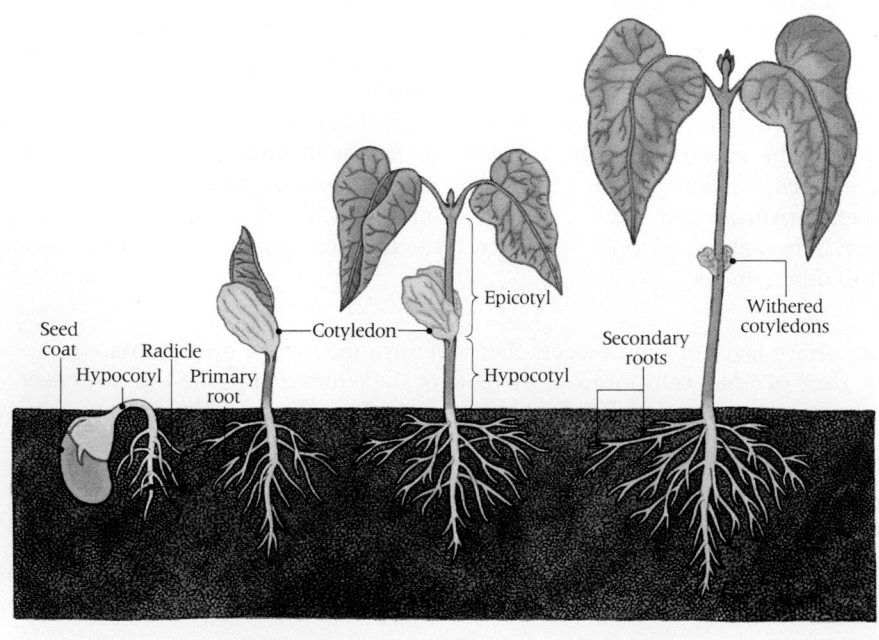

21-21 Germination and early development of a bean seed
Stages in the germination of the common bean, Phaseolus.

21-22 Development of gymnosperm embryo
In contrast to the angiosperm embryo, the early stages of gymnosperm embryonic development are characterized by the appearance of many nuclei within one cytoplasm. Later in development individual cells form and begin to differentiate into meristematic tissue of the root and shoot.

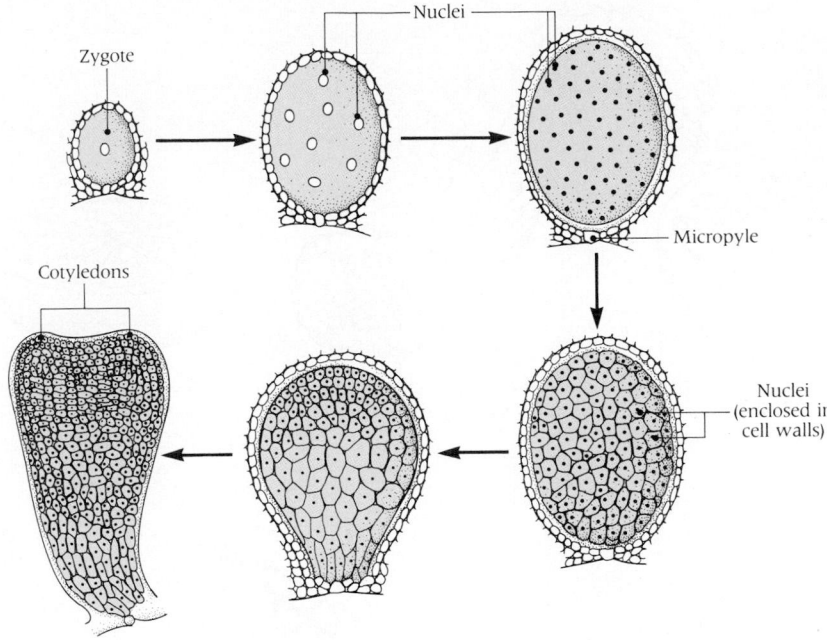

later into the shoot meristem. Aspects of the later development of plants will be discussed in Chapter 28.

Seeds and Fruit

Seeds were a remarkable evolutionary innovation that protected embryos from the dry conditions of open air and contributed to the dispersal of the seed plants. Thus seeds have contributed to the great diversity and abundance of land plants.

After fertilization most of the flower withers. Only the ovary is left with its ovule tissues around the young embryo (see Figure 21-19). These tissues harden to form a tough coat. The final product, a seed, is thus an encased sporangium containing an embryonic sporophyte, usually with a triploid endosperm as a developmental byproduct.

In flowering plants, seeds are typically retained within the ovary, which undergoes further growth to become a **fruit.** Thus, fruits are modified or ripened ovaries, sometimes with other flower parts, in which seeds are embedded. Their growth often produces the succulent sweet tissues so palatable to animals. The bitter taste of unripe fruits discourages animals from eating them before the seeds are fully matured. As they mature, the taste becomes sweet as sugar content rises. Hungry animals eat the fruit and help disperse the seeds by passing them unharmed through their digestive tracts.

There are many kinds of fruits in addition to the fleshy ones encountered in the supermarket. In fact, many of the foods called "vegetables" are really fruits. Examples include green beans, squash, and tomatoes. Technically, any plant structure (derived from the ovary) that encloses seeds is a fruit. Some fruits become hard and split open to release seeds. Other fruits have hooks, spines, wings, or hairs that aid in their dispersal by animals or wind. The diversity of fruits is illustrated in Figure 21-23.

SOME IMPLICATIONS OF PLANT IMMOBILITY

Plants are usually rooted in place. The fact of their immobility has affected many features of their reproductive biology.

Getting the Sexes Together

A problem is posed by the inability of two immobile plants to come together to reproduce in the fashion of mobile animals. Often this immobility is counteracted

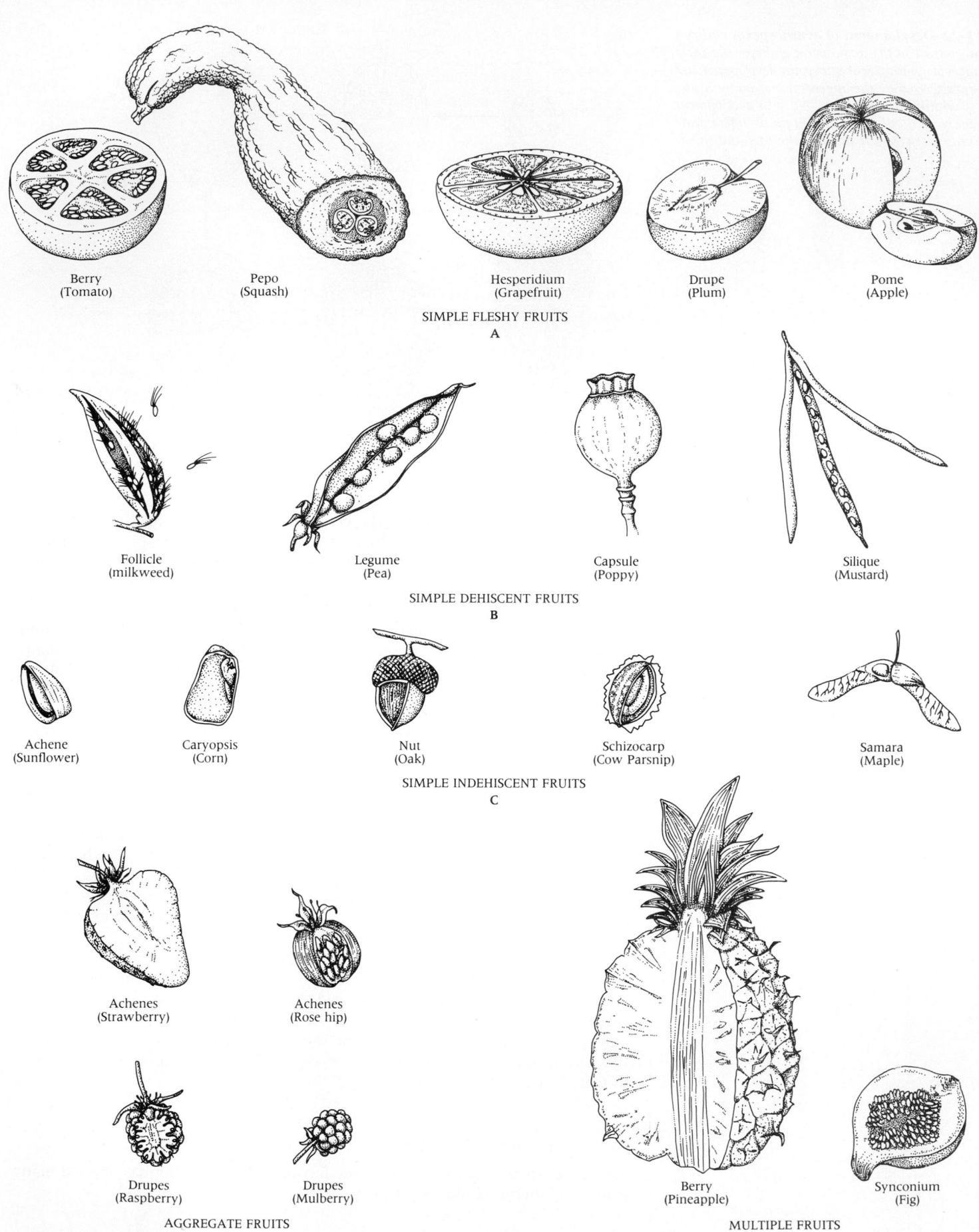

Berry
(Tomato)

Pepo
(Squash)

Hesperidium
(Grapefruit)

Drupe
(Plum)

Pome
(Apple)

SIMPLE FLESHY FRUITS

A

Follicle
(milkweed)

Legume
(Pea)

Capsule
(Poppy)

Silique
(Mustard)

SIMPLE DEHISCENT FRUITS

B

Achene
(Sunflower)

Caryopsis
(Corn)

Nut
(Oak)

Schizocarp
(Cow Parsnip)

Samara
(Maple)

SIMPLE INDEHISCENT FRUITS

C

Achenes
(Strawberry)

Achenes
(Rose hip)

Drupes
(Raspberry)

Drupes
(Mulberry)

AGGREGATE FRUITS

D

Berry
(Pineapple)

Synconium
(Fig)

MULTIPLE FRUITS

E

21-23 A fruit sampler
(A) *Simple fleshy fruits.* **(B)** *Simple dehiscent fruits.* **(C)** *Simple indehiscent fruits.* **(D)** *Aggregate fruits.* **(E)** *Multiple fruits.*
Dehiscent fruits open at maturity; indehiscent do not. In aggregate fruits, the separate carpels remain together as a unit.

by the mobility of gametes. Examples are the water-borne male gametes of attached algae, the sperm of ferns, and the highly mobile pollen grains of seed plants.

Many plants, including most seed plants, are monoecious—that is, both sexes occur in the same plant. In spite of the many devices that transmit pollen from plant to plant, many plants can pollinate and fertilize themselves when such devices fail. If continued for many generations, self-pollination would lead to a decrease in genetic diversity. Many plants have therefore evolved elaborate ways to ensure cross-pollination. For example, the anthers may mature before the stigma, minimizing a tendency to pursue the "last-resort" of self-pollination. Pollen in such a case is likely to go to another plant that is far enough developed to have a ripe stigma.

Among the most interesting and engaging facets of plant biology are the remarkable relationships between flowering plants and the animals they attract as pollinating agents. Hummingbird-flowers such as fuchsia, red columbine, and hibiscus, often a brilliant red color that attracts hummingbirds, are long and tubular so that only the long-tongued hummingbird can get the nectar. *Salvia* has a flower with a trigger that pulls down a stamen and dusts a nectar-seeking visitor with pollen (Figure 21-24A). Some orchids have flowers that look like female wasps that live in the vicinity (Figure 21-24B). They produce flowers yearly—precisely at the time when male wasps are out and flying but female wasps are not yet developed. The flower mimics the female so accurately the amorous but deceived male wasp mounts the flower and tries to copulate with it. The male wasp's genital organs become dusted with pollen which the insect carries to the next orchid that seduces him.

Geographic Dispersal

The geographic dispersal of some plants, algae, and fungi depends on the capacity of spores to move about. For example, ferns produce numerous lightweight spores that travel far and wide on currents of air. Indeed, air is seldom free of spores.

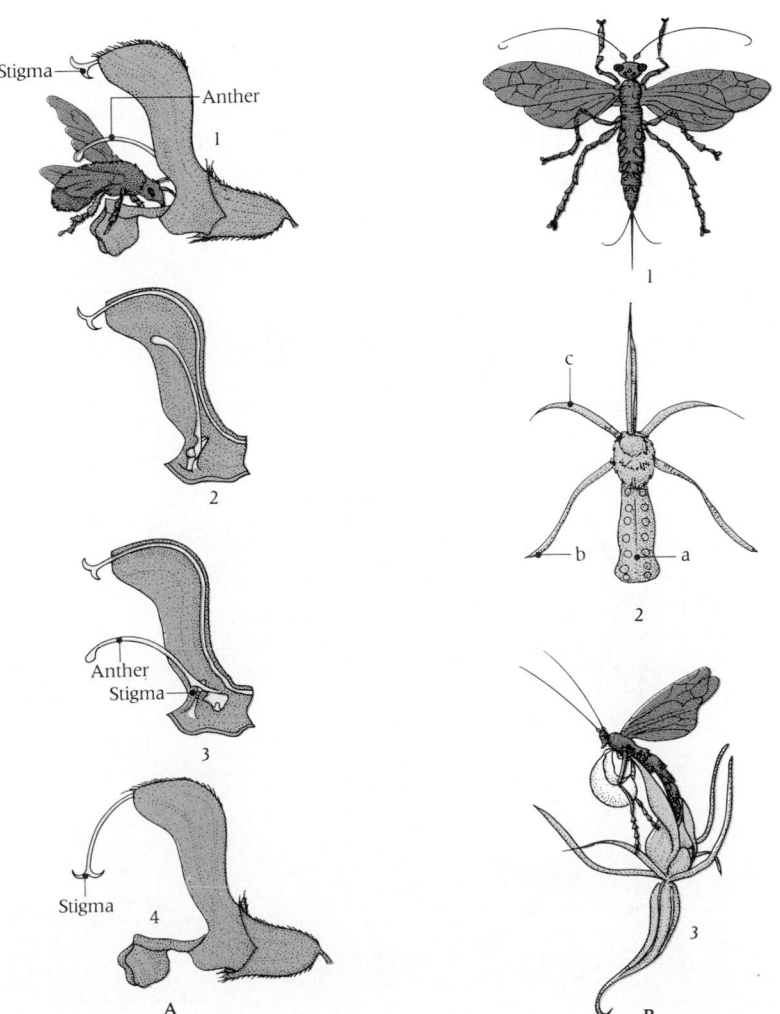

21-24 Insect-flower relationships
(A) *In sage* (Salvia), *the male organ (stamen and anther) in the flower matures earlier than the female organ (stigma). 1, In young plants, the style has not completed growth and so the stigma does not reach the bee's abdomen, although the anther does do so. 2, The anther is shown in the rest position. 3, Exploitation of a bee as cross-pollinating agent. The stamen is short. The anther is asymmetrically shaped, so that the short arm is pushed down by the bee as it searches for nectar. This results in a downward movement of the long arm of the anther so that the bee is brushed with pollen. 4, In older flowers the style has completed its curved growth and the stigma touches the bee, receiving pollen from other flowers that the bee has visited previously.* **(B)** *The ichneumon wasp* (Nemeritis) *cross-pollinates the Australian orchid. Part of the flower resembles the wasp's abdomen, and other parts simulate legs and antennae (1, 2a–2c). 3, The male wasp attempts to copulate with the flower. When he leaves, he carries away pollen to the next plant he mounts.*

In seed plants, pollen flies free of the parent organism, but pollen alone cannot produce a new plant. Therefore, geographic dispersal depends on the durability and portability of seeds. Here is another repertory of tricks and devices as varied and cunning as those that pollinate the flowers. Dandelions and similar plants give their seeds little parachutes and then send them out on the winds. Burred seeds stick to the fur of animals—and sometimes to the sweaters of small children. The hard seeds of many fruits pass unharmed through the digestive systems of fruit-eating animals. Coconuts and mangroves have floating seeds well suited to long sea voyages. Vetches have pods that open suddenly and scatter seeds like shrapnel from a grenade. In Russian thistles, the "tumbleweed" of country music, the globular plant breaks away from its roots and rolls along the prairie, scattering seeds as it goes.

We see, then, that despite the immobility of a developed individual plant, its methods of spore or seed dispersal and the large number of spores and seeds produced, may effectively cause populations of plants to spread even more widely and rapidly than populations of mobile animals.

REPRODUCTION IN FUNGI

The fungi are as different from protists and plants as they are from animals. Indeed some of their many biological peculiarities are just beginning to be understood. They are discussed in Chapter 40. Here we summarize major features of their reproductive cycles.

Two kinds of reproductive structures occur in fungi: **sporangia,** which are involved in spore formation, and **gametangia,** structures in which gametes are formed in a single fungal phylum (Oomycetes). Meiosis immediately follows the formation of the zygote in most fungi (zygotic meiosis). Nonmotile spores are a major reproductive device in many fungi, which reproduce asexually by releasing into the air or onto the bodies of insects the spores produced by a single parent. Haploid fungal spores are generally produced in vast numbers—a single giant puffball of the fungus *Calvatia gigantea* may contain several trillion spores and released spores may be caused to travel great distances. If they land in an adequately moist spot, they start to grow and divide by mitosis to produce long threadlike filaments, one cell wide, called **hyphae** (singular, hypha). The cells within a hypha may contain one, two, or many nuclei, depending on the fungus species. A mass of hyphae is called a **mycelium.**

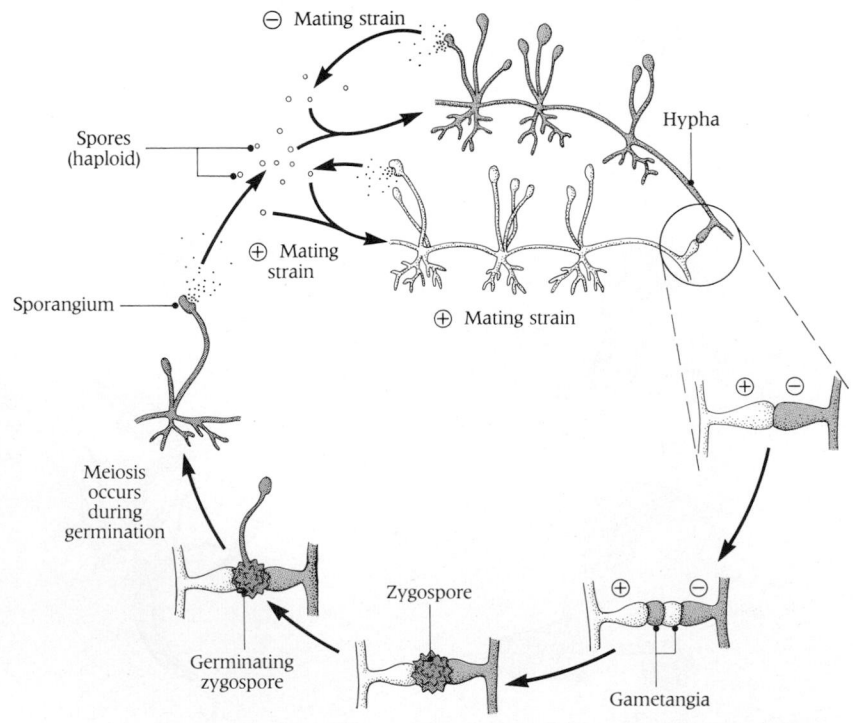

21-25 Reproductive cycle of a fungus
The species shown as an example is the bread mold, Rhizopus. *See also Figure 40-5.*

Some fungi are also capable of a form of sexuality, termed **parasexuality,** that differs from animal and plant sexual reproduction. In broad terms, parasexuality is a mechanism of genetic recombination that is based on the mitotic cycle rather than the meiotic (sexual) cycle. Another genetic feature that is unique among the fungi is **heterokaryosis.** This occurs when two parental hypha that are genetically dissimilar combine. The hyphae resulting from their fusion is said to be **heterokaryotic.** This is in contrast to **homokaryosis,** which occurs when hyphae of two genetically similar mating types come together and fuse. In heterokaryosis, genetically different nuclei are present in a single cell, a heterokaryotic cell, which contains at least one haploid nucleus from each "parent" hypha. For much of the life of the organism, the nuclei simply coexist. The new cell divides mitotically for a time to produce hyphae that are still composed of heterokaryotic cells.

In some of the heterokaryotic cells, two nuclei that have descended from different parents eventually fuse to form a diploid zygote nucleus. The zygote nucleus immediately undergoes meiosis to form haploid spores—and the life cycle starts again (Figure 21-25; see also Figure 40-5).

Heterokaryosis in some ways resembles conjugation occurring in other diploid organisms. Often, the two parental cells are indistinguishable in outward appearance. But if they are genotypically different they can produce phenotypically distinctive hyphae as readily as Mendel's pea plants produced phenotypic variants.

REPRODUCTION IN ANIMALS

The story of reproduction in animals also has many subplots, but fewer perhaps than the reproduction of plants, protists, and especially fungi. One concerns similarities and differences between the generalized reproductive cycle of Figure 21-6B and the many distinctive reproductive cycles of individual animal species. A second would be about the various ways and means by which animal species reproduce. A third would consider reproductive stratagems from an evolutionary viewpoint. And fourth is the story of mammalian reproductive biology, including that of human beings.

The last story, predictably, has subplots of its own. For example, "getting the sexes together" includes the whole range of courtship and mating behaviors. In the case of human reproduction, it includes the many psychological and emotional factors covered by the term "human sexuality," which in large measure detach the sex drive from the conscious desire to reproduce.

FEATURES DISTINCTIVE TO ANIMAL REPRODUCTION

We begin our survey of features distinctive to animal reproduction with one that is not so distinctive: the dominance of the diplophase.

Absence of Gametophytes

Multicellular animals appear to have evolved from animallike (that is, nonphotosynthetic) flagellated protists. As far as we know, the animallike flagellates are haplo-

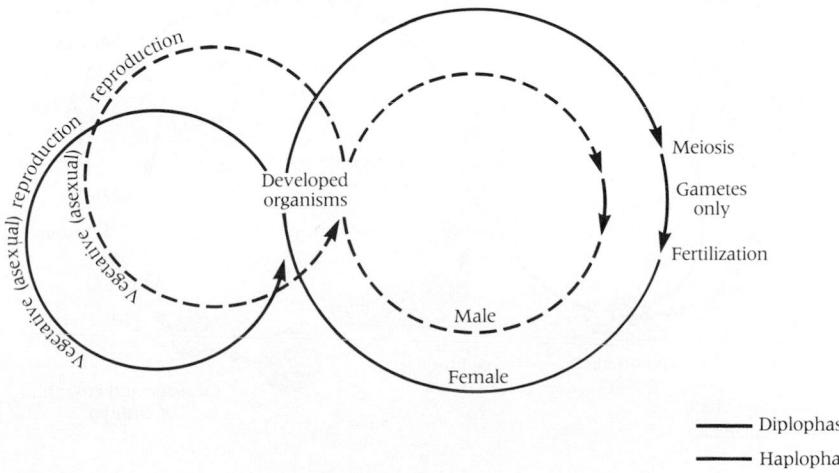

21-26 Generalized reproductive cycle of animals

―――― Diplophase

―――― Haplophase

phase organisms, diplophase being restricted to the zygote itself. Therefore, haplophase may have had some early emphasis. But, once animal life evolved beyond the protistan level it became exclusively diplophase. From then on, haplophase was represented only by gametes.

The animal reproductive cycle most familiar to us is the human cycle. It is characteristic of vertebrate animals in having no haplophase comparable to the plant gametophyte (Figure 21-26). (Compare this figure with Figures 12-37 and 12-38 which display the three major life cycle patterns.)

Variations of the animal cycle exist, but these relate to such features as the regression of sexuality and other curiosities—for example, the addition of asexual reproduction to the diplophase portion of the main reproductive cycle. Other striking aspects of animal reproduction are concerned not so much with the reproductive cycle as with the varieties of anatomical equipment and physiologic mechanisms that have arisen in the name of reproduction and development.

Sexual and Asexual Reproduction

Many invertebrates and tunicates (animals like sea squirts, which are chordates distantly related to the vertebrates) reproduce asexually; but even so, asexual reproduction is not the sole means of reproduction.

Asexual reproduction in animals is usually by budding (as in *Hydra*) or by fission (as in planarians). Among the anatomically simple animals called coelenterates, asexual and sexual reproduction alternate regularly. In *Obelia*, a familiar example of this group of jellyfish-like creatures (Figure 21-27), the zygote develops into an attached larva which then grows into a highly branched colony of **polyps.** This is the result of a budding process that is not followed by separation. Parts of the colony finally do form buds that detach and swim away. Because many tentacles hang from these swimmers, they are called **medusas** (after the mythological woman whose hair turned to snakes). Medusas are male or female and produce eggs or sperms that unite in fertilization to give rise to new attached colonies. The complex life cycle thus includes developed organisms of two diverse types: (1) attached, sexless, colonial polyps, and (2) free-swimming, sexed, noncolonial medusas. Both life forms are diploid. Hence, do not confuse the alternation of sexual and asexual reproduction in coelenterates with the alternation in plants of diploid sporophyte and haploid gametophyte generations.

Profoundly important biological questions are raised by this pattern. *Obelia* lives for a time as a colony of many organisms. If the colony is viewed as a highly integrated society, one may ask when a society becomes so well integrated that it is no longer a society but an organism.

21-27 Reproductive cycle of **Obelia**
(**A**) *Both polyp and medusa stages are in diplophase, with the haplophase stage represented only by the gametes. Note the "alternation of generations" (medusa–polyp–medusa–polyp, etc.).* (**B**) *A living* Obelia. *Note the feeding polyps.*

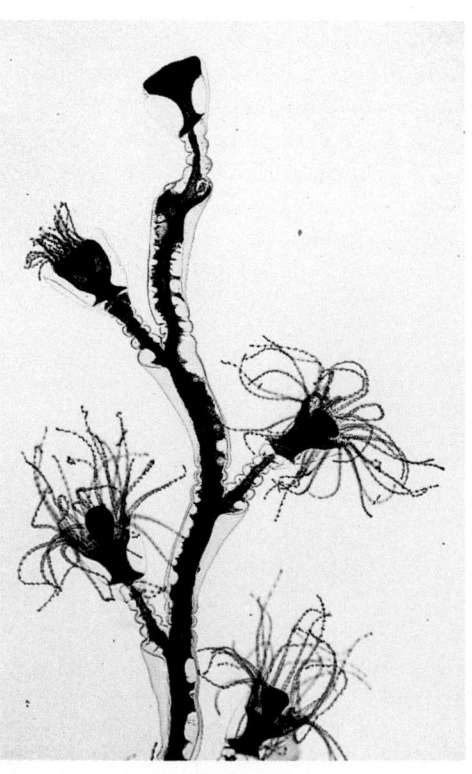

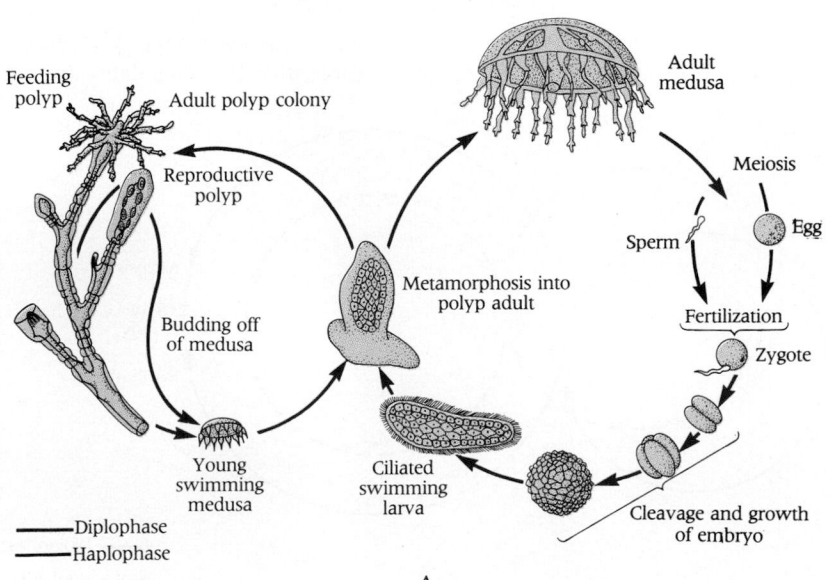

Feeding polyp

Adult polyp colony

Reproductive polyp

Budding off of medusa

Young swimming medusa

Ciliated swimming larva

Metamorphosis into polyp adult

Adult medusa

Meiosis

Sperm

Egg

Fertilization

Zygote

Cleavage and growth of embryo

——— Diplophase
——— Haplophase

A

B

Regression of Sexuality

In all organisms except some bacteria, cyanobacteria, and a few fungi, genetic recombination is achieved through sexual reproduction. As we have seen, the fully developed sexual process includes meiosis and fertilization. However, the sexual cycle is modified or absent in a number of organisms. Since the evolutionary ancestors of most of these organisms presumably did engage in sexual reproduction, these organisms must represent examples of lost or regressive sexuality. Regressive sexuality is manifested in several ways.

- Hermaphroditic self-fertilization is a sexual process, but from an evolutionary viewpoint it must be considered regressive. In the terminology of evolution, this refers to situations in which a species lacks functions or structures possessed by its ancestors.
- Parthenogenesis, the production of an embryo from an unfertilized egg, occurs sporadically in animals and commonly in plants.

Some organisms can be both parthenogenetic and fully sexual (biparental). Sometimes there is a fairly regular alternating pattern. This is the case with many aphids, or plant lice. In bees, males arise from unfertilized eggs and are thus haploid. Females arise from fertilized eggs and are thus diploid (Figure 21-28).

We saw that both sexual and asexual reproduction can occur in the same organism. If in the course of evolution asexual reproduction is lost, reproduction becomes exclusively sexual. This evidently happened among the ancestors of the vertebrates. Similarly, reproduction can become entirely asexual. This has happened in a number of species here and there, but not to the whole of any large group.

The Remarkable Story of Fertilization

In one of the earliest modes of fertilization, eggs and sperm were shed directly into the water. This remains the pattern typical of most aquatic invertebrates. For example, the sperm of a male American oyster are ejected into the surrounding water in such numbers the water appears milky. When a free sperm encounters an ejected egg, fertilization occurs.

Many modifications of this simple process are known. In European oysters, the eggs are retained in the mother, where they are fertilized. In some freshwater clams, the eggs leave the ovary and are retained under the gills. To minimize the waste of gametes that occurs when sperm are shed into vast watery surroundings, animals, like plants, developed a large repertory of tactics and devices to bring egg and sperm together. One stratagem synchronizes sexual activity into seasonal periods that parallel the seasonal flowering of plants. Another employs internal timing devices—sometimes called "clocks"—that synchronize sexual activity with events of a certain time of the month or day.

Many marine animals release their gametes on a 28-day cycle in relation to a particular moon phase. The palolo worm (*Leodice*) of South Pacific coral reefs goes

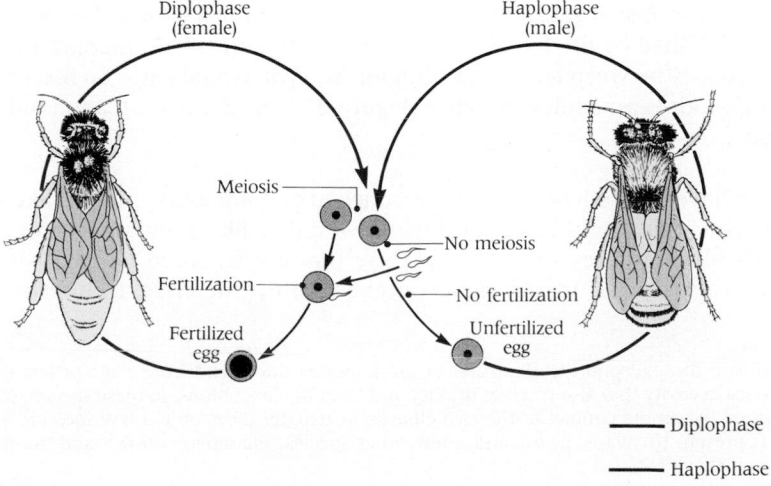

21-28 Reproductive cycle of the bee
The bee, and other members of the Hymenoptera, produce haploid males, while all fertilized eggs produce females. Since males are haploid, they produce normal haploid sperm without meiosis.

through an extraordinary reproductive ballet. As the moon rises in its last quarter, the worm pinches off the rear end of its body, which contains the gametes. These sections rise to the surface in vast numbers and there—only at full moon—they make the sea milky with eggs and sperm. Meanwhile the front part of the worm survives at the bottom. Next season it develops a new tail.

Aside from the physiological processes that synchronize sexual activity, motile animals have evolved intricate behavioral adaptations that synchronize gamete release. Some males and females may expel gametes only in each other's presence. Or an elaborate courtship display may produce the same result (Figure 21-29).

The evolution of reproductive devices in animals gradually freed gametes from unpredictable conditions in surrounding waters. But despite these many developments, animal gametes have continued to depend on a watery medium. Sea water was simply replaced with body secretions.

The egg shell is an interesting adaptation of animal reproduction that helped to solve the problem of incubation outside of a watery medium or a maternal body. If an egg were not surrounded by fluid, it would dry out and die. The development of a tough, outer, waterproof shell made it possible for eggs to survive in a dry environment. But, while protecting the egg, the shell excluded the sperms. What to do? Several evolutionary developments provided answers. Insect eggs developed a shell with a small hole through which a sperm could enter. The eggs of reptiles and birds developed shells that form after fertilization.

In mammals no shell forms at all. Land animals are mobile and evolution placed the crucial event of fertilization within the mother's body. The male takes his sperm to the female and injects them into her in the act of copulation. In all animals adapted to life on dry land—insects, reptiles, birds, and mammals—this became the favored mode. The body must be equipped for this act. The male must have a penis that can be inserted into the female to supply sperm.[4] The female must have a vagina or other receptacle from which deposited sperm can move toward eggs through a fluid medium. In insects, females have sacs within which deposited sperm are kept alive for long periods of time. Sperm are then released from the sacs whenever eggs are to be fertilized. The queen bee copulates only once and stores the received sperm for the rest of her life, sometimes as long as 17 years.

Development of Larvae

Life cycles of several different types are found in the animal kingdom.

- In placental mammals (like ourselves), a fertilized egg develops in the mother's body. When the developing individual reaches a stage at which it can be self-maintaining, it becomes a more or less independent organism resembling its adult parents.
- In most reptiles and birds, embryonic development occurs in an egg that provides food and protection until the young animal hatches.
- In most species of invertebrates (including insects), many fishes, and many amphibians—groups that collectively embrace the great majority of animals—the embryo does not develop directly into an adultlike organism. A free, postembryonic life begins in a form quite unlike that of the adult. This creature, known as a larva, is able to get about, find food, and eat, which it often does voraciously. As exemplified by the life cycle of *Drosophila* (Figure 14-6), there is a notable dissimilarity between larval and adult forms. Who would guess on first encounter that a moth is an adult caterpillar (Figure 21-30) or a frog an adult tadpole (Figure 21-31)?

Larvae have two functions. First, for attached and immobile animals like sponges and corals, larvae are a means of dispersal—rather like plant seeds. Even mobile animals—like starfishes—may have larvae that are far more mobile than adults. Second, heavy feeding marks the larval phase of the life cycle. In insects, most or

21-29 Courtship of the great crested grebe Courtship serves to synchronize the release of gametes. **(A)** Mutual head shaking. **(B)** Female displaying before a male who has dived and appears before her in the water. **(C)** Two views of the male immediately after surfacing from the dive, showing the "collar" of feathers. **(D)** Both sexes have dived and brought up weeds, which they display to each other.

---

[4] Birds are the exception to this rule. In most species ducts lead from male or female gonads to the cloaca, a cavity that also receives urinary and intestinal excretions. In these species, copulation is performed by simple contact of the two cloacae. It usually takes only a few seconds. However, a penis is present in swans, geese, and a few other species. Flamingos, storks, and herons have a vestigial penis.

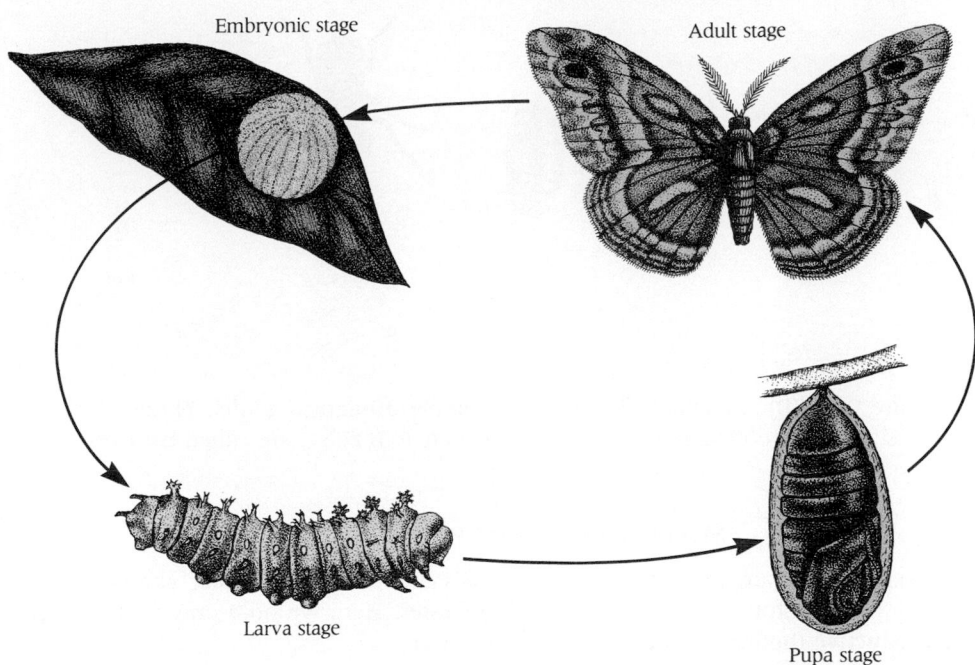

21-30 Metamorphosis of a moth
Details are given in the text.

Embryonic stage

Adult stage

Larva stage

Pupa stage

all feeding and growth occur in the larvae. Some adult insects eat nothing, depending entirely on nutrients they acquired as larvae. In most insects, both larvae and adults eat, but they eat different foods.

The change from larval to adult form, called **metamorphosis,** may be gradual or abrupt. In grasshoppers, the wingless larva acquires adult form gradually by repeated shedding of the external skeleton, a process termed **molting.** Dragonfly metamorphosis occurs in a single molt. In butterflies and most other insects, a special inactive phase, called a **pupa** or **chrysalis** depending on the insect type, occurs between larva and adult (Figure 21-31). The pupa becomes enclosed in a case called a **puparium.** There it undergoes radical metamorphosis. Almost all the organs of the larva are transformed, and a virtually new organism develops from small cell buds or clumps—a pattern reminiscent of the life cycle of *Obelia.* In insects that metamorphose, the adult is an **imago.** Thus, the full life sequence is as follows:

$$egg \rightarrow larvae \rightarrow pupa \rightarrow imago$$

This sequence, known as **complete metamorphosis** (Figure 21-32A), occurs in more than 90% of all insects. Some insects undergo **simple metamorphosis**

Fully metamorphosed frog

Young frog metamorphosing from larva

Adults spawning in water

Older larvae with legs

Fertilized egg

Early cleavage of the egg

Later embryonic stages

Young larva (tadpole)

21-31 Life cycle of the frog

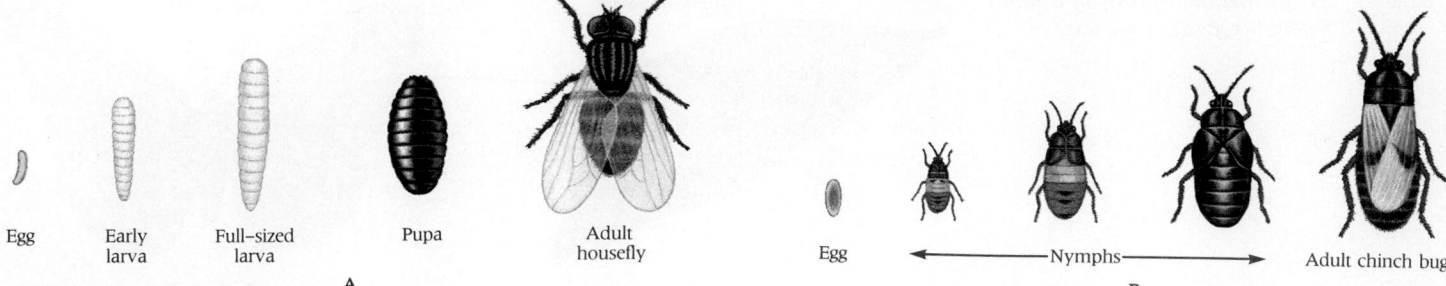

Egg Early Full-sized Pupa Adult
larva larva housefly

A

Egg ←—————Nymphs—————→ Adult chinch bug

B

(Figure 21-32B), in which the young resemble miniature adults. These youthful stages, which are usually wingless and distinctive in color, are called **nymphs.**

MAMMALIAN REPRODUCTION

Human beings are mammals and mammals are vertebrates. Later chapters will show how evolution gave rise to the vertebrates. Here we note only that certain reproductive themes characterize their evolution.

Evolution of Mammalian Reproduction

The early reptiles were the first wholly terrestrial vertebrates. Total emergence from the water required many changes. The most important adaptations for reproduction were the perfection of copulation, internal fertilization, and the ability to develop shells around fertilized eggs before they were laid.

Within the shells of such eggs, a membrane called the **amnion** developed from the tissues of the embryo (Figure 21-33). The amnion formed a fluid-filled sac that enclosed the embryo in amniotic fluid. Thus the embryo managed to develop its own self-contained aquatic environment. Other membranes, the **allantois** and **chorion,** formed other sacs and surfaces that facilitated respiration, food absorption, and the further protection of the embryo.

In the long transition from early reptiles to advanced mammals, developing eggs came to be retained in the tubes leading from the ovaries to the exterior. In the course of evolution, these areas became enlarged and thickened. The resulting adaptation was the organ known as the **uterus.** In earlier forms, uteri were paired. As shown in Figure 21-34, mammals displayed a progressive tendency for uteri to fuse. Thus there developed a duplex uterus with two chambers (found in marsupials, many rodents, and bats), a bipartite or bicornuate uterus containing a small central chamber (found in most mammals), and a simplex uterus with complete fusion (found only in humans and other primates). A duplex, bipartite, or bicornuate uterus has ample room for the development of several embryos at once. In a simplex uterus, only a single embryo ordinarily develops per pregnancy.

With the evolution of the uterus, the egg shell was lost. Membranes of the embryo then began to come in direct contact with the wall of the uterus where nutrients, gases, and wastes could be exchanged with the wall by diffusion. The efficiency of such exchanges was vastly improved by the evolution of the **placenta.** This complex structure, located between the embryo and uterine wall, permits small capillaries from the maternal and embryonic circulatory systems to come close enough together to permit free exchanges of dissolved substances (see Chapter 19). Special

21-32 Two types of metamorphosis in insects
(**A**) In complete metamorphosis, the younger stages differ markedly from the adult. An intermediate resting stage, the pupa, is important for the structural changes that occur, as shown for the housefly. (**B**) In simple metamorphosis, represented by the chinch bug, the younger stages closely resemble the adult.

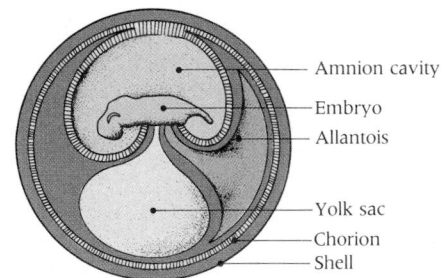

Amnion cavity
Embryo
Allantois
Yolk sac
Chorion
Shell

21-33 The amniotic egg of a terrestrial vertebrate
Also see Figure 25-1.

21-34 Progressive fusion of the uterus in placental mammals
(**A**) Duplex uterus. (**B**) Bipartite uterus. (**C**) Bicornuate uterus. (**D**) Simple uterus.

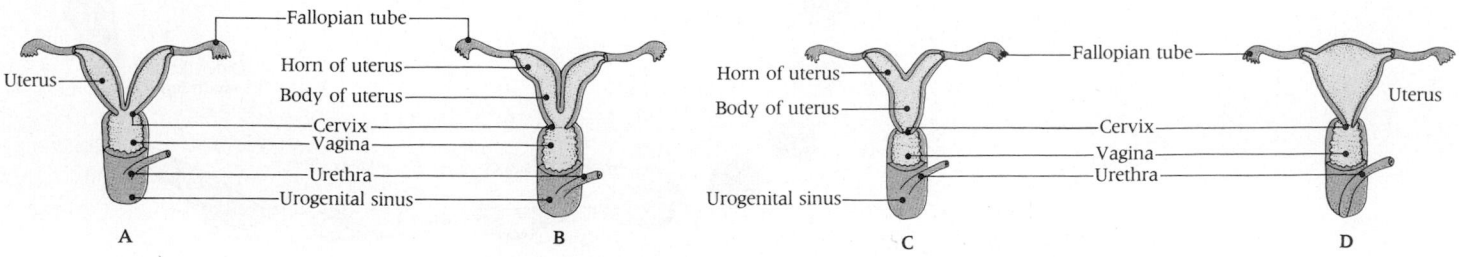

Fallopian tube
Uterus — Horn of uterus
Body of uterus
Cervix
Vagina
Urethra
Urogenital sinus

A B

Horn of uterus Fallopian tube
Body of uterus Uterus
Cervix
Vagina
Urethra
Urogenital sinus

C D

blood vessels, made by the embryo, connect the embryo with the placenta and shut down at birth.

Other arrangements for the sustenance of newborn offspring also arose. Even before eggs were first retained in a uterus, mammary glands evolved that provided nourishing milk to the newborn. In humans, even after weaning, long periods of parental care with mental and social training became a necessary adaptation that enhanced survival during a dangerous period of the child's life.

Issues Unique to Human Reproduction

Any study of human reproduction brings out factors made unusual by our own perceptions of them and attitudes toward them. These include such themes as the increasing discrepancy between physiological maturity—the age at which people can reproduce—and social maturity—the age at which they should reproduce. It also includes other aspects of human sexuality that play significant roles in the emotional and mental health of individuals and societies. It includes, too, such biologically unique considerations as the human desire to plan parenthood and its increasingly important correlates, contraception and abortion.

Male Reproductive System in Humans

The chief functions of the reproductive system in human males are as follows:

- Spermatogenesis
- Participation in copulation
- Production of male sex hormones, or androgens, which control the development of primary and secondary sexual characteristics.

The major structural components of the male reproductive system are the penis, testes, epididymides (the plural of epididymis), vasa deferentia, seminal vesicles, and prostate (Figure 21-35).

As we learned in Chapter 19, spermatogenesis takes place in the testes. Each testis descends from an abdominal position before birth and comes to lie in a skin-covered sac called the **scrotum.** A **testis** is an ovoid body to which is attached a crescent-shaped **epididymis.** The combination of testis and epididymis is commonly called a testicle.

Both testis and epididymis have complex systems of ducts. They join to form tubules that drain eventually into a **vas deferens,** which extends upward from the testis through the **spermatic cord.** The cord runs a lengthy course into the **ejaculatory duct.**

The **penis** is the copulatory organ. Its body consists of three longitudinal columns of erectile tissue termed the **corpora cavernosa** (Latin for "cavernous bodies"). Erectile tissue is full of vascular spaces (caverns). Sexual excitement causes blood to pour into these spaces faster than it can be drained out. As a result the corpora become distended, and the penis becomes hard and erect. Only in this condition can it be inserted into the vagina. After sexual excitement has passed, blood leaves the erectile tissue and the penis becomes soft again.

Sensations of touch, thoughts, and many other kinds of stimuli can initiate sexual arousal. Indeed, dreams can cause sexual excitement that culminates in ejaculation. Erection is the first effect of male sexual stimulation. Accompanying nervous impulses then cause the glands in the system to secrete fluids that lubricate the urethra. As stimulation intensifies, nervous reflexes produce rhythmic contractions that expel semen into the urethra. Further nervous impulses then produce the rhythmic contractions that ejaculate the semen from the urethra to the exterior.

Ejaculated semen is composed of fluids from the vasa deferentia, the **seminal vesicles, prostate,** and mucous glands. Prostatic fluid gives it a milky appearance; fluids from the seminal vesicles and the mucous glands give it a thick consistency. These fluids contribute sugars, nutrients, buffers, and other factors that activate sperm motility. Immediately after ejaculation, the spermatozoa are relatively nonmotile, possibly because of the viscosity of the semen. Within 30 minutes of ejaculation, the mucoid consistency is gone owing to the action of proteolytic enzymes in semen. The sperm then become highly motile.

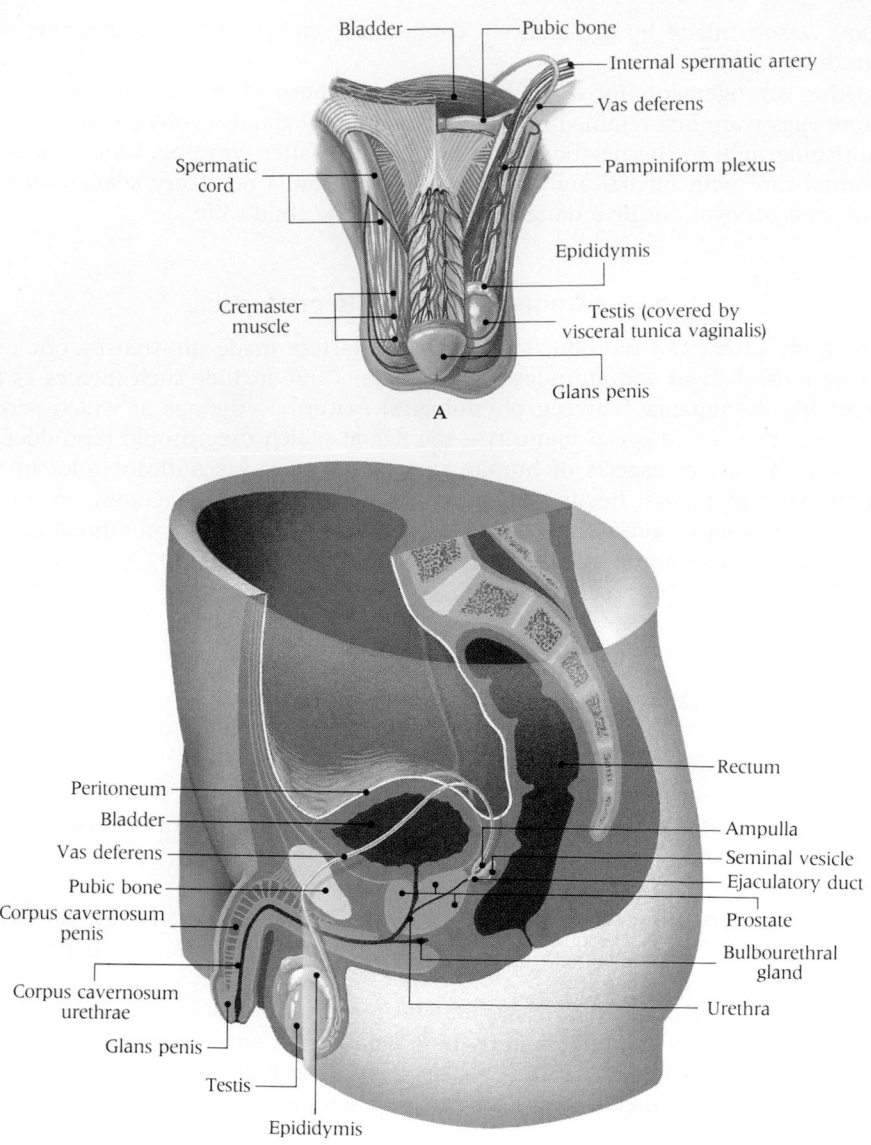

Bladder — Pubic bone
— Internal spermatic artery
— Vas deferens
Spermatic cord
— Pampiniform plexus
Epididymis
Cremaster muscle
Testis (covered by visceral tunica vaginalis)
Glans penis
A

Peritoneum
Bladder
Vas deferens
Pubic bone
Corpus cavernosum penis
Corpus cavernosum urethrae
Glans penis
Testis
Epididymis
— Rectum
— Ampulla
— Seminal vesicle
— Ejaculatory duct
— Prostate
— Bulbourethral gland
— Urethra
B

Female Reproductive System in Humans

The reproductive system of the human female appears in Figure 21-36. The internal organs are the ovaries, Fallopian tubes (oviducts), uterus, and vagina.

The paired **ovaries** lie on either side of the uterus below the Fallopian tubes. Internally, an ovary consists of an inner medulla containing blood vessels and nerves and an outer cortex containing numerous eggs and **follicles** in various stages of development. The cortex is concerned mainly with the maturation of eggs and the secretion of ovarian hormones (Figure 21-37). The surface of the ovary is covered by a delicate membrane that breaks whenever follicles rupture in the monthly process called **ovulation.** This is defined as the maturation of a single egg within an ovary and its expulsion through the ovarian surface into the abdominal cavity.

Following ovulation, the egg passes down one of the **Fallopian tubes** to the uterus. The tubes that perform this function in other animals are referred to as **oviducts.** Only in humans are they called Fallopian tubes. As shown in Figure 21-36A, they are slender structures lying horizontally above the ovaries. The ends near the ovaries flare out in a funnellike arrangement of fringed processes, called fimbriae, that guide an egg into the tubes. Ciliated epithelium lining the tubes, together with the contractions of smooth muscle in the tube walls, propel the egg toward the uterus.

The **uterus** is a thick-walled muscular organ in the upper pelvis. It is usually

21-36 Human female reproductive system
(A) Anterior view. (B) Sagittal section.

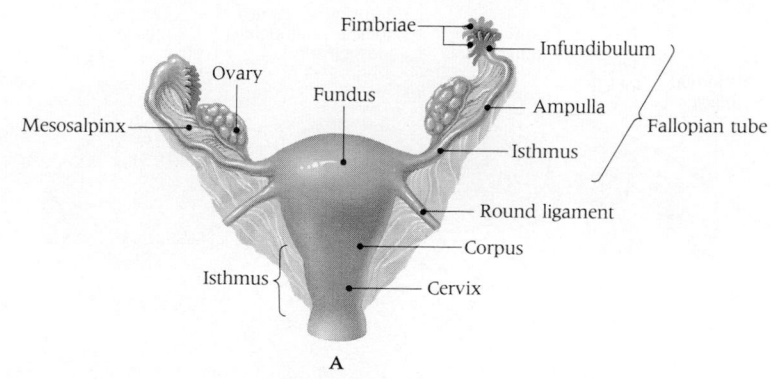

A

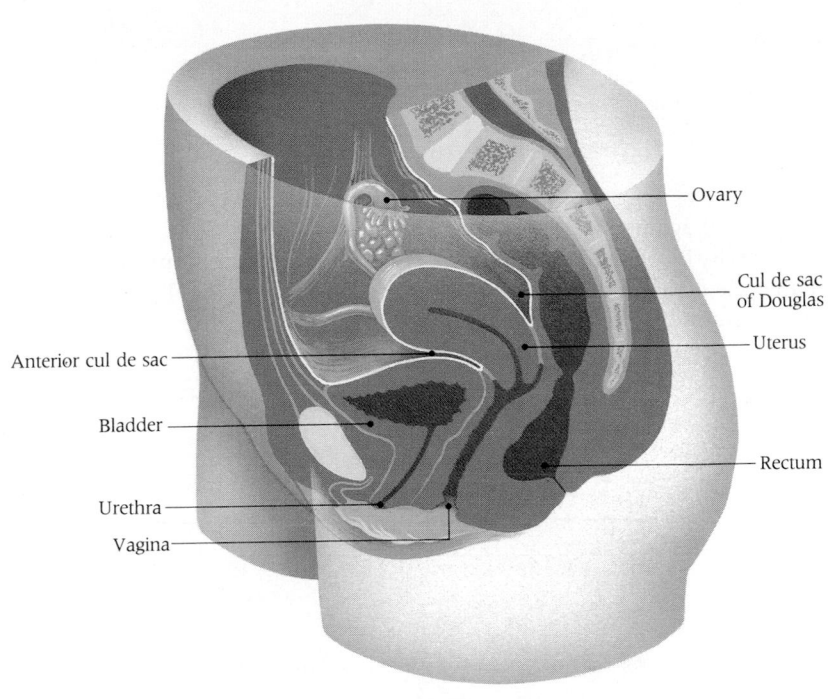

B

TABLE 21-3
THE UTERINE CYCLE

Days	Events
1–4	Menstruation
5–12	Proliferative phase Preovulatory period Development of follicle (follicular phase) Growth of endometrium High estrogen level
13–15	Interval phase Ovulation
16–22	Secretory phase Migration and breakdown of unfertilized ovum Development of corpus luteum (luteal phase) High estrogen and progesterone levels
23–28	Premenstrual phase Regression of corpus luteum Fall of estrogen and progesterone levels Deterioration of endometrium

tipped toward the bladder, but its position may vary. Its lowermost portion, the **cervix,** extends into the upper part of the **vagina.**

The external opening of the vagina is partially covered by a membranous fold called the **hymen,** which varies considerably in size, shape, and extensibility. Ordinarily it is no obstacle to sexual intercourse, either because it is flexible or because earlier athletic or other activity has ruptured it.

Ovarian cycle
The two main functions of the ovaries are production of eggs and secretion of the hormones, **estrogen** and **progesterone.** The most distinctive feature of ovarian function is its elegantly regulated periodicity. This pattern is termed the **ovarian cycle** (Figure 21-38). It includes the uterine cycle, and because its average period is 28 days, it is also called a **menstrual** (Latin, monthly) **cycle.** A timetable of the major events of the human uterine cycle appears in Table 21-3.

The human ovarian cycle has two significant results. First, it causes a single mature egg to be released from one of the two ovaries each month. Second, by secreting progesterone, it effectively prepares the lining of the uterus (the **endometrium**) for implantation. In the absence of a fertilized egg, the endometrium breaks down and sloughs away in the process called **menstruation.**

As shown in Figure 21-38, estrogen secretion steadily rises during the first half of the menstrual cycle (the so-called follicular phase) until just before ovulation, which is usually on about the 14th day of the cycle. Shortly before ovulation occurs, progesterone production begins to rise sharply. During the second half of the cycle

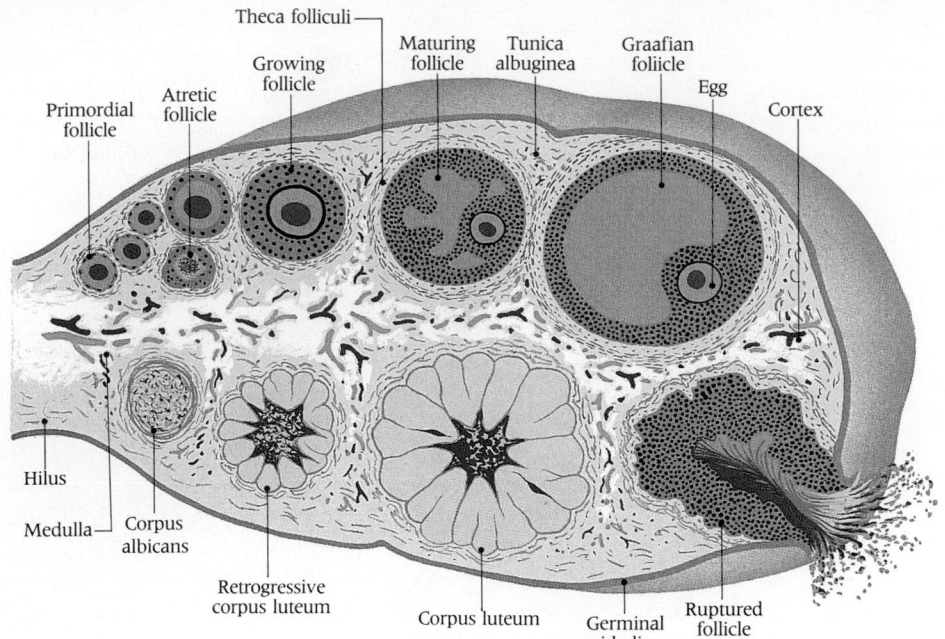

21-37 Structure of the human ovary
*Follicles in various stages of development are
present. The most mature stage, the Graafian
follicle, releases the egg at ovulation. The empty
follicle then degenerates to a corpus luteum.*

(the so-called luteal phase), estrogen and progesterone levels are both relatively
high, but they fall again as menstruation approaches.

Ovarian activity is controlled almost entirely by complex events outside the ovaries.
The human cycle depends directly on the production of hormones by the hypothalamus and the anterior portion of the pituitary gland, called the adenohypophysis

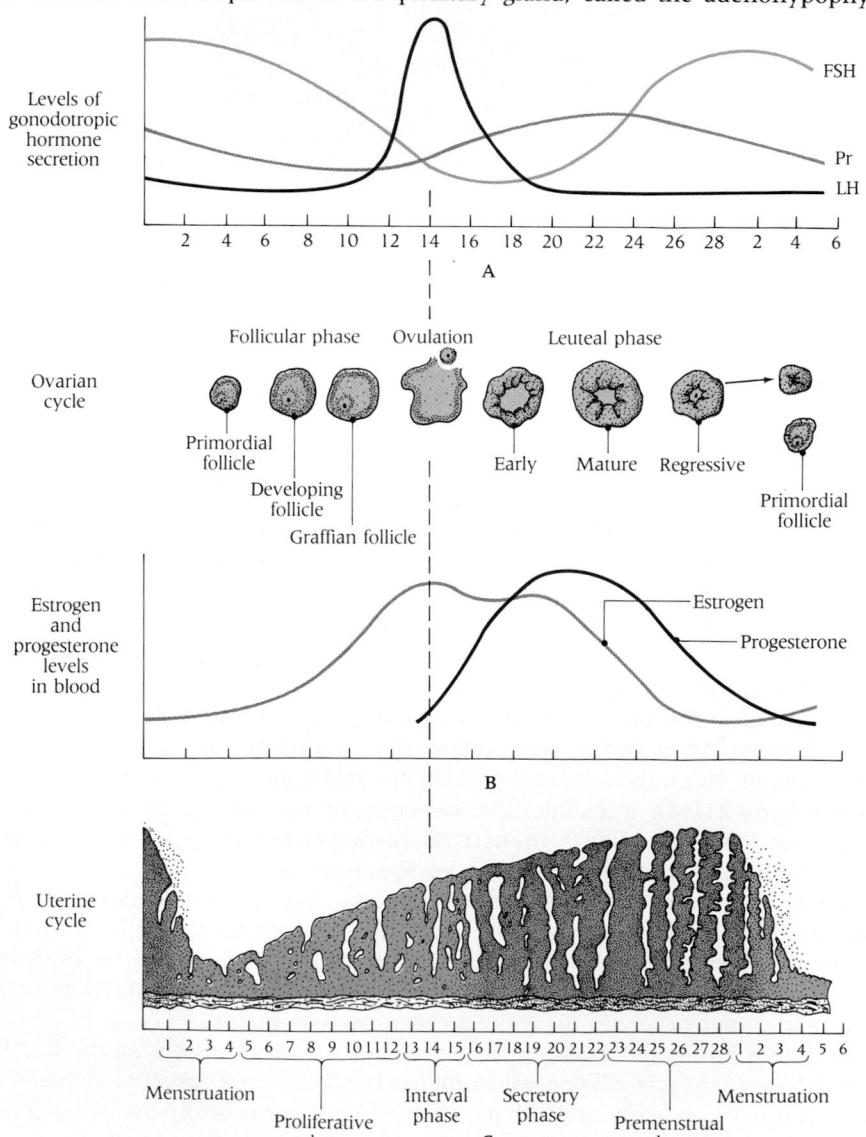

21-38 The ovarian cycle
(A) *The levels of the hormones FSH (follicle
stimulating hormone), Pr (prolactin), and LH
(luteinizing hormone) vary throughout the
ovarian cycle.* **(B)** *Changes in the follicles in
the ovary are accompanied by changes in estrogen
and progesterone levels in the blood.* **(C)** *The
endometrium increases in thickness throughout
the uterine cycle, and is sloughed off during
menstruation if fertilization does not take place.*

(see Chapter 28). Under the cyclical influence of these gonadotropic hormones, an ovarian follicle matures, and ovulation takes place. The three major gonadotropic hormones are **follicle-stimulating hormone (FHS), luteinizing hormone (LH)** and **prolactin (Pr)** (also called **luteotropic hormone** or **LTH).** The adenohypophysis sharply increases its LH output around the time of ovulation. This stimulates the release of progesterone.

Estrous cycle

Most mammals breed seasonally. In them, ovulation does not occur on such a short monthly schedule. Rather, it occurs only during the periods called **estrus.** In most female mammals, eggs mature only one or a few times a year. Secretion of ovarian hormones increases simultaneously and is greatest when the egg is discharged from the ovary. Only then are females willing to copulate. They are then said to be "in heat."

When copulation does take place, nervous stimuli associated with it cause the adenohypophysis to release a burst of LH, which directly leads to ovulation. During the part of the year when she is not in estrus, the female is decidedly unreceptive to mating activity. The whole process is termed the **estrous cycle.**

In a sense, the human ovarian cycle is an aberrant estrous cycle. In the human male, maturation of sperm and secretion of sex hormones go on continuously from early adolescence into old age, though the vigor of both decreases in later years. Female humans, like other female mammals, have an estrous cycle, but it is an unusual one with regular periods of 28 days, though normal cycles commonly range from 24 to 36 days. Fertilization in humans can occur only during a brief period after release of an egg from its follicle. As we have noted, this occurs when the ovum is in a Fallopian tube.

The repetitive uterine cycle in women occurs continuously from early adolescence until the late forties or early fifties, except during pregnancies and for a time during lactation. The ending of the cycles of ovulation and menstruation and of the ability to reproduce marks the **menopause.** This event reflects a normal waning of ovarian activity. Temporary physical and emotional disturbances due to changing hormone levels may occur at this time until the adjustment is completed.

Pregnancy and birth

Pregnancy, or gestation, begins with the implantation of the fertilized egg in the lining of the uterine wall (Figure 21-39A). The duration of gestation in the human species is normally 280 days.[5] The course of embryonic development is illustrated in Figure 21-40E. At full term the embryo (now termed a **fetus**) is head down in 95% of the cases. The **umbilical cord,** which connects it to the placenta, is coiled around its body.

Membranes begin to surround the embryo early in pregnancy (Figure 21-39B,C). Immediately surrounding the embryo is the **amnion.** The closed space within the amnion is filled with **amniotic fluid.** Amniotic fluid, a protective cushion permitting fetal movement, is completely replaced every 24 hours. Its source is the fetal kidneys and bladder—that is, it is fetal urine.

During the first months of pregnancy, the mother often loses a few pounds, sometimes as a result of nausea, but over the entire period she gains an average of 24 lb, most of it in the last 6 months. Approximately 7 lb is the baby; 4.4 lb is fetal membranes and amniotic fluid. The remaining 13 lb arises in the mother— the uterus gains 2 lb, the breasts 3 lb, and the rest of the body 8 lb.

The act of giving birth is called **parturition.** As the end of pregnancy approaches, uterine muscles become more excitable until the onset of **labor,** a series of strong rhythmic muscular contractions the force of which finally expels the baby. The increase in the excitability of the uterus occurs gradually over a period of weeks. Labor is its dramatic climax.

Although the precise mechanism for triggering parturition remains unknown, it is clear that hormonal, nervous, and mechanical factors lead to the culminating contractions. The rising ratio of estrogens to progesterone is one factor inducing

---

[5] The gestation period varies widely among the mammals. Typical average durations are as follows: opossum, 13 days; mouse, 21 days; squirrel, 35 days; cat and dog, 62 days; lion, 106 days; chimpanzee, 250 days; human, 266 days; horse, 355 days; camel, 395 days; rhinoceros, 510 days; and elephant, 620 days.

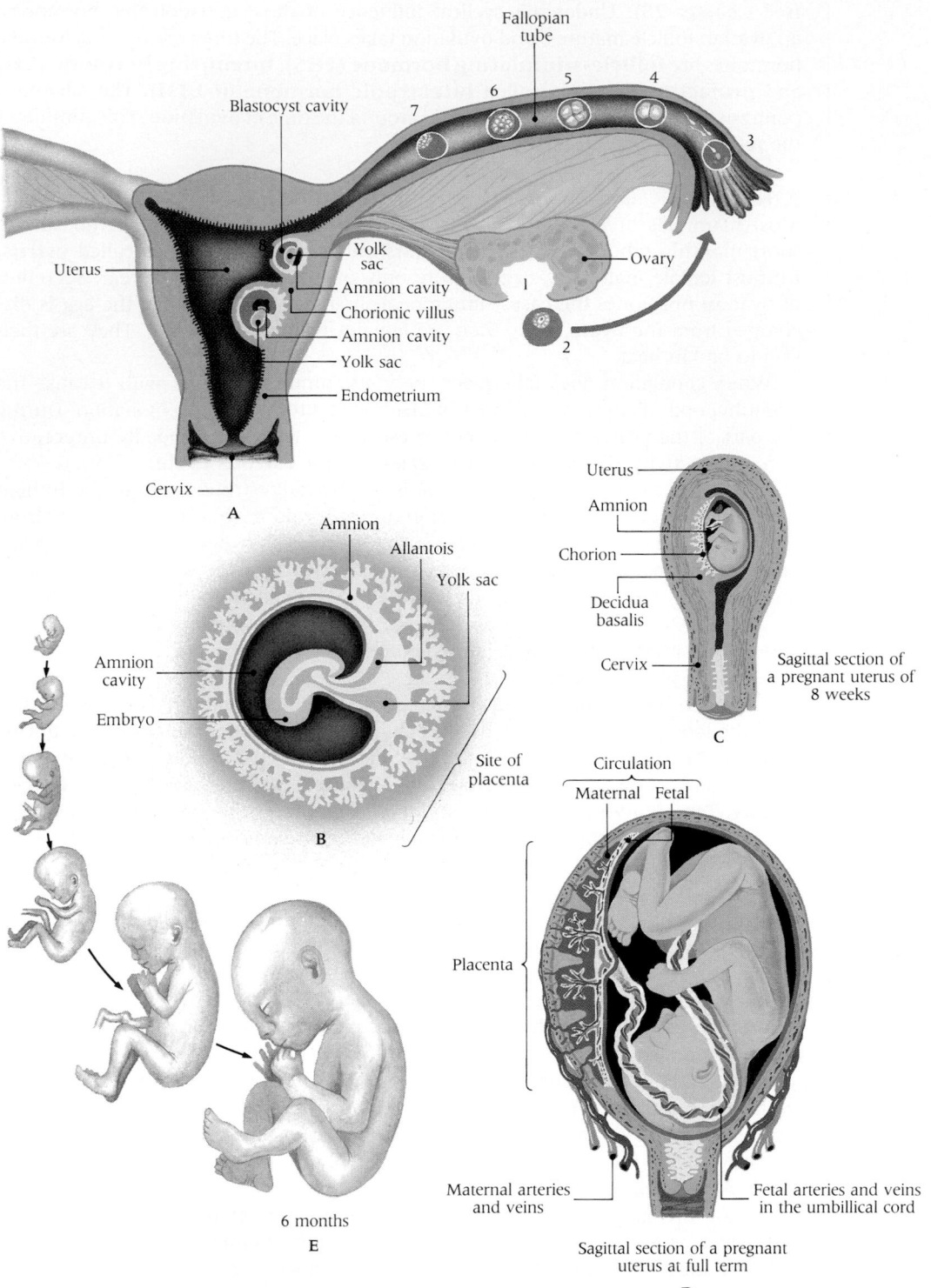

Fallopian tube

Blastocyst cavity

7 6 5 4

3

Uterus

Yolk sac
Amnion cavity
Chorionic villus
Amnion cavity
Yolk sac
Endometrium

Ovary

1

2

Cervix

A

Amnion
Allantois
Yolk sac

Amnion cavity

Embryo

Site of placenta

B

Uterus
Amnion
Chorion
Decidua basalis
Cervix

Sagittal section of a pregnant uterus of 8 weeks

C

Circulation
Maternal Fetal

Placenta

Maternal arteries and veins

Fetal arteries and veins in the umbillical cord

Sagittal section of a pregnant uterus at full term

D

6 months
E

increased uterine contractility. Another is the simple stretching of the uterus by the growing baby. Stretching of the cervix also elicits uterine contractions. Indeed, the obstetrician often can initiate labor by simply dilating the cervix and rupturing the membranes. This allows the descending head of the baby to dilate the cervix even more forcibly. Finally, **oxytocin,** one of the hormones secreted by the posterior portion of the pituitary, the neurohypophysis, specifically promotes uterine contractions (see Chapter 28). When amniotic fluid pressure reaches a critical level, regular contractions begin. Intense contractions of the abdominal muscles soon follow. These help to expel the baby.

Within minutes of the baby's birth, the uterus begins to shrink. The result is a shearing effect between the uterine wall and placenta. This separates the placenta from its implantation site and permits it to be expelled. In one week, the weight of the uterus drops to less than half its weight at the time of parturition. In 30–40

21-39 Human reproduction
(**A**) *The ovary is near the open end of the Fallopian tube. The numbers represent sequential stages in the movement of egg from ovary, through the Fallopian tube (where fertilization occurs), to implantation in the uterine wall.* (**B**) *The implanted embryo is surrounded by membranes and the amniotic cavity.* (**C**) *Sagittal section of a pregnant uterus at 8 weeks.* (**D**) *The fetus in an advanced stage of development is shown in the uterus, with the fetal circulation in the umbilical cord leading to and from the placenta.* (**E**) *Appearance of the embryo at 8 weeks, 9 weeks, 10 weeks, 3 months, 4 months, and 6 months.*

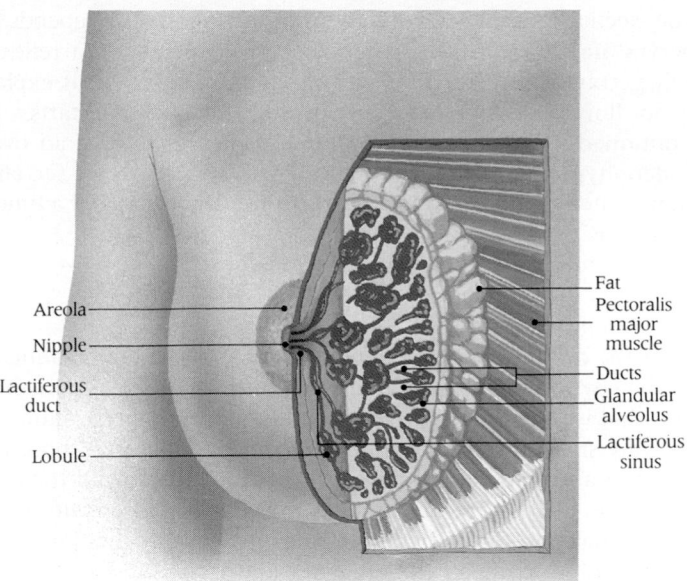

days, the endometrium is restored and ready for a normal uterine cycle. The uterus returns completely to its nonpregnant state in 4 or 5 weeks.

Breasts and lactation

Mammae, or breasts, are distinguishing features of mammals. Their milk-secreting portions, the **mammary glands,** are modified skin glands that develop along two rows called milk lines on the trunk. In animals that have large litters, each milk line gives rise to as many as six pairs of mammary glands—and in humans, occasional accessory breasts or nipples appear, sometimes as far down as the groin.

The development of breasts in men and women is controlled by hormones that are more plentiful in females, where they arise in the ovaries. Female breasts remain small until puberty, when an accumulation of fat enlarges them. The glandular portion of the breast (Figure 21-40) consists of 12 to 20 irregular lobes, arranged radially around the nipple. Each lobe is subdivided into lobules, and each lobule, in turn, into small glandular alveoli. Each lobe has a main duct opening into a collecting space beneath the pigmented area around the nipple, the **areola.** Each main duct receives many smaller ducts. The ducts from the lobules contain sphincters that are under nervous control. Breasts undergo additional growth and development during pregnancy, owing to the influence of estrogens and progesterone that arise in the placenta. Estrogens cause the ducts to proliferate and estrogen and progesterone together stimulate alveolar growth.

The breasts become capable of secreting milk at or soon after the fourth month of pregnancy, but little milk is secreted until after parturition. This is probably because the process of secretion is blocked by the high levels of progesterone present during pregnancy. At birth, progesterone levels in the blood fall sharply as the placenta is expelled (Figure 21-41). However, the fully developed breasts for a time discharge only a little fluid each day. This watery, yellowish protein-rich material is **colostrum.** It differs from milk in that it lacks lipids. It is useful to the newborn because it contains maternal antibodies which help to protect the child from infectious diseases during its early life.

The formation of milk, or **lactation,** suppressed during pregnancy, begins soon after parturition. A curious nervous phenomenon has a role in this process. The act of suckling by the baby (or "milking" in the case of dairy animals) stimulates the adenohypophysis to secrete prolactin (see Figure 21-41). Within 2 or 3 days, the breasts are engorged with milk instead of colostrum.

The continued production of milk depends on continuation of suckling. When the suckling stimulus is absent, secretion of prolactin and other essential hormones stops. The capacity for milk production is then lost in a week or two. Lactation is thus controlled by the demand for milk. Prolactin secretion declines gradually over a period of 7 to 9 months despite continued milking (though it sometimes continues for years), and milk production eventually terminates. Lactation may be deliberately stopped by administration of estrogens, which inhibit prolactin secretion.

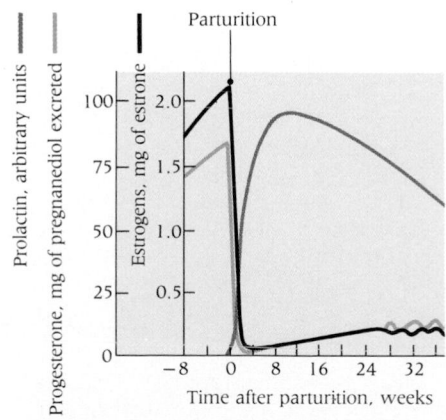

21-41 Effect of parturition on estrogen, progesterone, and prolactin secretion

The ejection of the secreted milk from the breast also depends on the interaction of nerves and hormones. This process, too, depends on a reflex that begins with suckling. Oxytocin secretion is part of the mechanism. This explains why the milk does not flow freely until a minute or two after suckling starts.

Continued lactation for a time prevents resumption of an ovarian cycle—as if the adenohypophysis were so preoccupied with prolactin secretion it neglects secretion of the other gonadotropic hormones. However, after a time it again produces enough FSH to reinstate the cycle.

Human milk and cow's milk are compared in Table 21-4.

Contraception and abortion

Perhaps the only way in which the size of the human population can be controlled, other than by war, disease, and famine, is the increased use of methods of **contraception**—that is, methods that prevent conception, or fertilization.

Broadly speaking, these methods may be divided into those that require strong motivation and forethought before sexual intercourse and those that do not. They may also be divided into techniques that employ essentially mechanical methods of preventing conception and those that employ techniques for altering the physiological functioning of one or the other sex partner.

Mechanical-type methods, in brief, are as follows: **coitus interruptus,** or withdrawal of the penis before ejaculation, an unsatisfactory and often unsuccessful method; periodic abstinence (the so-called **rhythm method**), which depends on the fact that in women with regular cycles ovulation rarely occurs outside the 9th and 19th days after the beginning of the menstruation and which is also unsatisfactory and in many cases ineffective; the rubber **diaphragm,** which fits into the vagina over the cervix, and the **sponge** (each often combined with spermicidal jelly or foam), and the **condom** over the penis (an advantage of which is its value in preventing transmission of AIDS). The diaphragm, sponge, and condom trap sperm before they can enter the cervix.

Sterilization of a woman by the tying off of her Fallopian tubes or of a man by the severing of his vasa deferentia (vasectomy) are mechanical methods in that they prevent, respectively, delivery of the egg to the uterus or sperm to ejaculate. Both are highly reliable but require surgery and are usually not reversible.

Contraceptive methods that depend on techniques for disturbing the physiology of reproduction are gaining in acceptability. As we have seen, reproduction in humans (and other animals) has two aspects: (1) mechanisms concerned with the maturation of the germ cells; and (2) mechanisms that safeguard the mature germ cells and the embryo. Each of these aspects is governed by specific hormones.

Thus, normal fertility is a complex of processes that are subject at almost every step to a balanced program of hormonal controls. Infertility may arise from a defect in any of the steps. Though it may occur spontaneously, it may also be induced artificially in deliberate attempts to prevent conception. The most promising method of inducing infertility involves the continuous administration of progesterone-like hormones. The result is an artificial duplication of the infertile conditions existing during the late luteal phase of the ovarian cycle (Figure 21-38) and during early

TABLE 21-4
COMPOSITION OF MILK

Component	Concentration in Human Milk (approximate)	Concentration in Cow's Milk (approximate)
	g per 100 ml	g per 100 ml
Water	88.5	87.0
Lactose (milk sugar)	7.0	4.8
Lipid	3.3	3.5
Casein	0.9	2.7
Lactalbumin and other proteins	0.4	0.7
Ash	0.2	0.7
Potassium	0.041	0.150
Calcium	0.030	0.120
Phosphorus	0.013	0.095
Sodium	0.011	0.050

pregnancy. At these times, natural estrogen and progesterone levels are high, and consequently the levels of ovulation-stimulating hormones, FSH and LH, are low. Ovulation does not occur; hence there is temporary, or anovulatory, infertility. In practical terms, a hormonal **pill** is taken every day for three weeks and then withdrawn for a week during which an artificial, scanty menstrual flow occurs. The method is 98–99% effective and is popular because it requires no thought or action immediately before sexual intercourse. Despite intensive research, a male contraceptive "pill" has not yet been developed that assures male infertility.

It should be noted that for some years birth control was achieved by the insertion into the uterus of a metal coil or other irregularly shaped object, which had the effect of preventing implantation of an embryo into the uterine wall. These so-called **intrauterine devices,** or **IUDs,** caused many women to experience severe cramps and occasional bleeding—and acquired infections that led ultimately to infertility or sterility. For this reason, IUDs have fallen into disfavor.

After abstinence, **abortion** is surely the most effective of all methods of preventing childbirth. It is also the least desirable, since it requires the destruction of a fetus that is already under development. At what point a fetus becomes a human child is a controversial and contentious biological and ethical question. The moral dilemma is further complicated sometimes by knowledge that a given fetus is defective (see Box 15-2) or unwanted.

Several methods of producing abortions are available.

- The uterus can be vacuum suctioned to remove the embryo.
- Dilatation and curettage empties the uterus by scraping the endometrium.
- "Salting out" involves the injection of a 20% NaCl solution that induces labor and delivery within 24 to 48 hours.
- Drugs that poison the embryo and promote its expulsion are a reliable but hazardous method.
- Hormones called prostaglandins (see Box 28-1) are now the most commonly used method. These powerful stimulators of uterine muscle contraction may be introduced either by injection into the amniotic fluid or by vaginal insertion of a suppository.
- In any woman whose menstrual period is a few days overdue, a high dose of estrogens (the "morning after pill") will stimulate uterine contractions and menstruation and thereby interrupt a pregnancy in its first few hours.
- A new "pill" called RU486, a progesterone antagonist available in Europe but not the United States, produces complete nonsurgical abortions in the first two months of gestation in most (but not all) pregnant women.

SUMMARY

Reproduction is a self-evident necessity for the continuance of life, and organisms have evolved asexual and sexual reproductive methods to achieve this end.

Asexual reproduction, whether by fission, budding, sporulation, or parthenogenesis, yields offspring that are genetically identical to their parents. Often (though not always) uniparental, it is simple and rapid and is the primary mode of reproduction in unicellular organisms, both prokaryotic and eukaryotic.

Sexual reproduction, in contrast, involves the fusion of two haploid nuclei from different gametes. Therefore it has the advantageous capability of introducing genetic diversity among offspring. A species able to produce genetically diverse offspring has a better chance of generating variants that can better survive the challenges of a changing and hostile environment. The existence of self-fertilizing hermaphrodites shows that sexual reproduction is not always biparental. However, sexual reproduction by male and female parents better promotes genetic diversity. It also allows a useful male-female division of labor in reproductive tasks such as egg laying and territorial protection. A generalized life cycle of sexually reproducing species reveals alternation of haploid and diploid forms, but the relative duration of haplophase and diplophase varies greatly among different species.

The life cycles of protists vary widely. Green algae such as *Chlamydomonas* spend most of their life cycle in haplophase, but other algae demonstrate alternation of generations in which haploid gametophyte and diploid sporophyte organisms are both well developed.

In plant reproduction, gametes arise not from meiosis, as in animal reproduction, but from mitosis in haploid gametophytes, which developed originally from spores produced meiotically by sporophytes. Sporophytes, in turn, arise from the zygote formed by the union of two gametes. Three representative plant categories, mosses, ferns, and seed plants, differ in the relative size and prominence of the gametophyte and sporophyte generations: in mosses, the familiar plant form is the gametophyte; in ferns and seed plants, it is the sporophyte. Seed plants have two kinds of spores and thus can form male and female gametophytes. While female gametophytes remain parasitically within the parent sporophyte, male gametophytes make up the pollen grains, which fertilize the female gametophytes.

A seed is a ripened ovule with an embryo inside of it. Seed plants are nonflowering (gymnosperms) or flowering (angiosperms). Conifers are gymnosperms—and pines, the most common conifers, are monoecious plants bearing male and female strobili (cones) on the same plant. Each microsporangiate (pollen) cone has several microsporophylls bearing two microsporangia, each of which produces many microspores that become dispersible pollen grains. Each macrosporangiate (ovulate) cone has seed-scale complexes bearing two ovules, each of which contains a megasporangium bearing a haploid megaspore that gives rise to a female gametophyte which remains embedded in the sporangium and ultimately bears egg-producing archegonia. The combination of sporangium, female gametophyte, and outer covering is an ovule. Pollen reaching an olvulate cone fertilizes its egg cells. The resulting zygote forms a tiny embryo—still within the female gametophyte which is within the sporangium. The entire ovule is shed as a seed.

Flowering plants (angiosperms) cluster their male and female reproductive organs within the flower. Zygote formation is accompanied by a second fertilization, in which a haploid sperm nucleus fuses with a diploid cell, forming a triploid endosperm, which nourishes the growing diploid embryo within the seed. Plant embryonic development differs from that of animals in that morphogenetic cell movements do not occur, and cytoplasmic determinants are not present. The fruit that in many flowering plants surrounds the seeds develops from the ovary and facilitates seed dispersal. The many mechanisms by which immobile plants disperse pollen and seed reflect the difficulty they face in getting the sexes together and then moving offspring away from the parental plant.

In fungi, meiosis follows immediately upon the formation of the zygote. Hence, these organisms exist primarily in haplophase. Many fungi produce spores asexually. Some are also capable of a form of sexuality, called parasexuality, in which gametes from two genetically dissimilar types combine, but their nuclei may not fuse. For much of the life of the resulting organism, the two nuclei simply coexist in each cell (heterokaryosis).

Animal reproductive cycles are marked by the absence of a multicellular haplophase. Asexual reproduction alternates with sexual reproduction in some invertebrates, but is not widespread. In most invertebrates, the embryo does not develop directly into an adult organism, but into a larva. Larvae aid in dispersal of relatively immobile species, and are also specialized for feeding, and thus growth.

Metamorphosis is the process by which larvae become adults.

Most animals have distinct male and female gametes. In species employing external fertilization (fishes, amphibians, etc.), gametes released directly into water find each other either by accident or by synchronization of gamete release. More complex animals make use of a variety of anatomical, physiological, and behavioral adaptations to bring their gametes together.

Early land animals (reptiles and birds) depended on internal fertilization and the production of eggs with shells. With the development of the uterus in placental mammals, shelled eggs were no longer essential. Instead, a placenta permitted exchange of nutrients, gases, and wastes between the developing embryo and the bloodstream of its mother. Milk from mammary glands and more elaborate parental care evolved to sustain the newborn offspring.

As in other mammals, the chief functions of the human male reproductive system are spermatogenesis, participation in copulation, and production of male sex hormones, the androgens. In human females, the ovaries promote egg maturation and secrete the ovarian hormones estrogen and progesterone.

Under the influence of hormones secreted by the hypothalamus, human ovaries secrete estrogen and progesterone in a pattern (the estrous cycle), which repeats approximately every 28 days. Estrogen levels rise during the first half of the cycle. A single egg is then released into the Fallopian tube from one of the ovaries (ovulation). Progesterone levels then rise as well. Then, at the end of the month, both estrogen and progesterone levels fall. These steroid hormones maintain the uterine lining; when their levels fall, menstruation occurs. Most other mammals breed seasonally, and ovulate, not monthly, but only one or a few times per year.

Pregnancy begins with the implantation of a fertilized ovum in the uterine wall and ends when a combination of mechanical and hormonal factors triggers uterine contractions, which expel the infant. After birth, mechanical and hormonal factors stimulate milk production (lactation). The story of human reproduction is, of course, complicated by issues centering on contraception, abortion, and human emotions and needs.

KEY TERMS

abortion
allantois
alternation of generations
amnion
angiosperm
anther
archegonium
asexual reproduction
binary fission
budding
carpel
chorion
cotyledon
dioecious
endometrium
endosperm
estrogen
estrous cycle
Fallopian tube
fetus
fission

follicle
fruit
gametic meiosis
gametophyte
gymnosperm
hermaphrodite
heterokaryosis
heterosporous
homokaryosis
medusa
megaspore
microspore
monoecious
ovarian cycle
ovary
oviduct
ovulation
ovule
parasexuality
parthenogenesis
pistil

placenta
pollen grain
polyp
progesterone
pupa
radicle
seed
sepal
sexual reproduction
sporangia
spore
sporic meiosis
sporophyte
stamen
stigma
style
syngamy
testis
tracheophyte
vegetative reproduction
zygotic meiosis

QUIZ QUESTIONS

1. Which of the following is *not* a function of asexual reproduction?
   A. increasing genetic diversity
   B. replacing injured individuals

C. permiting growth of populations
D. replacing diseased individuals
E. All of the above are functions of asexual reproduction.

2. In angiosperms,
   A. the sporophyte grows upon the gametophyte.
   B. vascular tissues are absent.

C. reproductive structures are located in flowers.
D. the female gametophyte is motile.
E. A and D

3. _____ is the part of the generalized life cycle that occurs after meiosis but before fertilization.
A. Diplophase
B. Haplophase

C. Metaphase
D. Pupation
E. Germination

4. The immediate products of meiotic division of a microspore mother cell are
A. tube cells.
B. generative cells.
C. megaspores.
D. pollen grains.
E. microspores.

5. The _____ is a key feature that allowed the invasion of land by vertebrates.
A. mycelium
B. amnion
C. heterokaryon
D. endometrium
E. corpora cavernosum

ESSAY QUESTIONS

1. Why is the reproduction of organisms a necessary part of life on Earth? What is the relation between the reproduction of organisms and such cellular activities as mitosis, meiosis, and fertilization? Why are gametes needed in the reproduction of organisms?

2. Contrast asexual and sexual reproduction. Characterize the four major modes of asexual reproduction. What are the advantages and costs of sexual reproduction?

3. What is the primary purpose (function) of the estrus cycle in female mammals? Why doesn't it occur in reptiles or birds?

What aspects of the human female ovarian cycle distinguish it from estrus in other mammals?

4. What is a flower; a seed; a fruit? What role do they play in reproduction in plants? To what structures do they compare in the animal kingdom?

REFERENCES AND SUGGESTIONS FOR FURTHER READING

Bell, G. (1982). *The Masterpiece of Nature: The Evolution and Genetics of Sexuality.* University of California Press, Berkeley.

A summary of theories on why sexual reproduction evolved as it did.

Cook, R. E. (1983). Clonal plant populations. *American Scientist 71,* 244–253.

A good discussion on the role of asexual reproduction among plants in their natural habitats and communities.

Cornish, E. C., Anderson, M. A., and Clarke, A. E. (1988). Molecular aspects of fertilization in flowering plants. *Annual Review of Cell Biology 4,* 209–228.

An informative review.

Galston, A. W., Davies, P. J., and Satter, R. L. (1980). *The Life of the Green Plant,* 3rd ed. Prentice-Hall, Englewood Cliffs, N.J.

A useful introductory text containing excellent discussions of plant reproduction and development.

Harvey, P. H., and Read, A. F. (1988). When incest is not best? *Nature 336,* 514–515

A stimulating note pointing out that inbreeding is sometimes beneficial and sometimes deleterious.

Katchadourina, H. A., and Lunde, D. T. (1980). *Biological Aspects of Human Sexuality,* 2nd ed. W. B. Saunders, Philadelphia.

A beginning text on human sexuality with emphasis on its biological aspects.

Kevles, B. (1988). *Females of the Species: Sex and Survival in the Animal Kingdom.* Harvard University Press, Cambridge, Mass.

The author argues that in the animal kingdom, females have taken a far more active role in the evolution of species than many have believed.

Koller, D. (1959). Germination. *Scientific American 200,* April, 75–84.

A somewhat dated but interesting review of the biology of seeds, their dormancy and the factors affecting their rates of germination.

Kondrashov, A. S. (1988). Deleterious mutations and the evolution of sexual reproduction. *Nature 336,* 435–440.

The prevalence of sexual reproduction, in spite of its costs, implies evolutionary advantages over asexual reproduction. Here a Soviet biologist argues that sex more efficiently eliminates deleterious mutations.

Hartmann, J. F. (Ed.) (1983) *Mechanisms and Control of Animal Fertilization.* Academic Press, Orlando, Fla.

A scholarly review of a topic of intense current interest.

Longo, F. J. (1988). Reorganization of the egg surface at fertilization. *International Review of Cytology 113,* 233–270.

New developments in an interesting area of research.

Mayer, A. and Poljakoff, A. (1975). *The Germination of Seeds.* Pergamon Press, New York.

A comprehensive treatise on all aspects of seed biology.

Maynard-Smith, J. (1978). *The Evolution of Sex.* Cambridge University Press, New York.

Incisive thinking on the evolutionary origin of sex.

Segal, S. J. (1975). The physiology of human reproduction. *Scientific American 233,* September, 52–62.

The author discusses the interactions between the nervous and endocrine systems in regulating the sexual cycle.

Ulmann, A., Teutsch, G., and Philibert, D. (1990). RU486. *Scientific American 262,* June, 42–48.

A fascinating story of discovery by the French developers of RU486.

Waring, P. F., and Phillips, I. D. J. (1981). *Growth and Differentiation in Plants,* 3rd ed. Oxford University Press, New York.

A good review.

Woman walking, throwing scarf over shoulder

Dog in gallop. one stride

Acorn to Oak

PART 3

HOMEOSTASIS

Whatever complexity of organization we may infer from electron micrographs or genetic studies or metabolic data from single cells pales in magnitude when we consider whole multicellular organisms. From their study, we have learned how living systems integrate and maintain their organization—how, in short, they maintain homeostasis. Part 3 is devoted to these matters.

Two major systems that defend the body against adversity—the blood and the immune system—are discussed in Chapters 22 and 23. Both are areas in which progress is occurring at a spectacular pace. Although a large body size has advantages, it also creates difficulties that are overcome only by such mechanisms as a transport system, a cardiovascular system in animals and a vascular system in plants (Chapter 24). Our review of excretion (Chapter 25) and respiration (Chapter 27) shows interesting and revealing differences that relate to an animal's mode of life and evolutionary history. The procurement and processing of needed materials and energy are reviewed in Chapter 26.

Attention then shifts to the integration of constituent body parts into an ordered and smoothly functioning whole. Such coordination demands communication within an organism. Chapter 28 describes the chemical messengers (hormones), which are transported from place to place in animals via the bloodstream and from cell to cell in plants. Chapters 29 through 31 are devoted to the second coordination system in animals, the nervous system. In addition to effecting communication among internal parts, the nervous system obtains and handles information about the environment and regulates an organism's behavior. Chapters 32 and 33 deal with the behavior of organisms, so profoundly important in the struggle for survival.

SECTION 1

DEFENDING AGAINST ADVERSITY

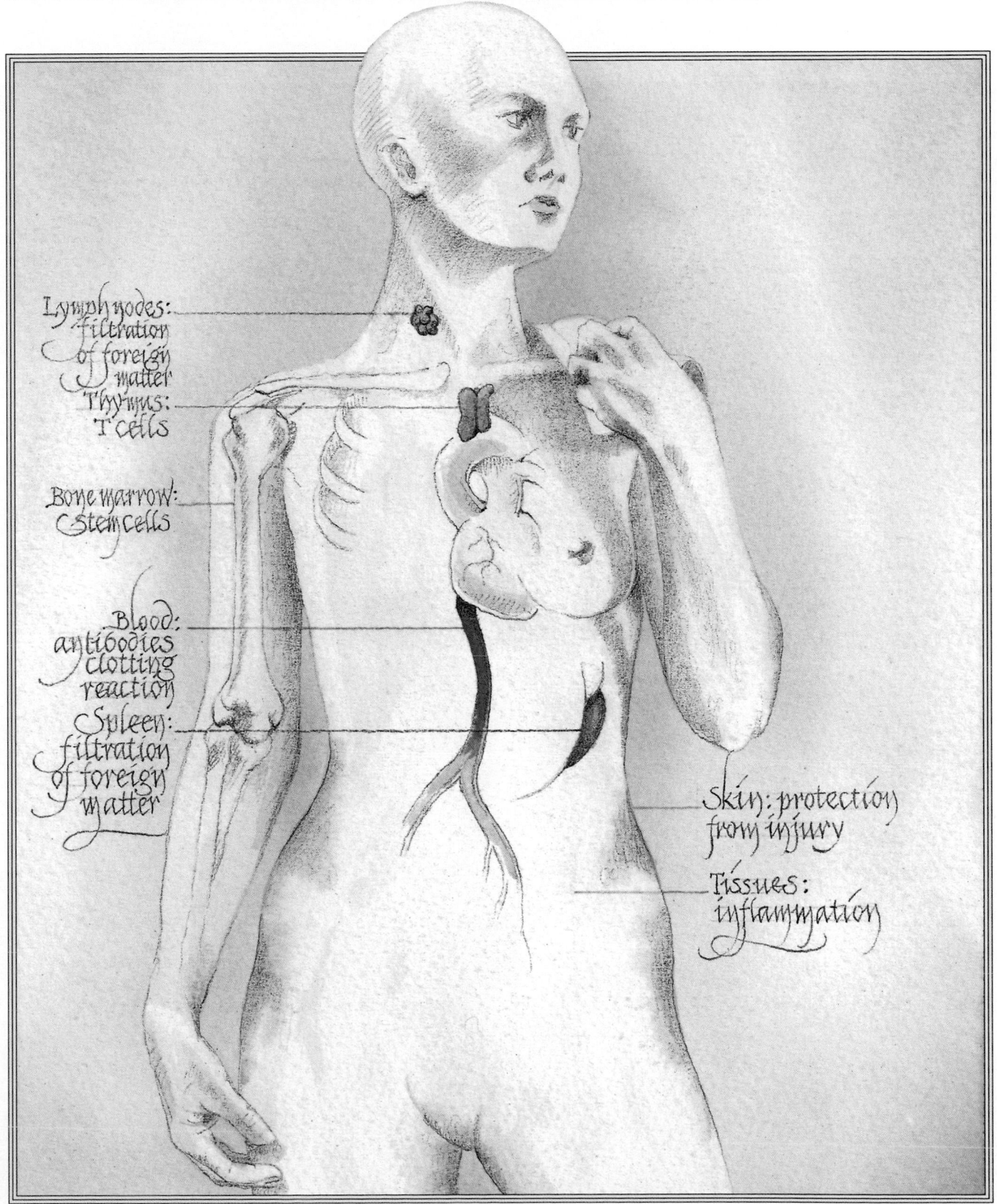

Lymph nodes: filtration of foreign matter

Thymus: T cells

Bone marrow: stem cells

Blood: antibodies clotting reaction

Spleen: filtration of foreign matter

Skin: protection from injury

Tissues: inflammation

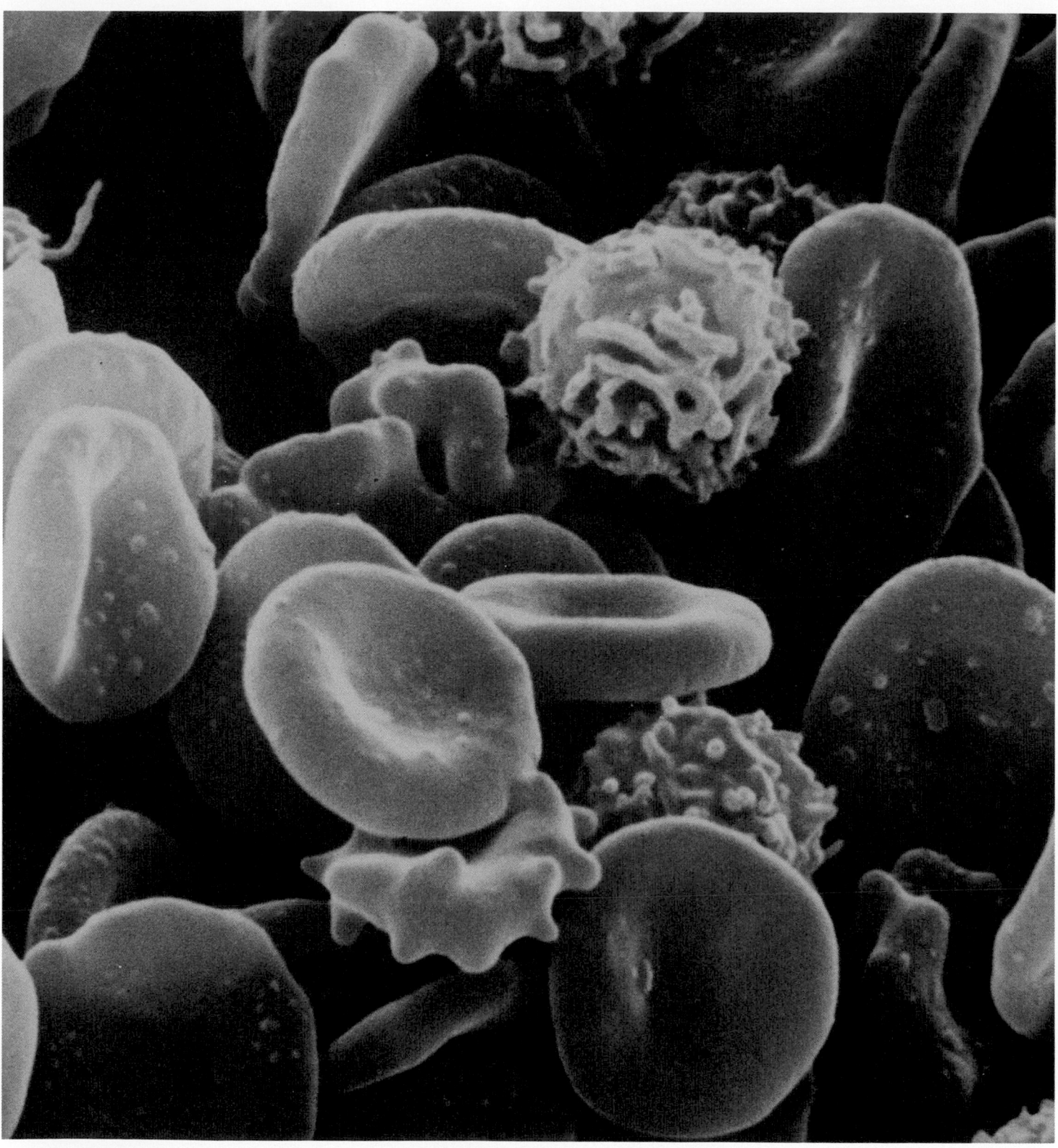

BLOOD

Organisms living in a hostile and capricious external environment must somehow preserve the constancy of their internal chemical environments.

In this and the next chapter, we are concerned with systems that defend organisms against adversity of one type or another. We learned in Chapter 1 that **homeostasis** refers to an organism's tendency to maintain a stable internal environment. Disturbances trigger sensitive mechanisms, which set about the restoration of balance. Major responsibility for the achievement of homeostasis rests upon the extraordinary fluid called blood.

BIOLOGICAL SIGNIFICANCE OF BLOOD

Blood has three major functions.

- It is the major vehicle of transportation within animals.
- It harbors several important defense mechanisms that include, in whole or in part, the clotting system which prevents fatal bleeding and the immune system which destroys or inactivates invading organisms and other foreign matter.
- It has a leading role in the regulation of salt and water distribution, acid-base balance, and temperature.

In sum, blood is engaged in transport, defense, and regulation.

As we shall see, blood participates, directly or indirectly, in every body function.[1] If blood is considered a tissue, it is the body's most voluminous. Its volume is 5 or 6 liters in an average-sized man or woman.

PHYSICAL CHARACTERISTICS OF BLOOD

Whole blood is, in fact, a **plasma** suspension of three kinds of blood cells. Their informal (and scientific) names are: **red cells (erythrocytes); white cells (leukocytes);** and the curious fragmentlike **platelets (thrombocytes)** (Figure 22-1A). So unusual are red cells and platelets—neither contain nuclei—some have questioned whether they deserve to be called cells. That is probably why the evasive terms "corpuscle" and "formed element" are sometimes used instead. Interestingly, the red cells of birds, amphibians, fishes, and reptiles, unlike those of humans and other mammals, do have nuclei (Figure 22-1B-D).

When blood is shed, it coagulates to form a **clot.** Blood drawn from the body into a glass test tube also forms a clot which, if undisturbed, contracts, leaving a

Blood
This scanning electron micrograph displays two of the three major cell types of blood: red cells (the biconcave disks) and white cells (bodies with irregular surfaces). Along with platelets, these cells, in great numbers, are suspended in the complex fluid called plasma. (× 8600)

---

[1] Although this discussion refers primarily to humans and other vertebrates, most of the general principles apply to all animals. We shall deal later with the fluids of plants.

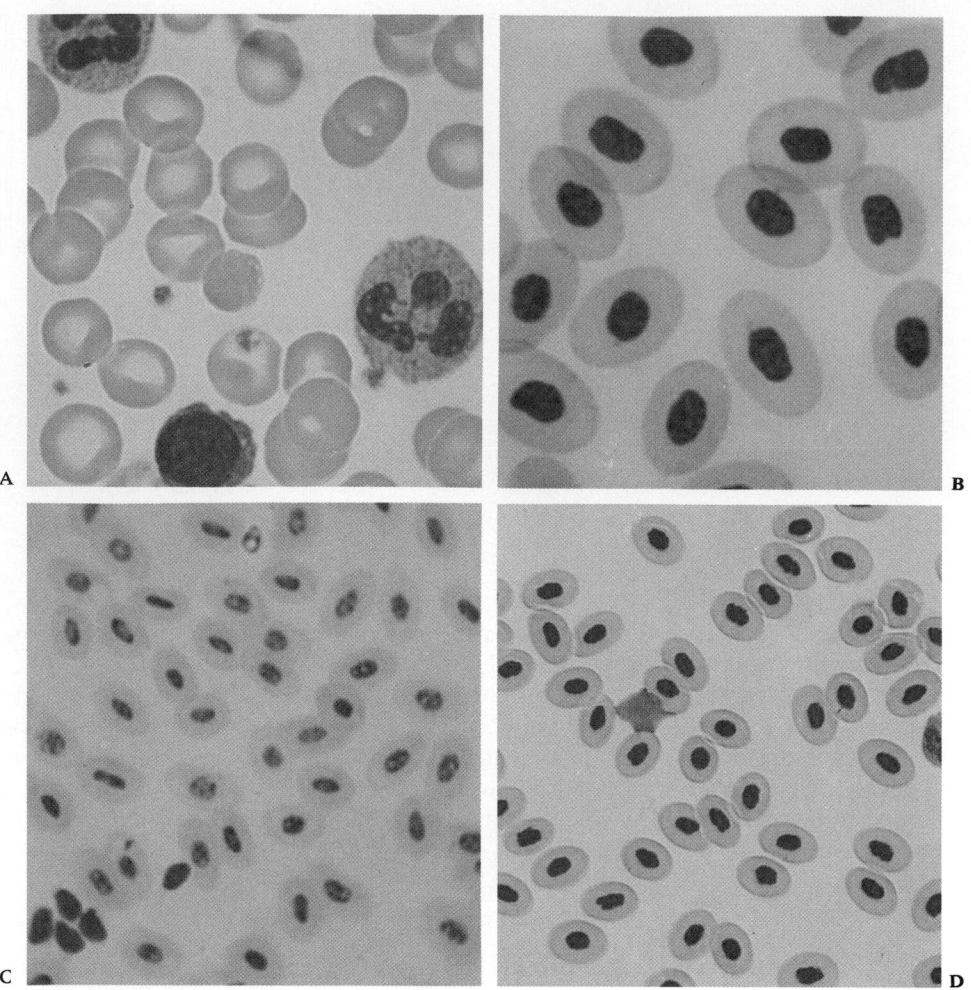

22-1 Stained blood smears showing various cell types
(A) A smear of human blood showing many red blood cells, a few white blood cells with prominent nuclei, and tiny platelets. (× 1200)
(B) In nonmammalian blood, red cells are nucleated. This is a stained smear of pigeon blood at high magnification. (× 300) **(C)** Fish (carp) blood. (× 900) **(D)** Frog blood. (× 400)

clear yellow fluid called **serum** (Figure 22-2). We can obtain whole blood in the fluid state without the formation of a clot only by adding a substance that prevents clotting. This is called an **anticoagulant.**

A clot forms because a soluble plasma protein, **fibrinogen,** has been converted to the insoluble protein, **fibrin.** The major difference between serum and plasma is that serum contains no fibrinogen or other clot-forming proteins. Serum is the fluid remaining after a clot forms. It cannot clot. Plasma does contain fibrinogen and can clot. Plasma is the fluid found at the top of a test tube after whole blood containing an anticoagulant has been centrifuged. A clot does not form in the test tube because an anticoagulant is present. Clotting does not normally occur within the blood vessels either, despite the fact that within the body plasma contains no anticoagulant. If clotting did normally occur within the blood vessels, life would not be possible. In the normal body, the fibrinogen in plasma is kept from becoming fibrin by mechanisms we will examine later.

The word **hematocrit** was originally applied to an instrument for determining the relative amounts of blood cells and plasma in whole blood. The quantity measured is the volume of packed cells per 100 ml of blood. "Hematocrit" is now synonymous with this percentage.

The hematocrit is determined by centrifugation of a blood specimen in a glass tube until the red cells are firmly packed in the bottom of the tube. Above them lies a white layer of white cells and platelets, and above this lies clear, cell-free plasma (Figure 22-3). The volume of red cells—that is, the red cell hematocrit— averages around 45% in normal men and 42% in normal women. In other words, there are 45 (or 42) ml of red cells per 100 ml of blood.

The white cells and platelets settle on top of the red cells because their specific gravities are lower than that of red cells and higher than that of plasma. They form a little white cap on the column of red that is called the buffy coat. The size of the buffy coat varies with the number of circulating white cells and platelets.

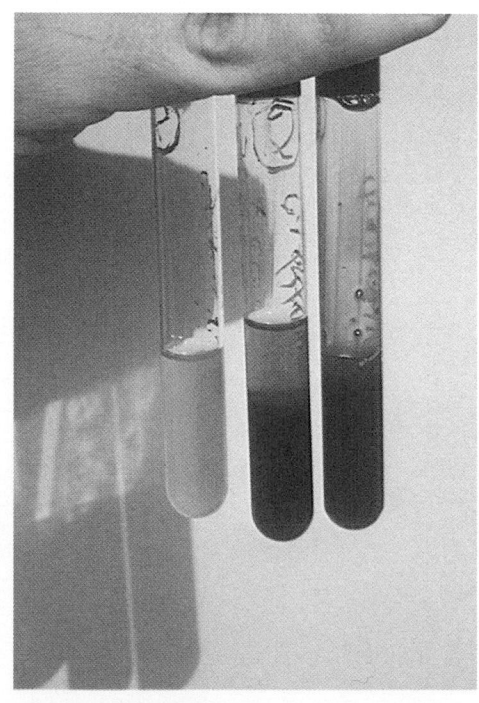

22-2 Formation of a clot
On the right is a tube of clotted blood. The clot is at the bottom. Serum is on top. The serum can be poured off, as shown in the tube on the left, leaving the clot in the bottom of the middle tube.

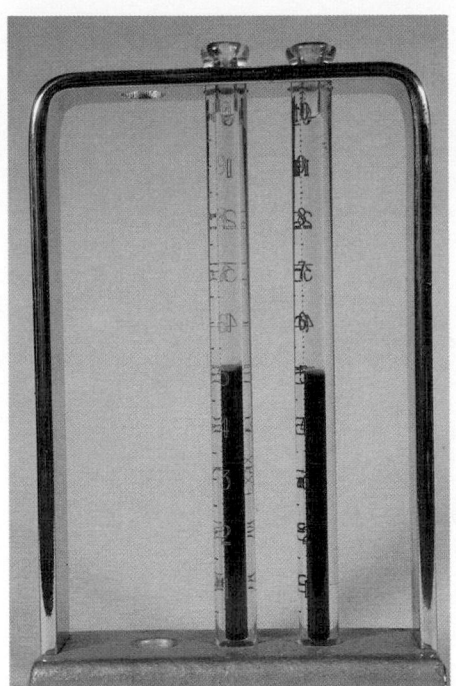

22-3 Appearance of blood after centrifugation in a hematocrit tube
Centrifugation packs red cells and white cells into the bottom of a hematocrit tube. The tube is graduated, therefore the volume of packed red cells per unit volume of blood can be measured as a percentage. Note the thin layer of white cells on top of the red cells.

CELLS OF THE BLOOD

SITES OF FORMATION

The embryonic mesoderm (Table 19-2) is the precursor of both the heart and blood vessels and the blood cells—as well as of the connective tissues, bones, and muscles. In parts of the early embryo's body, mesoderm takes the form of unorganized masses of actively migrating cells called **mesenchyme.** Mesenchyme forms a meshwork of tissue and then develops into the circulatory system and blood cells as well as the lymphatic system.

Early in embryonic life, certain mesenchymal cells begin to form the tubes of a primitive circulatory system. At the same time, and for a brief duration, some mesenchymal cells differentiate into free-floating elements that are progenitors of the stem cells of adults. These are soon swept along by a mounting stream of plasma. When they divide, some of the daughter cells remain stem cells and preserve the stem-cell pool. Others acquire hemoglobin. These are the body's first precursors of red blood cells.

Later in embryonic life, the liver becomes the major site of blood cell formation, or **hematopoiesis.** The spleen, lymph nodes, and thymus also manufacture blood cells during embryonic life. Curiously, hematopoiesis continues in these organs only until soon after birth. At the midpoint of an embryo's life, **bone marrow** becomes the major site. At birth and for the rest of the life span, it is the only site of hematopoiesis—except in abnormal conditions.

In the human adult, the hematopoietically active portion of bone marrow is largely restricted to the flat bones (skull, pelvis, and sternum) and parts of the vertebrae. Some blood cells are produced in the ends of the long bones of the legs and arms, but few are produced in their shafts, the marrow of which is quite fatty and thus termed yellow marrow in contrast to active marrow, which is red. Yellow marrow is transformed into active red marrow in times of need—when the body requires additional red cells or white cells. Collectively, the total bone marrow (including red and yellow marrow) of a human adult weighs 1500–3000 grams. It is thus an organ of formidable size. Why is all of this necessary?

The answer is straightforward. Red and white cells in the circulating blood are short-lived and must be constantly replaced. Hence, the process of hematopoiesis is not only enormous in scale, it is also complex. Cells of nine distinct lineages are tightly packed together in the marrow. Furthermore, hematopoiesis must be capable of rapid but controlled fluctuations to meet a variety of emergencies ranging from blood loss to infection.

This remarkable system is currently under intense scientific scrutiny. One of its novel aspects is that all blood cells originate from a very small population of multipotent **stem cells** that, as noted earlier, were formed during a brief interval in early embryonic life. Ever since, this population of precursor cells has maintained hematopoiesis by virtue of its extensive capacity for self-generation. These stem cells produce the unipotent precursors of each of the three blood cell types: red cells, white cells, and platelets—as well as several other cell types (Figure 22-4). These committed precursors then undergo maturation in the marrow. Mature blood cells are ultimately released into the blood.

RED BLOOD CELLS

The **red cell,** or **erythrocyte,** is the blood cell produced in greatest numbers. It is the primary transportation vehicle for hemoglobin, which carries oxygen.

Mature red cells in mammals are shaped as biconcave discs (Figure 22-5) that differ from most other cells in several ways. As the red cells of mammals have no nuclei, they cannot divide or reproduce. They arise in bone marrow from precursors that do have nuclei and do divide; but before they are released into the blood, they lose their nuclei. The nucleus is extruded—literally banished. In the human bloodstream, red cells live an average of 120 days. They then become senescent and are destroyed by specialized cells of the spleen and liver.

The number of red cells in the adult body is very large indeed. The blood of an average-sized man contains about 5 million (5×10^6) red cells per cubic mm (μl). Assuming a total blood volume of 5 liters, this is equivalent to a grand total of 25×10^{12} circulating red cells. (The normal red count is 5–10% lower in women.)

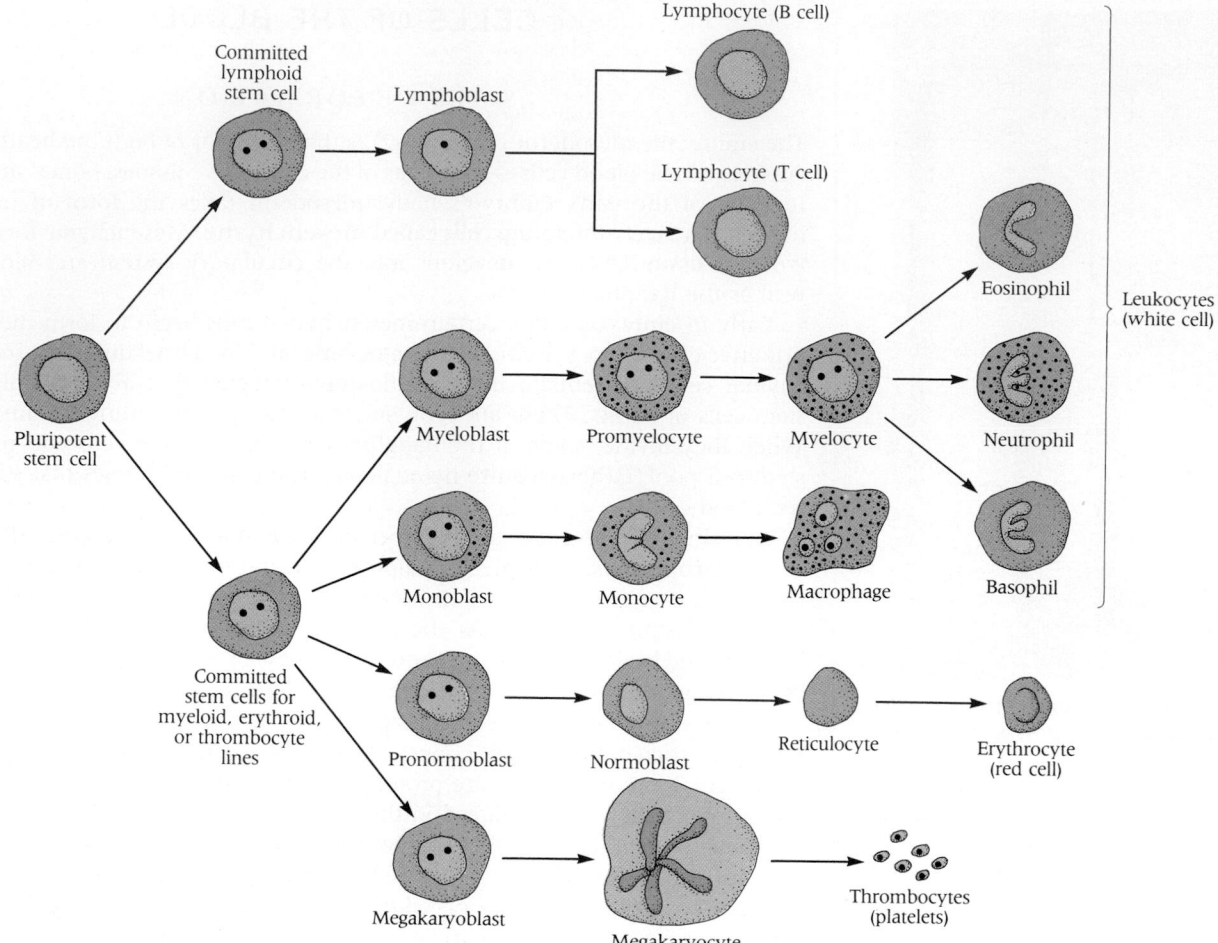

Every day 1/120th of this number (about 2×10^{11}) are formed and the same number are destroyed. That means that every second about 2.3 million new red cells pass from the marrow into the blood and about 2.3 million old red cells are recognized and destroyed. If there is inadequate formation of red cells or if destruction occurs too rapidly, **anemia** (defined as a deficiency of red cells) will ensue.

Hemoglobin

Hemoglobin is a remarkable red protein that travels within red cells and transports oxygen (and carbon dioxide). Why is it there? It is fabricated in only the immature precursors of red cells. Why did such a complex method of carrying hemoglobin molecules from place to place arise in evolution? Why does hemoglobin not travel about dissolved in the plasma, thus making red cells quite unnecessary?

22-4 Stem cells give rise to differentiated blood cells
Different maturation sequences give rise to different types of blood cells, but all derive from the same pluripotent stem cell.

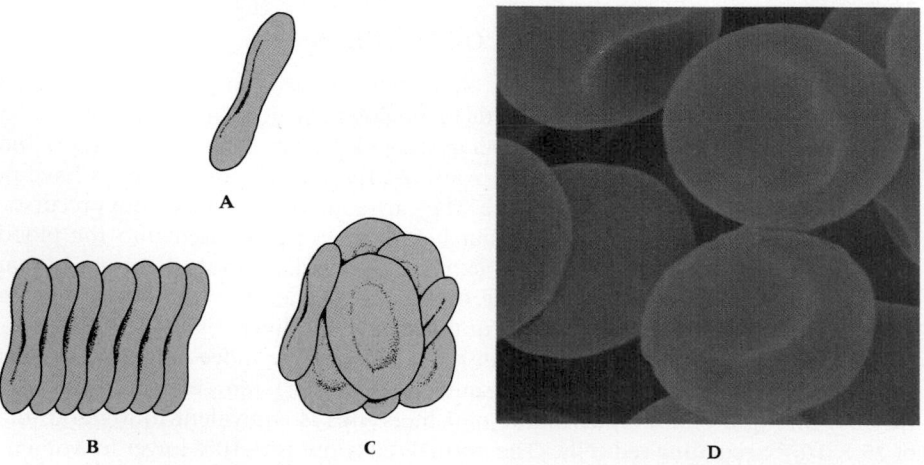

22-5 Shapes of red cells
(A) *Individual erythrocyte.* **(B)** *A rouleau of red cells.* **(C)** *A clump of agglutinated cells.* **(D)** *Scanning electron micrograph of red cells displaying their biconcave shape. Compare with the photograph at the beginning of this chapter.*

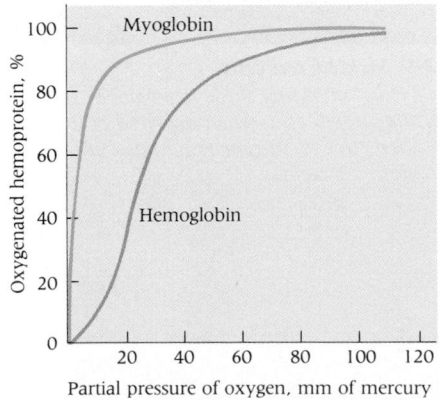

22-6 Structures of pyrrole and its porphyrin derivatives
(A) *Pyrrole.* **(B)** *Porphin.* **(C)** *Protoporphyrin.*
(D) *Heme. Compare with the discussion of tetrapyrrides in Figure 10-7, and see Box 10-1.*

22-7 Oxygen dissociation curves
The curve for hemoglobin is sigmoidal; that for myoglobin is not. This accounts for a major difference in the ability of each protein to carry and release oxygen.

A hemoglobin-like pigment, called **hemocyanin,** is in fact dissolved in the plasma of crayfish, crabs, and other crustaceans. But if free hemoglobin were present in circulating human plasma, some of it would be oxidized to **methemoglobin,** a form that can no longer transport oxygen. Some would be tightly bound to **haptoglobin,** a specific hemoglobin-binding protein in the plasma. Some would be altered chemically. And some would be excreted by the kidneys.

Moreover, if a solution of hemoglobin in plasma even approached the concentration achieved by the hemoglobin within red cells (about 33%), it would fatally increase the viscosity and osmotic pressure of plasma. The placement of hemoglobin within red cells was a compromise of sorts. It may not transport oxygen as rapidly in a red cell as it would in free solution, but its life span in red cells is the same as that of the cells—120 days—and there it is continuously and efficiently protected by red cell enzymes that act to prevent its oxidation.

Molecular structure
Hemoglobin belongs to the category of conjugated proteins called **chromoproteins** (see Chapter 4), which are of wide biological distribution. In these chromoproteins, a specific protein is linked to a porphyrin derivative. For that reason, they are also called **hemoproteins** (Figure 22-6; see also Chapters 9 and 10).

We saw in Box 10-1 that porphyrins readily combine with a metal atom that serves as the active site in the several kinds of reactions in which hemoproteins participate. The metal, iron, is the site of oxygen binding in hemoglobin and of electron transfer in the cytochromes. The protein of hemoglobin is **globin.** The porphyrin is **heme.** Hemocyanin, the respiratory pigment of crayfish and other crustaceans, contains copper instead of iron. It is blue in color.

Role in oxygen transport
Hemoglobin binds oxygen reversibly. This means that it takes up oxygen when the surrounding oxygen pressure is high (as it is in the oxygen-rich blood of the lungs) and releases oxygen when the surrounding oxygen pressure is low (as it is in the oxygen-depleted blood near body cells).

This behavior is represented graphically by the **oxygen dissociation curve** in Figure 22-7. This capability of hemoglobin accounts for the transfer of oxygen from the lungs to the interior of body cells (see Chapter 27). As we will see, the sigmoidal (S-shaped) character of hemoglobin's oxygen dissociation curve is a critical prerequisite for the proper execution of this task. **Myoglobin,** a hemoglobin-like pigment of muscles, lacks hemoglobin's tetrameric (four subunits) structure. As a result, its curve is nonsigmoidal and it could not function as exquisitely as does hemoglobin as an oxygen-transporting agent.

Hemoglobin is bright red when saturated with oxygen. Blood in arteries is highly oxygenated. Therefore, arterial blood has a distinctive red color. Blood in veins is returning from the tissues and is low in oxygen content. Venous blood is a darker and more purplish red.

Hemoglobin has another interesting property, one that takes on added significance in today's polluted environment. The iron in hemoglobin combines with carbon monoxide (CO) even more avidly than with oxygen. Having done so, it becomes incapable of binding oxygen. Carbon monoxide is produced in vast quantities by gasoline engines, gas ranges, heaters, and charcoal grills. Severe CO poisoning can be rapidly fatal because it denies oxygen to body cells.

Variants
In early embryonic life, human embryos contain two distinctive hemoglobins that are called hemoglobin Gower 1 and hemoglobin Gower 2. These are replaced by the end of the third month of embryonic life by a new hemoglobin, called **fetal hemoglobin (hemoglobin F).** Fetal hemoglobin is present in high concentrations (75 to 95%) in blood at birth but gradually disappears during the first year of life. Fetal hemoglobin differs in significant ways from the hemoglobin of adult blood **(hemoglobin A).**

■ Hemoglobins F and A have slightly different amino acid sequences. This has several important consequences. One provides the basis of a reliable old assay method: hemoglobin A is rapidly denatured in alkaline solution; hemoglobin F is alkali-resistant.

- Like hemoglobin A, hemoglobin F contains α chains. But instead of β chains, it contains a unique polypeptide termed γ chain. Thus if normal hemoglobin A is formulated $\alpha_2\beta_2$, hemoglobin F would be $\alpha_2\gamma_2$. The β and γ chains have amino acid sequences that differ in 39 of the 146 residues in each chain.
- Hemoglobin F has a greater affinity for oxygen than hemoglobin A, though this is more difficult to demonstrate in humans than in other mammals. This is the basis of the important functional difference between the two hemoglobins—and of the advantages for the fetus of having hemoglobin F. In the low oxygen tension found in the placenta, hemoglobin F competes very effectively with hemoglobin A (the mother's major hemoglobin) for the available oxygen.

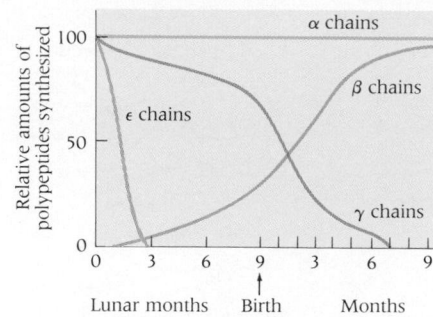

22-8 Hemoglobin production before and after birth
The synthesis of four different globin molecules takes place at different times during human development.

The embryonic hemoglobin Gower 2 also contains α chains, and chains termed ε replace the β chains. Thus Gower 2 is formulated $\alpha_2\epsilon_2$. Gower 1, it turns out, is ϵ_4. As shown in Figure 22-8, α chain synthesis occurs from the beginning of embryonic life, ε chain synthesis ceases early, and γ chain synthesis stops during the first year of life, giving way to β chain synthesis.[2]

For many reasons interest in the abnormal hemoglobins is intense. More than 400 variants are now recognized. The best known, **sickle hemoglobin** or **hemoglobin S,** causes red cells to sickle (Figure 22-9; Box 22-1). In **thalassemia,** a genetic disorder of hemoglobin synthesis common in peoples of certain regions, the defect is a quantitative one (Figure 22-10): in α-thalassemia (common in Asia) there is impaired synthesis of α chains. In β-thalassemia (common in the Mediterranean basin), a more common disorder, the defect is of β-chain synthesis. In both forms of thalassemia, the hemoglobin that is synthesized is qualitatively normal in its amino acid sequences. This is one genetic disorder that can be diagnosed before birth (see Box 15-2). In its homozygous form, thalassemia is a serious illness, causing severe anemia and related problems (such as impaired growth).

The hemoglobin molecules of animal species differ in the details of their chemical structures. The main differences are in the amino acid sequences or in the number of the polypeptide chains. Of the four polypeptides in a molecule of human hemoglobin, the two α chains contain 141 amino acids each and the two β chains contain 146 amino acids each. By studying these differences carefully, molecular biologists have been able to infer the evolutionary history of these species.

Most animals possess hemoglobin molecules of several kinds. In humans, hemoglobin A predominates, comprising about 96% of the hemoglobin (Table 22-1). In other species several major variants may be present in nearly equal amounts.

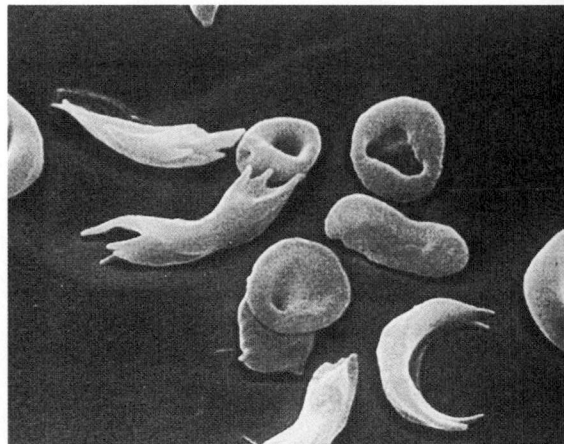

22-9 Sickled red cells
Red cells containing sickle hemoglobin snap into bizzare shapes only when deprived of oxygen—in a test tube or in remote portions of the circulatory system.

Red Cell Metabolism

Despite its deceptively simple appearance, the red cell has a fairly complex program of metabolic activities. Interestingly, some of its enzymes have no known physiological functions. For example, several proteolytic enzymes occur in mature red cells although no detectable breakdown of protein occurs there.

Adult red cells of mammals lack mitochondria.[3] For this reason they are obliged to derive most of their energy from the inefficient oxygen-independent process of glycolysis (see Chapter 9). Once pyruvic acid has been produced in red cells, it undergoes almost no subsequent oxidation. Hence, there is virtually no metabolic utilization of oxygen.

We learned in Chapter 8 of the critical importance of ATP production in all cells. The red cell has four major energy-requiring functions that are wholly dependent upon the cell's ability to generate ATP.

- Initiation of glycolysis requires ATP.
- Transport of ions across the cell membrane, which maintains a distinctive pattern of ion concentration inside the cell, is ATP-dependent.

---

[2] We referred to this fact in Chapter 18 in describing recent successes in stimulating hemoglobin F production in adults with thalassemia by administration of the drug 5-azacytidine.

[3] However, the red cells of birds, fishes, amphibians, and reptiles retain mitochondria, as well as ribosomes and nuclei (see Figure 22-1).

22-10 The molecular basis of thalassemia
In the α-thalassemias, the rate of α-chain synthesis is reduced. This decreases synthesis of Hb A, Hb F, and Hb A₂. Because of this, tetramers of unpaired β chains or γ chains form. In β-thalassemias, the rate of β chain synthesis is reduced. This decreases synthesis of Hb A and increases Hb A₂ relatively. In some cases, synthesis of δ chains can be similarly affected.

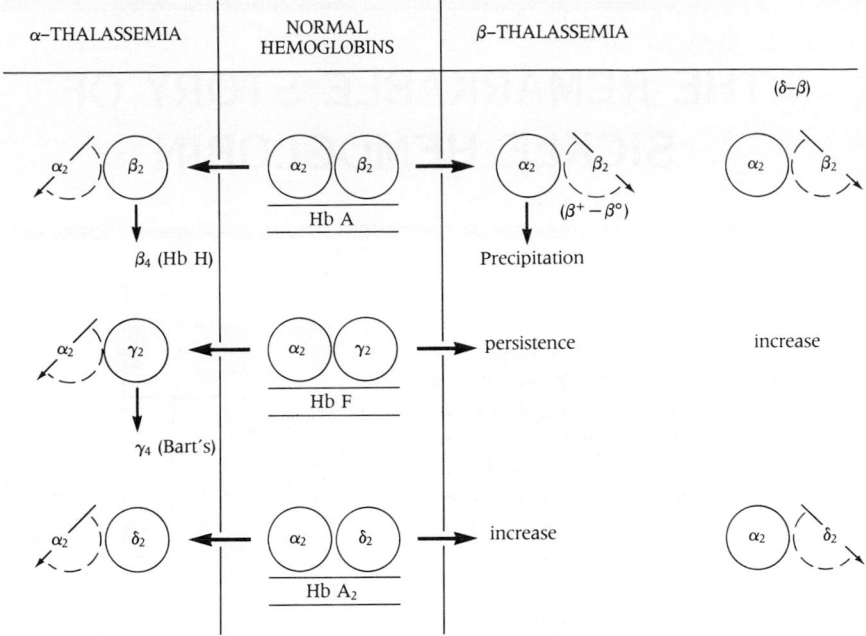

- Transport of nutrients across the cell membrane requires ATP.
- Maintenance of hemoglobin in reduced form is energy-dependent. Without the reducing power generated by glycolysis, hemoglobin would be oxidized to methemoglobin. Normally, the iron in hemoglobin is kept in the form of Fe^{2+}, rather than its oxidized form, Fe^{3+}, which is unable to bind oxygen.

WHITE BLOOD CELLS

Unlike red cells, white cells, or leukocytes, do contain nuclei (see Figure 22-1). White cells are scattered about in body tissues and they occur in blood, where they circulate about on their own almost like parasitic protists.

Leukocytes are one of the body's major defenses against bacterial invasion, or infection. (Antibodies, the other defense mechanism, are discussed in the next chapter). Leukocytes are also capable of movement, a property that has earned them the name "wandering cells." Their ability to migrate through the walls of capillaries into surrounding tissues represents a crucial difference between them and red cells. Red cells remain in the blood throughout their life span, performing their major function (oxygen transport) there. It would be a difficult assignment for them if they wandered about in tissues unconfined by blood vessels. White cells, on the other hand, occur in the blood but do not do most of their work there. Those tasks are performed elsewhere. When in the blood they are, so to speak, on their way to work. In sum, white cells are most useful when they escape from the bloodstream. Red cells are useless if they escape.

TABLE 22-1
STRUCTURE OF NORMAL HUMAN HEMOGLOBINS

Hemoglobin	Structure	Comments
A	$\alpha_2\beta_2$	Comprises 92% of adult hemoglobin
A_{1c}	$\alpha_2(\beta\text{-NH-glucose})_2$	Comprises 5% of adult hemoglobin; increased in diabetes
A_2	$\alpha_2\delta_2$	Comprises about 2% of adult hemoglobin
F	$\alpha_2\gamma_2$	Predominant hemoglobin in fetus from the 3rd through 9th month of gestation; facilitates transfer of oxygen across placenta
Gower 1	ϵ_4	Present in early embryo; function unknown
Gower 2	$\alpha_2\epsilon_2$	Present in early embryo; function unknown
Barts	γ_4	Trace present in newborns; may comprise 100% of hemoglobin in certain forms of thalassemia; nonfunctional

In 1910, Dr. James B. Herrick of Chicago examined a 20-year-old black college student and discovered peculiar elongated and sickle-shaped red cells "no duplicate of which I have ever seen described" (see Figure 22-9). Physicians later discovered many cases of the condition, which was named **sickle cell anemia.** It proved to be a genetic disorder to which blacks are particularly susceptible. Not until 1949 was it discovered—by Linus Pauling, Harvey Itano, and coworkers—that the disorder is due to the presence of an abnormal hemoglobin molecule that was called hemoglobin S to distinguish it from normal adult human hemoglobin, or hemoglobin A.

The molecular abnormality produced by the mutant gene is a subtle one. When the oxygen supply is adequate, red cell shape is normal, but when the partial pressure of oxygen is reduced, the sickle shape appears suddenly and dramatically. Sickle cell anemia aroused much interest, for it involves a genetic trait that is manifested in the peculiar properties of a specific molecule. It was thus the clearest illustration to date of the one gene, one polypeptide theory of gene action (see Chapter 16) and the prototype of the so-called molecular diseases. Indeed, this descriptive term was first used in referring to sickle cell anemia.

Investigators then set about trying to determine the nature of the abnormality in hemoglobin. S. Pauling and coworkers found that hemoglobin S differs from hemoglobin A in its electrophoretic mobility, moving

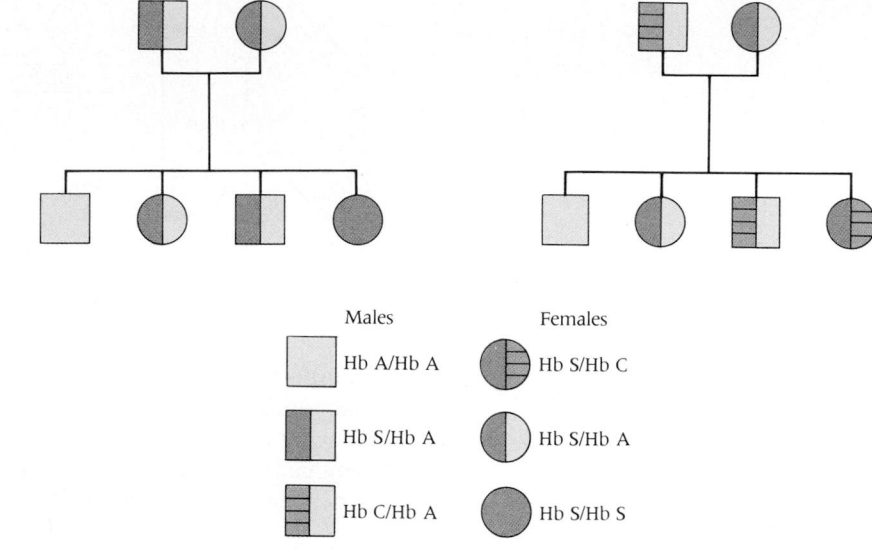

Males

Hb A/Hb A

Hb S/Hb A

Hb C/Hb A

Females

Hb S/Hb C

Hb S/Hb A

Hb S/Hb S

22B Family tree of individuals with hemoglobins S and A (Left); and hemoglobins S, C, and A (Right).

more slowly because it has two fewer electrical charges (Figure 22A). In severe sickle cell anemia, nearly all the hemoglobin is of the slow-moving variety, whereas in milder anemias the proportions are roughly half hemoglobin S and half hemoglobin A. James Neel then demonstrated that the hemoglobin S molecule is transmitted as a simple Mendelian allele, with homozygotes (SS) exhibiting severe sickle cell anemia and heterozygotes (SA) an asymptomatic state called sickle cell trait. The family tree of a typical individual with sickle cell anemia is shown in Figure 22B.

The charge difference between hemoglobin A and hemoglobin S established that the chemical difference between them is small. Investigators wondered: can it be an abnormal amino acid? A whole hemoglobin molecule contains 574 amino acid residues, but amino acid analysis was accurate only to within 3%. Hence single amino acid substitutions would be hard to recognize. The problem was ingeniously solved by Vernon Ingram in 1957. Following digestion with trypsin, each of the hemoglobin molecule's two identical halves yields 26 short peptides. By a simple method that has come to be known as fingerprinting, Ingram found that 25 of these peptides are the same in hemoglo-

bin A and hemoglobin S (Figure 22C). The difference was in the remaining one, which contains only 8 amino acid residues. Ingram then showed that 7 of the 8 amino acids

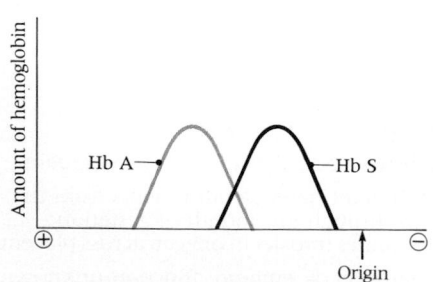

22A Electrophoresis of hemoglobin molecules
Normal hemoglobin (Hb A) and sickle-cell hemoglobin (Hb S) exhibit different mobilities when placed in an electric field. This indicates a difference in the net charge of the proteins in each molecule.

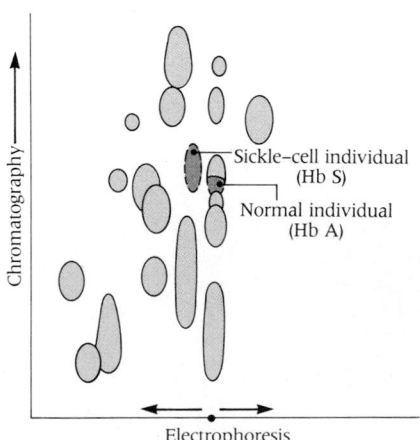

22C Two-dimensional fingerprint of normal and abnormal hemoglobin molecules
Each spot represents a peptide. In comparing normal and sickle-cell hemoglobins, only one peptide (color) distinguishes these two molecules. This peptide migrates differently following electrophoresis and chromatography, indicating a difference in net charge.

are identical in the normal peptide and its abnormal counterpart. The other amino acid—the sixth in order—was found to be glutamic acid in the hemoglobin A peptide and valine in the hemoglobin S peptide (Figure 22D). As it happens, the abnormal peptide is located at the N-terminal end of the β chain, and so the sixth amino acid in the peptide is the sixth amino acid in the β chain. Since a glutamic acid molecule carries one negative charge, the net effect of the substitution of valine for glutamic acid in hemoglobin is the deletion of two negative charges, one from each half of the hemoglobin molecule, precisely as predicted from the electrophoretic pattern.

Hemoglobin C, another abnormal hemoglobin, has an electrophoretic mobility approximately half that of hemoglobin S. Hence, its net electric charge is more positive than that of hemoglobin S. Its abnormality is a substitution of lysine for glutamic acid at the same locus as the hemoglobin S abnormality. Instead of being sickle-shaped, a cell carrying hemoglobin C is abnormally thin, with a central dot of hemoglobin, its appearance having suggested the name target cell.

The amino acid substitutions determined so far are in perfect harmony with the genetic code (see Table 16-1). One of the triplet code words for glutamic acid (GAG) differs from one for valine (GUG) and one for lysine (AAG) by only a single nucleotide. Thus the mutation responsible for the replacement of hemoglobin A by hemoglobin S or hemoglobin C may be tentatively visualized as having altered only a single nucleotide in the DNA sequence comprising the hemoglobin A gene. The interesting fact that sickle cell anemia appears to protect its victims against malaria (which explains the prevalence of the sickle cell gene in the malarial belts of Africa) is discussed in Box 35-1.

More than 400 abnormal hemoglobins have been discovered so far. Some have abnormalities in the α chain. Others have an abnormal β chain. Genetic studies have revealed numerous gene combinations in afflicted families.

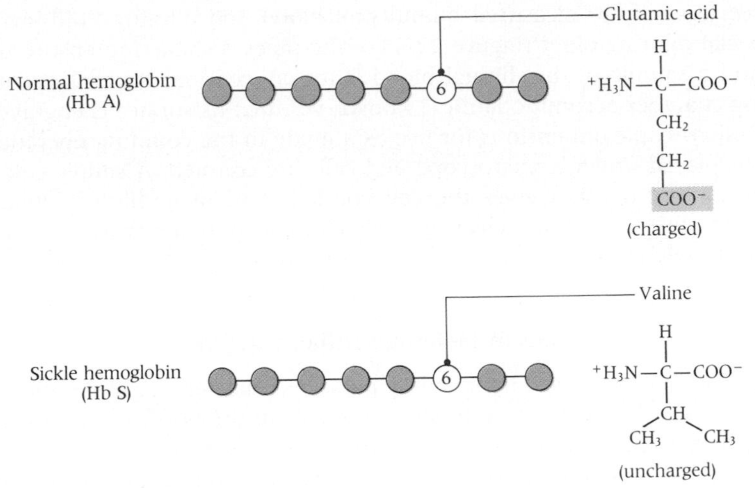

22D Analysis of charge difference between Hb A and Hb S
In this peptide, only the amino acids at position 6 in the sequence are different from each other. Hb A contains glutamic acid (negative charge), Hb S has the uncharged amino acid valine in this position.

Classification

We classify white cells on the basis of structure, source, and function.

One structural classification depends on the presence of cytoplasmic granules and divides leukocytes into **granulocytes** (cells that contain prominent cytoplasmic granules) and **nongranulocytes.** Nongranulocytes include **lymphocytes** and **monocytes.** Granulocytes are usually divided into **neutrophils, eosinophils,** and **basophils,** depending on the color of the granules when stained.

Another structural classification, based on nuclear shape, separates white cells into mononuclear and polymorphonuclear groups. A **mononuclear cell** has a single rounded or slightly indented nucleus that occupies most of the cell volume. A **polymorphonuclear cell** has a polymorphous nucleus that assumes many different shapes with many filaments and lobules (see Figure 22-1). Lymphocytes and monocytes are mononuclear cells. Mature granulocytes are polymorphonuclear cells.

Leukocytes are also categorized as **myeloid cells** if they are produced in bone marrow or **lymphoid cells** if they are produced in lymphoid tissues, though the pluripotent stem cell that gives rise to both is located in the bone marrow (see Figure 22-4). Immature lymphoid cells migrate to lymphoid tissue and there produce lymphocytes in abundance (see Chapter 23). Finally, we class leukocytes as phagocytic and nonphagocytic depending on their capacity for phagocytosis.

In the normal adult, the total white cell count is 5,000 to 10,000 cells per μl. The distribution of major cell types is roughly: neutrophils, 55–65%; eosinophils, 2–3%; basophils, 1–2%; lymphocytes, 20–30%; and monocytes, 4–7%.

The traditional methods for counting cells in a given volume of blood are well known, but the large error inherent in these methods is not sufficiently appreciated.

In practice, a carefully measured quantity of blood and diluting fluid are drawn into a special diluting pipet (Figure 22-11). The pipet is shaken to ensure uniform mixing and a sample of the diluted blood is introduced into a **hemocytometer,** a counting chamber accommodating a known volume; its surface is engraved with a grid of microscopic dimensions for use as a guide in the counting operation. The chamber is placed under a microscope and cells are counted. A simple calculation correcting for the dilution gives the cell count per μl of undiluted blood. Most medical laboratories today use electronic cell counters that are vastly more accurate and astonishingly fast.

Role in Defense: Inflammation

Infectious diseases are caused by invading parasites—bacteria or other microorganisms—that infect the host and nourish themselves in the host's body. Diseases are caused by bacteria, protozoans, fungi, worms, and viruses.

Animals have several lines of defense against infection. Some do not involve immunity. The first and most important line of defense is purely mechanical. It is the skin, which excludes many organisms that would find the moist, nutrient-rich tissue just below the skin a luxurious place to live. This is dramatically and often horrifyingly confirmed by the overwhelming susceptibility to infection of burn victims who have lost large areas of skin. That is why surgeons urgently attempt to transplant skin grafts from unburned areas to burn sites.[4] Any invaders that get past the skin barrier, or the epithelium of nasal passages, mouth, bronchi, and urogenital canals, which are covered with protective mucus, and into a tissue of the body, are promptly attacked by an army of white blood cells.

A normal reaction to an injurious event—such as a bacterial infection, a cut or burn—is the defensive process called **inflammation.** A general reaction to all forms of injury, it is a characteristic pattern of localized changes. Medical students have been taught for centuries that inflammation has four cardinal signs: swelling (Latin, *tumor*), pain (*dolor*), redness (*rubor*), and local warmth (*calor*).[5] These changes are initiated by the release of **histamine** and other biochemicals from **mast cells** in damaged tissues. Histamine enhances blood flow locally and increases the permeability of nearby blood vessels. As a result, fluids and clotting proteins leak into the area. Vasodilation reduces resistance to blood flow, so blood accumulates in the affected area (thus the swelling), carrying with it proteins that have various beneficial effects. Meanwhile, a variety of leukocytes are attracted into the area.

Neutrophils and other granulocytes also play a major role in the inflammatory response. When an invasion of bacteria occurs, white cells move out from the capillaries in massive numbers. By the process of phagocytosis, they engulf, kill, and digest invading bacteria (Figure 22-12). Although microscopists had long ago noticed that granulocytes can devour solid objects, it was the Russian biologist Elie Metchnikoff (1845–1916) who, in 1882, established the importance of phagocytosis in body defenses. In a famous experiment, performed by the sea in Messina while his family was at the circus, he placed a rose thorn beneath the skin of a starfish larva in an area possessing no blood vessels or nerve fibers. The next day the thorn was surrounded by mobile phagocytic cells.

Granulocytes are not the only phagocytic body cells. The **macrophages** are another large class of phagocytic cells. These cells, which as their name suggests are relatively large, are found in the blood and out of it. In blood, a less mature form of macrophage is the monocyte. Macrophages in various tissues are given various names based on differences in form and function.

A witty colleague has said that macrophages are sitters or runners. The mobile monocytes in blood and the wandering macrophages in tissues are the runners.

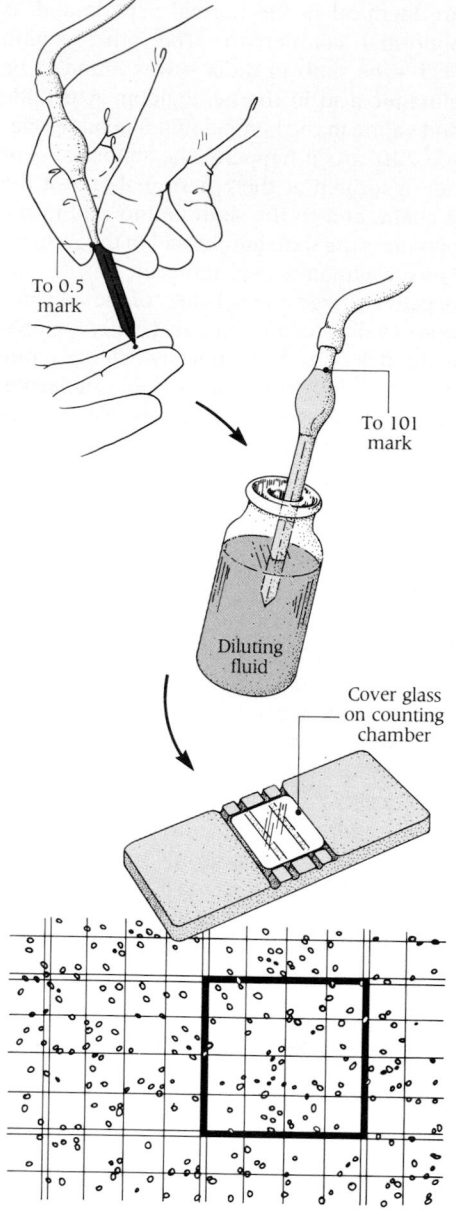

22-11 Counting blood cells with a hemocytometer
By appropriately diluting a sample of blood, one can count the number of cells seen in the microscope in a given volume and can calculate the undiluted blood cell count.

---

[4] Until recently, burn victims with large areas of destroyed skin presented a difficult and discouraging problem because of the shortage of normal skin area capable of donating grafts. In the following chapter, we will discuss the immunologic rejection mechanisms that make it virtually impossible to transplant skin from one person to another. An exciting discovery in the 1980s suggests a possible solution to this dilemma. Investigators have been able to make sheets of skinlike material from cultures of the victim's own skin cells mixed with pure collagen, a major protein of skin. The material "takes" like real skin and thus may be the long sought "artificial skin."

[5] This classic description of inflammation was put forth by the Roman physician Celsus in the first century A.D.

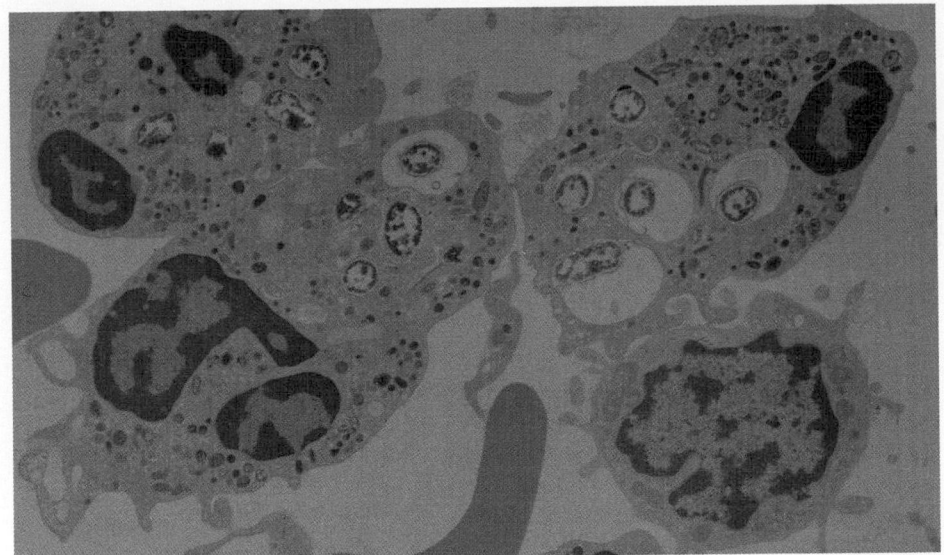

22-12 Phagocytosis by white blood cells
In this electron micrograph, white cells are ingesting bacteria, which are encapsulated in membrane-bound compartments in the leukocyte cytoplasm. (× 2800)

But many fixed cells in the body are also phagocytic—"fixed cell" being the antithesis of "wandering cell." These are the sitters. The major fixed phagocytic cells are the fixed macrophages (reticulum cells), most of which are in the spleen, liver, and lymph nodes, and the endothelial cells, which line the blood vessels, which only recently were revealed to be phagocytes. In 1924, all of these cells were collectively named the reticuloendothelial system. A more contemporary name is the **mononuclear phagocyte system.** The name change followed an important discovery: all of the macrophages in tissues derive from the blood monocyte (see Figure 22-4).

Phagocytosis

Phagocytosis has three phases: chemotaxis, ingestion, and digestion. **Chemotaxis** is the directional migration of white cells from small blood vessels to sites of bacterial invasion—rather like the cavalry in old westerns. Bacteria elaborate many substances that attract phagocytes. So do injured tissues, as well as the complement system discussed later in the chapter. Some of these factors cause the pain of acute inflammation. Chemotaxis is readily demonstrated in a plasma suspension of granulocytes when a clump of bacteria is added. The granulocytes move at once toward the microbes. This would be **positive chemotaxis.** When certain substances (e.g., kaolin) are added instead of bacteria, the granulocytes move away. This is **negative chemotaxis.** The actual attractive or repellent forces of chemotaxis are currently under study. At inflammatory sites in tissues, neutrophils arrive quickly; macrophages come after several hours.

Granulocytes actively ingest bacteria and other particles. This activity is affected by the nature of the particles. For example, some bacteria are protected against phagocytic ingestion by heavy capsules. Conversely, certain substances in plasma strongly stimulate phagocytosis. These agents, called **opsonins,** act like a seasoning that makes the bacteria a "tastier" dish to the white cells.

Sometimes the tide of battle goes the other way. Highly virulent bacteria may produce toxic substances that destroy white cells. When this happens on a large scale, the area soon becomes strewn with living and dead bacteria and white cells, dead tissue cells, digestive enzymes, cell fluids, and cell fragments in all stages of disintegration. The resulting semifluid mass is called **pus.** An **abscess** is a deeply embedded, walled-off accumulation of pus.

The role of granulocytes in inflammation depends upon two properties: their ameboid motility, which enables them to pass through capillary walls at sites of inflammation, and their capacity for phagocytosis. White cells also play a key role in causing fever, the elevation of body temperature that occurs in infection. When white cells are activated to phagocytize bacterial invaders, they release a substance called endogenous pyrogen, which acts on the temperature regulating center of the brain.

Three Cascades Initiated by Inflammation

Several dozen specific proteins of plasma are activated in emergencies by the inflammatory reaction. Activation evolves in three different reaction "cascades." A cascade is a reaction sequence in which each activated protein sequentially activates the next protein, as in this example of a simulated cascade.

$$\text{Protein X} \xrightarrow{\text{Catalyst: outside agent}} \text{activated protein X}$$

$$\text{Protein Y} \xrightarrow{\text{Catalyst: activated protein X}} \text{activated protein Y}$$

$$\text{Protein Z} \xrightarrow{\text{Catalyst: activated protein Y}} \text{activated protein Z}$$

Such patterns are common in nature.

The first cascade leads to the release of peptides known as **kinins** (Figure 22-13), notably bradykinin (Figure 22-14). Kinins act like histamine, causing increased vascular permeability and vasodilation, but they are activated more slowly. The kinin cascade begins with the activation of a protein known as Hageman factor, or factor XII, which is also the first reactant in the blood clotting cascade. Hageman factor is probably activated in tissues by contact with collagen. This would be the "outside agent" in the foregoing scheme. Activation converts Hageman factor from a **proenzyme** to an enzyme, which initiates a reaction sequence that in the end produces powerful kinins. It is worth noting that the venom of one wasp species (see Figure 22-14) is a kinin not unlike bradykinin. Another potent kinin (phyllokinin) occurs in the skin of a Brazilian toad that is poisonous to the touch.

The second cascade of the inflammatory reaction activates a series of 18 plasma proteins, known collectively as **complement.** This complex sequence of reactions seems designed primarily to protect the host against bacterial infection. It does this in a variety of ways.

- It promotes the lysis of certain bacteria.
- It promotes opsonization of bacteria.
- It releases chemotactic factors, which attract neutrophils into areas of injury, and other factors, which increase vascular permeability.

Unfortunately, these strenuous reactions sometimes damage the host's own cells. This system has an important role in immunity.

The third cascade is the clotting reaction, which acts to plug bleeding blood vessels following an injury to those vessels.

PLATELETS

Platelets, or thrombocytes, are fragmentlike bodies that lack nuclei (Figure 22-15). Since they are only 2 to 4 μm in diameter, they are smaller than red cells, which are 7 to 8 μm across. Normally there are some 150,000 to 350,000 platelets per μl of blood. Platelets are sometimes clumped together in blood. This fact led shrewd early observers to suspect that these cells may play a role in blood clotting, a part of the overall process called **hemostasis.** The definition of hemostasis and hemostatic mechanisms should be clearly understood. They refer to the entire process by which bleeding from a damaged blood vessel is stopped. Not until the 1920s was it confirmed that platelets participate in hemostasis.

Functions

Platelets are essential to hemostasis, and bleeding occurs at many sites when there are too few platelets or when platelets are present but defective. Their role in hemostasis depends on the following remarkable properties.

- They adhere to injured blood vessel walls.
- They clump together to form a temporary hemostatic plug at a bleeding point that helps to stop the flow of blood from a damaged vessel.
- Their surfaces promote activation of the clotting proteins in plasma.
- They help maintain the integrity of the endothelium in blood vessels.

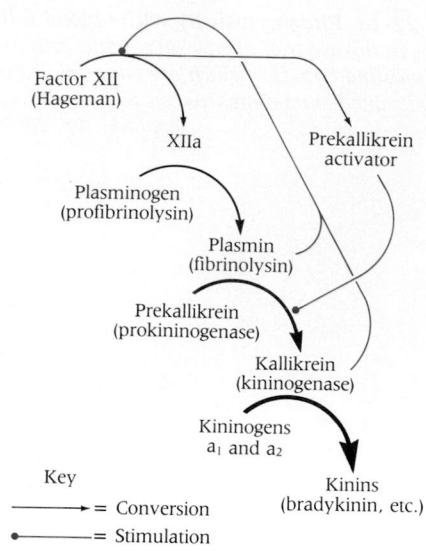

22-13 The production of kinins
Kinin molecules are formed in a cascade of reactions.

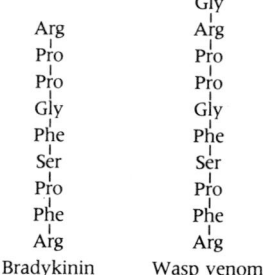

Bradykinin	Wasp venom
Arg	Gly
Pro	Arg
Pro	Pro
Gly	Pro
Phe	Gly
Ser	Phe
Pro	Ser
Phe	Pro
Arg	Phe
	Arg

22-14 The similar structures of bradykinin and a component of wasp venom

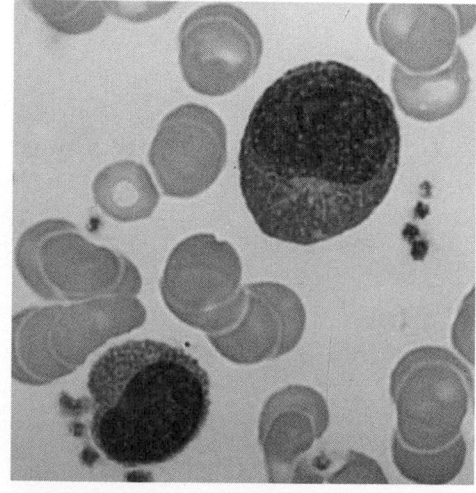

22-15 Blood smear showing platelets
The platelets are the small dark bodies. This blood is from a person with infectious mononucleosis.

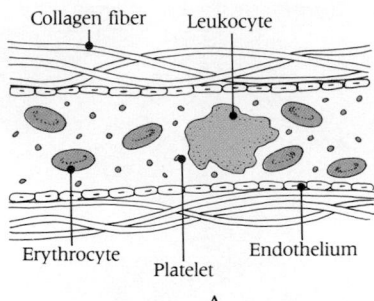

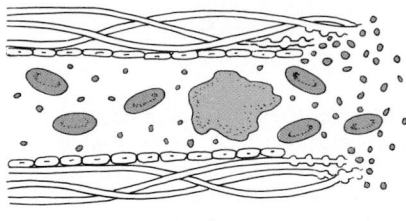

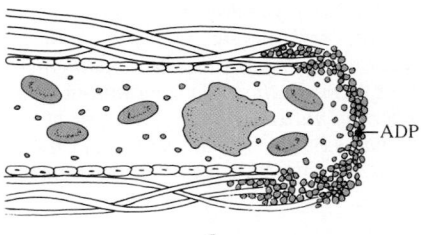

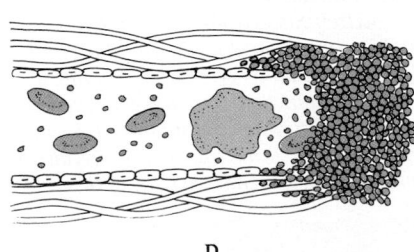

22-16 The role of platelets in the formation of the primary hemostatic plug
(A) An intact blood vessel in longitudinal section shows red cells, white cells, and platelets within the bloodstream. (B) Platelets begin to accumulate at the site of injury on the right. (C) Platelets discharge ADP and thereby promote platelet aggregation. (D) An aggregate of platelets forms the primary hemostatic plug.

■ They give rise to mediators that initiate repair of the vessel wall and regulate inflammatory reactions. These include **serotonin,** a substance that constricts small blood vessels thereby slowing bleeding, and **platelet-derived growth factor (PDGF),** a potent vasoconstrictor that tends to stop bleeding, and a stimulator of the growth of various types of mesenchymal cells.

The most important platelet function is the formation of a hemostatic plug (Figure 22-16). This capability depends on both platelet adhesiveness and platelet aggregation. Platelet adhesiveness is defined as a tendency to stick to foreign surfaces, especially exposed collagen fibers in blood vessel walls. Platelet aggregation involves a tendency of cells to stick to each other, or clump together. Aggregation is greatly enhanced by ADP (adenosine diphosphate), a nucleotide that resembles ATP but has one less phosphate group (see Chapter 8). Platelets actively release ADP as they start to form clumps. Therefore clumping promotes further clumping.

Evolutionary Aspects

Human beings and other vertebrates possess an elaborate hemostatic mechanism involving both platelets and various soluble clotting proteins. Mechanisms of hemostasis appear less complex in some simpler animals. The ancient arthropod *Limulus polyphaemus*, the horseshoe crab, possesses a single type of blood cell, the **amebocyte,** which is a combination phagocyte-platelet and a source of coagulation proteins. When a foreign particle enters the circulation, amebocytes migrate to the area and adhere to the particle, forming cellular aggregates. The amebocyte also secretes a clottable protein, roughly equivalent to mammalian fibrinogen, which immobilizes the invader and repairs tissue damage.

The two functions—formation of intercellular aggregates and production of clotting proteins—greatly increased in complexity in the course of evolution. Eventually, they were segregated into two separate systems—cellular elements and soluble plasma proteins called clotting factors.

Production and Structure

The platelet is a small fragment of a giant polyploid precursor cell, the **megakaryocyte.** Megakaryocytes are found mainly in bone marrow. Their most distinctive feature is their large size: diameters are 30 μm or more (Figure 22-17A). Within their large, irregular, multilobed nuclei is a huge number of chromosomes, ploidy averaging 8 to 64 times the haploid number. This comes about for an interesting reason. Megakaryocyte development is distinguished by an unusual process called **endomitosis,** in which nuclear units replicate again and again without simultaneous cytoplasmic division.

Platelets arise by fragmentation of megakaryocyte cytoplasm (Figure 22-17B). Mature platelets have complex internal architecture that cannot be appreciated by light microscopy (see Figure 22-15). However, electron microscopy reveals a rich internal structure (Figure 22-18).

PLASMA AND THE PLASMA PROTEINS

Before we can explore the other contributions of blood to defense, we must introduce its noncellular components.

COMPOSITION OF PLASMA

The components of plasma are divided into three categories, each with a different function.

■ Water is the essential vehicle of transport.
■ Myriad substances are transported to and from the tissues.
■ Plasma proteins maintain osmotic pressure and regulate the distribution of water between the intravascular and extravascular compartments. Included in this group are the clotting proteins, antibodies, and other components of the immune system.

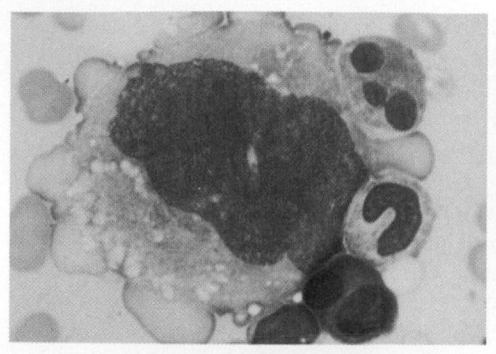

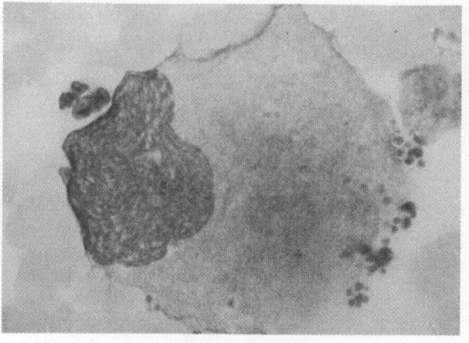

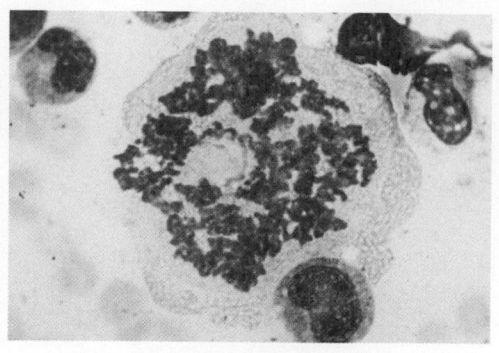

A   B   C

Some of the major components of plasma are listed in Table 22-2.

FUNCTIONS OF PLASMA

All plasma constituents are continuously transported. However, some are passengers and others are vehicles. As shown in Table 22-2, the passengers are the nutrients absorbed from the intestinal tract, metabolic intermediates and waste products (collectively called metabolites), and an assortment of hormones and other chemical messengers.

The nutrients transported by plasma include amino acids from the digestion of dietary proteins, glucose from the digestion of higher carbohydrates, fatty acids from the digestion of lipids, and plasma proteins themselves, which to an extent are utilized by the tissues as foods. More or less constant concentrations of nutrients are maintained through adjustments in the balance among the following processes: storage in body depots, removal from storage, intestinal absorption, excretion, and metabolic utilization by body cells.

Since plasma is a carrier of metabolites, it is the medium of exchange for the many biochemical intermediates. Interestingly, some of these small molecules are free in plasma, while others are bound to proteins or other large molecules. For example, **bilirubin,** a breakdown product of hemoglobin, travels attached to plasma albumin. Binding to such so-called **transport proteins** is best illustrated by the trace metals in plasma, among them iron and copper. In this way, quantities of substances that would be toxic in the free state are moved in a state of physiological inertness. The iron-binding protein of plasma is **transferrin,** and the copper-binding protein is **ceruloplasmin.** Cobalamin (vitamin B_{12}) has two transport proteins called **transcobalamins.**

Plasma transports the waste products of metabolism to the chief excretory organs—the skin, intestines, liver, lungs, and kidneys. It carries the breakdown products of protein metabolism, such as urea, creatinine, uric acid, and phenol; the breakdown products of hormone and hemoglobin metabolism; and others. Many metabolic end products from one tissue—e.g., lactic acid, an end product of glycolysis—may be utilized in other tissues.

22-17 Megakaryocytes give rise to platelets
(A) A megakaryocyte is a huge bone marrow cell containing a multilobed nucleus. Note its large size compared to nearby red cells and white cells. **(B)** Platelets are formed by the pinching off of megakaryocyte cytoplasm.
(C) Megakaryocytes are large because the nucleus repeatedly replicates by mitosis, while the cytoplasm does not divide. Hence, daughter cells do not form. In this mitotic metaphase of a replicating nucleus, the large number of chromosomes (8–32 times the haploid number) reveals the polyploidy typical of these amazing cells.

22-18 Platelet structure
(A) Electron micrograph of a platelet. (×20,000)
(B) Platelet sectioned at the equator of the cell showing that its cytoplasm contains a variety of organelles. Note the encircling band of microtubules which maintains platelet shape.
(C) The same cell in cross section reveals the canalicular system characteristic of this cell type.

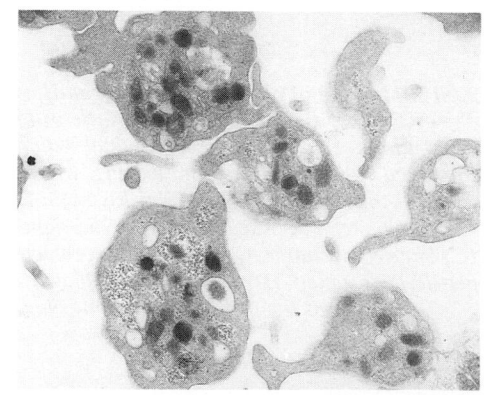

A

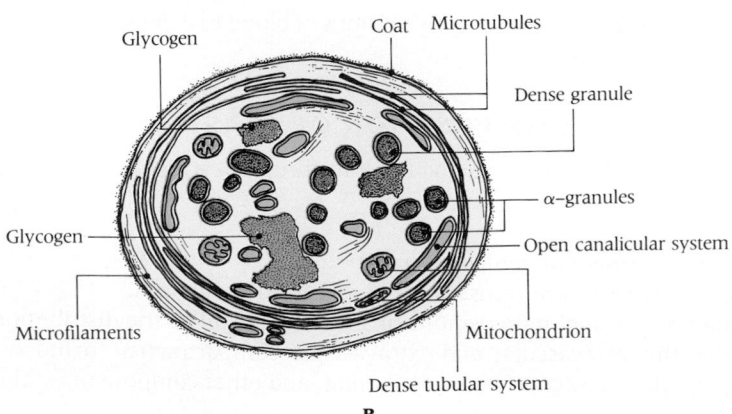

B

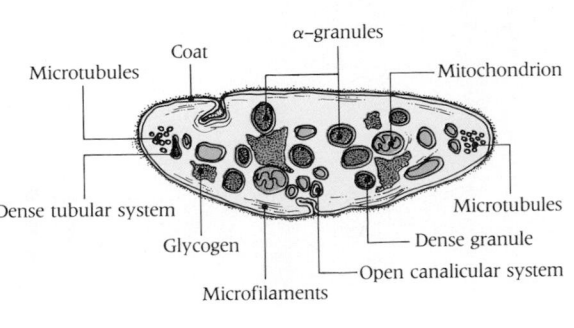

C

TABLE 22-2
MAJOR COMPONENTS OF HUMAN PLASMA

Component	Amount[†]	Component	Amount[†]
Electrolytes and metals		Aspartic acid	1.0
Bicarbonate	152–190	Citrulline	0.5
Calcium	8.5–10.5	Cystine	1.4
Chloride	355–380	Glutamic acid	0.8
Cobalt	0.4–0.7 μg	Glycine	1.8
Copper	88–124 μg	Histidine	1.4
Iodine	5–6 μg	Isoleucine	1.6
Iron	50–150 μg	Leucine	1.9
Magnesium	1.8–3.6	Lysine	3.0
Manganese	4–6 μg	Methionine	0.5
Phosphate	3.0–4.5	Phenylalanine	1.4
Potassium	15–20 μg	Proline	2.5
Sodium	315–330	Threonine	2.0
Sulfate	0.5–1.5	Tryptophan	1.1
Zinc	250–350 μg	Tyrosine	1.5
		Valine	2.8
Metabolic intermediates		Glucose	70–100
Bile acids	0.2–3.0	Vitamins	
Choline	26–35	Ascorbic acid	0.6–2.0
Citric acid	2.0–3.0	Biotin	1.0–1.7 μg
Lactic acid	6–16	Carotenoids	60–200 μg
Lipids	360–820	Folic acid	3–21 ng
Cholesterol	150–280	Niacin	200–300 ng
Fatty acids	200–450	Pantothenic acid	100–180 μg
Phospholipids	135–170	Riboflavin (B_2)	2.5–4.0
Triglycerides	100–250	Vitamin A	20–50 μg
Pyruvic acid	1.0–2.0	Vitamin B_{12}	20–80 ng
		Vitamin E	5–20 μg
Waste products		Hormones	
Acetone	0.3–2.0	Adenohypophyseal hormones	
Ammonia	40–70 μg	Adrenocorticotrophic hormone	
Bilirubin	0.4–0.7	(ACTH)	
Creatine	0.8–1.0	Follicle-stimulating hormone	
Creatinine	0.7–1.5	(FSH)	
Glutamine	5–12	Luteinizing hormone (LH)	
Urea	25–52	Prolactin (Pr)	
Uric acid	3–6	Thyrotrophic hormone (TSH)	
		Adrenal cortical steroids	
Proteins, g		Adrenal medullary hormones	
Albumin	4.6	Epinephrine	
Clotting factors	—	Norepinephrine	
Enzymes	—	Insulin	
Fibrinogen	0.38	Neurohypophyseal hormones	
Globulins	2.6	Vasopressin	
α_1-globulin	0.3	Oxytocin	
α_2-globulin	0.6	Ovarian hormones	
β-globulin	0.8	Estrogens	
γ-globulin	0.9	Progesterone	
Metal-binding proteins	trace	Parathyroid hormone	
		Testicular hormones	
Nutrients and vitamins		Testosterone	
Amino acids		Thyroid hormone	
Alanine	4.0		
Arginine	2.3		

[†] Values based on 100 ml plasma. Unless otherwise noted, amounts are given in mg.

The **osmotic pressure** of plasma proteins is profoundly important in body fluid balance and transport. If plasma were separated from distilled water by a membrane permeable to water but impermeable to all plasma solutes (including electrolytes), an osmotic pressure of 7.2 atm per kg of water (1 atm = 760 mm of mercury) would develop.[6] Such a pressure does not arise in the capillaries because their walls are impermeable only to the large protein molecules dissolved in plasma. These alone exert an osmotic force across the walls of the capillaries (see Chapter 24).

As noted above, a major defense system of plasma is the **clotting system** (Figure 22-19).

[6] This figure is derived as follows. The osmolality of normal human plasma is about 287 milliosmols per kilogram of water. One milliosmol is equivalent to an osmotic pressure of 19 mm Hg. Therefore, 287 × 19 = 5453 mm Hg. One atmosphere equals 760 mm Hg. Therefore, 5453/760 = 7.19 atmospheres.

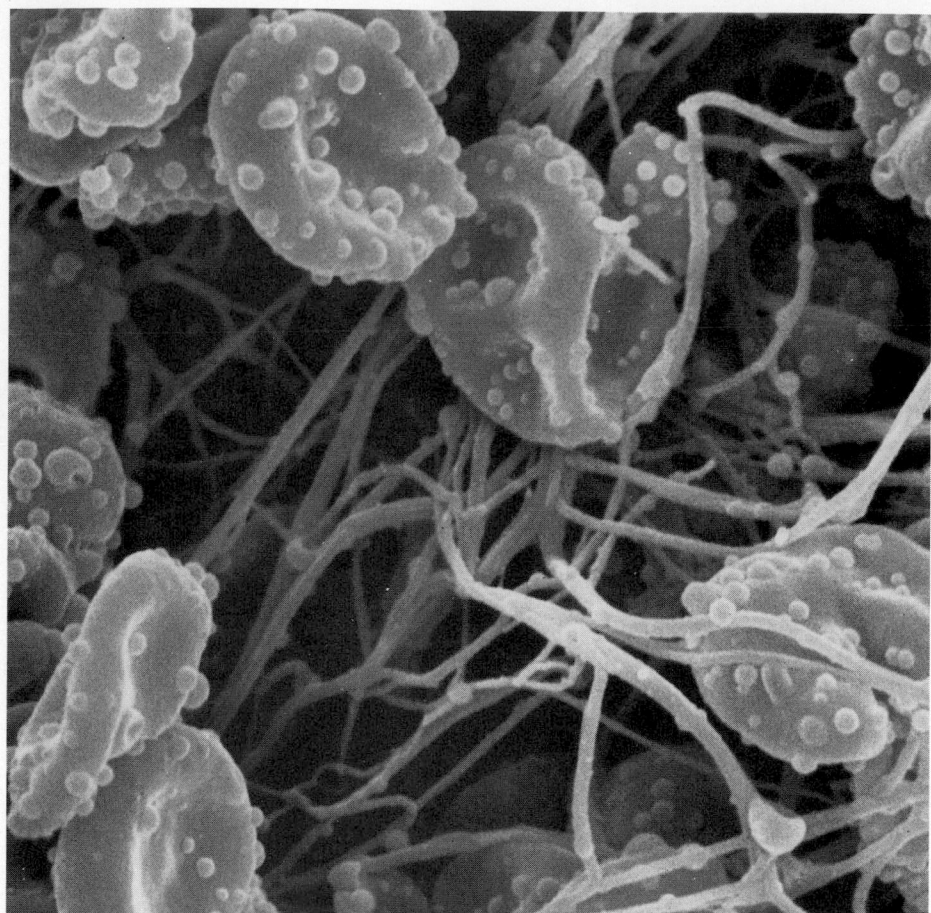

22-19 Scanning electron micrograph of a blood clot
In a clot, a complex web of tough fibrin strands enmeshes red cells. The internal environment of the clot soon deforms the trapped red cells. They develop surface blebs and spikes, some of which break off. (× 6000)

A second major defense system of plasma depends largely on **antibodies,** another group of plasma proteins. As we will see, a foreign antigen stimulates production of antibodies that antagonize the foreign substance. Since their production continues long after the antigen is gone, antibodies often provide immunity for years or a lifetime. The "memory" of the immune system is astonishing.

PLASMA PROTEINS

The diversity of the **plasma proteins** is suggested by the electrophoretic patterns of normal plasma. Figure 22-20 shows the pattern of classic electrophoresis and the patterns of the more recent techniques.

In diagrams of the type shown in Figure 22-20A, each protein or class of proteins of a singular electrical charge has a discrete peak. An ideally homogeneous protein has a peak of knifelike sharpness. The relative breadth of a plasma protein peak is a measure of heterogeneity. Each peak represents not an individual protein but a class of proteins, the concentration of the class being proportional to the area under the peak. Thus plasma contains only six major electrophoretic groupings (under the usual conditions of analysis), although there are hundreds of individual plasma proteins. Concentrations of the plasma proteins are given in Table 22-2.

Albumin, with the highest electrophoretic mobility, is the most abundant protein. Accordingly, its peak is largest. It constitutes 50 to 60% of the total plasma proteins and accounts for 80% of the plasma osmotic pressure. Therefore, it is critical in the maintenance of blood volume. In diseases causing decreased plasma albumin (e.g., certain liver ailments), water moves from the blood into the tissues (edema).

The **globulins** comprise several groups of proteins of high molecular weight, among them various glycoproteins, lipoproteins (carriers of plasma lipids that would be insoluble without them), metal-binding proteins, clotting factors and antibodies, also called immunoglobulins.

Fibrinogen constitutes only 4 to 6% of the total plasma proteins and may not be seen in the electrophoretic pattern(s) of normal plasma.

The large-scale efforts initiated during World War II to purify the plasma proteins

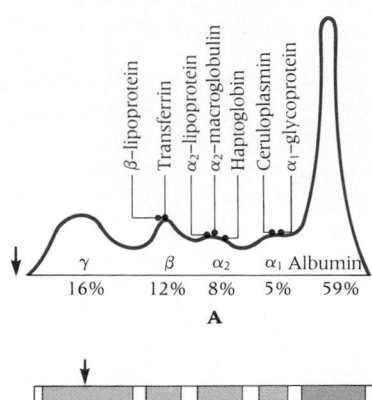

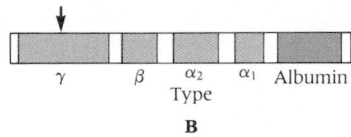

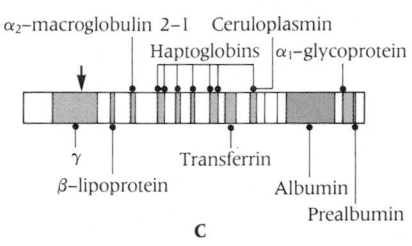

22-20 Electrophoretic patterns of normal serum proteins
The vertical arrow represents the starting point in each case. **(A)** *Tiselius method.* **(B)** *Paper electrophoresis.* **(C)** *Starch gel electrophoresis.*

led to much of our present knowledge. Large quantities of blood were needed for the treatment of battle casualties. Because of the perishability of whole blood, the possibility was considered of using plasma or certain purified plasma proteins as a blood substitute. Since the main consequence of blood loss is shock due to decreasing osmotic pressure and shrinking blood volume, infusions of osmotically active plasma protein(s) instead of whole plasma or blood were used to maintain blood volume.

BLOOD CLOTTING

When blood coagulates, it rapidly solidifies. Under normal circumstances, the only role of clotting is to halt bleeding. One of the most interesting self-regulatory systems in all of biology, clotting is obviously essential to the survival of most animals. The fact that blood clots when shed is as remarkable as the fact that it does not (normally) clot within blood vessels.

PRIMARY AND SECONDARY HEMOSTASIS

Hemostasis has two stages: a primary stage, in which bleeding is stopped by a combination of vasoconstriction and platelet aggregation; and a secondary stage, in which clotting produces a seal that is maintained until healing is complete. The bulwark of secondary hemostasis is a **clot,** a tough, solid meshwork of fibrin strands and entrapped red cells (see Figure 22-19).

The hemostatic sequence begins instantly following any injury that lacerates a blood vessel.

- The first event is local constriction of vessels near the injury. This reaction, which depends in part on nerve reflexes, slows blood loss but does not stop it.
- Disruption of the lining of the injured vessel exposes collagen fibers (see Figure 22-16B). Passing platelets are attracted to these fibers, adhere, and begin to accumulate.
- A remarkable transformation then takes place. Platelets swell and release into the area serotonin, ADP, and various platelet proteins. Serotonin prolongs local vasoconstriction and ADP causes additional platelets to aggregate. The result is the formation of a fragile plug that temporarily seals the ruptured blood vessel (see Figure 22-16D).

In the secondary stage, the clotting reaction produces a fibrin clot, which makes up the bulk of the familiar, crustlike scab over a wound. **Fibrin** is an insoluble protein, not present as such in blood. It arises from a precursor, a soluble plasma protein called **fibrinogen.** The clotting reaction is a series of precisely regulated events that begins with the release of platelet lipoprotein and ends with the conversion of fibrinogen to fibrin.

Former Confusions

It might enhance understanding of the clotting reaction if we first take brief note of the major sources of past confusion. Four problems long impeded our understanding of clotting.

The first was the puzzling fact that so many factors participate in the clotting of blood. Nearly 20 different proteins and other agents are needed for clot formation. Most are enzymes that sequentially activate one another. The net effect of this arrangement is to ensure that clotting occurs when it should and does not occur when it should not.

Second, confusion arose because many of these factors were given different names by different investigators. That problem was partially solved when an international commission assigned noncommittal Roman numerals to the factors. Now they are called factor I, factor II, and so on up to factor XIII (with a few numbers missing owing to scientific mistakes).

A third problem was the steady stream of newly discovered factors. The mode of discovery was most unusual: new clotting factors were invariably discovered by finding them missing—that is, by identifying people who lacked them and thus had a bleeding tendency.

Consider an example. Hemophilia, a disease with a prominent place in history (see Chapter 15), is due to a hereditary lack of factor VIII.[7] When factor VIII-deficient plasma is allowed to clot in a test tube, the clot forms very slowly. Addition of a small amount of normal plasma promptly restores the clotting time to normal because it supplies the missing factor VIII. However, addition of plasma from another factor VIII-deficient individual does not supply the missing factor and fails to correct the clotting time.

One day in 1952 physicians at Oxford University examined an individual who seemed to have typical hemophilia—that is, factor VIII deficiency. The clotting time of the patient's plasma was very prolonged. But addition of an equal volume of bona fide factor VIII-deficient plasma corrected the clotting time! That meant that the added factor VIII-deficient plasma must have been supplying a factor missing from the patient's plasma. Hence the missing factor could not have been factor VIII. Likewise, the patient's plasma must have been supplying factor VIII to the VIII-deficient plasma that had been added. The conclusion was clear. This patient's plasma lacked something other than factor VIII—and investigators were on their way to the discovery of a new clotting factor.

The principle is simple and straightforward: whenever one mixes two plasmas with abnormal clotting behavior and finds that the behavior is thereby improved the two plasmas must lack different factors. When mixing them produces no improvement, they must both lack the same factor.

A fourth issue puzzling to students of clotting was evidence indicating that clotting can be initiated by two mechanisms (Figure 22-21).

- The **intrinsic system** begins with exposure of blood plasma to an injured blood vessel wall and release of various platelet factors. This leads to a series of reactions that in the end forms an activated clotting factor called factor Xa. (Note: "a" means the factor is activated.) The intrinsic system is fully active in the test tube when whole blood is present.
- The **extrinsic system** is initiated by a glycoprotein in tissue fluids called tissue factor, which is not active until and unless tissues are injured. The extrinsic pathway also leads to the formation of factor Xa.

Both the intrinsic and extrinsic systems operate in the body—the intrinsic system when activation is entirely dependent on factors within the bloodstream, the extrinsic system when tissue factor is liberated by injury.

How the Clot Forms

Now we examine the clotting reaction itself (Figure 22-21).

As we have seen, normal hemostasis begins when vascular endothelium is damaged. Exposed collagen attracts platelets and induces their loose aggregation. These components in turn initiate the clotting reaction, which results in the generation of **thrombin,** a protein that leads to the laying down of a clot, a platelet-fibrin network.

While all of this is going on, various limiting processes are activated that work to confine clotting to the site of injury. Without these processes, the entire blood compartment would clot. Finally, lysis of the clot occurs when vascular endothelium is regenerated.

The critical final event in clot formation is the conversion of soluble, plentiful, circulating **fibrinogen** to tough, insoluble, fibrous **fibrin,** which subsequently shrinks and, under the influence of factor XIII, forms a tight-knit clot. That key conversion is catalyzed by thrombin, a powerful proteolytic enzyme that arises from **prothrombin** (factor II) under the influence of factor Xa, the activator that can arise from activity of the intrinsic pathway or the extrinsic pathway.

In the intrinsic system the various clotting factors combine in a definite sequence to form the prothrombin activator. In each step, a clotting factor is activated—

---

[7] Hemophilia has also been described in horses and many dog breeds. Thus it occurs in at least three orders of placental mammals: the Perissodactyla, the Carnivora, and the Primates (see Appendix). That means either that the mutation responsible for hemophilia is of great antiquity, since these groups have been distinct for some 65 million years, or that the mutation has recurred independently many times in the three orders, since it must be all but lethal in the wild.

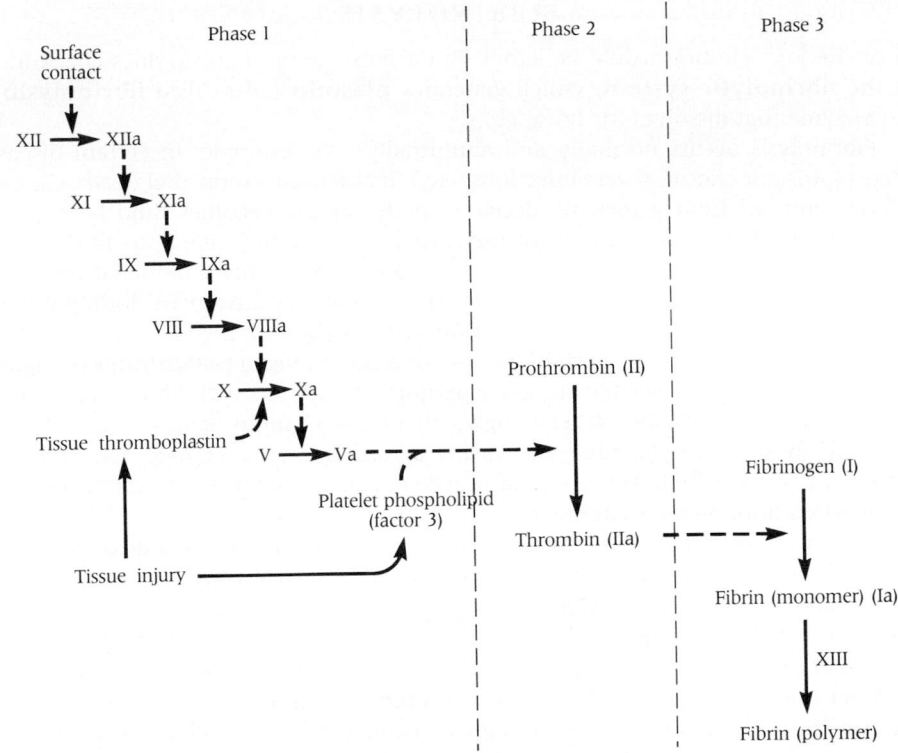

22-21 The clotting reaction
Clotting of blood depends on a cascade of reactions that eventually leads to the production of a fibrin clot. In this scheme, black arrows refer to chemical changes that activate a molecule (an activated molecule is indicated by ''a''). Colored arrows identify catalytic events. Note the three phases of the clotting reaction.

and we then add a small *''a''* to its name. Activated factors are also proteolytic enzymes or proteases that resemble the enzyme trypsin. Prior to activation, the clotting factors are enzymatically inactive: they are enzyme precursors (zymogens) that are plentiful in the blood but lack catalytic activity. Each is transformed to a trypsin-like protease (i.e., it is activated) in one of two ways: either by a conformational change or by scission of peptide bonds via action of a converting enzyme.

In the course of evolution homologous sets of zymogens arose through gene duplication and mutation. This resulted in the development of a linked series of reactions in which a zymogen is converted to a **serine protease** (i.e., a protease that contains serine at its active site) that then catalyzes a subsequent precursor-protease transition. In fact, activation of the first factor in the clotting sequence is due a conformational change induced by exposed collagen. But in later stages, activation occurs by proteolytic cleavage of the next zymogen in the sequence. Early reactions tend to occur on surfaces exposed by injury; later reactions take place on phospholipid surfaces of aggregated platelets. This linked, multistage system permits both amplification and modulation of the initial stimulus that sets the hemostatic mechanism into action.

The first step in the intrinsic pathway is activation of **factor XII** by a collagen fiber—or by the glass surface of a test tube. **Factor XIIa** then activates the next factor and so on down the line in a cascade of precursor-protease transformations until **factor X** is activated. **Factor Xa** converts **prothrombin** to **thrombin**— yet another precursor-protease transformation.

Such a cascade has two properties of biological importance.

- It amplifies small effects. Only a few molecules of activated factor XII are needed to set the sequence in motion.
- It affords many opportunities for the scope of the clotting reaction to be limited and modulated once it has begun.

Once formed, fibrin molecules polymerize and the polymers gradually grow and become insoluble. The resulting clot is held together only by hydrophobic and electrostatic bonds. Therefore it is a weak clot that could come apart and cause bleeding to start again. However, the next step introduces covalent bonds that cross-link polymerized fibrin molecules. This is catalyzed by factor XIIIa. The cross-linked clot is much stronger mechanically and thus better able to withstand the brunt of collisional events—both within the vascular system and at the body surface. It is also less likely to be rapidly dissolved by the fibrinolytic mechanism.

FIBRINOLYSIS

If unchecked, clotting would be lethal. But a sensitive mechanism does check it. It is the **fibrinolytic system,** which generates **plasmin** (also called **fibrinolysin**), an enzyme that dissolves, or lyses, clots.

Fibrinolysis occurs normally and abnormally—for example, in certain disease states (prostatic cancer, severe infections, etc.). It also occurs soon after death. Clotted blood removed from a recently deceased body rapidly becomes fluid because of fibrinolysis. The normal fluidity of menstrual blood is attributable to fibrinolysis, which keeps this blood fluid and allows it to be more easily eliminated from the body. In general, fibrinolysis tends to prevent or minimize unwanted clotting within the body, which is the condition called thrombosis (Box 22-2).

A key component of the fibrinolytic system is the zymogen **plasminogen** (Figure 22-22). Its unusual molecular structure permits it to interact with fibrin and become concentrated within clots. Interestingly, an unusual amino acid, ϵ-aminocaproic acid, avidly binds to plasminogen causing a conformational change that limits its ability to be activated. Thus, ϵ-aminocaproic acid is useful as an antifibrinolytic drug when fibrinolysis is excessive.

The successful production of tissue plasminogen activator by genetic engineering was a major achievement of biotechnology in the 1980s. By activating fibrinolysis in the body, it can lyse the clots causing coronary thrombosis (and other forms of thrombosis) and thereby save lives. However, it costs about $2000 per dose.

Under normal conditions, fibrinolysis is precisely regulated. Diffusion of plasminogen activators into the clot transforms plasminogen into plasmin, which remains bound to its fibrin substrate as it lyses it. Leukocytes in the clot also contribute proteases that help to hydrolyze the fibrin meshwork. Beyond the clot, fibrinolytic activity is feeble, largely because **antiplasmin** neutralizes any newly formed plasmin escaping into the plasma.

In summary, the hemostatic system rests ultimately upon two cascades yielding discrete proteolytic enzymes. The first is the clotting, or coagulation, system, the end product of which is thrombin. The second is the fibrinolytic system, the end product of which is plasmin. Thrombin initiates formation of a fibrin clot by cleaving specific peptide bonds in plasma fibrinogen. Plasmin breaks down the clot by hydrolyzing different peptide bonds in the fibrin molecule. These two serine proteases may also interact with plasma components that inhibit their activities. These inhibitors contain peptide sequences similar to those in their natural substrates. When thrombin or plasmin interacts with these inhibitors, they form stable complexes and chemical bond hydrolysis is hindered. We see in this pattern a satisfying symmetry: similar biochemical mechanisms are responsible for both the initiation and suppression of hemostasis.

A remarkable feature of the hemostatic system is the many regulatory interactions which coordinate the activities of its diverse pathways. At least two links (and probably more) connect the clotting and fibrinolytic systems (Figure 22-23). Not only does activation of factor XII catalyze the activation of factor XI (which then cascades down the intrinsic system), it is also linked to the conversion of plasminogen

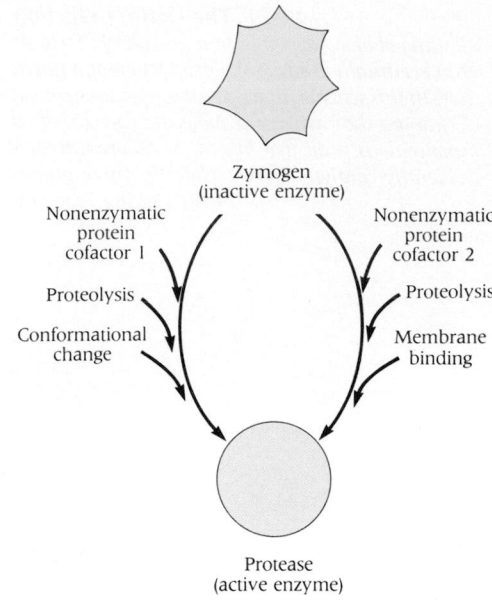

22-22 Activation mechanism for hemostatic zymogens

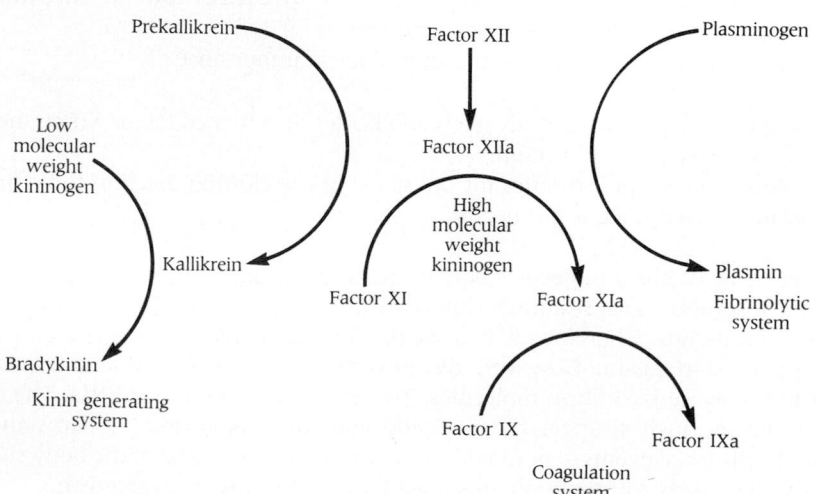

22-23 Consequences of factor XII activation
Activation of factor XII results in simultaneous activation of the kinin-generating system, the clotting cascade, and the fibrinolytic system.

BOX 22-2

THE PROBLEM OF THROMBOSIS

There can be no doubt about the danger to human life of the disorder called **thrombosis.** It is, in one form or another, one of the commonest of human ills. Yet its causes in many cases remain unknown.

In thrombosis, an unwanted clot (or thrombus) occurs within the bloodstream—either in an artery or a vein. The result is an obstruction of blood flow.

The size of the problem is readily apparent in various statistics. In 1980 in the United States, 1 million people died of myocardial infarction, in most cases a result of thrombosis in a coronary artery. Many thousands have also suffered and died from pulmonary embolism (an injury to the lung that occurs when a piece of a thrombus breaks off, travels to a branch of the pulmonary artery, and plugs it up). More than 80% of the elderly people who have fractured their hips have severe thrombosis of the leg veins at death—and of these some 40% have pulmonary embolism. Thrombosis is also one of the main causes of stroke. Indeed, thrombosis—the process in which fluid blood clots within a blood vessel—is a more significant killer than cancer.

The fact that a victim of thrombosis may nonetheless survive for years indicates that local and perhaps transitory abnormalities are critical in determining whether or not a disaster takes place. Factors such as sex, smoking, some blood groups, and oral contraceptives do predipose individuals significantly, if slightly, to various kinds of thrombosis.

Since the work of the Viennese pathologist Rudolf von Virchow more than a century ago, thrombosis has been variously ascribed to impeded blood flow, to endothelial injury, and to alterations in the composition of blood that foster intravascular coagulation, the so-called hypercoagulable state. What actually constitutes a hypercoagulable state is only vaguely understood. A number of alterations have been speculatively postulated.

- Increased amounts of clotting factors, platelets, or plasmin inhibitors
- Decreased concentrations of inhibitors of coagulation or of plasminogen or its activators
- The presence in blood of an excess of activated clotting factors
- Biosynthesis of abnormal clotting factors with increased ease of activation, or of abnormal plasminogen with decreased ease of activation
- Impaired generation of plasminogen activators
- Impaired removal of activated clotting factors from the blood

The problem is: Can any laboratory tests predict thrombosis? If so, what kind? At this writing, no simple blood tests can reliably identify individuals who are likely to develop thrombosis. However, some promising new procedures, sensitive radioimmunoassays, measure either released platelet membrane or granule proteins, small activation peptides released from coagulation factors, or coagulation factor inhibitor complexes that are formed during thrombosis.

It has been found, for example, that individuals with deep venous thrombosis often have increased circulating levels of fibrinopeptide A (which is removed from fibrogen by thrombin), certain prothrombin fragments, and thrombin-antithrombin complex—all the result of thrombin generation and clot formation.

These abnormalities revert to normal when a suitable anticoagulant is administered. Similar abnormalities have been noted in people with active thrombosis (or embolism), as well as those with so-called "prethrombotic" states—that is, people who are asymptomatic but who develop thrombosis later.

Although the cause of thrombosis is still uncertain in the vast majority of cases, three well-defined genetic abnormalities are known to produce prethrombotic or hypercoagulable states.

- The first is antithrombin deficiency, an autosomal dominant disorder that occurs in one of every 2000 individuals. Affected individuals have recurrent venous thrombosis and pulmonary emboli. They require anticoagulant treatment at times of surgery or trauma and lifelong anticoagulation after the onset of symptoms.
- A second genetic disorder that leads to venous thrombosis and embolism is protein C deficiency.
- Third, several families have an abnormal form of plasminogen that cannot be readily activated to plasmin and a high incidence of venous thrombosis and embolism.

In some manner, various congenital and acquired factors alter the blood in a way that makes people prone to thrombosis. Thus there is a spectrum extending from mild abnormalities detectable only by sensitive laboratory tests, to frank disease and death. If most of these events occur silently in otherwise healthy individuals, it becomes difficult therefore to know at what stage or stages in the evolution of a thrombus preventive treatment should be aimed. Clearly there are many gaps and much work to do.

Since platelets have a fundamental role in arterial thrombus formation, inhibition of platelet function by drugs is a potentially important way of preventing arterial occlusion. Several promising antiplatelet agents are known that inhibit the platelet release reaction. Aspirin, for example, inhibits the membrane-bound cyclo-oxygenase that produces prostaglandin endoperoxides for thromboxane formation, thereby impairing platelet aggregation (see Box 28-1). Aspirin also decreases the release of ADP, a mediation of platelet aggregation (see Figure 22-16).

to plasmin and to the conversion of prekallikrein to the serine protease, kallikrein. Kallikrein then cleaves the polypeptide chain of low-molecular weight kininogen, releasing the short peptide bradykinin, which we have seen arises locally following injury. There it participates in the inflammatory reaction, dilating capillaries and increasing their permeability. Kallikrein itself exhibits chemotactic activity for neutrophils and monocytes, recruiting them into sites of tissue injury.

SUMMARY

The major functions of blood include (1) transport of nutrients and waste, (2) defense against infection and bleeding, and (3) regulation of salt and ionic balance, acid-base balance, and temperature.

Whole blood consists of fluid plasma and the several types of cells that are suspended in it. The major cell types are red cells (erythrocytes), white cells (leukocytes), and platelets (thrombocytes).

Blood cells, and the cells comprising the blood vessels and heart, develop from mesenchymal cells. In adult vertebrates, blood-cell formation (hematopoiesis) takes place only in bone marrow, but in embryonic forms hematopoiesis occurs in the liver, spleen, and other tissues.

Mammalian red cells, which are produced in greatest numbers, lose their nuclei before they are released into the blood from the bone marrow. Because mature red cells lack mitochondria, their energy metabolism consists solely of glycolysis. The hematocrit is a measure of the volume of packed red cells per 100 ml blood.

Red cells in all vertebrates but mammals are nucleated and continue to synthesize hemoglobin throughout their lifespans. In mammals red cells function primarily as containers for hemoglobin, an oxygen-carrying molecule that contains four hemes (prosthetic groups composed of porphyrin and iron). Two advantages arise from the presence of hemoglobin in red cells rather than in plasma solution: (1) blood viscosity is lowered, and (2) red cell glycolysis can keep hemoglobin iron in the reduced form (Fe^{2+}) rather than the oxidized form (Fe^{3+}). Hemoglobin binds oxygen only when its iron atoms are reduced.

Hemoglobin binds oxygen reversibly. The curve tracing the affinity of hemoglobin for oxygen as a function of oxygen pressure has a sigmoidal shape. Hence, hemoglobin takes up oxygen where it is plentiful (as in the lungs), and releases it when it is scarce (as in actively metabolizing tissues).

Genetic mutations are responsible for many hemoglobin variants (such as sickle hemoglobin), which contain abnormal amino acid sequences. The hemoglobins of animals of different species have differing amino acid sequences despite their considerable similarity. The extent of the similarity is a measure of the degree of evolutionary relatedness of two species.

White cells (leukocytes) do contain nuclei and, unlike red cells, are found in both tissues and blood. Indeed, many of the defensive functions of white cells are performed outside of the vascular system. Leukocytes are classified according to their morphology, function, and tissue of origin.

Lymphocytes are largely produced in lymphoid tissues (lymph nodes and spleen). Granulocytes (neutrophils, basophils, and eosinophils) are produced in the bone marrow (hence they are termed myeloid cells). Granulocytes are distinguished from lymphocytes by their multilobed nuclei and cytoplasmic granules (lysosomes).

All leukocytes contribute to the body's defenses against infection and injury. In part, they do this by participating in the process called inflammation. At sites of tissue damage, blood vessels dilate, with resulting swelling. Leukocytes leave the capillaries and migrate to the site. There granulocytes remove dead cells, bacteria, and other foreign matter by phagocytosis. Monocytes also migrate to sites of injury where they become actively phagocytic macrophages. Lymphocytes produce antibodies that bind and render harmless many foreign substances. Inflammation also activates three enzymatic cascades: the kinin system, which causes vasodilatation; the complement system, which promotes the destruction of bacteria; and the clotting system, which stops bleeding.

Platelets have an essential role in hemostasis (the prevention of bleeding). They clump together to form a temporary or primary plug at bleeding sites. They also trigger the activation of inactive clotting proteins in plasma. As a result, the plasma protein fibrinogen is converted to the insoluble protein fibrin. The clot forms a permanent or secondary plug at bleeding sites.

Only mammals have true platelets. In other vertebrates, cells resembling leukocytes perform similar functions. Platelets are not typical nucleated cells. Rather they arise from fragmentation of the cytoplasm of the giant precursors in bone marrow called megakaryocytes.

Plasma, the fluid in which blood cells are suspended, transports nutrients, wastes, and chemical messengers. Some of these substances travel freely in solution; others are bound to transport proteins or other large carrier molecules. Albumin, the most prevalent plasma protein, has a role in the transport of several otherwise poorly soluble compounds. Other important plasma proteins are transport proteins, antibodies, and clotting factors. Plasma proteins are large molecules that cannot diffuse through blood vessel walls. Hence, they help in maintaining the osmotic activity of blood, which in turn sustains fluid balance throughout the body.

More than twenty proteins and various cofactors are required in the clotting reaction, a complex process that in the end forms a fibrin clot. Two independent enzymatic cascades—the intrinsic and extrinsic pathways—can trigger clot formation. Both cause a critical proteolytic enzyme called thrombin to convert fibrinogen to fibrin at the site of blood vessel injury. If unchecked clotting would be lethal, but it is normally regulated by the fibrinolytic system, an enzymatic cascade which generates plasmin, a dissolver of clots.

KEY TERMS

albumin
anemia
antibody
anticoagulant
bone marrow
chemotaxis
clotting system

erythrocyte
fetal hemoglobin (Hb F)
fibrin
fibrinogen
fibrinolytic system
globulin
granulocyte

hematocrit
hemoglobin (Hb A)
hemoprotein
inflammation
leukocyte
lymphocyte
macrophage

monocyte
mononuclear cell
myoglobin
opsonin
osmotic pressure
oxygen dissociation curve
plasma

plasma protein
platelet
red cell
serine protease
serum
sickle cell anemia

sickle cell hemoglobin (Hb S)
stem cell
thrombosis
transport protein
white cell
zymogen

QUIZ QUESTIONS

1. Which of the following is *not* a function of blood and associated blood vessels?
   A. transport of hormones
   B. transport of heat
   C. maintenance of osmotic balance of fluids in tissues and bloodstream
   D. protection of the body against invading microorganisms
   E. All of the above are functions of blood and associated blood vessels.

2. Which of the following does *not* occur as part of the inflammatory response?
   A. release of histamine by mast cells
   B. aggregation of platelets at the site of inflammation

   C. vasodilation of blood vessels at the site of inflammation
   D. leakage of fluids into the inflamed tissues

3. Which of the following substances stimulates phagocytic activity of certain types of leukocytes?
   A. bradykinin
   B. serotonin
   C. adenosine diphosphate
   D. opsonins

4. Megakaryocytes
   A. develop by the process of endomitosis.

   B. fragment to produce granulocytes.
   C. lack a nucleus.
   D. contain hemoglobin.

5. Which of the following is *not* involved in blood coagulation?
   A. fibrin
   B. thrombocytes
   C. plasmin
   D. factor X
   E. tissue factor

ESSAY QUESTIONS

1. What are the major functions of blood?

2. Why is hemoglobin carried within red blood cells rather than in solution in plasma? What makes hemoglobin red? Why is arterial blood brighter red than venous blood? Contrast the structure and function of hemoglobin and myoglobin.

3. What are the roles of the major classes of white blood cells in fighting infection? What is pus?

4. What are the functions of platelets? How do they form? Aspirin (acetylsalicylic acid) has the potentially harmful side effect of inactivating platelet function. What problems might result from excessive aspirin intake?

REFERENCES AND SUGGESTIONS FOR FURTHER READING

Beck, W. S. (Ed.) (1990) *Hematology*, 5th ed. The MIT Press, Cambridge, Mass.

Lectures on the physiology of human blood in health and disease.

Benesch, R., and Benesch, R. E. (1969). Intracellular organic phosphates as regulators of oxygen release by hemoglobin. *Nature 221*, 618–622.

Although it had long been known that red cells contain high concentrations of 2,3-DPG, its significance was completely unknown until this contribution by the Benesches.

Bunn, H. F., and Forget, B. G. (1986). *Hemoglobin: Molecular, Genetic, and Clinical Aspects*. W. B. Saunders, Philadelphia.

An excellent survey of a fascinating subject.

Dickerson, R. E., and Geis, I. (1983). *Hemoglobin: Structure, Function, Evolution, and Pathology*. Benjamin/Cummings, Menlo Park, Calif.

Beautifully illustrated review of important aspects of hemoglobin biology.

Linzen, B., and 16 other authors (1985). The structure of arthropod hemocyanins. *Science 229*, 519–524.

Hemocyanins are huge O_2-transporting proteins in many arthropods and mollusks. Unlike hemoglobin, they contain copper. This fascinating survey argues that hemocyanin evolved more than 600 million years ago.

Metcalf, D. (1989). The molecular control of cell division, differentiation commitment and maturation in haemopoietic cells. *Nature 339*, 27–30.

The author, discoverer of the colony-stimulating factors that regulate white cell production, here gives a broad summary of a major scientific advance.

Pauling, L., Itano, H. A., Singer, S. J., and Wells, I. C. (1949). Sickle cell anemia, a molecular disease. *Science 110*, 543–548.

A classic paper that demonstrated a molecular abnormality in sickle cell anemia and thereby opened a new era in genetics.

Perutz, M. F. (1978). Hemoglobin structure and respiratory transport. *Scientific American 239*, December, 92–125.

Hemoglobin structure—and the effects upon it of oxygen binding—are clearly explained by the Cambridge University investigator who elucidated the structure.

Ratnoff, O. D. (1987). The evolution of hemostatic mechanisms. *Perspectives in Biology & Medicine 31*, 4–33.

A delightful essay.

Shemin, D. (1982). From glycine to heme. In Kaplan, N. O., and Robinson, A. (Eds.). *From Cyclotrons to Cytochromes*, pp. 117–129. Academic Press, Orlando, Fla.

A stimulating account of an early and successful use of isotopes in unraveling the biosynthetic pathway of porphyrins and heme.

Wintrobe, M. M. (Ed.) (1980). *Blood Pure and Eloquent: A Story of Discovery, of People, and of Ideas*. McGraw-Hill, New York.

An impressive history of hematology with individual essays by many key contributors. Highly recommended.

IMMUNITY

Immunity
The immune system is a highly versatile adaptation that, by the actions of specific antibodies and cells, defends animals against an immense variety of molecules, viruses, and bacteria. In this unusual scanning electron micrograph, a macrophage (one type of defensive cell) extends its pseudopods toward several E. coli prior to ingesting them. Note the complex lobulations of the macrophage surface.

Organisms possess an array of defense mechanisms that enables them to withstand many forms of adversity. This chapter explores immunity, a defense system notable for its specificity—and complexity.

INTRODUCTION

An animal body invaded by foreign materials or microorganisms defends itself by developing immunity to the invader. In this process, the body "learns" from its experience with past invasions to deal effectively—and specifically—with future ones. **Immunity** is defined as a state of heightened resistance or accelerated reactivity toward microorganisms, transplants, or any other "nonself" substances that gain access to the body. It is a homeostatic mechanism of the first importance.

It has been said that immunology is both the most ancient and the most modern of sciences, the most theoretical and the most practical. Indeed, the broad concepts of immunology are today at the heart of biological thought. Let us consider why.

HISTORICAL NOTES

Thucydides, writing during the time of the Peloponnesian Wars, observed that the sick should be treated by those who had recovered since they were "free of fear." Presumably, that meant fear of again acquiring a contagious disease. Other ancients actually recorded that those who recovered from an infectious disease such as chicken pox, measles, and mumps seldom suffered from that disease again.

The notion that immunity might be induced deliberately was established by Edward Jenner (1748–1823) in the late eighteenth century with the introduction of a vaccine for smallpox (Figure 23-1). There was no knowledge then of viruses or other microbes, but Jenner listened when a young milkmaid told him that she could never have the disfiguring, and often lethal, disease smallpox (variola) because she had contracted the mild disease cowpox (vaccinia). His concern for the milkmaids' smooth complexions—with faces unscarred by smallpox—led to the notion that vaccination (a term derived from vaccinia) with cowpox might protect against smallpox. Jenner's vaccine virtually eliminated smallpox from Europe—and later from the world.[1]

By the mid-nineteenth century, it was accepted that an induced mild form of an infection could protect against a natural virulent form. With the development of medical bacteriology by Pasteur, Koch, and others, this principle was soon applied to many infectious diseases. Pasteur laboriously worked out methods for the attenua-

[1] In fact, the idea of inoculation against smallpox did not originate with Jenner. In 1734, Voltaire published an amusing essay (in his *Letters on the English*) titled "On Inculation." In it, he described how inoculation against smallpox was brought to England during the reign of George I by Lady Mary Wortley Montague on her return from Constantinople, where her husband had been ambassador. She learned that it had been practiced there for hundreds of years by parents anxious to keep their daughters from being disfigured by smallpox before they could be sold to the seraglios of Turkey and Persia.

tion (weakening) of organisms responsible for anthrax, rabies, and other diseases. It was to honor Jenner that Pasteur called his treatments "vaccinations." **Vaccination** stimulates immunity to an infectious disease by exposing the body to small amounts of a microorganism or its components in natural or altered form.

During the 1880s, two seemingly conflicting ideas emerged. One group of workers, having determined that blood serum frequently kills certain bacteria, concentrated on this bactericidal power of serum. They learned that it is enhanced after infection with the bacteria. Sometimes serum destroys the offending organisms, and sometimes it agglutinates them (clumps them together) without destroying them. These properties were ascribed to a class of substances, at first wholly hypothetical, called **antibodies.** Further work revealed their astonishing specificity. Those appearing after streptococcal infection, for example, attack only streptococci. These striking results led many to consider antibodies the sole basis of immunity to infection. Because antibodies are soluble components of blood—as are all body "humors" according to the ancients—this view became known as the **humoral theory** of immunity. In humoral immunity, antibody synthesis is stimulated by **antigens,** which are foreign molecules bearing specific chemical groupings called **antigenic determinants.**

Others, inspired by Metchnikoff (see Chapter 22), gave paramount importance to **phagocytosis,** the capacity of neutrophils, macrophages, and other cells to ingest bacteria or attack them in other ways, and advanced a competing idea—the **cellular theory** of immunity (Figure 23-2). As is so often the case, elements of both theories were correct.

- Humoral defenses employ highly specific, soluble antibodies that circulate in the plasma and react with the antigens that evoked their appearance.
- Cellular (or cell-mediated) defenses require the participation of certain body cells in reactions that are immunological in character. The cells responsible for cell-mediated immunity are lymphocytes. An operational distinction between humoral and cellular immunity is that humoral immunity can be passively transferred from one individual to another with blood serum; transfer of cellular immunity requires a transfer of cells. Granulocytes and macrophages also play a role, since phagocytosis is a form of cellular immunity. However, phagocytes lack the high specificity of the lymphocyte-mediated immune reactions.

Another impetus to progress was the discovery in the late nineteenth century that some bacteria, notably those causing diphtheria and tetanus, cause illness and death by liberating soluble, diffusible **toxins.** In 1890, it was found that injection of toxin in small nonlethal amounts often elicits a neutralizing antibody, then called an **antitoxin.** Evidently, antibody production is elicited by some bacterial products, as well as by intact bacteria. The studies of diphtheria toxin and antitoxin by Paul Ehrlich (1854–1915) provided the first unified theories of immunity. Because he could titrate one against the other, Ehrlich concluded that antitoxin neutralizes toxin molecules by combining with them chemically.[2]

Countless experiments have shown that many substances besides bacteria and their toxins elicit the production of antibodies. These include the serum or red cells of another species, such proteins as egg albumin, and bacteria and viruses. All of these substances are antigens.

While some investigators probed the problem of how immune responses protect against infection, others found that these responses can sometimes cause unpleasant and dangerous **hypersensitivity** (later called allergy) to a provoking antigen. Because immune reactions underlying hypersensitivity appeared to do little to protect the body against infection it became clear that immune responses can sometimes harm the body. Reflecting on this paradox early in this century, C. F. von Pirquet saw that immunity can no longer be viewed only in medical terms—a body system that protects against infectious disease. Rather, it has broader meaning: an individual exposed to an antigen displays altered reactivity—in some cases, a beneficial immunity, in others, a potentially harmful hypersensitivity state. The feature common to both is a process of recognition and specific response to the foreign nature of a wide range of substances.

23-1 Edward Jenner (1748–1823)

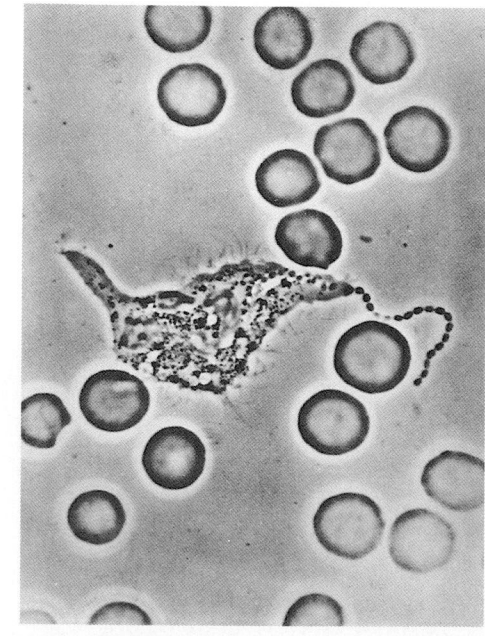

23-2 Phagocytosis
The photomicrograph was taken through a phase microscope, which visualizes living cells without preliminary staining. The white cell is ingesting a chain of bacteria (Streptococci). (×1400)

---

[2] The famous Ehrlich side-chain theory stated that cells making antitoxins had protoplasmic extrusions that would bind to the toxin and that this binding would induce the appearance of more side chains, which would then be shed, appearing as serum antitoxins. Ehrlich's theory came too early to be tested. Indeed, no cellular work of significance was done until recently.

We will describe the several types of immunity and then see how they work.

Humoral Immunity

Humoral immunity is based upon the activities and properties of antigens and antibodies. It is often termed "classic" or "orthodox" immunity, because antibodies were among the earliest discoveries in immunology.

Humoral immunity can be active or passive.

- **Active immunity** means that antibodies were produced in an animal's own tissues following its own exposure to an antigen.
- **Passive immunity** is produced secondhand by a transfer of serum from another individual who is immune. It may occur naturally, for example, by transmission to the fetus of maternal antibodies through the placenta or in the early milk (colostrum). Or it may be conferred artificially by injection of antibody-containing serum (antiserum) from another human or animal body.

The body tissues of a passively immune individual take no part in the production of antibodies. Since all antibody molecules have a certain lifespan in the body, immunity of this type is inevitably temporary. Artificially induced passive immunity is medically useful because it is immediate; active immunity must be built up over

23-3 Delayed and immediate hypersensitivity
(A) In delayed hypersensitivity, tuberculin sensitivity can be transferred from a sensitized to a nonsensitized animal only by lymphoid cells, not by cell-free serum. The inflammatory lesion takes 2 days to appear. (B) In immediate hypersensitivity, sensitivity to egg albumin is transferable with serum, but not cells. Within 5 minutes of an injection of egg albumin, the nonsensitive animal develops anaphylactic shock.

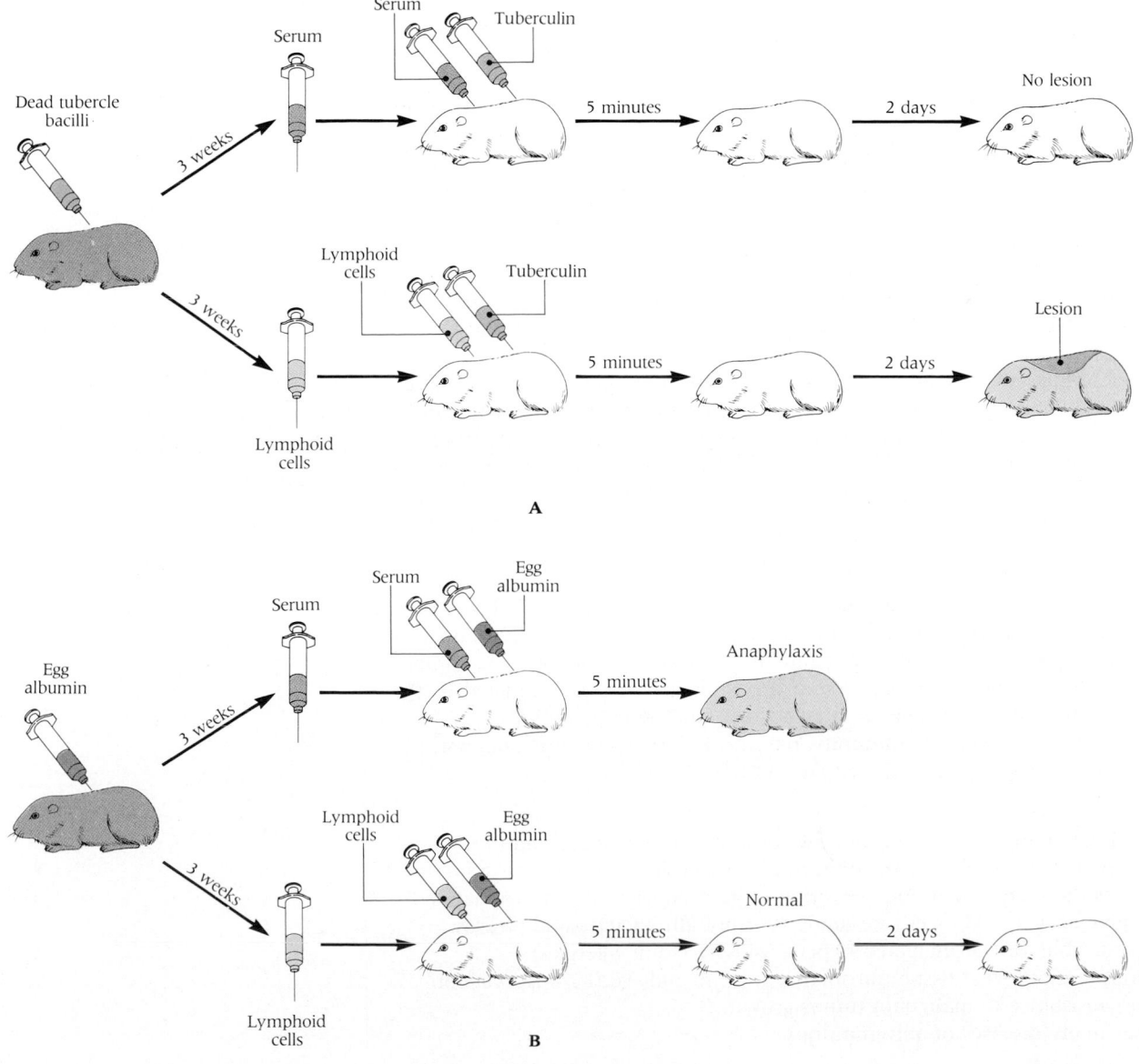

A

B

days or weeks. The use of tetanus antitoxin is a familiar application of passive immunity.

The antibody elicited by an antigen binds it with great specificity, often with the involvement of complement (see Chapter 22). This is termed **complement fixation.** Usually the antigen complex is then phagocytized by neutrophils.

Allergy is a miscarriage of the immunologic process in which overreaction of an essentially protective mechanism results in deleterious effects that range in severity from mild itching to sudden death. Its severest form occurs in one type of hypersensitivity, the nonbeneficial state studied by von Pirquet: the severe immediate hypersensitivity called **anaphylaxis.** As in all hypersensitivity states, first contact with a foreign substance is apparently harmless, but later exposure produces harmful, sometimes fatal, results. In anaphylaxis, the difficulties occur when an antigen-antibody complex triggers the release of various harmful substances (notably histamine) by a class of body cells called mast cells. Most forms of allergy are also antibody-mediated, but they do not display the sudden severity of anaphylaxis. Unlike delayed hypersensitivity (to be discussed later), immediate hypersensitivity is transferable to another animal with serum (Figure 23-3B).

In summary, antibodies have the following capabilities:

- They prepare bacteria for phagocytosis by neutrophils or macrophages. This is called **opsonization** (from Greek, "to prepare food").
- They neutralize bacterial toxins.
- They kill and lyse bacteria in serum.
- They can passively induce allergic and anaphylactic reactions.

Cell-mediated Immunity

The hypersensitivity responses studied long ago by von Pirquet were originally divided into two classes: **immediate** and **delayed.** The former occur in minutes; the latter take several days. We still use those terms but they now have different meanings based on underlying mechanisms. Many writers prefer the term *cell-mediated hypersensitivity* to delayed hypersensitivity. The corresponding term for immediate hypersensitivity is *antibody-mediated hypersensitivity.*

One of the earliest clear examples of cellular immunity to be studied was the delayed hypersensitivity of the tuberculin reaction. **Tuberculin** is a mixture of substances extracted from tubercle bacilli (which cause tuberculosis). When a small amount of it is injected into the skin of a human or guinea pig previously infected by tubercle bacilli, local redness and swelling appear within 24 hours (Figure 23-3A). Its cellular basis is indicated by the fact that tuberculin sensitivity can be transferred to a nontubercular animal by inoculation of lymphoid cells from the spleen or lymph nodes of a sensitive animal. Serum does not transfer tuberculin sensitivity. We conclude that prior exposure to tubercle bacilli stimulates and modifies lymphocytes so that henceforth they are hypersensitive to the foreign material.

In this immune response, cells of various kinds accumulate at sites of inflammation and there by various means harm invaders and cause their destruction. Cell-mediated immunity has a *specific* component, which is represented by certain lymphocytes (T cells) that recognize antigens and by certain macrophages, and a *nonspecific* component, represented by the granulocytes and macrophages that are attracted to the site, activated, and attack invaders by phagocytosis and other activities.

In summary, cell-mediated immunity has the following helpful and harmful functions (some of which are discussed later in this chapter).

- It aids in defending against infection by certain bacteria, viruses, and fungi, many of which are curiously resistant to phagocytosis.
- It involves the recruitment and activation of granulocytes and macrophages, which phagocytize and inactivate many (but not all) bacteria and viruses.
- It is responsible for cell-mediated hypersensitivity (some allergies).
- It promotes rejection of tissue transplants and the graft-versus-host reaction.
- It fosters resistance to malignant tumor growth.
- It causes many diseases of autoimmunity.

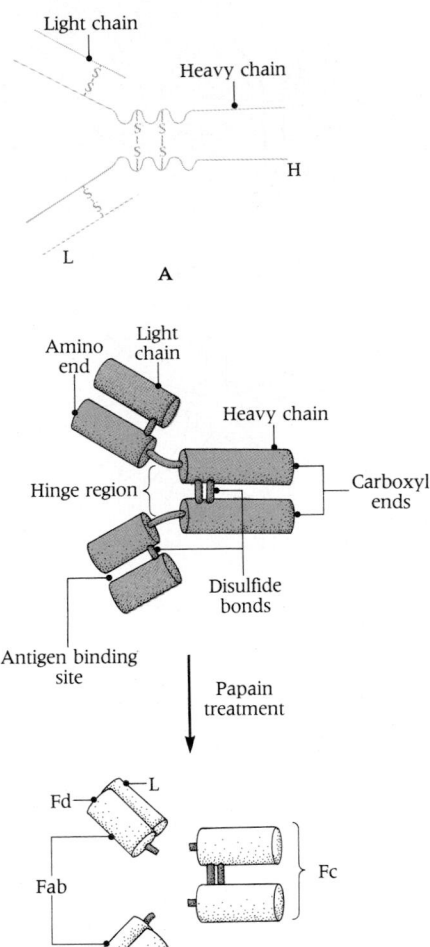

23-4 Structure of immunoglobulin G
(A) *Disulfide bonds covalently attach the light chains (L) to the heavy chains (H), and the two heavy chains to each other.* **(B)** *The polypeptide chains making up the light and heavy chains are here drawn as cylinders. The antigen binding site is at the amino ends of the L and H chains. Papain digestion of IgG splits the molecule into Fab, Fc, and Fd fragments.*

HUMORAL IMMUNITY

Antigens, as we have noted, are substances that stimulate antibody formation. Antibodies are molecules that are synthesized upon stimulation by an antigen and that combine with the antigen with great specificity. All antibodies belong to a class of plasma globulins known collectively as **immunoglobulins** (abbreviated **Ig**).

NATURE OF ANTIBODIES AND ANTIGENS

Properties of Normal Immunoglobulins

The five human immunoglobulin classes are termed **IgG, IgA, IgM, IgD,** and **IgE.** Table 23-1 summarizes their properties. Note the differences in molecular weights and sedimentation constants (which denote their behavior in an ultracentrifuge). Those with high sedimentation constants (14–19*S*) are called **macroglobulins.**

All immunoglobulins contain carbohydrates—hence they are glycoproteins. The macroglobulins are especially carbohydrate-rich. The most plentiful immunoglobulin is IgG, a γ-globulin (this Greek letter prefix signifies electrophoretic mobility) with a molecular weight of 150,000 and a sedimentation constant of 7*S*. It is a symmetrical molecule consisting of two identical subunits (Figure 23-4). In each, a small **light chain** (L) is linked through a disulfide bond to a large **heavy chain** (H). The two H subunits are themselves joined through two disulfide bonds. The IgG subunits are folded into a roughly cylindrical form some 35 Å in diameter and 280 Å in length—about one-millionth the volume of a typical bacterium. The end portions of each L and H chain have unique structural features. They confer the specificity upon each immunoglobulin, whereby it binds only the correct antigen.

An immunoglobulin molecule is readily split into fragments of several kinds by various treatments. L and H chains are easily dissociated by a reducing agent that cleaves disulfide bonds (e.g., mercaptoethanol). Treatment of the molecule with the proteolytic enzymes papain or pepsin yields fragments that are named **Fab** (antigen-binding), **Fc** (crystallizable), and **Fd.** For example, papain treatment yields an Fc fragment, which contains the combined terminal portions of the H chains—and two Fab fragments, each an intact L chain plus the remaining portion of an H chain. This portion alone is an Fd fragment (see Figure 23-4).

Such techniques led to the discovery that each class of immunoglobulins has a distinctive H chain. These are designated by the Greek letters corresponding with their Roman letter names: gamma or γ, alpha or α, mu or μ, delta or δ, and epsilon or ε, respectively, in IgG, IgA, IgM, IgD, and IgE (see Table 23-1). In contrast, L chains are of only two types, designated kappa (κ) and lambda (λ). Both types of light chain are associated with each class of heavy chain. In humans, interestingly, two-thirds of all the immunoglobulin molecules contain κ-type L chains and one-third contains λ-type L chains. Therefore, a precise formulation of the several types of immunoglobulin molecules would be as follows:

$$\text{IgG, } \gamma_2\kappa_2 \text{ or } \gamma_2\lambda_2$$
$$\text{IgA, } \alpha_2\kappa_2 \text{ or } \alpha_2\lambda_2$$
$$\text{IgM, } \mu_2\kappa_2 \text{ or } \mu_2\lambda_2$$
$$\text{IgD, } \delta_2\kappa_2 \text{ or } \delta_2\lambda_2$$
$$\text{IgE, } \epsilon_2\kappa_2 \text{ or } \epsilon_2\lambda_2$$

TABLE 23-1
PROPERTIES OF NORMAL HUMAN IMMUNOGLOBULINS

	IgG	IgA	IgM	IgD	IgE
Sedimentation constant (*S*)	7	7 (9–15)	19	7	8
Molecular weight ($\times 10^3$)	150	150–300	900	180	190
Carbohydrate content (%)	3	7	12	12	12
Electrophoretic mobility	γ to α₂	Slow β	Fast γ	Slow β	Slow β
Concentration in plasma (mg/100 ml)	1100	250	100	3	0.01
Relative abundance (% of total)	76	17	7	0.2	trace
Biological life span ($T_{1/2}$, days)	21	6	5	3	2
Heavy chain type	γ	α	μ	δ	ε
Activation of complement	+	−	+	−	−

IgG and IgD molecules are considered monomers, with molecular weights of 150,000 or 180,000. IgM molecules are pentamers and IgA molecules are either monomers resembling an IgG or IgD molecule or dimers or trimers (Figure 23-5). IgA is found not only in plasma, but in the secretions of various glands—in saliva and milk, for example. Secretory IgA is a dimer that differs from plasma IgA in possessing an additional polypeptide chain of molecular weight, the so-called secretory piece, which enhances the stability of the molecule.

A great deal of immunity can reside in a small quantity of immunoglobulin. Therefore, an increase in the amount of a single type of antibody in an immune serum need not be accompanied by a detectable increase in its total immunoglobulin content. Methods of demonstrating antibodies in the laboratory are summarized in Box 23-1. Serum containing a specific antibody to a given antigen is often called **immune serum,** or **antiserum.**

Antigens

Many antigens are proteins, though DNA and carbohydrates may also be antigenic. An antigen may be part of a virus, a bacterium, or a foreign cell—or it may be a soluble derivative of some such structure. It also can be a tumor cell or an environmental chemical.

One of the intriguing issues of immunology lies in the fact that there is no particular chemical feature that antigenic substances share. We have no way of telling in advance whether a substance is antigenic and—if it is—the degree of its antigenicity. Indeed, we define antigenicity in a highly operational way: a substance is an antigen if when introduced into the body it is antigenic—that is, if it elicits an immune response.

Only a portion of each antigen's molecular structure is critically involved in stimulating antibody formation. This small region, the antigenic determinant, is usually no more than a few amino acids long. The antigenic determinant is therefore the site to which the antibody binds. Interaction of antibody and antigen is strictly determined by both the three-dimensional shape and the chemical structure of the antigenic determinant and the antigen-combining site of the antibody. If a useful interaction is to occur, the fit between combining site on antibody and corresponding antigenic determinant must be very precise.

Most natural antigens possess several antigenic determinants—and each may be capable of eliciting several types of antibodies. Hence, there is usually heterogeneity in the antibody response to a given antigen. The population of antibodies produced against a single antigen may in fact be heterogeneous, not only with respect to the antibody's Ig class but also to binding affinity, a reflection of the primary amino acid sequence of the combining site of the molecule.

As one might expect, antibodies produced against two different molecules with related antigenic determinants will, to a considerable extent, cross-react (i.e., the antibodies react with a *class* of related molecules and not just the one that served as antigen). Even so, a small population within a class of cross-reacting antibody molecules can usually be shown to have absolute specificity for the immunizing antigen. The degree of cross-reactivity depends upon the structural similarities of the antigens.

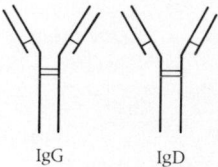
IgG   IgD

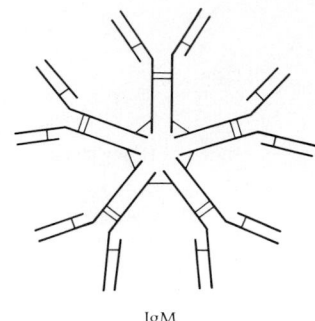
IgM

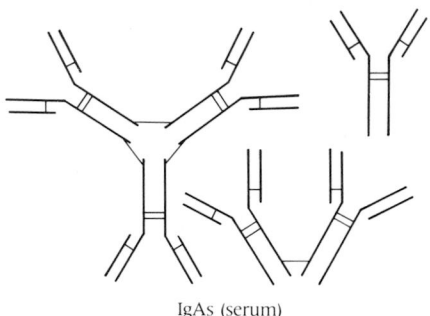
IgAs (serum)

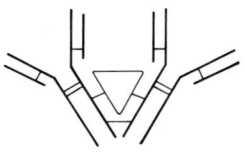
IgA (with secretory piece)

23-5 Structures of different immunoglobulin classes
IgG and IgD are monomers. IgA and IgM form aggregates. Secretory IgA includes an additional polypeptide chain.

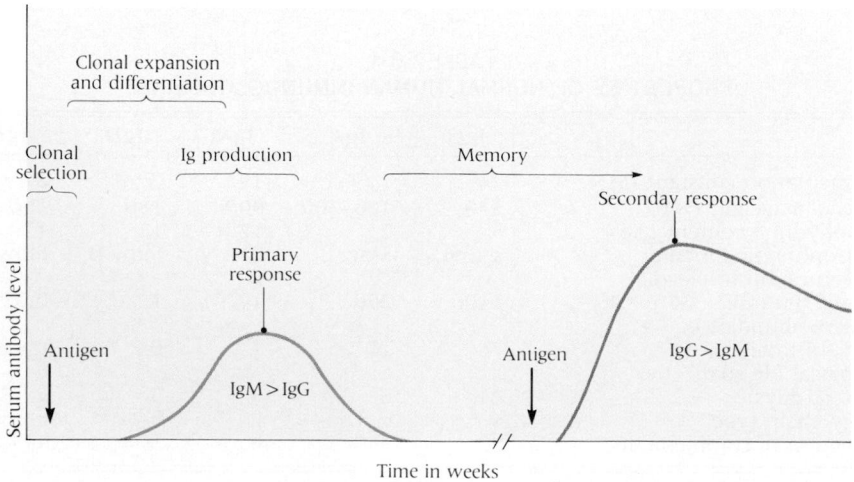

23-6 Antibody production in response to an antigen
The level of antibody in the serum increases and then falls after exposure to an antigen. In this primary response, IgM molecules predominate. If antigen is later reintroduced, a more intense secondary response occurs in which IgG molecules predominate.

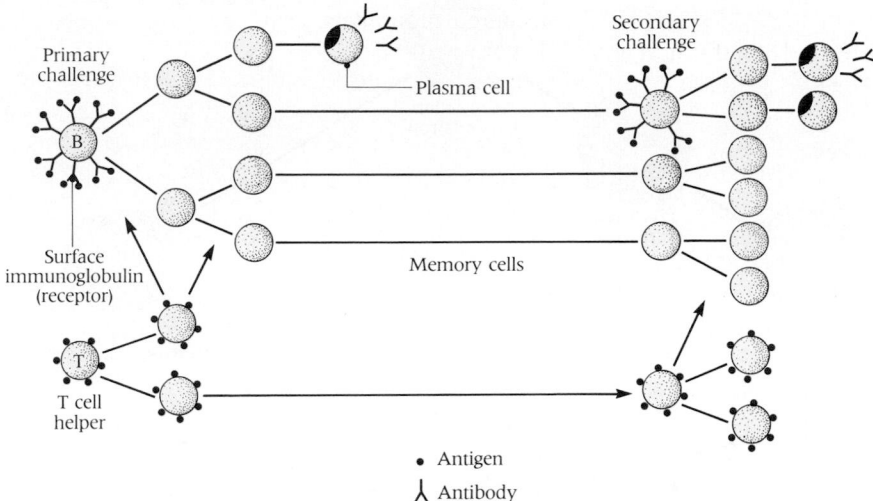

Primary challenge

Secondary challenge

Plasma cell

B

Surface immunoglobulin (receptor)

T cell helper

T

Memory cells

• Antigen

ʎ Antibody

One of the major advances of modern molecular immunology was the development of a simple technique for producing **monoclonal antibodies,** which are (1) pure and thus not heterogeneous, (2) of virtually any desired specificity, and (3) available in unlimited quantities. This monumental achievement by César Milstein and George Köhler is described in Box 23-2.

PHYSIOLOGY OF ANTIBODY PRODUCTION

Primary and Secondary Responses

Following exposure to a conventional antigen not previously encountered, the animal body does not produce antibody instantly. Nor does it become immune immediately. Rather, there is a latent period, during which the antibody-synthesizing apparatus prepares itself to elaborate antibodies. Typically, antibody appears abruptly within a week after a single initial injection of antigen (Figure 23-6). This is the **primary response.** It usually consists of IgM antibodies, with a small IgG component appearing later.

If the animal rests until the serum antibody concentration has declined to undetectable levels and then a similar or smaller amount of the same antigen is given, the second antibody response is much more rapid and stronger, the latent period is shorter, and the antibody produced is of the IgG class. This is the **secondary,** or **anamnestic, response.** After reaching its peak, the antibody levels sag much more slowly than in the primary response. This more rapid, intense, and qualitatively different response is the result of **immunological memory.** All medical immunizations (and their "booster shots") are based upon this concept.

Cells of the Immune Response

Many kinds of cells participate in the immune response, which begins with an interaction of an antigen with lymphocytes bearing specific immunoglobulins that serve as an antigen receptor on the plasma membrane (Figure 23-7). Lymphocytes all look nearly alike under the microscope (Figure 23-8), but they are in fact a heterogeneous group made up of two major classes, T and B lymphocytes (or cells).[3]

T lymphocytes
T cells develop to maturity in the thymus gland, a curious structure beneath the sternum (breast bone) in humans, which is quite large in childhood but then shrinks over the decades until it is virtually gone. As shown in Figure 23-9, their precursors, pre-T cells or stem cells, originate in the bone marrow and then migrate to the thymus. The maturation and differentiation of immature T cells are profoundly influenced by early interactions with the stromal (epithelial) cells of the thymus (Figure 23-10). Someone has said that a T cell "get its education" in the thymus from these interactions.

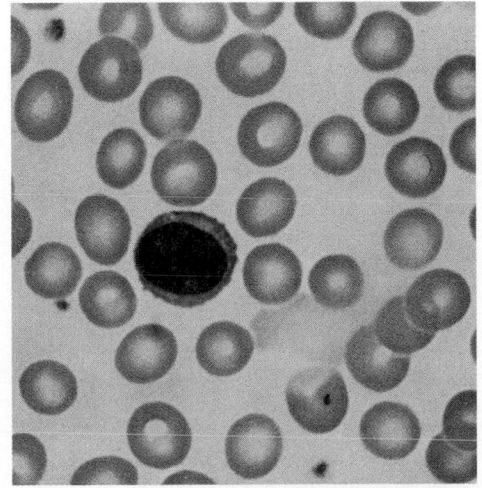

23-8 A typical lymphocyte
The lymphocyte is surrounded by red cells and a few platelets. (×1000)

[3] There is a third group of lymphocytes that lacks the characteristics of either B or T cells. These are called *null cells.* The evidence favoring their existence as a definite category is still fairly shaky.

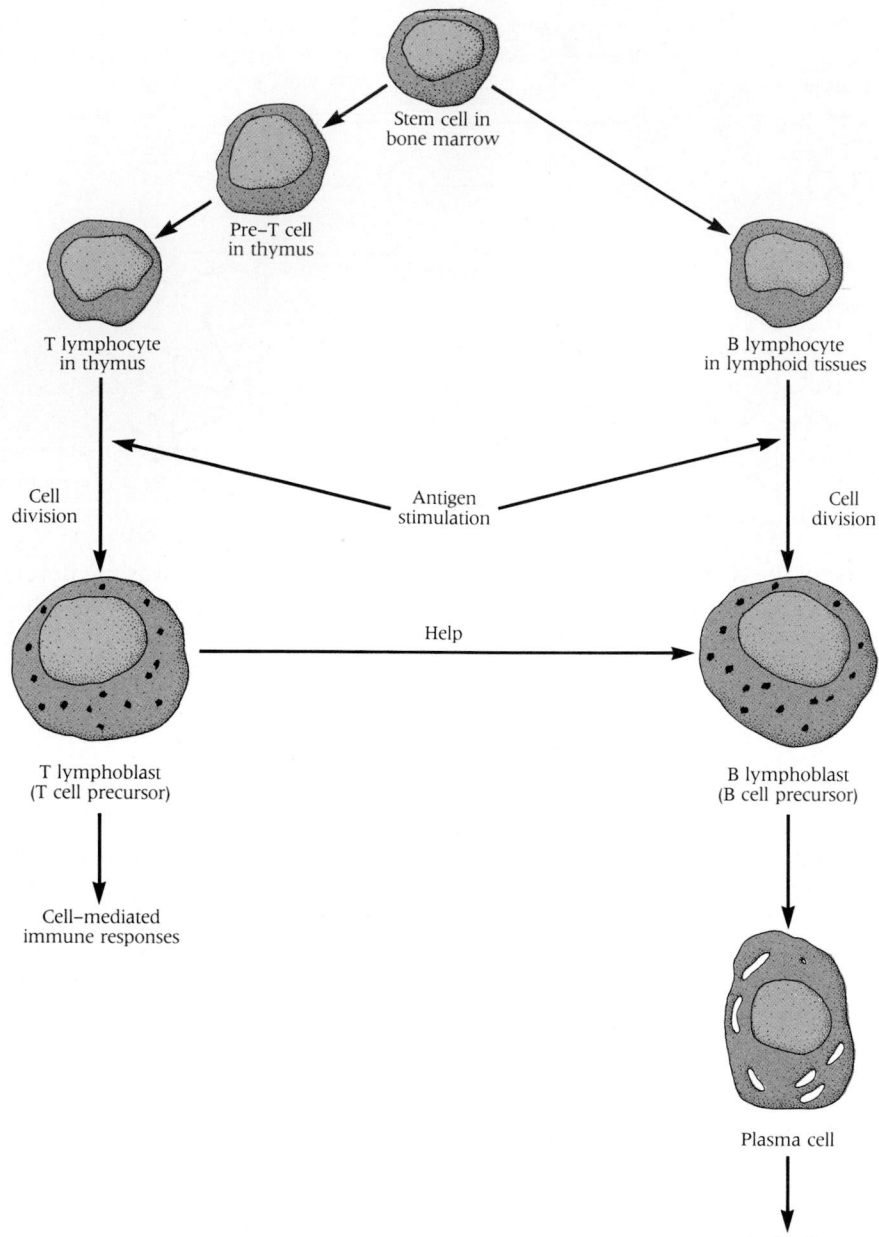

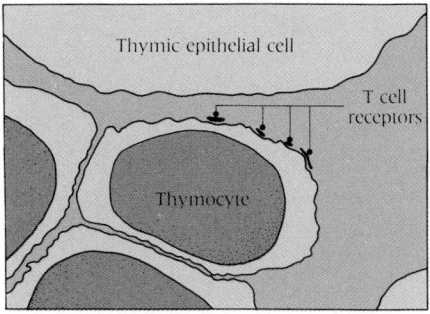

23-9 Development of T cells and B cells
Pre-T cells arise from stem cells in the bone marrow, migrate to the thymus, and differentiate into mature T cells. Following exposure to antigen, T cells begin to divide, producing a population of cells involved in cell-mediated immune responses. B cells also arise from stem cells in the marrow and begin cell division following antigenic stimulation. Many B cells differentiate into antibody-producing plasma cells. A subset of T cells "helps" in this process.

23-10 The "education" of a T cell
The thymocyte, an immature T cell, is surrounded in the thymus by epithelial cells, which confer on the T cell the ability to distinguish self from nonself. The T cell's surface receptors bind to "self"-defining MHC proteins on the epithelial cells' surface.

Some mature T cells take up residence in the lymph nodes. Others leave the thymus gland and continuously circulate from blood to lymph nodes and back to blood. **Lymphoid tissues** (lymph nodes and spleen) are tissues where blood lymphocytes of all kinds are formed, where they function, and through which they move. Lymphoid tissues are collectively termed the **lymphatic system** (see Chapter 24).

T lymphocytes (Figure 23-11) can be recognized in the laboratory by their ability to form "rosettes" with sheep red cells (Figure 23-12) and by the presence on their surfaces of certain protein "markers." These are specific proteins in and on the cell surface. Though other cells are of great importance, much of the important cellular business of the immune system is conducted by T cells.

There are three major T cell categories.

- **Helper T cells** regulate—or help—other cells. For example, they help B cells to differentiate into antibody-secreting cells, cytolytic T cells to become functional, and macrophages to become highly activated.
- **Suppressor T cells** act to limit or terminate the immune response. The intensity of their influence reflects the balance existing between helper and suppressor functions. Extreme shifts in this balance may result in abnormalities of immune function. One example of such an abnormality is the devastating disease called **acquired immune deficiency syndrome,** or **AIDS** in which some T cell types fail to function. (Box 23-3).

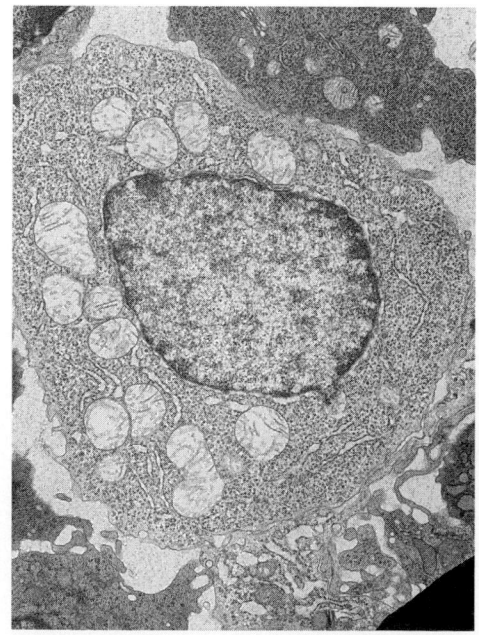

23-11 Electron micrograph of an immature T cell
This lymphoblast from a rat lymph node reveals fine structure and organelles (mitochondria, endoplasmic reticulum, ribosomes, etc.) like those in many cell types. However, when a cell of this type is tested under the conditions displayed in Figure 23-12, it is shown to be a T cell. (×3500)

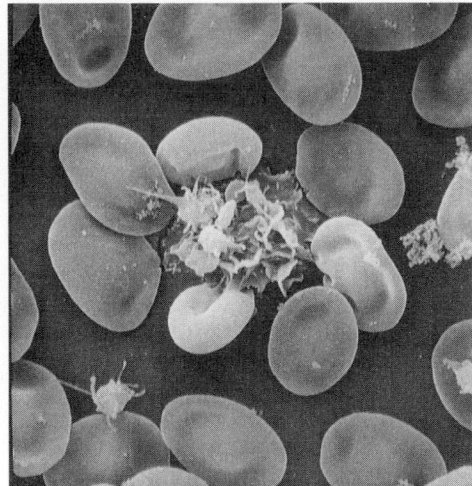

23-12 Rosette formation of T cells and sheep red cells

Sheep red cells attach to a human T cell, forming a rosette configuration. (×2000)

- **Cytolytic T cells** recognize specific antigens on the surfaces of abnormal cells (e.g., tumor cells, cells infected with viruses), attach to them, and directly destroy them, without the intervention of antibodies. Hence, they have been called **killer T cells.**

The immune response is heavily dependent on T cells, which have the following functions:

- They regulate (modulate) the activities of B cells (and other T cells). This is the responsibility of helper T cells and suppressor T cells.
- They are involved in cell-mediated immunity (delayed hypersensitivity, immunity to certain bacterial infections, etc.). T cells do these chores by elaborating a remarkable group of soluble, hormone-like proteins called **lymphokines,** which influence the behavior of lymphocytes, macrophages, and other cells of the inflammatory reaction. Lymphokins are remarkable because they are potent, numerous, and by dint of genetic engineering, available for experimentation and in some cases for medical uses.
- They are killers of certain virus-infected cells and tumors. This is the function of cytotoxic T cells. This phenomenon is now being studied by medical scientists actively pursuing a novel (and promising) cancer treatment.
- They are major participants in the immunology of tissue transplantation.

B lymphocytes

The precursors of **B cells,** like pre-T cells, arise in the bone marrow (see Figure 23-9). Some of these cells then migrate to lymphoid tissues; others remain in the bone marrow. B cells arise from their immature precursors in these tissues.

B lymphocytes were named for the **bursa of Fabricius,** their major source in birds (Figure 23-13). No single organ has been identified in mammals that corresponds to the bursa in birds. B cells (Figure 23-14) are identified in the laboratory by the presence on their surfaces of membrane-bound immunoglobulin molecules (abbreviated SIg) (Figure 23-15). B cells, like T cells, have other distinctive surface markers. The major role of B cells is in antibody production (Figure 23-16).

When mature, B cells appear in lymphoid tissues in clusters called **lymphoid follicles.** With the help of the antibody molecules bound to their surface membranes, which act as receptors, they bind antigens (see Figure 23-7). In the language of immunology, this is called **antigen trapping.** Antigens are also trapped in lymphoid tissues by macrophages and T cells. Which trapping mechanism predominates depends on the nature or state of the antigen. After interacting with antigen (and with helper T cells), B cells differentiate into active antibody-forming cells called **plasma cells** (Figure 23-17). Plasma cells secrete antibody with the same antigen specificity as that of the surface receptor of the B cell from which it is derived.

In sum, B cells have two major functions.

- They produce antibodies.
- With the help of their surface immunoglobulins, they trap antigens.

23-13 Location of the bursa of Fabricius in a chicken

Birds alone possess the bursa, a saclike structure that gives rise to the B cells which make antibodies.

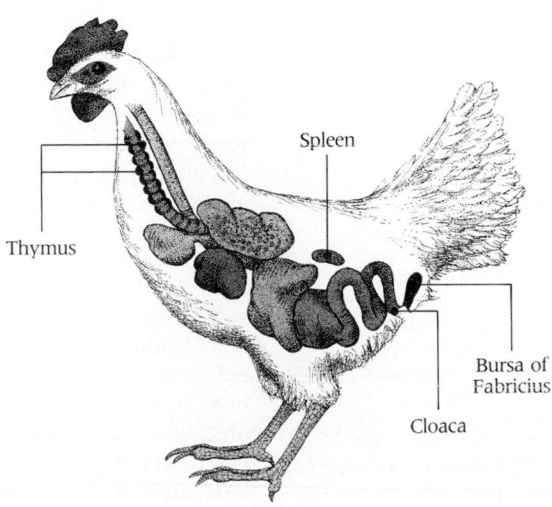

Spleen

Thymus

Bursa of Fabricius

Cloaca

Until recently, an antibody was recognizable as such only through the visible consequences of an antibody-antigen reaction. In the absence of such a reaction, we cannot infer that an antibody is present. It might be supposed that antibodies could be easily demonstrated by their power to destroy or inactivate infectious agents. That is true of many antibodies. For example, if a rabbit is inoculated several times with killed dysentery bacilli, its serum in a few weeks causes

suspensions of the bacilli to agglutinate, or clump, visibly.

Antibodies that agglutinate cells (e.g., bacteria, red cells) are called **agglutinins.** The antibodies called **lysins** cause the cells against which they act to dissolve. The antibodies called **precipitins** act against soluble antigens. When mixed with their antigen in appropriate concentrations, the antigen-antibody complex forms a visible precipitate (Figure 23A).

The important points are these.

- Antibody combines with antigen.
- The mode of combination depends upon the nature of the antigen and antibody.
- Some antigen-antibody reactions produce visible results which can be exploited as a means of antibody detection.
- Specificity remains a cardinal attribute of antibody-antigen reactions.

Among the categories of methods used to demonstrate the presence of an antibody are classic precipitation techniques, diffusion techniques, fluorescent antibody techniques, and complement fixation techniques.

CLASSIC PRECIPITATION TECHNIQUES

Because of its simplicity, the **precipitin reaction** has been extensively studied. Consider its features.

Figure 23A illustrates the precipitate formed when a rabbit antibody to purified bovine serum albumin (BSA) is mixed with BSA. Note that the mere combination of antibody with antigen does not ensure formation of a precipitate. An insoluble precipitate forms only when the two reactants are mixed in the proper proportions. The bulk of the precipitate increases at higher and higher ratios of antigen to antibody, but when excess antigen is present, much less precipitate forms.

Determination of the antigen to antibody ratio yielding maximum precipitates provided important clues about how antigens and antibodies interact. Unlike chemical reactants, which combine in definite proportions, the ratio of antibody to antigen in a precipitate varies with the amount of the two reactants present. The reasons: antibody is **bivalent** and antigen **multivalent**—that

is, each antibody has two combining sites and each antigen has many antigen determinants.

When antigen and antibody are present in correct proportions, antibodies attach to more than one antigen at a time and thus build three-dimensional lattices which become too large to remain in solution (Figure 23B). When antigen is present in excess—

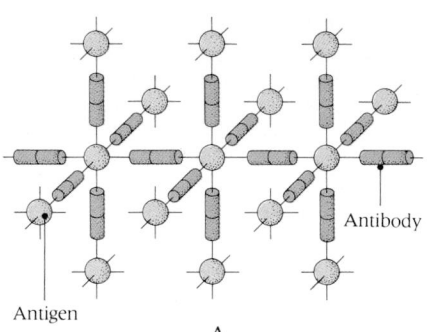

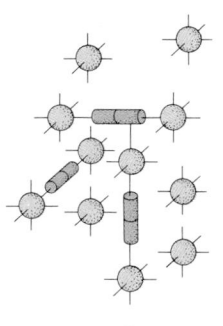

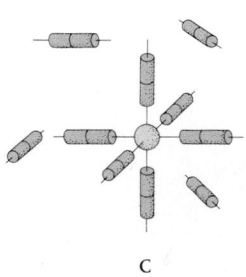

23B Lattice model of antigen-antibody precipitate formation
(A) Antibodies are shown as rods with two binding sites; antigens are spheres with six binding sites. When they are present in a ratio of one antigen to three antibodies, an insoluble lattice forms, causing a visible precipitate. *(B)* When antigen is in excess (prozone), only small aggregates form and these are mostly soluble. *(C)* The same is true when antibody is in excess (postzone).

23A The precipitin type of antigen-antibody reaction
(A) The tube at the left contains a rabbit antibody to bovine serum albumin (BSA). Addition of BSA produces a cloudy precipitate of antigen-antibody complex, which settles out in a few minutes. *(B)* The amount of antigen-antibody complex formed is influenced by the proportions of antigen and antibody. Each tube in the row contains the same amount of antibody to BSA, but the amount of added BSA increases from left to right. When BSA is present in great excess (in the so-called postzone), no precipitate forms.

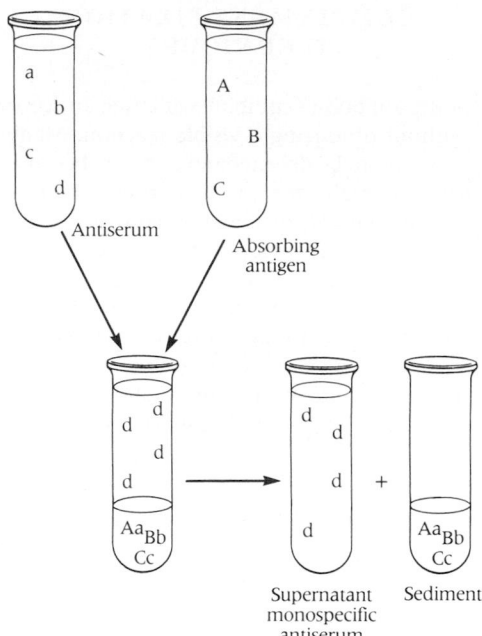

23C Absorption technique

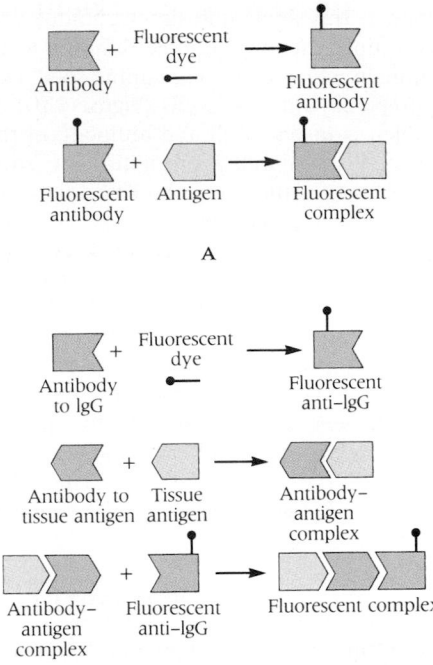

A

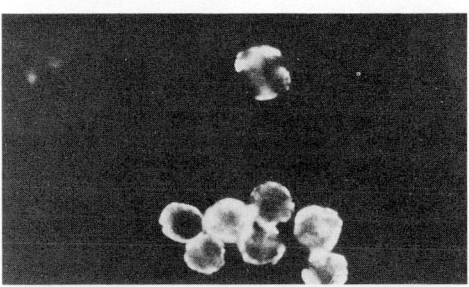

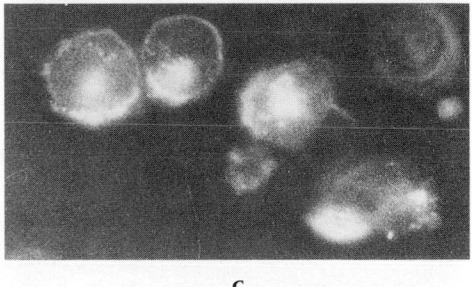

B

may be illustrated by an example. Suppose we wished to determine which of four different antigens—*A, B, C,* and *D*—is present in a solution. First we would immunize individual rabbits with pure samples of each antigen and obtain antisera containing specific antibodies against *A, B, C,* and *D*—respectively, *a, b, c,* and *d.* If a rabbit antiserum containing antibody *d* precipitated the unknown antigen, we would conclude it is *D.*

In practice, an antiserum containing antibody *d,* for example, might react weakly with *A, B,* or *C.* Either the rabbit immunized with antigen *D* formed small amounts of antibodies *a, b,* and *c* or it could reflect the presence of multiple antigenic determinants on the single antigen (*D*). If *A, B, C,* and *D* had one or more antigenic determinants in common, *a, b, c,* and *d* would cross-react to an extent. To obtain an antibody of absolute specificity—to prepare antibody *d* so that it is free of *a, b,* and *c*—one employs the **absorption technique** (Figure 23C).

in the prozone (see Figure 23A)—small aggregates form because there are not enough extra antibodies to crosslink together the small soluble complexes (composed of one antibody and two antigens) to form the larger insoluble complexes. These "nonlatticed" antibodies remain in solution. When antibody is present in excess—in the postzone—another type of water-soluble aggregate develops and lattice formation is again blocked.

Classic precipitation techniques are useful in two areas: (1) in the quantitative assay of antibody; and (2) in the qualitative identification of antigen. The strength, or **titer,** of antibody in a solution has traditionally been assayed by determining the degree to which the antibody solution can be diluted before it just fails to cause visible precipitation. Dilution methods permit relative, not absolute or quantitative measurements of antibody content.

The usefulness of precipitin reactions in the identification of an unknown antigen

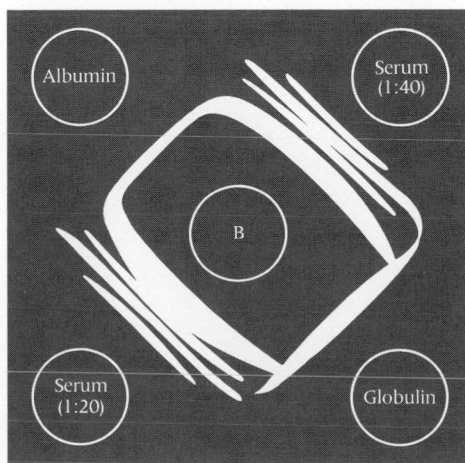

23D Double-diffusion agar technique

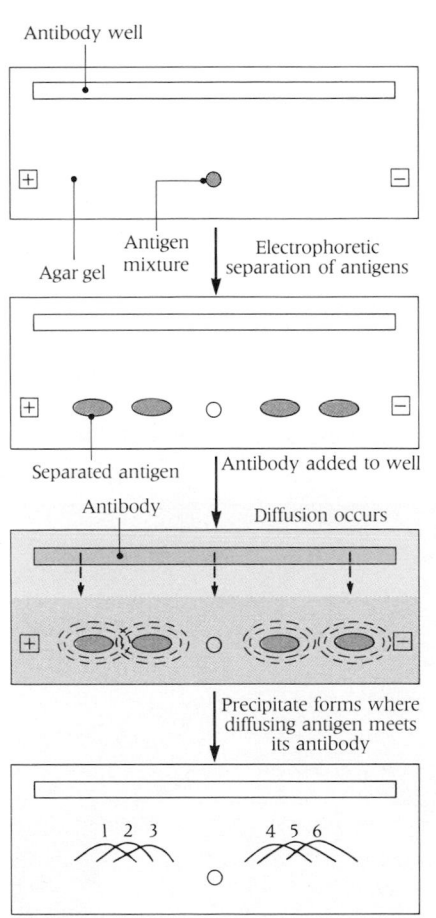

23E The principle of immunoelectrophoresis
A mixture of antigens (six in this example) are placed in a small well and separated by electrophoresis in a slab of agar gel. When the current is turned off, the antigens have clustered into four groups on the basis of electric charge. Antibodies specific for the antigens are then added to a long trough in the gel and they and the antigens begin to diffuse toward each other through the gel. A heavy white precipitation arc forms wherever an antigen meets its specific antibody.

C

23F Fluorescent antibody techniques
(*A*) *In direct methods, a fluorescent antibody, prepared by combining an antibody with a fluorescent dye, reacts directly with its antigen.* (*B*) *In indirect methods, a fluorescent antibody is prepared that reacts only with other IgG molecules. When allowed to react with a tissue element containing IgG antibody (as do plasma cells and B lymphocytes) or coated with an IgG antibody (as occurs in diseases of autoimmunity), a fluorescent complex forms from the fluorescent probe—an "antiantibody"—and the antibody in or on the tissue.* (*C*) *When these probes are applied to tissues or cells, and the slide is viewed under a fluorescent microscope, locations of antigens are clearly seen as light-emitting zones of fluorescence.*

DIFFUSION TECHNIQUES

When a solution of antigen is layered over a semisolid gel (agar) containing antibody, antigen diffuses into the gel and a sharp band of precipitate forms at the point where they interact. This single-diffusion technique was soon superseded by a double-diffusion technique, in which a dish containing solid agar is prepared with two wells (Figure 23D). If antigen is in one well and antibody in the other, a sharp milky precipitin line soon forms between the two wells. This is the antigen-antibody precipitate.

In the technique of **immunoelectrophoresis.** Antigens are first separated by electrophoresis in a gel (Figure 23E). Antiserum containing antibodies to one or more of the antigens is then placed in a well running the length of the agar gel. Antibodies and antigens diffusing through the gel encounter each other and each antigen reacts with its specific antibody to form a visible arc of precipitation.

FLUORESCENT ANTIBODY TECHNIQUES

Antigens in tissues are detected with antibodies combined with a fluorescent dye such as fluorescein isocyanate. Such **fluorescent antibodies** can be employed as a tissue stain (Figure 23F). Specific antigens in the tissue react with and locally bind fluorescent antibodies. When illuminated by light of certain wavelengths they display a bright fluorescence.

COMPLEMENT FIXATION TECHNIQUES

Some antibodies combine with their antigens without producing a visible reaction. Many of these can be detected nevertheless because serum contains the group of proteins known as complement, or the complement system, which combines avidly with many antibody-antigen complexes. Thus, the occurrence of an antibody-antigen reaction may be demonstrated, even though unaccompanied by visible consequences, if it can be shown that complement fixation—elimination of complement by binding or inactivation—has occurred.

THEORIES OF IMMUNITY

The immune system has two major properties upon which the animal body relies in dealing with infectious microorganisms.

- The system can recognize, and react to, a wide range of antigens.
- The system is highly specific and selective in coping with the diversity of microbes and parasites.

If the immune system is to function effectively, however, it must repel only foreign materials. If it reacted with parts of the host's body it would be a self-destructive body system. In a word, it must recognize self and be able to distinguish it from nonself. How does the immune system do it? This is one of the most interesting scientific questions of our time.

Two Theories

Attempts to explain the great diversity in antibody specificities were of two kinds, the instructive theory and the selective theory. The latter proved to be correct.

The **instructive theory** holds that an antigen impresses its own new and foreign

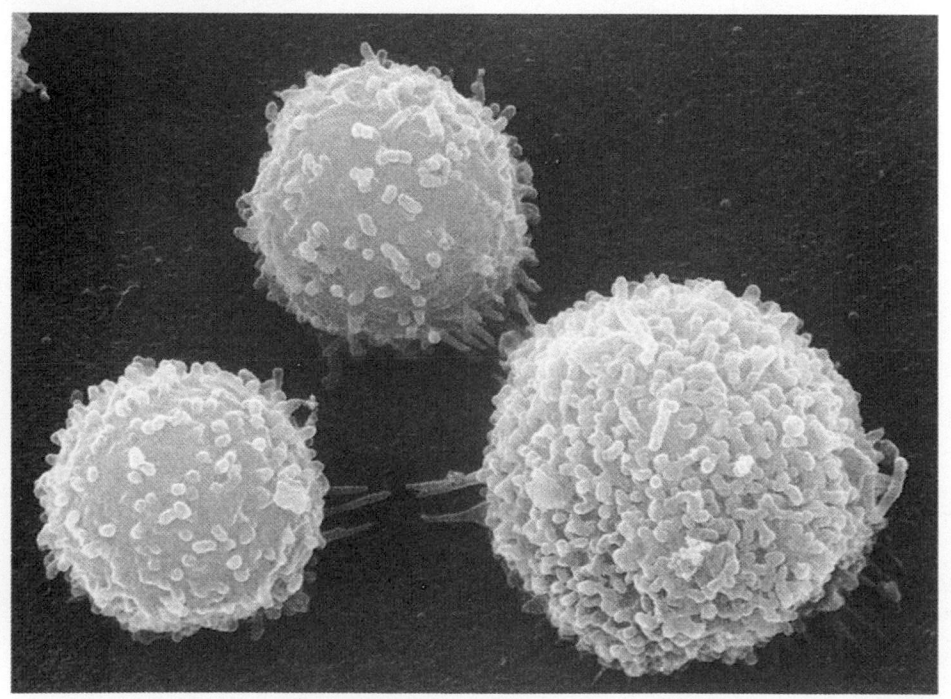

23-14 Scanning electron micrograph of lymphocytes
Large cell on the right is a B lymphocyte. The other two cells are T lymphocytes. (×4000)

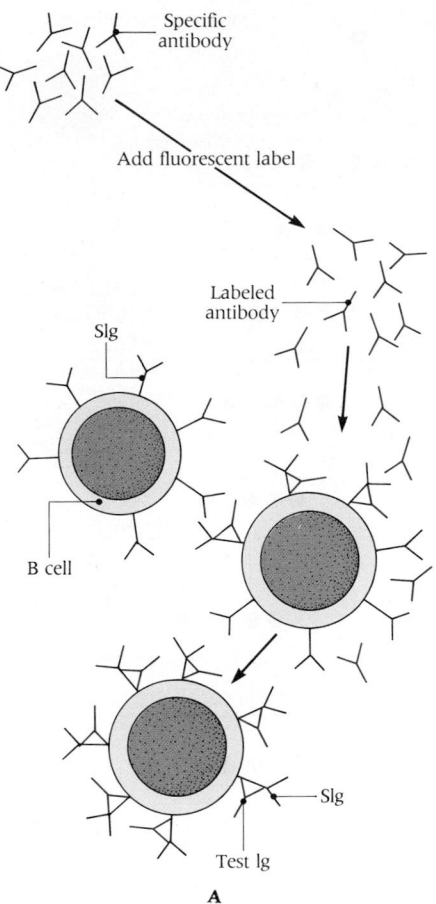

23-15 Surface receptors on B cells
(A) *To detect surface immunoglobulins (SIg) on B cells, one attaches a fluorescent label to antibodies against these proteins.* **(B)** *When the test antibody binds to the surface marker, a rim of fluorescence appears around the periphery of the B cell.*

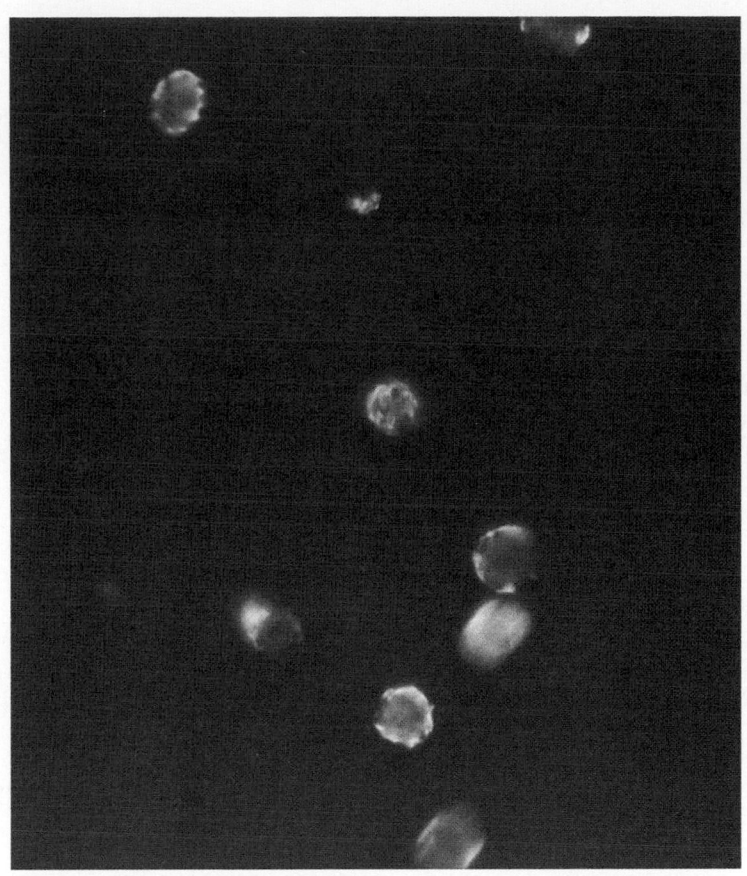

pattern upon the body's antibody-synthesizing apparatus (Figure 23-18A). The antibody combining site is somehow shaped as it is synthesized by the direct influence of a nearby antigenic determinant. The latter is thus visualized as an "instructor" that tells the protein-synthesizing machinery what to do and how to do it. Instructive theories were widely accepted for many years—despite their failure to explain how a plasma cell "learns" to synthesize antibody molecules bearing a specific combining site complementary to the antigenic determinant of an entirely foreign substance.

The **selective theory** holds that an antigen stimulates the synthesis of an anti-

23-16 Life history of T cells and B cells
In the development of T cells and B cells there is a progression from an antigen-independent state to an antigen-dependent state. Macrophages also play a critical role in the functions of both lymphocyte types.

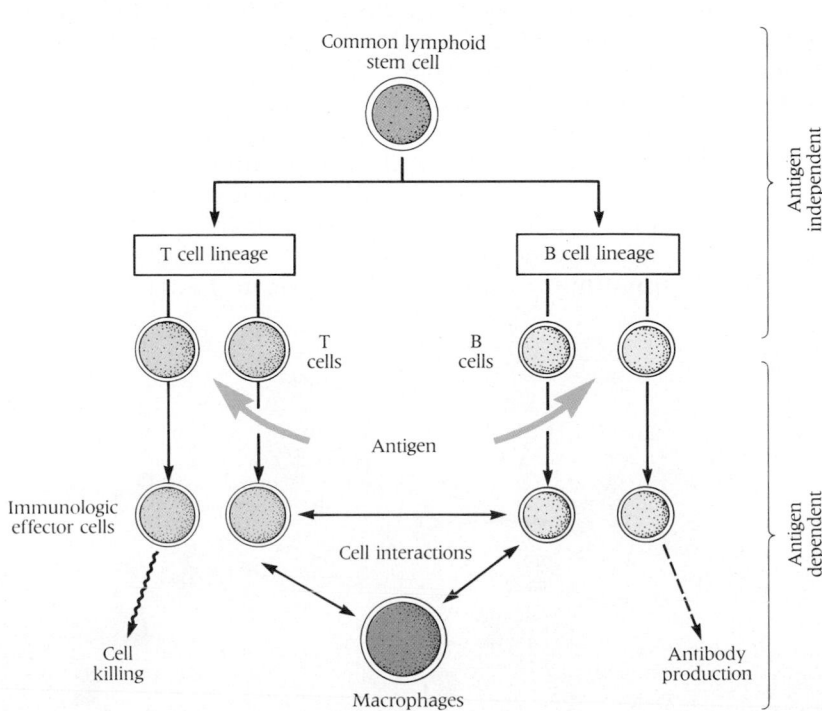

body, the specificity of which preexists in the body's genetic information (Figure 23-18B). In this view, antibody molecules are synthesized like all other proteins—according to genetic instructions contained in the nuclear DNA—uninfluenced by information from the outside. For each of the thousands of possible antigens, the body already contains a small group of cells that already "knew" how to make a specific antibody before the body ever encountered the complementary antigen. The antigen is viewed as a "selector" of these cells and a stimulator of their proliferation and subsequent antibody production.

Selective theories imply that a mature organism possesses an elaborate "dictionary," which it consults in deciding whether or not a "word" (i.e., the chemical configuration of a substance) is foreign. This conception, first stated by Niels K. Jerne and later by Joshua Lederberg, envisions a "dictionary" containing only foreign words (nonself), all words of its own language (self) having been purged. Moreover, it lists foreign words without ever having seen them or heard them!

Understanding in this area was enhanced by Sir McFarlane Burnet (1900–1985), who envisaged selection at the cell level and proposed a **clonal selection theory** of immunity. The basic tenets of this brilliant conception are as follows:

- Each animal carries a large population of lymphocytes capable of binding to a vast array of antigens.
- The immunologic specificity resides in the structure of the receptor molecules on the B cell surface (Figure 23-19). Each newly generated B cell (and its progeny) expresses a novel species of immunoglobulin on its surface with a unique set of binding specificities.
- The repertoire of B cell specificities is present prior to antigen exposure.
- B cells differentiate into plasma cells that synthesize and secrete antibody molecules with the same specificity and affinity as the surface receptors of the original B cell.
- Memory cells are also derived from original B cells and have the same specificity.
- A comparable mechanism is involved in the stimulation of T cells.

Mechanisms

Both B cells and T cells pick out and recognize specific foreign antigens, each using a protein on their surface called an **antigen receptor.**

- The **B cell receptor** is an immunoglobulin. All are monospecific (i.e., have a single specificity) and each B cell is stimulated to divide and differentiate only by a unique antigenic determinant (Figure 23-20). B cells acquire antigen receptors after migrating to the lymphoid organs (spleen, lymph nodes), where they make contact with circulating antigen. The receptors are specific for only one or a few related antigenic determinants.
- The **T cell receptor** was an elusive entity for years, in part because T cells react only with antigen on a cellular target, not with free antigen. (A cytotoxic T cell, for example, kills virus-infected cells, not free viruses.) Not an immunoglobulin, the receptor according to recent work is a single protein—containing two polypetides (α and β)—that recognizes both an antigen (or a peptide derived from it) and a second protein encoded by a gene cluster called the **major histocompatibility complex (MHC)** (Figure 23-21). These proteins,

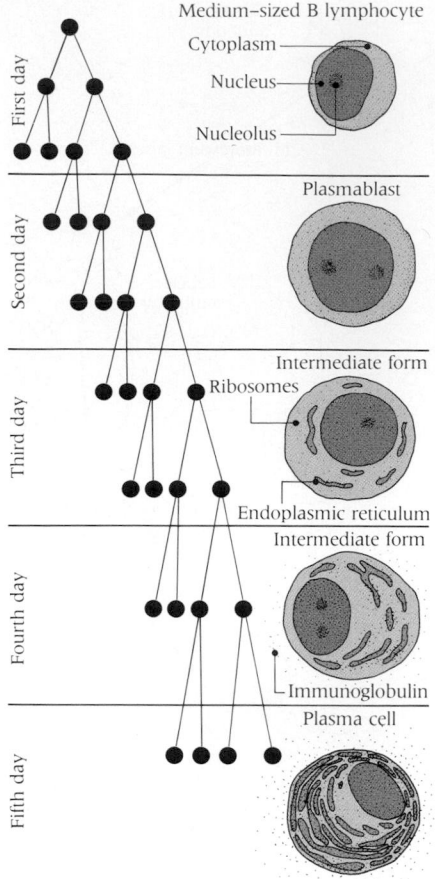

23-17 Clonal development of plasma cells
B lymphocytes begin to divide after contact with antigen, producing a population of plasmablasts. As division proceeds through the fifth day, the amount of endoplasmic reticulum increases in the cytoplasm, and newly synthesized immunoglobulin (antibody) is produced in large amounts.

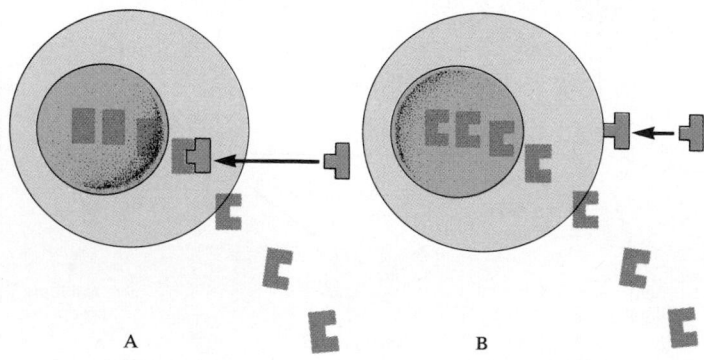

23-18 Two theories of how antibody diversity arises
(A) In the instructive theory, an antigen enters the cytoplasm of the plasma cell and molds the antibody so that its shape conforms to that of the antigen. (B) In the selective theory, the antigen contacts the surface of a selected cell and thereby stimulates production of the appropriate antibody.

23-19 The clonal selection theory
Different determinants of an antigen stimulate different lymphocyte clones to divide and differentiate into plasma cells. Each clone bears antigen receptors for one of the three determinants. An antigen with a fourth type of determinant does not induce proliferation of lymphocytes.

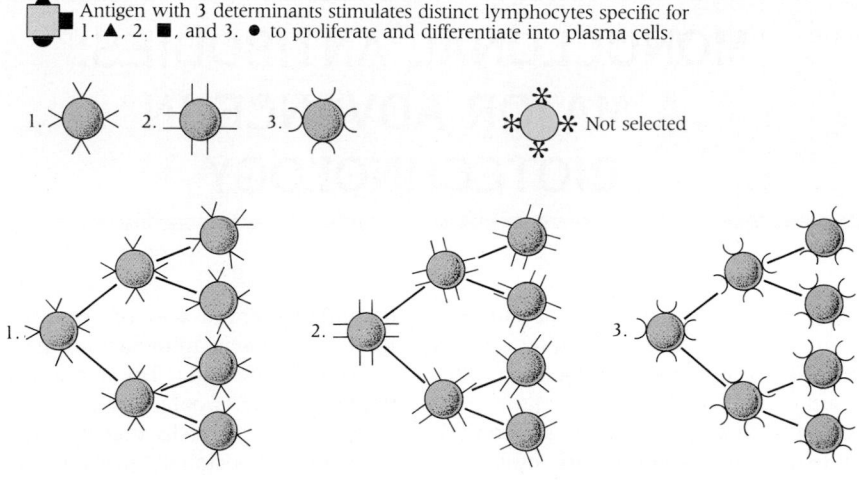

Antigen with 3 determinants stimulates distinct lymphocytes specific for 1. ▲, 2. ■, and 3. ● to proliferate and differentiate into plasma cells.

* Not selected

to be discussed later, are found on the surfaces of every body cell. They are markers indicating that that cell belongs to this body—and they are critically involved in transplantation immunity. The T cell receptor binds its antigen only when it can also bind a "self" MHC-encoded protein.

Thus, as predicted by Burnet, the immune system is triggered by the binding of antigens to those lymphocytes possessing unique complementary receptors of high

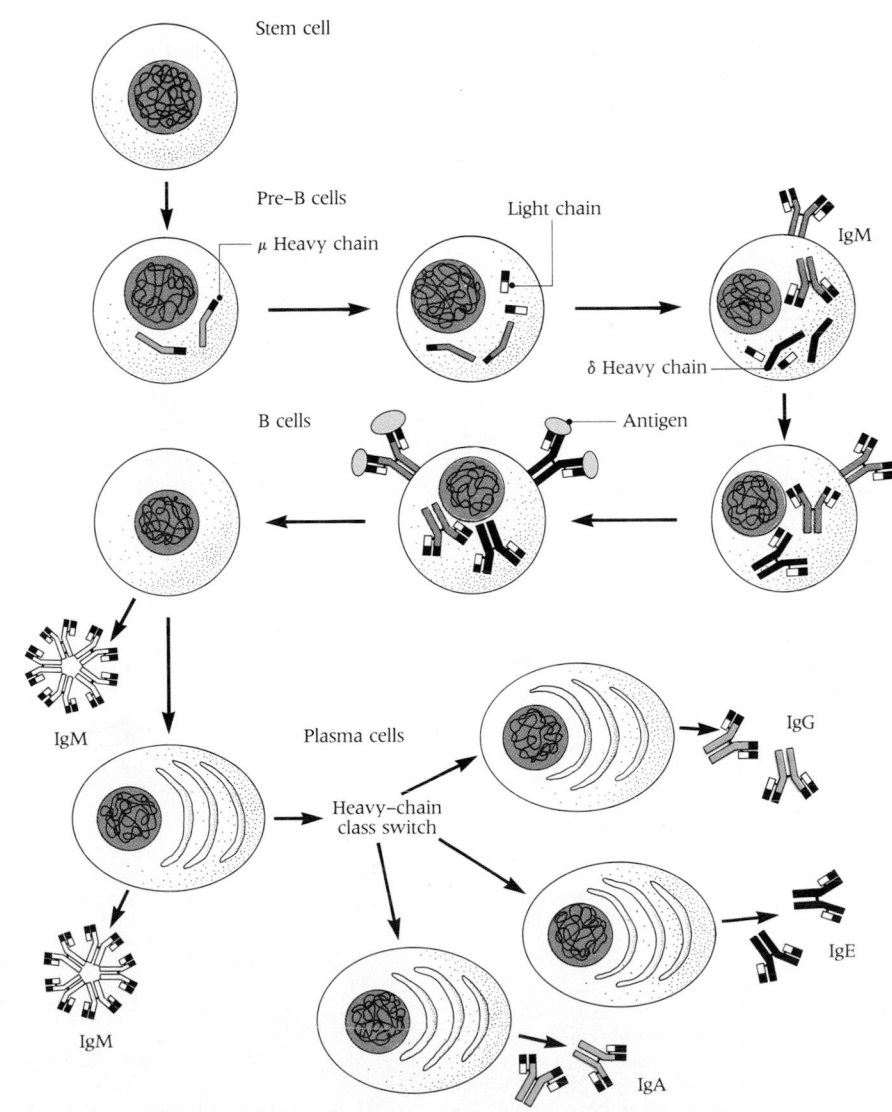

23-20 Differentiation of a B cell
Hematopoietic stem cells give rise to pre-B cells. The first step in B-cell differentiation is the synthesis of μ heavy chains and light chains. These form cytoplasmic and then membrane-bound IgM molecules. IgD molecules are then synthesized and the cell becomes a mature B cell. Following encounter with an antigen, IgM and IgD molecules in the membrane disappear. As discussed later, these cells mature into IgM-secreting plasma cells or alternatively undergo a heavy-chain class switch and express IgG, IgA, or IgE molecules on their cell surfaces (see Figure 23-24). Such cells can then differentiate into IgG-, IgA-, or IgE-secreting plasma cells.

BOX 23-2

MONOCLONAL ANTIBODIES: A MAJOR ADVANCE IN BIOTECHNOLOGY

Suppose that you could raise a potent antibody that was (1) specific for any antigen of your own choosing, (2) available in unlimited supply, and (3) completely pure (that is, uncontaminated by other antibodies). What would you do with it? Such antibodies, termed **monoclonal antibodies,** must be considered useful by someone because a multi-billion dollar a year industry to make and market them has sprung up since 1975. That was the year Georges Köhler, a new Ph.D., and César Milstein, his research mentor, discovered the **hybridoma technique** at Cambridge University. Their application of pure science to the solution of practical problems was a momentous advance in the history of biotechnology. It was recognized with a Nobel Prize in 1984.

Although the immune system produces specific antibodies against antigens, the antibodies are often of many kinds. Indeed, the immune system of a mouse can produce a thousand different antibodies against a single antigen (Figure 23G). The hybridoma technique is a way of obtaining a tissue culture clone of identical cells that manufacture a single or "monoclonal" antibody of desired specificity. Such a technique could not have emerged until earlier workers had shown how to raise antibodies, how to keep cells going in culture, and how to make two cells fuse into one (see Chapter 18).

The original aim of Köhler and Milstein was to study the possible importance of mutations in the genes that specify immunoglobulins. Obviously it would simplify the work if a clone of cells were cultured that produced only a single antibody against a known substance. But no such cells existed. The nearest thing to them were **myelomas,** malignant tumors of plasma cells that arise spontaneously and are somehow locked into producing a single antibody type in enormous quantities.

A simple idea began to take shape: Since normal B cells do not grow well in culture, why not fuse a normal antibody-producing B lymphocyte with a myeloma cell and thus create a hybrid cell combining the desired properties of its two parents? These are (1) the ability to produce a single antibody to a selected antigen like its lymphocyte parent (which alone is short-lived in culture), and (2) the ability to grow permanently and rapidly like its malignant myeloma cell parent.

In the spleen, B lymphocytes proliferate rapidly after antigen injection. Each lymphocyte produces a single immunoglobulin antibody that binds a particular antigenic determinant. The idea was to expose a mouse to a chosen antigen and then remove its spleen and fuse the lymphocytes with myeloma cells. With luck, a hybrid cell might form that produced a single antibody to the injected antigen and survive in culture where it could amply produce quantities of that antibody (Figure 23H). It was easy to think of reasons why this approach would fail. One was that lymphocytes are known to be bad fusers. Another was the extreme unlikelihood of producing a hybrid that made only the desired antibody. It seemed that it might be necessary to make and screen about a thousand hybrids before one might have even a chance of success. Köhler decided to try it anyway.

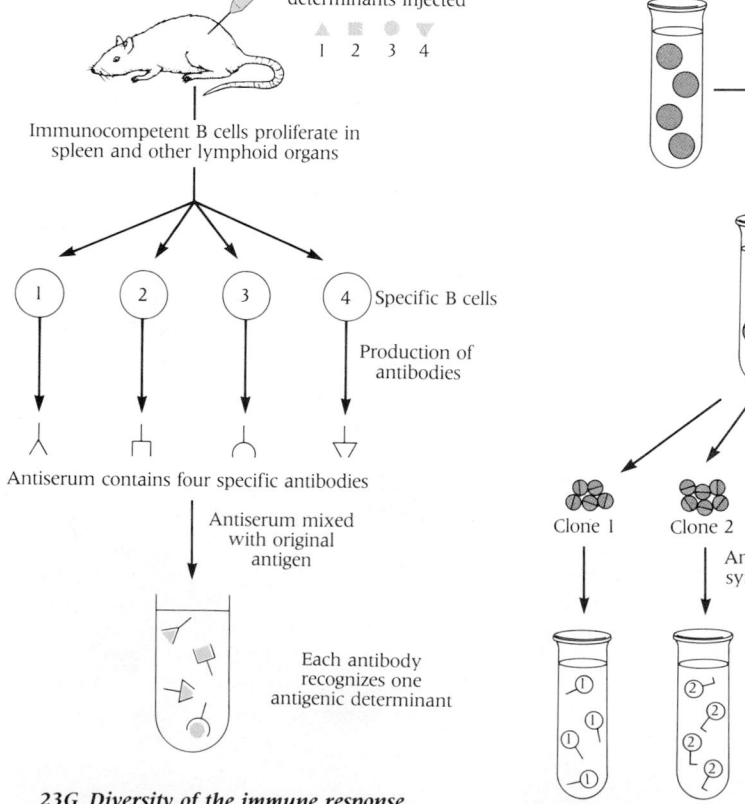

23G Diversity of the immune response
When an antigen bearing multiple antigenic determinants (four in this example) enters the animal body, multiple antibodies are elicited—each with a specificity for one of the antigenic determinants.

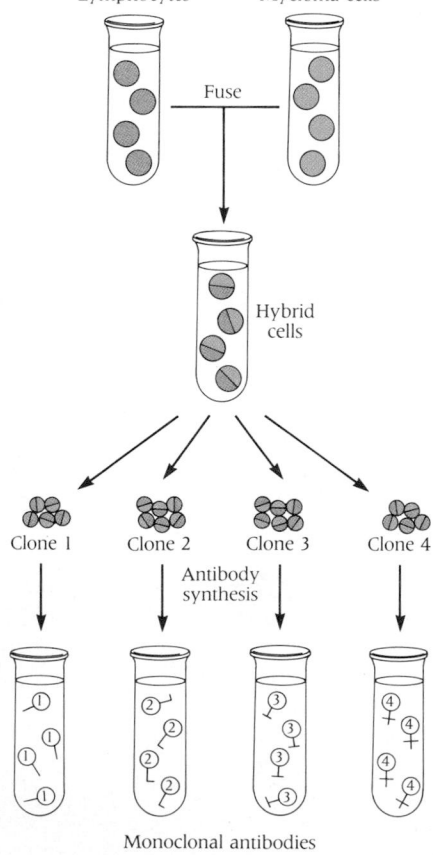

Monoclonal antibodies

23H Hybridoma technique
Details are given in the text.

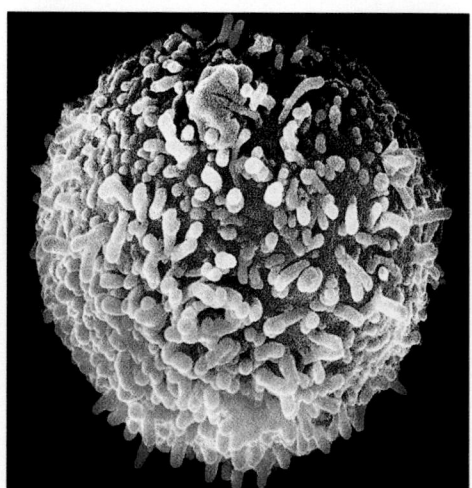

23I Hybridoma cell
Scanning electron micrograph of a hybridoma cell producing a monoclonal antibody to a cytoskeleton protein. (×3000)

In the first experiment, he injected a mouse with sheep red cells. He then removed the spleen, isolated its lymphocytes, and mixed them with an established tissue culture line of myeloma cells. A few days later, he could tell just by looking at the bottles that hybrids had formed. The critical test, however, was to see if any of these hybrids produced antibody against sheep red cells.

In a personal account of this episode, Köhler wrote: "I looked at the hybrids growing in the bottles and felt happy with myself for growing them. I was reluctant to test them for their specificity because I thought I probably didn't have specificity yet. So I waited 7 weeks before testing." By then it was around Christmas of 1974. He decided to do plaque assays, which take several hours. In this test, hybrids making antibodies

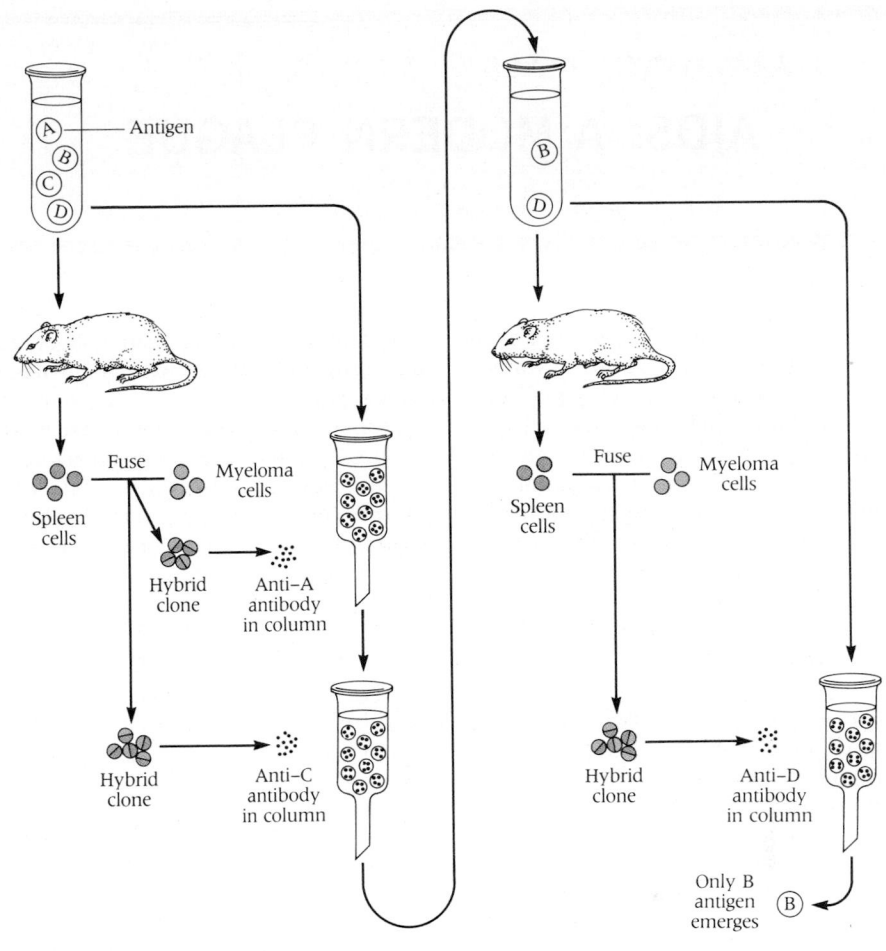

23K Purification of proteins using monoclonal antibodies

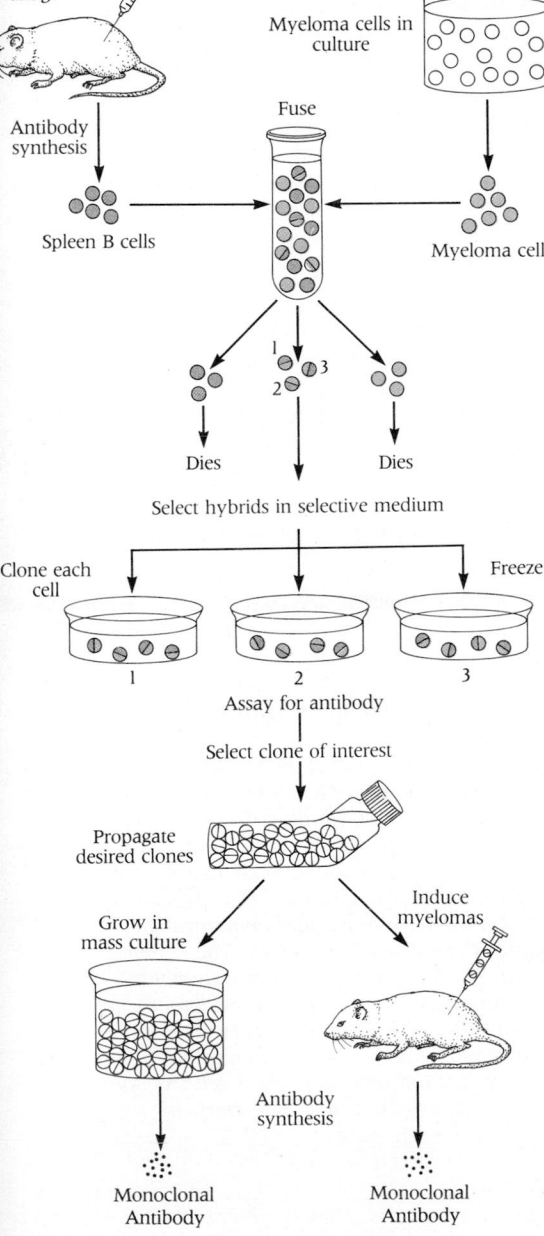

23J Details of hybridoma procedure

against sheep red cells form halos in the surrounding solid medium (Figure 23I). He started the assays at 5 P.M., went home, and returned later in the evening with his wife "because it would be so boring to score a negative result. We went down in the basement of the institute, which has no windows. I looked at the first two plates. I saw these halos. It was fantastic . . . the best result I could think of."

The myeloma-lymphocyte hybrids came to be known as **hybridomas.** The procedure routinely used today to produce monoclonal antibodies is sketched in Figure 23J. Note that the procedure employs a special selective medium and mutant myeloma cells that can grow in it only when hybridized. Since the parent lymphocytes from a mouse immunized with a chosen antigen produced a range of antibodies, the resulting hybridomas must be sorted according to the particular antibody they produce. This is easily done. When a hybridoma producing the desired antibody is finally isolated, it can be cloned into sources of a single antibody—either by growing them in mass cultures or by injecting them into animals where they induce myelomas that secrete the desired antibody.

Today monoclonal antibodies are finding countless uses (since each antibody will only recognize a particular molecule).

- Labeling procedures are employed, ranging from studies of cell membrane structure to the search for tumor viruses.

- Purification methods can accurately dissect unknown mixtures (Figure 23K).

- Novel therapeutic methods have been designed to attack infectious agents and cancer cells. For example, by developing an antibody to offending cancer cells and attaching to it a lethal drug, one can deliver drug exclusively to its target.

It was assumed that human-human hybridomas could be produced as easily as the mouse-mouse variety and that specific antibody probes would soon become available for the study of human tissues and diseases. However, it was not until 1980 that Lennart Olson and the late Henry Kaplan of Stanford University produced human-human hybridomas, which still did not grow well.

In the late 1970s, a disease of unknown cause began to attract attention. The disease is called acquired immune deficiency syndrome or AIDS. Defective immunity increases susceptibility to infection by microorganisms of all kinds. The most common is the protist *Pneumocystis carinii*, which frequently causes severe pneumonia in AIDS victims and almost never does so in individuals with normal immunity. Other common infectious agents are various viruses (such as cytomegalovirus and *Herpes simplex*), fungi (*Candida* and *Cryptococcus*), protists (*Toxoplasma*), and bacteria (tubercle bacillus). Experienced physicians consider these infections typical of those occurring in the immunosuppressed host.

The infections often take a relentlessly aggressive course. A few can be controlled with antibiotic drugs, but eventually they recur—or another infection overwhelms the patient. About a third of AIDS victims develop a hitherto rare form of cancer called Kaposi's sarcoma. Other kinds of cancer are also turning up with surprising frequency. The underlying disorder, impaired immunity, apparently induces vulnerability to infection—and, to cancer. The number of cases of AIDS is rising ominously (Figure 23L). More than 130,000 Americans have been diagnosed with the disease since 1981. More than half of these are dead, and no one with AIDS has yet been cured.

Humoral immunity is unimpaired in AIDS victims. Antibody levels in blood are normal or even elevated. However, the blood lymphocyte count is sharply depressed. It is the T cells, the ones needed for cellular immunity, that are low in number and abnormal in composition. In particular, helper T cells (of the type called T4) are depleted or missing or defective; killer or suppressor T cells are less diminished.

These findings give support to the immune surveillance theory, which holds that the immune system seeks out and destroys cancer cells before they can grow into life-threatening tumors. In immunosuppressed subjects, so goes the theory, cancer cells are encouraged to grow by the absence of normal restraints.

What causes AIDS? In 1983 Robert Gallo of the National Cancer Institute was able to clone DNA copies of the RNA genome of a virus called HTLV-III (for human T cell leukemia virus), which he suspected to be a cause of AIDS (Figure 23M). Gallo was not alone in the field. In May 1983, Louis Montagnier of the Institut Pasteur in Paris announced that AIDS is caused by a virus he termed lymphadenopathy-associated virus or LAV. LAV was soon found to be identical to HTLV-III. In 1986 both groups agreed upon a more noncommittal name, HIV-1 (for human immunodeficiency virus).

Recovery of HIV from the saliva or semen of some apparently healthy homosexual male volunteers (and of people suffering

23L Incidence of AIDS
Cumulative number of cases in the U.S. and other countries.

Chart legend:
- ▲ United States
- ▼ All but United States
- ● Total

Y-axis: Number × 1000, from 0 to 80
X-axis: 1979, 1983, 1987

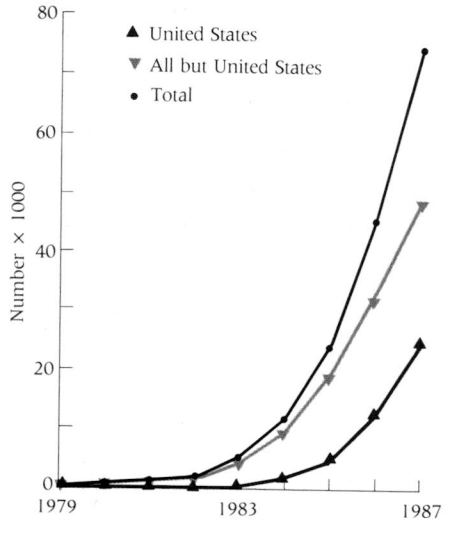

23-21 Recognition of proteins by the T cell receptor
The associated-recognition theory states that each T cell has a single receptor, which recognizes and binds the combined antigen-MHC protein complex. To test the theory, a hybrid T cell was constructed by fusing two normal T cells. One parent T cell was from an animal having F-strain MHC proteins, after immunization against the antigen hemocyanin. The second parent T cell was from a K-strain animal immunized against the antigen, ovalbumin. In accord with the theory, the hybrid cell recognizes hemocyanin only on F-strain cells and ovalbumin only on K-strain cells, since the T cell receptor binds both proteins at the same time.

Figure labels: Receptor for F–strain protein and hemocyanin antigen; Antigen-presenting cell; F-strain MHC-encoded protein; Hemocyanin antigen; T cell; Hybrid T cell; Ovalbumin antigen; K–strain MHC-encoded protein; Receptor for K–strain protein and ovalbumin antigen; F strain; Hemocyanin antigen; K strain; Ovalbumin antigen

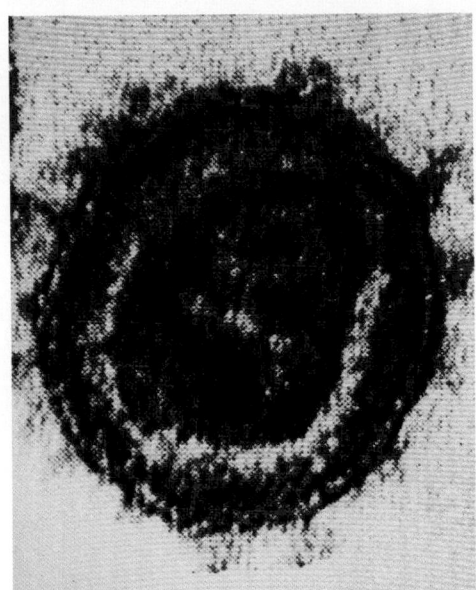

23M *HIV, the retrovirus that causes AIDS*
($\times 75,000$)

from early forms of AIDS) confirmed a suspicion that apparently healthy people can be virus transmitters or carriers. Further confirmatory data showed the presence in many apparently healthy people of an antibody against HIV. This antibody, rather than the virus itself, is what is measured in the much discussed "AIDS screening test" now being done. Because a long time interval is needed for the disease to develop in virus carriers, we cannot yet predict how many antibody-positive people are future victims (see Figure 23L). It is estimated that between 1 and 2 million individuals in the United States are infected with HIV and are at present without symptoms.

Statistics vary in different studies, but all suggest certain conclusions.

■ AIDS constitutes a worldwide public health issue, but distinct patterns have emerged in different regions. In the United States and Europe, the great majority of victims are homosexual (or bisexual) men or intravenous drug abusers. In Africa, most cases of AIDS are acquired heterosexually. Heterosexual transmission is increasing in Latin America and possibly in the U.S.

■ Affected homosexuals in this country (about 75% of all AIDS victims) tended to be very active sexually—and very promiscuous. More than 50% of all homosexual men in some urban areas may carry the AIDS virus.

■ Some 12% of AIDS victims (as opposed to virus-carriers without overt disease) are heterosexual users of drugs like heroin, which are taken intravenously (IV). Some 70% of IV drug users are virus carriers.

■ A third group of AIDS victims—about 6%—are Haitian immigrants to the United States. No one knows how these individuals acquired the disease.

■ Another group appearing in every survey are people with hemophilia, who acquired the virus from transfusions.

■ Other groups of victims include female sexual partners of members of the groups mentioned above; occasional recipients of routine blood transfusions (though antibody screening tests have greatly reduced this risk); the newborn babies of infected mothers; and, finally, a small

number of AIDS victims (and a somewhat larger number of virus carriers) who do not fit into any of these groups.

The evidence is overwhelming that AIDS is *not* transmitted by casual contact, insect bites, air droplets, and so on.

The ultimate test will be in the effort to control the HIV virus in human beings. Biologists recently succeeded in transplanting the human immune system from fetuses into experimental mice. Such mice may be useful in future AIDS research—especially in the testing of anti-AIDS vaccines. Much work lies ahead if a vaccine is to be produced. If, as it appears, the causal agent is a retrovirus (see Box 18-2), formidable hurdles may block development of a vaccine. One major obstacle is the virus's "hypermutability," which results in marked genetic instability. This has been attributed to the uniquely high error rate of its reverse transcriptase, which inserts the wrong nucleotide once in 2000 times. At this writing the drug AZT has lengthened survival and improved quality of life, but few useful drugs have yet been developed against any virus infection.

Meanwhile it is necessary to rely on such measures as understanding how the virus is transmitted, vigilance in the testing and handling of blood and other body fluids, and sexual prudence (including the use of condoms) to limit spread of the disease.

The emergence of AIDS and its rising incidence have had a profound impact on medicine, biological science, and society. It has also forced the public and its leaders to address many controversial issues in girding for the fight against both HIV and the factors that promote its spread.

affinity. Remarkably, these lymphocytes possess specific binding capacities prior to their first encounter with their antigens.

Binding of antigen to selected B cells stimulates them to divide and differentiate, producing daughter cells of the same specificity. In the absence of further stimulation, these B cells become nondividing—or resting—lymphocytes. Binding of antigen to selected T cells also stimulates clonal division and differentiation, but does not lead to antibody production. Instead, different kinds of T cells arise—helpers, suppressors, and killers. Helpers bind to antigen on the surface of a B cell that has already bound an antigen (Figure 23-22) and releases lymphokines that enable the B cell to do its job. Suppressors inhibit immune responses for a time.

These mechanisms are both fascinating and complex. Figure 23-22 summarizes in schematic form the events that follow the antigenic stimulation of killer (cytotoxic) T cells and helper T cells. The figure also shows how B cells are stimulated to make antibodies. An antigen presenting itself is first engulfed by a macrophage (designated *1a* and *1b* in Figure 23-22) and then processed into smaller pieces (*2a, 2b*) which bind to the macrophage surface (*3a, 3b*) to one of two classes of MHC proteins (see Figure 23-27). A T cell—killer or helper—recognizes these complexes (see Figure 23-21), and thus is selected for clonal expansion (*4a, 4b*). While all of this is going on, free antigen molecules select a B cell whose surface

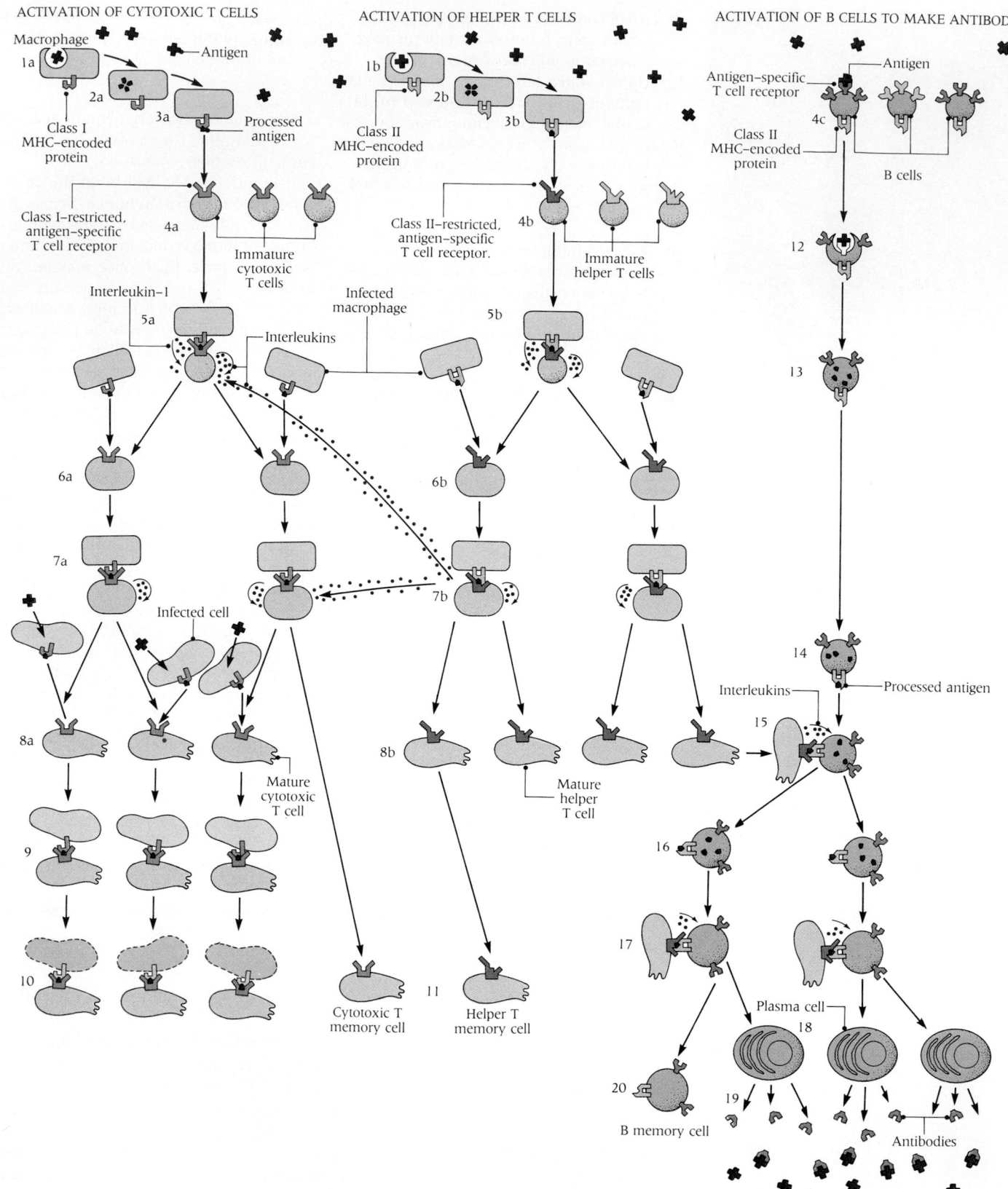

ACTIVATION OF CYTOTOXIC T CELLS

ACTIVATION OF HELPER T CELLS

ACTIVATION OF B CELLS TO MAKE ANTIBODY

Macrophage
1a
2a
3a
Antigen
Class I MHC–encoded protein
Processed antigen
Class I–restricted, antigen–specific T cell receptor
4a
Immature cytotoxic T cells
Interleukin–1
5a
Interleukins
Infected macrophage
6a
7a
Infected cell
8a
9
10

1b
2b
3b
Class II MHC–encoded protein
Class II–restricted, antigen–specific T cell receptor.
4b
Immature helper T cells
5b
6b
7b
8b
Mature cytotoxic T cell
Mature helper T cell
Cytotoxic T memory cell
11
Helper T memory cell

Antigen–specific T cell receptor
Antigen
4c
Class II MHC–encoded protein
B cells
12
13
14
Interleukins
Processed antigen
15
16
17
Plasma cell
18
19
20
B memory cell
Antibodies

receptor fits the antigen and the two become bound together (4c). T cells destined to become killer cells bind antigen with class I MHC proteins (5a); those destined to become helper cells bind antigen with class II MHC proteins (5b). The bound T cells stimulate the macrophage to release interleukin-1, which in turn stimulates T cells to divide and differentiate (6a, 6b). T-cell division and differentiation continue as long as cells are present that present surface antigen (7a, 7b). Only in the presence of cell-bound antigen can mature T cells (8a, 8b) "do their thing." If the mature T cell is cytotoxic, it can either bind to an antigen-presenting infected cell (9), or it can kill the cell (10), or it can remain in the blood and lymph as a memory cell

23-22 Antigen-specific responses of the immune system
The responses of cytotoxic T cells, helper T cells, and B cells are compared. Details in text.

(*11*) that can deal rapidly with similar future challenges. A mature helper cell can also become a memory cell (*11*). Helper T cells stimulate activated B cells to proliferate. A B cell, having engulfed its bound antigen (*12*) and processed it (*13*), also presents a piece of the antigen on its surface to a class II MHC protein (*14*). A helper T cell (*15*) then binds to the antigen-protein complex on the B cell. This releases interleukins, which stimulate B cell division and differentiation (*16*), which continues as long as it is stimulated by helper cells (*17*). The mature plasma cells that form (*18*) release antigen-specific antibodies, which bind and destroy free antigen (*19*). Some mature B cells remain as memory cells (*20*). Extremely complicated, yes—but highly effective!

How Can the Diversity of Antibodies Be Explained?

According to selective theories, the immune system has the potential to produce about 10^7 different antibodies. If the one gene, one polypeptide theory is correct (see Chapter 16), how can we account for such tremendous diversity?

Variable regions in the amino acid sequences of antibody proteins are found near the antigen-binding sight at the amino (N-terminal) ends of an immunoglobulin molecule (see Figure 23-4). The presence of variable sequences seems at first to conflict with everything we have come to understand about protein specificity and the role of genes in ensuring its error-free preservation.[4] Yet here we have proteins bearing antibody-combining sites with amino acid sequences that differ from antibody to antibody. This region consists of a sequence of 100 amino acids (more or less) at the N-terminal end of each heavy chain and a sequence of 100 amino acids at the N-terminal end of each light chain (Figure 23-4). Variations in these sequences account for the antigen specificity of each antibody type. In contrast, the rest of the heavy and light chains of the immunoglobulin molecule—that is, the **constant regions**—are invariant within an immunoglobulin class.

Faced with the paradox of variable regions in an otherwise invariant protein structure, W. J. Dreyer and J. C. Bennett in 1965 advanced a radical hypothesis that turned out to be substantially correct. They proposed that the variable region must be constructed according to the genetic information in any one of many genes, while the constant region is the product of a single invariant gene. This arrangement would, of course, free the immune system from a rigid set of germ line-encoded responses. It would also provide each animal with the potential to create an almost unlimited set of different antigen-binding specificities.

This idea implied that a single constant region must be capable of being joined to any one of many variable regions. In this view, immunologically competent cells must have a unique pattern of genetic behavior. The genetic material in B cells, they suggested, must be *rearranged* at some point.

The 1970s brought rapid gains, largely in the laboratories of Philip Leder, Leroy Hood, and Susumu Tonegawa. It became clear (see Figure 23-20) that the B cell derives from a stem cell that is fully committed to the steps that eventuate in an antibody-secreting plasma cell (see Figure 23-17). Moreover, its genetic information will limit its responsiveness to only one (or very few) antigens. The immune response involves some sort of antigenic selection of discrete B cell clones—i.e., clonal selection—each restricted to the expression of a homogeneous set of immunoglobulin receptors.

Immunoglobulin formation during B cell differentiation begins with the appearance in the cytoplasm of heavy chains of the μ type. Next, there appear light chains, which interact with the heavy chains to form IgM molecules, expressed first in the cytoplasm and subsequently on the surface, where they are embedded in the cell membrane. Immediately thereafter, there occurs in the cell, now a pre-B lymphocyte, a parallel interaction between δ heavy chains and light chains that yields cytoplasmic and membrane IgD. The IgM and IgD have the same variable region sequences. The cell has now become a specific immunocompetent B cell (Figure 23-20).

By the ingenious use of recombinant DNA technology, Leder found that during development, **V** (for **variable**) and **C** (for **constant**) **regions** are encoded in separate

---

[4] Nobel Prizes were awarded in 1972 to Gerald Edelman and Rodney Porter who illuminated these questions. They were the first to determine the amino acid sequence of the polypeptide chains of several antibodies. These Rockefeller scientists became known as "The Chain Gang" to fellow research workers.

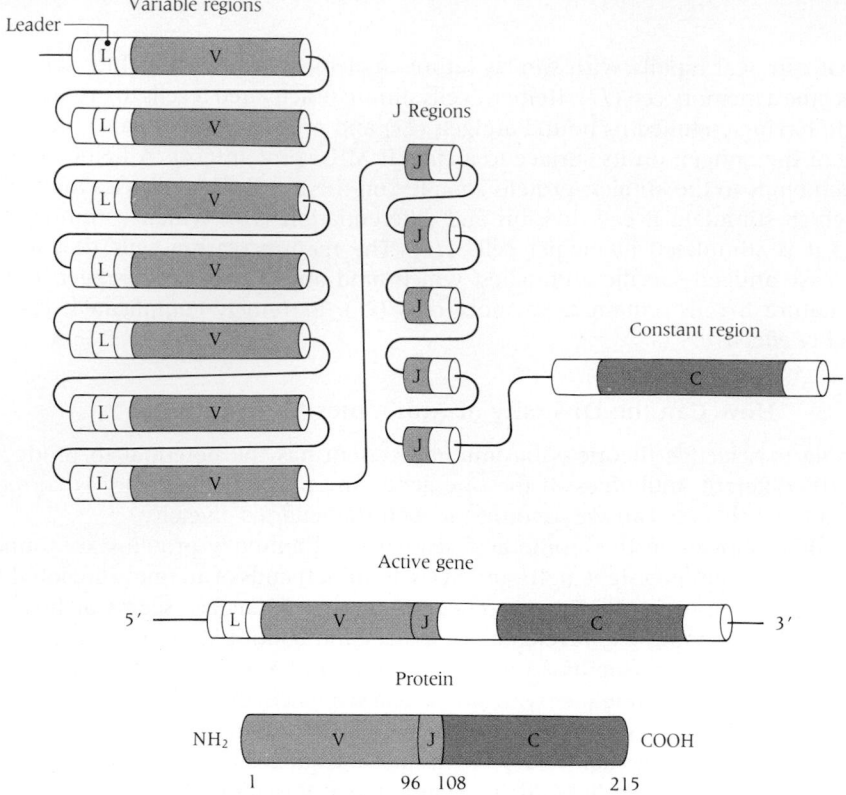

Variable regions

Leader

J Regions

Constant region

Active gene

Protein

NH_2 V   J   C   COOH

1   96 108   215

areas of the chromosome and that light-chain gene formation involves a unique somatic cell recombinational event. The so-called V segment of the κ-chain gene is itself encoded by two separate gene elements termed **V** and **J** (for **joining**) **segments.** (Figure 23-23)

In the formation of an active immunoglobulin light chain gene, a recombinational process joins one of many variable region sequences to one of these four or five J sequences. In other words, the V and C genes become rearranged during the course of plasma cell differentiation. There are some 100 to 200 different V-region sequences and five J-region elements.

In sum, the major events occurring during the assembly of light-chain genes are as follows:

■ The variable region of a light chain is encoded from three segments, the leader, V, and J regions of the DNA.

■ The constant region is encoded from a C segment located nearby on the same chromosome.

■ The noninformational region of DNA between the V and C regions is incorporated into the RNA formed to transmit the code for protein synthesis. This precursor RNA is then processed to remove the intervening sequences and the structural sequences are rejoined to form an intact light-chain code. (RNA splicing was discussed in Chapter 17.)

■ Note that the gene elements, V, D, and J, are selected at random and spliced together before intervening sequences are spliced out and lost.

Simple calculations show that 100 V-region sequences and 5 J-region sequences that can be connected in every possible way would yield 500 different combinations for the light chain alone. Yet another source of diversity is provided by point mutations known to occur in the V region. Indeed, "hypermutation" is a special mechanism found only in B cells.

. We have thus far been discussing light chains. Although the structure and rearrangements of heavy-chain genes and light-chain genes are similar in many ways, certain differences add interesting complexities to the heavy-chain system. The heavy-chain variable region is constructed by joining not two, but three, distinct segments of DNA: a V segment; a D, or diversity, segment; and a J segment. Heavy-chain J segments are separated by sequences of approximately 300 to 350 nucleotides occurring in regular tandem array (Figure 23-24). D-region segments, which encode only about 10 amino acids, are separated by about 10,000 bases. How these tiny islands of genetic information required for heavy-chain formation came to be placed

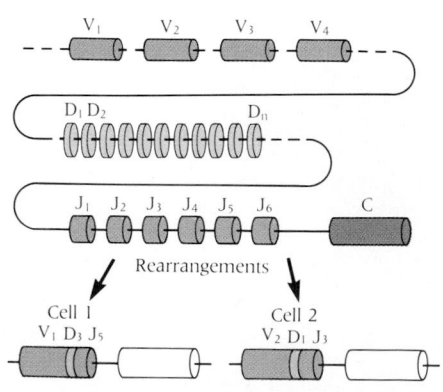

23-24 Rearrangements of heavy-chain genes
Rearrangements of variable (V), diversity (D), and joining (J) regions create unique heavy-chain genes. These end up next to constant (C) region genes.

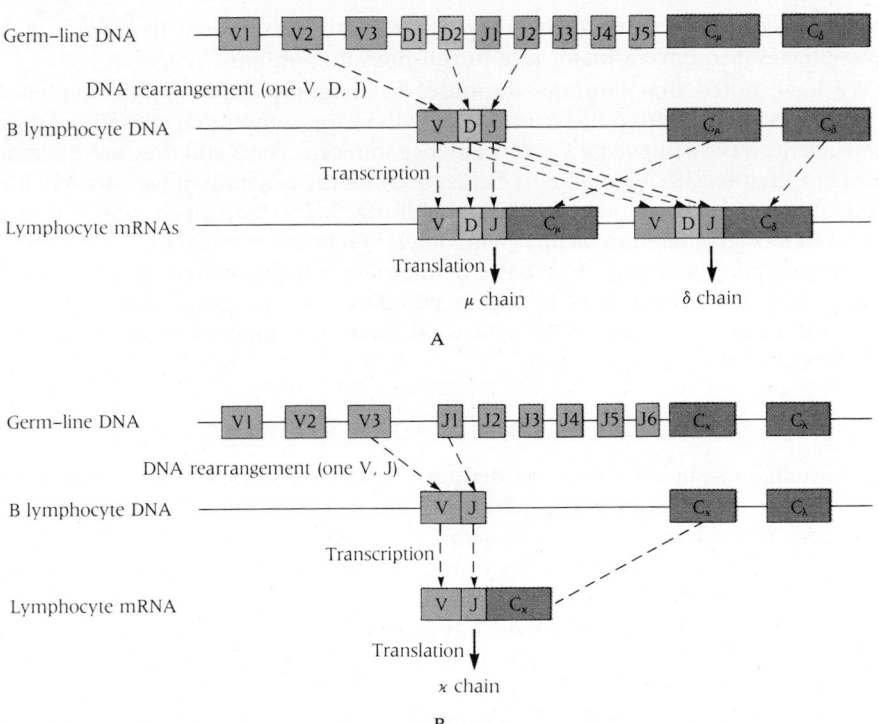

23-25 Production of heavy and light chains compared

A summary showing how heavy-chain and light-chain genes are generated by DNA rearrangements. (A) Production of the μ and δ heavy chains of two antibodies. (B) Production of a κ light chain.

adrift in a sea of nucleotide sequences remains unclear. There are probably 6 active copies of J-region sequences and a reasonably large family of D and V segments, each of the latter having a leader sequence. If, hypothetically, we projected 100 V, 15 D, and 6 J segments, we would have the potential to create 9,000 (100 × 15 × 6) different combinations. Since each combination allows for splicing error variations and since there is a high frequency of single-point mutations (especially in the gene encoding the V region of the light chain), it seems apparent that hundreds of thousands of different genes can be created. This clarifies the mathematical side of the puzzle. A system that can combine hundreds of thousands of different heavy-chain genes with thousands of different light-chain genes is clearly capable of constructing millions of immunoglobulins, each with a unique antigen-binding site.

On first encountering its "intended" antigen, surface IgD and IgM on an immuno-complement cell disappear, and the cell begins to synthesize and secrete IgM. The cell also undergoes division and differentiates into a plasma cell. This clone of cells can continue to produce IgM. Alternatively, it may switch over to one of the other heavy-chain classes. This mechanism, through which a single variable region can occur in association with several different heavy chains, is known as the **heavy-chain class switch.**

In sum, we have learned at last how a single organism can produce so many specific immunoglobulins. As shown in Figure 23-25, a large number of genes encode different portions of the immunoglobulin molecules. A great deal of the diversity arises from recombinational events that begin in the lymphocytes of the developing embryo. This process continues as new B cells are produced throughout life. Clearly, the DNA of the immune system has a unique ability to rearrange itself in different ways in different cells. Antibody gene assembly, which takes a little DNA from here and a little DNA from there, has been aptly compared to ordering dinner in a Chinese restaurant.

CELL-MEDIATED IMMUNITY

We have already summarized the essential features of cell-mediated immunity. Mechanisms of its several major antigen-specific responses are shown schematically in Figure 23-22.

It is clear that T lymphocytes serve as both effectors and regulators of the immune system. Unlike B lymphocytes, which produce antibodies that neutralize soluble antigens, T lymphocytes bear membrane-bound receptors that detect perturbations of cell surfaces caused by infection or other factors. As shown in Figure 23-22, macrophages also have a major role in cell-mediated immune reactions.

We have noted that immune responses to antigenic challenge are modulated by soluble factors. Some could be termed growth factors, others differentiation factors. These agents have achieved great prominence in recent years and they are discussed further in Chapter 28. As in all new fields based on the isolation of various activities, nomenclature remains both chaotic and shifting. More than once, purification of several supposedly different factors (with different names) revealed them to be identical. Most would agree with two basic definitions: **lymphokines** are produced by lymphocytes and **monokines** by monocytes and macrophages. Collectively, these factors form a network that amplifies the weak incoming signals of antigenic challenge and stimulates the immune system.

LYMPHOKINES AND MONOKINES

The existence of such factors was deduced in the early 1970s from experiments showing that media from tissue cultures of antigen-stimulated T cells contain factors—then termed T-cell replacing factor(s)—that could substitute for T cells in T cell-depleted lymphoid tissues by restoring production of specific antibodies. Within a short time, so many factors were discovered in comparable experiments, that the general result was bewilderment. Only after the development of a reliably specific monoclonal antibody to each individual factor was real progress made in sorting them out.

Among the many currently recognized lymphokines and monokines are various classes of low molecular weight peptides: the **interferons,** the **interleukins,** the **colony-stimulating factors,** and so on. Several general points about these factors can be made.

- These factors are numerous and their effects are diverse.
- They are generally soluble proteins of low molecular weight.
- Some have given evidence of multiple functions, including tantalizing hints of potent anticancer effects.
- Some may turn out to be identical (e.g., lymphotoxin and TNF, perforin and cytolysin).
- In part because they are prospective "cancer cures," some lymphokines and monokines have been produced by biotechnology companies by the methods of genetic engineering. It happens often that the experiences of detached observers dim the enthusiastic (and much publicized) claims of early investigators—as they did after interleukin 2 (IL-2) was held to convert T cells to active killer cells, which then destroy malignant growths. However, there are still reasons for optimism. Investigators have modified their approaches and now rely on more potent IL-2-stimulated killer cells isolated directly from tumor tissue. Other factors such as α-interferon do seem to have some value to certain cancers. Future work, one hopes, will make this approach more rewarding.

TISSUE TRANSPLANTATION

The Concept of "Self" and "Nonself"

In modern terminology, a graft of tissue from one person to another is termed an **allogeneic** graft, or **allograft.** It was earlier called a **homograft.** Terms currently used in transplantation immunity are summarized in Table 23-2. Note that the

TABLE 23-2
TRANSPLANTATION TERMINOLOGY

Current Nomenclature	Older Nomenclature	Relationship of Graft Donor and Recipient
Syngeneic graft	Autograft	Same individual
Isogeneic graft	Isograft	Same species, genetically identical
Allogeneic graft	Homograft	Same species, genetically nonidentical
Xenogeneic graft	Heterograft	Different species

prefix *syngene* means same individual; *isogene-*, same species or genetically identical; *allogene-*, genetically nonidentical member of same species; and *xenogene-*, another species.

An allogeneic skin graft survives only briefly, no matter how competently performed. The same is true for allogeneic grafts of other tissues. We say that such grafts have been *rejected*. Ordinarily, the only type of graft that can be relied upon to succeed permanently is a syngeneic graft, or **isograft**—one in which the donor is also the recipient—as in the transplantation of skin from one area of the body to another. The only major exception to this generalization is that grafts interchanged between identical twins are permanently accepted.[5] This fact has saved the lives of many members of identical-twin pairs who are afflicted with severe burns or advanced kidney disease. Tissues interchanged between ordinary brothers and sisters do not survive unless measures are taken to suppress the recipient's immune system.

An allogeneic skin graft heals satisfactorily at first. However, it soon becomes inflamed and infiltrated by macrophages and lymphocytes, its blood vessels become congested, and the graft sloughs away. The entire rejection process takes about two weeks. This failure of allogeneic grafts is not peculiar to humans: it is typical of all vertebrates and many invertebrates.

In these cases, the body is rejecting tissue that it recognizes as foreign to itself. The singular fact that animals of the same species cannot accept grafts from one another implies that along with the genes that determine species, individuals carry complex determinants that are responsible for the uniqueness of each individual.

Immunological Tolerance

A clue to the mechanisms underlying graft rejection came from experiments in which an organism was "tricked" into accepting as parts of itself cells or substances that, genetically speaking, had no right to be there. The most impressive examples of this phenomenon come from rare cases in which genetically dissimilar human twins share a common placental circulation in the uterus. Each twin continuously receives a large variety of cells from the other, including stem cells capable of surviving in the bone marrow where they colonize, multiply, and produce blood cells. Twins of this kind become and remain chimeras, each containing two blood groups: their own and one that is genetically appropriate to their twin.

Such fraternal twins differ in another way from ordinary dissimilar twins. Fraternal twins with double blood groups (chimeras) accept skin grafts from each other, as if they were genetically identical. To F. M. Burnet (who shared the Nobel Prize with Medawar), this suggested that a foreign antigen introduced early in embryonic life is henceforth regarded as "self" by the body's immune system, and does not excite antibody formation. That would imply that the ability to recognize "self" develops sometime during (or shortly after) embryonic life. Any configuration present during this period would, in the future, be regarded as self by the immune system and thus be nonantigenic.

In an experiment first performed by Medawar (Figure 23-26A), cells from the spleen, lymph nodes, or kidney of a mouse embryo of strain *B* were inoculated into a fetal or newborn mouse of strain *A*. The mouse developed normally. When a piece of *B* skin was later grafted to the inoculated mouse *A* (when sufficiently grown), the graft "took" and persisted in healthy condition.

In another experiment, R. E. Billingham and L. Brent inoculated a mouseling (fetal or newborn) of strain *A* with lymphocytes of an adult *B* mouse, not an embryo. Depending upon the number of cells and the particular pair of mouse strains, the mouseling either died within two or three weeks or developed into an undersized, scruffy-looking individual suffering from what was called runt disease (Figure 23-26B). An injection of plasma from an adult *B* mouse had no such effect.

The results of Medawar's experiment suggested that host *A* had become tolerant of the *B* cells implanted in its tissues before or just after birth. Note that the implanted cells are as immunologically competent as the host cells. If an equilibrium is to be reached, implanted cells must become tolerant of their foreign host as well as vice

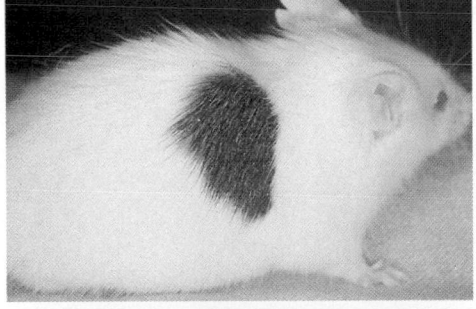
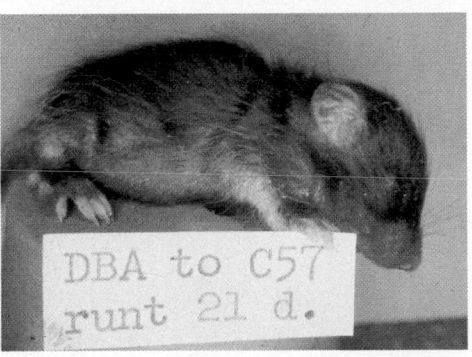

23-26 Immunological tolerance and the graft-versus-host reaction
(A) A white A mouse with tolerance to B mouse tissues, which has accepted a skin graft from a black B mouse. (B) A "runted" mouse which has been injected with lymphoid cells from a mouse of a different strain.

[5] Recall that human twins are of two types (see Box 19-2). Fraternal or nonidentical twins result from the simultaneous development of two ova. These have separate placentas (except for the rare phenomenon to be described later) and they are as genetically dissimilar as separately born brothers or sisters. Identical twins develop from a single ovum.

versa. As noted above, embryonic *B* cells do become tolerant. But in the second experiment, the adult *B* cells set up their own immune reaction against their host, producing a *graft-versus-host reaction* that causes runt disease or death. The host is too young to attack the grafted cells (of which it becomes tolerant) but the engrafted adult lymphocytes attack the host.

In sum, it appears that at some point in embryonic or neonatal life, the body's immunocompetent cells take inventory of the cells and substances then present and classify them as ''self.'' From then on, all other materials will elicit immune responses. ''Self'' forever receives the dispensation of **immunological tolerance.** Clearly, the function of tolerance is to prevent the body from reacting against its own constituents—though, as we shall see, this tolerance sometimes breaks down. Autoimmune reactions do sometimes occur.

The Basis of Allogeneic Graft Rejection

Rejection of an allogeneic graft is an immune response of the host to foreign antigens present in the grafted tissue. It is mediated, not by circulating antibodies, but by sensitized lymphoid cells.

Once elicited, the **sensitized state** is present in every part of the recipient body that is served by blood vessels. It is also long lasting. Later homografts from the original donor transplanted to a host that has already been exposed to tissue cells from that donor, undergo an accelerated rejection termed a **second-set reaction.** Indeed, this reaction provides an important biological test for allogeneic graft sensitization.

Tissue transplantation experiments revealed that all species contain a gene cluster termed the **major histocompatibility complex (MHC),** which encodes the cell surface antigens responsible for graft rejection. MHC loci encode highly polymorphic (that means with many forms or varieties) surface glycoproteins of two major classes—I and II—that are distinct in their chemical structure and biological effects (Figure 23-27). Antigens are presented to T cells by these proteins (see Figure 23-21). Class I MHC proteins are essential for *self-recognition.* Class II MHC proteins are involved in the *lymphocyte interactions* shown in Figure 23-22.

The so-called **histocompatibility antigens** are numerous. They are also strictly determined by Mendelian laws of dominant inheritance (see Box 34-5). The major factor determining the tempo and intensity of graft reactions is the degree of genetic disparity between donor and recipient. If they differ only by a single weak factor, a graft may live for many weeks. The number of MHC genes in humans is evidently large enough to permit an immense number of possible combinations. Hence, there can be little optimism about achieving a perfect match with respect to histocompatibility antigens. Nonetheless, donor selection by degree of MHC compatibility does increase the chances of success.

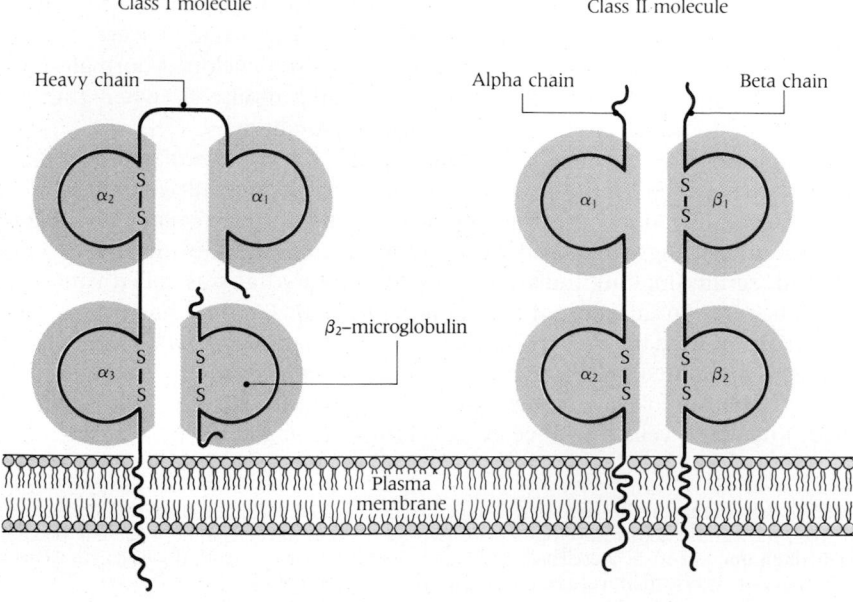

23-27 Structures of class I and class II MHC proteins

Histocompatibility antigens are individual-specific, not tissue- or organ-specific: one cannot distinguish between an individual's thyroid, spleen, or skin on the basis of these antigens.

The fundamental similarity of allogeneic graft sensitivity and tuberculin-type delayed hypersensitivity is indicated by the following evidence:

■ Both are transferrable between individuals by a transfer of lymphocytes.

■ A variety of serum antibodies do reject allogeneic grafts, but claims that they are the agents that destroy normal solid tissue grafts are unconvincing, and serum antibodies will not transfer allogeneic graft sensitivity.

■ Medawar demonstrated that guinea pigs, previously sensitized by skin grafts, respond to subsequent intradermal injection of cells or extracts from homologous donor tissues with a delayed-type inflammatory response virtually indistinguishable from the reaction to tuberculin in a sensitized individual.

■ The histologic appearance of skin undergoing rejection resembles that of other delayed-type responses in skin.

The weight of evidence indicates that transplantation immunity is provoked when histocompatibility antigens are emitted from an allogeneic graft. The development of new blood vessels in the graft facilitates its infiltration by sensitized or activated lymphocytes and macrophages from the recipient. Surgeons hoping to eliminate this barrier to free transplantation of tissues from body to body must devise a reliable method of eliminating an intended host's ability to react against the histocompatibility antigens of an intended donor (Box 23-4).

IMMUNOLOGICALLY MEDIATED DISEASES

Some diseases are caused by an abnormality of the immune process. The immune system is a balanced network of competing signals and responses. Abnormalities are caused by an imbalance of the same mechanisms that normally protect the individual—antibodies, complement, macrophages, T cells, and the rest. Such defects as impaired development of B or T cells, lack of various complement proteins, and disordered regulatory interactions are found.

Diseases of **autoimmunity** are an especially interesting category in which the immune system reacts to self—to normal (autologous) body components. They can be mediated by B cells, T cells, or both. The diseases produced by immune responses to autoantigens, the so-called autoimmune diseases, can manifest themselves in three ways.

■ An agent, such as a drug, attaches itself chemically to a cell component and elicits an immune response directed against it and the cell. This sometimes happens in people receiving penicillin.

■ An immune reaction occurs against a normally inaccessible antigen. For example, breakdown of cells of a damaged heart muscle leads to autoantibodies to muscle proteins that are normally "hidden."

■ Autoantibodies may be directed against many widespread normal body components—as in the disease called lupus erythematosus.

Irrespective of the possible role of autoantibodies in disease production, a successful theory of immunity must account for their occasional occurrence. It seems likely that many of these conditions represent anomalies of immunocompetent cells rather than abnormal antigens. What is abnormal is the presence of such reactive T cells or antibodies—and the breakdown of the homeostatic processes that should have prevented their emergence and pathogenic activity. Table 23-3 lists some of the diseases of autoimmunity.

BLOOD GROUPS AND BLOOD TRANSFUSION

We learned in Chapter 15 that human blood can be classified into groups. **Blood typing,** the method for identifying groups, depends on reactions between antigens on the red cell surface and antibodies in testing sera. Nearly 100 years ago, while

Methods of tissue transplantation used in humans seek to depress the immunological responses responsible for allogeneic graft rejection. Two have been successful. One involves the administration of drugs (e.g., azathioprine and cyclosporine). Unfortunately, such immunosuppressive drugs depress immunity to infectious organisms as well (Figure 23N). The second employs antilymphocyte globulin (ALG), an antiserum raised in one species (horse) against human T lymphocytes. ALG abolishes T cell-mediated immune responses, while sparing humoral antibody responses.

Today surgical organ transplantation has progressed to a mature and scientific technology available at virtually every major medical

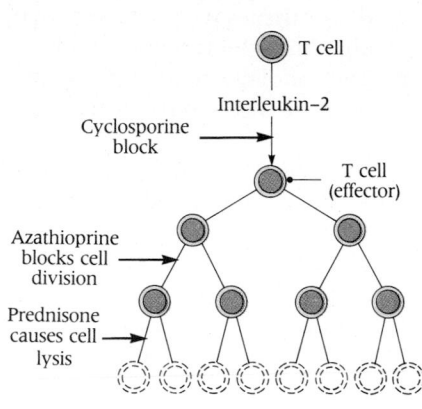

23N *Immunosuppressive drugs depress immunity*

center. In some measure, progress can be attributed to improved surgical techniques. However, most gains have been achieved by our ever-improving ability to control and blunt graft rejection.

We can here cite only two brief examples of the current situation. One is kidney transplantation. When the donor is a living sibling, the probability that a functioning graft will survive for 5 years, and in most cases for 10 years or more, is 90 to 95%. Even without such a close genetic relationship, such long-term transplant survival is achieved in well over 70% of cases (Figure 23O).

Unquestionably, the key factor behind this success is careful donor selection by

TABLE 23-3
SOME OF THE DISEASES CAUSED BY AUTOIMMUNITY

Disease	Target of Autoantibody	Major Sign or Symptom
Graves disease	Thyroid hormone receptors	Excessive stimulation of thyroid hormone production (hyperthyroidism)
Thyroiditis	Various thyroid proteins	Hypothyroidism
Hemolytic anemia	Rh locus in red cell membrane; other loci	Anemia
Lupus erythematosus	Native DNA, single-stranded DNA, nucleoproteins	Formation of antigen-antibody complexes that damage many organs
Myasthenia gravis	Acetylcholine receptors in neuromuscular synapses	Extreme weakness
Nephritis	Basement membrane of kidney glomeruli	Kidney failure and uremia
Rheumatoid arthritis	Proteins in synovial membrane lining joints	Inflammation in joints

searching for possible differences among individual human bloods, Karl Landsteiner (1868–1943) mixed serum from one normal individual with red cells from other normal individuals. He found that certain combinations of cells and serum caused marked clumping of the red cells, while other combinations showed no clumping (see Figure 15-16). By the use of these reactions, Landsteiner showed human beings to be classifiable into clearcut blood groups: A, B, AB, and O. This discovery made possible safe blood transfusion for the first time. As shown in Table 15-2, the ABO system depends upon the presence on red cells of A or B antigens. It is defined by anti-A and anti-B antibodies.

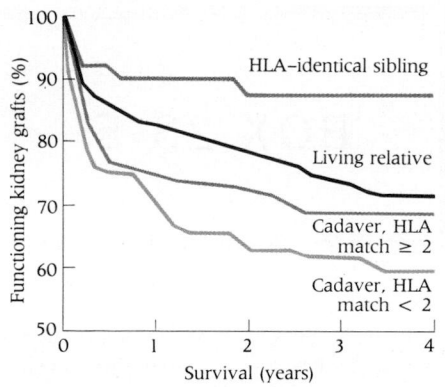

230 Source of an allograft influences transplantation outcome
The graph shows the survival in years of transplanted kidneys obtained from different sources. As expected, grafts from HLA-identical siblings fare best. The moderate (but respectable) survival of grafts from recently deceased bodies varies with the extent to which donor and recipient are HLA-matched. In recent years, the likelihood of rejection has been significantly reduced by rise of the drug cyclosporin.

matching of donor and recipient at the major histocompatibility complex (MHC), plus the skillful use of immunosuppressive drugs.

Prospective kidney donors are tested in a preliminary cross-match—the so-called mixed lymphocyte reaction—in which donor and recipient lymphocytes are mixed in culture and observed for compatibility.

If transplantation from a living donor is not possible, the patient then becomes a candidate for a kidney from a donor who has recently died. Transplantation surgeons call these cadaver kidney grafts. In such circumstances, it is often difficult to judge the degree of alloimmunity, but it is attempted nonetheless. The intended recipient, who it must be remembered has failed kidneys and is on dialysis, is tested against a panel of cells from 20 randomly selected normal people. In that way, the immune status of the recipient can be roughly evaluated on a percentage basis. If the recipient cells are reactive with one of 20, he or she is rated at 5%; with four, at 20%; and so forth. The lower the percentage, the better the chance of avoiding rejection of a cadaver kidney—though the correlation often fails. When a cadaver kidney becomes available, the cadaver donor is tested for MHC compatibility if conditions permit. Often, they do not.

The transplantation of bone marrow is also on the increase—for many disorders. Marrow is much more likely to be rejected than kidney, and graft-versus-host disease is more common and more severe. Marrow, unlike kidney, resembles a cell suspension. This fortuitous circumstance has encouraged researchers to attempt to "purge" donor marrow of unwanted T cells by the use of various monoclonal antibodies against T cells that are mixed with the marrow prior to its injection into the recipient. This approach is now being evaluated.

In a sense, mammals throughout evolution have constantly been exposed to "allogeneic grafts" during pregnancy. Every fetus is an allogeneic graft and possesses, even at an early stage, antigens inherited from its father that may be lacking in the mother. The biological success of mammals is strong evidence that special dispensations must apply. A cushion of immunologically inert cells (the trophoblast layer) as well as the vascular quarantine of the fetus is probably the basis of its exemption from immunological rejection. Its histocompatibility antigens may never cross the placental barrier into the mother, and even if they do, there is no pathway for immunocompetent lymphocytes of maternal origin to enter the fetus and cause it harm.

The use of novel test antisera of nonhuman origin soon uncovered new and different blood groups. When human red cells were mixed with antisera from rabbits that had been injected with human red cells, antibodies in the rabbit serum clearly distinguished two new human blood groups, arbitrarily called M and N. Their simple hereditary patterns made them extremely useful in the investigation of disputed paternity. An important new set of blood groups was found when human red cells were tested with sera from rabbits into which red cells from a rhesus monkey had been injected. This was how, in 1940, Landsteiner and Alexander Wiener discovered the **Rh (Rhesus) factor.** The rabbit antibodies agglutinated not only monkey red cells but also the red cells of about 85% of the human blood samples tested. Human red cells thus were first classified as Rh positive (Rh$^+$) if they contained the factor or Rh negative (Rh$^-$) if they lacked it. The medical implications of the Rh system were profound (Box 23-5).

PRACTICAL APPLICATIONS

Dozens of blood group systems are now known. Therefore, many millions of blood group combinations exist. Indeed, the complete pattern of blood group antigens is as distinctive in an individual as a fingerprint. This fact is of obvious importance to medical examiners and police laboratories investigating crime scenes.

As discussed in Chapter 15, the genetics of the ABO system and MN groups follow clear Mendelian principles. The presence of antigens is an uncomplicated dominant trait. No known blood group antigen is recessive. Like other genes, blood group genes occasionally undergo mutation. This is presumably the basis of their great variety. Most of the mutations accounting for present patterns probably occurred very long ago.

Because of their hereditary nature, blood groups are indispensable tools of the anthropologist. Unlike other phenotypic traits such as height and skin color, a blood group is totally uninfluenced by environment. Hence it reflects only the genotype. Study of the geographical distribution of blood groups shows that human beings fall into a number of large divisions, each with a different set of average frequencies of the various blood genes, and each occupying an area of subcontinental

The dangerous illness called **Rh disease**—properly **Rh hemolytic disease of the newborn**—occurs when an Rh-negative mother develops antibodies against Rh factor present in the red cells of her Rh-positive fetus. When an Rh-negative mother produces such antibodies, we say that she has been sensitized. Sensitization is caused by any of these three occurrences.

- A pregnancy (current or previous) with an Rh-positive fetus
- An earlier mismatched blood transfusion in which the Rh-negative mother was given Rh-positive blood
- The unlikely case in which the mother's own mother was Rh-positive, and a few Rh-positive red cells entered her body during fetal life

Note that Rh disease occurs only when the mother is Rh-negative and has been exposed to Rh-positive cells that stimulated her to form anti-Rh antibodies.

The number of women sensitized by blood transfusion is small compared to the number who become sensitized as a result of an Rh-incompatible pregnancy. The chance of sensitization with a single Rh-positive pregnancy is about 1 in 6. Hence, the risk to an Rh-positive baby of an Rh-negative mother is proportional to the number of existing Rh-positive siblings.

The Rh-positive baby that initiates the formation of anti-Rh antibodies in the mother escapes harm because these maternal antibodies rarely appear until some weeks after delivery. Anti-Rh appearing in the mother's serum for four to six weeks after delivery is proof that sensitization was initiated by that pregnancy. However, in many mothers anti-Rh levels are so low that they are undetectable until a subsequent Rh-positive pregnancy raises their concentration. (This is an example of the anamnestic reaction.) Consequently, the first substantial appearance of antibody occurs only during second and later pregnancies.

The mother makes antibodies because she is exposed to a fetal antigen—Rh factor—that her body has not previously encountered. This occurs when fetal red cells in small numbers pass across the placenta into the mother—a normal event. Indeed, fetal red cells may be found in the mother's blood as early as eight weeks after conception. However, sensitization is most likely to occur at the time of delivery since the passage of fetal red cells into the mother's blood is most active then.

The disease affects the fetus in several ways. The maternal anti-Rh antibody crossing the placenta destroys many of its Rh-positive red cells and anemia develops. This can result in heart failure. The fetus compensates for the destruction of its red cells by increasing erythropoiesis. Since its bone marrow is undeveloped, this occurs chiefly in the liver and spleen, both of which are greatly enlarged. Liver function is usually impaired as a result. Finally, massive destruction of red cells generates excess **bilirubin,** a yellow breakdown product of hemoglobin. Since the newborn is not biochemically able to handle such a load of bilirubin, there ensues the yellow skin discoloration called jaundice. High bilirubin levels can also cause brain damage with neurological impairment or even death.

It was observed in the 1960s that deliberate administration of anti-Rh antibody (an Rh immune globulin called Rhogam) to an Rh-negative mother during pregnancy or after delivery can eliminate the fetal Rh-positive cells in her body and thereby decrease her chances of becoming sensitized to Rh factor. This discovery, a major achievement of modern medicine, yielded an effective means for preventing Rh disease of the newborn.

Interestingly, the risk of Rh disease is much lower in Rh-negative mothers delivering ABO-incompatible Rh-positive babies. The reason is simple. Consider a group O, Rh-negative mother with a group A, Rh-positive fetus. Fetal red cells entering the mother's body will be destroyed by her naturally occurring anti-A antibody (see Table 15-2) before they can elicit anti-Rh antibody in the mother.

The widespread availability of Rhogam should lead to the virtual elimination of this once severe public health problem within a generation. Recent statistics clearly show that the trend is in this direction.

---

size. For example, the well-known gradient from a high incidence of type A in the south of England to a high incidence of blood group O in Scotland is thought to be due to the progressive retreat of peoples with a high incidence of group O before the onslaught of immigrants with a high incidence of blood type A.

BLOOD TRANSFUSION

It was obvious to the ancients that blood is associated with life and health and that loss of blood led to weakness and death. The Egyptians added blood to baths to rejuvenate the aged and sick. Ancient Romans drank the blood of dying gladiators in hopes of acquiring their strength and courage.

In 1667, blood was first transfused from animals to humans, but the recipients died and the practice was banned. Even the transfusion of blood from human to human caused deaths. Not until Landsteiner's discovery of the ABO system at the turn of the century was human blood transfusion placed on a rational basis. At that time, only direct transfusion from donor to recipient was possible, but with the advent of reliable refrigeration and anticoagulants, blood banking finally became possible. The first blood bank was established in 1937. About 12 million pints of blood were used in the United States in 1989.

An "incompatible" blood transfusion is a transfusion in which donor and recipient have different blood types, or antibodies, to blood groups. The main hazard of incompatible transfusion is the destruction of donor red cells by antibody in the recipient's serum. If severe enough, death may result. For example, if group A cells are transfused into a person whose blood contains B red cells, the A cells are rapidly destroyed by the anti-A antibody in the recipient's serum. Since type O red cells lack both A and B antigens, they cannot be attacked by anti-A or anti-B antibody in the recipient's serum. Thus O blood can usually be safely transfused into any member of any ABO group. For this reason O individuals are sometimes called **universal donors.**[6] Occasionally, however, donated O blood contains unusually high amounts of anti-A and/or anti-B antibody. In such cases, the donor's serum antibody destroys the recipient's A or B red cells. Donors of such O blood are known as "dangerous universal donors."

Other hazards of blood transfusion are related to the transmission of such viral diseases as hepatitis and AIDS. Diligent research has produced screening tests that detect antibodies to many of these viruses. Unfortunately, the tests are not perfect and a small but finite risk remains.

---

[6] We call O individuals universal donors because their red cells will not be attacked by the large amount of anti-A and anti-B antibodies in the whole blood volume of a prospective recipient. The opposite proposition—that O individuals can be universal recipients (of any blood type)—is not true. The reasons relate to the small volume of transfused blood and the large volume of blood already present in the recipient. If an O individual (who normally has anti-A and anti-B serum antibodies) received a transfusion of A blood, the recipient's anti-A antibodies would destroy the transfused blood. But if an A individual (who normally has anti-B serum antibodies) received a transfusion of O blood, the recipient's anti-B antibodies would not attack donor O cells, and the small amount of anti-A antibody in donor serum would be diluted out in the recipient's large blood volume and would have little effect.

SUMMARY

The responses of the animal body to invasion by foreign substances or cells include immune reactions of two major types: humoral and cell-mediated. Humoral immunity is based upon antibodies, which circulate in plasma and bind with high specificity to the foreign substances (antigens) that stimulated their production. Some antibodies directly neutralize toxins. Others coat foreign particles and induce white blood cells to engulf them. Still others kill bacteria through the mediation of the complement system.

Humoral immunity is most effective against bacterial invasion. Cell-mediated immunity is involved in neutralizing viral and fungal infections and in the rejection of transplanted tissues. Immune reactions that are excessive or inappropriate can harm the host by inducing hypersensitivity reactions. Unlike the antibody-mediated humoral system, which can cause immediate hypersensitivity as in anaphylaxis, the cell-mediated immunity system causes delayed hypersensitivity, as for example in the tuberculin reaction.

Most but not all antigens are proteins, but only small portions of each antigen's molecular structure are critically involved in eliciting a given antibody. These structures, termed antigenic determinants, are also the sites to which the antibody binds. Because a single antigen usually has multiple antigenic determinants, one antigen will elicit more than one antibody, each with an affinity for one antigenic determinant. The antibody response to a single antigen is therefore said to be polyclonal. Various laboratory techniques, however, can readily resolve the antibodies elicited by a single antigen into separate, pure monoclonal antibodies.

Antibodies are produced by B cells, one of two major lymphocyte types. It was once thought that antigens somehow "instruct" B cells to construct antibodies with the necessary complementary structure. However, we now know that antigens select lymphocytes with a pre-existing capacity to produce a single specific antibody, and then stimulate them to divide and differentiate into a clone of plasma cells which actively produce an antibody specific to that antigen. This implies that the body possesses a large number of different antibody-producing B cells, each bearing a surface receptor with a more or less unique specificity. When a B cell encounters an antigen for which it has specificity, it undergoes a series of transformations that results in the production of antibodies of the same specificity. Following an animal's first exposure to an antigen, production of specific antibodies increases and then falls off. If the animal is again exposed to that antigen, a large population of B cells specific for that antigen already exists (memory cells), and a vigorous secondary reaction increases antibody levels to new highs. This is the basis of medical "booster" immunizations.

If a different gene were necessary to encode every possible antibody, most of an organism's genome would be concerned with antibodies. This is not the case. Instead, unique somatic cell gene-splicing events rearrange the genes that govern the synthesis of different parts of the immunoglobulin molecule—the C (constant) and V (variable) regions, and other regions (J and D). By rearranging multiple nucleotide sequences, these recombinations can generate many thousands of different antibodies. In addition, a special mechanism leading to a high rate of point mutations after V(D)J splicing during each B cell's development greatly enhances diversity.

T cells, the second major lymphocyte group, are the prime agents of cell-mediated immunity. They also regulate B cell function in humoral immunity. Both types of immune responses are amplified and controlled by soluble molecules called lymphokines (made by lymphocytes) and monokines (made by monocytes and macrophages). T cells develop in the thymus early in life. Later, they proliferate in the spleen and lymph nodes. The three T cell categories are helper cells which promote activity in other cells of the immune system, suppressor cells which limit the immune response, and cytolytic (killer) cells which recognize and kill abnormal cells (foreign, infected, or cancer cells).

T cells like B cells are stimulated to divide when they bind specific antigens. They bind antigens only when they are found on cells in association with a particular group of surface proteins called histocompatibility antigens, which are encoded by genes of the MHC complex. These proteins are found on every cell within an individual, but they have so many allelic forms that each individual has a unique set. Thus they serve as critical markers, indicating to T cells whether a cell belongs to "self" or "nonself."

The tolerance of T cells to self-antigens occurs during fetal development. Indeed, tolerance will develop to foreign cells if they are transplanted into an animal early in its development. If an organism is tolerant of cells transplanted into it, but the engrafted cells are not tolerant of the organism, there occurs a graft-versus-host reaction in which engrafted cells reject the host. Tolerance to self-antigens breaks down in autoimmune diseases in which an individual rejects his own tissues.

The most commonly performed tissue transplantation is blood transfusion. The well-known A, B, and O antigens are but one of many antigenic systems that characterize each individual's red cells. If an individual receives red cells containing foreign antigens to which he has developed antibodies, the antibodies may destroy the transfused cells. The presence of maternal antibodies against Rh factor may harm an Rh-negative newborn.

KEY TERMS

agglutinin
AIDS (acquired immune deficiency syndrome)
allergy
allograft
antibody
antigen
antigenic determinant
antiserum
autoantibody
autoimmunity
B cell receptor
B lymphocyte (cell)
cell-mediated immunity (cellular immunity)
clonal selection theory
complement fixation

constant (C) region
cytotoxic T cell
Fab fragment
Fc fragment
heavy chain
heavy-chain switch
helper T cell
histocompatibility antigen
humoral immunity
hypersensitivity
immune system
immunity
immunoglobulin (Ig)
immunological memory
immunological tolerance
isograft
killer T cell

light chain
lymphatic system
lymphoid tissue
lymphokine
lysin
major histocompatibility complex (MHC)
monoclonal antibody
monokine
plasma cell
Rhesus (Rh) factor
second-set reaction
suppressor T cell
T cell receptor
T lymphocyte (cell)
vaccination
variable (V) region

QUIZ QUESTIONS

1. All of the following are classified as cell-mediated immunological responses or activities except for
   A. production of lymphokines.
   B. recognition of antigens as self or nonself.
   C. production of antibody molecules.
   D. production of monokines.

2. Which of the following statements about antibody molecules is not true?
   A. All are proteins.
   B. All contain carbohydrate molecules.
   C. All are immunoglobulins.
   D. All have light and heavy chains.
   E. All are produced by T cells.

3. _____ is the key feature of the genetic mechanism that allows production of the huge diversity of antibody molecules.
   A. Somatic recombination
   B. The presence in the genome of a huge array of structural genes, each coding for a different antibody protein
   C. The operation of single-point mutations
   D. The operation of differentiation factors

4. Proteins encoded by the _____ define "self."
   A. major histocompatibility complex

   B. J segments
   C. V segments
   D. leader sequence

5. Production of antibody molecules by plasma cells requires interactions of _____ and _____.
   A. stromal cells; B cells
   B. T helper cells; B cells
   C. T suppressor cells; plasma cells
   D. macrophages; plasma cells

ESSAY QUESTIONS

1. Contrast humoral and cellular immunity. What is the difference between active and passive immunity? What is allergy?

2. What is a monoclonal antibody?

3. Where do B-cells originate? What distinguishes B and T cells? What is a plasma cell?

4. Trace the steps of an immune response to an invading bacterium.

5. What are histocompatibility antigens and what role do they play in tissue-graft rejection?

6. Account for the disappearance of Rh disease of the newborn.

REFERENCES AND SUGGESTIONS FOR FURTHER READING

Ada, G. L., and Nossal, G. (1987). The clonal selection theory. *Scientific American 257*, July, 62–69.

A stimulating discussion of a fundamental concept of immunology.

Benjamini, E., and Leskowitz, S. (1987). *Immunology: A Short Course*. Alan R. Liss, New York.

A well conceived introductory text.

French, D. L., Laskov, R., and Scharff, M. D. (1989). The role of somatic hypermutation in the generation of antibody diversity. *Science 244*, 1152–1157.

The latest information on the mechanisms by which the body generates an enormous repertoire of antibodies before its first contact with antigen.

Gallo, R. C., and Montagnier, L. (1988). AIDS in 1988. *Scientific American 259*, October, 40–48.

A summary of where we stand—by the two investigators who established the viral cause of AIDS.

Gillis, S. (1987). *Recombinant Lymphokines and Their Receptors*. Marcel Dekker, New York.

Data on lymphokines accumulate so quickly no book can be entirely current, even on the date of publication. Nonetheless, this book gives excellent summaries of knowledge on the interleukins, interferons, TNF, lymphotoxin, and the rest of the players.

Golub, E. S. (1987). *Immunology: A Synthesis*. Sinauer/Blackwell Scientific, Boston.

An attractive and highly readable summary, said by one reviewer to be as compelling as an adventure novel.

Haseltine, W. A., and Wong-Staal, F. (1988). The molecular biology of the AIDS virus. *Scientific American 259*, October, 52–62.

A discussion of the three regulatory genes that direct the enzymatic machinery of body cells to make a new AIDS virus.

Heyward, W. L., and Curran, J. W. (1988). The epidemiology of AIDS in the U.S. *Scientific American 259*, October, 72–81.

Since 1981, more than 130,000 people in the U.S. have contracted AIDS. By 1992, there may be more than 300,000 cases. These and other sobering epidemiological data are discussed in this article.

Jerne, N. K. (1973). The immune system. *Scientific American 229*, July, 52–60.

Jerne was the imaginative investigator whose ingenious experiments provided direct support for the clonal selection theory by demonstrating that different lymphoid cells do indeed make different antibodies.

Kimball, J. W. (1986). *Introduction to Immunology*, 2nd ed. Macmillan, New York.

An excellent text by a popular teacher.

Kindt, T. J., and Capra, J. D. (1984). *The Antibody Enigma*. Plenum, New York.

The authors review the great debate surrounding the question: how can there be a separate gene for each of the many thousands of antibodies? The story is interesting, in part, for the failure of so many workers to foresee the harmonious solution later provided by molecular biology.

Köhler, G. (1986). Derivation and diversification of monoclonal antibodies. *Science 233*, 1281–1286.

This Nobel Prize lecture by a codiscoverer of monoclonal antibodies includes a personal account of how the discovery was made.

Leder, P. (1982). The genetics of antibody diversity. *Scientific American 246*, May, 102–115.

A clear account of the answer to the question discussed by Kindt and Capra (see above)—by the brilliant investigator who helped find the answer.

Marrack, P., and Kappler, P. (1986). *Scientific American 259*, February, 36–45.

Summary of the experiments which identified the receptor molecule enabling T cells to function.

Milstein, C. (1980). Monoclonal antibodies. *Scientific American 243*, October, 66–74.

Milstein, C. (1986). From antibody structure to immunological diversification of immune response. *Science 231*, 1261–1268.

Two articles by a codiscoverer of monoclonal antibodies. The second is his remarkable Nobel Prize address.

Multiple authors (1988). Special issue on AIDS. *Science 239*, 573–622.

A special issue of *Science* containing 8 authoritative articles on all aspects of the AIDS problem—from the properties of the virus to legal and ethical issues arising in this twentieth-century scourge.

Nisonoff, A. (1985). *Introduction to Molecular Immunology*, 2nd ed. Sinauer, Sunderland, Mass.

A concise and perceptive introduction.

Rosen, F. S., Steiner, L. A., and Unanue, E. R. (1989). *Macmillan Dictionary of Immunology*. Macmillan, New York.

Our colleagues have performed a valuable service in preparing this useful dictionary.

Sabo, V. L., and Gefter, M. L. (Eds.) (1981). *Cellular Immunology: Selected Readings and Critical Commentary*. Addison-Wesley, Reading, Mass.

Helpful comments by the editors set these reprints of classic papers into a coherent context.

Tonegawa, S. (1985). The molecules of the immune system. *Scientific American 253*, October 122–131.

A splendid article by a winner of the Nobel Prize in Medicine.

Unanue, E. R., and Benacerraf, B. (1984). *The Antibody Enigma*. Williams & Wilkins, Baltimore.

Admirably brief review of a dynamic field by distinguished authors.

Vitetta, E. S., and 3 other authors. (1989). Cellular interactions in the humoral immune response. *Advances in Immunology 45*, 1–106.

A scholarly review of an important and complex system.

Yancopoulos, G. D., and Alt, F. W. (1988). Reconstruction of an immune system. *Science 241*, 1581–1583.

A special article—called a "perspective"—that briefly summarizes the workings of the immune system and discusses various immune deficiency states, including AIDS.

SECTION 2
MAINTAINING INTERNAL CONSTANCY

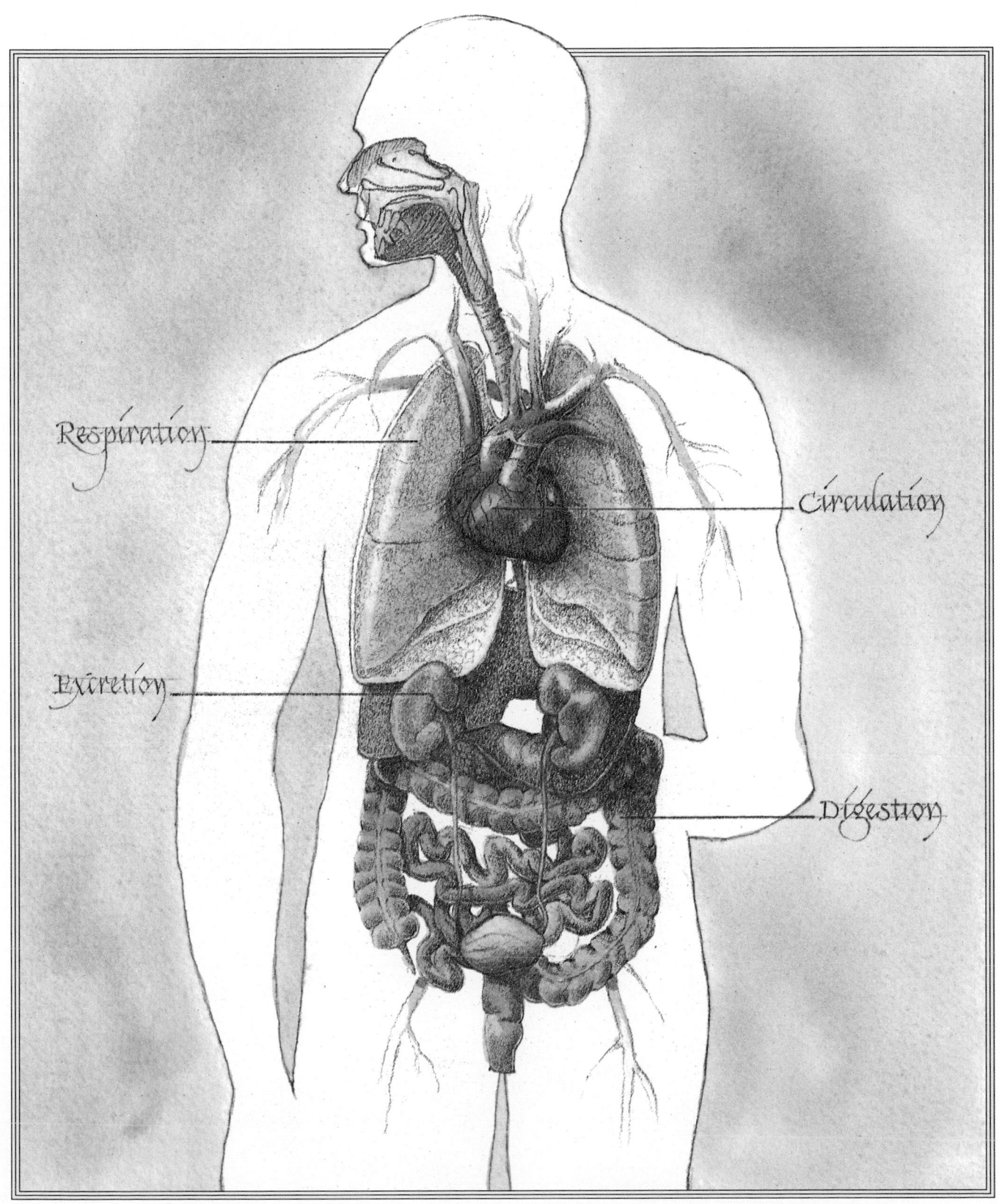

Respiration

Circulation

Excretion

Digestion

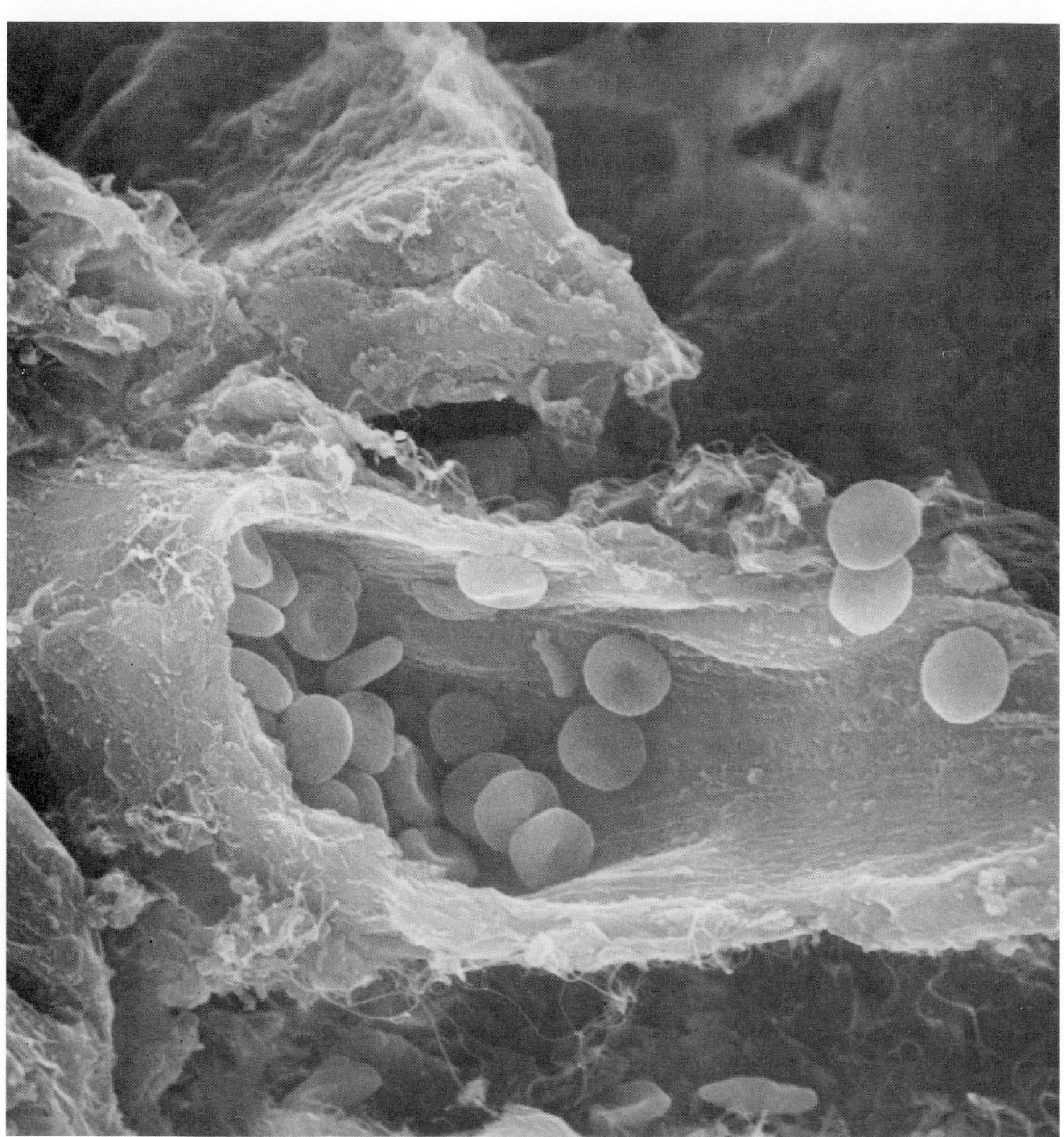

TRANSPORT

Consider the following mechanisms, all of which were discussed in earlier chapters: receptor-mediated endocytosis, oxidative phosphorylation, photosynthesis, mitosis, and crossing-over. What do they have in common? The answer: an orderly movement of elements—substrates, electrons, chromosomes, whatever—from one location to another.

In protists and other single cells, **diffusion** plays a major role in moving materials (see Chapter 6). Solutes within such cells exhibit ordinary random thermal agitation, which distributes them evenly unless intracellular membranes get in the way. Diffusion also facilitates the movement of molecules into cells from the outside and between cells within colonial and multicellular organisms.

However, diffusion is a slow process, and other mechanisms must help in moving materials around in multicellular organisms. Special transport systems are necessary when materials needed by a multicellular organism are not to be found in the vicinity of the cells where they will be used, or when materials produced by certain cells are needed by other cells that are far away from points of release. The various transport systems that evolved at both the cellular and multicellular levels nicely meet these needs.

One prevalent mechanism is the bulk movement of cytoplasm termed **cytoplasmic streaming** (Figure 24-1). Another is the movement of digestion products around the cell within food vacuoles. These mass flow systems, which move substances rapidly and in large quantities, depend upon the movements of the actin-containing microfilaments of the cytoskeleton (see Chapter 7).

In the relatively small group of plants described as nonvascular because they lack transport vessels, movement of materials is almost entirely by diffusion and cytoplasmic movement. Most of their cells synthesize foodstuffs from molecular precursors that enter directly from the environment. Hence, the cells of these plants are relatively independent of one another. Materials transported between cells go across cell walls and membranes or through **plasmodesmata,** tiny cytoplasmic threads that join cells (see Chapter 6).

Because of the relatively slow transport of materials by diffusion and other processes at the cellular level, nonvascular plants like mosses and algae are restricted in both habitat and size. These organisms generally inhabit aquatic or moist environments (see Chapter 41). They survive in dry habitats only because they are able to become dormant until moisture is available. And, although some can become relatively large, they are large in only one or two dimensions, so that each cell is within a short distance of the environment. The evolution of vascular tissue (xylem and phloem) allowed vascular plants to colonize the land more effectively than other mosses or algae.

Transport
All living organisms, whether unicellular or multicellular, possess systems that move materials from one place to another. This scanning electron micrograph shows a small human vein, cut open to reveal red blood cells within it.

TRANSPORT IN VASCULAR PLANTS

The evolutionary emergence of complex plants had to await the invention of more effective methods of transport—a network of tubular structure or ducts.

A plant's ability to transport throughout its body diverse organic and inorganic materials—along with vast quantities of water—is critical to its survival. Although the root endings of a large tree may be far removed from its topmost leaves, the water and solutes that move from the soil into these root tips course rapidly upward. Plants have no hearts to drive the sap upward. Nevertheless, in tall trees sap rises hundreds of feet. This is astonishing when one considers that ordinary atmospheric pressure can raise a column of water only 32 feet. Obviously, a vast amount of energy must be expended in moving water through a plant. But the tissue elements composing the vessels are dead cells incapable of generating active energy. How does the system operate?

As we saw in Chapter 7, a vascular, or duct, system connects the roots of plants to the leaves. In **vascular plants,** two kinds of ducts are arranged in specialized bundles in roots, stems, and leaves. These are the **xylem** and the **phloem.** Together they form a continuous system of pipes that penetrates nearly every part of the plant. Xylem carries water and inorganic solutes upward from roots to stems and leaves. Xylem sap also includes a variety of organic solutes (e.g., water-soluble plant hormones such as cytokinins) in low concentrations. Products from the leaves—mainly consisting of sucrose manufactured in photosynthesis—move through the phloem from the leaves to the roots, flowers, and buds (Figure 24-2). Vascular tissue forms the veins in leaves and the wood in the stems and roots that support the plant.

How do materials move these long distances to the various parts of a tree, defying gravity as they do so? To answer this question we must consider events occurring in the leaves (transpiration), in the roots (active ion transport and water absorption), and in the xylem and the phloem. We must also examine many of the properties of water itself.

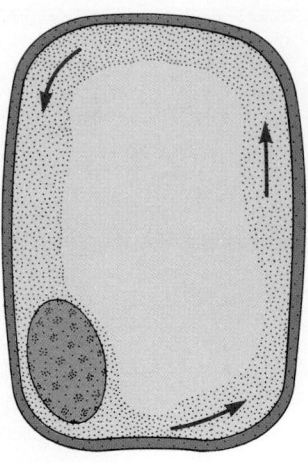

24-1 Cytoplasmic streaming
In a plant cell, the cytoplasm flows in one direction around the central vacuole. The cell nucleus is at the lower left corner.

TRANSPIRATION

What Is Transpiration?

In the early eighteenth century, the English clergyman Stephen Hales calculated that pound for pound one sunflower imbibes and eliminates each day 17 times more water than a human being does. Thus the total quantity of water absorbed by any plant is enormous—far greater than that used by any vertebrate animal of comparable weight. Why? Because a vertebrate recirculates most of its water throughout its body again and again in the form of blood plasma and other fluids. In plants, more than 90% of the water taken in by the roots is released by the plant into the air as water vapor. In one growing season, a single wheat plant, tomato plant, and corn plant loses, respectively, about 95, 125, and 206 liters of water.

Loss of water vapor by the plant body is termed **transpiration.** It can occur from any part of the plant, but leaves are by far the most important organs of transpiration. Hales found that transpiration depends on the exposure of leaves to the air (Figure 24-3).

Is there a good biological reason why plants lose such large quantities of water to transpiration? In part the answer is found in the structure and function of leaves. In part it relates to the nature of roots.

Role of Leaf Function

Leaf structure was described and illustrated in Chapter 10. There we saw how specialized **guard cells,** under the influence of osmosis, control the openings and closings of the **stomata** in the leaf's surface.

If photosynthesis is to proceed at maximal efficiency, a plant must expose the largest possible leaf surface to sunlight. But photosynthesis also depends on the availability of CO_2 from the air. In order for CO_2 to enter the plant cell, which it does by diffusion, it must be in solution. To go into solution, CO_2 must come in contact with a moist surface. But moisture evaporates whenever water is exposed to air. Water that evaporates from the surfaces of the spongy mesophyll cell layer lining the intracellular spaces (see Figure 10-26) is continuously replaced from the small leaf veins. This water loss by evaporation is transpiration.

Plants have developed a number of special adaptations that limit evaporation. Deciduous plants, for example, drop their leaves and become dormant during seasons

Phloem——
Xylem——

24-2 Transport in a vascular plant
Phloem channels carry sucrose and other nutrients upward and downward from the sites of photosynthesis. The xylem moves water and inorganic ions upward from the roots.

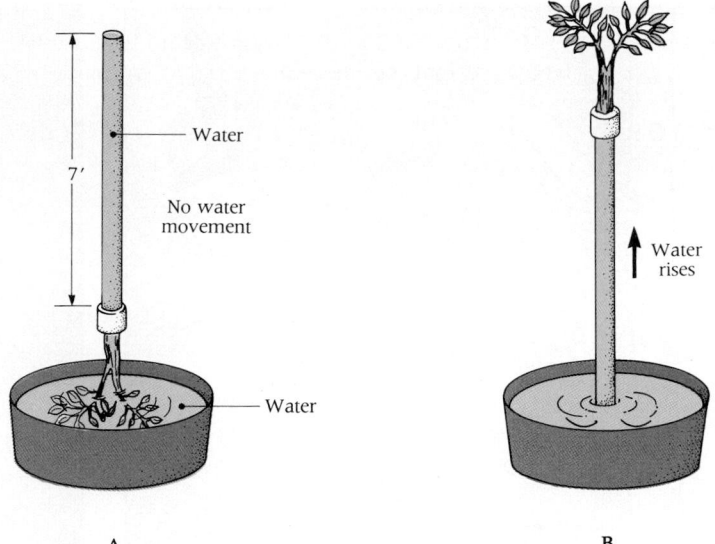

when water is either scarce or frozen. This stratagem conserves water, but it also prevents photosynthesis.

Acquiring CO_2 and controlling water loss are the two major functions of stomata. Hence, it is not surprising that CO_2 concentrations in the air in a leaf serve as cues that control stomatal opening. Low CO_2 concentrations and high relative humidity cause stomatal closing. (The mechanisms of stomatal opening and closing were discussed in Chapter 10.) Plants move and transpire much more water than would be necessary for the transport of minerals to the leaves. In a sense, this water loss is a detrimental side effect of the need to exchange gasses with the atmosphere. Not all water loss is useless, however. Water evaporation cools the leaves, especially on hot, sunny days when water loss is greatest. And, as we shall see, evaporation of water from leaf surfaces creates the forces that pull water from the soil into the plant's roots.

WATER TRANSPORT

Roots: Absorbers of Water and Nutrients

The root system serves to anchor a plant in the soil. It also collects from the soil nutrients and the water needed to meet the tremendous requirements of the plant's leaves. Thus, plants must have a root system with a huge surface area. The roots make contact with the surrounding soil, absorbing from it the nutrients that are present as very weak solutions of ions. Since this soil solution is unstirred, the replacement of ions absorbed from a given zone depends on diffusion of ions from elsewhere. This is a slow process and soil can readily become depleted of nutrient ions wherever it contacts a root. In striking contrast, a leaf makes contact with a rapidly shifting collection of mobile air molecules.

In response to these two features of the soil environment—extreme dilution and nutrient immobility—root systems have evolved that, by extensive branching and ramification, present an astonishingly large surface area to the surrounding soil. The contact area is expanded even further by the emergence of **root hairs,** fingerlike projections of a root that project into the soil at right angles to the root axis (Figure 24-4). Some plant species have as many as 2500 root hairs per square centimeter of root surface. In the 1930s, H. J. Dittmer convincingly showed how very large such a root system can be. He planted a winter rye plant (*Secale cereale*) in a box of soil 12 inches square and 22 inches deep. At the end of four months he carefully freed the root system from the surrounding soil and measured it. The total length, root hairs excluded, was 387 miles; the total surface area was 2554 square feet. When the length and surface area of the root hairs were added in, the totals for the entire system were nearly 7000 miles and 7000 square feet!

As discussed in Chapter 40, the roots of certain plants are associated with fungi that perform much of the work of absorbing nutrients from the soil. This association is mutually beneficial, the plant receiving water and minerals from the fungus and

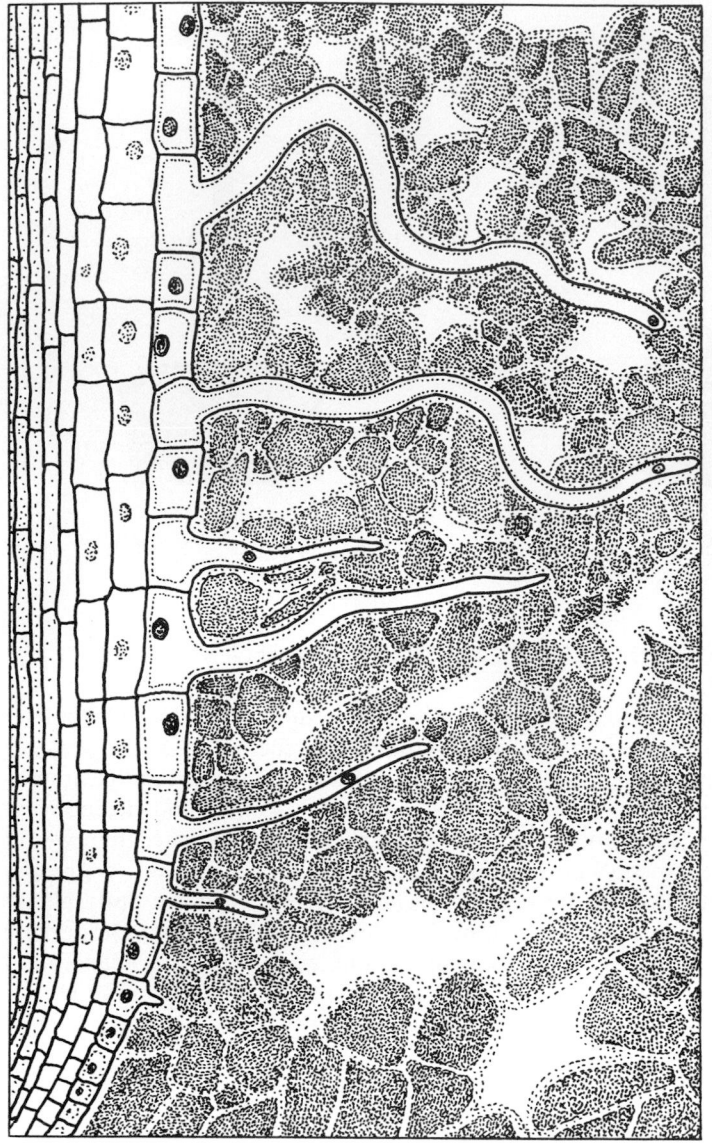

A

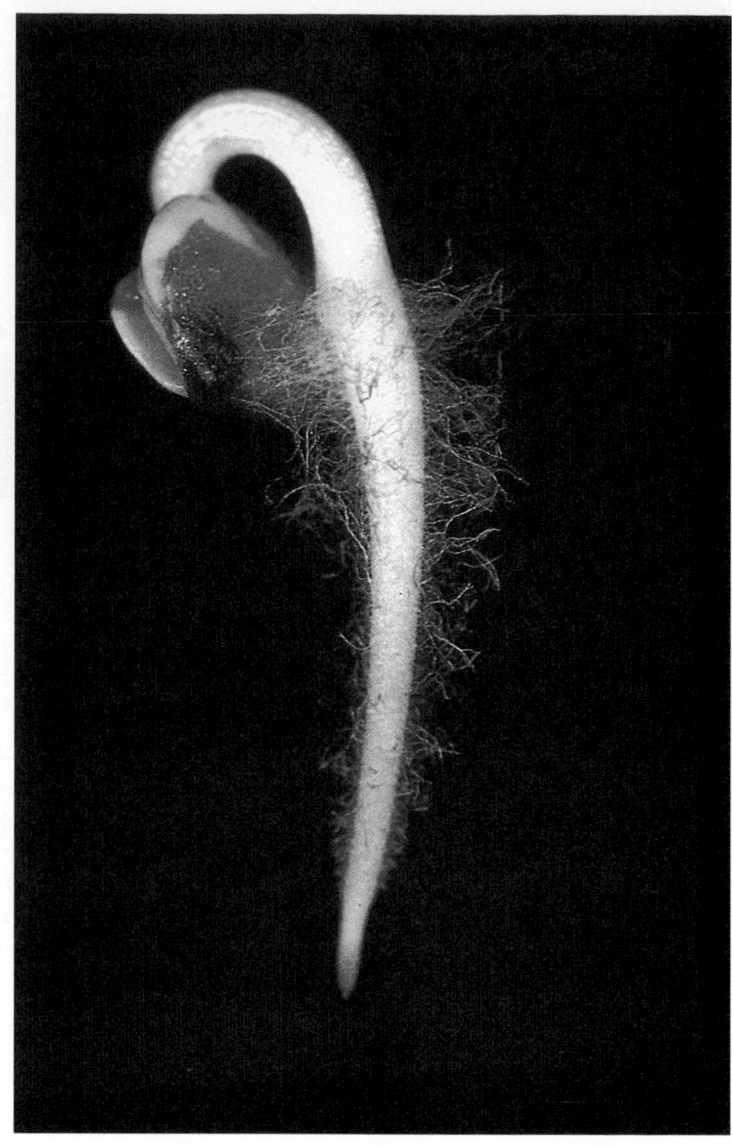

B

the fungus receiving energy from the plant. Such associations are called **mycorrhizae** (see Figure 40-20).

Like the stem, the root has an outer **epidermis** (Figure 24-5). However, the root has no waterproof cuticle like that of leaves—with the possible exception of young roots in some plants. Within the epidermal layer of a root—for example, of a dicot, one of two broad categories of seed plants—is a cylindrical **cortex** layer that is several cell layers thick. It is composed of parenchyma cells. The innermost layer is a cylindrical tube one cell thick, the **endodermis,** that does have a waterproof waxy zone called the **Casparian strip.**[1] Just inside the endodermis is another cylinder of cells, the **pericycle,** which is the site of origin of **lateral roots.** Within the pericycle—at the center of the root—are the vascular elements, the xylem and the phloem, which are continuous with their counterparts in the stem.

Almost all the water taken from the soil enters through the younger parts of the root. Absorption takes place directly through the epidermis of the root, outgrowths of which are the root hairs. These hairlike extensions of the epidermal cells have surface membranes containing a variety of ion transport channels that actively pump ions into the hairs.

As shown in Figure 24-5, two pathways are followed by water and its dissolved minerals.

24-4 Structure of root hairs
(**A**) *Root hairs extending between soil particles. The hairs increase the surface area of the root, thereby increasing uptake of water and ions from the surrounding soil.* (**B**) *A primary root of radish,* Raphanus sativus, *showing fine root hairs extending at right angles from the root.*

---

[1] This oddly named structure was first noticed by the German botanist J. X. Robert Caspary (1818–1887).

24-5 Water and mineral transport in a root
The apoplastic pathway (red) transports materials between cells. The symplastic pathway travels through the cells (black). The Casparian strip forces water and minerals through endodermal cell cytoplasm and prevents backflow into the cortex.

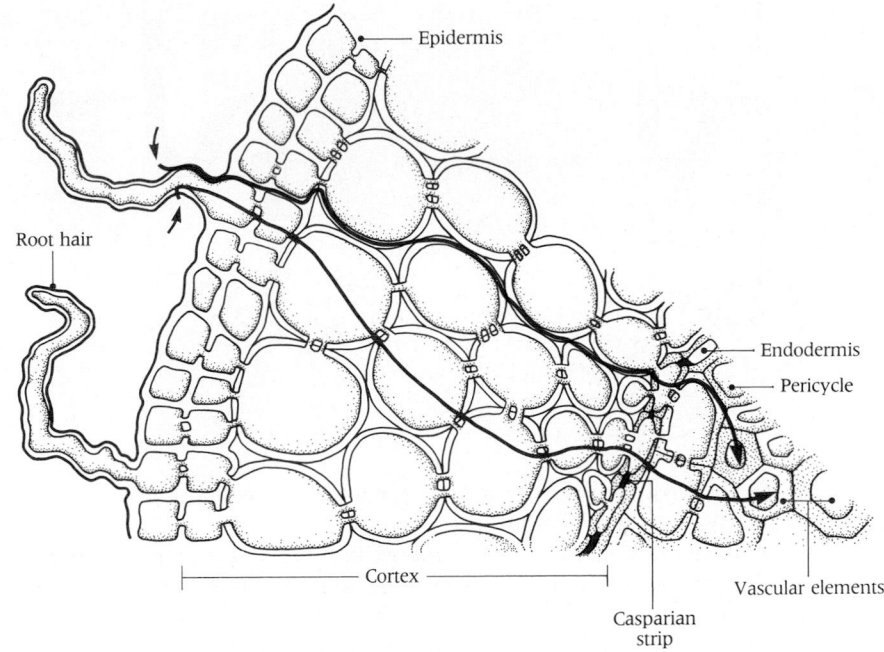

Epidermis

Root hair

Endodermis

Pericycle

Cortex

Vascular elements

Casparian strip

■ In the **apoplastic pathway,** water moves between cells in epidermis and cortex. When water reaches the endodermis, further movement is obstructed by the Casparian strips, which form a tight barrier between the cortex and vascular channels of the root (Figure 24-6). To reach the root interior, water and minerals must find a selective, semipermeable membrane they can cross. This is provided by the endodermal cells, which they pass through before entering the xylem.

■ In the **symplastic pathway,** a second pathway, water moves from cell to cell via plasmodesmata.

24-6 Movement of ions from root epidermis to xylem

Ions outside cells (open dots) are transported into the cytoplasm (closed dots) of epidermal and cortical cells. These ions bypass the Casparian strip barrier and eventually enter the xylem.

Active transport concentrates ions within the vascular tissue, which is interior to the endodermis. As discussed in Chapter 4, solute potential influences water potential by making it more negative. Therefore, the accumulation of ions lowers water potential. As a result of the lower water potential in the vascular region, osmosis moves water across the endodermis into the vascular tissue. This influx of

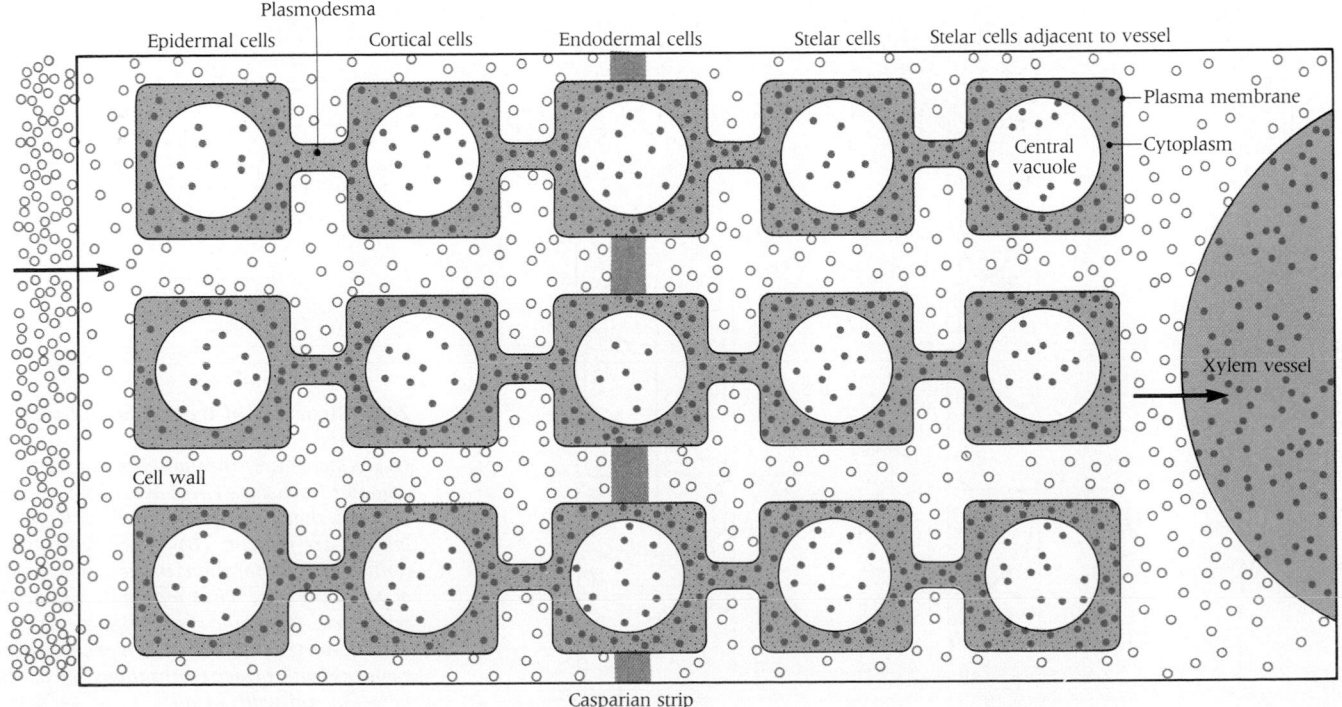

Plasmodesma

Epidermal cells   Cortical cells   Endodermal cells   Stelar cells   Stelar cells adjacent to vessel

Plasma membrane

Central vacuole

Cytoplasm

Xylem vessel

Cell wall

Casparian strip

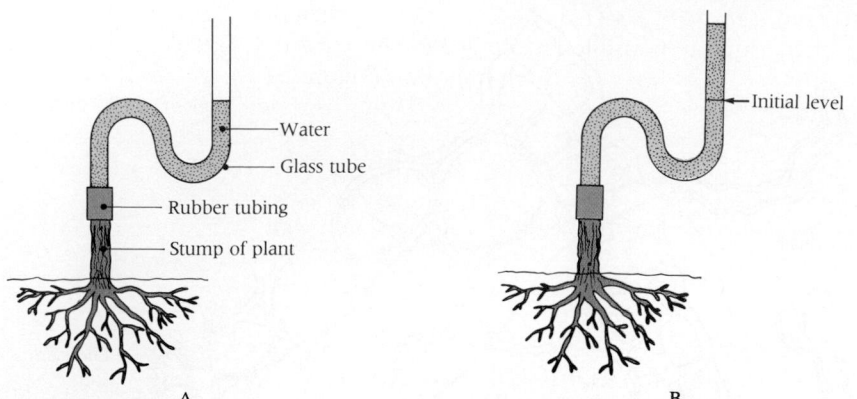

24-7 Demonstration of root pressure
(A) *A curved glass tube is sealed to a freshly cut plant and is then partially filled with water.* **(B)** *Sap rising into the tube from the roots raises the water level in the tube until the root pressure equals hydrostatic pressure.*

water raises the hydrostatic pressure in the xylem vessels, which thereby increases **root pressure.** Water then flows up the stem.

This is readily demonstrated in some plants by cutting the plant off near the ground and sealing the stump into a glass tube (Figure 24-7). Sap rises into the tube. The root pressure is measured from the height of the fluid in the tube. Due to the Casparian strip (see Figure 24-5), the endodermis surrounding the vascular elements of the root blocks the return of fluid from the root back into the soil. The resulting positive pressure drives the sap in the only direction available to it—upward. However, root pressure is not the main force moving water to the leaves of plants.

Roots use much energy for active transport and they need an adequate supply of oxygen. When the soil is wet and well aerated, root pressure is high. If the air

24-8 Guttation droplets on strawberry leaves
High root pressure forces water upward and forms droplets at the edges of leaves.

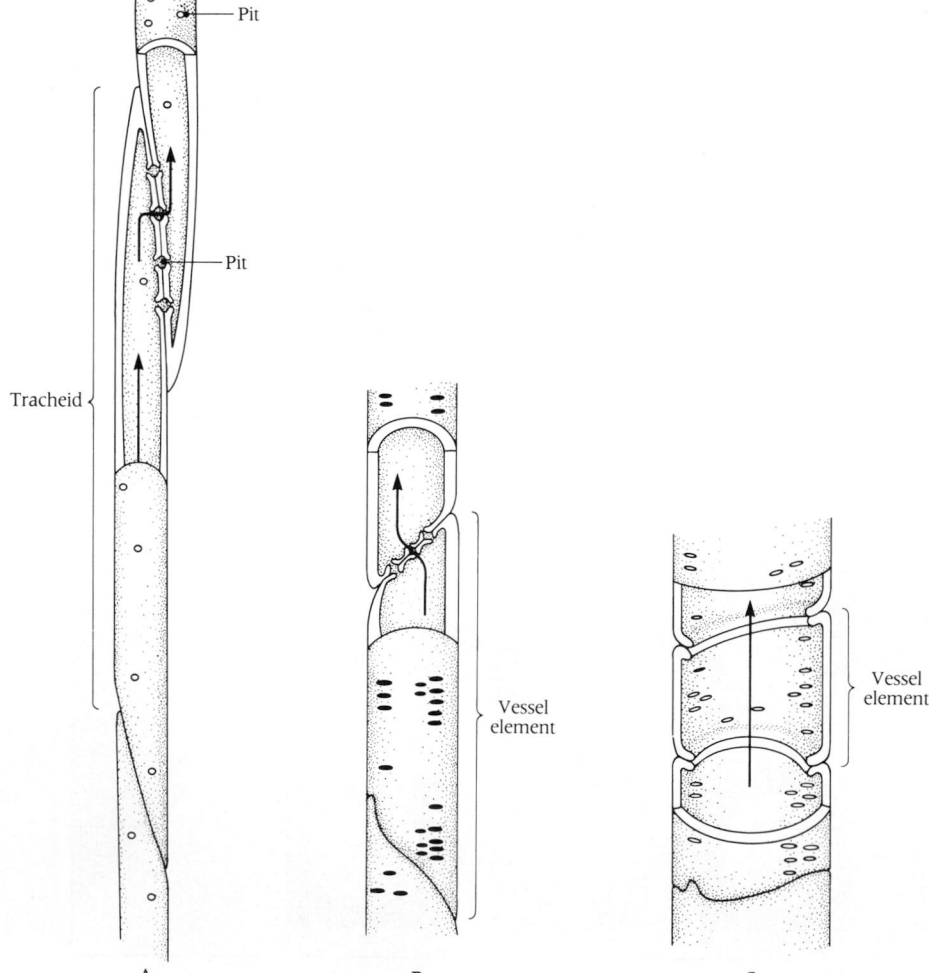

24-9 Structure of tracheids and vessel elements
(A) *Water passes through pits that connect adjacent tracheids. These areas lack a secondary cell wall, though the primary cell wall and middle lamella are present.* **(B)** *Vessel elements are connected by perforations in the primary cell walls of adjoining cells. Thus materials can move from cell to cell.* **(C)** *Some species display an evolutionary trend toward wider and shorter elements with larger perforations, or as shown here, no end walls at all.*

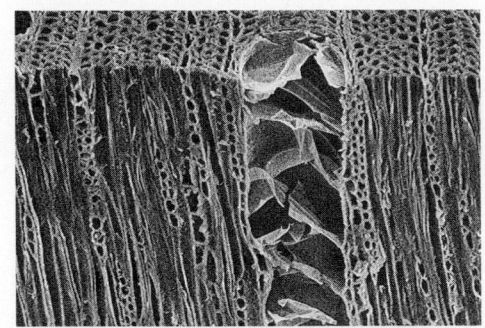

A

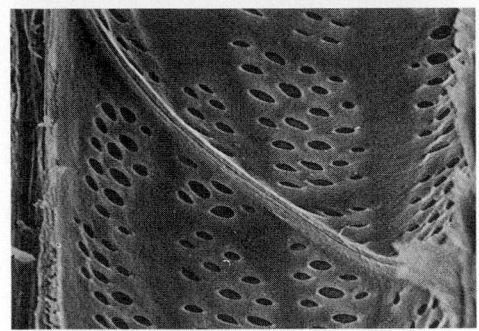

B

24-10 Appearance of tracheids and vessel elements in wood
(A) In this scanning electron micrograph of the heartwood of a walnut tree, Juglans, a prominent vessel element is shown in section, surrounded by many tracheids. (×200) (B) Unlike heartwood, the sapwood of walnut trees has open vessel elements that function in conduction. The end wall connecting adjacent cells appears as a perforated plate. Pits are found in the side walls. (×1000)

24-11 Structure of xylem
(A) Orientation of cuts to produce sections shown in (B) and (C). (B) The secondary xylem of pine shows many tracheids, which conduct sap. The rays participate in lateral conduction. (C) The secondary xylem of an angiosperm shows vessel elements as sap-conducting cells. The fibers strengthen the plant body.

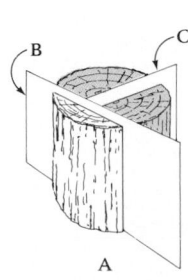

is also saturated with water vapor, water does not evaporate readily from the leaves. This situation is responsible for the dewlike water droplets at the tips of blades of grass and other leaves in the early morning (Figure 24-8). These droplets are not dew, which is water that has condensed from the air. Rather they come from within the leaf by a process known as **guttation.** The water of guttation is literally forced out of the leaves by root pressure. Guttation occurs only in short plants, at most a foot or two tall.

Root pressure is highest at night when transpiration is minimal, but even then it is never high enough to force water to the top of a tall tree. Indeed, many plants, such as conifers (among the tallest trees known) develop no root pressure at all. How then does sap rise? To answer this question, we must look again at the xylem and the phloem. It would be useful at this point to reexamine Figures 7-1, 7-2, and 7-3, which show the structural characteristics of xylem and phloem.

Structure and Functions of Xylem

Xylem has two functions—transport and support. Both conducting and nonconducting cells contribute to support. As discussed in Chapter 7, once the conducting cells of xylem have formed tubular elements, they die. When the cell contents disintegrate, a firm sap-filled hollow tube remains. These tubes eventually extend the length of the plant.

The two types of conducting cells in xylem are **tracheids** and **vessel elements.** The conducting cells of gymnosperms and the primitive vascular plants consist only of tracheids—long, thin, overlapping cells with tapering end walls (Figures 24-9 and 24–10). In a gymnosperm, such as pine, numerous **pits** in the walls of tracheids permit rapid passage of sap from cell to cell. Actually, these are not holes but only spots where the adjacent walls are relatively thin. Water and dissolved minerals pass readily through these thin places in the walls. Pits occur in both the side and end walls of tracheids. This permits adjustment of the water supply to different sides of the tree.

In more advanced angiosperms, there is further specialization of xylem cells, which results in the formation of holes. This allows sap to flow directly from one cell to the next. In more advanced angiosperms, the end walls are absent, and the cells function like sections of pipe. In addition, there is a tendency for the diameter of the "pipelines" to increase somewhat as they rise from roots to leaves. When this occurs, it reduces the frictional drag on water passing through xylem and permits it to travel faster. It is these hollow xylem tubes of angiosperms (and a few gymnosperms) that are termed vessel elements. Much of the wood of an angiosperm such as oak is made up of these conducting cells (Figure 24-11).

Structural Organization in the Stem

The term **stem** applies to all parts of the shoot system other than leaves. The tissues of the stem of a woody plant, such as ash or box elder are best studied in

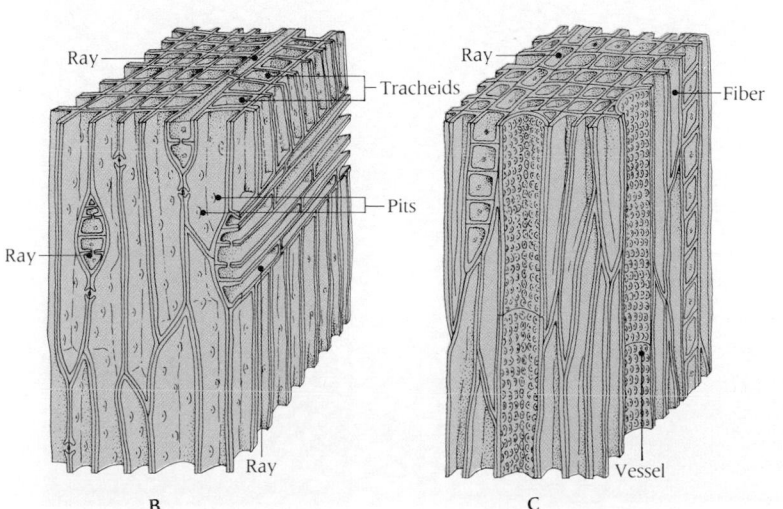

a cross-section of the stem. We see in Figure 24-12 that the tissues are arranged in concentric cylinders that surround a central mass of parenchyma cells called **pith.** The first cylinder around the pith is made up of layers called **primary xylem** and **secondary xylem.** Secondary xylem, or **wood,** can be differentiated into an inner, supporting cylinder of **heartwood** and an outer, water-transporting cylinder of **sapwood.** This is followed, in order, by cylinders of meristematic tissue (the **cambium**), secondary phloem, and primary phloem. Then there is the cortex, which consists of parenchyma cells, more meristematic tissue (the **cork cambium**), epidermis (or periderm), and in some cases, **cork.**

The cylinders of meristematic tissue cause the stem to grow in diameter as it gets older. The protective tissue of the cork consists of cells with walls containing a waxy waterproofing substance, **suberin,** that protects underlying tissues from water loss. However, it also keeps the cork cells themselves from obtaining water. They die and as they slough away they are replaced by new cells from the cork cambium.

The principal layer of meristematic tissue is the cambium, between the xylem and the phloem. In each growing season, new cells on its inner surface differentiate into new xylem while those on its outer surface become new phloem. Thus each season a new layer of xylem is added inside the cambium, nearer the pith, and a new layer of phloem is added outside the cambium nearer the cortex. Cells laid down in the spring are larger than cells coming later. As a result, phloem and xylem in cross section show annual growth rings (Figure 24-13). In climates without well-marked seasons, rings are faint or absent.

Cells of the xylem function for years. In time, however, they become infiltrated with gums, oils, resins, and pigments. Many of the xylem cells in the wood of a tree like the oak are dead vessel elements (heartwood) that have lost their conducting function and serve only as mechanical support. However, the actively growing cambium between the xylem and the phloem provides a continual supply of living cells that function in water transport as well as in support (sapwood).

As we shall discuss in detail in Chapter 41, the angiosperms are divided into two broad groups: the **dicotyledons,** or **dicots,** with some 200,000 species, and the **monocotyledons,** or **monocots,** with some 50,000 species. These terms refer to the presence in the plant embryo of either two or one **cotyledons** or seed leaves, respectively. The organization of the stem tissues we have been discussing is generally applicable to the dicots (Figure 24-14). In the monocots, which include the lily, bamboo, palm, corn, and various grasses, the same kinds of tissues are present, but they are not arranged in the series of concentric cylinders characteristic of the dicotyledons. Instead, vascular bundles of both xylem and phloem are scattered

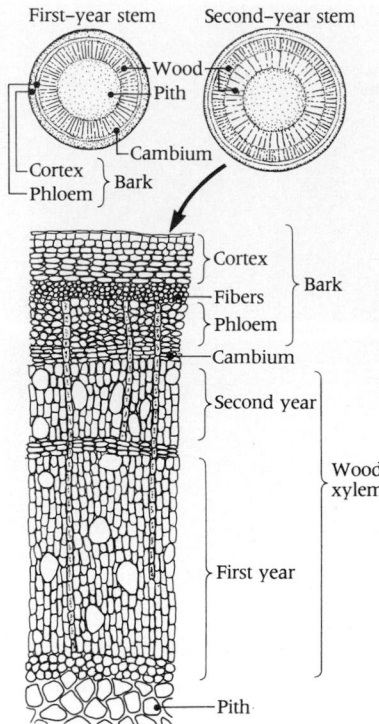

24-12 Tissue organization of a woody stem
The two cross sections at the top illustrate the concentric cylinders of tissues in first-year and second-year stems of box elder, Acer negundo. *The lower figure shows the cellular detail of the tissue layers in a stem.*

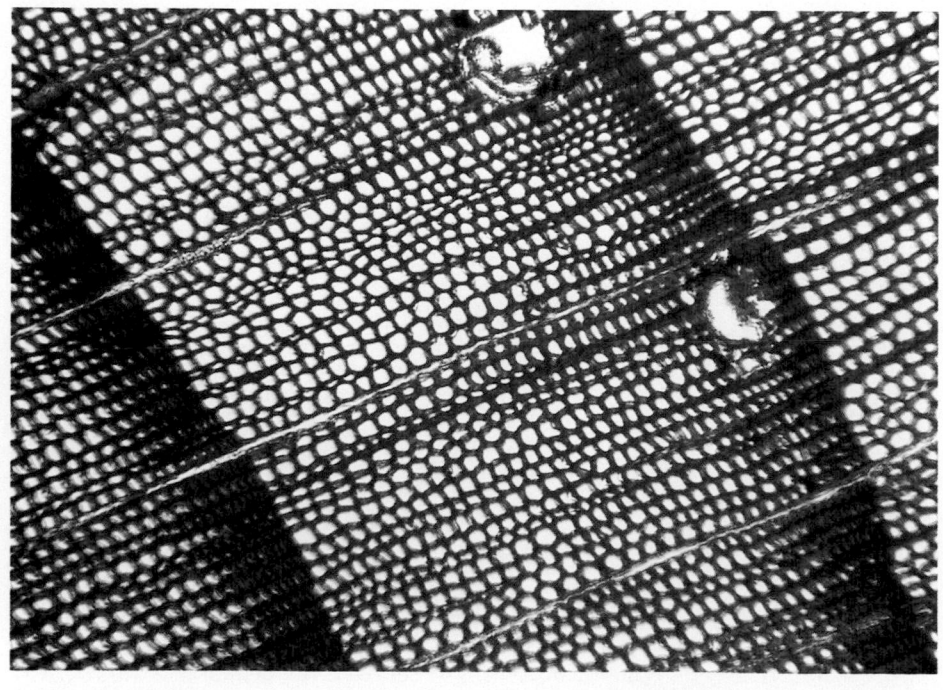

24-13 Annual growth rings
This cross section shows parts of three annual growth rings of a pine tree, Pinus. *(×80)*

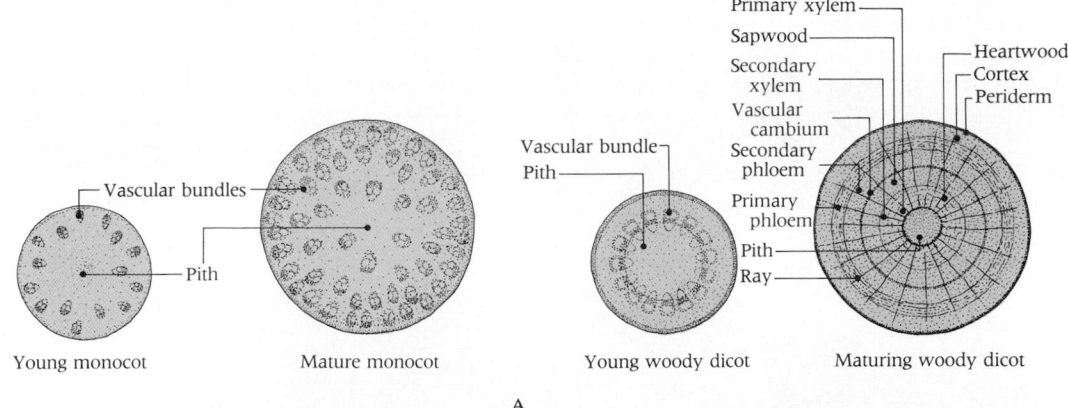

Young monocot   Mature monocot   Young woody dicot   Maturing woody dicot

A

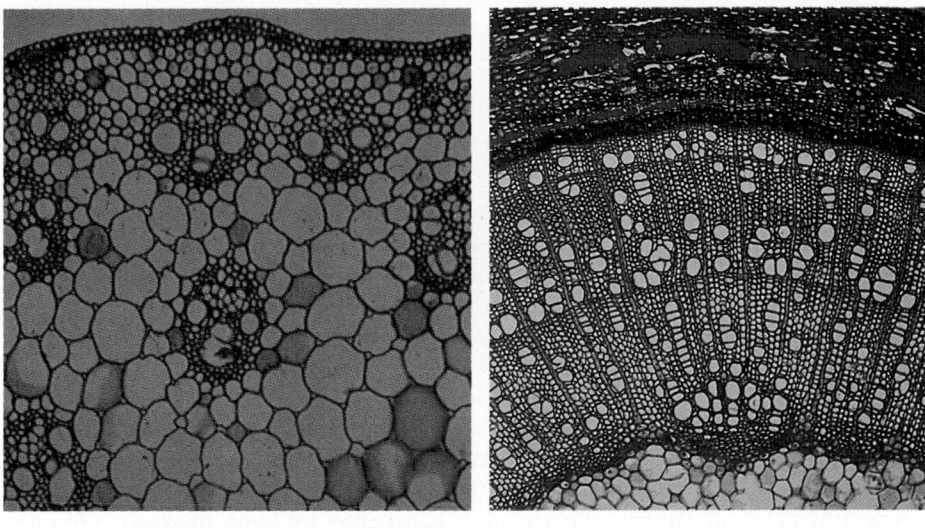

B

24-14 Structures of monocot and dicot stems (A) *As shown in the two cross sections on the left, the stem of a young monocot has vascular bundles arranged peripherally. These appear in the pith as the plant matures. On the right, the young woody dicot has peripheral vascular bundles as well. As the plant matures, the primary xylem and phloem are replaced by secondary xylem and phloem cells arising from the cambium, and the vascular bundles fuse to form a cylinder. (B) Photomicrographs of mature stems of monocot corn,* Zea *(left, ×20), and dicot maple,* Acer *(right, ×30).*

throughout the pith. Also, monocotyledons do not have continuous cambium layers. Consequently, most of them do not become continuously thicker.

How Is Transport of Sap Explained?

We have seen that water enters the plant via the roots and departs from the leaves. How it gets from one place to the other, often rising vertically over long distances, is a problem that has intrigued generations of biologists.

Children often color white flowers by placing their cut stems in a solution of vegetable dye. After the dye rises into the flower, by holding the stem of the flower to the light, one can see thin strands of color within it. Under a microscope, the dye appears inside large cells in xylem tissue; the rest of the cells in the stem keep their original color. This experiment clearly shows that xylem transports water upward. More sophisticated experiments using isotopes confirm that the water does indeed travel by way of the vessel elements (Figure 24-15).

In attempting to explain how the water rises, we can begin with simple logic. Water can be pushed from the bottom or pulled from the top. Root pressure, we noted earlier, does not develop in all plants. When present, it is insufficient to push water to the top of a tall tree. The simple cut-stem experiment just described also rules out root pressure as a crucial factor. So we are left with the hypothesis that water is pulled up through the plant body.

When water evaporates from surfaces lining the intercellular spaces within a leaf during transpiration, it is replaced by new water from within cells. This water diffuses outward across plasma membranes, which are freely permeable to water but not to the solutes of the cell. As solute concentration within a water-depleted cell increases and its water potential decreases, a gradient of water potential is established between this cell and nearby cells that contain more water. This gradient causes water to move from nearby cells into the water-depleted cells. Eventually

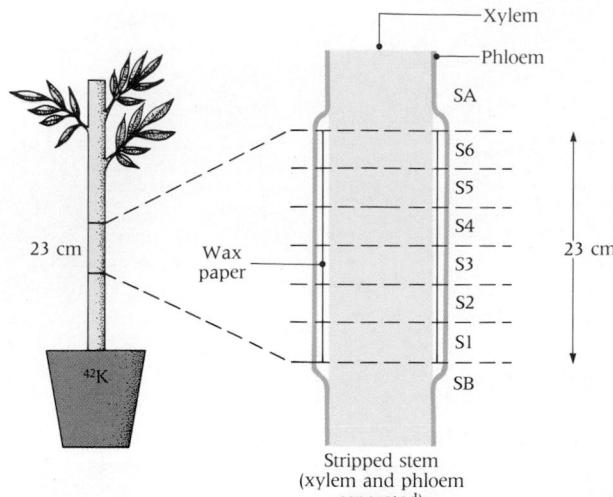

		^{42}K in phloem	^{42}K in xylem
Above strip	SA	53	47
Stripped section	S6	11.6	119
	S5	0.9	122
	S4	0.7	112
	S3	0.3	98
	S2	0.3	108
	S1	20	113
Below strip	SB	84	58

Stripped stem
(xylem and phloem
separated)

this brigade of water transfers reaches a vein. There a "pull" is exerted on the water column in the xylem.

The unique properties of water now become critical elements of the theory. Because of the extraordinary cohesion of water (see Chapter 3), the pull, or tension (the technical term for negative pressure) is transmitted all the way down the stem to the roots, so that water is withdrawn from the roots, pulled up the xylem, and distributed to the cells that are losing water to the atmosphere. This loss ultimately makes the water potential of the roots more negative, thereby increasing their ability to extract water from the soil (Figure 24-16).

This theory of water movement has been termed the **cohesion-tension theory,** because it depends on the cohesion of water, which permits it to withstand tension. However, some have suggested that the theory would better be called the **cohesion-adhesion-tension theory** (Figure 24-17), because adhesion of water molecules (by electrostatic bonds) to the highly hydrophilic (water-attracting) walls of the xylem and to the walls of leaf and root cells is as important for the upward movement of water as cohesion and tension. It also tends to prevent **cavitation,** the development of a vacuum above the water column (like that at the top of a barometer) due to the sheer weight of the water. The small internal size (in the range where capillary action can occur) of tracheids and vessels also helps prevent cavitation.

There is now no doubt that the tensile strength of water prevents pulling apart of water molecules under the tension required to move water up the xylem of tall trees. A column of water in a fine capillary tube is capable of withstanding a tension of 261 torr; the estimated tension required to move water to the top of a giant redwood is only about 20 torr.[2]

In sum, the flow of water in plants has a sound physical basis that is quantitatively described in terms of water potential, which is a measure of the free energy of water. As we saw in Chapter 6, water potential largely controls the movement of free energy in the soil-plant-atmosphere continuum.

Structure and Functions of Phloem

In addition to xylem, vascular plants have a secondary transport tissue, the phloem. Phloem is responsible for conducting various substances, especially organic compounds, throughout the plant.

Like xylem, phloem contains several cell types. In the more primitive vascular plants, including gymnosperms, the conducting cells of the phloem are the **sieve cells.** Angiosperms evolved more specialized phloem-conducting cells stacked on top of one another in the long columns called **sieve tubes.** Both the sieve cells of primitive vascular plants and the sieve tube members of angiosperms have special sieve areas—numerous pores in their cell walls through which substances can be

24-15 Demonstration that xylem conducts materials upward

In this experiment, the xylem and phloem layers are separated by wax paper; and the stem is placed in soil, to which is added water containing radioactive potassium (^{42}K). Lateral transport is prevented by the wax paper, and so upward movement of the isotope can be determined. A comparison of the amount of isotope in the xylem and phloem of stem sections shows that only the xylem channel conducts water and inorganic ions. The tabular data are expressed in parts per million.

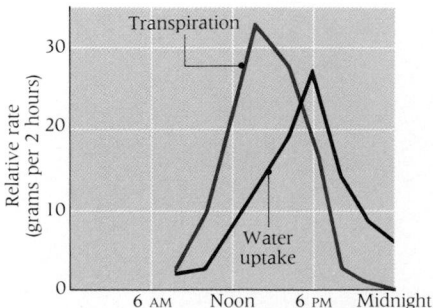

24-16 Relation between water uptake and transpiration

This experiment using ash, Fraxinus, *trees shows that an increase in water uptake follows a rise in transpiration. This suggests that evaporative loss of water generates the force responsible for water uptake.*

---

[2] A torr is a unit of pressure: 1 torr equals 1 mg of mercury in a barometer at standard conditions (0°C, sea level).

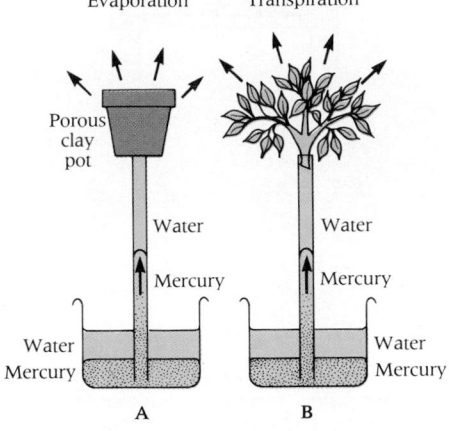

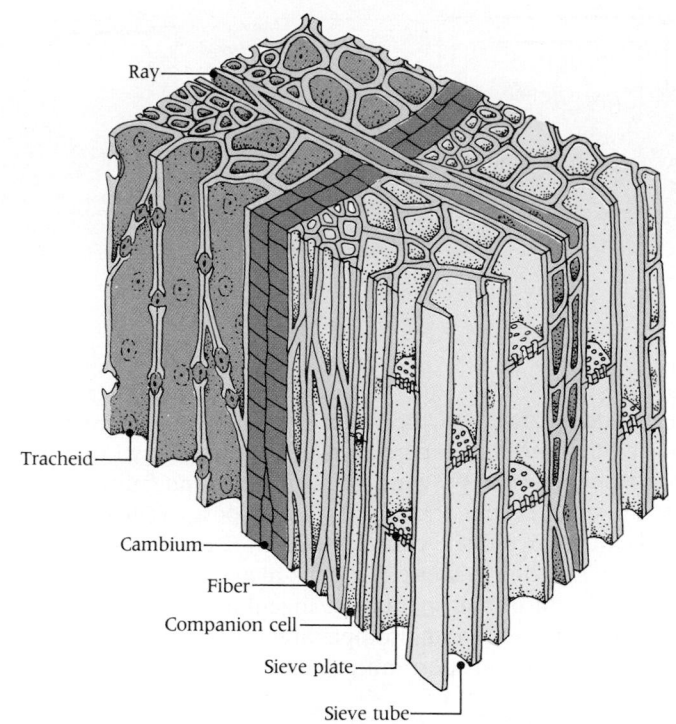

24-17 The cohesion-adhesion-tension theory of water movement (Above)
(A) A porous clay pot is filled with water and attached to the end of a long, narrow glass tube that is also filled with water. The end of the tube is placed in a beaker of mercury. As water evaporates from the pot, it is replaced by water in the tube, and mercury is pulled upward.
(B) If a plant shoot replaces the clay pot, transpiration from leaves results in sufficient water loss to create a similar negative pressure.

24-18 Structure of phloem (Right)
The xylem and phloem are separated by the cambium layer. The conducting cells of the phloem form sieve tubes. Their end walls form sieve plates.

24-19 Malpighi's girdling experiment
After removal of the bark (and phloem) encircling the tree, a sugary fluid exudes from the stripped area. However, the tree eventually dies, because the roots are unable to obtain nutrients. Girdling the tree severs the phloem channels and prevents nutrient transport to regions below the girdled area.

exchanged with neighboring cells. In angiosperms, these eventually evolved into **sieve plates** with somewhat larger pores (Figure 24-18).

The conducting cells of the phloem, unlike those of the xylem, are living, though they have lost their nuclei and most, if not all, of their mitochondria. Before a cell gains the capacity to conduct, the plasmodesmata strands passing between cells expand to form close connections between neighboring cells. In angiosperms, unequal cell division produces a large cell that degenerates to form a sieve tube member and a smaller cell that becomes an elongated **companion cell** for that sieve tube member (Figure 24-18).

In 1679, the renowned Italian physician Marcello Malpighi (1628–1694) tried to explain the functions of xylem and phloem by girdling trees (Figure 24-19). Girdling consists in the removal of bark in a complete ring around the trunk of a tree. This removes the phloem, which is attached to the inner bark, but leaves the secondary xylem, or wood, intact. After this treatment, he noticed a swelling of the bark just above the stripped area. Fluid exuding from it was sweet. We now know that it contains sucrose. The leaves remained unaffected for days or months, depending on the type of tree and its age. Eventually, however, they wilted and died, and death of the entire tree followed—usually in the next growing season.

Malpighi concluded that phloem transports nutrients throughout the tree. Without a food supply, the roots died after they had used the reserves stored below the girdle. Since the leaves remained healthy for a time, he concluded that the xylem of a girdled tree continues to transport water to them. Without a continuous water supply, the leaves would have wilted and died in hours.

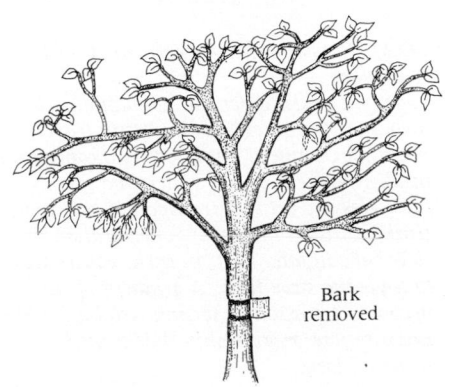

Bark removed

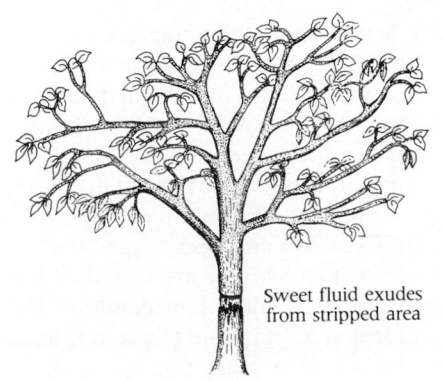

Sweet fluid exudes from stripped area

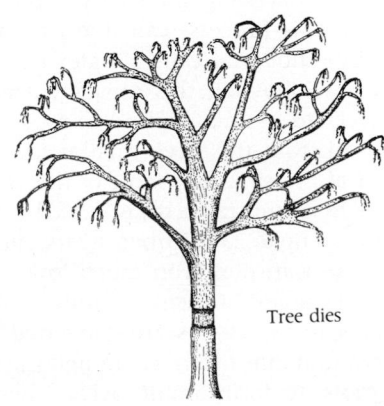

Tree dies

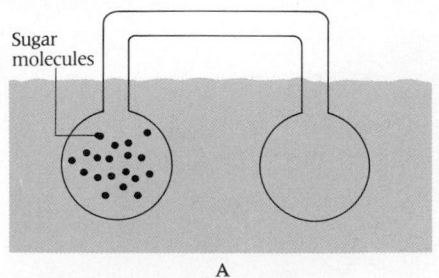

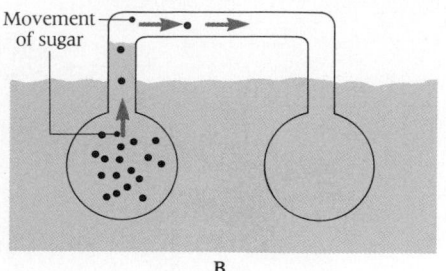

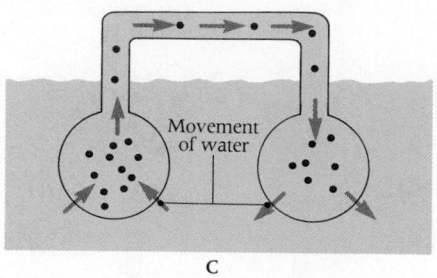

A   B   C

It thus appears that the vascular tissues divide the task of transport: the xylem moves water from the roots to the shoot system; the phloem moves food from leaves to roots. Other girdling experiments have since shown that the phloem also conducts food from leaves to growing buds, flowers, and fruits.

Biologists are still debating exactly how transport occurs in the phloem. The problem is difficult to investigate because phloem is extremely delicate; even slight disturbances to living phloem cause the sieve plates to become plugged. Transport then halts. For that reason, all manner of ingenious research methods have been devised, including some using radioisotopes and phloem-feeding aphids (Box 24-1).

Any theory of phloem transport must account for certain facts.

- The contents of conducting cells in the phloem are under pressure and a very large amount of fluid passes through each cell in a short time.
- In intact plants, substances may travel at speeds far greater than diffusion could explain.
- Direction of flow in a particular sieve element is sometimes reversed.
- Neighboring sieve elements may conduct in opposite directions at the same time.
- The small size of the pores in the sieve plates must surely hinder the flow of materials from cell to cell.
- The conducting cells contain living cytoplasm and, if cells are killed, transport in that part of the plant stops.

The **pressure-flow hypothesis,** first proposed in 1927 by Ernst Munch, proposes that solutes move as a result of turgor (or hydrostatic) pressure of osmotic origin. The principle can be illustrated by a physical model consisting of bulbs permeable only to water that are connected by glass tubes (Figure 24-20). The left bulb contains a sugar solution; the right one contains only water. When the interconnected bulbs are placed in distilled water, water enters the left bulb by osmosis. This increases the hydrostatic pressure within that bulb. This pressure forces the sugar solution into the other bulb, where it forces out water molecules. If this bulb is connected with another one containing water or sugar at a lower concentration, the solution will flow from the second to the third bulb, and so on down the line.

Transferring the model to a plant, it is postulated that the contents of sieve tubes are under higher turgor pressure in one part of the plant (the source) than they are in other parts of the plant (the sink). The result is a mass movement of sieve tube contents from source to sink. Sucrose produced by photosynthesis in a leaf cell is actively transported into the sieve tubes of the small veins (Figure 24-21). This process, called **phloem loading,** decreases the water potential in the sieve tube and causes water entering the leaf to move into the sieve tube by osmosis. Following this movement at the source, the sucrose solution is carried passively to a sink, such as a root cell, where the sucrose is unloaded (actively removed from the sieve tube at the sink). The sucrose may be either utilized or stored at the sink, but most of the water returns to the xylem and is recirculated.

The main substance transported by the phloem, sucrose, may be present in concentrations of up to 25%. Other sugars, such as hexose phosphates and small polysaccharides, are transported in much lower concentrations. When these sugars arrive at the roots, some of them are utilized in root cell metabolism, or are stored as food reserves; other sugars are converted to organic acids, which then combine with ammonium ions taken in through the roots or formed from nitrate taken in through the roots, to form amino acids. These amino acids then enter the xylem or the

24-20 Model of the pressure-flow hypothesis *(A) Two bulbs, permeable to water only, are connected and placed in water. The left bulb contains a sugar solution, the right bulb only water. (B) Water molecules enter the left bulb, forcing the sugar solution into the connecting tube and eventually into the right bulb. (C) Entry of sugar solution into the right bulb forces water out of the bulb.*

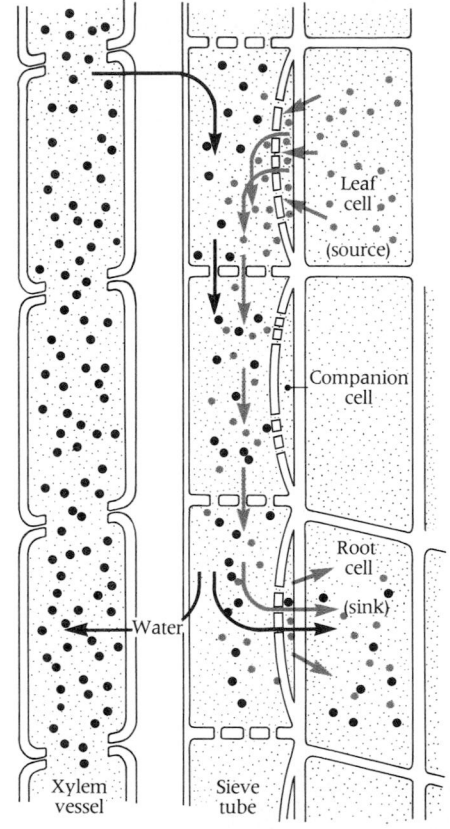

- Water molecule ——▶ Movement of water
- Sugar molecule ┈┈▶ Movement of sugar

24-21 Pressure-flow mechanism of phloem transport *Sugar molecules enter the companion cell from their site of synthesis (source) and traverse the sieve tube channels. The high sugar concentration inside these cells draws water into them. Sugar molecules are actively transported into root cells (sink), causing the sugar concentration in the sieve tube to fall. This, in turn, causes water to leave the sieve tube. A gradient of turgor pressure, from source to sink, establishes the pressure-flow relationship that transports sugar in the phloem.*

BOX 24-1

METHODS USED IN THE STUDY OF TRANSPORT IN PLANTS

Radioactive tracers have been used in a number of ways to study metabolism and transport in plants. In metabolic studies, a common approach measures rates of isotope incorporation into plant cells. Radioactive carbon, for example, will be taken up by a plant if its leaves are exposed to CO_2 that contains ^{14}C instead of ^{12}C (Figure 24A). Radioactive phosphorus will be taken up by a plant if the roots are exposed to a solution of phosphate ions containing ^{32}P.

The rate and location of isotope incorporation can be determined by **radioautography.** In this procedure, thin slices of plant tissues are placed in contact with a photographic film. Radiation from the isotope exposes the film in contact with the portions of the tissue section containing the radioactive material. Examination of the developed film under a microscope reveals that emanations of radioactive isotope expose (blacken) particles of photographic emulsion that were next to them. By comparing the black zones on the film with underlying tissue sections one can readily determine the exact location of the radioactive substance in the plant tissues (Figure 24B).

An ingenious approach to the study of phloem transport makes use of aphids, tiny insects that feed on the juices of plants (Figure 24C). Aphids have mouthparts called **stylets** that are arranged in the form of a long tube. An aphid inserts it stylets through the epidermis of a stem or leaf and extends the end of the tube until it punctures a single sieve element. Because the fluid contents of the sieve element are under considerable tur-

gor pressure, the sap is forced into the aphid's tube and on through its digestive tract. It flows with such force the feeding aphids often display a droplet of "honeydew" on their posterior ends. Since passage through the aphid changes the composition of honeydew, researchers seldom work with healthy intact aphids. Instead, they wait until the aphid has started to feed, anesthetize it, and sever its body from its mouthparts. The fluid that continues for hours or days to ooze out of the phloem through the detached stylets can be easily collected with a little pipette and analyzed. Fluid obtained in this way is the purest sieve tube sap obtainable. In most plants, it contains 10 to 25% dry matter, 90% or more of which is sugar—mainly sucrose.

By using several aphids on different parts of the plant, an investigator can introduce test substances at one point and measure how long it takes for them to move to other points, which direction they travel, and so forth. Such studies show that materials move longitudinally in the phloem with remarkable speed.

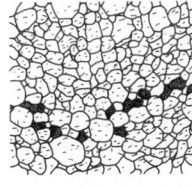

24A The use of radioactive isotopes to study plant metabolism
Two leaves of the broad bean, Vicia faba, *are enclosed in a plastic container and exposed to $^{14}CO_2$ generated by addition of acid (in syringe) to a radioactive carbonate. Radioactive carbon is later found incorporated into sugars that are transported to various parts of the plant.*

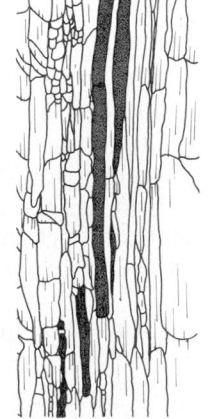

24B Radioautography localizes radioactive molecules
When film is placed in contact with the plant's stem for four weeks and then developed, radioactivity (dark grains on film) is seen to be confined largely to the sieve tubes.

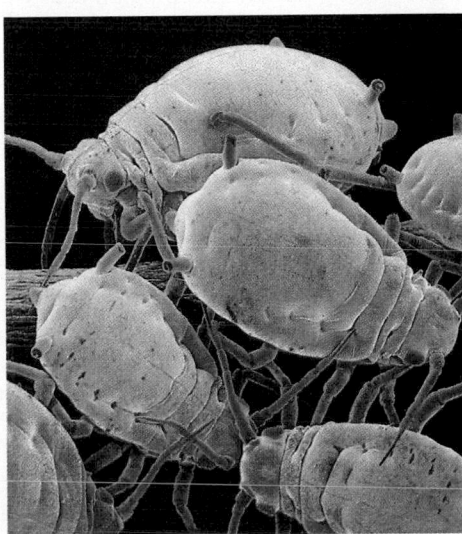

24C Scanning electron micrograph of aphids, Longistigma caryae, on a lemon plant (×32)

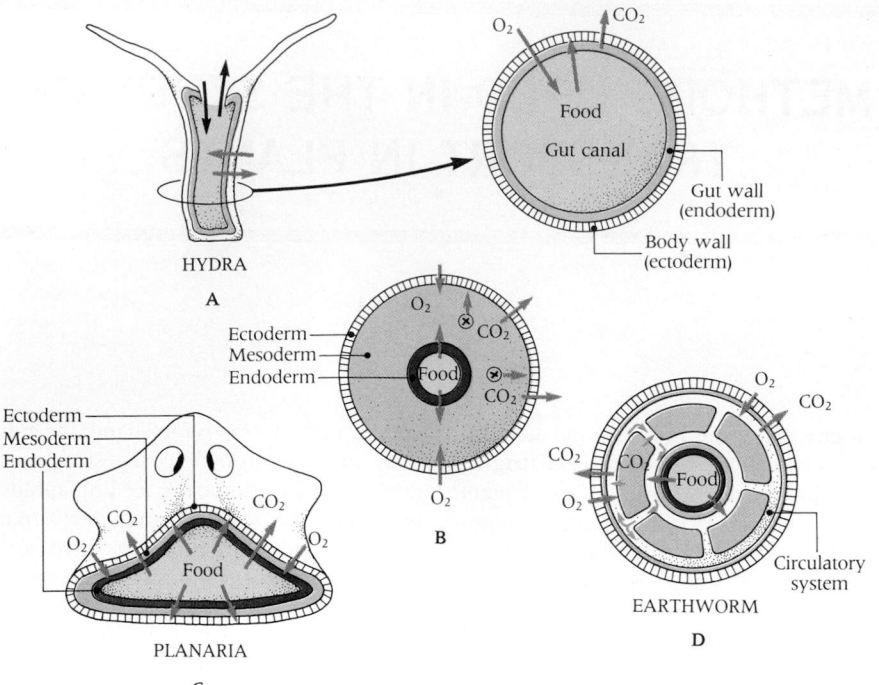

24-22 Fluid transport in multicellular animals
(**A**) In a coelenterate such as Hydra, water circulates into and out of the body cavity and diffusion accounts for transfer of O_2, CO_2, and food across the two cell layers making up the body. (**B**) Cross section through a hypothetical three-layered animal. Mesoderm cells marked ''X'' are too far from the gut and exterior to rely on diffusion to transport food and gases. (**C**) In the flatworm, the flattened body shape allows efficient transfer of food, O_2, and CO_2 by diffusion. (**D**) Earthworms and other animals more advanced than flatworms have evolved a circulatory system to distribute food and gases efficiently and over large distances.

phloem for transport. Phloem also carries nucleotides, hormones, and various other organic compounds. This is known as **translocation.**

TRANSPORT IN ANIMALS

Sedentary plants have less active rates of metabolism than mobile animals. Hence their transport can take place at a leisurely pace. Transport in animals must take place more rapidly.

ARRANGEMENTS IN SIMPLE ANIMALS

Smaller animals, particularly the simplest of the multicellular animals, sponges, corals, jellyfishes,[3] and others—resemble nonvascular plants in lacking vessels for the movement of internal fluids (Figure 24-22A). Instead, they have spaces into which fluids from a watery external environment move freely.

One group of animals, the flatworms (Figure 24-22C), lack a vascular system, but by being flat and keeping all cells close to the body surface, successfully diffuse O_2 and CO_2 into and out of body cells.

Internal spaces devoted to transport are found in most multicellular animals (see Figure 24-22D). In the simplest vascular systems, movement of fluid relies on enlarged regions in the vessel system, which have strong muscular walls and contract rhythmically. These organs, the **hearts,** force fluid through **vessels.** Systems of **valves** keep the fluid moving in one direction. Animal groups differ strikingly in heart number, arrangement, and structure. A squid has three hearts—one that pumps blood to the body and two that pump blood through the gills, from which it passes back to the body heart (Figure 24-23A). The common earthworm has 10 hearts (Figure 24-23C).

In animals, one or more hearts continuously pump blood in one direction, but there is no fluid loss comparable to transpiration in plants. Instead, blood makes a circuit of the body and returns to the heart. In other words, there is a true circulatory system.

---

[3] Jellyfishes, some of which are many feet long, are curious creatures. As much as 90% of their mass is a kind of "jelly," or mesoglea, within thin membranous sheaths. As in the *Hydra,* their coelenterate cousins, transport depends on external fluids passing in and out of a central cavity (see Chapter 42).

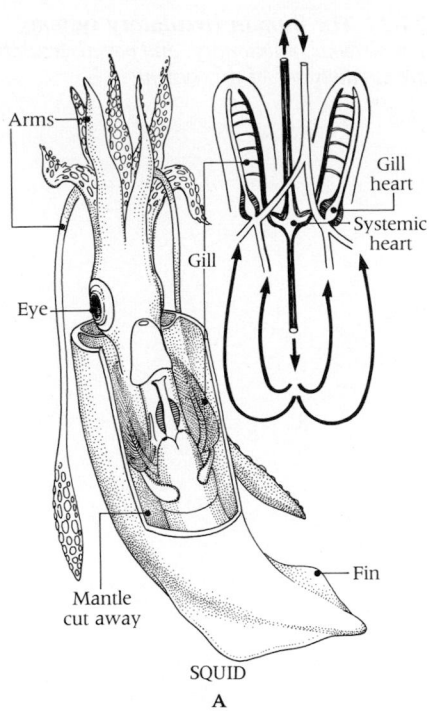

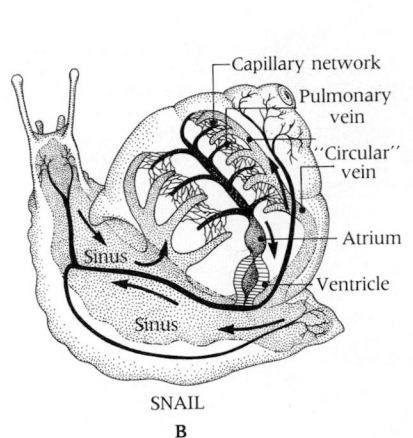

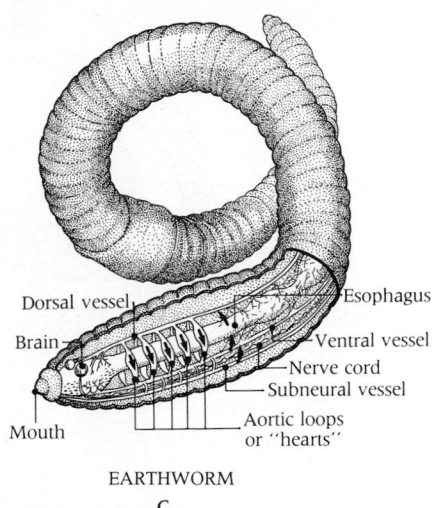

| SQUID | SNAIL | EARTHWORM |
| A | B | C |

24-23 Open and closed circulatory systems
(A) A squid has a closed circulatory system with three hearts, two for moving blood through the gills, and one for transporting blood through the body. (B) A snail has an open circulatory system with a single two-chambered heart. Blood is collected in sinuses and drains into veins on the surface of the lung, eventually reaching the heart again. (C) The closed circulatory system of an earthworm has 10 hearts.

In some animals, including insects, some worms, and many mollusks, the heart pours blood into irregular spaces called **sinuses** located among the tissues (see Figure 24-23B). The organs are bathed in blood that moves about sluggishly, exchanging materials with body cells and eventually seeping back into vessels that return it to the heart. This is an **open circulatory system.** Most worms and all vertebrates (fishes, amphibians, reptiles, birds, and mammals) have a **closed circulatory system.** The blood does not leave the vessels. Exchanges of materials between blood and the cells are accomplished in networks of the tiny, thin-walled vessels called **capillaries.**

THE CIRCULATORY SYSTEM

The circulatory system is similar in all mammals. At the center of the system is the heart, which maintains a constant flow of blood by applying the pumping pressure to drive it throughout the body. Let us take the plan of the human circulatory system as an example of a mammalian system.

The Human Circulatory System

As shown in Figure 24-24, oxygen-rich arterial blood leaves the **left ventricle** of the heart through the body's greatest artery, the **aorta.** Branches of the aorta extend to the head, arms, internal organs, and legs. In the periphery blood reaches the smaller arteries, the **arterioles,** and then it moves into and through the capillaries, supplying the tissues with O_2 and emerging as oxygen-poor venous blood, which is laden with CO_2 and other waste products. **Veins** from the lower portion of the body converge into the **inferior vena cava.** Those from the head and upper extremities converge into the **superior vena cava.** Both great vessels empty into the **right atrium** of the heart. This completes the **systemic circulation.**

Although blood leaves the heart freshly oxygenated and returns to it depleted of O_2, nowhere in this systemic circuit does it encounter the gas-exchanging surfaces of the lungs. To do that it must traverse a second loop, the **pulmonary circulation.** In that circuit, venous blood moves from the right atrium to the **right ventricle,** which then pumps it via the **pulmonary arteries** into the lungs. After discharging CO_2 and taking up O_2, the blood travels via the **pulmonary veins** to the left atrium and then to the left ventricle for another trip around the body. Structurally speaking, a single heart (of two cylinders) stands between two lungs, but functionally there is evidently one lung between two pumps (Figure 24-25): the right heart,

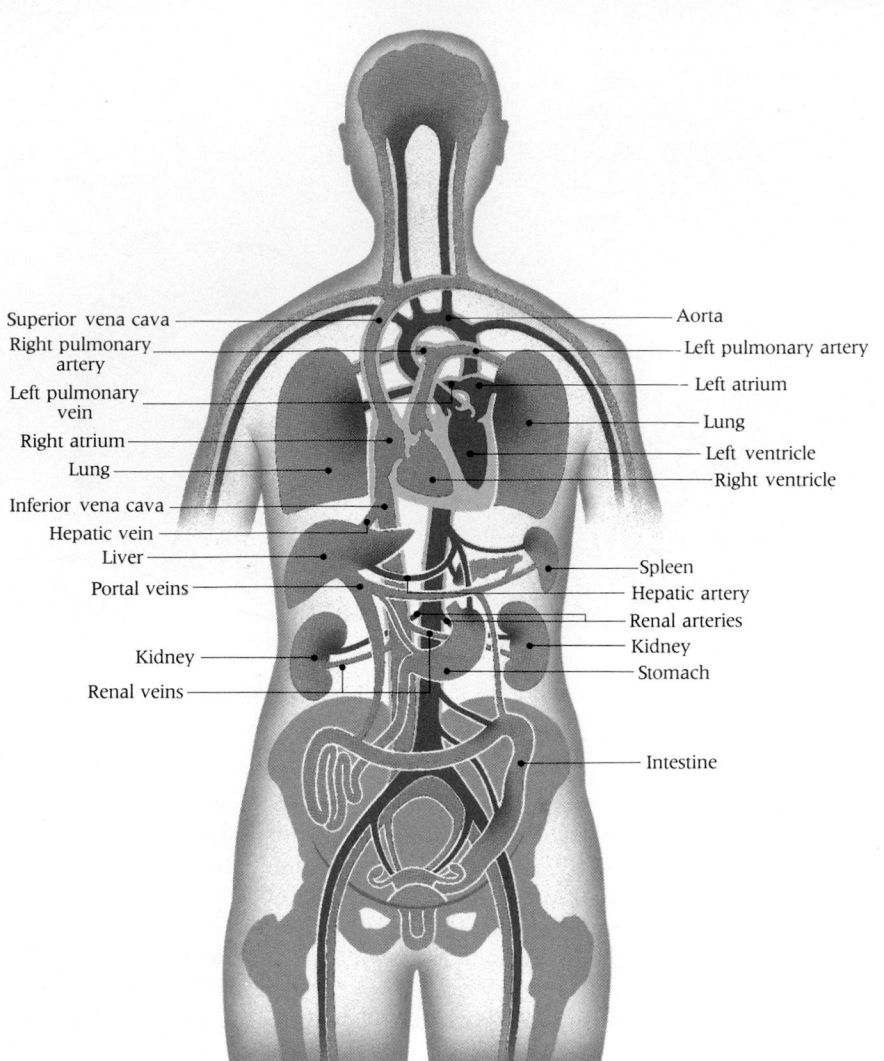

24-24 The human circulatory system
The systemic, pulmonary, and portal circulations are subdivisions of this system.

Superior vena cava
Right pulmonary artery
Left pulmonary vein
Right atrium
Lung
Inferior vena cava
Hepatic vein
Liver
Portal veins
Kidney
Renal veins

Aorta
Left pulmonary artery
Left atrium
Lung
Left ventricle
Right ventricle
Spleen
Hepatic artery
Renal arteries
Kidney
Stomach
Intestine

which receives venous blood from the body and pumps it to the lung; and the left heart, which receives arterial blood from the lung and pumps it to the body.

There is still another circulatory loop, called the **portal circulation.** As shown in Figure 24-24, this system begins in the intestines and other digestive organs and ends in the liver. The liver, in fact, has a double circulation. The portal circulation transmits nutrients absorbed from the alimentary tract to the liver. However, the bulk of the liver receives arterial blood from the hepatic artery. Both blood supplies drain into the hepatic vein, which runs directly to the inferior vena cava.

The heartbeat consists of a cycle with two major phases (Figure 24-26). If we begin at the point when the heart is relaxed—the phase called **diastole**—the first major event is an inward surging of blood from both vena cavas through the atriums into the large ventricles. The surge of blood into the ventricles is initiated by ventricular diastole, but it is completed by the contraction, or **systole,** of the atriums. The filled ventricle then contracts (ventricular systole), pumping blood out of the heart into the aorta and the pulmonary arteries. Flaplike valves between atriums and ventricles automatically close as ventricular systoles begin. This prevents the blood from moving back into the atriums.[4]

A cherished memory of every medical student is the first time he or she listened to the heart's sounds through a stethoscope. What is heard (or is supposed to be heard) is quite instructive. A dull "boom" marks the moment of ventricular systole.

[4] In 1982, after years of experimentation in laboratory and farm animals, surgeons successfully installed an artificial heart in a living human being. A plastic device that replaced the subject's own right and left ventricles (functionally, not anatomically), it was energized by an outside power supply to which the individual had to remain attached. This heart was regulated by dials and voltage regulators, not by the nervous system. Hence it could not respond to exertion or other demands as did the heart it replaced.

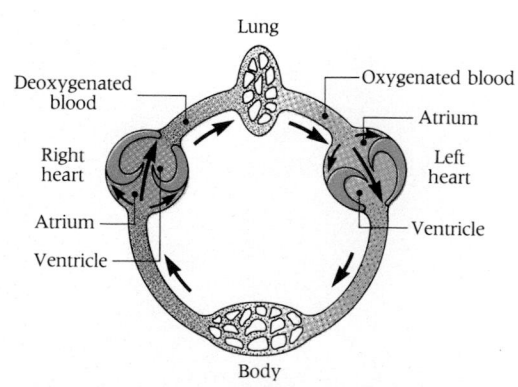

Lung
Deoxygenated blood
Oxygenated blood
Right heart
Atrium
Left heart
Atrium
Ventricle
Ventricle
Body

24-25 Functional relationship of systemic and pulmonary circulations
The right heart pumps deoxygenated blood to the lungs. The left heart receives oxygenated blood from the lungs and pumps it to body tissues.

24-26 Pumping cycle of the human heart
(A) The human heart has four chambers connected by valves that guide the flow of blood in one direction. The wall of the left ventricle is thicker because this side of the heart pumps blood through the body against higher peripheral resistance. **(B)** The pumping cycle begins with atrial diastole, in which the atria relax and fill with blood. In ventricular diastole, the ventricles relax and fill with blood from the atria. The cycle is completed when the atria contract (atrial systole). Ventricular systole pumps blood into the pulmonary arteries and aorta.

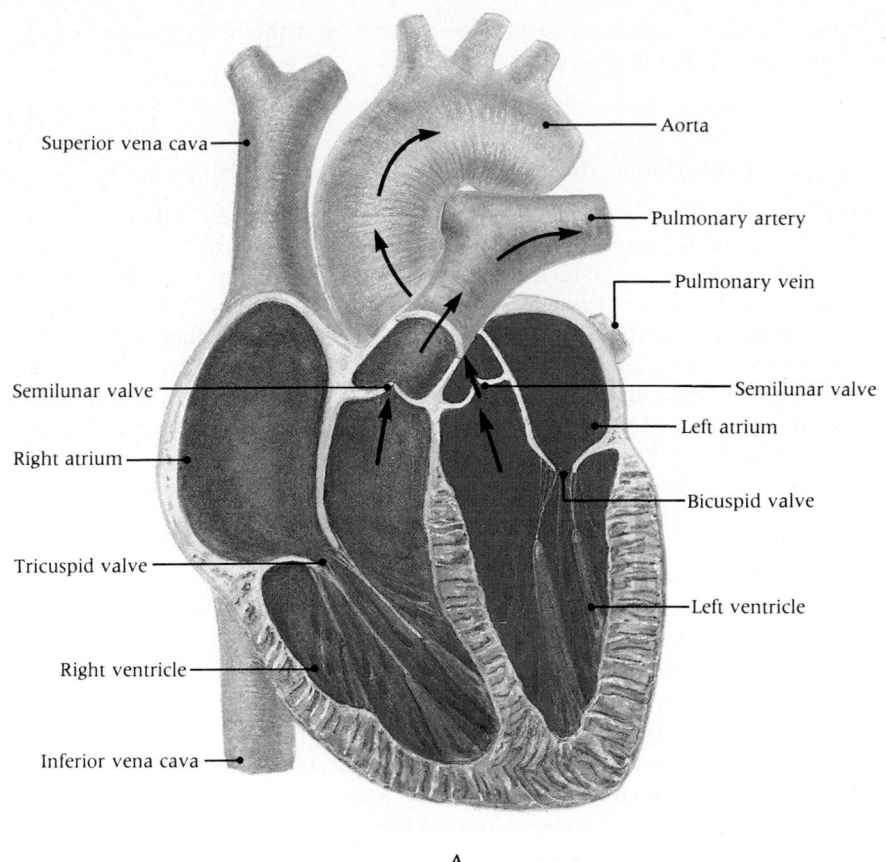

A

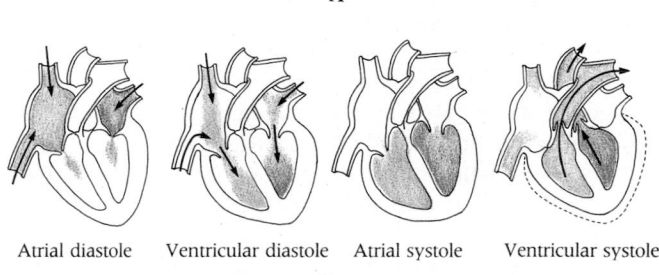

Atrial diastole   Ventricular diastole   Atrial systole   Ventricular systole

B

24-27 Evolution of the vertebrate circulation (Below)
(A) Fishes have a two-chambered heart, and oxygenated blood travels directly from the gills to the tissues. **(B)** In amphibians, the three-chambered heart accepts oxygenated blood from the lungs and deoxygenated blood from the body. Some mixing of blood occurs. **(C)** In reptiles, a septum partially divides the ventricle, foreshadowing the advent of a four-chambered heart. **(D)** The four-chambered heart of birds and mammals is connected to two separate circulations, the systemic and pulmonary, which separate completely the oxygenated and deoxygenated blood.

This is the sound caused by the vibrations of closing valves and contracting muscles. When the ventricles relax in their next diastole, crescent-shaped valves at the mouths of the great arteries prevent a retrograde flow of expelled blood back into the heart. The closure of these valves produces the second heart sound. Physicians often mimic the two heart sounds as "lubb-dup, lubb-dup." If the valves are defective and do not close completely, some of the turbulently flowing blood does leak back and a soft hissing sound may be heard. This is called a **murmur.** (Note that some murmurs can arise from abnormalities of the blood, such as anemia, in the absence of valve disease.)

In summary, the circulatory system consists of a pump and several circular loops

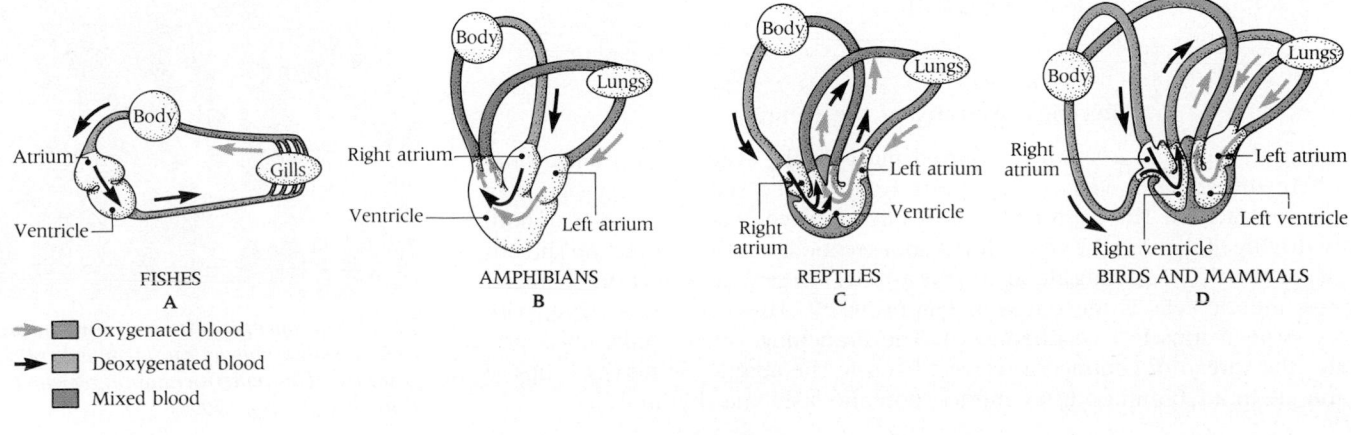

FISHES
A

AMPHIBIANS
B

REPTILES
C

BIRDS AND MAMMALS
D

→ Oxygenated blood
→ Deoxygenated blood
Mixed blood

CHAPTER 24   TRANSPORT

629

of tubing. By preventing backward flow, the valves of the heart and veins help to keep blood circulating in a forward direction.

Evolutionary History of the Vertebrate Circulation

As we have seen, the mammalian heart consists of two pumps that beat in unison and thereby circulate blood to the lungs and the rest of the body. The evolutionary origin of this arrangement, which took more than 100 million years to develop, is well documented.

It is often stated that the fish heart is composed of only two chambers, an atrium and a ventricle (Figure 24-27A). However, in addition to those two chambers the fish heart has a large **sinus venosus** and a well developed **conus arteriosus** (Figure 24-28), both of which have cardiac muscle in their walls. Deoxygenated (venous) blood from the body enters the sinus venosus and passes through the atrium, ventricle, and conus arteriosus to the ventral aorta and thence to the **gills,** where it is oxygenated. Continuous blood flow is a prerequisite for efficient O$_2$ uptake from the water. Oxygenated blood then passes through the dorsal aorta to the rest of the body. By this time, much of the propulsive force of the heartbeat has been dissipated by friction in the gill capillaries. Hence, the return of blood to the sinus venosus, after oxygen has been discharged into the tissues, is quite sluggish.

With the emergence of air respiration during the invasion of land by lungfishes and amphibians, an evolutionary novelty appeared—the entry of a pulmonary vein into the atrium carrying oxygenated blood from the lung (see Figure 24-27B). For the first time in the evolutionary history of vertebrates, the heart receives two kinds of blood—oxygenated blood from the lung and deoxygenated blood from the tissues of the body. The two bloodstreams are kept separated by a divided atrium. The pulmonary vein carries oxygenated blood to the left atrium, and the sinus venosus (now reduced in size) passes deoxygenated blood to the right atrium. Even though both atria discharge blood into a single ventricle, mixing of the two bloodstreams is minimized by a partial **septum** in the ventricle. In the conus arteriosus, separation of the oxygenated and deoxygenated blood is largely maintained by a **spiral fold,** which in turn divides the conus into two channels. The ventral channel dispatches O$_2$-enriched blood directly across to the dorsal aorta for direct distribution to the tissues, whereas the O$_2$-depleted bloodstream in the dorsal channel of the conus is conveyed to the lung for oxygenation via the pulmonary artery. Thus the amphibian evolutionary stage foreshadows the subsequent evolutionary trends in higher vertebrates.

In reptiles, the separation of the ventricle is nearly completed and the conus arteriosus becomes fully subdivided, to be incorporated in the bases of the aortic trunks (see Figure 24-27C). The sinus venosus is greatly reduced.

Birds and mammals have a completely divided heart (see Figure 24-27D). As a result, fully oxygenated blood is pumped into the systemic circulation under high pressure. The presence of a high-pressure double circulation system can be correlated with high metabolic rates. In the transition from reptiles to birds and mammals, the sinus venosus became further reduced and was finally incorporated into the atrial wall as the **sinoatrial node,** the heart's pacemaker.

In some ways, the embryonic development of the human heart appears to retrace this evolutionary sequence. Defects in embryonic development often result in abnormalities that mix the circulations (Box 24-2).

Electrical Activity of the Heart

The heartbeat begins in a sharply localized place in the wall of the right atrium near where the superior vena cava enters. Here can be seen the first sign of a muscular contraction. The contraction passes like a wave over the muscles of both atriums, driving blood into the ventricles. A split second later, the contraction spreads through the ventricles, ejecting blood into the arteries. The special branching character of cardiac muscle cells is illustrated in Figure 24-29. This contrasts sharply with ordinary skeletal muscle (see Chapter 6). The branching pattern links fibers and facilitates the spread of contraction waves. Even in the absence of nerves, strips of heart muscle in a laboratory dish contract spontaneously and rhythmically.

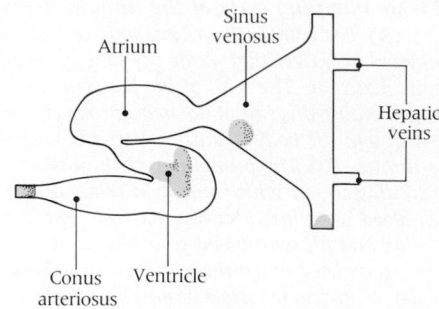

24-28 Structure of the fish heart
In this ventral view, the parts of the heart are drawn side-by-side, rather than superimposed as they occur normally. Blood flows into the sinus venosus, through the atrium and ventricle, and out the conus arteriosus to the gills.

24-29 Micrograph of human cardiac muscle
Unlike skeletal muscle, cardiac muscle cells are branched. This pattern enhances coordination of the heartbeat. (×5000)

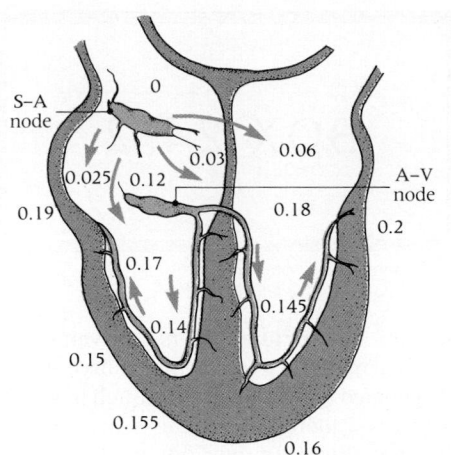

24-30 Impulse transmission through the heart

This cross section of the human heart shows the S-A node and A-V node, which initiate the heartbeat and transmit impulses to the heart muscle. Numbers indicate the time (in seconds) required for arrival of impulses from the S-A node to that point.

What controls the heartbeat? The impulse begins in a small mass of specialized cardiac muscle, which is an evolutionary remnant of the sinus venosus. This is the **sinoatrial,** or **S-A, node** into which various nerves lead (Figure 24-30). It is known as the **cardiac pacemaker,** since it initiates beats. Contractions beginning in the S-A node travel throughout the heart via specialized strands of cardiac muscle that are adapted to conduct impulses. This is the heart's conduction system. The result is a better synchronized and hence more efficient pumping action than might occur if excitation waves were propagated randomly.

Excitation waves that begin in the S-A node spread first through the atriums. Then they activate the **atrioventricular,** or **A-V, node,** another mass of specialized cardiac muscle low in the right atrium that serves as a second pacemaker for the ventricles. The A-V node sends impulses into the ventricles via the conduction system, which then divides into right and left branches, one running down into each ventricle. Each branch divides further into small branches, which spread over the inner surfaces of each ventricle.

The portion of the heart muscle that generates the most impulses per unit time imposes its own rhythm on the entire organ. Normally this is the S-A node; that is why the S-A node is ordinarily the pacemaker. If the S-A node is surgically removed (or damaged by disease), the heartbeat is drastically slowed. The atrial muscle or the A-V node then takes over as pacemaker, and the atria and ventricles now beat at a slower rate. The normal rhythm set by the S-A node, the so-called **sinus rhythm,** is 75 to 80 impulses per minute. On their own, the atrial muscle or the A-V node can produce only 40 to 50 impulses per minute, adequate to maintain the circulation.[5]

[5] If an individual's heart rate becomes too slow because of damage to the S-A or A-V nodes, it is possible to implant surgically a small battery-operated device that emits 75–80 impulses per minute and serves admirably as an artificial pacemaker. New nuclear-powered models make it unnecessary to remove the device yearly for a change of batteries.

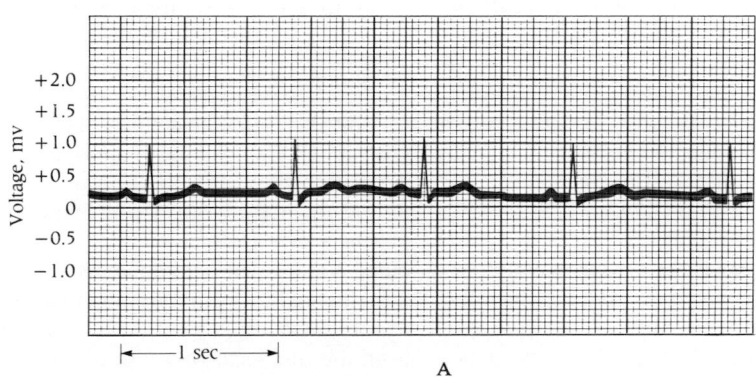

A

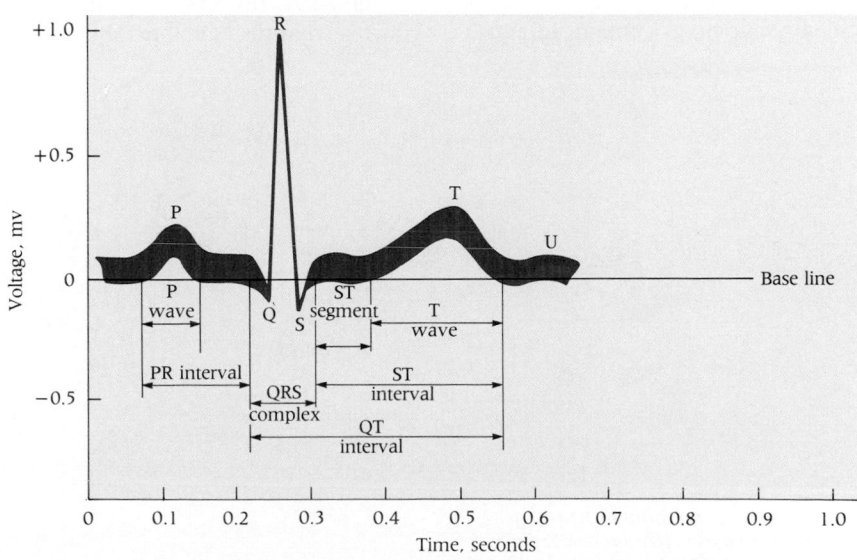

24-31 Electrocardiogram of a normally beating heart

(A) A four-second tracing. (B) A single heartbeat, magnified to show the various waves (P, Q, R, S, and T) and intervals.

B

The human heart and blood vessels arise during embryonic life in an integrated sequence of events that in some ways parallels the evolution of the heart. Most of this development occurs between the third and eighth weeks of embryonic life. As shown in Figure 24D, the heart in the early embryo is a mere tube that receives blood at its caudal (tail) end and discharges it at its cephalic (head) end. This tubular heart develops five sacculations, or bulges, which are as follows (reading from the caudal to the cephalic ends): sinus venosus, atrium, ventricle, bulbus arteriosus, and truncus arteriosus.

As the tubular structure lengthens, it forms a loop in which the two caudal segments, the sinus venosus and the atrium, lie behind (dorsal to) and above the cephalic segments. By the fifth week of embryonic life, a septum separates the right and left atria and begins growing within the ventricle. Next, the right atrium absorbs the sinus venosus. Then a septum appears in the truncus arteriosus, dividing it into pulmonary artery and aorta. Curiously, this septum takes a spiral course, producing a crossover between the pulmonary artery and the aorta in the adult (see Figure 24-24). In the eighth week, the ventricular septum is completed. Meanwhile, blood vessels are arising everywhere, lung buds are developing, and the liver and other organs are forming.

Two significant facts underlie the major structural and functional differences between fetal (Figure 24E) and adult circulations. First, embryonic lungs contain no air. Second, gases, nutrients, and waste products enter and leave the embryo via the placenta, where embryonic blood contacts maternal blood.

The embryonic heart is adapted structurally to the special requirements of embryonic life. Since the lungs do not yet exchange gases, blood flow to them is minimal. In fact, some venous blood is shunted directly from the right atrium to the left atrium through a special opening in the atrial septum, the foramen ovale. The remainder of the blood in the right atrium follows the conventional route—through the right ventricle into the pulmonary artery. Most of it traverses a second direct shunt from the pulmonary artery to the aorta, the ductus arteriosus. Clearly, this vessel has to close up and disappear at birth.

The blood reaching the left atrium proceeds to the left ventricle and the aorta, where it joins the blood from the ductus arteriosus. As might be expected, these bloods are poorly oxygenated.

How then does the embryo obtain its oxygen supply? Blood of the embryonic systemic circulation flows through the umbilical arteries and finally to the placenta where waste products are transferred to the mother's blood and oxygen is taken up. Blood returns from the placenta through the umbilical vein toward the liver. For a short time, it empties from the umbilical vein into the portal vein and thus passes through the liver on its way to the heart. However, development of the ductus venosus, another shunt pathway, provides a shortcut to the inferior vena cava that bypasses the liver. In the embryo, then, oxygenated blood, admixed though it is with vena caval blood, enters the heart through the right atrium. In the adult, the left atrium is the first heart chamber to receive oxygenated blood. Since both sides of an embryonic heart deliver blood to the body, the heart of the embryo resembles that of an amphibian or reptile (see Figure 24-27).

Dramatic and fundamental changes occur at birth. With great suddenness the maternal O_2 supply is literally cut off. An immediate exchange of O_2 and CO_2 through previously collapsed lungs is therefore essential. The circulation must also be rebalanced. Instead of one circulation, two—the systemic and the pulmonary—must now be established without delay, and further mixing of oxygenated and deoxygenated blood must be prevented.

When the umbilical cord is tied off, circulation through the umbilical vessels and the ductus venosus ceases. With the first few breaths, the lungs expand, so that the pressure within them decreases and the flow of blood to them through the pulmonary artery increases. The volume of blood returned to the left atrium via the pulmonary veins increases correspondingly, raising the pressure in that atrium and closing the foramen ovale.

A third passage must also close—the ductus arteriosus. With the onset of respiration, pressure in the pulmonary circulation drops below that in the systemic circulation, and the direction of flow through the ductus arteriosus is reversed. Soon the pressures in

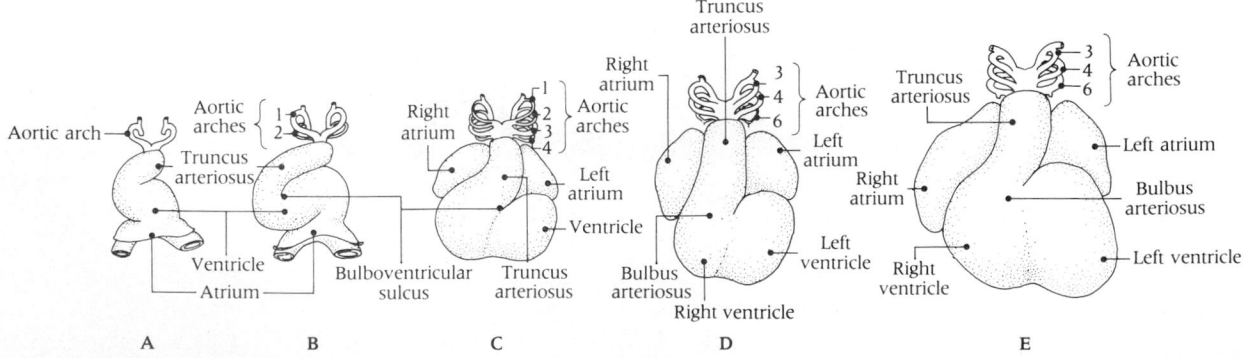

24D Development of the human heart
(**A**) *2.1 mm long embryo.* (**B**) *3.0 mm embryo.*
(**C**) *5.2 mm embryo.* (**D**) *6.0 mm embryo.*
(**E**) *8.8 mm embryo.*

the two sides of the heart are equal, and flow through the ductus stops, partly owing to the contraction of muscles in its wall.

Malformations of the heart result when development is arrested or flawed at some point in this train of events. Such developmental errors occur in 9 births in 1000. They may be genetic in origin or due to environmental assaults of one sort or another. For example, German measles (varicella) in early pregnancy commonly causes congenital heart lesions. Since the lesions ordinarily affect only one portion of the heart, the remainder of the circulatory system usually develops normally. Large books have been written on these interesting malformations, many of which can be diagnosed accurately and treated satisfactorily with aggressive surgical and medical methods. We can here mention only two important examples: patent ductus arteriosus and complete transposition of the great arteries.

In patent ductus arteriosus, one of the more common congenital heart abnormalities, the ductus fails to close after birth (Figure 24F). As a result, blood can freely pass back and forth between aorta and pulmonary artery. Since aortic pressure is usually higher, blood flow through the ductus is chiefly from aorta to pulmonary artery. This blood is distributed to the two left pulmonary arteries which pass it through the lungs. Then it returns via the pulmonary veins into the left ventricle and aorta. Consequently patency of the ductus overloads the left ventricle. This frequently results in heart failure. In the infant or young child, surgical correction can be accomplished with very little difficulty.

In transposition of the great arteries (TGA), which occurs in slightly more than once in 5000 births, the positions of the two great arteries are reversed. The aorta arises from the right ventricle and the pulmonary artery from the left ventricle (Figure 24G). Hence, the pulmonary and systemic circulations are arranged in parallel rather than in series with the systemic venous blood passing through the right heart chambers and then back out to the body and pulmonary venous blood traversing the left heart and returning to the lungs.

Before birth, TGA has few adverse effects on the fetus. After birth, however, survival depends completely on mixing between pulmonary and systemic circulation. For a short while the fetal pathways, the ductus arteriosus and foramen ovale, suffice. But after a few hours, pulmonary resistance is signifi-cantly lower than that in the systemic circulation. This facilitates shunting of deoxygenated blood from the aorta to the pulmonary artery. Since the pulmonary circuit cannot be overloaded, there must be shunting of blood returning in the pulmonary vein from left atrium to right atrium. This bidirectional shunting from aorta to pulmonary artery and left atrium to right atrium improves mixing. As the ductus arteriosus closes, however, the obligatory shunting is eliminated and the only site of mixing is the foramen ovale. Although some bidirectional shunting may occur that allows deoxygenated blood to get to the lungs and oxygenated blood to the systemic circulation, this is usually inadequate and a severe and dangerous lowering of blood oxygen results. Total surgical correction of the malformation is now possible and it is commonly done.

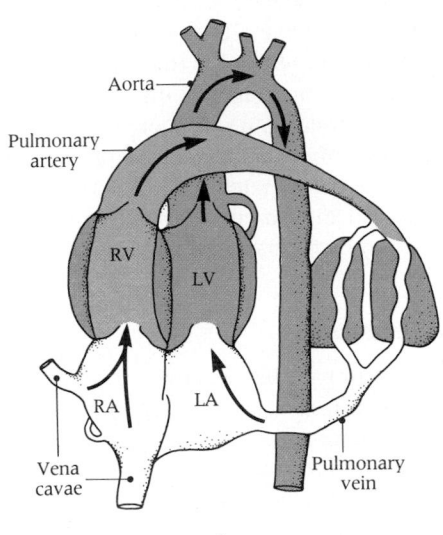

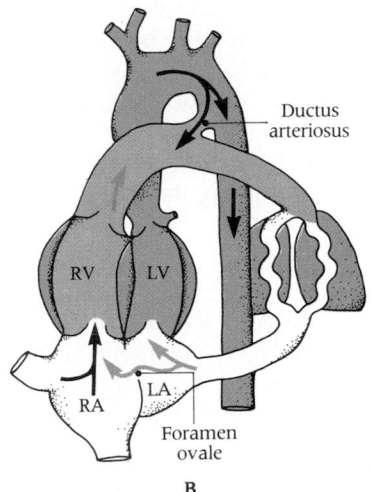

A

B

24F Patent ductus arteriosus

(A) *Normal human circulation at three days of age. The ductus arteriosus connecting the pulmonary artery and aorta is closed.* (B) *In patent ductus arteriosus, the channel connecting aorta and pulmonary artery* remains open. More blood enters the pulmonary circulation and the left side of the heart is overloaded. Blood also moves from left to right atria through a patent foramen ovale.

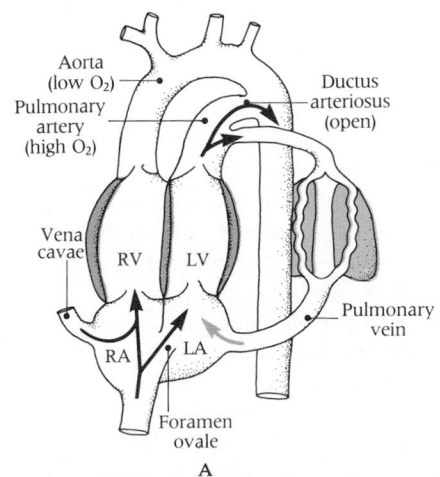

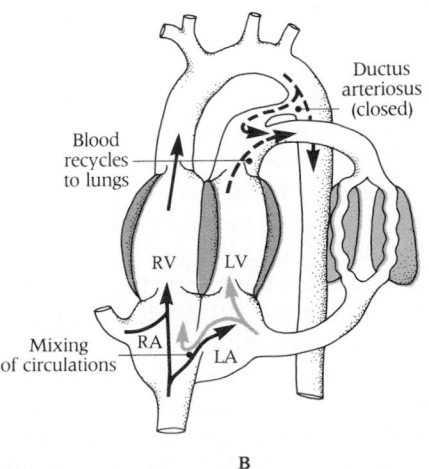

A

B

24E The embryonic circulation

The ductus arteriosus, foramen ovale, and ductus venosus are shunt pathways that are open and functional in the embryo, but are closed off at birth.

24G Transposition of the great arteries

(A) *Fetus with this abnormality has a patent ductus arteriosus and foramen ovale. Strikingly, the aorta leaves the right ventricle, rather than the left, and the pulmonary artery emerges from the left ventricle rather than the right.* (B) *The same situation in the* newborn shows that the circulations are arranged in parallel rather than in series. Blood on the left side of the heart circulates through the lungs, and that in the right side of the heart circulates through the body. Survival depends on mixing of the two circulations between the atria or through the ductus arteriosus.

What sort of mechanism could account for the emission of impulses with astonishing regularity 110,000 times in a day—and 2.8 billion times in the 70-year lifetime of a human being? In 1882, the English physiologist, Sidney Ringer, discovered that a solution containing sodium, potassium, and calcium ions in concentrations identical to those in blood sustains the beat of an excised frog or mammalian heart for a long time. Addition of glucose prolongs the performance. Thus the heartbeat—and the pacemaking "spark" in the S-A node—depends upon the availability of (1) a balanced mixture of various ions, (2) an energy-yielding compound such as glucose, and (3) oxygen.

Later research showed that the spread of impulses throughout heart muscle is associated with the spread of an electrically active area of the surface membranes of heart muscle cells. This does not mean that electricity flows as it does in copper wires. Rather it is the spread of a zone, which through stimulation has been caused to have a localized negative potential relative to the neighboring area.[6] When negativity spreads to a neighboring area, it subsides at its original locus.

The gross electrical activity of the heart is readily measured by the technique called **electrocardiography.** A typical **electrocardiogram**, or **ECG** (or **EKG,** in honor of its German origin) is shown in Figure 24-31. Its characteristic wave pattern depends upon the locations on the body of the positive and negative electrodes. If an excitation wave moves toward a positive electrode, the galvanometer needle is deflected in a positive direction (up); if the wave recedes from a positive electrode, the deflection is in a negative direction (down). By this method it is possible to study the heartbeat with great precision. For example, the ECG readily detects the disorder called **heart block** in which there is a delay or interruption in the conduction of excitation waves. It also detects abnormalities of the pacemaker and the condition in which the heart muscle has deteriorated locally—as it does in **myocardial infarction** (heart muscle damage) due to **coronary thrombosis** (interruption of blood flow to heart muscle through coronary artery).

Pulse and Blood Pressure

With each contraction of the left ventricle, a column of blood surges through the arteries under high pressure (Figure 24-32). Pressure drops sharply when the ventricle relaxes. We can detect these alternations of pressure when we feel the **pulse.** A pulse occurs in all arteries—not just in the radial artery at the wrist. A physician feels the pulse to assess the rate and regularity of the heartbeat.

In the familiar procedure for measuring blood pressure, the arteries in the arm are partially compressed until the systolic wave is no longer forceful enough to push blood past the obstacle. Compression is then relaxed until even the low diastolic pressure is enough to force the blood through. Thus the highest (systolic) and lowest (diastolic) pressures can be determined. The difference between them is the **pulse pressure.** A typical normal blood pressure reading would be designated "120 over 80" or 120/80. Many factors, emotional or physical, can cause the blood pressure to vary. Severe and persistent high or low pressures are dangerous and require investigation and treatment.

The average pressure at any given point in the systemic arterial system represents a balance between the rate at which the left ventricle pumps blood into the aorta and the rate at which the arterioles permit the blood to flow out into the capillary beds. Changes in the arterial pressure occur when there is a change in either the cardiac output or in the resistance to outflow from the arteries. The resistance to flow of blood in the systemic circuit is referred to as the **peripheral resistance.** Most resistance is met in the arterioles, and it is here that the pressure falls off rapidly (Figure 24-32). Wide and rapid changes in peripheral resistance can occur, since the diameter of the arterioles can vary with changes in the tension, or **tonus,** of the smooth muscle in the arteriolar wall (Figure 24-33). Variations in tonus make possible the shunting of blood to where it is most needed.

In the capillaries, pressure is radically reduced; hence there is no surging pulse. Blood returning in the veins has no pulse and is under little or no pressure from the heart. Blood flow in the veins depends in good part on breathing and other

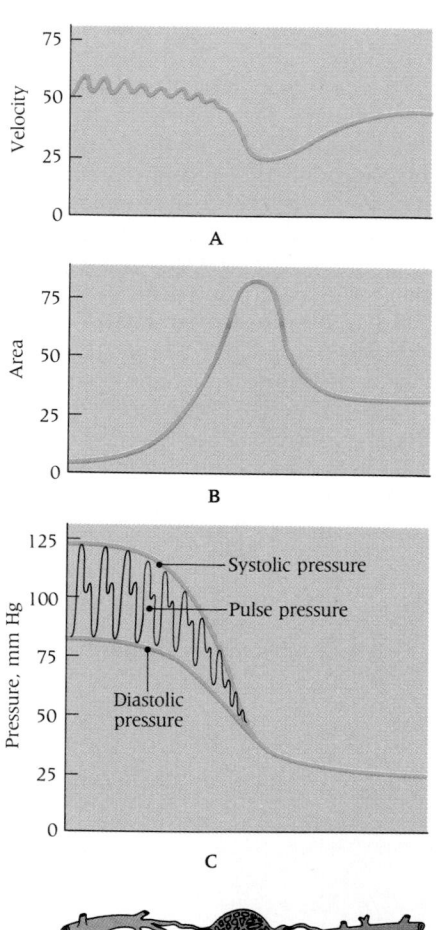

24-32 Functional attributes of different parts of the vascular system
The horizontal axis of each graph represents the different blood vessel types shown at the bottom of the figure. (A) The rate of flow is greatest in larger vessels. (B) The area of vessel walls is greatest in the capillaries. (C) Blood pressure in arteries and arterioles oscillates between systolic and diastolic pressures. Pulse pressure is the difference between them. Pressures are highest in vessels near the heart.

---

[6] This phenomenon resembles the spread of nerve impulses which we will discuss in detail in Chapter 29.

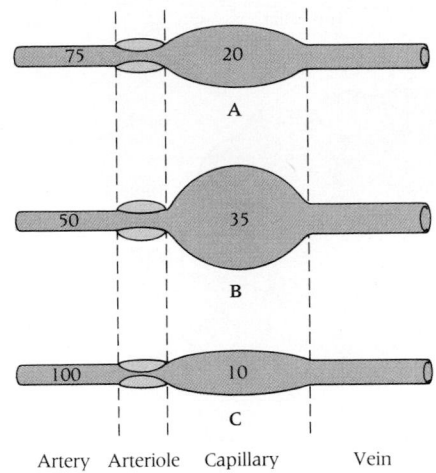

24-33 Effects of arteriolar tonus on capillary volume and pressure

(A) In the control, blood pressure is 75 on the arterial side and 20 on the capillary side of the arteriole. (B) Arteriole diameter is increased slightly, allowing more blood into the capillary. Arterial pressure decreases; capillary pressure and volume increase. (C) Constriction of arterioles increases arterial pressure and decreases capillary pressure and volume.

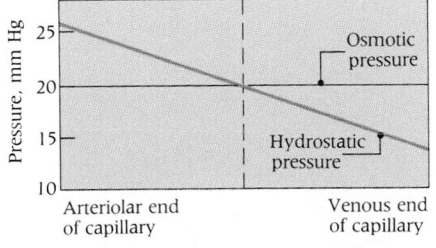

24-34 Relation of osmotic pressure to hydrostatic pressure in capillaries

Hydrostatic pressure exceeds osmotic pressure at the arteriolar end of the capillaries. This causes water to leave the capillary and enter the extracellular interstitial fluid. At the venous end of the capillary, osmotic pressure exceeds hydrostatic pressure, and water reenters the blood.

muscular movements. This is because the veins contain nonreturn valves. If a vein is compressed by muscular activity the blood is squeezed out and it has no alternative but to move toward the heart. Venous blood is kept moving toward the heart by this same series of valves.

Interchange of Materials Across Capillary Walls

The capillary may be regarded as the exclusive justification for the existence of the rest of the circulatory system, since it is in the capillary beds that most of the functions of the circulatory system are performed. A large part of the total volume of blood in the body is contained in the capillary beds and the capillary walls provide a very large surface or membrane separating the blood plasma from the fluid in the tissues, the so-called **interstitial fluid.** Blood moves slowly in the capillaries. This fact, along with the large surface available, makes possible the adequate exchange of substances between blood plasma and interstitial fluid. The basic mechanisms for the exchange are the same physical processes that transfer materials in solution across all membranes, namely, diffusion and osmosis.

The permeability of the capillary wall is such that all of the smaller dissolved particles can readily pass through. The plasma proteins are sufficiently large that they cannot pass through the capillary wall. Hence, all constituents of plasma other than protein diffuse freely back and forth between plasma and interstitium, while the proteins are retained in the plasma. Diffusible substances pass in both directions, but the net effect is always in the direction of equalizing concentrations on both sides of the capillary wall (see Figure 6-1).

Wastes, such as urea and CO_2, diffuse from body cells into interstitial fluid and thence into plasma because their concentrations in plasma are lower than they are in the interstitial fluid. However, the concentrations of waste products in plasma can never rise to equal those in interstitial fluid as long as blood in the capillaries keeps moving. Similarly, nutrients diffuse from plasma into interstitial fluid and thence into cells because their intracellular concentrations are continually reduced as the nutrients are used up. The concentrations of nutrients in interstitial fluid can rise to equal those in plasma only if cells cease to use these substances.

The net exchange of water across the capillary wall is in the direction of loss from plasma into interstitial space. To be sure, water molecules are free to diffuse in each direction. However, this net loss of water is not related to diffusion, since the concentration of water molecules is not greater in plasma than in interstitial fluid. Net loss of water from capillaries into interstitial spaces is caused by **filtration,** which occurs whenever there is a difference in pressure on two sides of a barrier that is permeable to the solvent.

Consider the diagram in Figure 24-34. **Hydrostatic pressure** in the capillary is traceable to the heartbeat. It is due to the pressure produced by ejection of blood from the left ventricle into the aorta. That pressure ranges from about 25 mm of mercury at the arteriolar end of the capillary to about 15 mm of mercury at the venous end. The major effect of this hydrostatic pressure—filtration of water and its solutes through the capillary wall—is in part counteracted by **osmotic pressure** acting in the opposite direction.

Since protein is present in much greater concentrations in plasma than in interstitial fluid the osmotic tension of plasma exceeds that of interstitial fluid by about 20 mm of mercury. Thus, at the arteriolar end of the capillary the hydrostatic pressure promoting filtration exceeds the opposing osmotic effect of plasma protein—by about 5 mm mercury. Here water is lost from plasma. At the venous end, in contrast, osmotic pressure can more than counteract hydrostatic pressure. Here water returns to plasma. Moderate arteriolar dilatation will cause a rise in pressure all along the capillary so that filtration may occur over the entire length of the capillary; arteriolar constriction will have the opposite effect.

Fluid that does not return to the capillary passes into the **lymphatic system,** which supplements the drainage function of the venous system.

LYMPH AND THE LYMPHATIC SYSTEM

Lymph is another name for the watery fluid of the interstitial spaces of body tissues. It derives from capillary leakage.

A remarkable drainage system exists to collect excess lymph and return it to

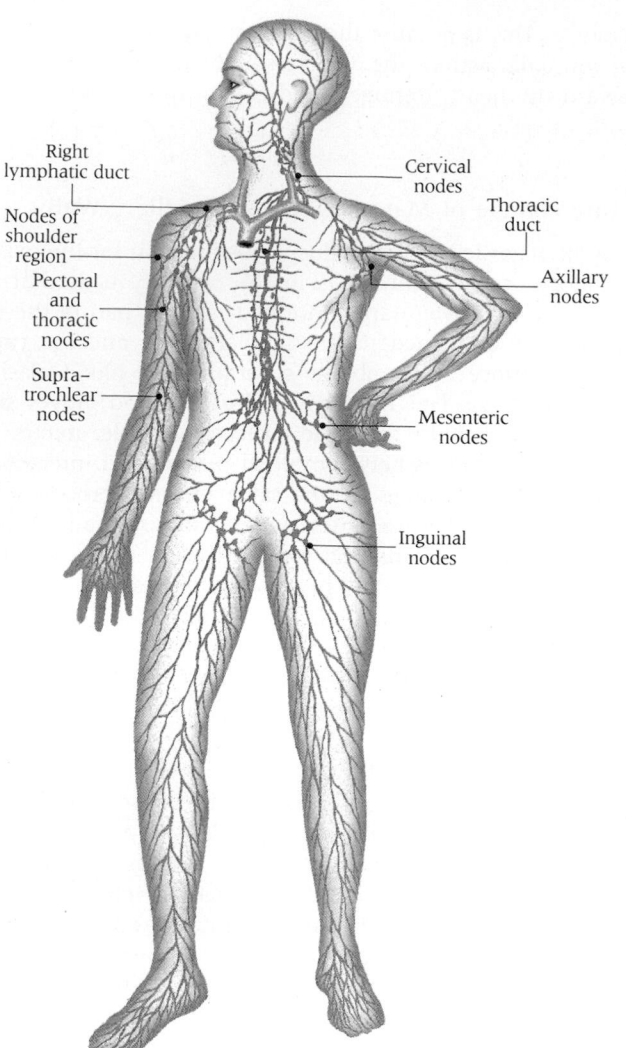

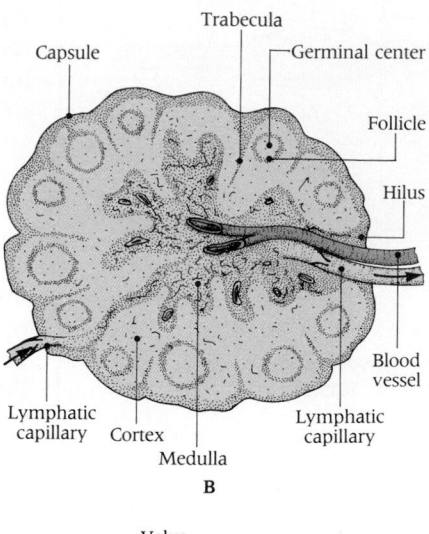

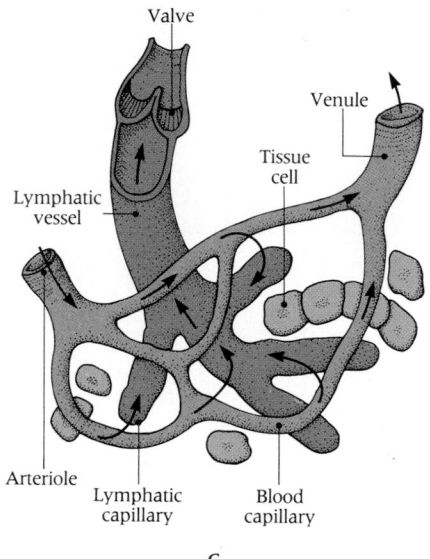

A

B

C

circulating blood. It is collected in the tissues by innumerable vessels that are separate and distinct from the blood-transporting capillaries (Figure 24-35A). These **lymphatic capillaries** form a complex network. Since their function is drainage, not perfusion or circulation, they have blind ends (or beginnings) in the tissues; thus this is a closed system. Fluid seeps into these vessels through their walls.

These capillaries unite to form larger and larger lymphatic vessels, which finally converge in two large lymphatic channels, the **thoracic duct** and the **left lymphatic duct,** which empty into the great veins as they enter the heart. Lymph vessels do have valves; hence returning fluids can flow against gravity. As in veins, flow through them depends largely on pressure and motion of surrounding tissues.

Lymph flows from and not to the body tissues; hence, it contains less dissolved food material than plasma. There is one notable exception to this statement. Lymph does pick up small globules of fat from the intestine and delivers them to the blood. Lymphocytes, which are formed in lymphoid tissues, enter the blood via the lymph.

At points along the lymph vessels there are lumpy enlargements, the **lymph nodes** (see Figure 24-35B). They are especially numerous in specific regions like the groin, armpit, and neck. One of their functions is lymphocyte formation. Another is to serve as filters for removing and destroying bacteria and solid particles arising in their drainage region. If, for example, there is an infection of the hand, the nodes in the corresponding armpit (axilla) will become painful and swollen. That indicates that they are doing their job—entrapping infectious microorganisms in their drainage region and preventing them from reaching the rest of the body. Similarly, lymph nodes near the lungs of city dwellers are often black with soot and dust particles.

24-35 The lymphatic system
(A) *The human lymphatic system.* **(B)** *Structure of a lymph node.* **(C)** *Detailed view of blood capillaries and lymph capillaries in the systemic circulation.*

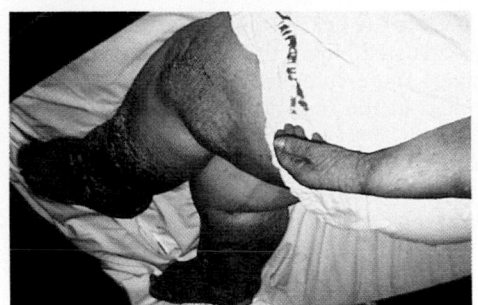

24-36 An unfortunate victim of elephantiasis

Sometimes lymphatic vessels become blocked. For example, obstruction by a parasitic African roundworm, *Wuchereria bancrofti,* produces a dreadful condition known as **elephantiasis.** The name reflects the fact that affected parts of the body often swell with accumulated fluid to elephantlike proportions. We see in Figure 24-36 what can happen to an unfortunate victim of this disorder.

In summary, the lymphatic system is a device for returning body fluid to the blood. It also filters out foreign particles, helps combat infection, and participates in the absorption of fats in the intestine.

One might suppose that these simple matters could be adequately managed by the blood and the circulatory system. It is not obvious, in other words, why a second system of vessels is needed. The fact is that the lymphatic system is an evolutionary newcomer that appeared for the first time in the vertebrates. Presumably, the increasing size of organisms and the increasing hydrostatic pressures in their circulatory systems enhanced the leakiness of the small blood vessels. When fluids began seeping out of the blood into the tissues, a drainage system became a necessity.

Interestingly, many vertebrates that lack lymph nodes have **lymph hearts** that pump the lymph along. Lymph hearts occur in amphibians (up to 100 of them!), reptiles (generally two), and in birds (a few).

EDEMA

It is essential that there be a proper balance between the rate of lymph formation and the rate of lymph removal by the lymphatic system. The famous British physiologist, E. H. Starling, proposed early in this century that this balance is determined by three factors.

- Lymph formation is promoted by the drop in hydrostatic pressure across the capillary wall.
- Lymph formation is reduced by the greater osmotic pressure of blood as compared with lymph (a result of the greater protein content of blood caused by the filtration effect).
- Lymph formation is increased whenever the leakiness of the capillaries is increased.

If any one of these factors changes, the rate of lymph formation changes. An example occurs when high venous blood pressure causes **edema,** the accumulation of water in the tissues. This is an important sign of a failing heart. Edema also occurs when starvation reduces the level of blood proteins. This lowers the osmotic difference between blood and lymph, consequently causing edematous swelling of much of the body. An extreme example is seen in Figure 24-36.

SUMMARY

In such simple organisms as primitive plants, transport occurs almost entirely by diffusion and by cytoplasmic streaming driven by actin filaments. More advanced vascular plants owe much of their success to an improved transport system wherein columns of fluid move in tubes called xylem and phloem. Most of the water absorbed through the roots travels upward via the xylem channels and is lost in the leaves by transpiration through stomata.

As water evaporates from the leaves, their solute concentration increases, and water from deeper within the plant diffuses into the leaves. Because of the extraordinary cohesion of water molecules and their adhesion to the walls of the narrow tubes through which they flow, tension results that can pull a thin column of water from the roots to the top of a tall tree.

Roots absorb water from soil in connection with the active transport of soil minerals. The resulting osmotic potential, called root pressure, pushes water and dissolved minerals up from the roots, while transpiration pulls them up. Root pressure, however, does not make a large contribution to the flow of water through plants.

The xylem transports water and dissolved materials from roots to leaves. In addition, the rigid xylem tubes provide structural support in stems. The walls of these tubes are formed by the walls of cells in which the cytoplasm has disintegrated. Hence, the tubes are not living tissue.

Tracheids, the only type of xylem in primitive vascular plants, consist of long, thin, overlapping cells. The end walls of tracheid cells are partially bridged by thin spots called pits, across which sap must move. Angiosperms also contain vessel elements. Because they have no end walls, these act as sections of pipe through which sap flows unimpeded.

Phloem, the second transport tissue in vascular plants, conducts water and dissolved sucrose and other substances throughout the plant. The conducting cells of primitive vascular plants are sieve cells. The more highly specialized phloem of angiosperms includes sieve tubes as well. Unlike tracheids and vessel elements, sieve cells and sieve tubes are living cells. Sieve tubes receive metabolic support from neighboring companion cells.

According to the pressure-flow hypothesis, fluid is driven through phloem by hydrostatic pressure of osmotic origin. Leaves actively load sucrose and other molecules into the phloem, increasing its osmotic potential and causing water to flow into it. As water enters, hydrostatic pressure increases and drives both water and solute molecules to other parts of the plant. Nutrients absorbed from the phloem are either utilized or stored.

Because animals have higher metabolic rates than plants, transport within them must take place rapidly. Only the simplest animals can rely on diffusion for transport. Most animals have one or more muscular chambers called hearts, which pump fluid through the body. Many have systems of vessels with valves that keep the fluid moving in one direction. Insects and some worms have an open circulatory system, in which blood can seep directly through tissues. Most worms and all vertebrates have closed circulatory systems, in which blood is contained within vessels.

The mammalian circulatory system is driven by two functionally independent pumps within the heart. The left ventricle pumps oxygenated blood through arteries to tissue capillaries, from which relatively deoxygenated blood flows back via veins to the right atrium. The right atrium empties into the right ventricle, the heart's second pump, which propels blood through the lungs. Oxygenated blood returns from this pulmonary circulation to the left atrium, empties into the left ventricle, and from there is pumped again throughout the body.

Fishes have only one atrium and one ventricle, and only a single circulation. All blood must flow past the gills and then through the body before returning to the heart. A dual circulation evolved in amphibians, with a pulmonary vein carrying oxygenated blood from the lungs directly back to the heart. To improve the separation of oxygenated and deoxygenated blood, the physical design of the heart became progressively more complicated in amphibians and reptiles until it culminated in the four-chambered heart of birds and mammals.

Blood is driven through arteries and arterioles by the heart's contractions, which generate the pulse and blood pressure frequently measured by physicians. By the time blood has reached the capillary bed, pulse pressure has been dissipated by the high resistance of blood vessels. Blood in the veins is driven back to the heart not by the heart's contractions but by those of the skeletal muscles, which compress the veins.

The capillary is the site of molecular transfer between plasma and tissue, which is the ultimate function of the circulatory system. Concentration gradients drive the net flow of nutrients from plasma to tissues and wastes from tissues to plasma. A net flow of fluid from plasma to tissues is also driven by the hydrostatic pressure produced when blood is ejected from the heart's left ventricle. The effect of hydrostatic pressure in forcing fluid from capillaries is counteracted in part by the osmotic activity of plasma proteins such as albumin, which move fluid in the opposite direction.

Fluid escaping the capillaries is free to pass into the vessels of the lymphatic system, which supplements the drainage function of the venous system, and eventually empties back into it. Lymph is lower in proteins and most nutrients than is plasma, but otherwise is generally similar in composition. Lymph nodes at intervals along the lymphatic vessels serve both as sites of lymphocyte formation as well as filters that trap bacteria and other foreign particles, and loci of antibody formation.

KEY TERMS

apoplastic pathway
atrium
capillary
closed circulatory system
companion cell
cortex
cotyledon
diastole
dicotyledon (dicot)
electrocardiogram (ECG, EKG)
endodermis
epidermis
guttation
hydrostatic pressure
interstitial fluid

lateral root
lymph
lymph node
lymphatic system
monocotyledon (monocot)
open circulatory system
pericycle
phloem
pith
plasmodesma (plasmodesmata)
portal circulation
pulmonary circulation
pulse
root hair
root pressure

sap
sieve cell
sieve tube
sinoatrial (S-A) node
suberin
symplastic pathway
systemic circulation
systole
tracheid
transpiration
vein
vena cava
ventricle
vessel element
xylem

QUIZ QUESTIONS

1. Which of the following is *not* a structure, substance, or process related to movement of sap upward from the root system to other parts of the plant?
   A. transpiration
   B. root pressure
   C. adhesion
   D. tracheids
   E. companion cells

2. Which of the following is a pathological condition directly related to inadequate functioning of the lymphatic system?
   A. suberin
   B. murmur
   C. edema
   D. coronary infarction
   E. diastole

3. Osmosis is involved in all of the following processes *except* for
   A. movement of water into roots.
   B. operation of guard cells.
   C. the apoplastic pathway.
   D. movement of water from the bloodstream into adjacent tissues.
   E. the pressure-flow hypothesis.

4. Organisms that lack specialized internal transport systems tend to have higher surface-to-volume ratios than do organisms with such systems.
   A. True
   B. False

C. Insufficient information is provided to determine the answer.

5. Valves are found in
   A. arteries.
   B. veins.

C. lymphatic vessels.
D. A and B
E. B and C

ESSAY QUESTIONS

1. What is transpiration? Where in the plant does it take place? How does leaf structure facilitate transpiration? What evolutionary adaptations limit evaporation of water from plants?

2. Certain small multicellular animals lack a vascular circulatory system. What aspects of the structure of sponges, corals, jellyfishes, and flatworms allow for this? What is an open circulatory system?

3. Why is it critical that the heartbeat entail a synchronized wave of muscle contraction? What would occur if the atria and ventricles contracted simultaneously?

4. What is the origin of the systolic and diastolic measures of blood pressure? What events in the heart give rise to the pulse felt in the wrist? Can a person's pulse be felt elsewhere?

5. What is lymph, and how does its composition differ from plasma? How do lymph vessels differ from blood vessels? What are the functions of the lymphatic system?

REFERENCES AND SUGGESTIONS FOR FURTHER READING

Adolph E. F. (1967). The heart's pacemaker. *Scientific American 216*, March, 32–37.

How the built-in rhythmicity of the remarkable S-A node regulates the heartbeat.

Clark, W. A., Borg, T. K., and Decker, R. S. (Eds.) (1988). *Biology of Isolated Adult Cardiac Myocytes*. Elsevier, New York.

Focuses on the isolated heart-muscle cell as a means of understanding heart function at the molecular, electrical, and physiological levels.

Crafts, A. S., and Crisp, C. E. (1971). *Phloem Transport in Plants*. W. H. Freeman, New York.

The authors present experimental evidence on how phloem functions as a transport mechanism in various plants. Includes a helpful comparative perspective of phloem transport.

Mayerson, H. S. (1963). The lymphatic system. *Scientific American 208*, June, 80–90.

How the lymphatic system recovers the water leaking from the vascular system.

Mossman, H. W. (1948). Circulatory cycles in the vertebrates. *Biological Reviews 23*, 237–255.

A classic paper.

Randall, D. J., and 4 other authors (1968). Functional morphology of the heart of vertebrates. *American Zoologist 8*, 179–229.

This symposium is the best comparative treatise on the evolution of the vertebrate heart. A stimulating discussion of the morphology and physiology of the hearts of lungfishes and amphibians reveals how the systemic and lung circulation remained separated in these forms.

Romer, A. S., and Parsons, T. S. (1977). Circulatory system. Chapter 14 in: *The Vertebrate Body*, 5th ed. W. B. Saunders, Philadelphia, pp. 402–449.

An informative chapter clearly describing the circulatory systems of the major vertebrate classes.

Russell-Hunter, W. D. (1969). *A Biology of Higher Invertebrates*. Macmillan, New York.

Many of these well written chapters discuss circulation and transport, among other matters.

Scrimshaw, N. S., and Young, N. R. (1976). The requirements of human nutrition. *Scientific American 235*, September, 50–64.

A readable discussion on the nutritive values of various food substances including trace elements.

Sutliffe, J. (1968). *Plants and Water*. St. Martin's Press, New York.

A good short account of the way plants relate to water.

Uribe, E. G., and Lüttge, U. (1984). Solute transport and the life functions of plants. *American Scientist 72*, 567–573.

An interesting review of the processes by which molecules are transported in plants, ranging in scope from short hops between organelles to long distance transport between tree roots to leaves.

Vander, A. J., Sherman, J. H., and Luciano, D. S. (1985). *Human Physiology*. McGraw-Hill, New York.

The chapter on blood circulation is especially good.

Wood, J. E. (1968). The venous system. *Scientific American 218*, January, 86–96.

Well illustrated summary.

Zimmerman, M. H. (1983). *Xylem Structures and the Ascent of Sap*. Springer-Verlag, New York.

The author is a pioneer in studies of the function of xylem in trees. Here he presents a comprehensive discussion of the structure of xylem, clearly showing its adaptive value as a mechanism for moving water.

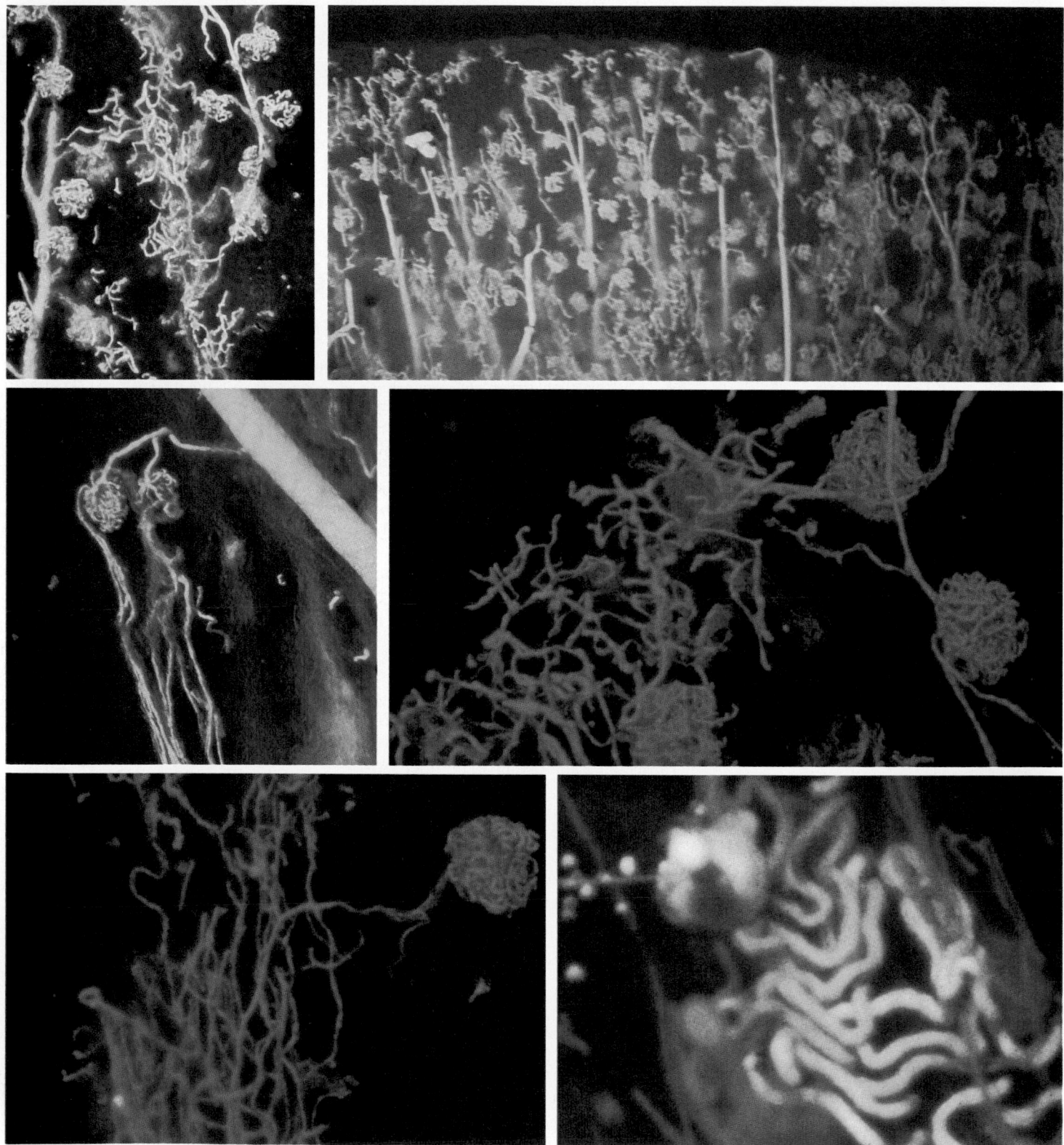

EXCRETION

Excretion
The animal kidney maintains the stability of the
fluids bathing body cells by determining which
molecules to eliminate and which to retain. Its
structure is an informative guide to animal
evolution, providing many clues to the
environmental challenges confronting our
ancestors. To study kidney structure, imaginative
investigators inject silicon rubber into tubules and
blood vessels of living kidneys, like these from
dogs. The technique clearly models the tubules,
glomeruli, and surrounding capillary networks.

Preceding chapters emphasized the need of all organisms to maintain the constancy of their body composition—a daunting challenge when nutrients, water, metabolic end-products, and toxic substances (poisons, drugs, alcohol, etc.) enter or form in the body at odd times and in unpredictable amounts. Major responsibility for this critical task is given in different species to a variety of excretory organs, of which the most familiar are the **kidneys.**

We normally think of the kidneys as organs of **excretion,** concerned only with eliminating wastes. As far as it goes, that generalization is correct, but it omits a significant truth: what the excretory organs retain is as important as what they eliminate. In the words of Homer Smith (1885–1962), a brilliant student of kidney function, "It is no exaggeration to say that the composition of the blood is determined not by what the mouth ingests but by what the kidneys keep; they are the master chemists of our internal environment, which, so to speak, they synthesize in reverse."

This chapter will consider the peculiar excretory needs of different categories of animals—and then demonstrate how those needs are met. The focus will be on protists and animals. Plants have no special organs of excretion. End products of metabolism diffuse from individual cells into the surrounding air or water—or they accumulate in vacuoles or in harmless, insoluble deposits elsewhere in the tissues of the organism.

WHAT MUST BE ELIMINATED

Because the end products of metabolism are useless and potentially harmful, they must be actively eliminated to prevent their accumulation in toxic amounts. The substances that must be excreted by animals include carbon dioxide, nitrogenous wastes, and a variety of other substances. In addition, the kidneys must eliminate excess water and various salts.

CARBON DIOXIDE, WATER, AND SALTS

Carbon dioxide (CO_2) is a gaseous byproduct of oxidative metabolism. It leaves the body through the same organs that acquire oxygen (O_2)—the gills or lungs (Chapter 27).

Water, in greater or lesser amounts, must be excreted by virtually all organisms. Some water is lost through the sweat glands and from the lungs by the familiar process of evaporation. Even when human beings are not actively sweating (and there is no visible water on the skin), they lose about 500 ml of water a day from the skin. The major function of perspiration is in the regulation of body temperature; it is not a major excretory path. The loss of body heat as sweat evaporates is an essential cooling mechanism. Similar statements could be made about plants—especially desert species with tough protective tissues that protect them from drought.

Salt concentration in body fluids can be regulated only if there is some ongoing excretion of inorganic salts. Marine (i.e., saltwater) bony fishes do this through their gills. Human beings do it with their kidneys through urinary excretion. Plants that live in salt or brackish water do it with special vacuoles that can accumulate large amounts of salt.

NITROGENOUS WASTES

These excretory activities do not account for a large and potentially toxic class of end products—those containing nitrogen. These are the end products of protein metabolism, since proteins are the major category of biomolecules that contains nitrogen (see Chapter 4). The major possible end products of protein breakdown are as follows:

- Urea, $CO(NH_2)_2$
- Uric acid, $C_5H_4N_4O_3$
- Ammonia, NH_3—or ammonium ions, NH_4^+

Various animal groups differ greatly in their most abundant nitrogenous end products—and this is an intimation of some very interesting biology. Most animals that are aquatic (live in water), whether invertebrates or vertebrates, excrete nitrogen as ammonia. Ammonia is highly soluble and diffusible, but it is also poisonous. Its excretion requires rapid solution in large amounts of water and a copious, continuous flow of urine.

Animals that are terrestrial (live on land) predominantly excrete urea or uric acid. The development of systems for converting toxic ammonia (or ammonium ions) to innocuous materials was an essential evolutionary adaptation to the inadequate water supply on land. The excretion of nitrogen as uric acid into the allantois of the amniotic egg of a reptile or bird (Figure 25-1) is apparently a method for taking advantage of the relative insolubility of uric acid. Since bird and reptile embryos live for a time within sealed eggs, deposition of solid uric acid is an ideal solution to the problem of nitrogen excretion.

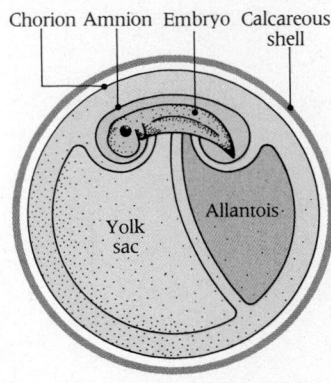

25-1 The amniotic egg of reptile or bird
The embryo receives nutrients from the yolk and excretes nitrogenous wastes into the allantois. The wastes are stored in this sac until hatching.

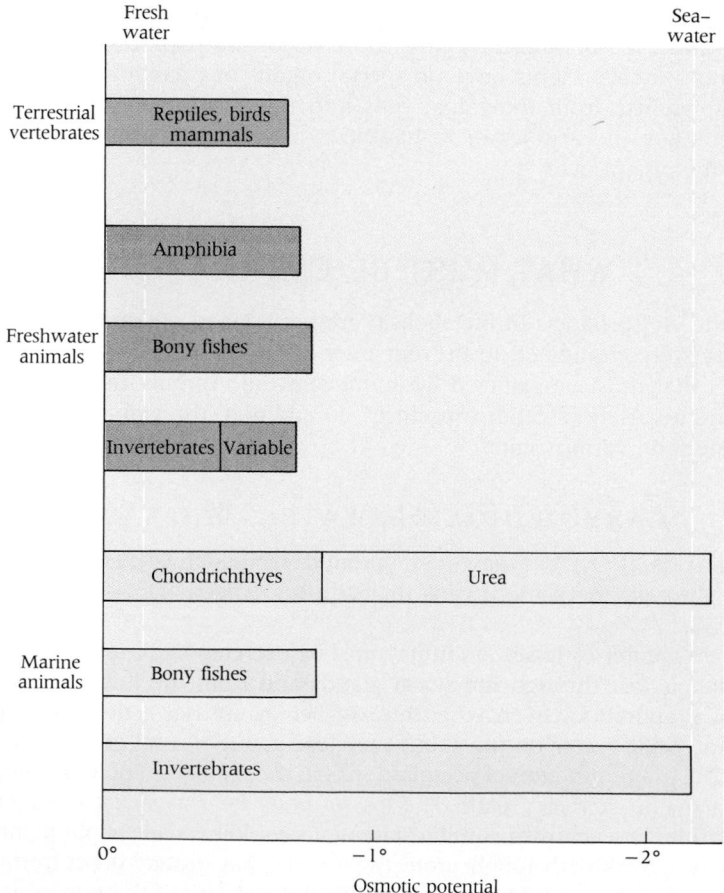

25-2 Osmotic potential of body fluids
The vertical dashed lines represent the osmotic potentials of fresh water and sea water. With the exception of the marine invertebrates, the osmotic potentials of body fluids in marine animals, freshwater animals, and terrestrial vertebrates are significantly less than that of seawater. Marine cartilaginous fishes, Chondrichthyes, increase osmotic potential by retaining urea. The fraction of osmotic potential attributable to urea in these fishes is shown in the graph.

The embryos of most mammals have no such problem. They are in close contact with the maternal circulation, which removes urea, a highly diffusible waste product. Urea is not seriously toxic except when present in unusually high concentrations.

OTHER SUBSTANCES

Other routes of excretion, to be discussed elsewhere, include the intestine, which we usually regard as an organ for absorbing nutrients, not excreting wastes. Feces are, of course, discharged from the intestine—but feces consist mainly of undigested food (or indigestible matter taken in with food) and masses of bacteria arising in the intestines.

However, feces do contain some substances that were excreted into the intestinal stream. The dark-colored bile pigments are such substances. They are excreted by the liver and flow into the intestine via the bile ducts (see Chapter 26). Bile pigments, which give feces their characteristic color, are the major end products of the breakdown of hemoglobin following the destruction of aged red cells in the spleen and liver (see Chapter 22).

DIFFERENT CHALLENGES CONFRONT DIFFERENT ANIMAL GROUPS

What we know about the course of early evolution persuades us that the excretory system is critically important in determining the course of events—especially in terms of its capacity for regulating body water. First, we consider the problems that faced major animal groups. Then we will show how these problems have been solved.

AQUATIC ANIMALS

Problems of Marine Invertebrates

Life began in the sea. Over long periods of time, organisms moved from seawater to fresh water and then to land. Those that stayed in the sea, the so-called marine organisms, had to cope with the gradual increase in its saltiness. At each stage of this saga, existing organisms confronted different problems.

Most invertebrates that live in the sea are isosmotic with the environment. Since their body fluids have the same osmotic pressure as the surrounding seawater, they are termed **osmoconformers.** This means that if the external, or environmental, osmotic pressure were to change, they would change their internal osmotic pressure too. In fact, this does not happen very often.

As we shall see, there is no tendency in such organisms for excess water to be forced into their bodies by a higher osmotic pressure in the surrounding water. Since they are bathed in water, they can simply excrete wastes from their entire body surfaces. That makes the maintenance of salt and water balance relatively easy. If the osmolality of body fluids falls, water is lost to the surrounding sea by osmosis and fluid osmolality returns to normal. If body fluid osmolality rises, organisms can gain water osmotically from seawater, and by dilution return fluid osmolality to normal.

Throughout evolutionary history, most marine invertebrates have had body fluids that are isosmotic with the surrounding seawater (Figure 25-2). Today the sea is considerably saltier—and therefore of higher osmotic potential—than the seawater of eons ago. Hence, the body fluids of many modern marine invertebrates are about three times as concentrated as those of freshwater invertebrates and most vertebrates—including bony fishes, which have body fluids that are hypotonic relative to sea water and hypertonic to fresh water (see Figure 25-2). As we shall see, of the vertebrates only sharks are osmoconformers. All of the other vertebrates are **osmoregulators**—that is, they maintain an internal solute concentration that does not vary, irrespective of their environment.

Problems of Marine Fishes

Fishes living in a salty sea—i.e., a medium that is hypertonic relative to their body fluids—face a rather different problem. Marine fishes and other marine verte-

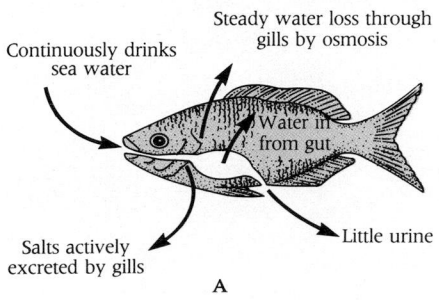

MARINE BONY FISHES

Continuously drinks sea water

Steady water loss through gills by osmosis

Water in from gut

Salts actively excreted by gills

Little urine

A

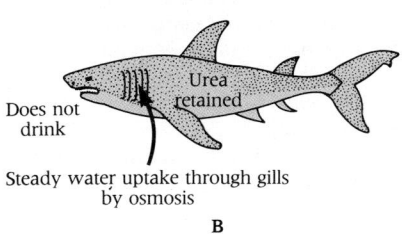

MARINE CARTILAGINOUS FISHES

Does not drink

Urea retained

Steady water uptake through gills by osmosis

B

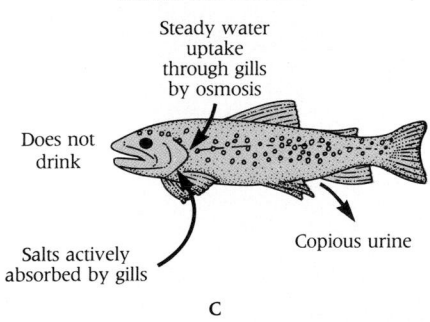

FRESHWATER FISHES

Steady water uptake through gills by osmosis

Does not drink

Salts actively absorbed by gills

Copious urine

C

25-3 Water balance in fishes
(A) A marine bony fish takes in large amounts of salt and actively transports the excess salt out of the body at the gills. (B) A marine cartilaginous fish does not take in salt water, but does retain urea, which gives its body fluids an osmotic potential slightly higher than that of the surrounding seawater. This draws water into the animal. (C) In freshwater fishes, water enters the body by osmosis, since body fluids have a higher osmotic potential than that of the surrounding water. Salts are absorbed and the excess water is excreted.

brates must protect their body cells from water loss, or dehydration. Wherever a semipermeable membrane in their bodies comes in contact with seawater—in the gills, for example—they tend to lose water by osmosis. Without an efficient compensatory mechanism, lethal dehydration would occur—even though the fishes were surrounded by water.

Most marine bony fishes must do two things (Figure 25-3A).

■ First, they must conserve body water. Although they lose some water through the gills and they excrete urine that is isotonic with their body fluids, they conserve some water by producing very little urine.

■ Second, they must eliminate excess salt in the large quantities of sea water they drink to replace the water lost osmotically. This is managed by the gills, which actively and continuously pump out salt against the osmotic gradient. The salt is moved by active transport from a solution of low concentration to one of high concentration.

Thus although marine bony fishes are surrounded by water, they actually live in something like a desert. They must constantly conserve water and excrete salt. Any engineer could design a more efficient system.

Marine cartilaginous fishes—sharks, skates, and rays—have a more efficient mechanism (Figure 25-3B). Like most other vertebrates, they have body fluids with a salt concentration about one-third that of sea water. But as we shall discuss more fully later, these animals cope with the high osmotic potential of the surrounding seawater by maintaining high concentrations of urea (or other nitrogen-containing compounds) along with other osmotically active materials in their body fluids (see Figure 25-2). This combination of salts and urea makes the osmotic potential of their body fluids slightly greater than that of seawater, so that their gills actually absorb a little water from the sea. This water is used to facilitate waste excretion.

We see, then, that organisms are not perfect. In many ways, they are not even efficient. However, they were not designed in a purposeful way. They are what a long, blind, unplanned evolutionary history has made them.

Problems of Freshwater Fishes

Freshwater fishes have the opposite problem. In migrating from salt water to fresh water, early organisms took with them an "internal sea" in the form of salt-containing body fluids. The osmotic pressure of their internal tissues is higher than that of the surrounding fresh water, and water tends to flow into the tissues. In this case, death would come from swelling if the tendency were not counteracted. Freshwater fishes must also prevent the loss of body salts by diffusion. A counter-strategy soon evolved (Figure 25-3C). These fishes drink little or no water, but they continuously form large quantities of dilute urine, thus eliminating the water acquired by osmosis. By itself this process would tend to flush out the body salts, but arrangements developed whereby fishes could reabsorb dissolved salts from urine before they leave the body. Special cells also developed in the gills, which absorb salt by active transport. Because they are capable of maintaining a stable pattern of internal solutes, these fishes (like all other vertebrates except sharks) are termed osmoregulators.

It has not been rigorously established whether the first vertebrates arose in salt water or fresh water, but the weight of evidence suggests the latter. Part of this evidence is found in the kidney itself. A discrete kidney first appeared in vertebrates, the early fishes. When vertebrates moved onto the land, the problems they faced were solved by devices that first appeared in freshwater fishes. These fishes needed and acquired an excretory apparatus capable of excreting water and conserving salt. Their kidneys, for the first time, controlled the composition of the internal environment.

TERRESTRIAL ANIMALS

Most amphibians live near fresh water. Their method of osmotic regulation is very similar to that of freshwater fishes. When a frog is in the water, there is an inflow of water that leads to the excretion of highly dilute urine by mechanisms and structures resembling those of the nephron of freshwater fishes.

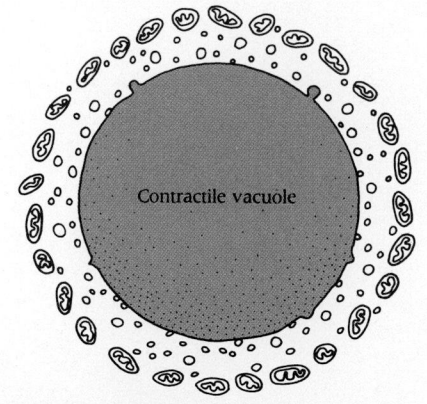

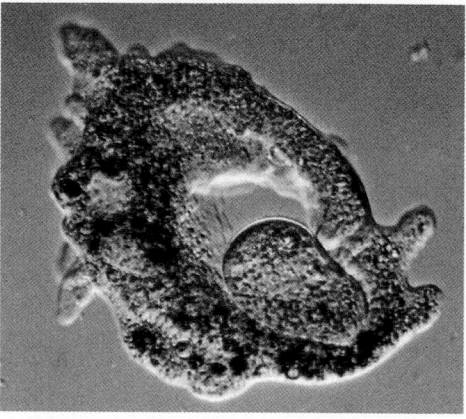

25-4 Contractile vacuole of Amoeba proteus
Within the cytoplasm of this cell, a contractile vacuole is surrounded by many tiny fluid-filled vesicles. After the ions have been pumped back into the cytoplasm, the vesicles fuse with the vacuole. The ring of mitochondria just outside the vesicles provides ATP required for pumping ions out of vesicles and for expelling vacuole contents from the cell.

The greatest physiological threat to life on land is dehydration. Terrestrial reptiles—remember there are also aquatic and marine reptiles—minimize this danger with dry and scaly skin that is relatively impermeable to water. The kidneys of terrestrial reptiles reabsorb most of the water in the filtrate and excrete a concentrated urine. In contrast, the kidneys of aquatic reptiles (turtles, tortoises, etc.) conform to the pattern of freshwater fishes—that is, they can produce dilute or isotonic urine, but cannot make urine that is more concentrated than blood plasma. Many marine reptiles (crocodiles, some turtles, lizards, etc.) eliminate salt by means of **salt glands,** located near the eyes or nose.

Birds and mammals are unique members of the animal kingdom in this regard; they can produce urine that is much more concentrated than the blood plasma. In birds, this ability is not very pronounced, but the mammalian kidney can produce urine up to 25 times as concentrated as plasma. As we will see, the capacity to form such urine depends upon the three-dimensional arrangement of renal tubules.

EXCRETORY MECHANISMS THAT SOLVED THE PROBLEMS

To comprehend the ingenious design of the vertebrate kidney, we must review some of the landmarks of its evolutionary development, bearing in mind that there are dangers in making inferences about the course of evolution from observations made on present-day organisms.

We earlier mentioned the fundamentally different challenges facing saltwater organisms with regard to salt and water conservation. To land-dwelling animals, always facing the threat of dehydration, water conservation is an absolute necessity.

CONTRACTILE VACUOLES

In protists and the simplest multicellular animals, too, excretion occurs mainly by diffusion from individual cells. In some (but not most) protists, the **contractile vacuole** has a role in excretion (Figures 25-4, 25-5).

The principal activity of the contractile vacuole is to control the water content of the cell. Elimination of waste products by these organisms is primarily by diffusion across the cell surface. However, some waste products are dissolved in the water expelled by the contractile vacuole.

Here we see in its most primitive form a phenomenon that is closely associated with the excretion of waste products all through the animal kingdom: organs of excretion also help to regulate the body's water content.

THE MEMBRANES OF AQUATIC INVERTEBRATES

As noted earlier, most aquatic invertebrates are osmoconformers. For them, the maintenance of salt and water balance is relatively simple. A need to conserve

25-5 Contracile vacuoles in Paramecium
*The schematic diagrams (**A–D**) and photomicrographs (**E–F**) show full and empty contractile vacuoles. The sequence of filling and emptying depends on the presence of radiating canals, which guide cytoplasmic water to the contractile vacuole. (**A**) The vacuole is full and the radiating canals are empty. (**B**) The vacuole contracts, releasing water to the outside. The canals begin to fill. (**C**) The vacuole is empty and the canals are full of water. (**D**) The canals contract, filling the vacuole with water. The cycle begins again.*

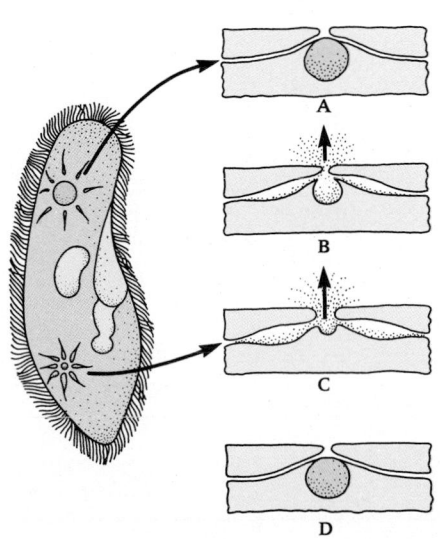

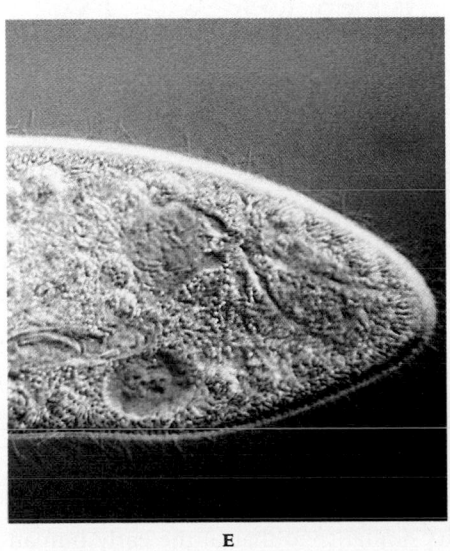

E

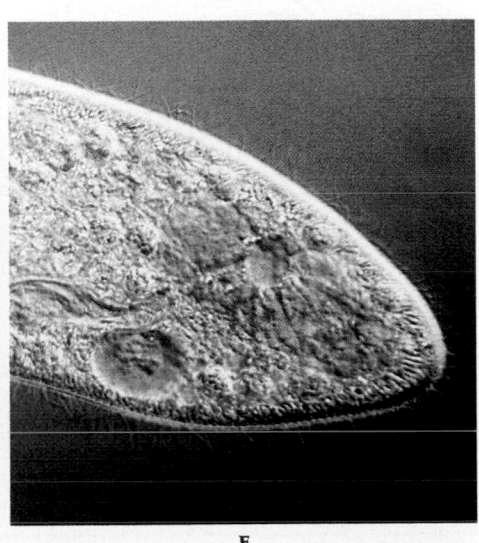

F

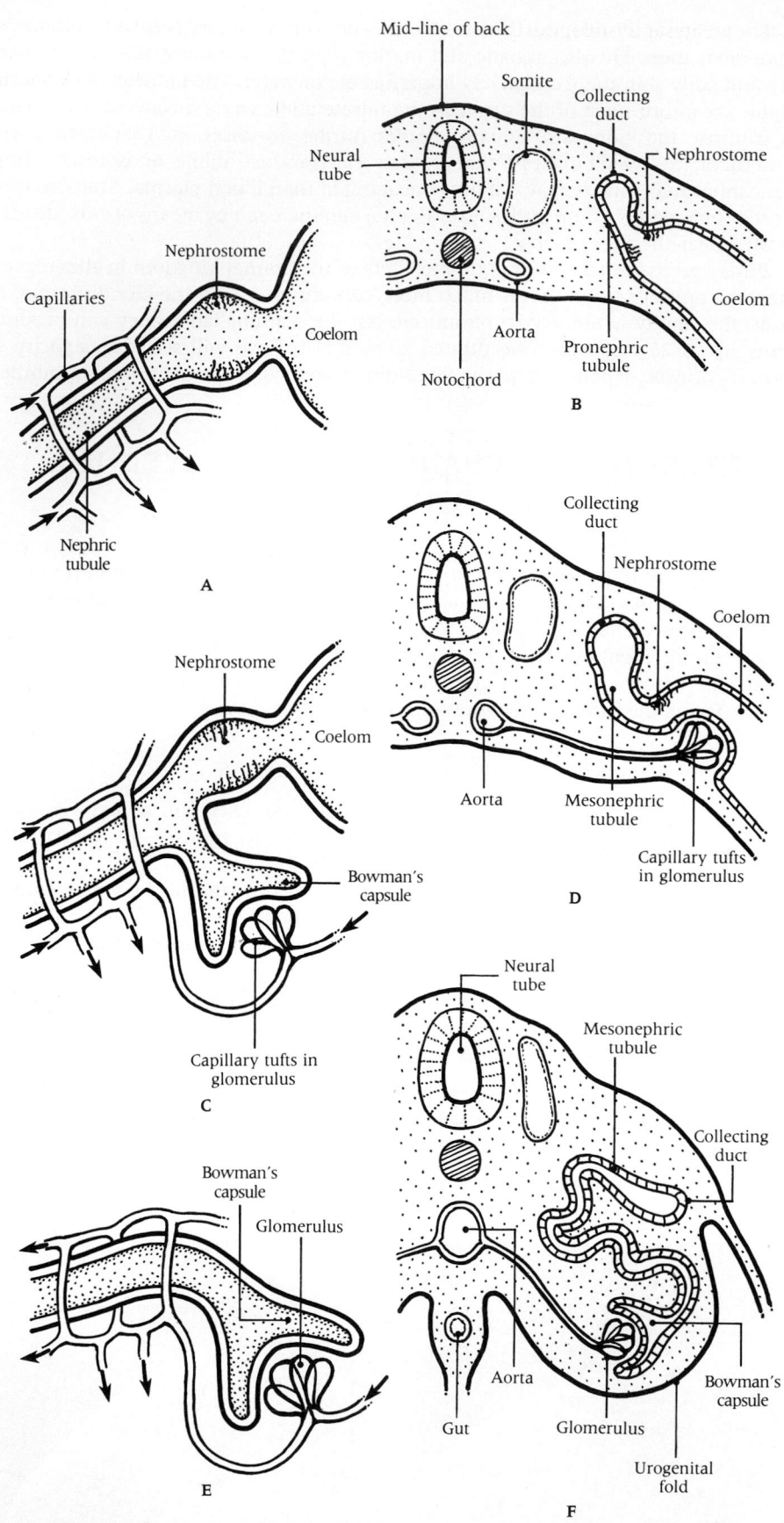

Mid-line of back

Somite

Collecting duct

Neural tube

Nephrostome

Nephrostome

Capillaries

Coelom

Aorta

Coelom

Notochord

Pronephric tubule

B

Nephric tubule

A

Nephrostome

Collecting duct

Nephrostome

Coelom

Coelom

Bowman's capsule

Aorta

Mesonephric tubule

Capillary tufts in glomerulus

D

Capillary tufts in glomerulus

C

Neural tube

Mesonephric tubule

Bowman's capsule

Glomerulus

Collecting duct

Aorta

Bowman's capsule

Gut

Glomerulus

Urogenital fold

E

F

25-6 Parallels between the evolutionary and embryological development of the kidney of the terrestrial vertebrate. Compare the evolutionary sequence (A, C, E) with the stages of embryogenesis (B, D, F). **(A)** In protovertebrates, nephric tubules drained the coelom via a nephrostome. **(B)** Human embryo at 3.5 weeks. **(C)** Early vertebrate with glomerulus and persisting nephrostome. **(D)** Human embryo at 5 weeks. **(E)** Nephrostome has disappeared leaving typical vertebrate nephrone. **(F)** Human embryo at 8 weeks.

water does not exist. If excess salt accumulates, it is excreted through the epithelium of the respiratory organs. Elimination of nitrogenous wastes, the only other significant excretory need, is managed by a meager set of membranes and segmented tubules that empty into the main body cavity, the coelom (Figure 25-6A). This type of primitive excretory equipment, striking in its similarity to that of the early human

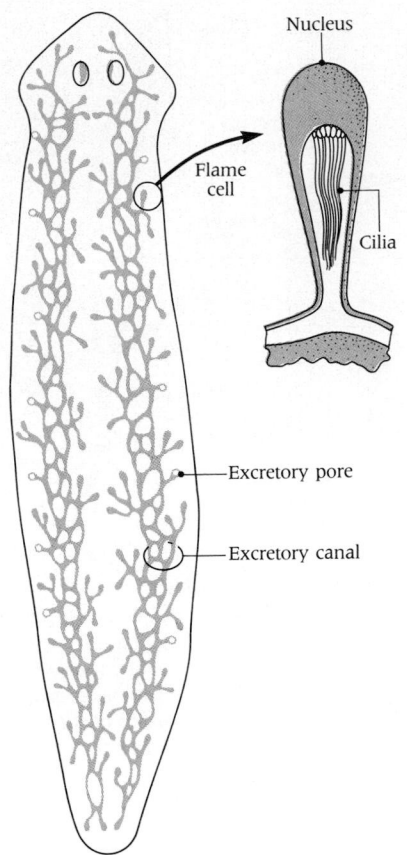

Nucleus

Flame
cell

Cilia

Excretory pore

Excretory canal

25-7 Excretion in flatworms
*The excretory system of a planaria is composed
of two excretory canals, longitudinal networks
of tubules that end in excretory pores or in
specialized flame cells. Cilia move materials
through the tubules and out the pores.*

embryo (Figure 25-6B), deals mainly with waste products incapable of leaving the body by simple diffusion.

Three of the major animal groups operate in this way—that is, they lack special organs for excretion or water regulation. They include members of the phyla Porifera (sponges), Coelenterata, or Cnidaria (corals, jellyfishes, etc.), and Echinodermata (starfishes, sea urchins, etc.). All are aquatic—that is, they live in water—and most are marine—that is, they live in seawater. All echinoderms are marine.

Freshwater flatworms (Platyhelminthes) do have organized excretory systems that make possible the excretion of excess water (Figure 25-7). In planaria and flukes, for example, body fluids are collected into specialized **flame cells** by the beating of cilia—which in a microscopic field resembles the flickering of a flame. Fluid (and some dissolved wastes) then passes by active transport into a series of tubules until it reaches an **excretory pore** at the body surface. This may represent the most primitive tubular excretory system.

EXCRETORY ORGANS OF EARTHWORMS

The earthworm has a closed circulatory system. It is therefore able to make use of small blood vessels in its excretory systems.

The body of an earthworm is composed of relatively discrete segments (Figure 25-8). Each has its own pair of excretory organs, the **nephridia.** A nephridium has a ciliated funnel, the **nephrostome,** that opens into the coelom and receives coelomic fluid into a long thin tubule. Materials needed by the organisms are reclaimed from fluid flowing through the tubule and these are transferred into a surrounding capillary network. The association of blood vessels with the coiled tubule is a major forward step in the evolution of excretory machinery.

THE MALPIGHIAN TUBULES OF INSECTS

As adults, the bodies of most insects are more than two-thirds water. Thus they must have extraordinary capacities to prevent water loss. They do, in part, with a hard, dry **cuticle** that often has an outer waxy layer. But this is not the whole story.

The evolutionary ancestors of insects probably resembled the ancestors of segmented earthworms, which had nephridia. But insects do not have nephridia. In them, feces and urine are eliminated through the same opening, the **anus.** Urine is formed by long, slender tubules, known as **Malpighian tubules,** which may number from two to several hundred. Each tubule has a blind end lying in the body cavity or coelom and at the other end opens into the intestine (Figure 25-9). Potassium and sodium ions are actively secreted into the lumen of the tubule and water follows passively by osmosis. As a result, large amounts of an ion-rich fluid are formed in the tubule, from which it enters the hindgut. In the hindgut, much

25-8 Nephridia of an earthworm
*Fluids are collected into the funnel-shaped
nephrostome, located in the segment anterior to
the rest of the nephridium. Materials move
through the system of coiled tubules, which are
closely associated with capillaries. A storage
bladder connects the tubule to the outside via a
nephridiopore.*

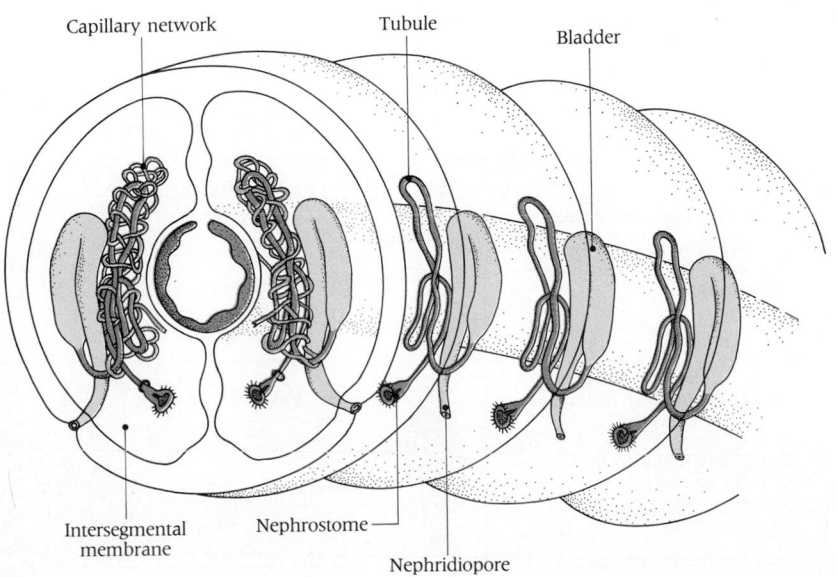

Capillary network

Tubule

Bladder

Intersegmental
membrane

Nephrostome

Nephridiopore

of the water and solutes are reabsorbed and uric acid (the major nitrogenous waste, which enters the tubule as water-soluble potassium urate) is precipitated. What remains in the hindgut is a mixture of solid uric acid and feces. In this way, insects can produce very dry excreta and may lose virtually no water in feces and urine.

THE KIDNEYS OF FRESHWATER FISHES

Discrete **kidneys** first appeared in mollusks and the earliest vertebrates, which were fishes. Freshwater fishes needed an excretory apparatus that could excrete water and conserve salt—and in addition excrete metabolic waste products. For the first time, water had to be eliminated. By elaborating dilute urine, these kidneys could keep salt loss to a minimum.

Freshwater fishes must eliminate water in large quantities to prevent self-dilution. Fortunately for them, their kidneys keep the osmotic concentration of body fluids constant in the face of external concentration differences and changes. Because they are osmoregulators, freshwater fishes can exploit habitats that would otherwise be inaccessible.

Five hundred million years ago the basic structural and functional unit of the vertebrate kidney, the **nephron,** made its first appearance in the most primitive vertebrates represented nowadays by the lampreys and hagfishes. Throughout subsequent evolution, in truly freshwater fishes the nephron underwent extensive development, laying down the basis for the elaborations that took place when vertebrates invaded the land (see Figure 25-6C). In contrast, marine fishes display drastically degenerative specializations.

One end of a single nephron is a cup-shaped device called **Bowman's capsule** (Figure 25-10A). Within it is a tiny tuft of capillaries called a **glomerulus.** Since the glomerulus was first described by Malpighi in 1666, it is also known as a **Malpighian corpuscle.** Hydrostatic pressure in the arteries forces fluid through the semipermeable membrane of Bowman's capsule. The membrane holds back large molecules (i.e., most proteins, but not some of the small ones). Water and small molecular solutes pass through freely. The product of **glomerular filtration** is an ultrafiltrate that now passes to the lower parts of the nephron.

The rest of the nephron is a long thin tubule that follows a complex course. The cells of the first portion, the neck, appear ciliated in tubular sections. They function as pumps and their cilia propel the tubular fluid (filtrate) down the next segment, the **proximal convoluted tubule.** Here macromolecules, glucose, and divalent ions, are actively reabsorbed from the filtrate into the fluids percolating through the tissues surrounding the tubule. Indeed, the proximal tubule plays a

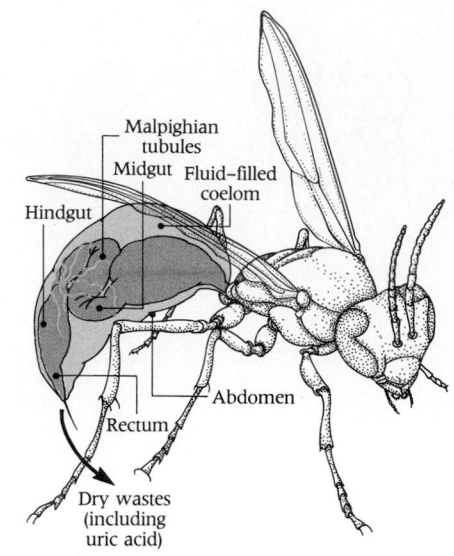

25-9 Malpighian tubules in insects
Malpighian tubules are blind sacs that float in the fluid of the coelom. They take up water, uric acid, and sodium and potassium ions from the fluid. All but the uric acid is resorbed into the coelom, leaving a dry deposit of nitrogenous wastes in the intestine, which is excreted through the anus.

25-10 Relation between nephron structure and the environment
There is a correlation between the environment and the structure of the nephron in vertebrates. **(A)** *Freshwater fishes.* **(B)** *Bony marine fishes.* **(C)** *Humans and other terrestrial vertebrates.*

A FRESHWATER FISHES			
Lives in water less concentrated than body fluids; fish tends to gain water, lose salt	Salt in (active transport by gills) / H₂O in / Salt out / No water-drinking	Glomerulus / Bowman's capsule / Very large glomerulus / Long proximal convoluted tubule / No distal convoluted tubule / No loop of Henle	Large volume of urine / Urine is less concentrated than body fluids
B BONY MARINE FISHES			
Lives in water more concentrated than body fluids; fish tends to lose water, gain salt	Salt out (active transport by gills) / Salt in / H₂O out / Drinks water	No glomerulus (or very small if present) / Large proximal convoluted tubule / No distal convoluted tubule / No loop of Henle	Small volume of urine / Urine is slightly less concentrated than body fluids
C TERRESTRIAL VERTEBRATES, HUMANS			
Terrestrial environment; tends to lose body water to air	Salt in (by mouth) / Salt out / H₂O out / Drinks water	Large glomerulus / Large proximal convoluted tubule / Distal convoluted tubule / Long loop of Henle	Moderate volume of urine / Urine is more concentrated than body fluids

key role in minimizing the loss of divalent ions (Ca^{2+}, Mg^{2+}, SO_4^{2-}, HPO_4^{2-}) into the urine.

The proximal tubule is followed in the nephron by a second "pump": this is the **intermediate segment,** which contains cilia that rapidly propel the fluid stream. Monovalent ions (Na^+, Cl^-, K^+) are actively reabsorbed from both the **collecting tubules.** These mark the nephron's terminus. Because of the pumping action of the neck and intermediate segment, copious amounts of urine are moved at a high velocity. Active reabsorption of glucose, macromolecules, bivalent and monovalent ions from the proximal, distal, and collecting tubules results in the production of very dilute urine.

"DEGENERATED" KIDNEYS OF SALTWATER FISHES

When surrounded by salt water, the ability of a fish to conserve water is at a premium. Since water had to be conserved, the flow of urine was greatly reduced—both in volume and velocity—by the elimination of the two ciliated nephric pumps, the neck and intermediate segment (Figure 25-10B). Glomerular filtration was also greatly decreased by reduction of the pressure in the capillaries and a thicker basement membrane in Bowman's capsule.

All marine fishes retain proximal tubules, which perform the vital function of active transport of salt ions from the tissue fluids surrounding the nephron into the urine of the proximal tubule. In most oceanic fishes, the distal tubule is lost, since Na^+ and Cl^- ions are excreted by the gills. As a result, the nephron of saltwater fishes produces scanty amounts of urine, thereby minimizing loss of water.

What we have here is termed a degenerative evolutionary trend in which a highly developed excretory organ begins to lose some of its capacities because such losses proved beneficial under existing circumstances. This trend climaxed with the complete loss of Bowman's capsule and its glomerulus in aglomerular marine fishes. Interestingly, the loss of these structures set the stage for the invasion of subzero waters by the fishes of Antarctica. The lack of a glomerulus prevented the loss into the urine of various low-molecular-weight glycoproteins that had distinctive antifreeze properties. Without glomerular filtration these glycoproteins never entered the tubular fluid. As a result, they remained in high concentration in the blood where they serve as an effective biological antifreeze.

A correlation between the structure of human kidneys and the terrestrial environment appears in Figure 25-10C.

THE KIDNEYS OF TERRESTRIAL VERTEBRATES: THE HUMAN MODEL

Let us briefly survey the stages of embryological development, as we noted in passing when discussing their evolutionary "counterparts." We urged caution (in Chapter 20) in interpreting such parallels to mean that "ontogeny recapitulates phylogeny." They do not. Nonetheless, they are striking and provocative.

EMBRYOLOGICAL DEVELOPMENT

The fact that the main arena of early vertebrate evolution was in fresh water rather than salt water had momentous consequences. It was the evolution of the kidney that conferred upon the animals destined to become vertebrates—the so-called **protovertebrates**—the stable internal environment necessary for the successful functioning of nerves, muscles, and glands. These early forms were segmented—and a degree of segmentation persisted in the skeletal muscles, nerves, and backbone of all vertebrates. The internal organs (viscera), however, were no longer segmentally organized. Instead they extended continuously from mouth to anus and were contained in an unsegmented body cavity, the coelom.

In the embryos of vertebrates, the viscera arise from endoderm; and the segmented muscles, nerves, and backbone arise from mesoderm. Except for its blood vessels, the kidney develops, not as one might expect from the endoderm, but from the mesoderm—and it does begin as a segmented structure (as did the kidney of early evolution).

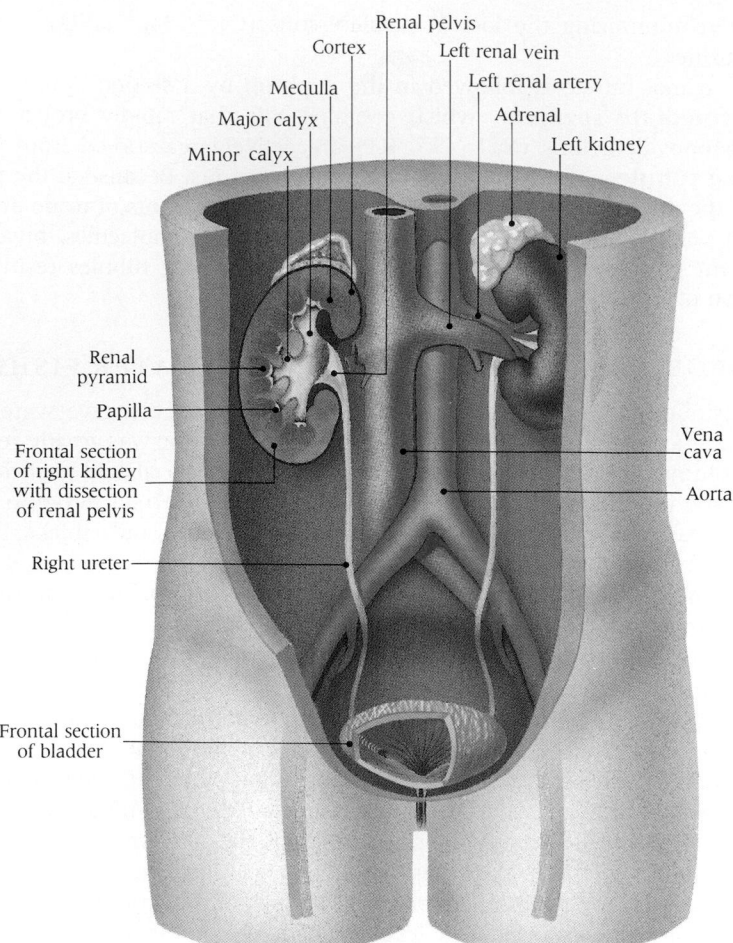

Renal pelvis
Cortex
Left renal vein
Medulla
Left renal artery
Major calyx
Adrenal
Minor calyx
Left kidney

Renal pyramid
Papilla
Frontal section of right kidney with dissection of renal pelvis
Right ureter

Vena cava
Aorta

Frontal section of bladder

25-11 Structure of the human urinary system

The kidneys connect to the circulatory system via the renal artery and renal vein. Blood is filtered in the nephron and urine formed from the filtrate descends through the ureter to the bladder where it is stored until discharged from the body.

In the early vertebrate embryo (see Figure 25-26B), short pronephric tubules on each side of the midline extend from the coelom into a collecting duct running the length of the body. A similar arrangement existed in the protovertebrate where nephric tubules communicated freely with the coelom via a mouthlike opening called a nephrostome. These drained to the exterior, either through separate vents in the body like the scuppers of a ship, or into a common longitudinal groove in the skin. These primitive nephric tubules also carried off eggs and sperm, which were shed into the coelomic cavity. Note the similarity of this model to that of the earthworm (see Figure 25-8).

We saw above that to increase the rate of water excretion, early freshwater fishes acquired glomeruli (see Figure 25-6C). With the tubule taking on a predominantly excretory, rather than reproductive, function, it gradually lost its connection with the coelom (see Figure 25-6E). As the body became encased in armor, the multiple external openings were lost and the tubules were bundled—first, in primitive fishes into the structure called a **pronephros,** then in higher fishes into a **mesonephros,** finally in vertebrates into a **metanephros,** or kidney, which then acquired a new external outlet, the ureter.

Comparable changes occur in the embryogenesis of the vertebrate kidney. As the tubules mature, capillary tufts are pushed into the expanded tubule end (Bowman's capsule) to form the glomerulus, while the tubule grows in length and becomes convoluted before it finally connects with the collecting ducts. In mammals, the pronephros and mesonephros are passing episodes—see the pronephric tubules in Figure 25-6B and the mesonephric tubules in Figure 25-6D,E—that culminate in the formation of an adult kidney.

STRUCTURE AND FUNCTION

In human beings, the two kidneys are solid, dark red, bean-shaped structures that lie at the back of the abdominal cavity (Figure 25-11). The dark brown exterior

25-12 Structure of the human nephron
Blood is filtered in the glomerulus, and the filtrate moves through a series of tubules where it is modified and concentrated. The resulting fluid, called urine, enters the collecting tubule, which empties into the renal pelvis, and then into the ureter. Note that in this sketch the several structures are not drawn to scale. Compare with the photographs at the beginning of this chapter.

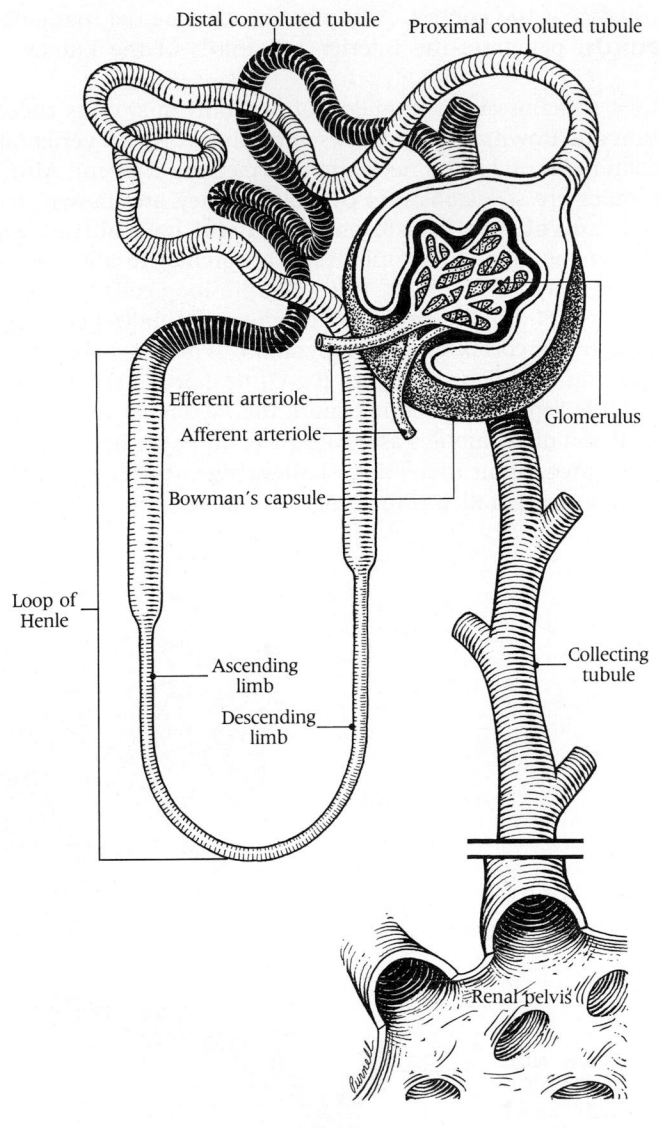

Distal convoluted tubule   Proximal convoluted tubule

Glomerulus

Efferent arteriole
Afferent arteriole
Bowman's capsule

Loop of Henle

Ascending limb
Descending limb

Collecting tubule

Renal pelvis

25-13 Structure of a human kidney
(A) *Blood enters the kidney via the renal artery, and traverses the medulla to reach the glomerular capillaries in the cortex. After passing through a second capillary bed, the blood (now cleared of wastes) leaves the kidney via the renal vein.* **(B)** *The nephrons (one is shown in color) extends from the cortex into the medulla. The glomeruli and convoluted tubules are in the cortex. The loop of Henle descends into the medulla.*

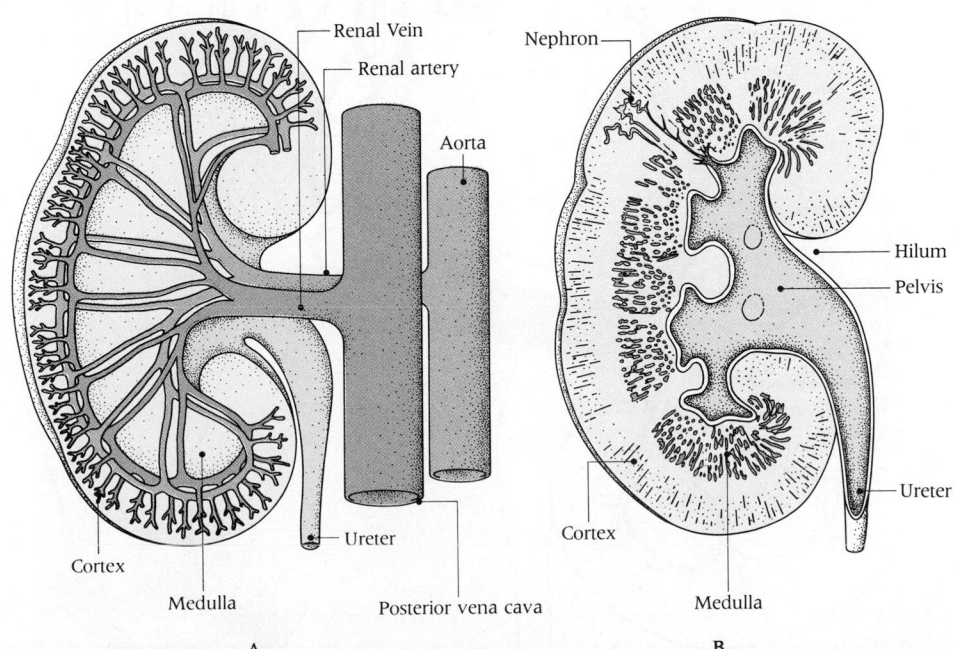

Renal Vein
Renal artery

Aorta

Nephron

Hilum

Pelvis

Cortex

Medulla

Ureter

Posterior vena cava

Cortex

Medulla

Ureter

A

B

third of the surface is the **cortex.** A number of separate red triangular masses, the **renal pyramids,** penetrate the interior two-thirds of the kidney, which is the **medulla.**

Each kidney contains about a million of the individual units called **nephrons.** The glomerulus and Bowman's capsule resemble those of other vertebrates. However, the mammalian nephron lacks a neck and intermediate segment. Also, the proximal and distal tubules are so extensively convoluted they are known, respectively, as the **proximal convoluted tubule** and the **distal convoluted tubule** (Figure 25-12). Between the convoluted tubules is an architecturally specialized part of the tubule found only in mammals. It is a hairpin loop called the **loop of Henle.** As seen in Figure 25-13, the proximal convoluted tubule becomes straight and descends through the cortex toward a pyramid, where it makes a sharp U-turn, proceeding back in a straight course to the cortical area near its own glomerulus. The loop of Henle represents nothing more than a spatial reorganization of parts of the proximal and distal tubules as we know them in the freshwater fish nephron. The distal convoluted tubule then joins a **collecting tubule,** which exits the kidney along a path that runs parallel to the limbs of Henle's loop.

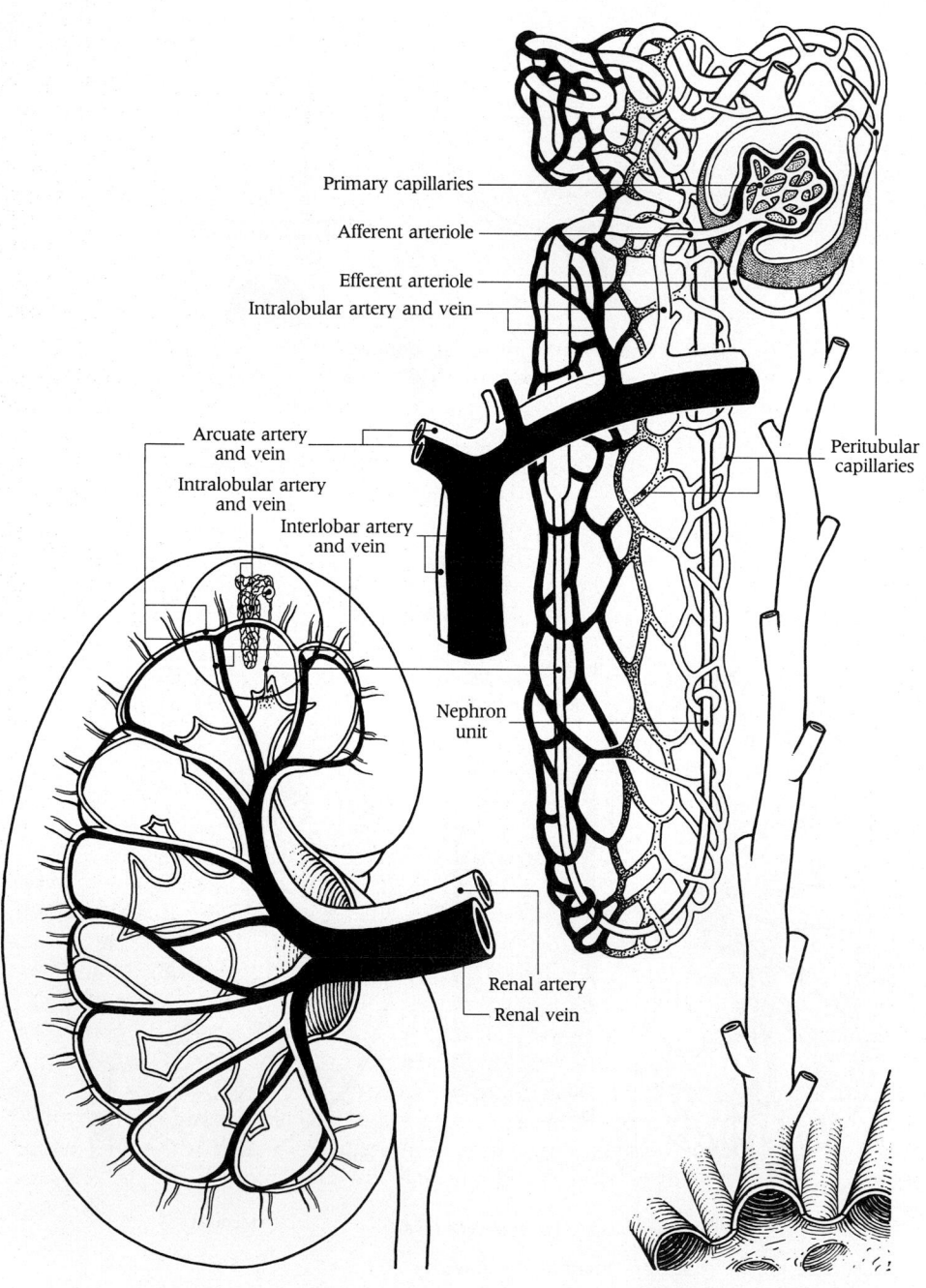

Primary capillaries

Afferent arteriole

Efferent arteriole

Intralobular artery and vein

Arcuate artery and vein

Intralobular artery and vein

Interlobar artery and vein

Peritubular capillaries

Nephron unit

Renal artery

Renal vein

25-14 Blood vessels of the kidney
Note the relations between the blood vessels and the nephron. There are two capillary beds, one in the glomerulus, and the other surrounding the tubules of the loop of Henle. Compare with the photographs of capillaries at the beginning of this chapter.

25-15 Scanning electron micrograph of a glomerulus
This view of a glomerulus in a monkey kidney shows the coiled capillaries that lie within Bowman's capsule. (×500)

TABLE 25-1

GENERAL CHARACTERISTICS OF HUMAN URINE

Property	Value*
Specific gravity	1.020 (1.002–1.030)
pH	5.5 (4.5–7.8)
Volume, ml	1200 (600–2500)
Titratable acidity, meq†	30 (20–40)
Total solids, g	50 (30–70)
Osmolality, mosmols/liter	960 (50–1400)

* Per 24 hr in a normal adult on an average diet. Average normal data (with ranges in parentheses).
† Meq = milliequivalents.

Urine made in a nephron (by mechanisms described below) drains into a treelike arrangement of collecting tubules that convey the urine to the **renal pelvis** and thence, by way of a **ureter,** to the **bladder** (see Figure 25-11). Urine is ejected from the bladder through the **urethra** by the voluntary process of **micturition,** or urination.

The kidney is rich in blood vessels (Figure 25-14). Each glomerulus arises from a branch of the **renal artery** called an **afferent arteriole.**[1] The **glomerular capillaries** then connect with an **efferent arteriole** (not the small vein into which ordinary capillaries lead), which passes out of Bowman's capsule. The outgoing arteriole then forms another network of capillaries. These are the **peritubular capillaries** that surround the rest of the tubule before they gather into the larger veins that drain into the **renal vein.**

The short renal artery that supplies the kidney branches directly off the aorta. Hence little of the pressure of the blood in the aorta is lost by the time blood enters the kidney. In fact, the blood pressure in the glomerular capillaries is twice as high as in all other capillaries. A high hydrostatic pressure is essential to the filtration step that initiates urine formation.

The total amount of blood pumped by the human heart each day is about 8000 liters. Since the entire blood volume of an average sized man is only 5 liters, the blood must be recirculated 1600 times a day. Of the 8000 liters pumped by the heart, 1900 liters enter the kidneys via the two renal arteries. In the glomeruli, about 180 liters of the fluid portion of the blood filters through the capillary walls. The high hydrostatic pressure developed by the heart literally forces everything in blood except blood cells and most plasma proteins across the glomerular capillary walls into Bowman's capsule (Figure 25-15) and ultimately into the tubules (Figure 25-16).

Interstitial fluid, the extracellular fluid of nearly fixed composition that bathes cells, passes back and forth between the tissue spaces outside the blood vessels and a compartment inside the blood vessels where it comprises the fluid of plasma. Capillary walls permit free passage of all soluble low molecular-weight blood components. That excludes the plasma proteins.

As the plasma filtrate—now termed the **glomerular filtrate**—passes down the tubule, tubule cells begin to reabsorb valuable ingredients of the filtrate, returning them to the general circulation. Most of the water is reabsorbed with it. The bulk of the water must be reabsorbed, since each day there is 180 liters of glomerular filtrate—but in the end only 1 liter of urine. Other substances reabsorbed by active transport in the tubule cells include glucose, salts, and amino acids. The kidneys reabsorb glucose so efficiently that the appearance of only a trace of glucose in the urine suggests the possible presence of a disease called **diabetes mellitus,** in which blood sugar levels are abnormally high. Oddly, half of the filtered urea is also reabsorbed. At the end of the tubule, the fluid, which only now can be called **urine,** is greatly altered in composition from the plasma (compare Tables 25-1 and 25-2 with Table 22-2). How does the nephron reabsorb water and various substances dissolved in it, and how does it do so with such precision?

The story of how a kidney can form concentrated urine is illustrated in Figure 25-17. First, consider the ascending limb of the loop of Henle. In its upper half, negatively charged chloride ions (Cl^-) are actively transported from the tubule

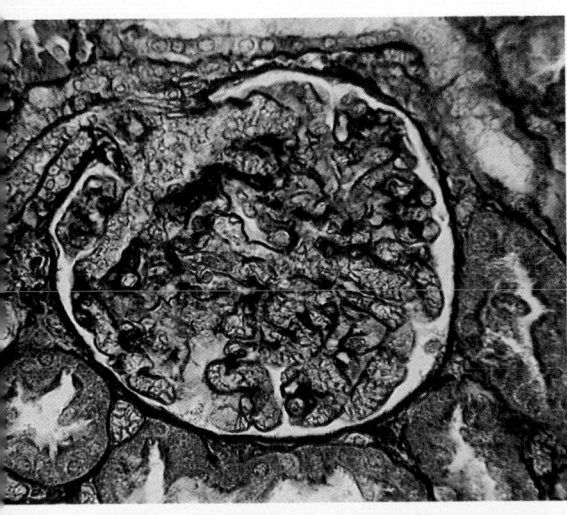

25-16 Light micrograph of glomerulus and surrounding tubules
The afferent arteriole is seen entering the glomerulus. Surrounding Bowman's capsule are tubules cut in cross section. (×320)

[1] Afferent means incoming. Efferent means outgoing.

into the surrounding tissue fluid that percolates among the nephrons; positively charged sodium ions (Na^+) follow passively by diffusion, thus maintaining electrical balance. Because the walls of the ascending limb are impermeable to water, large quantities of sodium and chloride ions accumulate in the interstitial fluid bathing both branches of the loop and the collecting duct.

Now consider the descending limb of the loop of Henle. Because its walls are freely permeable to water, large quantities of water move by osmosis from the tubule into the surrounding (interstitial) tissue. This tends to concentrate the fluid within the tubule as it travels down the descending limb. The tubular fluid that makes the hairpin turn thus is relatively concentrated as it presents itself to the active transport systems of the ascending limb, which pump out more chloride.

As the tubular fluid flows into the collecting duct, it must again pass through the zone of high salt concentration. The walls of the collecting duct are variably permeable to water. The degree of permeability depends on the presence or absence of antidiuretic hormone (ADH). Water moves by osmosis through the walls of the collecting duct, leaving within the duct a fluid that is isotonic with the briny interstitial tissue fluid that percolates in the duct, but hypertonic in relation to the body fluids as a whole. In this way, mammals are able to excrete urine that is far more concentrated than the plasma from which it originated. The water extracted from the urine is returned to the blood via the capillaries surrounding the nephric tubules.

Clearly, urine concentration in mammals depends on the hairpin structure of the loop of Henle. The fluids in the descending and ascending limbs flow in opposite directions. We call this a countercurrent flow. Such an arrangement helps to maintain differences in solute concentration from one end of the loop to the other (see Figure 25-17). When countercurrent flow is coupled with an active transport mechanism, it multiplies the differences and thus is termed a **countercurrent multiplier** system. The longer the loop, the greater the concentration differences.

The countercurrent multiplier model of Henle's loop elegantly explains the urine-concentrating ability of mammals. However, the role of urea remains unaccounted for. In the recently proposed two-solute model, the importance of urea is apparent (Box 25-1).

The permeability of the collecting tubules to water is controlled by **antidiuretic hormone (ADH),** also called **vasopressin,** which is formed by the hypothalamus and secreted by that portion of the pituitary gland called the **neurohypophysis.** Loss of water from the body stimulates ADH secretion, which slows excretion of water by the kidney (Figure 25-18). The body has two ways of detecting a decrease

TABLE 25-2
INORGANIC AND ORGANIC COMPONENTS OF HUMAN URINE

Component	Value* (meq)[†]	(g)
Inorganic		
Chlorine (as chloride)	170 (85−380)	10.0 (5.0−16.0)
Sodium	170 (85−300)	4.0 (2.0−7.0)
Potassium	50 (40−80)	2.0 (1.5−3.0)
Sulfur (as sulfate)	125 (45−22)	2.0 (0.7−3.5)
Phosphorus (as phosphate)	90 (60−120)	1.5 (1.0−2.0)
Ammonia	40 (17−60)	0.7 (0.3−1.0)
Calcium	7 (5−10)	0.15 (0.1−0.2)
Magnesium	12 (4−16)	0.15 (0.05−0.2)
Organic		
Nitrogen compounds	—	30 (25−35)
Urea	—	25 (20−30) (half total solids)
Creatinine	—	1.4 (1.0−1.8)
Uric acid	—	0.7 (0.5−0.8)
Creatine	—	0.10 (0.06−0.15)
Amino acids, ascorbic acid, coproporphyrins, hormones, oxalic acid, purines, and pyrimidines	—	Trace

* Per 24 hr in a normal adult on an average diet. Average normal data (with ranges in parentheses).
[†] Meq = milliequivalents

25-17 The loop of Henle is part of a countercurrent multiplier system

The transport of water, sodium ions, and chloride ions throughout the tubular system is shown by arrows. Note that both active transport and passive transport take place. Within the medulla, extracellular fluids become progressively more concentrated.

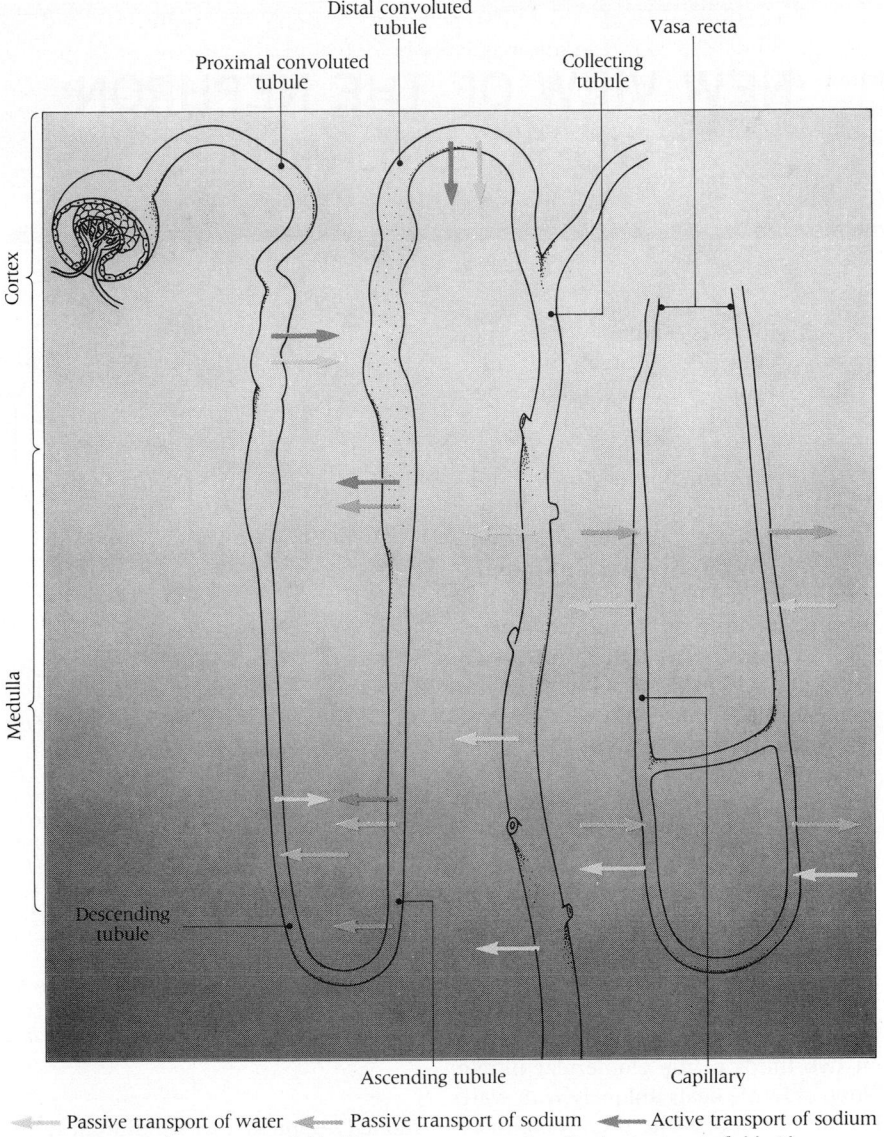

Passive transport of water ◄— Passive transport of sodium ◄— Active transport of sodium
Active transport of chloride ◄— Passive transport of chloride

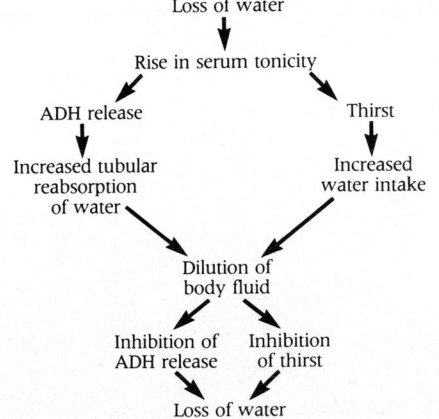

Loss of water → Rise in serum tonicity → ADH release / Thirst → Increased tubular reabsorption of water / Increased water intake → Dilution of body fluid → Inhibition of ADH release / Inhibition of thirst → Loss of water

25-18 Regulation of concentration of body fluids

Antidiuretic hormone (ADH) released in response to dehydration in the body, enhances water retention by the kidney. Increased thirst and water intake results. As body fluids reach the proper tonicity, ADH release is inhibited and thirst abates.

in its water level: it can sense a decreased blood volume, and it detects decreased osmotic potential of the blood.

The opposite processes also occur. Drinking a large volume of water (or beer) in a short time increases urine flow in less than 30 minutes. Flow reaches a peak in 1 to 2 hours and subsides in 3 to 5 hours. The increase results from slight dilution of the blood, a short-lived phenomenon the major effect of which is to suppress release of ADH by negative feedback (see Figure 25-18). Suppression of ADH secretion leads to the following:

- Decreased permeability of collecting tubules to water
- Decreased reabsorption of water back into the body
- Excretion of dilute urine

Increasing plasma osmolality stimulates the neurohypophysis to release ADH by affecting nerve endings in a brain center—the hypothalamus—that are sensitive to small charges in plasma osmolality. These **osmoreceptors** in turn set up nerve impulses that increase or decrease the secretion of ADH by the hypothalamus. ADH appears to work by increasing the permeability of the collecting duct to water. In the absence of ADH, the walls of this major duct are practically impermeable to water. Hence, little water is reabsorbed from the fluid emerging from the nephrons.

Lack of ADH production occurs in a disease called **diabetes insipidus.** Victims excrete vast quantities of dilute urine and experience extreme thirst. It is much

Renal physiologists—among the more imaginative students of organismic function—have firmly established three basic principles of kidney function.

- First, urine production begins with the formation of a protein-free ultrafiltrate at the membrane of the glomerular capillaries. This is **glomerular filtration.**
- Second, solutes and water are reabsorbed into the blood from the tubular lumen. This is **tubular reabsorption.**
- Third, there is a transfer of other solutes from the blood into the tubular lumen. This is **tubular secretion.**

Functionally speaking, it is useful to divide the renal tubule into three major segments (Figure 25A): the proximal tubule; the loop of Henle; and the distal nephron. Although physiologists subdivide these segments even further on the basis of structure and function, it is possible to assign certain general functions to each of them. The proximal tubule reabsorbs rather unselectively about two-thirds of the glomerular filtrate. The loop of Henle deals uniquely with water and solute transport. The distal nephron is the locus of a fine-tuning system that regulates both water and salt concentrations. These mechanisms are the targets of the hor-

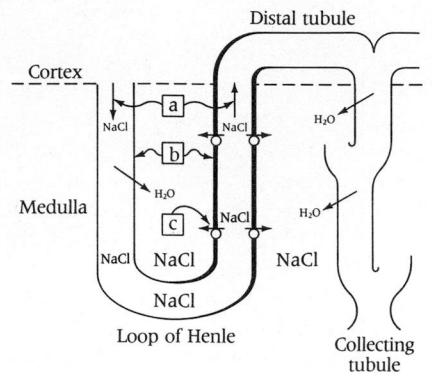

25A Countercurrent multiplier system of the loop of Henle
Three requirements must be met if the loop is to serve as a countercurrent multiplier. A countercurrent fluid flow must exist (a), there must be a difference in permeability of the epithelial cells lining the loops (b), and there must be a source of energy available for active transport (c). The thickened lining of the ascending limb and first part of the distal tubule signifies a water-impermeable lining. The descending limb can transport water. Energy is supplied by active reabsorption of NaCl. The progressive increase in NaCl concentration in the medulla (shown by variation in type size) extracts water from the descending limb and collecting duct.

mone (vasopressin, or antidiuretic hormone, or ADH) that controls body water levels.

The discovery of the countercurrent multiplier mechanism greatly clarified the puz-

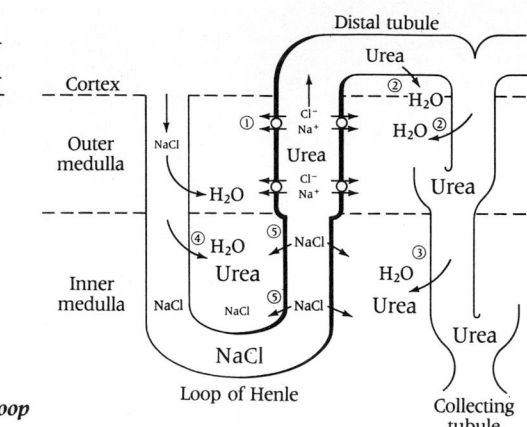

25B New ideas on the countercurrent multiplier hypothesis
The thickened lining indicates impermeability to water. (1) In the ascending limb, Cl^- is actively transported, and Na^+ is passively transported out of the tubule, making interstitial fluid hypertonic relative to dilute fluid in the tubule. (2) In the distal and collecting tubules, water is reabsorbed into the surrounding hypertonic environment, leaving behind urea, which is now more concentrated. (3) In the deep medulla, water and urea are both reabsorbed into the tissues, where the concentration of urea rises further. (4) The high urea concentration in the tissues draws water out of the descending tubule, thereby concentrating the NaCl left in the tubule. (5) As the NaCl-enriched fluid enters the water-impermeable section of the loop of Henle, NaCl moves down its concentration gradient, leaving the tubule and entering the tissues.

less common than an entirely different disorder associated with high urine volume, **diabetes mellitus.** Here, the large volume of urine is caused by the osmotic effect of glucose in the urine.

Aldosterone is a hormone produced by the **adrenal cortex,** which exerts its action by stimulating the formation of mRNAs and thereby the synthesis of specific proteins that appear to augment sodium transport in three ways: (1) by increasing the permeability of the distal convoluted tubule membranes to sodium ions (Na^+); (2) by increasing within the membrane the activity (or amount) of Na-K-ATPase, which may be a carrier in active Na^+ transport (see Chapter 6); and (3) by increasing mitochondrial oxidative phosphorylation and thereby supplying more energy to the Na^+ pump. The result of these effects is to inhibit the reabsorption of potassium ions (K^+) and stimulate reabsorption of Na^+ from the distal convoluted tubule.

Increased aldosterone production results in increased reabsorption of Na^+, whereas lowered aldosterone levels (as occurs in **Addison's disease**) cause the loss of NaCl and water in the urine.

zling problem of how the kidney concentrates urine. Among the consequences, however, was the uncomfortable need to assume that the descending limb of the loop of Henle is impermeable to water, solute, or both and that the ascending limb is impermeable to water and can pump salt. The fact was that no one was ever able to demonstrate convincingly that the thin ascending

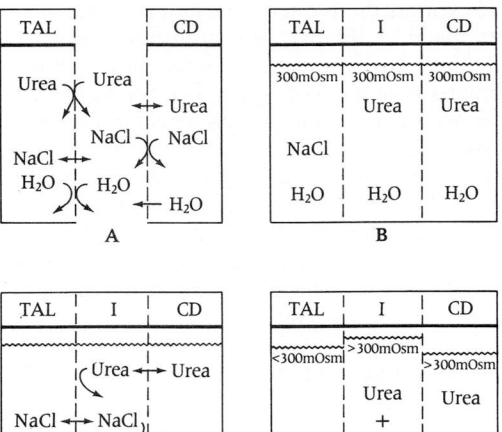

25C Generation of osmotic gradients from pure solutions by selective permeability
Three compartments are shown, the thin ascending limb (TAL), the collecting duct (CD), and the interstitial fluid (I). (A) TAL is permeable only to salt. CD is permeable to both urea and water. (B) The two compartments are separated by I, and each is filled with solutions of equal osmolarity (300 mOsm) but different composition (urea or NaCl). (C) NaCl can diffuse into the interstitial fluid, but cannot enter the collecting duct. (D) Since I now contains both urea and NaCl, its osmolarity is greater than 300 mOsm. Because the TAL is impermeable to water, water can enter only the I from the CD. This decreases the volume of fluid in the duct and concentrates the urea. The osmolarity of duct fluid now exceeds 300 mOsm.

limb has the capacity to extrude salt actively. This posed a problem for investigators.

The problem was: how can a continued rise in salt concentration in the thin ascending limb be explained without invoking an apparently nonexistent energy-dependent active process? The apparent answer came from an ingenious hypothesis—that the highly concentrated solution of urea in the interstitial spaces of the medulla is the source of additional free energy. This second solute, it was postulated, produces a concentrating effect that would not be possible in a single-solute model without a hydrostatic driving force or an active solute transport system, something that always seemed highly unlikely in the thin, mitochondria-poor cells of the ascending limb.

The elements of this resourceful model are summarized in Figure 25B. We see that urea accumulates in the deep (or inner) medulla as a result of active reabsorption of salt (sodium chloride) in the thick part of the ascending limb in the outer medulla. The high urea concentration in the medullary interstitium concentrates salt within the descending limb by osmotically removing water from it. As a result, a favorable concentration gradient is established for the passive removal of salt from the thin ascending limb. Since this structure is impermeable to water, the fluid within it becomes hypoosmotic relative to the surrounding interstitium.

A model using pure solutions nicely illustrates these concepts. In Figure 25C, a compartment is surrounded by a membrane that is permeable to salt but relatively impermeable to water and urea. Another compartment nearby is impermeable to salt but freely permeable to water and urea. The first compartment corresponds to the salt-permeable thin ascending limb (TAL); the second compartment corresponds to the water- and urea-permeable inner medullary collecting duct (CD). The space between them represents the interstitium (I). A pure NaCl solution at an osmotic concentration of 300 mOsm is added to the ascending limb compartment and a solution of urea at the same osmotic concentration is placed in the collecting duct and interstitial compartments. Because there is no salt in the interstitium and the ascending limb is permeable to salt, salt will slide down its concentration gradient, which runs from the TAL to the adjacent interstitium. The passive loss of salt from the TAL results in dilution of this compartment. In the interstitium, the added salt plus the urea originally present raises the total osmotic concentration of this compartment. Water might have been drawn into it from the TAL, but it is contained by a membrane that is impermeable to water. But the CD is permeable to water. Hence, its water enters the interstitium until its osmotic level equals that of the surrounding interstitium. In the end, the TAL compartment will have its original volume but will have lost salt. Thus, it has become diluted. The CD compartment has lost more water than solute. Thus, it has become concentrated. The water lost from the CD is now in the interstitium, which has increased its volume.

In sum, the **two-solute hypothesis** attempts to explain experimental data without invoking an active-transport system for the TAL. As long as compartment permeabilities are as postulated, thick limb transport of salt (first solute) is the only needed active system. Attractive as the hypothesis is, with its passive model of the TAL, a few hard facts remain unexplained by it. We can hope that these contradictions will be resolved soon by the energetic investigators now at work.

SUMMARY

Maintenance of relatively constant physiological conditions by excretory organs occurs only in animals. Constancy is achieved by simultaneous excretion of wastes and selective retention of water, ions, and necessary metabolites such as glucose. Substances that animals must excrete include carbon dioxide, salts, nitrogenous wastes, and greater or lesser amounts of water, depending on the animal's mode of life. Thus freshwater fishes are adapted to eliminate water, terrestrial animals to conserve water.

Nitrogenous wastes, the end products of protein and nucleic acid catabolism, consist chiefly of urea, $CO(NH_2)_2$, uric acid ($C_5H_4N_4O_3$), and ammonia (NH_3 or NH_4^+). Most aquatic animals excrete nitrogen directly as ammonia, the most toxic of these compounds. It is highly soluble and readily carried away by the watery environment. The internal environment of most marine invertebrates is isosmotic with the surrounding sea. These animals excrete wastes through the body surface and need no special excretory organs. Marine vertebrates have evolved with an internal environment that is hypotonic relative to seawater, and they have developed various mechanisms to excrete excess salt and conserve water.

Freshwater fishes have the opposite problem. Because their body fluids are hypertonic relative to the watery environment, they must prevent intake of too much water and loss of body salts. They accomplish this with excretory organs that constantly excrete a highly dilute urine and, to conserve body salts, actively reabsorb most of the salt from urine before it is excreted.

Terrestrial animals must constantly combat dehydration. They conserve water by excreting nitrogenous wastes in the form of uric acid or urea, instead of ammonia. Amphibians and reptiles have excretory organs resembling those of the freshwater fishes from which they evolved. They can make dilute or isotonic urine, but cannot produce urine more concentrated than plasma. Birds and mammals, on the other hand, are uniquely able to conserve water because they can produce urine that is up to 25 times more concentrated than plasma.

The organs of excretion vary widely in different species. Freshwater flatworms are among the simplest animals with an organized tubular excretory system, a key component of which is the flame cell. The excretory organs of earthworms, the nephridia, gain efficiency from their association with a closed circulatory system. Insects excrete water and wastes, primarily uric acid and potassium, through Malpighian tubules, which empty into the gut. There, water is reabsorbed, uric acid precipitates, and the resulting dry material is discharged with the feces.

Freshwater fishes have discrete kidneys, the functional unit of which is the nephron. A nephron is a tubule that begins as a cup-shaped device called Bowman's capsule, which encloses a tuft of capillaries called a glomerulus. The hydrostatic pressure of circulating blood forces some fluid through the glomerulus into Bowman's capsule. This ultrafiltrate contains only low-molecular-weight solute molecules (glucose, salts, etc.); larger molecules (notably proteins) cannot pass through the capsular membrane. As the filtrate passes down through the nephron tubule, certain solutes such as glucose and needed ions are reabsorbed into the body and thereby conserved. A highly dilute urine results.

Marine fishes have kidneys that, in the course of evolution, lost Bowman's capsule and its glomerulus and thus the capacity to form a glomerular filtrate. As a result urine production is scanty.

The embryological development of the kidney of terrestrial vertebrates in many ways reflects its evolutionary development. The human kidney, a typical mammalian kidney, has nephrons similar in many respects to those of freshwater fishes. Blood pressure forces solute molecules across glomerular capillary walls and into nephron tubules, where more than 99.5% of the filtered fluid is reabsorbed into the bloodstream by specialized active-transport systems in the tubules.

The striking ability of mammalian kidneys to reabsorb water from urine against an osmotic gradient derives from a combination of active tubular transport systems and the three-dimensional arrangement of the tubules within the kidney. Of particular importance is the hairpin turn made by the part of each tubule known as the loop of Henle. Filtrate flows in opposite directions in the descending and ascending parts of the loop, creating a countercurrent flow system. This arrangement, coupled with active transport in cells of the tubules, successfully maintains a steep concentration gradient of solutes between the start of the loop and its end, and establishes a hypertonic area within the kidney tissue. Urine in the collecting ducts, which pass through the highly concentrated zone, loses water to the surrounding tissue. Antidiuretic hormone (ADH) increases the permeability of collecting ducts to water. By allowing more water to return to body tissue from the ducts, ADH decreases water loss from the body. In diabetes insipidus, ADH is lacking. As a result, urine flow and thirst are enormous.

KEY TERMS

afferent arteriole
aldosterone
antidiuretic hormone (ADH)
Bowman's capsule
contractile vacuole
cortex
countercurrent multiplier
distal convoluted tubule
efferent arteriole
flame cell
glomerulus

isosmotic
kidney
loop of Henle
Malpighian tubules
medulla
nephridium
nephron
osmoconformer
osmoreceptor
osmoregulator
peritubular capillary

proximal convoluted tubule
salt gland
tubular reabsorption
tubular secretion
two-solute hypothesis
urea
ureter
urethra
uric acid
urine
vasopressin

QUIZ QUESTIONS

1. Which of the following is *not* a nitrogenous waste that must be excreted from the body?
   A. urea
   B. feces
   C. ammonia
   D. uric acid
   E. A and D

2. Which of the following structures is not related to osmoregulation or to maintenance of water balance?
   A. kidney
   B. gill
   C. contractile vacuole
   D. insect hindgut
   E. All of the above are related to these processes.

3. The two-solute hypothesis has been advanced recently to explain
   A. how the nephron functions.
   B. how contractile vacuoles operate.
   C. how the liver degrades hemoglobin into bile pigments.
   D. how excess salt is excreted at gill surfaces.
   E. how flame cells operate.

4. The urine of earthworms
   A. is released to the exterior through a pair of openings in each body segment.
   B. is formed from fluid that flows from the body cavity into a tubular excretory system.
   C. is formed by a process that involves a close association of blood vessels with the excretory tubule system.
   D. A and B
   E. A, B, and C

5. The mammalian excretory system involves all of the following *except*:
   A. hormonal control of permeability of walls in various excretory tubules.
   B. reabsorption of high molecular weight proteins from the glomerular filtrate.
   C. release of nitrogenous wastes into the intestines.
   D. active transport of chloride ions from the filtrate into the peritubular tissues.
   E. diffusion of sodium ions from the filtrate into the peritubular tissues.

ESSAY QUESTIONS

1. Explain why each of the following is the preferred molecule of nitrogen waste excretion in the species mentioned: ammonia (or ammonium ions) in aquatic animals, uric acid in reptiles and birds, urea in mammals.

2. Describe the structure of a human nephron. How does the glomerular filtrate compare in composition to plasma? Interstitial fluid? What are the functions of the proximal collecting tubule, the loop of Henle, the distal convoluted tubule, and collecting tubules?

3. Diabetes mellitus leads to a loss of glucose in the urine. One of the cardinal signs of the disease is increased urinary output. How might this be related to the high glucose level in the urine? The disease diabetes insipidus can result from an inability to produce the antidiuretic hormone ADH. What are the symptoms of this disease?

REFERENCES AND SUGGESTIONS FOR FURTHER READING

Brenner, B. M., and Rector, F. C., Jr. (Eds.) (1986) *The Kidney, Vols. 1 and 2,* 3rd ed. W. B. Saunders, Philadelphia.

A generous compilation of authoritative up-to-date essays.

Fraser, E. A. (1950). The development of the vertebrate excretory system. *Biological Reviews 25,* 159–187.

A highly regarded review article.

Hoar, W. S. (1983). Excretion. Chapter 15 in: *General and Comparative Physiology,* 3rd ed. Prentice-Hall, Englewood Cliffs, N.J.

This outstanding textbook of comparative physiology is informative in all branches of animal physiology. The chapter on excretion upholds the standard.

Jamison, R. L., and Kriz, W. (1982). *Urinary Concentrating Mechanisms. Structure and Function.* Oxford University Press, New York.

A meticulous review of the way hormones and varying permeabilities in the nephron permit the kidneys of terrestrial vertebrates to excrete urine that is significantly more concentrated than plasma.

Mommsen, T. P., and Walsh, P. J. (1989). Evolution of urea synthesis in vertebrates: the piscine connection. *Science 244,* 1152–1157.

A fascinating study.

Romer, A. S., and Parson, T. S. (1977). Excretory and reproductive systems. Chapter 15 in: *The Vertebrate Body,* 5th ed. W. B. Saunders, Philadelphia.

This chapter of a classic textbook jointly considers the evolution of excretion and reproduction in vertebrates.

Schmidt-Nielsen, K. (1959). Salt glands. *Scientific American 200,* January, 109–115.

An account of the structure and role in osmoregulation of the salt glands of sea birds, which permit them to drink seawater. Similar organs are found in marine turtles and iguanas.

Schmidt-Nielsen, K. (1981). Countercurrent systems in animals. *Scientific American 244,* May, 118–128.

Exchanges of molecules or heat between two fluids moving in opposite directions are basic to many systems that significantly enhance the adaptedness of animals to their environment.

Schmidt-Nielsen, K. (1984). *Animal Physiology: Adaptation and Environment,* 2nd ed. Cambridge University Press, New York.

This book includes among other things a review of excretion from a comparative perspective.

Seldin, D. W., and Giebisch, G. (Eds.) (1985). *The Kidney: Physiology and Pathophysiology, Vols. 1 and 2.* Raven, New York.

A comprehensive account of the state of renal physiology. The level is advanced but many chapters are accessible, brief, and well illustrated.

Smith, H. W. (1951). *The Kidney: Structure and Function in Health and Disease.* Oxford University Press, New York.

A classic work. A shorter version was issued in 1956 by the same publishers under the title *Principles of Renal Physiology.*

Smith, H. W. (1953). *From Fish to Philosopher.* Little, Brown and Company, Boston.

The story of the evolution of the kidney told by an eminent renal physiologist in an original and engaging manner. Highly recommended.

Taylor, C. R., Johansen, K., and Pollis, L. (Eds.) (1982). *A Companion to Animal Physiology.* Cambridge University Press, New York.

A collection of essays by various authorities that goes into greater detail than Schmidt-Nielsen's *Animal Physiology.* Includes lucid discussions of the one-solute and two-solute models of the mammalian nephron.

Valtin, H. (1983). *Renal Function: Mechanisms Preserving Fluid and Solute Balance in Health,* 2nd ed. Little, Brown and Company, Boston.

A concise and clearly written discussion of renal physiology. Excellent bibliographies.

NUTRITION AND METABOLISM

All organisms need nutrients. But methods of getting these nutrients and using them differ widely. The differences are of two broad kinds that are never completely separate.

The first difference relates to the organism's size. In an individual cell, the procurement, transport, and processing of raw materials pose problems drastically unlike those arising in a large multicellular organism like a primate or a tree. In large organisms, the distances to be traversed become so great that transport can no longer be entrusted to simple diffusion and cytoplasmic streaming.

The second difference is illustrated by the difference between a green plant (an autotroph) and an animal (a heterotroph). It has to do with an organism's degree of dependence on sources of energy and prefabricated materials. These differences in energy and nutrient sources were discussed in Chapter 9 (see Table 9-1).

These great differences among organisms raise many interesting problems. One relates to the means for obtaining food. In tiny organisms, such as protists, acquiring food is a relatively simple matter. However, large multicellular organisms must have special body parts devoted to feeding and, in the case of animals, digestive systems for converting food into usable forms. Special arrangements are also needed for the mass transport of molecules—blood and circulatory systems in animals, vascular systems in complex plants. Here we will be concerned with defining the nutritional needs of organisms of different kinds and how these needs are met.

The intake of food is termed **ingestion**—in animals we usually call it *eating.* Many ingested foodstuffs are useless to an organism until they are digested. By **digestion** we mean the breakdown, chiefly by hydrolysis, of the relatively large protein molecules and other biomolecules of food into simpler compounds. In the process of **assimilation,** these simple molecules are absorbed into the body for use in metabolism.

MODES OF NUTRITION

We learned in Chapter 9 that (1) the molecular building blocks needed by organisms may be simple, inorganic molecules like CO_2 and minerals, or complex ready-made organic molecules like carbohydrates, lipids, and proteins, and (2) the energy required by organisms is obtained either from sunlight or from chemical reactions that liberate the bond energy in organic molecules.

How organisms obtain energy is so important that we classify all living things in one of two categories. As we have seen, those that do not rely on organic molecules as an energy source are called **autotrophs.** Most autotrophs are photosynthetic organisms in one of several groups, as shown in Table 9-1.

Organisms that must rely entirely on organic molecules produced by autotrophs for their energy are called **heterotrophs.** These organisms are incapable of photosynthesis or chemosynthesis (the ability to use inorganic molecules such as H_2S as a source of energy for converting CO_2 into organic molecules).

Nutrition and metabolism
The need for food is universal. Here a green frog, Rana clamitans, *catches a fly, which it will ingest, swallow, and digest.*

AUTOTROPHIC NUTRITION

As previously mentioned, autotrophs need a source of energy and carbon for necessary biosynthetic systems. For most autotrophs, the energy comes from the sun and the means of capturing it is photosynthesis (Chapter 10). Photosynthetic autotrophs require CO_2, which diffuses into them from the surrounding air or water. Control of gas exchange in terrestrial plants was also described in Chapter 10. The water needed for photosynthesis is obtained from the environment of autotrophs, too, either by direct uptake (in aquatic autotrophs) or by absorption through roots (Chapter 24).

In addition, autotrophs require a number of inorganic nutrients, or minerals. Dissolved inorganic molecules dissociate to form ions and it is the ions that plants take up. The minerals required by plants and the form in which they exist in the soil is summarized in Table 26-1, along with information on deficiency symptoms. The remarkable properties of different soils, their composition, and physical properties, are discussed in Chapter 46.

The roots of plants are in intimate contact with soil (Figure 26-1), from which they absorb water and minerals. Water taken up by roots is accompanied by the minerals dissolved in it. Roots increase their absorptive surface area by growing numerous secondary roots. Root hairs, which are extensions of individual epidermal cells (see Figure 26-1), also greatly increase surface area. Yet, despite their large surface area, roots quickly deplete the nearby soil of many minerals. This depletion stimulates roots to grow and elongate so that they can "mine" new regions of soil. Roots may also secrete hydrogen ions that make some minerals more available, and use active transport to take up others.

As discussed in Chapter 40, some roots make use of the considerable absorptive abilities of soil fungi. During millions of years of occupying the same habitat, many mutual-benefit relationships—mycorrhizae—have evolved between fungi and plant roots. The fungus usually penetrates the root cortex (see Figures 26-1, 40-20) and the fungal mycelium (a network of filaments) becomes closely associated with the root cortex cells. This provides a large surface for interaction between the organisms. The plant typically derives minerals and water from the fungus, making use of the extensive fungal mycelium. The fungus in turn receives organic molecules from the plant. Many plants will not grow without a specific soil fungus.

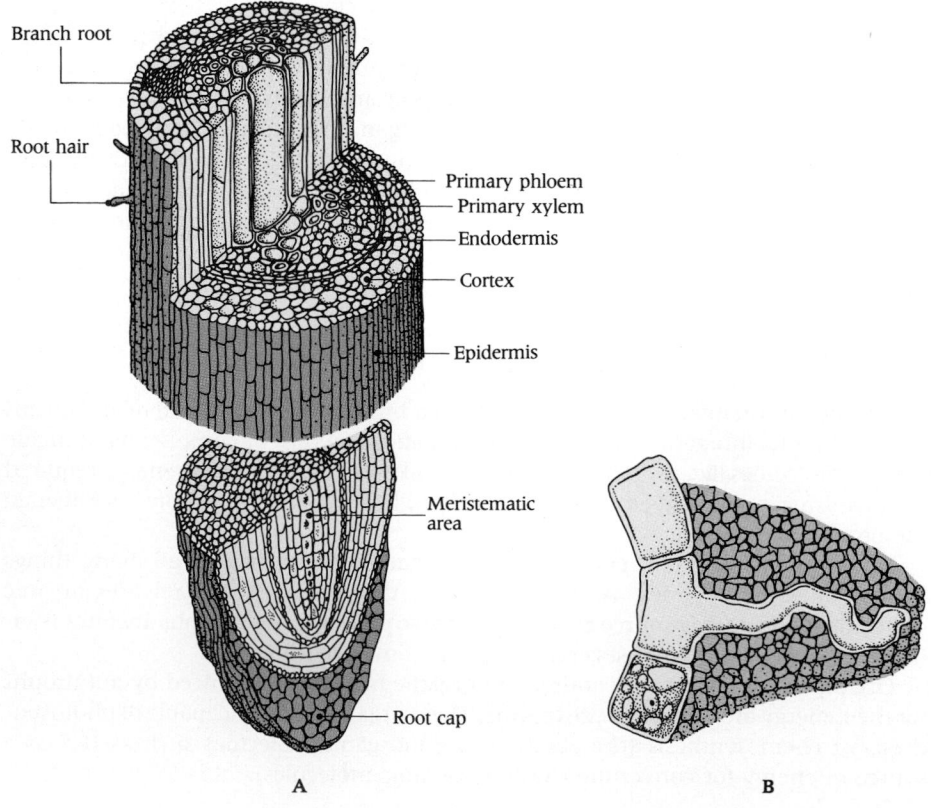

26-1 Root structure
(A) *Diagram of a typical plant root showing the meristematic (cell division) zone and the more mature region in which branch roots originate.* (B) *Epidermal cells of the root with root hairs growing among particles of soil. Compare with Figure 24-4.*

Table 26-1

MINERAL ELEMENT REQUIREMENTS BY PLANTS

Element	Symbol	Approx. Content as Dry Weight, %	Form in Which Usually Absorbed	Role	Typical Deficiency Symptoms
Macronutrients*					
Nitrogen	N	1.5	NO_3^- or NH_4^+	Constituent of proteins including all enzymes, constituent of chlorophyll, nucleic acids, various coenzymes	Leaves, especially in mature portions, light green (yellow or red in severe cases), slow growth
Phosphorus	P	0.2	$H_2PO_4^-$ or HPO_2^{2-}	Constituent of nucleic acids, phospholipids of membranes, coenzymes. Needed for phosphorylation of sugars. Involved in formation of high energy phosphate compounds such as ATP	Dark green to purplish leaves, slow growth, poor fruit yield, premature fruit drop
Potassium	K	1.0	K^+	Cofactor for some enzymes, regulation of stomatal opening	Mottled or curled leaves, slow growth, poorly developed root system
Calcium	Ca	0.25	Ca^{2+}	Cofactor for some enzymes, pollen germination and growth, maintenance of membrane selective permeability, wall constituent, necessary for root growth	Deformed or dead terminal buds, decreased root growth
Magnesium	Mg	0.2	Mg^{2+}	Constituent for chlorophyll, cofactor for some enzymes	Chlorosis (loss of chlorophyll) especially between veins of older leaves, eventually spreading to margins and tips
Sulfur	S	0.2	SO_4^{2-}	Constituent of proteins, constituent of coenzyme A	Similar to nitrogen deficiency, leaves light green, slow growth
Micronutrients					
Iron	Fe	0.1	Fe^{2+} or Fe^{3+}	Constituent of cytochromes and certain other enzymes, activator of some enzymes, essential for chlorophyll synthesis	General chlorosis in young leaves. Vein regions last to yellow
Boron	B	.0006 (6 ppm)	BO_3^- or $B_4O_7^{2-}$	Regulation of carbohydrate metabolism. Other functions poorly understood	Damage or death to tips. Distorted brittle leaves. Retarded root growth
Manganese	Mn	0.03	Mn^{2+}	Activator of some enzymes	Similar to iron deficiency. Chlorosis developing first between veins
Zinc	Zn	.0008	Zn^{2+}	Component of certain metalloenzymes (e.g., alcohol dehydrogenase), important in synthesis of auxin	Retarded leaf expansion, short internodes, abnormal arrangement root hairs
Molybdenum	Mo	.0001 (1 ppm)	MoO_4^{2-}	Component of certain metalloenzymes, important roles in nitrogen metabolism	Chlorosis between veins, leaves curl, similar to manganese or iron deficiency
Copper	Cu	.0006 (6 ppm)	Cu^{2+}	Activator of some enzymes	Variable, often chlorotic leaves with margins curling. Shoot tips often die back while lateral buds sprout giving a bushy appearance
Chlorine	Cl	0.1 (1,000 ppm)	Cl^-	Important for normal enzyme function in oxygen production associated with photosystem II, may be involved in regulation of stomatal opening, involved in osmotic regulation and ion balance	Young leaves appear bluish green and shiny, later reddish brown ("bronzing"). Eventually leaves die. Chlorine is so plentiful that it must be purposely withheld from the plant to induce these symptoms

* In addition to these six macronutrients the elements carbon, oxygen, and hydrogen are required in large amounts by plants and are considered macronutrients. Since they are widely available from water and carbon dioxide, they are not included in this table.

Source: *Plants/Their Biology and Importance* by Kaufman et al. Harper & Row 1989.

HETEROTROPHIC NUTRITION

Heterotrophic organisms may be divided into two categories based on where digestion takes place relative to the organism.

Those that release digestive enzymes into their environment and then absorb the products of digestion are **absorptive heterotrophs.** These include heterotrophic bacteria and fungi. When such organisms derive nutrition from living organisms, they are termed *parasites*. Absorptive heterotrophs that derive nutrition from dead organic matter are called *saprobes*. Saprobes play the important ecological role of decomposer (see Chapter 46). They are the main reason we do not trip over dead dinosaurs on the way to town.

A second group of heterotrophs are those that ingest ("eat") nutrients prior to digestion. These organisms are the **ingestive heterotrophs.** This group includes those nonphotosynthetic protists we lump together as protozoans, which digest food within food vacuoles, and animals, which have digestive systems of various kinds.

We should remember several facts about the nutrients needed by heterotrophs.

■ Their food includes the large organic molecules described in Chapter 40.
■ Such molecules are assembled from the same few simple subunits—in all organisms. Thus (with few exceptions), the proteins of a fungus contain the same twenty amino acids as the proteins of a bird, and lipids from a bacterium contain the same kinds of fatty acids as the lipids of a rose.
■ It takes energy to construct organic molecules, and energy is released when organic molecules are oxidized (see Chapter 9). These reactions always result in the loss of energy, usually as heat.

Given these facts, it makes energetic sense that a heterotroph digests its food into small organic molecules like the monosaccharides, amino acids, and fatty acids, and then uses those molecules to construct larger organic molecules. Complete oxidation of large organic molecules to CO_2 and H_2O, followed by complete molecular reconstruction would waste energy. Heterotrophs have been avoiding this waste for billions of years. Indeed, through retrogressive evolution, they have often lost the ability to manufacture some of the organic molecules that have always been available in their diet or surroundings.

Of course, heterotrophs also require inorganic minerals and water. In the process of assimilation, the products of ingestion and digestion are incorporated into the organisms for use in metabolism. Larger organisms like plants and animals have internal transport systems (Chapter 24) that move nutrients from places of abundance to places of need within the organism.

A few plants—such as butterwort (*Pinguicula grandiflora*), sundew (*Drosera intermedia*), and Venus flytrap (*Diomaea muscipula*)—have modified leaves that employ sticky secretions to trap insects (Figure 26-2). Enzymatic digestion occurs in it and digested food is absorbed from it. It is of interest that these insectivorous plants are, in fact, green photosynthetic autotrophs. The advantages they derive from ingesting insects is probably related more to the need for minerals than for calories or other kinds of nutrients.[1]

The remainder of this chapter is concerned mainly with aspects of heterotrophic nutrition in animals.

INGESTION OF FOOD

INGESTION BY PROTISTS

Protists absorb water and dissolved materials directly through the cell membrane. The process of **absorption** depends on diffusion and osmosis as well as active transport. Most protists also take in solid food. They eat other protists and even small multicellular organisms. Protists ingest solid food simply by surrounding and capturing it by various cellular devices. In *Paramecium*, food is taken in through the mouth and enclosed in a **food vacuole** in which digestion occurs (Figure 26-3). The food vacuole corresponds in function with our stomach and intestine, but the food vacuole is not a permanent part of the protist's anatomy. It is devised anew following each meal or "mouthful." In *Trichonympha*, a parasite of termites, bacteria are present in the cytoplasm. The bacteria digest cellulose in ingested wood particles to generate products useful to the termite (see Figure 26-3).

[1] The touch mechanism of these plants is discussed in Box 29-1.

26-2 Carnivorous plants
(A) *This sundew from South Africa,* Drosera capensis, *traps insects in its sticky secretions, here seen as glistening droplets on the plant body.* **(B)** *An unwary fly, having triggered the sensitive marginal bristles of a Venus flytrap,* Dionaea muscipula, *is trapped.*

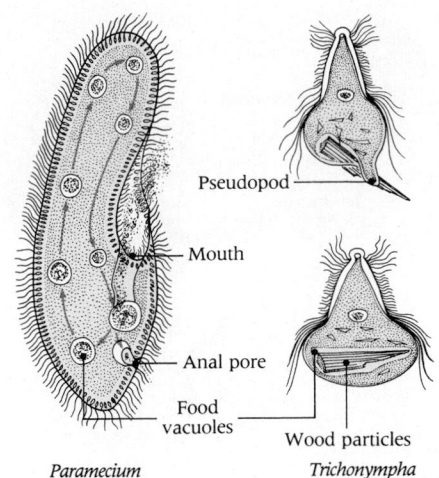

Pseudopod

Mouth

Anal pore

Food
vacuoles

Wood particles

Paramecium

Trichonympha

26-3 Digestion in unicellular organisms
In Paramecium, *a food vacuole engulfs bacteria and other microorganisms brought to the mouth by the beating of cilia. The food vacuole moves along a fixed path through the cytoplasm* (arrows). Trichonympha *inhabits the gut of termites. Wood particles eaten by the termite are engulfed and digested by bacteria in the cytoplasm of this protist.*

Some protists utilize solid food without actually taking it into their bodies. Instead, they secrete digestive fluid onto food particles outside the body and then absorb the dissolved products. Many more complex animals—for example, starfishes—do essentially the same thing: they hold their prey outside the body and surround it with digestive fluid. A starfish can actually stick its stomach between the shells of a clam. The extruded stomach secretes enzymes that digest the clam. In other words, starfishes digest their food before they eat it.

INGESTION BY ANIMALS

In animals, the usual method of ingesting food is by the introduction of solids and liquids into a tube called the **alimentary canal** that runs through the body (Figure 26-4). Within the canal, solids and large molecules are digested—hydrolyzed and dissolved—and then assimilated through the canal's wall into the body proper.

Animals take food into an alimentary canal in a number of ways.

- **Bulk feeders** eat sizable plants and animals, or chunks of them, at occasional meals. Such animals usually have a method of seizing food, a mouth, and teeth or other devices for breaking up the chunks. Seed-eating birds—relatively recent arrivals in the history of life because all but primitive seed plants were late to evolve—crack seeds with their powerful beaks. Alternatively, some fowl, quail, and pigeons have a **gizzard,** a specialized portion of the stomach that (often with the aid of small pebbles) helps grind up seeds (Figure 26-5). Many "chewing" insects have mouth parts that effectively tear the food into small pieces. Many insects also have a **proventriculus,** a crushing organ that performs the same functions as a bird's gizzard (Figure 26-6).
- **Filter feeders** (also called **suspension feeders**) chiefly include aquatic animals that eat food in the form of extremely small particles that must be recovered

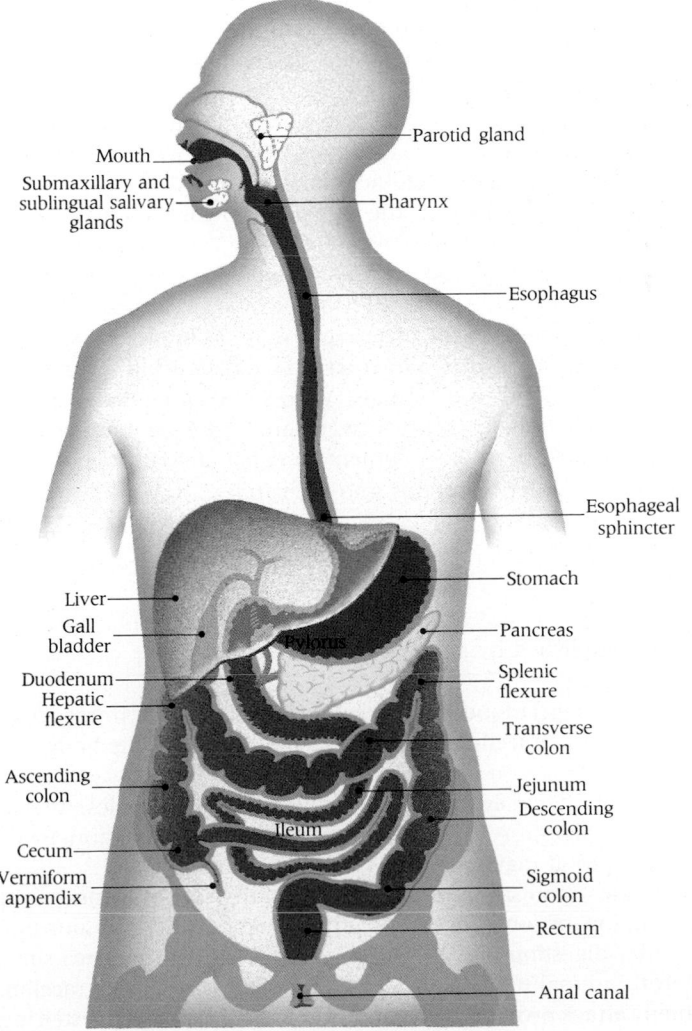

Parotid gland

Mouth

Submaxillary and
sublingual salivary
glands

Pharynx

Esophagus

Esophageal
sphincter

Stomach

Liver

Gall
bladder

Pancreas

Pylorus

Duodenum
Hepatic
flexure

Splenic
flexure

Transverse
colon

Ascending
colon

Jejunum

Descending
colon

Ileum

Cecum

Vermiform
appendix

Sigmoid
colon

Rectum

Anal canal

26-4 The human digestive system

from water. These animals almost continuously take in vast quantities of surrounding water from which they remove the suspended microscopic plants and animals, relying on anatomical strainers or filters to catch the food. Then they swallow the filtered masses (Figure 26-7). The variety of filter feeders is huge, ranging from clams and oysters to blue whales, flamingos, and aquatic insect larvae.

- **Fluid feeders** suck such fluids as blood or plant sap. Nutrients may be very dilute in these ingested fluids. Thus fluid feeders must be able to eliminate large amounts of excess water. Examples are blood-sucking leeches, vampire bats, and sap-feeding aphids (see Box 24-1).
- **Deposit feeders** are represented by the few animals that indiscriminately take in samples of their environment, digest any food within, and evacuate the rest. Earthworms are deposit feeders (Figure 26-8). They pass bits of whole earth through their alimentary canals, with no attempt to select wanted and unwanted portions. Such a system will not work unless the environment is fairly nutritious. There are no worms in sand dunes.
- **Suction feeding** occurs in aquatic habitats where teleost (bony) fishes—perch, bass, minnows, etc.—are the dominant vertebrates. Their skulls have mobile plates that can swing out, a forward-extending upper jaw and a mouth floor that can be lowered to an extraordinary extent. The effect of such an "exploding skull" is a sudden drop in the oral cavity as the mouth is opened (Figure 26-9). As a result, suction is generated—not unlike the operation of a powerful pipette. Because of the rapidity of this event—it takes most fishes from 0.001 to 0.025 second to convert the head from resting state to full expansion—prey is sucked into the mouth with accuracy, efficiency, and finality.

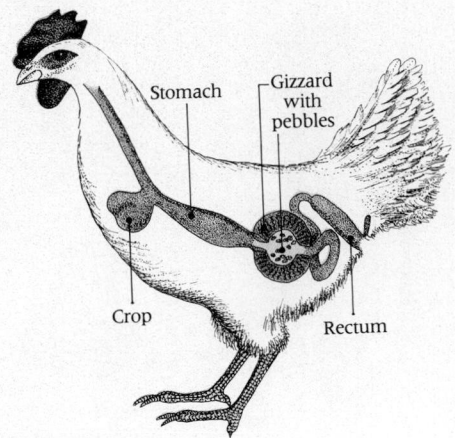

26-5 The bird gizzard
The digestive system of birds contains a specialized structure called the gizzard. Pebbles in the gizzard serve functionally as teeth, grinding up food.

THE HUMAN ALIMENTARY CANAL

The alimentary canal is about 30 feet long in the adult human. As shown in Figure 26-4, the major components of the system are the mouth, pharynx, esophagus, stomach, small intestine (which comprises the duodenum, jejunum, and ileum), large intestine (which comprises the cecum, ascending, transverse, descending and sigmoid colon), rectum, and anus.

To facilitate discussion, we divide the alimentary tract into upper and lower segments at the juncture of the stomach and small intestine. An inelegant but useful synonym of the lower alimentary canal is **gut.**

The chief accessory glands are the salivary glands, pancreas, and liver. These glands pour digestive secretions into the main tract through special ducts. In addition, the mucous membrane lining the tract is studded with tiny glandular invaginations that pour important digestive fluids directly into the stream.

Structures and functions in the digestive system, as in other systems, are remarkably coordinated. Although the alimentary tract is basically a tube, it is far from inert, displaying several forms of motility. Waves of coordinated contractions (called **peristalsis**) can propel its contents forward and, on occasion, backward. Further, though many glands open into the alimentary canal, it is capable of precisely timing the discharge of glandular secretions into the canal.

DIGESTION

Digestion is accomplished by animals in two ways.

- Individual cells ingest food by phagocytosis and digest it intracellularly.
- Food is digested extracellularly before being taken into the body.

Phagocytosis is obviously possible only with small particles. We may assume that extracellular digestion arose as an evolutionary adaptation to the need for breaking up large food masses before assimilating them.

In many simple organisms intracellular digestion is the only mechanism available for the assimilation of food. In more complex organisms, it is not uncommon to find intracellular digestion and extracellular digestion taking place simultaneously. In a carnivorous animal (e.g., in a coelenterate like *Hydra*), its extracellular digestive enzymes chiefly attack proteins while nonprotein materials are digested intracellularly

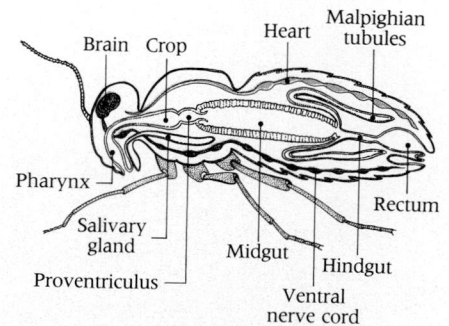

26-6 Insect digestive system
Although the structure of the insect's digestive system depends on the species and stage of the life cycle, the structures shown here are representative of most insects. The proventriculus crushes food before it enters the midgut.

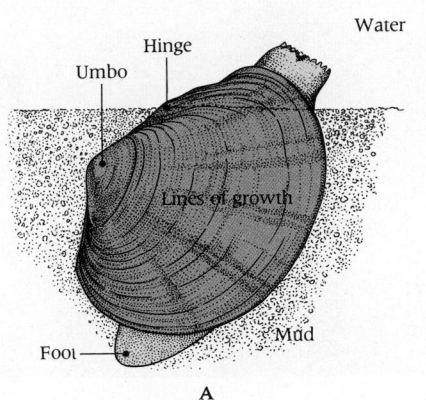
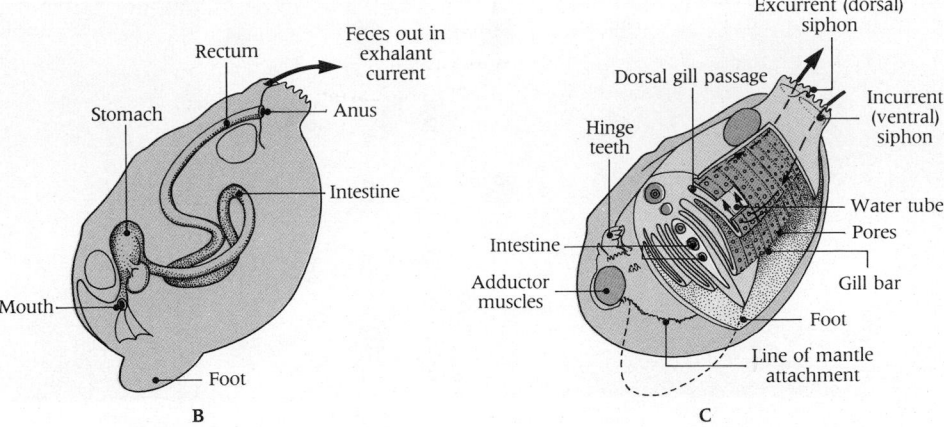

A Umbo, Hinge, Water, Lines of growth, Foot, Mud

B Rectum, Feces out in exhalant current, Anus, Stomach, Intestine, Mouth, Foot

C Excurrent (dorsal) siphon, Dorsal gill passage, Incurrent (ventral) siphon, Hinge teeth, Water tubes, Pores, Intestine, Gill bar, Adductor muscles, Foot, Line of mantle attachment

26-7 Filter feeding in a clam
(A) *A clam lives nearly submerged in the mud, with siphon exposed to the surrounding water.* **(B)** *The digestive system runs from mouth to anus.* **(C)** *Water is taken into the mantle cavity through the incurrent siphon and filtered through pores in the large gills. Minute food particles trapped in the mucus of the gill surfaces are kept moving in currents directed toward the mouth. Water freed of food particles is then ejected through the excurrent siphon. Mucus on the gills is like a continuously moving "flypaper" that enters the mouth; its motion is due to the organized beating of cilia.*

(Figure 26-10). If the animal is herbivorous, extracellular enzymes are produced that split carbohydrates. In vertebrates, the general disintegration of the food mass by extracellular enzymes is facilitated by mechanical movements in the walls of the digestive tract and digestion is largely extracellular. Thus the food mass is reduced to molecular species small enough to be taken into the body cells. However, an element of phagocytosis and intracellular digestion remains. It is confined chiefly to the fixed and wandering scavenger cells of the reticuloendothelial system (see Chapter 22).

One other digestive device that is unique to human beings has been introduced: cooking. Though many foods can be digested without cooking, the cooking process performs an initial digestive function without which the human body would find certain foods relatively indigestible. Cooking denatures food proteins and bursts the granules of natural starch, thereby facilitating attack by digestive enzymes. Without prior cooking, less than half of the raw starch of, say, potatoes is assimilated by the body, whereas 90% of cooked starch is digested.

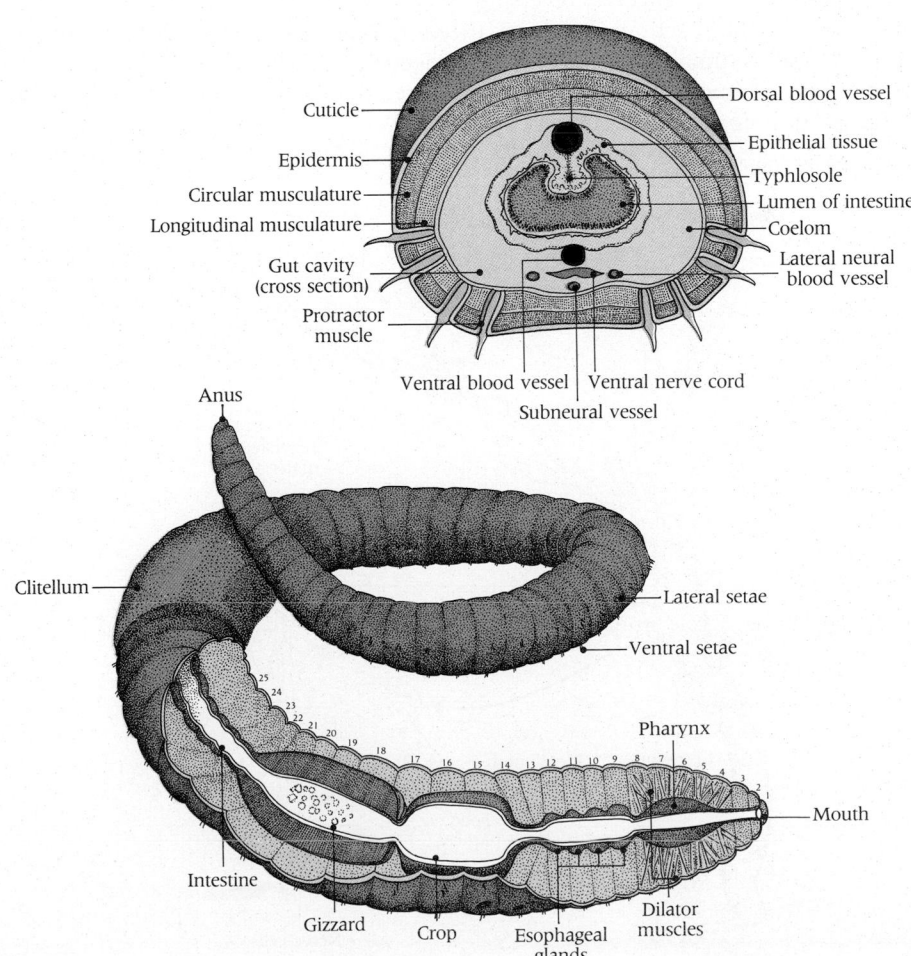

Cuticle, Dorsal blood vessel, Epidermis, Epithelial tissue, Circular musculature, Typhlosole, Longitudinal musculature, Lumen of intestine, Coelom, Gut cavity (cross section), Lateral neural blood vessel, Protractor muscle, Ventral blood vessel, Ventral nerve cord, Subneural vessel

Anus, Clitellum, Lateral setae, Ventral setae, Pharynx, Mouth, Intestine, Gizzard, Crop, Esophageal glands, Dilator muscles

26-8 Extracellular digestion in an earthworm
Earthworms are deposit feeders in which whole earth is passed from mouth to anus. A grinding organ, the gizzard, is present, as well as various glands that participate in food digestion. Indigestible substances are expelled through the anus.

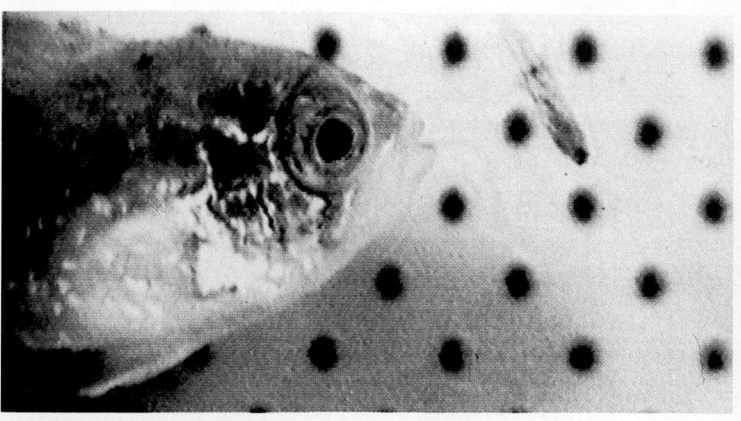 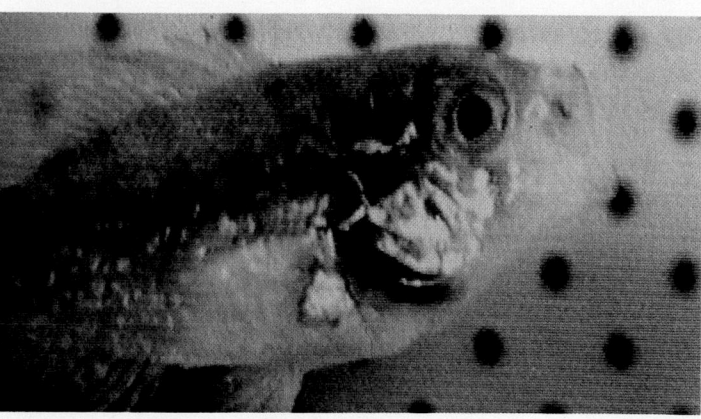

UNIFORMITIES OF DIGESTIVE FUNCTIONS

The human digestive apparatus is equipped in its different parts to do the following:

- Receive solid food
- Mince food mechanically into small fragments
- Secrete digestive enzymes from the intestines and accessory glands
- Transport absorbed material into the bloodstream for distribution to other parts of the body
- Remove most of the water from food residue
- Eliminate the solid remnant, freed of its useful nutrients

The chemical reactions occurring during digestion are substantially the same in all animals, whether digestion is intracellular or extracellular. They are enzyme-catalyzed hydrolyses. Hence, the digestive enzymes responsible for carrying out these reactions are all classified as **hydrolases.** The hydrolytic reactions occurring in digestion have the effect of reversing synthetic processes (see Chapter 11). The most fundamental reactions are summarized in Table 26-2. All such biosyntheses, we will recall, ultimately involve linking together molecules and the elimination of water. By adding back the elements of water (hence, its name), the hydrolyses of digestion undo the achievements of synthesis.

26-9 "Exploding skull" of a suction-feeding fish
A banded sunfish capturing a guppy within a time span of 0.015 second. The two photographs were made with a high-speed video recording camera. In the picture at left, the jaw is at rest as the fish contemplates its prey. In the right photo, various bony elements of the head and jaw expand to an amazing extent. Suction draws in the guppy as the jaws shoot forward to engulf it.

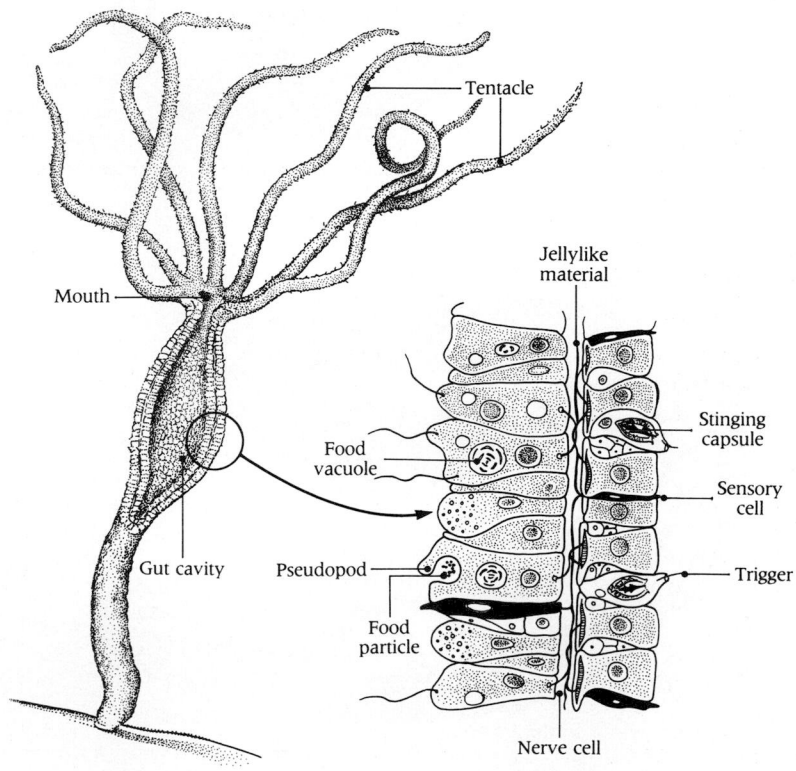

26-10 Digestion in Hydra
In Hydra, *as well as other coelenterates, digestion is partly extracellular and partly intracellular. The alimentary canal is a blind sac, with materials entering and leaving through the mouth. Partly digested but still solid food in the gut enters cells lining the cavity and is digested to completion in food vacuoles (as in protists).*

TABLE 26-2
COMPARISON OF THE NET REACTIONS OF BIOSYNTHESIS (ANABOLISM) AND DIGESTION (CATABOLISM)

Starch and other polysaccharides

$$n \text{ Monosaccharides} \underset{\text{digestion}}{\overset{\text{synthesis}}{\rightleftharpoons}} \text{polysaccharide} + (n-1)H_2O$$

Lipids

$$\text{Glycerol} + 3 \text{ fatty acids} \underset{\text{digestion}}{\overset{\text{synthesis}}{\rightleftharpoons}} \text{triglyceride} + 3H_2O$$

Proteins

$$n \text{ Amino acids} \underset{\text{digestion}}{\overset{\text{synthesis}}{\rightleftharpoons}} \text{protein} + (n-1)H_2O$$

It is important to recognize that enzymes catalyzing the many hydrolytic reactions of digestion are not the same enzymes that participate in the corresponding syntheses. Since digestion proceeds as a series of small steps, what is summarized as a single reaction may actually require a number of different enzymes. Unlike ordinary enzymes, which are found only within cells in minute quantities, digestive enzymes of the animal alimentary canal are poured into the gut from surrounding cells in relatively large amounts—and in fairly simple mixtures (Table 26-3).

Even though food passing along the digestive tract is acted upon by a series of hydrolytic enzymes, the digestive process is not divisible into clearly defined discrete steps, each yielding specific products. Though each enzyme, taken by itself, does carry out certain well-defined operations, food in the digestive tract is exposed to many enzymes at once. Digestion is a continuous process under the control of a *system* of enzymes, ordered to produce a complex *chain* of reactions.

PHASES OF DIGESTION

Chewing, or **mastication,** accomplishes several ends.

- By the tearing, cutting, and grinding actions of teeth, it reduces food to a size convenient for swallowing.
- It mixes ingested food particles with saliva.
- It increases the surface area of food particles and thus facilitates the attack of digestive enzymes.
- It participates in the stimulation of the salivary glands.
- It allows food to be tasted by taste receptors.

TABLE 26-3
MAJOR DIGESTIVE ENZYMES

Source	Enzyme	Optimum pH	Substrates	Products
Salivary glands	Salivary amylase	6.6–6.8	Cooked starches	Disaccharides (maltose)
Gastric glands	Pepsin	1.0–2.0	Proteins	Peptides
	Rennin	4.0	Milk casein	Clotted casein (curd)
Pancreas	Trypsin	7.9	Proteins, peptides	Amino acids
	Chymotrypsin	8.0	Proteins, peptides	Amino acids
	Carboxypeptidase	7.5	Peptides	Amino acids
	Pancreatic amylase	7.0	Starches	Maltose
	Lipase	8.0	Triglycerides (lipids)	Fatty acids, monoglycerides, glycerol
Small intestine	Ribonuclease	7.0–8.0	RNA	Ribonucleotides
	Deoxyribonuclease	7.0–8.0	DNA	Deoxyribonucleotides
	Aminopeptidase	7.8–9.3	Peptides	Amino acids
	Dipeptidase	7.3–9.0	Dipeptides	Amino acids
	Tripeptidase	7.5–9.0	Tripeptides	Amino acids
	Sucrase	5.0–7.0	Sucrose	Fructose, glucose
	Maltase	5.8–6.2	Maltose	Glucose
	Lactase	5.4–6.0	Lactose	Glucose, galactose

Chewing is less important in dogs, cats, and other carnivores, which bolt their food in large lumps, than it is in humans.

The sight, smell, and thought of food elicit the secretion of saliva and gastric juice. **Saliva** assists in the process of chewing in the following ways:

- It readily dissolves soluble food components.
- It partly digests a portion of ingested starch.
- It softens and lubricates the food mass.

The composition of saliva varies under different conditions. In fact, at least two distinct kinds of saliva are secreted by the three pairs of **salivary glands.** One is watery, the other rich in mucus, a lubricating mixture of glycoproteins. The main digestive enzyme in saliva is **salivary amylase.** It hydrolyzes starch into disaccharides, principally maltose, and a scattering of small polysaccharide fragments called **dextrins.**

Swallowing, or **deglutition,** is a reflex action that may be initiated voluntarily but usually begins involuntarily. In the first stage of swallowing, the mass (or **bolus**) of food is lubricated by saliva and passes into the **pharynx.** The second stage is an involuntary process in which the food bolus passes through the pharynx into the **esophagus** (Figure 26-11). As shown in Figure 26-4, food in the pharynx can exit in four directions, only one of which is desirable. Ordinarily it is forced into the esophagus by the closing of the other three routes. The mouth is shut off by the elevation of the tongue against the **hard palate,** the opening into the nasal cavity is closed by the elevation of the **soft palate,** and the opening into the larynx is closed by the vocal cords and by the elevation of the entire larynx—the familiar "bobbing of the Adam's apple"—though occasionally food slips into the larynx or trachea where it may cause choking. In the third stage of swallowing, the bolus passes along the esophagus and through the esophageal sphincter into the **stomach.**

When the semiliquid bolus of food reaches the stomach, it is converted into **chyme,** a thick grayish liquid. The stomach has three major functions.

- It serves as a temporary reservoir, which retains food until it has been reduced to chyme, the form acceptable to the small intestine.
- It digests some protein.
- It prevents bacterial growth through the acidity of its contents.

Although the stomach is not essential to survival, its surgical removal does impair digestive efficiency.

The most important components of **gastric juice** are pepsin and hydrochloric acid (HCl). In addition, gastric juice contains mucus, intrinsic factor (a protein necessary for the absorption of vitamin B_{12} in the small intestine), and a scattering of other digestive enzymes. Box 26-1 recalls a bit of the interesting history of how we first learned about gastric juice.

Pepsin, the principal enzyme of gastric juice, is stored in cells in the lining of the stomach in the form of its inactive precursor, **pepsinogen** (Figure 26-12). Pepsin acts in digestion by hydrolyzing protein molecules. It is the first of several proteolytic enzymes to be encountered by food in its passage through the alimentary tract. Acid—that is, a pH below 6—converts pepsinogen to pepsin, the conversion occurring most rapidly at pH 1.5. We say that this reaction proceeds autocatalytically. That means that free pepsin promotes the further transformation of pepsinogen to free pepsin. Therefore, as free pepsin arises from pepsinogen the rate of its appearance accelerates.

Rennin is the milk-curdling enzyme of gastric juice. The curdling or clotting of milk in an infant's stomach improves the utilization of milk by slowing its passage through the alimentary tract, thus keeping it in the stomach long enough for pepsin to act upon it.

The mucus that coats the stomach lining is secreted in the richly glandular **gastric mucosa** (Figure 26-12). It neutralizes some of the free HCl and also tends to inhibit the action of pepsin, thereby safeguarding the gastric mucosa against erosion by its own digestive secretions. Under normal conditions the mucosa of the stomach and duodenum is resistant to these enzymes, but when a local area undergoes

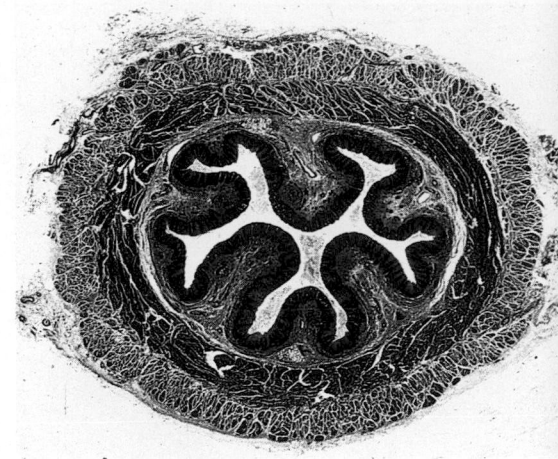

26-11 Scanning electron micrograph of the human esophagus in cross section ($\times 4$)

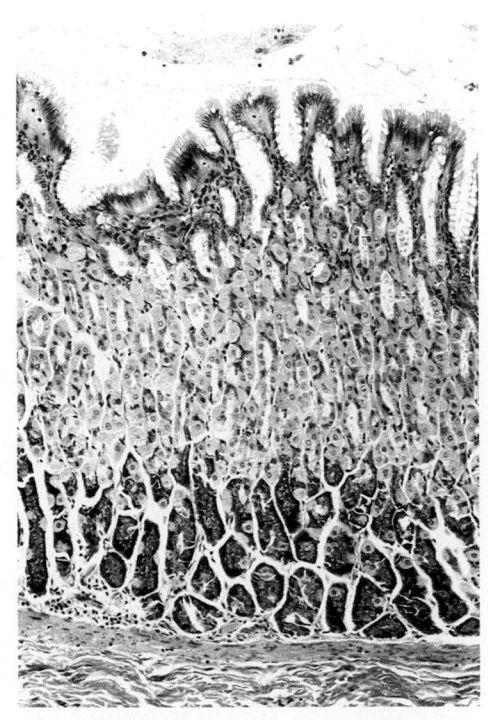

26-12 Photomicrograph of the human gastric mucosa ($\times 34$)

EARLY DAYS IN THE HISTORY OF GASTRIC PHYSIOLOGY

A remarkable chapter in the history of physiology concerns the evolution of methods for the collection of gastric juice and for the study of digestive processes in the stomach. Investigators seeking to determine whether the primary function of the stomach is to soften food mechanically or to attack it chemically early observed that gastric juice seemed to melt food away. The ingenious Lazzaro Spallanzani (1729–1799) swallowed small perforated wooden tubes containing food in small linen bags. He then recovered these tubes intact from his feces and found that the food was considerably digested despite the fact that the tubes were not crushed. Since the food was protected from mechanical manipulation, chemical attack must have occurred.

A classic contribution, the first involving direct, systematic observations of the human stomach, was made by William Beaumont, a United States Army surgeon. In 1822, Alexis St. Martin, a Canadian voyageur, was accidently injured by a discharging shotgun in a frontier trading post at Mackinac. Beaumont, called to attend him, found that the skin and muscles of the upper abdomen had been torn away, leaving a gaping wound in the outer wall of the stomach. He stitched the edges of the open stomach to the skin. St. Martin survived, but the opening into his stomach remained. Beaumont recognized a unique opportunity for investigating stomach physiology and arranged to keep St. Martin in his employ so that he could carry out a series of studies on him. Peering through the opening, he could watch the stomach move. He inserted a string with a piece of meat attached to its end and withdrew it two hours later to find that the meat had dissolved. He learned that gastric juice appeared only when food entered the stomach or while it was being chewed in the mouth. He withdrew pure gastric juice and, with the aid of Professor Dunglison of the University of Virginia, discovered that it contained hydrochloric acid.

Physiologists have devised clever methods for obtaining pure gastric juice from experimental animals. In order to collect gastric juice uncontaminated with food and other secretions, Heidenhain surgically isolated part of the stomach as a pouch (the Heidenhain pouch) that drained directly to the outside through an opening in the abdominal wall. The Heidenhain pouch and its later modifications have contributed much to our understanding of gastric function. In later years other accident victims became experimental subjects in the tradition of St. Martin.

injury, cellular breakdown occurs. The result may be a painful **peptic ulcer,** in which devitalized cells have been digested away.[2]

HUNGER AND APPETITE

For many years the dominant conception of hunger was a mechanical one. Hunger is a subjective sensation, but Walter Cannon in 1912 showed in studies on his students that hunger pangs occur simultaneously with gastric contractions.

In the modern era, hunger is viewed as a complex element in the regulation of food intake by the nervous system. **Hunger,** the bodily state arising from deprivation of food, manifests itself in so-called hunger behavior as well as in hunger sensations. Hunger behavior consists of a general restlessness and increased motor activity. Hunger sensations, the mental adjuncts of hunger, include feelings of generalized weakness, fatigue, irritability, emptiness, and tension, cramps, and pain in the upper abdomen.

With learning or conditioning (Chapter 32), humans and other animals come to associate the ingestion of food with a decrease in hunger sensations and behavior. Hence hunger sensations give rise to the conscious desire to eat. We call that **appetite.** The ingestion of food gradually abolishes the desire to eat, leading to satiety, a normal or physiological loss of appetite. **Anorexia** is an abnormal loss of appetite.

SPECIAL DIGESTIVE ARRANGEMENTS IN HERBIVORES

We commonly classify animals as **omnivores** (eating plants and animals), **carnivores** (meat-eating), or **herbivores** (plant-eating). Not surprisingly, the general plan of the digestive system in each of these groups is adapted to the kind of food it must deal with.

[2] The incidence of peptic ulcers is highest in those portions of the alimentary tract that are in closest contact with gastric juice: the upper duodenum is first (since it repeatedly receives jetlike pulses of stomach contents) and the stomach itself is second.

However, the three ways of classifying digestion in animals are not rigid. No animal can live on any or all kinds of food. The term **eurytrophic** has been applied to animals that can live on many kinds of food and **stenotrophic** to animals with a limited necessary diet. Humans are quite eurytrophic. Among the many stenotrophic animals are the numerous species that eat only ants or termites. The koala lives on only a few of the species of eucalyptus.

The term omnivore may be misleading, since animals in this group cannot digest plants that are rich in cellulose. Yet just such a diet is the mainstay of certain herbivores, nature having provided them with a remarkable set of special arrangements that permit them to subsist on forage that is rich in cellulose.

The key feature is a device for providing the necessary hydrolytic enzymes called **cellulases** in the alimentary tracts of these herbivorous animals. Such enzymes are produced in only a few unusual species. One is the famous woodboring shipworm *Teredo*, a bivalve that relies on cellulase to tunnel its way through timbers. The herbivorous animals we will now discuss are the **ruminants** (suborder Ruminantia), a huge group that includes many living and extinct genera and species.[3] The ones most familiar to us are the domestic farm animals, cattle, goats, and sheep. They cannot produce cellulase, but they have acquired a solution to the problem. They possess an extra stomach chamber called a **rumen** that is filled with dozens of varieties of bacteria and protists, many of which are capable of digesting cellulose on their host's behalf.[4] These organisms actively ferment the ingested cellulose-containing forage, hydrolyzing the cellulose into a variety of short-chain fatty acids (propionic acid, butyric acid, etc.).

Ruminants have large four-chambered stomachs consisting of a rumen, reticulum, omasum, and abomasum (Figure 26-13). The rumen and **reticulum** are microorganism-packed ''fermentation vats.'' To facilitate digestion, these animals periodically regurgitate the contents of the rumen and rechew them. This is the process we call ''chewing the cud.'' When finally digested the food mass is transferred, along with a large number of microorganisms, into the **omasum,** where it is concentrated, and then into the **abomasum.** Like the stomach of carnivorous animals, the abomasum contains HCl and various digestive enzymes. The microorganisms are now a significant part of the food mass and they are digested too. Thus they constitute an important source of food protein.

Rabbits, hares, and many rodents either do not have microorganisms to digest cellulose, or have fewer of them. Interestingly, the **cecum**, a blind sac projecting from the large intestine, contains many microorganisms. In some mammals, notably herbivores, bacteria and protists in the cecum are capable of digesting cellulose. For many species (e.g., horses), cecal fermentation is of little value because it occurs too far along in the alimentary canal to permit much absorption. In rabbits and some other species, however, an interesting adaptation exists. These animals evacuate two kinds of feces—soft fecal pellets that are dropped, and larger and lighter colored feces that are not dropped by the animal, but are eaten directly from the anus, a practice termed **coprophagy.** Food is therefore digested twice. If coprophagy is prevented, rats develop vitamin and mineral deficiencies.

DIGESTION IN OTHER ANIMALS

In the group of animals to which the hydras, corals, and jellyfishes belong (phylum Coelenterata, or Cnidaria) and also in the flatworms (phylum Platyhelminthes) (Figure 26-14), a body cavity opens freely to the outside through a single opening—a **mouth.** The mouth opens into a tubular **pharynx,** which is protruded through the mouth to touch the prey. Most of the animal's food is first brought into this cavity, which is filled with the water in which the animal lives, but which also contains some of the digestive enzymes secreted by the surrounding cells. Digestion begins in the cavity, but in most species it is not completed there. Usually small bits of food are taken directly into the cells lining the cavity, where they undergo intracellular digestion in food vacuoles, just as solid food is usually digested in

---

[3] The living ruminants include all the many deer and antelopes, pronghorns, giraffes, buffalos, bisons, muskoxen, domestic cattle, sheep, and goats.

[4] In some species, this microbial mix includes fungi (yeasts) that are cellulase producers. The cellulolytic capacity of rumen microorganisms is easily demonstrated in the laboratory by placing a sample of a culture directly on a piece of filter paper, itself a cellulose object. If cellulase is present, a hole soon appears in the paper.

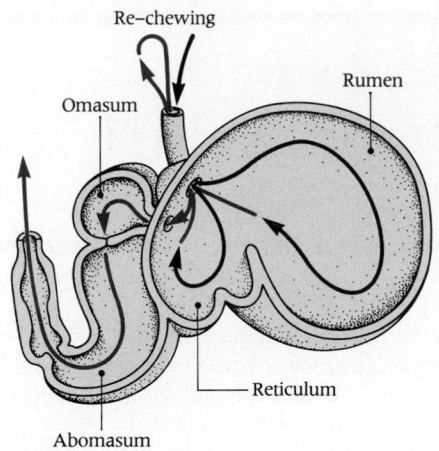

26-13 Stomach of a ruminant
The four chambers of a ruminant's stomach include the omasum, abomasum, reticulum, and rumen. The rumen contains a variety of bacteria and protists that digest the cellulose taken into the body. The abomasum corresponds to the true vertebrate stomach.

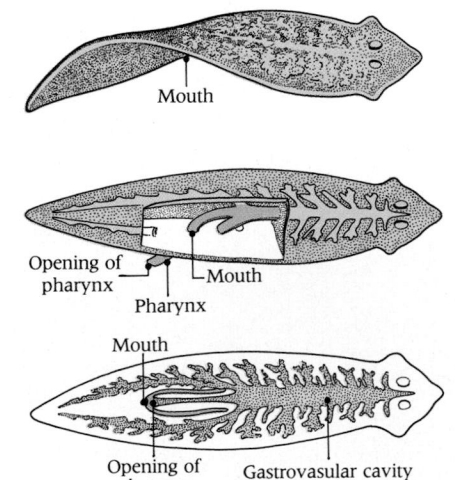

26-14 Digestion in flatworms
The alimentary canal, or gastrovascular cavity, is a highly branched blind sac that extends throughout the body. Food taken in through the mouth enters the gastrovascular cavity through the pharynx. Food vacuoles in cells lining the cavity take in partially digested materials and complete the digestive process.

protists. Digestion here is intermediate between the food vacuole system and the alimentary canal system.

The digestive system of insects (see Figure 26-6) resembles the human system in that it is complete, running from one of the openings, the mouth, to the other, the anus. Compared to a body cavity with one opening that takes food in and passes wastes out, such a system has obvious advantages: one-way flow and assembly-line processing of food permits separate and sequential digestive stages to take place simultaneously. Ingested food is chewed and mixed with saliva. The food mass is pulled into a **crop,** a pouchlike enlargement of the gullet, and then into a **proventriculus** that grinds the food before passing it into the intestine for digestion and absorption. Some insects do not swallow solid food but digest it in the mouth (or, more strictly, the pharynx), swallowing the solutions and spitting out the undigested parts.

Those fluid feeders that live on watery solutions, such as blood from other animals, or flower nectar, have digestive systems that are correspondingly modified.

ASSIMILATION

We interrupted our discussion of human digestion as the semifluid chyme was passing from the stomach into the duodenum. To this point, (1) the food mass had been broken up mechanically by chewing; (2) it had been liquefied by saliva and gastric juice; (3) starch and protein digestion had been initiated by salivary amylase and gastric pepsin, respectively; and (4) the pH had been sharply lowered by the addition of HCl.

COMPLETION OF DIGESTION

In the small intestine, digestion proceeds until complex food elements are simplified enough to be absorbed. The secretion of the pancreas is the main digestive juice. The volume of **pancreatic juice** produced in 24 hours varies from 700 to 2000 ml. The juice is a colorless liquid with a pH of about 8, consisting of water, electrolytes (including bicarbonate that accounts for its alkalinity), mucus, and digestive enzymes. The enzymes are grouped into four major types: proteolytic (protein-splitting **proteases**), amylolytic (carbohydrate-splitting **carbohydrases**), lipolytic (lipid-splitting **lipases**), and nucleolytic (nucleic acid-splitting **nucleases**). These enzymes are capable of digesting virtually all food without the aid of any other digestive agents.

The principal proteolytic enzymes are trypsin, chymotrypsin, and carboxypeptidase. Like gastric pepsin, pancreatic *trypsin* is secreted as an enzymatically inert precursor. The trypsin precursor is **trypsinogen.** Upon entering the small intestine via the **pancreatic duct,** trypsinogen is converted to trypsin by **enterokinase,** another enzyme. Once some trypsin is formed, it can itself transform trypsinogen to trypsin. Therefore, trypsinogen is converted completely to trypsin via two pathways.

$$\text{Trypsinogen} \xrightarrow{\text{enterokinase}} \text{trypsin}$$

$$\text{Trypsinogen} \xrightarrow{\text{trypsin}} \text{trypsin}$$

Chymotrypsin and **carboxypeptidase** (actually a family of enzymes) are also secreted as their respective precursors, **chymotrypsinogen** and **procarboxypeptidase.** In each case, conversion of precursor to active enzyme is catalyzed by trypsin but not by enterokinase. Chymotrypsin also participates in the conversion of chymotrypsinogen to chymotrypsin.

$$\text{Chymotrypsinogen} \xrightarrow{\text{chymotrypsin, trypsin}} \text{chymotrypsin}$$

$$\text{Procarboxypeptidase} \xrightarrow{\text{trypsin}} \text{carboxypeptidase}$$

Trypsin specifically hydrolyzes a peptide bond to which arginine or lysine contributes the carboxyl group (Figure 26-15). Like pepsin, chymotrypsin hydrolyzes a peptide bond involving an aromatic amino acid (i.e., tyrosine or phenylalanine). Pancreatic carboxypeptidase attacks the carboxyl end of a protein molecule, and

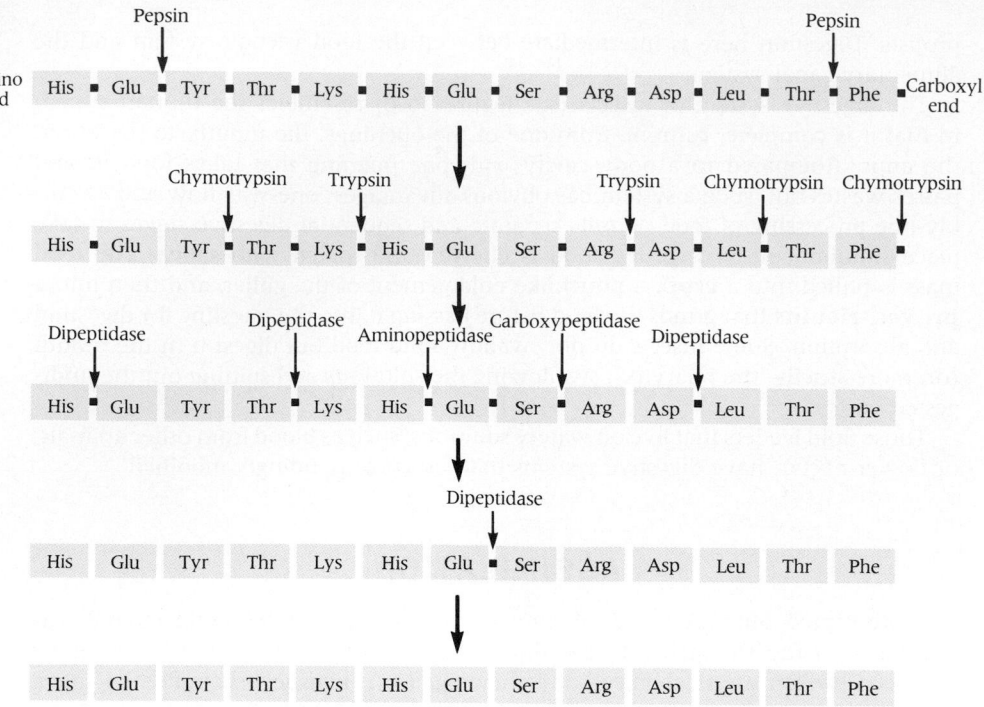

26-15 Protein breakdown by proteases
A protein containing a hypothetical sequence of 13 amino acids is shown. Pepsin in the stomach acts on the sequence to produce a single amino acid, a dipeptide, and a decapeptide (10 amino acids). In the intestine, trypsin and chymotrypsin cleave the molecule further at specific amino acids. Dipeptidase splits bonds between pairs of amino acids, amino-terminal bonds are broken by aminopeptidases, and carboxy-terminal bonds are hydrolyzed by carboxypeptidases. This completes breakdown to single amino acids. Amino acid abbreviations are given in Figure 4-14.

intestinal **aminopeptidase** the amino end, until only a dipeptide or tripeptide remains. In this way they complete the digestion of proteins begun by pepsin, trypsin, and chymotrypsin.

Pancreatic amylases break down large starch molecules to dextrins and disaccharides. Indeed, starch is not effectively digested until it is exposed to them. The disaccharide maltose must be cleaved into its two component glucose molecules to be absorbed. Pancreatic juice contains a weak enzyme, **maltase,** that splits a small fraction of the total maltose present. Hydrolysis of the rest requires intestinal juice enzymes. This illustrates that a given enzyme may act on only one kind of substrate molecule, but a given substrate may be attacked by more than one enzyme.

Pancreatic lipase breaks down lipids (i.e., triglycerides) to varying degrees, the major products being free fatty acids (and glycerol) and monoglycerides. Its activity depends on the presence of **bile,** a complex fluid produced by the liver. Bile is a dark-colored fluid, the principal components of which are: bile salts, bile pigments, fatty acids, cholesterol, various salts, and water. The **bile salts,** glycocholic and taurocholic acids, are formed by conjugation of the amino acids glycine and taurine, respectively, with cholic acid, a product of cholesterol breakdown. They are natural detergents that act to emulsify lipids in the intestine, thus making them more accessible to lipase action. The major lipase, pancreatic lipase, is extremely important in fat digestion. The pancreas appears to be the only source of a lipase that can function in the conditions prevailing in the alimentary tract.

The final enzymes in intestinal juice come from the glands of the small intestine itself. As shown in Table 26-3, these include the aminopeptidase just mentioned and a dipeptidase and tripeptidase which convert the dipeptides and tripeptides remaining from protein digestion to individual amino acids. There are also various disaccharidases that hydrolyze sucrose, maltose, and lactose.

ABSORPTION IN THE SMALL INTESTINE

The small intestine is by far the major site of **absorption,** by which we mean the transfer of digested food from the intestinal stream into the bloodstream. The lining of the small intestine has remarkable anatomical arrangements that maximize its surface area. The inner lining is draped into numerous **circular folds.** If you were to run a finger over the lining membrane, you would note that it feels very much like velvet. Examination of this "velvet" through a microscope reveals thousands of tiny fingerlike projections called **villi** (Figures 26-16 and 26-17). Digested food is absorbed by the villi. Higher magnification shows that each villus is covered in turn by many smaller fingers called **microvilli,** which increase absorption 1000-fold. The total area is more than 100 square feet.

The human small intestine is well adapted to absorb the molecules of digested

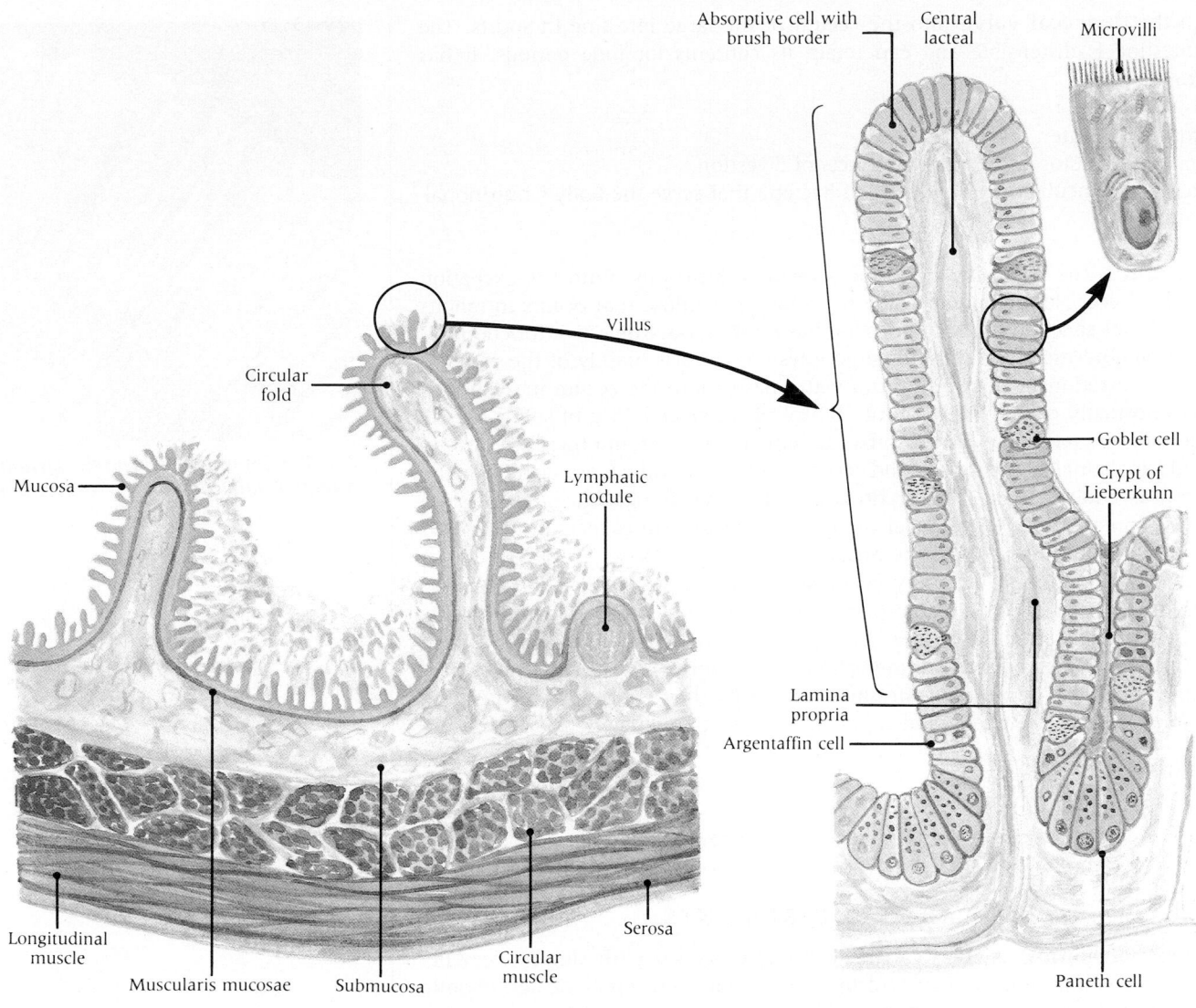

Absorptive cell with brush border

Central lacteal

Microvilli

Villus

Circular fold

Lymphatic nodule

Mucosa

Goblet cell

Crypt of Lieberkuhn

Lamina propria

Argentaffin cell

Longitudinal muscle

Muscularis mucosae   Submucosa   Circular muscle   Serosa

Paneth cell

A

B

26-16 Structure of the small intestine *(A) A longitudinal section through the intestine wall shows the villi, which increase the surface area for absorption. (B) A single villus is covered with many mucosal cells, each with multiple microvilli that enhance transfer of materials from the intestine to the blood and lymphatic capillaries (lacteals).*

food. In the adult, the small intestine is 20 to 25 feet long and about 1 inch in diameter. Its three major subdivisions are the C-shaped **duodenum,** approximately 10 inches long; the **jejunum,** about 6 to 8 feet in length; and the **ileum,** the remainder of its length.

In physical terms, intestinal absorption involves active transport of substances from the inside, or **lumen** of the intestine, across the microvilli of absorptive **mucosal cells** lining the intestinal surface, and into the blood-containing capillary network or the lymph-containing **central lacteal** of a villus (see Figure 26-16). This transfer proceeds against a concentration gradient. Thus it is an active process that requires the expenditure of energy derived from ATP.

The walls of the small intestine contain muscles that are responsible for several kinds of movements. These include rhythmic segmentation movements and peristalsis. **Segmentation** is the formation of pouches in the wall of the intestine by the alternate contraction and relaxation of circular bands of muscle running around the intestine. Contraction squeezes the intestine and its contents, making it resemble a string of sausages. **Peristalsis** is a wave of contractions that run down the length of the intestine and propel the intestinal contents forward. Segmentation movements blend the intestinal contents but do not push them along.

These movements mix and knead the intestinal contents, rub them across absorbing surfaces, and propel them toward the large intestine.

FUNCTIONS OF THE LARGE INTESTINE

The time required for the passage of food from stomach to the end of the small intestine varies widely, averaging several hours. Material from the ileum proceeds through the **ileocecal valve** into the cecum of the large intestine in spurts. The large intestine is distensible and can retain its contents for long periods. It has three functions.

- It conserves water.
- It temporarily stores the waste products of digestion.
- It acts as an incubator for a variety of bacteria that serve the body's nutritional needs.

The **colon** is the most important locus of these functions. With the exception of the role of cecal fermentation in the digestion of cellulose that occurs in rabbits and some other species, the large intestine has no important digestive functions.

The semiliquid mass entering the large intestine consists mainly of the residues of digestive secretions. About 500 ml of water passes into the cecum in 24 hours. The feces normally contain only about 100 ml of water and 35 g of solids for the same period. The large intestine absorbs the remaining water and thereby converts semiliquid intestinal contents to solid feces. Water is absorbed principally in the cecum and ascending colon. The secretions of the mucosal lining of the large intestine are scanty, containing no digestive enzymes. The **rectum** stores feces prior to their intermittent elimination through the **anus.**

Herbivores that subsist on bulky cellulose-containing vegetable fibers that are digested in part by colonic bacteria, have a long small intestine and an unusually long and capacious colon. Carnivores usually have a short, narrow colon and almost no cecum. Many frogs present an interesting case (Figure 26-18). As tadpoles they eat plant matter and have long intestines. As adults they eat animal matter and have short intestines. Humans and other omnivores are somewhere between the two.

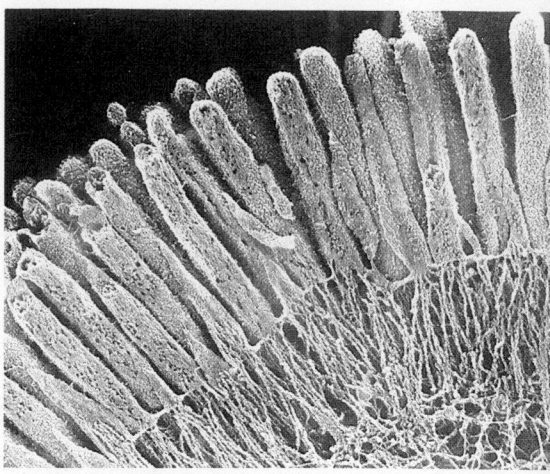

26-17 Scanning electron micrograph of the human small intestine in cross section ($\times 72$)

NUTRITION AND METABOLISM

DIETARY REQUIREMENTS

Autotrophic organisms can begin their synthetic processes with simple elements and inorganic compounds, but heterotrophs must start with prefabricated organic compounds such as sugars and amino acids. These are synthesized by organisms that are ingested as food.

Humans are heterotrophs that lack the ability to make many necessary organic molecules. Hence, their diet must include a wide variety of prefabricated molecules. We call these molecules dietary, or nutritional, requirements. In addition, humans thrive best on a diet containing all three major classes of food—carbohydrates, fats, and proteins—as long as there is some diversity in the foods of each class. Food from any of these classes can furnish the necessary energy. In this respect, the classes are largely interchangeable. However, they are not interchangeable as building materials or as sources for the synthesis of proteins (which include enzymes) and other compounds. For these needs, dietary proteins (or a number of amino acids) are absolutely indispensable.[5]

Another critically important nutritional requirement is for vitamins. A person who takes an apparently sufficient quantity of carbohydrates, fats, and proteins, could nonetheless develop serious nutritional deficiency disorders if needed vitamins were lacking. A deficiency disease is caused by a lack of substances essential for normal metabolism. One of the first of these diseases to be studied was **scurvy,** a potentially fatal disorder characterized by bleeding gums and painfully swollen joints. It frequently appeared among sailors on long voyages in the days when fresh foods were not available on shipboard. It was discovered by chance (during long voyages of British ships) that limes (and other citrus fruits) prevent this disease. (When a law insisted that British sailors drink lime juice, they acquired the nickname ''limeys.'') Later experiments and clinical investigations revealed that scurvy results from a lack of vitamin C (ascorbic acid). Other deficiency diseases result from shortages of other vitamins (Table 26-4).

[5] As a general rule, foods rich in carbohydrates are cheap, and those rich in lipids and proteins are expensive.

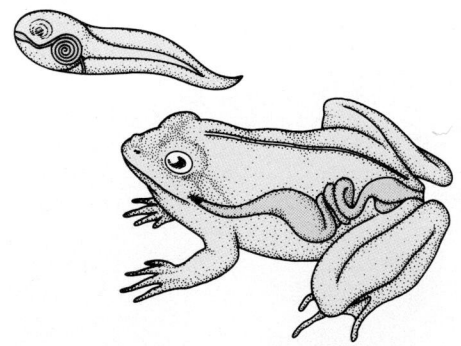

26-18 Differences in intestines of herbivores and carnivores
The tadpole is a herbivore that has a long intestine to increase the area for absorption of nutrients from plant material. This is needed because of the large amount of cellulose in its diet. The adult frog is a carnivore. Its diet lacks cellulose, and nutrients are more easily obtained. These animals have a shorter intestine.

TABLE 26-4
VITAMINS IMPORTANT IN HUMAN NUTRITION

Vitamins	Some Deficiency Symptoms	Some Food Sources
Fat-Soluble		
A (retinol)	Dry, scaly skin; night blindness	Milk, butter, liver oils, yellow and green vegetables*
D (calciferol)	Rickets (defective growth of bones and other hard tissues)	Milk, egg yolk, liver oils†
E (tocopherol)	Degeneration of muscles (also sterility in rats and possibly in other organisms)	Green vegetables, oils from seeds, egg yolk, meat
K (menaquinones)	Faulty blood coagulation and hemorrhage due to failure to complete synthesis of prothrombin	Green vegetables
Water-Soluble		
B_1 (thiamin)	Beriberi, polyneuritis (inflammation and degeneration of nerves)	Yeast, whole-grain cereals, lean meat
B_2 (riboflavin)	Soreness around mouth, inflammation of eyes	Vegetables, yeast, milk, liver, eggs
B_6 (pyridoxine)	Dermatitis, convulsions	Whole grain or enriched cereals, liver, fish, vegetables
B_{12} (cobalamin)	Anemia due to the inability of blood cell precursors to synthesize DNA, nerve damage	Liver, lean meat
C (ascorbic acid)	Scurvy	Citrus fruit, tomatoes, green peppers
Nicotinic acid or niacin	Pellagra (a disease affecting skin, alimentary canal, and nerves)	Yeast, meat
Folic acid	Anemia similar to that of vitamin B_{12} deficiency	Green vegetables
Biotin	Scaly skin	Liver, milk, eggs, whole-grain cereals

* Vitamin A is not required as such, since the human body can synthesize it from yellow pigments (carotenes). These are required if vitamin A itself is deficient.
† The human body can synthesize vitamin D, but the amounts synthesized may be inadequate.

All animals also require mineral elements (Table 26-5). Calcium and phosphorus are required in relatively large amounts because calcium phosphate is the principal component of bones and teeth. Calcium also has an essential role in nerve and muscle functions. Phosphorus is, of course, a component of nucleotides and nucleic acids and a key participant in energy transfer via the cycling of the ATP-ADP system. Iron plays an important role in cellular respiration and is the O_2-binding atom in both hemoglobin and myoglobin. Other needed minerals and trace elements were discussed in Chapter 3.

Vitamins are defined as organic compounds that are essential in small quantities for normal metabolism, but that are not synthesized by the organism. Therefore they must be present in the diet. This is another example of auxotrophism: in this case, too, an organism needs assistance from other organisms. This means that a substance required in the diet of one organism, for which it is a vitamin, must be synthesized by another organism, for which it is obviously not a vitamin. Most vitamins can be synthesized by one plant or another.[6]

These roles are now well understood for all but a few vitamins. As discussed in Chapter 8, almost every vitamin is converted in the body to a coenzyme. Until this transformation is accomplished, a vitamin cannot participate in cell metabolism. The vitamin nicotinic acid, for example, is converted to the coenzymes NAD and NADP. An individual deficient in nicotinic acid lacks these coenzymes in sufficient quantity and the metabolic pathways upon which they depend function subnormally. This is the ultimate cause of the disorder known as **pellagra.** Until the 1930s, persons with **pernicious anemia**—a disease in which the number of red blood cells steadily declines—eventually died because they could not transport about a millionth of a gram of cobalamin (vitamin B_{12}) across the intestinal wall each day. The protein called **intrinsic factor,** produced in the stomach, normally promotes the absorption of cobalamin across the mucosa of the ileum. If intrinsic factor is missing, as it is in pernicious anemia, cobalamin in the diet cannot be absorbed into the body and is useless. There is no cure for pernicious anemia, but patients can now live relatively normal lives with regular injections of cobalamin.

Most of the vitamins essential to humans, listed in Table 26-4, are classified as fat-soluble and water-soluble. The former, which include vitamins A, D, E, and K,

[6] An exception is cobalamin (vitamin B_{12}), which is synthesized only by certain bacteria. The most active cobalamin synthesizers include bacteria normally present in the rumens of cows, sheep, and other ruminants. This explains why meat, milk, and liver are major sources of cobalamin in the human diet.

TABLE 26-5
MINERAL ELEMENTS REQUIRED BY ANIMALS

Element	Source in Human Diet	Major Functions
Macronutrients		
Calcium (Ca)	Dairy foods, eggs, green leafy vegetables, whole grains, legumes, nuts	In bones and teeth, blood clotting, nerve and muscle action, enzyme activation
Chlorine (Cl)	Table salt (NaCl)	Water balance, digestion (as HCl) principal negative ion in fluid around cells
Iron (Fe)	Liver, meat, green vegetables, eggs, whole grains, legumes, nuts	In active site of many redox enzymes and electron carriers, hemoglobin, myoglobin
Magnesium (Mg)	Green vegetables, meat, whole grains, nuts, milk, legumes	Required by many enzymes, found in bones and teeth
Phosphorus (P)	Dairy foods, eggs, meat, whole grains, legumes, nuts	In nucleic acids, ATP, and phospholipids, bone formation, buffers; metabolism of sugars
Potassium (K)	Meat, whole grains, fruits, vegetables, legumes	Nerve and muscle action, protein synthesis, principal positive ion in cells
Sodium (Na)	Table salt, dairy foods, meat, eggs, vegetables	Nerve and muscle action, water balance, principal positive ion in fluid around cells
Sulfur (S)	Meat, eggs, dairy foods, nuts, legumes	In proteins and coenzymes, detoxification of harmful substances
Micronutrients		
Chromium (Cr)	Meat, dairy foods, whole grains, dried beans, peanuts, brewers' yeast	Involved in glucose metabolism
Cobalt (Co)	Meat, dairy foods	Vitamin B_{12}, formation of erythrocytes
Copper (Cu)	Liver, meat, fish, shellfish, legumes, whole grains, nuts	In active site of many redox enzymes and electron carriers, production of hemoglobin, bone formation
Fluorine (F)	Most water supplies	Improves resistance to tooth decay
Iodine (I)	Fish, shellfish, iodized salt	In thyroid hormone
Manganese (Mn)	Organ meats, whole grains, legumes, nuts, tea, coffee	Activates many enzymes
Molybdenum (Mo)	Organ meats, dairy foods, whole grains, green vegetables, legumes	Required by some enzymes
Selenium (Se)	Meat, seafood, whole grains, eggs, chicken, milk, garlic	Involved in metabolism of fats
Zinc (Zn)	Liver, fish, shellfish, and many other foods	Required by some enzymes, involved in physiology of insulin

are vitamins only in vertebrates. The latter, which include vitamin C and the vitamins of the B complex, are required by almost all animals.

The varied diet customarily eaten by most Americans—excepting the poor, and devotees of so-called junk food—usually provides sufficient amounts of all the needed vitamins, particularly if the diet includes vegetables and fruits in addition to meat and milk. When a vitamin deficiency exists, it should be corrected by administration of the lacking vitamin. Overdoses of some vitamins can be harmful.

ROLE OF THE LIVER

The amino acids and sugars that were absorbed into the capillaries of the small intestine are transported to the **liver.** This complex organ is the largest in the human body, weighing 1500 g in an average-sized man and a bit less in an adult woman. It is situated in the upper right of the abdominal cavity (see Figure 26-3) and from the front looks triangular.

As explained in Chapter 24, the liver has a unique circulatory system: it receives blood from two channels.

- One, the hepatic artery, brings arterial blood under the high hydrostatic pressure typical of arteries, but it furnishes only about 30% of the liver's blood supply.
- The other, the portal vein, carries a large volume of blood under low pressure from the capillary beds of the stomach, pancreas, and intestine.

Blood passing from intestine to liver via the portal vein is laden with all the nutrients resulting from intestinal digestion and absorption except the lipids, which, we recall, are carried by the lymphatic system (see Chapter 24). If these nutrients were to enter the systemic circulation directly, most would be rapidly eliminated by the kidneys. This arrangement permits the liver to store and transform the nutrients for use in the body. Major functions of the liver are summarized in Figure 26-19.

26-19 Functions of the mammalian liver
The liver carries out a variety of functions essential in the maintenance of homeostasis.

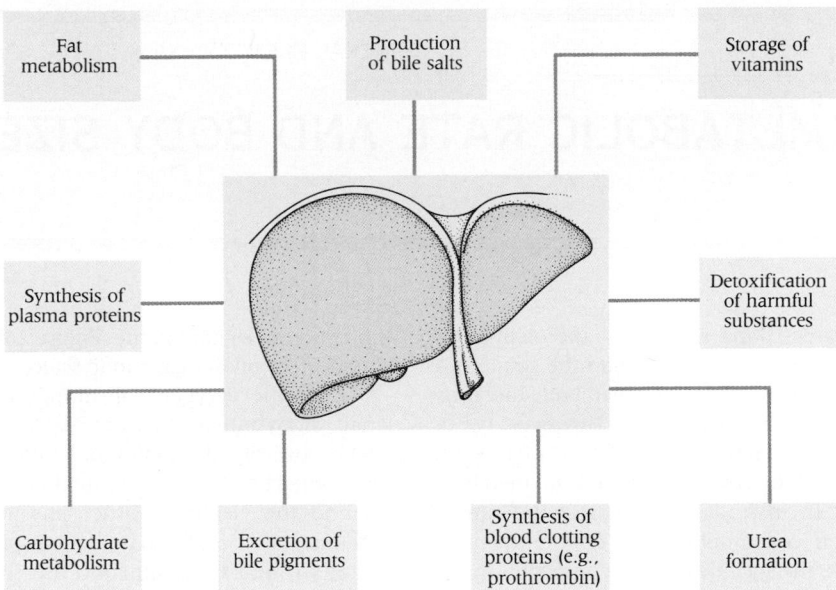

Fat metabolism

Production of bile salts

Storage of vitamins

Synthesis of plasma proteins

Detoxification of harmful substances

Carbohydrate metabolism

Excretion of bile pigments

Synthesis of blood clotting proteins (e.g., prothrombin)

Urea formation

Glucose is stored in the liver as **glycogen** (Figure 26-20). Excess glucose is converted to fat by the pathways summarized in Chapter 10. When the body needs fuel, the liver breaks down its stored glycogen into glucose, releasing it into the blood as needed. Animal cells cannot function without glucose. Yet the only vertebrate tissues that store glycogen in appreciable amounts are liver and muscle. Other tissues and organs, including the brain, must depend on the liver to release glucose into the bloodstream. The liver, then, is vitally important because it regulates the fuel supply that keeps the metabolic fires burning in the cells.

The liver has many other functions.

- It secretes the bile salts of bile and excretes the bile pigments.
- It stores amino acids used for growth and tissue repair.
- It removes ammonia from blood.
- It converts the nitrogen of excess amino acids into urea, the form in which nitrogen is excreted from the human body by the kidneys.
- Fatty acids absorbed in the small intestine are carried from the lacteals of the villi by lymphatic vessels, which ultimately pass them to the bloodstream. From there, fatty acids move to the liver, which synthesizes from them various plasma lipids, cholesterol, and other substances. These substances form part of the structure of cell membranes and the sheaths that insulate nerve cells.
- The liver detoxifies many potentially harmful substances by altering them chemically (by methylation, conjugation, oxidation, etc.). For example, many drugs and hormones are modified in the liver and thereby inactivated. Alcohol in beer and whiskey is broken down by the liver to acetaldehyde and then to acetyl CoA, relatively harmless substances. The alcohol itself causes the major effects of drinking.
- It synthesizes plasma proteins and many clotting factors.

The liver performs many other functions in addition. The importance of this vital "chemical factory" to the body is underscored by the fact that nature provides an excess of liver tissue—we could easily get by on about one-fourth of what we have. The liver can also regenerate quickly when damaged.

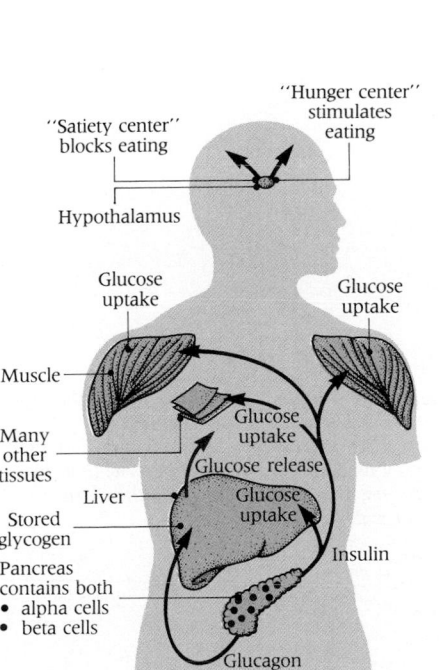

"Hunger center" stimulates eating

"Satiety center" blocks eating

Hypothalamus

Glucose uptake

Glucose uptake

Muscle

Glucose uptake

Many other tissues

Glucose release

Liver

Glucose uptake

Stored glycogen

Insulin

Pancreas contains both
- alpha cells
- beta cells

Glucagon

26-20 Control of blood glucose levels
The blood glucose level is controlled by a number of coordinated systems. Glucose is stored in the liver as glycogen. The pancreatic hormone glucagon stimulates release of glucose into the blood. The pancreatic hormone insulin promotes glucose uptake by tissues, including muscle. Feelings of hunger and satiety originate in the hypothalamus.

TOTAL BODY METABOLISM

The **metabolism** of the whole body is simply the sum of all the metabolic processes in all the cells of the body. Let us briefly consider certain quantitative aspects of total body metabolism.

We have seen that less than half of the potential energy of glucose is transformed into the energy-rich bonds of ATP. Energy yield from the catabolism of proteins is somewhat less than that from lipid metabolism, so that on average slightly more than half of the potential energy of foodstuffs is converted into usable bond energy. The remainder is dissipated as heat during ATP formation. Still more energy is

The largest land mammal—the elephant—is a million times larger than the smallest—the shrew—and its total metabolic rate must of course be much higher. However, we do not get an accurate assessment of the two mammals by comparing their total metabolic rates. If, instead, we calculate the rate of oxygen consumption per unit body mass, that is, the specific oxygen consumption, we get a very different relationship between body size and metabolic rate.

Figure 26A summarizes the measured rates of specific oxygen consumption of various mammals. We see that the rate of oxygen consumption per kilogram of body weight decreases consistently with increasing body size. One kilogram of shrew tissue consumes oxygen at a rate some 100 times that of 1

kilogram of elephant tissue. Figure 26B plots similar data on a logarithmic scale.

The higher oxygen consumption of the small shrew implies that the oxygen supply (hence the blood supply) per kilogram of tissue must be 100 times greater in the shrew than in the elephant. Other physiological variables such as food intake, respiration, excretion, and heart function are similarly affected.

Many physiological processes are functions of the body's surface area. As the size of an animal (or cell or organ) changes, surfaces increase (or decrease) in proportion to the square of linear dimensions, whereas volumes change in proportion to the cube of linear dimensions. Thus, the shrew has a disproportionately large surface area when

compared with that of the more voluminous elephant.

A great many physiological processes are critically governed by surface-volume relations. They include the following:

- The uptake of food in the intestine depends largely on the intestinal surface area.
- Heat loss is related to body surface.
- The uptake of oxygen in lungs or gills depends largely on the surface area of these organs.
- The diffusion of oxygen from the blood to the tissues takes place across the capillary walls, which again is a function depending on surface area.

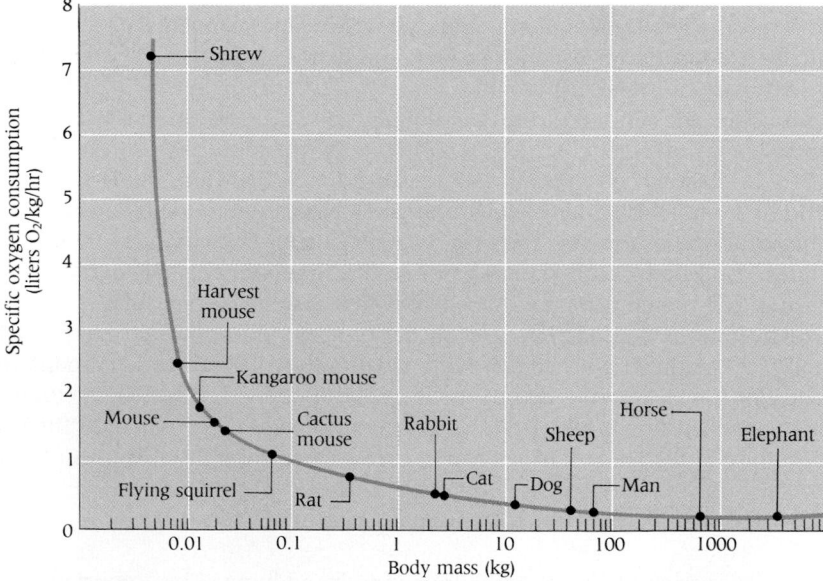

26A Semi-log plot of the rates of specific oxygen consumption of various mammals
Oxygen consumption per unit body mass increases rapidly with decreasing body size. Note that the x-axis has a logarithmic scale and the y-axis a linear scale.

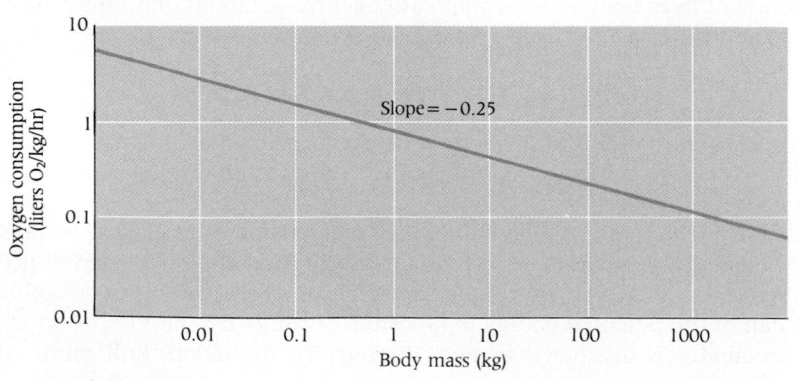

26B Log-log plot of the rates of specific oxygen consumption of various mammals
In a log-log plot, the data shown in Figure 26A fall on a straight line with a slope of −0.25.

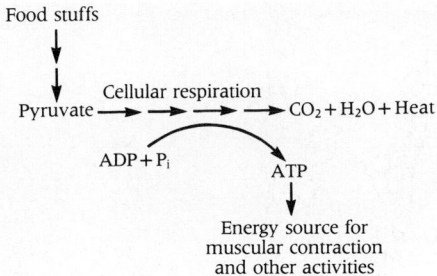

26-21 Review of how ATP is made available for muscular contraction
Carbohydrate taken in as food is catabolized to pyruvate, which in the presence of O_2 is converted to CO_2 and H_2O. Some energy is lost as heat. Recoverable energy is converted to ATP, which fuels muscular contraction and other activities. Compare with Figure 27-1 and see Chapter 9.

dissipated as heat during the utilization of ATP in such processes as glandular secretion, nerve conduction, muscular contraction, and biosynthesis, so that actually only about a quarter of the initial food energy participates directly in body functions. The rest is lost as heat.

However, with one exception, even this fraction is eventually converted into heat. Each of the many energy-requiring (endergonic) physiological processes using ATP is balanced by a simultaneous heat-producing (exergonic) process. For example, ATP energy is consumed in protein synthesis, but protein breakdown going on simultaneously elsewhere in the body releases as heat the energy stored in peptide linkages. Similarly, ATP energy is consumed in doing the work of muscular contraction (Figure 26-21), but the friction involved in this activity generates heat. The single exception arises when the muscles perform external work—creating either potential energy by lifting a mass against gravity or kinetic energy by operating a tool or turning a wheel. Since during rest practically all the energy of ingested foodstuffs takes the form of heat, the rate at which heat is liberated by the body is a measure of the overall **metabolic rate.** The relation of metabolic rate to body size is discussed in Box 26-2.

CALORIC VALUES OF FOODSTUFFS

A **calorie** (written with a small c and abbreviated **cal**) is the amount of energy needed to raise the temperature of 1 g of water 1°C. In discussing metabolism, we speak only of kilocalories (kcal) which are also called Calories (written with a capital **C** and abbreviated **Cal**). One kcal (Cal) equals 1000 calories.

Using a device called the bomb calorimeter, investigators have accurately measured the heat liberated during total combustion of a food sample. This gives a dependable measure of the sample's caloric content, since the end products of the complete calorimetric combustion of carbohydrate or lipid are CO_2 and H_2O, the same as those arising from the oxidation of carbohydrate or lipid in the body. Therefore, the heat produced in the bomb is equivalent to the energy yield in the body.

The situation is not as simple for protein. Because proteins contain nitrogen, the products of combustion necessarily include NO_2. Unlike the calorimeter, the body cannot oxidize protein nitrogen completely to NO_2. Rather it breaks protein nitrogen down to urea, which is $CO(NH_2)_2$. Hence, the energy available from protein equals the heat produced in the combustion chamber minus the heat that would be evolved if the body's nitrogenous end products were completely oxidized.

When a food's carbohydrate, lipid, and protein contents are known, the available energy can be calculated from known calorimeter values for the three constituents. We learn that for many organisms, 1 gram of carbohydrate yields 4.1 kcal, protein yields 4.1 kcal per gram, and lipid contains 9.3 kcal per gram. Allowing for the possibility that digestion or absorption is incomplete, these values are conveniently rounded off to 4, 4, and 9 kcal per gram, respectively.

The body's total energy output in any 24-hour period is the sum of the **basal metabolism,** which is the body's energy requirement at rest, plus a highly variable component related to muscular activity, plus a certain contribution attributable to the energy cost of food digestion. The total number of food calories needed per day varies considerably with age, sex, diet, and activity. The following is a reasonable estimate for an average (70 kg) man:

TABLE 26-6
DAILY CALORIC EXPENDITURE

	Energy Required	
Activity	kcal/hr	kcal/day
8 hr sleep	65	520
2 hr light work	150	300
8 hr moderate work	200	1600
2 hr evening chores	170	340
4 hr sitting	100	400
	Total	3160

The relationship between caloric intake and energy expenditure is the prime factor determining whether the body gains or loses weight (mass) over a period of time. Body weight is determined ultimately by the balance between food income and energy expenditure. When the balance is negative—either in calories or in essential nutrients such as vitamins and minerals—the body feeds upon itself, attacking first its energy reserves and eventually its tissues. The consequence is weight loss. When more food is ingested than is needed, the result is weight gain. The continued deposition of fatty tissue leads to obesity, a common affliction in both civilized and primitive societies.

In planning a reducing diet for ordinary mild to moderate obesity, we can conveniently calculate the desired daily caloric deficit on the basis that 1 lb of body fat is equivalent to approximately 3800 kcal. Thus a daily deficit of 500 kcal will produce over a long period an average weight loss of almost 1 lb per week; a daily deficit of 1000 kcal an average weight loss of almost 2 lb per week, etc. It is useful to increase the daily caloric requirement by adding, for example, an hour of walking—though an hour's walk increases energy expenditure by only 350 kcal. Rigorous restriction of the caloric intake accelerates weight loss, but most people find it difficult to maintain a daily intake of less than 1500 kcal for long periods. Indeed, it may be unsafe to ingest less than 1000 kcal per day for an extended time.

Overnutrition is far more prevalent than undernutrition in developed countries of the world, despite the fact that obesity decreases life expectancy. Obesity is ultimately traceable to a long-term caloric surplus, but the reasons why individuals habitually eat too much are complicated and controversial. We spoke above of hunger and appetite. Abnormalities of these mechanisms are doubtless involved, but whether they are genetic, emotional, cultural, or neurological in origin is unclear.[7] We do know that the satisfactions of eating are deeply rooted in human emotions.

26-22 *Thermophilic bacteria thrive in hot springs*

CONTROL OF BODY TEMPERATURE

Life can exist only within a relatively narrow range of temperatures. Particular types of organisms are at their best in an even narrower range. We know that the rates of chemical reactions in living cells are strongly influenced by temperature. As shown in Chapter 8, enzymes can catalyze biochemical reactions within a narrow range of temperatures. Such physical properties of protoplasm as fluidity and elasticity are also sensitive to temperature.

In the sea, the ranges of environmental temperatures are very much narrower than those in other environments. The range of temperatures at the surface is about −2 to 40°C (about 28 to 104°F), although there are local extremes in particular places. The upper limit, especially, is highly exceptional and can occur only in shallow rock pools in very hot climates. At any one place and depth, the range is usually only a few degrees. The distribution of marine organisms is markedly affected by water temperatures. Those living in warm seas are quite different from those of colder waters. In each place, however, the temperature changes are slight. Occasional incidents such as the upwelling of cold water from the depths may kill millions of animals adapted to warmer surface waters. Yet the usual fluctuations pose no severe problems, and a capacity for temperature regulation has never evolved in most sea animals.[8] In addition, water is an excellent conductor of heat (see Chapter 3). This makes it difficult for a marine organism to maintain a temperature different from that of its surroundings.

Land is the most difficult environment for life. Here temperature ranges are extreme, not only over a continent but also in single localities through the year and even in the course of a day. Some unusual organisms—for example, the thermophilic bacteria—have made a long-term evolutionary adaptation to a particular niche by tailoring their molecules to the stresses of a highly exceptional environment (Figure 26-22). These organisms, which are often found thriving in hot springs,

[7] Any attempt to establish the extent to which obesity is genetically determined encounters obvious difficulties. For example, one cannot do breeding experiments with human beings. Nevertheless, a recent study has ingeniously shown that the weight patterns of natural children correlate closely with those of their parents, whereas the weight patterns of adopted children do not correlate with those of their foster parents. This suggests (but does not prove) that obesity is a genotypic as well as a phenotypic trait.

[8] Fresh water has more variable temperatures than sea water, but still the variation is limited. Its temperature cannot drop much below freezing or rise above the temperature of the adjacent air. Organisms in waters with relatively great seasonal variation in temperature often pass the winter in an inactive state.

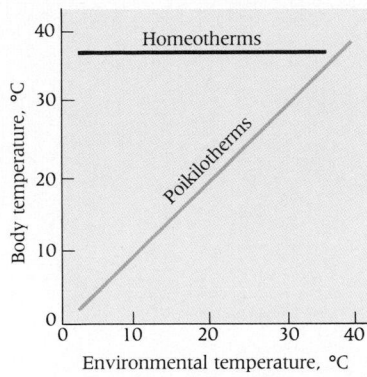

26-23 Relation between body temperature and environmental temperature in animals *Homeotherms maintain a constant body temperature in the face of changing environmental temperatures. In contrast, the internal temperature of a poikilotherm is directly related to that of its surroundings.*

can grow and reproduce at 90°C, 10°C below the boiling point! They manage this because they have transfer RNA molecules of high thermal stability. At the other extreme, certain Antarctic fishes live in ice-laden seawater at −1.9°C. They survive because they synthesize a unique glycoprotein "antifreeze."

Most animals live in less extreme conditions. In the Tropical Zone, temperatures change little from one season to another, but the daily rise and fall may exceed any seasonal change. Tropical animals, from mosquitoes to monkeys, commonly have a daily rhythm. They are active at the times of day or night when the temperatures suit them. They stay in shelter when temperatures are much higher or lower.

The most extreme seasonal changes occur in the Temperate Zone, where climates are the most intemperate. Residents of the central region of the United States often endure summer days hotter than any in the tropics and winter nights nearly as cold as those in the far north. During winter in this zone, land animals frequently either hibernate (as many mammals do) or migrate to warmer climes (as many birds do). Many animals and plants die when cold weather comes, their offspring surviving the winter as inactive, well-protected eggs or seeds. This is how insects and most leafy plants meet the problem of temperature range. Perennial plants and trees become quiescent in winter, and most shed all their leaves in the fall. Some shed all their aerial parts and grow them again from the subterranean parts when warm weather returns. In one way or another all plants suspend activity in the coldest weather.

Unlike birds and mammals, which have evolved sensitive means of controlling body temperature, two highly successful terrestrial groups—insects and flowering plants—generally lack mechanisms of temperature control.

Many insects have evolved behavioral patterns that control temperature to an extent (see Chapter 33), although no insects maintain constant body temperatures. Locusts move about so as to absorb more or less heat from the sun if they are too cool or warm. Ants move their larvae to warmer or cooler places in the nest. Honeybees keep the temperature in their hives within livable limits. If it drops to 13°C (about 55°F), the bees in the hive become very active and, by release of body heat, raise it to about 25°C (77°F). In summer they ventilate the hive with their wings.

A few flowering plants can raise the temperature of flower parts to 10–15°C above the surrounding air temperature, and can maintain this temperature for several days. Such **thermogenic plants** were first described by Lamarck in 1776. Generated heat volatilizes foul-smelling (to us) chemicals, attracting flies that can pollinate their flowers. Such heat production has a high energy cost and these plants have a correspondingly high metabolic rate. They also exhibit an interesting variation of electron transport called *cyanide-resistant respiration*. This process produces much less ATP than ordinary respiration. In its place, plants rely on the energy from NADH for heat production. Among the thermogenic plants are the skunk cabbage (*Symplocarpus foetidus*), found in swamps of northern United States and Canada, the common houseplant, *Philodendron,* and various *Arum* lilies.

Short-term changes may occur hours, days, or weeks after a chronic shift in environmental temperature. These alternatives, termed **thermal acclimation,** permit more effective body function at the new temperature. Although its mechanisms are not known, thermal acclimation occurs at many organizational levels, including changes in biochemical reaction rates and membrane lipid composition and viscosity.

Several terms are used to describe how animals react to environmental temperature changes. The unscientific terms "warm-blooded" and "cold-blooded" are often used as approximate synonyms, respectively, for the scientific terms homeothermic and poikilothermic. **Homeotherms** maintain a constant internal temperature at all times (Figure 26-23). Only mammals and birds are homeothermic. **Poikilotherms** are animals, like aquatic invertebrates and to some extent reptiles, whose body temperature varies with the environment, although it is often a little higher than that of the environment. Examples can be found of animals that occupy every bit of the range between complete homeothermy and complete poikilothermy. Interestingly, winter hibernation in homeothermic animals lowers body temperature far below its normal level—and body metabolism, heart rate, and so on are depressed. In a sense, this makes them temporarily poikilothermic. True hibernators include bats, rodents, and some other small mammals. Large animals like bears may spend the winter sleeping, but they subsist on body fat and their body temperatures decreases only slightly.

True poikilotherms do have some heat-regulating ability. A swiftly moving animal, and even some fishes, raises its body temperature above that of the surroundings, and this warming in turn helps to speed up the reactions that produce it. Reptiles often bask in the sun on cool days, and absorption of radiation can raise their temperatures above that of the air.

In regulating their body temperature, animals may operate as ectotherms or endotherms. **Ectotherms** obtain most of their heat from the environment. Their main mode of temperature regulation is behavioral: they migrate into areas where the environmental temperature is suitable.[9] In contrast, **endotherms** use metabolic means and other physiological adaptations to maintain a high and constant body temperature, even during rest.[10] Endothermy, results in a costly expenditure of great quantities of energy needed to regulate internal temperature and maintain body functions over a wide range of environmental temperatures. Practically the only terrestrial organisms fully active throughout the year in the Temperate and Frigid Zones are birds and mammals, which are endotherms.

All animals generate heat in metabolism. In ectotherms, this heat escapes into the environment. Endotherms usually have physiological arrangements for controlling the loss of heat from the body, and some are also covered with a layer of insulation. The distinction between ectothermy and endothermy is not always clearcut. Many animals fall on the spectrum between the two.

Only endotherms tend to maintain a constant temperature by entirely internal mechanisms. Variations of as much as 15°C in internal temperature may not be fatal in some birds, but under usual conditions the temperature range is kept within a degree. In humans, the body temperature is normally close to 37.0°C (98.6°F). A rise of less than 2°F may indicate serious illness.[11] The temperature of a healthy person remains nearly the same whether the air temperature is 115°F or −40°F. Obviously humans have some means of balancing heat production and loss.

As we have seen, heat is produced by oxidative processes in cells. The most important source of heat is the muscles, which produce much heat when exercised or tensed. We tend voluntarily to move about more when it is cold than when it is hot around us. Even if we do not, our muscles automatically become tense in the cold. If the tenseness is extreme, the muscles begin to quiver, and we shiver. This increases heat production.

Heat is lost by radiation and evaporation. The blood flows through a network of capillaries in the skin, where it is so near the surface that it loses heat by radiation. The more blood that flows here, the more heat is lost. The amount of flow is regulated by constriction or enlargement (dilation) of the arterioles, which changes the amount of blood in the capillaries. In cold air the arterioles constrict: less blood

---

[9] Experiments on Antarctic icefish nicely illustrate such behavior. If these interesting animals are placed in a two-chambered tank, with one chamber at 3°C (the temperature to which they had been acclimated) and one at 5°C, they are peacefully quiescent in the 3°C chamber. But in the 5°C chamber, they rapidly swim into the other chamber. This escape occurs so promptly—well before the fish's brain or internal body tissues could have changed—it appears that thermosensitivity at the body surface is great and that the animal places a high priority on remaining at the precise temperature at which it functions best.

[10] A controversy has arisen over whether the dinosaurs were endothermic or ectothermic. We will speak of it in Chapter 44.

[11] Although 98.6°F is marked as normal on clinical thermometers, this is not necessarily normal for all individuals. Perfectly healthy individuals may have somewhat lower or higher temperatures normal for them, and in the same individual, a slight temperature change need not signify illness.

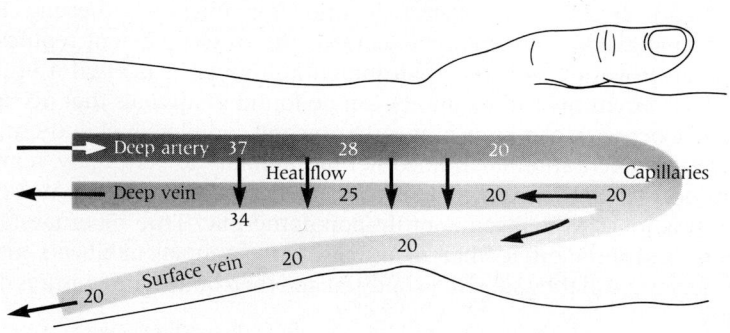

26-24 Countercurrent heat exchange in human circulation
Heat is lost in the capillaries of the hand. Blood running through the adjacent deep artery and vein exchanges heat so that arterial blood is cooled and venous blood is warmed. When heat conservation is unnecessary, most of the blood is returned to the body via a surface vein, where its temperature does not change. The nervous system controls the degree of blood vessel constriction and thus controls heat loss. The numbers are temperatures of the blood in °C.

flows near the skin, and less heat is radiated. In warm air, the arterioles dilate: more blood flows, more heat is lost. Evaporation causes heat loss and hence results in cooling of the body, as everyone knows who ever has felt a wet cloth on a fevered brow. Cooling evaporation occurs all over our skin, mostly from sweat glands. In summary, these mechanisms work as follows:

When it is cool,
  {
  skin arterioles constrict—
    and less heat is radiated.

  sweat glands are inactive—
    and less heat is lost by evaporation.

  muscles are more tense—
    and more heat is produced.
  }

When it is warm,
  {
  skin arterioles dilate—
    and more heat is radiated.

  sweat glands are active—
    and more heat is lost by evaporation.

  muscles relax—
    and less heat is produced.
  }

These diverse mechanisms are activated by impulses in nerves beyond conscious control. We cannot stop sweating when we want to. The nerve impulses involved originate in and are coordinated by a temperature-regulating center at the base of the brain (see Chapter 30).

Finally, we should note that the principle of countercurrent exchange operates to conserve heat in extremities such as arms, legs, ears, and nose. This is the same principle we found operating in the kidney in the long U-shaped loop of Henle of the nephron. As shown in Figure 26-24, heat in the warm arterial blood traveling out to an extremity flows freely to the returning blood in neighboring veins, which was cooled at the end of the extremity. As a result, blood entering an exposed limb is cooled so that it has little heat to lose; at the same time blood returning to the body is warmed.

Human beings differ in two important ways from most other homeotherms. First, few other animals have as many sweat glands. Birds have none, and dogs and other carnivores very few. In these organisms, heat loss is increased by more rapid breathing and by secretion of more fluid in the mouth. The second difference between humans and most homeotherms is that humans have little really useful natural insulation other than fat. Practically all other endotherms are covered with thick fur or feathers, highly effective insulation both for keeping heat out when the sun is hot and for keeping it in when the air is cool. A reflex raises hairs and feathers, increasing the thickness of the insulating layer in very cold air. Human beings still have this reflex, although we have lost the fur that makes it effective. In us, it just produces "goose flesh." We have made good our loss of fur, as we have many of our other deficiencies, by using our brains. In cold weather we wear clothes and burn fuel in furnaces.

SUMMARY

The ways in which organisms acquire nutrients are influenced by their mode of life, size, and degree of dependence on prefabricated organic compounds. Diffusion of nutrients across plasma membranes and phagocytosis adequately serve the needs of protists and some very small animals. Organisms are classified as autotrophs or heterotrophs depending on their degree of dependence on sources of prefabricated energy and materials. Green plants (photosynthetic autotrophs) can get along with only a few simple nutrients—small molecules like CO_2, water, and assorted minerals. The simple needs of plants and other autotrophs are met by diffusion and transport alone. Animals (heterotrophs) cannot manufacture all of their complex biomolecules and must acquire some of them ready-made, from plants taken in as food.

Some simple animals secrete digestive enzymes onto food particles external to their bodies. More complex animals have alimentary canals, within which extracellular digestion and nutrient absorption occurs. The alimentary canals of different animals differ significantly. Coelenterates and flatworms have digestive cavities with a single opening. Vertebrates have one-way canals that run the length of the body.

The accessory glands (salivary glands, pancreas, and liver) associated with the canal secrete digestive enzymes and lubricating substances. In all heterotrophs, both intracellular and extracellular digestion depend on essentially the same enzyme-catalyzed hydrolyses of ingested food molecules.

Animals are classified as bulk feeders, filter feeders, fluid feeders, or deposit feeders depending on their method of taking food into the alimentary canal. Bulk feeders generally chew their food, breaking it into smaller particles with a larger surface area, so that digestion by hydrolytic enzymes can proceed more efficiently. This process also mixes in saliva, which contains an enzyme (amylase) that begins immediately to break down starch.

In the stomach, food is acidified by HCl. This activates pepsin, a protease secreted by the stomach. The low pH also prevents bacterial overgrowth. Herbivores, such as cows and sheep, have alimentary canals that are appropriately adapted to diets rich in cellulose. Herbivores cannot make the enzyme cellulase. However, the stomachs of ruminants contain bacteria that excrete cellulase, which efficiently digests cellulose on behalf of their hosts.

In humans and other vertebrates, food leaves the stomach and enters the duodenum, where its pH is neutralized and it encounters a variety of hydrolytic enzymes secreted by the pancreas. Proteases, amylases, lipases, and nucleases are secreted as inert precursors that are activated in the alimentary canal by other proteases that cleave specific peptide bonds within the inactive precursor molecules. Bile, secreted by the liver into the gut, emulsifies dietary fat, thereby exposing it to the action of pancreatic lipase.

Food macromolecules are digested into small component molecules, which are absorbed into the body by transport across the intestinal mucosa, chiefly in the small intestine. Numerous projections called villi and microvilli greatly increase the surface area of the mucosa. Amino acids and carbohydrates are transported by specific transmembrane protein channels. Fats pass directly across the lipid bilayer of the plasma membrane and into the central lacteals which are lymphatic channels.

Peristaltic contractions move unabsorbed materials into the large intestine, which has little digestive or absorptive capacity. There much of the remaining water is reclaimed. What remains is expelled through the anus as feces.

The nutrient-rich blood leaving the capillaries of the gut enters the liver via the portal vein. This organ stores and transforms many nutrients. For example, glucose is stored as glycogen in liver cells. When blood glucose concentrations decline, this triggers breakdown of glycogen to glucose and release of this sugar into the blood. The liver also inactivates harmful substances that may have been absorbed through the gut, as well as those produced by the processes of body metabolism.

Food molecules entering the body are used either as structural building blocks or as fuel. Since much of the energy of ingested nutrients is eventually released as heat, the heat production of an organism is a measure of its overall metabolic rate. The energy yield of a foodstuff can be estimated in the laboratory by measuring the heat released during its complete combustion to CO_2, H_2O, and, if nitrogen was present, to NO_2. The kilocalorie (Cal) is a unit of heat released.

Heat released in the body by metabolism is utilized by mammals and birds, which are endotherms uniquely able to regulate internal body temperature. Many organisms (e.g., reptiles) are ectotherms that lack this ability, despite the potentially severe effects of changing temperature on enzyme reaction rates and molecular stability. They survive through a variety of biochemical and behavioral adaptations to heat and cold. Or, as in the case of many insects and plants, they do not survive, instead leaving large numbers of offspring to confront temperature extremes in the form of well-protected eggs, seeds, or spores.

To combat cold, humans increase heat generation primarily by increasing muscular activity. In hot weather, they increase heat loss by radiation from blood in skin capillaries and by evaporation of liquid from the body surface (sweat) and from the lungs. Few other endotherms sweat, relying instead on the insulation provided by fur or feathers.

KEY TERMS

abomasum	deposit feeder	microvillus
alimentary canal	digestion	nuclease
aminopeptidase	ectotherm	omnivore
amylase	endotherm	pancreas
anus	filter feeder	pellagra
assimilation	fluid feeder	pepsin
basal metabolism	gastric juice	pepsinogen
bile	gizzard	peristalsis
bile salts	herbivore	poikilotherm
bulk feeder	homeotherm	protease
calorie (cal)	hydrolase	rennin
Calorie (Cal)	jejunum	reticulum
carbohydrase	kilocalorie(kcal)	rumen
carboxypeptidase	lipase	ruminant
carnivore	liver	salivary gland
cellulase	macronutrient	suction feeding
central lacteal	maltase	trypsin
chymotrypsin	micronutrient	villus

QUESTIONS

1. _____ contains amylases, proteases, nucleases, and lipases.
   A. Saliva
   B. Peristalsis
   C. Pancreatic juice
   D. Gastric juice
   E. Chyme

2. All of the following are accessory structures of the vertebrate alimentary canal *except* for the
   A. larynx.
   B. liver.
   C. pancreas.
   D. tongue.
   E. salivary glands.

3. Absorption of nutrients and other materials occurs in the
   A. stomach.
   B. mouth.
   C. jejunum.
   D. A and C
   E. B and C

4. Each of the following is a way to cope with cold temperatures except for
   A. moving into the sunlight.
   B. vasodilation.
   C. increasing the thickness of the skin or body fat.
   D. shivering.
   E. growing a beard.

5. Plants are able to synthesize
   A. all needed amino acids.
   B. all needed carbohydrates.
   C. all needed water.
   D. all needed micronutrients.
   E. A and B

ESSAY QUESTIONS

1. Why are there no worms in sand dunes?

2. Most green plants do not feed on other organisms, but they must constantly take in raw materials for their synthetic processes. What are these raw materials and how do they enter the plants? Do plants have digestive systems? Do they perform the process of digestion?

3. How does chewing facilitate the digestion of food? What are the several functions of saliva? How do birds and other toothless animals grind up their swallowed food masses?

4. How have ruminants adapted to digest the cellulose in their diet?

5. Explain the observation that people who consume a diet sufficiently rich in calories to meet their energy demands may still be seriously malnourished.

6. What is the role of the liver in the transport, storage, and metabolism of glucose?

REFERENCES AND SUGGESTIONS FOR FURTHER READING

Beaumont, W. (1822). Experiments and observations on the gastric juice and the physiology of digestion. In: Baker, J. W. B., and Allen, G. E. (Eds.) (1970). *The Process of Biology: Primary Sources*. Addison-Wesley, Reading, Mass.

The original report of Dr. Beaumont on the gunshot wound of the stomach suffered by his patient Alexis St. Martin.

Cheung, W. Y. (1980). Calmodulin plays a pivotal role in cellular regulation. *Science 207*, 19–27.

Well written review by the discoverer of calmodulin.

Cossins, A. R., and Bowler, K. (1987). *Temperature Biology of Animals*. Chapman & Hall, New York.

This useful textbook provides a wide sampling of animal thermobiology. Engaging descriptions of such fascinating variants as the intermittent shivering of endothermic dung beetles, hawk moths, and sharks.

Davenport, H. W. (1972). Why the stomach does not digest itself. *Scientific American 226*, January, 85–93.

A stimulating discussion of gastric physiology.

Foster, D. W. (1984). From glycogen to ketones—and back. *Diabetes 33*, 1188–1199.

A lucid account of how anabolism and catabolism are sensitively controlled by the insulin: glucagon ratio.

Jennings, J. B. (1973). *Feeding, Digestion, and Assimilation in Animals*. St. Martin's Press, New York.

A comprehensive treatment of vertebrate digestion. The approach is comparative and functional.

Johnson, L. R., et al (Eds.) (1987). *Physiology of the Gastrointestinal Tract, Vol. 1 and 2*. 2nd ed. Raven, New York.

An authoritative multi-authored review of the entire field of gastrointestinal physiology.

Newsholme, E. A., and Start, C. (1973). *Regulation in Metabolism*. John Wiley & Sons, New York.

An excellent account of the control of carbohydrate and fat metabolism.

Newsholme, E. A., and Leech, T. (1983). *The Runner: Energy and Endurance*. Fitness Books, New York.

An interesting presentation of the biochemical and physiological basis of energy utilization in running.

Nordlie, R. C. (1984). Fine tuning of glucose concentrations. *Trends in Biochemical Sciences 10*, 70–75.

More on the hormonal regulation of the blood glucose level.

Ochs, R. S., Hanson, R. W., and Hall, J. (Eds.) (1985). *Metabolic Regulation*. Elsevier, New York.

A collection of interesting articles on metabolic control in both prokaryotes and eukaryotes.

Sanderson, S. L., and Wassersug, R. (1990). Suspension-feeding vertebrates. *Scientific American 262*, February, 96–101.

A stimulating survey of the diverse species that filter from large quantities of water prey or plants too small to be hunted individually.

Schmidt-Nielsen, K. (Ed.) (1984). *Scaling: Why is Animal Size So Important?* Cambridge University Press, New York.

Scaling is defined as the transitions in different animals, not only of spatial dimensions, but of surface area, volume, metabolic rate, respiratory rate, and so on. This lucid book attributes these phenomena to the physiochemical consequences of size. The most elegant correlations are seen in mammals and birds.

RESPIRATION

Respiration
Mountaineering physiologists have made many exhilarating contributions to our knowledge of respiration and gas exchange, especially those who dared Mount Everest and succeeded. In this remarkable photograph (taken on October 24, 1981 by Sherpa Young-Tenzing), we see Dr. Christopher Pizzo sampling his own alveolar gases while standing on the summit of Mount Everest—the first such measurements ever made.

Every cell in an organism's body produces energy by metabolic processes that in most cases continuously utilize oxygen (O_2) and evolve carbon dioxide (CO_2). Small quantities of these gases can cross cell membranes by diffusion. However, cells within organisms are not in direct contact with the surrounding atmosphere. For this reason, whole organisms must be equipped with specialized systems that can take up the O_2 needed by body cells and discharge the CO_2 they have produced as a waste product. In the language of physiology, this process of gas exchange is termed respiration.

THE MEANING OF RESPIRATION

Breathing air into the body is called **inspiration.** Breathing air out is **expiration.** Inspiration draws in O_2. Expiration expels CO_2. Oxygen is obviously not used in the lungs alone, any more than food is used in the stomach alone. In both cases, the organs named are receptacles for the intake of materials that are used throughout the body. Properly defined, the term respiration encompasses not only inspiration and expiration in the lungs, but also the consumption of O_2 and production of CO_2 by metabolic processes in an organism.

It is awkward that the same word has come to be used in more than one way. To avoid confusion in this book, we will restrict the term **respiration** to the processes of breathing, wherein gases cross a surface membrane and pass directly into (and out of) the blood. We will apply the term **cellular respiration** to processes of oxidative metabolism.

CELLULAR RESPIRATION

We have seen how solar energy is incorporated into simple sugars and other molecules by photosynthesis (see Chapter 10). Cellular respiration is the process that releases this bound energy. The basic strategy in most cases is oxidation of foodstuffs to CO_2 and H_2O, as discussed in biochemical terms in Chapters 8–11. It would be useful to review those pages at this point.

As we learned in Chapter 26, carbohydrates enter the metabolic cycles of the animal body as glucose from the intestine. Along with other items absorbed by the intestine, glucose is transported to the liver through a special circulatory channel, the portal vein. There it is converted into glycogen, an insoluble storage form of glucose. When needed, glucose is regenerated by hydrolysis of glycogen and distributed to all body cells in the blood. In the muscles, the body's chief consumer of glucose, it is reconverted to glycogen for temporary storage.

In the course of cellular respiration in muscles, glycogen is broken down via a now familiar pathway (Chapter 9). When glycogen reserves are exhausted, body fat is utilized as an energy reserve. Fat is broken down to fatty acids, which are highly competent fuels (see Chapter 11).

A great deal has been learned about cellular respiration from studies of what happens during severe muscular exertion. When the O_2 supply is sufficient cellular metabolism follows its usual pathway, with the carbon and hydrogen atoms of glucose becoming oxidized to CO_2 and H_2O. Released energy is trapped mainly in the form of high-energy phosphate bonds, which are passed on to ADP to form the ATP that ultimately supplies the energy consumed in muscular contraction (see Figure 26-21).

Running consumes large quantities of energy. ATP is needed continuously and is utilized at a great rate. In violent exercise, the rate of ATP consumption begins to exceed the rate at which it can be supplied. This occurs whenever oxygen is transported from lungs to muscles at a rate inadequate to meet the need. Fortunately, such limitations in the O_2 supply do not set an upper limit to the rate at which energy can be released in the muscle. Respiratory energy is released in oxidative phosphorylation by the removal of hydrogens, not by the addition of O_2. Oxygen acts as a "sponge" that "soaks up" the hydrogens stripped from sugar.

When, during the strenuous muscular exertion of, say, sprinting or marathon running, oxygen supply is insufficient to deal with the hydrogen atoms removed from sugar, some of these atoms are picked up by pyruvic acid, which is thereby converted into **lactic acid.** Lactic acid accumulates as long as heavy demands for energy continue. To an extent, the severely exercised muscle does some of its work using ATP formed by the partial metabolism of glycogen to lactic acid (rather than to CO_2 and H_2O). One result of an accumulation of lactic acid is an increase in the concentration of hydrogen ions in the blood—that is, a decrease in pH—which is a powerful stimulant of respiration. Another consequence is muscle fatigue: the runner has a cramp.

No cells carry a very large supply of free ADP and ATP. Instead, quickly withdrawable reserves of high-energy phosphate bonds are stored in animal cells in compounds called **phosphagens.** In humans and other vertebrates the main phosphagen is **creatine phosphate.** In Figure 27-1, the creatine molecule is designated simply as C, and creatine phosphate is shown as $C \sim$ Ⓟ. Of course, even when energy demands are low, aerobic muscle respiration continues, and the generated Ⓟ groups are passed, as usual, to ADP to form ATP. The ADP-ATP cycle becomes

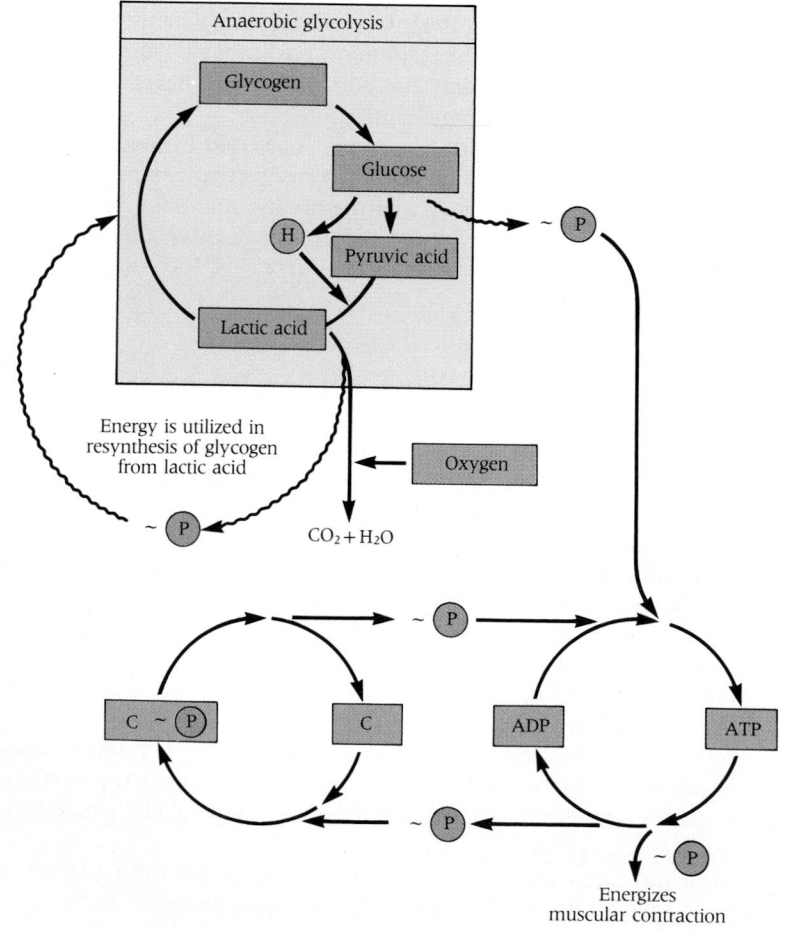

27-1 Release, storage, and utilization of energy in muscle
In anaerobic glycolysis, very little energy is derived from the breakdown of glucose to lactic acid. In aerobic respiration, energy is converted to ATP for use in muscle contraction. A reserve supply of high-energy phosphate bonds (~Ⓟ) exists in a pool of creatine phosphate molecules (C~Ⓟ), which can contribute Ⓟ to ADP to generate ATP when needed.

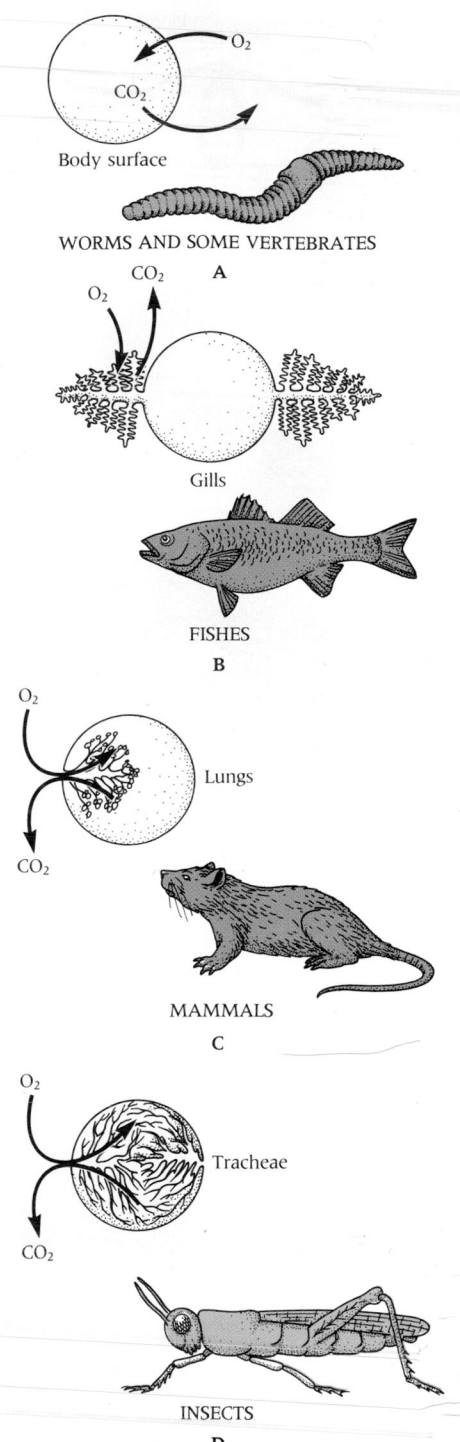

27-2 Respiratory surfaces in animals
In each of these drawings, the circle symbolizes the body surface. (A) In some animals, such as the earthworm and certain vertebrates, gases exchange across the body surface. (B) In fishes, the gills present a large surface area for exchange between water and the blood. (C) The lungs are protected within the body and present a large area for gas exchange. (D) Insects have a tracheal system that connects cells directly with the external environment.

saturated (when there is no more ADP) and ATP transfers some $\textcircled{P}$ to creatine, which is present in large amounts, for storage as $C \sim \textcircled{P}$. Later, when rapid exercise makes high-energy demands, these are met first from $C \sim \textcircled{P}$ reserves, but through the agency of the ADP-ATP cycle.

When heavy exertion stops, the animal breathes rapidly for a time—until the gas exchange system returns to normal. Panting accomplishes the following:

- Increases O_2 intake and does so as long as the ATP level is low
- Replenishes the ATP level
- Rebuilds glycogen and fat reserves
- Restores the creatine phosphate level
- Corrects the pH and restores ionic balance
- Completes the oxidation of a portion of the accumulated lactic acid, oxidizing it to CO_2 and H_2O and liberating energy as ATP
- Reconverts the remainder of the accumulated lactic acid to glycogen

Anaerobic metabolism occurs in the cells of some plants and fungi, where it may take a different course than it does in animals. In yeast cells, as in animals, anaerobic glycolysis proceeds to pyruvic acid, but this compound, instead of becoming lactic acid, is transformed into ethyl alcohol and CO_2 (see Chapter 9) in the process of **fermentation.**

Most organisms require oxygen to complete the oxidative phases of the metabolic cycle, but they can often get along without oxygen for shorter or longer periods. In humans, the period is very short, indeed, because our nervous system demands active ongoing aerobic respiration. Some animals and plants can live on anaerobic glycolysis for hours, days, or weeks, even though they do require aerobic respiration eventually. Some organisms, too, are normally anaerobic throughout their lives. This fact has medical importance, for most of these anaerobic organisms are bacteria or intestinal parasites.

RESPIRATION: THE RESPIRATORY SURFACE

In a unicellular organism, gases are readily exchanged across the outer cell membrane. The small amount of oxygen diffusing through the cell membrane is enough for the unicellular organism's needs because such small organisms have a high ratio of surface area to volume. A multicellular animal, however, cannot survive on the small amount of oxygen diffusing through its body surface because it has a relatively small surface area.

This problem was solved in the course of evolution by the development of special organs that provide large specialized respiratory surfaces. Four main kinds of respiratory surfaces are found in animals: the body surface, gills, lungs, and tracheae (Figure 27-2).

Body Surfaces in Respiration

The **body surface** provides all the oxygen needed by some animals, both aquatic and terrestrial. In most cases, this arrangement works only when specific criteria are met.

- The animal's size is small, with a high ratio of surface to volume.
- The metabolic rate is low, so that oxygen needs are minimal.
- The body surface is well ventilated.
- The body surface remains moist and protected from injury.

However, there are (as always) some surprising exceptions. Some vertebrates of considerable size can supplement or even replace the breathing they normally do through gills or lungs with skin breathing.

But most large animals require more oxygen than can be taken in by diffusion through the unmodified body surface. Moreover, many have skin that not only efficiently regulates body water content but also blocks gas diffusion. The most refined respiratory organs evolved among these animals.

Gills As Respiratory Organs

In aquatic forms the respiratory organs are usually **gills,** which arose independently in various groups and which are now highly diverse. Usually, gills are feathery outgrowths that are exposed to water and exchange gases across thin gill membranes (Figure 27-2B). Often, but not always, the gill extends outward from the body to form an appendage (Figure 27-3). In fishes, the gills are located behind the head (Figure 27-4A). Water enters through the mouth, passes from the pharynx across the gills and leaves via the opening behind the **operculum,** which covers the gills.

Most gills have in common (1) a filamentous or platelike structure, which compresses a large surface area into a small bulk, and (2) special mechanisms to pass a constant stream of water over the gill surfaces (Figure 27-4B,C). Blood circulation in fish gills is so arranged that it greatly increases the blood's efficiency in removing oxygen from water. The layout is another countercurrent exchange system—like that in the kidney—in which two fluids exchange substances with one another as they flow in opposite directions. (Figure 27-4C). This scheme continuously maintains an oxygen concentration gradient between the water and the blood.

How does it work? Water entering the gill area encounters blood that has already picked up some oxygen and is about to leave the gills. However, this blood is still able to hold more oxygen, which it picks up from the fresh, oxygen-rich water it encounters. As the water passes along, it loses more and more of its oxygen to the blood, but begins to encounter blood that is less and less saturated with oxygen. A gradient is maintained because the water always contains more oxygen than does the blood it encounters; hence the water continues to lose oxygen and the blood continues to gain oxygen. As shown in Figure 27-4D, if blood ran in the same direction as water, the blood would end up with a lower oxygen concentration than it does with countercurrent flow.

For the mechanism to work, the respiratory water current must move uninterruptedly in one direction. This occurs in fishes only so long as a constant pressure difference is maintained between the mouth and gill cavities. Throughout the respiratory cycle, pressure in the mouth cavity exceeds that in the gill cavities. Hence, water flows continuously in one direction—over the gills (Figure 27-5).

In certain mollusks and chordates, ventilation of the gills is tied to feeding. As water is drawn into the gill area, food is filtered out of it, and gas exchange occurs at the same time. Often it is difficult to determine whether the original function of the gills was respiration or feeding.[1]

Lungs and Respiration

True lungs occur in some fishes, in most adult amphibians, and in all reptiles, birds, and mammals. Human lungs are shown in Figure 27-6.

A **lung** is a respiratory organ that, unlike a gill, turns inward to form a cavity. Lungs are organs in which the bloodstream is brought in close contact with a continuously renewed supply of oxygen across a thin membrane of extremely large area (Figure 27-2C).

Actual gas exchange takes place in the lung in tiny air sacs called **alveoli,** which are clustered like grapes around the ends of the **bronchioles** (see Figure 27-6). Each alveolus, about 1 or 2 mm in diameter, is surrounded by a network of capillaries. The walls of the alveolus and capillary are one cell thick (Figure 27-7). Thus the barrier between inspired air and blood is extremely thin—only about 0.3 μm. Gases exchange freely across this barrier by diffusion. A pair of human lungs has some 300 million alveoli and a total respiratory surface area of 70 square meters— more than 45 times the total skin area of the adult body. In the common disorder called **emphysema,** the number of alveoli, and hence the total respiratory surface, decreases, often with disastrous results.

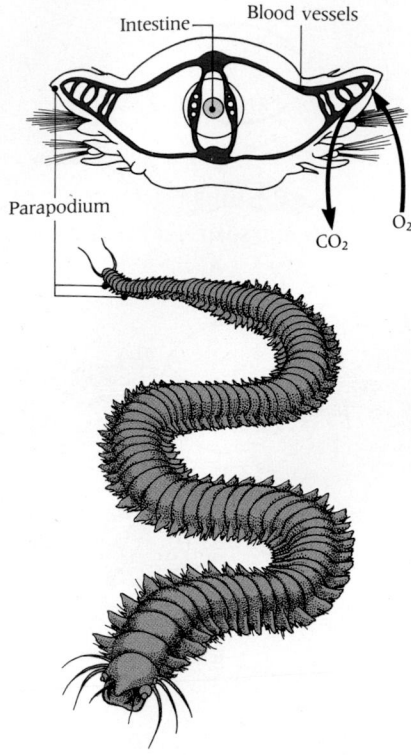

27-3 Gills of marine segmented worm
In this worm, Nereis, *each segment has two extensions called parapodia, which contain the gills. The cross section shows the extensive blood vessel network that ensures adequate exchange of O_2 and CO_2.*

---

[1] Some animals use respiratory water currents for locomotion. Water is forced out of the gill area, creating a current. If an animal does not have a way to keep itself steady, the body will inevitably move in the direction opposite from that of the outgoing current. Squid, for instance, eject water from the siphon with considerable force, creating a jet-propulsion stream that moves them rapidly forward or backwards depending on the direction in which the siphon is pointed (see Chapter 42).

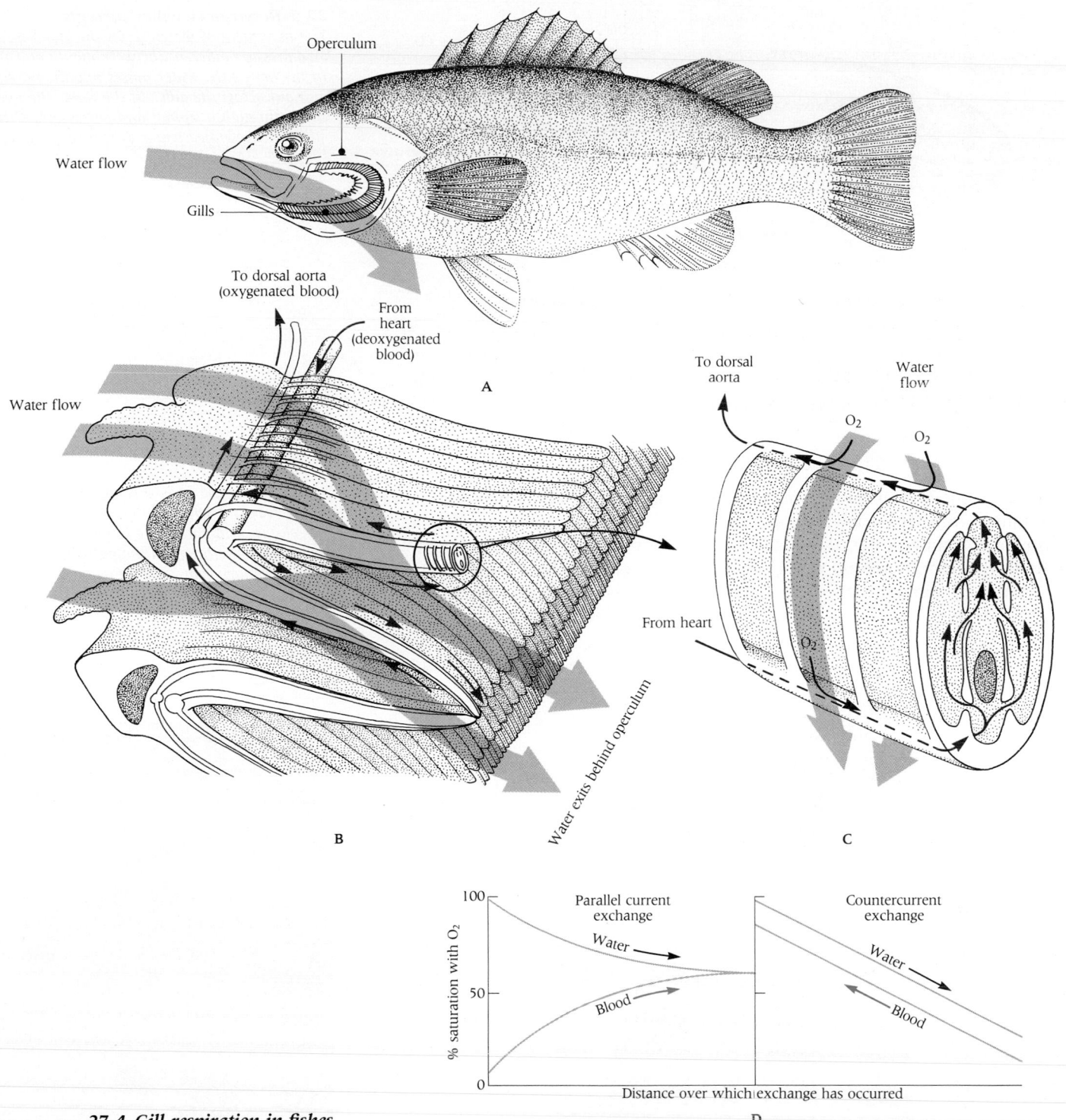

Operculum

Water flow

Gills

To dorsal aorta
(oxygenated blood)

From
heart
(deoxygenated
blood)

A

Water flow

B

To dorsal
aorta

Water
flow

O_2

O_2

From heart

O_2

Water exits behind operculum

C

Parallel current
exchange

Countercurrent
exchange

Water

Blood

Water

Blood

% saturation with O_2

Distance over which exchange has occurred

D

27-4 Gill respiration in fishes

(A) *Oxygen-rich water enters through the mouth, passes across the gills, and exits behind the operculum (within the dashed line).* **(B)** *In this enlarged view of the gill, the direction of water flow and blood flow is shown by arrows. Gas exchange takes place at sites called lamellae (circled).* **(C)** *This enlarged view of a lamella shows how gases are exchanged between blood and water, which flow in opposite directions. The countercurrent exchange principle causes blood to take up O_2 and give up CO_2.* **(D)** *In parallel current exchange, the oxygen gradient (color) between water and blood is reduced with exchanger length, so that less and less oxygen enters the blood. In countercurrent exchange, an oxygen gradient is maintained and up to 80% of the O_2 in water may enter the blood.*

Successful gas exchange in the lungs depends on the frequent periodic inflow and outflow of gases. **Ventilation** is the exposure of the respiratory surface to air. **Breathing** is the involuntary process of respiratory movements that ventilates the respiratory surface with air.

The basic functions of the lungs, collectively termed **pulmonary gas exchange,** may be divided into three separate processes (Figure 27-8).

- **Ventilation** is the flow of fresh air into the lungs and its distribution to all 300 million alveoli.
- **Diffusion** is responsible for the transfer of gases across the alveolar-capillary membranes.

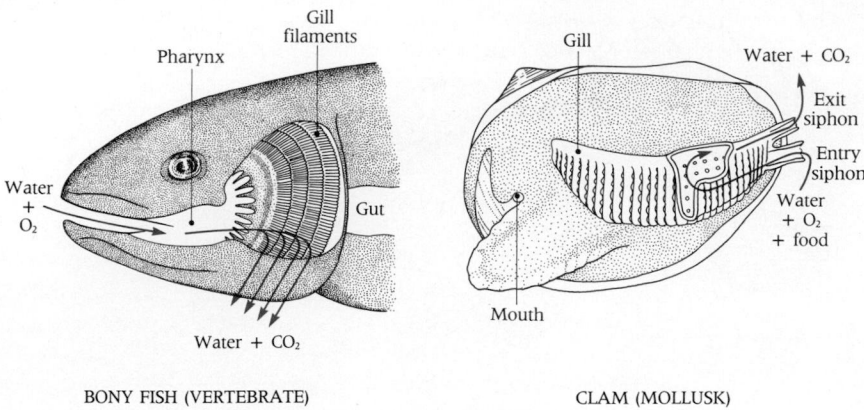

Pharynx
Gill filaments
Water + O₂
Water + CO₂
Gut
BONY FISH (VERTEBRATE)

Gill
Water + CO₂
Exit siphon
Entry siphon
Water + O₂ + food
Mouth
CLAM (MOLLUSK)

27-5 Respiratory water currents
The movement of water across the gills depends on a pressure difference between mouth and gills. In the bony fish, water passes into the mouth and out across the gills. In the clam, the water enters through a siphon and passes across the gills. Deoxygenated water exits through another siphon.

■ The **pulmonary circulation** contributes the flow of blood essential for the transport of gases to and from the lungs.

Tracheae and Insect Respiration

Finally, a unique system for air breathing has evolved in the other major group of land animals, the terrestrial arthropods—insects, centipedes, and many spiders. As shown in Figure 27-2D, they have a system of branched internal tubes called **tracheae** (singular, **trachea**) that extend throughout the body (Figure 27-9) and open to the air through paired openings in the body wall called **spiracles.** Air

27-6 Human respiratory system (Below left)
Air taken into the body through the mouth or nose traverses a series of channels to reach the sites of gas exchange in the lungs. Deoxygenated blood enters the lung via the pulmonary artery. The pulmonary vein carries oxygenated blood back to the heart.

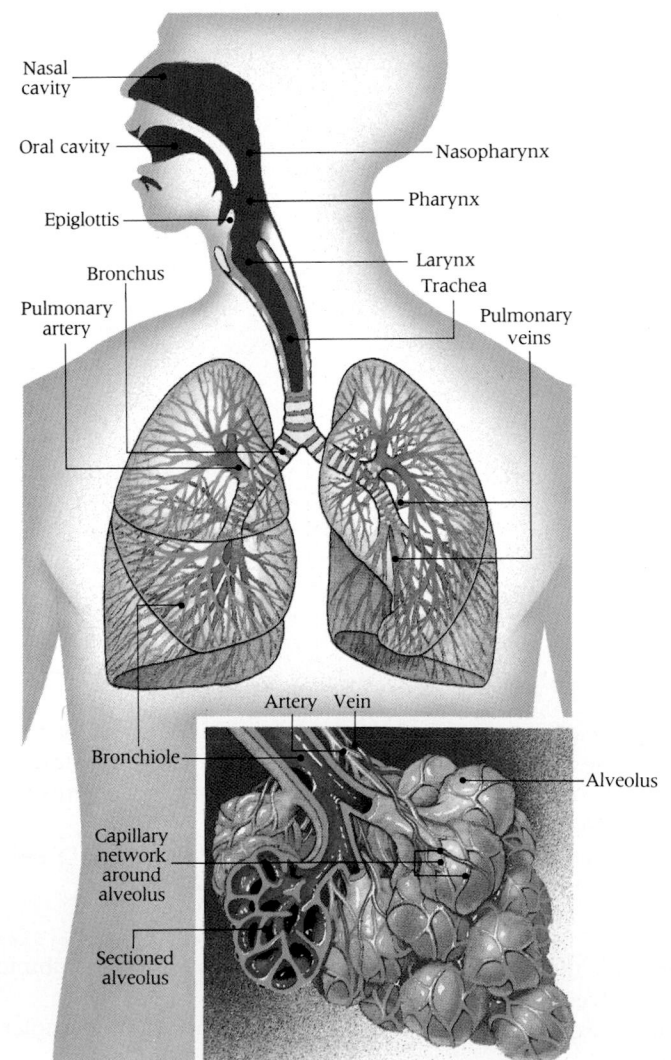

Nasal cavity
Oral cavity
Epiglottis
Bronchus
Pulmonary artery
Nasopharynx
Pharynx
Larynx
Trachea
Pulmonary veins
Artery Vein
Bronchiole
Alveolus
Capillary network around alveolus
Sectioned alveolus

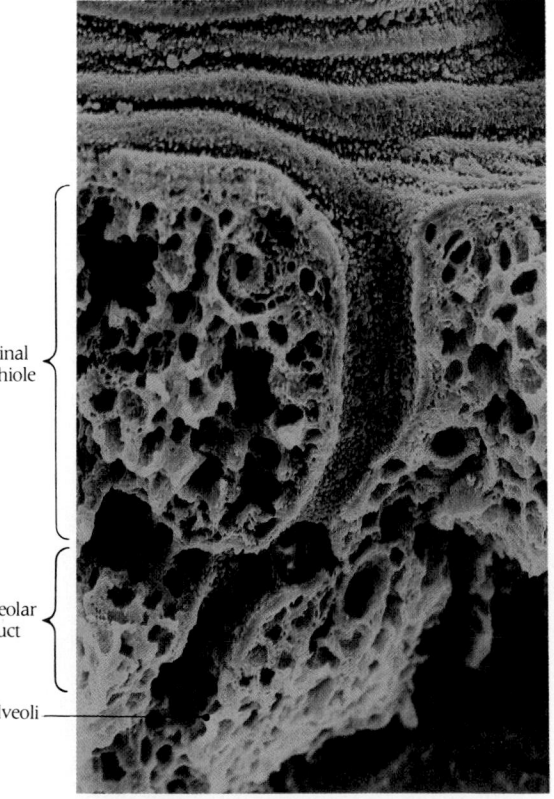

Terminal bronchiole
Alveolar duct
Alveoli

27-7 Structure of alveoli
Bronchioles lead to the alveoli, which are richly supplied with blood vessels. The large alveolar surface area permits efficient exchange of gases. The lacework of alveoli is connected to a narrow duct that leads to a bronchiole. This cross section shows how thin the alveolar walls are. (×30)

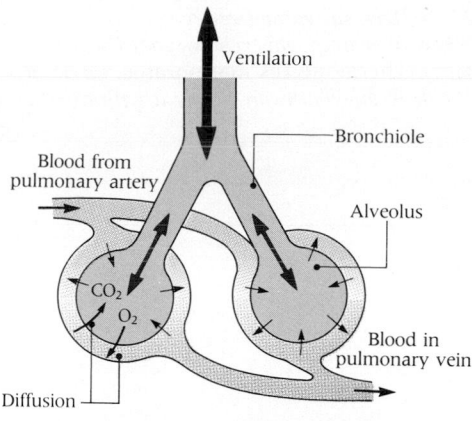

27-8 Relations between alveoli and pulmonary circulation

The alveoli are surrounded by capillaries that are part of the pulmonary circulation. Air enters the alveoli via the bronchioles, and gases diffuse across the alveolar membrane according to their concentration gradients, O_2 entering the blood and CO_2 leaving it. Exhaled air is thus higher in CO_2 and lower in O_2 than is inhaled air. Oxygenated blood returns to the heart via the pulmonary vein. Note that mixed venous blood from the systemic circulation is carried by the pulmonary artery. Blood in the pulmonary vein is destined to become systemic arterial blood.

27-9 Tracheal respiration in a grasshopper
(A) The main tracheal trunks are connected to the exterior by spiracles. Air sacs serve as reservoirs. (B) The tracheoles are highly branched and each terminus lies near a target cell. The termini may contain fluid that decreases in volume as O_2 requirements of the cell increase. This allows more air to contact the cell. (C) Nearly closed grasshopper spiracle. (D) Fully open ant spiracle.

flows into the tracheae, which branch repeatedly out to tiny terminal extensions about 1 μm in diameter called **tracheoles** and deliver air to all of the body's cells. Diffusion accounts for the actual exchange of gases.

In the course of evolution, insects remained small for several reasons. One was the inability of the tracheal system, which depends largely on diffusion, to supply oxygen to active cells located more than a few millimeters away from the external air. Unlike the situation in animals with gills or lungs, where it is the interstitial fluid that exchanges O_2 and CO_2 directly with the cells, insects do not employ body fluids to facilitate gas exchange. Instead, O_2-CO_2 exchange occurs directly by diffusion between air in the tracheae and the cells of the body. However, diffusion is a slow process. It is supplemented, especially in large insects and particularly during flight, by rhythmic pumping movements of the abdomen and synchronized openings and closings of the spiracles—apertures in the body wall—and effectively closed by valves.

An insect with an especially interesting respiratory mechanism is the diving water beetle, *Thermonectus marmoratus*. Whenever it dips beneath the surface of the water this creature carries an oxygen supply with it in an air bubble trapped beneath its wings (Figure 27-10). As oxygen from the bubble enters the spiracles, more oxygen diffuses into it from the water, replenishing the O_2 in the bubble.

VENTILATION

Lungs themselves have no muscles. Instead (in mammals) their filling and emptying cycles depend on a bellowslike motion of the chest wall (Figure 27-11). Lung expansion and contraction, then, result from movements of the **ribs,** which form a rigid but movable cage around the lungs, the **diaphragm,** the muscular partition separating the chest cavity from the abdominal cavity, and the other **respiratory muscles.**

The human respiratory apparatus has three components

- The **air passages,** which inspired air traverses on its way to the lungs, and which include the **nose, pharynx, larynx, trachea,** and **bronchi**
- The **lungs,** which include the **bronchioles, alveoli** and the veins, arteries, and capillaries of the **pulmonary circulation**
- The **skeletal and muscular respiratory apparatus,** which includes the **ribs, diaphragm, intercostal muscles,** and other muscles

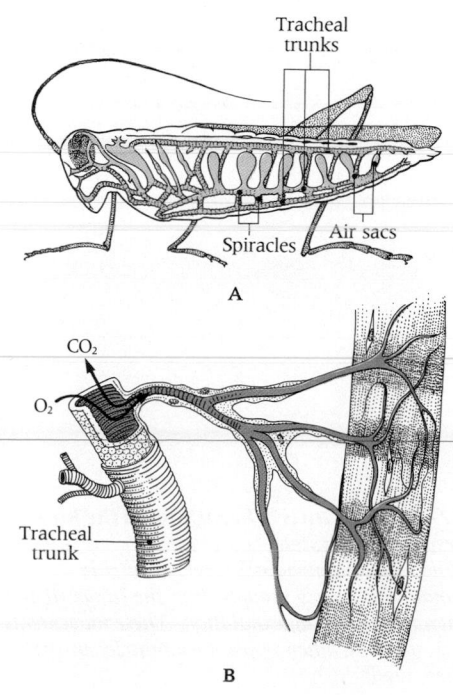

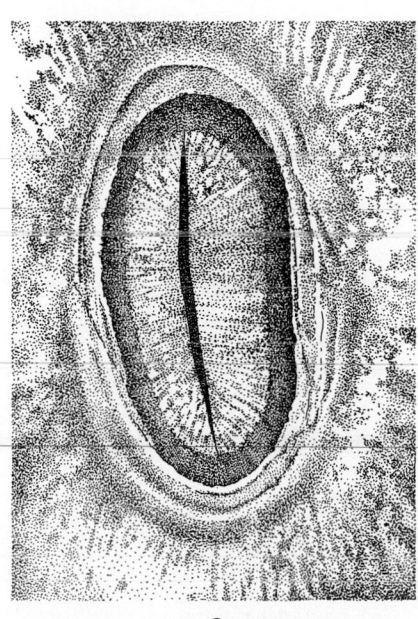

27-10 Diving water beetle
When submerged, this yellow-spotted water beetle, Thermonectus marmoratus, carries a bubble of air which can supply it with oxygen.

Humans draw in fresh air and expel impure air 16 to 20 times per minute.[2] Breathing can be illustrated by a model involving one balloon within another (Figure 27-12). The lungs are the inner balloon. The chest wall is the larger outer balloon. The narrow space between the balloons, the **pleural space,** is filled with fluid. The opening of the smaller inner balloon at the top corresponds with the trachea, which is in free communication with the atmosphere. In a breathing cycle, the following sequence of events takes place:

[2] It is not surprising to find that respiratory rates increase greatly with muscular activity in animals. A butterfly's respiratory rate is more than 150 times higher during flight than during rest. The rate in a running mouse is eight times that of a resting mouse. In humans and many other animals, the rate tends to be higher in larger individuals, in younger animals, and in males.

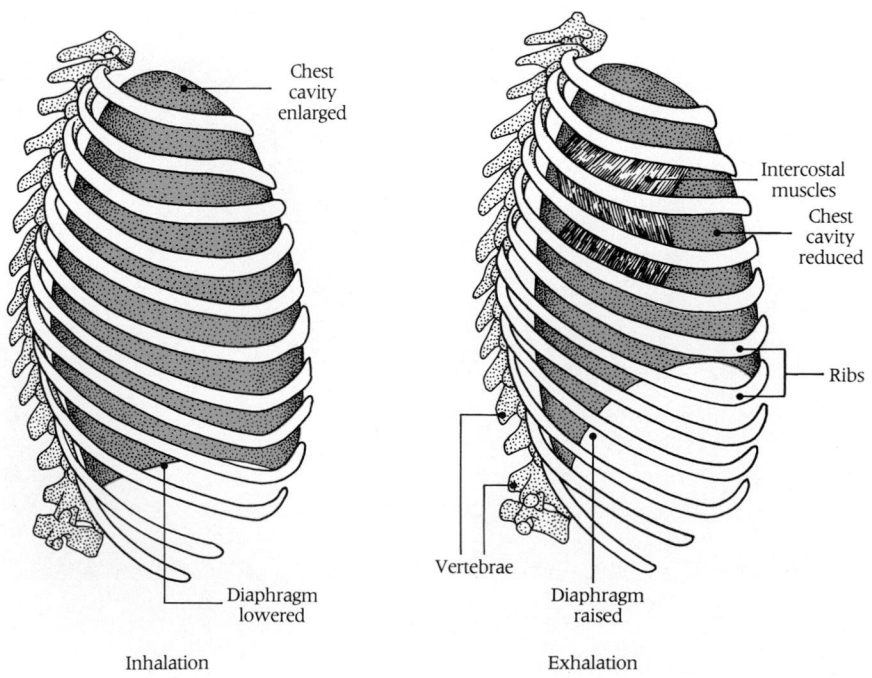

Chest
cavity
enlarged

Intercostal
muscles

Chest
cavity
reduced

Ribs

Vertebrae

Diaphragm
lowered

Diaphragm
raised

Inhalation

Exhalation

27-11 Mechanics of breathing in the human respiratory system
When the diaphragm is lowered during inhalation, air is brought into the lungs. It is then raised during exhalation. These movements are aided by other respiratory muscles in the chest wall.

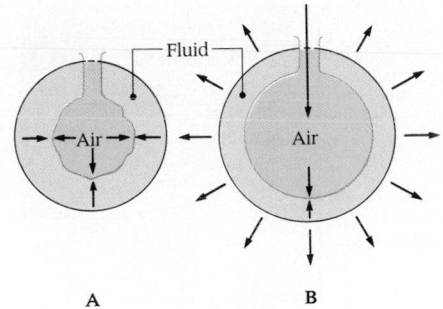

27-12 Balloon model of respiration
Fluid between two balloons causes the inner balloon to expand whenever the outer one does. **(A)** *Inner balloon is unstretched. Air pressure within it equals the pressure of the surrounding fluid.* **(B)** *The outer balloon is expanded by an outside force. Expansion of fluid space causes fluid pressure to be lower than air pressure within inner balloon. Hence, inner balloon expands. The resulting decrease in its internal air pressure draws in outside air.*

- At the start of a cycle, equal forces act upon the wall of the inner balloon. The pressure of the air within the balloon and the pressure of the fluid surrounding it are equal.
- During inspiration, the outer balloon (chest cavity) enlarges (see Figure 27-11). This sharply decreases intrapleural fluid pressure, since fluid is indistensible—that is, any change in volume of a fluid-filled space causes a corresponding change of fluid pressure within the space. Thus expansion of the outer balloon (chest cavity) causes the pressure in the pleural fluid to be lower than that in the inner balloon (lung). This difference expands the inner balloon and causes its internal air pressure to fall below atmospheric pressure. Atmospheric air is drawn in and the pressure difference is neutralized. Since pressure in the pleural space remains lower than that of the inner balloon, the latter expands until the elasticity of its wall balances the forces caused by the pressure differences across the wall of the inner balloon.
- In expiration, these changes occur in reverse.

In summary, breathing consists of periodic expansions and contractions of the chest cavity brought about by intermittent contractions of the respiratory muscles and passive recoils of the elastic lungs—except during forced exhalation, as in strenuous exercise. The advantage of the aspiration pump is that large volumes of air can be moved with minimal muscular effort and relative independence from the size of the mouth or nostrils.

In mammals, the subatmospheric pressure of the intrapleural fluid is maintained throughout life. If the chest wall is ever pierced—by a stab wound, for example—atmospheric air rushes into the pleural space, eliminating the pressure difference across the wall of the lung. This causes that lung to collapse. The condition is known as a **pneumothorax.**

This event, which often results from the rupture of a blisterlike bleb on the lung surface, is always potentially dangerous. In the fortunate victim, the huge volume of air in the pleural space is slowly reabsorbed as the "leak-holes" heal and normal pressure levels are restored. In the unfortunate victim, a situation may arise in which air is forced into the empty chest cavity and high positive pressure develops. Under these conditions, the uninvolved lung cannot expand. Without urgent treatment (in which a tube is inserted in the chest and negative intrathoracic pressure is reinstated), death soon follows.

REGULATION OF RESPIRATION

Movements of the respiratory muscles are exquisitely controlled by the nervous system. The rhythm of breathing is maintained, not in the lungs nor in the muscles of ribs and diaphragm, but in a region of the brain called the **respiratory center.** The center is located in the **medulla oblongata,** which we will learn in Chapter 30 is the lower part of the brain at the top of the spinal cord.

The respiratory center in the central nervous system has connections to many other parts of the body. Nerve impulses from these regions can modify the rate of breathing and coordinate it with other bodily activities. When a person purposely holds his breath, an impulse travels from the brain and stops the rhythmic impulses from the respiratory center to the muscles of breathing. Breathing also stops when a person swallows or quickly catches his breath—as when suddenly smelling irritating fumes.

The main factors influencing the respiratory center are as follows:

- When the level of CO_2 in the blood rises even slightly, the center sends signals to the respiratory muscles of the diaphragm and ribs that cause deeper and faster respiration. As a result, CO_2 is expired more rapidly and the blood CO_2 level falls.
- Falling O_2 levels and rising CO_2 levels in the blood also stimulate certain chemoreceptor cells in the aorta and carotid arteries (see Chapter 23), causing them to signal the brain's respiratory center.[3]
- Finally, the respiratory center is stimulated by rising H^+ levels (decreasing pH)

[3] We will further discuss chemoreceptor cells in Chapter 30.

in blood. This occurs in a variety of circumstances, one of which is severe muscular exercise (Figure 27-13). This condition, **acidosis**, is a major cause of increased rate of respiration.

Hyperventilation is a common occurrence in anxious or emotional states in which the ventilation rate is in excess of need. Too much CO_2 is expired. As a result, the pH rises in arterial blood **(alkalosis).** Since a major buffer system of plasma depends on the ratio of bicarbonate ion (HCO_3^-) to carbonic acid (H_2CO_3), and since H_2CO_3 is readily converted to CO_2 and H_2O, any excessive "blowing off" of CO_2 by a hyperventilator will hoist this ratio by decreasing its denominator. According to the Henderson-Hasselbach equation (see Box 3-2, Equation 1), that must raise the pH.

An interesting consequence of a rising plasma pH is a lowering of the blood concentration of free ionic calcium (as opposed to protein-bound calcium). As a result, there is increased nerve excitability and twitching or spasm of muscles—often in the hands and face—and feelings of numbness or tingling. Grandmother's advice was correct in this situation: a hyperventilating person can correct the disorder by breathing in and out of a closed paper bag.

Failure of respiration is one of the most dangerous catastrophes that can confront an individual. When this occurs, **artificial respiration** or **cardiopulmonary resuscitation (CPR)** is necessary.

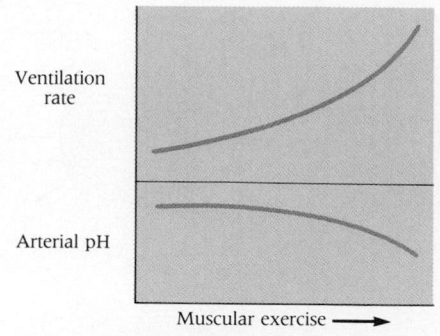

27-13 Ventilation rate and blood pH during exercise
As muscles contract repeatedly during exercise, the ventilation rate rises to enhance gas exchange in the muscle. During exercise CO_2 levels in the blood rise. This leads to an increase in the concentration of H^+ and thus a decline in arterial pH.

GAS EXCHANGE

The pulmonary circulation is a major factor in controlling the supply of O_2 to tissues, just as ventilation is a major factor in regulating CO_2 elimination from tissues. How does blood convey these vital gases?

GAS TRANSPORT IN THE BLOOD

Blood carries O_2 in two forms.

- In physical solution in plasma
- In chemical combination with hemoglobin in the red cells

Oxygen diffusing into the capillaries surrounding the pulmonary alveoli dissolves in the blood plasma, where it remains chemically uncombined. Most of it, however, diffuses farther into the red cells, where it combines with **hemoglobin** to form **oxyhemoglobin.** The presence of hemoglobin enormously enhances the capacity of blood to transport oxygen. Only about 0.3 ml of O_2 can be dissolved in 100 ml of plasma. In contrast, about 20 ml of O_2 is carried by the hemoglobin in 100 ml of blood.

Blood moves from the lungs to the heart, which vigorously pumps blood to the capillaries that perfuse the tissues of the body. Here oxyhemoglobin loses some of its O_2, which diffuses from the blood into tissue cells that have less O_2 than the blood.

Blood also transports CO_2 from the tissues to the lungs. CO_2 diffuses from the tissues where it is formed into the blood arriving from the lungs because this blood is low in CO_2. Blood transports CO_2 in three forms.

- A small amount of CO_2 is dissolved in plasma.
- About 25% of the transported CO_2 forms **carbaminohemoglobin,** a chemical derivative of the amino groups of hemoglobin.
- The balance is in the form of bicarbonate ions (HCO_3^-).

Bicarbonate is formed in two stages. First, CO_2 combines with water to form carbonic acid, H_2CO_3.

$$CO_2 + H_2O \underset{\text{carbonic anhydrase}}{\rightleftharpoons} H_2CO_3 \rightleftharpoons HCO_3^- + H^+ \qquad (27.1)$$

27-14 Dissociation curves of O_2 and CO_2
The dissociation curve of oxyhemoglobin is S-shaped. In the physiological range, small changes in pO_2 can produce large changes in total blood O_2 in the steep region of the curve. In contrast, the CO_2 curve is not S-shaped in the physiological range. Hence, small changes in pCO_2 produce small changes in total blood CO_2.

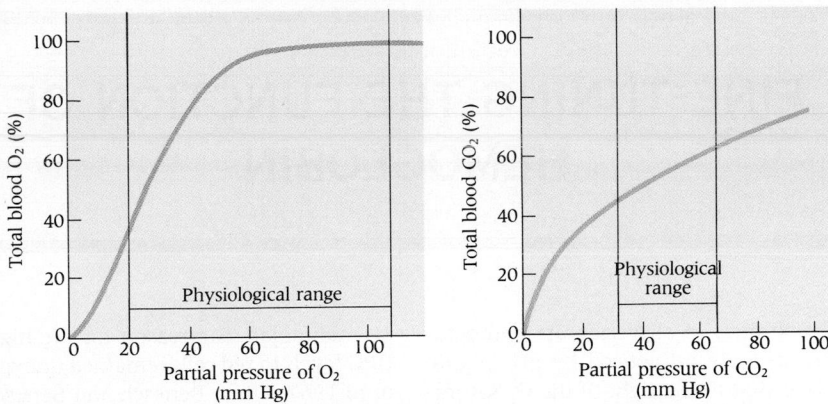

This reaction is catalyzed by a critically important red cell enzyme called **carbonic anhydrase.** H_2CO_3, a weak acid, dissociates to form HCO_3^- and H^+. As more CO_2 enters the blood, the hydrogen ion concentration rises and the pH falls. As discussed in Box 27-1, this in turn causes hemoglobin to release O_2 more readily (see Chapter 22). When blood reaches the lungs, CO_2 diffuses into the alveoli, because the CO_2 concentration in blood from the tissues is higher than it is in the alveoli. CO_2 then leaves the lungs in the expired air.

SIGNIFICANCE OF THE GAS CURVES

In discussing the curious sigmoidal (S-shaped) character of the O_2 dissociation curve in Chapter 22, we noted that oxygenation of hemoglobin is facilitated when the local pressure of O_2 is high. Compare the O_2 dissociation curve with the CO_2 dissociation curve shown in Figure 27-14. The differences are of crucial importance.

Within the range of CO_2 pressures occurring in the body—the zone marked *physiological range*—the CO_2 dissociation curve is nearly a straight line. This means that changes in CO_2 pressure produce nearly proportional changes in total blood CO_2. In contrast, the O_2 dissociation curve has a steep portion and a less steeply sloped portion in the physiological range. Small changes in O_2 pressure produce large changes in total blood O_2 on the steep portion, and large changes in O_2 pressure produce small changes in total blood O_2 on the flatter portion. The symbol **pO_2** denotes the local **partial pressure** due to O_2 alone. It is measured in **torr** units.[4] The flatter portion of the O_2 curve (above a pO_2 of 70 mm of mercury, or 70 torr) ensures a relatively constant arterial blood O_2 content. Hence, a human being can live in the thin air of high altitudes without much reduction in the O_2 carried by his or her blood.

The steep portion of the O_2 curve ensures ready delivery of O_2 to the tissues. The pO_2 of a metabolically active tissue is in this range. Hence oxyhemoglobin in the capillary blood serving such tissues surrenders much of its O_2, the tissues receiving a large shipment of O_2 despite the relatively modest decrease in pO_2.

With rising acidity, the O_2 dissociation curve shifts to the right, since oxyhemoglobin is sensitive to pH changes in the blood. In exercise, when CO_2 release is rapid, the resulting local acidity causes oxyhemoglobin to release O_2 to the tissues. Another influence on the dissociation of O_2 from hemoglobin is 2,3-diphosphoglycerate or 2,3-DPG. The story of this remarkable molecule, and its several relatives, is told in Box 27-1.

Typical normal values for gas pressures in arterial and venous blood are summarized in Table 27-1. Note that the pO_2 of venous blood is 62 torr lower than that of arterial blood and its pCO_2 is 6 torr higher. The shapes of the curves in Figure 27-14 account for the large decrease in total gas pressure as arterial blood becomes venous blood.

---

[4] Partial pressure is generally equivalent to local concentration. We will follow the now accepted practice of expressing pressure in torr units, named after Toricelli, coinventor with Galileo of the barometer in 1643. One torr is equal to the barometric pressure equivalent to 1 mm of mercury.

FINE-TUNING THE FUNCTION OF HEMOGLOBIN

BOX 27-1

The O_2 affinities of all vertebrate hemoglobins are strongly influenced by pH (Figure 27A). The shift to the right of the O_2 saturation curve when the pH decreases was first reported in 1904. The phenomenon was called the **Bohr effect** after its discoverer, Christian Bohr, father of the nuclear physicist Niels Bohr. Physiologists quickly recognized its importance: organisms bind and release O_2 and CO_2 reciprocally.

- In the lungs, O_2 is taken up as CO_2 is released.
- In the tissues, the reverse process occurs.

As we have seen, gaseous CO_2 plus water generates H^+ ions, thus lowering pH. Because of the Bohr effect, CO_2 exchange shifts the curve to the right and thus facilitates O_2 exchange.

Another important intracellular factor regulates hemoglobin function, but many years were to pass after the work of Bohr before it was recognized. In 1925 Isidore Greenwald of New York University discovered that a byproduct of the glycolytic pathway, **2,3-diphosphoglyceric acid** (abbreviated **2,3-DPG**), is present in an unusually high concentration in pig red cells. Later work showed that in most mammals, 2,3-DPG is by far the most abundant organic phosphate compound within the red cell. In some species, its concentration exceeds

10 millimolar. The reason for the high 2,3-DPG levels in red cells remained unexplained until 1967, when Benesch and Benesch discovered that this molecule is an allosteric modifier of hemoglobin function. By combining preferentially at a specific binding site with hemoglobin in (nonoxygenated) deoxyhemoglobin (see Figure 35J, Box 35-3), 2,3-DPG decreases the affinity of hemoglobin for O_2—that is, it shifts the curve to the right.

When biologists sought to determine how generally this phenomenon occurs in the animal kingdom, they found high levels of red cell 2,3-DPG only in mammals. However, other organic phosphates play a comparable role in some nonmammalian classes. In birds and some reptiles, another organic phosphate—inositol pentaphosphate—is the regulator of hemoglobin function. In fishes, amphibians, and most reptiles, it is adenosine triphosphate—ATP itself. When investigators studied a large number of mammalian species, they found high levels of 2,3-DPG in the red cells of all species except cats, sheep, cows, goats, and a few others, which contain barely measurable amounts of 2,3-DPG (Figure 27B).* Species with high 2,3-DPG levels have hemoglobins of intrinsically

* This group of exceptional species includes a portion of the order Carnivora (cats, hyenas, and civets) and a portion of the order Artiodactyla (ruminants, deer, giraffes, etc.).

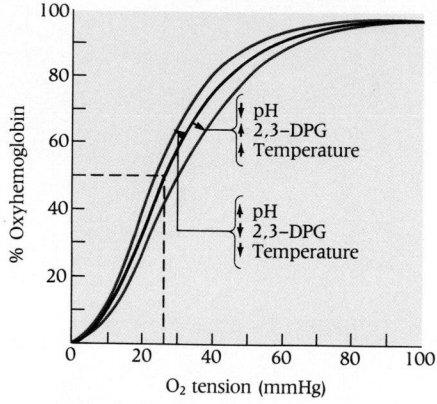

27A Factors influencing the hemoglobin-O_2 dissociation curve
Variations in arterial blood pH, temperature, and level of 2,3-diphosphoglyceric acid (DPG) in red cells all act to shift the normal hemoglobin-O_2 dissociation curve (black curve) to the left or to the right (colored curves). Upward arrows refer to increasing levels and downward arrows to decreasing levels of each factor.

high O_2 affinity, which is lowered by the addition of 2,3-DPG. In contrast, the carnivores and ruminants with low red-cell 2,3-DPG levels have hemoglobins of intrinsically low O_2 affinity that fail to interact with 2,3-DPG. Evidently, animals can get along without an intracellular hemoglobin modifier if their hemoglobins have a low enough O_2 affinity to permit optimal physiologic unloading of O_2 in tissues.

TABLE 27-1
GAS PRESSURES IN NORMAL ARTERIAL AND
VENOUS HUMAN BLOOD

Gas	Partial Pressure*	
	Arterial Blood	Venous Blood
Oxygen	100	38
Carbon dioxide	40	46
Water vapor	46	46
Nitrogen	574	574
Total	760	704

* Measured in torr

Hemoglobins of different types exist in the body. Their differences depend ultimately on variations in their amino acid sequences. As discussed in Chapter 22,

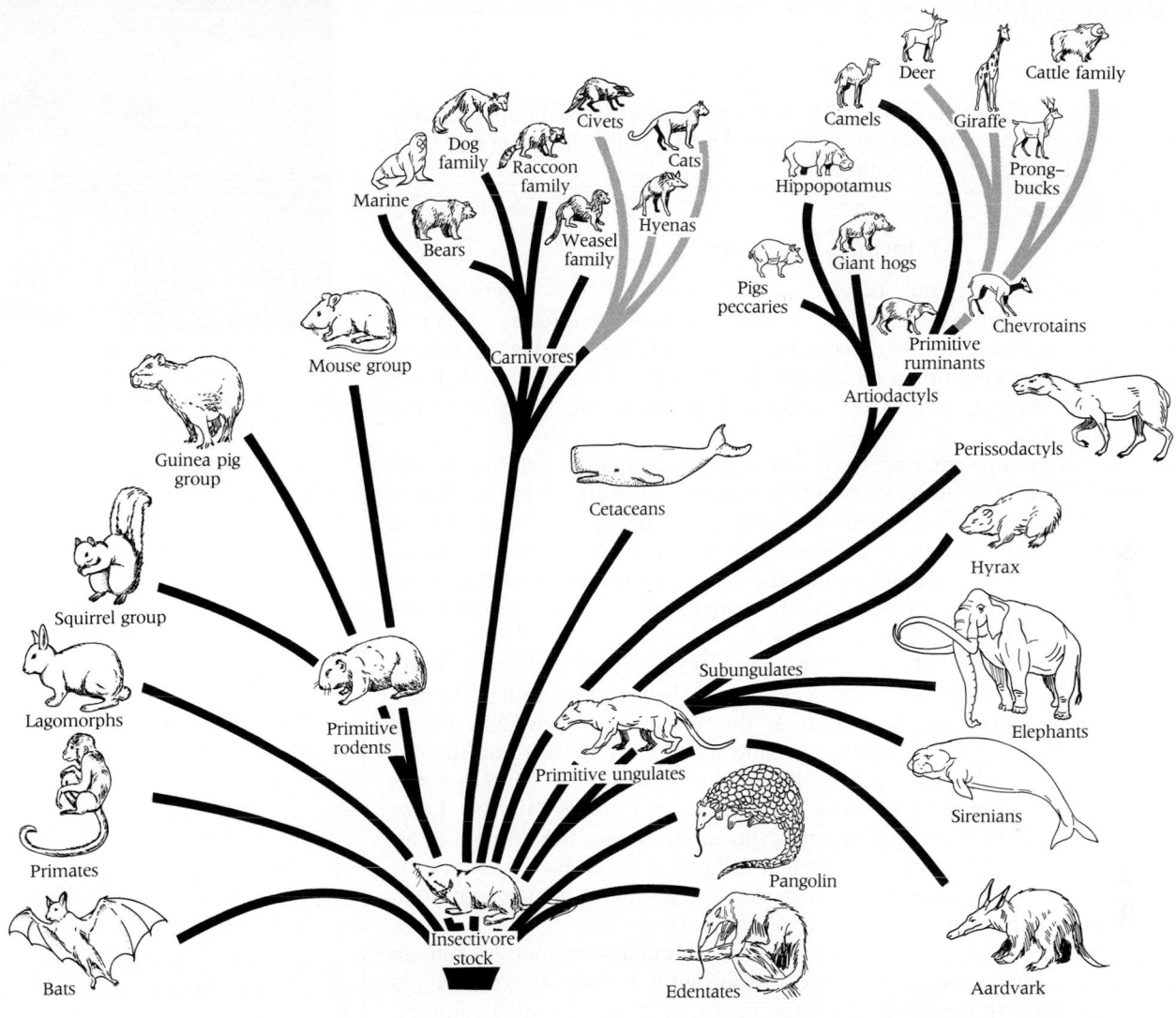

one interesting variant of adult hemoglobin, or hemoglobin A, is the hemoglobin
of the unborn fetus, hemoglobin F, or fetal hemoglobin. Its function is to extract
enough O_2 from the maternal blood percolating through the placenta to support
the rapidly growing fetus. Fetal hemoglobin has a notably higher affinity for O_2
than does maternal hemoglobin. Thus a fetus's O_2 dissociation curve is to the left
of the adult curve. The same is true of hemoglobin in the llama, which lives in
the thin air of the high Andes. In this sense, fetal life is like life on a high mountain.
During gestation, fetal hemoglobin gradually gives way to adult hemoglobin. Soon
after birth, fetal hemoglobin is totally replaced by adult hemoglobin.

The hemoglobins of various animals show many fascinating adaptations to different conditions of life. Small animals like rodents need to exchange O_2 rapidly and
their hemoglobin molecules surrender O_2 more readily than human hemoglobin—
that is, the rodent O_2 dissociation curve lies to the right of ours. Other variations
are discussed in Box 27-1.

VARIATIONS ON A THEME

A singular combination of human ingenuity, curiosity, and courage has in this century extended the frontiers of exploration to ocean depths, mountain peaks, and outer space. In each realm, major constraints have been the limits of human endurance and the capacities of men and women to survive and function in hostile and alien environments. A brief survey of the problems faced by those who go far below or far above the Earth's surface strikingly illuminates the biology of respiration.

DEEP DIVING

Interest in deep diving grew rapidly after the development of a compact self-contained underwater breathing apparatus, or scuba (Figure 27-15). Scuba divers receive their air supply from a tank they carry with them.

Principles of Diving

Air is compressible. Water is not. In Figure 27-16, we see that a weight resting upon a cylinder containing water and an air-filled balloon compresses the air in the balloon but not the water, which retains its original volume. An identical weight on an attached cylinder containing air compresses the air and restores the balloon to its original size. Pressure within the balloon again equals the pressure of the water surrounding it.

These relationships imply that pressure within the body's air-containing cavities must be raised to equal that of the surrounding water. If pressure within the lungs, sinuses, and middle ears—or even beneath faulty dental fillings—is not equalized with that of surrounding water, a relative vacuum exists. Of course, the lungs could collapse until the air in them attained the same pressure as the surrounding water. However, the frontal sinuses, which have rigid walls, cannot collapse and thus they might painfully fill with fluid or blood.

Air pressure at sea level is 760 torr or 14.7 pounds per square inch (psi). Water pressure increases by about 23 torr with every foot of depth. At a depth of 33 ft, a diver must withstand pressure twice that at the surface. As shown in Figure 27-16, when the pressure in the balloon (as controlled by the pressure on the air-containing cylinder) remains unchanged, the volume of air in the balloon varies inversely with the pressure of the water surrounding it. That is **Boyle's Law.** When the pressure on the water-containing cylinder decreases, the volume of air in the balloon increases. Therefore, a lung half inflated at a depth of 33 ft is fully inflated at the surface, and a lung fully inflated at a depth of 33 ft may rupture at the surface unless it expels air on its way up.

There is a stringent upper limit on the ability of human swimmers to remain submerged without special equipment. Expert and highly trained divers of Korea and Japan can remain underwater at depths up to 80 ft for almost 3 minutes. Their physiological adaptations include much larger lung volumes than those of nondivers and an ability to slow their heartbeat from about 100 beats per minute just before a dive to about 60 beats per minute as a dive nears its end.

These exploits seem paltry compared to those of another mammal, the Weddell seal (*Leptonychotes weddelli*) (Figure 27-17). This animal may plunge 1600 ft and stay submerged more than 70 min. How does this creature manage to do it? The answer is found in a remarkable pattern of functional adaptations to the challenges of deep diving.

27-15 Diver with SCUBA gear
The self-contained underwater breathing apparatus (SCUBA) permits divers to descend to great depths.

27-16 Demonstration of the compressibility of air and incompressibility of water
(A) *Two chambers of equal volume are set up, one with air and the other with water.*
(B) *Compressing the left chamber reduces the volume of air in the balloon without changing the water volume.* **(C)** *and* **(D)** *Varying the compression of air changes the balloon volume and increases the total volume in the left chamber.*

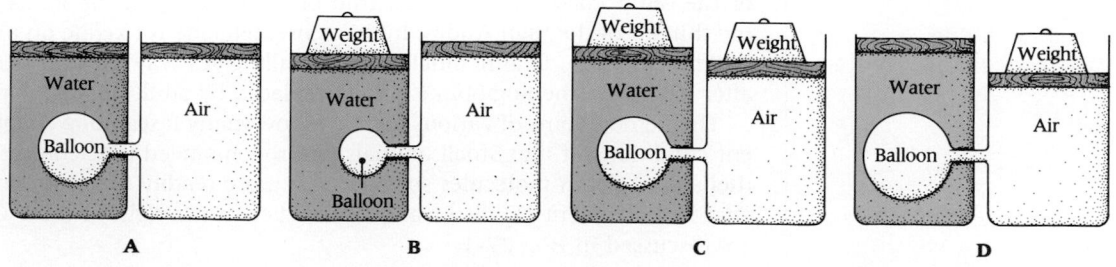

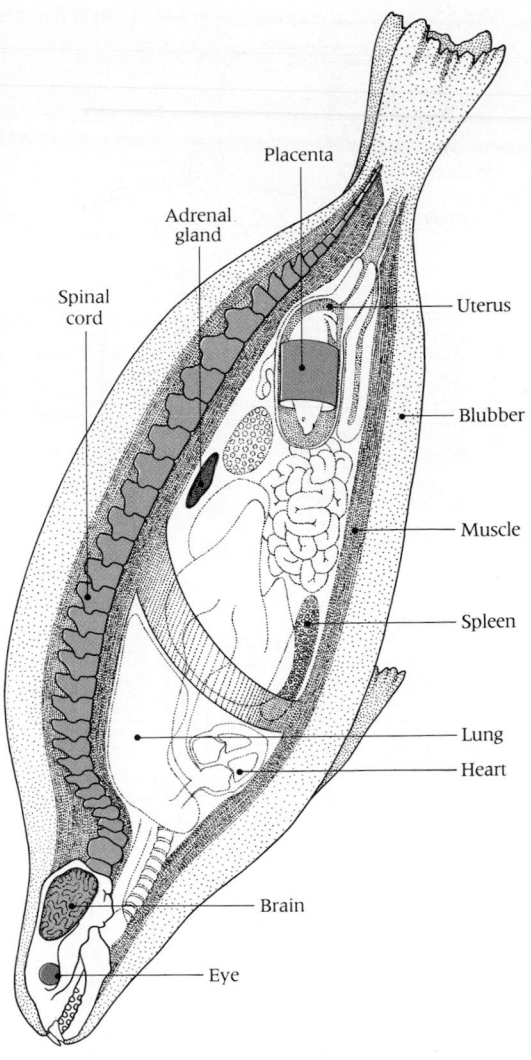

27-17 Diving adaptations in the Weddell seal

To prepare for dives of varying durations, the Weddell seal diverts O_2-rich blood to the eye, brain, spinal cord, and adrenal gland. In pregnant seals, this blood is also shunted to the placenta. This adaptation dramatically reduces the amount of O_2 reaching the muscles and other tissues. The spleen stores O_2-rich red cells and pumps them back into the blood as the seal dives.

Labels on figure: Placenta, Adrenal gland, Spinal cord, Uterus, Blubber, Muscle, Spleen, Lung, Heart, Brain, Eye

- The seal stores an abundant amount of O_2. Seal blood has 50% more red cells per milliliter than does human blood. Hence a large amount of hemoglobin is available to bind oxygen. Seal muscles contain huge amounts of **myoglobin,** the hemoglobin-like O_2-carrying pigment that delivers O_2 to muscles needing it (see Chapter 22).
- The distribution of body O_2 differs in human and seal (Figure 27-18), a seal storing twice the O_2 per kilogram of body weight as a human. Much more oxygen is in blood and muscles than in lungs.
- During a dive, arteries constrict, blood flow to abdominal organs is shut off and redirected to nervous system and eyes (the navigation system), even as heart rate and breathing slow drastically.
- The muscles switch from aerobic to anaerobic metabolism.
- The seal's large spleen collects oxygenated red cells before the dive and pumps them back into the blood when the animal submerges (Figure 27-19).

Hazards of Diving

The pressure on a diver at a depth of 50 ft is 1910 torr—two-and-one-half times the pressure at sea level (760 torr). At 100 ft it is 3060 torr! Humans can withstand pressure quite well. Indeed, some have descended to 600 ft without severe consequences. At this depth, pressure is 19 times that at the surface. Nonetheless, human divers do encounter certain hazards.

- At the bottom nitrogen narcosis and oxygen poisoning can occur.
- During ascent from deep water gases expand as pressure decreases and nitrogen escapes from solution, the condition known as "the bends."

In deep water, pressure is high; hence, gases dissolve more readily in liquids. The extra N_2 that dissolves in a diver's blood and tissue fluids produces dizziness. This is "rapture of the deep" or *nitrogen narcosis*. Its precise mechanism remains unknown.

When present in excess, O_2 itself is potentially harmful, sometimes causing convulsions and other nervous system aberrations. At sea level pressure, only about 0.3 ml of O_2 is dissolved in 100 ml of plasma. In contrast, about 6.5 ml dissolves in the same volume at a depth of 300 ft—and this is in addition to the 20 ml of O_2 in red cells per 100 ml of blood. Since the brain can utilize only 6 to 6.5 ml of O_2 per 100 ml per minute, its needs in deep water are largely met by the O_2 in plasma. That has an interesting consequence: blood leaving the brain under these conditions contains a large amount of oxyhemoglobin and an abnormally small amount of deoxyhemoglobin. Because of the Bohr effect (see Box 27-1), oxyhemoglobin has a relatively poor capacity for CO_2 transport. Therefore, plasma pCO_2 increases— by about 4 torr at a depth of 300 ft. This stimulates the brain's respiratory center and this causes hyperventilation and decreases arterial blood pCO_2 by about 6 torr. In severe cases, plasma pCO_2 may rise despite hyperventilation. The net result

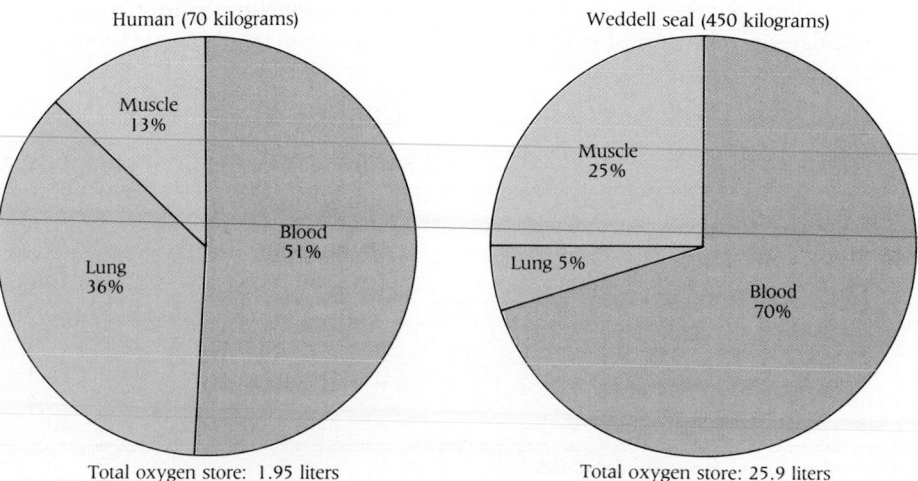

27-18 Oxygen storage and distribution in humans and Weddell seals

The amount of oxygen stored per kilogram of body weight is 0.058 liter (25.9/450) in the seal and 0.028 liter (1.95/70) in the human. Thus the seal not only stores twice the amount of O_2 per kilogram of body weight, it also concentrates this O_2 in the blood and muscles, rather than in the lungs.

Human (70 kilograms): Muscle 13%, Blood 51%, Lung 36%
Total oxygen store: 1.95 liters

Weddell seal (450 kilograms): Muscle 25%, Lung 5%, Blood 70%
Total oxygen store: 25.9 liters

is added delivery of O_2 to the brain and an enhanced tendency toward O_2 poisoning.

During ascent from the depths, the greatest hazards result from pressure reduction, or **decompression.** For example, lung air expands. A diver must exhale continuously during a rapid ascent. If he does not, the alveoli progressively distend, and some may rupture. The resulting introduction of air into the bloodstream is called an air **embolism.** Air embolism to the brain may be fatal.

The escape of gaseous N_2 during decompression, well known as "the bends," varies in severity with the depth and duration of immersion and the rapidity of ascent. The greater the depth, the greater the amount of N_2 going into solution. The longer the immersion, the greater the opportunity for it to do so. Depths to 30 ft can be tolerated indefinitely without decompression difficulties, but depths below 60 ft are safe for progressively shorter time intervals. More than 30 min at 100 ft dissolves so much N_2 that slow decompression is imperative. The greater the difference between the partial pressure of dissolved N_2 and atmospheric pressure, the larger and more numerous the N_2 bubbles escaping during decompression and the greater the chances of symptoms (Figure 27-20).

Once gas bubbles have formed, the severity of symptoms depends upon where they lodge. Pain, the chief symptom, mainly affects the joints and poorly distensible tissues such as tendons and ligaments. Symptoms appear in a few severe cases within minutes, in 85% of cases in 4 to 6 hours, and in the remaining 15% within 12 to 24 hours.

The only method of relieving decompression pain is recompression. The victim is placed in a special chamber in which air pressure can be reduced gradually so that excess gas may be eliminated slowly.

HIGH ALTITUDES

Much has been learned of the biology and physiology of respiration from studies of humans and other species living at high altitudes. An especially rich lode of information came from those who challenged the world's highest mountain peaks—and, more recently, outer space. Stories of the conquest of both domains are rich in history and high adventure.

Lessons of Mountaineering

About 25 million human beings live and work at altitudes greater than 10,000 ft above sea level. Very few live above 17,000 ft. Only a few birds, spiders, and insects live between 18,000 and 20,000 ft. What we call **altitude sickness,** the illness that overtakes lowlanders arriving at high altitudes, was first described in 1532 during the Spanish conquest of the Inca empire in Peru—but Pizarro and his men failed to attribute this often fatal malady to high altitudes so unfamiliar to Europeans. Gradually, it was recognized that air is "thin" at high altitude and in 1878 Paul Bert first proved that the low O_2 content of inspired air is the cause of altitude sickness.

In this century, investigators learned that when the body is deprived of O_2, two major adaptive responses occur.

■ The kidney secretes larger amounts of the hormone **erythropoietin,** which stimulates production of extra red cells in the bone marrow. These adjustments

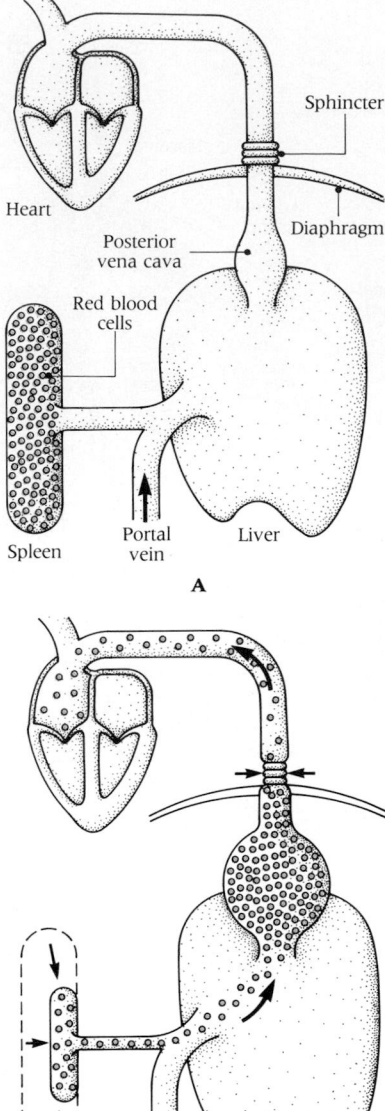

27-19 Circulatory adaptations in the Weddell seal
(A) *When the seal is breathing air at the surface, the large spleen collects O_2-rich red cells.*
(B) *When the seal dives, these red cells enter the circulation via the portal vein, liver, and posterior vena cava. A sphincter muscle near the diaphragm controls release of cells from the vena cava.*

TABLE 27-2
CHANGES IN ARTERIAL BLOOD AT VARIOUS ALTITUDES*

Altitude (feet)	Barometric Pressure (torr)	Hemoglobin (gm/100 ml)	Arterial O_2		
			Capacity (ml/100 ml)	Content (ml/100 ml)	Saturation (%)
0 (sea level)	760	15.0	20.1	19.3	96.0
4,800	650	17.1	22.2	20.9	93.8
12,000	550	19.4	25.2	22.1	87.6
14,700	420	21.1	27.5	22.3	81.0
17,400	390	23.2	30.2	23.0	76.0

* From studies of A. Hurtado in Peruvian Andes.

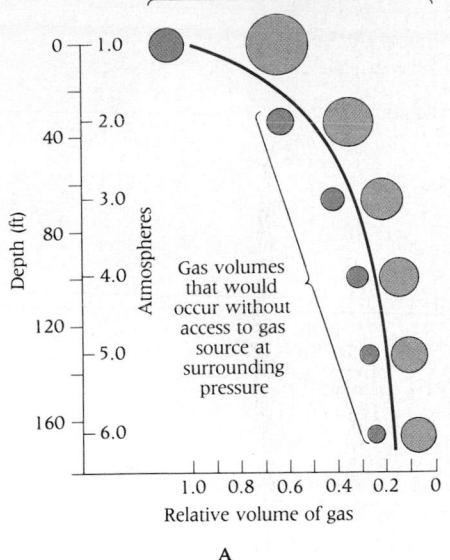

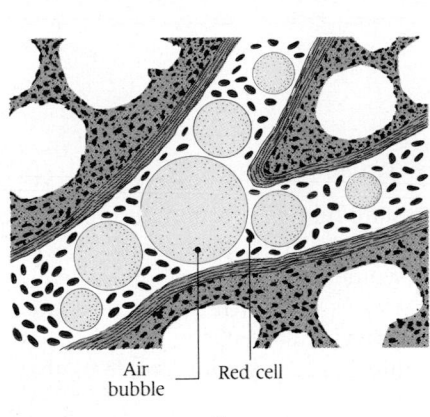

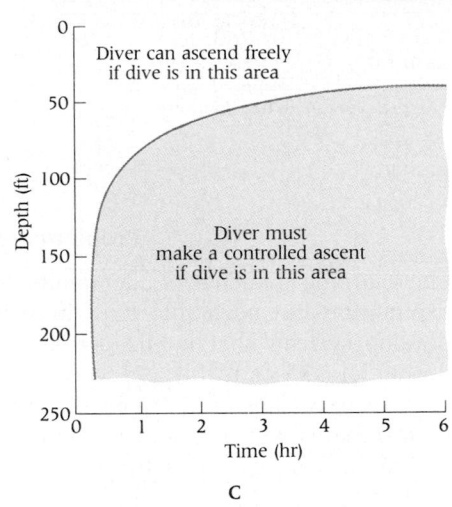

A

B

C

27-20 Physiological basis of the bends
(A) *Relative volumes of a quantity of gas as various pressures at and below the water surface.*
(B) *Gas bubbles form in the blood if decompression occurs too rapidly.* **(C)** *A "guide" for divers showing depths and durations at which ascent must be carried out.*

take time—days or weeks—and newcomers cannot function normally until they are completed. Despite the elevation of blood hemoglobin at high altitudes, there is insufficient O_2 to saturate the hemoglobin (Table 27-2).

■ Another adaptive response is a rise in the level of 2,3-DPG in red cells (see Box 27-1). This has the seemingly perverse effect of shifting the O_2 dissociation curve rightward (Figure 27-21), which raises the pO_2 level at which hemoglobin is half saturated. However, a simultaneous rise in blood pH—so-called respiratory alkalosis—causes a compensatory leftward shift of the curve. This is highly advantageous at very high altitudes because it enhances the uptake of O_2 by hemoglobin in pulmonary capillaries. Interestingly, humans that normally live at high altitudes and high-altitude dwelling animals—llamas and vicuñas, for example—have strikingly left-shifted curves. Red cell counts are high in humans, but this becomes disadvantageous beyond a certain point because it excessively increases blood viscosity.

The saga of the conquest of Mount Everest, the world's highest mountain (altitude 29,200 ft, or 8,848 m)[5], by daring mountaineers is intriguing for biologists. The first successful ascent was by Edmund Hillary and his Sherpa guide, Tenzing Norkay, on May 29, 1953, but they carried supplementary oxygen. In 1960, Hillary tried to scale the summit without supplementary O_2 after spending six months at 18,750 ft. Despite this intensive acclimatization, he failed.

In 1978, two mountaineers (Messner and Habeler) did reach the summit without supplementary O_2. It was a feat that greatly stimulated interest in the effects of extreme anoxia. A recent scientific expedition produced data that clearly show what an enormous challenge is presented at extreme altitudes (Table 27-3). These studies reveal that human beings can tolerate these conditions only by an enormous increase in ventilation, which results in an alveolar pCO_2 of 7.5 torr at the summit (compared to 40 at sea level) and an arterial pH of over 7.7 (compared to a normal value of 7.4).

These are conditions that press human physiology to its limits. Yet one recalls a report by George Lowe, who saw a perfect echelon of bar-headed geese flying far above the summit of Mount Everest. These birds start their migratory flight at sea level near the lakes of India and somehow effortlessly adapt to extremes of altitude in Tibet. How they do it is not known.

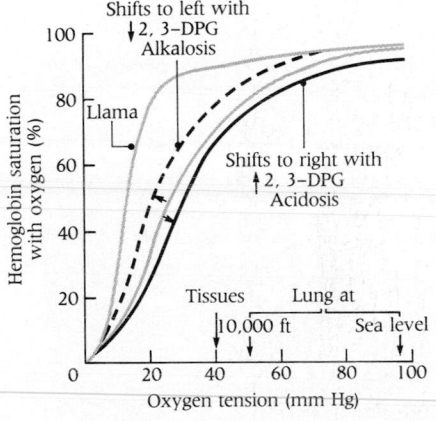

27-21 Oxygen dissociation curves of human and llama

In humans at sea level, the O_2 tension (partial pressure) in the lung is high enough to nearly saturate hemoglobin with O_2. At an altitude of 10,000 ft, hemoglobin is about 75% saturated. O_2 tension in the tissues is maintained at a certain level, regardless of altitude, by increased blood flow, or by shift of the curve to the right. Shifts in the curve can also be induced by pH changes in the blood. Hemoglobin in the llama binds O_2 more avidly at low O_2 tensions.

[5] A report published in 1987 challenges the conventional wisdom that Mount Everest in the world's highest peak. George Wallerstein, a University of Washington astronomer, made new measurements with equipment that received signals from a space satellite and concluded that another Himalayan peak, known only as K2, is higher than Mount Everest by 36 ft. The world now awaits confirmatory measurements of both peaks.

TABLE 27-3
ALVEOLAR AND ARTERIAL GAS PRESSURES ON THE SUMMIT OF MOUNT EVEREST*

Altitude	Barometric Pressure	Inspired pO$_2$	Alveolar pO$_2$	Arterial Blood		
				pO$_2$	pCO$_2$	pH
Sea level	760	149	100	95	40	7.40
Summit	253	43	35	28	7.5	> 7.70

* All pressures are in torr. From data of John B. West.

Problems of Space Travel

The daunting challenge of maintaining life and useful human function during space exploration has been met most impressively. The challenge, in a word, was to develop systems that would permit men and women (1) to work productively during long space flights, (2) to withstand the stresses of launch and reentry, and (3) to survive and function in the course of extravehicular activity, termed *EVA* by astronauts and *space walking* by laymen (Figure 27-22). Among problems that had to be dealt with were the following:

- Consequences of weightlessness, among them psychological disturbances, muscle atrophy, motion sickness, and loss of bone density
- The lack of a life-supporting atmosphere[6]
- Adverse effects of inertial or rotational forces
- Adverse effects of radiation of several kinds
- Danger of collision with small objects in space (micrometeoroids)

Our present concern is with the exchange of gases. On Earth, the atmospheric pressure is 14.7 pounds per square inch (psi), or 760 torr, at sea level. The gas mix is 20.9% O$_2$, 78.0% N$_2$, and 0.04% CO$_2$. Over much of the Earth's surface, the average temperature ranges from 22 to 27°C (72 to 81°F), an ideal temperature for a lightly clothed person. In space, the pressure approaches that of a perfect vacuum, there are no gases, and temperatures are determined solely by unshielded solar radiation or the cold blackness of space. Human survival requires a container, an encapsulated little volume of living space that provides an atmosphere at the proper pressure, gas concentration, and temperature. When astronauts step out of the spacecraft to engage in EVA, they require a spacesuit with similar capabilities.

Formidable issues faced engineers developing the needed equipment. Many failures marked the successive missions of the Mercury, Gemini, and Apollo programs—and many unforeseen problems necessitated difficult tradeoffs. Consider an example.

Concentrations of O$_2$ in spacecraft must be high enough to prevent hypoxia but not high enough to risk fire. Atmospheric gas pressure has to be sufficient to cool gas-cooled electronic equipment. It was not notably difficult to maintain cabin pressures at a comfortable 760 torr but gas pressure within the many-layered leak-proof spacesuits necessary for EVA could not be kept this high because suits containing gas at this pressure in vacuum of space, become as stiff as overstuffed sausages.

One solution was to lower spacesuit pressures to 212 torr (4.1 psi), just a bit lower than the barometric pressure at the summit of Mount Everest (see Table 27-3). This loosened up the spacesuits, but it created some other problems.

- Obviously, a more acceptable combination of cabin and suit pressures could be achieved by lowering cabin pressure or raising suit pressure. But if cabin pressure was reduced while keeping pO$_2$ constant (by increasing O$_2$ percentage), flammability increased dangerously. If cabin pressure was reduced and O$_2$ percentage was held constant, pO$_2$ was reduced excessively and hypoxia was threatened. If suit pressure was increased then suit mobility was decreased.

[6] Manned flight in near-Earth orbit usually takes place at altitudes of about 240 km (150 miles). This is below the region defined as "true space," but in a zone of negligible air resistance well beyond the Earth's atmosphere. The problem is even greater in lunar or interplanetary flight.

27-22 Space walking
Astronaut engaging in extravehicular activity (EVA). The black sky of outer space is seen in the background.

27-23 Lenticels and plant respiration
Lenticels appear as white areas on this young stem of the paper birch, Betula papyrifera. *In these regions, oxygen diffuses into the tissues below the bark.*

- With spacecraft pressures uncomfortably higher than spacesuit pressures, astronauts performing EVA risked decompression sickness (bends) on leaving the spacecraft. At first they planned to "denitrogenate" themselves before entering space by prebreathing pure O_2 for 3 hours. But this took too long and it failed too often to prevent bends.
- Upon reentry from space, astronauts have to breathe pure O_2 for 3 or 4 hours prior to suit decompression in order to equilibrate gas partial pressures in body tissues prior to cabin reentry. All of these requirements seriously reduce EVA time during a reasonable workday.

The ideal solution would be to develop a high pressure spacesuit (8.0 psi) of good mobility. Pending that development, a compromise solution in the Shuttle program was to prepare for EVA with a 12 hr exposure to a cabin pressure of 10.2 psi, a 40 min breathing session with pure O_2 and then decompression to 4.3 psi in the spacesuit. This program seems to work but others are being tested.[7]

RESPIRATORY SYSTEMS IN VARIOUS ORGANISMS

Plant Gas Exchange

In land plants, a process of diffusion occurs between cells and surrounding air. Many of these forms also develop air-filled spaces among the leaf's cells that open to the outside through stomata, which are controlled by guard cells. We considered these arrangements in Chapter 10. In a word, guard cells surrounding the stomata open these pores in response to abundant moisture, light, and low levels of CO_2 within the leaf; they close them in response to a lack of water, darkness, and high levels of internal CO_2. Since stomatal openings and closings regulate the extent to which mesophyll cells within the leaf are exposed to the atmosphere, it is appropriate to consider these changes as devices that regulate respiration. Within the leaf, of course, the actual diffusion of gases into and out of mesophyll cells takes place by diffusion through thin layers of surface moisture.

Respiratory gas exchange also occurs in structures other than leaves. An anatomical arrangement that in many plants permits O_2-CO_2 exchange in woody stems and roots consists of **lenticels,** small regions of the epidermis in which the cambium develops many intercellular spaces (Figure 27-23). These facilitate gas exchange through bark.

Other Invertebrates

The respiratory systems of a large group of invertebrates—the terrestrial arthropods—were discussed above.

Many worms exchange gases by diffusion—not only through the outer skin but also through cells lining the gut. Some worms have an enlarged, thin-walled hind gut and rhythmically take in and expel water through the anus. For them, this part of the gut is a true respiratory organ. In sea cucumbers (Figure 27-24), this form of respiration is developed further. Water, pumped in and out through the anus, circulates through highly branched respiratory trees. Animals living in water or in a moist environment often receive some of their O_2 by simple diffusion through the cells of the skin. This is still true in most frog species, even though they have gills as tadpoles and lungs as adults.

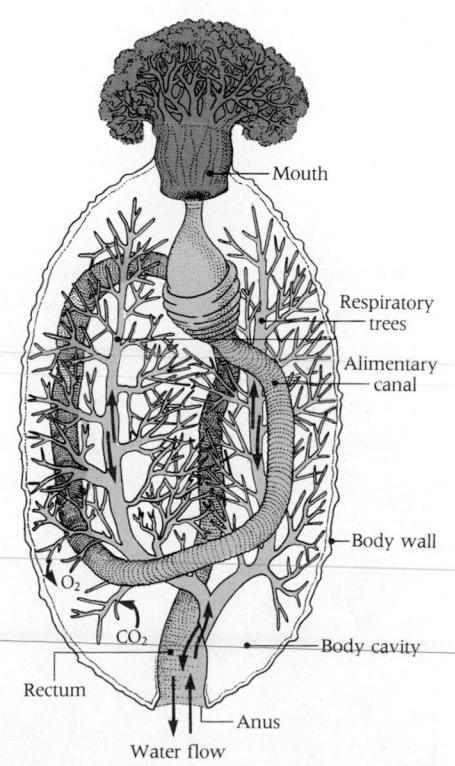

27-24 Respiration in the sea cucumber
Water enters the body through the anus and is pumped through the respiratory tree, where gas exchange takes place. It then goes back out through the anus to the exterior.

Mouth
Respiratory trees
Alimentary canal
Body wall
O_2
CO_2
Body cavity
Rectum
Anus
Water flow

[7] The spacesuit, known in space program jargon as an EMU or Extravehicular Mobility Unit, also includes an in-suit drink pack containing enough drinking water for a 7 hr EVA, a urine collection device, a communications system permitting voice communication with spacecraft and ground, a self-operated cooling unit, a visual display for various warning signals, and multiple visors to protect the crewmember from glare and ultraviolet radiation while permitting unobstructed vision out of the spacesuit. Finally, there is an emergency pack providing a 30 min O_2 supply, CO_2 washout, and cooling facility. A special facet of the biology of space flight—in spacecraft or spacesuit—concerns the need to minimize the use of consumables requiring resupply from Earth. Such items would include: gases (O_2 and nitrogen); CO_2-absorbant materials; water (for cooling and ingestion), and food.

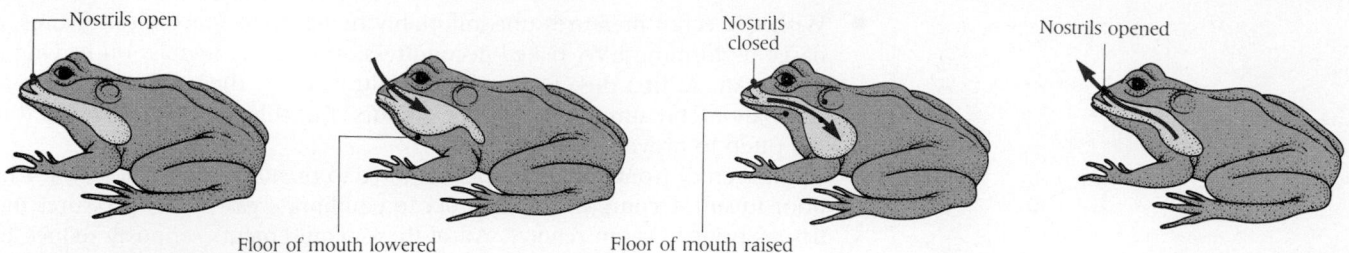

Nostrils open

Floor of mouth lowered

Nostrils closed

Floor of mouth raised

Nostrils opened

Frogs: Positive Pressure Breathers

Many adult frogs have a method of ventilating their lungs called positive pressure (or pulse) breathing (Figure 27-25). This is in contrast to the **negative pressure** (or **aspiration**) **breathing** in mammals, described earlier in this chapter, which arose with the evolution of reptiles.

Positive pressure breathing, which appeared before negative pressure breathing in vertebrate evolution, is in many ways related to the ventilatory movements of fishes. Many frogs breathe by opening their nostrils and lowering the floor of the mouth. Air rushes into the mouth through the nostrils. Gears are then reversed: nostrils close, the mouth floor rises, and air is forced into the lungs. During exhalation, air is drawn out of the lung by lowering the floor of the mouth aided somewhat by lung elasticity. Air moves into the mouth and then to the outside atmosphere.

Compared to negative pressure breathing, this system is mechanically inefficient because the amount of air that can be moved depends on the size of the mouth. It also requires significant muscular effort.

Air Sacs of Birds

The respiratory system of birds differs strikingly from that of mammals. A clue to this difference was discovered in 1758 by John Hunter, who showed that a bird with a blocked trachea can still breathe if a connection is established between one of its bones and the outside air.

Like mammals, birds have a pair of lungs to which a trachea leads. The lungs are small and compact and have a distinctive fine structure, the parabronchi, that enables air to enter one side of the lung, pass through it, and leave from the other side (Figure 27-26). In addition, they have a number of thin-walled air sacs throughout the body cavity that are much larger than the lungs (Figure 27-27). The air

27-25 Positive pressure breathing in a frog
As the floor of the mouth is lowered, air is drawn into the nostrils and forced into the lungs when the mouth and nostrils are closed and the floor of the mouth is raised. Air flows out of the lungs when the nostrils open again.

27-26 Comparison of the avian and mammalian lung structure
(A) In the avian lung, the parabronchi permit air to pass through the lung from one side to the other. (B) In contrast, air passes into and out of (rather than through) the alveoli of the mammalian lung. (C) Scanning electron micrograph of avian lung showing parabronchi.

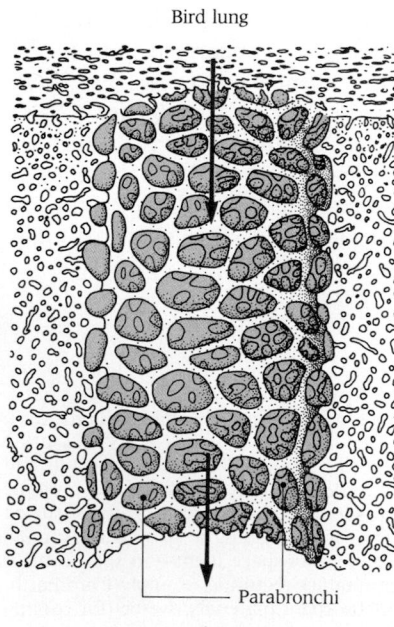

Bird lung

Parabronchi

A

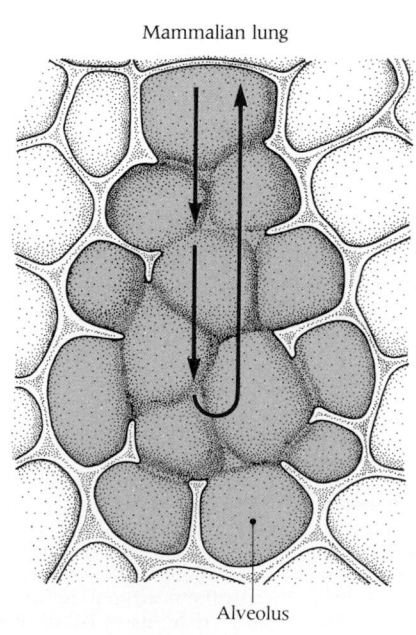

Mammalian lung

Alveolus

B

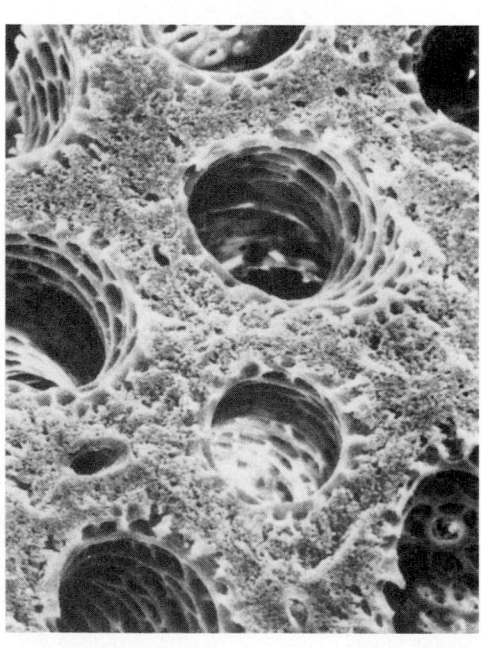

C

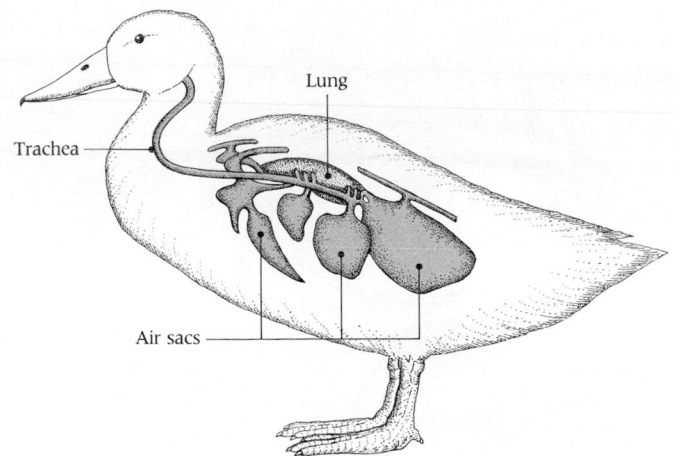

27-27 Avian respiratory system
In birds, inhaled air is passed into the lungs, as well as into air sacs, which are also connected to the hollow bones. The air sacs dissipate heat generated during muscular activity.

Trachea

Lung

Air sacs

sacs are connected to the air spaces in the hollow bones. They dissipate the large amount of heat generated during flight or other activities. Since they are not generously supplied with blood, their ventilation causes heat loss through the fluids of the body cavity. Little gas exchange occurs across their surfaces.

The trachea, bronchi, and other air-carrying tubes of birds are so arranged that air may or may not pass through the lungs as it enters the air sacs from the outside. The pathway taken depends on the bird's oxygen needs and its burden of heat. A pigeon in flight uses one-quarter of the air it inhales for gas exchange and three-quarters for evaporative cooling, which occurs mainly across the surfaces of the air sacs. Birds are notably more efficient at extracting O_2 from air than mammals.

Inhalation is the passive phase of respiration in birds. Air enters the lungs and air sacs when the breastbone (sternum) is lowered, the abdominal muscles relax, and the body cavity enlarges. In a remarkable study, Farish Jenkins and coworkers recently took high speed x-ray movies of European starlings as they flew inside a wind tunnel. Thus they were able, for the first time, to track the movements of the sternum, the wishbone (or furcula), and surrounding bones *during* flight. Among the surprising results was the observation that the wishbone widens to 1.5 times its normal width with each beat of the wings. This movement, they concluded, assists the birds' breathing and is probably related to an air sac between the two halves of the wishbone.

In sum, birds ventilate their lungs by means of negative or aspiration breathing. Movements of the abdominal and flight muscles (and the wishbone) forces air in and out of the lungs and air sacs.

Swim Bladders of Fishes

Most teleost (bony) fishes have a **swim bladder** (sometimes termed a **gas bladder** or **air bladder**) that stores gas and regulates the density of the fish's body and thus provides neutral buoyancy. Just as a submarine submerges or ascends by varying the amount of air in its ballast tanks, a fish can manage to float, rise, or sink. Bottom-dwelling fishes usually lack swim bladders.

The swim bladder evolved as an outpocketing of the pharynx much like the lungs (Figure 27-28). In some fishes, a connection still can be found between the swim bladder and pharynx. These fish can increase their density by "spitting out air." They do that when they swim into deeper water. Or, they can decrease their density when swimming upward by swallowing air bubbles to fill the swim bladder. Some gas also enters the bladder from a gas gland in the bladder wall.

Other fish species lack a connection between the swim bladder and pharynx. In these fishes, gases to be exchanged are delivered to or withdrawn from the bladders by a complex countercurrent arrangement of blood capillaries going to and from the bladder (Figure 27-29). The arrangement is yet another example of the countercurrent exchange principle at work in a living organism. In this case, it operates to keep the volume of gas in the swim bladder constant.

In many fishes, the swim bladder is connected to the ear, either by a chain of bones or by an extension of the bladder. Because air is more compressible than

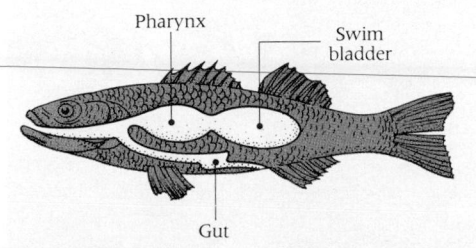

Pharynx

Swim bladder

Gut

27-28 Evolution of the swim bladder
Swim bladders evolved as an outpouching of the dorsal part of the pharynx.

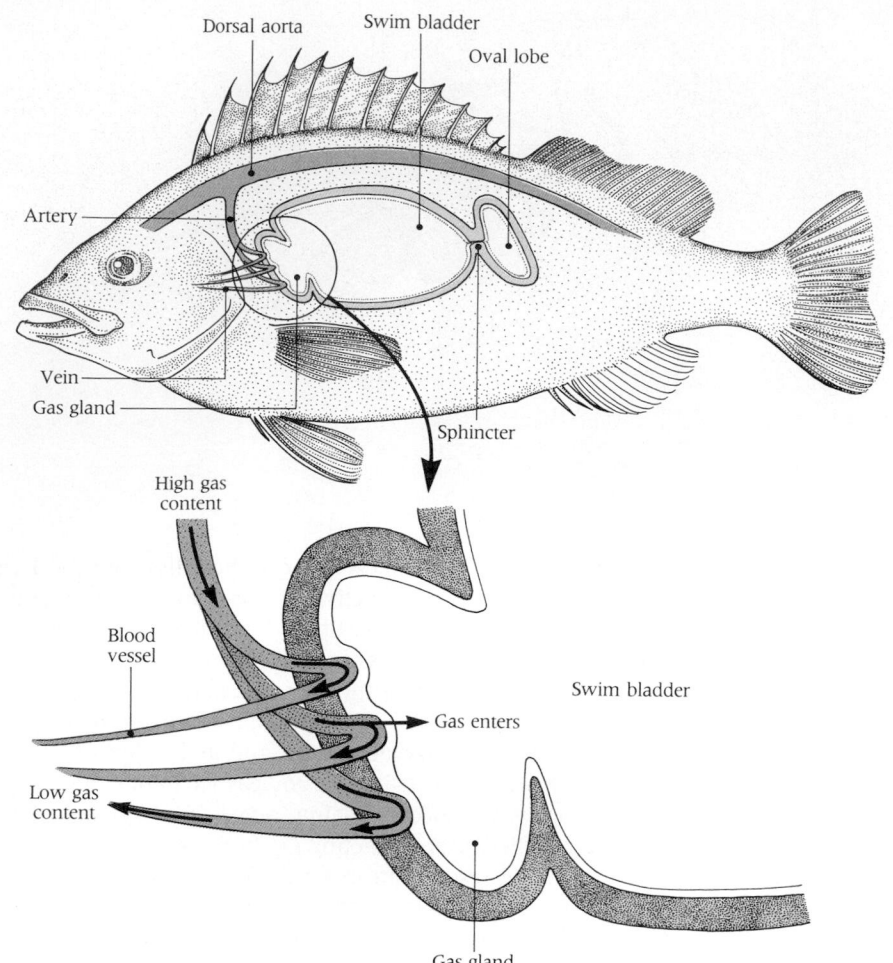

Dorsal aorta   Swim bladder
                      Oval lobe

Artery

Vein

Gas gland

Sphincter

High gas
content

Blood
vessel

Swim bladder

Gas enters

Low gas
content

Gas gland

27-29 Swim bladder function
Gas is added to the swim bladder via the gas gland, which has a countercurrent exchange system in the capillaries of the bladder wall. Gas travels down a concentration gradient from the outgoing to incoming capillary, as shown by arrows and numbers. The volume of gas in the bladder controls buoyancy of the fish in water.

water, the swim bladder can act as an effective resonator that enhances the underwater hearing capacity of these fishes.

Skin Breathing in Vertebrates

Recent work has shown that skin breathing is more widespread in nature than most biologists had realized. Early in this century, August Krogh, Denmark's Nobel Prize-winning physiologist, found that when he obstructed airflow to the lungs of quiescent wintering frogs the skin could supply enough O_2 to maintain life. During summer, when they were much more active, lung breathing was necessary.

We now recognize a long list of skin-breathing vertebrates. Three are shown in Figure 27-30. This type of respiration is limited, as far as we know, to amphibians, fishes, reptiles, and a few unusual mammals. Bats, for example, can eliminate up to 12% of their total CO_2 burden across huge, thin, highly perfused wing membranes.

27-30 Skin-breathing vertebrates
(A) The skin is the sole respiratory organ in the adult of this lungless salamander, Ensatina. (B) The plaice, Pleuronectes, takes in 27% of its O_2 through the skin. (C) The sea snake, Pelamis, absorbs more than one third of its O_2 and excretes more than three-fourths of its CO_2 through its skin. This mode of respiration is important in many other animal groups.

A

B

C

(A) O_2 diffuses rapidly across the alveolar membrane in the mammalian lung, so that hemoglobin is rapidly oxygenated. Arrows indicate the direction of blood flow. Deoxygenated hemoglobin is indicated by open circles and oxygenated hemoglobin by closed circles. **(B)** O_2 uptake through the skin is limited by the skin's thickness. Diffusion is slower than in the alveoli, and the blood is never fully saturated with oxygen, as indicated by the half-filled circles.

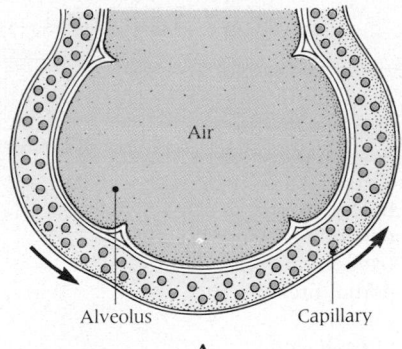
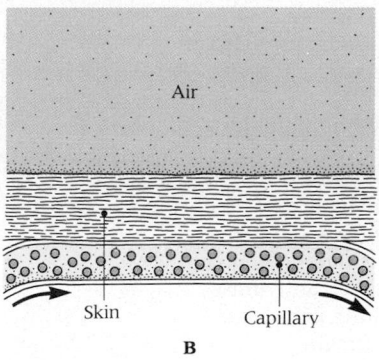

Air

Alveolus   Capillary

A

Air

Skin   Capillary

B

Skin breathing is not important in most mammals (including humans) and birds. All amphibians tested (frogs, salamanders, and caecilians) are substantial skin breathers.

As shown in Figure 27-31, skin breathing depends on the diffusion of gases across a thin layer of skin into small capillaries near the surface. Although it can be a major respiratory organ in some vertebrates, it is more commonly a supplementary system, one of several gas exchangers in the body. It is energetically costly, but its widespread occurrence suggests that it has proved useful.

SUMMARY

In biochemical terms, cellular respiration encompasses the reactions of oxidative metabolism by which individual cells generate energy. In physiological terms, respiration refers to the exchange of molecular O_2 and CO_2 between an animal and its environment. The rate of that exchange is limited by the area of the body surface across which the exchange can take place. Gills, lungs, and tracheae are three elaborations of the body surface that are specialized for gas exchange in different species. The gills of fishes utilize the principle of countercurrent exchange to extract O_2 from water. The tracheae of insects rely on diffusion alone to transfer O_2 from air to body cells. The lungs of vertebrates depend on the muscles of the chest wall to bring air into contact with the respiratory surface. The bloodstream transports O_2 from that surface to all body cells. Respiratory muscles in vertebrates are regulated by a center in the brain, the firing rate of which is influenced by receptors in blood vessels that respond to CO_2 levels in the blood.

Blood transports O_2 in two ways. A small amount of O_2 travels in physical solution in the plasma. Most of it, however, is carried in chemical combination with hemoglobin as oxyhemoglobin. Blood transports CO_2 in three ways. A small amount is dissolved in plasma. Some of it is bound to amino groups of hemoglobin, and most of it combines with H_2O to form carbonic acid (H_2CO_3), which dissociates to bicarbonate (HCO_3) and H^+. When blood CO_2 rises, as it does during exercise, H^+ is generated and blood pH falls. Decreasing blood pH in turn affects the affinity of hemoglobin for O_2, causing it to release more O_2 to the tissues.

The relationship between the concentration of a gas in the blood and the partial pressure of a gas in the body can be defined graphically to yield a characteristic dissociation curve. The dissociation curve of CO_2 is linear in the physiological range of partial pressures, whereas that of O_2 is sigmoidal, or S-shaped. As a result, where O_2 pressure (pO_2) is high, as in the lungs, the blood remains saturated with O_2, even if lung pO_2 drops significantly. Where pO_2 is low, as in muscle, even a small drop in pO_2 causes the blood to release large amounts of O_2 to the tissues.

The respiratory challenge confronting deep-diving mammals, from seals to human scuba divers, centers on the increased pressure in deep water and the decreased availability of O_2 there. Humans solve the latter problem with air tanks; seals combat it with high hemoglobin and myoglobin levels. Increased air pressure causes excess N_2 and O_2 to dissolve in blood, with resulting N_2 narcosis and O_2 toxicity. Upon returning to surface pressures, the excess dissolved gases come out of solution, forming gas bubbles in the blood that are painful and dangerous. This condition is known as "the bends."

Adaptation to high altitudes requires compensation for the low pO_2 that accompanies decreased air pressure. First, the kidney secretes erythropoietin, a hormone that stimulates red blood cell production. Second, an increase in 2,3-diphosphoglycerate (2,3-DPG) in red cells, along with a rise of blood pH, causes O_2 to be released to the tissues more efficiently.

Diverse strategies for gas exchange have evolved in different species. In plants, gas diffusion takes place through leaf stomata, bark lenticels, and root-hair membranes. Some aquatic worms, by pumping water in and out through the anus, use a thin-walled hindgut as a respiratory organ. Unlike mammals, which breathe by negative pressure, frogs ventilate their lungs by positive pressure breathing. In this process, air is drawn into the mouth and then (with the nostrils closed) is pushed into the lungs by a rising mouth floor.

The lungs of birds differ from those of mammals in that air entering one side of the lung from the mouth can leave through the other side. In addition, avian airways are connected to large air sacs and hollow spaces within the bones. The sacs help dissipate heat generated during muscle activity. Most bony fishes have swim bladders, thought to have evolved from primitive lungs, which increase buoyancy. Finally, skin breathing plays a role in respiration in some vertebrates, especially amphibians, supplementing the more complex lung-related respiratory systems. This process, however, is energetically costly.

KEY TERMS

alveoli
aspiration breathing
Bohr effect
Boyle's Law
bronchiole
bronchus
carbaminohemoglobin
carbonic anhydrase
creatine phosphate
diaphragm
2,3-diphosphoglyceric acid
(2,3-DPG, DPG)

expiration
gills
inspiration
lenticel
myoglobin
negative pressure breathing
oxyhemoglobin
partial pressure
pharynx
phosphagen
pleural space
positive pressure breathing

respiration
respiratory center
respiratory muscles
spiracle
swim bladder
torr
trachea
tracheoles
ventilation

QUIZ QUESTIONS

1. Diffusion is an important process in respiratory gas exchange in
   A. protists.
   B. aquatic plants.
   C. birds.
   D. A and B
   E. A, B, and C

2. During strenuous muscular activity, hydrogens removed from sugar molecules are accepted temporarily by ____ to form ____.
   A. myoglobin; carbaminohemoglobin
   B. pyruvate; lactic acid
   C. oxygen; carbon dioxide
   D. oxygen; water

E. none of the above

3. Which of the following causes a leftward shift in the oxygen dissociation curve?
   A. an increase in hydrogen ion concentration
   B. a decrease in pH
   C. increased carbon dioxide concentration
   D. A and B
   E. A, B, and C

4. Which of the following is *not* a respiratory exchange surface?
   A. air sacs of birds
   B. alveoli of mammals

C. gills of larval amphibians
D. skin of amphibians
E. All of the above are exchange surfaces.

5. Which of the following is *not* a means by which oxygen and/or carbon dioxide are transported in the bloodstream of vertebrates?
   A. in solution in the plasma
   B. as oxyhemoglobin
   C. as bicarbonate ions
   D. erythropoietin
   E. as carbaminohemoglobin

ESSAY QUESTIONS

1. During strenuous exercise, muscles function under anaerobic conditions. Explain the concept of "oxygen debt." Why does heavy breathing persist after exertion stops?

2. What is alkalosis, and how is it caused by hyperventilation? Why does one briefly hyperventilate after holding ones breath for an extended period?

3. What causes "the bends" in divers? How is it treated?

4. A person takes a plane from sea-level to the airport in Denver (the mile high city). As the person steps from the plane, what immediate changes help the person to accommodate to the "thin air"? If the person stays for several weeks, how will the body adapt to the high altitude?

5. Breathing eliminates CO_2 from the body because high CO_2 concentrations in tissues are harmful. If it were possible, would it be desirable to decrease the concentration of CO_2 in blood to zero? What would be the consequences of such a decrease?

REFERENCES AND SUGGESTIONS FOR FURTHER READING

Comroe, J. H., Jr. (1966). The lung. *Scientific American 214*, February, 57–68.

This helpful review is by an authority whose writings have illuminated the study of pulmonary physiology.

Feder, M. E., and Burggren, W. W. (1985)

Skin breathing in vertebrates. *Scientific American 253*, October, 125–139.

A demonstration that skin breathing can supplement or replace lung or gill breathing in several vertebrate classes.

Jenkins, F. A., Jr., Dial, K. P., and Goslow,

G. E., Jr. (1988). A cineradiographic analysis of bird flight: the wishbone in starlings is a spring. *Science 241*, 1495–1498.

X-ray movies made of European starlings as they flew in wind tunnel uncovered some surprising information about the role of the wishbone in bird respiration.

Jones, F. R. H., and Marshall, N. B. (1953). The structure and function of the teleostean swim bladder. *Biological Reviews 28,* 16–83.

A classic discussion.

Liem, K. F. (1988). Form and function of lungs: the evolution of air breathing mechanisms. *American Zoologist 28,* 739–759.

An overview of the evolution and physiology of the lung in vertebrates.

Schmidt-Nielsen, K. (1971). How birds breathe. *Scientific American 225,* December, 72–79.

A prolific comparative physiologist adds his insights to the story of bird respiration.

Shulman, R. G., Hopfield, J. J., and Ogawa, S. (1975). Allosteric interpretation of hemoglobin properties. *Quarterly Review of Biophysics 8,* 325–420.

A biophysical interpretation of the structural changes undergone by hemoglobin when it binds and unbinds oxygen.

Weibel, A. R. (1984). *The Pathway for Oxygen: Structure and Function in the Mammalian Respiratory System.* Harvard University Press, Cambridge, Mass.

This book links respiratory physiology and morphology and compares them in different species. An outstanding work.

West, J. B. (1984). Human physiology at extreme altitudes on Mount Everest. *Science 223,* 784–788.

The ability of humans and other animals to adapt to extreme altitudes has always intrigued biologists. In this fascinating paper, a respiratory physiologist reports data obtained at the summit on members of a research expedition, which included himself.

Zapol, W. M. (1987). Diving adaptations of the Weddell seal. *Scientific American 256,* June, 100–105.

A demonstration that collapsible lungs and a large splenic reservoir of oxygenated red cells are the features that enable a seal to dive deeper and hold its breath longer than most other mammals.

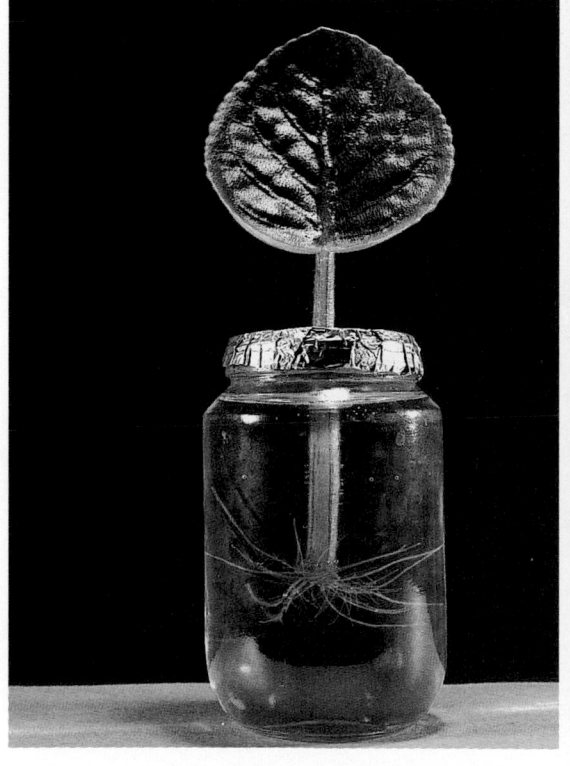

CHEMICAL COORDINATION

Chemical coordination
Production of the hormones controlling growth by the plant tissues themselves is implied in the top photograph showing Boston lettuce, Lactuca sativa, *being grown by hydroponics (in a solution containing only minerals). However, growth can be stimulated by the deliberate addition of hormones. At lower left, root development of an African violet,* Saintpaulia ionantha, *is enhanced in a hormone-enriched medium. At lower right, added gibberellin elongates the stems of mustard,* Brassica rapa. *For scientific and practical reasons, plant and animal hormone research is a lively field today.*

Coordination, or self-regulation, is a fundamental property of life. It ensures the maintenance of homeostasis, reproduction, normal growth and development, and effective adaptation to stress.

Coordination in any organization, living or inanimate, depends ultimately on transfers of information from one place to another. Some call it networking. This chapter and the two that follow are concerned with the mechanisms of biological coordination, and with the questions: what component systems do organisms need to coordinate and how do they do it?

TWO MAJOR COORDINATING SYSTEMS

As organisms grew larger in the course of evolution, two great coordinating systems emerged. One, the **nervous system,** relies on transfer of information via nerves. The other, the **endocrine system,** employs as messengers chemical agents called **hormones** and **growth factors.**[1] Both systems work to preserve homeostasis in the face of numerous complex challenges.

Most animals possess a system of peripheral nerves and central structures that coordinate the information supplied by those nerves and respond to it. It is this combination of peripheral and central structures—and their electrical and metabolic properties—that comprises a "true" nervous system. We consider that system in the next two chapters.

The other major mechanism of coordination, which as far as we know operates in all organisms, utilizes chemical messengers that travel from a place in the body where they originate to a place where they cause an effect. The messenger metaphor is an apt one. Specific molecules are dispatched to specific destinations. There is a measure of information in their very presence. There is further information in their quantity. Complications arise, however, when we try to define terms in this field—and scientific advances have made them more perplexing.

Hormones, the best studied chemical messengers, were traditionally defined as specific compounds that are produced in endocrine glands and transported in the bloodstream to target tissues elsewhere in the body where they induce specific effects. Today this definition is not adequate. Consider the following propositions and the questions they raise.

■ Some chemical messengers act on the cell that released them and do not travel at all. Sometimes they influence a neighboring cell. How far must a chemical messenger travel before we can call it a hormone?

[1] The word "hormone" was first used by the distinguished British physiologist E. H. Starling in his famous Croonian lecture to the Royal College of Physicians in 1905. In describing the discovery (made jointly with Sir William Bayliss) of the intestinal hormone secretin, Starling generalized the idea of humoral control of body function with these words, "These chemical messengers, or hormones as we may call them. . . ." With that discovery, the science of endocrinology was launched.

- Chemical messengers are known to arise in tissues or organs not previously regarded as endocrine glands, for example the brain, liver, intestine, kidney, and even the heart, while others arise in all cells rather than in specific endocrine glands or organs.
- Agents called growth factors share many of the properties of hormones, and the distinction between them is not always clear. These interesting substances include the lymphokines and monokines discussed in Chapter 23.
- Agents called neurotransmitters convey nerve impulses across tiny nerve junctions and behave like hormones (see Chapter 29). Yet they do not enter the bloodstream. Are they hormones?
- Cell-to-cell interaction between neighboring cells depends on gap junctions and plasmodesmata, bridgelike communications in which the cytoplasm of one cell directly connects with the cytoplasm of an adjacent cell. Such connections facilitate exchanges of ions and small molecules, some of which are chemical messengers. Are they hormones?
- Chemical coordination within the organism takes many forms. For example, CO_2 produced by body cells enters the bloodstream. As we learned in Chapter 27, blood CO_2 is monitored by the brain's respiratory center, which responds by regulating breathing. Is CO_2 a hormone?

We will apply the term hormone to those agents about which there is no ambiguity. Perhaps more appropriate terms would be *classic hormone* and *classic endocrinology*. A classic hormone is synthesized in an **endocrine gland.** It specifically influences certain **target cells** in a target organ. These cells are predisposed to respond to a particular hormone by altering their metabolism. In this definition, CO_2 is a chemical messenger, not a hormone.

CHEMICAL AND NEURAL COORDINATION COMPARED

It is instructive to compare biology's two great systems of coordination.

- In both the endocrine and nervous systems, the process being controlled is located at some distance from a control center.
- The nervous and endocrine systems differ in their mechanisms of action. To a large extent, both systems operate by **negative feedback** control. The nervous system exerts control by sending impulses over anatomically defined cables (nerve fibers). Its messages are nerve impulses that activate a limited number of cells near endings of the nerve fibers. The endocrine system controls by chemical messengers (hormones or their precursors). Its messages are expressed in the amounts and types of hormones released into the bloodstream, which disperses them.[2]
- The nervous and endocrine systems are not entirely separate. For example, a part of the brain, the hypothalamus, connects with the "master endocrine gland," the adenohypophysis (or anterior pituitary). Indeed, hormones are produced by the brain itself. In the wake of that discovery, neuroendocrinology emerged as an important new discipline.
- In general, nerve impulses control rapidly changing activities such as muscle movements, whereas hormones influence the rates of cellular metabolic systems. Some hormones are amazingly potent, tiny amounts producing maximal responses within minutes, or seconds (for example, the increase in heart rate after an abrupt increase in epinephrine secretion). Others continue for days, months, and years (for example, the hormonal regulation of growth).

BASIC ELEMENTS OF HORMONAL SYSTEMS
Nature of Hormones

The foregoing definition of hormones reflects the pattern in vertebrate animals, in which hormones have been studied most profitably. However, hormones are found

---

[2] As we have seen, blood-borne substances are sometimes called humors. Hence, the endocrine system is said to exert humoral control. The term recalls humoral immunity (Chapter 23), which is based on blood-borne antibodies.

28-1 Molecular structure of a steroid
Steroids are made up of four rings. The chemical properties of the R group and other substituents distinguish one steroid from another.

28-2 Steps in the synthesis of a typical peptide hormone
Many peptide hormones are synthesized as a large precursor molecule termed a preprohormone. This is often cleaved to produce a prohormone, which requires further proteolytic cleavage to yield the biologically active molecule.

in all five kingdoms and undoubtedly they arose early in evolution. We will later see examples of hormones from nonmammalian vertebrates, plants, and insects. Their evolutionary antiquity is suggested by the similarities of many hormones in different kingdoms. For example, a mating pheromone of yeast is structurally similar to mammalian chorionic gonadotropin. A hormone of the tobacco hornworm has features in common with glucagon. Similar statements could be made about hormones from earthworms, sea anemones, protists like *Tetrahymena*, and *E. coli*.

- In protists, hormones diffuse through the single cell from one part of the protoplasm to another.
- In nonvascular multicellular organisms (which lack circulatory systems), including plants and animals, such substances diffuse from one cell to another. They may also be carried by the water of the external environment in the immediate vicinity of the organism.
- In vascular organisms (which have circulatory systems), movements of hormones occur within and between cells, tissues, and organs via internal body fluids. These are the classic systems with which we are mainly concerned in this chapter.

At least 50 hormones—probably many more—are produced by animals. Chemically, they are of all sorts. Molecular structures of hormones are highly diverse and are usually grouped into four classes.

- **Steroids,** which are chemically related to cholesterol, are molecules of the general structure shown in Figure 28-1, where *R* can be any of several chemical groups. The identity of *R* as well as the other groups attached to the ring system determine the actions of steroid hormones.
- **Amine** or **amino acid derivatives** and analogues are simpler molecules with diverse functions.
- **Peptide hormones** may contain as few as 3–4 amino acids, while others have molecular weights of 20–30,000. Some are glycoproteins with multiple subunits.
- Other organic molecules can also function as chemical messengers.

This diversity makes it rather difficult to offer a summary description of the biochemical steps in hormone biosynthesis.

In the case of peptide hormones, what is often released into the blood is not the hormone itself but a chemical precursor, a **prohormone** or even a **preprohormone,** which is a precursor of a precursor (Figure 28-2). In these cases, the final hormone, the effector molecule that actually influences the target tissue, may be formed in the bloodstream, in the target tissue, or even in another organ.

Hormone Receptors

Hormones act on specific target cells or target tissues. In fact, target cells rely on receptors to recognize each specific hormone. A **receptor** is a specific and unique macromolecule that selectively binds a hormone and then mediates its effects upon the cell. More precisely, the two functions of a receptor are (1) to distinguish a particular chemical signal from the jumble of hormones and other molecular passersby constantly impinging on the cell, and (2) to relay a signal to the cell that causes an appropriate cellular response. Discovery of receptors and their modes of action answered many long-unresolved questions.

We saw in Chapter 6 that membrane proteins are of two types: tightly bound intrinsic (or integral) proteins and loosely associated extrinsic (or peripheral) proteins. Hormone receptors are a class of intrinsic proteins that contain a specific binding site (or sites) for a hormone. A finite number of receptors dot the surface of each target cell. Many receptors can glide through the lipid bilayer and arrange themselves in groups and clusters. Each receptor in a cluster tends to influence the binding behavior of its partners—much as the individual hemes in a tetrameric hemoglobin molecule influence one other (see Box 27-1).

Receptors differ in the particular hormones they recognize, in the method of information transfer after binding, and in their location within the cell.

Three types of hormone receptors are recognized based on their cell location.

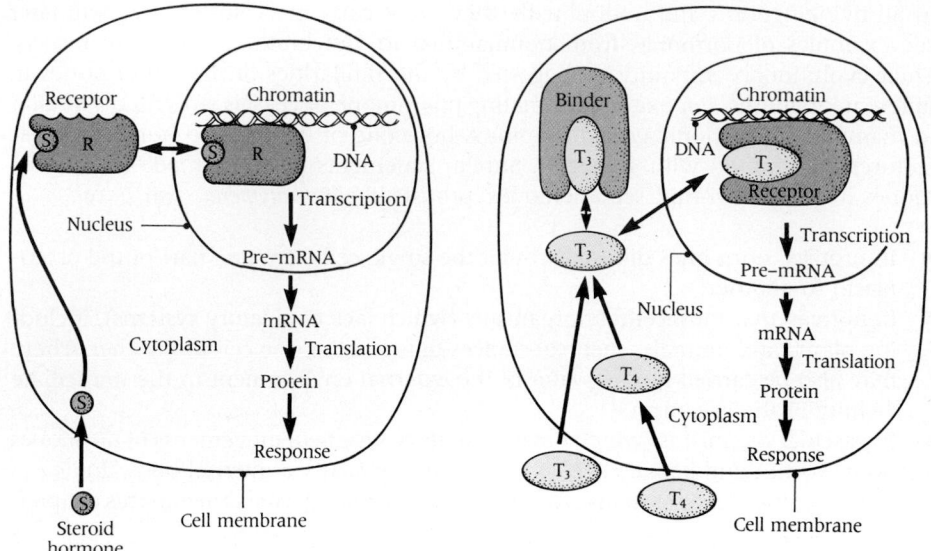

28-3 Mechanism of steroid hormone action (Left)

After penetrating the cell membrane, steroid hormone (S) binds to its cytoplasmic receptor (R), thereby altering its shape. The S-R complex enters the nucleus and binds to specific regions of chromatin, altering the types of mRNAs produced. As a result, new proteins are produced that have specific effects on the cell's metabolic processes.

28-4 Mechanism of thryoid hormone action (Right)

T_3 and T_4 are thyroid hormones; T_4 is converted into T_3 by enzyme action. T_3 enters the nucleus and binds to its receptor. The complex then associates with DNA, changing rates of transcription and thereby modifying the types of proteins in the cell. Some T_3 is temporarily held by binders in the cytoplasm.

- Some receptors lie within the cell rather than on its surface (Figure 28-3). Hormones bound by these receptors (typified by **steroid hormones**) are lipid soluble and hence pass readily through the lipid-rich plasma membrane. Inside the target cell, a steroid hormone binds to its specific receptor, which is promptly modified by the bound hormone. The **hormone-receptor complex** then migrates into the nucleus and attaches to chromatin, where it stimulates (or represses) production of a specific mRNA.

- Some receptors lie within the chromatin of target cells (Figure 28-4). To bind to these receptors, a hormone must be able to cross the cell membrane without difficulty. Intracellular receptors of **thyroid hormones** are in this class.

- Some receptors are located on the cell surface of the target cell (Figure 28-5). This is characteristic of the receptors for **peptide hormones, amine hormones,** and the hormones called **releasing factors.**

MECHANISMS OF HORMONE ACTION

We learned in Chapter 8 that mechanisms for regulating cellular metabolism include systems that control the rate of enzyme activity (feedback inhibition) and systems that control the rate of enzyme synthesis (repression and induction). Hormones are external agents that can work at both of these levels as well as several others. They influence cells in the following ways:

- By altering rates of activity of critical enzymes
- By activating transcription and translation of mRNAs encoding enzymes[3]

---

[3] This phenomenon was discovered in 1961 by P. Karlson and J. Edstrom in studies of ecdysone, the molting hormone of insects. This mechanism of hormone action was later found in many vertebrate systems.

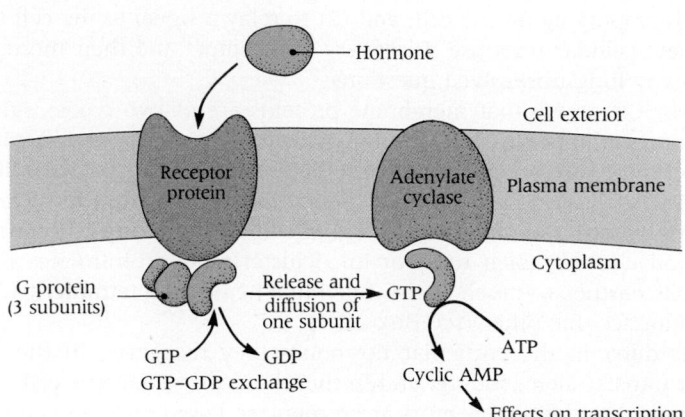

28-5 Mechanism of action of peptide hormones

Peptide hormones do not enter the cell. Instead, they bind to a specific receptor protein in the plasma membrane. The hormone-receptor complex activates a G protein which binds GTP in place of GDP and splits off a GTP-bearing subunit which migrates to a nearby molecule of adenylate cyclase and activates it. This enzyme converts ATP to cyclic AMP, the molecule that actually effects changes in cell functions.

TABLE 28-1
SOME MECHANISMS OF HORMONE ACTION

Class of Hormone	Example	Mechanism of Action
Steroid	Corticosterone (adrenal cortex), testosterone (testis)	Hormone enters cell, binds to receptors, moves into nucleus, stimulates transcription of specific mRNA
Amino acid derivative	Thyroxine (thyroid)	Hormone enters cell, binds to receptors in chromatin, stimulates transcription of specific mRNA
Amine	Epinephrine (adrenal medulla)	Hormone binds to receptors on cell surface, stimulates activity of adenylate cyclase, increases level of cAMP
Peptide	Parathyroid hormone (parathyroid)	Hormone binds to receptors on cell surface, stimulates activity of adenylate cyclase, increases level of cAMP
	Growth hormone (adenohypophysis), insulin (pancreas)	Hormone binds to receptors on cell surface, complex effects possibly involving other hormones, mechanism unknown

- By altering permeabilities of cell membranes
- By replacing natural coenzymes and thus stimulating the activity of target enzymes
- By directly or indirectly influencing the physical state of a multisubunit enzyme, causing it, for example, to disaggregate into catalytically inactive monomers.

By working together, in combinations befitting the challenge, hormones influence target-cell metabolism in ways that benefit the whole body. Table 28-1 lists the major mechanisms of hormonal action and cites examples.

How do hormones perform their tasks? How does the mere binding of a hormone to a hormone receptor located at the cell surface or within a cell initiate events that wrench the cell into a different metabolic posture? How can a tiny amount of hormone trigger profound effects in a target cell without contributing significant amounts of energy or matter to it?

One answer, long sought by investigators, is that there are one or more **second messengers.** This term is self-explanatory. The "first messenger," the hormone, brings a signal from another part of the body to the target cell's receptor. Its message is: "Do this or that." The second messenger brings "word" from the hormone-receptor complex to centers inside the cell that are capable of altering metabolism. Its message is: "Hormone has arrived. Launch game plan."

The Cyclic AMP System

In the earliest second-messenger system to be recognized (see Figure 28-5), the binding of a peptide hormone to a surface receptor results in the activation of **adenylate cyclase,** an enzyme that in the presence of Mg^{2+} catalyzes the conversion of ATP to **cyclic adenosine monophosphate (cyclic AMP or cAMP).** This molecule is structurally similar to ordinary AMP, yet decisively different from it (Figure 28-6).[4]

Note in Figure 28-5 that an important step precedes the formation of cAMP. The nucleotide GTP is an essential participant in the hormonal activation of adenylate cyclase. Associated with the receptor are two membrane-bound GTP-binding proteins that are often referred to simply as **G proteins.** The function of the G protein is to activate adenylate cyclase. This protein, which has three subunits, interconverts between a GDP form and a GTP form. In the absence of hormone, the G protein

---

[4] The discovery of this system by Earl W. Sutherland, Jr., of Washington University, was recognized by the 1971 Nobel Prize in Physiology and Medicine.

28-6 Structures of cyclic AMP and AMP

The structure of AMP is given as in Figure 16-9, except that the symbol Ⓟ is here replaced by the full structural formula of phosphate (in red). Note that in cAMP, a "cycle" is formed by the phosphate group, which is here attached to carbon 5' (as in AMP) and to carbon 3' via ester linkages.

Cyclic AMP

AMP

is in the inactive GDP form. The binding of hormone to its receptor triggers an exchange of GTP for bound GDP. The GTP-bearing subunit breaks away, diffuses to an adenylate cyclase molecule nearby, and activates it. ATP is then converted to cAMP, which acts as a second messenger.

In some hormonal systems, the adenylate cyclase is geared to respond to a stimulatory hormone (Figure 28-7). Depending on which hormone is bound to the receptor, a stimulatory or inhibitory G protein is activated—and GDP is concurrently replaced (in greater or lesser degree) by GTP. Again, the binding of GTP to adenylate cyclase determines its level of activity—and thus determines the amount of cAMP formed.

The work of recent years has revealed that there is a family of G proteins that serves to regulate a number of different intracellular signalling pathways—among them, ion channels in the plasma membrane, adenylate cyclase, and some of the enzymes of the phosphoinositide system. An interesting puzzle confronting investigators of the G protein transmembrane signalling system is to explain how the system achieves specificity. All of the G proteins so far studied have almost identical amino acid sequences. Yet, the activation of each by its specific external hormonal—such activating signals are called **agonists** to contrast them with inhibitory agents that are called **antagonists**—produces a distinctive and specific result.

Cyclic AMP is an ideal second messenger because it is formed only within cells and acts only intracellularly. Its major effect is to stimulate certain cAMP-dependent **protein kinases,** which phosphorylate various enzymes at specific sites. It does

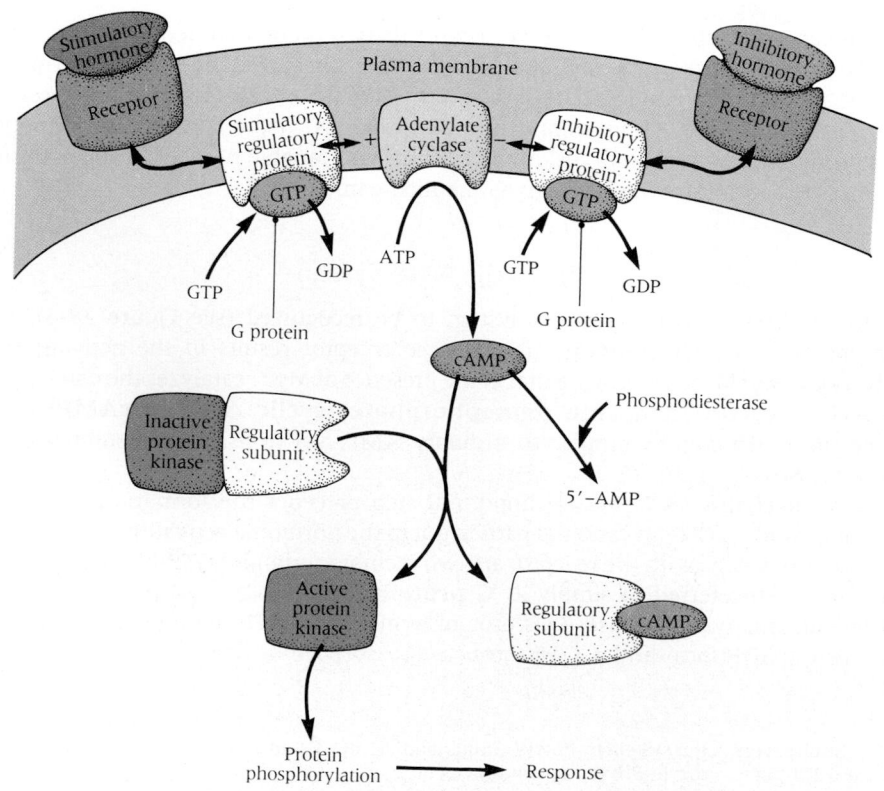

28-7 Mechanism of action of stimulatory and inhibitory hormones

Cyclic AMP is a second messenger molecule that exerts its effect by activating a protein kinase, which adds phosphate groups to enzymes, thereby altering their activity. When a stimulatory or inhibitory peptide hormone binds to the receptor in the membrane, G protein, in the presence of GTP, either stimulates or inhibits adenylate cyclase, with corresponding changes in the levels of cAMP.

this by dissociating their subunits. This, in turn, stimulates their activity. This system is notably important in the actions of polypeptide and amine hormones.

Adenylate cyclase and related enzymes occur widely in nature and are of surpassing importance. It is, for example, cAMP that causes individual cells of the slime mold *Dictyostelium discoideum* to aggregate into a patterned multicellular form.[5] In bacteria, cAMP is a "hunger signal," signifying an absence of glucose and leading to the synthesis of enzymes that can exploit other energy sources. Clearly, cAMP has a long evolutionary history as a regulatory molecule.

Calcium As Second Messenger

As long ago as 1883, Sydney Ringer found that an isolated frog heart beats only when Ca^{2+} is present in the fluid medium. Later workers found that intracellular Ca^{2+} concentration is the crucial element and that Ca^{2+} regulates many—perhaps all—physiological activities, including muscle contraction, endocytosis and exocytosis, cell motility, the movement of chromosomes prior to cell division and perhaps cell division itself, nerve impulse transmission, and many metabolic systems—among them glycogen metabolism, blood clotting, and cobalamin (vitamin B_{12}) metabolism.

Work in the early 1960s showed that Ca^{2+} does not act alone. It is a regulator only when embraced by a calcium-binding protein. In 1970, Wai Yiu Cheung of the University of Tennessee discovered a new protein that activates certain enzymes. The protein was **calmodulin,** but it was not so named until 1978, after its relationship to Ca^{2+} was recognized. That relationship can be briefly stated: calmodulin binds Ca^{2+}—it is, in fact, an intracellular calcium receptor—and is activated by it. In turn, the calmodulin-Ca^{2+} complex activates certain enzymes. The presence of calmodulin in all eukaryotes—plant and animal—is remarkable and hints that it is an

---

[5] This organism's unique life cycle makes it a useful experimental object. In the presence of a food source (usually bacteria) it exists as single-celled amoebae. When the bacteria are removed, the amoebae aggregate to form a multicellular mass, the pseudoplasmodium or slug, which includes about 10,000 cells. See Chapter 39.

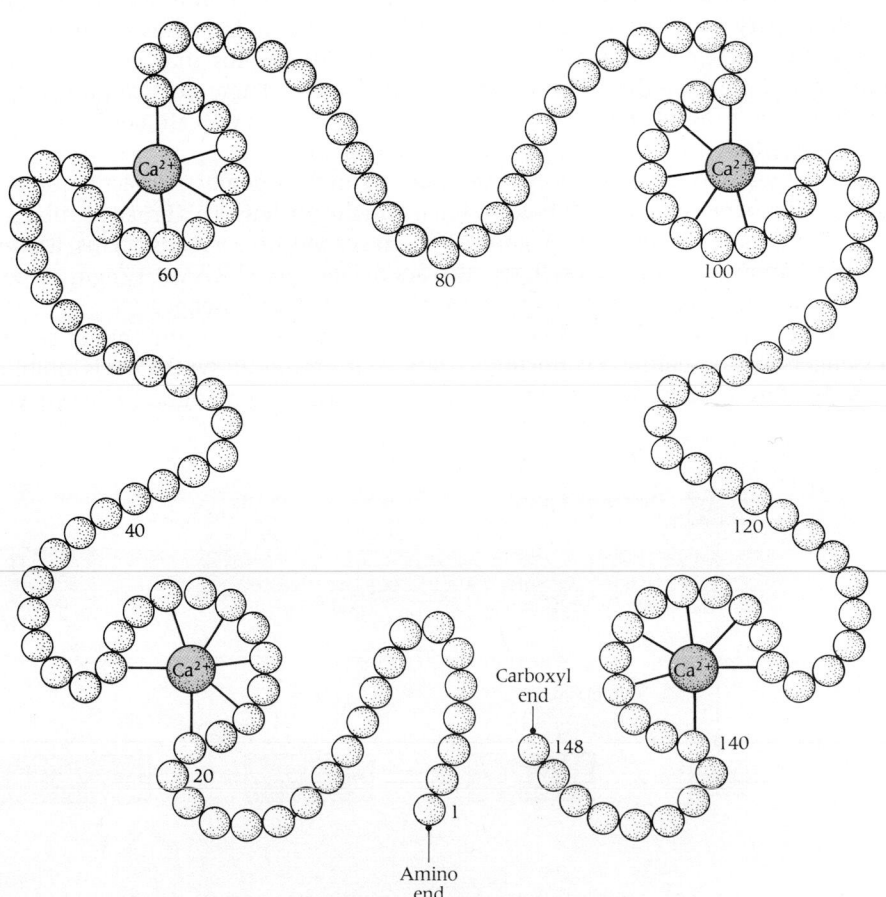

28-8 Calmodulin is a regulatory calcium-binding protein
Calmodulin is a single peptide chain of 148 amino acids which are represented as circles. The amino and carboxyl ends of the chain are shown. The four sites that bind Ca^{2+} are in color.

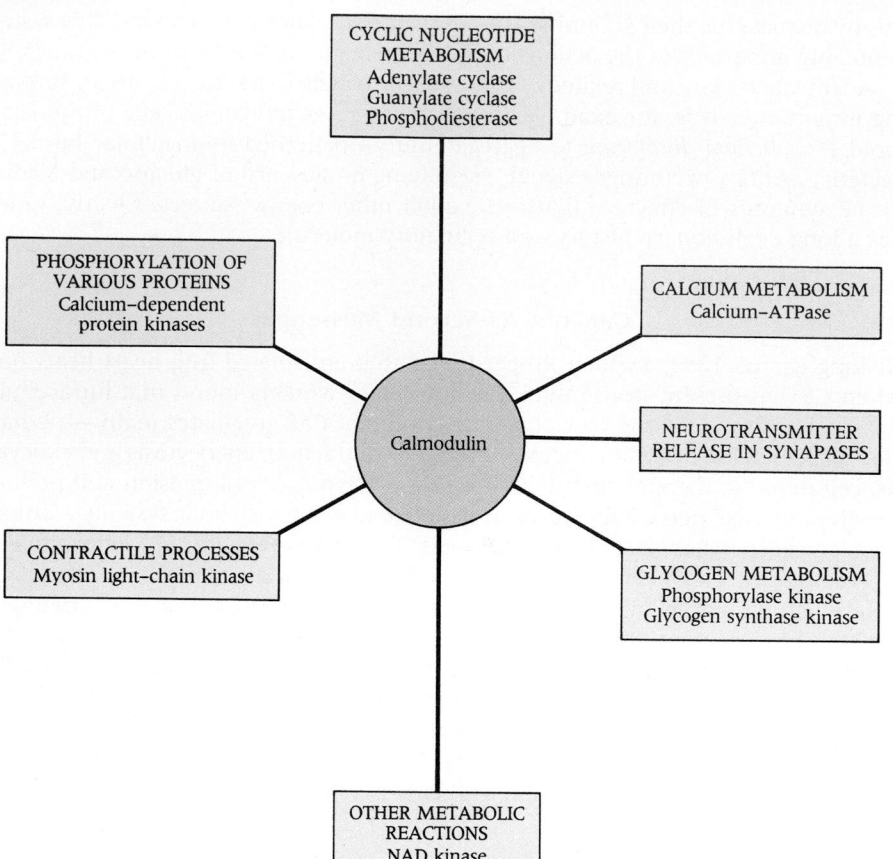

28-9 The many functions of calmodulin
Calmodulin affects a variety of important
processes by controlling Ca²⁺ availability in the
cell and thereby modifying the actions of
Ca²⁺-dependent enzymes.

ancient protein. In view of its ubiquity and profound importance it is surprising
that it eluded observers for so long.

Calmodulin contains 148 amino acids (mol. wt. 17,000) and has an unusually
flexible tertiary structure (Figure 28-8). A third of the amino acids are glutamate
or aspartate, acidic amino acids with —COOH side groups that ionize to form
negatively charged —COO⁻ groups (see Figure 4-14). These are the groups that
bind four Ca^{2+} ions into four binding sites. When Ca^{2+} is bound, calmodulin assumes
a helical conformation that makes it biologically active.

Calmodulin is found in many cytoplasmic structures—membranes, actin-containing microfilaments, and microtubules—but not in the nucleus. This broad distribution
befits calmodulin's many roles (Figure 28-9). In second-messenger systems, it interacts with another second messenger, the adenylate cyclase-cAMP system (Figure
28-10). For that reason, some have called Ca^{2+} a "third messenger."

Glycogen synthesis and breakdown are elegant examples of the role of second
messengers in modulating cell functions. The enzyme glycogen synthetase is inhibited
by cAMP-stimulated phosphorylation. This decreases glycogen synthesis. The enzyme

28-10 Mechanism of action of calmodulin
In the brain, for example, Ca^{2+} activates both
adenylate cyclase (which synthesizes cAMP from
ATP) as well as phosphodiesterase, which
converts cAMP to the inactive form, 5'-AMP.
In this way, calmodulin regulates the level of
cAMP in the cell.

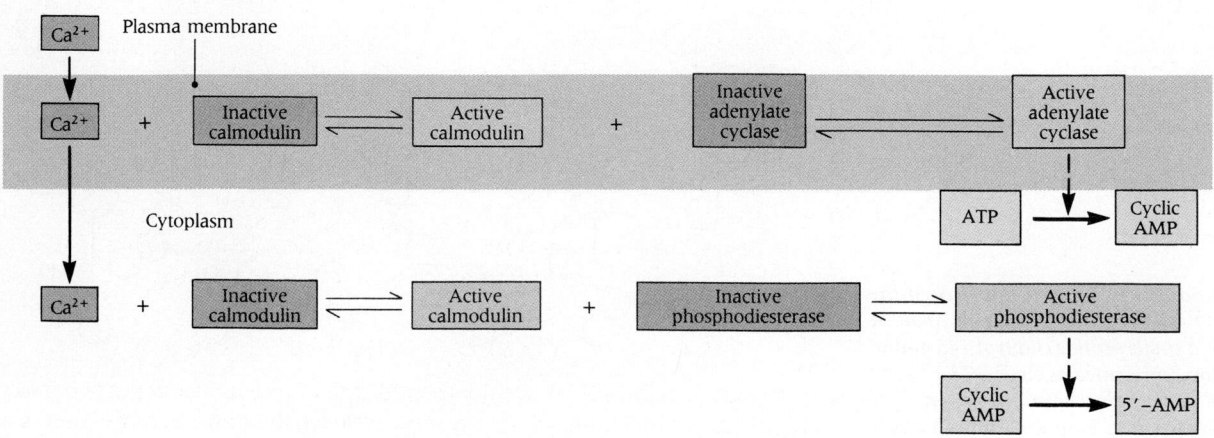

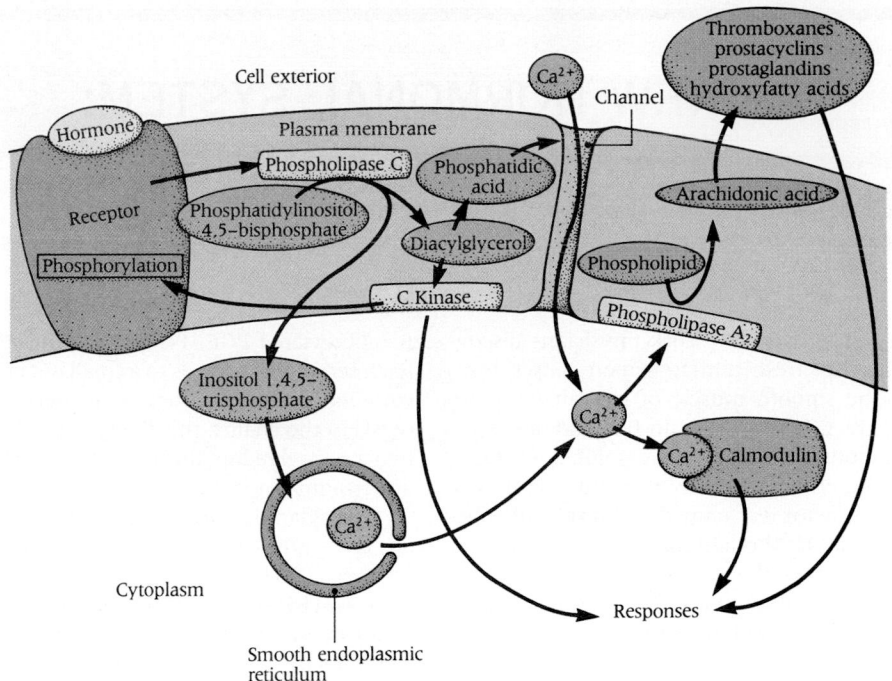

28-11 A new second messenger system: phosphoinositides
The binding of a peptide hormone to its membrane receptor activates membrane phospholipase C, which converts membrane phosphatidylinositol 4,5-bisphosphate to diacylglycerol and inositol 1,4,5-trisphosphate. These two molecules regulate Ca^{2+} levels in the cell by the pathways shown. Ca^{2+} modulates cell responses under the regulatory influence of calmodulin, and enhances synthesis of prostaglandins and related regulatory molecules (see Box 28-1).

glycogen phosphorylase, which catalyzes a breakdown of glycogen to glucose 1-phosphate, is strongly activated by Ca^{2+}.

Other Second Messengers

The repertoire of second messengers, though limited in size, continues to grow as work in the field proceeds. Those discovered to date are listed in Table 28-2. Let us note them briefly, for they hold high promise of clarifying many mechanisms of hormone action—and may provide methods for manipulating hormone action in cells.

A second-messenger system now at the forefront of interest involves membrane phospholipids, which mediate hormone action in two ways. One method depends on their role in stimulating increased turnover and/or synthesis of **phosphoinositides,** also called **inositol polyphosphates**—phospholipids that, according to older texts, were believed to play only a minor role in cell function. These phospholipids are present in the plasma membrane. Under the influence of various hormones and growth factors, they break down, liberating various messenger molecules that powerfully affect cellular activities. (Figure 28-11).

A phosphoinositide within the cell membrane (called phosphatidylinositol 4,5-bisphosphate) is broken down to **inositol 1,4,5-trisphosphate (IP₃)** by a series of steps involving a G protein and an enzyme called **phospholipase C.** Another product of phospholipase C action is **diacylglycerol.** IP₃ releases Ca^{2+} (from the smooth endoplasmic reticulum), which in collaboration with diacylglycerol activates a protein kinase, which in turn activates various enzymes. Thus IP₃ and diacylglycerol both serve as messengers capable of mediating a wide variety of regulatory extracellular signals. The result is an unusually versatile mechanism that controls many short-term cellular responses—contraction, secretion, metabolism—and perhaps some long-term ones, such as growth and information storage in the brain.

A second role of membrane phospholipids depends on their ability to supply **arachidonic acid** for the synthesis of **prostaglandins,** which directly or indirectly influence many target cells (Box 28-1). In most cases these compounds act as hormones. Recent work, however, implicates an intermediate in the biosynthesis of prostaglandins from arachidonic acid as a "second messenger" in the snail *Aplysia.*

GROWTH FACTORS

Growth factors (GFs) are polypeptides that, like hormones, are bound to specific, high-affinity receptors. They are also called **mitogens** because they stimulate cell

In 1930, two New York physicians discovered that fresh human semen causes strips of the smooth muscle of human uterus to contract. Investigators in England and Sweden soon found that a lipid-soluble substance in semen is responsible for this effect. The active factor was named **prostaglandin** because it was thought, mistakenly, to be present only in the secretions of the prostate gland. In the 1950s, Sune Bergstrom of Sweden showed that there is a family of prostaglandins.

Analysis of the first few milligrams of purified material revealed that the prostaglandins are organic acids possessing 20 carbon atoms and a unique structure (Figure 28A). Further work disclosed three important facts.

- The prostaglandins occur in many tissues outside of the reproductive system.
- Purified prostaglandins have extraordinarily potent effects of a wide variety.
- Naturally occurring prostaglandins are synthesized in the body from precursor molecules, the fatty acids, which are long-chain hydrocarbons with two or more double bonds (see Chapter 4).

Structurally, these molecules consist of a 5-carbon (cyclopentane) ring with two hydrocarbon chains attached to two neighboring carbon atoms (Figure 28A). Beyond this, nonchemists are quickly lost in a welter of substitutions, side chains, and mirror-image patterns that are named by a seemingly irrational system of roman letters, greek letters, numbers, subscripts, and superscripts. At the moment, the most important subdivision is into four types—prostaglandins E, F, A, and B (abbreviated PGE, PGF, PGA, and PGB). Each series corresponds to a small but critical variation in the 5-carbon ring. Because of twists in the chains, the differences between two prostaglandins are best seen in three-dimensional models.

Prostaglandins are not stored in tissues, but are synthesized as a result of membrane perturbations that release free fatty acids, notably **arachidonic acid.** This can be brought about by various hormones, acting directly or indirectly, as well as by inflammatory or immunological stimuli and mechanical agitation. As shown in Figure 28B, cyclooxygenase converts free arachidonic acid to the endoperoxide intermediate, PGG_2, which is then converted to other biologically active products, the nature of which depends on the enzymes present in that tissue. For example, platelets make thromboxane A_2 (TXA_2). Blood vessel walls form prostacyclin (PGI_2).

If the chemistry of the prostaglandins seems complicated, the number and diversity of their reported actions may leave the uninitiated gasping. Prostaglandins cause contraction of the human uterus. They also cause the vasa deferentia and seminal vesicles to contract. Brain tissue is rich in prostaglandins, several of which powerfully affect reflex activity and behavior. Prostaglandins also function like such chemical neurotransmitters as acetylcholine (see Chapter 29). Following nerve stimulation, several prostaglandins are released in the gastrointestinal tract, where they inhibit gastric secretion. Prostaglandins are nowhere so evident as in the kidneys. Clearly, prostaglandins are not hormones of the usual type. They are distributed and released far more widely.

28B Biosynthesis of prostaglandins E_2 and F_{2a}
Arachidonic acid is transformed by a series of enzymes into PGE_2 and PGF_{2a}. Arachidonic acid and the versatile enzyme cyclooxygenase occur within membranes

Possibly, they are normal constituents of all cells.

Though we speak loosely of prostaglandin target organs or tissues, these are so far-flung and varied that the term seems almost meaningless. Indeed, the prostaglandins more resemble vitamins. They derive from certain essential foods, the unsaturated long-chain fatty acids; and they play vitaminlike roles in many biochemical reactions. On the other hand, vitamins are trace substances that when lacking lead to disorders such as scurvy or beriberi. If such a disorder follows prostaglandin deficiency, it is unrecognized. The primary action of the prostaglandins may be to regulate adenyl cyclase, the enzyme that catalyzes formation of cAMP. PGI_2 and TXA_2 both influence that enzyme and thus influence intracellular metabolic pathways.

28A Prostaglandins
The basic structure of prostaglandins is shown above. Prostaglandins E, F, A, and B differ from one another in the structure of the ring. Thromboxane (TXA_2) and prostacyclin (PGI_2) are important chemical derivatives.

division and mitosis. They are highly active at extremely low concentrations—often as low as 1 picogram (1 pg = 10^{-12} g) per ml.

Some argue that growth factors differ from classic polypeptide hormones such as insulin and ACTH in their mode of synthesis and their mode of delivery from secreting cell to target cell. Most GFs are stored in vesicles or granules of the cells that synthesize them and are released in a steady stream, to diffuse to their target cells. We now recognize that chemical messengers have three modes of secretion and delivery (Figure 28-12).

- **Endocrine secretion** is the classic mode in which a secretory cell produces regulatory substances and discharges them into the blood.
- In **paracrine secretion,** the hormones or GFs are discharged by a secretory cell and travel to a target cell by diffusion.
- **Autocrine secretion,** in which the secretory cell produces factors that bind to receptors on that cell and influence its own behavior, is mainly found in a number of tumor cell varieties. It does suggest one possible reason for the autonomous growth of malignant tumor cells. By stimulating themselves, these cells are able to escape feedback control systems that sensitively regulate external or exogenous chemical regulators.

There has been an explosion of interest in polypeptide GFs in recent years. This is partly due to their possible connection with the proliferation of cancer cells and partly because they have turned up as key participants in many biological systems. These include both normal processes, such as immunity, wound healing, and blood vessel development (angiogenesis), as well as pathological states, such as atherosclerosis, or hardening of arteries, and cancer. Most of the lymphokines and monokines are, in fact, polypeptide GFs.

Here we can only sample the heady atmosphere in this field today. Some of the polypeptide GFs now crowding the journals of cell biology are listed in Table 28-3. Some were isolated very recently. Their amino acid sequences—and those of their receptors—were established with breathtaking quickness and many GFs have already been cloned and produced in industrial quantities by genetic engineers.

A number of generalizations have emerged from this new work.

- The polypeptide GFs comprise a large family of regulatory agents that includes many subgroups displaying structural similarities and common evolutionary ancestries.
- All GF receptors are on cell surfaces. None are within cells. As with some hormone-receptor complexes, many GF-receptor complexes are internalized by receptor-mediated endocytosis (see Chapter 6).
- GFs control and determine the growth behavior of cells. In the body (as opposed to the tissue-culture dish), growth patterns of tissues and organs are usually regulated by many interacting GFs and hormones.
- The actions of many GFs include both stimulation and inhibition of cell proliferation, as well as effects unrelated to cell growth.
- Almost all GFs have awkward names that in varying measure misrepresent function. For example, nerve growth factor (NGF) was discovered—and named— on the basis of its stimulatory effect on the growth of nerve cells. We now know that it affects fibroblasts, leukocytes, and other cell types. Yet the old name lingers.
- Like many classic polypeptide hormones, some GFs are synthesized as large precursor molecules that are converted to active forms by proteolytic enzymes.
- Detailed biochemical knowledge of how a GF stimulates cell division is still lacking. The immediate consequence of GF binding to its cell surface receptor is protein phosphorylation, as occurs with certain polypeptide hormones. Some GFs are bound to a protein kinase that specifically phosphorylates the amino acid tyrosine. Indeed, the cell surface receptor of EGF is itself a tyrosine-specific protein kinase.
- The relation of GFs to malignant growth is now under intense study. Evidence of a connection rests on the startling discovery, made simultaneously in two laboratories in 1983, that the product of *v-sis*, the transforming gene of simian sarcoma virus (a cancer virus of woolly monkeys), has the same amino acid sequence as one of the peptide chains of platelet-derived growth factor (PDGF).

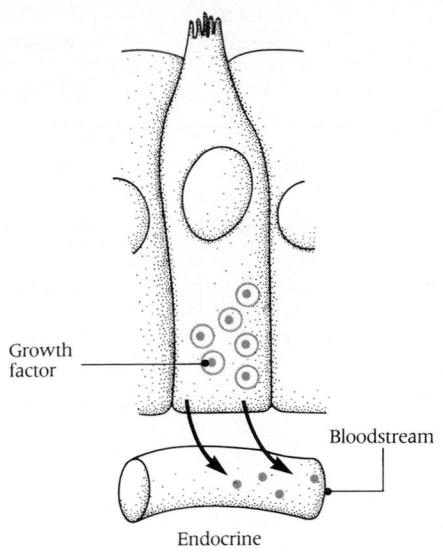

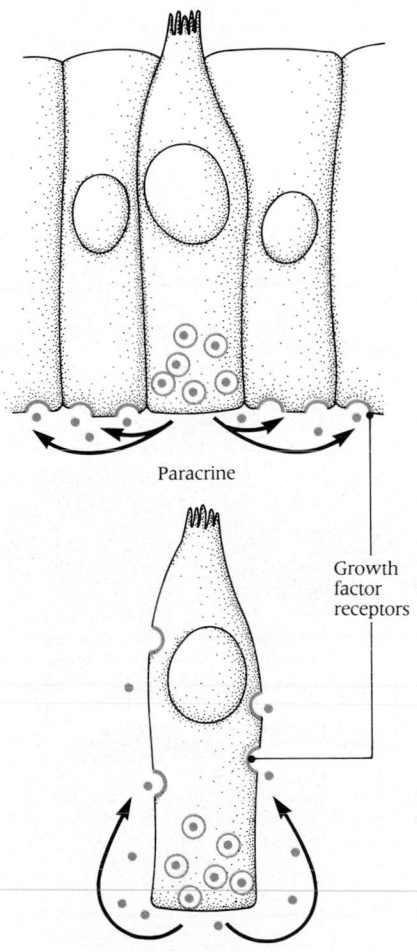

28-12 Three modes of action of peptide growth factors

Growth factors synthesized within a cell can act in three ways. Endocrine secretion transports these molecules to distant places via the blood stream. In paracrine secretion, they bind to receptor sites on nearby cells. In autocrine secretion, receptors are on the same cell in which the growth factor is synthesized.

TABLE 28-3
MAJOR POLYPEPTIDE GROWTH FACTORS

Growth Factor	Abbreviation	Source	Activity	Molecular Weight	Receptor Molecular Weight
Epidermal growth factor	EGF	Mouse submaxillary gland; human urine	Mitogen, keratinization	6,200	175,000
Platelet-derived growth factor	PDGF	Platelets (α-granules)	Growth of various cells	30,000	180,000
Nerve growth factor	NGF	Mouse submaxillary gland	Growth of various cells	28–35,000	98–190,000
Transforming growth factor β	TGFβ	Platelets, placenta	Growth of various cells	22–25,000	175,000
Colony stimulating factors; interleukins	CSF-GM, IL-2, IL-3, etc.	Macrophages, lymphocytes, many cells	Stimulates production of granulocytes, macrophages, lymphocytes	23–35,000	55–165,000
Tumor angiogenesis factor	TAF	Tumor cells, embryonic cells	Stimulates capillary proliferation	18,000	—
Insulinlike growth factors I and II	IGF-I, IGF-II	Plasma	Insulinlike metabolic effects	7,650	210,000

Similarity also exists between the epidermal growth factor (EGF) receptor and the transforming protein produced by a known oncogene (see Box 18-2). This homology between GFs and oncogenes hints at future revelations—perhaps that oncogenes encode GF-like mitogens which induce malignant growth.

Turning to the major GFs themselves (see Table 28-3), **platelet-derived growth factor (PDGF),** like other GFs, stimulates cell division in specific target cells in concert with other mitogens. PDGF, misnamed since it is derived from many cell types, stimulates proliferation of fibroblasts, smooth muscle cells, and certain cells of the nervous system. Critically involved in wound healing (Figure 28-13), PDGF released from platelets at sites of injury attracts neutrophils, monocytes, fibroblasts, and smooth muscle cells and stimulates the proliferation needed in healing.

Epidermal growth factor (EGF) and **nerve growth factor (NGF)** stimulate the cells indicated by their names—and many others. These, too, participate in the remarkable events of wound healing. Curiously, NGF is secreted from salivary glands (among other places), and saliva is rich in NGF. This may explain the tendency of animals to lick their wounds. Wound licking clearly promotes wound healing.

So-called **transforming growth factor-beta (TGFβ),** a product of most cells, controls many aspects of cell growth, differentiation, and function. It works not in isolation, but through the intervention of other GFs. Thus it has been termed a "master growth factor." **Colony-stimulating factors (CSFs)** are a family of factors that promote the production of leukocytes and macrophages by bone marrow (see Chapter 22). Some have been cloned and are now being tested—promisingly—as therapeutic agents in various diseases including leukemia and AIDS. The many **interleukins (ILs)** stimulate B cells, T cells, and macrophages in immune reactions (see Chapter 23).

Finally, **tumor angiogenesis factor (TAF)** was found to be a product of tumor cells after its discoverer, Judah Folkman, reasoned that such a factor must exist. Unless a solid tumor is provided with blood vessels by its host, it remains small and dormant. Aggressive tumor growth demands a rich supply of new blood vessels and, as shown in Figure 28-14, this is brought about by TAF, which stimulates nearby capillaries to grow and extend into complex new networks. This phenomenon, termed **angiogenesis,** is also necessary in embryonic growth. Discovery of TAF precipitated a search for TAF inhibitors that might be capable of slowing tumor expansion. That search, now in full cry, has begun to yield results. Heparin, which when attached to cell surfaces may mediate TAF actions (Figure 28-15), can inhibit TAF when free.

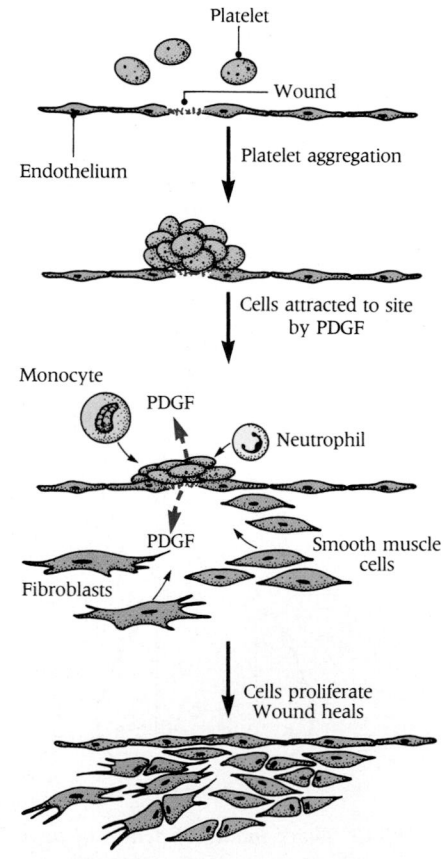

28-13 Role of PDGF in wound healing
Platelets aggregate at the site of a wound and release PDGF, which stimulates the migration of monocytes, neutrophils, smooth muscle cells, and fibroblasts to the site of injury. Smooth muscle cells and fibroblasts multiply and organize to heal the wound.

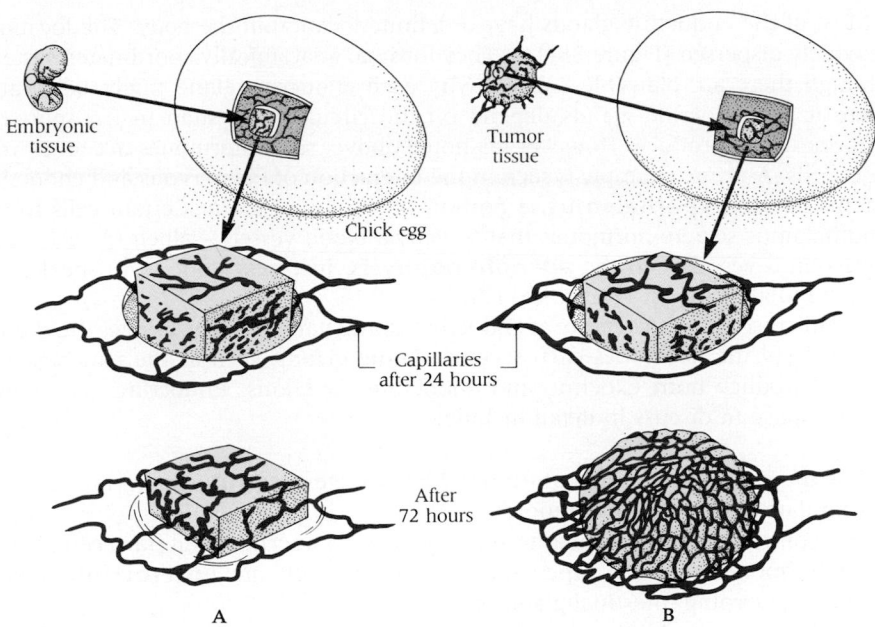

28-14 Action of tumor angiogenesis factor (TAF)

(A) When a piece of normal embryonic tissue is transplanted into a chick embryo, the capillaries of the transplant fuse with those of the embryo within 24 hours. Thereafter no significant change in the blood supply occurs. **(B)** When the same experiment is carried out with tumor tissue which secretes TAF, the implant acquires a rich supply of capillaries within 72 hours. Other growth factors also have some angiogenic activity.

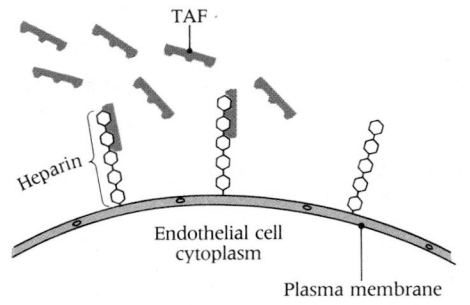

28-15 Modulation of TAF activity by heparin

Many believe that heparin molecules on the surfaces of the endothelial cells lining blood vessels may mediate the angiogenesis induced by TAF.

METHODS OF INVESTIGATION

Historically, the study of hormones passed through a long initial phase that encompassed the discovery of the hormones, the identification of their glandular sources, and the description of their gross effects. During this period, investigators were content to conclude that a hormone—for example, growth hormone—is necessary for growth because a child lacking it becomes a dwarf. Most knowledge was derived from clinical observations of unfortunate individuals.

Later on, endocrinologists were experimentally inclined. Typically, they traveled the following path:

- An endocrine gland was surgically removed from an anesthetized experimental animal (or, alternatively, glandular action was suppressed by drugs) and the effects of removal (or suppression) were observed.
- Next, a gland extract was administered to the deprived animal. If it restored normal physiological function, an attempt was launched to isolate the active hormonal principle from the active but impure extract.
- Proof that the gland in question secretes the identified hormone required demonstration of the hormone in the blood leaving a normal gland.
- Similar experiments were then performed on human beings afflicted with disorders leading to hypersecretion (too much hormone) or hyposecretion (too little hormone) by the gland.

Present endocrinology is concerned with the chemical nature of hormones and molecular details of their mechanisms of action. Today, we speak of a peptide hormone's amino acid sequence, the structure of its receptor, and the intracellular messengers that make things happen.

HORMONES OF THE HUMAN BODY

The following discussions focus on the hormones of the human body. The best understood hormones are those of vertebrates, particularly mammals, and the largest body of information concerns human hormones.

ORGANIZATION OF THE ENDOCRINE SYSTEM

Endocrine glands have no ducts. Their secretions are generally deposited not into the neighborhood of the gland (as are the products of *exocrine* glands) but into a swiftly moving bloodstream. The word *endocrine*, which derives from Greek roots meaning "inside" and "to separate or secrete," means internal secretion.

Most of the endocrine glands have a definite location in the body. The locations are widely dispersed (Figure 28-16). They form no anatomically coordinated system, although there are plausible reasons why each endocrine gland might be located where it is. Endocrine glands depend on the circulatory system to transport the hormones they produce. However, as noted above, some hormones act on nearby target cells. A prime example is seen in the interaction of the nervous and endocrine systems in the **hypothalamus,** a portion of the lower brain. Certain cells in the hypothalamus secrete hormones that travel via blood vessels—albeit special short-range portal vessels—to the **adenohypophysis,** just next door so to speak, and there stimulate it to act (Figure 28-17).

The microscopic architecture of endocrine glands varies widely, as does the chemical nature of the hormones each secretes. Some glands, such as the pancreas and gonads, produce both exocrine and endocrine secretions. Endocrine glands that we lack space to discuss in detail include:

- The mucosa of the pyloric antrum, which secretes **gastrin,** the hormone that stimulates parietal cell secretion
- The epithelium of the small intestine that secretes **secretin** and **pancreozymin,** the hormones initiating pancreatic exocrine secretion, and **enterocrinin,** a hormone governing intestinal gland secretion
- The kidneys, notably its juxtaglomerular apparatuses, which secrete **erythro-poietin,** a hormone controlling the rate of erythrocyte production, and **renin,** an enzyme that indirectly elevates blood pressure and influences the adrenal cortex to secrete the hormone aldosterone (and that some feel is not a hormone because it acts on a blood protein, not a target cell); and the prostate, which along with other tissues secretes the prostaglandins, a class of hormones with many diverse effects (see Box 28-1).

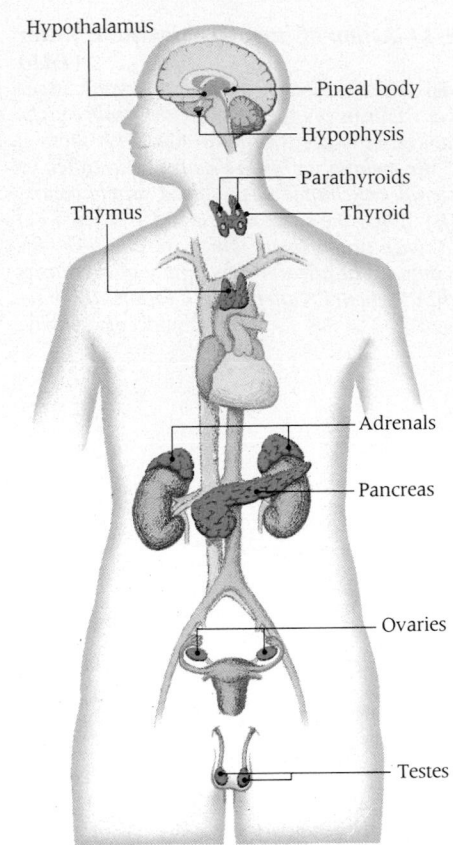

28-16 Locations of major endocrine glands in the human body

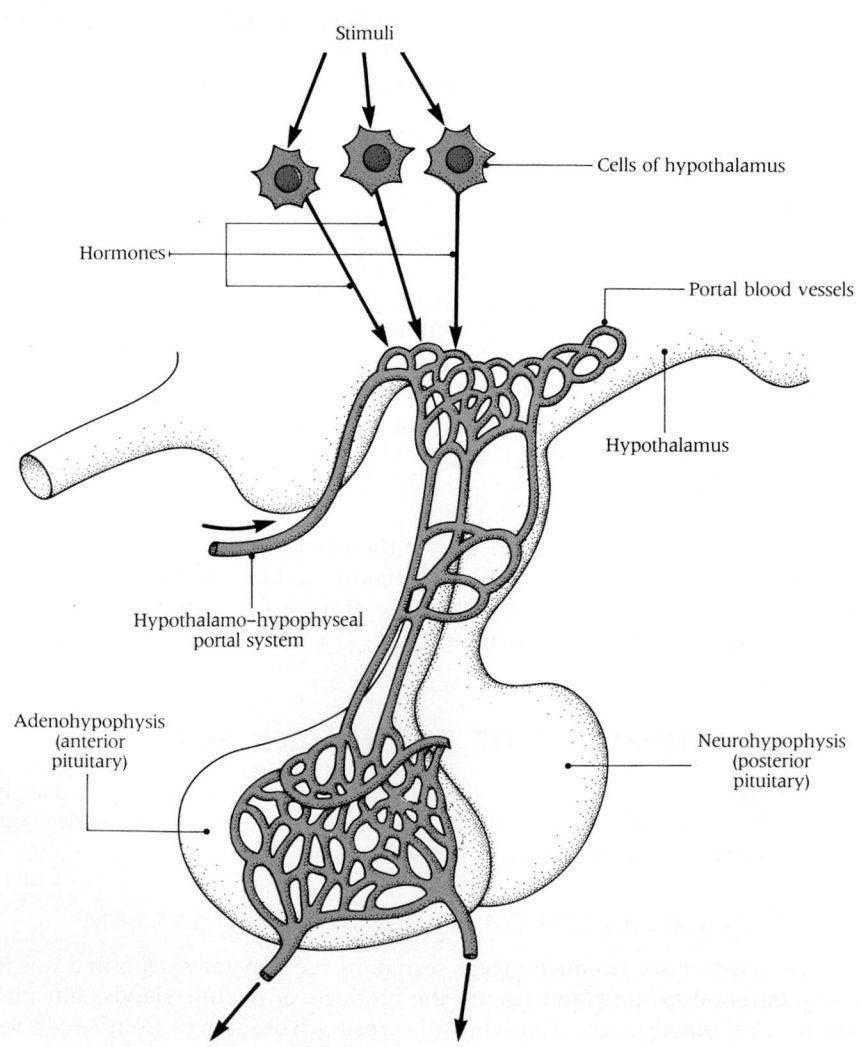

28-17 How the hypothalamus regulates the adenohypophysis
Releasing and inhibiting hormones synthesized in cells of the hypothalamus enter the short portal blood supply (hypothalamo-hypophyseal system), which transports them to the adenohypophysis, where they determine its level of hormone release.

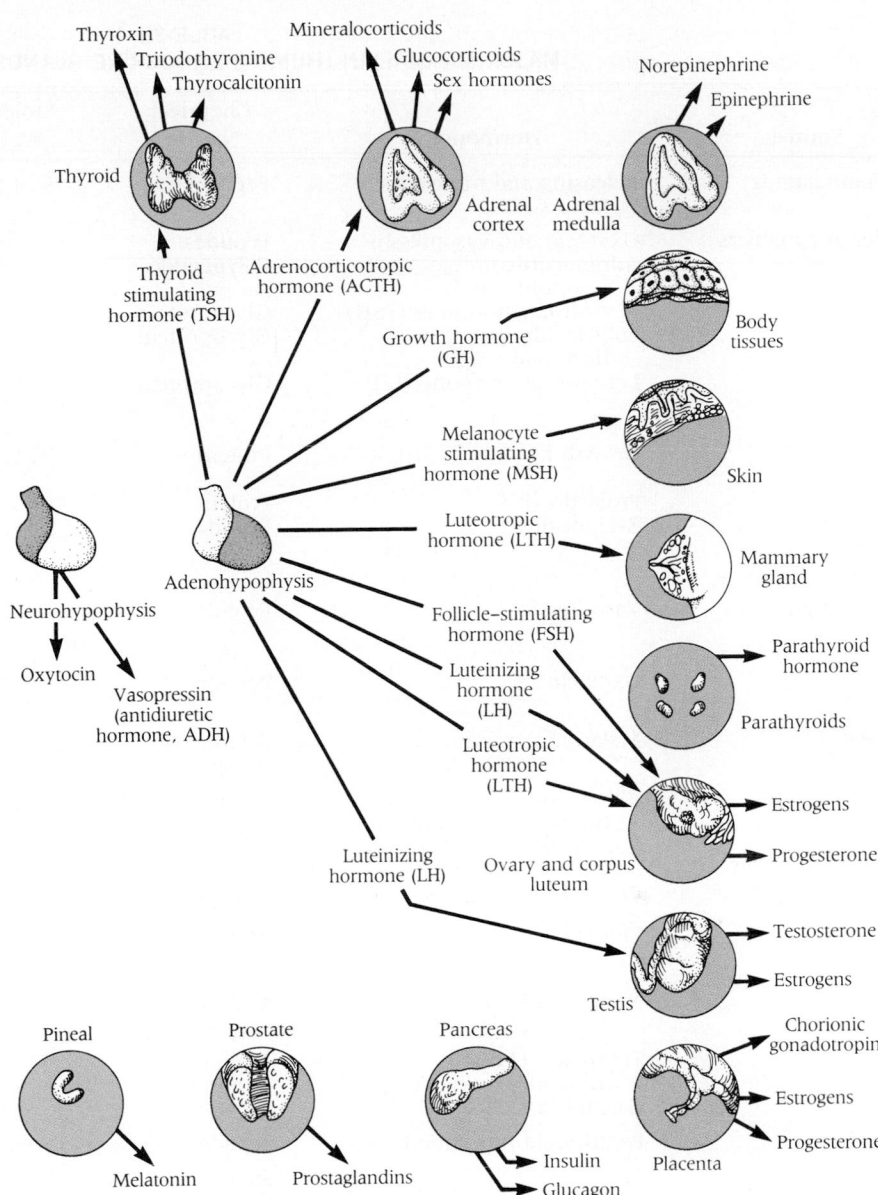

28-18 Scheme of the human endocrine system
A summary of the sites of hormone synthesis.

The major endocrine glands to be discussed in detail are the following:

- Hypophysis (or pituitary), which includes the adenohypophysis (or anterior pituitary) and the neurohypophysis (or posterior pituitary)
- Adrenal cortex
- Adrenal medulla
- Thyroid gland
- Parathyroids
- Pancreas
- Pineal body
- Gonads (testes in the male and ovaries in the female)

Table 28-4 lists the human endocrine glands, their hormones and effects on target organs. Figure 28-18 displays these interactions diagrammatically.

HYPOPHYSIS

The **hypophysis,** or **pituitary,** is an exceedingly small gland weighing only 500 mg that lies protected in a bony depression in the skull near the base of the brain. The bone pocket is the sella turcica (Turkish saddle).

The hypophysis has two components, the **adenohypophysis** (or **anterior pituitary**) and the **neurohypophysis** (or **posterior pituitary**). They are completely separate endocrine glands.

TABLE 28-4
MAJOR MAMMALIAN (HUMAN) ENDOCRINE GLANDS AND HORMONES

Source	Hormone	Chemical Nature	Molecular Weight	Processes Controlled/Functions
Hypothalamus	Releasing and inhibiting factors	Peptides	<1,200	Hormone release by adenohypophysis
Adenohypophysis	Oxytocin and vasopressin	Peptides	1,000	Stored and released by neurohyophysis
	Adrenocorticotropic hormone (ACTH)	Polypeptide	4,500	Activity of adrenal cortex
	Thyrotropic hormone (TSH)	Glycoprotein	29,000	Activity of thyroid
	Follicle-stimulating hormone (FSH)	Glycoprotein	29,000	Growth of ovarian follicles in females and of seminiferous tubules in males
	Luteinizing hormone (LH)	Glycoprotein	29,000	Conversion of ovarian follicle in corpus luteum; production of sex hormones by ovaries and testes
	Growth hormone (GH)	Protein	21,800	Growth; blood sugar level and other metabolic effects
	Prolactin (Pr)	Protein	23,000	Breast development and lactation
	β-Lipotropin	Peptide	11,200	Precursor of enkephalins and endorphins; pain, mood and behavior
Neurohypophysis	Vasopressin	Peptide	1,000	Excretion of water, contraction of smooth muscle in walls of arterioles (blood pressure)
	Oxytocin	Peptide	1,000	Release of milk; contraction of uterine muscles during childbirth
Pineal	Melatonin	Amino acid derivative	232	Day-night circadian rhythm of body functions; gonadal functions
Adrenal cortex	Aldosterone (mineralocorticoid)	Steroid	360	Salt balance
	Corticosterone and cortisol (glucocorticoids)	Steroids	346	Metabolism of carbohydrates, lipids and proteins
	Cortical sex hormones	Steroids	~280	Secondary sexual characteristics
Adrenal medulla	Epinephrine	Amine	183	Conversion of glycogen to glucose; effects similar to actions of sympathetic nervous system
	Nonepinephrine	Amine	169	Reactions to stress; elevates blood pressure
Thyroid	Thyroxine (T_4) and triiodothyronine (T_3)	Amino acid derivatives	777 and 650	Level of metabolism, oxidation rate
	Calcitonin	Peptide	9,500	Blood calcium levels; release of calcium
Parathyroids	Parathyroid hormone (PTH)	Peptide	9,500	Calcium metabolism; bone growth
Testis	Testosterone	Steroid	228	Male sexual development and functions
Ovary	Estrogens	Steroids	~270	Female sexual development and functions
Ovary (corpus luteum)	Progesterone	Steroid	314	Maintenance of uterine endometrium and development of breasts during pregnancy
Placenta	Chorionic gonadotropin	Glycoprotein	46,000	Maintenance of all body functions, maintains gestation
	Placental lactogen	Glycoprotein	21,800	Lactation and fetal growth
Pancreas	Insulin	Protein	5,800	Glucose metabolism, glycogen storage
	Glucagon	Protein	3,500	Conversion of glycogen to glucose
Stomach	Gastrin	Peptide	2,200	Promotes HCl secretion in stomach
Kidney	Erythropoietin	Protein	34,000	Promotes erythropoiesis
	Renin	Protein	40,000	Converts angiotensinogen to angiotensin, which elevates blood pressure
Intestine	Secretin	Peptide	3,100	Promotes pancreatic secretion
Heart	Atrial natriuretic factor (ANF)	Peptide	<5,000 (varies)	Promotes renal sodium excretion; blood pressure

Hormones of the Adenohypophysis

In spite of its small size, the adenohypophysis is the most complex of the endocrine glands. It produces at least nine different hormones—probably more. Its special role in the endocrine system is implied by the term master endocrine gland. The adenohypophysis is special because it produces **tropic hormones** (*trophic* in older

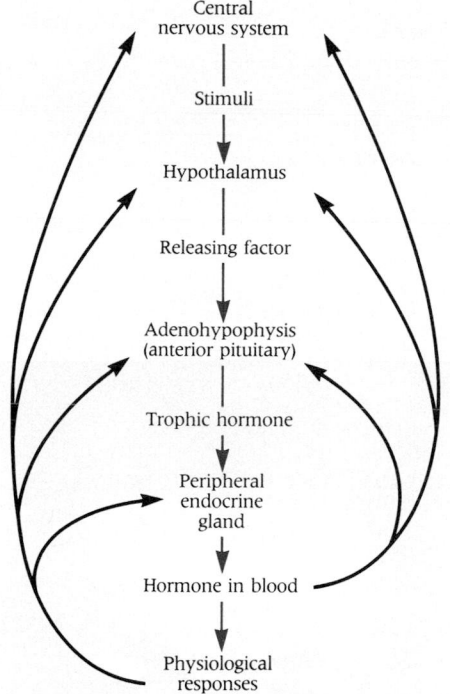

28-19 Feedback regulation in the endocrine system
This generalized scheme shows how the various endocrine glands can be regulated by negative feedback mechanisms (arrows). All of these regulatory influences may not operate at any one time.

Central nervous system

Stimuli

Hypothalamus

Releasing factor

Adenohypophysis (anterior pituitary)

Trophic hormone

Peripheral endocrine gland

Hormone in blood

Physiological responses

texts), which regulate other endocrine glands. The adenohypophysis also produces several nontropic hormones that appear to influence target tissues directly.

The four major tropic hormones are as follows:

- **Corticotropin,** or **adrenocorticotropic hormone (ACTH),** stimulates the adrenal cortex to secrete adrenocortical hormones.[6]
- **Thyrotropic hormone,** also called **thyroid-stimulating hormone (TSH),** stimulates the thyroid gland to secrete thyroid hormone.
- **Follicle-stimulating hormone (FSH)** stimulates hormone production by the gonads. Hence, it is a gonadotropic hormone (one of several). It is essential in females for the development of egg-bearing follicles in the ovary and in males for spermatogenesis in the testis.
- **Luteinizing hormone (LH),** another gonadotropic hormone, stimulates secretion of sex hormones by ovary and testis. The FSH and LH molecules of males and females are chemically identical.

Some adenohypophyseal hormones seem to act directly on their target cells.

- **Growth hormone (GH)** is essential for the normal growth of a child.
- **Prolactin (Pr)**, also called **luteotropic hormone (LTH),** in the human female stimulates breast development and lactation; in other animals it has widespread effects on metabolism and osmotic balance.[7]
- **β-Lipotropin,** the most recently discovered hormone of the adenohypophysis, plays a key role in the field of neuroendocrinology.

The molecular structures of TSH, FSH, and LH are of interest. All are glycoproteins consisting of two polypeptide chains or subunits. One subunit is identical in all three hormones. Target gland specificity is conferred by the other subunit. To understand the endocrine system, one begins with the tropic hormones of the adenohypophysis, which directly stimulate the growth and activity of specific target glands.

Hypothalamic Regulation

If the adenohypophysis coordinates other endocrine glands, what controls it. Or does it function in splendid isolation?

- First, the target endocrine glands themselves help to regulate the adenohypophysis by feedback mechanisms (Figure 28-19). For example, when the plasma concentration of a circulating hormone such as thyroid hormone rises, release of its tropic hormone—in this case, TSH—by the adenohypophysis slows down. When blood reaching the adenohypophysis contains a lower level of thyroid hormone, TSH production accelerates.
- Second, the hypothalamus, a portion of the brain just above the hypophysis, exerts important regulatory influences.

The **hypothalamus** secretes hormones that travel to the adenohypophysis through portal blood vessels, the **hypothalamo-hypophyseal portal system** (see Figure 28-17), and influences its cells to release hormones into the blood or to withhold them. Some of these regulatory hormones are called **releasing factors**

---

[6] Endocrinology is littered with abbreviations that have become so familiar they have replaced the terms they denote in ordinary communication. No one says corticotropin or adrenocorticotropic hormone in conversation. They say ACTH and everyone knows what is meant. We recommend that you learn these abbreviations. If you read further in this field, you will find that some authors have proposed competing names for some of these hormones. (We just saw an example earlier in this footnote.) To avoid confusion we have omitted most of these alternatives. However, it is necessary to mention them in some cases. For example, the hypothalamic releasing factor for ACTH is known as corticotropin-releasing factor or CRF, rather than ACTH-RF.

[7] Recent work has shown that breast cancer is much less common in women who have borne children than in women who have not. It is known that administration of prolactin to rats along with certain chemical carcinogens (cancer inducers) drastically increases the frequency of breast cancer in these experimental animals. This suggests (but does not prove) that long-term suppression of prolactin secretion by pregnancy in humans may protect against later breast cancer—an interesting hypothesis that awaits investigation.

(RFs) because they promote the release of tropic hormones by the adenohypophysis. Others are **release inhibiting factors (RIFs)** because they oppose hormone release. Someone has written in jest, "These factors tell the pituitary when and how much to pitu."

The RFs and RIFs are small peptides (see Table 28-4). Many have been characterized and synthesized. The story of how persistent investigators tracked down these elusive molecules is a fascinating tale of scientific ingenuity, risky and sometimes mistaken decisions, and human fortitude.[8] Key elements of the story were, first, the development of a novel assay for hypothalamic factors suspected to be present despite widespread skepticism and, second, bold efforts to purify these factors from massive quantities of brain tissue. One of the initial purifications of **thyrotropin-releasing factor (T-RF)** from the hypothalamus in 1968 yielded only l mg of the hormone from 300,000 sheep! Surprisingly, T-RF turned out to be a simple chain of three amino acids: *glutamate-histidine-proline*. This material was readily synthesized by chemists and the synthetic peptide was found to be potent in all vertebrates. It is now used to test adenohypophyseal function. When injected into the body, T-RF stimulates release of TSH from a normal adenohypophysis but not from an abnormal one.

Other RFs soon followed. In addition to T-RF, the list now includes **growth hormone releasing factor (GH-RF), luteinizing hormone releasing factor (LH-RF), follicle-stimulating hormone releasing factor (FSH-RF),** and **prolactin releasing factor (Pr-RF).** In addition, separate RIFs have been isolated— among them, **GH-RIF** (also called **somatostatin**) and **Pr-RIF.** The actions of these factors explain how external environmental conditions such as cold and stress can influence the adenohypophysis through the nervous system. The brain monitors these conditions and acts on its information by causing the hypothalamus to control release of hormones by the adenohypophysis. This system also accounts for seasonal changes in sex hormone activities.

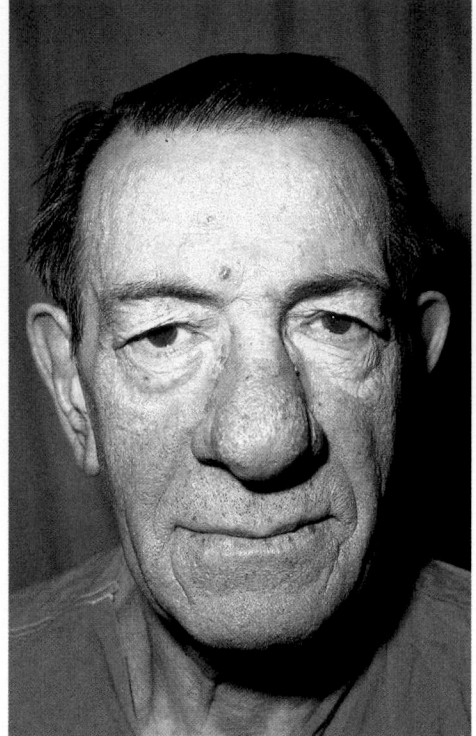

Growth Hormone

Of the adenohypophyseal hormones that act on their own, one of the most interesting is **growth hormone (GH),** which promotes and controls normal body growth. A child may never achieve full growth if for some reason there is a deficiency of GH.[9] Gross underproduction of GH produces midgets, and overproduction during early life produces giants. Martina de la Cruz, who lived to the age of 74 without topping 1 foot 9 inches, was a famous midget. Robert Wadlow, a famous American giant, was 8 feet 10 inches tall and growing when he died at the age of 22. Except for their size, giants and midgets are usually well proportioned. If GH secretion becomes excessive after growth is completed, **acromegaly** may develop, with gross enlargement of hands, feet, jaw, and facial bones (Figure 28-20).

GH is a polypeptide of 191 amino acids that differs only slightly from prolactin. Its release by the adenohypophysis is stimulated by GH-RF, a 44-amino-acid peptide from the hypothalamus, (Figure 28-21).[10] Its actions are opposed by GH-RIF (somatostatin) and evidently mediated by small peptides (mol. wt. 6000—9000) called **somatomedins.**

Many thousands of children in this country suffer from growth failure due to GH deficiency. This condition can be treated with human GH, but until recently supplies were severely limited because GH could be obtained only from human pituitary glands—hardly a convenient source. This situation spurred intense efforts

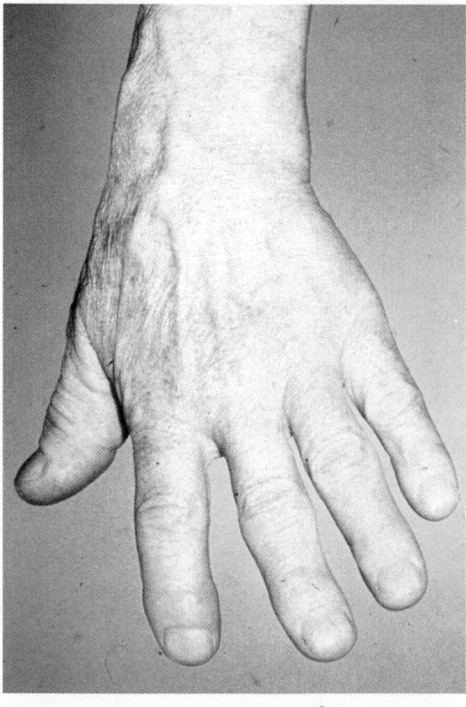

28-20 A man with acromegaly
Note the enlarged hands, jaws, and facial bones, which are due to excessive secretion of growth hormone that began after growth was completed.

---

[8] It led, among other things, to the award of the 1977 Nobel Prize in Physiology and Medicine to Andrew V. Schally and Roger Guillemin, long-time rivals in the pursuit of these brain hormones.

[9] The frequency of growth failure due to hypopituitarism is not known since the disorder is often undiagnosed. In a recent survey of 48,000 children between 6 and 9 years of age, one case of severe GH deficiency was found per 4,000 children. Mild to moderate deficiencies are surely far more frequent—and it is now necessary to distinguish the rare cases in which GH is lacking (which do respond to GH) from the more common one with defective GH receptors (which do not respond to hormone treatment).

[10] As noted above, somatostatin (a synonym of GH-RIF) arises in the hypothalamus, where it was first identified. But we now know it is widely distributed. The pancreas is another of its sources. It is the smallest pancreatic hormone, containing only 14 amino acids. It inhibits the release of insulin and glucagon.

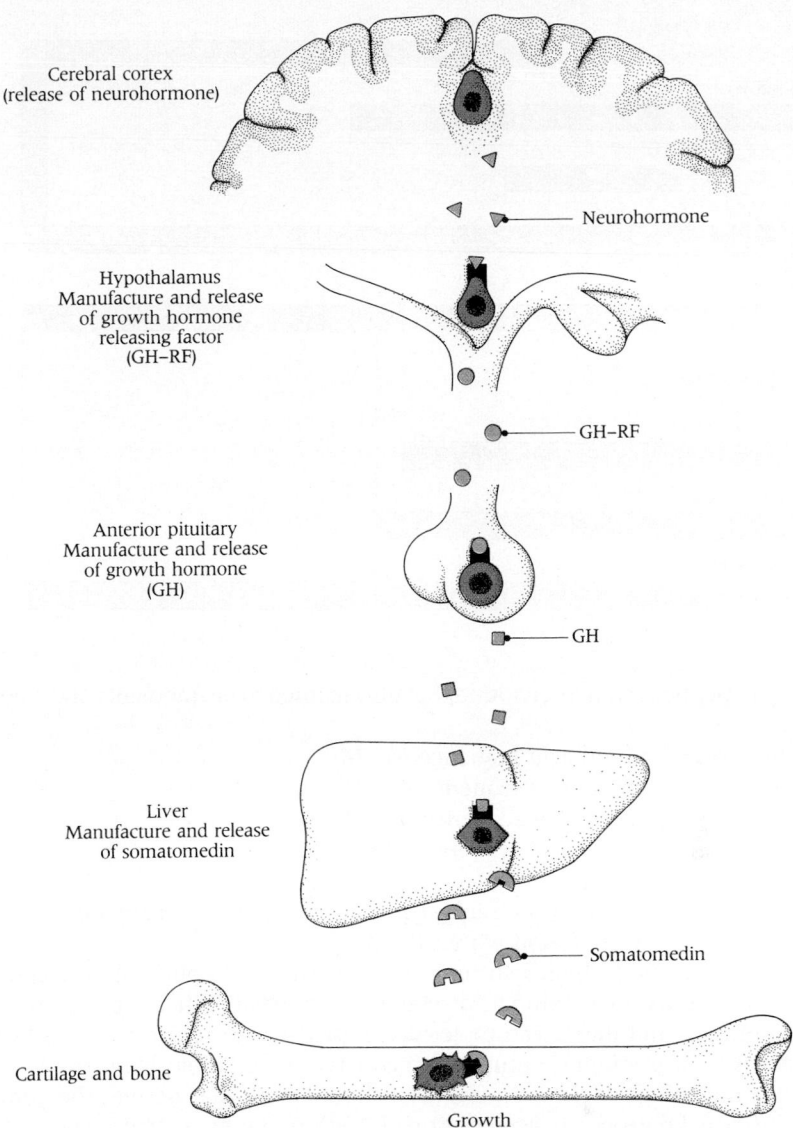

28-21 How growth hormone stimulates growth

The diagram illustrates the sequence of events that leads to growth. Release of a neurohormone from the brain leads to release of GH-RF from the hypothalamus. This hormone travels by the portal system to the adenohypophysis, which releases GH into the bloodstream. In the liver, GH binds to receptors and stimulates synthesis of somatomedin, which causes bone and cartilage to grow.

Cerebral cortex (release of neurohormone)

Neurohormone

Hypothalamus Manufacture and release of growth hormone releasing factor (GH–RF)

GH–RF

Anterior pituitary Manufacture and release of growth hormone (GH)

GH

Liver Manufacture and release of somatomedin

Somatomedin

Cartilage and bone

Growth

to incorporate the gene for GH into bacteria by recombinant DNA methods. This effort succeeded in 1979.

The Endorphins and β-Lipotropin

After the isolation in 1968 of T-RF, the first hypothalamic releasing factor to be fully characterized, neuroendocrinology expanded rapidly. Almost every peptide first isolated from the hypothalamus was soon found in every part of the nervous system. The newest brain peptides are the **endorphins** and **enkephalins.** It was their morphinelike effects in reducing pain that led to their discovery.

The endorphin-enkephalin story began in 1964 with the isolation from hypophysis of **β-lipotropin,** a relatively large polypeptide containing 91 amino acids (Figure 28-22). β-lipotropin had no obvious biological role. Work in the early 1970s revealed that the vertebrate brain contains receptors for the familiar pain-killing drug morphine and other opium derivatives. Investigators argued that these receptors had not evolved solely to bind products of the opium poppy—and the search began for body molecules (as opposed to drug molecules) that normally bind to receptors that also happen to bind morphine.

The successful search revealed two new peptides, each containing only five amino acids, of which four are the same in both. They were named methionine-enkephalin and leucine-enkephalin and were shown to have the following sequences:

methionine-enkephalin: Tyr-Gly-Gly-Phe-Met

leucine-enkephalin: Tyr-Gly-Gly-Phe-Leu

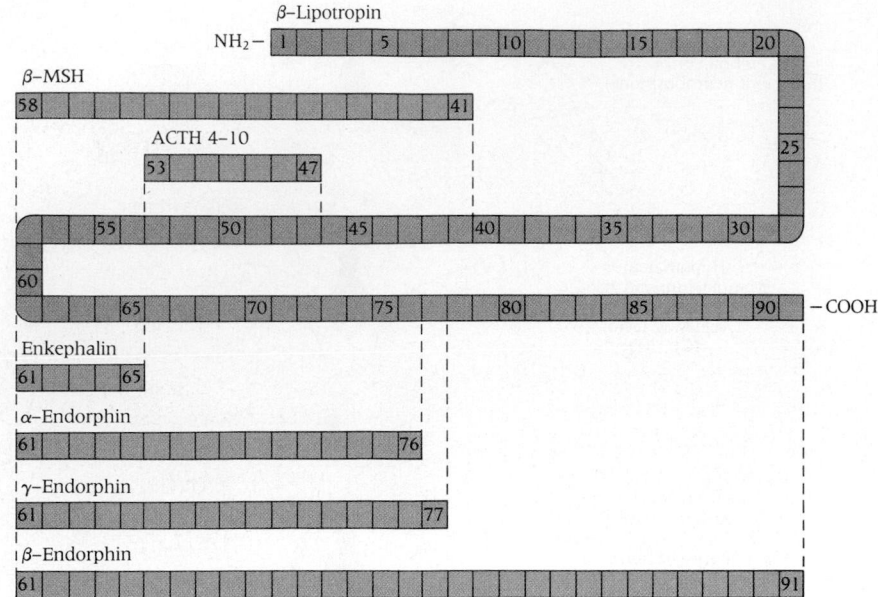

28-22 β-Lipotropin is a precursor of other hormones
Within the 91 amino acid sequence of β-lipotropin are several biologically active peptides, which are released by proteolytic cleavages. MSH, melanocyte stimulating hormone; ACTH, adrenocorticotropic hormone.

Four more peptides were then found that also seemed to be fragments of β-lipotropin.

- α-Endorphin, amino acid sequence 61–76
- γ-Endorphin, amino acid sequence 61–77
- δ-Endorphin, amino acid sequence 61–87
- β-Endorphin, amino acid sequence 61–91

Finally (and surprisingly), amino acid sequences of several other apparently unrelated peptide hormones were found within the β-lipotropin sequence.

The fact that the β-lipotropin molecule contains endorphin and enkephalin sequences within its own amino acid sequence suggests that it is a precursor of these two hormones, and that it can be termed a **prohormone.** However, β-lipotropin was the first prohormone found that gives rise to multiple breakdown products with different physiologic functions. We should note that despite the compelling notion that β-lipotropin is a precursor of methionine-enkephalin, some evidence suggests that methionine-enkephalin and leucine-enkephalin (whose sequence does not appear in β-lipotropin) are produced by an entirely different pathway.

It now appears that β-lipotropin itself derives from an even larger prohormone containing 132 amino acids. The precursor molecule of ACTH is also the precursor of β-endorphin and other hormones. This precursor, which also contains the sequences of **α-** and **β-melanocyte stimulating hormone (MSH),** was named **pro-opiomelanocortin.** Most of these peptides become biologically active only after they are cleaved from their precursor molecule.

Current evidence indicates that the endorphins profoundly affect mood and behavior. They are secreted along with ACTH in response to stress. They also aid in establishing homeostasis throughout the nervous system and the gastrointestinal system, and they may promote hormone secretion by the hypophysis and other glands. They are probably also involved in pain perception and may be part of the mechanism by which morphine and other pain-killing drugs act.

The physiological roles of methionine-enkephalin and leucine-enkephalin are not yet known. Perhaps they too promote hormone production.

Hormones of the Neurohypophysis

The neurohypophysis (posterior pituitary) releases two principal peptide hormones: **vasopressin** (also called **antidiuretic hormone,** or **ADH**) and **oxytocin.** Vasopressin and oxytocin are not synthesized in the neurohypophysis. They are synthesized in the hypothalamus and travel down to the neurohypophysis, where they are temporarily stored.

The intensively investigated hypothalamic-hypophyseal portal system (see Figure 28-17) consists of nerve cells (neurons), the cell bodies of which are in the hypothala-

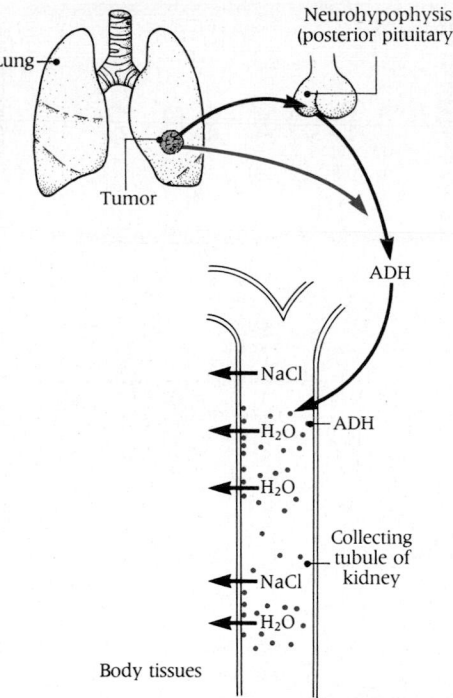

28-23 Inappropriate ADH secretion by a malignant lung tumor
Inappropriate ADH secretion occurs in certain kinds of lung cancer. ADH is synthesized by cancer cells (color), as well as by the neurohypophysis (posterior pituitary). ADH promotes water retention by the collecting tubules of the kidney. Compare this diagram with Figures 25-17 and 25-18.

mus and extensions of which (axons) travel to the neurohypophysis and terminate directly upon its capillaries. Hormones released from these terminals immediately enter the general circulation.

The synthesis of vasopressin and oxytocin within neurons resembles secretory processes in other cells. The small peptides are synthesized in the endoplasmic reticulum and carried to the Golgi apparatus where they are packaged into small granules which travel down the axons to the neurohypophysis.

As we learned in Chapter 25, vasopressin influences the kidney to reduce water excretion, an **antidiuretic effect** (see Figure 25-18).

- If body water content is low, as it is in a thirsty laborer on a hot day, blood plasma becomes slightly concentrated. This is detected by an "osmostat" in the hypothalamus, which causes the neurohypophysis to increase its output of vasopressin. This hormone is carried in the blood to the kidney, where it acts to reduce the urinary loss of water.
- If body water is in excess, as it is after drinking too much water, blood becomes slightly diluted. This is also detected by the hypothalamus, which reduces the amount of vasopressin secreted, thereby allowing the kidney to excrete a large volume of dilute urine.

If the hypothalamus or neurohypophysis is damaged, vasopressin is no longer produced and vast amounts of dilute urine may be excreted. This condition, **diabetes insipidus,** causes perpetual thirst.

Vasopressin also raises blood pressure in dogs or cats, but not in humans. This action, termed the **pressor effect,** explains the name of this hormone, which was applied before it and ADH were found to be identical.

Physicians have long been aware that malignant tumors of all kinds often produce peptide hormones of all kinds. Such ectopic hormone synthesis results in inappropriate secretion of hormones that can mimic the overactive states of true endocrine glands. For example, inappropriate GH secretion by a pancreatic tumor causes acromegaly resembling that of true hyperpituitarism. We raise the matter at this point because inappropriate ADH (vasopressin) secretion occurs commonly, especially with malignant lung tumors (Figure 28-23). One result, predictable from the known effects of ADH, is severe water retention.

Oxytocin was believed to have only two important functions—until recently (see Chapter 21).

- During the birth of a baby, sensory nerves are activated in the wall of the uterus and, in particular, the cervix. These nerves send impulses to the hypothalamus, which then increases output of oxytocin from the neurohypophysis. Oxytocin causes uterine contractions to proceed more vigorously. This is called the **oxytocic effect.**
- Second, oxytocin is essential for the ejection of milk from the breasts. When the baby sucks, sensory nerves in the nipple are activated. They send impulses to the hypothalamus, which then stimulates oxytocin secretion. Oxytocin travels in the blood to the breasts and makes the contractile myoepithelial cells contract, thus forcing milk out of the breast. This is the **lactagogic effect.**

Recent work shows that oxytocin has a role in hastening the onset of the estrus cycle in animals.

Chemical manipulation of the amino acid sequences of these two short peptide hormones has interesting and instructive effects (Box 28-2).

THE STRANGE PINEAL

The **pineal body** (also termed the **epiphysis**), is a pea-sized outgrowth of the base of the brain that was named for the pine cone it resembles. The seventeenth-century thinker René Descartes considered the pineal the seat of the soul. Physiologists of our century showed that the pineal synchronizes the body's responses to light. It now appears that the day-night cycle mediated by the pineal is the dominant factor, other than food, activating endocrine and behavioral rhythms.

The relations of light, the pineal, and the organs it controls have become clearer in recent years. We once thought that pineal function was controlled solely by

The discovery that antidiuretic-pressor and oxytocic-lactagogic effects are produced by two separate peptides led Vincent du Vigneaud and his associates at Cornell University Medical College to a remarkable series of discoveries in the 1940s and 1950s. By diligently purifying natural extracts of the neurohypophysis, they isolated and ultimately characterized the hormones vasopressin and oxytocin.

Although each of these peptide hormones contains only nine amino acids, their amino acid sequences differ in only two places (Figure 28C). These two differences are the following: phenylalanine and arginine in vasopressin are replaced, respectively, by isoleucine and leucine in oxytocin. Oxytocin has a large oxytocic effect and a small pressor effect; vasopressin has a large pressor effect and a small oxytocic effect.

Using novel procedures, du Vigneaud managed with difficulty to synthesize both peptides by the methods of organic chemistry. He had little difficulty in preparing "abnormal" peptides—that is, peptides with the "wrong amino acid" here and there in the chain. He was also able to make hybrids of vasopressin and oxytocin.

He soon learned, for example that substitution of a single residue in the oxytocin molecule—especially in the ring, which bears an interesting resemblance to part of the insulin molecule—greatly diminishes oxytocic activity. A strongly basic amino acid in the eighth position seems essential for pressor activity.

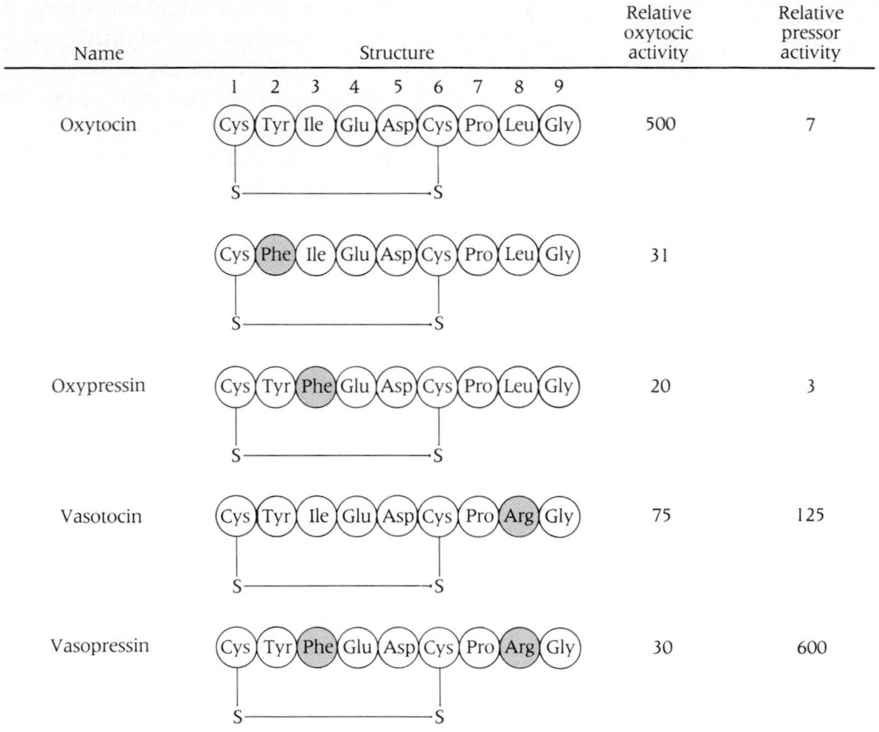

When investigators looked at these hormones in other organisms they found that their peptide structure is generally well conserved across a wide range of species. They also found variants rather similar to du Vigneaud's deliberately prepared analogues and hybrids. For example, in most vertebrates less advanced than mammals the peptide family is represented by a single substance,

Name	Structure	Relative oxytocic activity	Relative pressor activity
Oxytocin	Cys-Tyr-Ile-Glu-Asp-Cys-Pro-Leu-Gly (S—S)	500	7
	Cys-Phe-Ile-Glu-Asp-Cys-Pro-Leu-Gly (S—S)	31	
Oxypressin	Cys-Tyr-Phe-Glu-Asp-Cys-Pro-Leu-Gly (S—S)	20	3
Vasotocin	Cys-Tyr-Ile-Glu-Asp-Cys-Pro-Arg-Gly (S—S)	75	125
Vasopressin	Cys-Tyr-Phe-Glu-Asp-Cys-Pro-Arg-Gly (S—S)	30	600

28C Structure of oxytocin, vasopressin, and some synthetic analogues
Each of the short peptides differs from the others in only one or two amino acids (colored). Yet the differences in their oxytocic and pressor activity are striking.

called arginine-vasotocin because of the presence of arginine at position 8 in its amino acid sequence. Mammals have the hormones oxytocin and vasopressin. Oxytocin is identical with arginine-vasotocin except that in it leucine takes the place of arginine at position 8. Vasopressin is slightly more variable. In all mammals except the pig it is identical with vasotocin except that phenylalanine substitutes for isoleucine at position 3. In the pig one finds two substitutions: lysine takes the place of arginine at position 8 and phenylalanine takes the place of isoleucine at position 3.

In any case, it is clear that a minor change in the structure of the peptide alters its hormonal activity. Probably each peptide acts on different target cells. Perhaps some further minor modifications of the molecules will yield synthetic peptides whose specificity of action is greatly enhanced. Such substances may ultimately serve as valuable drugs.

environmental light, and that it secreted its chief hormone, **melatonin,** when the animal was exposed to darkness, and stopped secreting it when in sunlight or artificial light. We now know that the pineal has a glandular portion and a light-sensitive (photoreceptive) portion.

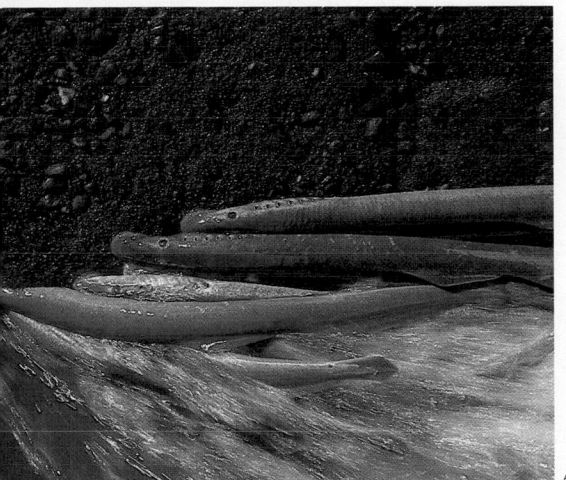

28-24 The pineal is a photoreceptor in certain species

Certain jawless fishes, like the lamprey, as well as frogs and reptiles have pineal glands that are responsive to light stimuli. (A) Pacific lampreys from Willamette, Oregon, Lampetra tridentata. *See also Figure 43-4. (B) A tree frog from Mexico,* Phrynohyas venulosa. *(C) A ringneck snake from Southern California,* Diadephis punctatus.

In lizards and certain other vertebrates, both portions of the pineal are present and functional. In others (notably mammals), the photoreceptive portion has been lost. While light controls melatonin secretion by the photoreceptive part of the pineal (in those species in which both parts of the pineal are functional), it now appears that the glandular part is controlled by an endogenous (internal) "clock" that causes the gland to function rhythmically even in individuals kept continuously in darkness.

We saw earlier that two criteria must be met to establish that a particular structure is a gland: (1) removal produces a noticeable change elsewhere in the body; and (2) injection of an extract of the gland or of its products reverses this change and reestablishes original function. These standards of proof were too restrictive for the pineal, for they failed to encompass the notion of a neuroendocrine transducer— that is, an endocrine gland that secretes its hormone in response to local neural influences. Pineal cells convert a neural input into a hormonal output (melatonin).

That the pineal might itself have photoreceptive capacities, at least in some vertebrates, was first postulated about 70 years ago. Noting that frog pineal cells strikingly resemble certain cells of the retina of the eye, anatomists speculated that the frog pineal is a photoreceptor, or "third eye," that detects light but cannot form images. But because pineal cells are not photoreceptors in mammals, some investigators concluded that the gland serves no biologic purpose in any species and is merely a vestige. This view gained support from the observation—well known to physicians who read skull x-rays—that the human pineal has a tendency to become calcified with age. However, later data showed that the frog pineal is, in fact, photoreceptive and does function as a third eye. Indeed, the pineal serves as a photoreceptor in a number of living animals—including certain cyclostomes ("jawless fishes"), frogs, and reptiles (Figure 28-24).

Our own ancestors up to the reptilian stage doubtless had a light-detecting pineal.[11] But it is not a mere vestige in human beings. Its role in mammals, including humans, is now thought to be as a mediator of the neuroendocrine system that converts information entering via the nervous system into factors controlling the output of melatonin. In this indirect role, the pineal reacts to light stimulating nerve cells in the eye. Nerves carry impulses from eye to brain, and then to certain nerve cells in the neck, which then travel to the pineal, which responds by secreting melatonin in inverse proportion to the external light level.

The hormone melatonin is a nonpeptide indole derivative that influences the rhythm of many physiological processes. Its major influence is antigonadal, suppressing gonadal growth and function. This explains how seasonal changes in day length influence reproductive behavior. When the days grow short in autumn, increased melatonin secretion tends to suppress gonadotropic hormone secretion and thus gonadal function. Melatonin also lightens skin color in frogs and other amphibians (see Figure 28-12).

Unlike the pattern of other endocrine glands, removal, loss, or dysfunction of the pineal does not cause abrupt termination of any one function. The reason is instructive. The pineal, a neuroendocrine transducer, possesses a degree of redundancy not widely present in the endocrine system. The pineal evidently shares responsibility with other neuroendocrine structures for regulating the timing and amplitude of biologic rhythms. Hence, it is never alone in controlling other glands or physiological processes.

ADRENAL GLAND

The **adrenal glands,** resting atop the kidneys (and thus termed suprarenals by earlier writers), contain an inner medulla and an outer cortex (Figure 28-25). Like the hypophysis, the glands contain two distinct regions that represent different and unrelated endocrine glands.

Adrenal Cortex

The **adrenal cortex** secretes a number of steroid hormones, or **corticosteroids.** Their synthesis and release is stimulated by ACTH. The actions of the several **adreno-**

[11] It is interesting that many fossil skulls of early amphibians and reptiles show large openings at the top where the third eye is found in living forms today.

cortical hormones are so numerous and diverse it is often difficult to tell whether their effects are primary or secondary. If, for example, adrenocortical secretion were to fail, functioning of kidneys, liver, muscle, and circulatory system would be profoundly altered. Important changes would also occur in the metabolism of carbohydrates, proteins, lipids, amino acids, electrolytes, and water. These hormones are vital in resistance to stress, a complex pattern that involves every part of the body.

Three generalizations concerning adrenocortical hormones can be made.

- Over 50 compounds can be found in the adrenal cortex, but only a few enter the bloodstream. The rest are biosynthetic intermediates.
- All compounds are steroid derivatives, hence the designation corticosteroids.
- The hormones can be grouped into three classes: (1) **mineralocorticoids,** which act upon extracellular Na^+, K^+, and Cl^-, and other ions; (2) **glucocorticoids,** which influence the metabolism of carbohydrates, proteins, and lipids; and (3) **sex hormones.** Most sex hormones come from the gonads; however, the adrenal cortex contributes some of them, including adrenocortical androgens, which mimic the testicular hormone, testosterone.

Among the mineralocorticoids and glucocorticoids, three are of particular importance: **aldosterone, cortisol,** and **corticosterone** (Table 28-5). In humans and many other mammals, aldosterone accounts for about 95% of the gland's mineralocorticoid activity and cortisol for about 90% of the glucocorticoid activity. Corticosterone accounts for a small percentage of both activities.

Deficiency of aldosterone and other mineralocorticoids leads to excessive excretion of Na^+ by the kidney—and with it an osmotically equivalent amount of water—and excessive retention of K^+. Administration of the mineralocorticoids reverses the pattern, causing the following effects:

- Increased renal retention of Na^+ and water
- Increased renal elimination of K^+ and H^+, presumably because the hormones stimulate Na^+-K^+ and Na^+-H^+ exchanges in the distal renal tubules (see Figure 25-17)

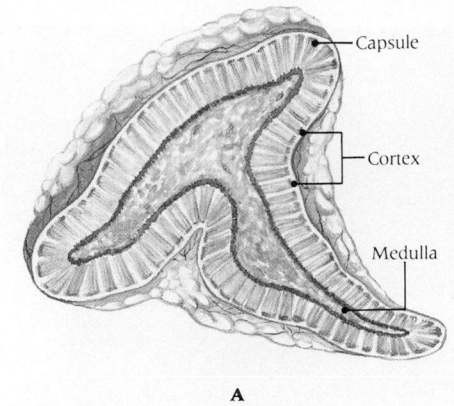

A

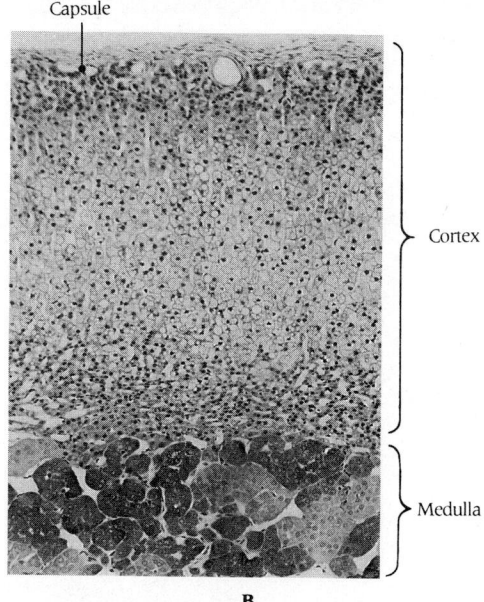

B

28-25 Structure of the human adrenal gland
(A) *The outer cortex and inner medulla are separate endocrine glands within the same organ.* **(B)** *A low power microscopic view of the adrenal gland in cross section.* (× 60)

TABLE 28-5
THREE MAJOR CORTICOSTEROIDS

Name	Structure	Properties
Cortisol		Main human glucocorticoid, produced in zona fasciculata of adrenal cortex. Daily output is 10 to 25 mg; little mineralocorticoid activity. Normal plasma level is 10 to 20 μg per 100 ml. Production is stimulated by ACTH.
Corticosterone		Produced in zona fasciculata of adrenal cortex; daily output is 2 to 5 mg; ratio of cortisol to corticosterone secretion 7:1; glucocorticoid activity, 30 to 50% that of cortisol. Normal plasma level 0.4 to 2.0 μg per 100 ml. Production is stimulated by ACTH.
Aldosterone		Main human mineralocorticoid, produced in zona glomerulosa of adrenal cortex. Daily output is 50 to 200 μg per 100 ml. Glucocorticoid activity insignificant; production stimulated by renin-angiotensin system.

PART 3 HOMEOSTASIS

These primary effects lead to many secondary effects. For example, large doses of aldosterone sharply increase extracellular fluid volume, cardiac output, and blood pressure. Occurrence of the last two effects, suggests that aldosterone acts on blood vessels and perhaps other body cells as well as those of the renal tubules.

The glucocorticoids are so named because they elevate the blood glucose level. The major glucocorticoid effects are the following:

- Increased protein catabolism and gluconeogenesis
- Increased deposition of liver and muscle glycogen
- Elevation of blood glucose
- Increased total body fat
- Increased erythrocyte production by bone marrow
- Decreased antibody formation
- Suppressed ACTH secretion
- Depressed thyroid secretion as a result of suppressed TSH secretion
- Anti-inflammatory effects, with decreased migration of leukocytes from blood to inflammatory sites and which underlie the use of glucocorticoids (cortisone or prednisone) in the treatment of inflammatory diseases like arthritis

Few of these effects occur visibly in normal individuals. Only when some abnormality causes increased or decreased hormone levels do they become apparent. In hyperadrenocorticalism, one observes elevated blood glucose levels, insulin insensitivity, and decreased inflammatory responses. In contrast, victims of hypoadrenocorticalism typically have low blood glucose levels and low blood pressure along with other abnormalities.

Adrenocortical activity rises strikingly in response to nonspecific stress—trauma, extreme temperatures, and the like. This is an important part of the body's response to environmental challenge: an adrenalectomized individual is incapable of withstanding stress. Precisely how glucocorticoids benefit stressed organisms is not clear. Perhaps they furnish additional substrates for energy metabolism. Glucocorticoid activity also rises under the influence of ACTH. Within minutes of the release of ACTH by the adenohypophysis—or of its medical administration—glucocorticoid secretion increases demonstrably. The molecular basis of ACTH's effects upon the adrenal cortex is incompletely understood. It is clear that ACTH induces the rapid utilization of adrenocortical cholesterol (in the synthesis of corticosteroids) and ascorbic acid.

Adrenal Medulla

It was known before 1900 that injection of extracts of **adrenal medulla,** a gland that is completely sheathed within the adrenal cortex, sharply elevates blood pressure.

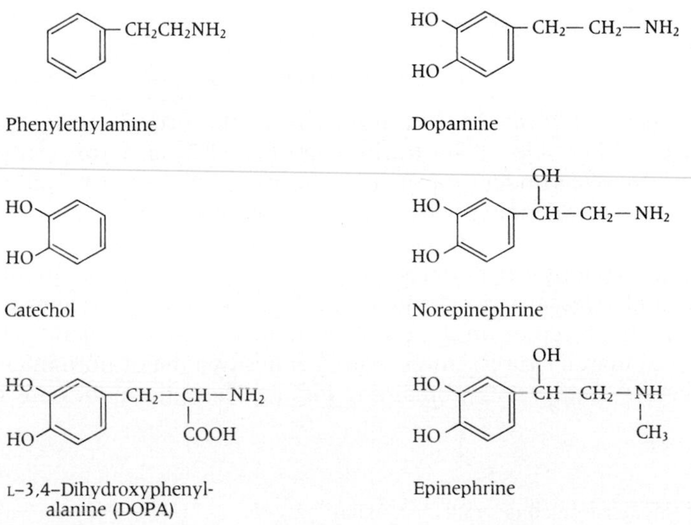

28-26 Molecular structures of various catecholamines
Note that epinephrine and norepinephrine differ only in the presence of a methyl group (CH_3) on the amine group of epinephrine.

Phenylethylamine

Dopamine

Catechol

Norepinephrine

L-3,4–Dihydroxyphenyl-alanine (DOPA)

Epinephrine

L-Tyrosine   L-DOPA   Dopamine   Norepinephrine   Epinephrine

In 1902, J. J. Abel isolated epinephrine,[12] an active crystalline compound, from the adrenal medulla.

The adrenal medulla secretes two hormones: **epinephrine** and **norepinephrine.** The two compounds differ structurally in only one respect: a methyl group is present in epinephrine (Figure 28-26). Both are chemically related to 1,2-dihydrobenzene, or catechol, and each has a short side chain bearing an amine (NH_2) group. Accordingly, they are both termed **catecholamines.** Epinephrine derives from norepinephrine (Figure 28-27). Both arise from the amino acid tyrosine.

Epinephrine and norepinephrine are secreted by chromaffin cells of the adrenal medulla. As we shall learn in Chapter 29, norepinephrine is also secreted in the sympathetic ganglia of the autonomic nervous system and in sympathetic postganglionic neurons. It is generally regarded as the chief transmitter substance at sympathetic neuroeffector junctions.

A study of the functions of the adrenal medulla is a study of the effects of epinephrine and norepinephrine. Although norepinephrine has a key role in sympathetic nerve function, the adrenal medulla is not indispensable. Adrenalectomized animals and humans thrive when given a sufficient supply of supplementary adrenocortical hormones. Yet no supplement of adrenal medullary hormones is required. Despite loss of the medulla, a major source of epinephrine, sympathetic nerves under proper excitation continue to liberate norepinephrine.

Adrenal medullary cells are modified sympathetic ganglion cells. The effects of epinephrine and norepinephrine mimic those produced by stimulation of the sympathetic nervous system. As we will see, these are the functions that ready the body for "fight or flight." Among them are blood pressure elevation due to arteriolar constriction, dilation of the pupils of the eyes and bronchioles of the lungs, accelerated heart rate, and inhibition of intestinal peristalsis.

Injection of either epinephrine or norepinephrine causes a sudden brief rise in blood pressure. Epinephrine increases the basal metabolic rate, the rate of glycogen breakdown in liver and muscle, and the blood glucose level. The last effect is attributable in the dog and probably in humans to the stimulation of adenylate cyclase and the cyclic AMP-induced stimulation of glycogen phosphorylase via a protein kinase. Norepinephrine has the same effects to a lesser degree. Glycogen synthase, the glycogen-forming enzyme, is simultaneously inhibited. Thus the catecholamine hormones of the adrenal medulla rapidly increase the production of glucose in the liver by stimulating glycogen breakdown and inhibiting glycogen synthesis. Both hormones also promote the release of fatty acids from fat stores, which raises the blood fatty acid level.

THYROID GLAND

The thyroid gland is located in the front part of the neck (see Figure 28-16). Its two lobes, on either side of the trachea, are joined by a narrow strip called the isthmus. The thyroid is under the primary control of thyrotropic (thyroid-stimulating) hormone (TSH), an adenohypophyseal tropic hormone, whose own release is controlled by thyrotropin-releasing factor (T-RF), which travels in the blood via the hypothalamic-hypophyseal portal system. Nearly every step of thyroid hormone production and release is accelerated by TSH. As shown in Figure 28-28, TSH binds to a specific receptor on the cell membrane of a thyroid gland cell.

The thyroid manufactures, stores, and liberates two major hormones, distinctive in that they are amino acids containing iodine. These hormones are **thyroxine,**

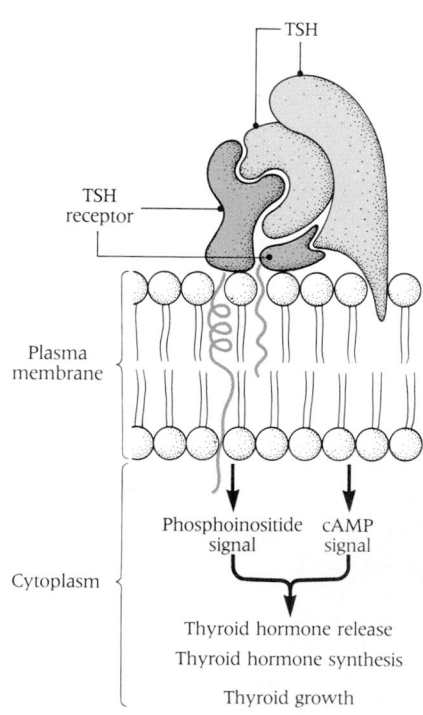

28-28 Mechanism of action of TSH
Thyroid-stimulating hormone (TSH) is composed of two polypeptide subunits. TSH binds to its receptor (which also has two subunits) in the plasma membrane of a thyroid cell. Release of thyroid hormone from the gland is mediated by cAMP and by signals from a phosphoinositide derivative.

[12] This hormone is sometimes called adrenalin.

also called **tetraiodothyronine (T₄),** and **triiodothyronine (T₃).** In addition, the thyroid secretes **calcitonin,** a hormone that regulates calcium metabolism.

T_4 and T_3 are both transported in the blood attached to plasma proteins. In the tissues, they alter the rate of practically every facet of intracellular metabolism. Consequently, a condition of hypothyroidism (absence of thyroid hormone) depresses the basal metabolic rate (BMR), whereas hyperthyroidism (excess of thyroid hormone) may sharply raise it. The thyroid is perhaps the most familiar of the endocrine glands (Figure 28-29).

Thyroid function involves a number of discrete steps or processes.

- Intestinal absorption of dietary iodide into the blood plasma
- Transport of plasma iodide to the thyroid gland
- Concentration of plasma iodide by thyroid tissue
- Enzymatic conversion of accumulated inorganic iodide that is "trapped" in the gland to organic iodide derivatives of the amino acid tyrosine, including the major thyroid hormone, thyroxine or T_4
- Retention of these compounds in follicles within the gland in the form of a protein called **thyroglobulin**
- Enzymatic digestion of thyroglobulin, under stimulus of TSH, with release into the blood of the iodine-containing thyroid hormones T_4 and T_3
- Binding of hormones to certain plasma proteins
- Transport of protein-bound thyroid hormones to body tissues
- Metabolic actions of thyroid hormones in body cells, possibly including a transformation to more active forms
- Chemical breakdown of thyroid hormones in the tissues
- Excretion of thyroid hormone degradation products by liver and kidneys

Thyroid hormones stimulate virtually every process of metabolism. The most prominent actions of thyroid hormones are the following:

- Elevation of the BMR
- Maintenance of normal growth
- Maintenance of normal functioning of the nervous system
- Increased glucose utilization and mobilization of glycogen stores
- Accelerated synthesis and breakdown of proteins
- Augmented epinephrine and norepinephrine effects

TSH stimulates both iodide transport and hormone synthesis in the thyroid. Inasmuch as TSH secretion by the adenohypophysis depends upon the hypothalamic secretion of T-RF, a means exists by which the brain, acting through the hypothalamus, can influence the level of TSH secretion. This influence accounts for the stimulation of thyroid hormone secretion by such factors as intense emotion and cold. TSH secretion is regulated by the levels of T_4 and T_3. Thyroid hormones in the plasma also exert a measure of direct control on the thyroid.

PARATHYROID GLANDS AND CALCIUM METABOLISM

As their name suggests, the **parathyroid glands** are next to the thyroid gland. However, they are entirely separate and distinct from the thyroid. Usually four in number, they are the smallest of the known endocrine glands, and for many years their presence was overlooked.

Parathyroid Hormone

Parathyroid hormone (PTH) is a single-chain polypeptide hormone composed of 84 amino acids (Figure 28-30). It is first synthesized in the parathyroid glands as a 115-amino acid precursor called **preproparathyroid hormone (preproPTH).** The first 25 amino acids are rapidly cleaved off to yield the 90-amino acid **proparathyroid hormone (proPTH),** which is then converted to PTH. Once again we see that an active peptide hormone is derived from an inactive prohormone that must be enzymatically cleaved. In this case, as in the case of insulin, a "pre" sequence

28-29 Endemic goiter
Hypothyroid individuals suffer from a deficiency of thyroid hormone. In some cases, this is associated with a thyroid gland enlargement called goiter. This is extremely common in certain parts of the world—for example, Bangladesh and Equador.

28-30 Biosynthesis of parathyroid hormone
Preproparathyroid hormone is the parent molecule; 25 amino acids are removed from it to generate proparathyroid hormone. Another 7 amino acids are cleaved off to produce biologically active parathyroid hormone.

called the signal peptide sequence is lopped off, leaving a final product with a tail. This is the "pro" sequence. The final hormone—the "pro" sequence minus its tail—is secreted by the gland after passing through the Golgi apparatus where it is glycosylated (Figure 28-2).

The essential role of PTH is to regulate blood levels of ionic calcium (Ca^{2+}), which is critical to the functioning of many organs (heart, brain, skeleton) and many metabolic events in the cell. The best known calcium-requiring systems are the following:

- Calmodulin-dependent control of many enzymes
- Nerve impulse transmission
- Stability of cell membranes
- Muscular contractions
- Blood clotting
- Manufacture of bones and teeth

The normal amount of calcium in plasma is 9–10 mg per 100 ml. Among the factors that keep it constant are the level of intestinal absorption of calcium, the renal excretion of calcium, bone metabolism, dietary calcium, and PTH actions.

Calcium in blood plasma takes three forms. About 46–50% of it exists as free calcium ions (Ca^{2+}), 46–50% is bound to the plasma proteins albumin and globulin, and 4–8% is complexed to phosphate. Only free Ca^{2+} is physiologically active. The ratio of ionized (free) to bound calcium depends largely on blood pH. If the pH rises, more calcium is bound.

PTH increases plasma calcium levels and decreases the concentration of phosphate in at least four ways.

- It causes release of calcium from bone into the blood.
- It conserves calcium by reducing its excretion by the kidneys.
- It promotes excretion of phosphate by the kidneys. The relation between serum Ca^{2+} and phosphate is poorly understood. Nonetheless, the product of $[Ca^{2+}] \times [PO_4^-]$ tends to remain constant (with well-known exceptions). Hence, if $[PO_4^-]$ decreases, $[Ca^{2+}]$ tends to rise.
- It increases calcium absorption from the intestine by mechanisms requiring the participation of vitamin D.

Output of PTH from the glands depends on the level of Ca^{2+} in the blood. If plasma Ca^{2+} falls, hormone output increases, returning Ca^{2+} to normal. If plasma Ca^{2+} rises, hormone secretion is suppressed and Ca^{2+} levels fall back to normal.

Calcitonin and Vitamin D

Several other substances play a critical role in the regulation of calcium metabolism. These include the peptide hormone **calcitonin,** the steroid **vitamin D,** and the ubiquitous protein **calmodulin.**

A rise in the level of blood Ca^{2+} stimulates secretion of calcitonin by the thyroid gland, which tends to lower Ca^{2+} levels (Figure 28-31). Calcitonin, a straight-chain peptide containing 32 amino acids, was discovered in 1962. However, in spite of nearly 30 years of active investigation its exact role remains controversial.

Vitamin D, another powerful regulator of calcium metabolism, is unique among the vitamins for several reasons. One is its peculiar relation with sunlight. A **provitamin D** is normally present in the skin. Under the influence of the ultraviolet component of sunlight, it is converted to a **previtamin D.** A slightly altered form of this molecule is transported to the liver, where it is chemically modified by the addition of —OH groups. The product is 1,25-dihydroxyvitamin D_3 (Figure 28-32). Its role is to promote the intestinal absorption of calcium. With parathyroid hormone, it maintains blood Ca^{2+} levels. Vitamin D deficiency profoundly impairs calcium metabolism and leads to the bone disease called rickets. The need for sunlight in the synthesis of previtamin D in skin explains why people living without sunlight (e.g. submarine crews) urgently need vitamin D supplements.

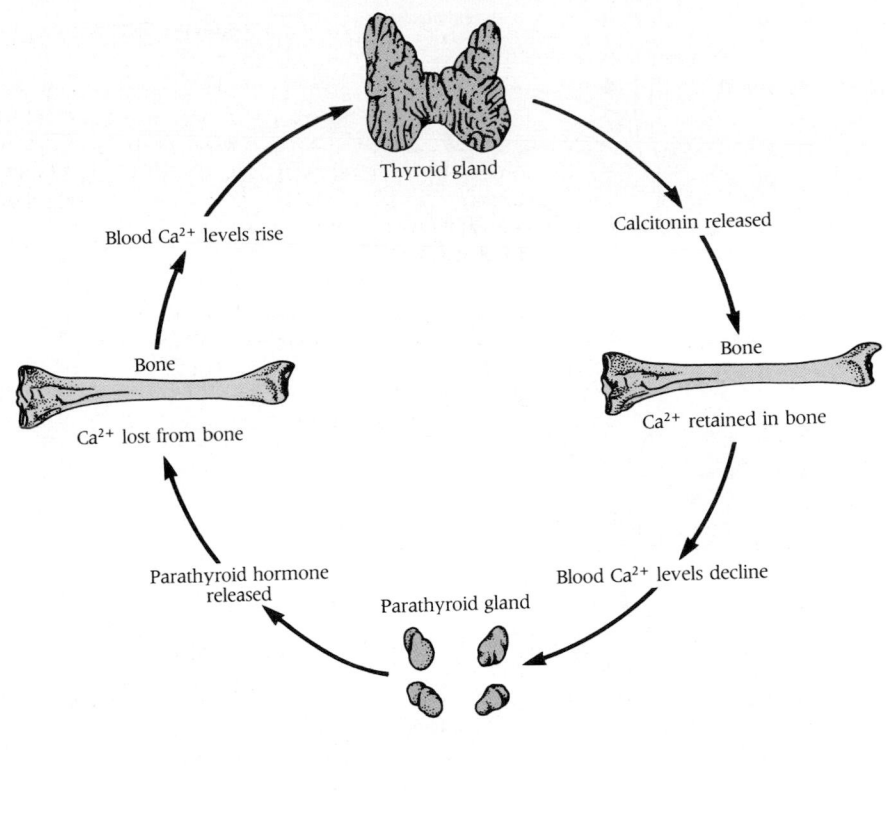

28-31 Regulation of Ca²⁺ metabolism

When blood Ca²⁺ is high, the thyroid secretes calcitonin which prevents resorption of bone. As a result, bone Ca²⁺ retention remain high and blood levels fall. This stimulates release of parathyroid hormone which enhances Ca²⁺ loss from bone, causing blood levels to rise.

Thyroid gland

Blood Ca²⁺ levels rise

Calcitonin released

Bone

Bone

Ca²⁺ lost from bone

Ca²⁺ retained in bone

Parathyroid hormone released

Blood Ca²⁺ levels decline

Parathyroid gland

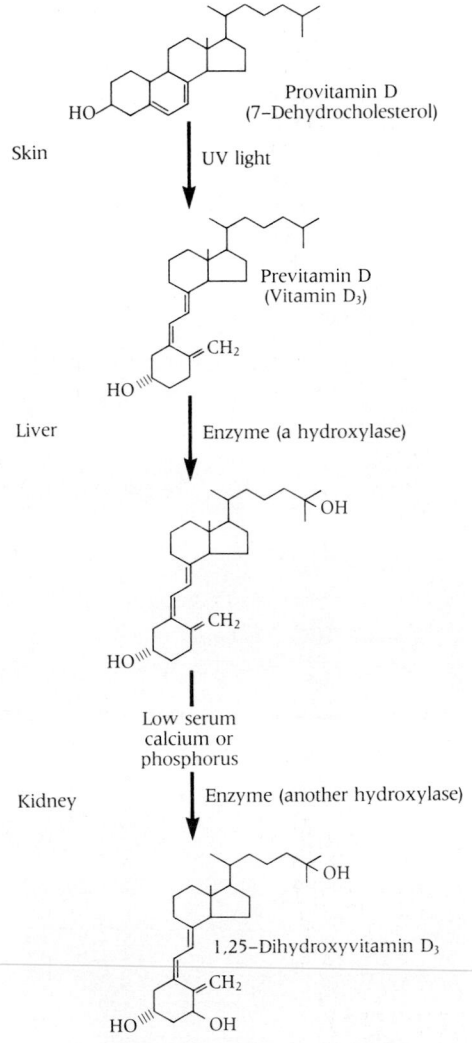

Skin

Provitamin D
(7–Dehydrocholesterol)

UV light

Previtamin D
(Vitamin D₃)

Liver

Enzyme (a hydroxylase)

Low serum
calcium or
phosphorus

Kidney

Enzyme (another hydroxylase)

1,25–Dihydroxyvitamin D₃

28-32 Biosynthesis of vitamin D₃

The active form of vitamin D, vitamin D₃ (1,25-dihydroxyvitamin D₃), is synthesized in an unusual pathway. The first step occurs in skin exposed to the ultraviolet component of sunlight. A cholesterol derivative, 7-dehydrocholesterol, is converted (by photolysis) to vitamin D₃. In the liver, one hydroxylase then adds an —OH group. In the kidney, another hydroxylase adds the second —OH group.

GONADS

The gonads—testes in the male and ovaries in the female—secrete several hormones important in determining the functions of other sex organs. These hormones include testosterone and estrogens in the male and progesterone and estrogens in the female. (Chapter 21).

Sex Hormones: Estrogens and Progesterone

The **estrogens** and **progesterone** control the development and function of the reproductive organs and the development of the secondary sexual characteristics in the female. The main sources of these hormones in the nonpregnant female are the ovaries. In pregnancy, the placenta secretes enormous amounts of both estrogens and progesterone. In addition, the membrane that forms from the placenta, the chorion, produces a gonadotropic hormone called **chorionic gonadotropin** and a lactogenic hormone called **placental lactogen.**

The estrogens have three sorts of effects on the body: (1) direct effects upon the female reproductive organs functioning as specific growth hormones, (2) effects upon the adenohypophysis, suppressing FSH release and promoting LH release, and (3) general metabolic effects.

In summary, estrogens have the following specific effects:

- They induce the changes of puberty, including development of the genital organs, breasts, pubic and axillary hair, and the feminine body contour with broadening of the pelvis.
- Cyclic production of estrogens causes the proliferative phase of the uterine cycle.
- The mucus secreted by glands of the cervix is increased, has a reduced viscosity, and an increased pH, which greatly favors migration, motility, and longevity of spermatozoa.
- FSH release is suppressed and LH release promoted by the adenohypophysis.
- The basal metabolic rate (BMR) is increased to a small extent.
- Sodium and water retention by kidney is promoted.
- Bone formation is accelerated and protein anabolism is enhanced.

The last effect is related to osteoporosis (loss of bone calcium) in postmenopausal women in whom estrogen secretion is diminished. Synthesis of the protein matrix

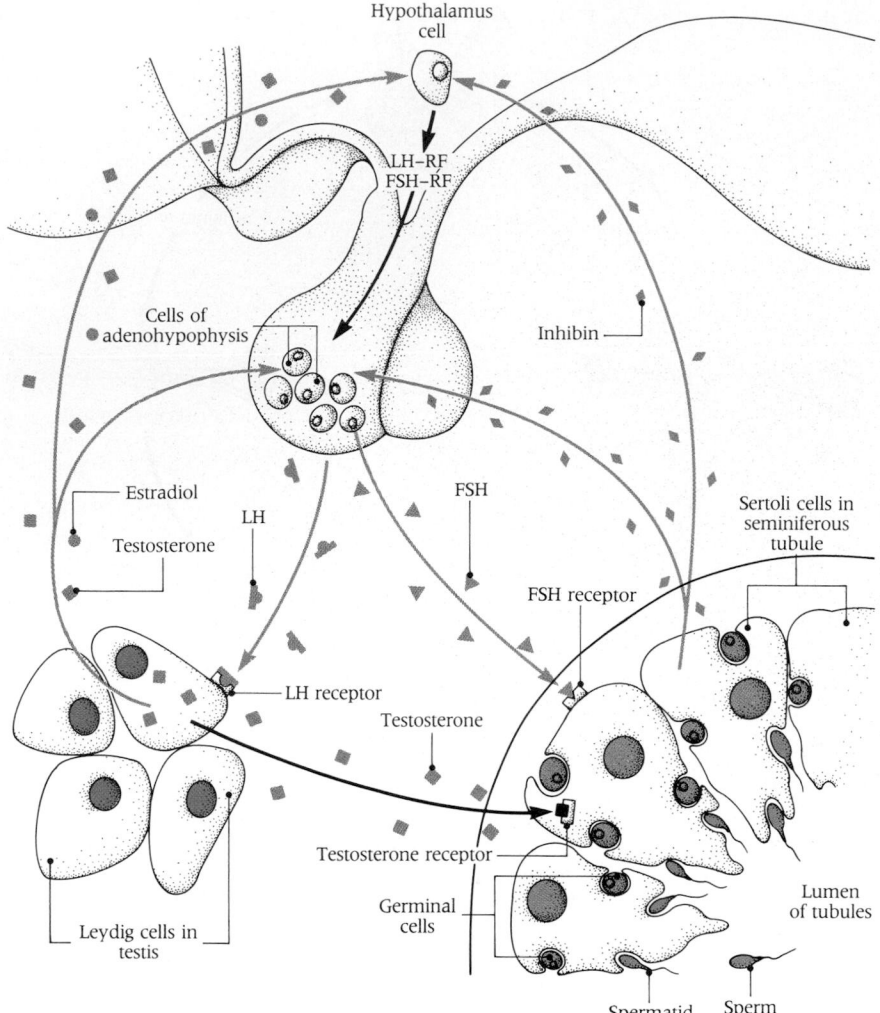

Hypothalamus
cell

LH-RF
FSH-RF

Cells of
adenohypophysis

Inhibin

Estradiol

Testosterone

LH

FSH

Sertoli cells in
seminiferous
tubule

FSH receptor

LH receptor

Testosterone

Testosterone receptor

Germinal
cells

Lumen
of tubules

Leydig cells in
testis

Spermatid   Sperm

28-33 Regulation of testicular function by LH and FSH
The hypothalamus and adenohypophysis control testis functions. LH-RF and FSH-RF, respectively, regulate LH and FSH release from the adenohypophysis. Binding of LH to receptors on the surfaces of Leydig cells leads to secretion of testosterone and estradiol into the blood. These sex hormones control the release of LH-RF and FSH-RF by negative feedback. Binding of FSH and testosterone to Sertoli cells of the seminiferous tubule stimulates spermatogenesis. Inhibin, a product of Sertoli cells, inhibits FSH secretion from the anterior pituitary.

of bone slowly diminishes. As a result, calcium salts lack a framework on which they can be laid down as bone salts.

In general terms, progesterone (as its name implies) acts to prepare the uterus for pregnancy. Specifically, it has the following effects:

- Stimulates growth of a secretory type of endometrium
- Diminishes muscular contractions of Fallopian tubes and uterus
- Promotes glandular development of the breasts

Most of these effects occur only after prior estrogen action. Progesterone is also believed to inhibit LH secretion and to stimulate FSH secretion in the next cycle. Because progesterone in high concentrations also inhibits ovulation, it is used in oral contraceptives.

Male Sex Hormones: Androgens

Like ovaries, testes are influenced by two adenohypophyseal gonadotropic hormones: **follicle-stimulating hormone (FSH)** and **luteinizing hormone (LH)** (Figure 28-33). However, their ultimate effects are different in the male. They stimulate, respectively, the germinal cells of the seminiferous tubules of the testes to produce spermatozoa and the Leydig cells to produce **androgens,** the main one of which is **testosterone.** As shown in Figure 28-33, secreted androgens complete a feedback loop that influences the hypothalamus to decrease the release of gonadotropic hormones by the adenohypophysis. A hormone called **inhibin** is released from Sertoli cells; it decreases FSH secretion.

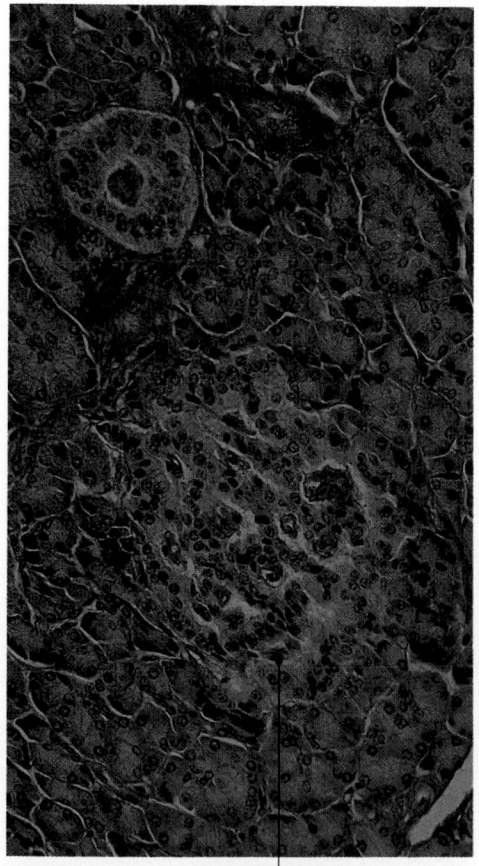

Islet of Langerhans

28-34 Islet of Langerhans in the pancreas
This microscopic view of a section of pancreas shows a cluster of cells known as an islet of Langerhans. Islets secrete insulin and glucagon into the bloodstream. (× 620)

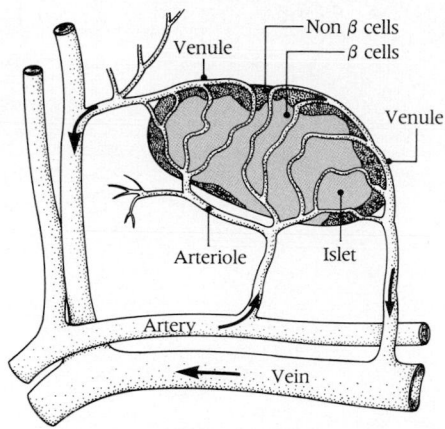

28-35 Blood vessels serving an islet of Langerhans
The diagram shows an extensive capillary bed surrounding the islet region containing beta (β) cells. Blood flows from arterioles into the capillaries which receive the hormones. It is then collected by venules.

Testosterone and other androgens have the following effects:

- Cause development of primary and secondary sexual characteristics, including growth of the genital organs, distribution of the body hair in a male pattern and occasional baldness, deepening of the voice, and alteration of the skin with darkening of pigment and stimulation of sebaceous glands, frequently resulting in acne
- Cause calcium retention and growth of long bones and causes bone growth to cease when correct bone length has been attained
- Stimulate protein anabolism with increased gluconeogenesis, acting somewhat in the manner of growth hormone
- Increase BMR
- Promote sodium and water retention
- Stimulate red cell production by bone marrow
- Cause aggressive behavior in many mammals

Like estrogens, androgens act by promoting selective gene transcription.

It should be understood that male and female sex hormones are both present and active in both sexes. However, one or the other predominates in each sex. In the course of aging, secretion of the predominant sex hormone slows down—estrogens in females, androgens in males. The other, nondominant hormone may then manifest itself more visibly.

PANCREAS

The **pancreas** is a compound organ that lies behind the stomach. As discussed in Chapter 26, certain exocrine glands of the pancreas produce various digestive enzymes that pass through the pancreatic duct to the duodenum. However, the organ is studded with numerous **islets of Langerhans,** little endocrine glands secreting the hormones that regulate carbohydrate metabolism (Figures 28-34 and 28-35). The β cells of the islets secrete **insulin.** The α cells secrete **glucagon.** The islets are about a million in number and together constitute about 1 g of tissue (about 15% of the weight of the pancreas).

The concentration of glucose in blood is normally 70–90 mg per 100 ml of plasma. Maintenance of a constant glucose level is one of the most important integrative functions of the endocrine system. Only in this way are all organs of the body able at all times to obtain adequate supplies of energy-yielding nutrient. Almost all body cells, apart from those of the liver and brain, place a barrier to the free entry of glucose from the blood. The hormones regulating glucose metabolism act in part by altering the effectiveness of this barrier. The exception is the cells of liver and brain, which are freely permeable to glucose at all times.

Humans—and most organisms—do not eat continuously. Hence, they need mechanisms for keeping the blood glucose level constant. Immediately after a meal, a large amount of glucose enters the blood and the plasma glucose concentration

28-36 Proinsulin is converted to insulin
Cleavage of proinsulin at two sites in the amino acid sequence (arrows) yields a C-peptide and an insulin molecule, which is composed of an A-chain and B-chain that are held together by disulfide bonds. Compare with Figure 4-16.

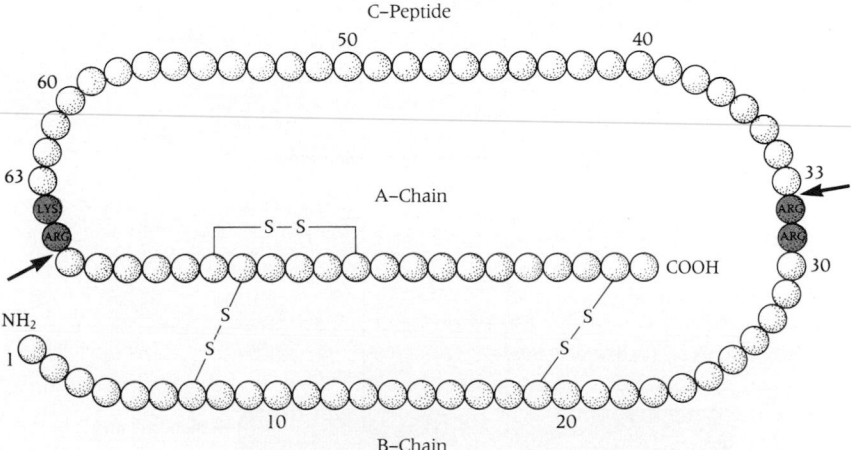

rises. But in a short time, absorption of the meal is completed and the blood glucose level falls. The aim of any regulating system is to damp these swings.

The liver is the first line of defense. All portal vein blood from the intestine goes to the liver. Just after a meal, when glucose is being absorbed rapidly, much of it is taken up by the liver and converted to liver glycogen. As a result, blood leaving the liver via the hepatic veins contains less glucose than blood entering the liver via the portal vein. Glycogen, the principal storage form of carbohydrates in animals, functions much like starch in plants. That is why glycogen has been called "animal starch." Between meals, when intestinal absorption has been completed and no glucose is entering the portal blood from the intestine, the liver calls up some of its stored glycogen and releases it into the blood as glucose. In this way, a sharp fall in blood glucose is avoided during times when no carbohydrate is being ingested.

This "smoothing-out" action of the liver is facilitated by the actions of **insulin,** a small polypeptide hormone (mol. wt. 6000). As we learned in Chapter 4, insulin has two polypeptide chains, the A-chain and the B-chain. However, it is synthesized in the β cells of the islets of Langerhans as a large single-chain precursor called **preproinsulin.** This molecule is then converted to **proinsulin** by deletion of 23 amino acids at the N-terminus of the B chain. The A-chain and B-chain sequences of the insulin molecule are linked within proinsulin by a 35-amino acid sequence called the connecting peptide or C-peptide (Figure 28-36). Deletion of the C-peptide in β-cell cytoplasm yields equal amounts of insulin and free C-peptide, both of which enter the blood (Figure 28-37). We still do not fully understand the mechanism of insulin action. Its most conspicuous effect is a lowered blood glucose level. It does this in three ways.

- It binds on cell surfaces to a specific insulin receptor, a molecule whose structure remarkably resembles that of immunoglobulin (Figure 28-38). There it alters the properties of cell membranes, causing enhanced transport of metabolites (glucose, fatty acids, amino acids, etc.) in most body cells. One immediate result is a sharp increase in glucose uptake. In skeletal muscles and in the heart, much of the glucose taken into the cells is converted into a glycogen store.

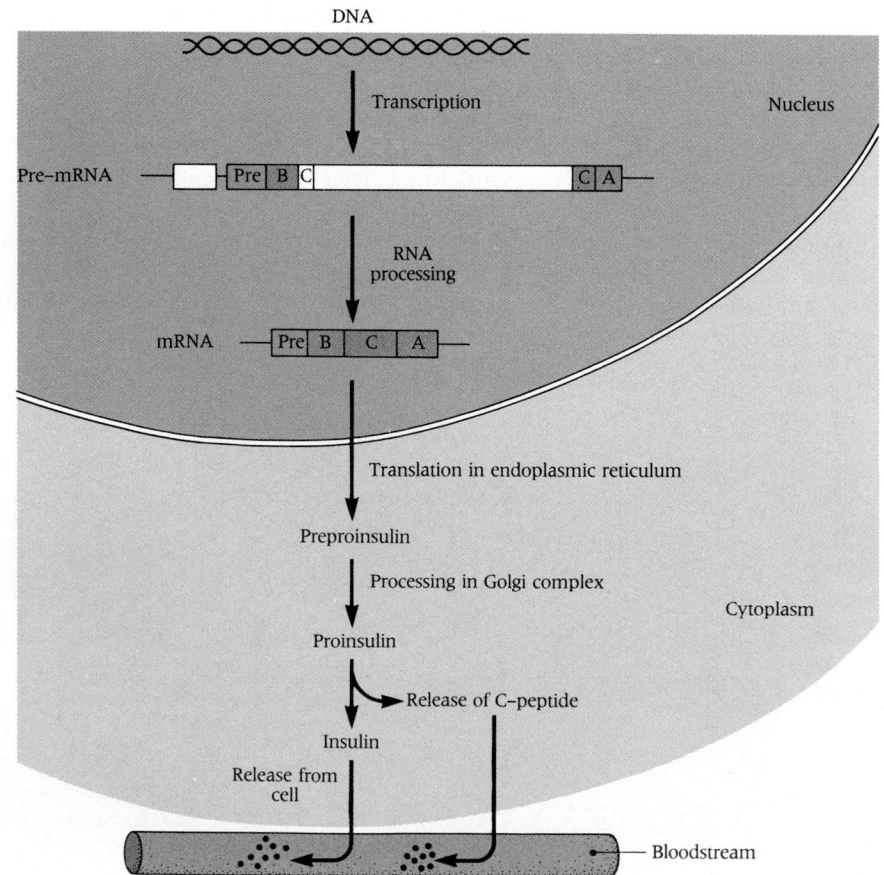

28-37 Overview of insulin synthesis and release from β cells
The product of mRNA translation is a large precursor, preproinsulin. Successive cleavages of the polypeptide chain generates insulin and C-peptide, both of which leave the β cell and enter directly into the bloodstream.

28-38 Structure of the insulin receptor
The receptor is composed of A and B subunits which are held together by disulfide bonds. Carbohydrate chains extend into the cell exterior. The activated insulin receptor is an enzyme (a protein kinase) that catalyzes the phosphorylation of tyrosines in target proteins.

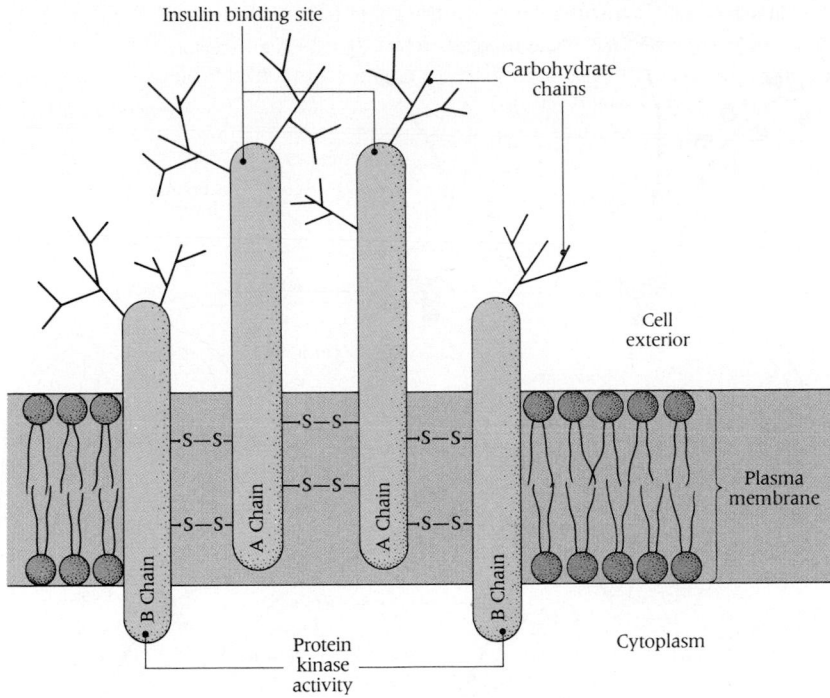

- It directs cell metabolism into biosynthetic activities (anabolism). All systems are affected, including the biosyntheses of glycogen from glucose, fatty acids from pyruvate, triglycerides from fatty acids, proteins from amino acids, and DNA and RNA from nucleotides.[13] These actions of insulin tend to increase the availability of carbohydrate inside cells and to reduce the availability of fat.
- Insulin also enhances cell growth, operating in tandem with other growth factors, including NGF, EGF, PDGF, and IGF-I and II (see Table 28-3).

Insulin release is controlled by blood glucose level. When blood glucose rises, insulin output rises and the blood glucose tends to return to normal. When blood glucose falls, insulin secretion falls; this decreases use of glucose by body cells so that blood glucose levels rise.

Diabetes mellitus is a state in which insulin levels are too low to cope with the intake of carbohydrate. As a result, blood glucose levels rise and glucose is excreted in the urine (glycosuria). Some cases begin in childhood, others in later life. Diabetes mellitus is probably a cluster of diseases of diverse causes. For example, a viral illness or an autoimmune process. Certain combinations of HLA alleles strongly influence susceptibility to diabetes in affected families. In most cases, diabetes is correctable in part by reducing the intake of carbohydrates, by administration of insulin, or by administration of drugs that mimic insulin or stimulate its production.

Interestingly, three cases have been found in which a mutation has replaced an amino acid involved in the binding of insulin to its receptor.[14] In each case, the individual's insulin has reduced activity and an illness exists that resembles diabetes mellitus. Normal insulin readily reverses the pattern. Figure 28-39 summarizes some of the recognized defects that can cause diabetes.

---

[13] These effects are mediated by indirect influences of insulin on the relevant enzymes. Most involve the phosphorylation of a biosynthetic enzyme by an insulin-activated kinase which thereby activates the enzyme.

[14] In the defect known as Insulin Chicago, phenylalanine is replaced by leucine as amino acid 25 in the B chain. In Insulin Los Angeles, serine replaces phenylalanine at position 24 in the B chain. In Insulin Wakayama, valine is replaced by leucine at position 3 in the A chain. The general similarity of these mutant proteins to the familiar mutations of hemoglobin (see Box 22-1) is striking.

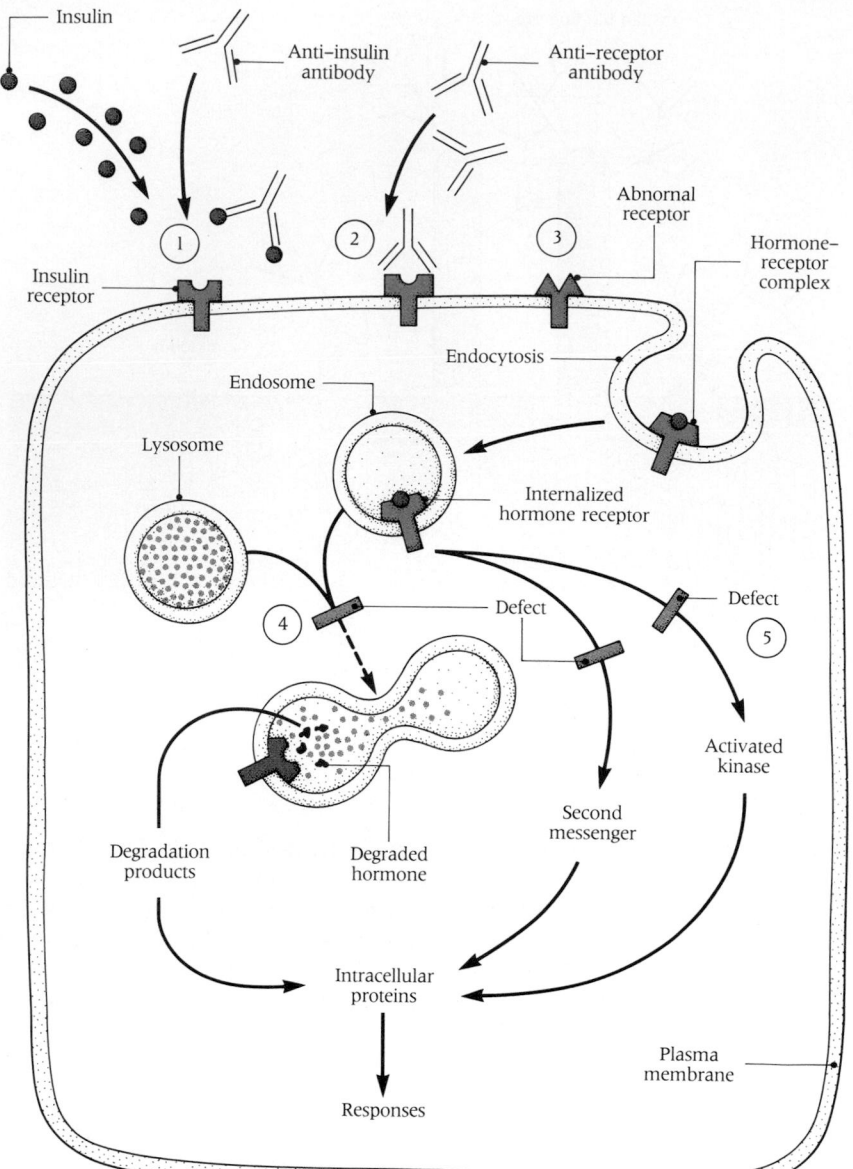

Glucagon, a polypeptide chain of 29 amino acids, is also produced by the islets of Langerhans, but its actions are opposed to those of insulin. By stimulating cAMP production, it depresses glycogen synthesis and raises blood glucose levels. It even stimulates the breakdown of protein to amino acids, which are then converted to glucose by gluconeogenesis. Such an effect is useful in coping with starvation.

For many years, insulin stood alone as the solitary member of its hormonal class. Yet its actions were counterbalanced by a host of regulatory hormones, including glucagon, epinephrine, growth hormone, and cortisol. Lest anyone consider this a simple system, we must record that insulin is now considered one of a family of structurally similar agents that includes two of the growth factors listed in Table 28-3—insulinlike growth factor 1 (IGF-1) (also called somatomedin C) and insulin-like growth factor 2 (IGF-2) (also called somatomedin A). Both share insulin's metabolic effects.

Significantly, these molecules form in the liver under the influence of growth hormone (GH), which does some of its work indirectly via the mediation of these chemical messengers. Two more peptides that may be classified as hormones arise in specific cells of the islets of Langerhans. One, simply termed **pancreatic peptide,** is a 36-amino acid peptide that appears to act on the gastrointestinal system and may produce satiety—a feeling of fullness and satisfaction—after eating.

Insulin has an illustrious role in the history of biology. It was the first protein hormone to be isolated and identified—by Frederick G. Banting and Charles H.

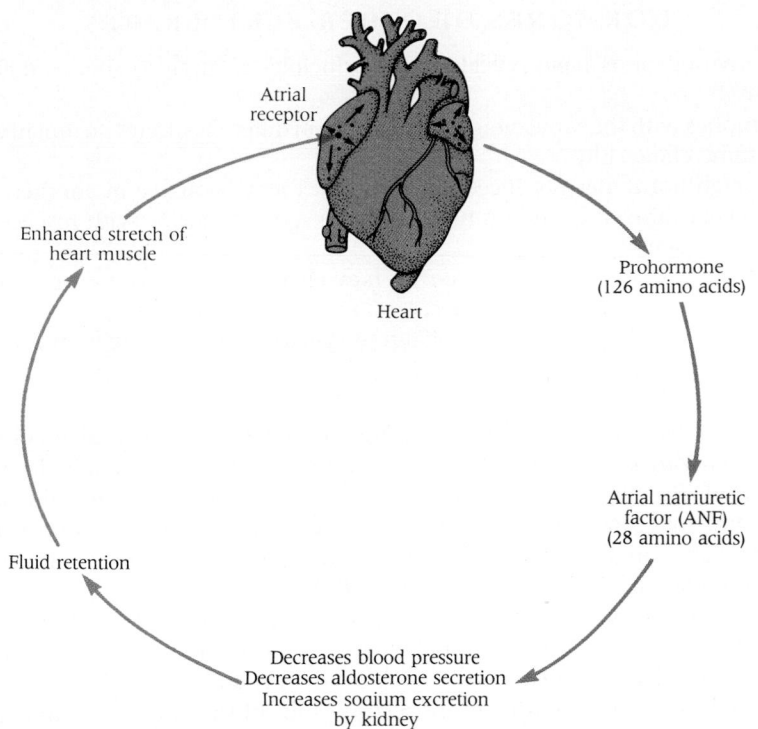

28-40 The atrial natriuretic factor system
ANF released from the atrium of the heart modulates fluid balance, thereby regulating the degree of contraction of cardiac muscle. One of its major effects (for which it is named) is increased sodium excretion by the kidney.

Atrial
receptor

Heart

Enhanced stretch of
heart muscle

Prohormone
(126 amino acids)

Atrial natriuretic
factor (ANF)
(28 amino acids)

Fluid retention

Decreases blood pressure
Decreases aldosterone secretion
Increases sodium excretion
by kidney

Best of the University of Toronto in 1922.[15] It was the first protein of any kind to yield its amino acid sequence. And it was also the first protein to have its crystal structure determined. Three Nobel Prizes were awarded for these contributions.

A HORMONE FROM THE HEART

As we have seen, homeostatic control of body sodium and water—and of blood pressure—involves a complex web of interacting hormonal and neural mechanisms. Major determining factors include central and autonomic nervous systems, cardiac output, blood vessel tonicity, renal function, aldosterone, the renin-angiotensin system, catecholamines, and ADH (vasopressin).

It was found in 1981 that the heart atrial muscle produces a small polypeptide hormone—**atrial natriuretic factor (ANF)**—that interacts with several of these factors. ANF, whose name alludes to "natriuresis" (sodium excretion by the kidney), potently promotes sodium excretion, lowers blood pressure, and inhibits aldosterone secretion (Figure 28-40). A substance with the properties of ANF was long sought on theoretical grounds. Its actual discovery, however, derived from morphological studies of cardiac muscle cells.

One facet of the ANF story is especially noteworthy: the swiftness with which this new hormone was postulated to exist, identified in its "prepro" and "pro" forms, isolated, sequenced, and cloned. Now the biological meaning of ANF is being pondered. ANF must have appeared early in evolution. This suggests that it may have different functions that vary in accordance with each species' environment.

HORMONES OF OTHER ANIMALS

The endocrine systems of most mammals resemble those of the human body. Vertebrates other than mammals also have somewhat similar endocrine systems, but significant differences are found.

[15] The story of this discovery is an inspiring tale. Banting and Best carried out the definitive experiments in the summer of 1921. In October 1920, Banting outlined the experimental plan in his notebook: "Ligate [tie off] pancreatic ducts of dog. Keep dogs alive until acini [exocrine glands] degenerate leaving islets. Try to isolate internal secretions of islets and attempt to relieve glycosuria [renal excretion of glucose by diabetic]." The point here is that if the exocrine glands had not been caused to degenerate, their potent proteases would have hydrolyzed insulin, which is, after all, a protein. The experiment was performed on July 30, 1921. Injection into a normal dog of an extract of duct-ligated pancreas caused a rapid drop in the blood glucose level. By December 1921, the two young scientists administered their extracts to human diabetics—with stunning results.

HORMONES OF OTHER VERTEBRATES

Recent investigations have revealed four principles relating to species differences.

- Hormones with the same biological function in different species are not necessarily the same chemically.
- Although hormones produced in one species may be active in another, the converse is also true: certain hormones are species-specific with respect to their biological activity.
- A hormone derived from one species may elicit a different response in the same target organ of a different species.
- The corresponding hormone in different species may have different functions in each species.

28-41 Hormonal control of frog skin color
The light-colored frog has been immersed in water containing melatonin. The dark-colored frog has received an injection of melanocyte stimulating hormone (α-MSH).

The last of the four is especially interesting. Human prolactin stimulates production of milk following childbirth. Prolactin has the same role in all mammals, as would be expected. But the corresponding hormone in pigeons stimulates the secretion of "pigeon's milk," which is not milk at all, but a secretion in the crop—a part of the alimentary canal. In hens, prolactin produces broodiness (an inclination to sit on eggs). Prolactin is even present in some fishes.

In many fishes, amphibians, and reptiles, hypophyseal hormones control variations in skin coloration. When injected into other species, those vertebrate hormones can induce dramatic color changes (Figure 28-41). There is close chemical similarity in some hormones despite the distant relationships of the animals producing them. Interestingly, color is also controlled by hormones in most species but by nerves in some. These hormones also occur in mammals, including humans; however, no mammals have **melanophores,** special skin cells that produce color variations. In other words, humans have the hormonal part of the mechanism for matching the color of the surroundings, but lack the effective part—the skin melanophores. Whether these hormones are only useless baggage inherited from remote ancestors or whether they have other roles in humans is not yet known.

Different but related processes may be regulated by the same hormone in different animal groups. Metamorphosis from tadpole to frog is initiated prematurely with injected thyroxin (T_4), or can be totally prevented by removal of a tadpole's thyroid gland. This remarkable transformation is unlike anything in human life; yet T_4 enhances human metabolism, too.

Going back still further, there is solid evidence that the thyroid gland in the earliest vertebrates evolved from grooved pouchlike structures that carried food particles into the front end of the digestive tract. This explains why the thyroid gland is in the neck. Why and how did a feeding mechanism turn into an endocrine gland? A reasonable hypothesis holds that as the developing pouches gradually lost all connection with the pharynx they became independent of the digestive system, functionally and structurally. As a result a functionally novel structure arose from an ancestral structure with an unrelated function. It is an evolutionary pattern we will see again and again.

Comparison of more ancient groups of vertebrates with more recent ones reveals another tendency. Older groups seem to have simpler endocrine systems. For example, some of the hormones now present in mammals are absent in fishes, their hypophyseal extracts lacking TSH, ACTH, and even GH. But despite the lack of GH, fishes do grow—and in a well-regulated and coordinated way. It is hard to imagine that hormonal regulation is not part of the process.

HORMONES OF INVERTEBRATES

Endocrine glands and their hormones have been recognized among some of the more active mollusks (octopuses, cuttle fishes) as well as in crustaceans and insects— all animals with well-developed circulatory and nervous systems. In several kinds of insects, metamorphosis from the larval to the adult stage is regulated by hormones.

The best studied example of hormone-controlled metamorphosis is in the silkworm moth, *Hyalophora cecropia.* The life cycle of cecropia and other insects includes three stages: larva, pupa, and adult (Figure 28-42). The processes of molting and metamorphosis depend on three hormones. In a gland that is closely associated with the brain, neuroendocrine cells secrete a hormone, once simply termed **brain hormone**

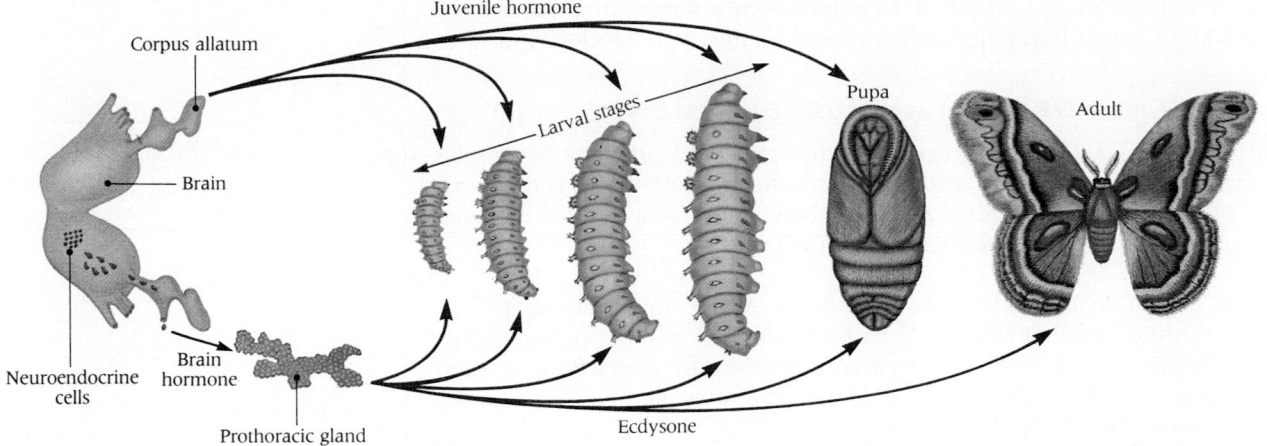

28-42 The role of hormones in the metamorphosis of a moth
The interactions of juvenile hormone (JH), brain hormone, and ecdysone control the molting sequence in a moth. If JH is abundant when the insect molts, it will molt into another larval stage. If little JH is present, the larva molts into a pupa. If JH is absent, the pupa molts into an adult.

and now called **prothoracitropic hormone (PTTH),** that stimulates metabolism and secretion in the **prothoracic gland.** This gland, which is farther back in the thorax (the three body segments to which legs and wings are attached), produces a hormone mixture including **α-ecdysone** and **β-ecdysone** (20-hydroxyecdysone). The ecdysones are molting hormones that induce the events associated with each molt.

A pair of glands behind the brain, the **corpora allata** (singular, corpus allatum), produce a third substance called **juvenile hormone (JH).** JH inhibits metamorphosis. High levels of JH at the time of the first few molts leads to the development of yet another immature form after the molt. In other words, metamorphosis has been temporarily suppressed. The level of JH decreases just before the pupal stage. It can no longer suppress metamorphosis and the pupa emerges. In the absence of JH, ecdysone can promote transcription of the genes for adult characters. In the presence of JH, this action of ecdysone is suppressed. Among other things, study of these hormones promises future pesticides that are specific and unlikely to generate resistant species (see Box 42-2).

Some insects, mollusks, and crustaceans have hormones that cause temporary changes of skin color. These, too, are secretions of definite endocrine glands in the head region.

It is instructive to compare insect endocrine glands and hormones with those of vertebrates. In both, master glands are located in the head in close association with the brain. In both, hormones from these glands have widespread influences on body metabolism and stimulate other endocrine glands. Yet the two systems certainly arose separately in evolution. The ancestries of insects and vertebrates became separated well over 500 million years ago—and at a stage when there may not have existed a definite endocrine system.

Even in insects with the most complex endocrine systems yet found among invertebrates, systems appear simpler than those in vertebrates. In invertebrates other than insects and crustaceans (which are relatives of insects), arrangements are simpler still. Indeed, no definite endocrine glands have been identified in the vast majority of invertebrates. Thus, in most of these animals we cannot speak of endocrine systems at all. This does not imply that in these species other forms of chemical coordination are lacking.

HORMONES OF PLANTS

Thus far, we have dealt exclusively with the hormones of animals. Plants are also coordinated by hormones. These substances—sometimes termed **phytohormones**—are members of a larger group of substances called plant growth regulators. These are defined as organic molecules, other than nutrients, that in small concentrations affect the physiological processes of plants.

In general terms, hormones act to control growth and development as well as tissue metabolism and function. Animal hormones actively participate in all aspects of life. Indeed, virtually every function of the animal body is under hormonal control. As far as we know, plant hormones are concerned mainly with development.

PLANT GROWTH AND DEVELOPMENT

We learned in Chapter 21 that the cardinal processes of development in animals are (1) differential gene expression, (2) morphogenetic movements, (3) cell interactions, and (4) embryonic induction. For the following reasons, this list does not accurately summarize the processes occurring in plants.

- Plant cells do not divide the way animal cells do. At the end of mitosis in animals cells, the daughter cells are free of one another. In plant cells, complex intercellular connections, consisting in part of cellulose microfibers, form between the new opposing walls of daughter cells. Elaborate arrangements for intercellular communication then develop.
- Unlike animal cell growth, which depends on the extensive synthesis of new protein and new membrane, plant cell growth is heavily dependent on the uptake of water, most of which ends up in the large central vacuole that presses the cytoplasm against the cell wall. Indeed, turgor pressure (see Figure 6-6) is critically important in plant growth and development. Much less protein synthesis takes place in plant growth than in animal growth.
- Plant cells do not migrate, so organ shape is determined largely by the orientation of the cellulose microfibrils. In a rapidly elongating stem, for example, the fibrils are in a predominantly horizontal pattern. When water enters, the walls cannot bulge sidewise so they expand in the longitudinal direction (Figure 28-43).
- Plants also have a distinctive method of forming organs from undifferentiated tissues. This is accomplished in animals largely by cell migrations. Since cell movement does not occur in plants, cell proliferation must be relied upon instead. But it must be confined to highly specific loci in the plant if development is to proceed in an orderly fashion. Such loci are the meristematic tissues found at the tip of each shoot or root and within stems and branches.

ROLE OF PLANT HORMONES

As we have noted, plant hormones are mainly concerned with growth and development. A striking feature of plant hormone action is the way different hormones interact in various combinations to produce a given effect—like budding, fruit drop, growth, and so on.

Two major classes of developmental hormones have been found in most, if not all, plants—the **auxins** and the **gibberellins.** In addition, several other plant hor-

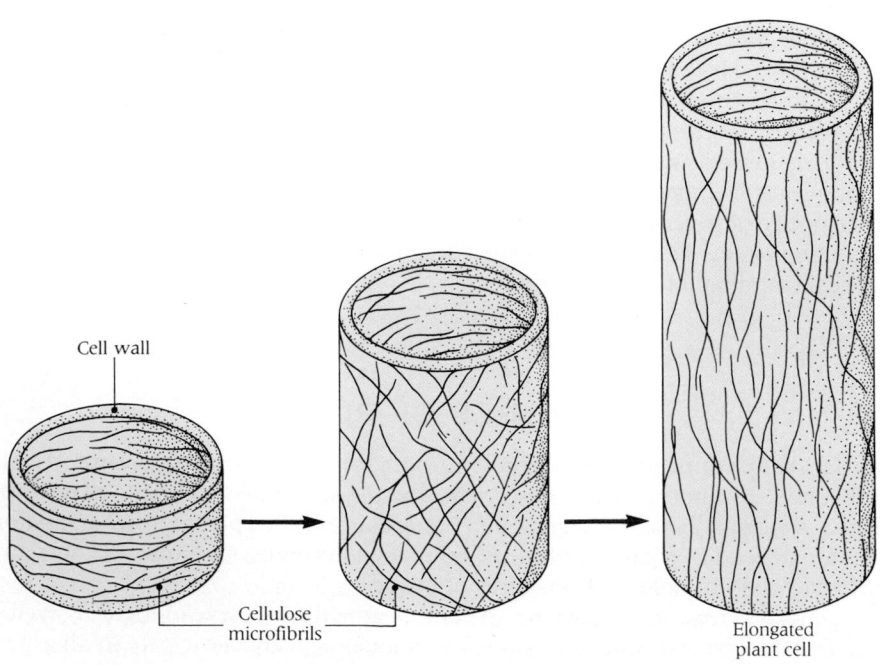

Cell wall

Cellulose microfibrils

Elongated plant cell

28-43 Reorientation of cellulose microfibrils during plant growth
As plant cells elongate, cellulose microfibrils move from a horizontal to a vertical position in the cell wall. New horizontal fibrils are added to the inner wall, as the older outer wall acquires fibrils in a vertical orientation. This cross-ply pattern strengthens the cell wall.

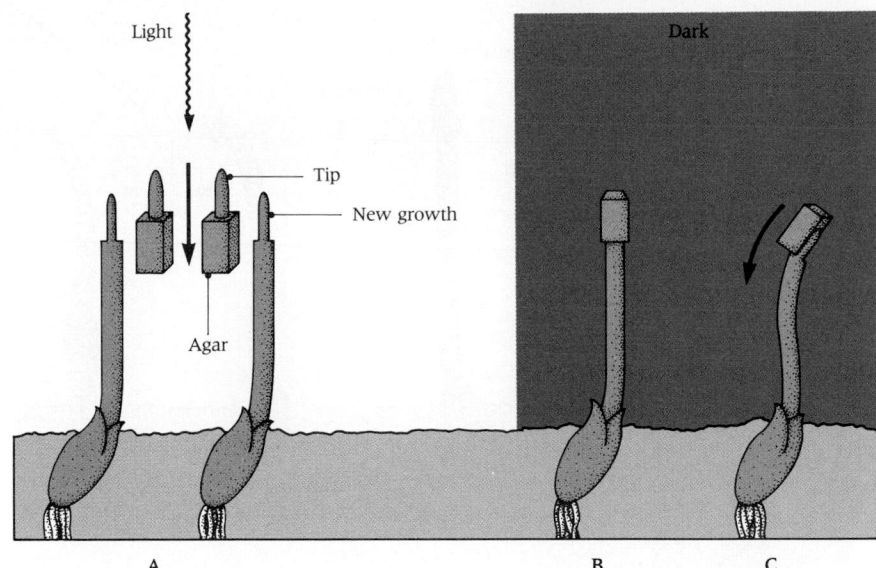

28-44 Went's experiment demonstrating the existence of auxins
(A) *A shoot tip is cut from a young seedling and placed on a block of agar. After a short time the tip is removed.* **(B)** *If the agar is placed on another decapitated seedling, the seedling continues to grow, even in the dark.* **(C)** *If the agar is placed to one side of the tip of the seedling, it bends away from the side on which the agar is placed. A hormone that diffused into the agar from the shoot tip is responsible for growth in the dark as well as the bending.*

mones have been discovered—among them, the several **cytokinins, ethylene, and abscisic acid.** These naturally occurring substances, the phytohormones, along with various synthetic compounds, comprise the plant growth regulators. The synthetic chemicals, many of which are widely advertised, kill weeds (herbicides) and stimulate the growth of plants considered desirable by human beings.

Plant hormones display at least one major difference from animal hormones: all plant hormones are secreted by unspecialized tissues. In plants, few or no other specialized messenger systems have evolved. Nonetheless, the mechanisms by which plant hormones act on their target cells parallel those of animal hormones, including enhancement of enzyme synthesis or activity and effects on ATP levels, proton pumps, and membrane permeability.

Since plants with few exceptions have no other system of coordination and yet are well-coordinated organisms, their hormones are of notable importance. Each plant grows at a particular time of the year, each grows in its own characteristic way, each flowers at a definite time, and the fruits and leaves of most perennials also drop at definite times. Nothing is haphazard about these activities. They are coordinated chemical processes.

Auxins

The **auxins** are a group of substances produced by the cells of growing tips of stems and roots that migrate to other zones of the plant, where they act to promote the process of cell elongation.

A classic experiment, first performed in 1926 by Fritz Went, established their existence (Figure 28-44). Went found that if the stem tip of a rapidly growing seedling is amputated, growth in the region below the cut slows down and stops. If the removed tip is replaced, growth continues normally, even though the tip is still detached. This shows that some "influence" emanating from the tip is conducted across the wound to growing cells. If a removed tip is placed on a block of agar for an hour or so and the agar block is then held to the decapitated stem, growth resumes. If the agar is placed on one side only, the resulting growth pattern causes the stem to bend. Agar blocks exposed to stems cut at lower levels had no effects. From such results, Went concluded that a substance—he called it *auxin* (from a Greek word for "grow")—moved from the tip to the agar block and from the block into the cut stem.

Later work revealed that many substances have auxin activity. The most common is the simple molecule, **indoleacetic acid** (Figure 28-45), which resembles the amino acid tryptophan and which occurs in many living organisms. We now recognize that in controlling the length of the stem and its curvature, auxins control the direction of plant growth. In the familiar phenomenon called **phototropism**, plants grow toward the light (Figure 28-46). One of its earliest students was Darwin himself, who with his son Frances performed some useful experiments. We now

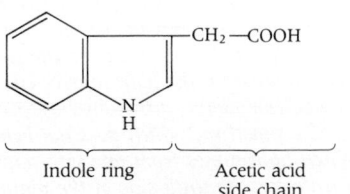

Indole ring   Acetic acid side chain

28-45 Indoleacetic acid

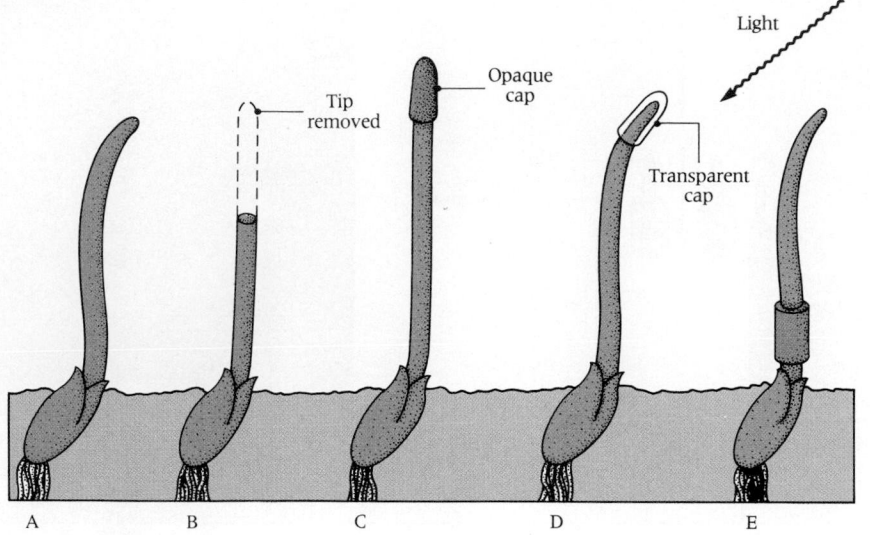

28-46 Phototropism experiment performed by Darwin
(A) *A growing tip bends toward the light.*
(B) *No bending occurs if the tip is removed.*
(C) *If the tip is covered by an opaque cap, the shoot remains straight.* (D) *Bending occurs if the tip is covered by a transparent cap.*
(E) *Bending occurs if the base is covered by an opaque sleeve.*

believe that light-sensitive structures in the plant tip, consisting perhaps of light-sensitive participants in photosynthesis such as β-carotene, respond to incoming light by redistributing auxin within the shoot (Figure 28-47). If it collects on the side away from the light, cells on that side elongate and the shoot bends toward the light.

Auxins have many other functions.

- Auxins determine which buds develop—the ones at the tip of the stem (apical buds) or the ones on the sides (lateral buds). In the intact stems of most plants, only the apical bud can grow. Removal of an apical bud—as in the pruning of a shrub—leads promptly to the growth of one or more lateral buds. If an apical bud is removed and auxin is then smeared on the cut surface, lateral buds do not grow. Conclusion: auxins inhibit lateral bud growth. However, more work needs to be done. It is not clear, for example, why lateral buds seem to be inhibited by auxin while apical buds are not.
- Auxins stimulate fruit development. Fruit forms from the ovary of a flower if pollination has taken place. Auxins released from pollen grains stimulate fruit formation. This is exploited commercially in the production of certain seedless fruits, e.g., the tomato. If auxins are applied artificially, the fruit will develop even though pollination, fertilization, and seed development have not taken place.
- Auxins regulate the fall of leaves (termed leaf abscission) from plants. Leaves of plants sprayed with auxins fall prematurely.
- Auxins are involved in the healing of plant wounds.
- The weed killers (herbicides), 2,4-D and 2,4,5-T, are synthetic auxins. Armies have used them to defoliate forests.

Gibberellins

The **gibberellins** are another intriguing family of plant growth hormones. In the late nineteenth century, Japanese farmers observed bizarre elongated seedlings in their rice paddies. The abnormally tall plants did not mature and rarely flowered. The farmers named the disease bakanae or "foolish seedling disease." In 1926, the same year Went did his agar block experiments in Holland, the botanist E. Kurosawa found that these seedlings were infected with a fungus called *Gibberella fujikuroi*. When the fungus was grown in a flask, a substance accumulated in the medium that produced the same overgrowth symptoms when transferred to a plant (Figure 28-48). The substance, named gibberellin for its source, was later shown to be a mixture of substances, some inhibiting and some stimulating growth. The chemical structure of one gibberellin is shown in Figure 28-49.

Not until the 1950s, well after the secrecy of World War II had ended, was this work noticed in the Western world. In 1956, gibberellins were isolated from a plant (the bean, *Phaseolus vulgaris*) rather than a fungus. We now recognize over

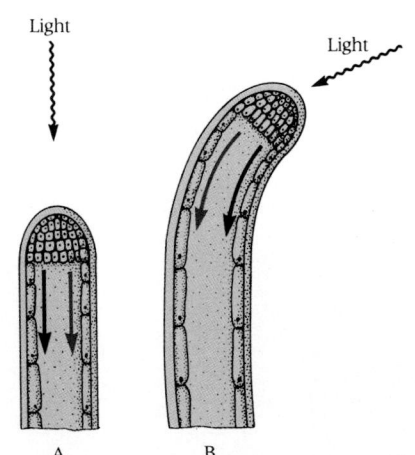

Light   Light

A   B

28-47 Auxin-dependent response to light
Light-responsive elements in the plant shoot tip respond by redistributing auxin to the side of the plant away from the light source. (A) When light comes from above, auxin distributes equally around the shoot tip, which does not bend. (B) When light comes from one side, auxin concentrates on the unlit side of the plant, enhancing elongation of cells on that side. As a result, the shoot bends towards the light.

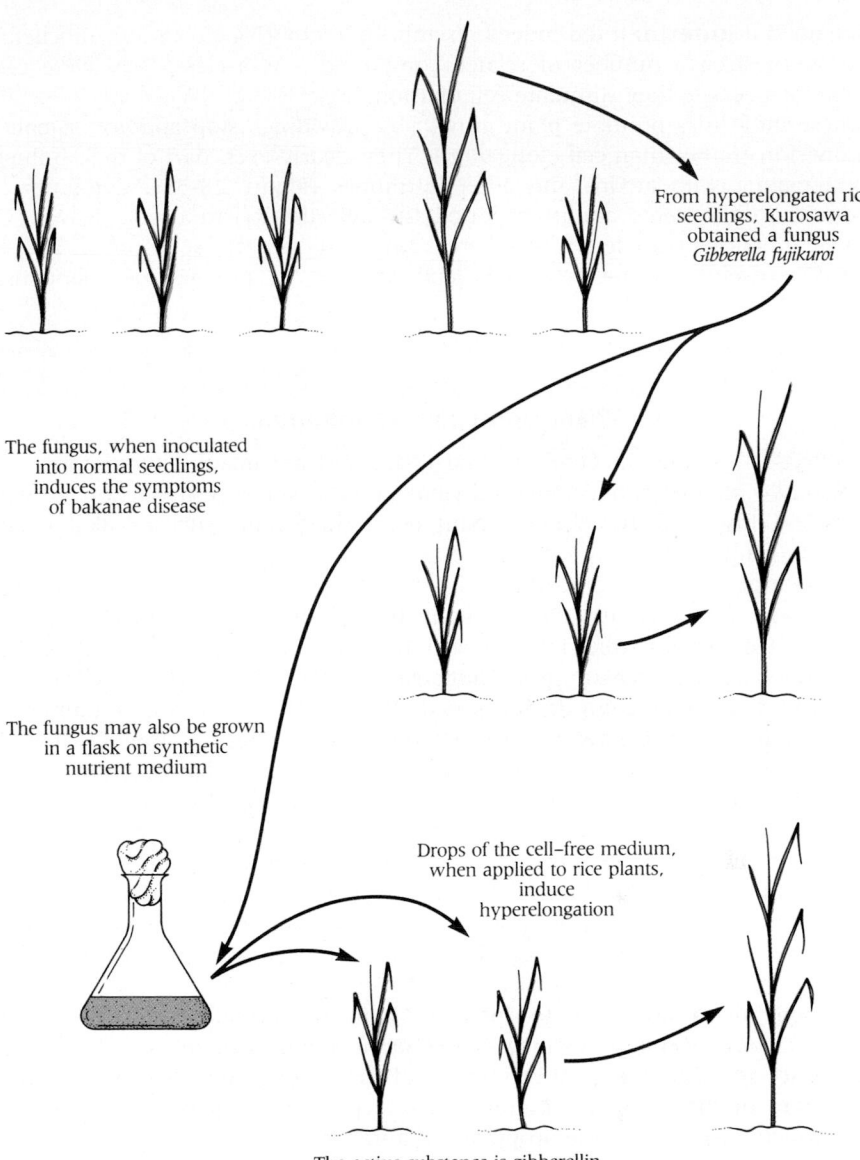

28-48 The discovery of gibberellins
The sequence of events in the discovery of the gibberellins is shown. This family of hormones stimulates growth of the plant.

From hyperelongated rice seedlings, Kurosawa obtained a fungus *Gibberella fujikuroi*

The fungus, when inoculated into normal seedlings, induces the symptoms of bakanae disease

The fungus may also be grown in a flask on synthetic nutrient medium

Drops of the cell-free medium, when applied to rice plants, induce hyperelongation

The active substance is gibberellin

28-49 Gibberellic acid
GA₃, one of the many gibberellins.

60 gibberellins. In some ways, their metabolic origins parallel that of steroid hormones in animals.

Recent investigation has shown that gibberellins activate certain genes, promoting synthesis of certain mRNAs and enzymes. These in turn alter the metabolism of plant cells in ways that induce physiological responses. For example, gibberellins alter the balance between elongating growth and leaf development. The result is a form of growth that best suits the plant's needs at different seasons.

The whole field of plant development remains one of the most fascinating of biological problems. Its relevance to agriculture and its implications for the problem of crop yields are obvious.

Cytokinins

The discovery of **cytokinins** by Folke Skoog and his group at the University of Wisconsin followed the observation that the purine adenine promotes bud formation in isolated plant tissues.

In searching for natural purine-containing substances that might have a similar effect, they tested the material found in an old bottle labeled "herring sperm DNA." It caused plant cells to divide. However, fresh new preparations of herring sperm DNA were quite inactive. Returning to another old bottle, they found activity again. Activity was also discovered in coconut milk, which soon became a standard additive in laboratory cultures of plant cells. In time, Skoog isolated the active principle and named it **kinetin;** it did indeed resemble adenine (Figure 28-50). Biochemists soon synthesized a number of related compounds. As a class, they were called cytokinins because they stimulate cell division.

These molecules promote plant growth by providing "stop and go" signals for cell division (rather than cell elongation). They clearly exert part of their influence by interacting with auxins and other hormones (Figure 28-51). Cytokinins are produced in the roots and transported through the xylem to the leaves. One of their functions is to retard plant breakdown and prevent senescence. They are especially useful as research tools and as aids in propagating plant tissue cultures.

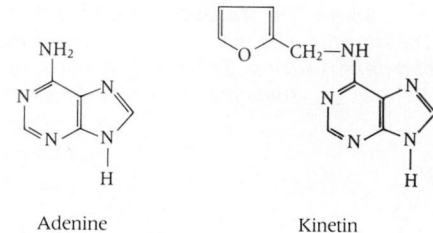

28-50 Molecular structures of adenine and kinetin compared

Plant Hormones As Inhibitors

The effects of auxins, gibberellins, and cytokinins are mainly stimulatory. There are also two important hormones, ethylene and abscisic acid, the actions of which are largely inhibitory. In addition, many other compounds such as alkaloids may function as inhibitors of growth and development.

Ethylene, a simple hydrocarbon with the formula $CH_2{=}CH_2$, is a gas at normal temperatures and pressures. In the early nineteenth century when illuminating gas was used to light street lamps it was regularly noticed, first in Germany and then elsewhere, that leaking gas defoliated the shade trees lining the streets. In 1901, a Russian investigator demonstrated that the active ingredient in illuminating gas was ethylene. We now consider ethylene to be a plant hormone. It retards stem elongation and increases stem radial expansion by changing the shape of stem cells. The formation of adventitious roots is stimulated by ethylene in some plants, and ethylene may also cause leaves to senesce and drop off. Ethylene also causes **epinasty,** an elongation of upper leaf or stem cells that results in a downward curvature of those organs. (Epinasty is most evident when dicot seedlings are emerging from the soil.) All of the above seem to be responses to stresses of various kinds that cause an increase in ethylene synthesis or an accumulation of ethylene in or near the plant.

Ethylene also causes fruits to ripen, and is in turn produced by ripening fruits. This may be a mechanism which causes large numbers of fruits to ripen at the same time, and thus the greatest number of fruit-eating animals can be attracted. With more of these animals present, there would be a greater chance of seeds being widely dispersed in the droppings of animals.

Abscisic acid was discovered in the 1960s as one of a group of compounds, occurring in leaves and fruits and accelerating leaf abscission. Its structure is shown in Figure 28-52). Abscisic acid (ABA) generally increases in plants in response to various kind of stress and causes growth inhibition and stomatal closure. ABA may also contribute to bud dormancy and the inhibition of seed germination in some plants. This helps to keep buds from opening and seeds from germinating at inappropriate times. Many botanists hope this molecule may prove useful someday in enhancing the ability of crop plants to survive stressful conditions.

The study of plant hormones is still in its adolescence. Nevertheless, its practical importance for agriculture and horticulture could hardly be exaggerated. Certain effects occur regularly.

■ Cytokinins and auxins stimulate cell division during early plant growth.
■ Cell division depends on the balance between cytokinins and inhibitors.
■ Cell elongation, a feature of later plant growth, is controlled by auxins and gibberellins.
■ Plant hormones determine patterns of growth, budding, flowering, fruit formation, and leaf drop.
■ Plant ripening and aging is implemented by ethylene and other agents.

Although little is known of the biochemical basis of plant hormone action, a major advance was achieved in 1978 with the discovery of several types of hormone receptors in plant cells that are entirely analogous to those in animal cells. Clearly the next major tasks are to establish the importance of these receptors in hormone action and to characterize the events between hormone binding and physiological response.

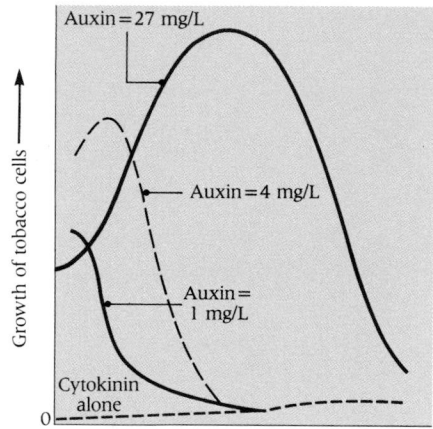

28-51 Auxin-cytokinin interactions
Cytokinin alone has little effect on the growth of tobacco cells in tissue culture. Auxin alone, at varying concentrations, causes the cells to attain a weight of about 10 g. When both hormones are present, growth increases to an optimum beyond which it declines.

28-52 Abscisic acid

SUMMARY

Multicellular organisms rely on both neural and hormonal systems in coordinating their body functions. This chapter considers hormones and related modes of chemical coordination. Neural coordination is discussed in Chapters 29–31.

Chemical messengers of various molecular types transmit information in various coordination mechanisms. Hormones, which are secreted into the bloodstream by endocrine glands, represent one class of messenger molecules. Although hormones come in contact with all or most of the cells in the body, they only influence the metabolism of certain specific target cells. Hormone receptors on or in a hormone's target cells account for this specificity. In many target cells, the binding of a hormone by its receptor triggers the release of "second messenger" molecules, which directly initiate hormone-dependent metabolic events within the cell. A single second messenger system may be triggered by different hormones. It may also have diverse effects in different cell types.

The best studied second messengers are calcium, certain phosphoinositides arising in plasma membranes, and cyclic AMP (cAMP)—a compound that resembles AMP except that its phosphate group is bound to two ribose carbons (thereby forming a "cycle"). Synthesis of cAMP is catalyzed by adenylate cyclase. The major effect of cAMP is stimulation of various cAMP-dependent protein kinases, which modify the activity of other cellular enzymes by phosphorylating them.

Growth factors are hormonelike polypeptides that stimulate cell division in various types of cells. The possible relation of growth factors to cancer is now under active study.

The major endocrine glands in humans and other mammals are the pituitary (also called the hypophysis), adrenals, thyroid, parathyroids, pancreas, gonads, and pineal. The anterior part of the pituitary, the adenohypophysis, secretes hormones called adrenocorticotropin (ACTH), thyroid stimulating hormone (TSH), and the gonadotropins which, respectively, control hormone secretion in the adrenal cortex, thyroids, and gonads. The adenohypophysis also secretes growth hormone, which is necessary for bone and tissue development in childhood, and β-lipotropin, a prohormone that gives rise to endorphins and enkephalins. The adenohypophysis is itself regulated by the hypothalamus and by feedback from adrenal, thyroid, and and gonadal hormones.

The neurohypophysis, the posterior part of the pituitary, does not synthesize hormones. It does, however, secrete two hormones, vasopressin and oxytocin, which are synthesized in neurons of the hypothalamus. The pineal gland secretes melatonin, a hormone that mediates body responses to daily light and dark cycles.

The adrenal glands, like the pituitary, contain two unrelated endocrine regions. One, the adrenal cortex, secretes steroid hormones of three classes: mineralocorticoids, sex hormones, and glucocorticoids. Mineralocorticoids, such as aldosterone, affect extracellular ion balance. Sex hormones influence the reproductive system. Glucocorticoids, such as cortisol, control carbohydrate, lipid, and protein metabolism. They are released in response to stress and have anti-inflammatory effects. The second portion of the adrenal gland, the adrenal medulla, secretes two hormones: epinephrine and norepinephrine. Released in response to stress or excitement, they increase heart rate and blood glucose.

Thyroid hormones, notable for their iodine content, stimulate virtually every phase of metabolism. Parathyroid hormone regulates blood levels of ionic calcium in a number of ways—for example, they cause release of calcium from bones.

The sex hormones arising in the gonads include estrogens and progesterone in the female and testosterone in the male. Estrogens induce female pubertal changes and cause proliferation of the uterine lining, whereas progesterone maintains the uterine lining and prepares the uterus for pregnancy. Testosterone causes male pubertal changes and maintains testicular function.

The pancreas acts as both an exocrine gland, secreting digestive hormones into the intestine, and an endocrine gland, secreting insulin and glucagon into the bloodstream. Insulin release, triggered by the rise of blood glucose, lowers it in turn by causing cells to take up glucose from the blood. In contrast, glucagon causes a release of glucose into the blood.

Atrial natriuretic factor (ANF), the most recently discovered hormone, is made in the heart. It stimulates excretion of salt and water.

Many mammalian hormones are present in other vertebrates, although there they control other functions. Thyroid hormone, for instance, triggers metamorphosis from tadpole to frog. Metamorphosis is also under hormonal control in insects. Insect juvenile hormone (JH) inhibits metamorphosis; ecdysone stimulates it.

Hormones also play critical roles in plants, especially in their growth and development. Plant development differs from that in animals in that it displays no cell migration; all shape and function changes result exclusively from localized differentiation. Two major classes of plant hormone regulating this differentiation are auxins and gibberellins.

All plant hormones are secreted by unspecialized tissues rather than specialized glands. Auxins are made in the meristematic tissue of root and stem tips. They promote cell elongation and bud and fruit development. Gibberellins, by activating certain genes, alter such aspects of metabolism as the balance between stem and leaf growth. Cytokinins, ethylene, and abscisic acid are other important chemical messengers that influence plant development.

KEY TERMS

abscisic acid
adenohypophysis
adenylate cyclase
adrenal cortex
adrenal gland
adrenal medulla
adrenocorticotropic hormone
agonist
aldosterone
androgen
antagonist
antidiuretic hormone (ADH)

atrial natriuretic factor (ANF)
autocrine secretion
auxin
calcitonin
calmodulin
corticosteroid
cyclic adenosine monophosphate
   (cyclic AMP, cAMP)
cytokinin
ecdysone
endocrine gland
endocrine system

endorphin
epidermal growth factor (EGF)
epinephrine
erythropoietin
estrogen
ethylene
follicle-stimulating hormone
   (FSH)
G protein
gibberellin
glucagon
growth factor

growth hormone (GH)
hormone
hypophysis
hypothalamus
indoleacetic acid
insulin
islets of Langerhans
juvenile hormone
luteinizing hormone (LH)
melatonin
nerve growth factor (NGF)
norepinephrine
oxytocin
pancreas
paracrine secretion

parathyroid gland
parathyroid hormone (PTH)
peptide hormone
phosphatidyl inositol
phototropism
phytohormone
pineal body
pituitary
platelet-derived growth factor
    (PDGF)
preprohormone
progesterone
prohormone
prolactin (Pr)
prostaglandin

protein kinase
prothoracic gland
receptor
release inhibiting factor
releasing factor
second messenger
steroid
steroid hormone
testosterone
thyroid hormone
thyrotropic hormone
thyroxine
tropic hormone
vasopressin
vitamin D

QUIZ QUESTIONS

1. The receptor molecules of _____ hormones are located _____ .
   A. amine; on the cell surface
   B. polypeptide; within the chromatin
   C. steroid; on the cell surface
   D. A and B
   E. A, B, and C

2. Which of the following is *not* involved in calcium metabolism?
   A. calcitonin
   B. parathyroid hormone
   C. calmodulin
   D. atrial natriuretic factor
   E. All of the above are involved in calcium metabolism.

3. Which of the following conditions results from an excess of growth hormone?
   A. diabetes insipidus
   B. acromegaly
   C. hyperthyroidism
   D. high blood pressure
   E. gluconeogenesis

4. Which of the following is *not* a secretion of the adrenal glands?
   A. androgens
   B. epinephrine
   C. aldosterone
   D. melatonin
   E. glucocorticoids

5. Phytohormones
   A. are producd in meristematic tissues.
   B. influence development.
   C. are responsible for the phenomenon of phototropism.
   D. B and C
   E. A, B, and C

ESSAY QUESTIONS

1. What is a hormone? In what ways do the following molecules fulfill (or fail to fulfill) that definition: lymphokines, neurotransmitters, growth factors, prostaglandins.

2. How might one prove that a gland produces a hormone? Would this approach work for the hypothalamus or the pineal?

3. The hormones of the adrenal medulla are said to mediate the "fight or flight" response of an organism to an environmental challenge. Explain.

4. How can the same tropic hormones, specifically FSH and LH, regulate sex hormones in both males and females? How do FSH and LH act in females, males?

5. Invertebrate animals have some hormones that function in ways that have no parallel in vertebrate animals. Identify these functions.

REFERENCES AND SUGGESTIONS FOR FURTHER READING

Barnes, D. M. (1988). Cells without growth factors commit suicide. *Science 242*, 1510–1511.

   On the absolute necessity of growth factors.

Beato, M. (1989). Gene regulation by steroid hormones. *Cell 56*, 335–344.

   A review of the mechanism of steroid hormone action.

Bentley, P. G. (1976). *Comparative Vertebrate Endocrinology.* Cambridge University Press, New York.

   A useful treatment of endocrine function in vertebrates that discusses the relations of endocrine functions with ecology and evolution.

Cantin, M. and Genest, J. (1986) The heart as an endocrine gland. *Scientific American 254*, February, 76–81.

   Summary of the surprising recent discovery that the cardiac atrium secretes a hormone, the atrial natriuretic factor, that interacts with other hormones in regulating blood pressure and volume. This is the first evidence that the heart is more than a pump.

Carafoli, E., and Penniston, J. T. (1985). The

calcium signal. *Scientific American 253*, November, 70–78.

An explanation of a major second-messenger system.

Casey, P. J., and Gilman, A. G. (1988). G protein involvement in receptor-effector coupling. *The Journal of Biological Chemistry 263*, 2577–2580.

This compact and up-to-date review is one in The *Journal's* outstanding "minireview" series.

Crozier, A., and Hillman, J. R. (Eds.) (1984). *The Biosynthesis and Metabolism of Plant Hormones.* Cambridge University Press, New York.

A detailed reference work.

Folkman, J., and Klagsbrun, M. (1987). Angiogenic factors. *Science 235*, 442–447.

Folkman argued in 1963 that tumors must depend on prior angiogenesis. Despite difficulty, he isolated TAF in 1971 and more recently other angiogenic factors. This remarkable story is reviewed here.

Levi-Montalcini, R. (1988). *In Praise of Imperfection: My Life and Work.* Basic Books, New York.

An inspiring and thought-provoking autobiography of the codiscoverer (with Stanley Cohen) of nerve growth factor. The two investigators received the 1986 Nobel Prize for work done 30 years earlier.

Majerus, P. W., and 5 other authors (1988). Inositol phosphates: synthesis and degradation. *The Journal of Biological Chemistry 263*, 3051–3054.

Another "minireview." This one is at a fairly advanced level.

Needleman, P., Turk, J., Jakschik, B. A., Morrison, A. R., and Lefkowith, J. B. (1986). Arachidonic acid metabolism. *Annual Review of Biochemistry 55*, 69–102.

A scholarly review for afficionados. Though chemically complex, these metabolic pathways have profound physiological effects.

Nishizuka, Y. (1986). Studies and perspectives of protein kinase. *Science 233*, 305–312.

A clearly written review.

Norman, A. W., and Litwack, G. (1987). *Hormones.* Academic Press, Orlando, Fla.

A stimulating textbook of modern endocrinology that emphasizes cellular and molecular aspects.

Roesler, W. J., Vandenbark, G. R., and Hanson, R. W. (1988). Cyclic AMP and the induction of eukaryotic gene transcription. *The Journal of Biological Chemistry 263*, 9063–9066.

This excellent "minireview" begins, "There is no more important or ubiquitous a regulatory molecule than cAMP. It goes on from there to explain how it works.

Schally, A., Castan, A., and Arimura, A. (1977). Hypothalamic hormones: the link between brain and body. *American Scientist 65*, 712–719.

A readable account of the interactions between hypothalamus and hypophysis (pituitary).

Weiss, E. R., and 6 other authors (1988). Receptor activation of G proteins. *FASEB Journal 2*, 2841–2848.

A good review.

Zeevaart, J. A. D., and Creelman, R. A. (1988). Metabolism and physiology of abscisic acid. *Annual Review of Plant Physiology and Plant Molecular Biology 39*, 439–473.

New insights on an important plant hormone.

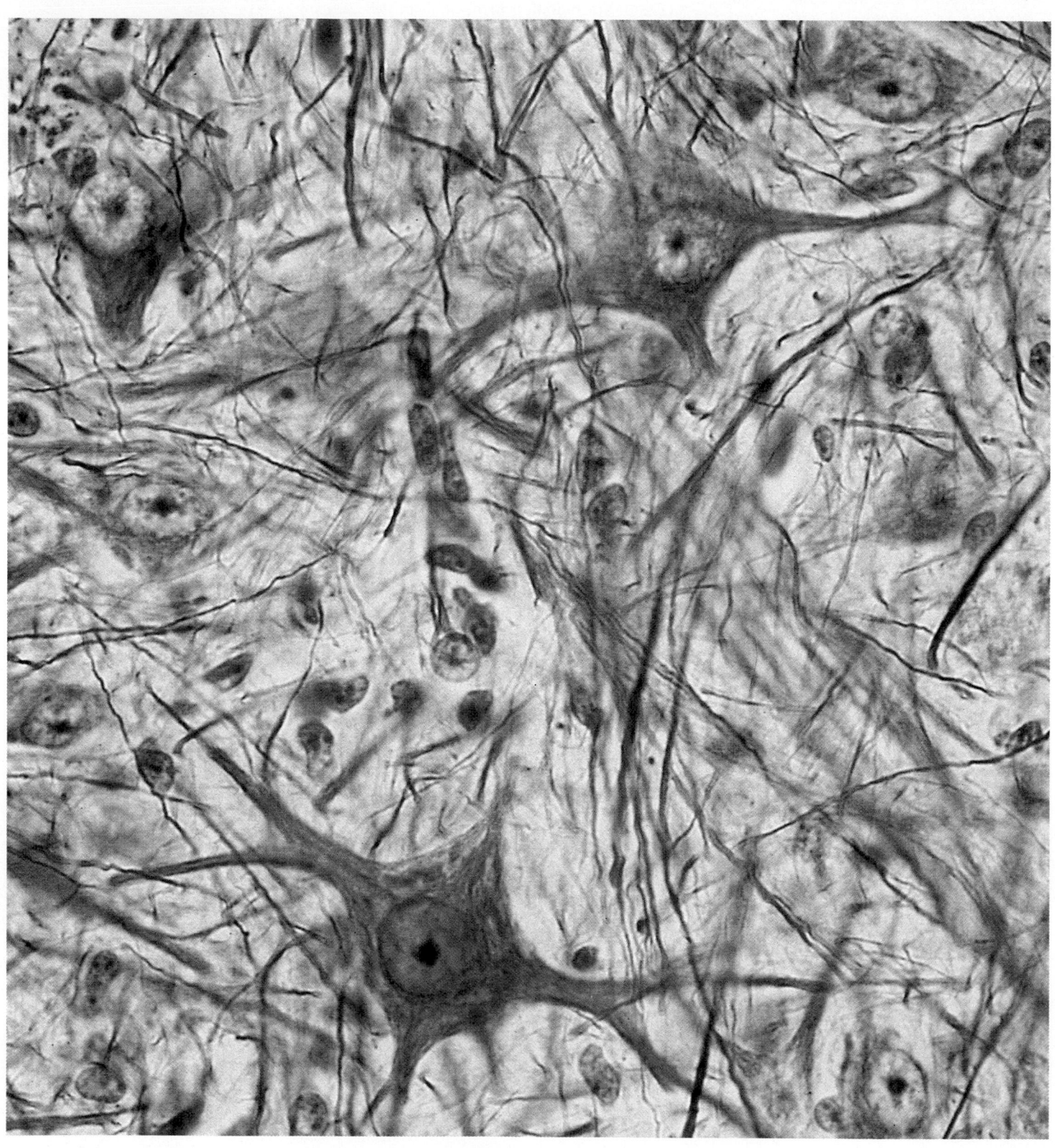

NEURAL COORDINATION

Neural coordination is the second major integrating system of living organisms. With greater or lesser refinement, depending on an organism's complexity, this system performs three functions.

- Collects information from the external and internal environments
- Coordinates, processes, and stores this information
- Initiates appropriate responsive actions

These functions are carried on by all organisms, but in animals, which perform them best, they are the business of the nervous system. It is the nervous system that gives our lives their singular human qualities—our feelings and sensibilities, our capacities to think, learn, reason, and remember. In a word, the nervous system gives us our minds.

RESPONSIVENESS

An amoeba touched with a needle turns away (Figure 29-1). This unicellular organism has no specialized organs (or organelles) for the perception of touch. Yet it withdraws from the noxious stimulus.

We cannot say that the amoeba actually felt the stimulus in the conscious way that we feel the prick of a needle. We believe it did not. We also know that the conduction of the effects of this contact did not occur, as in animals, along nerves. Nor did the amoeba's movement depend on muscles. Still, the chain of events clearly parallels what happens when a person is stuck with a needle. What can we learn from this primitive response of a unicellular organism?

REACTIVITY OF PROTOPLASM

In the harassed amoeba, three events occurred in sequence.

- First, a stimulus activated some sort of receptor.[1] This step involves the collection of information.
- Second, there was conduction of a signal of some kind from the locus of the stimulus to other parts of the cell.
- Third, there was an appropriate response.

Neural coordination
The role of the nervous system is to adjust the activities of an individual to the environment in which it lives and to coordinate its body functions. In its web of interconnected neurons lies the machinery of conscious experience and behavior. The complexity of that network is suggested by this photomicrograph of the motor cortex of a macaque monkey brain. (×1800)

[1] We are here using the same term we used for the part of the cell that binds a specific hormone but with a different meaning. Clearly, the two meanings of "receptor" have something in common—a biochemical structure that receives something (a hormone or a stimulus) and initiates a specific response. We should be careful not to confuse the technical definitions of the two receptors. To make things more complicated, we will soon see that the nervous system is full of receptors of the kind that bind specific chemical messengers—just like hormone receptors.

The capacity to react to environmental changes is a general property of all proto-plasm—whether in a single-celled protist or in a cell of a multicellular plant or animal. It is termed **irritability.**

In a single cell, surface area is large in relation to total volume. Hence, most internal structures are close to one another and to the surface. Communication over these short intracellular distances operates through simple biophysical and biochemical mechanisms and does not appear to require a specialized communica-tions network. The evolution of multicellular animals, however, produced a clear separation of these three phases of reaction to a stimulus. In all but the simplest animals, each process depends on the activities of highly specialized cells, tissues, and organs.

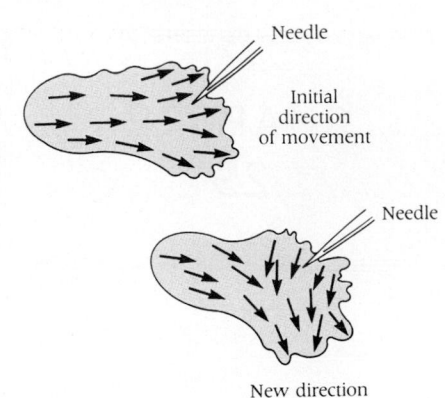

- First, there are **receptors,** each of which is specialized to receive a particular kind of stimulus. These are the sense organs.
- Then there are **conductors,** usually nerves, which carry signals from the receptors to the effectors.
- Finally, there are the special organs of response, termed **effectors** because they carry out, or effect, a response by the organism.

29-1 Responsiveness in the amoeba
If a needle prods at the forward-moving end of an amoeba, its direction of movement changes to avoid the stimulus.

A receptor-conductor-effector network of greater or lesser complexity is a distinctive characteristic of animals.

Compared to animals, plants have a notably small, slow, and unsophisticated repertory of responses to environmental change, but they do respond and make adjustments to a wide range of external influences. They can turn to the sun, but they cannot flee from danger. Although plants can transmit certain electrical signals, they lack a specialized communication network. Sensory receptors definitely exist in plants (Box 29-1). Many plants show periodic leaf movements associated with day-night cycles. Indeed, most plants keep track of day length, and recent work has revealed a pigment in many bud surfaces, a phytochrome, that is sensitive to transitions between light and darkness and that promotes blooming when the ratio of daylight to darkness reaches a certain value—in other words, when spring arrives (or fall, for some flowers).

The form and lifestyle of the animal body necessitates a system that transmits information inward, sorts it, and effects appropriate coordinated actions in response. Only a nervous system can expeditiously manage such complex functions. Together with the endocrine system, the nervous system controls and integrates almost every aspect of body function.

STIMULUS AND RESPONSE
Kinds of Stimuli

A **stimulus** is a change in an organism's environment that conveys information about an external event and eventually elicits a response. Certain kinds of stimuli affect some organisms but not others. Other kinds of stimuli affect all protoplasm.

- All cells are affected by changes in environmental pressure, temperature, and certain kinds of radiant and electrical energy.
- Discrimination of different chemical stimuli is also highly developed in most organisms. An amoeba moves toward some chemical stimuli associated with food and away from various noxious stimuli. So do many algae and fungi. Animals have the remarkable chemical senses of taste and smell.
- Both plants and animals also react to the force of gravity. Even motile bacteria seem to "know" which direction is up (see Box 5-2). In most plants, the main stem turns upward, against the pull of gravity. Many animals, humans among them, have organs that indicate the position of the body or of the head with respect to the pull of gravity.

There are also striking differences in the abilities of organisms to discriminate among incoming stimuli and to react to particular stimuli.

- Fishes have organs for detecting pressure changes in water currents. Humans do not.

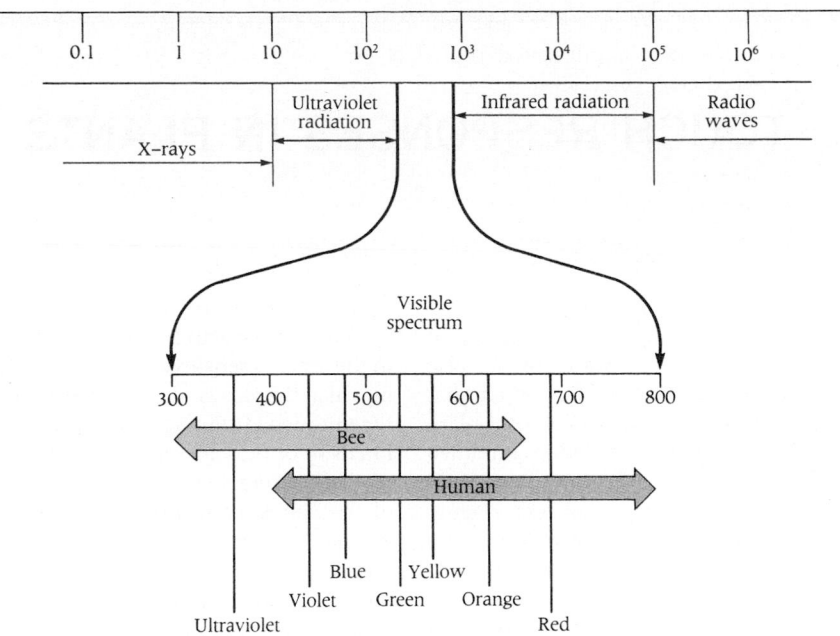

29-3 How colors appear to bees and humans
(A) *Silverweed as seen by human eye.* **(B)** *Under ultraviolet light, silverweed has this appearance to bees.*

- The rhythmic waves in air pressure we recognize as sound are stimuli for many organisms.
- Vision discriminates among the radiant energies of the electro- magnetic spectrum by direction, intensity, pattern, and frequency, the last of which is perceived as color. All organisms react to radiant energy but sensitivities differ (Figure 29-2). The eye is a specialized receptor that reacts to a limited range of wavelengths. Our eyes can see a spectrum ranging from shorter wavelengths (violet) to longer wavelengths (red). Many species see wavelengths we cannot see. Bees, toads, lizards, and hummingbirds can see ultraviolet light, invisible to us because the lenses and retinas of our eyes filter out wavelengths shorter than 400 nm (Figure 29-3). But, bees do not see the red rays visible to us.
- We once believed that magnetism is not a stimulus for living organisms, but recent work on homing behavior in pigeons revealed particles of iron-containing magnetite in a small structure (1–2 mm in diameter) on one side of the skull. This suggests that pigeons can sense magnetic fields, that they may have a "compass sense." Some bacteria also respond to magnetic fields (Box 29-2).

Finally, there are some stimuli to which no organisms respond. As far as we know, no organism has receptors for (or any reaction to) radio waves, which are of much longer wavelengths than light.[2] Humans can perceive them only by using external devices that turn them into other kinds of stimuli: sound (in radios) or visible light (in television sets). It is a striking fact that organisms do not react to many stimuli that conceivably could provide information about the environment. Probably they are not greatly deprived for a simple reason: the information they do not receive would not be all that useful. The only radio waves arising in nature come from the stars or from lightning. But radio signals from the stars or information about lightning after it has struck would hardly be useful to a plant or animal. Organisms generally can sense information that is necessary to their ways of life. Few, if any, sense more than that.

Stimulus Intensity

For each kind of stimulus and for each organism there is a stimulus intensity below which no reaction occurs. This is the **threshold** of stimulation. If one takes a glass of water and adds a drop of dilute salt solution, the water will still be tasteless. Keep on adding and tasting, drop by drop, and finally at a particular concentration,

---

[2] Although rare individuals with dental fillings of certain kinds can receive radio programs in their teeth.

Animals are not unique in their ability to respond when touched. Some plants also have this capacity. Perhaps the most familiar example is seen in the **tendrils** of some plant species. Tendrils, which are slender modified leaves, manage to entwine themselves around any object they encounter (Figure 29A). As a result, the plant is able to cling and climb. The response of a tendril to contact with an object can occur with amazing speed. Indeed, it may wrap itself around a solid support several times in one hour. This capacity for twisting is caused by differences in growth rates of the cells on the two sides of the tendril. Cells touching the support shorten slightly; those on the other side elongate. Auxin may play an initial role in this phenomenon.

Recent experiments show that when the tendrils of pea plants (*Pisum sativum*) are kept in the dark for three days and then touched, they do not coil until they are subsequently illuminated. Moreover, tendrils kept in the dark for as long as two hours after being touched still coil immediately upon being illuminated—as though they had been storing "sensory information." Why the motor function does not proceed in the dark is unknown. Perhaps ATP needed for coiling is consumed during the dark period and then restored in the light by photosynthesis. Perhaps a coiling inhibitor accumulates in the dark and is quickly removed in the light.

A more spectacular touch response is seen in the famous sensitive plant, *Mimosa pudica*, in which the leaves or leaflets abruptly droop when touched (Figure 29B). The speed and intensity of the movement and the quantity of plant tissue affected vary with the intensity of the shock, movement occurring first in the part touched and then spreading with diminishing force to other parts. Recovery takes 10 to 20 minutes. If a leaflet is heated, all of the plant's leaves may quickly droop. These movements are a result of sudden changes in turgor pressure in certain cells of the jointlike thickenings called **pulvini** at the bases of leaflets and leaves. We still lack full understanding of the mechanism. Loss of water from these cells follows the migration of K^+ ions from them. Two different mechanisms, one electrical and one chemical, may be involved in the spread of the stimulus throughout the sensitive plant.

There has been considerable argument about the survival value of this response to touch in *Mimosa pudica*, which often grows on dry, exposed ridge-tops where it may be subjected to drying winds. A strong wind may strike the leaves hard enough to make them droop and fold up, thus reducing transpiration and conserving water. Another suggestion is that the wilting response makes the plant unattractive to large herbivores. Conceivably, the folding response startles herbivorous insects. It has been claimed that

nonsensitive species of *Mimosa* growing near *Mimosa pudica* are more freely attacked by insects.

The triggering of turgor changes by touch is also involved in the capture of prey by the carnivorous Venus flytrap (*Dionaea muscipula*) which we encountered in Chapter 26. As shown in Figure 29C, each leaf half is equipped with three sensitive hairs. When an insect is attracted by the nectar on the leaf surface and walks across the leaf surface, it brushes against the hairs and the two halves of the leaf blade fold together and close like a trap. As the toothed edges mesh, the insect is pressed against digestive glands on the inner surfaces of the leaf.

The trapping mechanism is so fastidious it can distinguish between living prey and inanimate objects, such as pebbles and small sticks, that fall on the leaf by chance: the leaf will not close unless two of its three hairs are touched in succession or one hair is touched twice.

Studies in which microelectrodes were placed within the hairs have shown that the rapid response of the outer hairs results from an abrupt change in the electrical potential across the plasma membrane, much as occurs in neurons. The stimulus is received in the enlarged tip of the hair and transmitted down the stalk, which responds by bending.

29A Tendrils of a catbrier plant
Twisting results from differing growth rates on the inside and outside of the tendril.

29B A touch-sensitive plant, Mimosa pudica
The normal position of leaves and leaflets is shown. When touched, the leaves collapse into a folded position.

29C The carnivorous Venus flytrap, Dionaea muscipula
Its sensitive hairs control the trap.

the water will begin to taste salty. This threshold varies greatly among different people and in the same person at different times. Most commonly it occurs when the strength of the salt solution is about 0.1%. The threshold for detecting that *some* stimulus is present is sometimes lower than the threshold for accurately identifying the stimulus. For example, at very low concentrations, salty water may taste sweet.

Above the threshold intensity, the main factors determining whether a reaction is produced are the duration of the stimulus and its rate of change. An extremely brief stimulus—for example, a projected picture that is flashed on a screen for only one millisecond (1/1000 second)—may not be perceived. A rapidly changing stimulus, such as a flickering light, may be perceived as a continuous light when the flicker frequency is above a particular value. There may be thresholds for duration and rate of change of stimulus as well as for intensity, and the actual threshold of stimulation in an organism depends on all three. There are trade-offs. If time is represented as t and intensity as i, $t \times i = k$, where k is a constant. Thus at high illumination (high i), a flashed picture will be perceived at briefer durations or exposure times (low t). At longer exposure times (high t), illumination (intensity) need not be so great (low i).

Responses

An organism's response to a stimulus may consist of anything an organism can do—and one stimulus may elicit multiple responses. For example, the major response to an increase in temperature is acceleration of metabolism, while other responses activate cooling mechanisms (e.g., panting). The response built into the nerve of a tooth is a pain message directed to an appropriate part of the brain. A raw, exposed nerve gives a pain reaction whether you poke it, eat a lemon, drink cold water, or shock it electrically. When hit in the eye, one "sees stars"—an illusory flash of light. That is because a hard blow stimulates the optic nerve—and optic nerves "know" only how to produce a sensation of light. Determination of response by the organism is not exclusive to the nervous system. Some kinds of pressure, chemicals, radiation, and electricity may all produce the same response in an amoeba—contraction of the cell into a ball. Muscular tissue responds to a variety of mechanical, chemical, and electrical stimuli in the same way—by contracting.

As a rule, a response does not begin instantly after stimulus. There is a **latent period** before the response begins. This period varies in different tissues. Usually it is short, often only a millisecond. Nerve and muscle tissue also have the property of not responding again immediately after a response. For a short time, the so-called **refractory period,** no stimulus is strong enough to produce a new response.

- In heart muscle, this period is relatively long, up to 1/5th second or so. This has important and useful consequences. No matter how persistently the heart is stimulated, the muscle cannot remain contracted. If it could, its pumping action would cease and death would follow.
- In ordinary skeletal muscle, the refractory period of the whole muscle is shorter, usually around a millisecond. The reason is interesting. While some muscle cells are in the refractory period, other muscle cells are able to contract. Hence, one can flex a finger faster than the heart can beat and one can keep the finger muscles contracted indefinitely.

NERVE CELLS AND NERVE ACTION

As we have seen, the nervous system is the essential machinery whereby stimulus elicits response in all but the simplest organisms. Receptor organs in animals transform stimuli into **nerve impulses,** signals transmitted along the fiberlike extensions of nerve cells, or **neurons. Nerves** are organized bundles of nerve fibers that conduct impulses from receptors to effectors or from one part of the nervous system to another part.

Information coming in from peripheral receptors is analyzed centrally by the brain, which stores all or part of this information as memory, and issues motor commands. The brain does these remarkable things with its own neurons.

How information flows into, through, and out of the nervous system is diagrammed in Figure 29-4. These nerve networks and their connections provide the mechanisms

Many bacteria exhibit **chemotaxis**—that is, they move toward higher or lower concentrations of particular substances. Others display **phototaxis**—that is, they accumulate in illuminated regions. A novel tactic response has been arousing scientific attention since 1975, when it was found that certain bacteria perform a U-turn and swim in the reverse direction when a switch reverses a local magnetic field. These bacteria exhibit **magnetotaxis.** That means they tend to swim along the lines of a magnetic field.

This phenomenon was first reported by Richard P. Blakemore, then at the University of Massachusetts at Amherst and now at the University of New Hampshire. While observing bacteria from the mud of brackish marshes near Woods Hole, Massachusetts he noticed that some microorganisms persistently swim in one direction across the field of view. The direction is unaffected by the position of the light source illuminating the microscope slide. Therefore, it is not an example of phototaxis. Even when the microscope was covered with a box, turned around or moved to another room, the bacteria continued to swim in the same geographic direction. The direction was always north! Since a nearby hand-held bar magnet would alter their direction—the bacteria swimming now in the direction of the new magnetic field—it appeared that they had been under the influence of Earth's magnetic field. In other words, these bacteria were acting as biological compass needles or magnetic dipoles.

These results interested Richard P. Frankel at the Massachusetts Institute of Technology's National Magnet Laboratory. With an electron microscope, he found that each of these bacteria contains a linear string of as

many as 20 cuboidal particles, each about 50 nm on one edge (Figure 29D). Spectroscopy revealed the particles to be crystals of magnetite ($FeSO_4$), also called lodestone. Each is permanently magnetized and

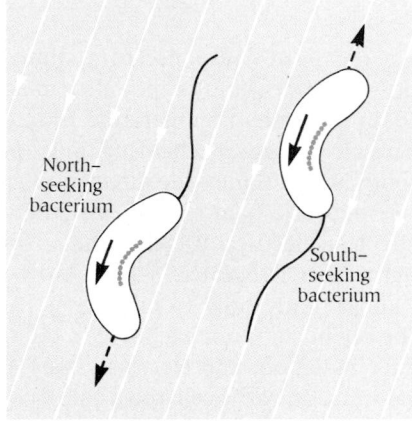

Woods Hole, Mass.

North–seeking bacterium

South–seeking bacterium

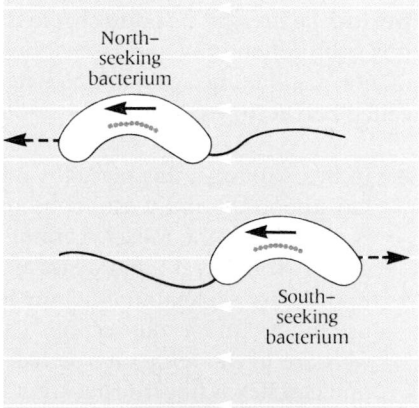

Fortaleza, Brazil

North–seeking bacterium

South–seeking bacterium

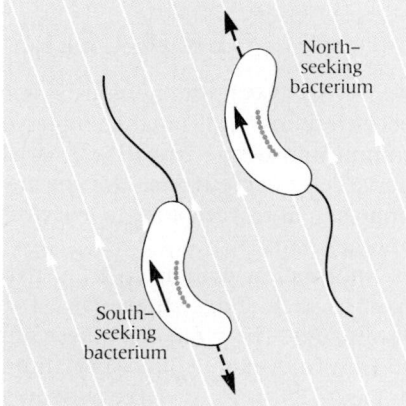

Christchurch, New Zealand

North–seeking bacterium

South–seeking bacterium

sheathed in a membrane. Blakemore and Frankel called them **magnetosomes.**

The magnetosome chain is both long enough and magnetic enough to position the bacterium in a magnetic field. As a result, swimming bacteria migrate along magnetic field lines. Even dead bacteria continue to orient along these lines, although they obviously do not swim. In the Northern Hemisphere, magnetic field lines incline downward so that bacteria swimming toward the North Pole will tend to swim downward in their medium. These bacteria happen to be bottom-dwellers that prefer an oxygen-poor environment. Magnetotaxis, it appears, might usefully serve to direct these bacteria downward toward the oxygen-poor sediments where they thrive best.

To test this hypothesis, the investigators did a simple experiment. They predicted that if such bacteria exist in the Southern Hemisphere, they should have the opposite magnetic polarity—if the same biological purpose (i.e., downward swimming) is to be served. That is precisely what was found. In 1980, they collected bacteria from sediments in New Zealand and Tasmania, regions with the same magnetic declination as New England, but with the opposite sign (Figure 29E). There the bacteria were predominantly south-seeking. This allowed them to navigate toward the bottom—just as their north-seeking cousins do. However, at the magnetic equator near Fortaleza, Brazil, where magnetic field lines are parallel to Earth's surface, they found equal numbers of north- and south-seeking bacteria. When a magnetic field was artificially imposed on the magnetic bacteria of Fortaleza, both groups—north-seekers and south-seekers—executed U-turns and swam in the opposite direction.

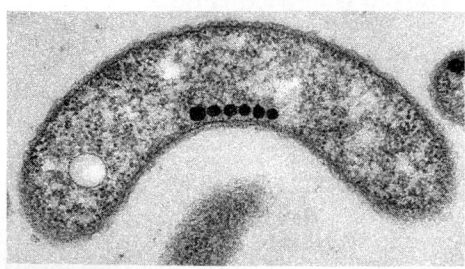

29D Electron micrograph of a magnetotactic bacterium
This thin section of Aquaspirillum magnetotacticum shows six magnetosomes aligned along the cell axis.

29E The importance of magnetotaxis to bottom-dwelling bacteria
The tops of these diagrams represent shallow water; the bottoms, deep water. In the Northern Hemisphere (Woods Hole), a north-seeking bacterium is directed downward toward sediments, whereas a south-seeking one moves upward toward the oxygenated surface. In the Southern Hemisphere (Christchurch), south-seeking bacteria swim downward and north-seeking ones upward. Near the geomagnetic equator (Fortaleza) both types of bacteria move horizontally. The direction of the geomagnetic field is indicated by background arrows.

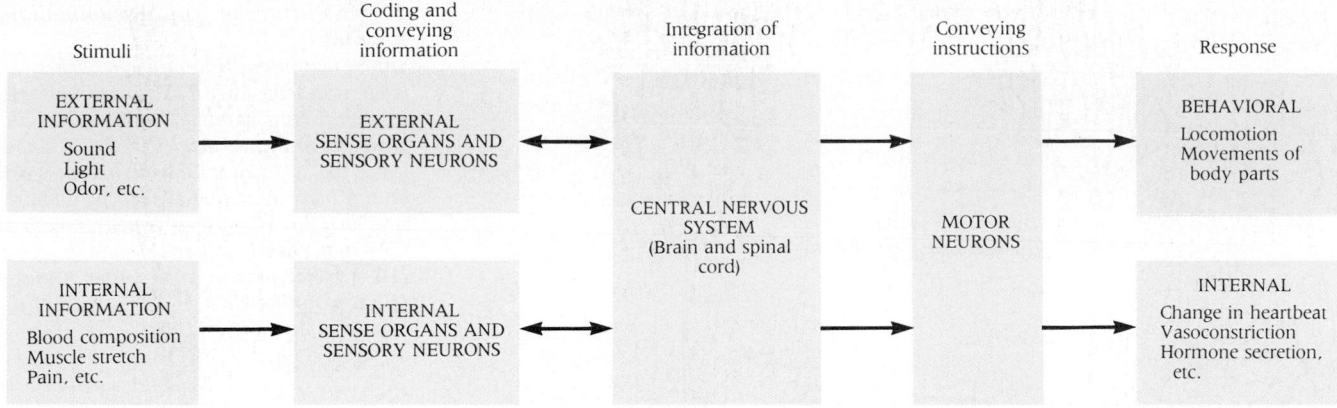

| Stimuli | Coding and conveying information | Integration of information | Conveying instructions | Response |

EXTERNAL INFORMATION
Sound
Light
Odor, etc.

EXTERNAL SENSE ORGANS AND SENSORY NEURONS

CENTRAL NERVOUS SYSTEM
(Brain and spinal cord)

MOTOR NEURONS

BEHAVIORAL
Locomotion
Movements of body parts

INTERNAL INFORMATION
Blood composition
Muscle stretch
Pain, etc.

INTERNAL SENSE ORGANS AND SENSORY NEURONS

INTERNAL
Change in heartbeat
Vasoconstriction
Hormone secretion, etc.

29-4 Flow of information into and out of the nervous system

that coordinate motor actions, memory, perception, association, and finally the mind. We shall soon recognize the many parallels between parts of the nervous system and a modern computer.

CELLS OF THE NERVOUS SYSTEM

The nervous system is made up of two basic classes of cells: the nerve cells, or neurons, and the neuroglial cells, or glia (or glial cells).

Neurons

The function of **neurons** is to transmit nerve impulses from one end of the neuron to the other and from one neuron to another. The same signaling mechanisms occur in all nerve cells. Nerve impulses are electrical disturbances of the same sort in all neurons, whether involved in vision, hearing, jogging, or composing music. It is not the separate neurons themselves, but their varying arrangements, connections, and effectors that determine the differences in their actions. All protoplasm displays irritability, and nerve cells are not unique in this regard. Their uniqueness lies in the fact that they maximize the conduction of impulses and direct them along specific anatomical pathways.

Our knowledge of neuron structure began in the nineteenth century with the remarkable insights of the Italian anatomist Camillo Golgi and his Spanish contemporary, Santiago Ramón y Cajal (Box 29-3). Their studies showed that these building blocks of the nervous system are fundamentally alike in structure and function in all animals that have them, although no two are identical. A typical neuron contains four distinct regions: a cell body, numerous short dendrites, a single long axon, and axonal terminal fibers (Figure 29-5A). Each region has a distinctive function.

- The spherical or pyramid-shaped **cell body** contains a nucleus, ribosomes, endoplasmic reticulum, and biochemical machinery for synthesizing proteins and other essential macromolecules. These are assembled into organelles and membranes and sent to other regions of the neuron.
- The delicately branching **dendrites** provide a large surface for receiving signals from sense organs or from the axons of other neurons.
- The **axon** emerging from the cell body provides a long pathway—in humans sometimes longer than 1 meter, in giraffes longer than 10 meters—for conducting nerve impulses to other parts of the nervous system or to an effector. Their diameters range from less than $0.2~\mu m$ in certain nerves of the brain to 1 mm in the well-studied giant nerve fibers of the squid. Within an axon are filaments made up of (in order of decreasing diameter) microtubules, intermediate filaments, and microfilaments (also called **neurofilaments**) that run the full length of the axon and carry proteins and other molecules from cell body to terminal fibers. This is called **axonal transport.**
- Near its end the axon divides into many fine branches, each of which ends with a specialized **terminal fiber.**

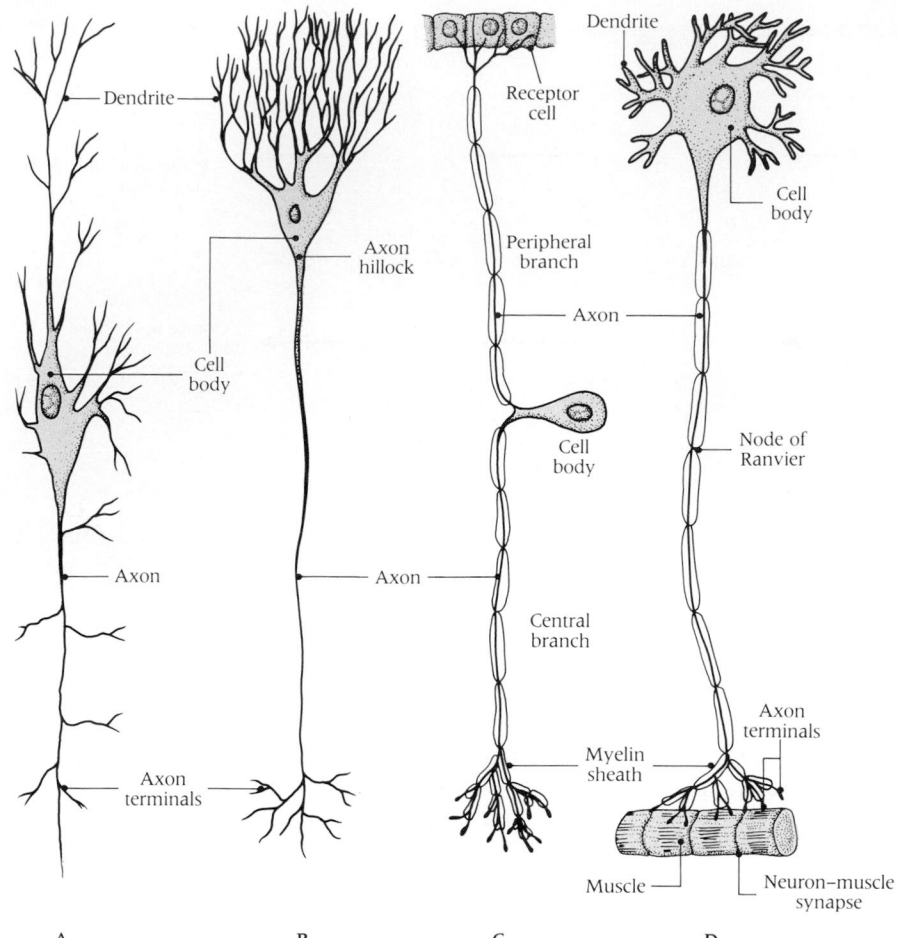

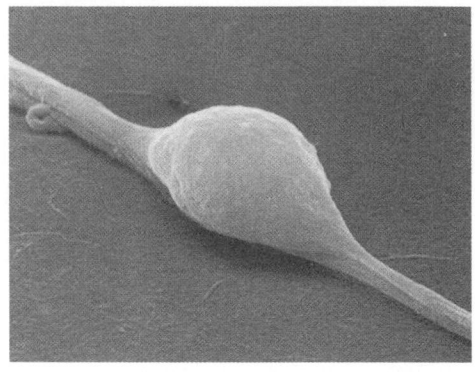

29-5 Structure of typical mammalian neurons
(**A**) *A neuron with multiple dendrites, a single axon, and some lateral axon terminals.* (**B**) *This neuron from a mammalian brain has an array of highly branched dendrites, and a single axon leaves the cell body at the axon hillock. Variations on this basic anatomical theme distinguish other neuron types.* (**C**) *A sensory neuron with a single fiber that branches after it leaves the cell body.* (**D**) *A motor neuron innervating a muscle cell.* (**E**) *A scanning electron micrograph of the cell body of a neuron.* (×1000)

A   B   C   D   E

Neurons differ in structure depending on their location and function. Brain neurons (Figure 29-5B) have profusely branched dendrites which connect with hundreds of other neurons. Sensory neurons (Figure 29-5C) connect to a sensory receptor. Motor neurons connect to an effector cell muscle (Figure 29-5D).

Impulses are transferred from one neuron to another at contact points called **synapses.** A neuron may have from 1,000 to 10,000 synapses and may receive information from perhaps 1,000 other neurons of different kinds (Figure 29-6). Neurons can conduct impulses in either direction (though retrograde conduction does not occur normally). Chemical synapses are the gates of the nervous system because they can conduct impulses in only one direction, although another type of specialized synapses (electric synapses) can conduct in both directions (see below). Hence neurons in a chain demonstrate **polarity** of impulse transmission. The **myelin sheath** that encloses some axons in a kind of insulating layer (Figure 29-7) is not essential to impulse conduction; but it speeds up the conduction of nerve impulses.

Certain aspects of neuron fine structure are of special interest. **Schwann cells,** named for their discoverer (the same Theodor Schwann who with Schleiden was cofounder of the cell theory) lie along the axon. The cytoplasm of the Schwann cells enfolds both myelinated and unmyelinated axons, although each myelinated axon has its own Schwann cell (Figure 29-7A). Groups of unmyelinated axons share the same cell (Figure 29-7B).

The Schwann cell plasma membrane is enriched in phospholipids and a variety of specific proteins. The plasma membrane wraps itself around the axon in multiple layers, creating the myelin sheaths around the axons of peripheral nerve fibers (Figure 29-7C). This explains why electron micrographs show that myelin exists in dense concentric layers. Other cells form the myelin sheaths of axons in the central nervous system. The myelin sheath appears to have evolved as a means of speeding impulse propagation by neurons.

Schwann cells and myelin have another interesting function. If a nerve is cut, there is degeneration of all the fibers distal to the cut—that is, all fibers detached from the cell bodies. As fibers and their myelin sheaths degenerate, Schwann cells

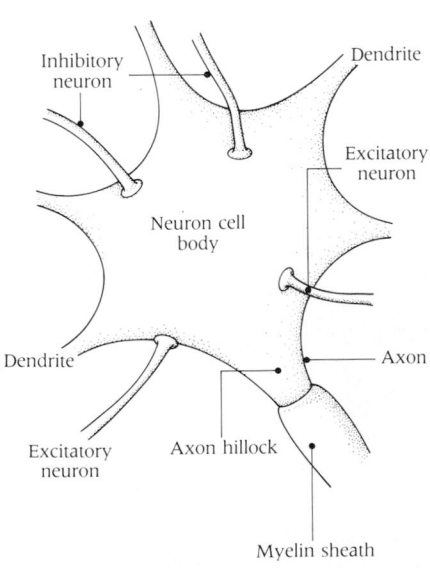

29-6 Multiple synapses on a single neuron
A single neuron may receive both stimulatory and inhibitory signals from thousands of neurons that synapse with it. Only a few are shown here.

BOX 29-3

STAINING RESEARCHES AND THE HISTORY OF NEUROANATOMY

Perhaps the most important single advance in the early years of neuroanatomy (after the microscope itself) came from the work of two Europeans in the late 19th century. One was the same Italian histologist, Camillo Golgi (1844–1926), whose name became attached to the Golgi apparatus. In 1873 he described a new method of metal impregnation, which stained a small proportion of the cells in a given region at one time and stained them in their entirety. The method required immersion of fresh tissues, first, in a solution of potassium bichromate, and then in silver nitrate. When free metallic silver was deposited in whole cells they appeared intensely black against a light yellow background (Figure 29F). It was an ingenious procedure, which had great value because it revealed only a few neurons in each microscopic field—and it revealed them completely, down to the smallest branches.

As often happens with important scientific discoveries, investigators of the time paid little attention. Not until the Golgi stain was championed by a Spanish contemporary, Santiago Ramón y Cajal (1852–1934), did it come into prominence. He improved the stain and then used it to investigate a fundamental question then disturbing neuroanatomists: Is there direct continuity between

nerve cells or are nerve cells completely separate entities that are joined into networks only by their synapsing branches? In one view the nervous system was seen as a special type of cellular **syncytium** (defined as a structure containing multiple nuclei within one membrane), in which nucleated "centers" were linked by uninterrupted fibers. In the other view, the cell theory held sway and nervous tissue was seen to consist of individual cells—like all other tissues. Using the Golgi stain, Ramón y Cajal proved that nerve cells are discrete units. For their contributions, Golgi and Ramón y Cajal shared the 1906 Nobel Prize in Physiology and Medicine (Figure 29G).

Ramón y Cajal made a second contribution of even greater importance. He compiled massive evidence to show that the incredibly complex interconnections among neurons are highly specific and that the organizational pattern is the very opposite of random. He exhaustively described the architecture of many kinds of structures in the brain, identified the various cells present, and showed how they are interconnected.

For a long time neuroanatomists had to be content with descriptions based on light microscopy using the Golgi stain and later the Nissl stain, another stain that reveals in

neurons the structures once called **Nissl bodies** and now recognized as clumps of ribosomes (see Chapter 7). More recently a new powerful tool was developed by Walle J. H. Nauta, now of the Massachusetts Institute of Technology. It exploits the fact that when a neuron is destroyed by mechanical or electrical means, or by heat, the nerve fiber coming from it degenerates. Just before disappearing, however, the fiber takes stain differently than does its undisturbed neighbors. This made possible the first accurate mapping of connections between different brain structures—for example, the different layers of the cerebrum. The result was an enormous expansion in the detail with which we can map the brain's architecture. Another new and powerful methodology exploits the tendency of injected horseradish peroxidase, which is easily detected with antibodies under the microscope, to outline the positions of nerve cells.

Many other methods are now available for the study of single neurons, their structure and connections—among them, sophisticated labeling procedures using radioactive isotopes or specific antibodies. Remarkably, neuroanatomists using these and other techniques are still discovering new connections in the brain.

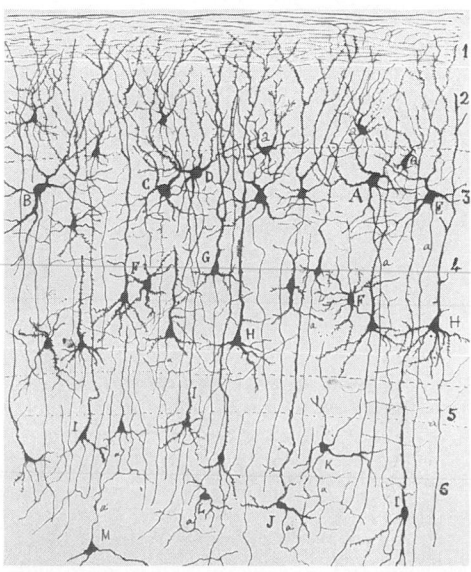

29F Golgi-stained nerve tissue as sketched by Ramón y Cajal in 1888

29G Pioneers in the development of neuroanatomy
(A) *Camillo Golgi.* **(B)** *Santiago Ramón y Cajal.*

multiply as if to form cellular tubes. When the two cut ends are close together, the fibers still connected to their cell bodies grow into the tubular sheath and replace the degenerating nerve fibers. This process, called nerve regeneration, proceeds at a rate of 1 to 4 mm per day. This sort of regeneration does not take place when the cell body is injured or when a neural tract of the central nervous system (brain and spinal cord) is cut.

In vertebrates, most of the cell bodies of nerves are in or near the brain and spinal cord. That is why some fibers run through nerves for such long distances. In humans, some fibers extend to the toes from cell bodies in the small of the back; in giraffes, they extend from spinal cord to hoof.

Nerve cells have very high rates of energy metabolism. They cannot store glucose, so they are highly sensitive to variations in levels of blood glucose, their principal source of energy. That is why temporary cessation of the heart beat threatens serious brain damage. The brain and other parts of the nervous system need blood-borne glucose continuously. When brain activity increases, the need for glucose increases. Thinking really is work, not just because some of us are reluctant to indulge in it, but because it uses energy. It requires less energy than muscular activity, however.

The surface membranes of nerve cells contain receptors, enzymes, and structural proteins. All are synthesized in the cell body of the neuron and exported in small membrane-sheathed vesicles.

Neuroglial Cells

The second class of cells in the nervous system are the **neuroglial cells,** or **glia,** as they are often called. The central nervous system contains some nine times as many glial cells as neurons. Glia have four general functions: (1) they are supporting elements—much like connective tissue cells elsewhere in the body—and physically segregate different groups of neurons from each other; (2) they provide nutrition to neurons; (3) they modulate neuron function; and (4) some of them make myelin, this being the function of two of the five classes of neuroglia.

The neuroglia are divided into five major classes.

■ **Oligodendrocytes,** which form the myelin sheaths of central nervous system axons (Figure 29-8A)
■ **Astrocytes,** which are numerous in areas of the central nervous system containing many myelinated axons (these regions are called **white matter** because of their whitish appearance in freshly cut brain slices) and in areas containing many cell bodies, dendrites, and synapses (these are the **gray matter**) (Figure 29-8B)
■ **Schwann cells,** which form the myelin sheaths of peripheral nerve axons
■ **Microglia,** tiny cells that are capable of acting as phagocytes
■ **Ependymal cells,** which line the inner surfaces of the brain

We should remember that glial cells are not directly involved in nerve impulse transmission. That is the function of neurons.

NERVE IMPULSES

It had been recognized for decades that the basic processes of nerve conduction are electrochemical—yet there was a discouraging scarcity of promising ideas about how an electrical impulse could be generated by biochemical means.

We can now say a great deal on this stimulating subject. Nerve impulses are electrical, but they are associated with and produced by fleeting changes in the highly specialized plasma membranes of nerve cells. However, impulse transmission is not the same as the simple conduction of an electrical current through a wire. There are two important differences.

■ In wires, current is carried by electrons traveling at the speed of light. In neurons, impulses are carried by bulky ions—mainly Na$^+$—which diffuse very slowly through solution.
■ An impulse in a wire involves a massive flux of particles that travel along the wire. A nerve impulse consists of a zone of changing differences in electrical

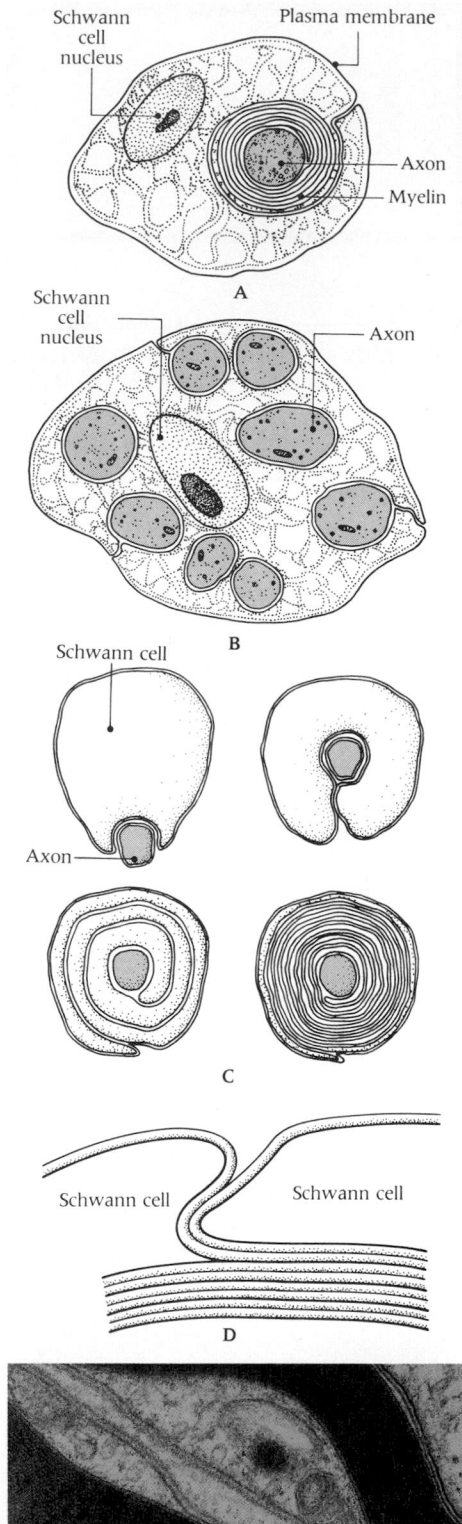

29-7 Relation of Schwann cells to myelinated and unmyelinated nerve fibers (**A**) A myelinated axon surrounded by a Schwann cell. (**B**) Several unmyelinated axons within a single Schwann cell. (**C**) A myelin sheath is formed as multiple layers of the Schwann cell plasma membrane wrap around the axon. (**D**) An enlarged view of the myelin sheath showing the plasma membrane layers that form myelin. (**E**) Myelinated axon. (×900)

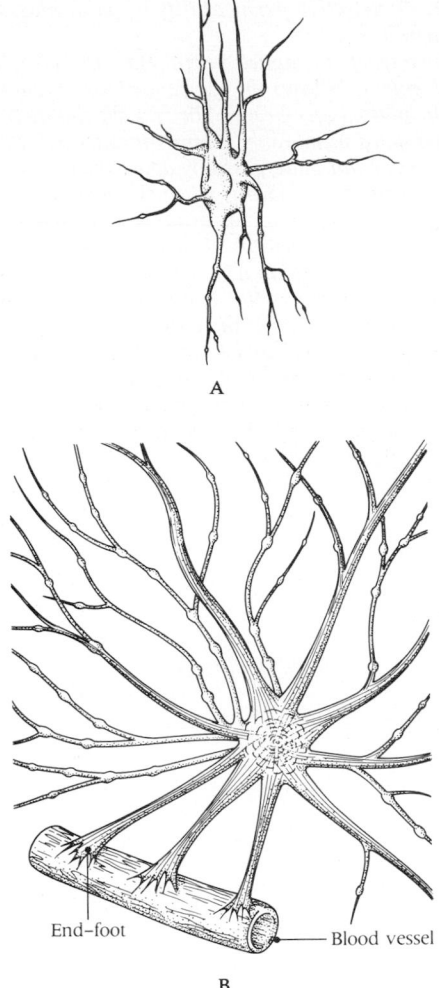

29-8 Examples of glial cells
(**A**) *Oligodendrocyte in white matter.* (**B**) *Fibrous astrocyte. These are the principal glial cell types in the brain and spinal cord. They are closely associated with neurons and form end-feet on blood vessels.*

Labels on figure B: End-foot, Blood vessel. Label A.

potential between the outer and inner surfaces of the plasma membrane. It is the zone that moves along the neuron.

This zone moves along the nerve fiber as a flux of small, purely local electrical currents resulting from movements of Na$^+$ and other ions that create ionic gradients across the membrane.

Despite the insensitivity of their instruments, workers in the early years of the twentieth century did manage to detect a difference in electrical potential across the plasma membrane of a resting neuron. The inside of the membrane is negatively charged; the outside of the membrane is positively charged. Thus the membrane is polarized. When a nerve cell is undisturbed, the separation of these charges by the membrane is responsible for a resting potential difference of about 60 millivolts (mV). Since we are here interested in potential differences, it is customary to arbitrarily set the outside charge equal to zero. Therefore, we say that the **resting potential** is -60 mV. This charge difference is of fundamental importance because propagation of a nerve impulse is associated with changes in the resting potential across the membrane. As we shall see, this change usually consists of a reversal in the polarity of the membrane.

It has been known since 1902 that the membrane of a resting neuron is selectively permeable to potassium (K$^+$) and impermeable to sodium (Na$^+$) (Figure 29-9). Negatively charged ions such as Cl$^-$ and amino acids cannot pass through the membrane. Because of this selective permeability, the K$^+$ concentrated inside the cell is able to diffuse outward and some NA$^+$ can diffuse in, but Na$^+$ and other positive ions are unable to enter in quantities sufficient to correct the deficiency of positive charges. We saw in Chapter 6 that the concentration of K$^+$ inside all metazoan cells is about 10 times its concentration in extracellular fluids, while Na$^+$ concentrations are much higher outside the cell than inside.

These gradients of ion concentrations are maintained by specific membrane proteins known as **pumps.** The most important is the **sodium-potassium pump,** an ATP-dependent mechanism in the membrane that drives Na$^+$ from the nerve fiber against a concentration gradient (that is why it is called a pump) in exchange for K$^+$ (Figure 29-10). A model of the Na-K pump in red cells, which expels three Na$^+$ ions for every two K$^+$ ions it takes in, was described in Chapter 6. Most neurons have 100–200 Na-K pumps per square micrometer of membrane surface.

These two properties of the neuronal plasma membrane—selective permeability and the capacity to maintain ionic concentration gradients—produce the difference in electrical potential between the inside and outside of the membrane.

During the passage of a nerve impulse, there is an abrupt increase in membrane permeability to Na$^+$ (Figure 29-11A). This permits Na$^+$ to rush into the fiber from the surrounding fluid. As the inner surface of the membrane becomes locally more positive, the potential difference drops to zero and then reverses, becoming positive. A reduction in Na$^+$ permeability and an outflow of K$^+$ not balanced by the passive efflux of Cl$^-$ restores the negative potential of the resting cell. The voltage reversal resulting from the sequential **depolarization** (change in resting potential from a negative to positive value) and **repolarization** in a highly localized region of the nerve cell membrane initiates a local **action potential,** which is propagated in the immediately adjacent portion of the membrane. In this way, it travels down the fiber (Figure 29-11B). In other words, the impulse migrates in a specific direction. Only after a brief **refractory period** is it possible for a second impulse to follow (Figure 29-11C).

The nerve-cell membrane does not lose its ionic selectivity during impulse transmission. Rather, it becomes specifically permeable to Na$^+$ for about half a millisecond. The movement of only a few Na$^+$ ions will generate an action potential. We saw in Chapter 6 that charged molecules do not pass freely through membranes which are lipid rich. Indeed, clusters of specific proteins form **membrane channels**—specific openings through which only certain ions can pass. Thus there are sodium channels and potassium channels. These channels possess so-called gates. A **gated channel** is one that can be opened or closed by a conformational change in a critical transmembrane protein, which serves as the "gate." As a result, a specific ion (in this case Na$^+$ or K$^+$) is permitted to move through the channel—or, if the gate is closed, is prevented from doing so (see Figures 29-9 and 29-10).

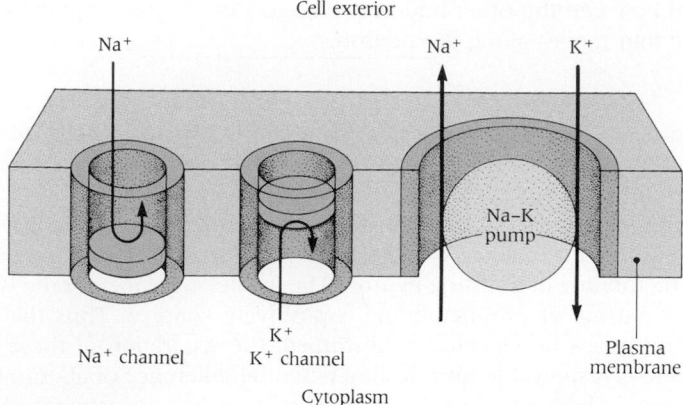

Cell exterior

Na⁺   Na⁺   K⁺

Na⁻K pump

Na⁺ channel

K⁺ channel

Plasma membrane

Cytoplasm

29-9 Selective permeability of axon plasma membrane
Two types of channels permit Na⁺ and K⁺ ions to enter and leave the axon cytoplasm through the plasma membrane. The two left channels are gated membrane channels through which the ions can move passively down their concentration gradients when the gates are open—K⁺ moving outward and Na⁺ inward. In this resting neuron, the gates are closed. In addition, there is an ATP-dependent Na-K pump that actively moves Na⁺ outward and K⁺ inward and tends to balance the passive fluxes, or leakiness, of Na⁺ and K⁺ by active transport in the opposite direction. The pump shown here (and in the following figures) is highly diagrammatic in style. A more realistic picture appears in Figure 6-18.

The factors that open and close these gates in nerve cell plasma membranes are nearby voltage shifts called **gating currents** (Figure 29-12). Do not confuse sodium gates with sodium pumps.

- Sodium gates open and close as an impulse moves along the axon, allowing Na⁺ to rush into the fiber, depolarizing the membrane. The subsequent opening of potassium channels repolarizes the membrane.
- Sodium pumps are relatively slow-working energy-dependent systems that correct the Na⁺-K⁺ gradient whenever it is disturbed.

An interesting sodium-channel protein has recently been isolated and purified. One of the stratagems that made this possible has involved the use of deadly nerve poisons, or **neurotoxins.** An example is **tetrodotoxin,** which occurs in the internal organs of the puffer fish. Another is **saxitoxin,** a product of some of the marine dinoflagellates that cause "red tides." Eating contaminated shellfish from red-tide waters can fatally damage the nervous system (see Box 39-1). One molecule of either toxin binds to each sodium channel with extraordinary specificity and high affinity. This property, which explains their toxicity, made possible the isolation of a part of the **sodium-channel protein,** a single polypeptide subunit.

One consequence of the way action potentials are propagated down a nerve is that the impulses move more slowly than does electric current in a wire. In the fastest nerve fibers of mammals, including humans, the rate is about 10 meters per second (22 miles per hour). In some invertebrate fibers, the rate is only 5 cm per second (0.7 mile per hour). A tortoise walks faster.

The velocity at which the action potential travels in a stimulated neuron is affected by the diameter of the nerve fiber, larger diameters being associated with faster impulse transmission. Many invertebrates have giant neurons with axons 0.5 to 1 mm in diameter. These are popular experimental objects of neurophysiologists because of their large size. The best known is in the squid, which has a jet propulsion system that must operate quickly if the animal is to escape danger (see Chapter 42). The nerve activating this organ has giant axons that transmit impulses at 10 times the rate in average-sized axons.

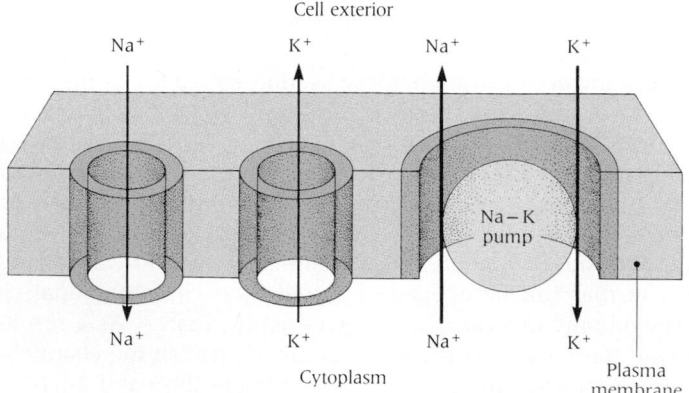

Cell exterior

Na⁺   K⁺   Na⁺   K⁺

Na—K pump

Na⁺   K⁺   Na⁺   K⁺

Plasma membrane

Cytoplasm

29-10 Ions pass freely when membrane channel gates are open
When the resting potential decreases, gates open in the membrane channels and ions pass through them freely. The Na⁺ channel opens quickly. The K⁺ channel opens slowly. The main result of these ion movements is an action potential.

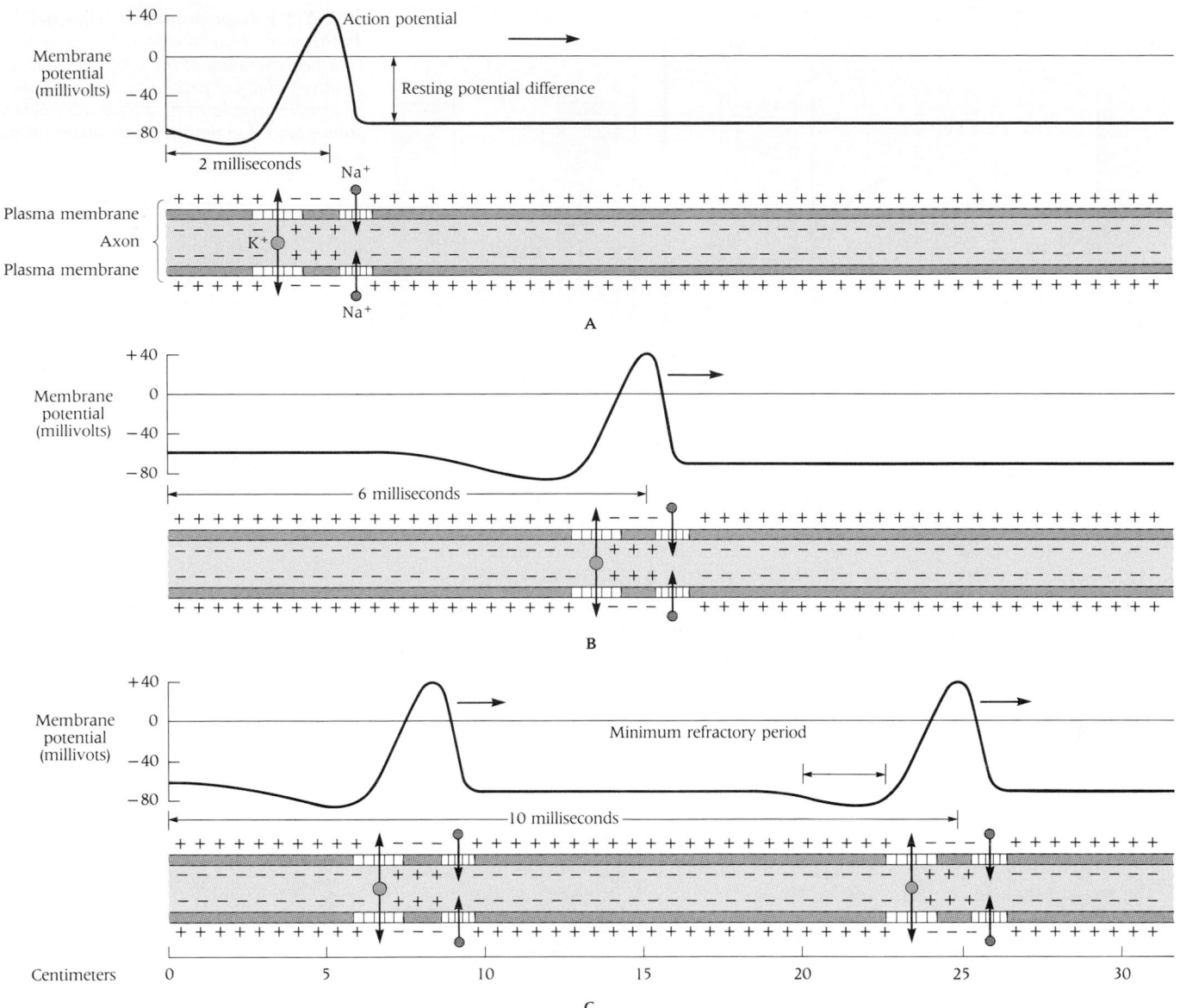

29-11 Propagation of an action potential
(A) *The action potential begins with a reduction in the membrane potential, which becomes positive as Na$^+$ enters the axon. This reversal in polarity closes the Na$^+$ channels and opens the K$^+$ channels, restoring the negative potential.* **(B)** *The action potential is propagated down the axon by sequential openings and closings of the gated channels in the membrane.* **(C)** *As the inpulse moves along the axon, it is trailed by a brief refractory state. After it passes, a second impulse can follow.*

The mechanism of nerve impulse conduction also implies that the strength of an impulse does not vary like current in a wire. In a wire, the resistance of the wire to the passage of current through it reduces the current so that less current emerges from the end of a wire than entered it at the beginning. There is no such drop-off in the conduction of nerve impulses. An impulse starting at full strength remains that way. Its passage more resembles the ignition of a sprinkled line of gunpowder than the activation of an electric wire. When the powder is lit at one end, a flash travels to the other end with a speed and strength that is independent of the heat of the igniting match or of the length of the line. In a similar way, the strength of a nerve impulse depends on energy generated by the nerve itself and on local events at each point.

For that reason, we call these events **all-or-none** reactions. The neuron either fires or it does not. Stimuli below a threshold strength will not fire a neuron. Stimuli above the threshold will fire it. The action potential in a stimulated nerve has a fixed size no matter how strong the stimulus.

How then does it happen that stimuli of varying intensities lead to effects of varying intensities? The answer is that a stimulus almost never starts with just a single impulse. Unless a stimulus is excessively brief, the impulses generated by it pass along the fiber in a volley, one after the other. Strengthening the stimulus increases the frequency of impulses, so that more of them arrive at the effector per second. This is what enhances the response. Nerves, therefore, are an FM (frequency modulation) rather than an AM (amplitude modulation) system.

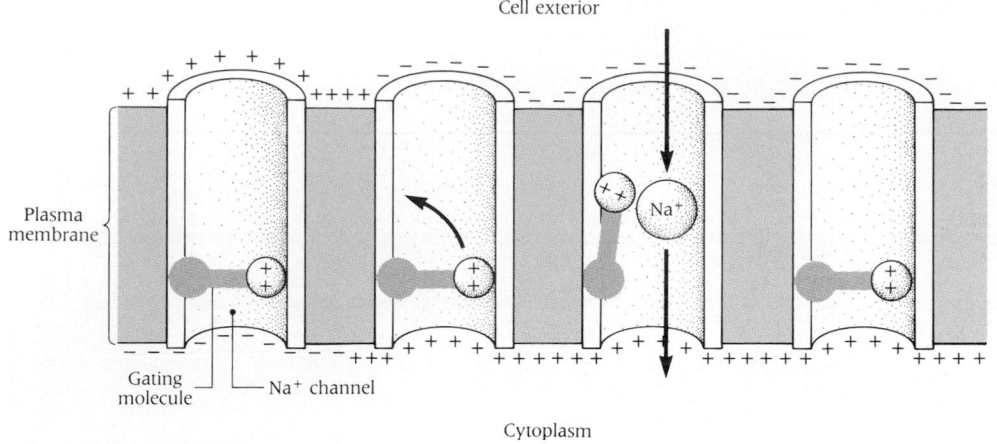

29-12 A voltage-gated Na⁺ channel
In this more detailed model of a gated Na⁺ channel, operation of the gate, which is a positively charged protein molecule within the channel, is regulated by a change in polarity from negative to positive in the axon cytoplasm.

The **refractory period** is the period immediately after stimulation or impulse transmission (see Figure 29-11C). During the absolute refractory period, the initial 0.5–1.0 millisecond of this phase, nervous tissue is impossible to excite—no matter how great the stimulus—because voltage-dependent channels are temporarily inactivated. For a period after the absolute refractory period (2–3 milliseconds), nervous tissue is less excitable than normal. This is the relative refractory period in which a stronger stimulus than normal can initiate an action potential. The refractory period accounts for the fact that nerve fibers normally carry impulses in only one direction. Impulses traveling in the opposite direction would stop and be extinguished at their meeting point, for each would have left in its wake a refractory region that could not be traversed by its advancing opponent. Before returning to its resting state, the neuron may pass through a supernormal phase of excitability. The refractory period occurs during **repolarization,** the restitution of the normal Na⁺ and K⁺ permeability of the axonal plasma membrane.

The myelin sheath has an interesting role in nerve impulse conduction. Its presence greatly increases electrical resistance across the axonal plasma membrane. As a result, the transmission of impulses down the axon is largely the responsibility of the **nodes of Ranvier** (Figure 29-13), bare junctures occurring every 1–2 mm at which myelin is missing and the axonal membrane is in direct contact with the extracellular fluid. The nodes are the only sites on the axon where voltage-activated sodium channels can be found.

Ionic currents cannot flow freely across the membrane where myelin is located, but the impulse is kept from dying out at the nodes, which rapidly reestablish the action potential (Figure 29-14A). The resulting current is conducted swiftly to the next node where it is again regenerated. We call this unusually fast form of node-to-node conduction **saltatory conduction** (Figure 29-14B). It is useful also because it requires less metabolic energy for the Na-K pump. Diseases resulting in the loss of myelin (demyelination) have devastating effects on nerve impulse transmission. Multiple sclerosis is one such disease.

29-13 Node of Ranvier
(A) *A node of Ranvier is a region of the axon plasma membrane that is not covered by the myelin sheath.* **(B)** *Scanning electron micrograph of a nerve showing nodes of Ranvier. (×600)*

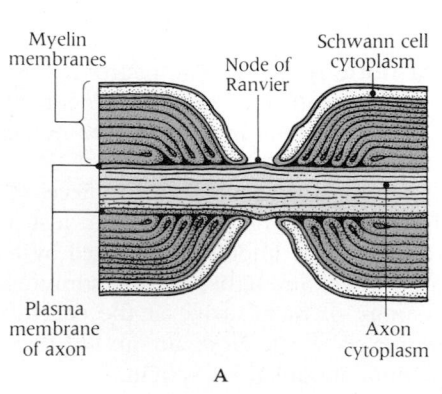

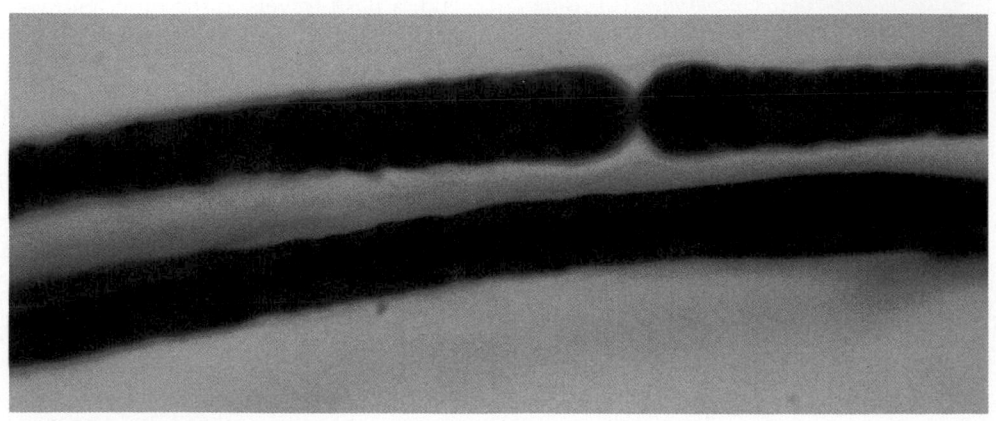

A   B

(A) *Ionic membrane current is much higher at the nodes of Ranvier than at regions between the nodes.* (B) *The action potential skips rapidly from node to node in myelinated axons.*

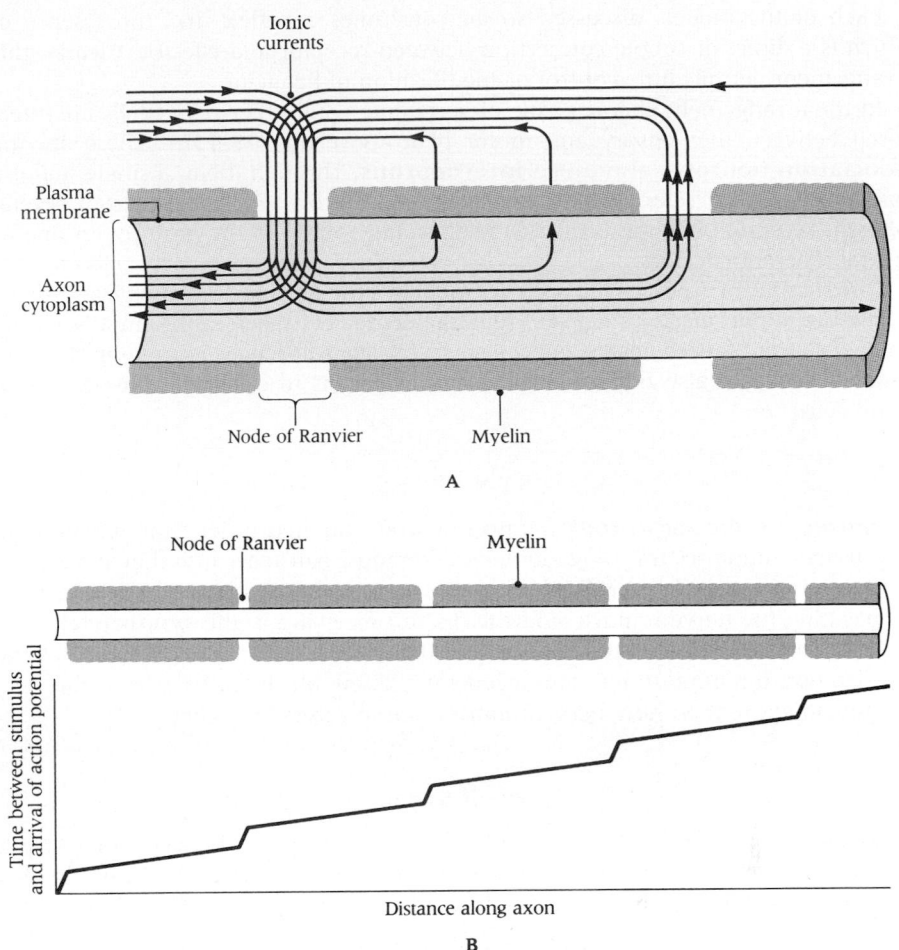

NERVE CIRCUITS: MODEL NERVOUS SYSTEMS

If we regard the nervous system as an enlarged version of an individual cell's primitive receptor-conductor-effector system, it is easy to devise an instructive model of the simplest of all possible nervous systems. Knowing that neurons tend to become organized into circuits, we can build upon this model in stepwise fashion.

In the simplest possible nervous system, a fiber extending directly from a receptor, or **sensory receptor cell,** in the body surface makes direct contact with an **effector cell** within (Figure 29-15A). The stimulated receptor cell transmits an impulse directly to the effector cell, which responds appropriately—for example, a muscle cell contracts. Systems of this type occur in simple animals such as sea anemones.

At the next stage of complexity (see Figure 29-15B), a receptor cell initiates an impulse in a separate nerve cell, a conductor that conveys the impulse to an effector cell. Systems of this type also occur.

In both of these models, reaction patterns are necessarily simple and entirely inflexible. When the stimulus is of adequate intensity, the receptor cell is "turned on" like a switch, and the impulse is invariably conducted to one particular effector, whose response can vary only in duration.

In more complex animals, receptor cells rarely make direct contact with effector cells. In vertebrates, they almost never do. Instead, there is almost always more than one nerve cell between the receptor and effector cells (Figure 29-15C). In some vertebrates, a separate **sensory neuron** receives an impulse from a receptor cell and then passes it on to a separate **motor neuron.** The motor neuron in turn conducts the impulse to an effector cell.

Though this system is still a simple one, the neural chain provides for a greater effectiveness and flexibility of reaction than does a direct connection between receptor and effector. Since a sensory neuron can stimulate more than one motor neuron (Figure 29-15D), a single stimulus can arouse an extensive response. Continuous stimulation can involve more and more effectors.

Each of the models discussed so far constitutes a **reflex arc,** the essence of which is a direct or simple connection between receptor and effector. Clearly, this arrangement permits little control or modification of behavior.

In the arrangement characteristic of vertebrates, still other nerve cells are interposed between the sensory and motor neurons (Figure 29-15E). These are the **association neurons,** also called **interneurons.** Through them, a single impulse may be passed on selectively to one or more effectors, or impulses from several different receptors may be brought together and routed to the brain or to one or several different effectors.

One significant aspect of this series of models is the introduction and progressively increasing importance of synapses, remarkable devices wherein the impulses from one neuron are ferried across a gap so they can continue their journey in the next neuron. Synapses also transmit signals from neurons to effectors (muscle cells or gland cells).

SYNAPSES

Synapses are the microscopic neuron-neuron and neuron-effector junctions at which nerve impulses traveling along an axon must pause because they have come to the end of the axon's terminal fibers. The space between the end of the neuron propagating the nerve impulse and the neuron receiving it, the **synaptic cleft,** is variable depending on the type of synapse. If a nerve impulse is to continue beyond the synapse, the impulse must be regenerated anew on the other side of the cleft. We must inquire then how nerve impulses "jump across" the cleft.

Types of Synapses

A critical issue in the 1950s was whether communication between presynaptic and postsynaptic cells requires chemical or electrical mediation. The answer, as it turned out, is that both processes occur.

- In **chemical synapses,** the more common type of synapse (Figure 29-16A), a chemical substance released into the synaptic cleft by the presynaptic neuron diffuses across the gap and changes neural activity in the postsynaptic cell. The synaptic cleft is only about 20 nm wide.
- In **electrical synapses,** the neuron communicates with the postsynaptic cell through gap junctions (Figure 29-16B). This permits more rapid and direct conduction of nerve impulses without a lag period.

Chemical neurotransmission was discovered in 1921 by Otto Loewi (1873–1961), then a professor of pharmacology at the University of Graz in Austria. While studying the behavior of isolated frog hearts, he passed saline solution through one beating heart into another. When he applied an electrical stimulus to the vagus nerve attached to the first heart, he observed the expected slowing of the heart beat. But to his delight, the second heart, which had been detached from all nerves, also began to beat more slowly. The salt solution had apparently transferred an inhibitory substance.

Loewi gave the mysterious substance the German name *Vagusstoff*. It was soon identified as the low molecular weight compound, **acetylcholine** (Figure 29-17).

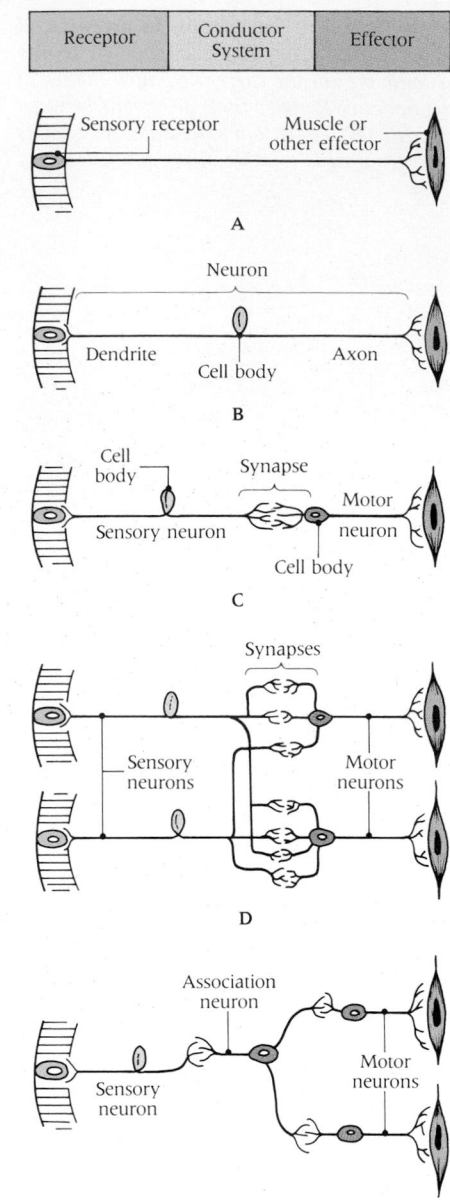

29-15 A hierarchy of receptor-conductor-effector systems
(A) In the simplest system, sensory cell connects directly to effector. (B) A single neuron connects sensory cell to effector. (C) More than one neuron connects receptor to effector. (D) More than one receptor innervates more than one effector via synapsing with multiple connecting neurons. (E) In this most highly developed system, association neurons connect sensory and motor neurons.

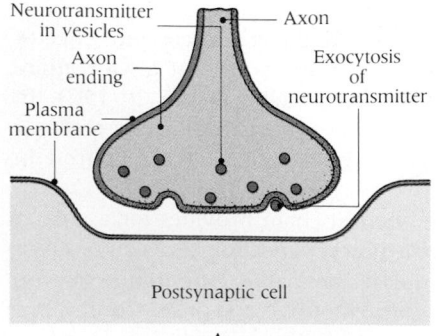
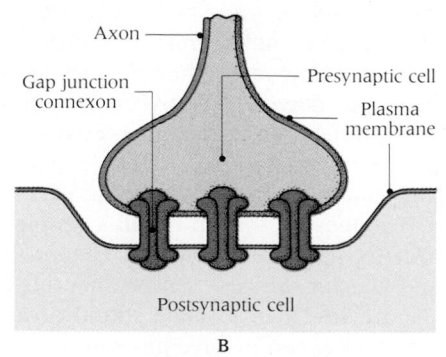

29-16 Two types of synapses
(A) In a chemical synapse, the synaptic cleft is about 20 nm wide. The terminus of the presynaptic axon contains synaptic vesicles filled with neurotransmitter molecules that are released into the space by exocytosis and diffuse to the postsynaptic cell surface. (B) In an electrical synapse, transmembrane channels called gap junctions link the presynaptic and postsynaptic membranes. The flow of ions through these channels transmits impulses from one cell to another.

$$CH_3-\overset{\overset{\displaystyle O}{\|}}{C}-O-CH_2-CH_2-\overset{\overset{\displaystyle CH_3}{|}}{\underset{\underset{\displaystyle CH_3}{|}}{N^+}}-CH_3$$

$\underbrace{}_{\text{Acetyl}}$ $\underbrace{}_{\text{Choline}}$

29-17 Acetylcholine

29-18 Structure of a chemical synapse
(A) *The knoblike axon ending is enriched in neurotransmitter-filled synaptic vesicles and mitochondria. When an action potential reaches the presynaptic knob, neurotransmitter molecules diffuse across the synaptic cleft and bind to receptors on the postsynaptic membrane, eliciting in it permeability changes and a new action potential on the other side of the synapse.* **(B)** *Electron micrograph of synaptic vesicles in the olfactory cortex of a rat* ($\times$ 82,000).

He concluded that it had been liberated by the vagus nerve, had crossed the synaptic cleft and, like a true chemical messenger, had triggered a new nerve impulse on the other side of the synapse. Many refused to accept the idea of chemical transmission of impulses across synapses, insisting that only electrical impulses could make the crossing so rapidly.

Since this early work, acetylcholine has been firmly established as the **neurotransmitter** in many, but decidedly not all, synapses in vertebrate and invertebrate animals. Our present view of how it works draws heavily on mechanisms we have already encountered in other areas of cell biology.

How Chemical Synapses Work

The chemical synapse is made up of a **presynaptic knob,** or **bouton,** which is the terminus of the presynaptic neuron, the **synaptic cleft,** and the specialized **postsynaptic membrane** of the postsynaptic neuron (Figure 29-18A). The presynaptic terminal contains many small **synaptic vesicles** and a few mitochondria. Each synaptic vesicle contains a single type of chemical transmitter. (We will here discuss acetylcholine as an example; other transmitters will be described later.) Acetylcholine is synthesized in the cytoplasm outside the synaptic vesicles and then taken up and stored by the vesicles, which are concentrated near the synaptic cleft. They are vividly revealed by electron microscopy (Figure 29-18B).

The key event of chemical neurotransmission, the release of a package of neurotransmitter molecules, is stimulated by an arriving voltage pulse in the presynaptic neuron. How is this brought about? The answer became clear—or clearer—only recently. What determines the release of neurotransmitter from the presynaptic cell is a short-lived influx of Ca^{2+} in the neighborhood of the synapse. An impulse arriving at the presynaptic knob makes the membrane more permeable to Ca^{2+} by opening voltage-dependent Ca^{2+} channels. Calmodulin in the presynaptic knob interacts with the membranes of synaptic vesicles and causes the vesicles to discharge their contents into the cleft (Figure 29-19). Vesicles migrate to the membranes of the knob, probably under the influence of a specific phosphoprotein called **synapsin.** They fuse with the synaptic membrane and then release transmitter substance. This is an example of exocytosis (see Chapter 6). It explains experimental results

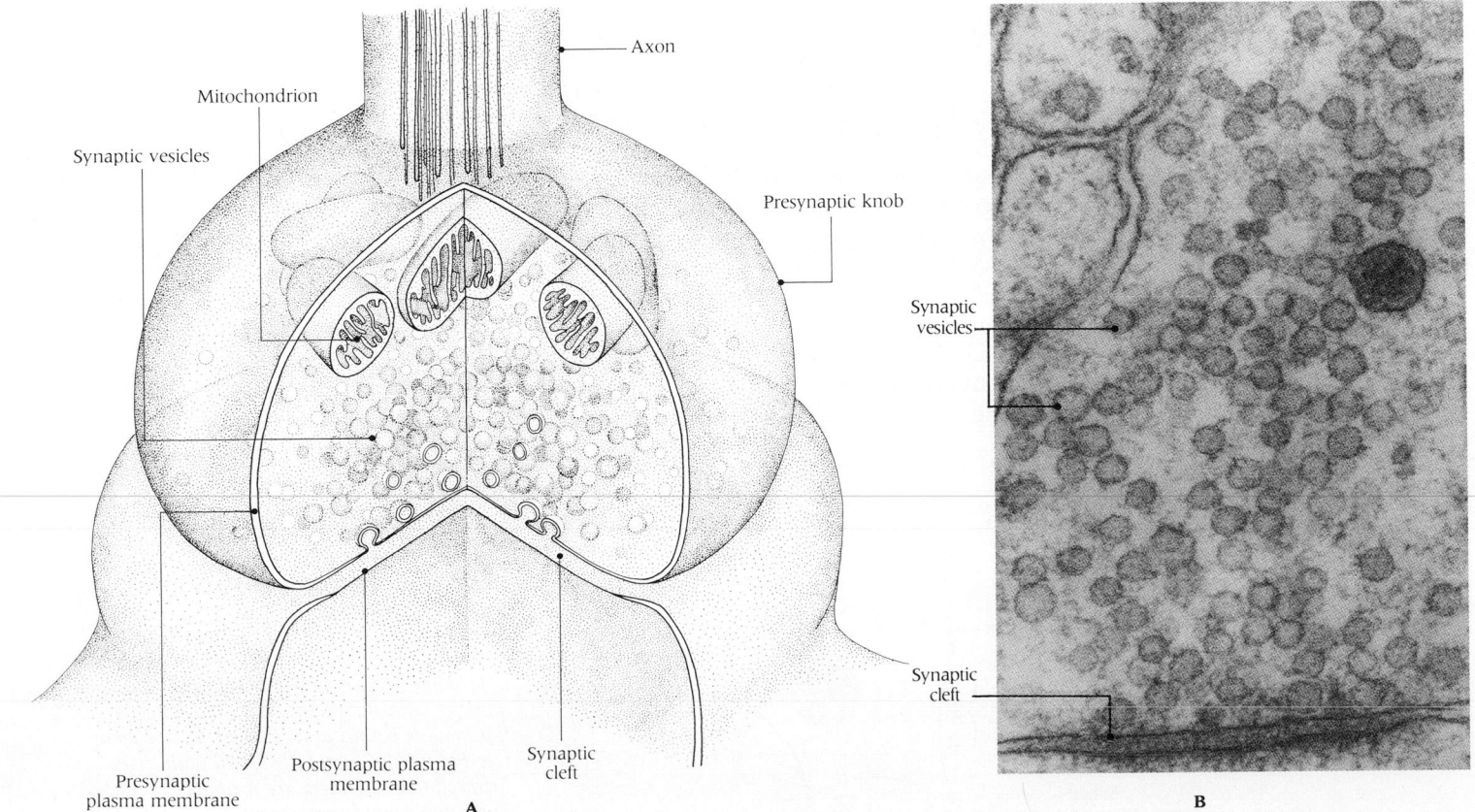

A

B

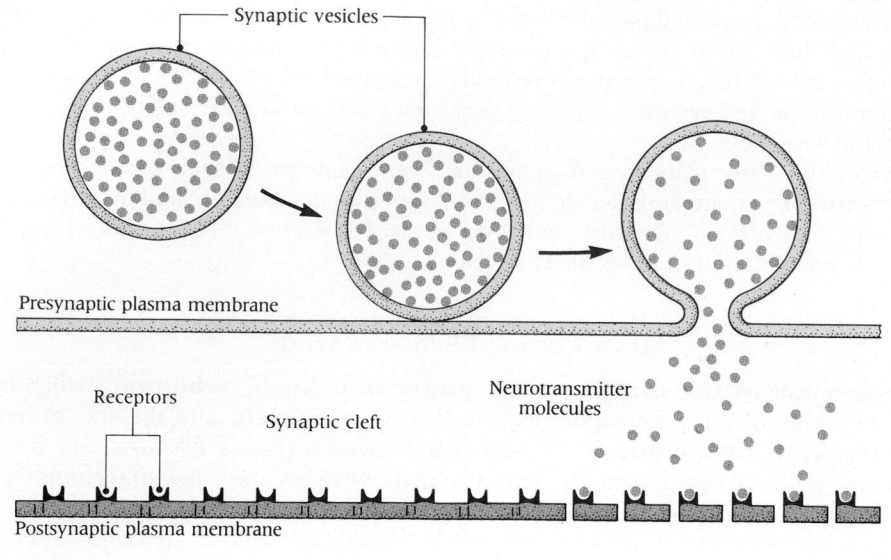

Synaptic vesicles

Presynaptic plasma membrane

Receptors

Synaptic cleft

Neurotransmitter molecules

Postsynaptic plasma membrane

A

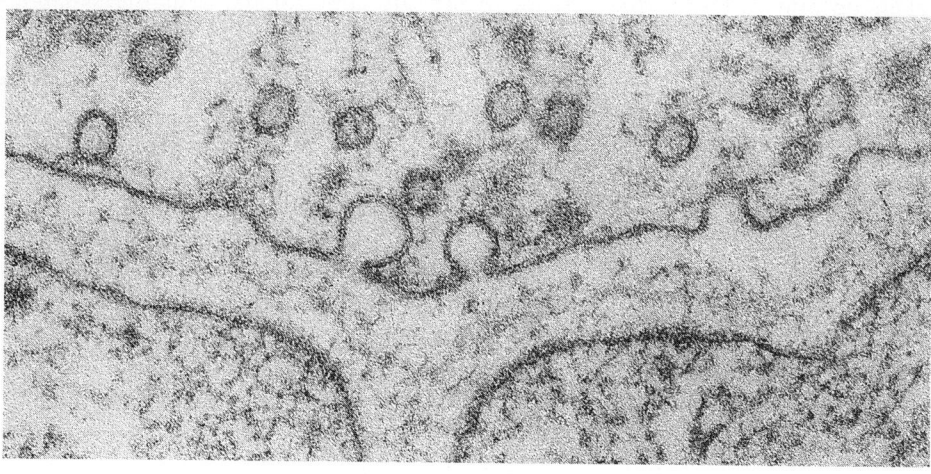

B

29-19 Exocytocis and the release of neurotransmitter molecules
(A) *Synaptic vesicles clustered near the presynaptic membrane fuse with the membrane. In this final step of exocytosis, they release neurotransmitter molecules into the synaptic cleft. The postsynaptic membrane bears receptors that bind the neurotransmitter. The synaptic cleft is narrower than it appears here, relative to the diameter of the membranes shown.* (B) *Electron micrograph showing fusion of synaptic vesicles with the presynaptic membrane.* (×160,000)

suggesting that neurotransmitter molecules are released in discrete bursts of a few thousand molecules. The released substance then attaches to special receptors on the subsynaptic membrane of the postsynaptic neuron.

Neurotransmitters bind to specific receptors on the surface of the postsynaptic neuron (Figure 29-20). In some cases, the receptor is also an ion channel. Binding a neurotransmitter causes it to open or close. In other cases, the binding event causes the receptor to stimulate adenylate cyclase, so that ATP is converted into cAMP. This in turn activates a protein kinase in the membrane that locally alters the membrane's permeability by opening or closing channels. This is the same second-messenger system participating in the actions of many hormones, among them norepinephrine, epinephrine, and various peptide hormones. Still other neurotransmitter receptors do not activate adenylate cyclase, but act through other second messengers.

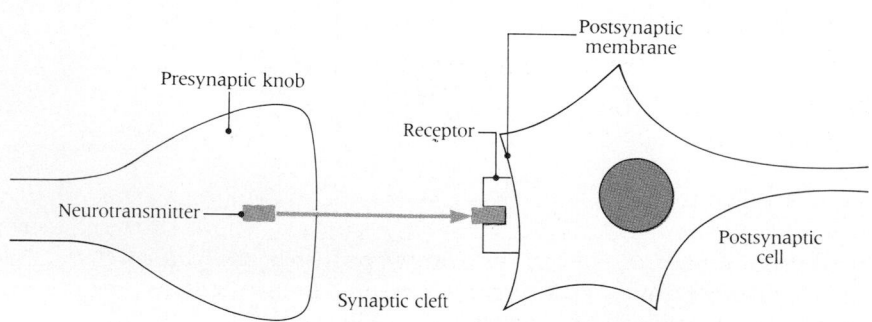

Presynaptic knob

Postsynaptic membrane

Receptor

Neurotransmitter

Synaptic cleft

Postsynaptic cell

29-20 Binding of neurotransmitter to postsynaptic receptor
Receptors on the postsynaptic membrane bind a specific neurotransmitter and are caused, thereby, to respond in a specific manner. Note the clear analogy with the events occurring when a hormone is bound to a cell's surface receptor (see Figures 28-5 and 28-7). In both cases, intracellular responses are mediated by cAMP and other second messengers.

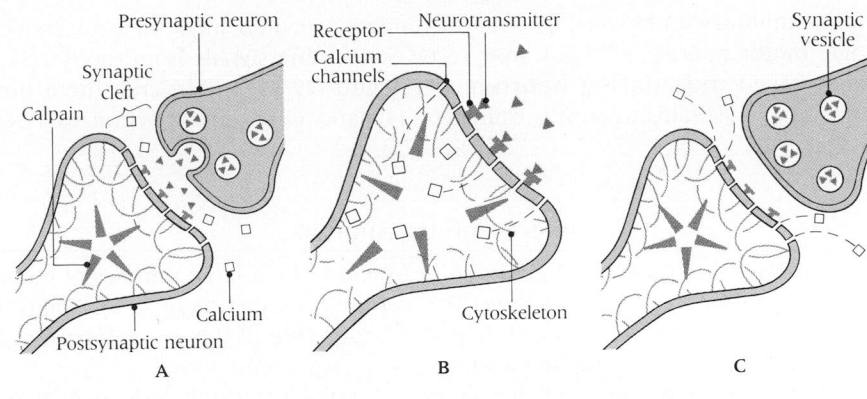

29-21 The role of calcium ions in nerve impulse transmission

(A) Neurotransmitter molecules released from the presynaptic neuron bind to receptors, which opens Ca^{2+} channels in the postsynaptic membrane. (B) Ca^{2+} ions enter the neuron and activate the enzyme, calpain, which attacks the cytoskeleton. (C) A reorganized cytoskeleton is reflected in a new shape of the terminus, which makes it more sensitive to subsequent nerve impulse transmission. Removal of Ca^{2+} from the terminus completes the process.

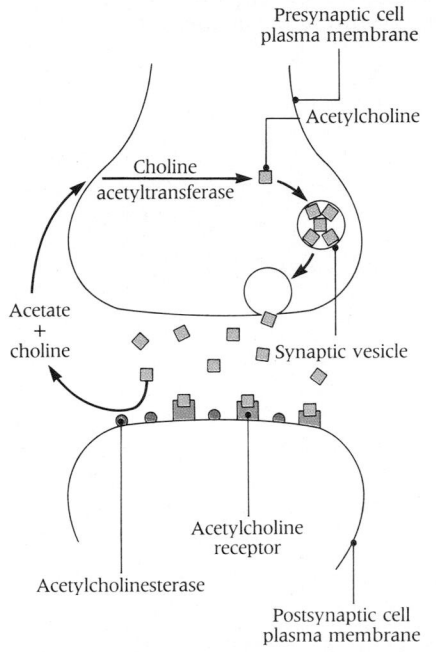

29-22 Role of acetylcholinesterase

In chemical synapses in which the neurotransmitter is acetylcholine, the postsynaptic membrane contains the enzyme acetylcholinesterase which removes acetylcholine from the synaptic cleft by hydrolyzing it to acetate and choline. These molecules can reenter the presynaptic axon ending and can be recombined by choline acetyltransferase to generate new acetylcholine molecules.

Transmitter binding to the postsynaptic cell stimulates ion channels to open or close. Recent work has revealed yet another aspect of synaptic transmission. Postsynaptic neurons contain a cytoskeleton—like that of other cells, which includes a cross-linking protein named **fodrin,** a molecule closely resembling spectrin, the major protein of the red cell cytoskeleton (see Figure 6-12). The postsynaptic knob also contains **calpain,** a proteolytic enzyme that can be activated by Ca^{2+} to attack fodrin and break up the cytoskeleton. This is postulated to occur in nerve impulse transmission as the binding of acetylcholine to its receptors opens calcium channels in the postsynaptic membrane. This permits Ca^{2+} to enter the cell and activate calpain (Figure 29-21). Ca^{2+} is then flushed out and the cytoskeleton reforms, but its shape—and that of the postsynaptic knob itself—may now be subtly different. As a result, the postsynaptic cell may be more sensitive to subsequent nerve impulses. This phenomenon may be one basis for memory and learning.

Meanwhile, the neurotransmitter must be gotten rid of. In the case of acetylcholine, this is accomplished by local hydrolytic enzyme, **acetylcholinesterase.** This clears the synaptic cleft of neurotransmitter after it has stimulated the postsynaptic neuron (Figure 29-22). If acetylcholinesterase did not perform this function, a single impulse would be enormously amplified within a synapse, and nervous system activity would be much harder to regulate. Just such nervous "storms" occur when large quantities of acetylcholine—or acetylcholinesterase-inhibiting chemicals—are injected into the body.

In summary, five biochemical steps occur in the course of chemical synaptic transmission (Figure 29-23). To be accepted as a neurotransmitter, a candidate molecule must be shown to go through the following steps:

- Synthesis in the presynaptic cell
- Release into the synaptic cleft
- Binding to a postsynaptic receptor
- Removal or destruction in the synaptic cleft
- Induce changes in membrane permeability of the postsynaptic cell, which in many cases causes it to fire

A possible sixth step, the importance of which is still being weighed, involves the inflow of Ca^{2+} through the postsynaptic membrane, activation of reversible proteolysis of the cytoskeleton, and structural reshaping of the neuron.

Modulators of Neurotransmission

Two major classes of synapses have been identified in the nervous system: excitatory and inhibitory (Figure 29-24).

- In **excitatory synapses,** which include most neuron-neuron and neuron-effector chemical synapses, the arrival of a neurotransmitter depolarizes the postsynaptic membrane and causes a local **excitatory postsynaptic potential,** or **EPSP.**
- In **inhibitory synapses,** the synapse secretes a transmitter that produces an **inhibitory postsynaptic potential,** or **IPSP.**

Both inhibitory and excitatory synapses may occur on a single neuron. Indeed, a single motor neuron cell body may receive incoming signals from thousands of these so-called **modulating neurons** (see Figure 29-6). In the end, the actions of many body mechanisms are determined by a balance between competing excitatory messages.

Other Neurotransmitters

Nerve impulses are transmitted across some synapses by a large number of neurotransmitters other than acetylcholine. The common denominator is an ability to influence ion channels in the postsynaptic neuron. One of the most important of the other transmitters is norepinephrine.

For many years, it was believed that all or most synaptic transmissions are mediated by either acetylcholine or norepinephrine. Thus, synapses were classified as **cholinergic** and **adrenergic.** More recently, neurophysiologists came to realize that a sizable number of low-molecular-weight substances serve as neurotransmitters (Table 29-1). Some of the more important are the following:

■ **Serotonin** (5-hydroxytryptamine) in the sea slug *Aplysia* also activates an adenylate cyclase-cAMP-protein kinase sequence—like those described in Chapter 28—which alters the postsynaptic membrane.

■ **γ-Aminobutyric acid,** known as **GABA,** operates exclusively in inhibitory synapses and is notably active in the brain. It affects Cl⁻ channels.

■ **Dopamine** mimics the actions of norepinephrine and, like norepinephrine, is a catecholamine.

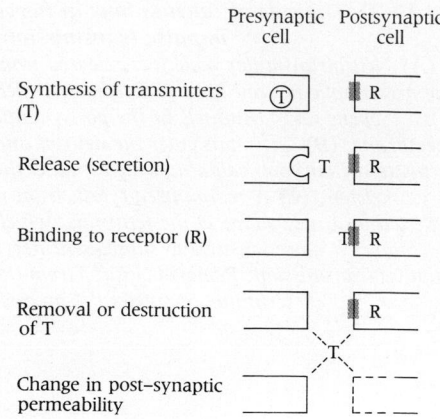

29-23 The five steps in synaptic transmission in summary

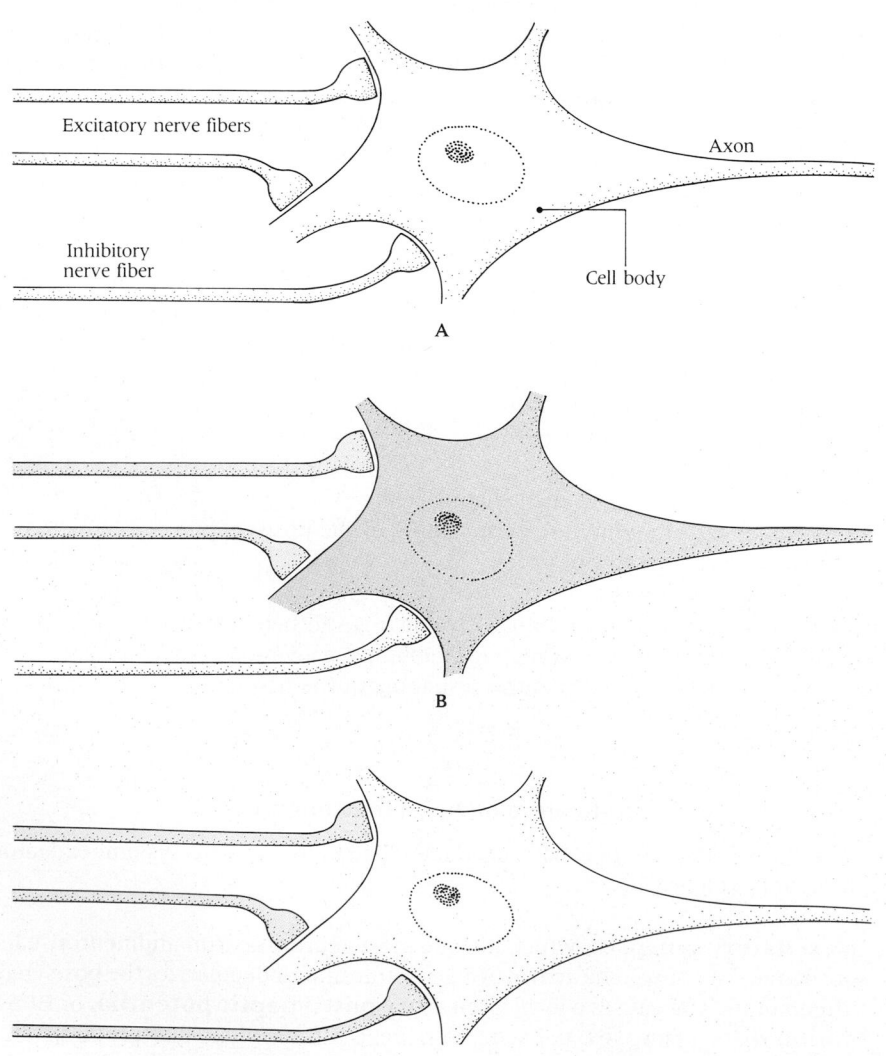

29-24 Excitatory and inhibitory synapses
A neuron cell body synapses with both excitatory and inhibitory nerve fibers. The balance between these two fiber types determines whether the neuron will fire. (A) Impulses received from only one excitatory fiber are insufficient to fire the motor neuron. (B) Impulses are received from both excitatory fibers, but not from the inhibitory fiber. The neuron fires. (C) The addition of inhibitory impulses restores the neuron to a subthreshold condition and it does not fire.

As discussed in Chapter 28, some two dozen short peptides have been discovered in recent years that (1) are localized in neurons, and (2) are very active, causing inhibition, excitation, or both, when applied directly to neurons. Some of them were already familiar as hormones. Some of those of current interest are listed in Table 29-1.

Many of these peptides probably serve as both neurotransmitters and neurohormones. Again we see a striking analogy between the mechanisms that complete communication in the endocrine system and in the nervous system. In both, chemical messengers (hormones or neurotransmitters) released by one cell traverse an extracellular gulf and bind to surface receptors of a second cell, thereby modifying its activity by means of the cAMP system.

Other Properties of Chemical Synapses

Impulses are conducted through synapses only from presynaptic terminals to postsynaptic neurons and never in the reverse direction. The synaptic cleft gives synapses an opportunity to alter impulses and even to determine which ones get through and which do not.

For example, isolated impulses arriving at a synapse may fail to activate the postsynaptic cell. As a result, the nervous system is protected from responding to random "noise" in the neurophysiological circuitry. Impulses approaching presynaptic terminals may summate, that is, come together for a single transmission. When impulses arrive simultaneously in many synapses on a neuron, transmission may be enhanced through **spatial summation,** also called multiple fiber summation. This means that many fibers are stimulated simultaneously in a nerve trunk, rather than just a single fiber. Thus, many action potentials travel together and "summate" at the end of the nerve to produce a greatly intensified effect. If impulses arrive in rapid succession in a single presynaptic terminal, they may be enhanced through **temporal summation**—that is, reactions may be intensified by the rapidity of arriving impulses. Because of summation, a volley may cross many synapses in a nerve network and the resulting effector response may be strong and widespread. More time is consumed in the transmission of an impulse across a synapse than down an equivalent length of axon. The extra time is called **synaptic delay.**

When a presynaptic terminal is continually and repetitively stimulated, the number of impulses transmitted by the postsynaptic neuron may progressively decrease owing to fatigue of synaptic transmission. The opposite situation also occurs. When a series of impulses has crossed a synapse, subsequent impulses cross more readily. This effect is known as facilitation. Continual crossing of a particular synapse at

TABLE 29-1
SUBSTANCES KNOWN OR SUSPECTED TO BE NEUROTRANSMITTERS

Nonpeptides	Neuroactive Peptides
Biogenic amines	Hypothalamic releasing factors
Epinephrine*	Thyrotropin releasing factor
Norepinephrine*	Luteinizing hormone releasing factor
Serotonin*	Growth hormone releasing factor
Dopamine*	
Histamine*	Hypophyseal peptides
Tyramine	ACTH
Octopamine	β-Endorphin
	α-MSH
Acetylcholine*	
Amino acids	Gut-brain peptides
γ-aminoisobutyric acid (GABA)	Methionine-enkephalin
Glycine	Leucine-enkephalin
Aspartate	Insulin
Glutamate	Glucagon
	Substance P
	Others
	Vasopressin
	Oxytocin
	Carnosine
	Bradykinin

* One of the major known neurotransmitters.

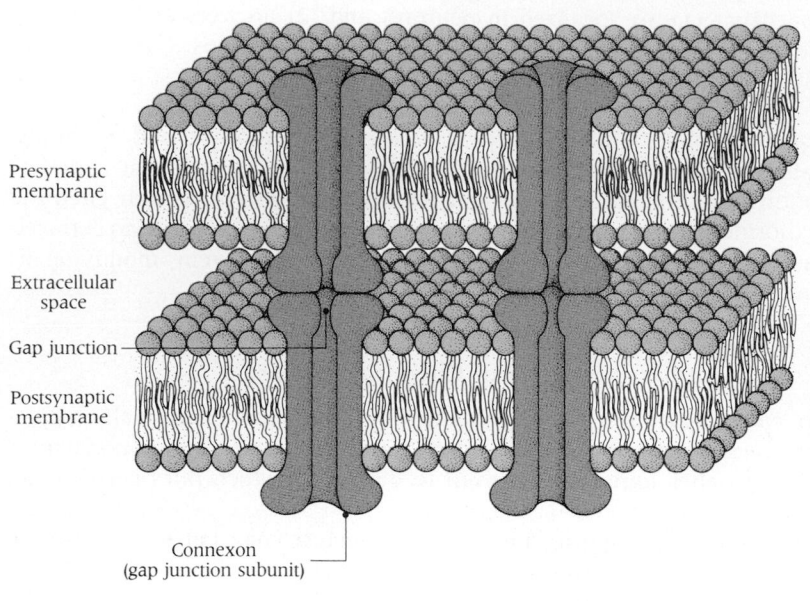

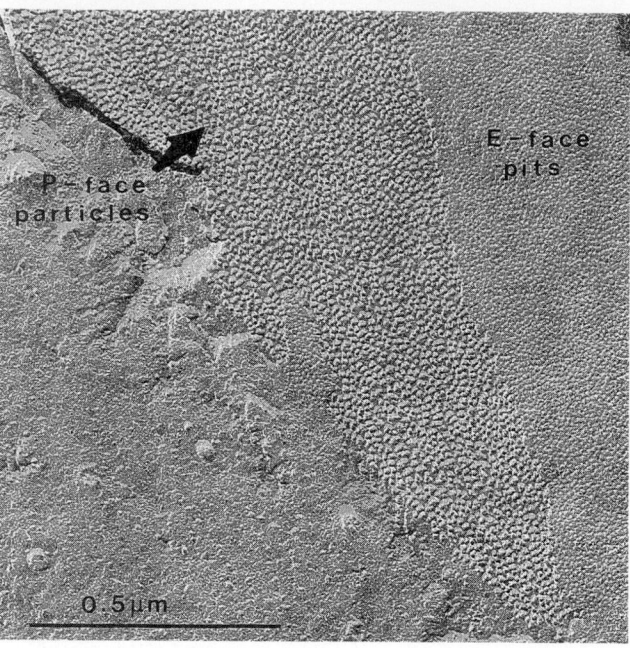

A

B

appropriate intervals can maintain its facilitated condition for a lifetime. The mechanism is not yet clear.

Electrical Synapses

Chemical neurotransmission is not the only method of conveying impulses across synapses (see Figure 29-16). A number of synapses transmit nerve impulses directly by means of low-resistance gap junctions (Figure 29-25), which we encountered in Chapter 6. We call these **electrical synapses** because positively charged ions move through the gap junctions and directly depolarize the postsynaptic membrane, as though the two neurons were electrically coupled.

These synapses, unlike chemical synapses, can transmit impulses in both directions. They also transmit impulses very rapidly. They are prominent in fishes and doubtless account for their uncanny ability to dart swiftly away from a threatening predator. Electrical synapses also operate efficiently at low temperatures that in cold-blooded animals would slow the metabolic synthesis of a chemical neurotransmitter. Although warm-blooded animals also have electrical synapses, it appears

29-25 Structure of an electrical synapse
(A) *Model of an electrical synapse, showing how gap junctions (subunits of which are called connexons) permit direct transfer of ions between the presynaptic cell and postsynaptic cell.* **(B)** *A large synapse from the brain of a hatchetfish. In the electron micrograph, the fracture plane is between the layers of the plasma membrane. The gap junction channels (connexons) are seen as large particles in the P-face membrane (outwardly facing inner layer) of one neuron, or as pits in the E-face membrane (inwardly facing outer layer) of a second neuron. (×90,000)*

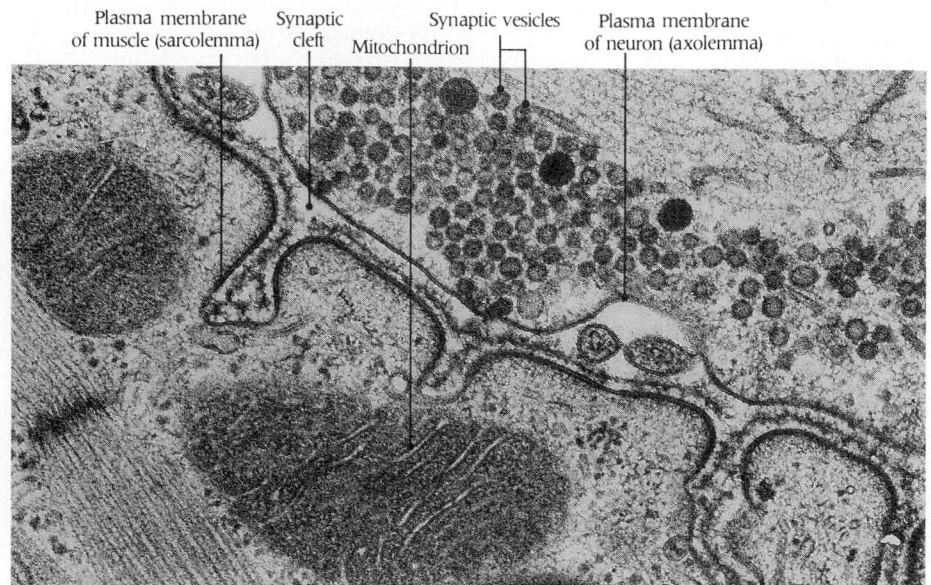

29-26 Electron micrograph of the motor endplate of a frog neuromuscular junction
Many synaptic vesicles can be seen in the terminus of the innervating neuron. A small synaptic cleft separates the neuron from the muscle cell plasma membrane. (× 40,000)

that evolution has favored chemical synapses because they are more susceptible to sensitive control.

Junctions Between Conductors and Effectors

The **conductor-effector junction** is a synapse of a particular kind. Much of what we know about synapses was first learned in studies of this neuron-effector junction, especially the junction between a motor nerve and striated muscle in the frog (Figure 29-26). This neuromuscular junction, called a **motor endplate,** is one of the best studied of all synapses. We regard it as a model of all such structures.

Unlike neuron-effector junctions associated with glands and smooth muscle, the junction with striated muscle is notable for the rapidity with which nerve impulses crossing it can stimulate muscular contraction. Both structurally and functionally, this junction has much in common with a typical synapse between two neurons.

The axon of the motor neuron, which has no myelin sheath near its end, enters the muscle fiber, expanding and ramifying over a disklike endplate of muscle substance containing many muscle cell nuclei. The narrow space of the synaptic cleft lies between the axon ending and the muscle. As in other synapses, the end of the axon is packed with small vesicles and mitochondria. The postsynaptic membrane of the muscle cell exhibits a feature not seen in other synapses: the membrane forms many **junctional folds,** or **palisades.**

Neuromuscular junctions are cholinergic. Hence, their chemical transmitter is acetylcholine. Impulses arrive at the axon ending. Acetylcholine is discharged into the space between it and the endplate from the vesicles in the axon ending. In less than 100 microseconds, released acetylcholine diffuses across the cleft, binds to specific receptor molecules in the endplate membrane and initiates depolarization by altering the membrane's permeability. The action potential thus produced induces the muscle fiber to contract. Remarkably, all of these things happen whenever we so much as wiggle a finger.

SUMMARY

In multicellular animals, a nervous system coordinates the various responses to changes in external and internal environments. The highly specialized cells that comprise the nervous system consist of receptors (sense organs), conductors (nerves), and effectors (muscles, glands, and so on). Although some stimuli can be detected by cells without specialized receptors, the detection of most stimuli requires their presence.

For each of the major types of stimuli, receptors for that stimulus will not respond until the intensity of the stimulus exceeds a certain threshold value. Above that threshold, the extent of a response is influenced by the duration of the stimulus and its rate of change, as well as by its intensity. The period between a stimulus and a response is called the response latency. After the response, a tissue is usually unable to respond again for a brief span of time called the refractory period.

The nervous system is made up of neurons and the glial cells that support them. The long thin processes that branch off along the cell body of each neuron include many dendrites and a single axon. Dendrites transmit impulses arriving from other neurons towards the cell body, which contains most of the metabolic machinery of the nerve cell. Axons transmit these impulses away from the cell body towards other neurons. Many axons are myelinated—through the activity of Schwann cells. This condition speeds nerve impulse transmission.

Each neuron maintains an electrochemical gradient across its plasma membrane of Na^+, K^+, and other ions. Nerve fibers at rest are negatively charged inside with respect to the surrounding extracellular fluid. This charge, called the resting potential, is maintained by Na-K pumps in the neuronal membrane. A neural impulse is not a flux of electrons like an impulse in an electric wire. Rather, it is a small region of membrane depolarization—a zone in which the neuron is positively charged—that travels down the cell from synapses on dendrites to those on the axon. Membrane depolarization is initiated by the brief opening of gated ion channels in the neuronal membrane. Ions then flow through these channels and temporarily reverse the membrane polarity. Firing in most neurons is associated with the all-or-none response known as an action potential. A neuron, thus, cannot encode stimulus strength by varying the size of the impulse generated. It can do so only by varying the frequency rate of impulse generation.

Nerve impulses are essentially similar from neuron to neuron. Therefore, the anatomical pattern of neural connections—the "circuit diagram"—is as important as firing patterns in encoding the information stored in the nervous system. Impulses are transferred from one neuron to another at contact points called synapses. In the simplest neural circuit, a receptor connects directly to an effector. In more complicated nervous systems, varying numbers of interneurons interposed in the circuits allow greater flexibility of integration and response. Some synapses are electrical—that is, ions flow directly from the presynaptic cell to the postsynaptic cell through gap junctions. The majority are chemical synapses, in which a neural impulse arriving at an axon terminal of the presynaptic neuron causes the release of a chemical neurotransmitter such as acetylcholine, norepinephrine, dopamine, or GABA. The transmitter diffuses across the synapse, binds to a receptor on the postsynaptic neuron, and opens ion channels. Depending on the synapse, the ion channels opened may have either an excitatory or an inhibitory effect on the postsynaptic neuron.

A single neuron may receive incoming excitatory and inhibitory impulses from thousands of synapses. Their effects are summated in the cell body; and, if net depolarization is large enough, it generates an action potential that travels down the axon to synapses with other neurons. Synapses are often sensitive to the pattern of impulses arriving on the presynaptic side. At some synapses, repetitive stimulation causes the number of impulses transmitted by the postsynaptic neuron to decrease progressively, owing to fatigue of synaptic transmission. Other synapses are facilitated by repetitive stimulation, so that the percentage of impulses transmitted increases. Neuron-effector junctions are specialized synapses. The best-studied of them, the neuromuscular junction, called the motor endplate, uses acetylcholine as a neurotransmitter. This junction is notable for the rapidity with which it transmits nerve impulses.

KEY TERMS

acetylcholine
acetylcholinesterase
action potential
all-or-none reaction
axon
axonal transport
chemical synapse
conductor
conductor-effector junction
dendrite
depolarization
dopamine
effector
electrical synapse
excitatory synapse
excitatory postsynaptic potential (EPSP)
facilitation
γ-aminobutyric acid (GABA)

gated channel
glia
gray matter
inhibitory synapse
inhibitory postsynaptic potential (IPSP)
interneuron
irritability
latent period
membrane channel
motor neuron
motor endplate
myelin sheath
nerve
nerve impulse
neuron
neurotoxin
neurotransmitter
node of Ranvier

presynaptic knob
presynaptic membrane
reflex arc
refractory period
repolarization
resting potential
saltatory conduction
Schwann cell
sensory neuron
sodium-potassium pump
spatial summation
stimulus
synapse
synaptic cleft
synaptic vesicle
temporal summation
threshold
white matter

QUIZ QUESTIONS

1. Which of the following is *not* involved in transmission of a nerve impulse across a chemical synapse?
   A. calmodulin
   B. acetylcholinesterase
   C. synapsin
   D. calcium ions
   E. All of the above are involved in the process.

2. Which of the following is *not* true of a chemical synapse?
   A. Signals can move in either direction across the synapse.
   B. The neurotransmitter substance secreted is not the same for all synapses.
   C. Neurotransmitter molecules are removed from the synaptic cleft after

they bind with receptor molecules.
   D. A minimum number of neurotransmitter-receptor complexes must form before the impulse is propagated in the postsynaptic neuron.
   E. All of the above are true statements.

3. Electrical synapses
   A. rapidly transmit impulses.
   B. occur in some fishes.
   C. are not as sensitive to temperature as are chemical synapses.
   D. allow impulses to pass in either direction.
   E. All of the above.

4. When a neuron is at rest,
   A. there are more sodium ions outside

of the cell than within.
   B. the inner surface of the cell membrane is negatively charged.
   C. the cell membrane is permeable to movement of all cations.
   D. A and B
   E. A, B, and C

5. For which of the following stimuli are there *no* known sensory structures in living organisms?
   A. magnetic fields
   B. radio waves
   C. vibrations
   D. pressure
   E. heat

ESSAY QUESTIONS

1. What factors affect the perception of a stimulus? What is meant by the "threshold" of stimulation? What factors affect the response to a stimulus? What is a latent period? A refractory period?

2. Why do we regard nerves as systems that

are regulated by frequency modulation (FM) rather than amplitude modulation (AM)?

3. What molecules function as neurotransmitters in the nervous system? What criteria must be satisfied for a

molecule to be considered a neurotransmitter?

4. Describe the fundamental cellular unit of a nervous system. How is it similar to other cells? How does it differ?

REFERENCES AND SUGGESTIONS FOR FURTHER READING

Frankel, R. B., Blakemore, R. P., and Wolfe, R. S. (1979). Magnetite in freshwater magnetotactic bacteria. *Science 203*, 1355–1356.

A fascinating account of a remarkable phenomenon.

Kandel, E. R., and Schwartz, J. H. (Eds.) (1985). *Principles of Neural Sciences*, 2nd ed. Elsevier, New York.

A comprehensive text and reference book on everything from molecular neurobiology to the study of intact nervous systems. Though its 62 chapters are written by a number of experts, the book is remarkably readable.

Keynes, R. D. (1979). Ion channels in the nerve-cell membrane. *Scientific American 240*, March, 126–135.

A clear exposition of the basic importance of ion channels.

Kimelberg, H. K., and Norenberg, M. D. (1989). Astrocytes. *Scientific American 260*, February, 66–76.

Stimulating discussion of one kind of glial cell.

Krueger, B. K. (1989). Toward an understanding of structure and function of ion channels. *FASEB Journal 3*, 1906–1914.

The latest word on a topic of growing importance.

Kuffler, S. W., Martin, A. R., and Nicholls,

J. G. (1984). *From Neuron to Brain: A Cellular Approach to the Function of the Nervous System*, 2nd ed. Sinauer, Sunderland, Mass.

This masterful summary of the mechanisms of nerve excitation and transmission is a fine bequest from the late senior author.

Lefkowitz, R. J., and Caron, M. G. (1988). Adrenergic receptors: models for the study of receptors coupled to guanine nucleotide regulatory proteins. *The Journal of Biological Chemistry 263*, 4993–4996.

This "minireview" elegantly demonstrates the similarity of adrenergic receptors in neural synapses and cell surface receptors for the hormones, epinephrine and norepinephrine. Both depend on G proteins.

Llinás, R. R. (1982). Calcium in synaptic transmission, *Scientific American 247*, October, 56–65.

In fascinating studies of a synapse of the squid's giant nerve cell, the author describes how a current of calcium ions triggers the passage of signals from one nerve cell to another.

Multiple authors. (1983). Molecular Neurobiology. *Cold Spring Harbor Symposium on Quantitative Biology*, Vol. 48.

A wealth of fine articles on membrane channels and sensory systems.

Multiple authors. (1979). Special issue on

the brain. *Scientific American 241*, September, 44–232

A cluster of 11 essays on different aspects of modern neurobiology by the leading investigators in the field. A useful introduction to the state of current research.

Snyder, S. H. (1985). The molecular basis of communication between cells. *Scientific American 253*, October, 132–140.

The author discusses chemical messengers that mediate long-range hormonal communication and short-range communication between nerve cells. The two systems differ in many ways, but some chemical messenger molecules are common to both.

Stevens, C. F. (1979). The neuron. *Scientific American 241*, September, 54–65.

A readable discussion of the structure and function of nerve cells.

Stroud, R. M., and Finer-Moore, J. (1985). Acetylcholine receptor structure, function, and evolution. *Annual Review of Cell Biology 1*, 317–351.

A thorough discussion of current knowledge.

Thompson, R. F. (1985). *The Brain: An Introduction to Neuroscience*. W. H. Freeman, New York.

A useful introductory text.

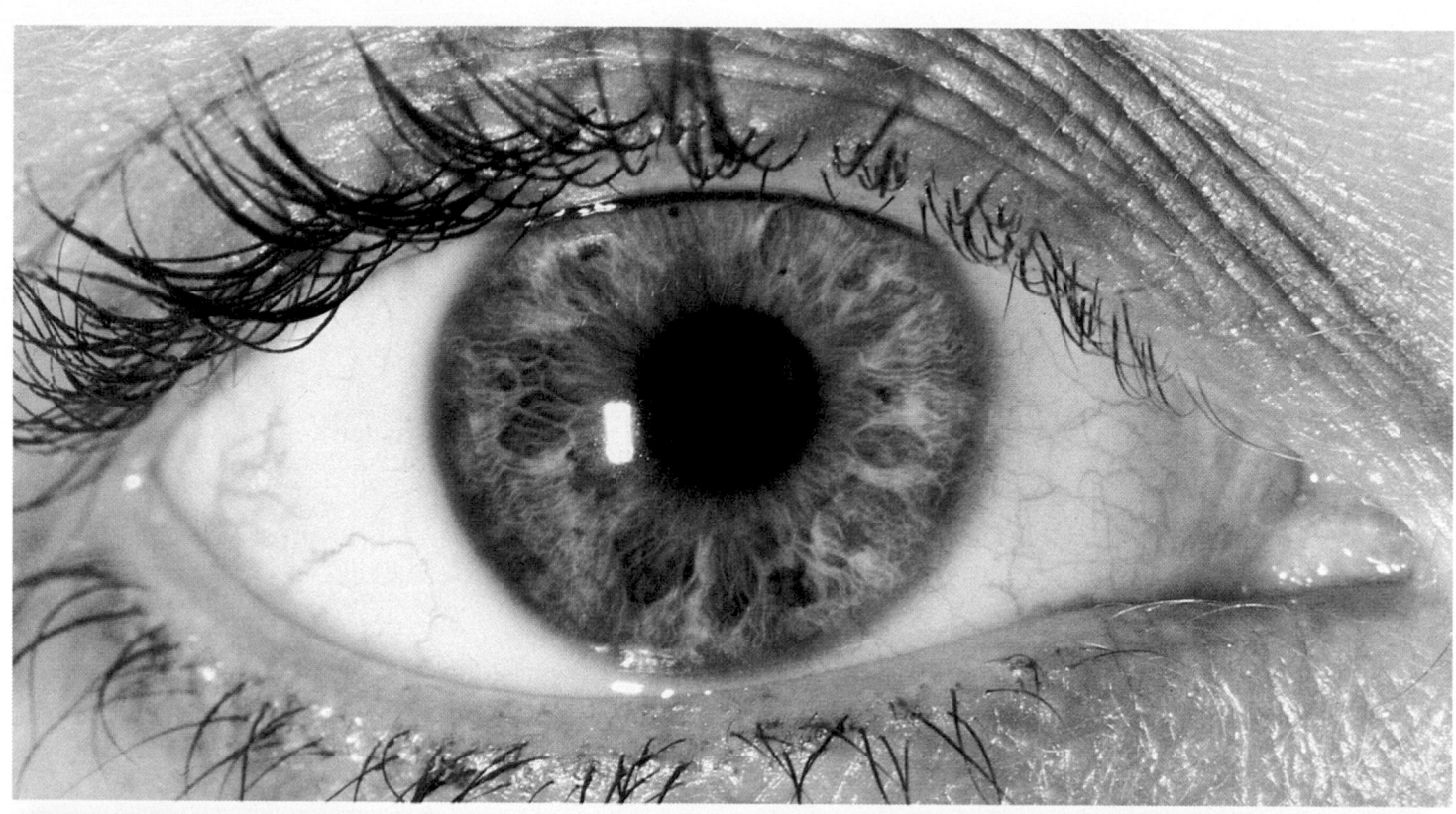

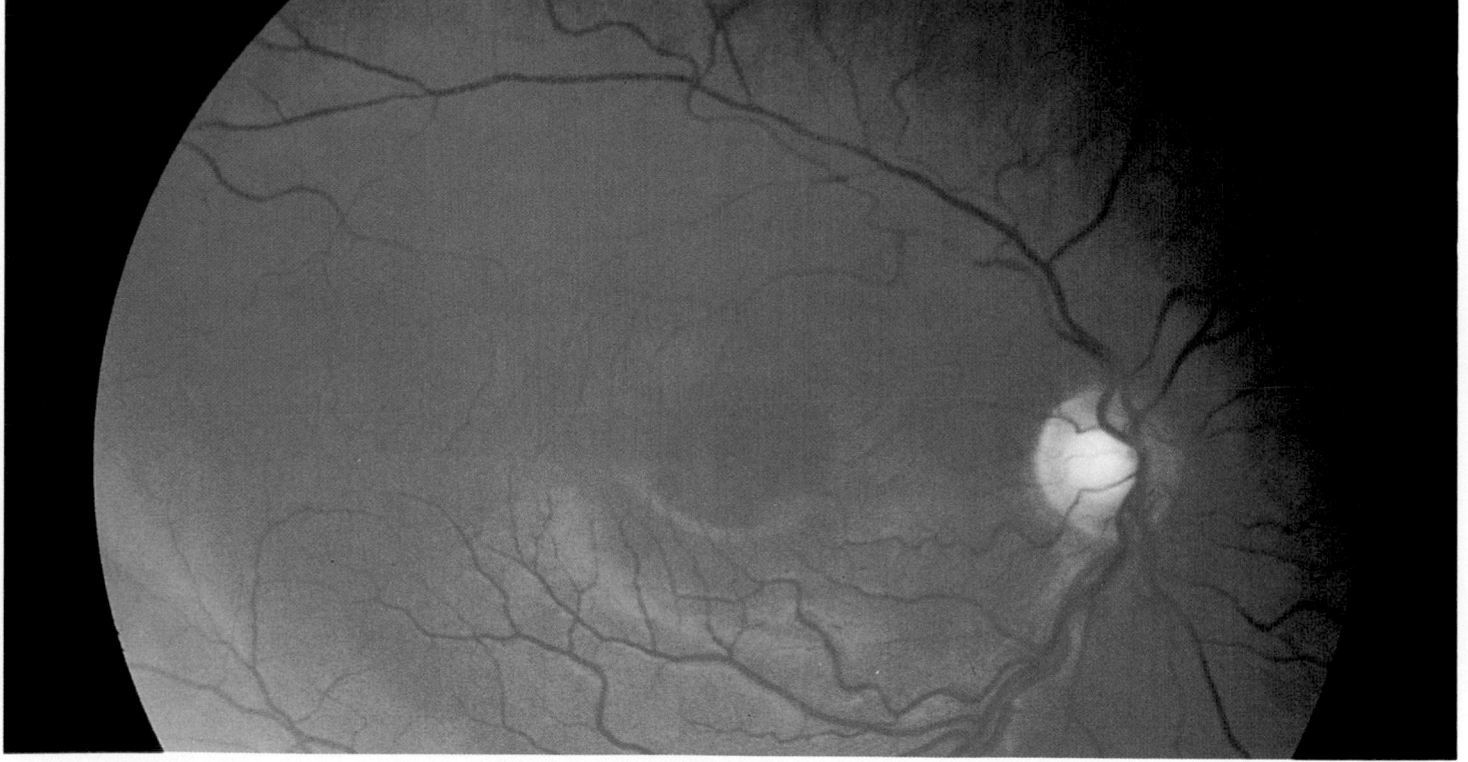

SENSORY RECEPTORS AND THE BRAIN

Sensory receptors and the brain
Many of the properties of sensory receptors—and of the brain as well—are displayed by the human eye, one of evolution's most remarkable constructions. Note the delicate structure of the iris muscles and the patches of brown pigment against a blue background. The retina, photographed in a normal eye, shows the red color of the underlying choroid. Note the disk, where the optic nerve exits, and the arteries and veins. These are the only blood vessels in the body that can be viewed directly.

The fundamental feature of all nervous systems was outlined in the last chapter. It is the receptor-conductor-effector network. In this chapter, we examine receptors and the primary organ of association, the brain.

SENSORY RECEPTORS

Nervous systems respond to changes in the environment. They sort out incoming stimuli, store information for future responses, and determine responses in accordance with existing conditions. We have seen that the specialized neurons (or neuronal derivatives) that are stimulated by external stimuli are the receptors. A **sensory receptor** is the part of the receptor-conductor-effector pathway that faces the outside world. It acts by transforming an external stimulus into a nerve impulse. Thus it is a **transducer.** That term, which derives from physics, denotes a device that transforms one form of energy into another. In linear physical transducers, output energy is proportional to input energy. Such devices are essential components in communication and signalling systems. A familiar example is the transducer in a telephone, which converts sound waves to electrical energy and vice versa. In biological transducers (sensory receptors), impulses initiated in a receptor are conducted away—in proportion to the intensity of the stimulus—through attached nerve fibers to synapses where they make contact with a succeeding neuron (see Chapter 29).

From antiquity, a popular misconception has held that human beings have but five senses: vision, hearing, smell, taste, and touch. It is true that the receptors of four of these are localized in the head and that we usually consider them separately as "special senses." But curiously, we have rarely included in this exclusive category another sense organ in the head, the organ of equilibrium in the inner ear. In addition, countless small, anatomically simple receptors are scattered throughout the body. They vary so much in structure, in stimulus, and in sensation it is hard to say how many senses they represent. To be sure, many of these receptors are adapted to receive specific stimuli. This is easily demonstrated by the prick of a pin or the touch of small hot and cold rods. Yet certain more complex sensations seem to ensue because more than one type of specific receptor has been activated. For example, the sensation of "wetness" may result from simultaneous stimulation of a combination of receptors, i.e., cold and pressure receptors.

The skin is especially rich in these scattered receptors. It has at least four kinds, on stimulation producing sensations of warmth, cold, pain, and touch-pressure. Other receptors in the body, many in number, doubtless account for feelings of muscular tenseness, joint motion, hunger, thirst, internal pain, nausea, sexual gratification, and other distinct sensations that arise from within.

Biologists have long wondered how an external physical or chemical event is converted into an internal physiological event of proportional intensity. The answer lies in the proportionality between stimulus intensity and the *frequency* of nerve impulses that are generated (Figure 30-1). Biological transducers have not yet been discovered for many stimuli to which physical transducers respond—for example, radio waves. As mentioned earlier, the list of stimuli provoking responses in organisms grows longer. Recent additions to it include magnetism, polarized light, wind, gyroscopic deflections, electric fields, infrared radiation, and ultrasound. The human nervous system is entirely insensitive, or relatively so, to many of these stimuli (Table 30-1).

PHOTORECEPTORS

All protoplasm is sensitive to some forms of radiation including the visible radiation we call light. Comparisons of simple and more complex photoreceptors (light receptors) in different organisms show that the more elaborate the receptor the more information it can extract from the same stimulus.

Evolution of the Eye

The simplest photoreceptors—one would hardly call them eyes at this stage—occur in some unicellular organisms as well as in simple multicellular animals (Figure 30-2). These are light-sensitive spots of pigment that give an organism no information from a distance other than whether or not there is light in the vicinity and whether it is increasing or decreasing in intensity. They do not perceive images. Such eyespots are found in *Euglena*, a unicellular protist described in Chapter 39 (Figure 30-2A).

In time, more complex eyes evolved among the invertebrates—for example, in starfishes, flatworms, and spiders (Figure 30-2B, C). These structures, termed **compound eyes,** display great diversity. One or both of two features are usually present: (1) a **lens,** and (2) few or many **photoreceptors,** each a light-sensitive cell that responds to incoming light by initiating nerve impulses. The lens concentrates light onto these sensitive cells. Hence, compound eyes can react to weak light intensities and can discriminate between different intensities. Such eyes also inform organisms about the direction from which light is coming—an evolutionary first. Given this

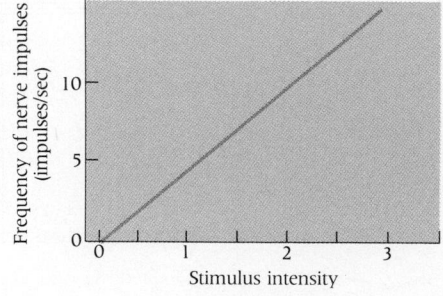

30-1 Relation between stimulus intensity and discharge frequency
As stimulus intensity increases, the frequency of nerve impulse propagation increases proportionally.

TABLE 30-1
SENSORY RECEPTORS

Stimulus	Type of Receptor (and subtype)	Location of Typical Receptor
Mechanical energy		
Touch	Mechanoreceptor	Skin (Meissner's corpuscles)
Pressure	Mechanoreceptor	Skin (Pacinian corpuscles)
Gravity, motion	Mechanoreceptor (statoreceptor)	Vestibular organ
Gravity, motion	Mechanoreceptor (proprioceptor)	Joints, muscle, spindles
Sound	Mechanoreceptor (phonoreceptor)	Organ of Corti of ear (hair cells)
Blood pressure	Mechanoreceptor (pressoreceptor)	Carotid sinus
Heat		
Environmental	Thermoreceptor	Skin (Krause's end bulbs); pit organs of pit vipers
Internal	Thermoreceptor	Hypothalamus
Light		
Specific wavelengths	Photoreceptor	Retina of vertebrate eye; ommatidia of arthropods
Chemical substances		
Volatile	Chemoreceptor (smell)	Olfactory epithelium
In solution (except O_2)	Chemoreceptor (taste)	Taste bud
O_2 in solution	Chemoreceptor	Carotid body
Electrical energy		
Specific voltages	Electroreceptors	Skin of some sharks, skates, and rays (elasmobranchs); bony fishes

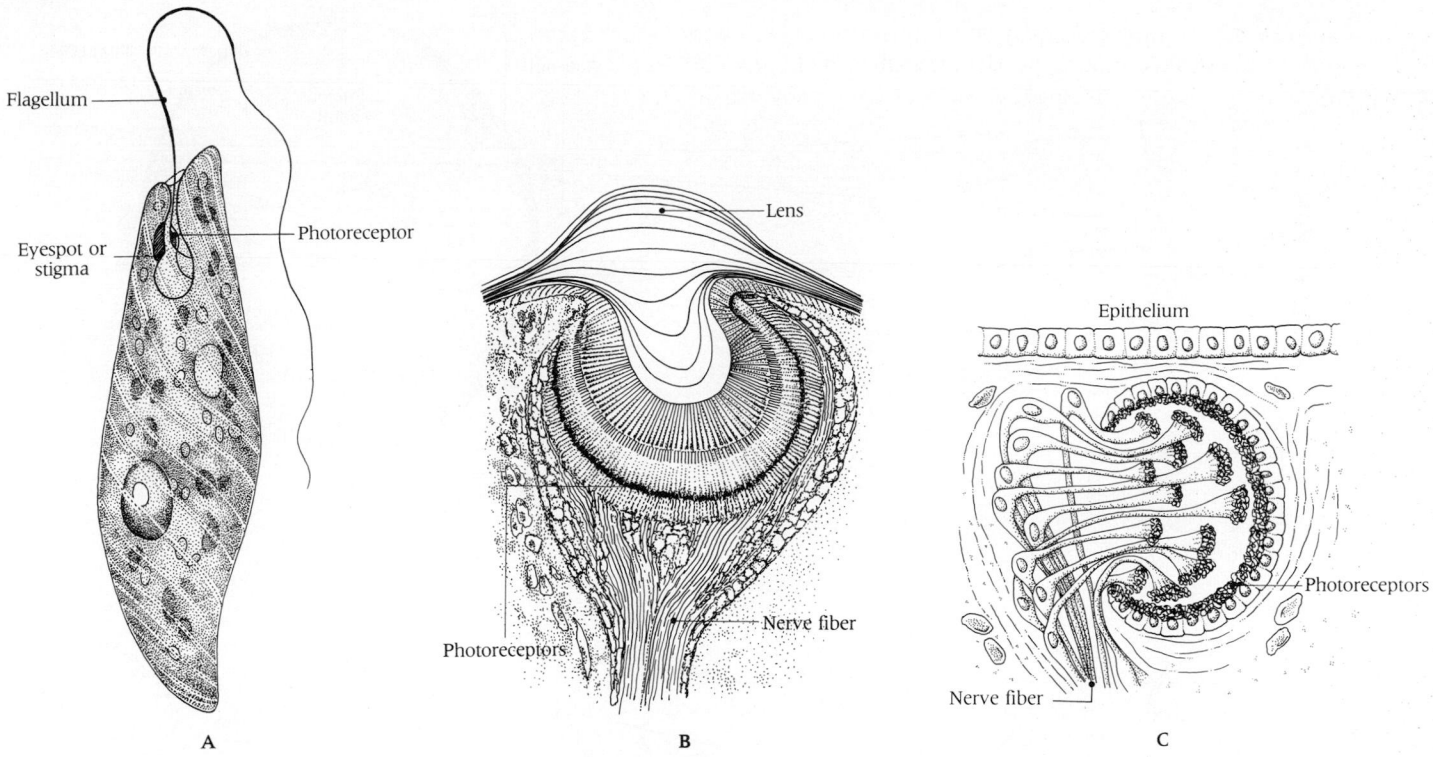

A   B   C

30-2 A variety of photoreceptive organs
(A) The protist Euglena *has a simple eyespot (stigma) containing pigments that absorb light energy. (B) The spider has a compound eye containing a lens that focuses light and a retina with photoreceptors connected to nerve fibers leaving the eye. (C) In the flatworm's eye, light-absorbing pigments are concentrated in a layer of cells in front of the nerve cells they stimulate. There is no lens.*

valuable information, animals can move more discriminatingly toward or away from perceived light.

As sensory cells behind the lens increased in number, even finer responses to light and its changing intensities became possible. Eventually, several different cells were stimulated in succession when light or dark objects moved before the lens. Animals with simple eyes cannot tell what is moving in front of them, since no true image is formed, but they can tell that motion has occurred and act accordingly. Compound eyes, by revealing happenings at some distance and in some cases by forming true images, greatly increased the amount of information available to an organism.

Lenses first served not as image-formers but as mechanisms for concentrating light coming in from different directions. However, lenses can also form images by focusing light on a surface. Indeed, the well-developed image-forming eye is exactly like a camera (Figure 30-3). Someone has remarked that if the eye were marketed as a camera, it could boast the most sophisticated technology available: it is fully automatic, it can adjust its shutter to changing light intensity, its ''film'' can replenish itself again and again as it records millions of images in the course of a lifetime!

The eye-as-camera analogy is enlightening. In both, a convex lens brings an inverted image to focus upon a light-sensitive surface. In both, an iris diaphragm controls the amount of admitted light,[1] a dark interior absorbing stray light that would otherwise be reflected back and forth, obscuring the image. The lens assemblies

---

[1] However, not all eyes have irises. The eyes of all vertebrates have irises, but some invertebrate eyes lack them.

30-3 Comparison of eye and camera
The eye has an iris that regulates the amount of light entering it, a lens and cornea that focus light, and retinal photoreceptors that form an image. Stray light is absorbed by the dark choroid layer. The camera is similarly designed, though its efficiency in recording images is inferior to that of the eye.

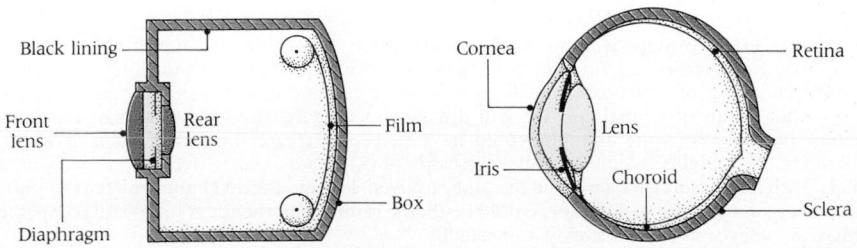

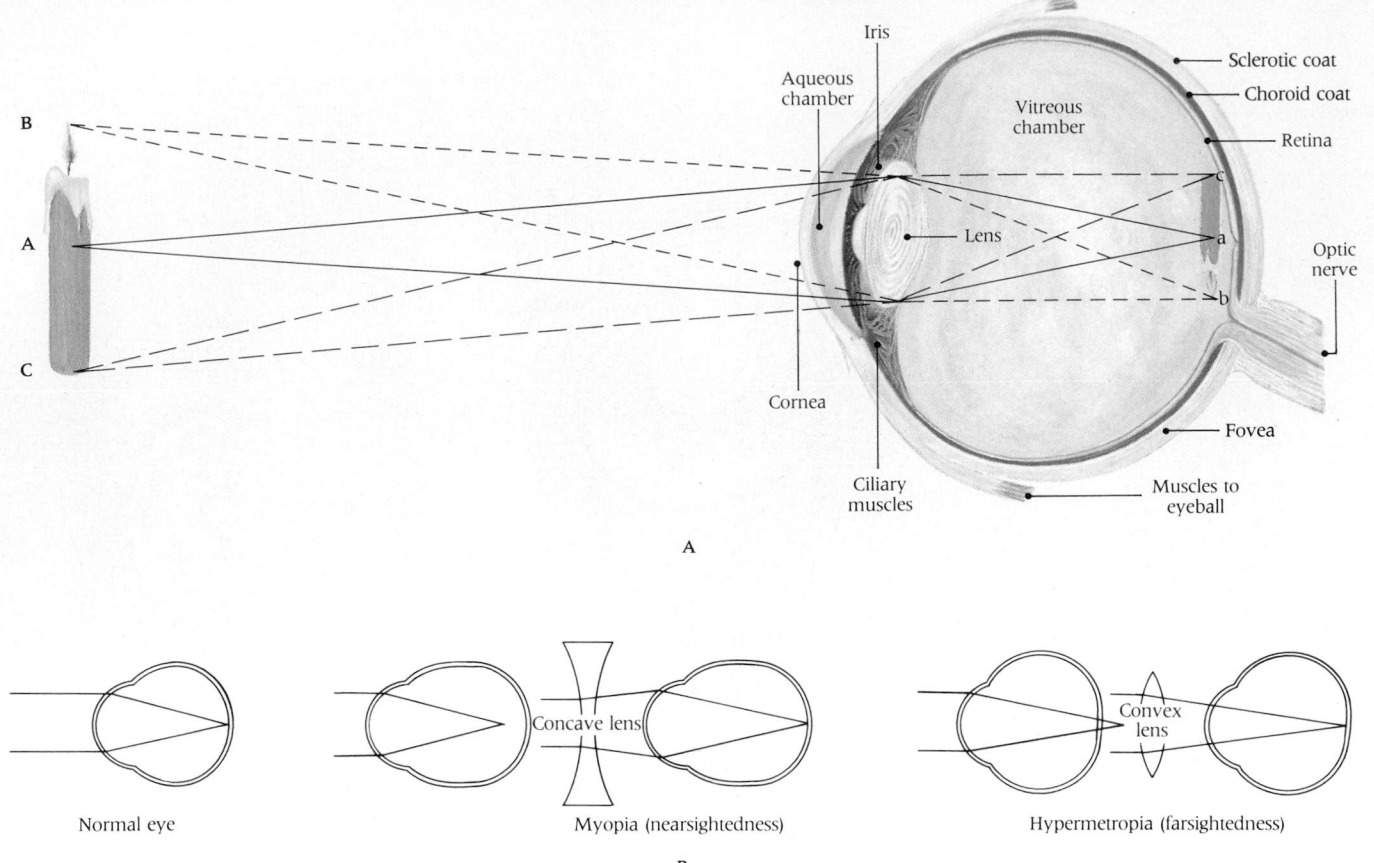

A

Normal eye   Myopia (nearsightedness)   Hypermetropia (farsightedness)

B

of many cameras include a front lens and a rear lens. In the eye, other structures (cornea, fluids, etc.) play the role of second lens.

The eye functions best when (1) the lens is of appropriate shape, (2) the number of light-sensitive cells is great enough, and (3) these cells are arrayed on an optimally curved surface upon which the lens focuses an image. Under these conditions, the eye can receive a sharp and detailed image (Figure 30-4A) and then translate it—or transduce it—into a pattern of nerve impulses. As shown in Figure 30-4B, nearsightedness (myopia) occurs if the eyeball is too long. If it is too short, the eye is farsighted (hyperopic). These conditions are corrected, respectively, by the concave and convex lenses of our eyeglasses.

Structures that focus light, discriminate it from dark, and form images developed separately and independently at least three times in the course of evolution: in some of the more active and complex mollusks (such as squids), in some spiders, and in the early vertebrates.[2]

Evolution has devised image-producing mechanisms other than the single-lens eye. Insects and some of their relatives evolved remarkable image-forming eyes that do not operate on the single-lens principle. They have compound eyes that consist of a large number—as many as 20,000—of densely packed tubes or **ommatidia,** each a partially independent light-sensitive unit with a lens of its own (Figure 30-5A, B). Fruit flies have compound eyes with about 800 ommatidial units per eye (Figure 30-5C). In the insect eye, each unit aims in a slightly different direction and receives a narrow beam of light from a different part of the insect's surroundings. The sum of all ommatidial reactions is a crude but recognizable mosaiclike image

30-4 The path of light in a human eye
(A) *The image of a candle is sharply focused on the retina. Focus is maintained by the curved corneal surface and by a lens that can adjust its own curvature by contracting or relaxing its attached ciliary muscles.* **(B)** *In myopia (nearsightedness) the eyeball is too long for its lens and cannot focus on distant objects. Concave lenses are corrective. In hyperopia, the eyeball is too short. Convex lenses are corrective.*

---

[2] Almost certainly there was no brief episode when eyes suddenly appeared and began to receive images. Increased discrimination of light intensity, direction, and movement probably led to the development of structures capable of producing a vague image. Since variation constantly occurs in all groups of animals (as we will discuss in surveying the mechanisms of evolution in Chapters 34 and 35), some animals would have perceived clearer images of their environment than others. This ability would give its possessor obvious advantages over those without it. If animals with clearer images fared better and survived longer, the likelihood increased that the genes passed on to future generations would be theirs, rather than their less successful competitors'. In this way, images improved slowly and steadily.

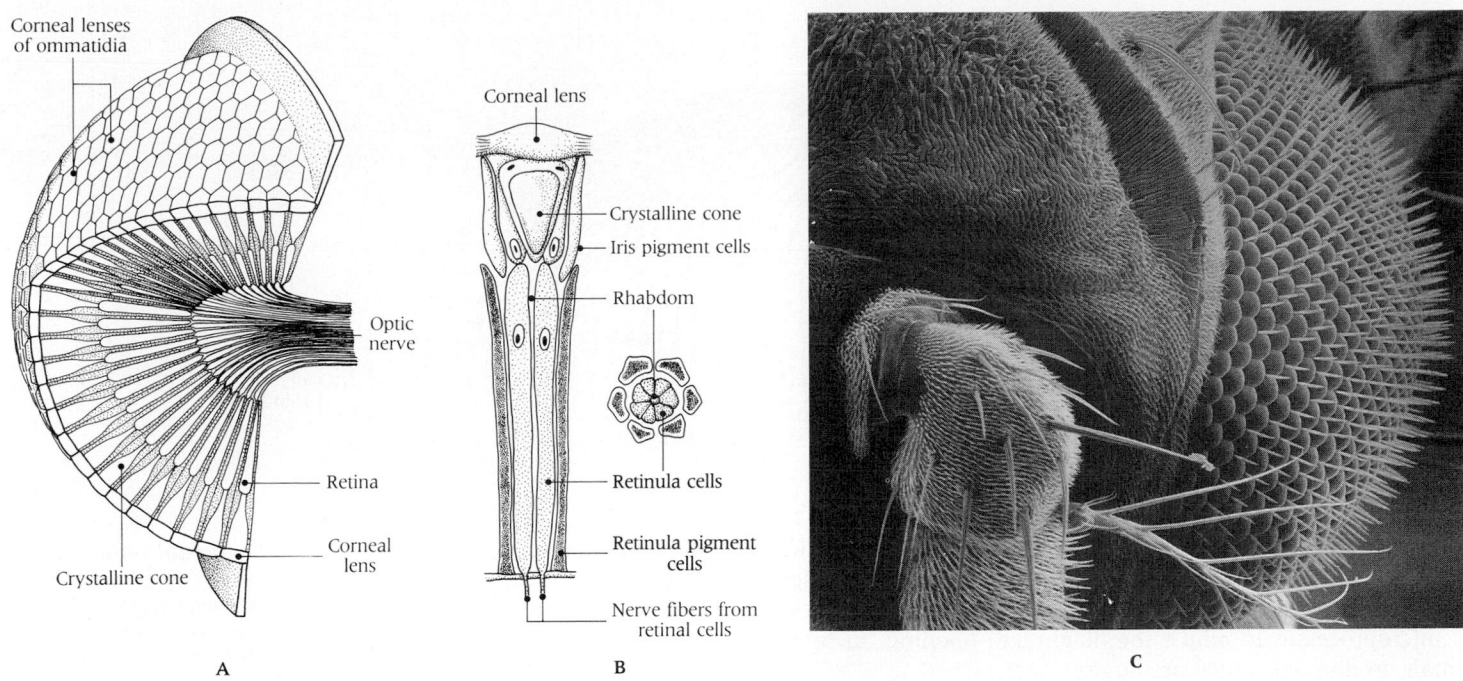

Corneal lenses
of ommatidia

Optic
nerve

Retina

Corneal
lens

Crystalline cone

A

Corneal lens

Crystalline cone

Iris pigment cells

Rhabdom

Retinula cells

Retinula pigment
cells

Nerve fibers from
retinal cells

B

C

30-5 Structure of an insect's compound eye
*(A) This cut-away view of an insect's eye reveals
many individual ommatidia. (B) A single
ommatidium. Each contains a corneal lens and
light-sensitive retinula cells that connect to the
optic nerve via nerve fibers. (C) Scanning electron
micrograph of the eye of a fruit fly showing
hundreds of ommatidia.*

of the outside world. Nonetheless it permits detection of rapid movement with five times the sensitivity of the human eye.

The more complex the eye's photoreceptors and the finer their discriminating power, the more elaborate is the range of possible responses by an organism's effectors. A simple eye might receive information sufficient to signal, "Something moved." But a compound eye would permit the more discerning observation, "A yellow cat a foot long is slowly coming toward me from 10 feet away a little to my left." Such detail is of no use to an organism that cannot then act on it in a variety of ways. An amoeba's list of possible responses, for example, does not go beyond rolling into a ball, moving one way or another, or projecting a pseudopod to engulf a particle. In sum, then, discrimination in receptors did not arise in evolution unless accompanied by a correspondingly complex association system and appropriately varied effector responses.

Color, which depends on the wavelengths of light energy, is another kind of information that may be extremely useful in identifying and discriminating among objects. Many animals see very well without discriminating differences between the wavelengths. In a real sense, they are color blind. Color vision, too, has evolved several times. Some crustaceans and many insects have it. Some insects, including honeybees, can see far into the ultraviolet spectrum. Indeed, they see these wavelengths as a distinctive color (see Chapter 29). Until recently, we had no more idea of how ultraviolet light looks to a bee than a bee has of how red looks to us. New work suggests that red appears as black to a bee—an absence of light—as does ultraviolet light to us (see Figure 29-3).

Curiously, color vision may have been lost and then regained in our own ancestry. Many fishes and reptiles and most birds have color vision, but most mammals do not. Primitive mammals probably lacked color vision and it is slight or absent among most present-day mammals, except humans and their close relatives, the apes and some monkeys. Dogs, cats, horses, and most other animals in some cases may have a feeble capacity to discriminate colors. More probably they have no color vision at all.

Structure of the Eye

The whole visual apparatus is enclosed in a tough ball of connective tissue called the **sclera.** The globe is turned by its own muscles. Its exposed portion is protected by eyelids and constantly washed by tears, which contain antibacterial agents. A camera lens is of fixed focal length and it is focused by moving it toward or away from a film. In contrast, the eye **lens** is elastic. It focuses by changing shape and

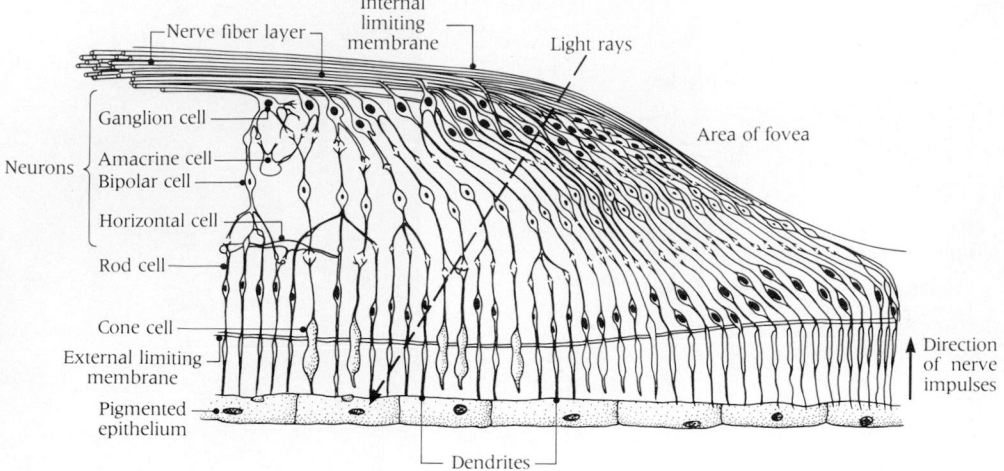

Internal
limiting
membrane

Nerve fiber layer

Light rays

Area of fovea

Ganglion cell

Neurons

Amacrine cell

Bipolar cell

Horizontal cell

Rod cell

Cone cell

External limiting
membrane

Pigmented
epithelium

Direction
of nerve
impulses

Dendrites

therefore focal length. At the same time, a special neural reflex automatically adjusts the variable diaphragm, the **iris,** in response to varying light intensity—much as the photocell system in a modern automatic camera measures light and then prompts a microprocessor to adjust the diaphragm opening. In bright light, the pupils are small; in dim light, they are large.

The sensory cells of the eye are arranged in a remarkable membrane called the **retina** (Figure 30-6). The retina contains photoreceptors of two kinds: the **rods** and **cones,** named for the shapes of their outermost segments (Figure 30-7). In a human eye, some 100 million rods are scattered uniformly over the retina—except in the **fovea,** a small area in the center of each retina. About 3 million cones are distributed irregularly and asymmetrically but are heavily concentrated in the fovea (see Figure 30-4).

30-6 Structure of the human retina
The nerve cells of the retina form several layers. Surprisingly, the actual photoreceptor cells (rods and cones) are in the back of the retina. Thus light must pass through layers of nerve cells to reach them. The interconnections of these nerve cells (ganglion, amacrine, bipolar, and horizontal cells) are shown in Figure 30-14. Nerve impulses initiated in the photoreceptor cells pass through them into the optic nerve which runs from the eye to the brain.

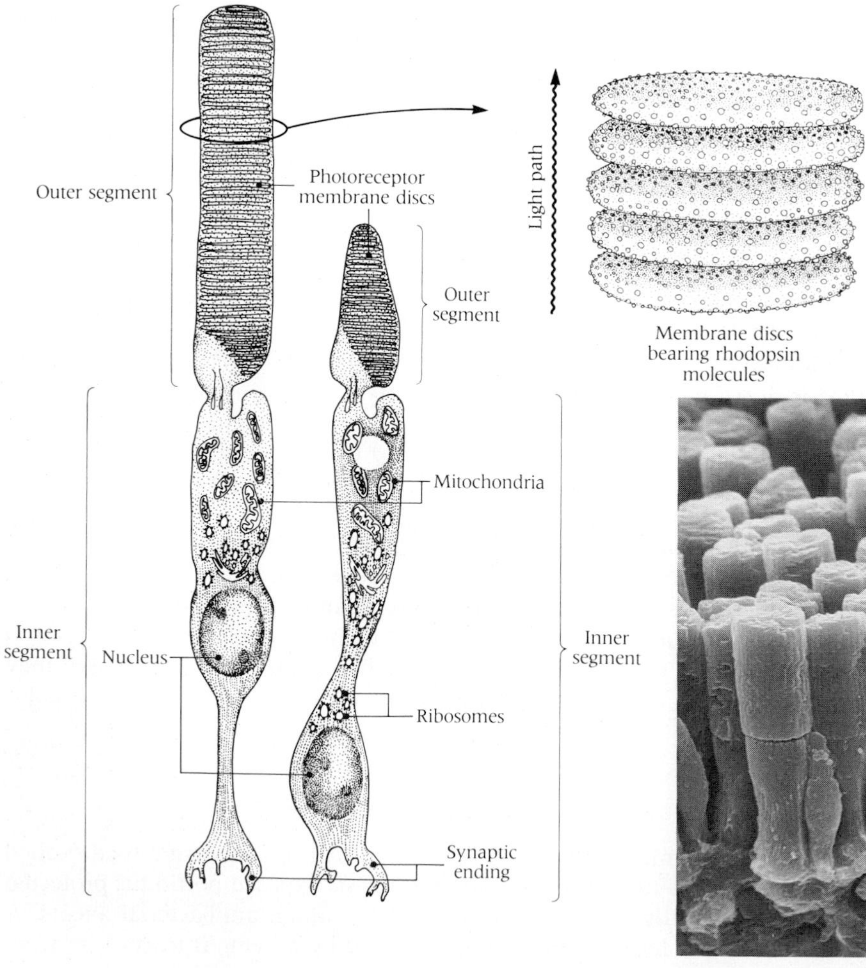

Outer segment

Photoreceptor
membrane discs

Outer
segment

Light path

Mitochondria

Inner
segment

Nucleus

Inner
segment

Membrane discs
bearing rhodopsin
molecules

30-7 Fine structure of rods and cones
(A) *Both rods and cones have a light-sensitive outer segment containing stacks of membrane disks studded with rhodopsin molecules. The synapse is at the opposite end of the inner segment.* **(B)** *Scanning electron micrograph of a tiger salamander retina. (× 900) The large cylindrical cells are the rods, the smaller conical cells the cones. These cells are four times thicker than human photoreceptor cells.*

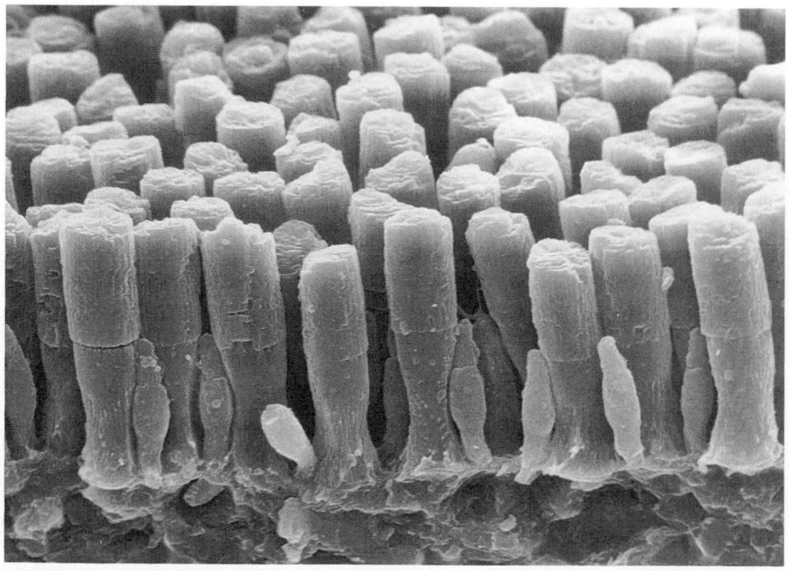

Ribosomes

Synaptic
ending

A

B

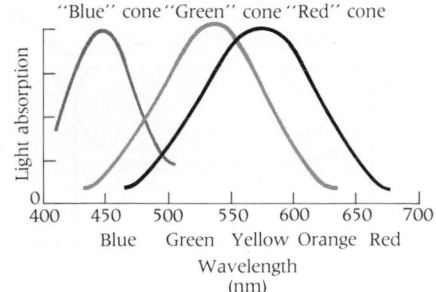

30-8 Basis of human color vision
The human eye contains three types of cones, each containing a pigment that is sensitive to a different part of the visible spectrum.

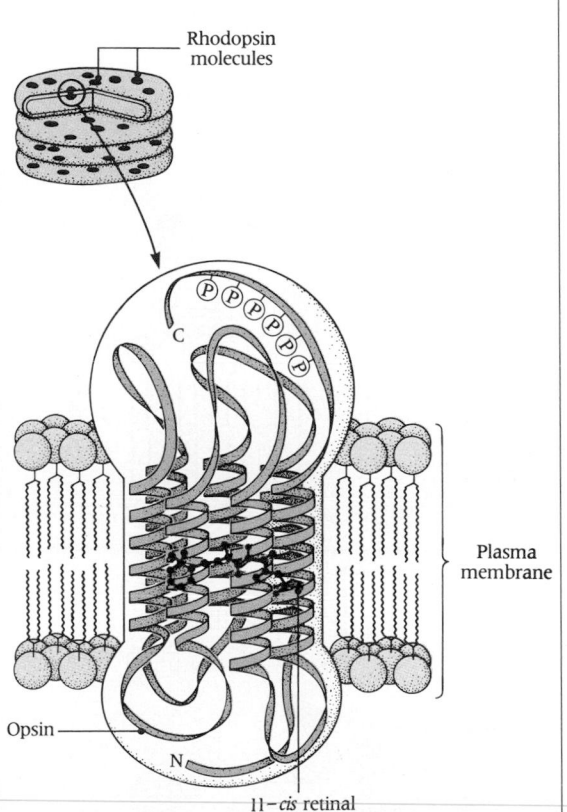

30-9 Model of rhodopsin
Rhodopsin is a transmembrane receptor protein containing opsin with 11-cis retinal near its center. Seven helical regions of the protein traverse the lipid bilayer. This unusual pattern occurs in other sensory and hormone receptors. The carboxyl end (C) of opsin is on the cytoplasmic side of the membranes; the amino end (N) is on the membrane surface facing the disk interior. Phosphorylations near the C end deactivate photoexcited rhodopsin. Each membrane disk contains many rhodopsin molecules.

These photoreceptor cells are located at the far end of the light path and various nerve cells and fibers lie in front of them between the lens and the receptors. This means that nerve fibers must find a common exit from the eye as they course into the optic nerve. This site lacks photoreceptors and is thus responsible for the eye's **blind spot.** In other animal groups—mollusks, reptiles, etc.—the photoreceptors are in front of the nerve cells. Hence, there is no blind spot.

Stereoscopic Vision

Both human eyes focus on the same point. This permits **stereopsis,** a three-dimensional effect that gives clues to distance. That is why a one-eyed person has difficulty judging how far away objects are—and may find it awkward to play tennis or drive a car. But the brain adjusts to this situation and in time finds ways to use other distance cues, as do other animals.

That insects also have stereoscopic vision was long suspected from the fact that one-eyed beetles, praying mantises, and water scorpions are rarely able to catch prey. Imaginative experiments in which praying mantises were forced to see through prismatic lenses—spectacles, really—confirmed that in estimating the distance to prey they do rely on binocular stereopsis and triangulation. (The latter is a technique for estimating the distance between two points or their relative positions by visualizing a triangle, two vertices of which are the points in question, and then inferring the position of the third point from assumptions on the triangle's sides or angles.)

The Role of Rods and Cones

Rods are more sensitive than cones in dim light. The relatively insensitive cones are not stimulated at all until light is about a thousand times brighter than the lowest intensity to which the rods respond. However, rods distinguish only different light *intensities;* cones distinguish different *wavelengths* as well as intensities. For that reason, cones are the cells of color vision.

There are three classes of cone cells in human eyes (Figure 30-8), each maximally sensitive to one primary color—blue, green, or red. (Other animals have as many as five cone types.) In 1986, a remarkable study demonstrated the existence of three different genes that direct retinal cells to synthesize three different light-trapping pigments—blue, green, and red.[3] (A fourth pigment, contained in rod cells, underlies colorless vision in dim light.) Interestingly, red-green color blindness, a sex-linked trait occurring primarily in males, was shown to be due to mutations in one or more of these genes. Predictably, the genes encoding red and green pigments were shown to be located on the X chromosome.

How Light Excites the Retina

Camera film and the retina in an eye both contain photosensitive molecules that are chemically transformed by light. In the retina, these photopigments are found within the rods and cones, in membranes of flattened disks that are piled like stacks of coins in the **outer segment,** the end of the photoreceptor furthest from the eye lens (see Figure 30-7A).

Light induces reversible chemical reactions in these molecules, generating nerve impulses that are ultimately transmitted to the brain. This photochemical effect of light is a key event in the visual process. The revelation of its biochemical basis by George Wald in the 1950s was a major discovery. It is a process that depends upon a reddish-purple molecule called **rhodopsin,** a conjugated protein consisting of the colorless protein, **opsin,** and a colored vitamin A derivative, **retinal.** As noted in Chapter 10, the vitamin A derivatives known collectively as **carotenoids** are synthesized only in plants. Animal vision thus depends on a continuing supply of dietary carotenoids.[4] We mentioned above that basic image-forming eyes appeared

---

[3] The study was by Jeremy Nathans, then a Stanford University medical student.

[4] Some individuals have abnormally poor vision in dim light. This defect is called night blindness. When it is due to a dietary deficiency of vitamin A, it is rapidly reversed if vitamin A is added to the diet. When vitamin A deficiency is too prolonged, however, supplemental vitamin A will not reinstate rhodopsin synthesis.

independently in evolution at least three times. How interesting then to realize that eyes from these three phyla—arthropods, mollusks, and chordates—contain almost identical visual pigments!

Many rhodopsin molecules are densely embedded in the disk membranes of rods (Figure 30-9). The role of rodopsin in vision depends upon its long side chain, which can be "bent" into several shapes. Opsin itself does not absorb visible light. As shown in Figure 30-10 light temporarily alters the shape of retinal from the *cis* form, which is bent, to the *trans* form, which is straight. It is this *cis-trans* isomerization of the visual pigments by light, termed bleaching, that leads to the initiation of a nerve impulse.

Rhodopsin must be replenished so that a response to new light stimuli can occur. Otherwise vision would be a once-in-a-lifetime experience. Regeneration of rhodopsin is accomplished by the enzyme **retinal isomerase,** which helps to reconvert inactive all-*trans* retinal to the active 11-*cis* isomer, a form that can bind to opsin. There are striking parallels between this system and the cycles that regenerate light-absorbing pigments in photosynthesis.

How Light Energy Is Transduced into Nerve Impulses

Finally, there is the issue of how bleaching rhodopsin by light leads to the initiation of the nerve impulse that enters the optic nerve.

Light leads to **hyperpolarization** of the plasma membrane of a rod cell. Hyperpolarization is quite different from depolarization, which we learned in Chapter 29 is associated with action potential generation in typical neurons. Like neurons, rod cell plasma membranes have specific ion channels for Na^+ and K^+. However, *these* channels are closed by light. In the dark, the Na^+ channels are open and Na^+ ions flow freely inward in the rod's outer segment where a very large Na^+ gradient is maintained across the membrane by ATP-dependent Na-K pumps located in the inner segment (see Figure 30-7A). Light closes these channels in the outer segment. As a result, the inside of the cell becomes *more* negative than usual—

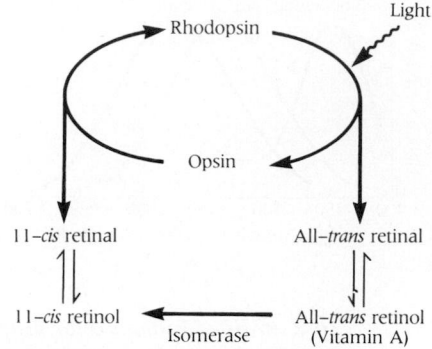

30-10 The visual cycle (Above)
*Absorption of light splits rhodopsin into opsin and all-*trans* retinal. Regeneration of rhodopsin depends upon an isomerase, which helps to convert all-*trans* retinal to 11-*cis* retinal.*

30-11 The cyclic GMP cascade of vision (Below)
A resting disk membrane (upper left) contains rhodopsin, phosphodiesterase (PDE), and the G protein transducin (T). Light activates rhodopsin, which now binds T, causing GTP to replace GDP on T. The subunit containing GTP dissociates from T, binding to and activating PDE by removing one of its subunits. As a result, PDE cleaves cGMP in the cytoplasm to GMP. Since cGMP opens membrane channels, its elimination closes them, hyperpolarizing the membrane and generating a nerve impulse. Rhodopsin, PDE, and T are regenerated to begin a new cycle.

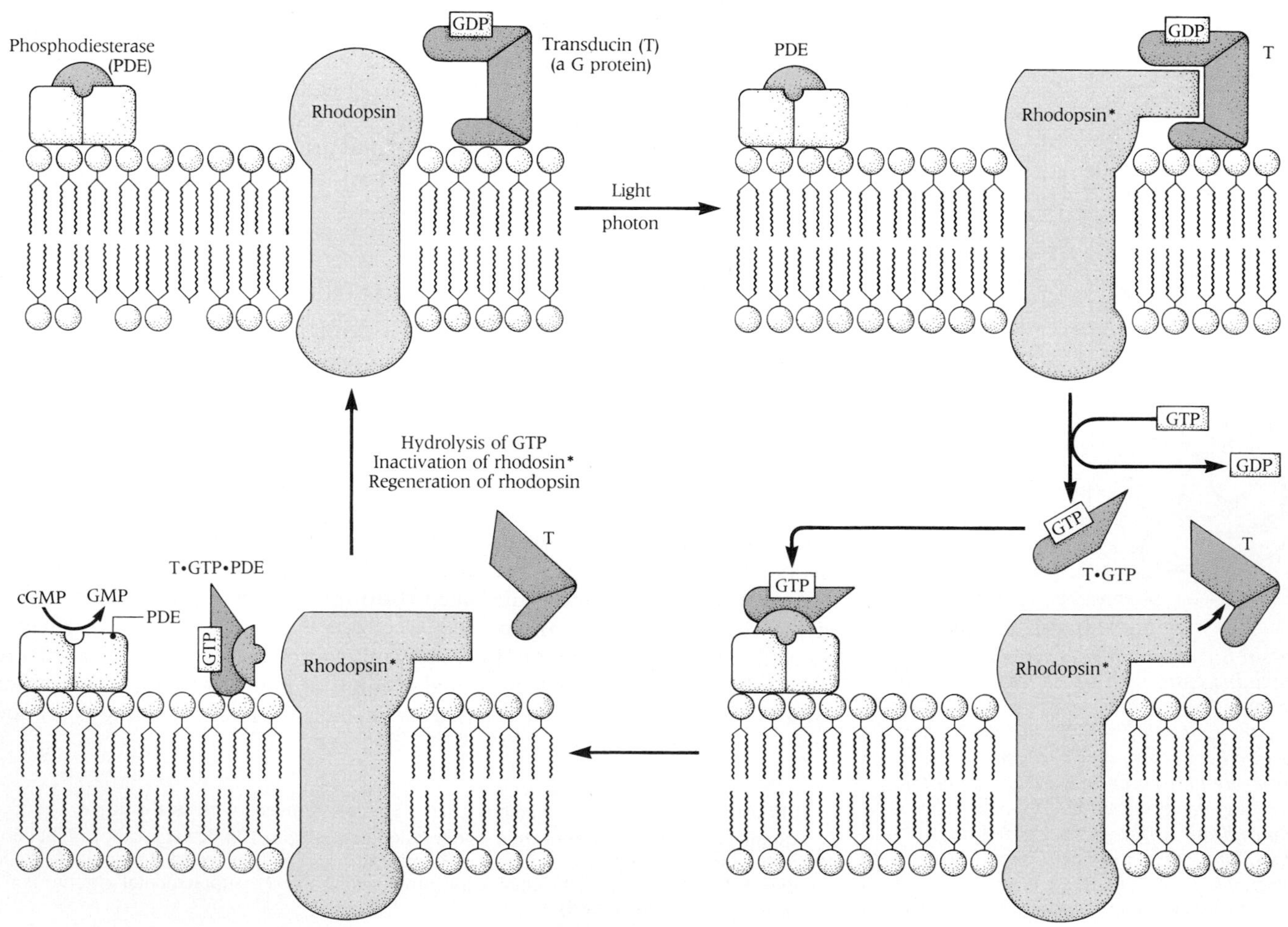

Phosphodiesterase (PDE)

Rhodopsin

Transducin (T) (a G protein)

GDP

Light photon

PDE

Rhodopsin*

GDP

T

GTP

GDP

Hydrolysis of GTP
Inactivation of rhodosin*
Regeneration of rhodopsin

T

GTP

T·GTP

T

cGMP   GMP

T·GTP·PDE

PDE

GTP

Rhodopsin*

GTP

Rhodopsin*

and the potential difference across the membrane increases. This zone of light-induced hyperpolarization travels along the membrane until it reaches the synapse. The speed and intensity of hyperpolarization both depend on the intensity of the incident light.

There is another notable difference between rod cells and neurons: rod cells do not produce all-or-none action potentials. Instead, they give a *graded* response that is proportional to light intensity. Thus, the signal sent to the synapse depends on the number of light photons absorbed.

Neural responses in the retina occur immediately after light arrives; thus they appear to be triggered by an early step in the bleaching of rhodopsin. An allosteric modification of the molecular conformation of opsin probably initiates the response. Remarkably, this occurs when a dark-adapted rod absorbs only a *single* photon of light! One photon could bleach no more than one molecule of rhodopsin. Hence, a thousand- to million-fold amplification must occur between the photochemical event and the first neural response.

It is clear that a local messenger molecule transmits information over the very short distance from the disk membrane, where light is absorbed, to the rod's plasma membrane, which becomes hyperpolarized. The messenger molecule that signals this event is **cyclic guanosine monophosphate** (cyclic GMP or cGMP), a compound with an obvious resemblance to cAMP. In sum, photoexcited rhodopsin triggers an enzymatic cascade that results in the hydrolysis of cGMP by an enzyme called phosphodiesterase (PDE). The lowering of cGMP levels closes sodium channels in the cell's plasma membrane, thereby initiating a graded response. As shown in Figure 30-11, **transducin,** a newly discovered G protein in the disk membrane, binds GDP when in one conformational state and GTP when in another. Light causes changes that favor GTP binding. This in turn leads to formation of cGMP—in contrast with the G proteins discussed earlier, which stimulate cAMP formation. Newly formed cGMP molecules directly open sodium channels in the cell's plasma membrane. Indeed, the gates of these channels are directly operated by cGMP.

Note the remarkable similarity between this cascade and the one responsible for the actions of peptide hormones such as epinephrine (Figure 30-12). It also recalls the GTP-dependent step in protein synthesis described in Chapter 17. Some believe that this mechanism first arose as a regulatory device in protein synthesis. It worked well and several billion years later was applied to other tasks, namely,

30-12 Parallels between the cascades of vision and hormone action
The binding of a hormone (H) to its receptor (R) activates a G protein, which binds GTP and then splits, one subunit of which activates adenylate cyclase (AC). This enzyme converts ATP to cyclic AMP, which induces a change in cell metabolism by regulating enzyme activity. Compare this cascade with Figure 28-5. It strikingly resembles events in the rod cell, where rhodopsin is activated by light. It then interacts with the G protein called transducin (T), which binds GTP and splits, one subunit activating phosphodiesterase (PDE), which lowers cell cyclic GMP levels by converting cyclic GMP into GMP. The absence of cyclic GMP closes channels and hyperpolarizes the cell.

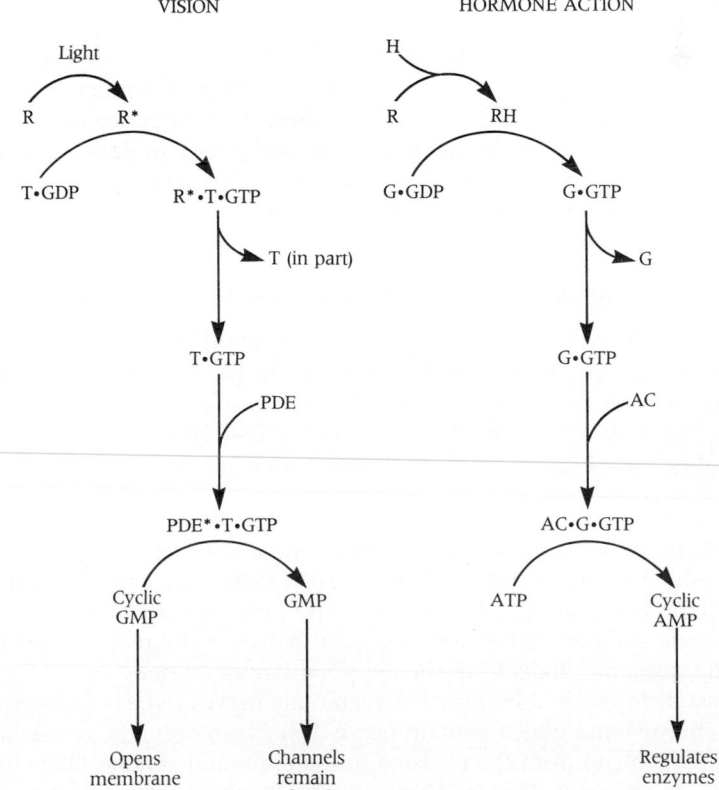

the transduction of hormonal and sensory stimuli. Presumably other sensory receptors have comparable transducer mechanisms.

A generalized scheme of the role of G proteins in signal transduction is shown in Figure 30-13. In this figure, the effector molecules may be adenylate cyclase, which produces cAMP, or guanylate cyclase, which makes cGMP, depending on which G protein is present.

Dark and Light Adaptation

An interesting problem relates to the range of light intensities the eye must perceive—how, in a word, it sees by starlight and by the light of a noonday sun, brighter by 10 billion times. The eye's versatility is vastly extended by **adaptation** to dark and light. Dark adaptation occurs as the eye gradually accommodates to darkness. The pupil dilates and the rods gain in sensitivity. Light adaptation occurs as the eye accommodates to normal or bright light.

Interactions Among Retinal Receptors

The retina is much more than a collection of individual receptor-transducers of light energy with passive connections to the brain. It also sorts and processes data in a way that stresses significant features and permits useful interpretations. The retina is in fact a highly organized structure containing elements that participate in many integrative functions—summation, inhibition, and temporal, spatial, and chromatic interactions—before transmitting visual information to the brain. Indeed, many neural cross-connections and integrating devices clearly reflect the fact that the retina develops embryologically as an extension of the brain. Colleagues who study this aspect of the retina say they are doing so because it is a "little brain" that may tell us how the "main brain" works.

The arrangement of the variously shaped neurons in the retina fosters interaction (Figure 30-14). This layout, so characteristic in brain architecture, permits fine tuning of several kinds. Photoreceptors (rods and cones) synapse with neurons called **bipolar cells,** which in turn synapse with **ganglion cells,** the axons of which become the **optic nerve.** Other cells in this layer, **horizontal cells** and **amacrine cells,** permit lateral transfers of information. Note that bipolar cells connect to more than one ganglion cell—and to amacrine cells as well.

"Data processing" in the retina has a number of interesting consequences. The first step in the processing of visual data occurs the instant an image spreads across the mosaic of photoreceptors. Stimulation of a receptor cell inhibits its immediate neighbors. As a result, the response of a single receptor is influenced both by the stimulatory light shining on it and by the inhibitory influence of its immediate neighbors. Together these systems provide another mode of dark-light adaptation that complements the better known one, which is based on photochemical pigment changes.

Transferring Visual Information to the Brain

Visual data processed in the retina travel to the brain via two optic nerves, one from each eye (Figure 30-15), which converge in the X-shaped **optic chiasma.** Here there occurs a remarkable sorting of nerve fibers that is best understood by examining the way images project upon the retina. Note in Figures 30-15 and 30-16 the difference between optic nerves and optic tracts.

- When both eyes are fixed on a single point (Figure 30-16A), the field of vision has a right and a left half—and an upper and a lower half.
- Light coming from the right (Figure 30-16B) strikes the temporal (outer) half of the left retina and the nasal (inner) half of the right retina.
- In the optic chiasma, fibers from the nasal halves of the retinas cross to the opposite sides, but fibers from the temporal halves do not cross.
- We must distinguish, therefore, between optic nerves (which run from the eyes to the chiasma and which contain nerve fibers from only one retina) and **optic tracts** (which run from the chiasma to the brain and contain fibers from both retinas). As shown in Figures 30-15 and 30-16, the right optic tract carries left

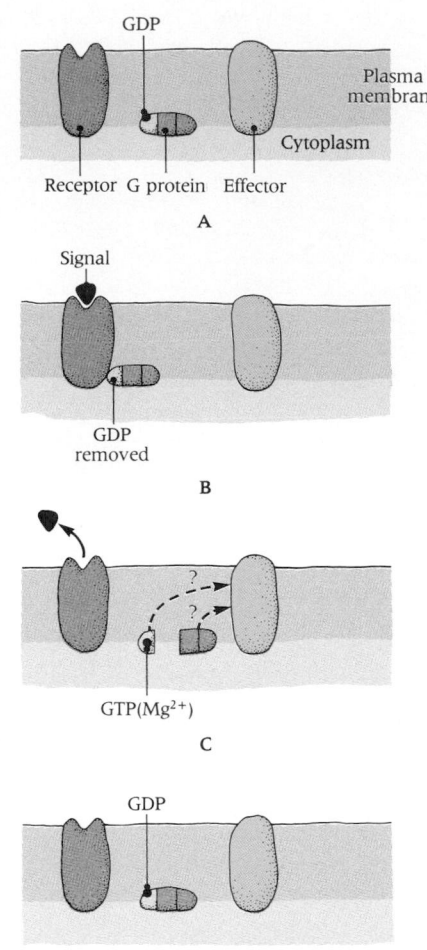

30-13 Summary of the role of G proteins in signal transduction
(**A**) In the unstimulated state, the receptor protein, G protein, and effector are not associated. GDP is bound to one of the three G protein subunits. (**B**) When the signal molecule (hormone, neurotransmitter, etc.) binds to the receptor, the G protein associates with the receptor. (**C**) GTP and Mg^{2+} exchange for GDP, and the subunits of the G protein separate and interact with the effector to cause a physiological response. The signal molecule then dissociates from the receptor. (**D**) An enzyme in the G protein converts GTP back to GDP and the membrane regains its former unstimulated state. Compare with Figures 28-5 and 30-12.

30-14 The cells of the human retina
The retina is composed of two types of photoreceptor cells (rods and cones) and four types of nerve cells (bipolar, ganglion, horizontal, and amacrine). The photoreceptor cells synapse at their base with bipolar cells. Bipolar cells then synapse with ganglion cells, the axons of which form the optic nerve. Processing of information occurs within the retina because several photoreceptor cells may synapse with a single bipolar cell and several bipolar cells may synapse with a single ganglion cell. There is also lateral transfer of information via horizontal cells (each of which receives synapses from many photoreceptors and synapses on many bipolar cells and on other horizontal cells) and via amacrine cells (which both receive synapses from and synapse on bipolar cells, and also synapse on many ganglion cells).

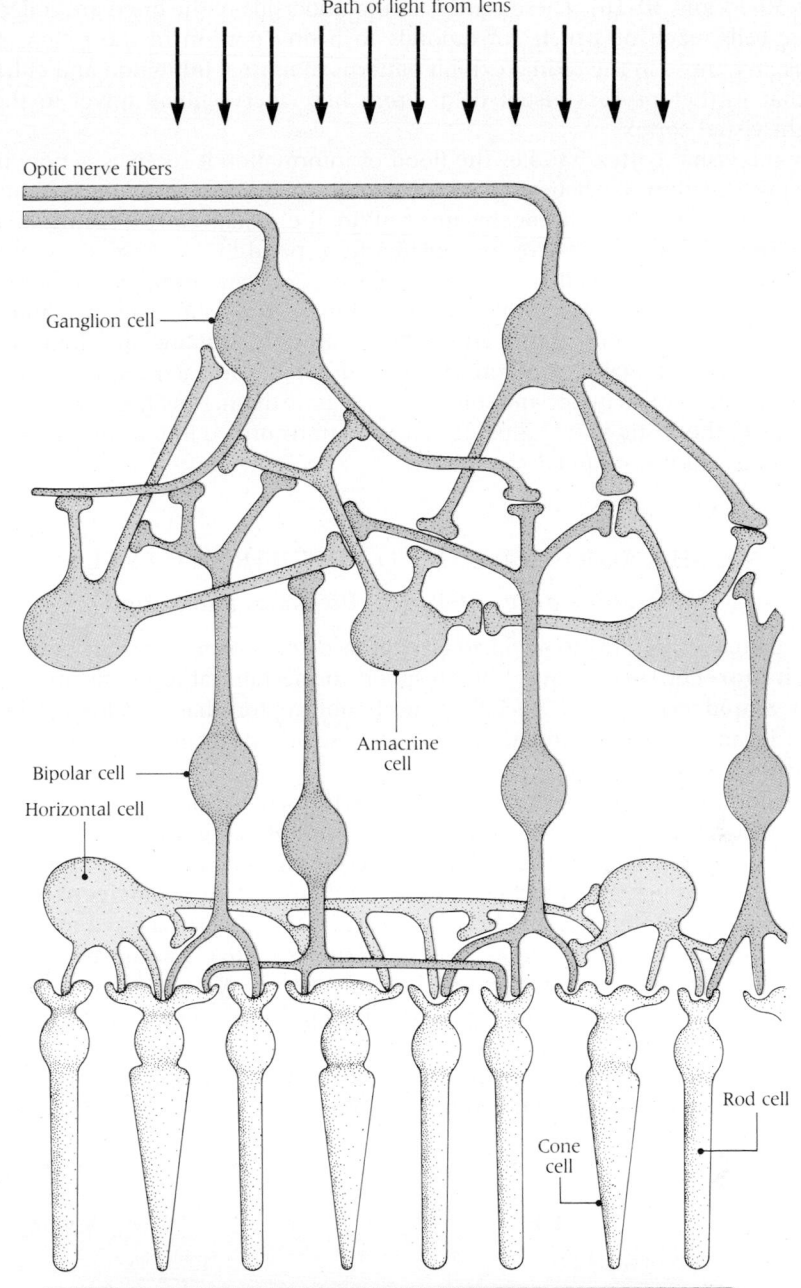

nasal and right temporal fibers; the left optic tract carries right nasal and left temporal fibers.

- If the left optic nerve is cut, the left eye is totally blinded. If the left optic tract is cut, the right half of the visual field in both eyes is eliminated through loss of the rightward-looking right nasal and left temporal retinal receptors. The left half of the visual field remains intact. Thus, all fibers from the left sides of both retinas (the rightward-looking sides) are bundled together and sent to the left side of the brain. Similarly, fibers from the right sides are sent to the right side of the brain—and fibers from the upper and lower halves of the retinas travel to the corresponding upper and lower halves of the visual cortex.

The chiasmic crossing results in good coordination between eyes and brain because images from both retinas are sent to each side of the brain and each retina has connections to both sides of the brain. In the brain, another set of integrative and associative operations begins.

After leaving the retina, impulses are propagated along the optic nerves and optic tracts to the left and right **lateral geniculate nuclei** in the thalamus (see

Figures 30-15 and 30-16). These centers in the underside of the brain are collections of nerve cells, each of which corresponds to a tiny portion of the retina. These neurons, like those in the retina, exhibit patterns of mutual inhibition and enhancement that further process visual data. From here, nerve fibers travel to the left and right visual cortex.

How the visual cortex handles the flood of information it receives is now under intense investigation. We believe that each fiber in an optic nerve connects ultimately to cells which form specific aggregates within the visual cortex. Some of these cells are specialized feature detectors, geared to respond to the presence or absence of a visual spot such as a line or moving object. Cells are arranged in alternating strips or columns and rows (like a three-dimensional spreadsheet), and each module or compartment of this gridlike organization contains specific cells that (with some redundancy) correspond to limited patches in each retina. These cells detect the features of an image and somehow integrate them. How these computerlike operations in the sorting and editing of data are transformed into a conscious visual perception is still unexplained.

MECHANORECEPTORS OF DIFFERENT TYPES
Common Features of Touch, Pressure, and Sound

Both receptors responding to *sound* and receptors of *touch* or *pressure* may be classified as **mechanoreceptors** because they respond to mechanical forces of one kind or another. Sound receptors are specialized mechanoreceptors that react to the vibratory changes in air (or water) pressure that comprise sound. They are also known as **phonoreceptors.**

In addition to phonoreceptors (which are housed in the ears), animals may have simpler receptors of other kinds for touch and pressure. In humans and other mammals, the skin contains a variety of mechanoreceptors (Figure 30-17). The various nerve endings that mediate the sense of pain are termed **nociceptors** (from the Latin *nocere,* to injure). Skin also contains several highly specialized mechanoreceptors. The best known of them is the pressure-sensitive **Pacinian corpuscle,** celebrated for its large size and onionlike architecture. A weight pressing on the skin stimulates mechanoreceptors to dispatch impulses to higher nerve centers—and the greater the weight, the more mechanoreceptors are stimulated. Since several receptors are innervated by the termini of a single sensory nerve fiber, the impulses from each receptor converge and summate. When a strong stimulus increases the

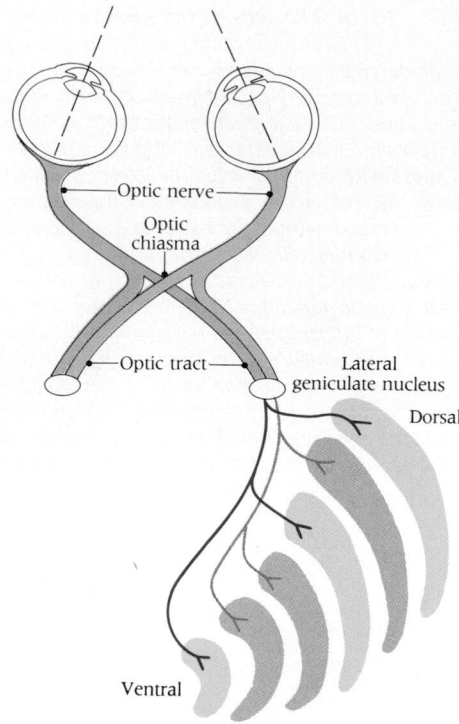

30-15 Path of visual images from the two retinas
Images received by photoreceptor cells in the retinas travel along the optic nerves, which converge at the optic chiasma. There, fibers from the nasal (inside) halves of each retina cross, while fibers from the temporal (outside) halves remain on the same side. Thus, the optic tracts which run from the optic chiasma to the brain each contain fibers from both retinas. These fibers eventually reach the lateral geniculate nuclei in the brain, which receive, in orderly array, fibers from the eye on the same side (ipsilateral) and from the eye on the opposite side (contralateral).

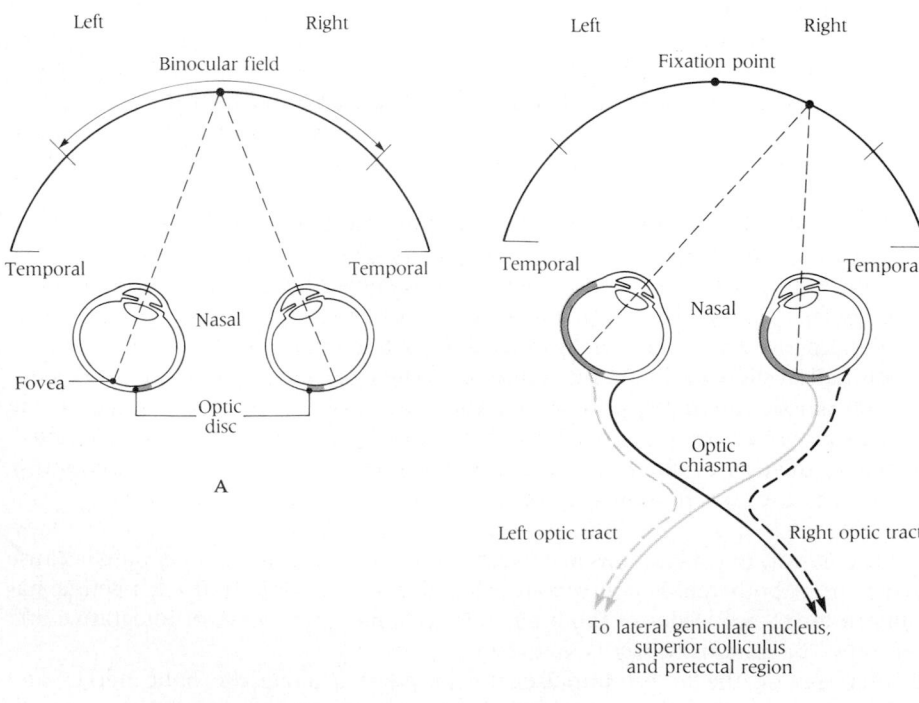

30-16 Organization of the visual field
(A) *Light from a single point in a binocular field strikes both eyes and is absorbed by receptors in both foveas.* **(B)** *Light from the right binocular field strikes the left temporal retina and the right nasal retina. Thus, fibers in the left optic tract contain a complete representation of the right side of the visual field.*

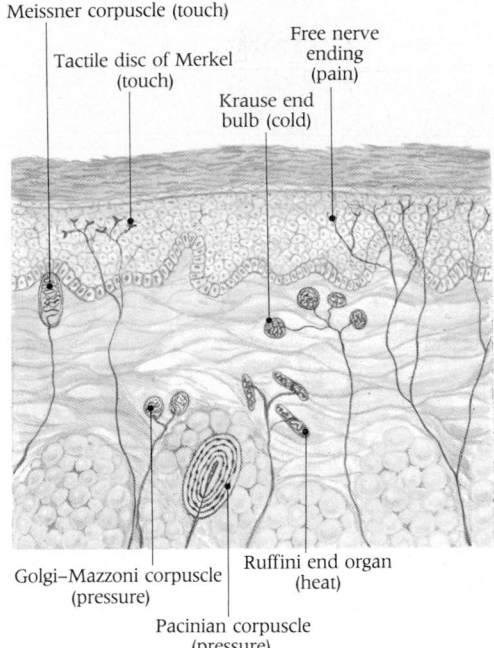

Meissner corpuscle (touch)

Tactile disc of Merkel (touch)

Free nerve ending (pain)

Krause end bulb (cold)

Golgi–Mazzoni corpuscle (pressure)

Ruffini end organ (heat)

Pacinian corpuscle (pressure)

30-17 Mechanreceptors and other receptors of the skin
Mammalian skin contains the receptors for touch, pressure, pain, and temperature.

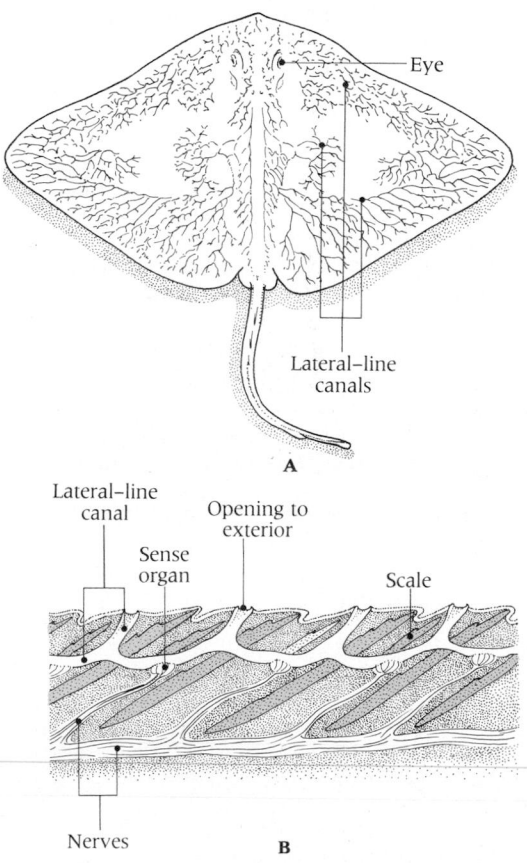

Eye

Lateral–line canals

A

Lateral–line canal

Opening to exterior

Sense organ

Scale

Nerves

B

30-18 The lateral-line system of fishes
(A) The lateral-line canals on the surface of this ray serve as a sensory system for detecting pressure changes in the surrounding water. In rays, the canals are highly branched. (B) Section through the surface of a perch. Note the scales embedded in the skin. A lateral-line canal has many openings to the exterior, which connect with numerous nerves.

number of activated receptors, the frequency of sensory nerve impulses increases correspondingly.

The Pacinian corpuscle has yielded interesting data on how the skin's highly specialized mechanoreceptors work. Many concentric connective tissue layers called *lamellae* enclose its nerve fiber ending. Almost all of the lamellae can be peeled away without impairing the receptor's transducer function. The membrane of this and other mechanoreceptors differs from those of other receptors in its ability to initiate impulses following mechanical deformation. Nerve impulses are propagated at a frequency proportional to the intensity of the stimulus.

In most aquatic invertebrates, touch and sound receptors, if present, are identical. Often these are sensitive hairs that vibrate in response to sound and that detect a touch. Insects receive sound energy via vibratory hairs, but they also have special small sound receptors that may be scattered throughout the body or grouped in special organs. These organs can discriminate stimulus intensities (loudness) and changes in frequency. They do not recognize the absolute frequency (pitch) of a sound.

Fishes and rays have a special sensory system, lacking in land-living vertebrates, that consists of a series of grooves or canals on the head and body containing clusters of sensory cells. These **lateral-line organs** (Figure 30-18) respond to changes in pressure and surrounding water currents. They also occur in the aquatic, larval stages of amphibians—for example, the tadpoles of frogs—but are usually lost in the adults. Lateral-line organs are absent in reptiles, birds, and mammals, including those, like whales, that have returned to the water.

Lateral-line organs are of particular interest because they seem to be the primitive pressure receptors from which organs of hearing and equilibrium evolved. In fishes, the ear is entirely internal and is mainly an organ of equilibrium. Some have debated whether fishes really hear. The answer is that some do. Some fishes also discriminate between different pitches of sound waves.

The Sense of Hearing

The organs of hearing in land vertebrates seem to have evolved from parts of the pressure and equilibrium receptors of fishes. Receptors responding to both pressure (sound) and equilibrium are still closely associated in the ear. The anatomical details of sound receptors differ widely in different animal groups, but the same principles operate in all of them.[5]

As shown in Figure 30-19, air vibrations (sound waves) strike an eardrum, or **tympanic membrane,** and cause it to vibrate with the same frequency as the sound wave. Vibrations from the drum then travel through a sequence of vibrating components to the phonosensitive apparatus of the inner ear. The major function of these components is to translate vibrations in air to vibrations in a fluid, which can be detected by specialized phonoreceptors. The sequence is as follows:

- Vibrations pass first through the middle ear—in mammals via a chain of tiny bones, the **ossicles,** in nonmammalian land vertebrates (including birds) via a single bone.
- The three ossicles—the smallest bones in the human body—are the **malleus** (hammer), **incus** (anvil), and **stapes** (stirrup). Their remarkable evolutionary history is summarized in Box 35-1.
- Vibrations of these bones propagate intermittent pressure waves through the oval window upon the fluid (perilymph) in the critical portion of the inner ear, the **cochlea.**[6]
- These pressure waves force a thin membrane covering the round window at the far end of the cochlear duct to bulge in and out, thus setting up additional small fluctuations in the perilymph that stimulate phonosensitive receptors in the **organ of Corti** (Figure 30-20).
- The organ of Corti contains phonoreceptor cells with projecting processes that are rich in actin filaments and resemble hairs. These are the **hair cells** (Figure

[5] We refer here to amphibians, reptiles, birds, and mammals. In most living amphibians the hearing organs are aberrant or degenerate.

[6] Strictly speaking, a true cochlea exists only in birds and mammals.

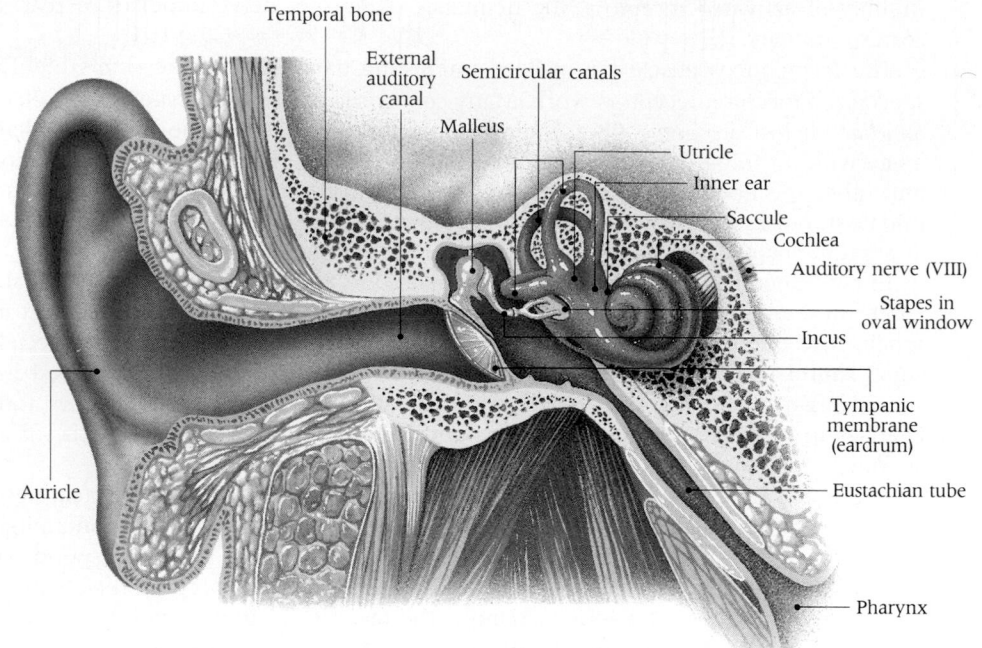

Temporal bone

External auditory canal

Semicircular canals

Malleus

Utricle

Inner ear

Saccule

Cochlea

Auditory nerve (VIII)

Stapes in oval window

Incus

Tympanic membrane (eardrum)

Eustachian tube

Auricle

Pharynx

30-21). Over them is the tectorial membrane, a gelatinous structure into which the hairs project. Sound waves are ultimately translated into vibrating movements of the **basilar membrane** of the organ of Corti. This movement is communicated to the hair cells, which are mechanically deformed because they cannot vibrate freely in the tectorial membrane. Deformation is transduced into an electrical potential, the so-called cochlear microphonic potential, which leads to liberation of a chemical transmitter that stimulates the afferent nerve fibers synapsing at the base of each hair cell (Figure 30-22).

Each hair cell is sensitive to only a limited frequency range, which is determined in part by its position relative to the basilar membrane. Therefore, if the ear is to receive useful information it must collect the output from thousands of hair-cell receptors. In fact, there are 15,000 or so hair cells in the cochlea of each human ear. The stimuli from them summate and excite all-or-none impulses in sensory nerve fibers. Impulses from each ear pass through several "way stations" and eventually reach the auditory cortex on both sides of the brain.

30-19 The human ear
Sound waves enter the external auditory canal and strike the tympanic membrane. The vibrations are transferred to the incus, malleus, and stapes and on to the cochlea of the inner ear, which contains the phonoreceptors that connect with the auditory nerve.

30-20 The cochlea
The cochlea is the organ of hearing. The model shows its coiled structure. A cutaway view shows the cochlear canal and the organ of Corti containing rows of sensory hair cells, which are stimulated by movements of the basilar membrane caused by pressure changes in the outer ear. Each hair cell is connected to nerve fibers that join to form the auditory nerve.

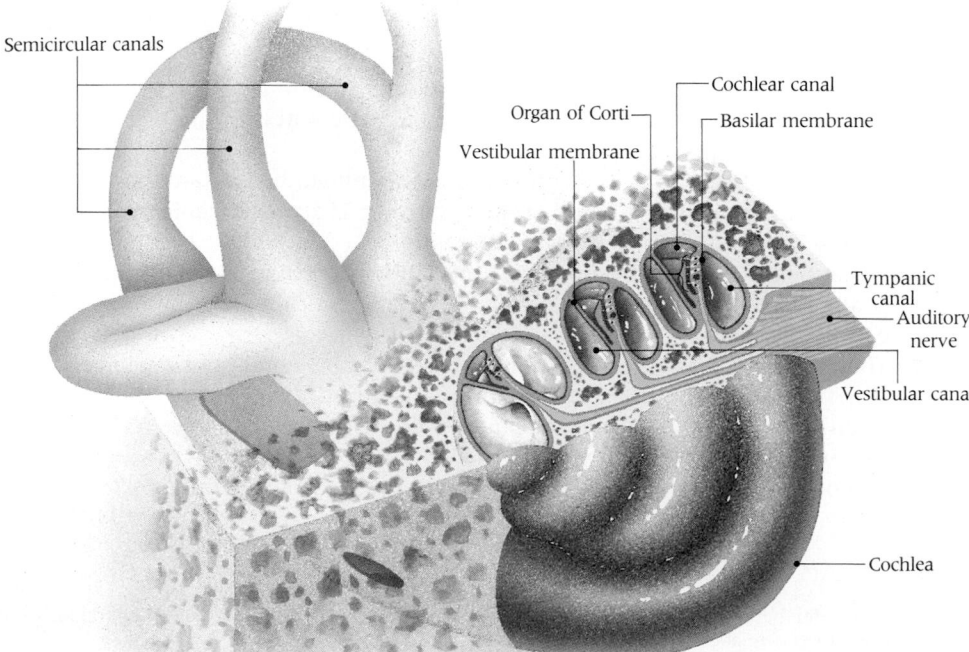

Semicircular canals

Organ of Corti

Vestibular membrane

Cochlear canal

Basilar membrane

Tympanic canal

Auditory nerve

Vestibular canal

Cochlea

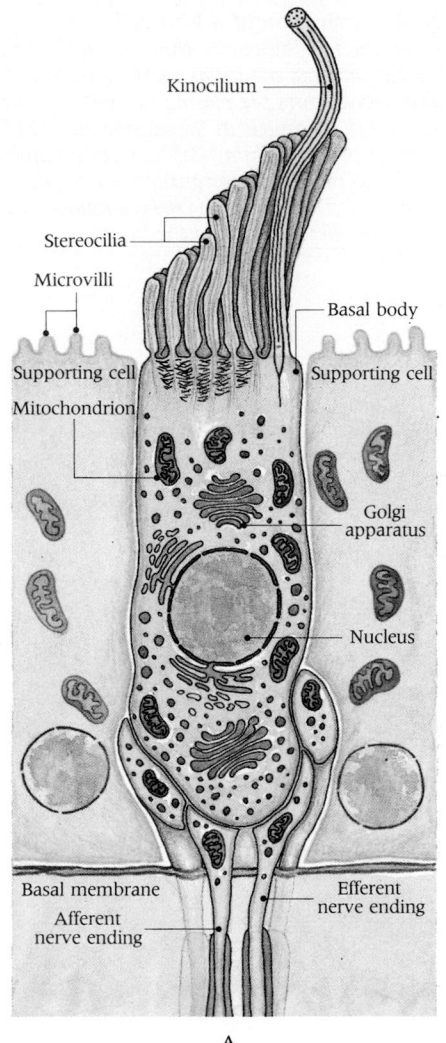

A

B

30-21 The hair cell of the organ of Corti
(A) Although hair cells resemble neurons, they lack dendrites and axons. The cylindrically shaped cell has a kinocilium, with a bulbous end, and a number of associated stereocilia. Bending of these structures causes the hair cell to send nerve impulses to the afferent nerve fiber that synapses at the other end of the hair cell.
(B) Scanning electron micrograph of hair cells from the inner ear of a bullfrog. Note the bundles of stereocilia and the single kinocilium in each cluster. (× 9600)

Sound is transmitted to the inner ear with remarkable fidelity. Because the cochlea is embedded in bone, vibrations of the entire skull can produce vibrations in the cochlear fluids. However, the cochlea's design tends to minimize the effects of bone conduction.

Hearing with two ears, called binaural hearing, does not have the same importance as does depth perception in binocular vision. However, it does contribute to a perception of spatial orientation of sound sources. That is the basis of stereophonic high-fidelity audio systems. Sounds heard through only one ear are difficult but not impossible to localize in space.

The Sense of Equilibrium

An animal swimming, walking, or flying, must know which way is up—in other words, it must orient itself in Earth's gravity field. Most animals have special organs that are sensitive to gravity.

In invertebrates, the organ is usually a **statocyst,** a sac containing a small stony ball called an **otolith.** Under the influence of gravity, the otolith stimulates sensory hairs or hair cells (Figure 30-23). It is curious indeed that insects rarely have a special gravity receptor.

Vertebrates have the same kind of receptor as invertebrates. In humans, each inner ear is the site of six small sensory organs that contain hair cells. One is the cochlea. Then there are two sacs, the **utricle** and the **saccule** (see Figure 30-19) and just beyond the cochlea are the three semicircular canals.

The utricle and saccule are two fluid-filled chambers, each of which contains floating crystals of calcium carbonate called **otoconia.** As shown in Figure 30-24A, a shift in the position of the head moves these solids, which then deform

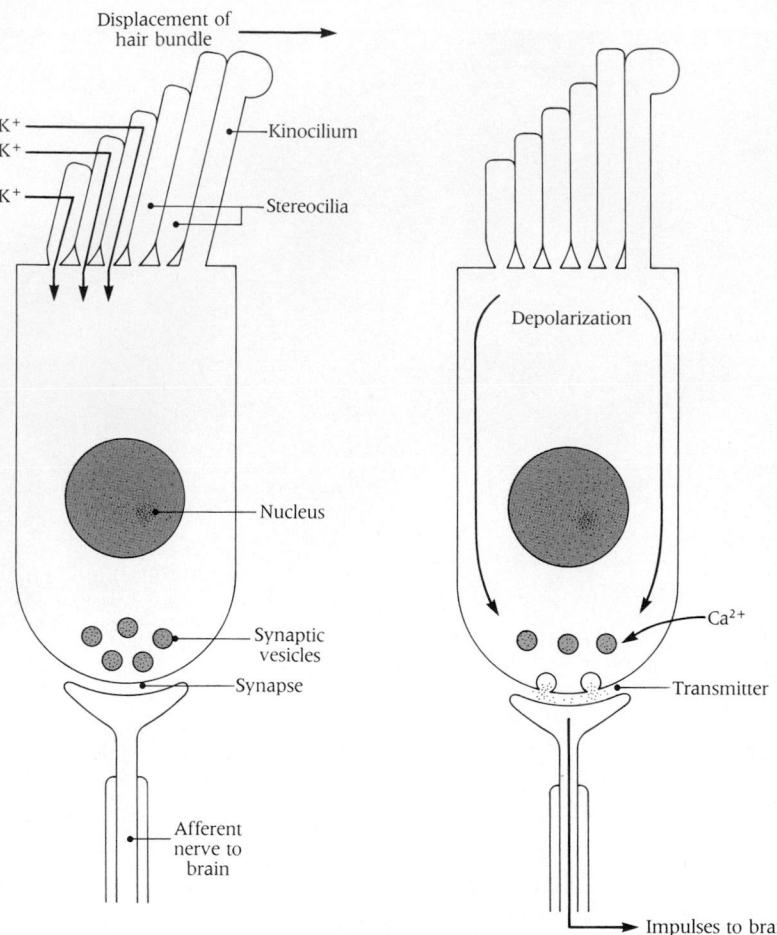

Displacement of hair bundle →

K⁺
K⁺
K⁺

Kinocilium

Stereocilia

Nucleus

Synaptic vesicles

Synapse

Afferent nerve to brain

Depolarization

Ca²⁺

Transmitter

Impulses to brain

30-22 Activation of a hair cell
When the hair bundle is displaced, K^+ enters the cell, making the inside more positive and thus depolarizing the plasma membrane. This opens Ca^{2+} channels in the membrane. The increase in intracellular Ca^{2+} causes synaptic vesicles to release neurotransmitter into the synaptic cleft. The resulting nerve impulse travels along an afferent nerve to the brain.

receptors in hair cells in tiny organs called **maculas.** This causes information to be sent to the brain concerning the body's orientation to linear acceleration.

Fluid-filled **semicircular canals,** arranged approximately at right angles to each other, respond to changes in rate and direction of motion, or to angular acceleration. Near the end of each canal is an expansion, the **ampulla,** which terminates in a tuft of hair cells called the **ampullary crest** (Figure 30-24B). Above the crest is the wedge-shaped **cupula** (Figure 30-24C). Any increase or decrease of motion causes fluid to shift in the canals. This stimulates hair cells in the ampullary crest and cupula, and in turn initiates nerve impulses. The pattern of fluid movement in the three canals depends on the direction of the acceleration. It is detected by hair cells called **statoreceptors.**

We see, then, that mechanical displacement of a bundle of hair cells is the critical stimulus in all of the sensory organs in the inner ear (Figure 30-25). This is also the basis of the lateral-line sense of fishes and the ability of certain animals to detect vibrations in the ground. In all of these situations, the common denominator is the hair cell.

One of the brain centers that collects and interprets information on rate and direction of motion from statoreceptors in the organs of equilibrium is the cerebellum. Another is the cerebral cortex. One way to put this system through its paces is to subject a volunteer to rotational acceleration. An individual being rapidly rotated in a swivel chair displays characteristic involuntary back-and-forth eye movements called nystagmus.[7]

■ If the head is erect during rotation, the horizontal semicircular canals are primarily affected, and nystagmus is horizontal, the eyes moving from side to side.

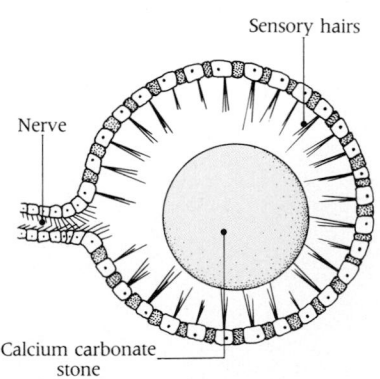

Sensory hairs

Nerve

Calcium carbonate stone

30-23 The statocyst as an equilibrium detector
In the mollusk, Pecten, the statocyst contains a central stonelike body composed of calcium carbonate. Gravity causes it to press on sensory hairs that line the hollow ball of cells. The animal's orientation cue depends on which hairs are touched.

[7] Nystagmus can also have other causes. In rotational acceleration, it is due to the vestibulo-ocular reflex. This is the principal mechanism that keeps visual images stable on the retina as we move our heads.

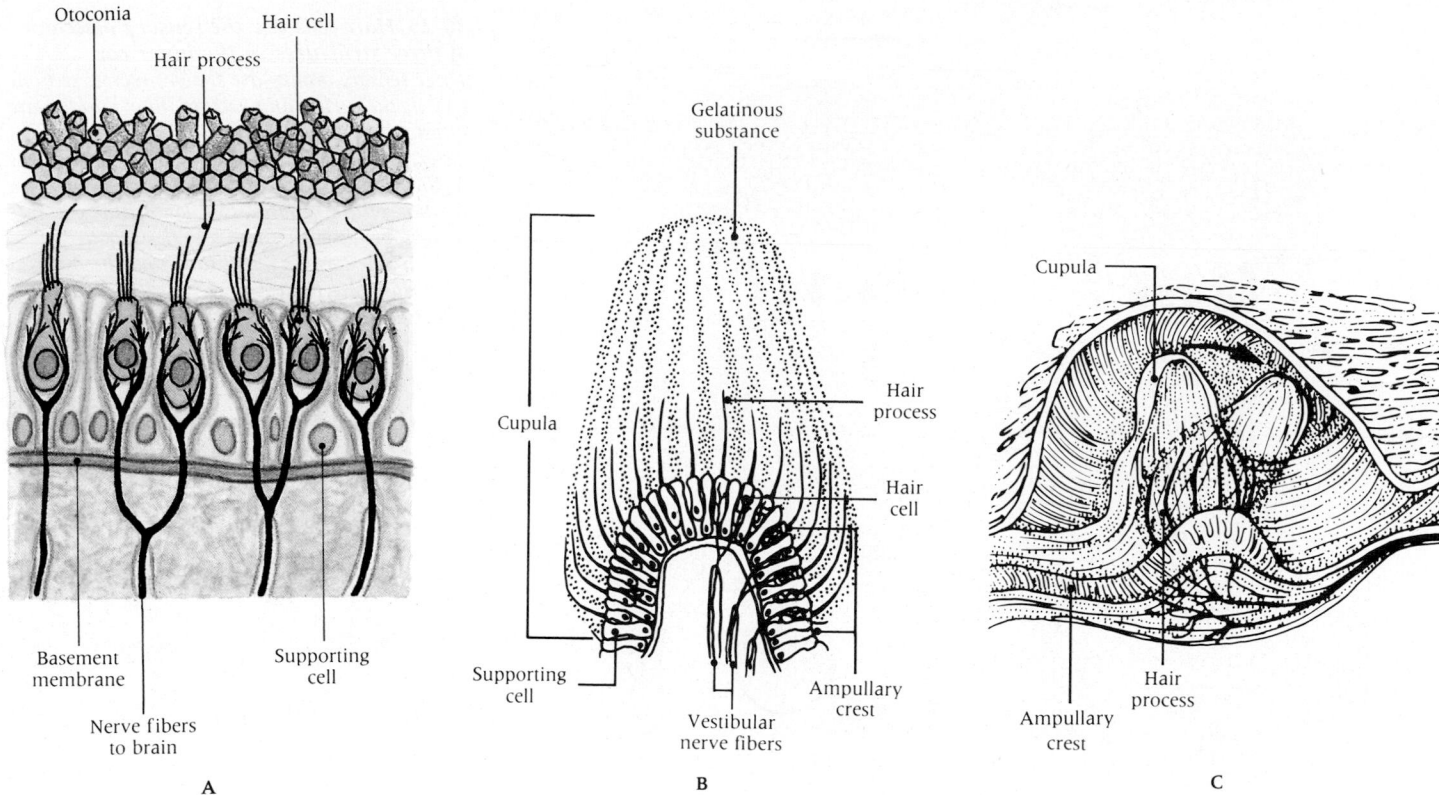

Otoconia   Hair cell

Hair process

Basement
membrane

Supporting
cell

Nerve fibers
to brain

A

Gelatinous
substance

Cupula

Hair
process

Hair
cell

Supporting
cell

Vestibular
nerve fibers

Ampullary
crest

B

Cupula

Hair
process

Ampullary
crest

C

**30-24 Human equilibrium receptors of the
vestibular apparatus**
(A) *The macula of a semicircular canal contains
hair cells as well as otoconia, which float in a
fluid and change position as the head moves.*
(B) *The ampullary crest of a semicircular canal
and its gelatinous cupula. **(C)** Response of the
cupula to movement of the surrounding fluid
(endolymph). In all of the structures, bending
of a hair process generates nerve impulses that
give information about the body's position.*

■ If the head rests on the shoulder in such a way that the anterior and posterior
canals lie in the plane of rotation, the nystagmus is vertical, the eyes moving
up and down.

Motion sickness is thought to be due to the continuous stimulation of parts of
these canals.

Much contemporary research concerns the equilibrium sense in aviation and
space flight. Acceleration is probably the most important of the several physical
stresses imposed upon a spacecraft pilot. Others are heat, vibration, radiation, bends,
weightlessness, and noise (see Chapter 27). Aside from its direct effects upon the
body at large, acceleration can cause sensory delusions. For example, a pilot deprived
of visual data may have false sensations of climbing or falling during acceleration
and deceleration.

CHEMORECEPTORS

All protoplasm is sensitive to chemical changes in the surroundings, but sensory
receptors evolved having greatly enhanced chemical sensitivity. These structures
are called **chemoreceptors.**

Even without such specialized chemoreceptors many protists and simple animals
can detect the presence of food or toxic substances through a general sort of chemical
sensitivity. Target molecules or ions diffuse through the water and act as stimuli.
The result is appropriate behavior. Harmful chemicals lead to defensive responses.
Nutrients stimulate "hot pursuit."

Most aquatic invertebrates are sensitive throughout their bodies to chemicals.
Many also have simple local chemoreceptors. Clams and other marine mollusks
have little patches of yellow cells that "taste" the water entering the gills. The
most elaborate chemoreceptors among the invertebrates are found in insects. There
are often small receptors on the antennae, and some insects even taste things with
their legs.

In our own bodies, *taste* and *smell* are the two major chemical senses. They are
called chemical senses because their primary stimuli are chemical substances in
the mouth and nose. Humans and other animals also have what we might call
the "common" chemical sense. This is a primitive sensitivity to *chemical irritation,*

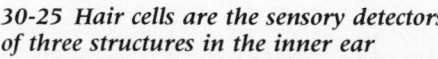

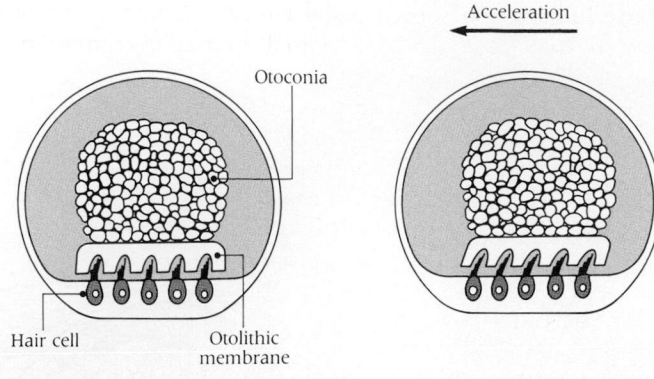

A

Acceleration

Otoconia

Hair cell

Otolithic membrane

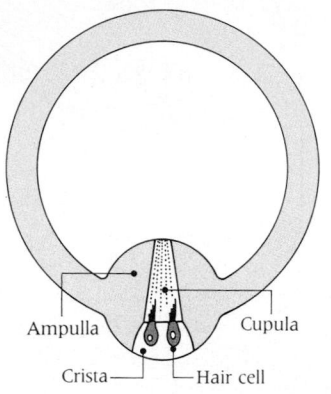

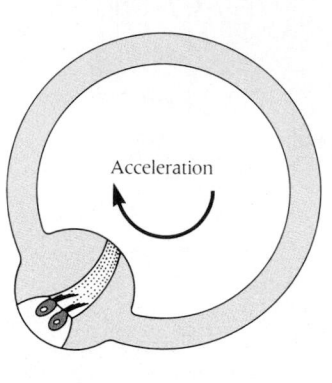

Acceleration

Ampulla

Crista

Cupula

Hair cell

B

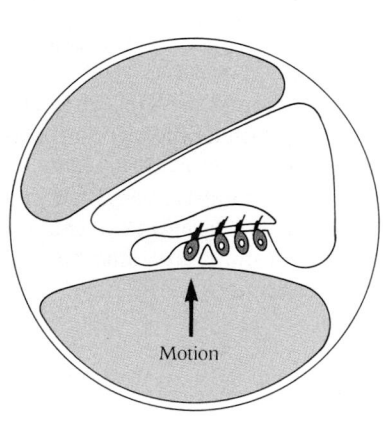

Vestibular membrane

Tectorial membrane

Vestibular canal

Middle canal

Tympanic canal

Basilar membrane

Organ of corti

Motion

C

30-25 Hair cells are the sensory detectors of three structures in the inner ear
Three sensory organs are shown in cross section. (A) In the utricle and saccule, hair bundles are inserted into the otolithic membrane, which is distorted by otoconia during acceleration. (B) In the semicircular canals, fluid moved by rotation of the head presses against the cupula, thereby bending the hair bundle. (C) In the cochlea, the basilar membrane and the tectorial membrane are hinged so that a deflection of the stapes bone bends the hair bundles with a shearing motion.

the kind caused by ammonia, acid fumes, and tear gas. Its receptors are present on large areas of the skin of many animals.

The senses of taste and smell have specialized receptors that differ from the common chemical receptors and certain other chemoreceptors within the body (e.g., those in the aortic and carotid bodies that respond to the blood oxygen level) in that they evoke conscious sensations.

The Sense of Taste

The word *taste* has had different meanings at different times. Originally, in most European languages it meant only *to sense*—to sense anything. Then it acquired the connotation *to appreciate*. In the biological context, some ambiguity remains because the term *taste sense* is often used interchangeably with the gustatory sense.

What we call taste in common speech really concerns *flavor,* a complex sensory pattern that combines gustatory, touch, thermal, and smell sensations. Here we are concerned primarily with the gustatory sense.

When smell is compromised by obstruction of the nasal passages, four primary gustatory sensations can be demonstrated: sweet, sour, salty, and bitter. The hundreds of tastes we perceive everyday are but combinations of these four sensations. We have become so accustomed to thinking in terms of these four basic tastes that we have tended to assume the same is true for other animals. Probably that is an error. Although fishes cannot smell or taste in the usual sense, they do have extremely sensitive and highly discriminating chemoreceptors for substances in the surrounding water. That is how spawning salmon recognize the chemical individuality of the one stream in which they were hatched.

The **taste buds** are small organs in the tongue which contain the chemoreceptors of the gustatory sense (Figure 30-26). The taste buds are located in small, rounded elevations called **papillae,** which form a V-shaped row at the back of the tongue and sparsely dot the rest of the tongue. Many also occur over the palate and the tonsillar region. A human adult has about 10,000 taste buds; after the age of 45 the number gradually decreases.

A taste bud is sensitive to many stimuli. Certain areas of the tongue react more strongly to one primary taste stimulus than to the others. The back of the tongue responds to bitter substances (for example, quinine and alkaloids); the edges of the tongue respond to sour substances (all acids) and salt (NaCl and KCl); and the tip of the tongue responds to sweet substances (sugars and alcohols). The center of the tongue contains few taste receptors.

To stimulate taste receptors, chemical substances must be present at a minimum concentration referred to as the *threshold* concentration. Threshold levels vary for different chemicals: HCl, about $0.0009M$ for stimulation of the sour taste; NaCl, about $0.02M$ for stimulation of the salt taste; sucrose, about $0.01M$ for stimulation of the sweet taste; and quinine, about $0.000008M$ for stimulation of the bitter taste. Since the bitter taste protects the body against many harmful agents, it is logical that it is the most sensitive of the four primary sensations.

30-26 Human taste receptors
The upper surface of the tongue is divided into areas responding to bitter, sour, salty, and sweet stimuli. Papillae at the back of the tongue contain some of the taste buds, one of which is shown in longitudinal section. Different types of papillae are found on the body of the tongue. Receptor cells respond to chemicals at the pore and send nerve impulses to the brain via synapses at the base of each cell.

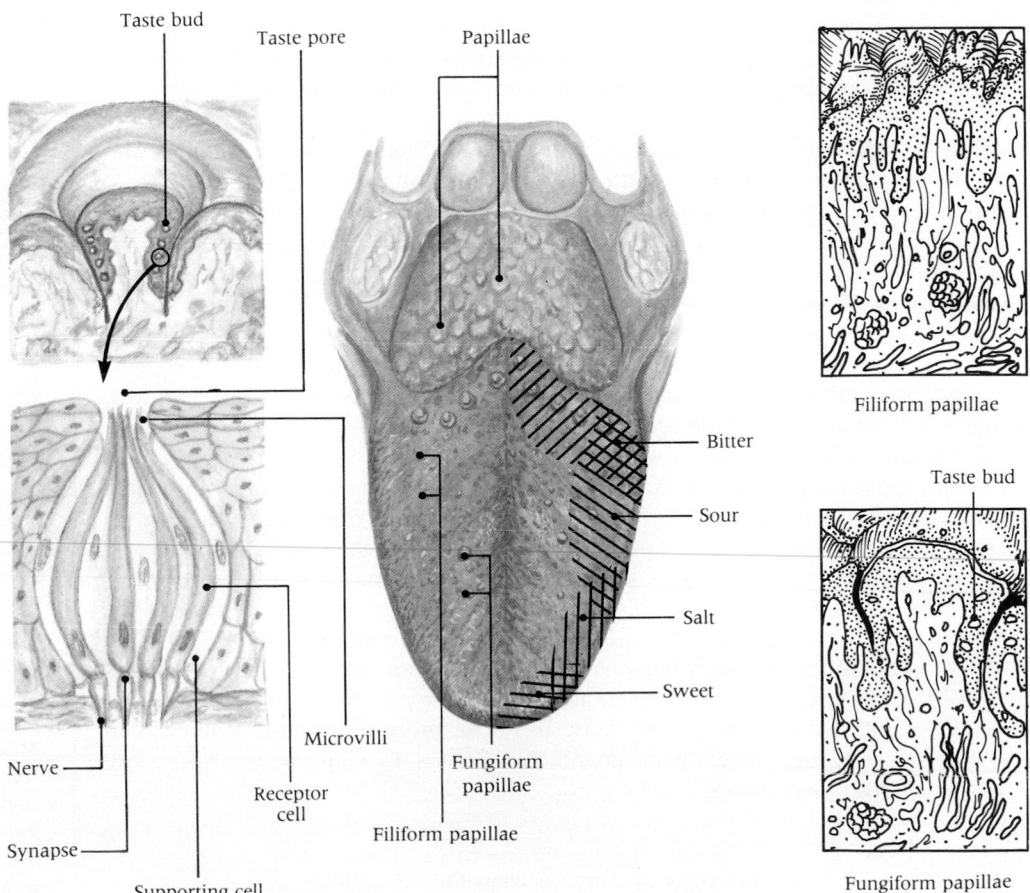

The ability to taste some substances is under genetic control. For example, phenyl-thiocarbamide tastes very bitter to about 65% of a tested population, but it is tasteless to the remainder.

The Sense of Smell

The sense of smell, the **olfactory sense,** plays so small a part in our own lives, relatively speaking, that we are inclined to underestimate its importance in other vertebrates. But to most fishes, amphibians, and reptiles, and to many mammals, this sense is of supreme importance. This is well known to dog fanciers. Dogs can see, hear, and feel very well, but for them smell is the main source of really reliable and meaningful information. They share this with most other vertebrates. Humans and most birds are among the conspicuous exceptions. Clearly dogs smell things that we do not.

Nostrils appeared first among early fishlike vertebrates as inlets through which water could reach specialized chemoreceptors. When air-breathing animals evolved, this chemical sense was retained in the same place and became sensitive to molecules in the air. The ability to use the nose for breathing, an incidental result, is still usually unnecessary. Most air-breathers can and do breathe through the mouth.

Even humans can detect remarkably small concentrations of some molecules in the air and can distinguish a very large number of smells. To have an odor, a substance must be volatile at ordinary temperatures and soluble in water or organic solvents. All known odorous substances are either gases, solids, or liquids with high vapor pressures. Most inorganic substances have low vapor pressures and hence no discernible odors.

The sense of smell is unusually acute in many organisms. The female silkworm moth, *Bombyx mori,* announces her presence by discharging into the air a substance called *bombykol.* This species of moth cannot fly, so the male cannot go very far in search of a mate. It does, however, have two remarkable antennae bearing olfactory hairs capable of detecting a single molecule of female sex attractant (Figure 30-27). A male can readily determine whether a female is near or far away—and then act. Such chemical compounds, which are secreted by an animal and elicit a specific behavior pattern, are called **pheromones** (Chapter 33).

Anatomically, odor-sensitive cells are simple structures (Figure 30-28). They occur in the **olfactory epithelium,** which in humans is an area of about a square inch in the upper part of the nose.[8] The olfactory receptor cells in the olfactory membrane are primitive neurons. Like the rods and cones of the retina, they are derived from the central nervous system. The olfactory receptors project into the nasal cavity where their exposed ends form knots bearing tufts of cilia called **olfactory hairs.** Their axons synapse with neurons in the olfactory bulb.

Investigation of the mechanisms of smell and taste has been a difficult challenge. The chemical substance being smelled is believed to alter the permeability of an olfactory receptor's plasma membrane, thereby initiating nerve impulses, perhaps through the intervention of a second messenger like cAMP or cGMP. Similar events follow light stimulation in rod cells and the recent discovery in olfactory tissues of specific G proteins and cAMP-gated (or cGMP-gated) ion channels is highly suggestive. It is provocative to realize how similar are the mechanisms of olfactory receptors, visual receptors, and hormone receptors.

The ability to discriminate smells remains baffling. Although olfactory receptor cells send nerve impulses that lack specificity, their plasma membranes probably contain specific chemically sensitive molecules, or "traps," that trigger (or in some cases inhibit) production of an action potential. Some believe that individual receptor cells respond selectively to various large groups of chemicals (e.g., benzenelike, acetonelike, etc.). If so, odor-specific elements are analogous to frequency-specific elements in the cochlea and color-specific elements in the retina.

There is no simple relation between a perceived odor and the structure of the molecule being smelled. Odor appears to be a property of the entire molecule, its size and shape. Although the smell sense has been more thoroughly studied than the taste sense, it has not yet been possible to discern a group of basic odors that parallels the four basic tastes.

30-27 Antennae of a male silk moth, Bombyx mori
More than half of the olfactory receptor cells in these remarkable antennae are adapted to respond to a single substance, bombykol, the sex attractant emitted by the female silk moth. Each hair on an antenna contains two chemoreceptor cells.

---

[8] Most mammals have much larger olfactory membranes relative to body size.

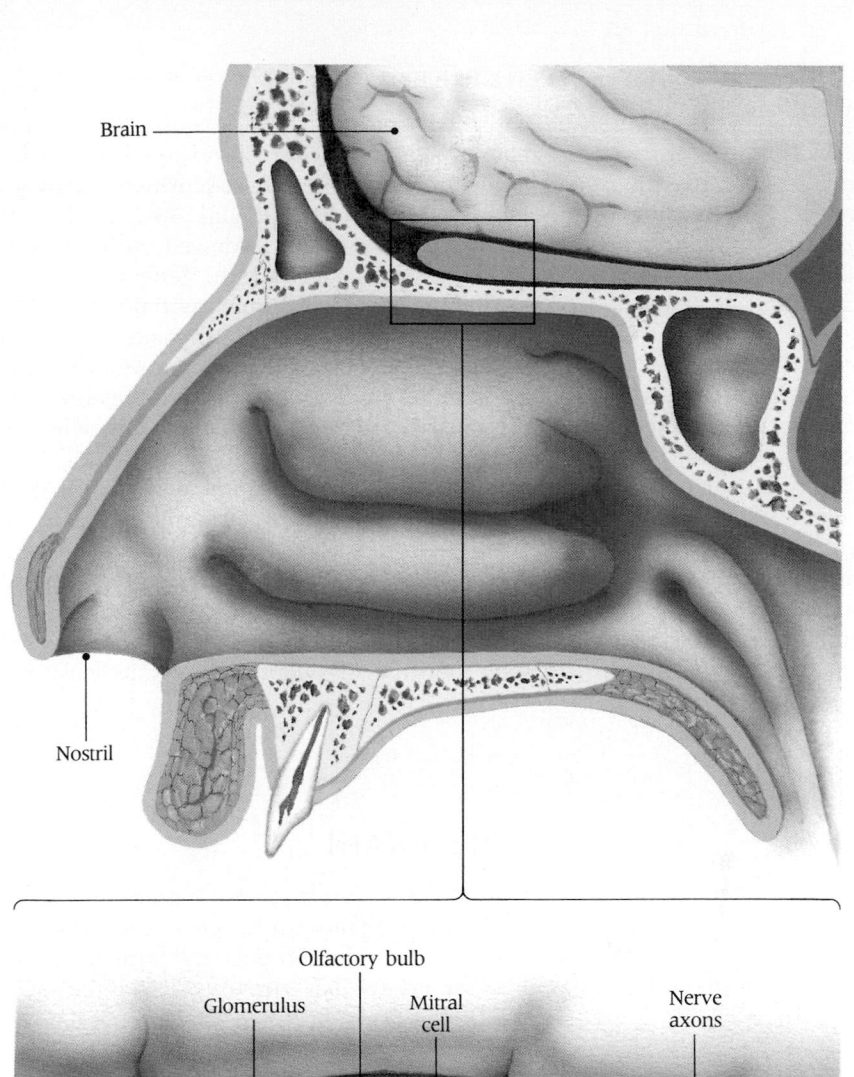

Brain

Nostril

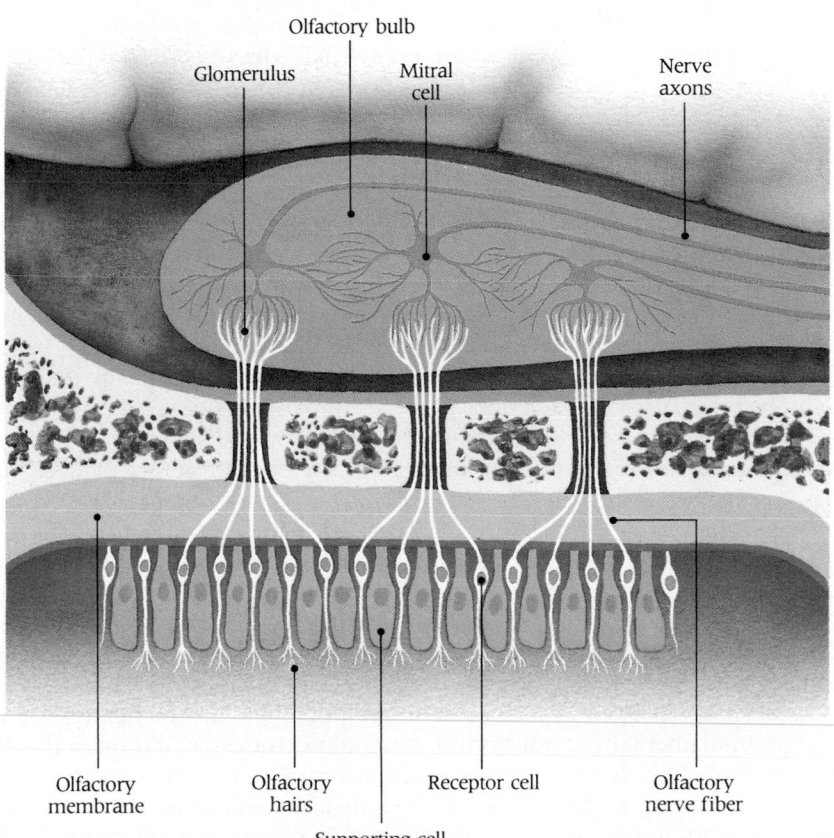

Olfactory bulb

Glomerulus   Mitral
cell

Nerve
axons

Olfactory
membrane

Olfactory
hairs

Receptor cell

Olfactory
nerve fiber

Supporting cell

Despite the biological importance of olfaction—and its practical importance to purveyors of food and perfume—scientific study of this sense has been slowed by unique problems, including a lack of objective methods for measuring the strength and quality of odors and the difficulty of obtaining truly pure test substances.

OTHER RECEPTORS

The sensation of warmth usually depends on simple local receptors. However, there are exceptions to that generalization. Two groups of snakes—pit vipers (subfamiliy Crotalinae) and pythons (subfamily Boidae)—have heat-sensitive **sensory pits** between eyes and nostrils on each side of the head (Figure 30-29). The pits are lined with thin, stretched membranes that are richly endowed with bushy nerve endings highly sensitive to infrared radiation, or warmth, generating volleys of nerve impulses in response to temperature rises as small as 0.003°C. With this capacity, sometimes called the "infrared vision" of snakes, rattlesnakes, water moccasins, and copperheads (the best known pit vipers in this country) can detect a small, motionless warm-blooded animal several feet away even in the dark. Military engineers have tried to develop optical instruments for night vision with even a fraction of this efficiency.

Many insects can sense and respond to changes in humidity, but it is doubtful whether this represents a special sense. In these insects, humidity sense organs occur either on antennae or in bristles on the back. These organs probably exploit sensitivities to temperature and pressure. Since the rate of water evaporation from the antennae varies inversely with the moisture content of air, humidity could well be registered by temperature receptors cooled by evaporation. Bending of the bristles on the back with changing moisture may flex the pressure-sensitive skin to which they are attached.

Pits

30-29 The pit organ is a thermoreceptor in many snakes
Sensory pits in the head of this reticulated python contain infrared-sensitive receptors which send nerve impulses to the midbrain, thus improving nocturnal detection of warm-blooded prey.

THE BRAIN

We conclude our survey of nervous systems with that most extraordinary product of evolution, the **brain.** It is best discussed against an historical background. How did the long and eventful chronology of evolution produce the human brain—the only thing in the universe (as far as we know) that tries to understand itself?

EVOLUTION OF THE NERVOUS SYSTEM

We have already noted some trends in the evolution of nerves and receptors, which are components of the nervous system. Let us now look at the history of the nervous system as a whole.

The only direct approach to such a history is an examination of fossil remains of organisms from the distant past. But for studies of the nervous system that approach is grossly inadequate. Nerve tissue is almost never preserved in fossils. Fossil vertebrates, however, do give useful data about one part of the nervous system, the brain, because the brains of these animals were encased by bony skulls. Bone is preserved in fossils—and skull shapes hint at the structure of brains they once contained. Still, the evolution of the brain from early fishes to modern human beings is only part of the story.

In such dilemmas a historian of life must fall back on present-day species, in which judicious comparisons of nervous systems can provide some evidence about their evolutionary history. A clam of a few hundred million years ago probably had a nervous system like that of a contemporary clam. By comparing animals that have changed little with animals that have changed more, one can make some reasonable inferences about their history. There are obvious risks in this method. All animals have changed in the course of evolution, and any comparison of living animals is fundamentally nonhistorical. It should be obvious (although this fact is sometimes forgotten) that no living animal is ancestral to any other (in an evolutionary sense). Moreover, no living animal is exactly like the ancestor of another. Above all, it is necessary to bear in mind that many sorts of ancestral animals have disappeared altogether and were not like any animal now alive.

Consider the nervous systems of vertebrates and insects. Vertebrates have more complex nervous systems than invertebrates, and they evolved from invertebrates. Insects have perhaps the most complex nervous systems among living invertebrates. It is therefore tempting to assume that the insect nervous system represents a stage that came directly before that of the vertebrates. However, that assumption is not warranted.

The ancestries of insects and of vertebrates diverged well over 500 million years ago, long before insects or vertebrates had themselves emerged. The common ancestors of both insects and vertebrates probably did have nervous systems. But these nervous systems were very simple and we cannot learn much about them from living insects. Unfortunately, no living animal is likely to resemble that common ancestor very closely, and we can visualize its nervous system only in general terms.

In any discussion of the evolution of anatomical structures or physiological processes, we must always keep in mind that comparative studies of *living* animals are not directly historical. They provide evidence that must be interpreted carefully and in the light of directly historical evidence.

Primitive Forms

It is likely that the nervous systems of present-day primitive animals resemble the nervous systems of forms that arose early in the course of evolution. We believe that the rising hierarchy of models in Figure 29-15 essentially recapitulates the early evolution of the nervous system. Some of the evidence tending to support that belief is the following.

The single-celled protist conducts impulses of a sort, since stimulation of the cell membrane at any point generates an action potential and may be followed by a response at some other point.

In sponges, the only major group of multicellular animals without nerves,[9] the protoplasm of each cell shows a similar capacity to conduct impulses within that cell. In most ways, each cell operates independently of the others, but a stimulus to one cell can be transmitted to the other cells so that a response is evoked in more distant cells. It is probable that some such stage as this existed among the earliest multicellular ancestors of other animals. Increasing cell specialization and tissue differentiation would permit the recruitment of intermediate cells as conductors. These would become primitive nerve cells.

Next simplest among major groups of living animals are the coelenterates (or cnidaria)—the corals, jellyfishes, and relatives—including the little freshwater hydra of pond water and biology classrooms. Present-day coelenterates also did not derive from evolution's major lines, but they are nearer to us than sponges, and it is likely that more complex nervous systems passed through an evolutionary stage like theirs. They have **nerve nets** composed of neurons, all nearly alike, all with several fibers of equal length (Figure 30-30A), all spread evenly in the body's outer layers. The short fibers synapse with fibers of adjacent neurons, roughly arrayed in a circle around each neuron.

Echinoderms, which include starfishes, sea urchins, and sea cucumbers, their associates, still have nerve nets, but of increasing complexity. Starfishes (Figure 30-30B) have three distinct nerve nets. One just under the thin skin has a circumoral ring and five sets of nerve cords running out to the "arms." Another serves the muscles between the skin plates called ossicles (see Chapter 42). The third connects to the tube feet. This degree of complexity permits locomotion, a variety of useful reflexes, and a degree of "central" coordination. For example, a starfish flipped on its back will right itself.

In nerve nets, the main response to stimulation is local contraction. A stimulus anywhere spreads through the whole net, and impulse conduction by nerve fibers is in either direction. Stimuli must be of sufficient strength and duration, however, to reach distant parts of the animal because fibers are short and many synapses must be crossed. Thus conduction is slow. Such nervous systems can mount and coordinate simple responses (see Box 32-1)—mostly contractions of the whole body or parts of it—but that is about it. They do permit simple reflexes, for there are scattered sensory cells and impulses go from them to effector muscle cells. But a nerve net can neither control nor coordinate complex reactions.

[9] We will discuss the major animal groups (phyla) in Chapters 42 and 43 (and they are presented in tabular form in the Appendix). Although they have not yet been formally introduced, we mention them here in general terms—more or less in a sequence of increasing complexity.

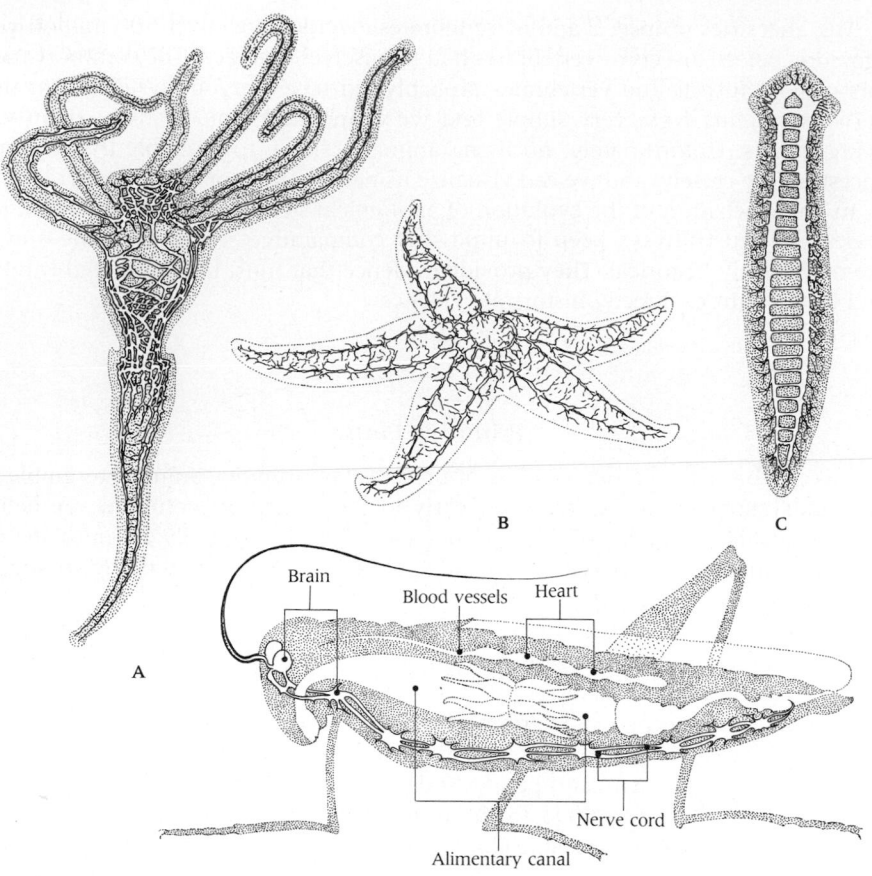

30-30 Nervous systems in different phyla
(**A**) In coelenterates, the nervous system is a simple network formed by the connected branches of individual neurons. (**B**) The starfish, an echinoderm, has a more complex nerve net with a ring and nerve cords extending into the arms. (**C**) In flatworms like planaria, two nerve cords extend the length of the animal, ending in the head as a primitive brain. (**D**) The arthropod's nervous system is represented by a double nerve cord and a brain containing ganglia surrounding the esophagus.

Evolutionary Trends in Invertebrates

Free-living flatworms, such as planarians (Figure 30-30C), show in simple form many of the fundamental features of more complex animals.

There is still an outer nerve net. Indeed, as just implied, this simple type of nervous structure may persist or reappear in all sorts of animals. There is, for example, a nerve net in the wall of the human intestine that controls its rhythmic contractions. Planarians also have the rudiments of a central nervous system. This includes a series of **nerve cords** and an enlargement of them that may be called a **brain,** although it is so rudimentary it barely deserves the name. The nerve cords are the main lines of conduction, each containing many neurons. The neurons have long fibers along the cord and impulses pass along the fibers in one direction only. From cell bodies in the nerve cords, fibers also connect with the nerve net and with other cells of the body.

Planarians have a definite front end. Here sensory cells (which also occur all over the body) are especially numerous, and here also are the eyes. The eyes are well-developed light receptor organs, sensitive to light intensity and direction, although they have no lenses and do not form an image. The nerve cords converge at the front end of the body and merge with an enlarged mass of nerve tissue, which as a courtesy we have called the brain. The eye and many sensory cells connect directly with this primitive brain, which contains associative cells and a fairly complex arrangement of synapses between the sensory nerve fibers and other fibers from the nerve cords.

This evolutionary development, known as **cephalization,** was important in the development of the nervous system. The front end of the organism, with several sense receptors concentrated there in close association with a central brain, became the focal point of more advanced nervous systems.

Simple as it is, the planarian nervous system permits control and coordination of particular responses throughout the body. It mediates between sense organs and effectors, and it permits association and some variability of responses beyond simple reflex actions. In a sense, the rest of the evolution of the nervous system

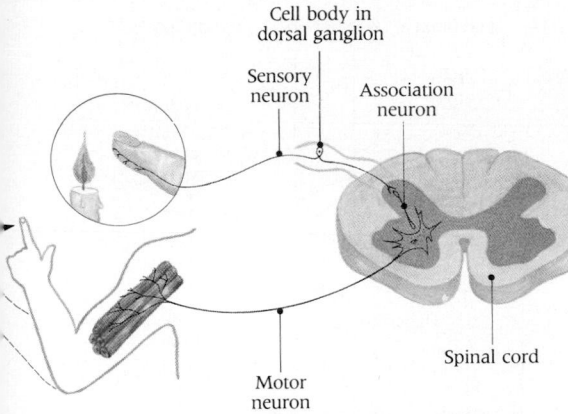

30-31 A simple reflex arc
A sensory neuron connected to the pain receptor in the finger is activated by the heat of a flame and sends impulses to an association neuron in the spinal cord. This neuron synapses directly with a motor neuron, which carries nerve impulses back to the finger, which contract muscles and withdraw the finger. No thinking or other higher brain activity is necessary.

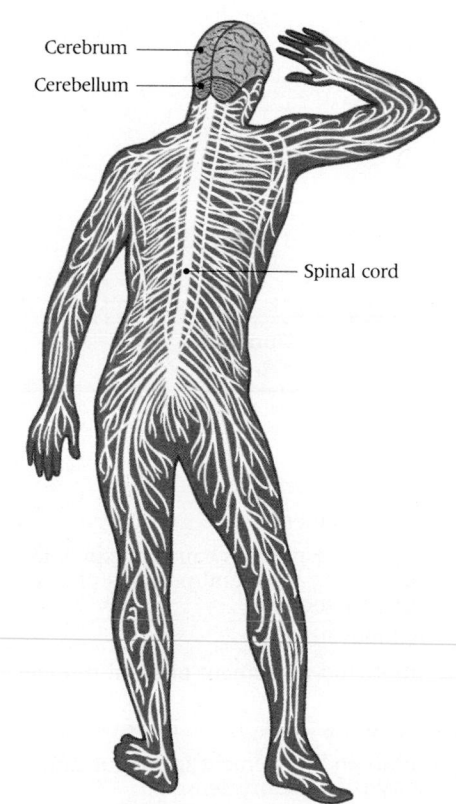

30-32 The human nervous system
The central nervous system in humans and other vertebrates consists of the brain, which contains the cerebrum and cerebellum and other structures; the spinal cord; and the peripheral nerves.

can be viewed as an elaboration of characteristics already present in a planarian. Note, for example, that the nervous system in the insect, an arthropod (Figure 30-30D), consists of a chain of segmental ganglia on the ventral (bottom) side.

Among the important trends already suggested in planarians and arthropods are the following:

■ Formation and concentration of a **central nervous system** occurs. Most cell bodies of neurons come to be concentrated in one or a few nerve cords, later the **spinal cord,** or in masses near the cords called **ganglia.** Connections from here to other parts of the body are by an elaborate **peripheral nervous system.** The peripheral system is mainly composed of nerves, bundles of long nerve fibers from the cell bodies in the central nervous system.

■ Afferent and efferent fibers and nerves in the peripheral nervous system become differentiated. Most nerve impulses are routed through the central nervous system and are carried in one direction only. Afferent nerves (''carrying to'') bring impulses to the central nervous system; those conducting impulses from it are called efferent nerves (''carrying away''). **Sensory nerves** are afferent; **motor nerves** are efferent. Because of this arrangement even the simplest **reflex arc** (Figure 30-31) passes through two or more neurons and is more flexible than the primitive one-neuron reflex.

■ There is an increased complexity of association among various parts of the nervous system. Accompanying the afferent-efferent system is an increase in associative neurons throughout the central nervous system and an increase in the number and complexity of nerve routes and connections. The central nervous system is, in essence, a complex of association neurons. Lying between receptors and effectors, it evidently derived from the primitive nerve net.

■ Development and complication of a brain (cephalization) takes place. The front end of the nerve cord (or cords) becomes enlarged, principally by the development of large numbers of associative neurons and tracts. Eventually most nervous impulses in the body are routed through this associative mass. Central coordination for complex responses is thereby provided.

■ There is an increase in number, complexity, and sensitivity of special distance receptors, or other sense organs. The complex sensory organs that developed at the front end of the organism—in the head—progressively became connected into the brain. But, at one time in their evolutionary history, these reflex centers consisted of massive collections of cell bodies, or ganglia. These ''eye brains,'' ''nose brains,'' and ''ear brains'' supplemented more primitive (older) ganglia representing the ''visceral brain.'' Because of the linear arrangement and segmental distribution of these older master ganglia, their part of the brain is generally called the ''brainstem.'' Because these master ganglia were well developed early in the course of evolution, it is also sometimes termed the ''old brain.''

As we learned in Chapter 19, the central nervous system of all vertebrates consists of a single, hollow **nerve cord** that runs along the back (or dorsal part) of the body. Such an anatomical arrangement is not found in present-day invertebrates, none of which retains those features of the ancient invertebrates that gave rise to vertebrates in evolution.[10] The brain of most modern invertebrates is a ring of nerve tissue surrounding the esophageal part of the alimentary canal (see Figure 30-30D). It consists of two ganglia that lie above the esophagus and are connected to two others lying below. Two solid nerve cords leave the subesophageal ganglia and extend backward along the length of the body under (or ventral to) the gut. The anterior expansion of the nerve cord into the brain is the principal feature linking the central nervous systems of the more complex living invertebrates with those of the vertebrates, exemplified by the human nervous system shown in Figure 30-32.

[10] A single, hollow, dorsal nerve cord does occur in some chordates that are not invertebrates, e.g., *Branchiostoma* (amphioxus). When we say ''invertebrate'' we almost always mean nonchordate and not simply nonvertebrate. This distinction is further discussed in Chapter 43.

TABLE 30-2
MAIN DIVISIONS OF THE VERTEBRATE BRAIN

Primary	Secondary	Adult
Forebrain	Telencephalon	Olfactory bulbs; cerebrum (cerebral hemispheres)
	Diencephalon	Thalamus; hypothalamus
Midbrain	Mesencephalon	Tectum; tegmentum; cerebral peduncles
Hindbrain	Metencephalon	Cerebellum; pons
	Myelencephalon	Medulla oblongata

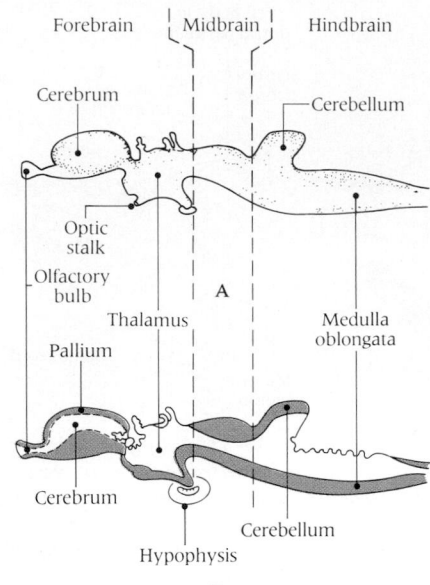

THE VERTEBRATE BRAIN

The evolution of the vertebrate brain is unquestionably the most important single factor underlying the success of the vertebrates. The most primitive vertebrate brain consisted of three irregular swellings of the hollow nerve cord, each with various thickenings of the walls. The three enlargements, the **forebrain, midbrain,** and **hindbrain** (Table 30-2, Figure 30-33), can still be distinguished in the human brain, even with its vastly greater complexity.[11] The secondary divisions of the brain are easily distinguished in the mammalian brain (Table 30-2), particularly during developmental stages of growth. Very early, even in primitive fishes, further subdivisions occurred in the structure and function of the brain.

- The forebrain became divided into three parts: (1) the **thalamus, hypothalamus,** and associated structures; (2) the cerebral hemispheres (or, taken together, the **cerebrum**), a pair of swellings farther forward and higher; and (3) the **olfactory bulbs,** which project as swellings from the lower front of each cerebral hemisphere.
- The midbrain developed swellings on its upper dorsal region, which became the **optic lobes,** centers associated with the optic nerves.

30-33 Parts of the vertebrate brain
(A) The forebrain, midbrain, and hindbrain are the three main divisions of the vertebrate brain. Each area is divided further into structurally and functionally distinct regions. **(B)** A longitudinal section showing the striking local differences in the thickness of the brain wall.

[11] There are, of course, many more parts than are named in Table 30-2. We have here used only the few terms necessary to an understanding of the brain's general functions and significance.

TABLE 30-3
THE TWELVE CRANIAL NERVES*

Nerve Number	Name	Function	Structure Innervated	Functions Served
I	Olfactory	Sensory	Olfactory epithelium	Olfactory sense
II	Optic	Sensory	Eye	Visual sense
III	Oculomotor	Motor	Four of six eye muscles	Eyeball movements
IV	Trochlear	Motor	One of six eye muscles	Eyeball movements
V	Trigeminal	Sensory and motor	Head and face muscles, lower jaw teeth and jaw muscles	Motion of face muscles causing expression, sensation in teeth and portions of jaw skin, jaw motion
VI	Abducens	Motor	One of six eye muscles	Eyeball movements
VII	Facial	Sensory and motor	Face	Sensation and movement of facial muscles
VIII	Auditory	Sensory	Inner ear	Hearing sense
IX	Glosspharyngeal	Sensory and motor	Tongue and pharynx	Sensation and movement in tongue and pharynx, gustatory sense
X	Vagus	Sensory and motor	Internal organs (heart, lungs, stomach, etc.)	Movement and sensation in heart and visceral organs
XI	Spinal accessory	Motor	Muscles of shoulder	Shoulder movements, gustatory sense
XII	Hypoglossal	Motor	Tongue muscles	Tongue movements, gustatory sense

*The hypoglossal is absent in fishes. In the lower vertebrates, the spinal accessory is incorporated into the vagus. Students of anatomy have for generations memorized the cranial nerves by remembering the first letters of the following couplet (or variations thereof): "On Old Olympus's Towering Tops, A Finn And German Viewed Some Hops."

- A large swelling developed on the forward, upper part of the hindbrain, which became the **cerebellum,** and the thickened lower wall of the hindbrain became the **medulla oblongata** (Figure 30-33B).

All parts of the brain connect directly (or through chains of neurons) with the spinal cord through the medulla oblongata, which merges with the spinal cord without an abrupt division. The brain also has a series of twelve paired (and numbered) nerves of its own (ten in some early fishes). These are the **cranial nerves,** sensory and motor nerves that connect the brain directly with some sense organs and muscles (Table 30-3).

- The forebrain connects with olfactory receptors in the nose (nerve I).
- The midbrain connects with the photoreceptors in the eyes (II) and, by two separate pairs of nerves, with the eye muscles (III and IV).
- The hindbrain has a series of nerves that connect with the more scattered receptors and with the muscles of the head (V through XII).

The main features of the vertebrate brain were already present, in rudimentary form, in the jawless fishes that are the earliest known vertebrates. Many important changes in details have occurred. The most striking later developments involved the forebrain, especially the cerebrum. As shown in Figure 30-34A, the cerebrum was at first outweighed by the rest of the brain. It was only a pair of small, smooth swellings involved mainly in smell associations. Even the most progressive fishes and amphibians have small, smooth cerebrums that are primarily olfactory. They are "smell brains." Their outer walls were a **pallium** of neurons (Figure 30-34B).

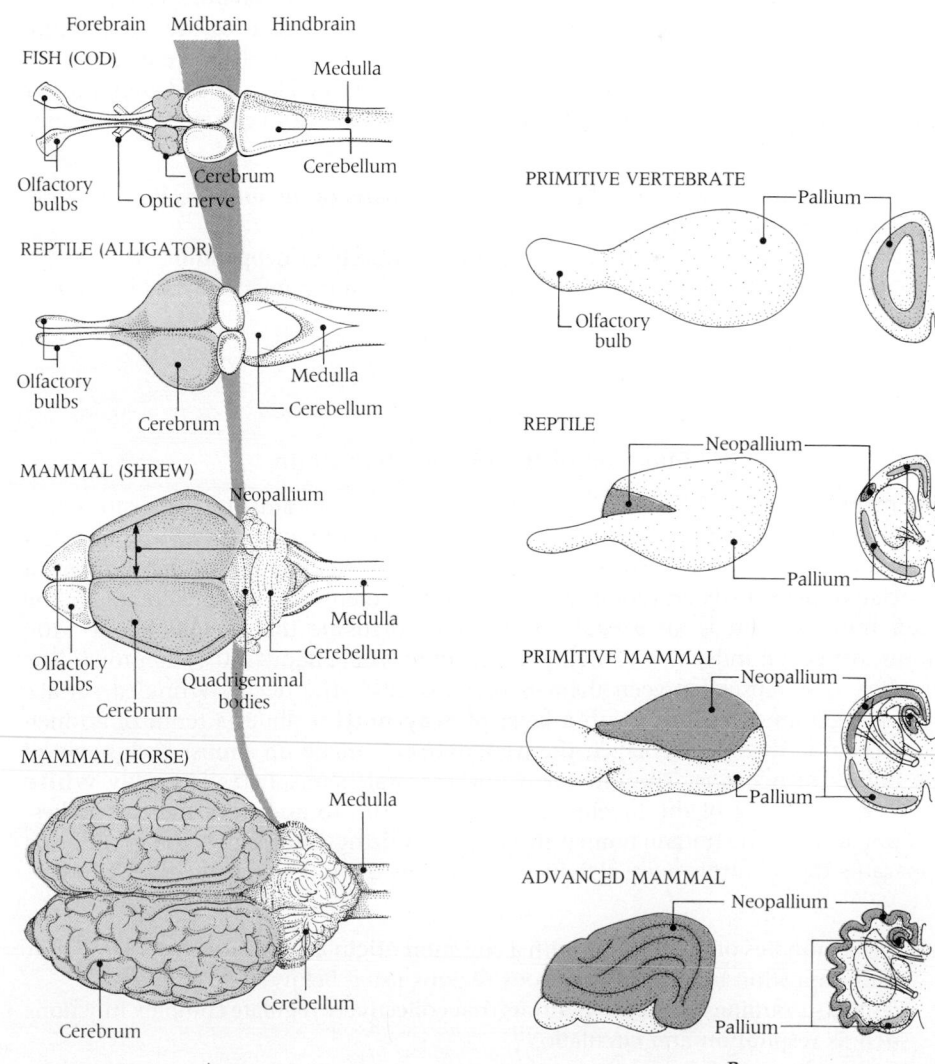

30-34 Evolution of the vertebrate brain
(A) *The brains of four classes of vertebrates are seen from above, showing the progressive increase in size of the forebrain and relative decrease in the size of the midbrain.* **(B)** *The neopallium evolved from a localized region of the brain in reptiles to the highly convoluted cerebral layer in advanced mammals.*

In early reptiles, the cerebrum, had enlarged. Most of it was still concerned with smell, but at the forward, upper part there appeared a new structure that associated and coordinated incoming impulses from other receptors and brain centers, and not primarily from olfactory receptors. This new part of the brain—new in the evolutionary sense—was the **neopallium,** which means "new cloak" (Figure 30-34B). In fact, it was a new sort of covering of gray matter on the cerebrum. (The older part of the brain is termed the pallium or archepallium, which means "old cloak.") The gray matter covering the cerebrum is the **cerebral cortex.** The neopallial part is called the **neocortex.**

Birds and mammals evolved separately from reptiles. In both of these groups, the brain became larger and the cerebrum became the largest part of the brain. But the expansion of the cerebrum in birds is almost entirely in the region near the base of the brain, which remains relatively small in all other vertebrates. This peculiarity is unique to birds, although its significance is not fully understood. It is of interest that the part of the brain where humans form more complex associative patterns—where, indeed, we think—is practically absent in birds. The epithet "bird-brain" thus has some justification. On the other hand, birds perform complex un-learned acts in courtship, nest building, and the like, and humans are wanting in brain areas that are well developed in birds. So a bird might be justified in denouncing a clumsy mate as a "human-brain."

In primitive mammals, the smell brain was not diminished. It remained large, as in reptiles; it even increased in size. Then, in a highly significant evolutionary development, the neopallium became separated by a furrow from the rest of the cortex and greatly expanded in most mammals, as did the number of its associative neurons. The increase came about in two ways: by actual expansion of the whole upper part of the cerebrum, and by the folding or convolution of its surface. A convoluted hemisphere has much more surface area than a smooth one. It is the surface area that determines the functional capacities of the cerebral cortex.

Increase in relative size of the cerebrum, in relative extent of the neopallium, and in its surface by convolution were trends in a number of different groups of mammals. But these trends in our own ancestors continued longer and went further than in any other group. Indeed, three trends were carried to extremes in humans.

- The cerebrum expanded right over the other parts of the brain so that nothing but cerebrum is visible from above.
- The cerebral surface came to consist almost entirely of neopallium.
- Only a small bit of smell brain remained visible in the middle of the bottom side of the cerebrum.

Function of the Mammalian Brain

The mammalian brain is primarily an associative and coordinating center for nerve impulses. It receives impulses from sensory receptors, organizes them, and transfers (or initiates) impulses to various effectors. In the vertebrate brain, the number of associative neurons is enormous and their arrangement is extremely complex. The adult human brain is an irregular solid globe of tissue that weighs about 1400 grams (three pounds) and contains more than 100 billion (10^{11}) neurons. The number of synapses between them is perhaps 10^{15}. The deeply wrinkled surface of the cerebrum is covered with a layer of **gray matter** about a tenth of an inch thick (Figure 30-35). This layer, the **neocortex,** is made up almost exclusively of the bodies of neurons. The interior of the cerebral hemispheres is mostly **white matter,** consisting of the myelinated fibers that run to and from the cell bodies.

Functionally, the human brain is the large organizing, integrating, and monitoring apparatus that controls body activities. It contains the following:

- Tracts, bundles of nerve fibers with a common origin and destination, that permit communication between the various regions listed below
- Centers, groupings of special neurons that collectively regulate complex functions such as respiration and circulation
- Nuclei, clusters of cell bodies, such as those the axons of which constitute the

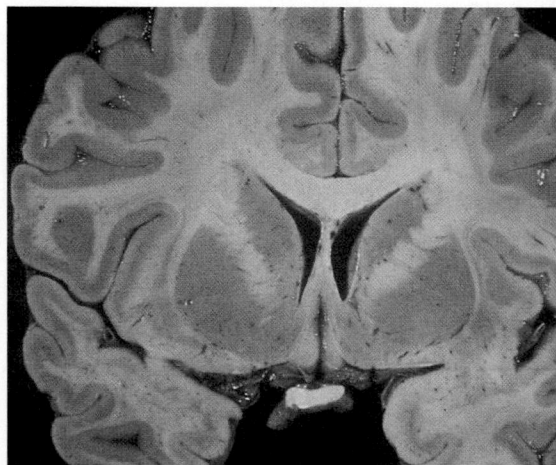

30-35 Slice of a human brain
Note the light colored white matter and darker gray matter.

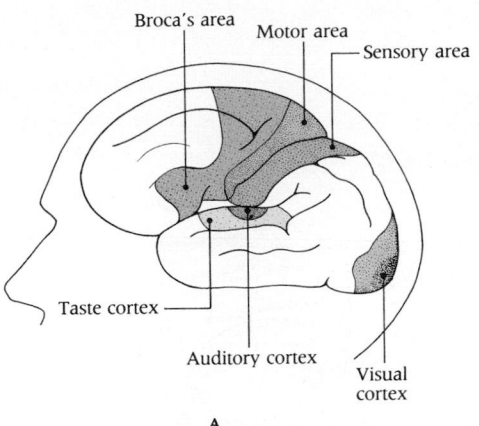

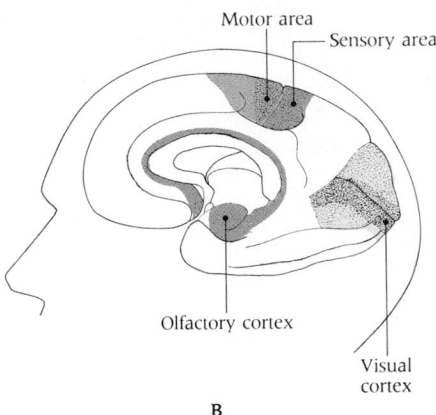

30-36 Functional areas in the human cerebral cortex
(A) *The brain is viewed from the left side, with only the major convolutions of the cortex drawn. The various functional areas are indicated.* (B) *The brain viewed from the middle, looking rightward.*

descending fiber tracts of the spinal cord and the twelve pairs of cranial nerves,[12] the brain's counterpart of the spinal nerves

- Motor areas that determine the body's motor responses to stimuli arriving through the sensory cranial nerves and the ascending fiber tracts of the spinal cord
- Sensory areas that interpret information reaching the conscious level
- Association areas concerned specifically with mental activity, memory, emotion, and learning

In simpler vertebrates, the hindbrain is a message center. Here begin regulatory responses such as automatic adjustments of posture, changes in rate of heartbeat or breathing, and the like. Messages involved in complex or modifiable responses are, however, passed on to parts of the brain farther forward. These relationships of the hindbrain are essentially the same in mammals. This is the part of the brain that changed the least in evolution.

The forebrain in simple vertebrates is also a message center, but a simpler one than the hindbrain. It receives information from a single sense, that of smell, makes preliminary associations, and passes on the organized information to the midbrain. In these animals, the midbrain is the main center of control and coordination. Here final associations are made with data from forebrain and hindbrain, and here the more complex and diversified responses are initiated.

Although a nervous system without a neocortex was adequate to meet the needs of simpler animals, central control later passed almost entirely from the midbrain to new centers in the neopallium of the forebrain. The midbrain then became a smaller, relatively unimportant reflex and secondary message center. Some secondary associations occur here. But most information is passed on to the neopallium.

In mammals, the older parts of the brain continue to relate to basic maintenance functions such as breathing and digestion. The neocortex has taken over and expanded more complex functions, such as coordination of sensory input (vision, hearing, and so on) and initiation of behavior appropriate to a perceived situation. Here, too, are "headquarters" for such functions as inquiry and planning.

We will soon see that mammalian behavior is endlessly varied and generally far more complex than the behavior of other animals. It is largely this complexity—carried to an historic pinnacle—that makes us human. To implement such behavior, even among mammals less complicated than ourselves and adapted in other ways, the neopallium itself had to become highly complex in structure, connections, and functions.

It was learned long ago that specific areas of the cerebral cortex are concerned with specific functions in different regions of the body. The search for cortical areas began as an attempt to relate certain brain areas with particular talents and moral qualities. In 1861, a young French surgeon, Paul Broca, reported the case of an old man who had lost his speech but understood all that was said to him. When the man died, Broca found a lesion in the left third frontal convolution of the cerebrum. Thus was born the dogma that the speech function resides mainly in this area, now known as Broca's area (Figure 30-36).

Experiments on the exposed brains of human beings[13] and other animals showed that electrical or chemical stimulation of certain cortical areas produced discrete and highly localized responses, such as sensations or movements of small muscle groups, without affecting other areas.

These data—and others collected following surgical removal of specific cortical areas—yielded a detailed functional map of the cerebral cortex showing specific motor and sensory areas (see Figure 30-36). There are also many large, ill-defined areas concerned with memory, reasoning, judgment, and other integrative processes.

---

[12] As shown in Table 30-3, some of the cranial nerves are exclusively motor, some are exclusively sensory, and some contain motor and sensory elements. Mixed cranial nerves have separate motor and sensory nuclei in the brain.

[13] There is a fascinating literature on the early experiments in this field. It is said, for example, that two Prussian medical officers, Fritsch and Hitzig, first demonstrated in 1870 that electrical stimulation of the brain causes muscular movements on the opposite side of the body, using the exposed brains of war casualties on the battlefield of Sedan. Another early worker, Bartholow, reported similar reactions in experiments performed in 1874 upon his servant girl, who had a huge scalp ulcer that exposed her brain. After communicating his observation to the medical society, he was promptly expelled for unjustifiable experimentation on a human being.

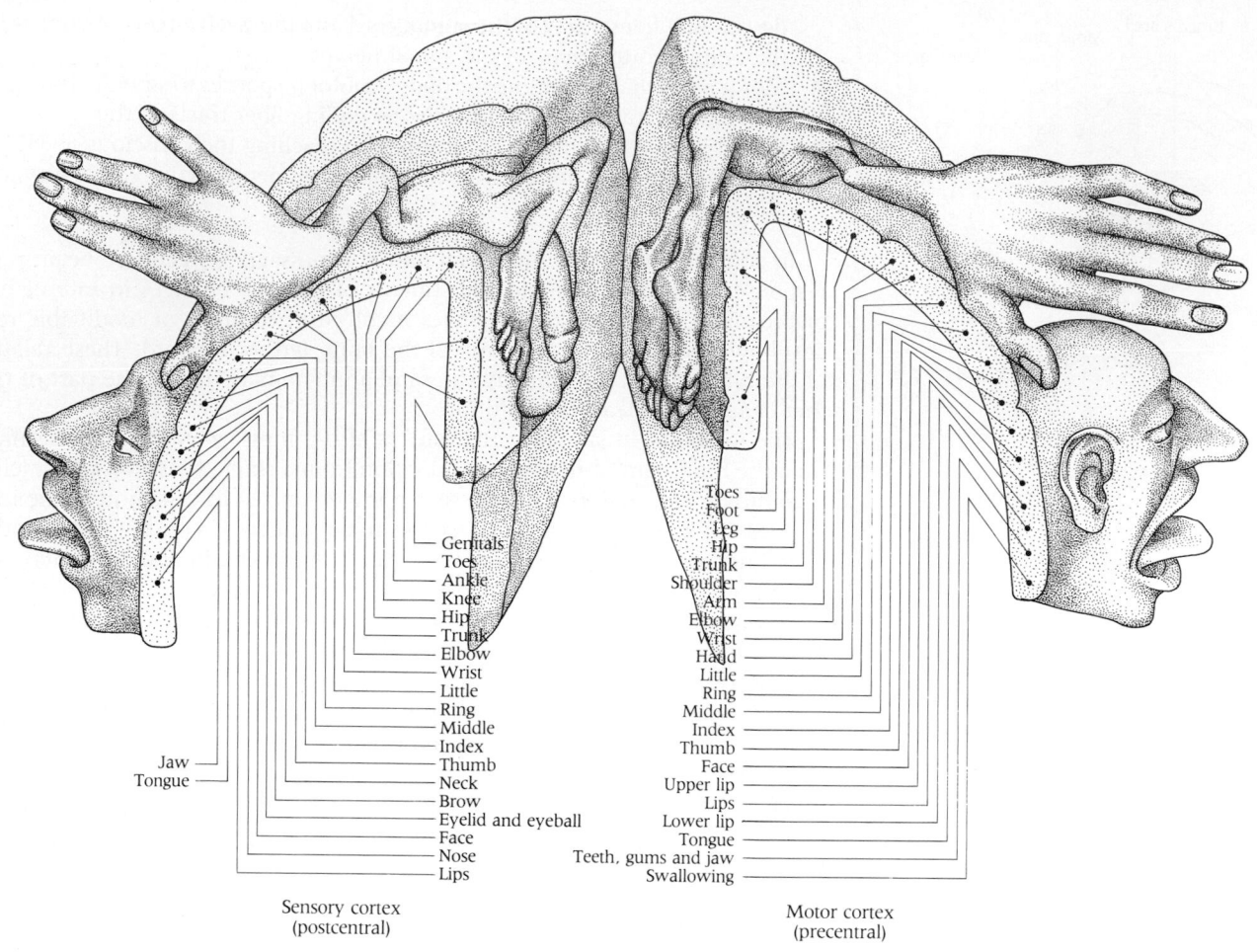

Genitals
Toes
Ankle
Knee
Hip
Trunk
Elbow
Wrist
Little
Ring
Middle
Index
Thumb
Neck
Brow
Eyelid and eyeball
Face
Nose
Lips

Jaw
Tongue

Toes
Foot
Leg
Hip
Trunk
Shoulder
Arm
Elbow
Wrist
Hand
Little
Ring
Middle
Index
Thumb
Face
Upper lip
Lips
Lower lip
Tongue
Teeth, gums and jaw
Swallowing

Sensory cortex
(postcentral)

Motor cortex
(precentral)

These are the **association areas,** the role of which resembles that of the association
neuron. Since this may not be true in all areas, some prefer the noncommittal
term **intrinsic sectors.**

The **motor areas** of the cerebral cortex send messages to muscles all over the
body. Various parts of the body are represented in discrete areas of the motor
cortex, as shown in Figure 30-37. The body is distorted to emphasize the relative
importance of each cortical area for a part of the body. The left side of the brain
represents the right side of the body. In general the lower parts of the body connect
with cortical areas along the brain's midline. Areas corresponding with the upper
part of the body are farther out. Those for the head are low on the sides of the
cerebrum.

The **sensory areas** immediately behind the motor areas receive information
concerning touch, pressure, temperature, position, and pain. The distribution pattern
along the convolution closely mirrors that in the adjacent motor area. Thus sensations
from the toes on the opposite side of the body are interpreted in the upper part of
the convolution near the great longitudinal fissure, whereas those from the trunk,
arm, and face are interpreted progressively nearer the lateral fissure.

Much that we know of these higher functions comes from studies of the consequences of removal of individual areas either experimentally or by injury or disease.
We have learned, for example, that the anterior portion of the frontal lobe, sometimes
called the prefrontal lobe, has an essential role in regulating behavior, though it is
without effect on reflexes, posture, and discrete movements. Removal of the prefrontal
lobe (or interruption of its connections with the rest of the brain) in an experimental
animal causes a characteristic alteration in behavior, with periods of hyperactivity,
aimless pacing, delayed reactions, and easy distractibility. Rightly or wrongly, observations such as these led in 1935 to the first surgical prefrontal lobotomy in an
emotionally disabled human being. The operation has been performed many times
since then but it is now judged to be of dubious value, often leading to loss of
certain higher sensibilities with a marked personality transformation, loss of judgment, and impaired sense of continuity and self.

*30-37 A map of sensory and motor areas
of the human brain*
*This cross section of the cerebrum shows the cortex
as a heavy line broken into fragments that
correspond to functional areas. The
corresponding body areas serviced by the cortex
are also shown as a ''homunculus,'' a little man
whose distorted body parts are drawn in
proportion to their importance in that brain area.
Although sensory and motor functions are each
labeled on only one side of the brain, they actually
exist on both sides. Compare the sensory and
motor functions of the body parts. Note that more
cortex is assigned to motor function of the hand
than to its sensory function. More sensory cortex
is assigned to the genitals.*

The Cellular Basis of Higher Functions

Consider learning and memory. The nature of the learning process has been a daunting problem. What parts of the nervous system participate in it? What is its mechanism? How does the brain encode, store, and retrieve information and memory? Even though various localized lesions are related to losses of certain functions, it is still not known whether learning occurs diffusely in the entire central nervous system, in localized cortical areas, or in combinations of lower and cortical areas.

In behavioral terms, **memory** appears to involve, first, one or more transitory or "holding" stages, then a period of consolidation with encoding of the transitory events, and finally the recording of a "permanent" memory. There must also be some arrangement for recalling that record. For any or all of these events, there need not be a single mechanism. Indeed, there seem to be two kinds of human memory: long-term or permanent memory, in which we store knowledge of language and most of everything else we know, and the more ephemeral short-term memory, which consists of the things we are paying attention to at a given time and which serves as a "gateway" to long-term memory.

When we speak of a computer's memory, we mean that it stores information—temporarily in random access memory, or permanently on a floppy disk, hard disk, or tape. Obviously, the memory of a machine differs in fundamental ways from the memory of a human or animal of another species. For example, one's long-term memory seems infinite in capacity (unlike one's hard disk). No one has ever been in possession of so much information that he or she couldn't learn something new.

Despite these differences, one feature *is* shared by both memories. It is the use of molecular or electrical patterns for storage of information. We understand the molecular basis of information storage in recording tapes and computer disks, but the basis of a brain's memory traces is more complicated. The best of the current theories on the basis of memory propose that memories are encoded as changes in the "wiring diagram" of the brain. The most important of these changes take place at synapses. They include facilitation and inhibition of preexisting synapses (Chapter 29). Interestingly, this idea has been around for a long time. In an unusual convergence of ideas, Ramón y Cajal, the Spanish neuroanatomist, and Sigmund Freud, founder of psychoanalysis, independently postulated in 1894 that learning might produce prolonged changes in the effectiveness of the synaptic connections between neurons and that patterning of these changes could provide a basis for memory. Surely, the synapses controlling one's fingers work differently—more, or perhaps less, effectively—in someone who has learned to play the piano than in someone who has not. A modern version of the notion of synaptic change as a mechanism of memory suggests that entirely new synapses may form as memories are encoded. Because memories originate from experiences, such synaptic changes must somehow be induced by the patterns of impulses being transmitted.

Much of our scientific knowledge of these processes comes from the study of elementary forms of learning in invertebrates. One body of work has focused on the gill and siphon reflex in the marine mollusk *Aplysia* (Figure 30-38A). So active has this field become, it has been dubbed "aplysiology." The learning capacities of *Aplysia* have been studied by observing the defensive withdrawal reflexes of external

30-38 Gill-withdrawal reflex in the marine snail, **Aplysia**

(A) Appearance of the snail; note that the gill lies under the protective mantle. (B) The neural circuits participating in the gill-withdrawal reflex. Some of the interneurons have been omitted.

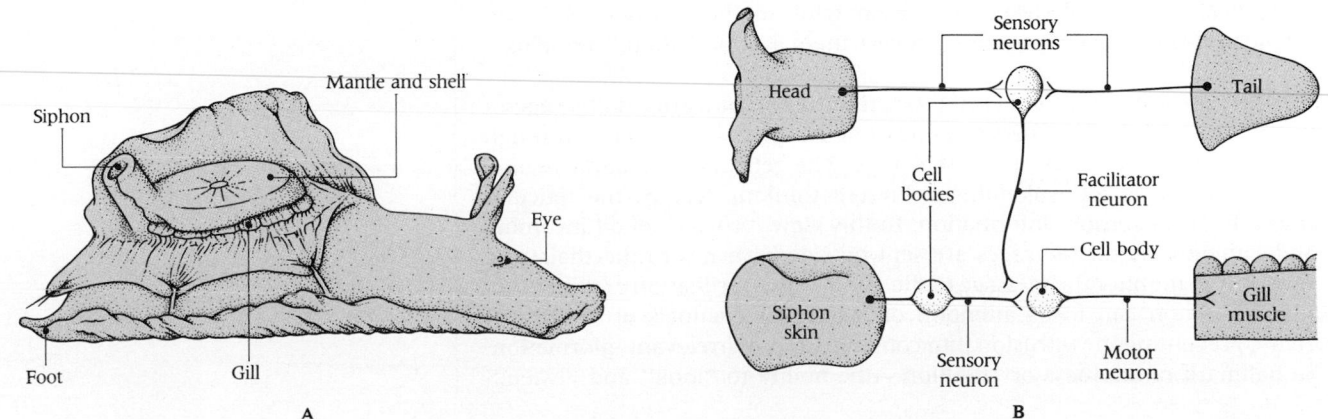

organs of the mantle cavity under various conditions (Figure 30-38B). This cavity, a respiratory chamber housing the gill, is covered by a protective sheet, the mantle shelf, which terminates in a fleshy spout, the siphon. When siphon or mantle shelf is stimulated by light touch, the siphon, mantle shelf, and gill all contract vigorously and withdraw into the mantle cavity. The reflex resembles vertebrate defensive escape and withdrawal responses, and can be modified in *Aplysia* by two kinds of learning experiences, habituation and sensitization.

The behavioral phenomenon of habituation (a dulling of a reflex by repeated stimulations) is produced by inhibition at a very specific set of synapses, those between the sensory neurons that transmit the touch stimulus and the interneurons and motor neurons that mediate gill withdrawal. The cellular basis of sensitization (an enhancement of a reflex by a strong or noxious stimulus), though more complex than that of habituation, depends on facilitation of the same set of synapses. Facilitation is mediated by a cAMP-dependent phosphorylation of proteins in the synapse. If this is a general molecular mechanism for short-term modifications of synaptic activity, some have wondered if it can also explain long-term memory. However, current evidence suggests that protein phosphorylation is too short-lived a phenomenon to underlie long-term memories.

Because long-term memory formation is accompanied by increased synthesis of mRNA and protein, many investigators suspect that it depends on the induction or repression of particular genes in the relevant neurons. The resulting changes in protein synthesis would presumably change synaptic structure as well as function. This hypothesis recalls the induction and derepression of gene function occurring in the early stages of embryonic development (Chapter 20).

We do not yet know how valid any of these proposals might be. It should be noted, however, that even if we knew all there is to know about the cellular and subcellular changes accompanying memory formation, we would still know very little of how actual, complex memories are stored in vertebrate brains. What we still need to know, for example, is how a single neural net, which somehow encodes one's knowledge of the best bicycle routes in Boston, can simultaneously encode thousands of specific, vivid memories of one's younger sister. Moreover, how do these memories persist even when individual cells within the net die?

Then there are the problems of consciousness, wakefulness, and sleep. It has been said that undefinable but intuitively recognized entities are a necessary foundation for most sciences. In neurophysiology, **consciousness** may be such an entity, for we can describe it only as the awareness of self and surroundings.

The unique value of consciousness is that it carries a hum of neural activity from one instant to the next, giving a semblance of continuity to what, in actual fact, may be extremely brief and isolated neural events. Although consciousness is necessary for the survival of organisms as they eat, fight, flee from threat, and so on, consciousness is not and never has been essential for the functioning of the nervous system. Rather, it is a supplement to the operations of the spinal cord, the brain stem, and the autonomic nervous system.

Investigation of the specific neural mechanisms of consciousness, wakefulness, and sleep has advanced in recent years. The surprising discovery was reported in 1982 by John Pappenheimer and Manfred Karnovsky of Harvard Medical School that the brain elaborates a specific peptide that is responsible for the loss of consciousness during sleep. Sleep has often been said to "rest the brain" and permit recovery from "wear and tear," but that is more poetic than scientific, and we must concede that the biological purpose of sleep, the reason why, in the words of Sir John Eccles, we must spend so much time in "abject mental annihilation," remains a mystery.

Some insight has come from studies of the **reticular formation**, a diffuse aggregate of cells and interconnecting fibers running from the spinal cord junction through the brainstem to the upper end of the thalamus. The reticular formation is now known to be responsible for wakefulness. Present thinking regards the reticular formation as a filter for sensory information. In this view, "copies" of all incoming sensory and outgoing motor messages are sent by side branches to the thalamus, which is thus able to monitor the message traffic. By selective facilitation or inhibition, the reticular formation can focus attention on particular channels of sensory or motor activity, preventing the intrusion into consciousness of irrelevant information. This is the hallmark of alertness or attention—the ability to "look" and "listen."

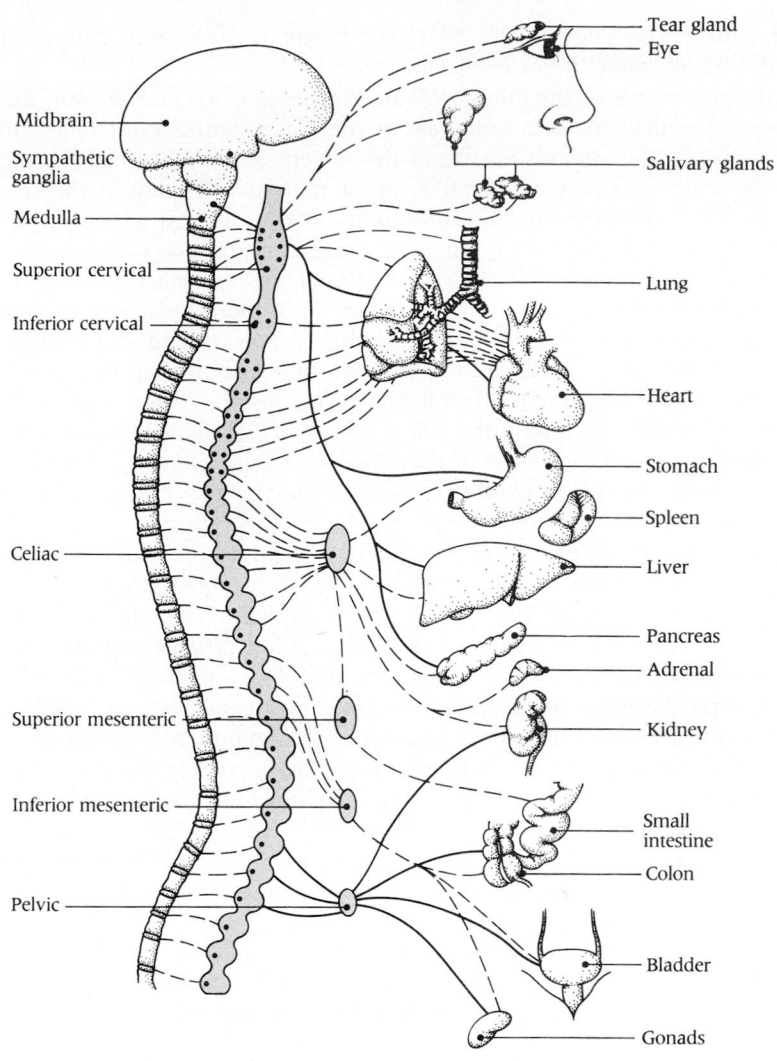

COORDINATION OF THE INTERNAL ENVIRONMENT

In considering control of the internal environment by hormones and other chemical messengers in Chapter 28, we noted many intimate connections between them and the nervous system. Clearly, coordination of this important domain is partly hormonal and partly neural, the two systems interacting closely.

Autonomic Nervous System

In the vertebrates—and in some of the more complex invertebrates (including insects)—nervous control of internal body functions is largely maintained by the so-called visceral nervous system. This portion of the nervous system controls a group of largely involuntary body functions—among them maintenance of blood pressure and body temperature, respiration, gastrointestinal motility and secretion, sweating, and pupillary movements in the eye. Their involuntary character is suggested by the synonymous, and now generally preferred name, **autonomic nervous system.**[14] As shown in Figure 30-39, this system is closely connected with the central nervous system. Yet it can act with some independence.

We should distinguish the anatomical pattern of the autonomic nervous system from that of the voluntary nervous system. In the classic spinal cord reflex arc (see Figure 30-31), the cell body of a sensory neuron lying in a dorsal ganglion outside the spinal cord, connects directly (or via an intermediary neuron) with a

---

[14] Involuntary nervous system, visceral nervous system, and vegetative nervous system are all used as synonyms for autonomic nervous system. In contrast, the voluntary nervous system is called somatic and nonvegetative.

motor neuron cell body in the cord. The motor neuron axon emerges from the cord and travels directly to the effector cell.

In the reflex arcs of the autonomic nervous system (Figure 30-40), the sensory neuron is identical to its counterpart in the classic spinal cord reflex arc except that it arises in the smooth muscle of the viscera (e.g., intestinal wall or arteriole wall). Its axon also connects with a motor neuron cell body in the spinal cord. Here the similarity ends. In the autonomic nervous system motor impulses from the spinal cord to an effector are conveyed along a two-neuron pathway. The synapse between the two neurons is in a special ganglion outside the spinal cord. The axon emerging from the ganglion innervates the smooth muscle of the viscera.

In sum, afferent nerves from the internal organs run to the central nervous system—to the spinal cord or the medulla oblongata in mammals—where they form reflex arcs with efferent visceral nerves that also have associative connections with higher centers including the brain. Much of the activity remains within the visceral reflexes—the processes of digestion, glandular secretion, and so on—that go on without our being aware of them or having any conscious control over them.

There are two complete sets of the efferent visceral nerves, which carry regulatory messages to the internal organs. These run to most of the same organs but have different connections with the central nervous system. The effects of the two sets of nerves are generally opposed. For example, one set, called **sympathetic,** accelerates the heart rate and speeds digestion by increasing intestinal motility. The other set, called **parasympathetic,** slows the heart rate and slows intestinal motility.[15]

In general, activity of the parasympathetic division primarily stimulates conservative and restorative processes such as slowing the heart rate, contracting the pupils to protect the eyes from light, and inhibiting the utilization of liver glycogen. It has a restricted distribution with local functions. Activity of the sympathetic division prompts expenditure of energy and defense in emergencies—widespread activities that require a diffuse distribution. Its functions, if discharged en masse, would gear a human being or other animal into an attitude of "fight or flight": acceleration of the heart rate, elevation of the blood pressure, stimulation of the breakdown of liver glycogen, and dilation of the bronchioles. Through balanced opposition, the autonomic nervous system coordinates the body's responses to widely varying internal and external conditions.

One of the considerations that originally led to the discovery of two divisions within the autonomic nervous system was the evidence of characteristic patterns in the responses to certain drugs. The differences arise because acetylcholine mediates impulse transmission in the neuroeffector junctions of the parasympathetic division and norepinephrine mediates it in the neuroeffector junctions of the sympathetic division.[16] Thus the parasympathetic division is cholinergic, and the sympathetic division is adrenergic. The transmitters are different only in the junctions between postganglionic neurons and effectors. Acetylcholine mediates impulse transmission in the synapses between preganglionic and postganglionic neurons in both divisions.

Many chemical agents have been discovered that, when administered as drugs, either mimic or block the actions of one division of the autonomic nervous system. Those causing behavior resembling that produced by the injection or endogenous formation of acetylcholine are termed parasympathomimetic. Those mimicking the actions of the sympathetic division are sympathomimetic. Agents inhibiting the actions of the parasympathetic and sympathetic divisions are, respectively, parasympatholytic and sympatholytic. Chemical agents inhibit the autonomic nervous system by combining with neurotransmitter receptor sites in synapses and by suppressing the synthesis and release of neurotransmitters. Such drugs have been indispensable tools in investigations of synaptic transmission and nerve impulse conduction.

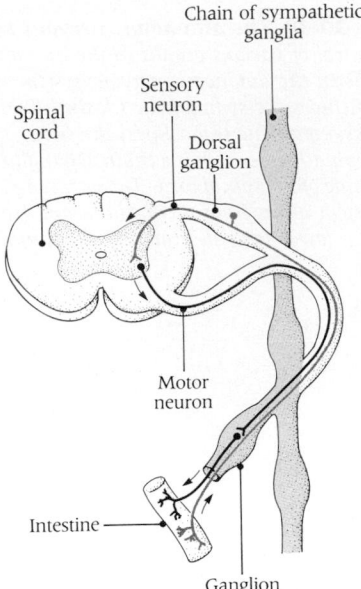

30-40 Reflex arc in the autonomic nervous system
Sensory information travels from the intestine to the spinal cord via a sensory neuron. A motor neuron receives these impulses, which travel to a ganglion outside the chain of sympathetic ganglia. There the impulses are transferred to a second motor neuron, which innervates smooth muscle in the intestine.

[15] Reciprocal relationships are not invariable, however. Many organs are controlled dominantly by one of the two divisions or similarly by both, so that active opposition does not occur. For example, parasympathetic stimulation excites the ciliary muscles of the eyes, whereas sympathetic stimulation is without effect. Both divisions excite the salivary glands.

[16] For many years, it was uncertain whether the adrenergic transmitter was norepinephrine or epinephrine. We now know it to be norepinephrine. We learned in Chapter 28 that both of these catecholamines are products of the adrenal medulla. Hence, these substances are hormones as well as neurotransmitters. When either is injected the response is a mass adrenergic response. The same thing occurs when one is frightened or excited. Its net effect is to stimulate the body and prepare it for instant action.

Role of the Hypothalamus

There is no pie in the vertebrate organism into which the hypothalamus does not dip its finger. We learned about control of endocrine glands by adenohypophyseal hormones and control of the adenohypophysis by the **hypothalamus.** This structure, which lies deep in the brain—under the thalamus, as its name implies—contains many nuclei, or brain cell clusters. The nuclei of the hypothalamus are primarily concerned with the regulation of visceral activities through the autonomic nervous system. Indeed, the hypothalamus has been called the "head ganglion" of the autonomic nervous system. Because of this arrangement many nervous reflexes arising within the body can find their way through the hindbrain and midbrain to the hypothalamus, which can take appropriate action by regulating the activities of the autonomic nervous system.

In all but the simplest animals, the role of the nervous system in bodily regulation is extensive. For example, when water becomes deficient in the body, a "thirst" signal reaches the brain centers. These centers make appropriate associations and send coordinated signals to the muscles. The organism then drinks and restores water balance. Here is a stabilizing mechanism, like the activity of a hormone or an autonomic reflex, but such responses are on a different level of activity. They are behavior, which is the subject of Chapters 32 and 33.

Finally, the hypothalamus has a key role in emotional behavior. The interplay between its activities and the neural activities of higher centers produces the emotional experiences we call fear, anger, pleasure, and contentment. Under conditions of stress or pain, the hypothalamus prepares the organism for appropriate action (attack, flight, etc.). These autonomic reactions, as we have seen, are involuntary. However, in the execution of such complex actions as attack, the forebrain must come into play. Perhaps consciousness arose in evolution because of the great complexity involved in dealing with the unpredictable environment. With the forebrain in operation, we have the luxury of delaying our actions and guiding them with images and plans.

SUMMARY

The senses of living organisms number well in excess of the five familiar ones (vision, hearing, touch, taste, and smell)—and there are many more than five kinds of sensory receptors. Each sensory receptor is a transducer that converts an external stimulus into volleys of nerve impulses, the frequency of which is proportional to the intensity of the stimulus.

Photoreceptors vary from the simple eye spots of protists, which contain little more than light-sensitive pigment molecules, to multicellular compound eyes, in which lenses allow the eye not only to detect light but to concentrate it and form images. Insects have compound eyes consisting of hundreds to thousands of independent ommatidia, each with a lens of its own.

Light entering the vertebrate eye is regulated by the diaphragm action of the iris. Light impinging on the retina there encounters sensitive photoreceptors (rods and cones). Rods are more sensitive to light than cones, but can distinguish only light intensities; cones can also distinguish different wavelengths of light. There are several kinds of cones—three in the human eye—each with a pigment responding maximally to a different color. Light affects photoreceptors by inducing reversible chemical changes in rhodopsin, a conjugated protein that contains retinal, a vitamin A derivative. Eyes adapt to dark and light by changing pupil size and modulating photoreceptor sensitivity and rhodopsin levels.

The retina does more than receive light and transduce it into nerve impulses. It also sorts and integrates the information it receives. Rods and cones respond to light by generating nerve impulses, utilizing G proteins and cyclic GMP as a second messenger to close specialized membrane Na^+ channels that remain open in the dark. Light closes them, hyperpolarizing the membrane in pro-

portion to the intensity of the light. The resulting nerve impulses are sent via the optic nerves to specialized brain areas—the lateral geniculate nuclei in the thalamus and the visual cortex—for further processing.

Mechanoreceptors in the skin include nociceptors (pain receptors) and touch-pressure receptors. Both generate nerve impulses in response to mechanical deformation of the receptor membrane. Sound receptors (phonoreceptors) are a special class of mechanoreceptors, inasmuch as sound vibrations stimulate phonoreceptors mechanically. Phonoreceptors are usually hair cells; the bending of their hair-like processes initiate nerve impulses. Because different hair cells respond to different sound frequencies, the ear can distinguish pitch. Equilibrium or gravity receptors are also hair cells.

Taste and smell are chemical senses. Mammals have four fundamental taste sensations: sweet, sour, salty, and bitter. No comparable set of fundamental odors is known.

The central nervous system evolved from the diffuse nerve nets like those of coelenterates. Cephalization (enlargement of the front end of the nerve cord to form a brain) began with flatworms like planaria. Other developments included the development of afferent and efferent nerves and specialized sensory receptors. Unlike invertebrates, vertebrates have a true brain and a hollow dorsal spinal cord.

The vertebrate brain has three regions: the forebrain (which includes the cerebral hemispheres and thalamus), the midbrain, and the hindbrain (which includes the cerebellum and medulla oblongata). The major evolutionary novelty in the mammalian brain was enlargement of the cerebrum. The vertebrate hindbrain controls such automatic functions as breathing and heartbeat. Forebrain regions have complex functions, especially in mammals, including specific cortical areas necessary for visual associations, fine motor control, and in humans, speech.

The mechanisms of learning and memory are not yet known with certainty. One current theory attributes the encoding of a memory trace to changes in the effectiveness of synapses between neurons. Short-term modulation of synaptic effectiveness may result from such biochemical events as receptor activation or inactivation. Long-term or permanent modulation may stem from the induction or repression of certain genes.

Involuntary body functions are controlled by the two independent subsystems of the autonomic nervous system—the sympathetic and parasympathetic nervous systems. The sympathetic system stimulates energy expenditure and "fight or flight" behavior. This effect is opposed in many ways to the effects of the parasympathetic system, which promotes processes that conserve body energy. Many autonomic reflexes are finally controlled by the remarkably versatile hypothalamus.

KEY TERMS

association area
autonomic nervous system
central nervous system
cerebellum
cerebral cortex
cerebrum
chemoreceptor
cochlea
compound eye
cone
forebrain
fovea
ganglion
hair cell
hindbrain
lateral-line system
mechanoreceptor
medulla oblongata

midbrain
motor area
nerve cord
nerve net
nociceptor
ommatidium
opsin
organ of Corti
ossicle
otolith
parasympathetic
peripheral nervous system
pheromone
phonoreceptor
photoreceptor
reflex
retina
retinal

retinal isomerase
rhodopsin
rod
saccule
semicircular canal
sensory area
sensory pit
sensory receptor
spinal cord
statoreceptor
statocyst
sympathetic
threshold
transducer
transducin
tympanic membrane

QUIZ QUESTIONS

1. Which of the following is *not* the site of transduction in some sensory structure?
   A. ganglion cells
   B. macula
   C. organ of Corti
   D. olfactory hair cells
   E. All of the above are sites of transduction.

2. Which of the following is *not* related to light perception?
   A. lens
   B. choroid
   C. cyclic GMP

   D. rhodopsin
   E. perilymph

3. Deformation of sensory dendrites is required to establish a nerve impulse in all of the following *except*
   A. gustatory hair cells.
   B. Pacinian corpuscles.
   C. lateral line organs.
   D. organ of Corti.
   E. macula.

4. The most extensively developed region of most mammalian brains is the

   A. cerebrum.
   B. midbrain.
   C. hypothalamus.
   D. optic chiasma.
   E. olfactory centers.

5. The ability to perceive pheromones is related to the sense of
   A. gustation.
   B. equilibrium.
   C. audition.
   D. olfaction.
   E. gravireception.

ESSAY QUESTIONS

1. Describe the organization of the human eye. What does each feature contribute to its functioning? What is the form and mode of functioning of the insect compound eye? Compare the ommatidia of fruit flies with a vertebrate eye.

2. How are sound waves (air pressure differences) converted into nerve impulses by the human ear? Why is the

middle ear cavity open to the throat? How does the inner ear provide information on position of the body and its movement?

3. Which sense organs in the inner ear are being taxed in the following situations:

attending a rock concert; riding a roller coaster; blasting off in a rocket; walking a balance beam?

4. What cellular changes are thought to underlie long-term memory in the brain?

How has this theory been applied to explain habituation and sensitization of the gill withdrawal reflex in *Aplysia*?

5. What is the role of the hypothalamus?

REFERENCES AND SUGGESTIONS FOR FURTHER READING

Alkon, D. L. (1988). *Memory Traces in the Brain*. Cambridge University Press, New York.

A discussion of the mechanisms of memory as revealed by the author's experiments with a marine mollusk. A nice illustration of what can be done with a simple system.

Bullock. T. H., Orkand, R., and Grinnell, A. (1959). *Introduction to Nervous Systems*. W. H. Freeman, New York.

A good introduction to the nervous system in the animal kingdom. The authors discuss both structural and functional aspects and they present a useful comparative perspective.

Changeux, J.-P. (1985). *Neuronal Man: The Biology of Mind*. Pantheon, New York.

A useful attempt to convey to nontechnical readers an interdisciplinary understanding of the human nervous system.

Coen, C. W. (Ed.) (1985). *Functions of the Brain*. Clarendon Press, Oxford, U.K.

Current doings in brain research. In the words of one reviewer, "a kind of pot-luck supper, to which a set of enthusiastic cooks each brings his own dish, be it spicy or a fluffy confection."

Denison, R. H., and 7 other authors (1966). The vertebrate ear. *American Zoologist 6*, 369–466.

The authors discuss the vertebrate ear from a comparative functional perspective. Includes an interesting discussion on the evolution of the hearing apparatus in the transition from reptiles to mammals.

Dowling, J. E. (1987). *The Retina: An Approachable Part of the Brain*. Harvard University Press, Cambridge, Mass.

The retina has two functions: it converts light into nerve impulses and it processes them with four layers of neurons before dispatching them to the brain. This book elegantly reviews what happens in between and why.

Gupta, A. P. (Ed.) (1987). *Arthropod Brain: Its Evolution, Development, Structure, and Functions*. Wiley-Interscience, New York.

The remarkable architecture of the insect brain was first pointed out by Ramón y Cajal in the honey bee. This multi-authored book brings us current, not only on the brains of insects, but of crustaceans as well.

Horn, G. (1986). *Memory, Imprinting, and the Brain: An Inquiry into Mechanisms*. Clarendon Press, Oxford, U.K.

A personal chronicle of the efforts of one research group to understand information storage in the brain. A stimulating account of how research is planned and how it is then carried out.

Hubel, D. H. and Wiesel, T. N. (1979). Brain mechanisms of vision. *Scientific American 241*, September, 150–162.

Two Nobel Laureates describe the processes of visual information in the cortex, presenting the experimental evidence briefly but convincingly.

Kleerekoper, H., and 4 other authors. (1967). Vertebrate olfaction. *American Zoologist 7*, 385–395.

This symposium on the comparative anatomy of the olfactory apparatus of vertebrates provides a comparative evolutionary perspective.

Lisberger, S. G. (1988). The neural basis for learning of simple motor skills. *Science 242*, 728–735.

Certain eye movements are used to study the neural basis of learning in monkeys. The results may apply to other forms of motor learning.

Llinás, R. R. (1988). The intrinsic electrophysiological properties of mammalian neurons: insights into central nervous system function. *Science 242*, 1654–1664.

Basic discussion of the electrophysiology of neurons.

Nathans, J. (1989). The genes for color vision. *Scientific American 260*, February, 42–49.

A review by the one who discovered the existence of different kinds of cone cells.

Nauta, W. J. H., and Feirtag, M. (1979). The organization of the brain. *Scientific American 241*, September, 88–111.

A concise yet complete discussion of brain anatomy, with special emphasis on the major pathways of information which flow within the brain. Most of the discussion concerns the human brain.

Nauta, W. J. H., and Feirtag, M. (1986). *Fundamental Neuroanatomy*. W. H. Freeman, New York.

A clear depiction of the central nervous system as a communication network and as a complex, three-dimensional architectural object.

Parker, D. E. (1980). The vestibular apparatus. *Scientific American 243*, November, 118–134.

Essay on the operations of otoliths and semicircular canals.

Schnapf, J. L., and Baylor, D. A. (1987). How photoreceptor cells respond to light. *Scientific American 256*, April, 40–47.

How light energy is changed into nerve impulses in an individual photoreceptor cell of the eye. A lucid account of the electrophysiology of rods and cones.

Snyder, S. H., Sklar, P. B., and Pevsner, J. (1988). Molecular mechanisms of olfaction. *The Journal of Biological Chemistry 263*, 13971–13974.

A stimulating "minireview" of a subject that only recently has begun to attract serious scientific interest, even though the olfactory system is of profound biological importance in the lives of most mammals and other animals.

Squire, L. R. (1987). *Memory and Brain*. Oxford University Press, New York.

An up-to-date survey of current understanding and interesting discussions of earlier research.

Stryer, L. (1987). The molecules of visual excitation. *Scientific American 257*, July, 42–50.

An account of the transducin story, written with gusto and authority, by a key contributor to our new understanding of sensory transduction.

Wald, G. (1968). The molecular basis of visual excitation. *Nature 219*, 800–807.

This Nobel lecture contains an interesting account of the discovery of the primary event in vision.

EFFECTORS

Effectors
At least in quantitative terms, muscle is the major effector in the animal body. The capacity of some muscles for sustained rapid-fire contraction is truly impressive. Here are two virtuoso performers: a honey bee, Apis mellifera, *and a graceful black chin hummingbird,* Archilochus alexandria.

Effectors are the components of organisms that carry out the many tasks of daily existence. The major effector in multicellular animals is **muscle.** Some effectors, we will find, are not shared by most animals and are unfamiliar to students of human physiology. These include electric organs, light-emitting organs, and others. Two principles should be noted at the outset.

■ Many effectors are under neural control. These systems operate in response to commands from the nervous system and may be considered the terminal elements of the receptor-conductor-effector sequence described in Chapter 29. They include endocrine glands, muscles, and others, and to a large extent are responsible for the outward manifestations of behavior.
■ Some effectors are not under neural control. These include cilia, flagella, and structural elements that provide mechanical support.

Some effectors under nervous control cannot perform without the help of other effectors that are not under nervous control. Muscle, the most familiar effector and from any viewpoint one of the most important, is widely distributed. Individual muscles under neural control account for body movement, heart action, breathing, peristalsis, speech, and much else. Yet in causing body movement, they are wholly dependent on another effector, the skeleton, which is independent of direct nervous system control.

MUSCLE

Movement underlies most manifestations of behavior. Indeed, it is often the most conspicuous sign that something is alive. In multicellular organisms, movement is invariably a result of muscle action.

SOME GENERAL PRINCIPLES

Motion that moves a whole organism from place to place is called **locomotion.** But motion may take place in only a small part of the organism. We now consider effectors that facilitate body movements of both types.

Muscle Cells and Nonmuscle Cells

We have studied the mechanisms responsible for movement in individual cells (see Chapter 7). Those, we recall, depend either on actin-containing microfilaments (along with myosin and related proteins) or microtubules (tubulin and microtubule-associated proteins). Those discussions would be profitably reviewed because much of what was said of these systems in single cells applies to the effectors of multicellular organisms now to be discussed.

Muscle, we will find, consists of actin-containing filaments and filaments of myosin, plus various regulatory proteins. It is an interesting fact of scientific history that studies of muscle contraction date back many decades, while studies of nonmuscle cell motility began only recently. Yet students of motility were astonished when the microfilaments they had discovered in nonmuscle cells turned out to consist largely of actin, working in concert with myosin in the cytoskeleton—the same proteins whose crucial role in muscle cell contraction had been appreciated so long ago. Soon thereafter, actin was recognized to be a universal component of all cells, in many cases comprising up to 15% of a cell's total protein.

Finally, several of the long list of known actin-binding regulatory proteins (Table 7-1), were found to be identical to certain proteins participating in muscle contraction. It is one more instance in which observations made in a few widely differing forms disclosed biological principles of global significance.

Muscle Types

The two main muscle types are **smooth** (or **visceral**) **muscle,** which is involuntary, and **striated** (or **skeletal**) **muscle,** which is voluntary (see Chapter 5). A third type, a variant of striated muscle is **cardiac muscle.** Only true striated muscle can carry out deliberate actions—although conditioning can sometimes impose a measure of voluntary control on involuntary muscles.

SKELETAL MUSCLES

There are more than 400 separate skeletal muscles in the human body. These muscle masses make up roughly 42% of the total body weight, although this figure varies somewhat.

In contracting, striated muscles generate mechanical forces which change the positions, relative to one another, of the two parts of the skeleton to which they are attached. The result with skeletal muscles is movement about a joint. Although each muscle is an independent unit, muscles almost always act in groups rather than singly.

Each skeletal muscle has a body and two attachments. The body is the fleshy part of the muscle; the attachments connect muscle directly to bone by a sheet of fibrous connective tissue or a **tendon.** When the precise locations of a muscle's attachments are known, its mode of action is usually predictable. As shown in Figure 31-1, most muscles can be arranged in antagonistic or opposing groups, which are named according to their actions: flexors and extensors, adductors and abductors, and internal rotators and external rotators. When, for example, the muscles that flex the forearm contract, the muscles that extend it relax.

The intricate organization of muscles and tendons is perhaps nowhere more vividly displayed than in the human hand, one of evolution's most remarkable products (Figure 31-2). A primary function of the hand is to grasp, firmly or gently,

31-1 Muscle-skeleton relations
(A) *In the human arm, two muscles, the biceps and the triceps, oppose each other. In this flexed arm, the biceps is contracted and the triceps is relaxed. When this pattern is reversed, the arm extends.* **(B)** *A similar system moves the limb joint of an arthropod's exoskeleton. Contraction of muscle a flexes the joint; muscle b extends it.* **(C)** *The beating of a bird's wing in flight requires the reciprocal action of paired muscles. Note the keel of the breastbone (sternum) and the wishbone (furcula).*

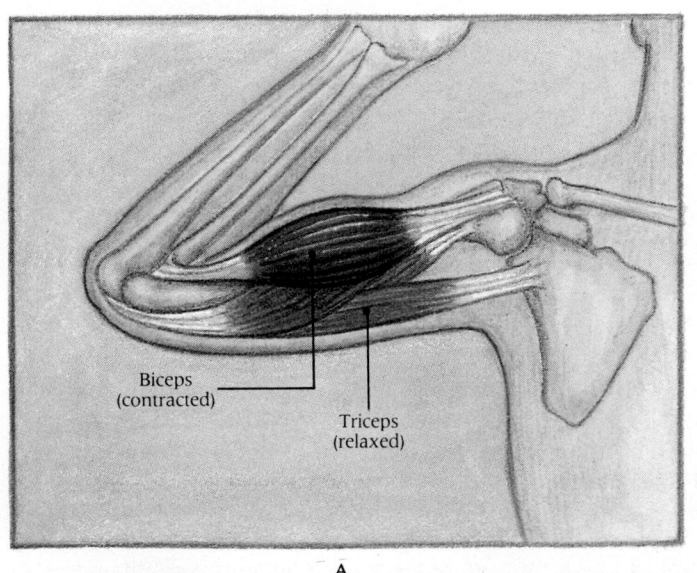

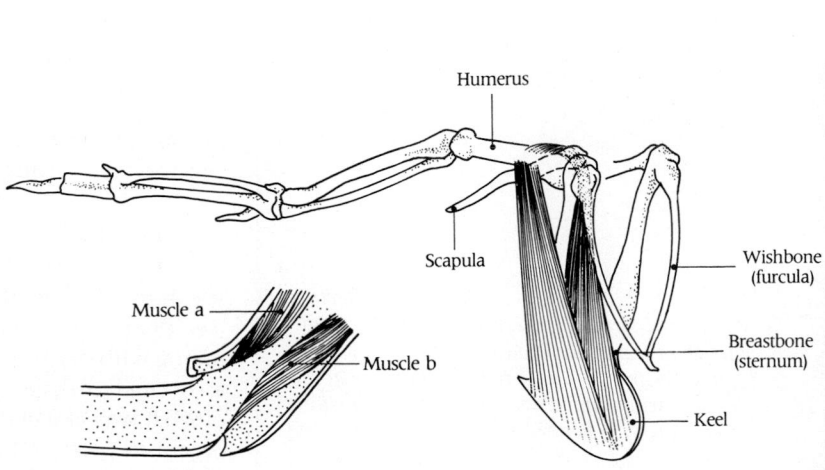

31-2 Design of the human hand
(A) *Flexion and extension of the fingers, and
rotation of the thumb using abductor and
adductor muscles.* (B) *This dissection of the hand
(left) and cross-section (right) reveals the intricate
positioning of the muscles and tendons required
for flexion, extension, and rotation.*

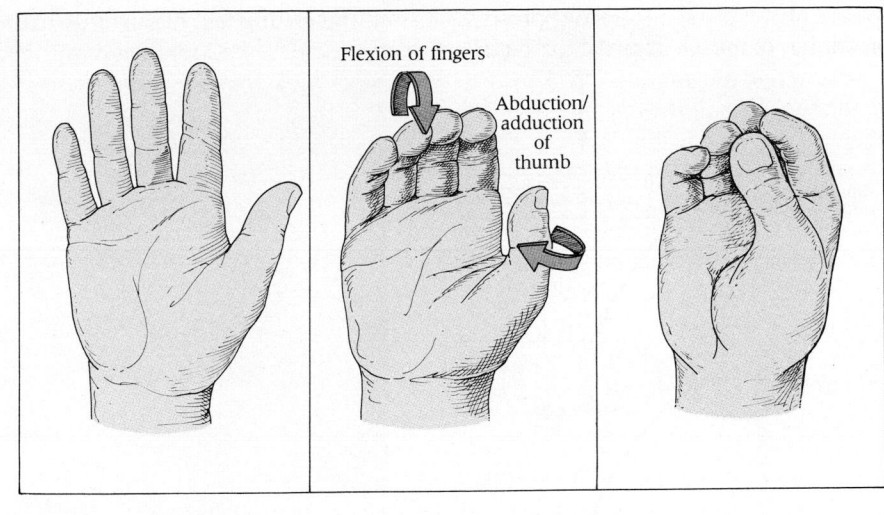

Flexion of fingers

Abduction/
adduction
of
thumb

A

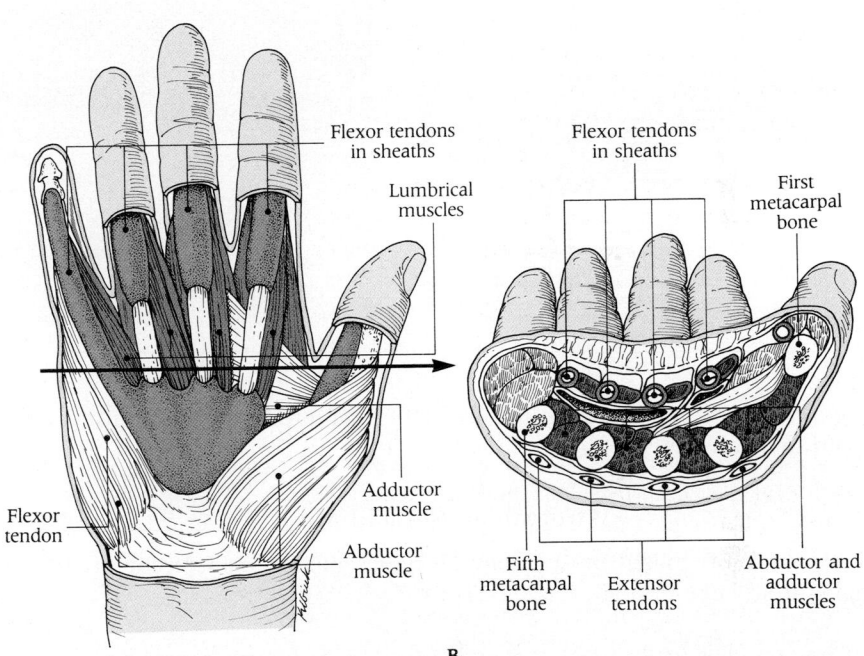

Flexor tendons
in sheaths

Lumbrical
muscles

Flexor tendons
in sheaths

First
metacarpal
bone

Adductor
muscle

Abductor
muscle

Flexor
tendon

Fifth
metacarpal
bone

Extensor
tendons

Abductor and
adductor
muscles

B

a wide range of objects of different sizes and shapes. The flexion and extension of
the fingers in grasping is supplemented by adduction and abduction and a special
movement of the thumb called opposition (in which the pad of the thumb is placed
upon the pad of another finger). Although numerous small muscles begin and
end within the hand itself, many hand and finger movements are controlled by
muscles in the forearm, the tendons of which descend to the hand in a complex
anatomical array.

MUSCULAR CONTRACTION

In the overall economy of the body, muscle contraction is a major consumer of
the energy-rich substrates and oxygen utilized in generating ATP. In earlier
chapters, we dealt with metabolic aspects of energy generation. The reader
should review those topics—especially the discussion of metabolic aspects of strenuous muscular exercise in Chapter 27. Recall that when demand outruns the oxygen supply, a fatigued muscle is forced to rely on two relatively inefficient methods
of ATP generation: the anaerobic conversion of glycogen to lactic acid, and a
standby store of readily withdrawable energy reserves called phosphagens. In humans and other vertebrates, the major phosphagen is creatine phosphate.

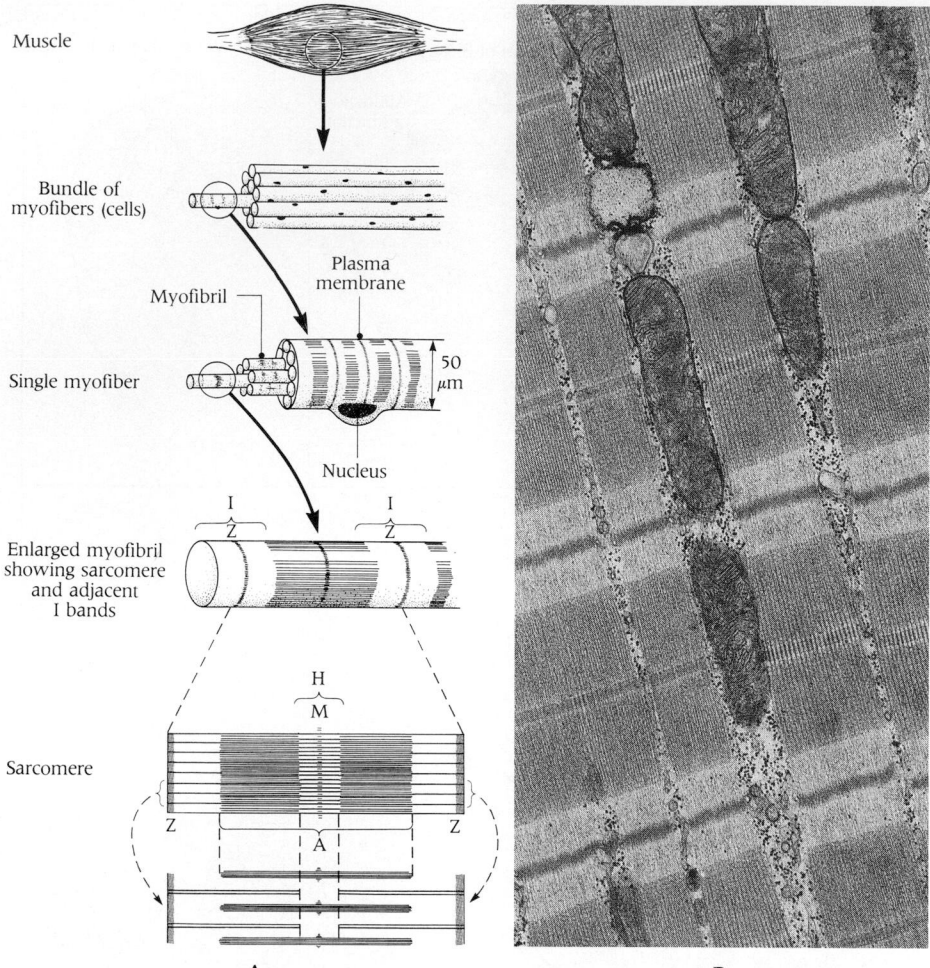

Despite recent progress, questions about the force-producing mechanism of muscle remain to be answered.

Structure of Striated Muscle

The structure of striated muscle may be represented as a hierarchy of structures, ranging from large to small (Figure 31-3A).

- Large bundles (or fasicles) of muscle fibers make up the muscle body.
- A typical muscle cell, termed a **myofiber,** is a long cylindrical fiber 10 to 100 μm in diameter and 1 millimeter to several centimeters in length.[1] This great length means that a single muscle cell may extend over a large part (or the entire length) of a single muscle. Note that each fiber is a **syncytium** (a mass of multinucleated cytoplasm) containing up to 100 nuclei.
- Muscle fibers contain bundles of parallel filaments, termed **myofibrils,** each 1 to 2 μm in diameter. Myofibrils are basic units of myofibers. Each myofibril consists of a sequence of repeating **sarcomeres,** each about 2 μm long (Figure 31-3B). The myofibrils are composed of two kinds of filaments: thin filaments and thick filaments.
- The **thin filaments** (6 nm in diameter and 1 μm in length) are primarily composed of **actin,** the roughly spherical molecules of which are aggregated like a twisted double strand of beads (Figure 31-4). Each filament also contains one molecule of a protein called **tropomyosin** and three **troponin** peptides.

---

[1] It is customary in discussions of muscular contraction to pay close attention to the sizes—lengths, widths, and diameters—of the various participating structural elements. We do this, in part, because these elements were first recognized by electron microscopy, a technique that lends itself to accurate size measurements and, in part, because judicious study of these dimensions has led to many highly successful guesses on their functions. For a refresher on these units of measure, review Box 3-1 and Figure 3-1. Meanwhile, recall that 1 μm equals 10^{-6} meters and 1 nm equals 10^{-9} meters.

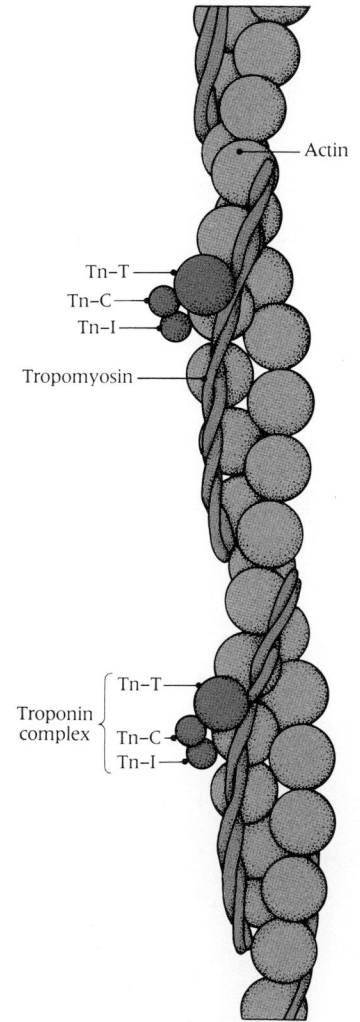

31-4 Structure of a thin filament
Thin filaments in vertebrae skeletal muscle are composed of individual actin molecules organized in a long filament. Tropomyosin and three troponin peptides termed Tn-T, Tn-C, and Tn-I complete the array of proteins.

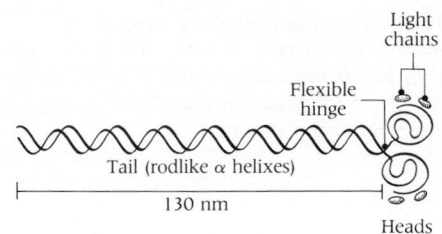

Light
chains

Flexible
hinge

Tail (rodlike α helixes)

130 nm

Heads

31-5 Structure of myosin
*Drawing of a myosin molecule showing two heavy
and four light chains. Note the two heads and
two tails and the flexible hinged joint.*

■ The **thick filaments** (16 nm in diameter and 1.5 μm in length) are composed almost entirely of **myosin,** protein molecules containing two heavy and two light chains with long rod-shaped tails and two globular heads (Figure 31-5). Myosin has three important properties: it binds actin; it polymerizes into filaments; and it is an enzyme that catalyzes the hydrolysis of ATP. As noted earlier, we call such an enzyme an ATPase.[2] This particular ATPase is active only in the presence of actin.

When muscle fibers are cut transversely, the filaments are seen to lie some distance apart in a regular hexagonal array (Figure 31-6). Each thick filament is surrounded by six thin ones; each thin filament is surrounded by three thick ones—and three thin ones.

Each of the two types of filament is aligned with others of the same kind and the two groups overlap for part of their lengths (Figure 31-7). The overlap gives rise to the characteristic striations or cross-banding pattern of the myofibril. Letter names are assigned to each band.

Thick and thin filaments are linked by an intricate system of cross-bridges that provide the sole mechanical connections between filaments (Figure 31-8). Cross-bridges are in fact the pear-shaped heads of the myosin molecules in the thick filaments. They project outward at right angles from each thick filament at regular 43 nm intervals over its entire length (except for a central region from which they are absent). The bridges around the axis of a thick filament, which form a helix, join a thick filament to each of the six surrounding thin filaments.

[2] ATPase is more properly termed adenosinetriphosphatase. It is an enzyme that catalyzes a hydrolytic cleavage of phosphate from ATP, thereby liberating (along with the phosphate) a molecule of ADP.

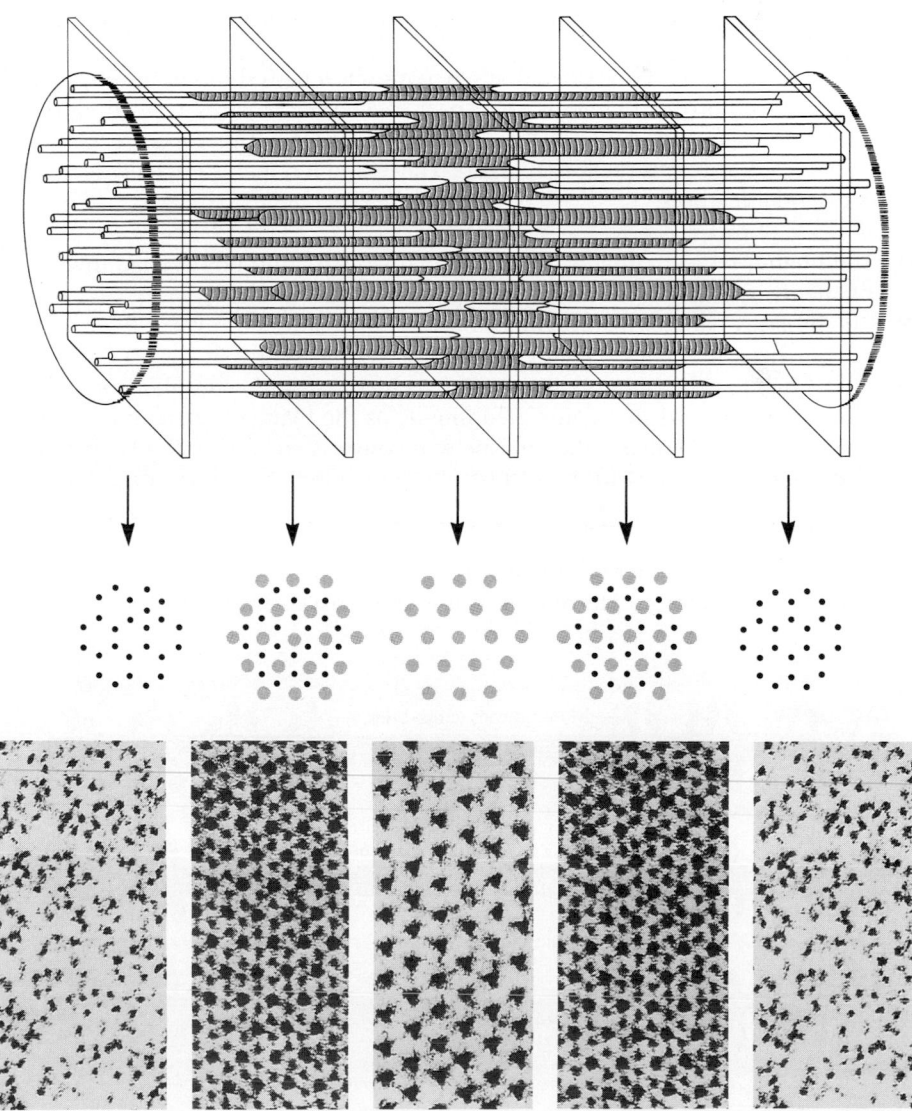

**31-6 Relationship between thick and thin
filaments in a sarcomere**
*A transverse section made at five different
positions in the sarcomere reveals hexagonal
arrays of thick filaments (color) and thin
filaments at each position.*

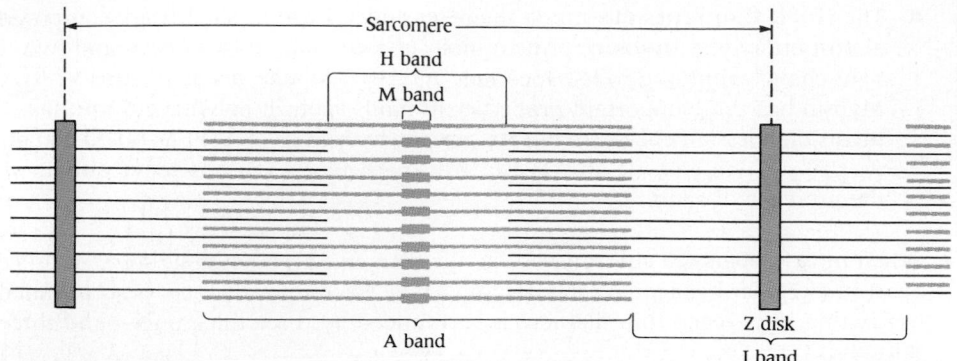

Sliding Filament Model

Careful study of the changing widths of the bands during muscle contraction suggested to H. E. Huxley that crucial events must be taking place at an organizational level above that of the individual protein molecules. Critical studies were carried out in England in 1953 by H. E. Huxley and Jean Hanson and independently by A. F. Huxley and R. Niedergerke. Observers were struck by the fact that the width of the A band remains constant when an isolated muscle fiber contracts (Figure 31-9). The I band and H bands narrow and virtually disappear and the Z lines (that mark the Z disks at each end of a sarcomere) come closer together. Since the width of an A band equals the length of a thick (myosin) filament, the length of these filaments must remain unchanged during contraction. Although the H band narrows with the I band, the distance from one H band to another (across a Z line) remains approximately constant. Therefore, thin (actin) filaments must also be of constant length.

The English workers concluded that when muscle length shortens during contraction the two groups of filaments slide past each other. In this way, they reduce their combined lengths, but not their individual lengths. This model explains the changes seen during contraction in Figure 31-9—and it remains unchallenged today in its essential features.

- The I band, which contains only actin filaments that do not overlap myosin filaments, is almost obliterated during muscle contraction.
- The broad A band, which includes overlapping actin and myosin filaments, approaches the Z lines, causing the I band to disappear.
- The H band at the center—the gap between two sets of actin filaments—also virtually disappears in a contracted muscle as the filament sets almost meet. Creatine phosphokinase, the enzyme that converts creatine and ATP to creatine phosphate, is attached to the M line of the myofibril, which is in the center of the H band.

Investigators then sought to identify the forces that induce the filaments to slide. What triggers this sliding? The answer was in the **cross-bridges** emerging at regular

31-8 Position of cross-bridges in a sarcomere
(A) The myosin heads extend at right angles from the core of the thick filaments which consists of myosin tails. Interaction of heads (or cross-bridges) with thin filaments is responsible for muscular contraction. (B) Electron micrograph showing actin-myosin cross-bridges in a striated insect muscle. (× 180,000)

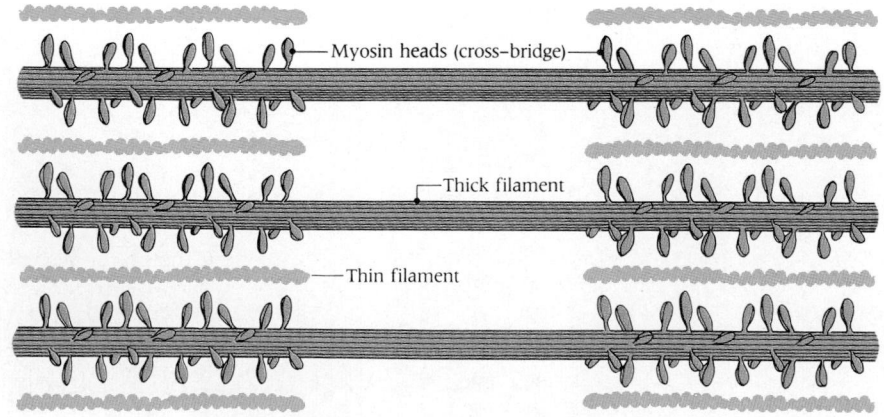

A

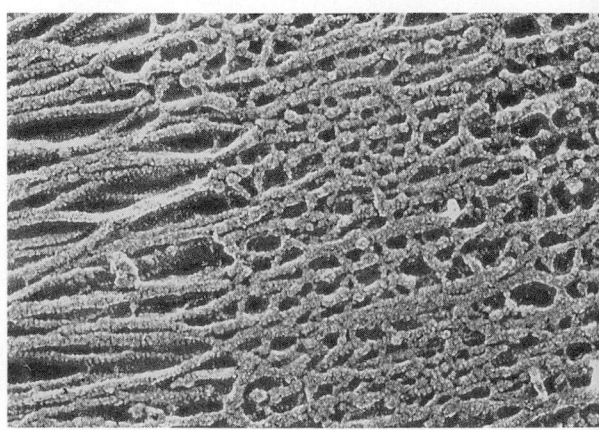

B

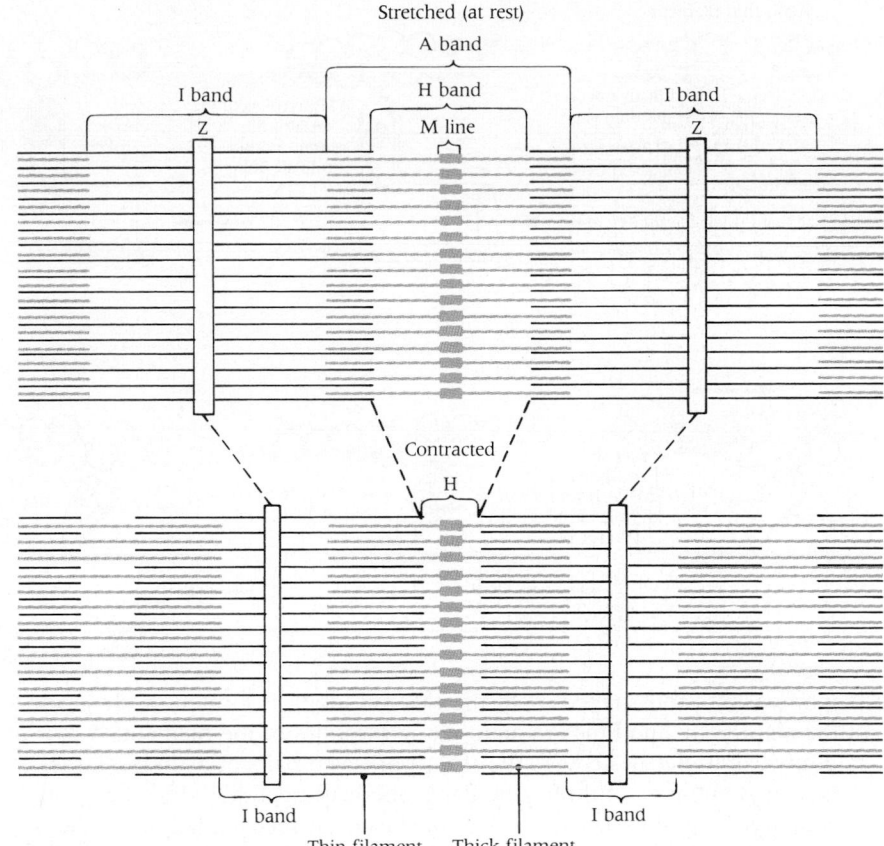

31-9 Change in band pattern during contraction of striated muscle
When skeletal muscle contracts, the A band remains constant in width. The I band and H band shorten.

intervals from the thick (myosin) filaments. We now realize that the bridges connect—reversibly—with specific sites on neighboring thin (actin) filaments (Figure 31-10). We noted above that myosin is an actin-activated ATPase. This enzymatic activity is located in a specific site on the myosin molecule—namely, the heads, which are the cross-bridges to actin and, predictably, have actin-binding sites as well. Myosin heads (cross-bridges) also have a high binding affinity for ATP and ADP.

The following properties of cross-bridges make possible the myosin-ATPase cycle that causes muscle contraction (Figure 31-11).

- When a molecule of ATP is bound, the connection between myosin and actin is weakened or broken.
- The ATPase of myosin hydrolyzes the ATP to ADP and inorganic phosphate (P_i), which momentarily remain bound to the head unit.
- This activates the actin-binding site. If Ca^{2+} is present, the head then binds to the actin filament.
- The myosin head then undergoes a conformational change in which it swivels around a "hinge" region in the headpiece until it is at an angle of 45°. As it does so, it moves the actin filament, causing it to slide by. The hinge can be seen in Figure 31-5A.
- During this step, P_i and then ADP are released. For a short time, there exists in that region a state of rigor, or stiff rigidity, in which the myosin head is locked onto the actin filament.
- The prompt binding of a new ATP molecule releases the head from actin and a new cycle can begin.

As energy is released from ATP at many points along the thin filaments, the filaments slide toward the center of the A band for a short distance, generating tension and movement. As a result, the sarcomere contracts. Eventually the filament returns to its original position. Each time a cross-bridge initiates such a cycle, one phosphate group is split from ATP. Under normal conditions, ATP is present in ample (millimolar) concentrations and the bridges continue to cycle as long as the Ca^{2+} concentration is adequate.

Two proteins that regulate muscle contraction are clearly related to proteins that regulate nonmuscle cell motility (see Figure 31-4).

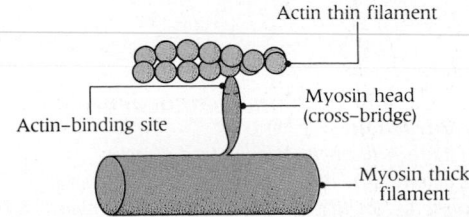

31-10 Relation between myosin and actin
The actin thin filament contains binding sites for myosin heads (cross-bridges). This association provides the fundamental basis for muscle contractions.

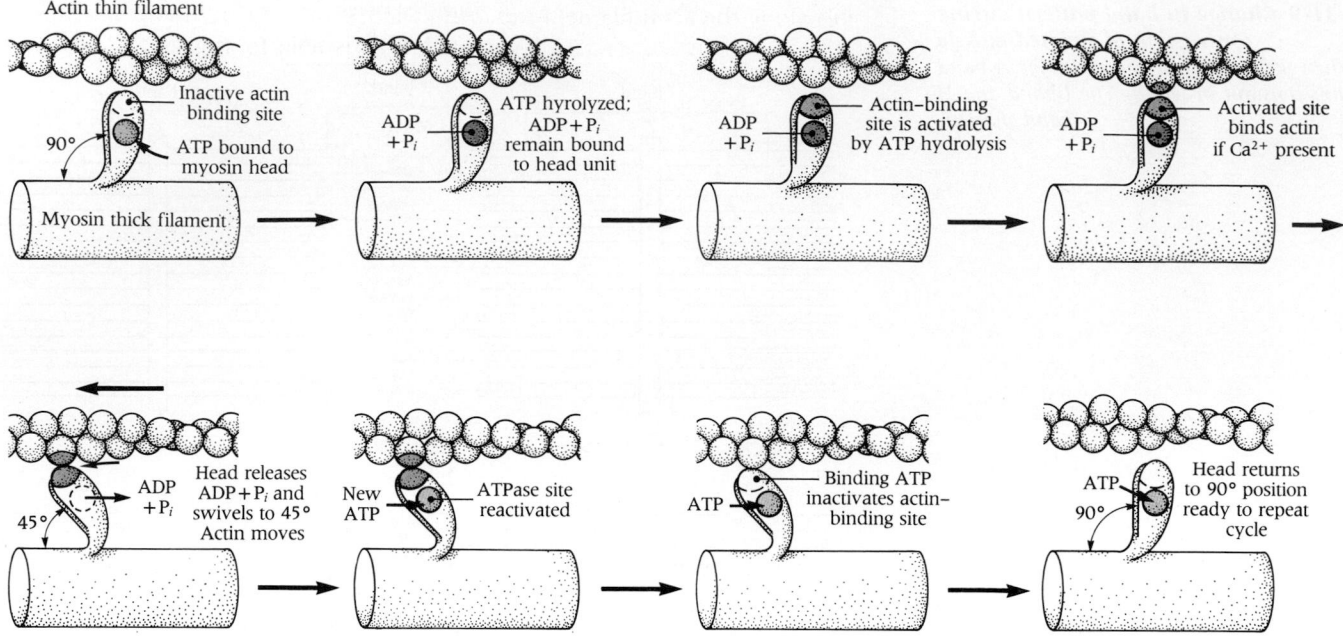

Tropomyosin, a rod-shaped molecule (40 nm long), is located in the grooves of the actin helix and binds to actin along the sides of the filament. A pair of tropomyosin molecules covers the length of seven pairs of actin monomers.

Troponin, a complex of three globular (spherical) peptides, binds Ca^{2+} (as well as actin and tropomyosin) and lodges in clusters at specific sites along the tropomyosin molecule.

Together these molecules control the regulatory effects of Ca^{2+}, and thus influence the binding of activated myosin heads (cross-bridges) to actin (Figure 31-12). These models postulate that the presence of Ca^{2+} is essential for contraction and that in striated muscle this critical role of Ca^{2+} is mediated by troponin and tropomyosin. As we will see, a nerve impulse entering a muscle alters the Ca^{2+} concentration around myofibrils. Troponin and tropomyosin are linked to that mechanism.

One question still to be answered is the problem of how an energy-liberating chemical reaction provides the motive force for the movements of contraction. As sketched in Figure 31-10, the cross-bridges joining the actin and myosin filaments are somehow caused to swivel to a different angle—from 90° to 45°—so that the cross-bridges pull the thin filaments past the thick filaments, just as if they were oars. After one stroke of the "oar," the bridges detach, swivel, and then reattach at a point farther along the thin filament. Huxley likens the sliding process to a common ratchet mechanism. In this familiar contrivance, a notched wheel is engaged by a toothlike device that permits the wheel to turn in only one direction. When bridges bind to specific sites on the thin filament, the two structures appear to engage each other like a tooth and ratchet, the bridge being the tooth, and binding

31-11 The sliding filament model of muscular contraction
A cycle of attachment-rotation-release is the basis of muscular contraction. It depends upon ATP hydrolysis by myosin ATPase and the association of myosin heads with actin in the presence of ADP and P_i. Rotation of a myosin head while it is bound to actin pulls the actin filament to the left (colored arrow). The binding of a new ATP molecule dissociates the cross-bridge from actin, thus preparing it for another cycle.

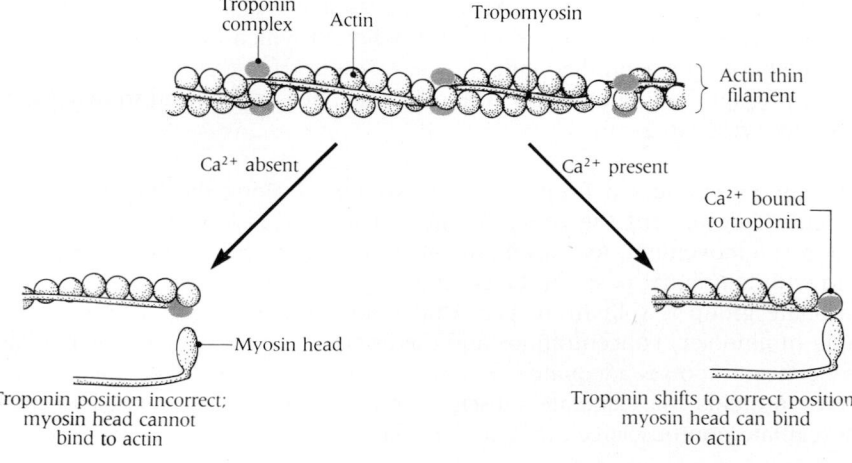

31-12 Role of calcium ions in muscular contraction
The thin filament is composed of actin, tropomyosin, and the three-peptide troponin complex. In the absence of Ca^{2+}, the myosin binding site of actin is blocked and cross-bridges cannot form. When Ca^{2+} binds to one troponin unit, the myosin binding site on the thin filament becomes available for cross-bridge formation and contraction occurs. This important mechanism is shown in greater detail in Figure 31-15.

sites along the actin filament the ratchet. When the bridge disengages momentarily from the ratchet, the swiveling cross-bridge is able to engage the next notch on the thin actin filament, which is moved toward the center of the sarcomere.

The formation of permanent links between actin and myosin in the absence of ATP may explain *rigor mortis,* the muscular rigidity that is present for a time after death. Muscles in rigor mortis rigidify in whatever position they were in when they ran out of ATP.

Links Between Excitation and Contraction

How does excitation by a nerve impulse initiate contraction? We saw in Chapter 29 that impulses arriving in a motor nerve fiber enter a striated muscle fiber via a neuroeffector junction, the motor endplate. Incoming nerve impulses are transmitted across the synaptic cleft by acetylcholine, which then depolarizes the muscle fiber membrane, the **sarcolemma.** Depolarization waves then sweep across the surface of the muscle fiber, find their way into the fiber and initiate a contraction. Even in a frog muscle cooled to 0°C, depolarization causes a contraction within 0.04 seconds. Calculations show that diffusion of a hypothetical chemical activator from the surface of a fiber to the fiber's interior would take too long to account for this rapid contractile response. What is going on?

It was A. F. Huxley and his coworkers who set themselves the task of explaining how neural excitation leads to muscular contraction through the widespread propagation of a zone of depolarization in the muscle fiber's plasma membrane. Exactly how depolarization proceeds into the muscle fiber from the plasma membrane was an enigma until electron microscopists detected the **sarcoplasmic reticulum,** a network of vesicles resembling the endoplasmic reticulum of other cells, and the **transverse tubular system** or **T system,** a ductwork of tiny transverse tubes that arise in the plasma membrane and extend perpendicularly into the muscle fiber in the region of the Z lines (Figure 31-13). The fluids within the sarcoplasmic reticulum are rich in Ca^{2+}. The fact that T tubules in different species open in the regions most sensitive to stimulation (Figure 31-14) bolstered the idea that the T system is involved in the inward spread of excitation.

Discovery of the T system provided the long-sought physical basis for the swift inward transmission of the depolarization waves triggered by arriving nerve impulses. Spreading throughout the T system, they lead to a release of Ca^{2+} from the sarco-

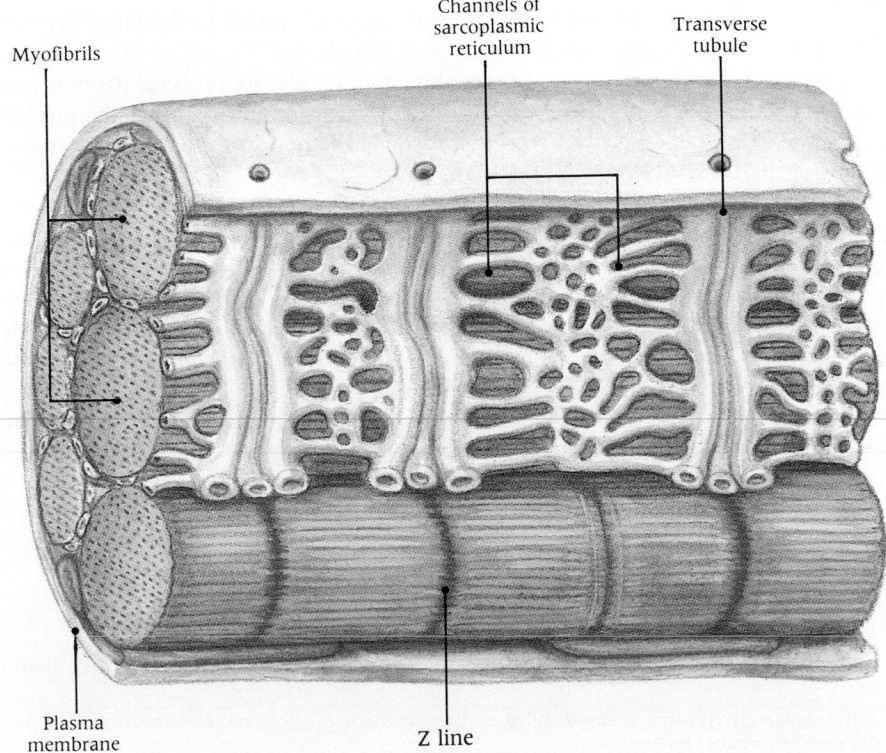

Myofibrils

Channels of sarcoplasmic reticulum

Transverse tubule

Plasma membrane

Z line

31-13 The transverse tubule (T) system of striated muscle

The T system of tubules is an extension of the plasma membrane of the myofibril. These tubules enter the fiber at each Z disk and extend deep into its interior. Within the fiber, they come in close contact with the sarcoplasmic reticulum.

plasmic reticulum. Recent work showing that the inositol polyphosphate system (Chapter 28) controls the release of Ca^{2+} from the endoplasmic reticulum of nonmuscle cells suggests that the same system may cause Ca^{2+} release from sarcoplasmic reticulum into muscle cells (see Figure 28-11).

As shown in Figure 31-15, released Ca^{2+} binds to troponin, thus changing the conformation of the troponin-tropomyosin complex. This frees the myosin-binding site of actin, which in the absence of Ca^{2+} is blocked by tropomyosin. Actin now binds myosin, forming cross-bridges that initiate a sliding movement of actin and myosin filaments. Energy for this process comes from hydrolysis of ATP.

When stimulation stops, the membrane of the sarcoplasmic reticulum is no longer permeable to Ca^{2+} and a calcium pump restores most of the Ca^{2+} to the inside of the reticulum channels. Tropomyosin again blocks the myosin binding sites of actin. ATP accumulates and old cross-bridges dissociate and soon disappear. In a word, the muscle relaxes.

There are variations in the basic patterns of muscle contraction. For that reason, skeletal muscle fibers are broadly divided into two types—though the muscles themselves are often mixtures of the two fiber types.

■ "Fast" muscle fibers contract rapidly in response to a nerve impulse and are used in vigorous movements.
■ "Slow" muscle fibers contract relatively slowly, are capable of a graded response, and are used in such activities as maintaining posture.

There is a good correlation between the speed of a muscle fiber's activity cycle and its content of sarcoplasmic reticulum and T tubules. The molecular basis of contraction is similar in fast and slow muscles, but there are biochemical differences between them. For example, the ATPase of slow muscle myosin functions at a lower rate than that of fast muscle.

Variations on a Theme

Certain animal groups face singular problems and they are equipped with specialized striated muscles that significantly enhance survival.

Bivalve mollusks (clams, oysters, etc.) have a strenuous burden: in the presence of enemies, they must keep their shells shut. This they do with an unusual type of muscle that can maintain tremendous tension for long periods of time.[3] Oddly, maintenance of this tension utilizes little more energy than a resting muscle. The phenomenon has been attributed to a "catch" mechanism. Once contraction is set, no more energy is needed to maintain it. In these muscles, regulation by Ca^{2+} operates differently than in vertebrate striated muscle. Ca^{2+} is active at much lower concentrations, it interacts with a different group of regulatory proteins (tropomyosin is present, troponin is absent), and it somehow causes the contracted muscle to "set", in which a change of state occurs, rather like the setting of an epoxy glue—except that it is reversible.

The unusual flight muscles of such flying insects as mosquitoes, flies, wasps, and bees are called "fast" because they contract far more rapidly than ordinary skeletal muscles, which are "slow" muscles. Weight for weight, these insect flight muscles generate more energy than any other tissue in the animal kingdom. Because these muscles contract and relax at an extraordinarily high rate—up to 1000 wing flappings per second—that greatly exceeds the frequency of incoming nerve impulses, they are termed **asynchronous muscles.**

In contrast, the flight muscles of other insects contract and relax in direct response to incoming nerve signals (as do striated muscles in vertebrates), averaging 35 cycles per second. Thus they are called **synchronous muscles.** Flying insects that buzz have asynchronous muscles. Those that do not buzz have synchronous muscles (Figure 31-16). Synchronous and asynchronous muscles have distinctive anatomical arrangements.

Two features distinguish asynchronous muscles.

■ Although a T system is present, the sarcoplasmic reticulum is practically absent.

[3] The delicately flavored seafood called scallops consists of these muscles from certain species of the shellfish genus, *Pecten*.

PART 3 HOMEOSTASIS

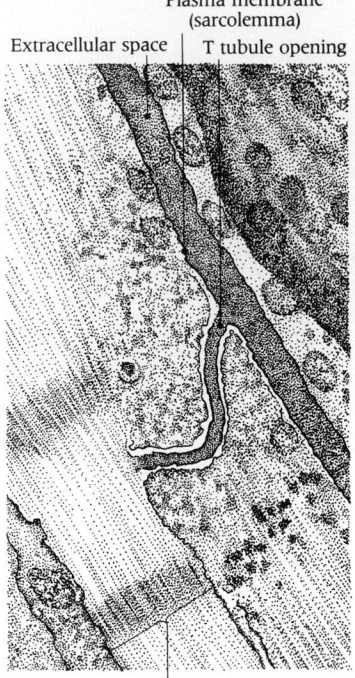

31-14 A T tubule entering a myofibril
This diagram shows a longitudinal section of the periphery of a frog muscle fiber. The T and extracellular space are filled with dense tracer molecules that make them more visible. Note that the T system and the extracellular space are continuous. ($\times$ 28,000)

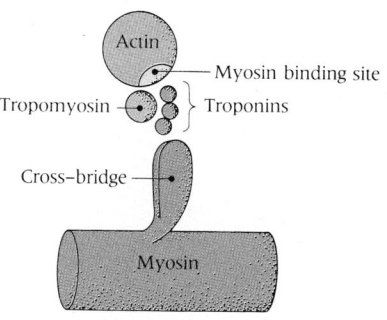

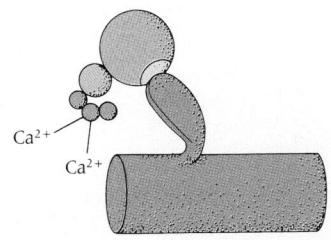

31-15 Control of cross-bridge formation by calcium ions
In the absence of Ca^{2+} tropomyosin blocks the myosin binding site on actin. When Ca^{2+} is released from the sarcoplasmic reticulum it binds to one of the three troponin units. This shifts tropomyosin away from the binding site so that myosin can bind to actin. Compare with Figure 31-12.

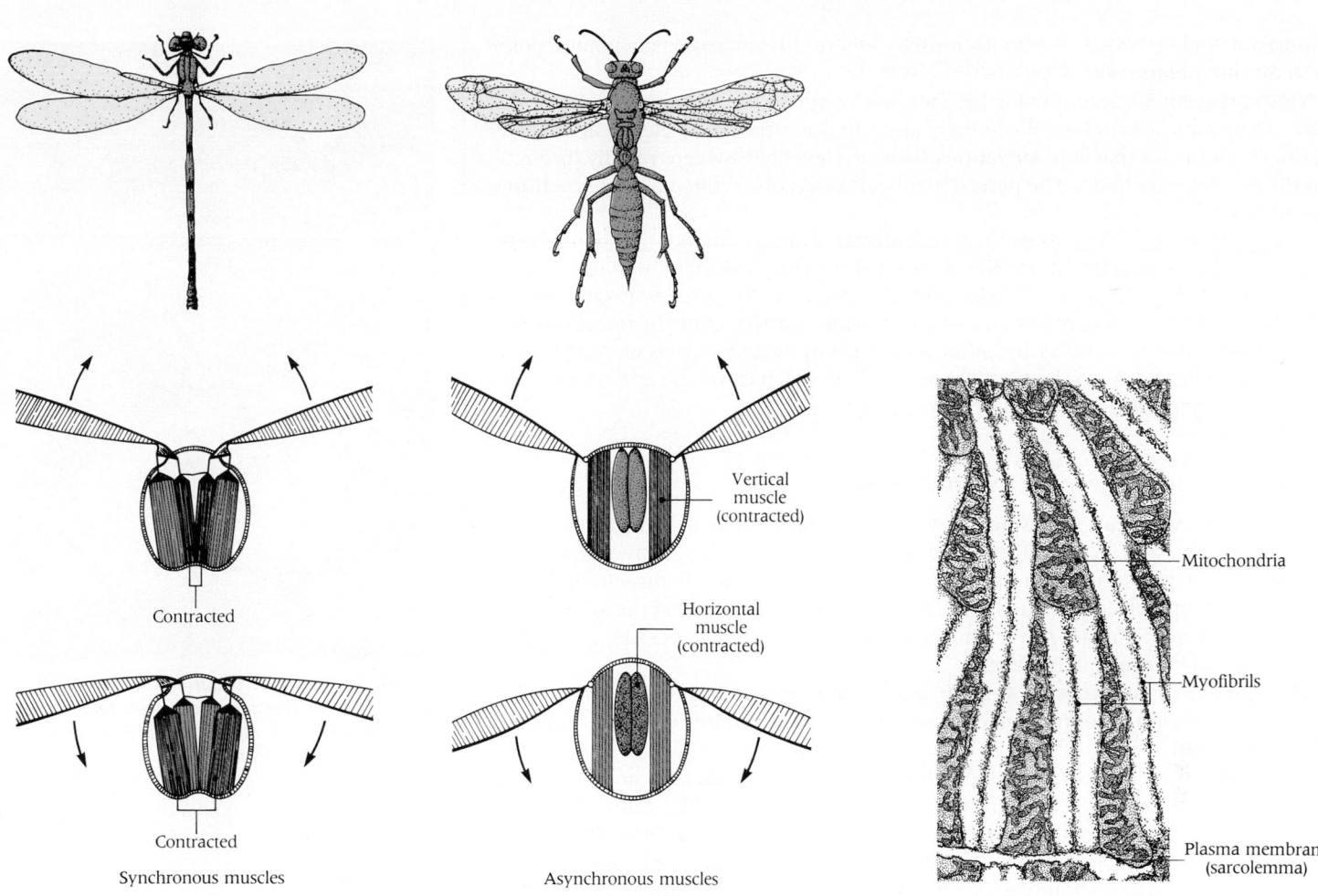

31-16 Synchronous and asynchronous muscles compared
(A) *In the damselfly,* Enallagma, *the flight muscles are synchronous. A section through the thorax shows that contraction of inner muscles lifts the wings. Contraction of the thicker outer muscles lowers them.* **(B)** *In the wasp* (Polistes), *the wings oscillate as antagonistic vertical and horizontal muscles alternately contract.* **(C)** *The flight muscle of a fly (as it appears in the electron microscope) showing numerous mitochondria among the muscle fibers.* (× 18,000)

- The mechanisms that permit asynchronous muscle to contract rapidly lie within the filaments themselves. When filaments or muscle are removed and suspended in a medium containing Ca^{2+} and ATP, they contract and relax in a rapid oscillatory fashion. Thus their activity seems not to be under the control of Ca^{2+} give-and-take.

In muscles of this type, the T system presumably turns on and turns off the cycle of contractions, instead of controlling each contraction by simply relaying information (as in synchronous muscle) from the reticulum. In this way, only an infrequent nerve impulse would be needed to keep the contractions going.

NONSKELETAL MUSCLE

Cardiac Muscle

Not all vertebrate muscle requires nerve impulses to initiate contractions. **Cardiac muscle,** a special kind of striated muscle in the hearts of vertebrates, contracts rhythmically in the absence of nerve stimulation. The contractions are **myogenic,** that is, spontaneous and automatic. The muscle's rhythmicity is said to be intrinsic. Interestingly, this tendency of cardiac muscle to contract is observed early in embryonic life, long before an actual heart has formed.

Unlike skeletal striated muscle, cardiac muscle cells are branched and they are mixed with noncontractile cells (Figure 31-17). Another distinctive feature is the close electrical contact established between neighboring fibers by structures termed **intercalated disks,** specialized regions of the plasma membrane which allow direct cytoplasmic contact between fibers. Actin filaments are anchored to the plasma membrane at the intercalated disk (Figure 31-18). This involves two actin-binding proteins. One, **vinculin,** we encountered in a discussion of actin-binding proteins in Chapter 7. In cardiac muscle it inserts into the plasma membrane and binds to **α-actinin.** This protein is also found in the Z disks of striated muscle and the cytoplasmic dense bodies of smooth muscle, where it is the major anchorage point for actin microfilaments.

This arrangement accounts for the fact that contractions in the ventricles of the heart occur almost simultaneously with those in the atria. Close electrical contacts among fibers means that depolarization of one or a few fibers sweeps rapidly throughout the neighboring fibers. The parts of the heart thus contract in a highly synchronized manner.[4]

Nerve impulses arriving from the central nervous system influence the spontaneous beating of the vertebrate heart. Nerves from the parasympathetic nervous system (particularly the vagus nerve) release acetylcholine and slow the heart rate. Nerves from the sympathetic system release norepinephrine and accelerate the rate. Epinephrine released into the blood by the adrenal medulla mimics the effects of norepinephrine. That explains why excitement or fear, both of which cause the adrenal medulla to secrete epinephrine, make the heart beat faster.

Smooth Muscle

Smooth (or **visceral**) **muscle** is found in many internal organs—in the walls of arteries and veins, digestive tract, bladder, and reproductive organs. Its role is to exert pressure on the space it encircles. Thus smooth muscle promotes peristaltic movements in the small intestine, maintains arterial blood pressure, eliminates urine from the bladder, ejaculates semen from the seminal vesicle, and expels babies from the uterus in childbirth.

The spindle-shaped cells of smooth muscle are arranged in sheets into which nerve fibers penetrate. Contraction of smooth muscle often continues for long periods without fatigue.

In smooth muscle tissues of vertebrates, component muscle cells are of variable length. They contain actin and myosin in quantity; however, the ratio of actin to myosin is 15 to 1 rather than 2 to 1 as in skeletal muscle. The arrangement of the two proteins is more similar to that of the motile nonmuscle cells discussed in Chapter 7. There are no striations. Electron microscopy shows that in smooth muscle cells thick and thin filaments are distributed throughout the cytoplasm, aligned with the long axes of the cells (Figure 31-19A). The thin actin filaments are anchored in electron-dense **dense bodies,** which are distributed beneath the sarcolemma. These have the same function as Z lines in skeletal muscle. Thin filaments contain small amounts of tropomyosin. Troponin has not yet been detected.

Between the thin filaments are the thick filaments containing myosin. Myosin molecules isolated from smooth muscle can be artificially induced to form thick filaments like those of striated muscle; however, the thick filaments in smooth muscle cells are organized differently. The projections thought to be myosin heads are distributed along the whole length of the fibers and no central bare zone is visible. Cross-bridges connect the thick and thin filaments of smooth muscle but do not occur in the repeated pattern typical of striated muscle. Perhaps the thick filaments of smooth muscle reflect an arrangement of myosin molecules in which the top and bottom sides of the filament have opposite polarity (Figure 31-19B).

During contraction individual smooth muscle cells undergo shortening, due to sliding of thin over thick filaments. The process is powered by a cyclic, ATP-dependent, cross-bridge linkage between actin and myosin. Because the pattern of thick and thin fibers in smooth muscle is less ordered, it has not yet been possible to observe the sliding directly.

Smooth muscle contraction is also controlled by Ca^{2+} ions, the concentration of which is regulated by sarcoplasmic reticulum vesicles like those of striated muscle. However, since troponin is absent in smooth muscle cells, Ca^{2+} cannot exert its control as it does in skeletal muscle. Instead, a key controlling role is played by **calmodulin.** Calmodulin closely resembles troponin, sharing about 70% of the amino acid sequence of one of the subunits of troponin in striated muscle. Like troponin, calmodulin binds four Ca^{2+} ions (see Chapter 28).

On binding Ca^{2+}, calmodulin undergoes extensive conformational changes as it

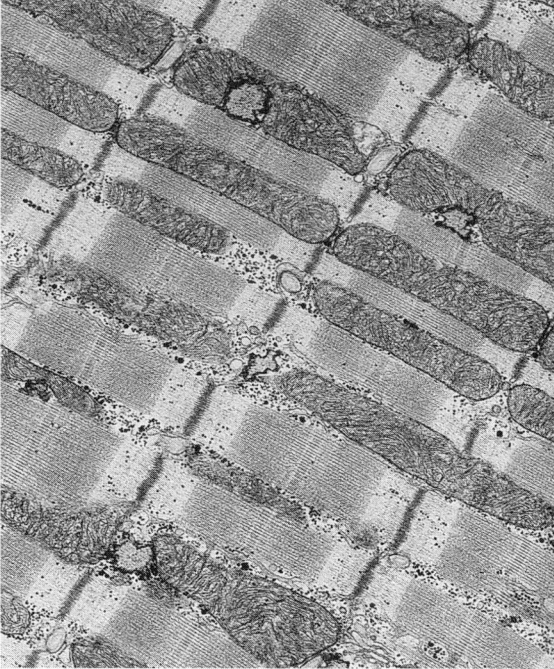

31-17 Structure of cardiac muscle
An electron micrograph of a longitudinal section showing numerous mitochondria between the sarcomeres. ($\times$ 12,500)

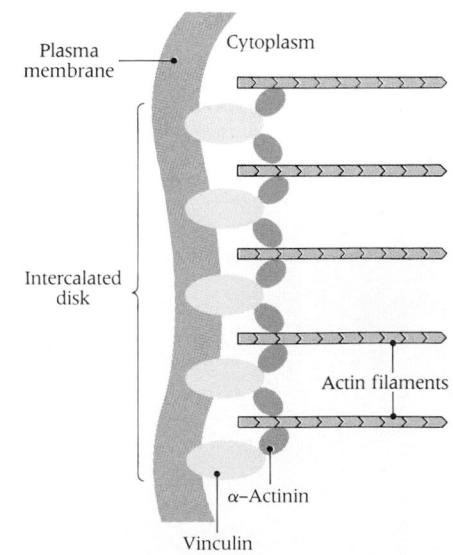

31-18 Structure of an intercalated disk
Actin filaments are anchored in the plasma membrane in the region of the intercalated disk through the intervention of two actin-binding proteins, vinculin and α-actinin.

---

[4] An interesting special feature in the heart is the pacemaker, a small group of modified muscle cells that spontaneously produce rhythmic depolarizations. The peculiar properties of the pacemaker are described in Chapter 24.

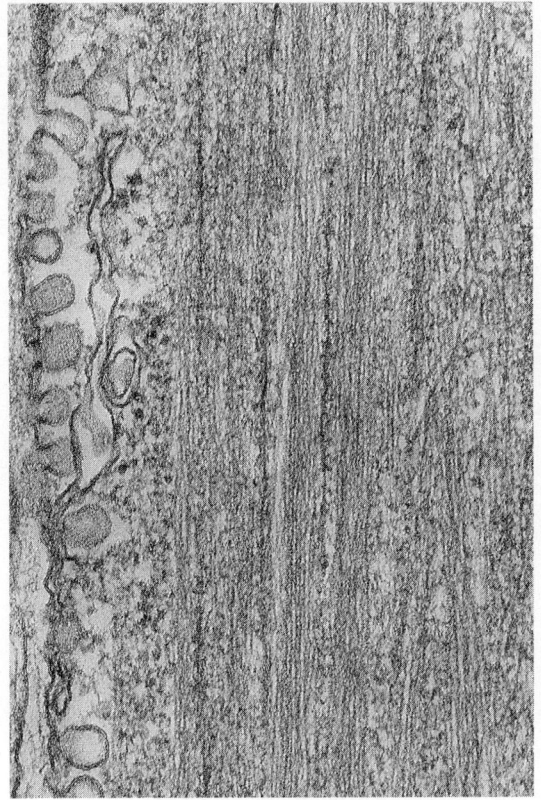

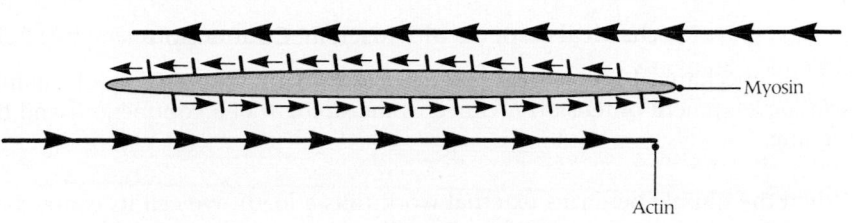

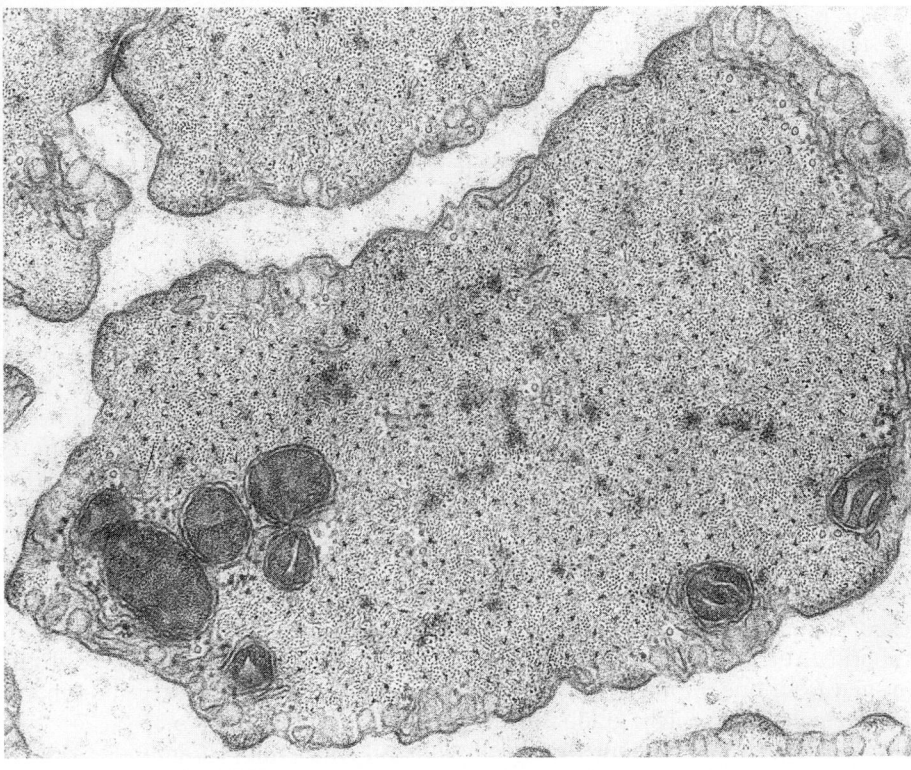

A

31-19 Smooth muscle
(A) Electron micrograph showing the dense bodies in smooth muscle to which actin filaments are anchored. (× 10,000) (B) Postulated arrangement of myosin heads (vertical lines) showing polarity of the heads on the two sides of the thick filament. (C) Electron micrograph of vascular smooth muscle cell in cross section showing actin and myosin filaments. (× 32,000)

converts from inactive to active form. In its active form, it regulates smooth muscle contraction by the following chain of events:

- Ca^{2+} is released by the smooth muscle sarcoplasmic reticulum vesicles in response to a nerve impulse reaching the smooth muscle cell. Smooth muscle lacks a T system.
- Calmodulin binds the Ca^{2+} released into the cytoplasm and is thereby converted to its active form, which stimulates the phosphorylating activity of an enzyme called myosin light-chain kinase.
- Phosphorylation of the myosin light chains releases inhibition of the myosin head group ATPase activity, thereby activating contraction. This is a much slower process than the ATPase-mediated hydrolysis of ATP in skeletal muscle.

Another familiar second messenger—cAMP—also modulates smooth muscle contraction by activating the cAMP-mediated protein kinase. This kinase phosphorylates the myosin light-chain kinase and inactivates it by lowering its binding affinity for the Ca^{2+}-calmodulin complex. Inactivation of this kinase causes the muscle to remain relaxed.

FUNCTIONAL PATTERNS OF MUSCULAR ACTIVITY

Each mammalian skeletal muscle fiber is innervated by only one motor neuron. However, one motor neuron innervates more than one muscle fiber. We call a motor neuron and the population of fibers it innervates a **motor unit.**

What is often referred to as the "classic" physiology of muscle is concerned with the properties of anatomically intact muscle, its neural excitation, and physical performance. The many interesting phenomena at this biological level, the level of motor units and intact muscle, ultimately relate to events at the molecular level.

Mechanical Aspects of Muscular Contraction

Two basic mechanical changes are associated with the contraction of an intact muscle: development of tension (defined as the strength of a contraction) and fiber shortening.

- When the muscle performs external work (lifts a load), we call its contraction **isotonic** because muscle tension remains constant while muscle length shortens. Isotonic contraction can be studied by hooking a muscle to one end of a lever and hanging weights near the fulcrum of the lever. The response to muscle stimulation is recorded by a stylus at the other end of the lever.
- When the muscle develops tension without performing external work, we call its contraction **isometric** because muscle length remains constant while muscle tension changes. If the contracting muscle is connected to a tension-measuring transducer that prevents shortening, tension development is found to be greatest when the muscle is set at the average length it has in the body (Figure 31-20).

Such studies yield a number of significant observations. In a muscle stimulated by a single maximum-strength volley of nerve impulses (or by a brief electrical pulse), a single sudden twitch occurs (Figure 31-21A). An isotonic twitch, interestingly, begins and ends more slowly than an isometric twitch, owing to the inertia of the load and the momentum of its return to rest, which causes it to overshoot.

If a second maximum stimulus is applied within a few milliseconds of the first stimulus, no additional response occurs. If it is applied slightly later but before the first twitch has ended, tension develops further (Figure 31-21B). This **summation** effect is due either to an increase in the number of motor units contracting simultaneously or to an increase in the rate of contraction of each motor unit.

If additional stimuli are applied at progressively shorter intervals, the summation effect is intensified. At rates between 50 and 100 stimuli per second, a tremulous response, termed subtetanus, occurs. At rates above 100 stimuli per second, a full response, **tetanus,** occurs. In tetanus, successive contractions cannot be distinguished from one another, and a maximum level of tension, far exceeding that of the original single twitch, gradually develops.

According to the eminent British physiologist, A. V. Hill of Cambridge University, a contracting muscle behaves as a two-component system, with a contractile component (muscle fibers) in series with an elastic component (noncontractile connective tissue and tendon). In isotonic contraction, the contractile component shortens and the elastic component maintains a constant length. The contractile component also shortens to some extent even in isometric contraction, and the elastic component lengthens. The development of tension depends upon these two actions.

A muscle may contract very rapidly when it has no load. As load increases, velocity of contraction decreases. The essential relationships are

$$V = \frac{(p_o - p)\, b}{p + a} \tag{31.1}$$

where V is the initial speed of shortening, p is the actual force acting on the muscle, p_o is the maximum tension that the muscle can develop, and a and b are constants with the dimensions of force and speed, respectively. Increased load also decreases initial speed of contraction and maximum degree of muscle shortening. Equation 31.1 fits all muscles investigated so far.

Heat Production

In another useful method for the study of muscle function, heat production is measured during the various phases of contraction. Stimulation of muscle causes its heat production to rise considerably above the resting level. This effect precedes and then accompanies the mechanical events of contraction (Figure 31-22). Hill showed that muscular heat production can be divided into initial heat and heat of recovery.

Initial heat is liberated during actual contraction. Most of it represents the energy expended by the muscle in keeping up the active state. In isometric contrac-

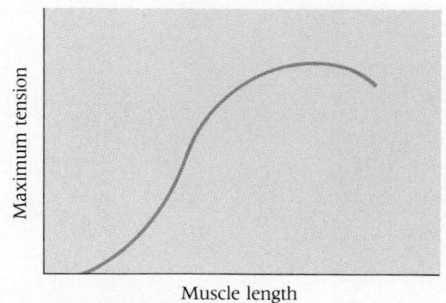

31-20 Relation between muscle tension and muscle length
Muscle tension peaks at a muscle length that is equal to its average length in the body.

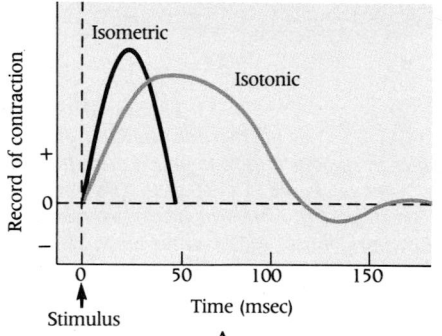

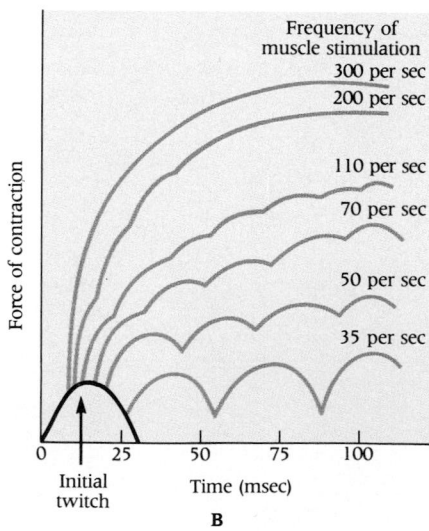

31-21 Properties of single and repeated muscle twitches
(A) *A muscle responds to a single stimulus with a single twitch, which is rapid when isometric or more prolonged when isotonic.* **(B)** *If the muscle is stimulated at progressively shorter intervals, contractions begin to add together (summate), until all the muscle fibers are contracting and maximum shortening occurs. This state is called tetanus.*

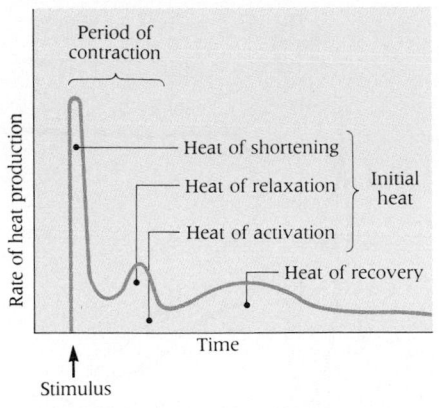

31-22 Heat production during muscular contraction
Initial heat is a direct consequence of muscle contraction. Heat of recovery is a post-contraction process that depends on ATP synthesis.

tion, virtually all the heat produced is of this type. In isotonic contraction, extra heat is produced in proportion to the muscle shortening. This heat is related to the energy expended by the cross bridges as the actin filaments slide over the myosin filaments.

Heat of recovery is liberated in the resynthesis of ATP by oxidative phosphorylation and glycolysis. When muscle is freely supplied with O_2, the heat of recovery is about equal to the initial heat. Hill found that the ratio of total energy (initial heat plus heat of recovery) to initial energy (initial heat plus work performed) is almost exactly 2:1.

Less than 25% of the food energy entering a muscle is converted to work. The remainder becomes heat. Maximum efficiency is realized only when the muscle contracts at moderate speed. If contraction is very slow, a large amount of heat of activation is liberated. If contraction is very rapid, too much energy is wasted in overcoming friction.

Muscular Fatigue

Prolonged contraction of a muscle leads to fatigue. Changes in fatigued muscle are induced by decreased availability of O_2 and accumulation of waste products (Chapter 26). Both of these conditions are offset by a free flow of blood. Hence interruption of blood flow to a muscle greatly hastens the onset of fatigue. There are mechanical problems in maintaining blood flow during prolonged contraction, since increased intramuscular tension tends to hamper flow.

An excessively fatigued muscle is likely to remain contracted (muscle "cramp"), apparently because the small quantities of energy needed to promote relaxation are unavailable. Contraction continues until new ATP has been formed.

MECHANICAL DESIGN IN ORGANISMS

In recent years, biologists have come to recognize that a critical determinant of evolutionary history was the mechanical design of organisms. They also conceded that this is an appropriate area of study for mechanical engineers, physicists, materials scientists, and what we now call bioengineers. Again and again, the viewpoint of the engineer has achieved interesting new insights at every level of biological organization, from molecules to ecosystems.

THE IMPORTANCE OF MATERIALS

All organisms, especially multicellular organisms, need structural support. They also need some sort of outer barrier to keep their fluids from trickling away and their "innards" from falling out. Structures of this type are best considered in light of basic engineering concepts such as tensile strength, compressibility, rigidity, deformability, pliancy, viscosity, elasticity, and permeability. These properties are best understood in terms of the interactions of atoms and molecules. That idea has led to new interpretations of the mechanical properties of the materials nature has used to construct support systems: bones, shells, skin, tendons, stems, and so on.

To do its job successfully, a structural element of an organism must be able to withstand patterns of stress that may be quite complex. All components must be structurally compatible. The bits must fit together—and stay fitted together. An element of the structure must be able to transmit a force over a certain distance and neither break nor deform excessively in doing so. It should also have a reserve of strength that will enable it to cope with unexpected overloads. Finally, it should use the least amount of materials to minimize the metabolic costs of synthesis and maintenance. These principles would satisfy any mechanical engineer designing a bridge, a computer, or a military weapon system.

SUPPORT SYSTEMS

The support system of every organism (excepting some aquatic forms) includes some sort of mechanical support device to hold up its weight and some sort of enclosure device to maintain its shape and to hold it together. Using a simple tent as an analogy, a central tent pole kept from bending or buckling by ropes or wires would be a support device. The canvas "membrane" of the tent would be an enclosure device.

How often we take for granted what is remarkable in our surroundings. For example, a person can stand in the middle of a room—that alone is an extraordinary feat requiring structural support, sensory and effector systems of maintaining balance, and so on—and then extend an arm horizontally to one side while raising a leg (Figure 31-23). An engineer would have a hard time designing a system that could carry out those motions while flawlessly maintaining positional stability.

We can do these things because we have a **skeleton,** the components (bones) of which are in engineering parlance *compression elements*. Like the girders in a building, they are stiff units that must resist collapse, buckling, or volume decrease. We must also have *tensile elements*—cords, strings, and tissues (tendons and muscles) that can withstand a stretching load without tearing. The elements of both living and nonliving structures are usually one or the other—compression or tensile elements—but one ingenious engineer, R. Buckminster Fuller, devised a principle of design he called **tensegrity** in which compression elements are greatly reduced and widely separated within a tensile web or membrane. This is the basis of his "geodesic dome" (Figure 31-24). We will see that nature "invented" it much earlier in the design of many insects.

Textbooks of biology have generally called the skeleton nature's foremost support system. We now realize that bones alone (compression elements) could not support the body without the aid of tensile elements (tendons, connective tissues, and muscles). Together, these comprise an organism's support system.

Hydrostatic Skeletons

The earliest organisms lived in the sea. We may visualize these primordial forms as having two parts: an outer membrane made up of organic molecules and an internal fluid containing various particulate inclusions. Irrespective of the composition of this fluid, we know that all fluids are noncompressible. Hence, these creatures were compression-resistant entities right from the start.

A watery sealike fluid proved useful and it remained a basic feature of virtually every species throughout evolution. But further evolution of the structural systems of both plants and animals necessitated the elaboration of flexible materials that would permit twisting, stretching, and other contortions, and of new kinds of stiff materials that could resist bending as well as compression. Only when this problem was solved was it possible for the numerous species of arthropods, vertebrates, and complex plants to propagate on land. Lack of success in elaborating new flexible and stiff materials in the later evolution of mollusks may account for their failure to exploit the land.

Organisms that remained in the sea had to withstand the pressure of the surrounding water. They met this challenge in a number of ways. Some (like jellyfishes and sea anemones) became jellylike forms. Some (like bivalves and snails) formed heavy rigid shells. Others (like corals) acquired rocklike accretions of calcium carbonate and other minerals.

An important advance was the development of the **hydrostatic skeleton,** so named because its support system depends in part on body fluids. These contrivances, which are found in both aquatic forms (squids, snails, etc.) and terrestrial forms (earthworms, caterpillars, etc.) (Figure 31-25) embody two essential features: (1) an internal fluid or semifluid volume that is incompressible, and (2) a body wall that consists of a remarkable combination of muscle fibers and other fibers that are wound around the body in a manner that pits them against each other. Engineers call such structures fiber-wound cylinders. Biologists refer to them as **stretched membrane hydrostats.**

Stretched membrane hydrostats are powered by sheets of muscle that can distribute forces and loads over large areas of the body wall. Animals utilizing them as their structural support system are generally cylindrical in shape. In their body walls,

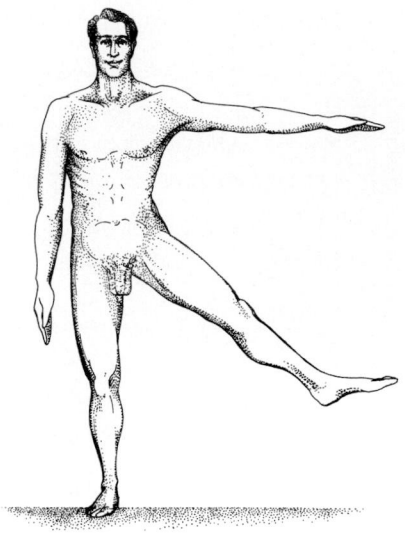

31-23 Support systems in humans
The remarkable ability of humans to maintain positional stability using internal support systems is here demonstrated. Readers are invited to try this themselves.

31-24 A geodesic dome
One of many designed by R. Buckminster Fuller.

31-25 A few animals with fluid compression support systems

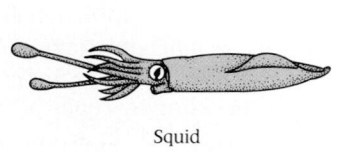

Squid

Snail

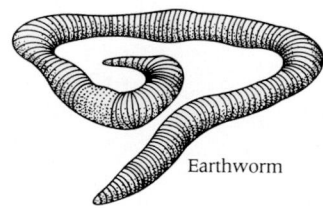

Earthworm

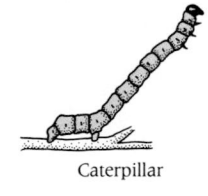

Caterpillar

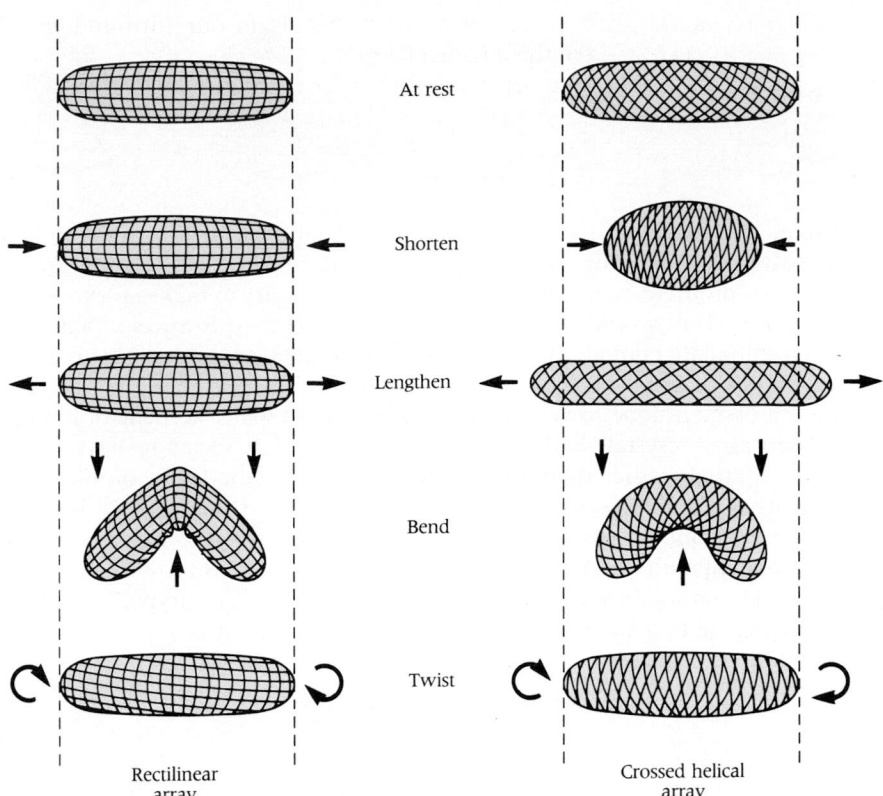

At rest

Shorten

Lengthen

Bend

Twist

Rectilinear array

Crossed helical array

multiple layers of fibrous connective tissue are wrapped in a crossed helical array around the body (Figure 31-26). Examples are seen in the body walls of many worms, in the hydraulic, contractile tube feet of echinoderms (starfishes, sea cucumbers, etc.), and in the mesogleal tissues of the jellyfishes called medusae. Such a body wall structure efficiently resists tensile forces. The fluid in their bodies makes these animals compression-resistant. This was the road taken by primitive animal life, which arose in the sea where incompressible water is abundant and cheap.

The helical array of collagen fibers in the body walls of hydrostatic animals allows their bodies to stretch, shorten, twist, and bend without kinking. This marvel of engineering evolved many millions of years before engineers learned to use strong, light, cheap fibers in crossed helical array to reinforce gas tanks and tires. Similar designs in animal intestines permit peristalsis and in arteries, permit stretching and resist tension.

Jointed Frameworks

Hydrostatic support systems are ideal for organisms living in or near water. But it became desirable to reduce dependence on watery fluids as a major support element—especially as plants and animals moved onto land, but also for those that remained near water. Terrestrial habitats subjected organisms to relatively greater gravitational forces than they experienced when surrounded by water. Those organisms that flourished had acquired (1) rigid support elements that could maintain body form, (2) jointed frameworks in their structural systems that permitted locomotion, running, and other actions, and (3) muscles of large cross-section that added force to body movements. The major phyla that achieved these advances were the mollusks, echinoderms, arthropods, and vertebrates.

There are two kinds of skeletons in the animal kingdom. Each became highly developed in one of the two groups in which the most elaborate behavior evolved: insects and vertebrates. In insects, the skeleton is a hollow shell on the outside of the body, and the muscles are attached to the inside of the hard outer shell (Figure 31-27), which is usually made of chitin, not bone. We call this an **exoskeleton.**[5] In vertebrates, the skeleton is inside the body and muscles are attached around it, fastening to the outer surfaces of bones. This is an **endoskeleton.**

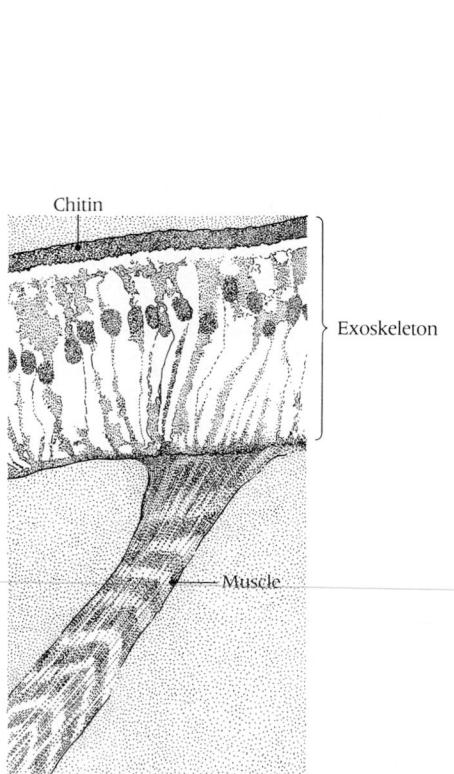

Chitin

Exoskeleton

Muscle

31-27 Muscle attachment to an insect exoskeleton
In insects and other arthropods, the musculature is attached to the inside of the exoskeleton.

[5] Many noninsect invertebrates have exoskeletons, the materials and arrangements of which are highly diverse. For example, most mollusks have carbonate exoskeletons. Brachiopods have both exo- and endoskeletons, as do some vertebrates.

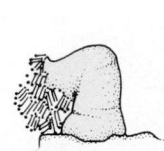 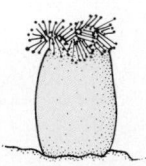 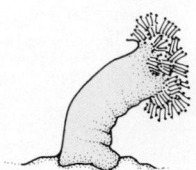 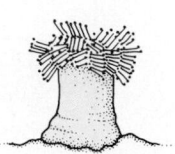 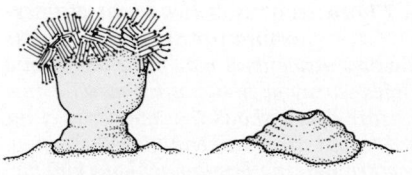

Mechanically speaking, the two systems are equally effective in small animals. In extremely small ones, an exoskeleton may be more effective. In larger animals, however, an inordinate increase in bulk would be necessary to make an exoskeleton strong enough. This would make it impossibly difficult to form workable joints and to accommodate enough muscles to budge the heavy apparatus. This is a major reason why insects are smaller than most vertebrates.

Bone is a tissue unique to vertebrates, a strong, hard material made of inorganic calcium phosphate crystals and organic collagen fibers. It seems to have evolved first as an external, rather than an internal, skeleton in the fishes of 400 or 500 million years ago called ostracoderms ("shell skinned"). One group of these, the osteostracans ("bone shelled"), had a bony covering over the head end of the body. The main hypotheses on why bone evolved concern its protective value against invertebrate predators—such as the giant water scorpions, or eurypterids (Chapter 44)—and its value as a storage site for calcium and phosphorus.

THREE EXAMPLES

Sea Anemones

Sea anemones are coelenterates with open extensible cylinders that are attached to a solid surface. One writer remarked that an open cylinder whose wall is flexible enough to collapse of its own weight sounds more like an empty stocking than it does an organism. Nonetheless, these remarkable creatures—which are classified as anthozoans (see Chapter 42)—use their mouths to control the ingress and egress of sea water and the muscles in their hydrostatic body walls to control body size. In this way, they can expand into tall vertical columns or shrink down into small wrinkled blobs (Figure 31-28).

All of these activities are made possible by an open cylindrical body with circumferential, radial, and longitudinal muscles in the walls, and long membranes (called mesenteries) that protrude inward from those walls. The main supportive material in the body walls is **mesoglea,** a viscous elastic composite of collagen fibrils in a fluid matrix that is capable of slowly recovering from mechanical deformation without the aid of muscular forces. There is also a crossed helical array of fibers in the outer layer and radially oriented fibers in the inner layer.

Earthworms

An earthworm, a member of the annelid phylum, is characterized by long cylindrical shape and extensive segmentation, with individual fluid-filled segments partitioned from one another and various organs—nerve ganglia, excretory organs, etc.—repeated in each segment (see Chapters 25 and 42). The design of annelids illustrates the benefits of a hydrostatic skeleton.

In addition, their well-developed, but still relatively simple, muscular system nicely illustrates the integrated nature of muscular activity. The striated muscles responsible for locomotion behavior form two nearly complete cylinders, one inside the other, throughout the body just inside the outer wall. In the outer layer, the fibers encircle the body; in the inner layer, they run lengthwise. Body segments in which outer layer muscles have contracted and inner layer ones have relaxed become long and thin (Figure 31-29). Where outer layer fibers relax and inner ones contract, that body section becomes short and thick. There are also paired bristles called **setae** down the sides of the body, each with its own small muscle fibers that erect the bristle or pull it back. The worm uses setae to grip the soil as it inches forward. Bracing itself with its bristles, the worm then sends long-thin and short-thick contraction waves along its body. The nerve cords (a fused pair) exercise central control over these muscular contractions, but what we call a brain only out of courtesy has little to do with the process. An earthworm with its front end cut off moves along just about as well as an intact worm.

31-28 The posturing of a sea anemone, Meritridium senile
This coelenterate can use its muscles to shrink down to a wrinkled blob, or it can take in sea water and increase its height to four times its width. It can do this because its hydrostat does not have a constant volume.

31-29 Locomotion in an earthworm
In forward motion, contraction of circular muscles squeezes forward segments into a long tube. Contraction of longitudinal muscles then shortens the tube. Setae anchor the worm to the soil. The longitudinal section of an earthworm shows the locations of muscle layers, setae, and the septa that separate segments from each other. More details of earthworm anatomy are shown in Figures 24-23C and 25-8.

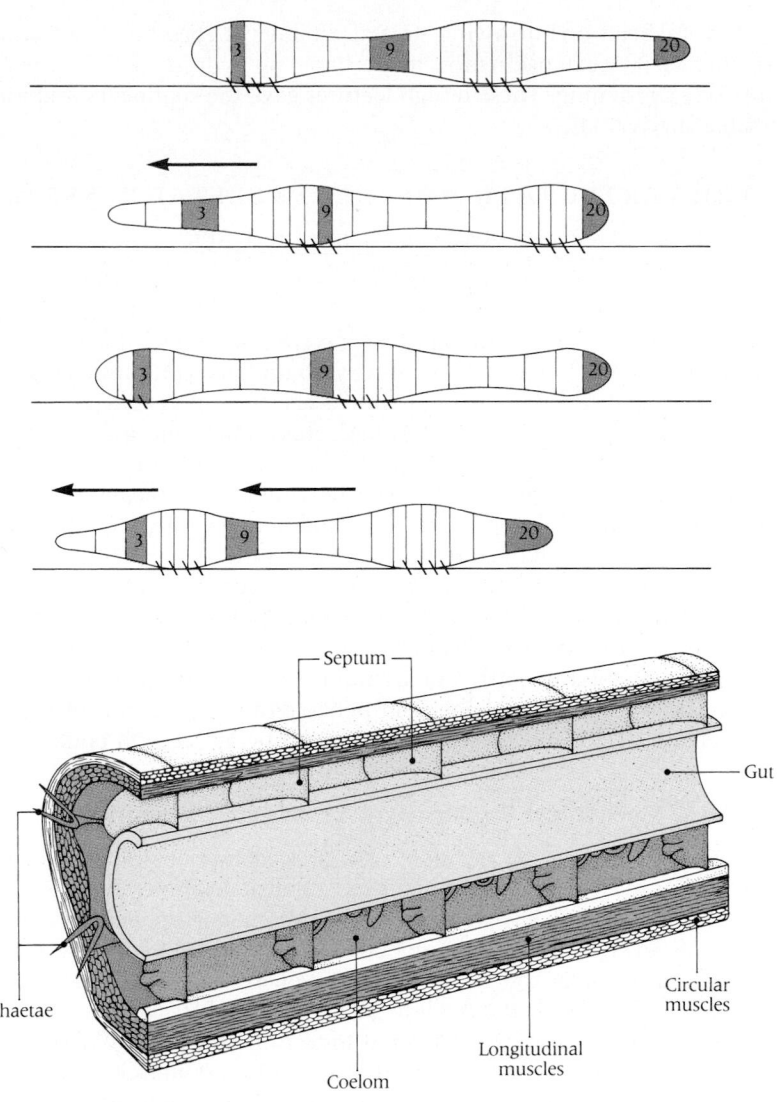

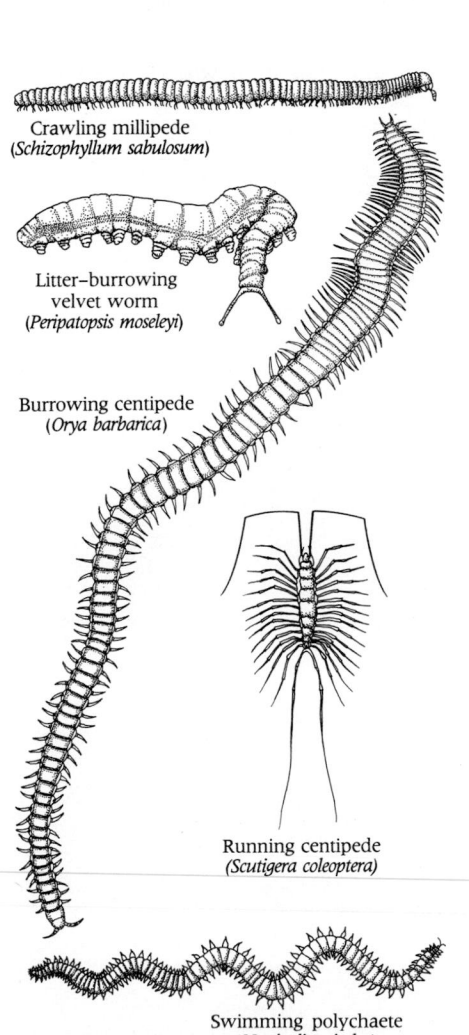

Crawling millipede
(*Schizophyllum sabulosum*)

Litter–burrowing velvet worm
(*Peripatopsis moseleyi*)

Burrowing centipede
(*Orya barbarica*)

Running centipede
(*Scutigera coleoptera*)

Swimming polychaete
(*Nereis diversicolor*)

31-30 Some segmented invertebrates
Segmentation of these wormlike creatures facilitates their running and burrowing. The burrowing centipede has just begun to move. Walking movements that began near the head have reached half way down the body.

As in vertebrates, a cylindrical sheet of muscular tissue around the worm's intestine moves food along by rhythmic contractions. There are also muscles in the walls of some blood vessels, especially in the worm's five pairs of hearts. The intestinal and circulatory muscles are visceral smooth muscles.

Centipedes

Centipedes (class Chilopoda) and millipedes (class Diplopoda) are arthropods with very wormlike, highly segmented bodies with one or two pairs of legs on each segment (Figure 31-30). A few centipedes have retained a feature of hydrostatic skeletons with single fluid-filled cavities, termed **haemocoels,** in each segment. But a major difference between them and soft-bodied hydrostatic forms is the inclusion in their outer cuticle of rigid, compression-resistant materials. To keep the many pairs of legs in running centipedes from tripping over one another, an extraordinary tensile suspension system has evolved in some forms in which the floating exoskeletal segments (called **sclerites**) are anchored to each other with muscles. The rigid sclerites of arthropods occur in a linear series along the body: those on the dorsal (back) surface are called **tergites;** those on the ventral (abdominal) surface are called **sternites** (Figure 31-31). These serve as attachment sites for body and leg muscles. In some forms, muscles in each segment go to three successive pairs of legs and to five successive sternites. In one centipede—the fleet *Scutigera*—there are 33 muscles running to the legs from specific sites on the exoskeleton. These are among the forms in which nature first evolved the principle of tensegrity.

These adaptations—a rigid exoskeleton, a segmented body (Chapter 20), a partial dependence on hydrostatic forces—made possible such functional advances as fast running, forceful head-on burrowing, great flexibility with bending based on segmentation (Figure 31-32), extension movements based on hydrostatic forces, and various modes of body shortening. These design features gave the centipedes and millipedes a formidable survival kit.

THE VERTEBRATE MUSCULOSKELETAL SYSTEM
Bones Are More Than Inert Rods

It is said that the musculoskeletal system consists of the skeleton for body support, and voluntary muscles, which permit body movement. That statement suggests a degree of skeletal inertness, as though bones were but a crutch for holding up soft flesh. In fact, bones are living tissues and they do undergo constant change. Indeed, one of their principal functions is to provide a reservoir of essential minerals, such as calcium and phosphorus. Thus bones have mechanical and metabolic functions and we speak of bones as organs because they are made up of many types of tissue (osseous tissue, cartilage, fibrous tissue, nervous tissue, vascular tissue, marrow, etc.). We reviewed the histology of cartilage and bone in Chapter 7, and we noted in discussing the parathyroid glands and vitamin D in Chapter 28 that bone tissue is in constant dynamic chemical equilibrium, individual bones being continuously remodeled during life. Here we are concerned with their mechanical role as effectors.

The human skeleton has two major divisions: the **axial skeleton,** which includes the skull, vertebral column, ribs, and sternum; and the **appendicular skeleton,** which includes limbs, the shoulders and pelvic girdle. A diagram of the skeleton appears in Figure 31-33. An average adult human being has 206 bones.

Mechanical Properties of Vertebrate Skeletons

The vertebrate skeleton is an example of a jointed framework, in which nearly linear muscles usually apply "point loads" to rigid skeletal members. Point load, a term from the physics of levers, denotes a force that is applied to a lever at a single point. A lever is a rigid bar or body that transmits forces by turning at a fixed pivot or fulcrum. In the skeleton, the pivot is a joint. Point loads on long, rigid skeletal elements allow the body to make use of leverage.

A machine is a mechanism that transmits force from one place to another, usually changing its magnitude in the process. Basically, all bone-muscle systems are machines. Some interesting calculations can be made about this machine's properties. Let us define an input force applied to a machine as the in-force (F_i) and the resulting output force from the machine as the out-force (F_o). In the vertebrate body, in-forces are applied by the pull of tendons, by gravity, and by external loads. Useful out-forces are realized at the teeth, feet, fingers, and elsewhere.

Locomotory mechanisms usually employ levers to transmit forces. In these biological levers (Figure 31-34), each force is spaced from the joint by a segment of the lever called the lever arm.

- The in-lever arm (l_i) extends from the in-force to the joint.
- The out-lever arm (l_o) extends from the joint to the out-force.
- The product of force times its lever arm is a moment.

Every functioning lever includes at least two moments, one for the in-system (M_i) and one for the out-system (M_o), which are defined as follows:

$$M_i = F_i l_i \quad \text{and} \quad M_o = F_o l_o \tag{31.2}$$

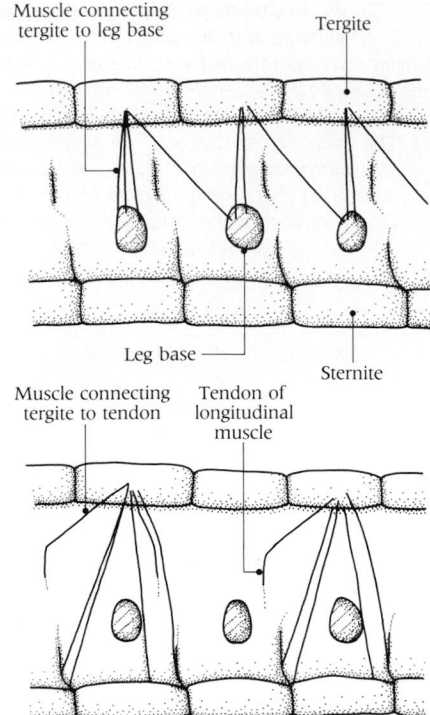

31-31 Attachments of muscles to tergites and sternites in a burrowing centipede
In this species, Haplophilus subterraneus, muscles connect tergites to the leg base or to tendons of longitudinal muscles.

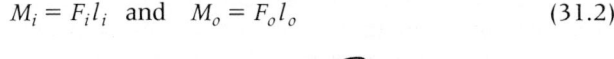
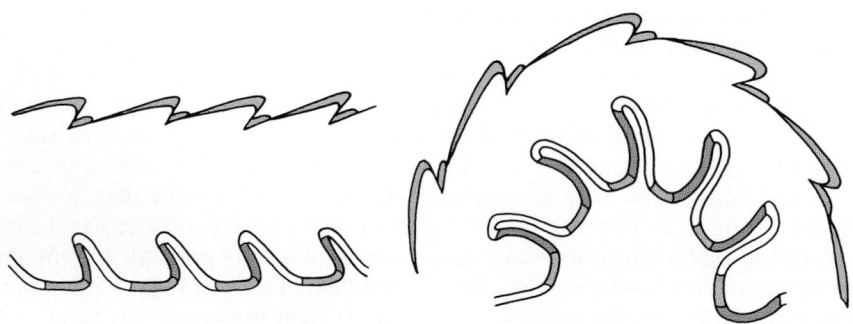

31-32 Bending is made possible by a segmented body plan
Sagittal sections through a burrowing centipede before and after bending show the importance both of segmentation and variability in cuticle rigidity. The most rigid cuticle extends along the top. Cuticle along the bottom is interspersed along with patches of intermediate and low rigidity.

31-33 The human skeleton

The axial skeleton includes the skull, vertebral column, ribs, and sternum. The appendicular skeleton includes limbs, shoulders, and pelvic girdle. The major bones are named.

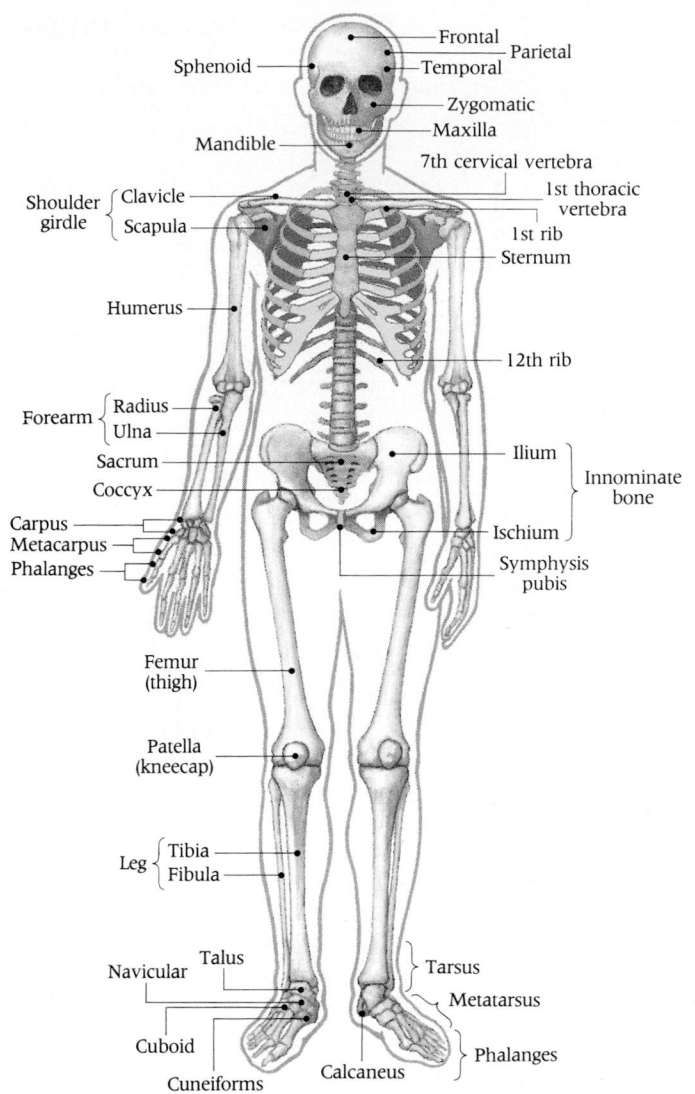

When $F_i l_i = F_o l_o$, a system is in equilibrium. In a standing horse, for example, postural muscles adjust so that the total sum of moments is in equilibrium.

Natural selection acts on limb design in various ways (see Figure 31-34).

■ In a digging mammal such as the mole, a premium is put on optimizing the out-force, F_o, which equals $(F_i l_i)/l_o$. This is accomplished by enlarging F_i by increasing muscle size, lengthening the in-force arm (l_i), and shortening the out-force arm (l_o). The design of the forearm of the mole clearly shows these mechanical adaptations.

■ In contrast, a habitually running mammal, such as a horse, deer, or cheetah, needs high velocity in a given direction rather than power. Thus a premium is put not on maximized out-force but on velocity, which on a lever is determined by its distance from the pivot. In-velocity and out-velocity (respectively, v_i and v_o) are thus related to the lengths of their respective lever arms, but in a reverse way.

$$v_i l_o = v_o l_i$$

Hence,

$$v_o = (v_i l_o)/l_i \qquad (31.3)$$

Since the velocity of muscle contraction (v_i) could not be increased very much in the course of evolution, all running mammals developed in the direction of an increased l_o and a decreased l_i.

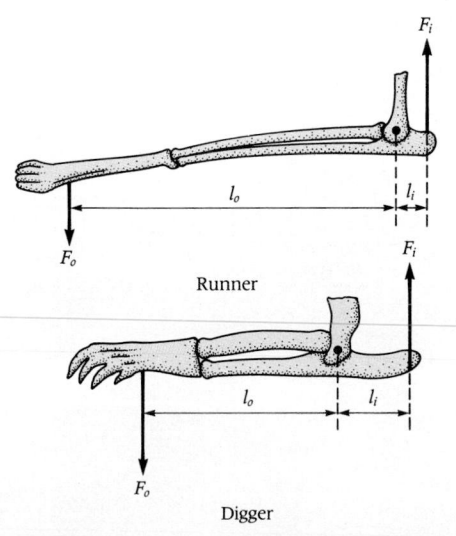

31-34 Mechanical adaptations in forearm structure

The forearm of a digging mammal is shortened and has a lower ratio of out-force to in-force than does the forearm of a running mammal.

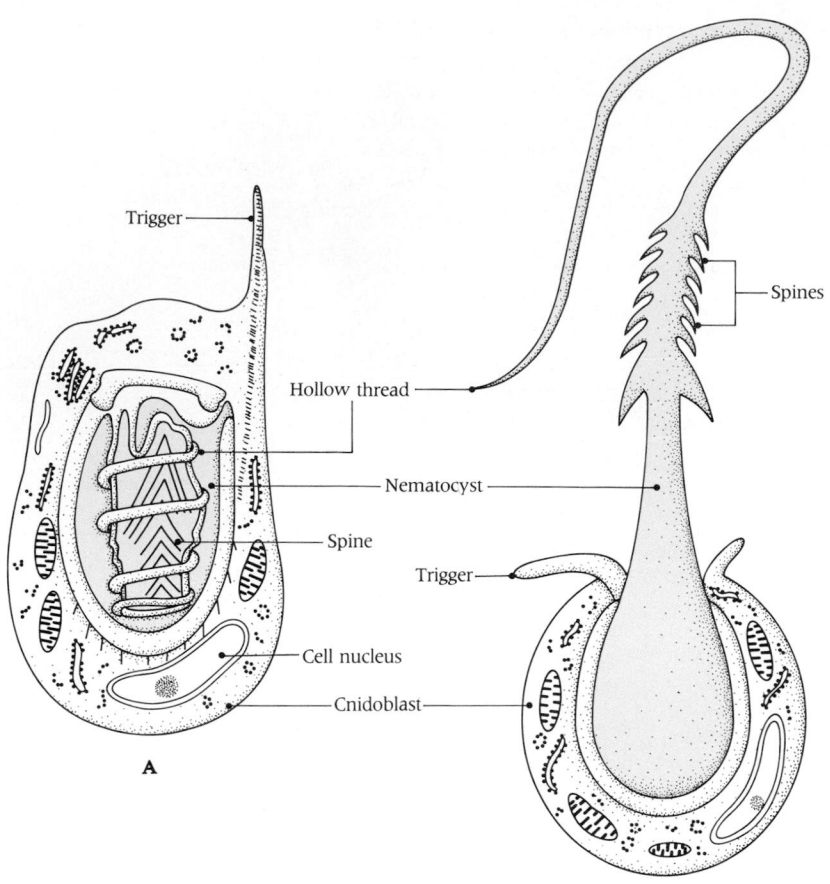

A

B

31-35 Missile-firing system in Hydra
(A) The nematocyst is a harpoonlike structure enclosed in special cells called cnidoblasts. Here the nematocyst is sequestered. **(B)** The nematocyst has been ejected by mechanical and/or chemical stimuli. Recent work has shown that long polymers of the amino acid L-glutamate are responsible for the generation and regulation of the high osmotic pressure within nematocysts. These molecules may be unique to coelenterates.

OTHER EFFECTORS

We mention briefly three other kinds of effectors of major biological importance.

MISSILE-FIRING SYSTEMS

Some organisms possess devices that fire defensive missiles or substances that are as effective in the battle for survival as the bullets, shells, guided missiles, and poison gases invented by the civilized human mind.

A remarkable system is found in corals, hydras, jellyfishes, and their relatives (the coelenterates). In these primitive animals, specialized cells called **cnidoblasts** are scattered over the surface of the body. Cnidoblasts contain miniature barbed

31-36 The defensive system of a bombardier beetle, Brachinus crepitans
(A) When stimulated by the nervous system, the defensive organ manufactures a highly toxic quinone that it releases explosively at temperatures approaching 100°C. The R group of the quinone is H or CH_3. **(B)** A bombardier beetle at rest. In Figure 8-2C, a bombardier beetle is shown in the act of discharging noxious secretions from its defensive organ with remarkably accurate aim.

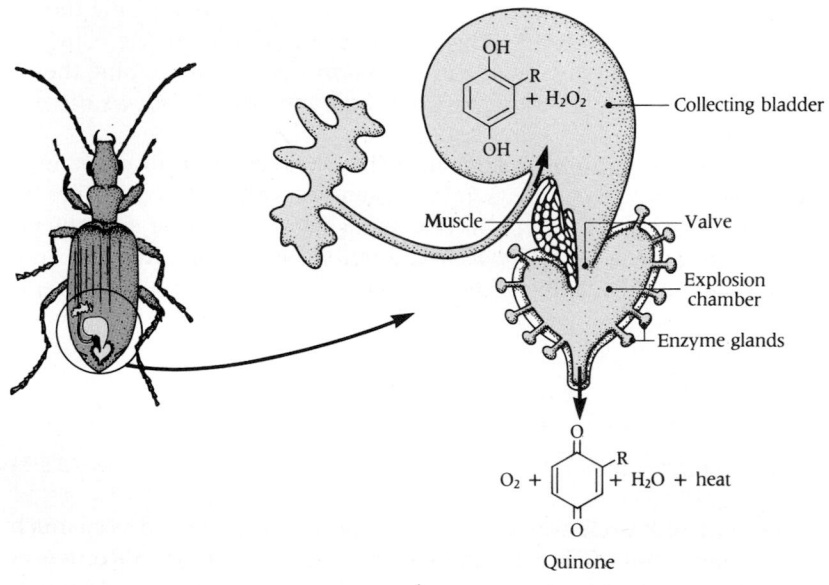

A

B

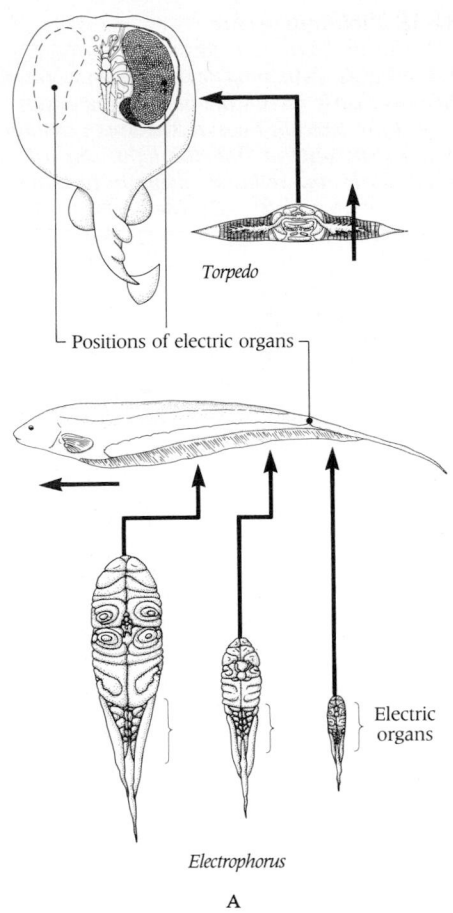

Torpedo

Positions of electric organs

Electric organs

Electrophorus

A

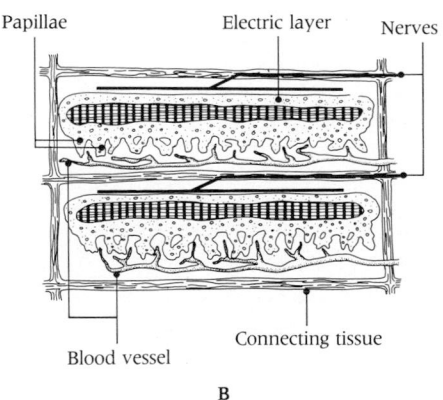

Papillae   Electric layer   Nerves

Blood vessel   Connecting tissue

B

31-37 Electric organs of two strongly electric fishes

(A) The electric eel, Electrophorus, *and the electric ray,* Torpedo, *showing the location of their electric organs in surface view and in sections at levels indicated by three arrows. Heavy arrows show the directions of active current flow through the organs. (B) Diagram of two neighboring electrocytes. Electrocytes (electroplates) are specialized skeletal muscle cells; their many nuclei are near the cell membranes. One face of the electrocyte is smooth with a myoneural junction (synapse); the opposite face is vascular and papillary.*

harpoons called **nematocysts** (Figure 31-35). The resting nematocyst is equipped with an external spinelike trigger and contains a coiled thread (Figure 31-35A). When touched, these cells literally explode, firing off dartlike poison-containing missiles. The thread uncoils, exposing tiny spines (Figure 31-35B). The nematocyst then penetrates or surrounds its victim, releasing poison at the point of contact. Anyone who has brushed against a jellyfish while swimming knows how painful this can be. Nematocysts are fatal to the usual prey of jellyfishes and have killed humans.

The organelle essential to this mechanism, the capsular nematocyst, is a complex secretory product of the cnidoblast's Golgi apparatus. Its firing is an extreme form of exocytosis. High-speed microcinematographic studies of the discharge of nematocysts in *Hydra* show that the firing results from a sudden increase in intracapsular pressure, a burst requiring only 3 milliseconds. The maximum velocity generated, 2 meters per second, requires a force of 40,000 times gravity! This is believed to be the fastest cellular process in nature. Its molecular basis remains unknown.

Many species of insects have evolved the capacity to discharge highly toxic substances. A group of beetles (*Brachinus*), known as bombardier beetles (Figure 31-36A), are common in Europe. They have a defensive organ in the rear of the body that is caused by nerve impulses to emit with explosive force a volley of discharges containing highly corrosive substances (quinones) in a hot cloud of fluid that is nearly boiling. This noxious material is cooked up in a collecting bladder (Figure 31-36B) just before it is discharged like a small machine gun. The biochemical reaction that makes the quinones involves hydrogen peroxide and a set of enzymes that are unusually heat stable.

Ciliated protists eject threadlike objects called **trichocysts** from their surface membranes. Propulsion is initiated by sudden elongation of the trichocyst shaft as it leaves the organism.

ELECTRIC ORGANS

Another unusual effector is found in the members of seven families of fishes that can generate electricity. The **electric organs** in many groups generate only weak voltages—in the microvolt range—and they are used in solving problems of species recognition, communication, and general orientation. Most of these so-called electric fishes live in murky water or other habitats where they cannot rely on vision—or they are active at night.

However, a number of species discharge a powerful pulse of electricity for offensive or defensive purposes. (Some of these, incidentally, also can produce weak electrical fields for peaceful purposes.) The more powerful electric fishes include: the giant electric ray (*Torpedo*), a north-Atlantic fish that puts out shocks of 50 amperes at 50 volts (Figure 31-37A); the African electric catfish (*Malapterurus*), which though smaller emits up to 350 volts; and the electric eel (*Electrophorus*) of South America, which can discharge 500 volts. Others are the electric skate and the freshwater knife fishes (*Gymnotus*). Representing widely separated groups, all of them have somewhat different methods of generating electrical energy. We may therefore assume that electric organs arose on several separate occasions.

The electric organs of these fishes evolved from muscle. They generate an electric potential in the same basic way a muscle does. Although they differ in structure and position in the body, all electric organs are made up of large disk-shaped cells called **electroplaques** that are stacked into columns like coins (Figure 31-37B). The eel, for example, has about 10,000 in each of about 70 columns. Their active cells, electrocytes, are specialized skeletal muscle cells. The electroplaques form an array that resembles a common voltaic cell. When discharged simultaneously as a result of sudden change in membrane permeability, they can produce a jolt large enough to stun a human being or light a panel of light bulbs.

BIOLUMINESCENCE

A third class of effectors includes those that produce the strange light of **bioluminescence**—a fire that burns without consuming—in the flickering of fireflies on summer nights, the glow of fox fire in dark forests, and the dazzling lights of some sea creatures. Many unrelated organisms possess it—among them bacteria, fungi, radiolarians, dinoflagellates, sponges, corals, coelenterates, nemerteans (bright-colored

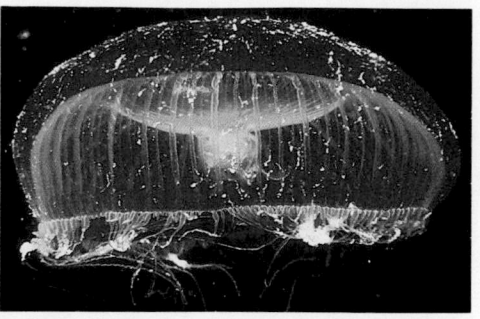

A   **B**

31-38 Bioluminescence
(A) *This flashlight fish,* Photoblepharon palpebratus, *swimming near the Grand Comoro Islands, clearly reveals the luminescent organ located just beneath the eye. The organ contains luminescent bacteria that emit light. The fish is small, dark, and reclusive, living in caves and recesses in deep water.* **(B)** *The Pacific jellyfish,* Aequorea aquorea.

marine worms), ctenophores (animals resembling jellyfish), crustaceans, bivalves, snails, insects, centipedes, millipedes, squids, and fishes.[6] More than half the major divisions (phyla) of the animal kingdom include bioluminescent forms.

The phenomenon is especially impressive among the luminous animals living in the unlighted depths of the sea (see Chapter 48). Consider the flashlight fishes (Figure 31-38A), a group of several types that light up the water with a large organ containing living luminous bacteria under each eye. The flashlight fishes use the light to see by, to communicate, to lure prey, and to confuse enemies.

Bioluminescence is also common in organisms at the sea's surface or that roam shallow waters. For example, the Pacific jellyfish, *Aequorea* (Figure 31-38B), is an organism whose umbrella, measuring three to six inches across, is rimmed with light organs that give off a bluish-green light in response to mechanical or chemical stimulation.

In the 2000 or so species of fireflies—really beetles (family Lampyridae)—of which a hundred inhabit the United States, bioluminescence is a method of communication between the sexes. Each species has a pattern and flashing rhythm of its own. Many of the American species are called roving fireflies because the males fly about singly in search of females. The male flashes his lantern rhythmically. When a female flashes a specific response rhythm, a courtship begins. An interloping male, by picking up the light-flashing rhythm of another amorous male, can trick the female into switching partners. Some mimic flashing patterns in order to lure in flies and then eat them. Certain species in Southeast Asia flash in precise unison hour after hour, a phenomenon that is beautiful to see on a dark night. This may be a form of group courtship behavior in which no male has an advantage over any other male. Many have speculated on how these fireflies synchronize their flashes.

The significance of bioluminescence in other kinds of organisms, particularly the bacteria, fungi, and protozoans, is still unknown. It may be merely an incidental byproduct of certain forms of oxidation, some of which may have evolved originally as protective mechanisms in organisms threatened by the increasing O_2 content of the atmosphere (see Chapter 44). Bioluminescent light is emitted when the complex molecule **luciferin** is oxidized under the influence of the enzyme **luciferase** in specialized cells. As shown in Figure 31-39, the reaction also requires the presence of free oxygen and ATP, the energy source.

---

[6] To be accurate, we should note that fishes are bioluminescent only by virtue of the fact that they have "adopted" bioluminescent bacteria. In fact, no amphibians, reptiles, birds, or mammals are known to be bioluminescent—despite some early claims to the contrary.

$$\text{Luciferin} + \text{ATP} + \text{Mg}^{2+} \xrightarrow{\text{Luciferase}} \text{Luciferin–AMP} + \text{PP}_i$$

31-39 Mechanism of firefly bioluminescence
The enzyme luciferase converts luciferin to reaction products and light. Note the essential participation of ATP.

$$\text{Luciferin–AMP} + \text{O}_2 \xrightarrow{\text{Luciferase}} \text{Bioluminescence} + \text{Reaction products}$$

EFFECTORS IN PLANTS

Plants typically have few, simple, and poorly differentiated receptors, conductors, and effectors. Corseted in cellulose, plant cells have limited freedom of movement. As a result, plant morphogenesis depends more on internal rearrangements of cytoplasm than on cellular migrations. Those plant motions that can be properly called behavior depend not on special effector cells or organs but on the ordinary activities of the ordinary plant cells. The famous experiment by the Darwins described in Chapter 28 showed that more elongation of cells on one side of the stem than on the other can cause a plant to bend, a distinct motion. This mechanism operates in most of the natural movements of plants.

Other mechanisms commonly involved in motions of parts of plants involve changes in the turgidity (stiffness or degree of inflation) of cells, which is mainly controlled by their water content. Leaves and stems droop or stiffen when their supportive cells are less or more turgid.

Among the few definite effector organs in plants is one that operates by the same sort of changes in water content and turgor pressure. As pointed out in Box 29-1, in so-called sensitive plants there are little swellings called pulvini at the base of each leaf attachment. When the leaf of a plant such as *Mimosa pudica* is touched, the cells on one side of the pulvinus lose water and become less turgid. The leaf then bends toward the side of less turgidity. As we have seen, the most behaviorlike movement in plants occurs in those like the Venus flytrap (*Dionaea muscipula*), which have "traps" that snap shut on a hapless fly.

Some movements of plant cells—notably those of cell division and cell enlargement—depend on microtubules and a microtubular cytoskeleton entirely comparable to those in protists and animal cells.

SUMMARY

Effectors are the structures in organisms that do things in response to stimuli. Most effectors (muscles, glands, firefly lanterns, etc.) are under neural control; some (cilia, flagella, etc.) are not. In vertebrates, skeletal (striated) muscle is the major effector responsible for body movements. Smooth muscle participates only in internal processes. Muscle cells (myofibers) are multinucleated fibers containing many myofibrils, each of which is packed with overlapping thick filaments (myosin) and thin filaments (actin and the regulatory proteins troponin and tropomyosin).

Intracellular Ca^{2+} triggers muscle contraction by binding to one of the three troponin peptides. This permits myosin to bind to actin. Energy released by the hydrolysis of ATP causes the crosslinks (myosin heads) between actin and myosin to swivel in a way that forces the filaments to slide past each other. As a result, muscle contracts. Neural excitation, which spreads into a muscle via T tubules, triggers muscular contraction by transiently causing release of Ca^{2+} from sarcoplasmic reticulum into muscle cells.

Cardiac muscle contracts spontaneously and rhythmically, requiring no nerve impulses to initiate contractions. Cardiac muscle fibers are in uniquely close electrical contact with one another. Hence, when one fiber contracts, all do.

Smooth muscle lines the walls of many hollow internal organs. Its contractions exert pressure on the contents of that organ. Smooth muscle differs from striated muscle in two major ways. The filaments of smooth muscle cells are much less regularly arranged than those of striated muscle—and nerve impulses induce contraction in smooth muscles through second messenger systems. Hence, their contractions are slower and more protracted than those of striated muscles.

Isotonic contraction occurs when muscle tension remains constant as muscle length shortens. In isometric contraction, muscle length remains constant as muscle tension increases. Increasing the load on a contracting muscle decreases its velocity of contraction. Measurements of heat production during contraction reveal a release of initial heat followed by the heat of recovery, which is generated during the resynthesis of ATP. Fatigue develops if a decreased oxygen supply and an accumulation of waste products prevent ATP resynthesis.

All multicellular organisms need structural support to keep them from collapsing under their own weight. This generally requires the presence of compression-resisting elements, such as a skeleton, and tensile elements, such as tendons and muscles. Many small sea creatures maintain their shapes by means of hydrostatic skeletons in which incompressible water fills a tension-resistant stretched membrane hydrostat made up of a crossed helical array of fibers in the body wall. Land animals, in contrast, commonly have a jointed skeleton—either an exoskeleton or an internal, bony endoskeleton. The bones of vertebrates are not inert rods but constantly changing structures with an active metabolism. In functional terms, bones are levers to which forces are applied at points along their shafts by muscles, gravity, and external loads. The fulcrum of the lever is the joint.

Among the many other effectors in the animal kingdom are the endocrine and exocrine glands, a variety of missile-firing systems, electric organs, and bioluminescence systems. Plants lack nervous systems, but they have some effectors that consist of systems capable of changing their water content and turgor pressure in response to certain stimuli.

KEY TERMS

actin
appendicular skeleton
asynchronous muscle
ATPase
axial skeleton
bioluminescence
calmodulin
cardiac muscle
cnidoblast
cross-bridge
effector
electric organ
electroplaque
endoskeleton
exoskeleton
heat of recovery

hydrostatic skeleton
initial heat
intercalated disk
isometric contraction
isotonic contraction
locomotion
luciferase
luciferin
motor unit
myofiber
myofibril
myosin
nematocyst
sarcolemma
sarcomere
sarcoplasmic reticulum

skeletal muscle
skeleton
smooth (visceral) muscle
stretched membrane hydrostat
striated muscle
summation
synchronous muscle
syncytium
tendon
tetanus
thick filament
thin filament
transverse tubular system (T system)
trichocyst
tropomyosin
troponin

QUIZ QUESTIONS

1. Which of the following occur(s) during muscle contraction?
   A. Filaments of myosin slide past filaments of actin.
   B. Z lines move closer together.
   C. Muscle fibers lengthen.
   D. Reduction of ATP liberates oxygen.
   E. A and C.

2. Anaerobic conditions
   A. do not occur in aerobic species such as humans.
   B. occur when oxygen demand exceeds oxygen supply.
   C. occurring in muscle tissue can be detected by an accumulation of lactic acid.
   D. A and B
   E. B and C

3. Which of the following is (are) not involved in muscle contraction?
   A. calmodulin
   B. calcium ions
   C. troponin
   D. creatine phosphokinase
   E. All of the above are involved in muscle contraction.

4. Operation of "effectors" in plants
   A. involves conduction of nerve impulses by vascular tissues.
   B. is restricted by the rigidity imposed by cellulose in cell walls.
   C. involves oxidation of luciferin.
   D. usually is related to changes in water content and turgor pressure.
   E. B and D

5. Electric organs develop from _____ tissue.
   A. muscle
   B. nervous
   C. epithelial
   D. vascular
   E. B and C

6. Which of these parts of the vertebrate brain increased greatly in proportionate size in evolution?
   A. olfactory lobe
   B. cerebrum
   C. thalamus
   D. midbrain
   E. medulla oblongata

ESSAY QUESTIONS

1. Why are muscles arranged in antagonistic groups (e.g. flexors and extensors)? What posture would the human arm assume if all of its extensors were paralyzed?

2. What are the major features of the sliding filament hypothesis of muscle contraction? What role do tropomyosin and tropinin play?

3. Compare the structural patterns and the functioning of the skeletal systems of vertebrates and insects. What are their respective advantages and disadvantages? How does the lever system of the bird leg or wing operate?

4. How does an earthworm move? What role do chaetae play?

5. Compare the design of the forelimb of a mole with the hindlimb of a horse. How do their differences reflect adaptation for the different tasks required of the two limbs?

6. Trace the fate of one molecule of ATP during formation of a cross-bridge between actin and myosin. Explain rigor mortis.

REFERENCES AND SUGGESTIONS FOR FURTHER READING

Alexander, R. McN. (1968). *Animal Mechanics*. University of Washington Press, Seattle.

Viewing animal design from an engineering perspective, the author discusses force, joint mechanisms, elasticity and viscosity, pressure, density and surface tension, motion and vibrations, and other matters.

Bagshaw, C. R. (1982). *Muscle Contraction*. Chapman & Hall, New York.

A concise and informative introduction.

Cohen, C. (1975). The protein switch of muscle contraction. *Scientific American 233*, November, 36–45.

An account of the critical interaction of calcium, troponin, and tropomyosin that controls muscular contraction.

Currey, J. (1984). *The Mechanical Adaptations of Bones*. Princeton University Press, Princeton, N.J.

In relating the mechanical and structural properties of vertebrate bone function, the author argues that natural selection produced optimum structures by compromising conflicting mechanical requirements.

Eisenberg, E., and Hill, T. L. (1985). Muscle contraction and free energy transduction in biological systems. *Science 227*, 999–1006.

Scholarly words on a stimulating subject.

Gans, C. (1974). *Biomechanics: An Approach to Vertebrate Biology*. University of Michigan Press, Ann Arbor.

This pioneering book on functional approaches to the study of animal structure shows how vertebrate structures are adapted to their environments and lifestyles.

Holstein, T., and Tardent, P. (1984). An ultrahigh-speed analysis of exocytosis: nematocyst discharge. *Science 223*, 830–833.

High-speed cinematography of nematocyst discharge in *Hydra* shows that it requires only 3 milliseconds. Thus it is one of the fastest cellular processes in nature. The film sequence in the paper is fascinating.

Huxley, H. E. (1965). The mechanisms of muscular contraction. *Scientific American 213*, December, 18–27.

A straightforward review by one of the architects of the sliding filament model of muscular contraction.

Korn, E. D., and Hammer, J. A., III (1988). Myosins of nonmuscle cells. *Annual Review of Biophysics and Biophysical Chemistry 17*, 23–45.

The latest word.

McMahon, T. A. (1983). *Muscles, Reflexes, and Locomotion*. Princeton University Press, Princeton, N.J.

A mathematical treatment of muscle function that attempts to answer the questions, "What generates force in a muscle?" and "How is this force controlled?"

Moore, J. A. (1988). Science as a way of knowing: form and function. *American Zoologist 28*, 441–802.

A comprehensive summary of effectors in vertebrates. This is one of a series of articles designed as supplementary reading for biology majors in their first or second years. This essay is highly recommended.

Nursall, J. R., and 7 other authors. (1962). Vertebrate locomotion. *American Zoologist 2*, 127–208.

This symposium on vertebrate locomotion in an evolutionary framework includes an important contribution on terrestrial locomotion without limbs and a good discussion on the evolutionary importance of bipedalism.

Rome, L. C., Funke, R. P., Alexander, R. M., and others (1988). Why animals have different muscle fiber types. *Nature 335*, 824–827.

Animal muscles have slow and fast fibers. This short paper shows that during slow locomotion slow fibers give peak power. Fast fibers power maximal movements because slow fibers cannot shorten rapidly enough.

Wainwright, S. A., Biggs, W. D., Currey, J. D., Gosline, J. M. (1982). *Mechanical Design in Organisms*. Princeton University Press, Princeton, N.J.

This stimulating discussion of the interface between mechanical engineering and biology reviews structural materials and their mechanical features. The authors show that function at any level of biological integration is dependent on structures at lower levels of integration.

Warrick, H. M. and Spudich, J. A. (1987). Myosin: structure and function in cell motility. *Annual Review of Cell Biology 3*, 379–422.

An up-to-date review.

BEHAVIOR

Behavior
One of biology's most challenging problems—
the basis of animal behavior and its remarkable
intricacy—is illustrated by this huge termite
mound in Kenya; an architectural masterpiece
providing a food-stocked nest in which
temperature, humidity, and gas exchange are
precisely regulated. What blueprints, one
wonders, guided the termite builders?

A motionless eagle perches high on a tall tree. Suddenly it spreads its wings, soars aloft, and dives at a fleeing rabbit. The eagle's cells and tissues, of course, have been functioning actively but flight is activity at a different level. The whole animal did something.

What it did was a product of body coordination—in the nervous and endocrine systems, in the muscles and other effectors. This different order of activity is called **behavior.** One of the things that makes animals so colorful and affecting is the behavior they display. It is also one of the factors that has influenced the course of animal evolution.

THE STUDY OF BEHAVIOR

WHAT IS BEHAVIOR?

Behavior is difficult to define with precision. As good a definition as any is that behavior is what an animal does. Some have defined it simply as a series of movements and postures. For these discussions, let us agree that behavior means any externally directed activity of an organism, however simple or complex it may be.

Most animal behavior induces change in an organism's relation with its environment. Some behavior is directed at survival or reproduction. It must find food and mates, avoid being eaten, react to members of its own and other species, and find its way around. An animal's behavior pattern is the sum of these responses.

Like other phenotypic traits, behavior is in part genetically determined—that is to say, a potential range of behavioral expressions is genetically determined. But specific behavioral patterns—the what, how, and how much of behavior—will always depend on the manner and extent to which a specific mode of behavior unfolds in the face of environmental events. It will also depend on an animal's past experiences.

Even though an animal's behavior may be a direct response to an external stimulus—a change in the environment, for example—its underlying mechanisms are, nonetheless, internal. We should bear in mind that behavior can be triggered by something other than an immediate external stimulus. The eagle may fly off, not because it saw a rabbit, but because it was hungry or weary of sitting. Even so, its behavior grew out of complex associative processes and memory.

Directly or indirectly, behavior stems from the combination of sensory systems, associative centers, and effectors described in Chapters 30 and 31—and it is influenced and determined by the functional attributes of receptors, conductors, associative mechanisms (when present), and effectors—the devices that "do the doing."

In this chapter and the next, we will first outline briefly some of the problems arising in the study of behavior. Second, we will survey various components of animal behavior. Third, we will mention genetic and evolutionary aspects of behavior, though we will return to this important topic in Chapter 34.

PROBLEMS OF INTERPRETATION

Behavior depends on many factors—and all are hard to isolate and identify. The problem of objectively interpreting the behaviors of widely differing organisms is so daunting that a rational approach to the study of behavior has emerged only quite recently.

The Problem of Anthropomorphism

We can ordinarily observe only the externals of behavior. We can recognize an environmental stimulus; we can note an effector's response. Until recently, we had no good way of examining what occurs between stimulus and response. Even now, information of that sort is very elusive. The best we can do as observers is infer the relations of visible external events to internal processes.

The habit of thought that interprets nonhuman actions in terms of human motivations is termed **anthropomorphism.** This way of thinking was one of two major obstacles to the development of a science of animal behavior. The other was the bothersome fact that when animal behavior is observed in real-world surroundings (as it usually is) the setting is often so complicated it is hard to identify the essential stimuli. Many "scientific" reports have been little more than anecdotes about what an animal was seen to do in complex natural circumstances, in which the critical stimuli were wholly unknown to the observer. The anecdotes were generally interpreted anthropomorphically—that is, as if the animal were human.

The Best Hypotheses Are the Simplest Ones

In 1893, the English psychologist Lloyd Morgan (1852–1936) made a famous pronouncement that became known as Morgan's canon[1]: never invoke a higher faculty if a lower one will do—in other words, interpret the behavior of an animal in terms of the simplest possible mental processes. This "law of parsimony" plays down conscious thought, decision, purpose, and foresight as explanations of animal behavior. It was proposed at the end of a century when a new theory of evolution was uppermost in everyone's mind.

Regard for Morgan's canon certainly did cut away vast amounts of nonsense. When an amoeba engulfs a food particle, we no longer had to say, in describing its behavior, that it smelled food, liked the smell, decided to eat, and therefore seized the particle.

Even before Morgan, few would have been quite that anthropomorphic about an amoeba. Yet many contemporary bird lovers still assume that a mother bird sits on eggs because she wants to have babies and that she cares for the young because she loves them and knows that they need food and warmth. What we should conclude instead is that the bird reacted to certain physiological needs within herself that were triggered by external stimuli—without the slightest idea that young would hatch from the eggs and with no conscious knowledge of the needs of the young. Experimental manipulation of the stimuli confirms this view. Birds react to pebbles or to cuckoo's eggs as they do to their own eggs.

It is possible, of course, to carry such thinking to extremes. The simplest hypothesis would be that nothing happens between stimulus and response except conduction from receptor to effector—in other words, that all behavior consists of simple reflexes—like those discussed in Chapter 30. After all, a reflex arc does allow for a variety of modifications (Figure 32-1). Indeed, some did argue that the study of behavior should boil down to correlations between stimuli and responses. This approach came to be called *behaviorism.* When applied to human beings, it led to a remarkable conclusion: our actions have nothing to do with our thoughts—beyond the fact that we may notice our own actions. It is true that we cannot get into the minds of others. However, we can examine our own minds (we call that process

---

[1] Morgan's canon is a corollary of a principle enunciated in the fourteenth century by William of Occam: if a phenomenon can be explained in several different ways, the simplest explanation is most probably the correct one. This idea, which became known as Occam's razor because it pared away many extraneous and nonsensical "explanations," clearly presaged Morgan's canon. A canon is a rule, principle, or standard for judgment.

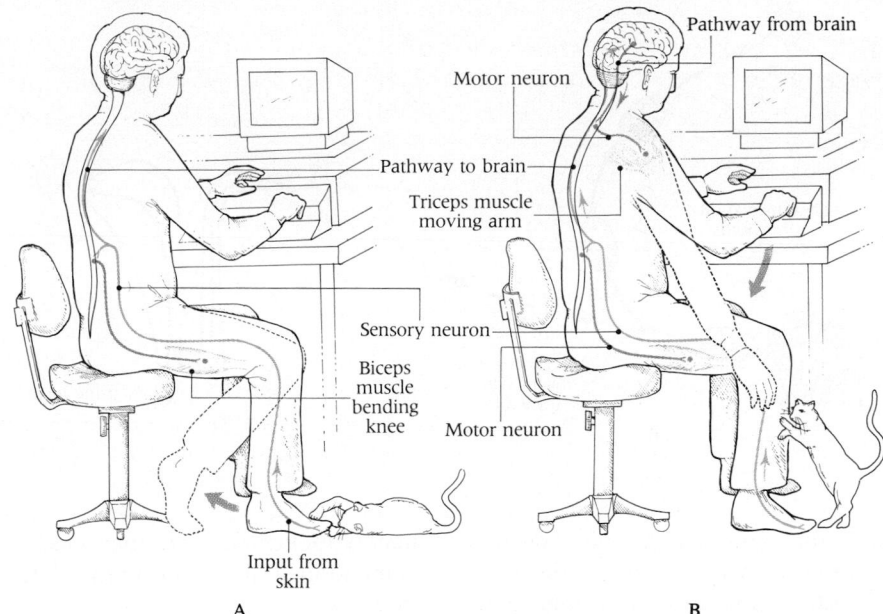

32-1 The simple modifiability of human reflex action
(A) A kitten touches the writer's toe. The stimulus travels along a sensory neuron to the spinal cord where it synapses with a motor neuron. The result is contraction of the biceps muscle in the leg and reflex bending of the knee. The stimulus also reaches the brain. (B) If the stimulus continues, simple reflex action is superceded by another reflex involving higher centers in the brain not including thought centers. The triceps muscle in the arm is now stimulated to contract, pushing the kitten away. See also Figure 30-31.

introspection) and we know that our own actions—not all, but a majority—are determined by complex mental processes and are not simple reflexes.

A question arises whether it is proper to insist that nothing of the sort happens in other animals. A dog sees its master and wags its tail. Strict followers of Morgan's canon would conclude that this is an established reflex:

stimulus (sight of master) → conduction (from eyes to tail muscles)
→ response (wag)

Is this a reasonable conclusion? It is possible now to determine objectively that when a dog sees its master, activity takes place in the cerebral cortex, rather than involving a simple reflex arc between muscle and spinal cord. This technical advance, called **electroencephalography,** employs sensitive detectors of electrical activity in the brain, the "brain waves" (or EEG pattern) that accompany activity (Figure 32-2). We also saw examples in Chapter 30 of how electrodes in nerves and nerve cells are successfully used in studying sensory receptors. Surgical manipulation of the nervous system also makes it possible to determine what parts are necessary for perception of a stimulus and production of a given response. Hence it is no longer true that behavior can be studied only by observing responses.

A reflex theory would not explain what we know of the dog's whole behavior pattern and it is simpler and more consistent with the data to conclude that the dog thinks—not like a human but like a dog, like this dog.

Modern behavioral research follows the same scientific principles as other areas of experimental science. Tests are made under standardized conditions that are kept as uniform as possible except for the one environmental factor that is varied.

32-2 Human encephalogram (EEG)
An electroencephalogram of brain waves is usually recorded from different points on the scalp. This recording from the back of the scalp shows that the waves are influenced by variations in the degree of alertness.

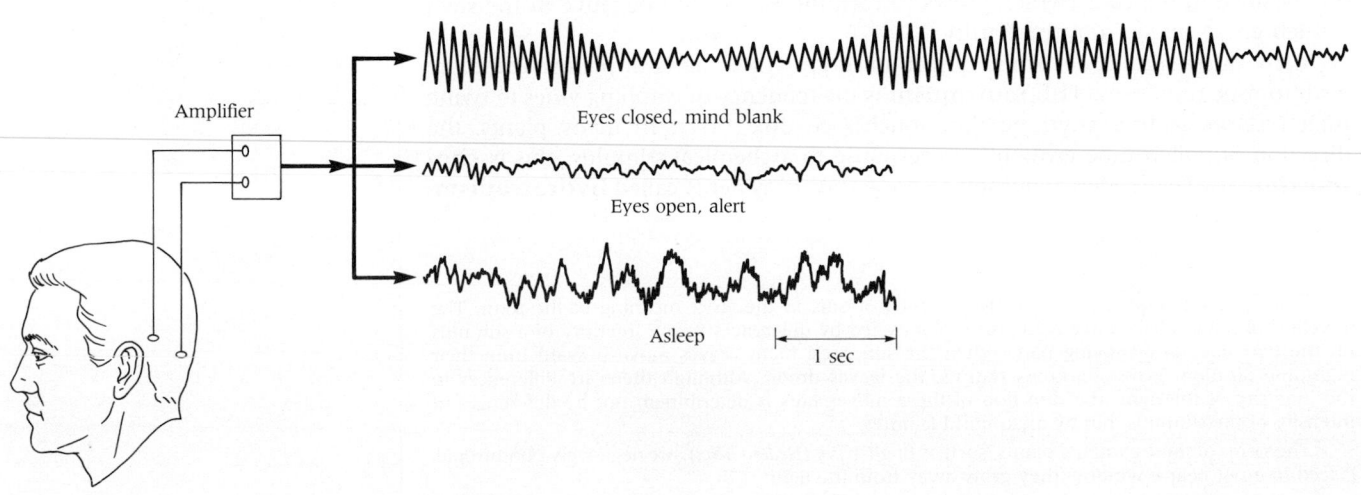

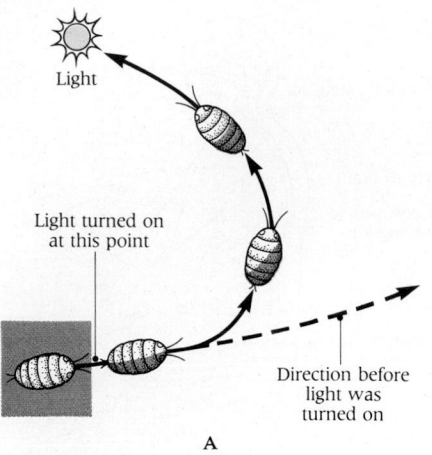

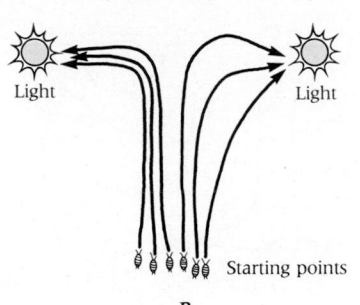

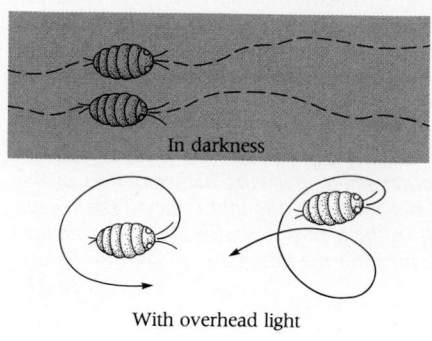

A   B   C

32-3 Forced movements of a pill bug
(A) *When a single light source strikes the pill bug from the side, one side is shaded (as indicated by body shading and eyes) and the bug turns until it is illuminated equally on both sides. It then maintains a straight path toward the light.*
(B) *The tracks of six animals exposed to two lights follow a path that produces equal light intensities on each side of the animal. Ultimately each animal turns toward one of the two lights.*
(C) *Two pill bugs have had their right eyes covered with black paint. In darkness, they hold to a straight path. Exposed to an overhead light, they both enter a circular path, moving toward the left, the side of the illuminated eye. These are all examples of phototaxis.*

That is the controlled stimulus. Responses, mainly movements, are then observed as stimuli are varied. Finally, the techniques mentioned above are used to study the mechanisms between stimulus and response.

TROPISMS AND TAXES

Tropism in Plants

Plant behavior is usually very simple. We have seen that a plant stem moves toward or away from light (see Chapter 28). The same stimulus invariably produces the same response—so memory or learning is not involved. The word **tropism** denotes such automatic plant responses. In its original sense, it had a narrow definition: the bending movements of plants in response to stimulation—by light or gravity—acting on the two sides of a growing plant part. Later, other tropisms were recognized and named for the directing stimulus, e.g., phototropism, gravitropism, etc. When some zoologists began to describe certain animal movements as "tropisms," the word's meaning became blurred.

The idea of tropism embodies two elements.

■ A tropistic movement has a definite direction that is determined by differences in stimulus intensity on the two sides of the moving organ.
■ The mechanism of the turning response is an innate and inflexible part of the organism's genetic endowment. It is not subject to modification or control by the organism.

Much so-called behavior in plants consists of tropisms, the most conspicuous being their responses to gravity and light.[2] Both are easily demonstrated in the garden. When planting seeds, one needn't put them in upright. No matter how a seed is set in the ground, roots will grow downward and stems will grow upward. There is a gravity (or Earth) tropism known as **gravitropism** (or **geotropism**): it is positive in the root (which grows toward the Earth) and negative in the stem (which grows away from the Earth).[3]

Many other stimuli determine direction of growth in parts of plants. These are also tropistic responses. **Thigmotropism** is the tendency of climbing vines to twine their tendrils around anything they touch (see Box 29-1). In many plants, the direction of pollen tube growth is in response to a chemical stimulus: this is **chemotropism.** The tendency of roots to grow toward water is called **hydrotropism.**

[2] However, many plant movements are not tropisms in the strict meaning of the term. The direction of many plant movements is not controlled by differences in the intensity of a stimulus on the two sides of a moving part. When the sun rises, many leaves move upward from their nighttime position. When darkness returns, the leaves droop. Although there are differences in the intensity of the light, the direction of these movements is determined, not by differences in intensity of the stimulus, but by anatomical features.

[3] The stems of some climbing plants, such as English ivy (*Hedera helix*), are negatively phototropic. Placed in a pot near a window they grow away from the light.

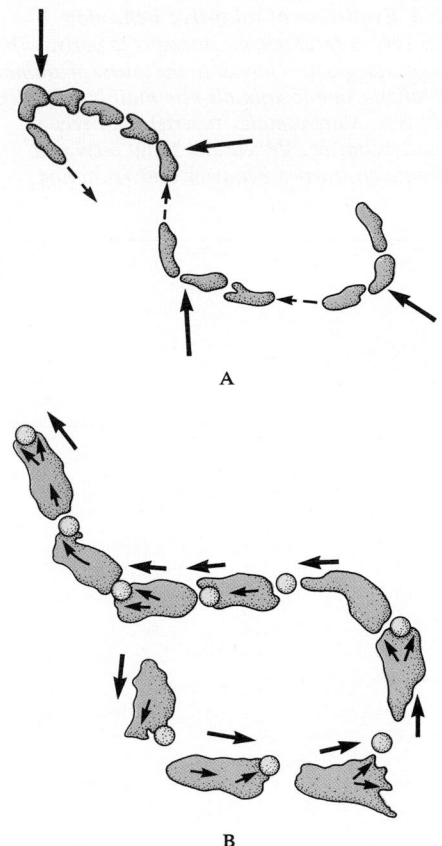

32-4 The behavior of an amoeba
(A) *Each time a light (solid arrows) is aimed at a moving amoeba, it turns away from the light source.* **(B)** *An amoeba in pursuit of a food particle (colored circle) can detect its position, presumably by chemical stimuli, even when it actually loses contact with the particle.*

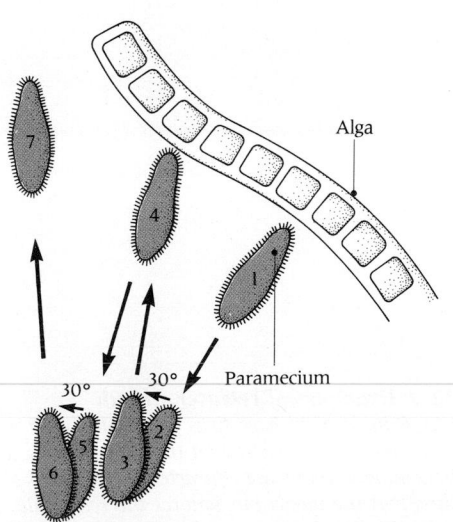

32-5 Trial-and-error behavior in the paramecium
On encountering an obstacle, such as an alga, the paramecium backs away (1–2), makes a turn of about 30° (2–3), and moves forward again (3–4). If the obstacle is still present, the maneuver is repeated (4–7).

Taxes: Animal "Tropisms"

Plant tropisms were reasonably well understood by the end of the nineteenth century, but the study of animal behavior was still bogged down by anthropomorphic thinking and anecdotal storytelling. Jacques Loeb (1859–1924), a German zoologist who worked in several American universities, tried to place the study of animal behavior on a sounder footing by extending the notion of plant tropism. His theory of animal tropism sought to show that patterns of movement in an animal—like those in a plant—were caused by differences of stimulation on its two sides. He called them *forced movements.*

Loeb's ideas may be illustrated by his explanation of the behavior of the pill bug (*Armadillium*), a tiny terrestrial crustacean that lives under rocks and decaying wood. When confronted with a single light source, a pill bug moves directly toward it (Figure 32-3A). With two lights, it moves along one of the paths shown in Figure 32-3B. In darkness, they maintain a straight path (Figure 32-3C). With one eye covered, they circle toward their "good" eye. In explaining these movements, Loeb drew a parallel between amounts of muscular activity on each side of the animal with amounts of growth on each side of a plant stem. Thus in Loeb's view, when the bug faces the light head on, it follows a straight path because its two sides are equally stimulated and equally active in response.

Loeb's forceful advocacy of what he saw as animal tropisms surely had the appeal of simplicity. However, few animal movements are like the motions of a pill bug facing a light.

The term **taxis** was introduced later in a hapless effort to eliminate confusion. Taxis (plural, "taxes") was applied to such directed movements, whether by bacteria, free-swimming unicellular organisms, and complex animals. Taxes, too, are named after the directing stimulus—for example phototaxis, geotaxis, chemotaxis, etc. "Tropism" was now reserved for plants.

We have already encountered several classic examples of chemotaxis. White blood cells move unerringly to sites of injury (see Chapter 22). Moths attracted by pheromones move toward their source (see Chapter 30). Many bacteria respond to chemical stimuli and, like the amoeba in Figure 32-4, move off in the direction of food, guided by chemical attractants.[4] Protists sometimes display what looks like trial-and-error behavior (Figure 32-5). If *Paramecium* encounters a "bad" object it backs up, turns approximately 30°, and tries again—and it does this again and again (Morgan and Loeb forbid us to say "in exasperation") until it finds something "good." This is a trial-and-error program, but it is hardly trial-and-error learning—to be discussed later—since the paramecium does not behave differently the next time out.

Adaptive Nature of Behavior

The harsh insistence that conscious purpose plays no part in animal behavior obscured the fact that "purpose" in another, quite different, sense is involved. For behavior in organisms is ordinarily adaptive. With few exceptions, behavior in any species has average results that are beneficial to the species. That is what we mean by adaptive, as discussed further in Chapters 34 and 35.

Behavior serves a purpose when it allows an animal to get food, avoid its enemies, and not only find but win a mate. Behavior serves purposes because it has to: an individual with maladaptive behavior would not last long. Species survived because the behavior of their members was adaptive. Indeed, differences in the behavior patterns of species now in existence hint at the evolutionary history of adaptive

---

[4] Early students of animal behavior defined a kinesis as a type of movement that, unlike taxes, which are directed movements, are indirect orientation movements of a random nature that become more active as the stimulus increases. The point was made in earlier texts that in this terminology, the phenomenon traditionally called bacterial chemotaxis is really a kinesis because it consists of random movements by an organism with no distance receptors that by random movements "climbs" a chemical gradient in step-by-step fashion. However, later developments toppled this interpretation. As discussed in Box 5-2, bacteria *do* have chemoreceptors—and, even though they lack nervous systems, there does exist a primitive receptor-effector link whereby the binding of a specific molecule by a chemoreceptor causes the flagellum (the effector) to propel them in the right direction. Studies in *E. coli* mutants show that four genes encode the chemoreceptors in its cell walls. Thus the term chemotaxis is appropriate for this behavior. Movements toward some chemicals and away from others are termed positive and negative chemotaxis, respectively.

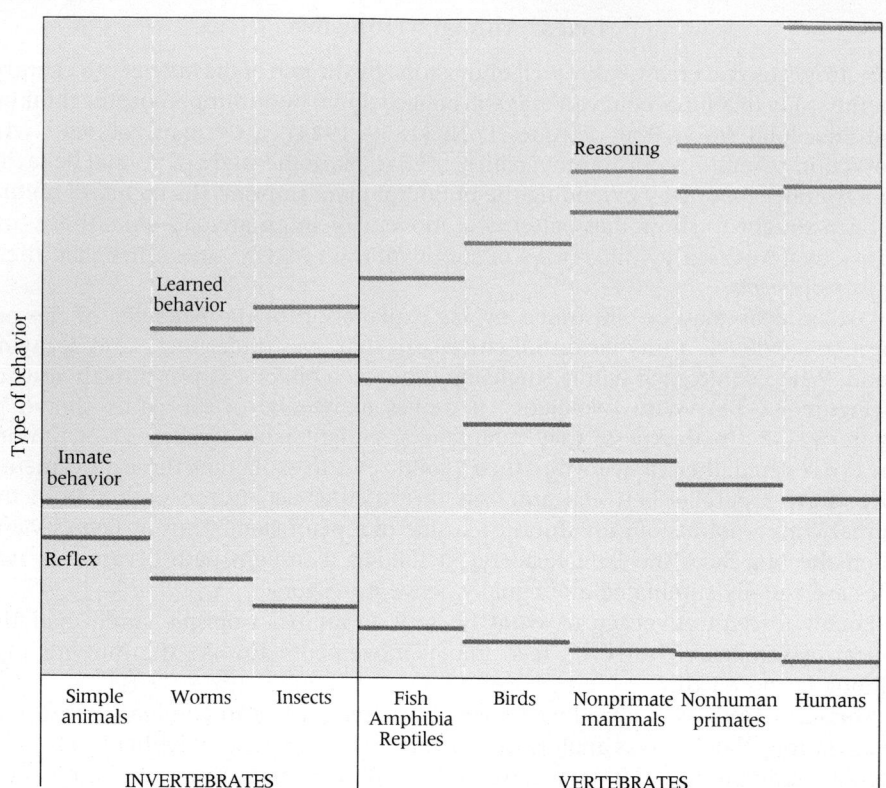

32-6 Evolution of adaptive behavior
This very general scheme attempts to portray the major changes in behavior in the course of animal evolution. Simple animals rely mainly on reflex behavior. More complex invertebrates rely on innate behavior. Vertebrates show increased reliance on learned behavior and reasoning.

behavior (Figure 32-6). Such information does not explain how particular sorts of adaptive behavior arose. That is a much harder question and we take it on in Chapter 34.

INNATE AND LEARNED BEHAVIOR

It is generally said that there are two modes of behavior: **innate behavior,** which is instinctive, built-in, and inherited; and **learned behavior,** which is modifiable by experience. In plants beyond any doubt, and in protists and coelenterates with only a whisper of a doubt, we are dealing with completely innate behavior. The behavior mechanism built into these organisms is rigidly fixed—that is, a certain stimulus always produces the same response. Other individuals of the same species (if in the same physiological condition) respond in the same way to the same stimuli.

In the echinoderms possibly, and in the flatworms and earthworms, primitive nervous systems have appeared. Though most of their behavior still seems innate or inherited, not all of it does (Box 32-1). Reactions can be changed by training. Responses to training are not the same in all individuals of the species. Differences occur that reflect an animal's past experience.

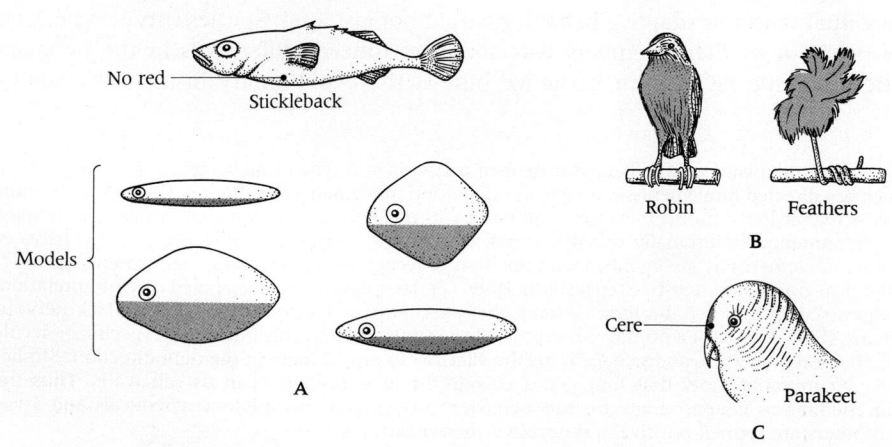

32-7 Examples of releaser stimuli
(A) *A male stickleback fish, confronted with a crude model painted red on the underside or a true male without the characteristic red belly, threatens the model but ignores the real male fish.* **(B)** *The European robin attacks a tuft of red feathers, but ignores an immature male robin lacking the red breast.* **(C)** *If the characteristic color of a female parakeet's cere (skin above the beak) is changed, her mate attacks her as if she were a rival male. The releaser stimulus in each case is the color of belly, breast, or cere skin.*

THE PROBLEM OF DISTINGUISHING BEHAVIOR MODES

Beginning in the 1920s, two major schools of thought developed in the field of animal behavior. One, based in experimental psychology, derived largely from the work of John B. Watson, an American whose book, *Behaviorism,* appeared in 1924. The other, mainly European, was founded in the 1930s by Konrad Lorenz, an Austrian, and Nikolaas Tinbergen, who moved from the Netherlands to Britain and for a time collaborated with Lorenz. Members of this school called themselves ethologists, defining **ethology** as the scientific study of behavior.[5] Their ideas stemmed from direct observation of animals in natural environments.

The two schools came into conflict, yet there was agreement on one point: even in animals with a flexible set of responses, even in those reared in isolation, there remain inherited patterns of behavior that are common to the species. Animals do display complex "goal-oriented" behavior patterns, such as nest-building and courting, without exposure to training and without having seen another animal perform them. The issue under dispute concerned the nature of these patterns, which in those days were called instincts. Ethologists were accused of underrating the role of learning and of viewing the nervous system as a "black box," the properties of which could be deduced only from environmental "inputs" and behavioral "outputs." Psychologists were said to be too concerned with animal behavior in laboratory settings and too little concerned with adaptation and evolution.

Early behavioral biologists defined an instinct as innate or built-in behavior that differs from learned behavior because it is predictable (given a definite stimulus) and thus is stereotyped. A given action was deemed to be either instinctive or learned: the experimenter's problem was to find out which.

In the 1930s, ethologists responded to this challenge by cataloging what they regarded as instinctive behaviors, painstakingly compiled after years of field observation. These listings came to be called *ethograms.* The problem, however, was that extensive variations in stereotypical behavior made it all but impossible to ascertain in a given case what behavior is innate and what is learned.

In time, Lorenz offered a more useful conceptual framework to account for such behavior. **Instincts,** he argued, are simple, stereotyped, inborn actions, which he called fixed action patterns, that are linked together by more flexible goal-oriented behavior, which is modifiable by experience—or learning. Although this concept is not without difficulties of its own, it is a basis (more or less) for much current thinking in behavioral biology.

BASIC COMPONENTS OF BEHAVIOR: CURRENT VIEWS

Most biologists today acknowledge that patterns of behavior issue from an intermingling of underlying mechanisms. These include sign stimuli, releasers, fixed action patterns, and motivations or drives.

Sign Stimuli and Releasers

Animals constantly receive stimuli of all sorts. Certain stimuli, however, have special significance.

■ The male three-spined stickleback (*Gasterosteus aculeatus*), a common fish of the north temperate zone, is a territorial animal. That means that individuals defend a particular area against intrusion by other individuals, especially those of the same species. In defending against other males, their behavior can be highly aggressive. Studies revealed that this behavior is triggered, not only by the red bellies of other territorial males, but by a wide variety of models, some bizarrely unrealistic, as long as the models have red undersides (Figure 32-7A). Models that more closely resemble the fish, but that lack a red color, do not trigger the aggressive behavior.

32-8 Prey-capture behavior of a toad (A) In this sequence, the toad responds to an elongated bar (stimulus) moving past it by turning toward it and striking at it with its tongue. The toad then swallows (eyes shut) and wipes its mouth, even though it hasn't caught anything. **(B)** This behavior is controlled by feature detectors in the brain that respond only to specific stimuli. Note that a rapid volley of nerve impulses is recorded in the toad's brain only when a wormlike bar crosses its field of vision.

---

[5] In 1973, Lorenz, Tinbergen, and Karl von Frisch (whose work is discussed later in this chapter) shared the Nobel Prize in Physiology and Medicine. The first ethologists to receive a Nobel Prize, the three were cited for "discoveries on the organization and elicitation of individual and social behavior patterns." Ethology, the hard-to-define science, had arrived.

The richness of behavioral repertory roughly parallels the intricacy of a nervous system. Consider the following generalizations.

- If behavior is to be complex, nerve conductors must have access to various alternative routes.
- Coordination of multiple effectors in a single response requires efficient associative mechanisms.
- Assembling many complex stimuli into information units that can elicit selective responses to different elements in the situation requires still more complex associative mechanisms.
- Centralized control requires a centralized message center containing an adequate combination of conducting and associating neurons.
- Delayed responses to stimuli require a storage mechanism, in other words, a memory.

Accordingly, as we learned in Chapter 30, the broad trends that followed diffuse nerve nets in the evolution of the nervous system included: centralization of the conducting pathways, increased numbers of association neurons, and the emergence of nerve masses—brains—at the front end of the body that could coordinate more and more effector activities, and thus could sensitively control behavior.

Each of these advances was accompanied by changes in behavior, which became more and more complex. Changes also occurred in the balance between innate, or instinctive, behavior, and learned or modifiable, behavior. Let us briefly retrace these evolutionary steps.

NERVE-NET BEHAVIOR

Among living animals (excepting sponges), the simplest nervous system and the simplest behavior is seen in coelenterates (hydras, jellyfishes, and corals), which have undifferentiated nerve nets (see Figure 30-30A). Though simple, coelenterate behavior is a good deal more complex than that of unicellular organisms, showing divisions of labor.

- If a tentacle of a coral touches a bit of food, that tentacle—not the whole organism—bends toward the mouth. If the food particle is large or struggling, other tentacles may join in landing the prey. There is no real coordination, however. The reaction simply spreads out in all directions from wherever the stimulus occurs.

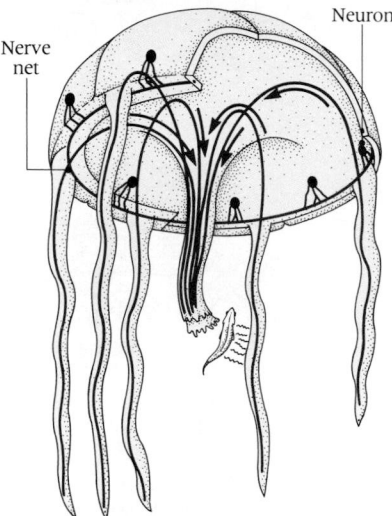

32A Control of swimming in jellyfishes
Neurons connect to a simple nerve net that encircles the body of the jellyfish. This circular ring helps to coordinate contractions that propel the animal through the water.

- The European red-breasted robin (*Erithracus rubecula*) is smaller and more pugnacious than the American robin, a large and peaceable thrush. During breeding season, male European robins establish territories from which they try to expel other male robins. However, they will threaten or attack any tuft of red feathers of a certain size (Figure 32-7B).
- Parakeets (*Agapornis*) have a patch of skin above the beak called the cere (Figure 32-7C). The color of a female's cere is highly significant to her mate. If its color is changed experimentally, the male will attack her as though she were a rival male.

The male stickleback fish, European robin, and parakeet all respond with aggressive behavior to only certain stimuli. These are sign stimuli, defined as the features of a stimulus that are of key importance in triggering a particular behavior pattern. Animals possess an exaggerated neural sensitivity to sign stimuli. One might say that their brains are "programmed" to respond to them. Because sign stimuli trigger or "release" specific behaviors, they are called **releasers.**

Such exaggerated neural sensitivity has been demonstrated in the visual system of the toad (*Bufo*). As discussed in Chapter 30, light entering a vertebrate eye generates nerve impulses that are sent to retinal ganglion cells via bipolar cells and then to the optic nerve. We saw that each ganglion cell receives input from many bipolar

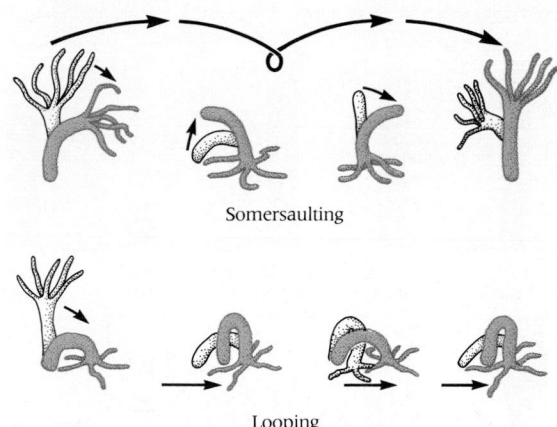

Somersaulting

Looping

32B Locomotion in Hydra
This coelenterate moves by both somersaulting and looping maneuvers.

- On the other hand, a jellyfish such as *Gonionemus* can swim because the simultaneous rhythmic contractions of the whole body eject jets of water (Figure 32A). This certainly is coordination, but of a simple sort and without local control. The nerve net, which forms a circular ring, gives rise to simultaneous rhythmic impulses throughout. If the ring is cut, coordination of these movements is lost.

- Perhaps the most complex (and most amusing) behavior in a coelenterate is the somersaulting locomotion of hydra (Figure 32B). Though not a highly elaborate maneuver, it is still a surprising one for an animal without central control of advanced associative mechanisms. The hydra is not pursuing anything. Rather it is moving to new and possibly better feeding grounds on a purely trial-and-

error basis. In spite of the odd method of locomotion, its behavior is not much more advanced than that of the struggling paramecium shown in Figure 32-5.

Echinoderms (starfishes, sea urchins, and their relatives) have a nerve ring and radial nerve cords in addition to a nerve net (see Figure 30-30B). They also have associative neurons and true reflex arcs. Predictably, their behavior is more highly coordinated and far more varied than that of coelenterates (Chapter 30).

BEHAVIOR WITH A PRIMITIVE BRAIN

We noted in Chapter 30 that earthworms have rudimentary brains—and we have seen something of their behavior. Highly devel-

oped reflexes operate through the central nervous system, but these occur locally in each of the body segments and scarcely involve the brain. Locomotion and most other behavior is hardly affected by removal of the brain. Nevertheless, the brain has some control over general sensitivity and muscle tension (tonus). Removal of the upper of the two anterior ganglia bends the worm's front end upward (because tonus is reduced in upper muscles); removal of the lower ganglion causes downward bending. A worm with its head cut off squirms more than an intact worm because inhibitory messages from the brain are lost. There is very little central control in an earthworm, though the beginnings of central control are there.

A flatworm (which is not closely related to an earthworm) has somewhat more central control. A planarian moves in several ways: by the beating of cilia on its lower surface, by rippling motions of the body, and by a looping crawl rather like an inchworm caterpillar's. Brain removal does not impair ciliary locomotion, but it interferes with rippling (without wholly preventing it). The worm stops crawling altogether. Thus, more complex behavioral reactions are coordinated and controlled by a brain.

Both flatworms and earthworms can be trained in simple ways—for example, to turn to the right or left in a Y- or T-shaped passage—by giving them an electric shock when they take the wrong turn. When a planarian brain is removed, some training is still possible, but it takes longer than in an intact animal. In other words, earthworms and flatworms can learn with their nerve cords alone, but a brain helps.

cells. When the toad is immobile and nothing is moving in the environment, electrical recordings from single cells in this visual system are relatively silent, indicating that no messages are being sent to the brain. Presumably the toad sees nothing (Figure 32-8A). The retina, we recall, is made up of various fields, each organized into a central inhibitory field surrounded by excitatory rings. Objects that cover the entire receptive field do not trigger a response, but images that move in and out of the center evoke action potentials in the optic nerve. To a toad (in a laboratory), the most powerful releaser by far is a bar-shaped object moving lengthwise across its visual field (Figure 32-8B). In a natural setting, this is precisely the cue provided by the worms and centipedes the toad feeds on.

Another example is the herring gull (*Larus argentatus*). The attribute that automatically stimulates a chick to peck at a parent's beak for food is a red spot on a downward-directed beak as it moves back and forth (Figure 32-9). It responds in this way even if it cannot perceive the rest of the parent. A highly unrealistic model with no head or a misplaced spot elicits an active response also. Interestingly, an unrealistic model with exaggerated spots and vertical bars is a stronger releaser than a normal gull head. When such exaggerated sign stimuli are superior to natural stimuli in triggering responses, we call them **supernormal stimuli.** Apparently the two releasers—a spot and a horizontally moving vertical bar—summate, or add up, to produce the exaggerated response. This is termed **heterogeneous summation.** Another supernormal stimulus is shown in Figure 32-9B.

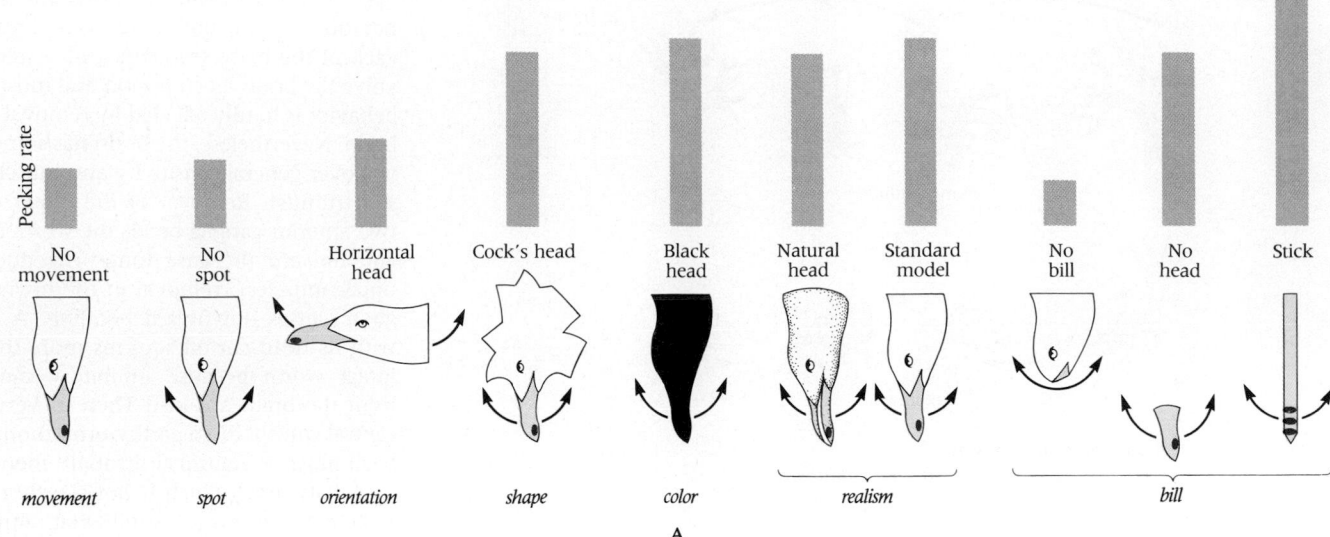

A

There is evidence that when the sign stimulus is correct, a behavioral threshold is lowered and a response becomes likely. This is what is meant when we say that signs release behavior.

The adaptive significance of releasers is that they can evoke specific critical behavioral responses automatically and quickly without risking the slow and error-prone processes of learning. But, as we have seen, the release phenomenon can have a major disadvantage: it can also be triggered by crude, inappropriate, and irrelevant stimuli. By natural selection, evolution has provided many animals with combinations of releaser-dependent and learning-dependent mechanisms.

Fixed Action Patterns

Releasers may trigger all-or-none responses that can be elaborate and highly orchestrated sequences of neuromuscular events. Many of these behavioral patterns are astonishingly unvarying. In the terminology of Lorenz, these are **fixed action** patterns in which stimulus and response are linked in a constant and predictable manner. In many cases, these response patterns are so firmly fixed in an organism's nervous system that they must go on to completion once released—even if the stimulus is no longer present.

- The greylag goose has an established action pattern for rolling a stray egg back into its nest (Figure 32-10). She proceeds through the whole pattern, even if the egg is taken away, continuing to roll an egg that is not there.
- A toad responds to a worm by turning toward it and striking at it with the tongue. Even when it has not caught the worm, the toad proceeds with "swallowing," shutting its eyes, and wiping its mouth (see Figure 32-8).
- A chameleon flips its tongue at a fly in less than 5/1000th of a second (Figure 32-11). Once the tongue flip is initiated, a high-velocity sequence of tongue and mouth motions proceeds unaltered and continues whether or not the fly moves.

Some of the most elaborate fixed action patterns are termed **chain reactions,** because each step is the stimulus for the next. Many courtship patterns consist of longer or shorter chain reactions (Figure 32-12).

As so often happens in science, a "simple" concept is blurred by additional research. This, alas, has been the fate of fixed action patterns. Nowadays, fixed action patterns, like walking or flying, are referred to as **motor programs** to indicate that they are virtually independent of sensory feedback and require little practice to develop normally. These are in contrast with motor activities that are controlled by sensory feedback loops—bird song, for example. Motor programs are important in initiating high-velocity activities.

B

32-9 Releasers and supernormal stimuli in herring gulls
(A) To determine the effect of releasers on the pecking rate of herring gull chicks, Tinbergen and coworkers confronted them with a series of realistic and inaccurate models. Bar heights denote pecking rates. A cardboard standard model and a real head were almost equally effective. A disembodied bill was almost as good. The head spot, pattern, and movement and bill orientations were crucial; but head shape, bill colors, and spot color were not. Evidently the two releasers for pecking are a vertical bar moving horizontally and a moving spot that contrasts with its background. Note the high pecking rate when a narrow stick bearing three spots is waved back and forth. That would be a supernormal stimulus. **(B)** A herring gull prefers to incubate a giant artificial egg (a supernormal stimulus) rather than one of its own smaller eggs.

32-10 Egg rolling by the greylag goose
When the goose sees the egg outside the nest, she touches it with her beak and then rolls it back in. This fixed action pattern is the same even if another object replaces the egg, or if the egg is removed after she has begun to reach for it.

32-11 Fixed action pattern in the chameleon, Chameleo jacksoni

Motivations or Drives

In many cases, the same animal responds to the same stimulus very differently at different times. The behavioral response to a given stimulus often depends on "forces from within" that motivate the animal to respond one way now and another way later. Such internal forces, termed **drives,** are more "push" than "pull." Everyone agrees that there are drives for hunger, thirst, and sex, but do drives exist for curiosity, freedom, aggressiveness, and so on? No one is sure. Moreover, many interacting factors appear to influence drives—among them, hormones, physiological states, habits, and such monitoring loops as the body system that keeps track of blood concentration and induces thirst when concentration increases.

A drive appears to operate by altering an animal's threshold to a stimulus, thus making it more or less likely to trigger certain behavior. Three phases occur in the unfolding of behavioral patterns involving drives.

- The first phase consists of appetitive behavior in which an animal seeks to satisfy bodily needs—with food for hunger, water for thirst. This type of behavior appears automatic or innate, requiring no prior experience.
- A second phase is the consummatory act (eating or drinking) that follows appetitive behavior and dissipates the drive that produces it.
- The third phase, quiescence, then follows.

When subjected to multiple simultaneous drives, an animal must "choose" among several possible behavior patterns. For example, its competing drives may be hunger or thirst, a desire to attract a mate, or a need to patrol territorial boundaries. At any moment, one of these may take precedence over the others, depending upon the circumstances. Shifting thresholds of response to various drives enable the animal to select the best response quickly. Carrying out the behavior with the highest priority at that moment lowers the urgency of the drive that motivated it.

When there are competing drives, the drive to escape from danger—and the ensuing behavior—almost always gets the highest priority. However, hunger and thirst can inspire consummatory behavior—a search for food or water—that is sometimes placed higher on the list of priorities than the need to escape from predators.

LEARNING

The mechanisms just discussed are so lacking in flexibility that they are capable on occasion of evoking inappropriate behavior. For many reasons, environmental conditions—including social conditions—usually demand flexible behavior. That can be provided only by a learning experience.

This argument does not apply to all animals—indeed, it does not apply to most animals, which are small and do not live long. We are referring to such creatures as insects and worms. In small, short-lived organisms, only meager learning abilities and a primitive nervous system evolved (see Box 32-1). The longer an animal lives, and the more varied its surroundings, the greater the survival value of learning is.

We define **learning,** in this context, as the modification of behavior by experience. As observers, we say that an animal has learned something when it behaves differently because of earlier experiences. Its new behavior is more likely to improve its reproductive success and survival. Through learning, animals progress from a total dependence on releasers, drives, and rhythms to a lesser dependence on them.

Learning can be demonstrated experimentally when one of two groups of animals of the same species acquires significantly different behavior patterns when it is given an experience denied to the other. We saw examples of learning in species as simple as snails (see Chapter 30). In that chapter, we also considered current views of the neurophysiological basis of learning. Here we examine the modes of learning at a higher organizational level, that of the whole organism.

Habituation

Habits are responses which, through long repetition, we learn to make after receiving a given stimulus. A habitual response becomes automatic in time. It is not innate. It is modifiable and remains so. Thus it is a simple form of learning. Habits die out if they are not repeated more or less frequently. One can purposely break a habit by avoiding the stimulus or by preventing or changing the response.

As discussed in Chapter 31, **habituation** is a form of "response-waning," a decrease in the probability of response to a stimulus upon repeated presentation of the stimulus. If snails are poked repeatedly and nothing bad happens afterward, they will no longer display a withdrawal reaction (drawing in of the tentacles). Even flatworms will cease to respond to vibrations if nothing follows. Habituation, one of the most widespread forms of learning, is especially important in the development of behavior in young animals. One of its chief values is its tendency to free an animal's brain of the need to respond to insignificant stimuli, while permitting it to concentrate on useful survival behavior.

Associative Learning

Many animals are capable of associating two or more stimuli with reward or punishment. This type of learning, termed **associative learning,** was first studied by the Russian physiologist Ivan P. Pavlov (1849–1936) (Figure 32-13). In experiments on the reflex that causes a dog to salivate when it smells food, Pavlov first presented dogs with food and measured the amount of saliva each produced. Then he added a sound—the ring of a bell—every time food was offered. After the two stimuli, food and sound, had been presented together six times, he found that when the dogs were presented only with the sound stimulus they produced saliva, even though sound alone did not initially trigger salivation (Figure 32-14). Pavlov termed this a **conditioned reflex,** which he defined as one in which the response to one stimulus becomes associated with another stimulus.[6]

Associative learning can be studied experimentally by methods like Pavlov's, which became known as Pavlovian conditioning, or classical conditioning, and many such experiments have been performed since then in animals of all kinds, from earthworms to chimpanzees and humans. Responses of every sort can be associated by conditioning. It became an extremely useful methodology.

Investigators now use the Pavlov technique in varied forms to develop what are called conditioned responses. An animal, for example, may be conditioned to respond differently to stimuli such as two tones. By making the tones more and more alike, the point at which the animal can no longer distinguish them is determined. Thus the technique permits study of both learning and discrimination.

If you wanted to find out whether bees recognize different colors, the use of conditioned responses makes it possible. The problem is to determine whether or not a bee can distinguish a color from that intensity of grayness to which the color would correspond if the bee were color-blind and recognized therefore only degrees of brightness in a series of grays. How the problem was solved by Karl von Frisch is shown in Figure 32-15.

Trial-and-Error Learning

True trial-and-error learning is common among animals, which are prone to make mistakes in dealing with environmental and social conditions. This process has been termed **operant conditioning** to contrast it with classical Pavlovian conditioning. Operant behavior is defined as behavior that is guided by its consequences. Not all behavior is influenced by the events that follow it. Some behavior, like that involving reflexes, is guided by the events that precede it. The word *operant* refers to an essential feature of goal-directed behavior: that it have some effect on the environment.

Suppose we repeat Pavlov's dog experiment with a few changes. When a conditioned dog hears the bell, it salivates. It also launches behavior likely to get it food—beg, search the room, try to get out of the room, etc. If one of these actions is successful—say, the dog finds a hidden bowl of food in a certain spot after all its trials and errors—next time the experiment is done the dog will skip the other

---

[6] Pavlov's major works appeared in English translation from 1902 to 1929. Along with John B. Watson, who came later, Pavlov is recognized as one of the founders of experimental psychology. However, he won the 1904 Nobel Prize for his work on enzymes! For a time after the publication of Pavlov's work, some enthusiasts believed that all behavior could be viewed as reflexes: innate behavior involved a simple reflex; learned behavior was a conditioned reflex.

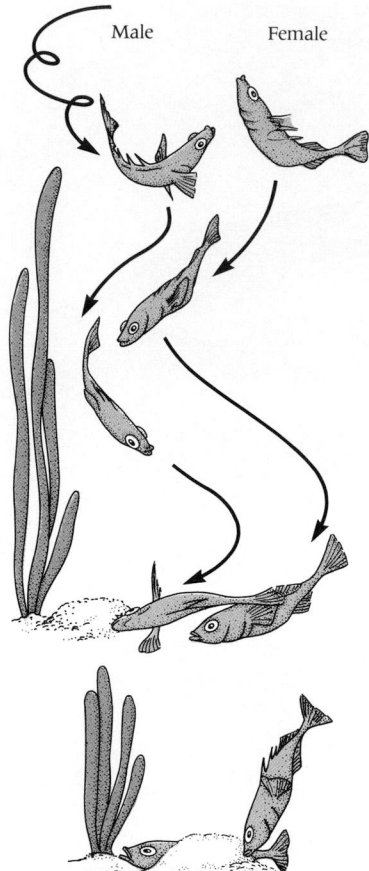

Male   Female

32-12 Chain-reaction behavior in stickleback fishes
In this courtship dance, each movement made by the male is a stimulus that elicits the next, fixed movement by the female. In turn, each female response stimulates the male to the next movement. First, the male performs a zigzag dance. The female then turns toward the male and takes an upright posture. The male then swims toward the nest. The female follows. The male shows the female the nest by putting his snout in the entrance and rolling on his side. The female enters the nest. The male nuzzles the female at the base of the tail. This stimulates the female to lay eggs. She then leaves the nest. The male then enters and fertilizes the eggs. (See also Figure 35-4.)

32-13 Ivan P. Pavlov (1849–1936)

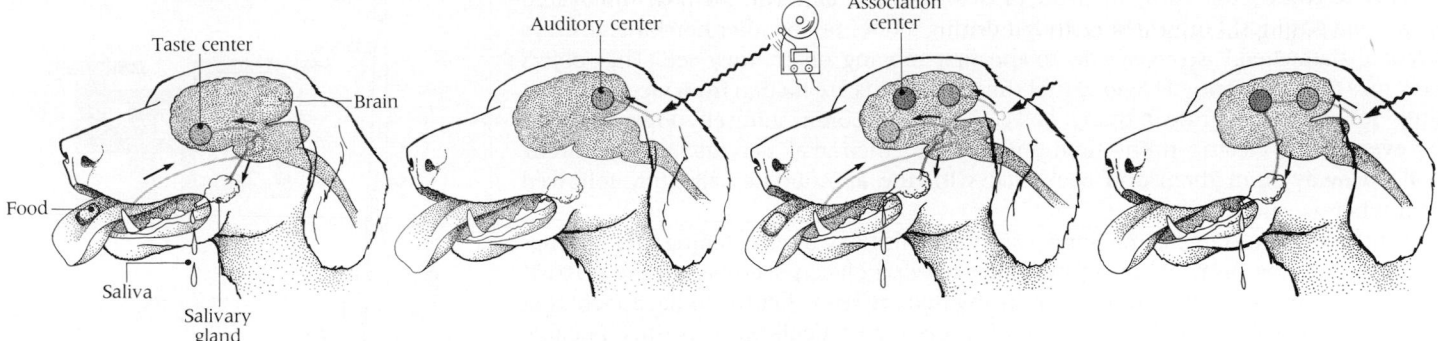

32-14 A simple conditioned reflex (Above)
When food reaches the dog's tongue, the cerebral taste center is activated and the salivary gland releases saliva. When a bell rings, the auditory center is activated, but no saliva is released. When food and the bell sound are presented simultaneously, salivation occurs via the taste center, but a cerebral association center is also activated by impulses from the taste and auditory centers. The dog thus learns to associate the bell sound with food. After this association is established, stimulation of the auditory center by a ringing bell elicits saliva flow.

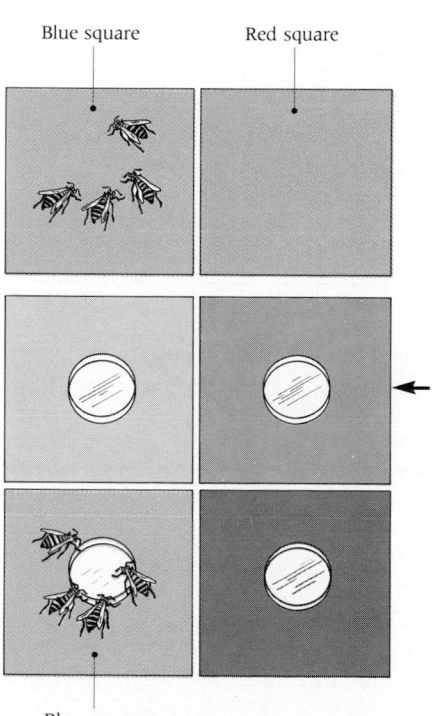

32-15 Using a conditioned response to demonstrate color vision in bees
Bees that are regularly fed on blue paper will settle on blue paper even though it lacks food, ignoring a blank red paper. Such bees are thus conditioned to blue paper. To show that the bees distinguish blueness, not grayness, a table is covered with one blue square and three grey squares each bearing an empty food dish. The bees settle on the blue paper, indicating that they recognize color and can distinguish it from equivalent shades of gray.

activities and go directly to the place where he previously found the food. The two phases of the dog's response involving (1) unorganized trials and errors that eventually prove successful, and (2) later repetition of the behavior that worked in a similar situation, are together called operant conditioning. In trial-and-error learning, an animal learns to eliminate behavior that produces no reward (or that produces discomfort) and to repeat a rewarded behavior. Nowadays psychologists speak of reinforcements rather than rewards.

Trial-and-error learning is a commonplace of our own lives. It is how we learned as children not to touch hot stoves. It is also how we learned to play a musical instrument (sour notes being a form of punishment), write, type, serve a tennis ball, and carry out many other highly coordinated activities.

Pioneering work in this field was performed by B. F. Skinner of Harvard, who coined the term operant conditioning. Skinner showed that a rat is able to learn a maze in a box with up to 30 "choice points." A maze is a series of pathways with one or more points where the animal must choose which way to go. With a wrong choice, the animal reaches a blind end, is punished, or fails to achieve a reward. Animals with complex nervous systems learn highly complex mazes. One often used with rats appears in Figure 32-16. Learning rates are scored by the number of wrong turns on successive trials. Note that the error rate decreases (the rat learns) in successive trials.

In a favorite maze study, a hungry rat (which takes readily to mazes) is rewarded by food for improving performance—either by giving a food reward for solving the path and coming out at the right end, or by requiring the rat to press a food lever. In a typical experiment of the latter kind, the maze has levers at various choice points nearer the exit that yield food pellets if pressed. The rat will first explore the maze feverishly. At first, it ignores the levers and moves about looking for a way out. Soon it finds by accident that pressing the lever yields food. Then it learns to operate the lever. But to get to the lever it must learn the maze. When hungry, it will learn the maze quickly—so it can spend its time pushing the lever. By varying rewards or other factors, an experimenter can evaluate different influences on learning rates.

As shown in Figure 32-16, ants tested in the same maze learn less rapidly. Some animals do very poorly in mazes.[7]

The rat's learning process involves trial-and-error. Here the game is rigged by the experimenter. In the real world, trial-and-error learning demonstrably increases the harvesting and foraging repertories of many kinds of animals. A capacity for such learning is clearly advantageous to animals since it allows them to solve a large variety of problems by acquiring new patterns of motor behavior.

Parental Imprinting

In some cases, certain vertebrates make strong associations between stimuli and motivations very early in life. Learning that takes place rapidly at a critical age is called parental **imprinting.**

---

[7] Perhaps because their natural habitat is an open plain, sheep become terrified in a maze and learn very little. If this explanation is correct, it would signify an important association of emotional and cognitive factors in learning.

Lorenz coined this term in 1935 to describe an innate mechanism discovered in greylag goslings. Lorenz showed that during a brief period after hatching goslings become "attached," so to speak, to the first moving object they see. That object becomes their "parent." Provided that the real parents are hidden from view, the goslings follow a substitute moving object—such as Lorenz himself (Figure 32-17) or even a toy electric train—and soon ignore their real parents. When Lorenz walked away from them and beckoned with the appropriate call, they followed him wherever he went.

If he waited to start his maneuver until the third day after hatching, the goslings would not follow him. He also found that moving objects seen before the **critical period** do not cause attachment, nor do any that are seen after the initial attachment has occurred. The particular object that is seen at a particular time is thus indelibly "imprinted." That is, the goslings must have quickly memorized the features of their parents—or what they take for parents—to ensure future recognition.

Imprinting does not depend on reward or punishment and, once learned, it is irreversible. Thereafter, the bird responds to that one stimulus and not (in the same way) to any other. In the wild, that object will almost always be the mother bird. The behavior is thus highly adaptive.

Imprinting is not limited to visual stimuli or to birds, but we know most about it in young birds. Goats are imprinted by olfactory cues. Pacific salmon are imprinted by olfactory cues in the stream where they hatched. While still in the egg, ducklings form a preference for the sound frequency of their mother's call.

Insight Learning

The highest form of learning, **insight learning,** is mainly the province of humans and other primates. In a famous example, a chimpanzee was placed in a room

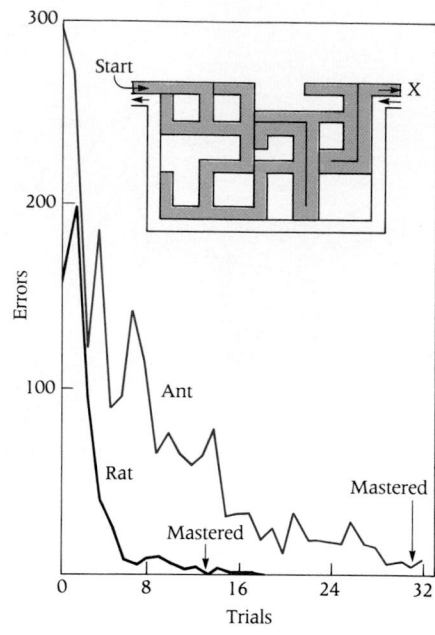

32-16 Maze learning
The inset shows the ground plan of a maze offered to rats and ants. The graphs show the progress of a rat and an ant in mastering the maze. The rat learned the maze more rapidly than the ant, mastering it by the thirteenth trial, whereas the ant required 31 trials.

32-17 Imprinted goslings following Konrad Lorenz
Imprinted goslings follow ethologist Konrad Lorenz as if he were their parent. Lorenz was the first living creature these young geese saw during the latent period immediately after hatching.

A   B

with boxes and a banana dangling from the ceiling out of its reach. Facing the problem for the first time, the chimp would often respond by stacking the boxes on top of one another, climbing up, and grabbing the banana (Figure 32-18A). However, the raccoon on a leash in Figure 32-18B fails to solve its problem.

Insight learning is characterized by reasoning. It consists of the recall of earlier experiences with stimuli unlike the ones currently being experienced, followed by the application and adaptation of what was learned and its utilization in the solution of a novel problem in a new context. Thus it involves a process of generalization.

One might reasonably ask if the chimp did not get the banana by trial-and-error approach. The answer is that insight learning is distinguished from trial-and-error approaches to a problem by the fact that not all the possible solutions are overtly attempted. Rather, it is as if the inadequate ones are tried and rejected mentally until, in a moment of sudden insight, the mind says "ah-ha."

PATTERNS OF ANIMAL BEHAVIOR

BIOLOGICAL CLOCKS AND CIRCADIAN RHYTHMS

Living organisms must correlate their activities with changes in the environment. Days alternate with nights. Tides ebb and flow. Wet seasons alternate with dry and cool seasons with hot. These cycles are the product of physical relationships of Earth with the moon and the sun.

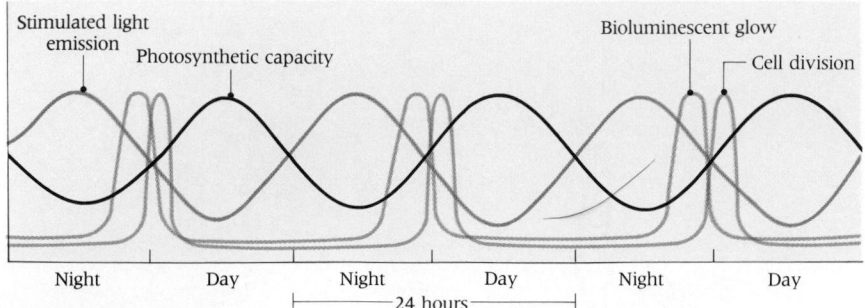

Rhythmic Patterns in Nature

Animals, plants, and protists behave differently at different times of the day or of the year. Many examples could be cited, including these.

- The unicellular alga, *Gonyaulax,* famed for its role in producing potent nerve poisons and red tides (see Box 39-1), displays at least four rhythmic patterns in the course of 24 hours: the photosynthetic fixation of CO_2; the timing of cell division; one form of bioluminescence that is referred to as a spontaneous glow; and another in which light flashes are emitted in response to various external stimuli (Figure 32-19). Although all four cycles have their own schedules, the evidence suggests that they are controlled by a single pacemaker, or "clock" mechanism. The bioluminescence patterns have been tempting to investigators because they depend on a known biochemical mechanism. Current work does indeed show a rhythmic pattern in the cell's content of the enzyme luciferin and its substrate luciferase at different times of the day (Figure 32-20). A 1989 study by J. W. Hastings and coworkers shows circadian rhythm in translation of the enzyme's gene.

- Cockroaches (*Periplaneta*) scurry about only at night. In the early 1950s, Colin S. Pittendrigh, then at Princeton University, carefully investigated the daily cycle of cockroach activity (and helped to spark new scientific interest in biological rhythms). The plot of the behavior pattern of a cockroach (Figure 32-21) shows its relative activity at different times of the day. The insect clearly becomes very active about 10 P.M. and quiets down about 4 A.M. Under a constant light, the cyclic pattern is unchanged. Evidently it is innate. Only after many weeks does it get out of phase with true time. After returning the cockroaches to their natural environments, daily rhythms are quickly resynchronized with the actual time.

- Populations of fruit flies (*Drosophila*) that have been synchronized with a normal daily solar cycle of light and dark—this necessary "training" experience is called **entrainment**—hatch their eggs just after dawn, even if they are kept in total

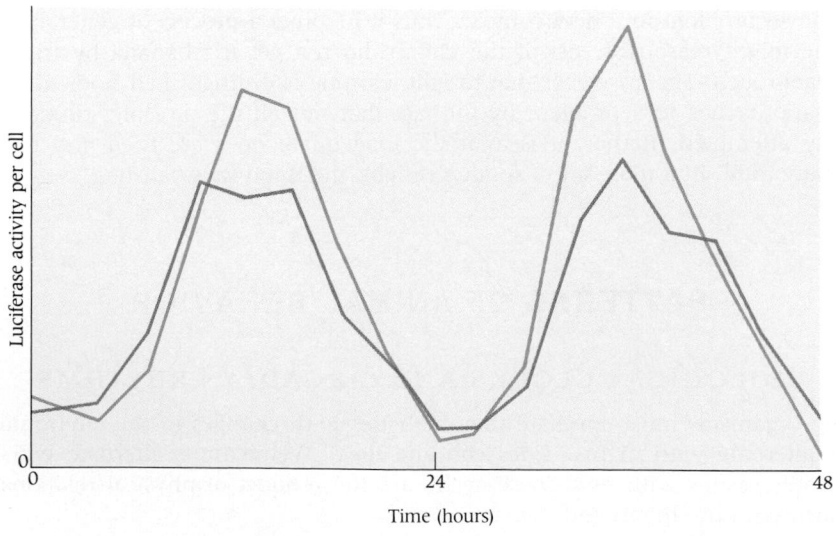

32-20 *Circadian variations in the luciferase content of* **Gonyaulax**
Culture of this unicellular alga grown in continuous light reveals a rhythmic rise and fall in the levels of luciferase in the cell over a 24-hour period. The results of two experiments are in excellent agreement.

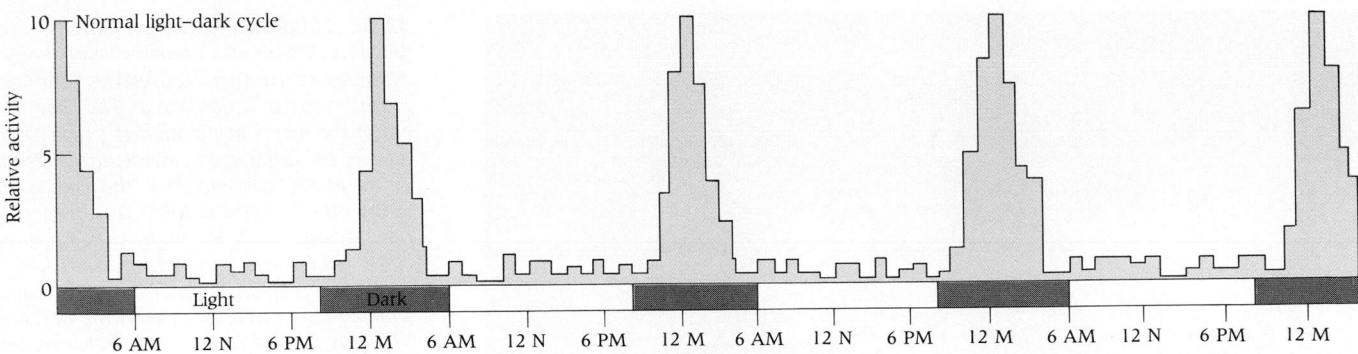

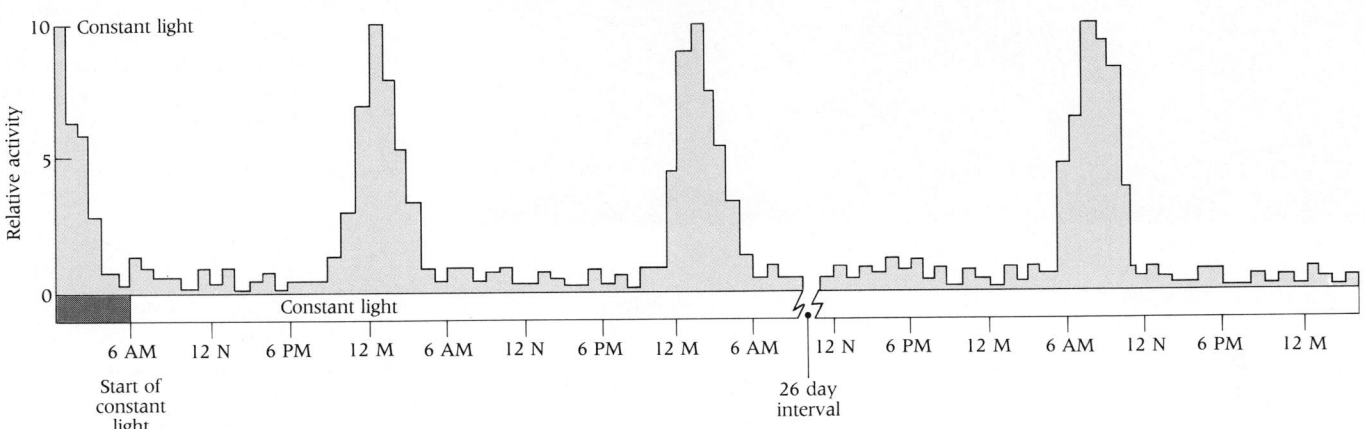

darkness. Recent work has revealed at least six different genes that are responsible for such rhythms in this species. The best studied is the gene called *per*. Some mutations of *per* completely abolish the rhythmic cycle. Others increase the day length from 24 to 29 hours. Remarkably, *per* mutations also affect another periodic activity—a vocal sound with which males try to attract mates.

- Many migratory birds of the north temperate region respond to the onset of autumn by flying south. This migration is believed to be initiated by a change in the length of nights.

- Breeding seasons of temperate birds are triggered by increasing day lengths in the spring and by cycles in the pituitary and gonads, which produce the sex hormones. These induce the development of such characteristic changes as brightly colored plumage and typical behavior patterns, including territorial defenses and courtship behavior. The term **photoperiodism** is applied to biological effects caused by changes in day lengths.

- Grunion, a species of small fish (*Leuresthes tenuis*), swim up on the beaches of southern California at the highest tide of the lunar cycle and lay their eggs in the sand between waves (Figure 32-22). If they fail to judge the tide correctly, their eggs are prematurely washed out to sea by a higher tide. This seasonal cycle is strictly cued by the phases of the moon.

- Colorful poinsettias emerge in the Christmas season in response to long nights—another example of photoperiodism. In New Zealand, where Christmas comes in midsummer, the plant symbolic of Christmas is the brilliant pohutukawa tree (*Metrosideros excelsa*). This species flowers in response to short nights. Hence, it flowers in December in those latitudes. The opening of many flowers is synchronized with the appearance of a pollinating insect. If flowers open too late, their chances of being pollinated are diminished or eliminated. If they open too early, they may be eaten or destroyed before being pollinated.

Circadian Pacemakers

Many, if not all, organisms possess what has been called an internal **biological clock.** For example, when animals normally displaying periodic behavior are kept

32-22 Tidal and lunar periodicity
(**A**) *Photograph of a remarkable phenomenon exhibited by grunion,* Leuresthes tenus, *on the southern coasts of California. The fishes wriggle out of the waves at the highest (night) high tides of the lunar cycle. Momentarily stranded on the beach between waves, the females wriggle down into the sand tail-first. In this position, they lay eggs while the male, curled around her body, ejaculates sperm, which travels over the female's wet flank to reach the eggs. Male and female both return to sea with the next wave. The eggs, buried in the sand, hatch in time to go to sea with the next high tide two weeks later.* (**B**) *The chart shows the patterns of grunion egg-laying in relation to tidal and lunar cycles at La Jolla, California. Heights of high tides (in feet) about 24 hours apart have been connected by smooth lines. The two tides each day yield the two curves. Night tide is indicated by the colored line. Intensity of runs is marked by short vertical bars below the graph. Moon phases are given below (black circles, new moon; white circles, full moon).*

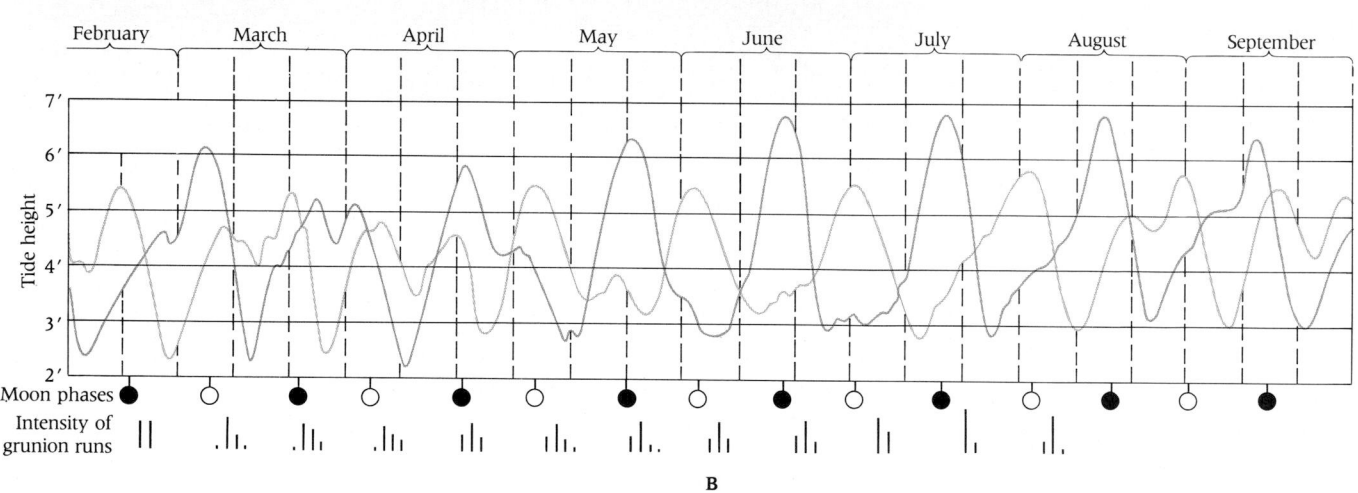

under constant conditions—say, constant darkness, light, or temperature—so that they receive no environmental cues, they will continue to show rhythmic patterns of activity. These cycles may persist for weeks without any indications of the time or season. Significantly, these oscillations are usually about 24 hours long. This fact led Franz Halberg to term them **circadian rhythms** (from the Latin *circa*, meaning "about," and *diem*, meaning "day"). The examples just cited were of circadian rhythms. (Circannual rhythms are related to the whole year.)

The model in Figure 32-23 has proved useful in the study of circadian systems. It proposes, among other things, that we must identify circadian pacemakers (what are they? where are they?) and the factors that regulate them.

The precision of these cycles is quite remarkable. Although we do not know what underlies most of them, some are yielding to investigation. We saw in Chapter 28 that the pineal gland may be a biological clock, serving in many vertebrate

32-23 Conceptual model of circadian timing system
The wiggles represent an oscillating system. The wiggle in a circle is an active cellular unit that can maintain a self-sustained oscillation with its own independent time periods. In the model, a photoreceptor system of some type receives an input signal and transmits it to the circadian pacemaker, which regulates (usually by chemical mediation) the metabolic activity leading to the output response.

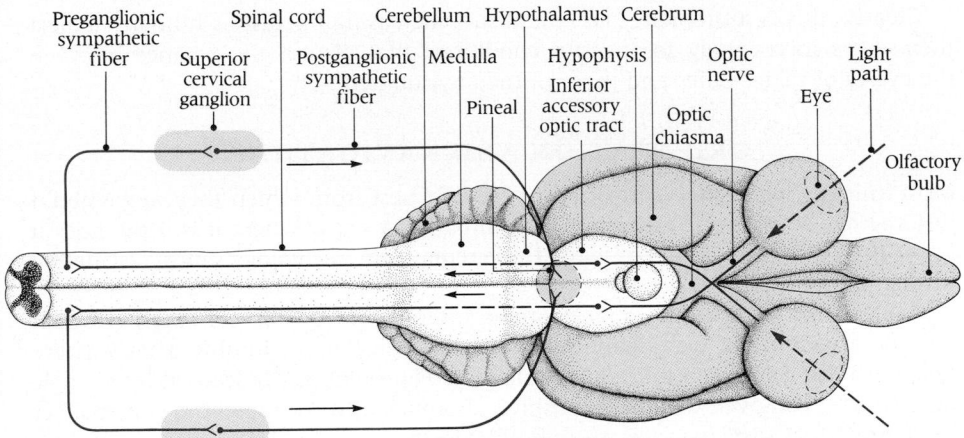

32-24 How light influences body rhythms in a rat
Light entering the eye by the usual route stimulates the retina, which sends impulses to both the visual cortex and the hypothalamus. There nerve fibers of the inferior accessory optic tract synapse with preganglionic sympathetic fibers (color) entering the spinal cord. Only after a second synapse with postganglionic sympathetic nerve fibers (color) do nerve impulses reach the pineal, which synthesizes melatonin.

species as a mechanism for keeping track of the light-dark cycle. In many birds, fishes, amphibians, and reptiles, the pineal serves as a "third eye."[8]

How does the pineal work? In mammals, nerve impulses elicited by sunlight entering the eyes travels by two paths to the brain (Figure 32-24): (1) the familiar path to the visual cortex (see Chapter 30), and (2) a less well known path to the hypothalamus. From the hypothalamus, impulses enter the spinal cord, exit through a nerve center in the neck, and travel to the pineal.

In all mammals, light stimulates a circadian cycle in the production of the pineal hormone melatonin. Despite its skin-lightening action in frogs (see Figure 28-41), this hormone has no such effect on mammalian melanocytes. It is synthesized from tryptophan through a short pathway, the third step of which is catalyzed by the enzyme *N*-acetyltransferase, which is found within the pineal (Figure 32-25).

Interest in this enzyme intensified when it was assigned primary responsibility for the rhythmic production of melatonin. In experimental rats, the enzyme's activity in pineal cells is significantly higher during the night than during the day (Figure 32-26). Even when the pineal bodies of some vertebrates are grown in culture, circadian rhythms in the activity of this enzyme persist for several days after the gland is removed from the body. This remarkable observation suggested that the pineal body may serve as a self-contained circadian oscillator. However, later work has implied that pineal activity may be driven by a component of the hypothalamus called the suprachiasmatic nucleus (SCN). For example, the pineal's ability to produce circadian rhythms in melatonin synthesis in constant darkness was lost when the SCN had been experimentally damaged. Melatonin, it appears, may be the hormonal link (especially in birds and lizards) coupling the endocrine system and various photoperiodic responses of reproduction.

Implications of Circadian Rhythms in Humans

Humans have several circadian rhythms. Body temperature reaches its highest daily level in the evening and its lowest in the morning. Similar daily cycles characterize many body functions—heart rate, blood pressure, waking and sleeping, hormone production, hemoglobin levels, and so on. There is ample evidence that these rhythms depend on an internal "clock" and are not simply responses to periodic changes in the environment. Although the nature of the clock has not been definitely established, it may well be the suprachiasmatic nucleus of the hypothalamus.

These considerations take on practical importance for human life in several areas. As people move very rapidly through Earth's time zones via jet travel, their biological clocks often get seriously out of phase with the local day-night cycle. This situation, termed jet lag, produces peculiar subjective feelings, with decreased alertness and disturbed eating and sleeping schedules. Several days are needed to readjust the body's circadian rhythms after such a flight.

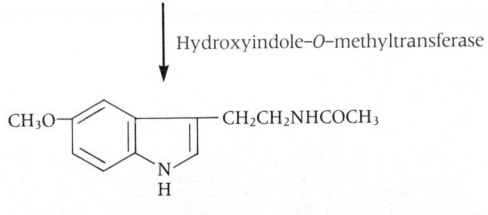

Tryptophan

↓ Tryptophan hydroxylase

5–Hydroxytryptophan

↓ Aromatic amino acid decarboxylase

Serotonin (5–Hydroxytryptamine)

↓ *N*–Acetyltransferace

N–Acetylserotonin

↓ Hydroxyindole–*O*–methyltransferase

Melatonin

32-25 Synthesis of melatonin from tryptophan in the pineal gland

[8] In some fishes, amphibians, and reptiles, the light-sensitive portion of the pineal is actually exposed at the top of the head. In others of these groups, a light-sensitive pineal is inside the skull.

Clearly, this is a problem that will attract increasing interest as humans subject themselves increasingly to stressful conditions that disrupt the balance between the cycles of their bodies and those of the environment.

ORIENTATION AND NAVIGATION

Most animals have some sort of home base or nest from which they move about in foraging. To return to its home, an animal must know where it is. How does it know this? How does it find its way? These remain among biology's most fascinating unsolved problems.

As discussed earlier, the simplest form of oriented movements are taxes, in which an animal assumes a well defined spatial relationship to a stimulus. That was the case with the celebrated pill bug described in Figure 32-3. Confronted by a single light source, it moved toward it (positive phototaxis). In contrast, the cockroaches in Figure 32-21 displayed negative phototaxis.

A more complicated form of oriented movement is the light-compass reaction displayed by many invertebrates. In orienting themselves, they make use of the sun, keeping constant the angle between the direction of movement and the direction of the light stimulus (Figure 32-27). Bees use the light-compass reaction in returning to the hive; so do ants when moving to and from their nests. Experimenters have interrupted the homeward journey of ants by putting them in a dark box for 2.5 hours. During that time the sun moved 37° across the sky. Upon release, the ants shifted their direction by 37° from the original path and missed their nest. Evidently, they had not allowed for the movement of the sun.

The capacity of animals to find their way over long distances is a phenomenon we call **navigation.**

- Many bird species make remarkable annual journeys between nesting grounds and distant feeding areas. The American golden plover (*Pluvialis dominicus*) breeds in northern Alaska, migrates in the fall to Argentina by way of Labrador, and returns in the spring over Central America heading north and west along the Mississippi River (Figure 32-28).
- Similar migrations occur among many insect species, including locusts, mosquitoes, monarch butterflies, dragonflies, and even certain aphids.

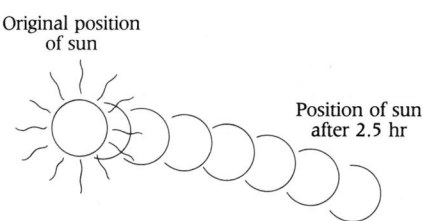

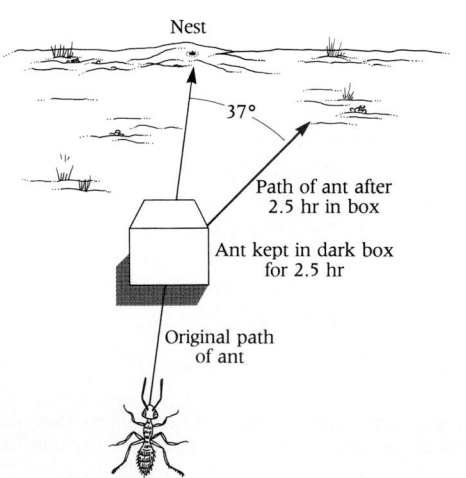

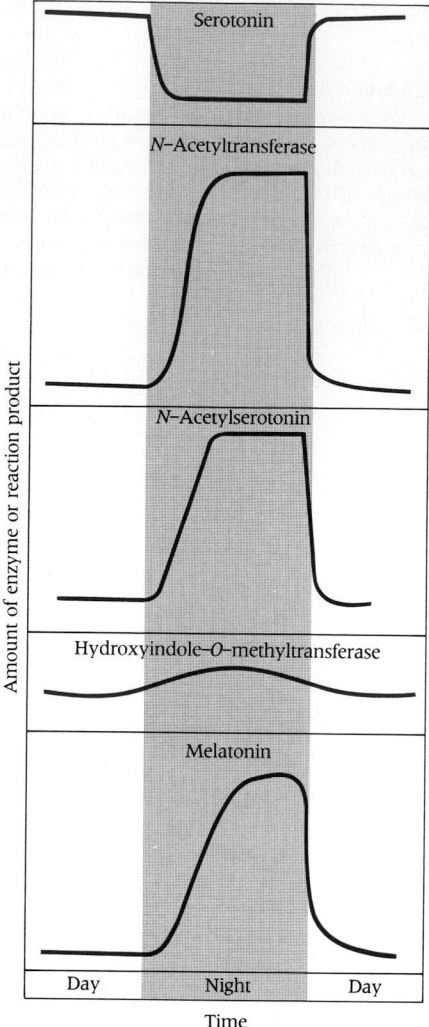

32-26 Periodicity of biochemical activities in the pineal
Activities of the enzymes that convert tryptophan to melatonin peak during the night. Serotonin is lowest during this part of the cycle and, as expected from the enzyme levels, the end product of the pathway, melatonin, is at a maximum during the night.

32-27 Light-compass reaction in ants
The reaction is described in the text.

32-28 Migration routes of the American golden plover

The annual migration of this remarkable bird spans many thousands of miles, from northern Alaska to Argentina.

32-29 Migrating salmon, Onchorynchus
(A) *The long migration routes of three varieties of Pacific salmon: silver (or coho); king (or chinook); and steelhead. Comparable routes are found in the Atlantic Ocean.* **(B)** *An Atlantic salmon leaps the falls as it heads upstream to its home waters.*

■ Some salmon are born in freshwater streams and then journey downstream to the ocean (Figure 32-29). Several years later, having fed and matured, they swim back upstream to spawn and in many cases to die. Remarkably, their journeys always end in the same stream in which they were born. They choose it from dozens or hundreds of equally suitable streams. In one study, 469,236 young salmon from a certain stream were marked and 11,059 were recovered from the same home stream several years later, after they completed their long return journeys—some up to 5,000 miles. Not a single marked salmon was ever recovered from an incorrect stream.

What guides animals in these long journeys in space and time? In many cases, the answer is known. Salmon, it seems clear, locate home rivers by smell. Like many fishes, they have an acute olfactory sense. They also appear to have a keen memory for slight differences in the chemical composition of water. It has been shown that the homing of salmon is attributable to an individual fish's memory of the distinctive "fragrance" or "bouquet" of its native stream. Salmon whose olfactory tissues have been destroyed cannot locate the right spawning stream.

What about the feats of long distance migration so common among birds and insects? Insect migrations are the more remarkable for the fact that most adult insects live for a year or less. The flights of many of these insects—some extending more than 2000 miles—are not annual or seasonal journeys, repeatedly undertaken by the same animals. Rather they are migrations undertaken once for the purpose of establishing new colonies at a considerable distance from their place of birth. On the other hand, many insect species migrate seasonally over very long distances toward predictable destinations. Several North American species—for example, leaf-hoppers (*Macrosteles fascifrons*) and harlequin bugs (*Murgantia histrionica*)—migrate north in the spring, ranging far up the Mississippi Valley. Once a year, billions of beautiful monarch butterflies (*Danaus plexippus*) fly from eastern Canada and the United States to a remote wintering place that remained unknown until 1976, when Fred A. Urquhart of the University of Toronto located it. Amazingly, it is a mere twenty acres of lofty wooded slope in central Mexico (Figure 32-30A). The near freezing temperatures there immobilize the insects, enabling them to conserve fat for their remarkable northward return trip in the spring (Figure 32-30B,C).

To an extent, the travel of these insects is aided by winds, but we still have no answer to the larger question: how are these migratory paths and destinations transmitted from generation to generation?

Unlike most insects, birds have life spans that generally extend over several seasons. In bird migrations, usually between northern and southern terminals, migrants typically follow a precise route that is characteristic for that species. Migrating birds probably rely in part on timing (by internal clocks keyed to light-dark cycles) and landmarks on the terrain below. In many cases, migration must depend on something other than landmarks. Many species fly over open ocean where there are no landmarks. Moreover, many migrating birds have never before made the journey and would not have become familiar with landmarks. Part of the answer appears to lie in the ability of birds to make use of celestial navigation.

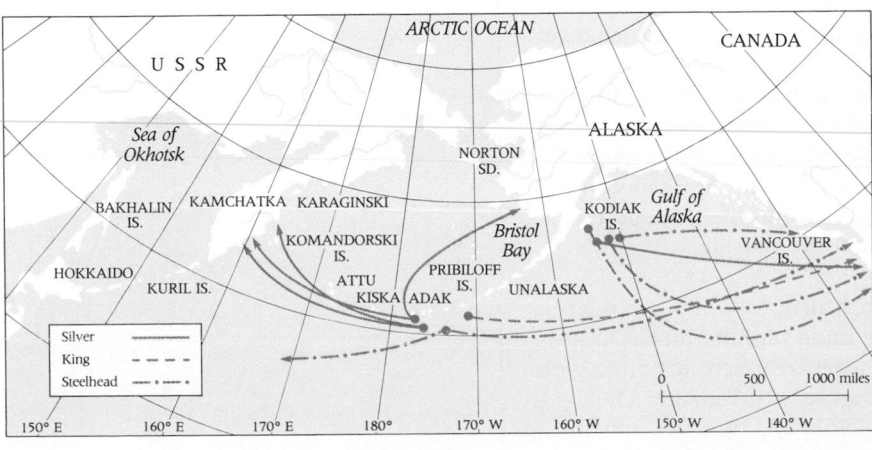

A

B

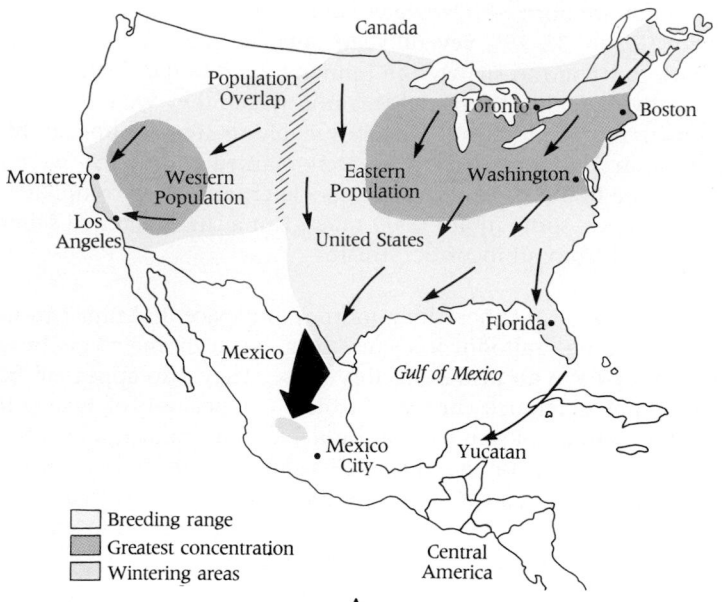

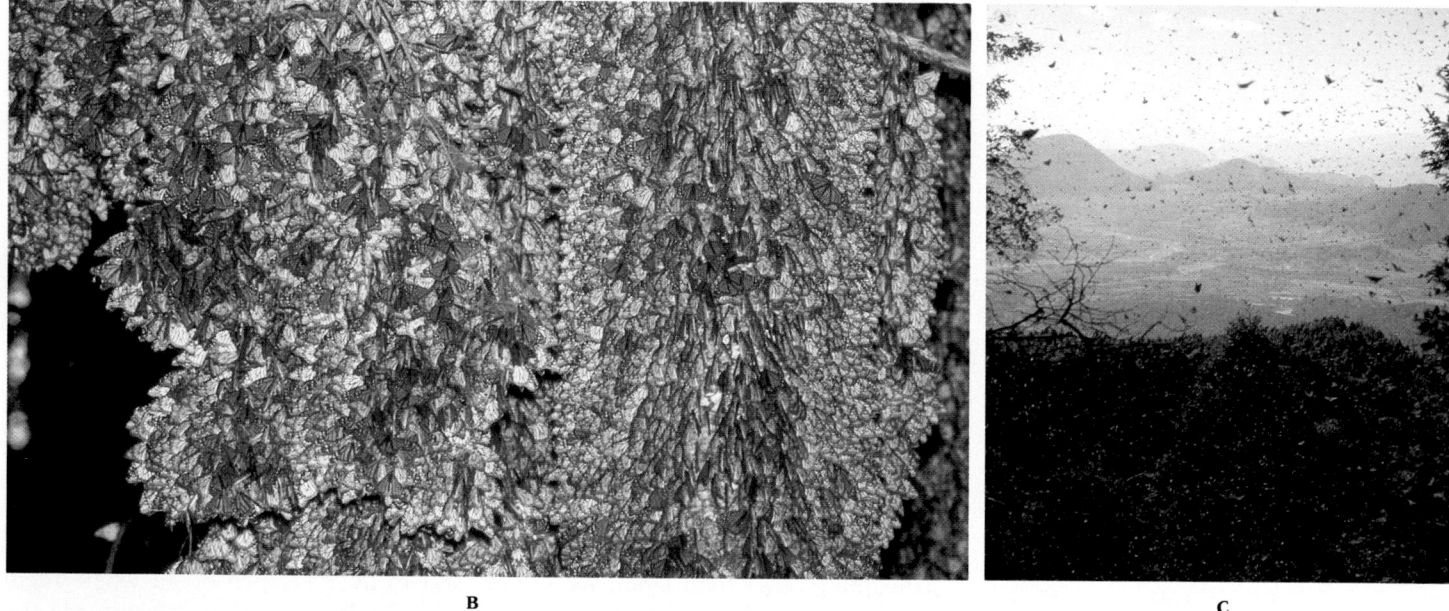

32-30 Migrations of the monarch butterfly
(A) *Migration routes to their Mexican wintering area.* **(B)** *Butterflies spend the winter clustered on trees.* **(C)** *In spring, they take flight.*

Map labels:
- Canada
- Population Overlap
- Toronto
- Boston
- Monterey
- Western Population
- Eastern Population
- Washington
- Los Angeles
- United States
- Mexico
- Florida
- Gulf of Mexico
- Mexico City
- Yucatan
- Central America
- □ Breeding range
- ▨ Greatest concentration
- □ Wintering areas

A

B

C

- At migration time, caged starlings (*Sturnus vulgaris*) become unusually restless. If permitted to see the sun, they begin to fly toward the side of the cage that lies in the direction of their normal migration route. When the sky is overcast, these movements persist but without direction. The importance of the sun compass is clearly demonstrated in experiments in which mirrors are used to "change" the sun's position.

- Night-flying migratory birds are believed to be guided by star patterns. In one imaginative experiment, North American indigo buntings (*Passerina cyanea*) were allowed to attempt flight under the artificial night sky of a planetarium. (Actually, they were placed in containers that recorded the orientation of their footprints.) The birds oriented themselves "correctly" with reference to the planetarium sky, which did not correspond with the actual night sky outside.

Finally, there is the impressive phenomenon of **homing** to which we referred briefly in Chapter 29. This is the process wherein homing pigeons (*Columba livia*) and other birds and animals return to their home when released at points great distances away, often hundreds of miles or more (Figure 32-31). After circling a few times near the release site, they fly homeward rapidly and directly, sometimes making a 600-mile flight from an unfamiliar release point to their precise home

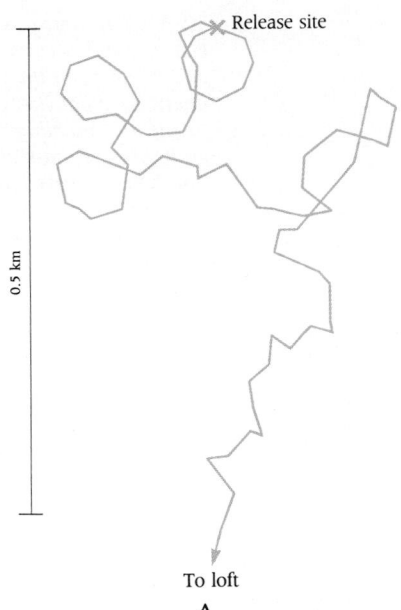

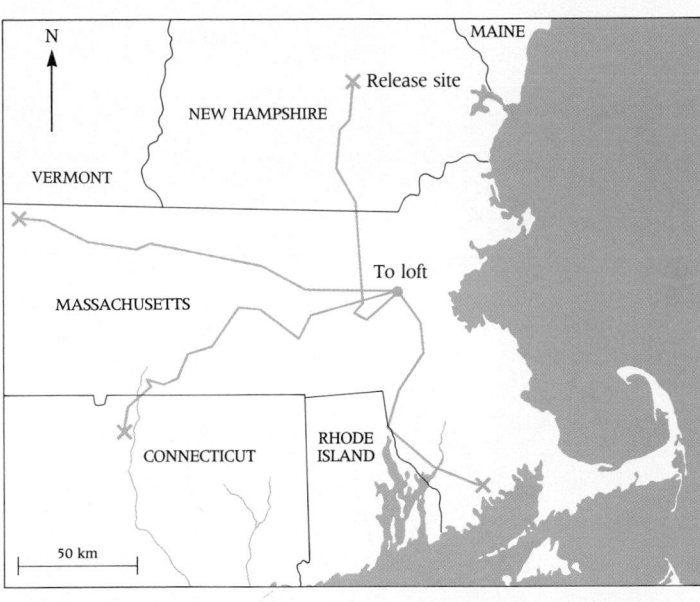

32-31 Homing pigeons
(A) Pigeons will circle a release site, and then take an irregular course home. (B) Routes are seldom direct. (C) A homing pigeon arriving home after a long flight.

location in a single day—a truly spectacular accomplishment. If they make the journey more than once, they fly a somewhat different route each time.

For a pigeon to home it must know where home is and its own position with respect to home, so that it knows in which direction to fly. Pigeons do use the sun or stars to take the right direction. However, the basis of the map sense by which pigeons and other birds determine their positions is still not known. There is evidence that part of the information is derived from Earth's magnetic field. We noted in Chapter 30 that pigeons have a magnetic body in their heads (or, according to some, neck muscles) to which nerve fibers attach. These might permit them in some way to gather information on the Earth's magnetic field. Pigeons do get disoriented in areas where there are magnetic distortions because of large iron deposits.

Sharks moving through Earth's magnetic field produce an electric current that can be detected by electroreceptors. This enables them to tell east from west.

Nature has evolved a number of astonishing systems by which animals orient themselves. Some seem the inventions of science fiction writers.

■ Some bacteria are capable of magnetic navigation (see Box 29-2).
■ Some bats orient themselves by means of a sonar system, with which they bounce their own signal off a target—the same system used by ships to locate

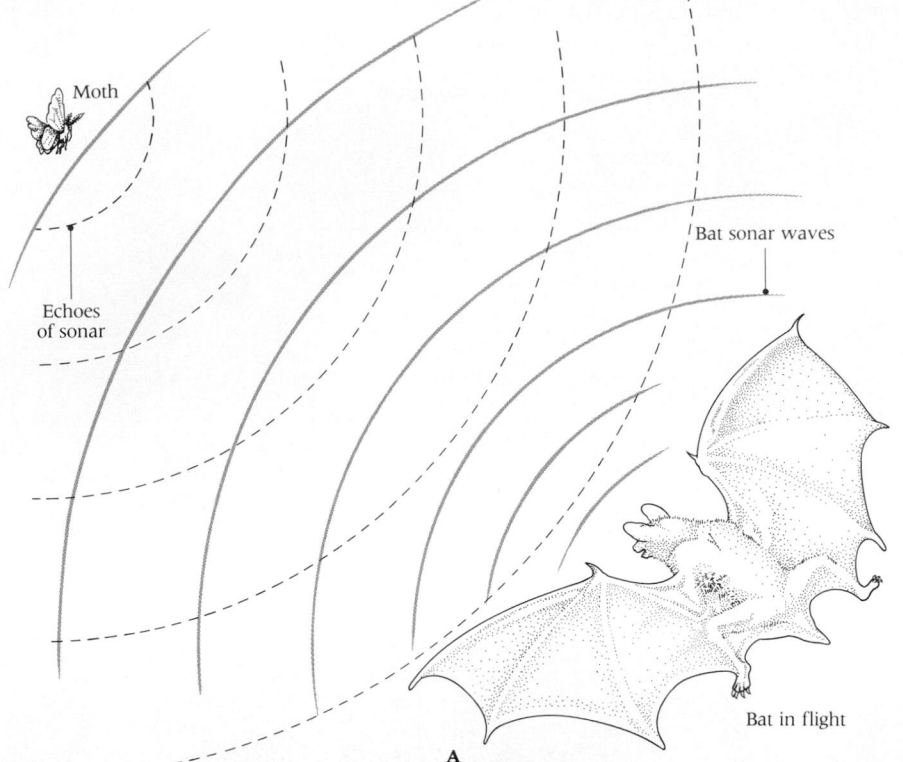

32-32 The sonar system of bats
(A) *Bats emit a series of sonar pulses (solid lines) and home in on the reflected signals (dashed lines). As a bat closes in, it emits pulses more frequently, enabling it to respond more precisely to the evasive maneuvers of the prey (in this case a moth).* (B) *Photograph of a bat closing in on prey.*

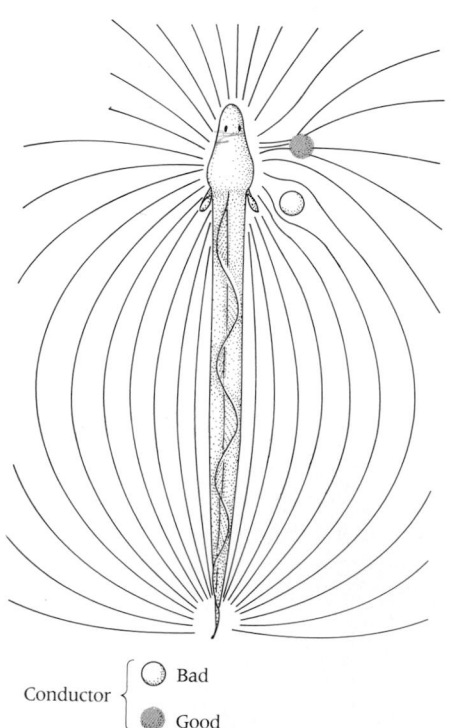

32-33 Electric eel uses electric field to detect foreign objects
The electric eel, Gymnarchus, *can generate an electric field around its body. Foreign objects, whether good or bad conductors, distort the field. The eel detects such distortions with receptors in its head.*

submarines (Figure 32-32). Many also have acute hearing that relies on sounds emitted by prey to guide them to it.

■ Porpoises and other cetaceans (and at least one bird species, the nocturnal Venezuelan birds called guácharos) also employ an echo-location system. Many blind persons are said to have developed this ability to some extent by making clicking sounds.

■ Certain electric fishes generate weak electric fields around their bodies to orient themselves in murky water (Figure 32-33). Nearby objects distort the field and the animal is able to detect the changes with specific receptors in its head.

These fishes also use electric fields to communicate with one another and to recognize species, a necessary prerequisite to mating. Different pulse patterns are used in threat, submission, and courtship.

We turn now to aspects of animal behavior that more directly concern their relations to each other, as well as to their environment. These matters, we will see in due course, are of fundamental importance in maximizing the survival of individuals and populations and in determining the course of evolution.

SUMMARY

The scientific study of behavior has long been hampered by two problems: the complexity of the relevant variables and the human penchant for anthropomorphic thinking. Nonetheless, many dramatic advances have deepened our understanding.

Tropisms in plants, and their animal counterparts, taxes, are simple forms of behavior in which an organism moves in a direction determined by a stimulus applied with differing intensities to each of its two sides. More complex behavior has traditionally been classified as innate or learned, but it is often difficult to distinguish innate from learned aspects of a behavioral action.

Complex behaviors often issue from an intermingling of relatively stereotyped releaser stimuli, fixed action patterns, and drives. Releasers are features of a stimulus to which members of a particular species respond with a specific behavior. Often the triggered responses are fixed action patterns—complex but highly stereotyped unlearned behaviors. Releasers and fixed action patterns make it possible for animals to make important behavioral responses automatically, without the need for a slow and error-prone learning process. To an extent, drives appear to operate by changing the

threshold of an animal to a stimulus, thereby making it more or less likely to respond with a particular behavior.

Learned behaviors are far more flexible responses to the environment than fixed action patterns. Habituation is perhaps the simplest form of learning. In associative learning, the association of reward or punishment with a stimulus changes the response of the animal. Classical (Pavlovian) conditioning and operant (trial-and-error) conditioning are two types of associative learning. Parental imprinting, a form of stereotyped, one-trial learning, occurs in some species within a few days of birth.

Behavior often exhibits circadian rhythms which, in the absence of environmental cues such as light-dark cycles, are directed by internal biological clocks. Seasonal behaviors such as migration and hibernation are generally driven by environmental cues: shortening or lengthening days, temperature changes, and so on. Animals exhibiting orienting and navigating behavior rely on environmental cues of extraordinary scope and subtlety: solar and celestial navigation, sonarlike echo location, magnetotaxis in the Earth's magnetic fields, and a large number of as-yet-undiscovered mechanisms.

KEY TERMS

associative learning
behavior
biological clock
chain reaction
chemotropism
circadian rhythm
conditioned reflex
critical period
drive
electroencephalography

entrainment
ethology
fixed action pattern
geotropism
habituation
heterogeneous summation
homing
imprinting
innate behavior
insight learning

learned behavior
learning
motor program
navigation
operant conditioning
photoperiodism
releaser
supernormal stimulus
taxis
tropism

QUIZ QUESTIONS

1. Which of the following involves learning?
   A. phototropism
   B. chemotaxis
   C. classical conditioning
   D. innate behavior
   E. reflex response

3. _____ is a behavior that is guided by its consequences.
   A. Thigmotropism
   B. A conditioned response
   C. Operant conditioning
   D. Reinforcement
   E. Circadian rhythm

2. Which of the following is an all-or-none response?
   A. fixed action pattern
   B. releaser
   C. appetitive behavior
   D. habituation
   E. Pavlovian conditioning

4. Behaviors of organisms correspond to changes in their environments. Which of the following environmental cues does *not* serve to regulate the behavior of some species?
   A. day-night light cycle
   B. ebb and flow of tides
   C. photoperiod
   D. seasonal wet and dry cycles
   E. None of the above

5. Which of the following is (are) *not* used, by some animals, as cues for orientation and navigation?
   A. magnetic fields
   B. position of stars
   C. chemical composition of the surrounding medium
   D. electricity
   E. All of the above are used as cues.

ESSAY QUESTIONS

1. What is behavior? Organisms at what levels of complexity exhibit behavior? Is a nervous system required? What is "behaviorism"? Does a computer exhibit behavior? Why might you consider that it does—or that it does not?

2. What is learning? At what levels of organismic complexity does learning occur? How can one demonstrate learning experimentally? What are instincts?

3. What is a sign stimulus? What is a releaser? How would one test the hypothesis that certain features of a stimulus indeed act as a releaser? What are some examples of sign stimuli in the animal world? What is a supernormal stimulus?

4. How does a conditioned reflex differ from a simple reflex? What distinguishes classical from operant conditioning? Provide examples of each. What is the role of reinforcement in conditioning?

5. What is imprinting? Why is the "critical period" critical? Can only visual stimuli be imprinted?

6. How can one experimentally demonstrate the presence of an internal biological clock in organisms? What are circadian rhythms? What neural structures might be important for synchronizing internal biological rhythms with environmental stimuli?

REFERENCES AND SUGGESTIONS FOR FURTHER READING

Adler, J. (1976). The sensing of chemicals by bacteria. *Scientific American 234,* April, 40–47.

A pioneer "bacterial psychologist" reviews the experiments that launched an important new branch of behavioral biology.

Alcock, J. (1987). *Animal Behavior: An Evolutionary Approach,* 3rd ed. Sinauer, Sunderland, Mass.

An excellent textbook that views animal behavior from an evolutionary point of view.

Binkley, S. (1979). A time-keeping enzyme in the pineal gland. *Scientific American 240,* April, 66–71.

A review of work on the role of the pineal in the body's circadian rhythms.

Boakes, R. (1984). *From Darwin to Behaviorism: Psychology and the Minds of Animals.* Cambridge University Press, New York.

This scholarly book offers a stimulating account of the various ideas on animal behavior from around 1870 to 1930. A useful guide to the historical background of current understandings.

Buck, J., and Buck, E. (1976). Synchronous fireflies. *Scientific American 234,* May 74–85.

Certain firefly species of Asia and the Pacific flash in unison, unlike the familiar fireflies of the Temperate Zone. The remarkable phenomenon of communal flashing is engagingly discussed in this article.

Buckley, K. W. (1989). *Mechanical Man. John Broadus Watson and the Beginnings of Behaviorism.* Guilford, New York.

An interesting biography of J. B. Watson, the "father" of behaviorism and America's first "pop" psychologist.

Dawkins, M. S. (1985). *Unraveling Animal Behaviour.* Longman, New York.

This unusual book includes ten stimulating essays on conceptual problems in animal behavior that have led to confusion and controversy.

Dyer, F. C., and Gould, J. L. (1983). Honey bee navigation. *American Scientist 71,* 587.

A first rate review by seasoned students of bee behavior.

Gould, J. L. (1982). The map sense of pigeons. *Nature 296,* 205–211.

A fascinating review which concludes that magnetic and olfactory cues account for the mysterious map sense.

Gould, J. L. (1982). *Ethology: The Mechanisms and Evolution of Behavior.* W. W. Norton, New York.

A textbook of animal behavior with a strong focus on animal senses from a behavioral perspective. It is especially successful in showing how important the sensory system is to the behavior of animals.

Gould, J. L., and Gould, C. G. (1988). *The Honey Bee.* W. H. Freeman, New York.

More on what have been called the *sapiens* of invertebrate evolution. A beautiful and compeling volume that is wonderfully readable.

Johnson, C. H., and Hastings, J. W. (1986). The elusive mechanism of the circadian clock. *American Scientist 74,* 29–36.

An excellent discussion of a tantalizing problem.

Keeton, W. T. (1974). The mystery of pigeon homing. *Scientific American 231,* December, 96–107.

Classical experimental evidence is presented on the mechanisms by which pigeons find their way.

Klopfer, P. H. (1974). *An Introduction to Animal Behavior: Ethology's First Century.* 2nd ed. Prentice Hall, Englewood Cliffs, N.J.

This well written book considers behavior from the viewpoint of ethology. It includes excellent discussions of early contributions to the field.

McFarland, D. (Ed.) (1981). *The Oxford Companion to Animal Behavior.* Oxford University Press, New York.

A mine of information about animal behavior. Although the editor claims it was written for laymen, not scientists, it is both useful and interesting.

McFarland, D. (1985). *Animal Behavior: Psychobiology, Ethology and Evolution.* Longman/Benjamin-Cummings, New York.

An excellent text that integrates the viewpoints of ethology and comparative psychology into a comprehensive treatment—and at the same time is easy to read.

Pauly, P. J. (1987). *Controlling Life: Jacques Loeb and the Engineering Ideal in Biology.* Oxford University Press, New York.

A lively study of Loeb and his remarkable influence on the history of biology. The author argues that Loeb, who vigorously sought reductionist explanations of organismic phenomena, saw biology in engineering terms.

Plotkin, H. C. (Ed.) (1988). *The Role of Behavior in Evolution.* The MIT Press, Cambridge, Mass.

A thought-provoking book about the evolutionary implications of behavior and the influence of organisms on their environment.

Ridley, M. (1986). *Animal Behaviour: A Concise Introduction,* Blackwell Scientific, Boston.

A well written beginner's book that is truly concise—and enjoyable.

Tamarkin, L., Baird, C. J., and Almeida, O. F. X. (1985). Melatonin: a coordinating signal for mammalian reproduction. *Science 227,* 714–720.

Discussion of one of melatonin's several roles.

Urquhart, F. A. (1976). Found at last: the monarch's winter home. *National Geographic 150,* 161–173.

A stimulating account by the biologist who spent 40 years unraveling a baffling mystery: where do monarch butterflies go in the winter?

Winfree, A. T. (1982). Human body clocks and the timing of sleep. *Nature 297,* 23–27.

Especially interesting discussion of jet lag, among other things.

CHAPTER 33

SOCIAL ASPECTS OF BEHAVIOR

As Darwin's theories on natural selection and evolution gained renown in the nineteenth century, what came to be called Darwinism was increasingly viewed as nature's authorization for a variety of ruthless social and economic practices rooted in the idea of "survival of the fittest." This in turn led to growing distortions in our view of the animal kingdom. With the supposed sanction of Darwinism, many began to regard competition between animals of different species (interspecific competition) or of the same species (intraspecific competition) as a brutal tumult of conflict and self-interest. For humans to be less, it was said, for them to give expression to altruistic impulses, was to deny their true nature. So believed many nineteenth century "social Darwinians" (Chapter 34).

But in this century, biologists and students of animal behavior began to grasp the importance of cooperation among living organisms. They also recognized the overall evolutionary trend toward social interdependence and organization. The social biology of animals is now known as sociobiology.

We will attempt in this chapter to survey, albeit briefly, certain aspects of behavioral biology that influence the social interactions of animals with members of their own and other species.[1]

SOME USEFUL PERSPECTIVES

GENETIC ASPECTS OF BEHAVIOR

Like all phenotypic traits, animal behavior has a genetic basis. Like other genetically determined traits, it has changed in the course of evolution.

The Evidence of Bird Song

In studies of song learning in birds, the following phenomena have been regularly observed.

- Males of most bird species sing species-specific songs to advertise their territorial sovereignty, sexual maturity, and readiness to accept a mate.
- Sometimes songs are modified to match those of neighbors, so that local "song dialects" arise.
- Some birds can mimic songs from other bird species as well as a whole range of other sounds (Box 33-1).
- In temperate zones, young males hear older males singing in the first summer of their lives, but they hear no songs during the subsequent fall and winter.

Social aspects of behavior
These graceful flamingos show what is meant by the proverb "Birds of a feather flock together." Group living, one aspect of social behavior, has many advantages and some disadvantages in the unending struggle to find food without at the same time becoming food.

[1] Because our discussions must be more concise than this interesting subject warrants, we strongly urge readers to be more diligent than usual in delving into some of the excellent books and articles cited at the end of this chapter.

A visitor to a rather large cage at the American Museum of Natural History may see a starling flitting from perch to perch—and be greeted with the words, "Hi Sam! Hi kid!" The voice is the bird's.

The uncanny ability of the starling (Figure 33A) and its relatives to imitate human speech—and a great variety of other sounds—has become a focus of study by many biologists. How, they are asking, can these creatures, whether in captivity or in the wild, display such virtuosity of sound with a primitive vocal apparatus and a brain the size of a grape? What permits birds of some types, notably the starling family (Sturnidae) and the macaw (Figure 33B), both famous examples, to speak more than 50 words in some cases and as many as 20 sentences? What enabled a well-known starling in Falmouth, Massachusetts not only to whistle "Michael Row the Boat Ashore" and the opening bars of Beethoven's Fifth Symphony, but also to talk on and on, saying such things as: "I go home. You go to sleep. He's a little baby boy. Yes he is! See you later. Bye bye"?

It seems clear that the complexity of a bird's vocal repertory is not related to the structure of its syrinx, the functional equivalent of a human larynx (Figure 33C). The syrinx, which is located at the bottom of the trachea, is simpler than the human vocal apparatus. Moreover, the syrinxes of most songbirds resemble one another. Sounds are produced by an airstream, the speed and volume of which is controlled by muscles in the trachea. Sounds are emitted through the mouth with little or no modulation. Human vocalizations, in contrast, originate in a larynx at the top of the trachea. The larynx produces relatively simple sounds, but they are significantly modified by the positions and movements of the tongue, cheeks, mouth, and lips, and the resonating effect of the hollow sinuses. The syrinx does have two resonating membranes and in many birds they can be independently controlled. This explains how birds can produce two different notes simultaneously. But this does

33A Bird mimicry
With tutoring, starlings can imitate human speech.

33B Macaw, Ara ararauna

They do not sing their own species-specific songs until the following spring when they start to breed.

Studies of the development of song in male white-crowned sparrows (*Zonotrichia leucophrys*) by Peter Marler, then at Berkeley, convincingly showed that this behavior is genetically controlled. Male sparrows reared in total isolation from sound fail to sing the species-specific song the following spring. If at the age of 10 to 50 days, they hear (if only briefly) the normal song of the white-crowned sparrow, they can sing that same song six months later, even if kept in total sound isolation after their brief audition of the species-specific song. Significantly, playing songs of other related species to young white-crowned male sparrows during the 10- to 50-day-old critical period does not cause the birds to adopt those songs. Clearly, what song the sparrow can learn is strongly influenced by its genes. Yet the ability to perform properly requires a learning experience—exposure to the father's song, which can be likened to a releaser.

Other Examples

Stevan Arnold of the University of Chicago demonstrated the genetic basis of feeding preferences in garter snakes (Figure 33-1). Coastal populations of garter snakes in

not account for the ability of parrots and mynas to imitate a human voice. The basis of that ability is found in the brain.

The portion of a bird's brain that controls its vocalizations corresponds roughly to the human cerebral cortex and is in the front of the skull. The males of most songbird species are the principal vocalizers and male forebrains are larger than female forebrains. The forebrain expands in size just before the mating and nesting season, when extensive singing and calling are necessary for identification and to attract mates, establish territories, and warn of possible dangers. For this and other reasons, we believe that in general the larger the forebrain, the larger a bird's song repertory.

As we have noted, most songbirds do not have innate or genetically derived song patterns, but learn their songs and calls at an early age from their parents or members of the same species. Most species in the wild do not respond to or learn the vocalizations of other species. Yet several species skillfully imitate the songs of many other kinds of birds. Some also imitate other environmental sounds. Birds that do not engage in mimicry appear to have a kind of "filter" in their brain that keeps them from learning alien vocalizations. In other words, their brains engage in selective learning. Much of current avian research focuses on locating such a mechanism and determining how it works.

Although we do not know why some birds are mimics and others are not, there are practical rewards for such behavior. A mockingbird may mimic a blue jay's call because blue jays are highly aggressive predators on the nests of many songbirds. By simulating their calls, a mockingbird may be able to exclude potential competitors from a nesting site. Another possible use for mimicry, especially in forests where birds may have difficulty seeing one another, could be to use the call of a more aggressive bird to establish territorial rights, protect food sources, and deter rivals from courting a mimic's mate. Oddly, parrots are not mimics in the wild.

33C Vocal apparatus of humans and birds
In birds, compression of the air sacs by chest muscles forces air from the lungs through the syrinx, where two simple tones can be produced at once. The human larynx is located at the beginning of the trachea, far from the lungs, where its sound is richly modulated by air spaces and flexible vocal organs.

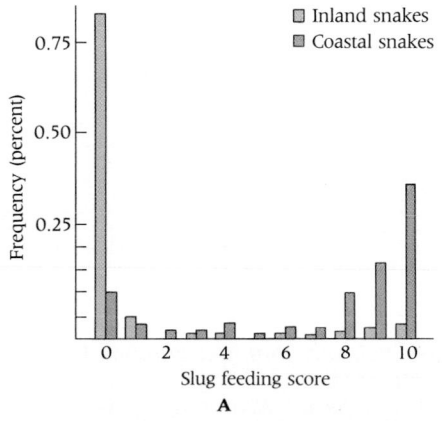

33-1 Feeding preferences in garter snakes
(A) *Data on feeding preferences, which are explained in the text.* **(B)** *A garter snake, Thamnophis.*

California feed primarily on slugs, while inland populations strongly prefer frogs and fish. Pregnant garter snakes from both localities were brought to the laboratory where they gave birth. Young snakes were then reared in the laboratory in isolation from their littermates. Each young snake was presented with a slug, the first prey

Peach-faced lovebird   Fischer's lovebird   Peach-faced lovebird–Fischer's lovebird
(*Agapornis roseicollis*)   (*Agapornis personata fischeri*)   hybrid

A

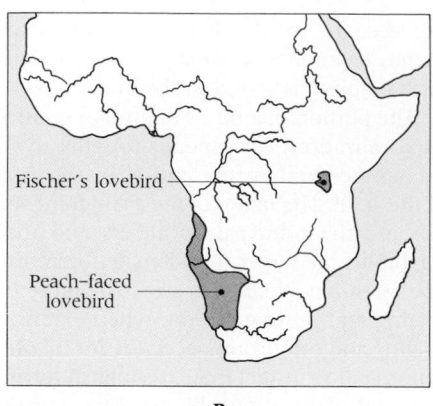

Fischer's lovebird

Peach-faced
lovebird

B

item it had encountered. Offspring of the coastal snakes without any prior experience with slugs ate the slug, while offspring of inland snakes did not. This remarkable difference in prey preference lasted into adulthood. By crossing experiments it was shown that the differences in feeding behavior in garter snakes are inherited.

Hybridization methods have been used to analyze the genetic control of complex behavior patterns. For example, two species of African lovebirds (*Agopornis*) have very different ways of carrying building materials for their nests (Figure 33-2). Peach-faced lovebirds (*Agapornis roseicollis*) carry them between the feathers of the lower back; Fischer's lovebirds (*Agapornis personata fischeri*) carry them in their bills. If the two species are mated, the resulting hybrids have great difficulty in carrying nest building materials—almost as though they had received two incompatible sets of genetic instructions about the way building materials are supposed to be carried (Figure 33-3).

Examples could be cited endlessly. All make it clear that the *potential* for an animal to execute certain behavior patterns is under genetic control. In most cases, however, there is a need for something more—an *experience* in the form of imprinting, releasers, learning, or combinations thereof.

33-2 Behavioral studies in Africa lovebirds
(**A**) *Peach-faced lovebird and Fischer's lovebird can interbreed to produce a hybrid.*
(**B**) *Geographical distribution of lovebirds in Africa. Each of the species inhabits a different area.*

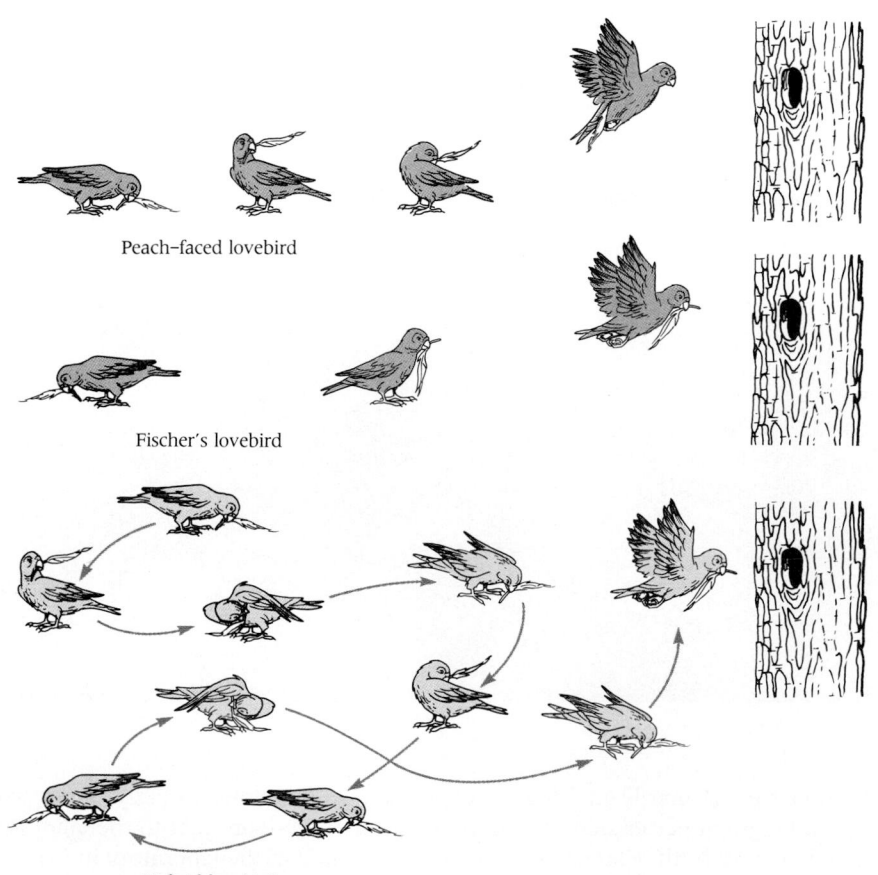

Peach-faced lovebird

Fischer's lovebird

Hybrid lovebird

33-3 Evidence that nest-building behavior is inherited
The peach-faced lovebird carries pieces of nest materials in the feathers on its back. Fischer's lovebird carries one bit of material at a time in its bill. The hybrid lovebird inherits both patterns from its parents, and thus acts in a complex and confusing manner when trying to build its nest.

EVOLUTION OF BEHAVIOR

The biologist Theodosius Dobzhansky once said, "Nothing in biology makes sense except in the light of evolution." This is surely true of animal behavior. It is an axiom of biology that behavioral traits improve the survival and reproductive success of the individuals or groups of individuals displaying them. Since the development of various behavioral patterns is influenced by genes, we can predict that natural selection (Chapters 34 and 35) will favor those gene combinations that enhance the survival of animals in their particular habitats and maximize the number of offspring they can produce. The branch of behavioral biology that seeks to explain the interaction of evolution and behavior is known as **behavioral ecology.**

Let us consider two major issues of behavioral ecology.

How Animals Communicate

Animals can manipulate objects in their environment (including other organisms) by the use of physical force or by communication. (Many biologists prefer the term *signaling* rather than *communication* because it clearly implies that the existence of a signal, an actor or signaler, and a reactor or signal receiver.)

Animal communication is a two-way process in which signals are used to influence the behavior of another animal. For communication to be effective—and if it were not effective it would not exist or evolve—the signal receiver must respond in a way that improves its own situation and at the same time avoids worsening the condition of the signaler. If as a result of communication, signalers were worse off than nonsignalers, the signaling system would not endure and could not evolve. The converse situation has no such strictures. If the signal receiver gains no benefit from the signal, while suffering no harm, the signal can be ignored.

Signals are of many kinds and can include externally visible features like crests or tail feathers and patterns of behavior such as "threat displays." Broadly speaking, animals can communicate via a number of channels.

Chemical channels are the most widespread and, in evolutionary terms, the oldest mode of communication. It is characteristic of chemical signals that they remain behind when the signaler has changed location. In a sense, then, they allow animals to be in many places at once.

- Pheromones are chemical signals between individuals of the same species that stimulate olfactory receptors and ultimately behavior. In this way, female insects often attract males from very great distances (see Chapter 30).
- Bees rely on odors to identify fellow hive members.
- Alarm substances produced in the skin are released by schooling minnows when an individual is attacked by a predator. These substances warn others, who then seek cover.
- Female rhesus produce sex attractants in their vaginas when they come into heat. These odd-smelling, to humans, substances, are mixtures of simple fatty acids.

Such chemical signals are of benefit because they are highly diverse, long-lasting, and capable of going long distances at little cost.

Visual channels depend on the visual sense, which is prevalent in nature. All sorts of visual signals have crucial roles in the lives of many species. Their advantages are many. They can be swiftly modified, and the identity of the signaler is always revealed—both to the intended signal receiver and to others—since it can usually be seen by one and all.

33-4 Body color and "mood" changes in fish
The African freshwater fish, Hemichromis fasciatus, *can change its body coloring rapidly. The changes are displays that supposedly reflect "mood" changes. Three of the eight recognized body color patterns are shown here.*

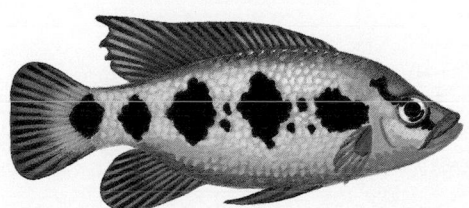
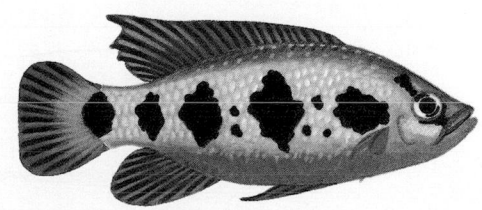

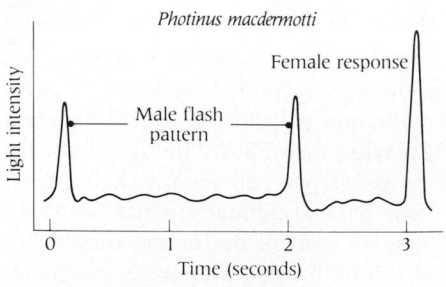

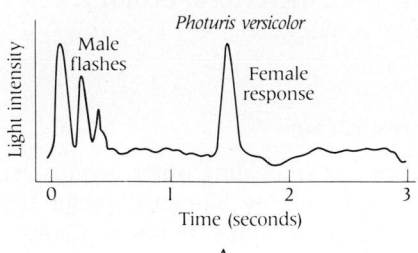

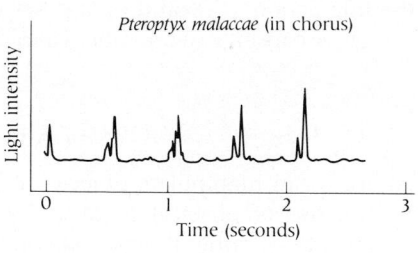

A

B

- The African fish *Hemichromis fasciatus* changes its body coloring with its "moods" (Figure 33-4).
- Fireflies produce pulsed flashes that can be seen by other fireflies at great distances. Males fly about flashing their lanterns according to a species-specific code (Figure 33-5). The females flash back from within the vegetation patch. The releaser that attracts a mate is the time interval between pulsed flashes.

Auditory channels have evolved almost exclusively for communication purposes and are widely used by mammals, which have a laryngeal apparatus and vocal cords and thus can vocalize.[2]

- Mammals use a **larynx** for roaring, speaking, and other vocal forms.
- Birds use their **syrinx** for singing (see Box 33-1).
- Fishes produce ultrasonic sounds with muscles in their gas bladders or by grinding their teeth.
- Many frogs invest almost as much energy in making species-specific calls as they do in locomotion.
- The singing of crickets, cicadas, and other insects is simpler than bird song, presumably because insects are relatively insensitive to variations in pitch.

Electrical channels are utilized by those vertebrates that can generate and receive electric signals. This capability appeared early in the evolution of fishes. Many modern fishes, especially those inhabiting very turbid waters in tropical Africa

33-5 The flash of a firefly is a sexual display
(A) *Using species-specific flash codes, male and female fireflies (family Lampyridae) identify and locate each other in the darkness. For example, a male* Photinus macdermotti *emits 2 flashes 2 seconds apart. The female waits 1 second and replies with 1 flash. A male* Photuris versicolor *flashes 3 times within 0.5 second and receives a 1-flash reply in 1 second.* Photuris *males may mimic the male* Photinus *code to attract mates. But* Photuris *females mimic the female* Photinus *response to attract* Photinus *males, which they eat.* **(B)** *In this Malayan tree, thousands of male fireflies,* Pteroptyx, *flash synchronously and rhythmically through the night. This communal sexual display is one of the great spectacles of the living world. The light is so strong, such trees along a shoreline are used as navigational beacons. The graph traces the synchronized flashes of a large number of* Pteroptyx. *An identical pattern is observed with a single firefly.*

[2] Vocal communication among mammals ranges from the barks and roars known to everyone to such vocal expressions as the low-frequency hum of elephants that is inaudible to the human ear. It also includes the phenomenon of language. Language is learned behavior acquired from those around us. Depending on who raises us we learn totally different languages. Nevertheless, there is an essential innate element in language. We are born with the nervous mechanism and effectors for talking and comprehending language. Without them, as when the left temporal area of the adult brain is damaged, we are unable to use language. Other animals are innately unable to acquire language. The point about linguistic behavior is that its mechanism is innate, but it is highly modifiable and can be expressed in many different ways. The ability to learn to talk is innate, but the particular language we talk is not.

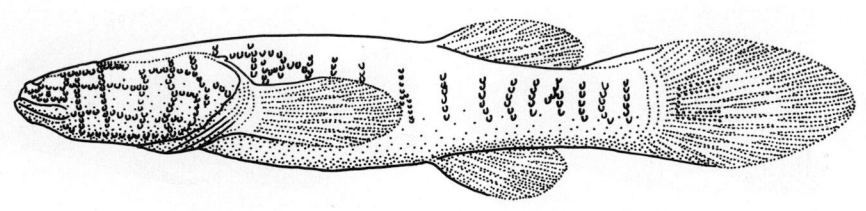

33-6 Cave-dwelling fish
The blind cave fish *Amblyopsis rosae.*

and South America, are capable of generating weak electric fields with their specialized effectors (see Figure 32-33).

Tactile channels are particularly well developed when darkness precludes visual communication. Cave-dwelling fishes (Figure 33-6) have elaborate mechanoreceptors in their skin for prey detection and communication, including sexual signaling.

A highly evolved communication system with an important tactile component is the famous "dance" of the honey bees (*Apis mellifera*). In 1945, Karl von Frisch of Munich painstakingly collected data showing that when a bee forages (searches for food) and finds a rich supply of honey (flower nectar), it promotes the efficient use of its fellow workers' time and effort by "telling" them both the distance and precise direction to the rich food source. The bee does this by performing a "dance" on the vertical faces of the honeycomb inside the dark hive. Within the movements of the dance is encoded information on the location of the flowers (Figure 33-7). The bee makes a circle and then transects it with a line that forms a figure-of-eight. The angle formed by the transecting line and the vertical (defined by gravity) corresponds with the angle formed by the lines joining the sun and the nectar source to the hive. As the bee's dancing transects the circle, it "waggles" its body. The number of waggles is inversely proportional to the distance between the source and the hive, each one denoting about 40 meters in von Frisch's bees. The richer the source of food, the more vigorous and longer the dance. Information in the dance is monitored by other bees that follow and touch the dancer.

Evolutionary Aspects of Communication

The various signals used in communication probably evolved from the elaboration and exaggeration of simple body movements. Some probably derived from autonomic nervous system activity. We have seen that stimulation of the sympathetic system can interfere with an animal's ability to do what it normally does, accelerating the heart rate and increasing breathing, heat production, and blood flow to muscles and brain. At the same time, the animal usually raises its fur or feathers—and

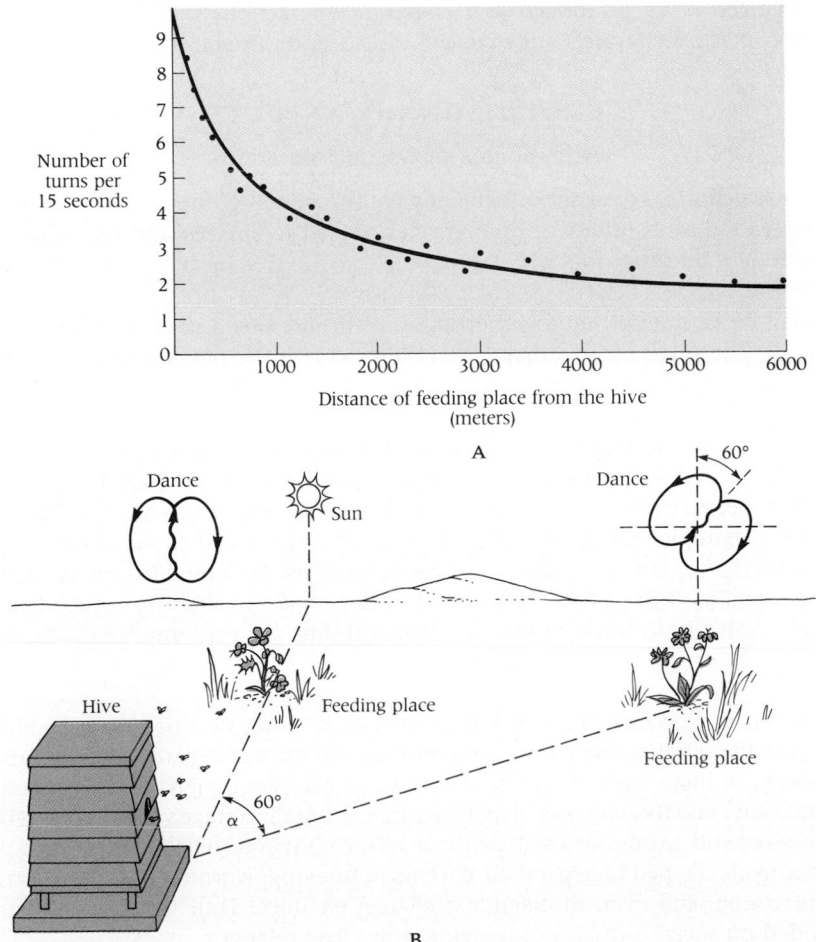

33-7 Communication in bees
Information is communicated by a dance, which indicates distance and direction of food from the hive. (A) The number of turns per unit time in the dance tells other members of the hive the distance to the food source. The closer the food, the faster the dance is performed. (B) The direction of food is defined by the angle between a straight line connecting the sun and the hive with a line connecting the hive with the food source. The angle is reproduced in the dance, which takes place on the upright portion of the comb in the hive.

this may have incidentally provided a behavioral display useful in communication. Various components of such movements could have been modified by natural selection into elaborated or exaggerated visual communicative displays.

Another common behavioral phenomenon is the performance of an irrelevant behavior when an animal's original intentions are thwarted by a motivational conflict. Even human beings do this at times of severe emotional conflict. Many people will light a cigarette during a crisis, attack an innocent bystander during a heated argument, or clean out a closet the night before an exam. An animal in such a situation will also redirect its actions toward nearby objects and individuals, a pattern termed **displacement behavior.**

Territorial male three-spined sticklebacks defend their territories feverishly, but at territorial boundaries there is an equal tendency to attack a neighbor or flee from a neighbor. Because of the motivational conflict during territorial boundary fights, a fish may suddenly engage in nest building behavior. From the evolutionary viewpoint, such displacement behavior reduces the high cost of fighting.

In the course of evolution, signals tended to become more and more elaborate. The adaptive significance is obvious: an elaborate or exaggerated signal is clear, unambiguous, and readily recognizable. But the selective benefits of signal clarity may carry a high cost. It is important, for example, for an aggressor to signal what a good fighter it is, but it should not advertise its intention to attack. If it is necessary to signal for a mate in an elaborate manner, the risk of attracting predators increases.

One of the amazing products of the evolution of interspecific communication (that is, communication between members of different species) is the repeated occurrence of behavioral deceit. Animals often benefit by posing as something other than what they are. We explore this phenomenon in Chapter 35.

SOCIAL BEHAVIOR

Though many animals lived solitary lives early in evolutionary history, they had to interact with one another to reproduce. From these brief encounters, we believe, there evolved more prolonged and elaborate interactions that in time led to the complex societies characteristic of social insects and primates.

EVOLUTIONARY ASPECTS
Advantages of Social Interaction

Natural selection favored social behavior within animal groups because individuals that interacted with others of their species tended to survive and reproduce more successfully than those that did not.

- Small birds often group together when confronted by a predator such as a hawk. The small birds gather into a tight flock. If the hawk attacks, its effort is diluted because the larger the flock's size, the smaller the chance that any particular individual will be a victim.
- Wood pigeons clearly exhibit the adaptive value of groupings. In an experiment, a trained predatory goshawk was most effective when attacking solitary pigeons. It became increasingly less successful as the number of pigeons in a flock increased.
- When taking a breath, air-breathing gourami fishes must come to the surface—but there they are vulnerable to fish-eating birds. To fend off these predators they tend to surface synchronously in a large and bewildering group. The birds have difficulty in tracking any one fish and thus are very much less efficient in capturing prey.

Social behavior may have arisen in evolution along two pathways, which are known as the familial and parasocial pathways. In the earliest stages of the **familial pathway,** animals bred solitarily and did not care for their young. Parental care emerged later and then increased in duration. Eventually a stage was reached wherein adults were still caring for their younger offspring when their older offspring were sufficiently developed to assist their parents in foraging, warning of predators, caring for the young, and even in defending against enemies. Thus the familial pathway depended on social groupings involving only close relatives.

33-8 Wolf pack devouring prey

33-9 Schools of fish
(A) *Wallin marine reef fish in soft coral.*
(B) *A fountain-effect tactic enables a school of small fish to outmaneuver a predator.*

A

The initial evolutionary stage of the **parasocial pathway** also consisted of solitary species. However, in later stages unrelated adults of the same generation formed groups centering around resources. Benefits were disproportionately available to cooperating groups. Cooperation intensified and divisions of labor emerged among the adults. This evolutionary trend culminated with the expansion of these cooperative groupings to include different generations. In parasocial groupings, the society involved active cooperation of nonrelatives.

Underlying Behavior Patterns

The organization of animal societies can be attributed to six principal behavior patterns. Except for the rigidly structured insect societies to be discussed below, social organizations may be classified by the degree to which one or the other of these patterns predominates.

■ **Cooperation** has evolved in many forms. By hunting in groups, animals can maneuver prey into places where they are easier to catch (Figure 33-8). Cooperative hunting was a key element in human evolution. In birds, predator avoidance is achieved by cooperative watching. The bigger the flock, the less time each bird has to watch and the more time it can spend feeding. Some animals breed communally. The banded mongoose (*Mungos mungo*) lives in packs of up to 40 individuals, but breeding is done only by several females in each pack, which produce litters simultaneously. The young are all the same age and suckle indiscriminately from any lactating female. Males help in caring for the young.

■ **Territoriality** is common where food and nesting sites are in short supply. Usually the male of the species controls and defends an area containing important resources, expelling all individuals other than mates or potential mates. However, territorial defense can be more subtle, some individuals maintaining exclusivity by mutually avoiding one another's scent, song, or other keepout signals. The topic of territoriality is further discussed in Box 45-2.

■ **Dominance** hierarchies or pecking orders occur within most social groups. In these hierarchies, every member knows which individuals it can defeat, and which individuals can defeat it. The more dominant males often take the lead in group activities. They are likely to have first whack at the food and they may monopolize the more desirable females. Such dominance hierarchies are established competitively, but once established they reduce competition and tension within the group and tend to eliminate fighting.

■ **Leadership** becomes an important phenomenon in mammalian societies that persist over long periods of time—unlike bird flocks, for example, which have a seasonal nature. The leader is the individual who sets the direction or pace of a group or who determines its mood, initiating alarm or feeding behavior. Leadership of a group may be divided so that different animals have different roles. Among the red deer (*Cervus elaphus*), males acting as leaders of their harems will go off by themselves when an alarm is sounded. Harem leadership then goes to another. Among primates, leadership is linked to dominance status. A change in one leads to a change in the other. In some species, leadership is unrelated to dominance status, spatial position in the herd, or apparent astuteness. In groups of ducklings and schools of fish (Figure 33-9), what seems to inspire leadership (or followership) is suddenness or directedness of movement. The animal that heads off most decisively draws the group along.

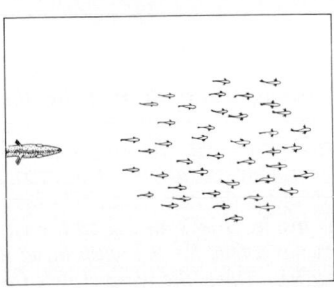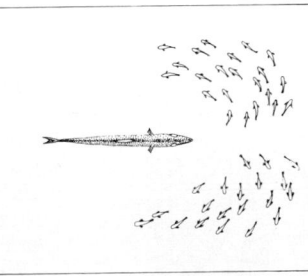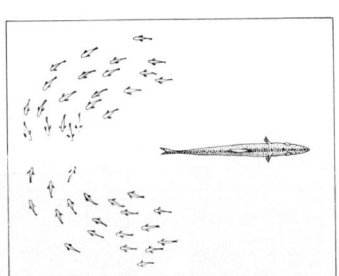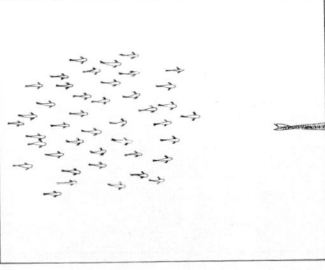

B

- **Parental care** and **mutual stimulation** are features of many diverse and unrelated species. Mutual stimulation includes mutual aid—for example, the grooming of one another by pairs of monkeys, as well as the visual and auditory displays birds and fishes provide one another. The prevalence of either parental or reciprocal care is closely associated with complex and long-lasting social bonds. Although primate and ungulate societies have been especially well studied from this point of view, observations of birds support the same conclusion.

INSECT SOCIETIES

Certain insects—notably ants, some bees and wasps, and termites (Figure 33-10)—form complex societies. Thus we call them the *social insects* in contrast with the majority of insects, which are nonsocial. Social insects are important for at least three reasons.

- Their ubiquity and enormous diversity have made them useful for studies of the evolution of social structure, exhibiting both the familial and parasocial pathways in bees and wasps.
- Societies such as theirs have evolved independently many times.
- They rank among the ecologically dominant animals of the land, turning more soil than all earthworms, and utilizing more energy than all birds and reptiles combined.

33-10 Termites are social insects

Insect Behavior

Insects have some of the most complex nervous systems of any invertebrates. They perform complex unlearned behavior. They also show more evidence of modifiable behavior and learning than do other invertebrates. They are at the end of a long evolutionary progression that has been diverging from ours for well over half a billion years. Compared to ours their intelligence is very different in kind.

Most insects are nonsocial and live quite alone, never seeing their parents or offspring, and seldom relating to other members of their species except during a brief mating. A few do not even mate: the females carry on the line without benefit of males. Yet each species has a distinct, often complex pattern of innate behavior that persists with little variation from one generation to the next.

A solitary female digger wasp (*Sphex ichneumoneus*), which has never seen her parent and need not have observed any other wasp doing the same thing, goes through the following remarkable sequence of actions (Figure 33-11). She hunts for an appropriate site and digs a nest. She then hunts prey, usually a caterpillar or katydid, paralyzes the victim, drags it to the nest, and shoves it in. The wasp performs this operation several times. Then she lays an egg in the nest, seals the opening, and smooths it over so that the location is disguised. When the egg hatches, the larva eats the food that was provided, but the wasp responsible for these provisions knows nothing of this.

In this sequence, there are several clear-cut responses to separate stimuli. The wasp's stinging of the caterpillar has all the marks of a reflex response. There is also learned behavior. The wasp's unerring return to the nest cannot be explained otherwise. But whether or not individual acts are learned or innate, the sequence as a whole is innate. Each act is the stimulus for the next and thus this is another example of chain reaction behavior. The wasp cannot vary the sequence, omit a step, or go back to repeat one. If, for example, the caterpillar is removed from the nest, the wasp lays an egg and seals it anyway.

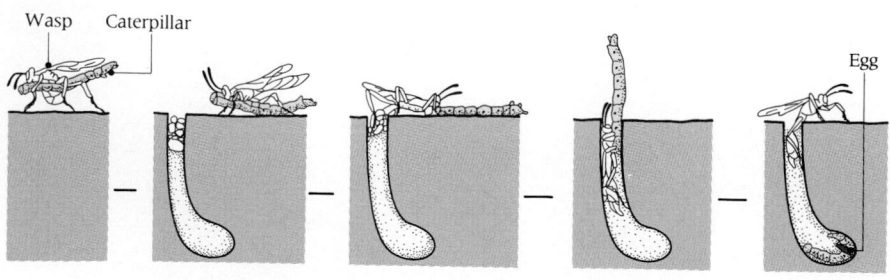

Wasp Caterpillar

Egg

33-11 A behavioral pattern of the digger wasp
Having dug out its nest, the wasp captures and stings a caterpillar, paralyzing it. It carries it to the nest, opens the nest, and packs the caterpillar inside. It lays an egg on the victim, which serve as a food for the wasp larva when it later hatches. Then, the wasp crawls from the nest, reclosing it, never to return.

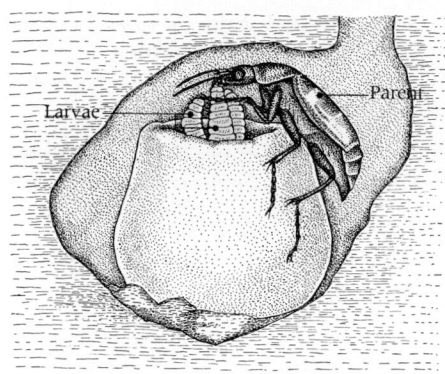

33-12 Feeding of larvae by a burying beetle
The parent beetle sips from the pool of fluid food in the top of a buried carcass and transfers the material to each larva, which instinctively rears up to receive the food like a nesting bird.

In sum, all insect behavior seems interpretable in terms of (1) innate structural patterns and reflexes, quite complex as to number and sequence, and (2) learning of the simplest conditioned kind with only very limited modifications of responses to stimuli. There is no evidence of foresight. An insect does not plan; it acts. The apparent foresight is built into the organism's genes, not decided on by the individual.

Social Patterns

To appreciate the social insects, we should note the various patterns that are seen among nonsocial insects.

- As noted above, many insect species are, in fact, solitary. Perhaps the most primitive pattern occurs in species in which females merely lay their eggs in a spot where they have some chance of survival. The offspring receive no further care.
- A few species are "subsocial," displaying parental care of the young but little else. The females make protective chambers, stock them with food, and lay the eggs on or in them. They provide no direct care of larvae.
- In still other species, the females stay with the eggs until they hatch and then protect and feed the larvae until they can shift for themselves (Figure 33-12). This trend may continue so that the female is still present when the next generation emerges. This may lead to cooperation by young females with their mothers in the rearing of additional broods.

The varying degrees of parental care seen among nonsocial insects suggest that some or many insect societies may have arisen through similar social pathways. It is estimated that less than 3% of present-day insect species are social or subsocial—some 24 different families of only 5 insect orders! As a matter of fact, advanced patterns of social organization have evolved repeatedly and at various times by parallel evolution at least 12 times among the insects: twice in wasps, once or twice in ants, at least eight times in bees, and once in termites.

All insect societies that we know about are actually extended and specialized families that begin in a familiar way—with parents and offspring. A key step in the rise of insect societies evidently occurred when the female parent and young began staying together for a time as an interacting family group. The final steps occurred when the young began to stay with each other and their mother as they matured, eventually taking over the task of feeding and protecting subsequent broods of young, and finally feeding the mother. In these stages, at least three distinct social roles are generally played by individuals that come to differ in structure and appearance (Figure 33-13):

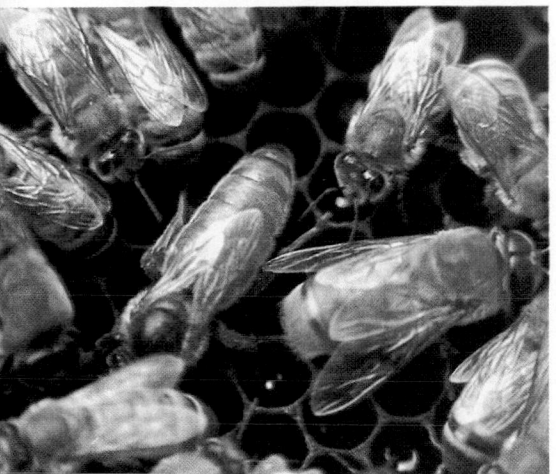

33-13 Bee castes
The queen is in the center (longer abdomen); stocky drones are on either side. The other bees are workers.

- The male whose role is to fertilize the female and who generally dies or leaves the group after the act
- The reproductive female, or queen, whose primary role is to lay the eggs from which the other members of the social group develop, but who may also rear the first brood of young
- The altruistic workers, who usually cannot reproduce

Altruism is defined as behavior that benefits another while exacting a price from the performer. These workers are altruistic because they have surrendered their own genetic fitness (measured as the survival and reproductive success of an individual and its kin), seemingly in order to enhance the well-being and fitness of others. The males, queens, and workers comprise the three basic castes. In ants and termites (but not in bees or wasps), there are other castes—notably soldiers, large creatures that protect the group with formidable weapons (Figure 33-14).

There are interesting and revealing differences among various species of social insects.

- In some bee and a few wasp species, females of one generation tend to group their nests in the same general area.
- In other species, there is a cooperative effort in building a large nest, a hive, with a single entrance. Within this nest, the female lays her eggs in cells she herself built and feeds her own larvae.

33-14 Soldier termite

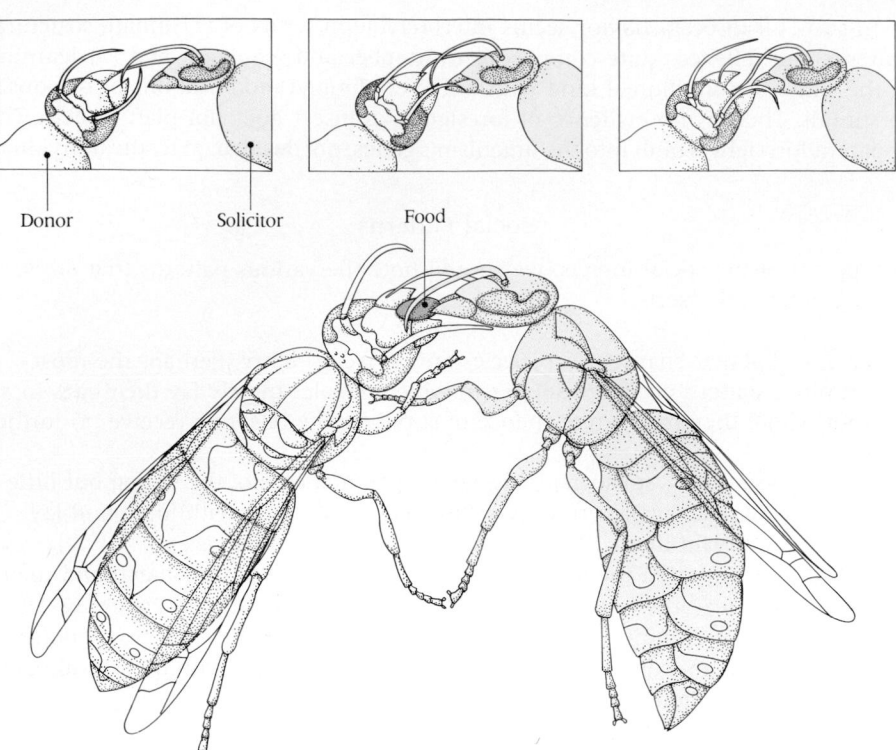

33-15 Trophallaxis between workers of the social wasp Vespula germanica
The solicitor on the right approaches the donor and places the tips of her flexible antennae on the donor's lower mouthparts. The donor responds by closing her antennae onto those of the solicitor who then begins to gently stroke her antennae up and down over the lower mouthparts. If this interaction continues, the donor will begin to regurgitate, and the solicitor is able to feed.

- In yet other species, there is cooperative rearing of the brood within the nest.
- In the most derived species, brood care is so elaborate that brooding is done by overlapping generations, which are not closely related. Selection for this parasocial grouping was probably based on improved protection from predators and improved reproduction.
- An elaborate form of behavior of importance to social insects is **trophallaxis,** the feeding of various liquid secretions to one worker by another. In a common form, it consists of the feeding of regurgitated liquid food from the donor's gut (Figure 33-15).[3]

Life in an Insect Society

Wasps, bees, and ants all belong to the order Hymenoptera. Only termites belong to the order Isoptera. Although termites are antlike and sometimes called "white ants," they are not related closely to ants. We believe that societies arose separately among ant and termite ancestors and only the social groups survived. Many solitary, nonsocial bees and wasps exist, and various species even now demonstrate most of the intermediate stages.

It is striking that the order Hymenoptera showed such a clear predilection for social evolution. A partial explanation may be that in most ants, bees, and wasps, the males are genetically haploid and thus homozygous, developing from unfertilized eggs (see Chapter 15). This is typical in Hymenoptera, which employ the haplodiploid mode of sex determination (see Figure 15-2D). Queens, workers, and soldiers, come from fertilized eggs, are diploid and genetically are females. This means that sisters are genetically very similar, each one having received identical genetic material from a homozygous haploid father. Thus they share three-fourths of their genes all having received the same haploid set of chromosomes from their father and another haploid set from their mother. Of the latter, half are likely to be the same in two sisters. Since they are genetically much closer to one another than ordinary sisters with a diploid father, it seems possible that their approaches to one another may be correspondingly more altruistic. In this situation, one's sister is almost (but

---

[3] This phenomenon was demonstrated by feeding a radioactive syrup to a single worker ant, which was then returned to the nest. Other workers were then removed from the nest at regular intervals and their stomach contents were tested for radioactivity. The data showed that the stomach contents of the first ant rapidly spread through the colony. In some cases, every single worker had some radioactivity in its stomach within 24 hours.

not quite) in one's own image and it may be more advantageous for hymenopteran females to care for their sisters than to rear their daughters.

In termites, the pattern is reversed: males are diploid and females haploid. Workers and soldiers may be either male or female genetically, although they rarely reproduce. The youngsters serve as workers and may grow up to be soldiers. Sometimes growth of certain individuals is arrested, apparently by something in the food, and such individuals remain workers throughout life.

Parental care is surely the most familiar example of altruism. Theorists point out that in evolution natural selection favors individuals who maximize their gene contribution to future generations. In those terms, a gene that causes parents to behave altruistically toward their young, which are likely to carry the same genes, will thereby become more numerous in the gene pool. It is possible to quantify the degree of relatedness between parent and offspring. Because of meiosis, any allele has a 50% chance of going into any one sperm or egg. Therefore, in a diploid species, the probability that a parent and its offspring will share a given allele is 0.5. This is termed the **coefficient of relatedness**—or r. In fact, individuals also share genes with other family members—siblings, cousins, grandchildren, and so on—and for these, r values range from 0.5 for a sibling to 0.125 for a cousin. Such sharing may account for displays of altruism toward other relatives. Thus what is maximized by natural selection is not individual fitness, but what W. D. Hamilton termed **inclusive fitness.** In this view, the fitness of an individual depends both on its own survival and reproductive success and on that of its kin.

The odd method of sex determination among the social hymenopterans tends to maximize their inclusive fitness. The queen is equally related to her sons and daughters ($r = 0.5$), but for her full sisters, $r = 0.75$. Worker females maximize their inclusive fitness by helping their mother to rear offspring rather than having offspring themselves, thus showing apparent altruism.

The important point is that an insect's social role is determined either in the egg or in early developmental stages. Moreover, the behavior appropriate for each caste is not learned. It is stereotyped and subject to little modification. Its elements are hereditary, and thus it can vary only within narrow limits. The number of distinct possible roles is also small—three, or rarely over four.

In insect societies, the individual has no choice of role. Changes in roles or increases in their complexity are matters of slow evolutionary change. A social insect has little chance of surviving if separated from its own group. Whether the group stands or falls depends on how well it copes with environmental demands.

NONHUMAN PRIMATE SOCIETIES

The next pinnacle of social evolution is that of the nonhuman primates. Among them, we find a truly momentous climax in the social evolution of vertebrates (Figure 33-16).

An important characteristic of most primates is **tribal behavior.**[4] Most species of monkeys, baboons, and anthropoid apes live in complexly organized societies that vary greatly in detail, notably in the degree of territoriality, group size, home-range size, and dominance behavior. Indeed, there is a spectrum of aggressive behavior with Rhesus monkeys and other macaques and baboons at the aggressive end of the spectrum. Some anthropoid apes, notably chimpanzees and gorillas, are at the other end. Perhaps these differences among the primates have something to tell us about the evolution of human social behavior.

There is an astonishing array of social organizations among the primates—and one aim of primate ethologists in recent years has been to seek meaning in these species differences. For example, gibbons (*Hylobates*) live in pairs and defend a territory of less than 1 square km, while baboons (*Papio*) wander over an undefended home range of 15–20 square km in groups of 40 to 80 in which **polygamy** (which means having many mates) is the norm. In many species, groups are organized internally and clear-cut social hierarchies or "pecking orders" exist, which serve to diminish intragroup fighting. Individuals recognize members of their own groups

---

[4] This is not true of all primates. The orang-utans, which live high in treetops, have less need for the kind of protection afforded by groups and they do not ordinarily live in tribes. Incidentally, all South American monkeys are highly arboreal, but some (e.g., howlers) usually live in tribelike groups.

Chimpanzee   Gibbon   Spider monkey   Howler

Gorilla   Orang-utan   Uakari monkey   White-throated Capuchin

33-16 The major nonhuman primates

and are suspicious and hostile toward members of other groups. However, fighting between groups is uncommon, since members tend to stay within their own range or to retreat to it when confronted. Several interesting primate social orders are described in Box 33-2.

Over the years primatologists have sought to link primate social structure with such ecological variables as food availability and predator pressure. Recently, T. H. Clutton-Brock of Cambridge University and P. H. Harvey of the University of Sussex classified all the primate species for which they could find data into three groupings.

- Nocturnal (night) or diurnal (night or day)
- Arboreal or terrestrial (tree or land dwellers)
- Insectivorous, frugivorous, or folivorous (insect, fruit, or leaf eaters)

Then they looked for consistent differences among these ecological categories. They discovered that nocturnal primates—all arboreal eaters of fruit or insects—are small, live in small groups, and have small home ranges. Both diurnal and nocturnal fruit eaters have large bodies and live in larger groups and home ranges than do leaf eaters. Finally, diurnal terrestrial primates live in larger groups and ranges and have more marked sexual dimorphism (anatomical differences between males and females) than do diurnal arboreal species.

These patterns relate to the ecological necessity of food gathering and predator avoidance. Nocturnal species need small bodies because they feed by crawling onto small twigs and hide to escape from predators rather than run away or fight. Hence they are solitary and inconspicuous. At the other extreme, conspicuous terrestrial monkeys living in open ground habitats have large bodies, partly to fend off predators and partly to give themselves mobility to cover wide areas in searching for scattered food. Group sizes differ widely among arboreal and terrestrial forms. Larger groups are advantageous for diurnal primates—perhaps an antipredator device—but competition for food limits group size. Species feeding on large food clumps can afford to live in larger groups than those eating dispersed foods. Home-range size is also related to group size and is probably set by the minimum area needed to support the group and the density and dispersion of food. Species feeding on evenly dispersed, dense foods have small, frequently defended home ranges. Those eating widely scattered foods tend to have large undefended ranges.

It is not so easy to account for the variations in mating systems and adult sex ratios. Some 14 primate species are **monogamous** (having a single mate). In contrast, **polygyny,** meaning having many female mates, is much more common. Presumably, monogamy evolved only when males were able to leave more offspring by helping the female rear them or when males could secure no more than one mate. The latter possibility may account for monogamy in gibbons. The distribution of their food requires territorial defense, and males may be unable to defend enough food for more than one mate. The ecological factors favoring one system over the other remain unknown.

It seems possible, in sum, to attribute differences in primate social structure to pressures associated with feeding and predation. The dimensions of social organization—group size, sex ratio, and the rest—may also be influenced by different ecological factors, but many of these are unclear.

HUMAN SOCIETIES

The societies discussed thus far are groups of individuals living together and reacting more or less as a unit toward other groups and outside forces. Each group has sufficient complexity of organization to permit the emergence of two basic qualities—elaborate integration and specialization of roles in different individuals. These features also typify human societies.

Beyond those broad resemblances, however, human society differs fundamentally from the societies of insects and nonhuman animals. This does not mean that human societies are fundamentally different from those of all other species—or free from ecological constraints. Most human societies face more or less urgent ecological constraints. What is still unclear is whether human social organization has been achieved through long-term genetic change and natural selection, through individual trial-and-error learning, or through a cultural transfer of the best strategies for utilizing resources.

We will survey human evolution in Chapter 44. We can note here, in brief anticipation, the current view that hominid (human) ancestors lived in social groupings as early as 7 million years ago, killing and eating small animals—such as baboons, porcupines, and lizards—and consuming seeds and berries. After 2 million years of rapid brain growth, *Homo erectus* emerged as the most effective hunter. Hunting in groups, these hominids could kill even large mammals. *Homo sapiens,* our own species, came next. Originally a hunter-gatherer, this creature exhibited typical human behavioral features: intraspecific war, heavy investment of labor in offspring, monogamous pairs with well-separated sexual roles, incest taboos, and body adornments. When domesticated herds and agriculture were developed, social groups settled in and expanded.

Humans may resemble other primates in many aspects of their social behavior, but they differ in having permanent male-female bonds. Human reproductive physiology and sexual behavior (which is virtually continuous, unlike that in nonhuman primates) evolved in a way that cements these bonds and makes the nuclear family a central element of most human societies. The essences of the human family include monogamy and cooperation of both sexes in caring for the young.

In human societies, the family is basic and persistent. In all mammals, the mother and her suckling young remain together and form a primitive social unit. Yet a group in which roles are differentiated only into father, mother, and young is not yet a society. A mammalian society is not an expanded family, but an aggregation of adults in which the children remain in the persisting but socially subordinate family units. Differentiation of roles within society goes far beyond the mere distinction of reproductive roles. Among nonprimate mammals, such differentiation seldom goes beyond dominance or leadership by some individuals over others. Among nonhuman primates (see Box 33-2), there is further differentiation. But in civilized human societies, with their butchers, bakers, and candlestick makers, differentiation is incredibly complex.

The organizational pattern of human societies—and of many mammalian societies—tends to override immediate male-female and parent-child relationships. We term such arrangements **associative societies,** which are not strictly familial like insect societies. Our societies evolve from aggregations of individuals and families rather than from the single family unit. Above all, the roles are less rigidly determined and—potentially, at least—more varied and changeable.

Heredity and development have some influence in determining human social roles, but not a great deal. Being male or female, constitutionally weaker or stronger,

Several recent writers have argued that much of human behavior is to be explained (or excused) on the assumption that it reflects a genetic heritage that cannot be altered. This, so it is claimed, is particularly true of those behavior patterns we call "aggression," "territoriality," and "mother love." To these thinkers, the evolutionary origin of such behavior patterns is understood as an extrapolation of patterns observed in other species. Ethologists now tell us, however, that a variety of behavioral mechanisms may serve similar ends, even among closely related species, and that the results of any particular "evolutionary extrapolation" may depend on the species at issue.

Consider the macaque monkeys. The dozen species of monkeys that comprise the genus Macaca include the rhesus macaque (*Macaca mulatta*), the bonnet macaque (*Macaca radiata*), the pig-tailed macaque (*Macaca nemestrina*), the Barbary macaque (*Macaca sylvanus*), and the Japanese macaque (*Macaca fuscata*). Despite the behavioral similarities observed in all macaques, there are many differences among the species. For example, motherless rhesus macaques will accept a surrogate mother much less readily than bonnet macaques.

The Japanese macaque (Figure 33D) has one of the best studied primate social orders. Males weigh 14.6 kg (32 pounds) and fe-

33D Japanese macaques

males 12.3 kg (27 pounds). These intelligent monkeys—terrestrial, diurnal fruit-eaters—travel in cohesive troops of 50 to 150 individuals. They confine themselves to home ranges of several hundred acres. They defend these ranges as roving bands moving through 8 to 10 acres a day and spending the nights in spaces of less than 2 acres. The ground they are occupying at the moment is defended with noisy threat behavior, tree shaking, ground slapping, and only occasional fighting. Their complex social organization embodies a highly developed social hierarchy based only partly upon aggressive behavior. A few adult males dominate all the other animals in the troop. Rigid dominance hierarchy, or pecking order, is maintained from moment to moment by a whole repertory of behaviorisms—voice, signals, posture, facial expressions, and so on. The top position—that of the leader, or in one terminology the "alpha monkey"—is always held by a mature male. Below him are five or

more or less intelligent, have a bearing of course, but a tremendous range of possibilities remains for everyone. Moreover, the precise roles taken are always learned. Individual humans are comparatively independent and self-reliant, prospering best in an accustomed social group but able to survive (temporarily, at least) without it, able to change roles within usually broad limits, and able to transfer from one group to another. All normal adult members of the society are able to reproduce. The lack of differentiation in this respect, so different from the insect society's situation, takes away from the family any really essential biological role in the organization of society beyond the function of producing and, to some extent, training the young. This further sharp contrast with insect societies is not contradicted by the fact that in human society the family remains an essential institution with profound psychological involvements and repercussions in most aspects of life.

There is little inherent limitation on the size or complexity of human social units. Among social insects, such units can be no larger than the number of progeny produced by one female (or a few) and they cannot attain even remotely comparable complexity. Both societies are biological adaptations that have been outstandingly successful in their different ways. Institutions and organizations in human societies may be viewed as biological adaptations meeting broad basic needs that also exist in insect societies. Human organizations typically depend on the emergence of a biologically versatile, all-purpose human individual—a founder, discoverer, inventor, leader. Ant organizations depend on the biologically restricted, specialized, and "depersonalized" insect individual.

In other words, if ants could write, it is doubtful that anyone would care to read more than one autobiography.

six lieutenants. Then come most of the adult females, many of whom are expert at chasing and threatening males. With them are their infants and juvenile offspring and below them in the hierarchy are the rest of the adult males who live at the periphery of the troop.

If a male fights with a female, other females will come to her aid. But adult males rarely assist each other. Female alliances thus are as important in regulating the social order as sexual attraction, which is seasonal and transitory. The dominant male who wins status has first access to food and to females of his choice.

An interesting aspect of the social biology of primates—and other animals—concerns the ability of parents and young to recognize each other as individuals. A good example was provided recently by H. Wu and G. Sackett of the University of Washington, who showed that young pigtailed macaques (*Macaca nemestrina*) can recognize their half-brothers or half-sisters without any previous contact with them. In their experiment, 16 test monkeys were related to their half-siblings only through their fathers, thus eliminating the possibility that some family-specific cue such as odor might have been acquired from the mother in utero.

The remarkable conclusion is that these monkeys have an inborn ability to recognize half-siblings, and presumably other close relatives.

The communication network that exists within macaque societies is enormously complex, far more so than that in more elementary societies, such as flocks, herds, and schools.

Then there are gorillas (Figure 33E), magnificent "gentle giants" of which several subspecies are threatened with extinction. Their numbers are estimated at 5,000 to 15,000 in the field and less than 700 in captivity. Male mountain gorillas weigh 160 kg (350 pounds). Females weigh about 93 kg (200 pounds).

In addition to mountain gorillas (*Gorilla gorilla beringei*), which range at altitudes from 6,500 to 13,500 feet in the region of several extinct African volcanos, there are two other recognized subspecies, the eastern lowland gorilla (*Gorilla gorilla graueri*) and the western lowland gorilla (*Gorilla gorilla gorilla*). From sea level to under 8000 feet, the eastern lowland gorillas are found almost exclusively in the rain forest of Zaire; the western lowland gorillas, at similar elevations, span an area from the west coast of Africa, to central Africa. Some 600 miles of Congo basin forest separate the western gorilla from the eastern populations including the mountain gorilla. The relatively minor differences among the three forms suggest that their divergence is recent.

The social and behavioral biology of gorillas arouses in one a sense of wonderment at the gorilla's biological and social similarities to humans. Leaf-eating herbivores, their diet and the ritualization of threat displays should dispel an unearned reputation for aggressiveness. Family groups usually consist of at least one adult male (more than 12 years of age), two or more adult females (10 years of age), and several young and juveniles, for a total of about 17. If one looks only at their social structure, revolving as it does around the competence, patience, and protection of the dominant male, mother-young interactions of care, play, and imitation during an infancy period of 3 to 5 years, dispersal of older juveniles of both sexes and so-called affiliative behavior—that is, contact and reassurance at all ages—the probability of the extinction of the gorilla, or of its continued existence solely in captive environments, is a cause for sorrow.

33E Lowland gorillas

SUMMARY

Interactions between animals—their methods of communication, cooperation, and social organization—are important aspects of behavior that crucially influence the survival and reproductive success of individuals and groups. Hence, they have profoundly influenced the course of evolution.

The methods of communication employed by animals are highly diverse, utilizing chemical, visual, auditory, electrical, and tactile channels. Communication, or signaling, always involves a signal, a signaler, and a signal receiver. Many of the signals sent by communicating animals probably arose as elaborations of simple body movement. Each communication system is a two-way process that persists through the generations when it benefits the participants—or avoids harming them, as they would, for example, if they attracted predators. Many species profit from mimicking the communications of other species.

Social behavior, in contrast, seems to have arisen along two evolutionary pathways. The familial pathway began with clusters consisting only of parents and close relatives, small groupings which at first shared only in the care of the young but later progressed into defense against predators, foraging, and other activities. In the parasocial pathway, associations of unrelated adults, by cooperation, made better use of resources. Regardless of their origin, the behavior patterns predominating in social groupings include cooperation, territoriality, dominance hierarchies, leadership, parental care, and mutual stimulation or care.

Certain insects—ants, some bees and wasps, and termites—form remarkably complex societies, which arose in evolution via the familial pathway. At least three social roles appear in most insect societies: the relatively parasitic male whose only role is to fertilize the female, the reproductive female (or queen), and the workers, which cannot reproduce. According to one proposed explanation of the forces that might have led to the selection of such complex societies, social insects commonly exhibit haplodiploidy: one sex is diploid, the other, haploid. Because a member of the diploid sex is genetically more closely related to its siblings than its offspring, such insects will be more likely to pass their genes on if they work altruistically to help their siblings than if they help their offspring, which have a lower level of relatedness.

Unlike the social behavior of insects, the behavior that maintains primate societies is largely learned. Tribal organizations are characteristic of primate societies. Their size and structure, however, vary widely among species depending on such variables as the nature and availability of food, foraging times in the 24-hour cycle, and the habitats of the species. Although humans are primates, human societies are drastically different from those of nonhuman primates, because learned traits and solutions to problems can be handed down from generation to generation by means of language and culture. Humans exhibit complex parasocial, or associative, organization—in addition to the nuclear familial organization, which is central to most human societies.

KEY TERMS

altruism
associative society
behavioral ecology
coefficient of relatedness
displacement behavior
dominance

familial pathway
inclusive fitness
monogamy
mutual stimulation
parasocial pathway
parental care

polygamy
polygyny
territoriality
tribal behavior
trophallaxis

QUIZ QUESTIONS

1. The ancestral form of communication between animals probably involved
   A. auditory channels.
   B. electrical channels.
   C. chemical channels.
   D. visual channels.
   E. A and B

2. _____ were probable evolutionary sources of elaborate visual displays.
   A. Irrelevant behaviors
   B. Preparatory behaviors
   C. Altruistic behaviors
   D. Tribal behaviors
   E. A and C

3. The scheme of tactile communications used by bees to indicate locations of food sources was discovered by
   A. P. H. Harvey.
   B. W. D. Hamilton.
   C. Theodosius Dobzhansky.
   D. Stevan Arnold.
   E. Karl von Frisch.

4. Which of the following is *not* true of an insect society?
   A. Each individual belongs to a particular caste.
   B. The role of the individual is determined by genetic and developmental factors.

   C. The size of a society is limited by the reproductive potential of one or a few females.
   D. Learning experiences enable an individual to alter its role in the society.

5. Mating systems found in societies of various primates include
   A. monogamy.
   B. polygyny.
   C. polygamy.
   D. A and C
   E. A, B, and C

ESSAY QUESTIONS

1. A newly hatched bird kept in quiet isolation will not learn to sing its species-specific song. Why can it be argued nonetheless that the bird's song is innate—that is, specified by its genes? What other animal behaviors are genetically determined?

2. How might communication behavior have evolved from simple body movements? Is all behavior under natural selection? Can some behaviors have a neutral effect on selection? What is displacement behavior?

3. How does a dominance hierarchy serve to reduce fighting within a group?

4. What is altruism, and how might this trait be evolutionarily selected? What is the coefficient of relatedness? How does this concept help to explain the evolution of altruism? What is meant by "inclusive fitness"?

5. In primate societies, what conditions favor monogamy?

REFERENCES AND SUGGESTIONS FOR FURTHER READING

Alexander, R., and Tinkle, D. (Eds.) (1981). *Natural Selection and Social Behavior.* Chiron Press, New York.

A collection of essays by specialists.

Axelrod, R., and Hamilton, W. D. (1981). The evolution of cooperation. *Science 211,* 1390–1396.

A stimulating review by two distinguished students of altruistic (and selfish and spiteful) behavior and its influence on inclusive fitness.

Caplan, A. L. (Ed.) (1978). *The Sociobiology Debate: Readings on Ethical and Scientific Issues.* Harper & Row, New York.

A collection of 42 essays at the interface of biology and ethics. The scholarly first half includes such luminaries as Darwin, Spencer, Lorenz, Tinbergen, and Hamilton. The heated second half traces what happened after the publication of Wilson's *Sociobiology* in 1975. Not to be missed.

Clutton-Brock, T. H., and Harvey, P. H. (1978). Mammals, resources and reproductive strategies. *Nature 273,* 191–195.

Breeding in mammals is strongly influenced by the density and dispersion of their populations. Using quantitative comparisons, the authors trace evolutionary relationships between these variables and animal behavior.

Dilger, W. C. (1962). The behavior of lovebirds. *Scientific American 206,* January, 88–98.

This classic study of the courtship and nest-building behavior of several species of African lovebirds and their hybrids establishes the genetic basis of behavior.

Ghiglieri, M. P. (1985). The social ecology of chimpanzees. *Scientific American 252,* June, 102–113.

A study of wild chimpanzees indicating that their social structure is shared only with that of human beings.

Jolly, A. (1985). The evolution of primate behavior. *American Scientist 73,* 230–239.

A fascinating analysis of primate behavior that proposes a progressive development of intelligence rather than a sudden emergence of human intelligence.

May, R. M., and Harvey, P. H. (1988). Behavioural ecology: Tampering with territories. *Nature 335,* 668–669.

A note on new work by John Maynard Smith that deals with the question: What criteria do males use in choosing a territory?

Nottebohm, F. (1989). From bird song to neurogenesis. *Scientific American 260* (February), 74–79.

The author has shown that when the adult canary needs to learn new songs, it grows some new brain neurons. This contradicts the old idea that brain neurons are never replaced or added to.

Poole, T. B. (1985). *Social Behavior in Mammals.* Chapman & Hall, New York.

A brief, if compressed, summary of mammalian social behavior.

Wilson, E. O. (1985). The sociogenesis of insect colonies. *Science 228,* 1489–1495.

Together with flight and metamorphosis, social organization was a major event in the evolution of insects, greatly enhancing their ecological success. This paper reports studies on the behavioral mechanisms that promote social development (sociogenesis) among the social insects.

Wilson, E. O. (1975). *Sociobiology: The New Synthesis.* Harvard University Press, Cambridge, Mass.

It was in this monumental work that the new field of sociobiology originated. An influential book that has catalyzed much new work and vigorous debate.

PART 4
BIOLOGY OF ORGANISMS AND POPULATIONS

In Part 4, for the first time we discuss organisms as groups—species, phyla, populations, and communities—rather than individuals. We begin with the mechanisms of organic evolution (Chapters 34 through 36), which can be understood only in terms of populations and the history of change in their genes and alleles. The process giving direction and order to evolutionary change is natural selection.

Evolutionary relationships are the basis of one method of classification of organisms (Chapter 37). Other methods use other approaches. In surveying time scales, Earth's history, and the nature of fossils, we find that for early stages of evolution the fossil record is meager. But those stages are not a book wholly closed to us.

We turn then to the immense diversity of life, attested to by the existence of well over a million living species and many more extinct ones. Chapters 38 through 43 survey the five living kingdoms. In Chapter 44, we trace the sweeping panorama of the history of life—beginning with the young planet Earth, devoid of life, and ending with the rich diversity of life that covers its surface today. Its most recent stage is marked by the emergence of human beings.

We consider in Chapters 45 through 48 the subject of ecology—the relations of populations and their communities to the environment and to each other, and the cyclic flow of materials and energy through communities and their environments. The geography of life is approached from two different viewpoints—ecological and historical—in Chapter 48.

Finally, in Chapter 49, we view the human species from a systematic and ecological point of view. We close with a survey of issues arising from human biology—the impact and dependence of this extraordinary society on natural communities, the problems of race, population size, and conservation.

SECTION 1
EVOLUTION

Miacis
Cynodictus
Cynodesmus

Tomarctus

Fennecs
Foxes
African Dogs

Cats
Hyenas
Civets

Raccoons
Weasels
Stoats

Coyotes, Jackals
Wolves
Dogs

CHAPTER 34

ELEMENTARY PROCESSES OF EVOLUTION

To understand life, we must examine not only the individuals, but the populations and communities to which they belong. A species cannot be understood by studying its individual members only. Species are understood by studying the dynamics of groups of individuals.

The balance of this book departs from the viewpoint of preceding chapters. From here on, we deal mainly with the biology of various groups of organisms.

INDIVIDUALS, POPULATIONS, AND OTHER UNITS OF LIFE

To facilitate our inquiry, we must now introduce and carefully define an organism, a population, a community, a deme, and a species. Each, we will find, is a distinctive biological unit at a particular level of integration.

INDIVIDUAL ORGANISMS AND POPULATIONS

The **individual** is a critical biological unit. Among unicellular organisms, cell and individual are usually one and the same. In multicellular organisms, they are obviously never the same. Whether made of one cell or many, an individual is the pertinent unit when we speak of physiological processes, metabolism, responsiveness, development, and related phenomena. But now that we have turned to an explicit discussion of evolution, individual organisms must move from center stage.

Just as processes in cells must be related ultimately to whole multicellular individuals, so must processes in individuals be related to populations. We define a **population** as all the members of a given species that occupy a particular area at the same time (Figure 34-1). Generally, members of a population breed together, actually or potentially, and produce fertile offspring. Examples of populations are the deer of New England, the rabbits of Australia, and the people of the United States.

Over long time spans, individuals are evanescent: they are born, they die. Only populations continue, their continuity depending upon transmitted genetic information. Evolution ordinarily occurs only in the course of many generations.

DEMES AND SPECIES

Individual paramecia in a pond all look alike. They reproduce for long periods without interbreeding.[1] They are without social organization and are as independent of one another as individuals can be. Yet, the whole *Paramecium* population is a unit with many interesting properties. The population also has a well-defined role

Elementary processes of evolution
Life's diversity, here symbolized by representatives of several kingdoms and phyla, is the product of evolution by common descent. Our understanding of its processes and mechanisms continues to deepen.

[1] We saw in Chapter 21 that strict and long-continuing uniparentalism is rare. *Paramecium* and other protists—indeed most organisms—do exchange genetic material from time to time. This is not strictly biparental or sexual reproduction, but it has some of the effects of interbreeding and it does produce genetic recombination. Thus, the discussion here applies as well to the few populations that reproduce uniparentally as the many that do not.

in the life of the pond it inhabits, and this population in this pond is to be distinguished from other *Paramecium* populations in other ponds.

Consider a grove of pine trees and the squirrels living within it. Each group is a population. These groups are defined in part by interbreeding relationships among their individual members. Future populations may derive from any or all members of the present local populations. Future populations inhabiting that particular locality are certainly less likely to derive from individual members of other populations of pine trees or squirrels.

A definable local unit or subunit of a population—like those of paramecia, pine trees, or squirrels—is called a **deme.** Demes are of several sorts. Their geographical limits and membership can differ widely. A deme defined in part by interbreeding among its individuals, as in demes of pine trees or squirrels, is also called a local genetic, or Mendelian, population. A social deme is united by a social organization. Usually, it is a breeding unit as well, so that it corresponds with the genetic population. However, this is not always true. Few individuals in a social deme of ants ever breed. When they do, they usually cross-breed with individuals from other social demes (anthills) and set up new ones. Thus the genetic population may in time include members of many social units. The real genetic population in a large area may consist of all the anthills.

Adjacent demes often merge into one another, or intergrade. For that reason, demes are characteristically vague in definition and fluctuating in numbers. It is likely, for example, that some pollen from the pine grove would reach a distant pine grove and vice versa, so that new genes would enter into the genetic material of each deme. If new pines should grow between two groves, the two demes could merge and become indistinguishable. As shown in Figure 34-2, adjacent demes exist in a certain geographic area called a **glade.** A group of glades constitutes a **cluster.** Clusters are grouped together to form a **region,** and various regions together constitute a **distribution range.** Thus a hierarchial fashion exists, ranging from the large distribution range all the way down to the individual.

Like the pine trees, some squirrels of one deme may occasionally interbreed with squirrels of a nearby deme. Genetic information thus does pass from deme to

34-1 Populations
A population is defined as the members of a species occupying a given area at the same time.

34-2 Hierarchical pattern of population distribution

The population structure of the plant Clematis fremontii *is shown, beginning with a large area, the distribution range, and ending with individuals in an aggregate.*

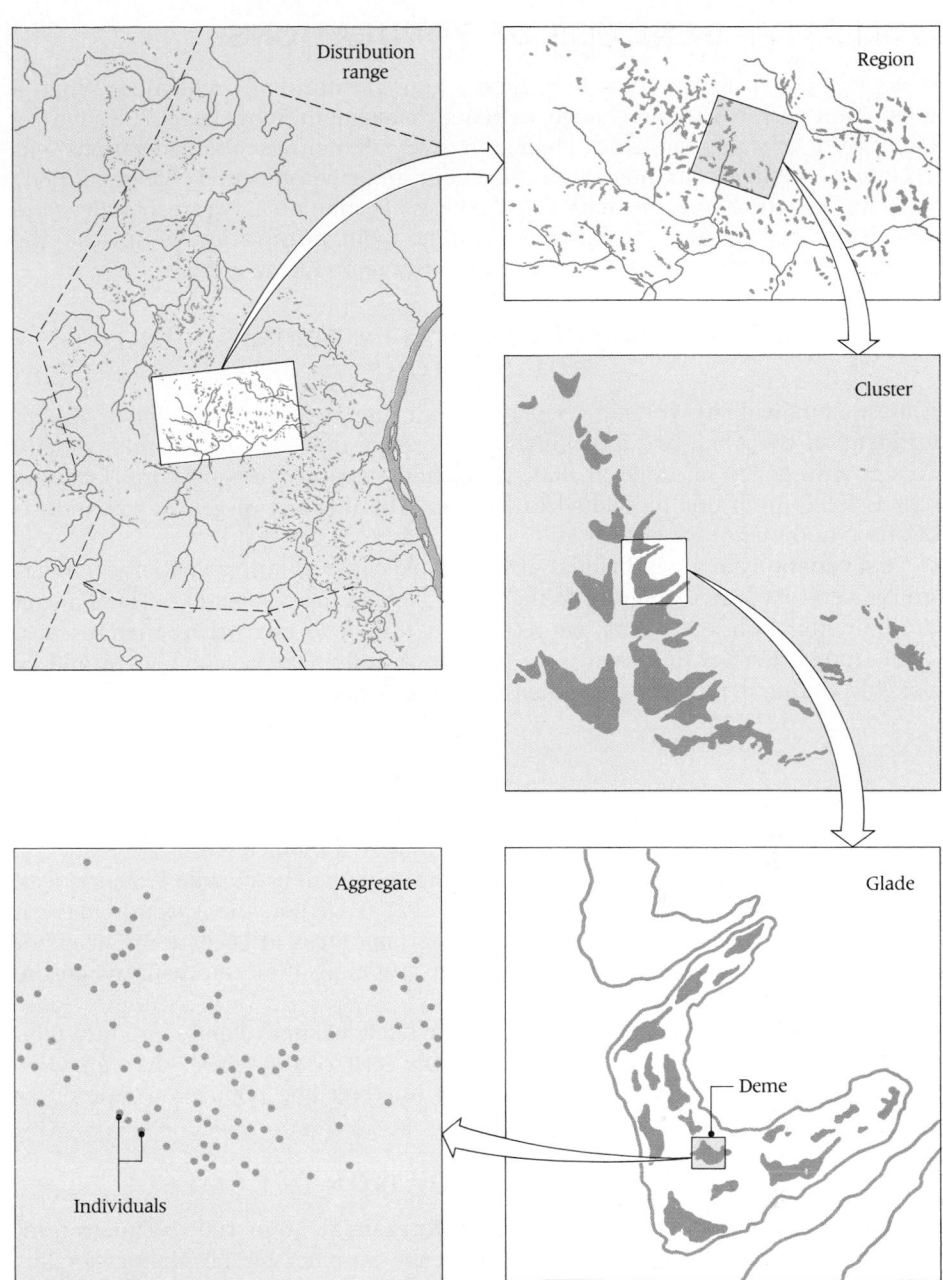

deme. If squirrels in adjacent demes interbreed freely, the demes fuse into one. If all the squirrels in one grove die out, a deme would cease to exist. But if squirrels from surrounding demes were to occupy the vacant territory, a new deme could arise—even though no permanent change in the squirrel population had occurred.

Though demes fluctuate and intergrade, certain larger population units are more permanent and clearcut. These are **species.** A species is a group of organisms so similar in heredity that their demes tend to intergrade, fuse, and replace each other without changing the ecological features of the group as a whole. A species is a series of populations within which a significant amount of gene flow occurs under natural conditions, but which is genetically isolated from other populations. A species includes many demes.

This definition of species applies only to biparental populations. Uniparental populations produce offspring with the same genetic makeup as their solitary parents; hence they lie outside this concept of species. They are classified as different species according to degrees of differences among their phenotypes and genotypes.

Biologists generally accept the above as a more or less suitable definition of species. It is difficult, however, to give a precise definition of species that applies to all organisms and all situations. This is the basis of the famous species problem to which we return in Chapter 36. Before delving into it, we must know more about the genetics of populations.

GENETICS OF POPULATIONS

A deme tends to persist for a long time—years, centuries, or millennia. When a population persists without change, its genetic makeup must be stable. The environment must be stable too, since change in a population over the generations—its evolution—results from interactions of genetic and environmental changes. Strictly speaking, genetic changes in demes are the basis of evolution. If we are to understand the history of life, we must know something about **population genetics**—the heredity both of individuals and of continuously reproducing groups.

VARIATION WITHIN SEXUALLY REPRODUCING POPULATIONS

Almost all individuals within a sexually reproducing population are unique. Except for identical twins, no two individuals have exactly the same genetic endowment. That is why police can rely on fingerprints and surgeons despair of transplanting organs freely from one individual to another. The ubiquity of genetic variation is of prime importance for evolution.

That variation among individuals exists should cause no surprise. A large population is very likely to exhibit more than one allele at many gene loci. These alleles arose by mutation. Each gene, we recall, may mutate to many more than just two allelic forms. Even if only two were present, *A* and *a*, three genotypes would be possible among the diploid individuals in a population:

AA, Aa, and *aa*

The number of possible genotypes increases sharply with an increasing number of alleles. This will be demonstrated mathematically in a moment.

Other factors increasing variation in natural populations include crossing-over, recombination, and gene mobility (Chapter 17). Together, these factors make it possible (if not always probable) for a gene at any locus to become the neighbor of any gene at any other locus. Along with mutation, these mechanisms ensure enormous variation within a population.

Individuals with useful variations are more likely to survive and reproduce than individuals without them—that means they have been *selected*. Hence, such variations tend to be preserved and propagated as an interbreeding population reproduces itself.

DISTRIBUTION OF VARIATION IN DEMES

One way to grasp the reality of variation is to examine 50 to 100 specimens from one deme of one kind of organism. Among the innumerable possibilities for this small experiment are full-grown flowers (each from a different plant of one species), shells, butterflies, field mice, or whatever. Next, we must measure or count some feature—petals of a flower, scales of a pine cone, ribs on a shell, and so forth. These are the principal kinds of observations used in studying variation and classifying diverse forms.

In any such sample, some features will be the same in all the individuals. It is, after all, a characteristic of demes that its members are similar. For instance, all the flowers of a single species one collects may have the same number of petals. But they may not. These and other characters may vary within the sample.

Such observations are best studied by tabulating them in the form of a **frequency distribution curve,** as in the example in Figure 34-3. Since such curves represent a standard measuring device, they are termed **normal curves** (Latin, *norma,* a carpenter's square). These curves have a characteristic bell-shaped pattern, and they can be analyzed mathematically.

- In the example shown, there is a **range** of values and a class that is most frequent. One might say that this class is "the fashion" among these animals. In fact, it is known as the **mode** (French, *la mode,* the fashion). It forms the peak of the frequency curve.
- Note the difference between mode, the single most frequent class, and **mean,** the arithmetic average of all the recorded values. Mode and mean may differ considerably. If students in a class were graded 4 (excellent), 3, 2, 1, and 0

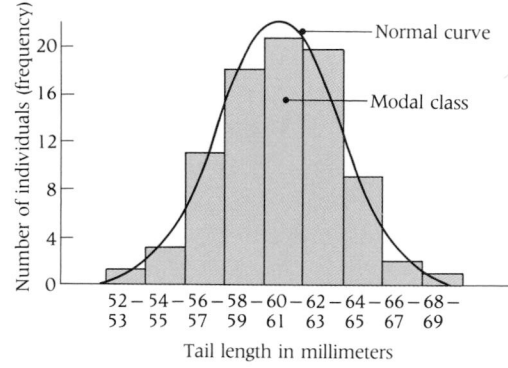

A

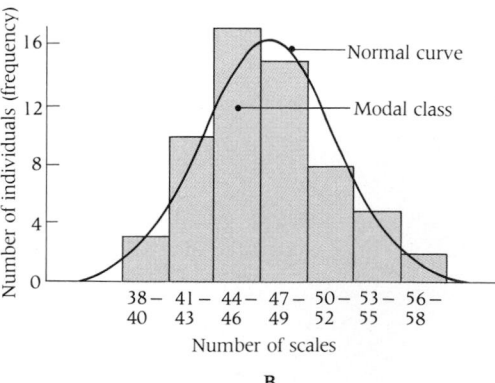

B

34-3 Examples of frequency distribution curve
The number of individuals exhibiting a particular characteristic often appears as a bell-shaped (normal) curve. The modal class forms the peak of the curve. (A) Data on tail length in a deme of deer mice. (B) Number of tail scutes in a deme of king snakes.

(less than excellent), the mode would be 2 if the largest number fell into that group, but the mean might be 2.75, a grade that was never given to anyone! In a normal curve, or normal distribution, the mode is equal to the mean.

■ On each side of the mode, frequencies fall off with fewer and fewer individuals in each class. The curve slopes down to form tails that extend to the ends of the distribution. Most real frequency distributions are somewhat asymmetrical—and there are measures of "skewness" and tests for its significance (that is, the probability that it could be attributable to the effects of random sampling from a normal distribution).

■ Many variable characters—that is, characters that can have any of many values in the individuals of a deme—have frequency distributions like that in Figure 34-3. This is important. The normal curve is said to be bell-shaped. When a curve departs from this shape, it may signify the presence within the group of two or more populations. A plot, for example, of the weights of all members of a species may not yield a normal, bell-shaped curve if there is a wide disparity between the mean weights of males and females. That curve would be *bimodal.*

The bell shape of the frequency distribution curve takes on meaning when we link it with biological principles. We know that variation results from (1) genetic differences within the members of the population, with environmental conditions remaining similar, or (2) interactions with differing environmental conditions, with similar genetically determined traits. Either or both may account for a bell-shaped curve. Let us think why.

If organisms with similar genotypes (Figure 34-4A), and hence similar ranges in their reactions to the environment, develop under similar conditions, their phenotypes will tend to be similar. Most will fall into a modal class that will reflect the usual interaction of genotype and environmental conditions. Conditions producing unusual phenotypes remote from the modal class will be rare. Thus, environmental factors alone are one cause of the bell-shaped curve.

On the other hand, genetic factors—especially in the case of characters influenced by many genes (polygenic inheritance)—can produce bell-shaped distribution curves within a deme, as we have seen. In an experiment discussed in Chapter 14, crossing large and small parents produced a broad, bell-shaped distribution in the F_2 generation (see Figure 14-17). In natural populations, the similar distributions result from interbreeding within a deme and tend to persist through the generations. In fact, when we see a symmetrical distribution attributable to genetic factors (Figure 34-4B), it is reasonable to postulate that the character under study is determined by multiple genes.

Environmental and genetic factors are both often at work, and it may be difficult to disentangle them. One way is to cross-breed the extreme variants from a deme—very long and very short, for example. If there were no significant environmental influence, the F_1 generation would be largely heterozygous and quite uniform in phenotype. In the F_2 generation, one would again see a normal distribution of phenotypes and genotypes. On the other hand, if extreme forms breeding among themselves (long × long and short × short) produced normal distributions of phenotypes in the F_2 generation, it would then appear that environmental factors were significant and genetic factors were less likely to be responsible for the variation observed. Another way would be to test the effect of selection on population variation (Figure 34-5), though obviously it is impractical to do breeding experiments on most of the countless demes in nature.

EVOLUTION AS A CHANGE IN GENE POOLS

A Population's Gene Pool

One definition depicts **evolution** as a change in a population's allele frequencies. Since individuals are almost always genetically unique, it follows that individual genotypes in a population constantly change from generation to generation. However, that is not what we mean by evolutionary change. Evolution occurs only when it affects the genetic characteristics—the genotypes and their frequencies—of the population as a whole. Populations evolve, not individuals.

If we knew every detail of a population's genetic makeup—all the alleles of all of the genes, their ratios or frequencies—we would know all there is to know about the genetic constitution of that population. Our information would have to

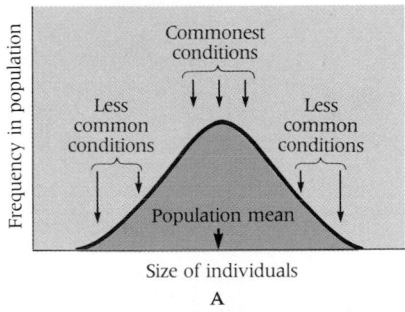

Absolutely constant genotype; variation caused entirely by environmental conditions

Frequency in population

Commonest conditions

Less common conditions

Less common conditions

Population mean

Size of individuals

A

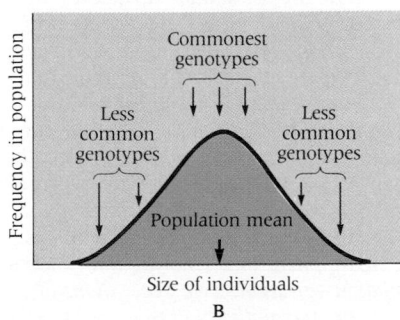

Absolutely constant environment; variation caused entirely by heredity

Frequency in population

Commonest genotypes

Less common genotypes

Less common genotypes

Population mean

Size of individuals

B

34-4 The bell-shaped curve of population variation

The distribution of phenotypic characteristics (such as size) in a population may theoretically be due to two factors. **(A)** *The bell-shaped curve may result from variation in the environmental conditions encountered by the population.* **(B)** *The curve may result from variation among the genotypes in the population. In natural populations, variation is nearly always due to both environment and genotype.*

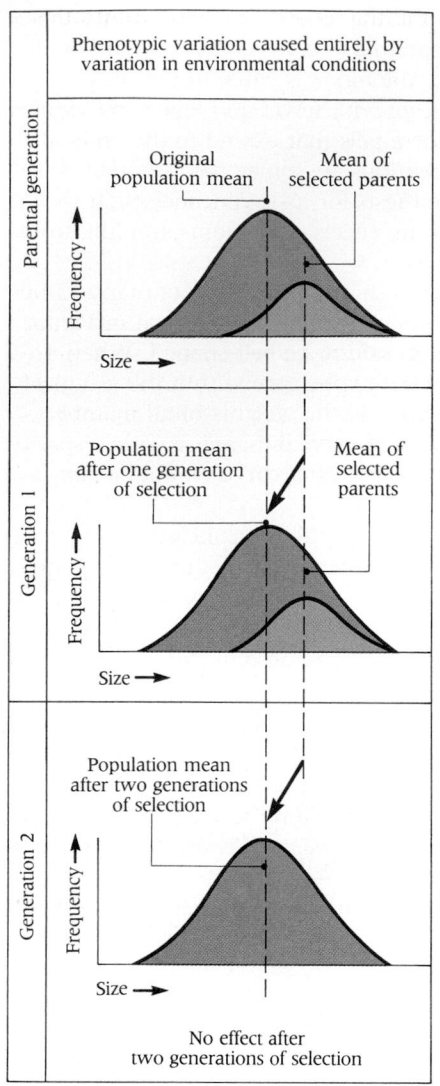

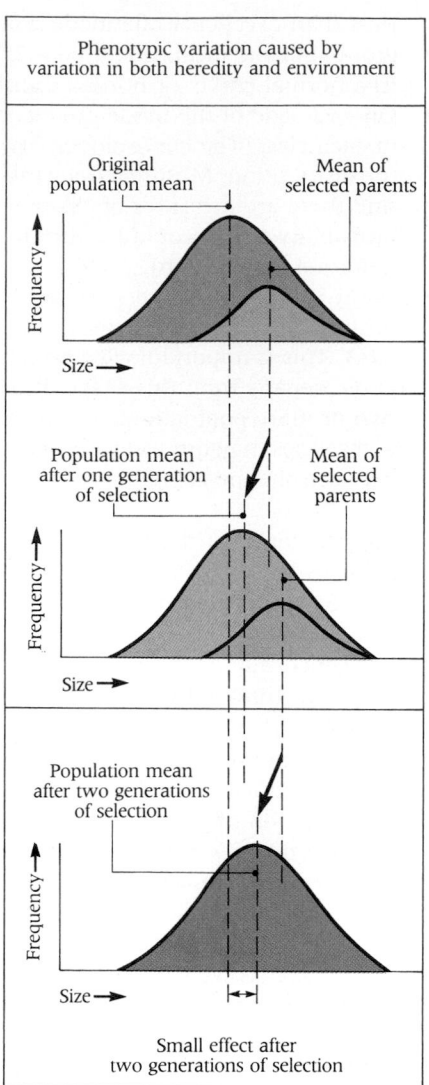

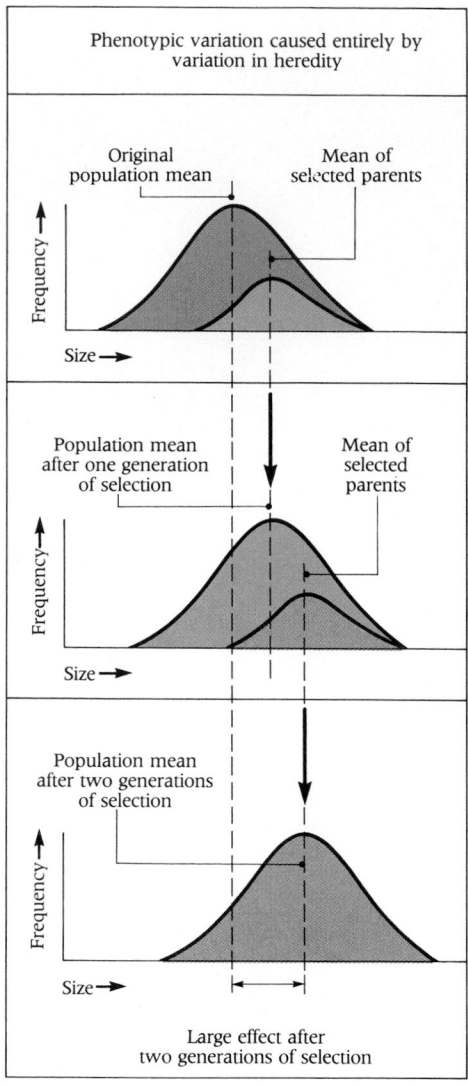

| Phenotypic variation caused entirely by variation in environmental conditions | Phenotypic variation caused by variation in both heredity and environment | Phenotypic variation caused entirely by variation in heredity |

No effect after two generations of selection

Small effect after two generations of selection

Large effect after two generations of selection

be that complete before we could claim to have fully characterized what is called the **gene pool** (or better, the allele pool) of a population. In reality, it is never that complete.

Processes That Can Modify Gene Pools

We are concerned now not with the mechanisms by which the genes of individuals change (which were listed earlier), but with the processes that can systematically modify the gene pool of a population.

These processes have classically been said to include mutation pressure, random drift (or genetic drift), gene migration (or gene flow), and natural selection. In addition, evidence accumulates for the unorthodox idea that some genes may be able to jump across the species barrier in eukaryotic organisms.

Alteration of the gene pool of a population by these processes has been called **microevolution.** The term denotes the collective accumulation of small changes that occur over the relatively short time span of two or more generations. This chapter is concerned with these processes. In contrast, **macroevolution** refers to larger events in the course of evolution, in which major groupings (individual species and categories above the level of species) evolved and differentiated. Macroevolution is the theme of the next two chapters.

The rest of this chapter is concerned with the mechanisms of microevolution, which may be regarded as the basic or elementary mechanisms of evolutionary change and that may operate alone or in combination. To facilitate discussion, we briefly define them at this point.

34-5 Effect of selection on variation within a population
Three sets of conditions are examined. The left panels show the population variation caused by environment alone. Selective breeding through two generations does not change the distribution. In generation 1, the offspring of individuals in the parental generation show the same distribution of phenotypes as that of the parental population. On the right, variation caused entirely by genotype leads, after two generations, to selection of larger individuals that approach the mean size of the parents. In the middle, both environment and genotype contribute to population variation. Two generations of selection produce individuals that are only slightly larger than the mean of the original parental population.

- **Mutation pressure** relates to the fact that mutations—both gene and chromosomal—tend to occur at a certain rate. As a result, there is an ongoing, ever present tendency towards variability.
- **Random drift** (or **genetic drift**) refers to processes by which gene frequencies in a gene pool are altered from one generation to the next by chance events often brought about by dispersal (migration).
- **Gene migration** (or **gene flow**) is the movement of genes from one population to another through interbreeding.
- **Natural selection** refers to the perpetuation in the next generation of preferential genotypes as a result of a shift in frequencies of certain phenotypes (and their underlying genotypes). The reason for the shift is that some phenotypes reproduce more effectively than others in the natural environment. Hence, they are selected.
- **Horizontal gene transfer** is a newly recognized mechanism wherein genes shift from one species to another (Box 34-1). It is too new to be considered an established mechanism of change.

GENETIC EQUILIBRIUM

While the processes just mentioned tend to alter the composition of a gene pool from one generation to the next, it should be recognized that other factors simultaneously operate to maintain the constancy of the gene pool, its allele frequencies or ratios. We term this tendency towards stability **genetic equilibrium.**

Mutational Equilibrium

Mutations produce hereditary variations and are the ultimate sources of variability and change. In one sense, they are not random events. External factors inducing mutations may act randomly, but mutations themselves occur repetitively and at predictable rates. Indeed, the same mutations occur again and again in different individuals. Significantly, they are to an extent reversible. A mutant allele can mutate back to its original allelic form—and this also occurs repetitively at a precisely predictable rate.

Consider a gene with two alleles, A and a. If at the start, all individuals in a population have only allele A, then this gene in the course of time will mutate to a in a few gametes. Rates of mutation vary widely. They are usually low and it is a fair approximation that most genes undergo mutation once every 10^6 to 10^8 replications. Despite the low rate, if we have a large enough population and a long series of generations, a alleles will accumulate, arising repeatedly and then propagating. The result is an evolutionary change.

Some a alleles, however, may mutate back to A. Thus, at this locus we would have the following situation:

$$A \xrightleftharpoons[\text{back mutation}]{\text{forward mutation}} a$$

As for the actual number of mutations, those from A to a will greatly exceed those from a to A at first, because at first the population had all A alleles. But as the process continues, the number of A alleles decreases as does the rate of their mutation to a. Meanwhile, the number of a alleles and their mutations back to A increase. A time comes when the numbers of mutations in each direction are equal. Change in the percentages of the two alleles ceases and we say that **mutational equilibrium** has been reached. At this point, evolution by mutation alone ceases.

Actually, such equilibriums are rarely reached in nature. Despite the logic of the case, their role in evolution is marginal for two reasons.

- Attainment of mutational equilibrium occurs so slowly that other processes may intervene before it is reached. For example, one allele or another may be so favored by selection that it spreads in the population regardless of the mutational equilibrium point.
- Back mutations are not common. A forward mutation, say from a to x, is likely to intervene before a mutates back to A. This further reduces the chance that descendants of a will mutate back to A.

Just as investigators began to ponder the evolutionary implications of mobile DNA within the genome of a given species, the first clear indications appeared that genetic mobility might extend across the species barrier in eukaryotic organisms—that genes can "jump" from species to species. The successes of bioengineers in deliberately transferring genes from one species to another have suddenly turned this into a plausible possibility.

If gene transfers between species actually do occur in nature, a new evolutionary dimension must be added to the phenomenon of jumping genes in eukaryotic organisms. Clearly, such a prospect would challenge many current views on the mechanisms of evolution and of the orderly transmission of eukaryotic genes.

Gene transfer between species is commonplace in prokaryotic organisms, where it is the basis of transformation, transduction, and conjugation in bacteria. However, interest in interspecific gene transfer (also known as horizontal gene transfer) is today very much focused on eukaryotic organisms. Anything as unorthodox as gene transfer between species necessarily raises several questions: Does it occur at all? If it does occur, how common and how important is it? And how does it occur?

Until recently, there were only a few clear examples of eukaryotic gene transfer. In one case, the common soil bacterium *Agrobacter tumefaciens* transfers some of its genetic mate-

34A Crown gall tumor growing on a sunflower

rial to the plant host, producing crown gall disease (Figure 34A). This well-studied tumor is caused by an unusual form of parasitism that amounts to a kind of "genetic colonization" of the plant's DNA by DNA from the bacterium.

It was the advent of reliable methods for DNA sequencing and gene cloning that finally permitted a systematic search for transferred genes. The following putative cases of transferred genes were fortuitous discoveries: (1) the genes for the enzyme superoxide dismutase in *Photobacter leiognathi*, a symbiotic bacterial parasite of the ponyfish; (2) the leghemoglobin gene in leguminous plants; (3) a family of histone genes in sea urchins; and (4) a subfamily of repeated sequences in sea urchins. Let us examine each of these cases briefly.

- Superoxide dismutase is a widely distributed enzyme that protects cells by eliminating harmful oxygen radicals (O_2^-). In eukaryotes, the enzyme contains copper and zinc. In prokaryotes it contains iron. A third form found in prokaryotes and mitochondria contains manganese. In 1974 researchers reported an anomalous finding—a copper-zinc enzyme in the bioluminescent bacterium *Photobacter leiognathi*, a prokaryote. When the amino acid composition of the bacterial enzyme was compared with dismutases from a range of eukaryotic and prokaryotic species, a clear-cut result emerged. The enzyme from *Photobacter leiognathi* fell unequivocally in the eukaryotic category, and by a number of other criteria closely resembled the enzyme from the ponyfish. It appears that the intimacy of the symbiotic relationship between fish and bacterium facilitated transfer of the copper-zinc superoxide dismutase gene from eukaryote to prokaryote. If this transfer did occur, it is far from clear what selective advantage it conferred on the bacterium.

- A second candidate for horizontal gene transfer concerns leghemoglobin, a myoglobinlike protein in leguminous plants that combines with oxygen. Leghemoglobin bears an obvious resemblance to globin of vertebrates. It even contains heme. The structure of the leghemoglobin gene is also similar to that of the vertebrate

globin gene, the resemblance extending to the precise location of two of its introns (see Box 35-4). It has been proposed that the le ghemoglobin gene came from *Rhizobium*, a bacterium that lives symbiotically with legumes. But the presence of introns (which do not occur in prokaryotes) makes it more likely that the gene was transferred to legumes from another eukaryote relatively recently in evolutionary history, possibly as a passenger on a virus. Such a mechanism would circumvent the rules of classical Mendelian genetics. It would also imply new mechanisms of evolution.

- The sea urchin, a popular experimental organism in developmental biology, provides two examples of possible gene transfer. Researchers recently reported interesting homologies in a superfamily of repeated sequences in several sea urchin species. A puzzling finding was the very close homology in one region between sea urchin species that had diverged almost 200 million years ago. Horizontal gene transfer has been proposed as a possible explanation.

- Finally, investigators sequencing clones of histone genes from different species of sea urchin to determine the mutation rate through evolutionary time were surprised to find that one minor clone from the North Atlantic species *Psammechinus miliaris* differed, on average, by only 1.3% of its bases from a homologous clone from *Stronglyocentrotus purpuratus*, a Pacific species from which it diverged some 65 million years ago. They postulated that the unusual homology might be explained either by an extreme and unusual conservation of sequences, which included nontranscribed sequences, or by horizontal gene transfer.

Nothing can yet be said about the mechanism of gene transfer in any of these cases, if indeed that is what they represent. Of several possible mechanisms, a viral vector is the current favorite and one for which there are precedents in other systems—and that is what investigators are now looking for. At present, it is important to search for additional examples. If horizontal gene transfer occurs, it might be a vastly important evolutionary mechanism.

TABLE 34-1
GENE FREQUENCIES IN AN EXPERIMENTAL POPULATION

Flies	Gametes		Totals
	A	*a*	
49 males are *AA* and produce	490	0	(490)
42 males are *Aa* and produce	210	210	(420)
9 males are *aa* and produce	0	90	(90)
The 100 males as a group produce	700	300	(1000)
Ratio of different gametes in the pooled population of sperms as a decimal fraction	0.7	0.3	(1.0)

Combinational Equilibrium: The Hardy-Weinberg Law

The most important factor promoting genetic equilibrium is the **combinational equilibrium** that occurs in sexually reproducing populations that breed or mate randomly. By random breeding, we mean that each genotype in a population is equally likely to mate with the other genotypes in the proportion in which they occur in the population. If in a population of birds the females of a species always chose to mate with males having yellow plumes, breeding would not be random and the genes producing that phenotype would increase in the population.

Consider an imaginary population of *Drosophila*. *A* and *a* are the two alleles of a given gene. We will introduce 200 *Drosophila* flies—100 males and 100 females—into a breeding cage arranged so that one generation can follow another without a break.

There are three possible genotypes in such a two-allele system.

AA, Aa, and *aa*

Of the 100 flies of each sex, we will arbitrarily choose 49 with the genotype *AA*, 42 with *Aa*, and 9 with *aa*, as follows:

100 females			100 males		
49	42	9	49	42	9
AA	*Aa*	*aa*	*AA*	*Aa*	*aa*

We then allow the males and females to mate at random. Can we predict the relative frequencies of the genotypes that will appear in successive generations? First guesses might be that (1) the less common allele, *a*, will gradually be lost from the population, and (2) it is hopelessly difficult to predict exactly what genotypes will result from indiscriminate matings among so many flies. Both guesses are wrong!

The prediction problem can be simplified if we regard all the females as though they were one female and ask: What kinds and frequencies of eggs will be produced by the population's "composite female?" Treating males similarly, we can determine what kinds of sperms the "composite male" will produce.

Let us begin with the males, who produce sperms in enormous numbers. To simplify we can assume the number of sperms per male to be 10, since the simplification does not affect ratios, which are our real concern. The 100 males produce gametes as shown in Table 34-1. We see that of every 10 sperms produced by the total population, 7 are *A* and 3 are *a*. The proportion of the total that can be assigned to a particular allele is termed the **gene** (or allele) frequency.[2] Thus, the gene (or allele) frequency of *A* in our population is 0.7. Since the initial females in our experiment had the same genetic constitutions as the males, it follows that in their pooled gametes (eggs) the frequency of *A* is 0.7 and of *a* is 0.3.

The next step in calculating what genotypes appear in the next generation is to set up a Punnett square like the ones in Figure 34-6. We used such squares in

[2] Recall that we express frequencies as a decimal fraction of one. In this usage, 100% equals 1.0 and a frequency of 50% equals 0.5 (see Chapter 13).

Chapters 13 and 14 to predict what genotypes could result from fertilizations by gametes from a single female and male. The method is just as valid when gametes from 100 females and 100 males are pooled.

When we crossed two heterozygous flies, gene frequencies in the gamete pool of each parent were 0.5 A : 0.5 a. The present population produces gametes in a ratio of 0.7 A : 0.3 a. The Punnett square in Figure 34-6A shows that, since matings are random, the genotypes of zygotes will be as follows:

$$0.49 \ AA : 0.42 \ Aa : 0.09 \ aa$$

The ratios are the same as those of the parental population.

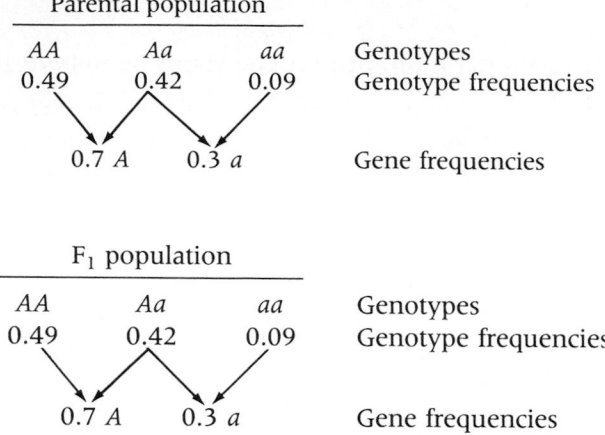

We can conclude that the F_2, F_3, and all subsequent generations would display the same gene and genotype frequencies. If more convincing evidence is needed, examine the Punnett square in Figure 34-6B which shows population gene frequencies at 0.4 A and 0.6 a.

Punnett squares are simply graphical expansions of the binomial expression $(p + q)^2$, where p is the frequency of the allele A in a population, and q is the frequency of the allele a. The expansion of $(p + q)^2$ describes the frequency of the three genotypes (AA, Aa, and aa) in the population. Taking our experimental fly population (Table 34-1, Figure 34-6A) as an example, we would have $p = 0.7$ (the frequency of A) and $q = 0.3$ (the frequency of a). Then it follows that:

(1) $(p + q)^2 = p^2 + 2pq + q^2 = 1.0$
(2) $(p + q)^2 = AA + Aa + aa = 1.0$
(3) $(0.7 + 0.3)^2 = 0.49 + 0.42 + 0.09 = 1.0$

Line 1 gives the expansion of the binomial; line 2 shows the zygote genotypes equivalent to the three terms in the expanded binomial; and line 3 shows how the frequencies of zygote genotypes are computed from the binomial. The solution is even clearer if expressed in equation form.

$$
\underset{(0.7 \times 0.7 = 0.49)}{p^2} + \underset{(2 \times 0.7 \times 0.3 = 0.42)}{2pq} + \underset{(0.3 \times 0.3 = 0.09)}{q^2} = 1.0
$$

The familiar Mendelian F_2 ratio for a single pair of alleles (1:2:1) is only a special case of this same general rule, in which $p = 0.5$.

These results were certainly not self-evident to early geneticists, who were concerned with two questions. How could both dominant and recessive genes persist in a population? Why do not dominant genes simply drive out recessives? A governing rule that answers these questions was proposed independently in 1908 in separate papers by two workers—G. H. Hardy, a distinguished English "pure" mathematician (1877–1947), and W. Weinberg, a German physician (1862–1937). Now known as the **Hardy-Weinberg law,** it is the fundamental law of population genetics.

The Hardy-Weinberg law can be formulated in different ways, but its general meaning is that in the absence of various disturbances—selection, nonrandom mat-

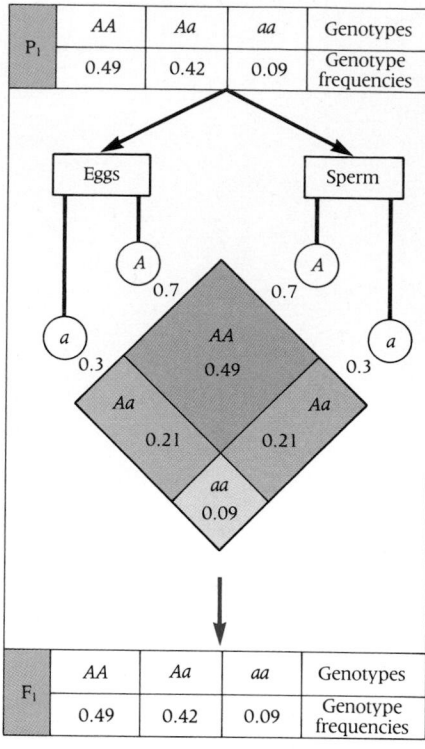

A

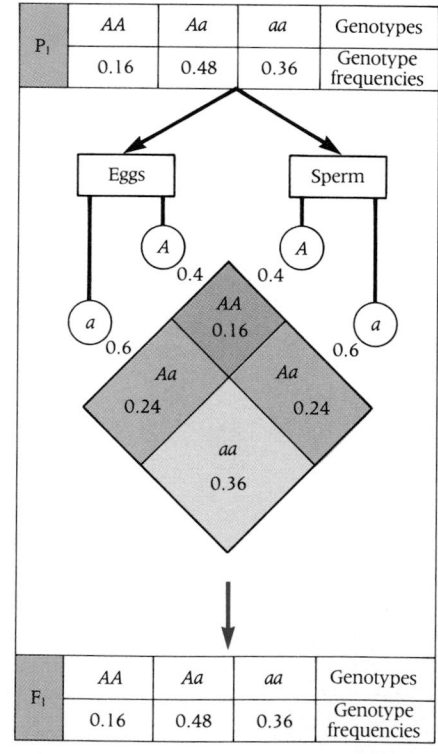

B

34-6 The Hardy-Weinberg law
Punnett squares clearly illustrate the Hardy-Weinberg law. (A) Genotype frequencies are 0.49 (AA), 0.42 (Aa), and 0.09 (aa). Gene (allele) frequencies are 0.7 (A) and 0.3 (a). Note that the genotype frequencies remain the same in the F_1 generation. (B) The same principle applies to different genotype frequencies in the parental population.

ing, migration, mutation, sampling errors, and accidents—the process of sexual reproduction does not by itself change the overall composition of the gene pool. On the contrary, it preserves whatever variability is present in the population. Under these conditions—we will state them more fully later—allele and genotype frequencies remain constant from generation to generation.

The Hardy-Weinberg law offers a useful approach to population genetics and its evolutionary implications. The formula accounts for the striking fact that a population as a whole, with its many variations and genotypes, may continue without change for many generations. Variation, which is necessary for evolutionary change, is preserved even when evolutionary change is not occurring. The importance of Hardy-Weinberg in the history of biology was considerable. In the nineteenth century, when most biologists accepted some sort of blending inheritance (see Box 13-1), it was hard to understand why variability was not simply halved in each generation and why rare characters were not diluted away to the vanishing point. This was a major problem for Darwin. Mendel resolved it with the concept of particulate inheritance and Hardy-Weinberg showed how variability is preserved.

Essential Conditions for Hardy-Weinberg Equilibrium

The Hardy-Weinberg law explains the stability of populations and species over a number of generations. Nonetheless, evolutionary change does occur! Therefore, Hardy-Weinberg is obviously not followed indefinitely. The reason is that conditions arise in populations that oppose the maintenance of genetic equilibrium. In fact, the Hardy-Weinberg law is strictly applicable only under the following conditions:

- No mutations: mutations either must not occur or must have reached a state of equilibrium
- No genetic drift: changes in allele frequencies due to chance or accident must be insignificant
- Isolation: gene migration either must not occur or, if it occurs, must reflect a precisely balanced interchange between populations
- Random reproduction: all aspects of reproduction—mating, fertility, and so on—must result from completely random forces
- No selection: no selective forces favor a particular genotype

We must now admit that in real life these rigid conditions rarely if ever exist. This means that, in reality, the Hardy-Weinberg equilibrium is a null hypothesis— that is, an hypothesis with implications that cannot be tested by reference to actual experience. Hardy-Weinberg thus does not describe the actual genetic trends of real populations. Why not?

EVOLUTION: A DEVIATION FROM GENETIC EQUILIBRIUM

Basic Causes of Genetic Change in a Population

The above list of necessary conditions for genetic equilibrium implies that opposite conditions could cause evolutionary change. These would be the following:

- Mutation occurs but without reaching mutational equilibrium.
- Genetic drift, or random changes in gene frequencies, are significant.
- Gene migration, or flow, are not in precise balance.
- Nonrandom reproduction occurs.
- Selection occurs that favors a particular genotype.

These disturbances may be considered the prime forces of evolutionary change. Let us briefly consider in sequence the first three of these forces. The rest of the chapter will then be devoted to nonrandom mating and natural selection—the most important and most complex of these forces.

Mutations and Evolutionary Change

Mutations are the ultimate source of the raw materials for evolution. As earlier chapters pointed out, mutations are of two types.

- Gene mutations are changes in individual genes that modify their function (see Chapter 17). They consist in alterations in the genetic code of DNA. As causes of genetic variation, they are the most important.
- Chromosome mutations are gross changes in the number or structure of an organism's chromosomes. We saw examples of chromosome mutations in the context of human genetics (see Chapter 15).

Table 15-5 and the concepts of haploidy (n), diploidy ($2n$) and polyploidy should be reviewed at this point. Mechanisms altering chromosome structure—rearrangements (inversions, translocations, and deletions) and duplications—would also be usefully reviewed.

Changes like these can have major phenotypic consequences, although one wonders why inversions or translocations should have such effects. The same genes are present: only their positions are changed. Sometimes, no difference in phenotype is apparent. Still, when chromosomes in a parent have undergone inversion or translocation, the pairing of the chromosomes in meiosis may result in a decrease in fertility or other peculiarities in inheritance. Sometimes the action of a gene is modified simply because it has new neighbors. In *Drosophila*, the size of the eye may vary (Figure 34-7) and is influenced by the identities of the genes adjacent to the gene specific for eye size. This position effect is part of the evidence that genes interact (see Chapter 14).

Thus, both gene mutations and chromosome mutations can generate variants in a population. If these mechanisms affect the phenotype, they may influence the population's evolutionary destiny.

Although mutation is necessary for continued evolutionary change, the fate of given mutations within a population depends largely on external factors. In exceptional cases—significant mainly in reference to some kinds of polyploidy, especially in plants (see Chapter 36)—mutation may produce a variant individual distinct from its parents that is capable forthwith of expanding reproductively into a new, genetically different population. In most cases, however, the individual mutant breeds back into the population in which it arose. The spread of the mutation in that population, if it occurs, is a gradual process over the course of generations. The actual outcome rests on any factors that might impede or promote that spread; the most important factor is natural selection. Contrary to the views of some early geneticists, mutation alone cannot determine a sustained direction of evolution and in itself rarely determines the nature of any evolutionary change.

Indeed, new mutations are not always necessary for change to occur. All natural populations have tremendous stores of genetic variation at any given time. Moreover, the potential number of possible allelic combinations almost always vastly exceeds the number of individuals in the population. It is this variation that is the material on which natural selection can act. These variations derive from earlier, ancestral mutations. Until exhausted, they can permit extensive evolutionary change.

Random Drift: Chance Alterations in Gene Frequencies

Random drift, also termed **genetic drift,** is defined as the change of allele frequencies attributable to chance alone. It is most likely to be important in very small populations.

Consider first what happens in large populations. When Mendel crossed two pea plants heterozygous for the flower color gene (*Cc*), 224 of 929 progeny in the F_2 were *cc* (white) homozygotes. This came to 24.1% instead of the theoretical 25% (see Chapter 13, Figure 13-5). We saw that such departures from "ideal" genetic ratios occur in experiments because of **sampling errors.** The gametes that produce the progeny are only a small sample of the total gamete population. This is as true for the progeny of individual parents. One hundred sperm taken at random from a population of males in which the gene frequency is 0.5 *A* : 0.5 *a* will contain approximately 50 sperm bearing the *a* allele. The actual number may be 53 or 48—or some other number around 50. Sampling error also accounts for chance departures from ideal or expected ratios that may be considerable in small samples and less serious in larger samples.

In the reproduction of large populations, sampling errors are usually negligible, and initial allelic ratios are accurately represented in the large number of gametes produced by each new generation. But in small populations—say, 50 individuals

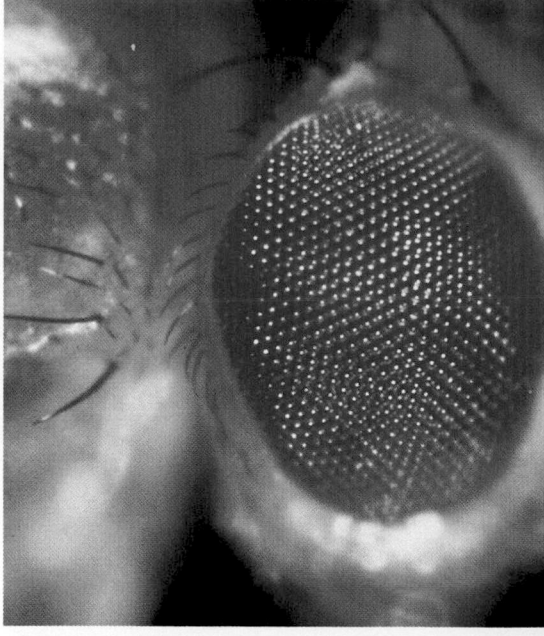

A

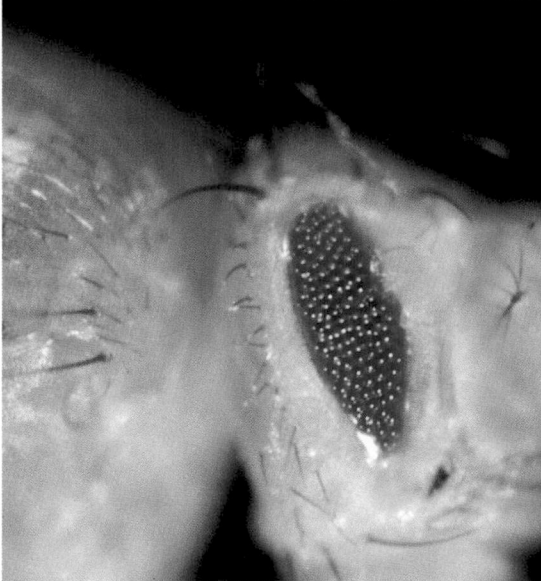

B

34-7 The position effect in Drosophila
A change in the position of certain genes results in a reduction in eye size, a condition known as Bar *eye.* **(A)** *Normal eye.* **(B)** Bar *eye.*

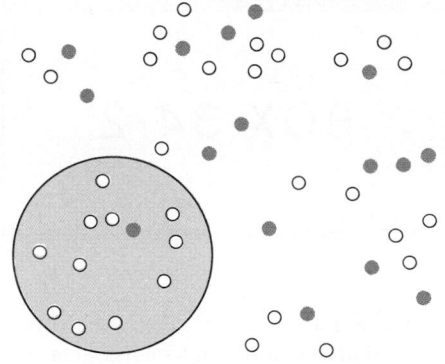

34-8 The principle of genetic drift
Individuals in a population are represented by small circles; those carrying a rare allele are in color. The large colored circle represents 25% of the population that breeds in a given year. If by chance, only one individual carrying the rare allele is in this breeding population, then the allele will be rarer in the next generation.

or fewer—gamete samples are small and sampling errors may accumulate (Figure 34-8) and the equilibrium of a gene pool can change (evolve) wholly by chance. This is termed **genetic** (or **random**) **drift** because when changes are due to chance, a genetic equilibrium is as likely to shift one way (for example, toward loss of allele *A*) as the other (toward loss of allele *a*). Genetic drift may influence gene and chromosome frequencies under certain conditions.

Because genetic drift can lead to changes in allelic frequencies without the influence of selection, it has been called **neutral selection.** In a small population, an allele can be lost simply by chance owing to genetic drift. It cannot reappear except by mutation, and once lost, the genetic variability of the population is reduced. When populations of endangered species become very small, they are often subjected to genetic drift and consequently lose their genetic variability.

Random drift often operates according to what has been termed the **founder principle,** a special case in which a very small number of parents strongly influences future gene frequencies of the population. When a species first colonizes a new area—say, an island—the pioneering individuals are usually few in number. These founders, the forebears of a whole new population, usually carry only a sampling of the original gene pool. Rarely do they include the whole genetic repertory of the species. New populations of this kind usually have a high incidence of homozygosity. Its members are also less likely to differ among themselves than are the members of the parental population.

Under these circumstances, the new population may start off with striking differences from the parental population as a whole. For example, if a large human population contained blue-eyed and brown-eyed members in a certain proportion, but the small group of founders of a new overseas colony were an exclusively blue-eyed sample, the new population, by chance alone, might be exclusively blue-eyed. A famous example of this phenomenon is discussed in Box 34-2.

This effect is most apparent on islands and in other secluded regions, which are invaded by small groups of nonrepresentative founders. Geography then isolates their descendants. It also occurs in temporary bodies of water—as, for example, when a river floods its banks and then recedes, leaving small ponds in which a few fishes are trapped.

In the continuous expansion of any species, even without island isolation, the marginal populations invading new areas are quite likely to differ genetically from the populations near the center of the region occupied by the species. Often, as in the case of migrants leaving a crowded area where food is scarce, the founders of a new population may have alleles that make them more vigorous than the average member of the parental population.

Gene Flow

In the wild, individuals often migrate between natural demes. A storm may disperse seeds far beyond the bounds of the local population. When new individuals move into an area we call it **immigration.**

When individuals immigrate successfully and survive to breed in their new home, new genes are added to the deme's gene pool. Such transport of new genes by an immigrant is termed **gene flow,** or **gene migration.** Often the immigration rate is so low—perhaps only one new individual per million per generation—that it is the equivalent of a mutation for the population, insofar as it adds some new genetic material without significantly altering allele frequencies. Sometimes the immigration rate is high—say, one immigrant per hundred in the host population. In such cases, gene flow can be a major source of new genetic variability. Clearly, gene flow is a prime evolutionary force.

In judging the possible quantitative importance of gene flow, a key factor is the degree of genetic difference between immigrants and the population at large. If the immigrants and indigenous population are genetically identical, no evolution occurs. But if the immigrants and indigenous population are genetically different, a few immigrants can have a major influence on the population—especially if the introduced gene proves adaptively superior in the new environment.

We see, then, that gene flow has consequences in common with those of both mutation and sexual recombination. If it introduces new genetic variation into a population, gene flow resembles mutation. If immigrant and host populations are closely related—for example, if they represent adjacent demes of the same species—then gene flow may resemble sexual recombination within a single deme.

The extreme case of gene flow occurs in **hybridization**, involving interbreeding between members of populations diverse enough to be considered different species.

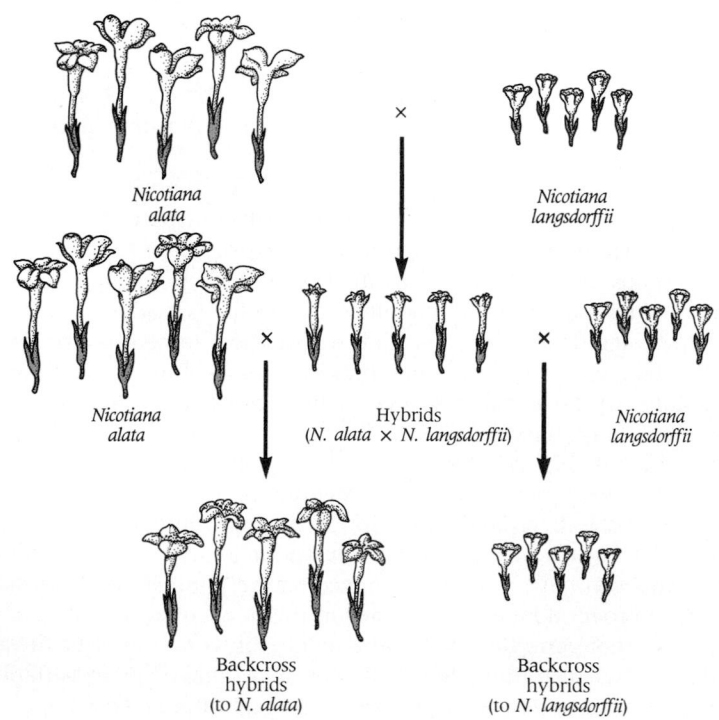

34-9 Introgressive hybridization
Interbreeding between two species of tobacco (Nicotiana alata *and* Nicotiana langsdorffii) *produces hybrids with intermediate characteristics. When the hybrids are backcrossed to each parent species, the offspring are more variable, and they create a bridge that enables genes to flow between species.*

Survey	Total population	Actual number of cases found	Expected number of cases by Hutterites norms	Expectancy ratio*
North Swedish area	8,651	107	94	1.14
Ethnic Hutterites	8,542	199	199	1.00
Bornholm Island	45,694	481	773	0.62
Baltimore Eastern Health district	55,129	507	822	0.62
Williamson County, Tenn.	24,804	156	271	0.58
West Swedish Island of Abo	8,735	94	186	0.51
Bavarian Villages, Rosenheim area	3,203	21	49	0.43
Thuringian villages	37,546	200	617	0.32

*Ratio of cases found to cases expected if the actual incidence had been determined by the incidence among Hutterites.

34B Incidence of mental illness among the Hutterites
The frequency of diagnosed cases of psychosis among Ethnic Hutterites is high, relative to that of six out of seven control groups.

of the colonies studied. Equally interesting were the differences observed between various colonies of Hutterites—even though all had descended from closely related founders.

- Blood group A gene varied from 32% in one colony to 52% in another.
- Among the Rh genes, R^1 (*Cde*) varied from 27 to 68% in different colonies. Its

incidence in England is 40.8%.

- Another Rh gene, the R^2 (*cDE*) gene, varied from 4 to 32% (14.1% in England).
- A third Rh gene, called *r* (*cde*), varied from 27 to 64% (38.9% in England).
- The frequencies of the M and N alleles (MN blood group) of all but one of the Hutterite colonies differ from those of nearly all known white populations, but

agree closely with those of the Dunkers, another genetically isolated population.

- The frequency of the Kell blood group in Hutterites differs from that of other human populations. That frequency is 0% in Negroes, Mongoloids, and Indonesians; it varies between 2 and 6% in all known white populations. In six colonies of Hutterites, it was 13, 20, 21, 22, 23, and 34%.

Many such studies from genetically isolated populations are now available. One such group is the Old Order Amish of Lancaster, Pennsylvania, a colony that long remained isolated after it was formed by three couples in the early 1770s. These people have a high frequency of an allele which in the homozygous state causes a combination of dwarfism and polydactylism (extra fingers). Since the group was founded, some 61 cases of this rare deformity have been reported—as many as have occurred in the rest of the world's peoples combined. Approximately 13% of all the Amish—some 17,000 people today—are believed to carry this mutant gene.

If the hybrid offspring are able to breed with one or both of the parent populations, the process is referred to as **introgression,** or **introgressive hybridization** (Figure 34-9). We shall later see that hybridization is both common and agriculturally important in the development of plant species.

NATURAL SELECTION

The course of evolution has two central features: it produces diversity among living things and it enhances their adaptation. A population's adaptedness may be considered from three different viewpoints.

- Its ability to survive and reproduce in its immediate environment
- Its ability to change as its environment changes
- Its ability to change as it moves into other environments

The evolutionary mechanisms mentioned so far are largely random forces that have no direct or causal link with adaptation. They are as likely to impair adaptation as enhance it. Indeed, under ordinary natural conditions, random changes are more likely to be nonadaptive than adaptive. That is a point of some importance.
In evolution, forces are nonrandom with respect to adaptation.

- **Nonrandom reproduction** is the mechanism that determines the relative frequency of genotypes among homozygotes and heterozygotes. It is important because it determines the amount of variation between genotypes in a population on which selection can act.
- **Natural selection** is the process in which two separate biological phenomena—differential survival and differential reproduction—cause population allele frequencies to change from generation to generation.

NONRANDOM REPRODUCTION AND NATURAL SELECTION

The Example of Hemophilia

The effects of nonrandom reproduction and differential survival on the frequency of mutant alleles in a population are well illustrated by the case of hemophilia (see Chapters 15 and 22). The disorder is an expression of a mutant allele we shall call h. Blood in subjects with the allele H clots normally.

The mutation $H \rightarrow h$ occurs about once in every 50,000 gametes, a fairly high mutation rate. If the mutational processes at the H gene locus were to reach an equilibrium, hemophilia (h) ought to be quite common. In fact, it is not common—about one h allele occurring for every 10,000 H alleles in human populations. Hemophilia is much rarer than it would be if mutation was the only relevant factor.

The causes of this discrepancy are easily found. One is that severe hemophiliacs may die while still too young to have reproduced (differential survival). In former days, even when they survived to sexual maturity, they were less likely to be chosen as marriage partners (nonrandom reproduction) or, if married, were less likely to have many children than people with normal blood-clotting mechanisms (differential reproduction).

Let us imagine a population (unlike any real population) in which H and h alleles have reached a mutational equilibrium at 0.3 H : 0.7 h. The mutant allele (h) is more common than the normal one (H) because in our example the forward-mutation rate exceeds the back-mutation rate. Thus in this population, hemophilia is common. A Punnett square tells us what zygote genotypes would appear in this population (Figure 34-10)—but assumes that all zygotes are equally competent reproducers.

If the values 0.3 H and 0.7 h represent the frequencies of alleles in the population's pool of gametes (Figure 34-10A), we should expect (based on the Hardy-Weinberg calculation) that zygotes would appear generation after generation with the following frequencies:

$$HH \ (0.09) : Hh \ (0.42) : hh \ (0.49)$$

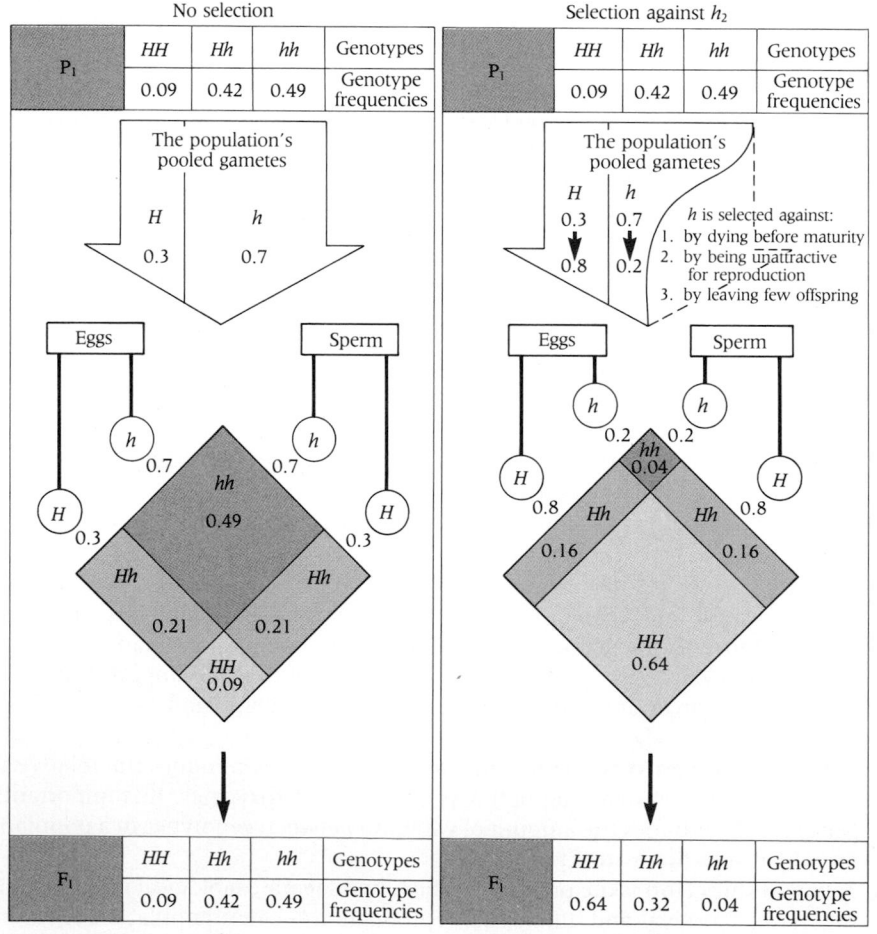

34-10 Effect of selection on genotype frequencies
Hypothetical alleles, H and h, are present in a population. (A) In the absence of selection, genotype frequencies in the parental and F_1 generations remain constant. (B) If h is selected against (for any of the listed reasons), gene frequencies change from 0.3 to 0.8 for H and 0.7 to 0.2 for h. This changes the genotype frequencies in the next generation and hh becomes rarer.

But the assumption of random or equal reproductive ability with respect to the two alleles in this hypothetical population is unjustified for the reasons just given. Hemophiliacs are more likely to die before sexual maturity than are nonhemophiliacs. They are less likely to reproduce. If they do, they are likely to have small families. Therefore, the sample of gametes that actually initiates the next generation contains a smaller proportion of h alleles than the initial value of 70% (Figure 34-10B). Zygote frequencies in the next generations change comparably and hemophilia becomes rarer.

In sum, if the gametes that make up a new generation are elaborated at random and are a proportionate sample of the parental population, the allelic ratio $H:h$ would not change. But, because a gamete containing an h allele is less likely to be passed on to the next generation, the sample is biased (nonrandom) and not representative (random). Therefore, the proportion of h alleles in the population would tend to decrease. Indeed, h might eventually be eliminated—but only after many, many generations and only if it were not being continually replenished by mutation. The actual allelic ratio is determined by the relationship between mutation rate, nonrandom reproduction, and natural selection.

Elements in Nonrandom Reproduction

If reproduction is truly random, each of three phases must occur randomly; that is, the whole process must "ignore" genotype.

- A female must be as likely to mate with one male as with another.
- Any two gametes must be as likely to produce a viable zygote as any two others, regardless of genotypes.
- Every individual zygote must have the same chance of developing into a sexually mature and reproducing organism as any other.

In reality, reproduction is almost never truly random—especially in plants, which are more likely to be fertilized by their neighbors than by individuals of faraway populations. Moreover, many plant species normally engage in self-fertilization; though, as we shall see in Chapter 41, most plant species have evolved mechanisms that tend to prevent self-pollination and favor cross-pollination.

The word "nonrandom," in the sense we use it, means "affected by genotype." When reproduction is affected by genotype, it is nonrandom and clearly tends to produce nonrandom changes in gene pools. Nonrandom or directional change *is* evolution.

Each of the three phases above affects the others. Let us look at them.

Nonrandom Mating

We can best explain nonrandom mating with an example. In an experimental population of *Drosophila* with a mutant allele causing white eyes instead of the usual red eyes, we observe immediate evolutionary change. Evidently, white-eyed males have less sex appeal than red-eyed males—and they are unsuccessful in courting both white- and red-eyed females. As a result, the white-eye allele disappears from the population in a few generations—even when white-eyed males outnumber the females initially. The relative attractiveness of red- and white-eyed males to females is easily measured. Hence, the rate of elimination of the white-eyed allele can be accurately predicted (Figure 34-11).

Mating customs within a species and patterns of courtship powerfully push reproduction in nonrandom directions. The result is an important evolutionary force. Selection will work against mutations causing deviations from the ordinary pattern of courtship and so these will have little chance of entering the next generation. Accepted patterns of courtship, on the other hand, act as a conservative force, eliminating deviants (like white-eyed flies). They are also an impetus to evolutionary change. In fishes, birds, and other animals, brightly colored or showy body parts may stimulate the opposite sex and become the inherited "password" for copulation. When, as in some fishes and birds, the courtship password is a color spot or a showy pattern, larger spots and showier patterns evolve. New mutations that make

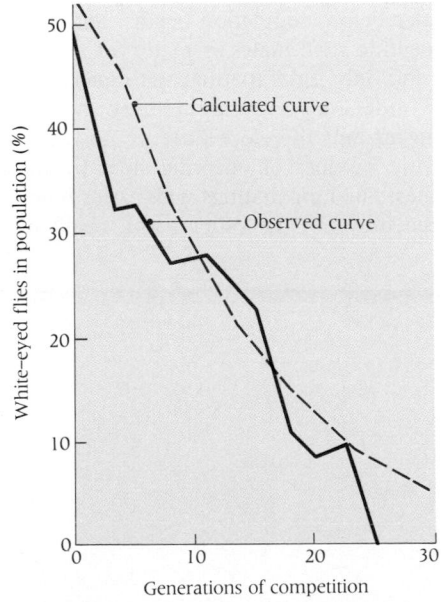

34-11 Natural selection due to nonrandom mating
Drosophila *females prefer red-eyed males to white-eyed males. The rate at which the white-eye gene is naturally selected from the population can be predicted (dashed line). When this prediction was tested experimentally for 30 generations, the observed loss of white-eyed flies from the population closely matched the calculated values.*

In *The Descent of Man and Selection in Relation to Sex*, Darwin conjectured how traits that occur in only one sex might have evolved—for example, the peacock's tail and the deer's antlers. He proposed that such traits resulted from one of two selective processes that favor mating success: (1) selection for weapons that can be used in combat between males competing for females (Figure 34C); and (2) selection for adornments with which males can induce females to choose them (Figure 34D). Darwin termed this process sexual selection, as opposed to natural selection. He noted that adornments are most vivid in males of those species—for example, birds of paradise—in which the females choose mates from a strutting group of displaying males. The importance of competition among males as a force in the evolution of male traits has never been questioned, but the importance of female choice in sexual selection has been very much debated.

Most often the existence of female choice can be demonstrated only indirectly—for example, by studies of the distribution of females around males of different endowments. However, recent work by Randy Thornhill has provided direct evidence on the question. While a graduate student at University of Michigan, he found on walking through the woods that females of the black-tipped hangingfly *Hylobittacus apicalis* do indeed choose their mates even though all males and females look very much alike. They make their choice on the basis of a gift presented during courtship: a prey insect (Figure 34E). Not all blacktipped hangingfly females choose among males in this way. Those that do, however, choose the males with the largest prey as mates. Since such choices increase a female's chances of survival and the number of eggs she lays, this pattern is one of the few known examples of adaptive female choice.

The mating sequence begins when a male either catches a prey insect or steals one from another male. Holding the prey with its rear legs, it catches onto a twig with its front legs and begins to eat. Not all the insects caught by males are offered as nuptial gifts. Indeed, studies of 42 marked males in Michigan forests revealed that they are highly discriminating about the prey they use for this purpose. Of the 345 insects caught by marked males, 110 (32%) were discarded after the males had fed on them briefly. All discarded prey measured less than 16 mm^2 in surface area, which suggests they were too small to be suitable offerings.

When a male has captured a prey insect that is large enough to be a gift and has fed on it briefly, he begins making short flights in search of females while holding on to the prey with his strong rear legs. The cues that initially attract a female to the vicinity are olfactory, not visual. At the end of a "prenuptial flight," the male hangs from a leaf or a twig, everts a pair of glandular sacs from his abdomen and begins releasing a pheromone, a message-bearing chemical that attracts females. Mating does not begin until a female is hanging by her forelegs facing the male. The male then presents his nuptial gift to the female and, while retaining his own hold on it, allows her to grip it with her hind legs. As the female begins to feed, the male attempts to mate with her. If the male's gift is sufficiently large, the female accepts him. The mating process is terminated by the male, and a struggle ensues as the male tries to disengage from the female and to pull the prey from her grasp. If, on the other hand, the male's gift is unpalatable (as it is when it is a ladybird beetle, which contains distasteful substances) or too small, the female will either refuse to mate with the male or, if mating is initiated, will terminate it quickly.

In sum, the female black-tipped hangingfly discriminates against males with small prey on two levels: (1) by rejecting such males before copulation begins, and (2) by accepting such males as mates for a limited period only. Brief matings with males holding undersize prey result in little or no sperm transfer and therefore little or no increase in the number of offspring sired by those males. The long matings with males holding large prey, on the other hand, result in a

---

the courtship password more readily perceived or otherwise more effective are more likely to succeed than their old alleles.

In our examples, there was sexual acceptance or refusal of one animal by another—what Darwin called sexual selection (Box 34-3). Darwin distinguished it from natural selection. We now regard it as only a special case of natural selection.

Nonrandom Fecundity or Differential Reproduction

Fecundity refers to the actual number of offspring produced—or more precisely, the number of viable zygotes produced. "Viable" in this context means capable of developing normally. We must distinguish two factors here: (1) how many gametes (potential zygotes) are produced; and (2) what proportion of them actually unite into viable zygotes?

Production of many offspring plainly influences natural selection. If individuals with allele *A* regularly produce 10 offspring for every one produced by those with allele *a*, the proportion of *A* in the population will tend to rise. It might appear

34C Weaponry as a means of sexual selection
Males competing for females may use horns as weapons. Successful males tend to have successful offspring.

34D Adornments as a means of sexual selection
Males may induce females to mate by parading colored plumage, like that of this bird of paradise.

34E Blacktipped hanging fly Hylobittacus apicalis presenting prey to prospective mate
The bigger the gift, the better the male's chances.

maximum insemination of the female, with obvious consequences.

Darwin was aware that his theory posed a difficulty that has not been fully resolved: the traits favored by females often appear to work against the survival of males by exposing them to predators. As we have noted, the peacock's male adornment attracts females but it also attracts predators. In other words, there is a trade off. Why should females choose to mate with those males whose endowments expose them to the greatest risks and who are therefore least likely to be favored by natural selection? Recent work has confirmed a model suggested in 1930 by R. A. Fisher, which shows that selection can favor female mating preferences for characters that actually lower male viability, even to the point of driving a population to extinction.

Fisher's model had a surprising outcome—male characters that impair survival could be selected for and maintained in a population simply as a consequence of female choice. He considered species that display heritable variation of a male character, say tail size, and that have females who choose mates on the basis of tail size. If ordinary natural selection led to an increase (or decrease) in male tail size, then male tail size would change over the generations. However, so would female preferences. The point here is that males with longer tails would tend to be mated by females who prefer long-tailed males. The male offspring of those females will not only inherit their fathers' long tails and be favored by natural selection but will also inherit their mothers' preference genes and pass them on to their daughters, thus causing an increase in the

frequency of those preference genes.

Eventually, natural and sexual selection would be expected to come into an equilibrium, but while there is a net advantage to further tail development, there will also be a net advantage in favor of making it more preferable sexually. Fisher termed this situation **runaway selection.** In the absence of severe counterselection, he argued, the rate of tail development should increase exponentially with time—to be checked only when males had tails so long they would be unlikely to survive and females preferring them would be left without mates. The idea that sexual selection can lead to decreased male viability and hence to extinction is an important one for evolutionary biology. Interestingly, a 1989 report described sexual selection in pheasants, in which males survived well—an exception perhaps.

that high-fecundity species would always be favored. There is, however, a balance between the fecundity factor and the survival factor.

If an offspring's chances of survival are low because of small egg size, then selection may favor high fecundity—as it does, for example, among fishes that produce millions of tiny eggs and suffer high mortality among the young. If, on the other hand, an individual's chances of survival are high because of large egg size, selection may not favor fecundity. As often happens in evolution, there are alternative solutions to the same problem: high fecundity, low survival is one; low fecundity, high survival is another. We return to this concept in Chapter 45 where we consider its ecological implications and introduce the terms *K* **strategists** (which have low fecundity but survive by superior competitive ability) and *r* **strategists** (which have high fecundity but manage to propagate through dispersal and rapid development despite low survival).

Differences in fecundity means that different numbers of eggs are produced from species to species. Indeed, these numbers vary enormously. A major selective factor is the need for zygotes to be fully viable. In cases of hybridization, gametes may come together but the resulting zygote fails to develop or does so abnormally. Or, the effect may be deferred to later generations. A hybrid like a mule develops into a vigorous adult, but it is infertile. Hence, its genetic characters are selected against. Many mutant alleles may do no harm in heterozygotes but may prevent normal development in homozygotes. Since a homozygote does not survive to reproduce, its alleles are selected against.

Survival to Reproduce

Most organisms must grow and undergo sexual maturation before they can enter into mating and fecundity. The nonrandomness with which they survive to reproductive age accounts for many adaptations. For example, the speed and strength of lions not only helps to keep them alive, but also preserves their ability to contribute to the next generation's gene pool. What counts in natural selection is the leaving of offspring.

Survival beyond the reproductive age is rare in nature. What happens to an organism after it has exhausted its ability to transmit genes to a new generation matters little to natural selection. In some social groups (including insects and human beings) nonreproducers may help to raise the next generation and so promote survival, but that is exceptional. The many ailments that plague older humans suggest that natural selection in our ancestors placed no value on postreproductive survival. Modern medicine is more and more devoted to combating the unpleasant but natural consequence of the way humans originated.

Environmental Pressures

An individual's ability to reach sexual maturity often depends on how competent it is to withstand the rigors of the physical and biotic environments. Hence, they too are determinants of natural selection.

Consider an example. Many tropical ponds and rivers undergo extreme seasonal fluctuations in their content of dissolved oxygen. Oxygen levels can even fall to zero when stagnant waters are warmed and decaying organic matter is abundant. Fishes with a capacity to breathe air can withstand such environmental rigors by staying in these waters and breathing air. A famous example is the walking catfish (*Clarias*), which can survive and reproduce in the harshest environments. In its natural environment in Africa—it has been accidentally introduced into Florida— these catfishes can survive when an entire riverbed has dried out by covering themselves with mud and switching from water breathing with their gills to air breathing with their lunglike organs (Figure 34-12).

The environment also includes powerful biological elements. The universal scramble to eat and not be eaten imposes selective pressures on the members of any community. An example of special interest concerns the relation between herbivores and plants, which is discussed in more detail in Chapter 47. Many plants have evolved protective devices, such as thorns or toxic chemicals, which can be considered defenses. However, herbivores often overcome these defenses with mechanisms called counterdefenses. It is generally assumed that an evolutionary "arms race" between herbivore and plant takes place, with each species evolving new ways of foiling the other.

For example, the vines of *Passiflora* produce toxic compounds that repel insects. The caterpillars of the butterfly *Heliconius* evolved counterdefenses that allow them to feed on *Passiflora* without ill effects. But some species of Passiflora evolved specific defenses against *Heliconius,* such as certain plant structures that mimic *Heliconius* eggs. As a result of their presence, butterflies are deterred from laying eggs on these vines because they appear to be occupied. One species, *Passiflora adenopoda,* has evolved hooked hairs that immobilize the *Heliconius* caterpillars, which then starve to death. These evolutionary "contests" between defenses of the prey and counterdefenses of predators illustrate the power of biotic environmental pressures.

NATURE OF NATURAL SELECTION

Selection and Adaptation

That natural selection is the directive factor in evolution is now generally accepted. Since natural selection rests upon a complex web of phenomena, no one definition

A

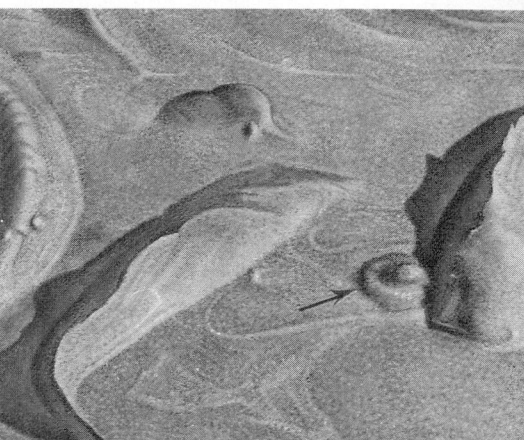

B

34-12 Catfishes, **Clarias macrocephalus**
These catfishes are in a dried out river bed in South Africa. (**A**) *They survive by breathing air through the mouth (see arrow).* (**B**) *They exhale spent air from underneath the gill cover (see arrow).*

of selection satisfies everyone. A definition acceptable to many is as follows: **natural selection** is the consistent differential survival and reproduction of two or more classes of entities. But which entities? Let us explore this definition further.

Any change in gene frequency requires that one allele be replicated more than another. Selection occurs when individuals of one genotype survive and reproduce better *on the average* than those of a different genotype. But if an average superiority is to be expressed, there must exist at least several copies of each gene—or several individuals of each genotype. In other words, there must exist classes of genes or genotypes. Thus, selection requires the presence of multiple copies of each gene and the hereditary transmission of that gene. The stronger its degree of heredity (that is, the more frequently it is represented in the next generation), the more effective and long lasting will be the results of selection.

Selection of whole genotypes in a sexually reproducing species occurs only rarely because whole genotypes are not precisely reproduced. Similarly, selection of whole chromosomes barely exists because a "superior" chromosome lasts only a few generations before it is broken up by crossing-over. Selection is most effective and has its most enduring effects when it acts on single genes, which are faithfully replicated and strongly inherited.

For this reason, the gene is undoubtedly the major unit of selection. Other biological entities—chromosomes, genotypes, populations, anything that is replicated and retains its identity from one generation to the next—also function as units of selection, albeit less efficiently. At the level of populations, **interdemic,** or **group, selection** can occur when populations of one kind emerge or die out at a different rate from populations of another kind. **Species selection** occurs when the species is the unit of selection. In this case, a given species with a given feature more rapidly leads to other species (or less often becomes extinct) than species with another feature.

Fitness is an important aspect of natural selection. Fitness is a consequence of the relationship between an organism's phenotype and the environment in which it lives. Thus, the same genotype could have different degrees of fitness in different environments. In other words, fitness of a genotype influences the contribution of that genotype to the genetic constitution of succeeding generations relative to the contribution of other genotypes. The more fit an individual, the greater its genetic contribution to subsequent generations. We say more about this in Chapter 35.

Clearly, an individual can influence the frequencies of the alleles it carries in future populations. It can do this in two ways.

- By producing its own offspring—that is, by **individual selection**
- By promoting the survival of relatives with the same alleles because they have descended from a common ancestor—a mechanism called **kin selection**

Combinations of individual and kin selection establish the inclusive fitness of an individual (see Chapter 33). In general, species that are solitary or that reproduce in pairs tend toward individual selection, while highly social species, such as primates and social insects, give prominence to kin selection.

Changing Concepts of Natural Selection

To many of Darwin's contemporaries, natural selection seemed a brutal struggle for survival in a carnage, in Tennyson's line, of "Nature, red in tooth and claw." Such phrases as "struggle for existence" and "survival of the fittest" were used by Darwin, but in a metaphorical sense.

From such concepts there developed a doctrine called social Darwinism, which interestingly was not supported by Darwin himself. As we mentioned in Chapter 33, natural selection was supposed to warrant as "right" all kinds of cutthroat competition, including wars between classes and nations, on the ground that in this way only the "fittest" would survive. The belief was unwarranted for several reasons. Natural selection is not an ethical principle that indicates what is right in human behavior. It is, like the law of gravitation, a fact about nature that is neither good nor bad in itself. As lucidly put forward by Darwin himself, the "struggle for existence" may go on in nature, but that does not mean it is a good thing. It may be natural for populations to have a high infant mortality, or for human beings or animals to be infested with parasites, but that does not mean we should happily embrace the situation.

The modern concept of natural selection, set forth earlier in this chapter, developed from Darwin's thought, yet it differs in some essential ways from nineteenth-century ideas on the subject.

Animals do sometimes fight and drive the weaker to the wall. Plants do compete for space, water, and sunlight. However, the competition has no bearing on populations and their genetic, evolutionary changes unless it leads to differential reproduction. That, and not the winning or losing of a struggle by individuals, is the point of natural selection. Moreover, the competitive aspects of nature are not the only ones that result in nonrandom breeding. An animal that gets along best with its neighbors may be the one that has the most offspring. In that case, selection by nonrandom breeding would favor absence of competition. Surely this accounts for the success of well-integrated plant and animal communities and animal (including human) social organizations that have arisen under the directive influence of natural selection.

Creative Selection

Some nineteenth-century critics argued that the effects of natural selection could not be creative. They conceded that natural selection could account for the elimination of the unfit, but denied that it could explain the origin of the fit, a more important problem.

It is easy now to see that natural selection is indeed creative. For one thing, the selective elimination of an allele from a population does not occur unless there is an alternative allele that, under existing conditions, is superior in promoting reproduction. Elimination of an ''unfit'' allele and retention of a ''fit'' allele are really two sides of the same coin. We cannot have one without the other. In a famous example to be discussed in Chapter 35, selection for darker color literally created a population of moths better fitted to survive and reproduce in the dark pine woods.

Natural selection is creative in a second way that is more complex, subtle, and important. An organism's many traits are, in reality, not determined separately and independently by individual genes. Usually each gene affects many traits and each trait is affected by many genes. Genes also interact, so that a given allele may have different effects depending on neighboring alleles. As noted above, natural selection acts not only on each allele but on the genetic system as a whole. It tends to produce gene associations and integrated genetic systems that would have little or no chance of arising and spreading through populations by any random process. This is a truly creative feature of natural selection.

Forms of Selection

Natural selection produces evolutionary change by acting on the variability within a population that is under genetic control. Evolutionists have described three forms of selection (Figure 34-13).

- **Stabilizing selection** occurs when those phenotypes in the middle of a range of phenotypes are favored. An early student of natural selection found after a storm that killed many birds that mortality had been highest among the largest and smallest birds and lowest among those of average size. Stabilizing selection works against individuals with traits at the extreme ends of the distribution of polygenic traits, slowing their spread in populations and thereby maintaining a steadily high ratio of ''better'' alleles at a given locus. This type of selection generally favors reproduction of the well-adapted individuals near the average, at the expense of deviants at the extremes. In a more subtle way, stabilizing selection seems to favor systems that tend to produce normal organisms despite varying environmental influences during development. This is how natural selection acts on most populations most of the time.
- **Directional selection** occurs when one phenotype—large size, for example— is favored over another phenotype, say small size. Its result is to shift the mean size in the direction of largeness in subsequent generations. Obviously, this

34-13 Three forms of selection

The effect of different types of selection on a polygenic trait such as weight. (A) The population before selection has a normal weight distribution. (B) The individuals eliminated by selection appear in color. Stabilizing selection removes very heavy or very light individuals, leading to a population with less weight variation. Directional selection eliminates lighter individuals, producing a population with a higher mean weight (dashed line) than in (A). Disruptive selection removes medium-weight individuals, producing two populations with different median weights (dashed lines).

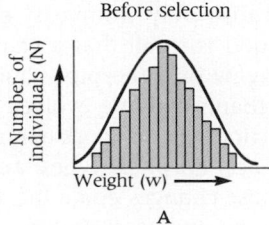

Before selection

A

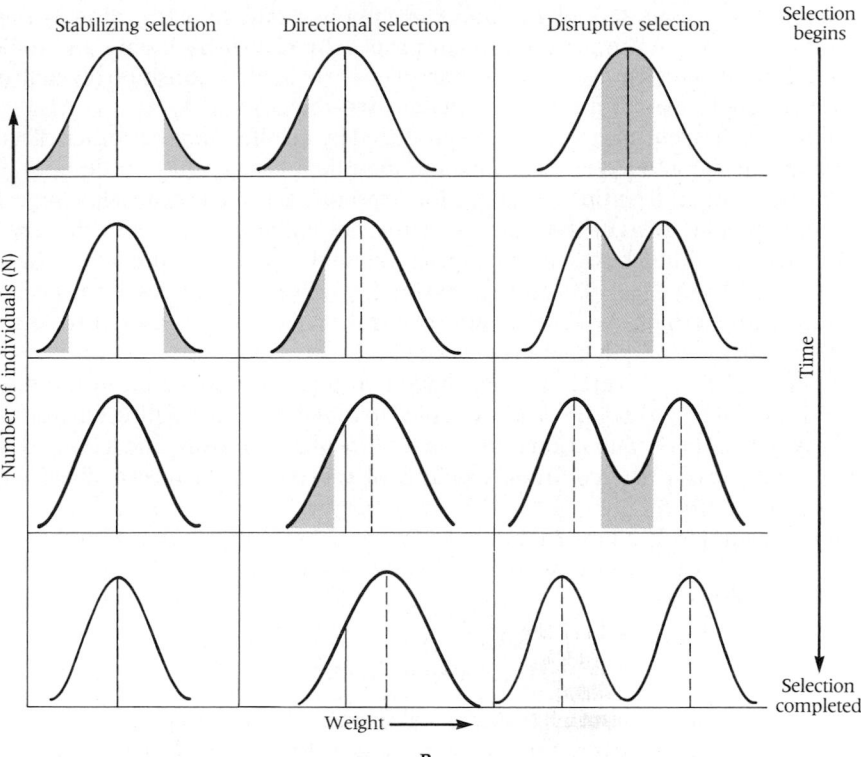

B

can continue only as long as underlying genotypic differences remain in the population for a given trait. Eventually, when the population becomes genotypically homogeneous with respect to that trait, selection can no longer alter the population mean for that trait.

- **Disruptive selection,** seemingly less common than the other two types of selection, occurs when both the extremes of a phenotype's distribution range are favored over the middle—or indeed over any intermediate type. Disruptive selection, we believe, determined the sizes and shapes of the bills of the famous ground finches of the Galápagos Islands (see Chapter 36). During dry spells, the only food types available were far more accessible to finches with extreme bill sizes and shapes than to birds with bills nearer the center of the size-shape range. By splitting the population into two or more types, disruptive selection also has a potential for promoting balanced polymorphism, to be described later in the chapter. As we shall find in Chapter 36, it can even divide a population into two or more new species.

Limitations of Selection

Even though natural selection often impels populations toward improved responses to environmental challenges, there are limits to what it can do. For one thing, it can operate only on genetic variations that are actually present in a population. Over a limited span of time, selection within a population can produce only slight modifications of what was already there. Mutations with really large effects are almost always so poorly integrated with the rest of the evolving organism that they yield "monstrosities" that are promptly eliminated by stabilizing selection. A dog cannot mutate into a viable cat. Of course, drastic changes do result from natural selection, but we believe that these occur only by the slow accumulation of lesser changes over many generations in the course of millions of years.

Possible future changes can arise only from past mutations, which are often nonadaptive. Surely, the likelihood is small that a chance mutation will improve something as intricate as an enzyme or a receptor. Quite the contrary, and this is undoubtedly why elaborate mechanisms have evolved for repairing altered DNA.

In other words, natural selection has to work with what is available. Also—and this deserves emphasis—it acts on phenotypes, and these can be many steps removed from the genotypes whose changes guide the course of evolution. Natural selection is a tinkerer rather than an engineer. As a result, we find many examples of "jerry-rigged" contraptions in nature. A famous example is the panda's thumb (see Chapter 35). Another is the human propensity to low-back pain and difficult childbirth, both of which can be blamed on a less than flawless transition of quadrupedal (four-footed) locomotion to the bipedal (two-footed) variety.

Adaptive characters may actually be produced by an allele that has other effects—including nonadaptive ones. The outcome may be a compromise in design rather than a crowning perfection. Consider, for example, the male peacock's large and brilliantly colored tail. On the one hand, the tail enhances its possessor's mating success. On the other hand, it jeopardizes his ability to avoid predators. Natural selection strikes some sort of balance between the two selective forces—and peacocks with tail feathers that are too long and colorful or too short and drab have given way to peacocks with tails that are "just right."

The limitations of natural selection have had a remarkable result in the history of life: the great majority of species have failed to remain adapted during environmental changes. Failure of adaptation is the usual cause of **extinction,** and it is interesting to note that most of the countless millions of species that have ever lived did in time become extinct.

Natural selection is a limited and blind process that cannot always produce adequate adaptation. The whole world of life attests nonetheless to its overall success.

POLYMORPHISM

For clarity, we have been simplifying natural selection by referring only to alternative alleles at a given gene locus. But, as we have just noted, the whole genetic system determines what an organism will be—and selection acts on the whole aggregate of genotypes. In so doing, selection does not simply increase the frequency of the "best" or "fittest" allele of each gene.

Variation, we have learned, has adaptive value. A completely homozygous population would have no variation from which natural selection could select. Hence, there could be no evolutionary change. When the inevitable environmental changes occur, such a population is no longer well adapted, and it cannot become so. The most rigorous of all the selective sanctions is then applied: the population becomes extinct. Populations that have survived these rigors have always generated and maintained enough variation to permit adaptive change.

Persistent Variations in a Population

Sometimes variations within a population or species persist over many generations. Such populations display **polymorphism** ("many forms"). By definition two or more distinct morphs, or forms, regularly occur in a polymorphic population.

Natural populations are highly polymorphic and it is easy to cite examples. For instance, within a deme of snails some shells may have different colors or coiling direction (Figure 34-14). Most human populations are polymorphic in their blood types—including the A, B, AB, and O system (Figure 34-15), the Rh system, and numerous other systems. There are two species of snow geese (*Chen caerulescens*), one blue and the other white, which were once believed to be distinct species (Figure 34-16A), and two patterns of the king snake (*Lampropeltis getulus*) of California (Figure 34-16B). In both cases, the two forms are often found in the same litter.

Differences in snail-coiling patterns and human blood groups are examples of phenotypic polymorphisms that arise from genetic polymorphisms in a way that is independent of their selective value. However, other phenotypic polymorphisms, such as coloration, may have selective value. Polymorphism has been observed in virtually every class of organisms from protists to vertebrates.

34-14 Polymorphism in shell colors and coiling direction
These members of a snail population are highly polymorphic, demonstrating variations in color patterns, size, and handedness of coiling direction.

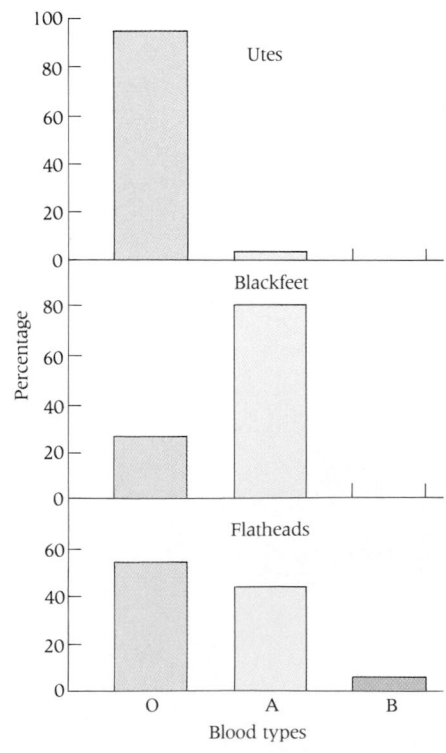

34-15 Polymorphism in ABO blood group alleles
Each of three distinct Native American populations in Montana have strikingly different frequencies for the O, A, and B alleles.

A

B

34-16 Polymorphic variation
(A) *Different color forms of the snow goose,* Chen caerulescens. *(B) Two patterns of the California king snake,* Lampropeltis getulus.

Phenotypic polymorphism is a consequence of the presence in a population of two or more alleles for a given character that express discontinuous phenotypic effects. What genetic mechanisms are responsible for its origin and continuance? That has been a difficult and controversial question in population genetics. The frequent occurrence of polymorphism must mean that mechanisms exist which actively promote it.

■ One proposed explanation of phenotypic polymorphism concerns the actions of allelic genes. We saw in Chapter 13 that the alleles of single genes may determine sharply dissimilar phenotypic effects. Mendel's flower colors and pea seed characteristics are characters of this kind. These are called **discontinuous phenotypes.** They are to be contrasted with **continuous phenotypes** like tall-short, which grade into one another, that are determined by the interactions of many genes, and that in fact represent the extreme ends of bell-shaped curves. Polymorphic characters in many natural demes closely parallel Mendel's flower colors. They are determined by one (or a few) genes, each with two (or a few) alleles that are distributed among the population.[3] Indeed, the view is now widely accepted.

■ A second proposed mechanism concerns the adaptive superiority of heterozygotes. According to the theory of natural selection, if one polymorphic form were more advantageous or adaptive than the other, that one would in time appear in all members of the population. The appearance of the advantageous trait in a heterozygote, a common occurrence, fosters hybrid vigor. We term that situation **heterosis.** However, highly fit heterozygotes cannot persist in a population unless the comparatively unfit corresponding homozygotes also appear regularly. When the most fit form is heterozygous, neither allele can eliminate the other because neither one is best by itself. Depending on the relative fitnesses of heterozygotes and both homozygotes, alleles will in time exist in definite ratios in the gene pool. The result is an optimum proportioning of heterozygotes and homozygotes. By "optimum" we mean "most effective for continued reproduction of the population." The optimum ratio leaves the most offspring (see Box 35-2). Heterosis is extremely important in the improvement of corn and other agricultural plant species.

A classic example of the second situation is offered by the genetic disease, sickle cell anemia (see Box 22-1). This disorder poses a perplexing question: why, if sickle cell anemia is so devastating, is the gene so plentiful in the population? Current views on that question are summarized in Box 34-4.

This steady-state preservation of variation in the population is called **balanced polymorphism.** One of its best-studied instances is seen in the land snail, *Cepaea nemoralis* (Figure 34-17), the shell of which may display any one of six banding patterns ranging from none to five bands. In addition, the snail's shell may be brown, yellow, or pink. The relative numbers of the various forms differ greatly from one place to another, even between localities less than a mile apart. It appears that certain color and banding patterns provide better camouflage defenses against attack. Which pattern is best depends on the situation and the season. Hence, selection has worked to preserve several different forms within the population.

Transitional polymorphism, in contrast with the more common balanced polymorphism, occurs during the period when a selectively advantageous allele is in the process of replacing other alleles. In Box 35-2 we enounter a famous example of this—the case of the melanic moths (*Biston betularia*).

In sum, phenotypic polymorphism is a type of variation that is not distributed continuously and smoothly along a bell-shaped curve. Rather it is a kind of variation in which a number of often strikingly different classes, types, or morphs exist. That number is usually (but not always) small. One seldom cited example is sex. Populations of animals and other sexually reproducing species include males and females. What more sharply differing phenotypic forms could there be?

---

[3] This distinction, like many in genetics, is not an absolute one. There are continuous scales of (1) gene action intensity, and (2) single-to-multiple gene effects on phenotypic characters. Thus, to some extent, polymorphic distribution curves (which always have two or more peaks) and bell-shaped distribution curves (which have one peak) intergrade in real populations. Nonetheless, the two types of distributions remain quite distinct in their more extreme or characteristic forms.

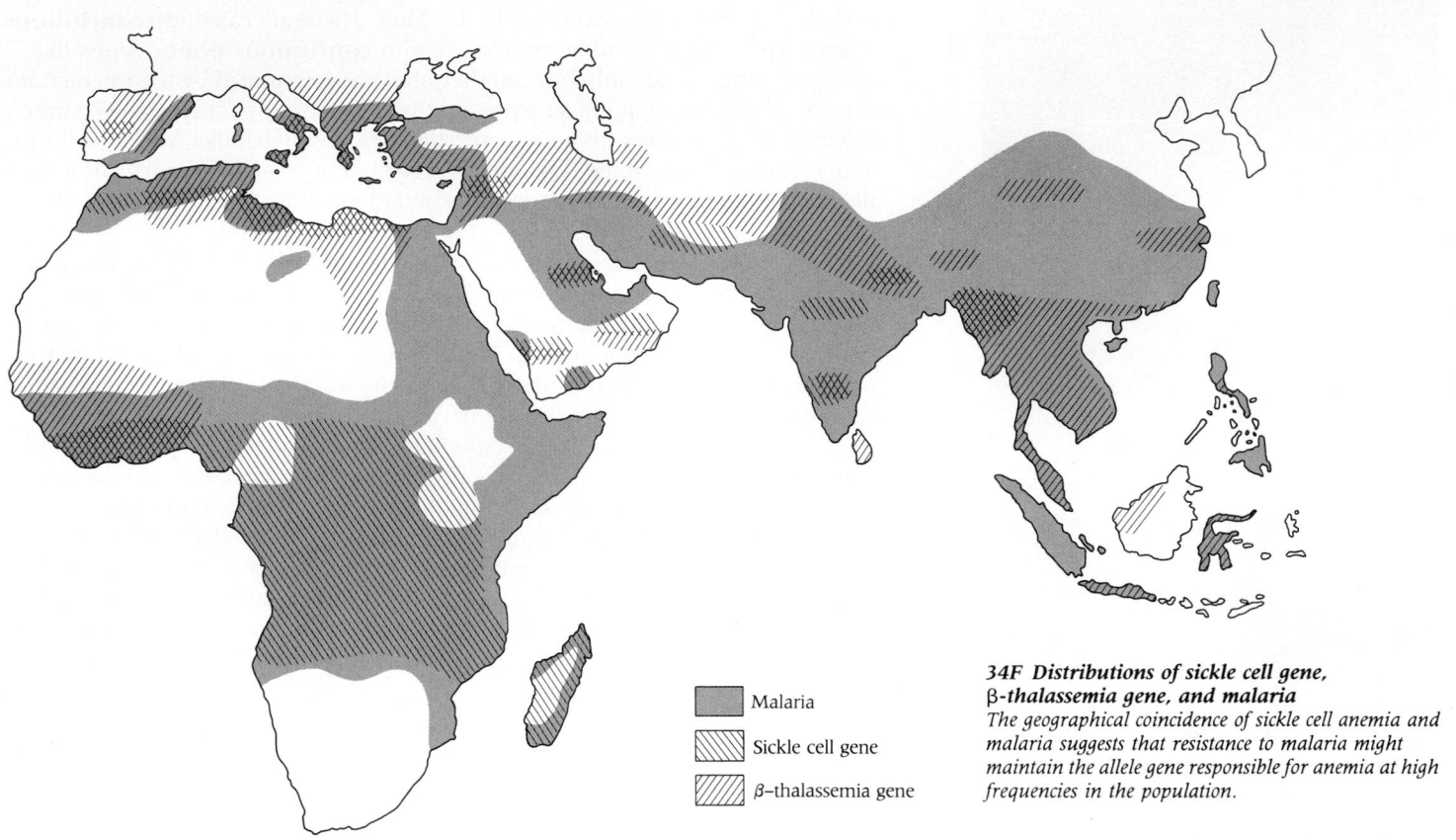

34F Distributions of sickle cell gene, β-thalassemia gene, and malaria
The geographical coincidence of sickle cell anemia and malaria suggests that resistance to malaria might maintain the allele gene responsible for anemia at high frequencies in the population.

Legend:
- Malaria
- Sickle cell gene
- β–thalassemia gene

Approximately 9% of black Americans are heterozygous for the sickle cell hemoglobin gene—that is, their genotype is *SA*—and 0.25% are homozygous, or *SS*. As many as 30% of the individuals in some African tribes have the sickle cell allele. Homozygosity leads to severe anemia and often lethal disorders of many organs owing to the obstruction of capillaries by sickled red cells. The fascinat-

Additional Explanatory Hypotheses

We have just considered two of the mechanisms that may produce polymorphism. Other proposed explanations of the polymorphism of various populations include frequency-dependent selection and disruptive selection—and a few other theories, which we omit.

34-17 Polymorphic forms of the land snail, Cepaea nemoralis
Polymorphism exists for shell coloration (yellow, pink, and brown) and for banding patterns in these snails from England.

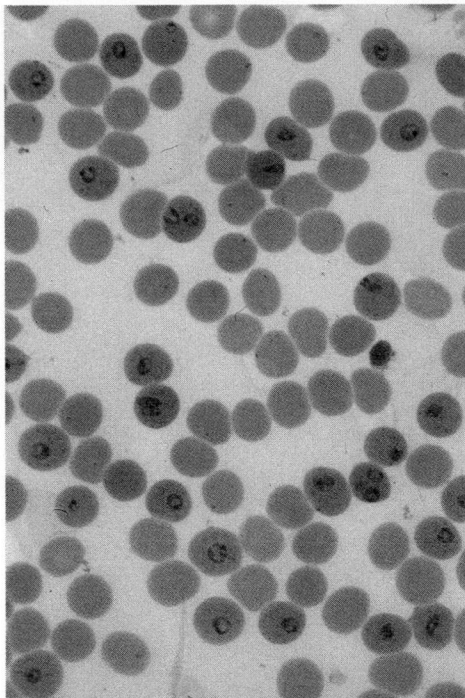

34G Plasmodium falciparum *parasites within red blood cells*
See also Figure 39-21.

ing problem that arises is: why if the sickle cell allele is so devastating is it so plentiful in these populations? The surprising answer is that the allele for sickle cell hemoglobin occurs at a high frequency for a very good reason: it protects against malaria.

The main clue to the explanation was the striking similarity of the pre-1930 geographical distribution of the sickle cell allele and the form of malaria caused by the parasite, *Plasmodium falciparum* (Figure 34F). (The genetically determined blood disease β-thalassemia has a similar distribution.) This coincidence raised the possibility that the sickle cell allele might somehow increase an individual's resistance to malaria. Abundant clinical data soon supported this concept. It was

shown, for example, that heterozygous (*SA*) children had much less severe malaria than did *AA* children, among whom the mortality rate was nearly 50%. Thus *AA* individuals suffer from their susceptibility to malaria and *SS* individuals from the severity of their sickle cell anemia. Any mutation that afforded protection against these hazards must clearly provide a selective advantage.

How does the allele for sickle hemoglobin protect against malaria, a disease that remains one of the major causes of death in large areas of the world? *Plasmodium falciparum*, the intracellular parasite that causes malaria in Africa, lives inside of red blood cells (Figure 34G). It was not possible to grow malaria parasites in the laboratory until 1977, when investigators then found that the parasites grow as well in *SA* red cells as in *AA* or *SS* red cells. However, when *SA* (or *SS*) red cells enter the low oxygen environment of tissues, they sickle. Potassium ions leak from a sickled red cell. As a result, the intracellular parasite dies (Figure 34H). Such a mechanism could confer protection against malaria even if all of the parasites were not killed. Any mitigation of numbers permits the immune system to gear up a more adequate antibody response and this can tip the balance between life and death for the host.

As noted earlier, β-thalassemia also protects against malaria. So do genetically determined deletions of red cell enzymes such as glucose 6-phosphate dehydrogenase. These, it appears, protect by a different mechanism relating to the damaging effects to the red cell's surface membrane of H_2O_2 generated by the parasite. However, the net effect is the same: a loss of red cell K^+, which is fatal for the parasite.

The delicately balanced contest between the selective effect of malaria on the one hand and of sickle cell disease and β-thalassemia on the other provides a classic example of

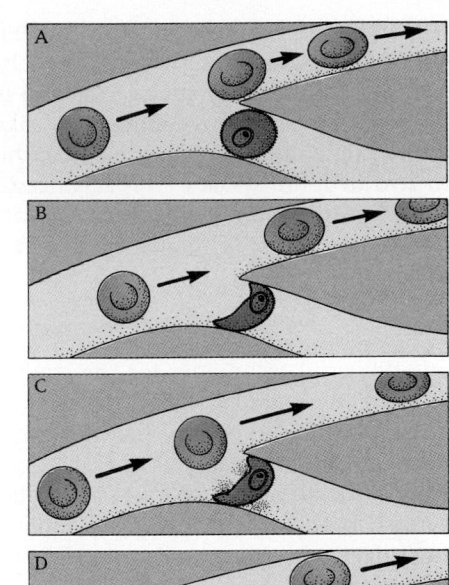

34H How AS *heterozygotes may be protected against malaria*
(A) *A parasite-containing red cell has an abnormal surface and a low intracellular pH.* **(B)** *The parasitic red cell occludes a capillary. Oxygen levels fall locally and the cell sickles.* **(C)** *Sickling causes the cell to leak* K^+. **(D)** *Lowering of* K^+ *within the cell kills the parasite. Meanwhile, the body mounts an immune response.*

balanced polymorphism—a situation in which a heterozygote advantage coupled with a homozygote disadvantage maintains a mutant allele at a low but consistent level in the population. This leads, as it always does when a heterozygote is favored over both homozygotes, to a stable gene-frequency equilibrium. Whenever this happens, all three forms—both homozygotes and heterozygotes—survive in the population. That is why we call it balanced polymorphism.

Frequency-dependent (or **apostatic**) **selection** occurs when the less frequent of two alleles is favored by natural selection. This is common in nature. For example, when a female *Drosophila* is offered a choice of males of two genotypes, she prefers to mate with the rarer male. In contrast, most vertebrate predators seek the most common form of prey and tend to ignore rare phenotypes (Figure 34-18).

At least two circumstances bring about frequency dependence. One is diversity of environments. For example, a predator may attack a disproportionate number of individuals belonging to the more common type and repeatedly shift its preferences. Or, some environments may favor carriers of one allele while other environments favor the competing genotype. Another circumstance involves interactions of carriers of different genotypes. For example, there may be a pattern of mating in which individuals preferentially select others with a different allele. This is called **disassortative mating.** As a result, the rare genotypes can reproduce at a higher rate and increase their abundance until they attain the frequency at which they are no longer scarce enough to gain any advantage.

Disruptive selection, a rarer phenomenon, was described above. As shown in Figure 34-13, polymorphism—or at least bimodality in continuously varying traits—can result if sufficiently strong selection is directed in a sustained manner against intermediate types. The resulting two peaks thus become two subpopulations within a larger population. In extreme cases, the two subpopulations may even split up into two or more species. This has been demonstrated in laboratory experiments but not yet in nature. It could occur in nature if individuals showed strong mating preference for those of similar phenotype. This is called **assortative mating.**

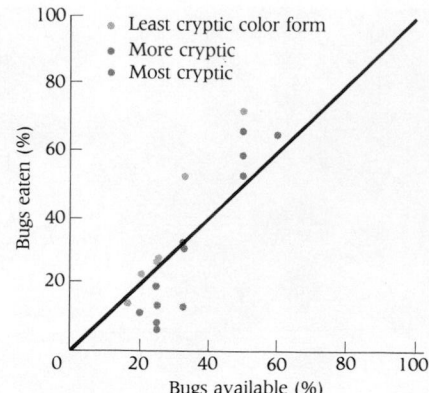

34-18 Frequency-dependent selection in vertebrates
Predation by a fish on three color morphs of the corixid bug, Sigara distincta. *Each morph suffers proportionately higher predation when it is common than when it is rare.*

Polymorphism at the Molecular Level

Many powerful techniques for the study of protein structure have recently come into wide use—among them, high-resolution electrophoresis (in which proteins are separated in an electric field on the basis of charge), isoelectric focusing (an even more sensitive method of resolution by charge), and various immunological methods, including monoclonal antibodies (see Box 23-2). These methods—and others such as DNA and protein sequencing (see Chapter 35)—have revealed a greater degree of genetic polymorphism at the molecular level than geneticists had earlier suspected.

Consider the ubiquitous enzyme alkaline phosphatase, a form of which is found in human placentas. When extracts from several thousand placentas were subjected to electrophoresis and stained to locate the migrating alkaline phosphatase bands, a remarkably large number of variants were found (Figure 34-19). Indeed, there was evidence for as many as 14 different alleles of the gene determining this enzyme's structure.

When similar studies were performed on the serum proteins of human blood, the enzymes of red cells, and various proteins of white cells and platelets, the evidence of polymorphism at a molecular level seemed overwhelming (Table 34-2). Interestingly, polymorphisms among serum proteins have provided so many new genetic markers that they have provided valuable tools for anthropologic and medicolegal studies. That is why when an FBI agent analyzes a blood stain at the scene of a crime and demonstrates a specific combinations of the alleles listed in Table 34-2, the agent has a better chance of identifying the source. The same is true for the curious and incredibly complex system of proteins found in all cells and tissues that derive from the major histocompatibility complex of genes (see Chapter 23) and determine histocompatibility (tissue graft rejection or acceptance)—the so-called HLA antigens (Box 34-5).

This phenomenon of extensive polymorphism at the molecular level was first examined in 1966 when R. C. Lewontin and J. L. Hubby reported interesting findings on the amount and degree of heterozygosity in several natural populations of *Drosophila pseudoobscura*. These workers set out to determine whether molecular polymorphisms were exceptional cases or were signs of a widespread phenomenon. This pioneering study surveyed the electrophoretic patterns of 18 different enzymes (Table 34-3). The results surprised classical geneticists, for they suggested that a single average *Drosophila* population is polymorphic at no less than 30% of all its genetic loci and that these polymorphic loci have so many alleles (2 to 6) at such high frequencies that an average fly in the population is likely to be heterozygous at about 12% of its loci! This revelation of extensive variation was so striking that it raised a sharp challenge to classical thinking. By conservative estimate, 30% of

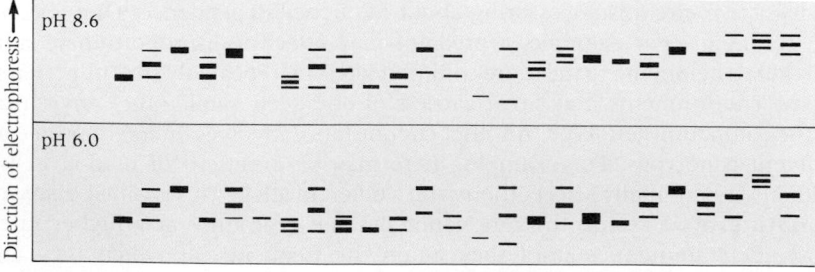

Individual samples of placental tissue

34-19 Polymorphism of alkaline phosphatase
Individual samples of placental tissue were separated by electrophoresis at pH 6.0 and pH 8.6. Each black band in a vertical array represents alkaline phosphate molecules of a distinct molecular weight. The data suggest that many different forms of the enzyme are present in the population.

TABLE 34-2
EXAMPLES OF HUMAN SERUM PROTEIN GENETIC GROUPS

Protein	Genetic Locus	Major Alleles	Method of Detection[*]
IgG1	G1M	a, f, x	3
IgG2	G2M	n	3
IgG3	G3M	b, g	3
Kappa chains	K1M	1	3
Haptoglobin (α chain)	HPA	1F, 1S, 2	1, 2
Gc-Globulin	GC	1F, 1S, 2	1, 2
β-Lipoprotein	AG	x, Y, A1	4
β-Lipoprotein	LP	a, x	4
α$_1$-Antitrypsin	PI	M1, M2, M3, S, Z	1, 2
Ceruloplasmin	CP	B, C, A	1
C3	C3	F, S	1
C6	C6	A, B	2
C8	C8	A, B, A1	2
Factor B	BF	F, S, F1, S1	1
Plasminogen	PLGN	A, B	2

[*] Methods of detection: 1, electrophoresis; 2, isoelectric focusing; 3, inhibition of specific agglutination; 4, immunodiffusion.

the loci in *D. pseudoobscura* means that at least 2000 loci are involved. How can enough selection occur to keep 2000 loci polymorphic?

Attempts at answering this question have generated much argument. One answer may be a selection process that favors heterozygosity per se. Many alleles at many different loci may interact in favorable or unfavorable ways (much as polygenes do) to influence the development of a single character. Instead of summing up thousands of selective processes as though they were independent events, perhaps the object of selection is the whole individual. In each generation (in this formulation), whole individuals are tested, so to speak, and only those above a certain phenotypic threshold survive (are selected) to be the parents of the next generation. If individuals heterozygous for a certain fraction of the loci are generally superior, a relatively small set of selective hurdles in each generation could sustain a large number of polymorphisms.

This process, termed **truncation selection,** occurs in two ways. In one (Figure 34-20A), the phenotypic value (truncation point) is fixed and constant over successive generations. In the other (Figure 34-20B), which is more common, a certain percentage of the population survives in each generation. As a result, the truncation point shifts as the population's frequency distribution moves.[4]

An important alternative hypothesis holds that genetic polymorphisms are maintained because they are in Hardy-Weinberg equilibrium. Such an equilibrium is

[4] Such mechanisms are often encountered in the real world of agriculture where forms (egg-laying chickens, corn plants, etc.) are selected that are more and more productive in successive generations. One must be careful in examining such cases. Part of the observed improvement may be environmental, the result of better fertilizer, insecticides, or whatever.

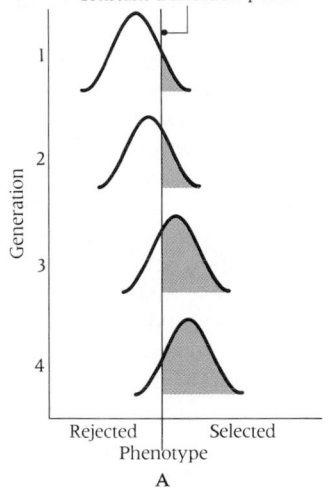

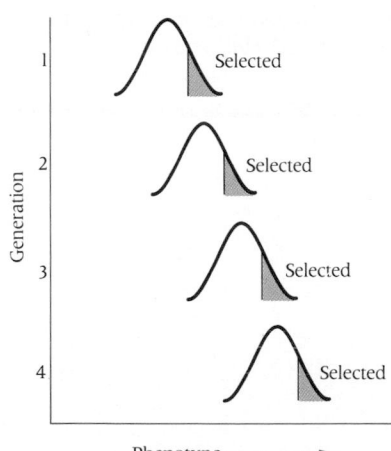

34-20 Two modes of truncation selection
(A) In constant truncation, a phenotypic value is fixed over successive generations. Therefore, fewer and fewer individuals are rejected. *(B)* In proportional truncation, a constant percentage is rejected in each generation. Therefore, the phenotype shifts over many generations.

TABLE 34-3
POLYMORPHISM AND HETEROZYGOSITY AT 18 ENZYME LOCI IN *DROSOPHILA PSEUDOOBSCURA*[*]

Population	Number of Loci Polymorphic	Proportion of Loci Polymorphic	Proportion of Genome Heterozygous Per Individual
Strawberry Canyon	6	0.33	0.148
Wildrose	5	0.28	0.106
Cimarron	5	0.28	0.099
Mather	6	0.33	0.143
Flagstaff	5	0.28	0.081
Average		0.30	0.115

[*] After Lewontin and Hubby, 1966.

We saw in Chapter 23 that the surfaces of human leukocytes carry a system of antigenic glycoproteins that have come to be known as the **human leukocyte antigen** (or **HLA**) system. For at least three reasons, study of this system in recent years has been prolific.

- HLA antigens are related to the transplantability of tissues from one human being to another. This is termed histocompatibility and hence the HLA system has also been called the **major histocompatibility complex,** or **MHC.**
- There is evidence that an individual's pattern of HLA antigens permits certain predictions regarding the diseases he or she is likely to experience.
- The complexity of the HLA system raises the old evolutionary problem—how do we account for such extensive polymorphism?

The HLA antigens are products of four, or possibly five, genes closely linked on a single chromosome—chromosome 6—and thus they are usually inherited as a unit (Figure 34I). The ingenious immunological (or serological) methods used to detect HLA antigens reveal five genetic loci: HLA-A, HLA-B, HLA-C, HLA-D, and HLA-DR. The A, B, and C antigens are expressed on almost all

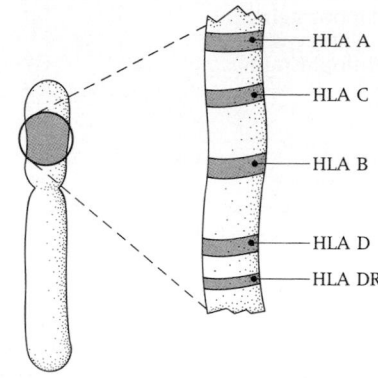

34I Chromosomal location of the major histocompatibility complex
Five loci have been identified on the short arm of chromosome 6 in humans.

human cells; DR antigens are restricted to B lymphocytes, monocytes, and possibly endothelial cells. (DR stands for "D-related," since there is evidence that the D and DR genes may be identical.)

Each locus can be occupied by one of many possible alleles. More than 90 are recognized (Figure 34J). Since 1980, many alleles have been found at each genetic locus. Since a person inherits genes from both parents, the number of possible gene combinations is enormous.

Some alleles occur only in certain ethnic groups. Some combinations of alleles occur more frequently than would be anticipated if HLA alleles segregated randomly. This is called linkage disequilibrium. In some people, only one antigen representing a particular locus can be detected. The usual reason is that the same allele is inherited from both parents. It can also mean that an allele is present for which no testing antiserum exists.

The combination of alleles inherited from each parent is called a **haplotype.** The alleles comprising an individual haplotype can usually be determined by comparing a person's HLA antigens with those of his parents. This is the basis of the highly successful application of the HLA system to investigations of disputed paternity. In one case, the system was sufficiently powerful to show that a pair of fraternal twins was sired by different fathers. In the absence of family studies, however, HLA typing assigns antigens to phenotypes, not to individual haplotypes. It is at this phenotypic level that most associations between HLA types and disease are studied.

The HLA system came into prominence in the late 1960s when researchers found that antigens on lymphocytes evidently determine compatibility between kidney transplant donors and recipients (see Box 23-4). The discovery in mice of an association between leukemia and genes of the mouse major histocompatibility complex led to inten-

---

possible because the variability of the individual alleles is neutral—that is, it does not contribute to the fitness of the organism in which it occurs. (Neutral characters are discussed in Chapter 35.)

- According to the "neutralist school," most variations (especially those of enzymes, which result in profusions of alloenzymes) are alike in the fitness they confer. Hence, they conclude, allele frequencies are determined largely by the rate of nucleotide substitutions in a population, which is determined in turn by the neutral mutation rate, genetic drift, and gene flow.
- According to the "selectionist school," in contrast, the variants do differ in fitness and thus variation is maintained by balancing selection. In this view, regions of the genome that are constrained by selection have a lower neutral mutation rate.

In sum, if the neutralists are right, evolution at the molecular level occurs largely by genetic drift and gene flow—and the observable variation is irrelevant to a

sive searches for association between diseases and antigens of the major histocompatibility complex in humans. The first hint was an apparent association between the skin disease psoriasis and the HLA-B13 antigen. People with that antigen were up to five times more likely to have psoriasis than people without the antigen. Then came reports of associations between HLA-DR3 and celiac disease, and between HLA-B8 and myasthenia gravis.

Figure 34K lists some of the HLA-associated diseases in the order of their relative risks. Their wide variety has spurred the use of HLA typing in the diagnosis of disease. Note that "associated with the disease" does not imply that the disease occurs only in people with the antigen, or that the antigen is found only in people with the disease. Only for relatively high prior probabilities (0.5 or more) does the test tend to support a diagnosis. A negative test, however, makes that diagnosis considerably less likely.

Why and how has this extraordinary degree of polymorphism been achieved and maintained in mammalian evolution? An answer may lie in the fact that the MHC antigens regulate immune responses. For example, killer and helper T cells (see Chapter 23) can recognize foreign antigens only in association with MHC molecules—and certain foreign antigens are more efficiently recognized in association with some MHC alleles than with others. Thus an animal heterozygous for MHC antigens may respond efficiently to a wider range of pathogenic organisms than a homozygote, and polymorphism may be maintained by heterozygous advantage. The polymorphism may be stable, or the MHC molecules may have to evolve rapidly to keep pace with pathogenic invaders that are continually evolving ways to evade recognition, so that advantageous new variants are continually selected.

A deeper understanding of the selective pressures at work on MHC molecules may come from current work in which cultured cells are transformed with cloned MHC genes. Such experiments may at last settle these and other fundamental questions that have baffled immunologists for years.

*The designation w refers to antigens provisionally identified at research workshops.

34J Multiple alleles at five loci for HLA antigens

Some Diseases Associated with HLA Antigens in Caucasians		
Disease	HLA Antigen	Relative Risk*
Narcolepsy	DR2	129
Ankylosing spondylitis	B27	82
Reiter's syndrome	B27	40
Sicca syndrome	DR3	19
Dermatitis herpetiformis	DR3	18
21–Hydroxylase deficiency	Bw47	15
Celiac disease	DR3	13
Subacute thyroiditis	Bw35	12
Acute anterior uveitis	B27	10
Addison's disease	DR3	9
Hemochromatosis	A3	7
	B14	4
Insulin–dependent diabetes	DR3	6
Rheumatoid arthritis	DR4	6
Chronic active hepatitis	DR3	6
Psoriasis vulgaris	B13	5
Graves' disease	DR3	4
Multiple sclerosis	DR2	3
Myasthenia gravis	B8	3

*The estimate of relative risk varies from study to study. The estimates given here are typical. A disease with no HLA association would have a relative risk of 1.

34K Some diseases associated with HLA antigens in Caucasians

population's response to selection. If the selectionists are right, natural selection is the primary force in molecular evolution, and the most highly varied populations would have the greatest evolutionary potential.

SUMMARY

Our survey of life turns now to the biology of groups of organisms, rather than individuals. Different groups are distinct biological units at different levels of integration.

Populations are defined as all the members of a given species occupying a particular area at the same time. Demes are local subunits of populations. Species are series of populations within which interbreeding occurs under natural conditions. A species is a group of organisms so similar in heredity that their demes tend to intergrade and replace each other without changing the features of the group as a whole. Demes, populations, and species vary in their degree of reproductive isolation. Members of different populations may sometimes interbreed, but members of different species do not.

Almost all individuals within a sexually reproducing deme are phenotypically unique. Their variation, of both genetic and environmental origin, generally traces a normal, bell-shaped distribution curve. Different demes often have differing gene pools—that is, different allele frequencies and, sometimes, different alleles—and thus different ranges of variation.

Evolution can be defined as a change in a population's allele frequencies. A population with stable allele frequencies is said to be in genetic equilibrium. Such stability is mandated by the Hardy-Weinberg law, which holds that in the absence of perturbation of allele frequencies by extrinsic factors, sexual reproduction will maintain a population's allele frequencies at status quo. The law also allows predictions of allele frequencies from phenotype frequencies (and vice versa) by use of a binomial equation.

The extrinsic factors, which in nature do perturb Hardy-Weinberg equilibria, and thus cause evolution in a population, include spontaneous mutations, natural selection, gene flow (allele migration) through breeding with other populations, nonrandom reproduction, and genetic drift.

Mutations are the ultimate source of all genetic variation, but they are too infrequent to drive evolution by themselves. Moreover, evolution can occur even if no new mutations occur within a population. A huge reservoir of variation in most populations provides materials upon which selection can act.

Genetic drift can cause evolution in small populations, where chance can produce large deviations from statistically predicted gene frequencies. The founder principle, a special case of genetic drift, applies when an isolated area is colonized by only a few species members whose own alleles may not fairly reflect the parent population's gene pool. Hence, populations descended from founders will exhibit an unusual pattern of allele frequency compared to the parent population. Genetic drift should be distinguished from gene flow, in which alleles added to a gene pool by immigrants to a population do change allele frequencies.

Natural selection is the only evolutionary mechanism that consistently causes adaptive changes in allele frequency. Nonrandom reproduction, one of its major manifestations, affects the gene pool in three ways—through nonrandom mating, nonrandom fecundity, and nonrandom survival to reproduce. Nonrandom mating, or sexual selection, changes the gene pool when some genotypes are more successful than others in attracting mates. Even when a mate is successfully found, different genotypes may exhibit different fecundity as well. Finally, irrespective of fecundity, descendants of particular genotypes may vary in their ability to survive long enough to reproduce. Fecundity and survival are, of couse, strongly influenced by environmental pressures.

Selection operates on any group of biological entities exhibiting variation so long as some of the variation is heritable. The fitness of an allele, defined as the likelihood of finding that allele in the genetic constitution of succeeding generations, influences future allele frequencies in two ways: through individual selection (passing the allele to one's direct descendents), and kin selection (promoting the survival of relatives with the same alleles).

There are three modes of selection. In stabilizing selection, the fittest individuals have phenotypes in the middle of a range. In directional selection, those at one end of the range are fittest. In disruptive selection, those at either end of a range are selected, while those in mid-range survive less well.

Evolutionary change is ultimately limited by the available variation or polymorphism within a species. Hence, mechanisms underlying polymorphism are of interest. They include: the actions of allelic genes in determining dissimilar phenotypes; cases in which heterozygotes are more fit than homozygotes (heterosis); frequency-dependent selection; and disruptive selection. The techniques of molecular biology show that there are many more polymorphic genes than had been suspected. Many polymorphic loci have so many alleles they can be used as genetic markers in legal as well as biological investigations.

KEY TERMS

allele frequency
assortative mating
combinational equilibrium
deme
directional selection
disassortative mating
disruptive selection
distribution range
evolution
founder principle
frequency-dependent selection

frequency distribution curve
gene migration
gene pool
genetic drift
group selection
Hardy-Weinberg law
horizontal gene transfer
hybridization
individual selection
kin selection
macroevolution

microevolution
mutational equilibrium
natural selection
neutral selection
nonrandom reproduction
polymorphism
population
range
species
stabilizing selection
truncation selection

QUIZ QUESTIONS

1. Which of the following is *not* a means by which genetic variation can occur within an individual organism?
   A. genetic drift
   B. recombination
   C. gene mobility
   D. crossing-over
   E. mutation

2. Which of the following normally occurs in most species?
   A. mutation
   B. genetic drift
   C. gene migration
   D. nonrandom reproduction
   E. All of the above

3. Which of the following is the form of selection that operates in most natural populations?
   A. disruptive selection
   B. directional selection
   C. stabilizing selection
   D. kin selection
   E. None of the above

4. Fitness is an important concept in natural selection. Fitness is most properly a property of
   A. a phenotype.
   B. a genotype.
   C. an individual.
   D. a population.
   E. a species.

5. Random breeding within a deme is referred to as
   A. introgression.
   B. hybridization.
   C. panmixia.
   D. heterosis.
   E. fecundity.

ESSAY QUESTIONS

1. What is a population? What constitutes a deme? How does one define a population and a deme in terms of geographic distribution and pools of genetic information?

2. What defines a species? Draw a diagram of intersecting circles showing the relationship among species, demes, and populations.

3. Explain the Hardy-Weinberg Law. What assumptions about microevolutionary processes underlie the Hardy-Wienberg Law? What conditions must be met before the Hardy-Wienberg Law would apply in the strictest sense?

4. How do organisms change over long periods of time? What role does genetic mutation play in evolution? Could evolution occur without genetic mutation?

5. Much has been written about the "survival of the fittest" and "nature red in tooth and claw." What is your view of the meaning of natural selection?

6. Widely varying human cultures all consider incest taboo. In microevolutionary terms, what are the deleterious effects of inbreeding?

REFERENCES AND SUGGESTIONS FOR FURTHER READING

Box, J. F. (1978). *R. A. Fisher: The Life of a Scientist.* John Wiley & Sons, New York.

Sir Ronald Fisher (1890–1962) applied the power of statistical mathematics to many areas of science—and above all to evolutionary biology. This biography by his daughter portrays his genius and complexity.

Bradbury, J. W., and Anderson, M. B. (Eds.) (1987). *Sexual Selection. Testing the Alternatives.* Wiley-Interscience, New York.

Report of a conference on sexual selection, a topic attracting renewed interest in recent years. Emphasis is on unsolved problems.

Clutton-Brock, T. H. (Ed.) (1988). *Reproductive Success. Studies of Individual Variation in Contrasting Breeding Systems.* University of Chicago Press, Chicago.

Interesting essays on current attempts to measure the actual or potential strength of selective forces in natural populations.

Eldredge, N. (1986). *Unfinished Synthesis: Biological Hierarchies and Modern Evolutionary Thought.* Oxford University Press, New York.

A controversial discussion of evolutionary biology that focuses on such philosophical questions as the nature of causation and explanation. Not for the faint of heart.

Endler, J. A. (1986). *Natural Selection in the Wild.* Princeton University Press, Princeton, N.J.

An eloquent defense of natural selection as a potent evolutionary force, based on examples observed and quantitatively studied under field conditions. Lucid and well written.

Futuyma, D. J. (1985). *Evolutionary Biology,* 2nd ed., Sinauer, Sunderland, Mass.

A balanced overview of current evolutionary theory—both at the population and higher taxon levels.

Grant, V. (1985). *The Evolutionary Process: A Critical Review of Evolutionary Theory.* Columbia University Press, New York.

The testament of a biologist steeped in natural history, this nice volume was portrayed by one reviewer as "a gentleman's book, which deals largely with the controversies of yesterday."

Greenwood, P. J., Harvey, P. H., and Slatkin, M. (Eds.). (1985). *Evolution: Essays in Honor of John Maynard Smith.* Cambridge University Press, New York.

A collection of balanced essays spanning all of evolutionary biology—from microevolution to macroevolution, from evolutionary stable strategies to sympatric speciation, and from population processes and patterns to models for the evolution of sex and behavior.

Kitcher, P. (1982). *Abusing Science: The Case Against Creationism.* The MIT Press, Cambridge, Mass.

A distinguished philosopher of science defines the issues at stake in the debate of science with creationism, delivers a mortal blow, and then illustrates his views with specific examples.

Lewontin, R. C. (1974). *The Genetic Basis of Evolutionary Change.* Columbia University Press, New York.

After reviewing the findings of the rapidly expanding investigation of population variation and evolution, the author suggests new directions for future study.

Mayr, E. (1982). *The Growth of Biological Thought: Diversity, Evolution, and Inheritance.* Harvard University Press, Cambridge, Mass.

Among many other things, this monumental work by an eminent evolutionist discusses the differences between the living and nonliving worlds—and notes that physicists do not grasp the uniqueness of living individuals.

Multiple authors. (1978). A special issue on evolution. *Scientific American 239,* September, 46–231.

This superb collection includes authoritative essays by Ernst Mayr, John Maynard Smith, Richard C. Lewontin, and other leaders.

Simpson, G. G. (1949). *The Meaning of Evolution.* Yale University Press, New Haven, Conn.

A remarkable book by one the century's great evolutionists.

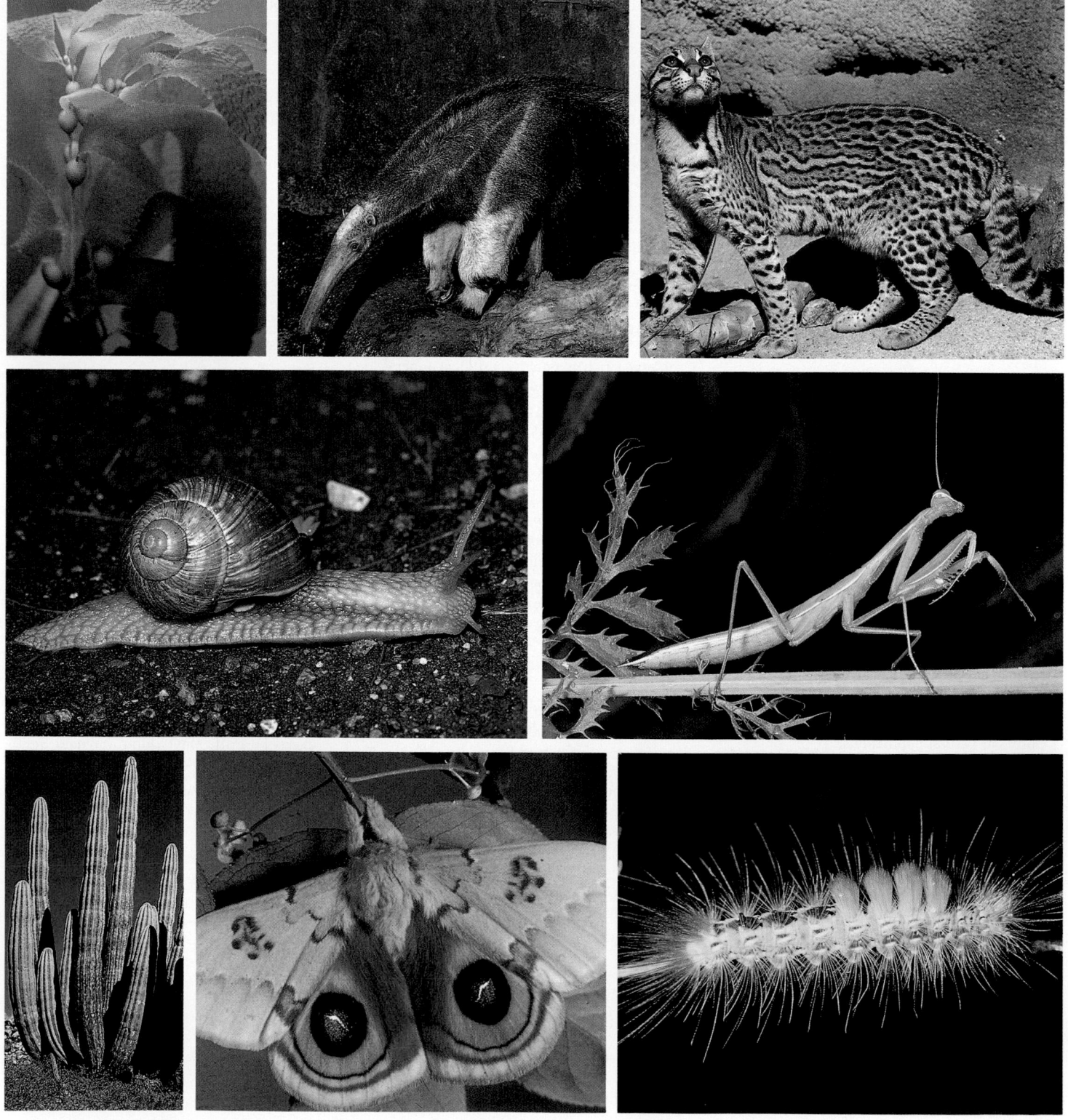

Adaptation

Adaptation, the adjustment of organisms to their environment in the course of evolution, has produced all sorts of structural, functional, and behavioral modifications. As these examples illustrate, they are remarkably appropriate to the lifestyle of each surviving species.

Contemplating the grand sweep of evolution and our own place in nature, Darwin closed *The Origin of Species* (1859) with the words, "There is a grandeur in this view of life . . . ."

In that seminal work, Darwin outlined his theory of evolution by natural selection. It can be summarized as follows:

- There are differences in the traits of individuals of the same species— the principle of variability.
- Many of the differences are heritable—the principle of heritability.
- Some of the differences improve the survival and reproductive success of a species in the environment in which it lives—the principle of adaptation.
- Some differences in adaptedness will result in the production of differing numbers of variant offspring either immediately or in later generations—the principle of natural selection.

Darwin's brilliant contribution was his recognition that **adaptation** is the means whereby organisms enhance their prospects of leaving more descendants in the next generation.

EVOLUTION AND THE PROBLEM OF PURPOSE

What forces underlie evolution? What is its purpose? We approach those issues as scientists, recognizing always that science rests on certain philosophical premises: that our perceptions can be reliably related to objective phenomena, that there is order in the universe, and that we can approach an understanding of this order only by observation and experiment.

We saw in Chapter 2 that there is a touchstone by which we can assess such inquiries. A question is scientific if its possible answers can be tested or verified by observation and thus empirically validated or invalidated. What mechanisms are responsible for evolution? That is a scientific question because the various answers proposed by biologists are testable. What is the meaning of the universe? That is a philosophical question because no objective observations could permit one to choose an answer. The point to be stressed is that though both kinds of questions are useful and provocative, we should try to distinguish between them. To put it another way, scientists speaking as scientists should take care to answer scientific questions scientifically.

The observed data plainly indicate that there is no single direction of evolution. As far as evolution has gone, humans may have been one "end-point"— or, if you prefer, pinnacle. But so is the tapeworm or mosquito or any of millions of other species, which do what they do very well, surviving and reproducing successfully. Certainly, there is a purposeful aspect to the world of life. Indeed, that world is infused with what could be called "purpose." The highest evolutionary purpose that is scientifically discernible is the universal one: survival to reproduce.

That is the long-term, or ultimate, purpose of living organisms. There are also more limited purposes, such as food-getting, escape from enemies, and so on. Most of these could be called short-term, or proximate, purposes.

Adaptations are the seemingly goal-directed features of living things that impress us with the notion that organisms do have purposes, even though we cannot assume for any organisms other than ourselves that these purposes are conscious or that they are predetermined beyond the universal goals of survival and reproduction. It is a major goal of evolutionary biology to answer the question: how do adaptations come about? We have already given the answer that biologists now accept: in general an adaptation is the result of natural selection acting within the whole complex of biological factors—genetic, populational, and ecological.

NATURE OF ADAPTATIONS

One outcome of evolution is that organisms adapt to their surroundings. Like many biological terms, "adaptation" is hard to define precisely. Essentially, **adaptation** is the process whereby members of a population become better suited over the generations to survive and reproduce. *An* adaptation is any aspect of an organism—structure, function, or behavior—that promotes its adaptedness.

When we speak of the adaptedness of an individual, we refer to that organism's success in obtaining food, avoiding enemies, and perpetuating its genes by leaving descendants.

35-1 Morphological adaptation of insect mouth parts
A comparison of the mouth parts of the cockroach, moth, mosquito, plant bug, and housefly shows differences that reflect specific adaptations that enhance the success of each individual in its environment. The cockroach bites and chews. The moth sucks nectar from a flower. The mosquito pierces skin and sucks. The plant bug pierces leaves and fruit. The housefly sucks juices from surfaces.

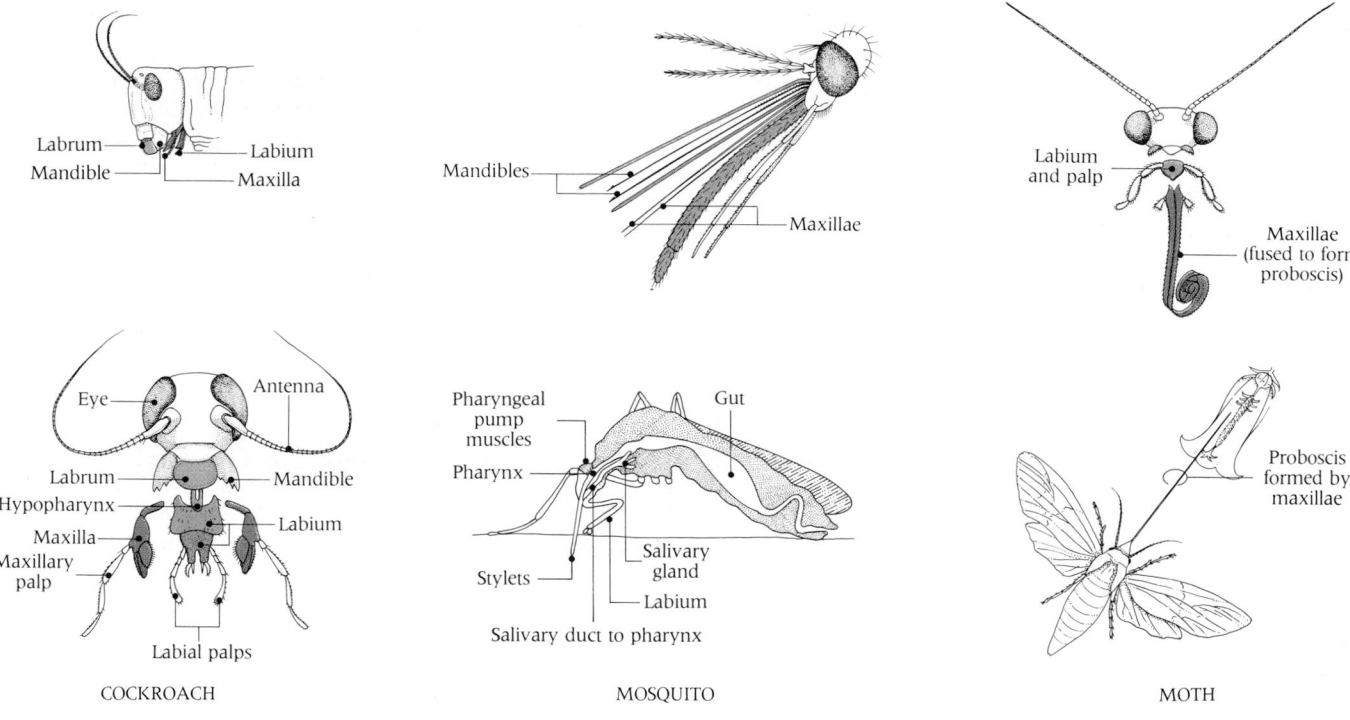

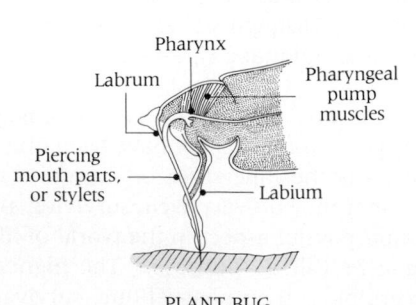

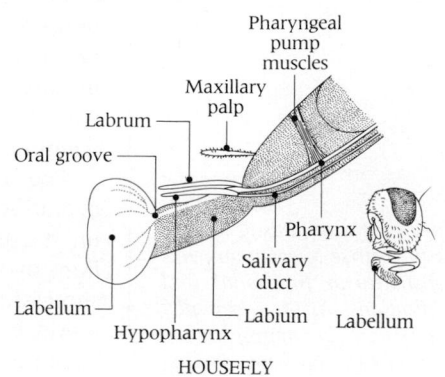

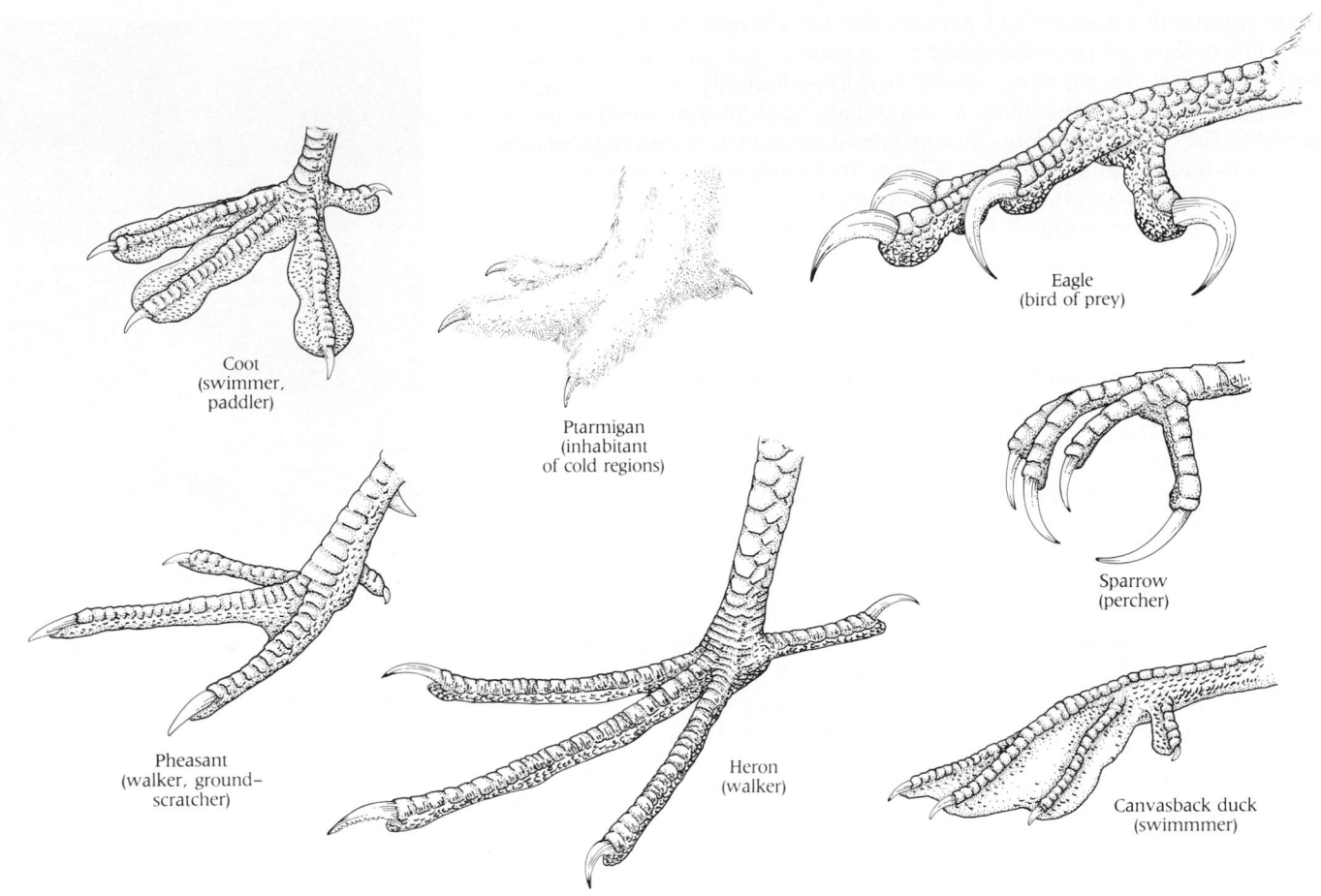

Coot
(swimmer,
paddler)

Ptarmigan
(inhabitant
of cold regions)

Eagle
(bird of prey)

Sparrow
(percher)

Pheasant
(walker, ground–
scratcher)

Heron
(walker)

Canvasback duck
(swimmmer)

35-2 Adaptive specialization of birds' feet
*The role of each bird species is indicated. Note
how the structure of their feet supports their
activities. For example, a heron is a walker,
but it must often walk on muddy terrain.*

**35-3 Asian leaf butterfly, Kallima
imachus**

DIVERSITY OF ADAPTATIONS

Specific adaptations may be structural, physiological, or behavioral, though the
three types often overlap. Let us illustrate the categories with examples and then
discuss them more fully.

- **Structural,** or **anatomical, adaptations** are the most obvious and familiar.
  A striking example is seen in the structural diversity of insect mouth parts,
  which promote efficient feeding (Figure 35-1), or bird feet, which improve walking, wading, perching or grasping (Figure 35-2). Compare the mouthparts of
  the cockroach, mosquito, and butterfly and the feet of the canvasback duck (a
  swimmer), the eagle (a bird of prey), and the other birds in Figure 35-2. All
  enhance the efficient functioning of organisms in their special environments.
  Other structural adaptations are the shapes and colors by which animals conceal
  themselves. A famous example is *Kallima*, an Indian leaf butterfly that looks
  like a leaf and is hard to recognize in foliage (Figure 35-3).
- **Physiological adaptations,** though less obvious perhaps, are omnipresent.
  Physiological adaptations are based upon biochemical processes. An example
  is the presence of digestive enzymes that permit the utilization of specific foods.
  Clothes moths possess special enzymes that digest wool. Other physiological
  adaptations are the various excretory systems that have evolved for handling
  salt and water (see Chapter 25). For example, in salt lakes like Great Salt Lake,
  shrimps have highly specialized mechanisms for regulating their internal osmotic
  pressure. A teleost fish (*Oreochomis alcalicus grahami*), the only fish living in an
  alkaline lake (pH 10) in Kenya, excretes urea instead of ammonia like most
  other freshwater teleost fishes.
- **Behavioral adaptations** are every bit as important as structural and functional
  adaptations. They include maternal behavior, courtship behavior, homing, and
  all the other behaviors discussed in Chapters 32 and 33. The stickleback remains
  a prime example (Figure 35-4). Termites also offer an interesting case. Like
  the shipworm *Teredo*, so injurious to wooden wharfs and ships, a termite can

live on wood, not because it can manufacture the enzymes needed to digest wood, but because its gut is inhabited by flagellated protists that can digest wood. A termite, like any insect, sheds its skin periodically in order to grow. In this process, it sheds the lining of its hindgut—and loses its wood-digesting flagellates. The ensuing problem—loss of digestive enzymes—is solved by a behavioral adaptation. As soon as its skin is shed, the termite eats it and thus reinfects itself with the flagellates it needs. A young termite freshly hatched from the egg case acquires the protists it needs by licking the anus of an adult termite.

BROAD AND NARROW ADAPTATIONS

These are hardly subtle examples. Too often biologists have thought of adaptations as collections of curiosities that permit organisms to succeed in unusual habitats or in unusual ways. That is an error. An organism in fact possesses a wide spectrum of adaptations that equip it to survive and reproduce in its usual environment.

The point is better made perhaps if we focus on the two opposite extremes of a continuum of adaptations: broad adaptations and narrow adaptations. Consider woodpeckers (Figure 35-5). Obvious narrow adaptations to their way of life include their posture, an ability to move up a tree trunk by means of specialized feet, powerful neck muscles that can move the head like a hammer, a large and chisel-like beak, and a long tongue that can probe insects from the crevices it cuts (Figure 35-6). These are the highly specialized ways in which woodpeckers are adapted to the narrowly defined task of extracting insects from tree trunks.

But woodpeckers are also broadly adapted to their way of life. They are birds, and they live an avian life. Both their wings and their bones, which combine lightness with strength, are adaptations to flying that are shared by almost all bird species. Their respiratory system, although specialized to the bird's way of life, is nevertheless shared in its general aspects by all birds, reptiles, mammals, and other forms. It is a fundamental vertebrate adaptation to life on land. Of even broader significance is the way in which the woodpecker's muscles and nerves relate to each other and to the bones to coordinate the movements on which its life depends. Broadest of all adaptations is the woodpecker's capacity to reproduce and thereby to copy and pass on the genetic message it inherited from its parents. This ability, shared by all organisms, is the broadest adaptation of all.

INDIVIDUAL ADAPTABILITY

A nearly universal adaptation is the ability of individual organisms to adjust to adverse or changed circumstances. A paramecium, ordinarily a freshwater creature, will die if it is exposed to a sudden large increase of salt in its water. But if the salt is added gradually it manages to adapt. Humans react similarly when exposed to arsenic and other poisons. Humans and mice raised in lowlands both have difficulty at high altitudes, where oxygen is scarce, but performance improves in time as an adaptive increase in circulating red cells enhances the oxygen-carrying capacity of the blood. There are many such examples. Even the capacity of animals to learn is an example of individual adaptability. In all of these instances, the adaptation occurs within an individual's life span.

We must therefore be careful in using the verb "to adapt." A student of human evolution might say, "Early humans adapted to life in trees by evolving modified forelimbs." That is a different process from the one implied in a sentence like "After adapting to the high altitude, he enjoyed his stay in Peru." The adaptation in the first case was associated with changes in genetic makeup that produced new instructions on limb structure. The second adaptation involved no change in genetic information. The flexibility of the body and its ability to respond to oxygen lack is already specified by heredity. What is inherited is a range of reactions.

Individual adaptability is itself an adaptation of a high order. Were a population to inhabit a rigorously stable environment, inherited information could be simplified by specifying an inflexible body organization appropriate to the enduring conditions. But no environments are completely stable. Biological environments change constantly. The time comes when, to survive, populations must meet the changing conditions where they are or move to new environments.

35-4 Adaptive behavior of stickleback
The male has entered the nest to fertilize the eggs. Compare with Figure 32-12.

35-5 Woodpecker, Melanerpes formicivorus, attacking a tree trunk

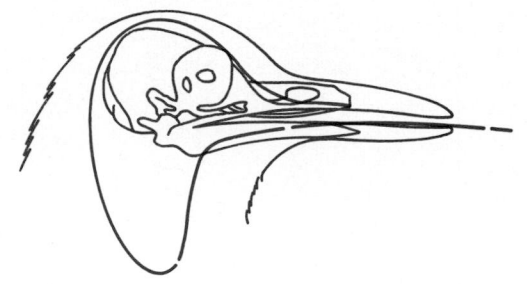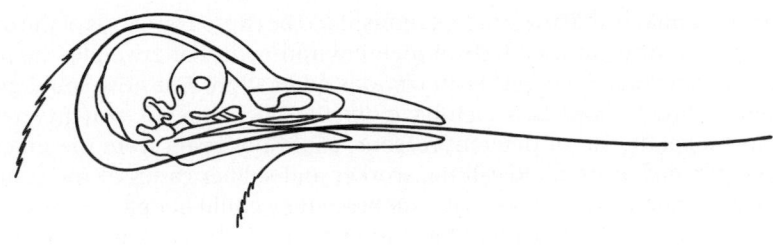

35-6 Adaptive specialization of hyoid cartilage in woodpeckers
In all higher vertebrates the tongue skeleton is the hyoid cartilage. In woodpeckers it is enormously long (color) and is adapted for probing deep crevices in trees. The tip of the tongue is equipped with barbs.

INHERITANCE OF ACQUIRED CHARACTERS

Two main scientific hypotheses have been proposed to account for adaptation. One, natural selection, has so successfully withstood testing across the years that it now ranks as established fact. The other, the inheritance of acquired characters, has not met such tests and has been discarded by most biologists. Still, it is worth discussing because it is a compelling lesson in the history and method of science. Moreover, some contemporary biologists have *not* discarded this hypothesis.

LAMARCKISM OR NEO-LAMARCKISM

According to the venerable **inheritance-of-acquired-characters hypothesis,** the results of individual adaptations are genetically transmitted to an individual's off-spring. Adaptations, in this view, are incorporations of such modifications into the genetic system through the generations. As mentioned earlier, the giraffe's long neck was seen as an evolutionary consequence of stretching as, generation after generation, giraffes reached for the leaves at the tops of trees (Figure 35-7). Another part of the hypothesis was the idea that environmental effects, including the harmful effects of injuries or malnutrition, also become heritable.

It is customary to call this hypothesis Lamarckian or Lamarckism, after the pre-Darwinian evolutionist, Jean Baptiste Lamarck. These terms are often used in discussions of the history of thought. However, both are unfair and inaccurate for the following reasons: the hypothesis of the inheritance of acquired characters was only a minor feature of Lamarck's theory of evolution; Lamarck was not its originator (and never claimed to be); and he himself did not accept the hypothesis in the same form or to the same extent as later "Lamarckians."[1]

When Lamarck published *Philosophie Zoologique* in 1809—and even when Darwin published *The Origin of Species* in 1859—the inheritance of acquired characters was a perfectly reasonable hypothesis in its time. It could be tested, and it was quite necessary that it be tested.

FAILURE OF AN HYPOTHESIS

Intensive testing in the latter nineteenth and early twentieth centuries revealed the weaknesses in the Neo-Lamarckian hypothesis. There were five principal objections.

■ The hypothesis demands that mutations in somatic cells would somehow have to be transferred to germ cells in order to be passed on to the next generation. As August Weissmann pointed out long ago, there is no way for such a transfer to occur.
■ All efforts to demonstrate the inheritance of acquired characters experimentally appear to have failed.

35-7 Giraffe eating the leaves of a tall tree

[1] Along with many other nineteenth century naturalists, Darwin did not accept the hypothesis in the way now labeled "Lamarckian" and misleadingly contrasted with "Darwinian." Hence, the terms Neo-Lamarckism and Neo-Darwinism are somewhat less objectionable in reference to the debate in Darwin's time between believers in inheritance of acquired characters and in natural selection of randomly occurring mutations as explanations of adaptation. These terms are now commonly used. Despite the vigor of the debaters, it was not really an "either-or" question as we shall see. Darwin himself accepted both of what later workers called Neo-Lamarckism and Neo-Darwinism.

- It seems unlikely that many adaptations could be caused by efforts of the organisms themselves or by direct effects of their environment. For example, the protective leafy green color of insects is an obvious adaptation. But how could green leaves influence insects, which have different pigments, to turn green?
- Some adaptations, for different reasons, could not result from the inheritance of acquired characters. Recall the worker and soldier castes of social insects. Since members of these castes do not breed they could not pass on any characters they acquire. Each individual must in fact inherit the capacity to develop its caste characters from parents that did not have those characters.
- If, as Neo-Lamarckism demands, effects of the environment were directly heritable as such, the nonadaptive harmful effects of, say, injury or malnutrition would accumulate along with the adaptive beneficial effects (assuming, of course, that the injured or malnourished managed to reproduce). Yet, though we must all have had ancestors who lost a finger or toe before producing offspring, five digits on each limb is still the normal endowment—as it was 300 million years ago.

More to the point, natural selection has been a sufficient explanation for observed data at each juncture where this hypothesis has failed.

Bacteria remain the last stronghold of Lamarckism. Unlike multicellular organisms, which have germ cells that are distinct from somatic cells, a single bacterial cell is at the same time the equivalent of a somatic cell and gamete in a multicellular organism—and it is fully exposed to the environment. Hence an inhibitory agent that enters it, such as an antibiotic, could conceivably cause mutations that would make the cell antibiotic-resistant, while at the same time selecting for such forms. To be sure, resistance appears in only a small fraction of the population. But the drug could have a directive influence with a low but steady level of effectiveness. In such a case, an external influence—a drug—would be producing the resistant forms that it was being used to detect.

In 1988, a surprising study by John Cairns and co-workers of Harvard University suggested that environmental factors could directly cause mutations in individual bacteria, which are of immediate benefit. For example, when *E. coli* that were unable to utilize lactose as an energy source (because of a chain-terminating codon in the structural gene for β-galactosidase) were placed in a medium containing lactose, they rapidly acquired the needed gene and survived. This effect was shown not to be due to enzyme induction or derepression, or some other mechanism that somehow called forth a pre-existing or cryptic gene. A critic of this important and much-discussed study wrote, "What's up? Can bacteria really direct their mutational processes?" It was his opinion that inheritance of acquired characters in bacteria has not quite been proved by this work and that various hypotheses could explain these results on the basis of familiar mechanisms, though these explanations would be complicated indeed. In any event, this work portends the new experiments needed to clarify underlying mechanisms.

MORE ON THE DIVERSITY OF ADAPTATIONS

Biologists have tended to be so impressed by successful adaptations that they sometimes forget organisms are not perfect. Some of their traits are **neutral,** or **nonadaptive, characters.** Such characters are without any beneficial effects that we can detect. They may even be harmful or **maladaptive.**

NEUTRAL (OR "NONADAPTIVE") CHARACTERS

Sometimes conflicting selection pressures working on a species render it less well adapted than it could be. Consider two examples.

- *Lotus corniculatus* is a plant that has evolved a highly effective defense against herbivores and infectious microorganisms: it produces cyanide, a powerful

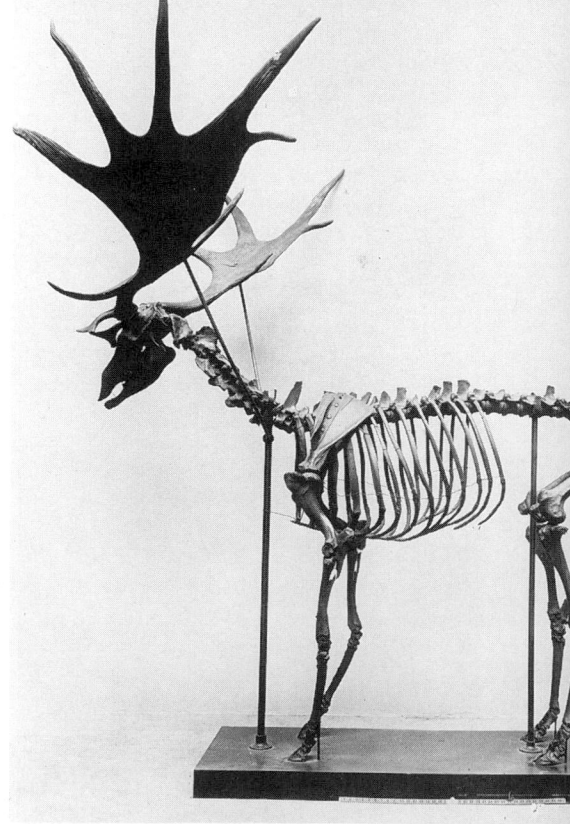

35-8 Disproportionate development of body parts
The Irish elk, Megalocerus, *or giant deer had antlers that were unusually large for its body. This may have led to its extinction.*

A

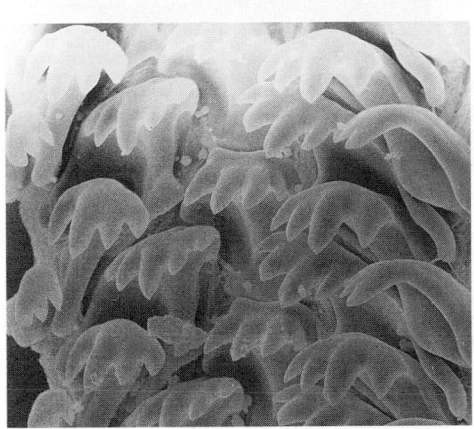

B

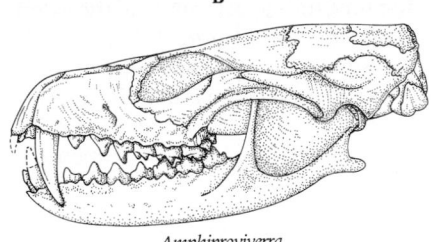

Amphiproviverra
(marsupial)

C

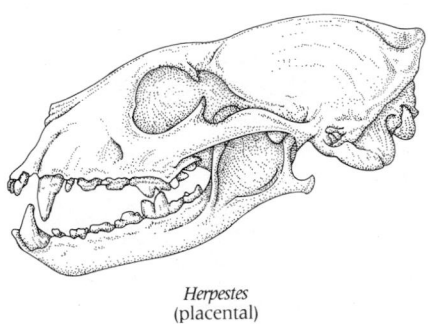

Herpestes
(placental)

D

**35-9 Similar selective pressures elicit
similar adaptations**
*Multiple rows of multicusped long stalked teeth
in a snail (A) and in a fish (B) appear as
adaptations to scrape algae from rock surfaces.
Adaptations for carnivory evolved independently
in marsupials (C) and eutherians (D).*

poison.[2] But in cold regions, the plant's cell membranes are disrupted by frost. As a result, in cold regions the plant poisons itself, thereby eliminating the selective advantage it has in warmer regions.

■ Many insect larvae construct silken threads which serve as parachutes that nicely catch the wind. This permits a highly effective method of dispersal. But because the larvae are unable to direct their movements to appropriate host plants, high mortality inevitably results. Sometimes the advantages of dispersal outweigh the disadvantages of high mortality. Sometimes they do not.

Neutral characters also can arise because two features are linked developmentally and their rates of growth may differ significantly. When increased body size is selected, other features that are not beneficial may be perpetuated. Often, for example, a specific body size is adaptive in itself—small animals can hide in small spaces, large ones are less susceptible to predators—but the disproportionate growth of a particular body part may not be adaptive. A famous case is the Irish elk (*Megalocerus*), now extinct, which was neither Irish nor elk (Figure 35-8). As its body size increased it became the largest species of deer. Yet its growth resulted in the disproportionate development of monstrously large antlers. It is argued that the double burden of carrying such a weight and of having to regenerate so much bone each year may have hastened its extinction, even though the large body size may well have functioned to reduce mortality by lowering the animal's susceptibility to predation.

Neutral features may also arise as a result of **pleiotropy**—that is, a gene having more than one phenotypic effect. Natural selection may operate to increase the frequency of a particular gene because one of the phenotypic effects is adaptive and thus has selective value, while the accompanying pleiotropic or unrelated effect is simply carried along. For example, an enzyme that detoxifies poisonous compounds by converting them into an insoluble pigment would be selected for this adaptive feature. But if as a side effect, the organism's color changes, this pleiotropic side effect would be a neutral character for which there could be no adaptive explanation.

Finally, even though evolutionary changes may be adaptive, they may simply represent alternative solutions to the same problem. The Indian rhinoceros has one horn. The African rhinoceros has two. Though there can be little doubt that horns are adaptations that protect against predators, it is not true that one horn is adaptive in Asia while two horns are in Africa. The difference can be explained by assuming that two different developmental systems existed and that the two systems responded to the same selective forces in slightly different ways.

These examples suggest that natural selection may not lead inevitably to adaptation. Still, the reality of adaptation is demonstrated by the indisputable fact that unrelated groups of organisms do respond to similar selective pressures with similar adaptations (Figure 35-9).

ROLE OF HISTORICAL OPPORTUNISM

What could happen in the course of evolution was always limited by various constraints. For change to occur, figuratively speaking, the environment had to offer an opportunity and a group of organisms had to possess the means of seizing it.

In many instances, nature seized an opportunity by exploiting an "accident," such as the ready availability of some structural elements or molecules for a new assignment. An illustration of this phenomenon at the molecular level appears in Box 10-1, which described how often biochemical adaptations have been based on tetrapyrroles. This chemical configuration in various guises and in various epochs took on many new and diverse functions in evolutionary history: chlorophylls in photosynthesis, cytochromes in electron transport, hemes in oxygen transport, and cobalamin in hydrogen transfer. In each case, the tetrapyrrole nucleus joined with different chemical materials to create a novel biological role.

---

[2] Hydrogen cyanide (HCN) is produced by many so-called cyanogenic plants as a defense against herbivores and infectious microorganisms. (Recall our discussion of cyanide-resistant respiration in Chapter 27.) When the tissues of such plants are experimentally heated or burned, cyanide forms from a class of tissue compounds called cyanogenetic glycosides. Interestingly, tobacco is such a plant, and tobacco smoke contains cyanide in a concentration of 6 parts per million. Almonds are also rich in cyanogenetic glycosides. That is why burnt almonds smell like cyanide, and why cyanide is often said to smell like burnt almonds.

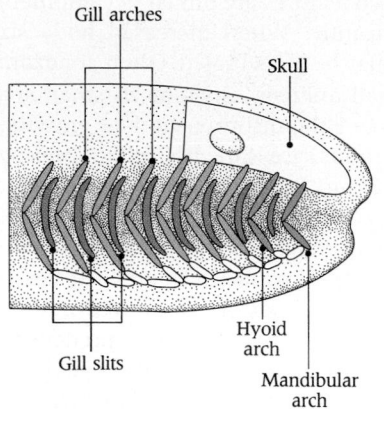

35A Jawless vertebrates
The evolution of the mammalian ear began with an ancestral jawless vertebrate possessing gill slits and gill arches.

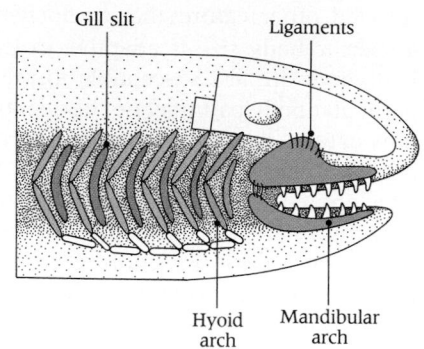

35B Placoderms
In these ancient fishes, the mandibular arch has increased in size and is attached to the skull by ligaments.

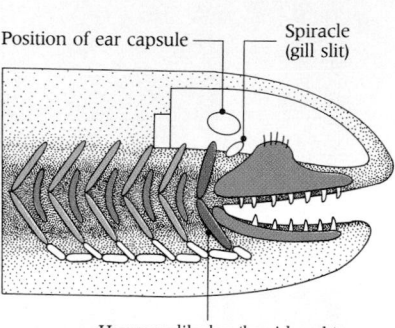

35C Cartilaginous fishes
The hyoid and mandibular arches present in earlier fishes have come together to form the hyomandibular arch.

The evolutionary history of the mammalian ear ossicles can be clearly delineated by comparing representatives of major stages in the evolution of mammals.

- In one of the earliest jawless vertebrates, an ancestral form, the wall of the throat, or pharynx, is perforated by gill slits and supported between the slits by a series of skeletal elements, the gill arches. The first two of these are the mandibular and hyoid arches, respectively. The jaws of all later vertebrates are evolutionary transformations of the first, or mandibular, arch (Figure 35A).
- In the ancient fishes called placoderms, the jaws are quite separate from the cranium, or brain case, and are attached to it only by ligaments (Figure 35B).
- In later fishes, like the shark, the second, or hyoid, arch has been moved forward. Its upper element (the hyomandibular) is now used as a brace for the jaw apparatus. At one of its ends, the hyomandibular attaches to the cranium as a point close to the ear capsule, the bony cavity housing the inner ear. At its other end, the

hyomandibular connects with the point of jaw articulation, firmly bracing it to the skull. In the hyoid arch's evolutionary movement forward, the first gill slit has been forced into a position above the jaws, near the ear capsule, and has been reduced in size (Figure 35C). It is now the spiracle and is clearly visible behind the eye in the ray, a close relative of the shark.

- A highly schematic cross section through one side of the skull of a fish shows how the jaw articulation is braced to the skull by the hyomandibular at the ear capsule and how the spiracle opens to the exterior above the jaws (Figure 35D).
- In amphibians and early reptiles, the principal innovation is that the upper jaw has fused with the cranium (Figure 35E). The main evolutionary consequence of

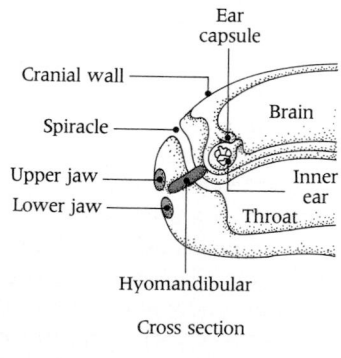

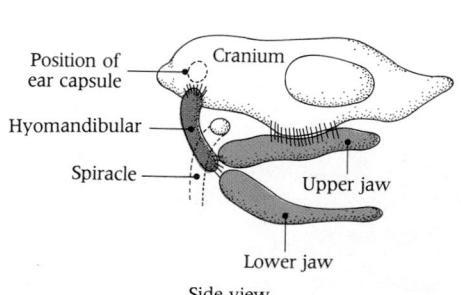

35D Bony fishes
Two views of the relationship between the hyomandibular and the upper and lower jaws.

The same sort of thing has occurred at the anatomical level. The opportunity to generate new adaptations of this sort derived from the physical proximity of anatomical parts. Chance variation and natural selection exploited the opportunity by joining

this was the liberation of the hyomandibular from its earlier function as a brace to the jaws. Its proximity to both the spiracle and the ear capsule was then exploited in a radical change of function. It came to lie in a middle ear cavity—a curious derivative of the early gill slit and the latter spiracle. The part of the spiracle communicating with the throat became the Eustachian tube. The middle ear cavity was closed off at the exterior by a tight membrane, the tympanic membrane or eardrum. Sound waves caused it to vibrate, and the vibrations were transmitted to the inner ear by the hyomandibular, now the stapes. The stapes lies in the middle ear, attaching at one end to the tympanic membrane and at the other end to a membrane that covers an opening (the oval window) into the inner ear; it is an efficient mechanical device to communicate sound vibrations from the surface of the skull to the auditory sense organ located deep in the head. The lower jaw includes bones called the dentary (which bear the teeth) and the articular (which articulates with a bone in the upper jaw called the quadrate). (Note that for simplicity, in Figures 35E, 35F, and 35G, only the articular and quadrate bones are given the characteristic shading showing their relationship with the whole of the lower and upper jaws in Figures 35A, 35B, 35C, and 35D.)

- In the mammallike reptiles, the actual ancestors of the mammals, the situation is substantially as it is in the amphibians and early reptiles (Figure 35F). However, two changes point the way to the ultimate condition in the mammals. First, the middle ear and tympanic membrane have shifted. The membrane is now very close to the point of quadrate-articular articulation. Second, the jaw itself is developing a new articulation with the skull. The end of the dentary bone is curved upward and is beginning to make contact with the cranium near the original quadrate-articular joint.

- The new articulation is complete in mammals, and the lower jaw consists solely of the dentary bone (Figure 35G). The other former jawbones assumed new functions or were lost. The articular bone became the malleus. The quadrate bone became the incus, and sound started being transmitted from the tympanic membrane to the oval window of the ear capsule by a chain of three bones, the malleus, incus, and stapes.

Thus, like the hyomandibular, the articular and quadrate bones were, in the course of time, liberated from one functional obligation and enjoined to perform a new one. In their history there was no sharply defined moment when the functional switchover took place. Rather, there was a period when, while still involved in jaw movement, the articular and quadrate bones also helped to transmit sound vibrations to the nearby ear mechanism. Their transformation to an exclusively auditory role was an exploitation of an anatomical opportunity—the fortuitous circumstance that they were located near the middle ear.

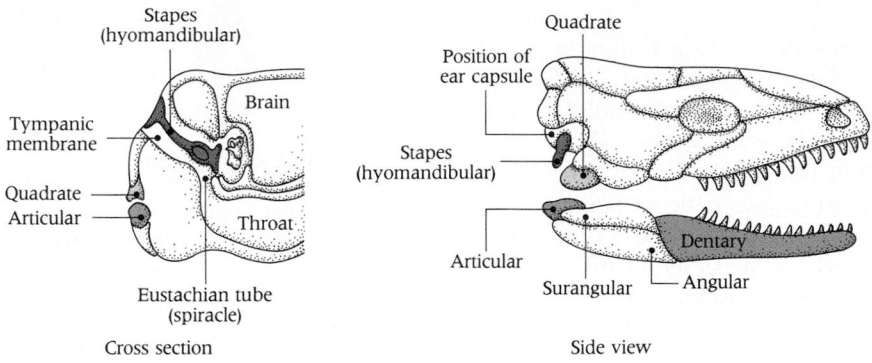

35E Amphibians and early reptiles
The hyomandibular becomes the middle bone of the inner ear. The quadrate and articular bones are highlighted to emphasize the relationships between upper and lower jaws.

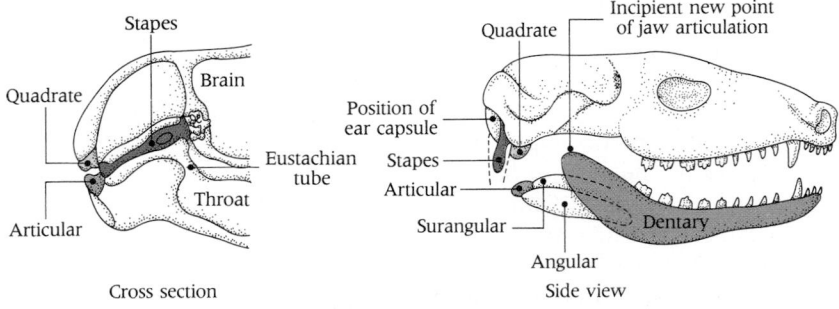

35F Mammallike reptiles
The lower jaw (dentary bone) has changed shape. The quadrate and articular bones become more closely aligned with the stapes (hyomandibular).

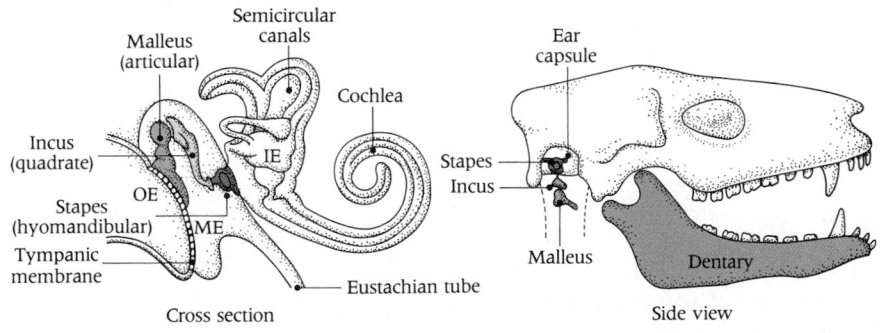

35G Mammals
The quadrate and articular become bones in the inner ear and connect with the stapes, forming the mechanism that transfers vibrations from the outer ear to the semicircular canals. The lower jaw also articulates with the upper jaw.

these elements in a different way and giving them an entirely new task. Two convincing examples concern the evolutionary origin of the mammalian ear and the panda's thumb.

The Mammalian Ear

The history of the ear is a clear case of evolutionary opportunity arising from anatomical happenstance. Today's mammalian ear is a complex of ducts, bones, and membranes, described in detail in Chapter 30. In earlier times, all (or most) of these parts had functions quite unrelated to hearing. The Eustachian tube was originally part of a gill slit serving the respiratory needs of fishes (Box 35-1). The three bones that now transmit vibrations from the eardrum to the inner ear also had functions unrelated to their present function. One (the stapes) was once a skeletal element in the throat. Later it propped up the jaw of the first biting vertebrates. The other two bones (incus and malleus) were basic elements of the reptilian jaw of 200 million years ago.

The present function of the three ear bones resulted from a fortuitous set of circumstances. First, their earlier roles happened, coincidentally, to bring them into close anatomical proximity to a pressure-sensitive structure (that would become the inner ear) in the brain case. While functioning as jaw components, these bones happened incidentally to transmit sound waves to this organ. Second, successive improvements in vertebrate jaw structure made unnecessary the earlier functions of the three bones. Released from one kind of selection, their future was determined by the selective premium placed on efficient sense organs. We describe this sequence in some detail in Box 35-1. It proves the value of careful observation.

The Panda's Thumb

Unlike other bears, which are omnivorous, giant pandas (*Ailuropoda melanoleuca*) subsist entirely on bamboo shoots. They deftly manipulate the stalks with their forepaws, passing them between an apparently flexible thumb and the other fingers. It had always been held that a dexterous, opposable thumb—one that can touch the small finger—is a hallmark of primate evolution, from whence came human beings. Most mammals lack this facility.

Investigation of this curious finding revealed that the panda has five fingers in addition to the thumb—and that the thumb is not a finger at all, but a wrist bone (called the radial sesamoid) that became so enlarged and elongated it came to resemble a thumb (Figure 35-10). Its muscles, too, are previously existing structures that were refashioned for a new function. According to D. D. Davis of Chicago's Field Museum of Natural History, the muscles operating this remarkable new mechanism required no intrinsic change from conditions already present in bears, the panda's closest relatives.

Both examples, the mammalian ear and the panda's thumb, tell an important story. In seeking a scientific explanation for the design of organisms, our task is to unravel a historical succession of needs met by transient opportunities. It is sobering to realize that the mammalian ear, long considered an anatomical masterpiece, arose like one of those cartoonist's inventions with pulleys, string, and tilting buckets. But it worked! And under the aegis of natural selection it worked better in each stage of its evolution.

ADAPTIVE COLORATION

Adaptive color and color patterns, brilliant or subtle, provide striking examples, some of which have furnished crucial evidence on the role of natural selection in producing adaptations. We here consider three kinds.

Cryptic Coloration

Many animals are hard to recognize in their natural surroundings because their colors or shapes blend them into the background (Figure 35-11). Such **cryptic coloration** is common among mammals, birds, and especially insects. Grasshopper species living in grassy meadows are green. Species living in dry prairies are the drab color of brush. The common moths of light birch trees and dark pine trees differ in their wing coloration. Each matches the surface upon which it usually rests. Though nearly invisible in their usual habitat, pale moths from birch trees are easily seen if they land on dark pine bark. Then they become visible to predatory

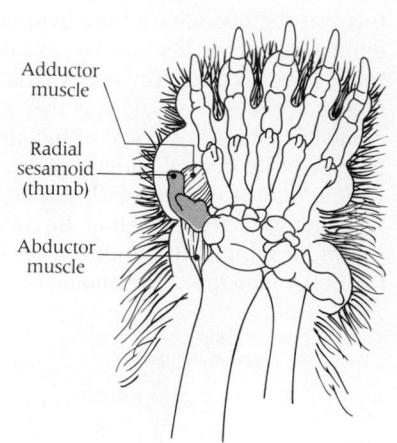

35-10 The panda's thumb
The radial sesamoid bone (color) became enlarged to the point that it could serve as another digit, a thumb. Adductor and abductor muscles give this digit agility in grasping and other actions.

BOX 35-2

THE CASE OF THE MELANIC MOTHS

The peppered moth, *Biston betularia*, one of 700 species of moths inhabiting the British Isles, provides biology with one of its most famous examples of cryptic coloration and its best documented case of microevolutionary change. Light in wing color with small dark markings, this creature flies by night and rests by day on such surfaces as lichen-clad tree bark. In the rural areas of Cornwall, Wales, and Scotland, its mottled appearance provides excellent camouflage against the bark. However, in smokey industrial areas, the great majority of moths are not light and mottled, but melanic—that is, nearly coal black. Significantly, dark moths became predominant since the mid-1800s, when factories began belching soot into the air. For that reason, this phenomenon has been called **industrial melanism.**

Its explanation is now clear. The color difference between the two forms is controlled by a single major gene locus. Thus this is a case of genetic polymorphism in which the light form, which once predominated, gave way in industrial areas to the dark or melanic form. In the Manchester district, the first black specimens of *Biston betularia* were caught in 1848. By 1895, 98% of the population in that area consisted of melanic moths, a remarkable turnabout.

In 1937, E. B. Ford of Oxford University offered an explanation for this striking pattern of change. While moths are resting in the daytime on tree trunks or rocks, their only protection from bird predators is their color, which matches the background color of their resting places. In the past, these tree trunks and rocks were light in color and resting light moths were hard to see, whereas melanic forms were highly visible (Figure 35H). Birds thus were able to capture melanic moths more easily than cryptically colored light ones—hence, survival of light forms was

35H Camouflage of light moth on light tree bark

35I Camouflage of dark moth on dark tree bark

favored. But when soot from local factories began blackening trunks and rocks, melanic moths came to resemble the background more closely than light moths (Figure 35I). Selection now favored the melanics.

Surprisingly perhaps, this hypothesis—which stressed the role of differential predation on two morphs (polymorphic forms)—

was opposed by many biologists who argued (without data) that birds are not that choosy about the color of the moths they capture. Between 1952 and 1956, H. B. D. Kettlewell of Oxford University decided to perform field experiments that were designed to settle the matter one way or the other. He released light and melanic forms of *Biston betularia* (marked with a spot of paint) into two different woody areas: one, in the trees of the badly polluted Birmingham district, the other in the light-colored, unpolluted trees of rural county Dorset, where 94.6% of the wild moth population was found to be light colored. Watching the resting moths with binoculars, he observed that insectivorous birds did prey on these moths, as expected. In the polluted woods near Birmingham, only 13% of the light-colored moths survived, while 27.5% of the dark-colored moths survived. In Dorset, 6.3% of the dark moths survived and were trapped, and 12.5%—more than twice as many—of the light forms survived. Clearly, light moths survived better in unpolluted woods and decreased predation was the key selective factor in their superior survival. These experiments, a classic of modern ecological genetics, have been said to prove, incidentally, that birds do hunt by sight. Not all agree.

Interestingly, some of the moth species that grew dark in nineteenth century England are now reverting to lighter colors in response to recent improvements in air quality. In one region where local "smokeless zones" were established in 1972, the frequencies of nonmelanic peppered moths in successive studies from 1961 to 1974 were 5.2%, 8.9%, and 10.5%. Other studies show that between 1970 and 1983 melanics had a 12% disadvantage in the areas studied. It appears that a reversal of industrial melanism is underway.

birds hunting food. A famous example of cryptic coloration in moths, the case of industrial melanism, is discussed in Box 35-2.

A simple theory seeks to explain the origin of cryptic coloration by natural selection. Most animals are subject to predation, that is, attack by enemies or predators. In any population, there is usually hereditary variation in color and pattern. Some variations are more likely to deceive predators than others. Thus, by lethal means

natural selection tends to preserve through the generations those genes that more adequately protect the population.

Critics of the theory of protective coloration argue that since some animal populations survive without such coloration, protective coloration is not essential and is not adaptive. Consider two similar but separate populations of light-winged moths on trees with dark bark. Both are conspicuous and heavily preyed upon by birds. Suppose now that in population A mutations occur that darken the wings slightly. It is incorrect to assume that the darker wings are not an adaptive improvement because they are only slightly less easily seen. Even if birds overlooked mutants only 1% more often, individuals with darker wings would increase in succeeding generations and the mutant allele would in time replace the original allele. Exactly this situation has been observed in nature, and studies such as those described in Box 35-2 have proved it a prime example of the selective pressure of predation.[3]

Meanwhile, population B retains uncryptic coloration and continues to flourish in spite of its lack of darkening mutations. However, this does not invalidate the conclusion that darker wings are protective. Population B might be nullifying predation pressure in some other way, perhaps by a fecundity rate that maintains a large population in the face of heavy losses. This, too, is an adaptation, one that often arises through natural selection. It nicely illustrates yet another facet of the situation—alternative means of adaptation may arise.

Warning (Aposematic) Coloration

An alternative adaptation is flashy coloration, the opposite of cryptic coloration. It, too, protects against predation, but in a different way. One argument against the theory of cryptic coloration held that concealment is not adaptive because its opposite, advertisement by color pattern, also occurs in thriving populations. Appropriately, this adaptation is called **warning coloration.** It is also known as **aposematic coloration** (after *apo*, away, and *sematic*, warning).

---

[3] Bacterial cultures clearly illustrate the consequences of small mutational advantages. Consider, for example, the effect of a mutation that causes strain A to grow only 1% faster than strain B in a mixed culture containing both strains. If 10 cells each of A and B are inoculated together into a flask and allowed to undergo 30 successive divisions, a process that in many bacteria would take less than 15 hours, the differential of 1%, compounded 30 times, yields a ratio of A to B in the flask of 1.35:1. If a small amount of this culture is transferred to fresh medium and the process is repeated every 15 hours, a differential growth advantage of 35% occurring with each transfer, not many transfers will be required before A completely outgrows B.

35-11 Cryptic coloration
Lizard, butterfly, moth, flounder, grasshopper, and katydid are hard to see in their natural habitats.

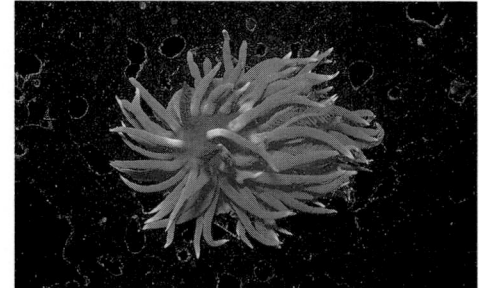

35-12 Warning (aposematic) coloration
Sea anemone, moth, and arrow frog display bright coloration.

Monarch

Viceroy

A

B

C

D

E

35-13 Batesian mimicry
(A) *The Monarch butterfly, which is distasteful to predators, is mimicked by the Viceroy butterfly, which is not distasteful. This protects the Viceroy from predation.* **(B)** *A blue jay eats a Monarch butterfly and* **(C)** *vomits.* **(D)** *Caterpillar with eye spot.* **(E)** *Syrphid fly mimics a yellowjacket.*

Many insects taste or smell bad, or they sting. These features, which make them disagreeable to predators, would profit them little if they were noted only after the insect is dead. When, however, the insect advertises its unacceptability with warning coloration, predators that have experienced it before learn to leave it (or its relatives) alone (Figure 35-12).

Aposematic coloration is especially effective against predators that can learn. Dogs, for example, will not attack a skunk more than once.

Mimicry

There are ways of avoiding being eaten other than by cryptic or aposematic coloration. Another, also involving coloration and pattern, is by **mimicry.** In one of several varieties of mimicry, called **Batesian mimicry** after Henry Walter Bates (1825–1913), the British naturalist who proposed this concept in 1862, the mimics are animals (especially insects) that would be acceptable as food but that look and act like another species. This species, termed the model, is obnoxious to predators and usually exhibits warning coloration. Survival of the mimic species is improved if it is mistaken by predators for an unpalatable model. Natural selection will then favor and reproduce the mimicry.

A classic example of Batesian mimicry is provided by the viceroy butterfly, the mimic, which looks like the monarch butterfly, the model (Figure 35-13A). During their larval stages, monarch butterflies (*Danaus plexippus*) absorb chemicals called cardiac glycosides from milkweed plants. These same substances exert powerful effects when administered to vertebrates, stimulating the heart and causing severe vomiting. As a result, monarch butterflies are highly disagreeable, even poisonous, to bird predators, which rapidly learn to reject them as prey after one or more unpleasant encounters (Figure 35-13B). The glycosides are highly concentrated in the monarch's wings, which in natural populations often reveal beak marks. The resemblance of viceroys (*Limenitis archippus*) to monarchs clearly protects them from bird predators.

Batesian mimicry systems work because gullible predators capture individuals of the unpalatable model species and thereby learn to avoid other individuals that resemble them. If, however, a predator captures a mimic, it finds it palatable and begins to associate palatability with prey of that appearance. This has an interesting result: when both mimics and models are present, individuals of the model species are attacked more often than they would have been had there been no mimics. As a result of this selective pressure, models gradually come to differ from their mimics. Thus Batesian mimicry systems are stable only when the mimic can evolve toward the model more rapidly than the model can evolve away from the mimic. That circumstance is most likely to occur when the mimic is much rarer than the model. In such a situation, each instance in which a predator kills a mimic decreases the total number of mimics by a higher percentage than would the killing of a model. This exerts a stronger selection pressure than would the death of a model. The opposite would be true if models were rarer than mimics.

In a second type of mimicry—**Müllerian mimicry,** described in 1879 by Fritz Müller (1821–1897)—one distasteful or venomous species may mimic another. In other words, different unpalatable species converge upon the same color pattern and each species is both model and mimic. What is the value of such mimicry to species already in possession of an effective weapon? The answer is that since predators have to learn by bitter experience which color patterns to avoid, Müllerian mimicry reduces the numbers of each species that are killed while predators learn to avoid a given aposematic color pattern. A bird, for example, will learn to avoid any insect with yellow and black stripes after it has been stung by two such insects. It is clearly advantageous for each striped prey species if, while learning, the bird kills only one member of one species and one of another.

We should remember that coloration may be adaptive without reference to predation. Showy colors may be recognition marks that help birds of a feather flock together. They may also be releasers or passwords that elicit sexual acceptance and appropriate mating behavior (see Chapter 32). Colors may also be adaptive in the physical environment. For example, dark skin may facilitate heat radiation from within and screen off damaging shorter frequencies of light (ultraviolet) from without.

The list of ways in which color is adaptive could be greatly extended.

BEHAVIORAL ADAPTATIONS

Behavior patterns are clearly adaptive and thus are important determinants of a species' reproductive success and evolutionary destiny. The field of behavioral ecology, among others, probes the ways in which behavior affects natural selection.

In Chapters 32 and 33, we discussed behavioral adaptations in the context of behavior. We now add a few more comments and examples in the context of adaptation and evolution.

Behavioral Mimicry and Deceit

One of the noteworthy products of the evolution of interspecific communication (that is, communication between members of different species) is the repeated occurrence of **behavioral mimicry** and deceit. Animals, it seems, benefit very often by pretending to be what they are not. So do plants.

■ In Lake Malawi (which lies between Tanzania, Mozambique, and Malawi in Western Africa), schooling cichlid fishes (*Tilapia*) display a characteristic size, body shape, and striped pattern. But another cichlid species (*Corematodus*) mimics the appearance of *Tilapia* (Figure 35-14A) and joins the school unnoticed. *Corematodus* then feeds on the small scales covering the tail of *Tilapia* by grabbing and holding it with jaws whose numerous small teeth give it the effect of sandpaper (Figure 35-14B). As the deceived prey struggles to escape, the deceiver gets a mouthful of nutritious scales.

■ Another cichlid fish of Lake Malawi carries its eggs into the nest of another species, which then cares for a mixed brood that includes some "wrong children." That is behavior typical of cuckoos. Some cichlid fishes in nearby Lake Tanganyika have their eggs incubated by an unsuspecting mouth-brooding catfish![4] The term **brood parasites** denotes species that deposit their eggs in the nests of other species, where they are incubated and reared by their host (see Figure 47-11).

■ The flowers of some plants that do not produce nectar—for example, the orchid *Epidendrum radicans*—closely resemble those of other plants that do produce nectar. Thus, they are able to attract butterflies and other insects and thereby promote pollination.

A

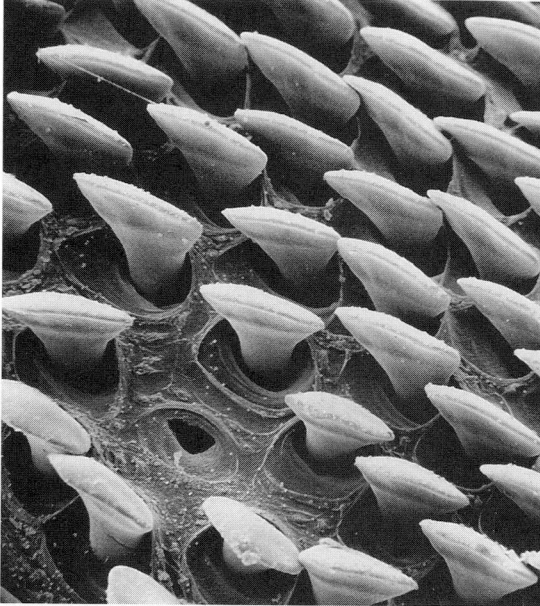

B

35-14 Cichlid fishes, **Tilapia** *and* **Corematodus**
(**A**) Corematodus (*bottom*) *mimics* Tilapia.
(**B**) *The teeth of a* Corematodus.

---

[4] Mouth brooding is a behavioral adaptation that evidently enhances reproductive success. The male cichlid fish possesses spots on the anal fins that resemble the eggs. The spots induce the female to snap at the spots that are near the male's ventral-posterior side. As a result, she takes sperm into her mouth. She also takes her eggs into her mouth, and by placing sperm and egg so close together ensures fertilization.

These are examples of interspecific deceit. There are also many instances of intraspecific deceit.

- In the bluegill sunfish (*Lepomis macrochirus*), a large male establishes and defends a territory in which he breeds. Small males mimic females and enter the territory safely disguised. There they deceive large males by signaling that they are female. They then proceed to inseminate the eggs of spawning females along with the large territorial male.
- Among garter snakes (genus *Thamnophis*), males normally recognize females by the pheromones they exude. In some strains, male-male rivalry is so intense that some males "pretend" to be females by producing the same substance. In a recent experiment, the mimic—the only one who knew who the real females were—was the successful copulator in 29 of 42 trials. Ordinary "honest" males were successful in only 13 of 42 trials.

Costs and Benefits

To assess the adaptive value of a behavioral trait, we must estimate its costs and benefits. **Benefits** are the sum of improvements in survival and reproductive success that an animal gains from executing a particular behavior. **Costs** are the negatives of benefits—the need to invest energy, the risks of danger or death, and so on. Both factors influence an animal's "choice" between different behaviors. How, for example, would an animal choose between these two options: easily accessible food in small quantities and large amounts of food obtainable only with difficulty and risk?

Costs are usually of three types.

- **Opportunity cost** is the benefit an animal loses by not being able to perform some other behavior during the same time interval. For example, when defending its territory, the animal must give up feeding.
- **Energetic cost** is the difference between the energy expended in performing the behavior and the energy expended during rest. For example, foraging may yield food, but it also requires extensive travelling and other activities that may consume so much time and energy it may not be worth it.
- **Risk cost** is the increased chance of injury or disease as a result of executing the behavior.

Even though it is difficult to measure costs and benefits directly, many highly reliable indirect methods are available. We can, for example, compare actual behaviors under different conditions set by an experimenter.

These considerations apply, not only to such "simple" behaviors as foraging and mating, but to highly complex behavioral capabilities. For example, obvious benefits attach to learning and the ability to form social groupings, but there are also costs. An animal that must learn a behavior may not learn the "right" behavior, and learning, we all know, is very arduous. Group living also has its costs, since it is conducive to the rapid spread of disease. Also, living in social groups may inhibit individuals in their reproductive activities.

If behavioral adaptations are to evolve by natural selection, the sum of all costs must be exceeded by the benefits attributable to that behavior. A complex behavioral repertory does not necessarily confer selective advantages upon its possessor. What matters is a positive balance in the benefit/cost ratio.

NATURAL SELECTION AS HISTORICAL PROCESS

An essential part of an organism's competence to leave offspring is its ability to find food and to make a living through the period of reproductive maturity. The natural selection that results from competition for food and living space has several interesting evolutionary consequences. Broadly speaking, the possibilities include the following:

- Further specialization and narrowing in competence
- Increased generalization and broadening in competence
- Change to a different mode of life

We now consider the factors that determine which of these possibilities is realized.

PRESSURE TO DIVERSIFY

All species are adapted to a particular environment that is more or less limited in size and resources. Eventually their reproductive capacity expands the population to a limit set by the availability of resources (see Chapter 46). This situation creates selective pressure wherein those individuals able to make use of otherwise unexploited environments and resources are at an advantage. Such a change puts them in a new environment where competition is lower, thus increasing the probability that they will leave progeny. The wide range of life's habitats (Figure 35-15)—from hot springs to arctic waters, from ocean floors to mountain streams, deserts, rain forests, and the air above, from the intestines of other animals to the pages of books—reflects the many instances in which opportunities were successfully seized by organisms seeking to escape from the competition of other more crowded environments or to occupy and exploit new habitats. In this way, the theory of natural selection explains the striking diversity of life.[5]

CONDITIONS FOR ENTRY INTO NEW ENVIRONMENTS

To enter a new environment or mode of life, an organism must be given an opportunity to do so in three different senses.

First, it must have the physical opportunity to enter. Conceivably, a butterfly species now limited to South American forests might be able to exploit a new environment in South Africa. But its evolutionary potential in this respect will remain unfulfilled so long as it lacks physical access to South Africa.

Physical access to a new environment, however, does not guarantee the successful evolutionary invasion by a species. For that, it must possess an appropriate design. That is the second condition. We define *design* in this context as the biological organization of structures in respect to a specific function. A species has the appropriate design for a new environment, a design opportunity, only when it already possesses some adaptation, however minimal, that will permit it to survive and reproduce while gaining a foothold. Once entry is established, selection should steadily raise the level of adaptation to the new conditions.

Physical and design access are insufficient in themselves to ensure an invasion of new habitats. The species must also have ecological opportunity—the third condition. That means that the competition in the new habitat must be slight enough to permit survival of the new invader during its initial phase, when its adaptation may still be relatively poor.[6]

ARE NATURAL SELECTION AND ADAPTATION TAUTOLOGICAL CONCEPTS?

Students of logic and the rules of language define a **tautology** as a statement that is necessarily true or true by definition. Tautologies thus are empty of empirical meaning (meaning relating to the data of experience). Rather, they are statements rooted only in the rules of language and the ways we use language. Examples of tautologies are: "It will rain tomorrow or it will not rain." "All bachelors are single." "Everyone who is 50 years old was alive 25 years ago."

In the same way, it would be tautologous to declare that "swift lizards run quickly." That statement is different in kind from a claim that "swift lizards escape predators." The latter statement can be tested by empirical observation and experi-

35-15 Range of habitats on Earth
Each of the major habitats, or biomes, will be discussed in detail in Chapter 48.

---

[5] We do not mean by "escape from competition" that a deliberate or conscious attempt is being made by a squirrel, bird, or bacterium to find a new way of life where the going will be easier. The structural and behavioral patterns that ensure random dispersal of a species are sufficient to ensure that some members of the species will sample new environments—and if they are easier, they will linger there.

[6] Even if all three of these conditions are met, an ancestral-descendant sequence will not split into two or more species unless part of the population becomes isolated, so that gene flow with the rest is diminished for a considerable period of time. Isolation and isolation mechanisms will be discussed in Chapter 36.

ment. The former cannot. Its truth rests on the fact that "swift" and "quick" are synonyms. Hence, that statement is necessarily true—and quite uninteresting. A statement is not a tautology if it is not necessarily true and thus has at least the potential of telling us something we did not know.

Some critics of evolutionary theory have argued that the principle of natural selection—and its often misused synoptic slogan "survival of the fittest"—is tautologous because the definition of "the fittest" is "those which survive." Like "swift" and "quick," they are synonyms and thus, according to these critics, it is empty of meaning to say that the fittest survive. This circularity, they argue, accounts for the supposed scarcity of testable hypotheses and predictions in evolutionary biology. Although this contention is mistaken, it is worth examining—not because it is wrong, but because the reason it is wrong teaches us something about the theory of evolution.

Evolutionary biologists, aware of these issues, have tried to keep "natural selection" from being a tautology by devising a definition of "fitness" that does not invoke the notion of survival. In this definition, fitness is grounded in good design. The fitness of a trait in a given environment is to be judged the same way an engineer evaluates a machine, or a biochemist judges the efficiency of an enzymatic pathway. For example, the muscles of a racehorse have higher fitness if the horse can run faster. By applying design standards to traits, we can determine their relative fitness without having first to measure survival rates. In this way of thinking, "survival of the fittest" and the concept of natural selection are not tautologous. A definition of fitness that is free of survival considerations allows biologists to distinguish cases where trait frequency increases by natural selection from those in which it increases by accident, by genetic drift, or whatever. If natural selection were truly tautologous, if fitness meant only an increase in prevalence, such cases would be indistinguishable.

Having aired that issue, we must now concede that in their daily work most biologists do use survival as a practical and accessible index of fitness or adaptation. Earlier in this chapter, we said that superior adaptedness results in the ability of certain variants to produce more viable offspring than do others. The workaday reliance on this alternate definition of fitness does not, however, disprove the point made above—anymore than does the tendency of biologists to ask "What is the purpose of this structure or that behavior?" disprove the principle that nature is devoid of teleological purpose (see Box 2-2). In fact, they are looking for *apparent* purposes—mainly because that habit of thought sometimes proves useful in guiding future experiments. It also can be misleading.

Whether or not it is justified to rely on survival data as a practical test of fitness depends on the important differences between cases in which biologists are testing the validity of the theory of evolution and cases in which they are investigating its down-to-earth operations. In an experiment designed to test whether natural selection occurs, it would beg the question to define fitness in terms of survival, since the goal in that experiment would be to show that those traits which confer fitness increase in frequency. Most experiments in evolutionary biology, however, assume that natural selection occurs. These experiments are designed to uncover the *effects* of selection in a particular case. Measuring fitness by survival in such experiments is therefore acceptable.

While much work in biology takes natural selection for granted in this way, this is not to say that there is no direct evidence that selection occurs. As we have seen, natural selection occurs when three conditions are present.

- A population exhibits individual variation in a particular trait.
- That trait is consistently transmitted from parents to their offspring.
- That trait is consistently associated with reproductive success and survivorship.

In other words, we can say that natural selection occurs if a population is *not* at equilibrium (a population is at equilibrium if a given trait has the same frequency in each generation) and the frequency of the trait among all offspring in the population is predictably different from that of all parents—over and above differences expected from the first two conditions alone.

All three conditions are necessary (and sufficient) for the process of natural selection to occur. When they are met, we can conclude that as a result of natural selection, the frequency of a trait changes in a predictable way over many generations. The simplest and most direct examples of natural selection are found in such cases as the industrial melanism of the moth *Biston betularia* (see Box 35-2), where in Karl Popper's words, "we can see selection occurring with our own eyes."

The concept of natural selection concerns the spread of new adaptations, not their origin. It explains why and how relatively better adaptations can increase in frequency. Why a given adaptation is superior to an alternative one depends on a design superiority that confers superior fitness—and that can be judged without recourse to survival data. When an adaptation is superior, it is consistently associated with improved reproductive success and survivorship.

ADAPTATION AND INHERITED INFORMATION

Again and again, biologists have had to confront profound questions posed by the complex organization of living organisms. How did life originate? When did DNA and RNA enter the picture? How did major changes—like the shift from anaerobic to aerobic metabolism—take place? How did genetic information, the specifications that underlie all traits, become so complicated? In view of the many biochemical similarities of all living organisms, how do we explain life's never-ending diversity?

Though we still lack comprehensive answers to all of these questions, recent developments are giving the study of evolution a new dimension.

EVOLUTION AT THE MOLECULAR LEVEL

In the 1930s and 1940s, students of genetics and natural history resolved their differences and joined forces in studying evolution. The result was the "modern synthesis," a term coined by Julian Huxley. It soon became a unifying theme in biology. Beginning in the 1960s and extending to the present, the flowering of molecular biology profoundly strengthened the meaning and scope of the modern synthesis: the selective inheritance of genetically determined variation had now become accessible to rigorous analysis.

Genetic variation was now seen to be rooted in the reassortments of genes during meiosis and sexual reproduction—along with such other mechanisms as gene transposition. With the discovery of the genetic code in the 1960s, gene mutations—a single changed nucleotide in a DNA codon within an allele—were recognized as basis of genetic variation, as were chromosomal mutations (Figure 35-16). Evolution then came to be viewed, at least in part, as a shift in the frequency of genes within a population.

The development of amino acid sequencing and recombinant DNA technology made it easily possible to dissect the genetic material and its protein products to a previously unattainable degree. When DNA sequencing was introduced, biologists were surprised at the amount of nonfunctional DNA, now known as "junk DNA," in the genomes of complex organisms and the extraordinary number of unexpected interruptions and redundancies (see Chapter 17). Genetics completed its change from an inferential science restricted to breeding experiments to a direct analytical science. Now it was possible to look at genes, determine their molecular and physical structures, and examine the proteins they encode.

Before the advances now to be summarized, there were plenty of clues to suggest the importance of molecular mechanisms in evolutionary biology. We saw in Chapter 16 how genetic information is actually carried in DNA, how the genetic code is translated into the specific amino acid sequences that comprise the proteins of each organism's body, and how the enzymes account for virtually every phenotypic trait. We also saw in Chapter 20 how DNA carries the specifications that govern development of embryos into complex adult organizations.

If we are what our enzymes make us—and if we think logically—we must come to an important conclusion. If evolution is caused by changes in gene pools, the DNA nucleotide sequences that comprise the genes must themselves have changed radically in the course of evolution. We have already seen how mutation alters the nucleotide sequences of DNA and how a single triplet—say UUG, a codon for the amino acid leucine—can be altered to make several new "missense" codons. As mutations modified inherited genetic instructions—first here, then there—novel amino acid sequences arose. The new sequences were in fact new proteins—new enzymes, perhaps, that could catalyze completely new reactions. As genetic recombi-

35-16 Change in chromosome number can alter phenotype
These daylilies, Hemorocallis, *demonstrate the phenotypic effect of polyploidy. From top to bottom the plants are haploid, tetraploid, and hexaploid.*

nation shuffled these variations among the individuals of a population, the presence of variant proteins in individuals led to variant individuals in populations.

If an altered mutant enzyme worked slightly less well or slightly better than an unaltered enzyme, the result would be a certain degree of selection pressure in a population for or against the individuals with the mutant enzyme. Eventually, a large part of the population would have the enzyme with the most advantageous amino acid sequence.

Thus we see that evolution proceeds at the molecular level as it does at the organismic level. A new variation arises by random mutation. Then there is selection from among the variants according to their efficiency in coping with the environment—their fitness. Out of the interplay of variation and selection comes evolutionary change. Therefore an important goal in the study of evolution is to try to reconstruct the past events that gave rise to the vast inventory of proteins in existence today. Even at the molecular level, natural selection is the agent that creates the patterns of coded information responsible for life's adaptations. In looking at an organism's genome, one is looking at history.

NEW APPROACHES TO THE STUDY OF EVOLUTION

Individual protein molecules are documents of evolutionary history, and their study led to a new field, **molecular evolution.** Its workers have set out to reconstruct the course of evolution at the molecular level. Its two general approaches involve the study of (1) the three-dimensional structures and amino acid sequences of individual protein molecules in members of different taxa (biological groups, such as phyla, orders, etc.), and (2) DNA and RNA nucleotide sequences in such organisms.[7] Both methods—really two sides of the same coin—seek to establish actual pathways of evolutionary change. Let us consider then, their successes and their problems.

Studies of Amino Acid Sequences

The premise of the school of evolutionary molecular biologists (known as "amino acid sequencers") is that it would have been far simpler in evolution to modify existing proteins through mutation or other modes of genetic change than to assemble completely new amino acid combinations for each new protein. One corollary of this view is that the great majority of proteins now found in living organisms have evolved from a very small number of **archetypal proteins.** (In evolutionary biology, an archetype is an ancient type that was forerunner to later forms.)

The main effort of this school has been to determine amino acid sequences of corresponding proteins in different organisms and then to compare them carefully in search of similarities or differences. For example, when corresponding proteins—say, hemoglobin—from two different organisms are found to display a pervading similarity in their amino acid sequences, we say they are structurally homologous even if some of their amino acids are different. Such **homology** is good evidence that the two proteins have a common ancestry. By making such comparisons systematically, the apparent pattern of evolution of the molecules can be traced.

Before proteins of similar amino acid sequence can be assigned a common evolutionary origin, two questions must be answered.

- Are similarities of primary structure due to chance?
- Are significant similarities of primary structure due to common ancestry or to **convergent evolution**—in other words, could proteins that arose independently have evolved similar structures via different lines of evolutionary descent?[8]

Computer analyses of protein sequences have been essential in the approach to both questions. One attempts statistical analyses of the likelihood that any given

---

[7] Review earlier discussions on the molecular structure of proteins (Chapter 4), modern methods for characterizing the amino acids in proteins (Box 4-2), the genetic code (Chapter 17), and methods for establishing DNA base sequences (Figure 16-15).

[8] We will return to the interesting issue of convergent evolution in Chapter 37. This concept was mentioned in Box 10-1 in connection with the possibility that tetrapyrroles arose several times in evolution.

sequence might have arisen on a random basis. Consider the proposition that, although all proteins consist of the same 20 amino acids (more or less) and an average protein contains about 100 amino acids, the number of unique proteins actually extant in the living world is far smaller than the number of possible proteins—which would be 20^{100} in this formulation. Indeed, the number of existing unique proteins is only about a million (give or take an order of magnitude), according to calculations based on the estimated sizes of genomes. For example, *E. coli* could make no more than 10^4 proteins—and, in fact, that much eukaryotic DNA is not translated. (In this tally, we count the same protein—e.g., hemoglobin or hexokinase—in different species only once even though they may differ in detail.) One million, then, is presumably the upper limit of what could arise by chance.

When comparing two corresponding proteins, such as hemoglobin from two different organisms, a statistical analysis is made of the similarities and differences between them. This analysis includes the percentage of identical amino acid sequences and the frequency of amino acid sequences whose deletion would make the proteins identical in the adjacent amino acid areas. These are called "gaps."

In many cases, this technique has permitted a high level of confidence that similarities are not due to chance, and has allowed reconstructions of the evolutionary events leading to the contemporary sequence. In other cases, the data have been either inconclusive or have suggested that similar sequences in two proteins of diverse origin did result from mere chance or, occasionally, by the workings of convergent evolution. A difficulty here is the fact that many proteins have not yet been sequenced. Also, many that have been sequenced are either small in size or do not come from organisms in evolution's "mainstreams" (see Chapter 44).

Nonetheless, this approach has given strong evidence of evolutionary relationships. It has also demonstrated the existence in the living world of a certain number of protein classes—about a thousand—that have been termed **protein families** (Figure 35-17). Proteins belong to the same family if statistical analysis of their amino acid sequences reveals authentic relationships. There are also **superfamilies,** which consist of proteins homologous in more than half their positions. New superfamilies are being recognized as data accumulate. For example, we once thought that the immunoglobulin family was restricted to antibodies (Chapter 23). We now know it to include many cell surface proteins.

Once it has been established that the amino acid sequences for a set of proteins have similarities greater than can be explained by chance, one attempts to assess phylogenetic relationships—that is, which form gave rise to which in evolutionary history—among the species carrying these proteins.

One basic approach has been to define the mutation or **replacement distance** between two corresponding proteins as the minimal number of nucleotides that would have to be altered in order for the gene for one protein to encode the other. In the example in Figure 35-18, an imaginary ancestral organism at the top apex splits into two descending lines, one of which splits again. The process ultimately produces three present-day species, *A, B,* and *C.* The number of nucleotide replacements that occurred in a particular gene since the *A* and *B* lines of descent diverged are represented, respectively, by *a* and *b.* The number of nucleotide replacements between the lower apex (where one of the original lines diverged) and the top apex (where the original divergence occurred) is represented by *c.* To obtain evidence that this "tree" sequence is correctly drawn, one can study amino acid or DNA sequences in molecules of the three species. Such studies in this case showed that the distance (number of nucleotide differences or replacements) between *A* to *C* is 28—4 less than the distance between *B* and *C.* This means that 4 more nucleotide replacements took place between the lower apex and the appearance of *B* than occurred in the descent of *A.* If $a + b = 24$ and $b - a = 4$, the calculated designated distances are $a = 10$, $b = 14$, and $c = 18$. We conclude that the pair with the shortest replacement distance, in this case *A* and *B,* are more closely related and more recently diverged.

The power of the method is illustrated by a particularly well studied protein, cytochrome *c.* Studies of the evolutionary history of cytochrome *c* reveal that it has spanned at least the 1.4 billion years from chemoautotrophs to human beings (Figure 35-19). It is an interesting molecule to examine because it probably arose at a time (about 1.5 billion years ago) when the evolution of enzymatic machinery gave organisms a capacity to move from a reducing to an oxidizing atmosphere—in other words, a capacity for aerobic metabolism. It was a major forward step in

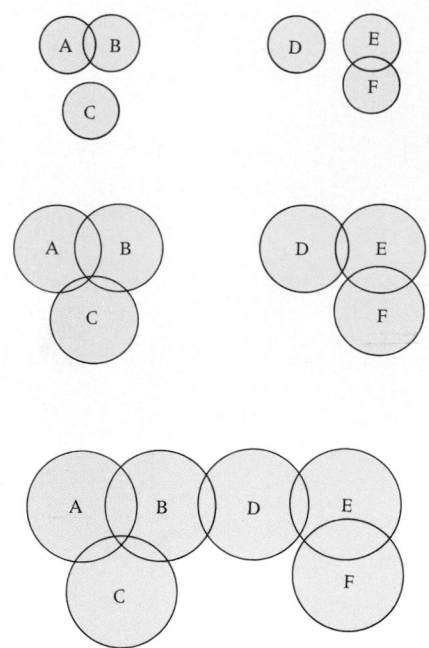

35-17 Protein families and superfamilies
Families of proteins that differ in less than half their amino acids are indicated by letters A through F. A superfamily contains two or more families. Superfamilies A–B and E–F are at the top. With more data, the superfamilies enlarge to A–B–C and D–E–F. Additional sequencing data shows that all six families belong to the single superfamily shown at the bottom.

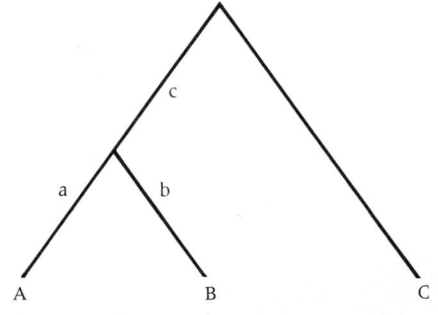

Nucleotide differences between two species		
	B	C
A	24	28
B		32

35-18 Determination of replacement distance
The evolutionary tree of a hypothetical organism. A, B, and C represent contemporary species.

Positions −9 to 40:

```
                          -9 -8 -7 -6 -5 -4 -3 -2 -1  1        5         10        15        20        25        30        35        40
Human, chimpanzee          G  D  V  E  K  G  K  K  I  F  I  M  K  C  S  Q  C  H  T  V  E  K  G  G  K  H  K  T  G  P  N  L  H  G  L  F  G  R  K  T
Rhesus monkey              G  D  V  E  K  G  K  K  I  F  I  M  K  C  S  Q  C  H  T  V  E  K  G  G  K  H  K  T  G  P  N  L  H  G  L  F  G  R  K  T
Horse                      G  D  V  E  K  G  K  K  I  F  V  Q  K  C  A  Q  C  H  T  V  E  K  G  G  K  H  K  T  G  P  N  L  H  G  L  F  G  R  K  T
Donkey                     G  D  V  E  K  G  K  K  I  F  V  Q  K  C  A  Q  C  H  T  V  E  K  G  G  K  H  K  T  G  P  N  L  H  G  L  F  G  R  K  T
Cow, pig, sheep            G  D  V  E  K  G  K  K  I  F  V  Q  K  C  A  Q  C  H  T  V  E  K  G  G  K  H  K  T  G  P  N  L  H  G  L  F  G  R  K  T
Dog                        G  D  V  E  K  G  K  K  I  F  V  Q  K  C  A  Q  C  H  T  V  E  K  G  G  K  H  K  T  G  P  N  L  H  G  L  F  G  R  K  T
Rabbit                     G  D  V  E  K  G  K  K  I  F  V  Q  K  C  A  Q  C  H  T  V  E  K  G  G  K  H  K  T  G  P  N  L  H  G  L  F  G  R  K  T
California grey whale       G  D  V  E  K  G  K  K  I  F  V  Q  K  C  A  Q  C  H  T  V  E  K  G  G  K  H  K  T  G  P  N  L  H  G  L  F  G  R  K  T
Great grey kangaroo         G  D  V  E  K  G  K  K  I  F  V  Q  K  C  A  Q  C  H  T  V  E  K  G  G  K  H  K  T  G  P  N  L  N  G  I  F  G  R  K  T

Chicken, turkey            G  D  I  E  K  G  K  K  I  F  V  Q  K  C  S  Q  C  H  T  V  E  K  G  G  K  H  K  T  G  P  H  L  H  G  L  F  G  R  K  T
Pigeon                     G  D  I  E  K  G  K  K  I  F  V  Q  K  C  S  Q  C  H  T  V  E  K  G  G  K  H  K  T  G  P  N  L  H  G  L  F  G  R  K  T
Peking duck                G  D  V  E  K  G  K  K  I  F  V  Q  K  C  S  Q  C  H  T  V  E  K  G  G  K  H  K  T  G  P  N  L  H  G  L  F  G  R  K  T
Snapping turtle            G  D  V  E  K  G  K  K  I  F  V  Q  K  C  A  Q  C  H  T  V  E  K  G  G  K  H  K  T  G  P  N  L  N  G  L  I  G  R  K  T
Rattlesnake                G  D  V  E  K  G  K  K  I  F  T  M  K  C  S  Q  C  H  T  V  E  K  G  G  K  H  K  T  G  P  N  L  H  G  L  F  G  R  K  T
Bullfrog                   G  D  V  E  K  G  K  K  I  F  V  Q  K  C  E  K  G  G  K  H  K  V  G  P  N  L  Y  G  L  I  G  R  K  T
Tuna                       G  D  A  K  G  K  K  T  F  V  Q  K  C  A  Q  C  H  T  V  E  N  G  G  K  H  K  V  G  P  N  L  W  G  L  F  G  R  K  T
Dogfish                    G  D  V  E  K  G  K  K  V  F  V  Q  K  C  A  Q  C  H  T  V  E  N  G  G  K  H  K  T  G  P  N  L  S  G  L  F  G  R  K  T

Samia cynthia              G  V  P  A  G  N  A  E  N  G  K  K  I  F  V  Q  R  C  A  Q  C  H  T  V  E  A  G  G  K  H  K  V  G  P  N  L  H  G  F  Y  G  R  K  T
Tobacco horn worm moth     G  V  P  A  G  N  A  D  N  G  K  K  I  F  V  Q  R  C  A  Q  C  H  T  V  E  A  G  G  K  H  K  V  G  P  N  L  H  G  F  F  G  R  K  T
Screw worm fly             G  V  P  A  G  D  V  E  K  G  K  K  I  F  V  Q  R  C  A  Q  C  H  T  V  E  A  G  G  K  H  K  V  G  P  N  L  H  G  L  F  G  R  K  T
Fruit fly (drosophila)     G  V  P  A  G  D  V  E  K  G  K  K  L  F  V  Q  R  C  A  Q  C  H  T  V  E  A  G  G  K  H  K  V  G  P  N  L  H  G  L  I  G  R  K  T
Baker's yeast        T  E  F  K  A  G  S  A  K  K  G  A  T  L  F  K  T  R  C  E  L  C  H  T  V  E  K  G  G  P  H  K  V  G  P  N  L  H  G  I  F  G  R  H  S
Candida krusei       P  A  P  F  E  Q  G  S  A  K  K  G  A  T  L  F  K  T  R  C  A  E  C  H  T  I  E  A  G  G  P  H  K  V  G  P  N  L  H  G  I  F  S  R  H  S
Neurospora crassa    h  G  F  S  A  G  D  S  K  K  G  A  N  L  F  K  T  R  C  A  E  C  H  G  E  G  G  N  L  T  Q  K  I  G  P  A  L  H  G  L  F  G  R  K  T
Wheat germ         A  S  F  S  E  A  P  P  G  N  P  D  A  G  A  K  I  F  K  T  K  C  A  Q  C  H  T  V  D  A  G  A  G  H  K  Q  G  P  N  L  H  G  L  F  G  R  Q  S
Sunflower seed     A  S  F  A  E  A  P  A  G  D  P  T  T  G  A  K  I  F  K  T  K  C  A  Q  C  H  T  V  E  K  G  A  G  H  K  Q  G  P  N  L  N  G  L  F  G  R  Q  S
Mung bean          A  S  F  N  E  A  P  P  G  D  S  K  S  G  E  K  I  F  K  T  K  C  A  Q  C  H  T  V  D  K  G  A  G  H  K  Q  G  P  N  L  N  G  L  F  G  R  Q  S
Castor bean        A  S  F  N  E  A  P  P  G  D  V  K  A  G  E  K  I  F  K  T  K  C  A  Q  C  H  T  V  E  K  G  A  G  H  K  Q  G  P  N  L  N  G  L  F  G  R  Q  S
Sesame seed        A  S  F  N  E  A  P  P  G  D  V  K  S  G  E  K  I  F  K  T  K  C  A  Q  C  H  T  V  D  K  G  A  G  H  K  Q  G  P  N  L  N  G  L  F  G  R  Q  S
```

Number of different residues:
1 3 5 5 5 1 3 3 4 1 4 3 2 1 3 3 1 1 2 4 3 4 2 3 4 2 1 4 1 1 2 1 5 1 3 3 2 1 3 2

Positions 45 to 104:

```
                    45        50        55        60        65        70        75        80        85        90        95        100    104
 G Q A P G Y S Y T A A N K N K G I I W G E D T L M E Y L E N P K K Y I P G T K M I F V G I K K K E E R A D L I A Y L K K A T N E
 G Q A P G Y S Y T A A N K N K G I I W G E D T L M E Y L E N P K K Y I P G T K M I F V G I K K K E E R A D L I A Y L K K A T N E
 G Q A P G F T Y T D A N K N K G I T W K E E T L M E Y L E N P K K Y I P G T K M I F A G I K K K T E R E D L I A Y L K K A T N E
 G Q A P G F S Y T D A N K N K G I T W K E E T L M E Y L E N P K K Y I P G T K M I F A G I K K K T E R E D L I A Y L K K A T N E
 G Q A P G F S Y T D A N K N K G I I W G E E T L M E Y L E N P K K Y I P G T K M I F A G I K K K G E R E D L I A Y L K K A T N E
 G Q A P G F S Y T D A N K N K G I I W G E E T L M E Y L E N P K K Y I P G T K M I F A G I K K T G E R A D L I A Y L K K A T K E
 G Q A V G F S Y T D A N K N K G I I W G E E T L M E Y L E N P K K Y I P G T K M I F A G I K K K D E R A D L I A Y L K K A T N E
 G Q A V G F S Y T D A N K N K G I T W G E E T L M E Y L E N P K K Y I P G T K M I F A G I K K K E E R A D L I A Y L K K A T N E
 G Q A P G F T Y T D A N K N K G I I W G E D T L M E Y L E N P K K Y I P G T K M I F A G I K K K G E R A D L I A Y L K K A T N E

 G Q A E G F S Y T D A N K N K G I T W G E D T L M E Y L E N P K K Y I P G T K M I F A G I K K K S E R V D L I A Y L K D A T S K
 G Q A E G F S Y T D A N K N K G I T W G E D T L M E Y L E N P K K Y I P G T K M I F A G I K K K A E R A D L I A Y L K Q A T A K
 G Q A E G F S Y T D A N K N K G I T W G E D T L M E Y L E N P K K Y I P G T K M I F A G I K K K S E R A D L I A Y L K D A T S K
 G Q A E G F S Y T E A N K N K G I T W G E D T L M E Y L E N P K K Y I P G T K M I F A G I K K K A E R A D L I A Y L K D A T A K
 G Q A V G Y S Y T A A N K N K G I I W G D D T L M E Y L E N P K K Y I P G T K M I F T G L S K K K E R T N L I A Y L K E K T A A
 G Q A A G F S Y T D A N K N K G I T W G E D T L M E Y L E N P K K Y I P G T K M I F A G I K K K G E R Q D L I A Y L K S A C S K
 G Q A E G Y S Y T D A N K S K G I V W N N D T L M E Y L E N P K K Y I P G T K M I F A G I K K K G E R Q D L V A Y L K S A T S —
 G Q A Q G F S Y T D A N K S K G I T W Q Q E T L R I Y L E N P K K Y I P G T K M I F A G L K K K S E R Q D L I A Y L K K T A A S

 G Q A P G F S Y S N A N K A K G I T W G D D T L F E Y L E N P K K Y I P G T K M V F A G L K K A N E R A D L I A Y L K E S T K —
 G Q A P G F S Y S N A N K A K G I T W Q D D T L F E Y L E N P K K Y I P G T K M V F A G L K K A N E R A D L I A Y L K Q A T K —
 G Q A A G F A Y T N A N K A K G I T W Q D D T L F E Y L E N P K K Y I P G T K M I F A G L K K P N E R G D L I A Y L K S A T K —
 G Q A A G F A Y T N A N K A K G I T W Q D D T L F E Y L E N P K K Y I P G T K M I F A G L K K P N E R G D L I A Y L K S A T K —
 G Q A P G F S Y T D A N I K K N V L W D E N N M S E Y L T N P X K Y I P G T K M A F G G L K K E K D R N D L I T Y L K K A C E
 G Q A Q G Y S Y T D A N K R A G V E W A E P T M S D Y L E N P X K Y I P G T K M A F G G L K K D K D R N D L V T Y M L E A S K
 G S V D G Y A Y T D A N K Q K G I T W D E N T L F E Y L E N P X K Y I P G T K M A F G G L K K D K D R N D I I T F M K E A T A —
 G T T A G Y S Y S A A N K N K A V E W E E N T L Y D Y L L N P X K Y I P G T K M V F P G L X K P Q D R A D L I A Y L K K A T S S
 G T T A G Y S Y S A A N K N M A V I W E E N T L Y D Y L L N P X K Y I P G T K M V F P G L X K P Q E R A D L I A Y L K T S T A
 G T T A G Y S Y S A A N K N M A V I W E E K T L Y D Y L L N P X K Y I P G T K M V F P G L X K P Q D R A D L I A Y L K S E T A
 G T T A G Y S Y S A A N K N M A V Q W G E N T L Y D Y L L N P X K Y I P G T K M V F P G L X K P Q D R A D L I A Y L K N A T A —
 G T T P G Y S Y S A A N K N M A V I W G E N T L Y D Y L L N P X K Y I P G T K M V F P G L X K P Q D R A D L I A Y L K E A T A —
```

1 3 3 6 1 2 3 1 2 5 1 1 2 6 3 3 2 6 1 7 3 5 2 2 5 3 1 1 3 1 1 1 1 1 1 1 1 1 1 3 1 5 1 2 2 1 6 9 2 1 7 1 2 2 2 2 2 2 7 4 4 5 5

35-19 Comparison of cytochrome c sequences among organisms

The organisms are indicated on the left, and amino acid position is given across the top. The colored areas represent regions of the protein that deviate in amino acid content at the position given. For example, at position 89, the sequence of amino acids in all the organisms tested is extremely variable, while position 17 amino acid is identical in all the organisms.

the efficiency of energy extraction from food molecules. A central component of the new machinery was cytochrome *c*. Relatives of this early protein can be found today in virtually every living cell.

From studies of the amino acid sequences of the cytochrome *c* molecules from many diverse organisms, investigators have tried to estimate the rate at which the protein has evolved since early times (Figure 35-20). The evolutionary tree in Figure 35-21 is no more than an approximation, yet it is remarkably close to classical phylogenies based on the fossil record (despite some notable differences)—and it is based on only a single protein's primary structure. Thus it shows that primates branch off the ancestral mammalian line before marsupials, the turtle is more similar to the birds than the other reptile (rattlesnake) in the set, and the shark appears more closely related to the lamprey than to the tuna. It is of interest that the

35-20 Evolutionary distance determined from cytochrome c sequence analysis
The degree of divergence among organisms is measured by comparing sequence divergence of cytochrome c. For example, the difference in sequence between horse and monkey is 11 amino acids, between sesame and monkey is 40. The mean differences in sequences of important categories of organisms are also indicated. Abbreviations: Thwm, tobacco horn worm moth; Sw fly, screw worm fly.

Lower-triangular matrix of amino-acid differences in cytochrome c (columns correspond, in order, to: Human/Chimpanzee, Rhesus monkey, Horse, Donkey, Cow/Pig/Sheep, Dog, Rabbit, California grey whale, Great grey kangaroo, Chicken/Turkey, Pigeon, Pekin duck, Snapping turtle, Rattlesnake, Bullfrog, Tuna, Dogfish, Silkworm moth, Tobacco hornworm moth, Screwworm fly, Fruit fly (Drosophila), Baker's yeast, Candida (a yeast), Bread mold (Neurospora), Wheat germ, Sunflower seed, Mung bean, Castor bean, Sesame seed):

Organism	Hu/Ch	RM	Ho	Do	Cow	Dog	Rb	Wh	Kg	Ch	Pg	Dk	Tur	Rat	Bf	Tu	Df	Sk	Th	Sw	Ff	Ye	Ca	Mo	Wt	Su	Mb	Cb
Human, Chimp																												
Monkey	1																											
Horse	12	11																										
Donkey	11	10	1																									
Cow	10	9	3	2																								
Dog	11	10	6	5	3																							
Rabbit	9	8	6	5	4	5																						
Whale	10	9	5	4	2	3	2																					
Kangaroo	10	11	7	8	6	7	6	6																				
Chicken	13	12	11	10	9	10	8	9	12																			
Pigeon	12	11	11	10	9	9	7	8	11	4																		
Duck	11	10	10	9	8	8	6	7	10	3	3																	
Turtle	15	14	11	10	9	9	9	8	11	8	8	7																
Rattlesnake	14	15	22	21	20	21	18	19	21	19	18	17	22															
Bullfrog	18	17	14	13	11	12	11	11	13	11	12	11	10	24														
Tuna	21	21	19	18	17	18	17	17	18	17	18	17	18	26	15													
Dogfish	24	23	16	15	16	17	17	16	20	19	19	17	19	26	20	20												
Silkworm	31	30	29	28	27	25	26	27	28	28	27	27	28	31	29	32	32											
Thwm	31	30	28	27	27	25	26	27	28	28	26	27	29	33	30	30	31	5										
Sw fly	27	26	22	22	22	21	21	22	24	23	23	22	24	29	22	24	25	14	12									
Fruit fly	29	28	24	24	24	23	23	24	26	25	25	24	24	31	22	25	26	15	14	2								
Yeast	45	45	46	45	45	45	45	46	46	46	46	49	48	47	47	48	47	45	45	45								
Candida	51	50	51	50	50	49	50	50	51	50	50	50	52	52	51	47	52	47	46	47	47	28						
Mold	48	46	46	46	46	46	46	46	49	47	46	46	49	48	49	48	49	47	46	41	41	41	42					
Wheat	43	43	46	46	45	44	44	44	47	46	46	46	47	48	49	49	45	42	45	47	47	50	54					
Sunflower	42	42	46	45	45	44	44	44	44	46	44	44	44	45	47	48	48	44	44	44	46	48	42	54	15			
Mung bean	45	45	48	47	47	46	46	46	46	48	46	46	46	45	49	50	50	46	47	47	49	50	52	53	16	11		
Castor	41	41	41	40	43	42	42	42	43	45	43	42	42	43	45	45	48	46	45	44	46	46	50	43	13	11	8	
Sesame	40	40	45	44	43	42	43	43	42	46	44	43	43	42	47	48	49	45	45	46	48	47	50	42	13	12	6	6

Mean differences among groups (shaded regions of the figure):
- Primates vs. other mammals: 10.1 ± 0.8
- (Other mammals): 5.1 ± 1.3
- Mammals vs. birds: 9.7 ± 1.3
- Reptiles vs. mammals: 14.8 ± 4.3
- Amphib. vs. higher vert.: 13.5 ± 2.8
- Fish vs. land vert.: 18.6 ± 2.2
- Insects vs. vertebrates: 26.4 ± 2.5
- Primitive plants vs. animals: 47.4 ± 2.0
- Complex plants vs. animals: 45.0 ± 1.8
- Birds vs. reptiles: 12.8 ± 5.1
- Yeasts vs. Mold: 41.5 ± 0.5
- Primitive vs. complex plants: 48.5 ± 3.2

cytochrome c molecules of human and chimpanzee are identical, both containing 104 amino acids in exactly the same sequence. On the other hand, human cytochrome c differs from that of birds and reptiles in 10 to 15 amino acids, from that of insects in 26, and from that of the red bread mold *Neurospora crassa* in 44.[9] The data on differences between the cytochrome c molecules of various groups of organisms has been labeled the "evolutionary clock" by its proponents.

Many regard these changes as strong evidence for the neutralist theory of evolution, discussed earlier in this chapter. This theory, which holds that most genic changes in natural populations are adaptively neutral or "nonadaptive," is widely accepted by most evolutionary biologists in one form or another.

The family tree of cytochrome c is not the only one that has been traced in such detail. Similar histories have been worked out for the hemoglobins, the fibrinopeptides (a critical protein in blood clotting), certain snake venom proteins, and other molecules. In each case, common or homologous sequences establish apparent ancestral relationships and, like bits of paper dropped by a hiker, mark a trail of molecular change that reflects the history of the emerging organisms.

Amino acid sequence patterns often tell us interesting things about the proteins themselves. When an enzyme always has certain amino acids in the same place, the location of these invariant components stimulates biochemists to look for functional explanations of such conservatism. It is no accident that the longest invariant segment of certain homologous enzymes includes features that are essential for their catalytic activity.

[9] Cytochrome c molecules from various bacteria differ significantly from the molecules of eukaryotic organisms in total numbers of amino acids, which vary from 82 (in *Pseudomonas*) to 134 (in *Paracoccus*).

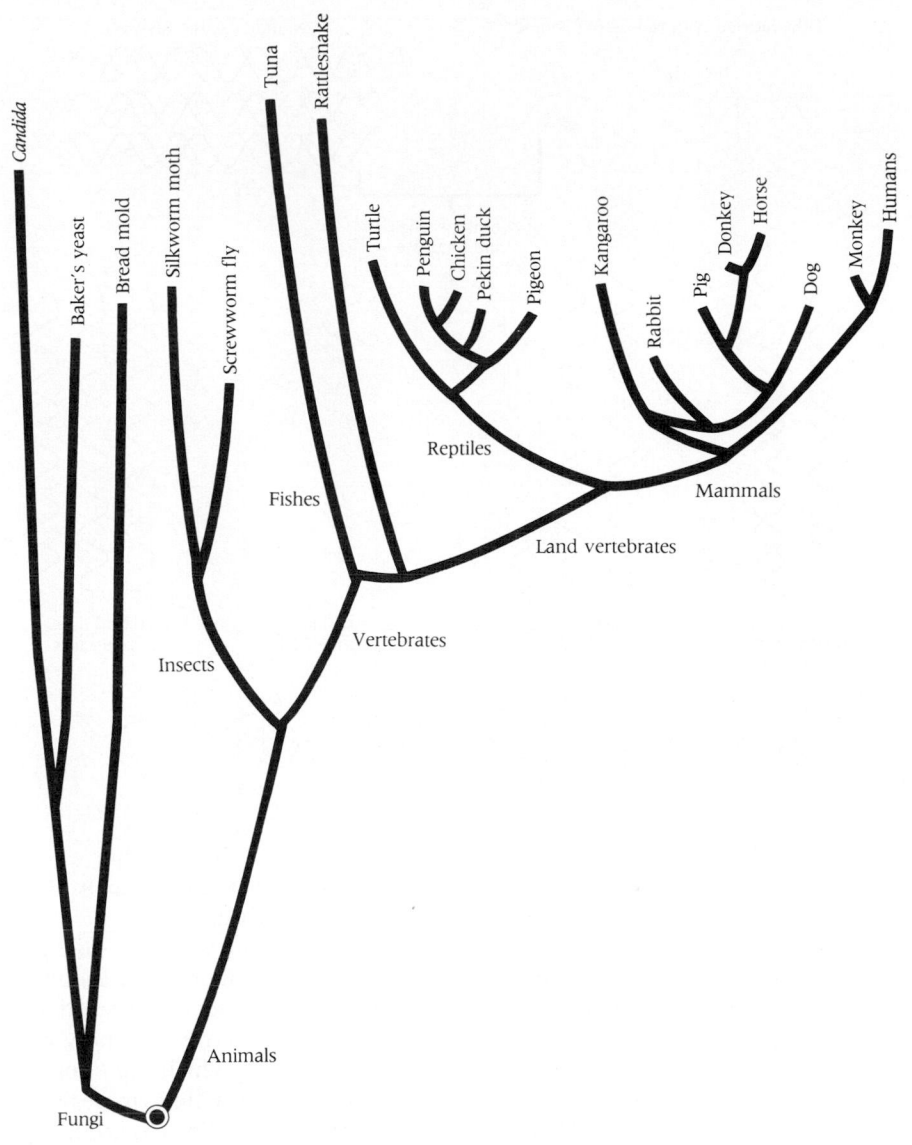

35-21 Phylogenetic tree based on cytochrome c sequence data
This partial tree is derived from an analysis of similarities and differences in the amino acid sequence of cytochrome c in the organisms indicated. The vertical axis represents the degree of difference among sequences over more than a billion years. Note that this scheme differs somewhat from those (shown later) that are based on the classical fossil evidence. For example, it appears here that rattlesnakes evolved before other land vertebrates and that placental mammals evolved twice. Neither of these conclusions, we believe, is correct.

Studies of DNA and RNA Sequences

Evolutionary studies of DNA sequences are far behind amino acid studies for a simple reason: methods for establishing DNA sequences were developed much later. However, DNA sequences are accumulating, and various computer programs are now being used in attempts to establish evolutionary trees by various methods. In these procedures, lineage trees are established on the same assumptions that were used to show protein lineages.

One relatively simple technique, DNA-DNA hybridization, has yielded notable successes and many controversies. In this procedure, the degree of homology and difference between two DNA molecules is established by allowing them to bind together (hybridize) in a test tube (Figure 35-22). They are then melted ("pried apart") by raising the temperature. The more bases that were complementary in the two strands, the higher the necessary melting temperature. Comparing the DNA of different species also reveals patterns indicative of evolutionary relatedness. It also reveals the remarkable mobility of nucleotide sequences within genomes, the multiplicity of gene copies, and vast stretches of untranscribed sequences, all of which point to a pool of potential genetic variation that is larger and deeper than had been previously envisioned. With this method, investigators have made interesting observations on the evolution of birds (Figure 35-23). In the example in Figure 35-24, DNA-DNA hybridization studies have demonstrated a previously unrecognized proximity in the evolutionary histories of starlings (tribe Sturnini) and mockingbirds (tribe Mimini).

Interestingly, the rates of DNA change seem to differ widely among different taxonomic groups. The slowest rates occur among the primates and some bird

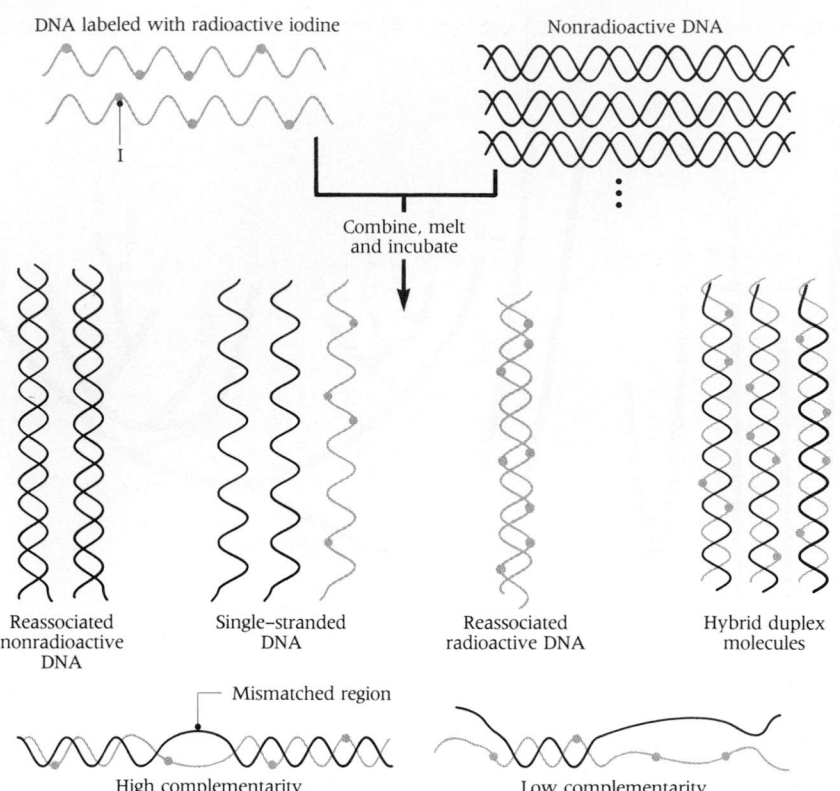

DNA labeled with radioactive iodine   Nonradioactive DNA

Combine, melt
and incubate

Reassociated   Single-stranded   Reassociated   Hybrid duplex
nonradioactive   DNA   radioactive DNA   molecules
DNA

Mismatched region

High complementarity   Low complementarity

35-22 DNA-DNA hybridization methodology
A small amount of radioactive DNA is mixed with a 1000-fold excess of unlabeled DNA and heated to separate the strands. The single-stranded DNA is permitted to reanneal. Among the double-stranded DNAs created is a hybrid, containing one radioactive strand and one unlabeled strand. To determine the sequence similarity of the two strands, the hybrid DNA is melted again, and the temperature at which the DNA returns to the single-stranded state is measured. The higher the amount of base pairing in the hybrid, the higher the temperature needed to melt the strands.

lineages. Much faster rates are observed in rodents, sea urchins and fruit flies. Perhaps improving mechanisms of DNA repair account for this.

As mentioned earlier, "DNA sequencers" had two stunning surprises—the previously unsuspected existence of introns (noncoding intervening sequences) and the remarkable mobility of genetic elements. Under the circumstances, genes and alleles could no longer be viewed in simple "beads-on-a-string" terms. We now accept that the genome is a mélange of replication units, some of which are "honestly" engaged in putting together an organism while others—20 to 100 times more plentiful on the average—are without known functions. Though the idea is troubling to many, the "junk" DNA may be the random detritus of gene duplications and meanderings in evolutionary time, the only remaining function of which is the "selfish" one of keeping themselves going. It is, of course, difficult to prove that a DNA sequence has no useful function.

The emerging view is that genetic variation draws upon a pool of variable genetic elements that is larger and deeper than anyone had supposed. With coding regions of split genes shuffling about, there is an obvious potential for modular change in the assembly of new proteins. Also, the propensity of some genes to make copies of themselves, which then shift to distant parts of the genome, generates raw materials that later might abet the processes of evolution.

What happens to these transposable elements?

■ Some may retain their function, but be expressed at different times in development or in different tissues, depending on where they have relocated. The allele encoding the γ globin chain (a subunit of fetal hemoglobin) is expressed before birth, that encoding β globin is expressed after birth.
■ Others might accumulate structural mutations in such high numbers they are inactivated. This converts them to so-called pseudogenes.
■ Still others might lose their noncoding regions on their way back to the genome, and thus become "processed genes."

In other words, these transposable elements can shift sections of DNA from one part of the genome to another, switching genes on or off as they do so. One survivor of such restless variability is the globin gene. Its remarkable usefulness in illustrating evolutionary history is discussed in Box 35-3. Another is the actin gene, the total numbers of which vary widely among species, for no known reason. The

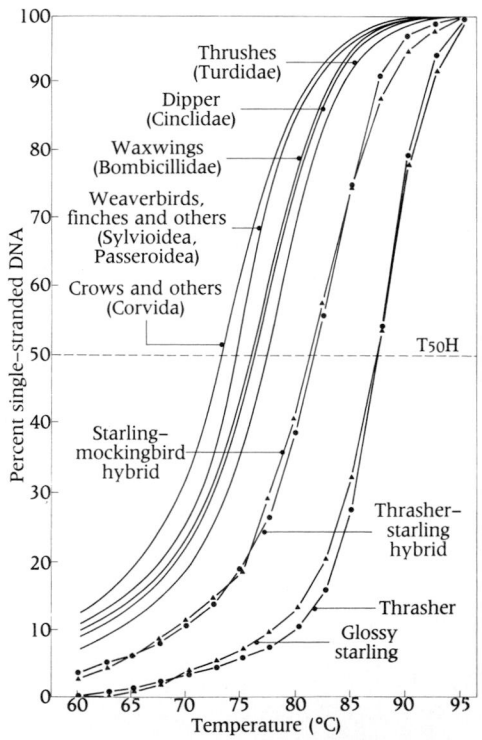

35-23 Evolutionary relationships defined by DNA melting curves
A measure of the degree of homology between two DNA strands is the $T_{50}H$, the temperature at which half the hybrids are converted to single-stranded DNA. The lower the $T_{50}H$, the poorer the match between the two DNA strands. Hybrids containing only glossy starling DNA or only thrasher DNA melt at a higher temperature ($87°C$) than do hybrids containing starling-mockingbird DNAs or thrasher-starling DNAs (about $82°C$). The $T_{50}H$ for hybrids containing DNAs of other bird groups and either starling or mockingbird DNA range from about $73–77°C$.

BOX 35-3

WRITING EVOLUTIONARY HISTORY IN GLOBIN GENES

The genes that encode the amino acid sequences of globin, the protein component of hemoglobin, have been more exhaustively studied than any other eukaryotic genes for at least three reasons: (1) they govern the synthesis of the most thoroughly investigated of all proteins, hemoglobin, (2) these genes have the interesting property of switching on and off at different times in an animal's development, and (3) knowledge of the globin gene family has illuminated the basis of the thalassemias, a group of widely distributed genetically determined blood diseases (see Chapter 22).

This scientific effort yielded an unexpected bonus. It is now possible to sketch out in considerable detail the 500 million year evolutionary history of the globin gene family. That exercise provides useful lessons about its evolution that are probably applicable to other eukaryotic genes. Although the eukaryotic genome is in a more dynamic state than was generally suspected until recently, the history of the globin gene family is marked by great stability for long periods of time. When change occurred, however, it seems to have occurred in large bursts.

Adult hemoglobin is a tetramer made up of two α-chains and two β-chains (Figure 35J). Each of the four polypeptide chains binds a heme group, which is an oxygen-binding locus. Humans make five different β-like globin chains at different times during development (see Table 22-1).

- ε in early embryonic life
- $^G\gamma$ and $^A\gamma$ in fetal life and early infancy
- δ and β in childhood and adulthood

There are three α-like globin chains.

- ζ in early embryonic life
- Two adult α-globin chains, α_1 and α_2.

These globin chains are gene products. As for the genes themselves, there are five functioning alleles in the β-globin family that correspond with the chains: ε, $^G\gamma$, $^A\gamma$, δ, and β, and there are four alleles in the α-globin family: ζ_1, ζ_2, α_1, and α_2. The α-globin gene cluster is half the size of the β-chain cluster (Figure 35K). In both, the alleles are placed along the chromosome in the order in which they are expressed in the course of develop-

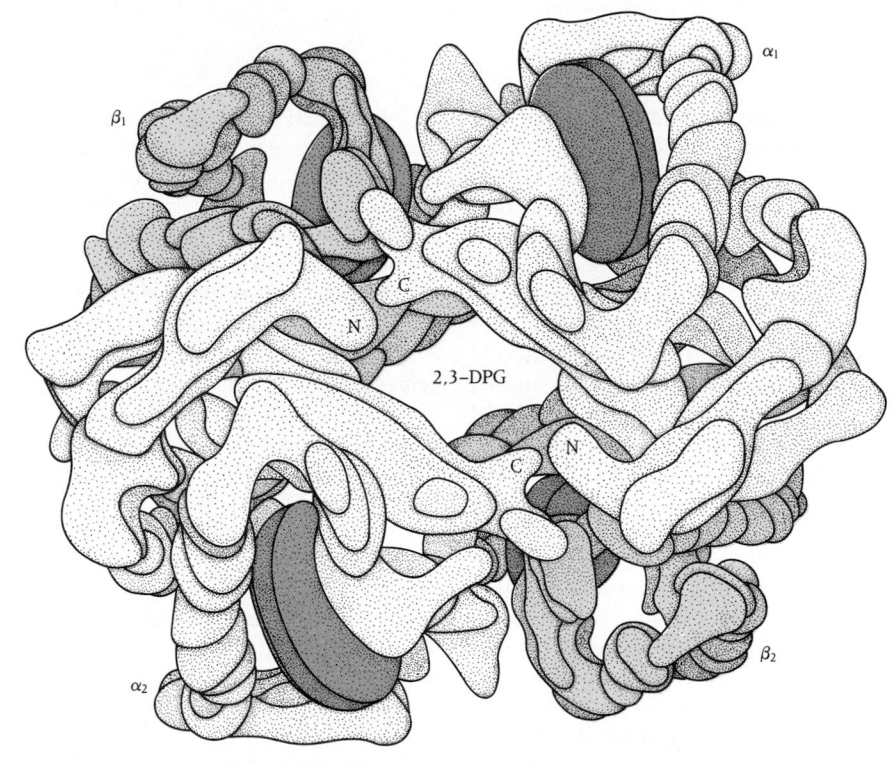

35J Structure of the human hemoglobin molecule
The diagram reflects the results of x-ray diffraction studies. Photographs of patterns at multiple planes within the molecule permit the construction of a three-dimensional model.

ment. The human α- and β-globin gene clusters are located on separate chromosomes (16 and 11, respectively). Questions about globin gene evolution concern (1) the structure of the genes (or alleles), and (2) the structure of the gene clusters.

The similarities between α- and β-globin genes imply that they arose from a single ancestral gene, probably by simple duplication. Studies of the differences between their DNA sequences date their divergence at around 500 million years ago—at the dawn of vertebrate history.

Like most eukaryotic genes, globin genes are interrupted by noncoding regions, or introns. Both the α- and β-globin genes of all vertebrates so far examined consist of three exons separated by two introns that are in corresponding, or homologous, positions. This, too, is consistent with their common origin. Unlike exons, introns accumulate mutations at a rapid rate. Thus intron

base sequences differ widely, though their sizes have changed relatively little. Despite the great antiquity of the early gene duplication that gave rise to α- and β-globin genes, the first intron is 116 to 130 base pairs long in all mammalian α- and β-globin genes so far studied. The second intron also varies little within the α- and β-globin gene families.

An intriguing story, now unfolding, suggests that during their long history vertebrate globin genes lost an intron. At one time there may have been three, not two, as we now see them. During globin synthesis, amino acids are assembled in four discrete regions, recently termed structural units F1, F2, F3, and F4. The F1 unit corresponds with the first exon or coding sequence, the last unit, F4, corresponds with the third exon. But the middle two units, F2 and F3, are encoded by a single exon, the middle one of the three, not by separate exons. Thus, selection pres-

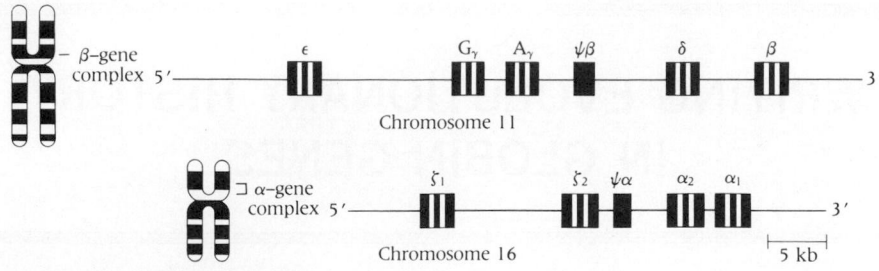

35K Organization and chromosomal locations of the human α and β globin gene clusters
The α and β globin genes each have two introns (the light bands in each gene).

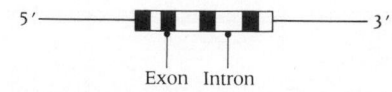

35L Structure of the leghemoglobin gene
The gene has three introns.

sure may have eliminated an intron that was present in an ancestral gene.

Corroboration of this view has come from an unexpected source—studies of the root nodules of leguminous plants. Nitrogen fixation, the prime task of root nodules (see Chapter 44), is facilitated by the presence of a protein that avidly binds oxygen. That protein is **leghemoglobin,** a plant molecule that is very similar structurally to animal myoglobin and hemoglobin. (We considered leghemoglobin and its interesting evolutionary implications in Box 34-1.) When the Danish investigators E. Ø. Jensen and colleagues sequenced the leghemoglobin gene of soybeans, it appeared that the similarity to hemoglobin was not a result of convergent evolution, as had been supposed. Although the animal and plant genes were closely homologous, the leghemoglobin gene had *three* introns (and therefore four exons), not two (Figure 35L). The deletion of leghemoglobin's extra intron would make the gene resemble that of animal globin.

It has been suggested that the presence of intervening sequences, or introns, in DNA can speed evolution by allowing novel proteins to be constructed from the pieces of existing ones. Perhaps far back in evolutionary history there was a separate small heme-binding protein that was encoded by a gene with two exons. Later, more exons may have been added to form the primordial globin gene of long ago, which then lost the intron that divides the heme-binding region. In this view, in the vast span of evolutionary time genes are assembled, like Lego toys, from "mini-genes" that correspond with structurally stable protein units or domains. The correspondence of the gene for part of the structure of certain cytochromes (which also bind heme) with globin's central exon supports the notion of an ancient heme-binding gene, which then traveled along several evolutionary paths to give a family of different heme-binding proteins.

Tracing the evolution of the β-globin cluster has shown that in addition to the five functioning structural globin genes, the 60-kilobase locus also contains a so-called pseudogene, designated ψβ (see Figure 35K). Pseudogenes bear a close resemblance to known genes, yet they are disabled through additions or deletions in their structure that prevent normal transcription and translation. Some think they may regulate the activity of neighboring genes. Others see them as diverged products of gene duplications—not necessarily relics of evolutionary change, but potential new genes and favorable sites for evolutionary change. Once a duplicated gene is released from selection pressure it is free to diverge extensively in its base sequence. Even though the β-globin gene complex contains a relatively large number of active genes, 95% of the locus is made up of noncoding DNA. The pseudogenes constitute a small fraction of the region. Some of the DNA is made up of representatives of well-known families of repetitive sequences. The remainder is DNA of no known function or meaning.

Gene evolution, as typified by the globin family, is a dynamic process, with major changes manifested in a stepwise manner. The principal source of the changes are gene duplications followed by sequence divergences that yield related, though distinctly different, members of the globin gene family.

slime mold *Dictyostelium discoideum* has 17 actin genes. The common yeast *Saccharomyces cerevisiae* has only one.

In addition to movements of genetic elements within the genome, it is clear that the evolution of cells in higher organisms has involved the migration of genes between the genome and such organelles as mitochondria and chloroplasts. The entangled functions of nucleus and organelles betoken a long and complex history.

Finally, yet another powerful methodology for the study of distant phylogenetic relationships among organisms rests on determining the nucleotide sequence of 18S ribosomal RNA (rRNA). This relatively simple technique guarantees that the sequences studied represent commonly transcribed RNA genes, not minor or inactive ones. It also focuses on the most conserved portions of the 18S rRNA molecule, which are most useful for broad phylogenetic comparisons. In Chapter 44, we shall point out specific contributions on the course of evolution that have come from rRNA molecular sequence data.

In sum, molecular biology has given us powerful—and controversial—new tools for studying evolution. By examining the amino acid sequences of proteins or the base sequences of genomic DNA or ribosomal RNA, we can begin to approach the mutational and transpositional history of single genes. Such studies have already shown that stepwise changes in these critical molecules are the foundation of evolutionary change. These powerful techniques have barely begun to scratch the surface.

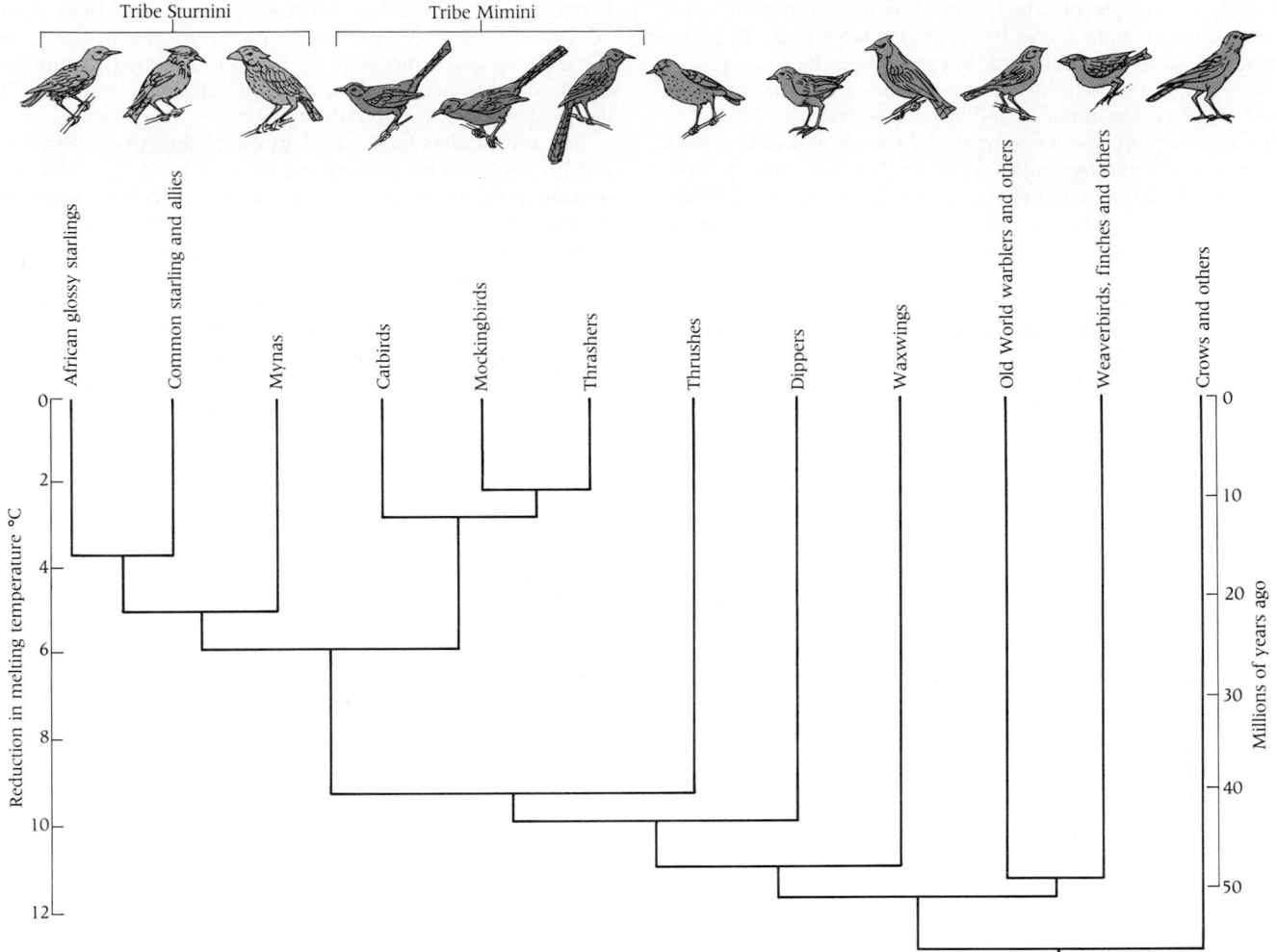

35-24 The avian phylogenetic tree
The evolutionary relationships between various bird groups can be determined by DNA-DNA hybridization. As discussed above, the greater the reduction in melting temperature of DNA-DNA hybrids, the larger the number of differences between the two DNA strands. Assuming that DNA evolves at the same average rate in different species, a decrease of 1°C in the melting temperature is equivalent to 4.5 million years. The data suggest that starlings and mockingbirds occupy adjacent branches of the tree, having diverged 25 million years ago. Both groups are related to the thrushes more closely than to other bird groups shown.

SUMMARY

Natural selection depends on the variability in traits among individuals of the same species, the heritability of some of the variations, and the differences they may produce in fitness or adaptedness. Variants that improve adaptedness enhance the production of viable offspring. Although adaptations may give an appearance of being purposive, it is important to remember that evolution is not goal-directed.

Adaptations range from narrow species-specific traits to the broad, fundamental traits shared by all members of a phylum or kingdom. Evolutionary adaptations must be distinguished from adaptive responses to specific situations which an individual may make in the course of a lifetime. The latter are not passed on to

offspring—mainly because such a phenomenon would require that changes in somatic cells be passed "backwards" to the germ cells of an organism. Nonetheless, the capacity to make individual adaptive responses is usually a heritable, and highly adaptive, trait.

Not all traits are adaptive. Conflicting selection pressures, environments that change faster than species can adapt to them, pleiotropic genes, and constraints on developmental mechanisms can all produce species characters that are neutral or even harmful. Even traits which are adaptive often arise from combinations of accidents that give new uses to old molecules, organs, or behaviors.

Protective coloration, which may serve to hide an organism (cryptic coloration) or warn predators away (aposematic coloration), is a well-studied example of adaptation at work. Coloration often involves mimicry, as do many behavioral adaptations. To assess the adaptive values of a trait, one must estimate its costs as well as its benefits. Costs include the opportunity cost, the energetic cost, and any increase in risk of injury or disease.

When two or more species compete for the same resource, each has the option of avoiding conflict by diversification: either by specializing, by broadening competence, or by changing to a different mode of life. They can change only if there is a new environment or ecological niche which is physically accessible to them and in which the competition is relatively low. Furthermore, they must be preadapted to some extent—that is, they must possess traits which allow them to survive in the new environment, even before natural selection causes them to adapt more successfully to it.

It is sometimes said that the concept of "survival of the fittest" is circular or tautological because "fitness" has been defined as "those who survive." However, a newer view of fitness invokes the independent concept of being well-designed to function in a certain environment, without making reference to survival. Hence, the proposition is not a tautology.

The advent of molecular biology established that natural selection acts on a species by influencing its gene pool. Heritable variation is nothing more or less than the variations in DNA nucleotide sequences that produce differences in the amino acid sequences of proteins. Amino acid sequence homology—the degree to which sequences for a single protein are the same in different species—permits assessments of the evolutionary relatedness of the species. Stretches of sequences that are conserved and do not vary from species to species often turn out to be essential for the protein's function.

KEY TERMS

adaptation
anatomical adaptation
aposematic coloration
archetypal protein
Batesian mimicry
behavioral adaptation
behavioral mimicry
brood parasitism

convergent evolution
cryptic coloration
energetic cost
homology
industrial melanism
maladaptive character
mimicry
molecular evolution

Müllerian mimicry
nonadaptive character
opportunity cost
physiological adaptation
pleiotropy
protein family
pseudogene
risk cost

QUIZ QUESTIONS

1. Evolution of the panda's thumb and evolution of the mammalian ear ossicles are examples of
   A. differential reproduction.
   B. Lamarckism.
   C. adaptive radiation.
   D. convergent evolution.
   E. historical opportunism.

2. Who of the following was most likely a proponent of the hypothesis of acquired traits?
   A. Charles Darwin
   B. Jean Baptiste Lamarck
   C. H. W. Bates
   D. August Weissmann
   E. None of the above

3. Industrial melanism illustrates the phenomenon of
   A. directional selection.
   B. concealing coloration.
   C. aposematic coloration.
   D. A and B
   E. A and C

4. For an adaptive radiation to be successful,
   A. the transitional form must face high levels of competition in order to refine the adaptations that allowed it to enter the new environment.
   B. the invader organisms must have a design appropriate for the new environment.

C. the transitional form must be provided physical access to the new environment.
   D. A and B
   E. B and C

5. Which of the following groups of birds is correctly matched with the islands on which these birds underwent an adaptive radiation?
   A. jays; Falkland Islands
   B. honeycreepers; Hawaiian Islands
   C. sparrows; Bahamas
   D. penguins; Aleutian Islands
   E. None of the above

ESSAY QUESTIONS

1. What are the four basic tenets of evolution outlined by Darwin in *The Origin of Species*? What is the most basic biological result of evolution?

2. Give some examples of structural adaptations. Physiological adaptations. Behavioral adaptations. Contrast broad and narrow adaptations of the typical woodpecker.

3. How does the evolution of the mammalian ear or the panda's thumb illustrate the role of historical opportunism in evolution?

4. What are the various means by which the coloration of moths and butterflies is adaptive? Contrast Batesian and Müllerian mimicry. Can coloration be adaptive without reference to predation?

5. Because of the redundancy of the genetic code, genes can sustain "silent mutations" which do not alter the sequence of amino acids. Are hemoglobins from related species likely to be more homologous in the amino acid or nucleic acid sequence? Why?

6. How do evolutionary biologists reconcile the supposed tautology in the phrase "survival of the fittest"? What is the biological definition of "fitness"?

7. To enter a new mode of life or environment, a species needs three types of opportunity. What are they?

REFERENCES AND SUGGESTIONS FOR FURTHER READING

Bendall, D. S. (Ed.) (1983). *Evolution from Molecules to Men*. Cambridge University Press, New York.

A stimulating collection of articles on molecular evolution.

Bonner, J. T. (1988). *The Evolution of Complexity by Means of Natural Selection*. Princeton University Press, Princeton, N.J.

Viewing the broad sweep of evolution, the author of this attractive book considers the evolution of complexity—notably size increase (relating overall size to organismic complexity), ecology (the diversity and abundance of organisms within communities), and animal behavior.

Cairns, J., Overbaugh, J., and Miller, S. (1988). The origin of mutants. *Nature 335*, 142–145.

This surprising paper claims that bacterial mutations are induced by environmental factors—seeming to support the inheritance of acquired characters. One reviewer termed it a "heresy in evolutionary biology."

Doolittle, R. F. (1981). Similar amino acid sequences: chance or common ancestry? *Science 214*, 149–159.

A well argued brief on the possible pitfalls of tracing evolutionary lineages on the basis of conserved amino acid sequences.

Felsenstein, J. (1988). Phylogenies from molecular sequences: Inference and reliability. *Annual Review of Genetics 22*, 521–565.

A thorough and critical review.

Field, K. G., and 7 other authors. (1988). Molecular phylogeny of the animal kingdom. *Science 239*, 748–753.

Having applied the sequencing of 18S ribosomal RNA to the study of evolutionary relationships, the authors present a scheme that is not entirely in agreement with traditional views of what happened in the remote past.

Gould, S. J. (1980). *The Panda's Thumb: More Reflections in Natural History*. W. W. Norton, New York.

The author's instructive essays on the panda's thumb are among many in this elegant collection.

Jeffreys, A. J. (1982). Evolution of globin genes. In: *Genome Evolution*. Dover, G. A., and Flavell, R. B. (Eds.), Academic Press, Orlando, Fla., pp. 157–176.

On the usefulness of globin genes as guides to the course of evolution.

Jensen, E. O., and 4 other authors. (1981). The structure of a chromosomal leghemoglobin gene from soybean. *Nature 291*, 677–679.

From comparative studies of the leghemoglobin and human hemoglobin genes, the authors attempt to explain their evolutionary history.

Ribbink, A. J. (1977). Cuckoo among Lake Malawi cichlid fish. *Nature 267*, 243–244.

An interesting description of cuckoo-like behavior by fishes. While diving in this African lake, the author found mother cichlids caring for mixed broods—offspring of their own species and young fry of another species. That is what cuckoos do.

Sibley, C. G., and Ahlquist, J. E. (1986). Reconstructing bird phylogeny by comparing DNAs. *Scientific American 254*, February, 82–92.

An impressive demonstration of the apparent success with which rates of DNA-DNA hybridization clarified the evolution and classification of various living bird groups.

Stebbins, G. L. (1977). *Processes of Organic Evolution*. Prentice-Hall, Englewood Cliffs, N.J.

A thoughtful synthesis of the evolution of plants and animals.

Wilson, A. C. (1985). The molecular basis of evolution. *Scientific American 253*, October, 164–173.

The discovery that mutations accumulate at steady rates over time in the genes of plants and animals has led to new insights into the course and tempo of evolution. The author shows how.

ORIGIN OF SPECIES

In many ways, the exhilarating appeal of biology lies in the diversity of living things. Evolutionary processes produced that diversity, which we exhibited in a symbolic way in Chapter 1. Diversity came in the wake of numerous adaptations to diverse environments. That, of course, is an important generalization, but students of evolution need to understand in detail how diversity arose. We consider some of the "details" in this chapter.

Individuals differ in any population. Still, in the natural setting of a reef or forest or polar icecap, organisms fall clearly into groups of nearly similar form. True, there are differences among individual buttercups in a meadow. But the similarities are so much more striking than the differences, that we recognize buttercups as a distinct kind of organism—a natural group of similar interbreeding individuals, a **species.**

VARIATION WITHIN SPECIES

Since no two individuals are ever exactly alike, we must distinguish between differences within species and differences between species. No matter how small the population, there is always variation among its members. As we seek to understand life's diversity, we shall encounter that fundamental proposition again and again. One has no difficulty in recognizing acquaintances in a crowd, because each person has a unique appearance. We are very familiar with human individual variation. It is, in fact, a characteristic of all sexually reproducing species.

SOURCES OF VARIATION

Heredity produces a range of reactions or capabilities. Throughout its whole life span, the specific circumstances of an organism's development determine where in that range an individual organism will actually stand. The differences that count most in evolution are the genetic differences. If an individual human or tree is taller than a neighbor solely because of better nutrition, that would be a difference between individuals, but it would ordinarily have little significance for populations of humans or trees. Such individual variation is nongenetic in origin and thus has no continuity in the population.

A list of causes of nongenetic variation is a long one and includes:

- Effects of the seasons (in the fall, for example, many mammals in cold regions molt into white winter fur)
- Generational variations (for example, in some rapidly reproducing insects, dry season individuals may be a different color than rainy season individuals)
- Other ecological variations, many of which we discuss in Chapter 45
- Social variations (for example, the caste system of social insects)
- Effects of disease, accidents, and parasites

Origin of species
Paleontologists exploring the diversification of life study fossils, true artifacts of Earth's history. Many fossils, like those shown here, have a haunting and evocative beauty.

In the face of phenotypic variability in nature, one can never say without further study how much of it is genetic in origin and how much nongenetic. Only careful breeding experiments can yield that information. It is of interest that genotypes may vary greatly with regard to possible individual ranges of the phenotypes they determine. For instance, blood group genes apparently permit no range at all, while the multiple genes affecting height often permit very wide ranges.

DIFFERENCES BETWEEN DEMES

As noted earlier, most species are divided into more or less distinct **demes**—also termed **local Mendelian populations.** Such species display two kinds of variation. One is the variations within demes that we have just been discussing. The other is seen as differences between demes. Two demes, in fact, are unlikely to be identical.

In sexually reproducing species, interbreeding between adjacent demes occurs occasionally and sometimes regularly. Offspring from one deme may wander off into the other deme. In many species, it is typical for males to move from the family's deme to other demes. For these reasons, adjacent demes usually intergrade. What does this mean? It means, first, that adjacent demes are seldom completely distinct from each other in any characteristic. It means, too, that it is rare for all the individuals of one deme to be of one form while all those of an adjacent deme are of a different form. If a character takes two forms, adjacent demes usually will have representations of both forms. When demes intergrade in a character in which they differ, the difference is usually recognized in the percentages of individuals of each form. We earlier saw variation of this sort in the variations in blood group frequencies in different tribes of Montana Native Americans (see Chapter 34).

With only rare (and short-lived) exceptions, two demes of the same species do not occur in the same place at the same time. If they did, the demes would fuse into one. Geography is often a relevant factor in these situations. Some geographic patterns of variation are quite irregular. Differences among demes of some species seem to be scattered about in checkerboard or kaleidoscopic fashion, without evident rhyme or reason (Figure 36-1). Such irregular patterns seem to have one of two causes, or a combination of them.

■ Demes may differentiate by selection and the resulting patterns may reflect local adaptations to irregularly distributed environmental conditions. In that case, an irregular pattern indicates an irregularly heterogeneous environment.

■ Demes may also differentiate into irregular patterns by genetic drift.

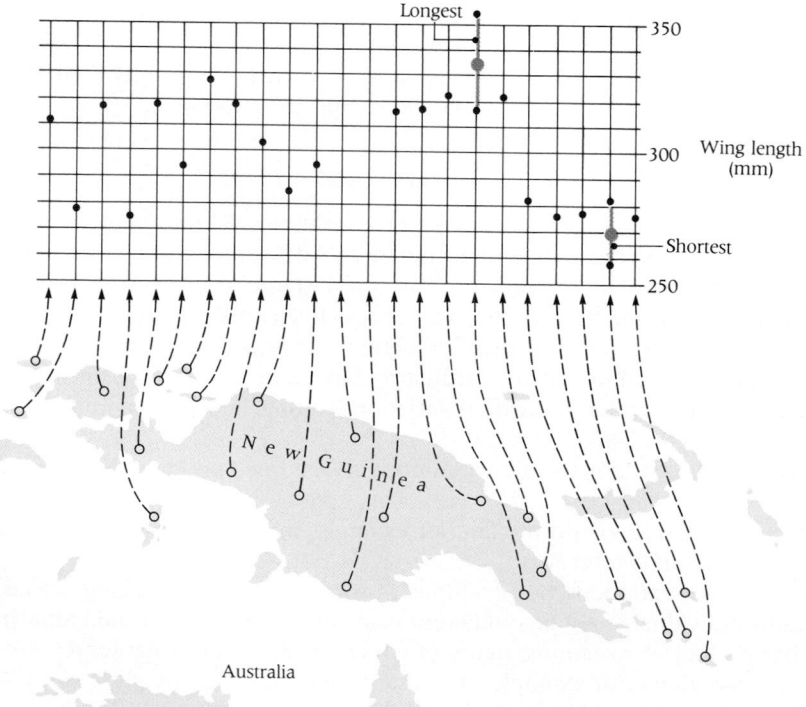

36-1 Chaotic interdeme variation
The mean wing lengths in 23 demes of the cockatoo, Cacatua galerita, *appear as solid dots above the map of New Guinea, which locates each deme. The variation is significant, as shown by the small ranges of variation (vertical bars) in the longest-winged and shortest-winged populations. There is no obvious trend or pattern in the variation.*

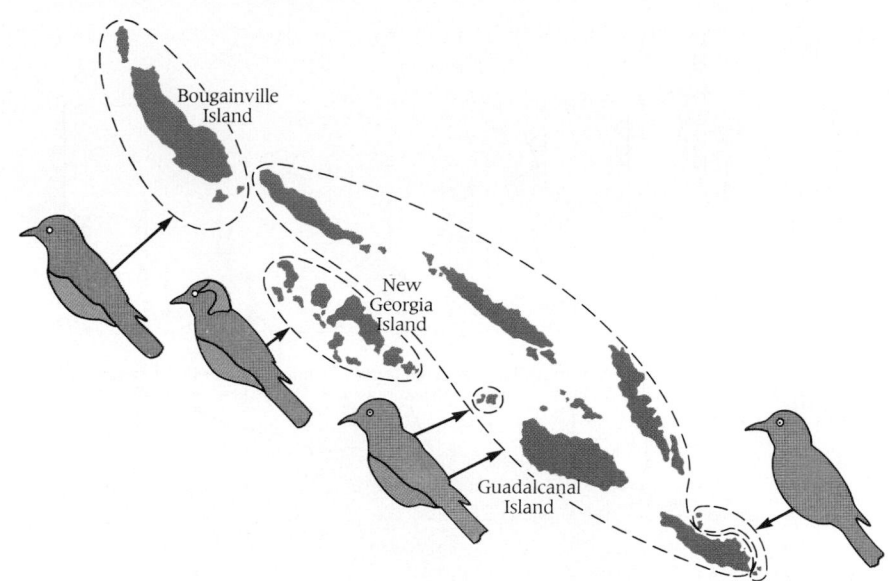

36-2 Subdivision of a bird species into subspecies
Five subspecies of the flycatcher, Monarcha castaneoventris, *exhibiting four different patterns, occupy different parts of the Solomon Islands (arrows). These are termed geographic subspecies.*

Usually all demes over a wide area are similar because they interbreed. Interbreeding over the area may be so common that there arises a large single deme or a complex of indistinguishable demes. This, we could say, begins to resemble a species. Putting it in other words, species are groups of actually or potentially interbreeding natural populations, or demes, which are reproductively isolated from other such groups. Before offering this as a definition, we must add the notion that interbreeding must lead to fertile and viable offspring. Horses and donkeys can interbreed but this combination does not not produce viable or fertile offspring.

Often there may be phenotypically distinct demes that are separated geographically. In biological classification, such deme complexes are called **subspecies,** or **races** (Figure 36-2). As discussed in Chapter 49, subspecies may be even more difficult to define rigorously than species. One reason is that where two subspecies of a given species are in contact, there is usually a zone of intergradation as a result of interbreeding—even though the subspecies are distinctive at a distance from this zone. Figure 36-3 illustrates such a pattern among two subspecies of snowbirds of the Northwest.

Zones of intergradation are not always narrow. The whole population in a region may intergrade from one end of its area to the other. In such cases, differences between forms may be slight at each point of the distribution, but forms may differ dramatically at the extremes. For example, populations may show gradual differences as they are followed from lower to higher elevations (Figure 36-4) or from wetter to drier locations. This pattern of continuity and complete intergradation implies

36-3 Intergradation of subspecies
The map shows the geographical distribution of two subspecies of the small snowbird, Junco oreganus, *and the intermediate population where contact between subspecies occurs. Symbols identify the localities studied. Frequency distributions of male wing length for J. o. shufeldtii, J. o. montanus, and the intermediate population reveal that the mean wing length (marked by arrow) for the intermediate population lies between the means of the two subspecies.*

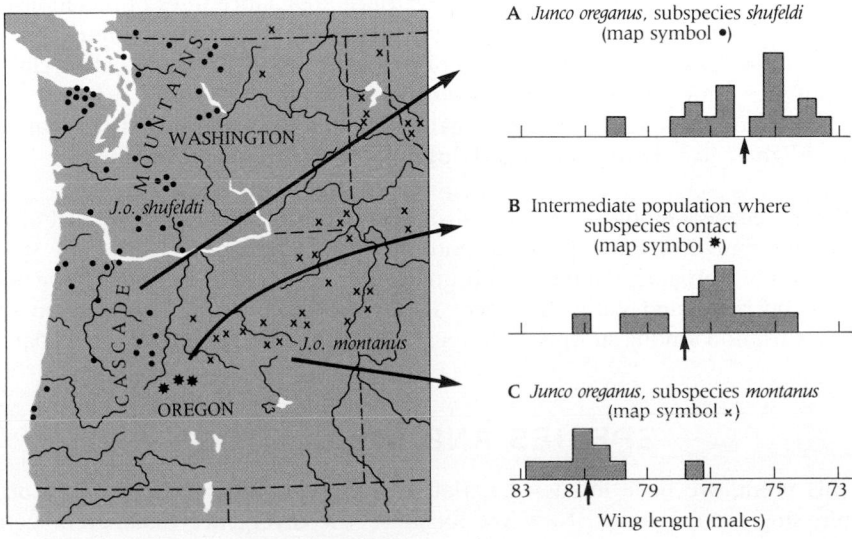

A *Junco oreganus,* subspecies *shufeldi* (map symbol •)

B Intermediate population where subspecies contact (map symbol *)

C *Junco oreganus,* subspecies *montanus* (map symbol ×)

83 81 79 77 75 73
Wing length (males)

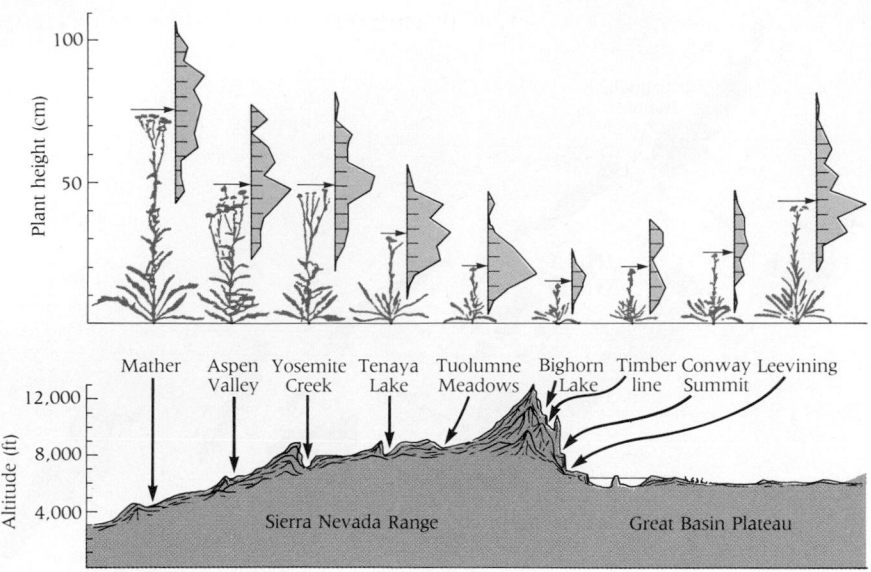

36-4 An altitudinal cline
The plant species Achillea lanulosa *occurs at all altitudes in the Sierra Nevada Range of California. Populations at lower altitudes are taller and those at higher altitudes are shorter. The range and mean (arrows) of each population at a given location are indicated. Considerable adaptive intraspecific variation exists in relation to the different environments at various altitudes.*

that interbreeding of adjacent demes is common and that the gradual change is due to gradually changing environmental conditions. In other words, there is an environmental gradient. A sequence of poorly separable demes with gradual change from one area to another is called a **cline.**

Many populations vary clinally in the frequency of different alleles at one or more loci. The clover *Trifolium repens* yields a higher ratio of cyanide-producing plants in the south than in the north. The frequency of cyanide producers apparently depends on a balance noted earlier between the advantage they derive from being distasteful to herbivores and the disadvantage they suffer when frost disrupts cell membranes, releasing cyanide into the plants' tissues.

Such environmental gradients are common. Every mountain range has gradients from bottom to top, from warm to cold, and from dry to wet. Plains have gradients from south to north. Lakes and seas have gradients of temperature, salinity, and light from the depths to the surface. These environmental gradients are commonly accompanied by gradients in the characteristics of the organisms inhabiting them. Sometimes gradients are similar in many different species. Three rules or generalizations, named for the nineteenth-century zoologists who posed them, apply (within any one species) to many mammals and birds.

- The average size of the individuals tends to be smaller in warmer climates and larger in colder climates **(Bergmann's rule).** These differences can be impressive. The usual explanation is that the volume of the body increases as the cube and the surface of the body as the square of any linear dimension. Large bodies thus will have a smaller surface to volume ratio. In cold climates, it would be advantageous to decrease the relative surface area, since rates of metabolism are more nearly proportional to body surface area than body volume.
- Protruding parts such as tails, ears, or bills tend to be shorter in colder climates than in warmer climates **(Allen's rule).**
- Colors tend to be darker (more heavily pigmented) in warm moist climates and lighter in cold dry climates **(Gloger's rule).** The reason is unclear.

These "ecogeographical" rules and others like them have many "real world" exceptions, but they are applicable more often than not. In fact, they are specific extensions of a broader and more profound generalization: Differences among demes tend to be correlated with differences in their environments. Putting it in other words, variation among subspecies or local populations of a species is often adaptive.

SPECIES AND SPECIATION

To this point, we have discussed variation and evolutionary change as it occurs within single populations. Now we focus on the divergence of different genetic lines. It is this divergence that leads to the emergence of higher taxonomic groups—and it begins with the differentiation of demes and their subsequent development into different species. We will see that speciation can be rapid and innovative.

THE SPECIES PROBLEM

Because organisms are classified by species,[1] the species has basic importance for biological studies. Ernst Mayr considers species as the real units of evolution and regards them as "the temporary incarnation of harmonious, well-integrated gene complexes." Yet biologists still have difficulty in rigorously defining "species," and for centuries they have debated what has come to be called "the species problem." In spite of millions of spoken and written words, the issue remains a lively one. Let us consider it briefly.

As we discuss more fully in the next chapter, biological classifications were long based on the most obvious features of organisms: structure. Organisms were classified according to how they are put together and what they look like. That approach relies on phenotypic differences that are taken as evidence of genetic individuality. For that reason—because it relies on phenotypic rather than genotypic criteria—biological classifications are not universally accepted. Since a phenotypically similar trait in two organisms may reflect very different genotypes, phenotypic structural features may be an unreliable basis for recognizing and defining species.

We might suppose, then, that the ability to interbreed is a defining characteristic of species. But aside from certain complications to be discussed in a moment, there is the problem of time and place. Populations that do not occur together in time and place cannot be put to the test. We will never know if Australian rabbits of long ago (and there were such rabbits) could interbreed with American rabbits of today. Must we say then that they are not of the same species?

Still another difficulty arises in asexual or self-fertilizing populations. Here a capacity to interbreed ordinarily cannot be tested and we are forced to rely on anatomical or other phenotypic traits. But what do we say about two such organisms—for example, two bacteria that look quite similar—when one is sensitive and the other resistant to penicillin?

We begin to see why species, the most fundamental persistent biological unit, cannot be defined simply. If structural similarity were the sole criterion, arbitrary decisions would be necessary on how rigid one wishes to be, since no two living beings are exactly alike. This would leave us with a concept that lacks precision. The situation is perhaps not as bad as we are making it sound. Two biologists may disagree on the strict definition of a species, but if they look at the same natural populations they will probably agree nine times out of ten on which groups are separate species and which are not. While squabbling over the tenth group, they often find they are saying the same thing in different words.

The tendency today is to take a middle position in defining species. Everyone agrees on certain elements of the definition: species are made up of populations of *similar* individuals, *differing* from other species, which breed *mostly* among themselves (and *rarely* with other species), are reasonably *stable* and yet are sufficiently flexible and plastic that they tend to *change* in the course of many generations. When we examine this statement (in which the vague words are italicized), we find that it says something about the (1) phenotypic character, (2) breeding behavior, and (3) status as evolutionary unit of a group of organisms.

Species Are Phenotypically Similar

A biologist speaking of a particular species has more in mind than **reproductive isolation,** an important aspect of the evolution of species. Species do have other attributes. Members of a species look alike—more or less. Consequently, we can say that species will differ from one another in phenotypic characters. These include many traits beyond looks. Defining characters will also involve physiology, biochemistry, and behavior. In these qualities, too, there is usually a gap or discontinuity between species. Therefore, if we find no intermediate populations, we can be more confident in calling a given group a species.

---

[1] The noun "species" can be singular or plural.

Thus, a species is one or more populations of which the individual members are rather closely related and therefore similar in many characters, although there is some variation among them. The kinds and degrees of variation are important.

Species Interbreed Freely

As long as the individuals in a group (1) can interbreed and thereby produce fertile offspring, and (2) can do so with some frequency, the group shares a gene pool and has the unity and continuity by which we define a species. Still, what is relevant is the extent of interbreeding under natural conditions, not experimental conditions.

The point is commonly illustrated by certain well-known **hybrids.** Horses can be bred with donkeys to produce mules. Lions can be bred with tigers to produce hybrids called "tiglons" (when the tiger is father) or "ligers" (when the lion is father). This proves that the mating forms are genetically related. It does not prove they belong to the same species. The critical question is: do the two forms breed freely when they have free access to one another in the field? Lions and tigers coexisted in large numbers in India until their populations were reduced by hunting. Still, no tiglons and ligers have ever been encountered in nature.

Nevertheless, some hybrids do arise in nature, with modest frequency, from some closely related species. Indeed, hybridization is relatively common among plants (Figure 36-5). Hence, our definition of species cannot demand that two distinct species *never* interbreed. The genetic distinctiveness of a species is preserved when breeding between species is markedly less frequent than breeding within species. Distinctiveness is also fostered by the fact that many hybrids are much less fertile than offspring of parents of the same species. However, some hybrids, natural and artificial, are completely fertile.[2] Hence, the term species is much less useful for plants than it is for animals.

Species Are Units of Evolution

Finally, we note the evolutionary significance of species. Natural populations, though they may be static for considerable lengths of time, are never permanently static. They are constantly changing, splitting up, often becoming less and less similar to each other (but sometimes converging). In a word, they are evolving.

We said before that a species is a group of organisms so similar in structure and heredity that their demes may intergrade and fuse. However, demes as such do not necessarily have evolutionary continuity. A species does. That is what makes species significant. We would say, then, that a species is a sequence of ancestral and descendant interbreeding populations that evolve independently of all others, that plays out its own evolutionary role. But how do new species arise?

SPECIATION

Obviously, two species with common ancestors in the remote past would differ only if one or both had changed. Changes occurring over time in a population would not of themselves increase the number of species unless circumstances arose that split the population. When that happens, speciation may take place. As mentioned earlier, **microevolution** denotes the processes producing genetic changes and variation within a population over a few generations. **Macroevolution** refers to the larger-scale processes of speciation.

A new species may arise in two ways.

- In one, a single lineage (succession of ancestral and descendant populations) may undergo so many changes affecting reproductive compatibility that the descendants become distinctive enough to be considered a new species when it appears that reproductive isolation has been established. Such a sequence is called successive, successional, or transient speciation. In biologist's jargon, it is referred to as **anagenesis,** or **anagenetic** (from a Greek root meaning *upward*) **speciation.**

36-5 New species arise by interspecific hybridization
This is the common potato plant, Solanum tuberosum, *which has a tetraploid chromosome number. We believe its diploid ancestors hybridized naturally.*

---

[2] Although the subject is disputed, some systematists believe that in some cases, at least, hybrids are not inherently or initially infertile, but become so as a result of negative selection on individuals and positive selection to make the parent species cease to hybridize.

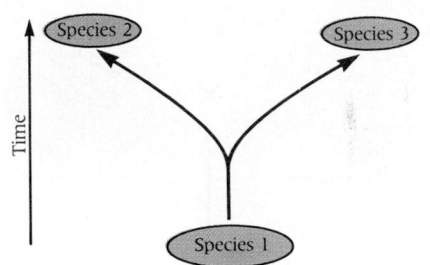

36-6 Divergent speciation
In this pattern of speciation, the ancestral species (1) gives rise to two (or sometimes more) descendant species (2 and 3).

- More commonly, speciation begins with the divergence of different genetic lines of an ancestral species. The divergence ultimately leads to formation of new species. This process is called **cladogenesis,** or **cladistic** (from a Greek root meaning *branch*) **speciation.**

Cladogenesis

In cladistic speciation, a group splits into two or more species somewhere along the line. In Figure 36-6, an ancestral or parental species (species 1) has given rise to two descendant species (species 2 and 3). In the opinion of many workers, macroevolution consists of a series of such dichotomies. Note that by definition the ancestral species ceases to exist following the dichotomous split.

How this type of speciation is believed to operate in evolution is sketched in Figure 36-7, which depicts a sequence of gradual evolutionary changes and illustrates the meaning of clades and grades. A **clade** is a group of species recently derived from a common ancestral species. (A clade is shown in Figure 36-6.) A **grade** was originally defined by Julian Huxley as a "unit of biological improvement."[3] That, of course, refers to improvement of adaptation. The point is that several evolutionary lines can reach the same adaptive state independently, but they do not necessarily share a recent common ancestor. Attainment of a new level of adaptedness—for example, lung breathing—often coincides with cladogenic branchings of lineages. But often it does not. Note in Figure 36-7 that members of a clade may belong to different grades because of differences in each line's rate of evolutionary change.

Speciation ordinarily takes place when parts of a population become isolated from one another. In many cases, isolation is a critical factor underlying speciation. As we will explain later in this chapter, one outcome of isolation is adaptive radiation.

Is Speciation Gradual or Punctuated?

There has been much argument on whether speciation occurs suddenly or gradually. Some early geneticists, notably the pioneer geneticist Hugo de Vries, and other biologists believed that sudden evolutionary leaps (saltations) could occur as a result of mutations. In their view, "big mutations" could abruptly yield new species— "hopeful monsters" in one worker's terminology—or even a whole new phylum. Of course, a new species is not a mutant or new form or type of organism. A new species is a new, reproductively isolated population of organisms and a single mutation rarely if ever leads to such a population. In fact, it is unlikely for it to do so. Therefore, most workers, especially after the emergence of population genetics, returned to the view held by Darwin: speciation is usually a gradual process.

We must record, however, that not all biologists today accept this gradualist model. Gradualists maintain that big changes occur by the accumulations of many small ones. In contrast, advocates of a model known as **punctuated equilibrium** hold that new species undergo most of their phenotypic modification as they first branch off from their parent species, and then change little. Proponents of the punctuated equilibrium model have advanced three major arguments.

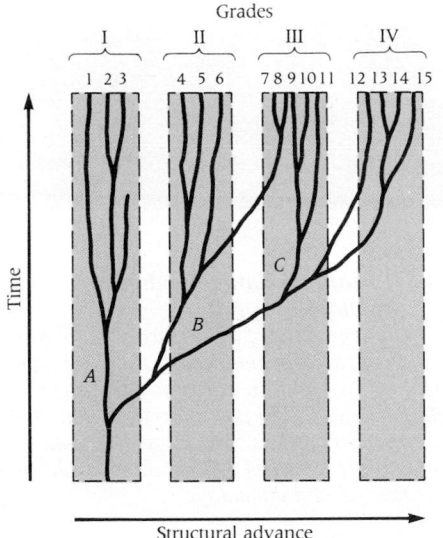

36-7 Grades and clades
A group of species with a recent common ancestor (for example, species 1, 2, and 3) forms a clade, that is, a branch of the evolutionary tree. A group of species at the same level of functional complexity (for example, species 7–11) forms a grade. Clade A forms Grade I. Clade B gives rise to Grades II and III. Grade III has been achieved twice by different clades, B and C.

- First, they have held that the fossil record reveals long periods of little or no change (termed periods of equilibrium, or **stasis**) that are punctuated by short periods of rapid change in which there is the appearance of new species, or macroevolution. Gradualists would respond that the fossil record is not complete and that speciation has often occurred without detectable differences in the fossil record.
- Second, punctuationalists have pointed to some actual observational data on specific evolutionary sequences that seem to support their position. One of the most impressive was a detailed 1981 study by P. G. Williamson of 21 species of fossilized freshwater mollusks from the eastern Turkana basin of East Africa. According to this "uncensored page of fossil history," as one editor described it, population measurements clearly revealed the pattern of evolution over the

---

[3] Be careful not to confuse *clade* and *grade* with *glade,* a term defined in Chapter 34.

last 5 million years. In 16 of the 21 species there were long periods of stasis. The other 5 species also changed little for most of the time, but then evolved rapidly during a "short" period of 5,000 to 50,000 years—that would be very short for a paleontologist—between long periods (5 million years) of stasis. This sequence, the first directly documented account of the evolution of one species into another, seemingly conforms to a pattern of punctuated equilibrium. But critics have argued, among other things, that these episodes of rapid change are not too rapid to be explained by conventional Darwinian natural selection. Stasis, they hold, reflects nothing more than stabilizing selection (see Chapter 34), in which selection favors typical members of a population relative to the extremes.

- Third, it is argued, recent work reveals that genomes are in a state of dynamic change. This provides the first plausible genetic mechanism for evolutionary leaps. As noted in the previous chapter (see Box 35-3), new evidence of noncoding intervening sequences (introns), "selfish" or "junk" DNA (repetitive sequences), and "jumping genes," suggests a previously unsuspected potential for large abrupt changes, in both chromosomal architecture and in the function of structural genes. There is a potential, to be sure, but key questions remain. How important is it? Could it lead to the establishment of a new species, which is what we mean by evolutionary change? Perhaps the answer will soon come from the accelerating convergence of two previously separate lines of research—evolutionary biology and molecular biology.

Instantaneous Speciation

Speciation based on accumulated changes in populations is a slow process. In some circumstances, however, speciation can occur instantaneously. This process, which involves changes in individuals, is an important exception to the rule that speciation does not occur by a single mutation or saltation.

By **instantaneous speciation,** we mean the production of a single individual (or offspring of a single mating) that for various reasons cannot interbreed with members of the species to which its parents belonged. If such an individual is to give rise to a new species, it must be able to survive and reproduce.

As far as we know, ordinary gene and chromosome mutations will not have such an effect (i.e., give rise to a new species) among biparental organisms, though they might do so in asexual groups. Mutations merely serve to increase heterozygosity in the population. However, when chromosome mutations and hybridization result in polyploidy (see Chapter 15), individuals will arise that cannot interbreed with their parental species but may produce populations among themselves. This process, a form of instantaneous, or abrupt, speciation, has occurred often enough among some plant groups to have had major effects on their diversification.

Polyploidy occurs when the number of chromosomes doubles, either within a species (autopolyploidy) or by combining chromosomes from two different species (allopolyploidy) (Figure 36-8). Polyploid individuals of either type usually cannot breed with either parent species because their chromosomes pair improperly in the first metaphase of meiosis. However, when such individuals can mate among themselves or self-fertilize, a new species can arise.

Because plants can engage in self-fertilization or reproduce asexually, speciation by polyploidy is seen most often in them. Indeed, about 47% of all flowering plant species are polyploid. Among animals, polyploidy is uncommon, the exception rather than the rule. However, it is found in the large group of ostariophysan fishes (minnows, carp, goldfish, catfish, etc.), isopods, dipteran flies, weevils, and some brine shrimp.

An example of such speciation is seen among the beautiful red, tubular-flowered gilias in the Mojave Desert of California. Five species are found there: three diploids and two tetraploids (with, respectively, two and four sets of chromosomes).

- One tetraploid, *Gilia malior,* arose by hybridization between *Gilia minor* and *Gilia aliquanta,* both of which are diploid.
- The other tetraploid, *Gilia transmontana,* is derived from *Gilia minor* and *Gilia clokeyei.*

These very similar looking species are reproductively isolated from each other.

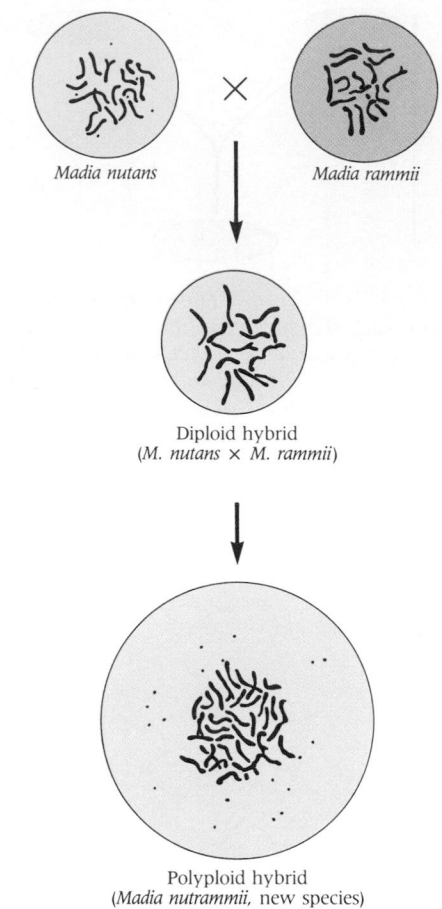

Madia nutans × Madia rammii

Diploid hybrid
(*M. nutans* × *M. rammii*)

Polyploid hybrid
(*Madia nutrammii*, new species)

A

B

36-8 Instant speciation by the development of polyploidy
(A) *The generation of a new species of California tarweed occurred spontaneously from two diploid species growing in an experimental garden.*
(B) *These tall "African" marigolds, familiar in American gardens, are tetraploid. Some varieties are produced by artificial hybridization of tetraploid and diploid types.*

TABLE 36-1
A CLASSIFICATION OF ISOLATING MECHANISMS IN ANIMALS

**Mechanisms that prevent interspecific crosses
(prezygotic or premating isolating mechanisms)**

Potential mates do not meet (*ecological isolation*)

Potential mates have different breeding seasons (*temporal isolation*)

Potential mates meet but do not mate (*behavioral isolation*)

Functional or metabolic factors prevent fertilization (*physiological isolation*)

Copulation is attempted unsuccessfully (*mechanical isolation*)

**Mechanisms that impair success of interspecific crosses
(postzygotic or postmating mechanisms)**

Sperm transfer takes place but egg is not fertilized (*gamete mortality*)

Egg is fertilized but zygote fails to survive (*zygote mortality*)

Zygote produces an F_1 hybrid but its viability is decreased (*hybrid unfitness,* or *breakdown*)

An F_1 hybrid zygote is produced and is fully viable but it is partially or completely sterile, or it produces a defective F_2 generation (*hybrid infertility,* or *sterility*)

REPRODUCTIVE ISOLATING MECHANISMS

Reproductive isolation serves to maintain the integrity of a species by protecting its gene pool and preventing its contamination by genes from other species. Any biological or physical factor that decreases interbreeding between different species is called an **isolating mechanism,** or **barrier.** Most species are protected by more than one isolating mechanism.

Reproductive isolating mechanisms prevent populations belonging to different species from interbreeding even if they live in the same area. As shown in Table 36-1, such mechanisms are usually based on (1) factors that prevent mating, called prezygotic (or premating) isolating mechanisms, or (2) factors that prevent reproduction if mating occurs, termed postzygotic (or postmating) isolating mechanisms.

Prezygotic Barriers

Prezygotic barriers or isolating mechanisms can be ecological, temporal, behavioral, physiological, or mechanical.

36-9 Reproductive (temporal) isolation in toads

These two common American toads breed at different times of the year. However, they are capable of interbreeding. (A) Bufo americanus. (B) Bufo woodhousei fowleri

A

B

Ecological isolation results from factors that may diminish interbreeding between otherwise similar populations in the same region for the simple reason that they have different habitats. Mountain climbers know that plants and animals tend to live at characteristic elevations. Interbreeding between similar populations may be reduced by their preferences for different altitudes. Florida has two groups of turtles that do interbreed to some extent but whose interbreeding is reduced by the fact that one group lives in lakes and the other in running rivers. Ecological factors often affect parasites, which are confined to certain plant or animal host species. Two species of parasites living on different hosts will not have many opportunities to mate.

Temporal isolation reduces interbreeding, as when animals have different breeding or pollinating seasons. These are determined by a meshing of behavioral patterns (see Chapters 32 and 33) and environmental conditions. In the eastern United States there are two groups of common toads that are usually regarded as distinct species (Figure 36-9). However, they occasionally interbreed, and when they do they produce fertile offspring. The populations as a whole are kept quite distinct, however, by the fact that one breeds early in the season and the other late. This type of isolation is common in plants.

Behavioral isolation is common among animals, especially insects and vertebrates. Males and females from different populations are simply not attracted to each other, or perhaps it would be more scientific to say they do not regard each other as potential mates. Such choices may be made by males, females, or both. Experiments with *Drosophila* suggest that males of one population are indifferent to females of another population while the females actively reject the "foreigners." Among guppies and other tropical fishes, the females do the selecting. Male guppies vary greatly in tail size, color, and color patterning—all heritable traits. Careful experiments by Bischoff, Gould, and Rubenstein showed that female guppies, like females of many other species, choose males on the basis of their conspicuousness. The larger the tail and the more flamboyant the color, the more attractive is the male to the female. Given a choice between males with tails of two different sizes (i.e, large and small, large and medium, or medium and small), females regularly choose to mate with males having larger tails.

Many animals have elaborate courtship rituals. In these species, a female does not breed unless she is properly stimulated by courtship behavior characteristic of males of her own species. A simple familiar instance is the strutting display of male turkeys. Some birds and fishes go through elaborate and complex courtships, displaying specific differences in breeding patterns (Chapter 34). Male sticklebacks (*Spinachia*) build nests attached to water plants and then, by zigzag swimming displays, induce females to lay eggs in them (see Figure 32-12). Differences in the choreography in different species are highly effective isolating mechanisms.

Interbreeding in plants is often limited by the tendency of pollinating insects to visit only one species of plant on a single foraging trip. This sort of behavioral isolation is carried to a high degree in orchids, for instance, which often have flowers that attract only one species of bee.

Physiological isolation includes various biochemical differences between populations. For example, the male and female gametes of interbreeding populations will combine, but those of different species will not. This may be due to biochemical features of the gametes themselves or of the seminal fluids. It has been directly demonstrated in one species of sea urchins, for example, that its sperm are much more capable of fertilizing the eggs of its own species than those of another species.

Finally, **mechanical isolation** is related to anatomical differences in sex organs, which by themselves can be an insurmountable barrier to interbreeding. Even within species, differences of size will tend to isolate subpopulations. St. Bernard dogs and toy poodles are both members of the species *Canis familiaris* but their size differences prevent interbreeding and thus tend to isolate them from one another. If there did not exist countless other intermediate dog breeds, which permit some gene flow between St. Bernards and poodles, they would probably evolve into two separate species.

Postzygotic Barriers

If mating does occur, a second set of barriers may act as an effective reproductive isolating mechanism (see Table 36-1).

36-10 Hybrid infertility
The mule is the product of a mating between a horse and a donkey. However, this beguiling animal (in Berks County, Pennsylvania) displays hybrid infertility.

■ **Gamete mortality** reduces the success of interspecific crosses. Even if gametes of different species are successfully brought together, they rarely form a zygote. Many aquatic animals release their gametes in the surrounding water where fertilization takes place. But even when two closely related species release their gametes simultaneously in the same place, cross-specific fertilization is rare. Gamete recognition is based on specific molecules on the surfaces of eggs, which adhere only to complementary molecules on sperm cells of the same species (see Chapter 21). Likewise molecular recognition mechanisms enable a flower to distinguish between conspecific pollen and that of other species.

■ When an egg of one species is fertilized by sperm of another species and a zygote does develop, development will not proceed beyond the embryonic stage because of the genetic incompatibility between the two species. This is called **zygote mortality.** Several frogs of the genus *Rana* may occasionally hybridize and form zygotes, but they generally do not complete development.

■ In a few cases, hybridizing species may produce F_1 hybrids that are viable and fertile. But when these hybrids mate with one another, or with members of either parental species, the offspring of the next generation is feeble or sterile. This process is termed **hybrid unfitness.**

■ Finally, even if a vigorous hybrid offspring is produced by a mating of two different species, reproductive isolation is often maintained because the hybrids are sterile. Such **hybrid infertility** or **sterility** is ordinarily caused by a failure of meiosis in the hybrid to produce normal gametes when chromosomes of the two parent species differ in number or structure. The best-known case of a sterile hybrid is the mule, a cross between a horse and donkey (Figure 36-10). Horses and donkeys remain distinct species because mules cannot backcross with either parental species.

Introgression

Genes may occasionally surmount all reproductive barriers. This could happen when fertile hybrids mate successfully with one of the parental species. We saw an example of this in Figure 34-9. Such passing of genes between species is called **introgression.** An example from agriculture is corn (*Zea mays*) which has some genes that came from a closely related wild grass (*Zea mexicana*). Introgression takes place when the two species hybridize and a fraction of the hybrids manage to cross with corn plants of the parental genotype. Such transfers of genes enrich the genetic variation that can then be exploited by breeders trying to produce new corn varieties by artificial selection. Because reproductive barriers do limit introgression to a trickle, the integrity of the two parental species is not seriously threatened.

MECHANISMS OF SPECIATION

The key processes in speciation are (1) genetic changes within populations, and (2) the splitting up of populations by sudden or gradual decreases in interbreeding between demes or other population groups. Several basically different kinds of mechanisms produce different sorts of speciation.

■ In allopatric ("other country") speciation, subpopulations are geographically isolated from one another.

■ In sympatric ("same country") speciation, populations are subdivided in the same general area without the establishment of physical barriers. Here the isolating mechanisms are biological (though biological isolating mechanisms operate as well in allopatric speciation).

■ In parapatric ("neighboring country") speciation, subpopulations are in the same geographic area but are isolated in different ecological niches or in some other manner.

Allopatric Speciation

The most effective reproductive isolation occurs when populations are segregated by geographical barriers. Such isolated groups are called **allopatric populations.** Allopatric populations often diverge to become new species by the process of **allopatric speciation.** It should be understood, too, that when populations are separated geographically, it is often the case that the different geographic regions have different environments and differing selective pressures. These circumstances—as well as mere physical separation—may lead to the biological diversity that is a prelude to speciation. This, we believe, is the most common mechanism of speciation.

Two nonmigratory animals a hundred miles apart obviously cannot interbreed. Nor can two plants at such distances, except in the case where winds, birds, or insects carry pollen unusually far. If plants or animals of the same species were regularly present throughout the intervening region, we could then not call the situation true isolation as far as the population is concerned. Although given individuals at opposite ends of the occupied region might not interbreed directly, they could pass their genes on to the rest of the population in the course of a few generations. Only when real spatial gaps appear in the populations' distribution does interbreeding across the gap diminish significantly. Then the populations on either side of the gap diverge by independent microevolution of their gene pools. In time, the kinds of variations differ as do the selection pressures operating on them. As gene frequencies diverge, new isolating mechanisms of the biological type are usually then added to the geographic isolation so that representatives of the two populations could no longer produce offspring. The likely result is the development of a new species.

A frequently cited example of allopatric speciation is the evolution of two populations of tuft-eared squirrels (*Sciurus aberti*) on the north and south rims of the Grand Canyon (Figure 36-11). Their many similarities suggest that they must have come originally from one ancestral population. We have always assumed that they never (or seldom) interbreed because they could not cross the canyon—though the situation may be more complex than previously suspected (Box 36-1). As a result, gene flow between them diminished and they became visibly different—for example, northern squirrels have darker underparts and whiter tails. Though the difference is not yet very great, the two populations—Abert squirrels on the south rim and Kaibab squirrels on the north rim—have been considered distinct species. The geographic gap in allopatric speciation need not be as wide as the Grand Canyon. A stretch of grassland between two forests may decisively separate the plant and animal populations of the forests. Such discontinuities, narrow and wide, occur on every side. If they persist, they often lead to the kind of separations that produce new species.

Geographic isolation alone seldom permanently decreases interbreeding. Even the Grand Canyon might not entirely prevent the squirrels from interbreeding; some might cross it, and populations can maneuver around the ends of a barrier. Also, in the broad view, even so tremendous a barrier as the Grand Canyon could in a remote time become less of a barrier—or human intervention might manage to circumvent the barrier. If, in those circumstances, the two populations were again to occupy the same geographical area, they could continue to coexist as two different species if biological isolating mechanisms had arisen in the two groups. In almost every instance of speciation, both kinds of isolating mechanisms—biological and geographical—eventually decrease interbreeding.

Separate species rejoined in this way may end up competing with one another so that advantages are conferred on variants that acquire different characteristics. As a result, the two species may diverge faster when they resume contact than they did when apart. This phenomenon is called **character displacement.** We mention it again in the context of ecology in Chapter 47.

Sympatric Speciation

In **sympatric speciation,** a new species evolves within the range of the parent population. Thus, reproductive isolation is accomplished by biological means and without geographical barriers. Instantaneous speciation by polyploidy in plants is an example. A single plant formed by hybridization between two diploid species can give rise to a fertile tetraploid offspring that is reproductively isolated from both parental diploid forms. The formation of new polyploid species is especially likely in annual plants that practice self-fertilization and in long-lived perennials, in which there is an opportunity for vegetative shoots to undergo somatic doubling of chromosomes and to produce unreduced gametes in the flowers.

However, in general, speciation is gradual, as initially incomplete barriers to gene flow become progressively more effective. One model of sympatric speciation

A

B

36-11 Allopatric speciation
*The Kaibab (**A**) and Abert (**B**) squirrel populations are said to be examples of allopatric speciation, since they are geographically separated from each other. Whether this is really true is discussed in Box 36-1.*

BOX 36-1

IS THE GRAND CANYON A REPRODUCTIVE BARRIER?

The story of the Kaibab and Abert squirrels (genus *Sciurus*) is an oft-quoted illustration of speciation following geographic isolation and, from a teacher's point of view, it is an excellent example. What deep chasm could be more isolating than the Grand Canyon? It is worth noting, however, that a closer examination of that case suggests a few complexities. As presented here (and in all previous editions of this book and in many other texts), the necessary implication is that in fairly late geological times a continuous and uniform population of *Sciurus aberti* was spread across the areas where the Grand Canyon now exists—and when the canyon developed it became a barrier that completely isolated the northern part of the population. That, however, may not be entirely correct. It now appears that the canyon barrier, in approximately its present form, existed before the northern population evolved into a different species. The former continuity of *Sciurus aberti* was probably not across the canyon but around it, to the north and east. Effective isolation probably resulted from a change in climate that produced deserts to the north and east and that eliminated the conifer forests on which *Sciurus aberti* squirrels depend for food. Then, since the preexisting canyon was also a barrier, the northern population became partially or perhaps almost fully isolated.

To add to the problem, some mammalogists now consider these squirrel populations to be subspecies, not species—that is, their differences are subspecific, not specific. Subspecies are denoted by a third term in the name. In this view, *Sciurus aberti aberti* is to the south (down to the Mogollon rim and far into New Mexico) and *Sciurus aberti kaibabensis* is to the north (on the Kaibab plateau only). To add further complications, the somewhat similar red squirrel, *Tamasciurus hudsonicus*, which is widespread and in Arizona largely sympatric with *Sciurus aberti*, ranges right across the Grand Canyon and occurs all over the Kaibab plateau without even a subspecific distinction. *Tamasciurus hudsonicus mogollonensis* is found on both north and south rims of the canyon.

One possibility is that Abert squirrels may indeed manage to cross the canyon (no one knows for sure) but seldom enough to permit much divergence of northern and southern demes. At present, they do not maneuver around the ends of that barrier. Although in exceptional circumstances one animal might get across the canyon in a day or two, it could not possibly survive a trek across the almost completely treeless area between the Colorado River and the Little Colorado River.

These considerations raise questions about the real effectiveness of the Grand Canyon as a reproductive barrier. When one makes a quick count of the species of mammals now or recently present on one or both of the main canyon rims (excluding bats, for which the canyon is obviously no barrier), it appears that the canyon is not a very significant barrier for mammals in general. Some of the identical mammals on both rims are very tiny insectivores (for example, *Sorex merriami*, a shrew) and rodents (for example, *Reithrodontomys emgalotis*, a little harvest mouse)—both with very limited migratory powers. One is left with the surprising notion that unless, improbably, *Scirus aberti kaibabensis* is a highly unusual example, there seem to be no pairs of related species that might have diverged *because* the canyon separated them.

None of this, of course, should weaken the validity of the principle that populations diverge following geographic isolation. However, this example may not offer the best evidence.

---

proposes that assortative mating may become intense enough to form a complete barrier between two phenotypic groups occupying the same territory. For example, blue geese are polymorphic in coloration and mate assortatively. If the fidelity of mate choice became perfect, the two color morphs would be reproductively isolated—and, it seems clear, a new species would arise. Mate choice may be based on imprinting (see Chapter 32), so that the offspring of two parents with like phenotype will choose only mates with that phenotype.

Parapatric Speciation

Parapatric speciation occurs when genetically distinctive organisms in a population gain access to some unoccupied ecological niche within the population's normal geographic range. In such cases, gene frequencies may change abruptly over short distances and individuals may become reproductively isolated from the rest of the population without the presence of a geographic barrier.

One example of a new species that appears to have arisen by parapatric speciation is the Old World mole rat, *Spalax ehrenbergi*. Mole rats and gophers are good candidates for parapatric speciation because of their burrowing habits and resulting immobility. Adult moles live in burrows during the day. They emerge only at night to forage, and even then they seldom stray very far. Consequently, a mole may spend most—or all—of its life in an extremely limited area. Four species of moles exist in an area of Israel that extends from the cool, humid Golan Heights in the north to the hotter, more arid region of Jerusalem and the Sinai Desert in the south. Since the ranges of several of the species overlap, some contact is possible. The populations differ in chromosome number: two northern species have 52 and 54 chromosomes; a species from central Israel has 58; and the southern species has 60. The four species have remarkably similar proteins: 96% of those tested were virtually identical in all four species. Hybrids with reduced viability can be produced in the laboratory, but these are not known to occur in nature. Fossil evidence suggests that today's four species descended from a single ancestral species (*Spalax mimtus*) that lived throughout the area a half-million years ago.

The probable explanation for these facts is that the four species of mole rats evolved parapatrically from the ancestral form. As different subpopulations of the ancestral population moved into different climatic areas (cool and humid or dry and arid), random chromosomal changes took place that in one or a few generations could have caused both reproductive isolation and metabolic differences. The mode of life also limited the contacts between subpopulations. This furthered the chance that small differences would accumulate and increase divergence between the groups until they became as different as we see them today. Now each area is homozygous for one chromosome type or the other and the zones between the different populations are very narrow.

Before leaving the topic of speciation we should note the view of some investigators that isolating mechanisms may not be as critically important as we have suggested. These workers have pointed to certain exceptional cases that seem to indicate either (1) that new species may fail to arise despite marked isolation with severely reduced gene flow, or (2) that new species can arise in the face of no isolation or poor isolation in situations in which considerable gene flow can take place.

There are many examples of the lack of speciation despite reduced gene flow. Many identical species of butterflies are found in regions as widely separated as the Northern and Southern hemispheres. Presumably, similar forces of selection preserve the uniformity of these organisms over wide areas despite the existence of physical barriers that appear to block gene flow.

Another proposed explanation is that most large populations evolve very slowly. New alleles and gene combinations that arise in a large gene pool are likely to be swamped by the huge number of existing genes.

ADAPTIVE RADIATION

Adaptive radiation is defined as the rapid development from a single ancestral species of many new species that spread out in a radiative pattern to fill multiple different niches. A **niche** is a functional concept having to do with the way an organism makes its living, including its way of life, habitats, foods and food sources, and so on (see Chapter 47). Each species in an area ordinarily has its own niche. Adaptive radiation into multiple niches is most vividly displayed on clusters of remote islands, though it may occur in any location.

It usually begins with the arrival of a few pioneers. A good example is the ancestral marsupials (pouched mammals), to be discussed in Chapter 43. They invaded the island continent of Australia some 60 million years ago over a fast-disappearing land bridge from the Asian mainland. The original pioneers, which resembled opossums, found a richly varied environment that was free of major competitors. The population grew and prospered, spreading throughout the continent and occupying all sorts of new environments. Each presented new and different selection pressures. As a result, new populations arose in different niches as invaders adapted to new conditions (Figure 36-12).

Darwin's Finches

Surely the most famous example of the evolutionary role of ecological opportunity is the history of a group of small land birds of the subfamily Geospizinae on the Galápagos Islands (Figure 36-13), which Darwin visited on the Beagle in 1832 (see Box 2-1). These birds came to be known as Darwin's finches.

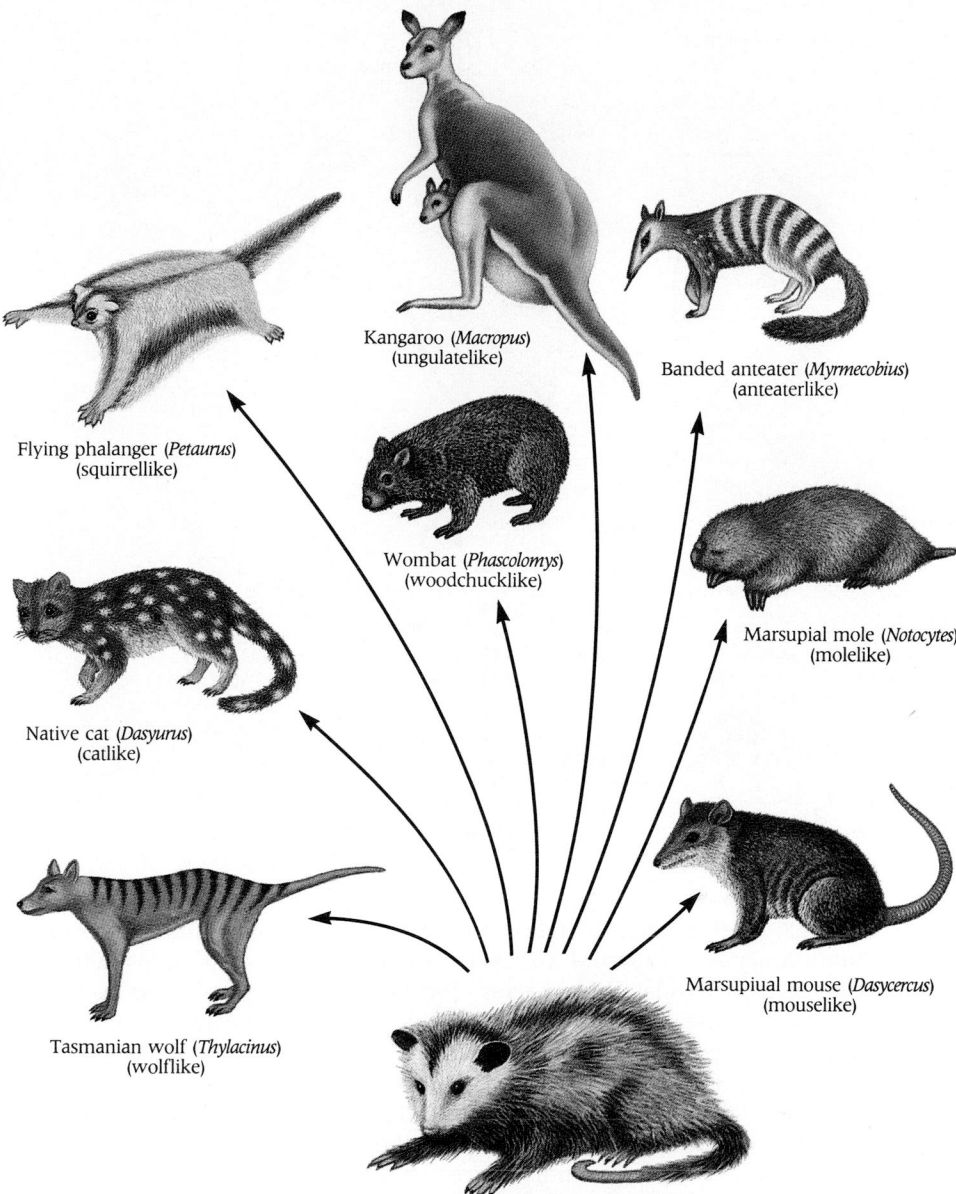

Flying phalanger (*Petaurus*)
(squirrellike)

Kangaroo (*Macropus*)
(ungulatelike)

Banded anteater (*Myrmecobius*)
(anteaterlike)

Wombat (*Phascolomys*)
(woodchucklike)

Marsupial mole (*Notocytes*)
(molelike)

Native cat (*Dasyurus*)
(catlike)

Marsupiual mouse (*Dasycercus*)
(mouselike)

Tasmanian wolf (*Thylacinus*)
(wolflike)

Original oppossumlike ancestor

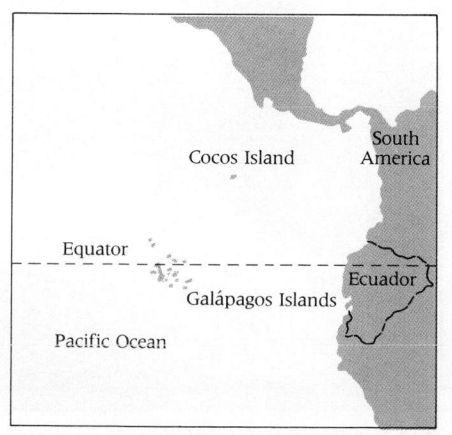

36-13 **Location of the Galápagos Islands**

The Galápagos are volcanic islands that arose from the ocean floor and were never connected with the mainland. Arising late in the history of life, only 3 to 5 million years ago, they initially provided an environment unoccupied by living organisms—a remarkable ecological opportunity for land organisms able to reach them. By examining the present-day flora and fauna on the islands, one can discern how the islands were colonized. In the absence of a land connection, only a few kinds of organisms ever reached the islands, and these were rare events brought about by chance movements of winds, currents, and floating debris. As soon as vegetation covered the islands, immigrants could enter any new environment for which they had the appropriate design.

Many land birds now living on the Galápagos are descendants of a small finch from the South American mainland. Since its arrival on Galápagos, the finch evolved into at least 14 distinct species, which constitute a subfamily found nowhere else in the world, not even in South America. The subfamily is said to be **endemic.** Each species specializes to some extent in exploiting the resources of the islands (Figure 36-14). The ancestral finches were evidently ground birds that fed mainly on seeds and other vegetation. The 14 species that evolved from this stock were still clearly finches, alike in their plumage, unmusical song, and nesting behavior. However, they differed in the kinds of food they ate.

The 14 species can be roughly grouped into two categories: six species of ground finches (genus *Geospiza*) and six species of tree finches (genus *Camarhynchus*) plus a pair of "warbler" species. In addition, there is an isolated species peculiar to

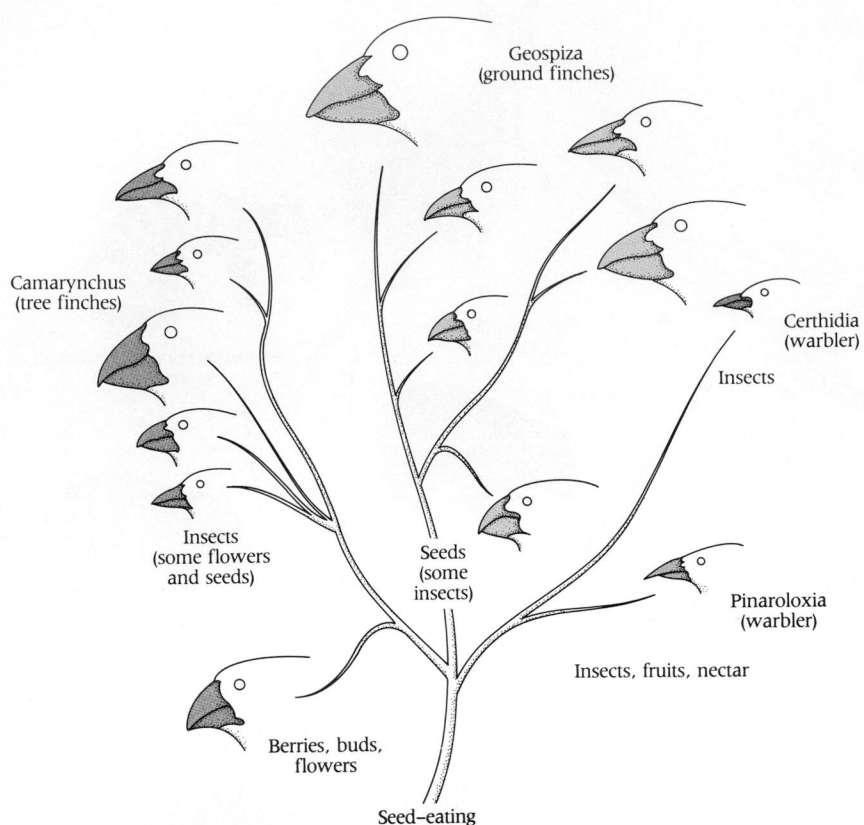

Cocos Island, 600 miles northeast, that originated from an unknown finch ancestor from the South American or Central American mainland. Of the six ground finches, three feed on seeds, two mainly on cactus, and one combines seeds and cactus. Of the six tree finches, the majority eat insects. Within these broad categories further specialization then developed as the species came to differ in beak size and structure, which reflected their food preferences. One tree finch (*Camarhynchus pallidus*) became what is essentially a woodpecker (Figure 36-15). Lacking the long tongue of true woodpeckers, it substitutes a remarkable behavioral trait. It snaps off a cactus spine and uses it to probe insect prey from a crevice it has chiseled with its beak. A tree finch that came to resemble a warbler in its habits is *Certhidia olivacea,* which lives on the insects it picks from bushes with its slender beak.

There were striking and instructive differences in the evolutionary futures of the mainland finches that early immigrated to the Galápagos and their fellows that stayed behind. Not one of the mainland finches underwent the extensive adaptive diversification that occurred on the islands, even though it had the necessary physical opportunities and design to do so. Mainland birds, it seems, lacked the ecological opportunities created by the vacant habitats of the Galápagos (Figure 36-16).

What promoted the evolution of new species in the new territory was the fact that the Galápagos are islands. The generally accepted model has three steps.

■ The first was the single over-the-water colonization of the Galápagos by a mainland finch.

■ Second, there was a population increase on the initial island. Island-to-island movement of these small birds, though limited, was adequate to populate all of the islands in time. Roughly the same ecological opportunities existed on most of the islands. However, the finches do not ordinarily fly across water for long distances. Thus a population on any one of the islands is essentially isolated from the populations on other islands. These openings were filled on every island, but similar niches were not always filled by the same species (Figure 36-17). Thus, the isolation of each new island population diverged into new species that to an extent became specialized to local conditions.

■ Third, if two new species evolved (call them *X* and *Y*), respectively, on the islands of Albemarle (Spanish name, Isabela) and Narborough (Fernandina), they could accidentally be reunited on one of the two islands with the aid of

36-15 Darwin's tree finch, Camarhynchus pallidus

36-16 Evolutionary opportunity and adaptive radiation

(A) On the South American mainland, the niches were filled and adaptive radiation could not take place. (B) On the Galápagos Islands, the finches had unlimited ecological opportunities and large-scale adaptive radiation took place.

Mainland (niches filled; no ecological access)   Galapagos Islands (niches open; full ecological access)

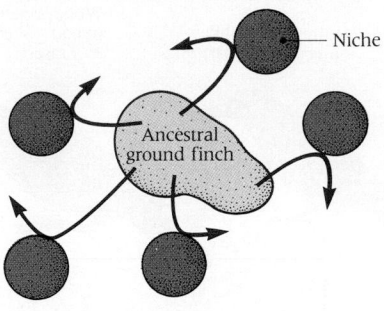

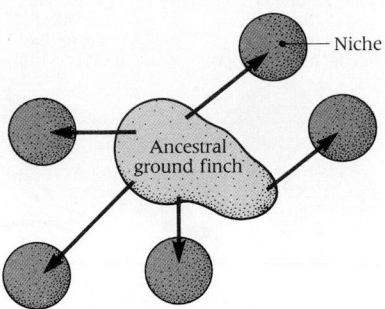

No adaptive radiation

A

Large–scale adaptive radiation

B

winds or other dispersal mechanisms. If *X* and *Y* have not diverged sufficiently for reproductive isolating mechanisms to occur, the two forms would interbreed and merge with each other. However if *X* and *Y* are reproductively isolated, the two forms would coexist on the same island, and competition between them would lead to rapid divergent evolution. This rapid evolution could lead to the formation of a third species—if, for example, species *X*, which originated from Albemarle, became more and more different in the presence of species *Y* on Narborough, until it diverged sufficiently to become a new species *Z*. The end result would be to have species *X* on Albermarle and species *Y* and *Z* on Narborough.

A similar but even more striking example of adaptive radiation is seen in the honeycreepers of Hawaii (Drepanidae). These small native birds must also have originated from a single pioneer species—in this case, one of the small goldfinchlike birds that colonized Hawaii from Asia or North America more than a million years ago. After colonizing the islands, they adaptively radiated into some 22 new species, many with distinctive food sources (Figure 36-18). Their early success was directly related to the abundance of Hawaii's insects and the variety of places where they can be found—as well as to an absence of competition. The first honeycreepers were probably curved-beaked feeders on nectar and insects. From this ancestry emerged an amazing range of beak sizes and shapes that perfectly matched feeding habits.

The Australian marsupials, the Galápagos finches, the Hawaiian honeycreepers, the cichlid fishes of East Africa, and many other plants and animals illustrate the recurrent evolutionary phenomenon of adaptive radiation.

36-17 Adaptive radiation of **Geospiza** species

(A) The Galápagos Islands, home of Darwin's finches. The map shows the English names of the islands. They also have Spanish names. Ancestral finches migrated from the mainland to these islands when they were unoccupied by other birds. (B) The exploitation of local evolutionary opportunities by Geospiza species. Some of the small outlying islands have never been colonized by some of the species from the central islands.

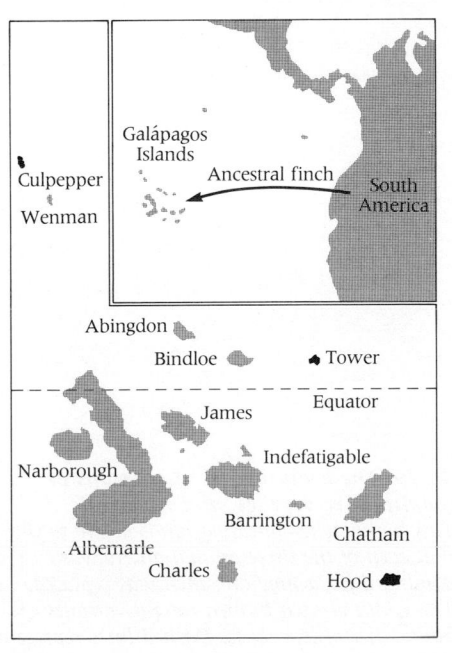

A

	Central islands	Small outlying islands		
		Tower	Hood	Culpepper
Large ground finch	*G. magnirostris*	*G. magnirostris*	*G. conirostris*	*G. conirostris*
Cactus ground finch	*G. scandens*	*G. conirostris*	*G. conirostris*	*G. difficilis*
Small ground finch (arid zones)	*G. fuliginosa*	*G. difficilis*	*G. fuliginosa*	
Small ground finch (humid woods)	*G. difficilis*	The outlying islands lack the moist woodland habitat occupied by G. difficilis in the central islands		

B

Psittirostra
kona
Finchlike
(seeds, a few insects)

Psittirostra
psittacea
Parrotlike
(flowers, fruit)

Hemignathus
procerus
Hummingbirdlike
(nectar, insects)

Hemignathus
wilsoni
Woodpeckerlike
(wood–dwelling
insects)

Pseudonestor
xenophrys
Unique
(longhorn beetle larvae)

Loxops virens
Warblerlike
(insects, nectar, berries)

Original colonizing species
Finchlike
(similar to *Loxops virens*)

36-18 Adaptive radiation of Hawaiian honeycreepers
These birds probably originated from a single species of finchlike bird that colonized the islands and later radiated by means of multiple invasion. Six of the more extreme adaptive forms are shown here. There are 16 other species.

Major Adaptive Radiations

The range of evolutionary opportunities afforded Darwin's finches was limited in two ways. First, the diversity of open habitats (ecological opportunities) was limited—for example, kinds of vegetation were restricted. Second, seed-eating ground finches had limited design opportunities. The finches lacked teeth and other prerequisites. Evolutionary opportunity of greater scope and significance could arise only when organisms with fewer design limitations confronted wider ecological opportunities.

A major adaptive radiation of great importance followed the first conquest of land by vertebrates. The earliest vertebrates that could emerge even temporarily on the land were fishes that (1) could breathe air with primitive lungs, and (2) could walk or wriggle with modified fins. These minimal prerequisites evolved as the fishes radiated. Once able to move and feed on land, the vertebrates had clear

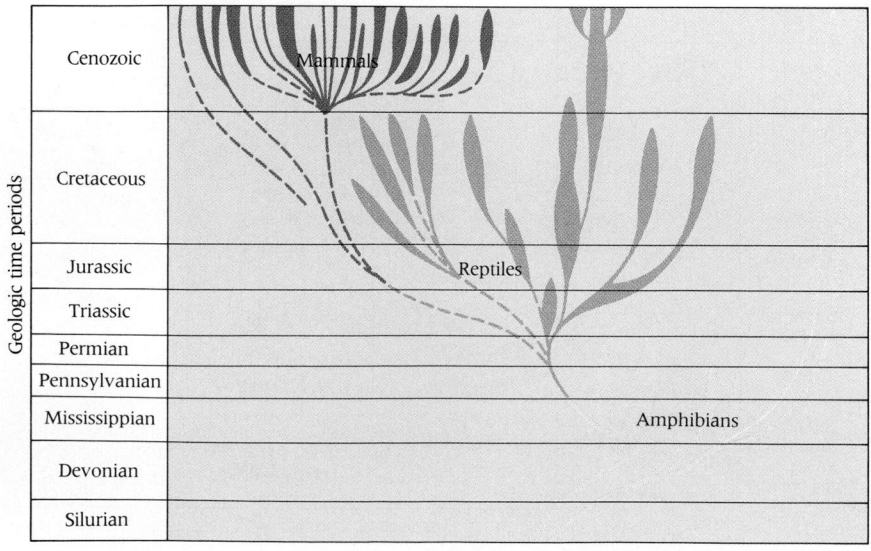

36-19 The evolutionary successions of amphibians, reptiles, and mammals
This highly schematic diagram is meant to show that each of the three animal orders was dominant for a time and then was replaced. The width of each branch roughly denotes the abundance of its species. Dashed lines represent unknown ancestral forms.

sailing. They were the first large terrestrial animals and they had little competition.

Amphibians, reptiles, and mammals then evolved in succession (Figure 36-19). Reptiles were the first fully competent land vertebrates. Equipped with greatly improved reproductive methods,[4] they were able to move inland and to exploit the rich array of wholly unoccupied land habitats. The reptilian radiation was a grand one. It produced herbivores in rich assortment, a diversity of carnivores, a host of flying forms, and even forms that successfully returned to water.

In examining Figure 36-19 and diagrams like it, one must be careful to recognize that contemporary amphibians, reptiles, and mammals are not the ancestral forms that gave rise to new branches on the evolutionary tree in an earlier time. Rather they are descendants of those distant ancestors. Today's amphibians are not precursors of today's reptiles, nor are today's reptiles precursors of today's mammals.

Two points are of interest in connection with the reptilian radiation.

- First, by establishing a competitive situation on land reptiles diminished the ecological opportunity earlier open to those fishes that were becoming amphibious (capable of living on land or in water). Most (but not all) agree that fishes evolved into land vertebrates on only one occasion (300 million years ago). That has always seemed puzzling. Why did they not do it again? The answer is simple: there was a transitional form in the evolutionary move from water to land that, in the absence of competitors, was able to succeed without being fully aquatic or fully terrestrial. A modern fish designed like this odd fellow would now meet numerous efficient competitors on land and would not stand a chance.

- A second point concerns the history of mammals. The fossil record indicates that one branch (some say more than one) of the early reptilian radiation produced animals (mammallike reptiles) that later gave rise to mammals (Figure 36-19). In fact, a few mammals of a sort were around throughout the prolonged heyday of the reptilian dinosaurs. For reasons still obscure, many reptiles became extinct about 65 million years ago. (We discuss this extraordinary event in Chapter 44.) Mammals, which had survived the mass extinction among the reptiles, were then free to radiate into the environments they had once filled. They radiated rapidly. Forms appeared with adaptations resembling those of most of the earlier reptilian groups. Later in the history of mammals came a whole series of more restricted radiations (see Chapter 44).

[4] We deal with this important event later in surveying the history of life (Chapter 44).

SUMMARY

Biological variation occurs, not in randomly distributed patterns, but in "clumps": members of a single species resemble each other much more than members of other species. Nonetheless, intraspecies genetic variation occurs, and it is essential for the formation of new species. Variation which has a purely environmental cause plays no role in speciation. Demes within a species often differ genetically, either because of adaptations to different environments or genetic drift.

Members of a species are phenotypically similar and reproductively isolated from other species under natural conditions. Because they interbreed, they function as units of evolutionary change. Although speciation occasionally occurs by successive changes within a single lineage, more new species arise through cladogenesis—the branching of a species into two or more descendant species.

Speciation is rarely caused by a single mutation. However, controversy surrounds the question of whether speciation occurs gradually or rapidly. In species that can self-fertilize or reproduce asexually, mutations leading to polyploidy cause reproductive isolation that can induce instantaneous speciation.

The gene pool of a species is protected from the disruptive effects of genes from other species, usually by more than one isolating mechanism. These mechanisms include factors that prevent mating—prezygotic barriers—and factors that prevent offspring if mating does occur—postzygotic barriers. Common prezygotic barriers include a tendency for two species to live in different habitats, breed at different times, have different criteria for mate selection, different methods of gamete dispersal, or mechanical or physiological barriers to cross-fertilization. Postzygotic barriers include zygote mortality, hybrid unfitness, and hybrid sterility. Despite all these barriers, however, genes are occasionally passed between species, a process called introgression.

Cladistic speciation requires the creation of reproductive barriers between formerly interbreeding populations. This process can be allopatric, sympatric, or parapatric, depending on the degree of geographic isolation of the populations. Speciation occurs most readily when populations are geographically isolated, or allopatric. They can then differentiate from each other—through genetic drift and the differing selective pressures that different geographic regions usually provide. In sympatric speciation, a new species evolves within the range of the parent population. Sympatric reproductive isolation can arise through polyploidy in some plants or, gradually, through such mechanisms as intense assortative mating. Parapatric speciation, intermediate between these extremes, occurs when variant organisms within a population gain access to a new ecological niche within that population's geographic range.

Adaptive radiation is the rapid development from a single species of many new species inhabiting different ecological niches. It occurs when a few founders colonize a new area in which few niches are occupied by competitors. This pattern occurred when Darwin's famous finches colonized and evolved into 14 species on the Galápagos Islands. Two of the major adaptive radiations in the broad scheme of evolutionary history were the colonization of land by vertebrates and the prolific speciation of mammals after the mass extinctions of reptiles 65 million years ago.

KEY TERMS

adaptive radiation
allopatric speciation
anagenetic speciation
behavioral isolation
character displacement
cladistic speciation
cladogenesis
cline

ecological isolation
hybrid
instantaneous speciation
isolating mechanism
mechanical isolation
parapatric speciation
physiological isolation
prezygotic barrier

punctuated equilibrium
reproductive isolation
species
stasis
subspecies
sympatric speciation
temporal isolation

QUIZ QUESTIONS

1. Which of the following is *not* a prezygotic isolating mechanism?
   A. mechanical incompatibility of genitalia
   B. chemical environment of female reproductive tract being inhospitable to sperm
   C. incompatible courtship rituals
   D. production of genetically inviable hybrids
   E. B and C

2. Evolution of a new species by events that involve its splitting from an ancestral species can occur via
   A. cladistic speciation.
   B. instantaneous speciation.
   C. allopatric speciation.
   D. parapatric speciation.
   E. All of the above

3. Which of the following is a way by which a species can arise and live in sympatry with the species from which it evolved?
   A. polyploidy
   B. hybridization
   C. macroevolution
   D. anagenetic speciation
   E. interspecific selection

4. _____ sometimes occurs in situations where individuals of two previously isolated demes come back into sympatry.

   A. Truncation selection
   B. Character displacement
   C. Frequency-dependent selection
   D. Heterosis
   E. Anagenetic speciation

5. Unrelated species that have achieved the same levels of biological improvement belong to the same
   A. deme.
   B. subspecies.
   C. clade.
   D. grade.
   E. None of the above

ESSAY QUESTIONS

1. What is reproductive isolation? What role does it play in defining a species? What barriers serve to reproductively isolate organisms? Which mechanisms tend to occur early in the separation of species? Which mechanisms evolve later?

2. What are examples of allopatric speciation? How is it different from sympatric speciation? Parapetric speciation?

3. What is a niche? How might the founder effect and geographic isolation combine to provide a strong force for speciation?

4. How do Darwin's finches epitomize the features of adaptive radiation?

5. What are the two major basic sources of variation in sexually reproducing populations?

REFERENCES AND SUGGESTIONS FOR FURTHER READING

Barton, N. H., Jones, J. S., and Mallet, J. (1988). Evolution: No barriers to speciation. *Nature 336*, 13–14.

An interesting brief note introducing three papers later in the issue on aspects of speciation and reproductive isolation.

Eldredge, N. (1985). *Time Frames: The Rethinking of Darwinian Evolution and the Theory of Punctuated Equilibrium*. Simon & Schuster, New York.

Another review of the much-publicized and much-disputed theory of punctuated equilibrium, by one of its two originators. The book is highly readable, though a recent reviewer asks, "What was all the fuss about?"

Grant, P. R. (1986). *Ecology and Evolution of Darwin's Finches*. Princeton University Press, Princeton, N.J.

Another book on Darwin's finches needs some defense, since as the author says "They have been done." This one offers some interesting new interpretations—for example, on the role of diet in shaping beak morphology. This book is well written and richly informative.

Grant, V. (1981). *Plant Speciation*. Columbia University Press, New York.

A comprehensive treatise on speciation in higher plants.

Mayr, E. (1963). *Animal Species and Evolution*. Harvard University Press, Cambridge, Mass.

An authoritative review of the various theories on speciation and a discussion of the most influential concepts. An abridgement of this book was published in 1970 under the title *Populations, Species, and Evolution*.

Simpson, G. G. (1944). *Tempo and Mode in Evolution*. Columbia University Press, New York.

A persuasive early work on the dynamics of selection pressure and speciation and the importance of time in determining the course of events.

Template, A. R. (1981). Mechanisms of speciation: a population genetic approach. *Annals and Review in Ecology and Systematics 12*, 23–48.

An up-to-date discussion of the genetic aspects of speciation.

Wilson, E. O. (1985). The biological diversity crisis. *BioScience 35*, 700–706.

A troubling discussion of the rapid rate at which living species are becoming extinct.

SECTION 2
THE DIVERSITY OF LIFE

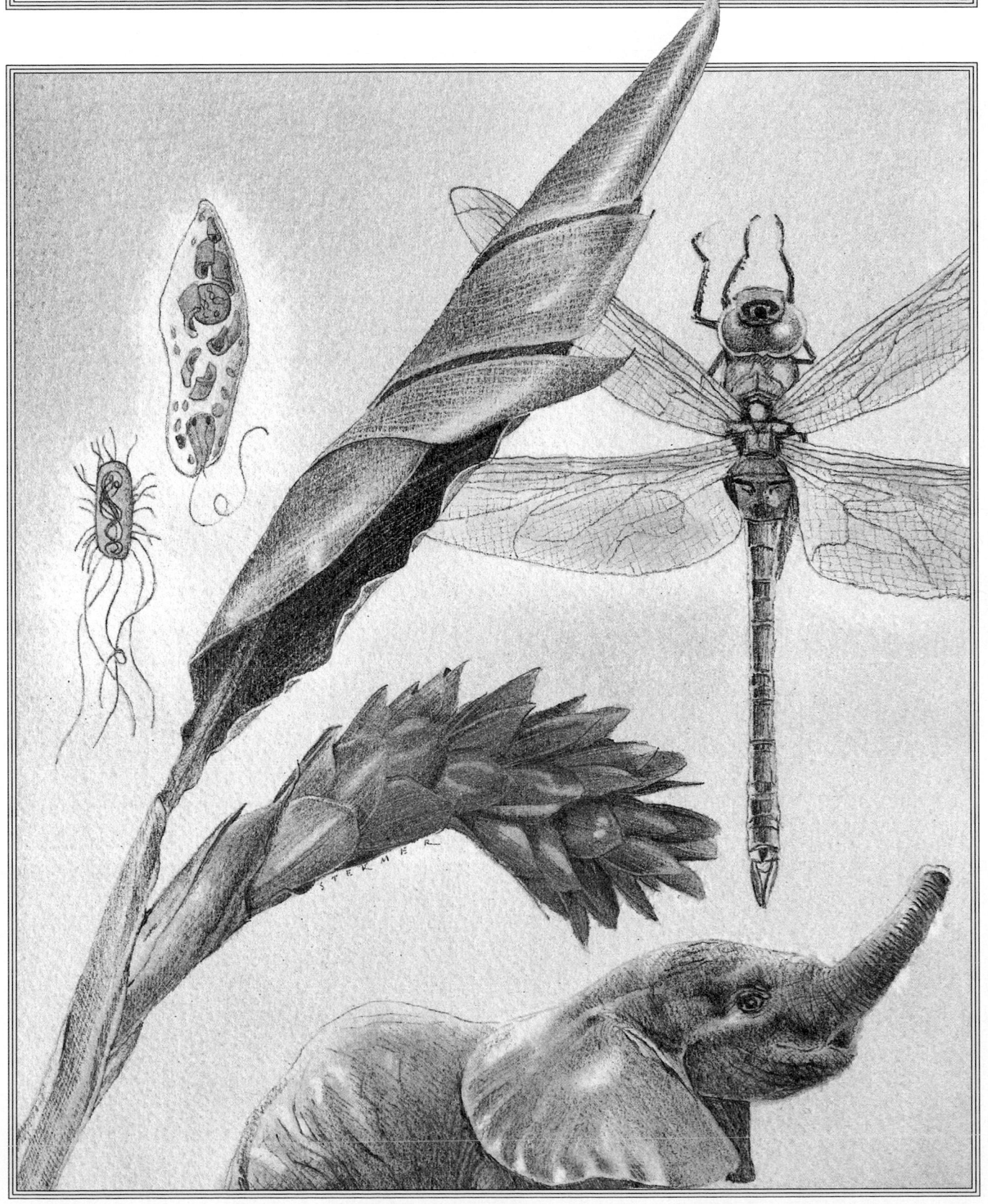

PRINCIPLES OF CLASSIFICATION

Principles of classification
Before living organisms can be understood, they must be arranged into an orderly classification system. This fascinating branch of biology examines both the diversity of organisms, such as those shown here, and their evolutionary relationships.

Biologists early realized that if they were ever to perceive meaningful relationships among a welter of forms and species, they must arrange them into a comprehensive and orderly system: in a word, to classify them. Classification, they argued, would serve three purposes.

- It would systematize knowledge about species and facilitate the collection, communication, storage, and retrieval of that knowledge.
- It would help determine whether and how new forms were still arising.
- It would make possible the discovery and generalization of evolutionary relationships.

As it turned out, classification of living organisms is not a simple matter. Any system of classification, whether biological or not, involves two procedural steps: (1) defining and describing the units to be classified while examining them for meaningful differences, and (2) arranging the units into a classification scheme. The best scheme, we will contend, is one that rests on sound biological principles. Our first step, then, will be to consider which principles to invoke.

Before delving into that problem, we must define certain terms as they are used in this branch of biology.

- **Systematics** is the scientific study of the kinds and diversity of organisms and their evolutionary relationships. A systematist examines organisms from many viewpoints with the aim of determining both their similarities and differences and their evolutionary history. Systematic study generally precedes classification.
- **Classification** is the ordering of living organisms into groups on the basis of associations by contiguity, similarity, or both.
- **Taxonomy** is the theoretical study of classification, including its bases, principles, procedures, and rules.
- **Nomenclature** is the application of distinctive names to each group, or **taxon** (plural, taxa), recognized in any biological classification.

RESEMBLANCE AND DIFFERENCE

The most noticeable thing about life's diversity is the fact that some organisms are more alike than others. Two daisies are enough alike to be indistinguishable. A daisy and an aster are much alike but they can be easily distinguished. Both differ strikingly from a pine tree. Red and gray squirrels are similar (Figure 37-1), but increasingly different from cats, frogs, fishes, and earthworms. The most straightforward goal of a classifier of organisms is to group together those that are alike and separate them from those that are different. Simple enough, but try to do it!

Two problems arise immediately. The first is that resemblances can vary in degree. This fact must be taken into account in any classification. A red squirrel resembles a gray squirrel more than it does a cat, it resembles a cat more than a frog, a frog

37-1 Resemblance
(A) *Red squirrel.* **(B)** *Gray squirrel.*

more than a fish, a fish more than an earthworm, and an earthworm more than a daisy. This problem is ordinarily not too daunting. The solution is simply to establish a **hierarchy,** an arrangement of groups of decreasing scope, one within another, with some subordinate to others. Consider the following example:

ORGANISMS:
everything that is alive, including daisies, earthworms, fishes, frogs, cats, gray squirrels, and red squirrels

ANIMALS:
no daisies, but includes earthworms, fishes, frogs, cats, gray squirrels, and red squirrels

VERTEBRATES:
no earthworms, but includes fishes, frogs, cats, gray squirrels, and red squirrels

TETRAPODS (four-limbed vertebrates):
no fishes, but includes frogs, cats, gray squirrels, and red squirrels

MAMMALS:
no frogs, but includes cats, gray squirrels, and red squirrels

RODENTS:
no cats, but includes all kinds of squirrels as well as rats, porcupines, mice, and others

SQUIRRELS:
divided into red and gray squirrels and others

This scheme is neither absolute nor arbitrary. Others could be devised that would be just as uncomplicated and useful as long as everyone accepted and understood them. We will shortly see that to some extent different classification methods have different purposes.

There is a second and more serious problem in classification. Organisms resemble each other in different ways. One kind of classification, which has been used since antiquity, is based on an organism's way of life, or **habit.** Plants, for instance, can be arranged as aquatic, herbaceous, shrubby, or arborescent. Animals can be grouped as swimming, walking, flying, and so on. Such classifications are also based on resemblances and differences of **habitat:** swamp or alpine plants, marine or desert animals, and so on. Classifications of this sort are in wide use, especially in the study of ecology (see Chapters 45–48), but are rarely used in the basic classifications of modern biology.

Even the ancients recognized that classification by habit and habitat can bring together organisms that are grossly different. It can also separate out those that are alike. For this reason, later classifiers began to rely on features of outward appearance, anatomy, biochemistry, and physiology. But, as we stated in the last chapter, these criteria are seldom clear-cut. Moreover, it would be difficult to classify a gray squirrel, for example, by concluding that it has 10,001 anatomical and physiological resemblances to a red squirrel, 8,346 to a cat, 3,921 to a frog, 2,754 to a fish, and 172 to a daisy (to cite some imaginary figures).[1] We have no idea what

[1] In fact, as we shall discuss later, there is now a school of systematists who call themselves numerical taxonomists (and are called numerical pheneticists by others) who do in effect exactly what we have here suggested is so difficult to do.

the correct figures would be or, indeed, how they could be established. We must also remember that outward appearances are phenotypic manifestations. Genotypic features could be just as important—perhaps more so—as a basis of classification.

Desert cactus nicely illustrates the problem. Genuine cacti and many South African euphorbias are succulent, spiny, flowering plants that are well adapted to the arid conditions of the desert. Cactus fanciers call both groups "cacti." To be sure, the anatomical and physiological resemblances between the two groups are numerous. However, the classifications long accepted by botanists tended to ignore those resemblances. Instead some of them sharply separated euphorbias from true cacti by placing them in a different family with, among other plants, the poinsettias. No one would think of calling *them* cacti.

Similarly, the Tasmanian wolf was called a "wolf" because it resembled a wolf in appearance and habit. But in today's zoological classification, the Tasmanian wolf is grouped with kangaroos. We shall explain this oddity later.

How do we account for such seeming arbitrariness? The answer is that classification by anatomy and physiology requires that certain characters be singled out and then carefully interpreted. Euphorbias and cacti came to resemble each other in their spininess and other ways because both responded similarly to the conditions of their environment. However, the fact that the ovary is superior (in position, that is) in spiny euphorbias and inferior in cacti (along with other differences) is for many reasons more important to classifiers than the fact that both plants have spines (Figure 37-2). The fact that Tasmanian "wolves" have pouches and bones like marsupials, while real wolves do not is what is emphasized in the preferred classification scheme. The fact that real wolves and Tasmanian "wolves" are both four-footed, running animals with flesh-cutting teeth is given secondary importance. What characters are selected, what meaning is assigned to them, and what classification is finally devised depends upon how we interpret and answer the deepest questions of biology—and perhaps of philosophy as well. What are the systematic units we use in classification and how do the characteristics of such units originate?

THE NATURE OF SYSTEMATIC UNITS

The form of classification of organisms still in use today dates from the work of the mid-eighteenth century Swedish botanist, Carolus Linnaeus (1707–1778), sometimes called the "father of taxonomy," and especially from his book *Systema Naturae*, which first appeared in 1735 and later appeared in many editions. His major legacy is the binomial (two-name) style now used in the scientific naming of organisms. He also devised some of the names themselves that are still in use today. Since the time of Linnaeus, the actual principles of classification have undergone two revolutionary changes in philosophy and a few changes in technique. As a result, the meaning of today's classification is wholly different from that of Linnaeus.

37-2 Cactus (Left) *and euphorbia* (Right) *both have spines*

THE LINNAEAN CONCEPT AND ITS SUCCESSORS

In Linnaeus's era, classification was based on a belief in the doctrine that species are fixed and unchanging units. That view was heavily reinforced by a theological dogma, that all species were created as such by God at the beginning of the world. Some slight changes within a species were admitted as possible, since it was well known that new races of domesticated animals could be deliberately developed. A few biologists at the time thought it possible that new species might arise. But these strange ideas had little influence at the time on systematics. The systematist's task in those days was to recognize and arrange all the units of divine creation—no more, no less. All units to be classified—like cats, dogs, pines, and maples—were supposedly sharply distinct. There was no "species problem." Linnaeus, for one, had little doubt that the 4235 animal species he listed in 1758 were the "kinds" provided by the Creator.[2] This list was not subject to further revision by systematists of the day.

For most systematists of that time, the characteristics of each species were fixed at its creation. Deviations, so evident in the real world as variation within species, were deemed accidental and irrelevant. The ultimate reality of a species was in fact not a tangible thing but a pattern, a divine idea, a **type**—or, as they often termed it then, an **archetype** (primeval pattern). The way for a systematist to look at organisms, then, was to ignore individuals, to brush aside variation and characteristics of populations as such, and to abstract the idea of what the individuals have in common. That abstraction was the ideal type or archetype of a species. The same concepts were applied to groups of wider scope than species.

The First Revolution: A Modern View of Species

The science of **systematics** was first erected on the principles that species are fixed and invariant units. Its first revolution followed the realization that species are not separate and unchanging creations, but have evolved, one from another, in the long history of life. This concept, profoundly unlike special-creationist and archetypal concepts, addressed the "species problem" for the first time. How, indeed, can one give a fixed definition to units that change and grade one into another in time and space?

Even more important were new discoveries about relationships among species. It was finally recognized that they are not abstract or metaphysical. Rather, species descend from other species in a perfectly material way. Moreover, many species may descend from a common ancestral species.

This revolutionary change in the principles underlying classification had no effect on its form and, at first, little effect on the practical methods of classifiers. Early preevolutionary classifications grouped organisms according to common anatomical and physiological characters, interpreting them as manifestations of a metaphysical idea or archetype. Later evolutionary classifications established their groupings in the same way—indeed, they appropriated most of the groups established by Linnaeus and other nonevolutionary biologists. However, common characters were now seen as evidence of the group's common ancestry.

Classification as it was practiced by the evolutionary systematists of the later nineteenth century did not completely do away with pre-evolutionary archetypes. The archetype was simply relabeled and differently interpreted. Each species was now visualized in terms of an individual specimen, a type, that was supposed to be a standard or model. Instead of being a disembodied or abstract archetype, it was embodied, a textbook picture of what a species member is thought to be.

---

[2] In 1753, Linnaeus knew of about 6,000 plant species and believed that the total might be about 10,000. He listed the 4,235 animal species he knew in 1758 and guessed that their total would also be about 10,000. In 1778, Linnaeus's contemporary, Zimmermann, made the far more realistic estimate of 150,000 plant species and 7 million animal species. Today we know of more than 250,000 species of seed-bearing plants alone, 1 million nematodes, 47,000 vertebrates, and according to one recent study, 751,012 insects! In reality, the true diversity of life is very much greater. The number of systematists working today is sadly inadequate and many species remain to be described. Some believe there are more than 30 million insect species. A recent reviewer, R. M. May, noting a proposal that a team of taxonomists catalogue all the species in one representative hectare of a tropical rain forest, wrote, "It would be better to census several such sites. Until this is done, I will not trust any estimate of the global total of species."

Individuals being classified were compared with these **type specimens** and placed in the species whose type they most nearly resembled—rather as one identifies stamps before placing them in a stamp album. That subjects were never exactly like the type in the book was simply a nuisance to be lived with. Variation within species was still not an integral part of the species concept. Rather it was an apparent imperfection of nature.

Groups higher in the hierarchy also relied in practice on archetypes—but in more or less veiled form. Although classification became evolutionary in principle after publication of Darwin's *The Origin of Species* in 1859, it remained in practice largely pre-evolutionary and typological.

The Second Revolution: The Significance of Populations

In the second major revolution after Linnaeus's time, the typological concept of systematic units gave way to a populational concept. This turning point cannot be associated with a particular name or a single date. Some early students (including Darwin) did grasp the idea of basing systematics on populations rather than types. Moreover, full comprehension of the importance of populations in systematics had to await the advent of Mendelian genetics, which finally led to the development of population genetics (Chapter 34). A systematics unambiguously based on populations and explicitly nontypological is an achievement mainly of the second and third quarters of the twentieth century.

MODERN CONCEPTS

We can now define systematic units in reasonably clear evolutionary terms. A **systematic unit** of organisms in nature is a population or group of related populations. Its anatomical and physiological characteristics are simply the sum of all characteristics possessed by the individuals making up the population. The pattern is not of a real individual, nor is it an idealized or abstract set of characters. Rather it is a frequency distribution of the existing variants of each character present at a given time.

Species, in this modern concept of systematics, are populations of individuals of common ancestry that live together in similar environments in a particular region and tend to have similar ecological relationships and unified, distinctive, and continuing evolutionary roles. In sexually reproducing biparental species, the distinctiveness and continuity of the group are maintained by (1) the extensive interbreeding that occurs within the species, and (2) the absence (or effective limitation) of interbreeding with members of other species. The next higher systematic unit—a **genus** (plural, *genera*)—is a taxon more inclusive than a single species. It is a group of species of common ancestry.

The only direct evidence that an individual belongs to a given species is not the individual's anatomy or physiology, but the fact that in a natural setting it is living and interbreeding with a specific population and functioning as a member of that population. In practice, such direct evidence may be unavailable for organisms not in their natural settings—for example, museum specimens. It may also be difficult or impossible to obtain good direct evidence that several species are of common ancestry. In such cases, judgment must rest on such indirect evidence as anatomy and physiology.

One other point needs to be made here. It is that the use of anatomy, for instance, as evidence that an organism belongs to a given species does not mean that species is *definable* in anatomical terms. Species (and higher systematic units) are defined in terms of populations and their biological, evolutionary relationships. These relationships have many consequences, anatomical ones among them, from which the status of a population as a species may be inferred by an observer. If two people look exactly alike, their resemblance is evidence that they may be identical twins. But they are not twins because they look alike. They look alike because they are twins. **Typological systematics** held that organisms belonged to the same systematic unit because they had the same anatomical features. **Populational systematists** hold, rather, that members of a systematic unit have certain anatomical and other features in common *because* they belong to a natural systematic unit that is defined in evolutionary terms.

We should note in passing that the existence of polymorphism—many phenotypic forms—can be a serious problem for classifiers relying on anatomical or other traits. An example is the arctic fox, which can have either white or silver-colored fur (Figure 37-3).

Anatomy, until recently, was the only kind of evidence available in the systematics of extinct species, which rests on fossils that chiefly preserve hard parts—shells, bones, and teeth. Today the identification and classification of species makes use of molecular evidence, as discussed in Chapter 35. This new approach goes well beyond DNA, RNA, and amino acid sequences. It now includes the immunological identity or nonidentity of proteins, a technique that has been successful in studies of a baby mammoth found in Siberia in 1977 that had been frozen 40,000 years ago. This approach has generally supported the conclusions of classical systematics and it has proved useful in a number of disputed cases.[3]

It is true that too little is known about some groups to permit their evolutionary classification, even though we use evolutionary principles to interpret what is known. Until recently, there was considerable doubt as to whether truly evolutionary classification is possible in a few groups, mostly among the monerans, notably the bacteria, and protists. The old typological method continued in use with members of those groups mainly because no one had come up with a better method. Then a few years ago a revolution began in the classification of bacteria and other prokaryotes. What had been a dry, esoteric, and uncertain discipline, in which accepted relationships were little more than officially sanctioned speculations (often based on the use of computers in estimating typological resemblance), suddenly became a fresh and exciting experimental field. For the most part, the transition reflects the probing insights made possible by the sequencing of proteins, DNA, and RNA.

37-3 *Polymorphism in fur color of the arctic fox,* **Alopex lagopus.**

INTERPRETATION OF FORM AND DESCENT

If classification is to have an evolutionary basis, the following major question always needs to be answered: Have resemblances between organisms been inherited from common ancestors? Before we can answer, we must review certain evolutionary principles and processes.

THE MEANING OF HOMOLOGY

The term **homology** denotes the correspondence between structures in different organisms arising from the fact that these structures were inherited from a common ancestry. Such structures are **homologs** and are said to be **homologous.**

The number and arrangement of bones in the human arm and the foreleg of a dog are remarkably similar (Figure 37-4). The resemblance extends to the way the limbs develop embryologically and, in some degree, to the way the muscles, blood vessels, and nerves are arranged in the limbs of the two species. The only reasonable explanation for these impressive similarities is that the forelimbs are

[3] One concerned an investigation of two shrunken heads in the British Museum that were assumed to be the remains of unfortunate Jivaran Indians of Ecuador or Brazil. After an expert (relying on anatomical evidence) had declared one and possibly both to be made of horsehide, precise testing with different antibodies specific for human and horse albumin and collagen revealed that both were authentic.

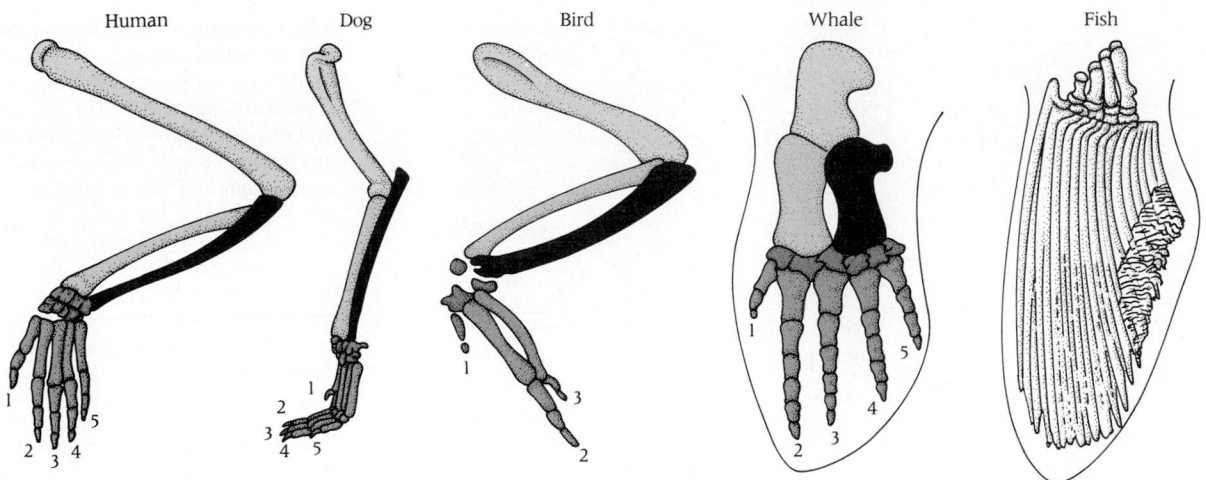

Human Dog Bird Whale Fish

37-4 Vertebrate forelimbs
Homologous bones in the limbs of human, dog, bird, and whale are indicated by color. Numbers refer to digits. The extent to which the fish fin is homologous is uncertain.

homologous. Significantly, they are homologous even though they are used mainly for manipulation in humans and for locomotion in dogs. We conclude from their homology that the limbs belong to species that descended from a common ancestry. The differences between them must have arisen *since* the evolutionary lines of descent leading to dogs and humans separated about 75 million years ago.

If we extend the comparison to a bird's wing, the similarities are less striking. Nevertheless, it is clear that a human's arm, dog's foreleg, and bird's wing are homologous—anatomically and embryologically. Indeed, some of the homologous bones are found in all three. Even more convincing, homology has been found in fossils of early mammals, of ancient birds, and of still older reptiles from which both mammals and birds arose. Separation of the reptilian ancestries of mammals and birds occurred at least 250 million years ago. But a fundamental resemblance remains.

When we add the front (pectoral) fin of a fish to the comparison, the resemblance appears even slighter. Now, we are comparing animals whose nearest common ancestors lived more than 350 million years ago. As a whole, the fish's fin is certainly homologous with the forelimbs of bird, dog, or human, but the separate evolutionary lines underwent changes so profound that the homologies of individual bones were all but obliterated.

Application of the concept of homology to anatomical structure is obvious. But biological structures contain cells and molecules. As we saw in Chapter 35, sequence studies reveal striking homologies of macromolecules (proteins or nucleic acids). In addition, homologies can be demonstrated at the molecular level in such functional traits as enzymatic activity, immunologic properties, electrophoretic behavior, and other properties.

Thus homology is based on anatomical or molecular evidence showing degrees of relationship among organisms. The degrees of homology in the examples given (both anatomical and molecular) permit inferences to be made concerning the origin of these species. We must consider such inferences in the classification of these organisms. Human and dog are more nearly related to each other than either is to a modern fish. An important problem, then, is to distinguish between resemblances that are homologous and those that are not.

HOMOPLASY, CONVERGENCE, ANALOGY, AND HOMOLOGY

Anatomical features in different organisms that resemble each other but do not share common ancestry are **homoplastic.** The phenomenon called **homoplasy** ("same-forming"), is associated with **convergence.** Anatomically similar structures in two animals or plants thus are either homologous or homoplastic. Homology is evidence of genetic relationship. Homoplasy is not. The terms are interpretive, of course, reflecting opinions based on available evidence. Use of the data of comparative anatomy in classification requires frequent decisions as to whether anatomical resemblances are homologous or homoplastic. The issue is crucial in evolutionary studies because different and completely unrelated species, exposed to the same environment, may in time develop strikingly similar phenotypic adaptations.

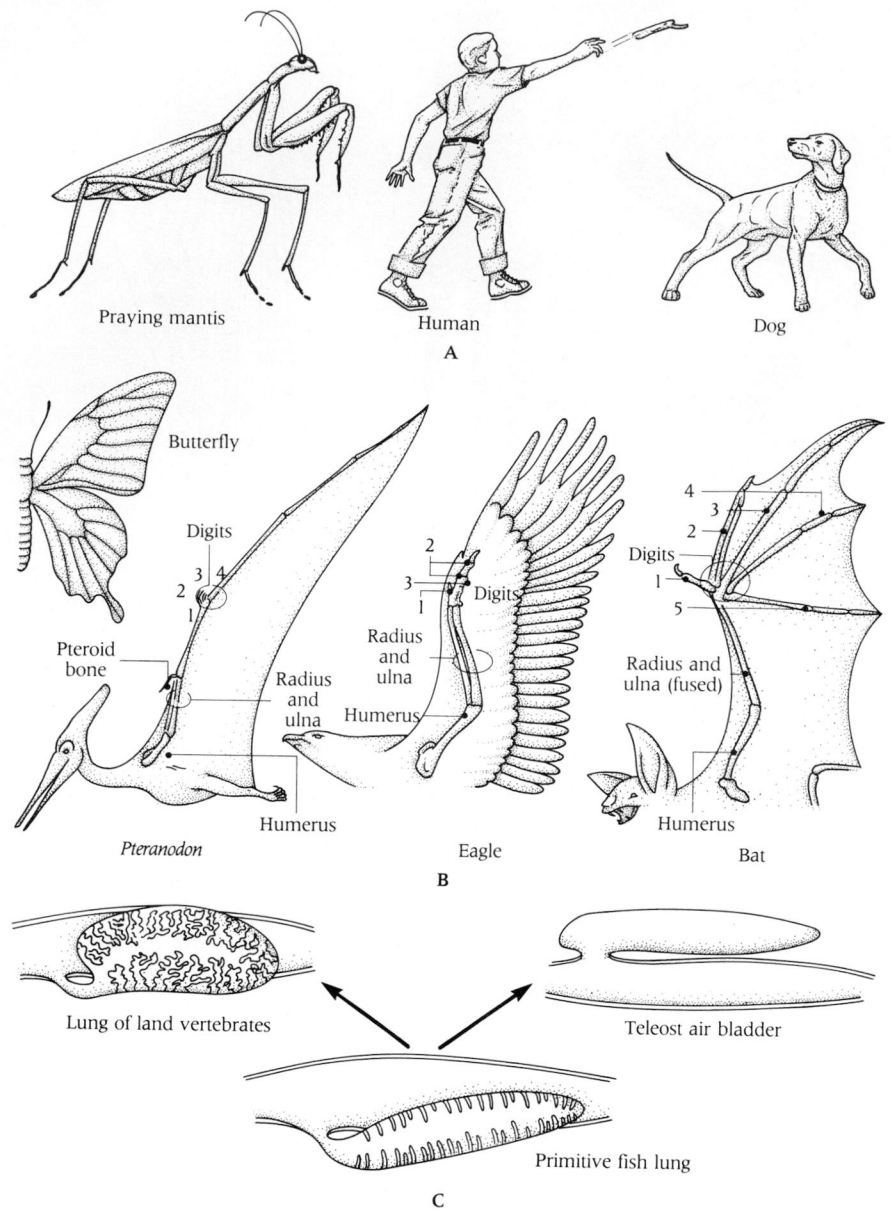

37-5 Homology, homoplasy, and analogy
(A) The wings and legs of insects are homoplastic to the limbs of human and dog, because they evolved independently. The forelimbs of human and praying mantis are homoplastic and analogous, because both are used for manipulation. The forelimbs of human and dog are homologous, but not fully analogous, since the dog's forelimb is used nearly exclusively for locomotion. **(B)** The wings of Pteranodon (an extinct flying reptile), eagle, and bat evolved separately from walking forelimbs and so are homoplastic as wings, yet they contain homologous bones. **(C)** Lungs of land vertebrates and air (or swim) bladders of fishes are homologous, but not analogous, because they have different functions. The fish gill is the analog of the vertebrate lung, because both are respiratory organs.

If we compare an insect's leg to the forelimbs just discussed, we see a good example of homoplasy (Figure 37-5A). In some ways, the front leg of an insect resembles a dog's foreleg or a human arm. But the limbs of insects and of mammals evolved independently. Presumably, genes determining their structure were not inherited from ancestors common to the two. Thus the limbs of insects and mammals are not homologous.

Another example of homoplasy is the vinelike habit that has evolved independently in many groups of plants. As shown in Figure 37-6, many different kinds of tendrils, holdfasts, and other structures have evolved along different evolutionary paths to facilitate the climbing of vines.

The homoplastic forelegs of insects and forelimbs of dogs do perform a similar adaptive function—walking. When structures that are not homologous perform similar functions, they are **analogous** and are said to be **analogs.** The wing of an insect is completely analogous with the wing of a bird (Figure 37-5B), but only to some extent are these structures homoplastic. These wings are certainly not homologous. Homoplastic structures are usually analogous as well. Structural resemblances not due to a common ancestry are usually associated with similarity of function.

However, structures may be analogous without being homoplastic—that is, without having any noticeable anatomical resemblance. The gills of a fish and the lungs of a mammal are anatomically so different they would hardly be called homoplastic, but both are organs of respiration and, in that respect, are analogous. In fishes, the homolog of the lungs is the swim bladder (Figure 37-5C).

37-6 Evolutionary modifications for climbing in vines

A variety of evolutionary paths have been taken by climbing plants. **(A)** Stipules modified into tendrils in Passifloraceae. **(B)** Leaflets modified into tendrils and suckers in Bignoniaceae. **(C)** Leaves modified into tendrils in Ranunculaceae. **(D)** Inflorescenses modified into hooks in Rubiaceae.

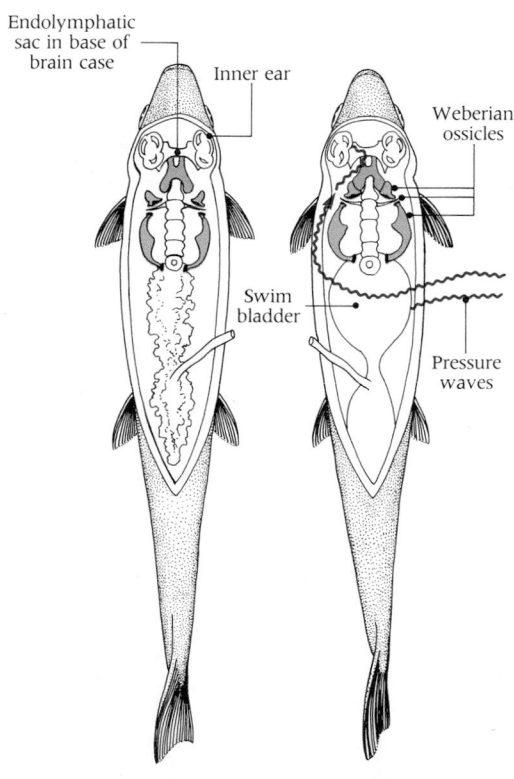

37-7 The swim bladder as a hearing device

In some modern bony fishes, the swim bladder functions in hearing. Bony processes of vertebrae close to the skull have evolved into Weberian ossicles, which are analogous to ear ossicles in land vertebrates. Pressure waves strike the swim bladder and are transmitted to the Weberian ossicles, then on to the endolymphatic sac in the brain case, which communicates with the inner ear. **(A)** The swim bladder is collapsed to show the ossicles. **(B)** When the swim bladder is inflated, the ossicles press on the endolymphatic sac.

TRANSFORMATION

Structures that are homologous sometimes show startling differences in both structure and function. For example, the swim bladder in most fishes is a gas-filled sac that regulates the fish's specific gravity, enabling the fish to swim at different depths while simultaneously maintaining neutral buoyancy. It has nothing to do with breathing. In one group of fishes (carps, minnows, catfishes, and characins), it has become even less lunglike and acts as a sort of sounding board or resonating chamber, the vibrations of which are communicated to the brain through a series of bones. The swim bladder is, in fact, analogous to an eardrum—fishes do not have true eardrums—although it is homologous with lungs (Figure 37-7).

It is odd that fishes should hear by means of a structure homologous with a part of our breathing apparatus. But it is just as odd that our hearing depends in part on structures homologous with portions of the reptile's jaw. As we have noted (see Box 35-1), two of the three little bones that transmit vibrations from our eardrums to our inner ears are homologous with the bones that form the joint between upper and lower jaws in reptiles (Figure 37-8).

It is fascinating that homologous structures can differ so much in both form and function. We must conclude that structures can change radically in their structure and operation. Such changes, termed **transformations,** have occurred commonly in the history of life. They have been especially prominent in the development of major new groups—for example, the rise of mammals from reptiles or of flowering plants from nonflowering plants.

The widespread occurrence of transformations tells us that new sorts of organisms, or new organs, or new adaptations (evolutionary adaptations) commonly evolve from what exists. Any good engineer could design a better reproductive apparatus than a magnolia flower, a better means of walking than a salamander's leg, or a better sound receptor than an opossum's ear. The point is that these structures

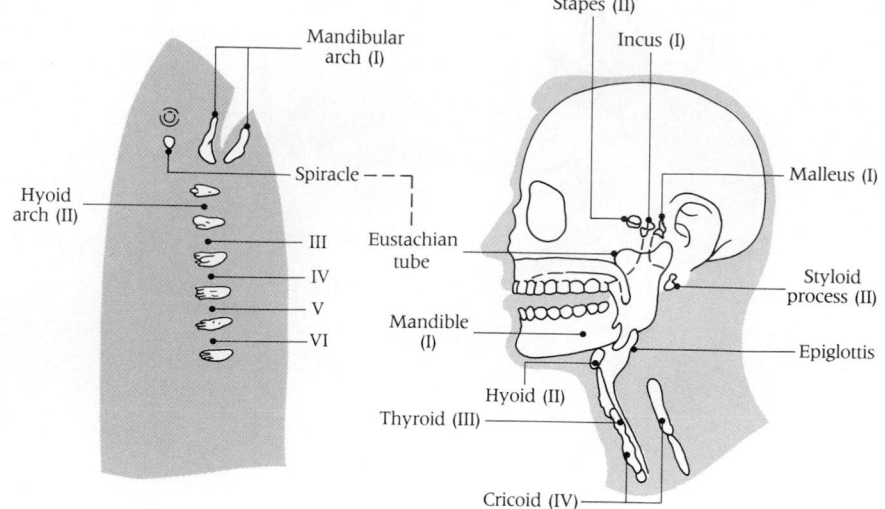

37-8 Evolutionary transformations of the primitive vertebrate gill arches
The evolutionary transformations of two of the primitive vertebrate gill arches into the mammalian ear ossicles were shown in Box 35-1. Here are seen the ultimate transformations of the remaining gill arches into other features of the skull and pharyngeal region. For simplicity, a human is compared with a contemporary shark, which retains ancestral features of the gill arches (but is not itself an ancestor of mammals). Homologous arches have the same Roman numerals. They have been transformed in the mammal into: (I) lower jaw and two ear ossicles (incus and malleus); (II) third ear ossicle (stapes), the hyoid cartilage (tongue skeleton), styloid process of skull, and styloid ligament; (III) thyroid cartilage (part of the larynx); and (IV) cricoid and epiglottal cartilages of the larynx.

were not designed by a mechanical engineer. They *evolved* on the basis of mutations affecting earlier structures that had different functions.

IRREVERSIBILITY OF EVOLUTION

Evolution builds on what exists—on the results of all previous evolution. This fact has a profound implication: evolution is irrevocable. The past cannot be undone by future events. However radical they may be, new evolutionary changes can never wholly erase the effects of previous evolution. If little mammalian ear bones had not been parts of the reptilian jaw, they could not have the relationships that they now have. If any of our ancestors, even those of the remote past, had been other than they were, we would now be different in some way from what we are.

This principle has another interesting facet. The past cannot be fully regained. Nothing like an earlier form of life ever arises again. Evolution, then, is not only irrevocable. It is irreversible.

The principle of irreversibility is important in systematics. An example will illustrate. Land vertebrates—reptiles and mammals—arose long ago from aquatic forms, the fishes. But some reptiles and mammals, like whales, became aquatic and fishlike in their habits and seemingly reversed the evolutionary trend. This did not make them fishes again, however. The irreversibility of their evolutionary past is plainly seen in their anatomy—for example, in their flippers (see Figure 37-4). The flipper functions like a fish's fin. But unlike a fin, it has passed through an evolutionary stage in which it was a leg. Moreover, even though the flipper became finlike, it has not lost traces of its land-living ancestry. The bones in a whale flipper are plainly homologous with those in the leg of a land mammal.

CONVERGENCE, PARALLELISM, AND DIVERGENCE

The story of whales and fishes shows that organisms of widely differing ancestry may be alike in many ways. Distantly related organisms may display such resemblances. Visitors to the American southwest may be astonished to see what appear to be large numbers of hummingbirds gathering nectar at dusk. However, a closer look shows that they are not hummingbirds but hawkmoths (*Sphingidae*)—insects that are almost identical with the birds in behavior, size, and superficial appearance (Figure 37-9).

We use the term **convergence** to denote what happens when features of two groups of unrelated organisms come to resemble each other in the course of time (Figure 37-10A). The resemblance between cacti and some euphorbias is convergent. So is that between Tasmanian and true wolves, and between the wings of insects, birds, and bats. In Australia, an island continent, the isolated evolution of marsupials has produced forms that converge toward many kinds of nonmarsupial (placental) mammals in the rest of the world (Figure 37-11). There are not only native "wolves" but also native "mice," "cats," "anteaters," "moles," and "sloths." These marsupials

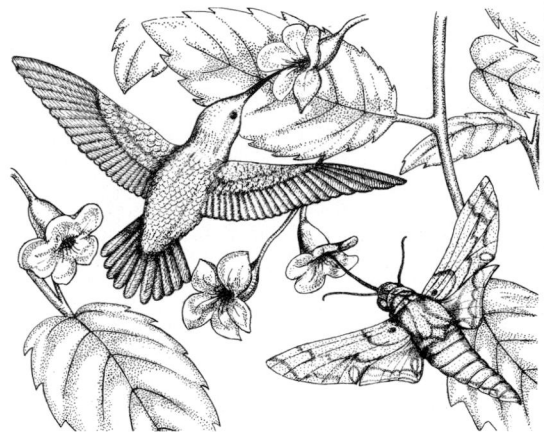

37-9 Convergent evolution
The hummingbird on the left and hawkmoth on the right have converged in form, flying habit, and feeding method in their common exploitation of the nectar in flowers as a food source. None of these resemblances existed in their remote common ancestry.

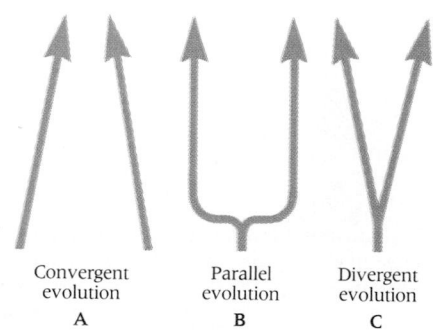

Convergent evolution
A

Parallel evolution
B

Divergent evolution
C

37-10 Three patterns of evolution
(A) *Convergent evolution occurs when two groups of unrelated organisms resemble each other more and more over time.* **(B)** *In divergent evolution, one ancestral group splits into two groups that become less alike as time passes.* **(C)** *In parallel evolution, two related species evolve in similar ways for long periods of time.*

PLACENTALS

MARSUPIALS

Wolf
(*Canis*)

Tasmanian wolf
(*Thylacinus*)

Ocelot
(*Felis*)

Native cat
(*Dasyurus*)

Anteater
(*Myrmecophaga*)

Anteater
(*Myrmecobius*)

Flying
squirrel
(*Glaucomys*)

Flying
phalanger
(*Petourus*)

Ground
hog
(*Marmota*)

Wombat
(*Phascolomys*)

Mouse
(*Mus*)

Mouse
(*Dasycercus*)

37-11 Convergent evolution of some placental and Australian marsupial mammals

are only distantly related to the placental true wolves, mice, cats, anteaters, moles, and sloths. The resemblances reflect convergent evolution. The fact that convergent forms had unlike ancestors means that they cannot become truly identical. Systematists usually have no trouble sorting out truly related organisms from merely convergent ones, though mistakes have been made. But no one classifying specimens today would mistake a hawkmoth for a hummingbird or a euphorbia for a cactus. The separate ancestries of these forms are so obvious that the convergent nature of their similarity is unmistakable.

Systematists' problems are more difficult when the ancestries of two forms are not very different. Even worse are cases of related ancestry in which later evolution has followed more or less the same course. This sort of evolution is not always clearly distinguishable from convergence, but it has been given a different name: **parallelism.** Parallelism (Figure 37-10B) is more common than convergence. Examples are the repeated evolution of female winglessness in the moth family Geometridae and other families and of colonial social behavior among bees, wasps, ants, and other hymenopterans.

Close parallelism may be impossible to distinguish from close community of ancestry unless the actual ancestors are known. For example, it was long assumed that American and Old World porcupines are closely related and that their spininess is homologous. However, later work suggested that this may be a case of parallelism and that a very remote common ancestor existed which was spineless. If this is so, the spines evolved independently in the two groups, presumably because both groups encountered similar selective forces. That means they are homoplastic, not homologous. The question is unsettled, but some systematists suspect that the history of porcupines does display parallelism.

In addition to convergent and parallel evolution, there is a third possible future for two evolving groups. They can become progressively less similar, that is, they can diverge (Figure 37-10C). **Divergence** is extremely common and is the obvious basis for the diversity of nature. Fluctuating and reversible divergence can also occur within a species. This occurs as demes and other groups develop differences from each other. As soon as speciation occurs, however, the divergence usually becomes irreversible and tends to increase as time goes on. A fourth possible future, not common but real, is "none of the above"—the two groups do not become either more or less similar.

SCHOOLS OF SYSTEMATICS

No well-defined schools of classification existed for nearly a century after the publication of *The Origin of Species* in 1859. As a result of the lack of accepted principles, there was a good deal of confusion. Different classifications were sometimes proposed for the same group of organisms—or new classifications would be proposed only because they were allegedly "better." Dissatisfaction with such arbitrariness in the 1950s and 1960s led to the development of three schools of classification: **numerical phenetics, cladistics** (both of which we mentioned earlier), and what is now termed **evolutionary systematics.**

NUMERICAL PHENETICS

We earlier cited a fanciful example in which a gray squirrel was classified by noting that it has 10,001 resemblances to a red squirrel, 8,346 to a cat, and so on. This is precisely what numerical pheneticists say should be done if we are to give every character equal weight and thereby make classification objective.

Numerical pheneticists use as many characters as possible, ignore issues of homology and homoplasy, and weight all characters equally. Their basic assumption is that if a sufficiently large number of characters is compared, there will be no need for subjective judgments. Pheneticists argue that any errors arising from their methods will be canceled out by the mass of their data.

Phenetic methods have attempted to substitute quantitative rigor and repeatability for the frequent subjective judgments that blemished the classical "natural history" of an earlier time. However, numerical phenetics faces serious problems when, as frequently happens, its two basic assumptions—that all characters are of equal weight in determining phylogeny, and that little convergent evolution has taken place—are invalid. Thus, if strictly applied, phenetics would classify the placental mole and the marsupial "mole" as close relatives, a conclusion pheneticists themselves recognize as erroneous. Today, the phenetic approach is mainly used as an auxiliary method of assessing similarities. Thus, even though we do not classify a placental mole and a marsupial "mole" together, despite their great similarity, phenetics can provide a quantitative measure of similarity that can be used in studies of function, design, and adaptation.

CLADISTICS (CLADISM)

Cladistics classifies organisms according to the sequential order in which branches (called clades) arise from a phylogenetic tree. At the same time, it ignores the degree of their divergence. A tree, constructed by cladistic methods, takes the form of a series of dichotomous branches: it is called a **cladogram** (see Figure 36-7). Branches arising at each branching point are defined by the new homologies that are unique to the various species on those branches.

Let us analyze a cladogram of five familiar vertebrates: lizard, cow, seal, dog, and cat. All possess five toes. Actually, the most primitive amphibians already have five toes. Thus, the presence of five toes is not useful in determining the branching points of the tree. The five-toed condition is regarded as a shared primitive, or **plesiomorphic,** character. Because cladistics defines each branching point by an evolutionary innovation, such as a derived, or **apomorphic,** character, it becomes necessary to seek a novel character to determine the branching point separating the lizard from the other four vertebrates (Figure 37-12). Hair is such a derived character, which is shared by the cow, seal, dog, and cat—but not the lizard. Thus we call it a **synapomorphic** character for mammals.

We must now identify the sequence of the subsequent branches. The dog, cat, and seal all share a new kind of cheek teeth called involuted teeth. The cow, however, does not possess this evolutionary innovation. The next branching point in the tree separates the seal from the cat and the dog on the basis of the emergence of carnassial teeth (specialized for cutting and shearing) in the cat and dog. Finally, cats branched off from the dogs because of a newly evolved character: retractable claws. We see, then, the major task in cladistics: It is to identify those features that characterize each branching point.

Emphasis on the sequential order in which branches arise from the phylogenetic tree without regard to the degree of divergence has led to novel hypotheses of evolutionary relationships. Consider, for example, the evolutionary relationships of the cow, lungfish, and trout, which are expressed in the cladograms depicted in Figure 37-13. When the cladistic method is applied, we find that lungfishes are

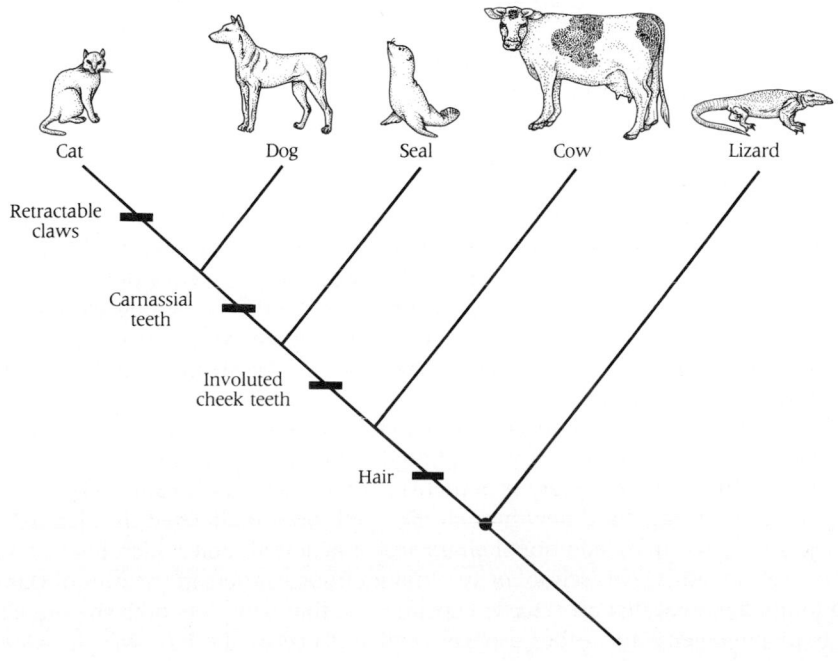

37-12 The cladistic method of classification *Cladistics identifies those features that characterize each branching point. In this example, lack of hair separates lizards from the other four vertebrates shown. Involuted cheek teeth distinguish cows, lack of carnassial teeth distinguish seals. In this group, only cats possess retractable claws.*

Cat   Dog   Seal   Cow   Lizard

Retractable claws

Carnassial teeth

Involuted cheek teeth

Hair

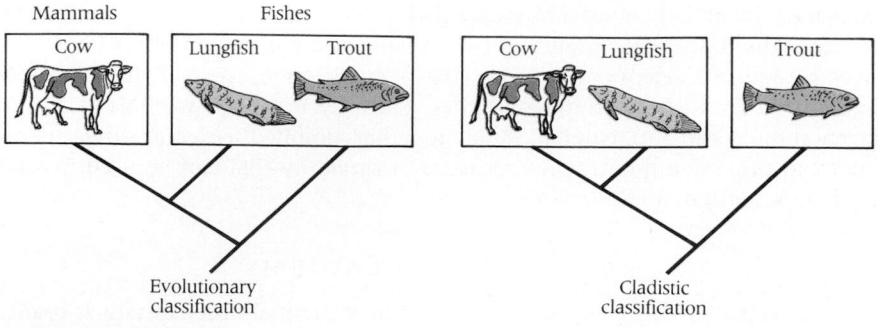

Mammals   Fishes
Cow   Lungfish   Trout

Evolutionary
classification

Cow   Lungfish   Trout

Cladistic
classification

37-13 Two phylogenetic trees of cow, lungfish, and trout
A comparison of evolution and cladistic approaches to evaluating the evolutionary relations between cow, lungfish, and trout. Note how the results differ.

more closely related to cows than the trout because, like cows, they share a novel feature: the internal nares (or choanae) in the roof of their mouth cavity. Thus, based on the branching sequence, lungfish and cow are sister groups because they share a more recent common ancestor than the trout. Evolutionary classification sees things differently. It gives weight to the fact that a lungfish and a trout are both fishes—and a cow is a mammal.

Remember, cladistics ignores degree of divergence. The fact that the lineage leading to cows diverged profoundly, passing through the structural grades of amphibian and reptile, while the lineage of lungfishes diverged far less is not relevant in the cladogram. Even though lungfish and trout both have fins and gills and live in water, their resemblances are based on primitive features. Based on the emergence of an evolutionary innovation, the internal nares, there is little doubt that lungfish and cow are more closely related to one another than to the trout. If one were to emphasize the degree of morphological divergence, one would consider the trout and lungfish more closely related to each other than either is to the cow.

Current debate focuses on the question: Should or should not classification express the fact that cows came to differ and lungfishes remained similar to their common ancestor? Cladists would answer: They should not because it is the sequence of branches that truly reflect relationships in the form of sister groups that share the most recent common ancestor in the genealogical tree.

EVOLUTIONARY SYSTEMATICS

Evolutionary systematists maintain that classifications ought to reflect the dual nature of evolutionary change.

- The splitting of phyletic lineages—that is, the branching in the phylogenetic tree
- The invasion of new niches and major new environments

In sharp contrast to cladists, evolutionary systematists do weigh the derived characters (apomorphies) that determine the branching points in the tree. An evolutionary innovation, for example, that signifies the successful invasion and exploitation of a major new environment is given more weight than a new character with a minor functional role.

In assigning weights to different novel characters, evolutionary systematists do engage in subjective judgments. Thus, an evolutionary systematist would recognize that cows and lungfishes are closely related because of the recency of their common descent (see Figure 37-13). But they would disregard the branching sequence and conclude that since cows have diverged far from their ancestral form and have invaded truly new major environments, cows should be given a status that clearly separates them from lungfish.

We see, then, that in evolutionary systematics major events in adaptive evolution are reflected in classification. By weighing evolutionary novelties in constructing phylogenetic trees, evolutionary systematists intend their classifications to say something about how rapidly a new branch diverged, how it changed in relation to its sister group, how many additional characters it acquired, and which new environments were invaded. This school of systematics finds its roots in theories of Darwin, who long ago advocated an eclectic classification that considers both the branching points in phylogeny and other aspects of divergence as the best way to generate biologically meaningful classifications that make possible the largest number of broad generalizations.

Although each of the schools of classifications insists that its methodology is "best," we would argue that the strength of the evolutionary approach lies in its intention to include and balance both major aspects of evolutionary descent: branching sequences and degrees of divergence in form and function. In contrast, phenetics and cladistics do not deal adequately with the evolutionary history of each taxonomic group. In practice, unfortunately, the results of studies based on branching and degree of divergence can be in conflict—and the conflict cannot be reconciled with mathematics. Therefore, systematic biologists who wish to construct classifications based on both branching sequences (cladogenesis) and divergence or similarity (anagenesis) are obliged to take refuge in subjective judgments.

THE PRACTICE OF CLASSIFICATION

Classification consists in essence of three operations.

- Describing and recognizing related groups of organisms according to the principles of populations and of phylogeny
- Fitting these groups into a formal hierarchy
- Naming the groups

Only the first of these operations depends on direct observation and interpretation of nature. The second and third put the results of the first into meaningful form and supply the names by which we can think and talk about the results. The second and third operations are necessarily subjective and more or less arbitrary.

RECOGNITION CLUES

The first operation of the systematist—describing and recognizing related groups of organisms—is the source of the data with which he or she answers the fundamental question: How does one distinguish between homologous structures reflecting common ancestry and analogous structures reflecting common environmental burdens? The modern systematist has access to a wide (and widening) range of helpful clues that were not available in the past.

First and foremost are the clues based on similarities of anatomical structure. Usually the systematist seeks homologies among structural characteristics that do not change very much during the evolution of large groups of organisms. The structure of forelimbs, as noted earlier, and the reproductive parts of flowering plants often provide useful clues to evolutionary relationships. Leaves vary too much in shape and display a good deal of convergence. In general, the number and arrangement of flower parts such as stamens and petals and the position of the ovary are the same in large groups of plants. As we shall see, the two great groups of flowering plants, the monocots and dicots, are distinguished by the number of seedling leaves (Figure 37-14).

Another type of clue that bespeaks relationship is found in the methods of reproduction and development. These processes are especially useful clues among the algae and fungi. They are so critical to the perpetuation of a species that there is small latitude within which changes can occur. Hence mutations affecting reproductive structures rarely survive. In animals, the study of embryos and larval forms may be helpful clues to relationships among major groups. For example, the embryos of all vertebrates display structures resembling gill slits. In fact, they are neither gills nor slits but are undoubtedly homologous with remote ancestral gill slits that have diverged enormously since that time. Similarly, gastrulation occurs in most animals, but not in plants. Clearly, such embryonic devices have withstood evolutionary modification because any mutation that altered them would be unlikely to survive.

Many other features—chemical, behavioral, and ultrastructural—offer useful clues. The fine structure of bacterial flagella differs from that of flagella of all other organisms. The very presence of mitochondria in all living organisms except prokaryotes (bacteria and blue-green algae) suggests that all animals, plants, fungi, and protists have a common ancestry. Clues are also found in the number and

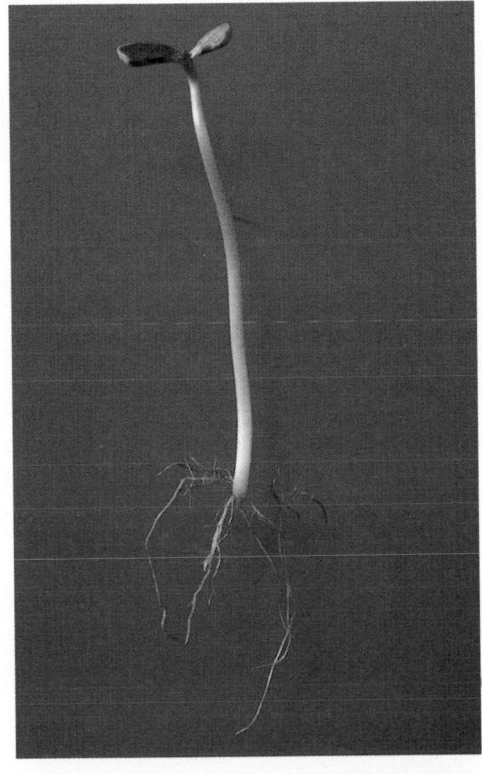

A

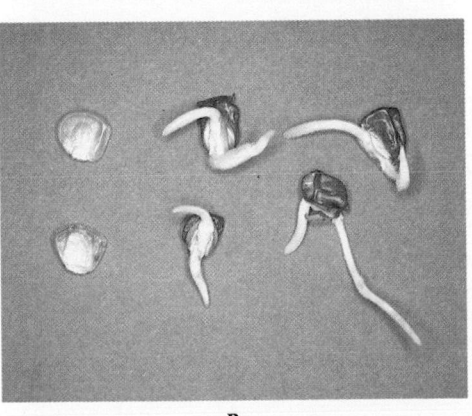

B

37-14 Dicot and monocot
(A) Dicot (bean) showing two cotyledons.
(B) Monocot (corn) showing one cotyledon.

shape of chromosomes—and even in behavior patterns such as the courtship displays of birds.

Finally, there are the intriguing clues that come from biochemical and molecular research. We spoke in Chapter 35 of new methods based on comparisons of amino acid sequences in similar proteins—for example, cytochrome c from different species—and of nucleotide sequences in DNA or RNA. Similarities (or differences) imply close (or distant) relationships and analysis of these molecules permits the establishment of a tentative timetable for the appearance of different species.

The Systematic Hierarchy

By general agreement our catalogers today use a form of hierarchy that is actually based on the one originally proposed by Linnaeus. A special term is applied to each level of the hierarchy.

Kingdom
 Phylum (plural, *phyla*)
  Class
   Order
    Family
     Genus (plural, *genera*)
      Species (plural, *species*)

There is no reason why the hierarchy must have seven levels. Groups of virtually any degree of inclusiveness could be recognized. As it happens, a hierarchy of seven levels became familiar through common usage. But with the tremendous increase in the known number of kinds of organisms, specialists found that seven levels were not enough. In turn, they began supplying additional levels with prefixes. Some classifiers (especially cladists) use many more than the 17 levels in the following example. We can see how the system works by examining a full formal classification of a human subspecies in which all the levels are named.

Kingdom Animalia
 Phylum Chordata
  Subphylum Vertebrata
   Superclass Tetrapoda
    Class Mammalia
     Subclass Theria
      Infraclass Eutheria
       Cohort Unguiculata
        Order Primata
         Suborder Anthropoidea
          Superfamily Hominoidea
           Family Hominidae
            Subfamily Homininae
             Genus *Homo*
              Subgenus *Homo* (*Homo*)
               Species *Homo sapiens*
                Subspecies *Homo sapiens sapiens*[4]

It should be noted that no two authorities agree on exactly how all organisms should be classified. There is, however, some agreement on the important features. A modern, but conservative, classification of plants, fungi, and animals is provided in the Appendix at the end of this book.

Classifying a Species

A species consists of groups of populations that may be subdivided in a number of ways but that have an essential unity and continuity in many spheres, including

---

[4] All living humans belong in the same species, but as discussed in Chapter 49, some investigators have considered the races of humankind to be subspecies. Others consider this a confusing and useless complication of classification.

their evolutionary role, geographic distribution, genetic relationships, and so on. Groups at higher hierarchical levels than the species level show less unity and continuity because they include more than one species. Groups at lower levels are less clearly defined because there is less decisive discontinuity between them. The species, then, is the systematist's fundamental population unit.

To classify a species, one must first determine a population's characteristics. The procedure resembles what would have to be done in characterizing the apples produced in a given orchard. One would look not at just one apple but at the whole crop—or more precisely at a fair sampling of the orchard's tens of thousands of apples. (Statistical analysis would provide a rigorous measure of how many apples comprise a sufficient sample.) This, in principle, is how the modern systematist estimates the characteristics of a natural grouping such as a species.

Higher Categories

The arrangement of species into genera, genera into families, and so on up the hierarchy does not depend upon judgments based on the sampling of populations. Rather, the basic population units, the species, are successively combined into larger and larger groups according to the way we interpret their evolutionary relationships. In principle (with certain exceptions), all the species of one genus have evolved from one ancestral species, all the genera of one family from one ancestral genus, and so on. Of course, our understanding of the natural relationships of all organisms is not yet detailed enough to permit a definitive and final classification that everyone would accept. Hence, classifications are only hypotheses about relationships. They are constantly changing as we learn more about the course of evolution.

Even when evolutionary groupings are well known, the easy shaping of higher categories does not follow automatically. A classification must be simple enough to be understandable and usable, and it may not be possible for the classification to take into account every intricacy of the relationships of species. All we can do is ensure that a classification is consistent with a reasonable theory of evolutionary affinities.

Nomenclature

Systematists are often chided because they do call a rose by another name—or they refer to an ordinary rat as *Rattus rattus rattus*. It is true that some among us enjoy being inscrutable. However, systematists were forced into a complex and stuffy nomenclature, as millions of species had to be named, not to mention all the other groups up and down the hierarchy.

In New Mexico alone, there are 13 species of wild roses. In the Old World, there are more than 560 species and subspecies of the genus *Rattus* and hundreds more are called "rats" that do not even belong to this genus. A less impelling but still important reason for not using common names is that these are different in every language. What is a *squirrel* in England is an *écureuil* across the Channel, an *ardilla* across the Pyrenees, an *egern* in Denmark, an *eichhörnchen* across the Rhine, and something else in other places.

In inventing an artificial system of names, systematists agreed to use the same names in every country, regardless of the native language. That is one useful holdover from the days when all scholars wrote in Latin. Later, many scientific names were derived from Greek; today they are derived from any language or from none, being invented for the occasion. However, all names are "latinized" and treated as if they were Latin words. We note also the custom of placing a dagger symbol (†) before the name of any group that is extinct (see Appendix).

The name of any group—from kingdom to genus—is designated by a single capitalized word. Genus names are usually printed in italics, but names of higher groups are not. The name of a species is two italicized words: The genus name (a noun) is followed by an adjective (not capitalized) distinctive to the species: for example, *Homo sapiens* (man, thinking). This is why the nomenclature is called binomial. For subspecies, a third italicized word (not capitalized) is added to the name of the species.

Even after systematists had agreed on a system of nomenclature, duplications and changes fostered disorder and confusion. Finally, international biological unions attached one distinctive name to each group and devised codes intended to avoid these problems that depend heavily on the rule of priority. The valid or accepted name of a group is the first published name applied to it. If the same name was given to two groups, the name belongs to the group to which it was first applied.

CLASSIFICATION SCHEMES

Understand that biological classification is both a theoretical and a practical necessity. The theoretical necessity, as discussed earlier, relates to the problem of establishing the course of evolutionary history. The practical aspect relates to a staggering fact: There are now at least three and perhaps ten million different species of living organisms.[5] In addition, some 130,000 fossil species have been distinguished, but these are thought to represent only a minute proportion of those species that have existed in the past. Surely an even greater number have become extinct than are now alive.

We have already identified the major taxa generally recognized by zoologists and botanists: kingdom (the largest and most inclusive taxon), phylum (called division by botanists), class, order, family, genus, and species. Let us now examine the highest taxa, the kingdoms and phyla.

The Kingdom Question

It was customary for many years to divide all living organisms into two obvious so-called kingdoms—plants (**Plantae**) and animals (**Animalia**) (Table 37-1 and Figure 37-15A). Plants did not move around, did not eat, and had cells with distinctive walls. Animals could move around, did eat, and lacked cell walls. But with the discovery of a wide variety of microorganisms, too many exceptions were encountered. Some organisms, like *Euglena*, seemed to fit both categories. Such difficulties

[5] The number of individuals per species varies from a few hundred or less for rare large vertebrates, the extinction of which may be near at hand, to meganumbers—for example, 10^{10} per ml—for small bacteria in nutrient-rich waters. The possible relation of species size and the differences between them has long been an issue of concern to ecologists (see Chapter 45).

TABLE 37-1
FOUR CLASSIFICATION SCHEMES

Two Kingdoms	Three Kingdoms	Four Kingdoms	Five Kingdoms*
	MONERA	**MONERA**	**MONERA**
	Bacteria	Bacteria	Bacteria
	Blue-green algae	Blue-green algae	Blue-green algae
		PROTISTA	**PROTISTA**
		Algae	Algae
		Protozoa	Protozoa
		Slime molds	Slime molds
		True fungi	
			FUNGI
			True fungi
PLANTAE	**PLANTAE**	**PLANTAE**	**PLANTAE**
Bacteria	Algae	Bryophytes	Bryophytes
Blue-green algae	Slime molds	Tracheophytes	Tracheophytes
Algae	True fungi		
Slime molds	Bryophytes		
True fungi	Tracheophytes		
Bryophytes			
Tracheophytes			
ANIMALIA	**ANIMALIA**	**ANIMALIA**	**ANIMALIA**
Protozoa	Protozoa	Multicellular	Multicellular
Multicellular	Multicellular	animals	animals
animals	animals		

* Modified from Whittaker

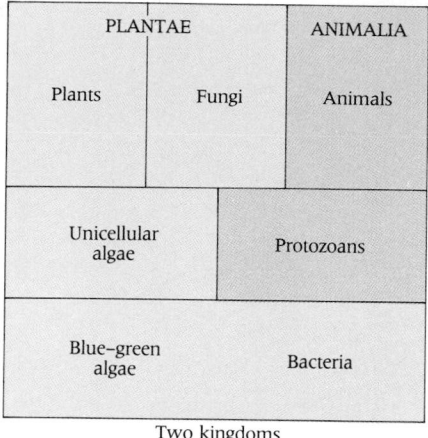

Two kingdoms
A

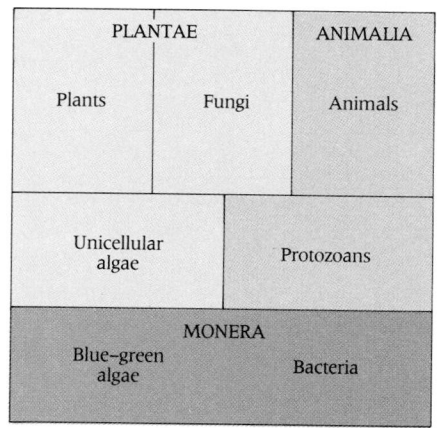

Three kingdoms
B

Five kingdoms
C

37-15 Three views of the kingdom question (**A**) *All organisms are classified as plants or animals.* (**B**) *The bacteria and blue-green algae are separated from plants and placed in a third kingdom, Monera.* (**C**) *The five-kingdom system.*

multiplied as other new fields and techniques began to lengthen the lists of characteristics that had to be taken into account.

A three-kingdom scheme was then devised (Figure 37-15B), in which a new kingdom, **Monera,** made up of bacteria and blue-green algae, was set apart from plants and animals (see Table 37-1). But new difficulties arose. Even a four-kingdom scheme left much to be desired. This scheme distinguished Monera and Protista. To **Protista,** the newest kingdom, were transferred the amoebas, flagellates, and other animallike protozoa (from Animalia), and the slime molds (from Plantae).

If an aim of classification is the grouping of organisms with shared or common ancestry,[6] it must be said that the evidence on the issue of relationship is still fairly equivocal among some of the great groups, notably fungi, and most of the algae. A five-kingdom scheme, shown in Table 37-1 and Figure 37-15C, recognizes the great differences between plants and fungi and gives **Fungi** the status of a separate kingdom. As we have mentioned earlier, we have adopted that scheme for this edition of our book.

Interestingly new organisms are occasionally discovered that do not fit into any of the existing kingdoms as currently defined. For example, the surprising recent discovery of a new group of microorganisms led to the claim that a new kingdom exists. The organisms, named **Archaebacteria,** are a group of prokaryotes as different from true bacteria as they are from eukaryotes (see Chapter 38). Some abhor O_2 and derive energy by converting CO_2 and H_2 to methane (CH_4). Thus they are *methanogenic.* Others are extremely *thermophilic* (preferring high temperatures). Other differences include a unique nucleotide sequence in certain regions of their tRNA molecules, and distinctive patterns of lipids and cell wall carbohydrates. Archaebacteria play a key role in the rumen stomachs of cows, elephants, and several other herbivores that utilize cellulose as a source of food. Whether or not they represent a new kingdom continues to be debated. We will deny them that rank in this edition of our book—but we may well promote them in the next.

An Introduction to the Five Kingdoms

Everyone agrees that the kingdom **Monera** includes the simplest known organisms and the ones from which the four other kingdoms evolved. It includes all prokaryotic organisms (except perhaps the archaebacteria), that is, the **bacteria** and the **blue-green algae** (also called **cyanobacteria**). Prokaryotes, we will recall, lack membrane-bound cellular organelles such as mitochondria, chloroplasts, or nuclei, and their genetic material is a naked chromosome consisting of a single circular molecule of double-stranded DNA without the associated proteins found in eukaryotes (see Table 5-1). Because all prokaryotes have these and other features in common, it is likely that they originated from some original founder-cell population and that they are thus a distinct evolutionary group.

The other four kingdoms are composed entirely of eukaryotes (Table 37-2). They have a definite or true nucleus bounded by a double membrane (envelope), and they contain chromosomes that are made up of DNA and proteins and that divide and segregate by mitosis. They also have other membrane-limited organelles— complex cilia and flagella with a "9 + 2" microtubular pattern, and mitochondria and vacuoles bounded by a single-unit membrane.

Eukaryotes, we should recall, have two other important features not found in prokaryotes: sexuality and (in all kingdoms but Protista) multicellularity with extensive development of tissues. The cells of multicellular eukaryotes also have mechanisms of intercommunication.

The kingdom **Protista** includes an extremely heterogeneous collection of single-celled eukaryotic organisms, among which are protozoa, algae, and slime molds. Some are plantlike; others are animallike. The former include those algae that

---

[6] It has become clear in recent years that a eukaryotic cell has a composite nature. As discussed in earlier chapters, it is generally assumed that mitochondria and chloroplasts are descended from an entirely different ancestral line from the rest of the eukaryotic cell, which therefore represents the end result of an essentially symbiotic (mutually beneficial) relationship between two cell types. Thus we may have to assume that at least two ancestral cell types were represented by the cell type that became the eukaryotic host cell and the prokaryotic cell types that became mitochondria and chloroplasts.

TABLE 37-2
SOME ADDITIONAL CHARACTERISTICS OF THE FIVE KINGDOMS*

	Monera	Protista	Plantae	Fungi	Animalia
Cell type	Prokaryotic	Eukaryotic	Eukaryotic	Eukaryotic	Eukaryotic
Means of genetic recombination	Conjugation, transduction, transformation, or none of above	Fertilization (syngamy) and meiosis, conjugation, or none of above	Fertilization and meiosis, alternating haploid and diploid phases	Fertilization (syngamy) and meiosis, or none, dikaryosis, haploid spore production	Fertilization, meiosis precedes gametogenesis
Mode of nutrition	Autotrophic (chemosynthetic and photosynthetic) and heterotrophic (saprobic and parasitic)	Photosynthetic and heterotrophic or combination of these	Photosynthetic mostly	Heterotrophic (saprobic and parasitic) by absorption	Heterotrophic by ingestion
Motility	Bacterial flagella, gliding, or nonmotile	9 + 2 cilia and flagella, ameboid, contractile fibrils	9 + 2 cilia and flagella in lower forms and in some gametes, none in most forms	9 + 2 cilia and flagella in some forms, none in most forms	9 + 2 cilia and flagella, contractile fibrils
Multicellularity	Absent	Absent	Present in advanced forms	Present in most forms (multinucleate)	Present in all forms
Nervous system	None	Primitive mechanisms for conducting stimuli in some forms	None	None	Present

* Table 5-1 summarizes cell structure (and some functions) in the five kingdoms.

manufacture their own food by photosynthesis; the latter include protozoans that ingest or absorb their food. As suggested in Figure 37-16, all protists may share a common ancestry in a primordial eukaryotic cell, which in turn evolved from a prokaryote. The protists, however, evolved in many directions.

The kingdoms **Plantae** and **Animalia** include multicellular eukaryotic organisms. The kingdom Plantae includes multicellular eukaryotic organisms whose cells have cellulose walls and photosynthetic pigments in chloroplasts. A few related organisms that have lost their chlorophyll are also considered to be plants. Plants appear to have evolved from several different ancestral populations of protists.

The kingdom **Fungi** could be considered to resemble plants—its members are nonmotile and have cell walls—but they lack photosynthetic pigments and are heterotrophs that grow in and through their food, absorbing nutrients as they go. They also differ in architecture and method of reproduction. They are usually composed of filaments containing many nuclei, and the filaments may or may not be divided by cross walls. Whether they arose from one or several ancestral populations of protists is still unknown.

The kingdom Animalia consists of multicellular eukaryotic organisms whose cells lack cell walls, chloroplasts, and photosynthetic pigments. They acquire nutrition mainly by ingestion, digestion, and absorption, and they move about with muscles that function by means of contractile fibrils. The animals appear to have evolved from two ancestral protists, one giving rise to the sponges and the other to all other animals.

Note that viruses do not fit into any of the five kingdoms. They are not cells, are not composed of cells, and are not independent organisms. They differ from all cellular organisms in possessing only DNA or RNA, not DNA and RNA, and bear very few genes. Their mode of life is parasitic.

Whatever other purposes they serve, the kingdoms—here characterized only briefly—provide immensely useful frameworks for the more detailed examination of individual organisms and groups of organisms, which will take place in the next few chapters. Monera and Protista will be discussed in Chapters 38 and 39, Fungi in Chapter 40, Plantae in Chapter 41, and Animalia in Chapters 42 and 43.

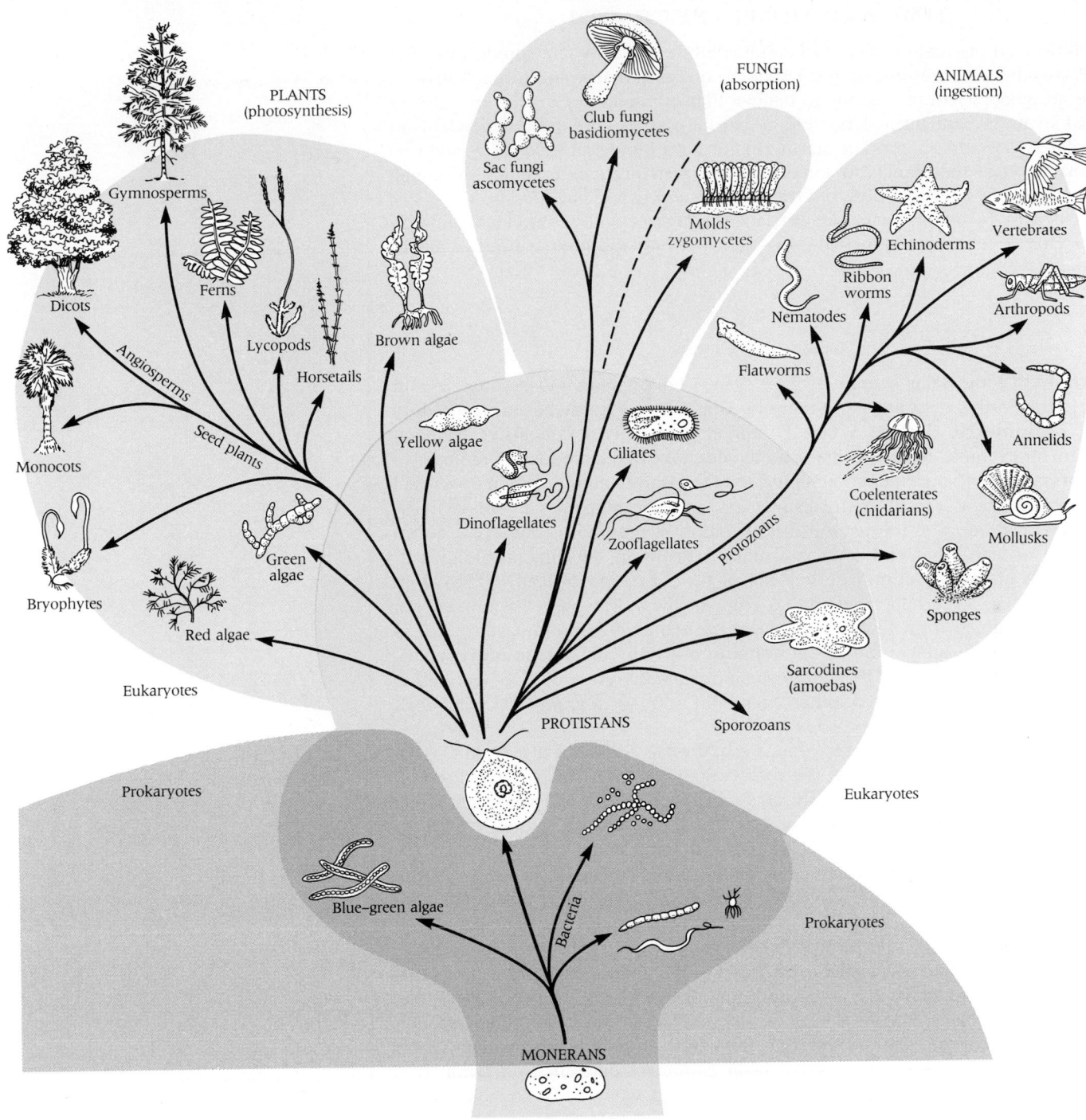

PLANTS
(photosynthesis)

FUNGI
(absorption)

ANIMALS
(ingestion)

Club fungi
basidiomycetes

Sac fungi
ascomycetes

Molds
zygomycetes

Vertebrates

Echinoderms

Ribbon
worms

Arthropods

Nematodes

Gymnosperms

Ferns

Lycopods

Brown algae

Flatworms

Annelids

Dicots

Horsetails

Yellow algae

Ciliates

Angiosperms

Seed plants

Green
algae

Dinoflagellates

Zooflagellates

Protozoans

Coelenterates
(cnidarians)

Mollusks

Monocots

Bryophytes

Red algae

Sarcodines
(amoebas)

Sponges

Eukaryotes

PROTISTANS

Sporozoans

Eukaryotes

Prokaryotes

Prokaryotes

Blue–green algae

Bacteria

MONERANS

**37-16 The five-kingdom system of
classification**

*The kingdoms are outlined by the oval groupings.
All organisms except those belonging to kingdom
Monera are eukaryotic.*

The appendix to this book offers a summary outline of one way in which living
things can be classified.[7]

STUDYING THE HISTORY OF LIFE

In order to discuss evolutionary origins of the major lineages of living organisms
and their distributional patterns over time, we must understand certain geological
principles, and must have some concept of what has been called "geological time."

[7] It is consigned to the Appendix, not because it is nonessential, but because it is reference
material. There are some who take little delight in such things, but those of orderly habits and
venturesome spirit will enjoy observing its complexity and discovering in its listings most of the
many organisms discussed in this book.

To follow such discussions, we need a suitable time scale. Where should our scale start? Was there a beginning of time? That is a question that science cannot answer. There are good scientific reasons to believe that our solar system has not always existed in its present form. Since life as we know it is absolutely dependent on the solar system, an appropriate starting point for a biological time scale might be the formation of the planets in something like their present condition. Although we do not know precisely how old the solar system is, we have reason to believe that it is 3 to 10 billion years old. Perhaps the answer will be provided someday by the Hubble telescope, launched into space in 1990, and now in need of repair.

Geological Clocks

We know that the solar system is more than 3 billion years old because certain rocks in the Earth's crust are about that old. That fact was learned from a procedure called **radiometric dating.** When a mineral containing a radioactive element, uranium for example, first crystallizes, it includes none of the products of radioactive transformation of that element. Such products then start to accumulate in the crystal at a constant rate as a result of the disintegration that accompanies all radioactivity. A stable end product of the natural disintegration of uranium is one of the forms (isotopes) of lead. That end product usually remains in the mineral along with what is left of the uranium. The half-life of the formation of lead (^{206}Pb) from uranium (^{238}U) is 7.6 billion years (7.6 eons, or byrs). The age of a mineral can thus be determined from the ratio between the remaining uranium and the lead produced by the disintegration of the uranium originally present. In other words, if we start with 1 gram of ^{238}U, in 7.6 billion years 0.5 gram of ^{238}U will be left and 0.5 gram will be lead. In another 7.6 billion years, 0.25 gram ^{238}U will be left and 0.75 gram will be lead, and so on.

Dating by this method requires good, fresh, crystals of radioactive minerals that were formed at the same time as the rock containing them. To be directly useful in the study of evolution, the minerals must also be associated with rocks containing fossils. Relatively few uranium minerals have both these qualifications. In recent years, another radioactive transformation—from radioactive potassium (^{40}K) to the inert gas argon (^{40}Ar)—has seemed more promising. Suitable potassium minerals are more common than uranium minerals and are more often associated with fossil-bearing rocks. An isotope of carbon (^{14}C), with a half-life of 5,730 years, is useful in dating materials less than 50,000 years old.

Some rocks have been estimated to be close to 4 billion years old by this method—and even older ones may exist. We know that 4 or 5 billion years ago Earth had a solid, cool crust and that processes of rock oxidation, weathering, and erosion were already going on. This means that there was already water on the surface and that the atmosphere cannot have been extremely different from what it is now. The biological significance is that life on Earth was aleady possible at that extremely remote date.

The subdivision of geologic time in years is now mainly based on the potassium-argon method, especially for the last 600 million years (600 myr), though uranium-lead and several other methods (rubidium-strontium) are also in use. The ages in years given in Table 37-3 accord with recently obtained potassium-argon dates. They are, however, still only approximations.

In discussing the history of Earth, geologists today speak of four main time divisions (Figure 37-17).

- **Hadean time** (before 3.9 billion years ago); no rocks had been identified from this period until 1990, when newly discovered rocks were shown to be 3.962 billion years old.
- **Archean eon** (from 3.9 to 2.5 billion years ago); life originated and simple prokaryotic organisms proliferated; identification of oldest known rocks and fossils (all of monerans); ended with the establishment of an oxygen-containing atmosphere.

TABLE 37-3
MAJOR DIVISIONS OF PHANEROZOIC TIME*

Era	Millions of Years Since Beginning	Period	Epoch	Some Important Events in the History of Life
CENOZOIC (Age of Mammals)	0.01	Quaternary	Holocene (Recent)	Repeated glaciations in north with Ice Age; increase, dispersal, and later thinning out of mammals
	2.5		Pleistocene	First true humans; rise of civilization
	10	Tertiary	Pliocene	Culmination of mammals; radiation of apes; flourishing of herbaceous angiosperms
	25		Miocene	Modernization of mammalian fauna; expansion of mammals and birds
	38		Oligocene	South America separates from Antarctica
	54		Eocene	Australia separates from Antarctica
	65		Paleocene	Early insectivores and primates
MESOZOIC (Age of Reptiles)	150	Cretaceous		Last of the dinosaurs and ammonites; great expansion of angiosperms; South America separates from Africa
	200	Jurassic		First mammals and birds
	250	Triassic		First dinosaurs; trilobites disappear
PALEOZOIC	290	Permian		Reptiles expand; mass extinction
	360	Carboniferous		Amphibians abundant; first reptiles; conifers appear; great coal-forming forests
	405	Devonian		First amphibians and insects; Age of Fishes
	430	Silurian		First land plants; first fishes with jaws
	500	Ordovician		Earliest known (jawless) fishes
	600	Cambrian		Appearance of abundant marine invertebrates
PRECAMBRIAN	>3000			First known fossils; origin of life in dim past
	?5000			Origin of Earth

* These divisions are shown in greater detail in the end papers at the back of the book.

- **Proterozoic eon** (from 2.5 billion to 600 million years ago); marked by formation of the large continents; single-celled organisms acquired sexuality and multicellularity, hallmarks of eukaryotic life.
- **Phanerozoic eon** (began 600 million years ago); marked by the appearance of many kinds of well-formed fossils of multicellular organisms.

In discussing the history of life, biologists are primarily interested in a system of time periods that builds upon the foregoing divisions (see Table 37-3). Because the first period of Phanerozoic time is the Cambrian, it is convenient to lump earlier eons together as Precambrian. This, the earliest time, covers the span from Earth's beginning 3 to 10 billion years ago to 600 million years ago when Phanerozoic time began.

We subdivide Phanerozoic time into three long eras of unequal length, that include eleven periods, some of which are divided into epochs.

- The **Paleozoic era** ("ancient life") covers the first 350 years of Phanerozoic time, a span from 600 million to 250 million years ago. It is subdivided into six periods, which correspond to critical geological events. The first of its periods is the Cambrian. Marked by expansions and extinctions of early archaic plants and animals, the Paleozoic era was terminated by the Permian mass extinction.
- The **Mesozoic era** ("middle life") covers the time from 250 to 65 million years ago. It has three periods and is often called the Age of Reptiles.
- The **Cenozoic era** ("recent life") extends from 65 million years ago to the present. Its two periods are divided into seven epochs. The Cenozoic era is known as the Age of Mammals. We are now living in the Holocene epoch (also called Recent).

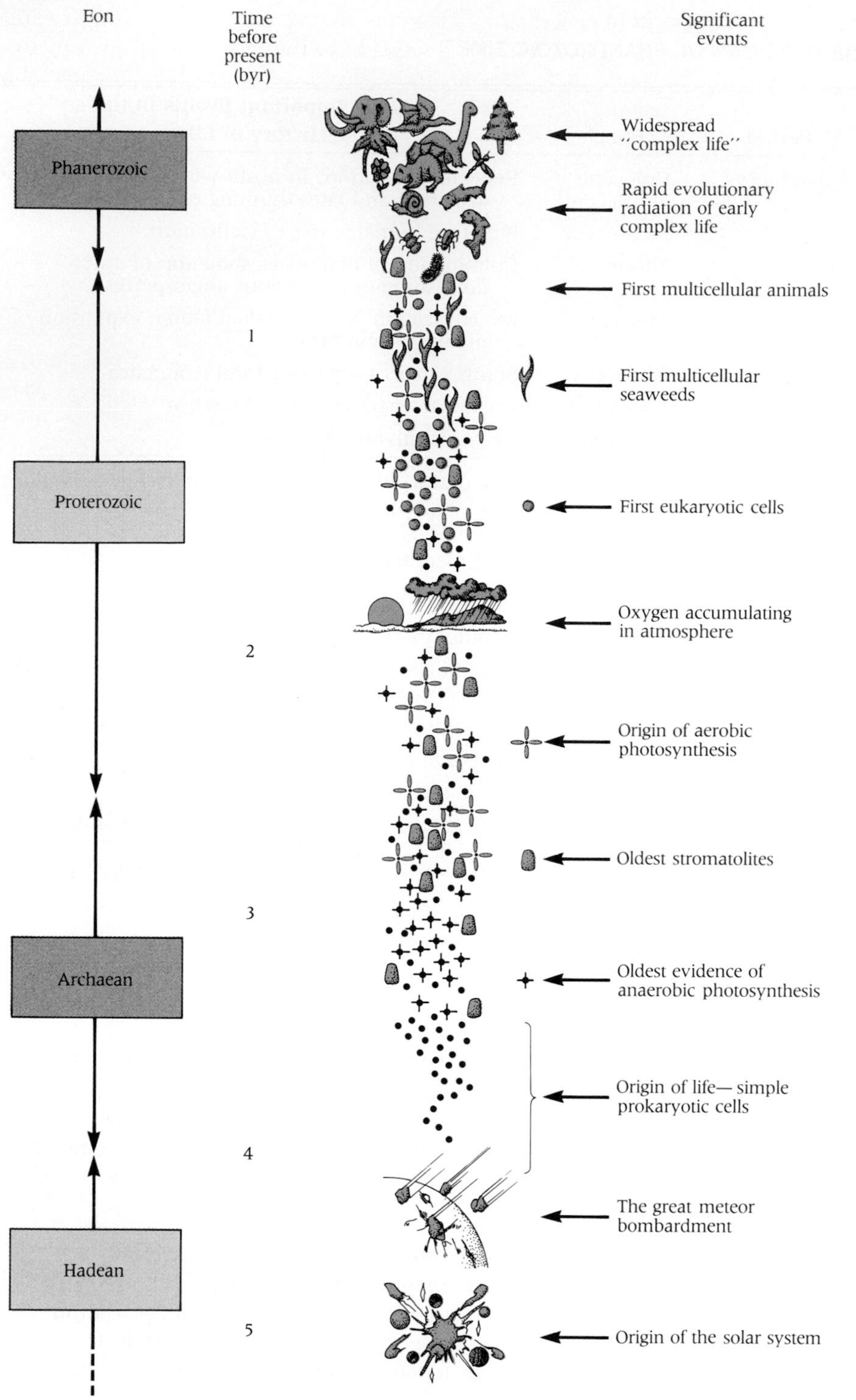
Eon
Time before present (byr)
Significant events
Phanerozoic
Proterozoic
Archaean
Hadean
1
2
3
4
5
Widespread "complex life"
Rapid evolutionary radiation of early complex life
First multicellular animals
First multicellular seaweeds
First eukaryotic cells
Oxygen accumulating in atmosphere
Origin of aerobic photosynthesis
Oldest stromatolites
Oldest evidence of anaerobic photosynthesis
Origin of life— simple prokaryotic cells
The great meteor bombardment
Origin of the solar system

37-17 The four time divisions in the history of the Earth
Some significant events are shown.

The Geological Time Scale

Radiometric methods of dating give useful approximations or orders of magnitude. However, the accuracy of the methods and the dates obtained from them are insufficient to warrant reliance on year dates alone.

Therefore, we must also use a different kind of scale, one that was devised long before radioactivity was recognized. The **geological time scale,** based on such evidence as the layering of sedimentary rocks (Figure 37-18), designates the sequence of rocks and events rather than the time intervals elapsing between them. One could, for example, comprehend the history of the United States by knowing the sequence of major events, even if the dates were not known. It would be handy in such a study to have names for successive periods. They could be designated arbitrarily or perhaps named for presidents or major events such as wars.

37-18 The Grand Canyon of the Colorado River

The layers of sedimentary rocks revealed in the walls of the Grand Canyon were deposited in sequence, one above another, over spans of hundreds of millions of years. By studying the superpositions of such rocks and the fossils contained within them, scientists have gathered evidence on the geological time scale.

37-19 The history of life on a 24-hour scale
For the purposes of this diagram, it has been arbitrarily postulated that life arose 5 billion years ago. The true figure is unknown and may be greater or lesser.

The time scale now in general use by biologists is given in Table 37-3. Approximate times in years are also given, but our discussion of the history of life will focus on the names and sequences of the eras and periods. We know much more about the later phases than about the earlier phases of the history of life, and so we use a more finely divided time scale as we approach the present.

It is difficult to grasp the vastness of these time spans—and to appreciate the increasing tempo of life's history as time went on. As mentioned in Chapter 1, we can better imagine relative time durations, at least, if we ponder a well-known analogy, which views the whole history of life as having occurred within 24 hours from one midnight to the next (Figure 37-19). Let us arbitrarily set the beginning, the first midnight, at 5 billion years ago. On that scale fossils did not become abundant until late afternoon—around 4:45 P.M. At 6:45 P.M. the invasion of land by plants was under way, and by about 7:50 insects and the first amphibians had joined them. The Age of Reptiles began about 11:10 P.M. It ended, and the Age of Mammals began at about 9:15 P.M. Modern humans appeared only a couple of seconds before midnight, and the whole span of recorded human history occupies about the last one-tenth second on the clock.

THE EVIDENCE OF FOSSILS
Principles

The history of life ceases to rest upon unsupported hypothesis or speculation when fossils are available for study. A **fossil** is a factual datum; it is a visible trace of an actual organism that lived in the geological past. That is the fundamental basis of the science of **paleontology,** the study of ancient life (Figure 37-20). It is the task of paleontologists deciphering the history of life to view these isolated facts, to consider their interrelationships to others, and to place them by inference in an explanatory context.

A fossil provides direct knowledge of an organism by showing in concrete form certain definite characteristics. It is then necessary to make inferences about characteristics and activities not directly preserved. In brief, the paleontologist must consider the following specific aspects of fossils:

- The geographic locality where a fossil was buried and therefore must have lived
- Its age, as judged from a range of geological and paleontological data
- Its associations with fossils of the same species, which is the basis of the systematic study of fossil populations
- Its associations with fossils of other species, which is the basis in part for the study of the community, the environment, and the period
- Characteristics of the rocks in which the fossil occurs and the position and mode of burial of the fossil in these rocks; these are also data for the study of environment and time period
- Relationships of the fossil population to other populations—earlier, contemporaneous, and later; these comparisons make possible useful inferences on their phylogeny and classification

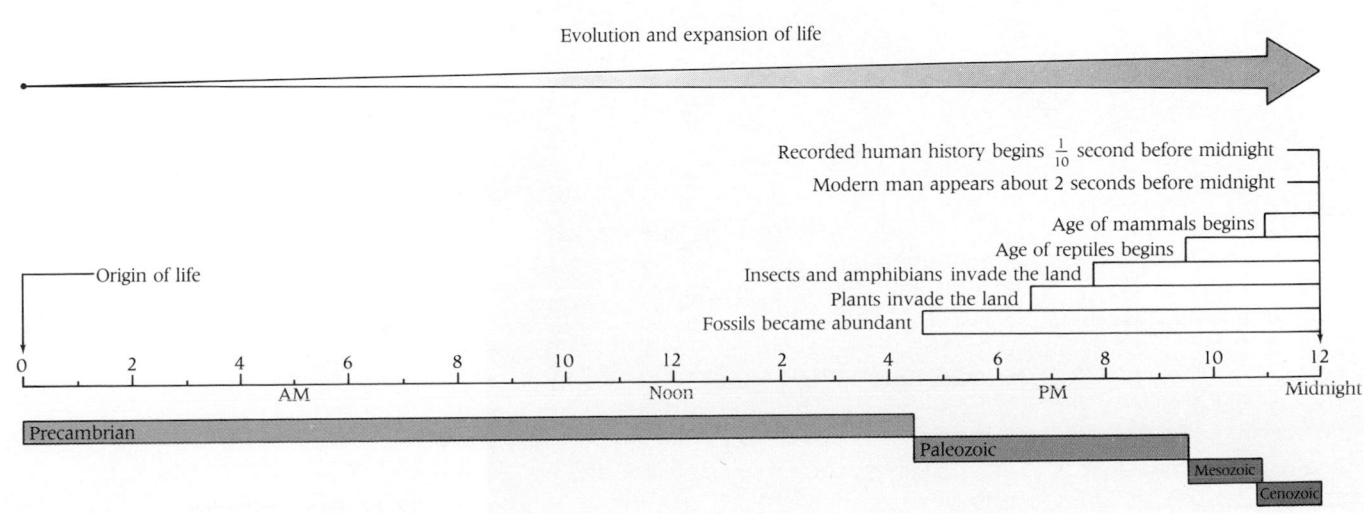

Evolution and expansion of life

Recorded human history begins $\frac{1}{10}$ second before midnight
Modern man appears about 2 seconds before midnight

Age of mammals begins
Age of reptiles begins
Insects and amphibians invade the land
Plants invade the land
Fossils became abundant

Origin of life

| 0 | 2 | 4 | 6 | 8 | 10 | 12 | 2 | 4 | 6 | 8 | 10 | 12 |

AM   Noon   PM   Midnight

Precambrian   Paleozoic   Mesozoic   Cenozoic

Uniformitarianism

An important principle of paleontology, geology, or any science with historical aspects is as follows: the present is a key to the past. That principle was the subject of bitter controversy a century or two ago, when it was endowed with a formidable name, the doctrine of **uniformitarianism.** Now it is widely accepted.

The doctrine holds that the fundamental properties of the universe, the nature and the modes of interaction of matter and energy, have not changed. They are independent of the passage of time. It is only the forms that they have taken and the status of the results of their past interactions that change. Water, for example, always has run downhill and a certain amount of water running at a given velocity over a defined bed always has eroded a certain kind of bed to a predictable degree. The amount of erosion that actually occurred and its effect on the shape of an eroded valley may change. Those aspects of erosion have a history, but the process of erosion itself was presumably the same then as it is now. Using this principle the geologist is able to interpret past changes in the sculpturing of Earth by processes that can now be studied.

Similarly the biologist interprets the fossil record in terms of processes still going on in living organisms—and which are subject to experiment. The timelessness of the properties and processes of the universe was by no means obvious to earlier thinkers. Establishment of that principle was a major advance in the history of thought.

What Are Fossils?

It is extraordinary that one can hold in one's hand the remains of an organism that lived hundreds of millions of years ago. Fossilization is a rare event. Most often dead organisms become unrecognizable in a few years at most, sometimes within hours. Yet many ancient organisms never wholly decayed.

The usual first condition for preservation of a fossil is burial before decay is complete. Natural burial occurs when a dead organism sinks into mud or sand or when these and other sediments are swept over the organism's remains by waves, streams, or winds. Organisms may be buried whole in what are termed sedimentary rocks, and sometimes fossils are found that are startlingly intact. More often, organisms are partly dismembered when buried. Fossils that are to serve as documents to us must then stay buried. An early Cambrian shell can end up in a laboratory only if it remained buried for some 600 million years and was never uncovered by erosion.

However, since fossils cannot be found and collected unless they are at or near the surface of the Earth, erosion must play a role in revealing them to the fossil hunter. Fortunately, rocks of all ages are now exposed at the surface, even though they may have been deep within the Earth's crust at some time in their history. In some areas, dramatic evidence of the geological strata is easily seen. The Earth's

37-20 Fossil hunting and preparation in field and laboratory

37-21 Fossil trilobites

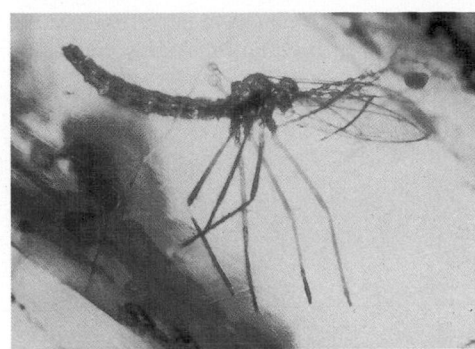

37-22 Fossilized insect preserved in amber

37-23 Fossilized fern in coal

crust has been continually rising in one place, sinking in another, buckling and breaking here and there. Thus some of even the oldest rocks are now at the surface.

The processes of decay are highly effective and may continue even after burial, if burial is not too deep and not in a naturally antiseptic environment (conditions which are unusual). Decay usually obliterates all the soft parts of an animal, leaving only the skeletal parts, which consist mainly of resistant inorganic materials. It is a common misunderstanding that a fossil is petrified, that the organism has "turned to stone." Instead, preserved hard parts usually consist of the same material as when the organism was alive, perhaps with some recrystallization or slight chemical change. But spaces left by the decay of soft parts, for instance, in the marrow of a bone or inside the cell walls of a tree trunk, are often filled with secondary deposits of a mineral, frequently silica (SiO_2) in various forms. After burial and hardening of the surrounding sediments, the hard parts of a fossil may be dissolved by percolating waters. Then a cavity—in effect a mold—may be left, or else mineral-laden waters may precipitate silica or some other mineral in the space, producing a mineral replica of the original fossil.

Even a cursory examination of some fossils of all ages affirms their great beauty and scientific fascination.

- Fossil protists are common, especially of those, like foraminifera, which had hard shells.
- Fossil invertebrates are usually represented by the shells alone. Examples are trilobites (Figure 37-21), a major arthropod throughout the Paleozoic, and snails.
- Fossil insects are uncommon because insects are unlikely to be buried, but those fossils that do occur are often unusually complete because of the insect's tough overall external skeleton. Many ancient insects have been preserved whole in amber (Figure 37-22).
- Fossil vertebrates seldom preserve anything but bones, teeth, and hard scales.
- Fossil plants also have usually lost all the soft parts, the protoplasmic cell contents. Tough leaf coatings, cell walls, spore skins, and such decay-resistant parts are, however, frequently preserved (Figure 37-23).

Looking at these objects, we sense very powerfully our links with the past.

SUMMARY

Modern biological classification groups organisms according to their evolutionary relations, which are inferred from similarities and differences of anatomy, physiology, and biochemistry. Early systematists regarded species as fixed, invariant units. In contrast, modern systematics sees them more flexibly as evolving, interbreeding populations of common ancestry. Their individual variation is essential to the processes by which species arise.

An important problem in classsification is to distinguish resemblances between species that are homologous from those that are homoplastic—that is, that arose by convergent evolution. Homologous structures resemble each other because they share a common evolutionary ancestry. Sometimes homologs have undergone evolutionary transformation, and thus show striking differences in structure and function. Nonetheless, irrevocable traces remain of their origin. Convergent or homoplastic structures resemble each other, not because of a genetic relationship, but because they have evolved in similar environments to do similar tasks. Analogous structures perform the same tasks but do not necessarily resemble each other. Parallelism describes cases in which closely related species show similarities which nonetheless arose independently. Therefore, they are homoplastic not homologous. In contrast, we describe cases in which related species became less and less similar as evolutionary divergence.

There are three modern schools of classification: numerical phenetics, cladistics, and evolutionary systematics. Phenetics provides a quantitative measure of similarity between species by counting shared traits. Because it weights all traits equally discounting the possibility of convergence, it is often an inexact guide to evolutionary relationships. Cladists classify species according to the sequence in which they branched off a phylogenetic line. Each branch point is identified by the appearance of a new distinguishing trait. Unlike evolutionary systematists, cladists do not consider the degree of divergence between groups, concentrating instead on the time when it began. Evolutionary systematists devise classifications that reflect the course of evolutionary change, emphasizing both the branchings of the phylogenetic tree and the innovative traits that allowed invasions of new environments.

The process of classification has three steps. First, species traits—e.g., similarities of anatomy, reproduction and development, and DNA sequence—are described and compared. Second, the species are placed into a formal hierarchy that traditionally contains seven major levels: kingdom, phylum, class, order, family, genus, and species. Third, the species is given a binomial name. Although not all agree, species are currently placed in one of five kingdoms: Monera, Protista, Plantae, Animalia, and Fungi. Viruses are not living organisms and do not fit in any of the kingdoms.

Evolutionary history is measured by a geological time scale determined by rates of radioactive decay and observable sequences of major geological events. Fossils provide the only direct evidence of the evolutionary history of modern species. In studying the fossil record, we rely on the doctrine of uniformitarianism, which holds that the fundamental geological and biological properties of the world have not changed. Fossils form when organisms are buried before they can decay completely. Only hard body structures are well preserved. Soft parts decay, but in some cases are replaced by informative mineral deposits.

KEY TERMS

analogy
Animalia
apomorphic
Archean eon
archetype
binomial nomenclature
Cenozoic era
cladistics
cladogram
class
classification
convergence
evolutionary systematics
family
fossil
Fungi

genus
geological time scale
habitat
Hadean time
homology
homoplasy
kingdom
Mesozoic era
Monera
numerical phenetics
order
paleontology
Paleozoic era
Phanerozoic eon
phylum
Plantae

plesiomorphic
populational systematics
Proterozoic eon
Protista
radiometric dating
species
synapomorphic
systematics
taxon
taxonomy
transformation
type specimen
typological systematics
uniformitarianism

QUIZ QUESTIONS

1. During which of the following intervals did life originate on Earth?
   A. Proterozoic eon
   B. Cambrian period
   C. Miocene epoch
   D. Archean eon
   E. Mesozoic era

2. The evolutionary importance of _____ was apparent to Darwin, but not to Linnaeus.
   A. typology
   B. cladistics
   C. variation

   D. numerical phenetics
   E. radiometric dating

3. A set of homologous characters cannot also be
   A. analogous.
   B. homoplastic.
   C. A and B.
   D. None of the above

4. A classification that purports to demonstrate phylogenetic relationships must be based on _____ characters.

   A. analogous
   B. homologous
   C. homoplastic
   D. None of the above

5. Photosynthesis can be conducted by at least some members of Phylum
   A. Fungi.
   B. Monera.
   C. Protista.
   D. Plantae.
   E. B, C, and D

ESSAY QUESTIONS

1. What principles can be used in classifying organisms? Why are these classification schemes imperfect?

2. What is biological homology? What are examples of homologous structures in nature?

3. What distinguishes cladistic and evolutionary systematics? On what grounds do evolutionary systematists assess the relatedness of the two organisms?

4. How are radioactive elements used in dating geological artifacts? What other methods allow for determination of the geological time scale?

REFERENCES AND SUGGESTIONS FOR FURTHER READING

Duncan, T., and Stuessy, T. F. (1984). *Cladistics: Perspectives on the Reconstruction of Evolutionary History.* Columbia University Press, New York.

The core of the book consists of seven chapters devoted to questions of cladogram construction, including parsimony and compatibility methods, and their interrelationships.

Frängsmyr, T. (Ed.) (1983). *Linnaeus: The Man and His Work.* University of California Press, Berkeley.

Essay by Swedish historians of science on aspects of the life and work of their distinguished countryman. Notes, among other things, that Linnaeus was the first to place humans in a system of biological classification.

Gee, H. (1988). Taxonomy blooded by cladistic wars. *Nature 335,* 585–585.

An interesting report of a recent meeting of systematists.

Hull, D. L. (1988). *Science as a Process: An Evolutionary Account of the Social and Conceptual Development of Science.* University of Chicago Press, Chicago.

In attempting to draw informative parallels between the evolution of life and progress in science, the author discusses the heated debate between cladists and pheneticists.

Margulis, L., and Schwartz, K. V. (1982). *Five Kingdoms: An Illustrated Guide to the Phyla of Life on Earth.* W. H. Freeman, New York.

A readable, well illustrated, and commendably brief catalogue of the major phyla according to these authors. Their scheme does not agree with those of some systematists, but their presentation has impact. An excellent overview of biological diversity.

May, R. M. (1988). How many species are there on Earth? *Science 241,* 1441–1449.

A review of the various factors affecting diversity, including the structure of food webs, the relative abundance of species, the number of species and of individuals in different categories of body size, along with other determinants of the commonness and rarity of organisms.

Ridley, M. (1986). *Evolution and Classification: The Reformation of Cladism.* Longman, New York.

The author argues that the leading schools of taxonomy have failed to live up to the objective standards required of science, with the notable exception of cladism. In his view, numerical phenetics and the evolutionary systematics of Simpson and Mayr are sadly deficient in this respect.

Stone, A. R., and Hawksworth, D. L. (Eds.) (1986). *Coevolution and Systematics.* Clarendon (Oxford University Press), New York.

A collection of authoritative essays celebrating the emergence of systematics from a period of eclipse.

Whittaker, R. H. (1959). On the broad classification of organisms. *Quarterly Rev. Biology 34,* 210–226.

Whittaker, R. H. (1969). New concepts of kingdoms of organisms. *Science 163,* 150–160.

Two papers by the distinguished Cornell biologist who developed and persuasively advocated the five-kingdom system of classification.

Wiley, E. O. (1981). *Phylogenetics.* John Wiley & Sons, New York.

An in depth and broad coverage of the theory of systematic biology. The book discusses how characters are selected and how phylogenies are reconstructed. It also discusses practical methods and rules of nomenclature.

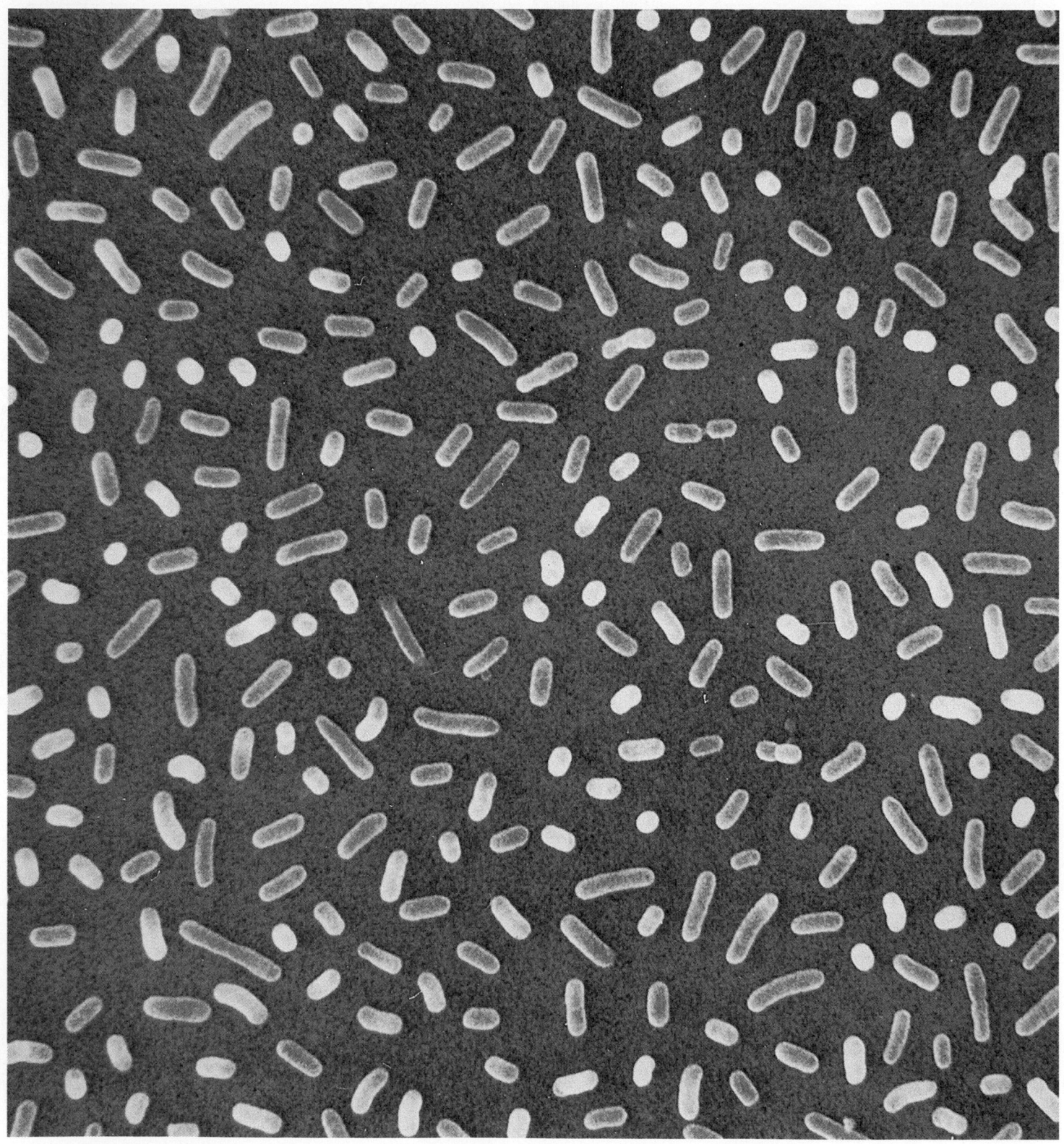

KINGDOM OF THE SMALL: MONERA

Kingdom of the small: Monera
One of modern biology's greatest insights is the universality of life's mechanisms. Bacteria, like the E. coli *in this electron micrograph (× 8000), are the smallest organisms capable of self-replication and growth. They are prokaryotes, yet their physiology has much in common with that of eukaryotes.*

Until relatively recently, human beings were entirely unaware of the existence of a vast subworld of microorganisms. Even now many people are uncertain of what microorganisms are, what they look like, and what niche they occupy in the living world. These uncertainties are reflected to an extent in the confusion that has surrounded the taxonomy of microorganisms.

The terms **microorganism, microbe,** and **germ** are all synonyms without precise technical meanings. They simply mean bacteria, some protists, and a few fungi. With few exceptions, most microorganisms are too small to be seen with the unaided eye (see Figure 5-3). However, when individuals pile up in large numbers they do make their presence known. Such accumulations of microorganisms occur frequently in nature. A good example is seen in the red tides, which sweep the ocean shore from time to time and are dangerous to human health. Other familiar and important collections of microorganisms are found in unpasteurized (and to a degree pasteurized) milk, intestinal contents and feces, saliva, and garden soil. A spoonful of garden soil contains about 10^{10} bacteria. The average healthy human mouth contains more bacteria than the number of people who have ever lived.

The discovery by early microscopists of a whole world of organisms too small to see profoundly stirred human thought. These organisms were soon found to include many varieties that cause infectious diseases in humans and other living organisms—diseases like cholera, syphilis, and tuberculosis that at times have spread across the world in spectacular plagues, decimating entire populations. In the years before the discovery of microorganisms, it was impossible to think rationally about the causes of these diseases. Outlandish ideas and superstitions flourished. Lepers, for example, were considered demonized and were punished or exiled. We now know that leprosy is caused by a specific microorganism, *Mycobacterium leprae*. Yet traces of such archaic ideas still tinge our thinking. A contemporary example is the widespread discrimination against victims of AIDS.

The invisibility of microorganisms to the naked eye was responsible in part for early fallacious ideas about spontaneous generation (see Chapter 4). One of the most important consequences of the discovery of microorganisms was a growing awareness that a world of organisms, too small to be observed directly, plays essential roles in the intricate ecological cycles of life. It is fair to say that much of biology could not be understood, even superficially, until investigators had recognized these smallest of life's forms and taken their measure. One interesting factor undoubtedly delayed this understanding. It was the fact that each kingdom has traditionally been studied by a different group of scientists (Figure 38-1)—and communication among them was hardly voluminous.

We shall speak in this chapter and the next of the kingdoms that include most of the microorganisms—the kingdom **Monera** and the kingdom **Protista.**

Note again that we omit viruses from this discussion, for reasons outlined in earlier chapters.

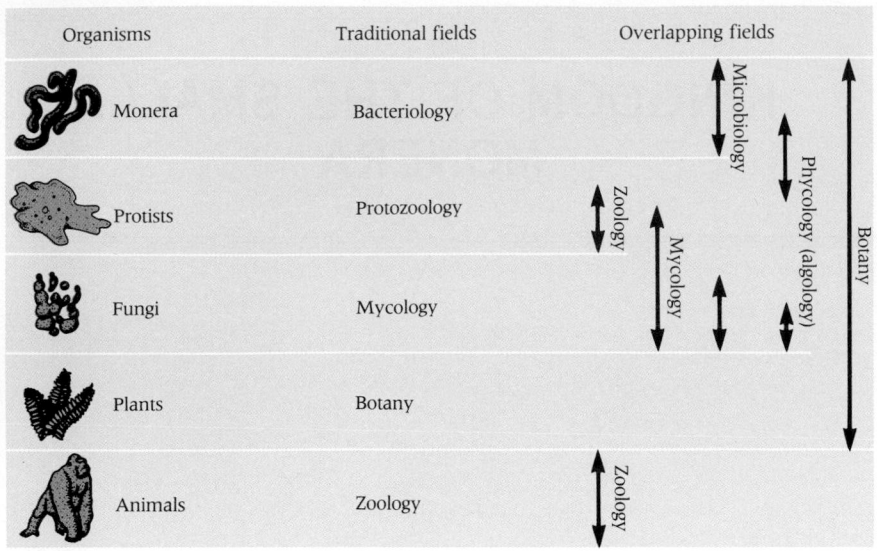

38-1 Kingdoms of living organisms and traditional fields of biology
The study of living things was originally divided into circumscribed fields. Today the fields overlap to a large extent. This reflects the universality of such underlying mechanisms and processes as molecular biology, evolution, and ecology. We call the inclusive field biology.

TAXONOMIC CONSIDERATIONS

In this text, the kingdom Monera is considered to consist of two large phyla—**Schizonta** (true bacteria and certain other forms) and **Cyanonta** (blue-green algae, or cyanobacteria)—and a third newly described phylum, **Prochloronta,** which superficially resembles Cyanonta, but which has been revealed to have distinctive pigment systems that separate it from the phylum of blue-green algae. Others have termed these phyla Schizophyta, Cyanophyta, and Prochlorophyta; but the *-phyta* suffix implies a relation to the kingdom Plantae. Still others list as many as 16 moneran phyla.

THE PROBLEM OF CLASSIFICATION

Classification of the monerans has proved unusually difficult and the difficulties are both fascinating and instructive. We saw in Chapter 37 that, in the case of animals and plants, traditional taxonomic schemes are based largely on morphological, physiological, and developmental homologies. The keystone of these schemes is the species. We also saw that the higher taxa (genus, family, order, and class) lack clean, operational definitions.

With monerans—especially the true bacteria—the situation has been different, and far more challenging. The number of morphological characteristics is smaller. There is no "ontogeny to recapitulate phylogeny"—a principle of development now no longer honored. The fossil record is scanty. Hence we cannot always reliably determine which characters of bacteria were acquired earlier in evolution and which later.

For all of these reasons, traditional classification of bacteria has relied to a great extent on certain morphological and physiological characteristics. When examined with ordinary (light) microscopes, bacteria turn out to have many and diverse shapes. Some are spherical, some are elongated or rodlike, and some are spiral. These morphological characteristics appear to be enduring traits developed early in the evolution of bacteria. The same might be said of the major energy-yielding or metabolic patterns, such as various specific fermentations, oxidative metabolism, or photosynthesis. Yet most of these metabolic patterns are found in many of the morphological groups.

For years, purely descriptive classifications continued in general use, mainly because nothing better was available. Such classifications served mainly as aides in the identification of an unknown organism or in comparisons of one organism to another. The decision to subordinate one such criterion to another was always quite arbitrary. Consequently, a major cooperative effort at classification, such as *Bergey's Manual of Systematic Bacteriology,* substantially changes its taxonomic conclusions in each edition. Happily, fresh new breezes are blowing in the field of prokaryote taxonomy. As mentioned earlier, the field has been transformed by molecular sequencing and other biochemical methods. Today many features of bacteria are used in classification. They include the following.

- Visible features (including shape, size, color, motility, flagellar pattern, capsule, colonial morphology, and staining properties—especially the Gram stain[1])
- Patterns of energy-yielding metabolism
- Patterns of nutrition (growth requirements and ability to utilize various foods)
- Production of characteristic chemical products
- DNA content of guanosine and cytosine—often expressed as "molar percent GC," or simply as "GC content"[2]
- DNA-DNA hybridization data providing clues to the degree of sequence homology that can support or deny close relationship
- Ribosomal RNAs (rRNAs), which are universally distributed, constant in function, easily isolated, and slow to change in the course of evolution; sequence studies especially useful in defining taxonomic groups above the species level[3]
- Other biochemical data from studies of proteins, their immunologic and catalytic properties; cell wall structure, including the presence or absence of characteristic surface macromolecules
- Ecological relations (including the ability to parasitize higher organisms and thus to cause disease)

In recent years, nucleotide sequence studies have invalidated some traditional taxa based on morphological characteristics. Interestingly, bacteria with any of three features—spherical shape, mode of cell division, and lack of cell wall—were most frequently placed in the "wrong" taxon.

By diligently cataloging the features of bacterial strains and establishing reference collections of "type cultures" (standard strains) in several nations,[4] bacteriologists have managed to arrange the bacteria into a coherent system. Following the Linnean tradition, each "species" is assigned an official binomial Latin name that includes a genus name (capitalized) followed by a second trivial (uncapitalized) name. This specific name is usually printed in italics. Unitalicized colloquial names may be derived from, or identical with, the official name. For example, "pneumococcus" denotes *Streptococcus pneumoniae*. "Salmonella" denotes various *Salmonella* strains. Unfortunately, these naming conventions developed late, are not always adhered to, and undergo frequent change. Hence, the same organism may be encountered in the literature of biology under several names—for example, *Bacillus typhosus, Bacterium typhosus, Eberthella typhosa,* and *Salmonella typhi* all refer to the same organism!

EVOLUTIONARY ORIGINS

The implications of these classification problems go far beyond the bounds of mere classification. The development of a prokaryotic phylogenetic tree is essential to our understanding of early events in cellular evolution. The conventional tree shown in Figure 38-2 was drawn before the discovery of archaebacteria, which some say belong in a separate kingdom. It suggests, first, that a common ancestral prokaryote, now known as **progenote,** gave rise to the major phyla of Monera and, second, that sublines of some of these evolved into an ancestral eukaryote (compare with Figure 37-16). This took place after the atmosphere acquired oxygen.

[1] This staining procedure, named for the Danish physician Hans Christian Gram who discovered it accidentally in 1884, divides bacteria into Gram-negative and Gram-positive groups. The difference is due to the presence or absence of a lipopolysaccharide layer in cell membranes. Gram-positive bacteria have this layer.

[2] This value varies widely—from 30 to 80%—in different bacteria, but it is constant for each species. When two species have a similar GC content, they may be closely related, or they may not be. But if these values differ widely, we can be certain that they belong in different taxa.

[3] The claim that archaebacteria represent a new kingdom rests in part on studies of rRNA sequences that distinguish them from all other bacteria and from eukaryotes (see Chapter 37).

[4] Unfortunately, there is a type culture—that is, an authentic specimen of the original "beast"— for very few of the 30,000 species now believed to exist. Hence there is nothing with which a proposed new species can be compared. This makes it difficult to identify, unambiguously, a supposedly new species.

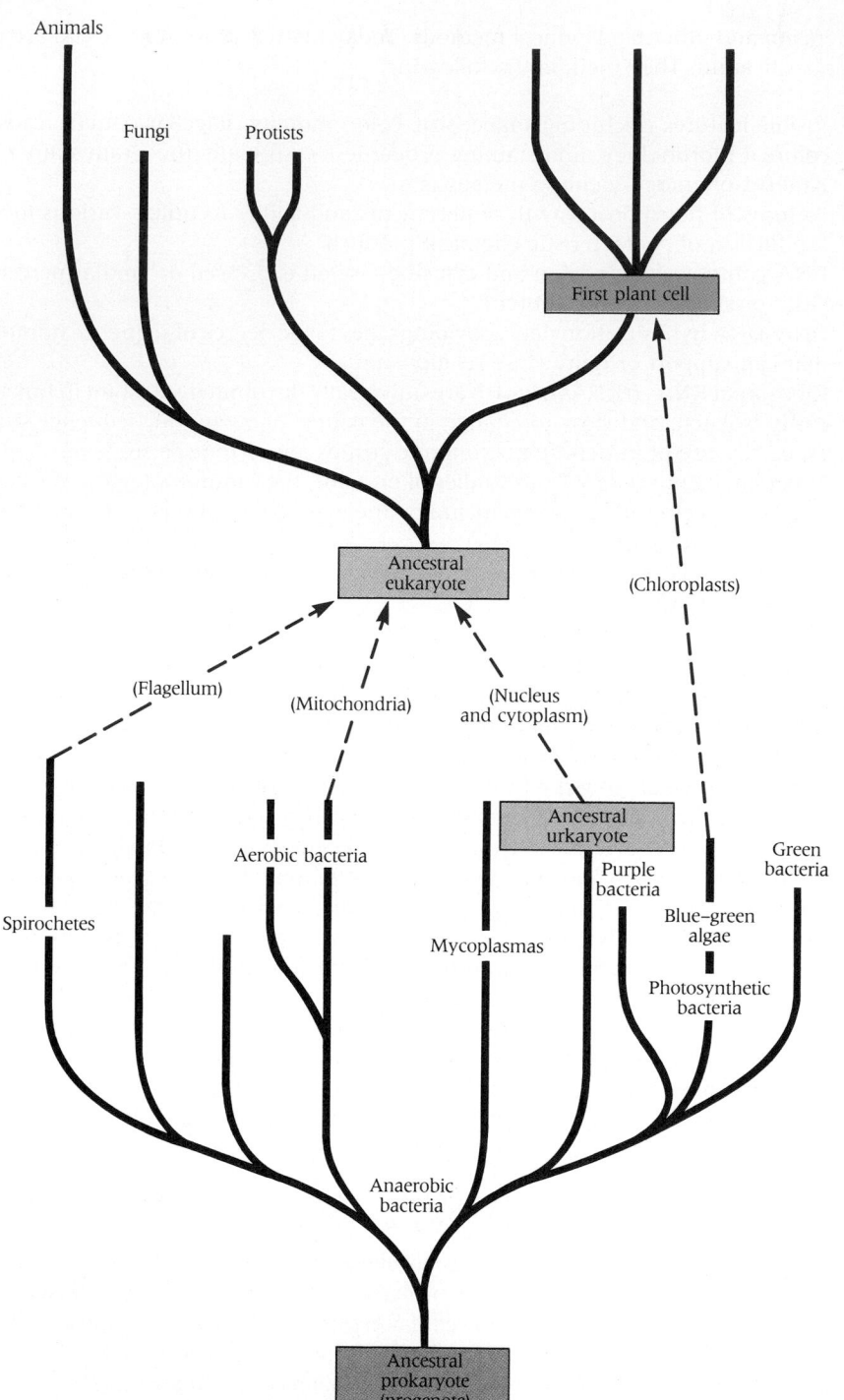

Animals

Fungi   Protists

First plant cell

38-2 The phylogenetic tree prior to the discovery of archaebacteria
Before the discovery of archaebacteria, phylogeny was thought to have followed two principal lines of descent to prokaryotes and eukaryotes. The ancestral prokaryote may have been an anaerobic bacterium that obtained energy by fermentation. The dashed lines represent the symbiotic events that provided the flagella, mitochondria, nuclei, cytoplasm and chloroplasts that combined to produce the ancestral eukaryotic cell. It in turn gave rise to the protists, fungi, animals, and plants.

Ancestral eukaryote

(Chloroplasts)

(Flagellum)

(Mitochondria)   (Nucleus and cytoplasm)

Aerobic bacteria

Ancestral urkaryote

Green bacteria

Purple bacteria

Spirochetes

Mycoplasmas

Blue–green algae

Photosynthetic bacteria

Anaerobic bacteria

Ancestral prokaryote (progenote)

After the recognition of archaebacteria, and after extensive study of rRNA sequences (and other biochemical data), the scheme shown in Figure 38-3 was proposed—despite rampant skepticism about dividing prokaryotes into true bacteria and archaebacteria. Today, many accept the view that the progenote, the first prokaryote cell, gave rise to three major lines—monerans, archaebacteria, and eukaryotes—and that none of these groups gave rise to the others.

The phylogenetic scheme in Figure 38-3 portrays this momentous three-way split. Note again the three emerging major lines of descent: (1) the *archaebacteria*, now proposed as a separate kingdom; (2) the forerunners of monerans, the *ancestral true bacteria*, often called eubacteria (though we choose to apply this term to a major class of true bacteria); and (3) the so-called *ancestral urkaryotes*, now extinct ancient forms that provided nucleus and cytoplasm to the forms that later became eukaryotes.

Two of the organelles of eukaryotes—mitochondria and chloroplasts—have their own circular DNA genomes (Chapter 18) and other elements that so strongly suggest

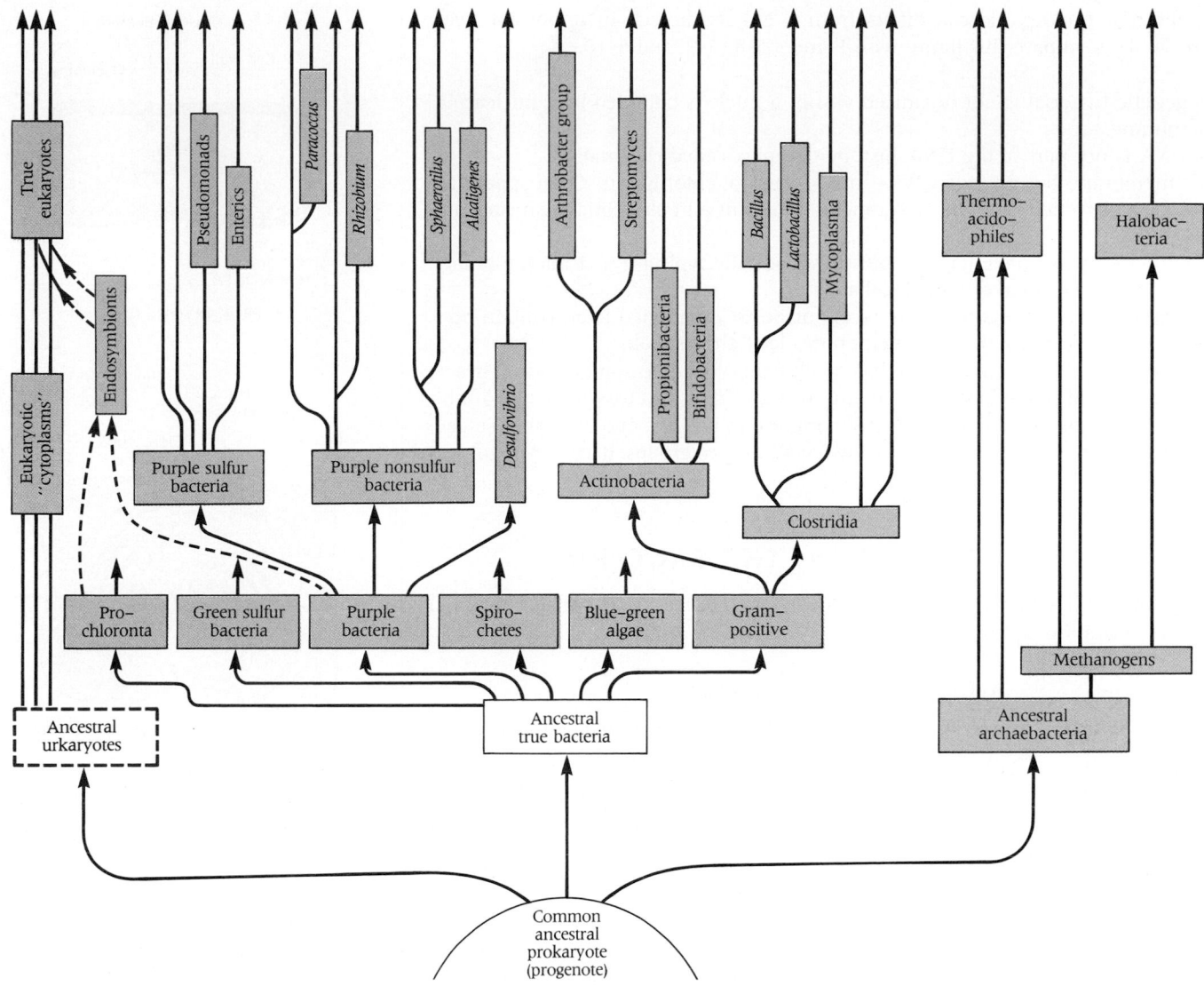

38-3 The major lines of prokaryotic evolution

Three lines arose from the common ancestral prokaryote: the archaebacteria, the eubacteria, and the urkaryotes. Each line then diversified, giving rise to many different types of cells.

a bacterial pedigree, it is now believed that the ancestral urkaryote engulfed some free-living bacteria. Once inside, these symbiotic (mutually beneficial) bacteria began supplying the host cell with energy by oxidative phosphorylation—and, in the case of plants, by photosynthesis. Hence, they are termed **endosymbionts.** Their arrival changed urkaryotes to true eukaryotes, which thus became chimeras—organisms composed of two or more genetically distinct components (see Box 20-1). As the promitochondria and prochloroplasts gradually turned into specialized organelles, some genes were transferred from their DNA to the host cell's DNA. Only a few genes remained behind in the organelle (see Figure 18-15). We now believe that mitochondria came from purple bacteria and chloroplasts from prochloronts or cyanobacteria.

All of the ancestral forms shown in Figure 38-3 were anaerobic. Aerobic metabolism evolved later, apparently arising independently a number of times. Photosynthetic forms also evolved early. Nonphotosynthetic heterotrophs came long afterwards. We will return to the questions of prokaryotic phylogeny after surveying the phyla of Monera.

GENERAL CHARACTERISTICS

All monerans are prokaryotic. They are, in a sense, one-celled organisms. In another sense, however, they are noncellular organisms. Most are organized like the single cells of multicellular organisms—that is, they contain a complete complement of genetic and protein-synthesizing machinery, including DNA, RNA, and all of the enzymes needed to translate the genetic code into various specific proteins. The prokaryotic cell also contains metabolic machinery for the generation of ATP.

However, a prokaryotic cell differs from a eukaryotic cell in important ways (Figure 38-4). Compare this figure with Figures 5-8, 5-9, and 5-10.

- Its genetic material is not organized within a nucleus bounded by a nuclear membrane.
- Its DNA is not part of the DNA-histone complex called chromatin.
- The membrane-bound organelles—mitochondria, chloroplasts, Golgi apparatus, endoplasmic reticulum, and lysosomes—found in all eukaryotic organisms are lacking.
- Some prokaryotes have flagella, but these lack the typical 9 + 2 microtubular structure found in eukaryotic flagella.
- Prokaryotes, with the interesting exception to be mentioned later, contain no microtubules or tubulin. Their membranes lack cholesterol.
- Finally, almost every prokaryote has a cell wall of unique composition. These walls consist of highly polymerized amino sugars that are cross-linked by amino acids. This unusual structural pattern, common to bacteria and blue-green algae, is an important item of the evidence linking those two groups. It also distinguishes both of them from the eukaryotes.

PHYLUM SCHIZONTA: THE BACTERIA

We will hold with a traditional classification of the monerans. The major bacterial groups are summarized in Table 38-1.

GENERAL PROPERTIES

Bacteria are found in almost every location that has not been deliberately freed of them, even in places where other forms of life are rare or absent. They are abundant in the waters of Earth—in ocean depths, in mountain streams, and in damp soils.

The most striking characteristic of bacteria is their small size and diversity of shapes (Figure 38-5). They are the tiniest of living things. Only certain viruses are smaller, but viruses do not fulfill the usual definition of a living organism. The smallest bacteria are 0.1 μm in diameter and length; the largest are about 6 μm across and 60 μm long. A typical bacterium like *Escherichia coli* is 1 μm^3 in volume and weighs about 10^{-12} gram. That means that one trillion (10^{12}) cells weigh only 1 gram. In contrast, a liver cell is 1000 times the size of an *E. coli* cell (see Figures 5-8 and 5-10).

Bacteria are a diverse group of organisms differing in size, shape, and habitat. They differ significantly in their metabolism, energy sources, and other necessary

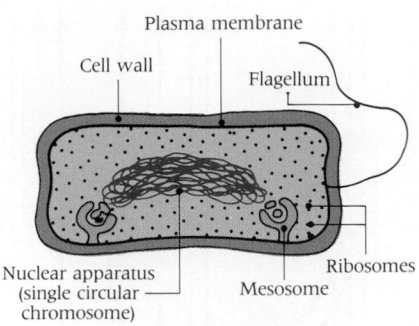

PROKARYOTIC CELL

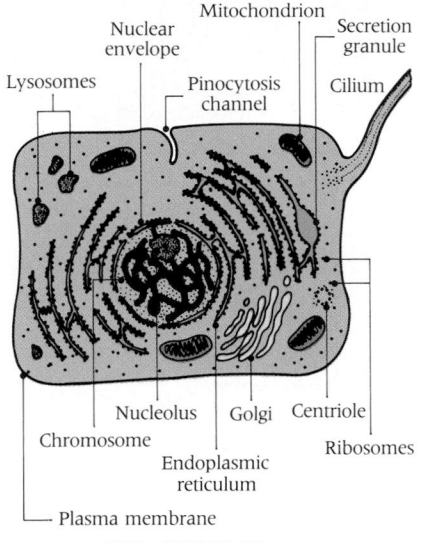

EUKARYOTIC CELL

38-4 Structures of generalized prokaryotic and eukaryotic cells
The two major cell types show striking differences in complexity and internal organization. Not all prokaryotic cells have flagella, nor do all eukaryotic cells have every organelle shown here.

38-5 The diversity of bacterial cell shapes
(A) Cocci are round cells that may grow singly, in pairs, or (B) as long chains. (C) Bacilli are rod-shaped and may exist singly or in chains. (D) Many bacteria are flagellated. (E) Some, such as spirillum, are helical.

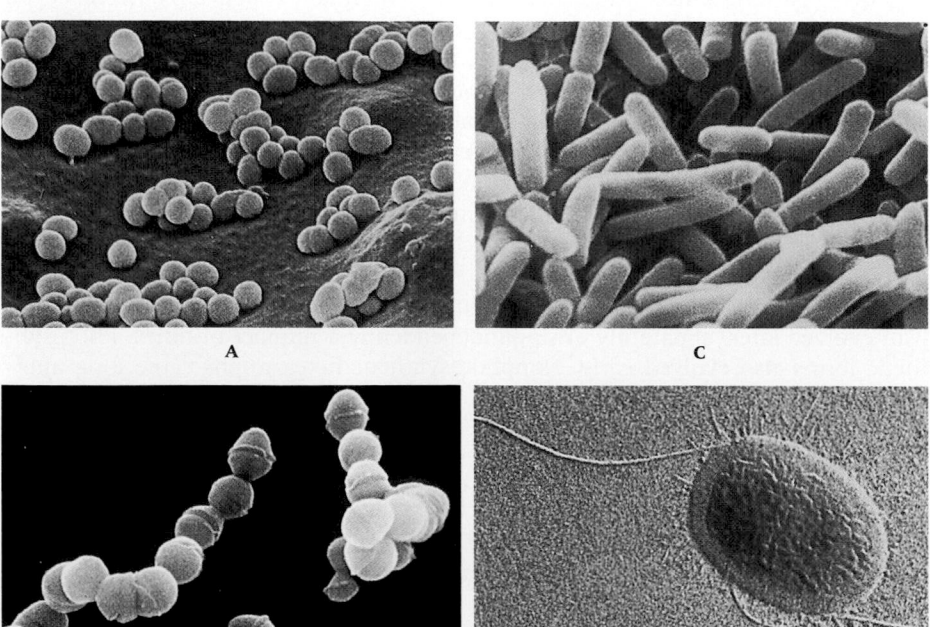

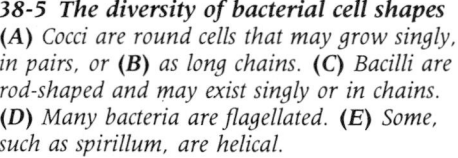

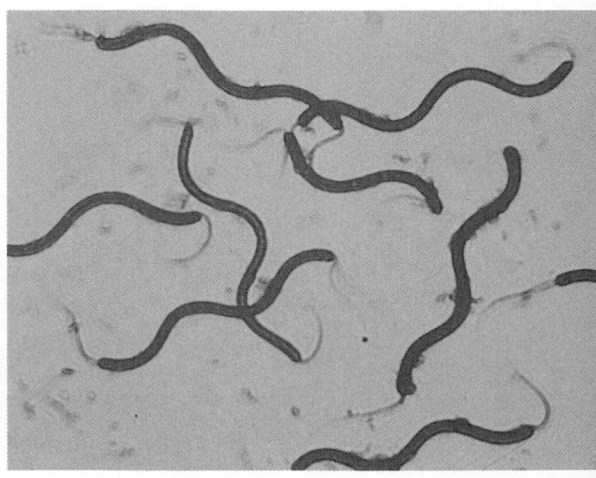

Table 38-1

SIX MAJOR CLASSES OF BACTERIA

Class	Examples	Morphology	Type of Motility	Mode of Nutrition	Distribution	Ecological Role	Diseases Caused
Eubacteria	*Escherichia, Streptococcus, Staphylococcus*	Rod-shaped bacillus, coccus spirillum (many curves), vibrio (one spherical curve)	Gliding, flagella	Chemoauto-trophs, photosynthetic autotrophs, heterotrophs	Soil, water, parasites	Decomposers, symbionts, pathogens	Tetanus, diphtheria
Myxobacteria	*Myxococcus, Cytophaga, Chondromyces, Stigmatella*	Rod-shaped, flexible, in slime	Gliding	Heterotrophs	Soil, some aquatic	Decomposers (especially of complex polysaccharides)	None
Actinobacteria	*Actinomyces, Corynebacterium, Nocardia, Streptomyces, Lactobacillus*	Rods and spheres, some unicellular, some linked in strands	Mostly nonmotile	Heterotrophs	Soil, parasites	Decomposers, pathogens, some produce antibiotics	Tuberculosis, leprosy, nocardiosis, diphtheria
Spirochaetae	*Spirochaeta, Treponema, Leptospira, Borrelia*	Extremely long, helical	Whirling (axial filament)	Heterotrophs	Aquatic (polluted water), parasites	Symbionts, pathogens, decomposers	Syphilis, infectious jaundice, relapsing fever
Mycoplasma	PPLO, *Mycoplasma, Bartonella, Acholeplasma*	Smallest free-living cells, no cell walls	None	Heterotrophs (many parasites)	Intracellular parasites, soil	Pathogens	Mycoplasma pneumonia
Rickettsiae	*Rickettsia*	Small	None	Heterotrophs (parasites)	Intracellular parasites	Pathogens	Typhus, spotted fever

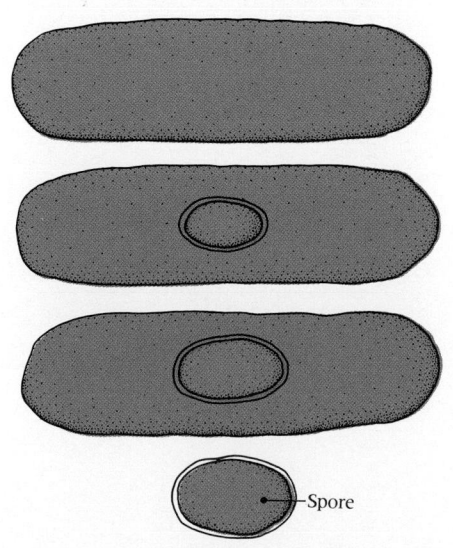

38-6 Sporulation in bacteria
A spore is formed from elements in the cell and is encased in a tough protective coat while still in the cytoplasm.

materials from their environments. We learned in Chapter 9 that organisms are classifiable by their energy sources. Let us look at the bacteria from this viewpoint.

■ Some bacteria are **heterotrophs**—that is, they depend on other organisms as sources of needed preformed organic molecules. Those heterotrophic bacteria that subsist on dead organic matter are called **saprobes.**

■ Other bacteria are **autotrophs** and thus are able to synthesize all the organic molecules they need, requiring carbon only in the form of CO_2. Some autotrophs derive energy from light, utilizing pigments that are closely similar to the chlorophyll of green plants. These are photosynthetic autotrophs. Unlike plants, their photosynthetic process produces no free oxygen. Other bacteria are chemo-synthetic autotrophs, deriving their energy from inorganic chemical sources, notably sulfur, hydrogen, or simple inorganic compounds of nitrogen.

Unlike animals and most plants, many bacteria—the **anaerobic** ones—can prosper in the absence of free oxygen. Some of these are actually harmed by oxygen. Others—the **aerobic** bacteria—tolerate or require oxygen. Different kinds of bacteria can live and reproduce at temperatures from 0 to 75°C. Quick-frozen bacteria can survive almost indefinitely at temperatures far below 0°C, and some pass through a stage that can briefly survive boiling (100°C at sea level)—though they do not reproduce under such extreme conditions. Each kind of bacterium has a temperature at which it grows best. These optimal temperatures range in different species from 12°C to a high of over 90°C (nearly boiling!)—the optimum in a curious heat-loving category of so-called **thermobacteria.**

Some bacteria form spores, reproductive cells that can grow into new individuals. The process of spore formation, termed **sporulation,** permits cells to survive unfavor-able conditions. It involves the production from cytoplasmic precursors of a **spore** that is sheathed in a tough impermeable outer coat (Figure 38-6). Within the spore is the cell's DNA and a bit of cytoplasm. Bacterial spores are remarkably durable, easily tolerating otherwise lethal conditions such as boiling temperatures. When favorable conditions are restored, spores can then give rise to new bacterial cells.

Many bacteria are useful, in that their activities are essential in maintaining the living communities of other organisms. Indeed, the entire scheme of life as it has evolved depends upon bacteria. Bacteria are the principal organisms of decay. Decay is a process in which organic compounds are broken down, usually by bacteria but also by some primitive plants. Proteins are broken down successively to amino acids and then ammonia. Bacteria are virtually the only organisms capable of **nitrogen fixation**—the conversion of atmospheric N_2 to ammonia (NH_3) (see Chapter 11). Other bacteria oxidize ammonia to nitrites (NO_2^-) and still others oxidize nitrites to nitrates (NO_3^-), the principal nitrogen source for green plants.

Some nitrogen-fixing bacteria are able to fix (utilize) free N_2 in the synthesis of their own proteins. After further metabolism and decay, those nitrogenous compounds eventually become available to other organisms. The nodules on the roots of many plants of the pea family (legumes) contain colonies of bacteria that obtain energy from the carbohydrates of the host plant and utilize part of that energy to fix atmospheric N_2.

Although most bacteria are useful in the sense that they help to keep the cycles of life going, many kinds of bacteria are parasites, which we define as organisms that live at the expense of other organisms. When they invade the body of a larger organism, we use the term **infection.** If infecting bacteria multiply rapidly (despite immune attack) or if they produce toxins (substances that are poisonous to the host), **disease** ensues.

THE MAJOR CLASSES

Table 38-1 lists the cast of bacterial characters. Let us briefly examine six of the best-known, here traditionally named, classes: (1) Eubacteria; (2) Myxobacteria; (3) Actinobacteria; (4) Spirochaetae; (5) Mycoplasma; and (6) Rickettsiae.

Eubacteria

The largest class, **Eubacteria,** is sometimes designated the "true bacteria." These are Gram-positive bacteria with thick and relatively stiff cell walls. Most eubacteria are nonmotile, but a few can move by means of flagella, whiplike filaments that extend singly or in tufts from one or both ends of the cell, or all around it. The eubacteria encompass an enormous number of species, structural forms, and means of producing energy. We have already considered some of these metabolic mechanisms in an earlier discussion of autotrophy and photosynthesis (see Chapter 9). Here we simply call attention to the diversity of organic molecules that can serve as both carbon and energy sources for bacteria.[5] Some groups of eubacteria can accomplish the difficult biochemical task of fixing free N_2. Evidently, the only important energy-related biochemical process beyond the capacity of these organisms is aerobic photosynthesis.

Eubacteria display three characteristic shapes (see Figure 38-5), which are major recognition clues. There are spheres (called **cocci**), rods (called **bacilli**), and spiral or helical forms (called **spirilla**). Cocci can occur singly, or in pairs (diplococci), clusters (staphylococci), or chains (streptococci). A major cause of pneumonia is a certain streptococcus (pneumococcus). Staphylococci are responsible for many serious infections characterized by boils or abscesses. Bacilli and spirilla occur singly or in chains. The chains consist of fully independent organisms. Because most bacteria reproduce simply by the fission of one cell into two, chains arise merely by the adhesion of cells after fission.

---

[5] Bacteriologists once argued that no carbon compound existed that could not be metabolized by at least one strain of bacteria. Sadly, we have learned in recent years that certain compounds—notably, chlorinated hydrocarbons like DDT and related insecticides, and chlorinated derivatives of benzene like the weed killer, 2,4-D—are in fact poorly utilized by bacteria or fungi. Thus they are not digested (i.e., decayed) by ubiquitous microorganisms, and accumulate in the environment in dangerously high amounts.

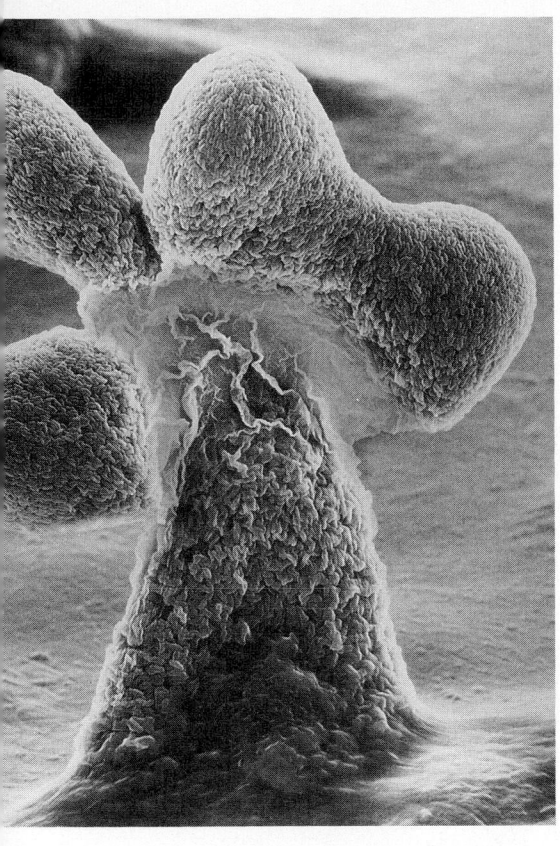

A

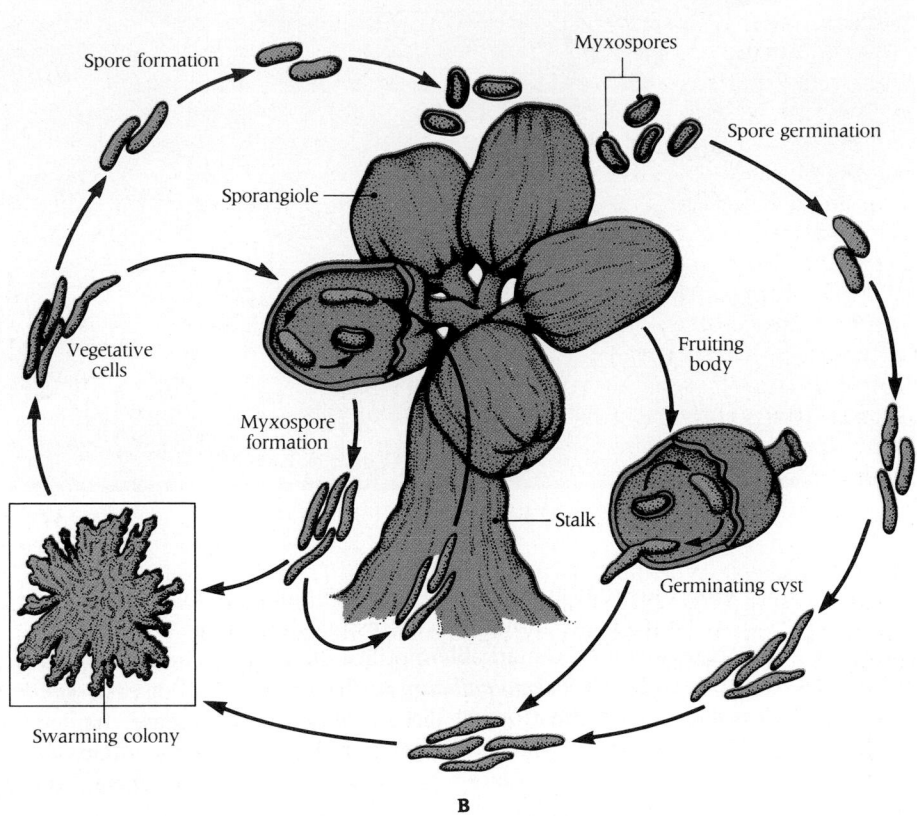

B

38-7 Myxobacterium
(A) Scanning electron micrograph of a fruiting body containing nearly a million cells. (× 500). **(B)** Life cycle of Stigmatella aurantiaca.

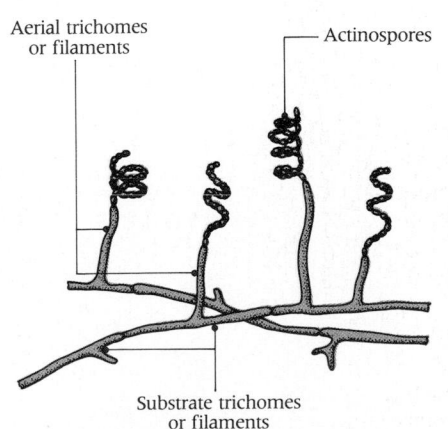

38-8 Mycelium of Actinobacteria
The matlike mycelium is formed from trichomes or filaments, some of which extend upward. Actinospores are located at their tips.

Myxobacteria

A second major class is the **Myxobacteria** (from *myxa*, Greek for "mucus" or "slime"). These bacteria are distinguished by their morphological complexity, their secretion of a slime track on which the cells glide, and by the fact that some types, when nutrients are depleted, form the complex reproductive structures known as **fruiting bodies** (Figure 38-7A). These are treelike forms, usually less than a millimeter high, that are topped by a branched globular structure. On the tips of the branches are large, colored membranous cysts, each containing thousands of bacterial spores that lie dormant during periods of insufficient moisture or nutrition. When conditions are favorable again, the cysts can germinate and release bacteria.

Individual myxobacteria are unicellular Gram-negative rods. They often aggregate into complex colonies that show distinctive form and behavior. They commonly inhabit soil, where they act as decomposers. The life cycle of one myxobacterium (*Stigmatella aurantiaco*) is shown in Figure 38-7B.

Actinobacteria

A third class, **Actinobacteria** (Greek root *aktis* meaning "ray"), includes bacteria that do not form internal spores. Some were originally mistaken for fungi because they form **actinospores** (Figure 38-8)—and they were misnamed actinomycetes. Actinobacteria are unicellular or linked in strands. Many form matlike mycelia (singular, mycelium). Many are pathogens causing serious human diseases (nocardiosis, actinomycosis, etc.). The organisms that cause tuberculosis (*Mycobacterium tuberculosis*) and leprosy (*Mycobacterium leprae*) are actinobacteria.

On the other hand, some actinobacteria—notably *Streptomyces*—are producers of streptomycin and other antibiotics that successfully combat infectious diseases in humans and domestic animals.

Spirochaetae

Members of the fourth class, **Spirochaetae,** look like coiled snakes. They are notable for (1) their great length (from 5 to 500 μm) and narrowness (about 0.5 μm), and (2) their unique **axial filaments,** which resemble internal flagella.

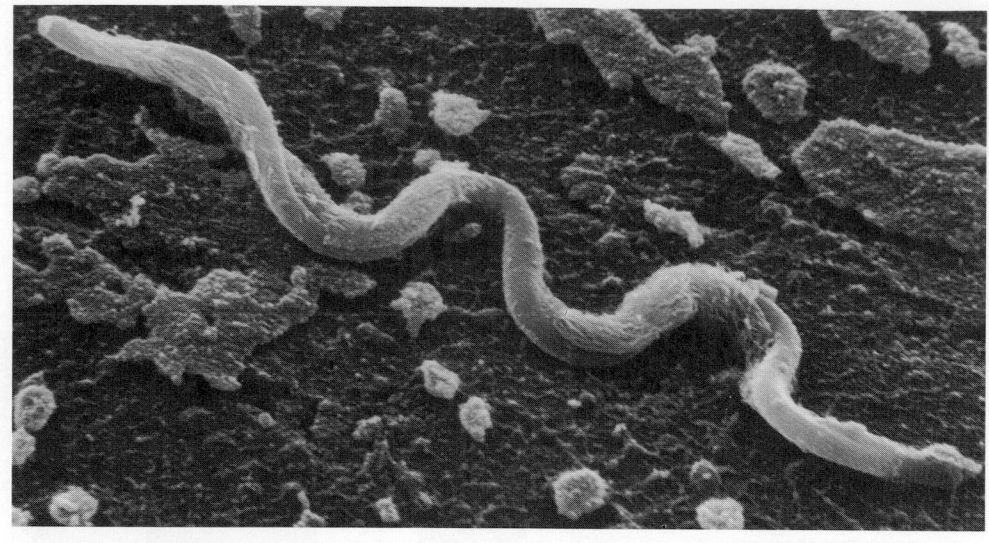

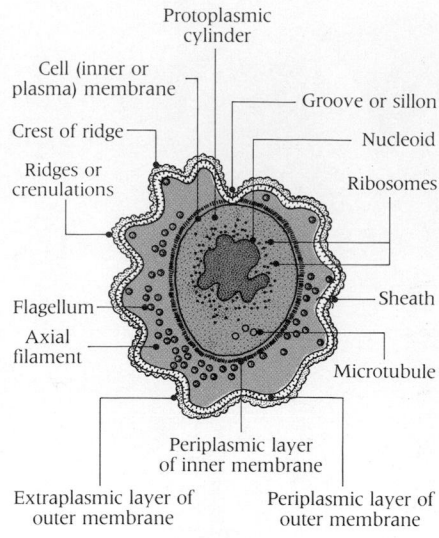

38-9 Spirochetes
(A) *Scanning electron micrograph of a spirochete,*
Cristispira. (× 25,000) (B) *Cross section of a*
generalized spirochete. Note the microtubules,
which are unique among monerans.

The cells are long rods that are coiled around the filaments in a helical pattern (Figure 38-9). The filaments are thought to be responsible for the whirling motion that gives these organisms their remarkable motility. Many spirochetes are parasites in humans. The spirochete *Treponema pallidum* causes the venereal disease *syphilis*.[6] Another, *Treponema*, causes the tropical disease *yaws*. Lyme disease, named for the Connecticut town where it was recently recognized, is a tick-borne inflammatory disorder of the joints, skin, and nervous system caused by a spirochete, *Borrelia burgdorferi*.

Spirochetes are an exception to the rule, given earlier, that prokaryotes lack microtubules. Microtubules have been found in certain spirochete species. Though of somewhat smaller diameter than eukaryotic microtubules, they appear to be made of tubulin. This finding has led to the suggestion that the flagellar microtubules of eukaryotic cells arose in evolution by an invasion of these cells by tubule-containing spirochetes, a mechanism resembling the one postulated to account for the origin of mitochondria and chloroplasts (see Figure 38-2).

Mycoplasmas and Rickettsiae

Mycoplasmas and **Rickettsiae** are the smallest prokaryotes known. They are intracellular parasites that have been cultured only with difficulty in the absence of the living cells on which they have become dependent. They were once considered to be evolutionary forerunners of the larger, more complex bacteria, but it is now agreed that, like other parasites, Mycoplasmas and Rickettsiae gradually lost many of their structures and functions.

Both groups are responsible for certain human diseases. Rickettsiae are the causative agents of Rocky Mountain spotted fever and typhus, which are frequently transmitted by ticks and fleas. They do not seem to cause any disease symptoms in their arthropod hosts. Mycoplasma can cause an odd and lingering pneumonia in humans.

METABOLIC FEATURES

The eubacteria discussed so far are nonphotosynthetic. We should note, however, that anaerobic photosynthetic bacteria include several kinds of photosynthetic eubacteria. We mention them only briefly, though they are of very great interest.

Three important, delightfully colored, groups are known as green sulfur bacteria, purple sulfur bacteria, and purple nonsulfur bacteria (see Figure 38-3). Like plants

---

[6] The geographic origins of treponemes, the spirochetes that cause syphilis and yaws, have long been a subject of controversy. A widely held view—that Columbus's ship brought these diseases to Europe from the New World—rests on evidence of classic syphilitic lesions in the bones of Central American Incas and Aztecs, which date back 1000 to 3000 years. A recent study gives support to a Western-Hemisphere origin by presenting new evidence of syphilitic bone disease (confirmed by immunological techniques) in the remains of a Pleistocene bear from Indiana that was dated at 11,500 years before the present.

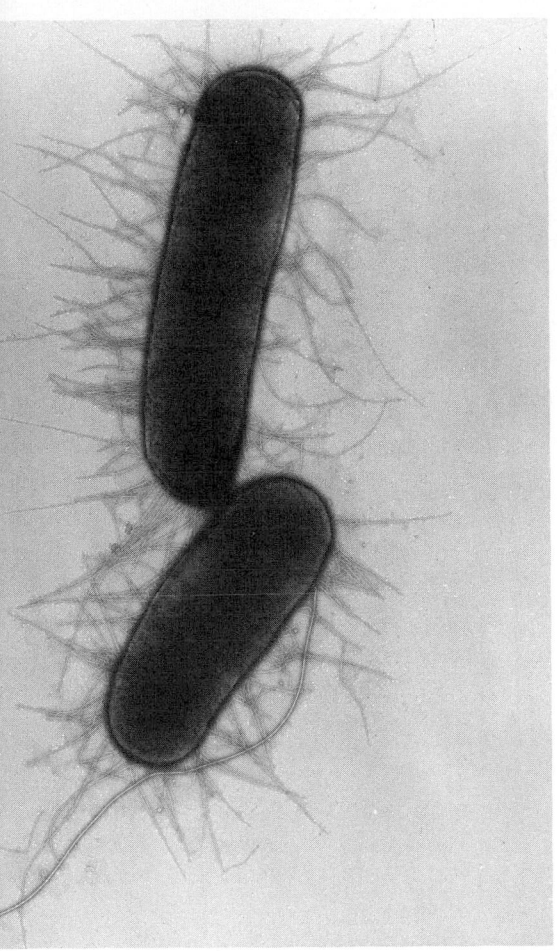

38-10 Electron micrograph of E. coli
($\times$ 40,000)

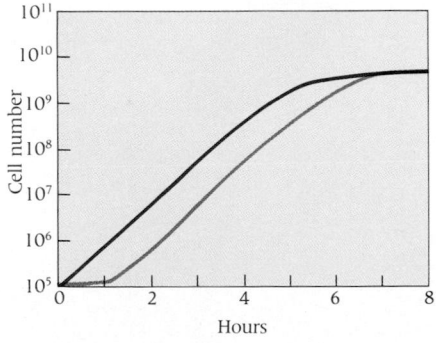

38-11 Growth curve of E. coli at 37°C
The black line shows an increase in cell number following inoculation of a sterile, nutrient-rich solution with 10^5 cells from an exponentially growing culture of E. coli. If cells from a slow-growing, nearly saturated culture, had been used as the inoculum, a lag period of nearly 1 hr would have preceded exponential growth (color).

In both cases, growth reaches a plateau in a few hours as a result of nutrient exhaustion and waste accumulation.

and blue-green algae, these photosynthetic bacteria contain chlorophyll, but it is not located within plastids. The diverse colors of these forms are attributable to their accessory photosynthetic pigments.

Some of these pigment combinations cause interesting variations in the photosynthetic process—for example, because of their presence many of these organisms react to light of much longer wavelengths than that used by other photosynthetic organisms. Photosynthetic sulfur bacteria assimilate CO_2 (as do plants). But they use hydrogen sulfide (H_2S) as a reducing agent rather than H_2O as in plant photosynthesis. Thus, in photosynthetic sulfur bacteria the following reaction occurs:

$$CO_2 \ + \ 2H_2S \ \xrightarrow{\text{light}} \ (CH_2O) \ + \ H_2O \ + \ 2S$$

This is to be compared with the pattern in photosynthetic plants.

$$CO_2 \ + \ 2H_2O \ \xrightarrow{\text{light}} \ (CH_2O) \ + \ H_2O \ + \ 2O$$

Since bacterial photosynthesis occurs under strictly anaerobic conditions, molecular O_2 is never a byproduct. In sulfur bacteria, the hydrogen donor is H_2S (hydrogen sulfide). In nonsulfur bacteria, it is a small organic molecule like lactic acid (CH_3—$CHOH$—$COOH$) or ethanol (CH_3—CH_2—$COOH$).

The green sulfur bacteria are all nonmotile rods. Most of the purple sulfur bacteria are motile rods or spirilla. All reproduce by simple fission but one purple bacterium reproduces by *budding*. In place of a "simple" cleavage that yields two daughter cells of equal size, such bacteria sprout at one end a small protuberance that increases in size and ultimately breaks away, a new and independent cell. Since green and purple bacteria can fix N_2, they occupy an important ecological niche in the life of a stagnant pool.

A CLOSER LOOK AT THE COLON BACILLUS

The story is told of a microscopist who pointed out to Pasteur that an organism that he had taken for a coccus was in reality a small bacillus. Pasteur replied, "If you only knew how little difference that makes to me!" The reason, of course, was that Pasteur was a bacterial physiologist not a taxonomist. For him, one species was as good as another in illustrating how bacteria function.

For comparable reasons, modern bacterial physiologists and geneticists have tended to concentrate their attention on the bacterium *Escherichia coli*, much as earlier geneticists concentrated on the fruit fly. *E. coli* is convenient, it is certainly available (being a plentiful denizen of the normal human colon), and it is safe to work with, rarely causing an infectious disease. Many other bacteria possess similar and equally favorable attributes. However, once serious work had started on *E. coli* it seemed a good idea to stick with it. Investigators were saying in effect, "Why switch to another organism when we can all work on this one and thereby deepen our understanding of it?" As a result, our knowledge of *E. coli*, its physiology, genetics, and fine structure, is much more complete than our knowledge of any other bacterium, though much of what we have learned of *E. coli* is true of other bacteria as well.

A typical *E. coli* cell is rod-shaped and about 2 μm in length and 1 μm in diameter (Figure 38-10). It is of the family Enterobacteriaceae (the intestinal or "coliform" bacteria), which are small, Gram-negative, nonspore-forming rods that are either nonmotile or, if motile, equipped with several flagellar filaments. *E. coli* is definitely motile, and recent studies of the way its flagella work led to the surprising and important discovery that this organelle rotates, thus propelling the cell in a given direction (see Box 5-2).

E. coli reproduces by increasing its length, and then undergoing a fission process that yields two cells of equal length. Growth occurs best at about 37°C. Generation time (doubling time in an actively growing culture) under optimal conditions is about 20 minutes (Figure 38-11).

Cell structure is readily studied under the light microscope or electron microscope. Such observations, however, cannot reveal whether a given cell is alive or dead. This can be established only by determining whether that cell divides to form daughter cells. This determination is usually made by spreading a small number of cells on

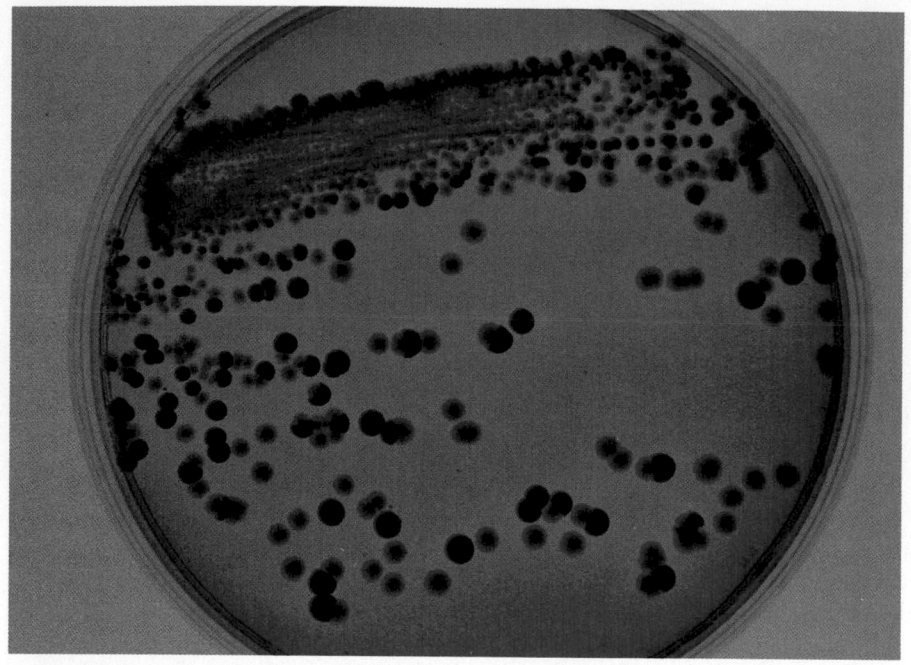

A

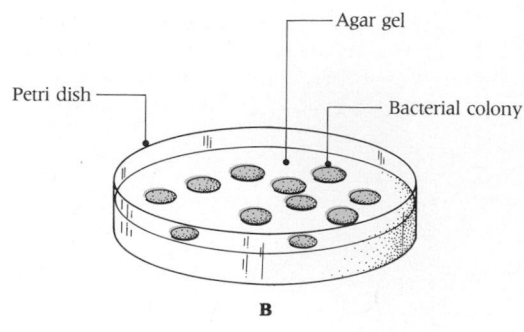

B

38-12 Single bacteria form colonies on a culture plate
(**A**) *Actual appearance of bacterial culture plate containing individual colonies.* (**B**) E. coli *are placed on the agar surface and permitted to divide. Each colony is formed by repeated divisions of a single original cell.*

top of a solid agar surface, which has been supplemented with the nutrients necessary for cell growth (Figure 38-12). If a cell is alive, it will divide to form two daughter cells, which in turn give rise to subsequent generations of daughter cells. The net result after 12 to 24 hours of incubation at 37°C is discrete masses, or **colonies,** of bacterial cells. Hence, each colony is considered a clone of a single bacterial cell.

As we have seen, the chromosome of *E. coli* is not enclosed within a nuclear membrane. There is no structural distinction between nucleus and cytoplasm. Nonetheless, bacterial cytologists often refer to the visible region occupied by the DNA as the **nucleoid** or nuclear region (see Figure 38-10). The current view of how the bacterial chromosome is replicated during the division process called "simple fission" is summarized in Figure 12-1A. The only essential similarity between mitotic divisions in higher organisms and simple fission in bacteria is that both result in the accurate duplication and partitioning of the chromosomes to daughter cells.

As discussed in Chapter 16, it was long believed that bacteria, including *E. coli*, had no sexual process involving cell fusion. Since 1947, however, it has been known that cells of opposite mating types do exist, and that a rare cell-fusion process takes place in which diploid cells are produced. What happens in this process of **conjugation** is an actual transfer of a chromosomelike body from a male to a female *E. coli*, so named because the former injects the body into the latter (see Figures 16-7 and 21-5). The process has been demonstrated by electron microscopy (Figure 38-13), and its genetic consequences have been traced. Such sporadic exchanges of genetic material within a large population that usually reproduces by fission has the effect of providing invigorating interminglings of genetic substance. Conjugation is beneficial because such distributions tend to prevent an accumulation in the population of altered and possibly debilitating genes.

An electron micrograph of a thin section of *E. coli* cut from a rapidly growing cell is a fairly drab landscape (see Figure 38-10). On the outside is the rigid cell wall, a 10 nm thick mosaic of protein, polysaccharide, and lipid molecules. Just inside the cell wall is a flexible, 10 nm thick cell membrane, composed largely of lipids and proteins. This membrane is semipermeable and it controls which molecules enter and leave the cell. Of vital importance is the ability of the membrane to maintain a concentration gradient, since most molecules, both small and large, are present at much higher levels inside the cell membrane than outside. This is true for inorganic ions such as K^+ and Mg^{2+} and, most important, organic molecules. The membrane must actively prevent molecules from diffusing into the outside area where concentrations are very much lower. As shown schematically in Figure 38-14, about one-fifth of the interior of the cell is occupied by DNA. Immediately surrounding the DNA are 20,000 to 30,000 ribosomes, usually in the form of the

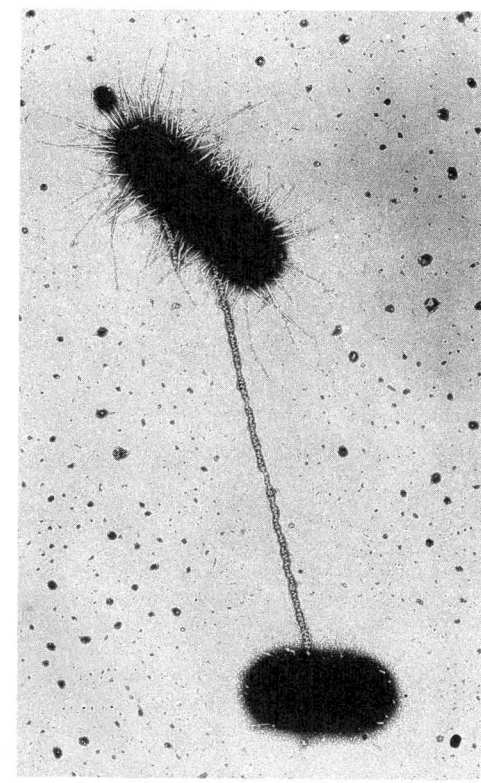

38-13 Conjugation in E. coli ($\times$ *17,000*)

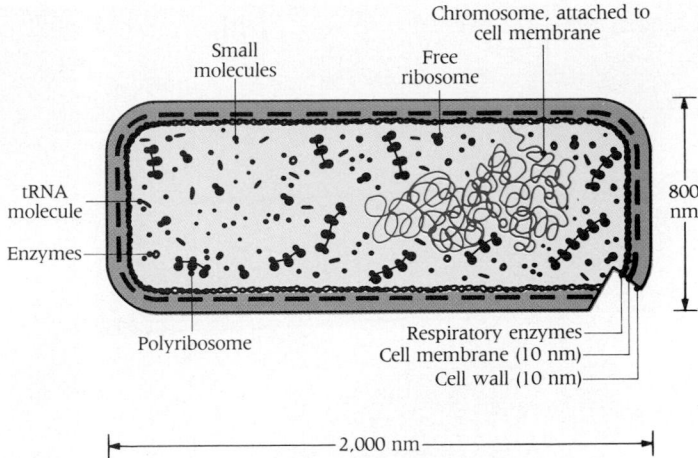

38-14 Cross section of an E. coli cell
The various components of E. coli *include a circular chromosome, which is attached to the cell membrane.*

Chromosome, attached to cell membrane

Small molecules

Free ribosome

tRNA molecule

Enzymes

Polyribosome

Respiratory enzymes
Cell membrane (10 nm)
Cell wall (10 nm)

800 nm

2,000 nm

aggregates called polyribosomes. They are the cellular sites of protein synthesis. The remainder of the cell's interior is filled with water, water-soluble enzymes, and a large number of various small molecules.

A remarkable thing about bacteria is that these minute and seemingly simple organisms are so complex in molecular composition and structure. They carry out all of the basic life processes and, in the complexity and general nature of their transformations of matter and energy, do not differ fundamentally from other organisms, including human beings.

PHYLUM CYANONTA: THE BLUE-GREEN ALGAE

Despite their name, only about half the **blue-green algae**—also known as cyanobacteria, cyanonts, cyanophytes, and blue-green bacteria—are actually blue-green in color. The rest of the several thousand species are blue, green, yellow, red, or colors in between, owing to the presence of **phycocyanin** (blue), **phycoerythrin** (red), or other pigments. Until recently, cyanonts were considered part of the plant kingdom because they are photosynthetic. It is now clear that cyanonts resemble other bacteria in that they have a rigid cellulose cell wall and contain no mitochondria, Golgi apparatus, or other membrane-limited organelles (Figure 38-15). Nor is there a nuclear membrane.

This phylum displays great diversity—indeed, it is the most diverse prokaryote phylum, ranging from simple unicellular forms to complex filamentous organisms. Some unicellular blue-green algae are solitary and free-living. Most blue-green algae consist of clumps or colonies of attached cells, with little further differentiation

38-15 The blue-green alga Anabaena cylindrica
(A) Blue-green algae (cyanobacteria) lack membrane-bound organelles. Photosynthesis takes place in chlorophyll-containing membranes within the cytoplasm. (B) Scanning electron micrograph of the Anabaena microfilaments within the leaf of a water fern, Azollo. A quarter of the filament cells are heterocysts, which fix atmospheric N$_2$ and convert it to ammonia. The two-cell hair is part of the fern. (× 930)

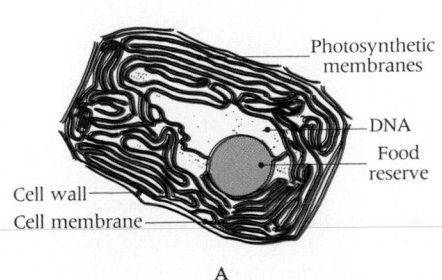

Photosynthetic membranes

DNA

Food reserve

Cell wall

Cell membrane

A

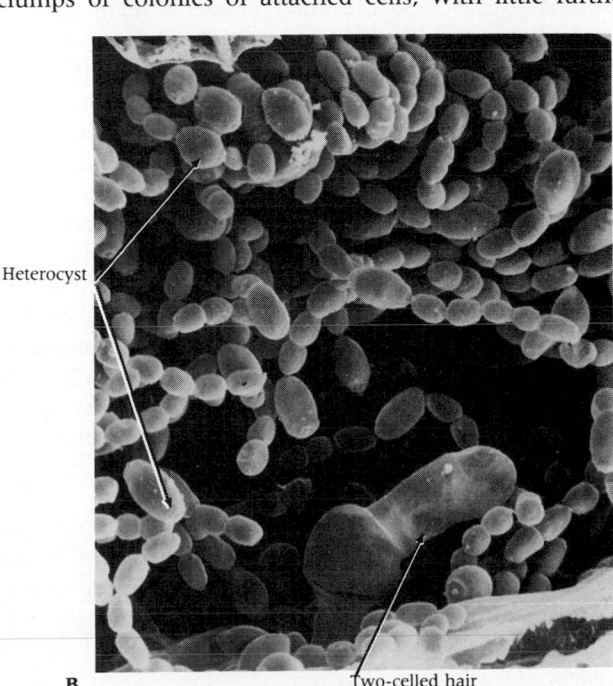

Heterocyst

Two-celled hair

B

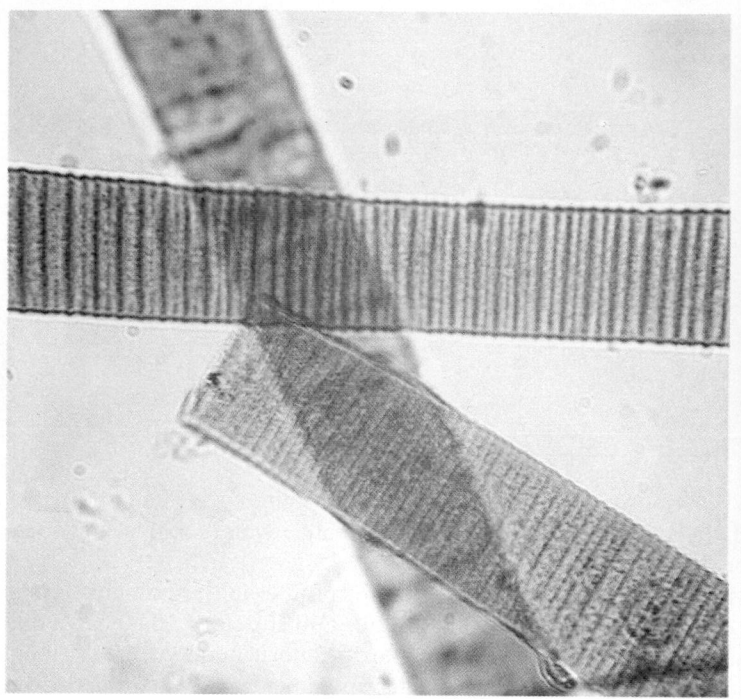

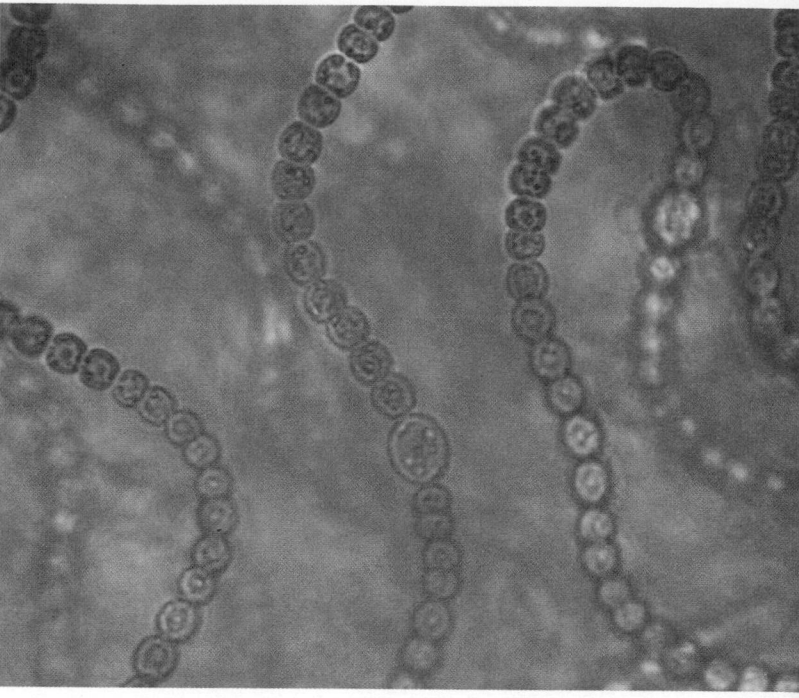

A

B

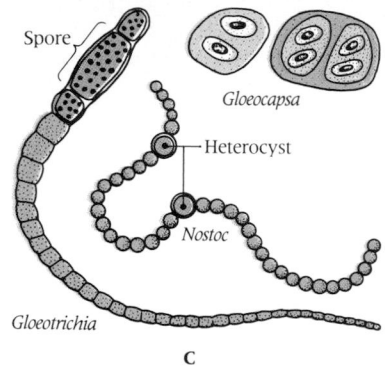

(Figure 38-16). The most characteristic cyanont form is a filament made up of cells attached end to end. In all of them, the cell (or colony) has a mucilaginous sheath or coating composed of pectic substances, and many of them engage in a gliding sort of motility when in contact with solid surfaces.

Some filamentous forms display a division of labor among colony member cells. Beside ordinary cells, a filament may contain cells specialized for attachment to a surface, spore-producing cells, and structures called **heterocysts,** which carry out N_2 fixation—the process that converts atmospheric N_2 to NH_3 (Figure 38-16C). Reproduction is vegetative or by fission. Sexual or parasexual processes have not yet been observed. Except for euglenoids, this is the only group of which this can be said.

Members of this phylum are totally independent nutritionally. They are photosynthetic oxygen-producers, using true chlorophyll (actually chlorophyll *a*)—unlike photosynthetic bacteria—as well as the other pigments mentioned earlier. They even have photosystems I and II like chloroplasts. However, because they are true prokaryotes, their chlorophyll is not in chloroplasts (as in plant cells) but an elaborate system of flattened photosynthetic vesicles or membranes called thylakoids—the same term as that used for chloroplast membranes. Blue-green algae respire aerobically, and many of them fix atmospheric N_2 on a wholesale basis. In many parts of the world, most notably southeastern Asia where fertilizer is hard to come by, blue-green algae are important providers of the fixed nitrogen needed for the growth of rice plants. In fact, they need (beyond a few minerals) only water, N_2, CO_2, and light. That is why they are able to thrive on bare rocks, sand, and soil.

Blue-green algae are abundant in the seas. The Red Sea may have been so named because of the presence of a red species of blue-green algae. They also abound in lakes and streams. Many contain gas bubbles that allow them to float on water. Many species prefer especially rigorous environments such as hot springs, hot deserts, bare rock surfaces, highly mineralized waters, and Antarctic pools where they form "algal peat." They also play an important role in the eutrophication of lakes (see Chapter 48).

It has been suggested that the blue-green algae are in fact advanced colonial bacteria, or that such bacteria are extremely primitive algae, which have remained more or less at the same very primitive evolutionary level where true algae, which are protists, arose. Algae arose very early in Earth's history, probably in Precambrian times. In any case, it is clear that the blue-green algae are not closely related to true present-day algae.

38-16 The phylum Cyanonta
(*A*) *The cells of* Oscillatoria *are arranged in chains or filaments.* (*B*) Nostoc *has a beadlike arrangement punctuated with heterocysts, which are centers of nitrogen fixation.* (*C*) *Members of these genera possess chlorophyll* a *and several other pigments, such as phycocyanin, which gives these cells their characteristic blue-green color.*

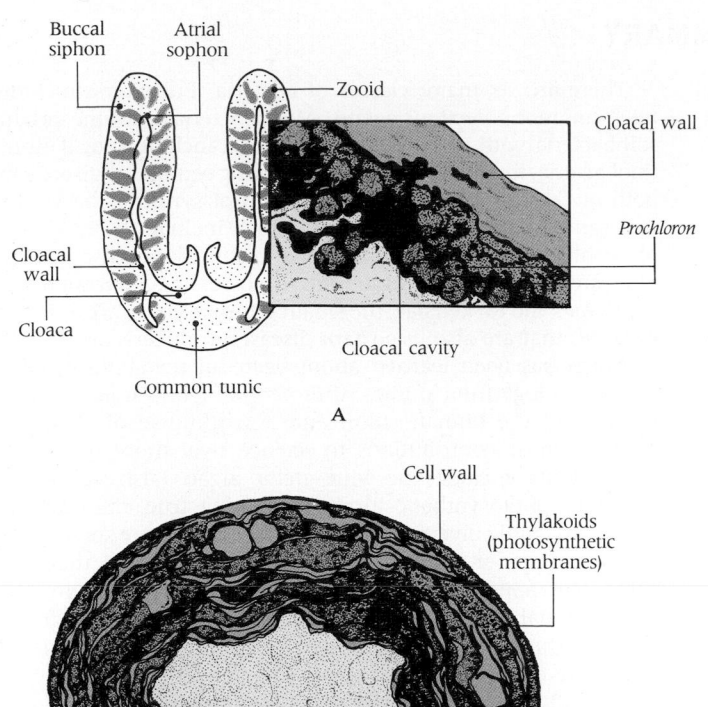

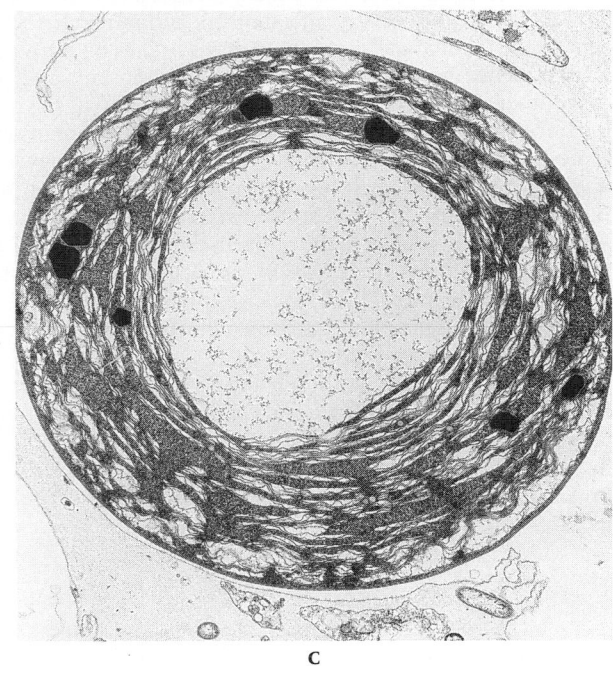

38-17 The phylum Prochloronta
(A) Prochloron, *lives in the cloacal wall of the tunicate,* Lissoclinum patella, *a native of the South Pacific.* **(B)** *Diagram of a cross section of* Prochloron *showing that it lacks a nucleus. Hence, it is a prokaryote.* **(C)** *Electron micrograph of* Prochloron *in cross section.*

Unlike bacteria, blue-green algae rarely cause decay or disease. This may be why university departments of cyanotology (if they existed) would be so much smaller than departments of bacteriology.

THE NEWEST PHYLUM: PHYLUM PROCHLORONTA

An interesting new microorganism was discovered in 1975 in Baja California, Mexico. Its remarkable properties indicated that it represents a new moneran phylum. It was at first thought to be a protist, resembling green algae, but electron microscopy clearly revealed it to be prokaryotic, with no sign of a nucleus (Figure 38-17).

The new phylum is **Prochloronta** (also called Prochlorophyta). The genus name assigned to the new form was *Prochloron*. For a time, *Prochloron* was the only genus in the new phylum, a strange situation indeed. But in 1986, it was joined by *Prochlorothrix*, a presumed relative. Metabolically, these organisms are photosynthetic O$_2$-producers, possessing both chlorophylls *a* and *b* (blue-green algae have only chlorophyll *a*), carotenoids, and xanthophylls—like green plants. Unlike blue-green algae, they lack the phycobilin pigments phycocyanin and phycoerythrin. Lack of these pigments is additional evidence that they are not blue-green algae.

There are two reasons for their considerable importance for evolutionary biology: (1) their intermediate position between blue-green algae (cyanobacteria) and green algae, and (2) the likely possibility that they are the evolutionary ancestors of plant chloroplasts, which we believe to have arisen through the endosymbiotic incorporation of prochloronts into the cytoplasm of ancestral urkaryotic cells. If this is correct, Prochloronta would be a critical early participant in the evolution of eukaryotes from prokaryotes (see Figure 38-3).

Investigators attempting to verify this hypothesis by comparing DNA and RNA sequences from chloroplasts and *Prochloron* were frustrated by conflicting data—and by their inability to culture *Prochloron*. *Prochlorothrix*, the new phylum member, *can* be cultured, but sequence data published in 1989 are still conflicting. There the matter stands.

SUMMARY

The kingdom Monera includes the prokaryotes. Its three phyla are: Schizonta (true bacteria); Cyanonta (blue-green algae, also called cyanobacteria and cyanophytes); and the recently discovered Prochloronta (which includes only two species).

Monerans reproduce asexually. Therefore, it has not been feasible to define moneran species as reproductively isolated units. Instead, systematists have been compelled to classify them according to similarities in morphology, metabolism, and other characteristics—a generally inadequate approach. Recent studies on DNA and amino acid sequence homology have added an important and promising new dimension to the taxonomy of prokaryotes.

Since the recognition of archaebacteria, which many believe to represent a possible sixth kingdom, it is widely believed that the ancestral prokaryotic cell (the progenote) gave rise to three evolutionary lines—monerans, archaebacteria, and the ancestral urkaryotes which provided the nucleus and cytoplasm of cells that became eukaryotes after invasion by bacterial endosymbionts destined to become mitochondria or chloroplasts. Mitochondria probably derive from ancient purple nonsulfur bacteria, and chloroplasts from prochloronts or blue-green algae.

Different bacteria have widely differing characteristics. There are heterotrophs, photosynthetic and chemosynthetic autotrophs, anerobes, and aerobes. Some bacteria are causal agents of infectious disease, but others are essential for the survival of multicellular organisms. Bacteria are among the few organisms capable of nitrogen fixation.

There are six major classes of bacteria. Eubacteria include the most common infectious agents. Myxobacteria are slime-producing soil bacteria with complex life cycles that include sporulation. Actinobacteria, filamentous forms resembling certain fungi, are capable both of causing infectious diseases and of synthesizing antibiotics that can cure them. Spirochaetes, which include the causal agents of syphilis and other diseases, have distinctive corkscrew shapes. They are unique among monerans in possessing microtubules. Mycoplasmas and Rickettsiae, the smallest prokaryotes, are intracellular parasites that are also important disease-producers.

Much has been learned about bacterial metabolism, genetics, and physiology from studies of *Escherichia coli*, a noninfectious inhabitant of the human colon and a workhorse of experimental biology whose contributions to science rival those of *Drosophila*.

Cyanonta include the blue-green algae (also called cyanobacteria), photosynthetic forms containing true chlorophyll. The phylum is highly diverse. Some blue-green algae are solitary. Others live as clumps or colonies of attached cells. Within the colonies, some cells appear to display rudimentary specialization. Despite the potential confusion arising from their name—their several names—the blue-green algae are not closely related to true algae, which are eukaryotic protists (see Chapter 39).

Like all monerans, prochloronts are prokaryotes. Yet they have characteristics intermediate between those of cyanonts and true algae. For example, they lack phycocyanin and phycoerythrin. They may be the evolutionary ancestors of chloroplasts.

KEY TERMS

aerobic metabolism
anaerobic metabolism
Archaebacteria
autotroph
bacillus
blue-green algae (cyanobacteria)
clone
coccus
colony

conjugation
cyanobacteria (blue-green algae)
endosymbiont
eubacteria
fruiting body
heterocyst
heterotroph
Monera
phycocyanin

phycoerythrin
Prochloronta
progenote
saprobelt
Schizonta
spirillum
spore
sporulation
thermobacteria
urkaryote

QUIZ QUESTIONS

1. Which of the following is *not* a feature of prokaryotic cells?
   A. ribosomes
   B. endoplasmic reticulum
   C. cell wall
   D. DNA
   E. A and C

2. Which of the following pigments is *not* found in at least some species of monerans?
   A. chlorophyll b
   B. carotenoids
   C. phycocyanin
   D. phycoerythrin
   E. All of the above pigments occur in Kingdom Monera.

3. Various anaerobic, autotrophic species of monerans use all of the following, *except* for _____, as reducing agents.
   A. water
   B. hydrogen sulfide
   C. lactic acid
   D. ethanol
   E. B and C

4. Select the false statement.
   A. Some monerans are important components of the global nitrogen cycle.
   B. Some monerans are important because they cause diseases.
   C. No monerans are known to be involved in decomposition of organic materials.

   D. Some monerans are pioneering colonizers of habitats that otherwise seem biologically sterile.
   E. All monerans are unicellular.

5. Which of the following eukaryotic structures is thought to have evolved via invasion of cells by various types of prokaryotic cells?
   A. microtubules of flagella
   B. mitochondria
   C. chloroplasts
   D. B and C
   E. A, B, and C

ESSAY QUESTIONS

1. What makes classification within the kingdom monera so difficult? On what grounds is classification within the kingdom monera based?

2. What features distinguish moneran cells from those of eukaryotes?

3. What are the major classes of bacteria? What are their properties?

4. What morphologic features of bacteria facilitate their classification?

5. Why is archaebacteria thought to represent a sixth kingdom?

REFERENCES AND SUGGESTIONS FOR FURTHER READING

Cooper, J. I., and MacCallum, F. O. (1984). *Viruses and the Environment.* Chapman & Hall, London, U.K.

A helpful account of the biology of viruses.

Davis, B. D., Dulbecco, R., Eisen, H. N., and Ginsberg, H. S. (Eds.) (1980) *Microbiology,* 3rd ed. Harper & Row, New York.

This masterful, multi-authored textbook of microbiology deals with basic aspects of bacterial physiology and with their role as infectious agents.

Eisenberg, H. (1988). Archaebacteria coming of age. *Trends in Biochemical Sciences 13,* 416–417.

The author argues that the archaebacteria must now take their rightful place in phylogenetic and taxonomic schemes.

Ferris, F. G., and Beverage, T. J. (1985). Functions of bacterial cell surface structures. *BioScience 35,* 172–177.

Bacteria are among the most successful forms ever to have lived. The authors attribute that success to their simplicity and adaptability. Both characteristics apply to their cell surface, which is here described.

Fox, G. E., and 18 other authors. (1980). The phylogeny of prokaryotes. *Science 209,* 457–463.

An important paper outlining current controversies (that is, current in 1980) concerning the ancestry and evolution of prokaryotes.

Holt, J. G., and Kreig, N. R. (Eds.) (1984). *Bergey's Manual of Systematic Bacteriology,* Vol I. Sneath, P. (Ed.) (1986). Idem. Vol. II. Williams & Wilkins, Baltimore.

This remarkable work, in print for many years, is the authoritative guide to the identification of bacteria. Packed with useful and interesting information.

Mortlock, R. P. (Ed.) (1984). *Microorganisms as Model Systems for Studying Evolution.* Plenum, New York.

This collection of authoritative essays offers a glimpse of how studies on the metabolism and genetics of bacteria bear upon the mechanisms of evolution.

Penny, D. (1989). What, if anything, is *Prochloron*? *Nature 337,* 304–305.

The discovery of Prochlorophyta in 1975 introduced a new phylum and a likely ancestor of chloroplasts. But new DNA and RNA sequence data raise questions. More work is needed.

Schwartz, R. M., and Dayhoff, M. O. (1978). Origins of prokaryotes, eukaryotes, mitochondria, and chloroplasts. *Science 199,* 395–403.

One of the papers that gave weight to the endosymbiont hypothesis of mitochondrial and chloroplast origin.

South, G. R., and Whittick, A. (1987). *Introduction to Phycology.* Blackwell Scientific, Boston.

Phycology, the study of algae, remains a difficult field because of algal diversity, which exceeds that of land plants. This introductory text on prokaryotic and eukaryotic algae is both interesting and attractive.

Woese, C. R. (1981). Archaebacteria. *Scientific American 244,* June, 98–122.

The author presents convincing arguments that the methanogens he calls archaebacteria represents an important and separate kingdom of living organisms.

Zinsser, H. (1955) *Rats, Lice, and History.* Little, Brown, Boston.

A famous investigator's popular account of the role of typhus and other plagues in human history.

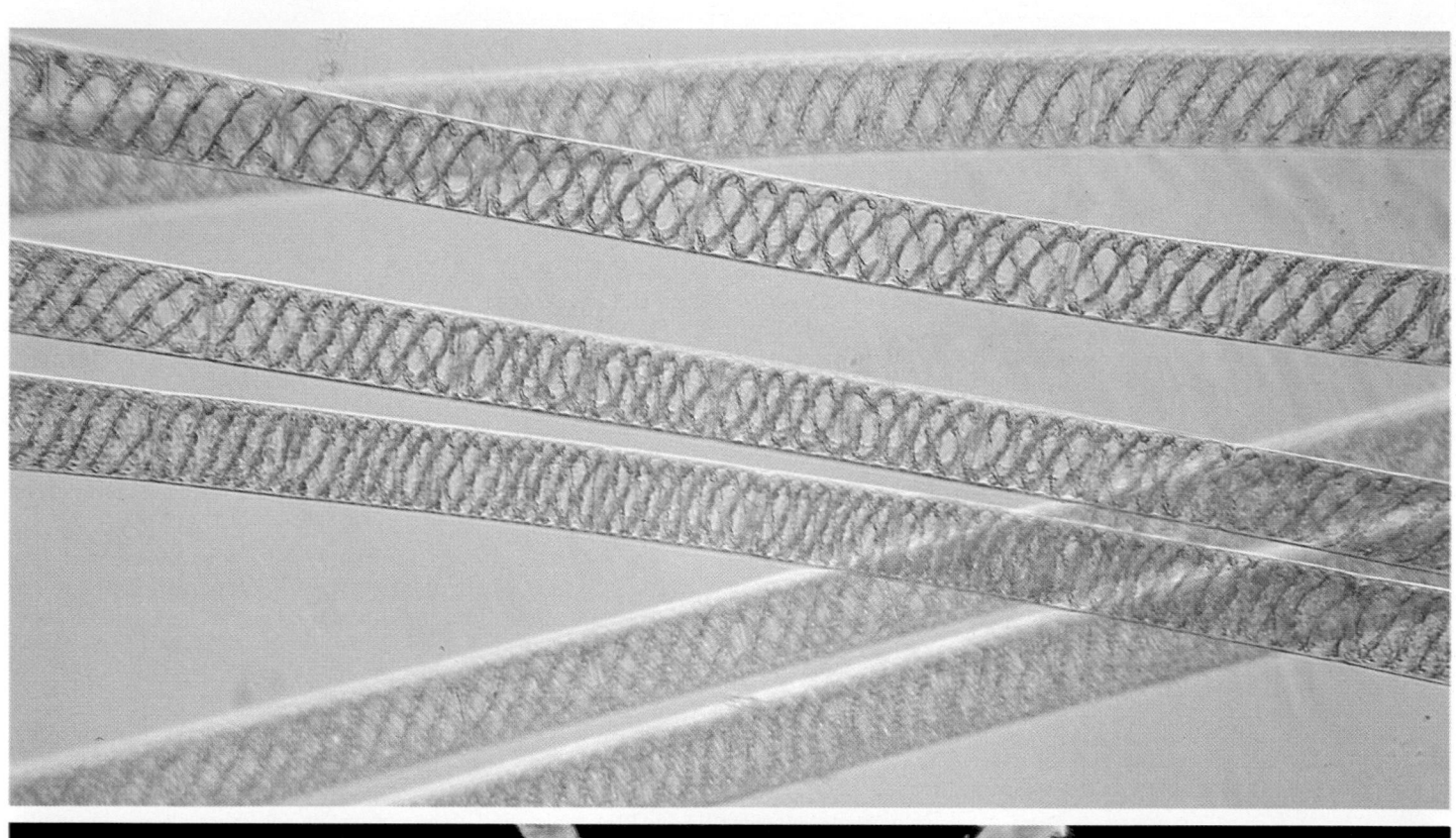

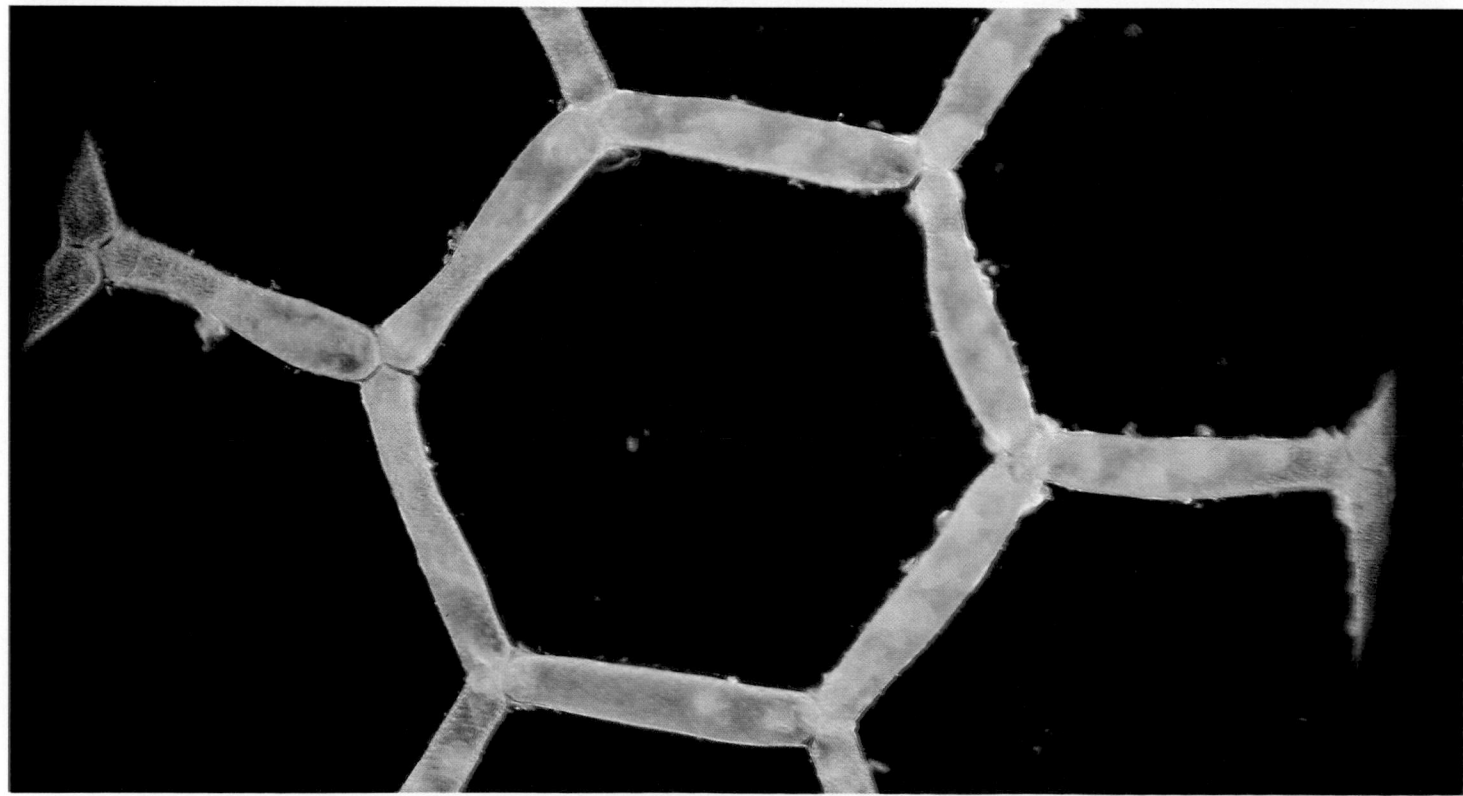

PROTISTS

Protista

The members of Kingdom Protista are aquatic organisms that live in the ocean, in fresh water, or in the body fluids of other organisms. These diverse forms show wide differences in cellular ultrastructure and organization, modes of cell division, and reproductive cycles. The living representatives shown here are both green algae (Chlorophyta). Top: Spirogyra (×100), *a filamentous form whose individual cells can reproduce sexually by conjugation with cells of neighboring filaments. Bottom:* Hydrodictyon (×20), *a common member of the class* Chlorococcales *that forms "water nets."*

Protista is a large kingdom of organisms that are both unicellular and eukaryotic. In Chapter 5, we spoke of the protist kingdom in a general discussion of differences between unicellular and multicellular organisms. This is the kingdom in which two major changes evolved, multicellularity and sexuality. It is also the kingdom that in the view of some includes the ancestors of each of the three kingdoms of multicellular organisms—fungi, plants, and animals. That, however, is not universally accepted.

EVOLUTION AND CLASSIFICATION: SOME UNRESOLVED PROBLEMS

What we do know is that eukaryotes evolved from prokaryotes in one of the most important events of biological history (see Figure 37-16). This conclusion rests on two lines of evidence: (1) the earliest prokaryote fossil, some 3.5 billion years old, antedated the earliest unquestioned eukaryote fossil by some 1.3 billion years; and (2) the genetic code, as we learned in Chapter 17, is the same in prokaryotes (such as *Escherichia coli*) and in eukaryotes (such as *Homo sapiens*). That, along with other biochemical similarities, suggests that the latter arose from the former.

POSITION OF PROTISTA IN EVOLUTIONARY HISTORY

As noted earlier, most protists are unicellular. However, though they lack the tissues and organs of multicellular organisms, many have subcellular structures that display remarkable functional specialization. Moreover, when they reproduce by cell division, daughter cells may stick together to produce a colony of individual organisms. Such colonies, we speculated, may have been steps in the evolutionary path from unicellular organisms to metazoans. In various stages of their lives, some metazoans—for example, sponges—and many true plants have cells that are almost undistinguishable from protists. It almost becomes a matter of personal taste, with present knowledge, whether we look upon such an aggregation as an advanced colonial union of individual protists or as a rudimentary grouping of cells in an individual of a higher order. We return to this theme in Chapter 42.

Clearly, there is some disagreement over protistan evolution. Figure 39-1 presents in simplified form three different interpretations of the available data, which have been proposed in recent years. None of them has been conclusively established.

- One scheme (Figure 39-1A) recognizes the fact that, morphologically, some protists are "funguslike," some are "animallike," and some are "plantlike." In this model, protists stood at a critical fork in the road and indeed were evolutionary precursors of the multicellular fungi, animals, and plants.
- In a second scheme (Figure 39-1B), Protista emerged as a separate evolutionary line prior to the rise of multicellular organisms.

- A third view (Figure 39-1C) gives kingdom status to Archaebacteria and Eubacteria (rather than Monera) and to slime molds. It then has Protista diverging later as a separate line—before a postulated common ancestral eukaryote (now called an **urkaryote**) gave rise to fungi, animals, and plants. In this model, neither bacterial group was a precursor of the eukaryotic kingdoms. The ancestral urkaryote was that precursor (see Figure 38-3).

Despite these uncertainties, it is indisputable that the evolution of eukaryotes from prokaryotes was a giant leap that encompassed the development of true nuclei (limited by a membrane and containing nucleoprotein) as well as other organelles, the evolution of mitosis with all of its machinery, and the appearance of other features distinctively eukaryotic. Metabolically, the Protista are highly diverse and include free-living heterotrophs, parasitic heterotrophs, photosynthetic autotrophs, and a few versatile types that can be heterotrophic or photosynthetic.

We have already discussed (in Chapters 5, 9, and 37) the widely accepted endosymbiont hypothesis, according to which mitochondria arose in evolution as a result of the "invasion" of urkaryotic cells by once free-living prokaryotes (probably purple nonsulfur bacteria) that remained there in a mutually beneficial relationship.[1] A similar event explains the evolutionary origin of chloroplasts.

Because chloroplasts differ in their pigment content in the plant and protist kingdoms—and in the various protist phyla as we note later—this theory of chloroplast origin explains the differences by postulating different bacterial invaders in different taxa. For example, the chloroplasts of plants, green algae, and euglenoids may derive from *Prochloron*, the newly discovered prokaryote described in Chapter 38. However, proof is still lacking. The chloroplasts of red algae, to be described later, probably derived from blue-green algae (cyanobacteria). If these ideas are correct, it would appear that chloroplasts arose in evolution not once but several times.

CLASSIFICATION

Authorities differ not only on the evolution of protists but also on their classification, disagreeing on which forms belong in the kingdom Protista, which belong in each of its phyla or divisions (the equivalent term), and which in each class. It is fair to say that when fungi, plants, and animals were raised to kingdoms, many systematists decided simply to place all remaining eukaryotes in the kingdom Protista—hence, its diversity.

One must choose, and, in this book, we divide the kingdom Protista into two major groups or subkingdoms.

- **Algae**—autotrophs that are plantlike in many ways
- **Protozoa**—heterotrophs that are animallike in many ways

Note that in earlier classifications (see Table 37-1) and Figure 37-15), these two groups were considered plants and animals. We are aware that the term "algae" has had a checkered history, meaning many things to many people. We alluded to that controversy in Chapter 21. Here we use the term to denote a large subgroup (subkingdom) of the protistan kingdoms and define its perimeter as best we can.

SUBKINGDOM ALGAE: AUTOTROPHIC PROTISTS

The "algae" discussed in Chapter 38 were blue-green algae, now called cyanobacteria. These organisms are prokaryotes. Hence, they are not protists, which are eukaryotes. The algae before us now are very different organisms—they *are* eukaryotes—despite the confusion of a once shared name.

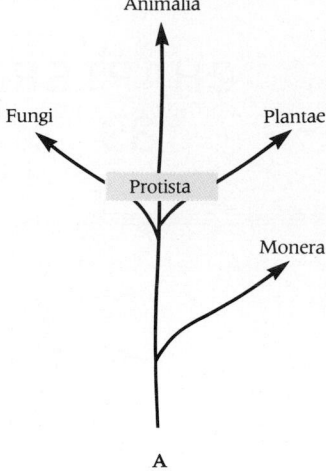

A

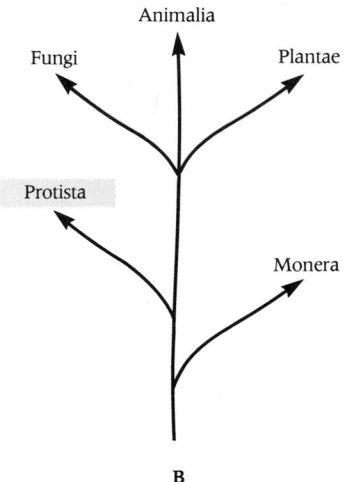

B

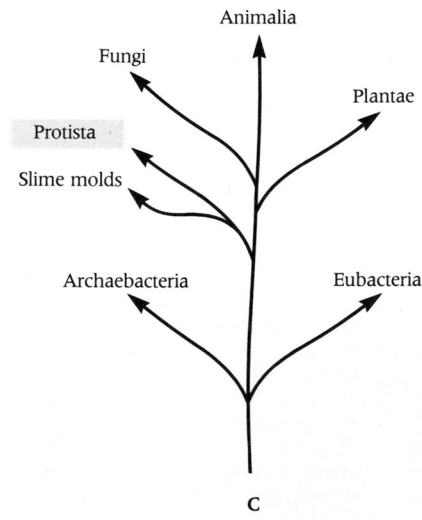

C

39-1 Three models of protistan evolution
The schemes are discussed in the text. Scheme C corresponds with the phylogenetic tree illustrating Moneran evolution in Figure 38-3.

[1] Not all eukaryotic cells have mitochondria. In fact, it is not the mitochondria, but the nucleus, internal membranes, and cytoskeleton that are the hallmarks of the eukaryotic cell. A great many species of protists, and a few fungi, have no mitochondria at all. Perhaps these are relicts of the ancestral urkaryotic cells.

The eukaryotic algae of the kingdom Protista are primary food producers. Marine algae carry on about 30 to 40% of all the photosynthesis that occurs on Earth.

We emphasize again that photosynthesis is the basic dynamic process of life. Although the situation may have been otherwise in the past, practically all of the energy utilized by living things today is first made available to them by photosynthesis. Photosynthesis is also a key mechanism in the building up of organic molecules. It occurs in fields and forests, but today, and surely in the past, a vast amount of photosynthesis occurs in water, especially in the world's oceans, by a variety of protists.

For this reason, algae have been called the "grass of the waters." They are the major energy-fixers of the "meadows" of the sea and they are the basic food source for most of the untold numbers of aquatic animals. For centuries, people in the Far East have eaten red and brown algae, but this greatest of all food sources has barely been touched as yet by most humans. The most significant human use—and it accounts for only a small fraction of the food consumed by humans—is indirect: the eating of fish and other seafood.

Nearly all algae are photosynthetic organisms that are largely adapted to life in water. There are about 30,000 known species of algae. Some are unicellular, but most have simple multicellular structures. Some are as small as bacteria. Others consist of filaments of cells, plates of cells, or large structures resembling vascular plants. Some—like the seaweed called **giant kelp**—are more than 200 feet long.

In multicellular algae, there is little evidence of differentiation into tissues or organs. In many species, all cells are practically identical. Others do have some

TABLE 39-1
CHARACTERISTICS OF THE PROTISTA

Phylum (examples)	Photosynthetic Pigments	Flagella or Cilia	Cell Wall or Shell	Food Reserve	Remarks
Subkingdom Algae (photosynthetic autotrophs)					
Euglenophyta Euglenoids (*Euglena*)	Chlorophylls *a* and *b*, carotenoids	1 to 3 (apical)	None	Paramylon	Mostly fresh water, no sexual reproduction
Pyrrophyta Dinoflagellates (*Noctiluca*)	Chlorophylls *a* and *c*, carotenoids (including peridinin)	2	Cellulose or absent	Starch, fats oils	Marine and fresh water, sexual reproduction rare
Chrysophyta Golden algae (*Ochromonas*) Diatoms (*Navicula*)	Chlorophylls *a* and *c*, carotenoids (including fucoxanthin)	1 or 2	Pectic compounds, siliceous shells	Leucosin, oils	Marine and fresh water, some multicellular
Chlorophyta Green algae (*Chlamydomonas*)	Chlorophylls *a* and *b*, carotenoids	0, 1, 2–8	Polysaccharides, cellulose in some	Starch	Mostly fresh water, some marine
Phaeophyta Brown algae (*Fucus*)	Chlorophylls *a* and *c*, carotenoids	2	Cellulose in some, alginic acids	Laminarin, mannitol, oils	Mostly marine, some multicellular
Rhodophyta Red algae (*Polysiphonia*)	Chlorophyll *a* and *d*, carotenoids, phycocyanin, phycoerythrin	0	Cellulose, pectic compounds, $CaCO_3$ in some	Floridean starch	Marine and fresh water, most multicellular
Subkingdom Protozoa (nonphotosynthetic heterotrophs)					
Zoomastigina Zooflagellates (*Trypanosoma*)	None	Many (varies)	None	Glycogen, lipid	May have evolved from euglenoids, some are pathogenic
Sarcodina (*Amoeba*)	None	None	None	Glycogen, lipid	Can pursue and engulf food particles.
Sporozoa (*Plasmodium*)	None	In some stages	None	Glycogen, lipid	Complex life cycle, parasitic, causes malaria
Ciliophora (*Paramecium*)	None	Many	None	Glycogen, lipid	Can "learn," sexual and asexual reproduction
Myxomycota Slime molds (*Physarum*)	None	0 or 2	None	Glycogen, lipid	Taxonomy disputed

distinctions of size, shape, and function among their cells but without differentiation of typical plantlike organs (leaves, roots, and so on).

The many species of algae are grouped into six major phyla, which generally appear to represent a series of parallel evolutionary lines.[2] The -phyta suffix in their phylum names implies their plantlike character.

- Euglenophyta (euglenoids)
- Pyrrophyta (dinoflagellates)
- Chrysophyta (golden algae and diatoms)
- Chlorophyta (the green algae)
- Phaeophyta (brown algae)
- Rhodophyta (red algae)

The first three phyla consist almost entirely of unicellular organisms. The last three include groups that are multicellular.

Most algae are fully aquatic, living on rocky shores, in seas, lakes, ponds, rivers—even in swimming pools. Others grow on or in soil, where they are abundant, or on the barks of trees. They also live in such unusual places as polar ice, the nearly boiling hot springs of Yellowstone National Park, in brine lakes of ten times the sea's salinity, and among the tissues of other organisms. They have even been found growing between the crystals of rocks. They always live in damp or wet places. Even in deserts they subsist on traces of dew.

The cell walls of algae generally consist of a polysaccharide matrix. Sometimes cellulose is present. A rich variety of storage products is found in each phylum—including carbohydrate food reserves and lipids. Almost all algae contain pigments of the chlorophyll family and are actively photosynthetic, but a few have lost their pigments and become organisms of decay. Although we usually classify the algae by their color—green, golden, red, and so on—other traits provide more rigorous taxonomic criteria. These include biochemical data on the nature of the pigments and the stored nutrients, details of anatomy and the life cycle. We will be concerned with the six main groups of algae, the characteristics of which are summarized in Table 39-1. Let us glance at them briefly.

PHYLUM EUGLENOPHYTA: EUGLENOID FLAGELLATES

The phylum Euglenophyta includes "whip-bearing" flagellates, a famous member of which is *Euglena gracilis*. Most **euglenoid flagellates,** of which there are hundreds of species, are unicellular and aquatic. Some are marine. Some are colonial and some are parasitic. Most euglenoids are photosynthetic autotrophs, but others lack chloroplasts and thus are heterotrophs. Euglenoids store their carbohydrates in the form of **paramylon,** a polysaccharide unique to this phylum.

Euglenophyta is an unusually interesting eukaryotic phylum because euglenoid flagellates are transitional between strictly plantlike and strictly animallike organisms. When *Euglena* is grown in the light, it is green and very plantlike—with chloroplasts and photosynthetic pigments, including chlorophylls *a* and *b* and several carotenoids (Figure 39-2). That is why botanists have generally claimed euglenoids as one of their own, placing them in a phylum (division) of the plant kingdom. However, when *Euglena* is grown in the dark, it loses its chloroplasts and photosynthetic pigments. The result is a cell that is henceforth heterotrophic and fully dependent on ingested food. Some consider these to be different organisms that are similar to *Euglena* and classify them as protozoa.

Leeuwenhoek, in the first description of this one-celled creature, wrote in 1675, "These animalcules had divers colours, some being whitish and transparent; others with green and very glittering little scales; others again were green. And the motion of most of these animalcules in the water was so swift, and so various, upwards, downwards, and round about, that 'twas wonderful to see." It still is. When grown under light in the laboratory, *Euglena* appears beautifully green as the organisms gather near the light source (see Figure 39-2).

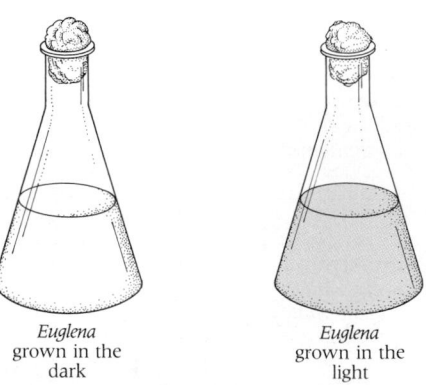

Euglena grown in the dark   *Euglena* grown in the light

39-2 Appearance of Euglena grown in dark and light
Each flask contains a culture of Euglena. *When grown in the dark, cells lack chloroplasts and are colorless. When grown in the light, cells contain chloroplasts and are green. Thus,* Euglena *can be a heterotroph and an autotroph.*

[2] Those omitted here include Xanthophyta (yellow-green algae) and Cryptophyta (cryptomonads). Many regard these are classes of Chrysophyta.

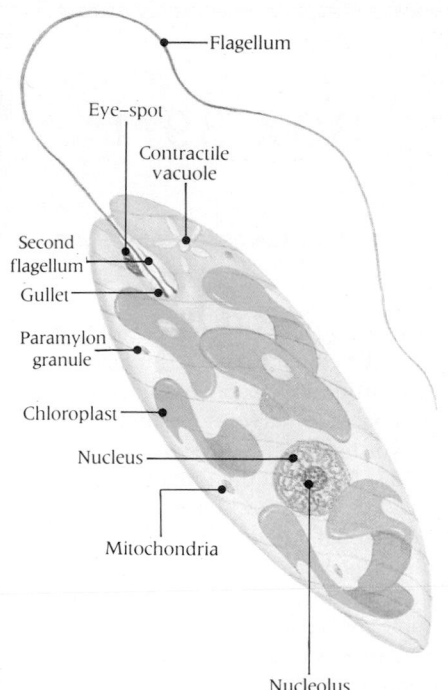

39-3 Structure of Euglena gracilis
In this light-grown cell, many chloroplasts are scattered throughout the cytoplasm. Two flagella are present, but only one is functional. The eyespot is a photoreceptor. (× 1400)

Labels on figure: Flagellum, Eye-spot, Contractile vacuole, Second flagellum, Gullet, Paramylon granule, Chloroplast, Nucleus, Mitochondria, Nucleolus

39-4 Dinoflagellates
(A) Ceratium. **(B)** Peridinium.
(C) Gonyalaux tamarensis. (*Approximate magnifications of scanning electron micrographs, × 1000*)

A swimming *Euglena gracilis* is an elongated cell, about 50 by 15 μm, with a whipping locomotor flagellum that protrudes from a **gullet** or **cytopharynx** (Figure 39-3). A short second internal flagellum does not emerge from the gullet.

E. gracilis has many disklike chloroplasts, a conspicuous nucleus (with a prominent nucleolus), numerous mitochondria, paramylon granules, and lipid inclusions. Also present are contractile vacuoles that regulate osmotic pressure by collecting and eliminating excess water, and a specialized structure, the **stigma** and its associated photoreceptor, which comprise the light-sensitive **eyespot.** They are close to the gullet near the bases of the two flagella.

In the dark, cultures of *E. gracilis* contain both motile and quiescent cells. Many are rounded up, reduced in size, and lacking in chlorophyll. The ability to assume the slender "gracilis" form is favored by light. Active swimming is resumed along with chlorophyll synthesis when organisms are furnished with light. The organisms reproduce by mitosis. Sexual reproduction is unknown among euglenoids. The growth of *Euglena gracilis* is totally dependent on the presence of cobalamin (vitamin B$_{12}$) in minute amounts. Indeed, this protist is used in the laboratory to assay cobalamin, which it can utilize in concentrations as low as 10^{-13} grams per ml.

PHYLUM PYRROPHYTA: DINOFLAGELLATES

Because many of Pyrrophyta's members are "fiery" red, this phylum of **dinoflagellates** takes its name from the Greek root *pyro,* meaning fire. Also known as Dinoflagellata, this phylum consists of about 1000 species of unicellular and colonial organisms. Some dinoflagellates occur in fresh water, but most are abundant in warm seas. There, as major elements of the free-floating marine organisms known collectively as **plankton,** they share with diatoms (phylum Chrysophyta) basic ecological roles in the turnover of organic materials. Some produce powerful poisons.

Some of the better-known genera of Pyrrophyta are *Gonyaulax, Ptychodiscus, Gymnodinium, Ceratium, Noctiluca,* and *Protopsis. Gonyaulax* (the circadian rhythms of which we discussed in Chapter 32), *Ptychodiscus,* and *Gymnodinium* are the producers of the strange toxic blooms, known as red tides, that have been a recurrent menace to human populations—not to mention the fishes and the shellfish they poison (Box 39-1).

Many dinoflagellates display impressive bioluminescence of extraordinary intensity which is often marked by periodicity. Some cause a peculiar shimmering phosphorescence that can be seen in seawater at night. One of these, *Noctiluca* ("nightshiner"), produces twinkling lights that are visible in the ocean waves at night.

Most dinoflagellates are bizarre in appearance, with stiff external cellulose plates that resemble a helmet or coat of armor (Figure 39-4). Almost every dinoflagellate has one or several whiplike flagella, which drive or pull it along. (The Greek root *dino* means "whirling," so these are "whirling whips.") The structure of the flagellum is the characteristic 9 + 2 pattern. The cell body is usually definite in form. Although

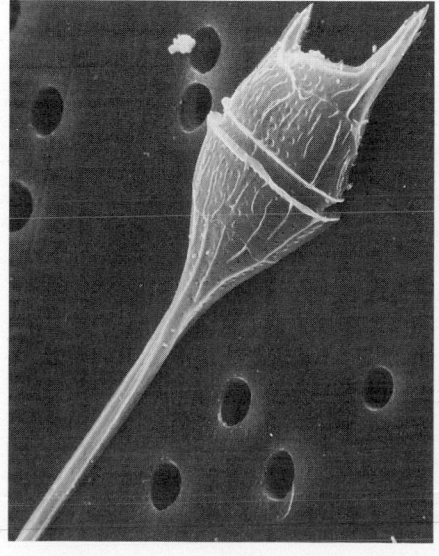

A

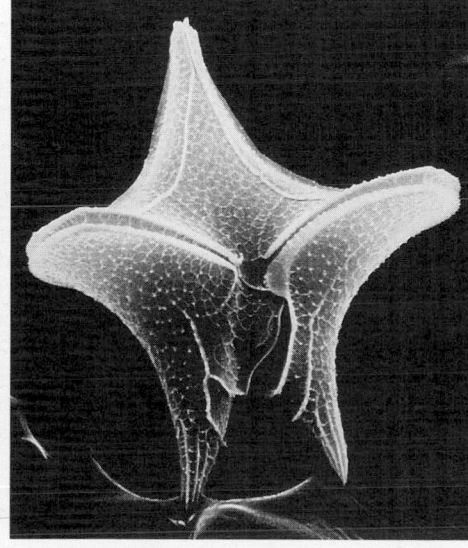

B

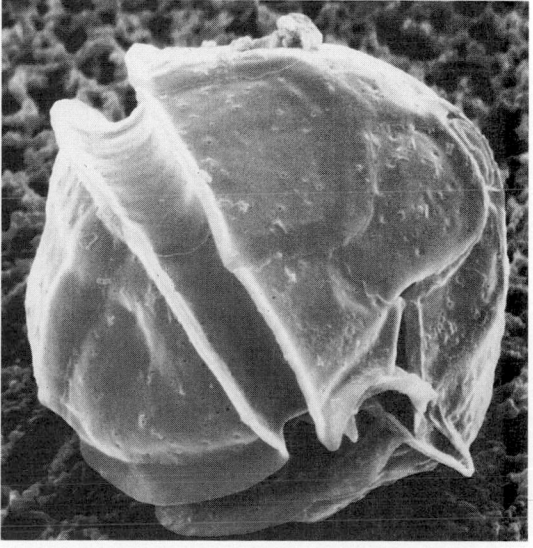

C

On a September day in 1972, a wildlife warden at Plum Island, Massachusetts, telephoned the state Department of Public Health and reported that within the past 14 hours he had found the bodies of 95 dead adult gulls and black ducks. At first he suspected that the deaths were due to aerial spraying of pesticides. However, autopsy revealed intensive hemorrhage in the birds' internal organs and mollusk shells in their stomachs. Later that day, a report came in from Gloucester of a peculiar reddish-brown color in a tidal pool near the University of Massachusetts Marine Biology Laboratory. When investigators examined the water microscopically, they saw large numbers of a microorganism that had never before been seen in quantity in Massachusetts waters. Suspecting that a potent toxin was present, the scientists prepared a filtered extract of the seawater and injected it into a laboratory mouse. The results confirmed their suspicions. The mouse was dead in eight minutes. A public health emergency was declared.

The organism was finally identified as the dinoflagellate *Gonyaulax excavata*. The infestation was an instance of a "red tide" (Figure 39A). A highly toxic marine dinoflagellate that infects shellfish, *Gonyaulax* produces a toxin that can be fatal within hours to humans who eat the shellfish—as it was to the unfortunate birds. This phenomenon has been called paralytic shellfish poisoning. Fatality can result from eating shellfish that contains 0.5 to 1.0 mg of the toxin per 100 g of shellfish meat. Researchers found up to 10 mg per 100 g in samples collected near Gloucester.

Officials were alarmed because there was no way of determining how much of the deadly shellfish had already reached markets, restaurants, or home refrigerators. Within hours, however, the entire 2000-mile Massachusetts coastline was closed to the harvesting of shellfish and stocks in markets and restaurants were confiscated. Studies showed that contamination was widespread and that the levels of toxin were extraordinarily high. The incident caused a general panic: 26 people were seriously poisoned by contaminated shellfish and public confidence was so shaken that, four years later, the Massachusetts shellfish industry had returned to only about two-thirds of its former level.

Red tides are now a recurring and increasing problem. During the winter and spring of 1974, the west coast of Florida was ravaged by its 25th major red tide since 1844. Hundreds of thousands of dead fishes littered the beaches, and millions of tourist dollars

39A Red tide
A red tide caused by large accumulations of the dinoflagellate, Gonyaulax excavata.

39B Fishkill due to Ptychodiscus brevis

were lost. In 1980, the entire coast of Maine was legally closed to shell-fishing from mid-August to mid-October. Commercial fishers were unable to harvest an estimated $7 million worth of clams, oysters, and mussels. In 1987, 200 dead dolphins washed ashore from Virginia to New Jersey.

In fact, red tides are produced by a variety of dinoflagellates. The term "red tide" is used to describe any bloom, toxic or not, even though the responsible alga may turn the waters brown, green, yellow, or other hues, depending on their pigments. In one Florida episode, the organism *Gymnodinium breve* turned the water blood red and fatally clogged the gills of fishes, but it was not toxic to human beings. In a devastating bloom off the coast of Naples, Florida, in August 1988, the offending organism was *Ptychodiscus brevis*. It turned the water yellow, caused major fish kills (Figure 39B), and affected 41 people with respiratory and neurologic illnesses. Two months later, the same dinoflagellate invaded the marine estuaries along much of the North Carolina coast.

The poisons produced by some dinoflagellates, such as *Gonyaulax cantanella*, are powerful nerve toxins. The chemical nature and biological effects of most of these toxins are now well understood. One of these toxins, saxitoxin, was described in Chapter 29. It binds specifically to a sodium channel protein in nerve membranes and is a useful laboratory tool. This effect of saxitoxin underlies its toxic effects on the nervous system.

On the other hand, the factors that cause the red tides themselves are poorly understood; levels of nutrients and certain trace metals, sewage runoff, ocean salinity and temperature, winds, light, and many other factors all seem to play some role. The 1972 surprise attack by *Gonyaulax* in Massachusetts, the first in the state's history, has never been adequately explained. Since then, red tide has occurred every year in Massachusetts—a fact that is equally perplexing to scientists. The periodic recurrence of red tides is compounded by the observation that once a "bloom" has occurred, if conditions shift for the worse, the dinoflagellates can lose their flagella and form the cysts that sink to the bottom and lie dormant until conditions are more favorable for their renewed activity.

some dinoflagellates are heterotrophs, many are photosynthetic autotrophs. Even the autotrophs require cobalamin, so they are not wholly autotrophic. In the ecology of the sea, and of coral reefs, the importance of dinoflagellates as primary photosynthesizers is second only to that of diatoms.

The life cycles of dinoflagellates resemble those of other algae (see Chapter 21), though many details relating to individual species remain uncertain. As in the complex life cycle of *Chlamydomonas,* a green alga (see Figure 21-7), the dinoflagellate reproductive cycle includes flagellated male and female haploid gametophytes that fuse to form diploid cells. Unlike the diploid forms of many green algae, those of some dinoflagellates develop into **cysts,** which fall to the ocean floor where they survive for long periods. When a cyst is later ruptured—by changes in water turbulence, temperature, light, or oxygen—a mobile cell emerges that divides into four daughter cells, each of which divides again and again. This probably accounts for the sudden "blooms" leading to red tides. Interestingly, fossilized bottom-lying dinoflagellate cysts, which go back to the Cambrian and even earlier, have proved useful in the dating of geologic seafloor strata.

Many dinoflagellates live symbiotically within other organisms—notably jellyfishes, mollusks, sea anemones, and corals—many of which are thereby rendered bioluminescent. In this situation, they lose their platelike armor and flagella, becoming the strange golden-brown forms that are known as **zooxanthellae.** As noted in Chapter 1, they are essential to the ecology of coral reefs, playing a vital role in the nutrition of the coral animals.

A distinctive feature of many dinoflagellates is the presence of large amounts of hydroxymethyluracil in place of some of the thymine (which, chemically, is methyluracil) in the DNA. Although dinoflagellates are eukaryotes, the organization of their nuclei, its DNA fibrils, and its chromatin is so unusual, that some authorities have termed them **mesokaryotes** (meaning between prokaryotes and eukaryotes).

PHYLUM CHRYSOPHYTA: GOLDEN ALGAE AND DIATOMS

The chrysophytes are a huge and diverse group of over 11,000 living aquatic species, the plastids of which contain yellow-golden carotenoids and other pigments. They are all photosynthetic and contain chlorophylls *a* and *c,* but lack chlorophyll *b.* They accumulate a distinctive polymer of glucose called **chrysolaminarin.** They also make alcohols and store energy in an oily lipid. Chrysophytes comprise a large fraction of the photosynthetic cells in plankton, the floating microorganisms of critical importance in marine food chains (Figure 39-5).

The phylum Chrysophyta includes **golden algae** (class Chrysophyceae) and **diatoms** (class Bacillophyceae), though some place them in separate phyla (divisions). Golden algae, about 1500 known species, are either unicellular, colonial, or filamentous (Figure 39-5).

The diatoms, a group of more than 10,000 living species and 40,000 (perhaps many more) extinct species, are remarkable, tiny, plantlike creatures that are usually unicellular but occasionally occur in small, colonylike or filamentous aggregations. Many diatoms and other golden algae secrete unique skeletal structures that are impregnated with silica (SiO_2) and often of complex and exquisite architecture (Figure 39-6). Usually the skeleton is constructed in two overlapping halves, like the bottom and overlapping lip of a porcelain pill box. They display a fascinating variety of shapes, but all seem to be either radially or bilaterally symmetrical. The oldest diatoms so far identified are early Jurassic in age (see Table 37-3).

The photosynthetic pigments of diatoms are chlorophylls *a* and *c,* but a rich endowment of carotenoids accounts for their golden color (see Table 39-1). Their storage products are oils rather than starches. This explains why fresh water containing many of these algae has a distinctive oily taste—or feel, or smell.

Many chrysophytes are multicellular. They usually reproduce asexually and many forms lack both germ cells and meiosis. However, diatoms do occasionally reproduce sexually, and unlike most protists are diploid in a part of their life cycle. In other chrysophytes, so-called **swarmer cells** may swim away from a colony and develop into a new colony. Alternatively, a colony may split in two, each half floating off and establishing itself elsewhere.

Diatoms are important because, despite their small size, they are often so abundant as to be the main photosynthetic organism over wide areas of seas, lakes, and streams. As noted above, they are a principal component of plankton. Thus, the

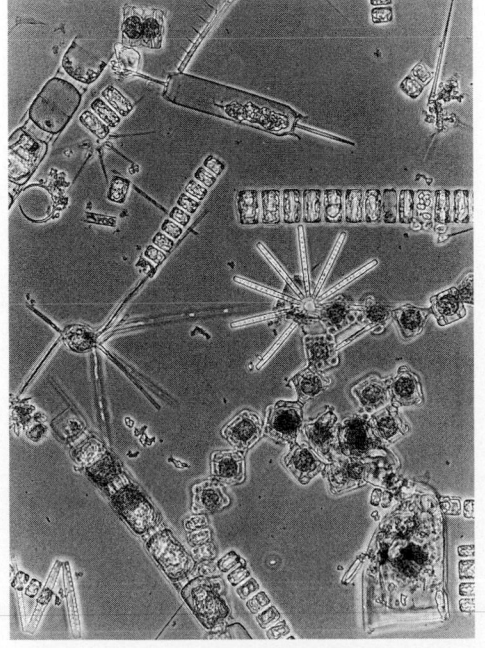

39-5 Phytoplankton (× 100)

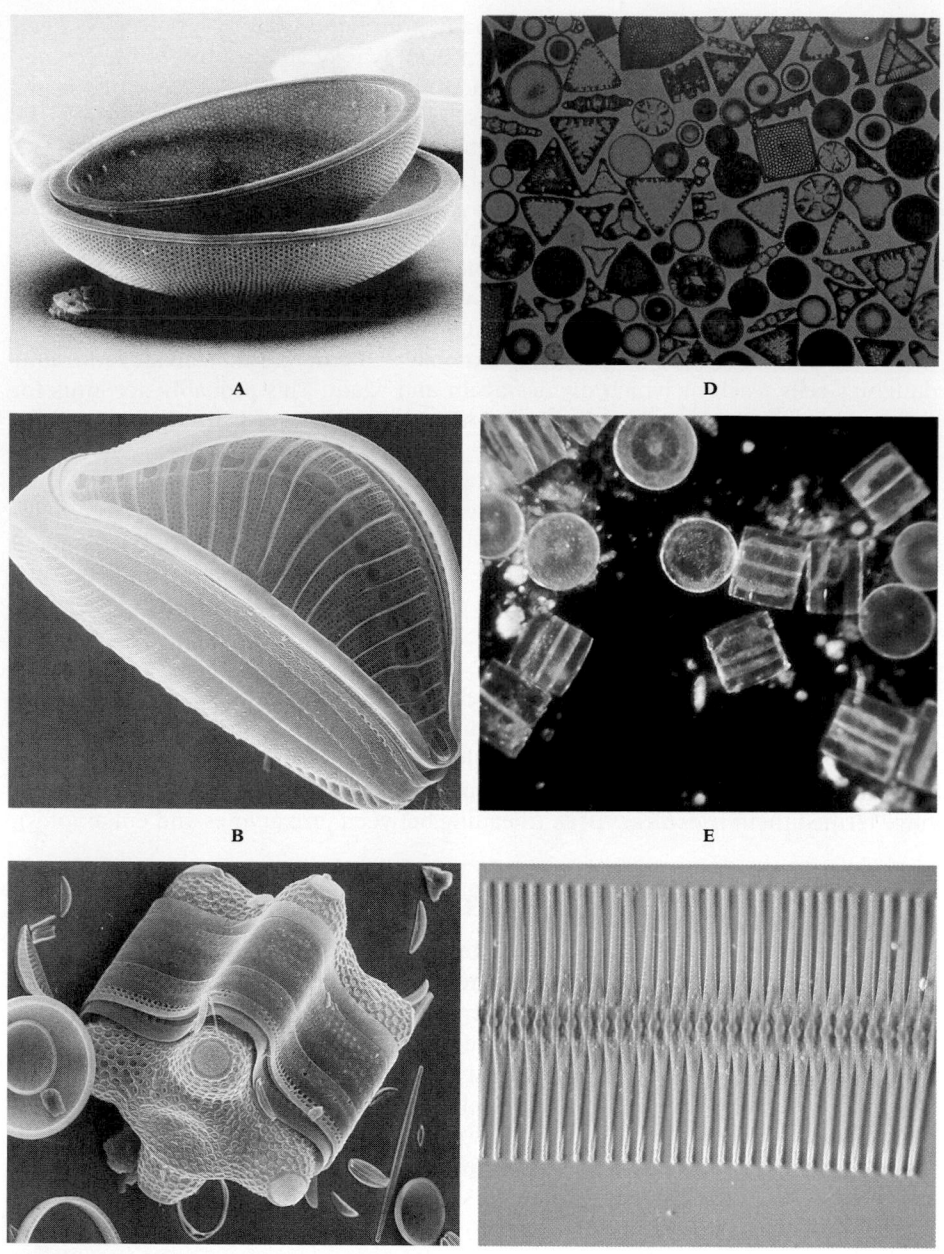

39-6 Marine diatoms
A selection of forms showing their extraordinary diversity and intricate structure.
(A-C) *Scanning electron micrographs.* (× 1500)
(D-F) *Light micrographs.* (× 110)

A

B

C

D

E

F

whole of marine life, from the tiniest animals to the great whales, depends to an extent on diatoms.

When diatoms die they settle on the sea bottom, where their rigid cell walls resist breakdown. This has created huge deposits of sedimentary rocks composed almost entirely of their siliceous skeletons. The material called *diatomaceous earth* is obtained in vast quantities from such rocks. Near Lompoc, California, more than 300,000 tons are quarried each year for industrial uses ranging from the polishing of gems to the filtering of freshly brewed beer. Diatomaceous earth is also good insulating material and is the light-reflecting ingredient in the painted center lines on highways.

PHYLUM CHLOROPHYTA: GREEN ALGAE

The **green algae** are divided into two major classes (plus several minor ones): Chlorophyceae, by far the larger and more diverse of the two, consisting of about 7000 species, and Charophyceae, the stoneworts. Most live in fresh water (or moist land regions), but some live in the sea.

The diverse forms of green algae (Figure 39-7) include the single-celled immobile (nonflagellated) *Chlorella,* the unicellular, largely haploid flagellated *Chlamydomonas, Pandorina,* which has about 16 cells, *Eudorina* with 32, the colonial *Volvox* (the flagellated individual cells of which collectively give mobility to a whole colony

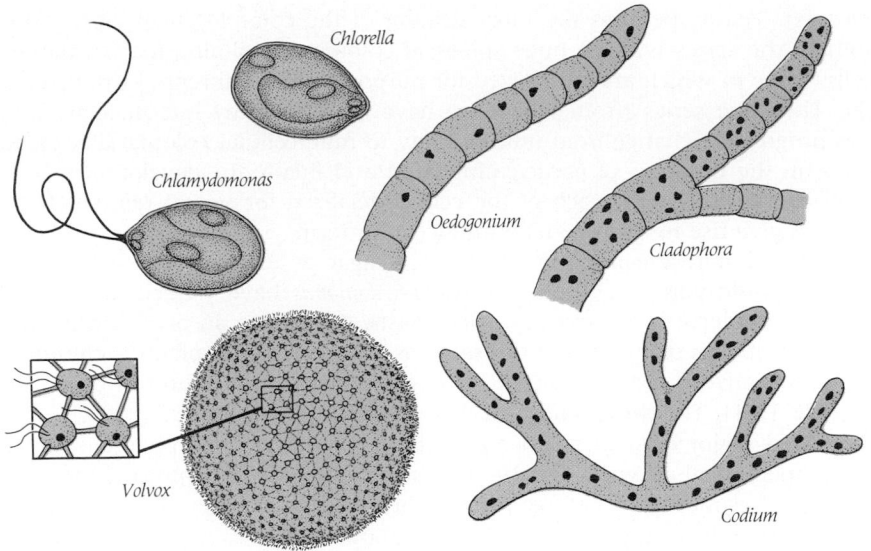

39-7 Green algae display a diversity of forms
Chlorella *is seemingly one of the simplest of green algae.* Cladophora *consists of multinucleated cells.* Codium *has no cell walls except during its reproductive phase.* Volvox *is a colonial alga composed of flagellated individual cells. Each of the forms is shown at a different magnification.*

containing 500 to 60,000 cells), the filamentous *Oedogonium* (each cell uninucleate), and the filamentous *Cladophora* (each cell multinucleate).

Then there is *Acetabularia*, familiar to us for its historic contributions to the study of differentiation (see Chapter 12)—an amazing giant single cell one and one-half inches in height with the unusual structure and shape that gave it the nickname ''mermaid's wineglass.'' Finally, there is *Ulva*—the sheets of cells commonly known as sea lettuce (Chapter 21).

We have already surveyed the various combinations of sexual and asexual reproduction in several of these forms. For example, we encountered in Chapter 21 the green alga, *Chlamydomonas,* and learned that its method of reproduction is an asexual mitotic cycle plus an occasional sexual union between haploid gametes (see Figure 21-7). This type of life cycle, perhaps the first form of sexual reproduction in a unicellular organism, may have been a highly significant evolutionary step in the origin of all metazoan life (see Chapter 44).

The point to be stressed here is that, collectively, green algae display an apparent transformation from unicellular to colonial life (see Figure 39-7). If we were to postulate that this is an evolutionary series, we would put *Chlamydomonas* at the unicellular end (Figure 39-8), and *Oedogonium, Pandorina,* and *Eudorina,* simple filamentous or colonial forms containing only 4, 16, or 32 cells, respectively, would

39-8 Chlamydomonas, *a green alga*
(A) *This unicellular green alga possesses a large chloroplast that takes up much of the cytoplasm. In addition, it has the other organelles typical of eukaryotic cells. Its two flagella provide motility.* (× 6500) **(B)** *Living* Chlamydomonas *in a pond.*

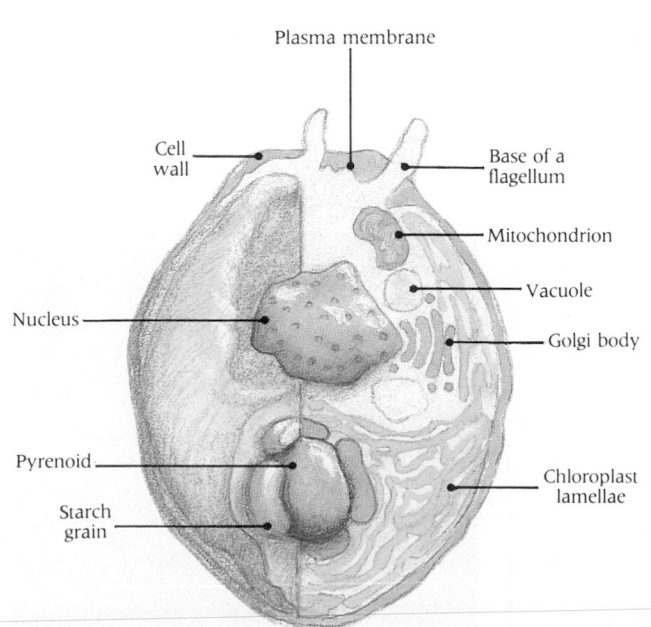

Plasma membrane
Cell wall
Base of a flagellum
Mitochondrion
Nucleus
Vacuole
Golgi body
Pyrenoid
Starch grain
Chloroplast lamellae

A

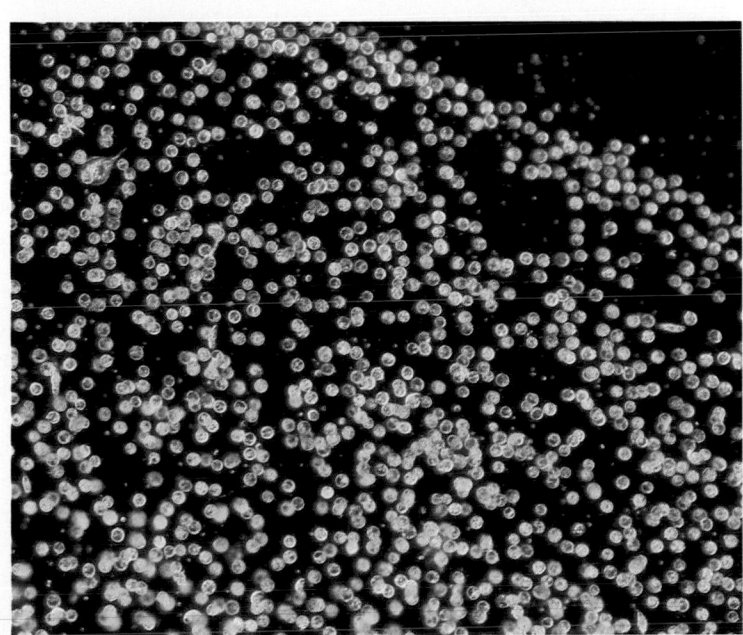

B

come next. *Volvox,* perhaps the most striking of the colonial green algae, would culminate the series with its huge spherical colonies containing many thousands of cells, some of which are specialized for purposes of sexual reproduction (Figure 39-9). Thus, the series (from which we have omitted many intermediate forms) shows progressive change from unicellularity to multicellular colonial life, gradual increase in the numbers of participating cells, and finally the development of cell specialization (in *Volvox*). Each of the cells specialized for sexual reproduction in *Volvox* can give rise to a whole new multicellular form.

Finally, we note the importance of green algae as experimental organisms in studies of photosynthesis. Mutants of *Chlamydomonas* have yielded much of our knowledge of electron transport in chloroplasts. *Chlorella* is an organism in which the photosynthetic pigments are the same as those in higher plants—chlorophyll *a* and chlorophyll *b,* which (except for euglenoids) are not found in other algae (see Table 39-1). The carotenoids are also typical of higher plants.

A second major class of green algae, the stoneworts (Charophyceae), are the most complex of all algae (Figure 39-10), though they are of relatively little importance in the economy of nature. (Predictably, perhaps, some authorities classify them separately from algae.) There are only about 250 living species, but they are well represented in the fossil record. Like several other algae, they have structures resembling roots, branching "stems," and "leaves," but, uniquely, Charophyceae have strangely marked "seeds." These structures are not true roots, stems, leaves, or seeds like those of higher plants, but they do serve similar purposes and show a remarkable degree of differentiation for algae. Stoneworts secrete calcium carbonate (lime or crystal) in their cell walls and may contribute to the formation of limestone deposits and marl. Thus stoneworts are among the calcareous algae.

PHYLUM PHAEOPHYTA: BROWN ALGAE

Brown algae, which include the kelps, are multicellular structures of large size, some spanning 300 feet or more. The phylum's 1500 species are almost all marine and are commonly found along rocky seashores at low tide. They also float in the open sea. A famous example is *Sargassum,* a seaweed that forms the dense floating mats of vegetation covering much of the surface of what is called the Sargasso Sea, an area of some two million square miles in the mid-Atlantic (Figure 39-11).

All brown algae are multicellular, often displaying extensive differentiation into stemlike stalks, leaflike blades, and gas-filled cavities or bladders. The body **(thallus)** may be a filament, or it may be a large and rather complex three-dimensional structure.

The characteristic color of brown algae derives from the carotenoid **fucoxanthin,** which is present in large amounts in the plastids. This yellowish-orange pigment combines with the green of chlorophylls *a* and *c* to yield a characteristic dirty brown color. The cell walls may contain as much as 25% alginic acid, a gummy polymer of mannuronic and glucuronic acids (both sugar derivatives). This substance, which cements cells and filaments together, makes good glue and is commonly used as an emulsifier in ice cream.

We saw the reproductive cycle of one brown algae, *Ectocarpus,* in Figure 21-10. Reproduction in brown algae is either asexual or sexual. The life cycle is marked in some genera by an alternation of gametophyte (haploid) and sporophyte (diploid) multicellular generations. Diploid forms have two kinds of sporangia. These are called **antheridia** if they produce male gametes and **oogonia** if they produce female gametes.

Kelps such as *Laminaria* show a complex alternation of generations (Figure 39-12). The haploid gametophytes are small and numerous (many are a food source for phytoplankton). The diploid sporophytes, produced in specialized structures called conceptacles, are large and prominent. In a few, such as the rockweed *Fucus,* reduction of the haploid stages goes so far there no longer exists a multicellular haploid gametophyte (Figure 39-13). Gametes are the only haploid cells in the reproductive cycle. This cycle rather resembles that of animals.

PHYLUM RHODOPHYTA: RED ALGAE

Most **red algae,** a highly distinctive group of some 4000 species, are multicellular marine seaweeds. Their carbohydrate reserves accumulate as small cytoplasmic gran-

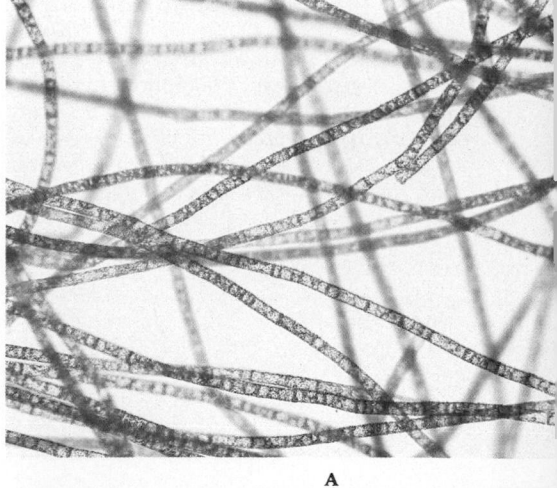

A

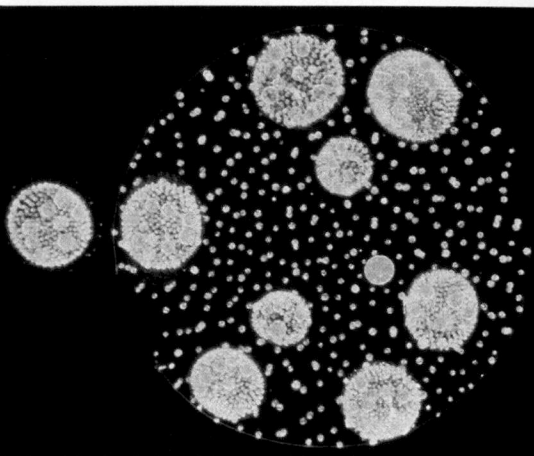

B

39-9 Colonial forms in green alga
(A) Oedogonium. (× *30*) **(B)** Volvox (× *80*)

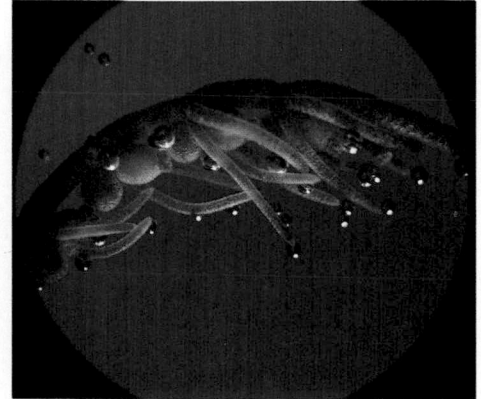

39-10 Stoneworts, **Chara**

39-11 Sargassum, *a brown alga*

ules of floridean starch, a branching polyglucan. Their cell walls contain cellulose and mucilaginous polysaccharides that are commercial sources of agar, carrageenan, and other gels used as stabilizers and moisturizing additives for puddings, ice creams, and cosmetics.

The characteristic color of the red algae is due to the pigment **phycoerythrin,** which occurs in large amounts in chloroplasts along with some phycocyanin, carotenoids, and chlorophyll *a*. Red algae are not always red. The relative amounts of their photosynthetic pigments depend on the light conditions under which they are growing. Red algae that grow in the lower part of the intertidal zone or wholly submerged in the ocean commonly have abundant phytocoerythrin and little or no phycocyanin—hence, they are red. It is because the phycoerythrin can capture blue-green light that they can penetrate deeper waters. But red algae from the upper intertidal zone and from fresh water commonly have large amounts of phycocyanin (and chlorophyll) and much less phycoerythrin. Hence, they do not have a red color.

Phycoerythrin and phycocyanin are the phycobilin pigments we mentioned in connection with the blue-green algae (cyanobacteria). Like the carotenoids, they absorb light energy and transmit some of it into the photosynthetic process. Phycoerythrin absorbs reasonably well throughout the green part of the spectrum, with an absorption peak near the green-yellow boundary. Only blue-green light penetrates water very well. Variation in pigmentation as a result of variation in the color of incident light is called **chromatic adaptation.**

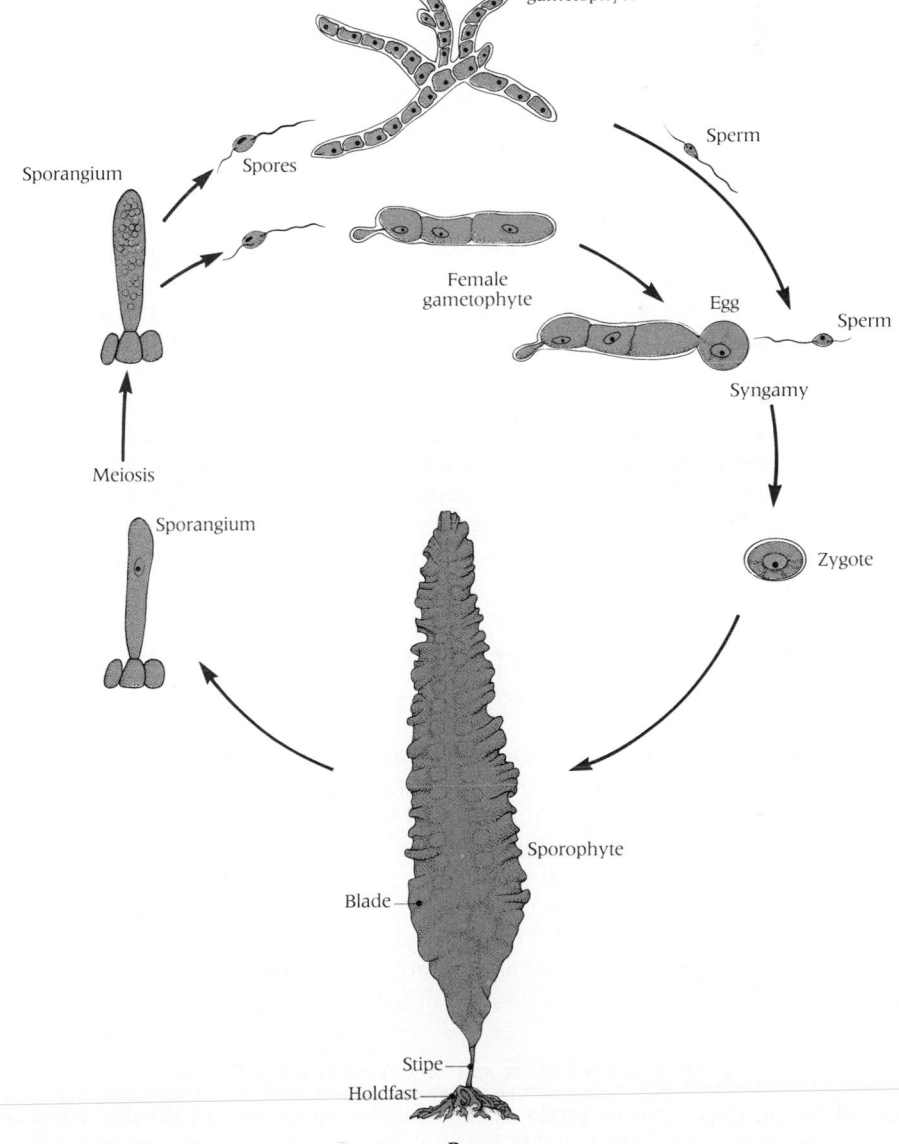

39-12 Laminaria, *a brown alga*
(A) Living kelp, Laminaria, *at low tide.*
(B) The reproductive cycle is characterized by a conspicuous sporophyte and much smaller gametophyte. This example of alternation of generations is more complex than that of other brown algae. Compare with Figure 21-11.

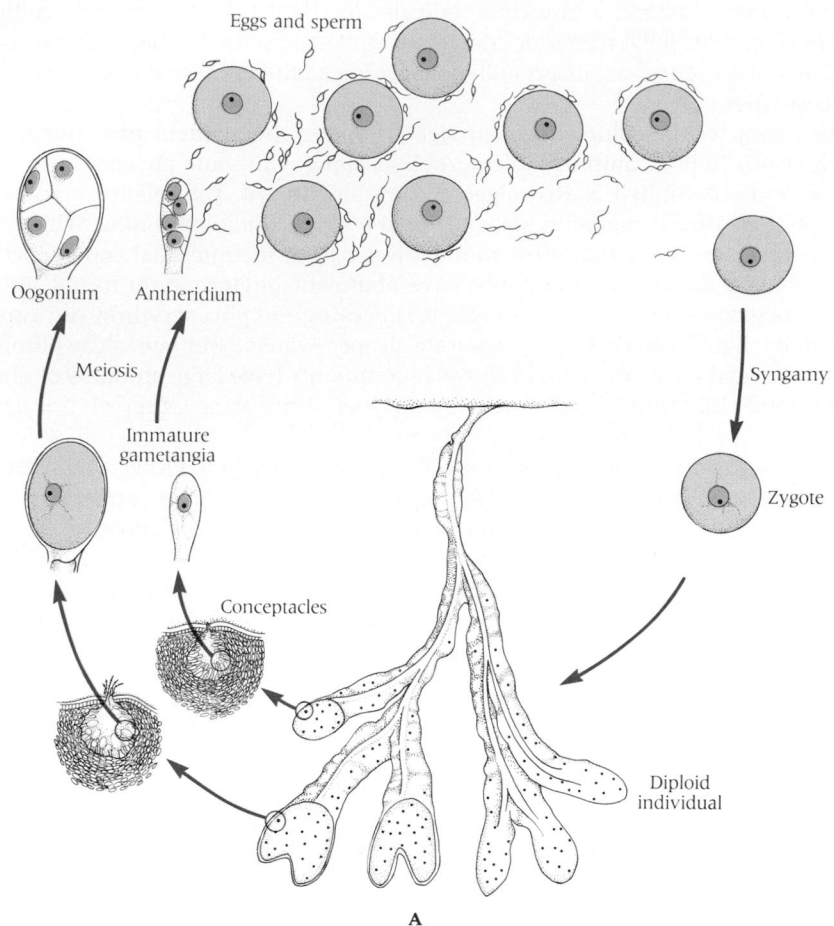

A

B

39-13 Fucus, *a brown alga*
(A) In the reproductive cycle, the gametophyte stage is absent, and the zygote develops directly into the diploid individual. Compare with Figure 21-12. (B) Living rockweed, Fucus, off the Maine coast.

The red algae are so different from other algae that they stand well apart taxonomically. They share certain features with blue-green algae—for example, phycobilins are important accessory photosynthetic pigments, and there is a complete absence of flagellated cells. Nevertheless, the red algae are definitely eukaryotic. In contrast, the organization of the blue-green algae is prokaryotic. If the eukaryotic condition evolved only once, as most authorities believe, then the red algae must be more closely related to the other eukaryotic algae than to the blue-green algae.

Some red (and green) algae have important roles in the formation of tropical reefs and sand beaches. Although this activity by algae of various kinds long antedated reef building by corals, red algae share with corals the biochemical machinery for depositing precipitates of calcium carbonate in and around their cell walls. After the death of red algae, the calcium carbonate persists, sometimes forming substantial rocky masses.

SUBKINGDOM PROTOZOA: HETEROTROPHIC PROTISTS

The status of *Euglena*, halfway between plants and animals, calls attention to an issue of some interest. What in fact is the difference between plants and animals? In T. H. Huxley's classic definition, animals are those forms of life that require preformed organic molecules in their nutrition and that possess such characteristics as locomotion, flexible body walls, and some sort of integrated "nervous" control. Photosynthetic plants, in contrast, depend very little on externally supplied organic substances; they can synthesize all or nearly all of their constituents from inorganic molecules. This implies that a hallmark of animality is the ingestion of particulate food. Where do the heterotrophic protists fit in?

THE QUESTION OF CLASSIFICATION

Several of the major protist phyla do seem unambiguously to resemble what we know as animals. In others, notably the slime molds, the resemblance is less clear.

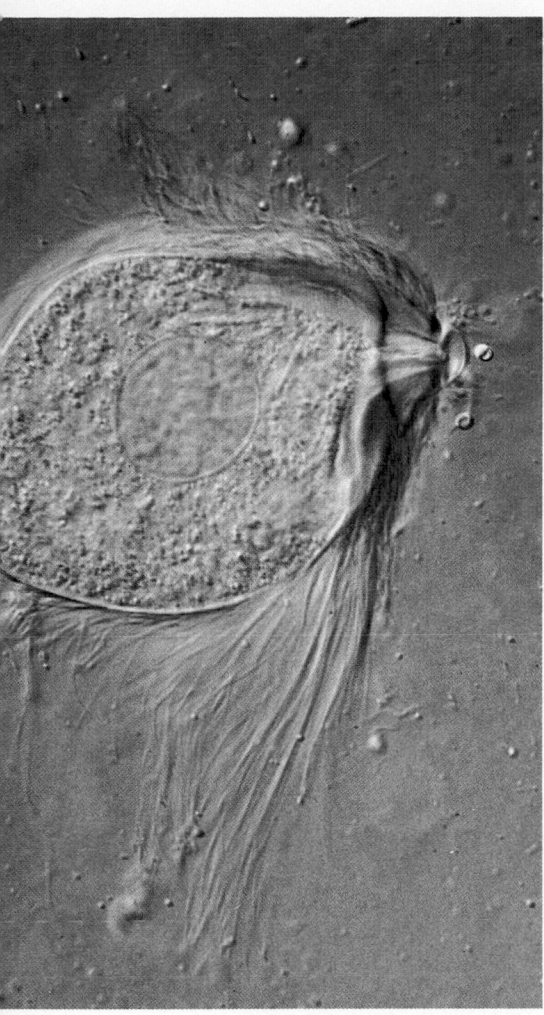

**39-14 The flagellated protozoan,
Trichonympha ampla (× 300)**

It is appropriate, therefore, to classify them as the subkingdom Protozoa, the animal-like protists. The **protozoans** include five major phyla (see Table 39-1).

- Zoomastigina (the zooflagellates)
- Sarcodina (the amoebas and their relatives)
- Sporozoa (parasitic forms such as *Plasmodium*)
- Ciliophora (the ciliates)
- Myxomycota (the slime molds)

Protozoa means "first animals." The term implies a belief that these creatures preserve those characteristics of the earliest organisms that, by any definition, were animallike. All "protozoa" now living have surely undergone profound evolutionary changes. It is unlikely that they resemble the first animals of early evolution.

The two most animallike features of the protozoans are the following: (1) they eat by ingesting chunks of food of diverse origin; and (2) most are quite active, moving about within their small worlds and scouring their surroundings for nourishment with a degree of purpose—or better, an orientation. They tend to move toward food or better living conditions and away from obstacles or poor living conditions. Thus protozoans exhibit the rudiments of animallike behavior. Plants, in contrast, have restricted movement. In addition the more complex of these protists have simple organs or organelles that are more animallike than plantlike. They also maintain an animallike food reserve—glycogen and lipid—rather than plantlike starches and oils.

All protozoa are small, some as little as 2 μm in diameter—as small as many bacteria. A few, the largest of all protists, may reach diameters of 3 or 4 cm. Most are between 100 to 300 μm, and can be easily observed with the light microscope. They have visible nuclei and divide by mitosis like the cells of plants and animals. They contain microtubules. Their genetic apparatus is fully established. However, there is an element in their heredity not present, at least not in the same degree, in multicellular organisms. The young protozoan generally begins life with a full set of parental chromosomes and with a body structure and substance directly derived from its parent. In most groups of these protists an exchange of nuclear material occurs from time to time between two individuals. This does not lead forthwith to reproduction, but over a sequence of generations it has the same genetic consequences as sexual reproduction in multicellular organisms.

PHYLUM ZOOMASTIGINA: ZOOFLAGELLATES

The **zooflagellates,** which are animallike protists in contrast to the plantlike phytoflagellates (euglenoids, dinoflagellates, etc.), are usually what is meant by the unqualified term *flagellates*. These primitive unicellular forms bear at least one flagellum. Some, like *Trichonympha,* bear hundreds (Figure 39-14). All zooflagellates are heterotrophic and lack chloroplasts. That is why we say that they are animallike. They are sexual or asexual, and are free living or, commonly, parasitic in the bodies of animals or plants.

The phylum Zoomastigina (also called Mastigophora and Zooflagellata) includes a number of unusually interesting forms. Some, like *Trichonympha,* live in the gut of wood-eating termites and roaches, where they assist their hosts in the digestion of the cellulose in the wood of the insect's diet. Without the parasite and the cellulose-digesting enzymes it contributes, the host could not subsist on a wood diet. We return to this interesting case in Chapter 47, where we consider the nature of parasitism.

One flagellate genus, *Trypanosoma,* is especially important in medicine and local agriculture, causing elephantiasis (American trypanosomiasis), sleeping sickness (African trypanosomiasis), and in South and Central American Chagas' disease in humans and domestic animals. Nagana, a widespread insect-borne trypanosomal disease of cattle, is so severe it has made farming virtually impossible in large areas of sub-Saharan Africa. Trypanosomes living in the blood of an infected host are transmitted to new hosts by the tsetse fly (*Glossina*), a blood-sucking insect. A modern research laboratory in Nairobi, Kenya, seeks to control trypanosomiasis and ameliorate its massive economic burden. So far progress has been meager, in part because frequent genic rearrangements in trypanosomes alter their surface proteins often enough to thwart the immune reactions necessary for a reliable vaccine.

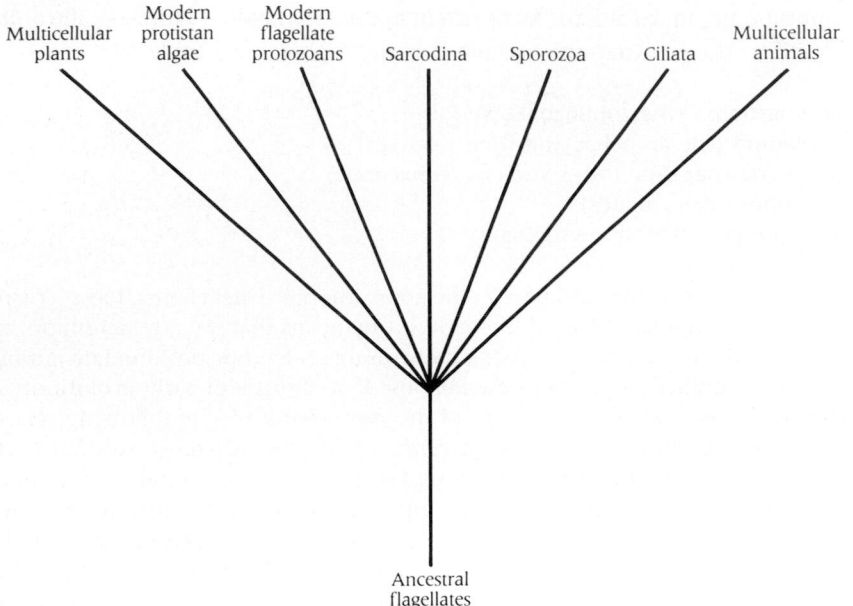

Multicellular plants
Modern protistan algae
Modern flagellate protozoans
Sarcodina
Sporozoa
Ciliata
Multicellular animals

Ancestral flagellates

39-15 Eukaryotic adaptive radiation is believed to stem from ancestral flagellates *Many consider that modern protists and multicellular plants and animals evolved from ancestral flagellates, beginning in Precambrian times nearly 1.5 billion years ago. These events may have occurred as the Earth's atmosphere became richer in oxygen.*

The zooflagellates are believed to be an important transitional form in the evolution of protistan (unicellular) to metazoan (multicellular) organisms. Although no living flagellate exactly resembles the ancient protists that were ancestral to multicellular organisms, today's flagellates are probably much more like their remote ancestors than are other present-day protists. One theory holds that zooflagellates arose when some euglenoids lost their chloroplasts. These in turn are believed to have given rise both to the other animallike protists (Sarcodina, Ciliophora, Sporozoa, and Myxomycota) and to all multicellular organisms (Figure 39-15). The present consensus is that multicellular plants and sponges, at least, were derived from colonies of protists somewhat like zooflagellates.

If all life on Earth except the zooflagellates were destroyed, they would provide a promising starting place for a renewal of evolutionary diversification of both the plants and animals. Like the prokaryotes, these organisms present a wide range of successful adaptive types.

PHYLUM SARCODINA: AMOEBAS AND THEIR KIN

The **sarcodines** include amoebas and amoeboid protozoa that lack a coat or wall outside their plasma membranes, although some species do have shells. They take their name from *sarcode*, an old word coined to describe the "simple, glutinous, and homogeneous jelly" of which simple cells were once thought to be composed. Tens of thousands of sarcodine species, living and fossil, have been described. It is

39-16 Amoeba
(A) *Appearance in phase microscope.*
(B) *Appearance in light microscope.* ($\times$ 400)

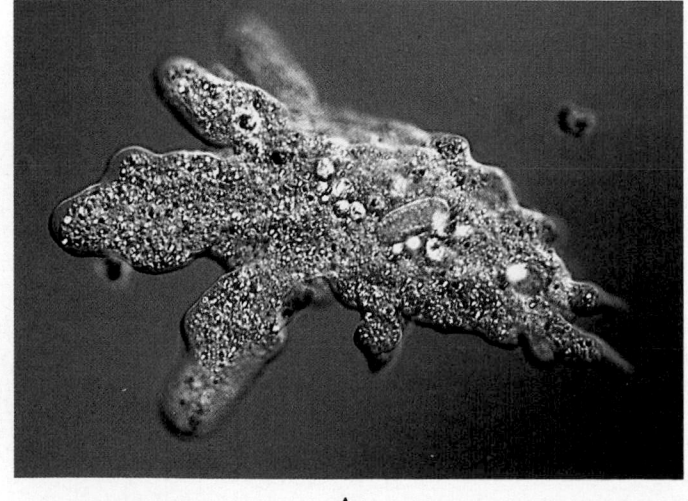

A

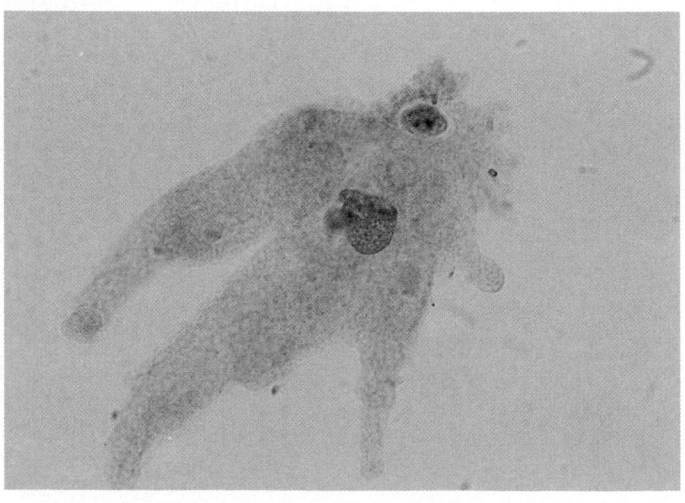

B

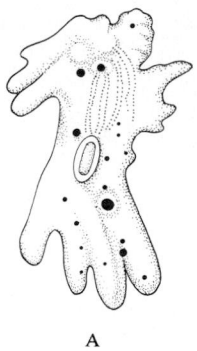
A

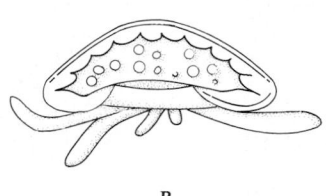
B

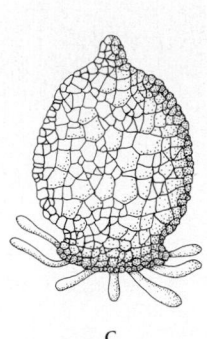
C

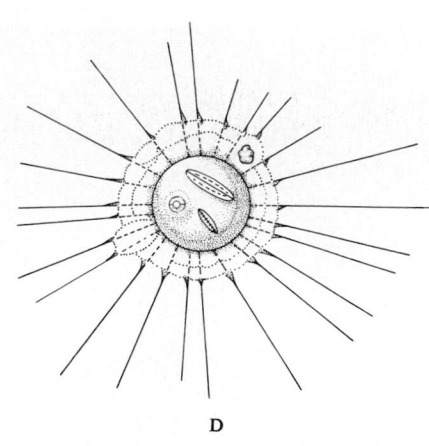
D

39-17 Amoebas and their kin
(A) Amoeba proteus *lacks a shell and moves by pseudopodia.* **(B)** Arcella discoides, *one of the sarcodines possessing a shell of chitinous material, through which pseudopods extend.*
(C) Difflugia ureolata *has a heavy outer chitinous shell that is supplemented with foreign particles.* **(D)** Actinosphaerium eichhorni *has pseudopods that are filamentous in shape.*

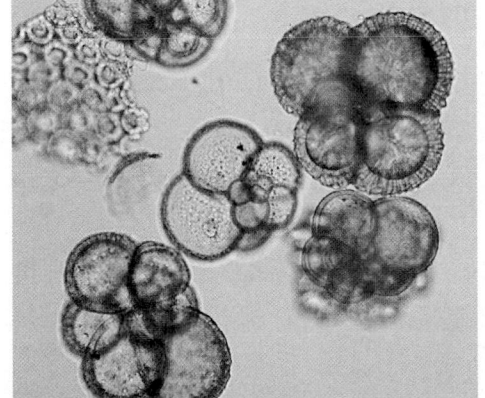

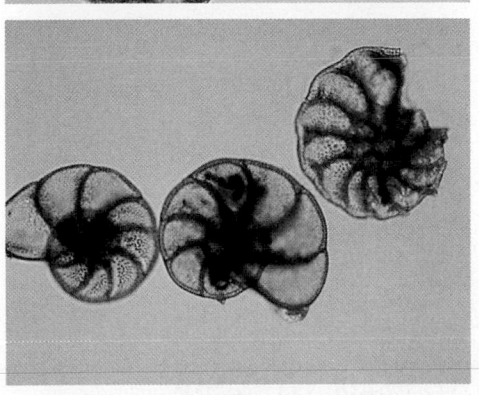

39-18 Foraminifera (× 500)

a curious historical fact that amoebas escaped the watchful eye of Leeuwenhoek. They were first observed by Rösel von Rosenhof in 1755.

An amoeba looks as simple as a fully developed organism can look and it is famous on that account. However, despite their uncomplicated appearance, amoebas are complex cells that are capable of complex behavior patterns, as when they sense and pursue prey.

Through the light microscope amoebas look like lumps of protoplasm, without front or back, and with little evidence of organelles or other specialization, except for the nucleus (Figure 39-16). Some species have within the granular cytoplasm small, clear bubblelike contractile vacuoles that pulse and expel water through the cell membrane, thereby controlling osmotic pressure. Freshwater amoebas take up more water than they lose through the cell membrane by osmosis alone. By forcibly expelling water, they prevent bursting. Marine protists, which do not take up excess water by osmosis, do not have this device in most cases.

We saw in Chapter 7 that an amoeba moves by a streaming motion of the cytoplasm, which pushes out an irregular bulge in the direction of movement and pulls in the bulges at the opposite end. The bulges are called **pseudopods** ("false feet"). When food is encountered, the pseudopodia surround and engulf it. The food is then taken into the body in a sac surrounded by a membrane—much like phagocytosis of particles by white cells. Digestive enzymes are secreted into the cavity thus formed, products are absorbed through the membrane, and undigested remnants are expelled from the amoeba's body.

Amoebas lack shells. However, many other sarcodines have shells and these forms are widespread (Figure 39-17). The most varied and most interesting perhaps are the **foraminifera,** known familiarly as **forams** (Figure 39-18), and Radiolaria, termed **radiolarians** (Figure 39-19), most of which secrete hard shells of calcium carbonate or silicates. When magnified, the shells are of great beauty and amazing complexity.

Forams are so abundant in the seas that much of the bottom ooze is made up of their shells. The white cliffs of Dover and similar chalky deposits throughout the world are the result of the long accumulation of these discarded shells. Forams are also common as fossils. These provide geologists with clues to the ages of rocks. When, for example, a well is bored in rocks originally laid down in the sea, forams are often found in abundance. For these reasons, they are carefully studied by petroleum engineers and by what are called micropaleontologists, which are not small paleontologists but paleontologists who study small fossils. Forams even contributed to the petroleum itself, which was probably derived from their soft parts.

The bottom ooze in deeper parts of the sea (>15,000 feet) is due not to the calcareous shells of forams, which dissolve at this depth, but to the siliceous shells of radiolarians. These sarcodines, surely the most elegant of all microorganisms (see Figure 39-19), elaborate skeletal designs of endless variety.

Sarcodine reproduction may be asexual or sexual. Asexual reproduction takes place by cell division accompanied by mitosis in which the nuclear envelope usually does not break down. In sexual reproduction, the diploid cells undergo meiosis, forming gametes that fuse to form zygotes. Some sarcodines, such as those that cause amoebic dysentery, are parasites.

The evolution of the sarcodines from flagellates elegantly illustrates two great evolutionary principles—overlapping functions and opportunism. The latter is illustrated in Figure 39-20, which shows that flagellates evolved into sarcodines and into ciliates and suctorians (described below)—and the other forms noted in Figure 39-15. Pseudopods evidently arose first as useful supplementary feeding devices in animal flagellates. Some living flagellates represent transitional stages that must have occurred in the evolution of pseudopods. *Oikomonas* is a zooflagellate in which pseudopods on a localized area of an otherwise firm cell surface engulf solid food particles. *Mastigamoeba* retains a flagellum but has pseudopods all over its surface. It is one of the protists without mitochondria (see footnote 1). The evolution in flagellates of a mobile cell surface useful primarily for feeding purposes created an opportunity for further evolution of a new mode of locomotion. Exploitation of this opportunity gave rise to the sarcodines, which are characterized by amoeboid or pseudopodial movement. Interestingly, many sarcodines (forams, for example) produce flagellated gametes, reminders of an ancestral mode of movement.

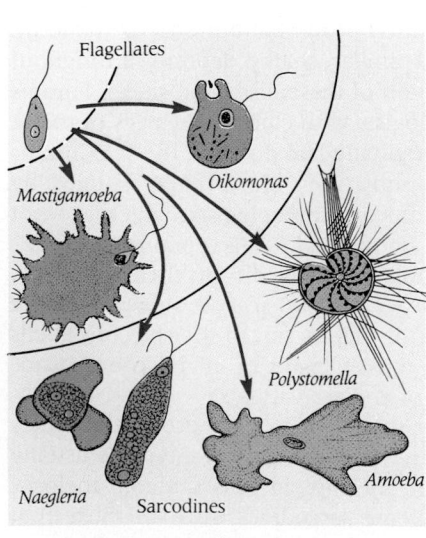

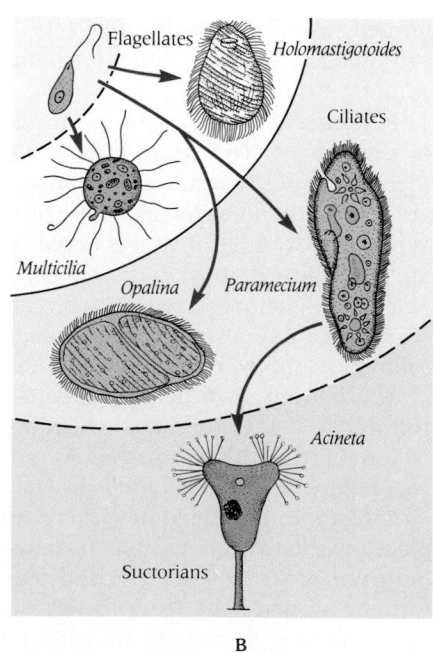

39-20 Descendants of flagellates
(A) *The relation of flagellate ancestors to sarcodines.* **(B)** *The relation of flagellates to ciliates, and to their descendants, the suctorians.*

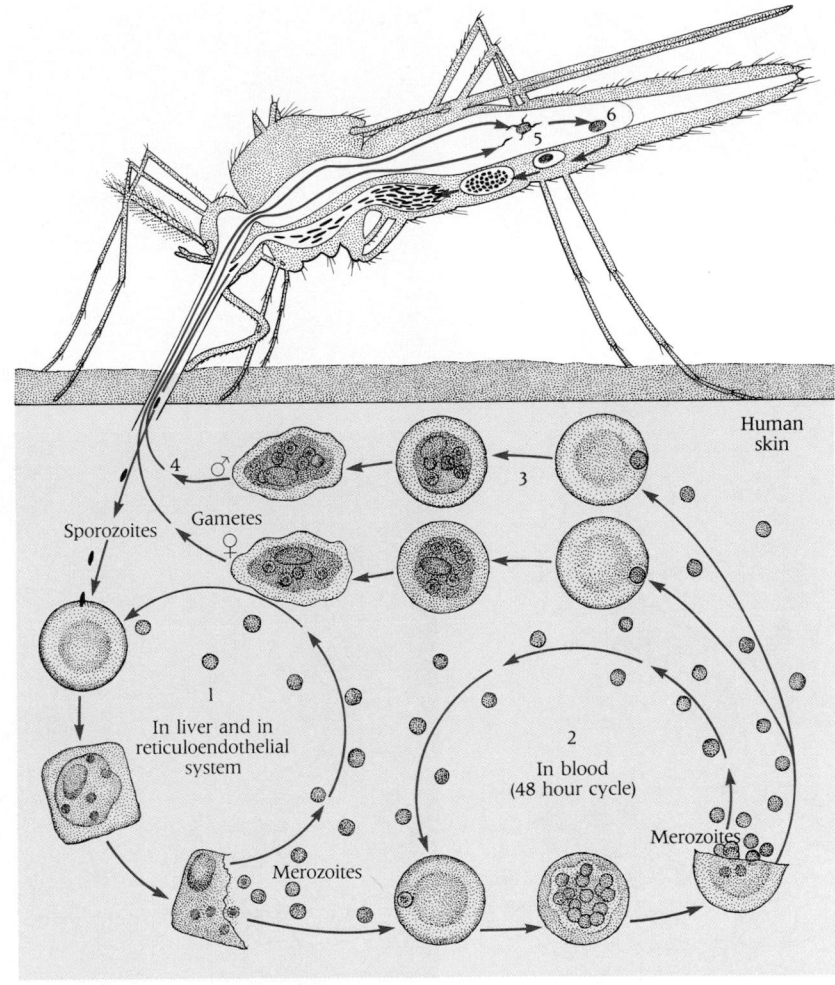

39-21 Life cycle of Plasmodium
The parasite that causes malaria has a life cycle that spans two hosts, humans and mosquitoes. Sporozoites enter the human bloodstream with the saliva of a biting Anopheles mosquito. They develop into merozoites in the liver and reticuloendothelial system (1). Released merozoites enter red cells where they reproduce asexually at regular intervals. Those that develop into gametes (3) return to another mosquito when the infected person is bitten (4). The gametes differentiate in the insect's stomach and fuse sexually to create zygotes (5). These form cysts in the gut lining (6). Sporozoites within the cysts proliferate. When the cells burst, they are released and enter the salivary glands, where the cycle begins anew.

Human skin

4 ♂

Sporozoites   Gametes ♀   3

1
In liver and in reticuloendothelial system

2
In blood (48 hour cycle)

Merozoites

Merozoites

PHYLUM SPOROZOA: THE PARASITIC PLASMODIUMS

The phylum Sporozoa illustrates two prominent characteristics of the animallike protists: (1) parasitism, and (2) complicated reproductive cycles that include both sexual and asexual phases. **Sporozoans** are characterized by their lack of flagella in adult forms and in most cases by an amoeboid (amoebalike) body form. The best known sporozoans are members of the genus *Plasmodium*, which cause malaria.[3]

39-22 Paramecium
The organism possesses cilia on its body surface and has a macronucleus and a micronucleus.
(× 850)

[3] Note that the word "plasmodium" is used in two ways: (1) when capitalized, it is the genus name of the sporozoans that cause malaria; and (2) uncapitalized, it is the noncellular, multinucleate, jellylike, amoeboid, assimilative stage in the life cycle of the slime molds (phylum Myxomycota).

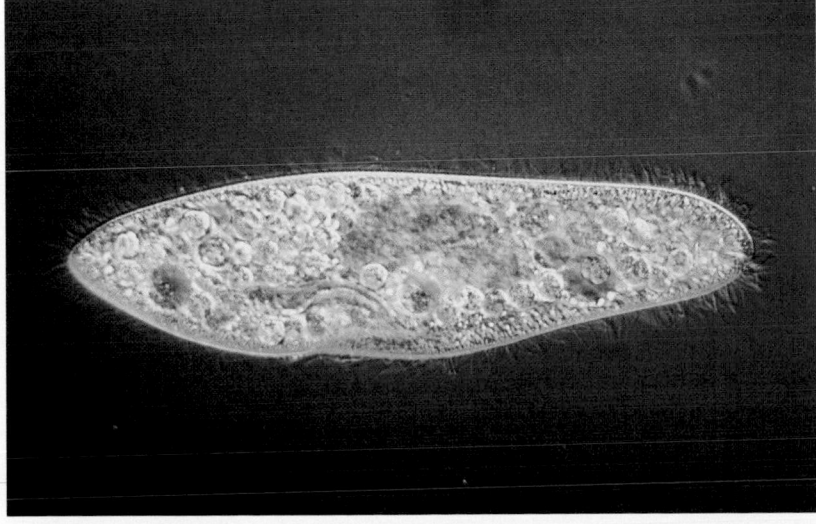

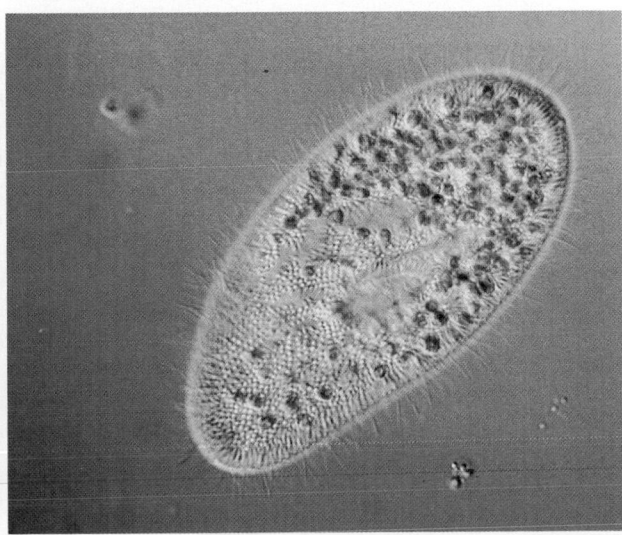

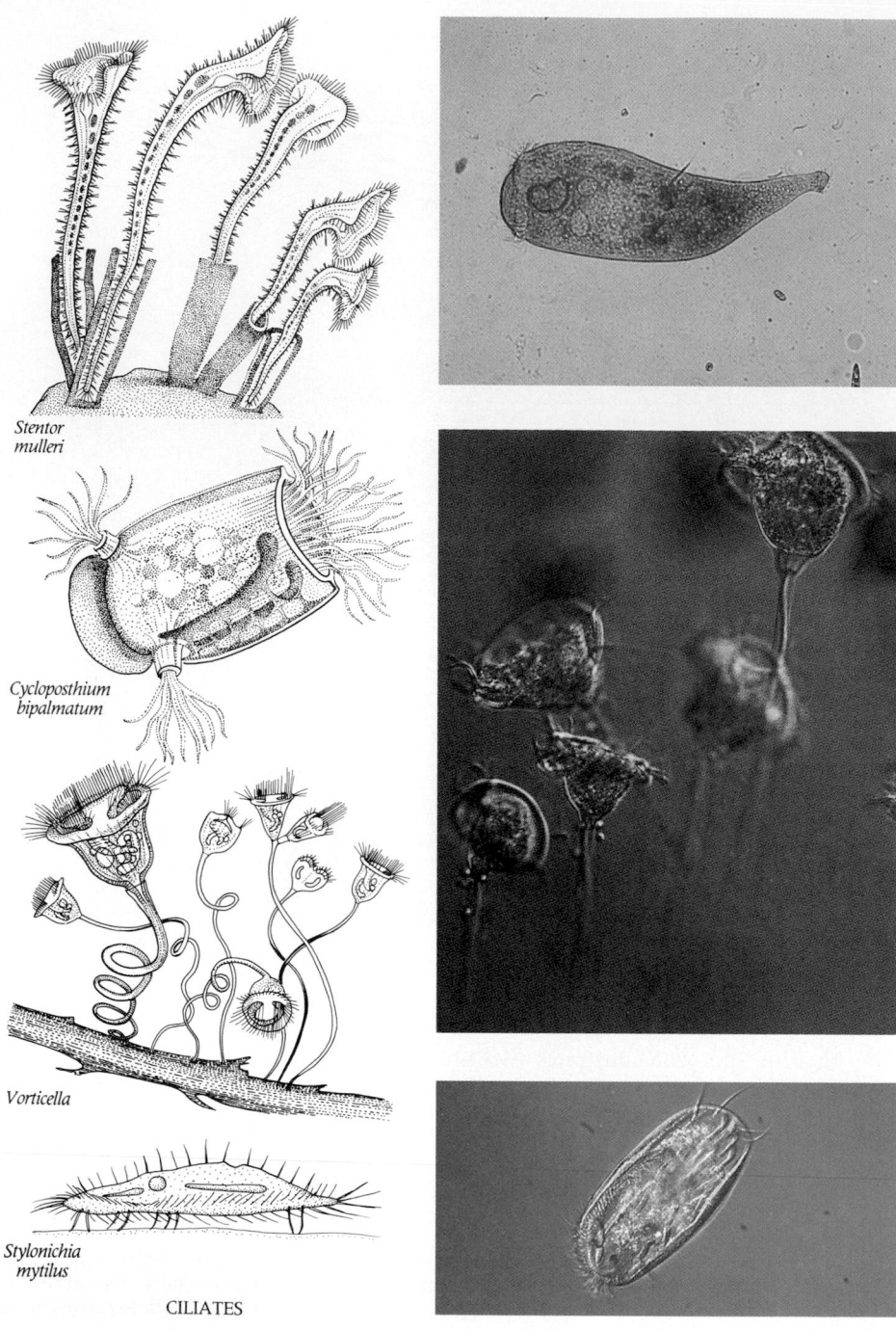

Stentor
mulleri

Cycloposthium
bipalmatum

Vorticella

Stylonichia
mytilus

CILIATES

Plasmodium vivax and *Plasmodium falciparum* are two well-known examples. Each causes a distinctive form of malaria, perhaps the most common of all human diseases.

The combination of parasitism and a complex reproductive cycle is clearly evident in the sporozoans that cause malaria. While living as parasites in *Anopheles* mosquitoes, the malarial sporozoans go through a sexual phase (Figure 39-21). Multiple fissions ensue, yielding **sporozoites** that migrate to the mosquitoes' salivary glands, from which they can be introduced into the human bloodstream during a mosquito bite. In the human body, sporozoites divide to form **merozoites,** which invade human red blood cells. Within the red cells, merozoites reproduce asexually at regular intervals—with a periodicity that causes the characteristic episodic fevers of malaria. Some merozoites develop into gametocytes. The reproductive cycle of *Plasmodium vivax* is completed when a second biting mosquito acquires individual gametocytes from the blood of an infected human. Fertilization occurs in the mosquito's gut. The resulting zygote produces sporozoites that migrate to the mosquito's salivary glands. We will speak again of this interesting cycle in connection with the general topic of parasitism.

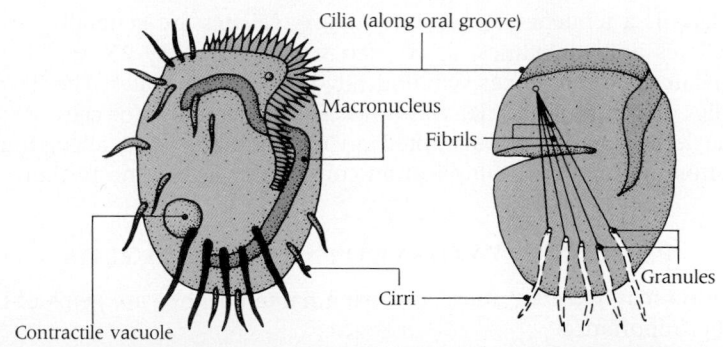

Cilia (along oral groove)

Macronucleus

Fibrils

Cirri

Contractile vacuole

Granules

39-24 Ciliate locomotion using cirri
This ciliate, Euplotes, *has a band of cilia around the oral groove, and locomotion is achieved via fused cilia called cirri. Coordination of cirri is controlled by a nervelike network of neurofibrils located inside the cell (right). A single neurofibril runs to each of the cirri. Cutting the neurofibrils (as shown) impairs their coordination.*

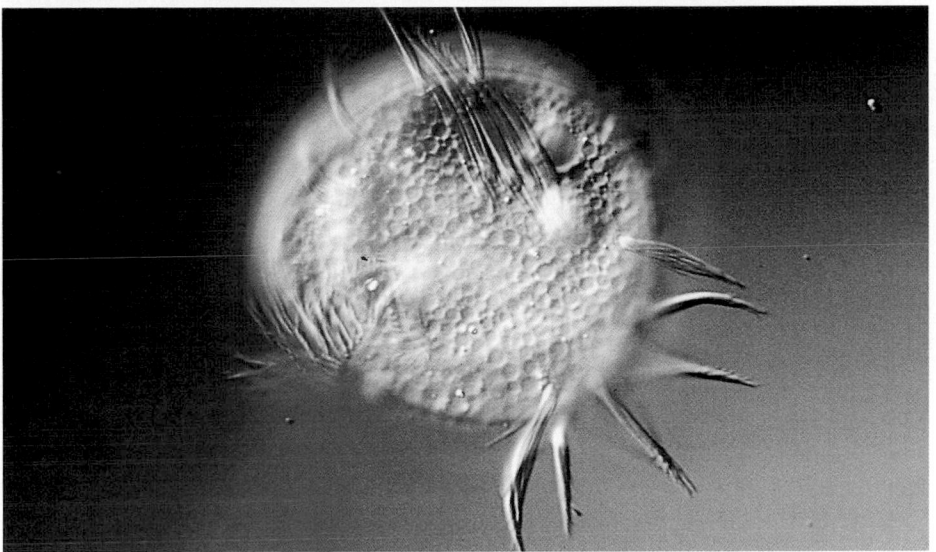

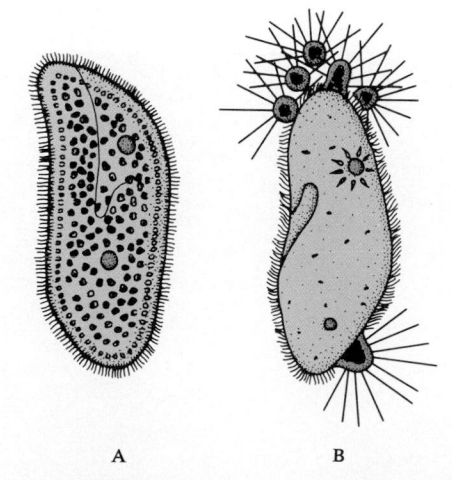

A   B

39-25 Relatively large ciliates often harbor small parasites
(A) *In* Paramecium bursaria, *symbiotic green flagellates live within the cytoplasm.* **(B)** *Smaller parasitic sarcodines on a paramecium.*

PHYLUM CILIOPHORA: THE CILIATES

The **ciliates** have numerous tiny hairlike projections, or **cilia** (singular, cilium), which beat rhythmically and drive these protists through the water in which they live. They are the largest and most complex of the protists. Ciliates match the flagellates in diversity and ecological importance.

The paramecium, one of the best known and abundant of the ciliates, occurs almost everywhere in fresh water. It is about 100 to 300 μm in length. Parameciums undergo conjugation, in which two cells of opposite mating type exchange genetic material. They can also self-fertilize. These processes are dependent on the presence of the second major characteristic of ciliates (the first is cilia). They possess two types of nuclei: a large **macronucleus** (which is highly polyploid) and, within the same cell, from one to as many as 80 small diploid **micronuclei** (Figure 39-22). The micronuclei are the carriers of the genetic information. The macronucleus arises from the micronuclei. Also called the vegetative nucleus, it is not essential in reproduction but it does control metabolism. Thus it plays a role in the expression of genetic information in the phenotype.

The ciliates represent some sort of pinnacle in the evolution of protists (Figure 39-23). Although unicellular, ciliates have well-developed structures and organelles that are entirely analogous to the organ systems of multicellular animals. There are locomotor systems (the cilia) that operate with elegant coordination. There are "muscular systems" in the contractile fibers of ciliates. The coordinating and conductive systems of ciliates are in every way incipient "nervous systems" that permit some ciliates to walk on surfaces using fused cilia called **cirri** (Figure 39-24). Ciliates have an alimentary canal with a "mouth," "gullet," and "anus." Stiffening plates resemble a "skeleton." As shown in Figure 39-20, ciliates gave rise to **suctorians,** a group of ciliates (class Kinetofragminophora) that illustrates the extent of complication possible at the protistan level.

Ciliates, like sarcodines, probably arose from primitive zooflagellates. Indeed, zooflagellates that are still living are more or less intermediate between the two phyla (see Figure 39-20). The ciliates illustrate several interesting evolutionary phenomena. One is a tendency to increase in size. Ciliates are generally much larger than flagellates and sarcodines, as we can see in Figure 39-25, which illustrates small flagellates and sarcodines symbiotically parasitizing ciliates. The development of organelles and multiple nuclei was a necessary adaptation for cells of such bulk. But in the long run, these constituted an evolutionary blind alley. Intracellular differentiation and nuclear multiplication could go so far and no further.

PHYLUM MYXOMYCOTA: SLIME MOLDS

Slime molds may have an unprepossessing name but they are gems of biological interest and importance.

As we have seen, systematists often disagree on which fundamental characters to make the basis of classification. Most protozoans have easily recognizable classification criteria: they are phytoflagellates (like *Euglena*) or ciliates (like *Paramecium*) or amoebas, or whatever. Some workers regard the slime molds as fungi, which we regard as a separate kingdom. Another school puts true slime molds in the protist phylum Myxomycota and cellular slime molds in the phylum Acrasiomycota. To others, the slime molds are a phylum (Myxomycophyta) of the plant kingdom— and to others still (like us) all of the slime molds may be fairly considered a phylum of the kingdom Protista. We chose to recognize some of its animallike characters.

Wherever they are placed, slime molds include several varieties. There are so-called **true slime molds** (such as *Physarum*), which are found in moist soil or decaying leaves or logs in a damp forest—glistening masses of viscous white (but sometimes red or yellow) slime. At the gamete stage in their life history, true slime molds are flagellated unicellular organisms that might well be classified as animallike flagellate protists (Figure 39-26). However, those that survive later lose their flagella and become thoroughly amoebalike. In both, they reproduce extensively by simple fission, with mitosis of the nucleus, which in these stages is haploid. Eventually many of the amoeboid individuals aggregate into large clumps. Separate cell membranes break down and the result is a single mass of protoplasm with hundreds or even thousands of nuclei. This vegetative mass, a multinucleate **plasmodium** (see footnote 3, page 1059), moves about and feeds like a gigantic amoeba that may reach a diameter of 25 cm or more (Figure 39-27).

So far the simple slime mold's life cycle resembles that of an animallike protist— solitary at first and later colonial in a peculiar way. But the colonial mass can undergo an extraordinary change that makes it plantlike in some ways. From the basal mass, stalks grow upward and bulblike expansions called **fruiting bodies** develop at their ends. These are sporangia. The basal cells (or nuclei), those in the stalk and those coating the bulb, do not further reproduce. They die, while the germ cells continue the line. In the slime molds, the inner nuclei of the bulb continue the line. They fuse two by two in a sexual process that produces a diploid zygote, which divides by mitosis to form a multinucleate mass called a **plasmodium.** Fruiting bodies form, meiosis then occurs, and haploid spores develop the plantlike walls of which contain cellulose. The spores are scattered, and the fortunate ones that germinate may each produce individual flagellated gametes, which (as the cycle continues) fuse to form zygotes that lose their flagella and become amoeboid.

Then there are the so-called **cellular slime molds** (such as *Dictyostelium discoideum*), the life cycles of which are quite different (Figure 39-28). Spores become, not flagellated gametes, but amoeboid cells with haploid nuclei that live free in the soil. As they feed, they divide mitotically. When the food supply dwindles, growth ceases and an active phase of differentiation begins. All the cells in the area aggregate to form a **slug,** also called a **pseudoplasmodium** (false plasmodium), in which the individual haploid cells retain their identity and do not fuse— unlike the events just described in the diploid multinucleate true plasmodium of the true slime molds. The slug can move about, but soon it settles down and one-third of its body forms an apex that becomes a stalked fruiting body whose coat is made up of an unusual polysaccharide. Note that all stages of the cycle are haploid, though not all agree. There are no sexual stages.

Cellular slime molds like *Dictyostelium discoideum* are easily grown in the laboratory and are fascinating objects of study, presenting many opportunities for the analysis of basic mechanisms. How, for example, does a dwindling nutrient supply initiate morphogenesis of the stalk and fruiting body? How are the individual amoebae recruited into a large aggregate? One theory suggests that they are attracted by

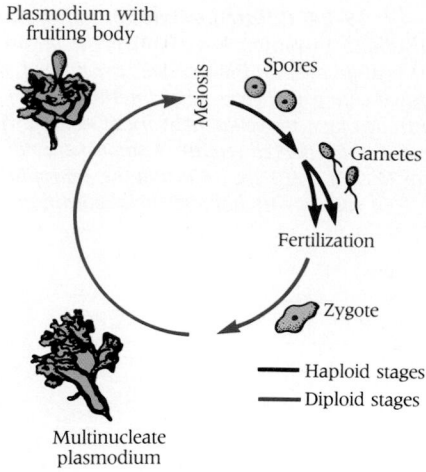

39-26 Reproductive cycle of a true slime mold
Haploid spores give rise to gametes that fuse to form a zygote. The diploid zygote forms a multinucleate plasmodium in which individual cells are indistinguishable. A fruiting body develops within which meiosis occurs, forming haploid spores. Note the regular alternation of haploid and diploid stages.

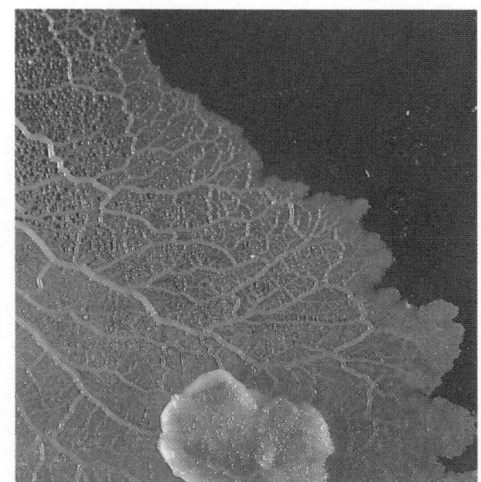

39-27 Plasmodium stage of true slime mold reproductive cycle
Physarum polycephalum on an oat flake.

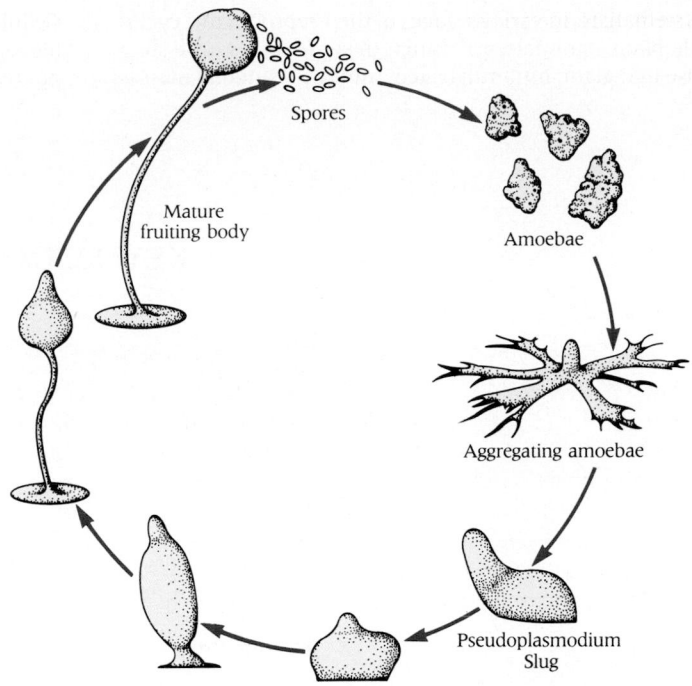

39-28 Reproductive cycle of the cellular slime mold Dictyostelium discoideum
When food is unavailable, individual amoebae aggregate into a slug (pseudoplasmodium), which moves to a new environment and elaborates a fruiting body, from which spores are released to begin the next generation.

Spores

Mature fruiting body

Amoebae

Aggregating amoebae

Pseudoplasmodium
Slug

cyclic AMP, the same "second messenger" nucleotide that participates in hormonal and nervous coordination in animals. Perhaps the best answer is that slime molds evolved from protists that somehow struck off on a peculiar line of specialization of their own, not like that followed by other protists. This can be viewed as another independent approach toward large size and more or less multicellular organization. It was successful enough to ensure survival of some forms, but it was another blind alley that led no further.

SUMMARY

Protists are unicellular eukaryotes. Hence, they contain the organelles typical of eukaryotic cells. Members of the kingdom Protista (flagellates) are believed to be the evolutionary ancestors of fungi, plants, and animals. There are two protist subkingdoms: Algae, which are plantlike autotrophs, and Protozoa, which are animallike heterotrophs.

Algae account for most of the Earth's photosynthesis. Thus, they are a major source of nutrient molecules for other organisms. Most algae are multicellular, though they lack differentiated tissues or organs. Whether aquatic or terrestrial, all algae need moist environments.

There are six major phyla. Phylum Euglenophyta includes flagellates. Because they are both motile and photosynthetic, they have both plantlike and animallike features. Some have photosensitive eyespots, which enable them to seek light.

Phylum Pyrrophyta (the dinoflagellates), along with the phylum Chrysophyta (diatoms), are the major constituents of plankton. Dinoflagellates have complex life cycles, which include a resting cyst stage that promotes rapid proliferation. These lead to the blooms responsible for toxic red tides. Diatoms have unique silica-impregnated skeletons that give "diatomaceous earth" many practical uses.

Chrysophyta include the golden algae. Chlorophyta, the green algae, range from unicellular organisms to differentiated colonies.

They also include the stoneworts, an unusual group with structures resembling the roots, stems, and leaves of plants. Brown algae, such as kelp, are found in phylum Phaeophyta. Red algae (phylum Rhodophyta) share certain features with the prokaryotic blue-green algae (cyanobacteria).

Members of subkingdom Protozoa (protozoans) are classified into five major phyla. Phylum Zoomastigina includes protists with flagella—the zooflagellates. Modern zooflagellates probably differ little from their remote ancestors, which are thought to have been the precursors of all other protists as well as multicellular animals and plants. Phylum Sarcodina consists of amoeboid protists, including *Amoeba*.

Phylum Sporozoa are the parasitic plasmodia, certain of which cause malaria. Sporozoans have complicated life cycles, in which different developmental stages occur in different host species. The malarial parasite, *Plasmodium*, has a sexual stage within the *Anopheles* mosquito and asexual stages in the mosquito-bitten human or animal host.

Ciliophora are the ciliates. Their members, which include *Paramecium*, are large cells, with well-differentiated structures that are analogous to the organs of animals.

Phylum Myxomycota, the true and cellular slime molds, present a difficult taxonomic problem and have been classified differently by various systematists. In various stages of their reproductive cycles, they resemble plants, animals, and fungi. In some stages, true slime mold cells fuse into giant, multinucleated, motile plasmodial masses.

Cellular slime molds, which form pseudoplasmodia, have complex life cycles including unicellular and multicellular stages that are of great current interest to students of developmental biology.

KEY TERMS

algae
antheridia
brown algae
chrysolaminarin
ciliate
cirri
colony
diatom
dinoflagellate
euglenoid flagellate
foraminifera
fruiting body

fucoxanthin
golden algae
green algae
macronucleus
merozoite
mesokaryote
metazoan
micronucleus
oogonia
phycoerythrin
plankton
plasmodium

Protista
Protozoa
pseudoplasmodium
radiolarian
red algae
sarcodine
slime mold
sporozoan
stigma
thallus
zooflagellate
zooxanthellae

QUIZ QUESTIONS

1. Which of the following is a material from which cell walls or shells of various protists are composed?
   A. glycogen
   B. alginic acid
   C. floridean starch
   D. leucosin
   E. paramylon

2. In which of the following taxa does photosynthesis *not* occur?
   A. Sporozoa
   B. Rhodophyta
   C. Pyrrophyta
   D. Euglenophyta
   E. B and C

3. Which of the following pigments do species of Chlorophyta share with higher plants?
   A. chlorophyll a
   B. chlorophyll b
   C. chlorophyll c
   D. A and B
   E. A and C

4. Which of the following protists do *not* have multicellular representatives?
   A. Chrysophyta
   B. Phaeophyta

   C. Rhodophyta
   D. Euglenophyta
   E. All of the above include multicellular representatives.

5. Which of the following major advances were achieved by members of phylum Protista?
   A. multicellularity
   B. autotrophy
   C. sexuality
   D. A and B
   E. A and C

ESSAY QUESTIONS

1. What evidence suggests that the eukaryotes evolved from the prokaryotes?

2. What properties define eukaryotic algae? What role do the algae play in the foodchain? What features distinguish the

huge multicellular algal forms (e.g. giant kelp) from vascular plants?

3. What microorganisms constitute plankton? Red tides? Diatomaceous earth?

4. Why are the protozoa considered "animallike"?

5. Why do slime molds present a particularly difficult problem of classification?

REFERENCES AND SUGGESTIONS FOR FURTHER READING

Bold, H. C., and Wynne, M. J. (1985). *Introduction to the Algae,* 2nd ed. Prentice-Hall, Englewood Cliffs, N.J.

A factual, concise, and well-written summary.

Fryxel, G. A. (Ed.) (1983). *Survival Strategies of the Algae.* Cambridge University Press, New York.

This book appeared just as we began to realize that the cyst stage in the life cycles of dinoflagellates is a reservoir for seeding blooms. What we know and don't know about this and other topics are well summarized.

Gall, J. G. (Ed.) (1986). *Molecular Biology of Ciliated Protozoa.* Academic Press, Orlando, Fla.

A stimulating general introduction to ciliate biology and genetics that highlights new understandings of their molecular properties.

Laybourn-Parry, J. (1984). *A Functional Biology of Free-Living Protozoa.* University of California Press, Berkeley.

A clearly written text on the protozoa.

Lee, J. J., Hutner, S. H., and Bovee, E. C. (Eds.) (1985). *An Illustrated Guide to the Protozoa.* Society of Protozoologist, Lawrence, KA.

A magnificently illustrated guidebook.

Noble, R. C. (1990). Death on the half-shell: the health hazards of eating shellfish. *Perspectives in Biology and Medicine 33,* 313–322.

After quoting Jonathan Swift ("He was a bold man that first eat an oyster."), the author tells a disturbing story of red tides, neurotoxins, and other contaminants in the shellfish we eat.

Perasso, R., and 3 other authors. (1989). Origin of the algae. *Nature 339,* 142–144.

New data based on nucleotide sequences of 28S ribosomal RNA helps to clarify algal evolution. Part of the interest lies in the power of the method.

Sleigh, M. (1978). *The Biology of Protozoa.* University Park Press, Baltimore.

A short general introduction that nicely covers classification and morphology.

Taylor, F. J. R. (Ed.) (1987). *The Biology of Dinoflagellates.* Blackwell Scientific, Boston.

A collection of authoritative papers on all aspects of dinoflagellates, made timely by rising worldwide concern with the hazards of red tides.

Vidal, G. (1984). The oldest eukaryotic cells. *Scientific American 250,* February, 48–58.

A survey of the oldest microfossils of eukaryotic cells.

Fungi
Fungi, a large kingdom of eukaryotes that are incapable of photosynthesis, are ubiquitous on Earth. Many are spread by spores like these from an ascomycete. The curious sculptured surfaces of this spore species is clearly revealed by scanning electron micrography. (× 7000)

To most people the word *fungus* evokes a vaguely unpleasant image: globs of primitive living matter that sometimes appear on spoiled foods, on basement walls, and in other damp places. Their true character is, in fact, rather different, for they are exquisite constructions of nature, with life cycles that are among the most complex to be found in any kingdom. Most of us have encountered fungi as curious denizens of our gardens, forests, and food cupboards (Figure 40-1). These include the mushrooms, edible and poisonous, of field and forest, the puffballs that give forth clouds of spores when kicked by hikers, and the bracket fungi on decaying logs and trees.

For a long time, fungi were considered to be plants (see Table 37-1). But they were unusual organisms to be so classified. The great majority of plants contain chlorophyll and are autotrophs that perform photosynthesis. However, all fungi lack chlorophyll, and they are indisputable heterotrophs. Together with heterotrophic bacteria and a few other heterotrophic forms, fungi are the great decomposers of the living world. Their major role in the ecosystems of Earth is to degrade dead organic matter.

GENERAL CHARACTERISTICS

Fungus (plural, fungi) comes from the Latin word originally applied to mushrooms. It has gradually been extended to cover molds, yeasts, and other forms, the total number of known species of which now exceeds 100,000.

MORPHOLOGY

Fungi can exist as single-celled organisms. The best known examples of these are the yeasts (Figure 40-2). The vast majority of fungi, however, are multicellular. Their basic structural unit is a threadlike filament called a **hypha** (plural, hyphae) (Figure 40-3). The elongated cells that comprise the hyphae contain one, two, or many nuclei. One cell in a hypha is separated from another by a perforated cross wall or **septum.**

The mass of hyphae that constitutes a fungal growth is referred to collectively as a **mycelium.** The high surface to volume ratio of a mycelium suits it well to the task of food absorption. A mycelium sometimes become organized into elaborate fruiting structures, such as puffballs or mushrooms. Their walls contain a variety of polysaccharides (sometimes including cellulose). But they also usually contain **chitin** (Figure 40-4), the same polymer that toughens the exoskeletons of insects and other arthropods. Note in the figure that chitin is a polymer of nitrogenous glucoselike units called *N*-acetylglucosamine. The presence of the *N*-acetylamino groups distinguishes this polymer from cellulose.

Variation in cell structure sufficient for purposes of classification does occur at the ultrastructural level among the fungi, but historically the phyla have been defined mainly by their method of reproduction and the characteristic structures associated

A **B** **C**

with it. Other criteria, such as the presence or absence of cross walls in the hyphae, are of limited use.

NUTRITION

All fungi are heterotrophs. Unlike plants, they cannot produce their own food by photosynthesis, nor can they eat like animals. Many fungi require only a simple sugar as the major organic food source. They then use the sugar as the starting point for all the other necessary syntheses—or they store it as glycogen, a capability that also sets them apart from plants. Some require certain more complex organic foods. In any case, all fungi must absorb needed sugars or other preformed organic molecules through their cell walls and cell membranes.

Fungi obtain their food either as **saprobes** (organisms that live on the dead remains of other organisms), **symbionts** (organisms that live in a mutually beneficial association with other living organisms), or **parasites** (organisms that live at the expense of other living organisms). In all cases, they first digest food substances with extracellular enzymes, which they secrete into the food mass; then they absorb them. These are the mechanisms that make fungi the world's major agents of decomposition. Some fungi (for example, yeasts) can release energy by anaerobic respiration. This is the process by which they ferment glucose to ethanol, a property exploited by brewers and vintners (Box 40-1).

IMPORTANCE FOR HUMAN LIFE

The impact of fungi on our daily lives is tremendous. We have already spoken of the molds, the mycelia of which spread visibly on old food in damp, dark places,

40-1 Three types of fungi
(A) Bracket fungus on a decaying tree.
(B) Amanita muscaria, *a poisonous mushroom.*
(C) Puff balls, Lycoperdon, *releasing spores.*

40-2 Other common fungi
(A) The mycelium of the mold Penicillium chrysogenum, *showing individual hyphae. This fungus produces penicillin.* (× 530) *(B) Cells of brewer's yeast,* Saccharomyces cereviseae, *in the process of budding.* (× 120)

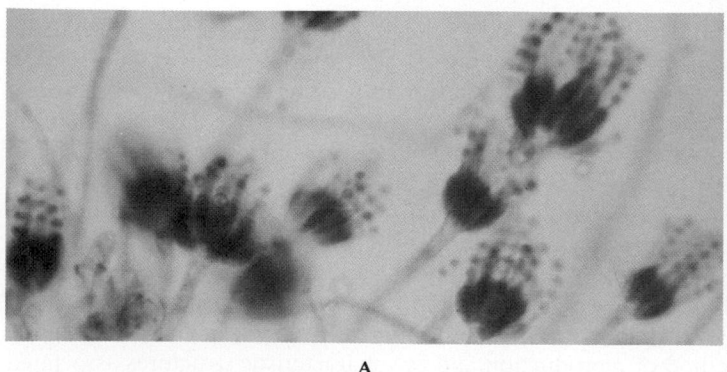

A

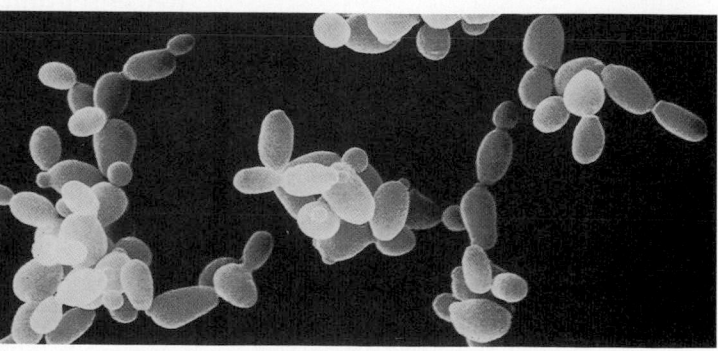

B

Spore

Hypha

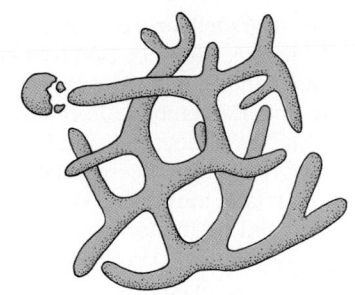

Mycelium

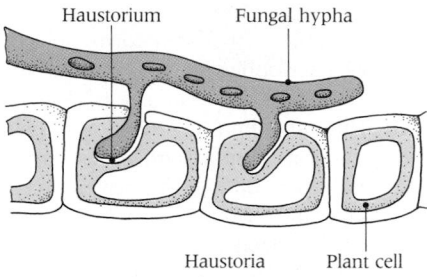

Haustorium   Fungal hypha

Haustoria   Plant cell

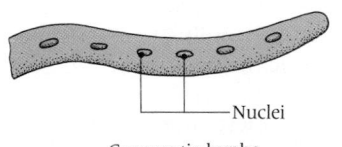

Nuclei

Coenocytic hypha

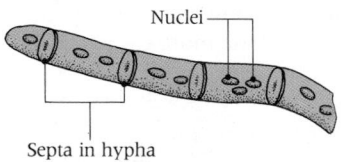

Nuclei

Septa in hypha

Septate hypha

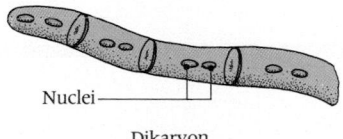

Nuclei

Dikaryon

40-3 Basic features of fungal anatomy
A mycelium develops from a single spore and forms a network containing many nuclei. In some species septa divide the hypha into segments. The haustorium is a specialized hyphal extension of some parasite and symbiotic species that forms a close association with the cells of a plant it infects and absorbs nutrients.

and sometimes on the surfaces of other plants or even of animals. Mushrooms are delicacies of the human diet, but certain varieties (sometimes inaccurately called toadstools) may be lethally poisonous. Food producers use baker's yeast to leaven bread and brewer's yeast to generate alcohol in beer, wine, and whiskey beverages. Farmers know only too well of such devastating fungus diseases as corn smut or black stem rust of wheat.

All of these familiar and visible organisms are fungi. In many fungi, such as mushrooms, what is visible to us are their reproductive structures, which appear as masses of tightly packed hyphae, or as in many parasite fungi, the spore masses. The vegetative mycelium often lies hidden in the soil or inside the tissues of a host plant or animal and it is this part of the fungus that is harmful to plant and animal tissues. Therefore, when spore masses become visible, as in smut and rust infections, the damage has already been done to the host.

The most widespread and destructive plant diseases are caused by parasitic fungi. The damage they do is reckoned in many millions of dollars. Even when an infected plant is not killed, it may be stunted, its productivity reduced, or its fruit blotched or scabbed. Fungi also cause a variety of human and animal diseases. Trichophytosis ("athlete's foot"), coccidiodomycosis (the endemic disease once called San Joaquin fever), candidiasis, cryptococcosis, and blastomycosis are human fungal diseases notable for their resistance to treatment.

Although some fungi do untold damage from the human point of view, many others are highly useful. As noted, brewing and all industrial operations involving fermentation depend on yeasts, which have the attractive property of metabolizing sugars into ethanol (ethyl alcohol) and CO_2. Bread raised with yeast utilizes the same reaction, but here it is the CO_2 bubbles we desire in the dough. Fungi are also essential in the making of cheese (Box 40-1). And, as discussed in earlier chapters and later in this chapter, these form mycorrhizae with the roots of certain plants that are essential to their growth.

Certain molds synthesize compounds poisonous to competing microorganisms. The term **antibiosis** is applied to the phenomenon in which organisms take the lives of others to preserve their own. The naturally synthesized antibacterial compounds are called **antibiotics.** They are metabolic products, formed regardless of any "need" the fungi may have for protection against invasion by or competition from bacteria. Some of them destroy bacteria that cause human diseases but are nontoxic or only mildly toxic to the human host. The first of these life-saving fungal products to be recognized, **penicillin,** was named for *Penicillium,* the mold that produces it. It has now been followed by cephalosporin, gentamicin, griseofulvin, and dozens of others.[1]

REPRODUCTION

In considering fungal reproduction, it would be useful to reexamine Figure 21-1, which summarizes the principal types of life cycles in living organisms.

Fungi reproduce both sexually and asexually—and their life cycles may be quite complex. We examined in Chapter 12 the three major types of eukaryotic life cycles and their distinctive patterns in the location and timing of meiosis. As shown in Figures 12-37C and 12-38C, one mode is characterized by alternation of generations and zygotic meiosis. This pattern is characteristic of fungi in which meiosis occurs immediately after zygote formation. Hence, we say that in fungi meiosis is zygotic. We would ordinarily conclude that all fungal cells except the zygote are haploid. However, because of their peculiar nuclear organization, the haploid-diploid terminology applied to other eukaryotes is not applicable to fungi.[2]

Fungal reproduction, whether sexual or asexual, depends on spores, which are discrete reproductive units that are capable of being widely disseminated. A spore is a cell that can develop into an "adult" organism without fusion with another cell (unlike a gamete, which must unite with another gamete to form a zygote, that then develops into an adult individual). Spores typically contain one or more

---

[1] Antibiotics are also produced by many bacteria. Examples include streptomycin and tetracycline.

[2] For example, although the nuclei in a fungus may be haploid, the fungal cells may be "haploid" (with one nucleus), "diploid" (with two nuclei), and so on. Further complexity is added by the presence of perforated septa and the potential for nuclear migration.

A fascinating aspect of microbiology concerns the practical uses to which microorganisms have been put, wittingly or unwittingly, by peoples of many cultures over the centuries. The methods used in making cheese and wine are instructive and diverting examples. In both, art, science, and technology come wonderfully together.

Cheese is made from milk. It is thought to have originated in southwestern Asia some 8000 years ago. The Romans introduced it to Europe between 60 B.C. and 300 A.D. and influenced its terminology: *caseus,* the Latin word for cheese, is a root of many terms in the lexicon of cheese and cheese making. Some 2000 kinds of cheese, most the products of specific countries or districts, can be subdivided into about 20 basic types.

Cheese production begins with that portion of milk that can be precipitated by acid, proteolytic enzymes, or heat. The precipitate, the **curd,** includes various lipids, casein (a characteristic protein of milk), calcium, and carbohydrates. When the curd is removed, a watery supernatant fraction, the **whey,** remains. This is drained off and discarded.

Curd can be precipitated by lowering the pH—with acid or by adding lactic acid-producing bacteria such as *Lactobacilli.* Curd can also be formed by adding **rennet,** a material containing rennin (a proteolytic enzyme) that is obtained from calf stomachs or from fungi such as *Mucor miehei* (a cheaper source). Rennin (when it works properly) alters the casein molecule so that it denatures, or coagulates. Interestingly, some cows produce rennin-resistant milk. This appears to be a genetically determined trait.

There are two major types of cheeses: fresh and ripened (or cured).

- ■ Fresh cheeses, which spoil if not eaten promptly, are made from milk coagulated by acid or heat. Examples are cottage cheese, cream cheese, ricotta, and mozzarella cheeses.
- ■ Ripened cheeses, which include the great majority of famous varieties, are made from milk coagulated by lactic acid-producing bacteria and rennet and then inoculated with various bacteria or fungi.

Cheeses are also commonly classified as hard, semisoft, or soft.

To ripen cheese, the producer adds to curd freed of whey a culture of appropriate microorganisms, either bacteria or fungi, whose actions are largely responsible for a cheese's distinctive flavor and texture. They are added in large amounts, usually one or two billion organisms per gram of curd. Cheese thus has the highest concentration of microorganisms of any basic human food. As cheese ripens, the microorganisms die and release their enzymes. These (along with milk lipase) act on curd lactose (which is thereby converted to lactic acid and ketones), lipids (which form free fatty acids), and proteins (which are degraded to free amino acids). Considerable gas may be evolved—CO_2 from Cheddar and Emmentaler cheeses, NH_3 from Brie and Camembert. Excessive NH_3 decreases quality, as does H_2S and H_2, which also may evolve. While cheese ripens, a process that may take months or years, little is required beyond regulation of temperature and humidity (Figure 40A).

Ancient cheese makers knew nothing of microbiology, yet they devised ingenious ways of achieving results. Often the correct organisms would come from the air and seed the curd. Today cheese factories carefully maintain cultures in frozen form, selecting them for bacteriophage-resistance and other properties, so that results can be reliably predicted. The aim, then and now, is to get a good curd that supports microbial growth and to maintain cultures of microorganisms that "cooperate." Doing that successfully is as much art as science.

Although cheese making most commonly begins with cow's milk, cheese makers around the world also rely on the milk of other animals, especially sheep and goats. Presumably, the milk of any mammal could be used. Since the milk source does influence the flavor of a natural ripened cheese, milks other than these three might well yield unique cheeses. However, most other mammals are either unavailable or the wrong size. Goat's milk gives a more sharply flavored cheese than cow's milk, primarily because its lipids include more caproic, caprylic, and capric acids (see Table 4-1). Sheep's milk has six times the caprylic acid of cow's milk and twice that of goat's milk. This does not affect the milk's flavor, but it does influence that of the cheese made from the milk.

Examples of the methods used in ripening some cheeses are shown in the table.

40A Curing of cheese

nuclei derived from either sexual or asexual reproduction. Spores differ in their motility. Some, called **zoospores,** are motile, swimming, flagellated structures. Others are transported by wind. Still others **(sporangiospores)** are nonmotile. Zoospores are produced by some fungi (Oomycetes) in structures called zoosporangia (singular, zoosporangium). **Oospores** are also distinctive spores in Oomycetes. Note that oospores are products of meiosis, which form zygotes if fertilized. Zoospores are products of mitosis.

CHEESE	MILK	METHOD OF RIPENING
Hard cheeses		
Cheddar	Cow	Cured with *Streptococcus lactis* for 2–12 months.
Cheeses rich in serine		
Emmentaler (Swiss)	Cow	Cured with *Streptococcus lactis* and then with *Streptococcus thermophilus* in high humidity for 2–10 months.
Stilton	Cow	Ripened with fungus *Penicillium roqueforti* for 2 weeks. Cured for 6 months.
Semisoft cheeses		
Bleu	Goat/cow	Ripened with *Penicillium roqueforti* for 3 months. Cured for 2–3 months.
Roquefort	Sheep	Ripened with *Penicillium roqueforti*. Salted and stored cool for 3 months.
Soft cheeses		
Brie	Cow	Ripened with fungus *Penicillium candida* for 10 days. Distributed in 14 days under refrigeration.
Limburger	Cow	Ripened by *Bacterium linens* and cured for 3 weeks in high humidity.

Cheese, they say, needs wine, and wine making is an equally ancient discipline. It begins with the juice of grapes (Figure 40B), yet it is similar in principle to cheese making. There are thousands of grape varieties, which vary in size, shape, color, flavor, and chemical composition. Their differences largely account for differences in wine flavor and style.

The making of wine is the subject of **enology.** About 20% of the contents of ripe grapes is a mixture of glucose and fructose, with a small amount of tartaric, malic, and other organic acids. As grapes ripen on the vine, sugar increases and acid decreases (Figure 40C). Grapes grown in cold climates often contain excess acid and yield hard-tasting wines. Grapes grown in warmer areas may be deficient in acid and yield wines that are flat and perishable.

The production of wine from grape juice depends on the process of **fermentation,** in which the enzymes in yeast convert a molecule of glucose to two molecules of ethanol and two molecules of CO_2 (see Chapter 9). Fermentation is complete when all of the sugar is converted to ethanol (Figure 40D)— or when the alcohol level reaches 15%.

It was Louis Pasteur who converted the folk art of wine making into modern enology by showing that living yeast cells convert grape juice to wine. Today vintners add their pedigreed yeast cultures to grape juice and hope for the best (Figure 40E).

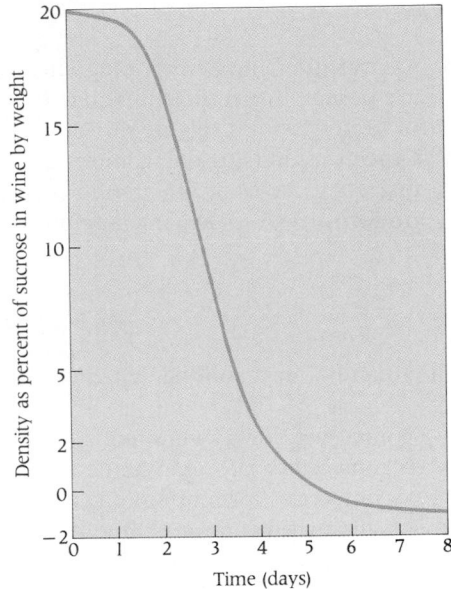

40D Fermentation in wine making
Sugar, the main soluble solid of ripe grapes, is converted to ethanol and CO_2. This process starts slowly when the yeast are few in number. It then accelerates, but slows down again when most of the sucrose has been converted to ethanol. This curve is for red wine fermenting at about 25°C. Density reaches a negative value because of the presence of ethanol.

40B Cultivated grapes and sunny vineyard

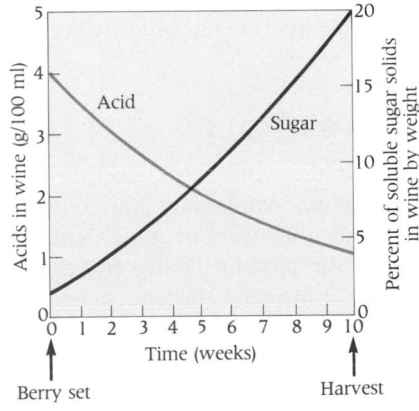

40C Acid and sugar content of grapes
In the time between grape set and harvest, the acid content of grapes declines and the sugar (sucrose) content rises. If the acid content drops too low, the grapes may be harvested early, or acid may be added to the juice to maintain quality wine.

40E Wine aging in barrels

Zygospores, ascospores, and basidiospores are, respectively, the spores produced by members of the phyla Zygomycetes, Ascomycetes, and Basidiomycetes. They arise by meiosis. Unlike zoospores, they are not flagellated and not motile. These nonmotile spores, the characteristic reproductive units of sexually reproducing fungi, are usually covered with a tough wall and can survive long periods of dryness at extremes of temperatures. They are sufficiently tiny to float freely in air, and the air around us is rarely free of them. If spores land on a suitable medium, they can develop into **vegetative bodies,** which in turn develop new spore-bearing organs that are often elaborate in shape and structure. This surely accounts for the wide distribution of many fungi. It is why, for example, a piece of bread left exposed to air will almost inevitably acquire a covering of mold. It is also why compost heaps often sprout mushrooms (that may grow with astonishing rapidity). All develop from ubiquitous air-borne spores.

In sexual reproduction, nuclei of two genetically different mating types come together and fuse. Subsequent events differ in the several phyla and are described below. Asexual reproduction occurs either by fragmentation of the mycelium, each fragment being capable of forming a new individual, or by the production of spores.

One of the many diverse life cycles of fungi is that of the familiar black bread mold, *Rhizopus* (Figure 40-5). Two kinds of reproductive structures are found in fungi.

- **Sporangia** (singular, sporangium), which are involved in spore production and are borne on specialized hyphae called **sporangiophores** that are the stalks of fruiting bodies (Figure 40-6).
- **Gametangia** (singular, gametangium) are differentiated cells that either produce discrete gametes or function in their place. A female gametangium is called an **oogonium.** A male gametangium is an **antheridium.**[3]

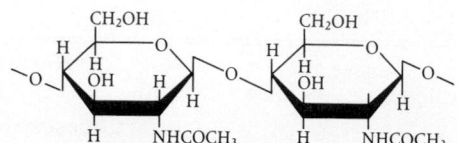

40-4 Structure of chitin
Chitin, a polymer of N-acetylglucosamine units, is a strong, compact polysaccharide. Chitin occurs in the cell walls of many (but not all) fungi. Compare with Figure 4-8 which shows the structure of glycogen, another polysaccharide.

MAJOR PHYLA

The fungi in our taxonomic scheme comprise five phyla.

- Oomycetes (form swimming zoospores)
- Zygomycetes (form nonmotile zygospores)
- Ascomycetes (form nonmotile ascospores)
- Basidiomycetes (form nonmotile basidiospores)
- Deuteromycetes, also termed Fungi Imperfecti (lack meiospores)

Major properties of these phyla are summarized in Table 40-1. Let us briefly examine each phylum.

PHYLUM OOMYCETES: WATER MOLDS

The **oomycetes** include organisms called water molds, white rusts, and downy mildews. They comprise a large heterogeneous phylum with many hundreds of species made up of filamentous or unicellular fungi, the walls of which consist mainly of cellulose. Indeed, a distinctive feature of the phylum Oomycetes is the presence in cell walls of cellulose (or celluloselike molecules) instead of chitin. Unlike the other fungal phyla, oomycetes produce motile zoospores bearing two unequal flagella. For this reason, some systematists regard them as protists and relate them to brown algae.

Typical members of this phylum are the water molds (*Saprolegnia*), aquatic forms that are distinctive because they produce flagellated, swimming spores. Their life cycle (Figure 40-7) is composed of two phases. In the asexual phase, zoospores emerge from zoosporangia, become encysted, and later germinate (Figure 40-8). In the sexual phase, two kinds of gametangia—a female oogonium and a male antheridium—produce gametes (eggs and sperm nuclei) that fuse and yield the thick-walled cells called oospores, which are zygotes that eventually germinate (Figure 40-9).

Some members of the phylum are parasites or saprobes. Others are pathogenic. Some saprolegnians attack fishes and their eggs. Certain of the pathogenic oomycetes have had economic consequences of historic magnitude. For example, *Plasmopara viticola* is the downy mildew so feared by wine growers (see Box 40-1). This mold nearly destroyed the French wine industry in the nineteenth century and in the wet season of 1963 it did dreadful damage to the famous vineyards of Bordeaux.

---

[3] Gametangia allow for direct contact of nuclei and thus obviate the need for specialized gametes of fungi to transport nuclei in the process of sexual reproduction.

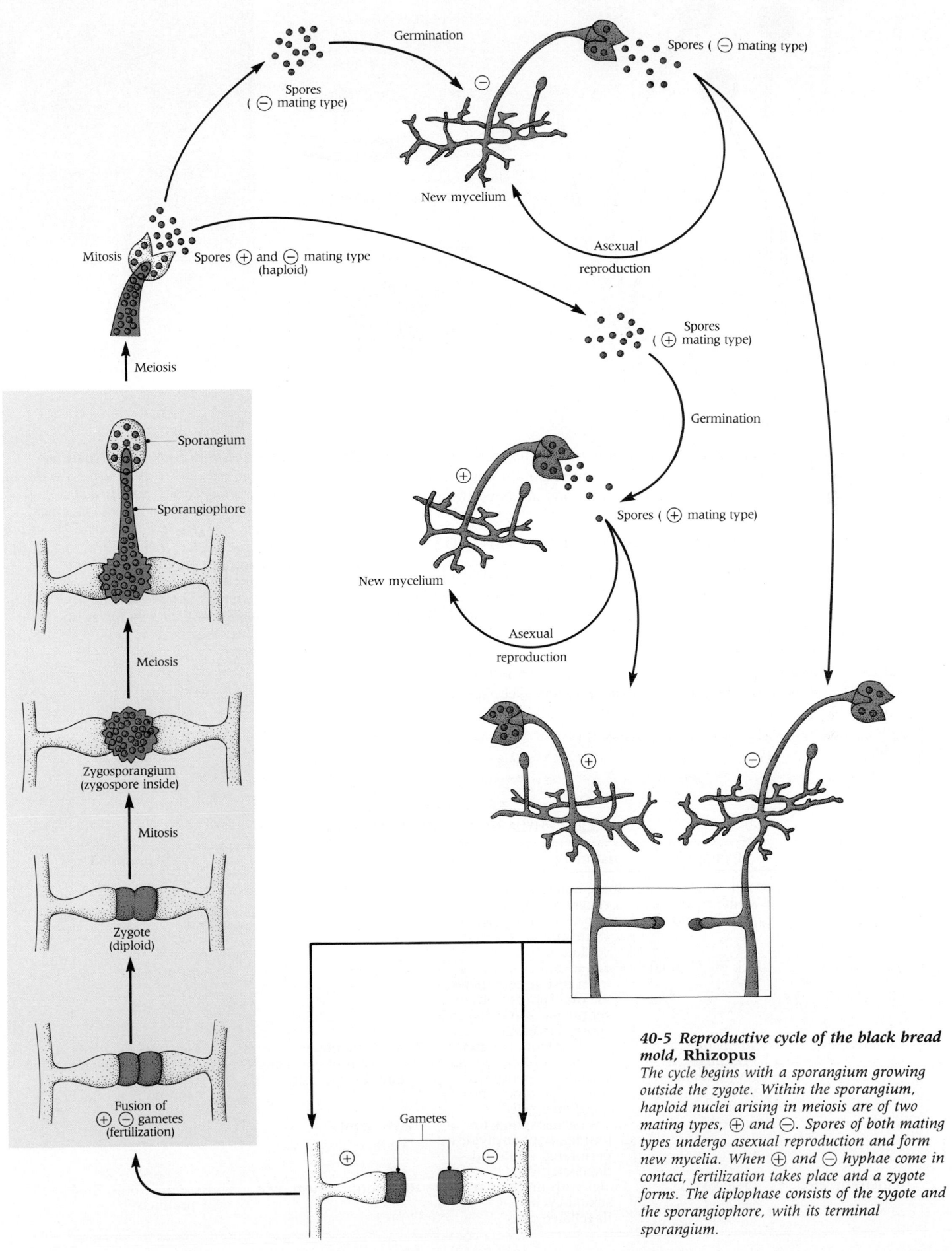

Germination

Spores
(⊖ mating type)

Spores (⊖ mating type)

Mitosis

Spores ⊕ and ⊖ mating type
(haploid)

New mycelium

Asexual
reproduction

Meiosis

Spores
(⊕ mating type)

Sporangium

Germination

Sporangiophore

⊕

Spores (⊕ mating type)

Meiosis

New mycelium

Asexual
reproduction

Zygosporangium
(zygospore inside)

⊕

⊖

Mitosis

Zygote
(diploid)

Fusion of
⊕ ⊖ gametes
(fertilization)

Gametes

⊕   ⊖

**40-5 Reproductive cycle of the black bread
mold, Rhizopus**

*The cycle begins with a sporangium growing
outside the zygote. Within the sporangium,
haploid nuclei arising in meiosis are of two
mating types, ⊕ and ⊖. Spores of both mating
types undergo asexual reproduction and form
new mycelia. When ⊕ and ⊖ hyphae come in
contact, fertilization takes place and a zygote
forms. The diplophase consists of the zygote and
the sporangiophore, with its terminal
sporangium.*

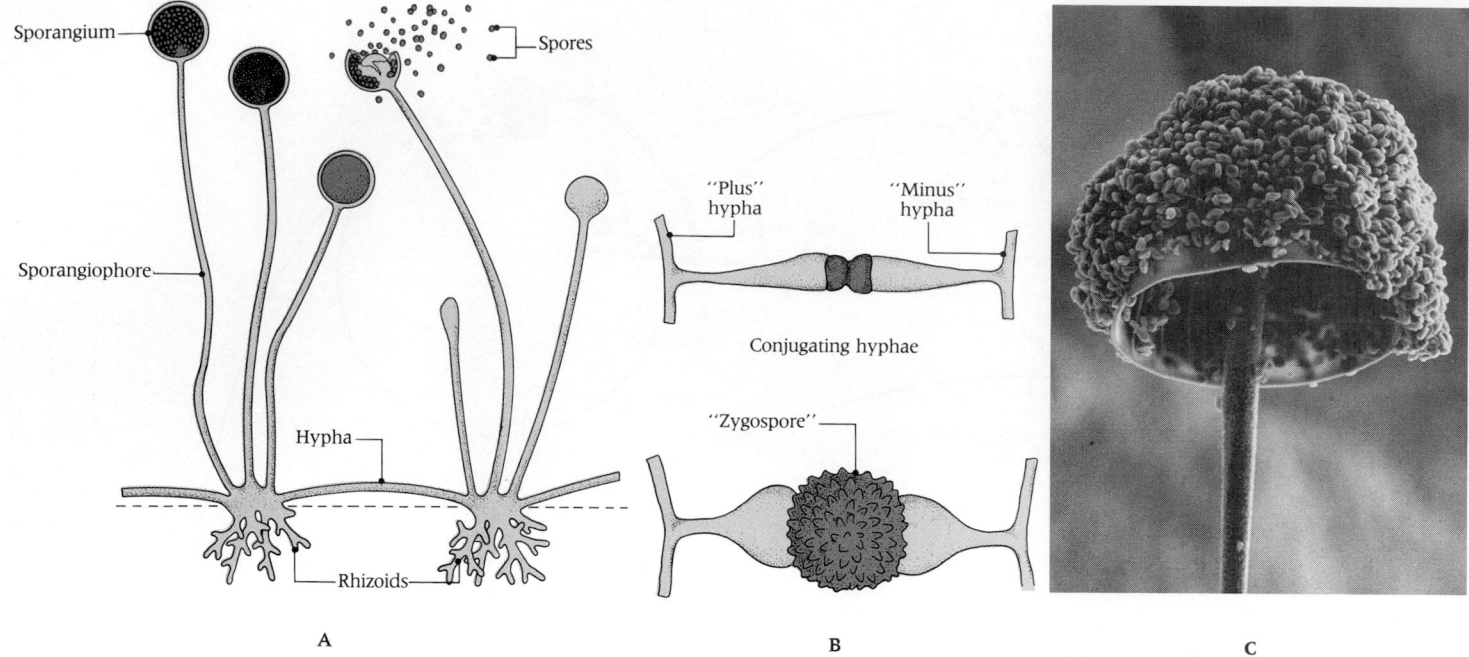

A   B   C

Another famous species of oomycetes, *Phytophthora infestans*, caused the "late blight" of potatoes, which in the 1840s destroyed the entire potato harvest of Europe and led to mass starvation and the migrations of millions of people from their homes.

PHYLUM ZYGOMYCETES: BREAD MOLDS

The **zygomycetes** are some 600 species of terrestrial filamentous fungi that have chitinous walls and lack cross walls or septa except those between the reproductive structures (sporangia and gametangia) and the rest of the mycelium. They reproduce asexually more often than sexually.

The term zygomycetes (from the Greek roots *zygon* "pair," and *mykes* "fungus") refers to their distinguishing characteristic—the production of sexual resting spores that are distinctive because they develop from a zygote (see Figure 40-6). For that reason we call them **zygospores.** They are thick-walled hardy structures. Unlike oomycetes, zygomycetes produce no flagellated spores at any stage of their reproductive cycle.

40-6 Fruiting bodies of Rhizopus
(A) Rhizoids anchor the mycelium to the soil. Each fruiting body is composed of a sporangiophore and a spore-containing sporangium. (B) Sexual reproduction occurs when two hyphae from different mating types come together, forming gametangia. These fuse to form a thick-walled zygote called the zygospore. (C) Scanning electron micrograph of Rhizopus sporangium covered with spores. (× 100)

TABLE 40-1
MAJOR PHYLA OF THE KINGDOM FUNGI

Phylum	Examples	Distinctive Characteristics	Diseases	Economic Uses
Oomycetes	Water molds, white rusts, downy mildew	Some aquatic; flagellated zoospores; eggs and sperm form in special gametangia; cell walls contain cellulose	Blights, mildews of plants, fish infections	None
Zygomycetes	Black bread mold	Form sexual zygospores from fusion of gametes; asexual spores from sporangia; some disperse spores forcefully	Few	Various drugs
Ascomycetes	*Neurospora, Penicillium,* a few *Candida* yeasts, morels, truffles	Form ascospores in asci; asexual conidia; hyphae divided by perforated septa; dikaryons; no flagellated cells	Powdery mildews of fruits, chestnut blight, Dutch elm disease, ergot	Food (morels, truffles), wine, beer, breadmaking (yeasts), cheeses, antibiotic production
Basidiomycetes	Mushrooms	Form basidiospores on basidia; hyphae divide by perforated septa; no flagellated cells	Rusts, smuts	Food (mushrooms)
Deuteromycetes	*Cryptococcus, Candida, Fusarium*	Fungi with no known sexual cycles; no flagellated cells	Ringworm, thrush, wilt disease	Research tools, industrial products

40-7 Reproductive cycle of **Saprolegnia,** an oomycete

Asexual reproduction occurs by means of flagellated zoospores. Sexual reproduction depends upon sperm and egg cells. In both cases, the resulting forms are the tubular filamentous growths called hyphae.

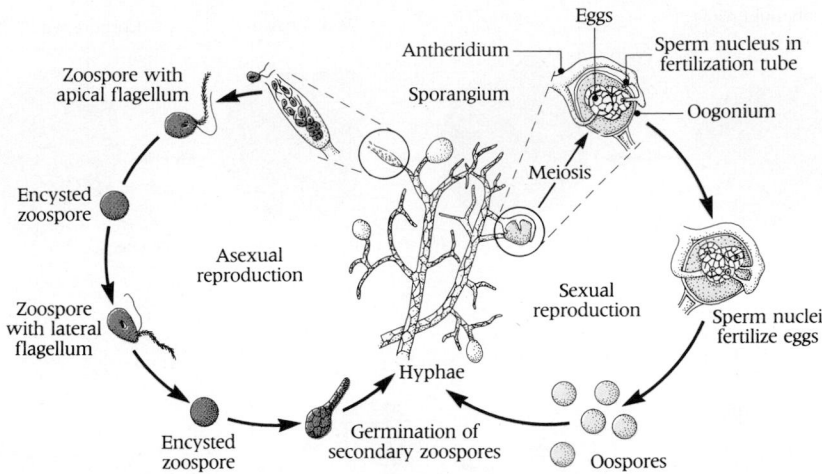

Zoospore with apical flagellum

Encysted zoospore

Asexual reproduction

Zoospore with lateral flagellum

Encysted zoospore

Hyphae

Germination of secondary zoospores

Eggs

Antheridium

Sporangium

Sperm nucleus in fertilization tube

Oogonium

Meiosis

Sexual reproduction

Sperm nuclei fertilize eggs

Oospores

This phylum includes *Rhizopus,* the common black bread mold illustrated in Figure 40-5. Many zygomycetes are saprobes—that is, they feed on dead plant or animal matter. However, some grow parasitically on living plants, insects, or small soil animals.

PHYLUM ASCOMYCETES: SAC FUNGI

The **ascomycetes** ("sac fungi") are the largest group of fungi, including about 30,000 described species of powdery mildews (*Microsphaera*), the red bread mold (*Neurospora*), which played an illustrious role in the history of molecular biology (see Chapter 16), yeasts, both baker's and brewer's (*Saccharomyces*), and the morels and truffles (*Morchella*) of haute cuisine renown.

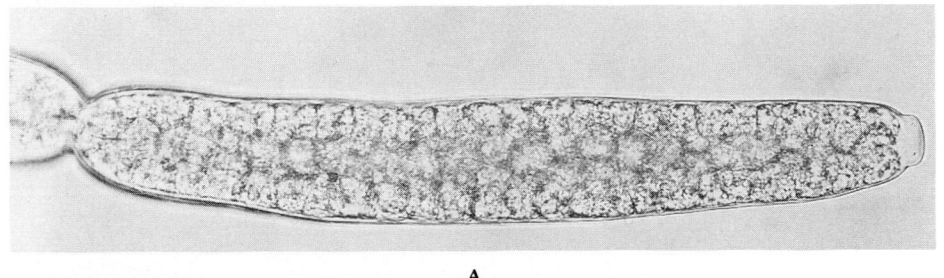

A

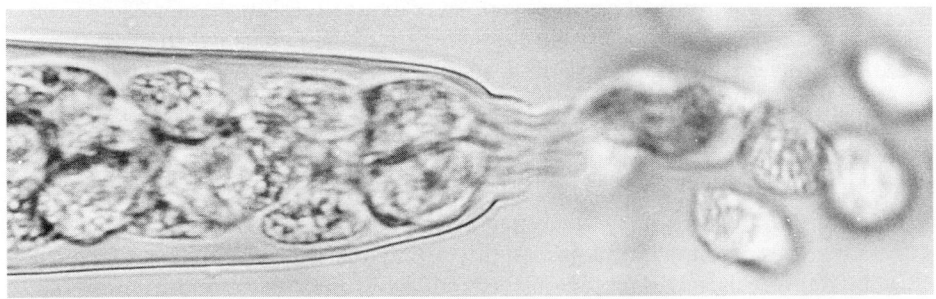

B

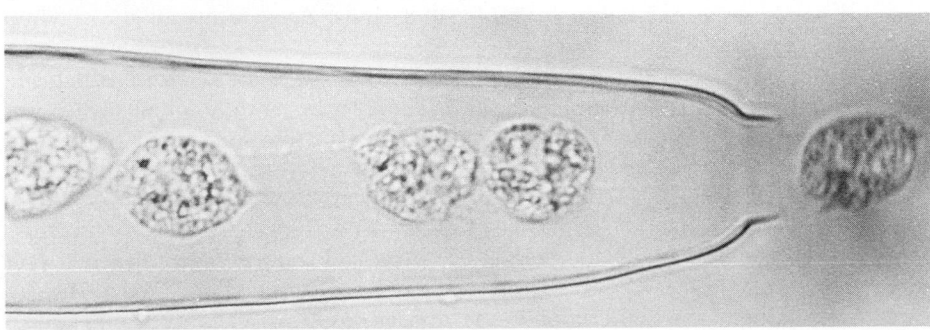

C

40-8 Zoosporangia and zoospores of Saprolegnia

(A) *Zoospores in a mature zoosporangium.* (× 600) **(B)** *Zoospores escaping through an apical opening.* (× 1800) **(C)** *Zoosporangium with few remaining zoospores.* (× 2700)

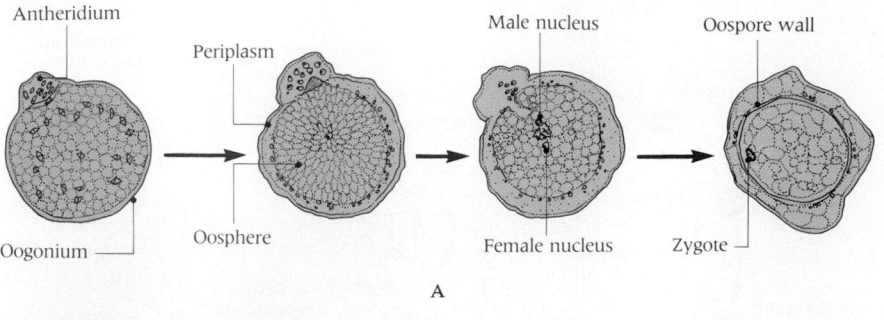

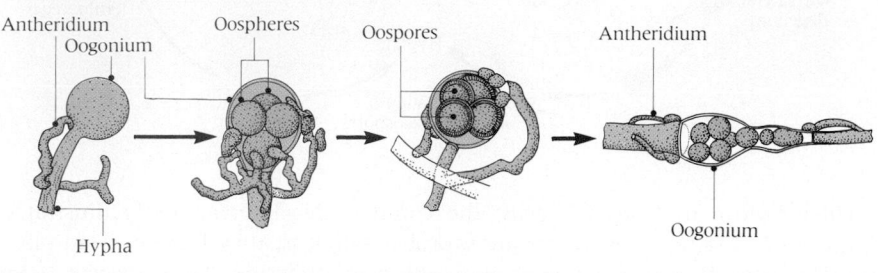

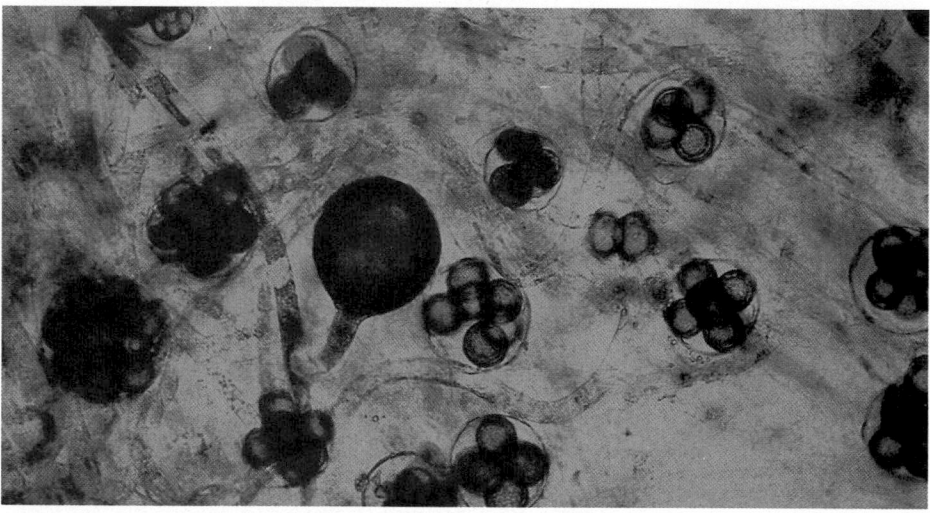

A

B

C

40-9 Sexual reproduction in oomycetes
(A) *In this sequence depicting fertilization in oomycetes, several eggs (oospheres) are produced inside an oogonium. An antheridium (male gametangium) around the oogonium transfers male nuclei, which unite with the eggs to produce diploid zygotes. Each zygote develops into an oospore.* **(B)** *Sexual reproductive structures formed by one of the best known oomycetes,* Saprolegnia literalis. **(C)** *Electron micrograph of* Saprolegnia literalis. *The large spherical structures are oogonia. The smaller spherical bodies within them are oospores.*

Members of this group of fungi are sources of many useful antibiotics. The ascomycetes *Penicillium* (Figure 40-10) produces the antibiotic penicillin. The two species, *Penicillium camemberti* and *Penicillium roqueforti,* are responsible for the characteristic flavor of Camembert and Roquefort cheeses (see Box 40-1). Someone who is hypersensitive to penicillin may have a serious allergic reaction upon eating one of these cheeses. Ascomycetes are also the causes of many diseases. The plant diseases they cause include chestnut blight (caused by *Endothia parasitica*) and Dutch elm disease (caused by *Ceratocystis ulmi*), which has destroyed many of America's stately elms. Fungal diseases of animals and humans are numerous. As noted above, they include coccidioidomycosis, a noncontagious respiratory disorder common in southwestern states, histoplasmosis, the most common fungal respiratory disease, and certain cases of candidiasis. As shown in Table 40-1, most *Candida* are of the phylum Deuteromycetes.

Ergotism, a fungus-produced disease, is caused by *Claviceps purpurea,* a parasite of rye grain. A very small amount in the grain supply may cause gangrene, delusions, convulsions, and death. Ergotism in one epidemic in 994 A.D. killed more than 40,000 people. In 1772, ergotism decimated both men and horses of the cavalry of Peter the Great as they prepared for battle against the Black Sea ports of Turkey. Ergot derivatives, which cause muscles to contract and blood vessels to constrict, have various medical uses. One derivative is the hallucinatory drug, lysergic acid diethylamide (LSD).

In many ascomycetes, the hyphae are divided by cross walls, or septa, unlike the hyphae of the oomycetes and most zygomycetes. Each compartment generally

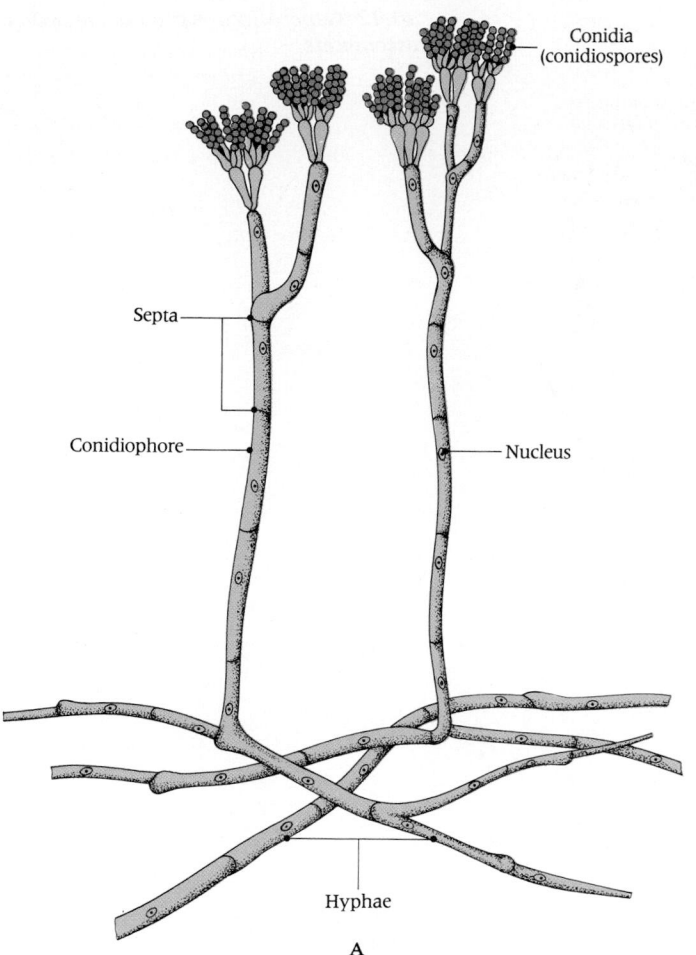

Conidia
(conidiospores)

Septa

Conidiophore

Nucleus

Hyphae

A

B

40-10 Penicillium
(A) In this ascomycete, the hyphae are divided into septa containing a single nucleus.
(B) Photomicrograph of Penicillium. (× 2000) Compare with Figure 40-2A.

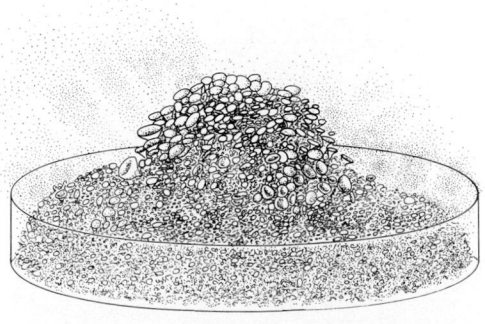

40-11 Visible ascocarps
These are the fruiting bodies within which asci are formed.

contains one or two nuclei, but the septa have openings in them through which the cytoplasm and often nuclei can pass. Spores are formed sexually and asexually.

■ Asexual spores, termed **conidia,** are formed either singly or in chains at the tips of specialized hyphae called **conidiophores.**

■ Sexual spores, termed **ascospores,** form within a specialized cell called an **ascus** ("little sac"), the structure that is characteristic of the group (Figures 40-11 and 40-12). The ascus arises within the ascocarp, a complex structure made of densely interwoven hyphae. The ascus is a cell that at first contains two haploid nuclei. These fuse within each ascus to form a diploid zygote, which divides immediately in the course of meiosis and then again by mitosis. Eight ascospores (or in some case four) form in each ascus. When mature, they are ejected onto the winds through an opening in the top of the ascus. Ascospores reach the ground, often a considerable distance away. There they germinate, each spore growing into a hypha, which becomes a mycelium.

This cycle involves the unique mechanisms of parasexuality (see Chapter 21), which are peculiar to fungi. One unusual aspect is that the hyphae of ascomycetes can be **homokaryotic** ("same nuclei") or **heterokaryotic** ("different nuclei"). In the former, all nuclei in the hypha are genetically identical. In the latter, the nuclei are genetically dissimilar. The hyphae of ascomycetes can be homokaryotic or heterokaryotic. A monokaryotic hypha, which has only one nucleus per cell, is always homokaryotic. A **dikaryotic** hypha has two nuclei, but the hypha can be homokaryotic or heterokaryotic.

In heterokaryosis, the sexual cycle of ascomycetes leads to the formation of a distinctive structure, a **heterokaryon,** in which both haploid parental nuclei remain independently present in the cytoplasm. As the hyphae grow and become septate, the nuclei migrate and divide synchronously with the host nuclei for a considerable period of time. Only with the formation of asci do the nuclei finally fuse.

The ascomycetes are divided into two broad groups, depending on whether or not the asci are contained within a specialized fruiting structure. Those species

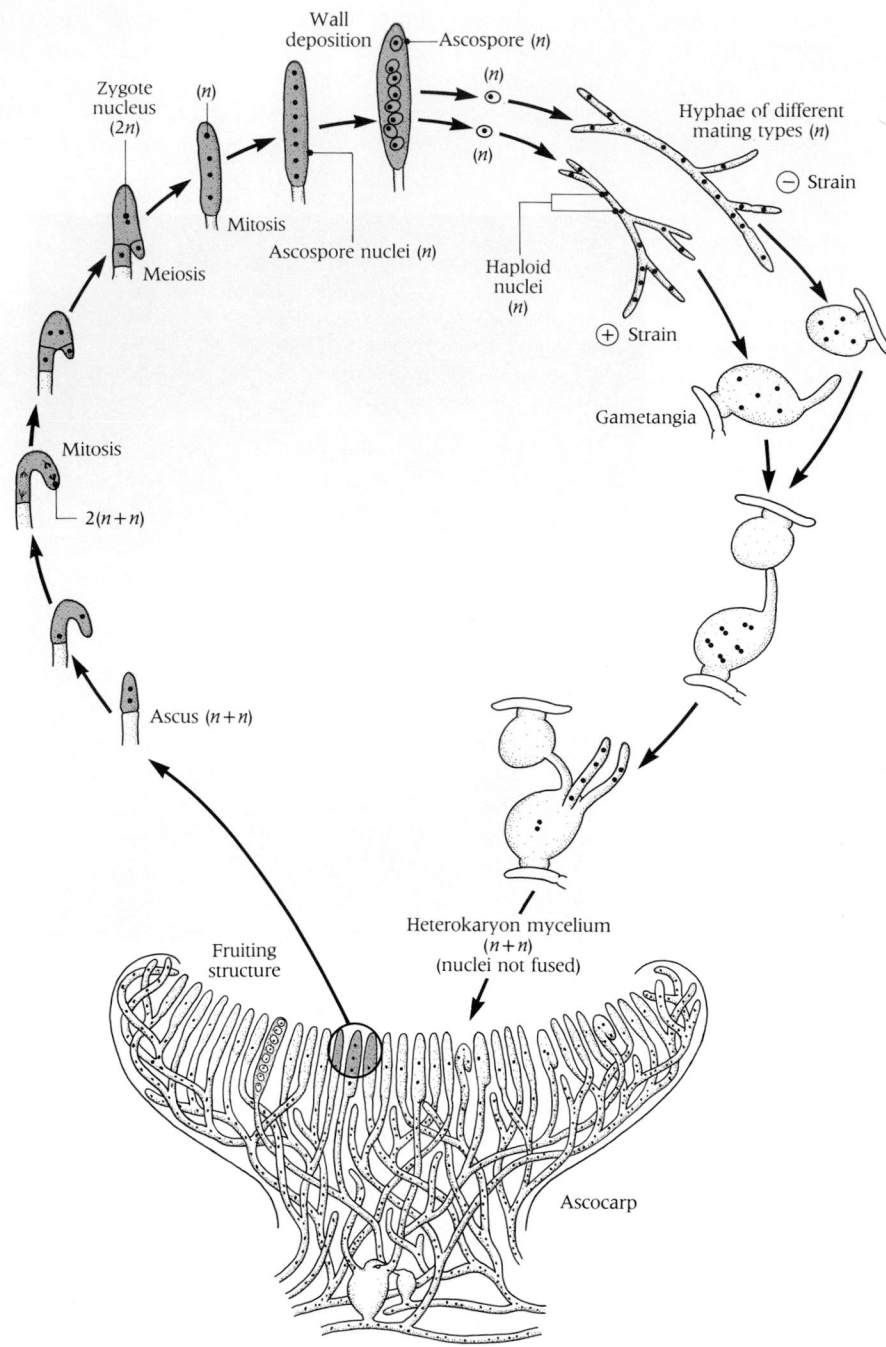

Zygote nucleus (2n)

(n)

Mitosis

Meiosis

Wall deposition — Ascospore (n)

(n)

(n)

Ascospore nuclei (n)

Haploid nuclei (n)

Hyphae of different mating types (n)

⊖ Strain

⊕ Strain

Gametangia

2(n+n)

Mitosis

Ascus (n+n)

Fruiting structure

Heterokaryon mycelium (n+n) (nuclei not fused)

Ascocarp

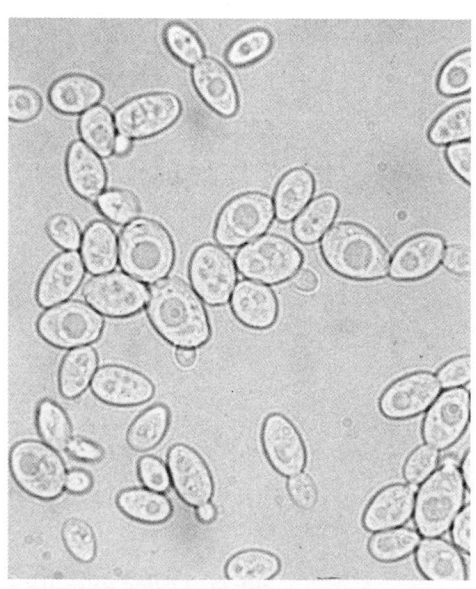

Bud

having a fruiting structure enclosing the asci are collectively called **euascomycetes,** while those without it are called **hemiascomycetes.** Ascus formation itself is a peculiar process. One might expect that a hyphal tip containing a pair of dissimilar nuclei would become walled off as its nuclei fused and underwent meiosis. But instead, a hook (crozier) forms at the tip of the heterokaryotic hypha, and the paired nuclei come to lie on either side of it (see Figure 40-12). Both nuclei then undergo mitosis simultaneously, their mitotic axes paralleling the hyphal axis. New walls are laid down and the tip, containing a single nucleus, fuses with the hyphal wall it touches, releasing that nucleus back into the lower portion of the hypha. The other two nuclei fuse, and meiosis begins ascospore formation.

Bearing this generalized life cycle in mind, consider the hemiascomycetes. These organisms are very small and frequently unicellular. The best known are the yeasts (*Saccharomyces cerevisiae*) (see Figure 40-2). Of all the fungi, the yeasts are the most important of the domesticated types. They multiply either by simple fission or, in the better known genera, by a process of budding analogous to that found in the budding eubacteria (Figure 40-13). Budding occurs in both haploid and diploid phases of the life cycle, so single cells can be haploid or diploid. Conjugation occurs

40-13 Budding cells of bread yeast, Saccharomyces cerevisiae (*Bottom × 180*)

only occasionally between two adjacent compatible cells, and nuclear fusion is followed immediately by meiosis and a single mitosis, so that the entire structure becomes an ascus. There is no extended heterokaryon stage.

Among the euascomycetes, the largest and the best known group, are *Neurospora* (so important to genetic research) and *Morchella* (so valued by French chefs) (Figure 40-14).

PHYLUM BASIDIOMYCETES: CLUB FUNGI

Basidiomycetes ("club fungi") are fungi in which sexual spores, here termed **basidiospores,** are formed externally on a specialized hypha called a **basidium,** from a Greek word for "club." The basidium plays the same role in basidiomycetes as the ascus in ascomycetes. It is the locus of nuclear fusion and meiosis. Most basidiomycetes go through three stages of development.

40-14 Morels, Morchella esculenta
The fleshy structure of this ascomycete contrasts with that of the capped mushroom, a basidiomycete.

40-15 Basidiomycete
(A) *Underside of the cap of the common edible field mushroom* Agaricus campestris. *Other members of the order Agaricales are poisonous or hallucinogenic.* **(B)** *Life cycle.*

A

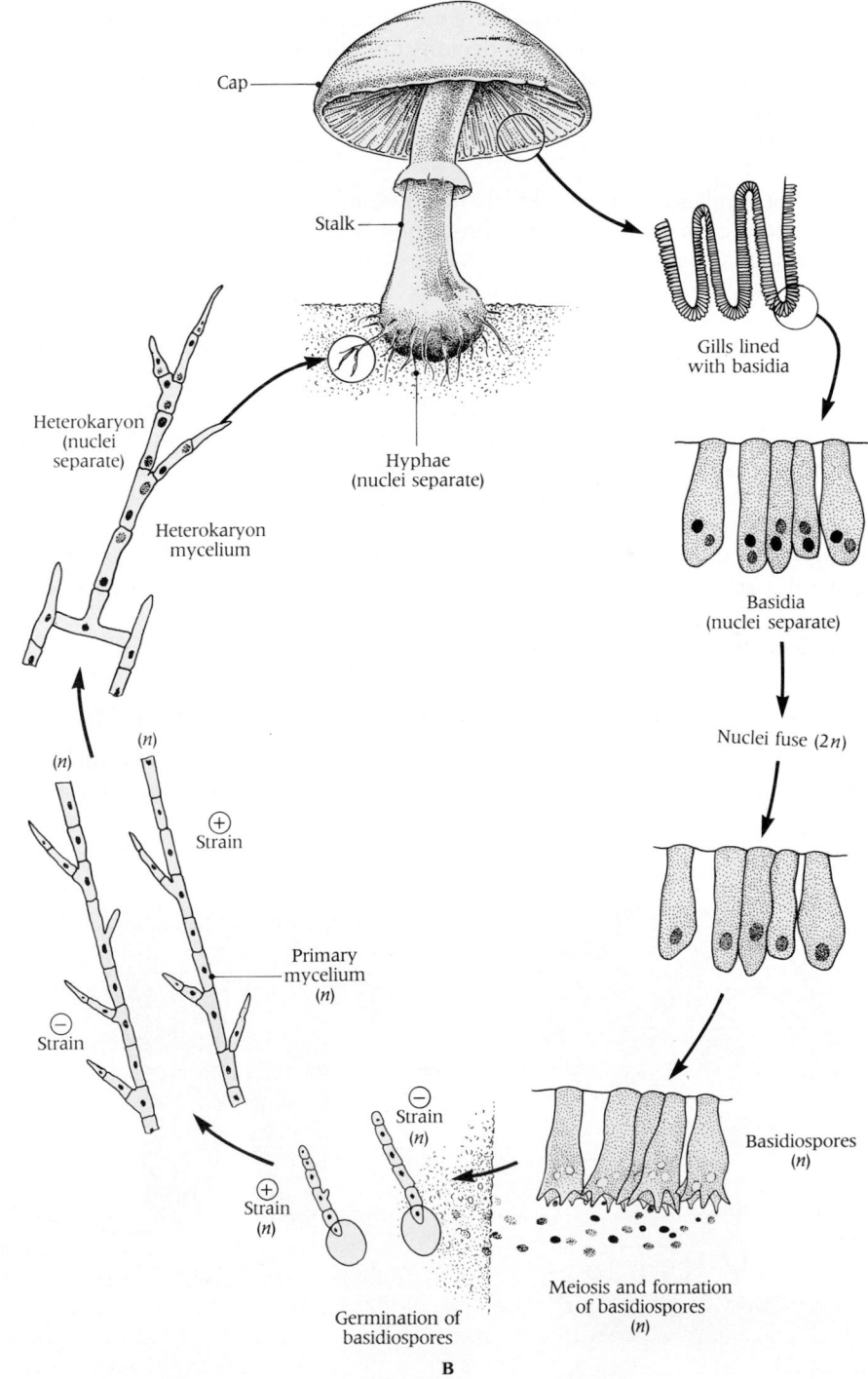

B

- First, a basidiospore germinates to form hyphae that become a **primary mycelium,** which is monokaryotic and lacks septa. Eventually, septa divide the hyphae into uninucleate segments (Figure 40-15). Though monokaryotic, hyphal cells contain different kinds of nuclei, one from each of the original dissimilar haploid partners. Sexual reproduction begins when hyphal cells containing differing nuclei fuse to form a heterokaryon.
- Second, the heterokaryon, which may persist for many years, forms an elaborate **secondary mycelium.** This is the dikaryotic stage. An extensive hyphal mass develops under the soil. This mycelium eventually sends up a fruiting body, the **basidiocarp,** familiar to us as the mushroom cap. Eventually some of the nuclei in the heterokaryon (usually a dikaryon with two nuclei, one of each mating type) fuse to form a diploid nucleus.
- Third, dikaryotic mycelia grow. Simultaneous meiotic divisions of the diploid nuclei produce four new haploid nuclei. Four small projections develop on the ends of a basidium and haploid nuclei enter into them. These tips become the basidiospores, which are then released. In Figure 40-16, we see the familiar ''gill mushroom'' in cross section. The gills beneath the cap are lined with prodigious numbers of basidia that release vast quantities of spores—many millions per hour—into the air. One can demonstrate this phenomenon by placing a mature mushroom cap on a sheet of paper for a few hours. It will release fine spores that trace a fine negative image of the gill pattern on the paper—in the spore's color.

The most familiar basidiomycetes are mushrooms, of which there are some 25,000 species. The mushroom is the spore-producing body. The best known are the **gill fungi,** which include *Agaricus campestris,* the common field mushroom, a cultivated variety of which one buys at the supermarket. Gill fungi also include most of the known poisonous mushrooms, the properties of which have provided a fascinating chapter of toxicology. Of the many toxic mushrooms, only a few are lethal. Those of the genus *Amanita* are the most poisonous. In Europe (where mushroom gathering is more popular than in the United States), 90–95% of all deaths from mushroom poisoning are due to one species of that genus—*Amanita phalloides.* But there are other toxic amanitas, including *A. verna,* a white variety. Curiously, many amanitas are nontoxic and thus edible (for example, the red variety, *A. rubescens* and *A. citrina*). Some genera of toxic mushrooms, such as *Psilocibe* (the source of psilocybin), *Panaeolus,* and *Stropharia,* were once eaten by Mexican and Central American Indians as part of a religious ritual. The hallucinogenic effects of these mushrooms begin within minutes. Clearly, those who would gather wild mushrooms need expert knowledge as well as a sense of adventure.

PHYLUM DEUTEROMYCETES: FUNGI IMPERFECTI

The fungi phyla just discussed (Oomycetes, Zygomycetes, Ascomycetes, and Basidiomycetes) are distinguished by their manner of sexual reproduction. But a large

40-16 Gill mushrooms are basidiomycetes
Gills on the underside of the fleshy cap are lined with basidia containing the diploid nuclei that arose from fusion of haploid nuclei. Meiosis produces four haploid basidiospores. These tiny structures are dispersed by the wind. The three drawings are at progressively higher magnifications.

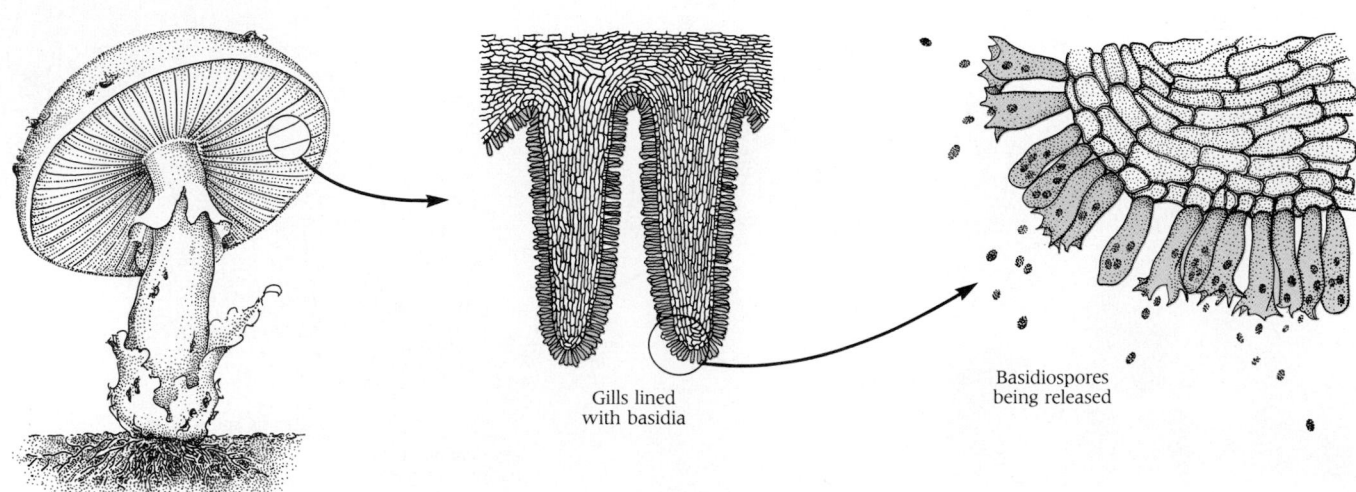

Gills lined
with basidia

Basidiospores
being released

number of fungi, both saprobes and parasites, have no sexual stages—or at least none so far observed. It thus becomes difficult to place them in any of the foregoing phyla, although many systematists have done so anyway. Such fungi, some 25,000 species, are therefore deposited in a taxonomic catch-all—a group of the homeless and unclassified—called **Deuteromycetes.** They are also termed **Fungi Imperfecti,** the word "perfect" being the old botanical term signifying a full capacity for sexual reproduction. In this terminology, sexual reproduction is a "perfect state."

Although they lack a conventional sexual cycle, some deuteromycetes exhibit parasexuality. In this process, as noted above, haploid nuclei in a heterokaryon may fuse to form diploid nuclei within a single cytoplasmic mass, or syncytium. Some of these nuclei are heterozygous (that is, derived from genetically different parental nuclei). The parasexual process, which gives rise to genetically distinctive true-breeding offspring and may involve crossing-over, does not involve specialized mycelia.

Despite their odd name, many fungi imperfecti are the causes of highly distinctive plant and animal diseases, including thrush (caused by some strains of *Candida*) and cryptococcosis, a disease resembling tuberculosis.

FUNGI AS SYMBIONTS

Some fungi regularly form mutually beneficial partnerships with other organisms, sometimes to the exclusion of an independent existence. We call such relationships *symbiosis* and the partners *symbionts.*

Here we consider two well-known examples of symbiosis, though many others exist: **lichens,** in which fungi relate with an alga, and **mycorrhizae,** in which fungi relate with the roots of vascular plants.

40-17 Lichen structure
(A) *The close relationship between the fungus and alga making up the lichen is apparent in the cross section of thallus and fruiting body.* **(B)** *Scanning electron micrograph showing algal cells being enveloped by fungal hyphae.* (× 10,000)

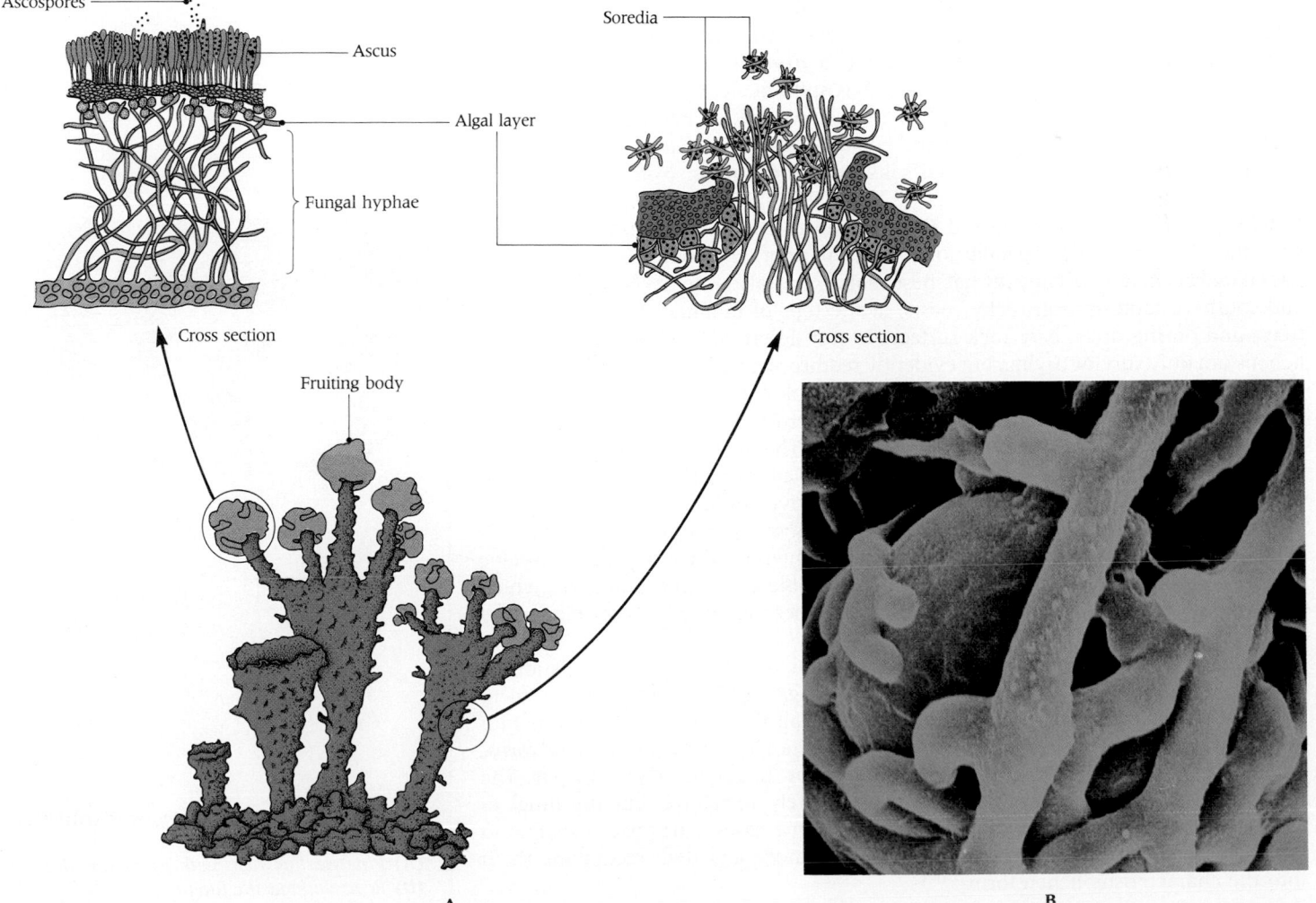

40-18 Lichens can be found on tree bark or stone.

LICHENS

The lichens are a familiar enough group of organisms, but their peculiar character has long confounded taxonomists, one of whom recently wrote, "The species concept in lichenology in the twentieth century is pre-Darwinian." The reason is that a lichen is a symbiotic association of a fungus and a photosynthetic alga—two quite distinct kinds of organisms—that grow together, always intertwined in a close obligatory association within a **thallus,** a new and distinctive morphological entity that cannot exist without both components. The term **phycobiont** refers to the photosynthetic partner. The fungal partner is a **mycobiont.** After two centuries of study, the relationship of the components is still not clearly understood.

The fungus portion of a lichen forms a dense web of hyphae within which the algae grow (Figure 40-17). The fungus obtains organic food from the algae, and the algae obtain water and dissolved salts from the fungus. Note the soredia in Figure 40-17A. These are fragments of algal cells and fungal hyphae by which lichens propagate. The strong and durable construction of the fungus, together with the photosynthetic capability of its algal partner, is a robust association that has enabled lichens to occupy the harshest habitats on Earth—in the farthest northern and southern latitudes, on rocky coasts, at the tops of mountains, on the bark of trees, and on the driest bare rock surfaces in hot deserts. Recent work shows that lichens not only survive drying, but evidently require alternating wet and dry periods. Lichens are familiar to us as the scalelike, varicolored patches on rocks or tree trunks (Figure 40-18). In Scandinavia and northern Canada, they form extensive ground covers where caribou and reindeer graze. The ability of lichens to grow in dry, harsh environments—often with no soil at all—exceeds that of any plant and provides a stunning example of nature's adaptive versatility.

Lichens are ecological pioneers that often play an important role in the first steps of breaking down rocks into soil. Those few lichens in which the photosynthetic partner is a blue-green alga (cyanobacterium) are able to fix atmospheric N_2, which then leaches into the environment where it becomes available to other pioneering organisms.

The fungal partner of a lichen is the dominant member of the association: it determines the type of thallus. Some 25,000 fungal species have been found in lichens. When the fungus is an ascomycetes we call the lichen an *ascolichen*. In a *basidiolichen,* the fungus is a species of basidiomycetes. There are also *deuterolichens.* The alga may be any of 12 genera of blue-green or 26 genera of green algae. The specific algae found in lichens may grow separately in nature, but the fungi in lichens are generally found only in lichens. In some cases it has been possible to grow the algae and fungi separately in the laboratory and then recombine them into the characteristic lichen form.

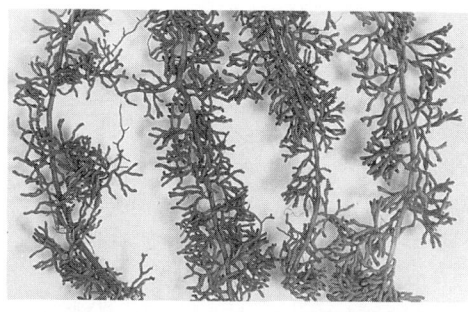

A

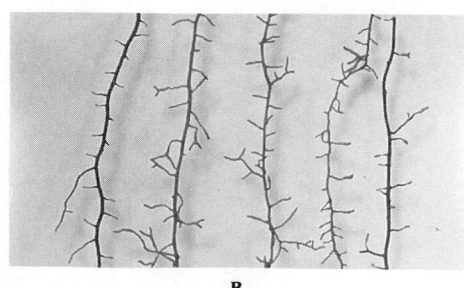

B

40-19 Ectomycorrhizal fungus Pisolithus tinctorius
(A) *Pine seedling roots with ectomycorrhizae.*
(B) *Roots without the fungus.*

40-20 Endomycorrhizae penetration of orchid root

In this cross section of the root of an orchid, fungal hyphae are penetrating root cells. This results in intracellular digestion of plant cells.

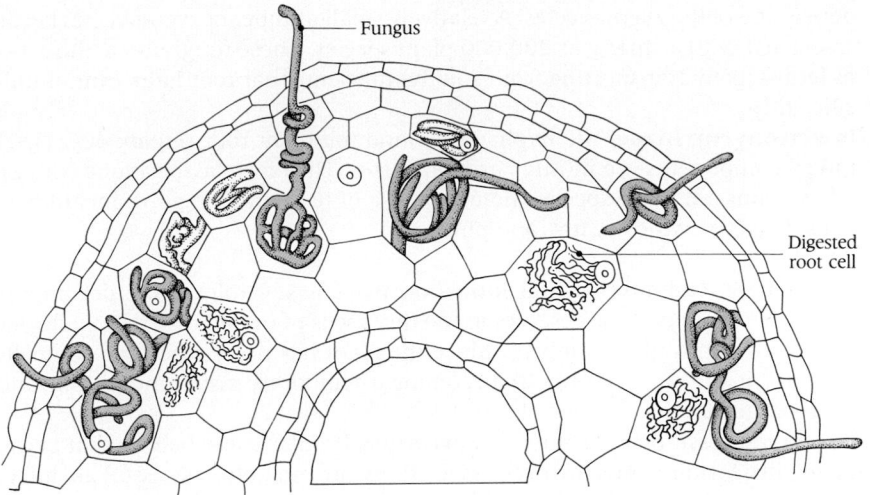

The unusual abilities of lichens to absorb substances dissolved in rain and dew have made them extraordinarily sensitive to airborne pollutants. Absorbed pollutants damage chlorophyll and thus impair photosynthesis. This unravels the delicate physiological balance between alga and fungus and degradation of the lichen follows.

Since air pollutants affect the distribution of lichens around industrial cities, lichens have been extensively viewed as indicators of air quality. Many lichen varieties are also used to monitor ozone in the air above the mountains surrounding Los Angeles—and to monitor radioactive isotopes in areas around uranium mines. It is a paradox that the very properties that allow lichens to colonize the harshest environments make them highly susceptible to man-made pollutants.

MYCORRHIZAE

The symbiotic association of a nonpathogenic fungus and a root of a vascular plant is called a **mycorrhiza** (Figure 40-19). It is estimated that the roots of about 80% of all plant species are associated in symbiotic relationships with certain species of fungi.

The fungus lives as an invader of the root and derives organic carbon and other nutrients from it. The plant also benefits. Fungi associated with roots act like root hairs, aiding in the direct transfer of phosphorus, zinc, copper, and other nutrients from the soil.

Two types of mycorrhizae are recognized. In both types, the mycelia may extend far out into the soil.

- In **endomycorrhizae,** the hyphae penetrate the outer cells of the plant root and also extend out into the surrounding soil (Figure 40-20). The fungal compo-

40-21 Ectomycorrhizae penetrating a root

This cross section of a plant root shows the sheath of ectomycorrhizae (color) with hyphae extending between the plant cells.

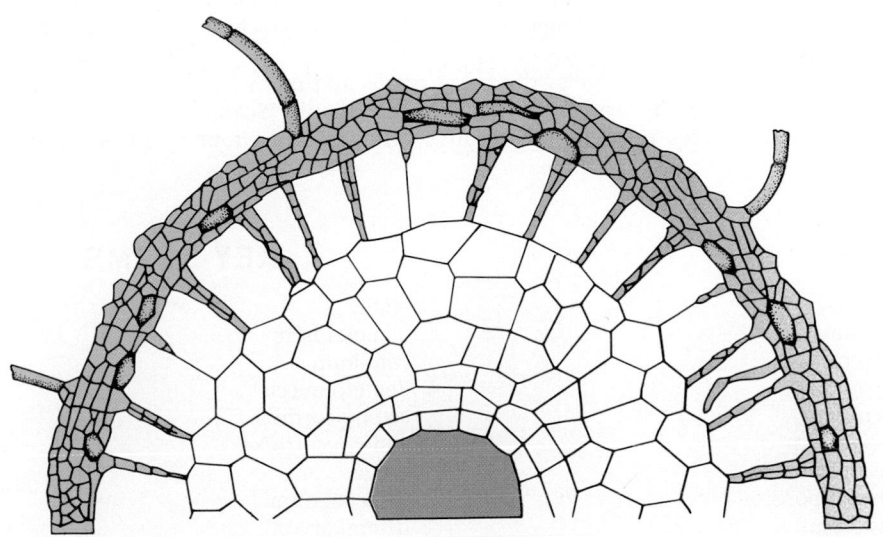

nent is one of the zygomycetes. A relatively small number of zygomycetes become associated with as many as 200,000 plant species. These fungi mimic the actions of fertilizers by transporting from soil the nutrients that root hairs cannot utilize efficiently.

- In **ectomycorrhizae,** the hyphae surround the plant root (Figure 40-21). The fungal components are mostly basidiomycetes. They form associations with only a few thousand plant species, notably those of temperate regions, including oaks, beeches, willows, firs, and pines.

Interestingly, endomycorrhizal formation plays a key role in the development of orchids. The minute food reserves in the tiny seeds of orchids are rapidly depleted after a few divisions of the embryo. Since the plant has not yet become completely autotrophic, it can survive only by becoming a temporary symbiont with a fungus (see Figure 40-20).

The earliest fossil plants have endomycorrhizal roots. Some believe that because recent plants forming mycorrhizal associations are notably successful in infertile soils, the possession of endomycorrhizal roots may have enabled the earliest plants to invade lands with soils that were lacking organic matter.

SUMMARY

Many fungi bear a superficial resemblance to plants (and fungi were once classified as plants), but they contain no chlorophyll and they are all heterotrophic. Some fungi—notably the yeasts—are unicellular, but most include the filamentous, multicellular structures called hyphae. Because of their broad morphological similarity, fungi are generally classified according to their method of reproduction. Fungi are also characterized by the common presence of chitin in their cell walls.

In both sexual and asexual phases, fungal reproduction almost always depends on airborne spores, which can germinate into hyphae. Hyphae collectively make up a mass called the mycelium. A homokaryotic hypha contains genetically identical nuclei. A heterokaryotic hypha contains nuclei of two or more genetic types. This phenomenon (heterokaryosis) is unique to the fungal kingdom. In all fungi, meiosis occurs immediately after zygote formation. Hence, the zygote is the only diploid form in the entire life cycle of fungi. The products of meiosis become spores.

Fungi play a critical ecological role in promoting the decay of dead organisms of all kinds. They also can cause many serious diseases of plants and animals (including humans). Fungi also have many useful properties from the human viewpoint. Some of them synthesize powerful antibiotics. They raise bread dough, ferment beer and wine, and ripen cheese.

Phylum Oomycetes includes the water molds, white rusts, and downy mildews. All form motile zoospores. Unlike other fungi, they are diploid during most of their life cycle. They also lack chitin in their cell walls.

Phylum Zygomycetes includes the bread molds and many soil saprophytes. Neither the spores nor gametes are motile. Sexual reproduction depends upon the conjugative fusion of cells from different hyphae. Zygomycetes form cell walls only during the formation of reproductive elements. Their hyphae are ordinarily multinucleate.

Sac fungi, so called because of the shape of their sporangia, comprise the phylum Ascomycetes, a diverse collection that includes yeasts, mildews, *Neurospora*, truffles, and the causal agent of Dutch elm disease. Indeed, Ascomycetes is the largest fungal phylum. Most of them (yeasts excluded) have in common the saclike ascus within which haploid spores are produced. Cells within the heterokaryotic hyphae of ascomycetes are multinucleate. The Ascomycetes genus *Penicillium* is responsible not only for the antibiotic penicillin, but for the flavors of Camembert and Roquefort cheeses.

Members of phylum Basidiomycetes typically have club-shaped sporangia. This phylum includes mushrooms, toadstools, puffballs, and bracket fungi. Cells within the heterokaryotic hyphae of basidiomycetes are dikaryotic, containing nuclei of opposite mating types (sexes).

Phylum Deuteromycetes (Fungi Imperfecti), unlike the other fungal phyla, are not known to reproduce sexually. However, haploid nuclei from different parents in a syncytium sometimes fuse, allowing gene transfer in the absence of true sexual reproduction. Deuteromycetes are classified according to the properties of their spores and spore-forming structures.

Fungi occasionally live with other organisms in symbiotic relationships so close that to the uninitiated the combination of two species may seem like one. This is the situation in lichens, in which fungi and algae live together, and in mycorrhizae, in which fungi live around and within the roots of various plants. There they act as root hairs that aid in nutrient absorption. About 80% of all plants have mycorrhizae.

KEY TERMS

antheridium	chitin	homokaryon
antibiotic	conidiophore	hypha
ascomycetes	conidium	lichen
ascospore	deuteromycetes	mycelium
ascus	ectomycorrhiza	mycorrhiza
basidiocarp	endomycorrhiza	oogonium
basidiomycetes	fungus	oomycete
basidiospore	gametangium	oospore
basidium	heterokaryon	penicillin

saprobe
septum
sporangium

spore
symbiont
thallus

zoospore
zygomycetes
zygospore

QUIZ QUESTIONS

1. Which of the following is a type of asexual spore?
   A. conidia
   B. zygospore
   C. oospore
   D. ascospore
   E. basidiospore

2. Sexual reproduction occurs in all of the following groups except
   A. Ascomycetes.
   B. Basidiomycetes.
   C. Deuteromycetes.
   D. Oomycetes.
   E. Zygomycetes.

3. Which of the following groups of fungi includes species that are autotrophic?

   A. Ascomycetes
   B. Basidiomycetes
   C. Deuteromycetes
   D. Oomycetes
   E. None of the above

4. Cell walls of various fungi are made of materials including
   A. cellulose.
   B. mycelium.
   C. chitin.

   D. A and B
   E. A and C

5. Formation of a crozier at the tip of heterokaryotic hypha is characteristic of sexual reproduction in
   A. Ascomycetes.
   B. Basidiomycetes.
   C. Deuteromycetes.
   D. Oomycetes.
   E. Zygomycetes.

ESSAY QUESTIONS

1. What characteristics distinguish fungi from plants? Why were they once classified as plants?

2. Do fungi contain cells? How do cells of fungi differ from the cells of plants?

3. What are hyphae? Mycelia? What is a spore, and what distinguishes a spore from a gamete; a plant seed? What is a heterokaryon?

4. Why do lichens pose a difficult problem for taxonomists?

5. In what ways can fungi be harmful to the interests of human beings? Provide concrete examples. How have the fungi benefited humankind?

6. What role do fungi play in Earth's ecosystems?

REFERENCES AND SUGGESTIONS FOR FURTHER READING

Ahmadjian, V. (1982). The nature of lichens. *Natural History 91,* March, 30–37.

   The author reports on experiments relating to the formation of lichens.

Botstein, D., and Fink, G. R. (1988). Yeast: an experimental organism for modern biology. *Science 240,* 1439–1443.

   The authors explain why the yeasts *Saccharomyces cerevisiae* and *Schizosaccharomyces pomb* have become popular and successful models for research in eukaryotic biology at the cellular and molecular levels.

Courtenay, B., and Burdsall, H. H. (1982). *A Field Guide to Mushrooms and Their Relatives.* Van Nostrand, Reinhold, New York.

   A useful manual that identifies more than 350 well-illustrated and clearly described species of basidiomycetes and ascomycetes.

Garroway, M. O., and Evans, R. C. (1984). *Fungal Nutrition and Physiology.* John Wiley & Sons, New York.

   A well written and attractive textbook on mycology with emphasis on fungal nutrition.

Hale, M. E. (1983). *The Biology of Lichens,* 3rd ed. University Park Press, Baltimore.

   A general treatise of all of the biology of lichens. Function, form, diversity, and ecology are summarized.

Kosilowski, K. V. (1985). Cheese. *Scientific American 252,* May, 88–99.

   A fascinating summary of the microbiological basis of cheese-making.

Lawrey, J. D. (1984). *Biology of Lichenized Fungi.* Praeger, New York.

   A review of current thought on the lichens and the biological roles of the fungal components of these unique symbiotic relationships with algae.

Moore-Landecker, E. (1982). *Fundamentals of the Fungi.* Prentice-Hall, Englewood Cliffs, N.J.

   An excellent general textbook of mycology.

Rayner, A. D. M., Brasier, C. M., and Moore, D. (Eds.) (1987). *Evolutionary Biology of the Fungi.* Cambridge University Press, New York.

   This authoritative multi-authored work appears at a time of rapid progress in the study of fungi. The major message of these occasionally technical papers: taxonomists can no longer ignore genetic, ultrastructural, and molecular data.

Ross, I. K. (1979). *Biology of the Fungi: Their Development, Regulation, and Associations.* McGraw-Hill, New York.

   One of the classic books on many aspects of the biology of fungi. Structure, development, physiology, reproduction, and ecology are all discussed.

Timberlake, W. E., and Marshall, M. A. (1989). Genetic engineering of filamentous fungi. *Science 244,* 1313–1317.

   New techniques have made possible the genetic engineering of some of the fungi important in medicine, industry, agriculture, and biological research.

Webb, A. D. (1984). The science of making wine. *American Scientist 72,* 360–367.

   The author interestingly explains how an ancient practical art became a scientific branch of microbiology, biochemistry, and engineering.

Wicklow, D. T., and Carroll, G. C. (Eds.) (1981). *The Fungal Community: Its Organization and Role in the Ecosystem.* Marcel Dekker, New York.

   Part of a series on all aspects of mycology. The emphasis in this volume is on the ecological roles of fungi, a broad field that the editors say has been neglected.

PLANTS

In scientific terms, we would describe the kingdom Plantae as a large group of multicellular, photosynthetic, eukaryotic organisms that are mainly terrestrial and possess chloroplasts containing chlorophylls *a* and *b,* and carotenoids and cell walls containing cellulose. In simpler terms, the green land plants we know best push up freely from the soil into the air and are conspicuous in our own environment.

Although plant classification is sometimes clear-cut, taxonomic uncertainties persist in this kingdom as in others, and again we will occasionally need to make difficult choices and arbitrary assignments.

In this book, we divide plants into two basic groups or subkingdoms—**Bryophyta** and **Tracheophyta.** In accord with botanical custom, we use the term *division* in place of *phylum.*

- The bryophytes are **nonvascular** land plants (liverworts, hornworts, and mosses). Lacking water-conducting and food-conducting tissues, they lack true leaves, stems, and roots.
- The tracheophytes, a vastly larger group, are **vascular** plants (club mosses, horsetails, ferns, and seed plants) that contain the conducting tissues xylem, which transports water and salts, and phloem, which transports sugars and other metabolites (see Chapter 7 and Chapter 24).

This chapter surveys the major plant groups and considers their evolutionary history. Other aspects of plant biology are discussed elsewhere.[1]

EVOLUTIONARY HISTORY

Both plant groups, as we have noted, probably descended from ancestral multicellular green algae more than 450 million years ago (Figure 41-1). Support for this idea comes from the remarkable similarity between the chloroplast genome in a flowering tobacco plant (*Nicotiana*) and the chloroplast genome of a primative bryophytic nonvascular plant (*Marchantia*).

Plants most likely invaded the land in the Silurian period of the Paleozoic era—about 415 million years ago. Land plants first appear in the fossil record of the Devonian period as rootless and leafless seaweedlike organisms that were nonetheless upright. The earliest fossil land plants were three sorts of tracheophytes—the extinct groups Zosterophyllophyta, Rhyniophyta, and Trimerophytophyta (Figure 41-2).

We now believe that Chlorophyta (green algae) early gave rise to ancestral groups that further evolved into the two major categories of higher plants. One became

Plants
The study of plants, their taxonomy and physiology, is among the most dynamic branches of modern biology. Their importance to human life is both practical and esthetic. Consider, for example, the beauty of this graceful ponderosa pine tree.

[1] The following topics are discussed in other chapters: plant cell structure (Chapter 5); plant cell walls (Chapter 6); tissue structure (Chapter 7); photosynthesis (Chapter 10); reproduction (Chapter 21); fluid transport (Chapter 24); nutrient absorption (Chapter 26); respiration (Chapter 27); growth, development and hormonal coordination (Chapter 28); effectors (Chapter 31); fungi (Chapter 40); evolution (Chapter 44); and ecology (Chapters 46 and 47).

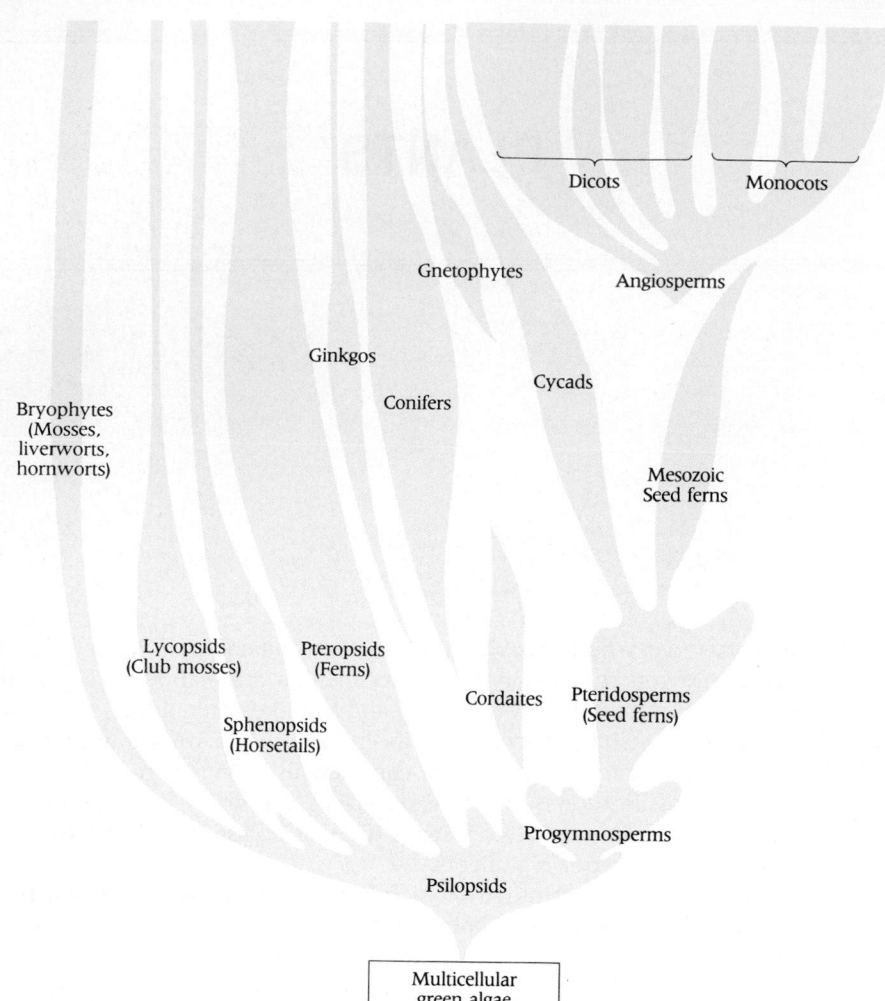

Dicots Monocots

Gnetophytes Angiosperms

Ginkgos

Cycads

Bryophytes (Mosses, liverworts, hornworts)

Conifers

Mesozoic Seed ferns

Lycopsids (Club mosses)

Pteropsids (Ferns)

Cordaites Pteridosperms (Seed ferns)

Sphenopsids (Horsetails)

Progymnosperms

Psilopsids

Multicellular green algae

the **bryophytes,** the other the **tracheophytes.** Together with fungi and one group of animals (insects), plants are the only major group of organisms that occur almost exclusively on land. It is likely that plants did arise on land (or shallow fresh water) and that the ancestors of the plants were the first forms to become terrestrial, arriving on land before the early arthropods, mollusks, and chordates. We shall trace some of the details of this fascinating story when we view the grand panorama of life's history in Chapter 44.

PROBLEMS CONFRONTING PLANTS ON LAND

It is interesting to consider what adaptational problems had to be solved before plants could survive on land. First, there was the need for keeping cells supplied with water when no longer immersed in water. This serious problem was solved when plants developed a number of physiological and structural adaptations.

- Dessication-resistant spores
- A waxy cuticle layer on their outer surfaces that retards water loss
- Stomata in leaf surfaces that permit sensitive control of water exchanges (see Chapter 10)
- Vascular tissues that make possible the transport of needed solutions to cells far from an external water source

Second, there was a need for specialized plant organs. A land plant needs water and dissolved minerals from the soil below and it needs sunlight and CO_2 from the air above. Hence, there had to occur a differentiation of specialized organs— **roots** to serve as an absorptive system and land anchor, and **leaves,** structures to house the photosynthetic apparatus. Between roots and leaves is the **stem,** a structural element containing the conduits of the conductive system. The stem had a

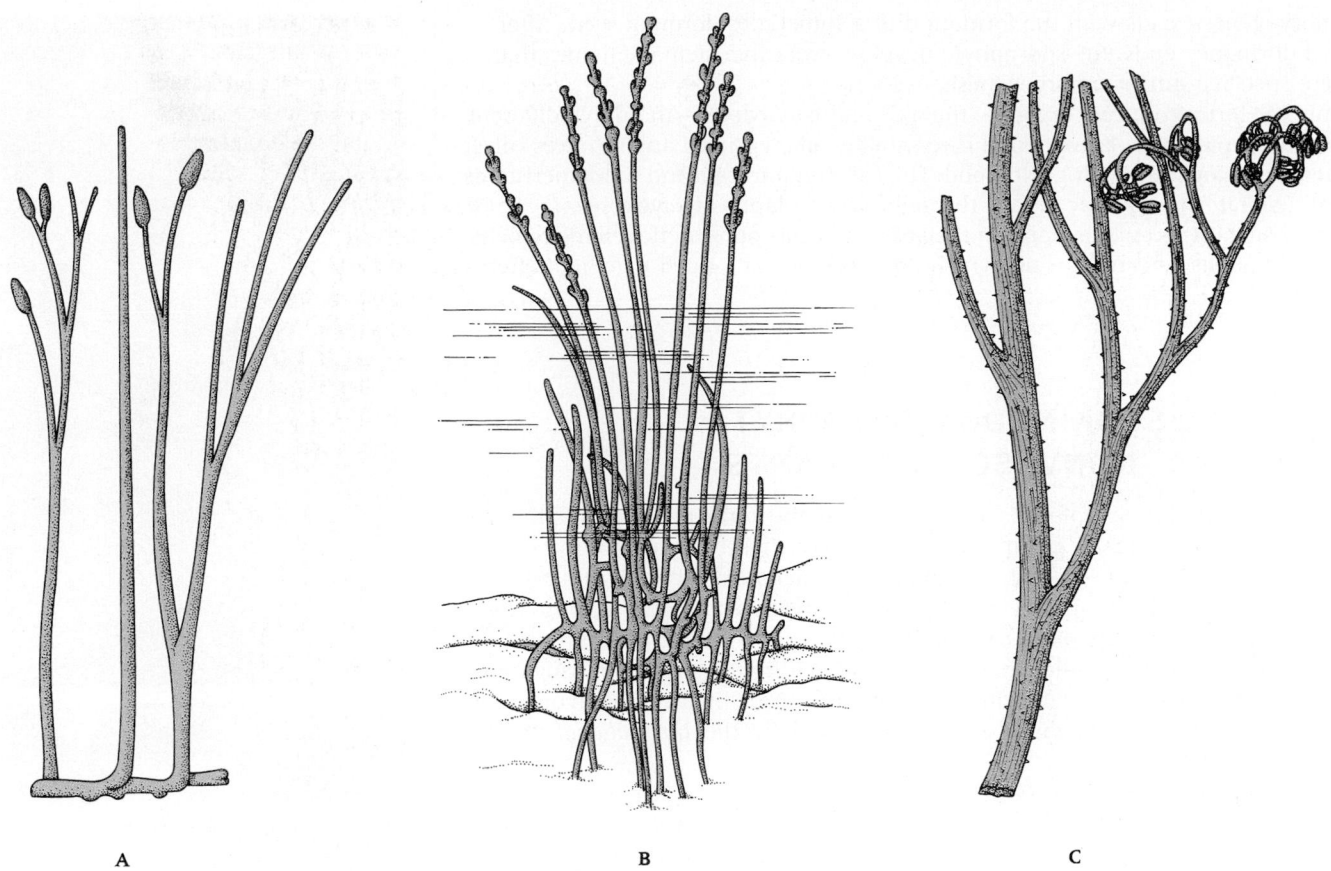

A	B	C

41-2 Early vascular plants
(A) *Members of the group Rhyniophyta, the simplest of the known vascular plants, have shoots that are leafless and a simple branching system.* Rhynia major *is shown here.* **(B)** *In the Zosterophyllophyta, the sporangia are aggregated at the end of the shoots, and the plants are simply branched. The example is* Zosterophyllum. **(C)** *Trimerophytophyta consisted of plants with a strong central axis, smaller side branches, and terminal sporangia, as in* Psilophyton princeps, *shown here. The terminal sporangia in these primitive forms (and on the tips of mosses) is evidence that their early evolution took place in shallow water.*

further requirement. Organisms living in water could simply float. But plants living on land needed strong stems, rigid enough to support both the stem and its attachments (leaves, in time fruits, and so on). Many land plants evolved highly specialized supporting tissues. In some divisions their stems became woody; in others they acquired specialized tissues such as collenchyma and sclerenchyma (see Figure 7-2C,D).

Third, there was a need to develop physiological adaptations that would meet certain specific challenges—among them the need for enough moisture on plant surfaces for efficient gas exchange (achieved by transpiration-driven transport of water from roots to leaf surfaces), and the need to withstand extremes of temperature (by evaporation of moisture from leaf surfaces), ultraviolet radiation (absorbed by pigments responsible for flower colors), wind (by stem strength), and other environmental hazards. Much of the story of land plant evolution must be understood in the context of these physiological adaptations.

Fourth, there was the need to carry on sex and reproduction under conditions in which spores and male or female gametes would not simply be shed into an enveloping watery medium. Conquest of this difficulty required a long series of specializations in land plants that culminated in **pollen** and **seeds.**

PLANT REPRODUCTION AND DEVELOPMENT RECALLED

The complexities of plant reproduction were discussed in Chapter 21. However, let us here summarize once again some of the principles of seed plant reproduction, most of which are unique to plants.

- Plant reproduction relies on both spores and gametes which function in alternating generations.
- Two independent processes trigger sporophyte development in angiosperms: **pollination** (involving a specific interaction between pollen grain and pistil) and **fertilization** (involving an interaction between sperm nuclei and three of the eight female gametophyte cells). In angiosperms, the ovule is sometimes covered with a protective carpel wall. In gymnosperms, it is exposed to the air at the time of pollination.

- Embryogenesis ends with the formation of a sometimes dormant seed. After seed dormancy ends, the sporophyte develops from meristematic tissues that were specified during embryogenesis.
- Embryos form from two organs—the axis and cotyledons—that have different developmental fates. A cotyledon is terminally differentiated and senesces (dies) after supplying the seedling with food. The axis contains root and shoot meristems that generate sporophytic organs throughout the plant's life cycle.
- Morphogenesis occurs in plants in the absence of morphogenetic cell movements because plants cells have walls which are more or less "glued together" after cell division.

SUBKINGDOM BRYOPHYTA: NONVASCULAR PLANTS

For reasons of tradition, botanists have preferred the term division to phylum. Botanists also employ many group names that differ slightly from those preferred by other systematists. The endings, or suffixes, of categories generally indicate the ranking of higher taxa. Divisions (phyla) end in *-phyta*, subdivisions in *-phytina*, classes in *-opsida* (though some systematists have used -opsida for divisions) or *-ae*, orders in *-ales*, and families in *-aceae*. A problem here, as in other kingdoms, arises from the wide use of informal or common group names. To avoid such difficulties, botanical names must now be approved by the *International Code of Botanical Nomenclature*.

The **bryophytes** are the liverworts, hornworts, and mosses. (The suffix *-wort* is an archaic term for "herb.") They are simple inconspicuous plants—some 24,000 living species of them—that mostly grow in moist places on land and thus might be called amphibious. They lack (or nearly lack) vascular systems and this limits their size and the range of possible land environments. But compared to green algae, bryophytes are well advanced in their adaptation to the land.

Bryophytes grow most luxuriantly in wet, shady places and in bogs. Some are aquatic, though none is marine. A few manage to live in Arctic wastes and arid deserts. Those hardy species survive cold and dryness by suspending vital activities until warmth and moisture revive them. Even on dry land they make for themselves a virtually aquatic habitat, the low tangled vegetation of mosses holding water like a sponge. Usually they form a dense mat that is inhabited by bacteria, algae, fungi, worms, insects, snails, and others (Figure 41-3).

We spoke in Chapter 21 of the reproductive cycle of the bryophytes. The familiar green plant that is readily visible to the eye is the haplophase gametophyte. It is the dominant stage. The sporophyte is a simple structure containing little or no chlorophyll, a virtual parasite depending on the gametophyte in most cases.

Although remarkable for the ability of many to revive after drying, bryophytes do require moisture for sexual reproduction. A flagellated bryophyte sperm must swim through a film of rain or dew in order to meet and fertilize an egg. The resulting zygote divides mitotically to produce a diploid sporophyte. With flagellated sperm cells capable of swimming, they were never fully emancipated from their ancestral aquatic environments.

Bryophytes are relatively primitive compared with vascular plants in terrestrial habitats. Without true vascular tissues they cannot ascend to gather light. Since they cannot transport fluids internally for long distances, they must be wet at some time to complete their life cycle. That means their gametes are highly vulnerable. Although the better developed among them have structures superficially resembling roots, stems, and leaves, these are not true vascular organs. Yet, despite these limitations, the bryophytes have survived and are relatively abundant. Indeed, they are the most abundant plants in Arctic and Antarctic regions.

One reason for this situation is that some environments exist for which bryophytes are better adapted than other plants. Peat bogs are such locations. They characteristically contain mosses of the genus *Sphagnum* (some 300 species), which in total actual mass far outweighs all of the other mosses of the world combined. Another reason bryophytes survive is their toughness. If they cannot compete for the "good things of life," they manage to make do with what they have. They are often pioneers, spreading into places not yet reached by other vegetation. When, in part

41-3 A variety of mosses

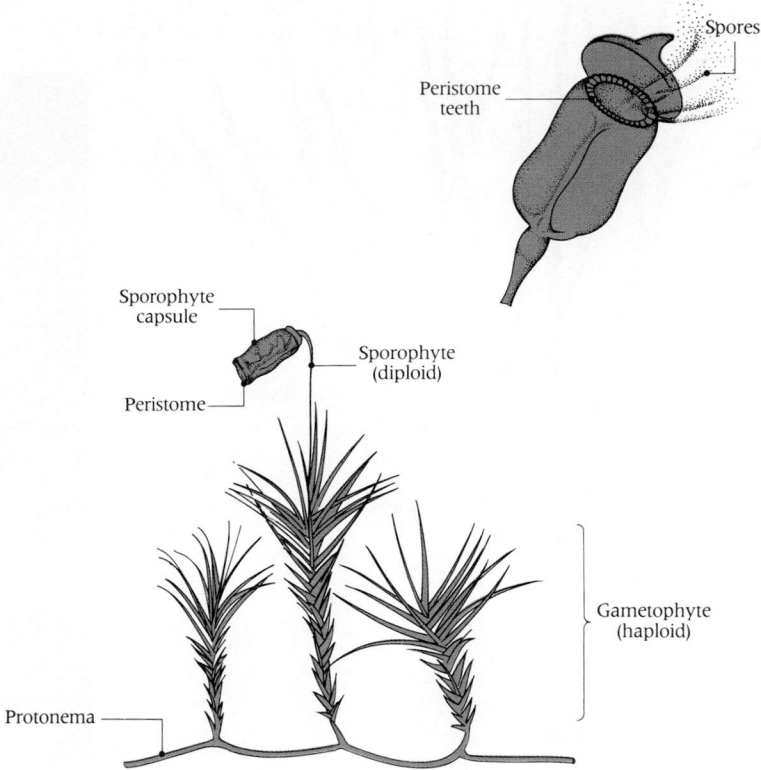

41-4 Reproduction in a common woodland moss

In Polytrichum, *the diploid sporophyte ends in a capsule that contains the spores. The enlarged view at the upper right shows release of spores through the peristome (opening) of the capsule.*

through their own activities, conditions are improved, other plants move in and the pioneers may be crowded out.

There are three classes of bryophytes.

- Class **Hepaticopsida** (also called Hepaticae), called **liverworts** and named for the liverlike shape of the thallus of some, includes about 9000 species of leafy and nonleafy forms.
- Class **Anthocerotopsida** (also called Anthocerotae) are the **hornworts,** about 400 species of worldwide distribution. The sporophytes of hornworts have a hornlike shape; hence their name. They are green and photosynthetic and thus, to an extent, nutritionally independent of the gametophytes to which they are attached.
- Class **Muscopsida** (also called Bryopsida), the largest class, includes the **mosses,** some 14,500 living species. Many vascular plants have traditionally been called mosses by nonbiologists. In fact, "reindeer mosses" are lichens, "Spanish mosses" are vascular plants, and "sea mosses" are algae. The reproductive cycle of a common moss species is shown in Figure 21-13. It is further displayed in Figure 41-4.

Bryophytes are well represented in the fossil record dating from the late Paleozoic, but in later times they were probably never dominant land plants.

SUBKINGDOM TRACHEOPHYTA: VASCULAR PLANTS

All groups of the plants still to be dealt with have vascular tissues. All had the evolutionary potential to rise into the air and become upright, an evolutionary path nearly opposite that of the bryophytes.

- In the bryophytes, the gametophyte became the principal vegetative stage in the life cycle. That development was successful in a limited way, but it seems to have taken those plants into an evolutionary blind alley from which no radical further progress was possible.
- In tracheophytes, the vascular plants, the gametophyte became reduced until in the latest and most progressive groups it is microscopic and transient. The

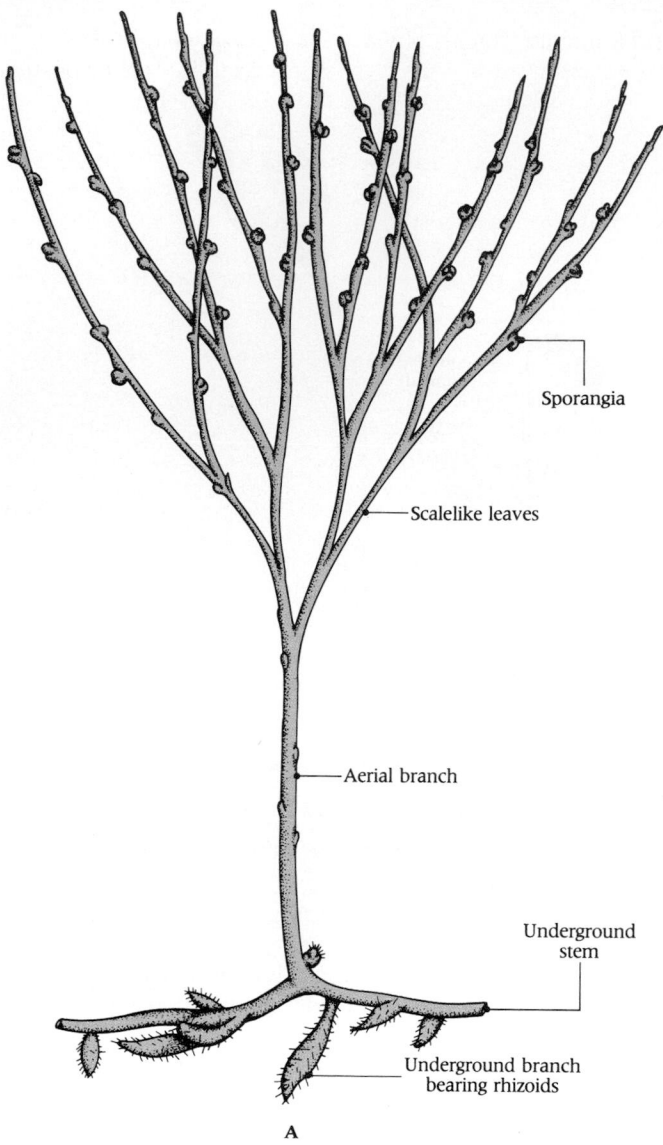

Sporangia

Scalelike leaves

Aerial branch

Underground
stem

Underground branch
bearing rhizoids

A

B

female gametophyte became a well-protected parasite on the sporophyte, and the sporophyte became in all vascular plants the principal vegetative photosynthetic phase in the life cycle.

Plants that had reached only the early phases of these important evolutionary changes were at first spectacularly successful in covering the land with vegetation. Later most of them were replaced by plants with more specialized reproductive cycles that involved seeds. A few relicts[2] of earlier groups live on, however, and one such group, the ferns, has survived in abundance.

The vascular plants (subkingdom Tracheophyta) developed countless adaptations that fitted them to all but the harshest terrestrial environments. In the course of this success, they diverged enough to form five divisions (or subdivisions if Tracheophyta is considered a division rather than a subkingdom—as some systematists prefer).

- Division Psilotophyta: whisk ferns
- Division Lycopodiophyta (also called Microphyllophyta): club mosses
- Division Sphenophyta (also called Anthrophyta): horsetails
- Division Pteridophyta: ferns
- Division Spermophyta: seed plants

[2] A relict is a species or other taxon of organisms that survives, sometimes in a limited region, after most of its close relatives have become extinct.

41-5 Psilophyte structure
(**A**) *The sporophyte of* Psilotum. (**B**) *The sporophyte of a living* Psilotum.

41-6 A lycopod

All members of the subkingdom Tracheophyta (save a few) have at least four major attributes that are fundamental adaptations for a life on land: (1) they possess specialized vascular conducting tissues—always xylem and usually phloem; (2) with exceptions, they produce multicellular embryos that are retained within the female reproductive organ, the archegonium, which is disrupted as the embryos expand into gametophytes; (3) they have an outer protective cuticle on exposed parts that prevents desiccation; and (4) their leaves have stomata surrounded by guard cells, which act to prevent water loss. It was largely these adaptions that allowed plants to spread over the land.

DIVISION PSILOTOPHYTA: WHISK FERNS

The earliest known vascular plants are the **psilotophytes,** a group for which we have few common names other than whisk ferns. In the Silurian and Devonian periods of the Paleozoic era, some 400 million years ago, they dominated the landscape. Today they are all but extinct, with only two relict genera surviving—*Psilotum*

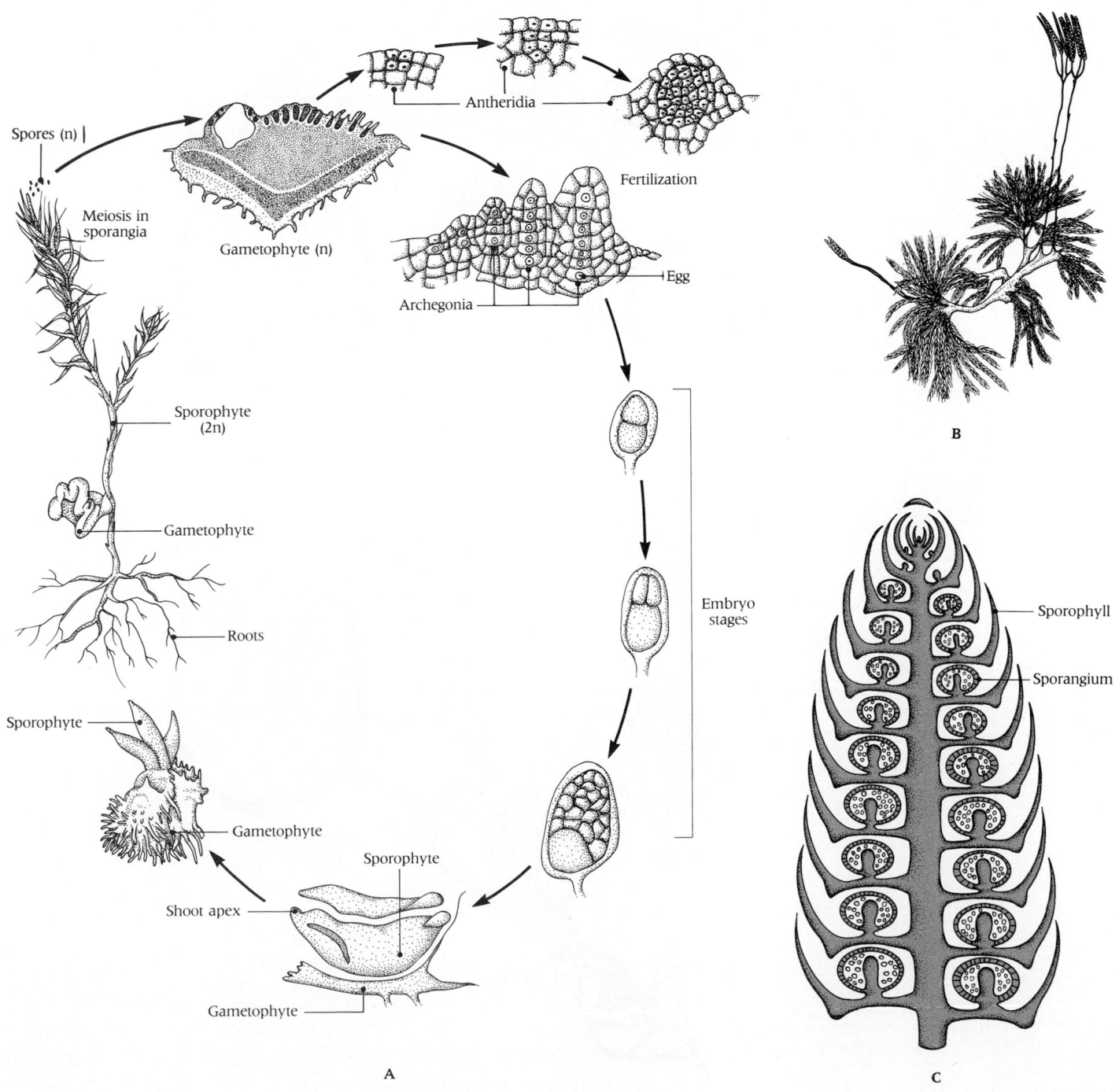

(Figure 41-5), which is mainly confined to the tropics, and *Tmesipteris*, which is confined to Australia, New Zealand, and other islands of the South Pacific.

The anatomy of *Psilotum* is almost diagrammatically simple. There are no true roots; a subterranean part of the stem serves the same needs. And there are usually no leaves. In the leafless plants, the aerial part of the stem carries on the necessary photosynthesis. Even the fossil record clearly shows that these were vascular plants—but were they sporophytes or gametophytes? The answer: they were sporophytes. Within sporangia, at the tips of some of the branches, meiosis produces haploid spores. These in turn give rise to subterranean gametophytes that produce both eggs and sperms. The gametes unite in fertilization to form diploid zygotes that develop into the prominent sporophyte plant shown in Figure 41-5A. In surviving forms, the gametophytes are quite in contrast—small, colorless, and subterranean—sometimes with symbiotic fungi. In the ancient psilotophytes, gametophytes are as yet unrecognized. They must have had them, but they were ill-suited for preservation as fossils.

The interesting suggestion has been made that the only living psilotophyte forms, *Psilotum* and *Tmesipteris*, may actually be primitive ferns or reduced ferns. There

41-8 Life cycle of Selaginella
In this heterosporous plant, two kinds of sporangia borne on the sporophyte give rise to two morphologically distinct kinds of gametophytes, both much smaller than the sporophyte. As in seed plants, the young sporophyte is enclosed by the tissues of the megagametophyte; and the major source of food for the developing embryo is material that was stored in the megaspore. However, there is no dormant period in the development of the embryo of Selaginella, as there is in most seed plants.

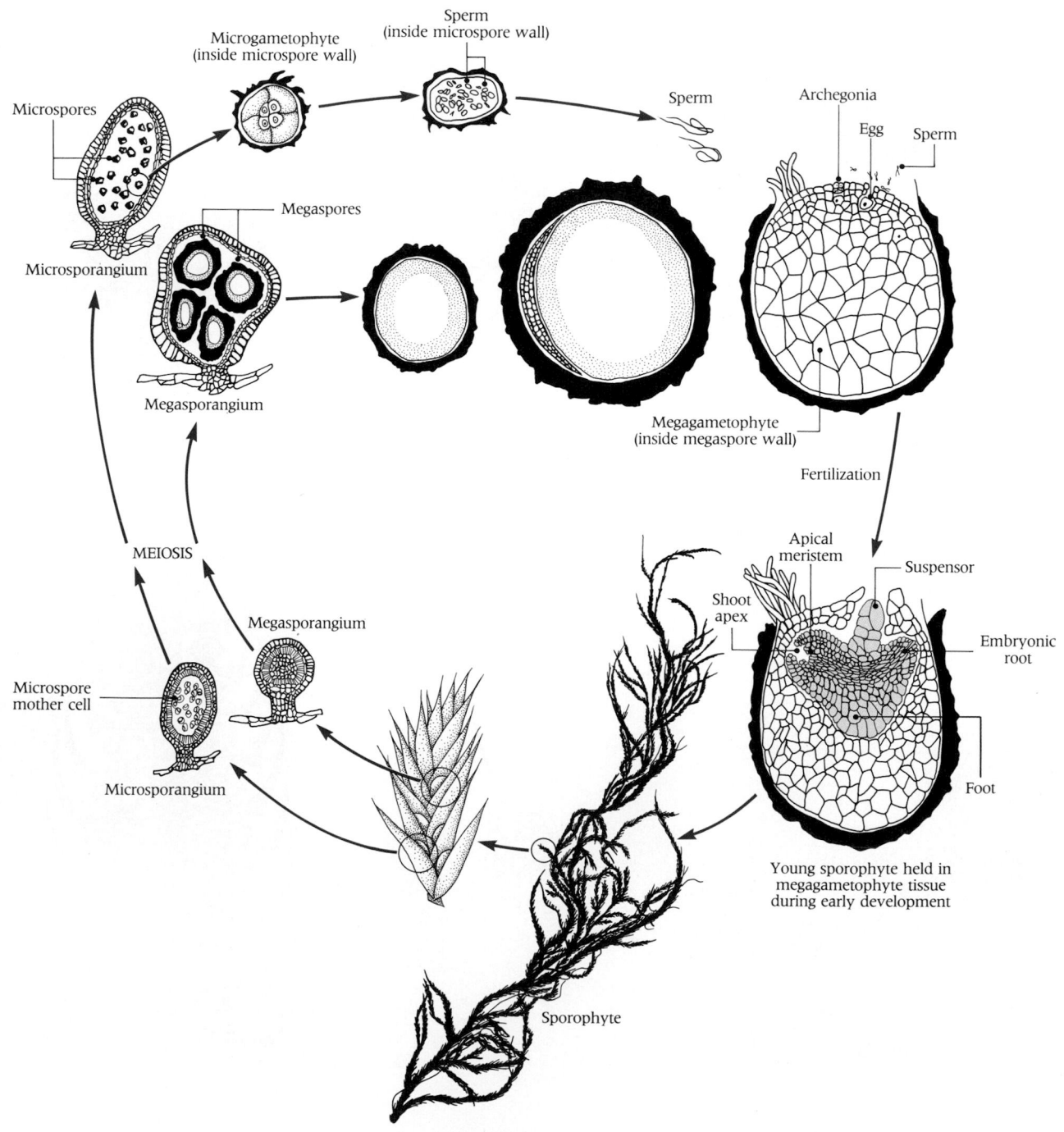

B

C

A

41-9 Sphenophyta (Above)
(A) The stalks of these plants bear reproductive cones (strobili). (B) Closeup view of Equisetum *strobili. (C) Fossil of an extinct sphenophyte,* Annularia, *from the Devonian. Note the whorls of leaves at stem joints.*

41-10 Ferns

are indeed no roots, but as we shall see soon, ferns may form roots only quite late in development. If correct, this thesis means that the psilotophytes are entirely extinct. In this book, we accept the two living psilotophytes as members of a separate phylum of simple, seedless vascular plants.

DIVISION LYCOPODIOPHYTA: CLUB MOSSES

The **lycopods** (division Lycopodiophyta) are another ancient and once dominant group that is now represented only by relicts. The five living genera of Lycopsida are *Lycopodium, Selaginella, Phylloglossum, Isoetes,* and *Stylites.*

These genera include some numerous and widespread species. Most living lycopods are low-lying herbs (Figure 41-6). Some of the vinelike species may be up to 20 meters long. In their time of glory, in the coal forests of the Carboniferous, some lycopods were great trees of the first forests, l to 2 meters in diameter and 30 to 40 meters high (see Figure 44-12). Some of our coal deposits were formed from these trees, which died out some 280 million years ago.

Unlike the psilotophytes, lycopods have true, if poorly developed, roots (Figure 41-7). They also have true stems and leaves and most are photosynthetic. Thus, their anatomy and reproductive cycle is more complex than that of the psilotophytes. Some—for example, *Selaginella*—produce two distinct kinds of spores—microspores, which develop into male gametophytes, and macrospores (termed megaspores by many botanists), which develop into female gametophytes. The gametophytes are tiny but independent organisms that live on organic debris in the soil. The life cycle of *Selaginella* is shown in Figure 41-8.

A few fossil lycopods show another remarkable advance. Their eggs developed in a well-protected case on the female gametophyte and were retained there even after they had been fertilized and had begun to develop into embryonic sporophytes. The result was something resembling seeds. (Some species of the living genus *Selaginella* exhibits the same property.) They are also an example of how independent but parallel responses to comparable selective pressures—the need for protection against dehydration—can lead to similar evolutionary changes.

DIVISION SPHENOPHYTA: HORSETAILS

All living **sphenophytes** belong to one genus, *Equisetum,* that includes only about 15 species (Figure 41-9). They are easy to find in bogs or in moist sandy soil near streams. Because of their appearance, some members of the genus *Equisetum* are known as "horsetails." Others are called "scouring rushes" because gritty silica in their stems made them useful for cleaning pots and pans in former days. Our native species are rarely over a meter tall, although some tropical equisetums reach greater heights. The equisetums, too, are relicts, well represented in the fossil record.

Nothing like a seed is known in the sphenophytes, and the spores are never today differentiated into microspores and megaspores. In degree of differentiation and life history, sphenophytes are like lycopods and ferns. They are, however, rather different in appearance and anatomical detail. The hollow stems are vertically ribbed and jointed with a whorl of small narrow leaves at each joint in the living genus. Oddly, the vestigial leaves contain little or no chlorophyll. Photosynthesis occurs mainly in the stems, which are green.

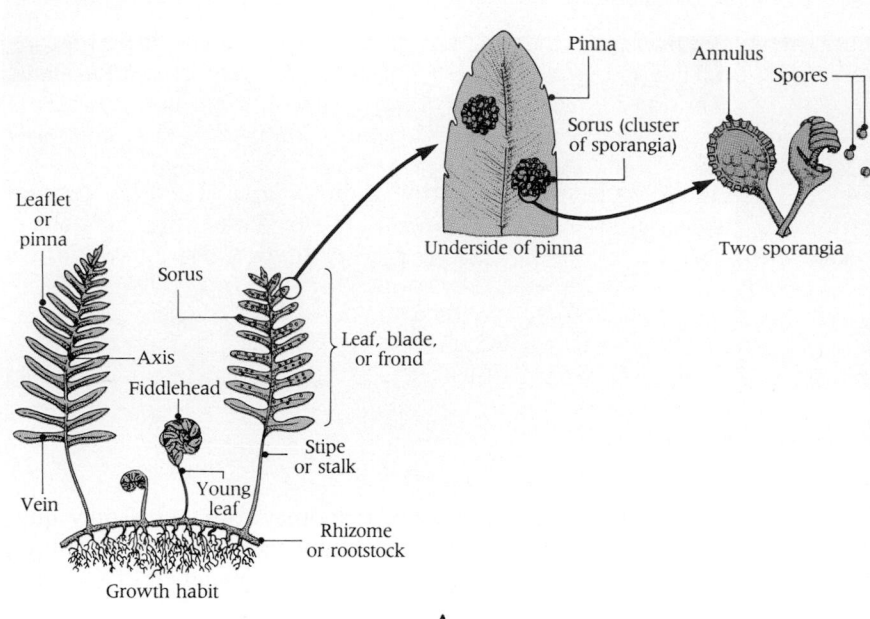

A

B

DIVISION PTERIDOPHYTA: FERNS

The fourth division of the vascular plant subkingdom (Tracheophyta) is **Pterido-phyta.** It includes the **ferns** (Figure 41-10).

Pteridophyta differs from Lycopodiophyta and Sphenophyta, other vascular non-seed plant divisions, in having a more specialized vascular system and large complex leaves. All vascular plants have leaves, but two basic types are discernible.

- The first type are small leaves with a single vascular channel that is an offshoot of the stem's vascular system. Such leaves, called **microphylls,** are typical of lycopods and sphenophytes.
- The second type, termed **megaphylls,** are the more familiar large-stemmed (petiolated) leaves with complex vein systems. They appeared first in the pterido-phyta, probably arising from a flattening of a branching stem system and develop-ment of a web of photosynthetic tissues between the branches.

Ferns, with their well-developed vascular systems, clearly illustrate this important evolutionary development.

All the major groups of nonseed plants reached their evolutionary climax at about the time of the Carboniferous coal forests. Two of these groups, the lycopods and sphenophytes, declined thereafter, though a few survivors remain. The still more ancient and originally ancestral psilopsids also straggled along even when overshadowed by their more exuberant descendants. The ferns were such a pro-gressive group. Although they have declined in importance since the Carboniferous, they are still so abundant and varied that they cannot be called relicts.

The leaves of a fern may be only a few millimeters long, but in some tropical climbing ferns they may reach the astonishing length of 30 meters or more. As shown in Figure 41-11, ferns have leaves called **fronds.** In most cases, the leaves are divided into leaflets **(pinnae),** which may in turn be subdivided. The last level above the "teeth" (serrations) of the leaf are called **pinnules.** Thus arises the lacy appearance so characteristic of familiar ferns. The stems have a well-developed vascular system, and true roots are usually present. In the familiar species of the temperate zone, the stems are usually horizontal, on or in the soil, with simple roots extending into the soil from the prostrate stem. In warm climates, ferns may have large, vertical stems or trunks which may develop "bark" and grow to 20 meters in height.

The reproductive and life cycle of ferns is similar to that of other vascular plants (see Chapter 21). To briefly recap the cycle, which is shown in Figure 21-14, haploid spores are produced by meiosis in the sporangia on the lower surfaces of some leaves (sporophylls). Shed spores develop into tiny, free-living haploid gametophytes (Figure 41-12). From the lower surfaces of these photosynthetic plants, rootlike

41-11 Fern sporophyte
(**A**) *Fern fronds are part of the sporophyte stage of the life cycle. Compare with Figure 21-14. They produce spores in clusters of sporangia called sori.* (**B**) *Note the many sori on the underside of the fertile frond of* Osmunda cinnamonea.

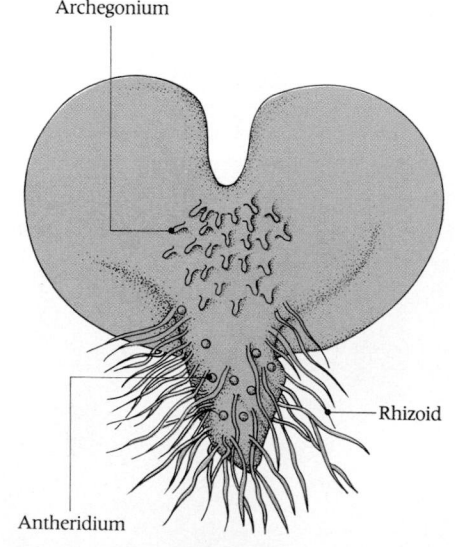

41-12 Fern gametophyte
The eggs in the archegonia are fertilized by sperm produced in the antheridia. Then the embryo develops into the sporophyte stage. The diagram is highly magnified.

41-13 Staghorn fern, Platycerium
This plant is an epiphyte. That means that it attaches itself to another living plant, but it is not parasitic.

41-14 Reproduction in angiosperms and gymnosperms
(A) *In angiosperms, the carpels form an ovary enclosing the seed, as shown in a magnified view in A_1.* **(B)** *In gymnosperms, the sporophyll is a flat, primitively leaflike structure bearing the seeds on its surface. This is seen in side view in B_1.*

filaments called **rhizoids** enter the soil. The gametophyte bears **archegonia** and **antheridia.** The former produces egg cells and the latter give rise to flagellated sperm cells that swim through an external watery medium to fertilize the eggs on the damp undersides of the gametophyte. The diploid sporophyte (the large leafy plant) develops directly from the zygote.

Most modern ferns are **homosporous**—that is, all spores are alike in appearance. However, there are two orders of aquatic ferns (Marsileales and Salviniales) which are **heterosporous**—that is, two types of spores (which germinate to form male and female gametophytes, respectively) are separate in the sporophyte. They are produced in different sporangia and male spores are always smaller and more numerous. In these orders, the spores are termed **megaspores** and **microspores.**

In some respects, ferns were no better adapted to life on land than were the ancient bryophytes. The vascularized sporophytes of ferns could survive in dry places and grow large, but the absolute dependence on moisture of their free-living gametophytes and sperms limited them to reasonably wet habitats. A few ferns are actually aquatic.

Many ferns manage to grow in cold or arid climates. But ferns are far more numerous and various in the more humid tropics. They culminate in wet tropical forests in a great profusion of forms, which include not only herbaceous species and tree ferns, but also creepers, vines, and **epiphytes.** (Epiphytes grow on other plants, using them for attachment, but they draw no nutrition from their hosts and are not parasitic.) In rain forests, such epiphytic ferns as the staghorn (Figure 41-13) contribute to the aerial gardens that grow profusely on the upper trunks and branches of the tall forest trees (see Chapter 48). In these situations, the forest floor is so dark that few photosynthetic plants can live there. The epiphytes are too small to receive the limited sunlight below, and so they utilize their tall competitors as bases—in that way growing far up to where they can share the light (see Figure 48-14).

We see, then, that abundant as they are, the ferns still do not represent a completely successful adaptation of plants to land life.

DIVISION SPERMOPHYTA: SEED PLANTS

The vulnerabilities of fernlike gametophytes and sperm were eliminated in plants that evolved **seeds.** In this group of plants, the **spermophyta,** the partially developed male gametophyte becomes a pollen grain, which is often highly resistant to drying and can float in air for long periods without dessicating (see Chapter 21). As shown in Figure 21-18, the female gametophyte is entirely parasitic and lives its whole life protected within the tissues of the parent sporophyte. The microspore becomes a male gametophyte while still enclosed within the microspore walls. The zygote also forms within parental tissues. The zygote then develops further into an embryonic

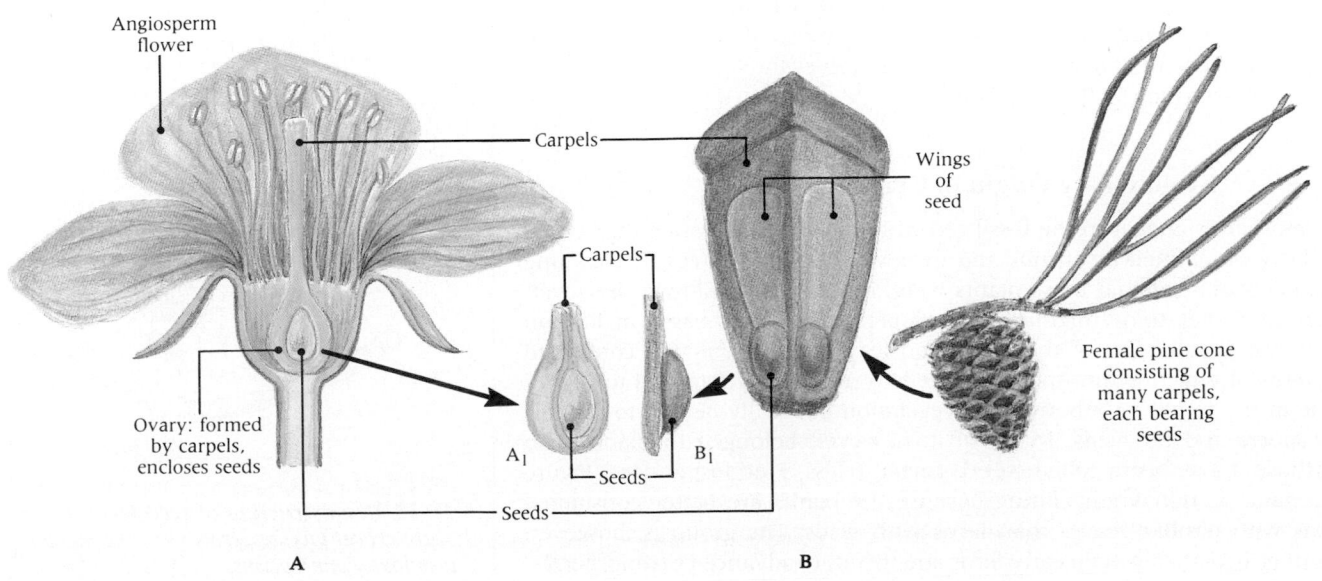

Angiosperm flower

Carpels

Wings of seed

Ovary: formed by carpels, encloses seeds

Carpels

Seeds

Female pine cone consisting of many carpels, each bearing seeds

A_1 B_1

Seeds

A   B

sporophyte, which is enclosed in protective tissue and provided with food before it is freed from the parental plant by some agency such as wind or rain.

The evolution of seeds was associated with the extensive and effective development of root and stem systems and of tissues and leaves. These factors made seed plants incomparably the most abundant and widespread of land plants. In most terrestrial environments, they have nearly replaced plants lacking seeds.

In the first seed plants to evolve, protection of the seeds was not yet perfected and flowers had not appeared. Spores, gametes, and seeds developed in organs, sometimes of considerable complexity, but without the still higher degree of organization seen in a true flower. The seeds arose on modified leaflike structures called **sporophylls** that did not enclose them.

Plants with such unprotected seeds are called **gymnosperms** (naked seeds). Plants with flowers and with seeds forming from ovules within structures called **carpels** are called **angiosperms** (enclosed seeds) (Figure 41-14). It is generally believed that carpels evolved from the enfolding closure of sporophylls around the ovules (see Chapter 21).

GYMNOSPERMS: NONFLOWERING PLANTS

Seven major classes of the division Spermophyta—four living and three extinct—have been traditionally grouped together as Gymnospermae, which has been assigned the rank of subdivision. Some recognize the following classes as divisions.

- Class Pteridospermopsida: seed ferns, now extinct
- Class Cordaitopsida: cordaites, now extinct
- Class Cycadeoidopsida: cycadeoids, now extinct
- Class Cycadopsida: cycads
- Class Ginkgopsida: ginkgos
- Class Gnetopsida: gnetophytes, "joint firs"
- Class Coniferopsida: conifers

The gymnosperms are a group of vascular plants, the ovules and seeds of which are exposed on the surface of sporophylls or analogous structures, without protective ovarian walls. Many botanists today hold that the differences between the four living gymnosperm classes are greater than their similarities, and there is some sentiment in favor of dropping the term gymnosperm and dealing with the classes directly.

Members of this cluster early underwent an extensive adaptive radiation (in the latter Paleozoic). Numerous groups became divergently adapted to various environments without much fundamental change in basic characteristics. Some of the main branches of this basic radiation—the seed ferns, cycadeoids, and cordaites—have become extinct, probably through competition with more progressive or efficient plants. Others survive—among them the ginkgos, cycads, and joint firs—but are so diminished in numbers, diversity, and geographic distribution that they are today mere relicts. Only one major branch of the early gymnosperm radiation has continued to the present in great abundance and diversity: It is the conifers or cone-bearers, the pines, firs, cedars, and their relatives. Let us briefly survey the several gymnosperm classes, of which fewer than 750 species survive.

Evolutionary Origin of Gymnosperms

The first gymnosperms appear in the fossil record in the early Carboniferous, when coal forests were dominated by shrubs and trees with fernlike leaves. It was long assumed by paleobotanists that these plants were, indeed, ferns. Slowly, however, the suspicion grew that some might be gymnosperms. This was based at first on minor anatomical peculiarities of the stems and leaves. Finally, it was confirmed by finding fossils in which gymnospermous seeds were actually attached to "fern" fronds. Some of the fernlike Carboniferous vegetation did really belong to ferns—plants with spores and no seeds. Much of it, however, belonged to plants with seeds and those have been called **seed ferns** (class Pteridospermae; Figure 41-15). The name is not wholly fitting because the plants are better considered gymnosperms with fernlike leaves than ferns with seeds. The group is, however, more or less intermediate between early ferns and the more advanced gymnosperms.

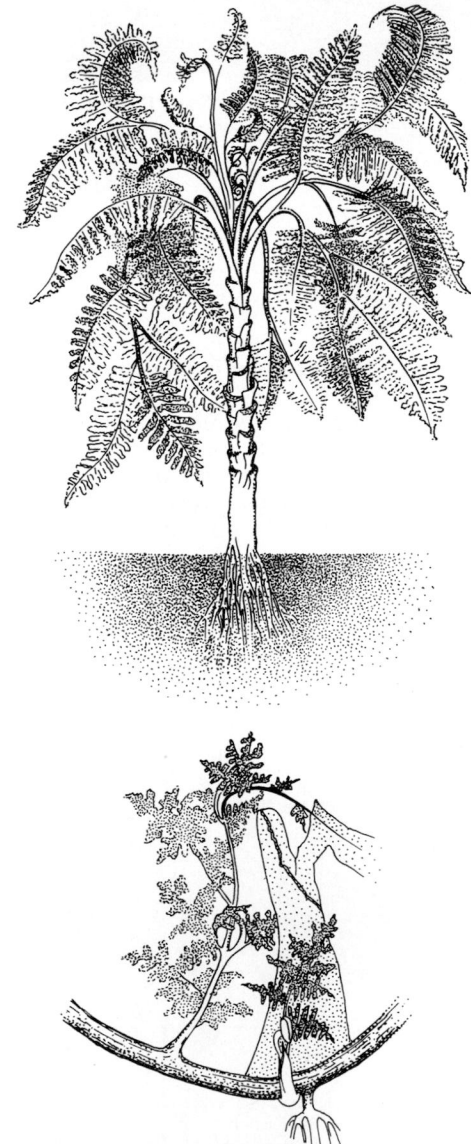

41-15 Reconstruction of seed ferns
These extinct pteridosperms were the most primitive gymnosperms.

The seeds of seed ferns grew singly and openly rather than in conelike structures, or strobili. They were attached along a frond or leaf stalk in various ways. Although for a time abundant, the seed ferns were soon replaced by more progressive groups. They are now extinct.

Cycads, Cordaites, Ginkgos, and Gnetophytes

The extinct **cycadeoids** (class Cycadeoidopsida) and the living relict **cycads** (class Cycadopsida) are similar in general appearance (Figure 41-16), but their reproductive organs are quite different. They have short, stumpy, palmlike stems, rarely exceeding a few meters in height. The foliage is also palmlike, and the cultivated plants, often raised in tubs in conservatories, are often mistaken for palms (Figure 41-17). (Real palms are not gymnosperms but specialized angiosperms.) The cycadeoids were common and widespread around the middle of the Mesozoic but became extinct soon thereafter. In them, the male and female organs were united in a single, complex structure resembling a flower. The detailed anatomy, however, is not flowerlike. Evidently, the evolution of these pseudoflowers and of the true flowers of angiosperms took place independently and in parallel. Cycads are still rather widespread in places with warm climates, including Florida and southern California, but they are nowhere abundant. Cycads are distinguished from cycadeoids by the fact that pollen and egg cells are produced in separate strobili. In this respect cycads resemble conifers.

The **cordaites** (class Cordaitopsida), now long extinct (since early Triassic), were another common group in the Carboniferous coal forests (Figure 41-18). They were tall trees with long, straplike leaves. Pollen and egg cells were produced in separate strobili. The special significance of cordaites is that they may have given rise to the conifers.

The **ginkgos** (class Ginkgopsida) were another abundant group but are now nearly extinct. Today it includes only one species. Therefore, that species, *Ginkgo biloba*—the maidenhair tree—is a "living fossil." It bears true gymnospermous seeds (Figure 41-19) and is a striking, woody, branched deciduous tree that has been spread all over the world by humans. It is now familiar along American streets and in parks, although few are aware of its long and unusual history. Ginkgos share with cycads a primitive feature lost in all other living gymnosperms and in

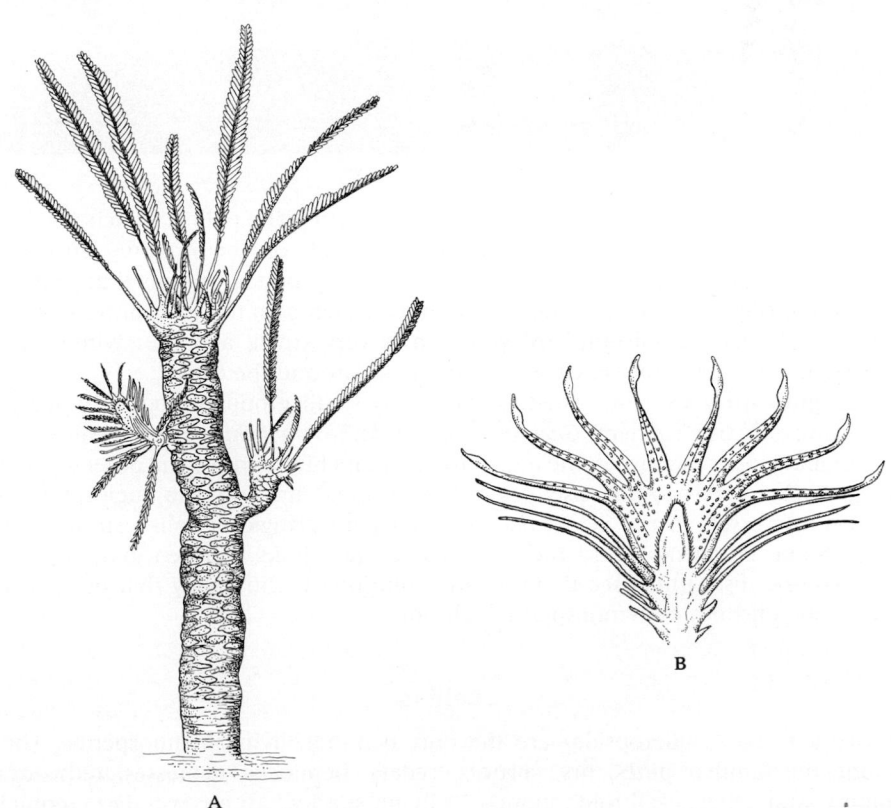

41-16 Fossil cycadeoids
(A) The extinct cycadeoid, Williamsonia sewardiana, *lived about 150 million years ago.*
(B) A median section through the so-called flower of the cycadeoid, in which a central conical structure bore ovules and the surrounding corollalike structure bore stamens. This "flower" is not related to that of the angiosperms.

41-17 Cycads
These living relicts have palmlike foliage, but are not palms. Cycads are gymnosperms. Palms are angiosperms (monocots).

angiosperms. They still have motile, flagellated sperms. The pollen, which you will recall is a partly developed male gametophyte, lands on the structure enclosing the egg cells. There the pollen develops further and produces sperms that swim to meet and fertilize the egg cells, much as in moss or ferns. In these primitive gymnosperms, however, the distance to be swum is very small, and the swimming is indoors; it is in a chamber enclosed by the tissue around the eggs.

   The gnetophytes (class Gnetopsida) are a small group of about 70 species containing only three genera: *Gnetum, Ephedra,* and *Welwitschia.* These include trees and climbing vines that live in the moist tropics, branching shrubs, and other vascular plants. Their interest to botanists lies mainly in their similarities to angiosperms—presence of vessels in the xylem, lack of archegonia, angiospermlike strobili, and so on. Some botanists regard them as evolutionary links between gymnosperms and angiosperms. Others see them as an evolutionary blind alley that brought to an end one offshoot of gymnosperm evolution.

Conifers

Conifers (class Coniferopsida) are the only common living gymnosperms. They include the familiar pines, firs, spruces, cedars, hemlocks, cypresses, redwoods, junipers, and others—all told, about 450 living species. All are woody perennials

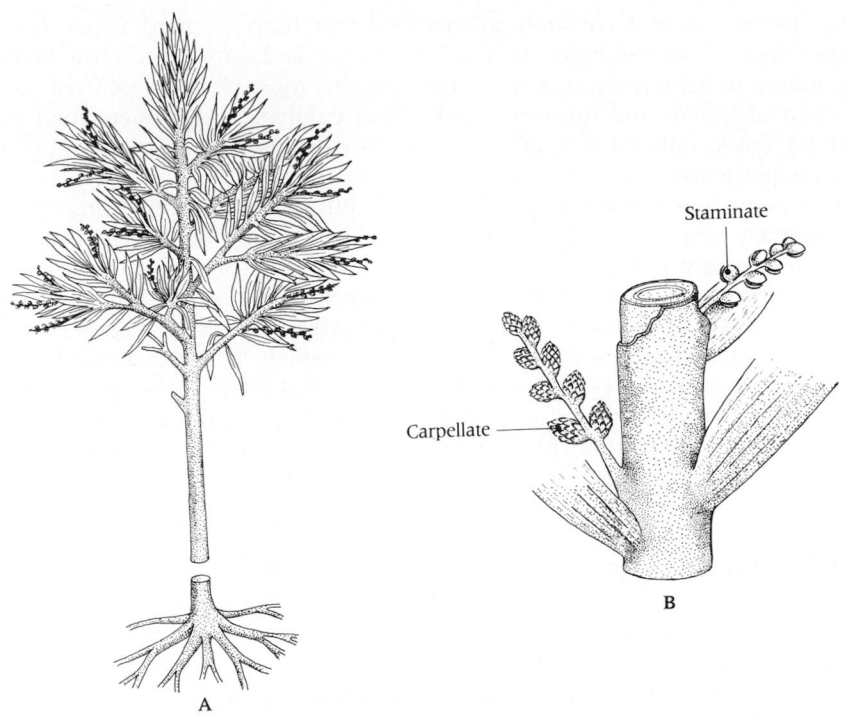

41-18 Cordaites, extinct gymnosperms
(A) The whole plant, with part of the stem removed, shows separate branches bearing strobili. **(B)** This expanded view of the main branch shows two minor branches bearing male (staminate) and female (carpellate) strobili.

Staminate

Carpellate

B

A

and most are trees, though some are shrubs. They include the largest and oldest (as individuals) of all living things—the closely related species of sequoias and redwoods, which may be 100 meters tall, some 4000 years old (by rough estimate) and have trunks 10 meters in diameter (Figure 41-20). The oldest authenticated age for a Sierra redwood is 3200 years.[3] Bristlecone pines are even older than redwoods, 4300, 4600, and 4900 year old species having been reported.

Conifers, the dominant trees over much of the Temperate Zone, often form great forests made up of only one or a few species. In spite of this wide and dense distribution, conifer forests are most common in relatively unfavorable situations: in the dry, cold, and windy upper slopes of mountain ranges, or in the poor sandy soil of many northern and southern regions. In more favorable situations, angiosperm forests predominate.

Most conifers have narrow, needlelike or scalelike leaves (Figure 41-21). A few have broad leaves; a very few have reduced leaves and perform photosynthesis mainly in their stems. Most conifers are evergreens. Evergreens do shed their leaves, as do all trees, but they do not shed them all at once, and so green leaves are always in place.

The life cycle of a conifer is an interesting example of the seed mode of reproduction (see Chapter 21). The large conifer tree is itself a diploid sporophyte. Microsporangia and macrosporangia are borne in separate cones, usually on the same tree. In

---

[3] The Sierra redwood ("bigtree") belongs to the genus *Sequoiadendron*, whereas the Coast redwood is of the genus *Sequoia*. The latter is the taller of the two; bigtree has the greatest mass. These two taxa are treated as separate genera because of relatively major differences when compared to other living and fossil members of the family Taxodiaceae, which includes *Sequoiadendron*, *Sequoia*, *Metasequoia* (the so-called dawn redwood), *Taxodium* (bald cypresses), and others.

41-19 Ginkgo biloba, *the maidenhair tree*
The two sexes occur in separate trees.
(A) *A male tree. Interestingly, the female ginkgo produces foul-smelling butyric acid and many communities have laws against planting them.* **(B)** *Mature strobili of the male tree.* **(C)** *A female sporangium with immature ovules.* **(D)** *Mature seeds.*

A

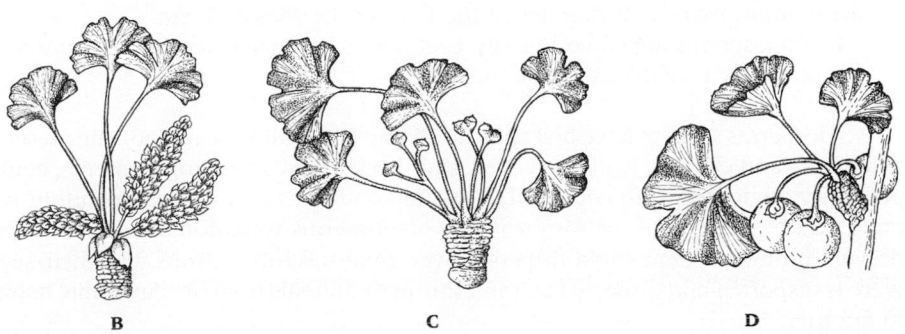

B   C   D

some species, cones have been so modified that they are hard to recognize. A juniper "berry," for example, evolved from a cone and thus is not a true berry.

Conifers are extremely important economically. Most of the wood used for construction of homes and furniture comes from conifers, and conifer wood is also used for posts, railroad ties, and musical instruments. Most paper is also made from conifer wood.

As a group, conifers are abundant enough, but in evolutionary terms they have an extremely long history and are not as abundant as they once were. Many types are extinct or are relicts. *Araucarias*, for example, were once worldwide but they now occur only in a limited part of South America, and in Australia and certain Pacific islands. The famous petrified forest in Arizona consists largely of ancient members of the araucaria group. Even more remarkable is *Metasequoia*, known as a fossil from many parts of the world and in the United States. It was long considered to have been extinct for tens of millions of years, but in 1944 a living *Metasequoia* forest was found in China. This "resurrected fossil" is now being cultivated.

ANGIOSPERMS: FLOWERING PLANTS

Angiosperm means "seed in a vessel"—that is, the seed is enclosed by an ovary wall. That arrangement is one of the differences between gymnosperms and angiosperms—the members of the subdivision Angiospermae. Another difference is the flowering habit of angiosperms.

By any criterion, angiosperms, our familiar flowering plants, are the most successful of land plants today—in terms of abundance, number of species (probably about 260,000), land area covered, and total metabolic activity. Much of our food—cereals, vegetables, and fruits—comes from angiosperms. Our food animals also live on angiosperms. Conifers are a more important source of wood, but angiosperm forests of maple, oak, or birch are also productive in this way. Many drugs (like digitalis, morphine, caffeine, marijuana, and quinine) and many poisons (like strychnine and the active agents of poison ivy) come from angiosperms. So do spices, rubber, and fibers. In a very real sense, angiosperms are a fundamental foundation of human existence and civilization.

The angiosperms are incredibly diverse in size, running from the barely visible duckweed (which is not always so small) to great trees. Their flowers may be microscopic in size or several feet across. They are of every imaginable hue. Some have color patterns invisible to our eyes, although none is black. A few angiosperms are aquatic. On land they grow wherever life exists. Some are parasites, some are organisms of decay, but the majority are photosynthetic. A few, like the Venus flytrap and sundew, are in part carnivorous (see Box 29-1).

Evolutionary Origin of Angiosperms

The origin of angiosperms is not clearly understood. Although the fossil record of early land plants is fragmentary, it appears that the oldest vascular land plants lived during the Silurian period—over 400 million years ago (see Table 37-3). We believe that there were four subsequent innovative episodes (Figure 41-22).

- Early vascular plants of simple morphology proliferated through the Silurian and Lower Devonian periods.
- A radiation of derived lineages in the Upper Devonian-Carboniferous included many pteridophytes (ferns).
- Seed plants appeared in the Upper Devonian and subsequently radiated into the gymnosperms that dominated the flora of the Mesozoic era.
- The angiosperms appeared in early Cretaceous and expanded down through the relatively recent Tertiary period.

Angiosperms seem to have first appeared some 135 million years ago. The presence of stomata and cuticles, leathery leaves, and specialized conducting elements, equipped early angiosperms to cope with stressful conditions of all sorts. Evolution was rapid and by the end of the Mesozoic era angiosperms were dominant. Specialized flowers promoted their rapid dispersal over great distances. Fruits and their seeds were transported and dispersed by (and through) animals from one favorable habitat to another.

41-20 *A redwood tree,* **Sequoia**

41-21 *Needlelike leaves of a conifer,* **Pinus sylvestris**

*41-22 Relative abundance of terrestrial
plants over geologic time*
*Following the invasion of land by early vascular
plants (I), three other groups successively became
dominant.*

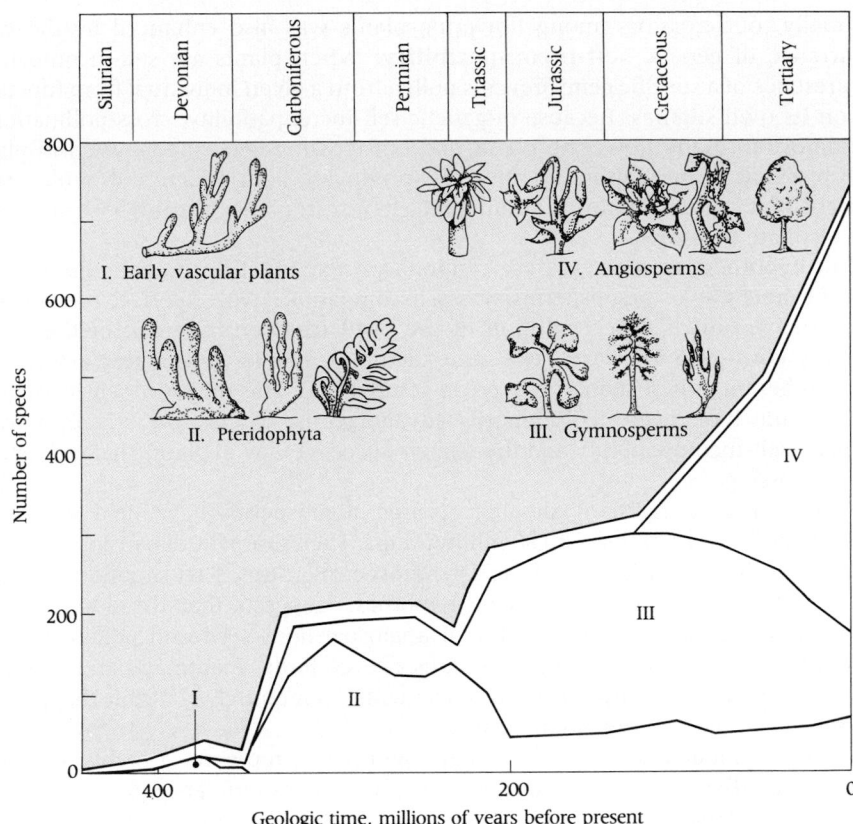

Angiosperms diversified for some 40 million years after their first appearance. So did new groups of insects. Because special relationships exist between flowers and insects (see Chapter 21), the two groups underwent **coevolution:** The more complex and varied the flowers became, the more diverse the insect groups interacting with them. Flowers have also evolved relationships with pollinating mammals and birds. We will see in Chapter 44 that in the history of life a number of mass extinctions took place. The dinosaurs are the best-known victims of such an episode. Yet, we know of no instance of mass extinction in the evolution of angiosperms. They seem to have evolved, steadily and continuously, even during the stressful times at the end of the Cretaceous when dinosaurs, certain types of marine plankton, and many other groups became extinct. A recent theory attributes these extinctions to a massive cloud caused by the impact of an asteroid or extensive volcanic activity that led to worldwide darkness. However, angiosperms continued their evolutionary diversification throughout this postulated time of darkness. We will return to this puzzle in Chapter 44.

The evolutionary radiation of angiosperms is vividly evident in the wide array of present-day flower morphology. Among the evolutionary trends that brought this about were probably factors promoting **out-crossing** (crossbreeding with flowers of other plants)—and thus the genetic diversity essential for the evolutionary development of new forms.

Among such trends were interesting variations in flower structure, including the aggregation or grouping together of various floral parts and various reductions or losses of floral components. If one first examines the structure of a typical flower, the nature of these trends will be clear. Most flowers, like the one in Figure 21-17, contain both stamens and carpels—and botanists term them **perfect.** However, the evolution of **imperfect** forms enhanced the transfer of pollen from one flower to the stigmas of other flowers, rather than just to the stigma of the same flower. In these imperfect flowers, either the stamens—collectively termed the **androecium** ("house of man")—or the pistils each consisting of one or a few fused carpels—collectively called the **gynoecium** ("house of woman")—were lost. When the androecium is lost, the flower becomes **pistillate** and produces only ovules and no pollen. When the gynoecium is lost, the flower is **staminate** and produces only pollen. Such changes made out-crossing necessary.

Finally, out-crossing among flowering plants was also enhanced by the wide occurrence of genetic **self-incompatibility.** When plants are self-incompatible, the presence of a specific gene prevents pollen from a given individual from functioning on its own stigmas. Because of genetic self-incompatibility, cross-pollination is mandatory in many flowering plants and out-crossing is ensured—even in plants with bisexual or hermaphroditic flowers (producing both pollen and ovules, with stamens and stigmas maturing simultaneously) or in plants bearing both staminate and pistillate flowers.

One hypothesis concerning the evolutionary origin of self-incompatibility assumes that the ancestor of angiosperms was self-compatible, with "perfect" or bisexual reproductive organs. The evolution of the pistil (the central organ of the flower typically consisting of ovary, style, and stigma)—may have provided a setting in which different pollen genotypes were in competition. This means of self-incompatibility would have been evolutionarily advantageous. Clearly, there is an analogy between self-incompatibility and the relative success today of plants that are capable of out-crossing.

Though the youngest of all plant groups, angiosperms have held worldwide dominance for most of the last 100 million years. Their dramatic expansion coincided with the decline or extinction of most gymnosperm groups. That situation undoubtedly involved cause and effect. One hypothesis suggests that the dominance of angiosperms related to the interplay of angiospermous seed and pollen dispersal systems—both of which made ingenious use of birds, mammals, and insects—and the overall superiority of these arrangements.[4] It is entirely plausible that gymnosperms "gave rise to" the angiosperms.

Whatever their origin, it is clear that angiosperms represent a highly successful phase of plant evolution—not only in extending their dominance, for older groups remain dominant in some environments, but in occupying environments farthest removed from the ancestral sea—and in thriving prosperously in those particular environments.

Classification of Angiosperms

In their general physiology, angiosperms are much like other green plants. In structure, they present many variations on the same themes found in other vascular

[4] The means of pollination and of seed dispersal in angiosperms are so numerous, some of them so peculiar, and all of them so interesting, that they merit additional discussion. This is given in Chapter 47.

41-23 Leaf structure in typical monocots and dicots
(A) *Tulips are monocots. Their leaves are elongated with parallel veins.* **(B)** *Roses are dicots. Their leaves display a network of veins arising from a central vein.*

A

B

41-24 Flower structure in typical monocots and dicots

(A) In Trillium, *a monocot, flower parts are grouped into threes.* **(B)** In Geranium, *a dicot, there are five petals in each flower.*

TABLE 41-1
COMPARISON OF GYMNOSPERMS AND ANGIOSPERMS

	Gymnosperms	Angiosperms
Dominance of gametophyte	0	0
Dominance of sporophyte	+	+
Roots, stems, leaves	+	+
Vascular tissues	+	+
Seeds	+	+
Flowers	0	+
Cones	+	0

Note: +, present; 0, absent

plants. Their manner of reproduction is typical of all seed plants with the addition of carpel-covered seeds and flowers. That material, discussed in Chapter 21, would be profitably reviewed, along with the discussion of embryological development.

The angiosperms are divided into two groups: Class Monocotyledonae and Class Dicotyledonae, more familiarly termed **monocots** and **dicots** (see Figure 24-14). Respectively, there are about 60,000 and 200,000 species in the two classes. The differences between them are the following:

- The bulk of an angiosperm seed is normally made up of the leaflike cotyledons that are packed with nutrients, especially starch, which the young plant needs to begin its development. Monocots have one such organ; dicots have two. The distinction is clearly seen, for example, on comparing a grain of corn (a monocot seed) with a dried bean (a dicot seed) (see Figure 37-14).
- A monocot leaf usually has many longitudinal parallel veins of equal prominence with short cross veins at right angles to the longitudinal veins; a dicot leaf usually has a vein network with a main central vein from which others arise by successive branching (Figure 41-23).
- In monocots, vascular tissue bundles are scattered throughout the ground tissue; in dicots, the bundles form a cylinder around the pith (see Figure 24-14A).
- Dicots usually have cambium and secondary growth; monocots do not.
- Floral parts (sepals, petals, stamens), when present in a definite number, typically occur in sets or multiples of three in monocots and of four or five in dicots (Figure 41-24).

The evolutionary origin of monocots and dicots is disputed. Some think that monocots are a more specialized group that derived from early dicots. In any event, monocots are now fewer in number than dicots.

Familiar monocots include grasses, corn, wheat, rye, sedges, cattails, daffodils, lilies, onions, tulips, irises, palms, and orchids. The more numerous dicots include magnolias, carrots, peas, beans, mints, morning glories, potatoes, mustards, squashes, dandelions, sunflowers, spinach, poison ivy, and gardenias, as well as almost all the other broad-leafed shrubs and trees—oaks, maples, elms, willows, and many others.

Most of our important angiosperm food plants have been cultivated since prehistoric times, although many varieties have been developed by human ingenuity only recently. Even early humans exercised vigorous selection, so that the older cultivated plants have been changed, sometimes beyond recognition, from their wild ancestors. For example, a much disputed problem of genetics and botany has to do with the wild plants from which pre-Columbian Indians selected corn (or maize). The ability of humans to develop new strains, even wholly new species, of cultivated plants (cultigens) is good evidence that nature has acted similarly, though more slowly and less systematically from the human point of view. Much has been learned about evolutionary principles from experiments with such plants. Knowledge of evolution has, in turn, fostered the development of desirable new cultivated plants.

We close this brief discussion of one of biology's most fascinating topics with two summary tables—one recapping the differences between angiosperms and gymnosperms (Table 41-1) and one summarizing the characteristics of the major plant groups (Table 41-2).

TABLE 41-2
MAJOR PLANT GROUPS

Subkingdom	Division (Examples)	Subdivision or Class (Examples)	Major Characteristics	Vascular Tissues
Bryophyta	Bryophyta	Hepaticopsida (liverworts)	Alternation of generations with gametophyte the conspicuous form; sporophyte lacks stomata and is short-lived, dying after spores mature; unicellular rhizoids or processes on epidermal cells	None
		Anthocerotopsida (hornworts)	Gametophyte conspicuous; sporophyte has stomata; sporophyte growth continues by addition of new meristem cells	None
		Muscopsida (mosses)	Gametophyte filamentous at first; later develops erect stem with leaves; sporophyte grows out of gametophyte; multicellular rhizoid	None
Tracheophyta	Psilotophyta	*Psilotum*	Dichotomously branching plants; many lack true roots or leaves; stems contain vascular tissue and carry out photosynthesis; spores produced at tip of some branches	Tracheids
	Lycopodiophyta (club mosses)	*Lycopodium*	Vascular plants with microphylls (little leaflike structures) in spirals; extremely diverse in appearance; all have motile sperm	Well-developed tracheids, cells with thin areas (pits) in their lateral walls
	Sphenophyta	*Equisetum*	Vascular plants with jointed stems marked by conspicuous nodes and elevated silicious ribs; sporangia in a strobilus at the tip of the stem; leaves are scalelike; sperm are motile	Primitive forms of tracheids, cells with thin areas (pits) in their lateral walls
	Pteridophyta (ferns)	Ferns	Vascular plants. Mostly homosporous. All possess a megaphyll; gametophyte, free-living and usually photosynethetic; multicellular, gametangia and free-swimming sperm	Tracheids, vessels; well-developed phloem
	Spermophyta	Gymnospermae	Vascular plants with seeds, lacking flowers and fruits	
		Pteridospermopsida (seed ferns)	Extinct ferns that bore seeds, true leaves, stems, and roots	Tracheids, vessels; well-developed phloem
		Cordaitopsida (cordaites)	Extinct trees with leaves, stems, and roots; pollen and eggs produced in separate strobili	Tracheids, vessels, phloem
		Cycadeoidopsida (cycadeoids)	Extinct trees with male and females organs united in a pseudoflower; mostly monoecious	Tracheids, vessels, phloem

SUMMARY

Two subkingdoms, Bryophyta and Tracheophyta, make up the plant kingdom. Bryophtes are nonvascular plants—or nearly so. They lack true leaves, stems, and roots. Tracheophytes are vascular and contain xylem and phloem. Tracheophyte characteristics, such as vascular tissue, a waxy cuticle layer, strong stems, and dessication-resistant spores, arose in response to the harsh environmental conditions plants faced when they first invaded land. Among the most important challenges on land was the need to conserve water and to stand upright.

Bryophytes are the liverworts, hornworts, and mosses. Though not as well adapted to survive in dry environments as tracheophytes, they can live in environments too shady or cold for many tracheophytes. The gametophyte generation is dominant in bryophytes; the dominant form in tracheophytes is the sporophyte.

There are five tracheophyte phyla. Division Psilotophyta, whose members lack roots and have at best rudimentary leaves, contains the oldest known tracheophytes. Division Lycopodiophyta includes the club mosses which have true roots, stems, and leaves. The horsetails make up the division Sphenophyta. Ferns comprise division Pterophyta, the first phylum to exhibit vascularly complex (megaphyllic) leaves. Division Spermophyta, by developing seeds, avoided depending on moist environments for reproduction, a re-

Subkingdom	Division Examples)	Subdivision or Class (Examples)	Major Characteristics	Vascular Tissues
		Cycadopsida (cycads)	Seed plants with slow cambium growth and pinnately compound, palmlike, or fertile leaves; ovules not enclosed; sperm are flagellated and motile, but are carried to the vicinity of the ovule in a pollen tube; dioecious	With sluggish cambium growth
		Ginkgopsida (ginkgo)	Seed plants with active cambium growth and fan-shaped leaves with dichotomous venation; sperm carried to the vicinity of the ovule in a pollen tube but are flagellated and motile	Active cambium growth
		Gnetopsida (gnetophytes)	Some are cone-bearing desert plants; many angiospermlike features including vessels in xylem, lack of archegonia, etc.	The only gymnosperms in which vessels occur
		Coniferopsida (conifers: pine, fir, spruce, hemlock)	Megaphylls often reduced to needles or scales; seed plants with active cambialum growth and simple leaves, in which the ovules are not enclosed and the sperm are not flagellated	More highly specialized, well-developed conducting tissue for the transport of water and organic materials and minerals
		Angiospermae (flowering plants)	Multicellular plants with vascular tissues and cambium; gametophyte small and dependent upon sporophyte, which is large, becoming herbaceous, shrubby, or woody; leaves usually broad; sperm without flagella, transferred to ovule by a pollen tube; seed enclosed within an ovary that develops into a fruit; possesses flowers; double fertilization to produce a triploid endosperm	Highly specialized, well-developed conducting tissue for the transport of water and organic material and minerals
		Dicotyledonae (dicots) (carrots, oak, maple)	Embryo inside seed with two cotyledons; flower parts mostly in fours or fives; vascular tissue in distinct strands or bundles arranged in a cylinder or circle; cambium commonly present; leaves net-veined	
		Monocotyledonae (monocots) (lily, iris)	Embryo with one cotyledon; flower parts mostly in threes or multiples thereof; vascular tissues usually in scattered bundles; leaves with parallel veins	

quirement that has constrained the other plant phyla. Seed plants are by far the most abundant—and thus successful—land plants.

Division Spermophyta includes the gymnosperms, a subdivision that has naked or unprotected seeds, and angiosperms, a subdivision with enclosed seeds.

Gymnosperms probably evolved from ferns. Living classes include the cycads and ginkgos, which are unusual in having motile sperm, the gneteae, which may be a link between gymnosperms and angiosperms, and the conifers. Conifers have male and female sporangia and gametophytes. The pollen and seeds are usually wind-dispersed.

Angiosperms are the flowering plants. Many angiosperms are pollinated by insects. Indeed, the relationship between angiosperms and insects is so intimate and essential that the two groups have coevolved. The extraordinary radiation of angiosperms was made possible by their genetic diversity, which resulted from mechanisms that promote out-crossing.

Angiosperms are divided on the basis of their early embryonic development into two groups, monocotyledons (monocots) and dicotyledons (dicots). Monocots generally have leaves with parallel veins, vascular bundles scattered throughout the pith, and flower parts which occur in multiples of three. Dicots have venous networks, vascular bundles around the pith, and flower parts in multiples of four or five. Dicots are more numerous than monocots.

KEY TERMS

angiosperm
antheridium
archegonium
bryophyte
carpel
coevolution
dicot
division
gymnosperm
heterosporous
homosporous
leaf

megasporangium
megaspore
microgametophyte
microspore
microsporophyll
monocot
out-crossing
ovulate cone
ovule
perfect flower
phylum
pollen

pollination
pteropsid
rhizoid
root
seed
self-incompatibility
spermopsid
sporophyll
stem
strobilus
tracheophyte

QUIZ QUESTIONS

1. Which of the following groups is *not* a tracheophyte?
   A. Dicotyledonae
   B. Sphenophyta
   C. Hepaticopsida
   D. Angiospermae
   E. All of the above are tracheophytes.

2. Motile sperm cells swim through an external film of water to reach female gametes in all of the following groups *except* for
   A. Pteridophyta.
   B. Spermophyta.
   C. Lycopodiophyta.
   D. Sphenophyta.
   E. Anthocerotopsida.

3. In which of the following groups are seeds present?
   A. Psilotophyta
   B. Ginkgopsida
   C. Lycopopdiophyta
   D. Bryophyta
   E. None of the above

4. Which of the following is *not* a trend followed during the evolutionary history of land plants?
   A. absence of seeds to presence of seeds
   B. simple leaf venation to complex leaf venation
   C. independent gametophyte to dependent gametophyte
   D. nonflowering to flowering
   E. dominance of sporophyte generation to dominance of gametophyte

5. A flower possessing both male and female structures is described as
   A. perfect.
   B. homosporous.
   C. strobilous.
   D. staminate.
   E. carpellate.

ESSAY QUESTIONS

1. What features characterize plants as kingdoms? What evidence suggests that the non-vascular and vascular plants evolved from a common ancestral multicellular green algae?

2. What environmental pressures confront land plants? What evolutionary adaptations allow plants to survive such pressures? In what environments are bryophytes better adapted to survive than tracheophytes?

3. Trace the evolutionary history of root and leaf development in the major classes of the tracheophyta. What are the two basic leaf types?

4. What advantages were gained in the evolution of the seed plants? What mechanisms account for seed dispersal?

REFERENCES AND SUGGESTIONS FOR FURTHER READING

Bell, P. R., and Woodcock, C. L. F. (1983). *The Diversity of Green Plants*. Edward Arnold, London.

An excellent, short survey of all plants (and fungi, algae, and cyanobacteria).

Blackmore, S., and Ferguson, I. K. (Eds.) (1986). *Pollen and Spores: Form and Function*. Academic Press, Orlando, Fla.

Papers from a symposium on modern and fossil pollen and spores and their implications for the evolution and relationships of plant groups.

Cleal, C. J. (1988). Questions of flower power. *Nature 331*, 304–305.

An interesting, brief discussion of the evolution of the early angiosperm groups.

Dahlgren, R. M. T., and Clifford, H. T. (1982). *The Monocotyledons: A Comparative Study*. Academic Press, Orlando, Fla.

The relation between monocots and dicots has important implications concerning the angiosperm phylogeny. Most discussions of that topic are "dicot centered." This handsome text surveys the monocots alone.

Dahlgren, R. M. T., Clifford, H. T., and Yeo, P. F. (1985). *The Families of Monocotyledons: Structure, Evolution, and Taxonomy*. Springer-Verlag, New York.

Another major effort to bring order into an area famous for its disorder—especially in the field of classification. Excellent line drawings illustrating the monocot families and various phylogenetic schemes.

Goldberg, R. B., Barker, S. J., and Perez-Grau, L. (1989). Regulation of gene expression during plant embryogenesis. *Cell 56*, 149–160.

A stimulating and up-to-date review.

Heywood, V. H. (1978). *Flowering Plants of the World*. Mayflower Books, New York.

A comprehensive guide to the families of world's flowering plants.

Kaufman, P. B., and 6 other authors. (1989). *Plants: Their Biology and Importance*. Harper & Row, New York.

A handsome new volume on all aspects of plant biology.

Kende, H. (1989). The state of plant biology: Views from the other side of the fence. *Cell 56*, 914–915.

Enjoyable comments.

Mapes, G., Rothwell, G. W., and Haworth, M. T. (1989). Evolution of seed dormancy. *Nature 337*, 645–646.

An interesting brief report on the "most ancient ever discovered with gymnosperm seeds."

They are from conifers and date to the Permo-Carboniferous.

Murray, D. R. (Ed.) (1987). *Seed Dispersal*. Academic Press, Orlando, Fla.

It has been said that one advantage to studying plant populations is that they just sit there waiting to be examined. Tracing the fate of dispersed seeds is a different matter. This book summarizes present knowledge of seed dispersal.

Radford, A. E. (1974). *Vascular Plant Systematics*. Harper & Row, New York.

An authoritative discussion of the systematics of the vascular plants.

Raven, P. H., Evert, R. F., and Curtis, H. (1986). *Biology of Plants*. Worth, New York.

A well-written textbook of plant biology.

Ray, P. M. (1972). *Living Plants*. Holt, Rinehart, and Winston, New York.

A relatively short but authoritative treatment of aspects of plant development and factors that influence plant growth.

Ray, P. M., Steeves, T. A., Fultz, S. A. (1983). *Botany*. W. B. Saunders, Philadelphia.

A general textbook in plant biology.

Richardson, D. H. S. (1981). *The Biology of Mosses*. John Wiley & Sons, New York.

A brief authoritative overview of the biology of mosses.

Salisbury, F. B., and Ross, C. W. (1985). *Plant Physiology*. Wadsworth, Belmont, Calif.

A review of the development, function, and structure of green plants.

Scagel, R. F., and 5 other authors. (1982). *Non-Vascular Plants: An Evolutionary Survey*. Wadsworth, Belmont, Calif.

The authors present a comprehensive treatment of the diversity of protists, plants, and fungi.

Schofield, W. B. (1985). *Introduction to Bryology*. Macmillan/Collier, New York.

Much of this excellent and abundantly illustrated text is devoted to the mosses—structure, classification, and even evolution, though little is known about it.

Stewart, W. N. (1983). *Paleobotany and the Evolution of Plants*. Cambridge University Press, New York.

A distinguished treatise on plant evolution.

Sussex, I. M. (1989). Developmental programming of the shoot meristem. *Cell 56*, 225–229.

Another excellent review in the *Cell* tradition.

INVERTEBRATE ANIMALS

Invertebrate animals
*About ninety percent of known animal species
are invertebrates. This small sampling suggests
their diversity and beauty.*

Plants live upon the energy that comes to them directly from the sun and are able to prosper even though rooted in one spot. Animals, on the other hand, cannot acquire energy directly from the sun. Instead, they feed on plants and other animals. This means that most must hunt for food and hence must have the power of movement. Much of animal structure and function is traceable to this single unceasing basic mission—the search for food.

OVERVIEW

We begin our study of the animal kingdom with a surprisingly difficult question—one we have asked before. What, in fact, is an animal?

We know that dogs, cats, and cows are animals. But what about ants, sponges, corals, worms, clams, and a vast array of other organisms? An evasive answer to our question might be that an animal is an organism that belongs by descent and common ancestry to any of the phyla that biologists agree to call "animals."

WHAT IS AN ANIMAL?

In fact, it is possible to develop some helpful descriptive generalizations about what distinguishes animals. Let us summarize the more important and uniform differences between animals and plants.

- *Mobility.* Most plants are sessile (attached to a surface), nonmotile as developed organisms, and dispersed only in their specialized reproductive phases, such as spores or seeds. Many animals are also sessile as adults, but most are mobile throughout life, and all have a mobile postzygotic stage (Figure 42-1A,B).
- *Metabolism.* Most plants are photosynthetic. Those that are not were probably derived from photosynthetic ancestors. Animals also may have had photosynthetic ancestors among the protists, but no animals are capable of photosynthesis. All eat other organisms: they are heterotrophs (see Figure 42-1C,D). Animal nutrition depends primarily on ingested food which, with certain exceptions, is digested in an internal cavity.
- *Structure and organization.* Most animals have fixed structures to which new elements are added only at limited, usually early, phases of development. Most plants have less fixed patterns, new elements adding on periodically throughout life, at almost any time. Tissue and organ differentiation in animals is often more definite and complex than in plants. Most individual plant cells have rigid walls; animal cells do not. Plant cells usually are vacuolated when mature; animal cells usually are not.
- *Maintenance.* In most animals, cells are either surrounded by sea water or are in a fairly constant internal environment that resembles sea water in containing NaCl (though at much lower concentrations) and other salts. In plants, internal fluids usually contain far less NaCl. Specialized systems for the maintenance of

a constant internal environment are much less evident in plants than in animals.

- *Responsiveness.* Most animals have nerves and muscles, conspicuously unlike any plant tissues. Almost all plants lack special sensory receptors, and the unusual plants in which these do appear are both rare and simple (see Box 29-1). Most animals have sensory receptors that are both numerous and complex. Moreover, animals display remarkable and characteristic patterns of behavior (Chapters 32 and 33). In these and many other regards, animal organization transcends that of plants in complexity.
- *Reproductive cycles.* Most plants have a sexual cycle with alternating haploid and diploid generations between meiosis and fertilization. In animals, development of haploid stages in other than the gametes is highly exceptional, occurring almost exclusively among the insects.

In Chapter 43, we will discuss these basic characteristics of animals in the context of animal evolution. First, let us get acquainted with the major animal phyla.

THE ANIMAL PHYLA

A detailed classification of animals appears in the Appendix of this book. As discussed in Chapter 39, protozoa are considered by some authorities as a subkingdom or phylum of the kingdom Animalia. For reasons stated above, we elected to place those phyla within the kingdom Protista.

The list in the Appendix includes 20 phyla, although as many as 35 have been proposed. Some systematists recognize more phyla and others fewer, preferring to combine and divide the classes in various ways. For example, in earlier days there was a phylum called Bryozoa, which included two major classes, Ectoprocta and Entoprocta. When new evidence indicated that these classes may have had different evolutionary origins, they were designated as separate phyla.

Some two million species of multicellular animals have been described. The rate at which new species are reported suggests that these two million are only about half of the extant species on Earth. In numerical terms, the great majority of known animals, both living and extinct, belong to only 12 of the 20 phyla of our classification. Those 12 appear in boldface type in the Appendix. The 10 major phyla (in order of decreasing species numbers[1]) are as follows:

- **Arthropoda:** spiders, crustaceans, insects (at least 1 million)
- **Mollusca:** snails, clams, octopuses etc. (110,000)
- **Aschelminthes** (roundworms): pinworms, hookworms, nematodes (80,000)
- **Chordata:** mainly vertebrates (45,000)
- **Platyhelminthes** (flatworms): planarians, flukes, tapeworms (15,000)
- **Coelenterata:** hydras, jellyfishes, sea anemones, corals (10,000)
- **Porifera:** sponges (10,000)
- **Annelida:** segmented worms, leeches (9,000)
- **Echinodermata:** starfishes, sea urchins, sea cucumbers (6,500)
- **Ectoprocta:** moss animals (5,100)

The other taxa include rarer forms of lesser importance today. For example, **graptolites** (previously placed in phylum Graptolithina) are known only as fossils and are now extinct. Once thought to be related to coelenterates, they are now viewed by many paleontologists as relatives of the chordates.[2] Conceivably, we would consider them a separate phylum if we better understood their evolutionary relationships.

The same is true of living animals belonging to several smaller phyla. If we better knew their history, we might relocate them in larger or better-known phyla. For example, the **ctenophores** or comb jellies (phylum Ctenophora) may be offshoots of Coelenterata (Figure 42-2). The **mesozoans** (phylum Mesozoa) may be degenerated members of the phylum Platyhelminthes, though some regard these

---

[1] These numerical estimates include only living species. Many extinct species are also known and many species, both living and extinct, are presumed to be still undiscovered.

[2] The authoritative text, *Treatise on Invertebrate Paleontology,* has placed them into a class in a subphylum called Stomochorda of the phylum Chordata. We have followed that scheme.

A

B

C

D

42-1 Animals and plants differ
(A) *Plants are rooted in place.* **(B)** *Animals are mobile.* **(C, D)** *Animals are heterotrophs. Plants are photosynthetic autotrophs.*

42-2 Comb jelly
The ctenophores are carnivorous, radially symmetrical marine animals that probably arose from the coelentrates.

tiny and obscure marine organisms with a distinctively small number of cells (20 to 30) as a separate subkingdom. Hence we put both phyla in lightface type.

The ideal way to learn the visible characteristics of the various animal phyla is to see them alive in their natural surroundings—or, failing that, in aquariums and zoos. Next, one should examine preserved specimens, and models in museums or laboratory collections, and if possible dissect some specimens. Finally, detailed descriptions may be found in libraries. Good starting places are the references cited at the end of this chapter. This is not a textbook of zoology, and we do not propose to describe the various phyla in great detail. Rather, we hope to examine the broad diversity of animal life, to call attention to the major features of the more important phyla—those of invertebrates in this chapter and of chordates (mainly vertebrates) in Chapter 43—and then to fit them into a broad biological scheme that charts their evolutionary origins and their roles in the economy of nature.

Surveying the phyla and some of their individual species, one should remember that the evolution of animals embodied certain major trends and innovations. It is our plan first to briefly note these major developments here. Then we will describe the major animal phyla in this chapter and the next. Finally, we will attempt to place the many changes we have encountered in the grand scheme of animal evolution. Only then will the following major descriptions acquire deeper meaning.

In summary, the following major evolutionary trends led to the emergence of animals and are reflected by them.

- Evolution from unicellularity to multicellularity
- Division of labor among cells with specialized cells (for example, muscle cells, nerve cells, etc.) functioning on behalf of the whole organism
- Development of tissues (groups of cells of mixed type) that operate together to perform specific functions—for example, loose connective tissue, bone, and cartilage
- Development of organs (groups of two or more tissues that are usually combined in a specific way to form an organ that can perform specialized tasks)—for example, heart, pancreas, and lung
- Development of organ systems (groups of organs that function together)—for example, cardiovascular system and alimentary system
- Evolution from an indistinct pattern of embryonic cell layers to forms with two and then with three layers
- Evolution from forms with no special digestive system, to forms with a pouchlike cavity and one opening, to forms with a tubular gut with both mouth and anus
- Evolution from forms lacking a true body cavity or coelom, the acoelomates, to those possessing such cavities, the coelomates, via intermediate forms with unlined cavities, the pseudocoelomates
- Development of external skeletons, exoskeletons, or internal skeletons, endoskeletons
- Development of a circulatory system
- Appearance of appendages and segmentation (termed metamerism)[3]
- Evolution, in many cases from radial to bilateral symmetry with the appearance of a right and left side of body that are approximately mirror images of one another
- Elaboration of complex nervous systems with brains and sense organs that are concentrated in the anterior end of the animal
- Acquisition of behavior patterns that can be modified by experience
- Increased specialization of the brain that ultimately permitted the rich repertory of capacities we find only in the human brain

Some of the characteristics of the major animal phyla are summarized in Table 42-1.

We begin our survey with the sponges.

---

[3] The essence of metamerism is a serial succession of body segments, each containing unit subdivisions of several organ systems, muscles, and nerves. Functionally, this body plan (along with hydrostatic skeletons) greatly improved locomotion in soft-bodied forms—for example, earthworms. Segmentation also acquired a critical role in embryogenesis and morphogenesis. Recall, for example, our discussion of homeotic genes in Chapter 20.

TABLE 42-1
SOME CHARACTERISTICS OF THE MAJOR ANIMAL PHYLA

Phylum	Embryonic Cell Layers; Symmetry	Digestive System	Circulatory System	Larvae	Other Features of Adults
Porifera	Indistinct; radial	No special organ; no coelom	Absent	Amphiblastula	Sessile; microscopic food ingested from flagella-produced currents
Coelenterata	Two; radial	Pouchlike; one opening; no coelom	Absent	Planula	Sessile; food captured by tentacles; stinging cells
Platyhelminthes	Two; bilateral	Pouchlike; one opening; no coelom	Absent	Trochophore-like	Motile; flattened, wormlike; passive or immobilized animal food
Aschelminthes	Two; bilateral	Tubular; two openings; pseudocoel	Absent	None	Motile; cylindrical, wormlike; becomes parasitic early
Ectoprocta	Two; bilateral	Tubular; two openings; true coelom	Absent	Trochophore	Sessile, bilateral; external skeleton; flagellated tentacles; early colonial
Brachiopoda	Three; bilateral	Tubular; two openings; true coelom	Present	Trochophore	Sessile, bilateral; dorsal, ventral shells; flagellated tentacles inside shells; noncolonial
Mollusca	Three; bilateral	Tubular; two openings; true coelom	Present	Trochophore	Motile creeping on ventral foot; shelled
Annelida	Three; bilateral	Tubular; two openings; true coelom	Present	Trochophore	Motile, cylindrical, wormlike; bristle appendages; segmentation
Arthropoda	Three; bilateral	Tubular; two openings; true coelom	Present	None or secondary	Highly motile; jointed legs; extended skeleton; segmentation
Echinodermata	Three; secondarily radial	Tubular; two openings; true coelom	Present	Pluteus or pluteuslike	Sessile or sedentary; heavy protective skeleton; noncolonial
Chordata	Three; bilateral	Tubular; two openings; true coelom	Present	Pluteuslike, none, secondary	Highly motile; internal skeletons aid movement; segmentation

PHYLUM PORIFERA: THE SPONGES

The word *porifera* comes from Latin roots meaning ''pore bearer.'' These ancient and primitive forms, in fact, resemble colonies of protists in that the separate cells appear to lead nearly independent lives within a higher organization. Indeed, they have been described as a ''republic of cells.''

All sponges are aquatic and most are marine (Figure 42-3). They are abundant in every sea. They vary in size from less than an inch to many feet in diameter. Their way of life is hardly tumultuous. What they do is lie there, immobile and sessile, drawing a steady current of water into their bodies through their pores, filtering out the microscopic food particles (diatoms, protozoa, etc.) in the stream, and ejecting the water out through a large opening at the top, the mouth or **osculum**

42-3 A marine sponge
Sponges are usually attached to the sea floor. They obtain nutrients by filtering seawater.

(Figure 42-4). Thus they are animated filters. They have no digestive system, no circulatory system, no organs of any kind, no real tissues. Even the division of labor among the cells is meager. One type of cell, the **collar cell (choanocyte),** lines the body cavity **(spongocoel)** and pumps water, captures and digests food, and forms sperms and eggs. A distinctive feature of these cells is the flagellum, which keeps the water flowing through the body cavity. Interestingly, if sponge cells are separated in the laboratory with a fine screen, they promptly reaggregate to form new sponges. If sponges are cut into small pieces, each piece will grow to full size. Sponge fishermen have known this for years.

Sponges have an internal skeleton that is secreted by amoeboid cells in their body walls. The skeleton is made up of crystalline **spicules** of silica or calcium carbonate and a network of elastic fibers composed of the protein **spongin.** The fibers grow in the living sponge much as horn or bone grows in a vertebrate animal. Many, perhaps most, sponges do not have the soft supple skeleton we are most familiar with. Rather, they have skeletons made rigid by crystalline deposits.

Sponge diving is a thriving business in the warm shallow waters off Greece, Florida, and Australia. The once-familiar bathroom sponge is in fact the skeleton of a sponge from which the living cells have been removed. Only when dried and cleaned does it emerge as the sponge we use for cleaning. Of the approximately 10,000 species of sponges, only a few are usable in this way. Sponge fishing has led to a decrease in the sponge population, but the advent of artificial sponges has decreased sponge fishing.

Sponges seem to have evolved from protists independently of the other animals. For this reason, they have been classified in a subkingdom of their own—Parazoa ("alongside the animals"). Fossil remains suggest that sponges have been numerous since early in the history of life and have not changed a great deal for some hundreds of millions of years. One can wonder why in so long a time sponges (and some other primitive organisms) have not evolved into something more elaborate. The answer is that sponges have not advanced evolutionarily because they were ecologically successful as they were. When life was young, they were already well adapted to a widely and continuously available way of life. Thus they are something of an offshoot from the main paths of animal evolution.

42-4 Structure of a sponge
Water entering the body cavity through pore cells is moved by flagella of the collar cells out through the mouth. Nutrients are filtered from this moving stream. Body support is provided by skeletal spicules.

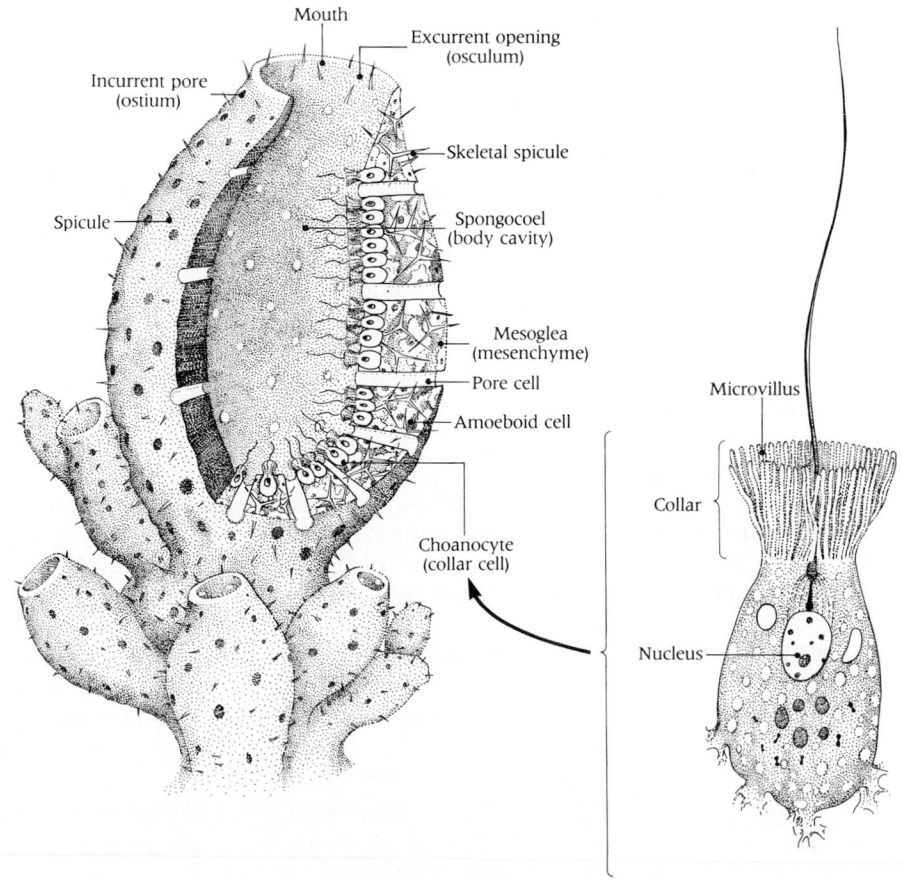

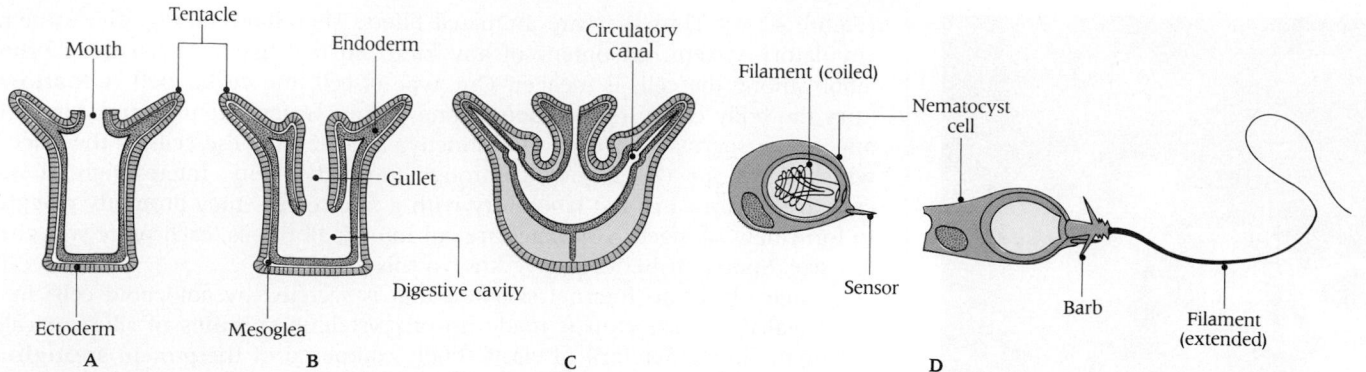

Mouth · Tentacle · Endoderm · Circulatory canal · Filament (coiled) · Nematocyst cell

Gullet · Sensor · Barb · Filament (extended)

Ectoderm · Mesoglea · Digestive cavity

A   B   C   D

42-5 The coelenterate body plan
The coelenterates have two cell layers, with a jellylike mesoglea in between. **(A)** *Simple polyp with mouth and digestive cavity.* **(B)** *More advanced polyp of sea anemone with gullet.* **(C)** *A medusa, the predominant body form in jellyfishes, has a circulatory canal that is continuous with the digestive cavity.* **(D)** *The nematocyte is a stinging cell used for prey capture. See also Figure 31-36.*

PHYLUM COELENTERATA: HYDRAS, JELLYFISHES, AND CORALS

Phylum Coelenterata (also called Cnidaria) includes some of the most beautiful animals in the sea. The word *coelenterate* comes from Greek roots meaning "hollow intestine." The significance of the term is that these animals have an internal digestive cavity. Like sponges, coelenterates are exclusively aquatic and predominantly marine, although there are a few freshwater types such as *Hydra*. The coelenterate's body shows **radial symmetry**—that is, from above it is seen to be organized in a circular wheellike plan around an axis. This is true only of coelenterates and one other phylum, Ctenophora. In other words, coelenterates have a top and bottom but no front and rear.

Coelenterate bodies are constructed of two cell layers, an outer epidermis of ectodermal origin and an inner layer of endodermal origin (Figure 42-5A-C). (A third middle layer of mesodermal origin, the mesoglea, is not well developed.) A major advance in coelenterates was the acquisition of specialized tissues. They also have organs. Thus they have a pouchlike digestive cavity with a single opening that serves both as mouth and anus (see Figure 26-9).[4] This cavity is surrounded by tentacles with stinging cells called **nematocytes** that discharge microscopic harpoons called **nematocysts** (Figure 42-5D; see also Figure 31-35). They also possess epithelial cells, muscle fibers, and nerve nets (see Figure 30-30A).

Coelenterates are carnivorous and can snare food in their tentacles, often after stinging it into paralysis. Their prey ranges from microscopic animals of all sorts to relatively large crustaceans, worms, or fishes. Digestion is partly in the central cavity (stomachlike) and partly within its lining cells (spongelike).

Coelenterate bodies are of two basic forms: the treelike polyp and the bell-shaped medusa (Figure 42-6). **Polyps** are long cylinders that are sessile, with one end attached to a firm surface. The other end has tentacles and a mouth. **Medusas,** flatter and broader, are usually free-swimming. As shown in Figure 42-6, the polyp and medusa may alternate in the life cycle of one coelenterate.

When the medusa is present in the polymorphous life cycle, it buds off asexually from the polyp (see Figure 21-27). It is male or female and is the stage that effects sexual reproduction (gamete production). The free-swimming nature of the medusa also guarantees dispersal.

In the evolution of coelenterates there has been a recurrent tendency for the polyp stage to form polyp colonies. Other polyps commonly arise by budding. But when budded polyps separate incompletely, a complex organism develops that is made up of many recognizably distinct but connected polyp units. Since all developed from a single zygote, they can in a sense be considered parts of one individual.

The phylum includes three major classes (plus an extinct class).

■ The **hydrozoans** or hydroids (class Hydrozoa, "water animals") include the freshwater hydroids, *Obelia* (Figure 42-6A) and *Hydra* (Figure 42-7), and such complex forms as the redoubtable Portuguese man-of-war (*Physalia*; Figure

---

[4] We usually reserve the term "anus" for the posterior opening of a tubular alimentary canal. Coelenterates, in fact, have no alimentary canal and no posterior opening.

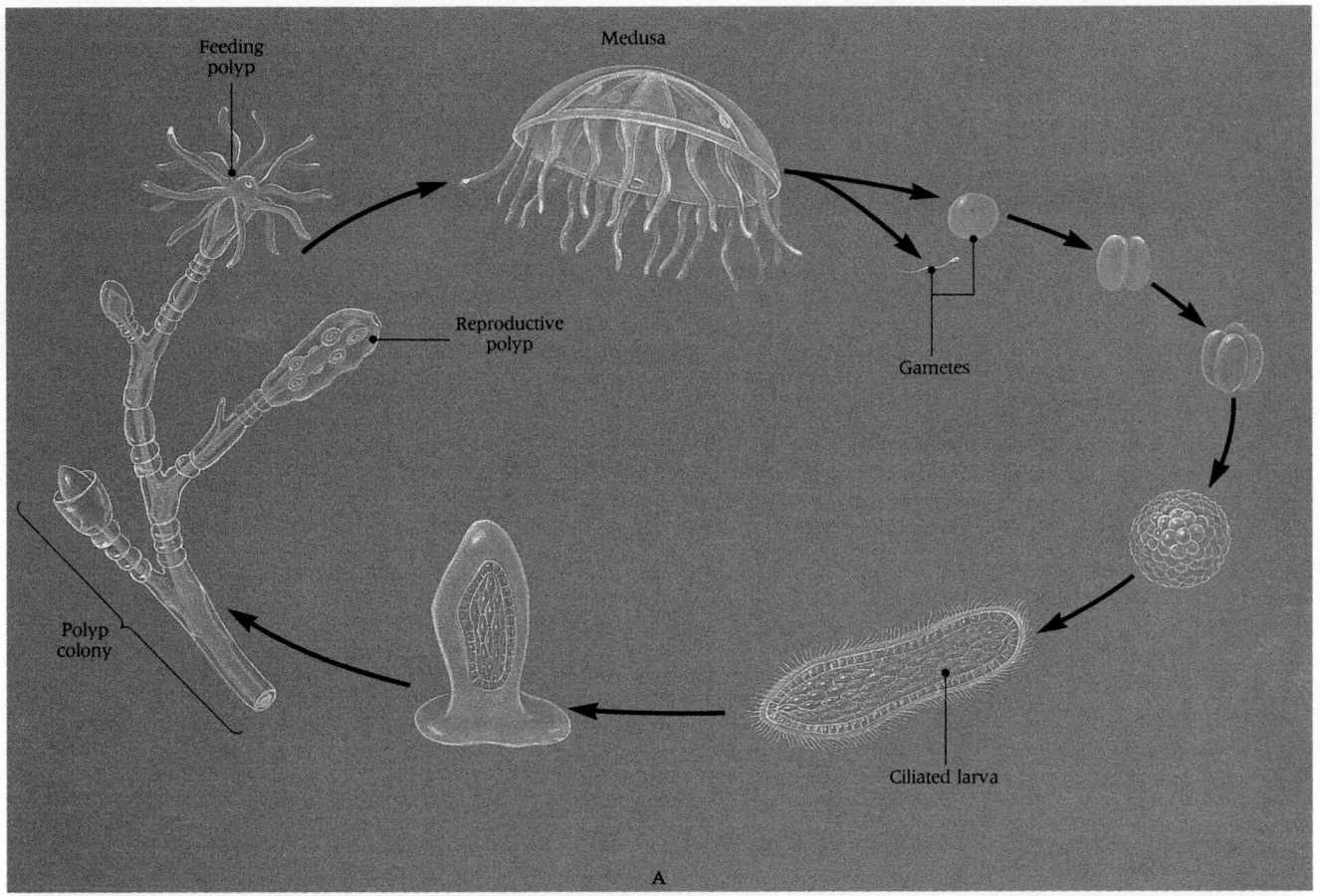

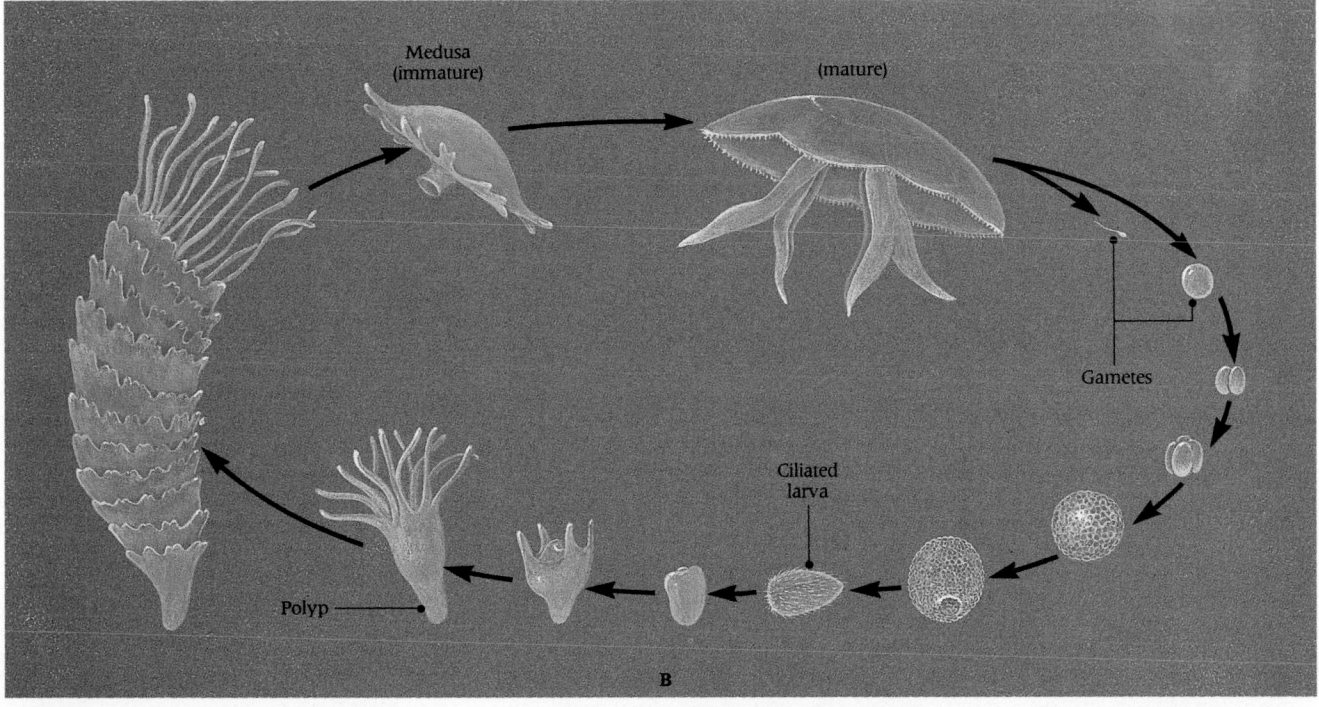

42-6 Life cycles of coelenterates
(A) In the hydroid, Obelia, *the principal stage is the polyp, which becomes a branched colony by budding. Some polyps gather food for the colony, while others are released as medusae for sexual reproduction. See also Figure 21-27. (B) In the jellyfish,* Aurelia, *the medusa is the dominant body form. The polyp is a transient stage in the life cycle.*

42-8). Many hydrozoans are small and soft-bodied (but not all: some form large masses with strong skeletons). Solitary or colonial polyps predominate, but medusas are often present in the life cycle. More complex forms like the floating Portuguese man-of-war are colonies made up of both polyps and medusas. Each of the colony of polyps is differentiated to perform special tasks such as locomotion, feeding, or reproduction. Medusas function as swimming organs or floats and reproductive organs.

■ The **scyphozoans** (class Scyphozoa, "cup animals") are the true jellyfishes. The familiar form of these animals that we can readily see is the free-swimming medusa stage. The polyp is a minute larval stage—or it does not occur at all. The 200 known species are all marine.

■ The **anthozoans** (class Anthozoa, "flower animals") are the sea anemones, sea fans, and corals, relatively large, solitary, soft-bodied creatures (with hydrostatic skeletons) that have only the polyp stage. Some 6000 species are known. All are marine. Anthozoans are the most complex of the coelenterates. Sea anemones (named for flowers they resemble) are solitary but quite large (Figure 42-9). As we saw in Chapter 1, corals form vast colonies of polyps that develop stony calcareous skeletons. When the old individuals die the skeletal supports remain. As layer forms upon layer, coral reefs, atolls, and islands emerge. The Great Barrier Reef off northern Australia beautifully illustrates how corals have affected geological history.

Coelenterata is another ancient phylum that became more diversified but changed only in minor ways for several hundred million years. Whether polyps are more primitive than medusas, or vice versa, is an interesting and disputed problem. Were the first coelenterates free swimming or attached? Fossils of the two forms are equally old and there are good arguments on both sides. Probably the characteristic coelenterate structure arose in sessile forms. Hence, the earliest coelenterates were either polyps, or alternately polyps and medusas. The radial symmetry of coelenterates and their hermaphroditic character support this conclusion. Both features are characteristic of sessile organisms.

PHYLUM PLATYHELMINTHES: FLATWORMS

To most people, a "worm" is any long, squirming animal—usually disagreeable— that lacks eyes, head, or legs. As a matter of fact, this body form occurs in many animal groups that may differ fundamentally in structure and origin. Early zoologists arbitrarily placed them all in a single phylum called Vermes (Latin for worms). We know now that worms belong to many phyla and display a wide range of anatomical arrangements.

■ Of the 20 phyla in our classification, the worm body form is prevalent in eight: Platyhelminthes, Mesozoa, Nemertina, Aschelminthes, Acanthocephala, Phoronidea, Chaetognatha, and Annelida.

■ Five other phyla include some groups or forms that are more or less wormlike: Ctenophora, Mollusca, Arthropoda, Echinodermata, and Chordata.

Many so-called worms (cutworms, inchworms, apple worms, etc.) are really insect larvae. Blindworms, or slow worms, are legless lizards. As we will see, worms are

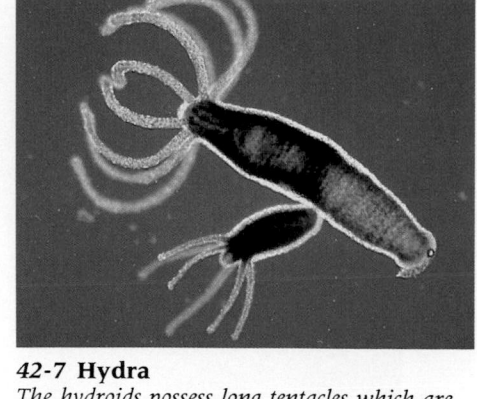

42-7 Hydra
The hydroids possess long tentacles which are used to capture prey and pass food into the mouth. This specimen is shown slightly larger than life size.

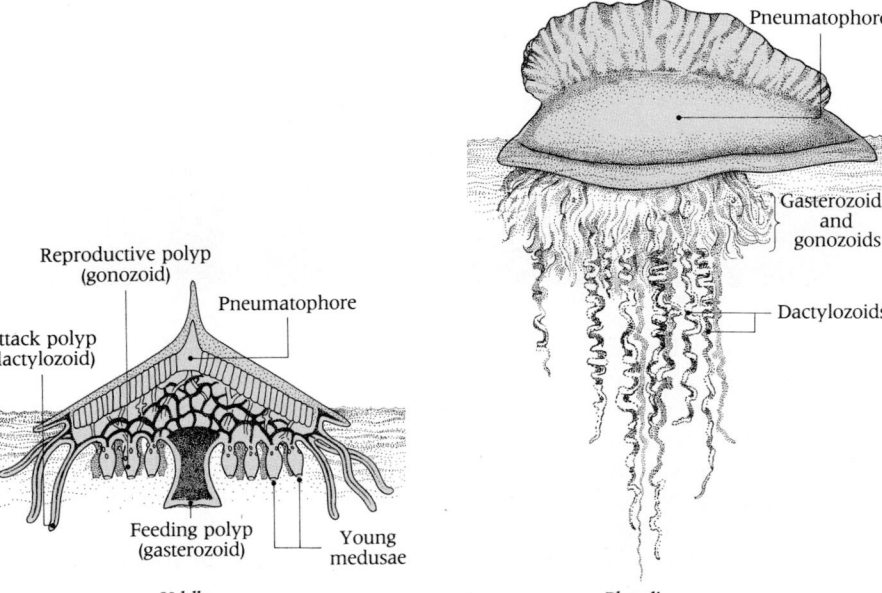

Reproductive polyp (gonozoid)
Pneumatophore
Attack polyp (dactylozoid)
Feeding polyp (gasterozoid)
Young medusae

Velella

Pneumatophore
Gasterozoids and gonozoids
Dactylozoids

Physalia

42-8 Colonial organization in coelenterates
Both Velella *and* Physalia *are highly evolved coelenterates with colonies of polyps that have become specialized to perform various functions, including buoyancy, feeding, reproduction, and prey attack.*

These anthozoans are polyps that lack a medusa stage. (**A**) Aiptasia pallida. (**B**) *Sea anemone,* Tealia lofotensis, *eating a starfish,* Patina miniata.

A

B

often parasitic, but many are free-living. The worm shape has evolved so often it is clearly a successful and advantageous way of life. Its advantages are generally related to its tendency to accelerate locomotion and facilitate burrowing and crawling—all of which improve the organism's responses to environmental danger or to predators.

The simplest of the wormlike animals are the **platyhelminths,** or flatworms (Figure 42-10). As their name implies they are flattened from top to bottom and have bilateral symmetry. Thus they sometimes resemble a piece of ribbon or tape.

The Platyhelminthes are bilaterally symmetrical. Like most such organisms, flatworms have a head end (anterior) and a tail end (posterior). Their bodies have three layers—ectoderm, endoderm, and mesoderm that is a well-developed solid mass in the space between gut and body wall. Thus there is no coelom or true

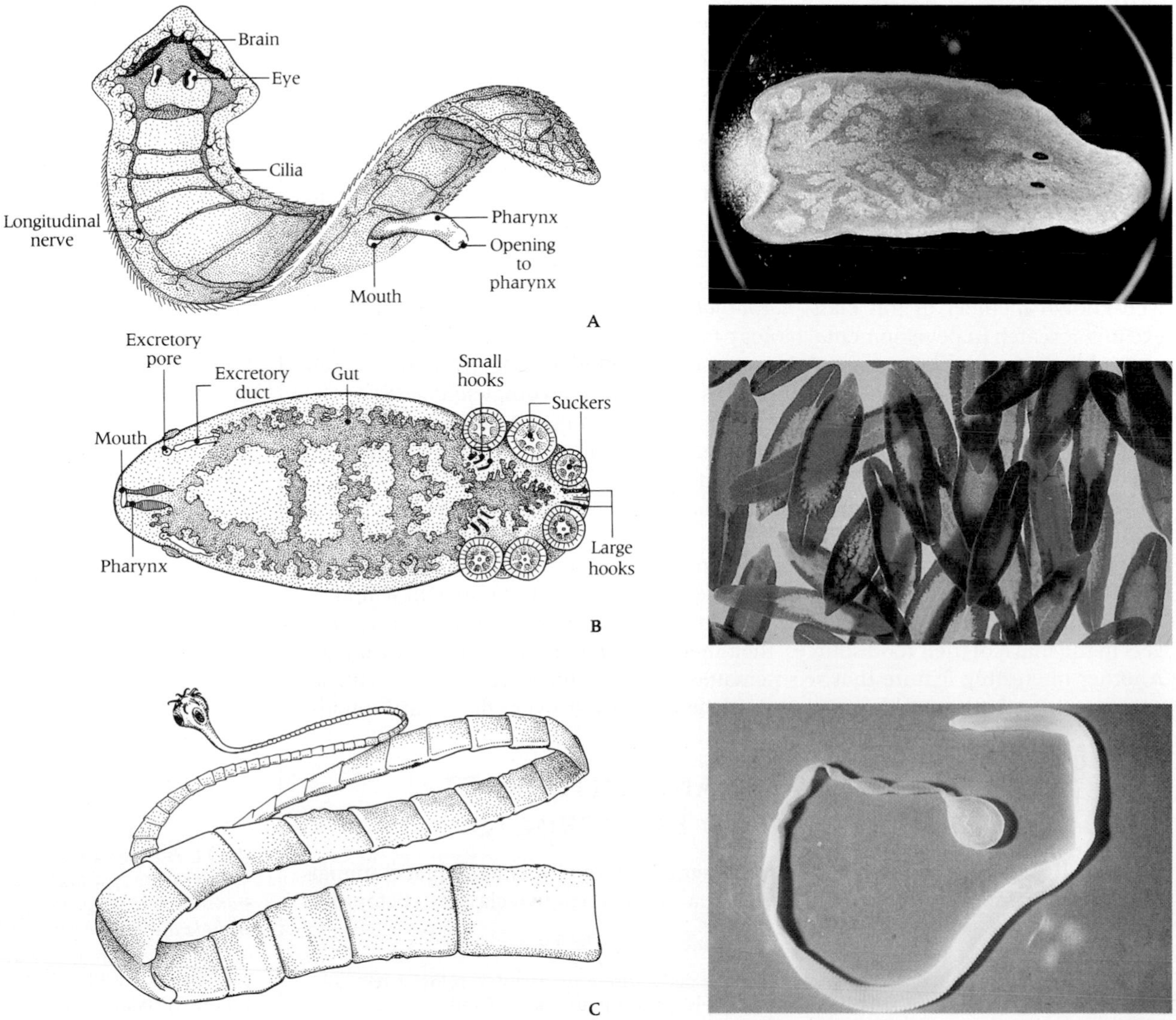

42-10 Platyhelminths

*The flatworm phylum includes three classes. (**A**) Turbellarians, like this planarian, are free-living organisms. (**B**) Trematodes, or flukes, are exclusively parasitic. (**C**) Cestodes, or tapeworms, live in the intestinal tracts of vertebrates.*

Brain

Eye

Cilia

Longitudinal nerve

Pharynx

Opening to pharynx

Mouth

A

Excretory pore

Excretory duct

Gut

Small hooks

Suckers

Mouth

Pharynx

Large hooks

B

C

body cavity in this space and members of this phylum (along with one other, Nemertina) are **acoelomate.**

The three classes of the phylum are **Turbellaria, Trematoda,** and **Cestoda.** The first includes free-living flatworms. The latter two classes include parasites that can cause serious diseases of humans and animals.

Turbellarians, belonging to the only nonparasitic class, are mostly marine but a few live in fresh water or moist habitats on land. The **planarians** (order Tricladida) nicely illustrate the class. Ranging in length from microscopic to several feet, they resemble coelenterates in having a digestive cavity with only one opening that serves as mouth and "anus." They have extensive powers of regeneration of body parts. But in other respects they are more like complex animals. They are fully motile and have several organs (for reproduction and excretion), a fairly complex nervous system including a brain of sorts (see Figure 30-30C), and a flame-cell excretory system (see Figure 25-7). They also have sensory receptors, including primitive eyes (that detect light but do not form images). The most primitive turbellarians are the tiny marine worms called acoels (order Acoela). These creatures even lack a gut (hence "acoel") and distinctive reproductive and excretory organs.

Tapeworms (class Cestoda) and **flukes** (class Trematoda), both parasites, are the two best-known classes of the phylum. They are carnivorous and feed on nutrients in the fluids, tissues, or intestinal contents of their hosts. Bathed in foods provided by their hosts, they absorb predigested nutrients through their body walls. Diffusion, plus some active transport, nourishes all body cells, none of which is far away.

PHYLUM ASCHELMINTHES:
ROUNDWORMS AND OTHERS

Phylum Aschelminthes (also called Nemathelminthes) consists of pseudocoelomate organisms. The number of known species, which has been variously estimated to be between 10,000 and 500,000 (we cited the compromise figure of 80,000 earlier), includes about 50 extremely widespread human parasites—for example, *Enterobius* or pinworms, the most common helminthic parasite of humans in the United States; *Ancylostoma* or hookworms, blood-sucking worms that cause widespread anemia and iron deficiency in many countries; *Trichinella,* the worms that can cause trichinosis, a disease that is transmitted by uncooked pork, one ounce of which can contain 80,000 worm cysts.

Roundworms, or **nematodes** (class Nematoda) are the most plentiful of the five classes of the phylum Aschelminthes and the only class we will discuss. A well-known member of this class, *Caenorhabditis elegans,* is currently an object of exciting research in developmental biology (see Chapter 20). Nematodes are superficially different from flatworms. They are actually rounder and less flattened. But there are also more fundamental differences. They have a complete digestive tube with two openings—a mouth and an anus (Figure 42-11). Roundworms possess the peculiar body cavity termed a **pseudocoelom,** which occupies the space between the mesoderm of the body wall and the endoderm of the gut. In the pseudocoelomate body plan, no muscle surrounds the gut and hence peristalsis cannot move food through the alimentary canal.

Indeed, a unique feature of nematode anatomy is the total absence of circular muscles. Longitudinal muscles account for the ability of some to swim with whipping movements. But most are inactive. Since nematodes almost always live like scavengers in the midst of their food sources, their inability to move well is no disadvantage. Another interesting feature that sets nematodes apart is their tough, elastic cuticle, an outer "skin" of several layers that they shed as they grow, as do the arthropods.

PHYLA ECTOPROCTA AND ENTOPROCTA:
ECTOPROCTS AND ENTOPROCTS

We noted earlier that taxonomists once assigned a group of small aquatic animals they called **bryozoans** to the phylum Bryozoa[5], which had two classes—Ectoprocta

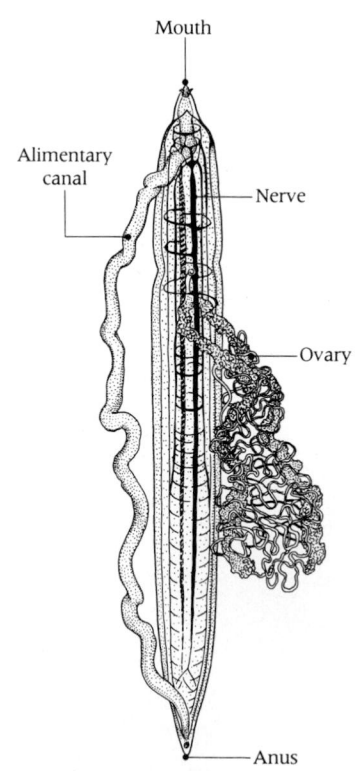

42-11 Body plan of the roundworm
The sketch is of a dissected female Ascaris, *a parasitic threadworm. The internal organs include a complete alimentary canal; a nervous system consisting of a nerve ring around the pharynx and ventral and dorsal nerve cords; and a reproductive system.*

[5] From roots meaning "moss animals." The name derives from the fact that colonies resemble moss. Some texts call them polyzoa ("multiple animals") because they are colonial.

A

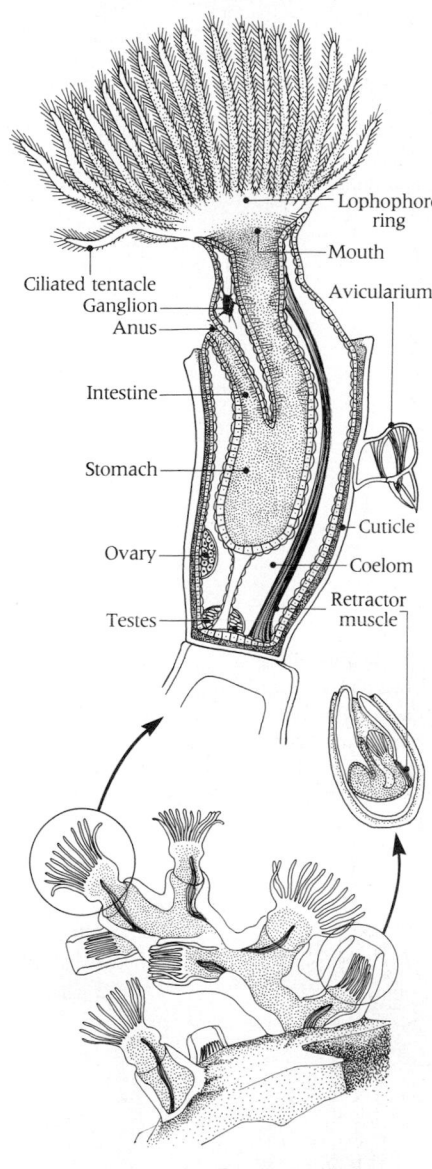

Lophophore
ring

Mouth

Ciliated tentacle
Ganglion
Anus

Avicularium

Intestine

Stomach

Cuticle

Ovary

Coelom

Testes

Retractor
muscle

B

42-12 Ectoproct anatomy and habitat
(A) *A colony of the ectoproct,* Amathia
convoluta. **(B)** *The freshwater ectoproct,*
Plumatella. *One individual is shown to illustrate
anatomical detail. The whole mouth region and
tentacles can be withdrawn by a retractor muscle.
The avicularium is a modified individual animal
whose "jaws" snap shut on moving prey.*

and Entoprocta. We now consider these separate phyla, though the members of Ectoprocta are still called bryozoans in the jargon of old-timers. Most ectoprocts are colonial, the colonies forming crusts on shells and rocks or mats or branching fans (Figure 42-12A). The colonies build supports of limy material, and the individuals live in small cups in the skeletal framework. They also have numerous ciliated tentacles around the mouth, but there the resemblance to coelenterates ends. Even the tentacles are different: in ectoprocts they have no stinging cells and do not capture food directly. Rather they are parts of a specialized food-gathering, funnellike organ called the **lophophore** (Figure 42-12B). In ectoprocts, cilia on the tentacles that make up the lophophore create water currents that carry plankton and other microscopic food particles into the mouth.

Ectoprocts are exclusively aquatic and mainly marine. Freshwater forms are not uncommon. Some 5000 species of them occur in all the seas and they are especially numerous on coral reefs, where they add materially to the stony matter. Fossil ectoprocts, including many reef-dwellers, are far more abundant and diverse than the living forms.

Their anatomy is complex: The digestive system is a complete tube that is doubled into a U-shape so that the anus is near the mouth—a necessary arrangement for those that live in a cup without a rear exit. There are nervous and muscular systems, and most ectoprocts have true coeloms. The abundant members of phylum Ectoprocta have a coelom and an anus that is outside the tentacle-bearing **lophophore ring.**

A much rarer group of animals superficially resembles Ectoprocta but lacks a coelom and has an anus inside the lophophore ring of tentacles (Figure 42-13). These are members of the phylum Entoprocta. About 60 species are known. Many form large populations that are "matted" along the bottom of shallow seas. Entoproct cilia beat in an uncoordinated way; ectoproct cilia beat in rhythmic waves.

This account would be incomplete without mention of the **avicularia,** one of the most peculiar products of evolution. As shown in Figure 42-12B, these strange creatures project from the surfaces of ectoproct colonies. Resembling birds' heads, their large beaklike jaws snap at anything that blunders by. By entrapping small animals, they keep the colony clean.

PHYLUM BRACHIOPODA: BRACHIOPODS

Brachiopods, the marine "lamp shells," so named because they resemble old oil lamps (like Aladdin's), belong to the phylum Brachiopoda ("arm-footed"). Their name is based on an elaborate error. The structures carrying tentacles look like arms and were thought to have something to do with walking. In fact, they have no functional resemblance to arms or feet, but the name was retained and is universally used.

Brachiopods are enclosed in two shells of approximately equal size (hence, they are literally bivalved). To that extent they resemble bivalved clams (of the class Bivalvia or Pelecypoda, phylum Mollusca), but the resemblance is superficial. In clams, the shells are located laterally on the right and left sides of the body. In brachiopods, they are dorsoventral, on the top and bottom of the body. Thus there is a difference in symmetry.

Two brachiopod classes are distinguished by the presence or absence of interlocking "teeth" that act as a tongue-and-groove hinge. Class Articulata has the hinge. Class Inarticulata lacks a hinge. The latter includes *Lingula,* probably the oldest living genus, dating from Cambrian times. Brachiopods are more complex than any phylum so far discussed. They have a coelom and well-developed nervous, muscular, digestive, circulatory, excretory, and reproductive systems (Figure 42-14). Like ectoprocts, they are lophophorate animals, possessing a conspicuous lophophore that encircles the mouth and is crowned with tentacles, bearing many vibrating cilia. As in the ectoprocts, the tentacles (or, strictly speaking, the tiny hairlike cilia on the tentacles) stir up water currents that sweep food particles into the mouth.

So-called lophophorate animals include a third phylum, called Phoronida, some 15 species of wormlike marine animals. Unlike ectoprocts, they have a closed circulatory system containing red blood cells and hemoglobin.

Brachiopods are noncolonial. An individual usually becomes attached by a short fleshy stalk that protrudes through the hind end. They are exclusively marine and always have been. The survivors are not very abundant or diverse. Only about

300 living brachiopod species are known. But they are remnants of a great line. Their fossil record is one of the best and more than 30,000 extinct species have been found by the millions in the rocks of many ages.

PHYLUM MOLLUSCA: MOLLUSKS

In no way does the list of modern animal phyla represent a steady and inexorable evolutionary march in the direction of progress. Still, it is clear that three of the phyla, each in its way, went farther than the others in complexity, diversity, and adaptational success. They are the mollusks, arthropods, and chordates.

Embracing more than 100,000 species, phylum Mollusca is second in size only to phylum Arthropoda. Its most familiar representatives are squids, octopuses, clams, oysters, and snails, which include the many exquisite forms so valued by the world's collectors of sea shells. The name Mollusca, which derives from a Latin word, *molluscus,* meaning soft, suggests that the animals are soft-bodied. In fact, most of them have shells, and the bodies within the shells are no softer than those of most other animals. But as with many systematic names, its inappropriateness soon ceased to be important.

The mollusk phylum is subdivided into seven major classes (Figure 42-15). Three are extremely large and diverse. These are the classes **Gastropoda** (snails and relatives), **Bivalvia** (clams and relatives), and **Cephalopoda** (octopuses, squids, cuttlefish, and nautili). The smaller classes are **Amphineura, Aplachophora, Monoplacophora,** and **Scaphopoda.**

The classes are so different and so old it is hard to generalize about the phylum. There is every reason to believe that evolution took seven different routes to arrive at the classes that have been designated. Nonetheless, there is enough of a basic uniform pattern to indicate common ancestry. The molluscan body (except in one small class) is unsegmented and typically consists of an anterior **head,** a large ventral **foot,** and a dorsal **visceral mass.** The viscera of most mollusks includes practically all of the organ systems found in other animals: digestive (sometimes with a unique rasping toothlike device, the radula), circulatory (with a heart), respiratory (usually with complex gills called ctenidia), excretory (with kidneys),[6] nervous (often with brainlike ganglia and sometimes with well-developed eyes and other sense organs), muscular, and reproductive systems. Above the viscera is a **mantle,** a fold of specialized tissue that more or less covers the body. With certain exceptions it includes glands that secrete one or more shells, within which the body is sheltered. The mantle sometimes overhangs the sides of the visceral mass, enclosing a mantle cavity in which gills may lie.

[6] The term ''kidney'' is used rather loosely to refer to the excretory organs in different metazoan groups. The molluscan kidney has no evolutionary relationship (or homology) with the vertebrate kidney. Rather it is a much elaborated version of the simple excretory tubules (nephridia) discussed in Chapter 25.

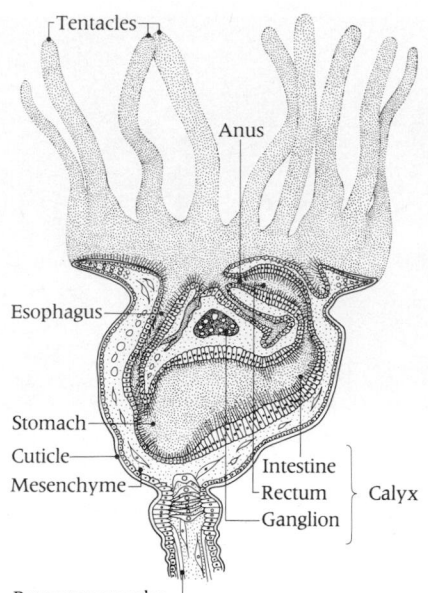

42-13 Entoproct body organization
In the entroproct Barentsia matsushimana, *the anus is within the lophophore ring of tentacles.*

42-14 Brachiopods
(A) *Anatomical organization showing a conspicuous lophophore within the mantle cavity. The ciliated tentacles generate water currents inside the two shells.* (B) *Brachiopods,* Liothyrella ura, *on a rock surface 20 feet under water near Antarctica.*

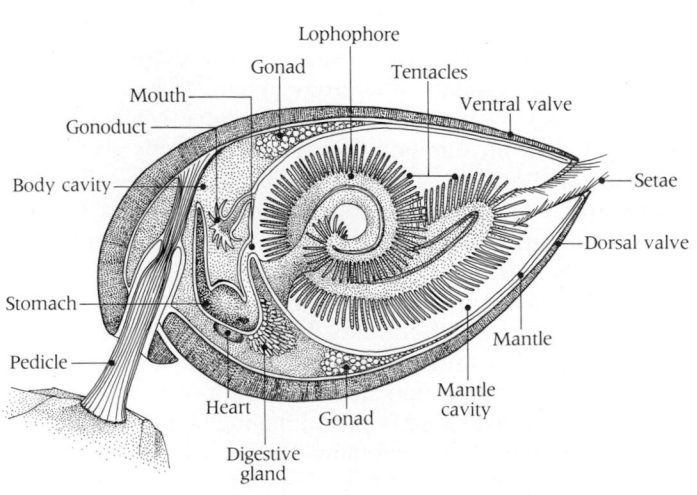

A

B

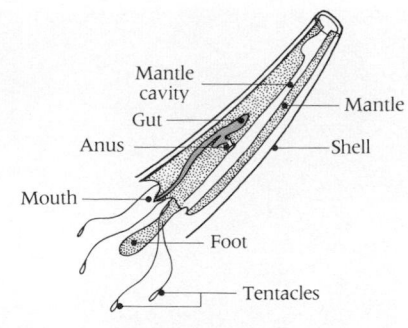

Scaphopoda (tusk shell)

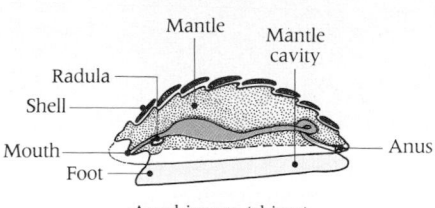

Amphineura (chiton)

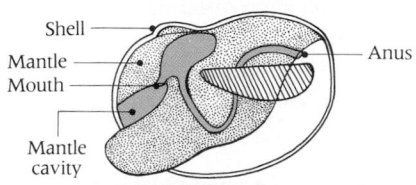

Pelecypoda (clam)

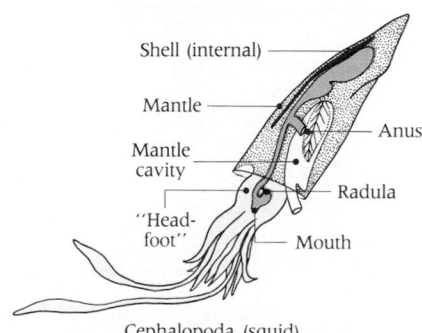

Cephalopoda (squid)

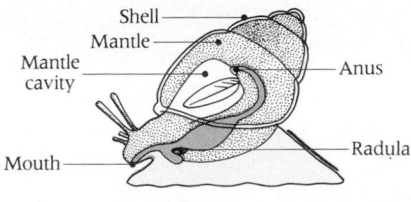

Gastropoda (snail)

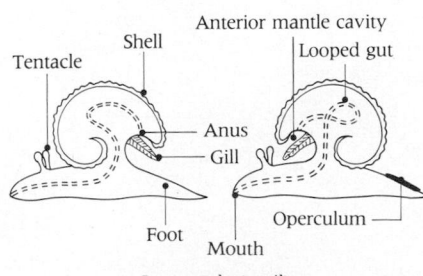

Gastropoda (snail)

42-15 The principal molluscan classes
The five major groups of mollusks have a generally similar body plan. A representative of each class is indicated in parentheses. All sketches are oriented with dorsal (back) body parts on top and ventral parts on bottom. Anterior aspects are at left and posterior at right.

CLASS AMPHINEURA: CHITONS

Most of the features just mentioned are clearly seen in the class Amphineura (also known as Polyplacophora).[7] Its best known members are the **chitons,** which are among the simplest mollusks, and probably one of the most primitive (Figure 42-16). The first mollusks to arise in evolution may have resembled the chitons, although much of the simplicity of chitons may have arisen secondarily. For example, the lack of a well-differentiated head and special sense organs is more likely due to degenerative loss than to primitiveness. In any event, chitons clearly illustrate certain basic molluscan characteristics. They have a bilaterally symmetrical ovoid body, an anterior mouth, a posterior anus, and eight overlapping dorsal shell plates that are joined by a fleshy girdle.

Chitons are herbivorous marine organisms that lead a sluggish, nearly sessile life, adhering to rocks and other hard surfaces. They slowly crawl about in shallow waters, feeding at night and using the horny **radula** to scrape algae and other food from rocks. The chiton foot develops tremendous suction, clamping tightly onto rocks, especially when they are disturbed. A chisel is sometimes needed to pry them loose.

There are about 600 living species of chitons, but they are not common fossils. They appear early in the fossil record, but are poorly preserved.

CLASS MONOPLACOPHORA: MONOPLACOPHORANS

The **monoplacophorans,** also primitive and simple, are internally segmented mollusks with single shells (univalves). They are frequently found in the fossil record and appear as early as any of the mollusks (early Cambrian).

The class was thought to have become extinct hundreds of millions of years ago, but on May 6, 1952 the grappling hook of the *Galathea,* a Danish oceanographic vessel, dredged up ten living specimens of an unknown limpetlike mollusk from the deep ocean bottom off the west coast of Central America. These "living fossils" (named *Neopilina galatheae*) sparked great excitement among biologists who wondered whether they were the "missing link" between wormlike molluscan ancestors and true mollusks. This proved not to be the case. It is now believed that *Neopilina* is merely an interesting survivor of an odd molluscan stock.

CLASS SCAPHOPODA: TUSK SHELLS

Scaphopods ("boat-footed") form a distinctive class of marine mollusks with a long, unchambered, tapered, tubelike shell that is open at both ends (see Figure 42-15). The shell resembles a little elephant tusk or a canine tooth; thus they are called tusk shells or tooth shells.

About 350 living species are known that range from shells an eighth of an inch long to five inches. Their empty shells, found at low tide on beaches all over the world are often used as ornaments. Scaphopods are an ancient class dating from the Ordovician period. Their 400 fossil species are much more common late in the Mesozoic era.

CLASS GASTROPODA: SNAILS AND THEIR RELATIVES

Gastropods ("stomach-footed") consist of the snails and slugs that seem to crawl on their bellies. This largest mollusk class includes the familiar land snails, freshwater snails, slugs, and marine snails—conchs, sundials, tritons, augers, murexes, whelks, limpets, and a multitude of others (Figure 42-17).

The snails and their relatives have a distinct head, usually with eyes on the tips of long tentacles, and a foot with which they crawl. Their nervous systems and learning ability has made at least one snail, *Aplysia,* something of a celebrity in recent years (see Chapter 30). Snails are motile—their unhurried gait is legendary—and searching for food is their principal activity. Characteristically, they have a single coiled dorsal shell, or **valve,** that has one opening. Most gastropods can

---

[7] We will note the molluscan class, Aplacophora, only in passing. These strange, poorly studied, shell-less organisms, commonly known as solenogasters, are wormlike things without a head, foot, mantle, or kidneys. They are covered with shelly spicules. Only about 250 species are known. They were once classified with the chitons.

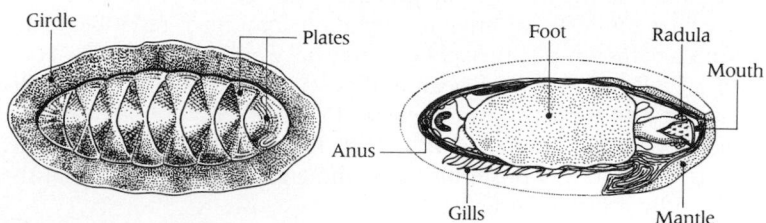

Girdle — Plates — Foot — Radula — Mouth — Anus — Gills — Mantle

42-16 Chitons
Chitons, Amphineura, though typical molluscans in most respects, are atypical in lacking special organs in the head and in their many plates (shells).

withdraw their bodies completely into their shells and in some groups a protective plate, the operculum, may be present to close the opening. Opercula may be beautiful and are often set in jewelry.

The viscera is typically contained in the shell. Early in embryonic life (the larval stage) the gastropod body is bilaterally symmetrical, but in the course of metamorphosis the visceral mass is coiled into a spiral. This process, called **anopedal flexure,** brings the anus from its position in the neck region behind the mouth to a point above the mouth. Then, in a process called **torsion,** the whole visceral mass is rotated by 180 degrees so it comes to lie dorsal to the head. Simultaneously the mantle cavity moves up and forward around the head (see Figure 42-15). However this peculiarity evolved, it is highly practical for an animal that lives in a house with only one entrance and that crawls about feeding on the bottom.

After the torsional twisting occurs in development, half of the paired visceral organs (gills and kidneys of one side) fail to develop. Further lopsided growth causes the mantle to become twisted spirally. Since the mantle secretes the shell material, a coiled shell results. The fossil record suggests that the coiling gastropod shell evolved before the phenomenon of flexure. The spiraling of the shell and torsion, therefore, reflect separate evolutionary events. It is noteworthy that in certain close relatives of snails, such as the unlovely garden slug, the viscera were straightened out again by further evolutionary events in which the shell was lost. But the organs missing on one side were not restored.

Gastropods that have lost their shells include the marine **nudibranchs** (order Nudibranchia). These forms illustrate the irrevocability of evolution. The young nudibranch has a spirally coiled shell but it loses the shell during development. Still the viscera show the torsional twisting and asymmetry that accompanied shell development in their ancestry. It is not clear what evolutionary forces led to loss of the shell. But its loss clearly led to the later evolution of protective devices, which substituted for the shell's protective function. For example, certain shell-less gastropods that feed on coelenterates evolved the remarkable ability of digesting all the victim's tissues except embryonic stinging cells. These cells are borrowed, eventually appearing in the gastropod's soft skin where they mature and help to protect their former predator. Such gastropods are brilliantly colored and ornamented. Some nudibranchs are luminous in the darkness and others give off flashes of light at the least touch.

Like all mollusks, gastropods were originally marine creatures. Those that still are derive their oxygen supply through gills that were placed in front of the heart by torsion. These are the **prosobranches** (subclass Prosobranchia)—the name means "front gills"—that are marine gill-breathers. However, some gastropods became almost fully terrestrial. They are usually found in moist locations.[8] Terrestrial gastropods are **pulmonate** (subclass Pulmonata)—that is, they have lost their

A

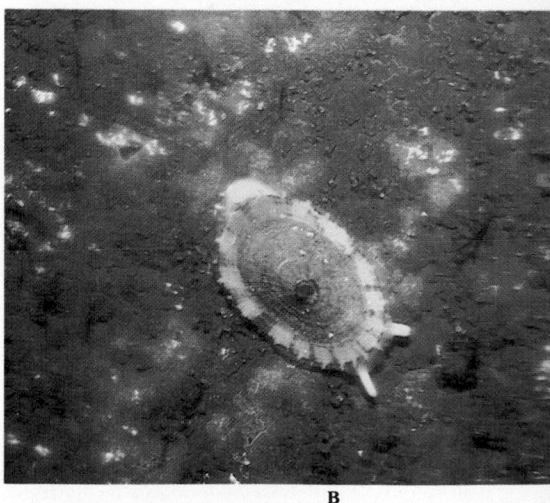

B

42-17 Gastropods
(A) *A snail of the family Trochidae.* **(B)** *A limpet of the family Patellae.*

---

[8] The silvery trails left by slugs and terrestrial snails are seen commonly in the garden. The trail consists of mucus secreted by glands in front of the foot. The animal skids along the mucus trail propelled by foot movements.

A

B

C

42-18 Bivalves
Two shells enclose the soft body parts. **(A)** *Pacific giant clam*, Tridacna gigas. **(B)** *Bay scallop*, Aquipecten irradians. **(C)** *Noble scallop*, Chlamys nobilis, *of the Western Pacific.*

gills and have in their place a pulmonary sac, a lunglike cavity in the mantle that enables them to extract oxygen from air. The number of pulmonate species is immense. Among the numerous freshwater snails, some have gills and some are pulmonate. Those with gills are of marine origin. Those with "lungs" had terrestrial ancestors but returned to the water. That they did not regain their long-lost gills is another example of the irrevocability of evolution. Their story recalls the return of air-breathing mammals, the whales, to the remotely ancestral seas, where they retained their lungs and did not regain the lost gills of their fish ancestors.

Gastropods have evolved into a tremendous array of different forms, many of which are beautiful, colorful, and bizarre (Box 42-1). There are about 80,000 living species, three subclasses, and many orders.

CLASS BIVALVIA: CLAMS, OYSTERS, AND RELATIVES

The **bivalves,** which include the clams, oysters, scallops, and mussels, are also known as **pelecypods.** Pelecypod means "hatchet-footed." The name refers to the hatchet- or wedge-shaped outline of the foot in some members of this class. The class Bivalvia once had another alternative name, Lamellibranchia ("plate-gills"), but this is now considered the largest of the three subclasses of Bivalvia. The other two—in most but not all classifications are Protobranchia and Septibranchia. Their exceptionally fine fossil record dates back to the Cambrian.

Bivalves, the second largest class of mollusks, retain bilateral symmetry and have two shells or valves that are usually nearly symmetrical and enclose the soft body. The valves remain open, but in the presence of danger, powerful adductor muscles clamp them tighter together.

These uncoiled animals are motile, but they usually lead sedentary lives, moving even more sluggishly than snails. They seldom move very far from where their embryonic larvae settled down. Some, like oysters and mussels, become permanently attached. A few swim fairly well; scallops (genus *Pecten*) swim by clapping their shells together. (The unusual character of their major muscle was described in Chapter 31.) By using their muscular foot, most of them make a specialty of burrowing. Many species of marine clams burrow in mud or sand, as do our common edible clams.[9] Others (like *Teredo*) burrow in wood and cause serious damage to old wooden boats and pilings, especially in salt water. Some (like *Pholas*) burrow in solid rock. The astonishing giant clams (*Tridacna*) of the South Pacific, which may be 6 feet across and weigh 600 pounds, burrow in living coral reefs.

In keeping with their sedentary life style, bivalves have lost a differentiated head region. Thus these strange animals have no head, pharynx, jaws, radula, or tentacles. However, many have numerous "eye spots," simple light receptors in the mantle along the opening between the shells, that can detect movement or shadows but cannot define images.

Like many sedentary animals, the bivalves feed by filtering microorganisms from their watery medium. In most bivalves, filtering is accomplished by the typical molluscan gill, initially a respiratory structure but here much enlarged and modified as a feeding sieve. The water it filters flows in and out of the mantle cavity through tubes called siphons.

Bivalves are less spectacularly diverse than gastropods, but they are still extremely varied (Figure 42-18). About 8,000 living species and some 126 genera are known— and they are certainly abundant. About two-thirds are marine and one-third (notably freshwater clams and river mussels) live in fresh water. None are terrestrial.

---

[9] Members of this group of mollusks have been prized foods since antiquity. The quahog, or little-neck, hard-shell clam, familiar to devotees of clam chowder and clams-on-the-half-shell, is *Mercenaria mercenaria* (formerly *Venus mercenaria*). The marine mussels popular as seafood all over the world belong to the Mytilus family. *Mytilus edulis* is among the most abundant and hardy species of edible mussels. Oysters of the genera *Crassostrea* (American) and *Ostrea* (European), like mussels, are not ordinarily cultivated commercially. Oysters of the genus *Pinctada* are cultured for the production of pearls, which develop when a foreign body penetrates the mantle. In the early part of this century, a Japanese research worker named Mikimoto discovered how to "seed" a pearl oyster deliberately.

CLASS CEPHALOPODA: OCTOPUSES, SQUIDS, AND NAUTILUSES

Cephalopods are "head-footed" because what is the foot in other mollusks is transformed in cephalopods into tentacles or "arms" around the mouth in the head region—8 in the octopus (hence, order Octopoda), 10 in the squid (hence, order Decapoda), and 70 or 80 in the chambered nautilus. Octopuses and squids make up the subclass Dibranchia. This highly specialized group of free-living mollusks is famous for locomotion, an activity that they can engage in with swift exuberance. The distinctive nautilus comprises the subclass Tetrabranchia.

Despite tales of alleged unpleasant behavior toward humans and despite their sharp beaks and waving tentacles, octopuses (other acceptable plurals: "octopodes" and "octopi") are shy creatures that have seldom if ever seriously harmed a human being. Divers often describe them as lovable. They and the squids are among the most complex and advanced products of evolution (Figure 42-19). They have well-formed heads with relatively large brains and they are capable of an astonishing degree of learned behavior.

The octopus is a highly mobile sea dweller that either creeps about on its "arms" or swims actively through open water by ejecting water through a tubular siphon below the head. Like squids, they are literally jet-propelled. They also resemble squids in having eyes remarkably like those of vertebrates. However, the evolution of these elaborate image-forming eyes, which work on the same cameralike principles as our own, occurred quite independently—a classic example of convergent evolution.[10]

It is a familiar fact that cephalopods have tentacles with many suction disks, or suckers. They use the tentacles and suckers to seize prey and convey it to the sharp, shearing jaws. (Octopuses also use them in clambering about.) Cephalopods are voracious predators. They actively pursue, kill, and devour fishes, crabs, clams, and many other kinds of animals (not humans) from holes, caves, and crevices. The octopus may sit for hours on a coral ridge awaiting a victim. When an unwary crab goes by, the octopus whips out an "arm" and immobilizes it with hundreds of suckers. Evidently the traits we admire most—brain development, keen senses, and skillful coordination—are best developed in predatory animals.

Octopuses, like squids, are mottled with pigment bodies, or **chromatophores,** in the skin with which they can change their color. An octopus can be blue-green one instant, brown the next, then reddish or nearly white. Some species even flaunt zebra stripes. Octopuses appear to have emotions that are closely linked to the color mechanism. It was once thought that the octopus mimics the colors and pattern it sees. Later work has shown that color patterning still occurs in blinded animals. It appears that there is a preprogrammed repertory of patterns and that incoming information, the nature of which we do not know, determines which one is displayed.

The cone-shaped squid has a large mantle cavity with muscular walls and 10 tentacles that include two extra-long ones with suckers at the tip. The mantle cavity opens near the head, admitting water to the gills for respiration. But the water does not leave the mantle cavity by the same opening. Instead it is rapidly expelled through a funnellike siphon. This permits the squid to move swiftly about by jet propulsion. If the tube is aimed backward, the squid darts forward in pursuit of prey (Figure 42-20). However, when it is pursued rather than in pursuit (the more usual situation), it can turn the tube forward, contract the powerful mantle muscles, and without turning suddenly dart backwards by reverse thrust!

Giant nerve fibers give them dazzling quick-start speed for capturing prey or evading enemies. Neurophysiologists have learned a great deal by studying isolated preparations of the unusual nerve nets of squids and their giant nerves, which can

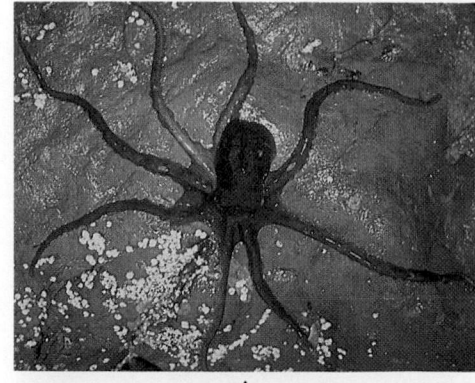

A

B

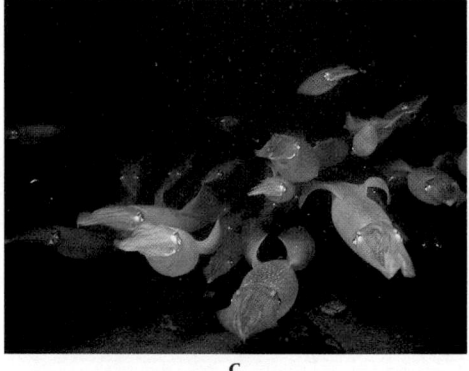

C

42-19 Cephalopods
(A) *Giant Pacific octopus*, Octopus dofleini. **(B)** *Squid*, Loligo opalescens. **(C)** *School of squid*, Loligo paelei.

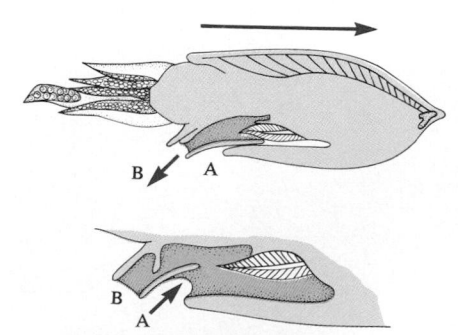

42-20 Cephalopod locomotion
The squid moves in the direction of the colored arrow by propelling water from the mantle cavity (color) through the siphon (arrow B). The animal can move in the reverse direction by directing the siphon backward (not shown). The mantle cavity is then recharged with water (arrow A).

---

[10] The convergence of their eyes with those of vertebrates has given cephalopods a curious place in the history of evolutionary thought. In the early nineteenth century, the idea of evolution had not yet been securely accepted. In 1830, the French naturalist Geoffroy Saint-Hilaire (1772–1844) engaged in a famous debate with the French zoologist Georges Cuvier (1769–1832), an opponent of evolution. Saint-Hilaire staked the case for evolution in part on the resemblance between cephalopod and vertebrate eyes. Cuvier had no trouble in demonstrating that vertebrate and mollusk eyes were quite unrelated in any sense (evolutionary or otherwise) and evolutionary thought suffered a serious setback from Saint-Hilaire's unfortunate choice of example.

42-21 Chambered nautilus
The coiled shell provides buoyancy for Nautilus pompilius. *It is the only cephalopod with an external shell.*

have diameters up to 1 mm (compared to 0.02 mm in humans). The squid has a large and complex brain, as befits its active life. Alone among invertebrates, it has a cartilaginous internal skeleton, including a skull-like case around the brain—yet another remarkable evolutionary convergence with the vertebrates.

We lack the space to discuss this remarkable invertebrate in greater detail and can note only that squids have a virtuoso collection of fantastic tricks and devices at their disposal. In addition to their color-changing ability, squids can squirt clouds of black-brown ink from an ink sac opening into the funnel.[11] Some species can blink myriad glowing lights and others can glide through the air for more than 100 feet after a burst of submarine jet propulsion. Squids inhabit every ocean and among the larger creatures only the fishes exceed them in abundance. Some species are as small as minnows. The giant squid, on the other hand, is the largest living invertebrate, measuring up to 60 feet overall and weighing two tons.

The cuttlefish differs from most squids by having a hard calcareous shell within its body. Cuttlefish bone (or cuttlebone) is the form in which we give calcium to our caged pet birds. The cuttlefish is the real chameleon of the cephalopods. It can deceive both its prey and its enemies by matching the color and pattern of almost any natural background.

The cephalopods of subclass Tetrabranchia are the one surviving group that is markedly different from Dibranchia, the subclass that includes the squids and octopuses. Tetrabranchiates are distinguished from dibranchiates by having numerous small arms lacking suckers and two pairs of gills instead of one pair. In the past there were many tetrabranchs. Those known as ammonites were especially conspicuous during the Mesozoic, more than 5000 extinct species from that era having been recognized. The genus *Nautilus* is the sole surviving representative. It is also the only living cephalopod with an external shell.

The nautilus lives in a coiled shell that is partitioned into chambers (Figure 42-21). From time to time as the animal grows, it moves outward, sealing off the old chamber and living in the most recent one. Because of the sealed chambers, the entire shell floats, even after death. The four or five known species of nautilus are limited to the western Pacific.

PHYLUM ANNELIDA: SEGMENTED WORMS AND LEECHES

The name *annelid* derives from Latin roots meaning "ring." The bodies of these worms do resemble a series of rings. This group is very different from flatworms or the so-called roundworms belonging to the phylum Aschelminthes. Annelids really are round worms, having a cylindrical tubular body with a mouth at one end and an anus at the other. As we saw in Chapter 31, their cylindrical surface is modified by rows of bristles called setae and often by more complex appendages (see Figure 31-30).

A striking feature of these worms is their numerous segments, which are visible externally as ringlike bulges (Figure 42-22A). Internally, they are separated by

42-22 Polychaete annelids
(A) This polychaete worm, Nereis, *exhibits characteristic bristle-bearing parapodia on each segment. (B) Many polychaetes, including this peacock worm,* Sabella, *are sedentary, bringing food to their mouths on currents set up by the tentacles. (C)* Chaetopterus

[11] Artists have used this ink, called sepia, for 2000 years. It was a forerunner of India ink. In 1817, Georges Cuvier used cephalopod ink to draw the illustrations in his classic work on the anatomy of Mollusca.

A

B

C

The sizes and shapes of mollusk shells fill a multifarious catalog. Yet, one sees in this diversity a common architectural plan. In essence, the gastropod shell is an elongated cone that has been wound into a spiral around a central axis. Three common shell types—sundials, cowries, and miters—are illustrated in Figure 42A.

The colors of marine shells range from red to violet, though bright greens and vivid blues are rare. The most common colors are shades of brown and green in combination with white. The most colorful seashells are generally found in shallow tropical waters. Temperate seas support less brightly colored species and species with largely white or pale-colored shells populate polar seas.

Large gaps exist in our knowledge of shell pigmentation. While shell substance can be deposited by the mantle throughout a snail's life, pigment production is usually intermittent. This may account for the variation in color pattern that is characteristic of each species. It has been suggested that some snails deposit pigments in the shells as a means of disposing of otherwise unmanageable waste products. Pigments of many types are found in the blood of mollusks and in their color-depositing cells. These pigments collect along the edge of the mantle and eventually produce a highly distinctive color pattern in the snail's shell.

Some of the most colorful shells, totally hidden by the fleshy mantle during the snail's life, are revealed only after the soft parts are gone. The colorful cowrie shells in Figure 42A were covered in life by an outer protective skin (periostracum). The brightly colored aperture and interior surface are also concealed in many species by soft parts, as in the abalone, or by a large horny or limy operculum.

In some gastropods, shell color doubtless plays a role in their adaptation to the environment, offering either cryptic coloration or mimicry (see Chapter 35). However, the direct evidence in support of this notion is less than adequate. Some of the most colorful species live at levels too deep for the penetration of light. For snails that live in the dark, a vividly colored shell would seem to serve no useful purpose. Hence, though we admire these elegant products of nature, we are still largely baffled by the complex systems that control the coloring of these creatures and the sculptural ornamentation of their shells.

Despite our ignorance of these matters, shell colors and their distinctive arrangements, which some consider a biological "code" of some kind, are indispensable in classification and are of immense value to collectors. The study and collecting of shells—the field of conchology—is among the oldest of human hobbies. There is evidence of shell jewelry from archaeological sites over 30,000 years old and diggings at Pompeii (A.D. 79) unearthed shell collections. Today, shell collecting appeals to people of all sorts—divers, biologists, and casual beach walkers. Some fear the possibility that indiscriminate collecting may deplete the world's store of sea shells. Others envision a world in which all the mollusks are on display in cabinets, museums, and private collections, while the seas, lakes, and streams are devoid of them. Something like this already has happened to the butterflies of various regions.

42A Variety in shell shapes
Miter, sundial, and cowrie shells.

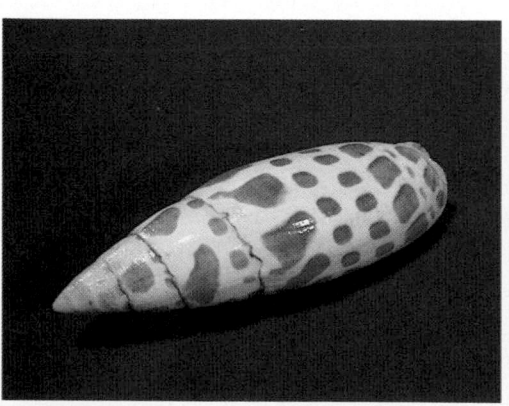

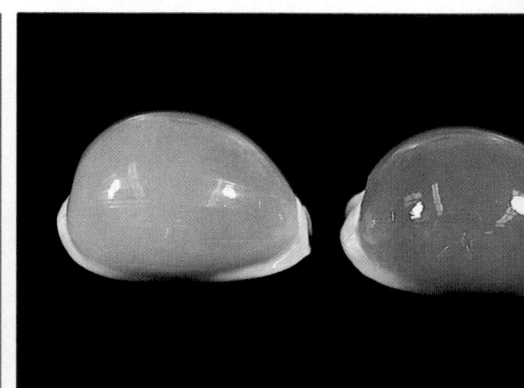

partitions and many of their well-developed organs—nerve ganglia, circular blood vessels, and excretory tubes, or nephridia—are repeated segmentally. However, the digestive tube, principal lengthwise blood vessels, and nerve cords, run through the segments. Control of the whole is ensured by a brain in the first fused segments that constitute a head. Indeed, these worms are highly integrated and are capable of quick, accurate, directed movements.

Earthworms are the annelids best known to us. However, annelids are found not only in soil, but in the sea and in fresh water. The more than 9000 species that have been described include many kinds of earthworms. Specialized for life in moist soil, these soft-bodied creatures literally eat their way through the ground, passing the dirt through the alimentary canal as they go. Organic matter is digested from the soil, and the residue is ejected from the anus in the form of familiar worm castings. They rework the soil so extensively, aerating it and reducing its acidity, Darwin doubted whether "there are any other animals which have played such an important part in the history of the world." Earthworm farms in the United States and Europe produce up to 500,000 per day. When the dikes of Holland collapsed in 1953, seawater destroyed the earthworms and it was necessary to reintroduce them.

The phylum Annelida has three major classes and a few minor ones.

- Class **Oligochaeta** (more than 3000 species) bristle-footed worms, includes earthworms (order Terricolae) and most freshwater annelids.
- Class **Hirudinea** includes some 300 species of leeches, semiparasites that live in the sea, in fresh water, and on land—largely in the tropics—subsisting on the blood of other invertebrates and vertebrates. Parasitism also occurs in the other annelid classes, but it is less common among annelids than among other worms. Most species are hermaphroditic, but each copulates with another individual. Leeches are distinguished by their two muscular suckers, one in front and one in back.[12]
- Class **Polychaeta,** also called paddle-footed worms, includes the majority of annelids, some 6000 species, which are marine and have little resemblance to earthworms. They live almost everywhere in the sea, are sometimes free swimming, but often live in burrows or tubes. They are mostly dioecious and spawn their eggs and sperms into the sea so that fertilization is external. Early development includes ciliated planktonic swimming larvae. One would not have to be a worm fancier like Darwin to recognize the beauty of the exquisite brilliantly colored polychaete plumes (see Figure 42-22B,C), which function in feeding and respiration.

Annelids appear to have evolved from flatworm ancestors in precambrian times.

PHYLUM ARTHROPODA: CRUSTACEANS, SPIDERS, AND INSECTS

The **arthropods** make up incomparably the largest of all animal phyla. They include, among others, the crustaceans, spiders, and insects. No one knows how many species of arthropods there are in the world. The number is at least a million (and that number would exceed all other animal species put together), but some estimates run to 10 million. Arthropods have been brought up from the deepest sea bottoms. They have been encountered by aircraft flying miles above the Earth. They are everywhere that life exists—in deserts, jungles, hot springs, and icecaps. Many live within the bowels of other animals as parasites. They are copiously eaten by other animals and in turn they eat other organisms of all kinds. Arthropods are humankind's chief competitors for food.

There is no doubt that the annelids and arthropods are very old, having shared a common origin more than 600 million years ago. In basic structure, the arthropods have several features in common with the annelids—notably bilateral symmetry and segmentation. To these inherited annelid characteristics, arthropods have added two major new features of great adaptive value: the jointed movable leg—the

---

[12] Leeches occupy a colorful place in the history of medicine. In former times, surgeons tried to cure many ailments by bleeding the patient, often by taking some "medicinal leeches" (*Hirudo medicinalis*) out of a bucket and applying them to the suffering victims who had the misfortune to be under their care. The practice was especially popular in Germany in the nineteenth century. It is astonishing how few of these physicians ever took the trouble to determine scientifically whether this practice did or did not have therapeutic value. In the act of bloodsucking, leeches deposit a secretion into the wound they have opened with their jaws. This substance is *hirudin,* an anticoagulant, which keeps the blood from clotting. It serves to keep ingested blood in the leech's alimentary canal in a liquid state for months. Although some physicians (notably, plastic surgeons) have begun using leeches again, the practice is of dubious value and a common cause of infection.

"jointed foot" which gives the phylum its name—and the hard external coating or **cuticle,** which serves as an almost impermeable external skeleton **(exoskeleton)** (Figure 42-23). These important inventions were discussed in Chapter 31.

Many of their other features are useful if less obviously spectacular improvements on annelid characteristics. Thus, arthropods have definite muscle groups that are mechanically related to specific movable parts. In contrast, the muscles of annelids form relatively simple sheets throughout the body. Arthropods are indeed segmented like annelids, but they generally have fewer segments, and there is a tendency for the segments of some regions, notably in the head, to fuse and to become strongly differentiated in structure. Arthropods have distinctly developed jaws that open side to side instead of up and down like ours and those of other vertebrates. Their nervous systems are usually more highly developed than those of annelids and are accompanied by elaborate sensory receptors, especially in the antennae and in their remarkable compound eyes.

The jointed appendages and exoskeleton have undoubtedly been one of the secrets of the arthropod's success story, although all of the arthropod characteristics just mentioned have improved their ability to react to environmental stimuli. It is the efficiency and adaptability of their behavior that accounts for their advantage over other invertebrates. We might consider our own kind of behavioral adaptation, with its larger element of flexibility in the individual, as "better" or "higher." However, arthropod species are from 10 to 100 times more numerous than vertebrate species, there are vastly more individual arthropods, and they are adapted to a much wider range of environments. They are also better at withstanding attacks by humans and other animals.

Our classification recognizes 13 classes of arthropods, of which two are extinct. The three outstanding arthropod classes are those of the crustaceans (class Crustaceae), the insects (class Insecta), and the spiders and their relatives (class Arachnida). We will discuss them in some detail.

Of the other eight, one class (Onychophora) includes only a few rather obscure living wormlike animals, about 70 species, that live mostly in the tropics and are known as "walking worms." This class is of interest because it includes the genus *Peripatus,* which is not fully arthropodlike and has some distinctly annelidlike characteristics (see Figure 42-23). Hence, it tends to link the two phyla.[13] *Peripatus* breathes by means of tracheae (like insects) and has an insectlike heart. But its body segments contained paired excretory organs like the nephridia of annelids. Also, its muscle is not striated like arthropod muscle, but is smooth like annelid muscle. For these reasons, onychophores have been termed proarthropods in contrast to euarthropods, which are fully developed arthropods.

The arthropod classes that are completely extinct include the trilobites and eurypterids. The trilobites (class Trilobita), the most primitive arthropods, are abundant and important fossils from Cambrian and Ordovician times that breathed their last in the Upper Permian period. They are interesting as probable forerunners of later arthropods. Their heavily armored, symmetrical, segmented bodies with three body regions (head, thorax, and tail or pygidium) have made their richly informative fossils prized specimens for collectors (see Figure 42-23; also see Figure 37-21). The eurypterids (class Eurypterida) were the remarkable giant water scorpions that in Ordovician times were much to be feared.

Two living arthropod classes of minor importance in the phylum are the familiar centipedes and millipedes. Centipedes (class Chilopoda) are flat-bodied "hundred-leggers," which despite their name may have over 350 legs, though 30 to 70—that is, one pair per segment—is more usual. Millipedes (class Diplopoda) are the familiar round-bodied "thousand-leggers" (an exaggeration) that are distinguished by having two pairs of legs per segment (see Chapter 31). There are about 3000 species of centipedes and 8000 species of millipedes. These two classes (along with two very minor ones) are known collectively as myriapods ("many legs") and in some classifications they are subclasses of a single class, Myriapoda (see Figure 42-23). The two other minor classes (class Pauropoda and class Symphyla) include centipedelike animals with fewer legs.

---

[13] We have called Onychophora a class of Arthropoda, but over the past two decades it has increasingly been considered a separate phylum by some taxonomists. It is probably an evolutionary offshoot of the line leading to arthropods from an ancient annelidlike ancestor. Onychophores are among the oldest of animals.

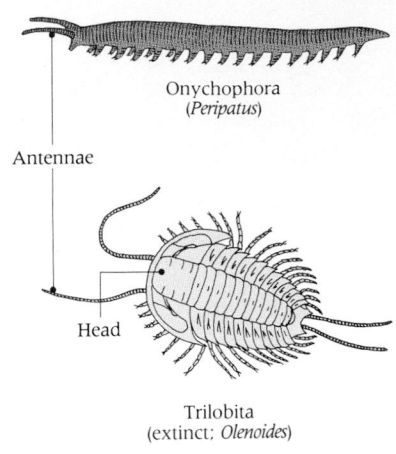

Onychophora
(*Peripatus*)

Antennae

Head

Trilobita
(extinct; *Olenoides*)

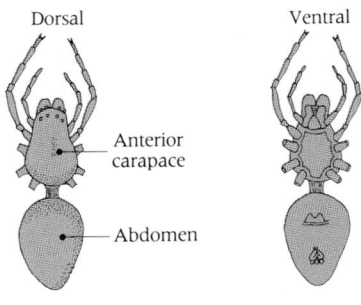

Dorsal   Ventral

Anterior
carapace

Abdomen

Arachnida
(spider, with posterior
three pairs of legs off)

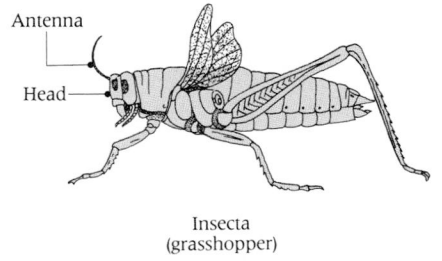

Antenna

Head

Insecta
(grasshopper)

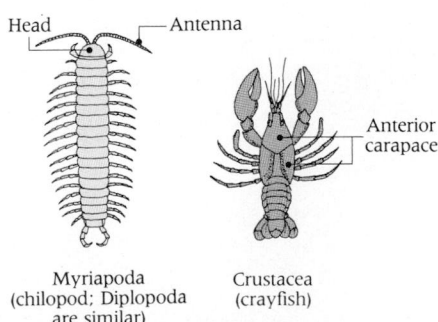

Head   Antenna

Anterior
carapace

Myriapoda   Crustacea
(chilopod; Diplopoda   (crayfish)
are similar)

42-23 Principal classes of arthropods
Arthropods have a segmented body, bilateral symmetry, an exoskeleton, and jointed appendages.

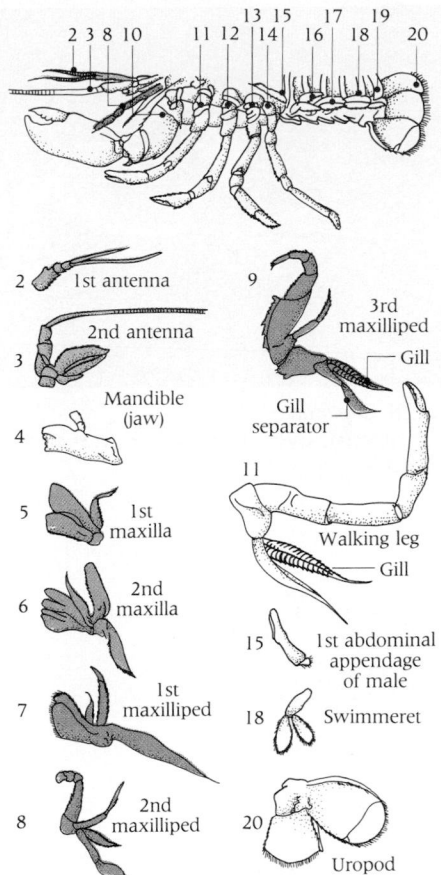

2 — 1st antenna
2nd antenna
3
Mandible (jaw)
4
5 — 1st maxilla
6 — 2nd maxilla
7 — 1st maxilliped
8 — 2nd maxilliped

9 — 3rd maxilliped
— Gill
Gill separator
11 — Walking leg
— Gill
15 — 1st abdominal appendage of male
18 — Swimmeret
20 — Uropod

42-24 Appendages of the lobster
The numbers represent segments bearing the appendages shown. Colors emphasize appendages of similar function.

A

B

42-25 Crabs
(A) Purple shore crab. *(B)* Sally Lightfoot crab.

Most of the remaining arthropods, including the members of the three principal classes, are traditionally subdivided into two major subphyla, Mandibulata and Chelicerata.

- **Mandibulata:** mandibles (jaws) and antennae serve as sensory structures; three distinct body parts (head, thorax, and abdomen); includes Crustacea, Insecta, Myrizpoda, and several minor classes
- **Chelicerata:** division of the body into a cephalothorax and an abdomen, feeding appendages called chelicerae present, and antennae absent; includes Arachnida, Merostomata, and three other classes (of which one is extinct)

SUBPHYLUM MANDIBULATA

Class Crustacea: Crabs, Lobsters, Shrimps, and Relatives

Crustaceans (whose name means "with a hard crust or shell") are familiar to us as our seafood delicacies—lobsters, crayfishes, crabs, shrimps, and prawns. Each of these common terms is in fact applied to a number of different genera and species. These few common groups are certainly well known, but there are some 28,000 species in this highly successful class of animals. As we will see, they include many unfamiliar forms that bear little superficial resemblance to the better-known ones. Indeed, crustacean diversity is so extraordinary that it has taxed the best taxonomists.

Crustaceans characteristically have two pairs of antennae, a pair of mandibles, and two pairs of maxillae. But the remaining appendages vary widely. Primitive crustaceans probably had many segments, each of which had similar appendages. Such a primitive condition is almost unknown among the true crustaceans, but it is approached in the older, extinct trilobites. However, in a crustacean like the lobster,[14] no two of the 19 pairs of appendages are exactly alike—even the two members of the same pair are not necessarily alike—and those of different regions are highly differentiated in forms and functions (Figure 42-24). It is a useful evolutionary generalization that, when primitive organisms have numerous parts similar in structure and function (like trilobite appendages), the analogous parts in descendants tend to be both reduced in number and locally specialized.

The crustacean exoskeleton is of special interest and it serves as a model for all arthropodal exoskeletons. It is composed of flexible chitin, hardened in many cases by calcium carbonate. Since exoskeletons are hard, they must be shed periodically and replaced if the animal is to grow. This process is called *molting*. Crustaceans do this throughout their lives, though at a slower pace as they grow older. This means that they grow ever larger without reaching a final characteristic size. The new shell takes some time to harden. Hence, crabs, lobsters, and other crustaceans are temporarily soft-bodied after molting. This is what is meant when the menu reads "soft-shelled crabs."

We can here pluck only a few of the major groups from different parts of the complex crustacean classification and discuss them briefly.

Decapods (order Decapoda), meaning "ten legs," are our edible crustaceans. They include shrimp (*Crangon*), spiny lobster (*Palinurus*),[15] lobster (*Homarus*), crayfish (*Astacus*), and crab (*Cancer*) (Figure 42-25). Most are marine, but they are also numerous in fresh water. Some crabs can survive for long periods in the air as long as they do not dry out. All are carnivores or scavengers.

Decapods are not the most numerous of crustaceans or the most important in the economy of nature. Those distinctions belong to the small, even microscopic, crustaceans that swarm by the billions in every sea and in most bodies of fresh water. **Krill,** the principal food of many marine animals including the whales, is composed of crustaceans (*Euphausia*) less than an inch long. Others, even smaller,

---

[14] The familiar big-clawed lobster of the frigid waters of the Northeast (*Homarus americanus*) may weigh more than 50 pounds (perhaps much more). Sadly, the numbers of these homarids have been so sharply diminished by overfishing and pollution that they may soon disappear from the seas and from our tables.

[15] These clawless lobsters are the *palinurids* that live off the coasts of South Africa, Australia, and in other warm waters. Their meat is marketed simply as "lobster tails." Closely similar crustaceans abound (or once did) off the Florida Keys where they are called rock crabs.

abound in salty seas (*Artemia*) or fresh ponds (*Daphnia*). There are many more specialized crustaceans. We list these few only to illustrate their diversity.

- Ostracods (subclass Ostracoda) are tiny crustaceans that secrete, in addition to the usual arthropod cuticle, two protective shells, much like miniature clam shells (example, *Cypris*).
- Copepods (subclass Copepoda) are probably the most abundant animals in the world, both in total numbers and in percentage of total animal biomass. Those that are free living, such as *Cyclops* (Figure 42-26), feed on algae, flagellates, and diatoms (phytoplankton) and comprise the zooplankton, an important fish food ("brit"). But many copepods are parasitic.
- Barnacles (subclass Cirrepedia) are sessile as adults and secrete a complex shelly cup within which they live. These ubiquitous squatters irritate bathers and frustrate boatmen, but they have fascinated scientists for centuries. Darwin filled his house with 10,000 of them, many gathered on the voyage of the *Beagle*. A shell-less relative of barnacles, *Sacculina*, is parasitic on crabs. Its life history (including a stage in which it attaches itself to a crab and shakes off its own thorax, legs, and abdomen, saving only its reproductive organs) is a classic example of extreme evolutionary regression (a concept that we will discuss in Chapter 44).
- Isopods and amphipods are small, bug-like crustaceans that live in salt water and fresh water or in damp places on land. They include "sow bugs," armadillo-like "pill bugs" or "wood lice" (see Figure 32-3), "beach fleas," and many others. They are entirely distinct from true bugs, lice, or fleas, all of which are insects.

42-26 Copepod
This copepod, Cyclops, *is carrying two external egg cases.*

Class Insecta: Insects

That **insects** are the most successful of all invertebrates is indicated by the fact that no other group of any kind—monerans, protists, fungi, plants, or other animals—comes close to matching their diversity. At least half, and probably more, of all living species and nearly three-fourths of all animal species belong to this one class. The number of individuals is also prodigious. According to one estimate, they make up 60% of all living organisms. According to another, 10^{18} or a billion billion insects inhabit the Earth. If true—and it would be hard to prove—that would be one billion per human being.

Insects occupy every imaginable habitat on land and in fresh water. Yet, there is one major exception to their near-universal preeminence. They failed to adapt to life in the sea, which was so well exploited by their fellow arthropods, the crustaceans. Insects are a rare and unimportant group in the ocean and few live out their entire lives there. Perhaps this was because insects first arose as terrestrial animals. Many of them subsequently adapted to life in fresh water, commonly in the larval stages only. There is perhaps only one genus (*Halobates*) that can be considered fully a creature of open oceans, and it is what is called a "water strider"—that is, a creature that lives *on* the water surface. They really live *in* the air. Water is the floor of their world. In fact, four major types of insects live on water surfaces:

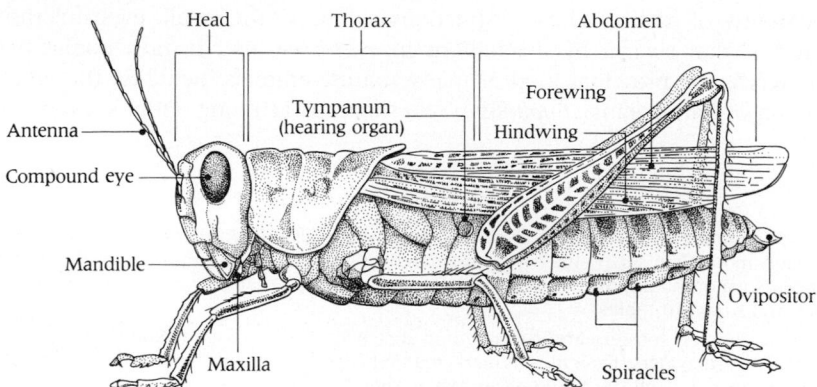

42-27 Insect organization
The grasshopper body exemplifies the basic insect body plan. Note the three major body parts—head, thorax, and abdomen. The grasshopper has 11 abdominal segments.

water striders (see Figure 3-16), whirligig beetles, backswimmers, and springtails. Each lives in a different relationship to the water surface and to the air above and water below. Only *Halobates* lives on salt water.

Insects probably arose from a millipedelike ancestor. The most primitive insect fossils may have been found in rocks from the Devonian period some 400 million years ago. But very soon after they first appeared, insects became dominant. Their greatest expansion was in the Carboniferous and Permian periods, when 21 of the 36 recognized orders first appeared in the fossil record. A period of great diversification occurred 200 million years ago, in the Triassic period, when one insect order arose. Flowering plants got under way in the late Jurassic when four insect orders (including the important Diptera and Hymenoptera) first appeared in the record. As noted in Chapter 41, flowering plants were first really dominant in the time immediately following the early Cretaceous.

Most insects are small and nonaquatic. All share certain key features. For example, the insect body is segmented. Chapter 20 summarized important recent work on the role of "homeo box" genes in the development of the segmented body plan of *Drosophila*.

Most features of the insect body plan are clearly seen in the common grasshopper (order Orthoptera). The grasshopper's body is divided into three regions: head, thorax, and abdomen (Figure 42-27).

The head consists of six segments that are fused together and practically indistinguishable in the adult. The paired appendages of one segment became the sensory antennae. The head also bears other sensory receptors, usually including compound eyes (see Chapter 30). Other ancestral appendages became three pairs of rather complex mouth parts. As shown in Figure 35-1, these are modified in different insects to satisfy their special feeding requirements—chewing, sucking, etc. For example, many mosquitos—of which there are some 3000 species—have a remarkable proboscis that can saw its way through a victim's skin, suck blood, and in some cases inject saliva containing anticoagulant substances. It is then withdrawn and, like a bloody dagger, resheathed (Figure 42-28).

The thorax comprises the central section of the body, separate from head and abdomin, and is made up of three segments, each bearing a pair of walking legs. In many insects, the second and third segment of the thorax each has a pair of stiff, membranous wings of similar size.

The abdomen has 9 to 12 segments and usually lacks appendages. However, remnants of the ancestral appendages of the most posterior segments have been specialized for mating and egg laying.

There is a tracheal breathing system, which was a key adaptation for life on land. In this system air-filled tubes carry air directly to body cells (see Figure 27-9). The circulatory system is open, without capillaries or veins. Oxygen transport by blood is not necessary because the tracheal branches oxygenate cells directly.

The nervous system is complex with two large ganglia, or "brains," in the head and a double ventral cord (see Figure 30-30D). Simple and compound eyes occur, as well as many other receptors.

The cuticle is made partly of chitin and partly of the tough, chemically unusual protein, sclerotin. It is covered with another layer, the epicuticle, a fragile film only 1 μm thick of varnishlike material, that completely waterproofs the cuticle, a critically important survival mechanism. This exoskeleton, unlike that of crustaceans, is not mineralized.

It is interesting to consider the rather narrow range of insect size. The smallest insects (like *Alaptus magnanimus,* which are tiny parasites in the eggs of other insects) are about 0.2 mm in length. Most insects are at least 2 mm long. Few insects are more than 40 mm long, despite rare extinct forms (like the giant dragonfly) which had wingspans of over two feet. The upper limit of body length in living insects is around 275 mm. Insects, in a word, are small. Why should that be?

For one thing, mechanical factors relating to the exoskeleton limit the size of insects. Although insect exoskeletons are amazingly light, there is a limit to the weight any skeleton can support. This limit is much higher for an internal skeleton than for one that is external. It is no accident that the only place large arthropods (lobsters and eurypterids, for example) have been found is in water, which buoys their weight. The respiratory and vascular systems also impose limits on the size of insects. These systems are very efficient for animals of small size, but they probably could not adapt to any considerable increase in bulk.

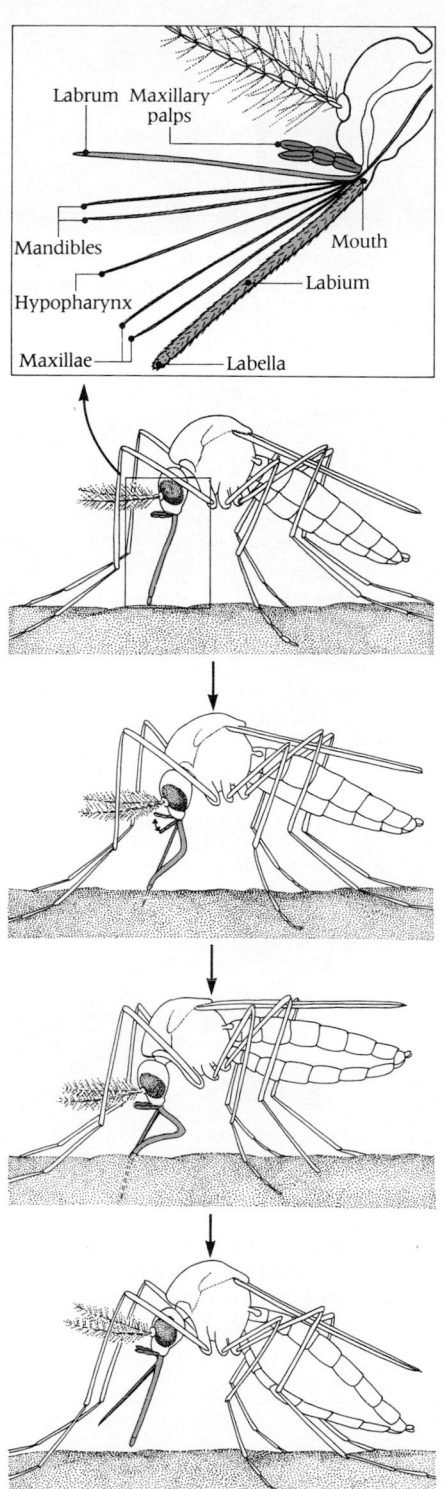

42-28 The mosquito proboscis
The parts of the proboscis are used for piercing prey and feeding. The finely toothed maxillae saw into the tissue and allow the insertion of the labrum, mandibles, and hypopharynx, while the labium folds back on the skin surface. The mosquito then draws blood. When feeding is complete, the withdrawn mouth parts snap back into the sheathlike labium. See also Figures 35-1 and 39-21.

Their small size has had many consequences for insects. It has helped them to find many remarkable niches to occupy. It has hindered them by limiting their ability to develop large nervous systems like those of vertebrates. Is not the small size of insects, and their small individual food requirement, one of the reasons why they are so numerous and diverse and such successful pests from the human point of view?

Specialists on insects (entomologists) divide the insect class into some 36 orders, 27 living and 9 extinct, the number varying with the authority's opinion. One can make some sense of the bewildering array of insect types by recognizing among the living insects two broad subclasses and their several subdivisions, each representing an evolutionary level.

At the most primitive level are the **apterygotes** or wingless forms (subclass Apterygota), which include among others the primitive springtails (order Collembola) and the familiar silverfish (order Thysanura)—large groups of small insects that live hidden lives in rock crevices, soil, and in association with various farm crops.

The second subclass represents the winged insects, or **pterygotes** (subclass Pterygota). These include truly winged forms as well as wingless forms, especially the lice, fleas, and other parasites that lost their wings secondarily. The pterygotes, which include 90% of all known insect species, are divided into three groups.

- The primitive Paleoptera cannot hold their wings backward against the body. They are good fliers but need plenty of open space. They include the mayflies (order Ephemeroptera) and the dragonflies and damselflies (order Odonatoptera) (Figure 42-29).
- In the Polyneoptera, the wings fold backward on the body at rest. These include termites (order Isoptera); grasshoppers, crickets, cockroaches, and locusts (order Orthoptera) (Figure 42-30); and earwigs (order Dermoptera).
- The Oligoneoptera have similar wing-folding habits, but are more advanced in many ways. They include the beetles (order Coleoptera), of which there are more than 300,000 species (including 25% of all living organisms); butterflies and moths (order Lepidoptera) (Figure 42-31A); flies, gnats, and mosquitos (order Diptera); bees, wasps, sawflies, and ants (order Hymenoptera) (Figure 42-31B); and another group, so little known it has no popular name (order Strepsiptera), in which only males have wings and only the hindwings are present. The females are parasites in other insects, from which they protrude their hind ends for fertilization by flying males. Insects commonly have equal-sized wing pairs. In many, notably the beetles, the front pair (the forewings), became a protective cover. In flies (order Diptera, or "two-wings"), the hind pair is reduced in size and is a balancing, not flying, organ.

Insects show two modes of development (see Figure 21-31). In hemimetabolous development, metamorphosis is incomplete. Nymphs that hatch from eggs become adults after growth and molting. Complete metamorphosis occurs in holometabolous development. Eggs hatch into feeding larvae which grow, molt, and enter an inactive pupa, often in a cocoon. The adult that emerges develops from disks of cells that were present in the larva. Most of the insects listed earlier, including the most successful and diverse orders (Hymenoptera, Coleoptera, Lepidoptera, Diptera, etc.) undergo holometabolous development. The thrips (order Thysanoptera), mayflies (order Ephemeroptera), dragonflies (order Odonatoptera), true bugs (order Hemipt-

42-29 Paleopteran insects
The wings of the dragonfly cannot be held close to the body.

A

B

42-30 Polyneopteran insects
In this group, the wings fold against the body.
(A) Prairie mole cricket, Gryllotalpa major.
(B) American cockroach, Periplaneta americana.

A

B

42-31 Oligoneopteran insects
These wings have more complex folding patterns.
(A) Painted Beauty moth, Vanessa virginiensis.
(B) Great golden digger wasp, Sphex ichneumoneus. See Figure 33-11.

era), grasshoppers (Orthoptera), and cicadas, aphids, scale insects (deadly enemies of fruit crops), and leafhoppers (order Homoptera) do not undergo metamorphosis, but are hemimetabolous.

Many of the curious and complicated life cycles of all these insect groups are still not fully understood, and entomologists the world over are diligently studying these strange and intricate life histories. In part, this study is stimulated by economic factors. A satisfactory insecticide is one of civilization's most pressing needs (see Box 42-2).

We note again that aspects of insect biology have been discussed in other chapters. These discussions cover insect reproduction and development (see Chapter 21), excretory physiology (see Chapter 25), respiration (see Chapter 27), neural coordination (see Chapter 30), and behavior and social tendencies (see Chapters 32 and 33). These topics would be usefully reviewed in connection with the study of this chapter.

SUBPHYLUM CHELICERATA

In most **chelicerates,** the first two body parts are fused into a single cephalo thorax. They have six pairs of appendages. But the first pair and often the first two pairs are so modified that they do not look or function like walking legs. Commonly, the first two pairs of appendages have become the specialized mouth parts called **chelicerae** and **pedipalps.** The chelicerae hold and manipulate the prey (and in some spiders serve as fangs for injecting poison). The pedipalps rip the prey apart (and in some males transfer sperm to the female).

Chelicerates have **book gills,** a series of leaflike lamellae named for their resemblance to the pages of an open book.

Subphylum Chelicerata includes four classes (three are living). One (Pycnogonida) includes the sea spiders or nobody-crabs, an unobtrusive marine group of browsing carnivores comprising about 500 species. The other two are Arachnida (the spiders) and Merostomata (the horseshoe crabs). One class (Eurypterida) has been extinct since the Paleozoic era.

Class Arachnida: Spiders and Their Kin

The **arachnids**—mites, ticks, scorpions, daddy longlegs, and spiders—are named from the Greek word for spider (which is based in turn on a myth about a girl named Arachne who was transformed into a spider). Like octopuses, they have an undeservedly bad name. Most spiders and ticks are poisonous, but few are dangerous to humans. Most spiders are beneficial to humans because they prey on undesirable insects. It is estimated that there are 50,000 spiders per acre in green areas and that each one destroys hundreds of insects per year. Daddy longlegs or harvestmen are especially helpful in this regard. Mites and ticks include parasites

42-32 Arachnids
Spiders have two body segments, the cephalothorax and abdomen. There are 4 or 5 pairs of legs.

Much has been written about insect pests and their impact on the world's food supply (see Box 46-2). In view of its importance and the underlying biological issues, let us briefly survey the techniques of insect control in use throughout the world.

Chemical insecticides are synthetic toxic substances that fatally impair the metabolism of insects. Examples are **chlorinated hydrocarbons** such as DDT (dichlrodiphenyltrichloroethane, $(ClC_6H_4)_2CHCCl_3$) and various **organic phosphates** (such as malathion and parathion). Their aim is to inflict lethal effects on insects without harming humans, other animals, or plants. Two major problems impair their usefulness. One is the propensity of pest insects to become resistant to these agents. A second problem relates to the persistence of these agents in the environment. Ideally, such agents should not persist. Unfortunately, DDT is a very persistent substance (Figure 42B). A study in Lake Michigan showed that although DDT concentrations were relatively low in lake sediments—0.0085 parts per million (ppm)—concentrations in aquatic invertebrates were 48 times higher (0.41 ppm). In the fishes feeding on these invertebrates, there was a further 20-fold increase in DDT concentration (reaching a level of 3 to 8 ppm). Predatory birds feeding on these fishes accumu-lated as much as 3180 ppm of DDT in their fatty tissues. These high concentrations in the uppermost consumer did have serious toxic effects. Several species of predatory birds had difficulty reproducing because DDT (and similar chemicals) interfered with eggshell formation. In sum, DDT is an excellent toxic agent, but it has low target specificity and it is a serious ecological hazard.

Bioactive chemical agents represent a second group of chemicals currently of great interest. These are termed "bioactive" or "biorational" to distinguish them from chemical insecticides. These agents act by modifying an insect's behavior in a way that affects its life cycle adversely. Examples include: the pheromones, which are secreted by some insects to influence the behavior of other insects; insect hormones such as insect growth regulators; and various hormone antagonists, attractants, and compounds that attract parasites. Although a pheromone is as much a chemical as DDT, its action has great species specificity, and it is active at very low concentrations. Also, pheromones are neither persistent nor toxic; nor is it likely that insects will develop resistance to them.

Microbial agents (bacteria) are used to inflict a specific infection on various insect species without affecting humans or other animals. Several such bacterial types are now commercially available. An example is *Bacillus thuringiensis*, which produces a toxin active against moth larvae; *Bacillus popilliae* causes "milky disease" of Japanese beetles. It is doubtful that these techniques will be widely used, but they may prove useful as supplements in specific situations.

Biological methods for the control of pest insects depend on the deliberate introduction of harmless competitors for ecological niches or natural predators (or parasites). (Biological control also involves the breeding of insect-resistant plant strains.) The process, which was used as early as A.D. 304, is illustrated by the use of the citrus ant, *Oecopylla smaragdina*, in the control of various insect pests of orange trees, now a common practice in China. Another example is the control of red scale, a citrus pest, with the wasp *Aphytis melinus*, its natural enemy. Unfortunately, these wasps survived least well in citrus districts where they are most needed. In another procedure, sterile male screwworm flies are introduced to disrupt natural life cycles. This is far from optimal because new batches of sterilized males must be released periodically. Further, it is subject to error and is likely to work only in large areas with natural geographical barriers (for example, islands) where continuous pest infestation from the outside is minimized. In princi-

dangerous to humans and domestic animals and they transmit organisms that cause diseases such as scrub typhus and Rocky Mountain spotted fever.

Most people think arachnids are insects, though they are really quite different. Insects have six legs (three pairs). Arachnids have eight or ten (four or five pairs). This is what systematists call a key diagnostic character. However, an animal is classified as an arachnid, not because it has four or five pairs of legs, but because we have judged it to have a common ancestry with other arachnids and a different ancestry from insects.

Spiders (order Aranae) are one of the two largest of the nine arachnid orders; more than 30,000 living spider species having been described (Figure 42-32). (The other large order is Acarina—mites and ticks.) All are carnivorous and have good sense organs. There are two body regions, a cephalothorax and an abdomen (see Figure 42-23). The former bears all the appendages and the eight eyes, which are simple. It never bears antennae. A fascinating feature of many spiders (and some mites) is the ability to produce silk. Spider silk is a complex polymer of a protein called *fibroin* that acquires tensile strength and elasticity when it hardens. It is not the same as silkworm silk, the material from which silk threads and fabrics are

ple, biological approaches offer nontoxic, nonpolluting, relatively inexpensive, long-lasting and self-perpetuating protection. Nonetheless, available procedures have drawbacks. For one, any introduction of beneficial parasites and predators cannot offer immediate control. Some degree of insect damage must be accepted.

The category of cultural control methods includes a complex array of management techniques that together keep the ecosystem optimal for food crops and inhospitable for insect pests. These methods include such procedures as quarantine, sanitation, crop rotation, pruning, irrigation, fertilization, weeding, mowing or grazing, scarecrows, and culturation.

Finally, we note that the production, packaging, and application of pesticides require significant amounts of energy. To produce the simplest pesticides, such as DDT, requires 11,000 kcal per pound produced. More complex pesticides require up to 12,500 kcal per pound. With an estimated 1.2 billion pounds of pesticide applied annually in the United States, some 3.0×10^{13} kcal in fossil energy is being expended in pesticide controls. This represents a consumption of 1 billion gallons of fossil fuel equivalents per year.

Intensive research all over the world now seeks new techniques in each of the major categories of control methods.

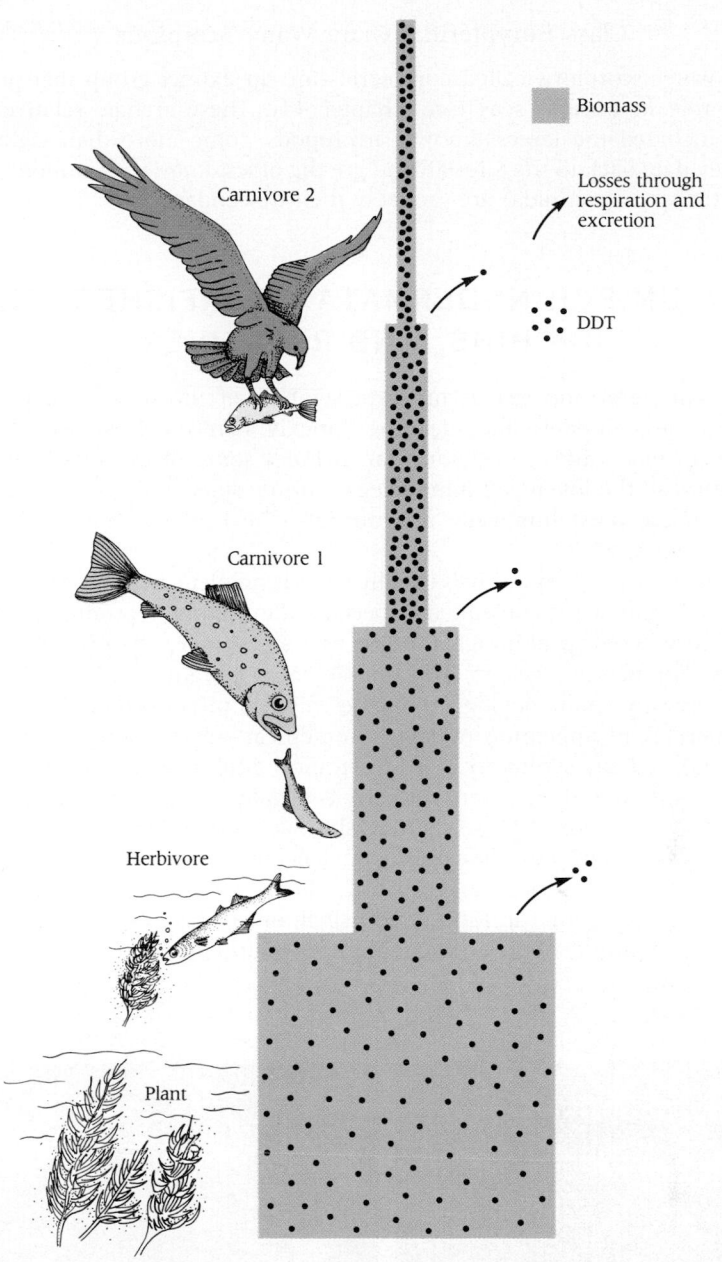

42B Increasing concentration of DDT at different levels of the food chain
As DDT moves through the food chain, from low levels in plants to higher levels in carnivores, its concentration in living tissue increases. Since this fat-soluble molecule is not excreted from the body, there are massive accumulations in the bodies of animals high on the pyramidlike food chain.

made. The spider's abdomen bears three or four pairs of tiny spigots called spinnerets that extrude a liquid containing fibroin, which hardens into silk. Silk fibers are useful for safety lines, for the construction of webs, and for many other purposes.

Class Merostomata: Horseshoe Crabs

Familiar creatures on ocean beaches, horseshoe crabs include only four genera. The most common, *Limulus* (Figure 42-33), is an interesting form with a broad protective dorsal shield (carapace) that appears crablike. But its legs (five pairs) and internal anatomy show clear-cut phylogenetic resemblances to spiders and decisive differences from crustaceans. This fascinating animal is a famous evolutionary relict. It is the last representative of an ancient class of arthropods that has changed very little since it lived in Ordovician seas hundreds of millions of years ago. It probably descended from the trilobites.

Horseshoe crabs are bottom-dwelling scavengers that feed on mollusks and annelids and that have few predators. They swim on their backs by moving their abdominal plates. They can also walk on their legs.

Class Eurypterida: Giant Water Scorpions

The giant water scorpions called eurypterids are an extinct group that played an important role in ancient seas (see Chapter 44). These archaic relatives of the arachnids included the largest known arthropods, some more than eight feet in length. They date back to the Silurian and are the oldest known arthropods. Today's scorpions (order Scorpionida) are probably their descendants.

PHYLUM ECHINODERMATA: STARFISHES, SEA URCHINS, AND RELATIVES

Starfishes (sea stars) and sea urchins are symbols of the sea and appropriately their phylum—Echinodermata (meaning "prickly skin")—is exclusively marine. It is made up of a varied collection that includes some of the most beautiful of living creatures—the lovely sea lillies, elegant brittle stars, and spiky sea urchins—and some of the most lumpishly unbeautiful—the leathery sausage-shaped sea cucumbers (Figure 42-34).

These relatively complex animals usually have a complete digestive tube (digestion is entirely extracellular), a coelom, and specialized excretory, reproductive, nervous, and circulatory systems, although the last two are simpler than in most animals so complex. The nervous system, for example, lacks a brain.

Though they are said to belong among the "higher" phyla, echinoderms resemble the coelenterates in appearing radially symmetrical—even though their internal organs are not radially symmetrical. At least, most adult echinoderms look radially symmetrical, although their outer symmetry is usually pentaradiate—which means five-sided. (Note in Figure 42-34B the starfish's five arms.) Interestingly, their larvae are bilaterally symmetrical. Characteristic of all echinoderms is the internal skeleton composed of calcareous ossicles that may articulate with one another (as in starfishes) or may be sutured together to form a rigid shell, or test (as in sand dollars).

The peculiar anatomy of the echinoderms has two unique features of great physiological importance.

42-33 Horseshoe crab
This horseshoe crab, Limulus polyphenus, *is evidently going somewhere.*

42-34 Echinoderms
(A) *Sea lily,* Crinoidea. **(B)** *Sea stars or starfish,* Asteroidea. **(C)** *Sea urchin,* Echinoidea. **(D)** *Sea cucumber,* Holothuroidea.

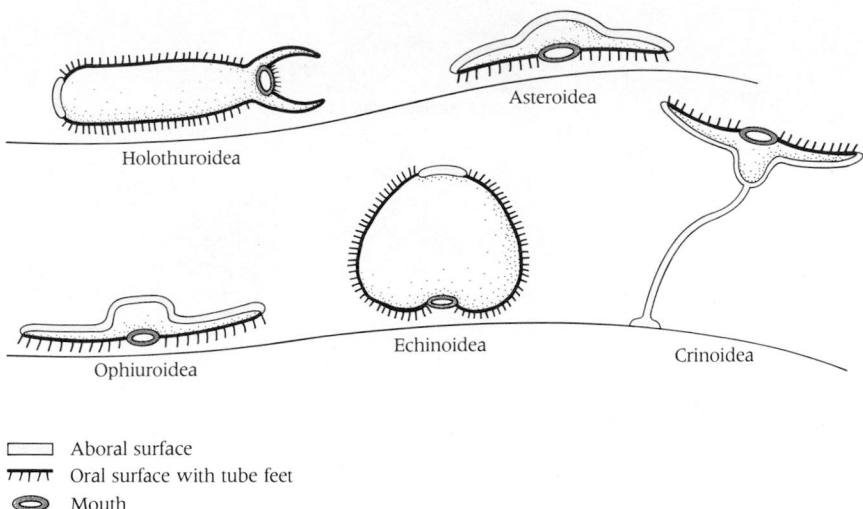

42-35 Body plans in living echinoderms
The posture of the members of the five major groups of echinoderms with respect to hard surfaces emphasizes the importance of tube feet.

Holothuroidea

Asteroidea

Ophiuroidea

Echinoidea

Crinoidea

▭ Aboral surface
⊤⊤⊤⊤ Oral surface with tube feet
◉ Mouth

- **Skin ossicles** (which give the phylum its name) are present. These small bony plates in the skin are separated by pockets of thin skin with a respiratory function. In some echinoderms, such as the sea urchins, these are united into a tough, protective box. (Sea urchins also have movable spines jointed to the outside of the box.) Besides functioning as a locomotory organ, this prickly exoskeleton has tended to discourage predators, thereby overcoming a major selective force in animal evolution. This may be why some echinoderms have changed very little in the past 250 million years.

- A well-developed **water vascular system,** a subdivision of the coelom that exists alongside a true (if poorly developed) circulatory system, has evolved. Its network of canals is connected to the outside by pores in a special bony sieve plate. Sea water enters the canals through the pores. In many echinoderms, the canals of the water-vascular system connect with numerous bulbed structures called tube feet. Water forced into the tube feet by muscle action operates them by hydraulic pressure (Figure 42-35).

The echinoderms are a very ancient and diverse group. Although the internal anatomy is for the most part essentially the same in all of them—some 6,500 living species and 20,000 fossil species—the five classes of living echinoderms are extraordinarily different in appearance and habits.

The sea lilies and feather stars (class Crinoidea) are enclosed in a rigid calcareous box, with slender, branching, flexible arms extending from around the mouth and anus (see Figure 42-34A). Sea lilies are usually sessile, attached to the sea floor by a stalk, and the flowerlike feather stars are free swimming. These animals lack tube feet. They are notable for their profusions of color. Crinoids are survivors of the most archaic stock of echinoderms. They are first known from the early Ordovician.

Starfishes or sea stars (class Asteroidea), of which there are four orders, have familiar star-shaped bodies that are stiff but flexible (see Figure 42-34B). They are free to move by crawling slowly. This they do by employing the suction of their approximately 1000 hydraulically controlled and well-developed **tube feet** (Figure 42-36A). They are predaceous carnivores, usually feeding on clams and other relatively large invertebrates. Their stomachs are completely eversible—that means the stomach can be extruded—and this occurs when they are feeding on small bivalves and when they are eliminating undigested waste. The everted stomach secretes digestive enzymes. When feeding on a clam, the prying force applied to the bivalve shell by the numerous tube feet of such starfishes as *Asteria* and *Pisaster* is equivalent to 5000 grams. Many species can slip folds of the everted stomach between pried-open bivalve shells and begin the digestive process.

Brittle stars and serpent stars (class Ophiuroidea) are surprisingly abundant creatures that somewhat resemble starfishes; however, they have long, slender arms

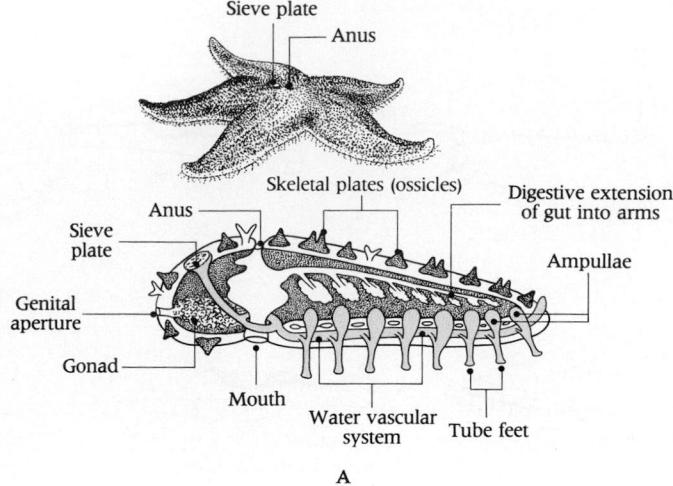

Sieve plate

Anus

Skeletal plates (ossicles)

Digestive extension of gut into arms

Anus

Sieve plate

Genital aperture

Gonad

Ampullae

Mouth

Water vascular system

Tube feet

A

42-36 Echinoderm body organization
The water vascular system is shown in color. Note the ossicles and the locations of mouth and anus. (A) Surface view of a starfish and one of its arms in cross section. (B) A sea urchin in surface view and section. Note the characteristic spines.

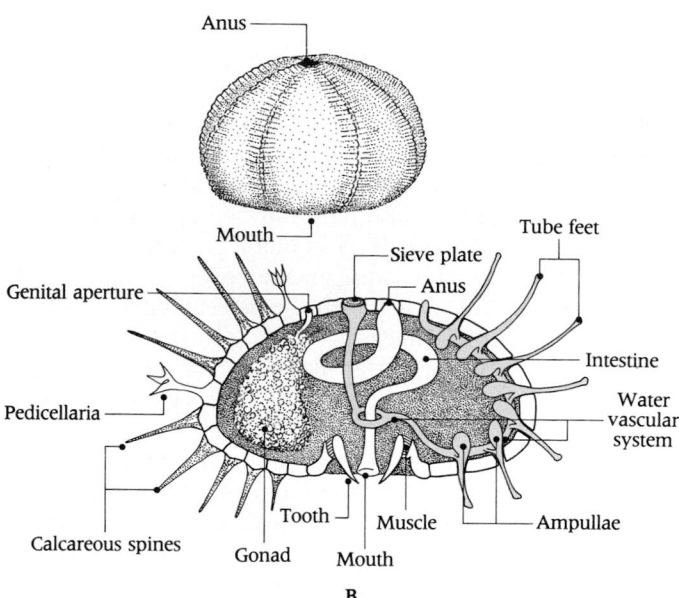

Anus

Mouth

Tube feet

Sieve plate

Anus

Genital aperture

Intestine

Water vascular system

Pedicellaria

Ampullae

Calcareous spines

Tooth

Muscle

Gonad

Mouth

B

that lash about rapidly in locomotion or in seizing small organisms as prey. These arms are solid and are made up of successive joints; they are not hollow and unjointed like the arms in starfishes.

Sea urchins and sand dollars (class Echinoidea) are sluggish animals that move slowly by means of spines and tube feet (see Figure 42-34D). Their skeletons are tightly joined into a rigid box (see Figure 42-36B). They feed largely on seaweeds and dead plant or animal matter and have elaborate digestive systems with food-crushing organs. Among the spines are curious structures with jawlike tips called pedicellaria, which can clamp down on a prey and release a poison that can cause a human victim severe pain and, in rare cases, death.

Sea cucumbers (class Holothuroidea) have long sausagelike bodies that lack arms or spines (see Figure 42-34D). Their leathery skin is reinforced by tiny scattered limy plates in the skin. They have tentacles around the mouth and move sluggishly, feeding on small animals or organic material in the mud or sand. Some are caught for food in the Far East.

The echinoderms have an exceptionally fine fossil record. It is noteworthy that most of the oldest forms were sessile. A feature that continues to puzzle biologists is their characteristic trait—five sidedness. Presumably, it has adaptive significance, but we do not know what it is.

SUMMARY

Animals differ from plants in a number of ways. Unlike plants, they are motile and heterotrophic. They maintain constant internal environments by means of complex homeostatic mechanisms, and they respond to the external environment by virtue of their nerves and muscles. Only a few animal species (mainly certain insects) display a haploid stage in their life cycle.

Phylum Porifera, the simplest animal phylum, includes the sponges. They seem to have arisen from protists via an evolutionary path independent of the lines leading to other animal phyla. The jellyfishes and corals of phylum Coelenterata are more complex than the poriferates. They have organs, two specialized cell layers, and greater motility.

The worm shape arose independently in many phyla. The most familiar are the phyla Platyhelminthes (flatworms such as planaria), Aschelminthes (roundworms such as nematodes), and Annelida (the segmented worms). Ciliated, funnellike food-gathering organs called lophophores are typical of several small phyla, including Brachiopoda (lamp shells), Ectoprocta (moss animals), Entoprocta and Phonorida (both so obscure as to lack common names).

Three dominant animal phyla, Mollusca, Arthropoda, and Chordata, represent culminations of differing evolutionary strategies. The several molluscan classes diverged from each other long ago and it is difficult to describe a stereotypic mollusk. Class Amphineura, the chitons, are the most primitive mollusks. Snails and slugs make up the largest class, Gastropoda. When they have shells, they are univalves with only one opening. This distinguishes them from members of class Bivalvia, which includes clams and oysters. Class Cephalopoda includes octopuses, squids, and nautiluses. Predatory and highly motile, members of this class possess the large brain and keen senses necessary in their demanding environments.

Arthropoda, by far the largest animal phylum, shares a common origin with Annelida, as well as such characteristics as segmentation and bilateral symmetry. To these, arthropods added two important innovations: the jointed leg and the exoskeleton. The three major arthropod classes are Crustacea, Insecta, and Arachnida. Crustacea are primarily aquatic animals with mineralized exoskeletons. They include the familiar lobsters, crabs, and shrimps, and also barnacles, terrestrial sow bugs, and copepods—tiny forms that exist both as parasites and in plankton. Unlike Crustacea, Insecta is primarily a terrestrial class. Hence, the weight of their exoskeletons limits their size. Insects, which probably arose from a millipede-like ancestor, generally have three body sections—head, thorax, and abdomen— each with multiple segments. They have tracheal breathing systems, open circulations, and nervous systems embodying ventral nerve cords. Class Arachnida, the spiders, ticks, and scorpions, is closely related to the relict class Merostomata—the horseshoe crabs.

The phylum most closely related to phylum Chordata is Echinodermata. Although adults approximate radial symmetry, their larvae are bilaterally symmetrical. Two unique anatomical features characterize echinoderms: small bony plates called skin ossicles, and a remarkable water-vascular system that functions alongside a true circulatory system.

KEY TERMS

acoelomate
amphineuran
annelid
anthozoan
apterygotes
arachnid
arthropod
bivalve
brachiopod
byrozoan
cephalopod
Chelicerata
chiton
chromatophore
coelenterate
collar cell
crustacean
ctenophore

cuticle
decapod
echinoderm
ectoproct
entoproct
exoskeleton
fluke
gastropod
graptolite
hydrozoan
insect
krill
lophophore ring
Mandibulata
mantle
medusa
mesozoan
mollusk

monoplacophoran
osculum
planarian
polyp
pseudocoelom
pteryogotes
pulmonate
radial symmetry
radula
roundworm
scaphopod
scyphozoan
tapeworm
torsion
tube feet
turbellarian
univalve

QUIZ QUESTIONS

1. Which of the following characteristics does *no* species of animal possess during some phase of its life cycle?
   A. mobility
   B. autotrophy
   C. responsiveness
   D. metabolism
   E. C and D

2. Which of the following phyla possesses all of these features: radial symmetry, a true coelom, a complete digestive tract, and lophophores?
   A. Aschelminthes
   B. Mollusca
   C. Porifera
   D. Arthropoda
   E. None of the above

3. Which of the following is *not* a trend seen in the evolution of animals?
   A. ancestral forms living in water → derived forms living on land
   B. ancestral forms with two openings in digestive tract → derived forms with one opening in digestive tract
   C. ancestral forms with open circulatory system → derived forms with closed circulatory system
   D. ancestral forms with one or two embryonic germ layers → derived forms with three distinct germ layers
   E. ancestral forms with radial symmetry → derived forms with bilateral symmetry

4. Division of Phylum Arthropoda into subphyla is based on
   A. presence or absence of mandibles.
   B. presence or absence of antennae.
   C. presence or absence of a cephalothorax.
   D. A and B
   E. A, B, and C

5. In which of the following taxa are there *no* terrestrial species?
   A. Echinodermata
   B. Arthropoda
   C. Mollusca
   D. Annelida
   E. B and C

ESSAY QUESTIONS

1. A team of marine biologists happens upon a novel organism in an underwater reef. Although it is firmly attached to the reef by a stalk, they agree to classify the organism as an animal. What features of the organism's structure, life cycle and behavior might have suggested that the organism was an animal?

2. What is a coelom; a pseudocoelom? Trace the evolution of body cavity form in the sponges, coelenterates, roundworms (aschelminthes), and annelids.

3. What are the major classes of arthropods? Why are insects small? Where are the largest arthropods found?

4. What distinguishes arachnids from insects? Why is a horseshoe crab more closely related to a spider than to a lobster?

5. Discuss the adaptive value of body segmentation. Name some of the major segmented animals.

REFERENCES AND SUGGESTIONS FOR FURTHER READING

Adams, S. L. (1988). The medicinal leech: a page from the annelids of internal medicine. *Annals of Internal Medicine 109*, 399–405.

This essay—amusingly titled in view of the journal name—gives a fascinating account of the rise and fall (and rise) of medical leech therapy.

Barnes, R. D. (1987). *Invertebrate Zoology.* Holt, Rinehart and Winston, New York.

A very good book for serious students.

Barnes, R. S. K., Calow, P., and Olive, P. J. W. (1988). *The Invertebrates: A New Synthesis.* Blackwell Scientific, Boston.

An attempt to explain invertebrates by combining knowledge of function and diversity. An enjoyable book.

Blum, M. S. (Ed.) (1985). *Fundamentals of Insect Physiology.* John Wiley & Sons, New York.

A comprehensive, up-to-date, and readable account of the principles of insect physiology.

Buchsbaum, R., Buchsbaum, M., Pearse, J., and Pearse, V. (1987). *Animals Without Backbones.* University of Chicago Press, Chicago.

A familiar old friend in a new edition. A non-technical delight.

Borror, D. J., Delong, D. M., and Triplehorn, C. A. (1980). *An Introduction to the Study of Insects.* Holt, Rinehart and Winston, New York.

A standard textbook on insect biology. Though its coverage is broad, the main focus is on morphology and systematics.

Cameron, J. N. (1985). Molting in the blue crab. *Scientific American 252*, May, 102–109.

A stimulating discussion of the exoskeleton. The author clearly explains how animals pos-

sessing one can grow.

Evans, H. E. (1984). *Insect Biology: A Textbook of Entomology.* Addison-Wesley, Reading, Mass.

A text with a functional approach to the biology of insects. It includes anatomy and systematics, but has a more physiological, behavioral, and ecological viewpoint than most texts.

Evans, H. E. (1985). *The Pleasures of Entomology: Portraits of Insects and the People Who Study Them.* Smithsonian Institution Press, Washington, D.C.

This book begins: "The insect never ceases to astound with its capacity to combine beauty and ugliness, power and frailty, familiarity and other-worldliness—in a package so small we can eliminate it with a stamp of the foot." These are the themes of this informative volume.

George, D., and George, J. (1979). *Marine*

Life: An Illustrated Encyclopedia of Invertebrates in the Sea. John Wiley & Sons, New York.

A refreshing and visually pleasing collection.

Hickman, C. P. (1973). *Biology of the Invertebrates.* Mosby, St. Louis, Mo.

An informative text on various aspects of the biology of invertebrates. The balanced discussion covers anatomy, physiology, and systematics.

House, M. R. (1979). *The Origin of Major Invertebrate Groups.* Academic Press, Orlando, Fla.

This authoritative discussion of the origins of the various invertebrate groups gives a clear account of the major adaptive specializations that emerged during evolution.

Lawrence, J. (1987). *A Functional Biology of Echinoderms.* Johns Hopkins University Press, Baltimore.

A water vascular system and ossicles distinguish echinoderms from other animal phyla. These and other aspects of echinoderm biology are interestingly discussed here.

Manton, S. M. (1977). *The Arthropoda: Habits, Functional Morphology and Evolution.* Clarendon Press, New York.

A classical work that integrates functional morphology with evolution and behavior.

Morton, J. E. (1979). *Mollusks.* The Hutchinson Publishing Group, London.

The author offers an extensive summary of all aspects of molluscan biology.

Richardson, J. R. (1986). Brachiopods. *Scientific American 255,* September, 100–106.

The author argues that these clamlike bivalves are not as scarce or evolutionarily static as many have thought.

Ward, P. D. (1988). *In Search of Nautilus.* Simon & Schuster, New York.

A historical, nontechnical account of scientific interest in the pearly nautilus. Delightfully lucid.

Winston, M. (1987). *The Biology of the Honey Bee.* Harvard University Press, Cambridge, Mass.

Another excellent work on a fascinating subject. Very readable.

CHORDATES AND ANIMAL PHYLOGENY

Chordates and animal phylogeny
The stunning diversity of vertebrate animals is
well illustrated by these representatives of several
classes of the vertebrate subphylum: (top)
Japanese macaque, Macaca fuscata, *and*
iguana, Conolophus subcristatus; *(bottom)*
turtle, Terrapene carolina, *and African*
elephant, Loxodonta africans.

In this chapter, we explore the phylum **Chordata** and thus complete our survey of the major animal groups. Members of this phylum, our own, include the animals we know best.

After we have examined the animal phyla, we will reconsider certain aspects of their evolution and phylogenetic relationships.

PHYLUM CHORDATA: THE CHORDATES

Most chordates—fishes, amphibians, reptiles, birds, and mammals—are vertebrates. Chordata is one of three phyla that have reached the greatest complexity of structure, the other two being Mollusca and Arthropoda.

Three key characteristics mark the chordate phylum.

- A **notochord** (Chapter 19) is present at some stage of life. The name "chordate" denotes this attribute.
- A single hollow **dorsal nerve cord** develops along the embryo's back. The notochord lies under it. The dorsal nerve cord, which represents the developing central nervous system, enlarges at the front end to become a brain in most forms.
- **Pharyngeal slits** appear during embryonic development. These cleftlike openings in the pharyngeal walls serve different functions in different chordate groups. In early chordates, their role related to filter feeding. In this process, seawater enters the body through an incurrent siphon, passes through the pharyngeal slits, and exits through an excurrent siphon. Food particles filtered from the water current by a mucous net are passed into the intestine by cilia (Figure 43-1). The slits became modified for gas exchange in fishes.

INVERTEBRATE CHORDATES

The phylum Chordata is usually divided into four subphyla, three of which are invertebrates. Some taxonomists regard these subphyla as separate phyla.

Subphylum Hemichordata: Acorn Worms

Hemichordates (subphylum **Hemichordata**) are "half-chorded." These wormlike marine creatures have some chordate characteristics (a dorsal nerve cord and gill slits). The **acorn worm** (Figure 43-2A) is an example.

Subphylum Urochordata: Sea Squirts

Tunicates (subphylum Urochordata) look even less like vertebrates as adults (Figure 43-2B), a stage in which most are sessile creatures and some are colonial. They are often found adhering to rocks, docks, and boats. The name Urochordata ("tail-chorded") signifies that the notochord is confined to the tail. This subphylum is also called Tunicata because all forms are encased in a tough covering or tunic.

The best known class of tunicates are the curious sessile **sea squirts** (class Ascidiacea), which shoot a jet of water through their excurrent siphon when disturbed. A variety of sea squirts was described by Aristotle. Some tunicate larvae superficially resemble tiny fishes, with tails supported by a notochord as swimming organs (thus the name urochordata). Thus, all three chordate characteristics are evident in the larval forms. However, even adult tunicates retain some chordate trademarks. We shall return to tunicates and their postulated place in the remote ancestry of vertebrates.

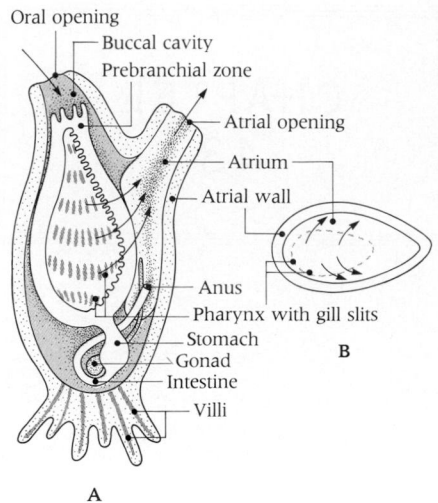

43-1 Water flow in sea squirts

Subphylum Cephalochordata: Amphioxus

Cephalochordates (subphylum **Cephalochordata**) are "head-chorded," the notochord extending to the tip of the head (Figure 43-2C), which it does not do in other chordates. These marine animals, of which some 30 species are recognized, are fishlike in appearance throughout life and resemble an idealized chordate, displaying a prominent notochord, a well-developed dorsal nerve cord, and numerous pharyngeal slits. Cephalochordates swim by coordinated serial contractions of muscle segments along the sides of the notochord, which develop from blocks of mesoderm called **somites.**

Cephalochordates have some of the basic vertebrate characteristics, such as segmentation, but they lack vertebrae, a true brain, and the appearance of neural crest cells during their development. The common cephalochordate usually studied is **amphioxus** (genus *Branchiostoma*), the lancelet—a tiny animal only a few centimeters long that wiggles backward into the sand, leaving only its anterior end exposed. It is a filter feeder, in which a food-laden water current is produced by ciliary action that draws water into the mouth and pumps it out of the pharyngeal slits.

Although the three groups of invertebrate chordates are represented today by a relatively small group of marine organisms, they cast some light, albeit an uncertain one, on a very knotty problem—the evolutionary origin of vertebrates.

SUBPHYLUM VERTEBRATA: ANIMALS WITH VERTEBRAL COLUMNS

In many respects, the fourth chordate subphylum, the vertebrates (subphylum **Vertebrata**) has one of the most prominent roles of all animal groups in the Earth's major ecosystems. However, vertebrates are much less diverse and abundant than insects, and their direct contribution to the total metabolic turnover of the living world is far smaller than that of several plant groups and also of several other animal groups. Of course, human activities have important indirect effects.

As a whole, vertebrates are characterized by a highly developed ability to receive environmental stimuli and to react to them flexibly. Moreover, they include human beings, which in most respects are incomparably the most potent and dominant influence on Earth today.

The vertebrates have a segmented backbone made up of **vertebrae** that surround the notochord and nerve cord. In the more derived forms, the notochord is present only during embryonic life. Its protective role is soon assumed by the vertebrae themselves. In addition, vertebrates have a closed circulatory system that contains in its fluid phase hemoglobin-filled red blood cells, motile white blood cells and platelets, and soluble clotting factors, which protect the vascular system's integrity by forming a clot that seals a blood vessel whenever it is punctured. There is also a four-chambered heart, a highly differentiated nervous system with a brain of great complexity, and improved sense organs. In other words, the trend toward cephalization is maximal.

From the evolutionary and developmental viewpoints, the single most important innovation that characterizes the vertebrates is the appearance of embryonic neural crest cells (see Chapter 19). It is now clear that all the shared and derived characteristics of vertebrates are either induced by, or otherwise associated with, tissues from

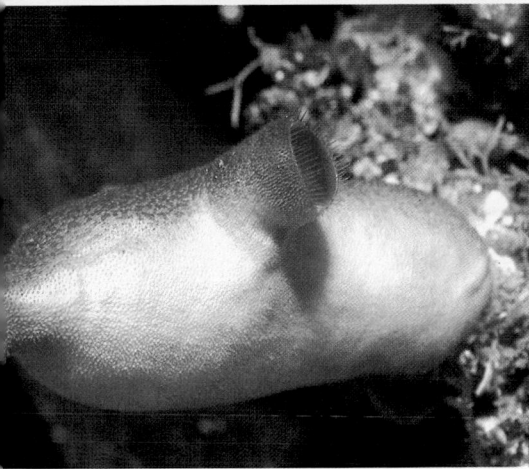

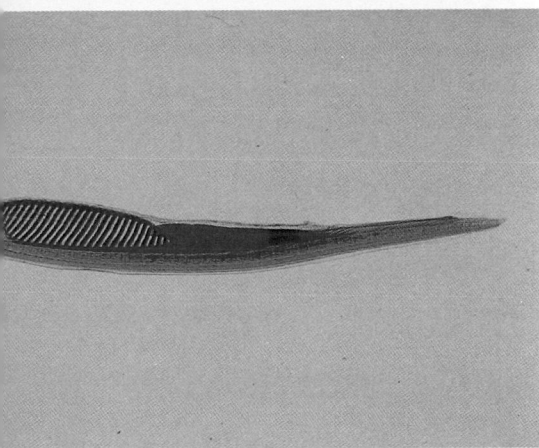

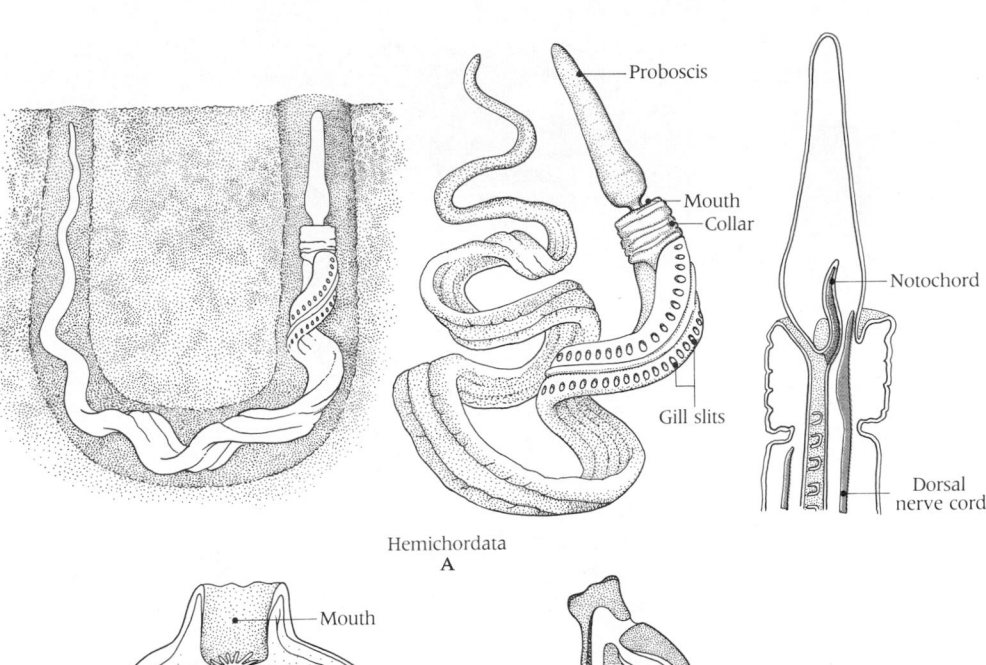

Hemichordata
A

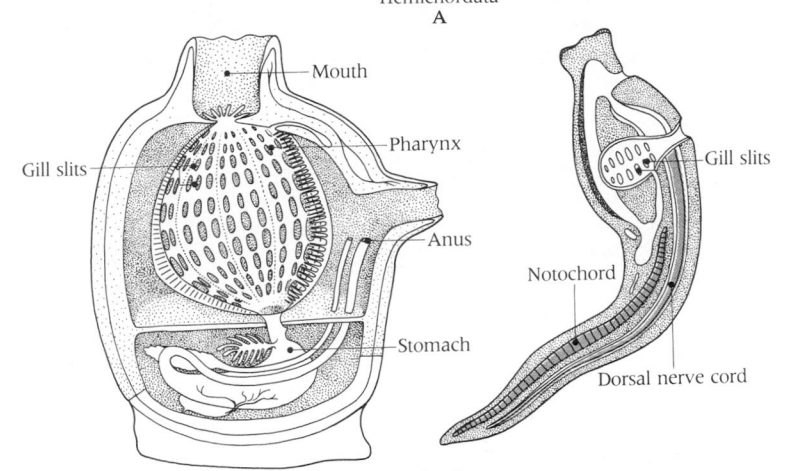

Urochordata
B

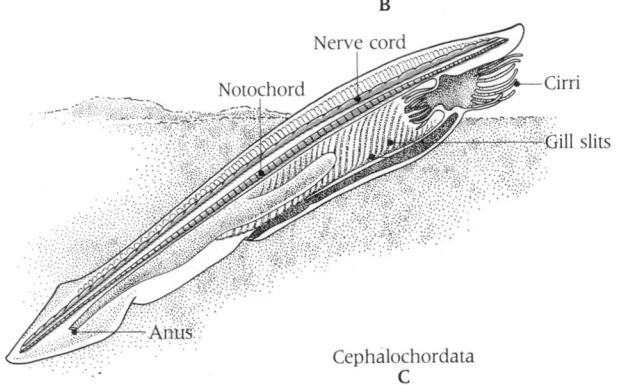

Cephalochordata
C

43-2 The invertebrate chordates

Note the location and extent of the notochord and dorsal nerve cord in each animal. (A) An acorn worm, Balanoglossus, is seen in its U-shaped burrow in the sand. Note its external features: the burrowing proboscis, collar, mouth, and gill slits. The detail of the head region shows the dorsal nerve cord and notochord, extending forward from the roof of the pharynx into the proboscis. (B) A sea squirt, Ciona, or tunicate. An adult is shown with its body wall cut away. Compare with Figure 43-1. The larva on the right is bilaterally symmetrical, elongate, and free-swimming. (C) Branchiostoma, an amphioxus, half-buried in sand where it lives a semisedentary existence as a filter feeder like the other invertebrate chordates.

the neural crest cells. As summarized in Table 19-2, these cells give rise to many bones (including the jaws, anterior skull, teeth, and skeletal supports of the gill arches), most of the peripheral nervous system, major components of the paired sense organs, and the adrenal medulla. Equipped with this combination of new structures, the vertebrates evolved into active and efficient predators, abandoning for the most part the filter-feeding habits so characteristic of the invertebrate (or protovertebrate) chordates.

The subphylum Vertebrata consists of eight classes (Figure 43-3). Four are lumped together in the superclass **Pisces,** the fishes, which are finned, usually gill-breathing aquatic vertebrates.

- Class **Agnatha:** jawless fishes
- Class **Placodermi:** placoderms, now extinct
- Class **Chondrichthyes:** sharks
- Class **Osteichthyes:** bony fishes

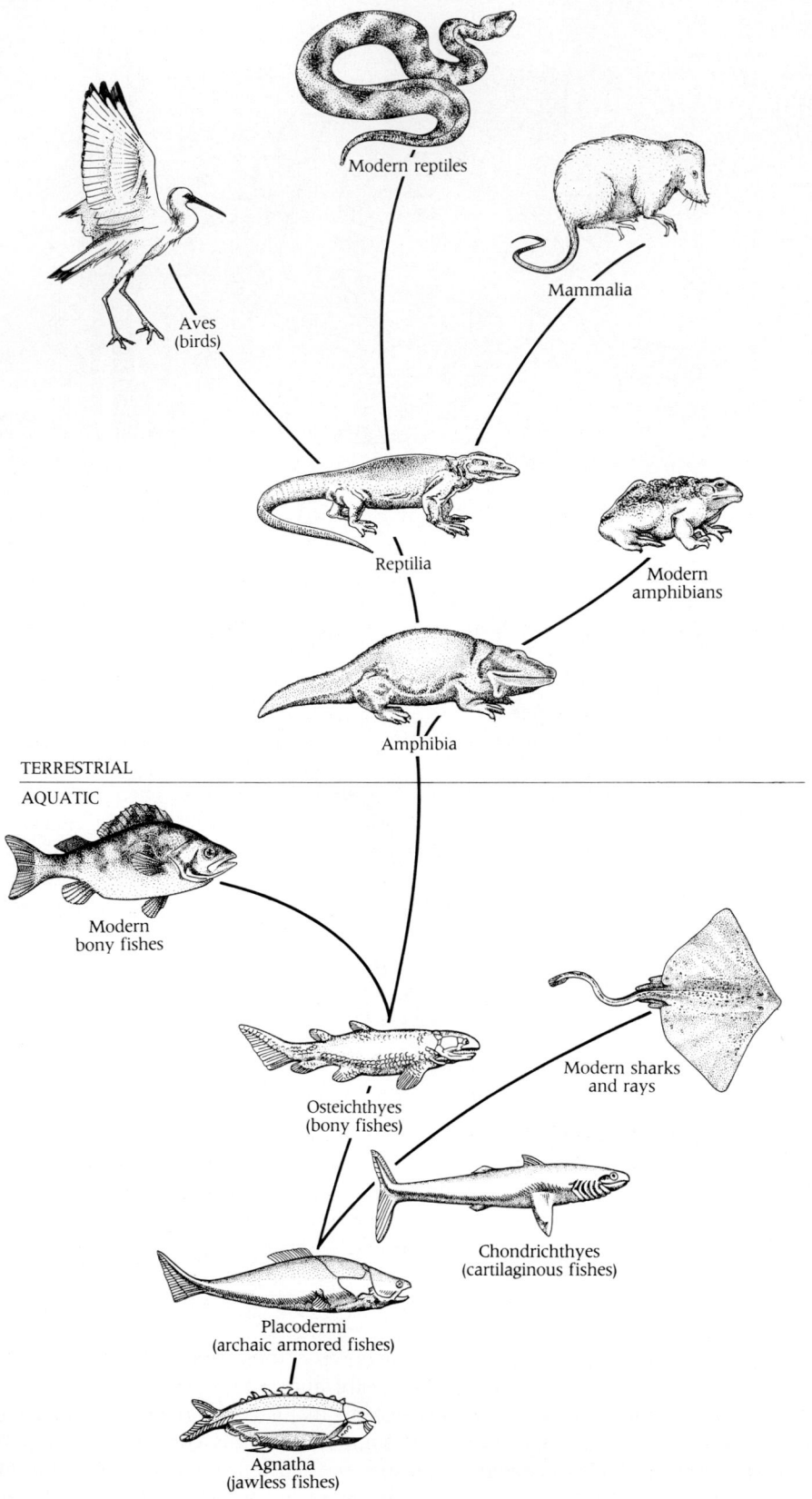

Modern reptiles

Mammalia

Aves
(birds)

Reptilia

Modern
amphibians

Amphibia

TERRESTRIAL

AQUATIC

Modern
bony fishes

Osteichthyes
(bony fishes)

Modern sharks
and rays

Chondrichthyes
(cartilaginous fishes)

Placodermi
(archaic armored fishes)

Agnatha
(jawless fishes)

The other four classes make up the superclass **Tetrapoda,** the four-limbed lung-breathing land vertebrates.

- Class **Amphibia:** amphibians
- Class **Reptilia:** reptiles
- Class **Aves:** birds
- Class **Mammalia:** mammals

Let us look at these classes and then turn to the interesting problem of their evolutionary origin.

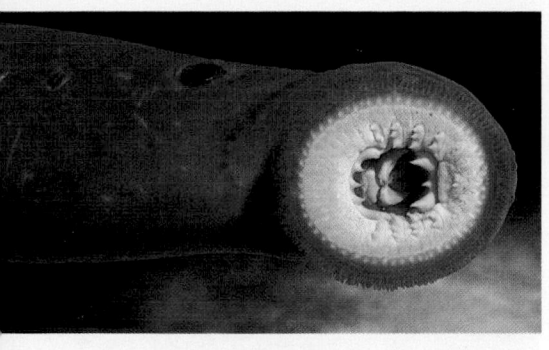

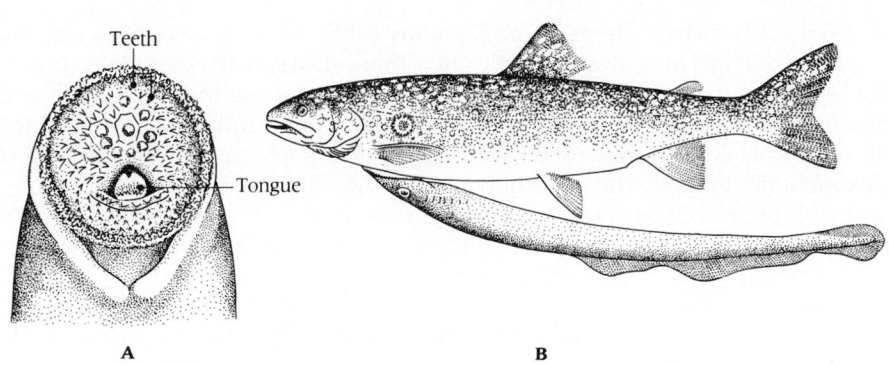

Teeth

Tongue

A   B

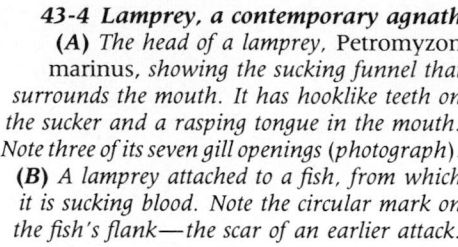

43-4 Lamprey, a contemporary agnath
(**A**) *The head of a lamprey,* Petromyzon
marinus, *showing the sucking funnel that
surrounds the mouth. It has hooklike teeth on
the sucker and a rasping tongue in the mouth.
Note three of its seven gill openings (photograph).*
(**B**) *A lamprey attached to a fish, from which
it is sucking blood. Note the circular mark on
the fish's flank—the scar of an earlier attack.*

Class Agnatha: Jawless Fishes

The first vertebrates, the jawless fishes or **agnaths,** date back at least to Ordovician times. These ancient fishes had highly differentiated bony skeletons but lacked jaws and paired fins. Fossils show that agnaths had skeletal armor that enclosed the head and protected the body. Their dominance of the realm of fishes was relatively brief—some tens of millions of years. Then they disappeared almost completely.

Only a few forms—31 species of lampreys (family Petromyzontidae; Figure 43-4) and 32 species of hagfishes (family Myxinoidei)—survive as relics. Today's agnaths have all the basic vertebrate characteristics, but lack the bony tissue and armor of their ancestors. Though jawless, they are highly specialized in certain respects—notably, in their feeding habits and elongate, eel-like body form. Such forms did not occur in the earliest vertebrates, which were otherwise diverse in shape. As for the feeding habits of living agnaths, they are predators that live either semiparasitically, sucking the blood of other fishes, or as scavengers.

Lampreys produce a distinctive larval form called an **ammocoete.** It lives up to seven years, far longer than the adult stage, and it feeds, as most early agnaths probably did, by sucking up mud that contains microorganisms and organic debris. After metamorphosis, it develops a round, sucking funnel lined with teeth. This organ is used to attach the larva to other fishes for the purpose of extracting their blood. There are both fresh water and marine lampreys.[1]

Hagfishes, which are not known to have larvae, also have a sucking funnel with which they eat their way into and through the bodies of other fishes, usually attacking the disabled or dead. All the hagfishes are marine.

Class Placodermi: The First Jawed Vertebrates

The archaic **placoderms** ("plate skins") of 400 million years ago were armored, jawed fishes with bony skeletons and paired appendages. Most of the early jawless fishes were probably mud-sucking filter feeders, extracting nourishment from the organic matter in mud. They were soon replaced by placoderm fishes with true movable jaws, clearly a more effective device for securing food than mere jawless openings. Indeed the evolution of jaws and teeth was a tremendous advance. Also, their paired fins contributed significantly to swimming and maneuvering.

The evolution of jaws from a pair of hinged arches supporting the gill region is an instructive example of evolutionary transformation (see Box 35-1). Placoderms became extinct and were replaced by other lineages of jawed fishes with more complex and mechanically more efficient jaws and fins. In the next chapter, we shall again mention placoderms and their important acquisition, hinged jaws—and its role in the history of life in the sea.

Class Chondrichthyes: Sharks, Rays, and Skates

Two large groups of jawed fishes evolved independently from placoderms at about the same time. Between them, they soon replaced the placoderms. A possible reason for this basic early subdivision of higher fishes into two main classes is that one of

[1] Sea lampreys, known as "vampires of the deep," have been increasing in the Great Lakes and seriously depleting stocks of salmon, trout, and other fish species, which they nearly wiped out in the 1940s and 1950s.

them—the **Chondrichthyes** ("cartilaginous fishes")—may have been originally specialized for life in seawater, while the other—**Osteichthyes** ("bony fishes"), which are discussed later—may have been adapted for life in fresh water. In the millions of years since these groups arose, a few chondrichthyans have wandered into fresh waters (but only a few), and they did so haphazardly. The group is still fundamentally marine. The osteichthyans, on the other hand, while continuing to dominate the fresh waters of Earth, spread early to the sea and are now by far the dominant marine fishes. However, their success did not competitively wipe out the chondrichthyans, which remain significant in the life of the sea.

The most obvious and universal anatomical feature of chondrichthyans is the one they are named for—a fully **cartilaginous skeleton.** Their placoderm ancestors had bony skeletons. The cartilaginous skeleton, which is usually a precursor to bone in ontogeny (the development of an individual) in other vertebrates, remains unossified in chondrichthyans. Apparently, chondrichthyans simply stopped replacing youthful cartilage with adult bone—in a sense, a retrogressive development.

The principal chondrichthyans are the sharks, rays, and skates (Figure 43-5). Placoid (platelike) scales with structures called **denticles** (resembling small teeth) give shark skin its characteristic rough texture. They are similar in form to the teeth in the jaws (Figure 43-6). Probably teeth evolved from such scales in the mouth region. All sharks and most rays have paired pectoral and pelvic fins, but in the males the posterior pelvic fins are modified with structures called **claspers** that facilitate the introduction of sperm into females. Indeed, internal fertilization was a notable specialization of this group. Another was the appearance of horny, leathery shells around individual eggs.

As we discussed in Chapter 25, an interesting feature of this class is the way it handles the osmotic challenge of salty water. The problem is to prevent loss of body water to the high osmotic pressure of the surrounding sea water. These fishes solve the problem by reabsorbing and retaining the urea arising metabolically (see Figure 25-3B). As a result, the urea content of their blood is 2000 to 2500 mg per 100 ml, or $0.45M$. This is to be compared to the blood urea concentrations in freshwater fishes and humans of $0.001M$. The high level of urea (and of trimethylamine oxide and other osmotically active salts) raises internal osmotic pressures to levels approximating that of sea water.[2] Thus, marine chondrichthyans, unlike other marine fishes, are in osmotic equilibrium with their environment. This adaptation in the chondrichthyans nicely illustrates the fact that physiological adaptations are as numerous and important as the anatomical ones we often tend to emphasize.

The oldest chondrichthyans, the sharks (order Selachii), are still common today—some 300 species of them. They are changed in many details but still of the same adaptive type: elongated, streamlined, swift-swimming predators that definitely have jaws. Only the two outer rows of pointed teeth are functional. The inner rows are replacement teeth that move forward as the outer teeth wear out or fall out. By shaking their heads from side to side, modern sharks use their teeth as saw blades to cut out chunks of flesh from large prey.

Their highly acute sense organs are adaptations that befit the carnivorous life style of sharks. Sharks have sharp vision, though they cannot distinguish colors. They also hear efficiently and have a well-developed sense of smell. The shark head also possesses special detectors of electric fields, the ampullae of Lorenzini. These electroreceptors are sensitive enough to detect the electric fields generated by the muscles or hearts of nearby fishes. Recent work has shown that sharks can also detect the Earth's magnetic fields, which they use for orientation in the oceans (see Chapter 32).

These and other characteristics of the class are well seen in the dogfishes (suborder Squaloidea), small sharks that are often used as laboratory animals. Later in origin and now abundant are the rays and skates (order Rajiformes)—broad flattened forms, most of which live on the sea bottom where they devour various invertebrates, especially crustaceans and some mollusks. Many rays have whiplike tails, which

---

[2] As every biochemist knows, urea in high concentrations denatures many proteins. How, then, do the chondrichthyans withstand this hazard? The answer was suggested by recent work showing that one protein (hemoglobin) of a shark, a ray, and a skate is uniquely resistant to this effect of urea, unlike the hemoglobins of other vertebrates. Evidently, the evolutionary adaptation that permitted these animals to deal with the high osmotic pressure of seawater included appropriate modifications in the molecular structures of their proteins.

A

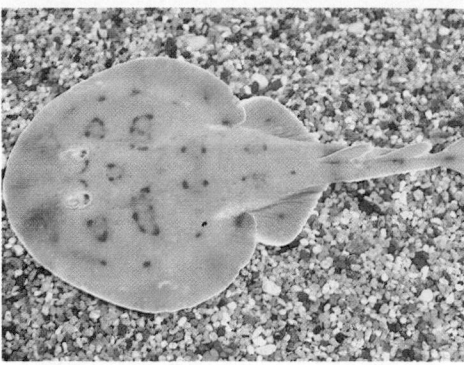

B

C

43-5 Cartilaginous fishes
(**A**) *Shark,* Notorynchus maculatus. (**B**) *Ray,* Narcine Grasiliensis. (**C**) *Skate,* Raja binoculata.

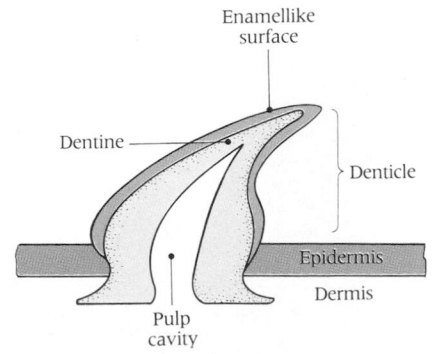

43-6 The placoid scale of a shark
These scales, called dermal denticles, are embedded in the dermis layer of the skin and also make up the teeth in the jaws.

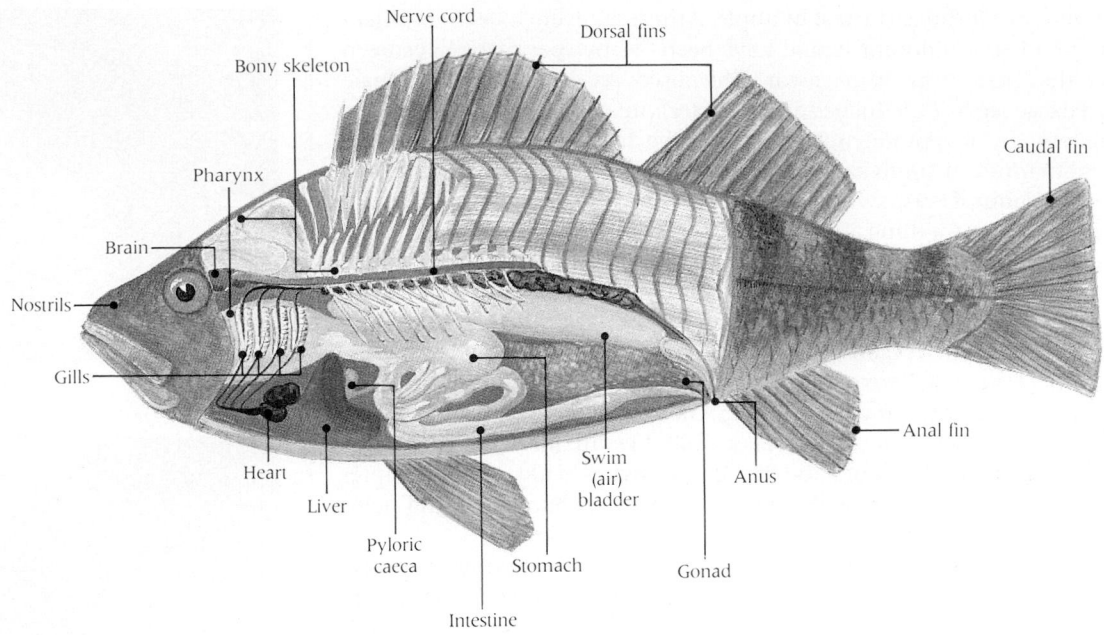

Nerve cord
Dorsal fins
Bony skeleton
Caudal fin
Pharynx
Brain
Nostrils
Gills
Heart
Liver
Pyloric caeca
Intestine
Stomach
Swim (air) bladder
Anus
Gonad
Anal fin

43-7 (Above) *Anatomy of a bony fish*
The nervous system and bony skeleton are highlighted in color. Note the multiple fins (dorsal, caudal, and anal), the gills, and swim bladder.

Barracuda

Cowfish

Butterfly fish

Blowfish

Sea horse

Moray eel

43-8 *Body shapes of some bony fishes*
The stippled silhouettes represent transverse sections through the various fishes. Their diversity is striking.

in some species (like the stingray) bear venomous barbs. Some 21 families of sharks and 15 families of rays are recognized, with about 600 species in both groups.

A third group of chondrichthyans, fossil and recent, are the chimaeras (or ratfish) and their relatives (subclass Holocephali), mostly deep sea fishes today with only a few known genera.

Class Osteichthyes: Bony Fishes

It is the bony fishes that usually come to mind when one thinks of fishes (Figure 43-7). They are the sardines, herrings, salmon, trout, and countless others. Indeed, nearly 30,000 living species are known and a large number remain to be discovered, particularly in the deepest part of the ocean. The abundance of some species is also remarkable: A billion billion herrings populate the world's oceans.

Bony fishes range in size from tiny creatures 0.6 mm in length to giants (like sunfishes) over 7.5 meters long. Their shapes, colors, and life styles are of every imaginable kind. Figure 43-8 merely hints at the variety of shapes they display.

In most osteichthyans, skeletons are of true bone, bodies are covered with scales, and there are two pairs of paired fins in addition to the medial fins. The earliest osteichthyans known are from the late Silurian. They were found in fresh water in the Devonian—the so-called Age of Fishes—but they spread to the sea early in their history. As we will learn in Chapter 44, they early became the dominant vertebrates in fresh and salt water and remain so to this day.

The osteichthyans, especially the large groups that are classed together as the superorder Teleosti, are the dominant aquatic animals today, having become adapted to virtually every watery environment—from frothy mountain streams to deep sea bottoms of unvarying and frigid darkness. Not only are these fishes dominant in both fresh water and salt water, there are some—salmon, eels, and others—that move from one to the other.

Some of the earliest osteichthyans were heavily armored with bony scales coated with a hard enamel-like tissue. A few modern fishes, such as the gars, have retained an armor, but in most of them it has evolved into more flexible scales. Some fishes (most eels, for example) have completely lost their scales.

There is disagreement on the classification of the living bony fishes. Our classification has two major subclasses—the **Sarcopterygii,** or **lobe-finned fishes,** and the **Actinopterygii,** or **ray-finned fishes.** Sarcopterygii includes a group of special interest—the crossopterygians (order Crossopterygii), also called **crossopts** (the name means "fringed fins"). These animals were abundant in the earliest days of osteichthyan history when, with the lungfishes, they were dominant. Thereafter, they dwindled and were long believed to have been extinct for about 75 million years. Then, in December 1938, a single living crossopt, a coelacanth (*Latimeria* *chalumnae*) was caught off the east coast of South Africa, an event as extraordinary as the discovery of a living dinosaur would have been. Many specimens have been caught since in the vicinity of Madagascar. The discovery in 1972 of a female coelacanth with large eggs (3.5 inches in diameter) in one ovary was the first evidence of their mode of reproduction. In 1987, the first motion pictures were made (from a submarine) of six living coelacanths in their natural environment in the Indian Ocean (Figure 43-9), swimming about with their paired and unpaired lobed fins. The most important thing about the crossopts is not their survival, amazing as it is, but the possibility that amphibians and, through them, all vertebrates evolved from an ancestral lobe-finned fish of the Devonian period (Figure 43-10).

Primitive osteichthyans had lungs as well as gills. The lungs were probably an adaptation to aquatic life in deoxygenated waters that provided supplementary respiratory support. One group of Sarcopterygii includes the few living osteichthyans that have retained lungs—the lungfishes (order Dipnoi), which survive as relicts in Australia, South America, and Africa (Figure 43-11). In most osteichthyans, the lung lost its respiratory function and became a swim bladder (see Chapter 27), a gas-filled sac that regulates a fish's density, modifies its buoyancy, and helps to maintain its position in the water.

The higher bony fishes (subclass Actinopterygii) include several primitive orders that some taxonomists place in a separate subclass called Chondrostei. In this group, represented today by the sturgeons and paddlefishes (order Acipenseriformes) and the bichirs (order Polypteriformes), bone has retrogressed in varying degrees and significant portions of their skeletons consist of cartilage. In these so-called **chondrostean** fishes, gill slits are unlike those in the chondrichthyans, in which each opens separately to the exterior. Instead, as in other teleosts, the gills open into a chamber covered by a common bony plate or flap, the operculum (Chapter 27).

The teleosts (superorder Teleostei) comprise a major adaptive radiation, with an estimated 23,000 species exceeding by far the diversity (and population sizes) of any vertebrate group. Indeed, this number accounts for more than half of all living vertebrate species. Their feeding apparatus is admirably suited for life in water. It can generate suction very rapidly, often within 0.025 second (see Figure 26-9). Suction draws water into the mouth and forces it out from the slit along the free edge of the operculum. In this way, a great variety of prey are drawn into the mouth at high speed. In addition, teleosts possess a symmetrical tail fin that greatly enhances the efficiency and precision of locomotion. Teleosts also have remarkably proficient respiratory mechanisms that extract up to 95% of the oxygen dissolved in the water entering their mouths.[3]

---

[3] The bony fishes are a major source of protein for the human diet and more than 50 million tons of fish are caught each year throughout the world. Unfortunately, overfishing and environmental pollution are damaging many of the world's major fishing grounds. The ocean is not as limitless as we once believed, and humankind's activities are changing its character ominously.

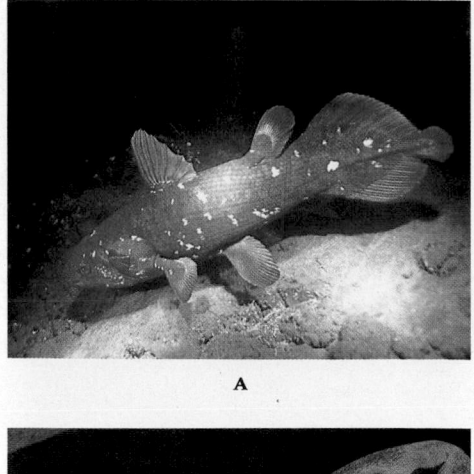

A

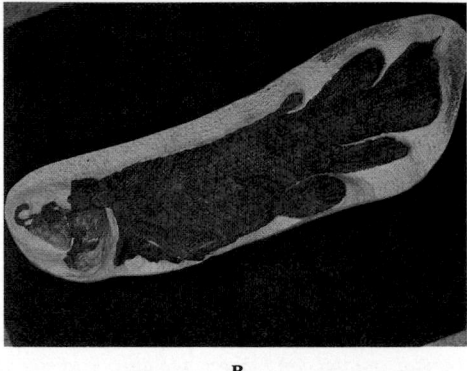

B

43-9 Coelacanths
(A) *A living coelacanth,* Latimeria chalumnae. *This remarkable photograph of a lobe-finned fish, long believed to be extinct, was taken from a minisubmarine in the Indian Ocean at a depth of 200 meters.* **(B)** *A fossil coelacanth dating from the Pennsylvanian epoch (late Carboniferous).*

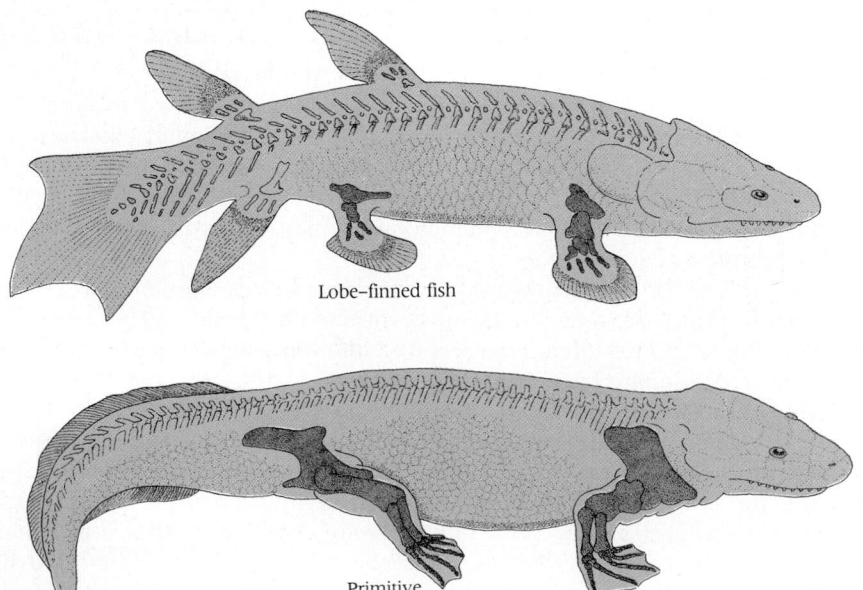

Lobe–finned fish

Primitive amphibian

43-10 Evolution of amphibians from lobe-finned fishes
The legs of amphibians evolved from the bony fins of lobe-finned fishes. The fishes were probably able to crawl between streams and ponds on their stubby fins.

A

B

43-11 Contemporary lungfishes.
These bony fishes have both gills and lungs. Hence, they can survive by air breathing when forced to do so by drought conditions. (A) An African lungfish, Protopterus aethiopicus, *which lives in eastern and central Africa. (B) The Australian lungfish,* Neoceratodus forsteri, *is native to Queensland.*

Bony fishes present us with some of behavioral biology's most provocative questions. Some of them we discussed in Chapters 32 and 33. Recall, for example, the ability of electric eels to develop charges of up to 550 volts with which they stun their prey. How do they do it? How do the fishes that live in the austerity of very deep sea waters (below 3300 feet) endure the cold and dark? We could go on.

Class Amphibia: Amphibians

Amphibian, meaning "with a double life," is the name of a class both aquatic and terrestrial. The key characteristics of amphibia are as follows: (1) they lack scales; (2) they respire through their moist, soft skin and through gills, lungs, or mouth linings; and (3) they lay eggs in water, where they spend at least their early life.

The amphibians did not evolve from later osteichthyans. The first land vertebrates arose from the sarcoptyerygians, the primitive forms that lived near the beginning of osteichthyan history. Usually when a radical adaptive change gives rise to a new major group, it originates from the primitive members of the ancestral groups. With a drift of adaptation toward one way of life or the other, a time often comes when the trend seems irrevocable.

A key factor in the fish-amphibian transition was the presence in primitive bony fishes of lungs, which enabled them to live in waters that at times were deficient in oxygen. However, it was not the emergence of lungs that marked the beginning of the vertebrate conquest of the land. Rather, it was the change of the paired fleshy fins of forms like crossopts into the walking legs of amphibians, which made them into tetrapods (see Figure 43-10). In other respects, the first amphibians were still quite fishlike. They probably spent most of their lives in the water.

Modern amphibians include about 2500 species in three orders.

- Order Urodela, about 300 species
- Order Anura, about 2100 species
- Order Apoda, about 75 species

Virtually all urodelans and anurans have four limbs that are adapted for walking or leaping. Apodans have no limbs (but their ancestors did). The body temperature of amphibians is variable. They lay jelly-coated eggs, and their skins are bare and moist. Unlike reptiles, birds, and mammals, most amphibians are obliged to remain near water for at least part of their life cycle.

Some modern amphibians are tied to the water by the ineffectiveness of their lungs as respiratory organs and a need to keep their skin moist. Air is inadequately forced into the lungs by the weak bellows action of the floor of the mouth cavity (see Chapter 27). The frog's skin, richly supplied with blood vessels, is its principal respiratory organ and, as such, must be kept moist (see Figure 27-31B). It is uncertain whether this dependence on the skin for respiration characterized ancestral Amphibia. Perhaps this was an evolutionary novelty in later Amphibia, which were confined to moist places.

A

B

Among modern amphibians, the salamanders and newts (order Urodela, "tail visible"), though specialized or degenerate in many respects, have most nearly retained the ancestral habits, fishlike body form, and even fishlike undulatory movements of the body (Figure 43-12A). They go through a larval stage that closely resembles the adult stage. However, the larvae have gills that are lost and replaced by functioning lungs when metamorphosis occurs. We saw in Chapter 20 that some salamanders display neoteny, in which larval features, such as external gills, are retained (see Figure 20-34). Some forms display progenesis when the animal spends its entire life in the larval stage but becomes sexually mature.

The anurans, or jumping amphibians (order Anura, "untailed"), are the frogs and toads, in almost all of which the familiar tadpole is the larval stage. Tadpoles, too, are fishlike—even more so than salamander larvae—lacking legs in early stages. Their metamorphosis is more drastic, yielding an adult frog or toad completely unlike a tadpole. The anuran family with the largest number of species is that of the Ranidae, the true frogs. The largest genus, *Rana*, is found almost everywhere.

Adult anurans are among the most distinctive and specialized of all vertebrates (see Figure 43-12B). The most obvious specializations are their lack of tails (that is one of our most obvious specializations, too, but we did not inherit it from the anurans) and their extraordinary capacity for leaping. In addition, the anatomy, physiology, and habits of anurans exhibit other unusual features. Some display brilliant and fantastic coloration. Some secrete a powerful nerve poison, such as batrachotoxin, that aborigines applied to blow-gun darts. Physiologists often use it in experiments on nerve function. (Its ecological role is discussed in Chapter 46.) Such features give anurans a way of life in which they have few serious competitors. Moreover, they live long lives—some as long as 30 years.

The caecilians (order Apoda, "footless") are a small and relatively unimportant group of wormlike, living amphibians—about 160 living species restricted to the tropics. They are blind (or practically so) and entirely limbless burrowers in wet soil. "Caecilian" is from the generic name of one of the apodans. It derives from the Latin *caecus*, for "blind."

Many modern terrestrial amphibians feed on insects, which they capture with long sticky tongues that can be flicked out a long distance with incredible speed (see Figure 32-11). The aquatic larvae of frogs and toads feed mainly on algae.

It is important to note that modern amphibians represent highly specialized lineages, in which some original amphibian features have degenerated. Early amphibians were large animals (up to 4 meters in length) that were often covered with bony plates. Indeed, they superficially resembled reptiles. Adaptive radiation of the earliest forms took place in the Carboniferous—the Age of Amphibians. Amphibians began to decline in the late Carboniferous period.

C

43-12 The three orders of amphibians
(*A*) *A urodelan: the California newt,* Taricha torosus. (*B*) *An anuran: Fowler's toad,* Bufo woodhousei fowleri. (*see also Figure 36-9*). (*C*) *An apodan (caecilian),* Dermophis thomensis, *from West Africa.*

A

B

C

43-13 Modern reptiles
(A) *Tortoise,* Testudo elephantopus.
(B) *Alligator,* Alligator mississipiensis.
(C) *Lizard,* Tarentola mauritanica. *Each of these is but one of many varieties now living.*

Class Reptilia: Reptiles

Following their emancipation from water, it was the reptiles that became the first true land vertebrates. They evolved from a branch of early primitive amphibians in the Carboniferous, some 310 or more million years ago.

New modes of reproduction and respiration were the key adaptations behind the great success of amphibians. The most significant was the evolution of eggs that are laid on land rather than in the water. These so-called **amniotic eggs,** which were discussed in Chapter 21, have limy or leathery shells that keep them from losing precious water. The eggs contain a large amount of yolk, which provides nutrition, and albumin, which provides a water reservoir. Full development, without a larval stage, occurs within the protective shell before the young hatch out— essentially as miniature adults.

Three embryonic membranes develop within the egg.

- The chorion, an external membrane, sheaths both embryo and yolk.
- The allantois forms a lunglike breathing mechanism that captures the oxygen penetrating the porous shell.
- The amnion, from which the egg type is named, encloses the developing embryo in a liquid-filled space, the amniotic cavity, which is a miniature imitation of an ancestral pond.

The advent of the amniotic egg freed reptiles of the need to return to the water to shed their sperm and egg cells. Mating was now possible on land and the wholly terrestrial reptiles achieved fertilization internally within the female body by means of copulation. The amniotic egg was an advance of such key importance in the evolution of land vertebrates that reptiles and their descendants, the birds and mammals, are referred to collectively as amniotes.

An improved respiratory mechanism was another adaptation that made reptiles better land dwellers than were amphibians. Lungs were more efficiently ventilated by a variety of bellowslike mechanisms that inflated and emptied them. Other improvements hinged on structural changes in the circulatory system, especially the heart, which functionally became four chambered. Of the other changes, some were barely perceptible in the first reptiles but became pronounced in their descendants. One was a tendency for limbs to become powerful and large, elevating the body enough to keep the belly from dragging on the ground and enabling a longer stride during locomotion.

The reptiles diversified into many different habitats. Lacking special mechanisms for controlling body temperature, they avoided the coldest climates—though there are exceptions: one lizard ranges inside the Arctic Circle. Some even returned to the sea, where they competed with fishes—and consumed quite a few. They were dominant during the Mesozoic era, the glorious Age of Reptiles that is famous for its dinosaurs. This dominance and their later decline are dramatic chapters in the history of life, to which we return in Chapter 44.

Systematists generally recognize 16 orders within the class Reptilia. Of these, only four survive today. One is on the verge of extinction. The other three are abundant, familiar, and successful (Figure 43-13).

- Order Rhynchocephalia ("peak-headed"): now represented by a single species, the tuatara (Sphenodon punctatus), which lives on 20 tiny islands in the Cook Strait off the coast of New Zealand[4]
- Order Chelonia (Greek, "tortoises"): turtles and tortoises
- Order Crocodilia: crocodiles and alligators
- Order Squamata (Latin, "scaly"): lizards and snakes

Lizards and snakes are the most abundant living reptiles. More than 2500 living species of each occur everywhere—from high seas (sea snakes) to deserts. Despite the apparent difference between the legged, swift-running lizards and the legless, crawling snakes, they are rather closely related. Snakes are descendants of an ancient

---

[4] Tuataras look like lizards, but their internal anatomy shows that they split off from lizard ancestors before lizards, as such, appeared.

43-14 Mesozoic reptiles
The Mesozoic Era was distinguished by an array
of reptiles with varying habitats and life styles.
Herbivores (such as Triceratops), terrestrial
carnivores (such as Tyrannosaurus), and a
variety of pterosaurs and such aquatic species
as ichthyosaurs and pleiosaurs dominated the
Earth during that era. Compare with Figures
44-16 and 44-17.

group of burrowing lizards. There are still some burrowing lizards, of later origin, that have also become legless and snake-like.

Lizards—not to be confused with salamanders, which are amphibians—have kept the body plan typical of land vertebrates. They have legs with five toes on each foot. Most are less than a foot long but some are huge, weighing more than 300 pounds. The spectacular Komodo dragon (*Varanus komodoensis*), which lives on certain islands near Java, is the largest of living lizards, and measures about twelve feet in length.

Several lizards are noted for an ability to discard their tails. Spontaneous self-excision of the tail, termed **autotomy,** occurs when the tail is seized by an enemy, but sometimes it results from stress or fear. It is due to violent, reflex (nonvoluntary) muscle contractions that press on one of the weak points present in each of the tail vertebrae. A circular muscle then clamps off the tail artery to suppress hemorrhage. Tail autotomy is an effective defense that allows the lizard to escape while its attacker is left distracted by a squirming segment of a detached tail. The tail regenerates—another instance of regeneration—but not in its original form. The new tail cannot undergo autotomy.

We will say more in Chapter 44 about the extraordinary success of the reptiles. We note here only that the 12 reptilian orders that became extinct included the Ornithischia (reptiles with a birdlike pelvis) and the Saurischia (dinosaurs). Other forms in this abundance were the following (Figure 43-14):

- The flying reptiles called **pterosaurs** (order Pterosauria)
- The **therapsids** (order Therapsida), ancestral to mammals
- The **plesiosaurs** (order Sauropterygia), which returned to the sea
- The **thecodonts** (order Thecodontia), which we will see gave rise to many other ancient forms as well as a living order, the birds

Class Aves: Birds

The latest and most progressive classes of land vertebrates are the birds (class Aves) and the mammals (class Mammalia). Both evolved from early reptiles, though they followed separate evolutionary paths.

43-15 Feather structure
(A) *Down feathers provide insulation.* **(B)** *Contour feathers cover the animal and contribute to streamlining and insulation.* **(C,D)** *The contour feather has a central shaft. Vanes projecting from it are attached to each other.*

Shaft

Vane

C

Aftershaft

Quill

A

B

D

By the late Triassic, or early Jurassic, a group of reptiles had arisen from the thecodonts that had acquired the power of flight. The success of this group was related to the emergence of wings that bore **feathers,** an entirely new invention. Surprisingly few details are available on the evolutionary origin of feathers, the remarkable structures that came to distinguish birds from reptiles. It is assumed that feathers came from skin scales. Their functions are both aerodynamic and heat conserving (Figure 43-15). What is so puzzling is the specific circumstance that may have led to the evolution of the first feathers. What selective pressures produced them?

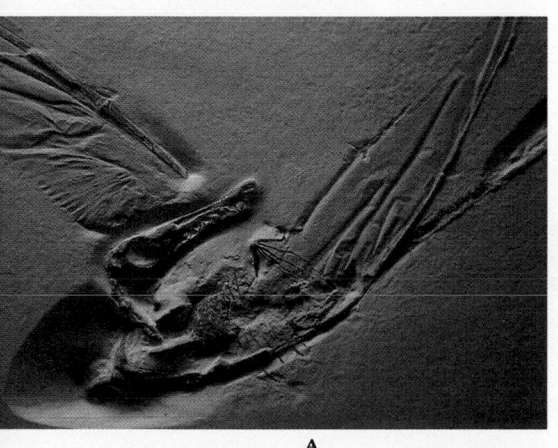

A

43-16 Fossil Archaeopteryx
(A) *Appearance of a fossil* Archaeopteryx. **(B)** *Differences between the fossil bird and a pigeon include presence or absence of teeth; the size of the brain case; the presence of separate or fused digits; and the size of the sternum, pelvic girdle, and tail.*

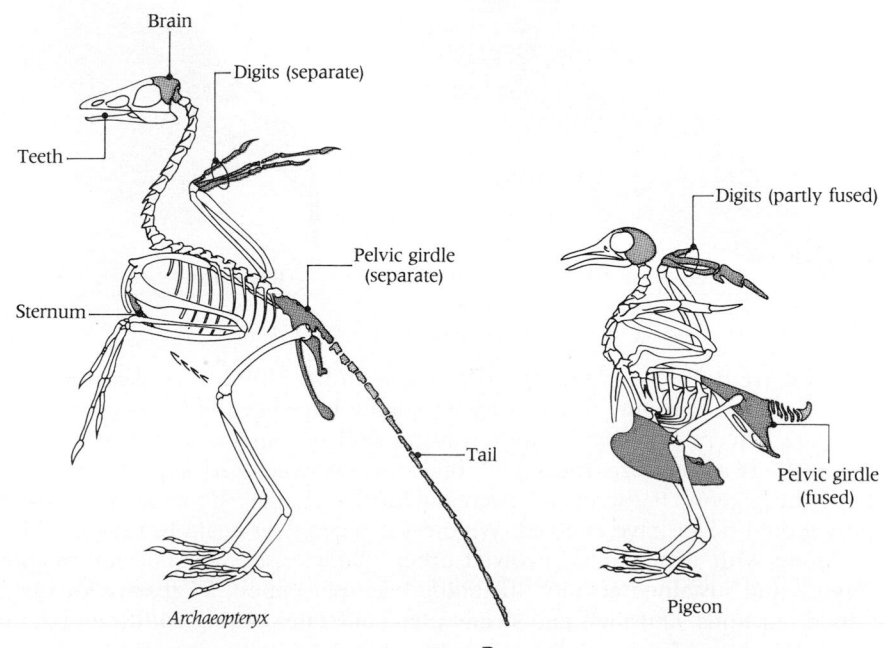

Brain

Digits (separate)

Teeth

Pelvic girdle
(separate)

Sternum

Tail

Archaeopteryx

Digits (partly fused)

Pelvic girdle
(fused)

Pigeon

B

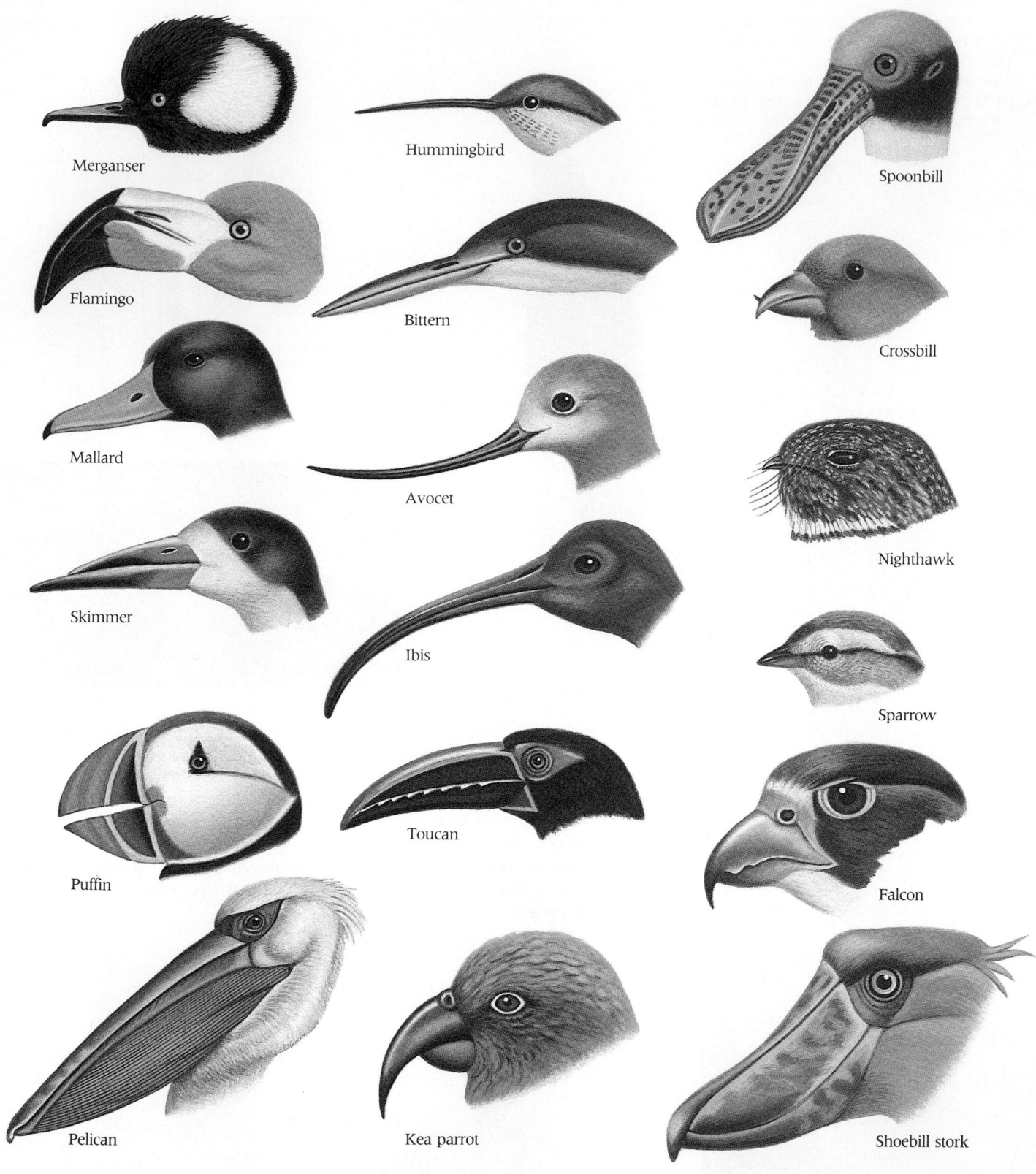

Merganser

Hummingbird

Spoonbill

Flamingo

Bittern

Crossbill

Mallard

Avocet

Skimmer

Nighthawk

Ibis

Sparrow

Puffin

Toucan

Falcon

Pelican

Kea parrot

Shoebill stork

43-17 Diversity in bird beaks

What we do know is that the oldest known fossil bird, *Archaeopteryx* (''ancient wing'')—of which only five of varying quality have been found—was still almost reptilian, except for a few minor features and one major one: feathered wings. Figure 43-16 summarizes the notable differences between *Archaeopteryx* and a modern bird, the pigeon. If there had been no further change, birds as we know them now would never have evolved. We meet *Archaeopteryx* again in Chapter 44.

Along with wings, birds evolved other characteristics suitable for creatures of intense and sustained activity, including keen perception, a capacity for rapid and varied reactions, and high and steady metabolic rates. Along with the mammals, a few reptiles, and fishes, birds are capable of maintaining a constant body tempera-

A

B

C

D

E

F

43-18 A few modern birds
(**A**) *Common puffin.* (**B**) *Tricolored heron.*
(**C**) *Common brown kiwi.* (**D**) *Bald eagle.*
(**E**) *Scarlet macaw.* (**F**) *American robin.*

ture—that is, they are **homeotherms** (see Chapter 26). Feathers assist in this regard. A sparrow, with its close-fitting feather "suit," can maintain a body temperature of 106°F on the coldest winter days. The circulatory system was upgraded by the emergence of a real four-chambered heart. This heart is completely divided into what are essentially two separate hearts, so that there is no mixing of oxygenated and deoxygenated blood (see Chapter 24). This was the final step in the evolution of an efficient circulatory system that can operate properly in relation to lung respiration. The hollow, air-filled bones are admirably adapted for flight and the structure of the sternum ("wishbone") permits the attachment of powerful flight muscles. These muscles, some of which are familiar to us as the breast meat of chicken and turkey, pull the wings downward during the main propulsive movements of flight.

Bird senses—especially vision, equilibrium, and hearing—are notably acute. Their brains are complex and distinctively specialized (see Chapter 30). Their behavior, though in considerable part stereotyped or instinctive, can be very complicated, displaying such phenomena as birdsong (see Box 33-1), nest building, parental care of the young, specific egg-laying patterns,[5] and uncannily precise migratory behavior. Many of these patterns were described in Chapters 32 and 33.

Ancient birds, even some that lived much later than *Archaeopteryx*, retained the simple teeth they inherited from the reptiles. Today we say there is nothing scarcer

[5] There are fascinating differences among birds' eggs and patterns of egg-laying behavior. For example, some birds are determinate layers. If one egg is taken away, the bird does not lay another. Albatrosses lay only one egg per breeding season. Woodcocks lay three or four eggs and no more. Other birds (woodpeckers, sparrows, etc.) are indeterminate layers and keep on laying eggs if eggs are taken from the nest. Fortunately for us, ducks and chickens are in this category. One duck is reported to have laid 363 eggs in 365 days!

than hen's teeth. Indeed, no bird has had teeth for tens of millions of years. Instead, birds have beaks of bone covered with horn. Beaks serve not only to obtain food, but as multi-purpose instruments useful for many things, from knot tying to wood boring. Beaks cannot grind seeds, but birds have an adequate functional substitute, the muscular gizzard, which contains abrasive grit or small pebbles.

In their basic anatomy, birds are remarkably alike. Perhaps that reflects the stringent mechanical demands of flight. But within the limits of the basic stereotype, they are fascinatingly diverse. The many forms of their beaks, most of which are clearly adaptive, illustrate this fact impressively (Figure 43-17). This was clearly demonstrated by Darwin's famous finches (see Chapter 35). The feet, too, have numerous adaptive patterns. And everyone, whether a "birder" or not, knows and enjoys the diversity of bird colors and patterns.

The systematics of recent birds has been worked out in greater detail than that of any other large animal group. Two superorders have been defined on the basis of anatomical details of jaw and palate structure. These are known as Paleognathae ("old jaws") and Neognathae ("new jaws"). Most present-day birds, of which there are some 8600 species, belong to the neognathan superorder. A few birds are illustrated in Figure 43-18. Modern classification recognizes about 27 different orders of living birds. More than 5000 living species (well more than half) are perching birds (order Passeriformes), which includes among many others the kingbirds, larks, swallows, crows, wrens, thrushes, warblers, blackbirds, and sparrows. The large and familiar groups are summarized in Figure 43-19.

Perhaps the most unusual birds (in anatomy and physiology) are those that can no longer fly—like the penguins (see Chapter 1), which still do "fly," but only under water, or the ostriches, of the superorder Paleognathae. Flightless birds are probably descendants of once-flying ancestors. Most are found on islands where they confront relatively few predators.

Like many vertebrates today, various bird species are seriously endangered. Nearly 80 species have become extinct in the last three centuries, and 100 or more are now threatened. However, most of the birds that have recently become extinct or are on the edge of extinction have been harmed by human activities. Some, such as the passenger pigeon, were hunted to extinction. Others, such as the whooping crane and the ivory-billed woodpecker, have had most of their habitats destroyed, and are drastically reduced in numbers. A few, such as the brown pelican, the bald eagle, and the osprey, are vanishing because insecticides like DDT have become concentrated in the tissues of the fish they eat. Living with humans remains the birds' greatest problem.

Class Mammalia: Mammals

We come, finally, to the most highly developed vertebrate class, the mammals. Like birds, they arose from the reptiles, but not as a result of the appearance of a new key characteristic like flight. Rather, they emerged following long-continued and gradual changes of many kinds. Two of the innovations that distinguished this class were **milk** for nourishing the young and **hair** for conserving body heat.

Mammals, named for **mammae** or **mammary glands,** the milk-secreting organs of females, now occupy much of the range that was once the domain of the reptiles—even though they went much further than the reptiles, a few of which still live successfully with the mammals. Thus, the mode of mammalian evolution contrasted with that of the birds, which exploited a new realm of life in consequence of a new key adaptation.

The line between mammallike reptiles and mammals cannot be drawn precisely. Nonetheless, we can say that six new characters evolved in the lineage now recognized as mammals and that all of these contributed significantly to their success (Figure 43-20).

■ Endothermy (warm-bloodedness), as occurs in birds, along with external insulation using hair or fur rather than feathers, usually with a layer of subcutaneous fat, was an important evolutionary development. Both characters appeared independently in the two classes and were related to the need for higher rates of metabolism and superior metabolic regulation, whatever the outside temperature.

43-19 The major bird orders

Rheiformes — Rhea
Struthioformes — Ostrich
Gaviiformes — Loon
Procellariformes — Albatross
Sphenisciformes — Penguin
Pelicaniformes — Pelican
Ciconiiformes — Stork
Anseriformes — Swan
Falconiformes — Eagle
Galliformes — Chicken
Charadriiformes — Gull
Columbiformes — Pigeon
Psittaciformes — Parrot
Cuculiformes — Roadrunner
Strigiformes — Owl
Apodiformes — Swift
Trochiliformes — Hummingbird
Coraciformes — Kingfisher
Piciformes — Woodpecker
Passeriformes — Tanager

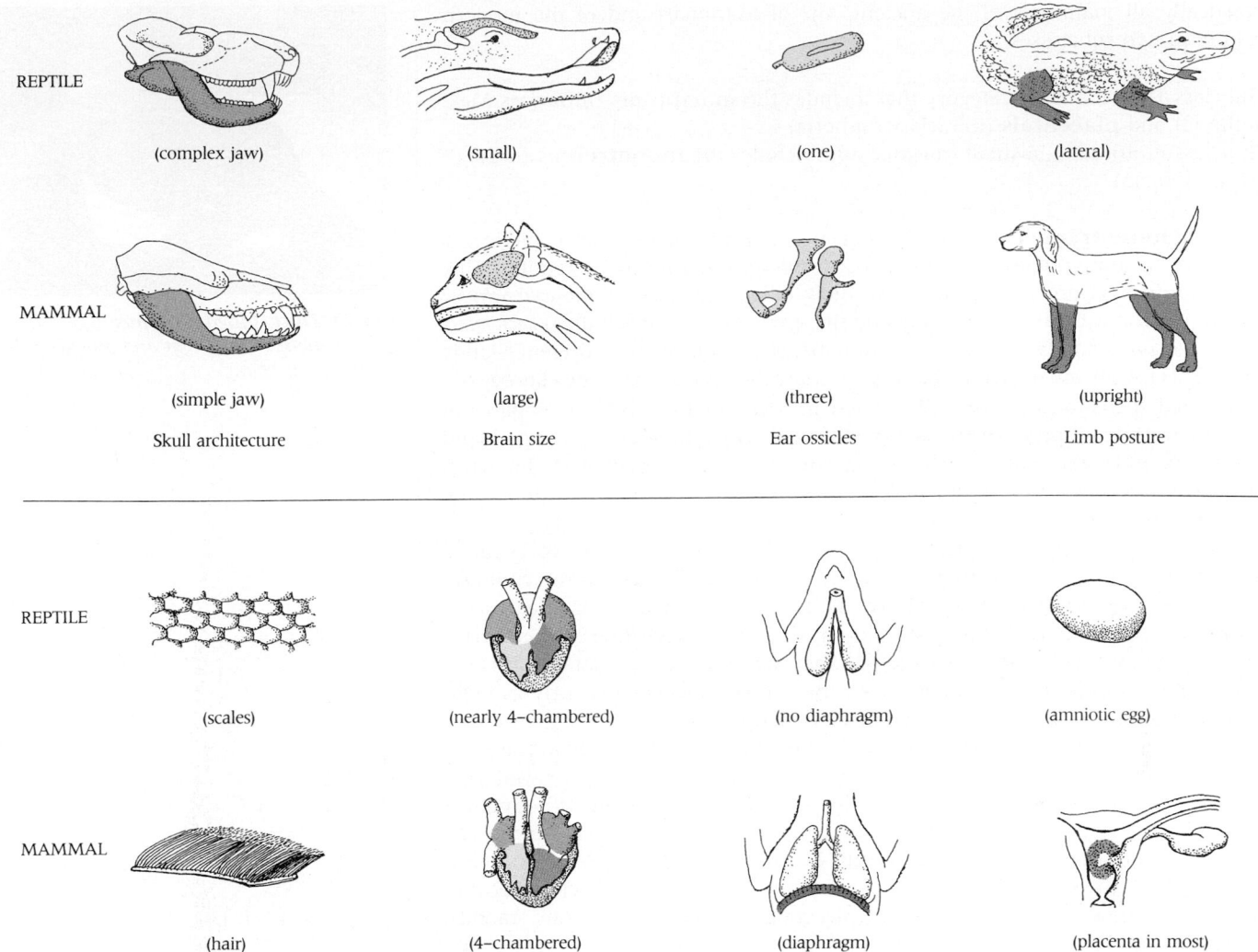

REPTILE	(complex jaw)	(small)	(one)	(lateral)
MAMMAL	(simple jaw)	(large)	(three)	(upright)
	Skull architecture	Brain size	Ear ossicles	Limb posture

REPTILE	(scales)	(nearly 4–chambered)	(no diaphragm)	(amniotic egg)
MAMMAL	(hair)	(4–chambered)	(diaphragm)	(placenta in most)
	Skin	Heart	Respiration	Reproduction

43-20 Comparison of the key features of a reptile and a mammal
Reptiles have a multiple-boned jaw, small brain case, single ear ossicle, lateral limbs, scaly skin, functionally four-chambered heart, no diaphragm, and amniotic eggs with no care of the young. Mammals have a single-boned jaw, large brain case, three ear ossicles, upright posture, hair, a true four-chambered heart, diaphragm, placental reproduction (in placental mammals), and parental care of the young.

■ Skull architecture came to include complex differentiated teeth and a new joint between the lower jaw and skull that facilitated chewing. Together these changes made food utilization more efficient and led to remarkable diversity in mammalian feeding patterns. Reptiles (and indeed all nonmammalian vertebrates) have several different bones in the lower jaw (mandible). Mammals have only one. The changeover also improved hearing. As discussed in Box 35-1, some reptilian bones were incorporated into the ear in mammals.

■ Limbs are more upright and more beneath the body, which thus is carried higher off the ground. Changes in the limbs (including changes in the joints, and in the manner of long bone growth) led to faster and more efficient locomotion, essential in both chase and flight.

■ Respiration was improved by three major anatomical changes: (1) increased, though still incomplete, separation (by the palate) of the respiratory and alimentary passages, which made it possible to breathe while eating; (2) complete separation of the chambers of the heart; and (3) the acquisition of a diaphragm, which led to more efficient ventilation of the lung.

■ Increased sustenance and protection of the young, both before and after birth, led to greater reproductive efficiency. Fertilization occurs internally and embryos develop within the mother's body—that is, mammals are viviparous. In most mammals, nourishment of the embryo occurs through a placenta. The exceptions (monotremes and marsupials) will be discussed shortly. After birth, offspring continue to receive parental care and a diet of mother's milk.

■ Changes in behavior included greater capacity for individual modification by experience and wider reaction ranges. Both reflect the increase in relative brain size, especially of the cerebral cortex.

Practically all mammals of the ancient Age of Mammals and of the modern world are in two subclasses.

- Subclass Theria, a large category that includes the **marsupials** (infraclass Metatheria) and **placentals** (infraclass Eutheria)
- Subclass Prototheria, a small category that includes the **monotremes** (order Monotremata)

The term **monotreme** ("one-hole") indicates that the anus and urogenital tract do not have separate openings. In this regard, monotremes resemble reptiles. This curious group of mammals includes the duck-billed platypus (*Ornithorhynchus*), found only on the Australian mainland, and the spiny anteater or echidna (*Tachyglossus*), which occurs mainly in Australia, Tasmania, New Guinea, and adjacent islands (Figure 43-21). They are intriguing because they are not viviparous—they lay eggs— yet they feed their young with milk, which trickles not from typical nipples but from numerous small pores in the abdominal wall. Thus, they are a strange blend of characters otherwise reptilian, characters otherwise mammalian, and characters unique to themselves. They represent a line that branched off from the main line of mammalian ancestry very long ago.

Marsupials were once considered to be an older group that was ancestral to the placentals. That now seems unlikely. Rather, they simply diverged from a common ancestry. The most striking of the many of the peculiarities of marsupials is the absence of a placenta. In many cases, developing embryos receive little or no nourishment from the mother while in her uterus. They stay there only briefly, subsisting on an ample egg yolk and emerging at a very immature stage—as early as eight days after fertilization. They then crawl into a pouch (marsupium) on the mother's belly, hang onto a nipple, and remain there for the several weeks needed to complete their development. (Although marsupials are named for this pouch, it is not very obvious in many.) The marsupials of Australia—the "land of marsupials"—include kangaroos, koalas, wombats, brushtails, and other exotic forms (Figure 43-22). Marsupials, mostly opossums, are increasingly abundant in North and South America. Unfortunately, many species in Australia are now threatened with extinction because of the introduction into that country of placental competitors like rats, rabbits, foxes, sheep, and cattle.

In placentals, the fetus is nourished, oxygenated, and cleared of wastes by a placenta, a flat organ derived from the embryonic membranes and certain uterine tissues (see Chapter 21). The placenta attaches to the uterus and connects to the fetus via an umbilical cord. This arrangement allows the young to reach an advanced

43-21 Duck-billed platypus
The monotremes lay eggs and provide milk to nourish the young.

43-22 Marsupial mammals
Marsupial mammals nourish the young with milk provided through nipples inside a pouch in the abdominal region. **(A)** *Koala.* **(B)** *Kangaroo.*

A

B

43-23 Modern placental mammals
(**A**) *Blue whale.* (**B**) *Musk ox.* (**C**) *Bengal tiger.*
(**D**) *Elephants.* (**E**) *Timber wolves.* (**F**) *Zebras.*

stage before birth. Some, like mice, are blind, hairless, and quite immature at birth. Others, like colts that walk minutes after their birth, are remarkably well developed. Although not as diverse as they were a few million years ago, placentals are still dominant vertebrates on land and have been throughout the Age of Mammals (Figure 43-23).

A modern classification divides placental mammals into 16 living orders and 13 extinct orders, more or less. The exact number is uncertain. The living orders, which include some 4500 species, are summarized in Figure 43-24.[6] Note that placentals burrow, fly, climb, run, and swim. They eat worms, fruit, grass, seaweed, insects, squids, crustaceans, bark, and each other. They live in the open sea, in treetops, on polar icecaps, in apartment houses, and in deserts.

The outstanding placentals in terms of numbers—both of species and of individuals—are the rodents. Rats, mice, squirrels, and other rodents swarm practically everywhere. Some, like the beaver, have an amphibious life style. In only one respect do they fall short of being the dominant mammals and the climax of vertebrate evolution: they are not as smart as human beings.

Intelligence is the reason we have named the order to which humans belong the **Primata** (Latin, "the first")—not that all primates are notably intelligent. Some living primates are below the average intelligence for mammals. So were the oldest primates (judging from the brain form known for a few of them). At first, there was little to distinguish primates beyond the use of the forefeet as hands and an increasing coordination of visual perception and manual response. Yet somehow

---

[6] We should remember that, as in marsupials, the name of a systematic group does not need to be, and often is not, appropriate for all its members. Some so-called insectivores do not eat insects. Most edentates (notably the armadillos) have teeth; some carnivores do not eat meat; and so on.

these primitive creatures had the potential to evolve the highest intelligence ever reached by any organisms.

The earliest primates lived everywhere except Australia and South America. They were the **prosimians** ("pre-monkeys"). Some prosimians (lemurs, bush babies, aye-ayes, tarsiers, lorises, etc.) still survive in modified form in Africa, Madagascar, and southeastern Asia (Figure 43-25). Three other major groups evolved from early prosimians: the ceboids (New World monkeys), the cercopithicoids (Old World monkeys), and the hominoids.

The New World monkeys and marmosets (superfamily Ceboidea) are confined to Central and South America (Figure 43-26). The vernacular name "ceboid" means "cebuslike." *Cebus* is the genus of common and well-known South American capuchin monkeys. The Old World monkeys, macaques, and baboons (superfamily Cercopithecoidea) live throughout the warmer parts of the Eastern Hemisphere (except Australia). Cercopithecoids, the name for the whole group of Old World monkeys, derives from *Cercopithecus*, the generic name of some common African monkeys. Rhesus monkeys (a variety of macaque much used in experimentation), mandrills, and many others belong to this group. There are several major differences between ceboids (New World monkeys) and cercopithecoids (Old World monkeys).

- Ceboids have three premolar teeth on each side of the upper and lower jaws; cercopithecoids (and hominoids with luck) all have two.
- Some ceboids have a prehensile tail that they use for grasping; cercopithecoids lack a prehensile tail.
- In ceboids, nostrils point laterally and forward; in cercopithecoids they point down.
- Ceboids lack the brightly colored buttock areas so prominent in Old World monkeys.

Originating somewhere in the Old World and soon spreading throughout it was the third and climactic group of the primates, the hominoids (superfamily Hominoidea, "human-like"). Formerly much more diverse, this group now includes the apes (orang-utans, gibbons, chimpanzees, and gorillas) and human beings.

CHARACTERISTICS OF THE ANIMAL PHYLA

ADAPTIVE RADIATION VERSUS THE LADDER OF LIFE

The ancient Greeks started the idea that organisms form a "ladder of life," with simplest organisms at the bottom and all the others in a sequence that culminates with human beings at the top. We learned in Chapter 2 that orthodox biological theory once envisioned a "ladder" of evolutionary progress, wherein natural selection gradually and progressively produced improved descendants.

It has also been argued that a listing of organisms from simple to complex is a representation of the actual course of evolution. That non sequitur has led to serious error. Even today, the old and mistaken notion of a ladder of life underlies much biological thinking and teaching. Some still view the animal phyla as an "evolutionary series"—with some dead ends and side branches, to be sure, but on the whole one major sequence. In this spirit, students are often asked to study the "evolution of the vertebrates" by dissecting a dogfish, a frog, and a cat.

The truth is not so simple. The dogfish, frog, and cat all live today, but no one of them is likely to be ancestral to any other one. Mammals (including cats) did arise by way of reptiles from amphibians (frogs are amphibians), and amphibians arose from fishes (including dogfishes). However, the fossil record clearly shows that a modern frog is unlike the amphibian ancestor of a mammal, and a dogfish is unlike the fish ancestor of an amphibian. In both cases, the ancestral forms were radically different.

The significance, then, of the differences between these animals is that each one has a decidedly different way of life from the others—and these differences are not usefully characterized as lower or higher than those of the others. (The term "progressive" is often used to refer to these features.) These differences plainly reflect adaptations to different ways of life in different environments. The history of how these adaptational differences arose is another question. Putting it another way, a nematode is as satisfactorily adapted to its environment as a human being

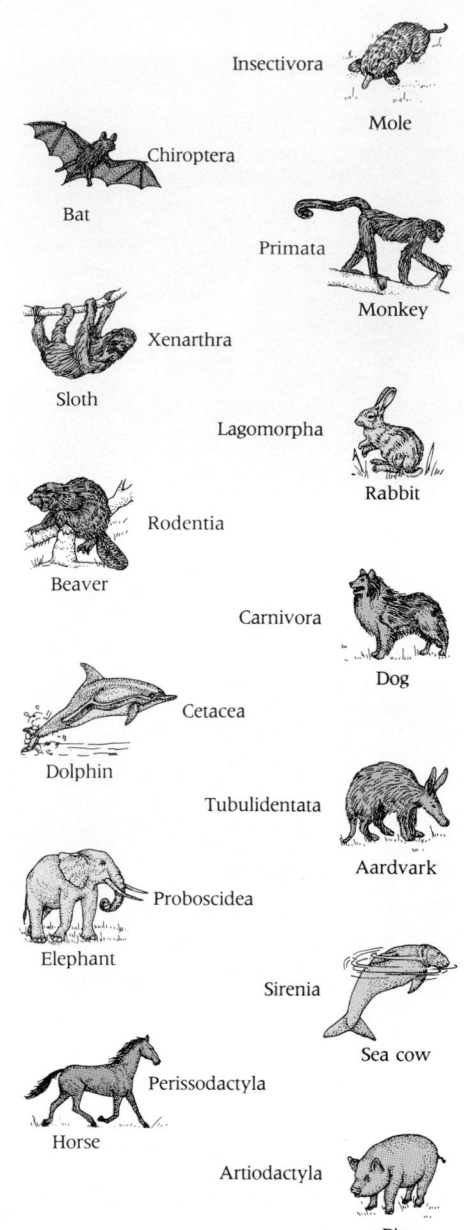

43-24 *Major orders of placental mammals*

43-25 **A prosimian primate**
The tarsier, Tarsius tarsier, *is the only surviving member of an ancient and primitive group of prosimian primates.*

43-26 New World monkeys
(A) *The capuchin monkey,* Cebus apella, *of South American forests.* (B) *The mandrill,* Mandrillas sphinx, *is an African monkey that lives on the ground and walks four-footed.*

is—and a nematode (or a bacterium, for that matter) has many capabilities that we lack. Nonetheless, a human is clearly more complex, more highly differentiated, and, in the sense we mean it, more progressive than a nematode.

The same considerations arise when we look at the differences between the animal phyla. Some appear simpler than others. but the point is that the various phyla are different. The differences arose as adaptations. They are related to the different ways of life of the early members of each phylum.

In reality, the fossil record gives no support to an idealized "scala naturae"—as we pointed out back in Chapter 2. It is true that the earliest creatures were very simple. The forms found in rocks nearly 3.4 billion years old were bacteria and blue-green algae—and for more than 2 billion years life remained at the prokaryotic level. Then, in a relatively short time span, virtually all of the major designs of multicellular life made their appearance at the beginning of the Cambrian about 600 million years ago. The evidence suggests, then, that complexity did not evolve sequentially by slow and steady addition of structures and functions.

Organisms evolve as they become better adapted in their local environments. But the trends that result do not necessarily reflect a general movement toward perfection of design. For each trend toward group complexity we can doubtless find a commensurate regression in another group—such as the tissue losses attending the evolution of highly adapted parasites. To repeat, improvements occur in evolution when organisms become better adapted to local environments.

Clearly, animal phyla do not form a single ladder or even several ladders in which one phylum after another arose in succession. A better model is a branching tree, the limbs of which are phyla originating from various stems and ultimately from a common trunk (see Figures 37-16, 43-3, and 44-9). The end branches represent taxa that arose by invasions of new environments. Variations upon basic body plans offer great potential for diversity. When a new body plan or design emerges, it often precipitates the vigorous phase of diversification we call adaptive radiation. As we saw in Chapter 36, adaptive radiations occur when organisms enter unoccupied environments or when they exploit new modes of life—such as occurred when insects and birds took to the air.

Only two points remain of the old idea of evolutionary sequence among the phyla. First, some phyla probably did arise from early members of other phyla (which in all are cases are unidentified), from which they inherited some of their complexities. Second, some phyla have changed more radically than others since they first arose.

BASIC RELATIONSHIPS AMONG THE ANIMAL PHYLA

For a broader understanding of animals and their diversity, we now seek an overall view of differences and resemblances among the phyla. Then in the following chapter, we will present a coherent historical view of the origins of these major animal groups.

Contrary to what we might have supposed, most (and perhaps all) of the animal phyla arose in the sea.[7] Most phyla include both marine and freshwater species. (The exclusively marine nonparasitic phyla include Graptolithina, Ctenophora, Brachiopoda, Phoronidea, Chaetognatha, and Echinodermata.) Both salt water and fresh water are aquatic environments, but they may require radically different physiological adaptations. Several groups, notably the nematodes and annelids, are abundant in damp soil, an environment resembling fresh water in its demands.

Life on land requires four major adaptations: (1) the ability to breathe air effectively; (2) the ability to overcome gravity during locomotion; (3) a means of preventing desiccation; and (4) a reproductive system that can function outside of an aquatic habitat.

Only three phyla include strictly terrestrial animals that have successfully achieved these capacities and thus are capable of carrying on all of their life activities on land: Mollusca (the fully terrestrial forms are the land snails); Arthropoda (insects, spiders, and some others); and Chordata (reptiles, birds, and mammals). These

---

[7] Two minor phyla (Mesozoa and Acanthocephala) are exclusively parasitic and of uncertain origin. Both include parasites of marine animals, and they may well have originated in the sea, but we are not certain. Flatworms (Platyhelminthes) and aschelminths (especially nematodes and rotifers) are more common in fresh water of damp soil than in marine waters but do include marine species. All the other phyla seem clearly to be of marine origin.

phyla also include many marine and freshwater animals. Although evidently of marine origin, some of the evolutionary lines of these phyla became adapted to the land, where life is so difficult that few phyla have been able to cope with it. Yet when these groups moved onto land no major changes occurred and no new phyla emerged. Significantly, the three phyla that did conquer the land were also extremely successful in the water.

The basic differences among phyla include anatomical features concerned with such processes as nutrition, internal transportation and maintenance, organ differentiation, coordination, and locomotion. They add up to a distinct overall pattern for each phylum, which is summarized in Table 42-1 and in Figure 43-27.

Tissue Layers

In Porifera, there is limited tissue differentiation in embryo or adult. In Coelenterata, there are only two distinct tissue layers (endoderm and ectoderm). The other phyla have three layers (endoderm, mesoderm, ectoderm) that differentiate into tissues and organs during development.

Digestive Systems

The Porifera have no special digestive organs. As we have seen, microscopic food particles are carried along on currents set up by flagellated cells, and these are engulfed and digested within the cells. The Coelenterata have a central cavity with only one opening, into which food is conveyed by the tentacles. Digestion is partly extracellular in the cavity (stomachlike) and partly intracellular within cells lining it (spongelike). The digestive system of the Platyhelminthes also has only one opening, and digestion is also partly in the cavity and partly in its lining cells. The cavity is, however, much more complex than in Coelenterata, with intricate branching throughout the body. Digested products thus are available near all other tissues in spite of the absence of a special transport system. The other major phyla have a digestive tube with openings at either end for food ingestion and waste elimination. The tube may be differentiated into regions or accessory organs.

Coelom

Porifera, Coelenterata, and Platyhelminthes have no cavity between the digestive cavity and the body wall. We call them acoelomates. In Aschelminthes, a pseudocoelomate, there is such a body cavity, but it does not have mesodermal lining and so is deemed a pseudocoel (or "false coelom") rather than a true coelom. The other major phyla have a true coelom with a cellular (mesodermal) lining.

Skeletal System

Most multicellular animals have evolved skeletal systems to support their otherwise flabby tissues, to facilitate locomotion, and in some cases to provide protection. As we have seen, many groups (coelenterates, bryozoans, brachiopods, mollusks, etc.) have an external skeleton that is a calcareous cup or "house." The exoskeletons of arthropods are elaborately jointed to permit locomotion. Echinoderm skeletons are developed from internal tissue, but are functionally external and calcareous with no real resemblance to bone. Vertebrate skeletons are internal and bony or cartilagenous.

Circulatory System

Porifera, Coelenterata, Platyhelminthes, Aschelminthes, and Bryozoa have no special circulatory systems. In them, cell layers must be thin (usually only one or two cells thick) and in contact with fluids or with the digestive cells. Brachiopoda, Mollusca, Annelida, Arthropoda, Echinodermata, and Chordata have vascular circulatory systems; these became necessary as cell layers and masses became thicker and complex.

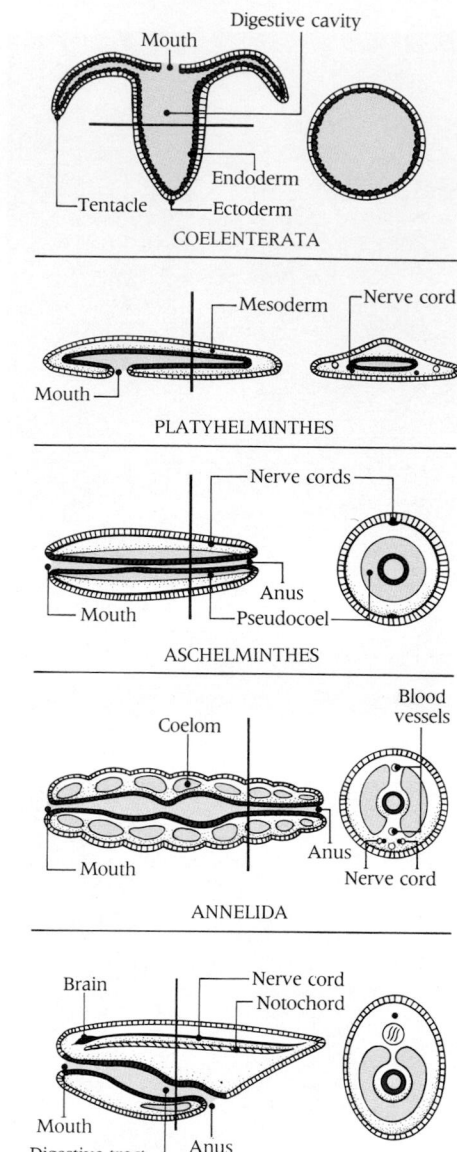

43-27 Characteristics of some animal phyla
The cross sections on the right side of each panel are taken at the level identified by the black line traversing the figure on the left.

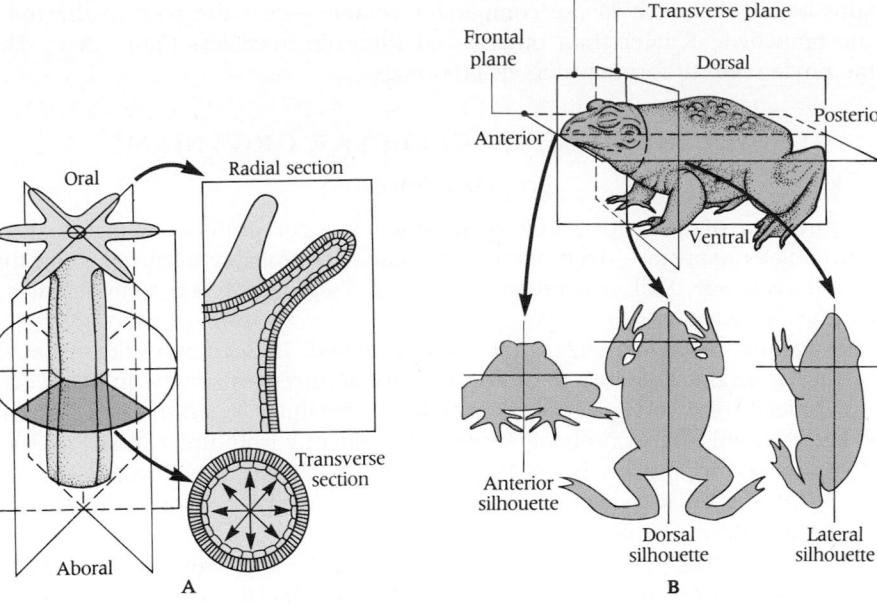

43-28 Animal symmetry
(A) *Radial symmetry, exemplified by the coelenterate polyp* Hydra. *The transverse section shows perfect symmetry in all radial directions.*
(B) *Bilateral symmetry, exemplified by a frog. Both the anterior and the dorsal silhouettes are bilaterally symmetrical about the median plane but not about the dorsal plane. The lateral silhouette is not symmetrical about any plane. Also see Figure 19-35.*

Segmentation

Three phyla, Annelida, Arthropoda, and Chordata, are characterized by some occurrence of successive body segments during development. In annelids many of the segments are similar, with repetition of the organs within them. Arthropod segmentation early became different from the annelid pattern by the specialization or reduction of segments and decreased repetition of organs. Chordate segmentation arose independently and was primarily related to locomotion. It is most evident in certain muscles, nerves, and bones (the vertebrae). It is obscure or absent elsewhere.

Symmetry

Most animals are more or less symmetrical, at least in their external form. Among those most obviously symmetrical, some are radially symmetrical around a central axis, like the pattern of a spoked wheel (Figure 43-28). Among the major phyla, the Porifera and Coelenterata tend to be radially symmetrical. The Echinodermata also usually have radial symmetry, but their larvae are bilateral, and traces of bilateral symmetry are present in adults. The other animal phyla are bilaterally symmetrical: the two halves are more or less mirror images of each other on each side of a central plane. Animals that seem to have other kinds of symmetry (or none at all) often are found to have radial or bilateral symmetry when their development and ancestry are examined. Coiled snails, for example, are not symmetrical in adult form, but their ancestors were bilaterally symmetrical and modern snails still are in the early stages of their development. Coiling arises from torsion resulting from the unequal growth of the two originally symmetrical sides.

ORIGINS AND RELATIONSHIPS OF THE ANIMAL PHYLA

Explaining the origins and relationships of the phyla is one of the most interesting and difficult challenges of biology. In illuminating their origins—what we have called macroevolution—fossils help a little, but not much. (On the other hand, the fossil record often documents the evolution of classes and other groups within phyla, as we will see in Chapter 44.) The earliest members of each phylum were extremely ancient, small, soft-bodied animals that have not in any case been preserved as fossils. At least, no paleontologist has yet found them. The earliest known fossil representatives of each phylum are already very different from those of any other phylum. Nonetheless, they are more primitive than any living representatives.

We have compared the animal phyla to the branches of a tree. Although the origins of all of the phyla are comparably remote—near the base of the tree—some branched off later than others, and some diverged less than others. Thus some phyla seem to form clusters or groupings.

ORIGINS OF MULTICELLULAR ORGANISMS

Protistan Ancestors

The transition of unicellular eukaryotic organisms, or protists, to metazoans or multicellular organisms—true plants and animals—probably occurred more than a billion years ago. Still, it is probable that today's protists retain some features of their ancient ancestors.

The earliest metazoans have not been identified. It is entirely likely that the great step from unicellularity to multicellularity occurred separately on many occasions. Indeed, one survey listed 17 simple multicellular organisms—autotrophic and heterotrophic—that probably arose independently from unicellular Protista. If this is true, and it probably is, the evolutionary advantages of multicellularity must have been very great indeed, since it emerged in so many different habitats and among such diverse modes of life.

It is usually said that the ancestral type for the metazoans was a flagellated protozoan. We noted in Chapter 39 that modern zooflagellates (phylum Zoomastigina) resemble the ancestors of both the multicellular organisms (plants and animals) and the other animallike protists, notably the sarcodines (amoebas and their kin) and the ciliates (parameciums and their kin).

Flagellates, surely, were not too specialized to have given rise to multicellular organisms. In fact, they seem to have most of the properties we would expect in that important immediate ancestor. Their peculiar method of locomotion is found at least in the male gametes of almost all animals and the more primitive plants. The flagellates, of course, include the phytoflagellates—members of subkingdom Algae that are photosynthetic autotrophs and hence physiologically "plants"—and the zooflagellates of subkingdom Protozoa that are nonphotosynthetic heterotrophs and hence physiologically "animals." Finally, a number of flagellates form colonies—and individuals in those colonies may be somewhat differentiated. This suggests a foreshadowing of multicellular organization in which individual cells have undergone differentiation and specialization.

An animal of protist size is largely at the mercy of waves and currents and cannot actively seek its food or a suitable environment the way a larger multicellular animal does. Hence, there must have been selection pressure toward greater size and complexity that would permit better sensory and behavioral adaptive advantages. These adaptations and others were achieved by the evolution of a multicellular pattern of organization.

One of the evolutionary steps that led to multicellular animals probably occurred when sponges arose. Sponges (phylum Porifera) are one of the three living multicellular animal groups with notably primitive features. The other two are jellyfishes and corals (phylum Coelenterata) and comb jellies (phylum Ctenophora). Sponges

43-29 The origin of metazoans
Three hypotheses have been offered to account for the evolutionary origin of metazoans.
(A) *Haeckel's blastaea-gastraea hypothesis.*
(B) *The planuloid hypothesis.* **(C)** *The ciliate hypothesis.*

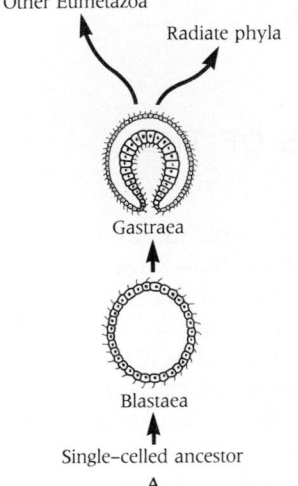

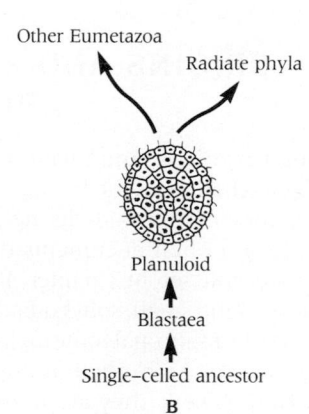

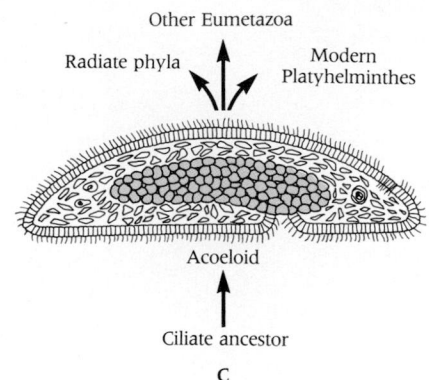

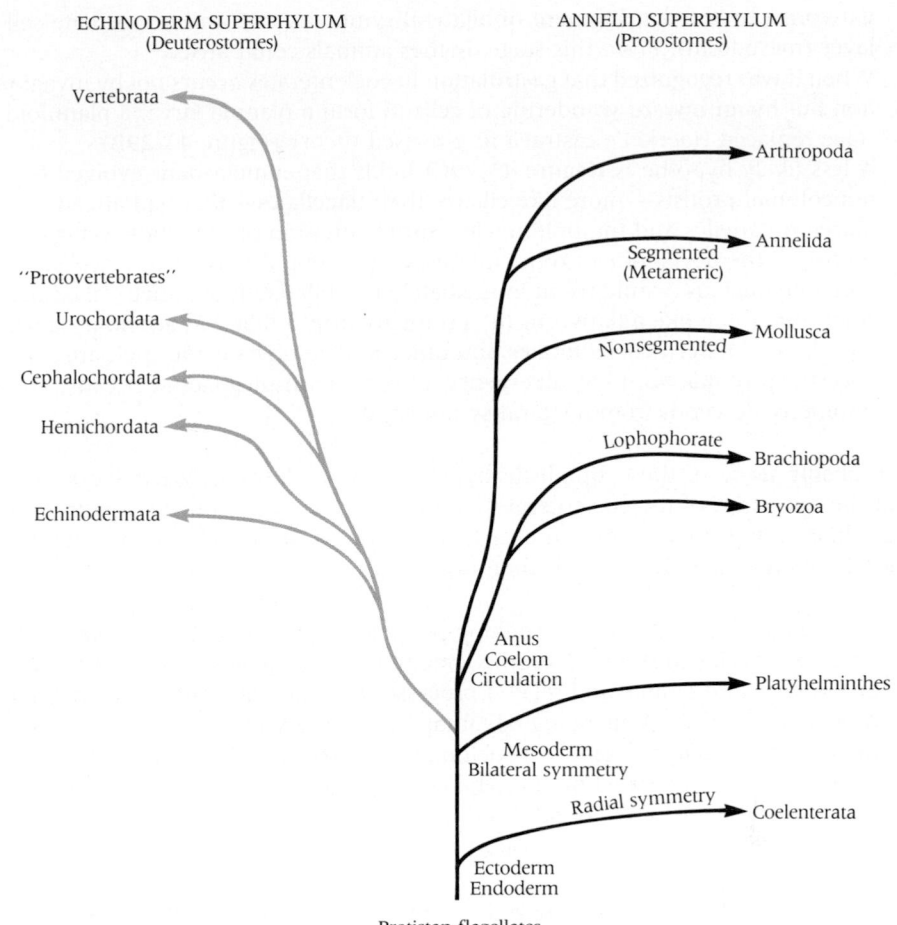

43-30 Phylogeny of animals
This phylogenetic tree is based on the morphology of adults and embryos. Note the two major superphyla. Based on embryological features, some would separate Brachiopoda from both protostomes and deuterostomes and place them in a separate group called Lophophorata. Mollusks are here seen as originating as acoelomates from flatworm ancestors. Some think they arose as true coelomates.

ECHINODERM SUPERPHYLUM
(Deuterostomes)

ANNELID SUPERPHYLUM
(Protostomes)

Vertebrata

Arthropoda

Segmented
(Metameric)

Annelida

"Protovertebrates"

Urochordata

Nonsegmented

Mollusca

Cephalochordata

Hemichordata

Lophophorate

Brachiopoda

Bryozoa

Echinodermata

Anus
Coelom
Circulation

Platyhelminthes

Mesoderm
Bilateral symmetry

Radial symmetry

Coelenterata

Ectoderm
Endoderm

Protistan flagellates

may have had a separate origin from all other animals. For that reason, they are sometimes placed in a subkingdom of their own (Parazoa), with the other multicellular animals gathered into subkingdom Eumetazoa (Metazoa in older classifications). It is even possible that sponges evolved long after eumetazoans, which appear in the fossil record some 100 million years earlier. Sponges almost certainly arose from flagellate colonies similar to those of the collar flagellates, and they still contain collar cells that are practically identical with them. Although successful at their own level, the sponges were in an evolutionary dead end.

Origin of the Eumetazoans

The ancestry of the multicellular animals, the **Eumetazoa,** was not part of the sponges' blind alley. The most crucial feature of that development was probably the differentiation of special contractile cells that eventually became muscles and special transmission cells that eventually became nerves. Of the primitive protists, some conservative lines changed very little. Some became other kinds of protists, such as sarcodines and ciliates. Several became plants. One became sponges. One—happily from our point of view—took a step in the direction of animal life.

The details are obscure and the time is unknown. There are three main hypotheses (Figure 43-29).

■ One hypothesis visualizes colonial integration (Figure 43-30A). This idea, proposed in 1874 by Ernst Haeckel, holds that metazoans evolved from colonial flagellates, as did sponges, but with more differentiation of individuals in the colony. Perhaps the original colony was a hollow globe like *Volvox* that resembled the blastula stage of embryonic development and thus was termed a *blastaea*. If differential growth pushed in one side so that it came in contact with the other side, the result would be a pouchlike, radially symmetrical body with two layers of cells in its wall, much like the gastrula stage of embryonic development. This pattern, termed a *gastraea*, superficially resembles that of the hydrozoan coelenterates. The primitive coelenterates may then have given rise to ancestral flatworms by the development of bilateral symmetry and an intermediate cell layer (mesoderm). From this stage, higher animals could arise.

- When it was recognized that gastrulation in coelenterates occurs not by invagination but by an inward wandering of cells to form a planula larva, a planuloid stage replaced Haeckel's gastraea in a revised theory (Figure 43-29B).
- A less likely hypothesis (Figure 43-29C) holds that eumetazoans evolved from noncolonial protists—more like ciliates than flagellates—that had already acquired organelles and multiple nuclei. Simple division of the whole syncytial protist by the development of membranes separating it into parts, or cells, each with one nucleus, would result immediately in a bilaterally symmetrical eumetazoan something like a flatworm (a "protoflatworm") that was acoeloid (lacking a coelom). Modern coelenterates and other radiate phyla could then arise from ancestral protoflatworms by divergent evolution (partly degenerate because radial symmetry descends from bilateral symmetry).

Probably none of these simplistic hypotheses is quite right. What we know is that the ancestries of the coelenterates and of the flatworms branched off from the main line of metazoan evolution at a very early stage in history. Within the main line, three important innovations then occurred.

- A circulatory system evolved, making possible greater bulk and complexity.
- An anus developed so that food movement in the alimentary canal began going one way, and digestive and related processes became regionally differentiated.
- A coelom evolved. Of its many advantages, a primary one was the ability of the coelomic cavity in primitive coelomates to serve as a hydrostatic skeleton allowing forces generated by the whole body wall to be focused in a small area. Also, the coelom permitted the free and lubricated motion of internal organs suspended in it, the transport of excretory products, and so on.

Having acquired these three new features, the Eumetazoa seem soon to have split into two main branches—the **protostomes** (also termed the annelid superphylum) and the **deuterostomes** (also termed the echinoderm superphylum). In this view, all higher phyla eventually arose within these two groupings, sometimes called *series* (Figure 43-30).[8]

THE TWO EUMETAZOAN SUPERPHYLA

Protostomes: The Annelid Superphylum

As shown in Table 42-1, four of the major animal phyla—Ectoprocta, Brachiopoda, Mollusca, and Annelida—have larvae called **trochophores** (Figure 43-31A). Despite differences in the different phyla, these larvae have certain fundamental resemblances. Every one is free living and bilaterally symmetrical, with a complete, regionally differentiated digestive tube and a ring of beating hairs or cilia situated in front of the mouth. The ring looks rather like a wheel, which explains the name trochophore ("wheel-bearer"). Other similarities in early development also suggest that the ancestors of these phyla all developed in the same way. For example, we will recall from Chapter 19 the importance of the blastopore in early development. As shown in Figure 19-37A, the blastopore eventually becomes the adult mouth in protosomes. (In deuterosomes, it becomes the anus.)

Hence, though the evidence is not conclusive, it seems very likely that the four phyla with trochophore larvae arose from the same ancient common ancestor, which also had trochophore larvae. This is far from telling us what the ancestor looked like. The adult, however, must have incorporated various trochophorelike anatomical features.

Arthropods do not have trochophore larvae. However, other evidence leaves no doubt that arthropods and annelids had a common ancestry. As we have seen, there is evidence that arthropods evolved from annelids.

---

[8] It is of interest that recent studies of animal phylogeny based on the sequence of 18S rRNA generally support this scheme, which is based on the morphology of adults and embryos. However, the 18S rRNA data suggest certain interpretive differences. For example, this scheme gives Coelenterata a separate protistan ancestry from that of Bilateria (bilaterally symmetrical animals).

Some flatworms (Platyhelminthes) have larvae that resemble trochophores except that their digestive systems lack an anus. That is also true of adult flatworms. They also lack a coelom, but there is a mass of mesoderm that arises much as it does in annelids. The consensus, therefore, is that flatworms probably are related to the phyla with trochophore larvae.

This group of phyla comprises the protostomes or Protostomia, which have also been called the annelid superphylum. A **superphylum** is a major branch of the phylogenetic tree. The branches of a superphylum are individual phyla. In the context of this discussion, these branches are related phyla. The striking peculiarities of the flatworms—for example, the absence of an anus and coelom—are probably explained by the branching off of their ancestors before these characteristics had evolved.

In sum, the annelid superphylum includes at least the major phyla: Platyhelminthes, Ectoprocta, Brachiopoda, Mollusca, Annelida, Arthropoda, and probably some of the lesser phyla. It is reasonable to imagine an ancestral protostome group that split or radiated into at least four basically different branches, as shown in Figure 43-30.

- One evolved into flatworms (and other acoelomates).
- Another split again and evolved into ectoprocts (bryozoans) and brachiopods.
- Another evolved into mollusks and annelids, the most primitive of which split into two groups—one evolving into later annelids and one, after more radical changes, into arthropods.

Deuterostomes: The Echinoderm Superphylum

The fossil record leaves no doubt as to the origins and relationships of the various classes within the subphylum Vertebrata (as we will demonstrate in Chapter 44). But fossils fail to clarify the remoter origin of the first vertebrates. The oldest known fossil vertebrates are extremely primitive and without suggestive resemblances to any phylum other than their own, Chordata.

We have seen that besides the Vertebrata, three subphyla of Chordata survive: Hemichordata, Urochordata, and Cephalochordata. These subphyla lack vertebrae and other features of vertebrates, but they seem surely to have had common origin with the vertebrates. They must have branched off from their vertebrate ancestry

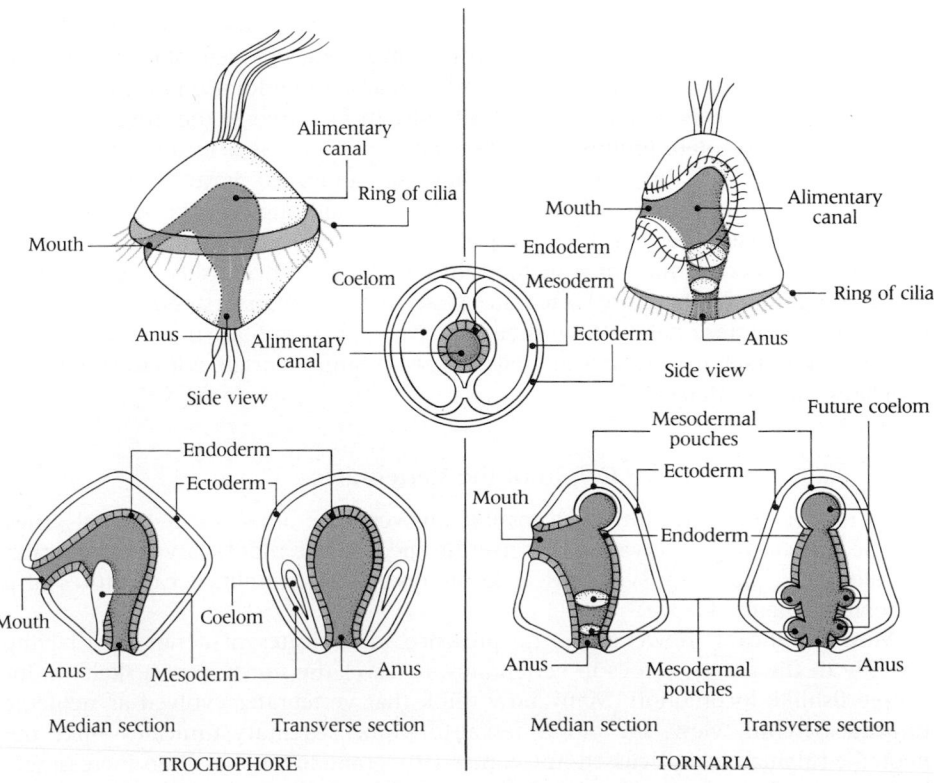

43-31 Larval forms and the origins of mesoderm and coelom
(A) The annelid superphylum is characterized by a trochophore larva. A side view, median section, and transverse section illustrate the relationship between tissues. The coelom arises as a split in the mesoderm. (B) A tornaria larva is found in members of the echinoderm superphylum. The mesoderm arises as pouches from the wall of the alimentary canal, and the cavities of the pouches give rise to the coelom. In the center is a generalized transverse section of both superphyla.

Alimentary canal
Ring of cilia
Mouth
Coelom
Anus
Alimentary canal
Side view
Endoderm
Mesoderm
Ectoderm

Mouth
Alimentary canal
Endoderm
Mesoderm
Ring of cilia
Ectoderm
Anus
Side view

Endoderm
Ectoderm
Mouth
Coelom
Anus
Mesoderm
Median section
Transverse section
TROCHOPHORE
A

Mesodermal pouches
Future coelom
Ectoderm
Mouth
Endoderm
Anus
Mesodermal pouches
Anus
Median section
Transverse section
TORNARIA
B

at a very remote time, before the vertebrates had originated. However, they might have retained some features of that ancestry that have been lost in the vertebrates. It is the patterns of development of some members of these groups that seem to offer the best clues to the origins and relationships of the vertebrates.

Vertebrates have no floating larvae. A larval stage of development probably occurred in their remote invertebrate ancestors, but it was lost in the course of evolution. In the vertebrates that do have larvae (some fishes, most amphibians), the larval stage is a new evolutionary development, not inherited from invertebrate ancestors. However, the hemichordates do have floating larvae that somewhat resemble trochophores. The striking difference is that the hemichordate larva (called a **tornaria**) has a twisted ring of cilia that encircles the mouth (see Figure 43-31B) instead of a single ring anterior to the mouth as in a trochophore. The tornaria possesses a true coelom. This larva is remarkably like the pluteus or bipinnaria of an echinoderm. This is strong, but not conclusive, evidence that hemichordates and echinoderms inherited their larval stages from the same marine invertebrate ancestor. Since the hemichordates seem to be an offshoot of the vertebrates, the latter are also probably derived from the same ultimate progenitor as the echinoderms. Because the hemichordate and echinoderm larvae are different from trochophores, the chordate-echinoderm ancestry probably became distinct from the ancestry of the annelid superphylum very early. Thus, chordates and echinoderms—both deuterostomes—are viewed as members of a second superphylum, the one known as the echinoderm superphylum (see Figures 43-28 and 43-30). How curious that we are so closely related to sand dollars and sea cucumbers!

Other evidence of relationship between echinoderms and chordates and a lack of relationship with the annelid superphylum is provided by the developmental mechanisms discussed in Chapter 19. Echinoderms and chordates are both deuterostomes. As shown in Figure 19-38B, that means that their embryonic blastopores become an anus (unlike the pattern in protosomes). A mouth arises later as a secondary opening. In both, pockets or folds arise from the endoderm of the developing digestive tract. Spaces in the pockets become the coelom, and their walls become mesoderm. This feature of development is much modified in most vertebrates, but it is present in primitive form in all chordates. It is also interesting that echinoderms and chordates contain the same phosphagen—creatine. The common phosphagen in the annelid superphylum is arginine.

It should be emphasized that this view does not envision the transformation of an echinoderm into a chordate. Rather the two phyla arose as widely diverging lines from common ancestors, now long extinct and of unknown adult structure, but with early developmental stages and larvae similar to those of echinoderms. From that remote ancestry, the echinoderms evolved as sessile, sedentary, or sluggish animals. One of the most distinctive chordate features (perforation of the pharynx by gill slits) is a likely adaptation to sessile life. In fact, two of the three surviving nonvertebrate chordate groups use a perforated pharynx as a feeding device (see Figure 43-2). One of these groups, Urochordata (tunicates), is still predominantly sedentary. The other, Cephalochordata (amphioxus), although capable of swimming, burrows in coarse sand, tail down, and lives an effectively sedentary life, drawing in water and filtering microorganic food through its pharyngeal slits. The evolution of the chordate (and therefore vertebrate) pharynx is easily understood as an adaptation to the sedentary habit. How, then, can we explain the fact that vertebrates—the most spectacularly motile of animals—have a common origin with the sedentary tunicates (urochordates)?

Origin of the Vertebrates

Some modern tunicates hint at a possible answer. Like other sessile animals, they rely heavily on motile larvae to disperse the species. The motile larvae of tunicates are elongate, bilaterally symmetrical forms resembling simplified caricatures of a vertebrate (Figure 43-32).

Most significant, however, is the presence in tunicates of a notochord lying dorsally to the gut as it does in vertebrates. It serves the function of a skeleton for simple fishlike locomotion. Many now think that vertebrates evolved as neotenic tunicates. In this view, the motile larvae of some sedentary tunicates—like the neotenic salamanders discussed in Chapter 19—gradually assumed an increasingly larger and more important part of the whole life cycle. Eventually they became

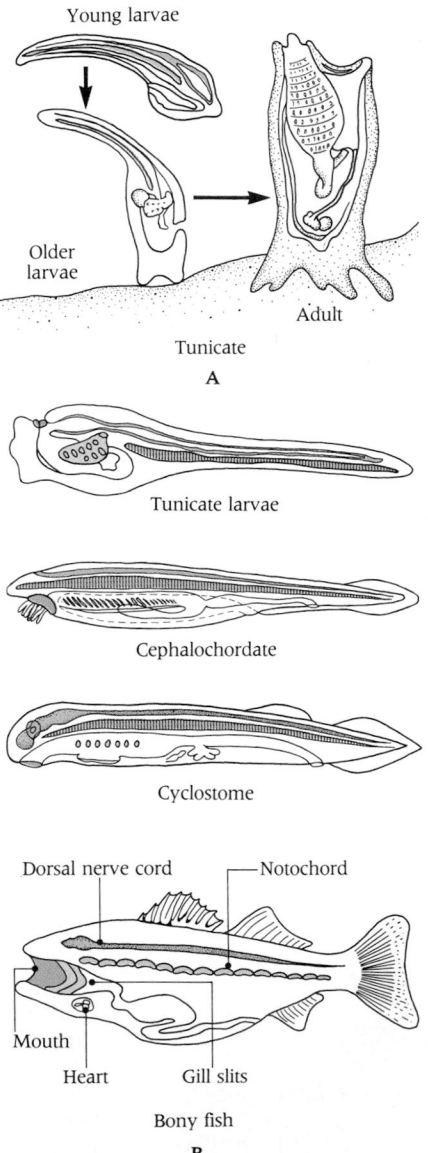

43-32 The tunicate origin of vertebrates
(A) *Young larvae metamorphose into older larvae and the adult stage.* **(B)** *A comparison of tunicate larvae with a cephalochordate (amphioxus), a cyclostome (agnathous fish), and a mature bony fish. Note that the tunicate is motile and bilaterally symmetrical, thus making plausible an evolutionary connection between the sedentary tunicates and the motile vertebrates.*

capable of their own reproduction and, so to speak, sloughed off the ancestral adult stages which were adapted to the sedentary habit.

An alternative hypothesis is that the common ancestor of tunicates and vertebrates was a free-swimming form somewhat like a tunicate larva. In this case, the ancestors of the vertebrates were not sessile and the sessile adult condition would be a tunicate specialization.

SUMMARY

Members of phylum Chordata display three key characteristics during at least one stage of their life cycles: (1) the presence of a notochord, a semirigid rod along the back; (2) the development of a tubelike dorsal nerve cord superficial to the notochord; and (3) the presence of pharyngeal slits.

Phylum Chordata is divided into four subphyla: Hemichordata, such as acorn worms; Urochordata (or tunicates), such as sea squirts (relatives of which are thought to have been the ancestors of all vertebrates); Cephalochordata, such as amphioxus; and Vertebrata. Many of the features associated with vertebrates—a segmented backbone, highly developed brain, peripheral nervous system, and sense organs—all derive from or are associated with embryonic neural crest cells.

Of the eight vertebrate classes, four are fishes. Agnatha are the jawless fishes; Placodermi is an extinct class containing the first jawed fishes. The sharks and rays of class Chondrichthyes have cartilaginous skeletons, unlike placoderms and the many members of class Osteichthyes, the modern bony fishes. Primitive osteichthyans had both lungs and gills. It was from this stock that the first vertebrates evolved. In most modern bony fishes, the lung became a swim bladder.

The other four vertebrate classes are all four-limbed, lung-breathing land vertebrates: Amphibia, Reptilia, Aves, and Mammalia. Reproductive, respiratory, and circulatory adaptations allowed reptiles to become the first completely terrestrial vertebrates. Both birds and mammals evolved from early reptiles.

The relationships between animal phyla is not a ladder linking "simple" ancestral phyla to complex, "advanced" phyla in a stepwise sequence. Rather, the evolutionary sequence has been a series of branching adaptive radiations. It is currently believed that certain tunicates were ancestral to the vertebrates.

Certain key characteristics are important in distinguishing and relating the phyla. These include the number of embryonic tissue layers, the structure of the digestive tube, the presence or absence of a coelom, the type of skeleton and circulatory system, the presence or absence of body segmentation, and the presence and nature of body symmetry.

Much evidence suggests that animals evolved from ciliated colonial protists. Subkingdom Eumetazoa, which includes all animals except sponges, can be divided into two superphyla on the basis of such features as larval characteristics and the early developmental patterns discussed in Chapter 19 (e.g., fate of the blastopore, pattern of cell cleavage, degree of morphogenetic cell movements during development, and so on).

Protostomes, which comprise the so-called annelid superphylum, includes many of the major animal phyla—Ectoprocta, Brachiopoda, Mollusca, Platyhelminthes, Annelida, and Arthropoda. In protostomes, the embryonic blastopore becomes the adult mouth.

Deuterostomes, the echinoderm superphylum, includes Echinodermata and Chordata. In deuterostomes, the blastopore becomes the anus and the mouth develops later as a secondary opening. Various biochemical features also mark the superphyla.

KEY TERMS

Aquatha

amniotic egg

Amphibia

amphioxus

Aves

Cephalochordata

Chondrichthyes

Chordata

crossopterygian

denticle

deuterostome

dorsal nerve cord

Hemichordata

mammal

Mammalia

marsupial

notochord

Osteichthyes

pharyngeal slit

Placodermata

placental

primate

protostome

Reptilia

swim bladder

teleost

thecodont

therapsid

tunicate

trochophore

Urochordata

Vertebrata

QUIZ QUESTIONS

1. Which of the following is *not* a feature of species of Phylum Chordata?
   A. presence of a notochord
   B. presence of pharyngeal gill slits
   C. presence of a ventral, hollow nerve cord
   D. presence of a coelom
   E. presence of three embryonic germ layers

2. Which of the following is *not* a correct common name with its taxon?
   A. lamprey; Placodermi
   B. lungfish; Osteichthyes
   C. lungfish; Dipnoi
   D. snake; Squamata
   E. dinosaur; Reptilia

3. Considering the entire theater of animal evolution, which of the following is *not* a derived trait or condition?
   A. closed circulatory system
   B. complete digestive tract
   C. bilateral body symmetry
   D. absence of body segmentation
   E. presence of a true coelom

4. Which of the following structures or traits is *not* related to homeothermic metabolism?
   A. a two-chambered heart
   B. feathers
   C. hair
   D. A and B
   E. B and C

5. Multicellular plants and animals probably evolved from
   A. solitary flagellated monerans.
   B. colonial flagellated protozoans.
   C. solitary protozoans.
   D. colonial parazoans.
   E. solitary parazoans.

ESSAY QUESTIONS

1. What are the cardinal features of the phylum Chordata? What is a notochord? Do adult humans possess a notochord? What are the several roles of the pharyngeal slits in chordates?

2. In what way do the distinguishing features of the vertebrates depend on the properties of neural crest cells?

3. What adaptations allowed the reptiles to become exclusively terrestrial?

4. What aspects of the anatomy of *Archaeopteryx* suggest an evolutionary link between reptiles and birds? What other structural developments allowed birds to better adapt? What aspects of the anatomy of monotremes are reptilian; mammalian?

5. What evolutionary developments characterize the class mammalia? Contrast fetal nutrition in monotremes, marsupials, and placental mammals.

6. Why is a tree a more appropriate metaphor for the course of evolutionary history than a ladder?

REFERENCES AND SUGGESTIONS FOR FURTHER READING

Alexander, R. McN. (1981). *The Chordates*, 2nd ed. Cambridge University Press, New York.

The author explains how animals are constructed, how they evolved, and how they work.

Blake, R. W. (1983). *Fish Locomotion*. Cambridge University Press, New York.

Evolution has adapted fishes to widely differing modes of life—and not all are streamlined fast swimmers. This lively book discusses the generation of propulsive forces and the functional significance of body form.

Carroll, R. L. (1966). *Vertebrate Paleontology and Evolution*. W. H. Freeman, New York.

A review of the fossil record of vertebrates in the context of evolutionary process and functional anatomy. Coverage includes early forms from the Paleozoic and Mesozoic.

Duellman, W. E., and Trueb, L. (1985). *Biology of Amphibians*. McGraw Hill, New York.

A valued textbook exclusively devoted to a fascinating animal class that deserves our attention.

Eisenberg, G. F. (1981) *The Mammalian Radiations*. University of Chicago Press, Chicago.

In synthesizing mammalian evolution and behavior, the author clearly explains the diversity of mammalian adaptation.

Fedducia, A. (1980). *The Age of Birds*. Harvard University Press, Cambridge, Mass.

An attractive book that thoroughly surveys the evolution of birds and their origin from reptilian ancestors.

Gans, C., and Huey, R. B. (Eds.) (1988). *The Biology of Reptilia, Vol. 16 Ecology, B: Defense and Life History*. Alan R. Liss, New York.

This new volume of a series concerns ecology and behavior. Topics include mimicry among snakes, the evolution of snake venom, invertebrates (centipedes, scorpions, etc.) that eat reptiles, autotomy, parental care, and many others.

Hanson, E. D. (1977). *Origin and Early Evolution of Animals*. Wesleyan University Press, Middletown, Conn.

An evolutionary biologist examines the mass of data bearing on the evolution of animals from protists and concludes that the origin of the Eumetazoa remains to be clarified.

Hildebrand, M., Bramble, D., Liem, K., and Wake, D. (1985). *Functional Vertebrate Morphology*. Harvard University Press, Cambridge, Mass.

The structure of vertebrates is clearly correlated with their functions.

Kemp, T. (1982). *Mammal-like Reptiles and the Origin of Mammals*. Academic Press, Orlando, Fla.

A engrossing scholarly review.

McFarland, W. N., Pough, F. H., Cade, T. J., and Heiser, J. B. (1985). *Effects of Pollutants at the Ecosystem Level*. Macmillan/Collier Macmillan, New York.

A stimulating in-depth review of the biology of vertebrates, covering systematic, ecological, and functional aspects.

Radinsky, L. B. (1987). *The Evolution of Vertebrate Design*. University of Chicago Press, Chicago.

A stimulating chronicle of the rise and fall of countless vertebrate species over the last 500 million years, this book looks at the major changes in body organization and their evolutionary importance.

Richard, A. F. (1985). *Primates in Nature.* W. H. Freeman, New York.

An admirable source book on primates, remarkable for its breadth and depth.

Romer, A. S., and Parsons, T. S. (1986). *The Vertebrate Body,* 7th ed. W. B. Saunders, Philadelphia.

Romer has made enormous contributions to the field of vertebrate paleontology and comparative vertebrate anatomy. This is an updated edition of a classic work on the vertebrate body and its evolutionary development.

Willson, M. F. (1983). *Vertebrate Natural History.* W. B. Saunders, Philadelphia.

This fine book on vertebrates illustrates the value of the branch of biology known as natural history.

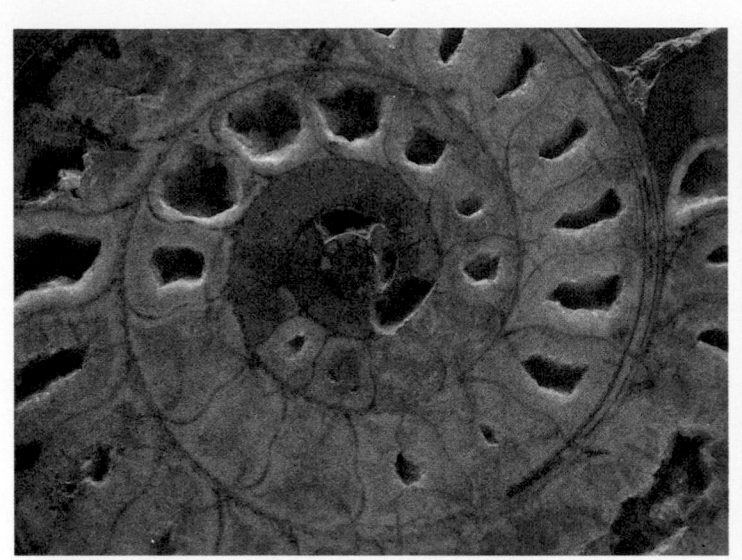

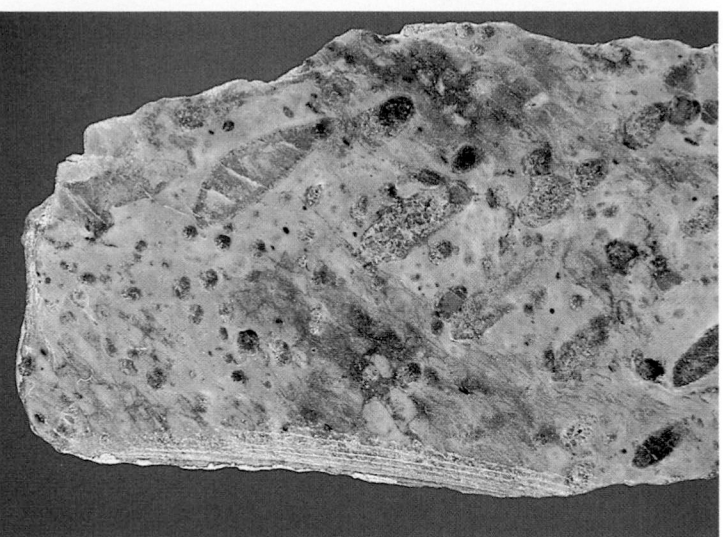

HOW DIVERSITY AROSE: THE HISTORY OF LIFE

How diversity arose: The history of life
Fossils, of different eras and evolutionary lineages, have shed brilliant light on the history of life. These beautiful specimens include: a sectioned ammonite (upper left) from the Lower Jurassic; protozoans (upper right) from the Permian; and a fish, Diplomystus, *from the Middle Cretaceous.*

The search for scientific understanding usually follows one of two broad avenues. One relates to *being,* the other to *becoming.* The physical sciences are studied mainly in terms of being. The law of gravitation is something that *is.* It will not become something different tomorrow or a million years hence.

But that approach does not yield an adequate understanding of living things. Groups of plants or animals were not always what they are now. They have become so, and how they became so is essential to our understanding of what they are. A tree grew from a seed. A squirrel developed from a zygote. The pursuit of biological principles must follow seed and zygote back through the generations to times when there were no seeds or zygotes, no trees or squirrels. Because biological entities change through time—at different tempos and in the differing contexts of ontogeny and phylogeny—we can best understand ourselves and other organisms by studying the long process of becoming that is evolution.

We have surveyed the five living kingdoms—and one by one the evolutionary origins of the major phyla. In this chapter, we consider the origin of life's diversity with a broader angle of vision. We will see that life as it is manifested in each living organism has a history. In reviewing that history, we will explore the peculiarly historical principles and contingencies underlying it.

EARLY ORGANISMS

The problem of how life began—biopoiesis or biogenesis—was introduced in Chapter 4. We saw that when Pasteur and other scientists of the nineteenth century succeeded in discrediting the notion that life is constantly arising by spontaneous generation they launched scientific discussion of biogenesis—the origin of life—for the first time. We saw also that *direct* information about the origin of life is totally lacking. There could be only two kinds of direct evidence, though neither is proof of what happened: (1) fossils of the first organisms; and (2) the repetition before our eyes of the early events leading to the appearance of living organisms from nonliving matter. No such fossils have been found, and as far as is known, no living things could be generated from nonliving matter today because modern environments rich in oxygen and competitors would prevent it.

Finally, we saw that modern biochemists have learned a good deal about the evolution of biological macromolecules from simple precursors and have accomplished many such syntheses in the laboratory under simulated primeval conditions. These portions of Chapter 4 might well be reviewed.

We consider it a well-supported hypothesis that simple and later more complex organic molecules were the first to arise in the ancient Precambrian seas—under an atmosphere of CO_2, N_2, and CO (with minor amounts of NH_3, H_2, CH_4, and H_2S).[1] We speak of these early events as abiotic formations of organic compounds, nucleotides, peptides, and catalysts.

[1] It was believed until recently that the atmosphere at this time was highly reducing and lacking in oxygen. New geochemical evidence now suggests that it had a nearly neutral position on the oxidation-reduction spectrum.

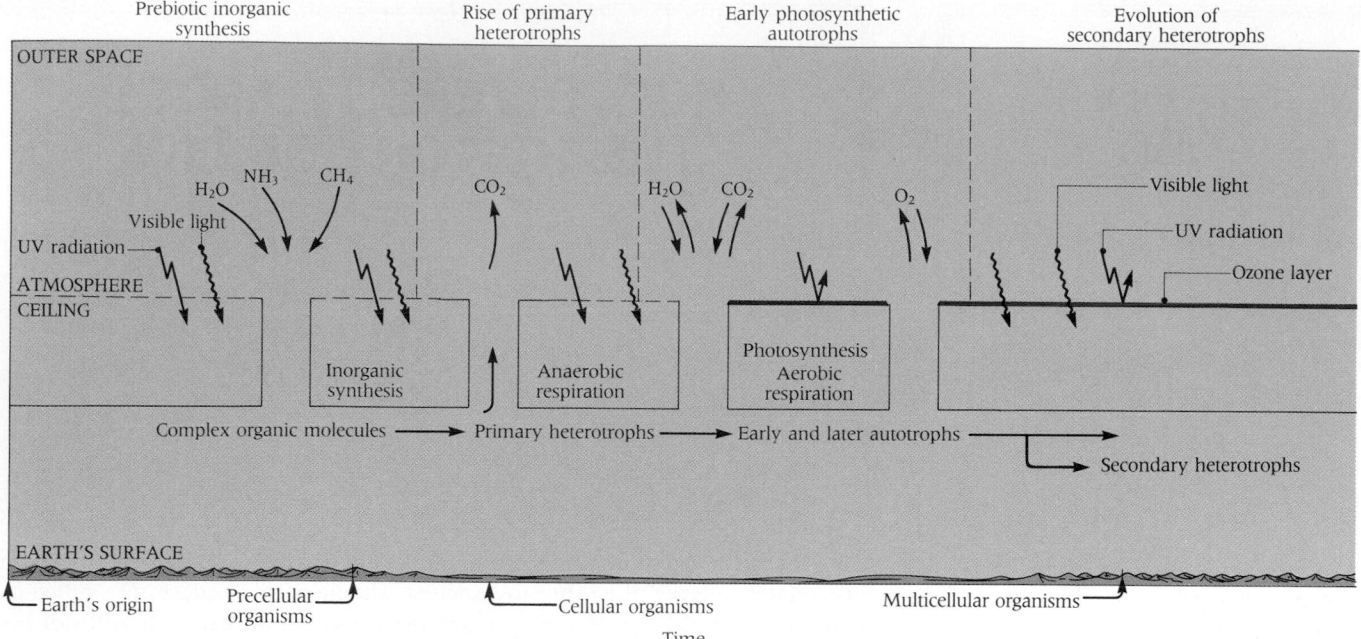

Prebiotic inorganic synthesis | Rise of primary heterotrophs | Early photosynthetic autotrophs | Evolution of secondary heterotrophs

OUTER SPACE

H_2O NH_3 CH_4 CO_2 H_2O CO_2 O_2 Visible light

Visible light

UV radiation

UV radiation

ATMOSPHERE

Ozone layer

CEILING

Inorganic synthesis | Anaerobic respiration | Photosynthesis Aerobic respiration

Complex organic molecules → Primary heterotrophs → Early and later autotrophs →

Secondary heterotrophs

EARTH'S SURFACE

Earth's origin | Precellular organisms | Cellular organisms | Multicellular organisms

Time

Whether such molecular systems were *alive* is a matter of definition. If they gradually became capable of self-reproduction and if that process in time allowed for modifications akin to mutations, then there may have been no exact point at which the nonliving became living—in which case the acquisition of life's full panoply occurred gradually. We believe that these molecular aggregations eventually interacted with each other and with their environment. The resulting groupings drew benefits from these associations. Probably, an association of nucleic acids, proteins, and the other compounds required for energy transfers occurred very early. These events—and some later ones—are sketched in Figure 44-1.

Once the formation of such interacting aggregates was regularized according to this speculative (but highly plausible and—given enough time—highly probable) scenario, they would tend to become more numerous and distinctive in their composition and organizational pattern. In this scheme, a complex aggregate of favorably interacting molecules would have notable advantages in attracting molecules compared to molecules that continued to "free-lance." Natural selection would favor changes that moved these aggregates in the direction of more effective and constant organization. The result in time would be a true multimolecular organism, a primitive protist that was alive by any definition.

The problem facing those seeking a rational account of the origin of life is a classic "chicken-and-egg" problem—that is, which came first? If all forms of life must carry distinctive sets of informational molecules (DNA) and functional molecules (proteins), then how could such a cycle have arisen? We have mentioned in earlier chapters the recent discovery that certain RNA molecules have true catalytic activity. If enzymes are defined as macromolecules that catalytically accelerate a specific reaction by lowering its activation energy (see Chapter 8), then these RNA molecules fully deserve to be called enzymes. This surprising phenomenon was first observed by Thomas R. Cech and coworkers at the University of Colorado, who found that an rRNA molecule of the ciliated protozoan *Tetrahymena thermophilia* catalyzes the excision of its own 413-nucleotide intron, that it does this elaborate splicing job all by itself in the total absence of proteins, and that it does it more efficiently than any known protein enzyme. Even more surprising was the later discovery that the excised intron can function as an RNA polymerase.

These amazing findings suggested an obvious way out of the "chicken-and-egg" dilemma. Such primitive RNA molecules, it seems, could have functioned as the first replicating (and therefore living) systems. A bold scenario now suggests that variants of this RNA polymerase first gave rise to a so-called **RNA world,** in which RNA was the only catalytic molecule. Then later, with the advent of protein synthesis, a more complex world evolved in which RNA and protein shared the catalytic tasks. This, in current jargon, was the **ribonucleoprotein,** or **RNP world.** We believe it was the immediate precursor of our own **DNA world,** a world in which information could be stored much more efficiently.

44-1 Early stages in the evolution of life
It is useful to divide early evolution into four major stages, beginning with the abiotic synthesis of complex organic molecules. This led in turn to primary heterotrophs, early and later autotrophs, and finally, secondary heterotrophs. Atmospheric accumulations of CO_2 led to photosynthesis, which in turn enriched the atmosphere with O_2. An ozone layer formed high in the atmosphere and screened out excessive ultraviolet radiation.

Aside from its logical attractions, support for this scheme can be found among the RNA viruses, which some now view as "living fossils" representing the ancient RNA world, and in present-day rRNA self-splicing mechanisms, which may be a vestige of a very old process.[2]

HOW EARLY ORGANISMS MAY HAVE DEVELOPED

It was long assumed that the earliest true organisms must have been autotrophs capable of building all their necessary organic molecules from simple starting materials, using solar radiation (or simple inorganic compounds) as energy sources. In this traditional view, the first organisms could hardly have been heterotrophs since there were no other organisms to feed on—another "chicken-and-egg" dilemma.

Many now question this assumption. Autotrophy always requires more complex organization than heterotrophy. Arguably, the very first organisms were simple rather than complex. Indeed, the process of biogenesis can be more easily visualized if the earliest organisms were heterotrophs. Of course, these primary heterotrophs could not feed on other organisms, since there were none. But they would be heterotrophic if they fed on preformed carbohydrates, amino acids, and other organic compounds that had been synthesized abiotically (not by organisms).

This state of affairs would not have persisted indefinitely. The earliest organisms, as heterotrophs, may have resembled modern fermenting bacteria like *Clostridium*, which are totally dependent on the breakdown of abiotically formed, energy-rich molecules. They were literally scavengers of the organic matter produced by electrical discharges (e.g., lightning) and ultraviolet radiation. If this is correct, the total amount of life that Earth could have sustained would have been sharply limited. As organisms multiplied, they would rapidly consume all available organic materials. The further synthesis of these compounds by abiotic processes would soon lag behind consumption. Moreover, conditions were changing. The organisms themselves would cause much of the change. They would, for example, tend to convert much of the carbon formerly in methane (CH_4) or other compounds into CO_2, from which more complex organic molecules were unlikely to be formed abiotically. Abiotic synthesis of foods must inevitably have fallen nearly to zero—where it is today. If nothing else had happened, life would have become extinct at a very early date.

As shown in Figure 44-1 what happened instead was the emergence of photosynthesis. There was, we presume, the first appearance of a mutant organism that could synthesize the compounds it needed from widely available simple precursors. When the mutant organism transmitted this tremendous advantage to its descendants, they soon became the dominant, perhaps the only, organisms extant. When additional mutations occurred—through a slow accumulation of changes due largely to the effects of radiation—there was a step-by-step trend toward self-sufficiency, toward a more complete autotrophy.

This stage of evolution presumably culminated in organisms that could use CO_2 as their sole external carbon source and solar radiation as their sole external energy source. Such organisms would be photosynthetic protists that eventually became plants. Evolution by natural selection was well on its way.

Meanwhile, conditions on Earth were changing radically. Earth's surface continued to cool. The NH_3 and CH_4 originally present in the atmosphere were incorporated into increasingly complex organic molecules and were transformed in part into N_2 and CO_2, which probably came to constitute most of the atmosphere soon after life arose. After the rise of photosynthesis, organisms themselves made the most important change of all in the atmospheric composition. They must have reduced the percentage of CO_2 in the atmosphere and increased the percentage of free O_2, previously very low.[3]

---

[2] Of course, those who envision a self-replicating RNA as the basis of the first living organisms must acknowledge that RNA is itself an evolutionarily advanced molecule that seems unlikely to have arisen *de novo* in the prebiotic soup. However, many imaginative ideas have been proposed to explain away that problem (for example, the pre-RNA evolution of various sugar-phosphate chains, etc.).

[3] That source may have been periodic bursts of atmospheric instability that were associated with photodissociation of an atmospheric component. According to interesting biochemical evidence, primitive prephotosynthetic Precambrian anaerobes had to protect themselves against occasional transient exposures to O_2 that could be lethal. This may have been done with such O_2-mediating enzymes as superoxide dismutase.

Regardless of how oxygen entered the atmosphere, it accumulated very slowly, gradually eliminating the reducing environments on Earth's surface. High in the atmosphere, the oxygen formed a layer of ozone (O_3), still present, which sharply decreased the amount of ultraviolet radiation reaching Earth's surface from the sun. Before this event, ultraviolet radiation streaming toward Earth had been an effective energy source for the abiotic synthesis of organic molecules. That ended when the ozone layer began to filter out ultraviolet radiation. Visible light, which is not significantly absorbed by ozone, became the chief source of energy through the vehicle of photosynthesis. These changes, largely derived from early living systems, made further spontaneous origin of such systems impossible. Living organisms now had a virtual monopoly on organic synthesis and a pattern of life driven by solar energy was fixed for all time on our planet.

PRECAMBRIAN FOSSILS

All of these hypotheses accord with modern biochemical concepts. Current evidence suggests that life was present 3.5 billion years ago and photosynthesis arose about 2.8 billion years ago. The major evidence dating the rise of photosynthesis is geological data showing that the early oceans were rich in soluble ferrous (Fe^{2+}) salts. About 2.8 billion years ago, this iron disappeared from the sea over a short span of several million years and entered sedimented rocks as insoluble ferric (Fe^{3+}) oxides. This marked the appearance in the atmosphere of plentiful oxygen—in direct consequence of the rise of photosynthetic organisms.

Eukaryotic cells evidently emerged at some time more recent than 2 billion years ago—in the wake of three important events: (1) the first appearance of free O_2 as a permanent component of the atmosphere; (2) the dwindling of CH_4 production as an important function in most living organisms; and (3) the onset of significant levels of sulfate reduction.

When evolution began to leave its traces in the fossil record the physical state of Earth and its atmosphere were much as they are today, and O_2-generating photosynthetic organisms were abundant. It was the existence of these autotrophs that made possible the evolution of secondary heterotrophs, organisms that fed on them and their by-products.

Although life doubtless existed 3.5 billion years ago, fossils become varied and abundant only with the beginning of the Cambrian, a mere 600 million years ago. Microscopic fossils (0.5–0.75 μm long) that are almost certainly bacteria occur in South African rock formations (called cherts) that are 3.2 billion years old. Similar rocks in Ontario, believed to be about 2 billion years old, also contain fossil bacteria— and what are almost surely blue-green algae (Figure 44-2A). From that time onward, fossils, even reefs, of blue-green algae occur the world over. Later Precambrian rocks, from 850 to 600 million years old, reveal the first promising traces of animals— impressions that resemble worm burrows and soft-bodied coelenterates, annelids, and various unidentified uncertain species (Figure 44-2B).

The oldest animals known with certainty are multicellular and already quite complex and varied.[4] So we cannot say whether their emergence was gradual or abrupt. Their age is not certain but it is late Precambrian and well over 550 million years. All of these fossils are of soft-bodied forms that must have been preserved under exceptional conditions. Some of these Precambrian animals are quite unlike any later ones that we know of, but coelenterates and annelids (almost certainly) and sponges (possibly) are among them. Indeed, the forms so far assembled include 26 metazoan species in 18 genera and 4 or more phyla.

THE PRECAMBRIAN-CAMBRIAN BREAK

Much more variety occurs in the earliest Cambrian rocks. The contrast is striking between Precambrian rocks, in which animal fossils are uncommon, and those of the Cambrian, in which there are abundant fossils of marine plants, worms, sponges, mollusks, lampshells, and trilobites. Indeed, almost all of the phyla appear as fossils

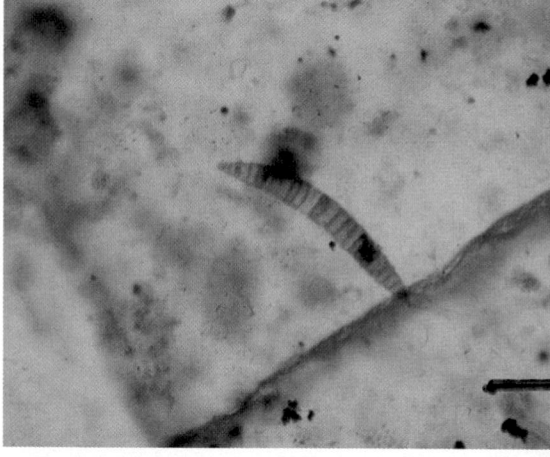

A

B

44-2 Microscopic fossils
(A) *Probable blue-green algae.* **(B)** *Layers of fossil sponges and other organisms.*

---

[4] They are best known in a rather large fauna named for the Ediacara Hills in South Australia. A few similar fossils are known from Namibia and England. The so-called Ediacarian period, which is characterized by the oldest known multicellular animal life, dates from 670 to 550 million years ago. The oldest known fossils (in millions of years) belonging to the "big three" phyla are as follows: mollusca (575), arthropoda (570), and chordata (540).

during this period. This relatively sudden diversification leading to the emergence of many new taxa was an immense event, perhaps the greatest single episode of increase in complexity in the history of life. Among other things, it shows clearly that the multicellular organisms had inherited the Earth. It also poses questions. Did this diversification of life occur abruptly? If it did, why? These are questions Darwin asked in *The Origin of Species*.

We do not yet know the answers. Some have argued that (1) an abrupt appearance of many new forms in the early Cambrian in "one enormous step" is not in keeping with our understanding of later evolutionary processes, and (2) these forms may have existed earlier during the Precambrian, but their fossils have not been preserved or found. The suddenness of the change from the barren Precambrian rocks is real enough. But it cannot be said that all metazoan phyla mysteriously arose at the same time. The Cambrian covered 100 million years, and the major animal groups straggle into the fossil record throughout that span.

MAJOR EVOLUTIONARY TRENDS

We propose now to examine in logical and chronological sequence some of the major events in the history of life since the early Cambrian.

A preliminary word is in order on connotations of the taxonomic categories in such discussions of the history of life. In defining phylum, class, order, and so on in Chapter 37, we said that evolutionary classification depends on the properties of organisms and, our current concern, their evolutionary history. We will find in the present discussion that the levels of the taxonomic hierarchy have special meaning. For example, a new class obviously could not arise within a phylum until that phylum had arisen. For that reason, the counting of the phyla, classes, and so forth represented in the fossil record of a period has been a useful tool in the study of evolution. Some of its conclusions are summarized in Table 44-1.

Let us begin our survey of major evolutionary tendencies by considering the history as a whole, as represented schematically in Figures 44-3 and 44-4. The figures reveal certain interesting broad tendencies that we should note before turning to the individual phyla.

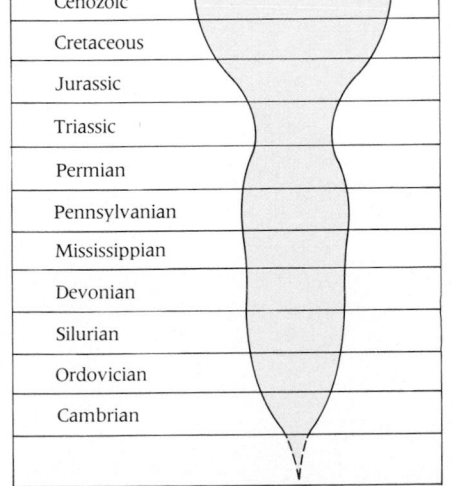

44-3 The expansion of life
The width of the pathway is approximately in proportion to the known diversity of plants and animals at various times in the past. Note the constriction of the pathway during the late Permian and Triassic.

TABLE 44-1
SIGNIFICANCE OF THE TAXONOMIC CATEGORIES IN STUDIES OF THE HISTORY OF LIFE, WITH ILLUSTRATIONS FROM ANIMAL EVOLUTION

Category or Taxon	Paleontologic Importance
Phylum	*Fundamental division of animal kingdom,* distinguished by unique aspects of basic "body plan." Most arose in the late Precambrian from small soft-bodied ancestors. Hence, there is no fossil record of intermediate forms. Evolutionary relationships are determined largely from the embryology and comparative anatomy of living representatives.
Class	*Major division within phylum,* distinguished by basic modifications of the body plan. Most marine classes with fossil records appear early in Phanerozoic. Fossil record contains intermediates between classes in some phyla (e.g., Mollusca).
Order	*Higher taxon,* new orders appearing throughout the Phanerozoic. Marine fossil record shows nearly constant numbers of animal orders from Ordovician to present.
Family	*Lower taxonomic category* for which comprehensive, accurate data is available for all marine and continental animals. Data on 5000 or so fossil families show such features of the history of life as major evolutionary radiations and mass extinctions. Many fluctuations in species, such as minor mass extinctions, are not reflected in families.
Genus	*Smaller taxonomic unit* than family showing more evolutionary detail. Accurate data available for only a few groups (e.g., nautiloid cephalopods). Between 25,000 and 40,000 fossil genera known.
Species	*Basic unit of evolution,* with more than 250,000 fossil animal species described in last 250 years, but no accurate listings available.

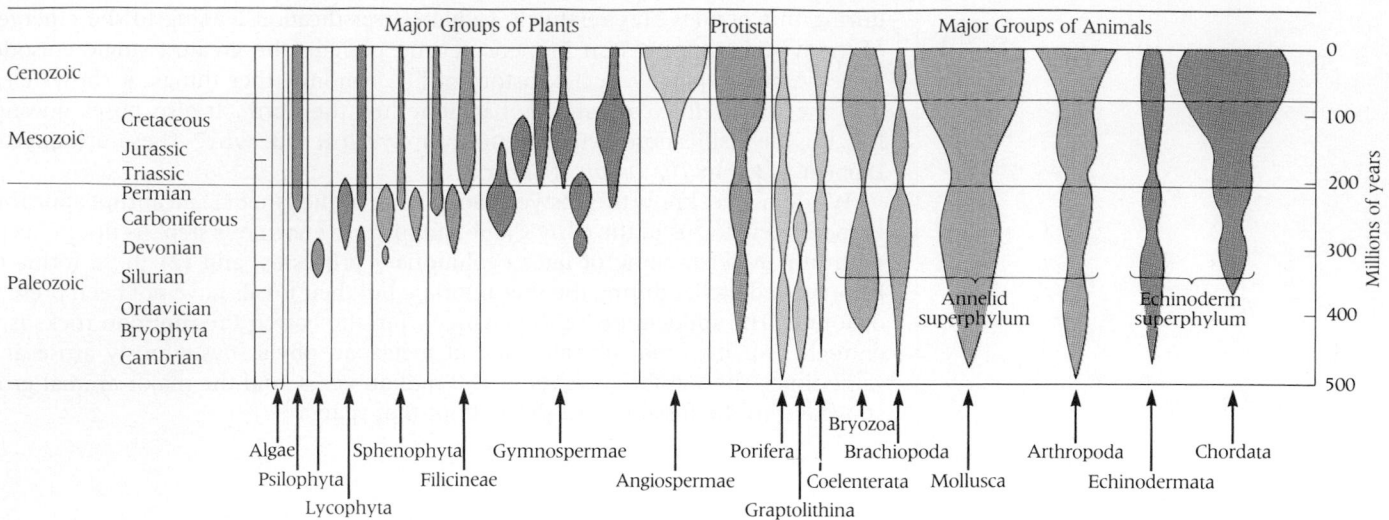

Major Groups of Plants | Protista | Major Groups of Animals

Cenozoic

Mesozoic — Cretaceous / Jurassic / Triassic

Permian / Carboniferous / Devonian / Silurian / Ordavician / Bryophyta / Cambrian

Paleozoic

Millions of years — 0, 100, 200, 300, 400, 500

Annelid superphylum   Echinoderm superphylum

Algae   Sphenophyta   Gymnospermae   Porifera   Bryozoa / Brachiopoda   Arthropoda   Chordata
Psilophyta   Filicineae   Angiospermae   Coelenterata   Mollusca   Echinodermata
Lycophyta   Graptolithina

EXPANSION

The first striking generalization is that the total number of living things has increased. So has the bulk of matter in living form, the number of different kinds, classes, and phyla of organisms, and the extent of their diversity. The divergence has continued. There are many more differences among a contemporary amoeba, maple tree, and horse than among any group of Cambrian organisms.

Expansion was not constant or continuous. Rather, there were great bursts of diversification. One occurred in the Devonian for plants and animals. There were also times when evolution seemed to lose ground—that is, diversity declined. Plants and animals both decreased in numbers and diversity between the Permian and Triassic. Nevertheless, a general expansionist tendency is plain.

Numbers of individuals, not directly discernible from the fossil record, have tended to be in inverse ratio to their size. The evolution of large plants, while predisposing to significant increases in the total number of plant species, may have caused a distinct decrease in the total number of photosynthetic organisms (if not in the total amount of photosynthesis going on). Several factors were responsible.

- The supply of solar energy and the efficiency of photosynthesis, are believed not to have increased significantly since the Precambrian. Were it not for these limitations, even a few primitive species in the Precambrian might have filled the environment with a great bulk of living matter.
- Both numbers and total bulk of organisms tended to increase with the invasion of new environments and the lengthening of food chains (see Chapter 46).

All of these factors persisted long after the Cambrian. Let us consider them briefly.

OCCUPATION OF NEW ENVIRONMENTS

The expansion of life was accompanied by increasingly extensive and intensive occupation of available environments. Indeed, this was a major reason for the general expansion. The most spectacular example was the invasion of the hitherto unoccupied land by plants and animals. Until Silurian times, life was confined to the water, including protists, simple plants, and a few worms.

Then a movement started into land environments that was not completed for hundreds of millions of years and that led to increasingly fine subdivision of ecological niches and increasing specialization of organisms in the niches. Lengthening of food chains facilitated the expansion and occupation of environments and their special niches. Each new species of mammal, for example, was a possible environment for a new species of parasite.

There was one more factor. New environments were created by the expansion itself and these were then occupied in turn. For instance, the "gardens" of epiphytes high in the crowns of forest trees exploit an environment that did not exist until forest trees had evolved.

44-4 Broad features of the fossil record
The width of each pathway is approximately in proportion to the known variety of organisms in the group through time. Note that diversity increases through time and that nearly every group constricts at the Permian-Triassic boundary. The more complex groups of plants and animals arise progressively later in the fossil record.

CHANGE: PERSISTENCE AND REPLACEMENT

Evolution is characterized by a constant state of flux. Nowhere in the record is there a time of completely static equilibrium. To biologists who think that such a final state has now been reached, we would argue that change has always occurred and there is no reason why it should stop now. The events of the next 10 million years should settle the point.[5]

It is true, of course, that expansion has not been constant in the past and cannot continue indefinitely into the future. There are presumably limits to the amounts and kinds of organisms that the Earth can accommodate at any one time. But can there be change without expansion? The fossil record is rich in case histories. Some groups of organisms are persistent for long times without notable change. However, there are more examples of the replacement of now extinct organisms by others better adapted to the same environments. Environments suitable for seed ferns, cycadeoids, cordaites, and certain other early land plants exist today, but those plants do not. The flowering plants (and to a smaller extent the conifers) have replaced them—though, of course, lycopods and equisetums still exist. Expansion and replacement are two of the main factors in the constant change among living things. If expansion is limited, replacement appears not to be.

COMPLICATION AND IMPROVEMENT

Increasing complication and functional improvement are the changes most often mentioned in discussions of broad evolutionary tendencies. We believe they are overemphasized and often misunderstood. Increasing complication has certainly been an important factor in various phases of the history of life, but it has not been a universal tendency, and it has not been particularly noticeable in the last few hundred million years of life's history.

The point to stress about improvement—or change for the better, or evolutionary progress—is that it is not inherent.[6] It is not built into the nature of the universe or of life. Improvement has occurred in the course of evolution, not because it is an inherent and general tendency. It occurred when it had a natural, immediate cause: selection.

There can be many definitions of improvement. What may reasonably be called improvement has occurred so frequently it may be considered a common, if not universal tendency in the history of life. It has taken three main forms: (1) increasingly precise and effective adaptation to a particular way of life; (2) marked change in structure and function, making possible ways of life, occupation of environments, and so on, increasingly far removed from the ancestral ways and environments; and (3) increasing perception of the environment and increasing complexity, flexibility, and appropriateness of reactions to environmental stimuli.

The first tendency is most widespread and least important in this overall view of the history of life, although of great importance in its details. An example of the second tendency is the rise of root-stem-leaf differentiation and of vascular tissue in plants, making possible true land life, or of the amniotic egg in reptiles. The third tendency is especially pertinent to the rise of human beings.

[5] Others hold that the rate of evolutionary change has tended to accelerate ever since life began and that flux is now at its greatest level.

[6] Though, in fairness, we must note that some biologists think that it is. That is, of course, an inference, not an observable fact, and therefore it is open to debate.

44-5 Major Cambrian faunas of the marine fossil record
The diversity curve represents the total number of families recognized in each period. The major Cambrian groups were the arthropods (such as trilobites), brachiopods, mollusks, and echinoderms.

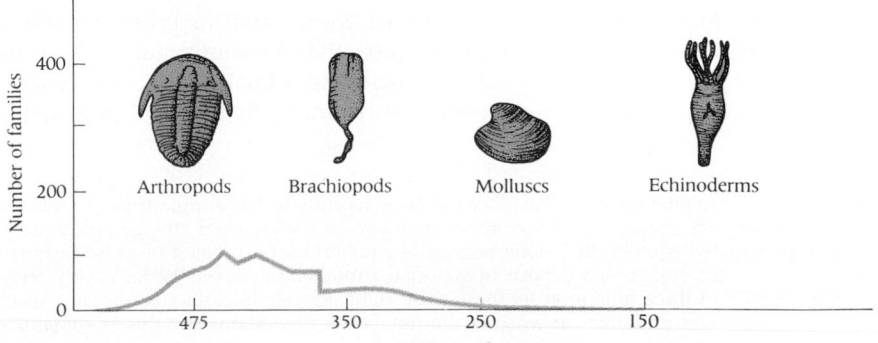

A   B

ANCIENT SEAS AND THE CONQUEST OF THE LAND

CAMBRIAN AND ORDOVICIAN SEAS

The world of 600 million years ago was very different from the world of today. The main ocean basins and continental masses may have already existed, but their outlines were different. There were deep seas and shallow seas, arms of the oceans, flooded far into the interior of the continental blocks, across what is now dry land. Such an extensive shallow water environment, we believe, was an essential precondition to the evolution of higher metazoan groups. The land of that time is hard for us to visualize because it was so utterly bare, lacking grass, shrubs, trees, insects, birds, or other kinds of life.

Then there was the sudden appearance of a diversity of multicellular organisms late in the Precambrian, which we have noted is the major enigma of the fossil record. Four major groups of organisms appeared in the Cambrian period (Figure 44-5):

- Trilobites (phylum Arthropoda, class Triblobita)[7]
- Lingula and other brachiopods (phylum Brachiopoda, class Inarticulata)
- Monoplacophorans (phylum Mollusca, class Monoplacophora)
- Eocrinoids (phylum Echinodermata, class Eocrinoidea), with features of classes Cystoidea (extinct) and Crinoidea (living)

The vertebrates do not appear until the Ordovician, the long 70 million-year period that followed the Cambrian.

Appearance of the Animal Phyla

The straggling into the record of major animal groups and the great expansion of life in these periods can be better told in figures than in words. Table 44-2 shows the numbers of phyla and classes of animals (including animallike protists) recognized at the stated times or earlier. Before the end of the Ordovician, all protistan and

---

[7] These small, crablike creatures have been at issue recently in the dispute between those who believe evolution has proceeded in a continuing steady pace of gradual change and those who believe in punctuated equilibrium—long periods of equilibrium punctuated by episodic bursts of change (see Chapter 36). A rich deposit of well-dated triblobite fossils in Wales clearly revealed that over a period of three million years there arose eight separate lineages (that became species), and in each there was a gradual increase in the number of ribs. Many saw this as supportive of evolutionary gradualism.

44-6 Fauna of early Paleozoic seas
(A) *Middle Cambrian.* **(B)** *Middle Ordovican.*
(C) *Middle Silurian.* **(D)** *Late Silurian.*

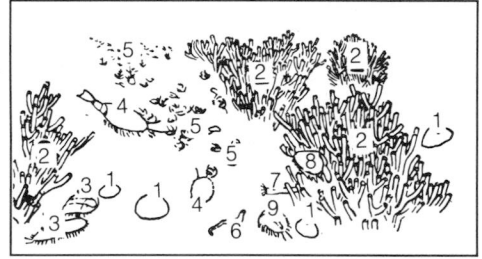

(A) *Middle Cambrian, western North America:*
(1) a jellyfish, (scyphozoan); (2) the spongelike Archeocyathus; *(3) a trilobite,* Ogygopsis; *(4) an arachnid,* Sidneyia; *(5) a crustacean,* Barrella; *(6) an annelid worm; (7) a holothurian (echinoderm); (8) a crustacean,* Hymenocaris; *(9) a trilobite,* Neolenus.

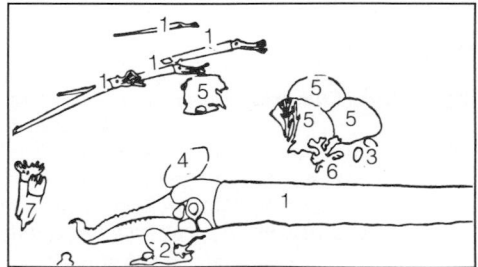

(B) *Middle Ordovician, central North America:*
(1) a straight-shelled (orthoconic) nautiloid cephalopod; (2) a gastropod; (3) a small trilobite, Calymene; *(4) a large trilobite,* Isotelus; *(5) massive coral; (6) branching coral; (7) two solitary corals.*

C

D

(C) *Middle Silurian, Illinois—a coral-reef community:* (1) *a stalked (sessile) cystoid echinoderm;* (2) *a cephalopod mollusk,* Phragmoceras; (3) *honeycomb coral,* Favosites; (4) *tube coral,* Syringopora; (5) *chain coral,* Halysites, (6) *a solitary coral;* (7) *a nautiloid cephalopod;* (8) *a trilobite,* Isotelus; (9) *a trilobite,* Actinurus; (10) *brachiopods,* Pentamerus; (11) *brachiopods,* Leptaena; (12) *a cephalopod mollusk,* Cyrtorizoceras.

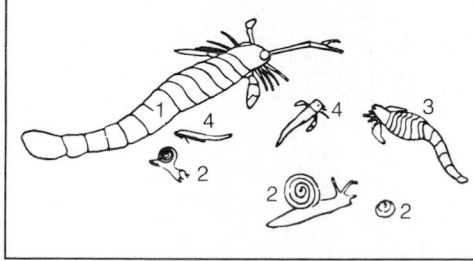

(D) *Late Silurian, New York;* (1) *a eurypterid,* Pterygotus; (2) *a snail,* Pycnomphalus; (3) *a eurypterid,* Carcinosoma; (4) *a eurypterid,* Hughmilleria.

animal phyla likely to be preserved as fossils were definitely present. With major body plans set, most of these forms were highly diversified. The land was still barren and lifeless, but the seas were full of organisms of great variety. Life in the sea has changed greatly since then, but its overall pattern was already well established over 425 million years ago.

Innumerable groups have died out since the Ordovician, but their places were taken over by other groups, generally of more recent origin. Thus, later changes were often replacements rather than expansions. Among animals and animallike protists likely to leave a fossil record, only 12 phyla and 31 classes are in the present seas—only a slight decrease from the 33 classes in late Ordovician seas. No species have survived from Ordovician to Recent, and perhaps only one genus, the primitive brachiopod, *Lingula* (class Inarticulata), has hung on, little changed. Indeed, *Lingula* may be the oldest living genus on Earth.

Life in the Late Ordovician

We cannot, even in a summary way, review the life of the seas through all the geological periods. A quick survey of Ordovician life shows characteristic marine floras and faunas, which we will recognize from descriptions in earlier chapters.

The major protists of the late Ordovician were flagellates, forams, and radiolarians (see Chapter 40), though other protist groups probably also existed. Eight major animal phyla were present at that time.

- Sponges
- Coelenterates, some of which built large reefs
- Graptolites, now extinct but plentiful then as floating forms

TABLE 44-2
NUMBER OF CLASSES OF ANIMALS KNOWN TO HAVE EXISTED IN CAMBRIAN AND ORDOVICIAN TIMES

Time	Number of Phyla	Number of Classes
Cambrian		
Early	8	12
Middle	10	20
Late	11	22
Ordovician		
Early	11	27
Middle	12	32
Late	12	33

- Ectoprocts, similar to those now living
- Brachiopods, much more numerous than they are now
- Mollusks, especially sea snails (gastropods) and nautiloids (cephalo-pods), related to the relict present-day chambered nautilus[8]
- Arthropods, including the extinct trilobites (Figure 44-6), which reached their climax in the Ordovician, and euryptids, large creatures (strange to us) that looked like crustaceans, but are related to spiders and scorpions (see Chapter 42)
- Echinoderms, an abundant group in late Ordovician seas

It was in the Ordovician that the first fossil vertebrates appear. This completes the roster of the animal phyla commonly preserved as fossils. Strange as it seems to think of a sea without fishes, that was its status before this time.

DEVONIAN: THE AGE OF FISHES

It is not certain whether the first fishes evolved in running fresh water or seawater. These fishes, the jawless agnaths, were primitive and scarce (Figure 44-7). Their emergence, however, was one of the most significant events in the history of life. A clear evolutionary pathway runs from these early agnaths to the amphibians and all other terrestrial vertebrates. Thus, they marked the rise of the phylum that was to become dominant in every sphere it invaded.

In the Silurian, the period following the Ordovician, graptolites declined, corals expanded, species and genera arose and died out. Though scanty in number, Ordovician and Silurian fossil vertebrates include not only jawless agnaths, but armored fishes (the placoderms)—the first fishes with jaws. Another major event was the development of paired fins. This, and the development of jaws (from gill arches), were the two most prominent structural adaptations of that time. Another event of the Silurian was the still feeble beginnings of the occupation of land.

In the Devonian which followed, evolutionary activity accelerated in many groups. Fishes expanded and developed significantly, their most basic differentiation occurring mainly in that period. That is why the Devonian is called the Age of Fishes. By its end, the agnaths were extinct except for the still unidentified lines that led to modern hagfishes and lampreys.

Here is another pointed example of the historical principle of replacement. Jawed fishes diversified quickly and soon replaced Devonian agnaths, which it would seem were ousted by fishes of greater efficiency, the now extinct placoderms (class Placodermi). The placoderms expanded through most of the Devonian, but toward its end they were replaced by two groups to which they had been ancestral, the cartilaginous fishes (Chondrichthyes) and the bony fishes (Osteichthyes).[9] These groups were highly successful, and are the dominant fishes today. Devonian chondrichthyans included sharks not unlike those of today and several other divergent lines, most of which became extinct. A branch that does live on are the comparatively uncommon **chimeras.** The flattened skates and rays, now quite common, evolved from sharks at a later date—in the Jurassic.

We saw in Chapter 43 that early chondrichthyans seem originally to have been adapted mainly to salt water and osteichthyans to fresh water. In the Devonian, the osteichthyans were still mostly freshwater fishes, although their eventually highly successful invasion of salt water was probably under way (Figure 44-8). Even then, the osteichthyans were differentiated into three basic groups with very different future prospects.

- The **crossopterygians** or **crossopts** gave rise in the later Devonian to the primitive amphibians (an evolutionary step in which these lobe-finned fishes developed legs and feet) and, through them, to all the vertebrates of the land and air (Figure 44-9). After the amphibians had appeared, other crossopt lines lingered on in dwindling numbers. They are represented today by the single known lobe-finned relict, *Latimeria*, the coelacanth (see Chapter 43).

---

[8] Ammonites, relatives of nautiloids with more complex partitions between chambers, had not yet appeared in the Ordovician.

[9] This may be oversimplified. There is evidence that the placoderms and the acanthodians (a class now placed between Chondrichthyes and Osteichthyes) may have arisen separately from agnaths. The ancanthodians, not the placoderms, are the probable ancestors of both Chondrichthyes and Osteichthyes.

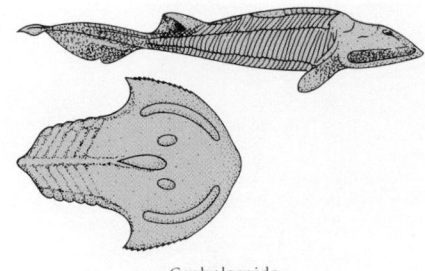

Cephalaspids

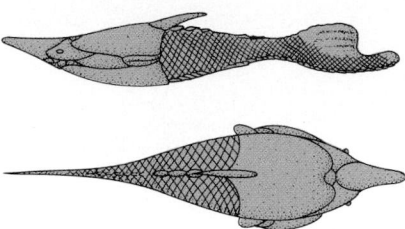

Pteraspids

44-7 Agnatha, the earliest jawless fishes
(*Top*) A side view of Cephalaspis *and a dorsal view of the head and trunk region of* Kieraspis, *showing the paired round eye sockets, the median nostril, and the pair of crescent-shaped apertures supposed to be electric organs like those of the modern electric eel.* (*Bottom*) *The side and dorsal views of* Pteraspis *reveal similar features.*

44-8 Some Devonian fishes
These fishes (Cheirolepis) *are paleoniscids, bony fishes* (Osteichthyes) *of a type that gave rise to all present-day fishes, including dominant teleosts.*

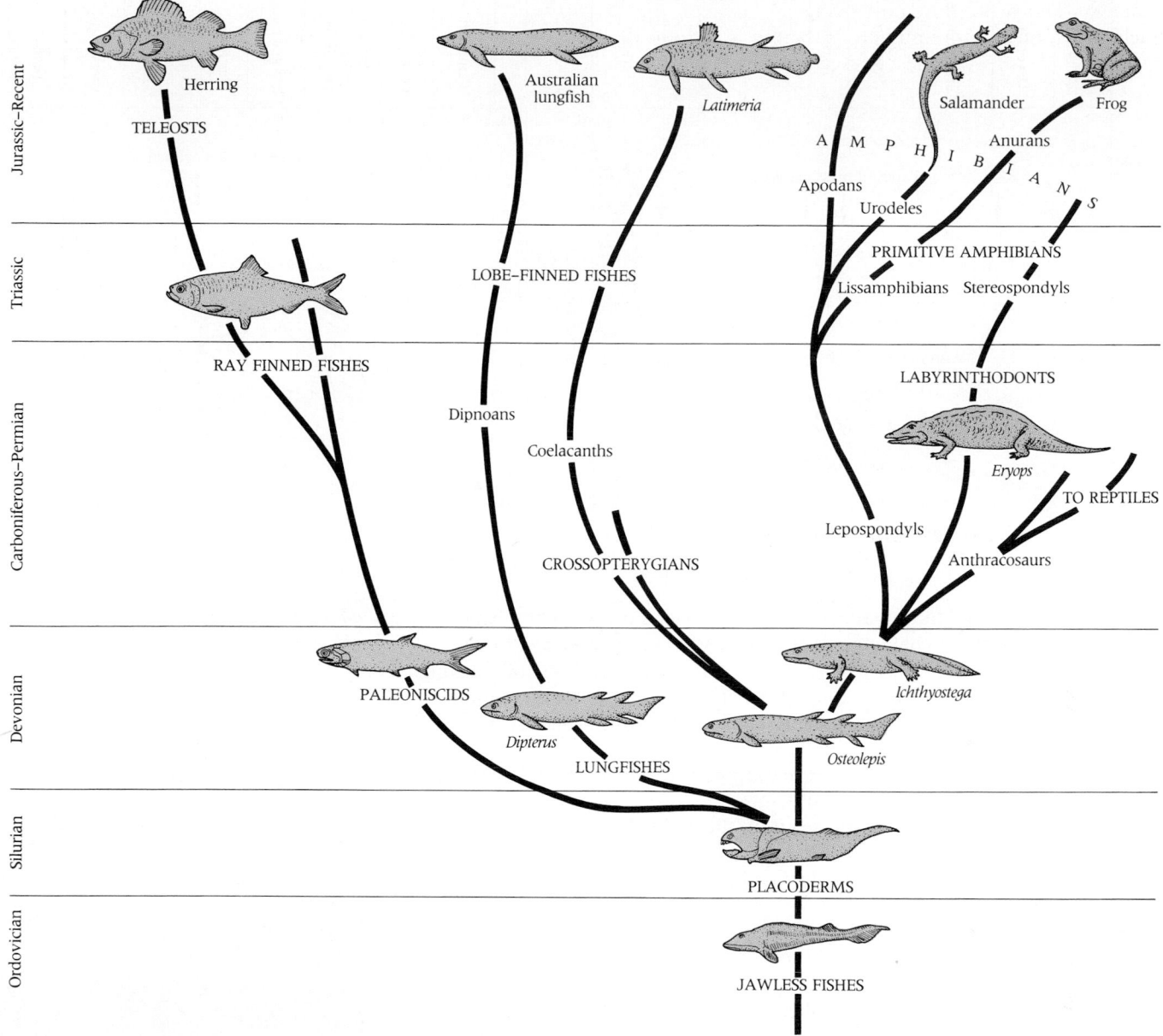

44-9 Phylogeny of lobe-finned fishes and amphibians

Devonian lobe-fins are classed in three groups. The dipnoans are represented today by lungfishes, the coelacanths by the living fossil, Latimeria, and the rhipidistians, here represented by Osteolepis, which had many amphibian features. Rhipidistians are extinct as fishes although all land vertebrates are among their descendants. Note the legs on Ichthyostega, an early amphibian.

- The **paleoniscids** (order Paleonisciformes), were the first bony fishes to originate from the ancestral acanthodians. They gave rise to a great radiation of fishes in both salt and fresh waters. That radiation, beginning in the Permian and continuing through the Mesozoic, produced nearly all our present-day fishes plus a number of extinct Mesozoic groups that were later replaced by the teleosts. Since the Cretaceous, teleosts have been the dominant fishes.
- The **lungfishes** (see Chapter 43), common in the later Devonian, dwindled steadily into a few surviving genera.

Apart from the fishes, an important event in the Devonian seas was the rise of the **ammonites,** cephalopods that were to be dominant marine mollusks in the Mesozoic. Their ancestral group, the nautiloids, continued, but it was dwindling in the Devonian. Most important of all, the movement of life onto the land was in full tilt during the Devonian. By its end, it was a fact.

OCCUPATION OF THE LAND

Problems to be Solved

We have spoken of the difficulties of life on land and the adaptations they made necessary in plants and animals.

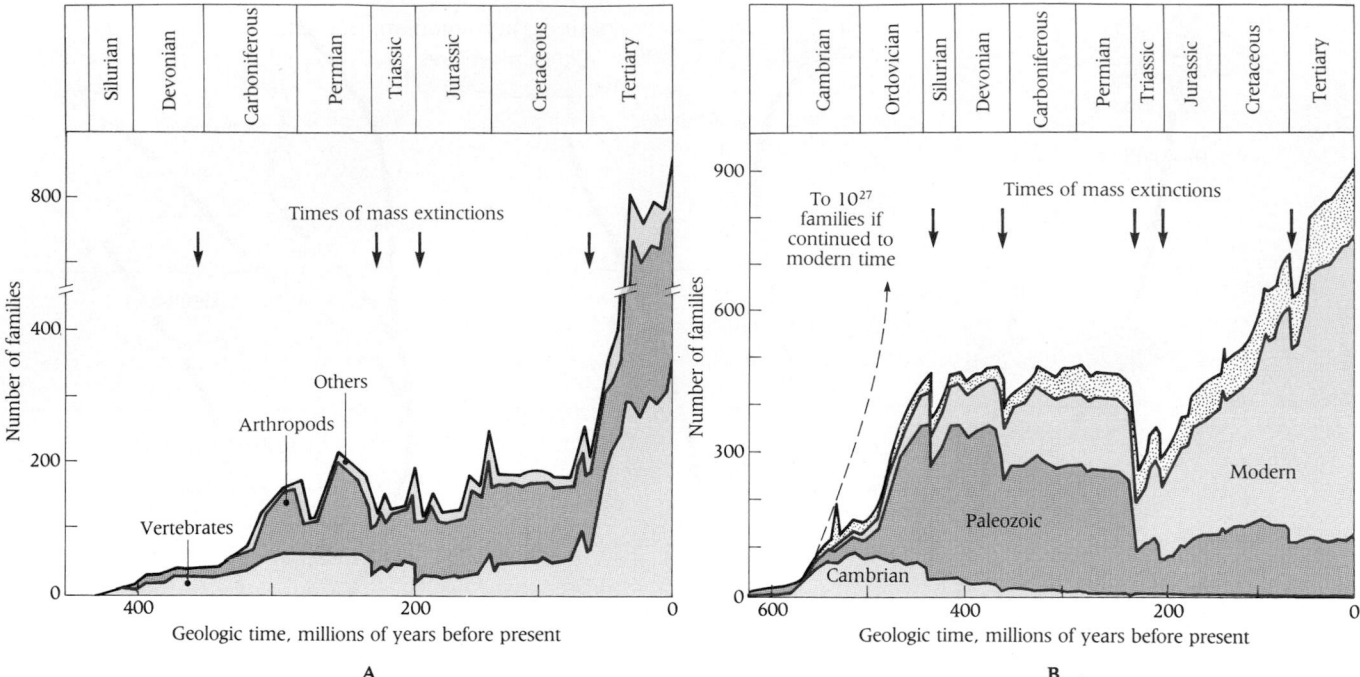

A

B

Unlike their ancestors, land organisms were no longer surrounded by water. This posed several major challenges. Water was still needed and had to be obtained from somewhere—by absorbing soil water through roots, by inhaling air-borne water vapor, by drinking liquid water, by eating plants and animals (which contain water), or by metabolic release of water from carbohydrates and lipids.

Once acquired, water had to be kept from evaporating. Hence, all fully terrestrial plants and animals needed external coverings.

Means were needed for extracting gases from the air (CO_2 and O_2 by plants, O_2 by animals), for getting these gases into solution in body fluids, and for discharging gaseous wastes into the air. Land plants developed stomata and associated structures. Land animals developed lungs, tracheae, or pouches or modified gills.

Gravity, no problem to an organism buoyed by water, became troublesome for all but the smallest land organisms. Many developed hydrostatic skeletons. Others developed strong jointed framework skeletons, external or internal. All true land plants have supportive tissues.

Extremes of temperature are far greater on land than in water.

Only four phyla came to include organisms that were fully and progressively adapted to land life: Tracheophyta, in the plant kingdom, and Mollusca, Arthropoda, and Chordata (Vertebrata) in the animal kingdom.

Earliest Land Plants and Animals

The fossil record of land biota[10] is considerably less complete than that of shallow marine forms. Nonetheless, the record confirms what seems a necessity of the historical sequence—namely, that plants should have led the way in the occupation of the land. Animals require food, and thus they are dependent directly or indirectly on plants and other photosynthesizers. Land plants did begin to appear before any land animals. The first certain and fairly common land-plant fossils appear in the third Paleozoic period, the Silurian (see Chapter 41). These were mainly small leafless **psilopsids,** the earliest vascular plants. The divergence of vascular plants was already underway.

Some wormlike phyla were probably early land dwellers, but the earliest fossilized land animals were arthropods. The oldest probable land animal, a fossil scorpion from the late Silurian, is enough like a fully terrestrial modern scorpion to suggest it had the same habitat. However, it also resembles aquatic forerunners.

[10] Biota is defined as all the living organisms, including the animal life (fauna) and plant life (flora) of a region or period.

44-10 Diversification of animal life
(**A**) *Increase in diversity of nonmarine animals with time. Note the times of mass extinctions.* (**B**) *Changes in the number of marine taxonomic families over geologic time. The upper curve depicts the total number of families.*

A

B

44-11 Vertebrates conquer the land
(**A**) *Devonian lobe-fin fish,* Eusthenopteron, *crawling out of the water.* (**B**) *Primitive Carboniferous-Permian labyrinthodont amphibians* Diplovertebron. *Note the legs.*

44-12 A model of a Carboniferous forest

Lycopsids: (1) *various species of* Sigillaria; (2) *two species of* Lepidodendron. *Sphenopsids:* (3) Calamites; (4) Sphenophyllum. *Ferns:* (5) Caulopteris; (6) Mariopteris. *Gymonosperms:* (7,8) *two species of* Neuropteris; (9) Cordaites. *Insects:* (10) *a primitive dragonfly,* Meganeura monyi.

It is in the following period, the Devonian, that animals were certainly terrestrial (Figure 44-10A). At first, all were arthropods. They included a mite, several forerunners of the spiders, and a creature that may have led to the insects. By the end of the Devonian, some of the crossopts acquired legs and evolved into amphibians (see Chapter 43). Though they were tetrapods, they still had fishlike tails and were probably as aquatic as fishes. Nevertheless, they were the first vertebrates to walk on land (Figure 44-11). Their descendants were to rise to dominance in this new environment.

During the Devonian, land plants became common (see Figure 41-22). Some reached the size of trees, and the first forests grew on the Earth. Psilopsids continued, but they were scarce by the late Devonian. The incoming, replacing groups were the lycopsids, sphenopsids, and ferns, all of which were abundant by the late Devonian. A few primitive gymnosperms (cordaites) had also appeared.

LATE PALEOZOIC: LIFE EXPANDS

The Carboniferous and the Permian followed the Devonian and were the last two periods of the Paleozoic era. They had much in common and are often considered together as the Permo-Carboniferous.

Rise of the Coal Forests

Carboniferous, meaning "carbon-bearing," is so named because many of our largest coal deposits, including those in the Appalachian regions, began in that age. *Coal* is the compressed remains of the plants that grew in the giant swamps and forests of the Carboniferous—among them, tall Lepidondron trees (lycopods), giant horsetails (*Calamites*), and great tree ferns, such as *Psaronius* (Figure 44-12).

Coal concists chiefly of carbon. Without long compression and sedimentation, partially decayed trees and plants form thick beds of the organic matter called *peat. Petroleum* and *natural gas,* which derive largely from dispersed organic materials in the sediments of inland seas and coastal marine basins, are mainly a mixture of hydrocarbons. It is obvious why all of these materials are called fossil fuels—but recent workers have asked: fossils of what? The answer seems to be plant remains and, surprisingly, lipids of the microorganisms responsible for their decay (see Chapter 45).

Despite deep burial for long periods of time, coal still retains traces of its original structure, and associated shales and sandstones often contain well-preserved fossil plants. For that reason, we are well acquainted with the compositions of forests and the structures of their individual plants (Figure 44-12).

We have already considered the main plant groups, the lycopsids (club mosses) and early gymnosperms—seed ferns, cordaites, and conifers, which did not become distinct until the late Carboniferous. In the late Permian and early Triassic, cycadeoids, cycads, and ginkgos appeared but they were not part of the earlier coal forests.

Land Invertebrates

Animal life, hitherto rare, literally swarmed in Carboniferous forests. Here appeared the first land snails, the only mollusks to complete the great transition from water to air. Undoubtedly, terrestrial scorpions and centipedes were present. The most important land invertebrates were the insects, the majority of which belonged to orders now extinct. But cockroaches were there, and so were forerunners of the dragonflies. Most Carboniferous insects were small, but a few were huge. One dragonflylike giant had a wingspread of nearly 30 inches. No later insect ever reached that size. It is obvious that the insects we now associate with flowers were absent in the Carboniferous, which lacked flowering plants. They begin to appear in the fossil record in the Jurassic and spread greatly during the Cretaceous and early Cenozoic, in parallel with the flowering plants.

Amphibians and Reptiles

As in the development of marine invertebrates of the Cambrian, many of the fundamental structural patterns of tetrapods developed early in their radiation.

44-13 Amphibians and reptiles of the early Permian
(**A**) Eryops, *a labyrinthodont amphibian.*
(**B**) Dimetrodon, *an early mammallike reptile.*

Amphibians were common during the Permo-Carboniferous, the Age of Amphibians, though they were already outnumbered by reptiles in the Permian before any of the modern groups had yet evolved. Most Permo-Carboniferous amphibians were labyrinthodonts ("labyrinth-toothed," so-called because sections of tooth tissues had a maze-like structure) (see Figure 44-9). Typically, they were clumsy brutes with four short sprawling legs, big flattened heads, and stubby tails. An example is *Eryops,* an early Permian labyrinthodont about 1.5 m long (Figure 44-13A).

The rise of the reptiles completed the transition of vertebrates to land life. The change was a gradual one. Some late Carboniferous fossils were already true reptiles, and reptiles were abundant in the Permian. The main lines of reptilian evolution are shown in Figure 44-14. Most of the Permian reptiles belonged to only two main groups, neither of which resembles today's reptiles.

- The primitive cotylosaursor are thought to have given rise to root reptiles, which rise to more or all of the higher land vertebrates.
- The mammallike reptiles were even more common. The earliest of them, the pelycosaurs ("basin reptiles," so-called because of their basinlike pelvis) included the ancestors of the whole reptile group (and hence of the mammals) and also some divergent, extinct lines of creatures like *Dimetrodon* (see Figure 44-13B), which had a fantastic fin on its back that was supported by long spines from the vertebral column. The later and eventually very diverse therapsids ("mammal-arched," so-called because the cheekbone or zygomatic arch was mammallike and not reptilelike) were the dominant land animals of the late Permian and early Triassic. We believe they were the immediate ancestors of the mammals.

THE AGE OF REPTILES

The Permian was in fact already an "age of reptiles," though that appellation is usually applied to the next three periods that together comprise the Mesozoic era—the Triassic, Jurassic, and Cretaceous.

There is growing awareness that evolutionary lines are often perturbed and that these disturbances may profoundly affect the course of events. As discussed in Chapter 36, some biologists have rejected a gradualistic model of evolution in favor of the view that evolution of species and higher taxa is a jerky or spasmodic process with periods of relative stability (stasis) that are occasionally interrupted, or punctuated, by brief episodes of great change. It is a position that is by no means unanimous.

There is ample evidence that episodes of abrupt change have occurred. The term *crisis* has been applied to them, though that term is defined in different ways.

44-14 The radiation of reptiles
The cotylosaurs, or root reptiles, were derived from the labyrinthodont amphibians. They gave rise to five major lines of reptiles: turtles, plesiosaurs, thecodonts (and snakes, lizards, and rhychocephalians), ichthyosaurs, and mammallike reptiles. The last line gave rise to mammals.

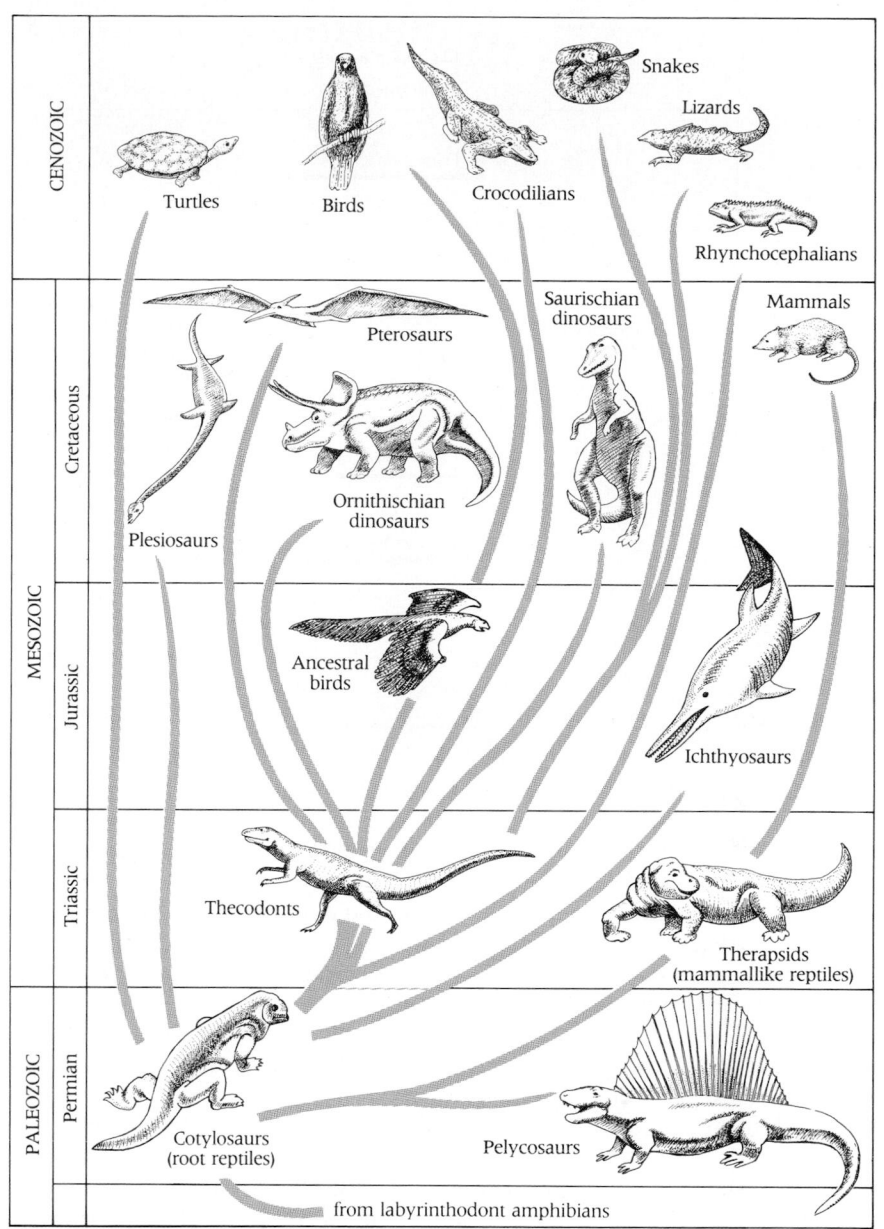

All agree that a crisis is a turning point, usually brief in duration and unpredictable, in which fundamental changes occur in the natural order of a system. However, a crisis may be good for some organisms and bad for others.

There have been a great many crises in the history of life. One, surely, was the addition of oxygen to the atmosphere in the Precambrian. Others were the five mass extinctions which took place in the late Ordovician, the late Devonian, the late Permian, the late Triassic, and the late Cretaceous periods. As shown in Figure 44-15, these crises are classified as major and intermediate. A number of others, about ten, were of "lesser" severity.[11] We are now concerned with the critical major extinction that took place in the late Permian and early Triassic. We will later consider the possible causes of these crises.

The Permo-Triassic Crisis

The Permian was an eventful time in the history of life. Evolution was proceeding rapidly in many groups. Decline or extinction of some groups was accompanied

---

[11] Many causes have been suggested for these fascinating episodes, in which large numbers of plant and animal species were wiped out in a relatively short time, but none has been proven. Among the causes are such major events as glaciation and sea level change in the late Ordovician and, at the end of the Cretaceous, a collision between Earth and a very large meteorite.

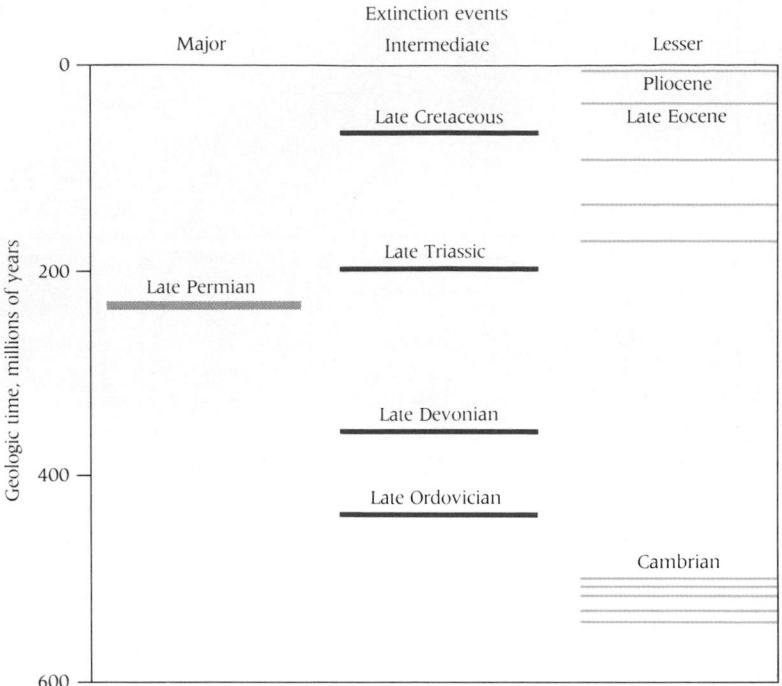

by sudden expansion of others. The ammonites (cephalopod mollusks) rapidly changed and expanded, but the brachiopods declined (some say because of competition from bivalved mussels) and were never again the common marine shells they had been through most of the Paleozoic. Other groups also came and went.

The huge scope of the extinctions of marine and nonmarine animals that occurred in the late Permian is shown in Figure 44-10. (The decline in plants is shown in Figure 41-22.) Among the casualties were the trilobites. Once plentiful, but in decline since the Ordovician, they now became extinct. The last of the placoderms disappeared. Most of the archaic chondrichthyans and osteichthyans declined, their places to be taken by more progressive offshoots that evolved and expanded steadily through the Mesozoic and on to the present.

On land, most of the great coal-forest trees became extinct, but the cycadeoids, cycads, ginkgos, and conifers expanded (see Chapter 41). The flowering plants probably originated later—in the Triassic. Not until the Cretaceous did they begin their remarkable expansion. By the end of that period they had replaced most other land plants.

Most of the amphibians were victims of the Permo-Triassic crisis. Even the labyrinthodonts, so common in the Triassic, had disappeared by its end. As far as we know, the only survivors were the immediate ancestors of the amphibians that exist today.

Dinosaurs

The reptiles also changed dramatically. The two main groups of the late Permian did survive and, for a time, expand in the Triassic (see Figure 44-14). But amazingly, they were nearly extinct at the end of that period. Only a few advanced mammallike reptiles continued into the Jurassic. During their great days in the Triassic, there was a tremendous radiation of new reptilian orders that set the pattern for the rest of the Mesozoic. The most impressive and renowned of the innumerable Mesozoic reptiles were the animals popularly known as dinosaurs (Greek, "terrible reptiles").

As we noted in Chapter 43, the dinosaurs comprised two orders of reptiles, the Ornithischia and Saurischia (Figure 44-16). Numerous and varied in the Jurassic and Cretaceous, their years of reign, they reached incredible physical dimensions. Indeed, no larger animals ever walked on land (though larger ones still swim the seas). For example, *Brontosaurus*, now called *Apatosaurus*, the genus that usually springs to mind when we think of dinosaurs, exceeded 85 feet (26 m) in length and 30 tons in weight. In 1979, a paleontologist digging in an ancient dry river bed on a windswept Colorado mesa found the bones of the largest dinosaur yet

44-16 History of the dinosaurs
The earliest dinosaurs of the late Triassic were saurischian and are represented by the small *Coelophysis. They were part of the radiation of thecodonts. The ornithischian dinosaurs diverged early as a distinct order.*

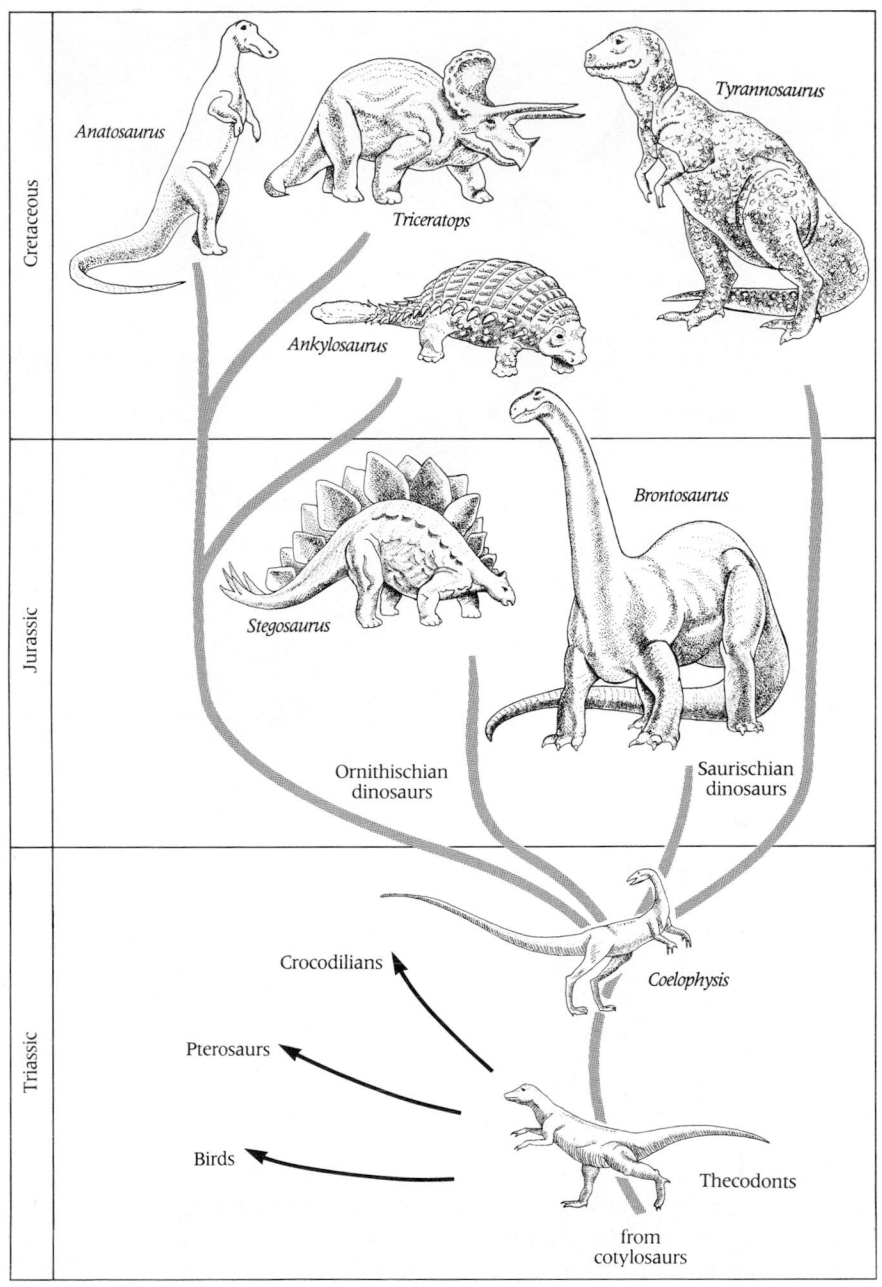

discovered. Nicknamed "ultrasaurus," it was some 85 feet (26 m) long and probably weighed 80 tons—a huge animal that could have looked into the top floor windows of a six-story building.[12] Of course, not all dinosaurs were large. They came in many shapes and sizes. Some were not much bigger than chickens.

Dinosaurs did not exist in the early Triassic. Their ancestry was represented at that time by a primitive group, the thecodonts, that later radiated widely. They gave rise not only to the two dinosaur orders but also to the crocodiles, flying reptiles (pterosaurs), and birds (see Figure 44-14). Indeed, a convincing argument holds that birds arose from early dinosaurs (the coelurosaurs).

Until recently, we regarded dinosaurs as typical reptiles with reptilian skeletons, cold-blooded (ectothermic) creatures with body temperatures that fluctuated with that of the environment. In 1965, it was proposed that some of them may have been warm-blooded (endothermic). The contention was based on evidence of intensive internal metabolism, patterns of blood vessels in bones and the presence of heat-conserving body coverings. While conceding its possibility, many reject this idea. There the matter stands.

[12] It is probably a species of *Brachiosaurus,* long-necked sauropod herbivores, but it will take years to classify it positively and name it formally.

The earliest true dinosaurs, appearing in the late Triassic, were slender, mainly bipedal (two-legged) forms such as *Coelophysis,* a coelurosaur (Figure 44-17). These soon gave way in the early Jurassic to a diversity of saurischians and about six genera of ornithischians. The saurischians included the **sauropods,** huge herbivorous, quadrupedal, long-necked and long-tailed relatives of *Brontosaurus*—and the **therapods,** a host of bipedal, mostly carnivorous forms.[13] Some, certain therapods, were tiny as dinosaurs go, but some, like the spectacular *Tyrannosaurus,* were very large, extending 50 feet. The first intact Tyrannosaurus skeleton was found in 1990.

All the ornithischians were primarily herbivorous. Four main groups arose.

- Plated dinosaurs, *Stegosaurus* and its relatives, ungainly quadrupeds with rows of strange upright bony plates down the back that some have suggested were both heat dissipators and solar energy panels
- Armored dinosaurs, such as *Ankylosaurus* (suborder Ankylosauria), animated four-footed tanks almost completely enclosed in a bony carapace
- Horned dinosaurs (suborder Ceratopsia), like *Triceratops,* hoofed creatures with facial horns
- Duck-billed dinosaurs, the hadrosaurs (family Hadrosauridae), including *Anatosaurus,* large herbivorous bipedal forms, many with extended skull bones over the neck and, in some forms, horns

[13] This is established by the discovery of the skeletons of small dinosaurs within the skeletal body cavities of larger dinosaurs. Sometimes the two skeletons were of the same species; thus cannibalism was practiced. However, this was not the exclusive diet.

44-17 Diversity of dinosaurs
The most prominent, Tyrannosaurus rex, *was the largest terrestrial predator. Its knee was 6 ft from the ground.* Coelophysis *was about the size of an ordinary dog.* Stegosaurus, *the plated dinosaur, was 30 ft long. The armored* Euoplocephalus *was 16 ft long. The horned* Triceratops *was 26 ft long, and* Anatosaurus *was 33 ft long. Aquatic reptiles included pleisosaurs, which had long necks; ichthyosaurs with prominent tail fins and small hindlimbs; and mosasaurs, which were up to 26 ft long. Among the winged reptiles were the slender pterosaur,* Rhamphorynchus *(wing span 6 ft),* Pteranodon *(wing span 26 ft), and little* Dimorphodon *(wing span only 5 ft). Note that all of these forms did not exist at the same time.*

About 250 dinosaur genera are known. They have been found on every continent except Antarctica.[14] The richest known deposits, however, are in North America. All of our examples are native to North America.

Much has been written about the puzzling questions raised by dinosaurs. How did they originate? How did creatures of such size generate the necessary metabolic energy to keep themselves going? Finally, the greatest mystery, what did them in? Why did they die? We return to that issue in a moment.

Other Mesozoic Reptiles

The dinosaurs occupied certain niches of the Mesozoic world, but all manner of other reptiles lived along with them. Land reptiles, some considerably smaller than dinosaurs, included even in the Triassic the rhynchocephalians (see Chapter 43). Four types of modern reptiles arose in the Mesozoic: lizards in the Jurassic, snakes in the Cretaceous, and crocodiles and turtles in the Triassic.

Several groups of fully aquatic and marine reptiles arose in the Mesozoic. The **plesiosaurs** ("nearly reptiles"[15]) were ocean-going reptiles with broad, flattened bodies, four paddles, and a long neck or tail, or both (see Figure 44-17). Someone has likened them to a snake threaded through an armorless turtle. The **ichthyosaurs** ("fish reptiles") looked something like porpoises or dolphins and indeed represented the most complete adaptation of the reptiles to an aquatic mode of life. It was a return to the sea and the life style of a hunter that fed on fishes and cephalopods.

The late Cretaceous saw the evolution of another widespread group of reptiles that had a brief but lively career: the large marine, fish-eaters called **mosasaurs** ("reptiles of the River Meuse"—first found there, but best known from later finds in Kansas). They were overgrown lizards that lived in the sea (see Figure 44-17).

There are no fully aquatic lizards today, though the Galápagos marine iguana is a lizard that finds its food in the sea, feeding on algae between the tides. Other present-day lizards typically escape predators by diving into the water. We would call them semiaquatic. Fully aquatic and marine snakes and sea turtles also exist today. They are further examples of successful marine adaptations among reptiles and are surprisingly numerous.

Appearance of Birds

Some reptiles took to the air. The only organisms up to this time that evolved a capacity to fly, other than insects, were two related but independent lines of late Triassic or early Jurassic reptiles (Figure 44-18). One group that enjoyed considerable but transient success were the pterosaurs ("wing reptiles"). They were the first true flying vertebrates. The other group, after a slow start, went on to become so distinctive, widespread, and diverse it could no longer be considered reptilian. Rather it became a new and more differentiated class: the birds (class Aves).

The first known bird, as we have seen, was *Archaeopteryx* ("ancient wing") from the middle Jurassic (see Chapter 43), of which six skeletons have been found— the sixth having been reported in 1988. All specimens come from the fine-grained limestones of Bavaria, renowned by printers for its excellence in lithography. For that reason, the species was named *Archaeopteryx lithographica*.

The teeth, a long, reptilian-type jointed tail, and other reptilian characters of archaeopteryx might have placed it among the reptiles, but by a rare chance, impressions of feathers, which usually do not fossilize, have been vividly preserved (see Figure 43-16). These remarkable fossils show that *Archaeopteryx* was about the size of a crow, though the largest of the six specimens is twice the size of the smallest.

[14] The southernmost dinosaur find has been in southern Argentina (latitude 48°S). The northernmost find (footprints) is on the island of Spitsbergen (80°N), well inside the Arctic Circle. As we will learn in Chapter 46, these were surely not the latitudes when the dinosaurs were living.

[15] We are not sure what the namer had in mind when he applied that name to a group completely reptilian, however queer in appearance and habits.

44-18 A Jurassic scene
On the ground is the small dinosaur Ornitholestes. *In the air and on the tree are pterosaurs and the ancestral birds* Archaeopteryx. *The trees are cycadeoids.*

For that reason, some think that two species are represented. It had feathered wings attached to an otherwise reptilian forearm and hand. Since a large breastbone (sternum) of the type needed for flight muscles was lacking, it is possible that *Archaeopteryx* lived in trees and used its feathers more for thermal insulation than for flight.

This fortunate discovery proves that a living, workable intermediate can exist between, say, a reptile and a bird, a fish and an amphibian, or a land mammal and a whale. As we mentioned in Chapter 36, it has been held that the really radical adaptive changes of evolution may have taken place all at once in one explosive mutational leap, or saltation. But *Archaeopteryx* is an obvious intermediate

44-19 (Below) *Reptile extinction in the late Cretaceous*
Of the 48 families shown, 24 died out at the end of the last stage of the Cretaceous, some 65 million years ago. A colored extension of the black bar in the Cretaceous indicates a record of the family elsewhere than in North America north of the Rio Grande. A colored bar in the Paleocene column indicates the survival of the family outside North America. Scarcely any reptilian order went unscathed.

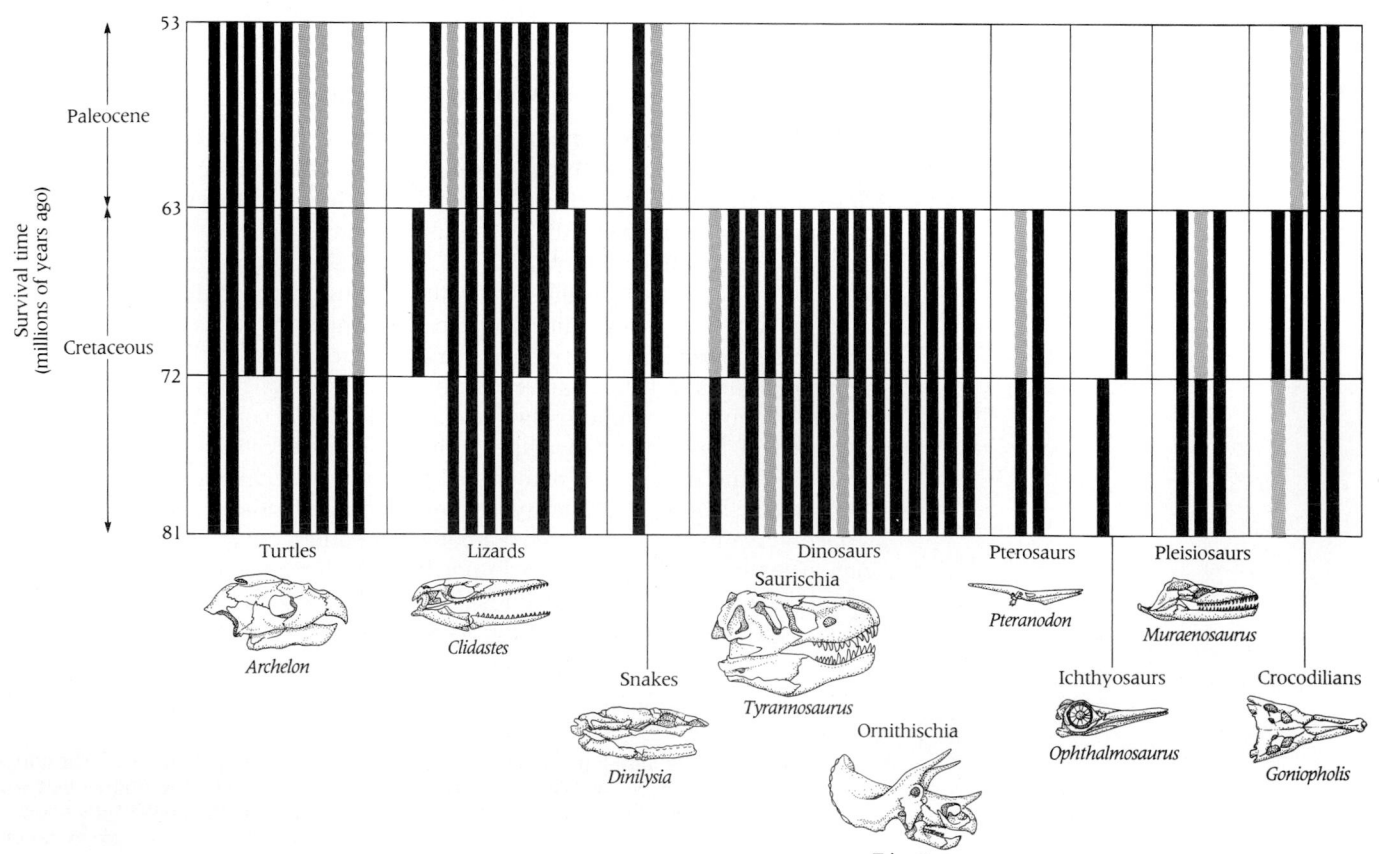

between a reptile and a bird. These characteristics suggest that reptiles did not become birds by saltation.[16]

In viewing the *Archaeopteryx*, an intermediate between reptile and bird, we do not suggest that it was halfway between them in all of the characters that distinguish these classes. Rather, it is a mixture, close to half-and-half, of reptilian and avian characters. This sort of thing is called **mosaic evolution.** Mosaics, which are quite common, display various combinations of ancestral and derived characters.

Mass Extinction in the Late Cretaceous: The "Great Dying"

One of the most striking events in the history of life was the mass extinction in the late Cretaceous, which witnessed the disappearance, at the end of the Mesozoic era some 65 million years ago, of many kinds of reptiles, certain kinds of marine invertebrates, and certain kinds of plants. Indeed, extinction was so widespread that it led to the destruction of 25% of all families, 13% of marine families, 50% of marine genera, and probably 60–75% of marine species. It also cleared the Earth of its great marine reptiles, plesiosaurs, ichthyosaurs, and mosasaurs—and of its dominant terrestrial animals, the dinosaurs and their kin (Figure 44-19). Marsupial mammals were hard hit, but they managed to survive. This event set the stage for the eventual dominance of mammals.

Of the many invertebrate groups that disappeared, some of the most interesting were the coil-shelled ammonites (Figure 44-20), cephalopod mollusks which are extremely abundant in most Cretaceous marine rocks. Yet, near the end of the period they disappeared completely.[17] As for the marine reptiles, nothing radical happened to the fishes on which they fed. Of the Mesozoic reptilian hordes, only four orders survived (see Chapter 43), and one (Rhynchocephalia) dwindled to relict status.

For generations, scholars have attempted to explain mass extinctions, and especially the extinction of the dinosaurs. We know there was a widespread environmental change—cosmic radiation temporarily increased, continents coalesced with a drastic reduction in the area of shallow seas, Earth temperature gradually declined, and as mentioned in the next chapter there is evidence of a temporary burst in the atmospheric CO_2 content at the end of the Mesozoic which would have briefly warmed the climate. There is also evidence that a layer of salt-free water "spilled over" from the Arctic and covered the world's oceans, lowering salinity at the surface and depleting oxygen below the surface, possibly destroying marine biota.

One recent theory is based on the discovery in late Mesozoic limestones of a layer of clay containing rich deposits of the heavy element *iridium* that is rarely found in the Earth's crust but is abundant in meteorites. This material evidently was deposited on Earth's surface by extraterrestrial meteoritic material arising in a gigantic stellar explosion, a supernova. Whatever its source, it is postulated that great clouds of dust-sized particles were injected into the stratosphere. The catastrophe blocked the entry of sunlight, drastically slowed photosynthesis—for a decade perhaps—and thus stressed all plants and animals. This idea has been weakened by new studies revealing changes in the flora and fauna *before* the catastrophic events that ended the age. These findings suggest that there were multiple, perhaps unrelated, causes for one of the greatest of mass extinctions. Despite an outpouring of imaginative hypotheses, we remain in doubt as to just what was the critical change.[18]

Whatever may have been its exact cause, the great dying did occur. It closed one era and opened the world to another era, the Age of Mammals.

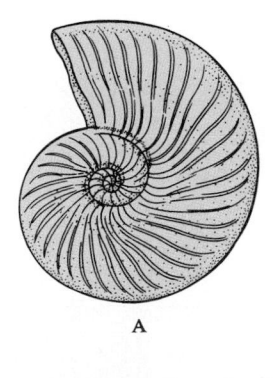

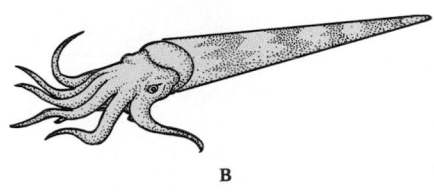

44-20 Cretaceous ammonites
(A) Oxytropidoceras acutocartinatum, *about ⅓ of its actual size.* *(B) Reconstruction of* Orthoceras. *The shells of ancient cephalopods were up to 3.8 m long. Most ammonites are coiled in plane spiral (see the ammonite fossil in the chapter opening collage), but some are straight. These cephalopods were enormously abundant.*

---

[16] Likewise, some rare late Devonian or early Carboniferous animals have turned out to be almost perfect intermediates between fishes and amphibians. The intermediates between land animals and whales have not yet been found, but who can claim they never existed?

[17] It is also interesting that they had almost become extinct long before, at the end of the Triassic, when only one small group, probably only a single genus, survived. From it, however, a tremendous new radiation occurred in the Jurassic and into the Cretaceous. The failure of even one genus to pull through the similar crisis at the end of the Cretaceous and to restock Cenozoic seas may have been due to competition from the abundant squidlike cephalopods.

[18] Anyone determined to find the answer should keep certain facts in mind. Groups that became extinct were of many different kinds, living in entirely different environments. Other groups living along with them in the same environments did not become extinct. Some did not even undergo evident change. Clearly, it is not necessary or probable that any one factor caused the extinction. Perhaps an asteroid arrived while the mass extinction was under way.

MODERNIZATION OF THE LIVING WORLD

By the end of the Cretaceous we are far along in the history of life, with only 70 million years to go to reach the present. What remains is the Cenozoic, the Age of Mammals. Only its brief last part includes humans, which are mammals too.

IN AQUATIC ENVIRONMENTS

We have seen that even in the late Ordovician, more than 425 million years ago, the seas swarmed with plants and animals about as diverse as those now living in the same environments. The aquatic phyla, the first to evolve, were the same then as now. Yet most of these groups became extinct and were replaced by others whose direct Ordovician ancestors were few and obscure.

The fossil record of aquatic plants is inadequate. Most of them are soft-bodied and, if they fossilize at all, they do so as a smear of carbon. Moreover, most of the algae present in the Precambrian were, as noted above, prokaryotic blue-green algae that are quite unlike later marine plants or plantlike protists. Probably most of the main groups of algae were present in the Paleozoic. They have never lost their dominance in aquatic environments, though many new species have evolved more recently. Thus aquatic floras seem already to have been modern several hundred million years ago. The most recent important change was the appearance of the diatoms, first found as fossils in the Jurassic and increasingly common since then. As we have seen, they account today for much of the photosynthesis in aquatic environments.

A few modern groups of aquatic invertebrates date from the early Paleozoic. Most, however, have been extensively replaced since then. Corals, ectoprocts, and clams, for example, had much the same roles in Paleozoic seas as they do now and looked about the same then. Still, these groups (or their major subgroups) were almost completely replaced.

Although they got started much later, the fishes rather closely paralleled the aquatic invertebrates in their modernization. As already noted, the archaic groups of fishes dominant in the Paleozoic dwindled during or before the Permo-Triassic crisis. The beginning of dominance of higher bony fishes was evident in the Triassic. The most progressive main group, the teleosts, had appeared and was spreading in the Jurassic. By the end of the Cretaceous it was fully dominant and modern in aspect.

The archaic amphibians also disappeared in the Permo-Triassic crisis. The rise of the modern groups is poorly documented by fossils, for some unknown reason, but frogs almost modern in appearance are known from the Jurassic. The salamanders may be equally old, but none is yet known before the Cretaceous.

The extinction of most aquatic reptiles in the Cretaceous left the turtles and the crocodiles (and their relatives the alligators) as the common amphibious or aquatic reptiles. Neither group changed very much during the Cenozoic.

The most striking change in aquatic environments during the Cenozoic, practically the only change of really deep significance, was the rise and spread of aquatic mammals, whales, seals, sea cows, and their relatives. They began to appear in the Eocene and were common and essentially modernized in the Miocene.

44-21 Early bird of the Cretaceous
The great-toothed diver, Hesperornis regalis, *was a large, wingless, swimming bird with teeth.*

44-22 Diatryma
A large, flightless bird of the Eocene.

IN LAND ENVIRONMENTS

The dominant, archaic vegetation of the Carboniferous vanished—largely during the Permo-Triassic crisis. In the Triassic and Jurassic, there was a sort of interim dominant floral type composed mainly of ferns, cycadeoids, cycads, ginkgos, and conifers. Most of those groups survive now, but in diminished numbers. Their dominance and the scarcity of flowering plants gave the plant life of the Triassic and Jurassic a decidedly unmodern aspect. By late Cretaceous, however, the aspect was fully modern. Plants known from late Cretaceous survive today.

Modernization of insects also began during the Permo-Triassic crisis, when most of the archaic groups that had appeared in the Carboniferous became extinct. No important groups that are now extinct evolved after the Permian. The whole picture of insect evolution during the Mesozoic was one of steady expansion by the evolution of new groups. Expansion went on also through the Cenozoic and is perhaps still going on. But most if not all of the main groups were present and modern in form in the late Cretaceous.

Birds rarely fossilize. However, patient study has revealed that they arose in the Jurassic and became fully birdlike, essentially modern in structure, by the end of the Cretaceous. Some of them still had teeth then, but their skeletons were no longer semireptilian. They were now avian. Moreover, they had undergone sharp divergence, for large, wingless, swimming birds existed at the time alongside ordinary flying birds (Figure 44-21).

Expansion and subdivision of the birds apparently occurred early in the Cenozoic and birds surely flitted in great numbers throughout the Cenozoic, which could as appropriately be called the Age of Birds as the Age of Mammals. Most bird orders were already present in the Eocene.

Large flightless birds evolved on all the larger land areas of the Cenozoic. Several groups have survived: ostriches in Africa and Arabia (and formerly over most of Eurasia), rheas in South America, cassowaries and the closely related emus in Australia and New Guinea, and penguins in Antarctica (unless it be allowed that penguins fly in water rather than air, since they propel themselves there with their wings rather than their webbed feet). More flightless birds have become extinct: *Diatryma* in North America (Figure 44-22), with relatives in Europe; moas in New Zealand; "elephant birds" (*Aepyornis* and relatives) in Madagascar: and several other kinds in South America.[19]

HISTORY OF MAMMALS

Mesozoic Mammals

In the late Triassic or earliest Jurassic the mammals gradually evolved from the once great mammallike reptiles, which then became extinct (Figures 44-23 and 44-24). The reptiles that gave rise to mammals were, of course, very different from

---

[19] It is an oddity of evolution that the birds, after acquiring flight and dominating a new environment, repeatedly gave rise to lines in which flight was lost. Nonetheless, these large running birds competed quite successfully with the mammals.

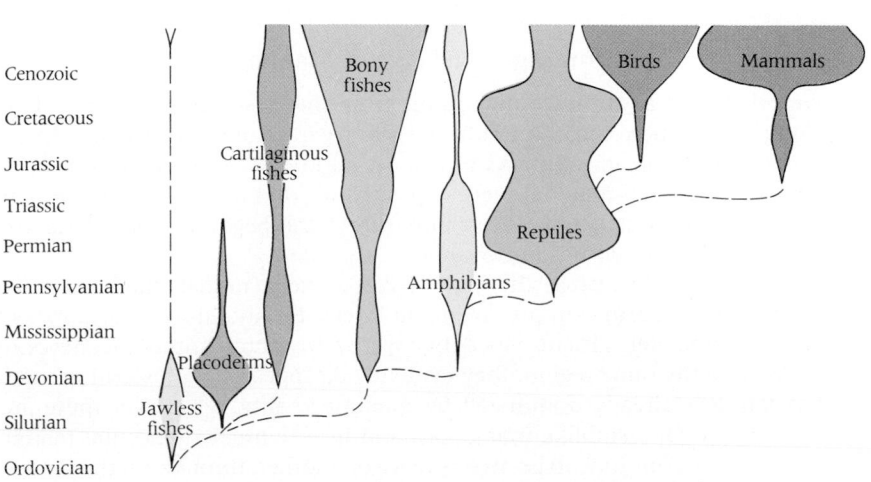

44–23 Historical record of the vertebrates
The widths of the pathways roughly approximate the relative numbers of known genera in the classes.

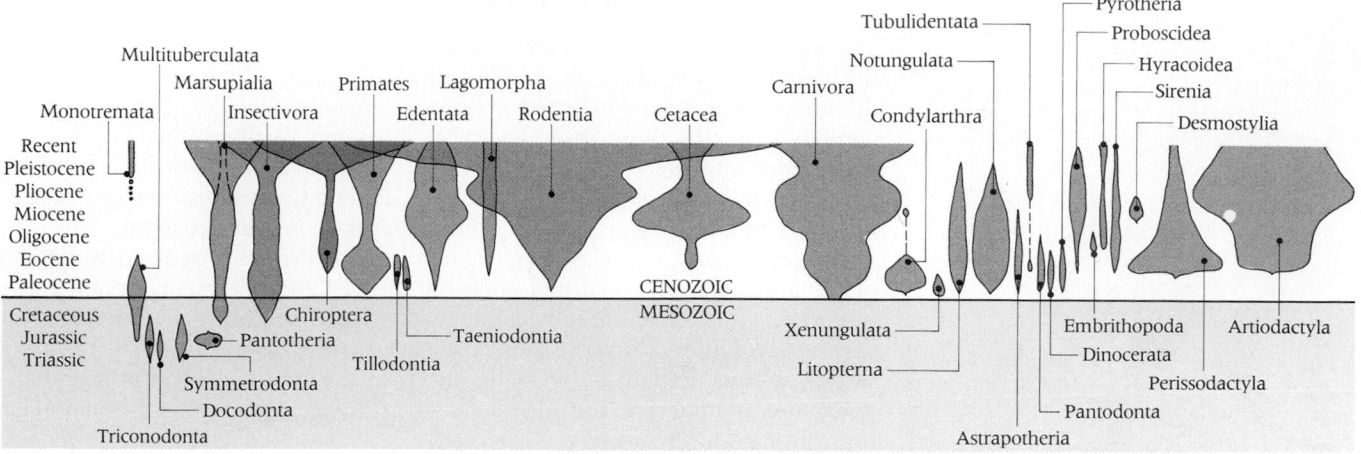

any reptiles that live today. Again a comparison of recent animals with ancestral animals misleads, for it gives no clue to the evolutionary changes that actually took place. The reptilian ancestors of the mammals were different and actually more mammallike than any later reptiles—rather as the fishes ancestral to amphibians were different and more amphibianlike than later fishes. Since the evolutionary process was gradual, some mammalian characteristics must have arisen in animals nominally classified as reptiles. Some, in fact, did not arise until after nominal mammals existed. It appears, too, that at the time when mammals evolved from reptiles the ancestors of today's living reptiles were committed to a radically different evolutionary trend.

Although they were the most progressive of organisms by the usual definitions, the mammals rose to dominance in their own environments more slowly than most other groups that were dominant in other environments at various times in the history of life. For more than 100 million years after they first appeared, mammals cut a small figure in the world. That is a considerably longer time span than the whole era of their later dominance, the Age of Mammals. Even at the end of the Cretaceous when the Mesozoic era came to a close, mammals were still small, mostly mouse-sized (Figure 44-25), and rare.

Why were the mammals so obscure for so long? We do not know, but we can offer a hypothesis. When amphibians arose, they had a new way of life with no competitors—so their expansion was rapid (speaking in terms of the geological time scale). Early reptiles also entered new environments empty of competitors. They expanded rapidly and soon nearly wiped out the amphibians, whose environments overlapped theirs. When mammals arose, the situation was different. The niches accessible to them were already occupied by well-adapted reptiles, which competed with them and preyed on them. Mammals may have become nocturnal as a result—and this circumstance may have generated selective forces favoring endothermy, a major factor in the later success of mammals.

Beginning of the Age of Mammals

The Cenozoic began at an unfortunate time from the fossil hunters' point of view. In the long seesaw between the uplift and the wearing down of the continents it was a time of widespread uplift. As a result erosion was widespread, but most of the sediments of eroded material were deposited beyond our reach. Consequently, few fossils of land animals have been found from the beginning of the Cenozoic. Much went on that we do not know about.

In spite of intensive search, the earliest Paleocene mammals have been found in only one part of Earth—in and near the Rocky Mountains. In these rocks the dinosaurs are completely absent. Since they were fairly abundant in the latest Cretaceous rocks from the same region, they evidently disappeared with startling suddenness. The fauna is already dominated by mammals, although as yet these are of only a few kinds. Opossumlike marsupials and insectivorelike placental mammals are present and will continue to be, in one place or another, throughout the Cenozoic,

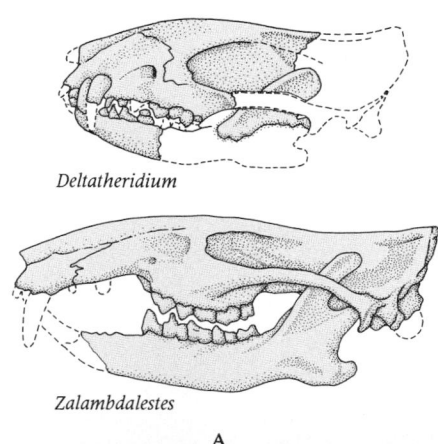

Deltatheridium

Zalambdalestes

A

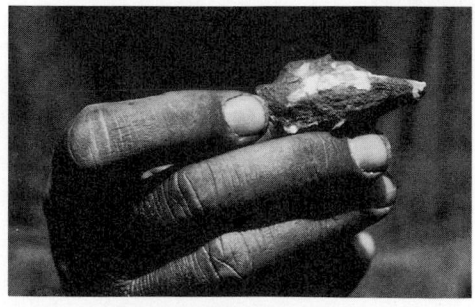

B

44-25 Mesozoic mammals
(A) *Partially restored skulls of two small Cretaceous mammals. Skulls are about 2 inches long.* (B) *An actual skull.*

44–26 *Ectoconus,* **a Paleocene mammal**

though in a small minority. The common mammals were somewhat larger (the largest about the size of a small collie) and much alike (Figure 44-26).

However, within the 15-million-year span of the Paleocene (which began 70 million years ago) the Age of Mammals was under way. Rapid expansion and divergence occurred among the placental mammals and by the late Paleocene there were many kinds: hoofed herbivores, some over a meter in height; various sizes and kinds of predacious carnivores; small forerunners of the monkeys; the first, still very rare, rodents; and others. Clearly, the mammals were finally inheriting the Earth. When environments once occupied by the Mesozoic land reptiles were emptied, the mammals in a world-wide adaptive radiation actively occupied them. The huge grazing dinosaurs were replaced by such grazing mammals as deer, buffalo, goats, and antelopes. Carnivorous dinosaurs were replaced by carnivorous mammals. More efficient and more capable of divergent adaptation, the mammals soon went further than the reptiles ever had in occupying and subdividing the environmental niches. All this only makes more mysterious and exasperating the problem of *why* the Mesozoic reptiles became extinct and left this opportunity to the mammals.

Modernization

For the geographical reasons to be discussed in Chapter 48, modernization of the South American mammalian faunas was long delayed until the late Pliocene and Pleistocene. Elsewhere, however, in all areas but South America, basic modernization occurred most prominently in the Eocene, during which all the main orders of living placental mammals were probably in existence.[20] Most of the modern families were dominant by the end of the third epoch, the Oligocene. It is significant that no family of placental mammals still living dates from before the Eocene and that most do date from before the Miocene. Expansion of modern groups at the expense of older ones is very evident through the Eocene (Figure 44-27). By the middle of the Oligocene the number of orders was almost down to the present level.

Eocene and Oligocene faunas look strange to us in spite of their rapid modernization. There are three main reasons for their exotic look.

[20] There are 16 living orders of placental mammals, of which 14 are definitely known from fossils in the Eocene. The other two (aardvarks and hyraxes) do not appear in the fossil record until later. However, both are small groups that originated in the Old World tropics, from which few Eocene fossils are known. Their apparent absence in the Eocene is almost surely due to lack of discovery.

44-27 Modernization of the mammalian fauna of North America
The percentages of North American mammals belonging to different groups changed through the Age of Mammals. Extensive turnover from archaic to modern groups is evident in the Eocene and Oligocene.

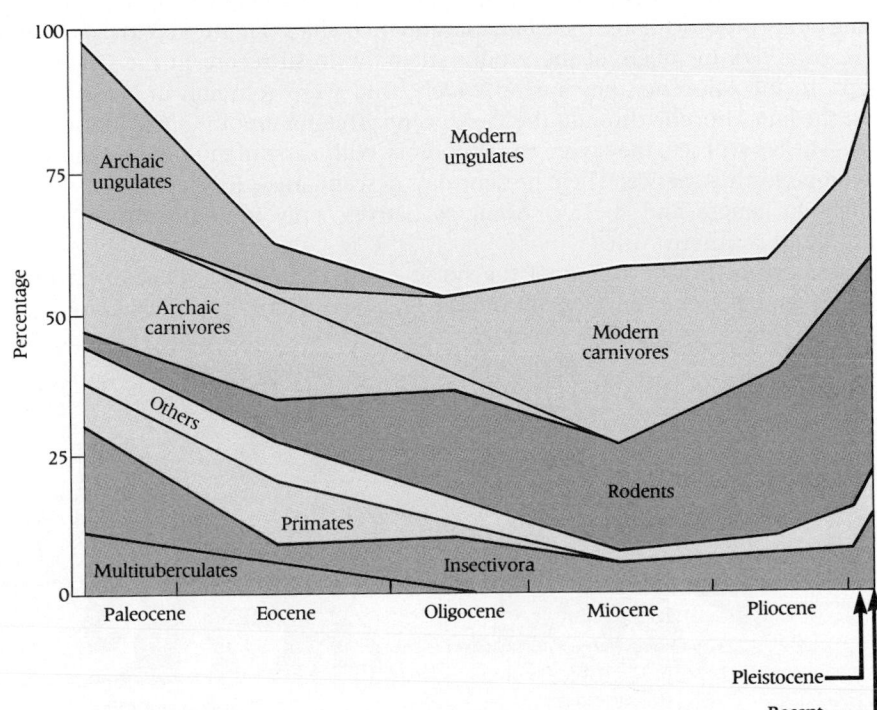

Uintathere

Oreodont

Titanothere

Eohippus

Camel

Oligocene sabertooth

- Some of the striking and peculiar ancient groups were not yet extinct.
- Some modern groups had different geographic distributions then and occurred in unexpected places.
- The direct ancestors of our present mammals had not yet reached just their present forms.

Ancient and now extinct groups of mammals prominent in the Eocene and Oligocene include among the herbivores the uintatheres, the titanotheres, and, in North America only, the oreodonts. Among striking extinct carnivores, the saber-tooths—frequently called "saber-toothed tigers" although they were not tigers—appeared in the earliest Oligocene and were prominent until the Pleistocene, when they became extinct. Startling later changes in geographic distribution are illustrated by the presence of abundant camels and horses in North America through much of the Cenozoic. (Figure 44-28).

Equally striking, especially at later dates, were the changes in areas occupied by the bulky proboscidians, the elephants and their allies (Figure 44-29). Mastodonts, more primitive members of the group, enter the fossil record in the Oligocene in Egypt. In the Miocene, they spread widely, and were common in North America from the late Miocene through the Pleistocene. The mammoths arose in the Pleistocene—as we will see, they were contemporary with early man—and spread widely, reaching North America. Their present-day descendants, the elephants (which include two genera and species) occur as natives, only in southeastern Asia and central and southern Africa.

The story of the evolution of the horse family is a classic illustration of how modern animal groups developed from the Eocene onward. Its earliest known mem-

44-28 Some Eocene and Oligocene mammals of North America

Moeritherium

Gomphotherium

Woolly mammoth

44-29 Some proboscidians
Moeritherium, *the earliest known proboscidian was amphibious and about the size of a pig.* Comphotherium, *a mastodont from the Miocene of North America, looks more familiar to us as a member of the elephant lineage. It had short tusks in both upper and lower jaws and an elephantlike trunk was present.* Mammuthus primigenius, *the woolly mammoth of the Pleistocene, was a fully evolved elephant.*

44-30 History of the horses

Miohippus, *a 3-toed browser, was only one product of an early radiation from* Eohippus. *Some of its descendents remained browsers, while others, such as* Merychippus, *made the step to grazing, exploiting the grasslands of the Miocene period. The line leading to the 1-toed* Equus *from* Merychippus *was one of many adaptive radiations that occurred. Thus the lineage from eohippus to* Equus *is the least direct of those that might be traced through the succession of adaptive radiations. This diagram shows only a few of the many genera known.*

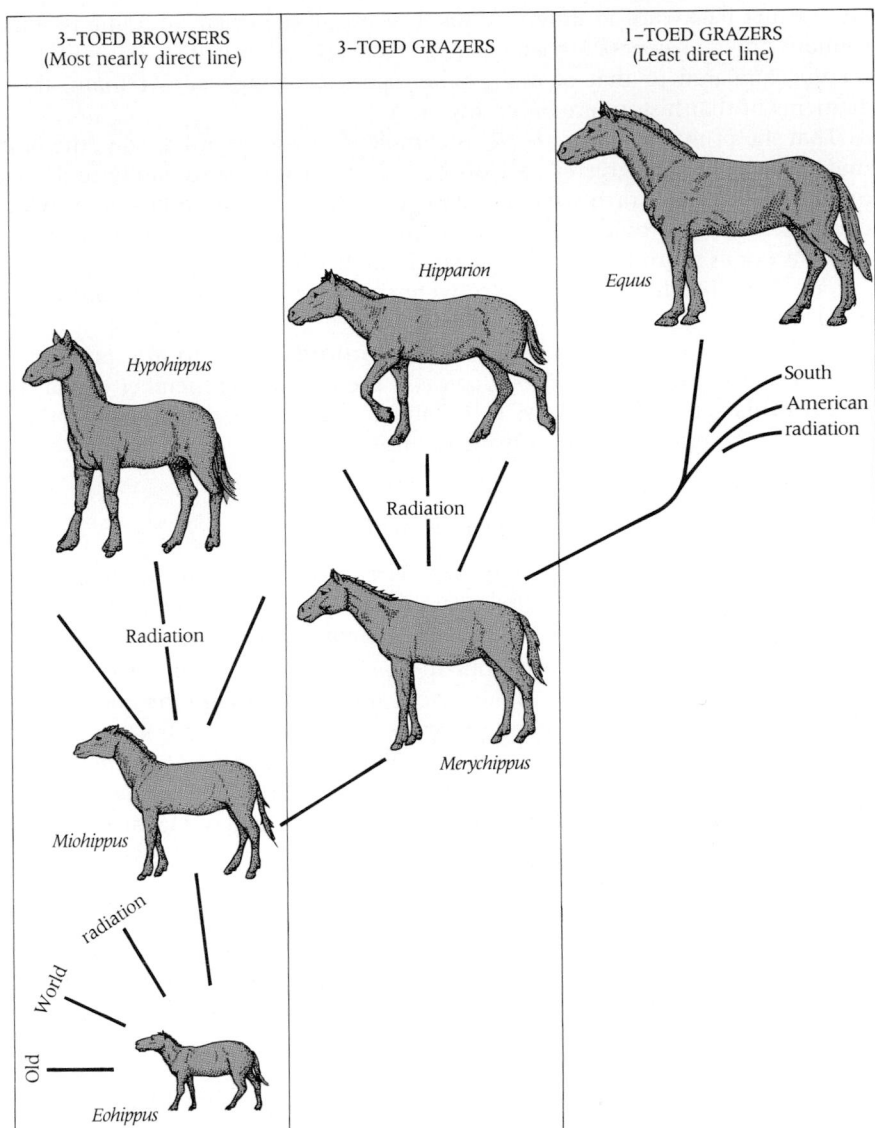

	3-TOED BROWSERS (Most nearly direct line)	3-TOED GRAZERS	1-TOED GRAZERS (Least direct line)

Hypohippus

Hipparion

Equus

South American radiation

Radiation

Merychippus

Radiation

Miohippus

Old World radiation

Eohippus

	PAD-FOOTED		SPRING-FOOTED	
	4-Toed	3-Toed	3-Toed	1-Toed

Equus

Hipparion

Hypohippus

Pliohippus

Merychippus

Mesohippus

Eohippus

44-31 Evolution of the forefoot in the horse family

*The single toe (stippled) of the modern horse (*Equus*) is the sole survivor of four toes present in the ancestral eohippus.* Hypohippus *and* Hiparion *show stabilization of the mechanism, without advancing toward the condition of* Equus. *The feet are of different sizes, but have been reduced to the same length for comparison.*

ber, the eohippus (*Hyracotherium*) lived in the early Eocene (Figure 44-30; see also Figure 44-28). Little eohippus, no bigger than a fox, had four toes on the front feet and three on the hind, each toe ending in a tiny hoof. The small head lacked the heavy muzzle of modern horses and had comparatively large eyes set near the mid-line, not as far back as in present-day horses. The teeth were simple, not fit for grazing but only for browsing on soft vegetation.

How little eohippus became big *Equus*, the large horses of today, can be followed with unusual clarity. Almost all the intermediate stages are known, a convincing demonstration that progressive change is a process of spread of small mutations and new combinations of genes and chromosomes in variable populations. We must emphasize that the picture of steady, gradual change from eohippus to the modern horse, as given in many popular discussions, is incorrect. The true history of the horse family does not show a lineage from eohippus to *Equus* involving a gradual increase in size, reduction in the number of toes (Figure 44-31), and development of more complex teeth. In fact, there was not one lineage, but many. The phylogeny is intricate and branched and all but a few of the branches have now become extinct. Moreover, increase in size and change in feet were not constant but sporadic. Necessarily, the complex phylogeny shown in Figure 44-30 is greatly simplified.

HUMAN FORERUNNERS

Studies of human evolution have increased in intensity (and discord) in recent years as new fossil discoveries and new techniques have pushed the history further and further backward in time. The fossil record of the origin of humans—and the primate order in general—is now more extensive than it was, but still relatively scanty compared to that of many other species. Though gaps remain, the main elements of that history are becoming clear.

That the primates did not leave as complete a fossil record as, say, the horses is understandable. The greater part of the history of the horse family took place in one area—central North America. Primate history was more far-flung, with vital episodes in several different regions—some in tropical areas neither rich in fossil deposits nor as thoroughly explored. The primate way of life also predisposes against a good fossil record. Most primates were tree dwellers unlikely to be buried and fossilized.

In the following discussion, the term **hominids** refers to any primate in the human family, Hominidae. *Homo sapiens* is the only living member of that family. **Hominoids** refers, collectively, to hominids and apes. Hominoids, together with monkeys, comprise the **anthropoid primates.**

Early Prosimians

Primate species were probably more widespread and numerous in the Eocene than at any later time. We surveyed the main primate groups in Chapter 43. Their history began with the most primitive of them, the **prosimians** or **premonkeys** (Figure 44-32). Their oldest fossils are from the late Cretaceous and early Paleocene of the western United States. Prosimians were abundant through the Eocene, in North America, and also in Eurasia and probably Africa, where the Eocene fossil record is skimpy. They are absent from the known North American record from the Miocene onward, but they lived on in parts of the Old World. Today they live only in tropical Asia and nearby islands, Africa, and the large island of Madagascar, comprising only a few species—tree shrews, tarsiers, and lemurs. One group, the lemurs, however, is common and diversified in Madagascar.[21]

Prosimians, even now, have poorly developed brains. Many have rather long, pointed snouts, markedly unlike the flattened faces and more localized noses of most monkeys and all apes and humans. The following are basic primate characteristics, present in early prosimians, that represented adaptations for their arboreal mode of life.

- They retained and further evolved grasping hands and, often, feet, with opposable first digits that make possible grasping.
- They possessed increased flexibility in the elbow and wrists—especially in rotational movements. Other traits shared by primates are flexible shoulder joints that permit arm rotation and collarbones (clavicles).
- Modifications in the sites of attachment of limbs to torso occurred.
- Eyes in the earliest primates are directed laterally so that visual fields overlap only slightly. Binocular vision, with overlapping visual fields arose later in the course of primate evolution. Together with grasping hands, this made possible superior hand-eye coordination.
- In locomotion and posture, primates are basically four-footed. In the larger arboreal monkeys and apes the trend was toward **brachiation** (swinging from tree branches by the arms instead of walking along the branches on four feet). This involved an adaptive straightening of the trunk and body, elongation of the arms and fingers, and significant regression of the thumb. As noted later, our own ancestors were never well-specialized brachiators like the modern apes (especially the gibbons), but some advance along this line may have facilitated an upright posture when our ancestors began to walk and live on the ground.
- Claws, primitive among placentals, began to evolve early among primates into flat nails and the inner surfaces of the hands and feet became sensitive hairless pads.

---

[21] This survival of so primitive a group is an interesting evolutionary phenomenon. Their ancestors gained access to the island early in the Cenozoic, before higher primates arose. The lemurs then underwent an adaptive radiation on the island and have been protected from effective competition. Later primates did not cross the sea barrier between Africa and Madagascar, so that the island became an asylum for prosimians.

A

B

44-32 Prosimians
(A) Notharctus *was an Eocene prosimian, with grasping hands and feet.* **(B)** *A living prosimian,* Loris tardigradus (*the slender loris*), *has eyes that are directed forward, permitting stereoscopic vision. The latter is an advanced character in which this particular prosimian has paralleled the higher primates.*

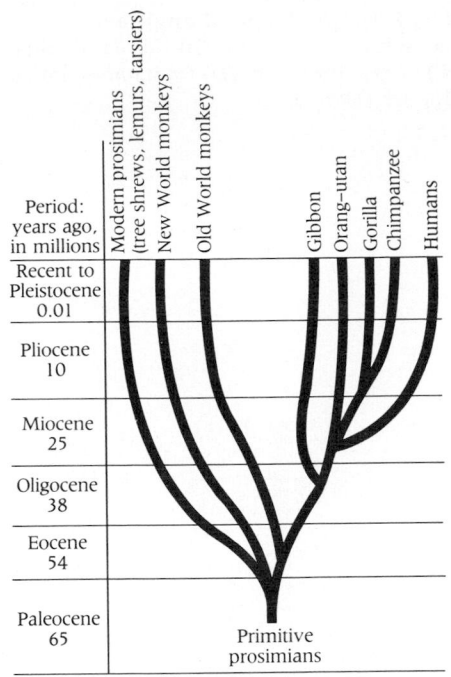

44-33 Primate phylogeny
All primates are believed to have evolved from a primitive prosimian ancestor, similar perhaps to the tree shrews of today. The primates originated from a branch of the mammals by moving into an arboreal environment.

There were innumerable divergent lines among Paleocene and Eocene primates. This is apparent in the tentative evolutionary scheme shown in Figure 44-33. It is agreed, however, that many early primates became specialized in peculiar ways and in time became extinct. Those that evolved further retained simple, primitive teeth and tended to show enlargement of the cranium, reduction of the snout, and enlargement of the eyes, which as noted moved forward in the head where they provided overlapping visual images affording a better three-dimensional view of the world.

Old World and New World Monkeys

The Old World monkeys, **cercopithecoids,** and New World monkeys, **ceboids** (see Chapter 43) are correctly portrayed in Figure 44-33 as fully distinct phylogenetic units. They have always had approximately the same geographic distribution that they have now: cercopithecoids are found in the warmer parts of Africa, Asia and Europe,[22] and ceboids in South and Central America. Both groups evolved from prosimians, but almost all of their evolution occurred separately—on the continents to which they are now confined.

Both groups first appeared in the Oligocene. Both differ from prosimians in having larger brains occupying a larger proportion of the head. The snout was reduced, and the characteristic flattened monkey face appeared. The muzzle shortened with a shortening of upper and lower jaws, and the number of teeth decreased. Hands and feet were grasping, but no more so than in some prosimians.

Apes

The name "ape" is sometimes applied to monkeys, but in strict usage it means a member of one particular primate family, the Pongidae (Figure 44-34), which is in turn part of the superfamily Hominoidea—which includes family Hominidae and subfamily Homininae, of which *Homo sapiens* is a member. Sometimes they are called *anthropoid apes* to make the distinction clear. Anthropoid means "humanlike." The German term is *Menschenaffen* ("men-monkeys").

The living apes of today are the **gorillas** (genus *Gorilla*) and **chimpanzees** (genus *Pan*) of central Africa, the **orang-utans** (genus *Pongo*) of Sumatra and Borneo, the **gibbons** (genus *Hylobates*) of southeastern Asia—though some primatologists place gibbons in a separate family, Hylobatidae. All are above average size for primates, strong, and intelligent. All gorillas lack tails.

All recent apes except gorillas are highly arboreal, spending most of their lives in trees. Gibbons, smallest of the apes, have tremendously long arms and fingers (with regressed thumbs) and they are astonishing acrobats. They swing (brachiate) by grasping overhead branches, and confidently jump great distances from branch to branch. On the ground, they walk on their hind legs, holding up their long arms as balancers. The other apes walk on all fours unless they have been taught to walk on their hind legs as a circus trick. Chimpanzees and orang-utans do not swing from limb to limb with the abandon of gibbons, but they are agile four-handed climbers and spend most of their time in trees.

The hominoids that we call apes have always been confined to the warmer parts of the Old World. The earliest known fossils occur in the Oligocene, along with the first Old World monkeys. Apes and cercopithecoids share certain characters absent in the ceboids and the prosimians. It appears that apes either were derived from a branch of the earliest cercopithecoids or that they and the cercopithecoids had an immediate common ancestry among the prosimians.

In the Miocene, apes became more varied than they are today (see Figure 44-33). Fossil apes have been known to paleontologists since 1856. Since then, diligent search has turned up fragments of many species in the Miocene and Pliocene of east Africa, Europe, and southern Asia. The family began a great expansion at the beginning of the Miocene. Some were ancestral to the modern apes and were already becoming specialized in similar ways. Still others, especially in the Miocene, were light, agile forms not yet strongly specialized for arboreal life and lacking

---

[22] The only European monkeys now are the famous "apes" (really monkeys, not apes) of the Rock of Gibralter. But monkeys once did occur more widely in southern Europe.

44-34 The great apes, Pongidae
(**A**) *Gibbon*, Hylobates. (**B**) *Gorilla*, Gorilla.
(**C**) *Chimpanzee*, Pan. (**D**) *Orang-utan*, Pongo.
Also see Figure 33-16.

other characters peculiar to the surviving apes. In the Miocene, *Proconsul* (named
for a captive ape in the London zoo) is the most completely known and is among
the less specialized (Figure 44-35). It is generally (but not unanimously) considered
identical with *Dryopithecus* ("pithecus" means ape), a genus widespread in Africa,
Europe and Asia.

Hominoids of the Miocene

The middle Miocene—9 to 14 million years ago—was a time of a widespread
shifting in the environment that led to profound changes in animal life. Until recently,
we lacked evidence of hominid remains of Miocene apes (hominoids), but recent
discoveries have greatly increased our knowledge of these groups and their radiations.
It now appears from dental structure and other data that *Sivapithecus* and its closely
allied (and perhaps identical) genus, *Ramapithecus*—as well as *Gigantopithecus* and
several other genera—occupied open woodland habitats of the middle Miocene—
between 8 and 14 million years ago.

For a variety of reasons, the hominoid *Ramapithecus* is of unusual interest. In
1961, Elwyn Simon gave strong support to proposals that it was the earliest hominid.
Though the fossil evidence was scanty, consisting at the time of little more than
two parts of a single broken upper jaw, the teeth and reconstructed face made
Ramapithecus the most hominidlike of the Miocene forms. Later work by Simon
and his student David Pilbeam even more firmly placed ramapiths first in the human
lineage, unequivocally distinguishing them from apes on the basis of tool use, bipedalism, and social life. In this scheme (Figure 44-36A)—we call it the traditional

44-35 Proconsul
A fossil ape from the Miocene of Kenya, Africa.

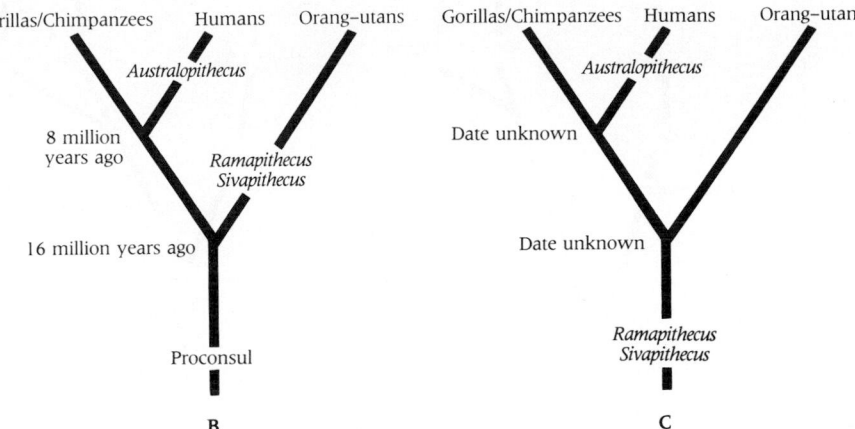

TRADITIONAL VIEW   PILBEAM'S SCHEME   WALKER'S PROPOSAL

44-36 Evolution of the hominoids
Three interpretations of the place of Ramapithecus *and* Sivapithecus *in the evolution of hominoids. In the traditional view, ramapiths are ancestral to humans. In Pilbeam's scheme, they are ancestral to the orang-utans. Walker and others proposed that ramapiths are ancestral to both humans and great apes.*

view—Prince Rama, as he was affectionately known, was first in the human line.

In 1978, however, new data persuaded David Pilbeam to change his mind and place *Ramapithecus* in the ancestry of orang-utans, not humans (Figure 44-36B). Then in 1984, Alan Walker, Richard Leakey, and others challenged this view, arguing that *Ramapithecus* and its close relative, *Sivapithecus*, were common ancestors of both humans and the great apes (Figure 44-36C).

The relationships of the living hominoid primates—apes and humans—have been clarified by molecular studies of recent years.[23] These data demonstrate several major cladogenic events (Figure 44-37):

- The early divergence of Old World monkeys from the common hominoid stock
- The divergence of the gibbon from the human lineage
- The subsequent divergence of the orang-utan from the great apes and human stock
- The most recent divergence between humans and apes

Comparisons of human and higher primate chromosomes and their banding patterns have generally supported this model. Molecular methods have dated these events as follows:

- Gibbon divergence, 12 ± 3 million years ago
- Orang-utan divergence, 10 ± 3 million years ago
- Human-ape divergence, 6 ± 3 million years ago

With *Ramapithecus* and *Sivapithecus*, thought to have been alive between 8 and 14 million years ago, these fossil genera may have arisen too early to be ancestral to humans. That is why, as noted above, some view *Ramapithecus-Sivapithecus* as part of the orang-utan clade (see Figure 44-36B). If correct, *Ramapithecus* can no longer be considered part of the human lineage. They were hominoids, but not hominids.

Pliocene and Pleistocene and Early Hominids

For years there was a long gap in the fossil record between *Ramapithecus* and the next younger known member of the human family, *Homo erectus*. The gap is now being filled in by an intense wave of new discoveries in the late Pliocene and early Pleistocene—2.5 to 3.5 million years ago—in Africa and southern Asia. The

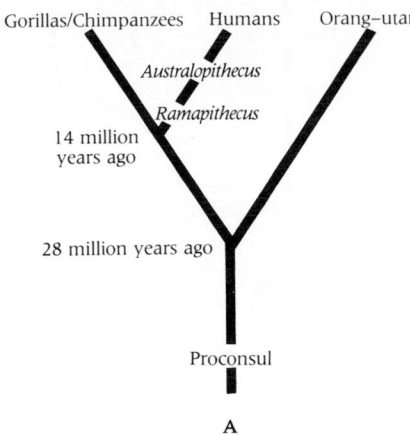

44-37 Hominid family tree according to molecular data
The DNA ''clock'' places the human-African ape split at 6 to 8 million years.

[23] As discussed in Chapter 34, these include studies of: (1) amino acid sequence changes in the proteins albumin, transferrin, hemoglobin, myoglobin (and others); (2) changes in mitochondrial DNA sequences: (3) sequences of nuclear DNA; and (4) sequences of ribosomal RNA. It is worth noting that in 1988 a sharp attack was mounted against published results based on the technique of DNA-DNA hybridization that claimed an unexpectedly close relationship between humans and chimpanzees (See Lewis in the reference list.) The debate rages on.

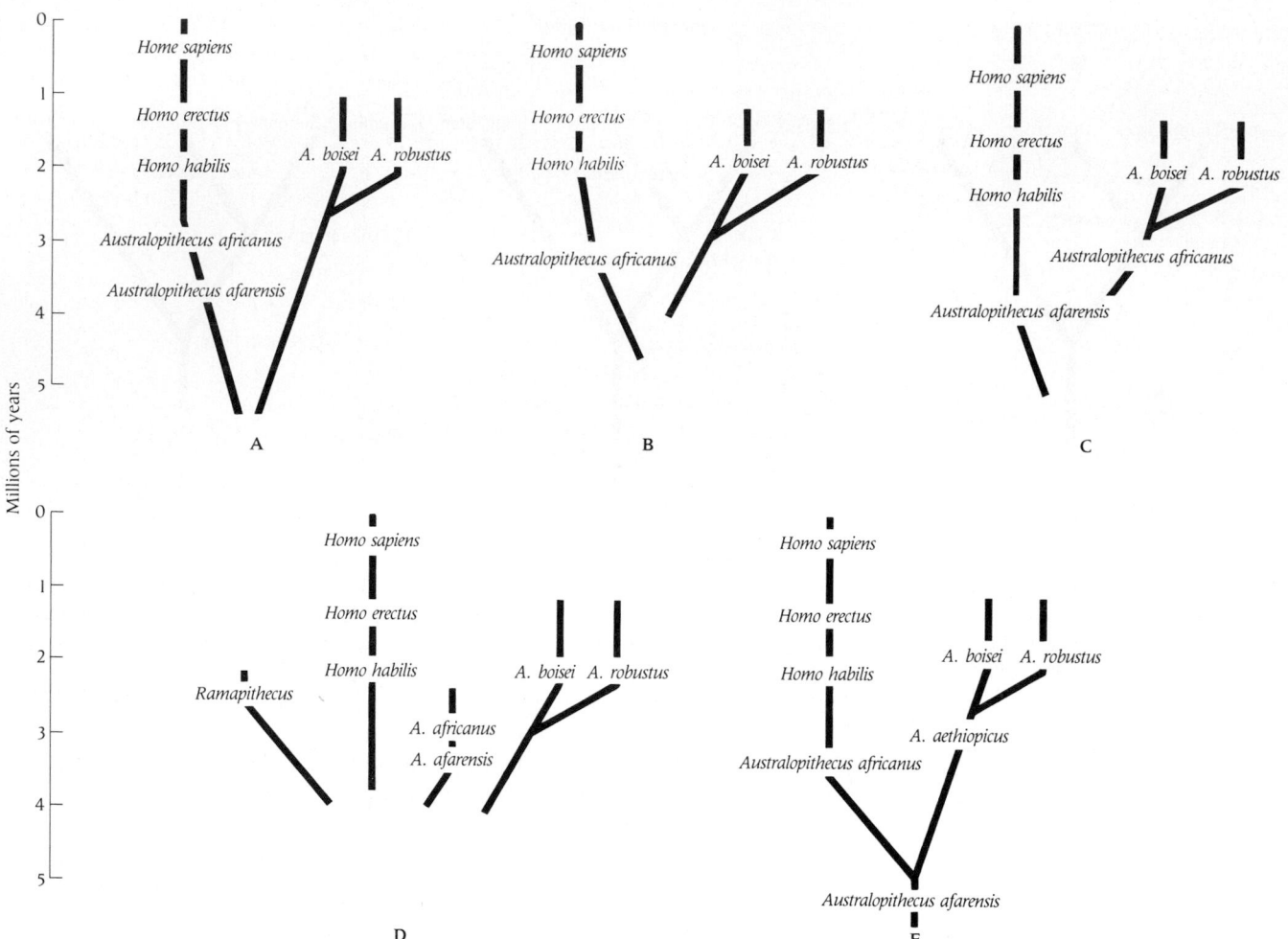

44-38 Alternative proposals on the major hominid taxa.
(A–I) *Although there are other proposed schemes, these five phylogenies are representative of current thinking. All concur that* Australopithecus africanus *appears early in the human line.* **(D)** *is the polyphyletic hypothesis of R. E. F. Leakey.* **(E)** *includes the recently discovered ''black skull''* Australopithecus aethiopicus. *See also Figure 44-40.*

first such discovery, made in 1925 in South Africa, was a juvenile skull of a small creature that was named *Australopithecus africanus* (the African southern ape). It is now agreed that it belongs in the human family, Hominidae. At the time of its discovery, it was the oldest known upright hominid.

Since then, many similar specimens have been found in South Africa, Tanzania, Kenya, and Ethiopia. They are known collectively as **australopithecines** and their positions in some of the alternative phylogenetic schemes are shown in Figure 44-38. Their considerable variation and the incompleteness of most of the specimens led to the too hasty naming of several genera and species among them, but better collections and further study have established that almost all can be put in just two distinguishable lineages or groups that were largely contemporaneous.

- The gracile australopithecines, which included *Australopithecus africanus*, are more like *Homo*, our own genus, and they are now considered an ancestor of *Homo*.
- The robust australopithecines included *Australopithecus robustus* (with its wide face and robust jaw) and *Australopithecus boisei* (protruding jaw, enormous cheekbones, and wide palate). An important specimen, discovered in 1987 in northern Kenya and called the ''black skull'' because it was darkened by nearby manganese-rich sediments, had hominid features suggesting it was *A. boisei*, but its age contradicted the general view that *A. robustus* was an ancestor of *A. boisei*. Many anthropologists now believe that the ''black skull'' is a new species—they call it *Australopithecus aethiopicus*—that is an ancestor of both *A. boisei* and *A. robustus*.

A still unsolved problem concerned the ancestry of the two australopithecine lines. The apparent answer came in 1978, with the discovery of the oldest good sample to that date of an undoubted hominid. This specimen, 2.9 to 3.5 million years old (the hominid fossil record is largely nonexistent before 4 million years ago), was unusually complete, including 40% of the skeleton (Figure 44-39). This

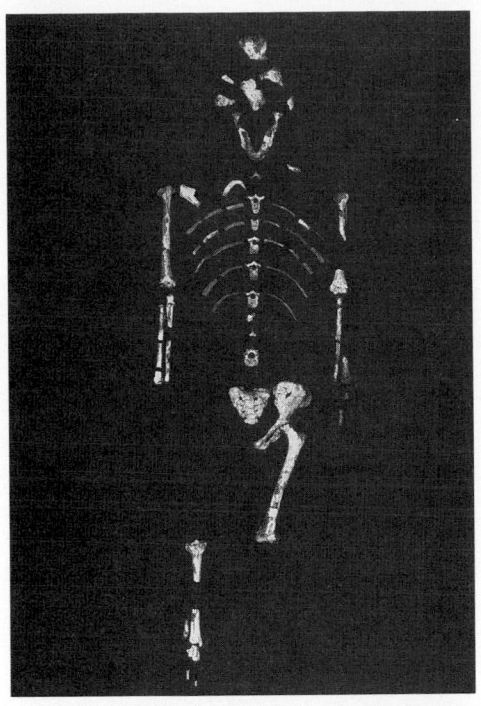

44-39 Lucy

The reassembled skeleton of Lucy, an approximately 2.9 to 3.3 million-year-old hominid skeleton uncovered in Ethiopia in the mid 1970s by Don Johanson and Tim White. The unusually complete bones of the arms and legs were useful in determining Lucy's posture and walking habits.

44-40 One phylogenetic scheme of hominid evolution

This scheme is based on that shown in Figure 44-38E.

individual, named *Lucy* and widely publicized, walked upright but had an apelike palate. It was named *Australopithecus afarensis,* having been discovered in the Afar triangle of Ethiopia. It (and similar fossils unearthed later) is now viewed as the common ancestor of the two main hominid branches. With the evidence of the "black skull," it now appears to many that *A. afarensis* is on the handle of an asymmetrical three-pronged fork (Figure 44-38E).

Paleoanthropologists carefully estimated the cranial capacity (which nearly equals the size of the brain) of these forms. For modern humans, the average figure is about 1360 cc (about 83 cubic inches). In gracile australopithecines, it is about 500 cc (about 30 cubic inches) or a bit more. In modern apes, the figure varies greatly but it rarely, if ever, exceeds about 570 cc (about 34¾ cubic inches) for a large male gorilla. Thus the australopithecines had brains in the size range of modern apes, a range far below that of modern humans. The size of the brain is not in itself a measurement of intelligence—indeed it has a complex relationship to body size—but with differences this great the intelligence of most australopithecines probably did not exceed that of apes.[24]

The australopithecines walked upright and their heads were set squarely on top of the backbone, not thrust forward as in apes. Their teeth were also more human than apelike. Erect posture and bipedal locomotion evolved before the acceleration of evolution of the brain. This should not be surprising, for selection for increased brain size would seem naturally to follow the freeing of hands from use in locomotion, making them available for more complex uses and manipulation. Some tools of human type have been found associated with australopithecines, but it is not clear what species made them or even whether the makers were or were not already members of the genus *Homo.*

We now consider the robust australopithecines a differently specialized, extinct side branch in the human family. The gracile line, it is widely believed, gave rise to *Homo.* The diagrams in Figure 44-38 are alternative current schemes of the

---

[24] A special index called the encephalization quotient, or EQ, takes the brain-body size into account. For gracile australopithecines, this averages about 3.86, with a range of 3.33–4.72. For one robust australopithecine it is about 3.25. For living apes it is about 2.10, and for modern humans it is about 7.60. These figures suggest, but do no more than suggest, that australopithecines were more brainy than apes but still far less so than humans. The formula for EQ makes the average EQ for living mammals in general equal to 1, so all these primates are more brainy than most other mammals.

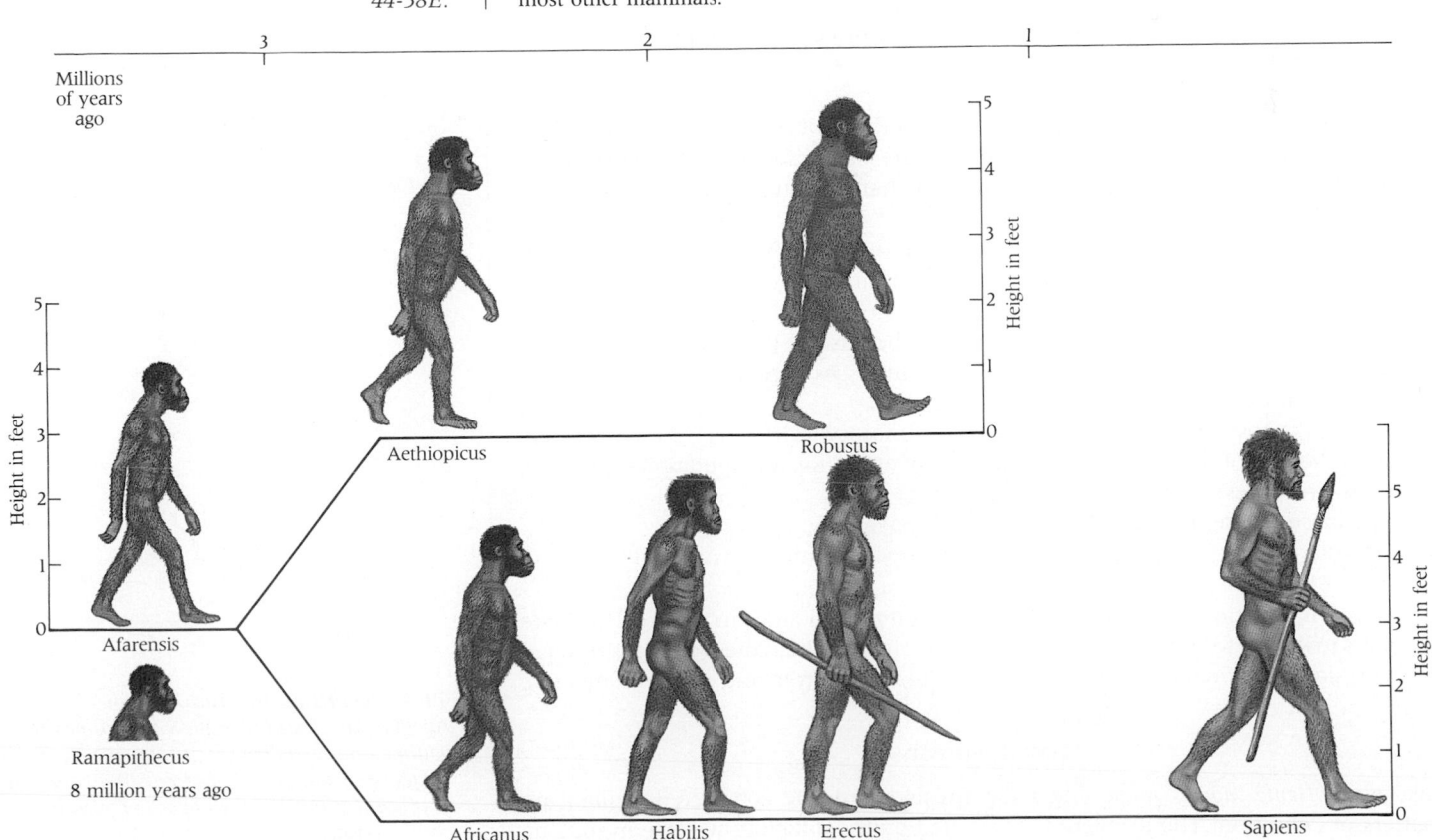

phylogeny of the major hominid taxa. The scheme in Figure 44-38E is shown in a different perspective in Figure 44-40.

MAN'S PLACE IN NATURE
Human Origins

Even the pre-evolutionary biologists recognized that human beings are animals, fundamentally like other animals but distinguished by higher intelligence. In the eighteenth century, Linnaeus, a pre-evolutionist, placed *Homo sapiens* in the order Primates with the prosimians, monkeys, and apes. That classification is still accepted by biologists. Knowledge of evolution makes it evident that the distinctive characteristics of human beings arose in the same way as those of other species of animals, by gradual change in varying populations. So stated Thomas H. Huxley (1825–1895) in his famous 1863 volume, *Man's Place in Nature*, written in defense of Darwin. That humankind evolved from other and (in intelligence, at least) lower animals is about as certain as a scientific conclusion can be.

The human position among the primates has clear phylogenetic validity. Precise relationships among the primates are still disputed in detail and, as shown in Figure 44-38, a variety of phylogenetic "trees" have been proposed to express various views on the details of human evolution. Nevertheless, there is a consensus on certain important points.

It is obvious that the apes are the most humanlike of living nonhumans. This is borne out by their gross structure, physiology, molecular makeup, and behavior. However, no living ape is a member of the actual human ancestry, or closely similar to it. Moreover, all living apes have specializations that surely did not occur in our ancestry. The apes undoubtedly share a common ancestry with us, but they have since diverged from the line of our later ancestors. Our common ancestor may well have been among the early Miocene dryopithecines (Pongidae).

The most distinctive characteristic of humans is the brain. But the origin of humans was dependent on many earlier primate adaptations, some of which were mentioned earlier. Three of the most important (in their order of appearance) were grasping hands, binocular stereoscopic vision, and upright posture with bipedalism. The australopithecines and possibly others already had grasping hands, stereoscopic vision, and erect posture, if not bipedalism. Some of the major anatomical changes marking the transition of these ape ancestors to modern humans are the following:

- The jaw became shorter, making the muzzle shorter. The long protruding face in front of the brain gave way to a short steep face under it.
- The teeth became smaller.
- The long bones of the legs, once shorter than those of the arms, became longer, as the arms became shorter. These are features of the so-called gracile skeleton. Others are the typical long-bone shape and shaft thickness and the relatively thin bone of the cranium and jaw.
- The tarsal bones of the feet, once short with long toes, became longer, as the toes became shorter. The feet flattened and an arch developed.
- The ape skull rested on the vertebral column at the back of the skull. The point of attachment shifted to the center of the base of the skull.
- Eyebrow ridges and other external cranial buttresses became reduced in size and muscles attached to them became smaller.
- An upright posture freed the hands from locomotory tasks and permitted efficient specialization of the limbs—fore limbs (now arms) for manipulation only, hind limbs (now legs) for locomotion only.
- Bipedalism arose from quadrupedal knuckle-walking.
- A substantial increase in brain size accompanied an increase in intelligence.

It was ramifications such as these that permitted our ancient arboreal primate ancestors to lay the foundation for human development. When these changes reached a certain stage, the final great expansion of the brain occurred and humans emerged.

Fossil Humans

Just when *Homo sapiens* arose and what fossils should be considered human are matters of definition. The problem is illustrated by the difficulty we face in judging

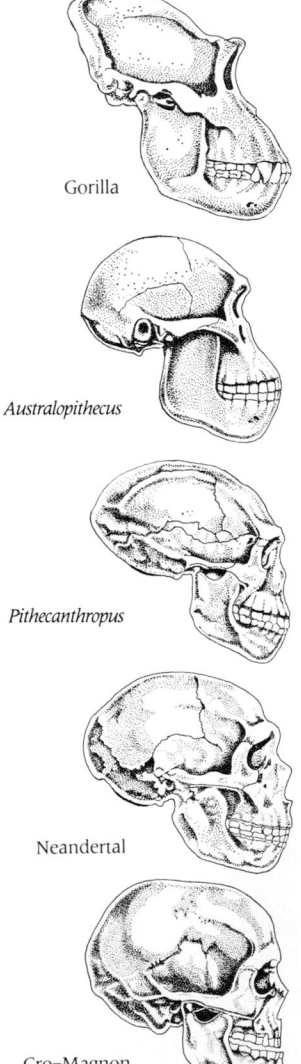

Gorilla

Australopithecus

Pithecanthropus

Neandertal

Cro–Magnon

44-41 Skulls of ape and human
Apelike skull features include a small brain-case volume and a shallow brain case, heavy brow ridges, protruding jaw, and receding lower jaw. Human features are the opposite for each characteristic.

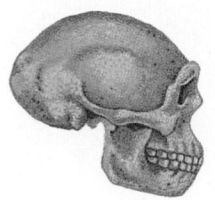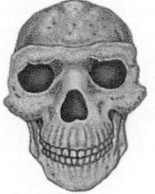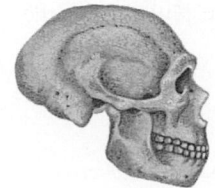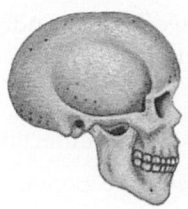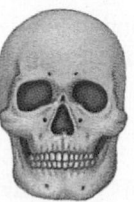

Homo erectus
1,500,000–200,000 years ago

Homo sapiens neandertalensis
100,000–35,000 years ago

Homo sapiens sapiens
90,000–present

44-42 *Skulls of the three major species of the genus* Homo

the evolutionary relationships of *A. africanus*. In some characteristics (cranial and dental structure, for example), it is the "perfect ancestor." In others (pelvic and femoral structure), it seems to belong to another group that developed in parallel with *Homo*. But not all agree. Some would call the australopithecines "human." Others would confine that designation to populations of *Homo sapiens* just like those now living.

We are concerned just now with the members of the human family that were more advanced than the australopithecines. All of these hominids are limited to the later Pleistocene. None is likely to be more than a million years old, and most must be under 500,000 years old. Dating is still inexact in this range.

Later in the Pleistocene there appear various fossil hominids that we do believe are archaic forms of *Homo sapiens*. The earliest and most controversial used tools. Hence they were called *Homo habilis* (Latin *habilis*, "able"). Probably direct descendants of *Australopithecus africanus*, they had long apelike arms (though limb remains are disturbingly scarce), but notably larger brains.

Abruptly, it appears, the primitive human underwent a spurt of evolution. The next stage was *Homo erectus*, a creature of modern size, who fashioned stone axes, cooked its food, and was first to migrate out of Africa. The *Homo erectus* story is interesting. The best-known fossil hominids (before the discovery of *Australopithecus*) were the forms originally assigned to the genera *Pithecanthropus* ("monkey man") and *Sinanthropus* ("China man"). These fossils, about 250,000 years old, are represented by a group of skulls, jaws, and other fragments that were found, respectively, in Java and China near Beijing (Figures 44-41 and 44-42). Both of these forms are so much alike that they are probably nothing more than different demes, or at most different subspecies, of a single specific population. The present view is that neither merits recognition as a genus distinct from that of modern humans. Hence, they were renamed *Homo erectus*.

Recently fossil remains of the same general type have been found in northern and central Africa, and members of the species probably occurred throughout most of the Old World. Some of the African finds referred to as *Homo erectus* are from the late Pliocene—much earlier than specimens of that species found elsewhere. This suggests that this species evolved in Africa and later, probably a million years ago, spread to Europe and Asia. None, it appears, ever discovered America.

44-43 Australopithecus *and fossil humans*.

Skulls of this archaic group retained apelike characters that are reduced or lost in modern humans (Figure 44-43). The brain case was small. Brain size varied greatly, but averaged about 1000 cc, which is intermediate between that of living

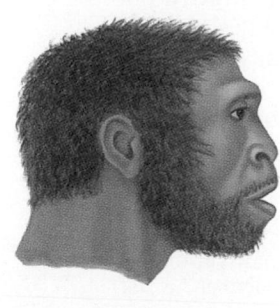

Australopithecus

Java Man

Neandertal Man

Cro-Magnon Man

apes and humans. There were heavy ridges above the eyes, the jaws and large teeth protruded in front, and the chin retreated.

Neandertal People

Next we encounter another large population more advanced than *Homo erectus*. Scattered finds in Europe, Asia, and Africa since the first find in 1856—some 300 in all—show individual, probably racial, differences, but they are enough alike to be considered as members of one variable population, the *Neandertals,* or Neandertaloids, which flourished in the Ice Age between 100,000 and 35,000 years ago (see Figures 44-41, 44-42 and 44-43).[25] Neandertals were the first fossil humans discovered. That may explain their wide, perhaps excessive, fame.

The Neandertals were short, stocky, powerful people with large, heavy-boned heads. Brow ridges were present and the chin was retreating, but neither character was as extreme as in the older Java and Peking men. The forehead was retreating and the brain low, but the total size of the brain, surprisingly, was like that in modern humans. We do not know how the Neandertals would perform in an intelligence test, but we do know that their intelligence was considerable in amount and human in quality. They constructed a large variety of well-made tools, and were successful hunters. They sometimes buried their dead, then put offerings or sacrifices in the graves, which implies that they had religion and rituals, and they produced various forms of art.

We do not know what happened to them. We also do not know if they were directly ancestral to modern humans. Recent reviewers clearly doubt it.[26] On the assumption that Neandertals and modern humans overlapped in their foraging territories, E. Zubrow of the State University of New York at Buffalo recently postulated that only a small disadvantage—1 or 2%—in life expectancy and mortality could have sealed the fate of the Neandertals in a relatively short time.

Early *Homo sapiens*

Several fossil skulls have been discovered that are somewhat older than typical Neandertals but that show greater resemblance to modern humans. Among these are the Swanscombe (England) skull (only a few fragments found), the more nearly complete Steinheim and Ehringsdorf (Germany) skulls, the Fontéchevade skull (France), and a cache of fossils from a cave in Qafzeh, Israel that has been accurately dated at 92,000 years. These and others are noted in Figure 44-44. They are not fully modern in appearance and do have some primitive and Neandertaloid traits. They suggest that our history in the Pleistocene was not a simple line.

There are two main possibilities. One is that Pleistocene man constituted a single species but was split up into highly varying demes, some more *Homo sapiens*-like, even at an early date, and others retaining or even accentuating more archaic characters. Selection within the species eventually eliminated archaic variants. This idea is supported by the fact that some fossil humans do seem to represent contemporaneous intermediates between the Neandertaloid and the more extreme modern types. The other main possibility is that the Neandertals were a separate branch, a species that became extinct and was subsequently replaced in competition with forerunners of *Homo sapiens*. The contrast between the two theories is not overwhelmingly clear-cut.

The intense recent debate on the origin of modern humans has been fueled not only by new fossil finds, but by interesting new molecular data—notably sequences of nuclear DNA and mitochondrial DNA (mtDNA). One of the most provocative recent studies rests on mtDNA, which as noted in Chapter 35 is a fast-ticking molecular "clock" that accumulates mutations rapidly and tracks only maternal inheritance (remember: sperm contribute no mitochondria to the zygote). Studies of mtDNA in 147 women from five different regions around the world suggested

---

[25] The term Neandertal comes from the Neander Valley ("tal" in German) near Düsseldorf, Germany, where the first described remains were found. An older spelling, "Neanderthal," is still in common use, but it is antiquated and encourages mispronunciation.

[26] We do know that only a small advantage is necessary for an advantaged form to thrive and a disadvantaged form to become extinct. A difference in mortality as trivial as 1% could lead to extinction in 30 generations—a single millenium.

44-44 *Temporal distribution of Neandertals*
Note that the time scale is logarithmic, which
expands the Middle and Upper Paleolithic
periods. Although many Neandertal sites in
Europe are not precisely dated, most are in the
75,000 to 35,000 year range. Those fossils more
than 80,000 years old are classified as early
Neandertals; those even earlier show varying
degrees of affinity with both Neandertals and
Homo erectus. The Upper Paleolithic sites more
recent than 35,000 years ago contain human
fossils of the modern type.

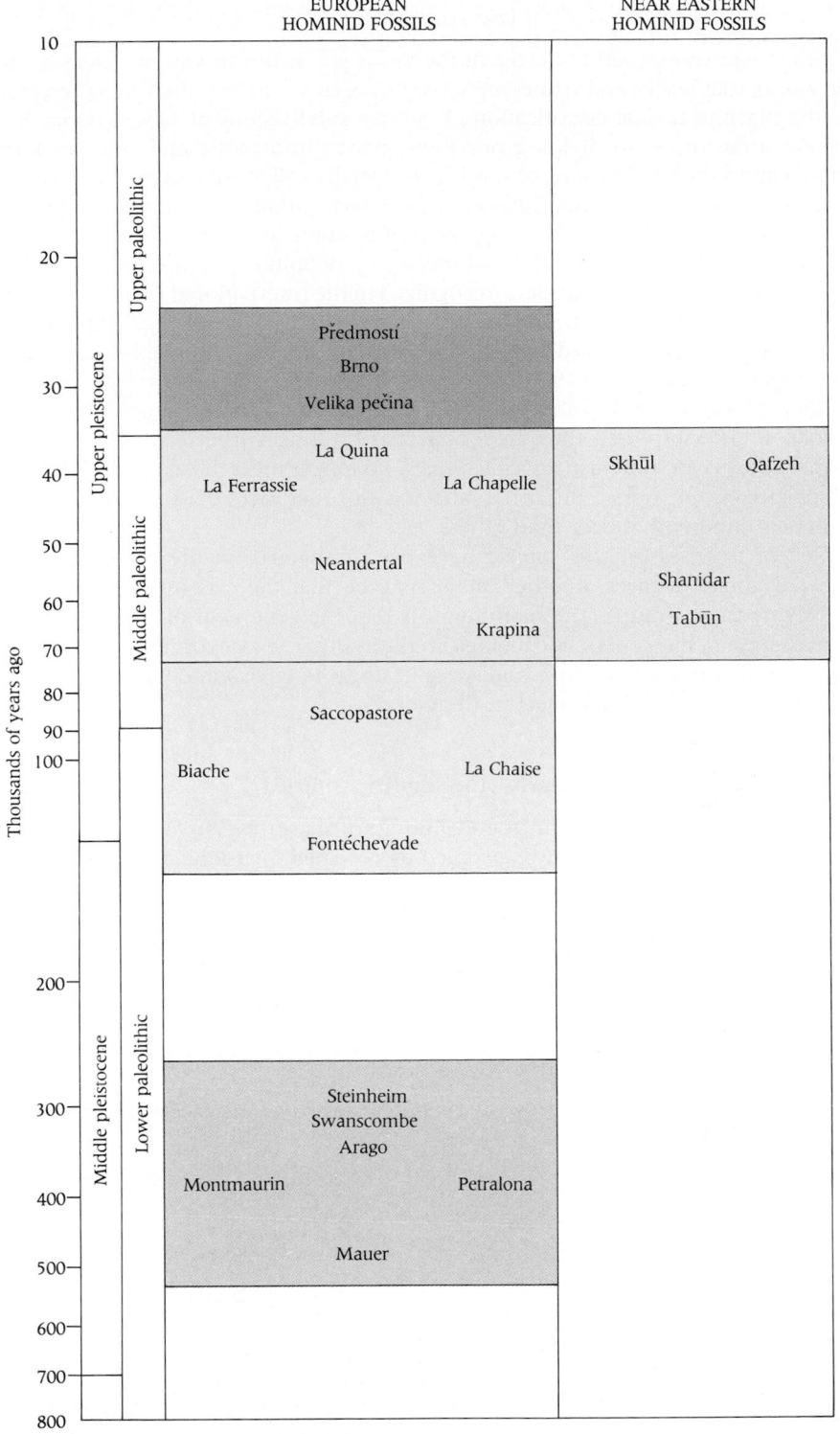

(1) that all were of relatively recent origin, and (2) that the African group was ancestral to the others. "Eve was an African!" said the press.

In sum, it appears that humans did not arise in many regions at once with subsequent extensive gene flow, but in a single locale—Africa—in the relatively recent early part of the late Pleistocene. Not everyone accepts this conclusion—some criticize the assumed mutation rate—and thus we do not know with certainty where modern *Homo sapiens* originated—or precisely when. What we do know is that the Neandertals of Europe appear to have been abruptly replaced. *Homo sapiens*, indistinguishable from modern humans, is first seen in the last glacial stage of the Pleistocene. Cro-Magnon man, first found in France, lived about 25,000 years ago (see Figure 44-43). All younger fossils appear to be true *Homo sapiens*.

Human modernization during the Pleistocene is thus a fact. By about 35,000 years ago, modern human beings had become the only hominid species extant.

The Human Races

We long ago recognized the error of the view, still heard in various backwaters of the world, that blacks and whites are separate species.[27] In fact, there is no agreement on the place in formal classification of various subdivisions of *Homo sapiens*. Since we are accustomed to dividing our own species into races, and since in formal classification the subdivisions of species are usually called subspecies, must we conclude that human races are subspecies of *Homo sapiens*? The answer is probably not—if only because the subspecies concept is vague and fading from use.

The concept is vague because subspecies by definition are not isolated entities. Nonetheless, this taxon is formally recognized in the International Code of Zoological Nomenclature and it continues in wide use despite some dissent. Human races obviously exist, they evolved as do subspecies, and they can reasonably be recognized and defined as subspecies. Subspecies within a species are always allopatric—that is, they do not occur together in one area or locality. That was doubtless true of human subspecies when they arose, but it is no longer true of them. Thus, these former subspecies, now usually called *races,* no longer meet the zoological definition of subspecies. Of course, the racial fallacy is not that races exist but that some are somehow inherently better than others.

Recent treatises on this subject have even disagreed on the number of races: one recognized 6 races, another 30. It may be that there is no scientific way of classifying these groups. The main point is that the gene pool of *Homo sapiens*, the human species, has genetic and biological continuity. The races or subspecies probably represent different adaptations and specializations to environmental conditions. We return to this interesting subject in Chapter 49.

Early Humans in America

No truly primitive humans (in the evolutionary sense) ever reached the Americas as far as we know. America was peopled by repeated invasions of modern humans from Asia, principally by way of Alaska. The oldest surely dated traces of humans in America are only about 10,000 years old, although some finds may be twice that age or more. In any case, it seems fairly certain that humankind did not reach America until the closing stage of the Pleistocene. Only *Homo sapiens,* mainly or entirely mongoloid in origin, ever reached the New World.

---

[27] This idea was accepted by no less a personage than Louis Agassiz, America's foremost biologist of the mid-nineteenth century, and it was a position that gave considerable comfort to the defenders of slavery.

SUMMARY

Life arose from self-replicating organic molecules that formed abiotically in the primordial seas (see Chapter 4). Evidence that some RNA molecules are enzymes strongly suggests that the earliest organisms belonged to an RNA world in which RNA was the only catalyst. With the advent of protein synthesis, RNA and protein molecules both served as catalysts in a ribonucleoprotein world. Only later did the familiar DNA world evolve. With the arrival of true organisms, life would henceforth arise only from preexisting life.

These first organisms were probably heterotrophs that rapidly depleted the seas of the abiotically-produced organic molecules on which they depended. The resulting selective pressure led to the emergence of photosynthetic autotrophs. The O_2 they produced increased atmospheric O_2 and formed an ozone layer that still protects the Earth's surface from the ultraviolet radiation that earlier had energized abiotic syntheses of organic molecules.

By the end of the Precambrian era, some 600 million years ago, eukaryotes had evolved; and in the succeeding Cambrian period, which began the Paleozoic era, a variety of animals (marine invertebrates) were abundant.

By the end of the next period, the Ordovician (430 million years ago), most of the major phyla were present. The trilobites (arthropods) were past their prime and the first vertebrates, fishes, had appeared. The Silurian period (ending 405 million years ago) saw the first land organisms: first plants, later worms and arthropods. The Devonian, the Age of Fishes, followed.

The end of the Paleozoic era was marked by the rise of conifers and the great coal forests (in the Carboniferous), the emergence and later expansion of reptiles, the first great insect radiation, the expansion and relative decline of amphibians, and the last of the world's trilobites.

The Mesozoic era, the Age of Reptiles, began 290 million years ago with the first appearance of dinosaurs (in the Triassic) and mammals (in the Jurassic). The first birds and flowering plants appeared during the height of the Age of Reptiles, which ended with mass extinctions at the end of the Mesozoic (65 million years ago). Despite these extinctions, whose multiple causes may have included the collision of a giant asteroid with Earth, there was a great concurrent expansion of angiosperms.

These remarkable events paved the way for the Cenozoic era, the Age of Mammals, which began 65 million years ago. By this time, aquatic organisms and terrestrial plant life had long been essentially modern, resembling the forms of today. The adaptive radiation of mammals occurred much more slowly after their origin than had that of amphibians or reptiles.

Prosimians arose near the start of the Cenozoic. Their adaptations to arboreal life, such as opposable thumbs and binocular vision, became basic primate traits. Hominoids—apes and hominids—separated from Old World monkeys early in primate evolution. Gibbons, orang-utans, and finally the hominid and ape families diverged from the common hominoid stock during the Miocene epoch. Australopithecines were hominids that were extant about 3 million years ago. Their erect posture and bipedal locomotion anticipated the later increases in brain size that have characterized the genus *Homo*. Its earliest known members, called *Homo habilis*, are less than a million years old and were evidently tool-users. *Homo erectus* followed them, and then between 75 and 35 thousand years ago came the Neandertals. Whether they were variant *Homo sapiens* or a kindred species is still not known. Modern *Homo sapiens* first appeared in the fossil record about 25 thousand years ago. Our species evidently arose in Africa.

KEY TERMS

australopithecines
brachiation
ceboid
cercopithecoid
crossopterygian
eocrinoid
gibbon

gorilla
hominid
hominoid
ichthyosaur
lingula
lungfish
mosaic evolution

orang-utan
plesiosaur
prosimian
psilopsid
sauropod
therapod
trilobite

QUIZ QUESTIONS

1. Of these groups of organisms, which one appeared on the Earth first?
   A. Reptilia
   B. Agnatha
   C. Trilobita
   D. Aves
   E. Crossopterygii

2. The invasion of land by vertebrates was accomplished by
   A. agnathans.
   B. placoderms.
   C. labyrinthodonts.
   D. dinosaurs.
   E. None of the above.

3. The crisis that drastically reduced the diversity of dinosaurs and other reptiles occurred
   A. at the end of the Permian era.
   B. at the beginning of the Mesozoic era.
   C. at the end of the Mesozoic era.
   D. during the height of the great coal forests.
   E. A and D

4. Which of the following groups evolved from reptiles?
   A. amphibians and birds
   B. amphibians and mammals
   C. birds and mammals

   D. mammals and crossopterygians
   E. trilobites and ammonites

5. Which of these sequences of names properly indicates an evolutionary sequence in family Hominidae?
   A. prosimian, australopithecine, *Homo habilis, Homo sapiens*
   B. prosimian, australopithecine, *Homo erectus, Homo habilis*
   C. *Homo erectus, Homo habilis,* australopithecine
   D. *Homo habilis, Homo sapiens,* australopithecine

ESSAY QUESTIONS

1. What biological macromolecules probably arose first in evolution? Which molecules came next? What class of organism likely arose first, heterotrophs or autotrophs?

2. What effect might early organisms, particularly those capable of photosynthesis, have had upon the environmental conditions of the Earth?

3. Which arose first on land, plants or animals? Explain. What roles did "crises" play in the evolution of life forms? In what periods did crises occur, and what were the results?

4. What kind of creatures were the inhabitants of the Paleozoic era? Within what set of environmental conditions did they live? What patterns of living did they follow? What is your imaginative view based upon?

5. What features of the reptilian structure and function make it possible to generalize that "the transition of the vertebrates to land life was completed when the reptiles arose"? As the first well-adapted land vertebrates, how did the reptiles exploit the land environment open to them?

6. What adaptations characterize the primates? What anatomic features distinguish modern humans from their apelike ancestors? Where on Earth did humans originate?

REFERENCES AND SUGGESTIONS FOR FURTHER READING

Bakker, R. T. (1986). *The Dinosaur Heresies: The New Theories Unlocking the Mystery of the Dinosaurs and Their Extinction.* William Morrow, New York.

The author views dinosaurs as dynamic, hot-blooded creatures—perfectly adapted masters of the terrestrial ecosystem that were so advanced they subjugated our mammalian ancestors for over 150 million years.

Berggren, W. A., and Van Couvering, J. A. (Eds.) (1984). *Catastrophes and Earth History: The New Uniformitarianism.* Princeton University Press, Princeton, N.J.

Stimulating essays on the great crises of Earth's history and their role in producing mass extinctions of living organisms.

Briggs, D. E. G., and Whittington, H. B. (1985). Terror of the trilobites. *Natural History 94,* December, 34–39.

A recently studied fossil, *Anomalocaris,* was an 18-inch "giant" by Cambrian standards. A swimmer with powerful jaws, it probably was a major predator of the trilobites.

Cavalli-Sforza, L. L., Piazza, A., Menozzi, P., and Mountain, J. (1988). Reconstruction of human evolution: bringing together genetic, archaeological, and linguistic data. *Proceedings of the National Academy of Sciences (USA) 85,* 6002–6006.

An attempt to clarify the human lineage.

Cech, T. R. (1986). RNA as an enzyme. *Scientific American 255,* November, 64–75.

An account of the discovery that certain RNA molecules can be as effective catalysts as enzymatic proteins. Another of Cech's papers is cited at the end of Chapter 18.

Delson, E. (Ed.) (1984). *Ancestors: The Hard Evidence.* Alan R. Liss, New York.

A useful update of the chronology of major East African hominid fossils.

Dickerson, R. (1978). Chemical evolution and the origin of life. *Scientific American 239,* September, 62–78.

A clear account of the most widely held view of how chemical evolution led to the rise of biomolecules and then of life.

Fox, S. (1988). *The Emergence of Life: Darwinian Evolution from the Inside.* Basic Books, New York.

The author, a pioneering experimentalist on the origin of life (see Chapter 4), gives a lively account of his ideas.

Hsü, K. J. (1987). *The Great Dying: Cosmic Catastrophe, Dinosaurs, and the Theory of Evolution.* Harcourt, Brace Jovanovich, San Diego, Calif.

An entertaining, popular book championing the idea that an asteroid hit the Earth and caused mass extinctions at the end of the Cretaceous.

Joyce, G. F. (1989). RNA evolution and the origins of life. *Nature 338,* 217–333.

The author argues that RNA played a key role in the early history of life, but it is doubtful that life began with RNA.

Kerr, R. A. (1988). Huge impact is favored K-T boundary killer. *Science 242,* 865–867.

A scientific discussion of the asteroid collision theory of mass extinctions at the Cretaceous-Tertiary boundary.

Leakey, M. (1984). *Disclosing the Past: An Autobiography.* Doubleday, New York.

This lively autobiography, written in the busy author's 70th year, is richly informative about the hunt for human ancestors and the Leakeys' part in it. It is a tale with many heroes and villains.

Lewin, R. (1982). *Thread of Life: The Smithsonian Looks at Evolution.* Smithsonian Books, Washington, D.C.

A well-illustrated and comprehensive survey of the history of life on Earth.

Lewin, R. (1987). *Bones of Contention: Controversies in the Search for Human Origins.* Simon & Schuster, New York.

An enthralling and even-handed account of the great "controversies in the search for human origins." These acrimonious disputes among major personalities have much to tell us about the nature of science.

Lewin, R. (1988). Conflict over DNA clock results. *Science 241,* 1598–1600 and 1756–1759.

Recent DNA–DNA hybridization data (*J. Mol. Evolution 20, 2,* 1984) showed a surprisingly close relation between chimpanzees and humans. This report details recent criticisms of the technique—or the way it was used in these studies.

Little, C. (1983). *The Colonization of Land: Origins and Adaptations of Terrestrial Animals.* Cambridge University Press, New York.

An outstanding monograph on the transition from aquatic to terrestrial life during vertebrate evolution.

Loomis, W. F. (1988). *Four Billion Years: An Essay on the Evolution of Genes and Organisms.* Sinauer, Sunderland, Mass.

An account of the evolution of life from prebiotic times to the relatively recent appearance of complex organisms.

Lovejoy, C. O. (1988). Evolution of human walking. *Scientific American 259,* November, 118–125.

How we got to walk around on two limbs rather than four, as most mammals do.

McMenamin, M. A. S. (1987). The emergence of animals. *Scientific American 256,* April, 94–103.

Well illustrated essay on the explosion of new life forms that took place at the beginning of the Cambrian.

Oxnard, C. (1973) *Form and Pattern in Human Evolution.* University of Chicago Press, Chicago.

The author shows how the application of statistical, physical, and engineering techniques to the study of living and fossil bones can illuminate the importance of shape and structure in human evolution.

Pilbeam, D. (1984). The descent of hominoids and hominids. *Scientific American 250,* March, 84–96.

An overview of modern ideas (including the author's latest views) on human evolution.

Russell, D. A. (1982). The mass extinctions of the late Mesozoic. *Scientific American 246,* January, 58–65.

An interpretation and analysis of the causes of one of the great extinctions during the history of the Earth.

Rutwick, N. J. S. (1972). *The Meaning of Fossils: Episodes in the History of Paleontology.* University of Chicago Press, Chicago.

A synthesis of our understanding of fossils, which clearly demonstrates how paleontology has influenced thought on the living world and the place of human beings within it.

Taylor, T. N. (1981). *Paleobotany: An Introduction to Fossil Plant Biology.* McGraw-Hill, New York.

A discussion of structural, functional, and ecological aspects of the evolutionary history of plants.

Tobias, P. V. (Ed.) (1985). *Hominid Evolution. Past, Present and Future.* Alan R. Liss, New York.

Record of a symposium celebrating the 60th anniversary of the discovery of *Australopithecus africanus.*

Walker, A., and Teaford, M. (1988). The hunt for Proconsul. *Scientific American 260,* January, 76–82.

Interesting discussion of the tree-living creasure of 18 million years ago that may have been the last common ancestor of humans and great apes.

ECOLOGY

POPULATIONS

Ecology is the study of the interactions, simple and complex, between living organisms and the environment, both physical and biological. We have seen that a hierarchy of organizational levels marks all biological systems (see Table 1-1). Ecology deals primarily with three levels of organization at the high end of the scale.

- **Populations** are groups of individuals belonging to the same species (Figure 45-1); the total assemblage of individuals of a given species in any ecosystem.
- **Communities** are composed of populations of one or more species living in a given area and interacting in diverse ways.
- **Ecosystems** are the communities and their environments considered together. Various ecosystems are interconnected by all manner of biological, chemical, and physical factors.

The entire Earth is an ecosystem. Only the thin shell of Earth's surface accommodates living organisms and their environments. This is the biosphere—a self-sufficient system in which water, oxygen, and nutrients are repeatedly cycled under the favoring influence of solar energy.

In this chapter, we will discuss the concepts of modern ecology at the population level. The two chapters that follow survey ecosystems and communities.

RISE AND FALL OF POPULATIONS

Each population is a system that has its own dynamics and interrelationships. Of these, population growth and decline have special importance. We consider them now, and then turn to particular intraspecific interactions that culminated in the evolution of societies.

BIRTH, DEATH, AND THE GROWTH RATE

What would happen if cells could divide without limit? If, for example, we took a single animal cell and let it and its progeny divide every 24 hours, how many cells would we have in a month—and then in a year? The answer may surprise. In a week, there would be 2^7 or 128 cells. In a month, there would be 2^{30} or 1,073,741,824 cells. And in a year, there would be 2^{365} cells. If the volume of the original cell was 5×10^{-10} cubic cm, the collective volume of the cells emerging from this hectic year would be 5×10^{100} cubic cm—or 5×10^{88} cubic kilometers, or 1.2×10^{88} cubic miles, which is a cube approximately 10^{30} miles in each direction. That is larger than the visible universe! Clearly then, cells that divide every 24 hours are not free to expand without limit.

These considerations teach two lessons: (1) unlimited population increase is *exponential*—that is, the larger a population the faster it grows; and (2) various constraints can and must limit this expansion.

Populations
Populations are groups of individuals of the same species in a given ecosystem. Here are some populations of bison, pine trees, star fishes, limpets, sea lions, impalas, grunts, gulls, green algae, and phytoplankton.

45-1 *More populations*
These are bees, shad, and pine trees.

Exponential growth may be formulated in this way:

$$\text{Growth in population} = \left(\text{average birthrate} - \text{average death rate}\right) \times \text{number of individuals} \qquad (45.1)$$

In mathematical form, this is expressed by the equation:

$$I = (b - d) \times N$$
$$\text{or} \quad I = r \times N \qquad (45.2)$$

where I is the increase in population, b is the birthrate (number of births per unit time—a year, for example—divided by the number of individuals in the population), d is the death rate (number of deaths per unit time divided by the same number), r is the usual symbol for $(b - d)$, and N is the number of individuals in the population. The difference between b and d—the term in parenthesis in the equation, or r—is the **intrinsic rate of increase.** More on the mathematical aspects of this concept appears in Box 45-1.

Let us consider an example of human population. Government statistics reveal the following interesting data about the population of the United States:

Year	1960	1965	1970	1975	1980	1985
Total population (millions)	179.4	193.2	203.8	214.9	226.5	238.2
Live births (millions)	4.31	3.80	3.74	3.14	3.61	3.75
Birthrate (per 1000)	23.8	19.6	18.2	14.6	15.9	15.7
Immigrants (millions)	0.33	0.37	0.44	0.45	0.85	0.58
Immigration rate (per 1000)	1.8	1.9	2.1	2.1	3.7	2.6
Deaths (millions)	1.71	1.83	1.93	1.89	1.99	2.08
Death rate (per 1000)	9.5	9.4	9.4	8.8	8.7	8.7

For our present purposes, we will add the immigration rate to the birthrate. This yields the following sums (which we will use in place of the birthrates in Equation 45.1).

	1960	1965	1970	1975	1980	1985
Birthrate plus immigration rate (per 1000)	25.6	21.5	20.3	16.7	19.6	18.3

Calculations using Equation 45.2 yield the following growth rates:

	1960	1965	1970	1975	1980	1985
Intrinsic rate of increase (per 1000)	16.1	11.9	12.8	10.0	11.3	9.4
(per cent)	1.61	1.19	1.28	1.00	1.13	0.94
Total growth per year (millions)	2.89	2.34	2.22	1.70	2.47	2.29

The intrinsic rate of increase has evidently been decreasing. Let us assume that the rate soon stabilizes at 8.5 per 1000, or 0.85%. In that case, a calculator pro-

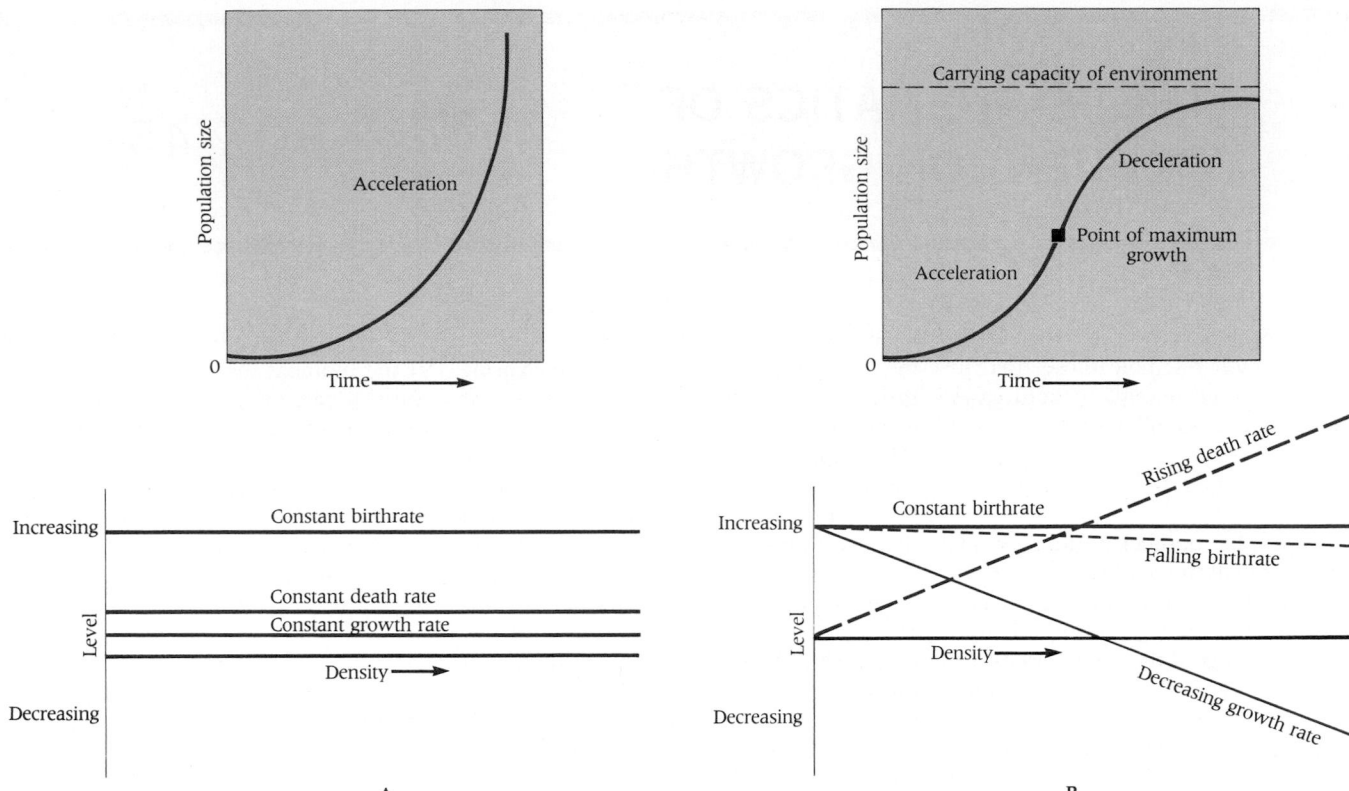

A

B

45-2 Population growth curves
(A) *Exponential (unrestricted) growth would occur if population expansion was unchecked and, as shown, birth, death, and growth rates remained constant despite increasing population density.* **(B)** *Logistic (restricted) population growth occurs when numbers plateau at the carrying capacity of the environment. In this case, growth rate and birth rate decrease with increasing population density.*

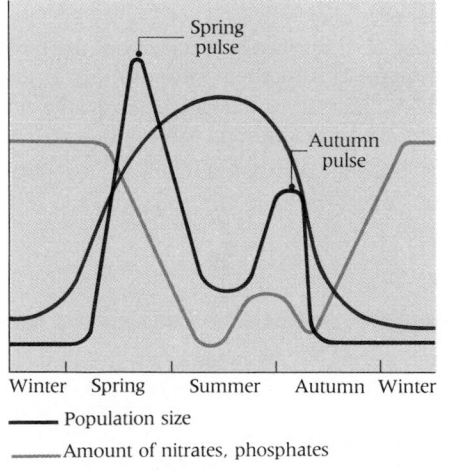

Population size

Amount of nitrates, phosphates

Amount of light

45-3 Seasonal fluctuations in diatom abundance in the North Atlantic
In winter, cold and poorly illuminated surface waters inhibit diatom growth. As a result, nitrate and phosphate levels are high. In spring, light intensity increases, waters warm up, and there is a rapid rise in diatoms. In summer, the diatom population declines because of lowered mineral levels and also predation. In autumn, a renewal of minerals results in a rise in diatoms. Growth slacks again in the winter.

grammed to do compound interest problems would quickly show that the population will be 334.2 million in the year 2025, 555.3 million in 2085, and 1 billion in 2155—that is, if the intrinsic rate of increase were to remain stable at 0.85% per year—a highly conservative assumption. Although this would be only a modest rate of increase, we begin to grasp the meaning of the term "population explosion."

The intrinsic rate of increase varies widely among different species and among different populations of the same species, depending on physiological aspects of the organism—and especially on environmental conditions. Obviously, if conditions were harsh enough to make reproduction difficult, the intrinsic rate of increase would approach zero. It reaches zero when the birthrate and death rate are equal— or when the population is zero—that is, extinct. It can even become negative if some individuals are dying while none are being born. If that were to continue such a population would soon become extinct.

We began this discussion with a projection of the fantastic consequences of an **exponential growth** pattern in which a population of cells doubled every 24 hours. The pattern of unrestricted growth is represented diagrammatically in Figure 45-2A. This is an exponential growth curve. It is fantastic because no population is likely to grow larger than the visible universe. Nevertheless, animal cells, for example, can and do double every 24 hours (and bacterial cells double much more rapidly), but they do so only under special conditions—in cell cultures, in tumors, in healing wounds, etc—and only for a limited time. Examples of exponential "population explosions" occur in nature under highly favorable conditions when, for a short time, the birthrate is much larger than the death rate. An example is seen in several species of diatoms in the North Atlantic (Figure 45-3), which at different times of the year undergo exponential increases that are triggered by variations in the environment and are always followed by "crashes" in which the populations are reduced by changing conditions.

A far more common pattern of population growth is one in which an initially exponential growth curve soon slows down, assuming a sigmoidal or S shape (Figure 45-2B). This is a **logistic growth** curve. This is the pattern of restricted growth that is typical of populations in which negative feedback control mechanisms relate the death rate to population density. Such a mechanism could be based on the limited availability of food or other more complex intraspecific and interspecific interactions. As we will see, some regulatory factors are density-independent.

In logistic population growth, the number of individuals in the population in-

The mathematical expressions that define population growth resemble those used to calculate compound interest. In both cases the basic formula is as follows:

$$(1 + i)^n$$

where n equals the number of time periods per year and i equals the yearly rate of interest per time period. When a bank pays *yearly* compound interest on an account, n equals the number of years. If the time periods are shorter than a year, say, months, i must represent the interest rate per month (not per year). Thus, if the bank paid 6% interest per year, the rate per month would be .06/12 or 0.005, or 0.5%.

Consider an imaginary bank that pays 6% compound interest yearly. At the end of the first year, each $1.00 in the account will be supplemented by $0.06. In the second year, 6% interest is paid on $1.06, and so on. According to the formula, at the end of 10 years each dollar originally in the account would be worth $1.79.

$$(1 + 0.06/1)^{10} = \$1.79$$

If interest were compounded *monthly*, the annual interest rate would have to be divided by 12 to obtain a monthly rate and n, the number of time periods, would become 10 × 12. After 10 years of monthly compounding, then, each dollar in the account would be worth $1.82.

$$(1 + 0.06/12)^{10 \times 12} = \$1.82$$

It begins to look as though the return on the account can be increased by increasing the number of times each year that interest is compounded. But how far can we push this? Suppose in this remarkable bank we decided to compound interest continuously—that is, at every instant throughout the year. How much would be in the account at the year's end? A colossal fortune? Not quite, for this process would generate an infinite series that converges to a limit. If one began with $1.00, the total on account after this boisterous year would be just short of $2.72. For if we take the trouble to expand the term $(1 + i)^n$, we find that as n, the number of time periods, becomes very large and indeed approaches infinity, i, the second term in the parenthetic expression becomes smaller, even though the whole expression is raised to an infinite power. The limit that is approached in this expansion is called e.

Mathematicians teach us that the exponential function

$$y = e^x$$

is the basic equation for describing growth of any kind. This is the basic equation that expresses the principles of exponential population growth that were presented in the text in words. In mathematical terms, it is written as follows:

$$\frac{dN}{dt} = rN$$

where N is the number of individuals present at the moment the count is taken; r is the intrinsic rate of increase (the interest rate); dN is the amount of growth in an infinitesimally small amount of time, dt; and dN/dt is the rate of increase in the number of individuals in the population. Differential calculus gives us a solution for this rate equation from which we can draw the curve in Figure 45-2A. It is as follows:

$$N = N_0 e^{rt}$$

This equation expresses N as a mathematical function of time. The term N_0 is the population at the starting point (time zero); t is the amount of time that has passed since time zero, and e is the term derived above (which equals 2.71828 . . . and is the base of natural logarithms). Since the independent variable, t, is in the exponent of this equation, the growth that occurs is exponential growth.

Logistic growth, in contrast, is expressed mathematically as follows:

$$\frac{dN}{dt} = \frac{r(K - N)N}{K}$$

where K is the carrying capacity of the environment. This mathematical rate expression is also identical to the one given in the text in words. Its solution, too complicated to present here, is the S-shaped curve shown in Figure 45-2B.

creases to a point at which their environment can no longer support them. The population density at that point is usually termed the **carrying capacity** of the environment. The carrying capacity is defined as the maximum number of individuals of a species that a given area can support indefinitely. At this point, the death rate equals the birthrate and the rate of increase in the number of individuals is zero. In the real world, there often are temporary fluctuations around this equilibrium density, with short-lived increases and decreases around the carrying capacity of the environment. It may seem surprising, but for most of its history the human species has had a growth rate close to zero. Only in the last few thousand years did technological and medical advances permit a necessarily temporary exponential increase (Figure 45-4).

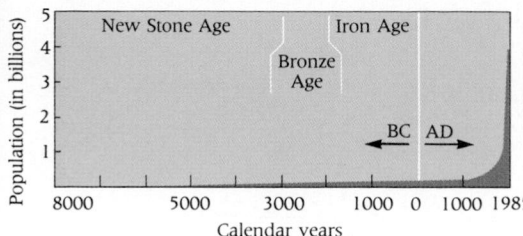

45-4 Human population growth
The growth of the world's human population was slow for many thousands of years, but in the last century it has accelerated exponentially.

The concept of environmental carrying capacity has important implications. The population density approaches the carrying capacity by a logistic or S-shaped curve, because as density increases, birthrates decrease. Birthrate and death rate come to equal each other when the carrying capacity is reached. The carrying capacity of the environment adds a new algebraic term to the population increase equation given earlier and thereby converts it to an equation for the logistic growth of a population.

$$\text{Growth rate} = \left(\text{average birthrate} - \text{average death rate}\right) \times \left(\frac{\text{carrying capacity} - \text{number of individuals}}{\text{carrying capacity}}\right) \times \text{number of individuals} \qquad (45.3)$$

Or, in mathematical terms:

$$I = r \times \left(\frac{K - N}{K}\right) \times N \qquad (45.4)$$

where I again equals the rate of population growth, r is the difference between birth and death rates (intrinsic rate of increase), K is the carrying capacity, and N is the number of individuals in the population. When the number of individuals is very small, the term in parentheses is virtually equal to one. Therefore, the growth rate is essentially as it was in the first equation—exponential. At the end of the curve, however, when the number of individuals is large and the population is equal to the carrying capacity, the term inside the parentheses approaches zero. At this point, we consider the environment to be saturated with individuals of a particular species. Growth rate of the population is then zero.

The rate of increase of a population is greatest at the inflection point (the point of transition from an accelerating to a decelerating rate), which is called the **maximum sustained yield.** In theory, one would expect that no damage is done to a population if harvesting is limited to the point of maximum sustained yield. However, if harvesting is allowed to reduce the population too far below the inflection point, the recovery of the population may be endangered. Other important variables, such as age and size distributions can play a key role in the maintenance of the population.

SURVIVORSHIP

The factors that determine birth and death rates are complex enough. Moreover, the study of population change, or population dynamics, is further complicated by the need to take into account how long individuals live. A population would achieve its minimum possible death rate only if all deaths were the result of "old age"— that is, if all individuals lived out their full life spans. In reality, the average length of life in a population is well below the theoretical maximum. Some individuals die at all ages from birth to senescence. The percentage of individuals that die at a given age—the death rate for that age—also changes markedly through the life span. Usually the death rate is high among the very young and the very old and reaches a low point somewhere in between. As shown in Table 45-1, this is as true of humans as it is of most other organisms. Note the higher death rates of males compared to females.

Premature death is more the rule than the exception in nature. It is presumably due not to anything programmed in the organism's genes, but to environmental incidents such as competition, infection, predation, or accident.

What we are considering is the striking contrast between the potential or physiological life span built into the organism and the shorter span that it usually manages to achieve. The traditional allotment of three score and ten years is too low an estimate for most humans. The potential span in *Homo sapiens* is the longest among mammals[1] and is longer than that of most other animals, but not all. Some mollusks have lived longer than any human, and other invertebrates may live longer than

[1] The belief that elephants live longer than humans is incorrect. The longest probably authentic survival for an elephant is 77 years. The record for human longevity is unknown, since extreme claims are unsubstantiated by evidence and are usually tall tales. Nevertheless, many cases are known of individuals who have lived beyond 110 years.

TABLE 45-1
DEATH RATES IN THE UNITED STATES, 1940 TO 1987†

Item	1940	1950	1960	1970	1980	1987*
Death rates per year	10.8	9.6	9.5	9.5	8.8	8.7
All ages						
Male	12.0	11.1	11.1	10.9	9.8	9.4
Female	9.5	8.2	8.1	8.1	7.9	8.2
Males, by age						
Under 1 year	61.9	37.3	30.6	24.1	14.3	11.2
1–4 years	3.1	1.5	1.2	0.9	0.7	0.6
5–14 years	1.2	0.7	0.6	0.5	0.4	0.3
15–24 years	2.3	1.7	1.5	1.9	1.7	1.5
25–34 years	3.4	2.2	1.9	2.2	2.0	1.9
35–44 years	5.9	4.3	3.7	4.0	3.0	2.9
45–54 years	12.5	10.7	9.9	9.6	7.7	6.5
55–64 years	26.1	24.0	23.1	22.8	18.2	16.5
65–74 years	54.6	49.3	49.1	48.7	41.1	37.0
75–84 years	121.3	104.3	101.8	100.1	88.2	83.5
85 years and over	246.4	216.4	211.9	178.2	188.0	181.9
Females, by age						
Under 1 year	47.7	28.5	23.2	18.6	11.4	8.8
1–4 years	2.7	1.3	1.0	0.8	0.6	0.4
5–14 years	0.9	0.5	0.4	0.3	0.2	0.2
15–24 years	1.8	0.9	0.6	0.7	0.6	0.5
25–34 years	2.7	1.4	1.1	1.0	0.8	0.7
35–44 years	4.5	2.9	2.3	2.3	1.6	1.4
45–54 years	8.6	6.4	5.3	5.2	4.1	3.6
55–64 years	18.0	14.0	12.0	11.0	9.3	9.1
65–74 years	42.2	33.3	28.7	25.8	21.5	20.9
75–84 years	103.7	84.0	76.3	66.8	54.4	51.5
85 years and over	227.6	191.9	190.1	155.2	147.5	143.0

From *Statistical Abstract of the United States*, 97th and 109th editions.
† Per 1000 population
* Preliminary data

any mollusk. However, most invertebrates have comparatively short lives—a few weeks or a few years. A few species of turtles live about 140 years.

Most plants also have short lives. Annual plants are so called because the potential life of the developed and reproducing plant is one year. The dormant seed, of course, can remain dormant, alive in a sense, for many years. Trees and some shrubs, however, may have spans counted in centuries.

The potential span for some protists may be a matter of hours or minutes. Of course, when a protist undergoes fission it becomes two new individuals. Hence, our definition of life span here is a little different.

A convenient way to represent the incidence of death in a population is by a **survivorship curve** (Figure 45-5), which shows the percentage of individuals still living at various times after birth. If most individuals live out the average potential life span, the curve is nearly horizontal until that span is reached and then it drops precipitously (curve A in Figure 45-5). At the opposite extreme, if most individuals die early in life and the survivors of that critical period have comparatively low death rates, the curve drops rapidly at first and then levels off (curve B). The first situation, practically speaking, does not occur in nature. The second is common, especially among plants and marine animals. Curve D, in which a constant percent of the original population dies at each age, is quite rare. Among animal populations, most survivorship patterns are intermediate between curves D and B. The curves for most plants are near the extreme of curve B. In other words, high mortality among the young is common in nature.

Human survivorship curves are of an intermediate type but they differ greatly in form in accordance with nutrition, sanitation, and medical care in a particular population. Where the level of public health is high (curve C), there is an approach to curve A; where it is low—as it was in early times and in primitive societies—the pattern approaches curve B.

One effect of this situation is an increase in the proportion of older people in the population. In the United States (see Table 45-1) and in most western countries, recent years have been generally characterized by a declining birthrate and a declining death rate. The net result of these two factors has been that the percentage of

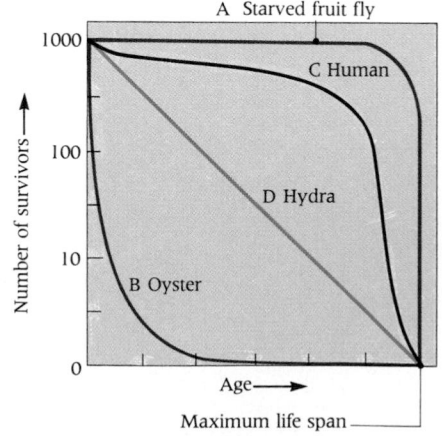

45-5 Survivorship curves
The number of survivors (survivorship per 1000 born) is plotted on a logarithmic scale. The curve for fruit flies (A) applies only to their life spans under the unnatural conditions of starvation in the laboratory. Note the striking differences between the pattern in fruit flies and humans and that in oysters.

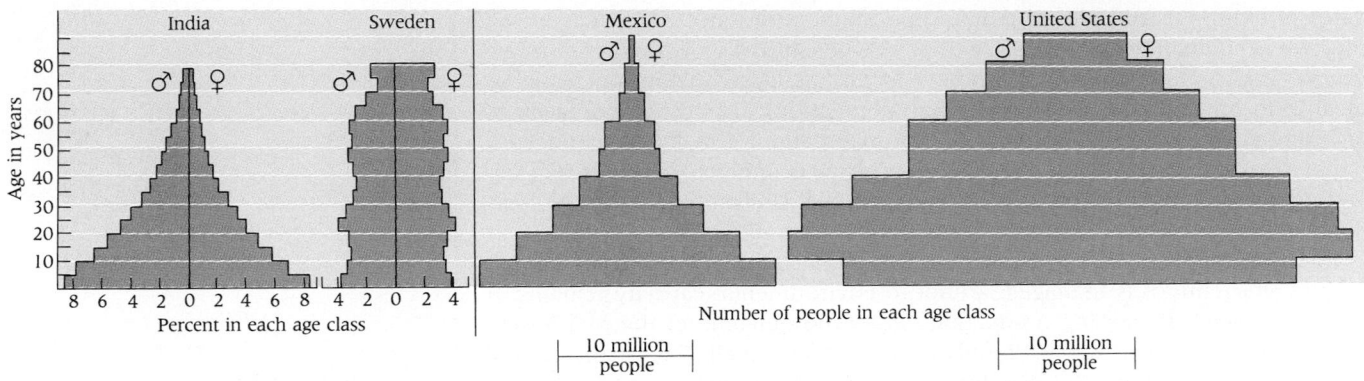

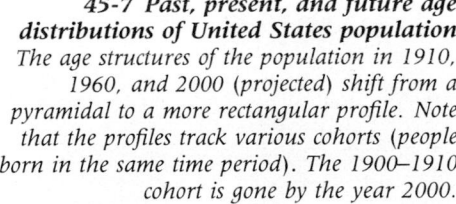

45-6 Age distribution in different countries
A pyramidal profile (India, Mexico) is characteristic of developing countries. A so-called rectangular profile (Sweden, United States) is typical of developed countries with a near-stable population.

Americans aged 20 to 45—the period of maximum rates of reproduction and productive labor—has stayed fairly steady, while the percentage of younger people, under 20, has decreased and that of older people—over 45—has correspondingly increased. Such patterns are readily seen in age distribution charts like those in Figures 45-6 and 45-7.

MALTHUS AND THE GROWTH OF POPULATIONS

As shown in the S-shaped expansion curve of Figure 45-2B, all offspring usually do not survive and breed. Only a small fraction of them do under ordinary conditions. The others are eliminated by competitors, within or outside their species, and hence by starvation, or by enemies, diseases, and other factors.

Thomas Robert Malthus (1766–1834), an English clergyman and economist, called attention to these grim facts in 1798. He extended the principle to humankind and concluded that if humans increase more rapidly than their means of subsistence, then famine, pestilence, or war must be the inevitable constraints on the expansion of human populations. Darwin saw that this Malthusian principle applies even more clearly to other organisms—and that the inevitable decimation of offspring may be a major factor underlying evolution by natural selection. Darwin was right on this point, and we now recognize that the sort of mortality envisioned by Malthus has been one of the important mechanisms of evolution.

The application of Malthus's unpleasant ideas to the human species provoked a storm of protest and many, including politicians, theologians, economists, and even a few biologists, objected that the Malthusian principle does not really apply to human beings. Indeed, subsequent events in the United States and other modern nations have seemed to refute Malthus's idea. In those countries, populations have increased because birthrates exceeded death rates and survival has increased. Yet

45-7 Past, present, and future age distributions of United States population
The age structures of the population in 1910, 1960, and 2000 (projected) shift from a pyramidal to a more rectangular profile. Note that the profiles track various cohorts (people born in the same time period). The 1900–1910 cohort is gone by the year 2000.

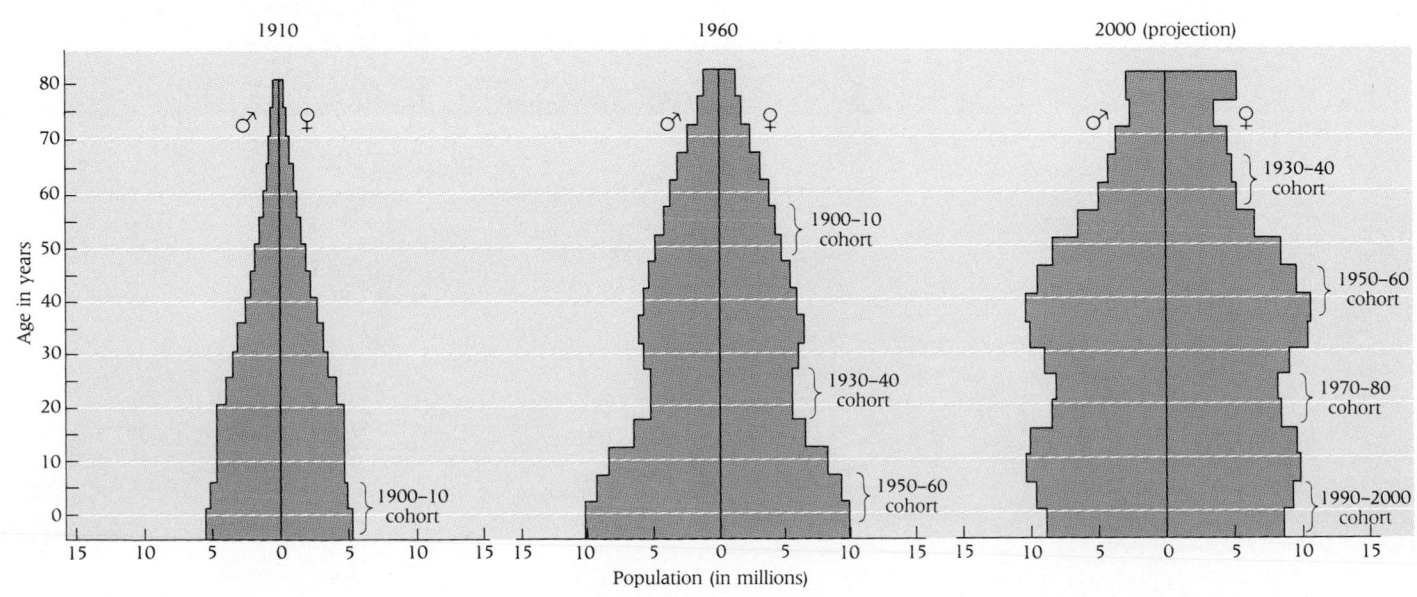

standards of living, instead of dropping to minimum subsistence levels, have risen sharply. The explanation, of course, is that the benchmarks of production have also risen—and population growth, though large, has lagged behind increases in production. In other words, for some human populations, the carrying capacity of the environment has increased faster than the population. This has encouraged a belief that the solution to human population problems is simply to continue increasing production—as broadly and as rapidly as possible. Unfortunately, it is biologically possible for that solution to work only temporarily and locally.

No matter how much of Earth's surface is brought into production for human use or how efficiently it is managed, a limit to environmental capacity remains. It cannot be removed. Increasing production delays the outcome of the Malthusian principle, but does not evade it. Populations must eventually be balanced, and birthrates cannot continue to exceed death rates. Whether balance is to be attained by decrease of birthrate or increase of death rate are economic and political issues of the greatest possible magnitude and urgency. They underlie our gathering ecological crises in energy and food production and environmental deterioration and we discuss them further in Chapter 49.

LIMITING AND BALANCING FACTORS

An important factor is **population density,** which is the number of individuals concentrated in a given space. The factors that limit population growth are either dependent on the density of the population or are density-independent (Figure 45-8).

- **Density-independent limiting factors** exert a constant influence regardless of the population density. These factors include weather and physical disruptions of the environment, such as floods or fires that occur regardless of population size. For example, a population of aphids in a field of alfalfa can grow at an exponential rate if the weather is cool and moist, but the population can crash suddenly when the weather abruptly becomes hot and dry. The sudden decline in the aphid population occurs regardless of population density.

45-8 Population limiting factors
Density-independent limiting factors include **(A)** *flood and* **(B)** *fire. Density-dependent limiting factors include* **(C)** *competition and* **(D)** *predation. All of these factors affect plants and animals alike.*

A

B

C

D

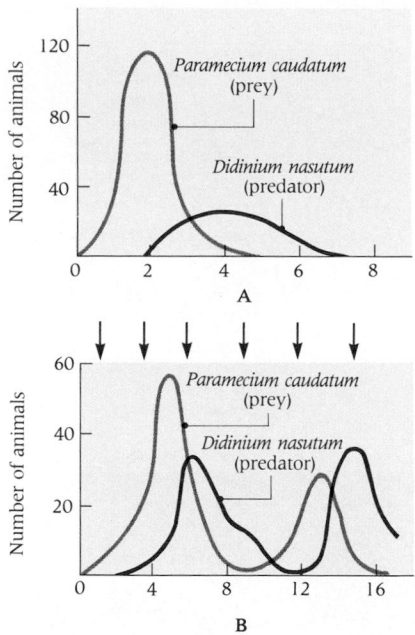

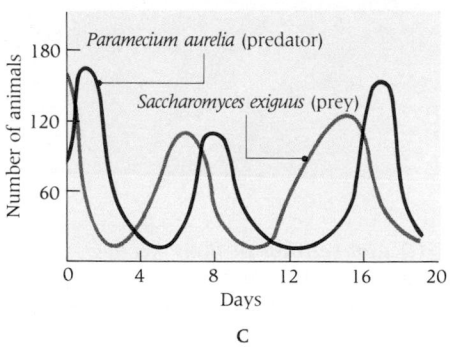

45-9 Prey-predator relationships
(A) *In the simpler case, the prey population* (Paramecium), *grows until predation by the ciliate,* Didinium, *commences. The predator population grows until the prey is exterminated. Now without resources, it declines to extinction.* **(B,C)** *In these cases of prey-predator relations, decline of the prey population causes a decline of the predator population and thus relieves pressure on the prey before it reaches extinction. Under temporary reprieve, the prey population again begins to grow, but the predator population—now with renewed resources— grows too. The system can persist for many oscillatory cycles. In* **(B)** *the prey population is preserved by additions of new individuals at times marked by arrows.*

- **Density-dependent limiting factors** are those whose effectiveness is determined at least in part by the density of the population being limited. The more crowded the population, the greater the impact of these factors. Biotic environmental factors—parasitism, predation, competition, etc.—are density-dependent limiting factors. Note that the second parenthetic term in Equation 45.3 is a mathematical expression of density-dependent growth limitation.

ENERGY AND MATERIALS

We have seen that the ultimate constraint on life is set by solar energy, only a small percentage of which is directly available for living organisms (see Chapter 10). This is explained by the low efficiency of photosynthesis and the massive amounts of energy needed to keep Earth livable, maintain its surface temperatures, and operate the water cycle on which all land organisms depend (Chapter 46). The materials available for organisms in any community are also limited, and no population can live beyond its budget.

The amounts of energy and materials available to the living world as a whole are independent of the density of any population. However, the amounts of these critically important commodities that actually become available to members of any given population in its environment are importantly related to population densities.[2]

FOOD CHAINS: PREDATION AND PARASITISM

We will see in Chapter 46 that each link in a food chain depends on the links that precede it. Hence, the size of a population at any link is limited by population sizes in all previous links. Population size in food chains is pushed in two directions: (1) the population size of a food species limits the population that feeds on it; and (2) the eating of a food species shrinks its population size.

- In **parasitism,** organisms live at the expense of other organisms but, though harmful to them, do not usually kill them. Thus there is a necessary balance between the numbers of parasites and the damage they do to their hosts.
- In **predation,** the eaten individual is killed. The predator population is usually less numerous than the prey population (except in transitory situations). As the density of the prey population increases, the number of predators usually increases also. This in turn tends to decrease the prey population.

These interactions can have very complex effects on the populations, but one of three reactions (or a combination of them) ordinarily ensues. These patterns are shown in Figure 45-9, which summarizes well-known studies carried out by G. F. Gause in the 1930s on *Paramecium caudatum.* In this study, *P. caudatum* in an aquarium is the prey of a ciliate predator, *Didinium nasutum.*

- First, increasing predation may destroy all the prey and wipe them out, after which the predator population, now without prey, starves to death and becomes extinct—or turns to other prey (Figure 45-9A). This cataclysmic short-range reaction obviously cannot persist in a balanced community.
- Second, increase of predation and decrease of prey may be followed so promptly by a decrease in predators that the prey survive and become more abundant with lessened predation; then predators and predation increase again, and so on (Figure 45-9B). This cyclic relationship can persist indefinitely, but it is delicately balanced and could lead to extinction of the prey, the predators—or to a more stable situation.
- Third, a stable balance—common in nature and probably the rule in established communities—may occur in which predators consume just enough prey to keep the prey population at or below the limit of environmental carrying capacity (Figure 45-9C). Here it could be said that predators are simply "cropping back" excess prey production. However, the balance can be readily disturbed by the exposure of the population to all sorts of adverse influences, among them toxic

---

[2] Examples are the fluctuations in the size of the diatom population (see Figure 45-3) and the events occurring when the level of O_2 in lakes is caused to diminish (see Figure 46-20).

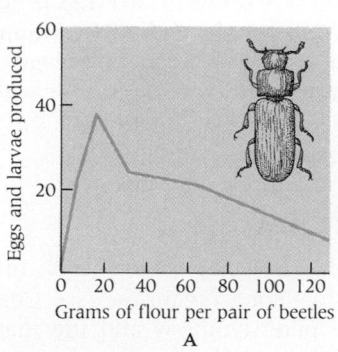

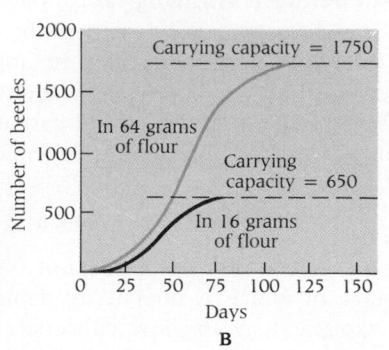

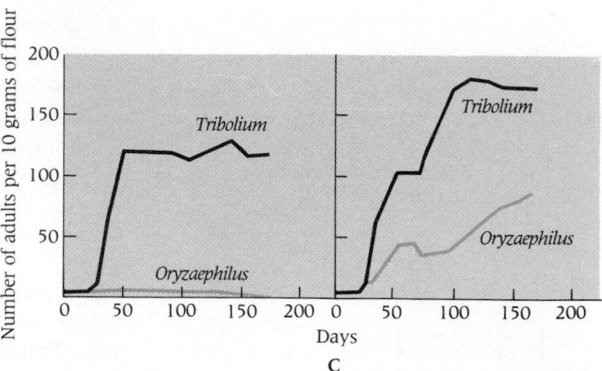

A

B

C

substances or poisons. Populations of predators are far more sensitive to poisons than populations of prey, because predator populations include fewer individuals. This drastically slows the evolution of strains resistant to poisons. In contrast, large prey populations can develop resistance faster because they reproduce at a higher rate.

These (and other) experimental results clearly implied that in stably balanced prey-predator populations, predation may actually be beneficial to the prey as a group, even though it destroys individuals. It has been suggested that in nature prey density is commonly controlled through predation. But recently it was found that this situation is not as common as once thought. The reason is that, in the wild, vertebrate predators often feed on young, old, or sick prey, thus eliminating those with the lowest reproductive potential in the population.

For example, it has been customary in biology textbooks (including earlier editions of this book) to cite the famous case of the mule deer (*Odocoileus hemionus*) on the Kaibab Plateau in Arizona—already familiar to us for its well-known squirrels (see Box 36-1). Before 1907, the plateau had a robust deer herd with a stable population kept well below the carrying capacity of the vegetation (and thus healthy and stabilized) by the heavy predation of pumas, wolves, and coyotes. A campaign of extermination was then waged against the predators with the idea of freeing the deer of their "enemies." The deer population was then reported to have increased enormously, from about 4,000 in 1907 to some 100,000 in 1924. The numbers then began to decrease, and by 1939 were down to 10,000. It was concluded that because the highest counts were far beyond the carrying capacity of the range, the deer had begun starving to death. We now realize that several variables (other than decreased predation and starvation) increased and then decreased the deer herd. Prior to 1907, cattle grazing and fires had decreased vegetation and thus limited the deer herd. When these factors were relieved in the 1920s, this change, as well as the shooting of predators, could have enlarged the herd. Ecologists now suspect that the earlier data were overinterpreted, because everyone had preconceived ideas and already "knew" that they reflected increased predation.

There are cases in which unforeseen and disastrous consequences did follow human intervention in natural communities. As noted in Box 42-2, uninformed efforts to control pests with insecticides sometimes have the unexpected result of disturbing natural predator-prey relationships, with dire consequences. When, for example, the nonpersistent organophosphate pesticide, Azodrin, was used against the boll weevil (*Anthonomus grandis*), the cotton farmer's nemesis, its effect was often the opposite of what the farmer intended. By devastating populations of insects that were predators of these pests, the pests made a big comeback in the absence of their natural enemies. The net result was an increase in the weevil populations— and more crop damage than before.

The spread of European peoples over the globe after the discovery of America soon led to the extinction of many species and local populations, including human populations. Among the best known were the dodo, passenger pigeon, sea cow, Tasmanian wolf, and Tasmanian human. Colonizing Europeans often encountered local hunting groups whose taboos effectively promoted conservation of the wildlife on which they depended. So arose the legend that primitive human hunter-gathers are prudent, and only civilized humans are exterminators. Yet accumulating evidence,

D

45-10 Competition in flour beetles, Tribolium confusum
(*A*) *The optimum density of flour beetles peaks at about 18 g of flour per pair of beetles.* (*B*) *The carrying capacity was varied by altering the amount of flour. After reaching the carrying capacity, the number of beetles in each population fluctuates only slightly.* (*C*) *Habitat complexity plays a role in survival of competing beetles* (Tribolium *and* Oryzaephilus). *In a uniform habitat of pure flour (left),* Tribolium *overcomes the second species. If pieces of glass tubing are added to the flour (right), both species survive and increase in number.* (*D*) *Four stages in the life cycle of flour beetles.*

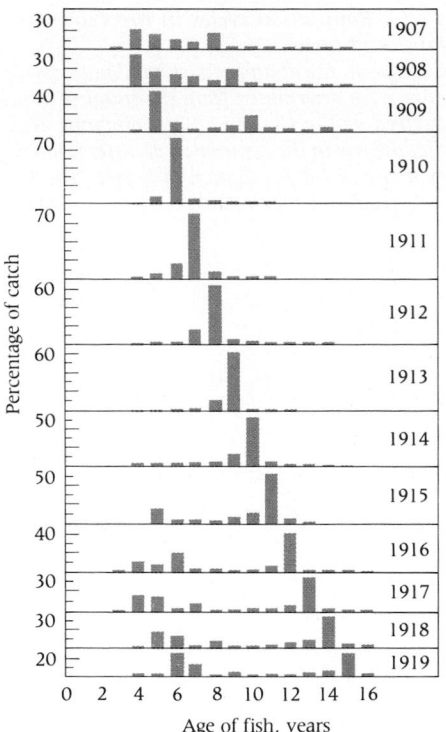

45-11 Population periodicity in the herring
The year 1904 was clearly a good one because herrings born that year made a major contribution to all catches through 1919. Note that in 1915, another good group (the "class of 1910") began to become prominent.

much of it recent, suggests that the truth may be very different. Hunter-gatherers and neolithic agriculturalists seem now to have played the part of exterminator whenever they occupied new areas of the globe.

COMPETITION

Competition is the interaction between organisms using the same essential resource, which is in short supply. The resource may be anything—a breeding site, a burrow, light, water, food, whatever. In general, competition between individuals of the same species—intraspecific competition—is far more intense than that between different species—interspecific competition. As first pointed out by Darwin, the greater the similarity between two species, the greater their competitive interactions. The frequent result of interspecific competition is the decline or even extinction of one of the competing species.

Consider a specialized feeder that depends on a single food source. It is obviously more vulnerable to extinction than one with a broader resource base—one that will eat anything, so to speak. This being the case, there will be more diversity in communities comprised of generalists than in communities where specialization is characteristic of the species involved.

When two species of *Paramecium* are cultured together, population growth of the successful competitor is slowed down, though it eventually reaches the size it would have reached without competition. The losing competitor's population grows normally at first but soon slows down and then gradually declines to extinction.

Population density is an important factor. High density leads to keen competition for food. Limitation by the environment is more often a problem of population density than of population size. Increased density slows population growth, eventually stopping it. When extreme, it decreases the size of the population. Density ordinarily affects a population by limiting the amount of food available, but it does not always do it in obvious ways. Experiments with growing flies under crowded conditions reveal that crowding reduces the number of eggs laid by a female. It was then found that the crowded females laid fewer eggs because they were undernourished. Hence, the density effect was due mainly, if not entirely, to limitation in food supply. Heavy densities also decrease populations by promoting the spread of disease. Forest succession and aspects of plant competition are discussed in Chapter 47.

For most species there is a range of densities at which the population does best, lower or higher densities impairing the maintenance and growth of the population. This is seen in Figure 45-10, which summarizes an informative study of flour beetles (*Tribolium confusum*), the same species that often cannibalizes its own larvae and pupae. Any population that becomes overcrowded is usually limited in size as a result.

A dense population may alter the temperature, humidity, and chemistry of its environment by muscular activity, respiration, excretion, and other processes. It may even poison the environment, as when too many fish are kept in one aquarium. On the other hand, high density may condition the environment favorably for the species. A goldfish grows better in water conditioned by other fishes than it does when living alone in clear water.

Undercrowding occurs when densities are too thin for population maintenance. It may be less obvious to an observer, but it happens nonetheless. There are several reasons. Isolated individuals may not find mates. Thinning populations may become increasingly inbred—with all the dire consequences of increasing homozygosity. There is often a loss of defensive strength in sparse numbers.

POPULATION PERIODICITY

It is well known that different plants and animals experience good years and bad years. Commercial fisheries provide many examples (Figure 45-11). Some fluctuations seem erratic; others range from vague to obvious cyclic patterns. We considered some of these periodicities and rhythms in Chapter 32. For many years, it was accepted that population cycles are caused by a time lag of predators to increasing density of their prey.

When investigators consulted the records of pelts processed by the Hudson Bay Company, which go back to 1750, they could plot the abundance of the so-called

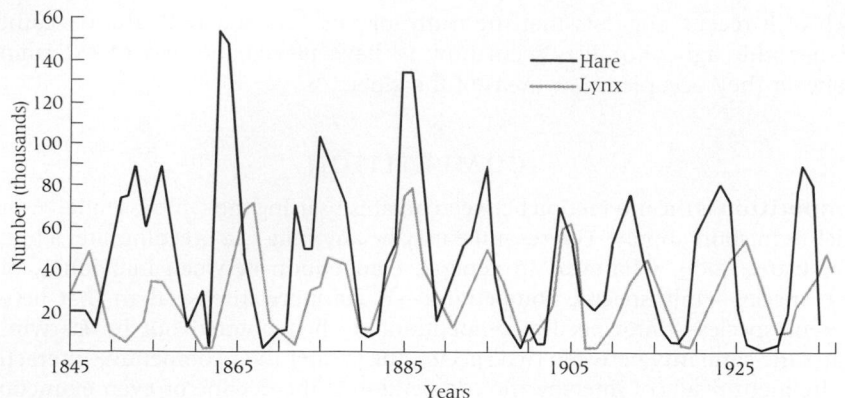

45-12 Population cycles in the snowshoe hare and lynx
Changes in the abundance of lynx and snowshoe hare were determined from the number of pelts received by the Hudson's Bay Company. Note that the rise in the population of hares is followed by a rise in the population of lynxes. The hare cycle reaches a maximum about every 10 years.

snowshoe or varying hare (*Lepus americanus*). These curves fluctuated in an irregular way but a cycle of about 10 years seemed to exist (Figure 45-12). Interestingly, records also revealed periodic fluctuations in the populations of predatory lynxes (*Lynx canadensis*) that closely followed those of the hares. This pattern recalls the experimental cycles in prey-predator relationships shown in Figure 45-9.

The hypothesis once generally assumed to explain these cycles went as follows. The lynxes prey on the hares, but as the hare population declines (because of overpopulation, not predation) the lynx population is regulated by its decreased food supply. Since reproduction is rapid among the hares, they build up anew to a state of overpopulation in a few years. Their rise is probably facilitated for a time by the decline in predation in the wake of decreased lynx populations.

However, when recent studies revealed that snowshoe hare populations show similar cycles on islands where there are no lynxes, an alternative explanation was offered for population cycles of the herbivorous snowshoe hare: high population density causes deterioration of food quality. When certain plants are damaged by herbivores, they produce increased amounts of defensive chemicals (see Chapter 47), which have adverse effects on the herbivore. Thus, the periodic crashes in the hare population may be due to changes in food sources brought about by overgrazing. It may be true that the density of the lynx population follows that of its prey, but the cycling hare population is regulated, not by its predator, but by the quality of its own food source.

The lesson to be learned here is that most population cycles are very probably caused by multiple factors that can be understood only after careful experimentation. The principal causes of regular population cycles are probably biological reactions within the communities themselves, as in the example of the hares and lynxes.

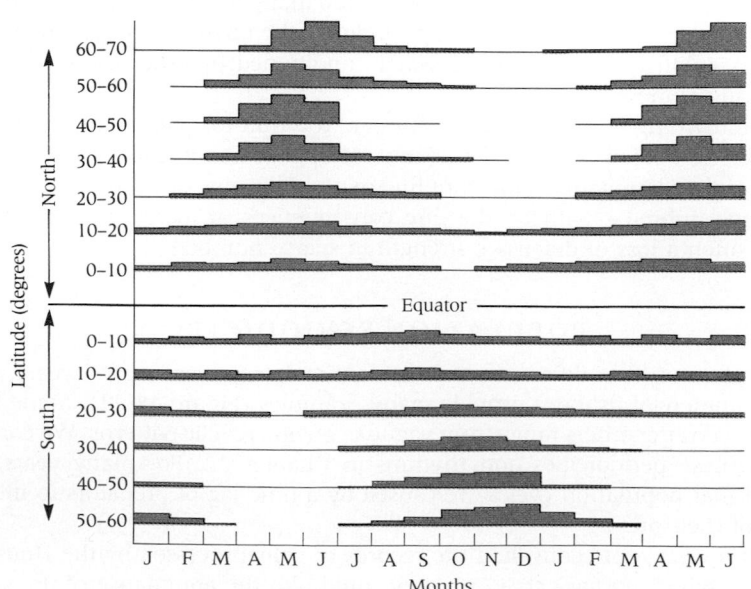

45-13 Annual breeding cycle of the English sparrow at different latitudes
The yearly cycle of breeding activity is most prominent at great distances from the Equator. Note that the cycles north and south of the Equator exhibit a 6-month phase difference.

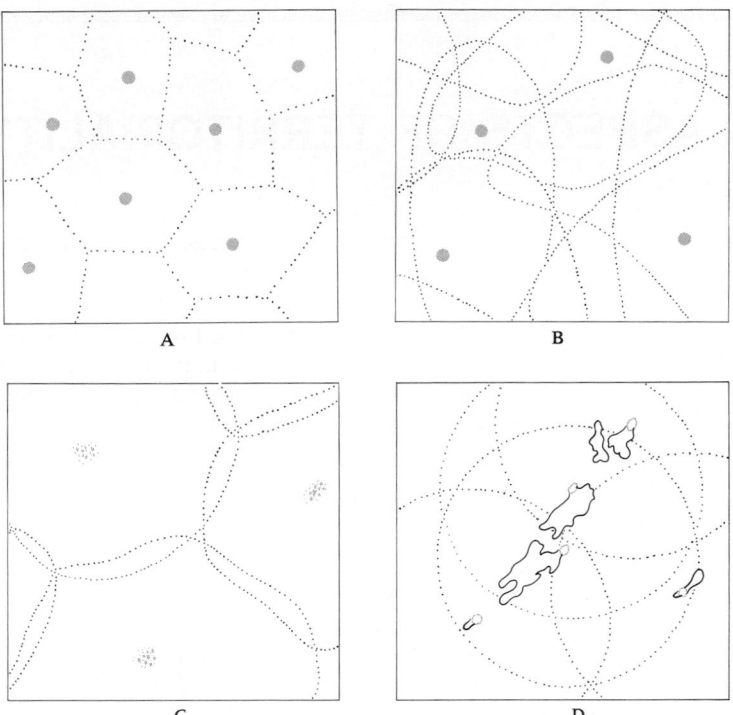

45-14 Basic types of "home ranges" in territorial animals
(A) Animals are solitary, and territories have sharp boundaries. Exclusion is virtually complete. (B) Animals are solitary, and territories overlap. The nest or territorial base, if any, may be defended. (C) Animals are gregarious, and territories of each group slightly overlap. (D) Animals are gregarious, and territories overlap extensively.

A

B

C

D

Irregular fluctuations, on the other hand, can have direct environmental causes such as disease or weather. Rhythmic environment-correlated breeding activities, such as those illustrated in Figure 45-13 (and in Figure 32-22), also influence population size.

INTRASPECIFIC INTERACTIONS

The dynamics of populations concern more than increases or decreases in population size. Each population also has some degree of internal organization, with characteristic patterns of interactions among its individuals. Competition, aggregation, and cooperation are at least as common within species as between species, but they tend to take different forms and have different evolutionary results. We now turn to such interactions.

COMPETITION WITHIN SPECIES

Members of the same species living in a given area essentially share the same basic requirements. Unless some other factor, such as predation, keeps the population at a level close to the carrying capacity of its environment, individuals must inevitably compete for the resources.

Many animals will "stake out" a specific area or volume of space for their activities which they defend. This is common among animals with more complex behavior, especially arthropods and vertebrates. It is called **territoriality.** Often the territory centers around a home, such as a hive, nest, or burrow. The animal may spend all its time there; if it does stray, it periodically returns to its home territory. The competitive background of such behavior is particularly evident when the territory

45-15 Variation in territories of Scottish red grouse
The sizes and shapes of territories were measured in 140 acres of moorland in the springs of 1958–1961. Territories marked by dots were held by males that did not mate. Average territory size varied from year to year, thus affecting the density of breeding. The number of territories ranged from 40 in 1958 to 16 in 1960; the number also correlated with food supply.

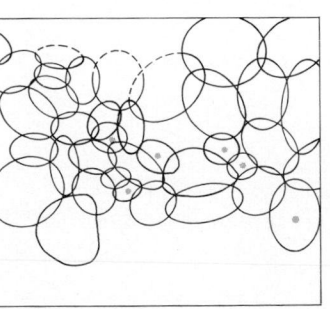
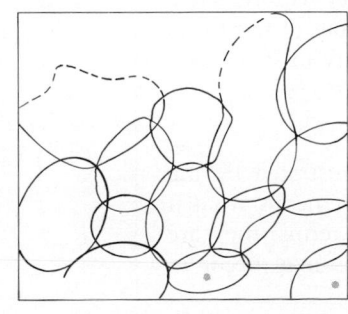
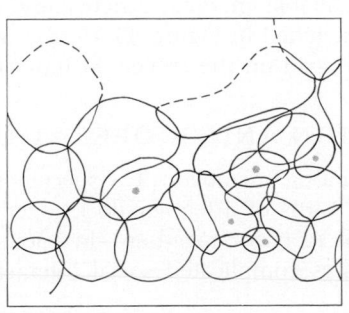
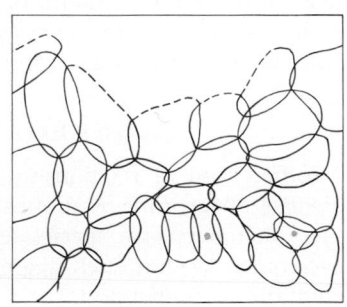

Many definitions have been given for the term **territory,** including "defended area" and "exclusive area." Most recently, it has been defined in terms of the spacing intervals between individual animals or groups. A territory is recognized when animals as individuals or groups are more spaced out than would be expected from a random occupation of habitats.

As discussed in Chapter 33, spacing out may be maintained by overt aggression (as in many birds) or by the sending of subtle "keep-out" signals, such as scents (as in many mammals). In the case of aggression, the defense of territory sometimes goes to the point of bloody combat. More commonly, the confrontation with invaders is ritualized, a threat of attack having more importance than the attack itself.

The size of the spacing out may vary widely (Figure 45A). A millimeter or two may exist between adjacent barnacles (*Balanus crenatus*) on a rock, 50 to 130 centimeters between tree frogs (*Hyla regilla*) near a pond, 20 to 90 meters between great tits (*Parus major*) in nesting boxes, and 5 to 10 kilometers between adjacent herds of buffalo (*Syncerus caffer*) in the Serengeti. Small perching birds may command territories of less than two acres. A golden eagle may be master of 250 square miles.

Territorial size is determined by the total number of individuals (or nesting pairs) that can be maintained in the area and by the ability of these individuals to defend them.

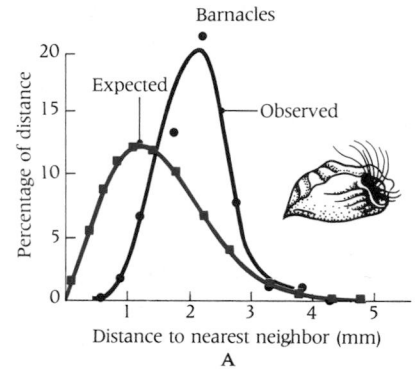

A

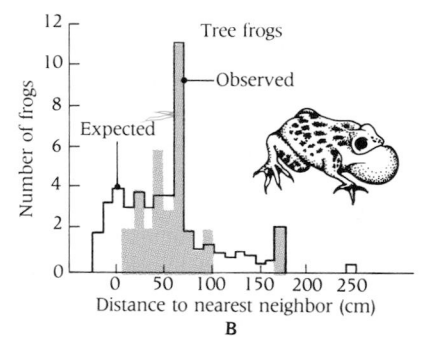

B

Strong individuals obviously will be able to hold on to larger territories than weak ones. Therefore, in species with a well-developed territorial instinct, those individuals strong

is defended against invasion by others of the same species. For example, chipmunks and many birds aggressively drive away trespassers on their preserves.

In essence, the territorial impulse divides available space into more or less discrete compartments. As shown in Figure 45-14, there can be some overlap of territories or no overlap. Each territory represents the amount of space needed for nesting or feeding by a single individual or nesting pair. Thus, the number of territories is related to the total amount of available energy, nutrients, and other requirements. If these change, as in the case sketched in Figure 45-15, the number of territories will change accordingly. More is said on the subject of territoriality in Box 45-2.

AGGREGATION AND COOPERATION

Another kind of relationship among members of a species is illustrated by the defense of territory by groups. "Birds of a feather flock together" means what it says: members of a species tend to occur together—in scientific terms, they are *aggregated*. Within such aggregates complicated social interactions may occur in addition to competition.

enough to defend their territories will probably be able to extract from them the food that they and their offspring require. This is an effective mechanism for ensuring the survival of the strongest individuals. The winners, as a rule, are those individuals most precisely and efficiently adapted to the niche. Competition, then, is one of the processes that leads to natural selection. It was the one most stressed by Darwin, but in the modern concept it is only one of many selective forces.

The nesting areas of songbirds are classic examples of territories. As the breeding season begins, male robins, thrushes and others spread over the landscape in territorial areas. Singing and visual displays by a male at the margins of a territorial area serve to determine the territorial limits (other males will see him and stay away). This behavior also advertises his readiness to breed.

Many kinds of animals have clear-cut feeding territories. Dragonflies have such areas and patrol them with diligence, chasing out other dragonflies. The ant horde in a single anthill fiercely attacks trespassing outsiders, whether of their own species or another. The howler monkeys of tropical South America live in bands and defend their territory from neighboring bands, although the defense is more likely to be a howling contest than a physical encounter. Wolf-pack territories, 50 to 120 square miles in area, are surrounded by buffer zones several miles deep, which separate them from surrounding territories. Interestingly, deer have learned to protect themselves from wolves by living in these "demilitarized zones." How has intraspecific competition been related to the evolution of territoriality? Can territoriality be regarded as an adaptation benefiting the individual family? The species?

The concept of home range is related to that of territoriality. A home range is the area that an individual or group uses in the course of its activities. Ordinarily they do not leave the home range. Even when chased by predators, animals may zigzag back and forth to avoid leaving the range or crossing its boundaries. All, none, or a portion of the home range may be defended against others of the same species. Only the defended area is the territory. In other words, the home range is the area traveled, which may extend beyond the territory. The home range has a number of fascinating aspects. Often an animal's range is that of its parents. Remarkably, migratory birds that travel great distances are likely to return year after year to the same home range.

45A Spacing out between different animals
In each case, the animals are spaced out more than would be expected from a random distribution of the available suitable habitats. (A) Barnacles on rocks. (B) Male tree frogs in breeding ponds. (C) Nestings of great tits. (D) Buffalo herds in the Serengeti.

In plants and in many animals, especially those of simple behavior, aggregation is of the passive variety. Often it is due to the response of individuals to some environmental cue—such as wind, tide, or water source. Individuals do not seek out each other's company, and there is little or no differentiation of roles within the group. Even so, the pattern of aggregation has biological significance. If nothing else, it facilitates interbreeding in biparental species. Thus, it has survival value for the individual.

Animals of more complex behavior are often solitary in habits. They live alone or briefly with a mate, and relationships within the species are dominantly competitive. However, an opposite tendency has appeared over and over again in the course of evolution. In many species, individuals live in *groups*. They may actively seek each other out, and actively maintain their aggregation. Many species of fishes gather into large schools or shoals, which may stay together over long periods of time. These groups are more than aggregations because the fish orient to one another, not the environment. Some birds live in flocks, and many others live in family groups on defended territories during nesting season and then gather into large flocks when the young take flight. Cattle, sheep, horses, and bisons, gather into

A

herds and flocks. Wolves hunt in packs, lions in prides. "Prairie dogs"—really a species of ground squirrel (*Cynomys ludovicianus*)—live in "towns," which frequently have populations of hundreds and may continue over many generations (Figure 45-16). Prairie dogs divide their towns into territories, the borders of which they guard zealously (Figure 45-17). Often, the legs of grazing cattle get stuck in these burrows. These events have important effects on the ecology of grasslands.

The size of animal aggregations is limited by the availability of food and, sometimes, space. Competition within the group is one of the factors that prevents overpopulation and determines which individuals will survive. But the relationships within the group as a whole are not predominantly aggressive. They are at least tolerant, and usually cooperative to some degree. Decrease of aggression within the group is advantageous in external competition. Shoaling is protective for small fishes, and it helps to promote breeding. Herrings, for example, shoal at mating time. Sentinels in prairie-dog towns give warning cries at the approach of danger, and the signal is passed quickly to the whole population.

It has generally been held that a major factor underlying the evolution of aggregative or gregarious behavior is the benefit gained from reducing the risk of attack by predators. One way in which individuals may gain protection from predators by joining a group is through a simple dilution effect. In other words, given a single predator attack, the larger the group of prey animals, the smaller the chance that any particular individual will be the victim.

DIVISION OF ROLES AND THE RISE OF SOCIETIES

In most protists and plants, and in many animals, all individuals play a similiar role among their **conspecific** ("belonging to the same species") associates and in their community. In species with individually separate sexes (most animals and some plants), the sexes do at least have different roles in reproduction. That difference

B

45-16 Labyrinth of burrows in a prairie dog town
(A) The mounds surrounding the burrow entrances keep surface water out of the system and serve as lookout vantage points. Many burrows have at least two exits. Bulbous excavations are nesting chambers. The passages are about 3 feet below the surface. **(B)** A prairie dog.

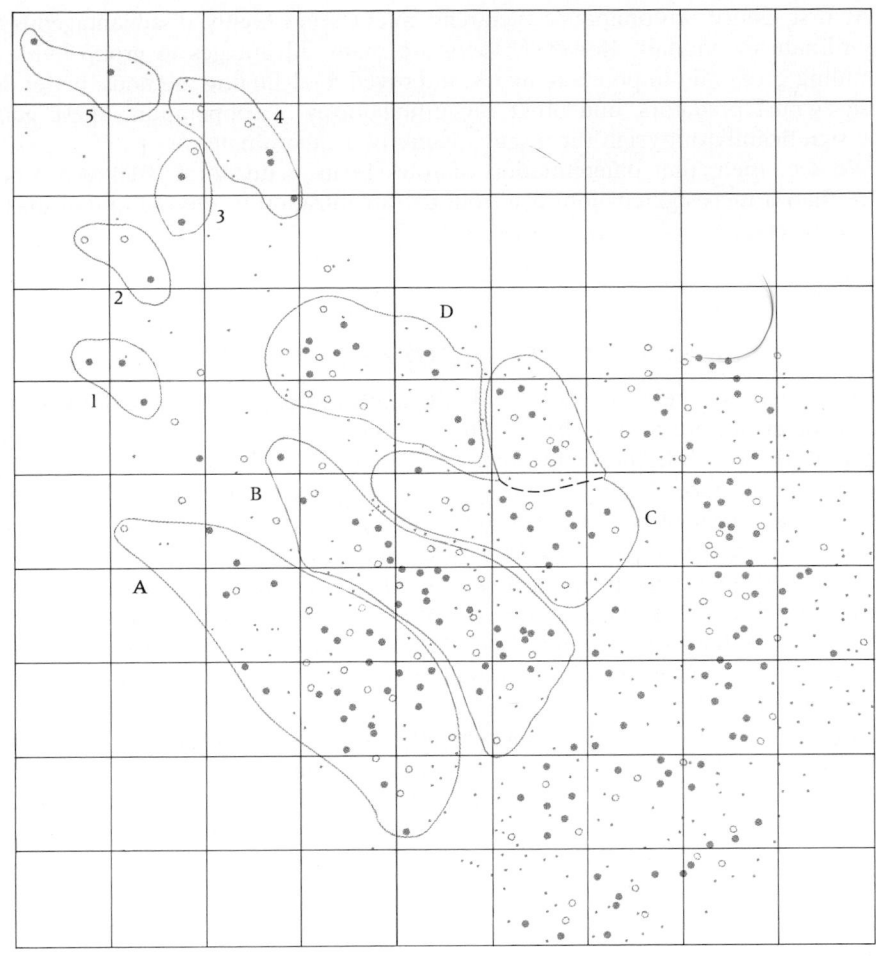

45-17 Coterie territories
Prairie dogs divide their towns into coterie territories (labeled A–D), the borders of which they zealously guard. Numbered areas at the upper left are new territories established by emigrating adults. Territory C is in the process of splitting. Solid circles indicate large, active burrows; open circles are smaller burrows; dots are holes without craters. Each square is 50 feet on a side.

is often accompanied by major differences in size, color, pattern, physiology, and behavior, especially in higher animals, with corresponding differences in roles even beyond the essentials of reproduction. Males and females may even belong to different food chains—like those mosquitoes whose males feed on juices of green plants and whose females feed on mammalian blood. In groups that defend territory, defense is usually by the male, though the territory may be shared with a female.

In cooperative groups, there is a further evolutionary tendency for a differentiation of roles that is not necessarily based on sex. Even in loosely organized animal aggregations, there is often an order of dominance, a "social scale" in which each individual has its place. As discussed in Chapter 32, such a dominance hierarchy is often called a "pecking order," because it is so clearly seen in the pattern of who pecks whom in a flock of fowls. Throughout a pecking order, each hen pecks those of lower status and is pecked by those higher in status. After the group is organized through competition into a pecking order, the amount of pecking decreases. Each hen, or each member of the sequence of whatever species, learns to know its place and gives way to its "betters" without further fuss. An interesting corollary is that these relationships require constancy of group membership and the ability of members to recognize each other individually.

Not all groups are organized into pecking orders. For example, ants, which live in rigidly organized groups, generally have none, though one ant species was recently shown to have a dominance hierarchy. There may, nevertheless, be a distinct division of roles. Sometimes the division is temporary. In a prairie-dog town or in a roving band of baboons, some individuals generally act as sentinels while the others go about their business, but sentinel duty rotates. On the other hand, there often is, in addition, a more or less permanent division of labor. When army ants are moving their base, the workers always carry the young, and the soldiers always protect the line. When a grazing herd is attacked, in most species it is the mature males that wheel into the first line of defense.

At first glance, a dominance hierarchy might seem highly disadvantageous to subordinate individuals. However, there are many advantages to group living—including proximity to potential mates, improved food-finding methods, better defense against predators, and other rewarding forms of cooperation. These gains may significantly outweigh the disadvantages of a subordinate role.

We see, then, that differentiation of roles in its many forms makes a group more than a mere aggregation. It introduces organization to the group and makes of it a society.

K STRATEGISTS AND *r* STRATEGISTS

Natural selection operating on individuals, and eventually on populations, in the course of evolutionary time has led to the evolution of a wide diversity of life histories. These also influence birth and death rates.

For example, most Pacific salmon reproduce only once in a lifetime—in a suicidal burst as three-year-olds. Annual plants reproduce only once. Many other species reproduce more than once. Most small birds, such as chickadees or great tits, breed in the spring following their birth, and continue to nest every year until their death. As adults, they have a 50% chance of surviving each successive winter. Oak trees have high adult survival rates, but take more than three years to produce their first acorns. Then they boost production until their acorns are numbered in the thousands each year.

Clutch size and offspring size vary greatly from species to species. So does the age at which fertility and first reproduction occur. The timing of reproduction in a life history can profoundly affect an individual's contribution to future generations. An organism with an ideal life history strategy would begin reproduction at birth and produce large clutches of well-endowed offspring many times throughout its infinite life span. Ecologists have termed such an imaginary evolutionarily ideal organism, a "Darwinian demon." We do not find such a life history strategy in nature, because some of its features are incompatible with other features. The real world requires compromises in life history features.

Among the organisms with diverse life history strategies, we distinguish two extremes that have been termed **K strategists** and **r strategists.** K and r will be recognized as terms in the logistic equation describing population growth (Equation 45.4)—r denoting rate of growth per capita, the intrinsic growth rate of a population, when its size is small, and K the carrying capacity of the environment.

Organisms operating at the extreme ends of the spectrum adjust to the carrying capacity of the environment—these are the K strategies—or adjust their intrinsic growth rate—these are the r strategists (Figure 45-19). Those that approach the carrying capacity must be K strategists to survive. Until a population reaches that point, it can get by as an r strategist.

- K strategists are usually at or near the carrying capacity (K) of their habitats. Their strategy is to maximize K and grow slowly in number. These species consist mainly of large mammals, such as elephants and whales. They also include many plants, e.g., redwood and sequoia. K strategists typically inhabit stable or predictable environments. Their fitness depends on such environments. It also depends on their ability to compete, escape predators, or effectively exploit limiting resources. K strategists tend to produce only a few large, slowly maturing offspring that receive extended parental care.

- Most r strategists have very fast rates of population growth. These include such species as bacteria, a number of insects, and annual plants. At this end of the spectrum, populations can grow almost exponentially, their growth rates approaching the intrinsic growth rate (r). Such organisms produce large numbers of small, quickly maturing offspring that get little or no parental care, live frenetic lives, and wear themselves out faster. They usually live in transitional, disturbed, or otherwise unpredictable environments. In the absence of severe competition, the fitness of r strategists is mainly dependent upon producing as many young as possible before some environmental catastrophe brings on a crash. The rapid growth of mosquito populations in springtime is typical of r strategists. Often, r strategists are good colonizers of new uninhabited areas because of their high

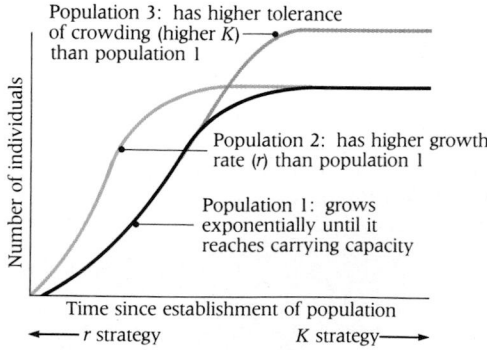

45-18 K *and* r *strategists*
K *strategists live near the carrying capacity of their stable environments. The* r *strategists live in transitional environments, have fast rates of growth, and their populations fluctuate extensively. The curves trace population growth in a benign but limiting environment. Populations grow up to the carrying capacity. The axes show when higher* r *or higher* K *values are adaptive.*

intrinsic rate of population growth. Examples are organisms such as dandelions, aphids, and mice.

K strategists and *r* strategists represent two extremes of a continuum. Most organisms are neither "pure" *K* strategists nor "pure" *r* strategists. Rather, their strategies lie somewhere between these two extremes or, as suggested in Figure 45-18, they can be adjusted according to their population density or to certain environmental conditions.

A benign and constant environment would allow populations to grow to the point at which *K* selection prevails. However, for *r* selection to persist, the population must be continually cut back to an early stage of population growth by predation or catastrophes of weather or climate.

We see, in sum, that in the game of life, an animal wagers its offspring against the capricious environment. The game is won when its offspring survive to play again. Evolution has given each species a strategy that is played out in terms of numbers and vigor of offspring, timing of reproduction, parental care (or its absence), dispersal (or aggregation), and other such adaptive tactics.

SUMMARY

Ecology concerns the interactions between living organisms and the physical and biological environment. It deals with three levels of organization: populations, which are groups of conspecific (members of the same species) individuals; communities, which are composed of populations of one or more species in a given area; and ecosystems, which are communities and their environments considered together.

The growth and decline of populations have important implications for all ecosystems. When birthrates exceed death rates, population sizes increase exponentially. However, this can occur for only a limited time. Various factors inevitably restrict population increases and their growth is ordinarily traced by a sigmoidal or S-shaped logistic growth curve. The maximum number of individuals of a given species that a given area can support indefinitely is termed the carrying capacity of the environment.

Population dynamics are sigificantly influenced by survivorship—how long individuals live. The Malthusian principle holds that because the resources available to a population are ultimately fixed and limiting, death rates must inevitably, in the course of time, come to equal birthrates.

Population limiting factors can be either independent or dependent on population density. The former include such factors as climate, flood, and fire; the latter relate to crowding and competition for resources.

Fluctuations in the population size of a link in a given food chain—as a result, for example, of predation—will influence the populations of other links in the chain. The relations of predator

and prey populations can be quite complicated. For example, a pesticide that initially diminishes an insect population may cause all of the insects' predators to starve. Their decline may in the end cause a net increase in the insect population. Predator-prey interactions often lead to cyclic fluctuations in both populations.

Competition between organisms is highly density-dependent. Because members of the same species have similar needs, intraspecific competition is generally more intense than interspecific competition. However, many animal species form intraspecific cooperative groups, within which competition is reduced. Species also decrease intraspecific competition by specialization of roles or requirements, as when males and females rely on different food sources.

Evolution has given each species a "life history strategy" that tends to optimize its success and survival and profoundly influences population growth. Key features of such strategies are variations in longevity, timing of reproduction in the life cycle, numbers of offspring, proficiency at escaping predation and finding resources, and so on. Two extreme modes of the many recognized strategies have been termed *K* and *r* strategies (after terms in the population growth equation denoting environmental carrying capacity and intrinsic growth rate). *K* strategists (generally large mammals) produce few offspring, but maximize their fitness with extended parental care and effective escape and exploitation skills. Conversely, *r* strategists (bacteria, insects, annual plants) typically produce large numbers of quickly-maturing offspring, which receive little parental care, but are numerous enough to survive "crashes" brought on by such catastrophes as heavy predation, drought, and adverse weather.

KEY TERMS

altruism
carrying capacity
community
competition
conspecific
density-dependent factor
density-independent factor

ecology
ecosystem
exponential growth
intrinsic rate of increase
K strategist
kin selection
logistic growth

maximum sustained yield
parasitism
population
predation
r strategist
survivorship curve
territoriality

QUIZ QUESTIONS

1. Which of the following is characteristic of a *K*-strategist species?
   A. occur in stable, predictable habitats
   B. produce relatively large numbers of offspring
   C. provide relatively little parental care to offspring
   D. populations often grow exponentially
   E. B and C

2. In the logistic growth equation, which of the following can yield a smaller increase (*I*) in population size?
   A. increased birthrate
   B. increased *r*
   C. decreased death rate

   D. increased death rate
   E. increased *N*

3. According to the logistic growth curve, the population level representing maximum sustainable yield
   A. occurs at the point of inflection, where rate of increase changes from accelerating to decelerating.
   B. occurs at carrying capacity.
   C. occurs early in the exponential phase of the curve.
   D. occurs at the value of *r*.
   E. None of the above

4. A plot of the proportion of a group of offspring still alive against time since birth/hatching is a

   A. logistic curve.
   B. exponential curve.
   C. survivorship curve.
   D. sigmoid curve.
   E. bell-shaped curve.

5. Which of the following species has *not* been used in experiments that provided us with an understanding of patterns of population growth and of patterns of interactions of individuals from the same and different species?
   A. *Tribolium confusum*
   B. *Paramecium caudatum*
   C. *Didinium nasutum*
   D. *Lynx canadensis*
   E. *Caenorhabditis elegans*

ESSAY QUESTIONS

1. What is an ecosystem? What are some examples of ecosystems?

2. What is the Malthusian principle as applied to populations of living organisms? What aspects of culture spare humankind (at least in part) this grim fate?

3. What are examples of density-dependent

and density-independent factors that act to limit population growth?

4. Rabbits are heavily preyed on by foxes, coyotes, etc. This situation is obviously advantageous to the predators—it provides them with a ready food supply. However, is there any beneficial value to the rabbit population from such constant

predation? What might happen to a rabbit population if all its predators were suddenly eliminated?

5. What biological values result from the tendency of many animals to remain in groups, such as a flock of birds or a school of fishes? What disadvantages occur from such group living? Do these features apply to human groups, such as the family unit?

REFERENCES AND SUGGESTIONS FOR FURTHER READING

Begon, M., Harper, J. L., and Townsend, C. R. (1986). *Ecology: Individuals, Populations and Communities.* Sinauer/Blackwell Scientific, Boston.

A highly recommended, clearly written text. Its fresh approach was called nouvelle cuisine by one reviewer.

Charlesworth, D. (1989). Why do plants produce so many more ovules than seeds? *Nature 338,* 21–22.

An interesting question receives a tentative answer.

Ehrlich, P. R. (1986). *The Machinery of Nature.* Simon and Schuster, New York.

A lively elaboration of the evolutionary approach to ecology.

Harper, J. L. (1977). *Population Biology of Plants.* Academic Press, Orlando, Fla.

This pioneering work on the population biology of plants is a fine introduction to an interesting field.

Hartl, D. L. (1988). *A Primer of Population Genetics,* 2nd ed. Sinauer/Blackwell Scientific, Boston.

New edition of a well regarded textbook that nicely integrates data and theory.

Huffaker, C. B., and Rabb, R. L. (Eds.) (1984). *Ecological Entomology.* John Wiley & Sons, New York.

A pioneering work on all aspects of the ecology of insects.

Hutchinson, G. E. (1978). *An Introduction to Population Ecology.* Yale University Press, New Haven, Conn.

A fine advanced text that is enriched with accounts of the historical development of modern ecological concepts.

Kingsland, S. E. (1985). *Modeling Nature. Episodes in the History of Population Ecology.* University of Chicago Press, Chicago.

A delightful account of the rise of population ecology that reads like a detective story.

Partridge, A., and Harvey, P. H. (1988). The ecological context of life history evolution. *Science 241,* 1449–1455.

Reproduction is so costly that fertility at all ages of the life cycle cannot be maximized by natural selection. Though reproductive modes evolved in response to different environments, they were constrained by many factors.

Shorrocks, B. (Ed.) (1984). *Evolutionary Ecology.* Blackwell Scientific, Boston.

Report of a British symposium that pays a great deal of attention to British contributions and all but ignores American work.

Stearns, S. (1976). Life history tactics: a review of the ideas. *Quarterly Review of Biology 51,* 3–47.

An in-depth discussion of the various factors that play important roles in the evolution of life history characteristics of both plants and animals.

ECOSYSTEMS

An **ecosystem** is a basic functional unit that is made up of a living community plus its physical environment. This ecosystem concept emphasizes the functional relationships among and between organisms and their environment.[1]

Energy flows through an ecosystem in a particular way. When energy input is balanced by energy output, the system is stable and it endures. The process we called homeostasis in the context of organismic physiology is equally relevant at the ecosystem level. By that we mean that perturbations in inputs and outputs are continuously modulated by various adjustments that tend to restore stability and equilibrium. This basic concept underlies much of contemporary ecology. As in organisms, homeostasis is maintained by various negative-feedback loops. As we have seen, limitations on increasing population size are imposed by other organisms through competition and predation and by various aspects of the physical environment. We will find that such factors have an important role in the regulation of ecological balance.

THE ENVIRONMENT

In earlier chapters, we repeatedly spoke of the "environment" without defining it. We do so now. The **environment** is the totality of physical and biological elements extrinsic to an organism that affect that organism. Like many definitions, this one blurs a bit when applied in specific situations. For example, "extrinsic" needs defining. It does not mean simply "external." A parasite in the stomach, intestine, or blood is an environmental condition. It is extrinsic to the essential structure of the host organism, yet it is not external to the host's body. This is a special case, but it gives emphasis to a concept of environment, which includes *everything* that can influence an organism but that is not an intrinsic part or state of that organism.

The environment so defined includes, first, the nonliving physical aspects of the place where an organism lives—its abiotic environment—and, second, all the living things that may affect the organism—its biotic environment.

THE ABIOTIC ENVIRONMENT

All organisms live either in water or in air—even if they live in the soil, for such organisms are effectively surrounded by water or air. Conditions of life are very different in the two spheres, and most organisms are confined to one or the other. To be sure, many plants live in both environments at once, with roots in the watery part of soil and stems and leaves in air (Figure 46-1), but the different parts are separately adapted to the different environments. Some animals spend part of their

Ecosystems
All organisms are parts of intricate webs of relationships with other organisms and the environment. Orchids, illustrated here by Stanhopea wardi, *often have distinctive relationships with at least three elements: mycorrhizal fungi, pollinating insects, and the physical environment. This makes their ecological roles unusually interesting.*

[1] In reviewing these ideas we are not actually dealing with new material. Rather we are viewing certain aspects of the living world from a different perspective. In recent years, these ideas have taken on new importance. We will attempt here to demonstrate why.

lives in water and part in the air, as do insects with aquatic larvae, but the organism is generally different in one environment than it is in the other.[2]

Solar Radiation, Climate, and the Water Cycle

Organisms that live in air are profoundly affected by weather and climate. Organisms that live in water are similarly affected by such physical factors as temperature and pressure. One of the major ingredients in the abiotic environment is the electromagnetic radiation coming from the sun (see Figure 10-4).

The importance of solar radiation cannot be overstated. With insignificant exceptions, it is the ultimate source of all of the energy required for all of the processes of life in all organisms. This is the income from which all life on Earth must be budgeted. This basic fact has special significance in particular local environments. Green plants, the primary converters of solar energy into biological energy, can grow only where solar radiation is received. Their activity is limited by, and roughly proportional to, the average amount of radiation reaching a given environment, provided that CO_2 and water are in adequate supply. The activities of other organisms are, in turn, limited by those of the green plants from which, directly or indirectly, their energy and materials must come.

Some environments receive little or no solar radiation. These locales have no green plants. They are inhabited only by animals, some nongreen plants, some fungi, and certain protists. The capacity of these organisms to thrive in lightless environments is due to their ability to live on foods that are imported from other environments that do receive solar radiation. One might suppose that environments without solar radiation are few and far between: caves, for example, have a sparse, though interesting, population but they are of limited importance. But in fact, lightless environments are more extensive than any others on Earth. They include the soil, below its topmost layer, and the vast reaches of the sea below the depths reached by daylight. Radiation is absorbed and scattered as it passes through water. Hence, there is no sharp line beyond which radiation cannot penetrate the sea. In very clear water, visible light may be evident at a depth of 2000 feet. However, solar radiation in the sea is seldom significant below about 600 feet, and deeper waters have few or no living plants.

What happens, in quantitative terms, to the solar energy reaching the surface of the Earth? A preliminary answer to this question appeared in Chapter 10 (see Figure 10-2). As we have seen, much less of that energy is converted directly into the energy of living organisms than is expended in heating the lower atmosphere and thus in maintaining the temperature of the environment. In an oft-cited study (Table 46-1), expenditures of solar energy were carefully calculated for an inland lake, Lake Mendota, in Madison, Wisconsin. These figures would be different in other environments. For example, a fresh snowfield reflects 75 to 90% of the incident

46-1 Water lilies, Nymphaea adorata
This fragrant water plant has a lovely flower with many free, spirally arranged parts.

---

[2] As always (or almost always) in biology, a generalization is never (or almost never) universally true. Penguins, for example, can reproduce only on land and can feed only in the water; whales can reproduce and eat only in water but require periodic contact with air; and so on. Impressively, there are hosts of plants and animals, especially along the shore, that alternate between the two environments. But in most of these cases, one environment is the vital one. The other is endured only briefly.

TABLE 46-1
RELATIVE EXPENDITURES OF SOLAR ENERGY IN AN INLAND LAKE*

Expenditure	Percent of Solar Energy Received
Reflected or otherwise lost	49.5
Absorbed and utilized in the evaporation of water	25.0
Used in raising temperatures in the lake	21.7
Used to melt the ice in the spring	3.0
Directly used by organisms	0.8

* From C. Juday, 1940, as corrected by R. L. Lindemann, 1942.

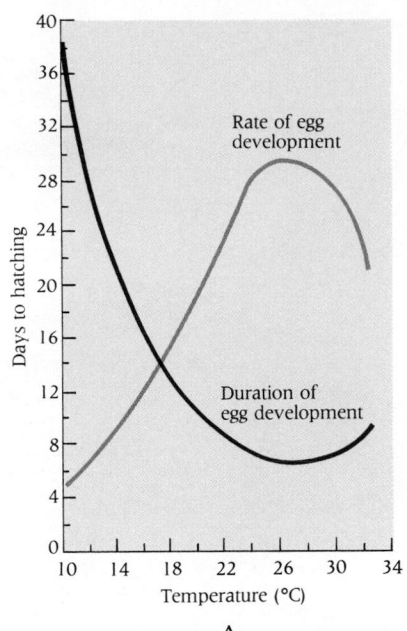

A

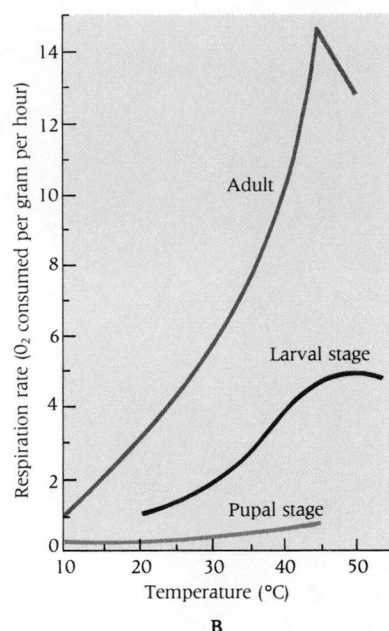

B

solar energy. Everywhere only a small fraction of incident radiation is used directly by green plants.

Nonetheless, much of the remaining incident radiation has biological meaning. A large part goes to warm the water or air. (At Lake Mendota most of the energy tabulated as lost heats the air above the lake.) It is true that maintenance of the environmental temperature is a necessity of life, which can exist only in the range of temperatures that is, in fact, maintained on Earth. Metabolic activities are strongly influenced by the temperatures of organisms (Figure 46-2), and in all plants and many animals the internal temperature depends largely on that of the environment. The way plants and animals adapt to their environment is also related to the temperatures in that environment, its averages and extremes. Orange trees require sustained warmth but apple trees thrive in regions with low winter temperatures. Polar bears live only in the northern cold and boa constrictors only in the tropical heat.

As Table 46-1 illustrates, enormous amounts of solar energy are expended in the evaporation of water. Without evaporation, life could exist only in the sea. Evaporation maintains the humidity of the atmosphere, it is the driving force for transpiration in plants, and it is the powerful mover in the **water cycle** (Figure 46-3), also known as the **hydrologic cycle.**

The water on Earth is distributed in *oceanic, atmospheric,* and *terrestrial* reservoirs (Table 46-2). Oceans are by far the largest reservoir, containing over 97% of the volume. Most of the remainder is in glacial ice. Relatively speaking, little water is found in lakes, rivers, the atmosphere, biosphere, or groundwater.[3]

---

[3] If the water volumes in Table 46-2 are added up, one finds that the total amount of water in the Earth's reservoirs is just under 4×10^{20} gallons or 1.5×10^{21} liters. A cubic meter contains 10^3 liters. Therefore, this amount of water is equal to a cube 1.15×10^7 meters (or 563 miles) on one side.

TABLE 46-2
EARTH'S RESERVOIRS OF WATER

Reservoir	Volume of Water (gal)*	Percent of Total
Oceans	$384,700.0 \times 10^{15}$	97.20
Glaciers	$7,700.0 \times 10^{15}$	2.15
Groundwater	$2,217.0 \times 10^{15}$	0.63
Lakes and inland seas	60.5×10^{15}	0.02
Atmosphere	34.1×10^{15}	0.01
Living organisms	3.4×10^{15}	0.001
Streams	0.3×10^{15}	0.0001

Source: B. J. Skinner, *Earth Resources,* Prentice-Hall, 1969.
* 1 gallon = 3.79 liters.

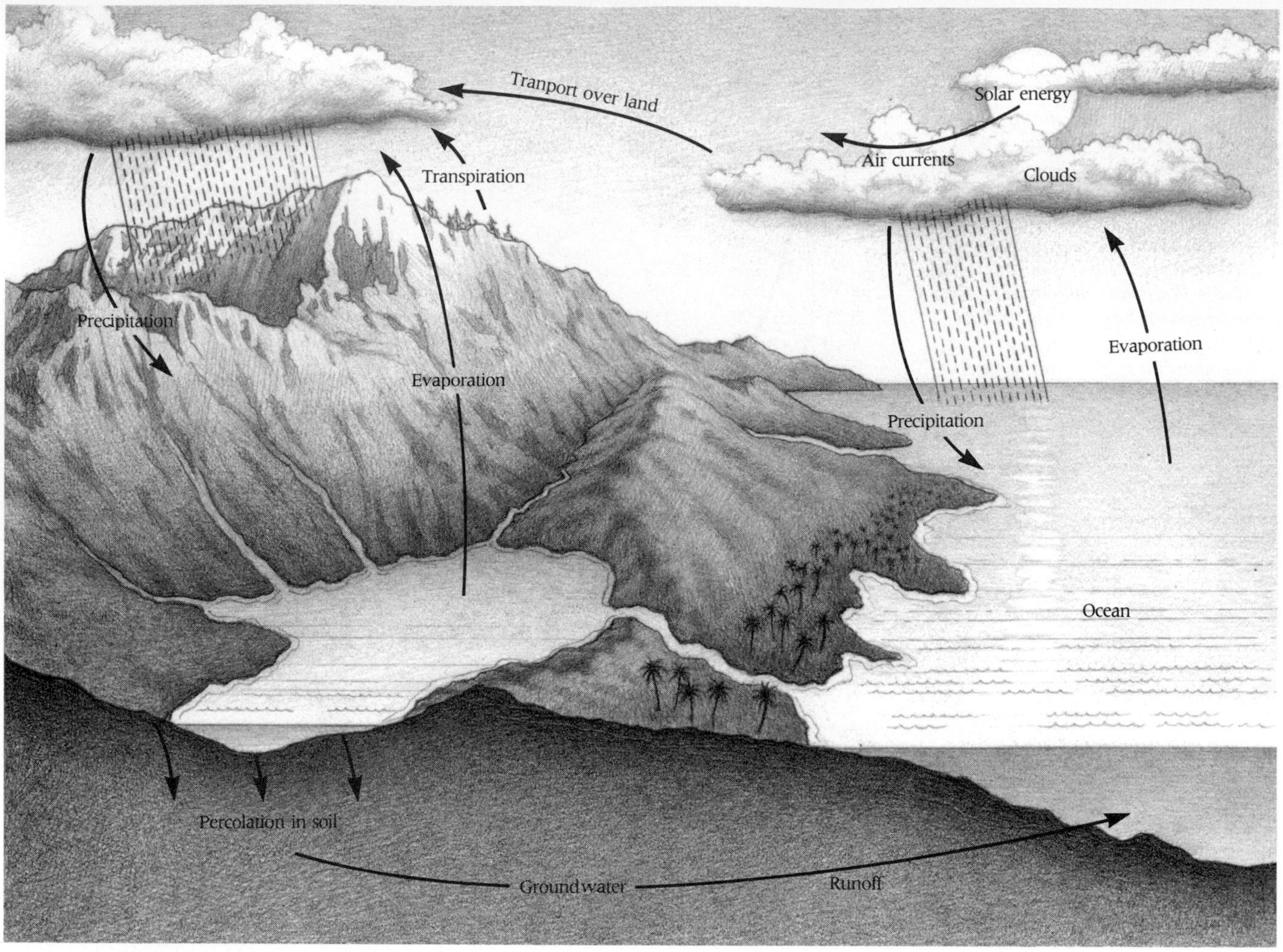

Figure labels (A):

Tranport over land

Solar energy

Air currents

Clouds

Transpiration

Evaporation

Precipitation

Precipitation

Evaporation

Evaporation

Ocean

Percolation in soil

Groundwater

Runoff

A

The water cycle is the ceaseless transfer of water among these reservoirs. Water evaporates from the sea to form clouds. Rainfall from clouds (which comes from condensation of evaporated water in the atmosphere) falls to Earth and maintains the streams, lakes, and groundwater (water beneath the surface of the ground), much of which returns to the sea. The cycle thus provides the enormous quantities of water required by terrestrial organisms.

Movement of water or air is another feature of the physical environment. All of these factors interlock. Air movement is a crucial element for rainfall and both water and air movement help to determine the distribution of temperatures. Wind and currents also influence organisms.

Microclimate, Niche, and Substratum

Climate is the average weather over a long period (Chapter 48). The weather data on which descriptions of climates are based must be gathered with care. Temperature is measured in a shady, ventilated, shelter high above the heat reflections or cold pockets of ground level. Similar precautions are taken in the measurement of wind velocity and humidity. But the weather, and hence the climate, of our immediate surroundings may be very different from that recorded by the Weather Bureau. For example, an organism a few inches under the soil will find the temperature different from and less variable than that on the surface. The climate on the floor of a forest is decidedly cooler, less sunny, more humid, and less windy than the climate in the tops of the trees (Figure 46-4).

Since weather variations can be highly local affairs, one should not expect to find the same kinds of organisms everywhere within a region. Thus we say that organisms live in **microclimates** of their own, which are diverse and often quite different from the idealized or average climate of the region. This means that though a particular organism lives in a region like a forest, it may really live not in the

B

46-3 The water cycle
(A) *Precipitation as rain or snow returns to the atmosphere through evaporation from streams, lakes, and oceans, as well as transpiration from plants. Drainage from streams and groundwater runs off to the ocean.* **(B)** *Thunder clouds over the Atlantic Ocean.*

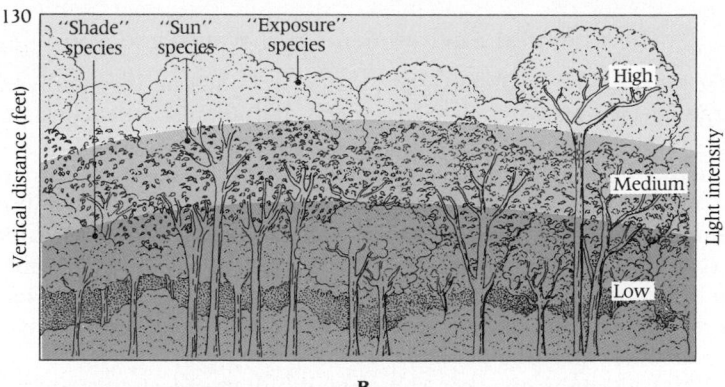

130

"Shade" species   "Sun" species   "Exposure" species

High

Medium

Low

Vertical distance (feet)

Light intensity

A

B

46-4 Stratification of forest microclimates
(A) Orchid growing on tree in a South American
rain forest. (B) The ecological significance of
differences in microclimates is shown by the
vertical distribution of epiphytic ferns, orchids,
and bromeliads that grow on trees. Each plant
has its own requirements for temperature,
moisture, and light.

regional "forest environment" but in a small environment of its own that is rather different.

The microenvironment of a particular species is one aspect of its ecological **niche.** We introduced this term in Chapter 36, but it requires further discussion because the concept is not a simple one. Ecologists have defined the niche of a living organism as "its place in the biotic environment, its relations to food and enemies" (Andrewartha), or "the constellation of environmental factors into which a species (or other taxon) fits: the outward projection of the needs of an organism, its specific way of utilizing its environment" (Mayr). Niche should not be confused with **habitat,** the actual physical locale in which the organism lives—although the habitat is one facet of the niche. The niche is not the microenvironment alone. Rather, it is a relationship between a species, or a local population of a species, and the environment.[4] Thus it is a complex of all the physical and biological factors and dimensions that typify an ecosystem. These include, among other things, the organisms that can be utilized for food, the types of competitors present, how and when it reproduces, and so on. Note that a niche must relate to an occupying population. Without that population, the niche concept is without meaning.

Every regional environment has a large number of different niches. Indeed, the environment is never precisely the same for any two individuals, although it tends to be similar for members of the same species at the same stages in their lives. It is also not exactly the same for any two species of organisms in a community. Every species has its own niche, which is a major determinant of the population's pattern of adaptations—structural, physical, and behavioral. Even two microorganisms of different species living side by side in the soil are influenced somewhat differently by their abiotic and biotic surroundings. Hence, while they may have the same habitat, they have different niches.

All sorts of organisms, from protists to whales, spend their whole lives suspended in water. Some spend much of their lives suspended in air. However, nearly all plants and most animals rest on a bottom of some sort from which they project into water or air. They may be firmly attached, like a tree or a coral, or may be highly motile, like a human being, but they are in contact with some surface most or all of the time. The surface is their **substratum.** This, too, is a part of the abiotic environment that strongly influences the organisms on it. The plants of clay, sand, and rocky substrata are usually strikingly different. So, in many cases, are the animals of sandy and of rocky stretches, even in the same region. The shore life of mud flats, sandy beaches, and rock pools has even greater contrasts. The influence of substrata is partly mechanical. Burrowing animals, for example, will generally be found in a soft substratum—though curiously quite a few animals, mostly mollusks, burrow in solids, even hard rock. Sessile animals are more common

[4] We encountered a famous and instructive example of the importance of niches and ecological opportunity in our discussion of the adaptative radiation of Darwin's finches on the Galápagos Islands (see Chapter 36).

on a hard substratum. The chemical nature of the substratum, and of the environment in general, is also of vital importance.

The Chemical Environment: Air, Water, and Soil

Of the three great ecological arenas—air, water, and soil—air is certainly the largest. It is the one affected least by living organisms and, thanks to its constant motion, it is the most uniform in composition (when dry): by volume about 78% N_2, about 21% O_2, and a very low concentration of CO_2 (about 0.04%).[5] Variations that can affect living organisms are those of water content, already mentioned, and local accumulations of gaseous pollutants—for example, oxides of carbon (CO), nitrogen (NO, NO_2) and sulfur (SO_2, SO_3), H_2S, HF, and others. Some of these arise from natural processes such as forest fires, volcanic activity, and wind erosion. But most, by far, are due to discharged exhausts from such human inventions as motor vehicles and industrial plants.

In contrast, soils and waters can show significant chemical differences that affect organisms profoundly. The most obvious are those between sea water and fresh water. We have earlier discussed some consequences of this difference (see Chapters 3, 6, and 25). Fresh water does contain mineral salts in large amounts, but they are in dilute solution. Their concentration in seawater is much higher, but seawater is not the strongest possible salt solution—that is to say, it is not saturated. Some salt lakes are saturated and have few and highly specialized organisms. The waters of the Dead Sea are ten times as concentrated as ordinary seawater.

Soils are of special interest to biologists because they influence the lives of nearly all land organisms, including human beings. The bulk of any soil is a mixture of grainlike particles of minerals, especially of silica, and clays. The mixture forms from the disintegration and decomposition of underlying rocks, or from silts and other sediments washed in by streams or blown by the wind. Between the soil particles are tiny spaces, together comprising a third to a half of the soil's volume, that are filled by air or water. The water in these spaces is the source of most of the vast volumes of water consumed by land plants. This water is, in fact, a complex solution of various ions (calcium, carbonates, phosphates, nitrates, etc.), which ultimately provides many of the materials eventually incorporated into land organisms of all kinds.

The uncertainty of the borderline between the physical and biotic environments is well illustrated by the soil, where these two facets of the environment interact. Surface soil is penetrated by roots, which change it physically (by loosening up its packed particles) and chemically (by withdrawing its minerals). As plants die, much of their organic matter becomes incorporated into the soil. Humus, which consists mostly of decaying vegetation (especially the breakdown products of cellulose and lignin), is the major repository of soil nitrogen and other organic materials. It also influences the physical conditions of the soil, giving it a porous, spongy texture that enhances efficient use of the nitrogen and organic nutrients released from it. These organic materials, along with parts of living plants in the soil, provide food for bacteria, algae, fungi, nematodes, earthworms, and other organisms. These further modify the physical and chemical properties of the soil, adding their excretions and dead bodies to its contents. We say more on this important subject in Chapter 49.

THE BIOTIC ENVIRONMENT

The biotic environment of an organism, by definition, includes all the living things that affect it. Living things that affect any one organism are also likely to be affected by it, either directly or indirectly. We have already encountered many examples.

---

[5] In addition, there are very small amounts of rare gases such as neon, krypton, and helium, which humans and their technology can extract and turn to their own uses but which are of little biological importance. We are, of course, speaking here of the present atmosphere. In its very early years, as we have seen, the Earth's atmosphere was composed of methane (CH_4) and ammonia (NH_3) and studies reported in 1987 of the composition of air bubbles trapped in amber over 40 million years ago revealed surprisingly high O_2 levels—in some cases approaching 30%.

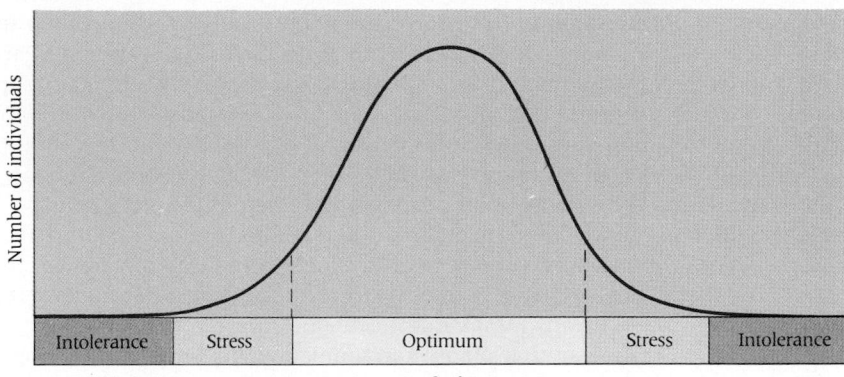

46-5 Tolerance and preference
Individuals survive best in the optimum tolerance range. On either side of that range, physiological stresses limit population size.

Number of individuals

| Intolerance | Stress | Optimum | Stress | Intolerance |

Limit of tolerance

So when we discuss how communities are organized and how their components relate to each other, we are discussing biotic environments.

Ranges, Limits, and Optimums

Every individual and every species must be able to live under a whole range of environmental circumstances. But the range within which a species can survive always has limits (Figure 46-5). Consider a simple example, in which only one factor—temperature—varies. It is obvious that a living population can exist only within a certain temperature range. The portion of the range of variation within which a species can survive is called the **tolerance range.** No plants and only a few endothermic animals can remain active long in environmental temperatures below their own freezing points; hence no species lives continuously in environments constantly below 0°C. At the other end of the scale, nearly all organisms die from heat at temperatures well below 100°C, the boiling point of water. Organisms cannot live at temperature extremes because many enzymes function suboptimally or not at all at these temperatures (see Chapter 8). That would obviously distort or halt intermediary metabolism.

Different species show drastic differences in the breadths of their tolerance ranges for temperature and in the positions of those ranges on the temperature scale. Pine trees survive both greater cold and greater heat than do banana trees; thus their total range is wider. Species with wide ranges for a given factor are called **euryecious.** Those with narrow ranges are **stenoecious.** The algae of warm springs cannot survive in the range of arctic algae. Thus the positions of the ranges are different. Similar differences in range exist for all environmental conditions—salinity of water, acidity or alkalinity of soil, light intensity, and the rest. We might note that it is the combined tolerance ranges for all of these factors that define the niche for a given species.

Within its range, each species and each organism generally has a point or a limited range at which it functions best. This is its optimum. We saw evidence of optimum temperatures for insect growth in Figure 46-2, and we saw that these temperatures are strikingly different for different species. Such optima exist for all environmental factors. This is an important reason why every species has a range, not merely an optimum, for each environmental condition under which it lives.

The existence of many different optima and the high unlikelihood of their simultaneous presence in a single environment underline the importance of systems within organisms for controlling their internal environment. Controls of this kind— homeothermy is a good example—keep the cells and tissues of a body within their ranges and near their optima. As a result, the ranges of external environmental conditions within which an organism can survive are widened.

Toleration and Preferences

A plant or animal *tolerates* the range of environments within which it can live, but its *preference* is for environments that come nearest to their optima. Although the words "tolerance" and "preference" are scientifically acceptable in this connection, they do savor of anthropomorphism. If we are careful not to assume they have the same meaning for plants and animals as they do for us, we can use them as vivid metaphors in discussing ranges and optima. In this sense, we wrote in Chapter 1, ponderosa pines are intolerant of shade, and horticulturalists *do* classify trees as shade tolerant or intolerant. Cacti are tolerant of dry, sandy soil; irises are not. Clams are tolerant of cool, brackish, muddy water; corals are not. Humans are tolerant of hot, humid summers; polar bears are not. Examples surround us.

Ecologists often use certain terms to describe tolerance and preference for particular conditions. Those with the suffix *-phile* (Greek, "lover") indicate preference; those that end with *-phobe* (Greek, "fearer") indicate intolerance. *Hydrophile* means preferring moisture. *Xerophobe* means intolerant of drought.

It seems a tautology to say that plants and animals occur only in environments they tolerate and predominantly in environments they prefer. In fact, the occurrence of many species seems more determined by tolerance than by preference. For example, sagebrush (*Artemesia*) grows best under conditions that are wetter than those in which it is usually found. Its optimum, then, would be in regions of more abundant and stable water supply. But these regions are exuberantly occupied by competitors that do not tolerate drier conditions. Hence, sagebrush grows not where it would prefer to grow, but where it has the least effective competition. Many other organisms, especially those in unusually rigorous environments such as deserts or the deep sea, live not where they would do best (as regards physical conditions) but where fewer competitors exist. They are, in fact, balancing two optima, and their abiotic optimum may not coincide with their biotic optimum.

Another important aspect of tolerance, one that is familiar to us in other contexts, is seen in the ability of many organisms to suspend animation in environmental conditions that could not otherwise be tolerated, and in certain features of their life cycles (see Chapter 21). Annual plants survive the severe "temperate" winter as seeds, and annual insects survive as eggs or pupae. Various protists survive similar vicissitudes by becoming encysted. Even some mammals hibernate, with greatly lowered metabolism, when lack of food and bitter cold would lead to death. Similarly, certain desert animals (rodents, some birds) cope with summer heat and dryness by entering a dormant state called estivation.

ECOSYSTEMS

CYCLES OF MATERIALS

The world is an open system with respect to energy, which continuously comes to it anew from the sun—in vast quantities, to be sure, but as we shall see, in strictly limited quantities. But the world is a closed system with respect to its elementary materials, which are both indestructible and finite in quantity. Carbon, nitrogen, minerals, water, and other basic materials move constantly in cycles—from the abiotic environment, into and through living organisms, and back to the environment again. The actual atoms—say, of carbon and nitrogen—that compose our bodies at this moment were present in the world long before we arrived in it—in the air, in the bodies of dinosaurs, in the trees of the Carboniferous coal forests, in the ancient trilobites. The cycling to which we refer is a repetitive sequence in which small molecules (like CO_2 or N_2) are built up into the large complex molecules of living organisms (like DNA or protein) only to appear again in simple form as a result of the death and decomposition of organisms. These circular pathways of elementary materials are called **biogeochemical cycles,** a term that emphasizes the participation of living organisms (*bio*) and the rocks, waters, soil, and air (*geo*) of the abiotic environment.

Perhaps as many as 40 of the 90-plus chemical elements participate in such cycles. Some are the well-known elements that are needed in bulk amounts—carbon, nitrogen, hydrogen, oxygen (see Table 3-2). Some are trace elements (like iodine or cobalt) which may enter living organisms in tiny amounts, but are nonetheless essential to their survival (see Table 3-3). Each of these elements has a cycle of its own in which it follows a distinctive pathway at a characteristic flow rate. All of these cycles have greater or lesser capacities for self-regulation—that is, for compensatory adjustments whenever flow along portions of the pathway is increased or decreased. This is important because compelling evidence suggests that, through

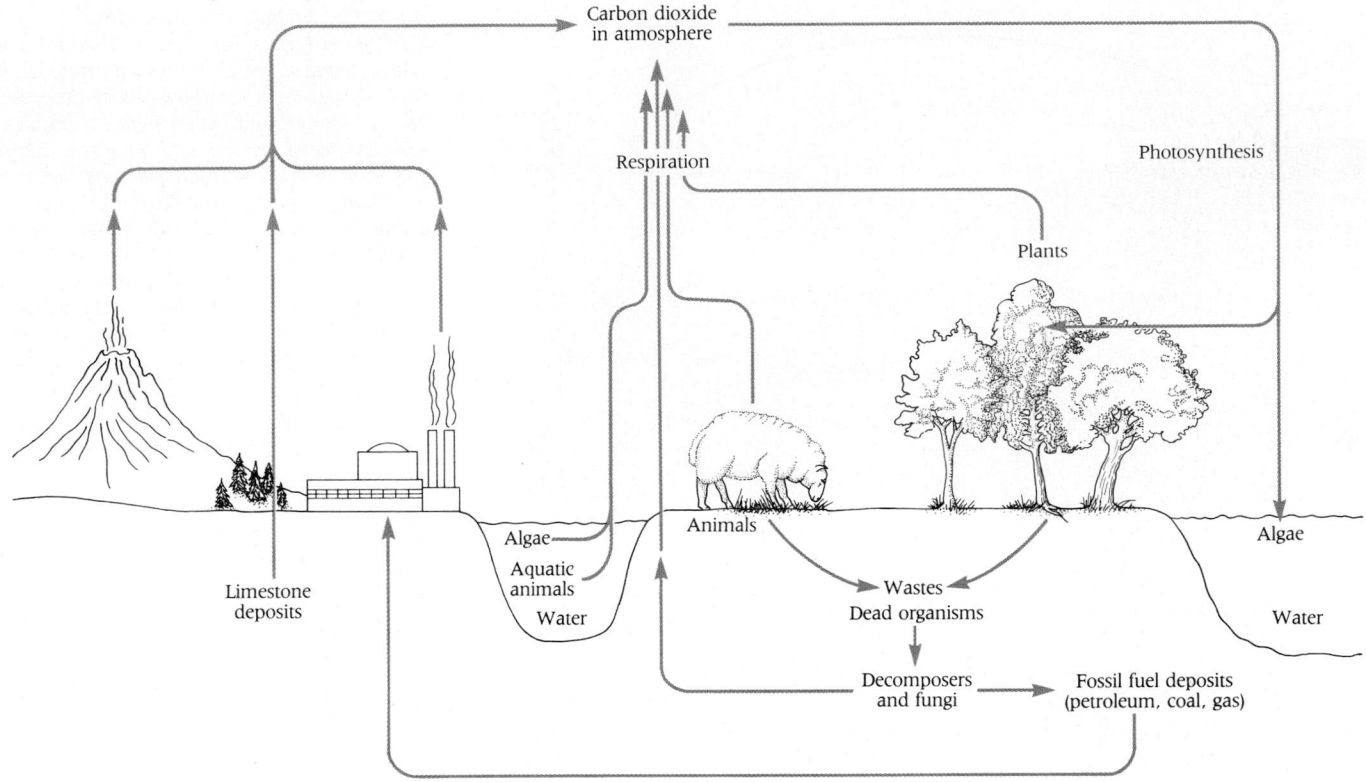

Carbon dioxide
in atmosphere

Respiration

Photosynthesis

Plants

Animals

Algae

Algae

Aquatic
animals

Water

Water

Limestone
deposits

Wastes
Dead organisms

Decomposers
and fungi

Fossil fuel deposits
(petroleum, coal, gas)

46-6 The carbon cycle
Carbon from the atmosphere, in the form of CO_2, is incorporated into carbohydrate as a result of photosynthesis and returns to the atmosphere following oxidative metabolism in living organisms. Some carbon is taken out of circulation for a time, appearing in carbon sinks such as fossil fuels or limestone deposits.

their multifarious activities, human beings are altering these cycles in significant and ominous ways.

We cannot here discuss all of the biogeochemical cycles in detail. Rather, let us consider a few of the major ones—the carbon, nitrogen, and mineral cycles—and also the newly understood cycles of environmental pollutants. We will find in these examples good illustrations of the general characteristics of biogeochemical cycles.

The Carbon Cycle

The major features of the flow of carbon in communities are summarized in Figure 46-6. The great reservoir of nonorganic carbon, and the source of almost all the carbon incorporated into living organisms, is the free CO_2 contained in the atmosphere and dissolved in the waters of the Earth. The first quantitatively important step of the **carbon cycle** in the utilization of carbon from this reservoir is photosynthesis, largely by photosynthetic bacteria, algae, and green plants (see Chapter 10). The carbon assimilated each year in this process—a little less then a tenth of the 700×10^9 metric tons of CO_2 in the atmosphere—first becomes parts of simple carbohydrates. These become polysaccharides, proteins, lipids, and other complex organic compounds. A few of these big molecules are broken down by respiratory metabolism of the plants themselves, and some carbon is released again as CO_2. But most of them remain in the plant until it dies or is eaten. When herbivorous animals eat plants, they eliminate some of these organic compounds in the urine and feces as waste products. They digest the rest and then reassemble them in their own tissues. They, too, release some respiratory CO_2. When carnivorous animals eat herbivorous animals, more digestion and resynthesis takes place.

As long as it is a vital part of a living organism, carbon occurs largely within organic molecules. But most of this organic carbon eventually reappears as CO_2 and returns to the inorganic realms of water and air. Indeed, the CO_2 dissolved in the waters of the Earth serves as an important reservoir or buffer in the carbon cycle. A bit of the CO_2 arises as a direct product of respiratory metabolism in plants, animals, and protists. The portion of carbon eliminated by animals in their waste products is still in the form of fairly complex organic compounds, and therefore does not immediately become atmospheric CO_2, which could be reutilized in

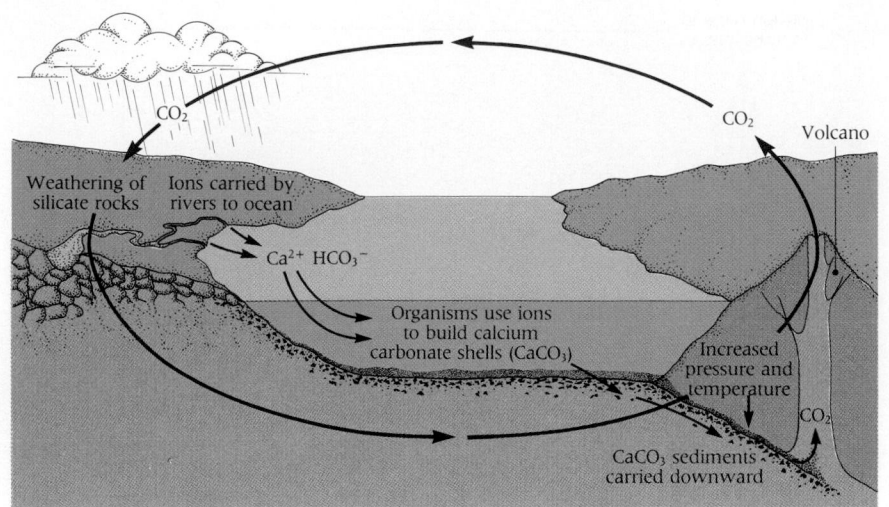

46-7 The carbon-silicate cycle
Carbonates form when CO_2 dissolved in rain reacts chemically with rocks containing calcium and silicate. Such reactions release Ca^{2+} and HCO_3^- into ground water, which transports the ions to streams, rivers, and the ocean. There, plankton and other organisms construct shells of calcium carbonate ($CaCO_3$). Their shells sediment on the sea floor when they die. With sea floor spreading, sediments slip under the continents. Exposed to high temperature and pressure, the sediment releases CO_2, which reenters the atmosphere via volcanic eruptions.

photosynthesis.[6] Substantial amounts of carbon remain in the tissues of organisms when they die.

If the carbon in animal wastes and dead bodies were not somehow reconverted into CO_2, all life would have ended long ago. Indeed, it has been estimated that atmospheric CO_2 would be exhausted in a year were not the atmosphere continually recharged with CO_2. Here is where the organisms of decay or putrefaction, mostly bacteria and fungi, play decisive roles in the carbon cycle. They attack, digest, and decompose the organic materials of dead plants and animals and of waste products, reducing them to the simpler and energy-poor compounds with which the various cycles can begin again. When these organisms have done their work, most of the carbon of organic compounds has become CO_2 again.

Other ways in which organic compounds can be broken down include fire, a rapid type of combustion, and oxidation and decomposition, slower types. When wood burns, the carbon of its cellulose and other constituents is rapidly converted to CO_2. But combustion that is not caused by humans is too intermittent and widespread to keep the natural carbon cycle going. This is an important principle: the continuity of cycles—especially those steps in which large organic compounds

---

[6] Urea, for example, is an abundant excretory product of protein catabolism in animals. Its formula is $CO(NH_2)_2$. Hence, it contains carbon. Uric acid and other excreted molecules contain carbon in even larger percentages.

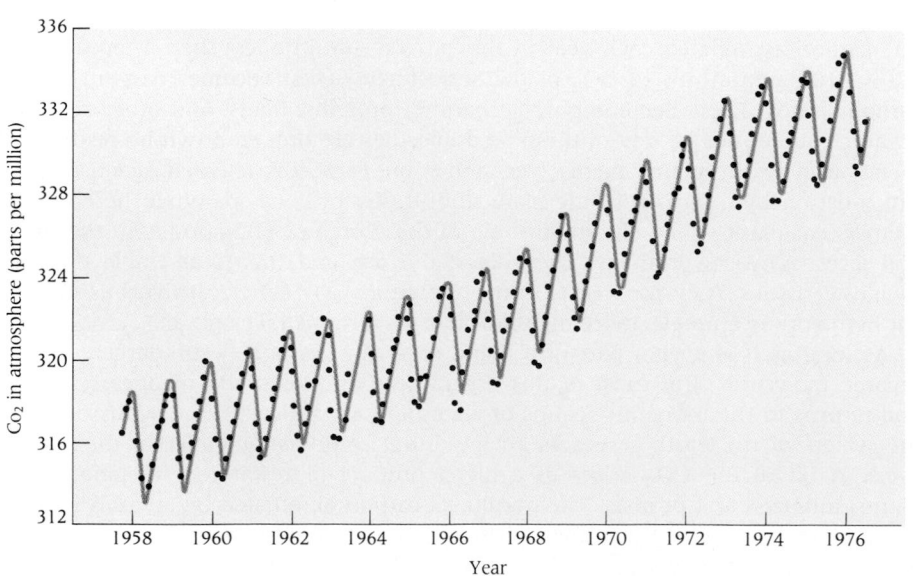

46-8 Trend in atmospheric CO_2
These data were collected by C.D. Keeling at the Mauna Loa Observatory on the island of Hawaii. The seasonal oscillations in atmospheric CO_2 are due to its removal by photosynthesis during the growing season in the Northern Hemisphere and its subsequent release during the fall and winter months. The average CO_2 content of the atmosphere has risen more than 5% since 1958.

46-9 The greenhouse effect

Water vapor, CO_2, and other gases in the atmosphere permit solar energy to reach Earth, but block the return of infrared radiation to space. This causes heat to increase on the Earth's surface. Compare with Figure 10-2.

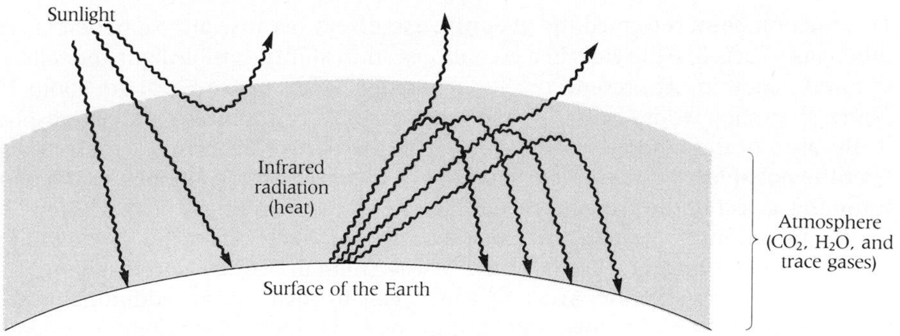

Sunlight

Infrared radiation (heat)

Surface of the Earth

Atmosphere (CO_2, H_2O, and trace gases)

are finally broken down into the small molecules with which new turns of the cycle begin—depends on the activities of living things.

Substantial amounts of carbon may be withdrawn from this cycle for long periods of time. Some of this carbon is stored in living organisms, especially trees. But a much larger amount is locked up in huge deposits of limestone, which is mainly calcium carbonate ($CaCO_3$). As we saw earlier, much but not all limestone is a result of the activities of organisms. In the sea, plankton and other organisms incorporate Ca^{2+} and HCO_3^- ions into their calcium carbonate shells. Coral reefs are one result of this process. Huge amounts of carbon are also taken out of circulation for varying periods as fossil fuels—petroleum, coal, and gas. In time, most of this will be returned to the atmosphere.

Recognition of these phenomena led to the discovery of another facet of the carbon cycle, the **carbon-silicate cycle,** a biogeochemical cycle that operates on a time scale in excess of 500,000 years (Figure 46-7). In this cycle, atmospheric CO_2 dissolves in rainwater and forms carbonic acid (H_2CO_3). As rainwater weathers and erodes rocks containing calcium and silicates, Ca^{2+} and HCO_3^- ions are released into the groundwater, eventually finding their way into rivers and oceans. There these ions are used by organisms to build calcium carbonate shells. The shells are eventually deposited as sediments on the sea floor and when the sea floor spreads (see Box 48-1) the sediments slip under the continental land masses. There heat and pressure cause calcium carbonate to react slowly with silica or quartz (SiO_2), with the reformation of silicate rocks and the release of CO_2, which reenters the atmosphere via midocean ridges or volcanic eruptions. Because of the leisurely pace of this cycle, vast quantities of limestone now lie deep in the Earth, where its carbon will remain unavailable for a long time.[7]

The Greenhouse Effect

The burning of fuel by humans has increased the CO_2 reservoir in the atmosphere by about 25% in the last 140 years—from 280 parts per million or less to about 350 parts per million (Figure 46-8). A good part of this increase has occurred since 1958. Today more than 1.1 tons of carbon (as CO_2) is released every year for every human being on Earth. Americans contribute an impressive 5 tons per person per year. Present estimates are that this release rate will quadruple by the year 2100.

This situation arouses concern because atmospheric CO_2, which is not readily decomposed by sunlight, plays a major role in regulating the temperature of the Earth's surface and thus its climate. It does this by acting as a one-way screen. It is transparent to solar radiation at visible wavelengths, where most of the energy of sunlight is concentrated. However, molecules of CO_2 in the atmosphere absorb and re-emit some of the longer wavelengths of this energy, the infrared radiation that would otherwise be transmitted back into space from the Earth's surface. As a result, much of this energy is trapped in the atmosphere as heat (Figure 46-9).

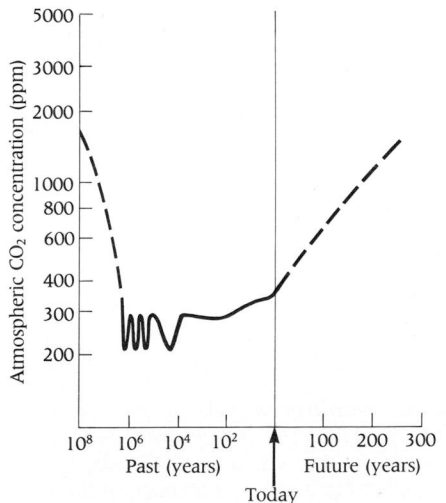

46-10 Atmospheric CO_2 past and future

The graph shows atmospheric CO_2 levels in the past and those projected for the future. If predictions are correct, the atmosphere will contain amounts of CO_2 exceeding the high levels of 100 million years ago. The dashed lines reflect estimates of CO_2 levels in the distant past and future. Note that CO_2 levels are plotted on a logarithmic scale.

[7] If all of this carbon were suddenly released as CO_2, it would exert a pressure of 870 pounds per square inch—60 times the normal atmospheric pressure at sea level, which is 14.5 pounds per square inch! The pressure exerted by present-day atmospheric CO_2 equals 0.0044 pounds per square inch.

This phenomenon is termed the **greenhouse effect** because atmospheric CO_2 (and other gases) acts like the glass in a greenhouse, transmitting visible light and reflecting infrared radiation. At present, the surface temperatures of the Earth are about 35°C higher than they would be if CO_2 and other gases were absent from the atmosphere.

Because of the tendency of CO_2 to warm the Earth's air blanket, there is alarm over the possible long-range consequences of increasing atmospheric CO_2. The most troubling aspect of the greenhouse effect is the global warming of 3° to 4°C predicted in the next century. Such an occurrence would likely raise the worldwide sea level by approximately 70 cm. If the West Antarctic ice sheet (see Figure 1-10) and the ice masses in the Arctic Ocean began to melt off, an additional 100 cm rise in sea level would follow, with disastrous consequences for the world's coastal regions.

If these predictions are fulfilled it is likely that humanity will face an historic dilemma. If human activities continue to increase atmospheric CO_2 (Figure 46-10) the results could be ecologically destabilizing. An increase in average world temperature would enlarge the area of arid lands and diminish agricultural production. But if humans change their ways—for example, by burning less fossil fuel or by preserving forests and replanting, the results might be equally destabilizing to social and economic patterns—although a few still consider the evidence on the issue inconclusive.

Environmentalists have argued recently that massive tree-planting programs might help—and they would be relatively inexpensive. Trees not only absorb CO_2 and store carbon as woody biomass, they also slow soil erosion, improve watersheds, and shelter a web of diverse species. In one study, a hectare of sycamores was shown to absorb 7.5 tons of carbon per year. Thus, we would have to plant 7 million square kilometers of trees to absorb 5 billion tons of carbon per year. That is an area about the size of Australia! The problem is still without a solution, particularly since recent evidence has indicated that the rise in CO_2 level may have a major cause beyond the burning of fossil fuels. It now appears that other gases, like chlorofluorocarbons (from aerosol cans), CH_4, nitrogen oxides, and ozone, could together be as important as CO_2 in augmenting the greenhouse effect.

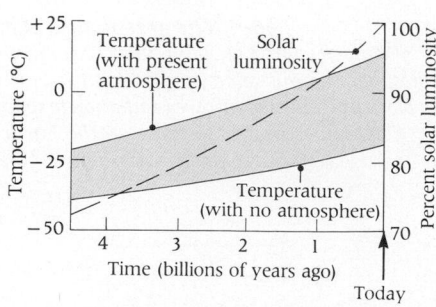

46-11 Earth's temperature throughout the course of geological time
Climate-model calculations indicate that, without the present atmosphere, Earth would have been frozen in its early years. At that time, the Earth received up to 30% less solar energy than it does today. The Earth's temperature with the present atmosphere is considerably higher due to the greenhouse effect. (The curve of solar luminosity is based on calculations by D. O. Gough of the University of Cambridge.)

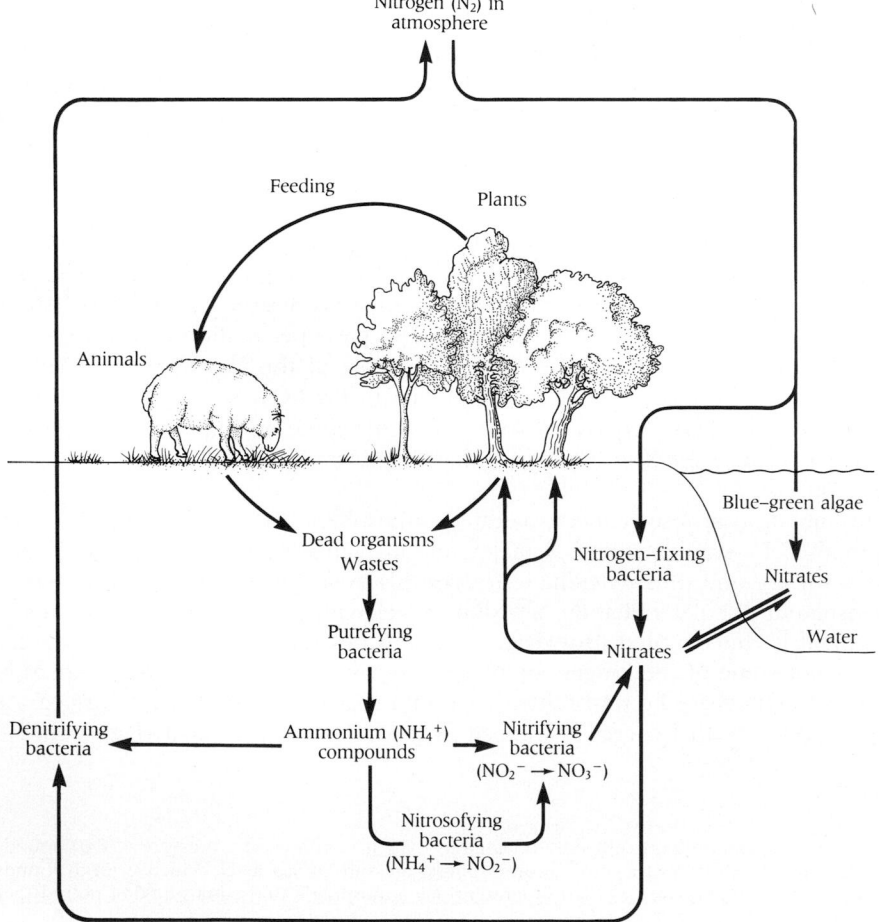

46-12 The nitrogen cycle
Gaseous N_2 from the atmosphere is converted into a biologically usable form (nitrate) by nitrogen-fixing bacteria in nodules on the roots of leguminous plants. Nitrates are used in the synthesis of amino acids by bacteria, green plants, and fungi. Most animals get their nitrogen by eating plant proteins, but corals depend on blue-green algae which live with them and provide a direct source of usable nitrates in the water. Nitrates can follow a number of metabolic pathways. Denitrifying bacteria convert nitrates and ammonium back into atmospheric nitrogen to complete the large cycle.

46-13 Nitrogen-fixing nodules on the roots of an alfalfa plant, Medicago sativa

Living organisms also release substantial amounts of CO_2, mainly through the destruction and harvesting of forests and the oxidation of humus. Hence, we do not really know whether the world's biota, its mass of living organisms, is a CO_2 sink (a net utilizer) or a source (a net producer). Almost certainly these considerations will loom large in the coming decades as nations struggle with decisions on new energy sources, optimal land use, and engineering counter measures. In addition, a substantial amount of methane from flatulent cows and other cattle (Chapter 47) adds to the greenhouse of CO_2.

An interesting implication of our understanding of the effects of atmospheric CO_2 on Earth temperatures concerns the role of this factor in evolution—particularly since the sun may have been 30% fainter in the distant past. As shown in Figure 46-11, the Earth's surface would have been frozen in its early years if the atmospheric composition was the same as it is today. This is evidence that atmospheric CO_2 levels were much higher in the past.

The Nitrogen Cycle

Nitrogen, an essential component of amino acids and of nucleotides, has a biogeochemical cycle somewhat like that of carbon, but certain of its features are distinctive (Figure 46-12).

We saw in Chapter 3 that gaseous nitrogen (N_2) comprises by far the greatest part of the atmosphere—almost 80%. But N_2 is quite inert biologically. Green plants use little or none of it in their biosynthetic pathways—in marked contrast with carbon, which plants actively extract from the air as CO_2. Green plants do in fact require nitrogen, but they can assimilate it only in the form of ammonia or ammonium ion (NH_3 or NH_4^+) or nitrate (NO_3^-). They can obtain these only through the activity of a small and remarkable group of microorganisms—a few bacteria, and blue-green algae—that have the capacity to combine gaseous N_2 with other elements to form useful substances such as ammonia and nitrate. This process is called **nitrogen fixation.**[8]

[8] A very small amount of nitrogen fixation can also result from atmospheric electrical discharges like lightning, cosmic radiation, and meteor trails, which momentarily provide the high energy needed to allow N_2 to react with O_2 or the hydrogen of water. But this is negligible compared with the activity of nitrogen-fixing bacteria. Interestingly, the metabolic machinery of these microorganisms permits them to get along with no nitrogen source other than atmospheric N_2, which they take up, "fix," and then transform into their own amino acids, nucleotides, etc. But only they know this trick. All other living organisms require nitrogen already "fixed" into chemical forms. This they must get directly or indirectly from soil microorganisms or applied fertilizer.

TABLE 46-3
RATES OF BIOLOGICAL NITROGEN FIXATION BY VARIOUS ORGANISMS OR SYSTEMS

Host Plant Group	Example	Microorganism (endophyte)	Estimated N_2 Fixed (kg/hectare/year)*
Legumes			
Soybeans	*Glycine*	*Rhizobium japonicum*	57–94
Clover	*Trifolium*	*Rhizobium trifolii*	104–160
Alfalfa	*Medicago*	*Rhizobium meliloti*	128–600
Nodulated nonlegumes			
Alder tree	*Alnus*	Actinomycete-like organism	40–300
Sea buckthorn	*Hippophaë*	Actinomycete-like organism	2–179
Tea Tree	*Ceanothus*	Actinomycete-like organism	60
Sumach	*Coriaria*	Actinomycete-like organism	150
Plant-algal associations			
Angiosperms			
Tropical herbs	*Gunnera*	*Nostoc*	12–21
Pteridophytes			
Aquatic ferns	*Azolla*	*Anabaena*	313
Fungus-algal associations			
Ascolichens	*Lichina*	*Nostoc, Calothrix*	39–84
		Free-living microorganisms:	
		Azotobacter vinelandii	0.3
		Clostridium pasteurianum	0.1–0.5

* 1 hectare = 10,000 square meters = 2.471 acres

BOX 46-1

IS THERE A LIMIT TO WORLD FOOD PRODUCTION?

One of the most interesting and least appreciated aspects of modern world history is the critical importance of an adequate supply of fixed nitrogen. Until this century, the major sources of fixed nitrogen for agriculture and also for gunpowder were the nitrate deposits of Chile. In 1893, Sir William Crookes warned that the Chilean deposits were becoming depleted. In World War I, the British blockade prevented Germany from importing Chilean nitrates, a possibility that had been completely overlooked by the German General Staff. The war would have been lost much earlier had it not been for the work of Fritz Haber in developing a process for converting gaseous N_2 and H_2 to ammonia. Never again was Germany troubled by shortages of nitrate. Haber, incidentally, was also the inventor of poison gas.

It was soon realized, however, that these methods for industrial nitrogen fixation have high energy costs. The H_2 used in the Haber process comes from natural gas or petroleum, and these materials are also being depleted. Thus the cost of synthetic nitrogenous fertilizers is affected by the price of fossil fuels. Clearly, other sources for agricultural nitrogen must be found.

In the years before World War II, leguminous crops were extensively used in the United States to add nitrogen to farmland. But this practice fell into disfavor as synthetic fertilizers became cheaper and crop lands became costlier. For this reason, investigators today are studying biological nitrogen fixation in search of ways of expanding it and thereby improving agricultural production.

Arguments for increasing our dependence on biological nitrogen fixation as a means of meeting part of the nitrogen requirements for production of food and fiber crops include the following:

- Biological nitrogen fixation takes place in fields, forests, and other places where nitrogen is utilized, in contrast to industrial ammonia synthesis and transportation of fertilizer nitrogen, which require substantial energy consumption.
- Increasing dependence on biological nitrogen fixation is a superior alternative in areas of the world where funds are limited for construction of ammonia synthesis factories and where energy sources are inadequate.
- Accumulations of ammonia, the product

of biological nitrogen fixation, help to regulate the fixation process and thereby minimize the possibility of accumulating excesses of nitrates in soils and water.
- Fixed nitrogen in the nodules of legumes and woody species is transported directly to the plant and is not immediately subject to losses.
- There is considerable evidence that the use of inorganic nitrogen fertilizers causes bacterial nitrogen fixation to decline and eventually stop. With heavy use, these fertilizers tend to kill off these bacteria or cause them to mutate to nonfixing forms.

In considering the probabilities of increasing biological nitrogen fixation, genetic engineers have proposed transferring nitrogen fixation genes to microorganisms that lack the ability to fix nitrogen. It is a difficult challenge because more than a dozen genes are involved. Alternatively, they have attempted to transfer some of these genes to the plants themselves—and there have been some successes amid many disappointments.

Part of the problem is the complexity of the process of root nodule formation in le-

Some nitrogen-fixing bacteria live independently in the soil (and in various other terrestrial and marine environments). But the primary nitrogen-fixing bacteria of importance in agriculture are those species that live in curious nodules that grow in the roots of certain plants, especially in the legumes—soybeans, cowpeas, lupine, clover, and alfalfa (Figure 46-13). The best known of these leguminous bacteria are members of the genus *Rhizobium*, a different species of which is associated with each legume (Table 46-3). A scanning electron migrograph of a root nodule appears in Figure 3-10.

The association between legumes and *Rhizobium* is the most highly developed and sophisticated system for biological nitrogen fixation, but it is not the only one. Most of the nitrogen needed for maintenance of forests and aquatic habitats comes from a variety of free-living nitrogen fixers—for example, see Figure 38-15B. As suggested in Table 46-3, the list of such organisms is a long one. Steps in the development of a legume nodule are shown in Box 46-1.

Wherever they are located, nitrogen-fixing organisms store very little fixed nitrogen. Instead, they furnish it directly to the cells of the host plants with which they live. They also release some of these compounds into the soil—both while they live and after they die and decompose. That is why legumes can grow well in soils that are poor in naturally available nitrogen, and why it has been the practice of farmers since ancient times to rotate crops each season. When fields are periodically

gumes (Figure 46A). The many steps include (1) attachment of soil rhizobial bacteria to root hairs, (2) encapsulation of the bacterium within a pocket of the cell wall, (3) dedifferentiation of cortical cells on the outside of the root to form a growing meristem that bulges out into a nodule, and (4) proliferation of the bacteria in the cytoplasm of host plant cells, where after they are covered with a special membrane synthesized by the host, they convert to so-called *bacteroids*. Within this protective environment, they finally become nitrogen-fixers by derepressing their genes.

In 1988, investigators at Cornell discovered a new bacterium which they called *Photorhizobium thompsonia*. It is the first known organism that is photosynthetic *and* can fix N_2. With other nitrogen-fixing organisms, plants pay a heavy price, providing up to 12 lbs of energy-yielding materials per pound of N_2 fixed. If the new organisms could be introduced to plant stems rather than roots (so they could receive sunlight), the plant would profit handsomely.

Experts have emphasized the urgent need for improving technical capabilities in this area. If the efforts of scientists to improve the nitrogen-fixing capabilities of nodulated legumes are successful, there would be enormous benefits to society in a short period of time.

46A Steps in the development of a legume nodule
Each step is marked by great specificity. For example, Rhizobium mellilota *grows on alfalfa but not on clover. The complexity of this system and its genetic control has posed a difficult challenge for genetic engineers.*

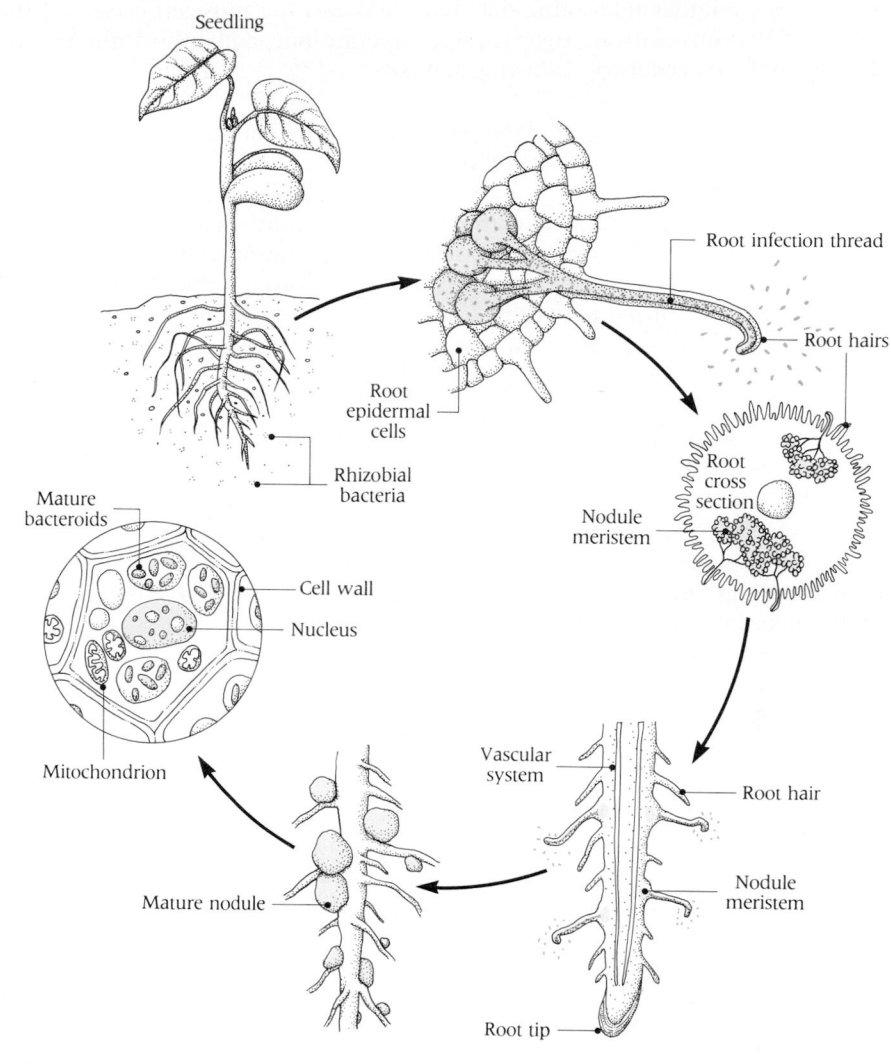

planted in legumes like clover or alfalfa, the fertility of soil is renewed and enriched by biologically fixed nitrogen.[9] This in some measure eliminates the need for constant addition of fertilizers (Box 46-1). In this way the nitrogen of green plants is acquired either from ammonia and nitrates (the preferred form for flowering plants) in the soil or from nitrogen-fixing bacteria. This nitrogen is incorporated into plant tissues—by far the largest share going into protein. It is then passed on to herbivorous animals, and then, like carbon, is passed from animal to carnivorous animal.

Again, like carbon, nitrogen is excreted by animals (for example, in urea). It also remains for a time in the tissues of dead organisms. Again it is mainly certain putrefying bacteria and fungi of decay that return to the cycle the nitrogen locked up in urea and other organic molecules. These bacteria, termed **decomposers,** produce ammonia from these molecules through a series of specific biochemical reactions. Then other bacteria, the nitrosifying bacteria, oxidize ammonia to nitrite (NO_2^-), and still others, the nitrifying bacteria, oxidize nitrite to nitrate (NO_3^-).

[9] As shown in Table 46-3, up to 600 kg of nitrogen may be fixed per hectare in a season by alfalfa. Biological nitrogen fixation by all types of microorganisms has been estimated to contribute 150 million metric tons of nitrogen per year to the whole Earth.

It is this remarkable two-step process of **nitrification,** the oxidation of ammonium ion to nitrite and of nitrite to nitrate, that restores to the soil inorganic nitrogen compounds that can be directly utilizable by green plants.

Here we see another interesting difference between the nitrogen cycle and the carbon cycle. Portions of the nitrogen cycle can operate independently of atmospheric N_2. For example, consider the following subcycle.

$$\text{soil nitrate} \rightarrow \text{green plants} \rightarrow \text{putrefaction} \rightarrow \text{nitrification}$$
$$\rightarrow \text{soil nitrate} \ldots$$

This cycle can continue indefinitely without the reappearance of nitrogen as gaseous atmospheric N_2. Some may reappear, however. When nitrogen does leave soil or water and return as N_2 to the atmosphere, it does so through the activities of still other bacteria, the denitrifying bacteria.

This brief survey of the carbon and nitrogen cycles has revealed several important principles.

■ First, each cycle includes a phase in which the molecular species is an atmospheric gas. This clearly facilitates its widespread distribution.

■ Second, a given atom of carbon or nitrogen can follow either of two routes. One, a speedy route, involves uptake and metabolism by living organisms. The other involves slow decay, long retention in geological deposits, and release only after long time intervals by weathering or by human intervention.

■ Third, green plants and putrefying bacteria must be present if these cycles are to continue indefinitely. For this purpose, animals are unnecessary.

■ Finally, we see that humankind has added several new pathways to both biogeo-chemical cycles.

Mineral Cycles

By drawing analogies with the carbon and nitrogen cycles and by recalling what we have learned in this and other chapters, we can readily work out the essentials of the water cycle (see Figure 46-3), the oxygen cycle, and other important cycles (sulfur, phosphorus, etc.). Since the mineral cycles have features of special interest (Figure 46-14), they merit brief further remarks. Before reading them, review the section in Chapter 3 on essential mineral elements.

With few exceptions, the sources of these elements are salts dissolved in water. Even fresh water contains many mineral salts in dilute solutions. These solutes are critical elements of two great cycles of nature, which they also link together.

■ One, the rock cycle (of which the carbon-silicate cycle is a part), is essentially an inorganic chemical system, though it is influenced by living organisms. The

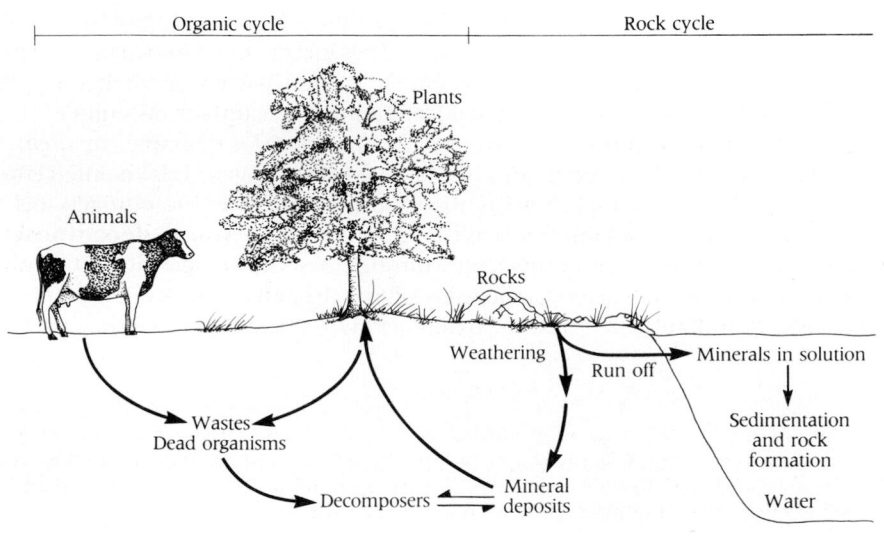

46-14 The mineral cycle
Minerals in rocks of the Earth are released and stored in the rock cycle. As mineral salts become available they are incorporated into organisms in the organic cycle.

mineral salts of the Earth came originally from the Earth's crust, and they continue to be liberated from rocks by weathering, disintegration, and decomposition.

■ Soluble salts released by weathering then enter the water cycle. With the flow of water, soluble salts move through soil, streams, and lakes, and eventually into the sea, where most salts will remain indefinitely. Along the way, salts are available to all kinds of organisms in all of life's environments. As noted earlier, some mineral salts do complete a cycle and return to Earth's crust through the process of sedimentation.[10] They are there incorporated into limestones, salt beds, and other sedimentary deposits. When weathering occurs, their mineral salts may enter the cycle again.

The acquisition of salts from salt solutions in the environment is most extensive in plants. All plants may return salts directly to the inorganic worlds of soil and water. Animals acquire salts from plants and from other animals eaten as food. Putrefying bacteria and other organisms of decay acquire salts from dead plants and animals.

It is worth noting that sedimentary cycles are more easily disrupted by humans (by mining, redirection of waterways, and so on) than are the carbon and nitrogen cycles. As we shall soon see, many ecologists now believe that humankind is accelerating losses of phosphorus and that this element will one day be a critical limiting resource in the biosphere.

Pollutant Cycles

Pollution, a conspicuous scientific term before the public today, has different meanings and implications for different people. Thus it usefully illustrates the relative nature of ideas and values. The issue is briefly stated as follows: what is one person's pollution may be another person's livelihood. Modern industry, agriculture, overpopulation, and lack of knowledge have combined to cause the release into the environment of vast quantities of chemicals that are new to most living organisms, ourselves included. Since there is no prior experience with most of them, there are in many cases no natural defenses against them.

Ecologists interested in this situation define pollution as a misplacement of resources that results in alterations of the biogeochemical cycles. This phenomenon has reached alarming proportions in the modern world. There are, perhaps, as many as half a million new chemical compounds in our environment that were not there 100 years ago, and hundreds of new ones are being created and used each year. Moreover, human society has only recently become aware of the potentially dire consequences of its influence on the cycles of nature. One thing that changed public attitudes on this problem was the space program, which yielded photographs of Earth such as the one in Figure 1-1. When we saw our planet, solitary and alone in the vast reaches of space, it was no longer difficult to understand that all living organisms are passengers on a "spaceship"—of limited resources, limited area, and delicate ecological balance.

One of the most disturbing effects of pollution is the depletion of the atmospheric ozone layer (see Figure 44-1) by such chemicals as chlorofluorocarbons, which are refrigerants and are used in aerosol cans. First observed in the late 1980s in the Antarctic atmosphere, it is now apparent in the Arctic atmosphere as well. Ozone protects the Earth's surface from excessive ultraviolet radiation. Its loss could have major genetic effects and could increase the incidence of cancer.

The pollution of the environment has vitally affected innumerable ecosystems, many of which are not fully understood. There is abundant evidence that entire ecosystems have changed in large areas such as forests, rivers, lakes, and seas. In some cases, familiar chemical compounds that were present on Earth long before the arrival of humans have reached unhealthy and menacing concentrations (for example, carbon monoxide). In other cases, alien chemical compounds have been incorporated into natural biogeochemical cycles where they either remained or

---

[10] They do not, to any great extent, travel as passengers in the return phase of the water cycle, which depends on evaporation. This process does not carry along minerals in appreciable amounts. There is, however, some return of salts from sea to land, sometimes many miles inland, by wind-blown particles of spray.

underwent slow degradation into relatively harmless, simpler substances. But even many of these are so different from anything that occurs naturally we have no understanding of their fate or possible impact on the biosphere.

We know that, with occasionally disastrous results, such compounds have entered the air we breathe (sulfur oxides, nitrogen oxides, etc.) (Figure 46-15), the water we drink (lead, detergents, phenols, pesticides, etc.), and the animals and plants we eat. Some of these materials are outright poisons. The chemical properties of many byproducts of industry and agriculture are still unknown.

Two principles need to be considered when viewing the possible impact of the entrance of pollutants into natural biogeochemical cycles. One is that a substance may not be harmful in the form in which it is released, but may be converted by bacterial or physical processes into substances that are harmful. A classic example is mercury, which has entered bodies of water near many chemical factories. It was once believed that released mercury was an insoluble, nontoxic substance that would settle to the bottom of the waters and there remain in stable chemical form, a harmless puddle of quicksilver. We then learned to our dismay that microorganisms can convert metallic mercury into methylmercury—interestingly, by means of a biochemical reaction that requires cobalamin (vitamin B_{12}). Methylmercury, it turned out, is a highly soluble compound that accumulates in fishes and other aquatic organisms and is toxic both to them and to the humans who may eat them.

Second, we should understand that the pollutants themselves undergo cyclical transformations. Eventually the rates of input and output of each of these substances will approach a balance. It is true also that the natural biogeochemical cycles will still take place. Nitrogen will still be converted to nitrates and nitrates back into nitrogen. It is significant, therefore, that the intensive use of nitrogen fertilizers has led to alarming increases of nitrate levels in the rivers and lakes that provide our drinking water. What we are concerned about here is the alteration of these cycles, both in rate and in character, and the possibility that the new balances will be unhealthy for humans, other animal species, and plants. Pollution alters a distribution of materials in two ways. It changes the rate of flow of basic materials and it introduces some materials in forms with which other species cannot cope.

Perhaps the best-known example of environmental pollution is that of DDT, the insecticide discussed in Box 42-2. Though it has drastically reduced such insect-borne human disease as typhus, for a time, and thereby saved many human lives and dramatically protected crops from insect pests, we began to realize several years ago (1) that overuse of DDT was damaging, even destroying, many valued species of plants and animals, and (2) that its ultimate effects on ecosystems were entirely unknown. Rachel Carson warned of this danger in her book, *Silent Spring,* which described in compelling and portentous terms the way that chemicals sprayed on croplands and gardens lie long in the soil and eventually enter the bodies of plants and animals, which pass them along in cycles of various kinds. Some pesticides pass into streams and well water and from them move into living organisms. DDT resists chemical breakdown and today is to be found in the tissues of nearly every living organism. Although DDT levels in human tissues appear to be decreasing,[11] many organisms now have extremely high DDT levels in their tissues, the implications of which are quite unknown.

How DDT is concentrated in food chains is shown in Box 42-2 (Figure 42B). Bermuda, which lies in the mid-Atlantic, more than 600 miles from North America, has never used DDT in local agriculture. Yet several species of endemic oceanic birds—for example, the Bermuda petrel—are now threatened with extinction. If DDT poisoning threatens Bermudian birds that occupy the top trophic levels in the long oceanic food chains, then all oceans of the world may well contain DDT in sufficient concentrations to pose a real threat to its carnivores at the top of the food chain.

Thus, DDT is a "persistent pesticide" that remains present for years as it becomes concentrated in higher links of the food chain. This process, termed biological magnifi-

46-15 Air pollution
Afternoon smog in the San Fernando Valley.

---

[11] Some encouragement, though not enough, may be found in the following data on the average concentrations of DDT (in parts per million) in human fat tissue in recent years: 1970, 7.95; 1975, 4.76; 1980, 2.82; and 1983, 1.67.

46-16 Biological magnification
The peregrine falcon, Falco peregrinus, *which is near the top of the food chain, has been reduced in numbers by accumulations of DDT in the tissues of the fish it eats.*

cation, has been found to have severely reduced the reproductive rates of predatory, fish-eating birds such as the bald eagle and the peregrine falcon (Figure 46-16). DDT, we now realize, interferes with the deposition of calcium in the eggshells of these birds. As a result, they lay thin-shelled eggs that break easily and are prematurely lost.

TRANSFERS OF ENERGY

The utilization and transfer of energy as fundamental features of the living world were stressed in Chapters 8 through 11. Our emphasis there was on the events taking place within cells and individual organisms. Here we are concerned with energy transfers in populations and communities.

Three great generalizations about energy, the capacity to perform work, should be recalled from earlier discussions: (1) energy may be potential or kinetic; (2) energy assumes many different forms (mechanical, chemical, electrical, radiant, and thermal), which are transformable from one to the other; and (3) transfers and transformations of energy are governed by the laws of thermodynamics.

The first law of thermodynamics, we should recall, relates to the conservation of energy. Whenever transfers or transformations of energy occur, there is neither gain nor loss in the total energy involved in the transaction. The significance of this first law for living things lies in their capacity to release the potential energy within the structures of organic molecules or to capture energy (such as that of sunlight) by transforming it into the potential energy of chemical structure.

The second law of thermodynamics, equally important for living organisms, was considered in rather formal terms in Chapter 8. According to this law, less and less of the total energy is available, as energy is transferred from one substance to another or transformed from one form to another. The total amount of energy cannot change, but the amount that can perform chemical, mechanical, or other

46-17 Net flow of energy and nutrients through a community
The flow of energy (colored arrows) and nutrients (black arrows) through a natural community depends on uptake by the primary producers and dissipation of energy as heat via cellular respiration. In contrast, almost all nutrients are recycled.

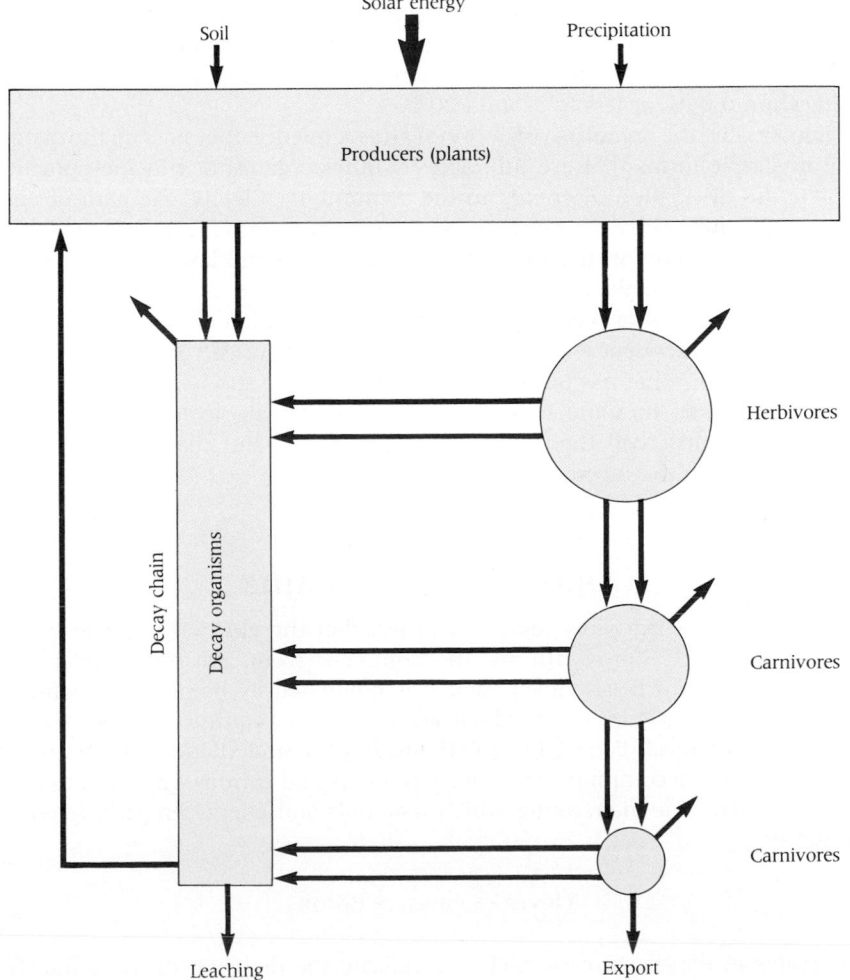

sorts of work, becomes steadily smaller. Hence, the usable energy in a sequence of transfers tends to run down, unless there is continuing input of energy from outside of the system.

We restate these laws of thermodynamics only to emphasize their importance for the activities of living things. In the cyclical movement of materials, such as carbon or nitrogen, nothing is lost. The system is *closed*, and all materials that enter the cycle are (or can be) returned to their original form after going through the cycle. There is no reason why such cycles could not go on forever without additions of anything from outside. This is not true of energy transfers, the transactions of which are those of an *open* system. The movement of energy is not cyclic but undirectional; energy flows from the producers to the consumers and to the decomposers.

According to the first law, an organism can utilize only the amount of energy that it receives. According to the second law, much of the received energy is made unusable by the organism's activities. The received energy cannot be destroyed (that would violate the first law), but it is dissipated (mainly as heat that is transferred to the environment) in forms that can no longer be used.

This means that organisms always pass less energy on to others than they received. When a herbivore eats a plant, it receives chemical energy in the form of molecules, but the energy derived is far less than the energy the plant received from the sun (Figure 46-17). A carnivore, in turn, acquires chemical energy by eating herbivores. In so doing, it ingests a plant's proteins (among other things), digests them, and reassembles their amino acids into its own proteins. However, it obtains from the herbivores less energy than the herbivores obtained from the plants they ate. To a large extent, this loss of usable energy reflects losses due to respiration and the performance of work—and also the storage of energy-rich molecules that may not be available to the next trophic level. For example, the cellulose in plants yields no energy to many herbivores.

In the past, a generalization proposed about this sequence—the so-called "10% law"—held that in this chain of energy transfers, only about 10% of the energy fixed by plants is transferred to herbivores; then about 10% of the energy entering the herbivore community is transferred to the first level of carnivores, and so on. Careful studies showed that this is approximately true for many freshwater ponds and laboratory aquariums, but not for much else. The actual fraction varies widely, ranging from 0.05% or less to around 20%.

When, finally, the organisms of decay end this sequence, they pass on the materials of life in simple forms that are utilizable by other organisms. But they practically complete the dissipation of energy in the community. Clearly, the path of energy flow in a community is not cyclic at all. It is a one-way sequence in which vital energy, like all energy in the universe, follows the second law and thus becomes continuously less available.

Since communities *do* keep going—they have been at it for over a billion years— it is obvious that new energy from an outside source must be continually acquired to compensate for what has been lost. That source is the sun, which is itself subject to the second law. In some remote day, its energy will no longer be in usable form. Life on Earth will then be possible no longer. But that event is perhaps billions of years in the future.

CHAINS AND PYRAMIDS

A contemporary of Darwin's suggested in jest that the glory of England was due to its "old maids." Sturdy Britons, the argument went, are nourished by roast beef from cows, which eat clover, which is pollinated by bumblebees, which are attacked in their nests by mice, which are kept under control by cats, which are raised by "old maids." Far-fetched? Perhaps, but the underlying point is valid: the different species in a community are linked in many and curious ways. The following sequence is part of a chain along which materials and energy are transferred in a community:

clover → cows → Britons

(The arrows in diagrams of such chains indicate the direction of these transfers.)

Another chain portrays material and energy transfers resulting from the relation among predatory species.

$$\text{bumblebees} \rightarrow \text{mice} \rightarrow \text{cats}$$

It is linkages like these that unite all living things into the single and seamless web of life.

The complexity of ecosystems is so enormous that it has proved difficult to analyze or describe them in any simple way that is both accurate and satisfying. However, the work of such pioneers as R. L. Lindeman called attention to the central importance of energy and material transfers in natural ecosystems and the quantitative relations that exist in communities.

Food Chains

The community is an organized association of different, interacting species, and the flow of materials and energy is its most fundamental feature. The term **food chain** refers to a given sequence of species through which materials and energy pass. Each species forms a link in the chain. The position of a given species in the chain is referred to as its **trophic level.**

As shown too simply in Figure 46-18, food web ordinarily start with **producers**—the photosynthetic green plants or algae that acquire energy from the sun and fix it in organic chemical forms. This is the first trophic level. The second trophic level consists of the **herbivores,** the consumers of green plants. The third level is the **carnivores,** which eat the herbivores, and the fourth the **secondary carnivores,** which eat the carnivores, and so on. These species, which are intermediate links

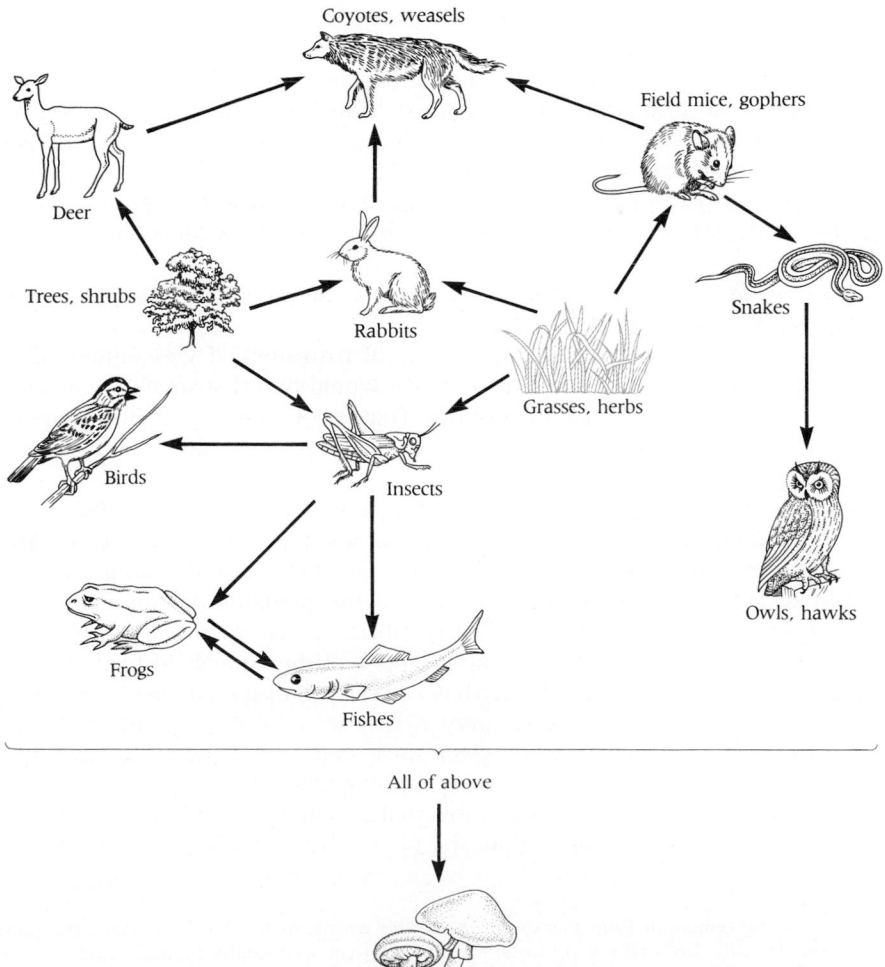

46-18 A food web
This food web in an American woodland community is only suggestive and is very incomplete. Arrows indicate the flow of energy and materials (consumption of food). This web is made up of many different food chains.

in the chain, are **consumers.** Almost all of them are animals. All chains end with **decomposers** or **reducers,** the organisms of decay. These include the fungi and bacteria that release CO_2, water, and heat.[12]

In a real community, the flow of materials and energy is far more complex than any single food chain would suggest. Most species may be consumed by more than one other species, and although some consumers and reducers do obtain food from only one species, most could consume numerous species. For example, one plant species may provide food for many species each of insects, birds, rodents, ungulates, worms, fungi, and bacteria. And one of the species of, say, the rodents, which live on those plants, may be consumed in turn by any of a dozen or more species of carnivorous snakes, birds, and mammals. Thus food chains converge and branch—so much so that the patterns of transfer in a whole community trace a **food web,** rather than a succession of distinguishable chains (see Figure 46-18). The extraordinary length and complexity of many food chains and food webs has long obscured detailed study of factors underlying the injury done to numerous species by pollutants.

Pyramids of Energy, Mass, and Numbers

In spite of the complexity of a food web, the portion of the chain running from producers through herbivore and carnivore consumers does take in a substantial segment of the food flow in any community. We will term this a *partial chain.*[13]

As noted earlier, each trophic level in a partial chain possesses less available energy than the previous level. This means that the total energy in the plants of a community is greater than that in the herbivores—which in turn is greater than that in the carnivores. The total energy of the reducers is less than that of all the rest of the organisms put together, but not necessarily less than that of any one link. Thus the distribution of **energy** in a community can be pictured as a pyramid, with the first trophic level (the producers) at the base and the last consumer trophic level at the apex (Figure 46-19).

Other attributes of ecosystems are also represented by pyramids. The total bulk or **mass** of living organisms—often termed **biomass**—also decreases from one link in the chain to the next. These decreases are not a necessary result of physical law as is the reduction in available energy. However, its occurrence is typical of all communities (except those temporarily in unusual circumstances). This means that the total mass of carnivores in a community is always less than the total mass of herbivores, which in turn is less than the total mass of plants. The examples in Figure 46-19 show that the producer level at the bottom of the pyramid has the most energy and mass.

Animal ecologists also speak of pyramids of **numbers.** If we counted all the animals in a community of different sizes, we would find that smaller animals are generally more numerous than larger ones. That is common experience, supported by the abundance of animals of the sizes of insects, mice, and cats or dogs in natural communities.

Pyramids of numbers often depend on pyramids of mass. Predacious animals, those that pursue and kill prey, usually eat animals smaller than themselves. Since these carnivores are higher in a food chain than their prey, the principle of the pyramid of mass holds that the total bulk of the predators is considerably less than that of their prey. The pyramid of numbers model often applies irrespective of predator-prey relationships or of the positions of the animals in food chains. A large animal requires more food (of whatever sort) than does a small one, and so may have to monopolize more territory if it is to survive. In addition, there is food for more small animals in any given area than there is for large ones. Large animals can also range more widely than can small animals.

Note that a pyramid of numbers is inverted in some communities. Plant-eating insects are usually much smaller than the plants they eat, and there are far more

---

[12] Among the omissions from Figure 46-18 are the worms in the deer's intestine, the pumas that eat the deer, the lice that live on owls, the birds that eat seeds and fruit, and so on.

[13] Reducers, although essential parts of all food chains and material cycles, somewhat confuse the picture because they take their food from every link in every chain and not only (in fact, least of all) from the last consumer link. Therefore, we omit them in most of this discussion.

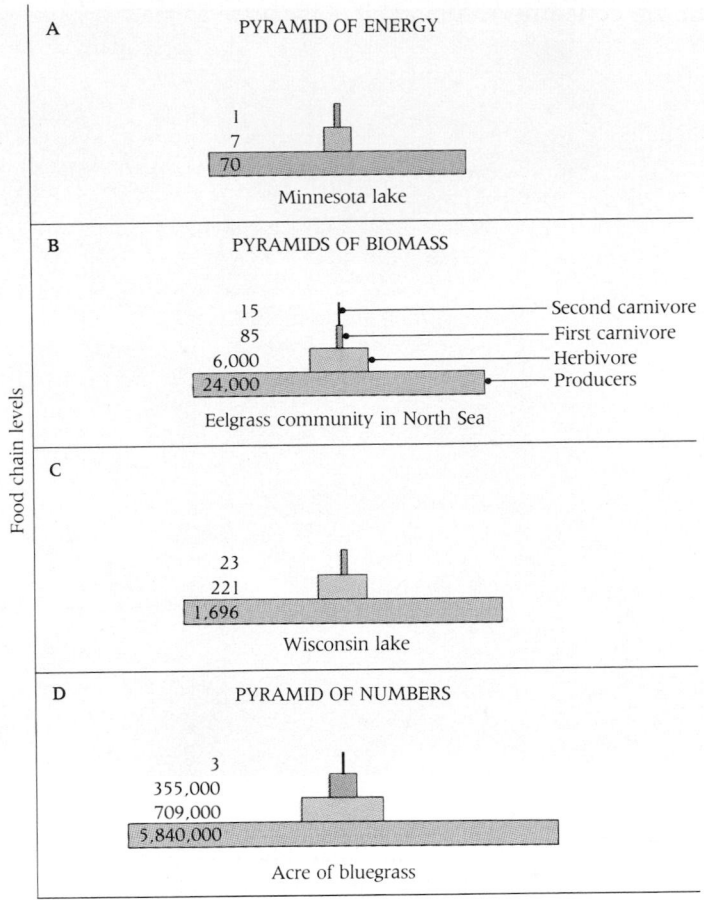

46-19 Pyramids of energy, biomass, and numbers
(A) Energy in a lake ecosystem is expressed as g-cal/cm². (B,C) Note the four levels in the food chain. Biomass in an eelgrass community in the North Sea is measured in thousands of tons; that of the Wisconsin Lake is in kg/hectare. (D) Numbers of individuals in each trophic level in an area of bluegrass. In each pyramid, numbers decline sharply at each higher level of the food chain.

A PYRAMID OF ENERGY

1
7
70
Minnesota lake

B PYRAMIDS OF BIOMASS

15 ————— Second carnivore
85 ————— First carnivore
6,000 ————— Herbivore
24,000 ————— Producers
Eelgrass community in North Sea

Food chain levels

C

23
221
1,696
Wisconsin lake

D PYRAMID OF NUMBERS

3
355,000
709,000
5,840,000
Acre of bluegrass

insect consumers (admittedly of lower total mass) than plants. In general, food chains including parasites have reversed-size relationships because the parasite is smaller and more numerous than the hosts. For these and other reasons, some ecologists caution that pyramids of numbers may misrepresent true community structures. How, for example, does one count the individual grass plants in a lawn? Also, some communities reflect extreme situations—for example, a single whale that eats innumerable tiny diatoms, hordes of parasitic aphids on a single tree (an inverted pyramid)—that are difficult to portray in a pyramid model.

In sum, two principles emerge from this discussion of trophic dynamics.

■ First, with certain exceptions, any complete food chain must begin with photosynthesis and end with decay. Thus a food chain or web may be represented in the following generalized form:

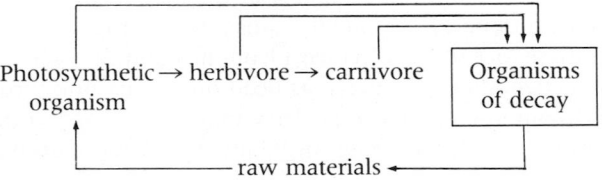

Photosynthetic → herbivore → carnivore   Organisms
organism   of decay

raw materials ←

There must be continual solar input to keep the food chain operating. A rare exception is noted in footnote 14.
■ Second, the shorter a food chain, the more efficient it is. Conversely, the more steps in such a chain, the greater the loss of usable energy.

Budgets: Factors Limiting Life

Most of the energy and materials in all living things must pass through photosynthetic green plants, but they are also a great bottleneck. The flow of energy through all

living organisms can proceed no faster than photosynthesis can fix that energy. We have already seen in Table 46-1 that only a small fraction of the sunlight reaching Earth is actually utilized by green plants in photosynthesis—about 1 part in 10,000. This supply of incoming solar energy, termed the **insolation,** cannot be increased, and photosynthesis, which can proceed no faster than solar energy permits, remains the only port of entry for all of the energy that sustains life on Earth.

The budget of available materials also sets important limitations, especially in certain localities. On land, the most obvious limitation may be in the water budget. Water is the material needed in largest quantities by all living communities and water supplies are notoriously variable, even in modern cities. The comparative scarcity of life in sun-drenched deserts is due to shortages, not of solar energy, but of water.

An important limiting factor in the soil is the nitrate supply. Phosphates are also limiting in the soil. On heavily cultivated land, these are the two principal materials that must be renewed by fertilization.

In aquatic environments, there is practically no limit to the water budget, but other materials, especially nitrates and phosphates, may be stringently limiting. Great quantities of these materials are washed into the sea from the land. That is why marine life is so abundant along shores and near the mouths of rivers. Farther out to sea, nitrates and phosphates are rapidly consumed in the upper levels of water where photosynthesis occurs. The return of nitrates and phosphates to the cycle by decay is more active at deeper levels, as dead organisms sink through the water before they are decomposed by bacteria. They may be returned to the sunlit surface waters by diffusion from deeper water or by the upwelling of deep currents along coasts.[14]

---

[14] Such a zone of upwelling along the coast of northern Peru is marked by an extraordinary richness of marine life. In this zone, great fissures in the ocean bottom serve as vents for the release of geothermal energy. Certain chemosynthetic bacteria can use this energy to synthesize carbohydrates and thus to start a food chain. This shows that some food chains need not start with photosynthesis. The fauna along this food chain—giant tubeworms, albino crabs, and the like—are consumers, ultimately dependent on chemosynthetic bacteria and their unusual geothermal energy source. Such hydrothermal vents are believed to have changed little over the past 4 billion years, and some have wondered whether life on Earth originated in their outflows, which were warm and were rich in minerals, H_2, N_2, CO_2, and possibly CH_4. However, a recent study concludes that they were too hot to allow this to occur.

Combined limitations of energy and material budgets are well illustrated by fluctuations in the abundance of diatoms in the North Atlantic (see Chapter 45). Diatoms are the most important photosynthetic organisms of the open sea. Hence the whole life of the ocean changes with fluctuations in diatom numbers.

An excess of a substance, surprisingly, can also have a limiting effect. Too much nitrate or phosphate, for example, is common in lakes and rivers polluted by sewage, fertilizers, or detergents. These elements "overfertilize" an ecosystem, which cannot dispose of them fast enough. The inadvertent result is termed **eutrophication** which means enrichment (Figure 46-20). Under these conditions, the situation in a lake or other body of water changes drastically. An excess of phosphate often stimulates a bloom of blue-green algae, which can fix atmospheric N_2 (see Table 46-3). The huge increase in algae causes the death of many organisms. Dissolved O_2 is used up by bacterial decomposers feeding on this sudden wealth of organic matter. With less O_2 available, some animals die, thereby generating more "food" for the bacteria, which further deplete the O_2 level. Other undesirable species with other toxic effects are also stimulated. For these reasons, the biogeochemical cycle is seriously compromised. Such lakes and rivers are said to be "dead."

SUMMARY

An ecosystem is a unit made up of a living community and its physical environment. The environments of organisms have both abiotic and biotic elements. Solar radiation, the most important abiotic element, makes possible the synthesis of food molecules by the photosynthetic organisms at the base of the food chain. It also maintains water and air temperatures and promotes the water evaporation that makes possible the Earth's important water (hydrological) cycle.

The niche of a species includes not only its environment, but also the way it interacts with that environment: what it eats, where it lives, and countless other facets of its mode of life. Thus a species' niche reflects the role it plays in its ecosystem. All species have a tolerance range for the variable factors within their environment and within those ranges they have preferences for certain optimal conditions. But because they must compete with other species, many species live not in optimal physical environments, but in environments where competitors are fewer.

The chemical elements in living organisms all participate in biogeochemical cycles, in which they pass continuously between organic and inorganic forms. The major reservoir of inorganic carbon is atmospheric CO_2, which becomes part of organic molecules only after being fixed by photosynthetic organisms. Organic carbon can again be recycled after it is released as CO_2 by cellular respiration or decay. Large amounts of carbon are trapped for long periods as limestone ($CaCO_3$) and fossil fuels. Release of CO_2 by combustion of those fuels has significantly raised atmospheric CO_2 levels in recent decades. The resulting greenhouse effect threatens to raise Earth's temperature in future years.

Nitrogen exists primarily in the form of gaseous N_2 in the atmosphere. Only a few microorganisms can fix N_2—that is, convert it to ammonia, nitrites, and nitrates, the forms in which nitrogen is required by other species. Complex systems recover and recycle organic nitrogen and make it available for biological reutilization.

Biologically essential minerals occur first as dissolved salts arising from the weathering of rocks. A slow-paced mineral cycle permits them eventually to reenter rocks through sedimentation. Pollutants, many unfortunately long-lived, have their own biological cycles. When they become concentrated in the tissues of animals at the top of the food chain, they may cause death or prevent reproduction.

Unlike the flow of materials, which is cyclic, energy flow in a community travels a one-way path. Thus, the biosphere is energetically an open system. Solar energy is transformed into chemical energy by photosynthesis, but at every stage in the food chain, much of that chemical energy is dissipated by various types of conversion to unusable forms.

At the base of the food chain are producers—photosynthetic plants. They are consumed by herbivores, which in turn are consumed by carnivores. These may be consumed by secondary carnivores. Eventually, organisms of decay consume the dead bodies of all organisms. Energy is lost at each level of the food chain; hence the biomass (total weight of organisms) of carnivores is less than that of herbivores, and that of herbivores is less than that of plants. In the end, the biomasses of species in a food chain depends on the limiting factors in that economy—whether it be sunlight in a rain forest, water in a desert, or phosphates and nitrates in the ocean.

KEY TERMS

abiotic environment
biogeochemical cycle
biomass
biotic environment
carbon cycle
carbon-silicate cycle
consumer
decomposer
ecosystem

environment
eutrophication
food chain
food web
greenhouse effect
habitat
homeostasis
hydrologic cycle
insolation

niche
nitrogen cycle
nitrogen fixation
pollution
producer
substratum
tolerance range
trophic level

QUIZ QUESTIONS

1. Which of the following is *not* involved in cycling of nitrogen?
   A. legumes
   B. *Artemesia*
   C. *Rhizobium*
   D. *Anabaena*
   E. *Azolla*

2. The primary reservoir of inorganic carbon that is accessible to organisms is
   A. cellulose.
   B. chitin.
   C. fossil fuels.
   D. carbon dioxide.
   E. calcium carbonate.

3. _____ is an integral component of mineral cycles.
   A. The water cycle
   B. Biological magnification
   C. The pollutant cycle
   D. The rock cycle
   E. A and D

4. For any ecosystem, the amount of energy present in the _____ level is greater than that present in any other trophic level.
   A. top consumer
   B. decomposer
   C. producer
   D. herbivore
   E. second-level consumer

5. Which of the following depict(s) the pattern of flow of energy and materials through an ecosystem?
   A. numbers pyramid
   B. food chain
   C. biomass pyramid
   D. food web
   E. B and D

ESSAY QUESTIONS

1. What is a niche? How is it distinct from a habitat; a substratum? What are the niche, habitat, and substratum of an earthworm?

2. How does the CO_2 content of the atmosphere affect the temperature of the Earth's surface? What is this phenomenon called? How might it be alleviated?

3. How is nitrogen assimilated into living systems? How does nitrogen return to the environment? What organisms are responsible for the two parts of this cycle? Does the nitrogen cycle necessarily involve gaseous nitrogen? If you were a farmer and could selectively encourage the growth of either nitrifying or denitrifying bacteria, while discouraging the growth of the other in your fields, which would you encourage? Why?

4. In an ecosystem, how does the biomass of herbivores compare to plants; to carnivores?

5. What is eutrophication? Describe the process as it might occur in a polluted lake.

REFERENCES AND SUGGESTIONS FOR FURTHER READING

Alcock, J. (1985). *Sonoran Desert Spring*. University of Chicago Press, Chicago.

A stimulating nontechnical book on the ecology of the Sonoran desert of Arizona.

Berner, R. A., and Lasaga, A. C. (1989). Modeling the geochemical carbon cycle. *Scientific American 260* (March), 74–81.

The authors present a model of a long-term cycle in which carbon is transferred between land, sea, and atmosphere.

Carpenter, S. R., Kitchell, J. F., and Hodgson, J. R. (1985). Cascading trophic interactions and lake productivity. *BioScience 35*, 634–639.

New approaches to the management of lake ecosystems in which food webs are altered by manipulating populations of consumer species.

Connell J. H. (1979). Diversity in tropical rain forests and coral reefs. *Science 199*, 1302–1309.

A comparative account of two of the richest and most diverse ecosystems. Similarities and differences are discussed.

Evans, H. J., Bottomley, P. J., and Newton, W. E., (Eds.) (1985). *Nitrogen Fixation Research Progress*. Martinus Nijhoff, Norwell, Mass.

What is new and what is true in an exciting field.

Grover, H. D., and Harwell, M. A. (1985) Biological effects of nuclear war II: Impact on the biosphere. *BioScience 35*, 576–583.

A sobering assessment that should be examined.

Kaufman, L., and Mallory, K. (1986). *The Last Extinction*. The MIT Press, Cambridge, Mass.

A readable discussion of the urgent threat of mass extinctions owing to the activities of human beings.

Long, S. R. (1989). Rhizobium-legume nodulation: Life together in the underground. *Cell 56*, 203–214.

Another excellent review from *Cell*.

Long, S. R., and Ehrhardt, D. W. (1989).

Nitrogen fixation: New route to a sticky subject. *Nature 338*, 545–546.

Interesting brief discussion of a daunting problem: the nature of the specificity inherent in root nodule formation.

Luoma, S. N. (1984). *Introduction to Environmental Issues*. Macmillan, New York.

A well integrated discussion of scientific principles underlying various environmental perturbations. Good discussions of ecosystems and biogeochemical cycles.

McIntosh, R. P. (1985). *The Background of Ecology: Concept and Theory*. Cambridge University Press, New York.

An outstanding introduction to current ecological thought and its historical development.

Mohnen, V. A. (1988). The challenge of acid rain. *Scientific American 259*, August, 30–38.

Acid rain is due mainly to oxides of sulfur and nitrogen that are emitted by power plants and automobiles. This article reviews newer technologies that have been proposed to control these emissions.

Schneider, S. H. (1989). The greenhouse effect: science and policy. *Science 243*, 771–781.

A sober analysis of the problem and possible strategies for dealing with it. The choice of tactics, says the author, is not a scientific question but a value judgment.

Seinfeld, J. H. (1989). Urban air pollution: State of the science. *Science 243*, 745–752.

Recent word on an important topic.

Sheehan, P. J., Miller, D. R., Butler, G. C., and Bourdeau, P. (Eds.) (1984). *Effects of Pollutants at the Ecosystem Level*. John Wiley & Sons, New York.

Humans have induced profound and disturbing changes in the biosphere. This informative book presents several case histories focused on how we decide if an ecosystem change is or is not due to a pollutant.

Woodwell, G. M., Hobbie, J. E., Houghton, R. A., and others (1983). Global deforestation: contribution to atmospheric carbon dioxide. *Science 222*, 1081–1086.

A discussion of the importance of forest loss in increasing atmospheric carbon dioxide.

COMMUNITIES

Communities
Communities are defined as interacting
assemblages of populations of different species
in a given area. Communities surround us on
all sides. Here are a few examples.

In this chapter, we further explore the relationships of living organisms in ecosystems. We will focus principally on the means by which different populations fit into communities. Just as the properties of a population are not the sum of the properties of all of its individual members, we will find that a community is much more than the sum of its constituent populations.

It has been said by critics that ecology, even the most theoretical kind, can be regarded largely as a formalization of common knowledge—an institutionalized walk through the woods, so to speak. To these skeptics, ecology is little more than natural history, even when computers and other technology are vigorously used. Others, ourselves included, believe that important ecological insights are being achieved outside of commonplace human experience. Some of those insights will be described in this chapter.

COMMUNITY ECOLOGY

A **community,** we recall, is an assemblage of populations of different species that are living and interacting in a given area. Here we explore the most important of the many interrelationships that bind members of a community into an organized system.

NICHES AND INTERSPECIFIC COMPETITION

An ecological **niche** was defined in Chapter 46. Restating the definition more simply, an ecological niche is the totality of what a population needs to survive and the resources it exploits. It is not to be confused with **habitat,** which is the physical environment an organism occupies, or its "address."

An organism's niche includes what nutrients it requires, what climate it can withstand, which predators and parasites it must deal with, where, when, and how it reproduces, how it responds to physical and biotic factors, and many other items of a utilitarian nature. Indeed, virtually every aspect of an organism's existence helps to determine its ecological niche. Thus a niche is a property of a population.

A field investigator can make a concrete assessment of a species' niche by establishing its **limits of tolerance** after each of the different conditions affecting the species is varied systematically. Results are then plotted graphically (Figure 47-1). After plotting many variables in this way, an investigator can define a three-dimensional volume within which every point corresponds to some combination of values for the several variables which would allow the species to survive. The result is a model in which the ecological niche of a species is seen as a multidimensional "hypervolume." The greater the similarity of two niches, the greater the overlap of the multidimensional hypervolumes for the two species. The more alike their overlapping niches are, the more intense is the competition between the species— that is, if essential resources are in short supply.

Relationships Among Niches

We can illustrate the relationships among different niches by recounting a natural drama that is played out regularly in some of the driest deserts of Arizona. There lives the gigantic cactus, *Cereus giganteus,* which is known locally as "saguaro" (or "sahuaro," in either case pronounced sah-*wah*-roh). It is so striking and picturesque that it has become a symbol of the American Southwest (Figure 47-2). Enter now two birds, a flicker (*Colaptes chrysoides*) and a common woodpecker (*Centurus uropygialis*), which cut round openings in the spiny, ridged stems of the saguaro, and excavate recesses that serve as nests in the soft internal tissues. They make more holes than they can use, and unused flicker or woodpecker holes are often occupied by elf owls (*Micropallas whitneyi*), dainty little creatures no larger than sparrows. In fact, elf owls rarely nest anywhere else.

Saguaros live only in deserts, and only in particular parts of them, and have their special niche among the plants. In turn, they provide one aspect of the otherwise differing niches of desert flickers and woodpeckers. These birds, in their turn, create the homes used by elf owls, and these homes are among the defining characteristics of the elf-owl niche. The interdependence of these four organisms is evident. So is the fact that their roles in the community are quite different. They all, in fact, have different niches.

Another owl, slightly larger than an elf owl but still small as owls go, also nests in woodpecker holes in saguaros: it is the saguaro screech owl (*Otus asio gilmani*). Owls of both kinds may be found in adjacent holes on the same cactus plant. Here you might think are two species occupying the same niche, but that is not the case. Elf owls feed mainly on small insects—beetles, ants, or crickets. Screech owls eat some insects, especially the larger grasshoppers and locusts, as well as scorpions, but their diet also includes mice and other rodents that are seldom attacked by elf owls. There are also differences in their other activities. Hence, these two owl species have distinctly different total relationships to their environment, or different roles in their community. In other words, they occupy different niches.

This appealing desert scenario illustrates a principle that operates in any community: each species in an established community has a distinctly different niche. Exceptions to that rule occur when one species invades the territory of another, but the duplication is only temporary. The basis of the rule is **competition.** When resources are in short supply, equal sharing by different species is unknown in nature. The principle can be expressed in another way: two species do not live together long if one of them can fully utilize an aspect of the environment or resource that is limited, and necessary to both of them. This generalization, known as **Gause's principle,** is also termed the **principle of competitive exclusion.**

Gause's principle has been verified again and again in different types of organisms. In a well-known experiment of G. F. Gause (Figure 47-3), different species of

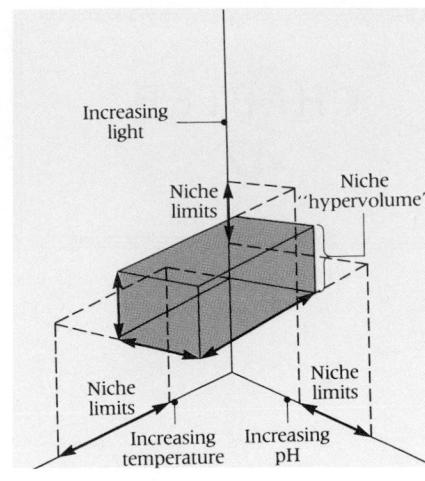

47-1 One method of defining a niche
An investigator attempting to estimate the niche of a given plant species might consider three variables: temperature, light, and soil pH. The more variables studied, the more precise is the definition of the niche. The "hypervolume" defining the niche is bounded by the tolerance limits for each variable.

47-2 Saguaro cactus

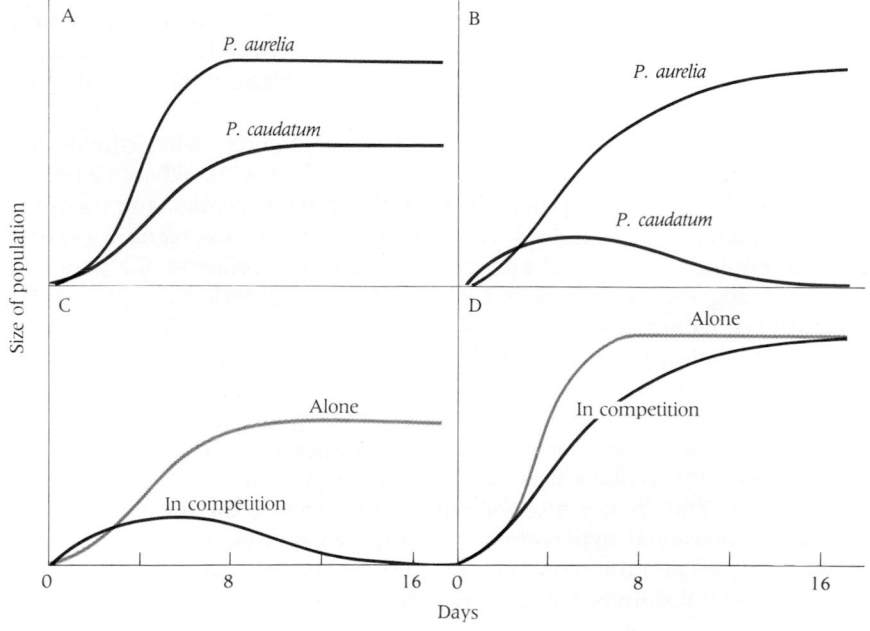

47-3 Interspecific competition
Two species of Paramecium *are grown in the same culture flask.* **(A)** *When the two species are grown separately,* P. aurelia *grows faster and to a higher density than does* P. caudatum. **(B)** *Competition between the two species occurs when they are grown in the same flask.* P. aurelia *survives and* P. caudatum *dies out. Regraphing the data shows the effect of living alone versus competition for each species separately.* **(C)** P. caudatum. **(D)** P. aurelia.

Paramecium were raised together in the same culture. When the species were *P. aurelia* and *P. caudatum*, only *P. aurelia* was found in the culture after 16 days. The niches of the two species are essentially identical, but the more rapid growth rate of *P. aurelia* causes it to outcompete *P. caudatum*. In contrast, when *P. caudatum* and *P. bursaria* were cultured together, both survived. Although their food requirements are the same, they managed to survive by living in different parts of the culture chamber. This simple physical separation is sufficient to provide each species with a separate habitat and hence to permit its survival.

Competition need not extend to all environmental necessities, and it seldom does. Elf owls and screech owls do not compete seriously for woodpecker holes. Different species of fishes in the sea do not compete for water. There is plenty of water in the sea and there are plenty of woodpecker holes in the desert. But species cannot exist together long if they continue to compete for *anything* that is in short supply and essential to them.

The requirements of two species in the same community may be so different that there is no real overlap. In that case there is no competition. Take the example of saguaros and woodpeckers. These two species affect each other in many ways. Aside from the fact that woodpeckers live in saguaros, both are parts of some of the same food chains: saguaros → insects → woodpeckers. Yet their niches are so different that there is no competition between them. Indeed, since woodpeckers eat insects and insects eat saguaros, woodpeckers are beneficial to saguaros. An environmental factor becomes crucial only when the competition for it could eliminate one or the other species. It is this factor that limits the populations of each species, keeping them down to the point where they do not compete for those things that they do, in fact, share.

This crucial factor is to be found among the things that are *different* in the overlapping niches of two species. For example, with the two desert owls, the crucial factor is evidently food: elf owls eat small insects and screech owls eat large insects and rodents. Even in food, there is some overlap because some kinds of insects are eaten by both. In times of abundance, this overlap does not matter. In times of scarcity, competition increases and they turn to different kinds of prey. The resulting differences in feeding efficiencies lead to a partitioning of resources. In other words, the two owls, driven to the wall, begin to eat different things. The elf owls are more efficient at feeding on small insects, which become their principal diet; the screech owls exploit rodents, which they capture with greater proficiency. Thus, in times of scarcity, there is a switching of food resources.

Among the most closely similar and most frequently competing species of a community there is always, so it seems, some such safety valve. Birds may willingly share the abundant foods of the late summer, and then, when the supply dwindles, take to different foods—or migrate to different regions.

Such escapes from competition during periods of environmental stress are common. Periods in which the availability of preferred foods may decrease sharply are called "ecological bottlenecks" because they put the species under intense selective pressure. Only species adapted to switch to alternate foods—so-called "refugium resources"—can survive the crunches of these periodic predicaments.

Consider the Central American cichlid fish, *Cichlasoma minckleyi*. Its feeding habits are related to the fact that it possesses two body forms, or "morphs." One morph, termed *papilliform*, has jaws with rather pointed teeth well suited for macerating the larvae of insects (Figure 47-4A). The second morph, the *molariform*, has stout blunt teeth that are effective in crushing snails (Figure 47-4B). When insect larvae are abundant, both morphs prefer to feed on them. But when the supply of insect larvae dwindles, the two morphs segregate their feeding activities. Only the molariform morph, with its specialized teeth, jaws, and associated muscles, is well adapted to exploit the less preferred refugium food resource.

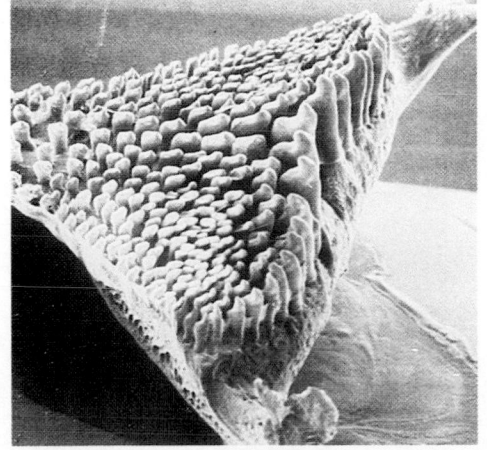

A

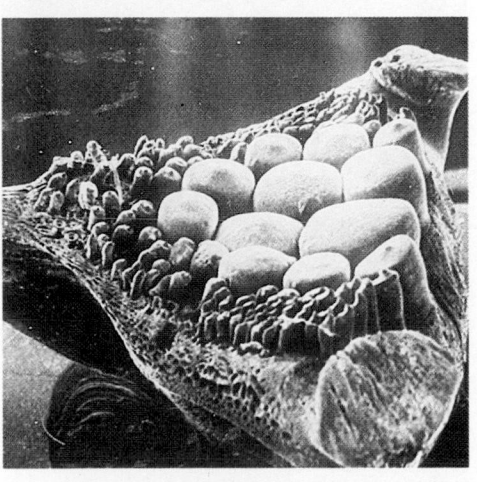

B

47-4 The teeth of cichlid fishes
(A) *The teeth of the papilliform morph.*
(B) *The blunt teeth of the molariform morph.*

Nature of Competition

Organisms may compete for almost anything from the environment that is needed and in short supply. Plants compete for water or sunshine, and sometimes for mineral salts—whatever is in shortest supply at a given time and place. In the desert, each creosite bush is surrounded by a zone of completely bare soil from which it has ousted its competitors for water. In the forest, a dense growth of seedlings thins out when a few more vigorous trees preempt the sunshine and cause others to die in their shade. In the sea, one species of photosynthetic organisms may drive out another when it wins a contest for nitrates or phosphates.

That food is the usual object of competition among animals lies behind the diversity and specificity of animals' food habits. These characteristics tend to reduce competition for food between different species.

When two species differing only slightly in their niches inhabit the same area and compete, they are likely to evolve toward greater niche differences in order to reduce competition. The process that accentuates differences between species to reduce interspecific competition is known as **character displacement.** We alluded to it briefly in Chapter 34. It is a major mechanism of evolutionary diversification.

Darwin's finches on the larger Galápagos Islands illustrate the process of evolutionary diversification. Character displacement has influenced the beak size of those that feed primarily on seeds. When small and medium-sized ground-finch species coexist on the larger islands, their beak sizes come to differ widely, with depths averaging about 8.4 and 13.3 mm, respectively (Figure 47-5). The pattern on the small islands where only one of the two species exists is in sharp contrast. There the finches have beaks of intermediate size, with depths averaging 9.7 mm.

To be sure, competition occurs just as frequently between phylogenetically unrelated species that both need one essential resource in short supply, while otherwise leading completely different lives. Competition between rabbits and sheep caused a crisis in the Australian wool growing industry. Humankind's severest competitors for food are the insects (Box 47-1).

In an instructive field study, J. H. Connell, of the University of California, Santa Barbara, showed that two competing species with partially overlapping niches can flourish side by side by occupying habitats that differ ever so slightly. Two species of barnacles live on the rocky shores of Scotland and northeastern United States. One (*Chthamalus*) occupies the upper part of the intertidal zone—the rim of the sea between high tide and low tide. The other (*Balanus*) keeps to a wider zone in the lower part of the intertidal zone—down to the low-water mark (Figure 47-6). The boundary between the two distributions hardly overlaps at all.

What keeps the two species apart? When Connell removed all *Balanus* from one area and all *Chthamalus* from another area, he found that free-swimming *Chthamalus* larvae would settle down for life and grow in lower parts of the intertidal zone in the absence of *Balanus*. In the reciprocal experiment, *Balanus* larvae did settle in the upper zones usually occupied by *Chthamalus*. But they did not survive because they lack the adaptations to desiccation and high temperature that allow *Chthamalus* to occupy this shallower zone. He then found that each species is capable of inhabiting an intermediate zone between the mean high neap tide (lowest tide of the lunar month) and the mean tide levels. Thus, it is competition that keeps *Balanus* and *Chthamalus* separate. The competitive interactions between them extend to the deeper intertidal zone, the niche *Chthamalus* might occupy if its competitor were absent. However, in its presence, *Chthamalus* has a much narrower niche and distribution.

This narrower niche is referred to as the realized niche. It is important to understand the difference between fundamental and realized niches when introducing organisms into new areas. A fish such as the herbivorous grass carp, which has a narrow realized niche in its Asian environment, turned out to be a real pest when it was introduced into North American waters. In the absence of its native competitors, its fundamental niche was fulfilled, making it a true omnivore in its adopted waters.

Competition and Evolution

If two species compete strongly, one of them usually becomes extinct. This is not the only cause of extinction, but it has been a common cause during the long history of life. An example occurred one to several million years ago when North and South America, each previously separate continents, were reunited (see Chapter 48). Each continent once had its own distinctive mammal species, including rodents, carnivores, and herbivorous ungulates (hoofed mammals). But North American species then invaded South America, and in the ensuing competition all the ungulates,

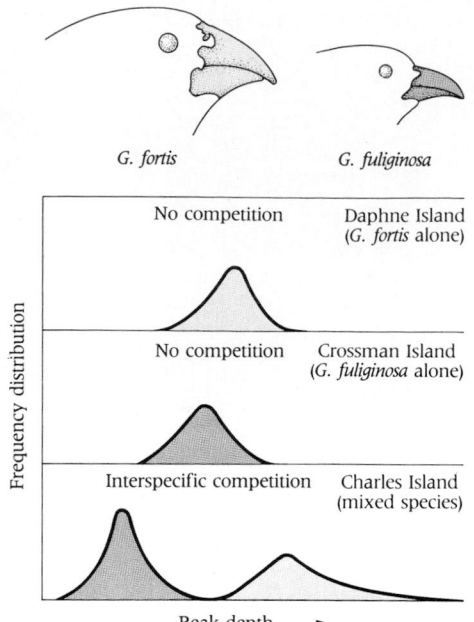

47-5 Evolutionary effects of interspecific competition
On Charles Island in the Galápagos, competition between two species of finches, Geospiza fortis and Geospiza fuliginosa, resulted in a shift in beak sizes relative to the sizes characteristic of each species living alone. Beak sizes reflect the insects they eat. Compare with Figure 36-11.

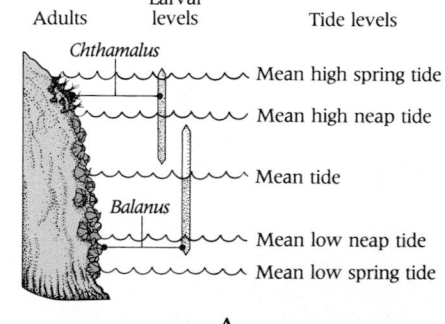

47-6 Interspecific competition between barnacles
(A) Though in the larval stages these two species occupy overlapping areas of the rock, by the time the adult stage is reached, Balanus is restricted to the lower regions below the tidal zone. Chthamalus is found only in the zone above the mean high neap tide. **(B)** Chthamalus are the light-colored rigid forms. Balanus is seen at the bottom of the photo.

BOX 47-1

HOW INSECTS COMPETE WITH HUMANS FOR FOOD

Expansion of the human population threatens people throughout the world with a serious food shortage. Contrary to popular belief, most arable land is now under cultivation. The finite supply of fossil fuels is also being rapidly depleted by the world's industrialized nations. Since fossil fuels are used in large quantities during food production—for fertilizer, pesticides, and many other activities—future prospects for an adequate food supply look bleak indeed.

It might be supposed that at least one of the major factors depressing food crop production could one day be eliminated—the problem of pest control. However, pests responsible for major crop losses in the United States include some 160 species of bacteria, 250 kinds of viruses, 8000 species of fungi, 2000 species of weeds, and 8000 species of insects. The many available techniques of insect control—their dependence on fossil fuels and their other drawbacks—were discussed in Box 42-2.

By 1945, the use of synthetic pesticides in the United States had grown to over 1.2 billion pounds annually. That total represented about 6 pounds per person per year. As shown in Table 1, recent crop losses due to insect pests, plant diseases, and weeds had a net worth of about $18.2 billion per year and consumed about 33% of the crops. These losses continue in spite of all efforts at control. As shown in Table 1, losses from weeds decreased somewhat in recent decades, while losses to insects and plant diseases increased. Although the overall percentage of crop losses to insects has increased despite the

use of insecticides, important advances have been made in reducing insect losses in certain crops. For example, losses in yield and quality from potato insects declined from 22% in 1910–1935, to 16% in 1942–1951, and 14% in 1951–1960. This reduction was perhaps to be expected in view of the recognized capacity of insecticides to control the major insect pests of the potato plant.

In contrast, losses of apples—caused primarily by the codling moth (*Carpocapsa pomonella*) and apple maggot (*Rhagoletis pomonella*)—have not declined despite increased use of insecticides. A 10.4% loss in yield was reported for the period of 1910–1935, a 12.4% loss for 1942–1951, and a 13.0% loss for 1951–1960. To an extent, this loss pattern is due to rising standards for marketable fruit. It also reflects a decline in the use of the sanitation methods that once controlled these pests in apple orchards. Losses to insects have also been increasing in corn, a major grain crop. A 3.5% loss in 1942–1951 rose to 18.0% in 1971–1980. Factors contributing to increased corn losses to insects include the continuous cultivation of corn on the same land year after year (a practice that increases the corn plant's susceptibility to rootworm) and the increased planting of insect-susceptible types of corn rather than insect-resistant types. Corn losses from birds and mammals are small—only about 1%. However, with certain crops and in certain locations (e.g., cherries and corn grown near water), losses of this kind may reach 10 to 50%.

Food losses to pests are also very high

in the rest of the world. World crop losses to pests are currently estimated at about 35%. That figure includes destruction by insects, disease-causing pathogens, weeds, mammals, and birds. Mammal and bird losses appear to be more severe in the tropics and sub-tropics than in the temperate region, but these losses are still low compared to the major pest groups: insects, disease-causing pathogens, and weeds.

Surprisingly, the widespread use of improved agricultural techniques—symbolized by the term "green revolution"—has intensified losses to pests (see Chapter 49). One reason is the increased susceptibility of many new high-yield crop strains to insects, pathogens, and weeds. New seed varieties are often planted uniformly over wide areas. This genetic uniformity provides an ideal ecological environment in which evolving pathogens can gain supremacy and decimate these particular genotypes.

Post-harvest losses in other countries average about 20%, ranging from about 9% in the United States to 40 to 50% in some developing nations in the tropics. The prime pests of harvested foods are insects, microorganisms, and rodents.

When pre-harvest losses are added to post-harvest losses, worldwide total food losses due to pests are estimated to be about 48% (35 + 13)! In other words, pest populations are consuming and/or destroying nearly half of the world's food supply. Surely that is a loss that human beings cannot afford as they face world food shortages and an ever increasing world population.

TABLE 1
ANNUAL AGRICULTURAL PEST LOSSES IN THE UNITED STATES IN RECENT YEARS*

Years	Insects ($)	Insects (%)	Diseases ($)	Diseases (%)	Weeds ($)	Weeds (%)	Total Loss ($)	Total Loss (%)	Potential Production ($)
1904	0.4	9.8	—	—	—	—	—	—	4.1
1910–1935	0.6	10.5	—	—	—	—	—	—	5.7
1942–1951	1.9	7.1	2.8	10.5	3.7	13.8	8.4	31.4	26.7
1951–1960	3.8	12.9	3.6	12.2	2.5	8.5	9.9	33.6	29.5
1974	7.2	13.0	6.6	12.0	4.4	8.0	18.2	33.0	55.0
1986	13.3	14.2	9.1	14.0	7.3	8.8	29.7	37.0	80.3

* $ symbolizes billions of current dollars

all the carnivores, and a great many of the rodents native to South America eventually became extinct.[1]

Competition has also played a less macabre part in evolution. Between (and within) species, it has tended to cause and maintain differences among populations. It has led to the occupation of particular niches and increased the numbers of niches in communities. Once a niche is occupied, competition becomes a conservative force tending to keep the population there, impeding its change or its spread into other niches.

It is easy to see how competition becomes an evolutionary force. Those variants in similar species that least resemble each other will encounter the least competition. Hence they are likely to be most successful in rearing offspring. Over many generations, the two species come to occupy niches that are increasingly different. On the other hand, in a well-integrated community with numerous occupied niches, variants of any species farthest from the usual or modal adaptation to its niche are most likely to encounter competition from the occupants of other niches. Natural selection will tend to go against these variants and thus favor maintenance of the status quo.

INTERACTIONS BETWEEN SPECIES: SYMBIOSIS

We have now considered two universal kinds of interactions occurring when species live together: *consumption* (including predation) and *competition*. However, other kinds of relationships exist within a community that spring from especially intimate modes of living together. Examples would be the relationships between a barnacle and the whale to which it is attached, between a dog and its fleas, and between a hermit crab and a sea anemone in and on an old snail shell (Figure 47-7). Such close associations between individuals of different species, occurring within the looser association of a whole community, are examples of **symbiosis,** a term with roots meaning "life together." We have mentioned symbiotic relationships before.

Types of Symbiotic Relationships

Symbiotic relationships are of three basic types, which are well illustrated by the examples just cited and those in Figure 47-8.

- If one species benefits from the association while the other is neither benefited nor harmed, the relationship is called **commensalism.** Assuming a whale does

---

[1] The native South American carnivores were marsupials. They, too, suffered large-scale extinctions that resulted in large part from competitive exclusion by invading placental mammals from North America.

47-7 Hermit crabs inhabit snail shells
These interesting crustaceans have soft abdomens. They make up for the lack of hard covering by seeking out gastropod shells of suitable size and placing their abdomens inside the shells. The shell is a temporary home, for they must find larger ones as they grow. The crab on the right has even found nicks in the shell to use as peepholes. Often, the shell becomes covered with sea anemones, which camouflage the shell and defend it with their stingers. The sea anemones make the relationship symbiotic.

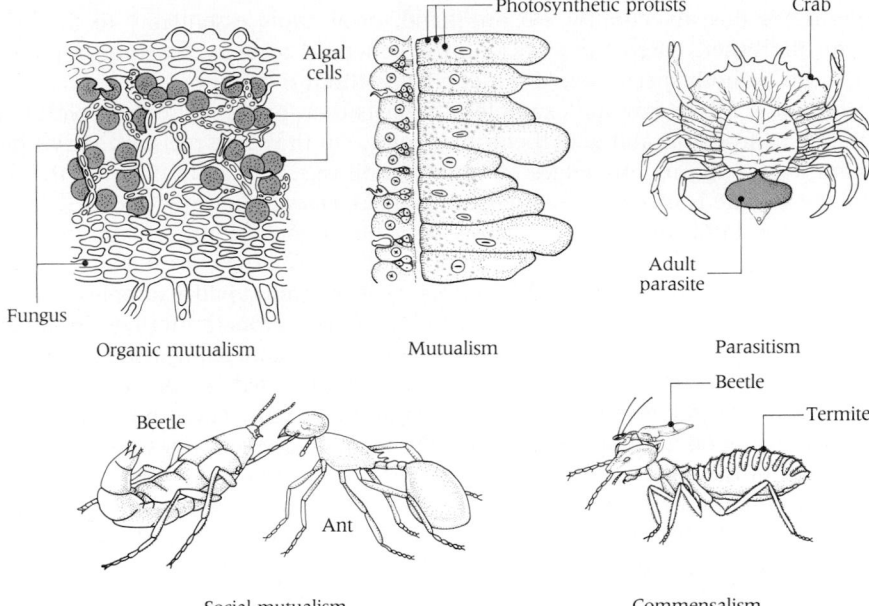

47-8 Examples of symbiotic relationships

Organic mutualism: a lichen which consists of nonphotosynthetic fungal hyphae and photosynthetic algal cells. Mutualism: photosynthetic protists in cells lining the body wall of the Hydra. Parasitism: the parasite Sacculina is represented in the adult crab by a pouch of reproductive tissue held between the crab's abdominal segments. Extensions from this pouch reach into the crab's tissues, digesting them and leaving a hollow shell. Social mutualism: a beetle that is reared and protected by ants for the sake of its secretions. Commensalism: a staphylinid beetle that lives in a termite colony as a tolerated scavenger. It rides on the head of a termite, collecting food fragments as they are passed between workers.

not mind a few barnacles on its hide it is neither helped nor harmed by their presence. The barnacle, on the other hand, benefits to the extent that it gets a place of attachment—and a free ride.

- If the relationship is beneficial to both, it is called **mutualism.** Hermit crabs and sea anemones, though devoid of altruistic motives, are nonetheless helpful to each other. The sea anemone uses its stinging cells to protect the crab. Sometimes it helps to make the snail shell fit the crab by actually remodeling it. The crab seeks out food and tears it to shreds, and the sea anemone lives on such bits as come its way.

- If one species is benefited at the expense of the other, the relationship is called **parasitism.** For fleas, the dog is food as well as home. But the dog, an unwilling host, receives only annoyance, and sometimes disease, in return.

The study of symbioses is as complex as it is important. Symbioses of one sort or another occur throughout the living world. They are diverse, often puzzling or obscure, and sometimes bizarre.[2]

Examples of Commensalism and Mutualism

Commensalism is quite common among marine invertebrates. Besides the barnacles on whales, there are barnacles that live only on the barnacles that live on whales. Several small fishes usually live among the tentacles of coelenterates (especially siphonophores and sea anemones), apparently immune to attack by stinging cells. For these fishes, the hosts provide protection and probably some food.

A similar relationship exists between a tropical fish (*Aeoliscus strigatus*) and a sea urchin (Figure 47-9). Some little crustaceans (a species of isopod) live in the mouths of fishes (menhaden), where they pick up bits of the fishes' food as it goes by. Examples of commensalism on land are the relationship between bryophytes and the trees on which they grow and the relationship between a tree and the birds that nest in it. For the bird and the bryophytes, the relationship is obligatory; the tree is unaffected.

As for mutualism, few are the plants or animals that are not inhabited by other plants or animals or by protists. Each of us is host to many bacteria and other protists that live on the skin and in the mouth, alimentary canal, and other cavities. Such relationships run the gamut—indeed two gamuts—from complete unimpor-

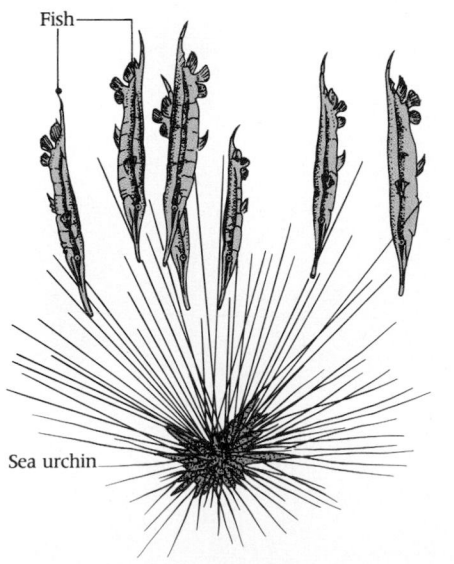

47-9 Commensalism

Symbiosis between the tropical fish, Aeoliscus strigatus, *and a sea urchin is an example of commensalism. The fishes are protected by sheltering spines without affecting the sea urchin.*

[2] It is sometimes difficult to distinguish between mere association in a community, commensalism, mutualism, and parasitism. Whether the partnership is completely unimportant to one species (making it commensal), more or less helpful (therefore mutualistic), or harmful (parasitic) may be impossible to determine or may be a matter of definition or point of view. Moreover, all sorts of intermediate relationships occur. In some cases, one sort of relationship has evolved into another.

tance to the host to lethality, on one hand, or absolute essentiality to the host's life, on the other. Often the association is beneficial to both, reaching a point at which one or both species could not survive without the other.

In Chapter 1, we saw the example of protists that live in reef corals, participate in their construction, and give them their color. On the same reefs, photosynthetic protists live in the mantle edges of various mollusks, brilliantly coloring them. A notable example is provided by the giant clam, *Tridacna* (Figure 47-10).[3] Relationships of this kind are common. Hosts may be animallike protists (protozoans), sponges, coelenterates (corals, green hydras, etc.), flatworms, or mollusks.

Mutualism is also common among plants. Especially vivid examples are seen in the tropics—nectar-producing plants and their pollinators (butterflies, hummingbirds, bats, bees, lemurs, and marsupials); obligatory ant-plant associations; and so on. Lichens, as noted in Chapter 40, may resemble single plants, but in fact are intimate, mutualistic associations of an alga and a fungus. The nonphotosynthetic fungus derives food from the alga. It also helps to maintain the water supply necessary for growth of the alga.

Similarly, a mycorrhiza is a mutualistic association of fungus and plant. In these structures, in which fungi form a feltlike coating around the roots of a green plant, the fungi are sometimes harmful. But many are beneficial. We saw in Chapter 40 that pines may die if deprived of their root fungi. Other plants also depend on them. The exact nature of the benefit to the host is unclear in some cases. More obvious is the benefit accruing from nitrogen-fixing bacteria in root nodules of peas and their leguminous allies (see Table 46-3).

Other implications of mutualistic interactions are now recognized. It appears, for example, that effective mutualism can arise indirectly in complex food webs. Suppose that species *A* and *B* appear to compete for prey species *C* and *D*, with species *B* concentrating mainly on *D*. Suppose, too, that species *C* is competitively superior to *D* and would eliminate *D* were it not for the effects of predatory species *A*. It could happen that removal of *A* would cause prey species *C* to displace *D*, thus decreasing the population of *B*, which is sustained largely by *D*. Thus, the apparently competing species, *A* and *B*, are, in fact, mutualistically related by these indirect interactions. When real-world time lags are also considered, the dynamics of mutualistic associations can become very complex indeed.

Another example of mutualism is seen in the wood-eating termites, which live on cellulose, a food that they cannot digest. As noted earlier (see Chapters 26, 36, and 39), the wood is in fact digested by symbiotic protists, a species of flagellates (*Myxotricha paradoxa* in Australian termites and other species in other termite species) that swarm in the termites' digestive tracts. The termites thus obtain their food from the protists. Without their protists, the termites cannot live—someone has said they "die of splinters"—and without their termites, which obtain the wood on which they subsist, the protists cannot live.[4]

Cows and other herbivorous mammals also indirectly acquire some food from cellulose, which they cannot themselves digest. But, unlike wood-eating termites, they are not wholly dependent on this process. Bacteria in the cows' alimentary canal digest some of the cellulose in grass or other forage. The cows, in turn, later assimilate some of the bacteria and their products (see Chapter 26). Indeed, many ruminants get up to 90% of their nutritional energy from bacterial fermentations in their rumens (Chapter 26). Methane and other gaseous products of this fermentation are emitted by cattle in large enough amounts to become important factors in causing the greenhouse effect (see Figure 46-9).

THE SPECIAL CASE OF PARASITISM

Nature of Parasitism

If the interspecific relationship we call "parasitism" were defined simply as living at the expense of other organisms, then the whole of the animal kingdom, including

47-10 Giant clam, Tridacna
See also Figure 42-18A.

---

[3] The photosynthetic protists that live in symbiosis with the tridacnids have been identified as a species of dinoflagellate, *Gymnodinium microadtriticum*.

[4] Interestingly, there is more to this story. Closer examination of *Myxotricha* revealed that many (but not all) of its "flagella" are not flagella at all but long, motile bacteria (spirochetes) that are attached at intervals to the protist's surface and beat as though they were flagella. Other bacteria also cover the protist and several of a third species are found *within* the protist. There they are believed to assist with the digestion of the wood particles taken up in the termite's gut.

47-11 Brood parasitism in birds
This hedge sparrow nest contains four sparrow eggs and a single cuckoo egg. Hedge sparrows cannot discriminate between their eggs and those of the cuckoo and so brood the foreign egg with their own. As soon as the cuckoo hatches, it will throw out the eggs of the sparrow. Other examples of brood parasitism are discussed in Chapter 35.

Homo sapiens, would be parasites. Even if we restrict the term to particular kinds of interactions between species, the relationship is not always clearly distinguishable from other relationships. A weasel kills a rabbit and sucks its blood. A fly sucks blood from a rabbit without killing it and then goes on its way. A louse spends much of its life on a rabbit's skin, taking blood when it cares to. A protist lives *in* a rabbit's bloodstream, within the fluid it feeds upon. Which of these are parasites?

Parasitism is distinguished from other relationships by three features.

■   A parasite lives on or in another organism for most of its life cycle.
■   It derives its food from its host.
■   It is more or less harmful to the host—at least, it is not beneficial.

The extent of parasitism and its diversity are astonishing. Few living things are free of parasites. Almost every part and tissue of the human body is a potential habitat for one parasite or another. Perhaps the only organisms not subject to parasitism are a few that are, you might say, the "last word" in parasites themselves. Many parasites, it seems, have parasites of their own. In the rhyme of Jonathan Swift, "A flea hath smaller fleas / that on him prey, and these / have smaller still to bite 'em, / and so proceed *ad infinitum.*"

Examples are without number. There are insects that begin life as parasites inside the eggs of other insects. The larval females of a small wasplike insect, *Coccophagus scutellaris*, are parasites in other insects, and the larval males live within the parasitic female larvae of their own species. In several species of annelids, crustaceans, and fishes, adult males are parasites on adult females. In brood parasitism, a parasite's young are habitually substituted for the host's young. The interloping brood then lives on food that was intended for the brood it dispossessed. This is common among birds—cowbirds and cuckoos (Figure 47-11)—and wasps and other insects.

Parasites are numerous among protists and most of the major plant groups. They are particularly common among fungi. As we noted in Chapter 42, several small animal phyla are wholly parasitic. The most noteworthy animal parasites are certain species of flatworms, nematodes, and arthropods, including some crustaceans, innumerable insects, and many arachnids (especially ticks and mites). Almost every phylum has at least a few partly parasitic species. True parasitism, however, is rare among vertebrates; indeed, except for a few partially or doubtfully parasitic fishes it is practically nonexistent.[5] Nevertheless, parasitism is one of the most widespread characteristics of all communities.

Parasitism leads to many peculiar adaptations. Internal parasites, for example, live in environments quite different from those of free and independent organisms. The peculiar conditions of life pose problems that are met by many strange adaptations. For one thing, the organisms must be comparatively small. Because the host unwittingly provides food, shelter, and a more or less stabilized environment, there is a tendency for parasites to evolve in the direction of simplification. This is the tendency usually labeled *degeneration*. A tapeworm in a mammal, for example, is bathed in predigested food that it need only to absorb through its body wall. It is also sheltered from the rigors of the outer world. Hence, tapeworms do not need or have special sensory organs, a developed brain, or a digestive system, although these occur in their free-living relatives.

Parasites live under conditions that require of them certain adaptations, so it should not be concluded that all evolutionary changes among parasites are degenerative. Again consider the tapeworm (Figure 47-12). At some point in its life cycle, it must be able to move from one host to another. Hence, it must be adapted both to its adult environment and to other rather different environments in the course of its life cycle. Accordingly, tapeworms, like other internal parasites, have evolved complex life cycles with a free-living stage, an aquatic embryonic or larval stage, one or more larval stages parasitic in intermediate hosts, and a final adult stage that settles down in a host of a different species. The final host usually acquires the parasite by eating the intermediate host. The commonest tapeworms in humans are two species that use domestic livestock as intermediate hosts; one occurs in cattle, the other in hogs. Humans acquire these tapeworms by eating undercooked

---

[5] Lampreys and hagfishes are often called parasitic, but this is a marginal case hardly distinguishable from ordinary predation. As mentioned earlier, a few fishes, such as the deep-sea angler, *Photocorynus*, have truly parasitic males, but the females are not parasitic.

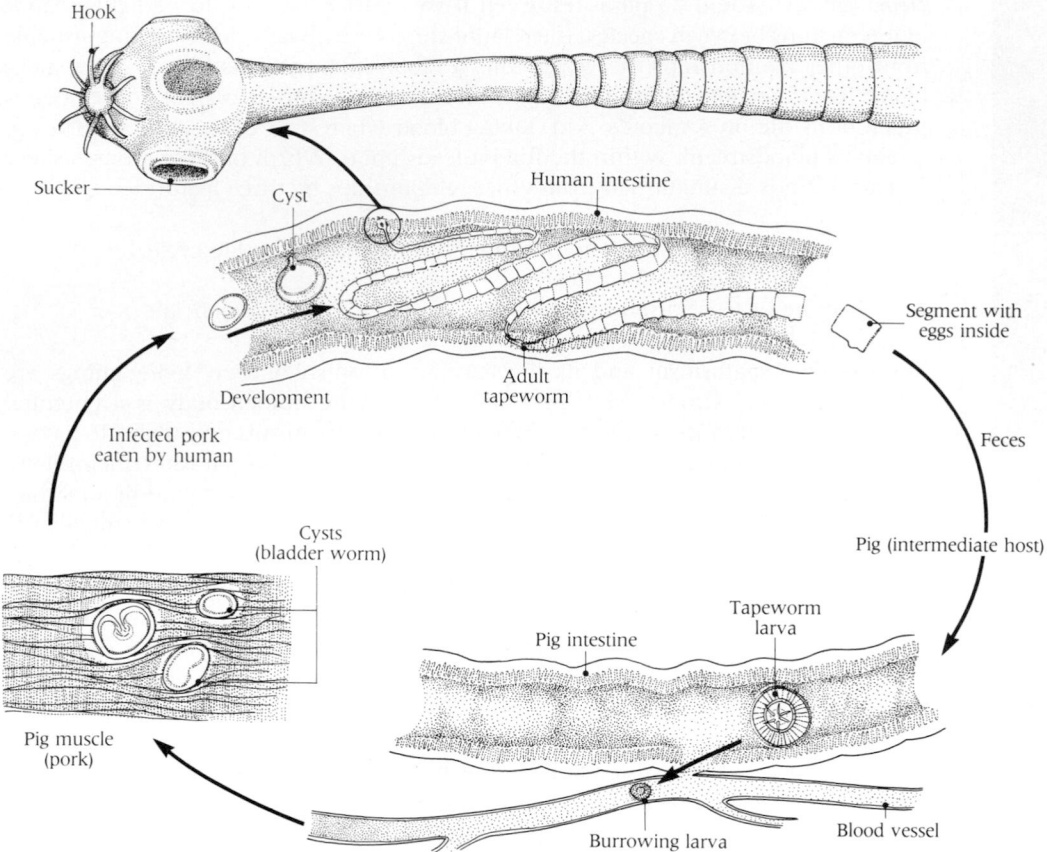

beef or pork. Another tapeworm in humans, a monster that may grow to lengths of over 60 feet, has two intermediate hosts and is acquired by humans from raw or undercooked fish.[6]

Other adaptations include various necessary mechanisms of subverting the defenses of the host. These include protective cuticles and lowered antigenicity that minimizes the formation of antiparasite antibodies. One of the most impressive adaptations to the parasitic way of life relates to its method of reproduction. Asexual reproduction—either as simple cell fission in protists or as budding in more complex parasites like tapeworms and flukes—is more prevalent in parasites than in any other animal group. Problems arise for the parasite if it is committed to sexual reproduction while in its host. Chances are poor enough that a larval parasite will encounter any suitable host; they are even poorer if the host is expected to harbor a potential mate.[7] Many crustaceans are parasites, and of these, the most successful are relatives of the common barnacle. They belong to a group (the Cirripedia) that is primitively sedentary and hermaphroditic.[8] In parasitic crustaceans, in which the sexes are separate, it is common for individuals of the opposite sex to join together permanently in their free-swimming larval stage. The female then becomes attached to a host, and the male—a midget—remains attached to the female near her genital opening.

Parasitism is a successful way of life, but in evolutionary terms it is a blind alley. The specializations and the degenerations that fit a parasite for life in its peculiar environment make it unfit for life in any other environment. They also tend to make improbable, if not downright impossible, further adaptation to anything

47-12 Parasitism by the tapeworm, Taenia solium
The adult tapeworm lives in the human intestine and obtains its nutrients by absorbing predigested food that had been ingested by its host. Segments of the tapeworm, rich in eggs, are released with the feces. The pig is the intermediate host in the life cycle, with the encysted larva found deep within the pig muscle. The cycle is completed when humans eat inadequately cooked pork containing living cysts.

[6] Having just learned that parasites must be comparatively small, a reader may be bemused to read that a 60 foot tapeworm can fit easily into a 6 foot human being. It definitely can—as one of the authors, having seen one, can attest. It does so by lying several times folded back on itself within the small intestine, which is itself 25 feet long.

[7] It is a striking fact that the only regularly hermaphroditic animal phylum—the Platyhelminthes—has been more successful than any other in the parasitic way of life.

[8] We should recall that the majority of plants are hermaphroditic (monoecious) and sedentary (see Chapter 41). One wonders if the early evolution of barnacles as sedentary animals may have had any bearing on their later evolution as parasites.

other than an even narrower parasitic niche. It is unlikely that any free-living organism has ever evolved from a parasitic ancestor.

Parasitism and Disease

When an organism invades the tissues of another organism, one of three things may happen.

- The invader may be killed off by one or another of the host's defense mechanisms—by far the most likely outcome.
- The parasite may survive but cause the host little harm.
- The parasite may survive and cause a disease in the host that ultimately may be fatal to both host and parasite.

Parasites cause the trouble we call *disease* in several ways. For one thing, they compete with the host for foods and other substances needed by the host's body. In some cases (e.g., tetanus, gas gangrene, botulism, etc.), parasitic microorganisms produce toxic substances that may be rapidly lethal to the host. Also, parasites are foreign biochemical systems with distinctive proteins and other substances which are foreign to the host and which stimulate immune reactions that may themselves harm the host.

These are the reasons parasites cause many diseases. We term such diseases *infectious* diseases. These are the diseases caused by various pathogenic parasites of microscopic size that are sometimes called "germs" or "microbes." Larger parasites also may cause disease.

Most infectious diseases are caused by viruses, bacteria, various parasitic protists, fungi, and small invertebrate animals.

- *Viruses,* whether or not we call them organisms, are foreign and parasitic to the host they infect. The damage they do to the host as they replicate in host cells is the ultimate cause of difficulty in a viral disease. Among the human diseases caused by viruses are AIDS (see Box 23-3), chicken pox, hepatitis, influenza, and poliomyelitis. In plants, they cause tobacco mosaic, leaf curl, and many other diseases.
- *Bacteria* cause most of the infectious diseases, among them meningitis, pneumonia, tuberculosis, diphtheria, bubonic plague, cholera, gonorrhea, syphilis, whooping cough, and brucellosis. The major bacteria that cause infectious disease in humans include the meningococci, streptococci, staphylococci, tubercle bacillus, gonococci, and others. Surprisingly, we are constantly encountering new infectious organisms. A 1976 outbreak of a serious lung infection that could not be attributed to any known organism led to the discovery of the infectious disorder named Legionnaire's disease—and its previously unrecognized causative agent, *Legionella pneumophila.* Certain bacteria (mycoplasmas) cause yellowing of many crop plants.
- *Rickettsias,* a group of parasitic organisms that are larger than viruses but smaller than all known bacteria (see Chapter 38), cause typhus, Rocky Mountain spotted fever, Q fever, and several other diseases.
- *Animallike protists* cause malaria, several forms of trypanosomiasis (see Chapter 39), including African sleeping sickness and amoebic dysentery.
- *Fungi* cause athlete's foot (trichophytosis), ringworm (which is not a worm), actinomycosis (a chronic infection resembling tuberculosis), and coccidioidomycosis (a serious lung infection that is also known as San Joaquin fever), and many other diseases (see Chapter 40). A fungal disease of rice plants led to the discovery of gibberellin (see Figure 28-48).
- *Invertebrates* are the larger of the parasites that produce human diseases. They include tapeworms, flukes (trematodes whose life cycle is shown in Figure 47-13), hookworms (nematodes), and trichinas or beef and pork worms (also nematodes).

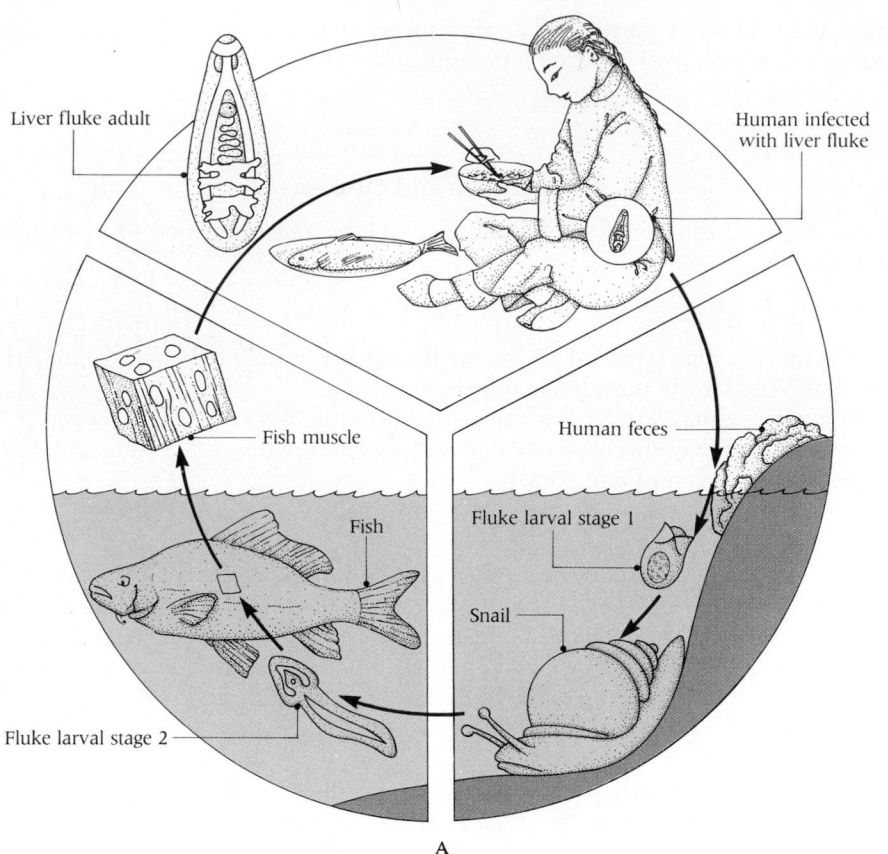

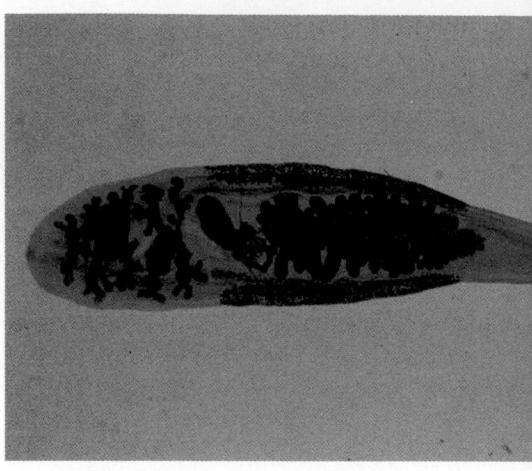

Liver fluke adult

Human infected
with liver fluke

Fish muscle

Human feces

Fish

Fluke larval stage 1

Snail

Fluke larval stage 2

A

B

47-13 The Chinese liver fluke, Clonorchis sinensis
(A) *The trematode liver flukes show some adaptations to parasitism that resemble those of the tapeworm. The life cycle requires three hosts (human, snail, and fish), and there is considerable asexual budding, which bolsters the total reproductive potential.* **(B)** *The adult stage of the liver fluke in the human.*

In the face of this long and melancholy catalog of infectious diseases, it may seem contradictory to say that the most successful, well-established parasites rarely cause serious diseases. However, such is the case and that fact illustrates an aspect of nature's dynamic balance. If a host dies rapidly after infection by a parasite, the parasite also dies. Therefore, when a host-parasite relationship continues for any long period, a balance is necessarily reached. The host species of a well-established parasite usually is not seriously damaged by the parasitism.

Most organisms, including humans, are hosts to innumerable parasites that do them no good but that also do them no great harm. Consider the case of bacterial parasitism. Healthy humans are infected with a variety of bacteria from infancy through old age. This is not surprising when one considers the ubiquitous distribution of bacteria in the environment. In fact, most of the air we breathe is heavily contaminated. Soon after birth, our body surfaces, external and internal, become colonized with bacteria. These we disseminate as we breathe, speak, cough, sweep the floor, sneeze, handle objects, and so on. The surprising fact is that we can live in peaceful coexistence with microorganisms capable of producing overt disease.

Two factors largely account for the development of this balance. When a species of parasite becomes adapted to a host it tends to evolve in such a way as to cause less serious disturbance; we say, then, that the parasite has become less *virulent*. At the same time, the host tends toward greater success in resisting disturbance; it develops *resistance* or *immunity*. Our tissues possess efficient mechanisms of antibacterial defense that restrict microbes to areas where they can be tolerated. When the defenses are penetrated and bacteria gain access to tissues not normally infected, disease usually results.

The distinction between infection and disease is of paramount importance to a physician. Confronted with the problem of establishing the cause of an illness, he or she must interpret laboratory tests with caution. The mere fact that a given microbe can be cultured from a patient's body may be totally irrelevant. For example, over 90% of throat cultures taken at random will be positive for a certain kind of streptococcus (viridans). A positive culture is therefore not indicative of disease. When, on the other hand, viridans streptococci are cultured from a patient's *blood*, which should normally be sterile, the diagnosis of subacute bacterial endocarditis (a heart valve infection) can be made with virtual certainty.[9] The presence in the throat of other strains of streptococci does signify disease.

Most bacterial species are incapable of penetrating the natural defenses of the host and are termed *nonpathogenic*. Others, which under the right circumstances are clearly *pathogenic*, often coexist with the host in a truce that is only occasionally broken. Still others are so highly pathogenic that they quickly overcome the natural defenses of the host and generally produce disease whenever they infect.

The varied effects of bacterial infection on the body economy of the normal host have been clarified by the study of axenic (germ-free) animals reared in a bacteria-free environment. These animals are unusually susceptible to a wide variety of infections. They show retarded development of antibody-forming lymphoid organs and are deficient in antibodies. Clearly, life-long bacterial infection influences the responsiveness of the animal host to its environment, including its susceptibility to bacterial diseases.

If host-parasite interactions do tend to become more or less balanced in time, why are there so many serious parasitic diseases? There are several hypotheses. One is that all species are capable of producing larger populations than their environments can support. Diseases that weed out the weakest individuals, especially among the very young and very old, without reducing the average population size and the vigor of the breeding adults, do not endanger survival of members of the species. They may even be beneficial to the species. In such a situation, natural selection does not tend to reduce the severity of the disease. Furthermore, parasites mutate, and mutations may produce newly virulent diseases that persist until the extinction of parasite, host, or both, or until establishment of a new balance. Parasites balanced with one host may spread to others with which they are not balanced and in which they then cause severe disease. Another explanation is that disease-producing organisms are selected for their own success.

PREDATOR-PREY INTERACTIONS

Plant-Herbivore Relationships

In the Earth's ecosystems, the interaction between plants and herbivores is a point at which one of the greatest transfers of energy occurs.[10] Some 2 million herbivore species the world over feed on some 300,000 plant species. Life on Earth is significantly influenced by the ways in which plants avoid being eaten—plant defenses—and the ways in which herbivorous animals evade those defenses and succeed in eating them anyway. We introduced this subject in Chapter 34.

The most obvious plant defenses against herbivores are morphological. Thorns, spines, and prickles definitely discourage browsers. Some plants, such as grasses, strengthen their leaves with deposits of silica. Plant hairs with glandular sticky tips play an important defensive role against insects.

Less obvious, but nonetheless very important defenses against herbivores are an incredible assortment of secondary chemical compounds. These are substances not involved in the plant's primary metabolic processes. Here are some examples.

- Members of the potato and tomato family (Solanaceae) are rich in alkaloids and steroids that can profoundly affect animal metabolism.
- Nicotine, an antiherbivore compound in tobacco leaves, powerfully affects the animal nervous system. To humans who smoke tobacco, it is an addicting drug.
- Digitoxin, a cardiac glycoside of foxglove (*Digitalis purpurea*), is an effective defense against herbivores (Figure 47-14). Purified digitoxin is used by physicians in the therapy of congestive heart failure because it increases the force of heart-muscle contraction without increasing oxygen consumption. However, an over-

47-14 Foxglove
This plant, Digitalis purpurea, *is the source of the cardiac stimulant called digitalis. Many plants contain similar cardiac glycosides, among them milkweed, whose toxic glycosides collect in the wings of Monarch butterflies (see Figure 35-13), and Oleander, whose compounds are lethally toxic to cattle and other animals that eat them.*

[9] Healthy persons infected with a given organism are often referred to as *carriers*. The percentage of individuals so infected in a population is known as the *carrier rate*. Not only does the existence of carriers often make it difficult for a physician to reach a definitive bacteriological diagnosis, but the occurrence of mixed infections, in which more than one potential disease-producer is recovered from a lesion, frequently requires that a distinction be made between primary and secondary invaders. The latter decision is usually based on a knowledge of the relative disease-producing potential of the microorganisms in question.

[10] The *greatest* transfer of energy takes place at the interface between organic matter and decomposers.

dose causes vomiting, visual disturbances, and cardiac disturbances that can be lethal if treatment is delayed.

- Caffeine is an ingredient of coffee beans, but its role in nature is not to fortify our morning coffee. Rather, it inhibits the larval maturation of various insect species. In other words, it is an insecticide.

Many alkaloids have been found in primitive angiosperms. That suggests that these compounds evolved early in the evolutionary history of the flowering plants and probably contributed to their success.

Recent studies show that angiosperms possess elaborate "medicine cabinets" that are richly stocked with chemicals capable of either killing many herbivores or preventing their reproduction or development. (One of them, hemlock, killed Socrates.) Consequently, most herbivores tend to avoid plants with chemical defenses. The presence of such chemical arsenals in plants has had major consequences on the evolution of both the plants themselves and herbivores.

Deployment of the Chemical Arsenal

Recent work revealed that several of the secondary compounds plants use in defense are not normally present, but are produced when needed. In a way, the system is analogous to the immune system of animals—and to the synthesis of antibacterial agents (antibiotics) by various fungi. For example, when a leaf is attacked by a herbivore, a chemical messenger can move rapidly through the plant, triggering the synthesis and accumulation of defensive chemicals that interfere with digestion in the herbivore's gut. Likewise, an infection of a plant by a fungus or bacterium may induce the plant to synthesize and deploy chemicals that slow down the spread of the fungus disease.

In this way, plants can conserve energy by delaying the synthesis of these chemicals until they are actually needed. Studies by David Rhoades at the University of Washington showed that trees such as willows and alders may communicate with neighboring trees. When willows are attacked by caterpillars, their leaves produce chemicals that reduce their nutritive value to the caterpillars. Surprisingly, the leaves of nearby unattacked trees also become less nutritious within four days. How the trees communicate is under investigation.

The Chemical Arsenal of Animals

Animals also produce a variety of defensive chemicals. Members of the frog family, Dendrobatidae, for example, produce *batrachotoxins* with the mucus covering their skin—toxins so potent that just a few micrograms injected into the bloodstream would kill a human. Several snakes, bees, wasps, scorpions, and spiders produce venoms and other chemicals not only for defense but for hunting prey.

Interestingly, some animals acquire their arsenal of defensive chemicals from the plants or animals they eat. Caterpillars of monarch butterflies feed on plants of the milkweed family, Asclepiadaceae, which contain cardiac glycosides that drastically affect heart action in vertebrates. However, monarch butterfly caterpillars do not break down these glycosides. As we saw in Chapter 35, they store them in fat bodies and pass them on to the adult stages and even to the eggs. As a result, all stages of the monarch butterfly are protected from predators. Any bird that eats a monarch butterfly containing cardiac glycosides becomes ill and quickly regurgitates its stomach contents (see Figure 35-13B,C). Many insects containing poisonous secondary compounds acquired from plants clearly advertise their poisonous nature by their unusually bright warning or aposematic coloration.

Many marine animals possess an enormous array of chemical compounds that hold promise as potential new drugs. Some marine invertebrates such as the sea slugs (nudibranchs) acquire their defensive weaponry from their prey. *Aplysia* grazes selectively on red algae, which contains *elatol*, an inhibitor of cell division. As a result, fish do not feed on *Aplysia*. Other sea creatures acquire stinging cells from the hydroids they eat.

Other Antipredator Defense Strategies

We already learned of the following strategies which animals use in self-defense.

- Electric shock, stinging cells, and corrosive chemicals
- Aposematic coloration and mimicry
- Various polymorphisms such as the banding, coloration, and whorl pattern in the shells of the snail, *Cepaea nemoralis,* which lead to apostatic selection
- The use of alarm calls at moments of potential danger.

Another defense strategy is **mobbing,** which has been defined as the ''hustling'' of predators by potential prey. It is seen among both vertebrates and invertebrates, but is most visibly manifested by birds and mammals. Mobbing may involve direct attacks upon the predator. Often the flurry and noise of such an encounter warns nearby members of the prey species of the predator's presence.

COMMUNITY PATTERNS

A community, we learned, is an assemblage of populations of one or more species that are living and interacting in a given ecosystem. The ecosystem is indeed a web linking all life and its external medium—air, water, and soil—into a common whole (Figure 47-15).

There is a common thread of interdependency between species that goes far beyond the obvious interactions between one population and another. The very existence of one population is often due to the fact that other populations exist to make available energy and nutrients, regulate population size, recycle wastes, and protect the community against the vagaries of the physical environment. Populations exist, in fact, because they are parts of communities.

In the balance of this chapter, we will discuss three aspects of the ecology of communities: changes in communities, the concept of climax, and the structure of communities.

CHANGES IN COMMUNITIES

Communities are dynamic systems that live in dynamic environments, changing ceaselessly from hour to hour, season to season, year to year, and epoch to epoch. As we have seen, most organisms have ''internal clocks'' that are adjusted to rhythms of the environment. The temperate zone has its dramatic cycle of the seasons and its familiar effects on the activities and compositions of local communities. Following each annual cycle, communities usually do not return to exactly the same conditions. As a result, variations accumulate as the years pass and there is gradual change.

The gardener finds his old enemy, the gypsy moth caterpillar (*Lymantria dispar*), more abundant some years than others. The New Englander sees the pastures of childhood overgrown with shrubs or merging into surrounding woods or paved with asphalt. The old swimming hole in the Midwest becomes a marsh or a prairie. The sod of the high plains, turned by courageous but injudicious pioneers, gives way to tumbleweeds and to desolate dust bowls, which may yet, with renewed rain and rational treatment, recover their verdure. In the Southwest, ruins of prehistoric Indian dwellings show that people once lived in wooded, fertile, watered valleys where desert now exists. Students of Earth's history find still greater changes. Turtles, alligators, and fishes once thrived where dry sagebrush flats of New Mexico now stand, and sands of barren dunes lie buried beneath fertile fields in the Mississippi Valley.

Within communities, the interactions of different species are changing and dynamic. A major question arises: Why are certain species found together in close associations within a community? Two theories have been proposed to answer this question.

One theory, first proposed by F. E. Clements at the turn of the century, views a community as a discrete superorganism undergoing development, in which the component species are integrated by interactions that make the community function as a single unit. This theory rests on the observation that certain species consistently live together in close association. Clements' theory would predict that plant species distributed along a gradient of environmental variables (e.g., temperature) should be clustered into groups, with relatively discrete boundaries, or disjunctions, between the communities.

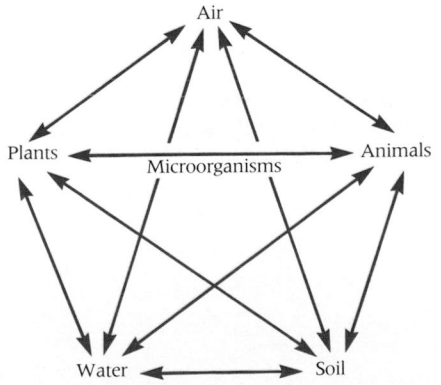

47-15 Interrelationships of biological and physical components in an ecosystem
Note the multiplicity of interactions

Another theory, proposed by H. S. Gleason in 1929, takes the opposite view and considers a community as a chance assemblage of species that occur in the same area because they happen to have similar environmental requirements. According to this individualistic theory, discrete boundaries would not exist between communities. Each species would have an independent distribution along an environmental gradient according to its tolerance range for the prevailing abiotic factors. According to Gleason, then, communities are loose associations without discrete boundaries.

In the real world, the composition of communities often changes spatially on a continuum, with each species independently distributed. This seems to support Gleason's concept of communities as relatively loose associations. However, there are cases where neighboring communities have discrete boundaries. Such sharp delineations are usually correlated with abrupt changes in important abiotic factors.

The Concept of Succession

Anyone old enough to read this book has surely seen a community change in character, perhaps with breathtaking speed where there was human intervention, more slowly perhaps without it. A striking feature of such change is that it is not so much a modification of a single community as a succession of more or less different communities in one area.

Succession, one of the most fascinating of natural phenomena and one of the oldest concepts in ecology, is the sequential replacement of one community by another that is better suited to that environment. In other words, it is a directional change in communities with time. It occurs whenever a new habitat is created, as by a geological process or some other disturbance (volcanic eruption, landslide, and so on). Pioneer species invade the the new habitat and then are gradually replaced by a sequence of other species until a relatively stable community is established.

- Primary succession is said to occur when previously lifeless or unoccupied terrain, such as a sand dune, a barren rock, or a new island risen from the sea, is gradually occupied by vegetation and other life.
- Secondary succession occurs in areas where previously existing plant communities have been disturbed or destroyed, as by fire or plowing.

Let us examine some examples of biotic succession.

Primary Succession

A classic example of primary succession—the changes in sand dunes at the southern end of Lake Michigan—was provided in the 1899 doctoral thesis of H. C. Cowles, a pioneer in ecology. These dry sand dunes are constantly shifting in the wind and therefore are unstable environments for life. Only a few species—mainly grasses—are able to colonize the bare, shifting dunes (Figure 47-16). But once populations of beach grasses are established, they act to stabilize the sand, allowing entry into the community of plants adapted to the low water levels but intolerant of the shifting sands. As individuals of these species die, some of their organic matter is incorporated into the sand as humus. This has two effects. It tends to bind particles, further stabilizing the dune, and it tends to bind water. As more water is retained in the upper layers of the dune, the soil becomes moister and new species of shrubs, including cottonwoods, can enter the community. The matted

47-16 Plant succession
Plant succession on the dunes at the southern end of Lake Michigan.

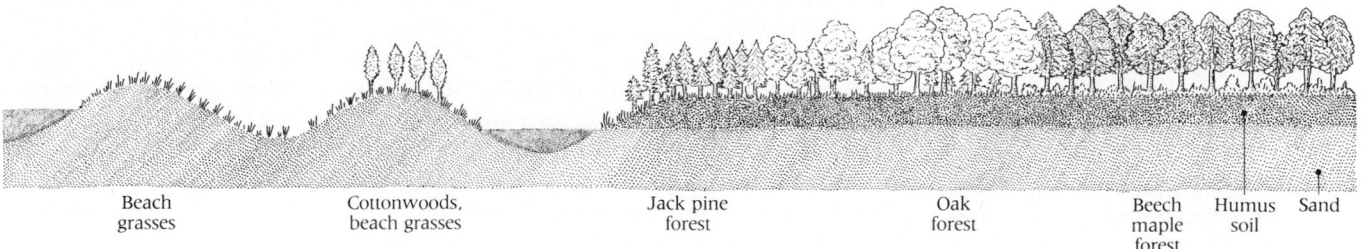

Beach grasses

Cottonwoods, beach grasses

Jack pine forest

Oak forest

Beech maple forest

Humus soil

Sand

roots of these plants further stabilize the dunes. Then come other shrubs, including juniper and jack pine. As productivity rises, so does the biomass of the plants and animals, as well as the amount of detritus (decayed organic matter) added to the sand. Detritus level, overall production, community biomass, and community diversity spiral upward until the dune is covered by a forest of oaks and blueberries, and the soil is brown humic sand rather than clean white sand. Eventually, water ceases to be a limiting factor, and the availability of nutrients gains importance—but as humus builds up, so do the nutrient levels. Finally, the soil is rich enough to support a forest of beech and sugar maple like those typical of this area.

The chief cause of change in the abiotic environment is the community. The stabilization of the dunes, the addition of humic material to the soil, and similar modifications of the abiotic environment are controlled by the organisms themselves. Thus the "driving force" is internal to the community. This is termed the **autogenic** aspect of succession. Factors that control succession from without constitute the **allogenic** aspect of succession.

Secondary Succession

When a forest is devastated by fire, it may become little more than a plot of bare ground. But soon secondary succession takes place. Some plants take root and grow and a simple community is established. Gradually, other species may enter the community and displace the original species.

Community succession occurs wherever change takes place. The change may be in climate; but succession also occurs without change in climate. Probably the most important changes are in the physical environment and very often these changes are induced by the community itself.

A coral reef grows toward the surface, its rich community changing as the species of the shallower water immigrate and become more numerous. Waves grind coral

47-17 Succession and the eventual obliteration of lakes
Initially a lake is surrounded by a beech and maple forest with a few conifers near the water's edge. Humus accumulates from lake plants and forms a marginal marsh. As the lake fills in, the surrounding forest takes over entirely.

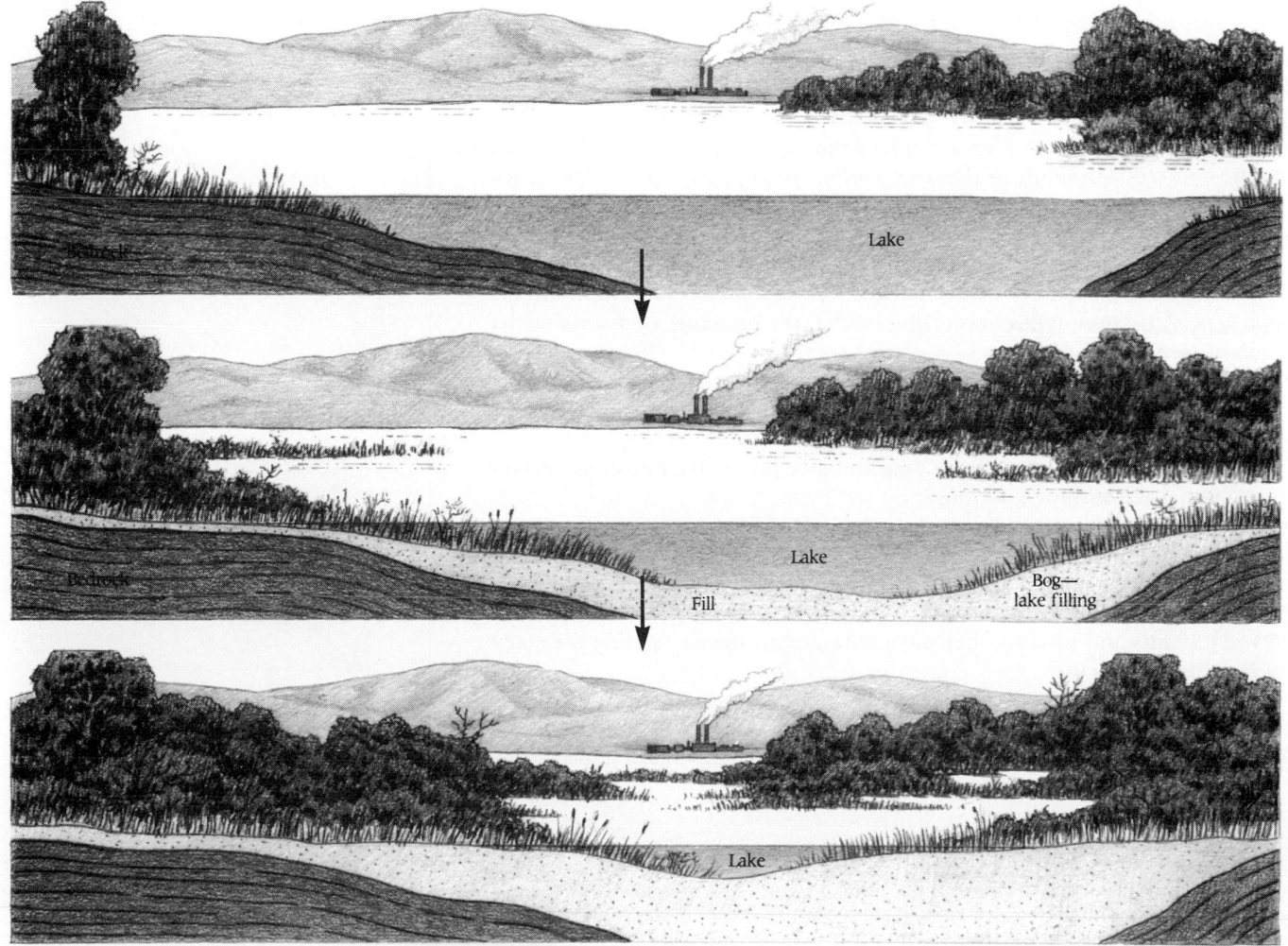

rock to sand; sand piles up in shifting dunes just above the tides. Vines take root, bind the sand in place, and contribute to it the humus of their dead tissues. Low trees, resistant to winds and spray, grow in the accumulating soil, to which they contribute in turn. Finally a copse of tall trees develops in the richer soil, and a more protected land environment gradually forms. Or perhaps the climate becomes progressively drier in a region where a pine forest flourishes. The pines die, starved for water, and scrubby junipers grow in their place. Still drier conditions see the junipers replaced by sagebrush and greasewood. Finally, a cactus and thornbush desert may emerge where a forest once stood. Very often successional communities alter the area in which they live in ways that make it less satisfactory for themselves and more favorable for others. In effect, a community may ruin things for itself.

Another instructive example is provided by a lake in the northeastern United States (Figure 47-17). The open waters of the lake have a community that includes protists, algae, small aquatic animals, and fishes, among other organisms. In shallower water near shore, are pond lilies, cattails, and other comparatively large plants that grow, so to speak, with their feet in the water and their heads in the air. Silt and soil wash in and are piled up by waves, and dead plant matter accumulates. As a result, the lake becomes shallow with marshy margins. The former shoreward rim of the lake becomes a damp bog in which grasses, herbs, and willows take root. The filling and the succession of new communities progress toward the center of the lake until finally open water and its aquatic community disappear. Meanwhile, the marshy marginal soil continues to build up as humus and silt accumulate. Eventually oaks sprout there, and finally the whole site of the vanished lake is occupied by a beech and maple forest, which follows upon the oaks.[11]

In all these examples, plants are emphasized because they are usually the clearest indicators of the situation *and* its biological keys. In each case, however, the association of plants is accompanied by a characteristic association of animals. The two together, along with all other organisms, make up the community.

Relationships in Succession

Ecological succession is an important subject about which ecologists have formulated the following generalizations (Figure 47-18):

- The number of species increases at first, but then stabilizes in time as the ecosystem matures. Succession tends to progress from large populations of a few species to smaller populations of many species.
- In general, pioneering species tend to be *r* strategists. Species associated with later successional habitats (termed equilibrium species to contrast them with pioneering species) tend to be *K* strategists. In fact, there is a continuum of strategies with "pure types" occurring only at the extremes of a successional gradient (see Figure 45-18).
- The total biomass in the ecosystem and the total amount of nonliving organic matter increases during succession until equilibrium is reached and the rate of net growth slows.
- Food webs increase in complexity and the interspecific relationships become more specialized as time goes on.
- The amount of new organic matter synthesized by producers (autotrophs) is high at the beginning of primary succession. Then it decreases to a stable, usually high, level but the percentage utilized at each trophic level rises.
- As succession progresses, interactions among the various species (for example, between plants and insects) alter the environment, thereby setting the stage for later developments.

If correct, these principles suggest that the tendencies of succession are toward more complex ecosystems—more species, larger food webs, increasing specialization,

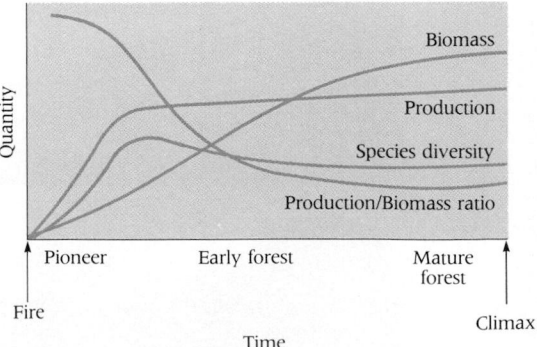

47-18 Evolution of community characteristics during succession
Following a fire in a generalized terrestrial environment, the various factors that define a community change in predictable ways over time until a common endpoint (climax) is reached.

---

[11] Similar successions of lake (or pond) to bog to forest can often be followed over periods of thousands, even tens of thousands, of years. Changes are recorded in layers of silt, peat, and humus that fill the lake basin and in the different kinds of pollen deposited in the various layers.

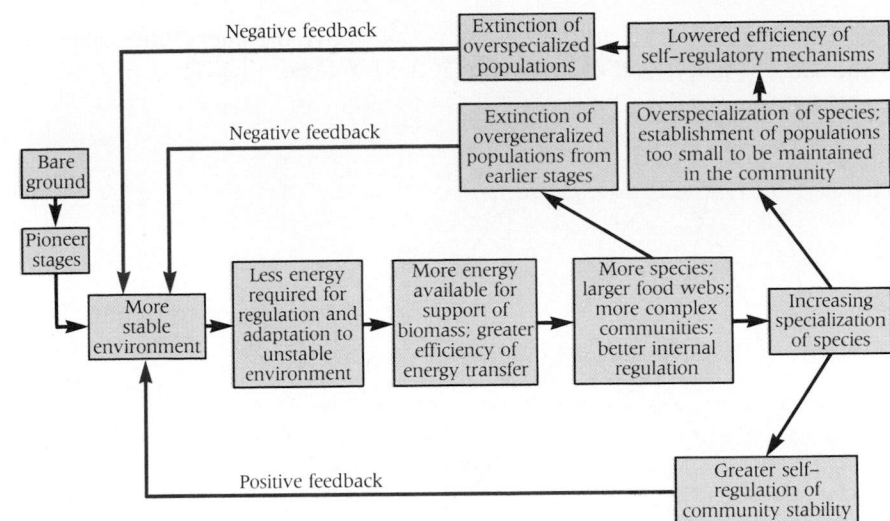

47-19 The progressive changes in communities during succession

A series of feedback loops defines the relationships among the changes occurring during community succession. During early stages, the positive feedback loop is dominant, leading to self-sustained increases in diversity and stability. In later stages, the introduction of more specialized species via the lower loop is counteracted by the loss of overgeneralized or overspecialized populations by the negative-feedback loops. The climax community is maintained by a dynamic equilibrium between the positive and negative feedback loops.

etc.—in which there is better internal regulation and sometimes community stability (Figure 47-19). Less energy is wasted in stable communities and larger biomasses are possible.

THE CONCEPT OF CLIMAX

Convergence and Climax

Let us return to the northeastern United States and view another example of primary succession: What happens to a bare rock near the same lake where we followed a community succession? The first living things to get a foothold on the rocks are small, scaly lichens, which grow during the brief periods when the rock is wet and lie dormant when it is dry. The lichens release acids and hasten the mechanical and chemical disintegration of the rock surface. Dust particles settle and bits of humus begin to develop in tiny crevices. More luxuriant lichens arise, and then come thickening mats of mosses and ferns. Soil develops along with increasing vegetation, and soon perennial shrubs spring up. As more and more plants survive and grow, the soil layer becomes thicker. Trees begin to grow, perhaps pines at first while the soil is still rather rocky and comparatively poor in organic matter. Oaks follow, and in their shade appear seedlings of beech and maple, which grow up to overshade the early trees. Finally a beech and maple forest flourishes and joins the beech and maple forest that grew where the lake had been earlier.

Different beginnings and different community successions evidently led to the same result: a beech and maple forest community. We say that the two successions

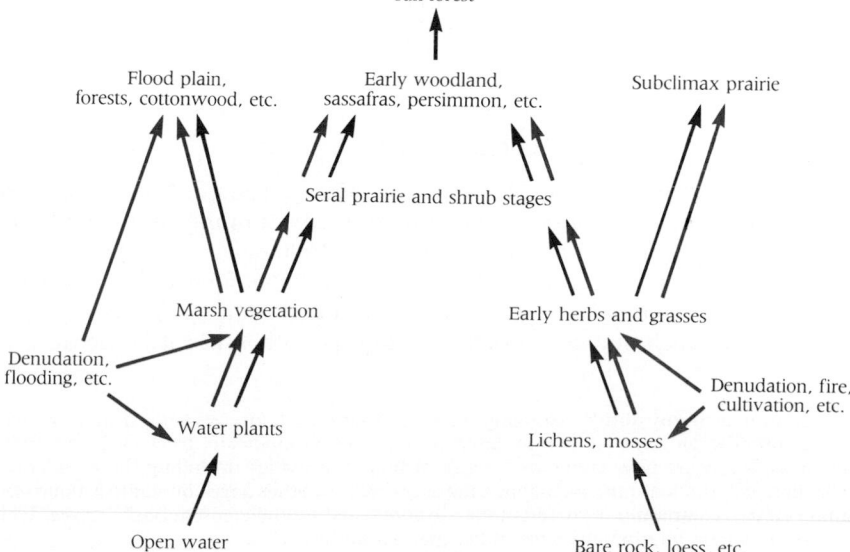

47-20 Succession from dissimilar starting points

The diagram traces successional pathways beginning at different points and moving toward a climax community in central Missouri. Convergence does not always occur, since one or more of the paths may stop short of the climax. Both primary (solid arrows) and secondary (broken arrows) paths are included.

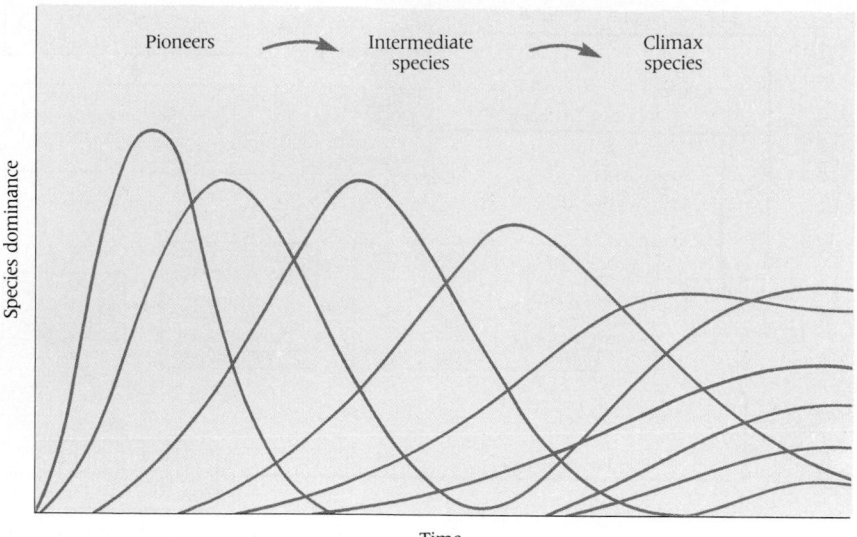

47-21 Successional development over time
As time passes, a succession of species inhabits a given area, starting with the pioneers and ending with the climax community.

converged and gave rise to the same kind of stable community—or perhaps to parts of the same community. This concept is depicted in Figure 47-20.

Such a forest is the usual culmination or **climax** of community succession in this region. Once such communities are established, they are self-perpetuating and there is little tendency for further progressive change in the community—that is, as long as the ecological conditions prevail. Seedlings of beech and maple are adapted to shade and do not grow in open sunlight. A mature beech-maple forest is so densely shaded that young oaks or pines cannot grow in it. An equilibrium state develops between the biotic community and the physical environment, and the forest community persists until some further geomorphic or climatic change occurs—or human intervention initiates a new community succession.

The progression of species in a generalized successional development is shown in Figure 47-21. This orderly sequence of communities, from the first or pioneer community to the terminal or climax community, is called a **sere.** The time required for the various seral stages and for their completion varies with climate and locality, but many years or even centuries may be needed. Successional processes occur most rapidly in warm, moist climates. Yet even in the tropics, 400 to 1000 years may be required for the completion of a sere that terminates in a mature rain forest.

Most regions have a type of climax community in which all community successions of the region usually end. Of course, the climax is not always a beech and maple forest (although that is a common one in the American northeast) or even a forest of any kind. On many mountains, the climax is a spruce and fir forest. In many more arid regions, mesquite communities are the climax, although in some places mesquite is dying out.[12] The climax over most of the high plains east of the Rockies is grassy prairie. The nature of the climax is ultimately determined more by climate than by any other factor.

Two Theories of Biotic Stability

Several basic concepts of biotic stability have been advanced. The oldest is the **monoclimax theory,** the principal proponent of which in the early years of this century was F. E. Clements, who published nearly 1000 papers describing different regions over a span of two centuries. This theory states that for each climatic region there is only one type of stable climax vegetation in which a sere could terminate. This implies that each successive plant assemblage establishes itself because the

---

[12] Mesquite is a thorny shrub belonging to the pea family. It survives only if its roots reach permanent groundwater. This limits it to desert regions with groundwater levels near the surface. In some areas, levels are now falling as a result of human activities (including the popularity of mesquite charcoal) and mesquite is disappearing as a result. In other areas, mesquite is flourishing. A common climax community in deserts today is dominated by the creosote bush, *Larrea*, which manages to survive on intermittent rains rather than groundwater.

preceding one has somehow modified the site, making it unsuitable for that assemblage but more suitable for the next. The sequence ends when the process stops and a climax community is established.

The oak-hickory forests of the Midwest would constitute such a climax community. However, certain prairie communities are known to persist under identical temperature and rainfall conditions. Supporters of the monoclimax theory call such prairie communities a **subclimax.** These are considered examples of arrested succession, in which development is terminated at a stage just short of the theoretical climax. They demonstrate ecological stability based on such local physical factors as soils and drainage.

A second basic concept of biotic stability is the **polyclimax theory.** This view regards all biotic communities displaying stability and functional permanence as climaxes. Thus the poorly drained soils largely responsible for the persistence of prairies in a forest climate provide the basis for a soil, or **edaphic,** climax. Other factors, including fire and grazing, imparting stability to a given community type are often relevant. The pine woods of the southeastern United States, for example, are considered in some cases to be fire-climax communities, since succession results in a forest of oaks, hickories, and other deciduous trees in cases where burning does not occur.

Climax and Change

If the communities in each region tend in time to converge toward the same type of climax, and if this has been going on for millions of years, how and why do communities undergo any further changes? Why are not all communities set and static at their climaxes? Two divergent views have emerged.

It is true that a climax community tends to persist, but it does so only as long as no internal or external disturbance affects the community and there is no essential geomorphic or climatic change in its environment. However, such disturbances and changes occur frequently. They include natural variations in the physical environment, broad-scale topographical variations, and all sorts of local heterogeneities, which, in the long run, are sure to happen. Indeed, change is so continuous and

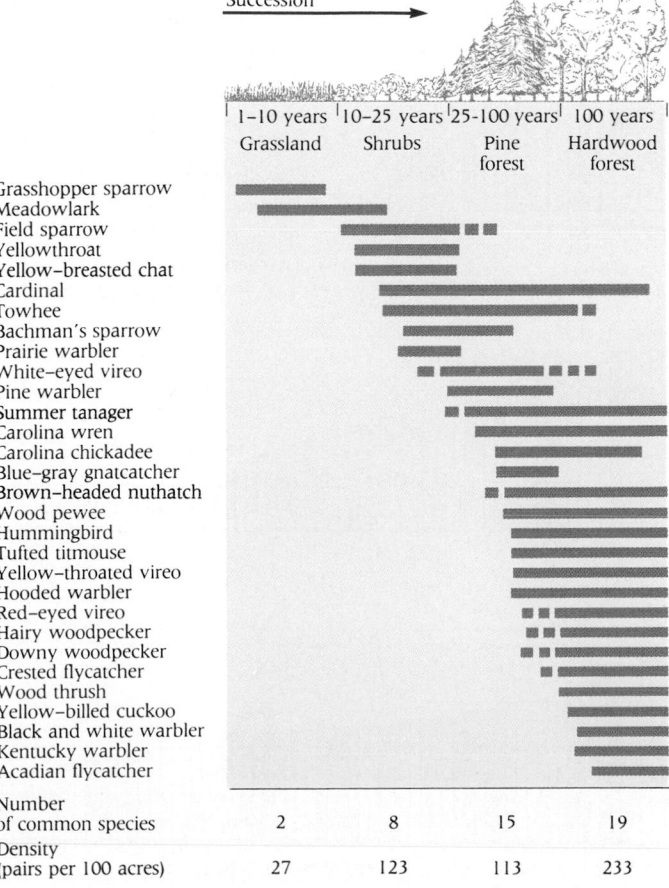

47-22 Succession of birds and plant life
The succession of bird species accompanies the progression from grassland to oak and hickory forest in Georgia. The climax forest harbors not only the greatest diversity of species, but also the greatest density of organisms.

persistent that a stable climax may be hard to identify—and, in fact, may not exist.

The notion that succession produces a predictable climax is rooted in the observation that species diversity reaches an equilibrium in the mature community. As a community matures, species interactions such as competition, symbiosis, and predation, become more extensive and varied, making increased diversity possible. According to the **equilibrium model,** succession reaches a climax when no additional species can be accommodated in the community unless niches become available as a result of some localized extinctions.

An opposing view, the **nonequilibrium model,** has also been offered. It suggests that most communities are in a continual nonequilibrium state, with the kinds and numbers of species changing during all stages of succession. The course of succession will depend on which species happen to colonize the area in the early stages. Disturbances such as fires and hurricanes affect species diversity of the climax stages. When disturbances are frequent and severe, the numbers of the community will be dominated by effective colonizing species, typical of early successional stages. If disturbances are rare and mild, then the most competitive species typical of late successional stages will be dominant. It is when disturbances are intermediate in nature and frequency that species diversity is greatest, because organisms of all the different successional stages will be represented.

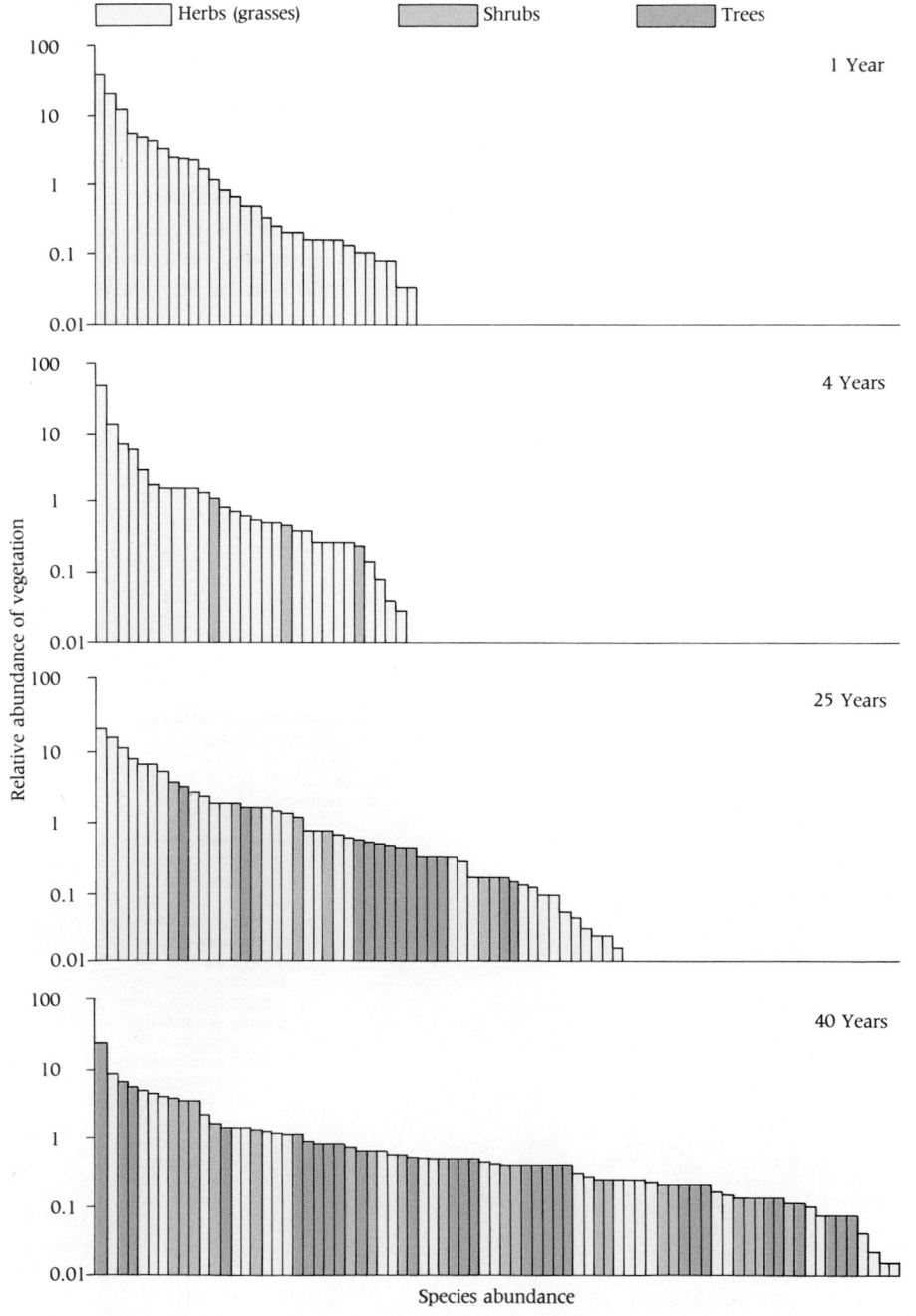

47-23 Plant succession over a forty-year period

The relative abundance of herbs (grasses), shrubs, and trees was measured in an abandoned field in southern Illinois over a forty-year period. Each bar represents a different species. Although in the first year, only herbs were present, by the fortieth year, herbs, shrubs, and trees were present in nearly equal abundance.

As shown in the real-life example in Figure 47-22, a striking succession of bird species accompanied the progression from the grassland of abandoned farmland in Georgia to an oak and hickory forest. The climax forest contains not only the greatest diversity of species but the largest number of individual organisms. In another example (Figure 47-23), the number of species of each of three vegetation types (herbs or grasses, shrubs, and trees) was studied over a forty-year period in an abandoned field in southern Illinois. Note that one year after abandonment the field contained only herbs. After only four years some shrubs appeared, and by twenty-five and forty years later there were trees present in increasing abundance. By the fortieth year each type of vegetation is present in more nearly equal quantities than at the beginning.

In concluding the subject of climax communities, we note that these generalizations have their limits and exceptions. Geomorphic changes, such as the filling in of lakes, do not always tend toward climax communities. New lakes are formed (otherwise all would be gone by now), and succession starts again in turn. Bare sands, clays, and harder rocks on which succession may have started can be laid bare by erosion, so that the process must begin again from the start. Not so long ago (10,000 years or so), almost the whole northern half of our continent was buried in ice, and new successions have occurred since then, not only in the glaciated areas but in most others as the climates changed.

Disturbances also occur within communities. Forest and prairie fires wipe out whole communities.[13] Population changes in climax or other well-established communities tend to fluctuate around an average, but they do not always do so. Great increases in the population of any one species may actually change the environment and the community as a whole, as will be shown in the next section. Decreases may go on to extinction, with a consequent long or permanent change in the composition of the community. New species frequently enter a community from elsewhere. Inevitably they change the interactions in the community, and they may profoundly change its nature.

Moreover, communities evolve not only by the extinction of species in them and the incursions of new species, but also by evolution within their populations. The species of which a community is composed are not static units. Over the generations each one of them changes, and so necessarily does the community of which it is a part.

THE STRUCTURE OF COMMUNITIES

Patterns in Space

All facets of communities vary in space. Most spatial community patterns are determined by abiotic environmental factors—temperature, rainfall, or degree of salinity—and they may be widespread or localized. In considering the spatial structure of a community it is useful to identify the factors that determine both its *vertical* and its *horizontal* structure.

The open waters of lakes and oceans contain loosely defined community structures. Vertically, they depend on the penetration of light and heat to different depths and on proximity to shore—as well as on oxygen, nutrients, and minerals. Horizontally, they depend on the influences of winds and currents. In terrestrial communities, by contrast, spatial structure is profoundly influenced by the organisms themselves—especially by plants. The immobility of plants gives the community a relatively permanent horizontal structure. High trees add a vertical dimension. Unlike many aquatic communities, terrestrial communities usually modify such aspects of their physical environments as the distribution of light and moisture, the formation of soil, and the cycling of nutrients.

Vertical structure is a prominent feature of forests. Canopy trees in temperate forests intercept more than half of the sunlight reaching the community; they also

---

[13] There is abundant evidence that natural fires antedated humans. For example, buried layers of charcoal testify to prehistoric fires. Tree trunks also contain records of past fires. A moderately intense fire often scars one side of a tree. If the tree survives, it may be possible to recognize the scar in the freshly cut stump and count off the years (rings) since the fire. From such evidence we know that there were about four fires a century during the 1100-year history of a stand of California redwoods. Nowadays humans are responsible for many of the fires in the world's forests and grasslands. In some cases, foresters have learned to make judicious use of fire as a tool in forest management.

change its spectral character. Below the canopy there are usually many smaller trees, including immature canopy trees and mature trees of other species. Less than 10% of the incoming sunlight may reach the next level down, the shrubs. Beneath them there is usually a layer of ground vegetation. Vertical structure continues down into the soil. The roots of different plants extend to different depths, and the soil itself is made up of various layers, or horizons, from the leaf litter on down. Here we have gradients, the most important characteristics of which are the availability, not of light, but of nutrients and water.

Animals, like plants, live a more or less vertically stratified existence within communities. Caterpillars are mostly found on leaves above the ground, ground beetles are in the detritus at the soil surface; and earthworms are at various depths below the ground. Some warbler species hunt for insects primarily in the canopy layer of the forest; others spend their time in the intermediate layers, and yet others, such as the ovenbird (*Seiurus aurocpillus*), are usually found hunting around on the ground. The horizontal structure of many communities is marked by patchiness—that is, organisms are not evenly distributed but are often clustered together in groupings of high density, separated by spaces of low density.

Patterns in Time

Community structure changes in time on several levels. The individual organisms of the community grow, interact, and die. These changes are influenced in the short term by the rhythms of the Earth's relationship with the sun, which causes daily and seasonal cycles, and with the moon, which influences tide cycles. As an example of a diurnal rhythm, different predators hunt the flying insects in a forest canopy at different times of the day: flycatchers and swallows by day, nighthawks at dusk, and bats at night.

Seasonal effects are especially prominent in communities exposed to long, severe winters or dry seasons. Most birds leave northern communities in winter, migrating south to join other communities and returning the following spring. Many mammals hibernate, and insects enter diapause (a resting state in which metabolism is subdued). Annual plants spend the winter as seeds, while deciduous shrubs and trees shed

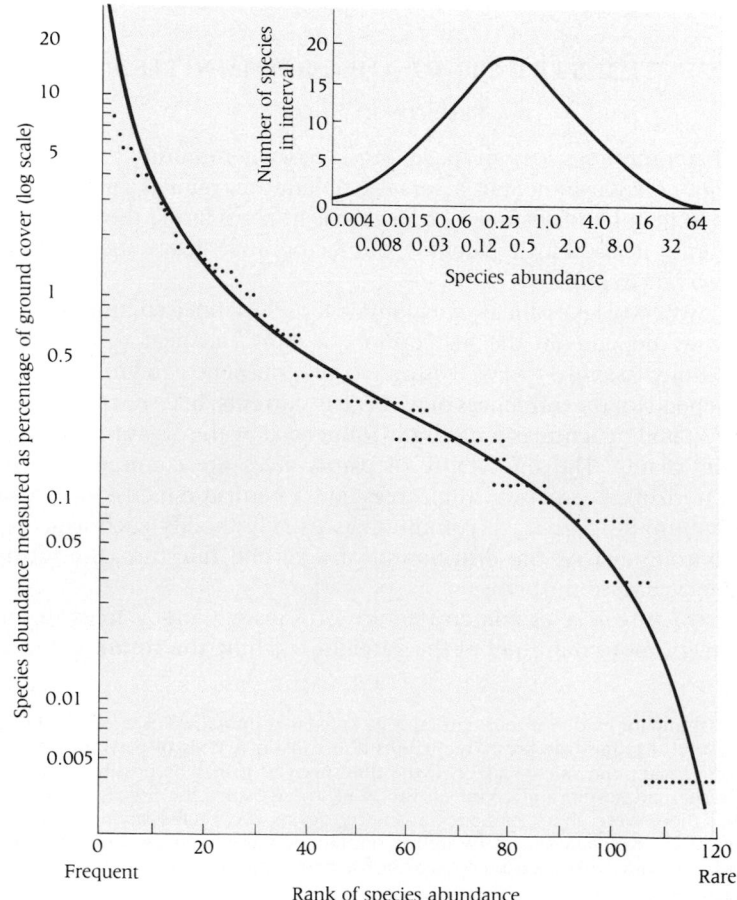

47-24 Species abundance of plants in Arizona
The abundance of plant species in the Santa Catalina Mountains of Arizona is measured in percentage of ground covered by each species. The insert shows the bell-shaped curve of species abundance. The semi-log plot reveals that a few species are frequent and many more are rare.

their leaves. Decomposer activities slow down or cease. The activity of the community is restricted to evergreen plants, which can continue photosynthesis, and foraging by a few hardy endotherms such as chickadees, woodpeckers, mice, and deer.

Species Abundance and Community Diversity

We spoke of the diversity of species in the course of succession. Some communities, such as a tropical rain forest or a coral reef, may contain thousands of populations of different species; others, such as the tundra or desert, may contain only a handful. The number of species in a community is termed its *richness*. It is not a particularly satisfying concept by itself, because it may be difficult to compare communities on the basis of their richness. For example, a community spread over a large area would be expected to be richer than a geographically restricted community, if for no other reason that the probability of finding a rare species in a large area is very low. Also, it is clear that two communities, each containing thirty species of organisms, are quite different in structure if each population of each species in the first accounts for 2–5% of the total, whereas two or three species in the second account for 95% of all the individuals in the community.

For this reason, ecologists use more complex measures of diversity that take into account, not only the number of species present, but also their relative abundance, or some other measure of their participation in the community (such as relative biomass or productivity). These measures of diversity usually reveal a steeply declining line or curve (Figure 47-24). There are few dominant species that are especially abundant or obvious, and there are a great many species that are rare or inconspicuous.

Island Biogeography

One of the methodological problems hampering ecologists is the difficulty—or near-impossibility—of manipulating ecosystems experimentally. Ecologists have thus been forced to examine "natural" settings as they are and more or less take what they can get in the way of meaningful and controlled observations. That is why ecologists have long been attracted to islands, especially when different islands in an archipelago turn out to have differing combinations of living organisms.

For ecological study, islands have many desirable features. They generally support fewer prey and predator species than comparable mainland habitats—and we often know when an island archipelago materialized from volcanic activity. Hence, it is regularly possible to date the beginning of an island's ecological history and to know with reasonable certainty that the history began with a blank slate—that is, with an uninhabited island. That is why classic studies such as those of Darwin's finches or the Hawaiian honeycreepers (see Chapter 36) were so consequential. They could only have been done in island settings.

Island studies have also contributed major insights to the topics of this chapter—the abundance and diversity of species in communities. For many years, islands were thought in some sense to be "impoverished" of species—perhaps because species must cross water barriers in colonizing them. However, when R. MacArthur and E. O. Wilson examined this situation in the 1960s, it appeared to them that islands might in fact be supporting as many species as possible.

MacArthur and Wilson assumed that species richness reflects a dynamic balance between immigration (the number of new species arriving) and extinction. When immigration and extinction are equal, they reasoned, species diversity should reach a state of equilibrium. Consider their argument. Each initially uninhabited island begins to receive colonists from the mainland. The mainland species that are potential colonists of the island are known collectively as the **species pool.** The initial arriving colonists represent new species on the island. As the number of colonizing species increases, a growing fraction of immigrants includes species already present. Thus, even if the same number of species continues to arrive as before, the rate of arrival of *new* species will decrease—until it reaches zero when the island contains all the species in the pool.

At the same time, extinction rates increase because such interspecific relationships as predation, competition, and parasitism will have a negative impact on populations. When the increasing extinction rate matches the decreasing immigration rate, an equilibrium is established (Figure 47-25). The equilibrium is of course dynamic, since continuing immigration and extinction may alter the species mix over time.

The theory further states that the two most important variables in determining species diversity are the size of the island and its distance from the source of the species pool. Smaller islands will reach a diversity equilibrium with fewer species than larger islands because smaller islands tend to have fewer habitats and more intense competition. This leads to a more steeply rising extinction rate than that of larger islands, which have many more available niches. As for the distance of the island from the mainland, the more remote an island from the source of the species pool, the lower the immigration rate.

In sum, the theory of island biogeography predicts that species diversity on islands is directly proportional to island size and inversely proportional to distance of the island from the source of the species pool.

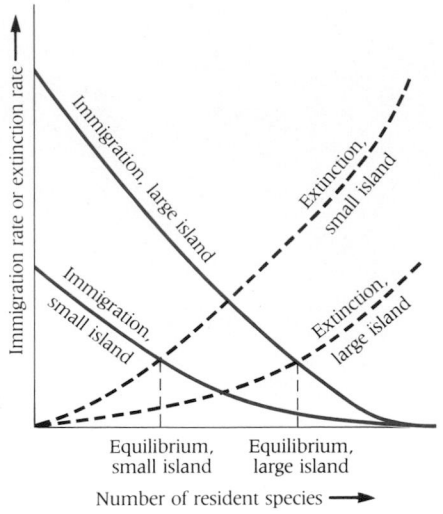

47-25 Extinction and immigration on islands

The Maintenance of Species Diversity

Diversity is a measure of the integration of a biological community that takes into account the number of constituent populations and their relative abundances. In comparing the community structures of different ecosystems, it is necessary to consider not only numbers of species in the community, but also the degree to which populations of different species approach equality in their relative abundance. This factor is termed the *equitability* of the community. The more equitable the community, the more nearly equal is the abundance of each of its species. However, because equitability is considered in measurements of diversity, the diversity of two communities may not be the same even though they are both equally rich in species, if the relative abundance of species in one is more equitable than that in the other.

The abundance of a species may not be the best basis for calculating community diversity. Abundance is only one reflection of a population's success, and it may be preferable to use some other aspect in the calculation. These aspects are collectively termed measures of *importance*. They include such factors as productivity, or the role played in cycling nutrient materials.

What maintains species diversity? What prevents one or more species in a trophic level from eliminating the others by competitive exclusion? One answer is that when coexisting species are in potential competition for a resource such as food supply, they subdivide this resource in some way. We saw earlier that different insectivorous warblers forage at different heights in a forest. Each warbler species is a specialist at feeding on insects at a specific range of heights above the ground. Flycatchers, nighthawks, and bats feed on the same general food resource (flying insects), but at different times of day or night. However, although they specialize in feeding on different foods, the niches of these different species may overlap in other ways. Members of all the warbler species, for example, feed by day, and they may form mixed foraging flocks on migration. Many species of oceanic sea birds may share the same cliff during the nesting season, but each specializes by using slightly different nesting sites on the cliff. An important reason for species diversity within a particular habitat, therefore, is specialization in the use of limited resources, which reflects niche differences among species.

A second factor influencing the overall species diversity of a region is the variety of habitats it contains. In a climax community, species diversity may vary with gradients of changing topography, elevation, soil type, and so forth. Another cause of diversity is the creation of different habitats within a region by periodic disturbances. Light gaps in a forest may be inhabited by species of birds and insects different from those in the nearby climax forest. The diversity of wildflower species in a prairie depends to a large extent on the frequency of fires.

Stable environments and climates tend to support a greater diversity of species than fluctuating ones, because a greater range of resources is consistently available. For example, the great lakes of Africa support innumerable species of fishes and aquatic invertebrates. In contrast, the continuously fluctuating conditions (including periodic deoxygenation) in swamps on the island of Java seem to limit species diversity.

According to the equilibrium theory, the numbers of species in communities are in equilibrium with ecological conditions that limit the capacities of environments to support species. Productivity is an important limiting factor. For example, there

47-26 Rocky intertidal community
Ochre sea stars, Pisaster ochraceus, *and purple sea urchins,* Stronglocentrotus purpuratus, *are among the many species in this complex community.*

47-27 Two well-known coral fishes
(A) *Parrot fish,* Scarus gibbus. *These interesting denizens of tropical and warm temperature seas have a massive parrotlike beak that contains strong pharyngeal teeth that can crush hard coral into a sandy paste. They play an important role in the wearing away of coral reefs and the formation of coral sand.* **(B)** *Butterfly fish,* Chaetodon xanthocephalus. *This and other brilliantly colored butterfly fishes have characteristic spines, fins, and protruding snouts that probe openings in coral in search of burrowing crustaceans and worms.*

is less usable energy at higher latitudes and thus less productivity. Therefore, at these latitudes we often see lower colonization and higher extinction rates than in tropical environments.

Diversity is also influenced by such biotic factors as predation and competition. Predation rates are often significantly higher in the tropics than in temperate climates. By removing competitors, high predation rates prevent the competitive exclusion of species, thereby permitting more species to live together. However, competition can also lead to more habitat specialization among species and therefore a greater diversity of species.

The intertidal marine communities provide a good example. There is intense competition for space because space is limited relative to the amount of food that is available (Figure 47-26). As different species came to require specialized habitats, it became possible in this small area for a great array of species to thrive together—among them, algae, sponges, coelenterates, mollusks, annelids, bryophytes, echinoderms, and chordates.

Finally, the structural complexity of the habitat may also influence the diversity of species within a community. The richest and most complex of all marine communities are the world's coral reefs. The structural complexity of the reefs is due to the corals. Like the trees of tropical rain forests, the corals provide the structures within which many different ways of life are possible. The richness of life in coral reefs is especially well exhibited by fishes, which have evolved an amazing radiation of body forms not found in fishes of any other community. Numerous jaws have evolved, from pipettelike elongated, tubular mouths of butterfly fishes to the coral-crunching beaklike jaws of parrot fishes (Figure 47-27). These feeding adaptations evolved in response to the great variety of the available prey—and because prey species found numerous shelters in which to hide in the complex coral-reef structure.

In sum, structural complexity and relative stability together with the processes of productivity, predation and competition have varying and profound effects on species diversity and contribute to the different patterns of species richness and diversity in various communities.

SUMMARY

A community is an assemblage of populations of different species that live and interact in a given area. Though there are many types of interactions between the members of a community, they all share, in some degree, overlapping ecological requirements.

Ecologists can partially characterize the niche of a given population by determining its tolerance ranges for many different variables. According to the principle of competitive exclusion, no two species occupy identical niches in a stable community. Hence, the selective advantages possessed by each of the species will never be identical. Sooner or later, one will win out—unless they diverge by the process of character displacement until they are no longer in serious competition.

Some species compete directly. Others do not compete because they have few needs in common. These may live together in symbiosis. Symbiotic relations are either commensal (only one species benefits), mutual (both benefit), or parasitic (one benefits at the expense of the other). Parasitic microorganisms are the agents that cause infectious diseases.

Parasites (by definition) spend most of their lives on or in their hosts. Hence, they commonly undergo evolutionary degeneration—losing advanced traits not necessary in the constant and relatively congenial environment provided by their host. They also acquire remarkable specializations—for example, mechanisms that allow them to subvert host defenses and hermaphroditism that allows them to breed when mates are unavailable (as they usually are within the host's body). Some parasites cause diseases, but selective pressures tend to modulate their virulence. If a disease they cause kills their hosts, they would at the same time be destroying their own habitats.

Nature is also abundant in the variety of its nonparasitic, predator-prey relations. Evolution has given prey species a wide array of morphological, chemical, and behavioral defenses against predator species.

Communities invariably undergo change. Newly formed habitats exhibit succession, a sequential replacement of one community by another which is better suited to that environment. The area is first colonized by a few rapidly reproducing species. These change the environment in ways that make it possible for other species to come in and displace the first colonists. Subsequent stages lead to more efficient use of resources, greater total biomass, and increased species diversity. Change continues until a stable state, known as the climax community, is reached. Although the succession of communities may vary, all successions in a particular physical environment tend to converge on a single climax community. If environmental disturbances are frequent, however, an area may never reach climax.

Diversity of species in any environment depends on many factors. These include the number of immigrating species; the rate of species extinction; the ability of the present species to specialize in the use of limited resources and to compete effectively for nutrients, habitats, and other resources; and the stability of the environment. Stable environments support more diversity. On islands, species diversity is directly proportional to island size and inversely proportional to the distance of the island from the geographical source of its colonizing species.

KEY TERMS

biome
brood parasitism
character displacement
climax
commensalism
community
competitive exclusion
competition

diversity
equilibrium model
habitat
limits of tolerance
monoclimax theory
mutualism
niche
nonequilibrium model

parasitism
polyclimax theory
sere
species pool
subclimax
succession
symbiosis

QUIZ QUESTIONS

1. Which of the following is(are) a factor(s) in vertical structuring of terrestrial ecosystems?
   A. plant structure
   B. penetrance of sunlight
   C. temperature
   D. A and B
   E. A, B, and C

2. Gause's principle is more likely to apply when
   A. population densities of competitive species are low.
   B. population densities of competitive species are high.

   C. shared resources are present in limited supply.
   D. A and C
   E. B and C

3. Which of the following contributed to evolution of the several species of finches on the Galapagos Islands?
   A. competition
   B. adaptive radiation
   C. character displacement
   D. natural selection
   E. All of the above

4. _____ is an interspecific association in which all species involved benefit.

   A. Commensalism
   B. Mutualism
   C. Competition
   D. Parasitism
   E. None of the above

5. All of the following increase as community succession progresses, *except* for
   A. relative amount of biomass produced by producers.
   B. number of species present.
   C. intricacy of trophic relationships.
   D. degree of specialization of species present.
   E. B and C

ESSAY QUESTIONS

1. How do investigators use limits of tolerance to define a species' niche? What is the principle of competitive exclusion?

2. What are the features of the three major types of symbiotic relationships? Provide examples of each. What limits the efficiency of parasitism?

3. What are the major mechanisms of plant defense against herbivores? What are some chemical defense strategies of animals?

4. What factors affect the stability of communities? Contrast the equilibrium and non-equilibrium models of community stability.

REFERENCES AND SUGGESTIONS FOR FURTHER READING

Barth, F. G. (1985). *Insects and Flowers: The Biology of a Partnership.* Princeton University Press, Princeton, N.J.

Insects and flowering plants are closely related species in many communities. This up-to-date survey of their diverse relationships is endlessly fascinating.

Brush, G. S. (1982). An environmental analysis of forest patterns. *American Scientist 70,* 18–25.

This survey of tree patterns in the forests of the eastern United States shows that they represent the larger communities to which they belong.

Cooper, C. F. (1961). The ecology of fire. *Scientific American 204,* April, 150–160.

After showing that the survival of some communities depends on periodic fires, the author argues that eradication of fires may be detrimental to forests.

Diamond, J., and Case, T. J. (Eds.) (1986). *Community Ecology.* Harper & Row, New York.

A stimulating summary of current approaches to the daunting problem of variability and scale in the populations of natural communities.

Finegan, B. (1984). Forest succession. *Nature 312,* 109–114.

An interesting review.

Harvey, P. H., and Partridge, L. (1988). Evolutionary biology: Of cuckoo clocks and cowbirds. *Nature 335,* 586–587.

A brief discussion of the known examples of brood parasitism among birds. The authors speculate on why female cuckoos might be nest predators as well as brood parasites.

Isack, H. A., and Reyer, H.-U. (1989). Honeyguides and honey gatherers: Interspecific communication in a symbiotic relationship. *Science 243,* 1343–1346.

In Africa, people searching for honey follow honeyguides to bees' nests. Old-timers say that the birds' flight patterns indicate the direction and distance to the nest and other information. Scientific study shows they are right.

Jackson, J. B. C., and Hughes, T. P. (1985). Adaptive strategies of coral reef invertebrates. *American Scientist 73,* 265–274.

Complex factors maintain the balance of highly diverse coral reef communities. The authors discuss the interesting ecological effects on that balance of such catastrophes as storms and predation.

Karr, J. R., Toth, L. A., and Dudley, D. R. (1985). Fish communities of Midwestern rivers: a history of degradation. *BioScience 35,* 90–95.

A persuasive essay on the decline of fish diversity and water quality.

Kikkawa, J., and Anderson (Eds.) (1987). *Community Ecology: Pattern and Process.* Blackwell Scientific, Boston.

Ecologists still ask whether communities are mere unstructured assemblages or tightly linked groups of interacting species. This collection reflects the views of some workers—and the problems they face.

Lande, R. (1988). Genetics and demography in biological conservation. *Science 241,* 1455–1460.

One of several papers in a special issue of *Science* on "perspectives in ecology," this essay clearly explains the importance of ecological and evolutionary data in the study of population dynamics.

MacArthur, R. H., and Wilson, E. O. (1967). *The Theory of Island Biogeography.* Princeton University Press, Princeton, N.J.

This short book significantly influenced thought on the ecology of species abundance and diversity, which is viewed in terms of population genetics. Part of the impact came from its imaginative reliance on experimental data and strict quantitative thinking.

Nabhan, G. P. (1985). *Gathering the Desert.* University of Arizona Press, Tucson.

One theme of this remarkable and often poetic book is the preservation of genetic diversity in crops and ecosystems. Its focus is the Sonoran Desert of Arizona and Southern California.

Pianka, E. R. (1988). *Evolutionary Ecology,* 4th ed. Harper & Row, New York.

A respected text reappears in a new edition. Useful discussions of most aspects of ecology with an emphasis on communities.

Rosenthal, G. A. (1986). The chemical defenses of high plants. *Scientific American 254,* January, 94–99.

Many plants produce chemicals that repel or kill herbivores. Some retard the growth of herbivorous insects. The author shows how some herbivores use plant-derived chemicals in their own defenses.

Roughgarden, J., Gaines, S., and Possingham, H. (1988). Recruitment dynamics in complex life cycles. *Science 241,* 1460–1466.

Another "perspectives" paper, this essay impressively illustrates the concerns of community ecology by analyzing the interactions of barnacles and other species (and their larvae) in a marine rocky intertidal zone.

Schoener, T. W. (1982). The controversy over interspecific competition. *American Scientist 70,* 586–595.

The author surveys current ideas on the nature of competition and predation and the influence they exert on the character of communities.

Tilman, D. (1988). *Plant Strategies and the Dynamics and Structure of Plant Communities.* Princeton University Press, Princeton, N.J.

Using a computer model that simulates the dynamics of plant communities, the author boldly attempts to identify the selective forces and constraints that dictate ecological specialization in plants.

THE GEOGRAPHY OF LIFE

The geography of life
The many landscapes of Earth are of obvious scientific interest. Many are also quite beautiful. We will see in this chapter how they influence life—in the present and throughout evolutionary history. The photographs are scenes from New Mexico; Baffin Bay, Canada; and Arizona.

No one can fail to observe the striking differences in plant and animal communities in different locales—meadows, forests, shorelines, rivers, and lakes. There are also broad regional differences. Along the road from Boston to Tucson, for example, a traveler passes first through regions in which the predominant natural community is a deciduous forest—though, in flatter areas, humans have replaced most of the forest with farmland. West of the Mississippi are open grassland communities that extend onto the high plains of eastern Colorado and New Mexico, but a botanist's eye would reveal different grass species. Then, depending on the route, there may be mountain forests and communities of sagebrush, tree junipers (locally called "cedars"), and piñon trees (nut pines). Finally, there are true deserts rich in cactus, mesquite, ocotillo, and greasewood.

Wider travel reveals more exotic differences in regional natural communities. The rain forests or "jungles" of South America include many unfamiliar species. Although the Argentine pampas resembles our own grasslands and much of Patagonia recalls our southwestern thornbush deserts, we will find, despite the superficial similarity of these surroundings, that many of the species present are as distinct from ours as those of the rain forest.

Clearly, the distribution of life on Earth lacks uniformity. Many geographic distinctions are very well known. Everyone knows that you must go north to see a polar bear, to Asia to see a wild tiger, and to Australia to see a wild kangaroo. Study of the distribution patterns of living things is the science of **biogeography.** Some biogeographic relationships take place within an area of only a few square meters. Others involve large regions, whole continents, or the entire Earth. It is the task of biogeographers to collect data on the distributions of species and to explain them on the basis of current knowledge of evolution, ecology, geology, climatology, and behavioral biology.

Biogeography has two major branches. **Ecological biogeography** is concerned with the ecological forces underlying the distribution of plants and animals and their associations and adaptations in communities in different locations. Where organisms live depends on the nature of the environment, and the adaptations—structural, functional, and behavioral—that have evolved to deal with that environment. If two deciduous forest communities—one in Pennsylvania and one in Missouri—were similar in climate, soil, and other environmental conditions, they could in principle support similar **biotas.** This term, important to biogeographers, is defined as the totality of organisms of a given place or region, its flora (plants) plus its fauna (animals).

Ecological factors do affect the geography of life. But they cannot account for it fully. Conditions in an Australian desert might closely match those in an African desert, or a Malayan jungle might match one in South America. Yet, despite their similarities, there are striking differences in their biotas that must reflect differences in the histories of the regions. Organisms arrived at different times and came from different places. Once there, they may have evolved differently. **Historical biogeography,** then, is the second aspect of the geography of life.

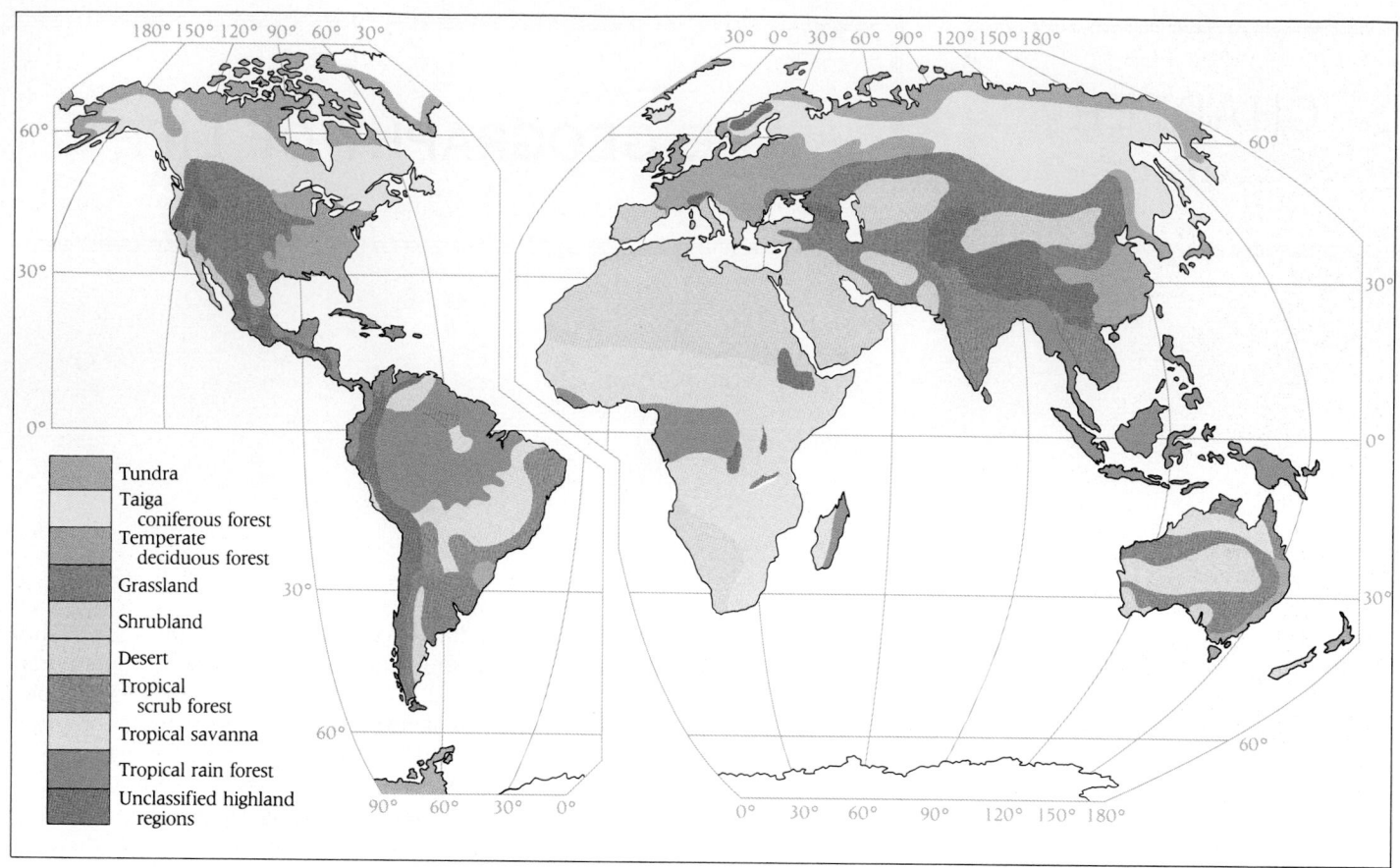

48-1 Global distribution of original native vegetation

The categories of vegetation on a map of this scale are necessarily broadly conceived. Many of them—like taiga (or boreal forest), temperate deciduous forest, and tropical rain forest— include within them a diversity of distinct subtypes. Within any one wide region there are also different types of vegetation in patches too small to show on this scale.

Map legend:

- Tundra
- Taiga
  coniferous forest
- Temperate
  deciduous forest
- Grassland
- Shrubland
- Desert
- Tropical
  scrub forest
- Tropical savanna
- Tropical rain forest
- Unclassified highland
  regions

ECOLOGICAL BIOGEOGRAPHY

Ecological biogeography rests on certain of the broad generalizations we have discussed in earlier chapters.

- Environments differ from place to place.
- Every organism is adapted to its particular environment.
- Every organism is a member of a community and is adapted to living with other members of the community, which are in fact part of the organism's environment.
- Interrelations within the community as a whole are adaptive and are such as to adapt the whole community to the conditions prevailing in its geographic position.

FACTORS REGULATING PATTERNS OF LIFE

Importance of Plants

Plants play a leading role in the geography of biotas, especially on land. Photosynthetic organisms are at the beginning of most food chains and their nature in a given place strongly influences the character of later links in the chains—and hence of the whole biota. Plants are sensitive to variations in the physical environment, especially of climate and soil. Such variations also affect animals, but their dependence on a given set of physical factors is usually less fussy.

Animals may seem to depend on a certain climate, though in reality they depend on a certain vegetation type that depends on climate. As a rule, grazing animals are most abundant where mean annual rainfall is 12 to 30 inches and rain occurs irregularly and in a relatively short season. Grazing animals probably do not do best in that sort of climate, though they are most common there. Many animal groups are restricted by the distributions of the plant communities they need. Maps of broad types of plant communities, called *plant formations* by botanists, are shown in Figures 48-1 and 48-2. To the extent that it is ecologically determined, the distribution of animals tends to follow the patterns of plant formations.

48-2 *The present-day distribution of original native vegetation in North America*

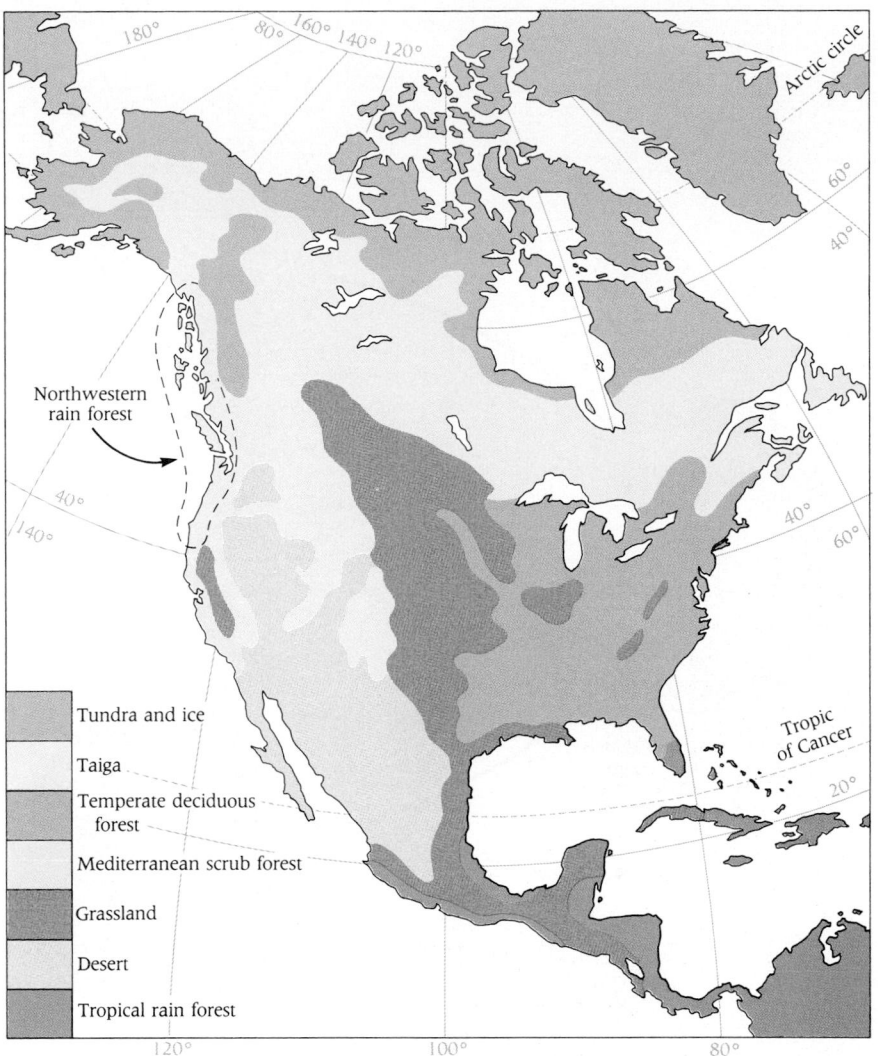

Northwestern rain forest

Tundra and ice

Taiga

Temperate deciduous forest

Mediterranean scrub forest

Grassland

Desert

Tropical rain forest

Plants are also basic to the whole ecology in aquatic communities. However, the more uniform aquatic environments generally have less sharply distinguished distributions of various kinds of plant communities.

Patterns of Climate

Air circulation, temperature, solar radiation, and precipitation (rainfall and snowfall) are the aspects of climate that control plant communities. Each affects vegetation directly or, through its effects on soil, indirectly. For each, there is meaning in both average intensity and distribution through the year. Climate varies with latitude for two reasons.

- First, there is an uneven distribution of solar energy at different latitudes. When the sun's rays strike the Earth at an angle of 90°, their heating effect is maximal because at that angle they spread very little on the Earth's surface. Because the Earth's axis of rotation is inclined at an angle of 23.5° with respect to the plane of its orbit around the sun, the Northern Hemisphere tilts toward the sun in summer and away from it in winter (Figure 48-3). Thus, in summer incoming rays are more nearly vertical—and daylight hours are longer. The angle of incidence of sunlight is most nearly vertical at the Equator (latitude 0°) during March and September, at the Tropic of Cancer (23.45°N) during June, and at the Tropic of Capricorn (23.45°S) during December. At very high latitudes, north or south, the sun's rays come in at a low angle and spread over wide areas. That is why polar climates are cold. Considerations such as these account for the **seasons** and for most aspects of regional climates.
- Second, there are significant variations in the pattern of air currents at different latitudes (Figure 48-4).

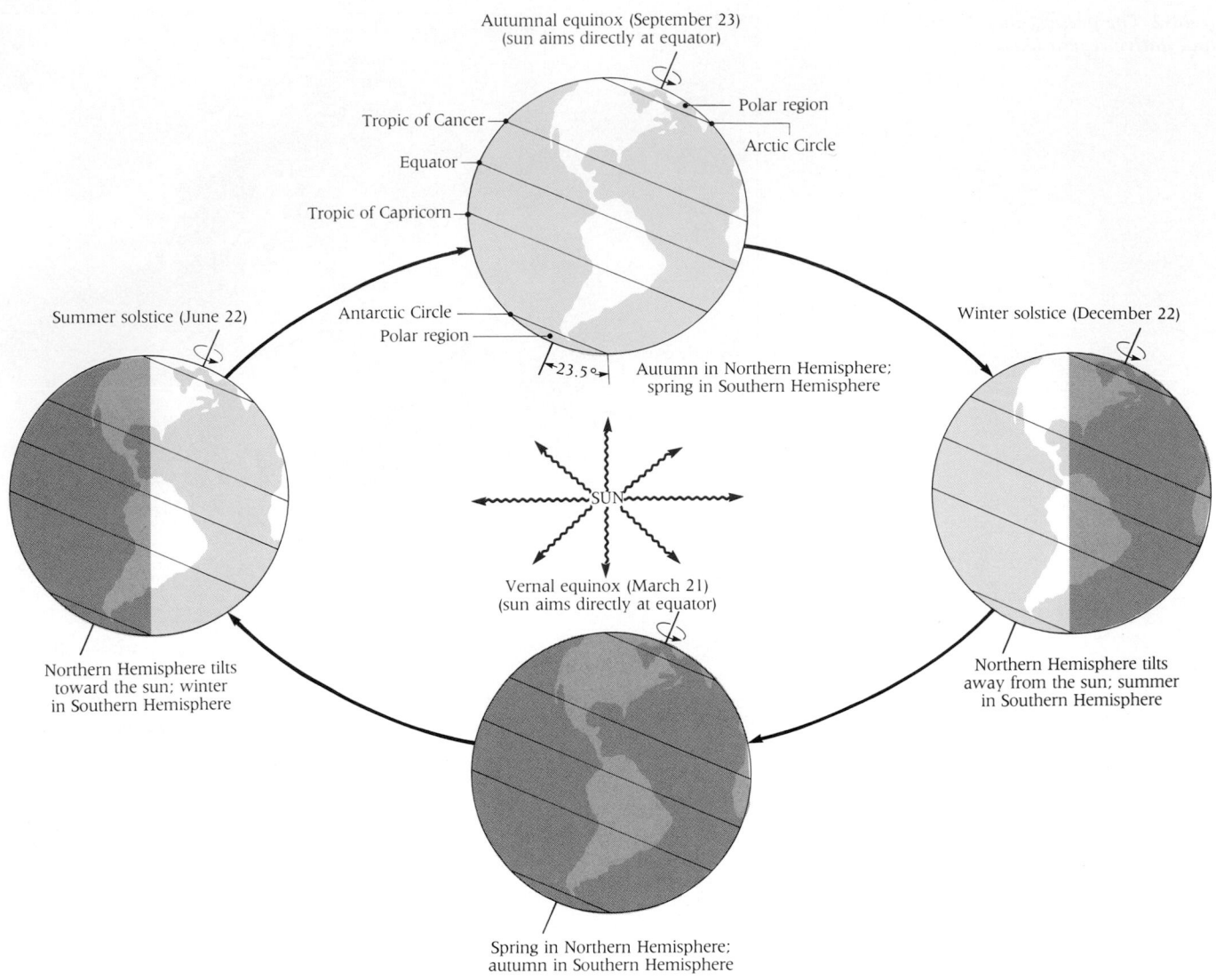

Autumnal equinox (September 23)
(sun aims directly at equator)

Polar region

Tropic of Cancer

Arctic Circle

Equator

Tropic of Capricorn

Summer solstice (June 22)

Antarctic Circle

Polar region

Winter solstice (December 22)

23.5°

Autumn in Northern Hemisphere;
spring in Southern Hemisphere

SUN

Vernal equinox (March 21)
(sun aims directly at equator)

Northern Hemisphere tilts
toward the sun; winter
in Southern Hemisphere

Northern Hemisphere tilts
away from the sun; summer
in Southern Hemisphere

Spring in Northern Hemisphere;
autumn in Southern Hemisphere

These movements of air masses hold to certain patterns. Near the Equator, heated air rises (forming a low pressure area) and flows toward the poles (Figure 48-5). Rising air cools and loses its water in profuse equatorial rainfalls. Air masses moving toward the poles gradually cool, increase in density, and descend, forming a subtropical high-pressure area at about 30° N or S—known for reasons lost in the mists of history as the Horse Latitudes. There the reheated air produces a zone of reduced precipitation and moves toward either the Equator or the poles. At about 60° N and S, air masses moving poleward encounter cold air coming from the poles. Where these air masses meet, there is an unstable zone of low pressure in which weather is changeable. At the poles themselves, cold dense air causes a zone of high pressure and low precipitation.

Beyond these bands of north-south circulation are three major prevailing air currents caused by interaction of the Earth's east-west rotation with patterns of world-wide temperature gain (see Figure 48-4). Between 30° N and S, the trade winds stream from the east-northeast in the Northern Hemisphere and from the east-southeast in the Southern Hemisphere, blowing steadily the year round toward the Equator from an easterly direction. The region where they meet is the **intertropical convergence.** Between 30° and 60° N and S, winds moving poleward become the westerlies that blow from west to east and dominate the climates of lands along the western marginal zones of the continents.

The heating of air increases its moisture-holding capacity; cooling does the reverse. At 30° N and S, the Horse Latitudes, sinking cool dry air produces an arid zone of low precipitation that is usually very warm. All of the world's great deserts are located along these latitudes. At 60°N and S, as relatively warm air rises and is

48-3 The source of the seasons
The Earth's axis is tilted 23.5° with respect to the plane of its orbit. Thus the sun's rays strike the Earth differently at different times of the year. This is responsible for seasonal climates.

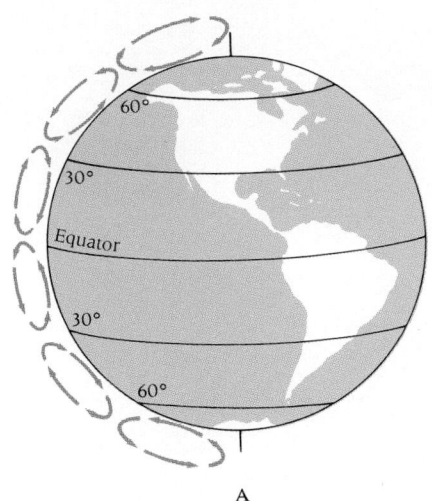

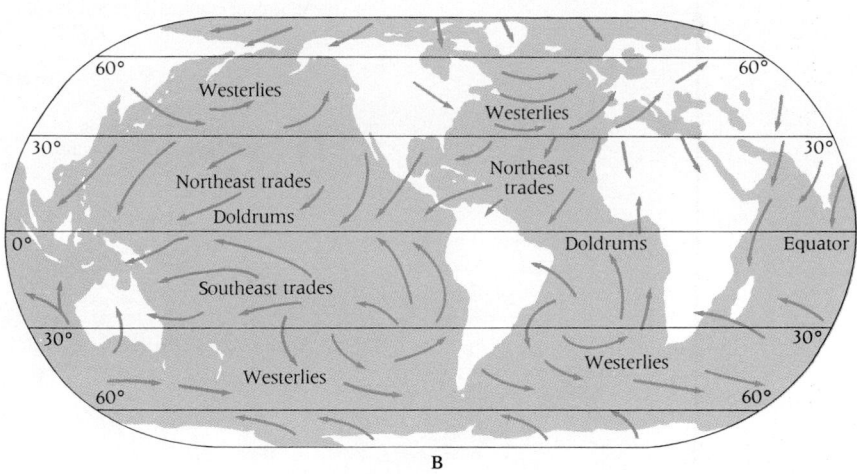

48-4 General patterns of air movement over the Earth
(A) The pattern out from and back to the surface.
(B) The major wind currents.

cooled where the great forests of the temperate zone are located, its moisture-holding capacity decreases. The result is precipitation.

Horizontal and Vertical Controls

We have seen that community structure usually has a horizontal and vertical dimension. Here we are concerned with horizontal and vertical dimensions, and their controls, on a much larger scale.

Again consider temperature and radiation. Their familiar north-south pattern clearly affects the vegetation map of North America. In the tropics, the lowlands are without frost. Daily temperature variation is greater than seasonal variation, and sunshine is intense and of about equal duration throughout the year. Moving northward or southward to the temperate zones, seasonal temperature variation becomes greater. Nearer the poles, frost occurs through a longer and longer part of the year. Summers may be no cooler than in the tropics—and may even be hotter—but periods of heat decrease in length. Radiation is not as intense as in the tropics, and its distribution through the year is increasingly uneven until the

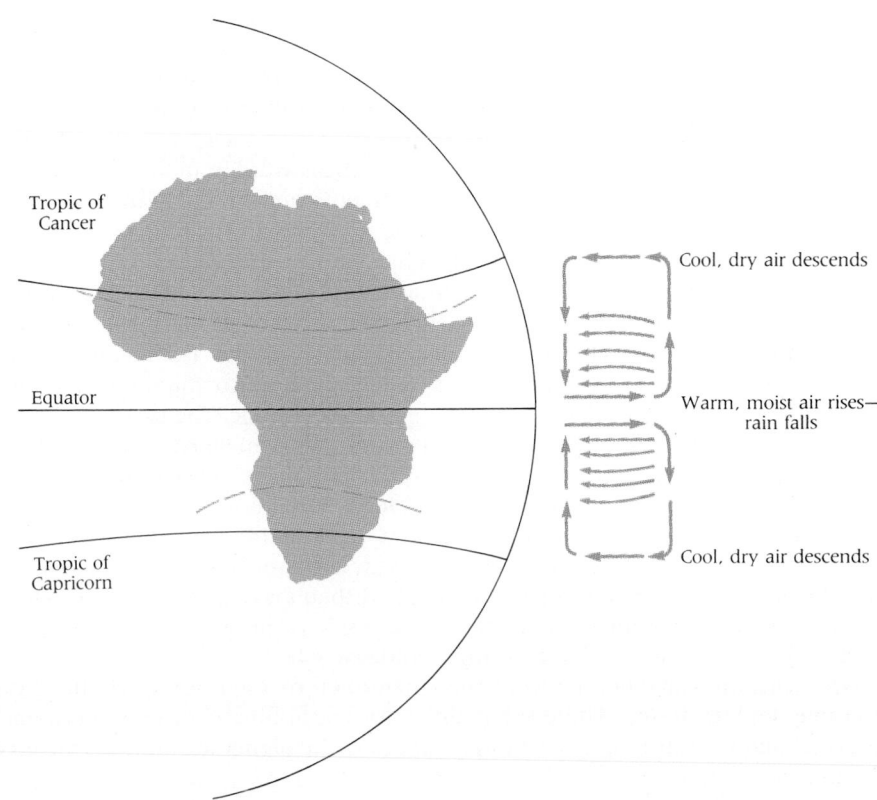

48-5 Vertical air movements and patterns of rainfall in different latitudes
Brown color indicates regions that are dry and barren.

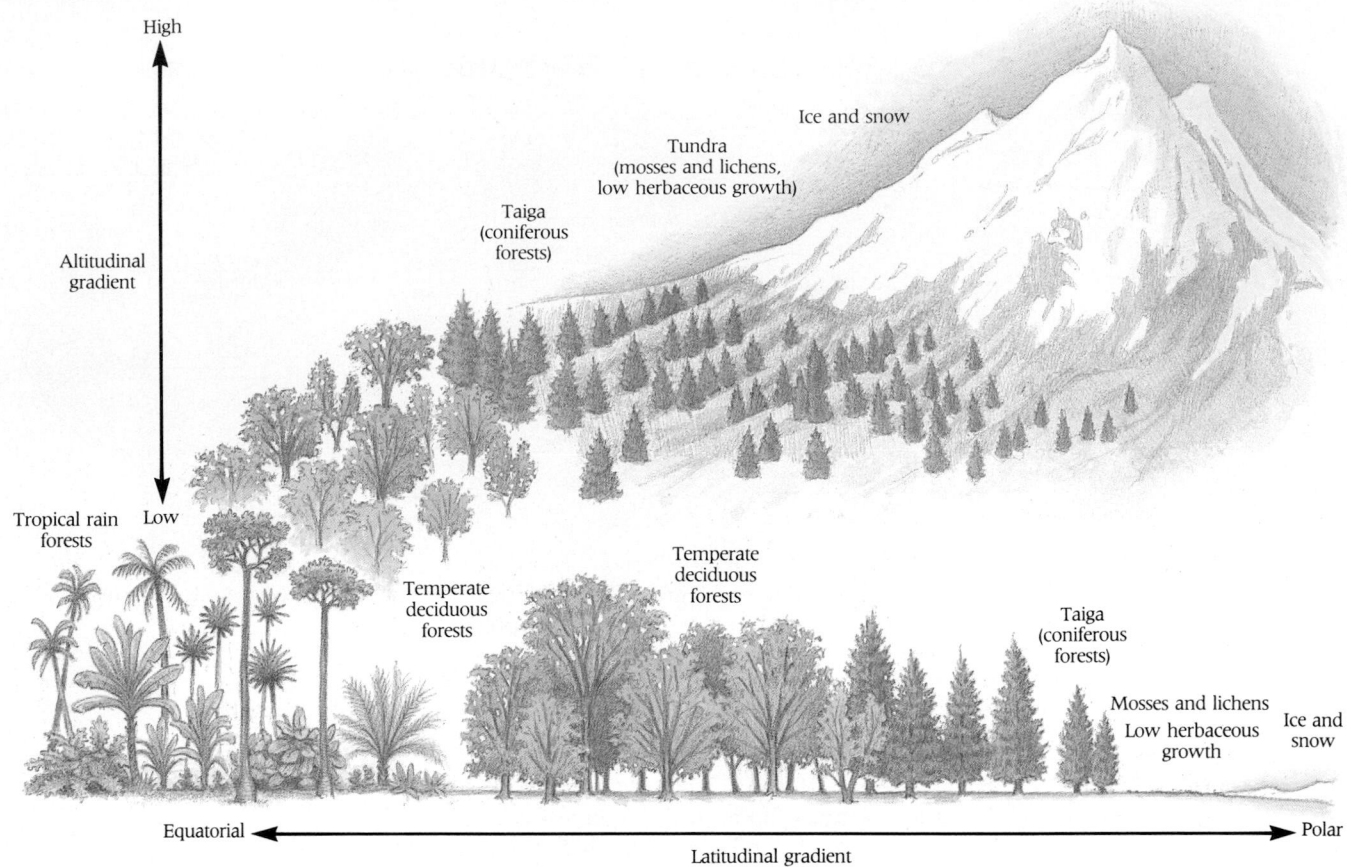

High

Altitudinal
gradient

Ice and snow

Tundra
(mosses and lichens,
low herbaceous growth)

Taiga
(coniferous
forests)

Low

Tropical rain
forests

Temperate
deciduous
forests

Temperate
deciduous
forests

Taiga
(coniferous
forests)

Mosses and lichens

Low herbaceous
growth

Ice and
snow

Equatorial

Latitudinal gradient

Polar

far north (or south) is reached. In these regions, there are months of continuous daylight—and other months of continuous darkness.

The same vegetation map shows notable east-west differences, especially in the temperate zone. The pattern is largely due to changes in precipitation and evaporation. Rainfall follows an erratic course—influenced by air movements, distance from the coast, and land topography. Rain is plentiful on the northwest coast and in the mountains of California. But air moving eastward from these areas loses moisture. The result is an arid belt of deserts and semideserts. Then, the Rocky Mountains again catch moisture and the plains immediately eastward of these arid regions are semideserts of high, dry grassland. From there on to the Atlantic coast, the general tendency (with many local exceptions) is for an increase in rainfall, mainly because of moist air that periodically moves northeastward from the Gulf of Mexico. Similar zoning of rainfall occurs on all continents and in all north-south climate zones of temperature and radiation. Deserts as well as lush rain forests occur in the tropics. Indeed, a desert may occur right next to a rain forest.

Besides these horizontal zonings of climates (and biotas), there is a vertical zoning. Climbing a high mountain, one notes conspicuous changes of climate and life on the way up. One climatic change is obvious: the air gets colder. This change resembles what occurs when one travels poleward from the equator. Conditions on a mountain are similar to those at sea level at a more northerly latitude. The tops of mountains, even at the Equator, have a considerable ecological resemblance to Arctic lowlands. The timberline, above which trees do not grow, becomes lower from the southern Rockies into Alaska and finally reaches sea level at the northern tree line.

The relationships between horizontal and vertical climatic zones are shown in Figure 48-6. The resemblance between an "arctic zone" on a high southern mountain and the actual Arctic is not exact. Factors other than average temperature—lengths of day and night, extremes of temperature, atmospheric pressures, etc.—are different in the two areas and may have distinct ecological effects.

Vertical zoning of biotas is vividly seen in our Southwest, from which the example in Table 48-1 is drawn. There the main factor controlling climate is precipitation, although temperature is important too. As a rule, the higher the altitude, the greater the precipitation.

48-6 Horizontal and vertical distributions of life
The parallel between these two distributions of life is evident. On the mountain, a rise of about 70 m in altitude is the equivalent of traveling 1° latitude northward of sea level.

TABLE 48-1
ALTITUDINAL LIFE ZONES IN NEW MEXICO

Altitude* (meters)	Name of Zone	Some Characteristic Plants	Some Characteristic Animals
800–1500	Lower Sonoran	Creosote bush, mesquite, ocotillo	Antelope squirrel, desert fox, road runner, diamondback rattlesnake
1500–2300	Upper Sonoran	Piñon pine, tree juniper ("cedar"), sagebrush, cottonwood, cholla cactus	Prairie dog, coyote, morning dove, meadow lark, plains rattlesnake
2300–2750	Transition	Ponderosa pine, narrow-leaf cottonwood, scrub oak, wild rose	Mule deer, Abert squirrel, mountain bobcat, black bear, wild turkey
2750–3800	Canadian	Spruce, fir, aspen, cinquefoil, gentian	Elk, spruce squirrel, marmot, lemming mouse, Steller's jay, junco
3800–4170	Hudsonian	Foxtail pine	Pika, mountain jay
4170–4500	Arctic-Alpine	No trees. Alpine forget-me-not, spring beauty	Mountain sheep, ptarmigan

* The stated altitudes are approximate averages. In different parts of the state and at different exposures on valley and mountain slopes, the elevation of a given zone may be as much as 320 m above or below the stated figure, but the sequence remains the same.

TERRESTRIAL COMMUNITIES

The study of ecological biogeography recalls those nested Chinese boxes or wooden eggs we enjoyed as children. Opening one always revealed a smaller one inside. The biggest "box" is the whole of life on Earth. Next in size is the totality of land communities on one hand and of aquatic communities on the other. So it goes until we consider the life of one thicket, one meadow, or one pool. To illustrate this representation and to bring out certain principles, we now discuss something in the middle of the series—broadly regional kinds of communities.

As noted in Chapter 47, ecologists refer to major climax formations as biomes. We will speak of the most important and widespread biomes that occupy major portions of the Earth (Figure 48-7). But our list is not exhaustive. There are distinctive geographically definable communities in salt lakes, marshes, or subterranean environments that we cannot consider here. You might enjoy working them out on your own. In each community, you should try to answer two questions. What are the special conditions of life here? How are they met by adaptations of the community and within the community?

Maps of the Earth's major terrestrial biomes (see Figure 48-1) show that a similar sequence of biomes goes northward or southward from the Equator. Many local

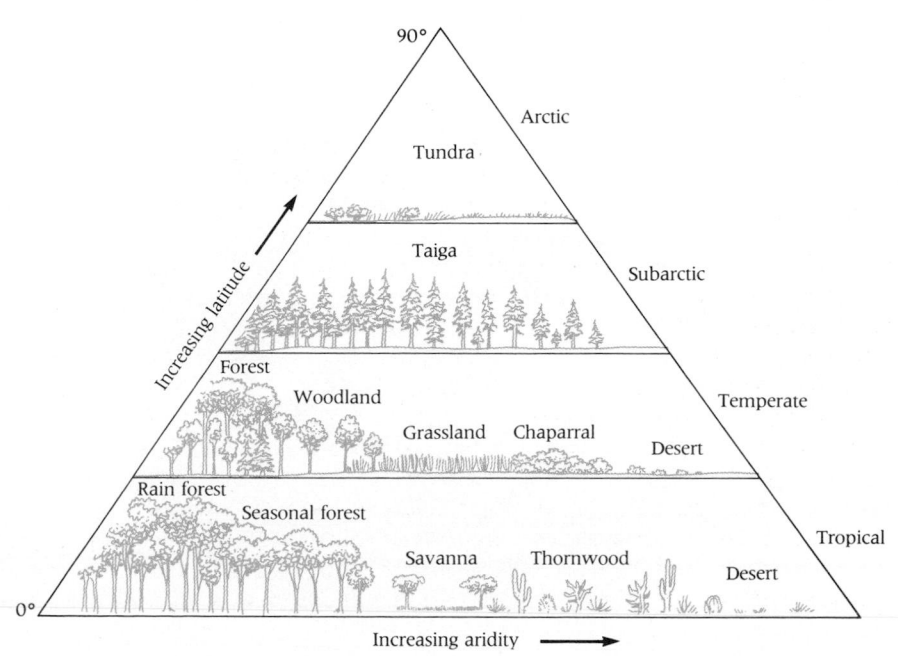

48-7 Major terrestrial biomes
Each of the principal terrestrial biomes is arranged along lines of increasing aridity at different latitudes. Note the influence of moisture and temperature on the structure of plant communities.

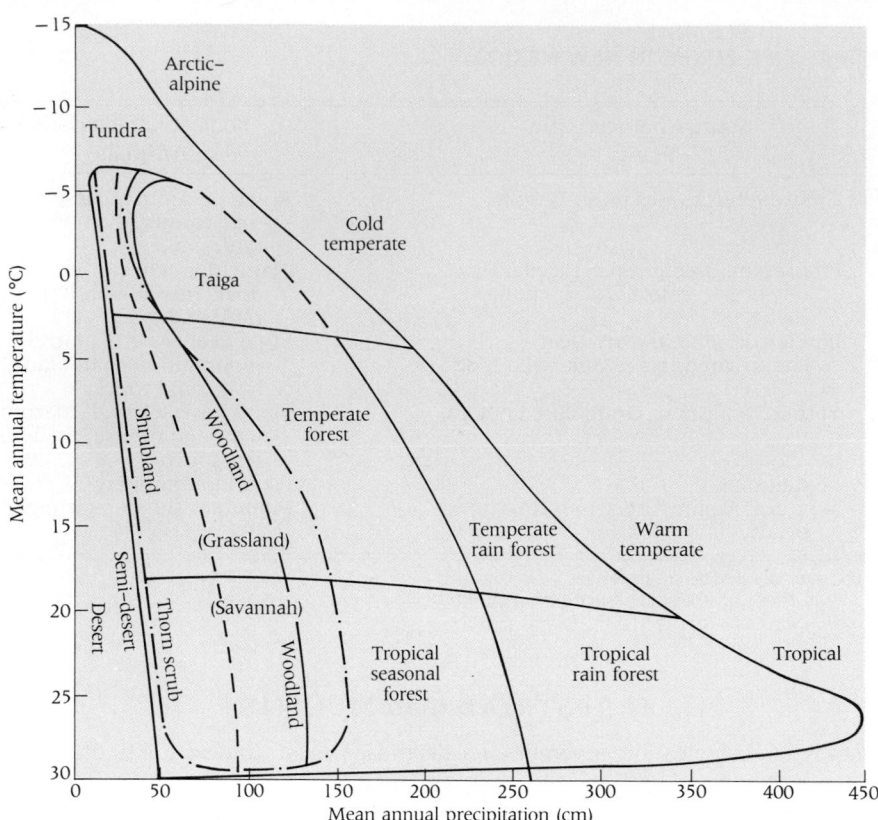

48-8 Biome distribution as a function of temperature and precipitation
The mean annual temperature and mean annual precipitation significantly influence the distribution of the major terrestrial biomes. Within regions outlined by the dashed line, other factors (oceanity, drought, human land use) may affect the biome type that develops.

variations are due to the ocean currents shown in the map. The diagram in Figure 48-8 relates the biomes to climate.

Tundra and Alpine Communities

Tundra—the word is from the Russian language—is a vast treeless zone of North

48-9 Tundra
The tundra rings the Arctic Ocean like a frozen collar atop the northern forests.

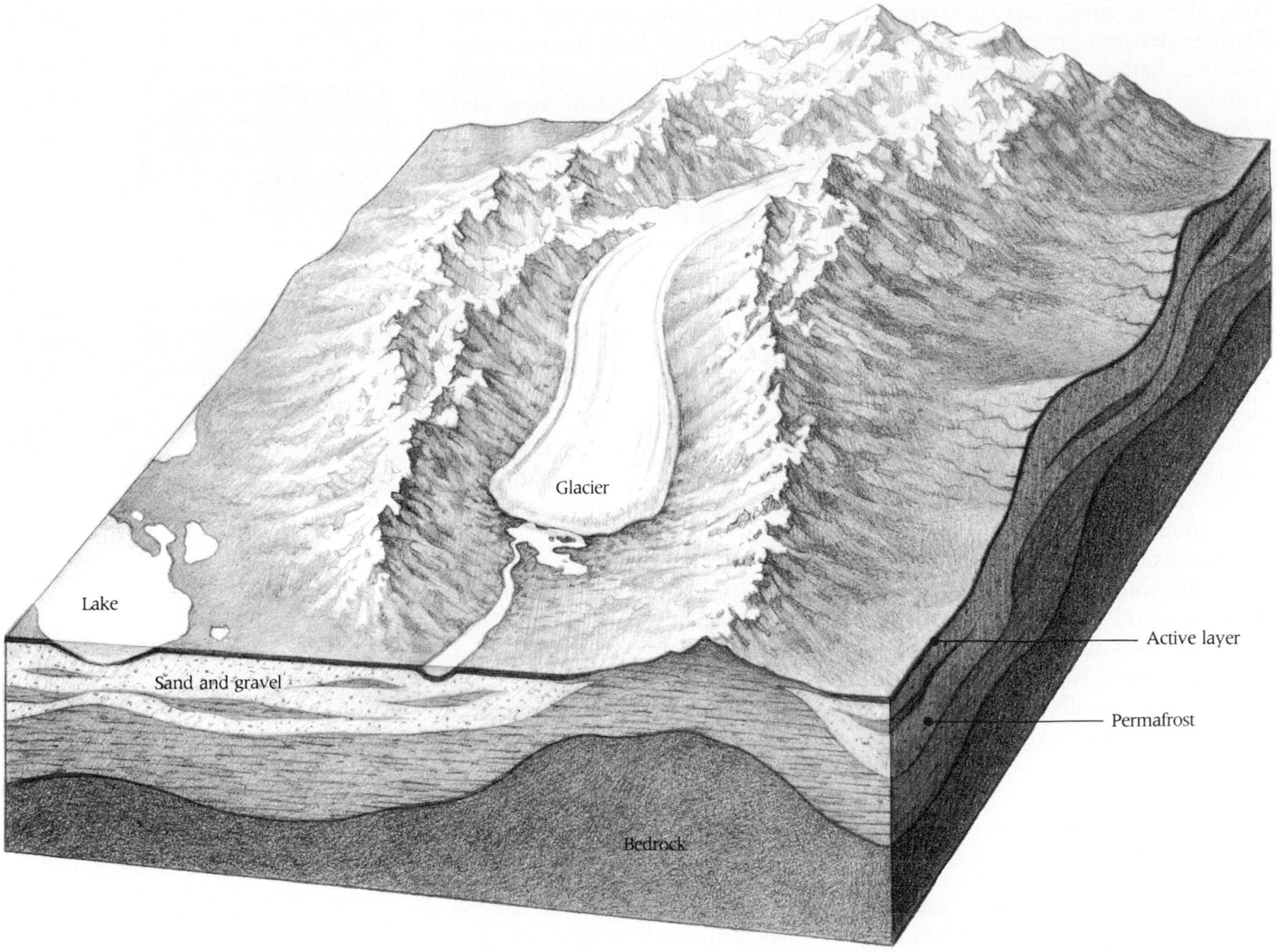

The permafrost lies just beneath the surface "active layer," which supports tundra life. The permafrost remains below 32°F, but the active layer thaws in the summer. The diagram is not drawn to scale.

Figure labels: Glacier, Lake, Sand and gravel, Active layer, Permafrost, Bedrock

48-10 Permafrost in the tundra
The permafrost lies just beneath the surface "active layer," which supports tundra life. The permafrost remains below 32°F, but the active layer thaws in the summer. The diagram is not drawn to scale.

America, Europe, and Asia that encircles the Arctic Ocean like a frozen collar (see Figures 48-1, 48-2, and 48-9), comprising roughly a twentieth of the Earth's land surface. A similar zone of comparable size does not occur in the Southern Hemisphere, where there is little land in corresponding latitudes (see Figure 1-9). The tundra begins where the northern forests end. Its climate is typical of the arctic—cold, with a long, dark winter and long or even continuous summer daylight. It receives little precipitation, and what water there is tends to be trapped near the surface of the ground, which is permanently frozen a few feet below the surface. This frozen layer is termed **permafrost** (Figure 48-10).

The ground surface of the tundra is notably fragile. Rock-hard in winter, like the hundreds of feet of frozen earth beneath it, it softens during summer thaws when the region becomes extremely wet. The flatness of the terrain limits water runoff. In winter, the topsoil freezes again. This repetitious freeze-thaw cycle is hard on the roots of plants and keeps them small. Driving winter snows and harsh dry winds further restrict plant growth. Temperature extremes are very wide, ranging from below −40°C in winter to moderate warmth (up to 30°C) in summer.

Because of these harsh conditions, the tundra is practically devoid of tall upstanding trees (Figure 48-11). Dwarfed, shrubby alders, birches, willows, and conifers cover much of the ground. Mosses (especially sphagnum), lichens (especially "reindeer moss," which is not a true moss), and a few grass species cover large ground areas with a springy carpet. Perennial herbs with large brilliantly colored flowers are conspicuous and quite beautiful. The growing season is very brief—June through August—so plants must mature without becoming large. Tundra plants spend most of their lives in a state of suspended animation, active only in brief periods of warming sunshine.

Vast hordes of birds, especially waterfowl, nest in the tundra in summer, but most desert it in winter. Permanent residents include a few birds and mammals, warm-blooded and well protected by feathers or fur. Some of the resident birds, like the ptarmigan (an Arctic grouselike bird), snowy owl, falcon, and raven, and

48-11 Barren landscape of the tundra

mammals, like the Arctic varying (snowshoe) hare, turn white in winter.[1] White fur or feathers give protective coloration in a snowy environment. They also minimize heat loss by radiation. Other mammals on the tundra are musk oxen and caribou (wild reindeer), large herbivores that depend mainly on the abundant moss and lichens, and numerous Arctic hares and lemmings (small ratlike rodents). Amphibious polar bears frequent the coasts and ice floes, but also wander inland on the tundra. Buzzing flies, mosquitoes, and other insects are so numerous as to be major nuisances from the human point of view. Their eggs and larvae are particularly cold-resistant, and the adults appear by the billions on warmer summer days. Still, the numbers of species permanently in residence there are smaller than in almost any other sort of community, even the deserts.

The areas on high mountains, between the timberline and heights where nothing can live, are called **alpine regions.** Alpine conditions resemble those of the tundra, but with some differences. Alpine heights are cold throughout the year, whereas summer nights are warm on the tundra. Alpine vegetation is similar to that of the tundra in general appearance. In mountains of the temperate zone, the flora and much of the fauna may consist of species related to those in the Arctic. Ptarmigan and varying hares extend far south of the tundra in alpine environments. Alpine insects are often Arctic species or closely related to them.

Recent times have brought the tundra into unaccustomed prominence as humans have feverishly sought to exploit its oil and mineral deposits. The effort to bring petroleum out of this region quickly and cheaply has posed at least two disturbing ecological threats—one, a long, above-the-ground pipeline that could block the complex north-south migration cycles of caribou herds, and the other, the ever-present danger of oil spills onto the tundra. We do need the oil below those slopes, but we also need to protect the desolate and fragile tundra.

48-12 *Taiga*

Taiga

Taiga (another word of Russian origin)—or **boreal forest**—is a still broader zone just south of the tundra that forms a belt across northern North America and Eurasia (see Figure 48-1). Like the tundra, and for the same reason, it is practically absent in the Southern Hemisphere. Winter temperatures may be as severe as in the tundra, but there is a well-defined summer growing season of three to six months. That is enough to thaw the subsoil and permit heavy growth of hardy trees. As a result, the taiga as a whole is a tremendous forest (Figure 48-12). In the typical taiga, monotonous coniferous forests extend across northern Canada and Siberia, mainly with spruce, though several other conifers occur (fir and tamarack). Alder, birch, and juniper thickets are also common. Burned areas of the coniferous forest are invaded by aspens and birches, which later are succeeded by conifers again.

Moose (called "elk" in Eurasia) occur throughout the taiga—where they have not been exterminated by humans—and are its most conspicuous animal. Smaller animals are much more varied than in the tundra. Black bears, wolves, and martens are more common here than elsewhere. Fishers, wolverines, and lynxes are practically confined to the taiga. So are some rodents, such as the northern vole, although most of the abundant rodents are subspecies of groups occurring farther south. Squirrels abound in these rich coniferous forests. So do many birds, most of which are summer breeders that migrate southward in the fall. The many insects and other invertebrates are of species that can lie dormant during the severe winters.

The coniferous forests of our western mountains have distinctive features of their own, but are essentially extensions of the taiga, occurring at increasingly high altitudes the farther south they are. Many of their plants and animals are those of the typical taiga. The hemlock-hardwood forest of southern Canada and down into the Appalachians is also an extension of the taiga.

Temperate Deciduous Forests

Temperate deciduous forests occupy regions with moderate, well-distributed precipitation and with cold winters and warm summers. These conditions occur

---

[1] Three species of varying (or snowshoe) hares are usually recognized—*Lepus americanus, Lepus arcticus,* and *Lepus otho.* They are so named because they doff their lightweight brown summer coat for fluffy winter white. The name snowshoe hare comes from the exceptionally large well-cushioned feet.

48-13 *Temperate deciduous forest*

in the temperate zones where the average annual precipitation is around 40 inches, without very well-defined dry and rainy seasons. These forests once covered most of the eastern half of the United States, which has such a climate (Figure 48-13). Northward the forests graded into the taiga and southward into the southeastern pine forests. The British Isles and practically all of Central Europe were also formerly occupied by temperate deciduous forests, and so was a large region in China and southeastern Siberia. There are similar forests in the temperate zone of South America, but they are not so widespread there because the precipitation is not suitable over such large areas.

"Deciduous" denotes trees that shed their leaves seasonally to conserve water in months when most water is frozen into the soil. Shedding leaves is characteristic of this climate and an obvious adaptation to it. Common trees of the deciduous forest are beech, tulip, sycamore, maple, oak, hickory, elm, poplar, and birch. Chestnuts were once common but have now been almost eradicated in the United States by blight. Stately elms seem likely to follow as they slowly die off from Dutch elm disease. The taiga and other coniferous forests include fewer species of trees, and locally a coniferous forest tends to be dominated by a single species. The deciduous forests have more varied local groupings, each of which commonly includes two or more species, as in the beech-maple and oak-hickory, elm-ash-maple, or willow-cottonwood-sycamore communities.

The most striking herbivores of the deciduous forests are the browsing deer, mainly the white-tailed or Virginia deer in North America (see Figure 1-17). In Eurasia, wild pigs (or boars) are also characteristic of this group of communities. The principal predators are large cats. Our variously named puma, mountain lion, cougar, or panther (all one species, *Felis concolor*) ranges into most of the environments of North and South America. It is now extinct in the eastern forests but was originally their commonest large carnivore. Wolves, though more characteristic of the taiga, once ranged widely into these forests, both in Eurasia and in North America. Recent reports have them returning to Canada and the northern United States after years of slaughter brought them to near extinction. Foxes are still common in these forests. Arboreal martens are locally as common here as in the taiga, and the racoon (absent in Eurasia) is especially abundant in our deciduous forests. These forests throughout the world are also rich in tree squirrels. Among mammals of the North American deciduous forests, over a third of the species are mainly arboreal. Tree-nesting birds are also abundant, and woodpeckers have the most obvious connection with the forest environment. The leaf and mold-covered forest floor is a world in itself, swarming with fungi and invertebrates.

Tropical Rain Forests

The lushest and most complex forest community, richest of all the biomes, is the **tropical rain forest,** which develops where there is an equable climate, an abundant and continuous water supply, and a long growing season that may continue through the whole year. Such rain forests can occur in the temperate zone—for example, in our northwest Pacific coast, where they have their own special characteristics and species. They are, however, most widespread and impressive in the tropics and subtropics (Figure 48-14). Fossil evidence suggests that the rain forest is one of the most ancient ecosystems, having existed since the Cretaceous.

Today, the rain forest is rapidly changing. It still covers much of Central America and northern South America, central Africa, southern Asia from India eastward, the East Indies and South Pacific islands from Sumatra through New Guinea, and small parts of northeastern Australia (see Figure 48-1). But it is no longer a continuous belt. Rather, it is fragmented and diminished in area, huge sections having been cut down for timber or farming. As discussed in Chapter 46, this is a major factor contributing to global warming from the greenhouse effect.

We refer to tropical rain forests when we speak of the jungle.[2] We all have romantic images of the jungle from travel brochures and movies, but these are usually inaccurate. Actual jungles include foliage of boundless green—in hundreds

48-14 Tropical rain forest

[2] The word "jungle" is of Sanskrit derivation. Originally it meant desert. It subsequently came to mean any wilderness. European travelers picked up the word in India and used it for the wilderness (as they considered it) of the rain forests there. Lately it has been so misused by explorers of the "Oh, how I suffered!" school, scientific explorers now avoid the word.

of shades—with a dense pattern of undergrowth that in some areas (e.g., the Amazon Valley) cannot be penetrated without chopping every foot of advance with a machete. But in other areas, the foliage on the forest floor is easily penetrated.

Tropical rain forests are highly productive, even though they spring largely from infertile soils. Only a thin layer of topsoil is present. Below the topsoil is virtually no organic matter—and nutrients are scarce. The roots of trees—spread out in the thin soil layer often only a few centimeters thick—are often involved in mycorrhizal associations (see Chapter 40), which efficiently transfer nutrients from fallen organic debris back to the trees. In view of the forest's high productivity, one might expect the soil layer to be very fertile and thick with humus. In fact, this soil is made as nutrient-poor as any soil on Earth by leaching, a process in which water removes soil nutrients. The lushness of the vegetation results from high rainfall and temperature, and efficient recycling of available nutrients.

A distinctive feature of the rain forest is its vertical stratification (Figure 48-15A). The conditions of life, the microenvironments, are very different at different elevations above the ground. Many of the organisms adapt accordingly. In a typical rain forest, the main trees all grow to about the same height, generally from 60 to 120 feet. At the top, their spreading leafy branches form a **canopy** (Figure 48-15B), which is continuously present throughout the year and that intercepts almost all the direct sunlight. In the leaves of the upper canopy, which receive full sunlight, photosynthesis is active and flowers and fruits are abundant. Still, the canopy is in some respects a difficult microenvironment. Water is at a premium, for no outside supply is available except when rain is actually falling. Variations in temperature and humidity are considerable. Perhaps two-thirds of the rain-forest species live aloft. Indeed, many plants and animals are born, reproduce, and die without ever touching the ground.

Moving down from the canopy to the forest floor, microenvironmental conditions change continuously and radically. The floor is dark, even at noon, and among green plants only a few with the most modest photosynthetic requirements can manage to grow. Rain is cut off by the canopy umbrella, but at lower levels there is high humidity of such unremitting constancy that things are usually dripping wet. The forest floor, in a sense, resembles a cave. The temperature is nearly constant throughout the day and throughout the year—usually around 25°C, and in different regions seldom less than 20°C or more than 30°C, well below maximum summer temperatures in all parts of the United States. The sustained warmth and humidity of the rain forest floor make it a natural hothouse that is unlikely to appeal to anyone accustomed to the Temperate Zone. Between the canopy and the forest floor are a range of microenvironments that provide for the animal denizens distinctive opportunities for concealment, a variety of available foods, and diverse facilities for swinging, climbing, flying, or running.

The most notable feature of the tropical rain forest is the richness of its biological diversity. It has been estimated that as many as half of all living species may live in the jungle canopy—most of them insects, many of them still unnamed. Nowhere is life more exuberant. A temperate or cold-climate forest frequently consists of one or two tree species, rarely a dozen. A tropical rain forest generally includes a hundred or more species of trees—though 500 have been counted in one such forest. Two trees of the same species seldom stand near each other.[3] The actual species present may be totally different in different rain forests, and they are sure to be different if the forests compared are in widely separated regions of the Earth. Always, however, there is an abundance of species in each forest, and the ecological make-up of a tropical rain forest is remarkably uniform.

Apart from the large forest trees themselves, two habits of vegetation are especially characteristic of tropical rain forests: lianas and epiphytes.

- **Lianas** (a word of French origin) are climbing vines. Rooted in the dark forest floor, they use standing trees as props on which they climb toward the canopy. There they spread their leaves in the light.

---

[3] Central American rain forests are estimated to contain 4000 species of trees—compared to about 700 in North America. Forest ecologists still have not identified a large number, perhaps a majority, of the plant species in rain forests. One problem is that many trees cannot be identified from the ground because their trunks look alike. Biologists often must climb (or be hoisted) to the top of a tree so they can identify leaves or blossoms.

48-15 Vertical stratification of the rain forest

(A) *Different flora and fauna inhabit ground level, lower canopy, and upper canopy regions of the rain forest. About two-thirds of the species are found in the upper canopy. The lower regions are more humid and receive less sunlight than the upper levels.* (B) Hymelobium *tree, reaching 120 feet into the canopy at La Selva, Costa Rica.*

Upper canopy

Lower canopy

Ground level

120

90

60

30

Feet

A

B

48-16 Epiphytic bromeliad growing in a tropical rain forest

■ **Epiphytes** are plants that grow on other plants without parasitizing them or deriving from them anything but a base on which to grow. Growing especially in the upper levels and canopy of the rain forest, epiphytes are well above the dark floor and are bathed in light, even though their own height is small. Orchids (some 20,000 species of them), ferns, and many other epiphytes form veritable aerial gardens among the high branches of the trees of the rain forests. Without roots reaching to a water supply in the soil, the epiphytes of a rain forest are paradoxically adapted to a dry climate. They include cacti, some of which store water in their pulpy tissues. The spread of cacti from desert soils to the rain forests is a remarkable example of what is called **preadaptation**—that is, an adaptation to one environment that turns out to be equally advantageous in another, often quite different, environment.

Other rain-forest epiphytes, notably the bromeliads[4] of South America, have their leaf bases arranged to catch rain and store it against future need (Figure 48-16). These little tanks form remarkable microenvironments of their own, in which insects, frogs, and other organisms develop and thrive. Like cacti, bromeliads largely owe their success as canopy epiphytes in rain forests to adaptations to poor water supply, which they acquired as desert plants early in their evolutionary history.

Most people picture jungles as teeming with animals. A visit to a rain forest is disappointing in that respect. Closer study reveals that animals are common enough, but probably no more so than in our familiar Temperate Zone forests and grasslands. Invertebrate density and abundance are very high, but vertebrate animals are rarely seen. They are more inconspicuous in the rain forest than in the temperate forest because they tend to be smaller than the animals of nearby grasslands, many are nocturnal animals, and of those that are active during the day—diurnal animals—most live high up in the canopy, where they are all but invisible from the ground. During the day, there is an oppressive silence in the jungle that is likely to be broken only by the chattering or howling of monkeys (most of which are diurnal) or the squawking of parrots overhead. At dusk, an ear-shattering chorus breaks out. Birds, mainly diurnal but foraging more quietly during the day, sound off as they settle for the night. Grasshoppers and allied insects make a din that is often clamorously joined by a tree frog chorus.

Insect species are incredibly abundant in the rain forest. In addition to ants, termites, flies, butterflies, beetles, scorpions, and other known insect species, a vast number of species are both undescribed and unnamed. Some insects, like the brilliant

[4] Named for a Swedish botanist, Olaf Bromel. Incidentally, this family of plants includes the Spanish "moss" (not moss at all) of our South, and also the pineapple.

A

B

blue morpho butterfly of tropical America, are found only in the rain forest. Frogs also reach a sort of pinnacle in this environment. Snakes are present but rare, contrary to the accounts of fiction writers. Mammals are less abundant in the forest than in adjacent grasslands, but they are still quite numerous. Arboreal forms include monkeys and rodents, especially squirrels. In the Old World rain forests, ground-dwelling herbivores include musk deer, small forest antelopes, and forest pigs. In South America, similar ways of life are represented mainly by terrestrial rodents and peccaries. In both hemispheres, herbivores are stalked by partly arboreal carnivores, especially cats, such as the Old World leopard and the New World jaguar. Here we see again that similar ecological roles exist in widely separated environments, but they may be filled by distinct species or even by animals of different families, orders, or classes in different regions.

C

Grasslands

In drier parts of the tropics, forests may still extend in narrow zones along the banks of streams and rivers where there is a good underground supply of water. These are the *gallery forests*.[5] Similar forest formations are seen in the temperate zone. Away from "stream galleries" are vast areas in both tropical and temperate zones where water supplies do not suffice for tree growth but do permit a heavy growth of grasses and other small herbs. These **grasslands** include the so-called plains and prairies of North America, the steppes of Russia, the savannas of East Africa, Australia, and South America, the pampas of Argentina, and the veldts of South Africa. All are ecologically similar (Figure 48-17).

The high plains east of the Rocky Mountains were once the location of the major North American grassland. Most of it has now been plowed under. It has been discovered, however, that the drier parts of the region were more productive as grassland than they were as farms, and an effort has been made to return some of them to grass.

The ecology of grasslands is dominated by a meager and intermittent water supply. Rainfall may be only 12 to 20 inches (300–500 mm) per year. On some grasslands it is far more than that, but the rainfall is unevenly distributed through the year. The irregularity of the rain, the porosity of the soil, or both factors, keep

D

48-17 Grasslands
The grasslands of the world look superficially similar, but they differ in their many species. (A) The great plains of North America (Wyoming). (B) Tall grass prairie in Kansas. (C) The Argentine pampas. (D) An arid veldt in southern Africa.

---

[5] Another type of tropical forest, which we shall not further describe here, are the *winter forests*. They occur in areas of strongly defined rainy and dry seasons. Their trees are deciduous, shedding leaves in the dry season and growing them again in the rainy season—"winter" for denizens of the tropics.

48-18 Desert
Scenes from southwest United States.

plant roots from receiving a continuous or ample supply of water. Wide differences in other environmental conditions give many grasslands their special characteristics. A savanna in the midst of the Venezuelan rain forest, and a high prairie in Alberta, have little in common except an unreliable water supply.

Different grasses are adapted to different special conditions of soil, precipitation, evaporation, and other factors. The eastern, wetter parts of the North American grasslands once had tall grasses that attained heights up to 3 meters: blue stems, Indian grasses, and slough grasses. But the tall-grass prairie was plowed under with rewarding results. It is now our richest agricultural area and includes the corn belt. In the arid western prairies short grasses predominated, with grama and buffalo grasses often growing among sagebrush. Between the extremes were mixed grasses, sod, and bunch grasses. C_4 plants are especially efficient under hot, dry conditions, and many grasses carry on this form of photosynthesis. Calculations show that in areas like the great plains of North America, the relative advantages and disadvantages of a C_4 system are roughly in balance (Figure 48-19)—and there is a 50-50 distribution of C_4 and C_3 grasses in such places. Further south, C_4 plants predominate and further north they slacken.

The grasslands swarm with animals more conspicuous and probably more numerous than in any other land community. Primary consumers are the large grazing mammals. Countless millions of bisons and pronghorns[6] once roamed North American prairies. Even now, African grasslands support large herds of zebras and of several species of grazing antelopes. Living in open country, these grassland ungulates are all fleet of foot, or **cursorial.** Hares and rodents are also common primary consumers. Some are also cursorial. Many rodents, like the prairie dogs and other ground squirrels or the pocket gophers, are burrowing, or **fossorial,** animals. Australian grasslands have herbivores very different in appearance and relationships but ecologically similar. Predators are adapted to herbivore prey. Wild dogs, lions, and the like prey on the ungulates; weasels, snakes, and others prey on the smaller herbivores. Herbivorous insects, such as locusts and grasshoppers, are also incredibly numerous, as are birds.

Deserts

Deserts are biomes that develop in the driest of environments. As aridity increases, grasslands grade into deserts without sharp lines of demarcation (Figure 48-18). Deserts have low precipitation, usually 10 inches (250 mm) per year or less, which usually falls in a few heavy showers at erratic intervals. They also have intense sunshine and very hot days, 35°C or 40°C and upward, at least during summer. Hence, evaporation rates are high. Nights are cold, even in summer, and daily variations in temperature reach extremes found in no other environment.

Water scarcity has led to numerous adaptations in various desert dwellers. For example, most annual desert plants are small. After a rain shower, they grow rapidly, bloom, and produce seed all within a few days. Among the most astonishing and beautiful scenes on Earth is a southwestern desert carpeted with brilliant and many-hued flowers a few days after a spring rain. In a few days, the desert is drab again, but scattered in it are millions of seeds waiting to perform the miracle anew.

Many perennial desert plants have small leaves or none at all. This circumstance decreases water loss. Some (usually shrubs) have tremendously long roots that reach deeply for buried water. Others, notably the desert cacti and other succulents, are shallow rooted and thus can absorb water rapidly after even a light rain and store it in spongy internal tissues. The fact that desert perennials are often spiny or thorny is doubtless related to the eternal struggle for water. Such weapons must discourage thirsty animals.

As we saw in Chapter 10, many plant species that live continuously in bright sunshine amidst a scarcity of water carry on crassulacean acid metabolism (CAM)— in which CO_2 taken up at night is fixed into such C_4 intermediates as malic acid, which is decarboxylated and then metabolized by the Calvin cycle during the day. To conserve water, these plants open the stomata of their leaves at night and close them during the day. In this way, they are able to cope with the intense heat of

---

[6] Usually called "buffalo" and "antelope," respectively, they are quite distinct from the Old World animals to which the same names were first applied.

desert days and still carry out photosynthesis. This mode of photosynthesis is common in succulent tropical plants, though certain desert cactuses have a mixed pattern of CAM and C_3 photosynthesis.

Desert animals are also adapted to a scarcity of water and extreme temperatures. Reptiles and insects produce dry, highly concentrated excrement and have waterproof outer coverings—with colors likely to reflect the heat. The cuticles and exoskeletons of arthropods are superb preadaptations to desert environments and an important reason why arthropods are so successful there.

Large mammals are rare, though some Old World antelopes are adapted to extreme desert conditions. Small rodents are numerous. Most live in deep, cool burrows. Many in different parts of the world have independently evolved a bipedal, leaping mode of locomotion. The kangaroo rat of the North American deserts is a good example. It lives exclusively on dry seeds and drinks no water, growing, reproducing, and nursing its young on a diet containing only 5 to 10% water by weight. Snakes and lizards are common, even though deserts sharply limit their activities.[7] Another adaptation by which desert animals cope with seasonal heat and dryness is **estivation**—a prolonged state of torpor. In contrast, **hibernation** is a condition of dormancy or torpor in certain small mammals—the insectivores, bats, and rodents—that gets them through cold winters by conserving energy reserves (see Chapter 26). By estivating, rodents and birds blunt the harsh seasons.

FRESHWATER COMMUNITIES

Land and aquatic habitats are as different as can be, and yet they do intergrade along shores, in coastal lagoons, and in swamps (Figure 48-20). Some plants and animals live habitually in such transitional zones. Many plants are rooted in water but rise from it into the air. Many animals spend part of the life cycle in water and part on land. Others alternate freely between them.

Freshwater habitats cover about 2.1% of the Earth's surface and are strikingly discontinuous. Organic and inorganic materials enter fresh water from the land, as plant and animal matter falls into streams.

Lakes and ponds are often stratified into three zones in which organisms live (Figure 48-21).

- A **littoral zone,** which extends from the beach to a point where the water is deep enough to be unaffected by the action of waves
- A **limnetic zone,** which is inhabited by plankton and other organisms that live in open water
- A **profundal zone,** which is below the limits of light penetration

Larger lakes in temperate regions are also thermally stratified. Since water is densest at 4°C, it sinks below waters that are either cooler or warmer. In winter, water at 4°C sinks below cooler surface water, which freezes when it reaches 0°C. Below the ice, water remains between 0°C and 4°C, and its organisms survive. In spring, surface waters warmed to 4°C sink, forcing cooler waters to the top with

[7] They are sluggish in the cold desert nights but quickly die of heat prostration in the sun. Consequently they are usually active only for short periods in the morning and evening and spend the rest of the time in burrows or crannies.

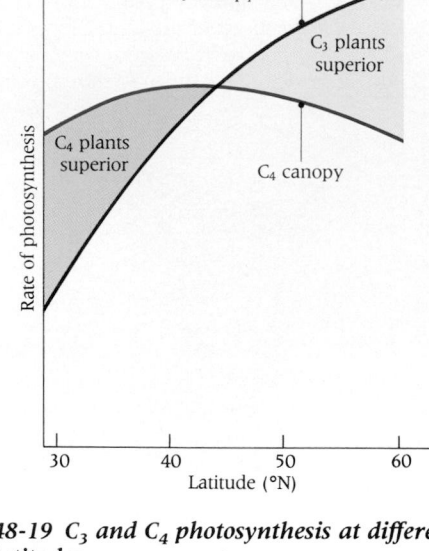

48-19 C_3 and C_4 photosynthesis at different latitudes
The predicted levels of photosynthesis in C_3 and C_4 plants over a range of latitudes in the Great Plains during July shows that C_4 plants are superior at latitudes below 45°N and C_3 plants are superior at higher latitudes. Of course, these curves would vary depending on season, moisture, and cloudiness. These types of photosynthesis are discussed in Chapter 10.

48-20 Some freshwater habitats
(A) A Manitoba lake where the water level is controlled by beavers. (B) A North Carolina millpond. (C) The waters of a clean creek join those of a muddy creek.

A

B

C

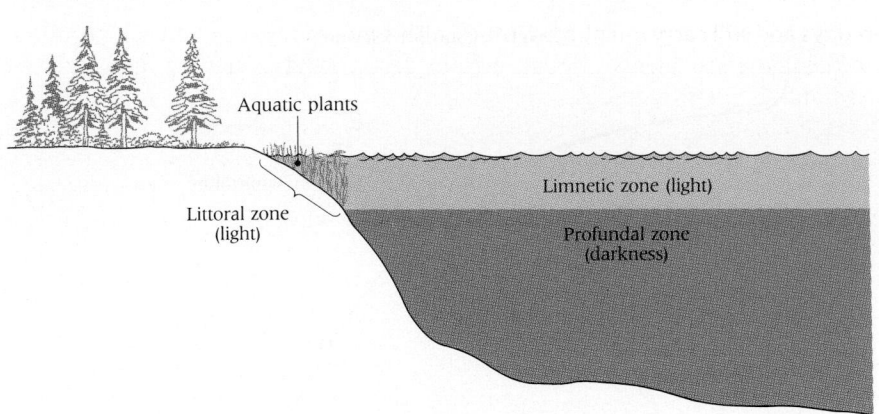

48-21 Three zones of ponds and lakes
The littoral, limnetic, and profundal zones are located in different parts of the freshwater pond. Each is characterized by a different environment for living organisms.

Aquatic plants

Limnetic zone (light)

Littoral zone (light)

Profundal zone (darkness)

nutrients that have accumulated below. This process is the **spring overturn.** As the hot summer sun raises the temperature of surface waters, this warmer water forms a zone, termed the **epilimnion** above a cooler 4°C zone known as the **hypolimnion.** Sandwiched between epilimnion and hypolimnion is a layer characterized by an abrupt change in temperature, the **thermocline.** Experienced swimmers often encounter it. In the fall, the temperature of the epilimnion drops until it equals that of the hypolimnion (4°C). When this happens, epilimnion and hypolimnion mix in the **fall overturn.** Spring and fall overturns play a key role in bringing up to the surface new supplies of dissolved nutrients from the lake bottom. Two types of lakes are recognized: eutrophic lakes, which have a generous supply of minerals and organic matter, and oligotrophic lakes, which are relatively sparse in organic matter and nutrients.

There are a good many flowering plants in freshwater communities: water lilies, duckweeds, water hyacinths, pickerel weeds, pondweeds, and others. Most of these, however, are only semiaquatic. They float on the surface or extend above it into the air. Algae are the predominant photosynthesizers of fresh water, as of all aquatic environments. Most common in fresh water are diatoms and cyanobacteria (blue-green algae), and green algae. They are the mainstay for photosynthesis in lakes and streams and may occur in such enormous numbers near the surface, where the light is strongest, as to form a scum of pea-soup consistency.

Freshwater faunas are rich in phyla and classes, more so than in any land communities. Only three animal phyla have members completely adapted to life in the open air, but almost all phyla have freshwater representatives. Among the commonest of freshwater organisms are monerans, protists, flatworms and several other groups of worms, rotifers, arthropods such as water "fleas," crayfishes, and insects (mostly as aquatic larvae), snails, and, of course, fishes.

Although we cannot discuss other freshwater communities further, we mention one extreme case—rapidly flowing water—to illustrate a peculiar environmental hazard and the adaptations it has led to. Torrential mountain streams would soon be swept free of organisms if their inhabitants did not have some means of staying in place in swift currents. For some fishes, notably trout, the situation is fairly simple. They can manage to stand still by swimming hard. Flatworms, leeches, snails, and many insect larvae in swift streams are flattened, streamlined, or limpet-like. They can cling to the bottom or to stones while the water flows by. Some snails, clams, and insect larvae moor themselves with spun fibers. Some caddis-fly larvae live in little cases to which they attach pebbles. The weight holds them on the bottom. Some fishes in swift streams lack swim bladders, and some salamanders lack lungs; both specializations make the animals heavier and better able to cling to the bottom. Other fishes and many tadpoles of swift water have sucking mouths with which they cling to rocks.

OCEAN COMMUNITIES

The oceans cover 71% of the Earth's face.[8] Over a century ago, when the wooden ship *H.M.S. Challenger* set sail from Portsmouth, England on the world's first oceano-

---

[8] The total area of the ocean is about 361 million square kilometers and that of the land about 149 million. The oceans occupy 81% of the total area in the Southern Hemisphere and 61% in the Northern Hemisphere.

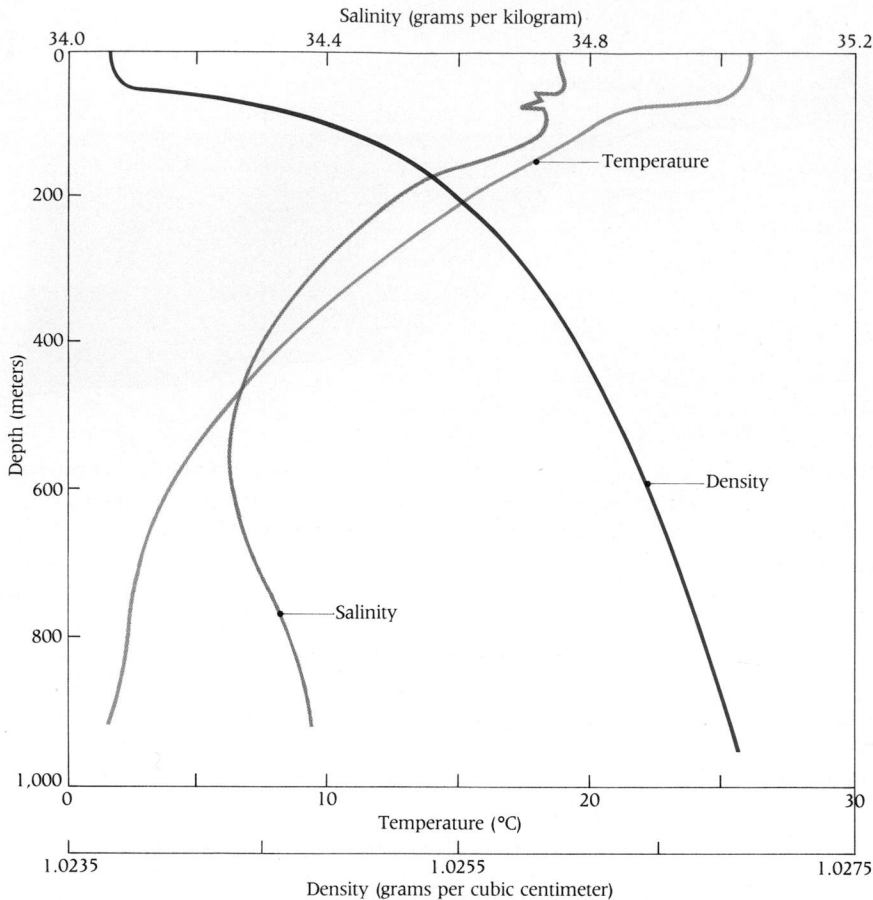

Temperature and salinity measurements were made with an instrument having a resolution of a few meters. In the top 60 meters the temperature and salinity are nearly constant due to mixing. Below this level, changes in the properties of water were as shown. The region of rapid temperature change extending down about 500 meters, is the thermocline. The region of maximum density change coincides with the thermocline.

graphic expedition, naturalists studied ocean waters by lowering bottles to prescribed depths, filling and sealing them, and then hoisting them aboard. Today, electronic devices can monitor temperature, salinity, density, and current velocity to depths of over a kilometer. Despite such advances, we have just begun to understand the ocean. Many simple-sounding questions remain unanswered, marine research having long been hindered by the difficulties of data-gathering, high costs, and various political problems.

The oceans have many features in common with lakes, but there are certain differences. At the top of the list is the salinity of the sea, which is remarkably uniform at 34 to 36 g per kg—that is, it contains 34 to 36 g of dissolved salt per kg of seawater.[9]

Unlike fresh water, seawater has no density maximum at 4°C. It becomes continuously denser as it gets colder—that is, until it freezes. Frozen sea water contains little salt and is lighter than seawater. The concentration of salt and the temperature of the water may vary from top to bottom (Figure 48-22). Another distinctive feature of the ocean is a world-wide system of warm and cold ocean currents—at the surface and in deep water. The best known—the Gulf Stream, the Labrador Current, etc.—are at the surface. Others are at great depths (see Figure 1-11).

Ocean currents trace huge surface gyrals that move clockwise in the Northern Hemisphere and counterclockwise in the Southern Hemisphere (Figure 48-23). By the ways in which they redistribute heat, they significantly influence not only life in the oceans, but also that on coastal lands. For example, the Gulf Stream in the North Atlantic swings away from North America near Cape Hatteras, North Carolina, and courses toward Europe. As a result, western Europe is much warmer than eastern North America at similar latitudes. In South America, the Humboldt Current brings colder water, rich in nutrients, northward up the west coast, making possible a great abundance of marine life that supports the fisheries of Chile and northern Peru.

[9] Oceanographers usually refer to the salinity of seawater as S ‰, which means grams of salt per kilogram of seawater. The symbol ‰ is read "per mil"—as % is read "percent."

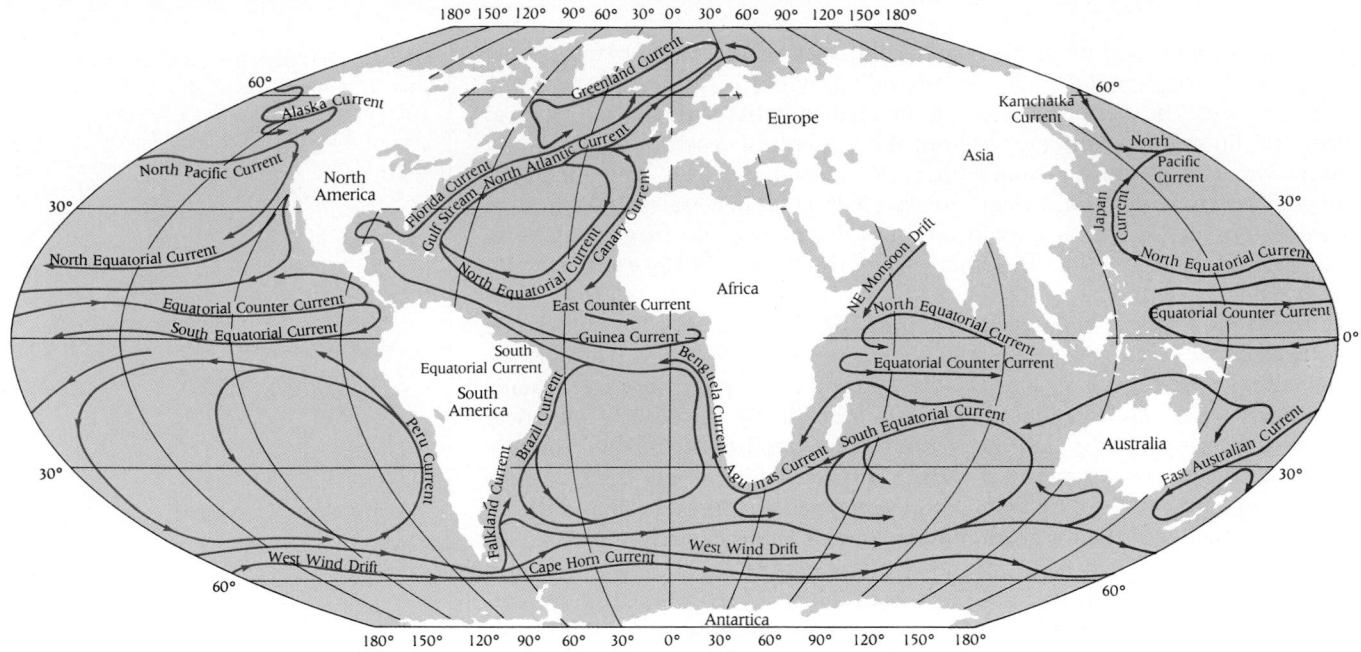

48-23 Surface currents of the world's oceans
Currents tend to form loops, with the strongest currents occurring on their periphery. Ocean currents move warm water poleward and cold water toward the tropics. This helps to distribute the incident radiation energy over the Earth's surface.

Despite the demanding character of many marine environments, most of the main groups of organisms among monerans, protists, fungi, plants, and animals—all five kingdoms—arose in the sea and are still abundant there. Among the phyla, only the Bryophyta and Tracheophyta are basically nonmarine. Even among the classes, the vast majority is dominantly marine, and nearly all have at least a few representatives in the sea. The only classes of animals that probably originated on land are those of the centipedes, millipedes, insects, and amphibians, and their descendants, the reptiles, birds, and mammals.

The brevity of this discussion of marine communities is obviously out of proportion with their overwhelming extent and richness. This mismatching of emphasis has three explanations: (1) we have already said much about marine life in earlier chapters; (2) for all their vastness and diversity, marine communities do not represent as radically divergent and sharply distinct ecological types as do those of land; and (3) the communities of land, our own environment, are more accessible for scientific study.

Marine Environments

From the biologist's point of view, there are two major provinces in the ocean: the **neritic** and the **oceanic** (Figure 48-24). The neritic province (from a Greek

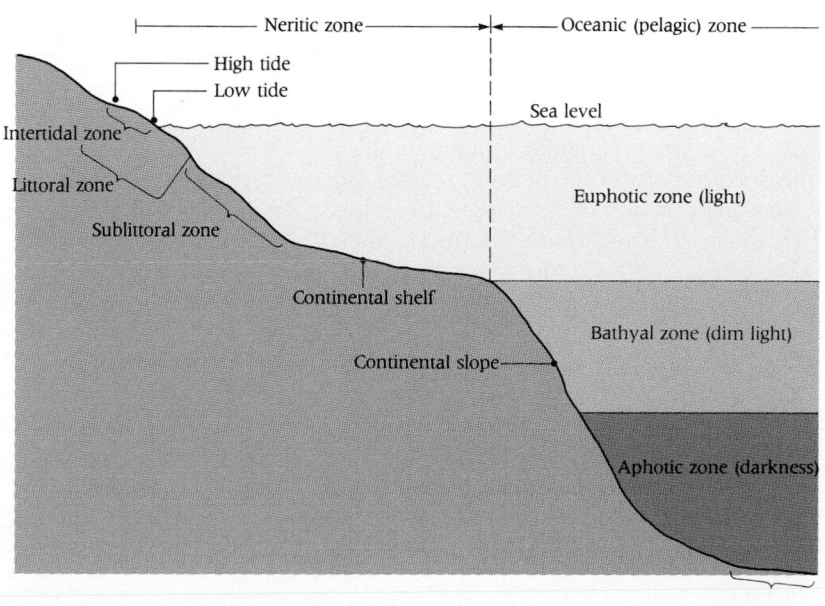

48-24 Profile of the oceanic zones
The neritic zone, which is the most productive part of the sea, constitutes less than 10% of its total area.

root denoting sea gods and nymphs) includes the shallow open waters from shore to the edge of the continental shelf. The oceanic province is the rest of the ocean and its main part. The neritic province is divided into **littoral** and **sublittoral** divisions. The littoral division extends from the uppermost reach of high tide out to shallow waters that are no longer stirred by tides and waves. The sublittoral division goes out to the edge of the continental shelf. Light penetrates to the bottom in both divisions of the neritic province. Some distinguish the **intertidal zone** between high tide and low tide from that part of the neritic province that extends beyond low tide.

Biologists also distinguish between the **benthic** division, which is the ocean bottom at all depths, and the **pelagic** (Greek, "the sea") division, which is the open water above the bottom. The pelagic division and oceanic province are virtually identical. Finally, biologists distinguish between a **euphotic** ("well-lit") zone near the surface and an **aphotic** ("lightless") zone in the depths. The aphotic zone, in turn, may be divided into a **bathyal** ("deep") zone, which is the depths below the euphotic zone that are still on continental slope, and an **abyssal** ("bottomless") zone, which is the depths below the euphotic zone in all other parts of ocean.

The littoral zone, a narrow band around all continents and islands, is as crowded with life as any place on Earth. It is also most accessible to investigation. Its outstanding environmental characteristic is the rhythmical ebb and flow of the tides, now totally covering the littoral zone, now leaving part of it exposed to air except for many shallow tidal pools. It is often argued that the intertidal zone, the part of the littoral zone exposed at low tide, may have been the home of the ancestors of those organisms that originally colonized terrestrial habitats. After all, the greater structural complexity needed to anchor an intertidal organism, and the waterproof covering needed to protect intertidal dwellers from the drying action of the air at low tide, may have served as key preadaptations for terrestriality. Solar radiation strikes the littoral zone harshly, and variations of temperature, turbulence, and salinity are far more drastic there than elsewhere in the sea.

Out beyond the low-tide mark, the penetration of radiation from the surface is the most important single factor affecting marine environments. Radiation is strongest near the surface. Here most of the photosynthesis in the ocean occurs. Temperatures at the surface vary with the seasons, as on land. The difference between summer and winter surface temperatures may be as much as 25°C though it is less than 5°C over most of the ocean surface. Temperatures are also zoned like the climates of the land, with a fairly constant surface temperature around 25°C in the tropics, grading to temperatures close to 0°C near the poles. Daily, seasonal, and climatic differences of temperature in the sea are less than on land.

As the sea grows deeper, there is no sharp point at which the penetration of radiation stops. Light becomes dimmer and dimmer until finally, below 1600 to 1900 feet (the aphotic zone), there is complete darkness. Light penetration varies with the clarity of the water and with latitude. It is greatest in the tropics and least in the Arctic and Antarctic.

Below the illuminated zone, there is eternal darkness, except for the eerie glow of those strange marine organisms that create their own light. The temperature at the top of the dark zone is usually 10 to 15°C. It grades down to near 0°C in the great depths of the oceans. At these depths, there is no daily change and little seasonal or latitudinal variation in temperature.

In the deepest parts of the ocean, pressure due to overlying water reaches nearly a thousand times the atmospheric pressure at the surface, and many kinds of organisms live under pressure more than 600 times that of the atmosphere.[10] It was once believed that high pressure made life in the great depths difficult or impossible. Now we know that pressure makes little difference to many organisms. Pressure inside the organism equals that outside the organism. As long as it remains constant, it is not felt any more than we feel the nearly 15 pounds per square inch of atmospheric pressure in which we normally live. Food supply and temperature are the more decisive factors in determining how many and what kinds of organisms occur at various depths below the lighted zone. Still it must be said that life in the ocean deep remains poorly understood.

---

[10] Compare this discussion with that of deep diving in Chapter 27.

A   B   C

48-25 Benthic forms
(A) *Pillar coral.* **(B)** *Trumpet fish in reef.*
(C) *Purple sea urchins.*

The organization of communities in the ocean depends not only on depth, temperature, light, and so on, but also on a special relationship between organisms and their surroundings. Many marine organisms simply float in the water, carried by currents and often sinking or rising with changes in radiation and temperature. These floating organisms—the diatoms, dinoflagellates, and algae—are planktonic. Together they make up the **plankton** (Greek, "wanderers"). Other organisms, especially fishes, swim freely in the water and can migrate, actively seeking what they may devour or evading what may devour them. They are nektonic (Greek, "swimming") and make up the **nekton.** Still others live on the bottom, where they may be attached or may crawl about, usually to a very limited extent. These plants and animals are benthonic or benthic (Greek for "depth" meaning, by implication, "living near the bottom") and are the **benthos** (Figure 48-25). Attached organisms in the sea face some of the same problems as the attached freshwater organisms described earlier that live in moving water. These forms—which include barnacles, stony corals, and sea anemones—display a remarkable array of adaptations to the forces generated by strong currents and crashing waves. Among them are tissue toughness, lacy or low-lying structure, and flexibility.

The kinds of communities in the sea are determined largely by physical zoning and by the various combinations of the life styles in the three zones. In the littoral division, most organisms are benthic, though some plankton and nekton come in with the tide and survive in tidal pools. In the well-lit sublittoral division of the neritic province, plankton, nekton, and benthos are abundant, and photosynthesis goes on throughout. Out in the oceanic province, the sea is vertically divided into three major environments and kinds of communities. Above, the euphotic zone is lighted, open water swarming with plankton, including abundant photosynthetic organisms and nekton. Below, in the bathyal and abyssal zones, are dark waters in which nearly all life is nektonic. At the bottom is the benthos, consisting of scavengers and organisms of decay, mostly bacteria.

Ecology of the Ocean

Many aspects of energy and material transfer and trophic levels in marine food webs are peculiar by terrestrial standards (Figure 48-26).

Consider first the entry into the system of energy and chemical building blocks. Beyond the littoral zone, marine food chains begin in the photosynthetic plankton (phytoplankton), source of 40% of the world's photosynthetic productivity. It consists largely of diatoms and dinoflagellates, although more conspicuous green or brown algae, such as the famous *Sargassum,* also occur. We now know that marine blue-green algae elaborate significant quantities of fixed nitrogen (see Table 46-3).

Large numbers of minute photosynthetic organisms, including blue-green algae (cyanobacteria) and algae, contribute half or more of the total productivity in marine

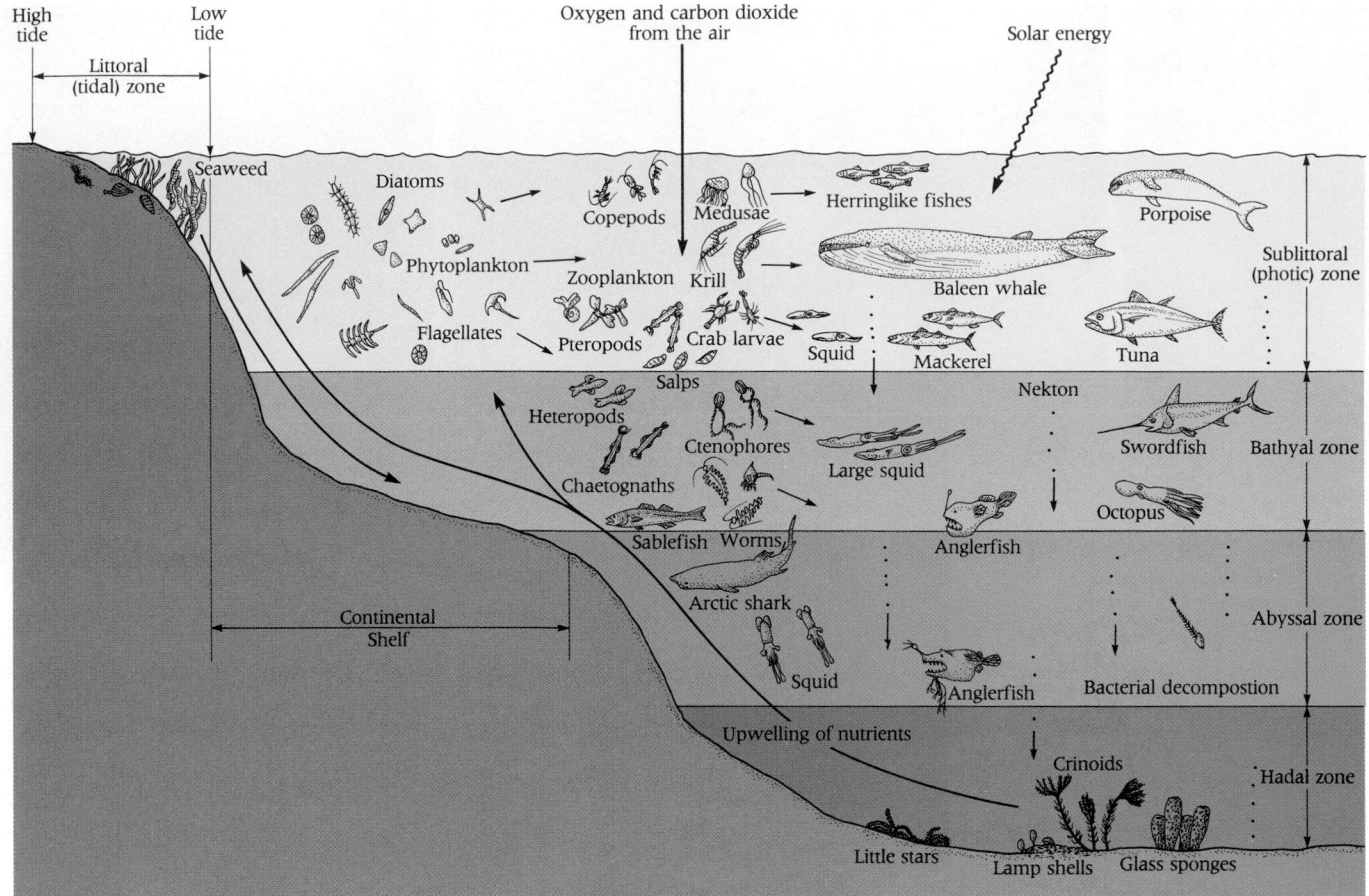

High tide — Low tide

Littoral (tidal) zone

Oxygen and carbon dioxide from the air

Solar energy

Seaweed

Diatoms

Copepods

Medusae

Herringlike fishes

Porpoise

Phytoplankton

Zooplankton

Krill

Sublittoral (photic) zone

Baleen whale

Flagellates

Pteropods

Crab larvae

Squid

Mackerel

Tuna

Salps

Heteropods

Ctenophores

Large squid

Nekton

Swordfish

Bathyal zone

Chaetognaths

Octopus

Sablefish   Worms

Anglerfish

Arctic shark

Abyssal zone

Squid

Anglerfish

Bacterial decompostion

Upwelling of nutrients

Hadal zone

Crinoids

Little stars   Lamp shells   Glass sponges

Continental Shelf

ecosystems. Blue-green algae are probably eaten by heterotrophic flagellated protists, which are estimated to consume a tenth or more of the total organic matter produced by all of the living world. Planktonic organisms are able to reproduce so rapidly, and the turnover of nutrients in the sea is so great, that productivity of the planktonic community has been underestimated.

Grazing, so to speak, on the grass of the sea's "sunny pastures" are many herbivores. These so-called zooplankton include various protists, medusae, copepods, and other crustaceans. Particularly abundant and important are the crustaceans called krill (*Euphausia*), which are microscopic in size but so numerous that they are the chief food of the largest animals that have ever lived, the blue whalebone whale (*Balaenoptera musculus*).[11] Zooplankton that eat phytoplankton produce fecal pellets that sink to the bottom and serve as one of the principal food sources of benthic forms.

Marine nektonic animals—baleen whales, toothed whales, sharks, squids, and many others—prey mostly on other carnivores. There are many long food chains of the "dog eat dog" variety in the nekton.

Dead or damaged phytoplankton cells that are not eaten by herbivorous zooplankton sink gradually to the bottom, where they nourish deep-sea (benthic) organisms. This is the fate of about 20% of the phytoplankton. In this way, communities below the illuminated zone depend on food that literally descends from above. Thus, a key element of benthic ecology is its link to the water above. Here, in the deep, no living photosynthetic organisms are to be found. Every animal within the community is potential prey for another. Some deep-sea fishes are among the most grotesque of all living things. A few can swallow whole fishes larger than themselves. Others feed on the fragmented carcasses of other marine animals.

48-26 Food web of the major ocean community
The directions of energy and material transfers within the community are indicated by arrows.

48-27 A deep-sea fish.
See also Figure 31-39.

[11] These whales, which as adults approach 100 feet in length, may eat over 8 tons of krill each day. As nursing calves, they may gain 200 pounds a day. Thin whalebone (called baleen) is attached to their upper palates. Plates of baleen, fringed at the edges, form sieves which retain the tiny krill from the vast quantities of seawater that enter the whale's mouth.

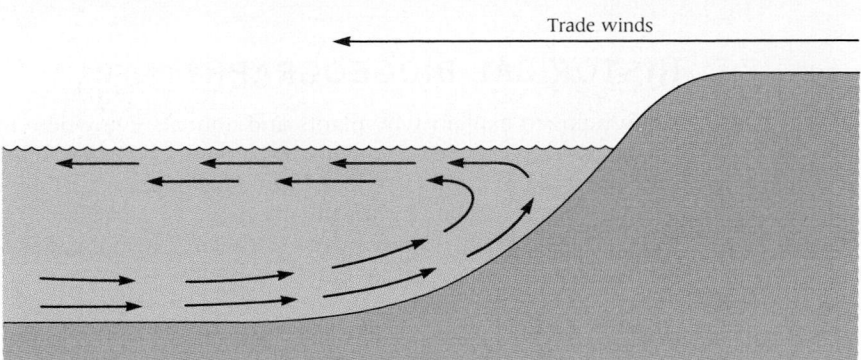

Trade winds

48-28 Zone of upwelling adjacent to a continent.
This phenomenon is associated with intense production of marine organisms because nutrient concentrations are high.

Many deep-sea creatures are luminescent, each light-producing species displaying distinctive colors and light patterns. Some are luminescent themselves; others harbor luminescent bacteria in various body cavities (Figure 48-27). The function of these animal lights in this dark setting is much disputed. It is odd that most deep-sea fishes have eyes capable of vision, though there is no light in their environment except that produced by other animals.

The rain of dead organic matter from above comes to rest on the dark sea bottom. Here, as on land, food chains come to an end with the scavenging bacteria and fungi that break down complex organic compounds into simpler molecules.[12] On land, the products of bacterial decay are largely available to green plants. Thus they start once more through the complex cycles and food chains. Since there are no photosynthetic organisms on the dark sea bottom, which is where most of the sea's organic materials eventually settle, this looks very much like a dead end. If that were so, the overall metabolism of the sea would be a one-way process, not a self-renewing cycle. In that case, life on Earth would tend to run down.

Fortunately, the total ecology of the sea is cyclical, but the cycle is completed in an unexpected way. At many places near western continental coasts, westward-blowing trade winds tend to push surface waters offshore. Their place is taken by an upwelling of water from the deep (Figure 48-28). With this water come the dissolved products of bacterial decay on the ocean floor. Wherever it occurs—sporadic in some places, almost constant in others—it has the effect of fertilizing the sea. Diatoms and other photosynthetic organisms increase enormously. The nutrient cycle is completed and another begins.

Recognition of this cycle helped to explain a remarkable natural phenomenon. When coastal surface waters are abnormally warmed by unusual currents, there arises the devastating oceanic and atmospheric condition known as El Niño, which can affect climatic patterns globally. El Niño—a Spanish term for the Christ child— was named by fishermen because it typically occurs around Christmastime in the waters off Ecuador and Northern Peru. Northward spreading warm waters of up to 22°C replace the usual cooler waters (at 16.5°C) that well up from the depths along the west coast of both North and South America. Since warmer waters are much poorer in nutrients than cool waters, there is a decrease in plankton populations—often to 5% of their normal abundance—with calamitous effects on commercial fish stocks. El Niños can launch a chain of events that influences weather in faraway places. For example, violent storms have hit the coast of California as a result of El Niños, and numerous bird colonies on islands throughout the central Pacific have starved to death as coral reefs succumbed off the Panamanian coast.

Despite great advances in our understanding of the land biomes—forests, grasslands, and the rest, we are only beginning to understand the life of the open sea. Great issues await the venturesome biologists and oceanographers of the future who are destined to reshape our notions of marine ecology with new data and imaginative insights. In these turbulent times, it is hard to imagine a field of study more important—and more daunting.

[12] We are now aware of the important trophic role of dissolved organic material known to marine biologists as *DOM*. This is undoubtedly the greatest of the organic reservoirs in the ocean, exceeding the total mass of phytoplankton and particulate organic material by several hundred times. Its average half-life of 3400 years means that it is utilized and removed slowly, perhaps only by certain deep-water bacteria that can utilize it. Wherever and however dissolved organic material is recycled, it may have a major role in the ecology of the sea that merits greater prominence than that given it to date.

HISTORICAL BIOGEOGRAPHY

Ecological biogeography helps to explain why plants and animals live where they do nowadays. Its insights, however, are never fully explanatory. Ecology tells us why monkeys live in forests and not in the deserts. It does not explain why monkeys in apparently identical ecological milieus in South America and Africa belong to different species, genera, and families—or why forests in eastern Australia, ecologically similar to those occupied by monkeys in South America, Africa, and Asia, harbor no monkeys at all.

All habitats present problems of that kind. Oysters living similar lives on both sides of the Atlantic are of different genera. Lungfishes in a few rivers of South America, Africa, and Australia are of different genera on the three continents—and completely absent in many seemingly suitable rivers. Large, spotted semiarboreal cats—leopards and their kin—prowl the jungles of South America and Africa, but the South American jaguar is a species apart from the African leopard.

These and many other problems of biogeography require explanations that are historical in nature. The Earth has changed during its long history; its floras and faunas have changed with it. Not only have they evolved into new species, they have changed in their distribution over the Earth's surface.

Facts like those just given are double problems requiring dual historical explanations. Consider the jaguar and the leopard. There is no land connection between South America and Africa today and thus no way in which the big cats could travel from one continent to the other. Yet the two species are closely related. Not long ago, geologically speaking, they had common ancestors in one place. Either those ancestors spread to both continents over routes no longer in existence, or the continents were not separated at that time. That is one aspect of the problem.

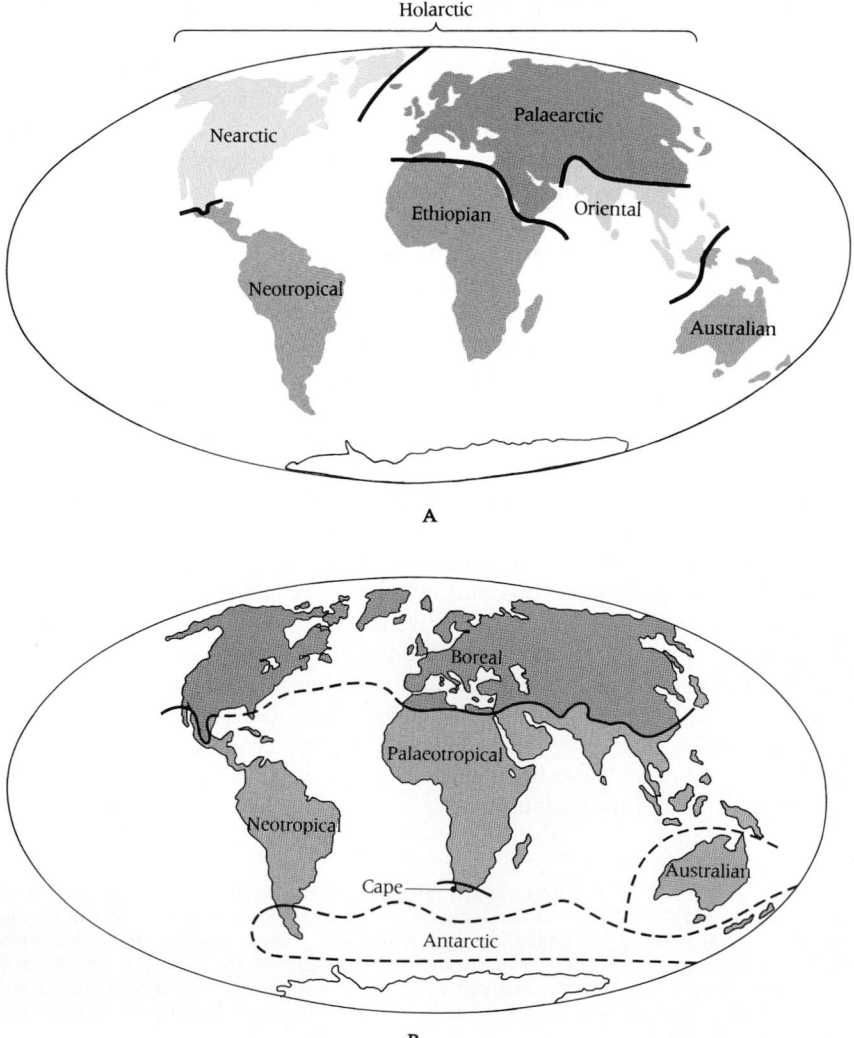

48-29 Faunal and floral regions of the modern world
(A) Faunal regions based on the distribution of mammals. **(B)** Floral regions.

The other concerns the fact that jaguars and leopards are different species. Given that there may have been a way for land animals and plants to spread to two continents (or that one original continent split into two), we must ask: How did their differences arise?

BIOGEOGRAPHIC REGIONS

Ecological and historical aspects of biogeography do, of course, interact and overlap. On the whole, however, ecological explanations apply to local questions—such as why a certain kind of community lives in one place and another kind a mile away. Historical explanations are most important for the large-scale questions, such as why the faunas of whole continents or seas have resemblances or differences.

Faunal and Floral Regions on Land

The observations from which biogeography arose began in antiquity. There was first an obvious fact: different plants and animals live in different places.[13] By the early twentieth century, the overall distribution of land mammals was known. When these facts were arranged and generalized, there emerged the familiar pattern of land **faunal regions** sketched in Figure 48-29A. The world's distinctive **floral regions,** discussed below, are shown in Figure 48-29B. The two maps show striking similarities.

The world, of course, consists of large continental land masses that are physically diverse and are separated by oceans, straits, or narrow land bridges. All of the regions have some faunal features in common. Each also has features that distinguish it from all other regions (Figure 48-30).

Division lines in nature are never sharp, especially where there is (or was in fairly recent geological times) a land connection between what are now viewed as separate regions. The zoogeographic map in Figure 48-29A is based primarily on the recent *continental* faunas of mammals and birds. Some islands—Sumatra, Java, and Borneo near Asia and the British Islands near Europe—have been connected

48-30 Mammals of the world's faunal regions

[13] Such patterns occur in the sea as well as on land, but they are not so clear-cut in the sea, nor are the marine patterns so well known. On land, regional patterns of plant and animal groups are better marked.

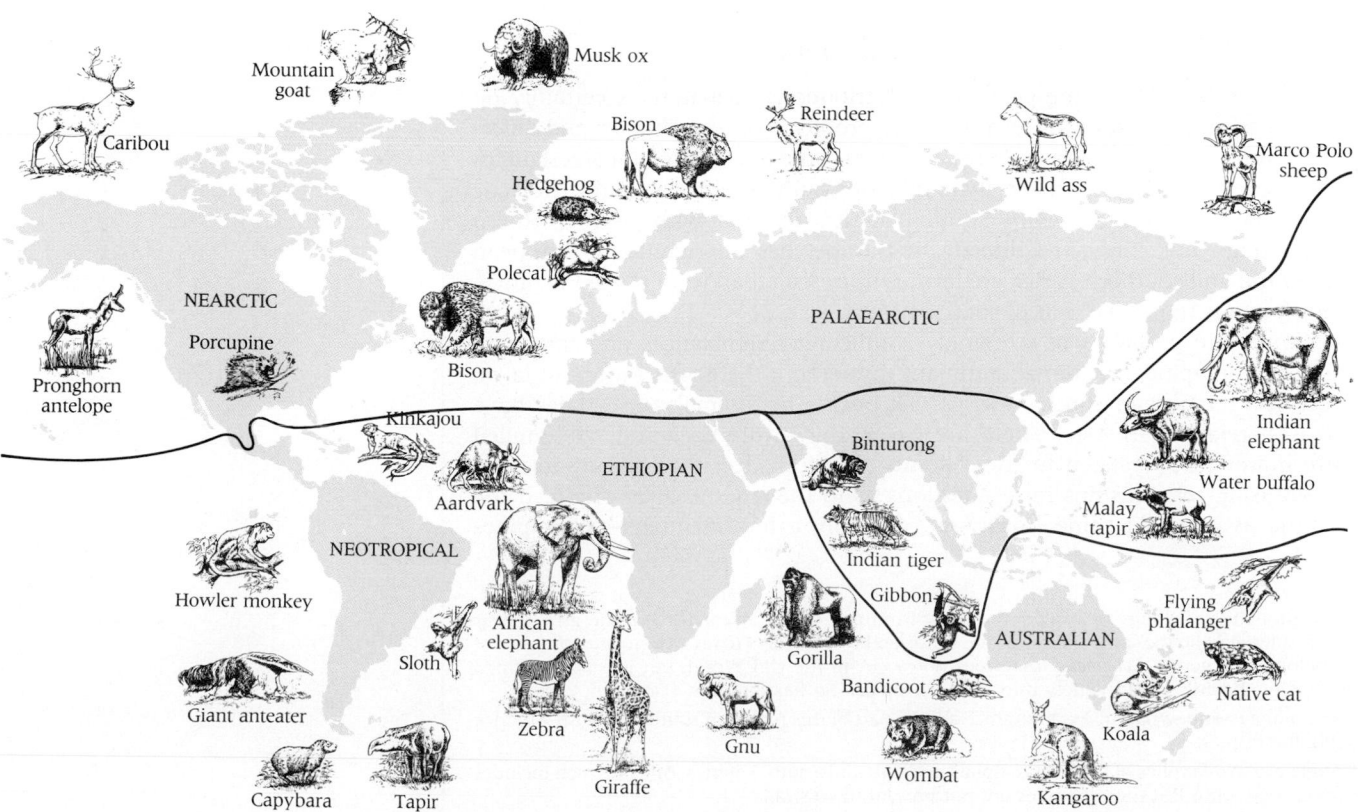

to the continents recently enough so that they still have essentially continental faunas and are here included in the continental regions. Other islands, such as Madagascar or New Zealand, lost such connections so long ago that they no longer have continental faunas. Still others—the Hawaiian Islands and the Galápagos—never had such connections or continental faunas. We leave islands without continental faunas outside the indicated continental regions. For continental faunas, the regions are real and definable. They have nevertheless been different in the past. For both historical and ecological reasons, the pattern for mammals and birds is not identical with that for some other animals—or for plants.

- The **Nearctic region** ("New [World] northers") and the **Eurasian Palearctic region** ("Old [World] northers") are often lumped together as the **Holarctic region** ("whole northers"). The timber wolf, hare, moose (called "elk" in Europe), and stag (called "elk" in America) range through most of it and only marginally elsewhere, if at all. The New World and Old World parts of the region are distinctive in a lesser way. Our commonest deer genus (*Odocoileus*) is absent in Eurasia. So is the Nearctic pronghorn antelope, mountain lion, and grizzly bear.[14] The wild boar, hedgehog, and wild ass of Palearctic Eurasia are absent here. The lists could easily be lengthened.
- The **Oriental region,** tropical Asia, is the haunt of the tiger, Indian elephant (not the same genus as the African elephant), gibbon, Malay tapir, two of the four living genera of rhinoceroses, and water buffalo.
- The **Ethiopian region,** which is Africa south of the Sahara, is especially characterized by the giraffe, zebra, gorilla, aardvark, African elephant, the other two genera of rhinoceroses, and African antelopes.
- The **Neotropical region**[15] ("New [World] tropical") includes tropical Mexico and Central and South America and is more distinctive than those just mentioned. Among the many mammals largely confined to this region are guinea pigs and related rodents, New World monkeys (ceboids), sloths, llamas, true anteaters, and armadillos.
- The **Australian region** has an even more unusual menagerie, which flabbergasted the zoologists of nineteenth-century Europe. Its mammalian fauna consists largely of marsupials, all belonging to families that do not occur elsewhere.[16] Those unusual monotremes—the platypus and the echidina or spiny anteater—are also confined to this region. There are also some native placental mammals—bats and rats.

Problems in Need of Solutions

The descriptive facts about the geographic distributions of mammals certainly do need explanations. Consider, for example, the problems of resemblance and difference. The fauna of North America north of Mexico—the Nearctic region—resembles that of northern Asia much more than that of South America. Yet the Nearctic region has land connections with South America and not with Asia. Northern Africa, although not connected directly to Europe, has an essentially European fauna. Central and southern Africa are farther from Asia than from northern Africa, but their fauna is more like that of southern Asia.

Then there are problems of seemingly conflicting resemblances and uncertain origins. The porcupine, an animal abundant throughout the North American taiga, has its closest relatives in South America. Yet most other animals of the taiga have their closest relatives in Asia. There are exceptions. Mule deer and white-tailed deer are more closely related to some South American deer than to any in Asia. These confusing relationships raise questions as to the true place of origin. If the taiga fauna as a whole came from Asia, did only a few members—porcupines

---

[14] The pronghorn belongs to a different family from any in the Palearctic, but the grizzly bear is now considered to be the same species as the Eurasian brown bears (*Ursus arctos*). Our commonest rabbits belong to a genus, *Sylvilagus*, that does not occur in the Old World, but it is not confined to the Nearctic, having spread widely into the Neotropical. So has the deer, *Odocoileus*.

[15] The name is somewhat misleading, since a large part of this region in southern South America is outside the tropics.

[16] There are two families of living marsupials in North and South America, one of which includes the familiar opossum. But these families are not present in Australia.

and deer—spread into South America? Or are the forms with South American affinities, the porcupine and the deer, of South American origin? Good historical research has shown that the first explanation accounts for the deer, the second for the porcupine.

Many of the classic problems of biogeography, including some still unsolved, arise from what are called **disjunctive distributions.** By disjunctive, we mean discontinuous or unconnected. We call two closely related groups of organisms disjunctive groups when they occur in widely separated regions with no closely related forms in between. Because they are closely related, they must have had a common ancestry in fairly recent times, geologically speaking. Three mechanisms can account for disjunctive distributions.

- Disjunctive groups may be evolutionary relicts—the scattered survivors of once widespread and dominant groups that failed to compete successfully with newer groups. The magnolias (*Magnolia*) have a modern distribution—some 80 species in southeast Asia and 26 in the Americas—that is explained in this way (Figure 48-31).

- Others are climatic relicts or habitat relicts that were greatly affected by past changes in climate, sea level, or other ecological factors. One category includes glacial relicts, plants or animals widespread in the Ice Age and now scattered disjunctively in the far north and on mountains. These are numerous. An example among the insects is the springtail (*Collembola*), which is common in Arctic locales—Iceland, Greenland, and northern Canada—but also appears in two other places—high in the Pyrenean Mountains between France and Spain and in the Tatra Mountains between Poland and Czechoslovakia.[17]

- Finally, some disjunctive groups, living and extinct, arose from the physical splitting of a once-continuous area as a result of continental drift. In such cases, ancestors of existing groups spread from one region to the other, or to both regions from a third. For the biogeographer the problem always is: what route did they follow and how? Famous examples among the mammals are the tapirs, which live only in Central and South America and in southeast Asia, and the camels, which live (as wild animals) only in South America and Asia. Both problems have been solved.

The problems of *how* disjunctive groups arose can be solved only by historical methods. Yet historical study raises problems of its own. It sometimes reveals that earlier faunal relationships were quite different from those of today. Hence, further explanations are required. The fauna of Honduras (in Central America, north of the Panama constriction) is now South American in its major affinities. But we know from fossils that a few million years ago the mammals of Honduras had nothing to do with those of South America and were all of northern affinities. They were, in fact, more nearly related to mammals of Eurasia.

Consider the world's distinctive floral regions. The distribution patterns of living angiosperms described by the British botanist Ronald Good (Figure 48-29B) include the Boreal, Neotropical, Palaeotropical, Cape, Australian, and Antarctic regions, many of which share historical backgrounds with the faunal regions. This is a more difficult field because (1) the evolutionary origin of many plant families is still not well understood, (2) there are far more living flowering plants (300 families and 12,500 genera have been described) than mammals (100 families, 1000 genera), (3) the fossil record is of little help, and (4) distribution patterns are highly diverse, perhaps because flowering plants are very successful at seed dispersal.

Of the six largest families of flowering plants, four occur almost everywhere: Compositae (daisies, sunflowers), Gramineae (grasses), Leguminoseae (peas, clovers), and Cyperaceae (sedges). Only 18% of the families are limited to one particular region, compared to 57% of the families of terrestrial mammals. Still, there are interesting regularities, as noted later.

A

B

48-31 Magnolia
(**A**) Magnolia stellata. (*Asia*). (**B**) Magnolia soulangeana (*North America*).

---

[17] Another example is the pikas (small-eared, tailless relatives of the rabbit), which occur disjunctively on various western mountain ranges and in the Yukon and adjacent parts of Alaska. These cold-climate animals became widely distributed during the Ice Age and now occur only where the climate resembles that of the Ice Age. They also occur on the cold northern steppes of Eurasia—not surprising for such a glacial relict. Disjunction between Old World and New World pikas is explained by the flooding of a former land bridge between Asia and Alaska.

Because the biogeography of land mammals is best understood, we stress their story as a paradigm of facts, problems, and derived principles. Bear in mind, however, that similar problems and principles apply to all groups of organisms.

Changing Fauna and the Changing Earth

As organisms developed, the Earth was constantly changing. Climates changed. Mountains arose and were worn down. Shallow seas have advanced and retreated where land now exists. Recent years have yielded major insights into how these changes took place (Box 48-1). The most important changes, in the context of this discussion, have been the connection and disconnection of major land masses.

Only during recent geological history—the last 15 or 20 million years—have the oceans and continents had substantially their present relationships, the only recent modifications having been in the land connections between North America and Asia on one side and South America on the other. However, with the important exception of the subcontinent of India, the plates of the Earth's crust bearing the continents have existed continuously as geographic units for more than 200 million years (see Box 48-1). The connections among them have changed, however, and that fact alone profoundly affected the distribution of organisms in the biogeographic regions.

Historical changes in any region and indeed in any community are of four kinds.

- Evolutionary change takes place within each species in a region or community.
- Shifts occur in the numbers of individuals of various species, some becoming more abundant, others less so.
- Some species disappear, either locally or by total extinction (this would be a special case of shifts in number in which proportionate numbers are reduced to zero).
- New species move into the region or community from elsewhere.

All of these changes are influenced by the geographic situation—and the map of the world has changed drastically (see Figure 1-2 and Box 48-1). The first kind of change is especially (but not solely) active when parts of a once widespread group of organisms become isolated geographically. Populations thus separated then continue to evolve independently in two or more separate regions. The last change mentioned is especially affected by geographic factors. Historical biogeography includes the spread of groups of organisms or of whole biotas into regions not previously occupied—this is called their dispersal—as well as the isolation of parts of groups or biotas previously united. Both of these processes have been connected with the shifting of continents or segments of them. For example, it appears that what is now India was once connected with Africa and then was isolated until it collided with Asia. The connection and severing of regional habitats also reflects the presence or absence of barriers apart from those directly involved in continental drift. For example, throughout the last 70 million years or more, the region now occupied by the Bering Sea (between Alaska and Siberia) was alternately *land,* which facilitated spread of land biotas between Asia and North America, or *sea,* which inhibited that sort of interchange but facilitated spread of marine biotas between the North Pacific and the Arctic oceans.

DISPERSAL AND ISOLATION

Means of Dispersal

To get from one place to another, species need a means of dispersal. Those means are most visible in animals that can fly, walk, crawl, or swim. The mobile category includes most vertebrates, insects and other arthropods, many worms, some mollusks (squids), some coelenterates (in the medusa form), most land animals (above microscopic size), and the nektonic swimming animals of fresh water and salt water.

Even among mobile organisms, dispersal is not a simple matter of packing up and going somewhere. Dispersal is not the same as migration, in which a whole population periodically moves to another region—a geographic movement, to be sure, but only to regions occupied on other occasions.

BOX 48-1

CONTINENTS ADRIFT: THE THEORY OF PLATE TECTONICS

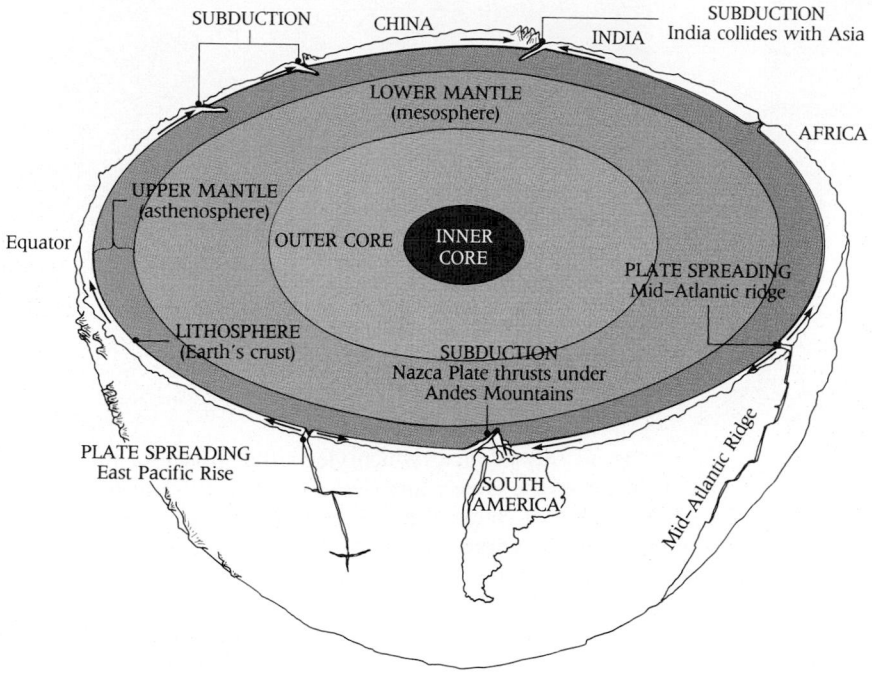

SUBDUCTION CHINA SUBDUCTION
India collides with Asia
INDIA

LOWER MANTLE
(mesosphere)

AFRICA

UPPER MANTLE
(asthenosphere)
OUTER CORE INNER CORE

Equator

PLATE SPREADING
Mid–Atlantic ridge

LITHOSPHERE
(Earth's crust)

SUBDUCTION
Nazca Plate thrusts under
Andes Mountains

PLATE SPREADING
East Pacific Rise

SOUTH
AMERICA

Mid-Atlantic Ridge

48A Cross section of the Earth

The science of geology was recently revolutionized by a new theory on the structure of the Earth. Early geologists realized that Earth's geography changed through the long span of geological time, that there were once seas where there now is land and land where now there are seas. Biogeographers in the nineteenth century concluded from the distribution of plants and animals that the relations of the continents must have been different in earlier times. However, they were uncertain as to how and where these relations had changed.

These notions remained idle speculation until 1908, when F. B. Taylor, an American geologist, proposed that the present continents are slowly drifting fragments of two

48B Patterns of ridges in the bottoms of the world's oceans
The crests of oceanic ridges are crossed by fracture zones. The dashed lines show the approximate limits of the oceanic rises. Regions of earthquakes are identified by color. Worldwide geological patterns provide evidence that the major land masses have been driven apart by a slow convection process that carries material upward from the mantle below the Earth's crust.

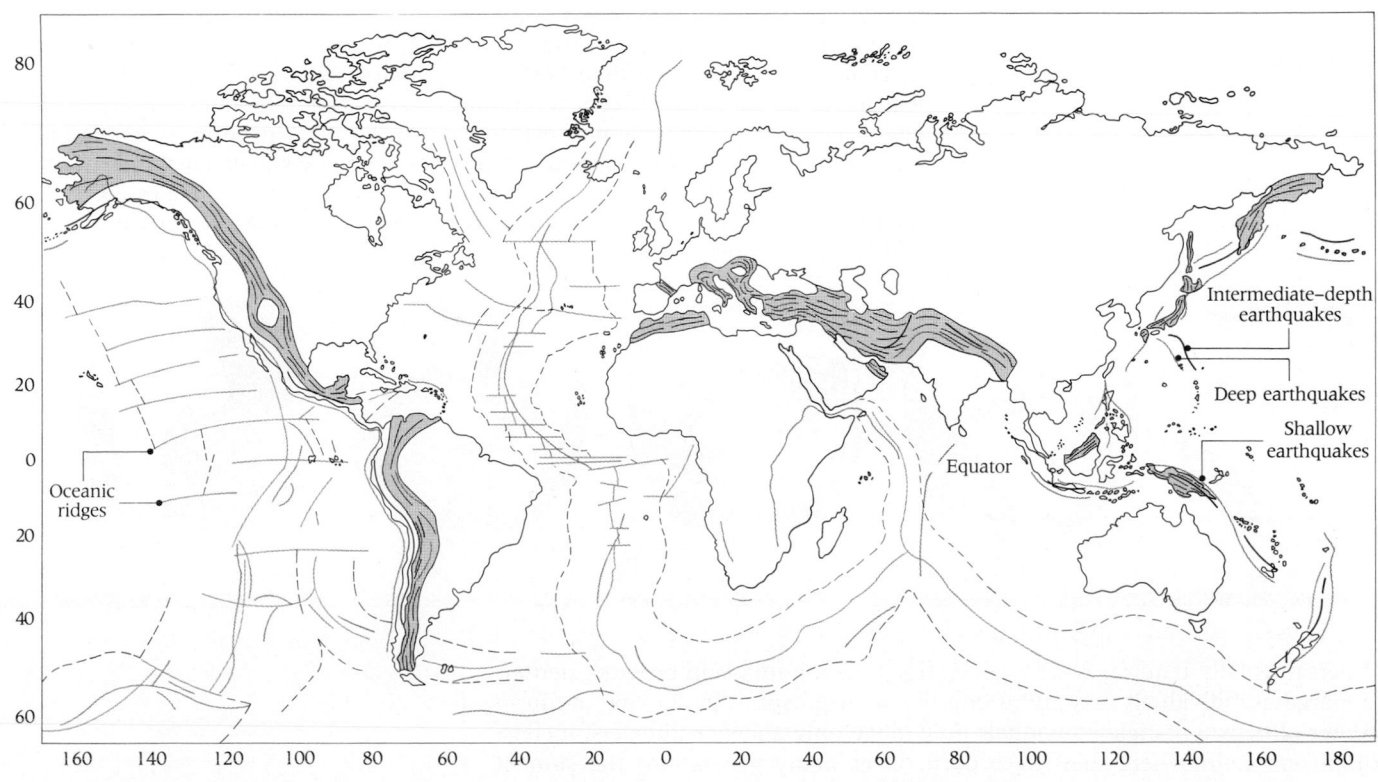

80
60
40
20
0
20
40
60

Intermediate–depth earthquakes

Deep earthquakes

Shallow earthquakes

Equator

Oceanic ridges

160 140 120 100 80 60 40 20 0 20 40 60 80 100 120 140 160 180

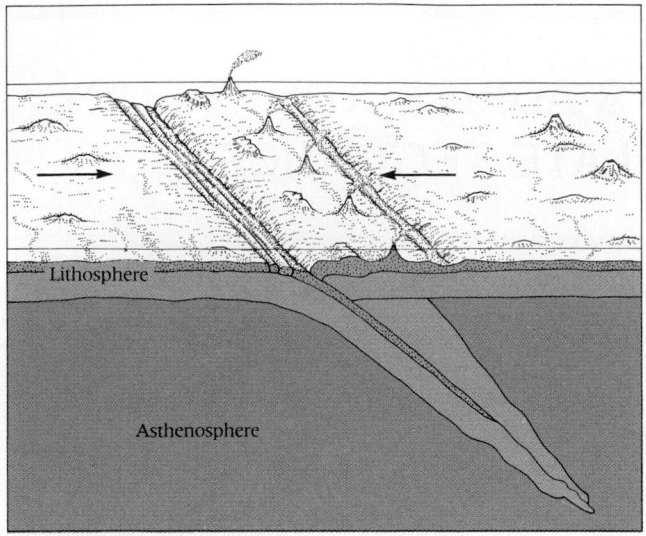

48C Subduction and movement of land masses
A trench is created where the leading edge of a plate that emerges from a fast-spreading center collides with another plate. The lithosphere plate plunges under the other and is destroyed in the asthenosphere, the hot layer below. The impact creates volcanoes, islands, and a deep trench.

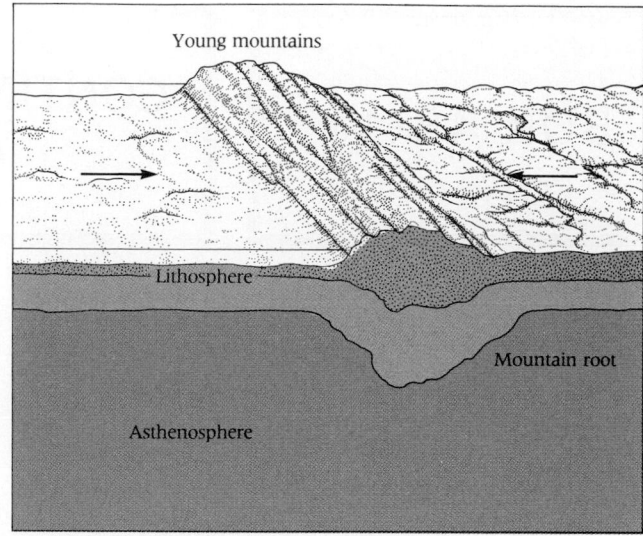

48D Formation of mountain ranges
When the leading edges of two plates come together at a rate less than 6 cm per year, both plates buckle, raising a young mountain range between them. Such ranges can be identified because they contain material typical of the ocean bottom.

great plates of the Earth's crust that had split up earlier. The large southern plate, he suggested, had split into South America and Africa. In this way he sought to explain a provocative fact, long obvious to mapmakers, that the sweeping curves of the east coast of South America and the west coast of Africa could fit together like the pieces of a jigsaw puzzle.

Taylor's ideas were ignored in his time. Even now he is seldom credited with what is correct in them. Instead, credit is given to Alfred Wegener (1880–1930), a German meteorologist, who without knowledge of Taylor's work published two papers in 1912

48E Formation of the continents
About 230 million years ago, the Earth had one ocean, now the Pacific, and one continent, now split into five. About 180 million years ago, the Sea of Tethys divided the land into two continents. Later a new split forms four continents plus Antarctica. Sixty million years ago, India heads for Asia while Australia moves laterally. The Earth is shown at the present. Compare with Figure 1-2. Details are given in the text.

and a book, *Die Entstehung der Kontinente und Ozeane,* in 1915, that advanced the same basic ideas—earlier fragmentation of continental masses and drift of the resulting fragments, our continents, into different positions. Wegener attracted ridicule, although he had a few supporters after 1920. In part, opposition rested on a reluctance to accept Wegener's idea that great continents plow about like icebergs through seas of solid material. In part, opposition stemmed from the inadequacy of the evidence. As it turned out, the critics were correct on one point—the continents do not move around through an ocean of bedrock. But they do move. A key question became: What makes them move?

Long after Wegener's death in 1930—he was found frozen on the Greenland icecap after leading an expedition—new evidence was obtained on the globe's *tectonics,* a term that refers to the nature and movements of the Earth's crust. This finally led to the now widely accepted theory of **plate tectonics,**

which views the surface of the planet as a massive mosaic, consisting of seven major and several smaller, thin rigid plates, comparable to the fragments of a cracked eggshell. Figure 48A shows a cross section of the globe with its several layers. Moving outward from the center, these are the *inner core* and *outer core,* the lower mantle or *mesosphere,* the upper mantle or *athenosphere,* and the Earth's crust or *lithosphere.* The lithosphere is a strong outer crust of rock that is about 80 miles thick. Presumably in response to forces generated in the athenosphere, the weak upper mantle layer just beneath the lithosphere, the outer crust broke up into separate plates. These have long been in slow but constant motion relative to one another and to the poles. The plates do the drifting, carried perhaps by forces in the upper mantle and pulled or driven by gravitational forces. The continents resting on the plates are thus rafted across the surface of the Earth.

The clearest evidence for this theory came

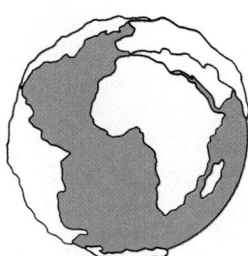

Dispersal usually requires several generations. As a population becomes denser, some marginal individuals may try to improve their prospects by moving out from the center of density. Single individuals may move only a few centimeters, meters, or kilometers from where they were born. Over many generations the sum of

from intensive efforts spurred by World War II to map the floors of the world's oceans. What was revealed was an alien landscape of deep mysterious trenches, endless plains, and strange mountains. Most startling of all was a continuous chain of ridges that meandered about the entire globe like the seams of a baseball (Figure 48B). New work revealed that the rigid floors of the oceans are continuously moving away from the ridges. In this phenomenon, termed **seafloor spreading,** new ocean bed is formed by upwelling of molten rock (magma) from the weak upper mantle layer into the ridge. There the magma hardens and moves away from the ridge in each direction, pushing a plate ahead as it is pushed from behind (Figures 48A, 48B). To make room for the new crust, old crust returns to the mantle in the deep trenches.

Investigators soon extended the idea of a rigid moving sea floor to the continents—and then to the plates themselves. Where the edges of the two plates come in contact, one may be pushed down under the other. This process, called **subduction,** creates a trench, the rock in the leading edge of a plate eventually melting to form new magma deep in the Earth (Figure 48C). This is happening, for example, along the west coast of South America, where what is called the Nazca plate (part of the floor of the Pacific Ocean), is being pushed under the edge of the plate bearing South America (see Figure 48A). In other instances, one plate may grind laterally along the side of the other. That process produces mountain ranges and earthquakes (Figure 48D).

This revolutionary theory had obvious significance for the history of life. Probably, the cores of what were to become the present continents were all incorporated within a single plate called Pangaea some 230 million years ago—toward the end of the Paleozoic era (Figure 48E).

Only in the 1960s did paleontologists begin to realize the importance of continental drift. For example, a puzzle was posed by the peculiar distribution of Cambrian trilobites. Two rather different types appeared—an Atlantic type that lived all over Europe and in eastern North America and a Pacific type that lived all over America and in the extreme western coast of Europe. Continental drift provides an explanation for these peculiar distributions. Europe and North America, separated in Cambrian times, drifted together (Figure 48F), joined, and then separated again. Portions of Europe retaining Atlantic trilobites remained with northern North America and a few bits of North America stuck to Europe. A similar theory—that Africa and North America were once separated, then joined, and then separated—accounts for the recent discovery of African-type trilobites in the American south along a broad belt stretching north from Alabama through Georgia and the Carolinas.

The theory of plate tectonics is giving us a new dynamic series of geographic settings in which the age-long evolution of organisms has occurred. As time goes on and research continues, the settings and their dates will become more clearly established than they are now. As time goes even further—say, 50 million years into the future—the map of the world will look different than it does now. The continents remain adrift.

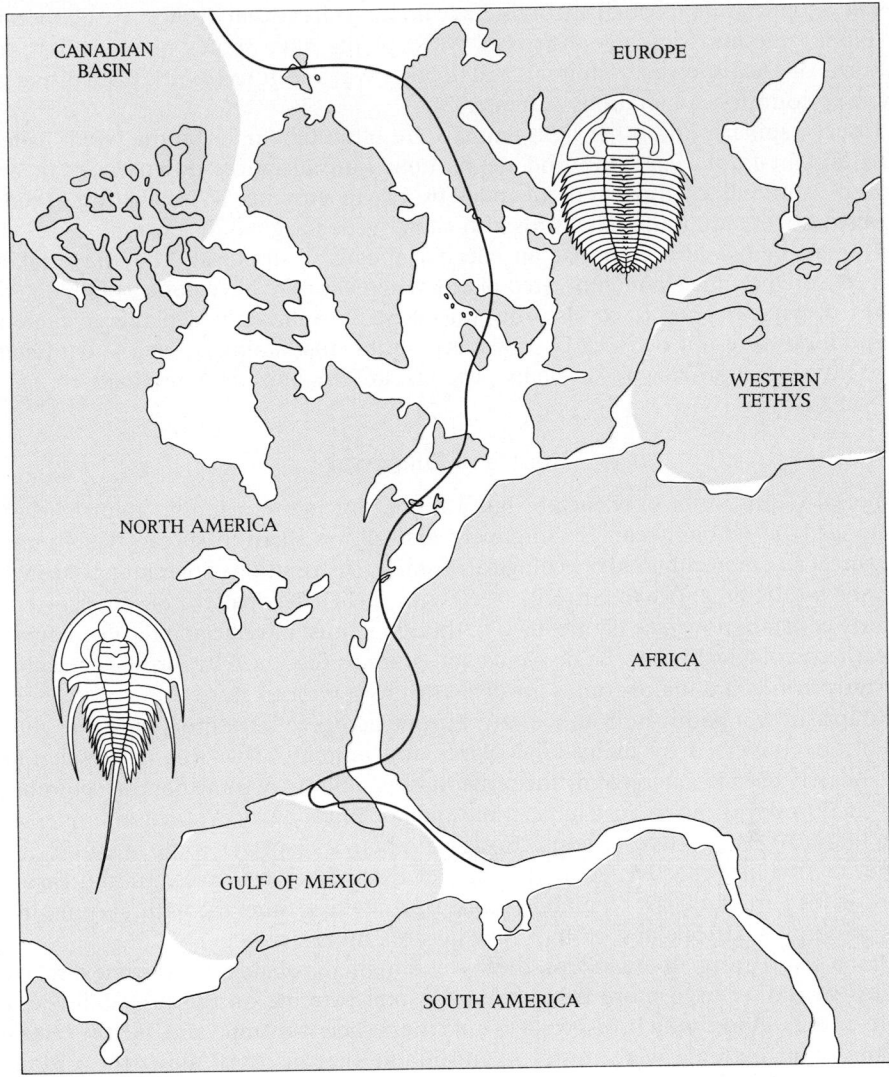

48F Distribution of Cambrian trilobites
The colored line divides regions that are connected today and marks the boundary between faunal realms that were distinctively different in Cambrian and Ordovician times. A fossil trilobite of each realm is shown: Paedeumias transitans in the Americas; Holma kjerul in Europe. As the Atlantic Ocean became smaller, differences between the two realms diminished. By the middle of the Paleozoic, there were no differences. Convergence of the two faunas suggests that what had once been a wide, deep ocean was swallowed up along zones of subduction as American and European continental plates came together.

such movements may spread the species (or species derived from it) over continents or oceans.

Protists, plants, and many animals are in a passive category as regards the dispersal of individual organisms. Plankton are dispersed by currents in which they float.

Sessile aquatic animals, such as corals, have floating larvae that in effect are temporarily planktonic and similarly dispersed. Most plants have spores or seeds that are airborne or are dispersed by animals and in other ways (Figure 48-32). Plant dispersal involves countless adaptations (Chapter 21).

Insects, spiders, and other light animals are often blown far by the wind. Fallen trees and mats of vegetation and debris float long distances down rivers or are carried hundreds or thousands of miles by ocean currents. With them go seeds, eggs, and even adult animals. Birds also carry seeds for great distances.

Human beings also comprise an effective means of dispersal. They have taken domesticated animals and cultivated plants wherever they have gone as they themselves dispersed. They have also introduced many animals and plants in regions where they were not native. Mice, rats, and other small animals, especially insects, have traveled as stowaways in boats, wagons, automobiles, and airplanes.

48-32 Dispersal of dandelion seeds by the wind

Routes of Dispersal

Dispersal requires not only means, but a route. For many nektonic animals of the open sea, the whole ocean is a highway that allows them to spread widely until they meet an environmental or ecological obstacle. Distribution of plankton is strongly affected by the great ocean currents (see Figure 1-11). The pattern of these currents is fairly constant now (see Figure 48-23), though it must have been radically different in earlier geological times.[18] For plants and animals dispersed by air, the prevailing westerly winds are major routes for dispersal from west to east.

Maps show a pathway from western Europe across to northern China that could readily be traversed by many land plants and animals. That this dispersal route was heavily used is suggested by the remarkable similarity of some natural communities in Europe to others in China, thousands of miles away. Yet communities are not identical in the two regions. However open a dispersal route may be, it is never completely effective. A marine barrier existed between western and eastern Eurasia for part of the last 65 million years. During those times the natural communities of western Europe and eastern Asia became more distinct.

For a given group of organisms, there was a higher probability that some dispersal routes would be used more than others. At one extreme on the probability scale were routes along which dispersal would have been prompt and nearly certain. At the other extreme were routes so unsuitable that dispersal along them would hardly have been possible. Such a "route" is better termed a barrier to the spread of a given group.

Overcoming the Barriers

Biogeographers recognize three types of dispersal routes.

- The easiest pathway is called a **corridor.** For whole biotas there is a continuous scale of probabilities of dispersal—or migration. In a corridor, the chances were good for the spread of many or most species of a biota (even though chances were poor or nil for some of its species). The Eurasian route just mentioned was a corridor for Holarctic floras and faunas. During the time of the latest Ice Age glacier, a 1000-mile-wide Bering land bridge between Asia and Alaska permitted many species of animals—and humankind itself—to cross into the Western Hemisphere from Asia (Figure 48-33).

- Routes are considered **filters** when they are selective—that is, when some species migrate along them while others do not because they pass parts of biotas and hold others back. The factor that makes a route a filter is usually its limited supply of habitats. There is no sharp distinction between a corridor, which is a filter for some individual species, and a filter, which is a corridor for certain species. It is merely a matter of what percentage of a whole biota follows the route. An example of a filter route is the Middle American connection between North and South America.

---

[18] For example, a current from the Gulf of Mexico once flowed through what is now Central America into the Pacific, instead of doubling back into the Atlantic as the Gulf Stream.

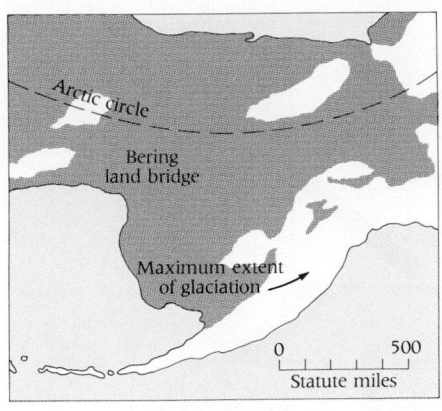

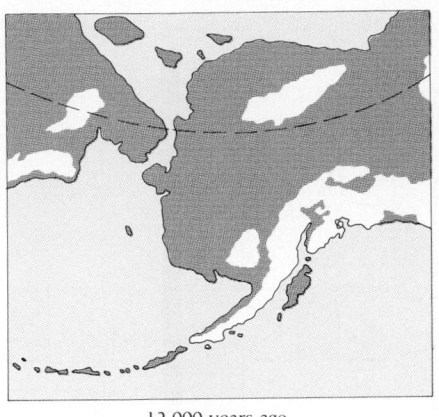

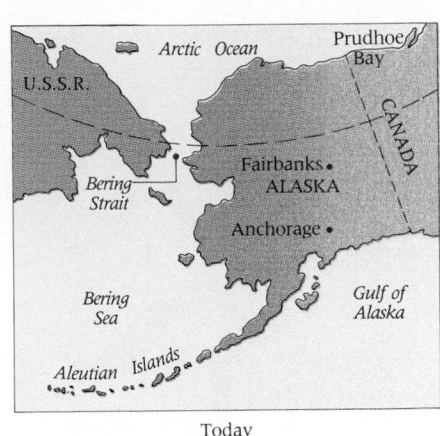

20,000 years ago

12,000 years ago

Today

48-33 Bering land bridge
Ice Age glaciers trapped so much moisture that the oceans receded. Across a 1000-mile-wide land bridge (left) wandered many species of animals, including humans. As time progressed, the glaciers melted and the seas rose, isolating the two continents from each other.

■ Some areas are surrounded by environments that are so wholly different in character that it would be extremely difficult for any organism to reach them. Chances of dispersal across this route are very low—and its occurrence is always due entirely to chance. We call this **sweepstakes dispersal** and its route a **sweepstakes route.** These are in fact barriers that are not merely difficult but downright impossible habitats for dispersing organisms. A plains animal is unlikely to cross a mountain range because it is a difficult barrier. But if somehow it did manage to find food and necessities on the way, it might make it. But a sea barrier of equal width is something else again, for the sea is an environment in which a land animal without a boat could not possibly survive. If the barrier is crossed at all, it must be by individuals and in one jump, so to speak—as a stowaway on a real boat or as a passenger on floating debris—not by gradual expansion of populations. Even the strongest barriers can be and have been crossed repeatedly in the long history of life by sweepstakes dispersal—so-named because an individual's chances of overcoming such barriers are as small as the chances of winning a sweepstakes. Yet it does occur (Figure 48-34).

When we consider whole biotas, almost any route could be a corridor for some species and a barrier for others. A filter route, of course, is a barrier for the species that are filtered out. A zone is a barrier when it is physically or ecologically unsuited for the organisms it impedes. For example, a mountain range is a barrier to species adapted to lowland conditions, and lowlands are barriers between mountain ranges.

Faunal regions are surrounded and delimited by major barriers (Figure 48-35). The change from the Nearctic to the Neotropical across Middle America is gradual, but it centers along a barrier formed by the change from temperate grassland and

48-34 Sweepstakes dispersal
A group of weevils, Cryptohynchinae, has island-hopped from west to east along a "biogeographic highway." Note the extent of diversity (number of genera) at each site. With increasing distance, fewer and fewer manage to follow the route. These insects are, even so, particularly good at sweepstakes dispersal.

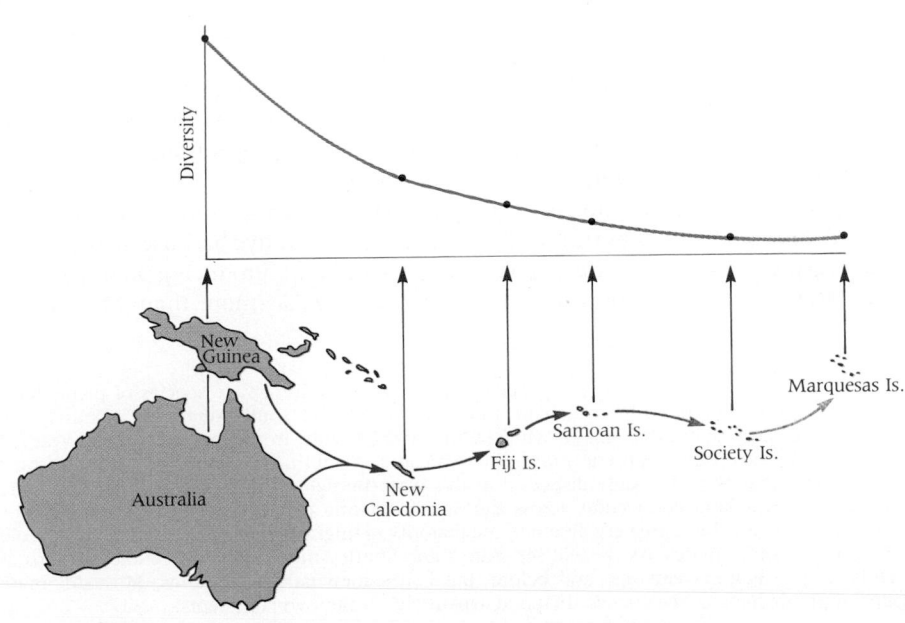

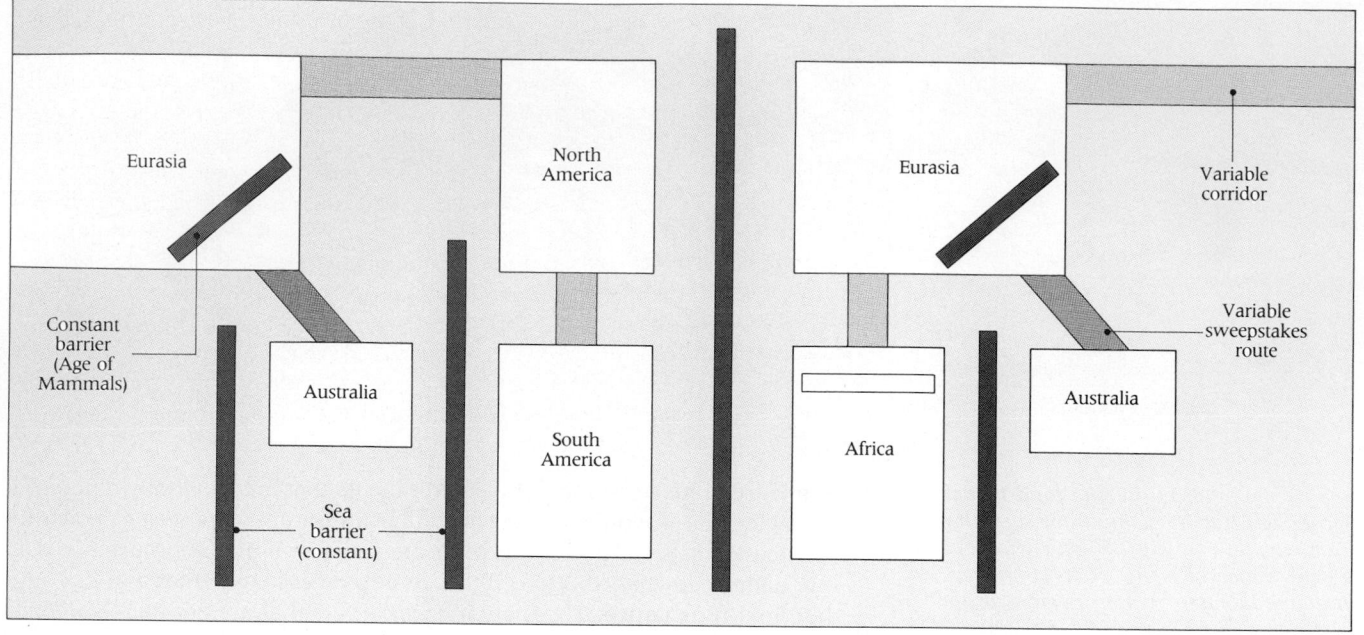

desert to tropical forests in Mexico. The Sahara and other deserts separate the Palaearctic and Ethopian regions in Africa. The Himalayas and other mountains are the barrier between Palaearctic and Oriental in Asia. These barriers are not absolute. They are filters, but strong ones, and they have not always been there.

Purely ecological barriers are among the strongest limitations on dispersal. Plant seeds may be wafted for hundreds of miles, even across an ocean. If they land on bare soil of suitable composition and in a suitable climate, the seeds will grow. Dispersal has occurred and the species has spread geographically to a new region. The catch is in the word "bare." Practically every environment is already occupied by well-adapted plants, so there is little chance that new seeds will land on bare soil. They will likely land in an established community where they must compete with well-adapted species already established there.[19]

As for sweepstakes dispersal, a great natural experiment along those lines occurred when the island volcano, Krakatoa, near Java in the East Indies, blew up on August 27, 1883. All traces of life on the island are believed to have been destroyed in the monumental explosion (though one earthworm species may have survived). The nearest island not destroyed by the eruption was over 11 miles away. Yet in only three years, 11 species of ferns and 15 flowering plants were growing on Krakatoa. Animals soon followed, and within 25 years 263 species of animals were living on the island. Most were insects, but there were 4 species of land snails, 2 of reptiles, and 16 of birds. In 1928, 45 years after the explosion, there were 47 species of vertebrates on the island, mostly flying forms (birds and bats), but including 2 species of rats. A visitor to the island today would never guess that it was little more than a steaming mass of lava a century ago. Sweepstakes dispersal also explains the origin of the biotas of the Hawaiian Islands and Galápagos Islands, two island groupings surrounded by tremendous oceanic barriers.

The 1980 volcanic eruption of Mount St. Helens in the state of Washington provided another rare opportunity for biogeographic study, because it destroyed dense evergreen forests and clear, cold streams and lakes, producing an apparently sterile landscape (Figure 48-36). Three years later, 90% (more than 230) of the

[19] There is reason to believe that the spores and seeds of innumerable species of plants have crossed the South Atlantic in both directions between Africa and South America. The number of such crossings, especially by means of winds, birds, and currents, in the last few million years of geological time must have been enormous. The African and South American floras do have some related species that were probably dispersed in this way (though a few botanists insist that there must have been a land connection across the South Atlantic in late geological times). On the whole, however, the floras are very different, the majority of migrants having failed to get a foothold in the foreign communities. As we will see, Africa and South America were in contact perhaps as recently as 135 billion years ago and before that had somewhat similar floras. Most dominant groups of plants have evolved since then and separately on the two continents.

48-35 World dispersal routes
Barriers, filters, and corridors are three major routes of world dispersal. The major features of the geographic history of faunas, especially mammals, are best accounted for by considering the continental blocks and the main sea barriers as constants and the three main filter bridges or corridors and one main sweepstakes route as variables.

48-36 Barren landscape on Mount St. Helens nine years after the volcanic eruption
Although vegetation has returned to some portions of Mount St. Helens, this view of the north face as it appeared in June 1989 shows that large areas of devastation remain. Erupting in a northerly direction, the blast stripped branches from trees, in many places leveling forests, and choked lakes with debris and mud. Forests on the south face was hardly damaged.

original plant species could be found, though many dominance patterns had changed drastically.

The possibility of sweepstakes dispersal is highest for plants and for small flying animals, notably insects, birds, and bats. Dispersal of strictly marine animals across a land barrier and of strictly land (nonflying) mammals across a sea barrier is least likely. There are, for example, no native land mammals (other than bats) on any of the Pacific Islands beyond those close enough for sweepstakes dispersal of land mammals. Some do have native rats, but these arrived in Polynesian canoes.

One reason why biogeographers have concentrated on land mammals is that they are little subject to sweepstakes dispersal and their migration routes were probably continuous land connections. Hence their geographic history is crucial in determining when and where variable earlier land connections existed. Nevertheless, it is all but certain that a few land mammals did undergo sweepstakes dispersal—and those rare instances markedly influenced mammalian biogeography in some regions. Recognition of these instances has cleared up some knotty problems.

Recent studies of continental drift, now to be discussed, have revealed yet another means of dispersal, which has playfully been called "Noah's ark." This would occur if a considerable segment of land broke away from one continent with a biota of that continent on board and eventually hit another land mass to which some of the biota was then transferred. Some believe that in this way India transported some originally African animals to Asia, but no example is known beyond all doubt. The known resemblances between African and Asiatic land mammals are almost certainly due to overland dispersal between northern Africa and southwest Asia. Yet there well may have been effective "Noah's arks" at work.

CHANGING BIOTAS AND GEOGRAPHY

The Land Mass Called Pangaea

We have seen that the fauna of northern North America is more like that of Asia, to which it is not connected by land, than of South America, to which it is connected— and that the Neotropical and Australian regions are more distinctive than the other major faunal regions. The historical basis for these facts is discussed in Box 48-1. In sum, all continents were once together (about 240 million years ago in the Paleozoic) in one big land unit, a supercontinent called **Pangaea.** This clustering of continents produced an enormous single interconnected "world ocean" called **Panthalassa.** The break-up of Pangaea, first into the two supercontinents **Laurasia** and **Gondwanaland** and then into smaller continental masses, occurred hundreds of millions of years ago. The evolutionary process was punctuated with great episodes of drift, movement, and geologic upheaval, with repeated isolations and alterations of the major land units and their faunal regions. By and large, the history of land faunas has followed the broad lines of continental change.

To the extent that a story of such mighty dimensions in space and time can be summarized, we have tried to do so in Figure 48-35, which shows major features of the geographic history of faunas, especially mammals. We see that Australia is an island continent. It has been for a long time. South America is not now an island continent, but it was during most of the Age of Mammals, the last 70 million years or so. Africa, Eurasia, and North America were separated from each other for various shorter times, but they were also connected periodically during the Age of Mammals. As a result, there has been frequent intermigration of faunas among the different major geographic provinces that have emerged most recently on the continents of today's world. For example, when a drifting Africa made its landfall with Eurasia in the Oligocene and Miocene epochs, a variety of Eurasian mammalian orders spread into Africa and crowded out some of the local forms. At the same time, some African mammals (notably the mastodons and elephant) went forth to conquer almost the entire world.

Fossils have shed much light on the relative intensities of these intermigrations during the past 60 million years or so. Estimates of the varying extent of two way dispersal of land mammals between Eurasia and North America appear in Figure 48-37. In general, dispersal was comparatively high when there was a land connection between the continents and low when there was a sea barrier. Other evidence indicates that the connection, when it existed, was between Alaska and northeastern Asia (see Figure 48-33).

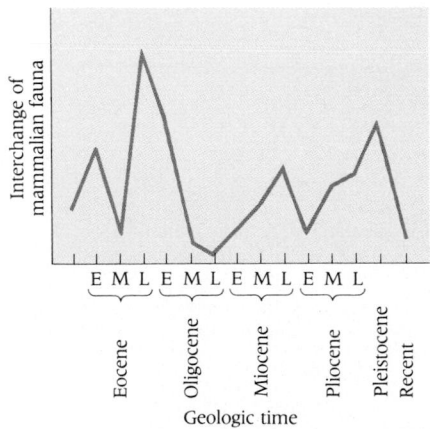

48-37 Interchange in the mammalian faunas through the Cenozoic
The interchange of North American and Eurasian mammalian faunas from the Eocene to the present is determined based on the number of items of evidence for the interchange. The intensity of the actual interchange doubtless followed rather closely the ups and downs shown here. E, early; M, middle; L, late. For details of the time scale, see Table 37-3.

The connections within the Old World kept the faunas sufficiently mixed so that they retained a broad similarity. Nevertheless, the connections were filters and new filters developed within the continents.

The differentiation of various regional faunas thus could and did occur in spite of repeated mixing by intermigration. Indeed, there did not develop a really uniform fauna, even where environmental conditions were closely similar. The alternating filter and barrier between Asia and North America caused a degree of isolation that is reflected now in their rather distinctive faunas. In the Old World, the desert filter in Africa and the mountain filter in Asia (see Figure 48-35) developed during the Age of the Mammals. Northern and southern faunas then became more sharply separated. Narrowing connections between Africa and Asia and extension of the desert filter finally resulted in the distinction we recognize now between the Ethiopian and Oriental regions.

Island Continents

Similarities between the faunas of two regions depend on the amount of intermigration between them. Regions long connected by corridors have similar faunas with only moderate local differentiation, largely on an ecological basis. Isolation, the interposition of barriers, enhances differences that on a regional basis are more historical than ecological. The longer the isolation, the greater the differences. Australia and South America, long island continents, illustrate this principle, which explains the regional peculiarities of their unusual and interesting floras and faunas.

Australia is surely the best illustration of the interplay of continental movements, climatic changes, and biotic dispersals. When flowering plants and the earliest marsupials first arose, radiated, and dispersed in the middle Cretaceous, Australia was still joined to both South America and Antarctica and had a warm, humid climate. Some 90 million years ago, it broke away and began drifting northward, reaching its present position and dry, cool climate in late Miocene times—before placentals had managed to reach it. Hence, in striking contrast with the rest of the world, which filled with placentals, most of the ecological roles (or ways of life) of Australian land mammals were filled by marsupials.[20]

Spreading over a whole continent, rich in varied environments and isolated by a strong barrier, the marsupials early produced a profusion of new species with many lines specialized in their adaptations to diverse ecological roles—an adaptive radiation on a grand scale (see Chapter 36). Since the roles assumed and niches occupied precisely paralleled those of the phylogenetically distinct placentals of the other continents, there were marsupial equivalents of squirrels, moles, badgers, rabbits, cats, and so on. The result was a striking convergence between many Australian marsupials and placentals elsewhere (see Figure 37-11).

Rats, of course, are placental rodents that had evolved on several other continents. Eventually, they reached Australia, became numerous, and evolved into many species and genera peculiar to the region. This can be explained only by sweepstakes dispersal. If the rats had come over a land connection, it is incredible that no other placentals accompanied them. Rats in fact are very good at overseas dispersal by island-hopping—via driftwood or as stowaways in boats, for example. The only other native placentals of Australia are bats, dispersed by flight and winds; and a wild dog, the dingo, whose ancestor was probably brought by aboriginal peoples some 3500 years ago. Of course, the aborigines themselves, we believe, came by boat some 30,000 years ago.

South America must have been connected with the other continents—including North America—very early in the Cenozoic, the Age of Mammals. Then the tenuous connection with North America was broken and South America remained isolated from the other continents during most of the Age of Mammals, not to be reunited with it until late in the Cenozoic. Thus the mammals of South America evolved in isolation for tens of millions of years. Although there was some sweepstakes dispersal in one direction or the other, South America started out the Age of Mammals

---

[20] Had there been a land route between Australia and the rest of the world during the Age of Mammals, there would surely have been early mixing of placentals and marsupials there. But Australia evidently started the Age of Mammals with only marsupials (and the now relatively insignificant monotremes).

with a less varied stock of land mammals than any other continent except Australia. Its three basic stocks were primitive marsupials, xenarthrans (armadillos, etc.), and ungulates. It was a strange mixture, but adaptive radiation occurred on a grand continental scale. As in Australia, there was extensive convergence toward the mammals of other lands. Placentals, evolving into families and orders peculiar to South America, took over most of the roles. The marsupials, however, became more diverse than they ever were elsewhere and took over various roles. Strikingly, all of the predacious carnivores of an island South America were marsupials. Virtually no marsupials evolved into predators on the other continents of the world. Only placental mammals did so.

Later on, around 40 million years ago (just about the middle of the Age of Mammals), two new groups—New World monkeys and rodents resembling guinea pigs[21]—appeared in South America as had the rats in Australia. Again, the newcomers probably got there by sweepstakes dispersal, island-hopping down from Central America—or across the Atlantic from Africa. Both types of newcomers expanded greatly in South America and are still characteristic of that continent.

Faunal Interchange

In the later part of the Age of Mammals, the mammalian fauna of an isolated South America had nothing in common with the faunas of North America or the rest of the World Continent. The major barrier isolating South America from North America was a sea channel or strait, the bottom of which was lifted above sea level in the late Tertiary to become the mountainous Panamanian land bridge. In this way, South America's isolation was abruptly ended some 3 million years ago— when the Andes were doubling their height and, by blocking moisture-laden winds from the Pacific, drastically altering the continent's climate and ecology.

The new "gateway" led to a dramatic historic event that is known as the Great American Interchange. At first in a trickle and later a flood, mammals from each continent poured onto the other (Figure 48-38). As noted earlier, such mixtures

[21] Recent discoveries indicate that monkeys and rodents probably arrived at about the same time. Although those rodents are called caviomorphs, "guinea pig-shaped," the first ones were not much like guinea pigs in general appearance and they became wildly diversified. They include the largest rodents now alive and among extinct ones the largest that ever lived. There are now many rodents in South America that are not descendants of these early arrivals.

48-38 North and South American fauna
(A) *Divergence of mammals occurred over millions of years when North and South America were not connected. Mammals found in the north and south were very different from one another. Those of the south included many that are now extinct.* **(B)** *Divergence of mammals began about the end of the Pliocene, after a land bridge was formed between North and South America. The armadillo, native to South America, migrated northward. Many mammals moved into South America, and some mammals indigenous to this area became extinct.*

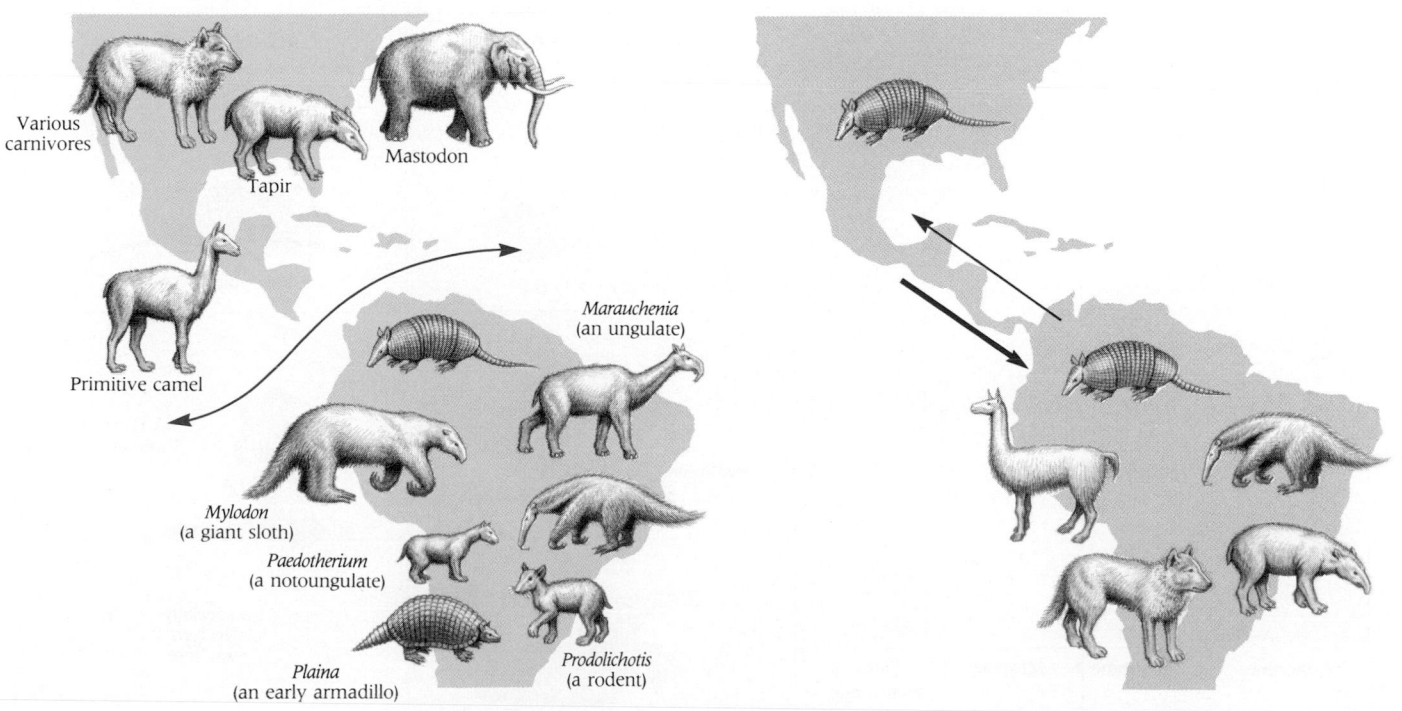

Various carnivores

Tapir

Mastodon

Primitive camel

Marauchenia (an ungulate)

Mylodon (a giant sloth)

Paedotherium (a notoungulate)

Plaina (an early armadillo)

Prodolichotis (a rodent)

A

B

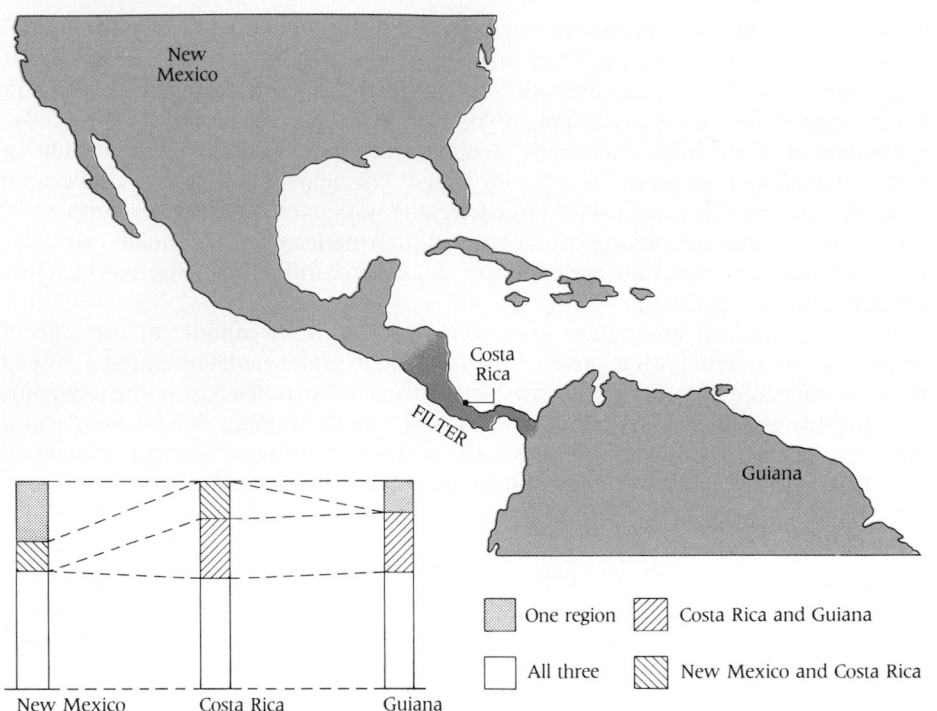

48-39 The Isthmus of Panama, a major filter route
The filter action of the isthmus is well illustrated by the graphs, which show the high proportion of families of mammals in both New Mexico and Guiana that either have entirely failed to cross the isthmus (shaded) or have gotten only halfway across (diagonal lines). The total height of the column for each of the three zones represents 100% of the local mammalian fauna.

New Mexico

Costa Rica

FILTER

Guiana

☐ One region   ◨ Costa Rica and Guiana

☐ All three   ◪ New Mexico and Costa Rica

New Mexico   Costa Rica   Guiana

of faunas have often occurred after the disappearance of a barrier, but this one is the clearest and best studied example (Figure 48-39).

The Panamanian connection was an efficient filter that passed only animals adapted to its tropical environment. Going from north to south in great numbers were wild dogs, raccoons, cats, weasels, field mice, peccaries, deer, tapirs, and many others. But other common North American mammals—beavers, pronghorns, and bison—did not make it. Going in the opposite direction, from South America into North America, came porcupines, capybaras (large, amphibious rodents, extinct here now but still present in South America), armadillos, glyptodonts (large extinct relatives of the armadillos), giant sloths (large extinct relatives of the living tree sloths), and opossums. Most South American mammals, however, failed to get through the filter.

48-40 Stratification of South American fauna
Three strata of South American fauna exist. The ancient stratum, present in the early Cenozoic, was found only in South America. The middle stratum, present in South America since the late Eocene or Oligocene, arrived by sweepstakes dispersal. The young stratum was present in South America since the late Miocene, and radiated to North America later after the land bridge was formed.

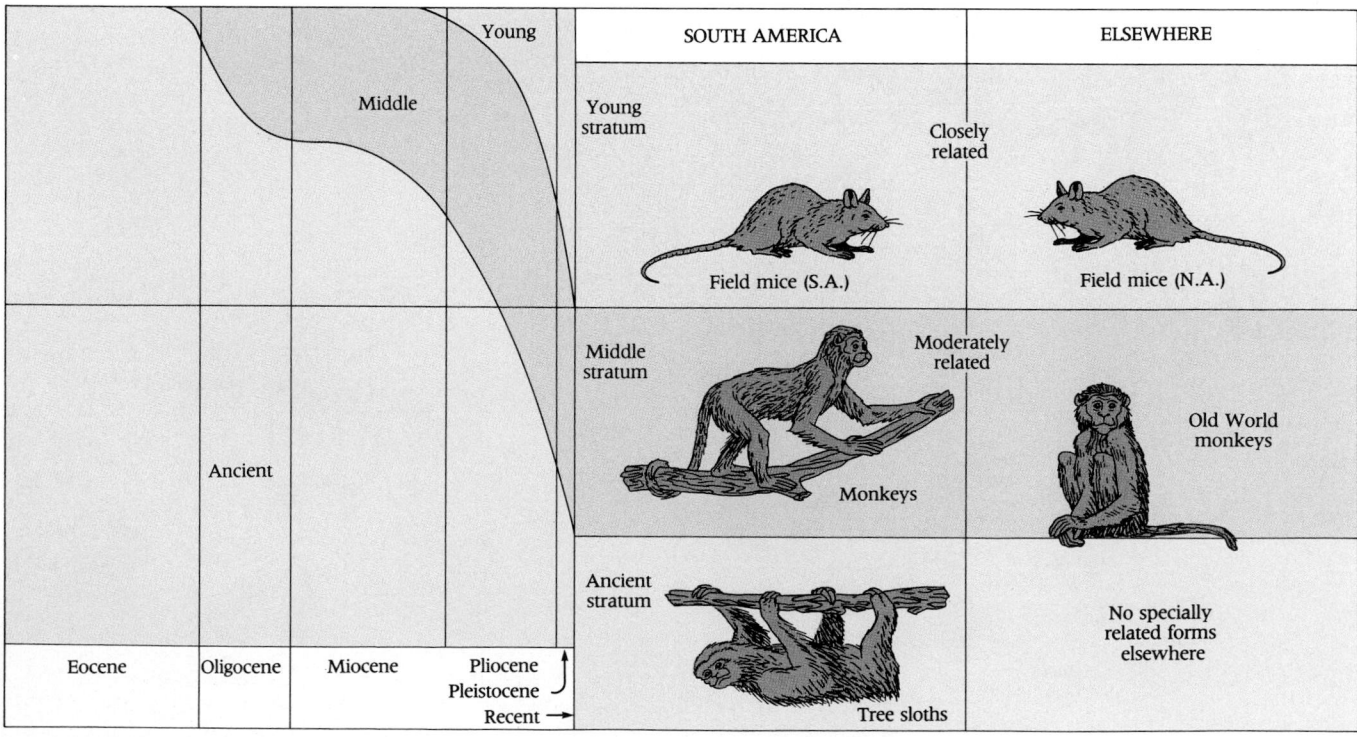

Both continents became temporarily richer in land mammals than they had been. Before the interchange, South America had 26 families of land mammals. The number rose to 39 after the interchange began and then fell to the present 35. An equivalent number of families experienced a comparable rise and decline in North America, which now has 33 families. Thus the interchange was both symmetrical and reciprocal.

The increase in diversity led to some duplication of ecological roles. Animals that had evolved convergently on the two separate continents now came into direct contact and competition. As we saw in Chapter 47, such a situation cannot last indefinitely. Ultimately, one of the competing forms wins out and one becomes extinct. Indeed, the Great American Interchange was followed by the widespread extinction of species, genera, and whole families. The mammals of North American origin were far more successful in South America than were those of South America in North America. Today in North America, only 29 (21%) of living land mammal genera are descended from South American stocks. But in South America, 85 (50%) of living genera derive from North American stocks. In other words, fewer immigrant northerners became extinct and there was greater diversification. The best-known mammals of known or probable southern origin still present in our fauna are the porcupine, armadillo, and opossum, and they do not loom very large in our fauna. In present-day South America, mammals that are descendants of comparatively recent invaders from North America include all of its native hoofed mammals (horses, llamas, peccaries), and many rodents. All of the predatory marsupials became extinct.[22]

Faunal Stratification

It has seldom if ever happened that a *whole* biota, a complete and integrated community, was dispersed all at once. There is always some filtering. Thus, it appears that the regional communities we have today consist of species whose ancestors spread into that region at different times. For example, in our now decimated grasslands fauna, the bison (or "buffalo") and pronghorn (or "antelope") were the major large herbivores, both equally at home on the high plains. But the ancestors of bison came here from Asia quite recently geologically speaking—some 500,000 years ago. The ancestors of the pronghorn, on the other hand, have been here for tens of millions of years.

The division of a fauna into different strata, depending on how long the various groups have been in the region, is particularly clear among the land mammals of South America (Figure 48-39). Three easily distinguished major faunal strata occur there. The oldest consists of descendants of animals dispersed to South America when it was connected with other continents around the beginning of the Age of Mammals. Armadillos, sloths, and anteaters are its prominent surviving members. The next stratum descends from animals that reached South America by sweepstakes dispersal around the middle of the Age of Mammals: monkeys and many native rodents such as the guinea pig. The youngest stratum consists of groups that invaded from North America when the continents were reunited toward the end of the Age of Mammals: field mice, dogs, cats, deer, and many others.

As a rule, with some exceptions, older faunal strata are more distinctive and peculiar to their region than are younger ones. In other words, the longer a group has been in a particular region the more likely it is to evolve along lines different from those of its relatives elsewhere. This is eminently true of the South American strata. There is nothing like a tree sloth anywhere else on Earth. South American monkeys do resemble their relatives in Africa and Asia, but they belong to a distinct superfamily (Figure 48-40).

[22] The distinctiveness of the present Neotropical region rests on three factors. Some old-timers that evolved in South America when it was isolated still survive there and have not crossed the filter to North America: armadillos (only one species is north of the filter), sloths, anteaters, many rodents allied to the guinea pigs, and monkeys. Some mammals of North American origin have become extinct there, but not in South America (the camels, llamas, and tapirs). Other groups are still common to the two continents, but in most cases they have diverged somewhat on the two sides of the filter. For example, the deer of South America are all of distinctive species, and several belong to distinctive genera that evolved there from North American ancestors but have never spread back to the north.

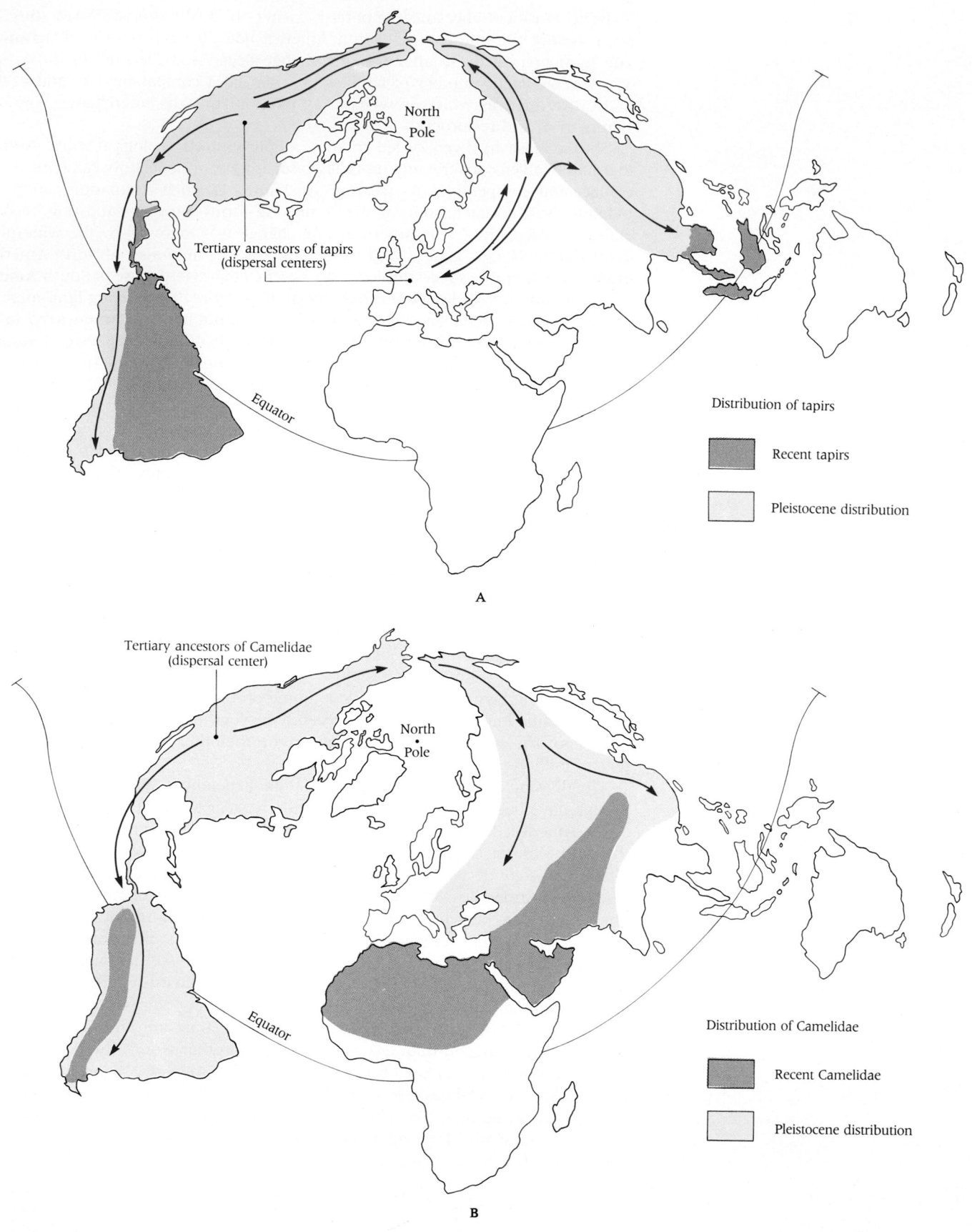

A

B

Disjunction

We are now ready to return to an aspect of disjunctive distribution (Figure 48-41). These are the cases, we recall, in which a species may occur, say, in South America, Africa, and southern Asia—and nowhere else.

The most striking and disputed instances of disjunctive distribution involve southern land areas. The tapirs and camels have already been mentioned as examples.

48-41 Disjunctive distributions
The strange disjunction in modern distributions like these is explained by their earlier distributions as revealed by fossils. Present-day disjunctive distributions are relicts of earlier continuous distributions. (A) Distribution of tapirs. (B) Distribution of camels.

Distribution of tapirs

Recent tapirs

Pleistocene distribution

Tertiary ancestors of tapirs (dispersal centers)

Equator

North Pole

Tertiary ancestors of Camelidae (dispersal center)

North Pole

Equator

Distribution of Camelidae

Recent Camelidae

Pleistocene distribution

Others include the marsupials in Australia and South America (absent in Eurasia), the southern beeches (*Northofacus*) and pines (*Araucaria*) in Australia and adjacent islands and in South America (but not in Africa or the northern continents), and a group of strictly freshwater fishes—the *characins*, often found in tropical aquariums—in South America and Africa (and nowhere in the north). Many other instances of disjunctive distribution exist among both plants and animals. We know from fossil evidence that many of these now disjunctive groups formerly occurred in northern lands. For many of these there is no reasonable doubt that they spread between the Old World and the New World across the Asia-Alaska land connection. Change in climate reasonably explains their present survival only in the southern parts of the two hemispheres. It is known that the northern lands now have much more severe climates than they had during most of geological time. They are only now emerging from an Ice Age. That explanation clearly applies to the tapirs (see Figure 48-41A), which formerly ranged all over the Holarctic region and hence spread southward into the Oriental and Neotropical regions. It also applies to the camels (see Figure 48-41B), which lived only in North America during most of the Age of Mammals and hence finally spread to Asia and to South America.

Fossils have demonstrated that many groups now disjunctive in the south were formerly northern. It is therefore a reasonable hypothesis, in the absence of contrary fossil evidence or of special considerations, that such disjunction involved the spread between Asia and North America. It is nevertheless likely that there are exceptions.[23]

[23] Continental drift has probably played a part in the disjunction of marsupials, a few other animals, and some plants between South America and Australia by way of Antarctica. Continental drift may also have been involved in disjunction between South America and Africa, but that could only apply to very old groups of plants and animals as so many of these have evolved since the two continents drifted apart. It is also possible that some groups, such as rodents and monkeys, did cross the Atlantic barrier by sweepstakes dispersal.

SUMMARY

Biogeography, the study of the distribution patterns of living things, has two branches. Ecological biogeography seeks to show how ecological forces define the distribution of plants and animals and influence their communities in different locations. Historical biogeography concerns the historical factors underlying plant and animal distributions.

Plants play a leading role in ecological biogeography, especially on land (though they are also important in aquatic communities). Plant life is strongly influenced by climate and soil. Animals are more dependent on certain vegetation types than on climate and other aspects of the physical environment.

The climates of Earth are determined by levels of solar radiation and patterns of precipitation and temperature. These in turn are influenced by such factors as latitude, altitude, and prevalent air currents as well as topography.

Ecologists studying terrestrial communities refer to the major climax formations of Earth as biomes. We have surveyed six major biomes, though there are many more. The tundra of the far north is typified by shrubs, mosses, and lichens similar to those found in mountainous (alpine) regions. Its permafrost layer and fragile surface exposes the tundra to many ecological hazards. Further south are the coniferous forests of the taiga (boreal forest) and then the temperate deciduous forests, each with its characteristic flora and fauna.

Tropical rain forests support a much greater diversity of plant and insect life than does any other region. Grasslands rather than forests predominate in areas where water is scarcer. Deserts arise in climates that produce less than ten inches of rain per year.

Because of the relative constancy of aquatic environments, freshwater communities show less variation than do terrestrial ones. However, they exhibit thermal stratification. Because fresh water is densest at 4°C, it sinks below waters that are cooler or warmer.

As a result, the thermal layers of a large lake undergo important overturns in spring and fall.

Ocean communities are not overturned in this way because seawater becomes progressively denser as it cools and freezes. The physical domains of ocean communities range from the rapidly changing intertidal zone to the harsh, unchanging abyssal regions, where organisms subsist on the debris that sinks from the euphotic surface regions. The products of this decay are inaccessible to other organisms until deep ocean currents return them to the surface. The chief nutrient producers of the oceans are photosynthetic planktonic microorganisms rather than plants. As a result, ocean ecology is dominated by the enormous numbers of tiny larval and juvenile animals that live on plankton.

Historical biogeographers seek to explain such phenomena as the presence of widely differing biotas in different regions with similar climates. Historical explanations also account for disjunctive distributions, in which related species are found in widely separated regions: they may be scattered survivors of once evolutionarily dominant groups, or they may have been separated by climatic changes or continental drift. Routes of dispersal are important factors in isolation and therefore speciation. Some are corridors, allowing species past the barrier unselectively. Others are filters, allowing only a few species to pass, and some are barriers so rigorous that only sweepstakes dispersal—the chance passage of a few individuals—is possible.

Dispersal of land species reflects the remarkable (and long unnoticed) drift of Earth's continents that followed the breakup of the supercontinent Pangaea about 230 million years ago. Because Australia and South America were island continents earlier and for a longer time than other major land areas, their fauna are appropriately distinctive. When faunal strata are present in a region, this often reflects successive waves of invaders.

KEY TERMS

abyssal zone
alpine region
aphotic zone
benthic
biogeography
biota
canopy
cursorial
desert
disjunctive distribution
epilimnion
epiphyte
estivation
euphotic zone

fossorial
Gondwanaland
grasslands
hibernation
hypolimnion
intertidal zone
intertropical convergence
Laurasia
limnetic zone
littoral zone
neritic province
oceanic province
Pangaea
Panthalassa

pelagic region
permafrost
plankton
plate tectonics
preadaptation
profundal zone
seasonal overturn
sublittoral
sweepstakes dispersal
taiga
temperate deciduous forest
thermocline
tropical rain forest
tundra

QUIZ QUESTIONS

1. Which of the following is *not* affected by solar radiation?
   A. wind currents
   B. precipitation
   C. temperature regimes
   D. rate of photosynthesis
   E. All of the above are affected by insolation.

2. Which of the following is *not* characteristic of tundra?
   A. vast coniferous forests
   B. permafrost
   C. large numbers of waterfowl during summer

D. few species living there year around
E. broad ranges of temperature

3. In which of the following is the diversity of species greatest?
   A. temperate deciduous forest
   B. tropical rain forest
   C. desert
   D. grassland
   E. taiga

4. Which of the following is *not* a zone within various aquatic ecosystems?

A. limnetic
B. euphotic
C. abyssal
D. neritic
E. All of the above are such zones.

5. A dispersal route that allows some species to pass but prevents passage of other species is called a
   A. barrier.
   B. corridor.
   C. filter.
   D. sweepstakes route.
   E. None of the above.

ESSAY QUESTIONS

1. What are biotas? Why do plants play a leading role in determining the geography of biotas? What are biomes?

2. Why does climate vary with latitude? Why do the tops of mountains at the Equator have an ecological resemblance to Arctic lowlands?

3. What accounts for the thermal stratification in freshwater lake

communities? Explain the spring and fall overturns. Does this occur in the oceans?

4. What are the major divisions of the marine environment? How does the penetration of radiation, depth, and temperature affect the composition of marine communities? How do freshwater communities differ from marine ones? What kind of environment is the largest abode of life and supports the greatest variety of fundamentally diverse

organisms—the ocean, fresh water, or land?

5. What sets a jungle apart from a grassland? A temperate deciduous forest from an arctic-alpine community?

6. Why is the fauna of northern North America more like that of Asia than of South America? Why are the fauna of Australia and South America so distinctive?

REFERENCES AND SUGGESTIONS FOR FURTHER READING

Attenborough, D. (1984). *The Living Planet: A Portrait of the Earth*. William Collins & Sons, London, U.K.

A popular account of the biomes of the world with fine illustrations.

Briggs, J. D. (1987). *Biogeography and Plate Tectonics*. Elsevier, New York.

A brief review of the interplay between dispersal of organisms and the shifting of continents as a result of plate tectonics.

Browne, J. (1983). *The Secular Ark: Studies in the History of Biogeography*. Yale University Press, New Haven, Conn.

A readable account of biogeographical thought from Linnaeus to Darwin, beginning with the seventeenth century conception that animals were distributed over the Earth from Noah's Ark.

Cox, C. B., and Moore, P. D. (1985). *Biogeography: An Ecological and Evolutionary Approach*, 4th ed. Blackwell Scientific, Boston.

The popularity of this respected text may deserve some of the credit for the recent renaissance of biogeography.

Dietz, R. S., and Holden, J. C. (1970). The breakup of Pangaea. *Scientific American 223*, October, 30–41.

This stimulating discussion of continental drift is one of 15 readings from *Scientific American* that appeared together in 1972 under the title *Continents Adrift* (W. H. Freeman, New York).

Finchman, A. A. (1984). *Basic Marine Biology*. Cambridge University Press, New York.

An excellent introduction to the biology of the sea.

Hallam, A. (1972). Continental drift and the fossil record. *Scientific American 227*, November, 56–66.

The fossil record of plants and animals are put in the conceptual framework of continental drift.

Isaacs, J. D. (1969). The nature of oceanic life. *Scientific American 221*, September, 146–162.

The author shows that marine environments have given rise to a food web in which the primary production of organic matter is carried out by microscopic plankton.

Jackson, D. D. (1989). Searching for medicinal wealth in Amazonia. *Smithsonian 19*, February, 95–102.

An interesting summary of the medicinal compounds derived from plants in the tropical rain forest. Most have been known for years to the Amazonian Indians.

Kasting, J. F., Toon, O. B., and Pollack, J. B. (1988). How climate evolved on the terrestrial planets. *Scientific American 258*, February, 90–97.

Earth cycles CO_2 into its atmosphere to promote greenhouse surface warming. This article discusses why the climates of other planets are inhospitable to life.

Molnar, P. (1988). Continental tectonics in the aftermath of plate tectonics. *Nature 335*, 131–137.

This 1988 article celebrates the 25th anniversary of the paper in *Nature* by F. J. Vine and D. H. Matthews that first connected seafloor spreading with the tectonics of continents and plates. A stirring review.

Nelson, G., and Platnick, N. (1981). *Systematics and Biogeography*. Columbia University Press, New York.

A useful synthesis of theory and observational data on organismic diversity that includes an account of the development of systematics and biogeography and an application of cladistic methods to biogeography.

Pears, N. (1985). *Basic Biogeography*, 2nd ed. Longman, New York.

A readable and lucid text on a fascinating branch of biology.

Perry, D. R. (1984). The canopy of the tropical rain forest. *Scientific American 251*, November 138–147.

Largely unexplored, the canopy of the tropical rain forest contains one of the most diverse plant and animal communities on Earth. The author's explorations using a new way of reaching the canopy allowed close observation of its ecology.

Rasmusson, E. M. (1985). El Niño and variations in climate. *American Scientist 73*, 168–177.

This discussion of the large-scale interactions between ocean and atmosphere shows how they affect weather patterns around the globe.

Simpson, G. G., and Roe, A. (Eds.) (1958). *Biogeography: An Ecological and Evolutionary Approach*. Yale University Press, New Haven, Conn.

A seminal work on ecological and historical biogeography.

Vermeij, G. J. (1978). *Biogeography and Adaptation*. Harvard University Press, Cambridge, Mass.

In outlining the forces underlying the diversity of forms in the world's great oceans, the author shows how natural selection influenced the shapes of living creatures and their patterns of distribution.

Whittaker, R. H. (1975). *Communities and Ecosystems*. Macmillan, New York.

A concise and authoritative overview of the features which characterize the biomes of the world.

Windley, B. F. (1984). *The Evolving Continents*. John Wiley & Sons, New York.

An authoritative text that includes interesting accounts of recent developments in this important field.

Hyla andersoni

Gymnogyps californianus

Vulpes velox hebes

oenothera deltoides ssp. howellii

HUMAN POPULATIONS AND ENVIRONMENTS

We now approach the end of a grand tour of the world of life. Fittingly, our tour traversed a great circle. It began with human achievement. It concludes with aspects of human biology. Human connections with biology transcend the obvious—the fact (sometimes overlooked) that the human species is but one of many species. It is part of nature. But it is more.

First, human beings are the source of biological knowledge. By their creative acts, data-gathering, marshalling of evidence, and recognition of past error, they have given themselves a perception of life that is a significant part of that human singularity, culture. Remarkably, evolution has produced an organism that is busily acquiring knowledge of itself.

Second, human culture has become a powerful directive force in the history of life. By altering the land, exploiting and overexploiting its resources, polluting the biosphere, and developing forces potentially capable of ending life, humankind has indeed become a force to be reckoned with. In part, we have acted in these ways as part of our struggle for survival. But with the evolution of culture, human motives have come to include what might be called a struggle for fulfillment, which operates in the realms of ideas, beliefs, purposes, loves, and hates.

LOOKING BACKWARD AND LOOKING AHEAD

These matters raise questions of great importance. As the human species faces its uncertain future, can the rapidly accumulating store of knowledge lead it to wisdom in the management of its affairs? How can science, particularly biological science, contribute to this wisdom? What will become of the human community in the distant future? Can biology help us to fathom its prospects? What will be biology's contribution to an understanding and solution of humanity's problems?

We close our survey of life with a glance at a few of the sectors in which biological science will almost certainly intrude on human life in the years ahead. We will confine ourselves to three areas of interest and concern: the nature of human races, the population explosion, and the preservation of the environment.

ORIGIN AND NATURE OF HUMAN RACES

Any traveler meeting the peoples in a rain forest of the Amazon Basin, or in a mountainous community of central Asia, or on an African savanna, or along the Norwegian seacoast, will attest to their differences (Figure 49-1). The human species is endlessly varied, and among its variations are sets of characteristics by which we distinguish local populations—groups that are entirely analogous to the demes and subspecies of plants and animals. However, despite differences in skin color, stature, hair form, and facial structure, and differences in more subtle features,

Human populations and environments
The crowd symbolizes the burgeoning of the human population and the problems it entails. One is the threat of extinction for many endangered species, of which four are shown in the painting on the preceding page, which opens Part 4, Section 4. As this chapter makes clear, humanity itself is an endangered species.

49-1 Some of the races of present-day humans
(**A**) *European.* (**B**) *African.* (**C**) *Australian aborigine.* (**D**) *American Indian.* (**E**) *Chinese.*

such as blood groups and enzyme systems, all humans are more alike than different. All share a set of specialized anatomical characters in the skull, brain, hands and wrist, and ankles and feet, not found in any other related primate species. All humans also share behavioral characteristics that include language, body adornment, property rights, sexual roles, and incest taboos.

Of all species, humans are the most widely dispersed. Like any widespread species, *Homo sapiens* has become subdivided into geographic races. As we have seen, small local populations (demes) tend to have gene frequencies somewhat different from those of adjacent demes, partly as a random result of mutations that are distributed and preserved by sexual reproduction, and partly as a result of natural selection under local conditions. Nevertheless, there is gene exchange among demes, and groups of demes tend to have genetic features in common that distinguish them, on the average, from other, more distant groups.

The human races have practically all differentiated on a geographic basis. In the early stages of human evolution, populations were sparse and widely separated. Passage of a gene by inheritance from one population, say, of central Africa to one in China would take a great many generations. It might well not occur at all.

Thus, human races differ genetically, but have no effective intrinsic reproductive isolating mechanisms. To be sure, there are barriers to interbreeding, but they are mainly cultural, social, religious, or political. In part, cultural barriers follow geo-

TABLE 49-1
TAXONOMY OF *HOMO SAPIENS* ACCORDING TO BLOOD GROUP FREQUENCIES

Early Classification*	Later Classification†
Early European	Early European
European	Lapps
	Northwestern Europe
	Eastern Europe
	Mediterranean
African	African
Asiatic	Asian
	Indo-Dravidian
American Indian	American Indian
Australoid	Indonesian
	Melanesian
	Polynesian
	Australoid

* Boyd, 1950
† Boyd, 1963

49-2 The value of mapping gene frequencies
(A) *Computer-drawn map of European gene frequencies of a principal component of 21 alleles of the human HLA-A and HLA-B loci (see Table 34-5).* **(B)** *Such studies, of which many are available, support the idea that farmers spread from the Near East to Europe 10,000 years ago.*

graphic divisions. However, cultural barriers often follow different lines. For example, in many countries, generations passed with little interbreeding between the wealthy and the poor. That barrier is now fading. In some regions, great social distances still play the role once played by geographic distances. The Amish of Pennsylvania and the Hutterites of South Dakota and Canada exist in rigidly defended enclaves that isolate them from surrounding communities. Proof of this isolation—and the resulting operation of the founder principle—is seen in the distinctive gene frequencies for such traits as blood groups (see Box 34-2). It is unlikely that such barriers could ever approach the effectiveness of the original geographic barriers. Today there are rarely distinct geographical divisions between the human races, since they generally intergrade over wide areas. It is probable that racial differences will decrease as time goes on.

Many of the phenotypic differences among races doubtless reflect adaptations to differing environmental conditions. There is fairly good but inconclusive evidence that differences in skin color were adaptive when and where they arose. Dark skin in tropical and subtropical regions would be a protective adaptation against damaging ultraviolet solar radiation. Lighter skin color was adaptive where such radiation (which is beneficial in small amounts) was deficient.

"How many races are there?" is a difficult question that is often asked. Authorities have recognized as few as three and as many as 30 races. One well-considered 1950 classification defined six races on the basis of blood type frequencies (Table 49-1). It is fair to point out that the three races recognized by all authorities are African (Negroid), Caucasian (Caucasoid), and Mongolian (Mongoloid).

Human variations embrace not only skin color but also head size and shape, facial form, nose form, blood groups, and biochemical traits such as lack of red cell glucose 6-phosphate dehydrogenase, various polymorphisms of plasma proteins (see Table 34-2), the presence of hemoglobin S, or sickle hemoglobin (see Boxes 22-1 and 34-4), and so on.

A more direct approach to the investigation of genetic racial differences, now actively under investigation, rests on the study of blood groups and histocompatibility (HLA) loci. Since all humans have a large number of blood group systems—the ABO system, the CDE (or Rh) systems, the MNS system, Lewis, Duffy, and dozens more—and many HLA genes, it is possible to evaluate alleles. In a recent study, gene frequencies of a large number of histocompatibility antigens yielded a map of Europe (Figure 49-2) that provided fascinating clues on the migrations of peoples in Neolithic times.

As peoples move about more and more, the old barriers to interbreeding—geographical, cultural, and social—are crumbling. Human cultural evolution proceeds

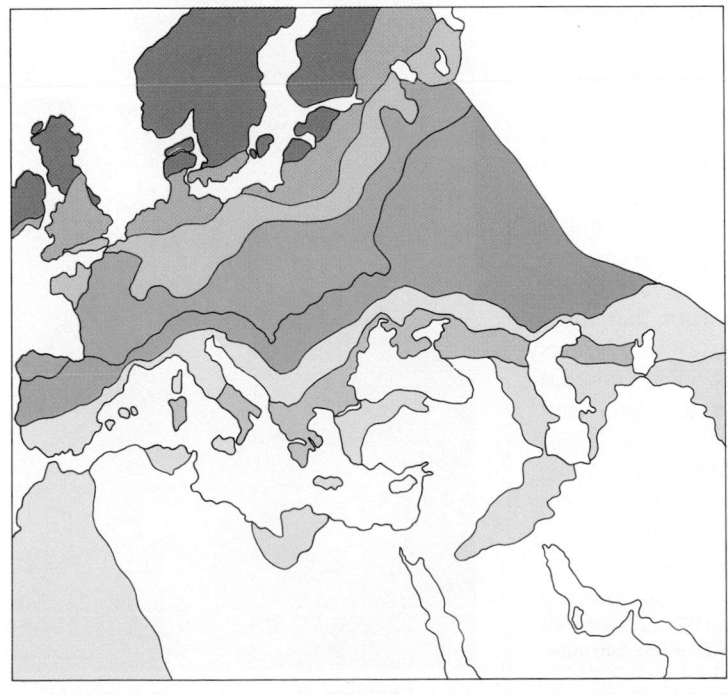

A

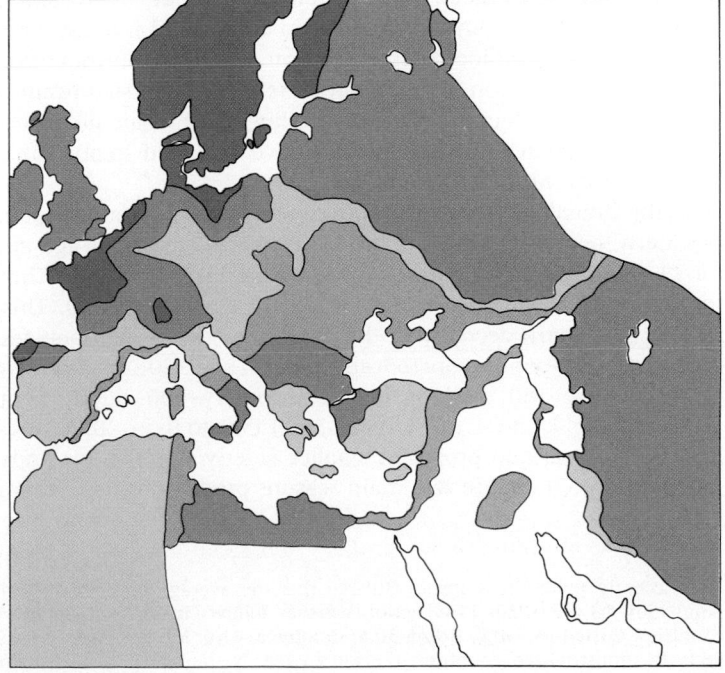

B

faster than its biological counterpart. Our civilization now takes rapid notice of and gives dominance to novel contributions of a few talented individuals. Human beings are unrivaled in their ability to alter their environments and to influence the evolution of other species. They have even developed basic tools that could alter the future evolution of their own species.

THE SIZE OF THE HUMAN POPULATION

The recent upsurge in the size of the human population constitutes a fateful problem for the future of humanity. Indeed, the severity and gravity of the threat it poses is of the highest order. It is estimated that about 5 million people lived at the time when agriculture developed. At the time of Christ, there were 130 million people. Computers somewhere in Washington officially informed us that on July 7, 1986, the birth of an unnamed person somewhere brought the world's population to 5 billion. Although our hypothetical calculations concerning future population trends in the United States assumed an annual rate of increase of 0.85%, the rate throughout the world in recent years has risen sharply, at times exceeding 2%. Significantly, rates of increase are highest in the poorest countries.[1] The expected increase in world population in 1990—90 million people—exceeds the total populations of France, Britain, and Germany.

Until recently, there was every reason to suppose that by the end of this century, world population could be twice its present size. That is a short doubling time, and the number being doubled is now a very large one. However, a worldwide survey conducted in 1980 in 61 countries revealed surprising recent decreases in fertility and birthrates in both developed and developing countries. Whether or not it is temporary remains to be seen.

The U.S. Census Bureau recently suggested that the United States will have a population of 309 million in the year 2050. Then, the Bureau predicts, the population may start slowly to decline. That contrasts sharply with the predictions in Chapter 45, where we calculated a U.S. population in 2025 of 302.7 million and in 2097 of 500 million based on the assumption that the intrinsic rate of increase would remain stable at 0.85% per year. If the Census Bureau is right, the intrinsic rate of increase will decline to zero by 2050. At that point, the population stops growing. What factors have accounted for the sharp increase in the nation's population in the two centuries since 1790? What factors will account for an expected population decrease in the future—the first time in our history that a decrease has been projected?

Many factors contributed to the rising population curve of past years. Some are more obvious than others. One is simple mathematics. As we saw earlier, populations would expand according to the laws of compound interest were it not for various limiting factors that block such expansion. Of the factors that have worked to expand the human population, medical progress has been a significant one. We now have the means for preventing and controlling most epidemic diseases. By improving hygiene, commerce and technology have also had important roles in decreasing the death rate. So has agriculture.

In the United States, demographers now foresee a sizable expansion in the number of elderly people (see Figure 45-7). People 65 years of age and older accounted for 11.4% of the population in 1980—and 11.9% in 1985. This group is expected to rise to 13.1% in the year 2000 and to 21.7% in 2050. This means that there should be a sharp decrease in the ratio of working-age population (18 to 64 years old) to elderly population (65 and older). In 1980, that ratio was 5.8 to 1—that is, there were 580 working people for every 100 elderly people. The projected ratios will be 4.7 to 1 in the year 2000, 3.0 to 1 in 2025, and 2.6 to 1 in 2050. Earlier forecasts had projected smaller elderly populations and thus higher ratios in the future. There are two main reasons predictions have changed: (1) the recent

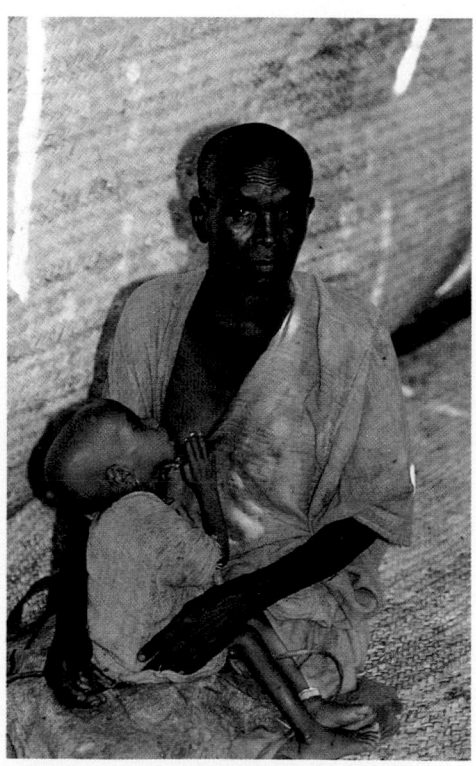

49-3 Evidence of starvation in Ethiopia (1987)

---

[1] Recently published statistics showed that the average number of births per year per 1000 population in the 1980–1985 period were as follows in selected Central and South American countries: Costa Rica, 30.2; Brazil, 30.6; Colombia, 31.0; Mexico, 31.7; Peru, 36.7; and El Salvador, 38.0. In industrialized countries, the rates were: Netherlands, 12.2; Britain, 13.0; France, 14.5; United States, 15.7; and Soviet Union, 19.1. Note that 20 births per 1000 population represents an annual birthrate of 2%. Of course, population increases depend on death rates as well.

49-4 Mediterranean fruit fly

decline in intrinsic rates of population increase, and (2) anticipated rapid improvements in our ability to extend human life.

At present, about 9.2% of the elderly population is 85 or older. In 2050, it is expected that 24% of the elderly will be 85 or older. That is why it is expected that deaths will exceed births for the first time in United States history in 2035. It is also a major reason why the Census Bureau predicts that the intrinsic rate of population increase will decline to zero by 2050. It is unlikely, however, that this pattern will be duplicated in poorer countries where the numbers of elderly are now relatively small (see Figure 45-6).

Let us now consider three matters that directly influence human population trends: food supplies, medicine and hygiene, and the environment.

WORLD FOOD PROSPECTS

A pressing (and depressing) preoccupation of most human beings is how to get enough food. Three related facts command our attention as we head toward the end of the twentieth century.

- Many of the world's peoples, especially in tropical and subtropical regions, remain undernourished (Figure 49-3).
- Population growth is accelerating most rapidly in the poorer sectors of the world.
- Agricultural production in many places is falling behind population growth.

The reserve provided by excess production capacity and surplus stocks, which the world enjoyed in recent decades, may be only a passing incident. Within a span of very few years excess production capacity and surplus stocks have dwindled, and much of the world's population today is living hand-to-mouth trying to survive from one harvest to the next, fearing always the impact of drought, plant disease, and pestilence (see Box 47-1). Surely this fear explains the intensity of the California effort to prevent the spread of the Mediterranean fruit fly (Figure 49-4). The North American drought of 1988 caused devastating crop losses and continuing dryness and high winds in early 1989, threaten the winter wheat harvest. With consumption rising and surpluses falling, the world's grain supply is severely threatened (Figure 49-5). This is reflected in the rising price of wheat.

To this bleak catalog may be added the fact that fertilizer, much of which is ultimately dependent on petroleum resources, has become scarcer and costlier (see Box 46-1). Thus we face a decreasing rate of expansion in food output and diminishing returns from the use of energy and fertilizer in agriculture. The necessary conclusion is that the era of cheap abundant food with large reserves of cropland may have passed.

A country of particular interest is Mexico, which undertook a change in agricultural techniques that has been termed the Green Revolution.[2] As a result, grain

---

[2] "Green Revolution" is the name given to a complex of radical improvements in agriculture that were based on local soil and water control, development of better seed strains, new techniques of animal husbandry, mechanical farm equipment, crop management, pest control, and fertilizer usage.

49-5 Evidence of a world grain shortage
These statistics, from the U.S. Department of Agriculture and Commodity Research, show clear evidence of a growing shortage of wheat throughout the world. (A) Total world consumption since 1973. (B) Total world reserves. (C) The rising price of wheat since 1987.

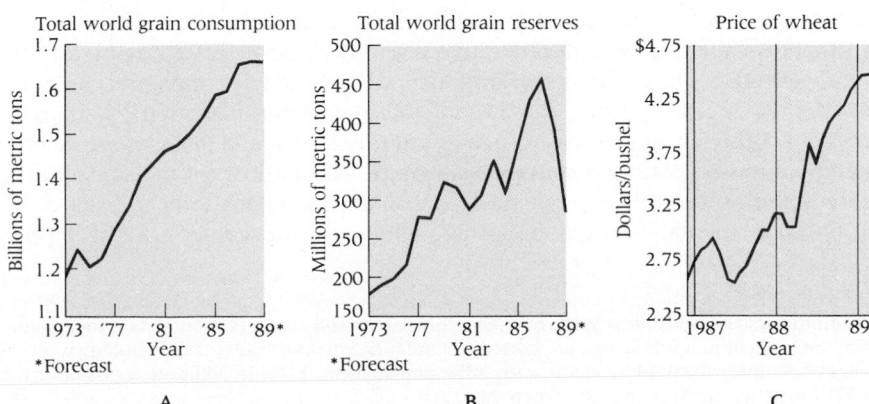

production doubled in the course of 25 years. But at the same time, the population doubled and Mexico in recent years was forced to import substantial amounts of grain. Until the mid-1970s, Mexico was without energy reserves and was importing oil. Although that situation changed dramatically in Mexico, there remains a vast gap between the wealth of Mexico on the one hand, with its advanced cities, automobiles, museums, human energy and intelligence, and its abject poverty on the other, with its unchecked human reproduction, and living conditions in many places resembling those of the distant past (Figure 49-6).

Even now limitations in energy and land resources make it impossible to feed the entire present world population of 5 billion a diet that is protein rich and generally well balanced. About 70% of the dietary protein in developed countries is of animal origin. About 70% of the dietary protein in developing countries is of vegetable origin. Vegetable proteins are of poorer quality because most are deficient in one or two essential amino acids. For example, the proteins provided by rice, wheat, and corn are low in lysine.[3]

Ocean fisheries contribute only about 5% of the total protein available in the human diet. Mismanagement of fisheries, principally through overfishing, local pollution, and destruction of fish breeding and feeding grounds, has led to alarming declines of up to 30% in the fish catches of recent years (Figure 49-7). About 3 billion head of livestock contribute 25% of the world's protein supply. About 29% of the world's protein supply is fed to the livestock. Most of the major plant crops now in world commerce were discovered thousands of years ago. Fewer than 200 species of the approximately 300,000 or more known plants have been harvested in agriculture since before 1800. Only rubber and oil palms have entered cultivation since 1800. A careful search for new crops out of the many known plant species would surely yield more in the future.

If we move toward greater consumption of animal protein, we would then face (1) shortages of land, water, and energy resources that will intensify as human populations increase, (2) further ecological degradation of land, water and other biological resources, which will tend to reduce the productivity of agriculture, and (3) the physiological limits of the feed crops and their abilities to respond to increased amounts of fertilizer and other production stimulants.

PROGRESS IN MEDICINE AND HYGIENE

In earlier societies, people were fatalistic about illness. There was little a physician could offer beyond providing comfort and minor palliation. Modern medicine, in contrast, commands a powerful arsenal of weapons that can often, but certainly not always, forestall death, relieve pain, cure disease, and rehabilitate the disabled. Today, the medical system of the civilized world is at its best in curing and controlling the infectious diseases that until about sixty years ago were untreatable and represented medicine's greatest challenge.

The sweeping change in the nature of human health problems during the twentieth century is evident when one compares the ten leading causes of death in 1900 with those in 1987. In 1900, three of the ten (tuberculosis, influenza-pneumonia, and diphtheria) were infectious diseases. Three more (gastroenteritis, nephritis, and diseases of early infancy) were indirectly related to infectious disorders. By 1987, of the first ten causes of death only influenza-pneumonia (and certain diseases of early infancy which ranked eleventh) were infectious diseases—and in both of these the mortality rate was far below the level in the year 1900, before the advent of antibiotics and immunization. Today the list is headed by heart disease, cancer, and other so-called chronic degenerative diseases associated with aging and with self-inflicted adversities such as smoking, alcoholism, and environmental contamination. In their mortality rates, diseases of this type have increased by many-fold since 1900. Other chronic diseases such as arteriosclerosis and diabetes are emerging as major causes of death. Of course, death rates tell only part of the story. Chronic disease significantly afflicts large numbers of the living for long periods of time with physical, mental, and emotional disabilities. An inevitable result of improved

[3] If amino acid compositions are expressed as grams per 100 grams of protein, and the quantities of each amino acid in a whole egg are taken as standards equaling 100%, then whole rice contains 52% lysine, whole wheat 44%, whole corn 38%, and soybean 11%. In addition, soybean contains only 53% as much methionine as a whole egg.

49-6 Poverty in Mexico City

A

B

49-7 Excessive fish catches
(**A**) *Small boats were once universally used.* (**B**) *This scow loaded with sockeye salmon employs high-technology fishing methods.*

TABLE 49-2
CHANGES IN LIFE EXPECTANCY IN THE UNITED STATES

Year*	Life Expectancy at Birth (yr)	Decade's Gain (%)	Life Expectancy at Age 45 (yr)	Decade's Gain (%)
1900	49.2	—	24.8	—
1910	51.5	9.7	24.5	−1.2
1920	54.1	5.0	26.3	7.3
1930	59.7	10.4	25.8	−1.9
1940	62.9	5.4	26.9	4.3
1950	68.2	8.4	28.5	5.9
1960	69.7	2.2	29.5	3.5
1970	70.8	1.6	30.1	2.0
1980	73.7	1.4	32.1	6.6
1986†	74.9	—	32.9	—

Source: U.S. National Center for Health Statistics
* Except for 1910 and 1986, the numbers given are three-year composites. For example, the 1970 data reflect changes occurring from 1969 to 1971.
† Preliminary data.

health in early and mid-life is the increased prevalence of such intractable forms of disease and disability in middle and later life.

The sharp increase in acquired immune deficiency syndrome (AIDS) since 1979 is a disturbing development of obvious importance (see Box 23-3). Although at the present time the numbers of victims remain small in the overall scheme of things—with total deaths in the thousands rather than the millions—its incidence is rising and there is so far no method of inducing immunity or curing established infections. Thus it is an ominous threat to future populations. At present, future trends in the incidence of AIDS depend largely on our ability to induce people to modify the kinds of behavior that spread infection.

Despite a widespread impression to the contrary, progress has been slow in recent decades in dealing meaningfully with chronic, degenerative, noninfectious human ills. Modern biomedical research has accumulated a formidable body of information about them—and there have been notable triumphs in the treatment of certain relatively uncommon chronic diseases—but the accumulated knowledge is not yet sufficient to ensure prevention or cure of any of the major diseases. Since physicians confronting the sick are obligated to act, to *do* something even if it is only to apply the best treatment of an unsatisfactory lot, we have a great many "halfway" medical technologies that, disturbingly, grow more and more costly.

Prolongation of life expectancy is perhaps the best overall measure of the success of modern medicine and hygiene.[4] Between 1970 and 1985, the life expectancy of a 45-year-old American increased by 8.6%, while that of a newborn increased by 5.5% (Table 49-2). These increases show that adults have shared equally in the gains in life expectancy occurring in the past generation—a reversal of an historical trend. For example, as the life expectancy of a newborn increased by 9.7% back in 1910, that of a 45-year-old actually declined by 1.2%. That is no longer true.

The aggregate data for the first 86 years of this century provides further evidence of the turnaround that has occurred. In the 50 years from 1900 to 1950, life expectancy at birth increased by 38.6% while life expectancy at age 45 increased by only 14.9%. However, in the 30 years between 1950 and 1985, life expectancy at birth increased by only 9.5%, as compared with an increase of 14.7% for life expectancy at age 45. Data from earlier periods are more difficult to obtain, but there is justification for the conclusion that the emerging trend is unprecedented.

The proportionately greater gain recently recorded in the United States in the life expectancy of adults relative to newborns—an important reversal of the historical trend—reflects the substantial capacity of biomedical science to identify health prob-

[4] Regrettably, infant mortality in the United States still exceeds that in more than a dozen other advanced nations. Our infant mortality is highest among economically disadvantaged segments of society, although it is still unacceptably high in the more affluent groups as well. The problem here is not lack of knowledge but lack of adequate medical care.

lems and develop solutions. But it also challenges us to mobilize our resources effectively to realize the further gains that are possible—and to prepare for the social and political implications of increased life expectancy.

Impressive as these numbers are, they pose difficult questions of a kind that previous generations had no need to ponder. Consider some of modern medicine's achievements. Effective contraceptives and safe legal abortions now allow fertile couples to choose the number of children they will bear. Advances in the technique of prenatal diagnosis are gradually permitting parents to prevent the birth of children with severe genetic defects. The possibility of controlling the quantity of children, their sex, and their genetic quality has obvious and enormous ramifications for patterns of child rearing, family life, composition of work forces, and sex roles.

At the other end of the birth-death spectrum, the capacity to transplant organs, to cure certain diseases, and to keep people alive demands that we ask what kind of life is worth living and how society's resources will be allocated to these increasingly costly efforts. The ability to sustain life artificially by machines has compelled more precise definitions of ''death''—and for that matter of ''life.'' If death can be forestalled, for how long and in what circumstances should it be forestalled? At what point does the rehabilitation of the crippled cease to be a real benefit to them? At what point does it cease to benefit society?

Between the extremes of birth and death, medicine is learning how to manipulate human behavior. Psychotropic drugs, psychiatry of certain kinds, and psychosurgery all offer the promise, albeit a controversial one, of medical intervention in the field of human emotion and cognition. But clearly this kind of intervention comes into conflict with our traditional ideas of autonomy and self-determination. It carries over, for example, into the legal system by attempting to define normality and deviancy. How far should medical technology go in trying to serve society's broad political and social needs?

As the frontiers of medical research expand, as new things are learned and new treatments are developed, more and more experimentation on human beings becomes necessary. In the move from theory to practice, human beings must necessarily be the final testing ground. But those often anonymous individuals experimented upon might be harmed by such undertakings. Yet if research is hampered by lack of experimental subjects thousands could be victimized by the failure to utilize promising treatments. If, on the other hand, the individual research subject is considered expendable for the good of future generations, then harm could be done to our hard-won values of human dignity and inviolability.

It is in the power to change the conditions of birth and death, to alter ways of life and behavior, to impose new dilemmas about the relationship between individual and social good, that medical research and clinical practice force a new confrontation with some very old human questions, many of which are ultimately philosophical and ethical. They are nevertheless of profound importance for the future of humankind. It will make a difference if we increase our numbers *and* lengthen human life—especially if increasing longevity prolongs only the infirmities of age without increasing the duration of productive working life.

Although limitations on such escalations are clearly inevitable, that fact has yet to seep into today's social or medical thinking. But soon we will be forced to harbor doubts about the nature of medicine's impact and to recognize conflicts between its aims and those of human society. Perhaps we will find it necessary to alter our notions of health in ways that will give more consideration to the communal good and the welfare of the species. Death, malaise, and suffering can never be banished, and future generations will have to decide how much of their energy and resources to invest in trying to hold them at bay.

Many now believe that advances in molecular biology and biotechnology will profoundly expand the capabilities of medical science in the twenty-first century. With full knowledge of the DNA sequence of the entire human genome, with an ability to alter DNA sequences at will, to amplify tiny DNA samples, to utilize so-called ''DNA diagnostics'' to detect genetic diseases pre- and postnatally, and to make use of automated, rapid, and inexpensive analyses for the DNA and RNA sequences associated with genetic, malignant, and infectious disease—physicians and biologists of the future will command incredibly powerful techniques for the prevention and therapy of human diseases. However exciting these prospects, this development is certain to raise new psychological, social, and ethical issues of the gravest importance.

HUMAN IMPACT ON THE ENVIRONMENT

When humans began to cut forests, burn fields, and plow lands, they disturbed the existing ecological balance (Figure 49-8). The results were far-reaching. For example, soil erosion rates, as we have seen, greatly increased. The growth of populations and the Industrial Revolution so accelerated these incursions that their effects—actual and potential—are apparent on all sides.

While overexploiting the Earth's nonrenewable resources, we have also been pouring into the environment waste products and other pollutants that are damaging to the environment and to human health and well-being. We have saturated the air with wastes from automobiles and factories and filled the water with chemicals and viruses. Our food contains chemical additives with a potential to cause cancer. Then there is **acid rain,** which is due largely to acidic air pollutants, such as sulfur dioxide and nitrogen oxide, that arise from combustion of fossil fuels. The increasing acidity of rain and snow throughout the world has been linked to sharp declines, sometimes to the point of extinction, in the number of fishes in many lakes and streams, and to extensive tree and plant damage. Soils bear the traces of fallout from atomic bomb tests.[5] And such physical hazards as environmental radiation, accidents, and noise have drastically increased.

What is to be done to preserve the environment? Where are we heading if we draw the balance between environmental and economic concerns too far on the side of commerce and production? To what extent should we preserve wilderness if it means denying ourselves promising new oil fields or shale and coal mines? To what extent should we drain scarce water into the production of new energy sources? What will be the implications of the ongoing destruction of wildlife? What can we do to preserve the nonmaterial or recreational resource of our environment, such as natural beauty and solitude, wild and interesting scenery, and opportunities for thought and adventure?

The answers to these questions are not known. At least scientists and political leaders have begun to discuss them, but a reading of such discussions shows how far we must travel before all can agree on a whole range of uncertainties. We have spoken of the greenhouse effect, a consequence of increasing levels of atmospheric level of CO_2—and CH_4 and other gases. And we have noted that it is not solely due to the burning of fossil fuel, but also to increasing deforestation (see Chapter 46). How can we reverse this trend?

These are questions of obvious and profound importance and the road to their solution is unclear. What *is* clear, however, is that increasing assaults on the biosphere have been generated by rapid population growth. Because no imaginable program of population control could significantly restrain our numbers in the next few decades, the urgent need is to achieve, as rapidly as possible, adequate levels of education

---

[5] Investigators have discovered a 2.5-fold increase in the leukemia death rate among children exposed to fallout from atomic bomb tests carried out in the Nevada desert in the 1950s. A notably large increase in the number of leukemia deaths occurred in children born about 1951—the beginning of the testing period. Leukemia death rates returned to their previous levels only after the testing was stopped in the early 1960s. Evidence has also emerged of an excessive incidence of leukemia among soldiers who took part in the tests and of cancer in people who lived downwind of the test sites. Although investigators are properly cautious about interpreting a correlation as indicating a causal relationship, these results convey few surprises to those of us who were associated with this work and who have long feared the consequences of a bomb-testing program uninformed by biological knowledge.

49-8 Plowed fields of the midwest

and significant progress in agriculture, science, and technology for *all* populations.

The basis for solving these problems will in the end be biological. Politicians and public alike must be informed about science, so that they may understand the basis for our survival. Biological literacy should be a prerequisite for all who hope to work at improving the world.

MODIFICATION OF THE ENVIRONMENT

When human culture alters the environment the results reverberate through the world of life and influence the environments of countless other organisms. If a "state of nature" is defined as one uninfluenced, directly or indirectly, by humans, then one would search far to find it anywhere on Earth today. Perhaps it still exists on some unscaled Himalayan height or unplumbed oceanic depth (though we have nearly run out of unscaled heights and unplumbed depths). But even there, some remote effects of human activity are likely to be felt.

All organisms influence their environments, and many have some small measure of active control over parts of their environments. That fact is obvious when a beaver builds a dam. Humans, however, control and modify the environment more extensively than do any other organisms.

LAND USE

In the most intensively developed parts of the world, the greatest changes have come from the destruction of natural communities by humans determined to convert them and the space they occupied to their own uses. Humankind has, in short, used nature's resources, exterminated bothersome species, cleared and cultivated the topsoil, and scattered refuse.

Native North American populations had already modified their environment before the European settlers arrived, but the changes they made were insignificant compared with what the Europeans did. In 1492, for example, almost the whole eastern half of North America was covered with forests. Some 300 million acres of forests have since been destroyed. Most of the early clearing was for farmland. In biological terms, forests were cleared to replace native communities with low carrying capacity for *Homo sapiens* with controlled communities more productive of food. Most of the good agricultural land was cleared long ago. The remaining forests mainly occupy land that is not suited for farming. They are now cut for forest products, mainly lumber. Destruction of forests still goes on, though in national forests and those controlled by private owners cutting is so managed as to draw on surplus growth and keep forests healthy. Even so, the controlled crop forest is a community unlike that of a virgin forest. The same is true of large tracts once cut for farmland and now gone back to forest by natural succession or deliberate reforestation. We are not, at this point, concerned with whether this is a "good thing" or a "bad thing" but only with the biological observation that natural communities have been destroyed or profoundly altered. It should be added that temporary destruction of forests still results from fires. Some of these are natural events, but some are of human origin.[6]

In addition to the vast conversion of grasslands to farmland or pastures, the introduction of new plants and animals, the draining of wet lands, and the irrigation of arid lands, we have also reshaped lands *not* used to produce food and raw materials. As noted in Chapter 1, cities and suburbs, industrial plants, airports, roads, and other such constructions wipe out the natural environment wherever they are built, and in the aggregate they cover a considerable and rapidly increasing

---

[6] It now seems well established that intermittent natural fires (which antedated humans) are essential for the conservation (repetitive restoration) of some desirable habitats. This issue came urgently to the fore in the summer of 1988, when unremitting fires burned more than one-third of Yellowstone National Park.

fraction of the surface of the land. Even areas not permanently inhabited, such as the high mountain country, or those deliberately kept "natural," such as national parks and national forests, are intensively used for recreation. No biologically informed person who joins the traffic jam in, say, Yellowstone Park, should imagine for a moment that he is seeing a community natural in the sense that it is free of radical human intervention.

The intensity of land use increases yearly—and it is unlikely that there is an acre of land in the United States that is not sometimes and in some way used by humans for their specific purposes and that has not been changed by that use. The same is true almost everywhere on Earth—and true to an even greater extent where human occupation has been of longer duration. At one time, it was possible to travel from Egypt to Morocco under a green canopy of luscious plant growth. Human intervention has so changed the environment that the journey today is exclusively through the desert.

DISTURBANCE OF COMMUNITIES

Certain specific things happen to communities as a result of human activities. The most radical, of course, is that a particular community is destroyed and replaced by another. This is by no means a simple occurrence. When grassland is plowed up and planted in corn, the result is not simply that we have corn instead of grass. Many food chains are cut off at the bottom when grass is destroyed. Some of the organisms in those chains die out or move elsewhere. It is our intention when we plant corn to start a food chain leading only to ourselves, but with all our controls we are usually unable to carry out that intention. Some members of the destroyed food chain switch to eating corn, and some of the old food chains continue. Other organisms that eat corn (any part of the plant) move in, and their populations increase. New food chains develop. A whole community organization is soon resumed on the new basis.

Wherever human beings break the soil, drain it, water it, or otherwise alter conditions, they create a new environment that is quickly exploited. Anyone who has lived in the West knows that tumbleweed (Russian thistle, *Salsola pestifer*) has great difficulty in breaking into an established community (Figure 49-9). It almost never grows anywhere but along roads, on plowed or abandoned fields, and on ranges so badly overgrazed that much of the native vegetation has been lost. Under those conditions of human origin, it rapidly spreads everywhere. Not only do new kinds of communities arise where humans have passed, but also new successions and new climaxes. Things are never quite the same again.

An unforeseen result of human interference has an ironic twist: use of insecticides often accelerates the evolution, by rapid selection, of strains of insects immune to the insecticides. Similarly, widespread use of antibiotics has led to the emergence of resistant bacterial strains.

Turning back to the tumbleweed, this, like many of our weeds, was unwittingly introduced by humans into North America (Hopi Indians call it "white man's plant"). All over the world such introductions have been made. If an introduction becomes established, one discovers how extensive the fundamental niche is as it unfolds in the adopted environment. It disturbs the native communities, and it frequently causes the extinction of native species in an unpredictable way. A policy of bringing in plants, birds, and mongooses considered more desirable than native life has decimated the native flora and fauna of the Hawaiian Islands and effectively destroyed what was a unique biological community. This is because the fundamental niche, constrained by competitors in the native environment, suddenly unfolded to its full extent in the new environments. In Australia, cacti and rabbits, also deliberately introduced, expanded explosively and almost ruined the economy of the country before coming under tenuous control.

Humans have in fact wiped out numerous species—examples are the passenger pigeon and the dodo—and they have brought others such as the bison and pronghorn antelope to the verge of extinction, before deciding to save a few for the world's zoos. Human overexploitation threatens the extinction of an estimated 15–20% of the millions of species on Earth within the next 50 years. It will occur during the lifetime of many of us. At this rate, the extinction rate may reach a few species an

49-9 Tumbleweed
Tumbleweed, Salsola pestifer, *is a familiar feature of the American west. Its presence is rarely an indicator of well-managed land.*

hour by the beginning of the twenty-first century. This rate would exceed the mass extinction rates that destroyed the dinosaurs 65 million years ago.

DEPLETION AND CONSERVATION

Humans are influenced by the same biological principles as any other species, and it is evident that their exploitation of the environment is reducing the capacity of that environment to support humanity. To that extent humanity is living beyond its biological income—and any species that does that faces eventual decimation or extinction. Humans, however, have the advantages of insight and foresight. We now recognize the possibility of disaster. It remains to be seen whether we will have the wisdom to avoid it.

The challenges posed by the interweaving of poverty, malnutrition, population growth, and land use are as much political as biological—and their full discussion is beyond the scope of this book. Suffice it to say that statistics depicting world patterns of resource depletion and poverty are themselves inadequate to illuminate the underlying causal forces, which to a major extent are political, historical, and cultural. Time will tell whether and how patterns of resource distribution change in the future—and if they change, whether global food production will then match population growth. If politics is defined as "who gets what," the answers we seek are rooted in its stern realities.

UTILIZATION VERSUS PRESERVATION

The usual first reaction of human beings to any widespread destruction of natural communities is to wish they could have been preserved. Many people, more sentimental than informed, understand "conservation" to mean opposition to any disturbance of nature. Biologists, more than others, are sensitive to the appeal of nature, but they have come to recognize that total conservationism is quite impossible.

Like all organisms, human beings utilize their environments. Indeed, human utilization of the whole of their environment is natural and inevitable. Even our destruction of nature is biologically unremarkable. Whether or not such utilization is wise depends upon whether, immediately and in the long run, it is beneficial for our species.

Wise utilization is that which increases (or does not decrease) the capacity of the environment to meet long-term human needs. Increase in capacity is inherently limited. Indeed, rapid increase may lead to ultimate decrease, as has happened in agricultural areas too intensively farmed. The ideal would be to put production on a cyclic basis, turning to human use such cycles as have kept all life going for some 2 billion years. To do that we must thoroughly understand biological and biogeochemical cycles, especially the fact that we cannot get more out of a cycle than goes into it and the fact that disturbance of any part upsets the whole.

Kinds of Resources

Resources that participate in cycles are *renewable*. The possible rate of use of a resource is limited by the rate of the cycle as a whole and by the rate of input of energy, a resource that is not renewable because of the second law of thermodynamics. As long as energy is available and resource utilization does not exceed the capacity of the cycle (and the cycle keeps going), renewable resources are inexhaustible. This is particularly relevant for such biological resources as food and raw materials derived from plants and animals.

Other resources are *nonrenewable*, or *finite*. Their production cannot be made cyclic. As everyone is now acutely aware, the world's present major sources of energy are nonrenewable. Most of our intensively used mineral products, notably the metals such as iron, copper, and many others, are nonrenewable—and about 90% of their use today is by the 20% of the world's populations in developed countries. Their atoms are not destroyed by use, but they are eventually so scattered as to be unrecoverable by any process economically feasible, now or in the foreseeable future. Increasing expectations throughout the world make it unlikely that the per

49-10 Solar energy is being used to heat these homes

49-11 Trees in a tropical rain forest
Its top soil is remarkably thin.

capita supply of these nonrenewable resources can be maintained in the developed countries in the near future.[7]

The wisest utilization of nonrenewable resources is clearly to use them as slowly as possible with the least possible waste. Energy is an especially severe problem. The sun stands out as the most promising renewable source of energy in the long run. It represents a clean and inexhaustible source, whether it is captured by solar collectors or fixed by photosynthesis. Solar energy would be especially suitable in areas where sunlight is abundant and energy most needed, such as the tropics and subtropics. It would reduce consumption of finite fossil fuel resources and alleviate the global CO_2 problem at the same time. Photosynthetic plants have captured solar energy for millions of years and serve as an excellent example for technologists seeking to solve this urgent problem in a biologically sound fashion (Figure 49-10).

The Importance of Soil

Like fossil fuels, arable lands are finite resources. The United Nations Food and Agriculture Organization estimated in 1973 that of all the world's land, 11%, or 3.7 billion acres, is suitable for cultivation. Nearly all of this is already under cultivation. About 22% of world land (7.4 billion acres) is now used for livestock, for pastures producing forage (grasses and legumes that feed livestock), ranges, and meadows. Forest covers 30% (10.1 billion acres) and the remaining 37% is too dry, too cold, or too steep for farming.

Conservationists are greatly concerned with soil utilization—and with good reason. We depend on natural soils for most of our food, directly as plant food and indirectly as animal food. It is possible to make excellent artificial soils or soil substitutes, but so far, this is so expensive that it has little promise as a large-scale source of human food. It is also unlikely that the aquatic environments, rich as they are, could economically be made the main source of plants and animals usable for human food.

It has been estimated that the United States once had 1517 million acres of usable farm and grazing land. Of this, by 1947, 282 million acres had been ruined and had become essentially unusable for the foreseeable future. The total acreage in farms in 1976 was 1085 million. It is estimated that continued use of traditional farming methods will soon ruin 775 million more acres.[8] Only 460 million acres are considered in reasonably good shape and not seriously threatened.

In the Soviet Union, there is no choice region like the U.S. corn belt that combines rich soil with dependable rainfall. In addition, there are severe winters and short growing seasons. In the tropics, where food production is needed most, the tropical forests are complex, productive ecosystems, but they do not produce much food for humans. How to replace them with other productive ecosystems that will support human beings remains an unsolved problem. When a tropical forest is cleared, we engage in a one-time only consumption of natural resources that will not be available again. The reason is that in the tropical rain forests the soils are very thin. Tree roots are restricted to only the top few centimeters of soil (Figure 49-11). Nutrients occur only in this thin top layer. The nutrients in the thin top layer of soil are exhausted very rapidly and are insufficient to sustain permanent agriculture.

[7] An example of this problem is seen in recent warnings that we are wasting helium, the lighter-than-air gas used to fill balloons and blimps. Helium is now recognized to be the unique and indispensable key to a future "supertechnology" in which helium-induced low temperatures (below 452°F) can induce a frictionless flow of electricity (superconductivity) that could vastly extend the usefulness of nonrenewable fuels, oil, coal, and uranium. However, helium is itself nonrenewable. When it is released into the air it is too light to be held by Earth's gravity and it is lost. It was formed in pockets of the Earth's crust over billions of years by the slow radioactive decay of uranium and thorium. We cannot, in any practical sense, make more than we have. The United States is estimated to have about 700 billion cubic feet of helium. We are dumping it into the atmosphere, mainly as an unwanted byproduct of natural gas production, at the rate of 13 billion cubic feet per year.

[8] The quality of arable land is degraded by erosion. Despite superlative agricultural technology, 14.5 metric tons of topsoil are lost annually per acre in corn production in Iowa. In the country at large, 1.5 billion metric tons of topsoil are lost annually. That is equal to 12.6 metric tons per acre under cultivation.

A   B

Soil is in reality a finite resource because it is formed by a very slow process. Under the best conditions, formation of a good agricultural topsoil takes centuries,[9] and in some places it may take thousands of years. When a farmer removes a crop from a field, he also removes material that came from the soil and that would, under natural conditions, have returned to it. Harvesting cuts off part of the cycle of materials and depletes the soil. The cycle can be kept going by interspersing harvested crops with planting that enriches the soil and by making good the losses by adding fertilizers. More serious, and practically impossible to compensate for, is actual physical loss of the soil by wind and water erosion after the protective natural plant cover is removed. A windstorm on May 12, 1934, swept up 300 million tons of soil from the plains east of the Rockies and began the "dust bowl" devastation that brought on an economic crisis and the migration of thousands of homeless farmers whose topsoil had blown away (Figure 49-12). In 1900, water dripping from a barn in Georgia started a gully in surrounding bare soil that eventually spread to 3000 acres and washed away whole farms. The millions of acres of soil already lost in such ways cannot be recovered. Further losses are being slowed down, at least, by contour plowing, terracing, and more care in maintaining a vegetation cover (Figure 49-13). On grazing land, heavy overgrazing has also laid bare the soil and promoted erosion (Figure 49-14), besides damaging the grass community and permitting invasion by undesirable weeds. The remedy is obvious though not easy: replanting of hardy grasses and reduced grazing.

Erosion is a natural process that went on for several billion years before human beings appeared, and that produced our badlands long before there were any grazing cows in North America. Wherever erosion has taken the soil from farmland and grassland, it would sooner or later have done so without human help. This does not alter the fact that humans have greatly hastened erosion and often localized it exactly where and when it does them the most damage.

Floods and Flood Control

Soil kept porous and covered with vegetation absorbs much rainfall. On cleared land the soil tends to pack, and more rain runs off over the surface. This is, of course, the immediate cause of accelerated erosion on cleared land, and it is also an interference with the water cycle that has extensive repercussions on living communities. Uplands from which forests have been cleared no longer retain water effectively, and the groundwater below them is also less steadily maintained. As a result, organic productivity is decreased by a lack of steady water supply. Urbanization itself has many effects on water runoff—partly as a result of the paving of land

[9] About 100 years are required to produce 25 mm of topsoil (or 1.2 metric tons per acre per year).

49-12 Wind erosion of farmland
(A) A dust storm in the Texas panhandle, March 1936. **(B)** A typical dust bowl scene on a South Dakotan farm, 1936.

49-13 Contour farming
By plowing furrows that parallel land coutours and terracing slopes, farmers can impede runoff and decrease soil erosion.

49-14 *The result of overgrazing of land*

49-15 *Flood control in the Netherlands is an engineering triumph.*

49-16 How dams trap silt
(A) *Gibraltar Dam was built in 1920 in the Los Padres National Forest in California to create a permanent reservoir for drought-stricken Santa Barbara. However, it is now two-thirds full of silt and will be completely filled within 20 years.*
(B) *Nearby Mitilija Dam and reservoir, also largely filled with silt, now has only a small storage capacity for water.*

(e.g., in huge parking lots), which decreases the soil surface needed for water absorption, and the roofing of houses and large buildings, which increases local runoff. Runoff, in turn, tends to produce floods more severe than those that would occur without human disturbance. A return to natural water conditions is generally desirable and can be achieved only by restoring vegetation on the upland.

Flood-control projects represent one of the most heroic efforts of humankind to control the intensification of natural forces resulting from human activities. The Dutch have doubled the size of their land by reclaiming it from the sea (Figure 49-15).[10] In such projects, some attempt is made to control floods where they come from, in the runoff in the uplands. But most of the effort seeks to delay floodwaters downstream by holding them behind storage dams. Water so impounded may also be used for the generation of electric power and for irrigation. Geologists note that the expedient is temporary, for all storage reservoirs will fill up with silt in a few generations at most and then be useless (Figure 49-16). Some have filled in a mere dozen years. Extension of a reservoir's usefulness demands that silting be reduced by the control of erosion upstream.

RENEWING RENEWABLE RESOURCES

When the whole situation is considered from our present vantage point, is evident that humans were, until recently, unaware that they have interfered with biological and biogeochemical cycles, thereby intensifying erosion problems, unwanted climatic changes, reductions in crop yields, and pollution in various forms.

However, there is reason for hope. Solar technology now being developed is likely to change the energy picture drastically as we replace nonrenewable resources with a renewable one. Genetic engineering is now being extensively applied to crop plants and farm animals. This technology holds great promise for the future. It has already produced new plants under laboratory conditions with very different characteristics. Thus in the not-too-distant future we may grow crop plants in areas that could not support growth before. Desirable features that genetic engineers may confer on plants include capacities to fix nitrogen, to mount new chemical defenses against disease and herbivores, to carry out C_4 photosynthesis, and so on.

The preservation of biological diversity should have a high priority on our agenda for the future. Agriculture as we know it today originally sprung from a wide diversity of biological forms. From these, humans selected those with desirable features by selective plant and animal breeding methods. We must ensure the preservation of this diverse genetic material in gene banks both in laboratories and in nature. From this, generations can develop new forms of agricultural plants and

[10] On the other hand, it can work the other way. Louisiana is losing 60 square miles of wetland per yer *because* of flood control levees.

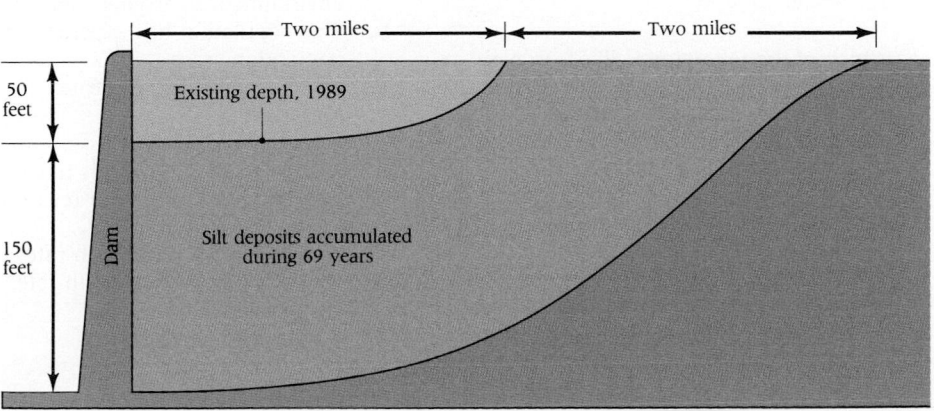

A

B

animals. We must design global schemes toward this end so that future generations will have the widest possible choice of options. It is the task of biologists to stamp out biological illiteracy if we are to maintain biological diversity and work toward the survival of the human species.

We end on a note of hope and affirmation. The greatest danger to our species will come from a tendency to forget that we are endowed with intelligence and curiosity and we must inquire, explore, and experiment. What we cannot do is assert that there are things we do not need to know. That, we believe, is a dependable path to disaster. That, we feel certain, is *not* the path humanity will follow in pursuing its future destiny.

SUMMARY

We complete our tour of the world of life by considering three selected topics which relate to human population now and in the future: race, population growth, and conservation.

Human races correspond to demes or subpopulations of the human species. Analysis of such genetic characters as blood groups can provide estimates of the relatedness of races—and of populations within races.

The future size of the human population is a matter of paramount biological and economic importance. A major factor limiting population growth in the near future is likely to be the restricted availability of food, especially protein sources of adequate quality. Expansion of food production is not now keeping pace with population growth, and such factors as fertilizer shortages and man-made global climatic changes threaten to limit food production in the future. One factor influencing population size has been the remarkable success of medicine and hygiene in increasing life expectancy over the last 100 years. Although many powerful medical techniques are in various stages of development, all raise many difficult issues. For example, many of these techniques are so costly that society must decide how much it is willing to pay for what may well be marginal improvements in the quantity or quality of human life.

The negative impact on the environment of human activities is immense, ranging from depletion of water and other resources to numberless forms of pollution. As a result of human activities, biological extinction rates threaten to exceed those of such natural mass extinctions as the one that ended the Age of Reptiles. Even seemingly innocuous activities such as clearing and plowing land for farming must inevitably damage or destroy various ecological communities.

The many nonrenewable resources currently being consumed include arable farmland. Present farming techniques in the United States frequently cause permanent losses of topsoil to erosion. In addition, cleared land absorbs rain water poorly and often causes damaging floods.

A combination of scientific advancement, public concern, and forceful intervention has the potential to solve these looming problems. We hope the forces of enlightenment will find people up to the challenge and will guide their leaders.

QUIZ QUESTIONS

1. Resources that operate in cycles are renewable. Which of the following is *not* renewable?
   A. energy
   B. sulfur
   C. soil
   D. phosphorus
   E. water

2. Which of the following is likely to play an important role in the future well-being of the human population on Earth?
   A. preservation of biological diversity
   B. solar technology
   C. genetic engineering
   D. biological and environmental education
   E. All of the above

3. Which of the following is true of growth of Earth's human population?
   A. This population is in the horizontal, slowly-increasing phase of the curve.
   B. This population is in the exponential phase of the curve.
   C. This population has exceeded the environment's carrying capacity.
   D. This population has experienced a population crash.
   E. None of the above are true statements.

4. Which of the following effects have medicine and hygiene had in regard to the human population?
   A. They have increased birth rate.
   B. They have increased death rate.
   C. They have increased life span.
   D. A and B
   E. A and C

5. The biological principles that apply to other species also apply to *Homo sapiens*.
   A. True.
   B. False.

6. Which of the following natural resources would not be considered renewable?
   A. salmon runs up coastal rivers
   B. gasoline for automobiles
   C. uranium for nuclear reactors
   D. fresh water
   E. timber forests

ESSAY QUESTIONS

1. What is meant by "race"? By what criteria are races distinguished?

2. What advantages do animal sources of protein have over vegetable? What problems are associated with an emphasis on cultivation of animals as sources of nutrition?

3. What is "acid rain"? What are its causes and consequences?

4. What factors serve to deplete fresh water and arable land? What practices can limit these losses or replenish the land?

5. What are the benefits of maintaining biological diversity through the preservation of species and their natural environments?

6. From your study of biology, what do you consider are the most critical biological problems facing humanity in light of its increasing utilization of the environment?

REFERENCES AND SUGGESTIONS FOR FURTHER READING

Blank. L. W., Roberts, T. M., and Skeffington, R. A. (1988). New perspectives on forest decline. *Nature 336*, 27–30.

This interesting commentary on the much publicized damage to German forests by acid rain notes that prophecies of disaster have not come to pass. Ecologists now view the situation more realistically.

Boyd, R., and Richerson, P. J. (1985). *Culture and the Evolutionary Process*. University of Chicago Press, Chicago.

Some believe that human culture can affect phenotypic characters and thus is an evolutionary force. Others think culture is too complex for analysis. This scholarly book addresses the issue. It merits attention.

Frost, P. (1988). Human skin color: a possible relationship between its sexual dimorphism and its social perception. *Perspectives in Biology and Medicine 32*, 38–58.

A stimulating essay on evolutionary, cultural, and sexual aspects of human skin color and hairiness.

Gasser, C. S., and Fraley, R. T. Genetically engineering plants for crop improvement. *Science 244*, 1293–1299.

An up-to-date review of an important topic.

Hansen, M., Busch, L., Burkhardt, J., Lacy, W. B., and Lacy, L. R. (1986). Plant breeding and biotechnology. *BioScience 36*, 29–39.

A provocative essay on the scientific, philosophical, and political problems that may arise as new technologies revolutionize agriculture.

Harwell, M. A., and Grover, H. D. A. (1985). Biological effects of nuclear war I: Impact on humans. Biological effects of nuclear war II: Impact on the biosphere. *BioScience 35*, 570–583.

A bleak assessment of the likely consequences for humans and the biosphere of a large-scale nuclear war. Beyond the direct effects in combatant countries is the certainty of starvation for noncombatant countries.

Holden, C. (1988). The ecosystem and human behavior. *Science 242*, 663.

News story on current studies of social scientists on how humans cause major environmental changes and how they are affected by them. Typical concerns: effects of global warming on land use; aftermath of large area going under water (e.g., Bangladesh), etc.

Huxley, A. (1985). *Green Inheritance*. Anchor Press/Doubleday, New York.

The author discusses the many ways in which humans depend on plants. The mutual interdependence of plants, animals, and humans requires a program of preservation of plant and animal diversity.

Kitcher, P. (1985). *Vaulting Ambition: Sociobiology and the Quest for Human Nature*. The MIT Press, Cambridge, Mass.

A stimulating critique of sociobiology and the claims of some of its proponents that evolution has determined human social behavior, a concept the author calls "pop sociobiology."

Landegren, U., Kaiser, R., Caskey, C. T., and Hood, L. (1988). DNA diagnostics—molecular techniques and automation. *Science 242*, 229–237.

Molecular biology has revolutionized our understanding of many diseases. This article describes developments in "DNA diagnostics"—the analysis of disease at the nucleic acid level—which will transform medicine in the future.

Moore, J. A. (1985). Science as a way of knowing: human ecology. *American Zoologist 25*, 1–155.

A thoughtful synthesis on the evolution of human beings and their multiple impacts on the global environments.

Pimentel, D., and Wall, C. W. (1984). *Food and Energy Resources*. Academic Press, Orlando, Fla.

A careful discussion of the problem of sustaining the world's food and energy production.

Simpson, G. G. (1969). *Biology and Man*. Harcourt Brace Jovanovich, New York.

Challenging essays.

Wilson, E. O. (Ed.) (1988). *Biodiversity*. National Academy Press, Washington, D.C.

The book discusses species diversity and the threats it faces in 57 chapters by specialists. Emphasis is on tropical rain forests, but many ecosystems (oceans, islands, grasslands, coastal zones, etc.) are covered.

A CLASSIFICATION OF ORGANISMS

In classifying organisms, authorities rarely agree. The following scheme presents either the general consensus or what we consider modest compromises in disputed cases.

A special problem arises in connection with Kingdom Monera, in which classification continues to be arbitrary (Chapter 37). Whether or not current classifications reflect meaningful genetic or evolutionary relationships remains uncertain. The species concept is especially vague when applied to bacteria and, according to some, indefensible. As for the major groups of protists, fungi, plants, and animals, most of the names in current use appear here.

We intend the classification to serve both as a glossary of systematic names and as an illustration of the tremendous diversity of living (and formerly living) things. These marvelous names and their interesting and evocative etymologies also add to the fun and romance of natural history.

The following ground-rules apply:

- All the **phyla (divisions)** we recognize are listed (sometimes with alternate names) and briefly described. Some systematists would add a few (or many) more. We show phyla and subphyla that are relatively obscure in the modern world in lightface type.
- Most of the **classes** are also listed. In some cases, these include groups called phyla by others. When classes are not designated within a phylum, either the phylum is small and has a single class, or a proposed division into classes is not satisfactory.
- In the very large classes—insects and fishes—we also list the major **orders** (though many are omitted). For the birds, mammals, and other large classes of special interest to us humans, we list *all* the orders we believe valid.
- For each phylum or class, at least one **genus** is named by way of example. More often several genera are cited. Many of them have been mentioned in the book, and we invite readers to locate some of these old friends in the scheme of things.
- The symbol † indicates that a group is *extinct*. Virtually all phyla and classes are known to include extinct smaller groups.

SUPERKINGDOM PROKARYOTA

KINGDOM MONERA

PHYLUM SCHIZONTA or **SCHIZOPHYTA,** the bacteria

Class Eubacteriae, true bacteria: *Bacillus, Escherichia, Azotobacter, Clostridium, Rhizobium, Pneumococcus, Streptococcus, Staphylococcus, Salmonella, Pasteurella*

Class Actinobacteriae: *Actinomyces, Corynebacterium, Nocardia, Streptomyces*

Class Myxobacteriae: *Myxococcus, Cytophaga, Chondromyces, Stigmatella, Lactobacillus*

Class Spirochaetae: *Leptospira, Spirochaeta, Treponema, Borrelia*

Class Mycoplasma: *Mycoplasma, Acholeplasma, Bartonella*

Class Rickettsiae: *Rickettsia, Coxiella, Chlamydia*

PHYLUM CYANONTA or **CYANOPHYTA,** blue-green algae (cyanobacteria)

Class Coccogoneaea, coccoid cyanobacteria: *Chamaesiphon, Chroococcus, Pleurocapsa*

Class Hormogoneaea, filamentous cyanobacteria: *Nostoc, Oscillatoria, Stigonema*

PHYLUM PROCHLORONTA or **PROCHLOROPHYTA:** *Prochloron, Prochlorothrix*

SUPERKINGDOM EUKARYOTA

KINGDOM PROTISTA

SUBKINGDOM ALGAE, photosynthetic autotrophs

PHYLUM EUGLENOPHYTA, euglenoids: *Euglena, Peranema, Astasia*

PHYLUM PYRROPHYTA or **DINOFLAGELLATA,** dinoflagellates: *Ceratium, Peridinium, Gymnodinium, Gonyaulax, Ceratodinium, Noctiluca, Protopsis*

PHYLUM CHRYSOPHYTA, golden algae

Class Chrysophyceae, golden-brown algae: *Synura, Dinobryon, Chromulina*

Class Bacillariophyceae, diatoms: *Navicula, Pinnularia, Tabellaria, Triceratium, Actinoptychus*

Class Xanthophyceae, yellow-green algae: *Vaucheria, Tribonema, Botrydium, Halosphaera*

PHYLUM CHLOROPHYTA, green algae

Class Chlorophyceae, grass-green algae: *Ulothrix, Oedogonium, Spirogyra, Eurodina, Closterium, Chlamydomonas, Ulva, Volvox, Tetraspora, Chlorella*

Class Charophyceae, stoneworts: *Chara, Nitella, Tolypella*

Class Siphonales, siphonous coenocytic green algae: *Valonia, Acetabularia, Codium, Hydrodictyon, Penicillus*

PHYLUM PHAEOPHYTA, brown algae: *Ectocarpus, Laminaria, Fucus, Sargassum, Dictyota*

PHYLUM RHODOPHYTA, red algae: *Porphyra, Batrachospermum, Nemalion, Lithophyllum, Thamnion, Polysiphonia, Dasya, Bangia*

SUBKINGDOM PROTOZOA,
nonphotosynthetic heterotrophs

PHYLUM ZOOMASTIGINA, (animallike) zooflagellates: *Trypanosoma, Polytoma, Chilomonas, Astasia, Oikomonas, Mastigamoeba, Hexamitus, Calonympha*

PHYLUM SARCODINA: *Amoeba, Globigerina, Lychnaspis, Entamoeba, Falbellula*

PHYLUM SPOROZOA: *Plasmodium, Gregarina, Babesia*

PHYLUM CILIOPHORA, ciliates

Class Oligohymenophora: *Paramecium, Vorticella, Epistylis, Tetrahymena*

Class Polyhymenophora: *Stylonychia, Stentor, Blepharisma, Euplotes*

Class Kinetofragminophora: *Entodiniomorpha, Tokophyra*

PHYLUM MYXOMYCOTA, plasmodial slime molds: *Physarum, Lycogala, Echinostelium, Ceratiomyxa*

PHYLUM ACRASIOMYCOTA, cellular slime molds: *Dictyostelium, Acrasia*

KINGDOM FUNGI

PHYLUM OOMYCETES, *Saprolegnia,* water molds; *Albuga,* white rusts and downy mildews; *Chytridium,* chytrids

PHYLUM ZYGOMYCETES, tube fungi: *Rhizopus,* bread molds; *Mucor, Phycomyces*

PHYLUM ASCOMYCETES, sac fungi: *Neurospora,* bread molds; *Saccharomyces,* yeasts; *Aspergillus, Penicillium,* blue and green molds; *Microsphaera,* powdery mildews; *Sclerotinia,* cup fungi; *Morchella,* morels

PHYLUM BASIDIOMYCETES, club fungi: *Psalliota,* mushrooms; *Amanita,* toadstools; *Fomes,* bracket fungi; *Puccinia,* smuts and rusts

PHYLUM DEUTEROMYCETES, fungi imperfecti: *Cryptococcus, Candida, Monilia, Histoplasma*

PHYLUM MYCOPHYCOPHYTA, lichens: fungus plus cyanobacteria (blue-green alga), e.g., *Nostoc;* fungus plus green alga, e.g., *Trebouxia*

KINGDOM PLANTAE

SUBKINGDOM BRYOPHYTA,
nonvascular plants

DIVISION BRYOPHYTA, bryophytes

Class Hepaticopsida or Hepaticae, liverworts: *Marchantia, Riccia, Conocephalum*

Class Anthocerotopsida or Anthocerotae, hornworts: *Anthoceros*

Class Muscopsida, mosses: *Sphagnum, Andreaea, Mnium, Funaria, Pottia*

SUBKINGDOM TRACHEOPHYTA,
vascular plants

DIVISION PSILOTOPHYTA, whisk ferns

Class † Zosterophyllopsida: † *Zosterophyllum*

Class † Rhyniopsida: † *Rhynia, Psilophyton*

Class † Psilopsida: *Psilotum, Tmesipteris*

DIVISION LYCOPODIOPHYTA, lycopods, club mosses: † *Lepidodendron,* † *Sigillaria, Lycopodium, Selaginella, Isoetes*

DIVISION SPHENOPHYTA, sphenophytes horsetails: † *Calamites, Equisetum*

DIVISION PTERIDOPHYTA, ferns: *Ophioglossum, Cyathea, Polypodium, Azolla, Osmunda, Platycerium*

DIVISION SPERMOPHYTA, seed plants

Subdivision Gymnospermae, gymnosperms.

Class † Pteridospermopsida: seed ferns

Class † Cordaitopsida: cordaites

Class † Cycadeoidopsida: cycadeoids

Class Cycadopsida, cycads: *Zamia*

Class Ginkgopsida: *Ginkgo*

Class Gnetopsida, gnetophytes: *Gnetum, Ephedra, Welwitschia*

Class Coniferopsida, conifers: *Pinus, Abies, Tsuga, Taxus, Sequoia, Metasequoia, Araucaria*

Subdivision Angiospermae, angiosperms, flowering plants

Class Monocotyledonae, monocots: *Pacnium, Stipa,* grasses; *Cyperus,* sedges; *Lilium,* lilies; *Tulipa,* tulips; *Yucca,* yuccas; *Sabal,* palms; *Cypripedium, Ophrys, Cryptostelis,* orchids

Class Dicotyledonae, dicots: *Magnolia,* magnolia; *Aristolochia,* snake root; *Eucalyptus,* eucalypts; *Quercus,* oaks; *Ulmus,* elms; *Acer,* maples; *Fagus, Nothofagus,* beeches; *Amygdalus,* peaches; *Cereus,* cacti; *Rubus,* blackberries; *Pisum,* peas; *Solanum,* nightshades; *Salvia,* sages; *Brassica,* mustards; *Taraxacum,* dandelions; *Ambrosia,* ragweeds

KINGDOM ANIMALIA

SUBKINGDOM PARAZOA

PHYLUM † ARCHAEOCYATHA: † *Archaeocyathus*

PHYLUM PORIFERA, sponges

Class Calcarea or Calcispongiae, chalky sponges: *Scypha, Sycon, Leucosolenia*

Class Hexactinellida or Hyalospongiae, glass sponges: *Hyalonema, Monoraphis*

Class Demospongiae, horny sponges, common bath sponges: *Spongia, Axinella, Dysidea*

Class Sclerospongiae, coralline sponges: *Astrosclera, Merlia*

PHYLUM PLACOZOA (only one known species): *Trichoplax*

SUBKINGDOM MESOZOA

PHYLUM MESOZOA: *Dicyema, Rhopalura*

SUBKINGDOM EUMETAZOA

PHYLUM COELENTERATA or **CNIDARIA,** coelenterates

Class Hydrozoa, hydrazoans: *Hydra; Obelia; Gonionemus; Physalia,* Portuguese man-of-war; *Velella, Tubularia*

Class Scyphozoa, scyphozoans, true jellyfishes: *Aurelia, Chrysaora, Charybdea*

Class Anthozoa, anthozoans, corals: *Astrangia, Diploria, Madrepora,* corals; *Corallium, Acropora, Actinia, Gonactinia;* sea anemones, *Metridium*

Class † Stromatoporoidea: † *Clathrodictyon,* † *Stromatopora*

PHYLUM CTENOPHORA, comb jellies, sea walnuts: *Cestus, Mertensia, Beroe*

PHYLUM PLATYHELMINTHES, flatworms

Class Turbellaria, planarians: *Planaria, Crenobia, Dugesia, Macrostomum*

Class Trematoda, flukes: *Fasciola,* liver flukes; *Clonorchis,* Chinese liver flukes; *Schistosoma,* blood flukes

Class Cestoda, tapeworms: *Taenia,* pork and beef tapeworms; *Diphyllobothrium,* fish tapeworm; *Dipylidium,* dog tapeworm

PHYLUM NEMERTINA or NEMERTEA, ribbon worms: *Lineus, Cerebratulus, Tubulanus, Nectomertes*

PHYLUM ASCHELMINTHES, roundworms

Class Nematoda, roundworms (the most important group of Aschelminthes and the class taken to represent the whole phylum in the discussion in Chapter 42): *Ascaris, Trichinella, Turbatrix, Ancylostoma, Nectar, Enterobius, Dracunculus*

Class Rotifera or Trochelminthes, rotifers: *Rotaria, Asplanchna, Hydatina*

Class Gastrotricha: *Chaetonotus, Macrodasys, Turbanella*

Class Kinorhyncha or Echinoderida: *Echinoderes, Semnoderes*

Class Nematomorpha or Gordiacea, horsehair worms: *Gordius, Paragordius, Nectonema*

PHYLUM ACANTHOCEPHALA, spiny-headed worms: *Giganthorhynchus, Echinorhynchus, Polymorphus*

PHYLUM ENTOPROCTA: *Urnatella, Pedicellina, Loxosoma*

PHYLUM ECTOPROCTA or **BRYOZOA:** bryozoans, moss animals

Class Phylactolaemata, fresh water: *Plumatella, Pectinatella, Lophopus*

Class Gymnolaemata, marine: *Paludicella, Bugala, Sertella*

Class Stenolaemata, marine: *Crisia, Hornera*

PHYLUM BRACHIOPODA, brachiopods or lampshells

Class Inarticulata: *Lingula, Glottidia, Crania*

Class Articulata: *Notosaria, Terebratulina, Leptodus, Spirifer*

PHYLUM PHORONIDEA, phoronids: *Phoronis, Phoronopsis*

PHYLUM CHAETOGNATHA, arrow worms: *Sagitta, Spadella, Eukrohnia*

PHYLUM MOLLUSCA, mollusks

Class Amphineura or Polyplacophora, chitons: *Chiton, Neomenia*

Class Aplachophora, solenogasters: *Nematomenia, Neomenia, Chaetoderma*

Class Monoplacophora, monoplacophorans, gastroverms: *Neopilina*

Class Scaphopoda, tusk shells: *Dentalium, Cadulus, Pulsellum*

Class Gastropoda, gastropods, univalves

Subclass Prosobranchia, front-gilled snails: *Buccinum, Ocenebra,* whelks; *Helix, Lymnaea, Planorbis,* snails; *Arion,* slugs; *Patella,* limpets

Subclass Opisthobranchia, shells reduced: *Archidoris,* nudibranchs (sea slugs); *Tethys,* sea hares

Subclass Pulmonata, lung-bearing snails: *Siphonaria, Latia, Physa*

Class Bivalvia or Pelecypoda: bivalves or pelecypods

Subclass Lamellibranchia, lamellibranchs (filter feeders) (The six orders of the subclass are usually used in place of the subclass name.): *Mercenaria,* Venus clams: *Anodonta,* river mussels; *Mya,* soft-shelled clams; *Pecten, Chlamys,* scallops; *Tridacna,* giant clams; *Pholas,* piddocks; *Teredo,* shipworms; *Solen,* razor clams; *Mytilus,* marine mussels; *Ostrea,* oysters

Subclass Protobranchia, protobranchs (deposit feeders): *Nucula; Yoldia; Xylophaga*

Subclass Septibranchia, septibranchs (pumping filter feeders): *Cuspidaria*

Class Cephalopoda, cephalopods

Subclass Dibranchiata: *Sepia,* cuttlefishes; *Loligo,* squids; *Octopus,* octopuses

Subclass Tetrabranchiata: *Nautilus,* nautilus

PHYLUM ANNELIDA, annelids, segmented worms

Class Oligochaeta, oligochaetes: *Lumbricus, Tubifex, Dendrobaena*

Class Hirudinea, leeches: *Hirudo, Haemodipsa, Macrobdella*

Class Polychaeta (including Archiannelida), polychaetes, sandworms: *Neanthes, Nereis, Aphrodite, Chaetopterus, Diopatra, Sabella, Eunice*

Class Gephyrea, sipunculid, echiuroid, and priapulid worms (often placed in one, two, or three separate phyla): *Sipunculus, Echiurus, Priapulus*

PHYLUM ARTHROPODA, arthropods (several other small classes often recognized)

Subphylum Trilobita

Class † Trilobita, † trilobites: † *Agnostus,* † *Neolenus,* † *Olenus,* † *Isotelus*

Subphylum Mandibulata

Class Crustacea, crustaceans: *Artemia,* brine shrimps; † *Barella;* † *Hymenocaris;* † *Daphnia,* water fleas; *Cyclops,* copepods; *Balanus, Lepas, Sacculina,* cirripeds (barnacles); *Armidillium,* pill bugs, wood lice; *Crangon,* common shrimps; *Caprella,* ghost shrimps; *Euphausia,* euphausids; *Leander,* prawns; *Palinurus,* spiny lobsters; *Homarus,* lobsters; *Cancer,* rock crabs

Class Insecta, insects

Subclass Apterygota, wingless insects

Order Collembola, springtails: † *Rhyniella, Isotoma, Anurida*

Order Thysanura, silverfish, bristletails: *Machilis, Lepisma, Thermobia*

Subclass Pterygota, winged insects

Order Psocoptera, book lice: *Liposcelis, Trogium*

Order Anoplura, sucking lice: *Pediculus, Phthirus, Linognathus*

Order Mallophaga, biting lice: *Haematomyzus, Colpocephalum, Trichodectes*

Order Ephemeroptera, mayflies: *Ephemera, Ecdyonurus, Caenes, Lestes*

Order Odonatoptera, dragonflies, damselflies: † *Dunbaria, Libellula, Calopteryx*

Order Dictyoptera, cockroaches, mantids: *Periplaneta, Blatta, Mantis*

Order Isoptera, termites: *Kalotermes, Amitermes, Syntermes, Cryptotermes*

Order Orthoptera, grasshoppers, crickets, katydids, locusts: *Schistocerca, Nemobius, Melanoplus, Locusta, Scudderia*

Order Dermaptera, earwigs: *Labia, Forficula, Hemimerus*

Order Coleoptera, beetles, weevils: *Calosoma, Copris, Phyllophaga, Melanotus, Scolytus, Coccinella, Leptinotarsa*

Order Lepidoptera, butterflies, moths: *Tinea, Pyrausta, Samia, Lycaena, Colias, Papilio, Parnassius, Tegeticula, Arctia, Morpho, Danaus*

Order Diptera, flies, gnats, mosquitoes: *Musca, Aedes, Asilus, Drosophila, Tabanus, Tipula, Anopheles, Glossina*

Order Hymenoptera, bees, wasps, sawflies, ants: *Scolia, Vespa, Apis, Formica, Bombus, Glypta, Pogonomyrmax, Coccophagus, Dorylus, Sphex*

Order Strepsiptera, parasitic stylopids: *Stylops*

Order Thysanoptera, thrips: *Heliothrips, Megalothrips, Franklinella*

Order Hemiptera, true bugs: *Cimex, Halobates, Lygus, Tingis, Ranatra*

Order Homoptera, cicadas, aphids, scale insects, leafhoppers: *Aleurods, Aphrophora, Chermes, Phylloxera, Aphis, Brachynus*

Order Siphonaptera, fleas: *Pulex, Xenopsylla, Tunga*

Order Trichoptera, caddisflies: *Philanisus, Enoicyla, Triaenodes*

Class Chilopoda, centipedes: *Lithobius, Scolopendra, Scutigera*

Class Diplopoda, millipedes: *Julus, Narceus, Glomeris, Siphonophora*

Class Pauropoda, centipedelike animals with 9–10 pairs of legs: *Pauropus*

Class Symphyla, centipedelike animals with 10–12 pairs of legs: *Scutigerella*

Class Onychophora: *Peripatus* (a separate phylum in many classifications)

Subphylum Chelicerata

Class Merostomata, horseshoe crabs: *Limulus,* † *Belinurus*

Class † Eurypterida, giant water scorpions: † *Eurypterus,* † *Pterygotus,* † *Carcinosoma,* † *Hughmilleria*

Class Arachnida: *Eurypelma, Theridion, Latrodectus,* spiders; *Vejovis,* scorpions; *Dermacentor,* ticks

Class Pycnogonida, sea spiders: *Pycnogonum, Nymphon*

Subphylum Pentastomida

Class Pentastomida, linguatulids: *Amillifer, Liguatula*

PHYLUM ECHINODERMATA, echinoderms

Class Crinoidea, crinoids, sea lilies, feather stars: *Antedon, Comactenia, Ptilocrinus*

Class Asteroidea, starfishes: *Asterias, Oreaster, Luidia*

Class Ophiuroidea, brittle stars, serpent stars: *Ophiura, Asteronyx, Ophioderma, Amphioplus*

Class Echinoidea, sea urchins, sand dollars: *Strongylocentrotus, Cidaris, Arbacia, Moira*

Class Holothuroidea, sea cucumbers: *Cucumaria, Caudina, Synapta*

Class † Cystoidea, cystoids: † *Caryocrinites*

Class † Edrioasteroidea, edrioasteroids: † *Edrioaster*

Class † Blastoidea, blastoids: † *Pentremites*

PHYLUM CHORDATA, chordates

Subphylum Stomochorda

Class Enteropneusta or Hemichordata, tongue or acorn worms: *Balanoglossus*

Class Pterobranchia, pterobranchs: *Rhabdopleura, Cephalodiscus*

Class † Graptolithina, graptolites: † *Didymograptus*

Subphylum Urochordata or Tunicata, tunicates

Class Ascidiacea, sea squirts, ascidians: *Ciona, Molgula, Clavelina*

Class Larvacea: *Appendicularia, Fritillaria*

Subphylum Cephalochordata, lancelets, amphioxus: *Asymmetron, Branchiostoma,*

Subphylum Vertebrata, vertebrates

SUPERCLASS AGNATHA, jawless fishes

Class Cyclostomata, cyclostomes: † *Cephalaspis;* † *Kieraspis;* † *Pteraspis; Petromyzon,* lampreys; *Myxine,* hagfishes

SUPERCLASS PISCES, aquatic vertebrates, fishes

Class † Placodermi, placoderms: † *Diplacanthus,* † *Coccosteus,* † *Dinichthys,* † *Pterichythyodes*

Class † Acanthodii: † *Acanthodes,* † *Climatius*

Class Chondrichthyes or Elasmobranchiiomorphi, cartilaginous fishes

Order † Cladoselachii: † *Cladoselache*

Order † Pleuracanthodii: † *Xenacanthus*

Order Selachii, sharks, dogfishes: *Squalus, Hexanchus, Alopias, Galeus, Charcharodon*

Order Batoidea or Rajiformes, rays, guitar fishes, skates: *Torpedo, Raja, Dasyatis, Chimaera*

Class Osteichthyes or Teleostomi, bony fishes

Subclass Sarcopterygii, fleshy-finned fishes

Order Crossopterygii, coelacanths, lobe-fins: † *Eusthenopteron,* † *Osteolepis, Latimeria*

Order Dipnoi, lungfishes: † *Dipterus, Neoceratodus, Lepidosiren, Protopterus*

Subclass Actinopterygii, higher bony fishes

Order † Palaeonysciformes: † *Paleoniscus*

Order Polypteriformes, bichirs, reedfish: *Polypterus, Calamoichthys*

Order Acipenseriformes, sturgeons, paddlefishes: *Acipenser, Scaphirhyncus, Polyodon*

Order Anguilliformes, eels: *Anguilla, Conger, Cyema*

Order Clupeiformes, herrings, anchovies: *Clupea, Alosa, Engraulis, Sardina*

Order Salmoniformes, salmon, trout, whitefishes: *Salmo, Osmerus, Onchorhynchus*

Order Cypriniformes, carp: *Charax, Agoniates*

Order Siluriformes, catfishes: *Ictalurus, Siluris*

Order Perciformes, perches, tunas, marlins: *Perca, Gramma, Roccus, Brama*

Order Lophiiformes, anglerfishes: *Photocorynus, Lophius, Ceratias*

Order Gadiformes, codfishes: *Gadus, Lotella, Merluccius*

Order Atheriniformes, flying fishes, grunion, garfishes, sandsmelts: *Exocoetus, Belone, Atherina, Leuresthes*

Order Gasterosteiformes, sticklebacks: *Gasterosteus, Culaea*

Order Solenichthyformes, pipefishes, seahorses: *Syngnathus, Hippocampus*

SUPERCLASS TETRAPODA, land vertebrates, tetrapods

Class Amphibia, amphibians

Order Urodela, salamanders, newts: *Necturus, Ambystoma, Trituris, Hynobius*

Order Anura, anurans, frogs, toads: *Rana, Bufo, Hyla, Xenopus*

Order Apoda, caecilians; *Gymnophis, Icthyophis, Sophonops* (plus at least 6 extinct orders)

Class Reptilia, reptiles

Subclass Anapsida

Order † Cotylosauria: † *Limnoscolis*

Order Chelonia or Testudines, turtles, tortoises: *Chelydra, Clemmys, Testudo, Aromochelys*

Subclass † Synapsida, mammallike reptiles

Order † Pelycosauria, pelycosaurs: † *Ophiocodon,* † *Edaphosaurus,* † *Dimetrodon*

Order † Therapsida, therapsids: † *Dicynodon,* † *Lycaenops,* † *Cynognathus*

Subclass † Ichthyopterygia

Order † Ichthyosauria, ichthyosaurs: † *Ichthyosaurus*

Subclass † Synaptosauria or † Euryapsida

Order † Sauropterygia, plesiosaurs: † *Nothosaurus,* † *Rhomaleosaurus*

Order † Placodontia, placodonts: † *Placodus,* † *Henodus,* † *Placochelys*

Subclass † Archosauria

Order † Thecodontia, theocodonts: † *Stegomus,* † *Phytosaurus*

Order † Saurishia, dinosaurs (with reptilelike pelvis): † *Tyrannosaurus,* † *Diplodocus,* † *Megalosaurus,* † *Plateosaurus,* † *Atlantosaurus*

Order † Ornithischia, dinosaurs (with birdlike pelvis): † *Ankylosaurus,* † *Iguanodon,* † *Triceratops,* † *Stegosaurus*

Order † Pterosauria, flying reptiles: † *Nyctosaurus,* † *Pteranodon,* † *Pterodactylus*

Order Crocodilia, crocodiles, alligators: *Crocodylus, Alligator*

Subclass Lepidosauria

Order † Eosuchia: † *Champosaurus*

Order Rhynochocephalia, tuatara: † *Pleurosaurus,* † *Homeosaurus, Shenodon*

Order Squamata, scaly reptiles (lizards, snakes): *Gerrhonotus, Crotophytus, Crotalus, Thamnophis, Iguana, Phyrnasoma, Chamaeleo, Hydrophis*

Class Aves, birds

Subclass † Archaeornithes, the most ancient birds

Order † Archaeopterygiformes: † *Archaeopteryx*

Subclass Palaeognathae, birds with primitive skulls and jaws

Order † Icthyornithiformes: † *Ichthyornis*

Order † Hesperornithiformes: † *Hesperornis*

Order Tinamiformes, tinamous: *Eudromia*

Order Rheiformes, rheas: *Rhea*

Order Casuariiformes, cassowaries, emus: *Casuarius, Dromaius*

Order Struthioniformes, ostriches: Struthio

Order † Aepyornithiformes, elephant birds: † *Aepyornis*

Order † Dinornithiformes, moas: † *Dinornis*

Order Apterygiformes, kiwis: † *Apteryx*

Subclass Neognathae, birds with specialized skulls and jaws

Order Gaviiformes, loons: *Gavia*

Order Podicipediformes, grebes: *Podiceps*

Order Procellariiformes, albatrosses, petrels, shearwaters: *Diomeda, Puffinus, Phoebetria*

Order Sphenisciformes, penguins: *Spheniscus, Pygoscelis*

Order Pelecaniformes, pelicans, gannets, cormorants: *Pelecanus, Phaethon, Fregata*

Order † Odontopterygiformes: † *Odontopteryx*

Order Ciconiiformes, herons, storks: *Ardea, Ciconia, Scopus*

Order Phoenicopteriformes, flamingos: *Phoenicopterus*

Order Anseriformes, ducks, geese, swans: *Cygnus, Anser, Anas*

Order Falconiformes, eagles, hawks, vultures: *Aguila, Falco, Vultur*

Order Galliformes, chickens, turkeys, partridges, peacocks: *Gallus, Tetrao, Meleagris, Pavo*

Order Gruiformes, cranes: *Grus*

Order † Diatrymiformes: † *Diatryma*

Order Charadriiformes, gulls, plovers, auks: *Larus, Pluvialis, Alca*

Order Columbiformes, pigeons: *Columba, Streptopilia*

Order Psittaciformes, parrots, lovebirds: *Psittacus, Agapornis, Kakatoe*

Order Musophagiformes, touracos: *Tauraco*

Order Cuculiformes, cuckoos, roadrunners: *Clamator, Geococcyx*

Order Strigiformes, owls: *Strix, Otus, Bubo*

Order Caprimulgiformes, nightjars, frogmouths: *Caprimulgus*

Order Apodiformes, swifts: *Apus, Chaetura*

Order Trochiliformes, hummingbirds: *Archilochus, Colibri*

Order Coliiformes, mousebirds: *Colius*

Order Trogoniformes, trogons: *Trogon*

Order Coraciiformes, kingfishers, hoopoes, rollers: *Halcyon, Upupa, Coracias*

Order Piciformes, woodpeckers, flickers, toucans: *Colaptes, Centurus, Rhamphastos*

Order Passeriformes, passerines, including sparrows, larks, swallows, shrikes, wrens, thrashers, thrushes, nuthatches, tanagers, starlings, crows, and many others: *Passer, Alauda, Hirundo, Lanius, Troglodytes, Mimus, Turdus, Sitta, Tangara, Sturnus, Zonotrichia, Corvus*

Class Mammalia, mammals

Subclass Prototheria

Order Monotremata, platypus, echidnas: *Ornithorhynchus, Tachyglossus, Zaglossus*

Subclass † Allotheria

Order † Multituberculata: † *Plagiaulax*, † *Ptilodus*, † *Taeniolabis*

Subclass † Eotheria

Order † Triconodonta: † *Triconodon*

Order † Docodonta: † *Docodon*

Subclass Theria

Infraclass † Patriotheria

Order † Symmetrodonta: † *Kuehneotherium*, † *Spalacotherium*

Order † Pantotheria: † *Amphitherium*, † *Paurodon*, † *Dryolestes*

Infraclass Metatheria, marsupials

Order Marsupialia, marsupials, including opossums, dasyures, bandicoots, phalangers, koalas, wombats, kangaroos, and others: *Didelphis, Dasyurus, Perameles, Phalanger, Phascolomys, Thylacinus, Petaurus, Myrmecobius, Macropus*

Infraclass Eutheria, placentals

Order † Leptictida: † *Palaeoryctes*, † *Ictops*

Order Insectivora, hedgehogs, shrews, moles: *Erinaceus, Sorex, Talpa*

Order Macroscelidea, elephant shrews: *Macroscelides*

Order Dermoptera, colugos: *Cynocephalus*

Order Chiroptera, bats: *Pteropus, Desmodus, Eptesicus*

Order Primata, lemurs, monkeys, apes, humans: † *Notharctus, Lemur, Cebus, Cercopithecus, Gorilla, Homo, Hylobates, Papio, Macaca*

Order † Tillodontia: † *Esthonyx*

Order † Taeniodonta: † *Ectoganus*

Order † Palaeanodontea: † *Epoicotherium*

Order Xenarthra, armadillos, glyptodonts, sloths, anteaters: *Dasypus, Mylodon, Bradypus, Myrmecophaga*

Order Pholidota, pangolins: *Manis*

Order Lagomorpha, rabbits, hares: *Sylvilagus, Lepus*

Order Rodentia, rodents, including squirrels, pocket gophers, kangaroo rats, beavers, rats, field mice, hamsters, porcupines, cavies, and many others: *Sciurus, Geomys, Heteromys, Castor, Rattus, Microtus, Cricetus, Erethizon, Glaucomys, Cavia*

Order Cetacea, whales, porpoises, dolphins: *Physeter, Eubalaena, Phocoena, Delphinus, Balaenoptera, Physeter, Tursiops, Phocoena*

Order † Creodonta: † *Oxyaena*, † *Hyaenodon*

Order Carnivora: dogs, racoons, weasels, civets, hyaenas, cats, seals, walruses, bears: *Canis, Procyon, Mustela, Viverra, Hyaena, Felis, Zalophus, Odobenus, Ursus*

Order † Condylarthra: † *Phenacodus*

Order Tubulidentata, aardvarks: *Orycteropus*

Order † Litopterna: † *Proterotherium*, † *Macrauchenia*

Order † Notoungulata: † *Thomashuxleya*, † *Scarrittia*, † *Toxodon*, † *Mesotherium*

Order † Pyrotheria: † *Pyrotherium*

Order † Astrapotheria: † *Astrapotherium*

Order † Trigonostylopoidea: † *Trigonostylops*

Order † Xenotheria: † *Carodnia*

Order † Pantodonta: † *Coryphodon*

Order † Dinocerata: † *Uintatherium*

Order Proboscidea, † mastodons, elephants: † *Gomphotherium*, † *Mammut*, † *Mammuthus, Loxodonta, Elephas*

Order † Embrithopoda: † *Arsinoitherium*

Order Hyracoidea, hyraxes: *Procavia*

Order Sirenia, sea cows (manatees, dugongs): *Trichechus, Dugong*

Order † Desmostylia: † *Desmostylus*

Order Perissodactyla, horses, † titanotheres, † chalicotheres, tapirs, rhinoceroses: *Equus*, † *Brontops*, † *Moropus, Tapirus, Diceros*

Order Artiodactyla, pigs, peccaries, oreodonts, camels, deer, sheep, giraffes, pronghorns, antelopes, goats, cattle, and others: *Sus, Tayassu, Camelus, Cervus, Giraffa, Antilocarpa, Gazella, Ovis, Capra, Bos, Bison*

GLOSSARY

The Glossary lists most of the boldfaced terms in the text. It includes only major taxonomic categories. Other taxa, species names, most units of measurement, and other terms will be found in the index, where boldfaced page numbers point to major discussions in the text.

The following points should make the glossary more accessible:

■ The number (or numbers) in brackets denote the chapter(s) in which the term was introduced, defined, or discussed.

■ We have alphabetized terms with Greek letter prefixes as though the prefix were spelled out. Therefore, the position of α-*glycoside* is that of *alpha-glycoside*, and β-*carotene* is alphabetized as *beta-carotene*.

■ Except for chemical compounds, terms beginning with a numeral are alphabetized as though the numeral were spelled out. Thus, *9 + 2* is listed as though it began with *nine*.

■ Definitions of terms with traditional conceptual associations (e.g., *hypotonic* and *hypertonic*) include "compare with" references to related terms.

Abdomen. In arthropods, the posterior portion of the body. [42] In mammals, the part of the body posterior to the thorax. [43]

Abiotic environment. All physical and nonliving environmental materials or forces (such as soil, temperature, etc.) that influence organisms. Compare with *biotic environment.* [46]

Abomasum. Fourth chamber of the ruminant stomach. Corresponds to a true stomach. [26]

Abortion. Expulsion from the uterus, spontaneous or induced, of a developing fetus before it is viable. [21]

Abscisic acid. A plant hormone that promotes various inhibitory effects, e.g., dormancy in seeds and buds, stomatal closing, leaf abscission, etc. [28]

Abscission. In vascular plants, the process by which leaves, petals, stems, and fruits drop from a plant after a growing season. [24,41]

Absolute temperature scale. The Kelvin (K) scale: 1°K has the same value as 1°C, but 0°K is −273°C, the temperature at which molecular motion ceases. [3,8]

Absorption spectrum. A graph plotting the amount of incident light of different wavelength (colors) absorbed by a given substance. Compare with *action spectrum.* [4,10]

Abyssal zone. A biogeographical realm, comprising the deep ocean beyond the continental shelf, which receives no light. Compare with *bathyal zone.* [48]

Accessory pigments. Secondary plant pigments, such as carotenoids or xanthophylls, that absorb light energy and transfer it to chlorophyll. [10]

Acetylcholine. The neurotransmitter molecule in vertebrate neuromuscular junctions and other synapses in the parasympathetic and central nervous systems. [29]

Acetylcholinesterase. The enzyme that catalyzes the degradation by hydrolysis of the neurotransmitter acetylcholine. [29]

Acetyl coenzyme A (acetyl CoA). A key metabolic intermediate that contains a high-energy bond and provides 2-carbon fragments for many processes (citric acid cycle, lipid synthesis, etc.). It arises from pyruvate and the degradation of fatty acids and certain amino acids. [9]

Acid. A substance capable of releasing a proton (hydrogen ion, H^+). Compare with *base* and *alkali.* [3]

Acidic. Having a pH of less than 7 (a hydrogen ion concentration greater than $10^{-7}\ M$). Compare with *basic.* [3]

Acoelomate. Animal, like a flatworm, that lacks a coelom. [19,42]

Acrosome. Caplike part of an animal sperm head. Releases hydrolytic enzymes and an actin filament that facilitates penetration of the egg during fertilization. [19]

Actin. A widely occurring protein that is essential for cell motility. It forms microfilaments in nonmuscle cells and thin filaments in muscle cells where with myosin it is responsible for muscle contraction. [7,31]

Actin-binding proteins. A class of proteins that regulate the polymerization and function of actin. [7]

Actinomycin D. A drug that inhibits mRNA synthesis and thus provides a tool for identifying mRNA-dependent mechanisms. [18]

Action potential. A neuronal impulse consisting of a transient, all-or-none wave of cell membrane depolarization that can travel along an axon and depends on voltage-activated sodium and potassium channels. [29]

Action spectrum. A graph of the effectiveness of different light wavelengths (colors) in causing a given biological effect, e.g., photosynthesis. Compare with *absorption spectrum.* [10]

Activating enzyme. An enzyme that transfers a high-energy bond (usually from ATP) to a low-energy molecule (e.g., an amino acid or fatty acid), thus enabling it to participate in exergonic biosynthetic reactions. [8,9,11]

Activation energy. The chemical bond energy that must be possessed by a molecule before it can enter into a given chemical reaction. [8]

Active immunity. Immunity generated by an animal after it is exposed to an antigen. Compare with *passive immunity.* [23]

Active site. The region of the enzyme surface that binds the substrate(s) of the reaction it catalyzes. [8]

Active transport. Energy-dependent transport of a substance across a membrane against a concentration gradient. Compare with *passive diffusion* and *facilitated diffusion.* [6]

Adaptation. (1) The adjustment of members of a population to their environment. (2) Any aspect of an organism's anatomy, physiology, or behavior that promotes its ability to survive and reproduce and thus improves its adaptedness. [35]

Adaptive radiation. Evolutionary divergence of members of a single phyletic line into a series of different adaptive regions or niches. Often occurs after a population moves to an area with many uninhabited ecological niches. [36]

Adenine (A, Ade). A nitrogenous purine base found in nucleic acids, ATP, ADP, and some coenzymes. Pairs with thymine in DNA. [16]

Adenohypophysis. The anterior pituitary gland. Compare with *neurohypophysis.* [28]

Adenosine triphosphate (ATP). An important nucleotide that stores energy in two high-energy phosphoanhydride bonds and serves as the universal "currency" of energy exchange in virtually all living cells. [9]

Adenylate cyclase. An enzyme, often coupled to a cell-surface receptor, which converts ATP to the "second messenger" cyclic AMP. [28]

Adhering junction. See *desmosome*. [7]

Adrenal gland. A two-component endocrine gland of vertebrates that secretes epinephrine (adrenalin) and norepinephrine from its medulla, and corticosteroid hormones (cortisol, aldosterone, etc.) from its cortex. [28]

Adrenalin. See *epinephrine*. [28]

Adrenocorticotropic hormone (ACTH). A hormone released by the adenohypophysis that stimulates the adrenal cortical hormone secretion. [28]

Aerobic. Denotes an organism or metabolic process that utilizes or depends on free oxygen. Aerobic metabolism can proceed only in the presence of oxygen. Compare with *anaerobic*. [9,38]

Afferent. Leading toward an organ. Compare with *efferent*. Afferent arterioles in the kidney carry blood from the renal artery to a glomerulus. [25] Afferent nerves carry impulses to the central nervous system. [29]

Agglutinin. A substance, often an antibody or lectin, that clumps cells of a particular type. [23]

Agnatha. A class of primitive vertebrates which lack true jaws; includes the jawless fishes. [43]

Agonist. An aid. An agonist of a hormone binds its receptor and has similar effects. An agonist of a muscle causes the joint to move in the same direction. Compare with *antagonist*. [28]

AIDS (acquired immune deficiency syndrome). A serious infectious disease caused by the human immunodeficiency virus (HIV). It induces susceptibility to secondary infections by impairing cell-mediated immunity. [23]

Albumin. (1) Any of a group of plant or animal proteins that are soluble in water or dilute salt solutions. [4] (2) A major protein of blood plasma that transports some molecules and osmotically maintains blood volume. A major protein of egg white. [22]

Aldose. A sugar containing an aldehyde group (—CHO). Compare with *ketose*. [4]

Aldosterone. A hormone secreted by the adrenal cortex that regulates body sodium and potassium levels by stimulating the kidney to retain sodium and excrete potassium. [25,28]

Alga. Any of a wide diversity of unicellular or simple multicellular organisms that are autotrophic and photosynthetic and thus more plantlike than animallike. [39]

Alimentary canal. The gut. A tube specialized for passage of food, digestion, and absorption. [26]

Alkali. A basic chemical, one that can accept a proton (hydrogen ion, H^+). Compare with *acid*. [3]

Allantois. An extraembryonic membrane that lies between the chorion and amnion. In bird and reptile eggs, it functions in excretion and respiration. In most mammals, it helps to form the placenta. [21]

Allele. One of two or more alternative forms of a gene. Found at corresponding sites (loci) on homologous chromosomes. [13]

Allele frequency. The rate of occurrence of a given allele in a gene pool. [34]

Allergy. An immune reaction marked by an exaggerated physiological response to a substance that causes no reaction in most individuals. [23]

Allogenic. A force external to a community with ecological consequences in the community (e.g., a change in habitat resulting from drought). Compare with *autogenic*. [47]

Allopatric speciation. Speciation occurring when populations diverge after becoming geographically isolated. Compare with *sympatric* and *parapatric speciation*. [36]

Allophenic animal. A chimeric animal displaying the phenotypic effects of more than one genotype. [20]

Allostery. Regulation of the activity of an enzyme by binding, at a site other than the catalytic active site, of an effector molecule that does not have the same structure as any of the enzyme's substrates. [8]

α-glycosidic bond. One steric form of the bonds linking mono- or oligosaccharide units to form glycosides. This type of bond links glucose units in glycogen. [11]

α-helix. A helical spatial configuration of polymeric chains. A common secondary structure of proteins. [4]

Alpine region. Biome on high mountains above the timberline. [48]

Alternation of generations. Succession of haploid and diploid phases in a sexually reproducing organism. In many plants and fungi, the diploid sporophyte produces a haploid gametophyte and vice versa. The more prominent phase may be haploid (as in mosses) or diploid (as in *Ulva* or vascular plants). In animals (male wasps and honey bees are exceptions), the haploid phase consists only of the gametes. [21]

Altruism. Actions by an animal that increase the evolutionary fitness of other members of its species without increasing its own fitness. [33,45]

Alveoli. Small, thin-walled sacs in the vertebrate lung, across which gases are exchanged. [27]

Amino acid. An organic compound of the general formula H_2N—CHR—$COOH$, where R is one of 20 or more different side groups. So named because it has both a basic amino group, —NH_2, and an acidic carboxyl group, —$COOH$. Proteins are polymers of amino acids. [4]

Amino acyl-tRNA synthetase. A family of enzymes that add the appropriate amino acid to each transfer RNA molecule. [17]

Aminopeptidase. A protease secreted by the pancreas that liberates an amino acid residue from the N-terminal end (amino end) of a protein. Compare with *carboxypeptidase*. [26]

Ammonia. NH_3. A nitrogen-containing end-product of protein metabolism in aquatic animals. Compare with *urea* and *uric acid*. [25]

Amniocentesis. A procedure for sampling amniotic fluid. Often done to determine fetal sex and health. [15]

Amnion. In birds, reptiles, and mammals (all amniotes), the innermost extraembryonic membrane. Filled with amniotic fluid. [15,21]

Amniote. A vertebrate (bird, reptile, or mammal) that develops an amnion and allantois around its embryo. [43]

Amniotic fluid. The watery fluid that suspends and cushions the fetus inside the amnion. [15,21]

Amoeba. Any of several unicellular animals of the phylum Sarcodina. Characterized among other features by its ability to change shape through the protrusion and retraction of cytoplasmic extensions called pseudopodia. [39]

Amoeboid movement. Movement of a single cell by pseudopodia (temporary extensions of the cell body). [7]

Amphibia. The first land-dwelling class of vertebrates; includes the orders Urodela (e.g., salamanders), Anura (e.g., frogs), and Apoda. [43]

Amphioxus. A tiny, fishlike animal; a member of the subphylum Cephalochordata. [43]

Amphipathic. A molecule that is soluble in both water and nonpolar solvents because it contains both hydrophilic and hydrophobic groups. Soaps are amphipathic. [4]

Amphoteric. A molecule containing both acidic and basic groups. Amino acids are amphoteric. [3]

Amylase. Any of a class of enzymes that digest (hydrolyze) starch into smaller polysaccharides. [26]

Anabolism. Synthetic reactions of metabolism, in which complex molecules are formed from simpler ones. Compare with *catabolism*. [8]

Anaerobic. An organism or metabolic process that can proceed in the absence of free oxygen (in some cases *only* in its absence). See *obligate* and *facultative anaerobe*. Compare with *aerobic*. [9,38]

Anagenetic speciation (anagenesis). Evolution occurring when descendents in a single lineage (a succession of ancestral and descendant populations) become different enough from their ancestors to be called a separate species. Compare with *cladistic speciation*. [36]

Analogy. Structures which perform similar functions in different species, but are not evolutionarily related (e.g., wings of birds and insects). Compare with *homology* and *homoplasy*. [37]

Anamnestic response. The rapid appearance of antibody in an animal after a second exposure to an antigen. [23]

Anaphase. A stage of mitosis and meiosis, in which chromatids separate from each other and begin to move apart. [12]

Anaphylaxis. An extreme, sudden allergic reaction in which antibody-antigen complexes trigger the release of possibly fatal amounts of histamine. [23]

Anatomical adaptation. See *structural adaptation*. [35]

Androecium. The male reproductive part of a flower, consisting of one or more stamens. [41]

Androgen. Any of a group of vertebrate male sex hormones, e.g., testosterone. [28]

Anemia. Deficiency of red blood cells. [22]

Aneuploidy. An abnormality of the chromosome number resulting from duplication or deletion of one or a few chromosomes of the set. Compare with *polyploidy*. [12,15]

Angiogenesis. The development of new blood vessels in a tissue. [28]

Angiosperm. A flowering plant. Member of the plant subdivision (subphylum) Angiospermae. A group of plants whose seeds are borne with a mature ovary (fruit). Includes monocots and dicots. [41]

Animal pole. The side of a zygote or early embryo which is characterized by the smallest cells and least amount of yolk. Compare with *vegetal pole*. [19]

Animal. A member of the kingdom Animalia. In general, a multicellular, free-living eukaryote that obtains food by ingestion. [42]

Animalia. A living kingdom that consists of multicellular eukaryotes with such distinctive features as nervous systems, food ingestion, etc. One of five phylogenetic kingdoms. [37,42,43]

Animallike protists. Nonphotosynthetic heterotrophic protists (protozoa), e.g., zooflagellates, sarcodines, etc. May have given rise to the first animals. Compare with *plantlike protists*. [39]

Anion. A negatively charged ion. [3]

Annelid. A member of the animal phylum Annelida; includes segmented worms, leeches, polychaetes, etc. [42]

Anoxygenic photosynthesis. A form of photosynthesis found in prokaryotes that does not produce oxygen. [10]

Antagonist. An opponent. (1) An antagonist of a hormone binds its receptor and has opposite effects. Compare with *agonist*. [28] (2) An antagonist of a muscle that causes the joint to move in the opposite direction. [31]

Antennapedia complex. A cluster of homeotic genes that in *Drosophila* affects head and thoracic segments. Mutations in this region can rearrange head and thorax so that antennae are replaced by legs. [20]

Antenna pigments. Pigments in a photosynthetic unit that transfer the energy they absorb to a single molecule of chlorophyll *a*. They include accessory pigments and other chlorophyll molecules. [10]

Anther. The terminal, pollen-bearing portion of a stamen in flowers. [21]

Antheridium (pl. **antheridia**). A multicellular sperm-producing organ in algae, fungi, and plants. Compare with *oogonium*. [39]

Anthozoan. A member of coelenterate class Anthozoa; polypoid forms that include sea anemones, sea fans, corals, etc. [42]

Anthropoid. A member of the primate suborder Anthropoidea; includes monkeys, apes, and humans. Compare with *hominid* and *hominoid*. [44]

Antibiotic. An antibacterial compound that is synthesized by a microorganism, usually a fungus or bacterium. [40]

Antibody. An immunoglobulin molecule, one of a group of plasma proteins produced by B lymphocytes, which specifically binds and inactivates the foreign substance (antigen) that elicited its production. Antibodies are the basis of humoral immunity. [22,23]

Anticoagulant. An agent that prevents blood clotting. [22]

Anticodon. A nucleotide triplet in transfer RNA that pairs with a complementary triplet (codon) in messenger RNA, thus aligning the two RNAs for protein synthesis. Each anticodon and codon specifies a given amino acid. [17]

Antidiuretic hormone (ADH). Vasopressin. A hormone made by the hypothalamus and secreted by the posterior pituitary, which increases water reabsorption in the kidney. [25,28]

Antigen. Any foreign substance that stimulates the body to generate antibodies against it. [23]

Antigenic determinant. A specific region of the chemical structure of an antigen that stimulates antibody synthesis and then is recognized by and bound by the specific antibody produced. [23]

Antiserum. Blood serum that contains antibodies to a specific antigen. [23]

Antitoxin. A neutralizing antibody that binds and inactivates a specific toxin. [23]

Anus. The opening at the posterior end of the gut through which digestive wastes are expelled. [26]

Aorta. The main arterial trunk leading from the heart to the systemic (as opposed to the pulmonary) circulation. [24]

Aphotic zone. The lightless zone in the depths of the ocean. Includes the bathyal and abyssal zones. Compare with *euphotic zone*.

Apical. Pertaining to the apex (the tip or highest point of a structure, such as a growing stem or root. See *apical meristem*. [24,41]

Apical meristem. The area of active cell division at the tip of a plant stem or root. Compare with *cambium*. [7]

Apoenzyme. The protein portion of an enzyme. An enzyme minus its coenzyme. [8]

Apogamy. In some plants (bryophytes and ferns), the direct, parthenogenetic development of sporophyte from gametophyte without fertilization. [21]

Apomixis. Vegetative reproduction without meiosis or fertilization. Progeny are identical to the parents. Compare with *automixis*. [21]

Apomorphic. A trait that is an evolutionary innovation not possessed by the population. Applied to derived characters in a cladogram that define evolutionary branching points. Compare with *plesiorphic* and *synapomorphic*. [37]

Apoplastic pathway. The active transport of solutes (and consequent passive flow of water) across plant cell walls from root surface to xylem. Compare with *symplastic pathway*. [24]

Aposematic coloration. A distinctive, usually species-specific appearance that intimidates potential predators. Also called *warning coloration*. Compare with *cryptic coloration*. [35]

Apostatic selection. See *frequency-dependent selection*. [34]

Apterygotes. Wingless insects, such as springtails, silverfish, etc. Compare with *pterygotes*. [42]

Aquatic. Living in or on water. [48]

Aqueous. Containing water, or dissolved in water. [3]

Arachnid. A member of the arthropod class Arachnida; includes spiders, mites, scorpions, etc. [42]

Archaebacteria. A group of prokaryotes that differ from bacteria and other monerans and are postulated to represent a sixth kingdom. [37]

Archegonium. The multicellular structure that produces eggs in bryophytes, ferns, and gymnosperms. [21,42]

Archenteron. The embryonic future digestive cavity of animals, such as sea urchins, whose blastula becomes a gastrula through invagination. [19]

Archeon eon. One of the four major time divisions in the history of Earth, it covers the period between 3.8 and 2.5 billion years ago during which life originated and simple prokaryotes proliferated. [37]

Archetypal protein. One of a small number of ancient proteins from which modern proteins evolved. [35]

Archetype. A pattern, or ideal design, that early systematists made the basis of classification, despite the many variations in observed individuals and populations. [37]

Arteriole. A small-diameter artery that ends in capillaries. [24]

Artery. A thick-walled muscular vessel that carries blood away from the heart. Compare with *vein*. [24]

Arthropod. A member of the animal phylum Arthropoda; includes spiders, crustaceans, and insects. [42]

Ascomycetes. Sac fungi. A phylum of the kingdom Fungi. [40]

Ascospore. A sexual spore that forms within an ascus (a sporangium). [40]

Ascus. In Ascomycetes (sac fungi), the club-shaped sporangium within an ascocarp which is the site of spore production by meiosis. [40]

Asexual reproduction. Reproduction of an organism without fusion of gametes. Compare with *sexual reproduction*. [21]

A-site. The site on a ribosome that binds the next amino acid to be added to a growing polypeptide chain during translation. Compare with *P-site*. [17]

Aspiration breathing. See *negative pressure breathing*. [27]

Assimilation. The absorption of nutrients into cells. [26]

Association area. An area of the cerebral cortex which is not directly concerned with sensory or motor function, but integrates them and controls higher cognition. [30]

Association neuron. See *interneuron*. [29]

Associative learning. Process in which an animal comes to associate a neutral stimulus with reward or punishment. Classical and operant conditioning are forms of associative learning. [32]

Associative society. Animal societies based on interactions of unrelated members of the same species. [33]

Assortative mating. A pattern of sexual selection in which individuals preferentially mate with others of similar phenotype. Compare with *disassortative mating*. [34]

Assortment. Random separation during meiosis of nonhomologous chromosomes and their genes. For example, if genes *A* and *B* are on nonhomologous chromosomes, meiosis of diploid *AaBb* cells will produce haploid cells in equal numbers of genotypes *AB, Ab, aB,* and *ab.* [12,13]

Aster. During mitosis, a group of microtubules radiating from a centriole, extending, between the spindle fibers which bind the chromosomes. [12]

Asymmetric carbon atom. A carbon atom with four different atoms or groups bonded to it. [4]

Asynchronous muscle. Muscle, such as the fast muscles of some insect wings, that contracts at a frequency higher than that of incoming nerve impulses. Compare with *synchronous muscle*. [31]

Atmosphere. (1) The layer of gases surrounding Earth. [1,48] (2) A unit of pressure equal to normal air pressure at sea level. [27]

Atom. The smallest unit of a chemical element. Contains a nucleus and one or more electrons. [3]

Atomic mass. The mass of a representative atom of an element. Approximately equal to the number of protons and neutrons in its nucleus. [3]

Atomic number. The number of protons in the nucleus of an atom. Also equal to the number of electrons around the neutral atom. Determines the chemical properties of the atom. [3]

Atomic weight. The average weight of an atom of an element. Depends upon the relative amounts of different isotopes of the element on Earth. [3]

ATP. See *adenosine triphosphate*.

ATPase. An enzyme that hydrolyzes ATP into ADP and inorganic phosphate. [31]

ATP synthase. Enzyme in membranes of mitochondria and chloroplasts that utilizes the proton-motive force of chemiosmosis in the synthesis of ATP from ADP and inorganic phosphate. [9,10]

Atrial natriuretic factor (ANF). A hormone produced by the mammalian heart. Promotes sodium excretion and lowers blood pressure. [28]

Atrium. A thin-walled antechamber in the heart, which receives venous blood and passes it on to the ventricles. [24]

Australopithecine. A member of the extinct hominoid genus Australopithecus. [44]

Autoantibody. An antibody that attacks a normal body component. The basis of autoimmunity. [23]

Autocrine secretion. Release of factors by a cell that act upon that cell. Compare with *paracrine* and *endocrine secretion*. [28]

Autogenic. A force internal to a community that drives ecological succession in that community. Compare with *allogenic*. [47]

Autoimmunity. Abnormal phenomenon in which an immunity develops to normal body components. [23]

Automixis. A form of parthenogenesis in which egg meiosis occurs, but then two sets of separated chromosomes rejoin and the ovum starts to develop into primarily homozygous offspring. Compare with *apomixis*. [21]

Autonomic nervous system. In vertebrates, the system controlling involuntary functions such as digestion and blood pressure. It contains the sympathetic and parasympathetic nervous systems. [30]

Autoradiography. The detection of a radioactive substance in a cell or organism by placing it in contact with a photographic emulsion. The location of the radioactivity in the cell is marked by darkened silver grains in the emulsion. [24]

Autosome. Any chromosome (in a eukaryote) other than a sex chromosome. [15]

Autotroph. An organism, such as a plant, that needs no preformed organic molecules and is capable of living on inorganic chemicals, CO_2, water, and an energy source such as sunlight or chemical reactions. Compare with *heterotroph*. [9,26,38]

Auxin. A class of plant hormones (derived from indoleacetic acid), which control growth, cell elongation, and other aspects of development and metabolism. [28]

Auxotroph. A mutant variant that requires one or more nutrients not required by the wild-type form of the organism. Compare with *prototroph*. [9,26]

Aves. The class of vertebrates which contains the birds. [43]

Axon. Fiber of a neuron that carries action potentials away from the cell body toward synapses with other neurons. Compare with *dendrite*. [7,29]

Axonal transport. A process in which molecules are carried from the cell body of a neuron down the axon to its synapses. [29]

Axoneme. A slender microtubular filament in the center of a cilium or flagellum. See *9 + 2 pattern*. [5]

Bacillus. A rod-shaped bacterium. [38]

Back mutation. A mutation in a mutant organism returns the progeny to the wild-type phenotype. Compare with *forward mutation*. [14]

Bacteriophage (also called *phage*). A virus that parasitizes (infects) a bacterium and may ultimately destroy it. Used in the laboratory to transfer genes between strains of bacteria and as model genetic systems. [16]

Bacterium. A prokaryote. Member of the kingdom Monera. An organism with chromosomes not contained in nuclear envelopes. [37,38]

Balanced polymorphism. Steady-state retention in a population of polymorphic variants from generation to generation. Usually occurs when the heterozygote is superior to both homozygotes. Compare with *transitional polymorphism*. [34]

Barr body. The darkly staining condensed chromatin in interphase somatic cells, which is the inactivated X chromosome in female mammals. [15]

Basal lamina. A sheet of extracellular matrix that separates two cell layers of differing embryonic origin. [20]

Basal metabolism. A resting organism's total energy output in a given period. [26]

Base. (1) A substance that can accept a proton (hydrogen ion, H^+). Compare with *acid*. [3] (2) In nucleic acids and nucleotides, a purine or pyrimidine. [16]

Base pair (bp). A unit of length in a DNA molecule or gene. Thus, a given DNA molecule might be 5,000 bp, or 5 kilobase (Kb) pairs, long. [12]

Base pairing. See *complementary base pairing*.

Basic. Having a pH greater than 7 (a H^+ concentration lower than $10^{-7} M$) Compare with *acidic*. [3]

Basidiocarp. A fruiting body produced by basidiomycetes, which bears basidia. A mushroom cap is an example. [40]

Basidiomycetes. Club fungi. A phylum of the kingdom Fungi. [40]

Basidiospore. A spore produced in and borne on a basidium. Results from nuclear fusion and meiosis. [40]

Basidium. In a basidiomycete, a structure bearing on its surface a species-specific number of basidiospores that are formed by nuclear fusion and meiosis and then shed. [40]

Basilar membrane. A membrane of the cochlea that mediates hearing by responding to sound energy and stimulating hair cells. [30]

Basophil. A white blood cell that plays a role in inflammation and allergic reactions. Related to mast cells. [22]

Batesian mimicry. A form of mimicry in which a palatable species (the mimic) comes to resemble an unrelated poisonous or dangerous species (the model) and enjoys protection from predators that mistake it for its model. Compare with *Müllerian mimicry*. [35]

Bathyal zone. The part of the deep ocean's lightless zone which is on the continental shelf. Compare with *abyssal zone*. [48]

B cell (B lymphocyte). Type of lymphocyte underlying humoral immunity. After recognition of an antigenic determinant, B cells develop into plasma cells which secrete antibodies. Compare with *T cell*. [3]

B cell receptor. A membrane-bound antibody on the surface of a B cell that allows it to recognize an antigen. Compare with *T cell receptor*. [23]

Behavior. Externally visible activity of an animal, in which a coordinated pattern of sensory, motor, and associative neural activity responds to changing external or internal conditions. [32]

Behavioral adaptation. An adaptation involving changes in behavior. [35]

Behavioral ecology. Study of the interactions of behavior and evolutionary forces. [33]

Behavioral isolation. A prezygotic isolating mechanism in which members of different populations exclude each other as potential mates. Compare with *ecological, temporal, physiological*, and *mechanical isolation*. [36]

Behavioral mimicry. A type of interspecific or intraspecific mimicry in which a deceitful member of one species gains advantage by ''pretending'' to be a member of another species. [35]

Belt desmosomes. Adhering junctions that girdle a cell, more for cell support than intercellular communication. Compare with *spot desmosome*. [7]

Benthic region. The ocean bottom. Compare with *pelagic region*. [48]

Benthos. Plants or animals that live on the ocean floor. [48]

β-carotene. One of the major accessory pigments in plant photosynthesis. A carotenoid. [10]

β-galactosidase. An enzyme that catalyzes the hydrolysis of lactose into glucose and galactose. [18]

β-pleated sheet. A secondary structure of proteins in which polypeptides aligned side-by-side are held together by interchain hydrogen bonds. [4]

Bilateral symmetry. Anatomical plan in which a body or part is divided by a plane into right and left halves that are essentially mirror images. Compare with *radial symmetry*. [19,42]

Bile. A fluid secreted by the liver into the intestine. It acts as a detergent, emulsifying dietary fats. [26]

Bile salts. Salts of glycocholic acid and taurocholic acid, found in bile, which help to emulsify fats. [26]

Bilirubin. A normal breakdown product of hemoglobin. [22]

Binomial nomenclature. The two-term Linnaean system of naming living things, in which the first name is the genus, and the second the species. [37]

Bioenergetics. The study of energy transformations in living organisms. [8]

Biogenesis. The concept that all living things come from pre-existing life, rather than from spontaneous generation. [4,12]

Biogenetic law. The principle of recapitulation (''ontogeny recapitulates phylogeny'') that concerns the supposed resemblance of early embryonic stages to phylogenetic ancestors. [20]

Biogeochemical cycle. The circulation of chemical elements through the living and nonliving portions of an ecosystem. [46]

Biogeography. The study of patterns of geographical distribution of organisms. Ecological biogeography deals with the habitats in which organisms live; historical biogeography with the geographic ranges of organisms and the historical and evolutionary factors underlying them. [48]

Biological clock. An internal timing mechanism that determines circadian rhythms in organisms even in the absence of environmental cues. [32]

Biology. The scientific study of life in all of its aspects and forms. [1]

Bioluminescence. Light production by living organisms. It is generated by biochemical systems, such as the oxidation of the luciferin by the enzyme luciferase. [31]

Biomass. The total mass of all the living organisms (or some group of organisms) in a given area. [46]

Biome. The community of organisms in a major geographic region, such as those of rain forests, grasslands, tundra. [47]

Biomolecule. The class of organic molecules found in living organisms. [4]

Biosphere. That part of the Earth and its atmosphere that is occupied by living organisms. The sum total of all organisms and their habitats. [1]

Biota. All the organisms of a given region. [48]

Biotechnology. The application of biology and biochemistry to the production of useful products and materials. [2]

Biotic environment. Portion of the environment consisting of living organisms. [46]

Bipedal locomotion. Walking on two feet. [44]

Bivalve. A member of mollusk class Bivalvia; includes clams, oysters, etc. [42]

Blastocoel. The hollow cavity within the blastula of vertebrate embryos. [19]

Blastocyst. In the embryogenesis of mammals, a modified blastula. [19]

Blastodisc. In the embryogenesis of birds, a disk of cells forming on the surface of the yolk mass of an embryo undergoing incomplete (meroblastic) cleavage. It is comparable to a blastula but occurs in forms in which a large yolk restricts cleavage to only one side of the egg. [19]

Blastomere. A blastula cell produced by early cleavage divisions of a zygote. [19]

Blastopore. In vertebrate embryogenesis, the opening of the archenteron cavity of a gastrula to the exterior of an embryo. A future mouth in protostomes. A future anus in deuterostomes. [19]

Blastula. An early stage of animal embryogenesis. Occurs after cleavage and before gastrulation. Usually a hollow single-layered sphere of cells that develops from the morula. [19]

Blue-green algae. See *Cyanonta*. [38]

B lymphocyte. See *B cell*. [23]

Bohr effect. Decreased affinity of hemoglobin for oxygen caused by decreased pH (usually due to increased CO_2 levels). It facilitates gas exchange in tissues and lungs. [27]

Bond. In chemistry, the linkage between atoms or ions in a compound. There are covalent, ionic, nonpolar, and hydrogen bonds. [3]

Book gill. A feature of chelicerates. A series of open, leaflike lamellae resembling open pages of a book. [42]

Boreal forest. See *taiga*. [48]

Bowman's capsule. In the kidney, a cup-shaped sac surrounding the glomerulus that collects fluid filtered from blood and delivers it to the tubule of the nephron. [25]

Brachiation. Movement by the grasping of overhead tree branches and swinging from branch to branch. Typical of many primates. [44]

Brachiopod. A member of the animal phylum Brachiopoda. Lophophorate marine forms known as lamp shells. [42]

Brain. In vertebrates, the enlargement of the central nervous system that is found in the head. In invertebrates, a cluster of anterior nerve ganglia. [30]

Bronchiole. The smallest air-conducting tubes in the vertebrate lung. Compare with *bronchus*. [27]

Bronchus. Paired air-conducting tubes in vertebrates that branch from the trachea and subdivide into bronchioles. [27]

Brood parasitism. The practice by a species of placing its eggs in the nest or habitat of another species, where they are incubated and reared by their host. [35,47]

Brown algae. Members of the protistan phylum Phaeophyta. [39]

Bryophyte. A member of the plant subkingdom Bryophyta. A nonvascular plant lacking true leaves, stems, and roots. [41]

Budding. (1) Asexual cellular reproduction in which a small new cell or organism is produced as an outgrowth of a larger parent. Compare with *fission*. [12,21] (2) The development in plants of a small mass of meristematic tissue in the angle between leaf stalk and shoot apex into a lateral branch. [21,41]

Buffer. Mixture of a weak acid and salt of that acid that tends to keep constant the pH of a solution when acids or bases are added. [3]

Bulk feeder. An animal that eats solid chunks of food, which must then be ground up to aid digestion. Compare with *filter feeder*. [26]

Bundle-sheath cell. A component of a layer or sheath surrounding xylem and phloem in plants. [10]

Bursa of Fabricius. The organ that produces B lymphocytes in birds. [23]

Byrozoan. A member of the animal phylum once called Bryozoa. Now split into phylum Ectoprocta and phylum Entoprocta. [42]

C₃ plants. Plants in which the three-carbon molecule, 3-PGA, is the first stable product of photosynthesis and RuBP is the CO_2 receptor. Compare with *C₄ plants*. See *Calvin cycle*. [10]

C₃–C₄ plants. Plants with the metabolic and structural characteristics of both C₃ plants and C₄ plants. [10]

C₄ plants. Plants which incorporate CO_2 into phosphoenolpyruvate thus forming oxaloacetate, a four-carbon molecule. This process adapts them to hot climates. Compare with C_3 plants. [10]

CAAT box. A promoter sequence found upstream of globin and other eukaryotic structural genes. [17,18]

Caecum. See *cecum*. [26]

Calcitonin. A hormone produced by the thyroid that lowers blood calcium levels. [28]

Calcium channel. An ion channel that permits passage of Ca^{2+} ions across a membrane. [6]

Calmodulin. A regulatory protein activated by binding four Ca^{2+} ions. It then binds to many enzymes and other proteins and modifies their activity. [28]

Calorie (Cal). A kilocalorie (1000 calories). Used as a measure of the energy released by the breakdown of a given amount of food. [26]

Calorie (cal). A calorie (small *c*) is the amount of heat needed to raise the temperature of 1 g of water 1°C (from 14.5° to 15.5°). A kilocalorie, or 1000 calories, is denoted as Calorie (capital *C*). The units used to express nutritional values are Calories, or kilocalories. [8,26]

Calvin cycle (Calvin-Benson cycle). Carbon fixation stage of photosynthesis in which CO_2 reacts with RuBP to form PGA, which is then reduced to a sugar (glucose) as RuBP is regenerated. Also called the reductive pentose phosphate cycle. [10]

Calyx. All of the (nonreproductive) sepals of a flower collectively. [21,41]

CAM. See *cell adhesion molecule*, or *crassulacean acid metabolism*.

Cambium. Lateral meristem. A cylindrical sheath of actively dividing cells which lies between the xylem and phloem and gives rise to both. Compare with *apical meristem*. [7]

cAMP. See *cyclic AMP*. [28]

Canopy. The ecological zone at the top of a rain forest. Consists of the top branches of trees and the organisms that live among them. [48]

CAP (catabolite activator protein) site. A regulatory site in the lac operon that binds a protein necessary for transcription. [18]

Cap (5′ cap). A post-translational modification of the 5′ end of eukaryotic mRNA that enhances translation. [17]

Capillary. The smallest blood vessels. They connect arteries and veins, and are the site of oxygen and waste exchange with the tissues. [24]

Carbaminohemoglobin. A complex of hemoglobin and CO_2 that participates in the transport of CO_2 from tissues to lungs. [27]

Carbohydrate. Class of organic molecules with the general formula $(CH_2O)_n$. Sugars, starches, glycogen, and cellulose are carbohydrates. [4]

Carbon cycle. The circulation and reutilization of carbon atoms through the world's ecosystems. [46]

Carbon fixation. The conversion of CO_2 into organic compounds. [10]

Carbonic anhydrase. An enzyme of red cells and other cells that converts CO_2 and H_2O into carbonic acid (H_2CO_3), which can then dissociate into bicarbonate (HCO_3^-) and H^+. Stabilizes blood pH and facilitates gas exchange. [27]

Carbon-silicate cycle. A worldwide biogeochemical cycle involving the circulation and reutilization of carbon (as carbonates) and silicon (as silicates). [46]

Carboxylation. The addition of a carboxyl group (—COOH) to a molecule. [10]

Carboxyl group. The acidic functional group of organic molecules (—COOH). [4]

Carboxypeptidase. A protease secreted by the pancreas. It hydrolytically removes amino acids from the C-terminal (carboxyl) end of a peptide. Compare with *aminopeptidase*. [26]

Cardiac muscle. Specialized muscle of the heart that produces the heartbeat. Although it is striated muscle, it is not under voluntary control and contracts even without neural stimulation. [31]

Carnivore. An organism that obtains food by eating the flesh of animals. Compare with *herbivore* and *omnivore*. [1,26]

Carotenoids. A group of red, orange, and yellow accessory pigments that are found in plastids and participate in photosynthesis. [10]

Carpel. Flower part that encloses one or more ovules. Female reproductive structure typically including stigma, style, and ovary. Also called *pistil*. [21,14]

Carrying capacity. The maximum number of organisms of a given species that can be maintained in a given environment. [45]

Cartilage. In vertebrates, a tough connective tissue found in joints, the outer ear, and elsewhere. [5]

Casparian strip. An impermeable layer containing suberin and lignin in the endodermis of vascular plant roots. Restricts water movement across the endodermis. [24]

Catabolism. Degradative reactions of metabolism that break down complex molecules. Compare with *anabolism*. [8]

Catabolite activator protein. A protein that must bind to the regulatory CAP site before transcription can be initiated. [18]

Catabolite repression. The inhibition of transcription by the presence of glucose breakdown products. [18]

Catalyst. A molecule that accelerates the rate of a reaction by lowering the activation energy without being consumed itself. Enzymes are biological catalysts. [8]

Cation. A positively charged ion. [3]

cDNA. See *complementary DNA*. [18]

Ceboid. A member of the primate superfamily Ceboidea, the New World monkeys. Compare with *cercopithecoid*. [44]

Cecum. A blind pouch opening into the large intestine near its beginning. In many nonruminant mammals, it contains bacteria that contribute to the digestion of food. [26]

Cell. The fundamental functional and structural unit of living organisms. Usually consists of cytoplasm containing a nucleus or other organelles, surrounded by a surface membrane. [5]

Cell-adhesion molecule (CAM). Molecules, mainly glycoproteins, on cell surfaces that foster specificity in cell recognition, adhesion, and migration. [7,20]

Cell cycle. The sequence of stages from one cell division to the next. [12]

Cell division. The reproductive splitting of a cell into two daughter cells. [12]

Cell-mediated immunity (cellular immunity). The part of the immune system that is based on the activities of T cells and and macrophages. Directed against cancer cells, parasites, fungi, intracellular viruses, and foreign tissues (grafts). Compare with *humoral immunity*. [23]

Cell plate. In dividing cells of plants and some green algae, a structure that forms at the cell equator in telophase. Precursor of the new cell wall separating daughter cells and in development of the middle lamella. [12,41]

Cell theory. The principle that all living organisms are made up of cells and their products and that all cells arise from pre-existing cells. [5]

Cellular immunity. See *cell-mediated immunity*. [23]

Cellulase. An enzyme that catalyzes the hydrolysis of cellulose. [26]

Cellulose. A polysaccharide made of glucose units. Chief constituent of plant cell walls where it functions as a supporting material. [4,41]

Cell wall. A rigid structure enclosing the cell membranes of plants, fungi, and many protists and bacteria. Animals do not have cell walls. [5]

Cenozoic era. A major division of Phanerozoic time extending from 65 million years ago to the present. Includes seven epochs. Often called the age of mammals. [37]

Central dogma. The principle that genetic information in a cell flows from DNA to RNA to protein. An exception occurs in retroviral replication. [17]

Central nervous system. The centrally located part of the nervous system that coordinates an animal's activities. In vertebrates, it is the brain and spinal cord; in invertebrates, it is nerve cords and ganglia. [30]

Centriole. Paired barrel-shaped organelles in animal and some protist cells that migrate to opposite poles of a dividing eukaryotic cell and organize the mitotic spindle fibers. [5,12]

Centromere. Point on a chromosome that assembles kinetochores, to which spindle fibers attach. Contains satellite DNA that bind functionally essential proteins and positions enzymes that separate intertwined DNA. [5,12]

Centrosome. A cytoplasmic area containing the centrioles. [5]

Cephalopod. A member of the mollusk phylum Cephalopoda; includes octopuses, squid, cuttlefishes, etc. [42]

Cercopithecoid. A member of the primate superfamily Cercopithecoidea, the Old World primates; includes apes and humans. Compare with *ceboid*. [44]

Cerebellum. An enlarged region of the vertebrate brain behind the cerebral hemispheres that coordinates movement and balance. [30]

Cerebral cortex. The layer of gray matter, containing neuronal cell bodies, which covers the surface of the cerebral hemispheres. It controls high-level sensory, motor, and cognitive processing. [30]

Cerebrum. The cerebral hemispheres of the vertebrate brain. It contains the cerebral cortex and controls high-level sensory, motor, and cognitive processing. [30]

Ceruloplasmin. The copper-binding transport protein of blood. [22]

Character displacement. What happens when niche competition or other interactions cause similar species to become less alike in regions they both inhabit than in regions where they are isolated from each other. [36,47]

Chelicera. The most anterior pair of appendages in some arthropods (Chelicerata). For example, the appendages used by arachnids to seize prey. [42]

Chelicerata. A class of arthropods possessing the feeding appendages called chelicarae. [42]

Chemiosmosis. Process powering ATP synthesis in mitochondria and chloroplasts. Protons pumped across a membrane generate proton-motive force (potential energy in the form of electrical and pH gradients) and then diffuse back again through a protein channel with ATP synthase activity. [9,10]

Chemoautotroph. See *chemosynthetic autotroph*. [9]

Chemoreceptor. A sensory structure, such as those in taste buds or the nose, that generates nerve impulses in response to certain chemicals. [30]

Chemosynthetic autotroph. An organism that uses energy released by specific chemical reductions to drive its metabolism. Compare with *photosynthetic autotroph*. [9]

Chemotaxis. Movement of a cell or organism under stimulus of a chemical gradient in the medium. [5,20,22,29]

Chemotroph. See *chemosynthetic autotroph*. [9]

Chemotropism. Growth of a plant toward or away from a chemical stimulus. [32]

Chiasma. The points at which members of a chromosome pair are joined during prophase I of meiosis, and at which recombination occurs. [12]

Chimera. An organism whose tissues are of two or more genetically distinct cell types. Often produced in the laboratory by fusing embryos. Also called a *mosaic*. [20]

Chitin. The tough, nitrogen-containing polysaccharide of the exoskeleton of insects and other arthropods. Also occurs in fungal cell walls. [40,42]

Chiton. A member of the mollusk class Amphineura. The simplest mollusks. [42]

Chlorophyll. A class of pigment molecules that absorb light energy in photosynthesis. [10]

Chloroplast. An organelle bounded by a double membrane containing the enzymes and pigments of photosynthesis. Occur only in eukaryotes. [5,10]

Choanocyte. A collar cell bearing cilia. [42]

Chondrichthyes. A class of vertebrates which contains the cartilaginous fishes, such as the sharks. [43]

Chordamesoderm. Embryonic tissue that develops into mesoderm and notochord. [19,20]

Chordata. The animal phylum that includes the hemichordate, urochordate, cephalochordate, and vertebrate subphyla. [43]

Chorion. The outermost membrane around amniotic embryos. In mammals it helps form the placenta. [21]

Chromatid. One of a pair of duplicated sister chromosomes, which are joined at the centromere. The pair separates in anaphase of mitosis. [12]

Chromatin. The DNA-protein complex making up eukaryotic chromosomes. [5,12]

Chromatography. A family of methods in which components in a mixture are separated by the different rates at which they diffuse into a moving medium from a stationary medium. [4]

Chromatophore. Pigment bodies in the skin of cephalopods. Under neural control, they can rapidly expand and contract, thereby changing skin color. [42]

Chromomere. Beaded thickenings in chromosomes that are visible during prophase I of meiosis. [12]

Chromoplast. Pigmented plastid, such as a chloroplast. [5]

Chromosome. In eukaryotes, a structure composed of DNA and proteins that conveys genetic information. Condenses and becomes visible only during meiosis and mitosis. Some also apply term to bacterial and viral DNA. [12,16]

Chromosome mutation. A change in the chromosomal complement of an organism caused by chromosomal nondisjunction, translocation, or other mechanisms. Compare with *gene mutation*. [14]

Chrysolaminarin. An oily substance within cells of members of the protistan phylum Chrysophyta. [39]

Chymotrypsin. A protease secreted by the pancreas. It cleaves peptide bonds containing phenylalanine or tyrosine. [26]

Chymotrypsinogen. The molecular precursor of chymotrypsin. [26]

Ciliate. A cell with cilia. Unicellular organisms that propel themselves with cilia. A member of the protistan phylum Ciliophora. [5,39]

Cilium (pl. **cilia**). Hairlike organelle used for locomotion or moving extracellular material. [5,39]

Circadian rhythm. A physiological or behavioral rhythm with a period of about 24 hours. [32]

Cirrus (pl. **cirri**). Flexible tentaclelike sensory appendage formed from fused cilia. [39]

Cisterna. A space between the membranes of endoplasmic reticulum or within a Golgi body. [5]

Citric acid cycle (also called *tricarboxylic acid cycle* and *Krebs cycle*). Cyclical series of reactions of oxidative metabolism, in which citric acid is formed from acetyl CoA and oxaloacetate, which is then regenerated. Its main products are CO_2 and hydrogen atoms in the form of $NADH^+$ and $FADH_2$. [9]

Clade. Group of species derived from a common ancestral species. [36]

Cladistics. A system of classification based on the sequential order in which branches (clades) arise from a phylogenetic tree. Compare with *numerical phenetics* and *evolutionary systematics*. [37]

Cladistic speciation (cladogenesis). The divergence of different genetic lines of an ancestral species that leads to the formation of new species. Compare with *anagenetic speciation*. [36]

Cladogenesis. See *cladistic speciation*. [36]

Cladogram. Graph of phylogenetic tree constructed according to cladistic principles. Points where populations branch off are defined by new traits unique to each branch. [37]

Class. A taxonomic category that lies between phylum (division) and order. A class contains multiple orders. [37]

Classical conditioning. Pavlovian learning in which a novel stimulus, repeatedly coupled with a stimulus triggering a behavioral response or reflex, in time also triggers that response or reflex. Compare with *operant conditioning*. [32]

Classification. The grouping of entities into categories based on phenotypic similarities, evolutionary relationships, or both. [37]

Clathrin. A protein that coats endocytic vesicles in receptor-mediated endocytosis. [6]

Cleavage. First divisions of the zygote into many cells with no overall increase in embryo size. [19]

Cleavage furrow. Groove appearing in the membrane of animal cells during cytokinesis that marks the site of imminent cell division. [12]

Climax community. The stable culmination of ecological succession in a given environment. [47]

Cline. Gradual phenotypic change from one area to another of poorly separated local populations of a species. [36]

Cloaca. Chamber at the end of the gastrointestinal tract in birds and other species. Also serves as exit for reproductive and urinary systems. [21]

Clonal selection theory. The theory that an antigen selects from a large, heterogeneous, pre-existing pool of T and B cells and activates only the few that are capable of making antibodies specific for that antigen. [23]

Clone. Genetically identical cells or organisms arising asexually from a common ancestor. One of a population of genetically identical individuals. [38]

Closed circulatory system. In vertebrates and other animals, a closed system of vessels through which blood flows without ever directly contacting other tissues. Compare with *open circulatory system*. [24]

Closed system. In thermodynamics, a system that does not exchange matter or energy with its surroundings. Compare with *open system*. [8]

Clot. A hemostatic plug constructed of tough fibrin strands formed from fibrinogen at sites of injury by the clotting system. [22]

Clotting system. A group of enzymatic plasma proteins that stops bleeding from an injured blood vessel by activating a reaction cascade to form a sealing clot. [22]

Cluster. A group of glades. [34]

Cnidoblast. Specialized cells on the surface of coelenterates (cnidarians) that expel harpoonlike stingers from nematocysts. [31]

Coadaptation. Traits that evolve together when natural selection causes genes interacting by epistasis to confer fitness on a species. [21]

Coated pit. Depression or invagination of a cell membrane that becomes coated with clathrin and pinches off to form a coated vesicle. [6]

Coated vesicle. A clathrin-coated vesicle that forms in the course of receptor-mediated endocytosis. [5,6]

Cobalamin (vitamin B$_{12}$). A vitamin for animals and many microorganisms. Converted to coenzymes essential in certain methylations and other reactions. [10]

Coccus. A bacterium of spherical shape.

Cochlea. A spiral tube in the vertebrate inner ear that contains the organ of Corti, whose phonoreceptors are the basis of hearing. [30]

Codominance. The ability of two alleles to produce a heterozygous phenotype possessing characteristics of both homozygotes. Compare with *incomplete dominance*. [13]

Codon. A triplet of three adjacent bases in DNA or mRNA that encodes a specific amino acid. Compare with *anticodon*. [17]

Coefficient of relatedness. The probability that two organisms within a species will share a particular copy of a gene. [33]

Coelenterate. A member of the animal phylum Coelenterata; includes hydras, jellyfishes, sea anemones, corals, etc. [42]

Coelom. A cavity, in the mesoderm of many animal phyla, in which the internal organs are suspended. [19]

Coelomate. An animal with a true coelom. Compare with *acoelomate*. [19]

Coenzyme. A nonprotein organic molecule that is usually bound to an enzyme and serves as an essential donor or acceptor of a chemical group in the reactions that enzyme catalyzes. Compare with *cofactor*. [8]

Coevolution. Evolution of two or more unrelated species with close ecological relationships, in which reciprocal selective forces make changes in one, e.g., a flower, that are strongly influenced by changes in the other, e.g., a pollinating insect. [41]

Cofactor. A nonprotein group, often a metal ion, which is essential for the functioning of certain enzymes. Compare with *coenzyme*. [8]

Cohesion. The mutual attraction (coherence) of like molecules within a substance. [3]

Collagen. A tough protein filament found in connective tissue, cartilage, and bone. One-third or more of body protein may be collagen. [20]

Collar cell. A specific cell in sponges that lines the body cavity (spongocoel) and pumps water, captures and digests food, and forms sperm and eggs. [42]

Collecting duct. Site of the last stage of urine formation in the kidney, in which urine is concentrated and its urea content increased. [25]

Collenchyma. Plant tissue, consisting of irregular thick-walled cells, that provides mechanical support in areas of primary growth. [7]

Colloid. A two-phase system in which fine particles of one material (e.g., a lipid) are stably suspended in a second phase (e.g., water). [6]

Colostrum. The protein-rich milk that a lactating mother secretes for the first few days after childbirth. [21]

Combinational equilibrium. The major factor promoting genetic stability in a population that breeds or mates randomly. See *Hardy-Weinberg law*. [34]

Commensalism. A form of symbiosis which is beneficial to one species but not to another, which is neither harmed nor benefited. Compare with *mutualism* and *parasitism*. [47]

Communicating junction. See *gap junction*. [7]

Communication. Action on the part of one organism (or cell) that alters the pattern of behavior in another organism (or cell) in an adaptive fashion. [33]

Community. All of the organisms of all species inhabiting a given area and interacting with each other. Compare with *population* and *ecosystem*. [45,47]

Companion cell. A plant cell that provides metabolic support to sieve tube cells in the phloem of some angiosperms. [24]

Competence. In embryonic development, the capability of a cell to differentiate in a number of ways. A manifestation of multipotence. [20]

Competition. (1) In ecology, the interaction between organisms which all need a resource in short supply. [45,47] (2) In biochemistry, the availability of multiple ligands for a single binding site on an enzyme or receptor, or multiple sites for a single ligand. [8]

Competitive exclusion. The principle that no two species can stably inhabit the same niche in an environment. Eventually only one will occupy the niche. [47]

Competitive inhibition. The inhibition of enzymatic activity by a molecule that binds to the active site of the enzyme thus displacing the substrate. Compare with *noncompetitive inhibition*. [8]

Complementary base pairings. The pairing of A and T (or A and U) and C and G between the two strands of DNA, the sense strand of DNA and the mRNA synthesized on it, and between portions of tRNA and mRNA with other portions and each other. [18]

Complement system. A group of proteins that participate in certain reactions of the immune system. They combine with antigen-antibody complexes and help destroy the antigen-carrying agents. They are not immunoglobulins. [23]

Complete metamorphosis. The process in most insect species in which a larva becomes a pupa and then an adult. Compare with *simple metamorphosis*. [21]

Compound. A molecule whose atoms are joined by covalent bonds. [3]

Compound eye. In arthropods and crustacea, an eye in which multiple lens-containing elements (ommatidia) each form a separate image on light-sensitive cells. The images are combined in the central nervous system. [30]

Concentration gradient. The change in concentration of a substance over distance. [6]

Conditioned reflex. An involuntary reflex produced by classical conditioning. [32]

Conditioning. See *classical* and *operant conditioning*. [32]

Cone. (1) In the vertebrate retina, a photoreceptor concerned with color discrimination and fine vision. [30] (2) In gymnosperms, a reproductive structure (strobilus) consisting of many tightly packed sporophylls or ovule-bearing scales on a stem. [21,41]

Conformation. The three-dimensional shape of a molecule or other structure. [4]

Conidiophore. A specialized hypha bearing one or more conidia (in ascomycetes). [40]

Conidium. An asexual spore that is borne on a conidiophore (in ascomycetes). [40]

Conifer. A member of the gymnosperm class Coniferopsida; includes the familiar cone-bearing pines, firs, cedars, hemlocks, redwoods, etc. Dominant trees over much of the Temperate Zone. [41]

Conjugation. (1) Formation of a bond between two dissimilar molecules. [4] (2) In the reproductive biology of unicellular organisms, a sexual union that transfers genetic material. [12,16,38]

Connective tissue. Animal tissue that surrounds, connects, and supports other tissues. Usually embedded in a collagen-containing matrix. [7]

Connexon. The protein channel comprising a gap junction; composed of the protein connexin. [7]

Consensus sequence. A short base sequence in DNA that is similar in many kinds of organisms and thus has been conserved by evolution. [17]

Conserved sequence. A DNA sequence, part of a gene or promoter region, which is the same even in distantly related species. This is taken as evidence that it has adaptive value for the species. [17]

Conspecific. Belonging to the same species. [45]

Constant (C) region. In an immunoglobulin molecule, a region similar in all antibodies. Compare with *variable region*. [23]

Constitutive enzyme. An enzyme that is synthesized at a constant rate whether or not its substrate is present. Compare with *inducible enzyme*. [8,18]

Contact guidance. The phenomenon in which certain surfaces control the movement of cells touching them. [20]

Contact inhibition. The inhibition of the growth or movement of a cell after it touches another cell. [20]

Continental drift. The slow movement of the land masses and continents that has taken place continuously throughout Earth's geological history. [48]

Contractile vacuole. A fluid-filled vacuole in some protists that pumps excess water out of the cell (especially in hypotonic media) and facilitates excretion by expelling its contents from the cell. [25]

Convergent evolution. An evolutionary process in which two unrelated and dissimilar species come to have similar (analogous) traits, often because they have been exposed to similar selective pressures. [35,37]

Cooperativity. A property of certain enzymes in which the binding of substrate at an active site increases the binding affinity of other sites on the enzyme. [8]

Cordaites. A member of the extinct gymnosperm class Cordaitopsida. A common group in the Carboniferous coal forests. [41]

Corepressor. A molecule that must bind to a repressor protein before it can function as a transcription inhibitor. [18]

Cork cambium. An outer layer of meristem in vascular plants that forms the periderm, producing cork. Compare with *vascular cambium*. [24]

Corolla. All of the petals of a flower, collectively. [21]

Corpus luteum. A structure formed from a ruptured follicle after ovulation that produces hormones essential to the maintenance of pregnancy. [21]

Cortex. An outer layer or rind. In plants, root or stem parenchyma, just under the epidermis. In animals, the outer tissue of certain organs, such as the adrenal gland and cerebrum. Compare with *medulla*. [24,25,28,30]

Cortical granules. In embryology, the material that forms the fertilization membrane to protect a fertilized egg from the entry of other sperm. [19]

Cortical reaction. A change in the fertilized egg that prevents entry of other sperm. [19]

Corticosteroid. A group of the hormones produced by the adrenal cortex. They regulate metabolism and response to stress. [28]

Cosmic rays. High energy electromagnetic radiations, with wavelengths of < 0.001 nm, that impinge on Earth from space. [10]

Cotyledon. An embryonic seed leaf in flowering plants (angiosperms) that provides food for embryos. [21,24]

Countercurrent multiplier. A system, important in kidney function and other body systems, in which a substance (water or O_2) or heat passes directly from one limb of a circulatory path to another running antiparallel to it, thereby increasing and maintaining concentration differences in the two limbs. [24]

Coupled reactions. The linking of an endergonic (energy-requiring) reaction to an exergonic (energy-yielding) reaction so that the latter drives the former. [8]

Covalent bond. A chemical bond arising from the sharing of electron pairs between atoms. [3]

Cranial nerves. Peripheral nerves that leave directly from the brain rather than from the spinal cord. They control sensory and motor functions of the head, and some autonomic functions. [30]

Crassulacean acid metabolism (CAM). A metabolic pathway in plants of the Crassulaceae and certain other families that allows them to incorporate large amounts of CO_2 into organic acids in the night and release CO_2 during the day. [10]

Creatine phosphate. See *phosphagen*. [9,27,31]

Crista (pl. **cristae**). Shelflike infoldings of the inner mitochondrial membrane that bear the enzymes necessary for oxidative phosphorylation. [5]

Crossing-over. The exchange of corresponding chromatid segments between homologous chromosomes during meiosis. It results in genetic recombination. [12,14]

Crossopterygian (crossopt). A group of primitive bony fishes, such as the coelacanth, thought to be related to the ancestors of the amphibians. [43,44]

Cross-pollination. The pollination of one plant by pollen from another plant. [21,41,47]

Crustacean. A class of arthropods; includes lobsters, crayfish, shrimp, etc. [42]

Cryptic coloration. External coloration or shape of an organism that blends with its background and permits it to escape visual detection. Compare with *aposematic* or *warning coloration*. [35]

Ctenophore. A member of the animal phylum Ctenophora; includes comb jellies. [42]

CURL (compartment of uncoupling of receptor and ligand). An endosome, formed by the fusion of several endocytic vesicles, in which the pH has dropped, causing ligands to leave their receptors. [6]

Cuticle. Hard waxy external coating of a plant or insect, usually composed of products of cellular metabolism, rather than cells themselves, that tends to retard water loss. [10,41,42]

Cyanobacteria (blue-green algae). See *Cyanonta*. [38]

Cyanonta (also called *Cyanophyta*). A phylum of the kingdom Monera which contains the prokaryotic cyanobacteria (blue-green algae). [38]

Cycad. A member of the gymnosperm class Cycadopsida. Short palmlike plants which produce pollen and egg cells in separate strobili. [41]

Cycadeoid. A member of the extinct gymnosperm class Cycadeoidopsida. [41]

Cyclic AMP (cyclic adenosine monophosphate, cAMP). A major second messenger in many eukaryotic cells that serves as an intracellular hormone. Also has a role in communication by cellular slime molds and in the transcription of catabolite-repressible operons in bacteria. [28]

Cyclic photophosphorylation. A mode of photosynthetic energy conversion that generates ATP but not O_2, NADPH, or carbohydrates. Compare with *noncyclic photophosphorylation*. [10]

Cytochromes. A class of iron-containing hemoproteins that serve as electron carriers in oxidative phosphorylation and photosynthesis. [9]

Cytokinesis. Cytoplasmic replication. Compare with *mitosis* and *karyokinesis*. [12]

Cytokinin. A class of plant hormones that regulate cell division, senescence, and other aspects of metabolism. [28]

Cytomatrix. The systems of filaments, including the microtrabecular lattice and cytoskeleton, which serve as scaffolding for the cytoplasm. [5]

Cytoplasm. The contents of a cell, excluding the nucleus. [5]

Cytoplasmic determinant. A molecule of egg cytoplasm that influences the development of the zygote. [20]

Cytoplasmic matrix. See *cytomatrix*. [5]

Cytoplasmic streaming. Bulk movement or circulation within a cell that distributes nutrients and wastes. [24]

Cytosine (C, Cyt). A nitrogenous pyrimidine base in RNA and DNA, where it pairs with guanine. [16]

Cytoskeleton. The supporting network of microtubules, microfilaments, and intermediate filaments within the cytoplasm of a cell. [5]

Cytosol. The fluid portion of the cytoplasm, excluding organelles and other solids. [5]

Cytotoxic T cell. See *killer t cell*. [23]

Decapod. Ten-legged crustaceans; includes shrimp, lobsters, crabs, etc. [42]

Decomposer. A fungus or bacterium that consumes dead organic matter, recycling it through an ecosystem. Also called *reducer*. [46]

Degenerate code. A code in which more than one codeword has the same meaning. In the genetic code, for example, most amino acids are specified by more than one codon. [17]

Deletion. In genetics, a mutation caused by loss of a segment of a gene or chromosome. [15]

Deme. A definable local unit or subunit of a population in which mating is random. Also called a *local, genetic,* or *Mendelian population*. [34]

Denaturation. Loss of the normal secondary or tertiary structure of a protein or nucleic acid, by exposure to heat, pH, or solvent changes. Usually leads to loss of biological activity. [4]

Dendrite. Fibers of a neuron that carry neural impulses from synapses toward the cell body. Usually much branched and relatively short compared with an axon. [5,29]

Density. Mass per volume. [3]

Denticle. A toothlike structure on the skin of sharks. May have been forerunners of true teeth. [43]

Deoxyribonucleic acid. See *DNA*. [4,16]

Deoxyribose. Five-carbon sugar of DNA which resembles ribose, the sugar of RNA, except that it lacks one oxygen atom (on the 2' carbon). [4,16]

Depolarization. In a neuron, the change of cell membrane potential from the negative resting potential to a more positive value. Action potentials are self-propagating depolarizations. [29]

Derepression. Reversal of repression occurring when cells are transferred to a medium lacking the molecule that repressed translation of the structural gene of an enzyme. [8,18]

Desert. Biome characteristic of environments which do not receive enough rainfall to support grasses and other vegetation. [48]

Desmosome. An adhering junction. The thickened site of attachment between adjoining cells. [7]

Determination. In embryology, the process by which a cell is irreversibly committed to a certain developmental fate. [19]

Deuteromycetes. A phylum of the kingdom Fungi. Also called *Fungi Imperfecti*. [40]

Deuterostome. One of two major lines of animal evolution. Characterized by radial cleavage, development of an anus from an embryonic blastopore, and other traits. Compare with *protostome*. [20,43]

Diabetes insipidus. Failure of water retention by the kidneys due to lack of antidiuretic hormone. Results in excessive urine flow and thirst. [25]

Diabetes mellitus. Impaired glucose metabolism due to lack of insulin. Resulting high blood sugar level leads to excessive urine flow and thirst. [28]

Diacylglycerol. A fatty acid derivative of glycerol that serves as a second messenger in the phosphatidyl inositol system. [28]

Dialysis. The separation of molecules in a solution by differences in their rates of diffusion through a semi-permeable membrane. [8]

Diaphragm. A barrier separating two cavities. (1) In vertebrates, it separates the chest cavity and abdomen and facilitates breathing. [27] (2) A contraceptive device that prevents sperm from entering the uterus. [21]

Diastole. The period during a heartbeat in which the heart chambers are filling with blood. Compare with *systole*. [24]

Diatom. A member of the protistan phylum Chrysophyta. [39]

Dicot (dicotyledon). A member of one of two angiosperm classes. Flowering plants with embryos bearing two cotyledons, net-veined leaves, 4- or 5-petaled flowers, and concentric vascular cylinders; includes woody or herbaceous plants, magnolias, tomatoes, cacti, etc. Compare with *monocot*. [24,41]

Dictyosome. A term for Golgi apparatus in plants. [5]

Differentiation. Increased specialization in the structure and function of cells or organs brought about in the development of once similar and less specialized cells or organs. [5,12,19]

Diffusion. Random movement of particles that causes them to be uniformly distributed. The net movement of a substance by diffusion is always from a region of high concentration to one of low concentration. [6]

Digestion. The hydrolytic breakdown of foods into small, soluble molecules that can be absorbed by cells. Compare with *assimilation*. [26]

Dihybrid cross. A genetic cross between two individuals which are identically heterozygous at two loci (e.g., $RrYy \times RrYy$). Compare with *monohybrid cross*. [13]

Dimer. A molecule composed of two similar subunits. [5]

Dinoflagellate. A member of the protistan phylum Pyrrophyta. [39]

Dioecious. (1) A plant species having its male and female organs in flowers on separate plants. (2) Any species in which the two sexes are in different individuals so that eggs and sperm are not produced in the same individuals. Compare with *monoecious*. [21]

Diploid. A cell or organism having two full sets of homologous chromosomes. A diploid individual (or cell) usually arises from the fusion of two haploid gametes, each with one chromosome set. [12]

Diplophase. In a sexually reproducing species, the diploid life stage between fertilization and meiosis. [21]

Directional selection. Natural selection for one extreme of a range of phenotypes, which shifts the phenotype of the population as a whole. Compare with *stabilizing* and *disruptive selection*. [34]

Disaccharide. A sugar, such as sucrose, consisting of two covalently linked monosaccharides. [4]

Disassortative mating. A pattern of sexual selection in which individuals preferentially select mates of dissimilar phenotype. Compare with *assortative mating*. [34]

Disjunctive distribution. Occurs when closely related populations inhabit widely separated regions with no intermediate forms in between. [48]

Displacement behavior. The performance of an irrelevant behavior when an intended behavior was thwarted by a motivational conflict. [33]

Disruptive selection. Occurs when both extremes of a phenotype's distribution range are favored over the middle range. Compare with *stabilizing* and *directional selection*. [34]

Distal. Situated far from the point of reference or attachment. Compare with *proximal*. [25]

Diversity. A measure of the number and abundance of different species within a community. [47]

Division. A taxonomic subgroup of the plant kingdom, further divided into classes. Equivalent to a phylum. [37]

Dizygotic twins. Fraternal twins. Twins arising from two zygotes. Compare with *monozygotic twins*. [19]

DNA (deoxyribonucleic acid). Double-stranded, helical polymeric molecule made up of deoxyribonucleotides (sugar is deoxyribose) whose sequence encodes the genetic information of living organisms and most viruses. Compare with *RNA*. [4,16]

DNA polymerase. A class of enzymes that in the presence of template DNA can synthesize DNA from its four component deoxyribonucleotides. [16]

Dominance. (1) The ability of one allele to determine the phenotype of a heterozygous individual carrying both it and another allele, which is recessive. Thus, if *A* and *a* are two alleles of a gene, *A* is dominant to *a* if *AA* and *Aa* are phenotypically identical and different from *aa*. [13] (2) A hierarchical set of relationships in some animal societies, in which one individual takes precedence over all others in eating, mating, and other activities. A second individual has precedence over all but the highest-ranking individual, and so on down the pecking order. [33]

Dorsal lip. An organizer in vertebrate embryonic development. In the blastula, the site of most active invagination. [19,20]

Dosage compensation. (1) Regulation at some autosomal gene loci that keeps homozygous dominants from producing twice as much gene product as heterozygotes. (2) Genes in female insects that decrease the levels of some X-linked gene products to levels seen in males. (3) Inactivation of X chromosomes in mammals so that no cell has more than one functioning X chromosome. [15]

Downstream. A location nearer the 5' end of a DNA molecule. Compare with *upstream*. [17]

Drive. An internal motivating force, such as hunger. [32]

Duodenum. In vertebrates, the first section of the small intestine. A major locus of food digestion. Compare with *jejunum* and *ileum*. [26]

Duplication. A mutation resulting from the introduction into the genome of an extra copy of a segment of a gene or chromosome. [17]

Dynein. A two-armed protein of flagellar microtubules with ATPase activity that causes movement by using ATP energy to "walk" one tubular doublet along its neighbor. [5]

Ecdysone. Invertebrate hormones that stimulate growth and molting. [28]

Echinoderm. A member of the animal phylum Echinodermata; includes starfishes (sea stars), sea urchins, sea anemones, corals, etc. [42]

Ecological isolation. A prezygotic isolating mechanism depending on habitat differences between otherwise similar populations in the same region. Compare with *behavioral, temporal, physiological,* and *mechanical isolation*. [36]

Ecology. The study of the interaction of organisms with their environment, including both the physical environment and other organisms living in it. [45]

Ecosystem. A community of organisms and its environment. [45,46]

Ectoderm. The outermost of the three embryonic cell layers. It gives rise to the skin, nervous system, and other structures. [19]

Ectomycorrhiza. Mutualistic association between a fungus (usually a basidiomycete) and a plant root around which it forms a sheath. Compare with *endomycorrhiza*. [40]

Ectoproct. A member of the animal phylum Ectoprocta. Moss animals with a coelom and an anus outside the lophophore ring. Compare with *entoproct*. [42]

Ectotherm. An organism that can regulate its body temperature by behavioral means, e.g., by moving into the shade from a sunlit forest clearing. Compare with *endotherm* and *poikilotherm*. [26]

Edema. An abnormal accumulation of extracellular fluid in body tissues. [24]

Effector. (1) A molecule that binds an enzyme at a site other than its active site (an allosteric site) and influences catalytic activity. [8] (2) A cell, organ, or tissue that translates neural activity into a useful bodily change. Muscles, bones, and glands are effectors. [29,31]

Efferent. Leading away from an organ. Compre with *afferent*. Efferent arterioles in the kidney carry blood from a glomerulus to the peritubular capillaries surrounding the renal tubules. [25] Efferent nerves carry impulses from the central nervous system. [29,30]

Egg. (1) In sexually reproducing organisms, a mature female gamete. [12] (2) In birds, reptiles, and some other vertebrates, a structure within which early embryonic development occurs. [21]

Egg activation. The immediate responses of an egg to fertilization. Includes the cortical reaction and the fusion of egg and sperm nuclei. [19]

Electric organ. Organs found in some fishes which can generate electric impulses used for communication, orientation, or to stun prey. [31]

Electrocardiogram (ECG, EKG). A record of the electrical activity of the heart. [24]

Electroencephalogram (EEG). A record of the electrical activity of the brain. [32]

Electrolyte. A substance, usually ionic, which when dissolved in water will conduct an electric current. Includes the major mineral ions of organisms, Na^+, K^+, Ca^{2+}, etc. [3]

Electromagnetic spectrum. The range of wavelengths of electromagnetic radiation from cosmic rays to radio waves; includes visible light. [10]

Electron. One of the fundamental particles of matter. Its mass is $1/1836$th (0.00055) the mass of a hydrogen nucleus and its charge is -1. [3]

Electron donor. A reducing agent. A compound that loses electrons in a chemical reaction. [10]

Electron microscope (EM). Powerful magnifying device that uses beams of electrons, rather than light photons, and focuses them with magnetic lenses. [5]

Electrophoresis. A procedure for separating mixtures of compounds in an electrical field on the basis of their charges. [22,23]

Electroplaques (also called *electroplax*). Diskshaped structural units (modified muscles) that generate electric charges in the electric organs of some fishes. [31]

Element. A single type of atom. [3]

Elongation factor. Protein necessary for the continued translation of an mRNA molecule. [17]

Embryo. A plant or animal in an early stage of development. Usually still contained within an ovary, egg, or uterus. [21]

Embryo sac. The female gametophyte of flowering plants. At maturity, consists of an egg nucleus and accessory nuclei. [21,41]

Endemic. Confined to or native to a particular region, thus having a comparatively restricted distribution and present at all times. [36]

Endergonic reaction. A reaction or process in which the end products possess more free energy than the reactants. Associated with anabolism. [8]

Endocrine gland. A ductless gland that secretes directly into the blood, hormones that influence target cells elsewhere in the body. Compare with *exocrine gland*. [28]

Endocrine secretion. Hormones that are passed from the gland that produces them into the blood. Compare with *autocrine* and *paracrine secretion*.

Endocytosis. Process by which cells engulf and take in particles or other material (including receptor-ligand complexes). Invaginated cell membrane pinches off to incorporate particles in an intracellular vesicle. See *receptor-mediated endocytosis*. Compare with *exocytosis*. [6]

Endoderm. The innermost of the three embryonic cell layers. It gives rise to such structures as the epithelium of the digestive tract and the bladder. [19]

Endodermis. In vascular plants, a specialized layer of cells just inside the cortex in roots and some stems, which often impedes the free diffusion of solutes. [24]

Endometrium. The membrane lining the mammalian uterus. [21]

Endomitosis. Nuclear or chromosomal replication without division of the surrounding cytoplasm. Results in multinucleated cells and polyploidy. [22]

Endomycorrhiza. Mutualistic association between a fungus (usually a zygomycete) and a plant root in which the fungus penetrates living outer root cells. Compare with *ectomycorrhiza*. [40]

Endonuclease. An enzyme that cleaves DNA or RNA within the polynucleotide strand. Compare with *exonuclease*. [12]

Endoplasmic reticulum (ER). A system of double membranes that divides eukaryotic cytoplasm into compartments. See *rough endoplasmic reticulum* and *smooth endoplasmic reticulum*. [5]

Endorphin. A class of naturally occurring neuronal peptides which, among other functions, have a morphinelike effect on pain perception. [28]

Endoskeleton. A skeleton, such as that of vertebrates, which is located within the body and surrounded by soft tissue. Compare with *exoskeleton*. [31]

Endosome. (1) An organelle created by the fusion of several endocytic vesicles. [6] (2) Intranuclear body, such as nucleolus. [5]

Endosperm. A food storage tissue in angiosperm seeds, which develops from the union of a haploid male nucleus and two haploid female polar bodies, and thus is triploid. [21]

Endosymbionts. Free-living bacteria that entered the cytoplasm of ancient urkaryotes, there elaborated a mutually beneficial relationship, and in time evolved into mitochondria or chloroplasts. [38,44]

Endotherm. An organism that regulates its body temperature by metabolic and physiological means. Compare with *ectotherm, poikilotherm*, and *homeotherm*. [26]

Endothermic reaction. A chemical reaction that requires the application of heat. Compare with *exothermic reaction*. [8]

End-product inhibition. See *feedback inhibition*. [8]

Energetic cost. The energy expended in carrying out a behavior or physiological process. Compare with *opportunity cost* and *risk cost*. [35]

Energetics. The study of energy and its transformations. [8]

Energy. The capacity to do work. [8]

Enhancer. A DNA sequence that increases the transcription of a particular gene. Can be quite distant from the gene on either its 3' or 5' side. Compare with *promoter*. [18]

Enkephalins. Two of the endorphins. [28]

Enterokinase. An enzyme secreted by the pancreas which converts trypsinogen into trypsin. [26]

Enthalpy (*H*). The heat content of a physical system. Change in enthalpy is ΔH. [8]

Entoproct. Member of the animal phylum Entoprocta. Lacks a coelum and has an anus within the lophophore ring. Compare with *ectoproct*. [42]

Entrainment. The process in which an animal's circadian or seasonal rhythms are synchronized with environmental rhythms. [32]

Entropy (*S*). A measure of the randomness of a system, i.e., of the amount of energy which has become evenly dispersed and is therefore no longer available to do work. The second law of thermodynamics states that entropy tends always to increase. [6,8]

Environment. The totality of physical and biological conditions extrinsic to an organism that affects its life. [1,46]

Enzymatic pathway. A series of enzyme-catalyzed reactions in which the product of one reaction is the substrate of the next. [8]

Enzyme. A biological catalyst. It was thought that all enzymes are proteins until the recent discovery of catalytic RNA molecules. [8]

Eosinophil. A white blood cell of the granulocyte class that is stained with the red dye eosin. Plays a role in inflammation and in allergic and parasitic disorders. [22]

Epicotyl. The portion of the seedling stem above the cotyledons. [21]

Epidermis. The outermost layer (or layers) of cells on the bodies of plants and the skin of animals. [7,24]

Epididymis. The convoluted tube that carries sperm from the testis to the vas deferens in mammals. [21]

Epilimnion. The warm surface waters of a lake (>4°C) that are mixed by wind and convection currents. Above the hypolimnion. [48]

Epinephrine (adrenaline). A hormone secreted by the adrenal medulla. Also a neurotransmitter. [28]

Epiphyte. A plant that depends on another plant for support but not nutrition. [41,48]

Episome. A genetic unit that can exist either as an integrated part of a chromosome or as a free element in the cytoplasm. Compare with *plasmid*. [16]

Epistasis. An interaction between genes in which an allele of one gene masks or modifies the effects of an allele of another gene. [14]

Epithelium. The layer of cells lining body surfaces and body cavities. [7]

Equational division. Meiosis II. See *meiosis*. [12]

Equatorial plane. A plane, roughly along the middle of a cell, on which the chromosomes line up during mitosis. [12]

Equilibrium (chemical). The point in a reversible chemical reaction at which net product synthesis ceases because the rate of the forward reaction equals that of the backward reaction. [8]

Equilibrium constant (K_{eq}). The ratio of product and reactant concentrations at equilibrium. Related to the free energy change of a reaction. [8]

Erythrocyte. A red blood cell. it contains hemoglobin and transports oxygen. In mammals, it lacks a nucleus. [22]

Erythropoietin. A hormone regulating red blood cell production. [28]

Esophagus. The part of the alimentary canal connecting the pharynx and the stomach. [26]

Essential amino acid. An amino acid that cannot be synthesized by an organism and thus must be a part of its diet. [11]

Estivation. Passing the summer or dry season in a dormant state, an adaptation that permits certain desert animals (rodents, some birds, etc.) to cope with seasonal heat and dryness. Compare with *hibernation*. [48]

Estrogen. Any of a group of mammalian steroid hormones that, with progesterone, regulate female sexual development, menstruation, and pregnancy. [21,28]

Estrous cycle. In most mammals, the cyclical pattern of female sexual receptivity which coincides with the ovarian cycle. Culminates in estrus. [21]

Estrus. The period of heat, or maximum sexual receptivity, in some female mammals. Ordinarily, the time of ovulation. [21]

Ethology. The scientific study of animal behavior in natural environments, stressing the adaptational and evolutionary origin of patterns. [32]

Ethylene. A simple gaseous hydrocarbon ($H_2C{=}CH_2$) produced in plants that stimulates growth and fruit ripening, and antagonizes auxins. [28]

Eubacteria. Largest class of the phylum Schizonta (bacteria). [38]

Euchromatin. Areas of diffuse, active chromatin in an interphase nucleus. Compare with *heterochromatin*. [15]

Eugenics. The control of procreation in an attempt to change the characteristics of future generations. [15]

Euglenoid flagellates. Members of the protistan phylum Euglenophyta. [39]

Eukaryote. An organism with membrane-bound organelles, most notably the nucleus, and DNA linked to histone proteins. [5]

Eumetazoa. A subkingdom of Animalia containing all animals except the sponges. [43]

Euphotic zone. The well-lit marine zone near the ocean surface. Compare with *aphotic zone*.

Euploidy. A state in which the chromosome number is normal. [12]

Euryecious. Pertains to species capable of surviving within a wide range of environmental conditions. Compare with *stenoecious*. [46]

Eurytrophic. An animal that can live on many different kinds of food. Compare with *stenotrophic*. [26]

Eutrophic. A lake with an abundant supply of organic nutrients. [48]

Eutrophication. Enrichment of a pond or lake with algae-stimulating nutrients. Algal growth then depletes pond water of oxygen, killing animals. [46]

Evolution. (1) Genetic change in a population of organisms over time; produced by the integrating agencies of natural selection and variation (genetic drift, gene migration, and mutation). (2) A change in a population's allele frequencies. [34]

Evolutionary systematics. A taxonomic school which classifies species according to phylogenetic lineage, but also weighs novel traits (evolutionary innovations) more heavily. Compare with *cladistics* and *numerical phenetics*. [37]

Excision repair. An enzymatic process of DNA repair in which an endonuclease removes a defective oligonucleotide, which is then replaced. [17]

Excitatory postsynaptic potentian (EPSP). A transient depolarization in the postsynaptic neuron of a synapse, induced by neurotransmitter release from the presynaptic neuron. [29]

Excretion. Release of metabolic wastes (unusable or excess materials) from an organism or cell. [25]

Excretory pore. In flatworms, the external openings of tubules that carry wastes from flame cells to the body surface. [25]

Exergonic reaction. A chemical reaction that releases free energy and thus yields products that possess less free energy than the reactants. Associated with catabolism. Compare with *endergonic reaction*. [8]

Exocrine gland. A gland with a duct which secretes externally (e.g., to the skin surface or gut) substances that act locally. Compare with *endocrine gland*. [28]

Exocytosis. Extrusion or secretion of materials from a cell by fusion with the cell membrane of the vesicle membrane surrounding materials being expelled. Compare with *endocytosis*. [6]

Exonuclease. An enzyme that catalyzes the hydrolytic removal of a single nucleotide from the end of a DNA strand. Compare with *endonuclease*. [12]

Exon. A portion of DNA in a gene which encodes a protein sequence. Compare with *intron*. [17]

Exoskeleton. A skeleton, such as that of insects and other arthropods, which takes the form of a hard external shell. Compare with *endoskeleton*. [13,42]

Exothermic reaction. A chemical reaction that releases energy as heat. Compare with *endothermic reaction*. [8]

Exponential growth. Growth, especially in the size of a population, which is a function of the size of the growing entity. The larger the entity, the faster it grows. Compare with *logistic growth*. [45]

Expression. Gene transcription and translation. [17]

Expressivity. The degree to which a penetrant gene is expressed phenotypically. Compare with *penetrance*. [14]

Extinction. The death of all members of a species or larger phylogenetic group that usually follows failure of adaptation to a changing environment. A common occurrence in evolutionary history. [34]

Extracellular matrix. Materials in multicellular organisms that lie between cells. [7]

Extrachromosomal inheritance. Transmission of genetic traits by mechanisms other than nuclear chromosomes (e.g., mitochondrial DNA). See *maternal effect*. [18]

Extrinsic (peripheral) protein. A protein found on the surface of a membrane. Compare with *intrinsic protein*. [6]

Extrinsic system. One of two mechanisms which can trigger blood clotting. Initiated by tissue factor, a protein released by injured tissue. Compare *intrinsic system*. [22]

Eyespot. See *stigma*. [39]

F_1 (first filial) generation. The offspring of a genetic cross. Compare with F_2 *generation*. [13]

F_2 (second filial) generation. The offspring of the F_2 generation of a genetic cross; the inbred grandchildren of the original cross. [13]

Fab fragment. Part of an antibody molecule (obtained by papain digestion) that contains one light chain and one heavy chain. Includes the antigen-binding site. Compare with *Fc fragment*. [23]

Facilitated diffusion. Diffusion of a substance across a membrane via a specific carrier protein or channel. Unlike active transport, it requires no energy and cannot proceed against a concentration gradient. Compare with *passive diffusion*. [6]

Facilitation. A simple form of learning, in which repeated presentations of a stimulus increase the probability that an organism or neuron will respond. Compare with *habituation*. [29]

Facultative anaerobe. An anaerobe that can live in the absence of free oxygen, but unlike an obligate anaerobe is not compelled to. [9]

Fallopian tube. The human oviduct. [21]

Familial pathway. The evolutionary development of social groups involving only close relatives. Compare with *parasocial pathway*. [33]

Family. The taxonomic category between order and genus. A single family includes several genera. [37]

Fate map. A chart that shows each early embryonic cell and the adult cells into which it develops. [20]

Fatty acid. A long hydrocarbon chain with a carboxyl group at one end. A component of many lipids. [4]

Fauna. All of the animals found in a given area. Compare with *flora*. [48]

Fc fragment. Part of an antibody molecule (obtained by papain digestion) that includes parts of two heavy chains and no antigen-binding site. Has many regulatory functions in the immune system. Compare with *Fab fragment*. [23]

Feedback inhibition. Metabolic regulation in which high levels of an enzymatic pathway's final product inhibit the activity of its rate-limiting enzyme by allosteric effects. Compare with *repression*. [8]

Fermentation. Anaerobic breakdown of a substance such as glucose to smaller organic molecules with the extraction of energy. Usually involves glycolysis. Sometimes used imprecisely for anaerobic metabolism. [9,40]

Fertility factor (F factor). A bacterial plasmid that confers "maleness" (the ability to donate genetic material to other bacteria during conjugation) on a cell that contains it. [16]

Fertilization. The union of male and female gametes to form a zygote. Also called *syngamy*. [19,21]

Fertilization membrane. A membrane that bars the entry of additional sperm into a fertilized egg. [19]

FeS proteins. Proteins containing iron and sulfur that participate in the electron transport chains of oxidative phosphorylation in mitochondria and photosynthesis in chloroplasts. [10]

Fetal hemoglobin (Hb F). A form of hemoglobin in the red cells of fetuses and infants that maximizes oxygen transport from placenta to fetus. [22]

Fetus. An embryo in its later stages of development. In humans, the embryo is called a fetus after about the second month of gestation. [21]

F factor. See *fertility factor*.

Fibrin. The fibrous insoluble protein that forms the structural framework of a clot. Forms when fibrinogen is acted upon by thrombin. [22]

Fibrinogen. A soluble plasma protein that is converted by thrombin to insoluble fibrin in blood clotting. [22]

Fibrinolysin. See *plasmin*. [22]

Fibrinolytic system. A system of plasma proteins that hydrolytically degrade fibrin and other clotting factors. [22]

Fibroblast. A cell of animal connective tissue that makes collagen and other components of the extracellular matrix. [7]

Fibroin. Protein secreted by spiders and silkworms that solidifies into the strong silk threads of webs and cocoons. [42]

Fibronectins. A class of glycoproteins that promote the adhesion of cells to a surface. [7,20]

Filter feeder. An aquatic animal that filters large amounts of water and feeds on the retrieved smaller organisms and particles. Compare with *bulk feeder*. [26]

Finalism. The once-prevalent doctrine that evolution has a preordained pattern that is purposefully directed toward a future goal. [2]

First law of thermodynamics. Principle that energy can be neither created nor destroyed [8]

Fission. A mode of asexual cellular reproduction in monerans and protists in which organisms split into two or more parts, each then becoming a complete new organism. Compare with *budding*. [12,21]

Fitness. (1) The ability of an organism owing to its genotype or phenotype to successfully meet environmental challenges. (2) The change of frequency of an allele over generations as a result of natural selection. (3) The change in frequency of an individual, the greater its genetic contribution to subsequent generations. [34]

Fixed action pattern. An instinctive, complex behavior invariably triggered by a simple stimulus. [32]

Flagellate. A member of the phylum Zoomastigina. Any unicellular protist that propels itself with flagella. [39]

Flagellum (pl. **flagella**). A long, threadlike organelle used in cell motility. Prokaryotic flagella differ from those of eukaryotes. Compare with *cilium*. [5]

Flame cell. A hollow excretory cell in certain flatworms and other groups. Uses cilia to collect waste materials. [25]

Flora. All of the plants found in a given area. Compare with *fauna*. [41,48]

Flower. The total reproductive structure of an angiosperm. Its parts include the calyx, corolla, stamens, and carpels. [21,41]

Fluid feeder. An animal that lives on nutrients available in a fluid such as blood or sap. [26]

Fluke. A member of the flatworm class Trematoda; parasitic forms. [42]

Fluorescence. Emission of light caused by a flow of light (or other energy) into the emitting body. Ceases abruptly when excitation stops. [10]

Fluorescent antibody. An antibody conjugated to a fluorescent labeling molecule (usually fluorescein). Used to label tissue or cells containing the appropropriate antigen. [23]

Follicle. A hollow chamber. In female mammals, an immature egg surrounded by nutritive cells. [21]

Follicle-stimulating hormone (FSH). A gonadotropic hormone produced by the anterior pituitary that stimulates follicle growth in ovaries and sperm production in males. [21,28]

Food chain. A portion of a food web, usually consisting of a sequence of prey species and the predators that consume them. Typically, plants are at the bottom of a food chain and carnivores at the top. [1,46]

Food web. A complete network of predatory-prey food links between and among species in a community. [46]

Foraminifera (forams). Hard-shelled members of protistan phylum Sarcodina. Abundant in the sea bottom. Common as fossils [39]

Forebrain. The area of the vertebrate brain which contains the cerebral hemispheres, thalamus, and hypothalamus. [30]

Forward mutation. A mutation in a wild-type organism which causes its progeny to differ from that type. Compare with *back mutation*. [14]

Fossil. Any structure deriving from an organism, or any impression from such a structure, that has been preserved from ancient times. [37]

Fossorial. Adapted for digging or burrowing. [48]

Founder cell. An early embryonic cell which is the progenitor of a specific cell type. [20]

Founder principle. Genetic drift following establishment of a new isolated population by a few pioneering individuals. Speciation may occur when descendants differ genetically from the rest of the species. [34]

Fovea. Small area in center of the vertebrate retina. Contains only cones (no rods), thus is specialized for color vision and fine discrimination. [30]

Fragile site. A chromosomal area that regularly breaks under certain conditions. [15]

Frame-shift mutation. Addition or deletion of a single base pair in a DNA gene sequence. The transcribed mRNA is translated normally until the point of the mutation. From then on, codons are read one base unit out of register. The resulting amino acid sequence is grossly abnormal. [17,18]

Fraternal twins. See *dizygotic twins*. [19]

Free diffusion. Movement of molecules across a membrane without involvement of carrier molecules. Not saturable and cannot cause net transport from a region of low concentration to a region of higher concentration. Compare with facilitated diffusion and active transport. [6]

Free energy (G). Also called *Gibbs free energy*. The energy available for doing work after allowance has been made for changes in entropy. The change in free energy (ΔG) of a reaction indicates whether or not it can occur spontaneously. [8]

Frequency-dependent selection. Natural selection for an allele whose fitness varies with its frequency in a population. Also called *apostatic selection*. [34]

Frequency-distribution curve. A plot of the value of a given variable versus the frequency with which that variable appears in a population. [34]

Fruit. In flowering plants, a ripened ovary (or group of ovaries) containing seeds. [21]

Fruiting body. Complex spore-bearing structures of myxobacteria and certain fungi. [38]

Fucoxanthin. Carotenoid pigment responsible for characteristic color of brown algae. [39]

Functional group. Substituent that confers characteristic chemical properties upon a molecule, e.g., carboxyl (—COOH) and amino groups (—NH$_2$). [4]

Fungi. One of the five kingdoms that includes distinctive eukaryotic heterrotrophs that are nonphotosynthetic and have characteristic structural features. A fungus is a member of the kingdom. [37,40]

Gametangium (pl. **gametangia**). Any plant or fungal structure that produces discrete gametes or functions in their place. [21,40]

Gamete. A haploid cell (egg or sperm) that must fuse with another haploid cell of the opposite sex or type to initiate the development of a new diploid organism. [12]

Gametic meiosis. Meiosis that occurs directly prior to gametogenesis. Occurs in most animals and fungi. Compare with *sporic* and *zygotic meiosis*. [21]

Gametocyte. The cell that gives rise to gametes. See *oocyte* and *spermatocyte*. [21]

Gametogenesis. Series of special cell divisions that leads to gamete production: spermatogenesis in males, oogenesis in females. [12,19]

Gametophyte. In plants with alternation of generations, the haploid gamete-producing phase in the life cycle. Compare with *sporophyte*. [21]

γ-aminobutyric acid (GABA). An inhibitory neurotransmitter in the central nervous system. [29]

Ganglion. A cluster of nerve cell bodies. In vertebrates, ganglia are outside of the central nervous system. In invertebrates, they comprise the central nervous system. [30]

Gap junction. A specialized junction between cells that contains tiny channels allowing passage of materials between two cytoplasms. [7]

Gastric juice. Fluid secreted by the stomach. Contains hydrochloric acid, pepsin, and other digestive enzymes. [26]

Gastropod. A member of the mollusk phylum Gastropoda; includes snails, etc. [42]

Gastrulation. The stage in early embryonic development in which blastula cells form three primary cell layers. [19]

Gause's principle. See *competitive exclusion*. [47]

Gel. A two-phase colloid system, either semisolid or solid, that has a large amount of liquid within the solid component. More solid than a sol. [6]

Gel-sol transition. See *sol-gel transition*. [6]

Gene. The unit of heredity. A site on a chromosome or a DNA sequence encoding a single polypeptide. [13,16,17]

Gene amplification. A selective increase in the number of copies of a particular gene in a cell. [17]

Gene cloning. The transfer of a gene from any organism into a microorganism, where it can be replicated. [18]

Gene duplication. A mutation which produces a second copy of a gene in the organism's genome. [17]

Gene flow. The movement of alleles between populations through interbreeding. Also called *gene migration*. [34]

Gene migration. See *gene flow*. [34]

Gene mutation. A change in the base composition of a specific genetic locus. Compare with *chromosome mutation*. [14,17]

Gene pool. All of the genes in a population or species, thus its genetic constitution; the alleles present and their relative frequencies. [34]

Genetic drift. See *random drift*. [34]

Genetic engineering. The branch of technology that produces new organisms or biological products by transferring genes between cells. [18]

Genetic equilibrium. The tendency in species toward constancy of the gene pool and its allelic frequencies. [34]

Genetics. The study of heredity. [12,13]

Genome. Complete set of genetic material of an organism in a haploid chromosome set. [13,16]

Genotype. The genetic makeup of an organism independent of its physical or functional traits. Also used to denote the genic basis of a single trait. Compare with *phenotype*. [13]

Genus. A taxonomic category. Multiple genera make a family. A genus includes multiple similar species. [37]

Geological time scale. A chronology of geological events tied to the relative sequence of rock formations, rather than to radiometric dating. [37]

Geotropism. See *gravitropism*. [32]

Germ cell. See *gamete*. Compare with *somatic cell*. [12,19,21]

Gibberellins. A class of plant hormones that influence growth and differentiation, stem elongation, seed germination, and flowering in certain plants. [28]

Gill. A respiratory organ of aquatic animals, consisting of thin-walled projections, richly supplied with blood, across which water passes and oxygen and CO$_2$ are exchanged. [27]

Ginkgo. A member of the gymnosperm class Ginkgopsida. Once an abundant group, today it includes only one species—the maidenhair tree. [41]

Gizzard. A muscular part of the alimentary canal in birds that grinds food (sometimes with the aid of pebbles) to facilitate digestion. [26]

Glade. Adjacent demes in a particular geographical area. [28]

Gland. An organ of secretion or excretion. [28]

Glia. Support cells in the nervous systems that provide nutritional support, make myelin, and scavenge debris. Do not conduct action potentials. [29]

Globulin. Any of several plasma protein classes, including the immunoglobulins (antibodies), clotting factors, and others. [22]

Glomerular filtration. The first step in urine formation in which an ultrafiltrate of blood forms as blood passes through the kidney glomerulus. [25]

Glomerulus. (1) In the kidney, a tuft of capillaries associated with each nephron from which fluid filters into Bowman's capsule. [25] (2) In the brain, a tuft of nerve fibers in the olfactory bulb. [30]

Glucagon. A peptide hormone produced in the vertebrate pancreas that stimulates glycogen breakdown and a rise in blood sugar. Compare with *insulin*. [28]

Gluconeogenesis. The production of glucose from noncarbohydrate precursors, such as amino acids and lipids. [11]

Glucose. The most common sugar, one of several monosaccharides with the formula C$_6$H$_{12}$O$_6$. [9,10]

Glycogen. A branched polysaccharide made of linked glucose molecules. The chief storage carbohydrate in animals. [4]

Glycolysis. The enzymatic breakdown of glucose to pyruvic acid, with a net synthesis of two ATP molecules. A widely occurring energy-yielding system. [9]

Glycophorins. A major class of plasma membrane glycoproteins. [6]

Glycoprotein. A protein to which a carbohydrate group has been attached. [6]

Glyoxylate cycle. A metabolic pathway in plant cells that converts fatty acids in stored lipids to succinic acid and then to sugars during seed germination. [11]

Glyoxysome. Plant cell organelle, usually found in cotyledons. Locus of the glyoxylate cycle. [5,11]

Gnetopsid. Member of the gymnosperm class Gnetopsida. Trees and climbing vines that may be an evolutionary link between gymnosperms and angiosperms. [41]

Golden algae. Members of the protistan phylum Chrysophyta. [39]

Golgi apparatus (or **body**). An organelle that sorts and packages substances (such as digestive enzymes) destined to be secreted from a cell. A stack of concentric membrane-bound sacs. [5]

Gonads. The male and female gamete-producing glands. The testis or ovary. [12]

Gondwanaland. A large land mass of the ancient world. One of two supercontinents produced by the breakup of the original supercontinent Pangaea. [48]

Gonial cell. See *spermatogonium*. [12]

Gonosomes. The sex chromosomes. [15]

G protein. A family of widely occurring membrane proteins that regulate second messenger systems. A signal transducer that binds GTP or GDP (G stands for guanyl nucleotide) and modifies the activity of adenylate cyclase and other enzymes. [28,30]

Granulocyte. A class of white blood cells containing cytoplasmic granules; includes neutrophils, eosinophils, and basophils. [22]

Graptolite. Extinct animals, once considered coelenterates, now viewed as relatives of chordates. Abundant in ancient seas. [42]

Grassland. Biome of small herbaceous plants. Found in temperate and tropical zones which lack enough water to support forests. [48]

Gravitropism. Growth of a plant toward or away from the gravitational pull of the Earth. Also called *geotropism*. [32]

Gray matter. Areas of the central nervous system with many neuronal cell bodies and few myelinated fibers. Compare with *white matter*. [29]

Green algae. Members of the protistan phylum Chlorophyta. [39]

Greenhouse effect. The gradual warming of Earth's surface by increasing concentrations in the atmosphere of CO_2 and other gases. [46]

Group selection. Natural selection favoring genes that increase the survival of the population as a whole, rather than the individual carrying the gene. Also called *interdemic selection*. Compare with *kin selection*, *individual selection*, and *species selection*. [34]

Growth. An increase in size of an organism, usually by an increase in the number of cells, but sometimes by an increase only in cell size. [19]

Growth factor. A molecule, usually a peptide, which triggers a specific type of cell to divide or to increase in size. [28]

Growth hormone (GH). A hormone from the anterior pituitary that is necessary for growth in young mammals. [28]

Guanine (G, Gua). A nitrogenous purine base found in RNA and DNA, where it pairs with cytosine. [16]

Guanyl nucleotide. GMP, GDP, or GTP. See *G protein*. [28,30]

Guard cells. Paired epidermal cells that flank stomata in leaves and open and close them. [10]

Gut. Digestive tract. [26]

Guttation. In plants, the exudation of drops of water from leaves or stem; a result of root pressure. [24]

Gymnosperm. A member of the plant subdivision Gymnospermae; includes four living and three extinct classes of the division Spermophyta. Has exposed seeds without ovarian walls on sporophylls or related structures. [41]

Gynandromorph. Abnormal organism in which part of the body is male and part female. [15]

Gynoecium. The female reproductive part of a flower, consisting of one or more carpels. [41]

Habitat. The place or environment in which an organism normally lives. [37,46]

Habituation. A simple form of learning in which repeated presentation of a stimulus decreases the probability that an organism

or neuron will respond to it. Compare with *facilitation*. [32]

Hadean time. One of the four major time divisions in the history of Earth, it covers the poorly defined period before about 3.9 billion years ago. [37]

Haemocoel. A blood-filled cavity. In some animals, it serves as a hydrostatic skeleton that maintains body shape. [31]

Hair cell. A mechanoreceptor of animals with pressure-sensitive, hairlike projections (stereocilia) that respond to sound energy in the cochlea and acceleration in the semicircular canals, saccule, and utricle. [30]

Half-life. The time necessary for half of a substance to undergo change. A measure of the rate of that change. [3]

Haploid. A cell or organism containing only one member of each homologous chromosome pair. The normal ploidy of gametes or spores produced by meiosis or of gametophytes that develop from such spores without fertilization. [12]

Haplophase. In a sexually reproducing species, the haploid life stage between meiosis and fertilization. [21]

Haplotype. The combination of alleles inherited from a parent. [34]

Hardy-Weinberg law. A law of population genetics stating that in the absence of selection, nonrandom mating, and other conditions, heredity tends to stabilize allelic frequencies in a population. It also predicts phenotype ratios from genotype ratios in a randomly mating population—in the absence of mutation, genetic (random) drift, migration, or selection. [34]

Heat capacity. The amount of heat required to raise the temperature of an object by 1°C without changing its phase. Compare with *specific heat*. [3]

Heat of fusion. The increase of heat (enthalpy) associated with the conversion of a mole (or given amount) of a solid to a liquid with no temperature change. [3]

Heat of vaporization. The amount of heat necessary to convert a mole (or given amount) of a substance from a liquid to a gas with no temperature change. [3]

Heavy chain. A subunit of an immunoglobulin. There are two heavy chains per molecule. Compare with *light chain*. [23]

Heavy-chain switch. Recombination of heavy chain genes in a B cell clone, which causes an antibody to change heavy chain class without changing antigenic specificity. [23]

Helper T cell. A class of T cells (lymphocytes); can activate B cells, killer T cells, and macrophages. [23]

Hematocrit. The volume of packed red cells expressed as a percentage of whole blood volume. [22]

Hematopoiesis. Blood cell formation. [22]

Heme. An iron-porphyrin complex associated with each polypeptide subunit of hemoglobin. [10,22]

Hemichordata. A subphylum of chordates; includes acorn worms and other marine animals. [43]

Hemocyanin. A blue, copper-containing, oxygen-carrying protein found only in the plasma of mollusks and arthropods other than insects. [22]

Hemoglobin. A red, iron-containing protein of red blood cells that transports oxygen and carbon dioxide. [22]

Hemoglobin S. Sickle cell hemoglobin. Because it differs from normal hemoglobin (hemoglobin A) by one of its hundreds of amino acids, it polymerizes at low oxygen pressure and causes red cells to sickle. [16,22]

Hemoprotein. A class of proteins, such as hemoglobin and the cytochromes, in which a protein component is conjugated to an iron-porphyrin prosthetic group like heme. [10,22]

Hemostasis. Mechanisms that stop bleeding. They include formation of a platelet plug (primary hemostasis) and clotting (secondary hemostasis). [22]

Herbivore. An organism that eats plants. Compare with *carnivore* and *omnivore*. [1,26]

Hermaphrodite. An organism possessing both male and female sexual organs. [15,21]

Heterochromatin. Areas of condensed, probably inactive, chromatin in an interphase nucleus. Compare with *euchromatin*. [15]

Heterochrony. Variations in the sequence of normal developmental processes that can give rise to evolutionary change. [20]

Heterocyst. A thick-walled, differentiated cell occurring at intervals in some cyanobacteria (blue-green algae), which can perform nitrogen fixation. [38]

Heterogametic sex. The sex in a species with two differing sex chromosomes and two or more kinds of gametes. Compare with *homogametic sex*. [12]

Heterogeneous nuclear RNA (hnRNA). Unprocessed RNA molecules of various sizes, some (but not all) of which are precursors of mRNA. [17]

Heterokaryon. A mycelial cell containing genetically different nuclei. Peculiar to fungi. Compare with *homokaryon*. [21,40]

Heteromorphic. (1) Having nonhomologous chromosomes, such as sex chromosomes, that differ in size or form. [14] (2) Having a different form at each stage of a life cycle. [21,40,41,42]

Heterosis. Hybrid vigor. The adaptive superiority of heterozygotes over homozygotes. [34]

Heterosporous. An organism producing two kinds of haploid spores, typically a microspore (which becomes a male gametophyte) and a megaspore (which becomes a female gametophyte). Compare with *homosporous*. [21,41]

Heterotroph. An organism that must obtain certain necessary organic molecules from other organisms (autotrophs) which provide them ready-made; includes parasites, carnivores, and others. [9,38]

Heterozygous (heterozygote). Having dissimilar alleles in regard to a given genetic trait. Compare with *homozygous*. [13]

Hexose monophosphate shunt. See *pentose phosphate shunt*. [11]

Hfr (high frequency of recombination) cell. A bacterium that has the fertility (F) factor gene integrated into its chromosome and thus frequently transfers its DNA to recipient (F−) cells. [16]

Hibernation. A state of winter inactivity of some animals (e.g., rodents, insectivores, bats, etc.) marked by decreased body temperature and energy conservation. Compare with *estivation*. [48]

Highly repetitive DNA (repetitious DNA). Regions of eukaryotic DNA that contain several types of highly repeated sequences. Compare with *moderately repetitive* and *simple sequence DNA*. [17]

Hindbrain. The part of the vertebrate brain between the midbrain and the spinal cord. It contains the cerebellum, pons, and medulla oblongata. [30]

Histone. Class of basic proteins in eukaryotic chromosomes. They form the nucleosome core about which DNA is coiled. Absent in prokaryotes. [12]

HLA (human leukocyte antigen) system. Complex system of polymorphic, antigenic, membrane glycoproteins that permit the immune system to identify cells of "self" and "non-self." It permits cell recognition by the immune system and influences histocompatibility in tissue transplantation. Also called the *MHC* (*major histocompatibility complex*). [23,34]

hnRNA. See *heterogeneous nuclear RNA*. [17]

Holoblastic cleavage. Embryonic cleavage that divides the entire zygote. Compare with *meroblastic cleavage*. [19]

Holoenzyme. An apoenzyme combined with its coenzyme. A complete enzyme. [8]

Homeo box. A DNA sequence of 180 nucleotides that is highly conserved in nature, appearing in widely diverse species. Associated with many homeotic genes, which determine the body plan in embryonic development. [20]

Homeostasis. The maintenance within an organism or group of organisms of a steady state, in such variables as temperature, chemical composition, or social structure, by means of physiological or behavioral feedback loops. [1,46]

Homeotherm. An animal which maintains a constant body temperature. Compare with *poikilotherm*. [26]

Homeotic mutation. Mutation in a homeotic gene that alters the developmental program, switching embryonic cells from one developmental fate to another. [20]

Hominid. A member of the family Hominidae, of which *Homo sapiens* is the only living species. Compare with *hominoid* and *anthropoid*. [44]

Hominoid. A member of the superfamily Hominoidea, which includes hominids and apes. Compare with *hominid* and *anthropoid*. [44]

Homogametic sex. The sex in a species that contains similar paired sex chromosomes in its cell nuclei and that produces identical gametes. Compare with *heterogametic sex*. [12]

Homokaryon. A mycelial cell containing nuclei of a single genetic constitution. Compare with *heterokaryon*. [40]

Homologous chromosomes. In diploid cells, chromosomes that are identical in size, shape, and gene content—and that associate during the first stage of meiosis. [12]

Homology. A similarity between structures (or molecules) in different organisms attributable to their inheritance from a common ancestry. Compare with *homoplasy* and *analogy*. [37]

Homoplasy. Correspondence between structures or organs in different organisms acquired as a result of convergent (or parallel) evolution. Compare with *analogy* and *homology*.

Homosporous. Organisms producing only one type of spore, either a microspore or a megaspore. Compare with *heterosporous*. [41]

Homozygous (homozygote). Having identical alleles in regard to a given genetic trait on both homologous chromosomes. Compare with *heterozygous*. [13]

Horizontal gene transfer. Transfer of genes between species without sexual reproduction. [34]

Hormone. A chemical messenger molecule (commonly a steroid or peptide), which is secreted by cells in one region of a multicellular organism and influences cell function elsewhere in the organism. [8,28]

Humoral immunity. Immunity provided by antibodies that are synthesized by plasma cells derived from B lymphocytes. Compare with *cell-mediated immunity*. [23]

Hybrid. The offspring of genetically dissimilar parents. [36]

Hybrid infertility. A term referring to the reproductive sterility of otherwise vigorous hybrids. [36]

Hybridization. (1) An extreme example of gene flow in which interbreeding occurs between members of different species. [34] (2) The fusing of cells from different species in the laboratory. [18]

Hybrid unfitness. A term referring to the feebleness (and sterility) of some hybrids relative to their parents. Opposite of *hybrid vigor*. [36]

Hydrogen bond. A weak chemical bond between a hydrogen atom bonded to an atom in one molecule and another nearby atom in the same or another molecule. The strongest hydrogen bonds form with oxygen, nitrogen, and fluorine. Important in proteins, nucleic acids, water, and other biomolecules. [3]

Hydrolase. An enzyme that catalyzes the rupture of covalent chemical bonds by hydrolysis. [8,26]

Hydrologic cycle. The solar-powered circulation of water molecules over the Earth and its atmosphere. Also called *water cycle*. [46]

Hydrolysis. Rupture of a covalent bond by the addition of components of a water molecule (—H and —OH) to the cleaved ends of the broken bond. [4]

Hydrophilic. A molecule that readily dissolves in water. Polar molecules are hydrophilic. Compare with *hydrophobic*. [3]

Hydrophobic. A molecule that does not dissolve in water. Nonpolar molecules (such as hydrocarbons) are hydrophobic. Compare with *hydrophilic*. [3,6]

Hydrosphere. The water portion of the Earth, as opposed to the solid part (lithosphere) or gaseous envelope (atmosphere). [1]

Hydrostatic pressure. The pressure at a point in a fluid due to the weight of the fluid above it. The pressure on the walls of a container from the fluid contained within it. See *pressure potential*. [24,31]

Hydrostatic skeleton. The maintenance by hydrostatic pressure of body shape in many unicellular organisms and invertebrates. [6,24,31]

Hydrotropism. Growth of a plant toward or away from water. [32]

Hydrozoan. A member of the coelenterate class Hydrozoa; includes hydra, Portuguese man-of-war, etc. [42]

Hymenopteran. A member of the insect order Hymenoptera, such as a wasp, bee, or ant. [42]

Hyperpolarization. An increase in charge separation across the membrane of a neuron, which makes the resting potential more negative. Generally makes the neuron less excitable. Compare with *depolarization*. [29]

Hypertonic. Having a higher concentration of solutes, and therefore a lower water concentration, than another solution. Compare with *hypotonic* and *isotonic*. [6]

Hypha. Structural unit of most fungi. A threadlike filament that may be multinucleate (zygomycetes, ascomycetes) or multicellular (basidiomycetes). [21,40]

Hypocotyl. The portion of the seedling stem below the cotyledons. [21]

Hypolimnion. The lower level of water in a lake. Has a uniform temperature and is cooler than the epilimnion layer. [48]

Hypophysis. See *pituitary*. [28]

Hypothalamus. A region of the vertebrate brain under the thalamus which regulates autonomic and endocrine functions, and drives such as hunger. [28,30]

Hypothesis. A proposition whose truth can be validated only by observation and experiment. [2]

Hypotonic. Having a lower concentration of solutes, and therefore a higher water concentration than some reference solution. Compare with *hypertonic* and *isotonic*. [6]

Identical twins. See *monozygotic twins*. [19]

Ileum. A segment of the lower small intestine. Site of food and water absorption. Compare with *duodenum* and *jejunum*. [26]

Imaginal disks. Clusters of cells in insect larvae that later give rise to specific adult structures. [20]

Imago. An adult insect that has undergone metamorphosis. [21]

Imbibition. Diffusion of water molecules into a dense material, such as gelatin, that causes the material to swell. [6]

Immune system. A defensive system that

recognizes foreign substances, cells, or organisms entering the body and attacks them with antibodies, macrophages, or other agents of cell-mediated immunity. [23]

Immunity. The condition of diminished susceptibility to an infectious agent or toxin. [23]

Immunoelectrophoresis. An analytical method that separates proteins in an electric field (electrophoresis) and then identifies them by exposure to antibodies that precipitate only specific antigens. [23]

Immunoglobulin (Ig). An antibody protein. [23]

Immunological memory. The production of a stronger and prompter immune response following a second exposure to an antigen. [23]

Immunological tolerance. Failure of the immune system to mount immune responses to antigenetic determinants of its own body components ("self"). [23]

Imperfect flower. A flower in which either stamens or carpels are lacking. [41]

Impermeable junction. See *tight junction*. [7]

Implantation. In mammalian embryonic development, the attachment of the early embryo to the wall of the uterus. [19]

Imprinting. Species-specific, rapid learning during a critical period early in life, in which social attachments are established. [32]

Inclusive fitness. Evolutionary fitness that takes into account not only a given organism's reproductive success, but also its influence on the success of closely related kin. [33]

Incomplete dominance. The ability of two alleles to produce a heterozygous phenotype differing from either homozygous phenotype. Compare with *condominance*. [13]

Individual selection. Natural selection favoring genes that increase the survival of the individual carrying them. Compare with *kin selection* and *group selection*. [34]

Indoleacetic acid (IAA). An auxin. A plant hormone that controls growth. [28]

Inducer. (1) A small molecule which, when added to a growth medium, sharply increases the level of some enzymes, usually by binding to repressor and changing its conformation so that it does not bind to the operator, thus increasing enzyme synthesis. [8,18] (2) In embryogenesis, a substance or cell group that causes another cell group to differentiate in a certain way. [20]

Inducible enzyme. An enzyme whose synthesis is regulated by changing levels of small molecule inducers (e.g., enzyme substrates). Compare with *constitute* and *repressible enzyme*. [8,18]

Induction. (1) Increased transcription of certain genes triggered by the presence of molecules (inducers) resembling the substrate of the enzyme. [8,18] (2) The influence of one cell group (inducer) over a neighboring cell group during embryonic development. [20]

Industrial melanism. An adaptive change in the coloration of a species that better matches it to an environment darkened by industrial smoke. [35]

Inflammation. A tissue response to injury or infection that usually causes local pain, redness, swelling, and heat. [22]

Informational macromolecule. A polymer, such as a protein or nucleic acid, that consists of a nonrepetitive, nonrandom sequence of monomeric units with potential informational content. [4]

Inhibitory postsynaptic potential (IPSP). A transient hyperpolarization in a postsynaptic neuron that is induced by neurotransmitter release from the presynaptic neuron. [29]

Inhibitory synapse. A synapse in which neurotransmitter released from the presynaptic neuron causes hyperpolarization in the postsynaptic neuron. Compare with *excitatory synapse*. [29]

Initiation factor. A protein that regulates initiation of the translation of an mRNA molecule. [17]

Initiator codon. AUG (for the amino acid methionine) and the nucleotide triplet at the start of an mRNA molecule. [17]

Innate behavior. An unlearned instinctive pattern of animal activity which is partly or wholly inherited. Compare with *learned behavior*. [32]

Inositol 1,4,5-triphosphate (IP$_3$). A second messenger released from phosphatidyl inositol in cell membranes. [28]

Insect. A member of a large class of arthropods, Insecta, which probably includes more than half of all living species. [42]

Insight learning. Learning characterized by reasoning rather than trial and error. [32]

Insolation. The supply of incoming solar energy to an environment. [46]

Instantaneous speciation. The production of a single mutant individual that cannot interbreed with members of its parents' species, but can nonetheless survive and reproduce. [36]

Instar. In insect larval development, a stage between two molts. [14]

Instinct. An unlearned, stereotyped behavioral response characteristic of a species. [32]

Instructive theory of immunity. The old theory that antigens directly shape the antibodies made in response to them. Superceded by the clonal selection theory. [23]

Insulin. A hormone produced in pancreatic islet cells, which stimulates glucose metabolism and glycogen synthesis. Compare with *glucagon*. [28]

Integument. A protective surface covering. (1) The skin in animals. (2) In gymnosperms and angiosperms, a layer of tissue around the ovule that will become the seed coat. [41]

Intercalary meristem. Plant meristem that forms below the apical meristem between regions of mature or permanent meristem tissue—for example, at the base of a leaf or in the nodes of grass stems. [7]

Intercalated disk. Specialized regions of cardiac muscle cell membrane which allow close electrical contact between neighboring cells, thereby synchronizing heart contraction. [31]

Interdemic selection. Occurs when populations of one kind emerge or die out at a different rate from populations of another kind. Also called *group selection*. Compare with *species selection*. [34]

Intermediate filament. A class of filaments intermediate in size between actin filaments and microtubules that are part of the cytoskeleton. [5]

Interneuron. A neuron that is neither sensory nor motor, but is concerned with higher-level processing and integration. [29]

Interphase. The period of the cell cycle in which a cell is not undergoing mitosis or meiosis. [12]

Intersex. An abnormal animal with sexual characteristics intermediate between those of males and females. [15]

Intertidal zone. The environment comprising beach regions between high and low tide. [48]

Intertropical convergence zone. The area between 30° and 60° N and S latitude where the prevailing trade winds from the two hemispheres collide. [48]

Intrinsic factor. A protein, secreted by the stomach, which is necessary for intestinal absorption of cobalamin (vitamin B$_{12}$). [23,26]

Intrinsic protein. A membrane protein that is embedded in the phospholipid bilayer of the membrane. Compare with *extrinsic protein*. [6]

Intrinsic rate of increase. The difference between the birth and death rates of a population. [45]

Intrinsic system. One of two mechanisms triggering blood clotting. It is initiated by platelet factors and the presence of an injured blood vessel wall. Compare with *extrinsic system*. [22]

Introgression. Interbreeding between the offspring of a hybridization and a member of one or both of the parent species. [34,36]

Intron. An intervening DNA sequence in a gene that does not encode a protein and after splicing is not represented in mature mRNA. Compare with *exon*. [17]

Inversion. A chromosomal mutation caused by the breakage, reversal (by 180°), and reinsertion, in the opposite orientation, into the same chromosomal locus. [15]

Invertebrate. Any animal that is not a vertebrate. One with a nerve cord that is not enclosed in a backbone of vertebral segments. [42]

Ion. An atom or molecule which, through gain or loss of electrons, has acquired a net electric charge. [3]

Ion channel. Protein channel complex in a membrane that allows ions of a specific type to diffuse through the membrane. May be opened and closed by chemical or electrical factors. [6]

Ionic bond. A bond in which ions of opposite charges are held together by electrostatic attraction. [3]

Iris. The colored ring surrounding the pupil of the vertebrate eye. It contains muscles which control pupil diameter. [30]

Irritability. The capacity of a cell or organism to react to an environmental stimulus. [29]

Islets of Langerhans. Cell clusters in the pancreas that secrete insulin and glucagon. [28]

Isoantibody. An antibody to antigens found in other organisms of the same species. [23]

Isogamy. A mode of sexual reproduction in which there is fusion of two flagellated gametes of similar size and morphology. Compare with *oogamy*. [21]

Isolating mechanism. Any biological or physical factor that decreases interbreeding between different populations or species. [36]

Isomer. One of two or more molecules that have the same numbers and kinds of atoms but differ in their spatial arrangement and therefore in certain properties. Glucose and fructose are isomers. Compare with *stereoisomer*. [4]

Isometric contraction. Muscular contraction in which muscle tension remains constant as muscle length shortens. Compare with *isotonic contraction*. [31]

Isotonic. (1) Having the same osmotic potential as another solution—or the fluid phase of a cell or tissue. Compare with *hypertonic* and *hypotonic*. [6,25] (2) Having uniform tension, as in the fibers of a contracted muscle. [31]

Isotonic contraction. Muscular contraction in which muscle tension changes as muscle length remains constant. Compare with *isometric contraction*. [31]

Isotope. One of two or more atoms having the same atomic number but a different atomic mass. Because they are radioactive or differ in mass but not in chemical properties, they are useful labels for tracer molecules in metabolic experiments. [3,4,9,10]

Jejunum. In vertebrates, the second section of the small intestine; a major site of food absorption. Compare with *duodenum* and *ileum*. [26]

Juvenile hormone. A hormone secreted by the corpora allata of insects. It inhibits metamorphosis and sustains larval growth. [28]

Karyokinesis. Nuclear replication. Compare with *cytokinesis*. [12]

Karyotype. The number, forms, and types of chromosomes in a cell. Sometimes denotes a photograph displaying the karyotype. [12,15]

Keratin. A tough sulfur-containing protein that is a major constituent of skin, hair, nails, feathers, and horn. [7]

Ketose. A sugar that contains a keto group. Compare with *aldose*. [4]

Kidney. A paired organ of excretion and water balance in vertebrates. [25]

Killer T cell. A cytolytic T lymphocyte that

recognizes and destroys abnormal cells of various types. [23]

Kilocalorie (kcal). A unit equal to 1000 calories. See *Calorie*. [8,26]

Kin selection. Natural selection favoring related individuals which increases their number and the frequency of alleles shared through common ancestry. Compare with *individual selection* and *group selection*. [34,45]

Kinesis. (1) General term for any physical movement. (2) Indirect orientation movements that become more active as a stimulus increases. Because they are random and undirected, faster or slower movements bring the organism nearer or farther from the stimulus source. Compare with *taxis*. [32]

Kinetics. (1) In chemistry, the study of reaction rates. (2) In physics, the study of the effects of forces on the motions of bodies. [8]

Kinetochore. A small structure near the centromere of a chromosome, to which the spindle fibers attach during mitosis. [12]

Kinetosome. The basal body of a cilium or flagellum, which anchors it. Similar to a centriole. [5]

Kingdom. The largest of the taxonomic categories, it is made up of multiple phyla (divisions). In this book, we recognize five kingdoms. [37]

Kranz anatomy. A leaf structure found in plants with C_4 photosynthesis. [10]

Krebs cycle. See *citric acid cycle*. [9]

Krill. Planktonic crustaceans that comprise the diet of many whales. [43,47,48]

***K* strategist.** Species whose members produce few offspring, but which survive by superior competitive ability. Compare with r *strategist*. [34,45]

Labrum. The upper lip of insects and crustaceans. [42]

***Lac* operon.** A well-studied gene cluster in *E. coli* that includes the structural genes encoding the enzymes of lactose metabolism and regulatory genes determining their rate behavior. [18]

Lactation. Milk production. [21]

Lagging strand. The DNA strand being replicated in the 5' to 3' direction. New nucleotides are added in discontinuous segments (Okazaki fragments). Compare with *leading strand*. [16]

Lamarckism. A theory of evolution by acquired characteristics, propounded in 1809 by Jean Baptiste de Lamarck. [1,34]

Larva. Immature form of an animal that is distinguishable from the adult in form and lifestyle. Develops into an adult through metamorphosis. [19]

Latent period. The period between a stimulus and the response. [29]

Lateral geniculate nucleus. In the vertebrate brain, the part of the thalamus that receives visual information from the eyes. [30]

Lateral-line system. Pressure-sensitive grooves along the sides of aquatic

vertebrates. They may have been evolutionary precursors of the ear. [30]

Lateral meristem. See *cambium*. [7]

Laurasia. One of two supercontinents produced by the breakup of the ancient supercontinent Pangaea. [48]

Leading strand. The DNA strand being replicated in the 3' to 5' direction. New nucleotides are added consecutively to this template. Compare with *lagging strand*. [16]

Leaf. Appendage of a plant stem. [5,10,41]

Learned behavior. Behavior that is modifiable by experience. Compare with *innate behavior*. [32]

Learning. The adaptive modification of behavior in response to experience. [32]

Lenticels. Small porous areas on plant surfaces that facilitate gas exchange through the periderm. [27]

Leukocyte. A white blood cell. [22]

Leydig (interstitial) cell. Testis cells that secrete testosterone. [19]

Lichen. A thallus composed of interdependent algal and fungal components. [40]

Ligand. A molecule that is specifically bound by another, usually larger, molecule, e.g., a small molecule (hormone) that binds to cell surface protein receptors. [6,28]

Light chain. A subunit of an antibody protein. There are two light chains per molecule. Compare with *heavy chain*. [23]

Light compass reaction. A reaction of many invertebrates in which the angle between their direction of movement and a line toward the sun is kept constant. [32]

Light reactions. Stage of photosynthesis that depends on light, i.e., the excitation of chlorophyll and the chemiosmotic synthesis of ATP. [10]

Lignin. The main noncarbohydrate component of wood. A complex polymer composed of higher alcohols, which with cellulose stiffens plant cell walls, especially in the secondary xylem. [5]

Limits of tolerance. In ecology, the range of a variable (e.g., temperature, moisture) within which a species can survive. [47]

Limnetic zone. An environment in a lake which comprises the upper regions of open water. Compare with *profundal zone*. [48]

Lingula. An extinct Brachiopod, abundant during the Cambrian period. [44]

Linkage. Lack of independent segregation. The tendency of two or more genes on a chromosome to segregate together during meiosis. [14]

Linkage group. A group of genes known to be physically associated on the same chromosome. [14]

Linkage map. A representation of the relative locations of genes on a chromosome. Distances between genes are estimated using the frequency of recombination between them. [14]

Linker DNA. DNA between nucleosomes in chromatin. [12]

Lipase. An enzyme that catalyzes the hydrolysis of fats into component fatty acids and glycerol. [26]

Lipid. A class of nonpolar organic molecules that are soluble in organic solvents and insoluble in water. Includes fats, phospholipids, waxes, and steroids. [4]

Liposome. Artificially produced membrane vesicle. [5,6]

Lithosphere. The solid crust of the Earth's surface. [1]

Littoral zone. A biogeographic aquatic environment between high- and low-water marks. Ranges from the upper limits of tidal action to the depths where the water is no longer stirred by wave action. [48]

Liver. In vertebrates, a large organ that stores glycogen, removes harmful substances from blood, secretes bile, and processes absorbed nutrients reaching it from the small intestine. [26]

Locus. A fixed location on a chromosome or a strand of DNA where a gene or one of its alleles is located. [13,14]

Logistic growth. The gradually slowing increase in a variable (e.g., population size) as to function of time. Compare with *exponential growth.* [45]

Loop of Henle. In the kidneys of birds and mammals, the hairpin-shaped portion of the nephron between the proximal and distal convoluted tubule segments. Site of water and salt reabsorption. [25]

Lophophore. Food-gathering, funnellike organ surrounding the mouth and bearing tentacles or filaments. [42]

Luciferase. An enzyme that produces bioluminescence by oxidizing luciferin. Found in fireflies and other organisms. [31]

Luciferin. A molecule that emits light when oxidized by the enzyme luciferase. [31]

Lumen. The hollow inner part of a tubular structure. [26]

Luminescence. See *bioluminescence.* [31]

Lung. An organ of air-breathing. In vertebrates, a cavity into which air is pumped. In mollusks, a part of the mantle. [27]

Luteinizing hormone (LH). A gonadotropic hormone, made by the anterior pituitary, which stimulates ovulation in the female, and the secretion of testosterone in the male. [21,28]

Luteotropic hormone (LTH). See *prolactin.* [21,28]

Lycopod. A member of the plant division Lycopodiophyta (club mosses). A once dominant group now represented only by relicts; they are vascular plants with true stems and leaves. [41]

Lymph. Clear fluid formed as a filtrate of blood. Collected by lymphatic capillaries and returned to the bloodstream via the lymphatic duct. [24]

Lymphatic capillary. Thin-walled vessel of the lymphatic system. [24]

Lymphatic duct. A large vessel of the lymphatic system that carries lymph into the great veins near the heart. [24]

Lymphatic system. A network of ducts throughout the body that carries lymph. [23,24]

Lymph heart. Small contractile organ that drives lymph circulation in amphibians, reptiles, and birds. [24]

Lymph node. An aggregation of lymphoid tissue within a fibrous capsule. Found along the course of lymphatic vessels. [24]

Lymphocyte. A large class of white blood cells important in the immune system. It includes B cells, which make the antibodies of humoral immunity, and T cells, which are partly responsible for cell-mediated immunity. [22,23]

Lymphoid cell. A lymphocyte or one of its precursors. [22,23]

Lymphoid tissue. A tissue rich in lymphoid cells. Includes lymph nodes, spleen, thymus, and scattered patches of tissue (tonsils, etc.) [23]

Lymphokine. A class of peptide growth factors made by T cells, which control immunity by regulating the activity of macrophages and other lymphocytes. Compare with *monokine.* [23]

Lyon hypothesis. The proposition that each cell in a mammalian female randomly and permanently inactivates one of its X-chromosomes. [15]

Lysin. A substance that binds to a cell and distintegrates it. [23]

Lysis. Disintegration of a cell by rupture of its membrane. [23]

Lysogeny. A phenomenon in which a bacterium is infected with a dormant virus (prophage) that remains inactive but has the potential of replicating and thus of lysing the bacterium. [16]

Lysosome. A membrane-bound organelle of eukaryotic cells containing hydrolytic enzymes that participate in intracellular digestion. [5]

Macroevolution. Large, long-term genetic changes in populations, leading to the development of new species. Compare with *microevolution.* [34,36]

Macromolecule. A large molecule (e.g., a nucleic acid or protein) made up of many smaller molecules joined by covalent bonds. [4,5]

Macronucleus. Larger of the two types of nuclei in ciliates. Site of mRNA synthesis. Required for asexual division. Compare with *micronucleus.* [39]

Macronutrient. An element needed by an organism in large amounts for normal growth and development. Compare with *micronutrient.* [26]

Macrophage. A large tissue cell that can engulf foreign particles by phagocytosis and produce monokines. It derives from the blood monocyte. [22]

Macrospore. See *megaspore.* [41]

Macula. A spot. (1) In the retina, the spot below the entrance of the optic nerve. (2) In the utricle and saccule of the inner ear, the area containing hair cells. [30]

Major histocompatibility complex (MHC). A complicated system of polymorphic genes encoding combinations of cell surface glycoproteins unique to individuals. Permits self-recognition, thus prompting graft rejection and other regulatory actions of the immune system. See *HLA.* [23,34]

Maladaptive character. A trait that decreases an organism's fitness. Compare with *neutral* and *nonadaptive character.* [35]

Malpighian corpuscle. The filtering unit of a kidney. A glomerulus and its Bowman's capsule. [25]

Malpighian tubules. Blind tubules (protonephridium) in arthropods that carry wastes from the coelom to the intestines. [25]

Maltase. An enzyme that hydrolyzes the disaccharide maltose into its component glucose molecules. [26]

Mammal. An animal of the class Mammalia. [43]

Mammalia. A class of vertebrates, possessing hair and milk glands; includes the monotreme, marsupial, and placental subclasses. [43]

Mammary gland. Glands within the mammalian breast that produce milk. [21]

Mandibulata. A class of arthropods in which mandibles and antennae serve as sensory structures; includes crabs, lobsters, shrimp, etc. [42]

Mantle. Soft part of the body wall of mollusks that secretes the shell, covers the visceral mass, and encloses the mantle cavity. [42]

Marine. Pertaining to or living in the ocean. Compare with *aquatic* and *terrestrial.* [46,48]

Marsupial. A mammal with no placenta belonging to the subclass Metatheria, such as the opossum and many Australian mammals. Most have a pouch (marsupium), containing milk glands, that serves as a container for the young. [43]

Mast cell. A tissue cell that releases histamine as a response to injury or allergic reaction. [22]

Maternal effect. The influence on a zygote's phenotype of factors, such as mitochondria or choroplasts, that are inherited from the female parent through the cytoplasm of the female gamete. [18]

Matrix. (1) In mitochondria, the material in the interior of the organelle. [5,9] (2) The intercellular substance of a tissue. [5]

Mature RNA. RNA that has undergone post-transcriptional processing. [17]

Mean. The arithmetic average of all recorded values. [34]

Mechanical isolation. A prezygotic isolating mechanism related to anatomical differences in sex organs. Compare with *behavioral, ecological, physiological,* and *temporal isolation.* [36]

Mechanism. (1) A term for the way a system works. (2) The universally accepted view life can be explained in terms of physical and chemical laws. Compare with *vitalism.* [2]

Mechanoreceptor. A sensory structure, such as the Pacinian corpuscles in the skin, that converts touch or pressure stimuli into nerve impulses. [30]

Medulla. Inner portion of a gland or other structure that is surrounded by cortex. Compare with *cortex.* [25,28,30]

Medulla oblongata. The most posterior part of the vertebrate hindbrain, before it joins the spinal cord. It controls basic functions such as heart rate and breathing. [30]

Medusa. A broad, bell-shaped, free-swimming sexual stage in the life-cycle of many coelenterates. May alternate with polyp stage. [42]

Megakaryocyte. Precursor of blood platelets. It is found in bone marrow and other hematopoietic tissues. [22]

Megaphyll. A large, broad leaf with a complex branched vein system containing several vascular channels that connect with stem vascular tissue. Compare with *microphyll*. [41]

Megasporangium. Megaspore-producing, multicellular structure in vascular plants. See *nucellus*. [21,41]

Megaspore (also called *macrospore*). In heterosporous plants, a haploid spore that develops into a female gametophyte. Usually larger than a microspore. [21,41]

Meiosis. A special form of one or two nuclear divisions that reduces the chromosome number by half (in meiosis I, the reductional division) and generates four haploid gametes from a diploid cell (in meiosis II, the equational division). Compare with *mitosis*. [12]

Melanocyte-stimulating hormone (MSH). A peptide hormone secreted by the anterior pituitary in mammals and the rhomboid fossa in lower vertebrates that causes melanin deposition in skin. [28]

Melanophore. A cell that controls skin pigmentation in fish, amphibians, and reptiles. [28]

Melatonin. Hormone secreted by the vertebrate pineal gland. It helps regulate circadian and seasonal metabolic rhythms. [28]

Membrane channel. Protein pores in the cell membrane through which specific molecules or ions can pass. Some channels are regulated by voltage or by ion concentrations. [29]

Membrane pump. A protein complex in a cellular membrane that actively transports ions or molecules across the membrane, thereby establishing a concentration gradient across the membrane. [6]

Mendelian population. See *deme*. [34,45]

Mendelian ratio. The ratio of phenotypes among progeny predicted to occur in a hybrid cross. [14]

Menstrual cycle. The human ovarian cycle. [21]

Meristem. Undifferentiated plant tissue containing actively dividing cells. The site of plant growth. [7]

Meroblastic cleavage. Embryonic cleavage dividing only part of the zygote. [19]

Mesenchyme. Embryonic or unspecialized cells derived from the mesoderm. [19]

Mesoderm. The middle of the three embryonic cell layers that is formed in gastrulation. Gives rise to such structures as connective tissue, the heart, and kidneys. [19]

Mesoglea. Gelatinous noncellular support material that maintains the shape of jellyfishes and other coelenterates. [31]

Mesokaryote. Term applied to forms supposedly intermediate between prokaryotes and eukaryotes. [39]

Mesonephros. One of the middle of three

pairs of embryonic kidney structures of vertebrates. It persists in adult fishes and is replaced by the metanephros in higher forms. See *pronephros*. [25]

Mesophyll. The largely photosynthetic tissue of a leaf between the upper and lower epidermis. [10]

Mesozoan. A member of the animal phylum Mesozoa; includes tiny marine organisms thought by some to represent a subkingdom. [42]

Mesozoic era. Covers the time from 250 to 65 million years ago; includes three periods and is often called the Age of Reptiles. [37]

Messenger RNA (mRNA). An RNA transcript of one DNA strand (the sense strand) that encodes a polypeptide and is translated by a ribosome. [17]

Metabolic pathway. See *enzymatic pathway*. [8]

Metabolism. All of the chemical transformations in an organism, including those of anabolism, catabolism, respiration, and photosynthesis. [26]

Metamere. A body segment in certain invertebrates. [20]

Metamorphosis. The change from larval to adult form in the development of certain species. [21]

Metanephros. The last of three embryonic renal structures in amniote vertebrates. It replaces the mesonephros and persists as the adult kidney in adult reptiles, birds, and mammals. [25]

Metaphase. A stage of mitosis and meiosis in which the chromosomes line up on the equatorial plane of the cell. [12]

Metazoan. A multicellular animal; excludes protists. [39]

Methanogen. Any of several bacterial groups that release methane as a metabolic product; includes the archaebacteria. [38,44]

Methemoglobin. Oxidized hemoglobin containing ferric (Fe^{3+}) iron, which is therefore incapable of binding oxygen. [22]

Microevolution. Small changes in a population's gene pool that do not lead to speciation. Compare with *macroevolution*. [34,36]

Microfilament. In cytoplasm, threadlike polymers of actin molecules that play a role in cellular motility. [5]

Microgametophyte. In heterosporous plants, the male gametophyte. [41]

Micronucleus. Smaller of two types of nuclei in ciliates. Participates in meiosis and autogamy but not asexual division. Compare with *macronucleus*. [39]

Micronutrient. An inorganic chemical, such as zinc or copper, needed by organisms in trace amounts. Compare with *macronutrient*. [26]

Microphyll. A small leaf with a single vascular channel that is an offshoot of stem vascular tissue. Typical of lycopods and sphenophytes. Compare with *megaphyll*. [41]

Micropyle. Flower part. The opening in an angiosperm ovule through which the pollen tube enters during fertilization. [21]

Microsome. Small vesicles produced from disrupted endoplasmic reticulum. [5]

Microsporangium. A plant sporangium that produces microspores. [21,41]

Microspore. A haploid spore that develops into a male gametophyte. In seed plants, a microspore is a pollen grain. See *heterosporous*. Compare with *megaspore*. [21,41]

Microsporophyll. A leaflike appendage that bears microsporangia. [41]

Microtrabecular lattice. A system of fiber (6 nm in diameter) forming a meshwork that connects and supports organelles in a eukaryotic cell. Part of the cytoskeleton. [5]

Microtubule. A tubular structure, a polymer of the protein tubulin, that is found in cilia, flagella, spindle fibers, and cytoskeleton. [5]

Microvillus (pl. **microvilli**). One of many hairlike extensions of the plasma membrane of a cell that serves to increase surface area. Compare with *villus*. [7,26]

Midbrain. Area of the vertebrae brain below the thalamus and above the pons. It contains the tectum and cerebral peduncles. [30]

Middle lamella. The layer of intercellular material that cements together the walls of adjacent plant cells. [5]

Mimicry. The adaptive similarity in the appearance or behavior of one species to another, which deters predators or attracts prey. See *Batesian* and *Müllerian mimicry*. [35]

Mineral. Any inorganic substance (other than water). [3]

Minimal medium. A material containing only the minimal number of nutrients that can support the growth in culture of wild-type cells but not cells with enzyme deficiencies (which require enriched medium). [16]

Missense mutation. A change in the genetic code that leads to synthesis of a protein containing an incorrect amino acid. Compare with *nonsense mutation*. [17]

Mitochondrion. An organelle of eukaryotic cells that contains in its complex membrane system the enzymes of the citric acid cycle, respiratory chain, and oxidative phosphorylation. Possesses its own DNA. [5,9]

Mitosis. Nuclear division in eukaryotic cells, that produces two identical daughter nuclei. Usually followed by cell division (cytokinesis). Compare with *meiosis*. [12]

Mode. In frequency distribution curves the class or value that is most frequent. [34]

Moderately repetitive DNA. A class of DNA containing long sequences of several types that are imperfectly repeated up to 100 times. Compare with *highly repetitive DNA*. [17]

Modifier gene. A gene whose alleles control the degree to which alleles of another gene are expressed. [14]

Mole. A quantity of a compound whose weight in grams is numerically equal to its molecular weight expressed in atomic mass units. Avogadro's number of molecules: 6.023×10^{23} molecules. [3,8]

Molecular evolution. An approach to evolutionary lineages based on homologies and differences in molecular structures. [35]

Molecule. Two or more atoms joined by covalent or ionic bonds. [3]

Mollusk. A member of the animal phylum Mollusca; includes snails, bivalves, cephalopods, etc. [42]

Molting. The shedding of all or part of an organism's outer covering. [21]

Monera. One of the five kingdoms; comprised of prokaryotes and includes bacteria, cyanobacteria (blue-green algae), and prochloronts. [37,38]

Moneran. A member of the kingdom Monera. A bacterium. [38]

Monochromatic light. Light of a single wavelength (and therefore color). [10]

Monocistronic mRNA. An mRNA molecule that encodes a single polypeptide. Compare with *polycistronic mRNA*. [17]

Monoclimax theory. The theory that for each climatic region there can be only one type of stable climax vegetation. Compare with *polyclimax theory*. See *sere*. [47]

Monoclonal antibody. A pure, homogeneous antibody made in cell culture from a single clone of antibody-producing cells. [23]

Monocot (monocotyledon). A member of one of two angiosperm classes. Flowering plants with embryos bearing only one cotyledon (seed leaf) having parallel-veined leaves, 3-petaled flowers, and scattered vascular bundles; includes palms, lilies, irises, grasses, etc. Compare with *dicot*. [24,41]

Monocyte. A white blood cell that engulfs foreign particles. Becomes a macrophage when it moves into tissues. [22,23]

Monoecious. A plant (e.g., corn) in which the male and female organs are found on the same plant but in separate flowers. Compare with *dioecious*. [21,41]

Monogamy. Having a single mate. Compare with *polygamy*. [33]

Monohybrid cross. A genetic cross between two individuals that are identically heterozygous for only a one gene pair, e.g., $Cc \times Cc$. Compare with *dihybrid cross*. [13]

Monokine. A class of polypeptide growth factors produced by monocytes and macrophages that mediate the activity of other cells. Compare with *lymphokine*. [23]

Monomer. A molecule that is the building block of a polymer. [4]

Monoplacophoran. Members of the extinct mollusk class Monoplacophora; bilaterally symmetrical segmented mollusks with single shells. The famous example is Neophilina. [42,44]

Monosaccharide. A simple sugar, such as glucose or ribose. [4]

Monotreme. A primitive, egg-laying mammal, such as the duck-billed platypus and spiny anteater. [43]

Monozygotic twins. Identical twins. Develop from the same zygote and thus have identical genes. Compare with *dizygotic twins*. [19]

Morphogenesis. The development of the form or structure of an organism. [19]

Morphogenetic movement. Cell and tissue migrations during embryogenesis that help to form the shape and structure of the adult. [20]

Morphology. The form and structure of an organism or any of its parts. [3]

Morula. An early stage of animal embryonic development. The solid mass of cells produced by cleavage of the zygote. [19]

Mosaic. (1) An organism whose cells have different genotypes. [15] (2) A chimeric animal created from the fusion of two or more embryos. See *chimera*. [20]

Mosaic evolution. The occurrence in an early member of a toxonomic group of a mosaic of traits, some characteristic of the ancestral group and some of the newly emerging group. [44]

Motor area. An area of the cerebral cortex that directly controls movement. Compare with *sensory area* and *association area*. [30]

Motor endplate. A neuromuscular junction. A synapse between a motor neuron and a muscle fiber it controls. [29]

Motor neuron. A neuron from the central nervous system that synapses on a muscle cell and controls its activity. [29]

Motor program. Behaviors, such as walking, that are virtually independent of sensory feedback and require little practice to develop normally. [32]

Motor unit. The unit of contraction of a muscle; a single motor neuron and the muscle fibers it innervates. [31]

mRNA. See *messenger RNA*. [17]

Müllerian mimicry. A form of warning coloration in which one dangerous species imitates the appearance of another. Compare with *Batesian mimicry*. [35]

Multipotent cell. See *pluripotent cell*. [19]

Mutant. (1) An organism produced by mutation. (2) A mutated gene. [14]

Mutation. A permanent, transmissible change in the genetic material. [14]

Mutational equilibrium. A condition reached when the rates of forward and back mutation of a given gene are equal. [34]

Mutation pressure. The occurrence at a certain rate of both gene and chromosome mutations. [34]

Mutualism. A symbiotic relationship that is beneficial to both participating species. See *symbiosis*. Compare with *commensalism* and *parasitism*. [47]

Mycelium. A mass of hyphae in a fungus. [21,40]

Mycobiont. The fungal partner in a lichen. See *phycobiont*. [40]

Mycorrhiza. Symbiotic association of a fungus and the roots of a vascular plant. See *ectomycorrhiza* and *endomycorrhiza*. [40]

Myelin. A fatty material forming a sheath around the axons of some neurons. It insulates them electrically and speeds their conduction of action potentials. [29]

Myeloid cell. A white blood cell arising in the bone marrow. [22]

Myocardial infarction. A heart attack in which heart muscle is damaged following interruption of local blood flow. [24]

Myofiber (muscle fiber). A vertebrate muscle cell. Often multinucleated, it contains myofibrils. [31]

Myofibril. Contractile unit of a myofiber. It consists of repeating sarcomeres containing overlapping actin and myosin filaments. [31]

Myoglobin. A hemoglobinlike pigment that helps vertebrate muscles store oxygen. [22,27]

Myosin. The protein that forms the thick filaments in muscle fibers. It interacts with thin actin fibers to produce muscle contraction. [5,7,31]

NAD (nicotinamide adenine dinucleotide). A compound found in all living cells. It has an oxidized form (NAD^+) and a reduced form (NADH). Serves as a coenzyme of oxidation-reduction enzymes. [8,9,10]

NADP (nicotinamide adenine dinucleotide phosphate). Resembles NAD, but has another phosphate group. Plays similar roles but with different enzymes. [8,9,10]

Natural selection. Change in allele frequencies in a population that is caused by differential fitness of different genotypes. The mechanism of evolution proposed by Darwin. [34]

Negative pressure breathing. The breathing mechanism of reptiles, birds, and mammals. The lung cavity expands, sucking in fresh air. Compare with *positive pressure breathing*. [27]

Nekton. Aquatic animals that can swim independent of water currents. Compare with *plankton*. [48]

Nematocyst. Capsule containing a poisonous harpoonlike projectile used by some coelenterates for predation or defense. See *cnidoblast*. [31]

Nematode. A member of the phylum Nematoda; a roundworm. [20,42]

Neoplasm. A tumor. A new and abnormal tissue growth in which cell multiplication is uncontrolled. May be benign or malignant. [5]

Neoteny. A form of paedomorphosis in which body development is retarded with respect to reproductive maturation. Results in sexually mature organisms with juvenile or larval characteristics. [20]

Nephridium. A tubular excretory organ in invertebrates. [25]

Nephron. The functional unit of the kidney, comprising a glomerulus, Bowman's capsule, and a tubule. It filters and resorbs material from blood. [25]

Nephrostome. Ciliated, funnel-shaped opening of a nephridium into the coelom in invertebrates. [25]

Neritic zone. A marine environment comprising the shallow open waters from the shore to the edge of the continental shelf. Compare with *oceanic zone*. [48]

Nerve. A bundle of nerve fibers (thin extensions of neurons) that carry neural impulses in the peripheral nervous system. Compare with *tract*. [29]

Nerve cord. A strand of nerve cells and fibers that forms part of the central nervous system of insects and some worms. [30]

Nerve impulse. An action potential. The

electrochemical signal that travels along neurons. [29]

Nerve net. The primitive nervous system of some invertebrates, in which neurons are evenly dispersed through the epithelium. It permits only a diffuse response to stimuli. [30]

Neural fold. Ridges on either side of the neural groove in a chordate embryo. [19]

Neural groove. A groove on the dorsal surface of a chordate embryo, under which the notochord forms. [19]

Neurohypophysis. The posterior pituitary. Compare with *adenohypophysis*. [25]

Neuron. A nerve cell. [7,29]

Neurotransmitter. A substance, produced in and released by a neuron, that diffuses across a synapse and excites or inhibits the postsynaptic neuron. [29]

Neurula. In embryonic development of a chordate, the stage at which the neural groove forms. [19]

Neurulation. The formation in the early embryo of a neural plate that subsequently forms a neural tube. [19]

Neutral character. A trait that does not improve or impair an organism's adaptedness. Also called *nonadaptive character*. [35]

Neutral pH. Refers to pH 7 at which a solution is neither acidic or basic. [3]

Neutrophil. A white blood cell that defends against infection by engulfing foreign particles by phagocytosis. [22]

Niche. The functional relationships of an organism or population to the environment it occupies. Compare with *habitat*. [46,47]

9 + 2 pattern. Cross-sectional pattern characteristic of cili and flagella, with an outer ring of nine microtubules surrounding two inner ones. [5]

Nitrification. The oxidation of ammonia to nitrite and nitrate ions, performed by certain soil bacteria. [46]

Nitrogen cycle. The worldwide circulation and reutilization of nitrogen. [46]

Nitrogen fixation. The incorporation of atmospheric nitrogen into chemical compounds of nitrogen (e.g., ammonia, nitrate) under the catalytic influence of the nitrogenase of certain microorganisms. [3,46]

Nitrogenous base. An aromatic, nitrogen-containing molecule, such as purine or pyrimidine, that will accept protons. [16]

Nociceptor. Free nerve endings in the skin that mediate the sensation of pain. [30]

Node of Ranvier. In a myelinated nerve, periodic regions without myelin. Each contains many sodium channels that help propagate action potentials to the next node. [29]

Nonadaptive character. See *neutral character*. Compare with *maladaptive character*. [35]

Noncompetitive inhibition. The inhibition of enzymatic activity by a molecule that binds at a site other than the active site of the enzyme. Compare with *competitive inhibition*. [8]

Noncyclic photophosphorylation. The photosynthetic light reaction of green plants that generates oxygen, ATP, and NADPH. Compare with *cyclic photophosphorylation*. [10]

Nondisjunction. An error in cell division that causes the chromosomal number to become abnormal. [12,15]

Nonpolar (hydrophobic) molecule. A molecule with no attraction to charged molecules, and therefore relatively insoluble in water. Fats and organic solvents are nonpolar. [3,6]

Nonrandom reproduction. Reproduction in which females are more likely to mate with some males than with others. Compare with *random reproduction*. [34]

Nonsense mutation. A change in the genetic code that blocks the translation of a gene. Compare with *missense mutation*. When such mutations change an amino acid codon to one of the codons (UAG, UAA, or UGA) signaling termination of translation, the resulting gene product is a shortened polypeptide that begins normally at the N-terminal end and ends at the altered codon. [17]

Normal curve. A Gaussian distribution. A bell-shaped frequency distribution characterized by a mean and a standard deviation. [34]

Notochord. A flexible dorsal rod of cartilage in embryonic chordates. Serves as the internal skeleton and is replaced by the vertebral column in most adult chordates. [19,43]

Nucellus. Tissue making up the main part of a plant ovule in which the embryo sac develops. Equivalent to a megasporangium. [21]

Nuclear envelope. The double membrane that surrounds the cell nucleus, the outer layer of which is usually continuous with the ER. [5]

Nuclear pore complex. An opening in the nuclear envelope through which large molecules may travel between the nucleus and the cytoplasm. [5]

Nuclease. An enzyme that hydrolyzes nucleic acids such as DNA and RNA into their component bases. [26]

Nucleic acid. A linear polymer of nucleotides whose nitrogenous bases (adenine, thymine or uracil, cytosine, and guanine) are attached to a "backbone" of alternating sugar and phosphate groups, e.g., DNA or RNA. [4,16]

Nucleoid. DNA-containing structure of prokaryotic cells. Not bounded by a membrane. [5,12]

Nucleolus. A structure within the cell nucleus in which ribosomal RNA is actively synthesized. [5]

Nucoeprotein. A complex of protein and nucleic acid. [16]

Nucleosome. A structure found in chromatin in which DNA is wound around histones, forming a "beads-on-a-string" arrangement. [12]

Nucleotide. The building blocks of DNA, RNA, and many important metabolic cofactors (e.g., ATP, NAD, etc.). Each nucleotide contains a nitrogenous base, a sugar, and a phosphate. [4,16]

Nucleus. (1) The dense central core of an atom, made up of protons and neutrons. (2) A membrane-bound organelle that contains the chromosomes in eukaryotes and is the site of DNA transcription. (3) A group of nerve cells in the central nervous system, usually contributing fibers to a particular nerve. [5]

Numerical phenetics. A method of biological classification based on the number of similar characters, which are given equal weight. Compare with *cladistics* and *evolutionary classification*. [37]

Nymph. A sexually immature and usually wingless stage of some insects that undergo simple metamorphosis. [21]

Obligate anaerobe. An anaerobe that can survive and grow only in the absence of free oxygen. Compare with *facultative anaerobe*. [9]

Oceanic zone. A marine environment comprising the deep waters between continents. Compare with *neritic zone*. [48]

Okazaki fragment. The discontinuous stretches of new DNA that are added to the lagging strand during DNA replication. [16]

Olfaction. The sense of smell. [30]

Olfactory membrane. In the nose, the site of odor-sensitive chemoreceptor cells. [30]

Omasum. The third chamber of the ruminant stomach. [26]

Ommatidium. The functional unit of the compound eye of arthropods. Each contains a lens and light-sensitive cells. [30]

Omnivore. An organism that eats both animal and plant material. Compare with *carnivore* and *herbivore*. [26]

Oncogene. A gene that can cause cancerous growth of the cells in which it is transcribed. Compare with *proto-oncogene*. [18]

One gene, one enzyme hypothesis. The view that each gene encodes an enzyme (or a single polypeptide) and that gene mutations are manifested by altered or deleted enzyme proteins. [16]

Ontogeny. The development of an individual of a species. Compare with *phylogeny*. [19]

Oocyte. The cell in female gametogenesis that undergoes meiosis to generate an egg cell (ovum). [12]

Oogamy. A mode of sexual reproduction in which gametes differentiate into motile sperm and large eggs. Compare with *isogamy*. [21]

Oogenesis. The process of egg formation and maturation in the ovary. [12,19]

Oogonium (pl. **oogonia**). (1) Unicellular egg-producing organ in algae, fungi, and plants. Compare with *antheridium*. [39,40] (2) Descendant of a primary germ cell that develops into an oocyte. [19]

Oomycetes. A phylum of the kingdom Fungi (water molds). [40]

Oospore. A thick-walled sexual spore that is characterisitc of the fungal phylum Oomycetes. [40]

Open circulatory system. In animals such as mollusks, a system of blood or fluid flow that is not contained in closed vessels. Compare with *closed circulatory system*. [24]

Open system. In thermodynamics, a system that exchanges matter or energy with its surroundings. Compare with *closed system*. [8]

Operant conditioning. Learning in which an animal increases the frequency of behaviors that produce a desired reward. Compare with *classical conditioning*. [32]

Operator. A site of gene regulation. A DNA sequence that binds repressor and thereby regulates transcription of nearby structural genes. [18]

Operculum. A lid. In fish, the gill covering. [27]

Operon. A cluster of genes, common in prokaryotes, that share the same promoter and operator regions. They are transcribed as a single mRNA molecule. [18]

Opportunity cost. The benefit an organism possessing a certain trait loses by the trait's preventing it from possessing other adaptive traits. Compare *energetic cost* and *risk cost*. [35]

Opsin. The protein component of the light-sensitive pigment rhodopsin. [30]

Opsonin. Proteins that bind to foreign particles in tissue and increase their chance of being engulfed by white blood cells. [22]

Order. A taxonomic category lying between class and family. Thus an order includes multiple families. [37]

Organ. Structure in a multicellular organism, such as heart, liver, brain, root, or leaf, that is specialized to perform one or more particular functions. Composed of different tissues. [7]

Organelle. Intracellular structure with specialized structure and function, e.g., mitochondria, lysosomes, etc. [5]

Organic. (1) Pertaining to any aspect of living matter, e.g., to its evolution, structure, or chemistry. (2) Any molecule containing carbon. [3]

Organizer. A region in an early embryo that induces the differentiation of nearby embryonic tissues. [20]

Organ of Corti. Part of the cochlea in the vertebrate ear. It contains the hair cells that transduce sound vibrations into nerve impulses. [30]

Organogenesis. Formation in the early embryo of the major organ systems. [19]

Osculum. The opening or mouth at the top of a sponge through which water is ejected. [42]

Osmoconformer. An organism that has the same solute concentration as its environment. Compare with *osmoregulator*. [25]

Osmoreceptor. Cells that are sensitive to small changes in surrounding osmolality. In mammals they include neurons in the hypothalamus. [25]

Osmoregulator. An organism that maintains an internal solute concentration independent of its environment. Compare with *osmoconformer*. [25]

Osmosis. The diffusion of water molecules across a semi-permeable membrane that allows the flow of water but not of solutes. If pressure is equal on both sides of the membrane, net water movement is toward the side of higher solute concentration. [6,22]

Osmotic potential (also called *osmotic pressure*). A measure of the lowering of water potential by solutes. Equivalent to the external pressure necessary to stop the flow of water molecules across a semi-permeable membrane from a solution of low solute concentration (and high water concentration) to one of high solute concentration. It is therefore an index of the difference in solute concentrations. [6,22]

Ossicle. (1) One of three small bones that amplify sound vibrations passing from the eardrum into the cochlea of the mammalian ear. [30] (2) Distinctive hard structures that form the exoskeleton of echinoderms. [31,42]

Osteichthyes. A class of vertebrates including the bony fishes. [43]

Otoconia. In vertebrate organs of balance, the equivalent of the invertebrate otolith. [30]

Otolith. A hard concretion that stimulates hair cells in an invertebrate statocyst when the animal moves. [30]

Out-crossing. Cross-pollination between different individuals of the same species. [41]

Ovarian cycle. In mammals, the regular pattern of egg and uterine development that prepares the female for fertilization and pregnancy. [21]

Ovary. The female gonad, in which eggs are produced. [21]

Oviduct. In vertebrates, a tube that conducts the egg from ovary to uterus. [21]

Oviparous. Having young that develop inside an egg rather than the maternal body. Compare with *viviparous* and *ovoviviparous*. [19]

Ovoviviparous. Having young that develop within the mother, but without nutrient support from a placenta. Compare with *viviparous* and *oviparous*. [19]

Ovulation. The release of an egg from a follicle in the ovary. [21]

Ovule. An immature seed consisting of the female gametophyte, the nucellus, and one or two integuments. Becomes a seed after fertilization. [21,41]

Ovum. An egg. A mature female gamete. [12,19]

Oxaloacetate. An intermediate in the citric acid cycle and other metabolic reactions. See *C_4 plants*. [9]

Oxidation. A reaction in which a molecule gives up electrons (or hydrogen atoms) to a molecule such as oxygen or through removal in ion-formation. Most biological oxidations are exergonic. Compare with *reduction*. [8]

Oxidation-reduction reaction. A reaction in which one reactant is oxidized and another is reduced. Every oxidation (loss of an electron) must be accompanied by a simultaneous reduction (gain of an electron). [8]

Oxidative phosphorylation. A process within the mitochondrial membrane that generates ATP by chemiosmosis, after transferring coenzyme-bound electrons (generated by oxidations of the citric acid cycle) to oxygen. [9]

Oxidizing agent. A compound that withdraws electrons from another molecule, thus oxidizing it and reducing itself. Compare with *reducing agent*. [10]

Oxygenase. An oxidative enzyme that catalyzes the incorporation of oxygen into its substrate. [10]

Oxygenic photosynthesis. The type of photosynthesis found in plants and green algae, that produces oxygen. Compare with *anoxygenic photosynthesis*. [10]

Oxyhemoglobin. The complex of hemoglobin and oxygen by which blood transports oxygen. [27]

Oxytocin. A hormone released by the posterior pituitary that stimulates uterine contractions during labor and milk secretion. [21,28]

P_{680}. The reaction center of photosystem II, a chlorophyll *a* complex that optimally absorbs light with a wavelength of 680 nm. [10]

P_{700}. The reaction center of photosystem I, a chlorophyll *a* complex that optimally absorbs light with a wavelength of 700 nm [10]

Pacemaker. (1) An excitatory region of the vertebrate heart that initiates the heartbeat. In mammals, the sinoatrial node. (2) A mechanical device that has the same function when implanted near the heart of someone with heart block. [24]

Paedomorphosis. A heterochronic process in which traits characteristic of juveniles in ancestral species are retained in adult descendants. [20]

Paleontology. The study of early life based on the evidence of fossils and other aspects of extinct organisms and species. [37]

Paleozoic era. The earliest era of the present (Phanerozoic) eon. It lasted from 600 to 250 million years ago, began with the Cambrian period, and ended with the Permian period. [37]

Palisade mesophyll (or **parenchyma**). A layer of leaf mesophyll with columnar, chloroplast-containing cells. Frequently found just below the upper epidermis. Compare with *spongy mesophyll*. [10]

Pancreas. In vertebrates, the composite gland that secretes most of the digestive enzymes (exocrine secretions), insulin and glucagon (endocrine secretions), and bicarbonate. [26,28]

Pangaea. The single supercontinent that existed during the Paleozoic era and broke apart to form multiple continents. [48]

Pangenesis. The ancient idea that particles from all over the body transmit parental attributes to offspring. [13]

Panthalassa. The single giant ocean that surrounded the supercontinent Pangaea during the Paleozoic era. [48]

Paracrine secretion. The release by cells of factors that influence nearby cells. Compare with *endocrine* and *autocrine secretion*. [28]

Parallel evolution. The evolution of organisms that were originally similar and had similar selection pressures. Hence, both evolved in the same direction but along different pathways. Compare with *convergent evolution*. [35,37]

Paramylon. A polysaccharide characteristic of euglenoid flagellates. [39]

Parapatric speciation. Development from a single parent population of two species after genetically distinctive organisms in a population gain access to an unoccupied ecological niche within the population's normal geographical range; a phenomenon intermediate between *allopatric speciation* and *sympatric speciation*. [36]

Parasexuality. Reproduction in which offspring have more than one parent but without meiosis or fertilization. Occurs in some monerans, protists, and fungi. In Fungi Imperfecti, for example, haploid nuclei in a heterokaryon fuse to form diploid nuclei, some of which are heterozygous recombinants. [21,40]

Parasitism. A symbiotic mode of life in which an organism lives on or in another of a different species, obtaining nutrients from it without killing it. Compare with *predation*. [45,47]

Parasocial pathway. The evolutionary development of social groups that involve the cooperative aggregation of unrelated animals of the same species around their resources. [33]

Parasympathetic nervous system. The portion of the autonomic nervous system that slows the heart rate, stimulates digestion, and slows the heartbeat. Compare with *sympathetic nervous system*. [30]

Parathyroid glands. Four glands near the thyroid that secrete parathyroid hormone, which raises calcium and lowers phosphate levels in blood. [28]

Parathyroid hormone (PTH). See *parathyroid glands*. [28]

Parenchyma. A plant tissue composed of relatively unspecialized cells without secondary walls. Composed of thin-walled, randomly arranged cells with large vacuoles. [7,10]

Parental generation (P). In a genetic cross, the stock whose offspring comprise the F_1 generation. [13]

Parthenogenesis. Asexual reproduction occurring when an egg develops without fertilization. Seen in lizards, insects, and others. [21]

Partial pressure. The pressure that would be exerted by one component of a mixture of gases (such as air) if it were alone in a container. A measure of the amount of that gas in a mixture. [27]

Parturition. In mammals, the process of giving birth. [21]

Passive diffusion. The spread of material from a region of high concentration by the random movement of its molecules without the intervention of active energy-dependent transport. Compare with *active transport* and *facilitated diffusion*. [6]

Passive immunity. "Second-hand" immunity produced in an unexposed animal by giving it immune serum from another animal that has been immunized. Compare with *active immunity*. [23]

Pasteur effect. The decrease in glycolysis when anaerobic cells are exposed to oxygen. [9]

Pectin. Complex carbohydrate found between plant cells and their cell walls. [5,12]

Pedipalp. Second appendage of an arachnid. [42]

Pelagic region. The open water above the ocean floor. Compare with *benthic region*.

Pellagra. Nutritional disease caused by deficiency of the vitamin nicotinic acid (niacin). Symptoms are due to tissue deficiencies of NAD and NADP, coenzymes containing nicotinic acid. [26]

Penetrance. The percentage of organisms carrying a gene that actually expresses the related phenotype. Compare with *expressivity*. [14]

Penicillin. An antibiotic of fungal origin. The first antibiotic to be recognized, isolated, and applied to the treatment of infectious diseases in humans and other animals. [40]

Pentose phosphate shunt. A pathway of glucose metabolism. An alternative to glycolytic pathway that produces ribose and reduced NADP (NADPH). [11]

Pepsin. The principle digestive enzyme of gastric juice. Hydrolyzes proteins. [26]

Pepsinogen. Precursor of pepsin. Secreted by the stomach. [26]

Peptide bond. The connecting bond in a protein, —CO—NH—. Formed by removal of water during linkage of amino acids by their —COOH to —NH_2 groups. [4]

Peptidyl transferase. An enzyme that adds new amino acids to the lengthening polypeptide during translation. [17]

Perfect flower. A flower in which both stamens and carpels are present. [41]

Pericycle. In plant roots, tissue just within the endodermis, but outside the root vascular tissue. Meristematic activity of pericycle cells gives rise to the lateral roots. [24]

Periderm. An outer layer of cork that replaces the epidermis in stems and roots undergoing secondary growth. [7]

Perinuclear space. The space between the outer and inner nuclear membranes. [5]

Peripheral nervous system. All the nervous tissue not contained within the central nervous system; includes sensory and motor nerves. [30]

Peristalsis. Wavelike muscular contractions proceeding along a tubular organ, propelling the contents along the tube. [26]

Peritubular capillary. Capillaries around the renal tubules. [25]

Permafrost. The permanently frozen subsoil of the tundra. [48]

Permeability. A measure of the ease with which molecules of a given kind can pass through a membrane. Depends both on the nature of the membrane and the molecule. [6]

Permease. A membrane protein that specifically transports a compound or family of compounds across the membrane. [18]

Peroxisome. An organelle in plant cells that contains enzymes catalyzing reactions of glycolic acid metabolism. Site of photorespiration in C_3 plants. [5]

Petal. In an angiosperm flower, a modified leaf, nonphotosynthetic and frequently colored, that attracts pollinating insects. [21,41]

pH. A measure of the acidity or alkalinity of a solution. Equals the negative log of the hydrogen ion (H^+, proton) concentration in a solution. Ranges from 0 to 14; pure water has a pH of 7. [3]

pK. The negative log of K, the ideal dissociation constant of a given acid; pK' is the apparent pK of an actual system. An indicator of the acidity of a compound, pK is the pH at which an acidic group is half-dissociated. [3]

Phage. See *bacteriophage*. [16]

Phagocytosis. Cellular ingestion of solid particles by endocytosis. Seen in protists, some digestive cells of invertebrates, and vertebrate white blood cells and macrophages. Compare with *pinocytosis*. [5,6,22,23]

Phanerozoic eon. The present eon, beginning 600 million years ago. It is divided into the Paleozoic, Mesozoic, and Cenozoic eras. [37]

Pharyngeal slit. Openings in the walls of the pharynx. Occurs in *chordates* during development. [43]

Pharynx. The throat. The cavity that connects the mouth, nasal cavities, and esophagus. [27]

Phenetic systematics. See *numerical phenetics*. [37]

Phenocopy. An organism whose phenotype has been altered by its environment so that it resembles the phenotype normally associated with another genotype. [14]

Phenotype. The physical or functional manifestations of gene activity in an organism as modified by environmental factors. Compare with *genotype*. [13]

Pheromone. A chemical substance released by organisms that influence other members of the same species, e.g., as a sex attractant, territory marker, or alarm signal. [30]

Phloem. A plant vascular tissue that transports dissolved nutrients through the plant. Composed of sieve tubes, fibers, and other specialized cells. In trees, the inner bark. [7,24]

Phloem loading. The secretion of food molecules from adjacent cells into plant phloem, a process that draws in water needed to transport molecules to other parts of the plant. [24]

Phonoreceptor. A structure, such as the hair cells of the vertebrate ear, that converts sound waves into nerve impulses. [30]

Phosphagen. Molecules, such as creatine phosphate, that furnish high-energy phosphate bonds for ATP synthesis when needed. [27]

Phosphate. The —PO_4 group (abbreviated P_i). Converted in metabolism to the high-energy phosphoanhydride bonds of ATP. Also has many other metabolic functions. [9]

Phosphatidyl inositol (also called *phosphoinositide*). A phospholipid in cell membranes that breaks down to release the second messengers inositol triphosphate (IP_3) and diacylglycerol. [28]

Phospholipids. Important components of cell membranes that contain phosphorus and are soluble in organic solvents. Differ from other lipids in which one fatty acid is replaced by a phosphate-containing compound. Examples are lecithin (phosphatidyl choline) and phosphatidyl inositol. [4,6]

Phosphorolysis. Rupture of a covalent bond by the addition of components of phosphoric acid (—H and —H_2PO_4, or P_i) to the cleaved ends of the broken bond. Typically occurs during glycogen breakdown. Compare with *hydrolysis*. [11]

Phosphorylation. Addition of a phosphate group to a molecule. [8,9,10]

Photolithotroph. An organism that uses inorganic compounds, rather than water, as electron donors during photosynthesis. [9]

Photon. A particle (quantum) of light energy. [10]

Photoorganotroph. An organism that uses organic compounds as electron donors during photosynthesis. [9]

Photoperiodism. Biological effects caused by changes in day length. [32]

Photophosphorylation. Photosynthetic reactions in which light energy trapped by chlorophyll is used to produce ATP. [10]

Photoreceptor. A specialized structure (e.g., retinal rod and cone cells) that contains light-trapping pigments and systems for converting light stimuli into nerve impulses. [30]

Photorespiration. Light-driven production of glycolic acid in chloroplasts of C_3 plants and its subsequent oxidation in perioxisomes and mitochondria. Generates no ATP but consumes oxygen and produces CO_2. Thus it competes with other photosynthetic reactions which do the opposite. [10]

Photosynthesis. A process in which light energy trapped by chlorophyll and other pigments is utilized to synthesize organic compounds (especially glucose) from CO_2 and water. [10]

Photosynthetic autotroph. An autotroph that derives energy from photosynthesis. Compare with *chemosynthetic autotroph*. [9]

Photosystems I and II (PSI and PSII). Interacting light-trapping units in the chloroplast membrane. PSI traps light energy for use in both cyclic and noncyclic photophosphorylation. PSII traps light energy and passes it to PSI for use in noncyclic photophosphorylation. [10]

Phototroph. See *photosynthetic autotroph*. [9]

Phototropism. Plant growth toward a light source. [28]

Phragmoplast. A large vesicle containing the precursors of the new cell wall laid down during the later stages of plant cell division. [12]

Phycobiont. The photosynthetic partner in a lichen. See *mycobiont*. [40]

Phycocyanin. A class of water-soluble blue pigments characteristic of blue-green algae (Cyanonta) and other algae. [38]

Phycoerythrin. A class of water-insoluble red pigments found in most blue-green algae and all red algae. [38]

Phylogeny. Evolution of a genetically related group of organisms. A schematic treelike diagram depicting such an evolutionary history. Compare with *ontogeny*. [20,34,44]

Phylum. A major taxonomic category that lies between kingdom and class. Thus a phylum includes multiple classes. The term *division* is used as a synonym in all kingdoms except Animalia and Protista. [37]

Physiological isolation. A prezygotic isolating mechanism in which biochemical differences between populations prevent fertilization. Compare with *ecological, temporal, behavioral,* and *mechanical isolation*. [36]

Phytochrome. A photoreceptor pigment regulating a large number of developmental and other phenomena in plants. Exist in two different forms, maximally absorbing light at 660 (red) and 730 nm (far-red). [10]

Phytohormone. Plant hormones or growth regulators. [28]

Phytoplankton. Free-floating photosynthetic organisms of microscopic size. Compare with *zooplankton*. [39,47,48]

Pineal body. An endocrine gland at the base of the brain that secretes melatonin and helps regulate circadian and seasonal rhythms. [28]

Pinna. (1) External portion of the ear. (2) Subdivision of a compound leaf or fern frond. [41]

Pinnule. Ultimate subdivision of a pinna. Applied mostly to fern fronds. [41]

Pinocytosis. A form of endocytosis in which a cell takes up liquid. Compare with *phagocytosis*. [6,21]

Pistil. See *carpel*. [21]

Pistillate. Having pistils and no functional stamens. [41]

Pith. The soft tissue at the center of a stem or root. Usually composed of parenchymal cells. [24]

Pits. In plants, thin spots in the walls of tracheid cells. [24]

Pituitary (also called *hypophysis*). Endocrine gland at the base of the brain in vertebrates. Most of its many hormones control the activities of other glands. Most of those released by the anterior pituitary (adenophyophysis) and posterior pituitary (neurohypophysis) are synthesized in the hypothalamus. [28]

Placenta. (1) In mammals, the organ joining the embryo to the wall of the mother's uterus through which nutrients and wastes are exchanged. (2) In flowering plants, the part of the ovary wall to which ovules or seeds are attached. [19,21]

Placental. Possessing a placenta. A characteristic of all mammals except the marsupials and monotremes. [43]

Placodermata. An extinct class of vertebrates containing the armored fishes. The first jawed fishes. [43]

Plankton. Free-floating, mostly microscopic, aquatic organisms. Includes photosynthetic forms (phytoplankton) and

nonphotosynthetic forms (zooplankton). Compare with *nekton*. [39]

Plantae. One of five phylogenetic kingdoms; includes plants (bryophytes and tracheophytes) that are multicellular, mostly photosynthetic eukaryotes. [37]

Plantlike protists. Photosynthetic protists, e.g., euglenoids, dinoflagellates, etc. Compare with *animallike protists*. [39]

Planula. The free-swimming, ciliated larva of coelenterates. [43]

Plasma. The fluid portion of blood that remains after cellular elements have been removed by centrifugation (not by clotting). Compare with *serum*. [22]

Plasma cell. An activated, antibody-producing B cell. [23]

Plasma membrane (plasmalemma). The cell membrane. A phospholipid bilayer in which proteins are embedded. Compare with *cell wall*. [5,6]

Plasma protein. A protein found in blood plasma. It has many functions including maintenance of the osmotic pressure of blood. [22]

Plasmid. A small piece of DNA, usually circular, that replicates independently of the chromosome. Used to transfer genes from one cell to an unrelated cell. [5,16,18]

Plasmin. An enzyme of the fibrinolytic system that dissolves clots. [22]

Plasminogen. Precursor of plasmin. [22]

Plasmodesma (pl. **plasmodesmata**). Fine cytoplasmic threads that pass through the cell walls of plant cells and form cytoplasmic links with adjacent cells. [6,12,24]

Plasmodium. Multinucleate mass of cytoplasm lacking internal cell boundaries. A stage in the reproductive cycle of myxomycetes (plasmodial slime molds) that moves about and feeds. [39]

Plasmolysis. The shrinking of cell contents and cell membrane away from a rigid cell wall, due to the osmotic loss of water from the cell. [6]

Plastid. Pigmented membrane-bound cytoplasmic organelle of plants. Includes chloroplasts and food-storage organelles. [5]

Plate tectonics. Geological model that views Earth's surface as a mosaic of rigid plates whose independent movements cause continental drift, volcanism, and earthquakes. [48]

Platelet. A membrane-bound anucleate body, that arises as a fragment of megakaryocyte cytoplasm in mammalian bone marrow and circulates in blood. Initiates blood clotting at sites of injury. [22]

Platelet-derived growth factor (PDGF). A peptide growth factor that is released from platelets and other cells. Stimulates proliferation of smooth muscle and connective tissue cells in blood vessel walls, constricts blood vessels, and promotes wound healing. [22,28]

Pleated sheet. See β-*pleated sheet*. [4]

Pleiotropy. The capacity of a gene to influence a number of phenotypic characters. [35]

Plesiomorphic. Term applied to a primitive character that is shared by all species in a cladogram. Compare with *apomorphic* and *synapomorphic*. [37]

Pleura. In vertebrates, the membrane lining the chest cavity (pleural space), the lung surface and inside of the chest wall. [27]

Ploidy. Chromosomal set number. A cell with one set of homologous chromosomes is haploid; one with two sets is diploid, etc. [12]

Pluripotent cell. A cell with many developmental pathways open to it. Compare with *unipotent cell*. [19]

Poikilotherm. A warm-blooded organism—one whose internal body temperature varies with the environment. Compare with *homeotherm*. [26]

Point mutation. A gene mutation in which only a single base is altered. Can give rise to wild-type revertants through reverse mutation. Compare with *deletion* and *duplication*. [17]

Polar body. A minute, nonfunctioning cell arising during the meiosis that produces the mammalian egg. Contains chromosomes but almost no cytoplasm. [12]

Polar (hydrophilic) molecule. A molecule with charged or partially charged groups. Polar molecules are water-soluble. Compare with *nonpolar (hydrophobic) molecule*. [3,6]

Polar nucleus. One of two nuclei derived from each end of the angiosperm embryo sac, both of which become centrally located. They fuse with a male nucleus to form the primary triploid nucleus that will produce the endosperm of the seed. [21]

Pollen grain. The immature male gametophyte in a flowering plant. [21]

Pollen tube. In flowering plants, a tube formed after pollen germination. It carries the male gametes into the ovule. [21]

Pollination. Process of transferring pollen from the anther to the receptive surface (stigma) of the ovary in plants. [21,41]

Polyandry. Having many male mates. Compare with *polygyny*. [33]

Polycistronic mRNA. An mRNA molecule that encodes more than one polypeptide. Compare with *monocistronic mRNA*. [17]

Polyclimax theory. The theory that any stable community is a climax community and that ecological succession can terminate in multiple climaxes. Compare with *monoclimax theory*. See *sere*. [47]

Polygamy. Having many mates. Compare with *monogamy*. [33]

Polygenic inheritance. Occurs when a phenotypic trait is influenced by more than one gene. [13,14]

Polygyny. Having many female mates. Compare with *polyandry*. [33]

Polymer. A molecule, usually large, composed of many similar or identical subunit molecules (monomers). [4]

Polymorphism. The presence in a population of multiple alleles of a gene that persist for many generations and are not due to random mutation. [34]

Polymorphonuclear cell. A white blood cell with a nucleus that has more than one lobe. Granulocytes are polymorphonuclear cells. [22]

Polynucleotide. A polymer of nucleotides. A nucleic acid. [16]

Polyp. The sedentary stage in the life-cycle of coelenterates. It is the result of asexual budding. May alternate with medusa as a stage in the reproductive cycle. Compare with *medusa*. [21,42]

Polypeptide. A molecule made of many amino acids joined by peptide bonds. Large polypeptides are called *proteins*. [4]

Polyploidy. The presence in a cell of multiple chromosome sets and thus multiples of the normal haploid chromosome number (n). Triploidy ($3n$) and tetraploidy ($4n$) are examples. Compare with *aneuploidy*. [12,15,36]

Polyribosome. See *polysome*. [5,17]

Polysaccharide. A class of molecules, including starches and cellulose, that are polymers of simple sugars. [4]

Polysome. A cluster of ribosomes that together translate the same strand of messenger RNA. Also called *polyribisome*. [5,17]

Polyspermy. The abnormal fertilization of one egg by more than one sperm cell. [19]

Polyunsaturated fatty acid. A fatty acid containing more than one double bond. [4]

Poly(A) tail. A sequence of adenosyl nucleotides that is added at the end of an mRNA molecule that is added after it has been transcribed. [17]

Population. (1) All members of a species that occupy a particular area at a given time. Compare with *deme* and *species*. [34,45] (2) In statistics, the group of entities under study from which samples are drawn. [14]

Population genetics. The heredity of both individuals and of continuously reproducing groups. [34]

Populational systematics. The evolutionary view that members of a systematic unit have anatomical features in common because they belong to the same systematic unit. Compare with typological systematics. [37]

Porphyrins. Widely occurring tetrapyrrole derivatives which give rise to heme, chlorophyll, and many other important molecules. [10]

Portal circulation. In vertebrates, the flow of blood from the capillaries of one organ (e.g., the intestines) to the capillaries of another organ (e.g., the liver) that does not go through the heart. [24]

Positive pressure breathing. The breathing mechanism of amphibians, in which air is pushed from the mouth into the lungs. Compare with *negative pressure breathing*. [27]

Postreplication repair. A mechanism that repairs DNA only after it has been replicated. [17]

Postzygotic barrier. A factor such as zygote mortality or hybrid unfitness that reproductively isolates two populations. Compare with *prezygotic barrier*. [36]

Preadaptation. An adaptation to one environment that turns out to be useful in a new environment as well. [48]

Precipitin. A substance that binds and precipitates a soluble antigen. [23]

Predation. A method of survival in which one organism kills and eats another. Compare with *parasitism*. [45]

Pre-mRNA. See *primary RNA transcript*. [17]

Pressure-flow hypothesis. The explanation of fluid flow in plant phloem based on the effects of osmotic pressure and turgor. [24]

Pressure potential. The actual physical (hydrostatic) pressure within a fluid due to the weight of the fluid above it. Compare with *osmotic potential* and *water potential*. [6,24]

Presynaptic knob (bouton). The termination of an axon in a synapse. It releases neurotransmitter on the postsynaptic neuron. [29]

Prezygotic barrier. A factor (ecological, behavioral, anatomical, etc.) that reproductively isolates two populations by preventing the union of their zygotes. Compare with *postzygotic barrier*. [36]

Primary growth. In vascular plants, growth at the lateral meristem that results in increased length. Compare with *secondary growth*. [7]

Primary immune response. The appearance of antibody several days after an initial exposure to antigen. Compare with *anamnestic response*. [23]

Primary RNA transcript. The immediate product of DNA transcription, before it is processed to become mRNA. [17]

Primary structure. The amino acid sequence of a protein. Compare with *secondary*, *tertiary*, and *quaternary structures*. [4]

Primary wall. The part of the plant cell wall secreted during cell expansion. [5]

Primate. A member of the mammalian order; includes humans, apes, monkeys, and lemurs. [43]

Primitive streak. A line running axially along the blastodisc of a bird or fish embryo from which cells migrate inwardly to form three layers. [19]

Prion. A minute infectious particle, believed to be the agent of kuru and other diseases, that may contain only protein and no nucleic acid. [5]

Prochloronta (also called *Prochlorophyta*). A recently discovered moneran phylum, whose members (so far only two in number) are believed to have evolved into the chloroplasts of green algae and plants. [38]

Profundal zone. An environment in a lake that is below the limits of light penetration. Compare with *limnetic zone*. [48]

Progenote. The postulated common ancestor of all organisms, including prokaryotes and eukaryotes. [38]

Progesterone. A mammalian steroid hormone that, with the estrogens, regulates female sexual development, menstruation, and pregnancy. [21,28]

Prokaryote. A simple cell without a nucleus or other membrane-bound organelles. Bacteria and other monerans. Compare with *eukaryote*. [5]

Prolactin. An adenohypophyseal hormone that affects metabolism and, in the human female, stimulates breast development and lactation. Also called *luteotropic hormone (LTH)*. [28]

Promoter. In an operon, a DNA sequence just upstream of a structure gene that binds RNA polymerase and thus must be present if the gene is be transcribed. The TATA and CAAT boxes are promoters. Compare with *enhancer*. [17,18]

Pronephros. The first part of the kidney to arise in vertebrate embryonic development. [25]

Prophage. Stage in the reproductive cycle of a temperate phage at which its DNA is integrated into its bacterial host's DNA and there is little other sign of it in the cell. Prophage can later enter a lytic phase to complete the cycle. See *lysogeny*. [16]

Prophase. The first phase of mitosis and meiosis, characterized by the condensation of the chromosomes. [12]

Prosimian. A primitive primate group containing the lemurs, bush-babies, and others. [43,44]

Prosobranch. Marine gill-breathing gastropods. [42]

Prostaglandins. A class of fatty acid derivatives that act as hormones or local mediators in mammalian cells. [28]

Prostate gland. A gland that secretes the fluids of semen. [21]

Prosthetic group. A nonprotein portion of a protein or enzyme. [4]

Protease. A proteolytic enzyme or peptidase; that is, an enzyme that hydrolyzes proteins into their component amino acids. [22,26]

Protein. A molecule composed of one or more polypeptide chains of amino acids, the sequence of which is specified by DNA. All enzymes are proteins, as are many of a cell's important structural elements. [4]

Protein family. A group of proteins from different species that differ in fewer than half their amino acids. In the living world, there are about a thousand protein families. [35]

Protein kinase. An enzyme that transfers a phosphate group from ATP to the enzyme's protein substrate. [28]

Proteolytic enzyme. An enzyme that catalyzes the hydrolysis of a protein or polypeptide. The digestive enzymes trypsin, pepsin, and carboxypeptidase are proteolytic enzymes. [26]

Proterozoic eon. The period from 2.5 billion to 600 million years ago, during which multicellular, sexual organisms arose. [37]

Prothoracic gland. A gland in some insects that secretes the ecdysones, hormones that stimulate molting. [28]

Prothrombin. A precursor protein that forms thrombin during blood clotting. [22]

Protist. A member of the kingdom Protista. [37,39]

Protista. One of five phylogenetic kingdoms. A highly diverse kingdom of eukaryotic, mainly unicellular organisms, it includes algae, protozoans, and (in the classification used in this book) slime molds. [37,39]

Proton. A positively charged component of atomic nuclei. A hydrogen ion. [3]

Protonephridium. (1) A primitive excretory tube in many invertebrates. (2) The duct of a flame cell. [25]

Proton gradient. A difference in proton (hydrogen ion) concentration across a membrane. In mitochondria and chloroplasts, it drives ATP synthesis by chemiosmosis. [6,9,10]

Proton pump. A transmembrane protein complex that pumps protons (hydrogen ions) across a membrane. Participates in chemiosmotic ATP synthesis. [6,9,10]

Proto-oncogene. A gene that is a precursor of a cancer-causing oncogene. [18]

Protoplasm. The cytoplasm and nucleoplasm of a cell. [5]

Protoplasmic streaming. The metabolically driven motion of the contents of a eukaryotic cell. The resulting "mixing" allows transport of materials over distances too great to allow mixing by diffusion alone. [5,6,24]

Protostome. An animal in which the embryonic blastopore develops into the mouth and there is spiral, determinate cleavage of the egg. One of two major lines of animal evolution. Compare with *deuterostome*. [19,43]

Prototroph. The nutritional wild type of an organism. Any deviant form that requires growth nutrients not required by the prototrophic form is said to be a nutritional mutant, or auxotroph. [17]

Protozoa. A unicellular or simple multicellular organism that is heterotrophic and thus more animallike than plantlike. See *protista* and *animallike protists*. [39]

Proventriculus. The gizzard of some insects. [26]

Provirus. See *prophage*. [16]

Proximal. Situated near the point of reference or attachment. Compare with *distal*. [25]

Proximal convoluted tubule. In the kidney, the part of the tubule before the loop of Henle, where macromolecules, glucose, and divalent ions are reabsorbed from glomerular filtrate. [25]

Pseudocoelomate. An animal that lacks a true coelom but has a body cavity in the ectodermal or endodermal layer. [19,43]

Pseudogene. A DNA sequence that resembles that of a protein-encoding sequence but does not itself encode a protein. [35]

Pseudoplasmodium. In the cellular slime mold such as Dictyostelium, an aggregation of single haploid amoeboid cells that do not fuse. Occurs prior to formation of a fruiting structure. Also called a *slug*. [39]

Pseudopod. A temporary extension of the cell body. Used in amoeboid movement. [7]

Psilophyte. A member of the plant division Psilophyta. The earliest vascular plant, it was dominant in the Paleozoic era. [41,44]

P-site. Site on a ribosome that catalyzes peptide bond formation between adjacent amino acids on the lengthening polypeptide chain. Compare with *A-site*. [17]

Pteridophyte. A member of the plant division Pteridophyta (ferns). It has specialized vascular systems and large complex leaves. [41]

Pterygotes. Winged insects; includes wingless forms (lice, fleas, etc.) that lost their wings secondarily. Compare with *apterygotes*. [42]

Puff. Visible enlargement of polytene chromosomes during development. They are sites of active mRNA synthesis. [18]

Pulmonary. Pertaining to the lungs. [24,27]

Pulmonary circulation. In mammals, the flow of blood through the lungs. Compare with *systemic circulation* and *portal circulation*. [24]

Pulmonate. Terrestrial gastropods with a lunglike cavity in the mantle. [42]

Pulse. The pressure wave that, following the heart's contraction, travels from the heart along the arteries. [24]

Pulse pressure. The difference between the peak and minimum blood pressures during the heartbeat. It measures the force generated by the heart's contraction. [24]

Punctuated equilibrium. A model of evolution according to which new species undergo most of their phenotypic modifications as they branch off from their parent species and then undergo little further modification. [36]

Punett square. A graphic method of determining all possible genotypes resulting from a genetic cross. [13]

Pupa. A stage between the larval and adult stages of some insects (the Holometabola), in which movement and feeding stop, but development proceeds. [21]

Purine. A class of nitrogenous base with a double-ring structure, found in DNA and RNA, and including adenine and guanine. [16]

Pyrimidine. A class of nitrogenous base with a single ring structure, found in DNA and RNA, and including cytosine, thymine, and uracil. [16]

Pyrrole rings. A molecular structure containing four carbons and a nitrogen. Porphyrins are tetrapyrroles (i.e., they have four pyrrole rings). See *porphyrin*. [10]

Pyruvate. A three-carbon acid. The end product of glycolysis and an initial substrate for the citric acid cycle. [9]

Quantum. An indivisible unit of energy. [8,10]

Quaternary structure. The aspects of a protein's shape that are determined by interaction between two or more polypeptide subunits. [4]

Race. See *subspecies*. [36]

Radial symmetry. Anatomical pattern in which body parts are arranged around one central axis. Any plane through the axis will divide the organism into similar halves. Compare with *bilateral symmetry*. [42]

Radicle. The embryonic root in a seed. [21]

Radiolarian. Hard-shelled member of protistan phylum Sarcodina. [39]

Radiometric dating. A method of estimating the age of a rock—and any fossils in the rock—by estimating the time during which radioactive crystals in the rock have been decaying. [37]

Radula. Horny toothed, tonguelike organ of mollusks. [42]

Random drift. The processes by which allele frequencies in a gene pool are altered from one generation to the next by chance events alone, most often due to dispersal (migration). Also called *genetic drift*. [34]

Random reproduction. Reproduction in which a female is as likely to mate with one male as another, any two gametes are as likely to produce a viable zygote as any two others, and each zygote has an equal chance to mature. Compare with *nonrandom reproduction*. [34]

Range. (1) In statistics, the span of values of a variable in a population. [13] (2) In ecology, the geographical spread of a species. [34]

Rate-limiting step. The reaction in an enzymatic pathway that proceeds at the slowest rate. Unless its rate is increased, increasing the rate of other steps will have no overall effect. [8]

Reaction range. In genetics, the potential range of phenotypes compatible with having a particular genotype. [14]

Reading error. A genetic mutation caused by incorrect reading of the genetic code during translation. Often caused by a shift in reading frame. [17]

Reading frame. The way in which a nucleotide sequence is divided into triplet codons. The nucleotide at which reading starts determines the reading frame. [17]

Receptor. (1) A macromolecule, usually a membrane-bound protein, that recognizes and binds a substance and mediates its effects. [6,28,30] (2) A cell that transduces a stimulus into a nerve impulse. [29,30]

Receptor-ligand complex. A receptor whose binding site is occupied by the substance (ligand) for which the receptor is specific. [6]

Receptor-mediated endocytosis. The import of specific proteins (ligands) into a cell by their binding to receptors in the plasma membrane and their inclusion into endocytotic vesicles. [6]

Recessive. Characteristic of an allele such that, in an organism possessing both alleles, it fails to determine the phenotype. Compare with *dominance*. [13]

Recombinant DNA technology. See *genetic engineering*. [18]

Recombination. The appearance in progeny of new gene combinations (recombinants) that differ from those of the parents. In eukaryotes, it is due mainly to random assortment and cross-over of homologous chromosomes. [12,13,14]

Red algae. Members of the protistan phylum Rhodophyta. [39]

Red cell. See *erythrocyte*. [22]

Reducing agent. A compound that donates electrons to other molecules, thus reducing them and oxidizing itself. Compare with *oxidizing agent*. [10]

Reduction. A reaction in which a molecule gains electrons. Compare with *oxidation*. [8]

Reductional division. Meiosis I. See *meiosis*. [12]

Reductive pentose phosphate cycle. See *Calvin cycle*. [10]

Reflex. A largely automatic response to a stimulus. It results from a neural circuit in which sensory neurons are connected relatively directly to motor neurons. [29]

Region. In the context of evolutionary biology and ecology, a group of clusters. [34]

Regulatory gene. A gene that controls (often by encoding a repressor) the rate of transcription of other genes. Compare with *structural gene*. [18]

Release factor. A protein necessary for the release of a completed polypeptide from a ribosome at the end of translation. [17]

Release inhibiting factor (RIF). A hormone secreted by the hypothalamus that inhibits release of another hormone from the pituitary (hypophysis). Compare with *releasing factor*. [28]

Releaser. A simple environmental stimulus that triggers a complex, instinctive behavior pattern in a species. [32]

Releasing factor [RF]. A hormone secreted by the hypothalamus that triggers release of a pituitary hormone. [28]

Renal. Relating to the kidneys. [25]

Rennin. The milk-curdling enzyme secreted by the stomach of young mammals. It aids in the retention and digestion of milk. [26]

Replacement distance. An index of sequence homology between two proteins. The minimal number of nucleotides that must be altered for the gene of one protein to encode the other. [35]

Replication fork. The point on a replicating DNA molecule at which the two parent strands split and nucleotides are added to form the daughter strands. [16]

Repressible enzyme. An enzyme whose synthesis can be decreased or prevented by the binding of a particular compound (effector) to a repressor protein. [8,18]

Repression. Metabolic regulation in which high levels of an enzymatic pathway's final product inhibit the synthesis of its rate-limiting repressible enzyme. Compare with *feed-back inhibition* and *derepression*. [8,18]

Repressor. A regulatory protein that binds to an operator DNA site and represses transcription of a nearby structural gene. The affinity of repressor for operator is enhanced by small molecules (effectors). [18]

Reproductive isolation. The inability of two populations to interbreed in their natural environment, whether because of anatomical, physiological, behavioral, or geographic barriers. [36]

Reptilia. A class of vertebrates, the first to have amniotic eggs; includes turtles, alligators, dinosaurs, lizards, and snakes. [43]

Respiration. (1) Oxidative breakdown of food molecules and release of energy from them. The terminal electron acceptor may be oxygen or, in anaerobic organisms, nitrate, sulfate, or nitrite. (2) The exchange of oxygen for CO_2. In vertebrates, it is mediated by lungs or gills. [9,27]

Respiratory center. An area of the vertebrate brain that controls breathing. [27]

Resting potential. The difference in electrical potential (usually about 70 millivolts)

between the inside and outside of a cell. In neurons at rest, the interior is negative to the exterior. [29]

Restriction enzyme (endonuclease). One of several enzymes, produced by bacteria, that cleave DNA only at sites with a particular short base sequence. Extensively used in recombinant DNA technology. [18]

Reticulum. (1) A fine network. (2) The second chamber of the ruminant stomach. It contains cellulose-digesting bacteria. [26]

Retina. The light-sensitive layer at the back of the veretebrate eye. It contains rods, cones, and neurons that transmit nerve impulses to the brain. [30]

Retinal. A derivative of vitamin A that in the eye is conjugated to opsin to make the photosensitive pigment rhodopsin. [30]

Retrovirus. A virus, such as that which causes AIDS, whose genetic material is RNA. Before replication, it must transcribe them into DNA, which integrates into host DNA. [18]

Reverse mutation. See *back mutation*. [14]

Reverse transcriptase. An enzyme found in retroviruses that catalyzes the synthesis of complementary DNA on an RNA template. Compare with *RNA synthetase*. [18]

Rh disease. An attack by the immune system of an Rh-negative mother on the red cells of an Rh-positive fetus. [23]

Rh (Rhesus) factor. A surface antigen on the red blood cells of humans (and Rhesus monkeys) that can trigger Rh disease. [23]

Rhizoid. (1) Rootlike filament in mosses, liverworts, and a few vascular plants that absorbs nutrients and water and anchors the plant. [41] (2) The branched, rootlike extensions of some fungi and algae. [38,39]

Rhizome. A special underground stem (as opposed to root) that runs horizontally beneath the ground. [41]

Rhodopsin. The light-sensitive pigment that transduces light stimuli in vertebrates, arthropods, and mollusks. It consists of the protein opsin conjugated to the vitamin A derivative retinal. [30]

Ribonucleic acid. See *RNA*. [4,16]

Ribonucleotide. The building-block monomer of RNA. A nucleotide containing the sugar ribose. [11]

Ribonucleotide reductase. The enzyme that converts ribonucleotides to deoxyribonucleotides. [16]

Ribose. A five-carbon sugar. The characteristic sugar of RNA. [4,11,16]

Ribosomal RNA (rRNA). A class of RNA that is a structural and functional component of ribosomes. [17]

Ribosome. A particle composed of protein and rRNA that carries out protein synthesis by the translation of mRNA. [5,16]

Ribulose 1,5-bisphosphate (RuBP). The molecule that reacts with CO_2 in the first reaction of the Calvin cycle. [10]

Risk cost. The increased chance of injury or disease that possessing a behavioral or physiological trait brings to an organism. Compare with *energetic cost* and *opportunity cost*. [35]

RNA (ribonucleic acid). A class of nucleic acids composed of nucleotides containing ribose. Involved in the transcription of genetic information and its translation into proteins. Compare with *DNA*. [4,16]

RNA maturation. The process of post-translational processing of primary transcript RNA, which produces mRNA. [17]

RNA polymerase. An enzyme that synthesizes RNA from its four component ribonucleotides. [16,17]

RNA primer. A fragment of RNA, complementary to a short stretch of DNA, that is necessary to initiate DNA replication. [16]

RNA replicase. See *RNA synthetase*. [18]

RNA synthetase. An enzyme found in some viruses that catalyzes the synthesis of a complementary strand of RNA on an RNA template. Compare with *reverse transcriptase*. [18]

Rod. A light-sensitive cell in the vertebrate retina; it is very sensitive to dim light. See *cone* and *fovea*. [30]

Root hair. A fine projection from a plant root that increases its surface area and its ability to absorb water. [24]

Root nodules. Spherical growths on the roots of leguminous plants. Contain nitrogen fixing bacteria. [41,46]

Root pressure. In vascular plants, the hydrostatic pressure that is produced in roots by osmosis of ground water and that causes fluid flow up the xylem. [24]

Rough endoplasmic reticulum (RER). ER that is studded with ribosomes and is a site of protein synthesis. [5]

r strategist. Organisms with low survival rates that manage to survive through high fecundity (many offspring) and dispersal. Compare with *K strategist*. [34,45]

RuBP. See *ribulose 1,5-bisphosphate*. [10]

RuBP carboxylase. The enzyme that fixes CO_2 to RuBP in photosynthesis. It also acts as an oxygenase in photorespiration. [10]

Ruffled membrane. In amoeboid movement, the delicate tip of an extended pseudopod. [7]

Rumen. The first chamber of the ruminant stomach. It holds cellulose-digesting bacteria and partly digested food (the cud) that can be returned to the mouth for further chewing. [26]

Ruminant. A suborder of herbivorous, cud-chewing mammals, including cattle, sheep, and deer, that can digest cellulose with the aid of bacteria in their rumens. [26]

Saccule. In the mammalian inner ear, an organ of balance in which hair cells sense linear acceleration. Compare with *utricle* and *semicircular canals*. [30]

Saliva. Fluid secreted into the mouth that lubricates and digests food. [26]

Salivary gland. A gland which secretes food lubricants and digestive enzymes into the mouth. [26]

Salt. A substance held together with ionic bonds. [3]

Saltatory conduction. The method by which an action potential travels quickly down a myelinated nerve. It is regenerated by

sodium channels at each node of Ranvier. [29]

Salt gland. An organ which excretes salt. In marine reptiles they are located near the eyes or nose. [25]

Sample. In statistics, a group chosen randomly to represent the population to which they belong. [14]

Sap. An aqueous solution of nutrients, minerals, and other substances that passes through the xylem of plants. [24]

Saprobe. A heterotrophic organism (usually a bacterium or fungus) that obtains its carbon and energy from dead organic matter. [38]

Saprophyte. See *saprobe*. [38]

Sarcodine. A member of the protistan phylum Sarcodina; includes amoebae, foraminifera, radiolarians, etc. [39]

Sarcolemma. The vertebrate muscle cell membrane. [31]

Sarcomere. The contractile segment of skeletal muscle that stretches between two Z bands. [31]

Sarcoplasmic reticulum. The specialized endoplasmic reticulum of vertebrate muscle cells. It stores calcium ions. When released, they trigger contraction. [5,31]

Satellite DNA. Nontranscribed DNA consisting of short, highly repetitive nucleotide sequences of unknown function. [17]

Saturated fatty acid. A fatty acid with no double bonds. [4]

Scala naturae. The old idea that there is a "ladder of nature" or hierarchy of living organisms in which some are more highly evolved than others. [2]

Scaphopod. A member of the mollusk class Scaphopoda; includes tusk shells. [42]

Schizonta. A phylum in the kingdom Monera; includes the bacteria. [38]

Schwann cell. A glial support cell in the peripheral nervous system which surrounds neuronal processes. It contains and synthesizes the myelin of myelinated nerves. [29]

Scientific method. A method of acquiring new knowledge in which testable hypotheses are validated or invalidated by observational and experimental data. [2]

Sclera. The tough white outer coat of the mammalian eyeball. [30]

Sclerenchyma. A support tissue in plants consisting of thick-walled cells that may or may not be living at maturity. [7]

Scurvy. The nutritional disease caused by vitamin C (ascorbic acid) deficiency. [26]

Scyphozoan. A member of coelenterate class Scyphozoa; includes true jellyfishes. [42]

Sea squirt. See *tunicate*. [43]

Secondary structure. The shape of a protein as determined by local interactions and hydrogen-bonding between amino acids. The α-helix configuration is a secondary structure. [4]

Second law of thermodynamics. The law stating that the entropy of the universe tends to increase and that, in any real (irreversible) process, free energy decreases and entropy increases. [8]

Second messenger. A molecule or system of molecules that mediates the intracellular response to a cell surface event, such as receptor binding. [28]

Secretin. A hormone secreted by the intestine that stimulates the pancreas to secrete digestive enzymes. [28]

Secretion. (1) Export of any molecule or class of molecules from the cell that produces it. (2) Also denotes the exported material itself. [26,28]

Secretory vesicles (granules). Membrane-bound collections of cell products to be secreted by exocytosis. [5]

Seed. A mature ovule (megasporangium) consisting of a protective coat, an endosperm containing stored nutrients, and an embryo. [21,41]

Seed fern. A member of the extinct plant class Pteridospermopsida. [41]

Segregation. In genetics, the principle that each haploid gamete receives only one allele of each parental gene. Mendel's first law. [13]

Self-incompatibility. Genetically determined state in which pollen of a given plant species is rejected by its own stigmas. [41]

Self-pollination (also called *selfing*). The self-fertilization of a plant in which pollen is transferred from the anthers of a plant to the stigmas on flowers of the same plant or another of the same genotype. Compare with *cross-pollination*. [21,41]

Semen. Complex nutrient fluid in which male gametes (sperm) are transferred to a female during copulation. [19]

Semicircular canal. One of three tubes in the inner ear of mammals that use hair cells to sense the animal's angular acceleration. Compare with *utricle* and *saccule*. [30]

Semiconservative replication. Established model of DNA replication, in which each partner strand is the template for a new partner strand so that after replication each double helix contains one old (parental) and one new strand. [16]

Semipermeable. A selectively permeable membrane, one which allows only some types of molecules to pass across it. [6]

Sense strand. The strand of a DNA helix that is transcribed during RNA synthesis. [17]

Sensory area. An area of the cerebral cortex that processes sensory information. Compare with *motor area* and *association area*. [30]

Sensory neuron. A neuron that receives sensory information, either from a sensory receptor cell or from another sensory neuron. [29,30]

Sensory pit. Heat-sensitive regions between the eyes and nostrils of snakes, which allow them to locate prey by the warmth of their bodies. [30]

Sepal. Flower part. Leaflike member of the outermost whorl (calyx) of a flower. [21]

Septum. (1) A cross-wall in a hypha or spore. [40] (2) The wall that divides the right and left halves of the heart. [24]

Sere. The orderly sequence of ecological communities in an environment, from the

pioneer community to the climax community. See *climax*. [47]

Serine protease. An enzyme that cleaves proteins. The amino acid serine is present in its active site. [22]

Serum. The liquid that separates from a blood clot after coagulation. Differs from plasma in lacking the clotting factors. [22]

Sex chromatin. See *Barr body*. [15]

Sex chromosomes. Chromosomes whose presence or absence determines the sex of an organism. Compare with *autosomes*. [15]

Sex-influenced character. A trait carried on an autosome whose expression is influenced by the sex of the organism. [15]

Sex linkage. The location of a gene on a sex chromosome. [15]

Sex-linked inheritance. Pattern of inheritance shown by X chromosome genes. Males are homozygous for such genes because they have only one X chromosome. [15]

Sexual reproduction. Generation of a new cell or organism by the fusion of two haploid cells so that genes are inherited from each parent. Compare with *asexual reproduction*. [12,15,21]

Sickle cell anemia. A disease caused by the abnormal gene for hemoglobin S, which in homozygotes polymerizes in the absence of oxygen, distorting red blood cells into sickle shapes that obstruct capillaries. [22,35]

Sickle cell hemoglobin. See *hemoglobin S*. [22]

Sieve cell. The elongated conducting cells of phloem in gymnosperms and some vascular plants other than angiosperms, which have specialized sieve tubes. [24]

Sieve plate. Porous cell wall areas separating sections of sieve tubes in the phloem of angiosperms. [24]

Sieve tube. A vertical series of cellulose-lined cells separated by sieve plates, which make up the phloem in angiosperms. [5,24]

Simple metamorphosis. The process in a few insect species wherein the nymph form becomes an adult. Compare with *complete metamorphosis*. [21]

Simple sequence DNA. The least repetitive type of repetitive eukaryotic DNA. It contains genes as well as noncoding sequences. Compare with *moderately* and *highly repetitive DNA*. [17]

Sinoatrial (S-A) node. The heart's pacemaker. A region of excitatory tissue that synchronizes and triggers the contractions of the heart. [24]

Sinus. A space. A blood sinus is an enlarged vein. A nasal sinus is a cavity in the bones of the face. [24]

Sinus venosus. An enlarged, muscular vein feeding into the single atrium of the heart of fishes, amphibians, reptiles, and embryonic mammals. [24]

Skeletal muscle. See *striated muscle*. [31]

Skeleton. Hardened tissues which form scaffolding of an organism. [31]

Slime molds. Heterotrophic organisms of the kingdom Protista. Some taxonomists consider them Fungi. [39]

Slug. See *pseudoplasmodium*. [39]

Smooth endoplasmic reticulum (SER). Endoplasmic reticulum that is not studded with ribosomes. SER is a site of lipid synthesis. [5]

Smooth (visceral) muscle. Vertebrate nonstriated muscle. The loosely structured muscle which lines the walls of internal organs and is under involuntary control. Compare with *striated muscle*. [31]

Social insect. A class of insects (termites, ants, certain bees and wasps) that form colonies with reproductive and worker castes. [33]

Sociobiology. The study of animal societies and communication. [33]

Sodium-potassium pump. Transmembrane protein complex that actively transports sodium ions across a membrane in exchange for potassium ions. Maintains the resting potential of neurons. [6,29]

Solute. The substance dissolved in a solvent to form a solution. [6]

Solution. A solvent and its dissolved solutes. [6]

Solvent. A liquid that can dissolve one or more substances (solutes). [6]

Somatic. Pertaining to the body, or body cells. Compare with *germ cells*. [13]

Sol. A liquefied gel. A system consisting of a colloid that is dispersed in a liquid medium. [6]

Sol-gel transition. Interconversion of a colloid, such as cytoplasm, between sol (liquid) and gel (semi-solid) states. [6]

Somatic cell. A differentiated, diploid body cell that does not give rise to haploid gametes. Compare with *gamete*. [12]

Somite. One of the segments into which an embryo becomes divided longitudinally, leading eventually to segmentation as illustrated by the spinal column, ribs, and associated muscles. [43]

Sorus. Cluster of sporangia or spores, e.g., on the underside of fern fronds. [41]

Spatial summation. The ability of multiple, subthreshold nerve impulses arriving simultaneously at different parts of a neuron to sum together and stimulate an action potential. [29]

Speciation. The processes of diversification of populations and the multiplication of species. [36]

Species. (1) A series of populations within which significant gene flow occurs under natural conditions, but which is genetically isolated. Includes many demes that tend to intergrade, fuse, and replace each other. [34,36,37] (2) The lowest category (except for subspecies) of the taxonomic hierarchy. [37]

Species pool. The different species which may potentially colonize an isolated environment such as an island. [47]

Species selection. Occurs when a given species with a given feature more rapidly leads to other species (or less often becomes extinct) than species with another feature. Compare with *interdemic selection* and *individual selection*. [34]

Specific heat. The heat capacity per gram of a material. [3]

Spectrophotometry. A determination of the amount of colored matter present in a

solution by measuring the amount of light it absorbs. [4]

Sperm (spermatozoa). Usually motile, male reproductive gametes arising from spermatid maturation. [12,19,21]

Spermatid. Immature sperm cells, generated by meiosis from spermatocytes. [12,19,21]

Spermatocyte. Cells that form spermatids by meiosis. [19]

Spermatogenesis. The process of sperm formation and maturation in the testis. [12,19,21]

Spermatogonium (pl. **spermatogonia**). Early cell in the maturation sequence of spermatogenesis, which produces primary spermatocytes. [12,19,21]

Spermatozoon. A sperm. [19,21]

Spermophyte. A member of the plant division Spermophyta. A seed plant. [41]

Sphenophyte. A member of the plant division Sphenophyta. Horsetails; vascular plants that are common in bogs and sandy soil. [41]

Spindle fiber. In mitosis and meiosis, the microtubules, radiating from a centriole, which attach to and separate the chromosomes. [12]

Spiracle. In arthropods, the external openings of the tracheae. [27]

Spirillum. A bacterium with a spiral or helical shape. [38]

Splice junction. The border between an exon and an intron in an unprocessed mRNA molecule. [17]

Splicing. In genetic transcription, the excision of introns from mRNA, leaving only exons that encode the gene product. [17]

Spongocoel. Central body cavity of a sponge. [42]

Spongy mesophyll. The lower layer of leaf mesophyll. Compare with *palisade mesophyll*. [10]

Spongy parenchyma. In leaves, a layer of loosely packed photosynthetic cells with extensive intercellular spaces for gas diffusion. Frequently found between the palisade mesophyll and the lower epidermis. [10]

Spontaneous generation. The old idea that life continuously arises from nonliving matter. Contradicted by the modern view that life arose from nonliving matter under primordial conditions early in Earth's history, but has since derived only from living organisms. [2,4,44]

Sporangiophore. Specialized fungal hypha that supports one or more sporangia. [40]

Sporangium (pl. **sporangia**). A saclike unicellular or multicellular fungal organ that produces and releases asexual spores. Borne on specialized hyphae (sporangiophores) which are the stalks of fruiting bodies. [21,40]

Spore. (1) An asexual reproductive structure that can survive for long periods and, under certain conditions, can give rise to new individuals. (2) In plants, a haploid cell which gives rise to the gametophyte generation. (3) In unicellular organisms, a resting form that can survive harsh conditions. [21,38,41]

Sporic meiosis. See *alternation of generations*. Compare with *gametic* and *zygotic meiosis*. [21]

Sporocyte. Specialized cells within a plant sporophyte that undergo meiosis and produce haploid spores. [21,41]

Sporophyll. A leaf or leaflike structure that bears sporangia; includes angiosperm carpels and stamens and sporangium-bearing leaves on ferns. [41]

Sporophyte. In species with alternation of generations, the spore-producing diploid phase of the life cycle. Compare with *gametophyte*. [21,41]

Sporozoan. A member of the protistan phylum Sporozoa; includes Plasmodia, the genus whose members cause malaria. [39]

Sporozoite. Life-cycle stage of sporozoans. Motile product of multiple mitoses (sporogony) of zygote or spore. [39]

Sporulation. The process of spore formation. (1) In plants, a stage in which the sporophyte generation reproduces asexually by forming haploid spores. (2) In unicellular organisms, the production of spore forms. [21,38,41]

Spot desmosome. A disklike site of adhesion between cells. Compare with *belt desmosome*. [7]

Stabilizing selection. Occurs when phenotypes in the middle range of a frequency distribution curve are favored. Compare with *directional* and *disruptive selection*. [34]

Stamen. The male, pollen-producing organ of a flower. [21]

Staminate. Having stamens and no functional pistils. [41]

Standard free energy change (ΔG°). The free energy change associated with a reaction taking place at standard temperature and pressure (0°C, 1 atm). [8]

Starch. A linear polymer of glucose, used to store energy in plants. [4]

Stasis. (1) Stoppage, as in blood stasis (slowed blood flow). [22,24] (2) Equilibrium state, as in evolutionary stasis (long periods of little or no evolutionary change, sometimes followed by short periods of rapid change in which new species arise). [36]

Statocyst. An organ of gravity detection and balance in invertebrates. It contains granules (otoliths) which stimulate hair cells when the animal moves. [30]

Stem. Cylindrical axis of a plant which contains conduits of the conductive system and which gives rise to leaf buds. [24,41]

Stem cell. A multipotent cell which can differentiate into a number of other cell types. [22]

Stenoecious. Pertains to species capable of surviving only within a narrow range of environmental conditions. Compare with *euryecious*. [46]

Stenotrophic. An animal that can live on only a limited number of foods. Compare with *eurytrophic*. [26]

Stereoisomer. An isomer that is identical in structure to another molecule, except that it is its mirror image. [4]

Stereopsis. Stereoscopic vision. The ability to perceive depth by fusing the slightly different images of two eyes. [30]

Steroid. A group of fat-soluble chemicals structurally related to cholesterol. They include various sex and adrenocortical hormones, vitamin D, and the active part of toad poisons. [4,28]

Stigma. (1) In flowers, the region of the female carpel that receives pollen. [21] (2) The light-sensitive photoreceptor of a euglenoid flagellate. Also called *eye spot*. [39]

Stimulus. Any agent or action which produces a response, i.e., a change in a receptor or irritable tissue. [29]

Stoma (pl. **stomata**). Small pores on leaf surfaces that are opened and closed by guard cells to facilitate gas exchange. [10]

Stomach. A muscular enlargement of the alimentary canal in which initial stages of digestion take place. [26]

Striated (skeletal) muscle. The vertebrate tissue which controls limb movements. It is under voluntary control. Compare with *smooth muscle*. [31]

Strobilus. The cone in gymnosperms. A cone-shaped mass of modified leaves (sporophylls) or ovule-bearing scales on a stem. [21,41]

Stroma. (1) The supporting tissue of an organ. (2) The contents of a plastid, such as a chloroplast. [5,10]

Stromatolite. Layered carbonate or silicate sedimentary rocks produced by early microorganisms, principally cyanobacteria. [9]

Structural adaptation. An adaptation involving change in body structure. Also called *anatomical adaptation*. [35]

Structural gene. A gene that encodes a protein. Compare with *regulatory gene*. [17,18]

Style. In flowers, the part of the carpel which supports the stigma. [21]

Subclimax. A stable community regarded by proponents of the monoclimax theory to be the result of arrested ecological succession. [47]

Suberin. A waxy material in plant cork that prevents water loss. [24]

Subspecies. A subdivision of a species. Phenotypically distinct demes that are separated geographically. Also called *races*. [36]

Substrate (substratum). (1) A substance upon which an enzyme acts. [8] (2) In ecology, the surface to which an organism is attached or relates, e.g., ground, rock, sea floor. [46]

Succession. The orderly changes in the species composition of an ecological community over time. In the absence of environmental disturbances or species mutation, it results in a climax community. [47]

Suction feeder. An animal, often a fish, that captures prey by abruptly sucking it into its mouth. [26]

Superfemale. A female with a higher than normal number of X chromosomes. [15]

Suppressor gene. See *modifier gene*. [14]

Suppressor T cell. A class of T lymphocytes that can inhibit immune responses to an antigen. [23]

Surface tension. A condition at the surface of a liquid where attractions (cohesion) of molecules of the liquid cause its surface to behave like an elastic membrane. [3]

Survivorship curve. A plot of the percentage of individuals still living as a function of their age. [45]

Suspensor. Tissue in plants that holds a developing seed embryo. [21]

Sweepstakes dispersal. Accidental transport of one or a few members of a species across a large migration barrier, as in the colonization of an island. [48]

Swim bladder. In bony fishes, a sac that stores gas, regulates body density, and aids in hearing. [27]

Symbiont. Organisms that live in mutually beneficial association with other organisms. [40]

Symbiosis. The close association of organisms of different species within an environment. Commensalism, mutualism, and parasitism are symbiotic relationships. [47]

Sympathetic nervous system. A division of the autonomic nervous system in vertebrates. It triggers responses to emergencies, for instance by accelerating the heart rate and arresting digestion. Compare with *parasympathetic nervous system*. [30]

Sympatric speciation. Speciation occurring without geographical isolation as the result of the acquisition of isolating mechanisms within demes. Compare with *allopatric* and *parapatric speciation*. [36]

Symplastic pathway. In plants, the flow of water and minerals through plasmodesmata from the root surface to the xylem. Compare with *apoplastic pathway*. [24]

Synapomorphic. Term applied to a character in a cladogram that is shared by multiple species and thus defines a larger taxon (e.g., hair defines mammals but excludes lizards). Compare with *apomorphic* and *plesiomorphic*. [37]

Synapse. The narrow gap between the axon terminal of one neuron and the dendrite of another. The point at which neural impulses cross from cell to cell, usually carried by the secretion of neurotransmitters. [7,29]

Synapsis. The specific point-for-point pairing alignment of homologous chromosomes during the first division of meiosis. [12]

Synaptic cleft. The narrow extracellular space between the two neurons joined by a synapse. [29]

Synaptic vesicle. A membrane-bound sac which stores neurotransmitter in the presynaptic terminal of a synapse. [29]

Synaptonemal complex. A complex that forms in the synapsis and crossing-over of homologous chromosomes during meiosis. [12]

Synchronous muscle. Muscle which contracts at the same frequency as the incoming nerve impulses that trigger the contraction. Compare with *asynchronous muscle*. [31]

Syncytium. A membrane-bound mass of cytoplasm with many nuclei. A product of cell fusion, or of mitosis without cytokinesis. Muscle fibers are syncytia. [29,31]

Syngamy. The union of two gametes to form a zygote. [19,21]

Synthase. An enzyme which catalyzes a synthetic reaction, e.g., ATP synthase. [9]

Syrinx. The sound-producing organ of birds, situated at the branching of the trachea into the bronchi leading to the lungs. [33]

Systemic circulation. The flow of blood through the body except for lungs or gills. Compare with *pulmonary* and *portal circulation*. [24]

Systematics. The scientific study of the kinds and diversity of organisms and their evolutionary relationships. [37]

Systole. The period during a heartbeat when the chambers, especially the ventricles, are contracting. Compare with *diastole*. [24]

Taiga. Boreal forest. A biome consisting largely of evergreen forests that lies south of the tundra in Eurasia and North America. [48]

Tapeworm. A member of the flatworm class Cestoda; a parasitic form. [42]

Taste bud. Specialized structure on the tongue which contain the chemoreceptors that mediate the sense of taste. [30]

TATA box. A promoter sequence found upstream to the transcription initiation site of virtually all eukaryotic structural genes. It correctly positions RNA polymerase. [17,18]

Taxis. Directed movement of a simple organism toward or away from an environmental stimulus. Plant phototaxis is upward movement toward light. Compare with *kinesis* and *tropism*. [32]

Taxon (pl. **taxa**). A general term for any category of biological classification, e.g., kingdom, phylum, class, order, family, genus, species. [37]

Taxonomy. The scientific study of classification, its principles, procedures, and rules. [37]

T cell (T lymphocyte). A class of lymphoid cells important in cell-mediated immunity. They are originally produced in bone marrow but proliferate and are conditioned in the thymus. Compare with *B cell* (*B lymphocyte*). [23]

T cell receptor. A T cell membrane protein that binds a specific antigen only when it is associated with a specific histocompatibility (HLA) antigen. Compare with *B cell receptor*. [23]

Telophase. A late stage in mitosis and meiosis, characterized by formation of two new nuclear membranes and the disappearance of the chromosomes. [12]

Temperate deciduous forest. The biome of temperate zones with moderate precipitation. Has trees that shed their leaves in winter. [48]

Temperate phage. A bacteriophage that does not ordinarily kill its host. Compare with *virulent phage*. [16]

Template. A pattern or mold; a strand of which specifies the synthesis of a complementary strand of DNA or mRNA, which in turn serves as a template for protein synthesis. [16,17]

Template strand. See *sense strand*. [17]

Temporal isolation. A prezygotic isolating mechanism based on differences in breeding or pollinating seasons. Compare with *ecological, behavioral, physiological,* and *mechanical isolation*. [36]

Temporal summation. The ability of multiple subthreshold nerve impulses arriving in succession on a single part of a neuron to sum together and stimulate an action potential. [29]

Tendon. A cord or band made primarily of collagen which connects muscle to bone. [31]

Terminator codon. A nucleotide triplet that signals the ribosome to stop translation of an mRNA molecule. [17]

Terrestrial. Of the land. Compare with *aquatic* and *marine*. [47,48]

Territoriality. A characteristic of some animal socities, in which each animal or group controls an area of land and expels intruders from it by aggressive behavior or display. [33,45]

Tertiary structure. The three-dimensional shape of a protein as determined by foldings of its secondary structure. Compare with *primary, secondary,* and *quaternary structure*. [4]

Test cross. A genetic cross of an individual of unknown genotype with a tester individual. For example, to determine if an unknown individual is *CC* or *Cc*, it is crossed with a *cc* tester and progeny are examined for *cc* individuals. [13]

Testis. The male gonad, the organ which produces sperm. [21]

Testis determining factor (TDF). The single gene found on the human Y chromosome which is supposedly responsible for male sexual development. [15]

Test of progeny. The determination of whether a novel phenotype is due to a mutation or is the result of other factors by examining its progeny. [14]

Testosterone. Male sex hormone secreted by the testis. [28]

Tetanus. Continuous muscle fiber contraction, in which nerve impulses arrive so fast the muscle does not relax between impulses. [31]

Tetraploidy. The state of having four times the normal haploid number of chromosomes. [15]

Tetrapyrrole. Large molecular ring structure composed of four pyrroles. Porphyrins that give rise to heme, chlorophyll, and many other biomolecules are tetrapyrrole derivatives. See *porphyrin*. [10]

Thalamus. A brain region under the cerebral hemispheres. It integrates the flow of information between all other parts of the brain and the cerebral hemispheres. [30]

Thallus. The body of a simple vegetative plant, fungus, or lichen that is not differentiated into roots, stems, or leaves. [39,40]

Thecodont. An extinct order of reptiles which gave rise to the birds. [43]

Therapsid. An extinct order of reptiles which gave rise to the mammals. [43]

Thermobacteria. A type of bacteria that grow best at high temperatures. [38]

Thermocline. An abrupt temperature transition (at about 4°C) occurring between the epilimnion and hypolimnion layers in a lake. [48]

Thermodynamics. The study of heat and energy and the laws governing their interconversion. [8]

Thick filament. Muscle myosin, a protein which interacts with actin (thin) filaments to produce contraction. [7,31]

Thigmotropism. In plants, growth stimulated by contact with a solid object, as when a vine twines around a pole. [32]

Thin filament. A polymer of the globular protein actin, which interacts with myosin (thick) filaments to produce muscular contraction. [31]

Threshold. The level of a stimulus that must be exceeded for an effect to be produced. [29,30]

Thrombin. A blood plasma protein that catalyzes the proteolytic conversion of fibrinogen to fibrin during clotting. [22]

Thrombocyte. See *platelet*. [22]

Thrombosis. Abnormal clot formation within blood vessels, which can lead to heart attacks and strokes. [22]

Thylakoid. A flattened membranous sac, stacks of which form the grana of chloroplasts. Thylakoids contain all of the chlorophyll in a plant and the enzymes of photophosphorylation. [5,10]

Thymine (T, Thy). A pyrimidine nitrogenous base found paired with adenine in DNA. [16]

Thymus. A ductless, glandular lymphoid structure with a fundamental role in the development of the immune systems of vertebrates. [23]

Thyroid. A two-lobed gland in vertebrates. Produces several hormones including thyroxin. [28]

Thyroid-stimulating hormone (TSH). See *thyrotropic hormone*. [28]

Thyrotropic hormone (TSH). A hormone released by the anterior pituitary (adenohypophysis) that stimulates the thyroid to produce thyroid hormone. [28]

Thyroxine (tetraiodothyronine, T4). The major thyroid hormone. An iodine-containing compound that stimulates development and metabolism. [28]

Tight junction. An area in which the membranes of adjacent animal cells are so tightly joined that there is no extracellular space through which molecules can diffuse. [7,20]

Tissue. A group of cells with similar appearance and function. Often integrated with other tissues to form an organ. [5,7]

Tolerance range. That portion of a range of variable environmental conditions within which a species can survive. [46]

Tonoplast. The limiting membrane of an intracellular vacuole. [5]

Tonus. Continuous contraction of a muscle. [24]

Tornaria. The free-swimming ciliated larva of hemichordates. Its resemblance to echinoderm larvae supports the claim that echinoderms and chordates have a close evolutionary relationship. [43]

Torr. A unit of pressure. Atmospheric pressure at sea level is normally 760 torr. [27]

Torsion. Twisting of the body during the development of gastropods, that results in formation of the mantle cavity, and situates gills and anus anteriorly. [42]

Totipotent cell. A cell with all the developmental capacity of a zygote. An entire organism can arise from it. [19]

Tracer. A label that permits a compound to be followed in metabolism or in chemical reactions. Radioactive atoms (isotopes) are often used as tracers. [3]

Trachea. A tube that permits gas exchange in land animals. In vertebrates, it is single and leads to the lungs. In arthropods, branching tracheae extend from multiple surface spiracles. [27]

Tracheid. One of two types of fluid-conducting cell in the xylem of vascular plants. Compare with *vessel element*. [24]

Tracheoles. In arthropods, the smallest branches of the tracheae. [27]

Tracheophyte. A member of the plant subkingdom Tracheophyta. A vascular plant containing xylem and phloem; includes ferns and seed plants. [21,41]

Tract. A bundle of nerve fibers in the central nervous system. Compare with *nerve*. [29]

Transcription. The biosynthesis of a molecule of RNA which is complementary to a piece of DNA, using the DNA as a template. Compare with *translation*. [17]

Transcriptional signal. A DNA sequence upstream or downstream of a gene that influences transcription of that gene; includes promoters and enhancers. [18]

Transcription factor IIIA. A protein that regulates the transcription of the 5S rRNA gene during embryonic development. [20]

Transducin. A GTP-binding protein (G protein) in the rods and cones of the eye that helps to convert light-induced bleaching of rhodopsin molecules into nerve impulses. [30]

Transduction. (1) The transfer via a viral carrier of genetic material from one cell to another. [16] (2) The conversion by a neural structure of one form of signal into another. [30]

Transferrin. The iron-binding transport protein of plasma. [6,22]

Transfer RNA (tRNA). A class of carrier molecules necessary for protein synthesis. Each contains a triplet nucleotide (anticodon) for one amino acid. [17]

Transformation. (1) Change occurring in homologous structures that alters their structure and function. [37] (2) Mechanism for transfer of genetic information in bacteria wherein pure DNA from bacteria of one genotype is taken in by bacteria of a different genotype and incorporated into its chromosome. [16]

Transforming growth factor-beta. A small protein that regulates growth and mediates inflammation. [20,28]

Transgenic animal. An animal derived from a genetic line to which foreign genes have been artificially introduced. [20]

Transitional polymorphism. Occurs during the period when a selectively advantageous allele is in the process of replacing other alleles. Compare with *balanced polymorphism*. [34]

Translation. The biosynthesis of a protein molecule by a ribosome, using an mRNA molecule as a template. Compare with *transcription*. [17]

Translocation. (1) A chromosomal mutation occurring when a piece of one chromosome breaks off and joins another chromosome. [15] (2) In plants, the movement of food molecules from the photosynthetic areas to other areas of the plant. [24]

Transpiration. The evaporative loss of water vapor from a land plant. Occurs chiefly through the stomata. Provides the motive force that raises water and solutes from the roots through the xylem. [24]

Transport protein. A class of plasma proteins which bind specific substances and carry them through the blood or across membranes. [6,22]

Transposable element. See *transposon*. [17]

Transposon. A DNA sequence flanked by insertion sequences which confer on it the ability to move from one DNA molecule to another. [17]

Transverse tubular system (T system). A network of tubes extending into a vertebrate muscle cell from the cell membrane. It conducts nerve impulses into the cell to trigger contraction. [31]

Tribal behavior. Primate social organizations, characterized by territoriality and dominance hierarchies. [33]

Tricarboxylic acid cycle. See *citric acid cycle*. [9]

Triglyceride. The usual form of stored fat in animals. It is composed by three fatty acids bound to a glycerol backbone. [4]

Triiodothyronine (T3). A thyroid hormone. [28]

Trilobite. An extinct class of Arthropoda, superficially resembling woodlice, that were abundant in the Cambrian to Silurian periods. [44]

Triplet. See *codon*. [17]

Triploidy. The state of having three times the normal haploid number of chromosomes. [15]

tRNA. See *transfer RNA*. [17]

Trochophore. A free-swimming larval form of mollusks, annelids, ectoprocts, and brachiopods. It is evident that these phyla are closely related. [43]

Trophallaxis. In insect societies, the feeding of liquid secretions to one worker by another. [33]

Trophic level. The position of a species in a food chain, indicating which species it consumes and which species consume it. [46]

Tropic hormone. A hormone that influences the growth or activity of another endocrine gland. [28]

Tropical rain forest. Distinctive biome found in tropical zones with abundant rainfall and a long growing season. [48]

Tropism. In plants, growth toward or away from an external stimulus such as light or gravity. Compare with *taxis*. [32]

Tropocollagen. Protein subunits of collagen. [20]

Tropomyosin. A protein in muscle fibers that interacts with actin and troponin to regulate muscle contraction. [31]

Troponin. A protein in muscle fibers that binds calcium and interacts with actin and tropomyosin to regulate muscle contraction. [31]

Trypsin. A protease produced by the pancreas that cleaves peptide bonds to which arginine or lysine contribute the carboxyl group. [26]

Trypsinogen. The inactive precursor of trypsin. [26]

T tubule. See *transverse tubular system*. [31]

Tube feet. Appendages of echinoderms, which are connected to the water vascular system and have locomotory and grasping functions. [42]

Tubular reabsorption. In the kidney, the active transport of substances back from the glomerular filtrate to the blood. [25]

Tubular secretion. In the kidney, the active transport of substances into the glomerular filtrate from the blood. [25]

Tubulin. The protein subunit from which microtubules are formed. [5]

Tundra. A treeless biome between the ice cap and the tree line of arctic regions, characterized by a permanently frozen subsoil and low vegetation. [48]

Tunicate. A member of the subphylum Urochordata; a sea squirt. It has larvae that may resemble the progenitors of subphylum Vertebrata. [43]

Turbellarian. A member of the flatworm class Turbellaria; free-living flatworms. [42]

Turgor. See *pressure potential*. [6,24]

Type specimen. Once regarded as an idealized pattern for a given species. See *archetype*. [37]

Typological systematics. The view that all organisms with the same anatomical features belong to the same systematic unit. Compare with *populational systematics*. [37]

Ultrastructure. Detailed structure of cells and tissues as revealed by electron microscopy. [5]

Ultraviolet radiation (UV light). Electromagnetic radiation with wavelengths shorter than visible light but longer than x-rays. [10]

Umbilical cord. In mammals, a tube containing blood vessels which connects the fetus to the placenta. [21]

Uniformitarianism. The doctrine that the fundamental properties of the universe have not changed over time, an assumption that justified the scientific study of geological history. [37]

Unipotent. A cell which has only one developmental pathway open to it; a determined or committed cell. Compare with *pluripotent cell*. [19]

Univalve. Refers to the single coiled dorsal shell of snails. Compare with *bivalve*. [42]

Unsaturated fatty acid. A fatty acid containing double bonds. [4]

Upstream. In a DNA molecule, a location nearer the 3' end of the molecule. Compare with *downstream*. [17]

Uracil (U, Ura). A pyrimidine nitrogenous base found in RNA. It is complementary to adenine. [16]

Urea. $CO(NH_2)_2$. The major nitrogen-containing end-product of protein metabolism in mammals. Compare with *uric acid* and *ammonia*. [25]

Ureter. The long tube which carries urine from the kidney to the bladder or cloaca. Compare with *urethra*. [25]

Urethra. The tube which carries urine from the bladder to the exterior of vertebrates. Compare with *ureter*. [25]

Uric acid. $C_5H_4N_4O_3$. The principal nitrogen-containing end-product of protein metabolism in some animals, notably those which must conserve water, such as birds, insects, and reptiles. Compare with *urea* and *ammonia*. [25]

Urine. The liquid waste excreted by the kidney and usually stored before elimination from the body. In birds, reptiles, and insects the principle excretory product. [25]

Urkaryote. The ancient cellular form postulated to be the evolutionary precursor of eukaryotes. Lacking cytoplasmic organelles, it acquired them when invaded by endosymbionts. [38,44]

Uterus. The muscular chamber in certain female mammals in which the embryo develops. [21]

Utricle. In the vertebrate inner ear, an organ of balance which uses hair cells to sense the animal's linear acceleration. Compare with *saccule* and *semicircular canal*. [30]

Vaccination. The technique of inducing immunity to a substance by injecting a vaccine (an antigenic substance in weakened form) into an organism. [23]

Vacuole. A large membrane-bound, fluid-containing sac, common in plants. [5]

Van der Waals force. A weak attractive force between two nonpolar atoms or molecules. [3]

Variable (V) region. In an antibody molecule, the region with a distinctive amino acid sequence that shapes the antigen-binding site and confers immunologic specificity. Compare with *constant region*. [23]

Vascular. Pertaining to tissues and organs that conduct fluid, e.g., blood vessels in animals, xylem and phloem in plants. [7,10]

Vascular cambium. In vascular plants, a sheath of meristem cells that produces secondary xylem and phloem. See *cork cambium*. [41]

Vascular ray. In vascular plants, radially oriented sheets of cells produced by the vascular cambium, carrying materials laterally between the wood and the phloem. [5,41]

Vasopressin. See *antidiuretic hormone*. [25,28]

Vector DNA. A DNA molecule that can replicate within a cell and thus can be used to transfer genes from cell to cell. [18]

Vegetal pole. The side of a zygote or early embryo containing the largest cells and greatest amount of yolk. Compare with *animal pole*. [19]

Vegetative reproduction. A mode of asexual reproduction by which organisms sprout offspring from their bodies as appendages. [21]

Vein. A thin-walled vessel which returns blood from the capillaries to the heart. Compare with *artery*. [24]

Vena cava. One of a pair of large veins that carry blood from the systemic circulatory system into the heart. [24]

Ventilation. In vertebrates, the repeated exposure of the lung to fresh volumes of air. [27]

Ventricle. A chamber. (1) In the heart, muscular chambers that pump blood. Compare with *atrium*. [24] (2) In the brain, they are filled with cerebrospinal fluid. [30]

Vertebrata. A subphylum of the phylum Chordata; includes the fishes, amphibians, reptiles, and birds. Animals whose nerve cords are enclosed in a backbone of bony segments (vertebrae). [43]

Vessel element. One of two types of tubular, fluid-conducting cells in the xylem of vascular plants. With tracheids, vessel elements conduct water and minerals in the plant. [24]

Villus. Fingerlike projections of a surface membrane. For example, the intestinal wall contains cells with microvilli that increase gut surface area. [26]

Viroid. An infective agent that is simpler than a virus in that it lacks a protein coat and consists only of nucleic acid. [5]

Virulent phage. A bacteriophage that reproduces in and kills its bacterial host. Compare with *temperate phage*. [16]

Virus. An infective particle consisting of a nucleic acid (DNA or RNA) core within a protein coat. Viruses can replicate only within living cells. [5,16,18]

Visible radiation. Light. Electromagnetic radiation with wavelengths between those of infrared radiation and x-rays. [10]

Vitalism. The discarded doctrine that living substance differs fundamentally from nonliving substance. Compare with *mechanism*. [2]

Vitamin. An organic molecule needed in small amounts by an organism but which it cannot synthesize itself. [26]

Vitamin D. A molecule that increases blood levels and intestinal absorption of calcium. In the presence of sunlight it is synthesized in skin. [28]

Vitelline envelope. A coating of eggs which helps regulate sperm binding. [19]

Viviparous. A mode of sexual reproduction in animals in which offspring develop inside the maternal body and are released either immature or as miniature adults. Compare with *oviparous*. [19]

Warning coloration. Conspicuous coloration that makes an organism highly visible. Common in species with disagreeable traits (stingers, bad taste, etc.). Also called *aposematic coloration*. Compare with *cryptic coloration*. [35]

Water cycle. See *hydrologic cycle*. [46]

Water potential. A measure of the potential energy of water, it equals the sum of osmotic potential (osmotic pressure) and pressure potential (hydrostatic pressure) and reflects the tendency of a solution to take up water from pure water through a semi-permeable membrane. Water flows toward a solution of more negative water potential. [6,24]

Water vascular system. Unique in echinoderms, a system of hydraulic canals that develops from the larval coelom and connects to tube feet. Involved in gas exchange, sensory reception, and food ingestion. [42]

White cell. See *leukocyte*. [22]

White matter. A region of the central nervous system containing few neuronal cell bodies and many myelinated fibers. Compare with *gray matter*. [29]

Wild type. With respect to a given trait, the genotype or phenotype that occurs in nature—as opposed to a mutant type. [14]

Wood. Accumulated secondary xylem cells in vascular plants. [24]

Work. The effect of energy transfer on a material. In mechanics, work equals the product of force and displacement. [8]

Xanthophylls. Red, orange, and yellow accessory pigments of photosynthesis. They are derived from carotenoids. [10]

X chromosome. A sex chromosome. [15]

X-ray diffraction. A technique in which molecular structure is deduced from the pattern of x-ray scatter produced when a crystal of a material is subjected to x-ray irradiation. [8]

XX-XO system. The sex chromosomes system of such animals as grasshoppers, in which females have two X chromosomes, males have one, and there is no Y chromosome. [15]

Xylem. Plant vascular tissue that conducts most of the water and minerals from the root to other parts of the plant. Composed of tracheids and other specialized cells. Constitutes the wood of trees and shrubs. [7,24]

Y chromosome. A sex chromosome found in the male of a species. [15]

Y-linked inheritance. The pattern of inheritance shown by the few Y chromosome genes. Only males possess Y-linked genes. [15]

Yolk. A nutrient storage pool of fat and protein that is found in most animal eggs. [19]

Yolk platelet. Small bodies in animal egg cytoplasm that contain yolk. [19]

Z-DNA. A form of DNA that spirals to the left rather than the right. [18]

Zona pellucida. Thick elastic envelope of the ovum. [19]

Zooflagellate. Member of the protistan phylum Zoomastigina; animallike protists. [39]

Zooplankton. The nonphotosynthetic organisms of plankton. [39,46,48]

Zoospore. A motile, asexually reproducing cell of fungal origin, bearing one or two flagella. A product of mitosis. [21,40]

Zooxanthellae. Unicellular golden-brown algae, often dinoflagellates (*Gymnodinium*), that live as symbionts within marine protists or animals. Important in the ecology of coral reefs. [39]

Z protein. In photosynthesis, a manganese-containing protein found in photosystem II that splits water molecules to replenish the electron supply of the photosystem. [10]

Zygomycetes. A phylum of the kingdom Fungi; bread molds. [40]

Zygospore. A thick-walled sexual resting spore produced by the fusion of two gametangia. Found in the fungal phylum Zygomycetes. Not flagellated. [21,40]

Zygote. A diploid nucleus or fertilized egg cell produced by the fusion of two haploid cells. Progenitor cell of a multicellular organism. [12,19]

Zygotic meiosis. A life cycle seen in some algae, in which meiosis takes place in the zygote immediately after it is formed. Compare *gametic* and *sporic meiosis*. [21]

Zymogen. A proenzyme. An inactive enzyme precursor which is activated by the action of an acid, another enzyme, or other means. [8,22]

ZZ-ZW system. The sex chromosome system of such aniamls as birds and butterflies, in which the male is the homogametic sex. [15]

QUIZ QUESTION ANSWERS

Chapter 1
1. D   2. A   3. D   4. B   5. D

Chapter 2
1. D   2. B   3. A   4. C   5. B

Chapter 3
1. A   2. B   3. D   4. D   5. B

Chapter 4
1. E   2. E   3. B   4. E   5. B

Chapter 5
1. D   2. B   3. D   4. A   5. C

Chapter 6
1. A   2. D   3. D   4. B   5. A

Chapter 7
1. D   2. B   3. A   4. E   5. E

Chapter 8
1. C   2. A   3. E   4. B   5. E

Chapter 9
1. C   2. E   3. B   4. C   5. D

Chapter 10
1. D   2. C   3. B   4. A   5. E

Chapter 11
1. A   2. A   3. B   4. C   5. E

Chapter 12
1. D   2. E   3. D   4. A   5. D

Chapter 13
1. C   2. B   3. D   4. B   5. E

Chapter 14
1. D   2. C   3. E   4. C   5. E

Chapter 15
1. B   2. C   3. E   4. D   5. E

Chapter 16
1. A   2. D   3. C   4. B   5. E

Chapter 17
1. A   2. C   3. E   4. D   5. C

Chapter 18
1. E   2. E   3. D   4. B   5. D

Chapter 19
1. B   2. H   3. B   4. C   5. A

Chapter 20
1. A   2. E   3. E   4. B   5. B

Chapter 21
1. A   2. C   3. B   4. E   5. B

Chapter 22
1. E   2. B   3. D   4. A   5. C

Chapter 23
1. D   2. E   3. A   4. A   5. B

Chapter 24
1. E   2. C   3. D   4. A   5. E

Chapter 25
1. B   2. E   3. A   4. E   5. C

Chapter 26
1. C   2. A   3. D   4. B   5. E

Chapter 27
1. E   2. B   3. E   4. A   5. D

Chapter 28
1. A   2. D   3. B   4. D   5. E

Chapter 29
1. B   2. A   3. E   4. D   5. B

Chapter 30
1. A   2. E   3. A   4. A   5. D

Chapter 31
1. B   2. E   3. E   4. E   5. A
6. B

Chapter 32
1. C   2. A   3. C   4. E   5. E

Chapter 33
1. C   2. B   3. E   4. D   5. E

Chapter 34
1. A   2. E   3. C   4. B   5. C

Chapter 35
1. E   2. B   3. D   4. E   5. B

Chapter 36
1. D   2. E   3. A   4. B   5. D

Chapter 37
1. D   2. C   3. C   4. B   5. E

Chapter 38
1. B   2. E   3. A   4. C   5. E

Chapter 39
1. B   2. A   3. D   4. E   5. E

Chapter 40
1. A   2. C   3. E   4. E   5. A

Chapter 41
1. C   2. B   3. B   4. E   5. A

Chapter 42
1. B   2. E   3. B   4. E   5. A

Chapter 43
1. C   2. A   3. D   4. A   5. B

Chapter 44
1. C   2. E   3. C   4. C   5. A

Chapter 45
1. A   2. D   3. A   4. C   5. D

Chapter 46
1. B   2. D   3. E   4. C   5. E

Chapter 47
1. E   2. E   3. E   4. B   5. A

Chapter 48
1. E   2. A   3. B   4. E   5. C

Chapter 49
1. A   2. E   3. B   4. E   5. A
6. B,C

CREDITS AND ACKNOWLEDGMENTS

ART CREDITS

General Studio: Vantage Art, Inc. (technical art)

Frontispiece: Barron Storey

Part Opener Art: John D. Dawson

Section Opener Art: Dugald Stermer

Artists: Ron DeFelice, Klarie Phipps, David Purnell (human and anatomical art), Kristine Rasmussen, Judy Skorpil, Dimitry Schidlovsky

PHOTO CREDITS

Chapter 1

page 2: (*tl,br*) Grant Heilman; (*tc,cl,c*) Dan McCoy/Rainbow; (*tr*) Breck P. Kent/Animals Animals; (*cr*) David Stone/Rainbow; (*bl*) Zig Lesczynski/Animals Animals; (*bc*) Doug Wallin/Taurus Photos. Figures 1–1 and 1–3: NASA/Science Source/Photo Researchers. Figure 1–4: Keith Gunnar/Bruce Coleman, Inc. Figure 1–5: Tom Brakefield. Figure 1–6: (A) Carl Roessler/Animals Animals; (B) John Lidington/Photo Researchers; (C) Peter Park, Oxford Scientific Films/Animals Animals; (D) G. I. Bernard, Oxford Scientific Films/Animals Animals. Figure 1–8: Carl Roessler/Animals Animals. Figure 1–12: G. L. Rooyman/Animals Animals. Figure 1–13: Roger Tory Peterson/Photo Researchers. Figure 1–14: M. P. Kahl/Bruce Coleman, Inc. Figure 1–15: Mike Holmes/Animals Animals. Figure 1–16: Jim P. Garrison/Rainbow. Figures 1–17 and 1–18: Tom Brakefield. Figure 1–19: Grant Heilman. Figure 1–20: Tom Walker/Stock, Boston. Figure 1–21: Cliff Fairchild/Taurus Photos. Figure 1–22: Ned Haines/Photo Researchers.

Chapter 2

page 20: The Bettmann Archive. Figures 2–1 and 2–2: The Bettmann Archive. Figure 2–3: UPI/Bettmann Newsphotos. Figures 2A, 2B, and 2D: The Bettmann Archive. Figure 2–4: NCI/Science Source/Photo Researchers. Figures 2–5, 2–6, and 2–7: The Bettmann Archive. Figure 2–8: Historical Pictures Service, Chicago. Figure 2–9: Stephen J. Krassemann/Peter Arnold, Inc.

Chapter 3

page 36: Alfred Pasieka/Taurus Photos. Figure 3–9: Wendell Metzen/Bruce Coleman, Inc. Figure 3–10: Jeremy Burgess, Science Source/Photo Researchers. Figure 3–16: Hermann Eisenbeiss/Photo Researchers.

Chapter 4

page 60: Computer Graphics Laboratories, Regents of the University of California. Figure 4–4: Clyde H. Smith/Peter Arnold, Inc. Figures 4–22 and 4–23: Historical Pictures Service, Chicago. Figure 4–25: Sidney Fox/Visuals Unlimited.

Chapter 5

page 88: W. L. Dentler, Univ. of Kansas/BPS. Figure 5–1: (A) The Bettmann Archive; (B) Omikron/Photo Researchers. Figure 5–5: (B) (*t,c*) Lee D. Simon/Photo Researchers; (*b*) Omikron/Photo Researchers. Figure 5–6: P. A. Merz/Institute for Basic Research in Developmental Disabilities/BPS. Figure 5–8: (B) Brian Eyden, Science Photo Library/Photo Researchers. Figure 5–9: (B) Biophoto Assoc./Photo Researchers. Figure 5–10: (B) T. J. Beveridge, Univ. of Guelph/BPS. Figure 5–11: Don Fawcett/Photo Researchers. Figure 5–12: (A) W. Rosenberg, Iona College/BPS; (B) Don Fawcett/Photo Researchers. Figure 5A: Courtesy, E. Leitz Co. Figures 5C, 5D, and 5E: Jim Solliday/Biological Photo Service. Figure 5F: J. F. Gennaro/Photo Researchers. Figure 5G: Science Source/Photo Researchers. Figure 5–13: (A) R. Bolender/Photo Researchers. Figure 5–16: (A) G. T. Cole, Univ. of Texas—Austin/BPS; (B) Don Fawcett/Photo Researchers. Figure 5–17: Courtesy, H. Loeb. Figure 5–19: Don Fawcett and R. Bolender/Photo Researchers. Figure 5–20: (A) Don Fawcett and Keith Porter/Photo Researchers; (B) Omikron/Photo Researchers. Figure 5–23: Don Fawcett/Photo Researchers. Figures 5–26 and 5–28: (C) David M. Phillips/Visuals Unlimited. Figure 5–29: (B) W. L. Dentler, Univ. of Kansas/BPS. Figure 5–30: (A) J. J. Wolosewick/Biological Photo Service. Figure 5–31: Omikron/Photo Researchers. Figure 5–34: (B) E. H. Newcomb and W. P. Wergin, Univ. of Wisconsin—Madison/BPS.

Chapter 6

page 122: Courtesy, John Hartwig. Figure 6–13: Don Fawcett/Photo Researchers. Figure 6–22: Biological Photo Services. Figure 6–24: Don Fawcett/Photo Researchers. Figure 6–26: (C) E. H. Newcomb, Univ. of Wisconsin—Madison/BPS.

Chapter 7

page 144: Courtesy, John Hartwig. Figure 7–5: (B) Courtesy, J. Heuser, Washington University. Figure 7–7: B. F. King, Univ. of California—Davis/BPS. Figure 7–12: Courtesy, Richard O. Hynes, MIT. Figure 7–13: (B) K. W. Jeon/Visuals Unlimited. Figure 7–14: (A) Don Fawcett/Photo Researchers. Figure 7–15: (A) Courtesy, Katherine B. Pryzwansky; (B) Science Photo Library/Photo Researchers. Figure 7–17: Don Fawcett and J. Heuser/Photo Researchers.

Chapter 8

page 164: Science Photo Library/Photo Researchers. Figure 8–2: (A) R. Wallace/Visuals Unlimited; (B) Bruce Coleman, Inc.; (C) Thomas Eisner and Daniel Aneshansley; (D) Howard Miller/Photo Researchers; (E) David M. Phillips/Visuals Unlimited; (F) M. Abbey/Visuals Unlimited.

Chapter 9

page 196: Biophoto Assoc./Photo Researchers. Figure 9–1: Gary Ladd/Photo Researchers. Figure 9–2: K. Pugh/Visuals Unlimited. Figure 9–11: K. G. Murti/Visuals Unlimited.

Chapter 10

page 220: Grant Heilman. Figure 10–3: Courtesy, C. J. Tucker, Goddard Space Center/NASA. Figure 10–11: Runck/Schoenberger/Grant Heilman. Figure 10–12: (C) E. H. Newcomb and W. P. Wergin, Univ. of Wisconsin—Madison/BPS. Figure 10–13: Courtesy, Keith Miller, Brown Univ. Figure 10–21: (B) Courtesy, W. M. Laetsch, Univ. of California—Berkeley. Figure 10–24: J. N. A. Lott, McMaster University/BPS. Figure 10–27: Courtesy, Joseph R. Thomasson, *Science*.

Chapter 11

page 248: Ivan Massar/Black Star. Figure 11–3: (B) Don Fawcett/Photo Researchers.

Chapter 12

page 262: Science Photo Library/Photo Researchers. Figure 12–1: (B) J. J. Cardamone, Jr., Univ. of Pittsburgh/BPS. Figure 12–2: Richard L. Moore/Biological Photo Service. Figure 12–3: (A,B,C,D) Peter J. Bryant/Biological Photo Service. Figure 12–4: David M. Phillips/Visuals Unlimited. Figure 12–5: The Bettmann Archive. Figure 12–6: Courtesy, J. Cairns. Figure 12–8: (B) J. F. Gennaro/Photo Researchers; (C) Science Photo Library/Photo Researchers. Figure 12–13: Daniel Mazia. Figure 12–17: K. Porter/Science Source/Photo Researchers. Figure 12–18: Don Fawcett/Photo Researchers. Figure 12–19: C. L. Rieder, N. Y. State Dept. of Public Health/BPS. Figure 12–20: (A) Biophoto Assoc./Science Source/Photo Researchers; (B) Omikron/Photo Researchers; (C) David M. Phillips/Visuals Unlimited. Figure 12–21: Courtesy, Ada L. Olins and Donald E. Olins, Univ. of Tennessee. Figure

12–25: (A) Don Fawcett/Photo Researchers; (B) Courtesy, U. K. Laemmli, Princeton Univ. Figure 12–34: Courtesy, David E. Comings, City of Hope Medical Center, Duarte, CA.

Chapter 13

page 294: Nordisk Pressfoto. Figure 13–2: Photo Researchers. Figure 13–3: (A) Larry Lefever/Grant Heilman. Figure 13–6: Charles Marden Fitch/Taurus Photos.

Chapter 14

page 308: John D. Cunningham/Visuals Unlimited. Figure 14–5: (A,B) Carolina Biological Supply Co.; (C) Peter J. Bryant/Biological Photo Service. Figure 14–11: J. N. A. Lott, McMaster University/BPS. Figure 14–22: Carolina Biological Supply Co.

Chapter 15

page 330: FourByFive. Figure 15–8: G. W. Willis, Ochsner Medical Institution/BPS. Figure 15–12: (l) Courtesy, J. M. Ortiz, Shodair Children's Hospital; (r) Courtesy, C. M. Moore. Figure 15–13: Courtesy, D. S. Borgaonkar, Medical Center of Delaware. Figure 15–15: (A) C. I. Scott/A. I. DuPont Inst., Wilmington; (B) Rose Fujimoto. Figure 15–20: Elizabeth Crews/Stock, Boston.

Chapter 16

page 358: K. G. Murti/Visuals Unlimited. Figure 16–2: Lee. D. Simon/Photo Researchers.

Chapter 17

page 388: Stock, Boston. Figure 17–1: (A) Courtesy, O. L. Miller, Jr., Univ. of Virginia; (B) Don Fawcett/Photo Researchers. Figure 17–30: Courtesy, Joseph G. Gall, Carnegie Institute.

Chapter 18

page 418: Stanley Cohen, Science Photo Library/Photo Researchers. Figure 18–6: Courtesy, B. Daneholt. Figure 18–8: Courtesy, Keith Yamamoto, Univ. of California—San Francisco. Figure 18C: Courtesy, Alexander Rich, Massachusetts Institute of Technology. Figure 18–19: John D. Cunningham/Visuals Unlimited. Figures 18–23 and 18–24: Courtesy, Henry Harris, University of Oxford.

Chapter 19

page 448: Petit Format, Nestle/Photo Researchers. Figure 19–1: Peter J. Bryant/Biological Photo Services. Figure 19–2: Alexander Tsiaras/Photo Researchers. Figure 19–3: William M. Harlow/Photo Researchers. Figures 19–4 and 19–5: Barka/William S. Beck. Figure 19–8: Biophoto Assoc./Photo Researchers. Figure 19–10: John Walsh, Science Source/Photo Researchers. Figure 19–12: Courtesy, Richard Kessel, Univ. of Iowa. Figure 19–17: G. Schatten, Univ. of Wisconsin/BPS. Figure 19–19: Courtesy, Everett Anderson, Harvard Medical School. Figure 19–21: (A,B,C,D) R. Yanagimachi, Univ. of Hawaii—Manoa/BPS. Figure 19–22: Omikron/Photo Researchers. Figure 19–26: David M. Phillips/Visuals Unlimited.

Chapter 20

page 476: David Scharf/Peter Arnold, Inc. Figure 20–4: Peter J. Bryant/Biological Photo Service. Figure 20–9: Clay-Adams, Inc. Figures 20A, 20B, and 20C: C. L. Markert and R. M. Petters, reprinted with permission, *Science, 202,* 56–58, 1978. Figure 20–12: Philip Leder and Richard Woychik, reprinted by permission from *Nature, 318,* 36–40, copyright © 1985, Macmillan Magazines Ltd. Figure 20–13: Courtesy, R. L. Brinster, University of Pennsylvania. Figures 20–14 and 20–16: J. E. Sulston. Figure 20–23: David Scharf/Peter Arnold, Inc. Figure 20–24: Walter Gehring, reprinted by permission from *Nature, 313,* cover,

copyright © 1985, Macmillan Magazines Ltd. Figure 20–27: David M. Phillips/Visuals Unlimited. Figure 20–28: R. D. Goldman, Carnegie Mellon Univ./BPS. Figure 20–29: D. W. Fawcett and J. Heuser/Photo Researchers. Figure 20–30: Jim Solliday/Biological Photo Services.

Chapter 21

page 506: (tl) Runk/Schoenberger/Grant Heilman; (tr) A. M. Siegelman/Visuals Unlimited; (cl) Biophoto Assoc./Photo Researchers; (c) D. J. Wrobel, Monterey Bay Aquarium/BPS; (cr) Petit Format/Nestle/Science Source/Photo Researchers; (bl) Tom Branch/Science Source/Photo Researchers; (br) Jeff Lepore/Photo Researchers. Figure 21–2: Jim Solliday/Biological Photo Service. Figure 21–3: Kenneth D. Whitney/Visuals Unlimited. Figure 21–4: Alford W. Cooper/Photo Researchers. Figure 21–5: Fred Hossler/Visuals Unlimited. Figure 21–8: Patrick Grace/Taurus Photos. Figure 21–9: (A) E. R. Degginger/Bruce Coleman, Inc. Figure 21–11: (B) D. P. Wilson, Eric and David Hosking/Photo Researchers. Figure 21–13: (B) Bruce Coleman, Inc. Figure 21–15: J. N. A. Lott, McMaster Univ./BPS. 21–16: (B) B. Ormerod/Visuals Unlimited. Figure 21–27: (B) Ed Reschke/Peter Arnold, Inc.

Chapter 22

page 552: David M. Phillips/Visuals Unlimited. Figure 22–1: (A) Fred Hossler/Visuals Unlimited; (B,C,D) William S. Beck. Figure 22–2: Stan Elems/Visuals Unlimited. Figure 22–3: Lester V. Bergman & Assoc. Figure 22–5: (D) Grant Heilman. Figure 22–9: Omikron/Photo Researchers. Figure 22–12: David M. Phillips/Visuals Unlimited. Figures 22–15 and 22–17: (A,B,C) William S. Beck. Figure 22–18: (A) Biophoto Assoc./Photo Researchers. Figure 22–19: Dennis D. Kunkel/Biological Photo Service.

Chapter 23

page 576: Lennart Nilsson/Bonnier Fakta. Figure 23–1: Science Photo Library/Photo Researchers. Figure 23–2: Omikron/Photo Researchers. Figure 23–8: William S. Beck. Figure 23–11: R. Rodewald, Univ. of Virginia/BPS. Figure 23–12: R. Albrecht, University of Wisconsin—Madison/BPS. Figure 23F: (C) Albert W. Coons. Figure 23–14: Don Fawcett/Science Source/Photo Researchers. Figure 23I: Jeremy Burgess/Science Photo Library/Photo Researchers. Figure 23M: Alfred Pasieka/Taurus Photos. Figure 23–26: Courtesy, Paul Russell, Massachusetts General Hospital.

Chapter 24

page 612: W. Rosenberg, Iona College/BPS. Figures 24–4: (B) and 24–8: John D. Cunningham/Visuals Unlimited. Figures 24–10: (A,B) and 24–13: J. N. A. Lott, McMaster Univ./BPS. Figure 24–14: (A) J. R. Waaland, Univ. of Washington/BPS; (B) John D. Cunningham/Visuals Unlimited. Figure 24C: David Scharf/Peter Arnold, Inc. Figure 24–29: Dennis D. Kunkel/Biological Photo Service. Figure 24–36: Lester V. Bergman & Assoc.

Chapter 25

page 640: Courtesy, Reinier Beeuwkes, III. Figure 25–4: (b) A. M. Siegelman/Visuals Unlimited. Figure 25–5: (E,F) M. Abbey/Visuals Unlimited. Figure 25–15: B. F. King, Univ. of California—Davis, School of Medicine/BPS. Figure 25–16: John D. Cunningham/Visuals Unlimited.

Chapter 26

page 660: Runk/Schoenberger/Grant Heilman. Figure 26–2: (A) Ken Lucas/Biological Photo Services; (B) Runk/Schoenberger/Grant Heilman. Figure 26–9: (A,B) Karel F. Leim. Figures 26–11 and 26–12: Biophoto Assoc./Photo Researchers. Figure 26–17: Fawcett and Heuser/Hirokawa Science Source/Photo Researchers. Figure 26–22: Gilda Schiff/Photo Researchers.

Chapter 27

page 688: Courtesy, J. B. West, Young-Tenzing/Christopher Pizzo. Figure 27–7: David M. Phillips/Visuals Unlimited. Figure 27–10: Peter J. Bryant/Biological Photo Service. Figure 27–15: Fred McConnaughey/Photo Researchers. Figure 27–22: Johnson Space Center/NASA. Figure 27–23: Ed Reschica/Peter Arnold, Inc. Figure 27–26: H. R. Duncker. Figure 27–30: (A) John D. Cunningham/Visuals Unlimited; (B,C) Tom McHugh/Photo Researchers.

Chapter 28

page 716: (t) Grant Heilman; (bl,br) Runk/Schoenberger/Grant Heilman. Figure 28–20: (t) Martin M. Rotker/Taurus Photos; (b) Photo Researchers. Figure 28–24: (A) Gary Gibson/Photo Researchers; (B) Tom McHugh/Photo Researchers; (C) Nathan W. Cohen/Visuals Unlimited. Figure 28–25: (B) Courtesy, Ira R. Telford. Figure 28–29: John Paul Kay/Peter Arnold, Inc. Figure 28–34: Manfred Kage/Peter Arnold, Inc. Figure 28–41: Courtesy, Aaron Lerner, Yale Univ.

Chapter 29

page 762: Biophoto Assoc./Science Source/Photo Researchers. Figure 29–3: (A,B) M. W. F. Tweedie/Photo Researchers. Figure 29A: Stephen J. Krasemann/Peter Arnold, Inc. Figure 29B: Charles Marden Fitch/Taurus Photos. Figure 29C: Nuridsany et Perennou/Photo Researchers. Figure 29D T. J. Beveridge, Univ. of Guelph/BPS. Figure 29–5: (E) Petit Format/Science Source/Photo Researchers. Figures 29F and 29G: Courtesy, Hospital Ramón y Cajal. Figure 29–7: (E) David M. Phillips/Visuals Unlimited. Figure 29–18: (B) R. Calentine/Visuals Unlimited. Figure 29–13: (B) Biological Photo Service. Figure 29–19: (B) D. W. Fawcett, J. Heuser, and T. Reese/Photo Researchers. Figure 29–25: (B) Courtesy, David H. Hall, Albert Einstein College of Medicine. Figure 29–26: Don Fawcett/Photo Researchers.

Chapter 30

page 788: Kit Johnson. Figure 30–5: (C) Dennis D. Kunkel/Biological Photo Supply. Figure 30–7: (B) Courtesy, Scott Mittman, University of California—San Francisco. Figure 30–21: (B) E. R. Lewis, Univ. of California—Berkeley/BPS. Figure 30–27: Peter J. Bryant/Biological Photo Service. Figure 30–29: R. Andrew Odum/Peter Arnold, Inc. Figure 30–35: I. Shoulson/Visuals Unlimited.

Chapter 31

page 826: (t) Kim Taylor/Bruce Coleman, Inc.; (b) Bob and Clara Calhoun/Bruce Coleman, Inc. Figure 31–3: (B) Don Fawcett/Photo Researchers. Figure 31–6: F. A. Pepe, Univ. of Pennsylvania/BPS. Figure 31–8: (B) Don Fawcett and J. Heuser/Photo Researchers. Figure 31–17: Don Fawcett/Photo Researchers. Figure 31–19: (A) Don Fawcett/Cooke/Photo Researchers; (C) Don Fawcett/Photo Researchers. Figure 31–24: John D. Cunningham/Visuals Unlimited. Figure 31–36: (B) Richard Parker/Photo Researchers. Figure 31–38: (A) Ken Lucas/Biological Photo Service; (B) R. Degoursey/Visuals Unlimited.

Chapter 32

page 854: John Gerlack/Visuals Unlimited. Figure 31–11: Kim Taylor/Bruce Coleman, Inc. Figure 32–13: The Bettmann Archive. Figure 32–17: Thomas McAvoy, *LIFE Magazine* © 1955 Time Inc./Life Picture Service. Figure 32–22: Peter J. Bryant/Biological Photo Service. Figure 32–29: (B) Leonard Lee Rue/Photo Researchers. Figure 32–30: (B,C) L. P. Brower. Figure 32–31: (C) Link/Visuals Unlimited. Figure 32–32: (B) Ross E. Hutchins/Photo Researchers.

Chapter 33

page 882: E. R. Degginger. Figure 33A: Stephen Collins, National Audubon Society/Photo Researchers.

Figure 33B: Milton J. Heilberg/Photo Researchers. Figure 33-1: (B) Tom McHugh, Steinhart Aquarium/Photo Researchers. Figures 33-5: (B) and 33-8: Bruce Coleman, Inc. Figure 33-9: (A) Taurus Photos. Figure 33-10: Bob Gossinger/Bruce Coleman, Inc. Figure 33-13: L. J. Conner/Visuals Unlimited. Figure 33-14: John D. Cunningham/Visuals Unlimited. Figure 33D: Akira Uchiyama/Photo Researchers. Figure 33E: Tom McHugh/Photo Researchers.

Chapter 34

page 906; (tl) Michael Colhier/Stock, Boston; (tc) David M. Phillips/Visuals Unlimited; (tr) John D. Cunningham/Visuals Unlimited; (cl) Coco McCoy/Rainbow; (c) Frans Lanting/Photo Researchers; (cr) D. Overcash/Bruce Coleman, Inc.; (bl) D. Neuman/Visuals Unlimited; (bc) Joe McDonald/Visuals Unlimited; (br) Owen Franken/Stock, Boston. Figure 34-1: (tl) Kim Taylor/Bruce Coleman, Inc; (tr) S. K. Webster, Monterey Bay Aquarium/BPS; (cl) John D. Cunningham/Visuals Unlimited; (c) Runk/Schoenberger/Grant Heilman; (cr) S. Rannels/Grant Heilman; (br) F. Gohier/Photo Researchers. Figure 34A: John D. Cunningham/Visuals Unlimited. Figure 34-7: (A,B) Carolina Biological Supply Co. Figure 34C: M. P. Kahl/Photo Reserachers. Figure 34D: J. P. Ferrero/Photo Researchers. Figure 34E: J. Alcock/Visuals Unlimited. Figure 34-12: (A,B) Karel F. Liem. Figure 34-14: John D. Cunningham/Visuals Unlimited. Figure 34-16: (A) Daphine Kinzler/Visuals Unlimited; (B) R. Andrew Odum/Peter Arnold, Inc. Figure 34G: Biophoto Assoc./Photo Researchers.

Chapter 35

page 940: (tl) John D. Cunningham/Visuals Unlimited; (tc,tr,bl) E. R. Degginger; (cl) Hans Pfletschinger/Peter Arnold, Inc.; (cr) Habicht/Taurus Photos; (bc) ESA/Ries/Visuals Unlimited; (br) J. Alcock/Visuals Unlimited. Figure 35-3: E. R. Degginger. Figure 35-4: National Audubon Society/Photo Researchers. Figure 35-5: Bob and Clara Calhoun/Bruce Coleman, Inc. Figure 35-7: E. R. Degginger. Figure 35-8: Royal Scottish Museum. Figure 35-9: (A,B) Karel F. Liem. Figure 35H: J. Alcock/Visuals Unlimited. Figure 35I: M. W. Tweedie/Photo Researchers. Figure 35-11: (tl) Dennis Paulson/Visuals Unlimited; (tc) J. Alcock/Visuals Unlimited; (tr) J. N. A. Lott, McMaster Univ./BPS; (bl) Ken Lucas/Biological Photo Supply; (bc) William Palmer/Visuals Unlimited; (br) T. Gula/Visuals Unlimited. Figure 35-12: (t) Ken Lucas/Biological Photo Service; (c) J. Alcock/Visuals Unlimited; (b) National Audubon Society/Photo Researchers. Figure 35-13: (B,C) L. P. Brower; (D) David Clayton/Visuals Unlimited; (E) J. Alcock/Visuals Unlimited. Figure 35-14: (A,B) Karel F. Leim. Figure 35-15: (A,B,C,D) E. R. Degginger; (F) Phil Degginger. Figure 35-16: Courtesy, W. Atlee Burpee & Company.

Chapter 36

page 970: (tl) J. N. A. Lott, McMaster Univ./BPS; (tc,cl) Barbara J. Miller/Biological Photo Service; (tr,cr,bl,bc,br) Runk/Schoenberger/Grant Heilman; (C) Bruce F. Molnia/Terraphotographics/BPS. Figures 36-5 and 36-8: (B) Courtesy, Peter B. Kaufman. Figure 36-9: (A) Bob Gossington/Bruce Coleman, Inc.; (B) David Overcash/Bruce Coleman, Inc. Figure 36-10: E. R. Degginger. Figure 36-11: (A,B) Pat and Tom Leeson/Photo Researchers. Figure 36-15: Miguel Castro/Photo Researchers.

Chapter 37

page 994: (tl,c) Tom McHugh/Photo Researchers; (tc,tr) E. R. Degginger; (cl) Ken Lucas/Biological Photo Service; (cr) Charles W. Mann/Photo Researchers; (bl) Klaus D. Francke/Peter Arnold, Inc.; (bc) Patrice/Visuals Unlimited; (br) Thomas Gula/Visuals Unlimited. Figure 37-1: (A) G. J. Chafaris/E. R. Degginger; (B) E. R. Degginger. Figures 37-2 and 37-3: E. R. Degginger. Figure 37-14: (A) E. R. Degginger; (B) D. Newman/Visuals Unlimited. Figure 37-18: Nolan Preece/Biological Photo Service. Figure 37-20: (A) Runk/Schoenberger/Grant Heilman; (B,C) Michael Collier/Stock, Boston. Figure 37-21: Barbara J. Miller/Biological Photo Service. Figure 37-22: W. B. Saunders, Bryn Mawr/BPS. Figure 37-23: Grant Heilman, Inc.

Chapter 38

page 1024: David M. Phillips/Visuals Unlimited. Figure 38-5: (A,B,C) David M. Phillips/Visuals Unlimited; (D) Fred Hossler/Visuals Unlimited; (E) Runk/Schoenberger/Grant Heilman. Figure 38-7: (A) K. Stephens, Stanford Univ./BPS. Figure 38-9: (A) P. W. Johnson and J. McN. Sieburth/Biological Photo Service. Figure 38-10: S. Abraham and E. H. Beachey, VA Medical Center, Memphis/BPS. Figure 38-12: (A) E. C. S. Chan/Visuals Unlimited. Figure 38-13: Charles C. Brinton, Jr. and Judith Carnaham, Univ. of Pittsburgh. Figure 38-15: (B) Courtesy, P. Dayanandan. Figure 38-16: (A,B) Runk/Schoenberg/Grant Heilman. Figure 38-17: (C) E. H. Newcomb and T. D. Pugh, Univ. of Wisconsin—Madison/BPS.

Chapter 39

page 1042: Runk/Schoenberg/Grant Heilman. Figure 39-4: (A,B,C) P. W. Johnson and J. McN. Sieburth/Biological Photo Service. (John Sieburth/Biological Photo Service. Figures 39A and 39B: Robert E. Pelham/Bruce Coleman, Inc. Figure 39-5: P. W. Johnson and J. McN. Sieburth/Biological Photo Service. Figure 39-6: (A,B,C) L. E. Roth, Univ. of Tennessee/BPS; (D) A. Copley/Visuals Unlimited; (E) William C. Joresensen/Visuals Unlimited; (F) J. R. Waaland, Univ. of Washington/BPS. Figure 39-8: (B) Kim Taylor/Bruce Coleman, Inc. Figure 39-9: (A) E. R. Degginger; (B) Dennis D. Kunkel/Biological Photo Supply. Figures 39-10 and 39-11: Runk/Schoenberg/Grant Heilman. Figure 39-12: (A) J. R. Waaland, Univ. of Washington/BPS. Figure 39-13: (B) Robert R. Carr/Bruce Coleman, Inc. Figure 39-14: M. Abbey/Visuals Unlimited. Figure 39-16: (A) M. Abbey/Visuals Unlimited; (B) Bruce Iverson. Figure 39-18: Jim Solliday/Biological Photo Service. Figure 39-19: Runk Schoenberger/Grant Heilman. Figure 39-22: M. Abbey/Visuals Unlimited. Figure 39-23: (t) Runk/Schoenberg/Grant Heilman; (c) Manfred Kage/Peter Arnold, Inc.; (b) E. R. Degginger. Figure 39-24: M. Abbey/Visuals Unlimited. Figure 39-27: Runk/Schoenberg/Grant Heilman.

Chapter 40

page 1066: Dennis D. Kunkel/Biological Photo Service. Figure 40-1: (A) Maslowski/Visuals Unlimited; (B) John D. Cunningham/Visuals Unlimited; (C) Barry L. Runk/Grant Heilman. Figure 40-2: (A) John D. Cunningham/Visuals Unlimited; (B) David M. Phillips/Visuals Unlimited. Figure 40A: Klaus Moller/Nordisk Pressefoto. Figure 40B: (t) W. A. Banaszewski/Visuals Unlimited; (b) Walt Anderson/Visuals Unlimited. Figure 40E: Robert Fried/Stock, Boston. Figure 40-6: (C) Dennis D. Kunkel/Biological Photo Service. Figure 40-8: (A,B,C) Carolina Biological Supply Co. Figure 40-9: (C) John D. Cunningham/Visuals Unlimited. Figure 40-10: Courtesy, Peter B. Kaufman. Figure 40-13: John D. Cunningham/Visuals Unlimited. Figure 40-14: John Serrao/Visuals Unlimited. Figure 40-15: (A) Forest W. Buchanan/Visuals Unlimited. Figure 40-17: (B) A. Ahmadjian/Visuals Unlimited. Figure 40-18: (l) J. N. A. Lott, McMaster Univ./BPS; (r) John D. Cunningham/Visuals Unlimited. Figure 40-19: (A,B) D. H. Marx/U. S. Dept. of Agriculture, Forest Service.

Chapter 41

page 1086: Grant Heilman. Figure 41-3: (t) Jeff Foott/Bruce Coleman, Inc.; (tc) Norman Owen Tomakin/Bruce Coleman, Inc.; (b) P. Gates, Univ. of Durham/BPS. Figure 41-5: (B) David S. Addison/Visuals Unlimited. Figure 41-6: Bruce Coleman, Inc. Figure 41-9: (A) J. R. Waaland, Univ. of Washington/BPS; (B) E. R. Degginger; (C) J. N. A. Lott/McMaster Univ./BPS. Figure 41-10: (t) Forest W. Buchanan/Visuals Unlimited; (b) Grant Heilman, Inc. Figure 41-11: (B) Charlie Ott/Photo Researchers. Figure 41-13: J. N. A. Lott, McMaster Univ./BPS. Figure 41-17: (tl,tr) John D. Cunningham/Visuals Unlimited; (bl) J. R. Waaland, Univ. of Washington/BPS; (br) J. N. A. Lott, McMaster Univ./BPS. Figure 41-19: (A) John D. Cunningham/Visuals Unlimited. Figure 41-20: E. R. Degginger. Figure 41-21: Runk/Schoenberger/Grant Heilman. Figures 41-23: (A,B) and 41-24: (A,B) E. R. Degginger.

Chapter 42

page 1110: (tl) Ken Lucas/Biological Photo Service; (tr) Peter J. Bryant/Biological Photo Service; (tcl) Frederik D. Bodin/Stock, Boston; (tcr) L. E. Gilbert, Univ. of Texas—Austin/BPS; (bcl) David S. Addison/Visuals Unlimited; (bcr) Thomas Gula/Visuals Unlimited; (bl) Ralph Oberlander/Stock, Boston; (br) Dan McCoy/Rainbow. Figure 42-1: (A,B) E. R. Degginger; (C) Tom McHugh/Photo Researchers; (D) John Mitchell/Photo Researchers. Figure 42-2: Jeff Rotman. Figure 42-3: R. Myers/Visuals Unlimited. Figure 42-7: Biophoto Assoc./Science Source/Photo Researchers. Figure 42-8: Fredrik D. Bodin/Stock, Boston. Figure 42-9: (A) Grant Heilman; (B) D. W. Gotshall/Visuals Unlimited. Figure 42-10: (A) Grant Heilman; (B) John D. Cunningham/Visuals Unlimited; (C) Larry Jensen/Visuals Unlimited. Figure 42-12: (A) C. P. Hickman/Visuals Unlimited. Figure 42-14: (B) Ken Lucas/Biological Photo Services. Figure 42-16: John D. Cunningham/Visuals Unlimited. Figure 42-17: (A) Ken Lucas/Biological Photo Services; (B) Glenn Oliver/Visuals Unlimited. Figure 42-18: (A) George H. Harrison/Grant Heilman; (B) Harold W. Pratt/Biological Photo Service; (C) Jeff Rotman. Figure 42-19: (A) Stan Elems/Visuals Unlimited; (B) William C. Jorgensen/Visuals Unlimited; (C) Harold W. Pratt/Biological Photo Service. Figure 42-21 (t) Ken Lucas/Biological Photo Service; (b) John D. Cunningham/Visuals Unlimited. Figure 42-22: (A) R. DeGoursey/Visuals Unlimited; (B,C) Tom McHugh/Photo Researchers. Figure 42A: (r) Ken Lucas/Biological Photo Service; (c,l) E. R. Degginger. Figure 42-25: (A) Stan Elems/Visuals Unlimited; (B) John Running/Stock, Boston. Figure 42-26: T. E. Adams/Visuals Unlimited. Figure 42-29: John Gerlach/Visuals Unlimited. Figure 42-30: (A) Richard Thom/Visuals Unlimited; (B) Ken Lucas/Biological Photo Service. Figure 42-31: (A) John Gerlach/Visuals Unlimited; (B) Peter J. Bryant/Biological Photo Service. Figure 42-32: (l) ESA-Ries/Visuals Unlimited; (r) Adrian Wenner/Visuals Unlimited. Figure 42-33: Runk/Schoenberger/Grant Heilman. Figure 42-34: (A) Ken Lucas/Biological Photo Service; (B) Glen Oliver/Visuals Unlimited; (C) Ted Clutter/Photo Researchers; (D) Runk/Schoenberger/Grant Heilman.

Chapter 43

page 1144: (tl) F. Gohier/Photo Researchers; (tr) Joe McDonald/Visuals Unlimited; (bl) Thomas Gula/Visuals Unlimited; (br) P. R. Ehrlich, Standford Univ./BPS. Figure 43-2: (A,C) John D. Cunningham/Visuals Unlimited; (B) Tom McHugh/Photo Researchers. 43-4: Rondi/Tani/Photo Researchers. Figure 43-5: (A,B) Ken Lucas/Biological Photo Service; (C) Patrice Ceisel/Visuals Unlimited. 43-9: (A) J. Schauer and H. Fricke, reprinted by permission from Nature, 329, cover, copyright © 1987, Macmillan Magazines Ltd.; (B) D. R. Schwimmer/Bruce Coleman, Inc. Figure 43-11: (A,B) Tom McHugh/Photo Researchers. Figure 43-12: (A,B) Ken Lucas/Biological Photo Service; (C) Carl Gans, Univ. of Michigan/BPS. Figure 43-13: (A) Bob Daemrich/Stock, Boston; (B) Leonard Lee Rue/Visuals Unlimited; (C) Ronald F. Thomas/Taurus Photos. Figure 43-15: (D) John Gerlach/Visuals Unlimited. Figure 43-16: (A) John D. Cunningham/Visuals Unlimited. Figure 43-18: (A) J. L. Lepore/Photo Researchers; (B) E. R. Degginger; (C,E) Tom McHugh/ Photo Researchers; (D) Stephen J. Krasemann/Photo Researchers; (F) Calvin Larsen/Photo Researchers. Figure 43-21: J. N. A. Lott, McMaster Univ./BPS. Figure 43-22: (A) D. & J.Heaton/Stock, Boston; (B) Davis Austen/Stock, Boston. Figure 43–23: (A) Daniel W. Gotshell/Visuals Unlimited; (B) Tom J. Ulrich/Visuals Unlimited; (C) Milton H. Tierney, Jr./Visuals Unlimited; (D,E,F) E. R. Degginger. Figure 43–25: Tom McHugh/Photo Researchers. Figure 43–26: (A,B) Tom McHugh/Photo Researchers.

Chapter 44

page 1176: (tl) John D. Cunningham/Visuals Unlimited; (tr,b) Runk/Schoenberger/Grant Heilman. Figure 44–2: (A) S. M. Awramik, Univ. of California/BPS; (B) Frank T. Awbrey/Visuals Unlimited. Figure 44–6: (A,B,C,D) Field Museum of Natural History. Figure 44–8: Thomas Lunt, American Museum of Natural History. Figure 44–11: American Museum of Natural History. Figure 44–12: John D. Cunningham/Visuals Unlimited. Figure 44–18: Field Museum of Natural History. Figures 44–25: (B) and 44–32: (A,B) American Museum of Natural History. Figure 44–34: (A,D) Tom McHugh/Photo Researchers; (B) Mike Mazzashi/Stock, Boston; (C) Joe McDonald/Visuals Unlimited. Figure 44–39: Cleveland Museum of Natural History.

Chapter 45

page 1218: (tl) John M. Burnley/Photo Researchers; (tc) Ken Lucas/Biological Photo Supply; (tr) Glenn Oliver/Visuals Unlimited; (cl) Tom Walker/Stock, Boston; (c) E. R. Degginger; (cr) Andrew Martinez/Photo Researchers; (bl) Gary P. James/Biological Photo Services; (bc,br) Runk/Schoenberger/Grant Heilman. Figure 45–1: (A) Scott Camazine/Photo Researchers; (B) Ken Lucas/Biological Photo Supply; (C) E. R. Degginger. Figure 45–8: (A) William R. Wright/Taurus Photos; (B) Kent and Donna Dannen/Photo Researchers; (C) Bruce Coleman, Inc.; (D) Tom J. Ulrich/Visuals Unlimited. Figure 45–10: (D) Bob Gossington/Bruce Coleman, Inc. Figure 45A: (A) Joyce Photo-

graphics/Photo Researchers; (B) John Gerlach/Visuals Unlimited; (C) A. Rider/Photo Researchers; (D) George Holton/Photo Researchers. Figure 45–16: John D. Cunningham/Visuals Unlimited.

Chapter 46

page 1240: Karl Weidmann/Photo Researchers. Figure 46–1: E. R. Degginger. Figure 46–3: Willie L. Hill/Stock, Boston. Figure 46–4: (A) Kenneth W. Fink/Photo Researchers. Figure 46–13: C. P. Vance/Visuals Unlimited. Figure 46–15: Ellis Herwig/Stock, Boston. Figures 46–16 and 46–20: E. R. Degginger.

Chapter 47

page 1268: (tl) V. Englebert/Photo Researchers; (tc) John D. Cunningham/Visuals Unlimited; (tr) Phil Degginger; (cl) Link/Visuals Unlimited; (c,bc) E. R. Degginger; (cr) J. Scherschel/Photo Researchers; (bl) L. Pyenson/Visuals Unlimited; (br) George Holton/Photo Researchers. Figure 47–2: Dan McCoy/Rainbow. Figure 47–4: (A,B) Karel F. Leim. Figure 47–6: William S. Ormerod/Visuals Unlimited. Figures 47–7 and 47–10: Ken Lucas/Biological Photo Service. Figure 47–11: W. B. Carr, reprinted by permission from *Nature, 335,* cover, copyright © 1988, Macmillan Magazines Ltd. Figure 47–13: (B) Arthur M. Siegelman/Visuals Unlimited. Figure 47–14: Farrell Grehan/Photo Researchers. Figure 47–26: Chris Luneski/Photo Researchers. Figure 47–27: (A) Fred McConnaughey/Photo Researchers; (B) Carl Roessler/Bruce Coleman.

Chapter 48

page 1298: (tl) Alan Pitcairn/Grant Heilman; (tr) E. R. Degginger; (b) Doug Sokell/Visuals Unlimited. Figure 48–11: Fran Kepler/Rainbow. Figure 48–12:

J. N. A. Lott, McMaster Univ./BPS. Figure 48–13: Grant Heilman. Figure 48–14: E. R. Degginger. Figure 48–15: (B) C. M. Pringle, Univ. of California/BPS. Figure 48–16: Grant Heilman. Figure 48–17: (A) John D. Cunningham/Visuals Unlimited; (B) Grant Heilman; (C) R. L. Ciochon/Visuals Unlimited; (D) David L. Pearson/Visuals Unlimited. Figure 48–18: (t) Christiana Dittman/Rainbow; (b) E. R. Degginger. Figure 48–20: (A,B) Barbara J. Miller/Biological Photo Service; (C) Grant Heilman. Figure 48–25: (A) Harold W. Pratt/Biological Photo Service; (B) S. K. Webster, Monterey Bay Aquarium/BPS; (C) T. W. Ransom/Biological Photo Service. Figure 48–27: Peter David/Photo Researchers. Figure 48–31: (A,B) E. R. Degginger. Figure 48–32: Barbara J. Miller/Biological Photo Service. Figure 48–36: Courtesy, Sohail H. Hashmi.

Chapter 49

page 1344: Owen Franken/Stock, Boston. Figure 49–1: (A) Cary Wolinsky/Stock, Boston; (B) Bob Burch/Bruce Coleman, Inc.; (C) Story Litchfield/Stock, Boston; (D) Terry E. Eiler/Stock, Boston; (E) Rick Smolan/Stock, Boston. Figure 49–3: Thomas England/Photo Researchers. Figure 49–4: Rick Brown/Stock, Boston. Figure 49–6: Owen Franken/Stock, Boston. Figure 49–7: (A) Sheryan P. Epperty/Visuals Unlimited; (B) John D. Cunningham/Visuals Unlimited. Figure 49–8: (l) Brandenburg/Bruce Coleman, Inc.; (r) John D. Cunningham/Visuals Unlimited. Figure 49–9: (t) David M. Newman/Visuals Unlimited; (b) E. R. Degginger. Figure 49–10: Bruce M. Wellman/Stock, Boston. Figure 49–11: E. R. Degginger. Figure 49–12: (A,B) Culver Pictures. Figure 49–13: (t) Dan McCoy/Rainbow; (b) Peter Menzel/Stock, Boston. Figure 49–14: J. Alcock/Visuals Unlimited. Figure 49–15: D. and J. Heaton/Stock, Boston. Figure 49–16: Courtesy, U. S. Dept. of Agriculture, Los Padres National Forest.

INDEX

Note: Page numbers in boldface indicate main discussion in text.

Crossing-over, 317–320
    chromosome mapping and, 320–322
        double cross over, **320**
        mechanisms in, **317–319**
        meiosis, **287**
Crossopterygians, 1151, **1186**
Cross-pollination, 299
Crustaceans, **1131–1132**
    characteristics of, 1131
    decapods, **1131–1132**
    types of, 1131, 1132
Cryptic coloration, **950–952**
Ctenophores, **1112**
Cuboidal cells, **149**
Cupula, **804**
Cursorial animals, 1313
Cuticle, **241**
Cutin, **241**
Cuvier, Georges, 22, 29, 1126–1127
Cyanobacteria, 516, 1026, 1044
Cyanonta, **1026**
    See also Blue-green algae.
Cyanonts, 1037, 1038
Cycadeoids, **1099**
Cycads, **1099**
Cyclic adenosine monophosphate
    (cAMP) system, **153, 721–728**
Cyclic guanosine monophosphate, **797**
Cyclic photophosphorylation, **234–235**
Cysts, algae, **1049**
Cytochromes, 188, **212**
Cytokinesis, **263**, 276
    animal cells, 277
    plant cells, 277
    process of, 277–278
Cytokinin, actions of, **755, 757**
Cytolytic T cells, **585**
Cytopharynx, algae, **1047**
Cytoplasm, 99, 102, **110–113**
    compartments of, 102
    cytomatrix, **111–112**
    cytoplasmic ground substance, **110**
    cytoskeleton, **111**
    molecular movement in, 123–124
    structure of, 111–113
Cytoplasmic determinants, **483**
Cytoplasmic streaming, **613**
Cytosine, **369**
Cytosol, **102**

Danielli, James F., 128
Dark-field light microscope, 100
Dark reactions, photosynthesis, **227**
Darwin, Charles, 21, 22, 24–25, 924–925, 984, 1181, 1272
Daughter chromosomes, **276**
Davenport, Charles B., 324
Davenport, Gertrude, 324
Davis, D.D., 950
Davson-Danielli model, of cell membrane, **128**
Davson, Hugh, 128
DDT, 1136
    pollution from, 1258–1259
Death, populations and, 1223–1225
Decapods, **1131–1132**
Deceit, intraspecific, 955

Deciduous forests. *See* Temperate deciduous forests
Decomposers, **1255, 1261**
Decompression, diving, **704**
Degenerate, genetic code, 397
Deglutition, **670**
Dehydration, 202
Dehydrogenases, **186**
Delayed hypersensitivity, **580**
Deletions, chromosomes, **348–349, 405**
Deme, **908–909**
    clines, **974**
    differentiation between, 972–974
    distribution of variation in, 910–911
    geographic influences, 972–974
    subspecies, **973**
Demyelination, 776
Denatured proteins, **78**
Dendrites, neurons, **769**
Dense bodies, **838**
Density, **50**
Density-dependent limiting factors, population and, **1227**
Density-independent limiting factors, population and, **1226**
Denticles, **1150**
Deoxyribonucleic acid (DNA). *See* DNA
Deoxyribonucleotides, **372**
Deoxyribose, 80, **369**
Depolarization, **773**
Depolarization event, **461**
Depolymerization model, chromosome movement in mitosis, **276**
Deposit feeders, **666**
Derepression
    enzyme synthesis, **193**
    gene regulation, **420**
Descartes, René, 26, 81
Deserts, **1313–1314**
    animals of, 1314
    plants of, 1313
Desmosomes, **151**
Determination, **450**
Deuterostomes, **471, 1170**
    echinoderm superphylum, 1171–1172
Development
    cellular mechanisms
        cell interactions, 498–499
        collagen, role of, **499**
        morphogenetic movements, 497–498
    embryonic induction
        competence, **496**
        organizer of early embryo, **494–496**
    exploration of genes during, 486–494
    molecular mechanisms
        commitment, **484–485**
        differentiation and cell division, 486
        gene expression and differentiation, 479–484
        gene transplantation, 482–484
        localization and cell division, 484
        nuclear transplantation, 477–481

ontogeny and phylogeny, **500–503**
    heterochrony, **501–503**
    principle of recapitulation, **500**
Developmental biology, **449**
Developmental research
    chimeras, **487**
    fate map, **490–491**
    homeobox, **491–494**
    transgenic mice, 486, 488–489
Development of zygote, **264**
Devonian period, 1087
    age of fishes, 1186–1187
    life in, 1186–1187, 1189
De Vries, Hugo, 296, 326
Dextrins, **670**
Diabetes insipidus, **655**
Diabetes mellitus, **653, 656**, 749
Diacylglycerol, **725**
Dialysis, 126, **186**
Diaphragm, **544, 695**
Diastole, **628**
Diatoms, characteristics of, **1049–1050**
Dicots, dicotyledons, **620, 1105**
Dictyosome, **104**
Dielectric constant, water, **52–53**
Differentiation, **450–451**
    beginning of, 465–466
    cell division and, 486
    cellular, **264**
    commitment and, **484–485**
    versus determination, 451
    gene regulation in, 479–484
    multicellular organism, **92**
    plants versus animals, 451
    regeneration and, 451
Diffusion, **613, 693**
    facilitated diffusion, **133–134**
    imbibition, **125**
    process of, **124–125**
    simple diffusion, **133**
    stages of, 125
Diffusion techniques, antibody detection, 587–588
Digestion, **661**
    cooking and, 667
    early study of, 671
    in flatworms, 672–673
    functions of, 668
    in herbivores, 671–672, 676
    hydrolytic reactions, 668–669
    in insects, 673
    large intestine, 675–676
    modes of
        extracellular digestion, 666–667
        phagocytosis, 666
    phases of, 669–670
    small intestine
        absorption, 674–675
        breakdown of molecules, 673–676
Digestive enzymes, types of, 669, 670, 673–674
Dihybrid crosses, Mendelian genetics, **304**
Dinoflagellates, **1047**, 1049
    red tide, 1048
Dinosaurs, 1192–1195
    areas found, 1195
    characteristics of, 1192–1193
    mass extinction of, 1197

orders of, 1192
    types of, 1192, 1194
Dinucleotide, 187
Dioecious, **512**
2,3-Diphosphoglyceric acid, **700**
Diploid cells, chromosomes of, 269
Diploid number, chromosomes, 280–281
Diplophase, 510, **513**
Direct communication, intercellular, 149–150
Directional selection, **928–929**
Disaccharides, **68**
Discontinuous phenotypes, **931**
Disease
    bacteria and, **1032**
    effects of medicine and hygiene on, 1350–1352
    versus infection, 1280–1281
    parasites and, **1279–1281**
    pathogenic bacteria and, 1279, 1280–1281
Disjunctive distributions, **1325, 1388–1339**
Dispersal
    means of, 1326–1330
    routes of, 1330
    sweepstakes dispersal, **1331–1332**
Displacement behavior, **890**
Disruptive selection, **929, 934**
Distal convoluted tubule, **652**
Distal nephron, 656
Distribution range, **908**
Dittmer, H.J., 615
Divergence, evolutionary, **1006**
Diversification, evolution and, 15–16
Diversity
    maintenance of species diversity, 1294–1295
    richness of species in community, 1293
Diving, 702–704
    hazards of, 703–704
    principles of diving, 702
Division furrow, **277**
Dizygotic twins, **467**
DNA, 61, 80, **361**
    B-DNA, **428**
    biosynthesis of, 372–377
    chloroplasts and, 116
    damage, types of, 407–408
    DNA-DNA hybridization, 963
    DNA gyrase, **377**
    DNA ligase, **377**, 378
    DNA methylation, 429–430
    DNA polymerase, **372–374**, 377
    DNA topoisomerases, **377**
    double-helix, 370, **371**, 373, 377
    events in flow of genetic information, 389–390
    histone interaction, 430–431
    intracellular DNA promiscuity, 415
    junk DNA, 958, 964, 978
    left-handed helix, 428–429
    linkage analysis, 352–353
    linker DNA, 279
    in meiosis, 283, 288, 289
    molecular components of, 368
    mitochondrial, 108
    organization of
        intervening DNA sequences, 410–413

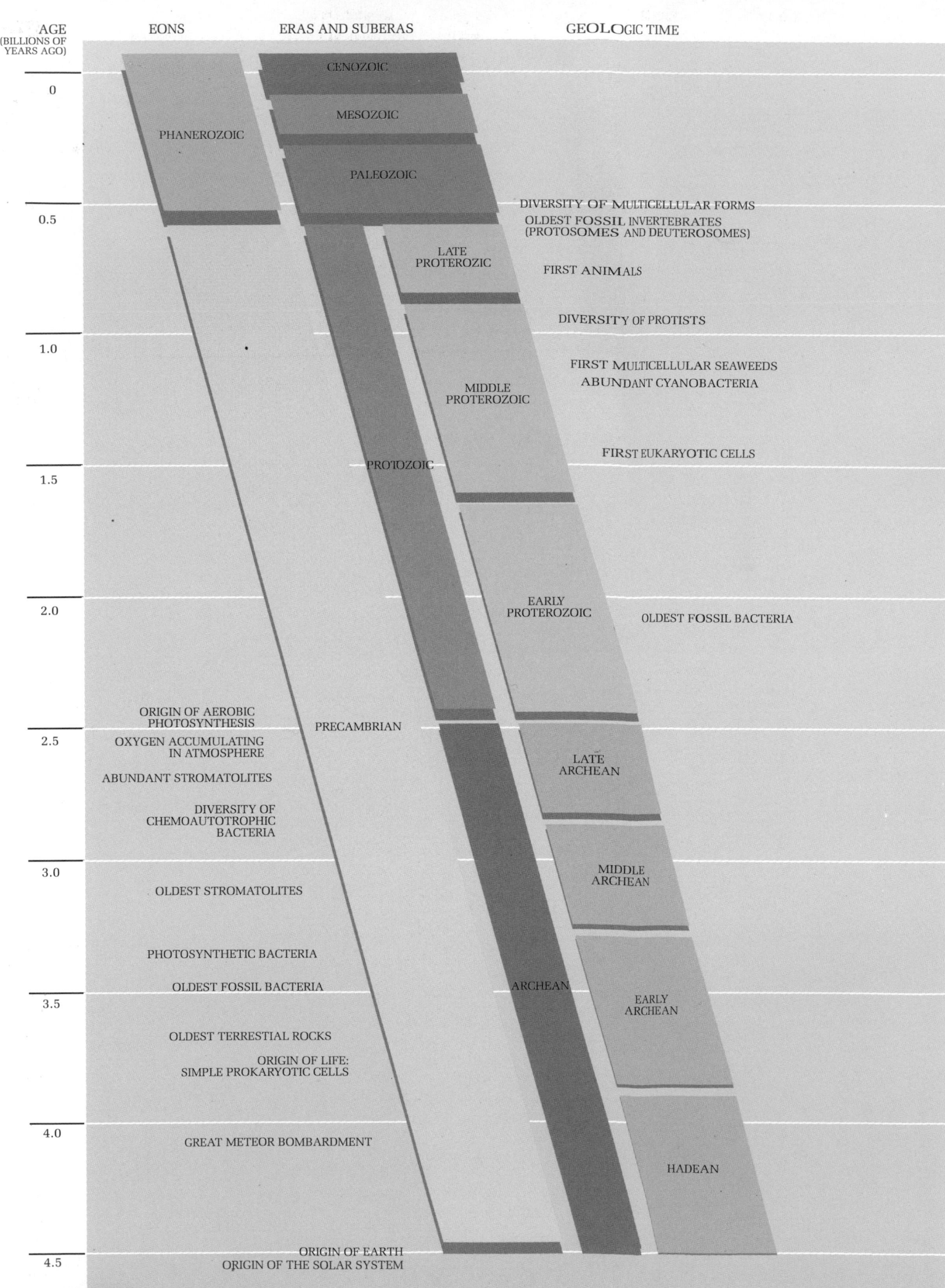